Biology
Life on Earth

Sunlight, here reflected in a dewdrop, provides the energy that powers nearly all life on Earth.

BIOLOGY
Life on Earth

Fourth Edition

Teresa Audesirk   Gerald Audesirk

University of Colorado at Denver

Prentice Hall, Upper Saddle River, New Jersey 07458

Library of Congress Cataloging-in-Publication Data

Audesirk, Teresa
  Biology: Life on Earth / Teresa Audesirk & Gerald Audesirk
  4th ed.

      p.   cm.
  Includes bibliographical references and index.
  ISBN 0-13-368150-5
  1. Biology. I. Audesirk. Gerald. II. Title.
  QH308.2.A935 1996   95-34325
  574--dc20   CIP

To our daughter, Heather, and to our parents, Lori, Jack, and Joe

Senior Editor: Sheri L. Snavely
Development Editor: Ray Mullaney, Nicole Gray
Production Editor: Debra A. Wechsler
Page Layout: Karen Noferi, Richard Foster, Molly Pike Riccardi
Art Director: Heather Scott
Art Manager: Patrice Van Acker
Editor in Chief: Paul F. Corey
Editorial Director: Tim Bozik
Marketing Manager: Kelly McDonald
Director, Production & Manufacturing: David W. Riccardi
Managing Editor, Production: Kathleen Schiaparelli
Formatting Manager: John J. Jordan
Creative Director: Paula Maylahn
Buyer: Trudy Pisciotti
Copy Editor: Margo Quinto
Cover Designer: Design W, Inc./ Wendy Helft
Interior Designer: Laura Ierardi
Art Retouching: Todd Ware, Marita Froimson, Rolando Corujo, Mumtaz Hussain
Illustrators: Rolando Corujo, Hudson River Studios, Howard Friedman,
        David Mascaro, Edmund Alexander, Page Two Associates
Production Support: David Tay, Jeff Harman
Project Support: Rhoda Sidney
Photo Research: Tobi Zausner, Beaura K. Ringrose
Cover Photograph: Mitsuaki Iwago

Printed in the United States of America
10 9 8 7 6 5 4 3 2 1

ISBN   0-13-368150-5

Prentice-Hall International (UK) Limited, London
Prentice-Hall of Australia Pty. Limited, Sydney
Prentice-Hall Canada Inc., Toronto
Prentice-Hall Hispanoamericana, S.A., Mexico
Prentice-Hall of India Private Limited, New Delhi
Prentice-Hall of Japan, Inc., Tokyo
Simon & Schuster Asia Pte, Ltd., Singapore
Editora Prentice-Hall do Brasil, Ltda., Rio de Janeiro

About the Authors

Terry and Gerry Audesirk both grew up in New Jersey, where they met as undergraduates. After marrying in 1970, they moved to California, where Terry earned her doctorate in marine ecology at the University of Southern California and Gerry earned his doctorate in neurobiology at the California Institute of Technology. While living in Southern California, both Terry and Gerry learned to scuba dive and were fascinated by the diversity and beauty of marine life. As postdoctoral students at the University of Washington's marine laboratories, they worked together on the neural bases of behavior using a marine mollusc as a model system. They are now professors of biology at the University of Colorado at Denver, where they have both taught introductory biology for over 10 years along with courses in neurobiology. In their research lab,

they investigate the mechanisms by which neurons are harmed by low levels of environmental pollutants such as lead. This program is funded by the National Institutes of Health.

Terry and Gerry share a deep appreciation of nature and the outdoors. They enjoy hiking in the Rockies, running near their home in the foothills west of Denver, and attempting to garden at 7000 feet in the presence of hungry deer and elk. They are long-time members of many conservation organizations. The birth of their daughter, Heather, has added another love to their lives, and they now spend much of their free time with her.

Biology: Life on Earth is the natural outgrowth of the authors' enjoyment of writing, their fascination with all aspects of biology, and their eagerness to share this fascination with their students.

Preface

Because we have a young daughter, we often hear the question "Why?" It's an exciting question, because it provides us with an opportunity to share some of what we know, or to learn more by looking up the answer to our daughter's question. The question "why?" begins a person's exploration of his or her world. For those fortunate enough to continue asking such questions, the exploration and the delight of learning become life-long and continuously enriching experiences. As teachers and scientists, we are continuously learning about life on Earth. In writing this textbook, we have had the good fortune both to learn more and to share our interest in biology with an extended classroom of thousands of students.

A course in introductory biology may be a student's first real exposure to the fascinating complexity of life. As teachers, we recognize how easily a student may become mired in the overwhelming number of new facts and terms, while losing sight of the underlying principles of biology and their relevance to daily life. *Biology: Life on Earth* has been carefully written and revised to integrate the necessary biological facts into a broader conceptual framework that stresses unifying themes and the ways in which an understanding of biology can enrich and enlighten day-to-day living. Why study biology? Maybe we're biased, but what can be more fascinating than learning about life on Earth?

Biology: Life on Earth

Focuses on Concepts

The subheadings within each major chapter section have been cast as sentences so that they introduce key concepts in each section of text. Our "At a Glance" section opens each chapter with an outline that brings these statements together, presenting an overview of the chapter and its most important ideas. Figure titles tell the student at a glance what idea the figure illustrates, before the details are described in the caption. Concepts are further stressed in highlighted in-text summaries and in end-of-chapter summaries.

Makes Biology Relevant

There is a danger that students will stop asking the question "why?" if the equally important question "so what?" isn't answered. In fact, there is no science that can tell us more about ourselves and our world than can biology. Knowledge of biology illuminates everyday life: how our bodies work, why our house plants bend toward the window, why a growling dog raises the fur along its back. A knowledge of biology is also crucial to understanding modern concerns and controversies such as population growth, the spread of diseases such as AIDS, the destruction of tropical rain forests, and the promise and perils of genetic engineering. In this text, we make a conscious effort to relate concepts in biology to everyday experiences and to important issues facing society. These applications are interwoven into the text, introduced in end-of-chapter essay questions, and highlighted in boxed essays such as:

 Earth Watch — a series of environmental essays addressing issues such as biodiversity, ozone depletion, and population growth.

 Health Watch — a series of essays investigating topics such as how ulcers form, how kidney dialysis works, and how bone fractures heal.

Communicates the Scientific Process

One cannot understand biology without knowing how science works. Biology is not just a compendium of facts and ideas; it is the outgrowth of a dynamic process of inquiry and human endeavor. In many places throughout the text, accompanying descriptions of biological facts and concepts, are descriptions of how those findings were achieved. The scientific process is further highlighted in Scientific Inquiry essays:

 Scientific Inquiry — students will learn, for example, how fossils are dated, how the structure of DNA was discovered, and how scientists have learned that a newborn baby prefers her own mother's voice.

Stresses Unifying Themes

Two themes are interwoven throughout the text: evolution and adaptation to the environment. As Theodosius Dobzhansky so aptly put it, "Nothing in biology makes sense, except in the light of evolution." Life on Earth has been forged in the crucible of evolution, and any modern biology text must reflect that fact. Many of the chapter introductions have an evolutionary theme, and many chapters end with a feature new to this edition:

 Evolutionary Connections — lively discussions that link chapter concepts into the broader perspective of evolution.

Adaptation to the environment flows naturally from the concept of evolution; the environment exerts selective pressure on organisms that helps shape their evolution. Nevertheless, not all structures are optimal solutions to environmental challenges. Many apparently unlikely biological designs are due to genetic drift, chance catastrophes, and the fact that the raw material for evolution consists of random mutations affecting previously existing structures. In the words of Sydney Brenner, "Anything produced by evolution is likely to be a bit of a mess," and we have tried to show that, too.

Our own concern for the environment also threads its way throughout this text. Whenever appropriate, we have tried to present students with the biological rationale for making sound environmental decisions.

Stimulates Exploration

For every "why?" that is answered, many more questions are raised. The knowledge and understanding of science that students gain from this text will allow them to continue to seek answers throughout their lives. Why does a flower have a particular arrangement of stigma and anthers? Watch what happens when a bee explores it! Thought-provoking questions end each chapter, along with interesting and informative references, where topics can be explored in more detail. A free supplement for students, *The New York Times Themes of the Times* brings together a collection of recent, topical, biological articles from the pages of *The New York Times*. The *ABC News/Prentice Hall*

Video Library for instructors features brief clips from programs like *20/20*, *World News Tonight*, and *Nightline*. Finally, the text challenges the student to enter the information age with its own home page on the World Wide Web. *Life on the Internet: Biology - A Student's Guide* is available as a supplement to help students navigate through the web to access the regularly updated resources on the *Biology: Life on Earth* home page:

 Net Watch — an icon at the end of each chapter highlights the specific address for each chapter's resources.

Is Flexible in Organization

There is no "correct" sequence of topics in biology; nearly every subject would be more understandable if all the other subjects preceded it! Our response to this dilemma has been to make the units as flexible as possible, using selective repetition of important points. For example, learning how plants became adapted to life on dry land is important in understanding their evolution, their diversity, and their physiology, so we have repeated key concepts separately in each discussion. Our first chapter sets the stage for coverage of topics in different orders by reviewing the diversity of life and the principles of evolution. Extensive cross-referencing throughout the text allows the student to find new related material or review previously covered areas.

By providing relatively complete coverage of all the major topics in biology, this text accommodates courses that provide a broad overview as well as those that delve into selected areas of biology. We recognize that some topics are harder than others for introductory students to grasp. When the details of more difficult topics are integrated into the text, students may lose sight of underlying concepts in the welter of detail. So we have written the text to provide a more general overview of complex subjects, providing the details in the boxed sections entitled A Closer Look:

 A Closer Look — These boxes focus on the more challenging details of topics such as chemiosmosis, the immune response, and urine formation in the nephron.

New in This Edition:

Our text has a new look and important new features which reflect the philosophy guiding the revision:

▸ Conceptual, full-sentence chapter subheadings highlight key concepts in each section. Figure titles now present the concept underlying each figure.

▸ The inheritance unit has been reorganized, allowing the classical experiments of genetics to be interpreted in the light of modern understanding of DNA and gene function.

▸ New essays have been added—some contributed by outside experts—covering topics such as DNA fingerprinting and the discovery of the gene for Huntington's disease.

▸ New end-of-chapter features include multiple choice questions, revised essay questions, and new questions that specifically relate chapter topics to issues in modern society and everyday life.

▸ The most creative, progressive supplements package available; including laser disc, CD-ROM, and Internet technology. Also a regularly updated *New York Times* supplement for students, and an ABC videos series. For a complete listing see page xi.

Acknowledgments

Biology: Life on Earth is truly a team effort, and we wish to acknowledge and thank our teammates.

The dauntingly complex task of putting together a book of this magnitude has been handled with unflagging dedication and zeal by the skilled development team at Prentice Hall. Ray Mullaney, Editor in Chief of Engineering, Science, and Mathematics development, and Nicole Gray, Assistant Development Editor, made sure the text was clear, consistent, and student-friendly. Ray continued to oversee the production of the text, contributing substantially to its design, photographic selection, and accuracy. Debra Wechsler, Senior Production Editor, patiently dealt with the minutiae of assuring accuracy in the art program and the translation from typed draft to final printed page. Tobi Zausner and Kathy Ringrose skillfully and doggedly tracked down excellent photos. Margo Quinto tackled the job of copyediting with meticulous care. Stephanie Hiebert did an excellent job in reading through the final typeset pages to help ensure accuracy in both text and art. We also wish to thank Heather Scott, our talented Art Director, for creatively guiding the text and cover design; Patrice Van Acker who coordinated the art program; John Jordan and his incredibly talented formatting team — Karen Noferi, Richard Foster, and David Tay. Mary Hornby has overseen the production of our print supplements, test bank, and transparencies. Linda Schreiber did a fantastic job of coordinating the multimedia program, especially our exciting new laser disc and CD-ROM.

Colleagues at other institutions have helped us enormously. Many, listed in the following pages, have stimulated us to rethink our presentation with careful, thoughtful reviews. Essays were contributed by Joseph Chinnici, Mark Shotwell, and Gisele Muller-Parker. A wonderful supplements package was engineered by Joseph Chinnici, Kristin Uthus, Rhoda Perozzi, Gayle Sauer, Linda Butler, Andrew Stull, Steven Brunasso, Mike McKinley, and Blanche Haning. We thank Bill Stark, Richard Mortensen, Gerald Summers, Rhoda Perozzi, Florence Juillerat, Jerri Lindsey, Scott Freeman, and Dan Doak for contributing thought-provoking end-of-chapter questions.

Marketing is being handled with talent and enthusiasm by Kelly McDonald, our Executive Marketing Manager. We also would like to thank Paul F. Corey, Editor in Chief of Science, for his insight and support through the past two editions. Finally, but most importantly, our editor, Sheri Snavely, has nurtured this project through thick and thin, smoothing rough spots and boosting our morale with her good humor, optimism, and energy.

Terry and Gerry Audesirk
Golden, Colorado

FOURTH EDITION REVIEWERS

W. Sylvester Allred, *Northern Arizona University*
William Anderson, *Abraham Baldwin Agriculture College*
Bill Barstow, *University of Georgia, Athens*
Colleen Belk, *University of Minnesota, Duluth*
Marilyn Brady, *Centennial College of Applied Arts & Technology*
Arthur L. Buikema Jr., *Virginia Polytechnic Institute*
Kathleen Burt-Utley, *University of New Orleans*
Linda Butler, *University of Texas at Austin*
W. Barkley Butler, *Indiana University of Pennsylvania*
Joseph R. Chinnici, *Virginia Commonwealth University*
Ethel Cornforth, *San Jacinto College, South*
Lee Couch, *Albuquerque Technical Vocational Institute*
Donald C. Cox, *Miami University of Ohio*
Patricia B. Cox, *University of Tennessee*
Peter Crowcroft, *University of Texas at Austin*
Carol Crowder, *North Harris Montgomery College*
Robert A. Cunningham, *Erie Community College, North Campus*
Jerry Davis, *University of Wisconsin, La Crosse*
Jean DeSaix, *University of North Carolina, Chapel Hill*
Ed DeWalt, *Louisiana State University*
Daniel F. Doak, *University of California, Santa Cruz*
Susan A. Dunford, *University of Cincinnati*
Mary Durant, *North Harris College*
Ronald Edwards, *University of Florida*
Joanne T. Ellzey, *University of Texas, El Paso*
Carl Estrella, *Merced College*
Rita G. Farrar, *Louisiana State University*
Marianne Feaver, *North Carolina State University*
Linnea Fletcher, *Austin Community College, Northridge Campus*
Scott Freeman, *University of Washington*
Donald P. French, *Oklahoma State University*
Teresa Lane Fulcher, *Pellissippi State Technical Community College*

Martin E. Hahn, *William Paterson College*
Blanche C. Haning, *North Carolina State University, Raleigh*
Alison G. Hoffman, *University of Tennessee, Chattanooga*
Florence Juillerat, *Indiana University-Purdue University at Indianapolis*
Thomas W. Jurik, *Iowa State University*
Jerri K.Lindsey, *Tarrant County Junior College, Northeast Campus*
D. McWhinnie, *De Paul University*
Karen E. Messley, *Rockvalley College*
Richard Mortensen, *Albion College*
Gisele Muller-Parker, *Western Washington University*
Harry Nickla, *Creighton University*
Jane Noble-Harvey, *University of Delaware*
Rhoda E. Perozzi, *Virginia Commonwealth University*
Bill Pfitsch, *Hamilton College*
James A. Raines, *North Harris College*
Mark Richter, *University of Kansas*
Paul Rosenbloom, *Southwest Texas State University*
Edna Seaman, *University of Massachusetts, Boston*
Linda Simpson, *University of North Carolina, Charlotte*
Shari Snitovsky, *Skyline College*
Benjamin Stark, *Illinois Institute of Technology*
William Stark, *Saint Louis University*
Gerald Summers, *University of Missouri, Columbia*
William Thwaites, *San Diego State University*
Sharon Tucker, *University of Delaware*
Lloyd W. Turtinen, *University of Wisconsin, Eau Claire*
Kristin Uthus, *Virginia Commonwealth University*
Jyoti R. Wagle, *Houston Community College,Central Campus*
Roberta Williams, *University of Nevada, Las Vegas*
Bill Wischusen, *Louisiana State University*
Colleen Wong, *Wilbur Wright College*
Wade Worthin, *Furman University*

MULTIMEDIA REVIEWERS

Consultant:
Michael McKinley, *Glendale Community College*

Photo Coordinator:
Blanche Haning, *North Carolina State University*

Advisory Panel:
Arthur L. Buikema Jr., *Virginia Polytechnic Institute*
Thomas W. Jurik, *Iowa State University*
Frank Schwartz, *Cuyahoga Community College*
Charlene Waggoner, *Bowling Green University*

Reviewers:
Sarah Barlow, *Middle Tennessee State University*
Guy Cameron, *University of Houston*
Susan A. Dunford, *University of Cincinnati*
Florence Juillerat, *Indiana University/Purdue University*
Arnold Karpoff, *University of Louisville*

Terry Keiser, *Ohio Northern University*
Donald Reinhardt, *Georgia State University*
Dan Skean Jr., *Albion College*
Lloyd W. Turtinen, *University of Wisconsin, Eau Claire*
Robin Tyser, *University of Wisconsin, LaCrosse*

INTERNET GUIDE AND WORLD WIDE WEB HOME PAGE REVIEWERS

Susan Brawley, *University of Maine*
Ronald Edwards, *University of Florida*
David Huffman, *Southwest Texas State University*
Jason Rosenblum, *Austin, Texas*
Patrick A. Thorpe, *Grand Valley State University*
Robin Tyser, *University of Wisconsin, La Crosse*
Roberta Williams, *University of Nevada, Las Vegas*

G. D. Aumann, *University of Houston*
Vernon Avila, *San Diego State University*
J. Wesley Bahorik, *Kutztown University of Pennsylvania*
Gerald Bergtrom, *University of Wisconsin*
Raymond Bower, *University of Arkansas*
Virginia Buckner, *Johnson County Community College*
William F. Burke, *University of Hawaii*
Nora L. Chee, *Chaminade University*
Joseph P. Chinnici, *Virginia Commonwealth University*
Dan Chiras, *University of Colorado, Denver*
Bob Coburn, *Middlesex Community College*
Martin Cohen, *University of Hartford*
Mary U. Connell, *Appalachian State University*
Joyce Corban, *Wright State University*
David J. Cotter, *Georgia College*
Donald E. Culwell, *University of Central Arkansas*
Jerry Davis, *University of Wisconsin, LaCrosse*
Douglas M. Deardon, *University of Minnesota*
Fred Delcomyn, *University of Illinois, Urbana*
Lorren Denney, *Southwest Missouri State University*
Katherine J. Denniston, *Towson State University*
Charles F. Denny, *University of South Carolina, Sumter*
Ronald J. Downey, *Ohio University*
Ernest Dubrul, *University of Toledo*
Michael Dufresne, *University of Windsor*
Wayne Elmore, *Marshall University*
Nancy Eyster-Smith, *Bentley College*
Gerald Farr, *Southwest Texas State University*
Charles V. Foltz, *Rhode Island College*
Douglas Fratianne, *Ohio State University*
Don Fritsch, *Virginia Commonwealth University*
Michael Gaines, *University of Kansas*
Farooka Gauhari, *University of Nebraska, Omaha*
David Glenn-Lewin, *Iowa State University*
Elmer Gless, *Montana College of Mineral Sciences*
Margaret Green, *Broward Community College*
Madeline Hall, *Cleveland State University*
Helen B. Hanten, *University of Minnesota*
John P. Harley, *Eastern Kentucky University*
Stephen Hedman, *University of Minnesota*
Jean Helgeson, *Collins County Community College*
Alexander Henderson, *Millersville University*
Laura Mays Hoopes, *Occidental College*
Michael D. Hudgins, *Alabama State University*
Donald A. Ingold, *East Texas State University*
Jon W. Jacklet, *State University of New York, Albany*
Arnold Karpoff, *University of Louisville*
L. Kavaljian, *California State University*
Hendrick J. Ketellapper, *University of California, Davis*
William H. Leonard, *Clemson University*

John Logue, *University of South Carolina, Sumter*
William Lowen, *Suffolk Community College*
Steele R. Lunt, *University of Nebraska, Omaha*
Michael Martin, *University of Michigan*
Margaret May, *Virginia Commonwealth University*
D. J. McWhinnie, *De Paul University*
Gary L. Meeker, *California State University, Sacramento*
Thoyd Melton, *North Carolina State University*
Glendon R. Miller, *Wichita State University*
Neil Miller, *Memphis State University*
Dr. Jack E. Mobley, *University of Central Arkansas*
Richard Mortenson, *Albion College*
Robert Neill, *University of Texas*
Daniel Nickrent, *Southern Illinois University*
James T. Oris, *Miami University, Ohio*
C. O. Patterson, *Texas A & M University*
Fred Peabody, *University of South Dakota*
Harry Peery, *Tompkins-Cortland Community College*
Ronald Pfohl, *Miami University, Ohio*
Bernard Possident, *Skidmore College*
Robert Robbins, *Michigan State University*
K. Ross, *University of Delaware*
Mary Lou Rottman, *University of Colorado at Denver*
Albert Ruesink, *Indiana University*
Alan Schoenherr, *Fullerton College*
Russel V. Skavaril, *Ohio State University*
John Smarelli, *Loyola University*
Jim Sorenson, *Radford University*
Mary Spratt, *University of Missouri, Kansas City*
William Stark, *Saint Louis University*
Dr. Barbara Stotler, *Southern Illinois University*
Marshall Sundberg, *Louisiana State University*
Bill Surver, *Clemson University*
Eldon Sutton, *University of Texas at Austin*
David Thorndill, *Essex Community College*
Professor Tobiessen, *Union College*
Richard Tolman, *Brigham Young University*
Dennis Trelka, *Washington & Jefferson College*
Gail Turner, *Virginia Commonwealth University*
Glyn Turnipseed, *Arkansas Technical University*
Robert Tyser, *University of Wisconsin, La Crosse*
F. Daniel Vogt, *State University of New York, Plattsburgh*
Nancy Wade, *Old Dominion University*
Michael Weis, *University of Windsor*
Jerry Wermuth, *Purdue University-Calumet*
Jacob Wiebers, *Purdue University*
P. Kelly Williams, *University of Dayton*
Sandra Winicur, *Indiana University at South Bend*
Chris Wolfe, *North Virginia Community College*
Tim Young, *Mercer University*

Supplements and New Media Tools

For the Instructor

Instructor's Examination Package (0-13-525734-4)

Instructor's Resource Manual
By Gayle Sauer, The Citadel
Contains a variety of lecture outlines and teaching tips.
(0-13-377987-4)

Transparency Pack
250 four-color, large type transparencies from the text. The balance of art is available as black line transparency masters.
(0-13-378076-7)

Slides
Same images as transparency acetates, available in slide format.
(0-13-378087-8)

Test Item File
By Joseph Chinnici, Rhoda Perozzi, Kristin Uthus, all of Virginia Commonwealth University
Contains over 2300 test questions. Makes use of a variety of question types so that users can customize tests and quizzes to incorporate conceptual, recall-oriented, applied, and critical thinking formats. (0-13-378035-X)

Prentice Hall Custom Test-IBM (0-13-378043-0)
Prentice Hall Custom Test-Macintosh (0-13-378068-6)
Available for Windows, Macintosh, and DOS, Prentice Hall Custom Text allows the educator to create and tailor the exam to their own needs. With the Online Testing option, exams can also be administered online and data can then be automatically transferred for evaluation. A comprehensive desk reference guide is included, along with online assistance.

ABC News /Prentice Hall Video Library for Biology
Coordinated by Linda Butler, University of Texas at Austin
This unique video library contains brief (5-20 minute) segments from award-winning shows such as 20/20, World News Tonight, and The American Agenda. This innovative resource shows biological principles at work and teaches students to critically analyze media messages based on their scientific knowledge.

The Prentice Hall Laser Disc for Biology
The Prentice Hall Laser Disc for Biology is a visual encyclopedia created specifically for use with Biology: Life on Earth. The disc features over 1000 still images from the text and other sources, 30 minutes of video, and 25 minutes of broadcast-quality three-dimensional animations. The animations focus on topics that are challenging to students and particularly difficult to visualize (e.g., cell respiration, enzyme function, photosynthesis). The video segments focus on applications of biology to students' everyday world (e.g., DNA testing, CT and MRI technology). (0-13-378100-3)

Prentice Hall CD-ROM Image Bank for Biology
This unique image bank makes available still images, animations, and video in a digitized format for use in the classroom or for students. It also includes a navigational tool to allow instructors to customize lecture presentations. Additional features include keyword searches and the ability to incorporate lecture notes based on custom presentations. (0-13-513987-2)

Multimedia Biology Presentation Manager 1.0
Providing an easy-to-use interface for creating dynamic multimedia presentations, this unique media manager is designed for use in conjunction with the Prentice Hall Biology Laser Disc.
(0-13-514134-6)

Director Academic
Authorware Academic
By Prentice Hall and Macromedia
Available exclusively through Prentice Hall, these educational adaptations of the leading multimedia authoring tools put the power of multimedia development into your hands at a fraction of the cost. Both products also contain templates designed specifically for academic uses. Available for both Macintosh and Windows platforms. For further information and/or sales, contact Prentice Hall Multimedia Group at 1-800-887-9998. Saleable product only ($150 for Authorware Academic/ $99 for Director Academic).

For the Student

Audesirk *Life on Earth* World Wide Web Home Page
Developed by Andrew Stull and Stephen Brunasso, California State University at Fullerton
Available January 1996, this unique tool is the first to harness the resources of the Internet for easy integration into the classroom. Designed to launch student exploration, resources are regularly updated and linked to text chapters (see Net Watch icons at the end of each chapter).

Life on the Internet: Biology—A Student Guide
By Andrew Stull, California State University at Fullerton
The perfect tool to help your students take advantage of our Life on Earth home page on the World Wide Web and open up the world of the Internet for exploration! Tied specifically to our home page (available January 1996), this unique resource gives clear steps to access our regularly updated biology resource area as well as an overview of general navigation strategies.
(0-13-244088-1)

Study Guide
By Joseph Chinnici, Virginia Commonwealth University
This essential study tool for students is written by the leading author of the test bank to ensure consistency in style. It includes a variety of study questions as well as chapter outlines and interactive exercises. (0-13-378027-9)

New York Times Themes of the Times Supplement
Coordinated by Linda Butler, University of Texas at Austin
This unique newspaper-format supplement brings together recent articles on dynamic biology applications from the pages of the world-renowned *New York Times*. This free supplement, available in quantity through your local representative, encourages students to make connections between the classroom and the world around them.

AIDS Update
By Gerald Stine, University of North Florida
This inexpensive paperback reader, updated annually, answers many of the questions students have about today's AIDS crisis.
(1996 version: 0-13-517939-4)

For the Laboratory

Explorations in Basic Biology, Seventh Edition
By Stanley Gunstream, Pasadena City College
This best-selling laboratory manual can be used with Biology: Life on Earth or any introductory biology text. It includes 40 self-contained, easy-to-understand experiments that blend traditional experiments with new investigative exercises.
(0-13-372939-7)

Instructor's Manual to Explorations in Basic Biology
(0-13-393927-8)

How to Use this Text: A Guide for Students

Biology: Life on Earth has been designed as one of the tools to launch your exploration of the exciting world of biology! Our goal is to help you master the basic concepts of biology and then begin applying them to the world around you. Take a few moments now to see how our text can help you succeed.

At a Glance

At a Glance overviews the chapter ahead and helps you organize your study.

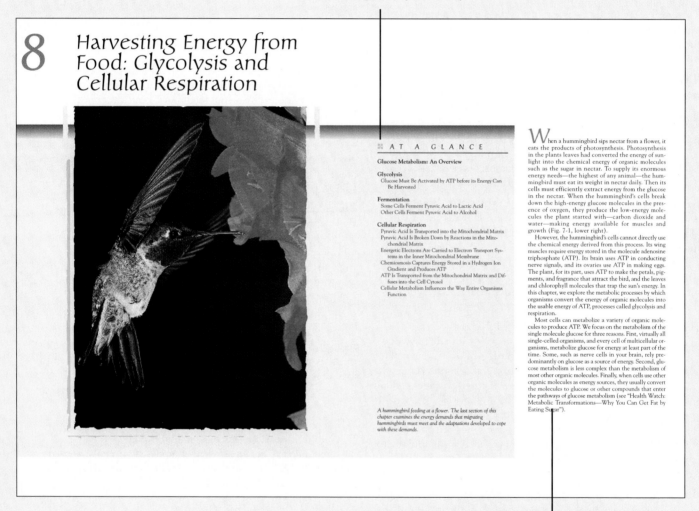

8 Harvesting Energy from Food: Glycolysis and Cellular Respiration

⁞⁞ AT A GLANCE

Glucose Metabolism: An Overview

Glycolysis
Glucose Must Be Activated by ATP before its Energy Can Be Harvested

Fermentation
Some Cells Ferment Pyruvic Acid to Lactic Acid
Other Cells Ferment Pyruvic Acid to Alcohol

Cellular Respiration
Pyruvic Acid Is Transported into the Mitochondrial Matrix
Pyruvic Acid Is Broken Down by Reactions in the Mitochondrial Matrix
Energetic Electrons Are Carried to Electron Transport Systems in the Inner Mitochondrial Membrane
Chemiosmosis Captures Energy Stored in a Hydrogen Ion Gradient and Produces ATP
ATP Is Transported from the Mitochondrial Matrix and Diffuses into the Cell Cytosol
Cellular Metabolism Influences the Way Entire Organisms Function

A hummingbird feeding at a flower. The last section of this chapter examines the energy demands that migrating hummingbirds must meet and the adaptations developed to cope with these demands.

When a hummingbird sips nectar from a flower, it eats the products of photosynthesis. Photosynthesis in the plants leaves had converted the energy of sunlight into the chemical energy of organic molecules such as the sugar in nectar. To supply its enormous energy needs—the highest of any animal—the hummingbird must eat its weight in nectar daily. Then its cells must efficiently extract energy from the glucose in the nectar. When the hummingbird's cells break down the high-energy glucose molecules in the presence of oxygen, they produce the low-energy molecules the plant started with—carbon dioxide and water—making energy available for muscles and growth (Fig. 7-1, lower right).

However, the hummingbird's cells cannot directly use the chemical energy derived from this process. Its wing muscles require energy stored in the molecule adenosine triphosphate (ATP). Its brain uses ATP in conducting nerve signals, and its ovaries use ATP in making eggs. The plant, for its part, uses ATP to make the petals, pigments, and fragrance that attract the bird, and the leaves and chlorophyll molecules that trap the sun's energy. In this chapter, we explore the metabolic processes by which organisms convert the energy of organic molecules into the usable energy of ATP, processes called glycolysis and respiration.

Most cells can metabolize a variety of organic molecules to produce ATP. We focus on the metabolism of the single molecule glucose for three reasons. First, virtually all single-celled organisms, and every cell of multicellular organisms, metabolize glucose for energy at least part of the time. Some, such as nerve cells in your brain, rely predominantly on glucose as a source of energy. Second, glucose metabolism is less complex than the metabolism of most other organic molecules. Finally, when cells use other organic molecules as energy sources, they usually convert the molecules to glucose or other compounds that enter the pathways of glucose metabolism (see "Health Watch: Metabolic Transformations—Why You Can Get Fat by Eating Sugar").

Opening Stories/Vignettes

Opening Stories/Vignettes in most chapters relate the opening photograph to the concepts that will be developed in the chapter.

Focus on Key Concepts

As you read, identify and focus on the key concepts.
Can you rephrase the main ideas in your own words?

Conceptual Headings

Conceptual Headings provide a context for the discussion to follow and give you a clear review of key concepts.

In-text Summaries

In-text Summaries help you check your understanding of complex processes, especially in the early chapters where the fundamental concepts of biology are laid out. Stop and read these summaries as you move through a chapter as a help in digesting new material.

---

see that cells have evolved elaborate mechanisms that ensure that each daughter cell inherits all the materials it needs to continue the flow of life.

The Essentials of Cellular Reproduction

When a cell divides, it must transmit to its offspring cells two essential requirements for life (Fig. 9-1): hereditary information to direct life processes and materials in the cytoplasm that the offspring need to survive and to utilize their hereditary information.

Cell Division Transmits a Complete Set of Hereditary Information to Each Daughter Cell

The hereditary information of all living cells is **deoxyribonucleic acid (DNA)**. Like many large biological molecules, a molecule of DNA consists of a long chain of smaller subunits (see Chapter 3). DNA has four different types of subunits, called nucleotides. Segments of DNA a few hundred to many thousand nucleotides long are the units of inheritance—the genes—that carry the genetic information to produce specific traits. The sequence of nucleotides in a gene encodes information for the synthesis of the RNA and protein molecules that are needed to build a cell and carry out its metabolic activities. We will see in Chapters 10 through 13 how DNA encodes genetic instructions and how a cell regulates which genes it uses at any given time.

For any cell to survive, it must have a *complete set of genetic instructions*. Therefore, when a cell divides, it can-

not simply split its set of genes in half and give each daughter cell half a set. Rather, the cell must first *duplicate its DNA*, much like making a photocopy of an instruction manual. Each daughter cell then receives a complete "DNA manual" containing all the genes.

Cell Division Transmits Essential Cytoplasmic Materials to Each Daughter Cell

Like the blueprints for a house, the i[n]
by DNA are useless without materials
newly formed cell must receive the m[e]
read its genetic instructions and to
enough to acquire new materials from
and to process them into new cellular
thermore, as we described in Chapter
thesize either mitochondria or chloro[p]
these organelles arise only by the div
existing mitochondria and chloroplas[ts]
cell divides, its cytoplasm is divided ab[ove]
the two daughter cells. This simple m[e]
provides both daughter cells with all [nu]
trients, enzymes, and other molecules

The Activities of a Cell from One [to]
the Next Constitute the Cell Cycle

Newly formed cells usually acquire n[utrients]
environment, synthesize more of their
grow larger. After a variable amount of
the organism, the type of cell, and the
the cell divides. This general descript[ion]
eukaryotic and prokaryotic cells. P[re]

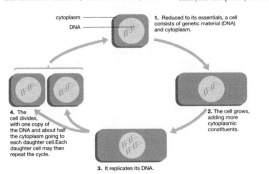

cytoplasm
DNA

1. Reduced to its essentials, a cell consists of genetic material (DNA) and cytoplasm.

2. The cell grows, adding more cytoplasmic constituents.

3. It replicates its DNA.

4. The cell divides, with one copy of the DNA and about half the cytoplasm going to each daughter cell. Each daughter cell may then repeat the cycle.

Figure 9-1   A generalized cell cycle.

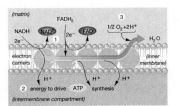

(matrix)
NADH   NAD⁺   FADH₂   FAD $1/2\ O_2 + 2H^+$ H_2O
$2e^-$ electron carriers   (inner membrane)
H^+ energy to drive   ATP   synthesis H^+
(intermembrane compartment)

Figure 8-6   The electron transport system of mitochondria

(1) The electron carrier molecules NADH and FADH₂ deposit their energetic electrons with the carriers of the transport system located in the inner membrane. (2) The electrons move from carrier to carrier within the transport system. Some of their energy is used to pump hydrogen ions across the inner membrane from the matrix into the intermembrane compartment. This movement out of the inner membrane creates a hydrogen ion gradient that can be used to drive ATP synthesis (for details, see "A Closer Look at Chemiosmosis in Mitochondria"). (3) At the end of the electron transport system, the energy-depleted electrons combine with oxygen and hydrogen ions in the matrix to form water.

inner membrane, into the intermembrane compartment (see the section on chemiosmosis, below).

Finally, at the end of the electron transport system, oxygen and hydrogen ions accept the energetically depleted electrons: Two electrons, one oxygen atom, and two hydrogen ions combine to form water. This step clears out the transport system, leaving it ready to run more electrons through. Without oxygen, the electrons would "pile up" in the transport system, with the result that the hydrogen ions would not be pumped across the inner membrane. The hydrogen ion gradient would soon dissipate and ATP synthesis would stop.

Chemiosmosis Captures Energy Stored in a Hydrogen Ion Gradient and Produces ATP

Hydrogen ion pumping across the inner membrane generates a large concentration gradient; that is, a high concentration of hydrogen ions in the intermembrane compartment and a low concentration in the matrix. The inner membrane is impermeable to hydrogen ions except at protein pores that are part of ATP-synthesizing enzymes. In the process known as **chemiosmosis**, hydrogen ions move down their concentration gradient from the intermembrane compartment to the matrix through these ATP-synthesizing enzymes. The flow

of hydrogen ions provides the energy to synthesize 32 to 34 molecules of ATP from ADP and inorganic phosphate in the matrix. The box "A Closer Look at Chemiosmosis in Mitochondria" examines chemiosmosis in more detail.

ATP Is Transported from the Mitochondrial Matrix and Diffuses into the Cell Cytosol

The ATP that was synthesized from ADP and inorganic phosphate in the matrix during chemiosmosis is transported across the inner membrane from the matrix to the intermembrane compartment. It then diffuses out of the mitochondrion to the cytosol through the outer membrane, which is very permeable to ATP. These ATP molecules provide most of the energy needed by the cell. ADP simultaneously diffuses from the cytosol across the outer membrane and is transported across the inner membrane to the matrix, replenishing the supply of ADP.

SUMMARY

ELECTRON TRANSPORT AND CHEMIOSMOSIS

Electrons from the electron carriers NADH and FADH₂ enter the electron transport system of the inner mitochondrial membrane. Here their energy is used to generate a hydrogen ion gradient. Movement of hydrogen ions down their gradient through the pores of ATP-synthesizing enzymes drives the synthesis of 32 to 34 molecules of ATP. At the end of the electron transport system, two electrons combine with one oxygen atom and two hydrogen ions to form water.

Cellular Metabolism Influences the Way Entire Organisms Function

Many students feel that the details of cellular metabolism are hard to learn and don't really help them to understand the living world around them. Have you ever read a murder mystery and wondered how cyanide could kill a person almost instantly? Cyanide reacts with one of the proteins in the electron transport system, immediately blocking the movement of electrons through the system and bringing cellular respiration to a screeching halt. Under more normal conditions, metabolic processes within individual cells have enormous impacts on the functioning of entire organisms. To take just two familiar examples, let's consider the migration of ruby-throated hummingbirds and track events in the Olympics.

Hummingbird Migration Requires Efficient Energy Storage and Use

As the days shorten in August, many birds in North America prepare to migrate to Central and South

Key terms

Key terms are boldfaced throughout.

Visualize Key Concepts

If you can visualize something related to a concept, you'll remember it more easily. Our art and photographic program will help you grasp concepts, so study the pictures as well as the text.

Micro to Macro Figures

To understand biology you must be able to visualize structures at several different scales. **Micro to macro figures** help you bridge this gap.

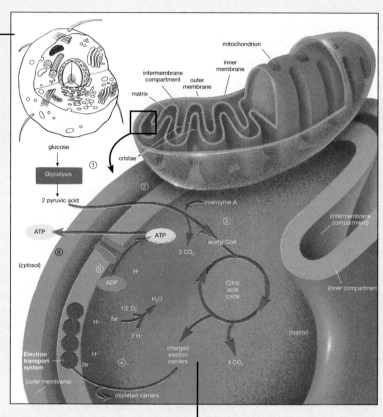

Compound Art

Compound art brings together different views of a topic.

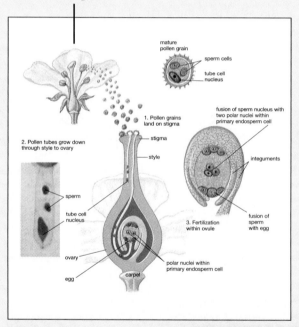

Process Art

Process art conveys key steps in processes, such as celluar respiration, as shown here.

Dynamic Photographs

Dynamic photographs provide vivid illustrations of concepts throughout.

Relate Concepts to the World Around You

Think about biology in the world around you. You have come into this course already knowing a lot about life. The concepts you are learning will give you a better understanding of your own life and of the world around you. Applications and stories are an important part of this text, and we hope that these features will spark your curiosity and make the study of biology as fascinating to you as it is to us.

EARTH WATCH

Has the Human Population Exceeded Earth's Carrying Capacity?

A glance at the age structure of less-developed countries, where most of the world's population resides, shows a tremendous momentum for continued growth. World population in the year 2010 is predicted and growing. A modest United Nation is that the human population may 2150 at 11.5 billion (Fig. E43-3). Ca twice its current population?

Earlier we defined carrying capacity ulation that could be indefinitely sus population requires that the ecosystem ways that lower its ability to provide no this definition we have already exceede pacity for people. The upper limit of capacity is determined by the ability o vest the energy from sunlight and prod ecules that organisms can use as food. biologist Peter Vitousek estimates that I already reduced the productivity of Ea lands by 12%. Each year, millions of a tive land are being turned into desert and deforestation, especially in less-o Earth's desert area is projected to inc year 2000 as a result of human activit

In the years between 1990 and 2010, the world population is projected to increase by 33%. During this same period, the available fish, cropland, pastureland, and

Earth Watch

Earth Watch essays present environmental issues such as biodiversity, ozone depletion, and population growth.

HEALTH WATCH

The Placenta Provides Only Partial Protection

When your grandparents were bearing children, doctors assumed that the placenta protected the developing fetus from most of the substances in maternal blood that could harm it. We know now that this is far from true. In fact, most medications and drugs and even some disease organisms readily penetrate the placental barrier and affect the fetus.

Infections May Cross the Placenta
The German measles virus can cross the placenta and attack the fetus, causing potentially severe retardation and other defects. As mentione ing genital herpes (during terium causing syphilis c defects in the developing may also cross the placent this incurable disease.

Drugs Readily Cross the
A tragic example of drugs c quilizer thalidomide, comr the early 1960s (Fig. E40-effects on embryos were di bies were born with missing More recently, in the late cutane was found to cause g to women using it (Accutan

Health Watch

Health Watch essays investigate issues such as prenatal diagnosis, sexually transmitted diseases, and osteoporosis.

A CLOSER LOOK

The Fate of Fats

Because digestion occurs in a watery environment using water-soluble enzymes, the digestion of water-insoluble lipids poses a particular challenge: Fats in chyme tend to aggregate into globs that resist the attack of digestive enzymes. Bile salts, secreted by the liver and released by the gallbladder, provide the first level of attack. If you have ever poured detergent into a dishpan with oily water and watched the oil disperse, you've witnessed an effect very comparable to the action of bile salts on lipids. Like your dish detergent, bile salts are molecules with both hydrophilic (literally, "water-loving," or lipid-soluble) and hydrophobic ("water-fearing," or lipid-soluble) portions. The hydrophobic end dissolves in the lipid while the hydrophilic end dissolves in the surrounding watery fluid. As illustrated in Fig. E32-1, this structure enables bile salts to disperse the fat globs into microscopic particles ①. These particles expose a much greater surface area to the surrounding fluid, which is rich in pancreatic lipases (lipid-

digesting enzymes). As the lipases break triglycerides into fatty acids and monoglycerides, these subunits are absorbed into the lipid-soluble portion of tiny clusters of bile salts called **micelles** ②. The micelles carry the fatty acids and monoglycerides to cells lining the intestine. There, the fatty acids and monoglycerides diffuse through the cell membranes into the intestinal cells, leaving the bile salts behind ③.

Within the intestinal cells, the fat subunits are reassembled into triglycerides, mixed with cholesterol and phospholipids, and collected into droplets coated with protein ④. These droplets, called **chylomicrons**, are packaged into vesicles by the Golgi apparatus and expelled by exocytosis into the interstitial fluid ⑤. Too large to enter capillaries, chylomicrons diffuse into the lacteal, or lymph vessel, that projects into each intestinal villus. The fat is then carried in the lymph to large veins into which the lymphatic vessels empty their contents.

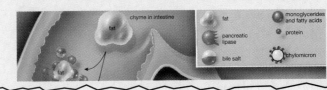

A Closer Look

A Closer Look essays treat selected topics in more depth, such as glycolysis, cycles in predator and prey populations, polymerase chain reaction.

The New York Times Themes of the Times

The New York Times Themes of the Times supplement brings together a collection of recent biology articles from this award-winning newspaper. Can you relate what you are learning in class to the issues discussed in recent articles?

The ABC News/Prentice Hall Video Library

The ABC News/Prentice Hall Video Library makes available brief clips from programs like *20/20*, *World News Tonight*, and *Nightline*. Will knowing more about the science behind the issues change the way you view the nightly news?

Understand the Process of Science

The processes by which scientists answer questions are as important as the questions themselves. *Life on Earth* emphasizes how we know what we know and gives you insight into the dynamic process of discovery.

Scientific Inquiry

Scientific Inquiry essays examine the people, the processes, and the methods of biology with topics such as DNA fingerprinting in forensics, Nancy Wexler on the trail of Huntington's disease, the use of isotopes in biology and medicine.

SCIENTIFIC INQUIRY

DNA Fingerprinting in Forensics—The Case of the Telltale Palo Verde

The polymerase chain reaction (PCR) procedure has caused a revolution in many fields of biology, but nowhere has the impact of this technique been felt more than in forensics, the collection of information to be used as evidence in court proceedings. The technique that is proving most useful has been nicknamed "DNA fingerprinting," because it may generate patterns of DNA fragments unique to each individual. This procedure involves the amplification of a set of DNA fragments of unknown sequence, a set that is likely to differ from one individual to the next. When these fragments are amplified and separated from one another on a gel, the pattern of bands is referred to as a DNA fingerprint. In theory, it represents a characteristic "snapshot" of the unique genome of an individual, allowing it to be distinguished from that of every other individual in the population. The quantity of DNA needed to analyze using PCR is exceedingly small; this amount of DNA can easily be obtained from a single spot of dried blood, from the cells clinging to the base of a hair shaft, from skin fragments under a victim's fingernails, or other small tissue samples collected at a crime scene.

Here is an example of the forensic use of DNA fingerprints that did not involve the suspect's DNA, but rather that of plants at the crime scene. On the night of May 2,

verdes show considerable genetic variation from one tree to the next. Otherwise, it would be impossible to unambiguously link a given pod to any particular tree. Fortunately, he found considerable differences in DNA between individual plants, allowing for discriminatory fingerprints.

Helentjaris then was given the two seed pods from the suspect's truck along with pods collected by investigators from 12 different palo verde trees near the factory. The investigators knew which of the 12 trees was near the body but did not tell Helentjaris. He extracted DNA from seeds taken from each pod and performed the PCR reaction, producing 10 to 15 distinct bands from each pod. The result of this experiment was unmistakable: The pattern from the pod found in the truck exactly matched the pattern from only one of the 12 trees, the one nearest the body. In an important additional test, Helentjaris found that this pattern was also different from that of 18 pods collected from trees at random sites around Phoenix. This result was admitted as evidence in the trial, the first time a DNA fingerprint of a plant had been used in a court case. This information was crucial in demolishing the suspect's alibi, and at the end of the 5-week trial the suspect was found guilty of first-degree murder. Riven the two seed pods from the suspect's truck along with pods col-

See the Connections

There are important patterns and themes that unify the discipline of biology. *Evolution* and *adaptation to the environment* are two related themes you'll revisit throughout the text. How do you see these themes at work?

Evolutionary Connections

Evolutionary Connections close many of the chapters, tying the subject matter of each chapter into the broader perspective of evolution. You'll also notice that many of the chapter introductions have an evolutionary theme.

lessly and frequently, such as leaves, songbirds, and geese. Predators are much less common, and the novel shape of a hawk still elicits instinctive crouching. Thus, learning modifies the innate response, making it more adaptive.

Trial-and-error learning can also result in more appropriate responses to releasers for fixed action patterns, as in the case of our bee-catching toad (see Fig. 41-14). The naive toad instinctively snaps at all flying insects of appropriate size, but from painful experience it learns to make certain exceptions!

EVOLUTIONARY CONNECTIONS

Why Do Animals Play?

Pigface, a giant 50-year-old African softshell turtle, spends hours each day batting a ball around his aquatic home in the National Zoo in Washington, D.C., to the delight of thousands of visitors and the puzzlement of behavioral biologists. Play has always been somewhat of a mystery. It has been observed in many birds and in most mammals, but, until zookeepers tossed Pigface a ball a few years ago, it had never been seen in animals as evolutionarily ancient as reptiles.

Animals at play are fascinating. Pygmy hippopotamuses push one another, shake and toss their heads, splash in the water, and pirouette on their hind legs. Otters delight in elaborate acrobatics. Bottlenose dolphins balance fish on their snouts, throw objects, and carry them in their mouths while swimming. Even baby vampire bats have been observed chasing, wrestling, and slapping each other with their wings. Solitary play often involves a single animal manipulating an object, like a cat with a ball of yarn, or the dolphin with its fish, or as seen on our cover, a macaque monkey making and playing with a snowball. Play may also be social. Often young of the same species play together, but parents may join them (Fig. 41-17a). Social play often includes chasing, fleeing, wrestling, kicking, and gentle biting (Fig. 41-17b, c).

What are the features of play? (1) Play seems to lack any clear goal. (2) Play is abandoned in favor of escaping from danger, feeding, and courtship. (3) Play seems to involve feelings of pleasure. (4) Young animals play more frequently than adults. (5) Play often involves movements borrowed from other behaviors (attacking, fleeing, stalking, etc.). (6) Play uses considerable energy. (7) Play is potentially dangerous. Young humans and other animals are frequently injured, and occasionally killed, during play. In addition, play may distract the animal from the presence of danger while making it conspicuous to predators. So why do animals play?

The only logical conclusion is that play must have survival value and that natural selection has favored those

Figure 41-17 Young animals at play

(a) A bonobo mother plays with her young. (b) Young polar bears wrestle on the ice of Hudson Bay, Canada. (c) Young warthogs in Kenya engage in a friendly shoving match.

Check Your Understanding.
Expand it. Explore!

✖ SUMMARY OF KEY CONCEPTS

The Composition of Chromosomes
Eukaryotic chromosomes are composed of DNA and protein. DNA is the molecule that carries genetic information. It is composed of subunits called nucleotides, linked together into long strands. Each nucleotide consists of a phosphate group, the five-carbon sugar deoxyribose, and a nitrogen-containing base. Four different bases occur in DNA: adenine, guanine, thymine, and cytosine.

DNA: The Hereditary Molecule
Transformation in bacteria and infection of bacteria by bacteriophages provided the first evidence that DNA is the hereditary molecule. Griffith discovered that living bacteria can acquire genetic material from dead bacteria, transforming the live bacteria. Avery and his colleagues showed that the genes taken up by living bacteria during transformation are composed of DNA.

A bacteriophage consists solely of DNA and protein. When it infects a bacterium, a phage injects its genetic material into the bacterium, where it directs the synthesis of more phages. Hershey and Chase demonstrated that phages inject DNA, but not protein, into bacteria. Therefore, DNA must be the genetic material of phages.

The Structure of DNA
The DNA of chromosomes is composed of two strands wound about one another in a double helix. The sugars and phosphates that link one nucleotide to the next form the backbone on each side of the double helix, while the bases from each strand pair up in the middle of the helix. Only specific pairs of bases, called complementary base pairs, can link together in the helix, held by hydrogen bonds: adenine with thymine and guanine with cytosine.

DNA Replication: The Key to Constancy
When a chromosome is replicated before mitosis or meiosis, the two DNA strands of each double helix unwind. DNA polymerase enzymes move along each strand, linking up free nucleotides into new DNA strands. The sequence of nucleotides in each newly formed strand is complementary to the sequence on a parental strand. As a result, two double helices are synthesized, each consisting of one parental DNA strand plus one newly synthesized, complementary strand that is an exact copy of the other parental strand. The two daughter DNA molecules are therefore duplicates of the parental DNA molecule.

✖ KEY TERMS

bacteriophage p. 181
base p. 178
complementary base pair p. 185
deoxyribonucleic acid (DNA)
    p. 177

DNA polymerase p. 188
double helix p. 184
helix p. 184
nucleotide p. 178
purine p. 178

pyrimidine
replication fo
semiconservat
    p. 186
transformatio

✖ THINKING THROUGH THE CO

Multiple Choice
1. If you wanted to label amino acids but not DNA, which of the following radioactive isotopes would you use?
   a. ¹⁸F   b. ³⁵S   c. ¹⁴C
   d. ³H   e. ³²P
2. Whose pioneering work on the gene used the bacteriophage to show that DNA was the part of this virus which went into the bacterial cell or replication?
   a. Watson and Crick   b. Avery   c. Hershey and Chase
   d. Griffith   e. Beadle and Tatum
3. Knowing what you now know about RNA, DNA, and DNA replication from this chapter and mitosis from the last chapter, sister chromatids
   a. each have one single strand of the two complementary DNA strands
   b. each have a fully formed double helix of DNA
   c. are the same as the homologous chromosomes of a diploid cell
   d. are present in bacteriophages, bacteria, and eukaryotes
   e. none of the above

4. How many different possible ba nucleotide chain 3 bases long?
   a. 1   b. 3
   c. 9   d. 64
   e. more than 64
5. Chargaff, in analyzing DNA, fo
   a. cytosine equalled that of guan
   b. cytosine equalled that of thym
   c. cytosine equalled that of aden
   d. each nucleotide was unrelated
   e. each nucleotide was equal to
6. Semiconservative replication re
   a. genetic information is contain molecule
   b. only the DNA of the bacteri cell
   c. certain bases pair with specifi
   d. one DNA strand remains whi
   e. point mutations sometimes in replication

Review Questions
1. Draw the structure of a nucleotide. Which parts are identical in all nucleotides, and which can vary?
2. Name the four types of nitrogen-containing bases found in DNA. Which bases are complementary to one another? How are they held together in the double helix of DNA?
3. What is transformation? In Griffith's experiments, why were the transformed bacteria able to synthesize capsules?
4. Diagram the life cycle of a bacteriophage.
5. Describe the structure of DNA in a chromosome. Where are the bases, sugars, and phosphates in the structure?
6. Describe the process of DNA replication.

✖ APPLYING THE CONCEPTS

1. Diagram and describe the Hershey-Chase experiment. Suppose they had found that *both radioactively labeled DNA and radioactively labeled protein* were injected into the bacteria. What would they have been able to conclude from this? (Note that if the bacteriophage chromosome was like the eukaryotic chromosome, that is, if it contained both DNA and protein, this is exactly what would have happened.)
2. A preview question for Chapter 11: Genetic information is encoded in the sequence of nucleotides in DNA. Let's suppose that the nucleotide sequence on one strand of a double helix encodes the information needed to synthesize a hemoglobin molecule. Do you think that the sequence of nucleotides on the other strand of the double helix would also encode useful information? Why or why not? An analogy to think about: Suppose that English were a "complementary language," with letters at opposite ends of the alphabet complementary to one another (for example, A is complementary to Z, B to Y, C to X, etc.). Would the complementary sentence to "All the world's a stage" make sense?
3. Suppose mammalian cells are grown and divide many times in a culture medium containing thymine made radioactive with a form of hydrogen (³H). The cells are then removed

from the radioactive medium and allowed to replicate several times in normal culture medium. Daughter chromosomes are tested each generation to determine whether they contain radioactive thymine. While in the radioactive medium, every strand of DNA of each chromosome has radioactive thymine. Assuming that each chromosome contains a single long double-stranded molecule of DNA, predict the radioactive status of daughter chromosomes after one, two, and three rounds of cell division in normal medium. Explain how your predictions are consistent with the Watson-Crick explanation of semiconservative DNA replication.
4. During the Inquisition, the church forced Galileo to disavow some of his scientific statements. Today, scientific advances are being made at an astounding rate, and nowhere is this more evident than in our understanding of the biology of heredity. Using DNA as a starting point, do you think there are limits to the knowledge people should acquire?
5. Historians debate the degree to which key people actually changed the course of history. From this chapter's historical perspective, discuss whether the pace or actual progress of knowledge is influenced by the individual scientists involved.

✖ FOR MORE INFORMATION

Crick, F. H. C. "The Structure of the Hereditary Material." *Scientific American*, October 1954. It is entertaining to consider a paper on one of science's most fundamental results while the finding was still new and before the author won the Nobel Prize for Physiology and Medicine.
Holliday, R. "A Different Kind of Inheritance." *Scientific American*, June 1989. Methyl groups attached to DNA may alter its use in a cell. What's more, patterns of DNA methylation may be passed down from parent to offspring.
Judson, H. F. *The Eighth Day of Creation: Makers of the Revolution in Biology*. New York: Simon and Schuster, 1979. A remarkably readable history of the beginnings of molecular biology.
Kimura, M. "The Neutral Theory of Molecular Evolution." *Scientific American*, November 1979. Kimura persuasively argues that some evolutionary changes can occur without selection.

Mirsky, A. E. "The Discovery of DNA." *Scientific American*, June 1968. The early history of DNA research.
Radman, M., and Wagner, R. "The High Fidelity of DNA Duplication." *Scientific American*, August 1988. Faithful duplication of chromosomes requires both reasonably accurate initial replication of DNA sequences and final proofreading.
Rennie, J. "DNA's New Twists." *Scientific American*, March 1993. Relevant to this and several subsequent chapters, this simple review covers some of the interesting modern contradictions which go beyond the standard textbook dogma.
Watson, J. D. *The Double Helix*. New York: Atheneum Publishers, 1968. If you still believe the Hollywood images, that scientists are either maniacs or cold-blooded, logical machines, be sure to read this book. Although scarcely models for the behavior of future scientists, Watson and Crick are certainly human enough!

✖ NET WATCH

On-line resources for this chapter are on the World Wide Web at:
http://www.prenhall.com/~audesirk (click on the table of contents link and then select Chapter 10).

Summary of Key Concepts
Summary of Key Concepts helps you begin your review.

Applying the Concepts
Applying the Concepts questions stretch your learning and ask you to apply your critical thinking skills to real-world scenarios.

Thinking Through the Concepts
Thinking Through the Concepts questions check your basic understanding. Both multiple-choice and short-answer questions are included.

Net Watch
Net Watch icons highlight the Internet's World Wide Web address of the *Life on Earth* Home Page where you can find resources and activities to aid your study and broaden your understanding of biology. If you have access to the World Wide Web (either though school or through a service like America Online) meet us on the Net! We'll be there as of January 1996.

For More Information
For More Information provides extra readings from books and the latest periodicals on chapter-related topics.

Brief Contents

1   An Introduction to Life on Earth   1

UNIT I   THE LIFE OF THE CELL   19

2   Atoms, Molecules, and Life   20
3   Biological Molecules   36
4   Energy Flow in the Life of a Cell   56
5   Cell Structure and Function   74
6   Cell Membrane Structure and Function   104
7   Photosynthesis: Capturing Solar Energy   126
8   Harvesting Energy from Food: Glycolysis and Cellular Respiration   140

UNIT II   INHERITANCE   157

9   Cellular Reproduction   158
10   DNA: The Molecule of Heredity   176
11   Gene Expression and Regulation   192
12   Gene Exchange, Meiosis, Eukaryotic Life Cycles   218
13   Patterns of Inheritance   234
14   Molecular Genetics and Biotechnology   260
15   Human Genetics   280

UNIT III   EVOLUTION   301

16   Principles of Evolution   302
17   How Organisms Evolve   322
18   The Origin of Species   346
19   The History of Life on Earth   362
20   Taxonomy: Imposing Order on Diversity   386
21   The Hidden World of Microbes   398
22   The Fungi   424
23   The Plant Kingdom   438
24   The Animal Kingdom   458

UNIT IV   PLANT ANATOMY AND PHYSIOLOGY   495

25   The Structure of Land Plants   496
26   How Land Plants Acquire and Transport Nutrients   518

27   Plant Reproduction and Development   534
28   Plant Responses to the Environment   556

UNIT V   ANIMAL ANATOMY AND PHYSIOLOGY   571

29   Homeostasis and the Organization of the Animal Body   572
30   Circulation   584
31   Respiration   604
32   Nutrition and Digestion   618
33   The Urinary System and Homeostasis   638
34   Defenses Against Disease: The Immune Response   650
35   Chemical Control of the Animal Body: The Endocrine System   678
36   Information Processing: The Nervous System   696
37   The Senses: Perception   726
38   Action and Support: The Muscles and Skeleton   746
39   Animal Reproduction   762
40   Animal Development   788

UNIT VI   ANIMAL BEHAVIOR   809

41   The Foundations of Animal Behavior   810
42   The Social Behavior of Animals   826

UNIT VII   ECOLOGY   847

43   Population Growth and Regulation   848
44   Community Interactions   870
45   How Ecosystems Work   890
46   Earth's Diverse Ecosystems   914

Essays

EARTH WATCH

Biodiversity Is Threatened by Human Activities — 14
Biotechnology Attacks the Chestnut Blight — 276
Natural Selection, Genetic Diversity, and Endangered
Species — 341
The Case of the Disappearing Mushrooms — 429
Amphibians in Decline — 486
Plants Help Regulate the Distribution of Water — 530
Disrupting Coevolutionary Relationships Threatens
Ecosystems — 550
Has the Human Population Exceeded Earth's
Carrying Capacity? — 864
Introduced Species Disrupt Community Interactions — 884
The Ozone Layer—We Have Punctured Our
Protective Shield — 920
Humans and Ecosystems — 942

HEALTH WATCH

Metabolic Transformation—Why You Can Get Fat
by Eating Sugar — 152
"Unnatural Selection"—The Evolution of
Drug-Resistant Pathogens — 409
Cardiovascular Disorders — 598
Smoking—A Life and Breath Decision — 614
Ulcers—When the Digestive Tract Digests Itself — 635
When the Kidneys Collapse — 647
Flu—The Unbeatable Bug — 672
Osteoporosis—The Hidden Crippler — 758
Sexually Transmitted Diseases — 784
The Placenta Provides Only Partial Protection — 804

EVOLUTIONARY CONNECTIONS

Natural Selection Has Produced Cell Membranes with
Diverse Compositions — 122
Kin Selection and the Evolution of Altruism — 342
Why Do Humans Walk Upright — 382
Classification Conundrums; or, Where Shall We
Put the Algae? — 393
Our Unicellular Ancestors — 420
Fungal Ingenuity — 435
Are Humans a Biological Success? — 490
The Evolution of Hormones — 693
Natural Selection Shapes Perception — 740
Why Do Animals Play? — 822
Altruism, Kin Selection, and the Selfish Gene — 843

A CLOSER LOOK

Protein Structure—A Hairy Subject — 48
The "Essentially Foreign Creatures" within Us—
The Evolution of Chloroplasts and Mitochondria — 92
Chemiosmosis—The Mechanism of ATP Synthesis in
Chloroplasts — 133
Glycolysis — 144
The Mitochondrial Matrix Reactions — 149
Chemiosmosis in Mitochondria — 151
Rube Goldberg Genetics—Making New Proteins
from Old Parts — 212
The Polymerase Chain Reaction—Hot Springs and
Hot Science — 268
The Hardy-Weinberg Equilibrium Population — 328
Viral Replication — 404
Adaptations for Seed Dispersal — 548
Rapid-Fire Plant Responses — 568
Gills and Gas Exchange—Countercurrent Flow — 608
The Fate of Fats — 632
The Nephron and Urine Formation — 644
Cellular Communication during the Immune
Response — 654
Sticks and Stones—The Repair of Broken Bones — 756
Homeobox Genes and the Control of Body Form — 803
Cycles in Predator and Prey Populations — 854

SCIENTIFIC INQUIRY

Does Life Arise Spontaneously? — 10
The Use of Isotopes in Biology and Medicine — 32
Viewing the Cell—A Gallery of Microscopic Images — 78
The Discovery of the Double Helix — 189
Cracking the Genetic Code — 198
DNA Fingerprinting in Forensics—The Case of the
Telltale Palo Verde — 274
In Search of the Huntington's Gene — 288
Collecting Fetal Samples and Analyzing Them Using
Biotechnology — 296
Modern Molecular Genetic Techniques Shed Light on
Evolutionary Relationships — 318
Radiometric Dating Traces Fossils through Time — 368
Scanning the Brain — 720
In Vitro Fertilization — 777
Robot Cricket Finds Her Mate — 814
Deciphering the Relationship between Ants and
Acacias — 881

Contents

Chapter 1 An Introduction to Life on Earth 1

The Characteristics of Living Things 1
Living Things Are Both Complex and Organized 2
Living Things Must Acquire and Use Materials and Energy 4
Homeostasis Maintains Relatively Constant Internal Conditions 5
Growth Is a Property of All Living Organisms 5
Living Things Respond to Stimuli 5
Living Things Reproduce Themselves 5
Living Things Have the Capacity to Evolve 6
The Diversity of Life 6
Members of the Kingdom Monera Are Prokaryotic Cells; Members of All Other Kingdoms Are Composed of Eukaryotic Cells 8
Members of the Kingdoms Monera and Protista Are Mostly Unicellular; Members of the Kingdoms Fungi, Plantae, and Animalia Are Primarily Multicellular 8
Members of the Different Kingdoms Have Different Ways of Acquiring Energy 8
Biology: The Science of Life 8
Scientific Principles Underlie All Scientific Inquiry 8
The Scientific Method Is the Basis for Scientific Inquiry 9
SCIENTIFIC INQUIRY: DOES LIFE ARISE SPONTANEOUSLY? 10
Science Is a Human Endeavor 11
Scientific Theories Have Been Thoroughly Tested 11
Evolution: The Unifying Concept of Biology 12
Three Natural Processes Underlie Evolution 12
Knowledge of Biology Illuminates Everyday Life 14
EARTH WATCH: BIODIVERSITY IS THREATENED BY HUMAN ACTIVITIES 14
Review Section 16

UNIT 1 THE LIFE OF A CELL 19

Chapter 2 Atoms, Molecules, and Life 20

Matter and Energy 22
Energy Exists in Several Forms That Can Be Interconverted 22
The Structure of Matter 22
Atoms, the Smallest Units of Elements, Are Composed of Still Smaller Particles 22
Atoms Will React with Other Atoms Only When There Are Vacancies in Their Outermost Electron Shells 24
Chemical Bonds: Joining Atoms to Make Molecules 25

Atoms Can Become Stable by Gaining or Losing Electrons, Forming Charged Ions; The Attraction between Oppositely Charged Ions Produces Ionic Bonds 25
Atoms Can Become Stable by Sharing Electrons with Other Atoms, Forming Covalent Bonds 26
Hydrogen Bonds Are Weak Electrical Attractions between the Polar Parts of Molecules 26
Organic and Inorganic Molecules 27
Water and Life 28
Water Interacts with Many Other Molecules 28
The pH of a Solution Is a Measure of Its Concentration of Hydrogen Ions 30
A Buffer Helps Maintain a Solution at a Relatively Constant pH 30
Water Moderates the Effects of Temperature Changes 30
Water Forms an Unusual Solid, Ice 31
SCIENTIFIC INQUIRY: THE USE OF ISOTOPES IN BIOLOGY AND MEDICINE 32
Water Molecules Tend to Stick Together 33
Review Section 34

Chapter 3 Biological Molecules 36

Synthesizing Organic Molecules: A Modular Approach 38
Biological Molecules Are Joined Together or Broken Apart by Adding or Removing Water 38
The Principal Types of Biological Molecules 38
Carbohydrates Are Sugars and Starches 40
Lipids Are Diverse Molecules That Are Insoluble in Water 42
Proteins Play Many Roles in Living Organisms 44
A CLOSER LOOK: PROTEIN STRUCTURE—A HAIRY SUBJECT 48
Nucleic Acids Form DNA and RNA, the Molecules of Heredity 52
Review Section 53

Chapter 4 Energy Flow in the Life of a Cell 56

Energy and the Ability to Do Work 57
The Laws of Thermodynamics Describe the Basic Properties of Energy 58
Living Things Use the Energy of Sunlight to Create the Low-Entropy Conditions Characteristic of Life 59
Energy Flow in Chemical Reactions 59
Exergonic Reactions Release Energy 60
Endergonic Reactions Require an Input of Energy 60
Coupled Reactions Link Exergonic with Endergonic Reactions 61
Chemical Reactions Are Reversible 61
Controlling the Metabolism of Living Cells 63
At Body Temperatures, High Activation Energy Makes Spontaneous Reactions Proceed Too Slowly to Sustain Life 64
Catalysts Reduce Activation Energy 64
Enzymes Are Biological Catalysts 65
Coupled Reactions and Energy-Carrier Molecules 68
ATP Is the Principal Energy Carrier in a Cell 68
Electron Carriers Also Transport Energy Within Cells 70
Review Section 71

Chapter 5 Cell Structure and Function 74

The Development of Cell Theory 75
An Overview of Cell Structure and Function 76
The Plasma Membrane Isolates the Cell But Allows Interactions with the Environment 76
The Genetic Material Provides a Cellular "Blueprint" 76
SCIENTIFIC INQUIRY: VIEWING THE CELL—A GALLERY OF MICROSCOPIC IMAGES 78
Cytoplasm Fills Each Cell and Surrounds the Nucleus 80
Cell Function Limits Cell Size 80
Types of Cells: Prokaryotic and Eukaryotic 81
Prokaryotic Cells Are Relatively Simple 81
Eukaryotic Cells Are More Complex 82
The Nucleus: Control Center of the Cell 82
The Nuclear Envelope Allows Selective Exchange of Materials 82
Chromatin Consists of DNA and Its Associated Proteins 82
The Nucleolus Is the Site of Ribosome Assembly 85
The Membrane System of the Cell 87
Membranes Consist of a Double Layer of Phospholipids in Which Proteins Are Embedded 87
The Cell's Membrane System Includes the Plasma Membrane, Endoplasmic Reticulum, the Golgi Complex, and Lysosomes 87
Chloroplasts and Mitochondria: Energy Capture and Extraction 91
A CLOSER LOOK: THE "ESSENTIALLY FOREIGN CREATURES" WITHIN US—THE EVOLUTION OF CHLOROPLAST AND MITOCHONDRIA 92
Chloroplasts Are the Site of Photosynthesis 94
Mitochondria Produce ATP Using Energy Stored in Food Molecules 94
Plastids and Vacuoles: Storage and Elimination 95
Many Types of Plastids Store Food in Plants 95
Vacuoles Serve Many Functions, Including Support and Storage and Elimination of Food and Wastes 95
The Cytoskeleton: Shape, Support, and Movement 95
Microfilaments Allow Cells to Change Shape and Guide the Movement of Organelles 96
Intermediate Filaments Provide a Supportive Framework 98
Microtubules Help Position, Anchor, and Move Organelles, and Alter the Shapes of Cells 98
Cilia and Flagella Move the Cell or Move Fluid Past the Cell 100
Review Section 101

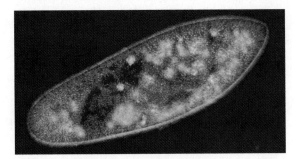

Chapter 6 Cell Membrane Structure and Function 104

Cell Walls 106
The Plasma Membrane 106
The Fluid Mosaic Model Describes the Structure of the Plasma Membrane 106
The Fluid Portion of Membranes Is Produced by a Double Layer of Phospholipids: The Phospholipid Bilayer 107
A Variety of Proteins in the Cell Membrane Form a Protein Mosaic 109
Transport across Membranes 109
Molecules in Fluids Move in Response to Gradients 110
Movement across Membranes Occurs by Both Passive and Energy-Requiring Transport 110
Passive Transport: Movement down Concentration Gradients 110
Molecules Diffuse from Areas of High Concentration to Areas of Low Concentration 110
Molecules Diffuse across Membranes Down Their Concentration Gradients 111
Osmosis Is the Diffusion of Water Across Membranes 113
Osmosis across the Cell Membrane Plays an Important Role in the Lives of Cells 114
Energy-Requiring Transport across Membranes 115
Active Transport Uses Energy to Move Molecules Across Membranes 115
Cells Engulf Particles or Fluids by Endocytosis 115
Transport across Intracellular Membranes 116
Vacuoles Help Regulate Cell Volume 118
Cell Connections and Communication 120
Desmosomes Attach Cells Together 120
Tight Junctions Leakproof the Cell 120
Gap Junctions and Plasmodesmata Allow Communication Between Cells 120
EVOLUTIONARY CONNECTIONS: NATURAL SELECTION HAS PRODUCED CELL MEMBRANES WITH DIVERSE COMPOSITION 122
Review Section 123

Chapter 7 Capturing Solar Energy: Photosynthesis 126

Photosynthesis: An Overview 128
Leaves and Chloroplasts Are Adaptations for Photosynthesis 128
The Light-Dependent Reactions: Converting Light to Chemical Energy 130
During Photosynthesis, Light Is First Captured by Pigments in Chloroplasts 130
The Light-Dependent Reactions Occur in Clusters of Molecules Called Photosystems 130
The Light-Independent Reactions: Securing Chemical Energy in Glucose Molecules 132
The C_3 Cycle Captures Carbon Dioxide 132
A CLOSER LOOK: CHEMIOSMOSIS—THE MECHANISM OF ATP SYNTHESIS IN CHLOROPLASTS 133
Carbon Fixed during the C_3 Cycle Is Used to Synthesize Glucose 134
The Relationship between the Light-Dependent and Light-Independent Reactions 134
Water, CO_2, and the C_4 Pathway 134
When Stomata Are Closed to Conserve Water, Wasteful Photorespiration Occurs 135
C_4 Plants Reduce Photorespiration Using a Two-Stage Carbon Fixation Process 135
C_3 and C_4 Plants Are Adapted to Specific Environmental Conditions 136
Review Section 137

Chapter 8 Harvesting Energy from Food: Glycolysis and Cellular Respiration 140

Glucose Metabolism: An Overview 141
Glycolysis 143
Glucose Must Be Activated by ATP Before Its Energy Can Be Harvested 143
A CLOSER LOOK: GLYCOLYSIS 144
Fermentation 145
Some Cells Ferment Pyruvic Acid to Lactic Acid 145
Other Cells Ferment Pyruvic Acid to Alcohol 146
Cellular Respiration 146
Pyruvic Acid Is Transported into the Mitochondrial Matrix 146
Pyruvic Acid Is Broken Down by Reactions in the Mitochondrial Matrix 148
Energetic Electrons Are Carried to Electron Transport Systems in the Inner Mitochondrial Membrane 148
A CLOSER LOOK: THE MITOCHONDRIAL MATRIX REACTIONS 149
Chemiosmosis Captures Energy Stored in a Hydrogen Ion Gradient and Produces ATP 150
ATP Is Transported from the Mitochondrial Matrix and Diffuses into the Cell Cytosol 150
Cellular Metabolism Influences the Way Entire Organisms Function 150
A CLOSER LOOK: CHEMIOSMOSIS IN MITOCHONDRIA 151
HEALTH WATCH: METABOLIC TRANSFORMATION— WHY YOU CAN GET FAT BY EATING SUGAR 152
Review Section 153

UNIT II INHERITANCE 157

Chapter 9 Cellular Reproduction 158

The Essentials of Cellular Reproduction 160
Cell Division Transmits a Complete Set of Hereditary Information to Each Daughter Cell 160
Cell Division Transmits Essential Cytoplasmic Materials to Each Daughter Cell 160
The Activities of a Cell from One Cell Division to the Next Constitute the Cell Cycle 160
Prokaryotic and Eukaryotic Chromosomes Compared 161
The Prokaryotic Cell Cycle 161
The Eukaryotic Chromosome 161
The Eukaryotic Chromosome Consists of a Single Strand of DNA Complexed with Proteins 161
Eukaryotic Chromosomes Often Occur in Pairs Containing Similar Genetic Information 162
The Eukaryotic Cell Cycle 163
Growth, Replication of Chromosomes, and Most Cell Functions Occur during Interphase 164

Cells Vary in the Duration of Interphase 164
Cell Division Consists of Nuclear Division and Cytoplasmic Division 164
Mitosis 166
During Prophase, the Chromosomes Condense and the Spindle Apparatus Forms and Attaches to the Chromosomes 166
During Metaphase, the Chromosomes Are Aligned along the Equator of the Cell 168
During Anaphase, Sister Chromatids Separate and Are Pulled to Opposite Poles of the Cell 168
During Telophase, Nuclear Envelopes Re-Form Around Each Group of Chromosomes 169
Cytokinesis 169
Cell Division and Asexual Reproduction 170
Review Section 172

Chapter 10 DNA: The Molecule of Heredity 176

The Composition of Chromosomes 177
DNA Is Composed of Four Types of Nucleotides 178
Is the Genetic Material DNA or Protein? 179
DNA: The Hereditary Molecule 179
Bacterial Transformation Indicates That the Hereditary Molecule of Bacteria Is DNA 179
The Hereditary Molecule of Bacteriophages Is Also DNA 181
The Sequence of Bases in DNA Can Encode Vast Amounts of Information 182
The Structure of DNA 182
The DNA of Chromosomes Is a Double Helix 184
DNA Replication: The Key to Constancy 185
The Steps of DNA Replication Are Catalyzed by the Coordinated Action of Several Enzymes 186
Base Pairing and Proofreading Produce Almost Error-Free Replication of DNA 188
SCIENTIFIC INQUIRY: THE DISCOVERY OF THE DOUBLE HELIX 189
Review Section 190

Chapter 11 Gene Expression and Regulation 192

The Relationship between Genes and Proteins 193
Most Genes Encode the Information for the Synthesis of a Protein 194
The Sequence of Bases in DNA Codes for the Sequence of Amino Acids in Proteins 195
Synthesizing Proteins from the Instructions in DNA 195
Genetic Information Flows in a Cell from DNA to RNA to Protein 196
The Genetic Code 196
The Genetic Code Uses Three Bases to Specify Each Amino Acid 196
RNA: Intermediary in Protein Synthesis 197
SCIENTIFIC INQUIRY: CRACKING THE GENETIC CODE 198
Transcription Produces RNA Molecules That Are Complementary Copies of One Strand of DNA 199
Three Types of RNA Cooperate in Protein Synthesis 199
Protein Synthesis 201
Mutations in DNA and Their Effects 204
Point Mutations Are Changes in a Single Base 205
Insertion and Deletion Mutations Result from Addition or Removal of Nucleotides 205

Mutations Differ in Their Effects on Protein Structure and Function 205
Mutations Provide the Raw Material for Evolution 207
Gene Regulation 207
Gene Regulation in Prokaryotes 208
Gene Regulation in Eukaryotes 209
Eukaryotic Genes Consist of DNA Segments That Code for the Amino Acid Sequence of Proteins Interrupted by Noncoding DNA Segments 209
A CLOSER LOOK: RUBE GOLDBERG GENETICS— MAKING NEW PROTEINS FROM OLD PARTS 212
Review Section 214

Chapter 12 Gene Exchange, Meiosis, and Eukaryotic Life Cycles 218

Genetic Variability, Genetic Exchange, and the Evolution of Sexual Reproduction 220
Mutations in DNA Are the Ultimate Source of Genetic Variability 220
The Effects of a Mutation Depend on the Nature of the Mutation, the Organism in Which It Occurs, and the Environment in Which the Organism Lives 220
Genetic Exchange May Combine Useful Mutations 220
Eukaryotic Organisms Use a Specialized Cell Division Process Called Meiosis to Combine Genetic Material from Two Separate Parents in a Single Offspring 220
"Permanent" Diploidy Protects against Some of the Harmful Effects of Mutations 221
Meiosis and Sexual Reproduction 222
Meiosis Separates Homologous Chromosomes in a Diploid Cell to Produce Haploid Daughter Cells Containing One Copy of Each Type of Chromosome 222
The Mechanisms of Meiosis 222
Meiosis I 222
Meiosis II 227
Mitosis, Meiosis, and Eukaryotic Life Cycles 227
In Haploid Life Cycles, the Majority of the Cycle Consists of Haploid Cells 228
In Diploid Life Cycles, the Majority of the Cycle Consists of Diploid Cells 228
In Alternation-of-Generations Life Cycles, There Are Both Diploid and Haploid Multicellular Stages 228
The Roles of Meiosis and Sexual Reproduction in Producing Genetic Variability 229
Review Section 230

Chapter 13 Patterns of Inheritance 234

Gregor Mendel and the Origin of Genetics 236
Doing It Right: The Secrets of Mendel's Success 237
Inheritance of Single Traits: The Law of Segregation 238

The Inheritance of Dominant and Recessive Alleles on Homologous Chromosomes Can Explain the Results of Mendel's Monohybrid Crosses 239
Mendel's Hypothesis Can Be Used to Predict the Outcome of New Types of Single-Trait Crosses 240
Inheritance of Multiple Traits: Independent Assortment 241
Inheritance Patterns of Genes Located on the Same Chromosome 242
Genes on the Same Chromosome Tend to Be Inherited Together 243
Crossing Over May Separate Linked Genes 243
The Frequency of Crossing Over Between Linked Genes Can Be Used to Map the Location of Genes on a Chromosome 246
The Inheritance of Sex and Sex-Linked Genes 246
Sex Linkage Is a Special Case of Linkage between Genes on the Same Chromosome 247
Variations on the Mendelian Theme 249
In Incomplete Dominance, the Phenotype of Heterozygotes Is Intermediate Between the Phenotypes of the Homozygotes 249
There May Be Multiple Alleles of a Single Gene 249
Many Traits Are Influenced by Several Genes 251
A Single Gene May Have Multiple Effects on Phenotype 251
The Environment Influences the Expression of All Genes 251
The Molecular Basis of Mendelian Genetics 253
Reflections on Genetic Diversity and Human Welfare 254
Review Section 255

Chapter 14 Molecular Genetics and Biotechnology 260

DNA Recombination in Nature 262
DNA Recombination Occurs Naturally Through Processes Such As Sexual Reproduction, Bacterial Transformation, and Viral Infection 262
Recombination Provides Raw Material for Evolution 262
There Are Both Similarities and Differences between Recombination in Nature and in the Laboratory 263
Recombinant DNA Technology 263
A DNA Library Consists of DNA from a Particular Organism Inserted into Bacterial Plasmids 263
The Genes of Interest Must Be Identified in the DNA Library 265
Selected DNA Sequences in the Library Can Be Amplified 265
Amplified Genes Can Be Used for Research, Industrial, Medical, or Agricultural Purposes 266
A CLOSER LOOK: THE POLYMERASE CHAIN REACTION— HOT SPRINGS AND HOT SCIENCE 268
Locating Genes 271
Restriction Enzymes Can Be Used to Provide Markers on a Chromosome 271
Restriction Fragment Length Polymorphisms Can Be Used to Locate a Gene 271
Sequencing Genes 272
Reflections on the Ethics of Biotechnology 273
Some Uses of Biotechnology Have Social Impacts 273
SCIENTIFIC INQUIRY: DNA FINGERPRINTING IN FORENSICS—THE CASE OF THE TELLTALE PALO VERDE 274
Bioengineered Organisms May Be Released into the Environment 274
Biotechnology May Eventually Allow Alterations in Human Genomes 275

EARTH WATCH: BIOTECHNOLOGY ATTACKS THE
   CHESTNUT BLIGHT 276
Review Section 277

Chapter 15 Human Genetics 280

Methods in Human Genetics 281
Single-Gene Inheritance 283
   Most Human Genetic Disorders Are Determined by Recessive
      Alleles 283
   A Few Human Genetic Disorders Are Determined by Dominant
      Alleles 286
   Many Human Traits Are Either Sex-Linked or Sex-Influenced
      287
   *SCIENTIFIC INQUIRY: IN SEARCH OF THE
      HUNTINGTON'S GENE 288*
Complex Inheritance 290
   Many Human Traits Are Influenced by Many Genes 290
   The Environment Influences the Expression of Genes 290
Chromosomal Inheritance 290
   Some Genetic Disorders Are Caused by Abnormal Numbers
      of Sex Chromosomes 291
   In Humans, the Y Chromosome Determines Maleness 292
   Some Genetic Disorders Are Caused by Abnormal Numbers
      of Autosomes 292
The Human Genome Project 294
**Medical Genetics Poses New Medical and Ethical
   Dilemmas 295**
   Phenylketonuria Illustrates a Biomedical Dilemma 295
   Knowledge of Human Genetics Poses Ethical Dilemmas 295
   *SCIENTIFIC INQUIRY: COLLECTING FETAL SAMPLES AND
      ANALYZING THEM USING BIOTECHNOLOGY 296*
   Review Section 297

UNIT III EVOLUTION 301

Chapter 16 Principles of Evolution 302

The History of Evolutionary Thought 304
   What Is a Species? 304
   The Philosophies of Plato and Aristotle Influenced Early
      Christian Beliefs 304
   Evidence Supporting Evolution Was Discovered Even before
      Darwin's Time 304
   Charles Darwin's Voyage on the *Beagle* Sowed the Seeds for His
      Theory of Evolution 307
   Darwin and Wallace Both Proposed That Evolution Occurs by
      Natural Selection 309
   Evolutionary Theory Arises from Scientific Observations and
      Conclusions Based on Them 311
The Evidence for Evolution 312
   The Fossil Record Provides Evidence of Evolutionary Change
      over Time 312
   Comparative Anatomy Provides Structural Evidence for
      Evolution 312
   Embryological Stages of Animals Can Provide Evidence for
      Common Ancestry 314
   Modern Biochemical and Genetic Analyses Reveal Relatedness
      among Diverse Organisms 314
   Artificial Selection Demonstrates That Organisms May Be
      Modified by Controlled Breeding 315
   Evolution by Natural Selection Occurs Today 316
   *SCIENTIFIC INQUIRY: MODERN MOLECULAR GENETIC
      TECHNIQUES SHED LIGHT ON EVOLUTIONARY
      RELATIONSHIPS 318*

A Postscript by Charles Darwin 318
   Review Section 319

Chapter 17 How Organisms Evolve 322

Evolution and the Genetics of Populations 323
   Genes, Influenced by the Environment, Determine the Traits
      of Each Individual 323
   The Gene Pool Is the Sum of All the Genes Occurring in
      a Population 324
   Evolution Is the Change of Gene Frequencies within a
      Population 324
   Mutation and the Recombination of Alleles During Sexual
      Reproduction Provide Sources of Variability 324
   The Equilibrium Population Is a Hypothetical Population in
      Which Evolution Does Not Occur 326
The Mechanisms of Evolution 326
   Mutations Are the Ultimate Source of Genetic Variability 326
   Migration Produces Gene Flow Between Populations 326
   Small Populations Are Subject to Random Changes in Allele
      Frequencies 327
   *A CLOSER LOOK: THE HARDY-WEINBERG EQUILIBRIUM
      POPULATION 328*
   Mating within a Population Is Almost Never Random 331
   All Genotypes Are Not Equally Adaptive 331
Natural Selection 332
   Natural Selection Acts on the Phenotype, Which Reflects the
      Underlying Genotype 332
   Natural Selection Can Influence Populations in Three Major
      Ways 333
   Natural Selection Takes Several Forms 335
Extinction 338
   Localized Distribution and Extreme Specialization Make Species
      Vulnerable in Changing Environments 338
   Interactions with Other Organisms May Drive Species to
      Extinction 339
   Habitat Change and Destruction Are the Leading Causes of
      Extinction 339
   *EARTH WATCH: NATURAL SELECTION, GENETIC
      DIVERSITY, AND ENDANGERED SPECIES 341*
   *EVOLUTIONARY CONNECTIONS: KIN SELECTION AND
      THE EVOLUTION OF ALTRUISM 342*
   Review Section 343

Chapter 18 The Origin of Species 346

Speciation 348
   Allopatric Speciation Occurs in Populations That Are
      Physically Separated 348
   Sympatric Speciation Occurs in Populations That Live in
      the Same Area 350
Maintaining Reproductive Isolation between Species 351
   Premating Isolating Mechanisms Prevent Mating between
      Species 352

Postmating Isolating Mechanisms Prevent Production of
Vigorous, Fertile Offspring 353
Phyletic and Divergent Speciation 353
The Genetics of Speciation 353
One Model of Speciation Stresses Gradual Accumulation of
Many Small Changes 354
A Second Model Stresses the Sudden Appearance of a Few
Major Changes 354
Which Model Is Correct? 354
Rates of Speciation 355
During Adaptive Radiation One Species Gives Rise to Many
under the Differing Selection Pressures of New Habitats 355
The Pattern of Evolution 355
Gradualism Explains Speciation As the Slow, Steady
Accumulation of Small Changes over Time 355
Punctuated Equilibrium Explains Speciation As Occurring
Relatively Rapidly amid Long Periods of Little Change 355
Gradualism and Punctuated Equilibrium Models Can Both Be
Applied to the Evolution of the Horse 356
Many Evolutionary Biologists Accept a Synthesis of Gradualism
and Punctuated Equilibrium 357
Review Section 359

Chapter 19 *The History of Life On Earth 362*

Origins 364
Prebiotic Evolution Was Controlled by the Early Atmosphere
and Climate 365
*SCIENTIFIC INQUIRY: RADIOMETRIC DATING TRACES
FOSSILS THROUGH TIME 368*
The First Life 368
Eukaryotes Evolved Membrane-Enclosed Organelles and A
Nucleus 369
Multicellularity 371
Multicellular Life in the Sea 371
Multicellular Plants Evolved Specialized Structures That
Facilitated Their Invasion of Diverse Habitats 371
Multicellular Animals Evolved Specializations That Allowed
Them to Capture Prey, Feed, and Escape More Efficiently
371
The Invasion of the Land 372
Some Plants Evolved Specialized Structures That Adapted
Them to Life on Dry Land 372
Some Animals Evolved Specialized Structures That Adapted
Them to Life on Dry Land 373
Human Evolution 376
Primate Evolution Has Been Linked to Grasping Hands,
Binocular Vision, and a Large Brain 376
Hominids Evolved from Dryopithecine Primates 377
Australopithecines, the First True Hominids, Could Stand
and Walk Upright 379
The Evolution of Human Behavior Is Highly Speculative 381
*EVOLUTIONARY CONNECTIONS: WHY DO HUMANS
WALK UPRIGHT? 382*
Review Section 383

Chapter 20 *Taxonomy: Imposing Order On Diversity 386*

Taxonomic Categories 387
The Origins of Taxonomy 387
Modern Criteria for Classification 388
The Five Kingdoms of Life 389
Taxonomy: An Inexact Science 393

Exploring Biodiversity: How Many Species Exist? 393
*EVOLUTIONARY CONNECTIONS: CLASSIFICATION
CONUNDRUMS; OR, WHERE SHALL WE PUT THE
ALGAE? 393*
Review Section 395

Chapter 21 *The Hidden World of Microbes 398*

Viruses 399
A Virus Consists of a Molecule of DNA or RNA Surrounded
by a Protein Coat 399
Viral Infections Cause Diseases That Are Difficult to Treat 400
Some Infectious Agents Are Even Simpler Than Viruses 400
How Did These Infectious Particles Originate? 402
The Kingdom Monera 402
Bacteria Are Difficult to Classify 402
Bacteria Possess a Remarkable Variety of Shapes and Structures
402
A CLOSER LOOK: VIRAL REPLICATION 404
Bacterial Reproduction Is by Cell Division Called Binary Fission
406
Bacteria Are Specialized for Specific Habitats 406
Bacteria Perform Many Functions That Are Important to Other
Forms of Life 407
Some Bacteria Pose a Threat to Human Health 408
*HEALTH WATCH: "UNNATURAL SELECTION"-THE
EVOLUTION OF DRUG-RESISTANT PATHOGENS 409*
Cyanobacteria Are Photosynthetic and Many Can Also Capture
Atmospheric Nitrogen 410
Archaebacteria Are Sufficiently Different in Cellular Structure
That Some Taxonomists Place Them in a Separate
Kingdom 410
The Kingdom Protista 411
Protists Are a Diverse Group Including Plantlike, Funguslike,
and Animal-like Forms 412
The Unicellular Algae, or Phytoplankton, Are Plantlike Protists
412
The Slime Molds Are Funguslike Protists 414
The Protozoa Are Animal-like Protists 416
*EVOLUTIONARY CONNECTIONS: OUR UNICELLULAR
ANCESTORS 420*
Review Section 420

Chapter 22 *The Fungi 424*

Fungal Form and Function 425
Most Fungi Have Filamentous Bodies 425
Fungi Obtain Their Nutrients from Other Organisms 426
Most Fungi Can Reproduce Both Sexually and Asexually 426
Economic, Ecological, and Health Impacts of Fungi 427
Some Fungi Form Symbiotic Relationships with Certain
Algae and Plants 427
The Classification of Fungi 428

EARTH WATCH: THE CASE OF THE DISAPPEARING
   MUSHROOMS 429

The Zygote Fungi (Division Zygomycota) Can Reproduce by
   Forming Diploid Zygospores 429

The Sac Fungi (Division Ascomycota) Form Spores in a
   Saclike Case Called an Ascus 429

The Club Fungi (Division Basidiomycota) Produce
   Club-Shaped Reproductive Structures Called Basidia 431

The Imperfect Fungi (Division Deuteromycota) Seem to
   Reproduce Entirely by Asexual Means 433

The Egg Fungi (Division Oomycota) Are Quite Different
   from Other Fungi 434

EVOLUTIONARY CONNECTIONS: FUNGAL INGENUITY
435

Review Section 436

Chapter 23 *The Plant Kingdom* 438

Evolutionary Trends in Plants 439
   The Plant Body Increased in Complexity As Plants Made
   the Evolutionary Transition from Water to Dry Land 439
   The Invasion of Land Required Protection and a Means
   of Dispersal for Sex Cells and Developing Plants 440
   Plants Have Both a Sporophyte and a Gametophyte
   Generation 440

Watery Origins—the Algae 440
   The Red Algae (Division Rhodophyta) Are Found Primarily
   in Clear Tropical Oceans 440
   The Brown Algae (Division Phaeophyta) Dominate in Cool
   Coastal Waters 443
   The Green Algae (Division Chlorophyta), Found Mostly
   in Ponds and Lakes, Probably Gave Rise to the Land
   Plants 443

Land—The New Frontier 443
   The Liverworts and Mosses (Division Bryophyta) Are Only
   Partially Adapted to Dry Land 444
   The Vascular Plants, or Tracheophytes, Have Conducting
   Vessels That Also Provide Support 445
   The Seedless Vascular Plants Include the Club Mosses,
   Horsetails, and Ferns 445
   The Seed Plants Dominate the Land, Aided by Two Important
   Adaptations: Pollen and Seeds 447
Review Section 454

Chapter 24 *The Animal Kingdom* 458

The Features of Animals 458
Evolutionary Trends in Animal Body Plans 458
   Over Evolutionary Time, Animals Have Increased in
   Complexity 458

Animal Phyla Show Trends toward Increasing Cellular
   Organization 461

Animal Phyla Show General Trends in Body Symmetry 461

Cephalization Increased over Evolutionary Time 462

A Coelem Evolved in More Complex Animals 462

Segmentation is First Seen in Annelid Worms 463

Digestive Systems Increased in Complexity 464

The Major Animal Phyla 464
The Sponges: Phylum Porifera 464
**The Hydra, Anemones, and Jellyfish: Phylum
Cnidaria 466**
The Flatworms: Phylum Platyhelminthes 468
The Roundworms: Phylum Nematoda 470
The Segmented Worms: Phylum Annelida 471
**The Insects, Arachnids, and Crustaceans: Phylum
Arthropoda 473**
   Insects Are the Most Diverse and Abundant Arthropods 474
   Spiders, Scorpions and Their Relatives Are Members of the
   Class Arachnida 476
   Crabs, Shrimp, Crayfish, and Their Relatives Are Members of
   the Class Crustacea 476
The Snails, Clams, and Squid: Phylum Mollusca 477
   Snails and Their Relatives Are Members of the Class
   Gastropoda 477
   Scallops, Clams, Oysters, and Their Relatives Are Members of
   the Class Bivalvia 477
   Octopuses, Squid, and Their Relatives Are Members of the
   Class Cephalopoda 479
**The Sea Stars, Sea Urchins, and Sea Cucumbers: Phylum
Echinodermata 479**
**The Tunicates, Lancelets, and Vertebrates: Phylum
Chordata 481**
   The Invertebrate Chordates Include Lancelets and Tunicates
   481
   The Vertebrates Have a Backbone and Other Adaptations That
   Have Contributed to Their Success 482
   Success on Dry Land Required Numerous Adaptations 484
EARTH WATCH: AMPHIBIANS IN DECLINE 486
EVOLUTIONARY CONNECTIONS: ARE HUMANS A
   BIOLOGICAL SUCCESS? 490
Review Section 491

UNIT IV PLANT ANATOMY AND PHYSIOLOGY 495

Chapter 25 *The Structure of Land Plants* 496

An Overview of Plant Structure 498
Plant Development and Growth 498
Plant Tissues and Cell Types 500
   The Dermal Tissue System Forms the Covering for the Plant
   Body 500
   The Ground Tissue System Makes Up Most of the Young Plant
   Body 501
   The Vascular Tissue System Consists of Xylem and
   Phloem 502
Roots: Anchorage, Absorption, and Storage 504
   Primary Growth Causes Roots to Elongate 504
   The Epidermis of the Root Is Very Permeable to Water 504
   Cortex Makes Up Much of the Interior of a Young Root 504
   The Vascular Cylinder Contains Conducting Tissues 506
Stems: Reaching for the Light 506

The Stem Has a Complex Organization Including Four Types of Tissue 506

Stem Branches Form from Lateral Buds Consisting of Meristem Cells 509

Secondary Growth Produces Thicker, Stronger Stems 509

Leaves: Nature's Solar Collectors 511

Leaves Have Two Major Parts: Blades and Petioles 511

Special Adaptations of Roots, Stems, and Leaves 512

Some Specialized Roots Store Food, Others Photosynthesize 512

Some Specialized Stems Produce New Plants, Store Water or Food, or Produce Thorns or Climbing Tendrils 512

Specialized Leaves May Conserve and Store Water, Store Food, or Even Capture Insects 513

Review Section 515

Chapter 26 How Land Plants Acquire and Transport Nutrients 518

A Comparison of Plant and Animal Nutrition 519

Plant Nutrition 520

Roots Acquire Minerals Using a Four-Step Process 520

Symbiotic Relationships Help Plants Acquire Nutrients 522

The Acquisition of Water 523

The Transport of Water and Minerals 524

Water Movement in Xylem Is Explained by the Cohesion-Tension Theory 524

Adjustable Stomata Control the Rate of Transpiration 525

The Transport of Sugars 528

The Pressure Flow Theory Explains Sugar Movement in Phloem 529

EARTH WATCH: PLANTS HELP REGULATE THE DISTRIBUTION OF WATER 530

Review Section 531

Chapter 27 Plant Reproduction and Development 534

Sexual versus Asexual Reproduction 535

Plant Life Cycles 536

The Evolution of Flowers 537

Complete Flowers Have Four Major Parts 538

Coevolution of Flowers and Pollinators 539

Some Flowers Provide Food for Pollinators 540

Sexy Deceptions Attract Pollinators 542

Some Plants Provide Nurseries for Pollinators 542

Gametophyte Development in Flowering Plants 543

Pollen Is the Male Gametophyte 543

The Embryo Sac Is the Female Gametophyte 544

Pollination and Fertilization 545

The Development of Seeds and Fruits 546

The Seed Develops from the Ovule and Embryo Sac 546

The Fruit Develops from the Ovary Wall 547

Seed Dormancy Helps Assure Germination at an Appropriate Time 547

A CLOSER LOOK: ADAPTATIONS FOR SEED DISPERSAL 548

EARTH WATCH: DISRUPTING COEVOLUTIONARY RELATIONSHIPS THREATENS ECOSYSTEMS 550

The Germination and Growth of Seedlings 551

The Shoot Tip Is Protected by the Coleoptile 551

Cotyledons Nourish the Sprouting Seed 552

Controlling the Development of the Seedling 552

Review Section 552

Chapter 28 Plant Responses to the Environment 556

The Discovery of Plant Hormones 557

The Work of Several Researchers Revealed the Hormonal Mechanisms of Phototropism 557

Plant Hormones and Their Actions 559

The Plant Life Cycle: Reception, Response, and Regulation 561

Abscisic Acid Maintains Seed Dormancy; Gibberellin Stimulates Germination 561

Auxin Controls the Orientation of the Sprouting Seedling 561

The Genetically Determined Shape of the Mature Plant Is the Result of Interactions between Hormones 563

Hormones Control the Differentiation of Xylem and Phloem 564

Daylength Controls Flowering 565

Hormones Coordinate the Development of Seeds and Fruit 566

Senescence and Dormancy Prepare the Plant for Winter 567

A CLOSER LOOK: RAPID-FIRE PLANT RESPONSES 568

Review Section 569

UNIT V ANIMAL ANATOMY AND PHYSIOLOGY 571

Chapter 29 Homeostasis and the Organization of the Animal Body 572

Homeostasis 573

Negative Feedback Tends to Maintain Constant Internal Conditions 573

Positive Feedback Events Are Self-Limiting 575

The Organization of the Animal Body 575

Animal Tissues Are Composed of Similar Cells That Perform a Specific Function 575

Organs Include Two or More Interacting Tissue Types 580

Organ Systems Consist of Two or More Interacting Organs 581

Review Section 582

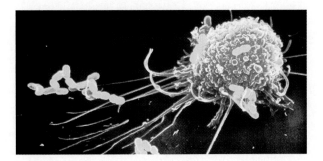

Chapter 30 Circulation 584

Types of Circulatory Systems 585

The Vertebrate Circulatory System 586

The Vertebrate Heart Consists of Muscular Chambers Whose Contraction Is Controlled by Electrical Impulses 587

Blood Transports Dissolved Nutrients, Wastes, and Hormones throughout the Body 591

The Blood Vessels Carry Blood to All Parts of the Body 594

The Lymphatic System 597

Lymphatic Vessels Resemble the Veins and Capillaries of the Circulatory System 597

The Lymphatic System Returns Fluids to the Blood 597
HEALTH WATCH: CARDIOVASCULAR DISORDERS 598
The Lymphatic System Transports Fats from the Intestine to the
    Blood 599
The Lymphatic System Helps Defend the Body against
    Disease 600
Review Section 601

Chapter 31   Respiration 604

The Evolution of Respiratory Systems 605
Some Types of Animals Exchange Gases without Specialized
    Respiratory Structures 606
A Variety of Respiratory Systems Have Evolved That Facilitate
    Gas Exchange by Diffusion 606
*A CLOSER LOOK: GILLS AND GAS EXCHANGE—
    COUNTERCURRENT FLOW 608*
The Human Respiratory System 610
The Conducting Portion of the Respiratory System Carries
    Air to the Lungs 610
Gas Exchange Occurs in the Alveoli 611
Oxygen and Carbon Dioxide Are Transported Using Different
    Mechanisms 611
Air Is Inhaled Actively and Exhaled Passively 612
Breathing Rate Is Controlled by the Respiratory Center of
    the Brain 613
*HEALTH WATCH: SMOKING—A LIFE AND BREATH
    DECISION 614*
Review Section 615

Chapter 32   Nutrition and Digestion 618

Nutrition 619
The Primary Sources of Energy Are Carbohydrates and
    Fats 619
Lipids Include Fats, Phospholipids, and Cholesterol 620
Carbohydrates, Including Sugars and Starches, Are a Source of
    Quick Energy 620
Proteins, Composed of Amino Acids, Perform a Wide Range
    of Functions within the Body 621
Minerals Are Elements and Small Inorganic Molecules
    Required by the Body 621
Vitamins Are Required in Small Amounts and Play Many
    Roles in Metabolism 621
Nutritional Guidelines Help People Obtain a Balanced
    Diet 624
The Challenge of Digestion 625
Digestive Systems Are Adapted to the Life-Style of Each
    Animal 625
Human Digestion 627
The Mechanical and Chemical Breakdown of Food Begins
    in the Mouth 627
The Esophagus Conducts Food to the Stomach, Where
    Digestion Continues 628
Most Digestion Occurs in the Small Intestine 630
Most Absorption Occurs in the Small Intestine 631
A CLOSER LOOK: THE FATE OF FATS 632
Water Is Absorbed and Feces Are Formed in the Large
    Intestine 633
Digestion Is Controlled by the Nervous System and
    Hormones 634
*HEALTH WATCH: ULCERS-WHEN THE DIGESTIVE TRACT
    DIGESTS ITSELF 635*
Review Section 635

Chapter 33   The Urinary System and
Homeostasis 638

The Functions of the Urinary System 639
Some Simple Urinary Systems 640
Flame Cells Filter Fluids in Flatworms 640
Nephridia in Earthworms Resemble Parts of the Vertebrate
    Kidney 640
The Human Urinary System 641
Urine Is Formed in the Kidneys 641
Blood Is Filtered by the Glomerulus 642
The Blood Filtrate Is Converted to Urine in the Nephron 642
The Loop of Henle Allows Urine to Become Concentrated 643
The Kidneys Are Important Organs of Homeostasis 643
*A CLOSER LOOK: THE NEPHRON AND URINE
    FORMATION 644*
HEALTH WATCH: WHEN THE KIDNEYS COLLAPSE 647
Review Section 648

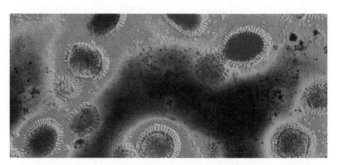

Chapter 34   Defenses Against Disease:
The Immune Response 650

Defenses Against Microbial Invasion 651
The Skin and Mucous Membranes Form Barriers 651
Nonspecific Internal Defenses Combat Microbes 652
*A CLOSER LOOK: CELLULAR COMMUNICATION
    DURING THE IMMUNE RESPONSE 654*
The Immune Response 657
A Successful Immune Response Recognizes, Overcomes, and
    Remembers 657
Recognition 658
Antibodies and T-Cell Receptors Recognize and Bind to Foreign
    Molecules, Triggering the Immune Response 658
An Antibody Contains Both Receptor and Effector
    Regions 658
T-Cell Receptors Bind Antigen and Trigger Responses 659
The Immune System Can Recognize Millions of Molecules 659
The Immune System Distinguishes "Self" from "Non-Self" 661
Attack 662
Humoral Immunity Is Produced by Antibodies in Blood 662
T Cells Produce Cell-Mediated Immunity 664
Memory 665
Medicine and the Immune Response 666
Antibiotics Slow Down Microbial Reproduction 666
Vaccinations Stimulate the Development of Memory
    Cells 666
Allergies Are Inappropriately Directed Immune
    Responses 666
An Autoimmune Disease Is an Immune Response against Some
    of the Body's Own Molecules 667
An Immune Deficiency Disease Results from the Inability to
    Mount an Effective Immune Response to Infection 667
AIDS 667

The Human Immunodeficiency Virus Is a Retrovirus That Infects and Destroys Helper T Cells 668
HIV Virus Is Transmitted by Exchange of Body Fluids 668
There Are Partially Effective Treatments, But No Cures, for AIDS 668
AIDS Is One of Many Widespread, Lethal Diseases 670
Cancer 670
Cancer Is Caused by Mutation, Activation, or Suppression of Genes That Control Cell Division 671
HEALTH WATCH: FLU-THE UNBEATABLE BUG 672
Review Section 675

Chapter 35   Chemical Control of the Animal Body: The Endocrine System   678

Animal Hormone Structure and Function 679
Hormones Have a Variety of Chemical Structures 680
Hormones Function by Binding to Specific Receptors on Target Cells 680
Hormones Are Regulated by Feedback Mechanisms 681
The Mammalian Endocrine System 683
Mammals Have Both Exocrine and Endocrine Glands 683
The Hypothalamus Controls the Secretions of the Pituitary Gland 683
The Thyroid and Parathyroid Glands Are Located in the Neck 687
The Pancreas Is Both an Exocrine and an Endocrine Gland 689
The Sex Organs Secrete Steroid Hormones 691
The Adrenal Glands Have Two Parts That Secrete Different Hormones 691
Many Types of Cells Produce Prostaglandins 692
Other Endocrine Organs Include the Pineal Gland, Thymus, Kidneys, Heart, and Digestive Tract 692
EVOLUTIONARY CONNECTIONS: THE EVOLUTION OF HORMONES 693
Review Section 694

Chapter 36   Information Processing: The Nervous System   696

A Comparison of Nervous and Endocrine Communication 697
The Functions and Structure of Neurons 698
Dendrites Receive Signals from Other Neurons or the Environment 698
The Cell Body Maintains the Neuron and Integrates Electrical Signals from the Dendrites 698
The Axon Carries Electrical Signals from the Cell Body to Their Destination 699
Synaptic Terminals Communicate with Other Neurons, Muscles, or Glands 699
Mechanisms of Neural Activity 699
How Is the Resting Potential Generated? 699
Action Potentials Can Carry Messages Rapidly over Long Distances 701
Neurons Communicate at Synapses 704
Building and Operating a Nervous System 705
Information Processing Requires Four Basic Operations 705
Neural Pathways Direct Behavior 707
Increasingly Complex Nervous Systems Are Increasingly Centralized 707
The Human Nervous System 708

The Peripheral Nervous System Links the Central Nervous System to the Body 709
The Central Nervous System Consists of the Spinal Cord and Brain 710
The Spinal Cord Is a Cable of Axons Protected by the Backbone 710
The Brain Consists of Several Parts Specialized for Specific Functions 713
The Nervous System Uses Many Neurotransmitters and Neuromodulators 716
Brain and Mind 717
The "Left Brain" and "Right Brain" Are Specialized For Different Functions 718
The Mechanisms of Learning and Memory Are Poorly Understood 719
SCIENTIFIC INQUIRY: SCANNING THE BRAIN 720
Insights on How the Brain Creates the Mind Come from Diverse Sources 720
Review Section 722

Chapter 37   The Senses: Perception   726

Receptor Mechanisms 728
Sensing Temperature: Thermoreception 729
Sensing Touch and Movement: Mechanoreception 729
Sound Perception: Hearing 730
The Structure of the Ear Helps Capture, Transmit, and Transduce Sound 730
Sound Transduction Is Aided by the Structure of the Cochlea 732
Gravity and Movement Perception: The Vestibular Apparatus 732
Sensing Light: Photoreception 733
Animal Eyes Range from Simple to Complex 733
The Mammalian Eye Collects, Focuses and Transduces Light Waves 734
Binocular Vision Allows Depth Perception 737
Smell and Taste: Chemoreception 737
The Ability to Smell Arises from Olfactory Receptors 738
Taste Receptors Are Found in Clusters on the Tongue 739
Sensing Pain 739
EVOLUTIONARY CONNECTIONS: NATURAL SELECTION SHAPES PERCEPTION 740
Review Section 742

Chapter 38   Action and Support: The Muscles and Skeleton   746

Muscle 747
The Intricate Structure of Skeletal Muscle Supports Its Active Function 748
Muscle Contraction Results from Thick and Thin Filaments Sliding Past One Another 750

Cardiac Muscle Powers the Heart 752
Smooth Muscle Produces Slow, Involuntary Contractions 752
The Skeleton 752
The Vertebrate Skeleton Serves Many Functions 753
Skeletal Tissues: Cartilage and Bone 753
Cartilage Provides Flexible Support and Connections 754
Bone Provides a Strong, Rigid Framework for the Body 754
A CLOSER LOOK: STICKS AND STONES-THE REPAIR OF BROKEN BONES 756
Body Movement: Muscle-Skeleton Interactions 757
Muscles Move the Skeleton Around Flexible Joints 757
HEALTH WATCH: OSTEOPOROSIS-THE HIDDEN CRIPPLER 758
Review Section 759

Chapter 39 Animal Reproduction 762

Reproductive Strategies 763
Asexual Reproduction Does Not Involve the Fusion of Sperm and Egg 764
Sexual Reproduction Requires the Union of Sperm and Egg 765
Mammalian Reproduction 767
The Male Reproductive Tract Includes the Testes and Accessory Structures 767
The Female Reproductive Tract Includes Ovaries and Accessory Structures 771
The Menstrual Cycle Is Controlled by Complex Hormonal Interactions 773
Copulation Allows Internal Fertilization 775
During Pregnancy, the Developing Embryo Grows within the Uterus 776
SCIENTIFIC INQUIRY: IN VITRO FERTILIZATION 777
Milk Secretion, or Lactation, Is Stimulated by Pregnancy 778
Reproduction Culminates in Labor and Delivery 779
On Limiting Fertility 780
Permanent Contraception Can Be Achieved through Sterilization 780
There Are Three Major Approaches to Temporary Contraception 780
Additional Contraceptive Methods Are under Development 781
HEALTH WATCH: SEXUALLY TRANSMITTED DISEASES 784
Review Section 785

Chapter 40 Animal Development 788

Differentiation 789
Gene Transcription Is Precisely Regulated during Development 790
Indirect and Direct Development 791
During Indirect Development, Animals Undergo a Radical Change in Body Form 791
Newborn Animals That Undergo Direct Development Resemble Miniature Adults 791
Stages of Animal Development 793
Cleavage Distributes Gene-Regulating Substances 793
Gastrulation Forms Three Tissue Layers 795
Adult Structures Develop During Organogenesis 796
Sexual Maturation Is Controlled by Genes and the Environment 796
Aging Seems to Be Genetically Programmed 797
Human Development 799
During the First Two Months, Rapid Differentiation and Growth Occur 799

The Placenta Secretes Hormones and Exchanges Materials between Mother and Embryo 801
Growth and Development Continue during the Last Seven Months 802
A CLOSER LOOK: HOMEOBOX GENES AND THE CONTROL OF BODY FORM 803
HEALTH WATCH: THE PLACENTA PROVIDES ONLY PARTIAL PROTECTION 804
Development Culminates in Birth 806
Review Section 806

UNIT VI ANIMAL BEHAVIOR 809

Chapter 41 The Foundations of Animal Behavior 810

The Genetic Basis of Behavior 811
Innate Behavior 812
Kineses Are Changes in the Rate of Random Movements 812
Taxes Are Movements toward or away from a Stimulus 812
A Reflex Is an Instinctive Movement of Part of the Body 813
SCIENTIFIC INQUIRY: ROBOT CRICKET FINDS HER MATE 814
Fixed Action Patterns Are Complex Innate Behaviors 815
Innate Behaviors Are Usually Adaptive 817
Learning 818
Imprinting Is a Genetically Programmed Form of Learning 818
Habituation Is a Decline in Response to a Repeated Harmless Stimulus 818
Classical and Operant Conditioning Are More Complex Forms of Learning 818
Trial-and-Error Learning Occurs through Experience 819
Insight or Reasoning Is the Most Sophisticated Form of Learning 820
The Instinct to Learn and the Learning of Instincts 820
EVOLUTIONARY CONNECTIONS: WHY DO ANIMALS PLAY? 822
Review Section 823

Chapter 42 The Social Behavior of Animals 826

Communication 828
Visual Communication Includes Active and Passive Signals 828
Communication by Sound Has Many Advantages 828
Chemical Communication Uses Pheromones 829
Communication by Touch Helps Establish Social Bonds 830
Social Behavior 830
Competition for Resources Underlies Many Forms of Social Behavior 830
Sexual Reproduction Requires Social Interactions between Mates 834

Social Behavior within Animal Societies Requires Cooperative Interactions 836

Human Ethology 840
Newborn Infants Exhibit Some Innate Behaviors 840
Innate Tendencies Can Be Revealed by Exaggerating Human Releasers 841
Simple Behaviors Shared by Diverse Cultures May Be Innate 842
Do People Respond to Pheromones? 842
Comparisons of Identical and Fraternal Twins Reveal Genetic Components of Behavior 843
EVOLUTIONARY CONNECTIONS: ALTRUISM, KIN SELECTION, AND THE SELFISH GENE 843
Review Section 844

UNIT VII ECOLOGY 847

Chapter 43 Population Growth and Regulation 848

Introduction to Ecology 849
How Populations Grow 850
The Biotic Potential of a Population Produces Exponential Growth If Not Restrained 851
Exponential Growth Sometimes Leads to "Boom and Bust" Cycles 853
A CLOSER LOOK: CYCLES IN PREDATOR AND PREY POPULATIONS 854
Environmental Resistance Limits Population Growth 855
Population Patterns in Space and Time 859
Populations Distribute Themselves in Different Ways 859
Populations Show Characteristic Patterns of Survivorship 860
The Human Population 861
The Human Population Is Growing Exponentially 861
The Age Structure of a Population Predicts Its Future Growth 862
The Population of the United States Is Growing Rapidly 863
EARTH WATCH: HAS THE HUMAN POPULATION EXCEEDED EARTH'S CARRYING CAPACITY? 864
Review Section 867

Chapter 44 Community Interactions 870

The Ecological Niche 872
Competition between Species 872
Interspecific Competition Helps Control Population Size 873
Predators and Their Prey 874
Predator-Prey Interactions Shape Evolutionary Adaptations 874
Keystone Predators Influence Community Structure 878
Symbiosis 878
Parasitism Harms, But Does Not Immediately Kill, the Host 878
Commensalism Benefits One Species without Affecting the Other 879
In Mutualistic Interactions, Both Species Benefit 880
Succession: Community Changes over Time 880
SCIENTIFIC INQUIRY: DECIPHERING THE RELATIONSHIP BETWEEN ANTS AND ACACIAS 881
There Are Two Major Forms of Succession: Primary and Secondary 883
EARTH WATCH: INTRODUCED SPECIES DISRUPT COMMUNITY INTERACTIONS 884
Succession Also Occurs in Ponds and Lakes 885
Succession Culminates in the Climax Community 886
Review Section 887

Chapter 45 How Ecosystems Work 890

The Flow of Energy 891
The Energy Captured by Photosynthetic Organisms Is Called Primary Productivity 891
Organisms Occupy Different Trophic Levels Based on How They Acquire Energy 893
Food Chains and Food Webs Describe the Feeding Relationships within Ecosystems 893
Energy Flows through Trophic Levels 894
Biological Magnification Occurs As Toxic Substances Are Passed through Trophic Levels 899
The Cycling of Nutrients 900
The Carbon Cycle Is an Atmospheric Cycle 900
The Nitrogen Cycle Is an Atmospheric Cycle 902
The Phosphorus Cycle Is a Sedimentary Cycle 902
Most Water Remains Chemically Unchanged during the Water Cycle 902
Human Intervention in Energy Flow and Nutrient Cycling 903
Acid Rain Is Caused by Overloading the Nitrogen and Sulfur Cycles 905
Interference with the Carbon Cycle Is Causing the Greenhouse Effect 908
Biodiversity and Ecosystem Stability 910
Review Section 911

Chapter 46 Earth's Diverse Ecosystems 914

Climate 915
Both Climate and Weather Are Driven by the Sun 916
Many Physical Factors Also Influence Climate 916
The Requirements of Life 918
Life on Land 918
Terrestrial Biomes Support Characteristic Plant Communities 918
EARTH WATCH: THE OZONE LAYER-WE HAVE PUNCTURED OUR PROTECTIVE SHIELD 920
Rainfall, Temperature, and Vegetation 933
Life in Water 934
Freshwater Lakes Have Distinct Regions of Life and Temperature 935
Freshwater Lakes Are Classified According to Their Nutrient Content 936
Marine Ecosystems Cover Much of Earth 937
EARTH WATCH: HUMANS AND ECOSYSTEMS 942
Review Section 944

Answers to Multiple-Choice Questions 948
Glossary G–1
Photo Credits P–1
Index I–1

1

An Introduction
to Life on Earth

"Viewed from the distance of the moon, the astonishing thing about the earth, catching the breath, is that it is alive. The photographs show the dry, pounded surface of the moon in the foreground, dead as an old bone. Aloft, floating free beneath the moist, gleaming surface of bright blue sky, is the rising earth, the only exuberant thing in this part of the cosmos."

LEWIS THOMAS in *The Lives of a Cell* (1974)

✴ AT A GLANCE

The Characteristics of Living Things
Living Things Are Both Complex and Organized
Living Things Must Acquire and Use Materials and Energy
Homeostasis Maintains Relatively Constant Internal Conditions
Growth Is a Property of All Living Organisms
Living Things Respond to Stimuli
Living Things Reproduce Themselves
Living Things Have the Capacity to Evolve

The Diversity of Life
Members of the Kingdom Monera Are Prokaryotic Cells; Members of All Other Kingdoms Are Composed of Eukaryotic Cells
Members of the Kingdoms Monera and Protista Are Mostly Unicellular; Members of the Kingdoms Fungi, Plantae, and Animalia Are Primarily Multicellular
Members of the Different Kingdoms Have Different Ways of Acquiring Energy

Biology: The Science of Life
Scientific Principles Underlie All Scientific Inquiry
The Scientific Method Is the Basis for Scientific Inquiry
Science Is a Human Endeavor
Scientific Theories Have Been Thoroughly Tested

Evolution: The Unifying Concept of Biology
Three Natural Processes Underlie Evolution

Knowledge of Biology Illuminates Everyday Life

Life on Earth is confined to a thin film encompassing Earth's surface: the biosphere. Earth, seen here from the moon, is an oasis of life in our solar system.

On your way across campus tomorrow morning, take a moment to look around you. Although you might not notice at first, an astonishing array of creatures thrives even in a place as domesticated as a college campus. Sparrows and squirrels scurry along branches, chirping and chattering. Trees, bushes, grass, and moss blanket the campus with green. Take a closer look. You may also see honeybees flitting from flower to flower, earthworms—or their winding tracks—at the edges of a small pond, a spider spinning its web, glistening mushrooms in the grass or beneath a bush. If you listen carefully, you may hear creatures you don't see; insects humming, birds calling in the distance, perhaps a frog croaking. But these are only the obvious organisms. Countless tiny ones, most too small to be seen with the naked eye, swim in the puddles left by last night's rain. A whole community of microscopic living things thrive in the soil. And living on, in, and around all these organisms, and the humans observing them, are billions of bacteria—simple, single-celled organisms that have survived nearly unchanged for billions of years.

How did such an astounding variety of living things evolve on Earth? In what ways do they interact with one another? How are these bacteria, fungi, plants, and animals alike and how do they differ? What processes must occur in each organism for it to survive in its environment and to reproduce? And perhaps the most basic question of all: How do the living and nonliving things differ? Common questions like these—asked by curious children and adults for centuries—form the basis of the science of biology. But they also relate to the very survival of the human species. For we too are organisms. We evolved in response to the same sorts of survival needs. Our bodies and those of all other living creatures consist of the same basic set of chemicals. The same processes allow us to survive and reproduce. As we journey through this book together, we hope you begin to share with us the excitement and awe that come from understanding the nature of Earth and its inhabitants and appreciating the beauty of it all.

The Characteristics of Living Things

To study the nature of life on Earth, we should begin at the beginning: *What is life?* If you look up *life* in a dictionary, you will find definitions such as "the quality that distin-

(a)

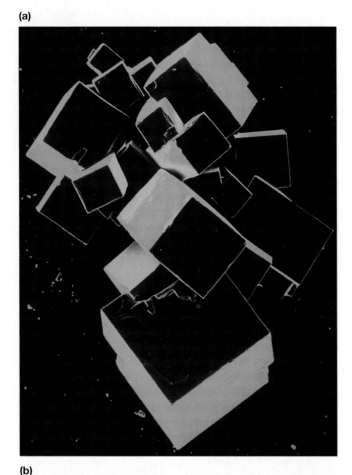

(b)

(c)

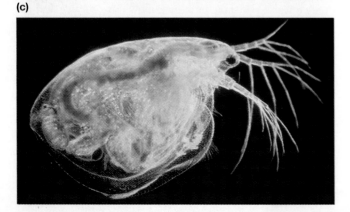

Figure 1-1   **Living organisms are complex and organized**

(a) Each crystal of table salt, sodium chloride, is a cube, showing great organization but minimal complexity. **(b)** The water and dissolved materials in the ocean represent complexity but very little organization. **(c)** Living things have both complexity and organization. The waterflea, *Daphnia pulex*, is only 1 millimeter (1/1000 meter) long, yet it has legs, a mouth, a digestive tract, reproductive organs, light-sensing eyes, and even a rather impressive brain considering its size.

guishes a vital and functioning being from a dead body," but you won't find out what that "quality" is.

We all have an intuitive understanding of what it means to be alive. Nevertheless, defining *life* is difficult, partly because living things are so diverse, and nonliving matter is sometimes so lifelike. A more fundamental difficulty in defining life is that living things cannot be described as the sum of their parts. The quality of life emerges as a result of the incredibly complex, ordered interactions among these parts. Because it is based on these **emergent properties**, life is a fundamentally intangible quality that defies any simple definition. We can, however, describe some of the characteristics of living things that, taken together, are not shared by nonliving objects. These characteristics are as follows:

1. Living things have a complex, organized structure based on organic (carbon-based) molecules.
2. Living things acquire materials and energy from their environment and convert them into different forms.
3. Living things actively maintain their complex structure and their internal environment, a process called homeostasis.
4. Living things grow.
5. Living things respond to stimuli from their environment.
6. Living things reproduce themselves, using a molecular blueprint called DNA.
7. Living things, taken as a whole, have the capacity to evolve.

Let's explore these characteristics in more detail.

Living Things Are Both Complex and Organized

Compared with nonliving matter of similar size, living organisms are highly complex and organized. A crystal of table salt (Fig. 1-1a), for example, consists of just two chemical elements, sodium and chlorine, arrayed in a precise cubical arrangement: The salt crystal is organized but simple. The oceans (Fig. 1-1b) contain some atoms of all the naturally occurring elements, but these atoms are randomly distributed: The oceans are complex but not organized. In contrast, even the tiny waterflea (Fig. 1-1c) contains dozens of different elements linked together in thousands of specific combinations that are further organized into ever larger and more complex assemblies to form structures such as eyes, legs, a digestive tract, and a brain.

Life on Earth is composed of a hierarchy of structures, with each level of the hierarchy based on the one below it and providing the foundation for the one above it (Fig. 1-2).

Biosphere	That part of Earth inhabited by living organisms; includes both the living and nonliving components	Earth's surface
Ecosystem	A community together with its nonliving surroundings	snake, antelope, hawk, bushes, grass, rocks, stream
Community	Two or more populations of different species living and interacting in the same area	snake, antelope, hawk, bushes, grass
Population	Members of one species inhabiting the same area	herd of pronghorn antelope
Species	Very similar, potentially interbreeding organisms	
Multicellular Organism	An individual living thing composed of many cells	pronghorn antelope
Organ System	Two or more organs working together in the execution of a specific bodily function	the nervous system
Organ	A structure within an organism usually composed of several tissue types that form a functional unit	the brain
Tissue	A group of similar cells that perform a specific function	nervous tissue
Cell	The smallest unit of life	a nerve cell
Organelle	A structure within a cell that performs a specific function	mitochondrion chloroplast nucleus
Molecule	A combination of atoms	water glucose DNA
Atom	The smallest particle of an element that retains the properties of that element	hydrogen carbon nitrogen oxygen
Subatomic Particle	Particles that make up an atom	proton neutron electron

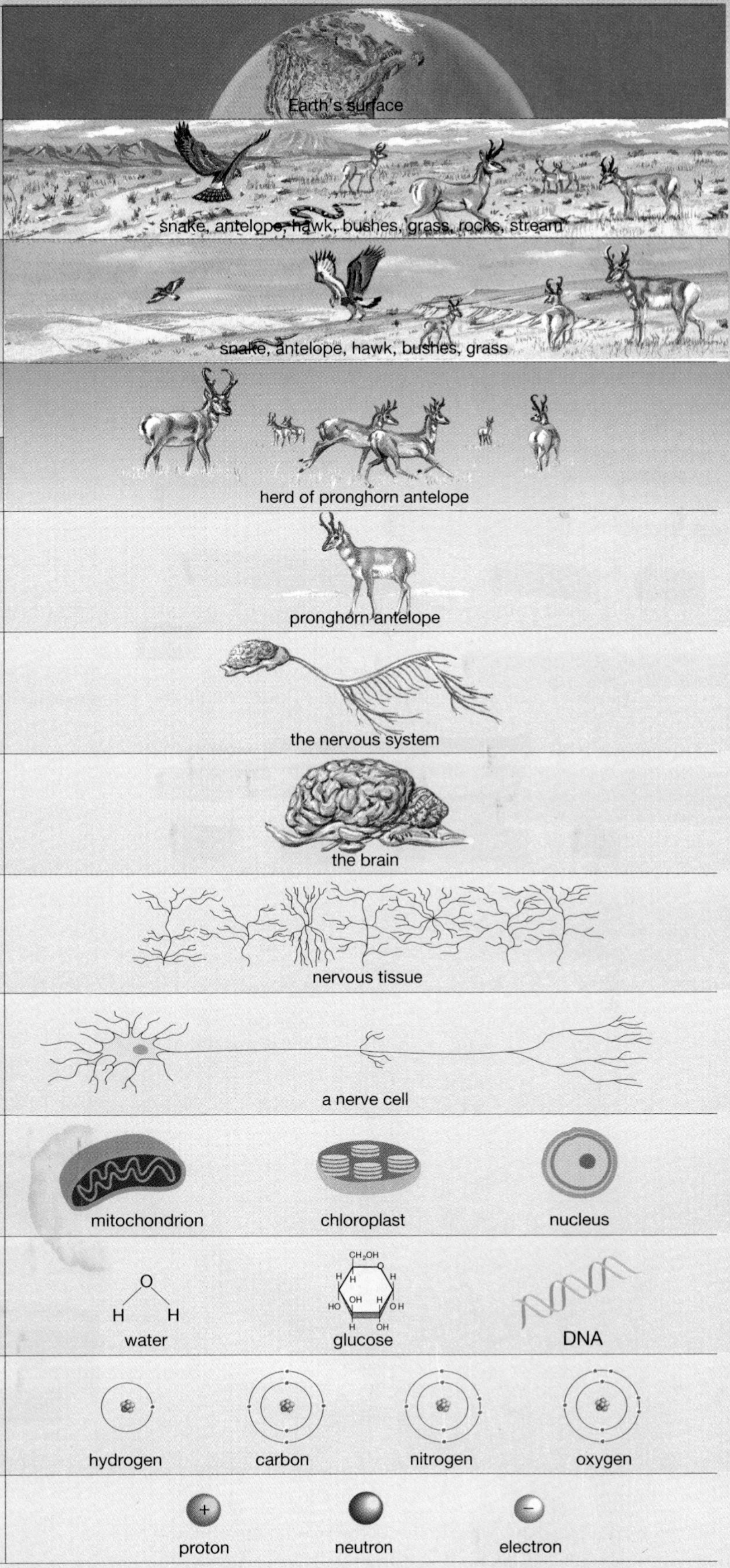

Figure 1-2   Levels of organization of matter on Earth

All life has a chemical basis, but the quality of life itself emerges on the cell level. Interactions among the components of each level and the levels below it allow the development of the next-higher level of organization.

All of life is built on a chemical foundation, based on **elements**, each of which is a unique type of matter. An **atom** is the smallest particle of an element that retains the properties of that element. For example, a diamond is made of the element carbon. The smallest possible unit of the diamond is an individual carbon atom; any further division would produce isolated **subatomic particles** that would no longer be carbon. Atoms may combine in specific ways to form assemblies called **molecules**; for example, one carbon atom can combine with two oxygen atoms to form a molecule of carbon dioxide. Although many simple molecules form spontaneously, extremely large and complex molecules are manufactured only by living things. The bodies of living things are composed primarily of complex molecules. The molecules of life, which are based on carbon, are called **organic molecules**.

Although the chemical arrangement and interaction of atoms and molecules are the building blocks of life, the quality of life itself emerges on the level of the cell. Just as an atom is the smallest unit of an element, so too the **cell** is the smallest unit of life (Fig. 1-3). The difference between a living cell and a conglomeration of chemicals illustrates some of the emergent properties of life.

Fundamentally, all cells contain **genes** that provide the information needed to control the life of the cell; subcellular structures called **organelles**, miniature chemical factories that use the information in the genes and keep the cell alive; and a thin **membrane** that encloses a watery medium (the cytoplasm) and separates the cell from the outside world. Some organisms, mostly microscopic, consist of just one cell, but larger organisms are composed of many cells whose functions are differentiated. In these multicellular organisms, cells of similar type form **tissues**, which perform a particular function. Examples of animal tissues are nervous tissue, composed of neurons (individual nerve cells), which produce electrical signals, and connective tissue, which often surrounds and supports other tissue types. Various tissue types combine to make up a structural unit, called an **organ** (for example, the brain). Several organs that collectively perform a single function are called an **organ system**; for example, together the brain, spinal cord, sense organs, and nerves form the nervous system. All the organ systems functioning cooperatively make up an individual living thing, the **organism**.

Beyond the individual organisms are broader levels of organization. A group of very similar, potentially interbreeding organisms constitutes a **species**. Members of the same species that live in a given area are considered a **population**. Populations of several species living and interacting in the same area form a **community**. A community plus its nonliving environment, including land, water, and atmosphere, constitute an **ecosystem**. Finally, the entire surface region of Earth inhabited by living things is called the **biosphere**.

The organization of this text will roughly follow the pattern of organization of life on Earth. We will begin with atoms and molecules, move to cells and principles of heredity, continue with the range of organisms and how they function, and conclude with the study of the interactions among organisms that are the focus of ecology.

Living Things Must Acquire and Use Materials and Energy

Organisms need materials and energy to maintain their high level of complexity and organization, to grow, and to reproduce (Fig. 1-4). The atoms and molecules of which all organisms are made may be acquired from the air, water, or soil or from other living things. Organisms extract these materials, called **nutrients**, from the environment and incorporate them into the molecules of their own bodies. The sum total of all the chemical reactions needed to sustain life is called **metabolism**.

Organisms obtain **energy**—the ability to do work, including carrying out chemical reactions, growing leaves in the spring, or contracting a muscle—in one of two basic ways. Plants and some single-celled organisms capture the energy of sunlight and store it in energy-rich sugar molecules, a process called **photosynthesis**. In contrast, neither fungi nor animals can photosynthesize, nor can most bacteria; these organisms must consume the energy-rich molecules contained in the bodies of other organisms. In either case, energy taken in is converted into a form that the organism can use or store for future use.

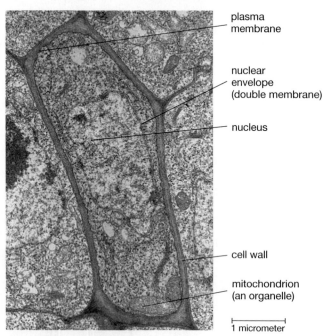

plasma
membrane

nuclear
envelope
(double membrane)

nucleus

cell wall

mitochondrion
(an organelle)

1 micrometer

Figure 1-3   **The smallest unit of life is the cell**

This electron micrograph clearly shows the plasma membrane that surrounds the cell, separating it from its environment; the nucleus that contains the cell's genes; and many other specialized structures, called organelles, that perform particular functions in the life of the cell. It also shows the cell wall present in many plant cells that surrounds and supports the cell while allowing free exchange of materials.

Ultimately the energy that sustains nearly all life comes from sunlight, captured by photosynthetic organisms and incorporated into energy-rich molecules. Organisms that cannot photosynthesize depend on photosynthetic life forms for food, either directly or indirectly. Thus, energy flows from the sun through nearly all forms of life and is eventually released again as heat.

Homeostasis Maintains Relatively Constant Internal Conditions

Complex, organized structures are not easy to maintain. Whether we consider the molecules of your body or the clothes in your closet, organization tends to disintegrate into chaos unless energy is used to sustain it (we will explore this tendency more fully in Chapter 4). To stay alive and function effectively, organisms must keep the conditions within their bodies fairly constant, a process called **homeostasis** (derived from Greek words meaning "to stay the same"). One of the many conditions regulated is body temperature. Among warm-blooded animals, for example, vital organs such as the brain and heart are kept at a warm, constant temperature despite wide fluctuations in environmental temperature.

Maintaining homeostasis is accomplished by a variety of automatic mechanisms; in the case of temperature regulation these include sweating during hot weather, me-

Figure 1-4 Living organisms acquire energy and materials (nutrients) from their environment

The plants in this Kenyan grassland capture energy from the sun and nutrients from the air, water, and soil. In contrast, the grazing Cape Buffalo extract both energy and nutrients from the plants. The lion will extract energy and nutrients from its animal prey, indirectly derived from plants. Without a continual supply of solar energy, nearly all life would cease.

tabolizing more food when it's cold, and behaviors such as basking in the sun or even (for modern humans) adjusting the thermostat in the room. In Chapter 29, and in Units IV and V, we will expand on the theme of homeostasis. Of course, not everything stays the same throughout an organism's life. Major changes, such as growth and reproduction, occur, but these are not failures of homeostasis. Rather, they are usually specific, genetically programmed parts of the organism's life cycle.

Growth Is a Property of All Living Organisms

At some time in its life cycle, every living thing becomes larger—that is, it grows. This feature is obvious for plants, birds, and mammals, all of which start out very small and undergo tremendous growth during their lives. Even single-celled bacteria, however, are small when they are first formed and grow to about double their original size before they divide. In all cases, growth involves the conversion of materials acquired from the environment into the specific molecules of the organism's own body.

Living Things Respond to Stimuli

Living organisms perceive and respond to stimuli in their internal and external environments. Animals have evolved elaborate sensory organs and muscular systems that allow them to detect and respond to light, sound, chemicals, and many other stimuli from their surroundings. Internal stimuli are perceived by receptors for stretch, pain, and various chemicals. For example, when you feel hungry, you are perceiving contractions of your empty stomach and low levels of sugars and fats in your blood. You then respond to external stimuli by choosing appropriate objects to eat, such as a piece of pie rather than the plate and fork. Animals, with their elaborate nervous systems and motile bodies, are not the only organisms to perceive and respond to stimuli. The plants on your windowsill grow toward the light, and even the bacteria in your intestine manufacture a different set of digestive enzymes depending on whether you drink milk, eat candy, or both.

Living Things Reproduce Themselves

The *continuity of life* occurs because organisms reproduce, giving rise to offspring of the same type (Fig. 1-5). The processes for producing offspring are varied, but the results—the perpetuation of the parents' genetic material—are the same. The *diversity of life* occurs in part because the offspring, though arising from the genetic material provided by the parents, are usually somewhat different from their parents, as explained briefly below, and in Chapters 16 and 17. The mechanism by which traits are passed from one generation to the next, through a "genetic blueprint" contained in molecules of DNA, produces these variable offspring.

Figure 1-5   **Living organisms reproduce**

As it grows, this baby orangutan will resemble, but not be identical to, its parents. The similarity and variability of offspring are crucial to the evolution of life.

DNA *Is the Molecule of Heredity*

All known forms of life use a molecule called **deoxyribonucleic acid**, or **DNA**, as the repository of hereditary information (Fig. 1-6). Much of Unit II will be devoted to exploring the structure and function of this remarkable molecule. For now, we will simply note that an organism's DNA is its "blueprint" or "molecular instruction manual," a guide to both the construction and, at least in part, the operation of its body. When an organism reproduces, it passes a copy of its DNA to its offspring. The accuracy of the DNA copying process is astonishingly high: Only about

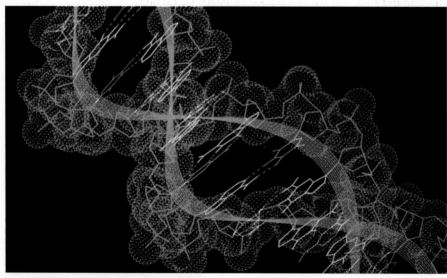

Figure 1-6   **DNA**

A computer-generated model of DNA, the molecule of heredity.

one mistake occurs for every billion bits of information contained in the DNA molecule. But chance accidents to the genetic material also bring about changes in the DNA. The occasional errors and accidental changes, called **mutations**, are crucial. Without mutations, all life forms might be identical. Indeed, there is reason to believe that, without mutations, there would be no life at all (see Chapter 19). Mutations in DNA are the ultimate source of genetic variations. These variations, superimposed on a background of overall genetic fidelity, make possible our final property of life, the capacity to evolve.

Living Things Have the Capacity to Evolve

Although the genetic makeup of a single organism remains essentially the same over its lifetime, the genetic composition of a species as a whole changes over many lifetimes. Over time, mutations and variable offspring provide diversity in the genetic material of a species. In other words, the species evolves. The most important force in evolution is **natural selection**, the process by which organisms with adaptive traits (those that help organisms cope with the rigors of their given environmental conditions) survive and reproduce more successfully than do others that lack those traits. The adaptive traits arising from genetic mutation that enhanced survival are thereby passed on to the next generation.

The Diversity of Life

Although all living things share the general characteristics discussed earlier, evolution has brought forth an amazing variety of life forms. In the following brief description of the features used to classify living organisms, we introduce you to the diversity of life on Earth. We describe the classification and structures of organisms in detail in Chapters 20 through 24.

Living organisms are often grouped into five major categories, called **kingdoms:** Monera, Protista, Fungi, Plantae, and Animalia (Fig. 1-7). Although biologists believe that the groupings reflect evolutionary relationships among organisms, we must classify living things according to the characteristics we can see and measure today. There are exceptions to any simple set of criteria used to define the kingdoms, but three characteristics are particularly useful: cell type, the number of cells in each organism, and the mode of acquiring energy (Table 1-1).

Figure 1-7 The five kingdoms

(a)

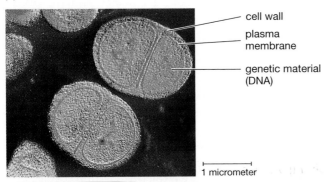

— cell wall

— plasma membrane

— genetic material (DNA)

|⊢————⊣|
1 micrometer

(a) The Kingdom Monera. A color-enhanced electron micrograph of a dividing bacterium. Monerans are unicellular and prokaryotic, and most are surrounded by a thick cell wall. Although some can photosynthesize, most absorb food from their surroundings.

(b)

oral groove ("mouth")   food vacuoles   contractile vacuole

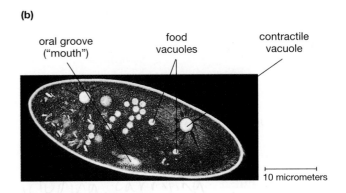

|⊢————⊣|
10 micrometers

(b) The Kingdom Protista. This light micrograph of a *Paramecium* illustrates the complexity of these large, usually single, eukaryotic cells. Some protists can photosynthesize, but others ingest or absorb their food. Many, including *Paramecium*, are mobile, moving with cilia or flagella.

(c)

(c) The Kingdom Fungi. An exotic mushroom found in Peru. Most fungi are multicellular, often with distinct cell types specialized for feeding or reproduction. Fungi generally absorb their food, which is usually the dead bodies or wastes of plants and animals. The food is digested by enzymes secreted outside of the fungal body. Most fungi cannot move.

(d)

(d) The Kingdom Plantae. This butterfly weed represents the flowering plants, the dominant members of the kingdom Plantae. Flowering plants owe much of their success to mutually beneficial relationships with animals, such as these pearl crescent butterflies, in which the flower provides food and the insect carries pollen from flower to flower, fertilizing them. Plants are multicellular, nonmotile eukaryotes with considerable specialization of cell types. Plants acquire nutrients by photosynthesis.

(e) The Kingdom Animalia. This clown fish and the anemone with which it lives are representative animals. Animals are multicellular eukaryotes that usually ingest their food. Animal bodies contain a wide variety of tissues and organs, composed of specialized cell types. Most animals are motile and can respond rapidly to stimuli. The anemone represents the largest major group of animals: the invertebrates, which lack a backbone. The clown fish is a vertebrate, an animal that (like humans) has a backbone.

(e)

TABLE 1-1 ▦ *Some Characteristics of the Five Kingdoms*			
Kingdom	*Cell Type*	*Cell Number*	*Major Mode of Nutrition*
Monera	Prokaryotic	Unicellular	Absorb or photosynthesize
Protista	Eukaryotic	Unicellular	Absorb, ingest, or photosynthesize
Fungi	Eukaryotic	Most multicellular	Absorb
Plantae	Eukaryotic	Multicellular	Photosynthesize
Animalia	Eukaryotic	Multicellular	Ingest

Members of the Kingdom Monera Are Prokaryotic Cells; Members of All Other Kingdoms Are Composed of Eukaryotic Cells

There are two fundamentally different types of cells: **prokaryotic** and **eukaryotic**. *Karyotic* refers to the **nucleus** of the cell: a membrane-enclosed sac containing the genetic material (see Fig. 1-3). *Eu* means "true" in Greek, and eukaryotic cells are recognized by the presence of a "true," membrane-enclosed nucleus. Eukaryotic cells are larger than prokaryotic cells and contain a variety of other organelles, many surrounded by membranes. Prokaryotic cells do not have a nucleus; their genetic material resides in the cytoplasm of the cell. They are small, only one or two micrometers long, and lack membrane-bound organelles. *Pro* means "before" in Greek, because prokaryotic cells almost certainly evolved before eukaryotic cells (and eukaryotic cells probably evolved from prokaryotic cells, as we will see in Chapter 19). Members of the kingdom Monera consist of prokaryotic cells; the cells of the other four kingdoms are eukaryotic.

Members of the Kingdoms Monera and Protista Are Mostly Unicellular; Members of the Kingdoms Fungi, Plantae, and Animalia Are Primarily Multicellular

Organisms of the kingdoms Monera and Protista are usually single-celled, or **unicellular**, although a few live in strands or mats of cells with little communication, cooperation, or organization among cells. Most fungi, plants, and animals are **multicellular**; their lives depend on intimate cooperation among cells.

Members of the Different Kingdoms Have Different Ways of Acquiring Energy

All organisms need energy to live. Photosynthetic organisms capture energy from sunlight and store it in molecules such as sugars and fats. These organisms, including plants, some monerans, and some protists, are therefore called **autotrophs**, meaning "self-feeders." Organisms that cannot photosynthesize must acquire energy prepackaged in the molecules of the bodies of other organisms; hence these organisms are called **heterotrophs**, meaning "other-feeders." Many monerans and protists, and all fungi and animals, are heterotrophs. Heterotrophs differ in the size

of the food they eat. Some, such as bacteria and fungi, absorb individual food molecules; others, including most animals, eat whole chunks of food and break them down to molecules in their digestive tracts (ingestion).

Biology: The Science of Life

Biology is probably the most diverse of all the sciences. While some biologists are creating new forms of life by manipulating genetic material, others are probing the workings of the brain, tracing the complex interactions within ecosystems, or seeking out new forms of life from the tropical rain forest to the ocean floor. Through the study of biology you will become familiar with how your body works. You will learn how some diseases are spread and how they are fought by our natural defenses and by medications. You will explore the intricacies of development from single cell to whole human being. You will learn how plants capture the solar energy that ultimately sustains nearly all life. By learning about the relationships between people and other forms of life, you will be prepared to make informed choices about land use, waste disposal, family size, and many other issues that affect the environment that sustains us.

Biology is a science, and its principles and methods are the same as those of any other science. In fact, a basic tenet of modern biology is that living things obey the same laws of physics and chemistry that govern nonliving matter.

Scientific Principles Underlie All Scientific Inquiry

All scientific inquiry, including biology, is based on a small set of assumptions. Although we can never absolutely prove these assumptions, they have been so thoroughly tested and found valid that we might call them scientific principles. These are the principles of natural causality, uniformity in space and time, and common perception.

Natural Causality Is the Principle That All Events Can Be Traced to Natural Causes

The first principle of science is natural causality. Over the course of human history, two approaches have been taken to the study of life and other natural phenomena. The first assumes that some events happen through the intervention of supernatural forces beyond our understanding. The ancient Greeks believed that Zeus hurled thunderbolts from the sky and that Poseidon caused earthquakes and storms at sea. In contrast, science adheres to the principle of **natural causality:** All events can be traced to natural causes that are potentially within our ability to comprehend. For example, until relatively recently, epilepsy was commonly thought to be a visitation from the gods. Today we realize that epilepsy is a disease of the brain, in which groups of nerve cells fire uncontrollably.

The principle of natural causality has an important corollary: The evidence we gather about the causes of natural events has not been deliberately distorted to fool us. This corollary may seem obvious, yet not so very long ago some people argued that fossils are not evidence of evolution but were placed on Earth by God as a test of our faith. If we cannot trust the evidence provided by nature, then the entire enterprise of science is futile.

The Natural Laws That Govern Events Apply Everywhere and for All Time

A second fundamental principle of science is that natural laws, laws derived from nature, are uniform in space and time and do not change with distance or time. The laws of gravity, the behavior of light, and the interactions of atoms, for example, are the same today as they were a billion years ago and will hold just as well in Moscow as in New York or even on Mars. Uniformity in space and time is especially vital to biology, because many events of great importance to biology, such as the evolution of today's diversity of living things, happened before humans were around to observe them. Some people believe that all the different types of living organisms were individually created at one time in the past by the direct intervention of God, a philosophy called **creationism.** As scientists, we freely admit that we cannot disprove this idea. Creationism, however, is contrary to both natural causality and uniformity in time. The overwhelming success of science in explaining natural events through natural causes has led almost all scientists to reject creationism.

Scientific Inquiry Is Based on the Assumption That People Perceive Natural Events in Similar Ways

A third basic assumption of science is that, as a general rule, all human beings perceive natural events in fundamentally the same way and that these perceptions provide us with reliable information about the natural world. Common perception is, to some extent, a principle peculiar to science. Value systems, such as those involved in the appreciation of art, poetry, and music, do not assume common perception. We may perceive the colors in a painting in a similar way (the scientific aspect of art), but we do not perceive the aesthetic value of the painting identically (the humanistic aspect of art). Moral values also differ radically among peo-

ple, often owing to their culture or religious beliefs. Because value systems are subjective, not objective, science cannot solve certain types of philosophical or moral problems, such as the morality of abortion.

The Scientific Method Is the Basis for Scientific Inquiry

Given these assumptions, how do biologists study the workings of life? Scientific inquiry is a rigorous method for making observations of specific phenomena and searching for the order underlying those phenomena. Ideally, biology and the other sciences use the **scientific method**, which consists of four interrelated operations: observation, hypothesis, experiment, and conclusion. All scientific inquiry begins with an **observation** of a specific phenomenon. The observation, in turn, leads to questions, such as "how did this come about?" Then in a flash of insight, or more often after long, hard thought, a hypothesis is formulated. A **hypothesis** is an educated guess that a certain preceding cause produces the observed phenomenon. To be useful, a hypothesis must be testable by further observations, or **experiments**. These experiments produce results that either support or refute the hypothesis, and a **conclusion** is drawn about its validity. A single experiment is never an adequate basis for a conclusion; the results must be repeatable not only by the original researcher, but by others as well.

Simple experiments test the assertion that a single factor, or **variable**, is the cause of a single observation. To be scientifically valid, the experiment must rule out a variety of other variables as the cause of the observation. So scientists design **controls** into their experiments, in which all the variables remain constant. Controls are then compared with the experimental situation, in which only the variable being tested is changed. In the 1600s, Francesco Redi used the scientific method to demonstrate that flies do not arise spontaneously from rotting meat (see "Scientific Inquiry: Does Life Arise Spontaneously?").

The scientific method can be used to solve everyday problems, as well as to generate new knowledge. Let's now consider an everyday situation in which you can apply the scientific method. Late to class, you rush to your car and make the *observation* that it won't start. Immediately, you form a *hypothesis*: The battery is dead. Quickly you design an *experiment*: You replace your battery with the battery from your roommate's new car and try to start your car again. The result seems to confirm your hypothesis; your car starts immediately. But wait! You haven't provided controls for several variables. Perhaps your battery was fine all along, and you just needed to try to start the car again. Or perhaps the battery cable was loose and simply needed to be tightened. Realizing the need for a good *control*, you replace your old battery, making sure the cables are secured tightly, and attempt to restart the car. If your car repeatedly refuses to

SCIENTIFIC INQUIRY

Does Life Arise Spontaneously?

The experiments of the Italian physician Francesco Redi (1621–1697) beautifully demonstrate the scientific method and also help to illustrate the principle of causality, on which modern science is based. Redi investigated why maggots appear in spoiled meat. Before Redi, the appearance of maggots was considered to be evidence of **spontaneous generation**, the production of living organisms from nonliving matter.

Redi *observed* that flies swarm around fresh meat and that maggots appear in meat left out for a few days. He formed a testable *hypothesis*: the flies produce the maggots. In his *experiment*, Redi wanted to test just one variable: the access of flies to the meat. Therefore, he took two clean jars and filled them with similar pieces of meat. He left one jar open (the *control* jar) and covered the other with gauze to keep out flies (the *experimental* jar). He did his best to keep all the other variables the same (for example, type of jar, type of meat, and temperature). After a few days, he observed that maggots swarmed over the meat in the open jar, but no maggots appeared in the meat in the covered jar. Redi *concluded* that his hypothesis was correct and that maggots are produced by flies, not by the meat itself (Fig. E1-1). Only through controlled experiments could the age-old hypothesis of spontaneous generation be laid to rest.

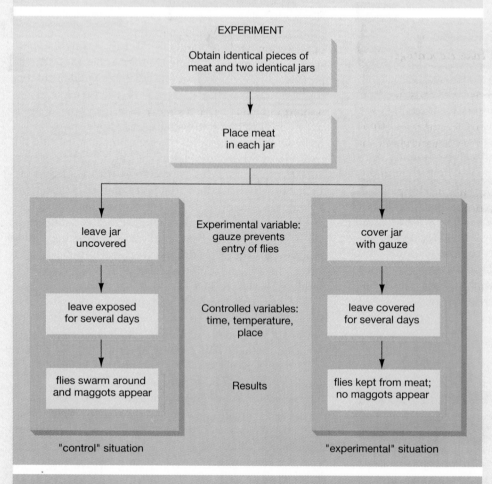

OBSERVATION
Flies swarm around meat left in the open;
maggots appear on meat.

HYPOTHESIS
Flies produce the maggots; keeping flies away
from meat will prevent the appearance of maggots.

EXPERIMENT

Obtain identical pieces of
meat and two identical jars

Place meat
in each jar

leave jar uncovered	Experimental variable: gauze prevents entry of flies	cover jar with gauze
leave exposed for several days	Controlled variables: time, temperature, place	leave covered for several days
flies swarm around and maggots appear	Results	flies kept from meat; no maggots appear

"control" situation   "experimental" situation

CONCLUSION
Spontaneous generation of maggots
from meat does not occur; flies are
probably the source of maggots.

Figure E1-1   The experiments of Francesco Redi

start with the old battery, but then starts immediately when you put in your roommate's battery, you have isolated a single *variable*, the battery, and you can safely draw the *conclusion* that your old battery is dead.

It is important to recognize the limitations of the scientific method. In particular, scientists can seldom be sure that they have controlled *all* the variables other than the one they are trying to study. Therefore, scientific conclusions must always remain tentative and are subject to revision if new observations or experiments demand it.

Science Is a Human Endeavor

Scientists are real people. They are driven by the same ambitions, pride, and fears as other people, and they sometimes make mistakes. As you will read in Chapter 10, ambition played an important role in the discovery by James Watson and Francis Crick of the structure of DNA. Accidents, lucky guesses, controversies with competing scientists, and, of course, the intellectual powers of individual scientists contribute greatly to scientific advances. To illustrate what we might call "real science," let's consider an actual case.

A Good Scientist Is Prepared to Take Advantage of Chance Events

When microbiologists study bacteria, they must use pure cultures—that is, plates of bacteria that are free from contamination by other bacteria, molds, and so on. Only by studying a single type at a time can we learn about the properties of that particular bacterium. Consequently, at the first sign of contamination, a culture is usually thrown out, often with mutterings about sloppy technique. On one such occasion, however, in the late 1920s, the Scottish bacteriologist Alexander Fleming turned a ruined culture into one of the greatest medical advances in history.

One of Fleming's bacterial cultures became contaminated with a patch of a mold called *Penicillium*. Before throwing out the culture dish, Fleming noticed that *no bacteria were growing near the mold* (Fig. 1-8). Why not? Fleming hypothesized that perhaps *Penicillium* releases a substance that kills off bacteria growing nearby. To test this hypothesis, Fleming grew some pure *Penicillium* in a liquid medium, filtered out the *Penicillium* mold, and then applied the liquid in which the mold had grown to an uncontaminated bacterial culture. Sure enough, the bacteria were killed by the liquid. Further research into these mold extracts resulted in the production of the first antibiotic: penicillin, a bacteria-killing substance that has since saved millions of lives. Fleming's experiments are a classic in scientific methodology, proceeding from observation to hypothesis to experimental tests of the hypothesis, followed by a conclusion, further supported by experimental evidence. But scientific method

alone would have been useless without the lucky combination of accident and a brilliant scientific mind. Had Fleming been a "perfect" microbiologist, he wouldn't have had any contaminated cultures. Had he been less observant, the contamination would have been just another spoiled culture dish. Instead, it was the beginning of antibiotic therapy for bacterial diseases. As the French microbiologist Louis Pasteur said, "Chance favors the prepared mind."

Scientific Theories Have Been Thoroughly Tested

In scientific terminology, a **theory** is a hypothesis that has been supported by so many cases that few scientists seriously doubt its validity. Take a trivial example: Each time an apple falls *down* to Earth rather than *up* into the sky, the theory of gravity is supported. Thus, a scientific theory is similar to what we would call a "law" or "fact" in common English usage. Scientists, however, must be open-minded about even their most cherished theories, and they must be willing to discard them if new observations render their theories obsolete.

Probably the foremost theory in biology is evolution. Since its formulation by two English naturalists, Charles Darwin and Alfred Wallace, in the mid-1800s, the theory of evolution has been supported by fossil finds, geological studies, radioactive dating of rocks, genetics, molecular biology, biochemistry, and breeding experiments. People who refer to evolution as "just a theory" profoundly misunderstand what scientists mean by that term.

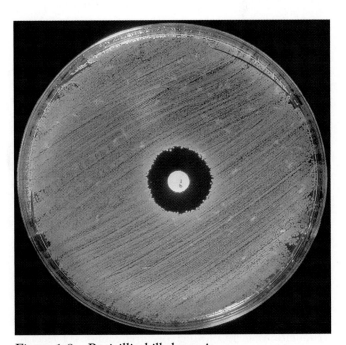

Figure 1-8   Penicillin kills bacteria

Penicillin diffuses outward from a penicillin-soaked disc of paper, creating a pronounced bald spot in the "lawn" of bacteria on this petri dish.

Evolution: The Unifying Concept of Biology

The most important concept in biology is evolution, a unifying theory that explains the origin of diverse forms of life as a result of changes in their genetic makeup. The theory of **evolution** states that modern organisms descended, with modification, from preexisting life forms. In the biologist Theodosius Dobzhansky's words, "Nothing in biology makes sense, except in the light of evolution." Why don't snakes have legs? Why are there dinosaur fossils but no living dinosaurs? Why are monkeys so like us, not only in appearance, but even in the structure of their genes and proteins? The answers to those questions, and thousands more, lie in the processes of evolution (which we will examine in detail in Chapters 16 through 18). Evolution is so vital to our understanding and appreciation of biology that we must briefly review its important principles before going further.

Three Natural Processes Underlie Evolution

In the mid-1800s, Darwin and Wallace formulated the theory of evolution that still forms the basis of our modern understanding. Evolution arises as a consequence of three natural processes: *genetic variation* among members of a population; *inheritance* of those variations by offspring of parents carrying the variation; and *natural selection*, the survival and enhanced reproduction of organisms with favorable variations.

Much of the Variability among Organisms Is Inherited

Look around at your classmates and notice how different they are. Although some of this variation is due to differences in environment and life-styles, it is mainly influenced by our genes: Most of us could pump iron for the rest of our lives and never develop a body like Arnold Schwarzenegger's. Where does genetic variation come from? The genetic instructions—the genes—of all living organisms are segments of molecules of deoxyribonucleic acid, or DNA. Occasionally, the DNA suffers an accident: Perhaps radiation strikes the DNA molecule just so, altering its information content. Mistakes in copying DNA during reproduction, although rare, also occur. These accidents and mistakes cause mutations, or changes in a gene, that may affect the organism's appearance or ability to function. Many mutations are harmless; some make the organism less able to function. But in rare cases mutations may actually improve an organism's ability to function. As a result of mutations, many of which occurred millions of years ago and have been passed from parent to offspring through countless generations, members of the same species are often slightly different from one another.

Natural Selection Tends to Preserve Genes That Help an Organism Survive and Reproduce

On the average, organisms that best meet the challenges of their environment will leave the most offspring. The offspring will inherit the genes that made their parents successful. Natural selection thus preserves genes that help organisms flourish in their environment. For example, a mutated gene providing instructions for larger teeth in beavers will be passed from parent to offspring. These offspring will be able to chew down trees more efficiently, build bigger dams and lodges, and eat more bark than "ordinary" beavers. Because these big-toothed beavers will obtain more food and better shelter than their smaller-toothed relatives, they will probably raise more offspring. The offspring will inherit their parents' genes for larger teeth. Over time, smaller-toothed beavers will become increasingly scarce, and after many generations all beavers will have large teeth.

Structures, physiological processes, or behaviors that aid survival and reproduction in a particular environment are called **adaptations**. Most of the features that we admire so much in our fellow life forms, such as the long clean limbs of deer, the wings of eagles, and the mighty columns of redwood trunks, are adaptations molded by millions of years of mutation and natural selection.

In the long run, however, what helps an organism to survive today may become a liability tomorrow. If environments change—for example, as ice ages come and go—then the genetic makeup that best adapts organisms to their environment will also change over time. When new mutations happen to appear that increase the fitness of an organism in the altered environment, these mutations in their turn will spread throughout the population.

Over millennia, the interplay of environment, variation, and selection results in evolution: the modification of the genetic makeup of species. In environments that are reasonably constant, such as the oceans, some well-adapted forms persist relatively unchanged and are sometimes called "living fossils." For example, sharks (Fig. 1-9) have retained essentially the same body form for tens of millions of years, as their sleek shape, powerful tail, acute sense of smell, and formidable teeth have made them superb predators. In changing environments, some species do not experience within an appropriate time frame the genetic changes that allow them to adapt. The rate of environmental change outstrips the rate of genetic changes, and those species become extinct. The petrified forests of Arizona (Fig. 1-10) are the remnants of giant trees that could not withstand the conversion of tropical lowland to high desert. Other species experience chance mutations that adapt them to meet new challenges. Fossil horses show gradual changes over time, caused by mutations that happened to produce adaptations to their changing environment (Fig. 1-11).

Figure 1-9 Ancient adaptations

This tiger shark, photographed in the clear waters of the Bahamas, possesses features that have characterized sharks for tens of millions of years: streamlined shape, a long, powerful tail, an acute sense of smell, and rows of sharp teeth.

Figure 1-10 Logs in Petrified Forest National Park

The petrified trunks scattered over the ground are remnants of vast forests of trees that lived here about 160 million years ago. As their habitat gradually changed from tropical lowland to high desert, the trees became extinct.

Figure 1-11 The evolution of the horse

The earliest horse, *Hyracotherium* (**a**; also called *Eohippus*) gave rise to several intermediate forms (**b, c**) culminating in the modern horse, *Equus* (**d**). The changing backgrounds through which the horses run depict the changing environments that molded horse evolution, from forests through open woodlands to hard, flat plains. The changing environments selected for larger, stronger hooves, stronger legs, and larger, harder teeth. See Chapter 16 for more on horse evolution.

The result of evolution is a tremendous variety of species. Within particular habitats, these species have evolved complex interrelationships with one another and with their nonliving surroundings. The diversity of species and the complex interrelationships that sustain them are encompassed by the term **biodiversity**. In recent decades, the rate of environmental change has been drastically accelerated by a single species, *Homo sapiens* (modern human). Few species are able to adapt to this change, and in habitats most affected by people, many species are being driven to extinction. This concept is explored further in the box "Earth Watch: Biodiversity Is Threatened by Human Activities."

Knowledge of Biology Illuminates Everyday Life

Some people regard science as a "dehumanizing" activity, feeling that too deep an understanding of the world robs us of vision and awe. Nothing could be further from the truth, as we repeatedly discover anew in our own lives. A few years ago, we watched a bee foraging at a spike of lupines. Lupines have a complicated structure, with two petals on the lower half of the flower enclosing the pollen-laden male reproductive structures (stamens) and sticky female pollen-capturing stigma (Fig. 1-12). In young lupine

---

E A R T H   W A T C H

Biodiversity Is Threatened by Human Activities

Ever since the United Nations' 1992 "Earth Summit" in Rio de Janeiro, Brazil, the word *biodiversity* has jumped out at us from magazines and news articles. What is biodiversity, and why should we be concerned with preserving it? In his article "Defining Biodiversity," biologist and author Peter Raven describes biodiversity as the sum total of all the living organisms in a particular area and all of the interactions among them. As you learned in this chapter, this definition is similar to that of an ecological community. Why should we be concerned with preserving biodiversity and maintaining biological communities?

Figure E1-2 Biodiversity threatened
Destruction of tropical rain forests by indiscriminate logging threatens Earth's greatest storehouse of biological diversity. Interrelationships such as those that have evolved between this *Heliconia* and its hummingbird pollinator sustain these diverse communities and are threatened by human activities.

Over the 3.5-billion-year history of life on Earth, evolution has produced an estimated 8 to 10 million unique and irreplaceable species. Of these, scientists have named only about 1.4 million, and only a tiny fraction of this number has been studied. Evolution has not, however, merely been churning out millions of independent species. Over thousands of years, organisms in a given area have been molded by forces of natural selection exerted by other living species as well as by the nonliving environment in which they live. The outcome is the community, a highly complex web of interdependent life forms whose interactions sustain one another. By participating in the natural cycling of water, oxygen, and other nutrients, by producing rich soil and purifying wastes, these communities contribute to the sustenance of human life as well.

The tropics are home to the vast majority of all the species on Earth, perhaps 7 to 8 million of them, living in complex communities. The rapid destruction of habitats in the tropics as a result of human activities is producing high rates of extinction of many species (Fig. E1-2). Most of these species have never been named, and others never even discovered. Aside from ethical concerns over eradicating irreplaceable forms of life, as we drive unknown organisms to extinction we lose potential sources of medicine, food, and raw materials for industry. The concept of biodiversity has emerged as a result of our increasing concern over the loss of countless forms of life and the habitat that sustains them.

Conservationists have come to understand that the fate of an endangered species is intimately linked to the fate of its community, and that communities must remain intact to sustain the functioning web of life. As biologist and author E. O. Wilson observed: "The loss of species is the folly our descendants are least likely to forgive us."

flowers, the weight of a bee pushing on these petals compresses the stamens, pushing pollen onto the bee's abdomen. In older flowers, the stigma protrudes through the lower petals, and when a pollen-dusted bee visits, it usually leaves behind a few grains of pollen.

Did our new-found insights into the functioning of lupine flowers detract from our appreciation of them? Far from it. Rather, we now looked on lupines with new delight, understanding something of the interplay of form and function, bee and flower, that shaped the evolution of the lupine. A few months later we ventured atop Hurricane Ridge in Olympic National Park in Washington, where the alpine meadows fairly burst with color in August (Fig. 1-13). As we crouched be-

side a wild lupine, an elderly man stopped to ask what we were looking at so intently. He listened with interest as we explained the structure to him; he then went off to another patch of lupines to watch the bees foraging. He too felt the increased sense of wonder that comes with understanding.

We try to convey to you that dual sense of understanding and wonder throughout this text. We also emphasize that biology is not a completed work, but an exploration that we have really just begun. Lewis Thomas, physician and natural philosopher, described our situation in an elegant way: "The only solid piece of scientific truth about which I feel totally confident is that we are profoundly ignorant about nature. Indeed, I regard this as the major discovery of the past hundred years of biology...but we are making a beginning."

We cannot urge you strongly enough, even if you are not contemplating a career in biology, to join in the journey of biological discoveries throughout your life. Don't look upon biology as just another course to take, just another set of facts to memorize. Biology can be much more than that. It can be a pathway to a new understanding of yourself and the living Earth around you.

(a)

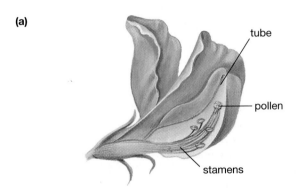

(b)

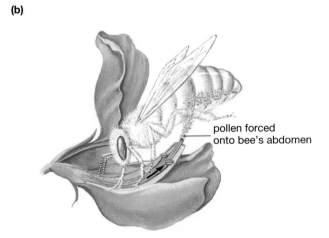

Figure 1-12   Complex adaptations help assure pollination

Lupines, like many other members of the pea family, have complex flowers. **(a)** The reproductive structures are enclosed within the lower petals. In young lupine flowers, the lower petals form a tube within which the stamens fit snugly. The stamens shed pollen within the tube. **(b)** When the weight of a foraging bee pushes on the lower petals, the stamens are thrust forward, and pollen is forced out the end onto her abdomen. Some pollen adheres to the abdomen and may come off on the sticky, pollen-receiving stigma of the next flower visited by the bee.

Figure 1-13   Wild lupines and subalpine fir

Thousands of people visit Hurricane Ridge in Olympic National Park each summer to gaze in awe at Mt. Olympus, but few bother to investigate the wonders at their feet.

✖ S U M M A R Y   O F   K E Y   C O N C E P T S

The Characteristics of Living Things
The quality of life emerges as the result of interactions of emergent properties. Living organisms usually possess the following characteristics: Their structure is complex and organized; they maintain homeostasis; they grow; they acquire energy and materials from the environment; they respond to stimuli; they reproduce; and they have the capacity to evolve.

The Diversity of Life
Living organisms can be grouped into five major categories, called kingdoms: Monera, Protista, Fungi, Plantae, and Animalia. Some features used to classify organisms into kingdoms are the type of cell the organism possesses, the number of cells in each organism, and the mode of acquiring energy.

Cell Type: Eukaryotic cells have their genetic material enclosed within a membrane-bound nucleus. Prokaryotic cells do not have a nucleus.

Cell Number: Organisms may consist of a single cell (unicellular) or of many cells bound together and working cooperatively (multicellular).

Energy Acquisition: Nearly all autotrophic organisms obtain energy by storing sunlight energy in energy-rich molecules through photosynthesis. Heterotrophic organisms obtain energy by eating energy-rich molecules

(food) synthesized in the bodies of other organisms. The food may be eaten in large chunks and broken down (ingestion) or may be absorbed molecule by molecule from the environment (absorption). The features of the five kingdoms are summarized in Table 1-1.

Biology: The Science of Life
Biology is based on the scientific principles of natural causality, uniformity in space and time, and common perception. These principles are assumptions that cannot be directly proved but that are validated by experience. Knowledge in biology is acquired through the application of the scientific method. First, an observation is made. Then a hypothesis is formulated that suggests a natural cause for the observation. The hypothesis is used to predict the outcome of further observations or experiments. A conclusion is then drawn about the validity of the hypothesis. A hypothesis becomes a scientific theory when repeated tests have confirmed it and none have refuted it.

Evolution: The Unifying Concept of Biology
Evolution is the theory that modern organisms descended, with modification, from preexisting life forms. Evolution occurs as a consequence of (1) genetic variation among members of a population, caused by mutation; (2) inheritance of those variations by offspring; and (3) natural selection of the variations that best adapt the organism to its environment.

✖ K E Y   T E R M S

adaptation   p. 12
atom   p. 4
autotroph   p. 8
biodiversity   p. 14
biosphere   p. 4
cell   p. 4
community   p. 4
conclusion   p. 9
control   p. 9
creationism   p. 9
deoxyribonucleic acid (DNA)   p. 6
ecosystem   p. 4
element   p. 4
emergent property   p. 2

energy   p. 4
eukaryotic   p. 8
evolution   p. 12
experiment   p. 9
gene   p. 4
heterotroph   p. 8
homeostasis   p. 5
hypothesis   p. 9
kingdom   p. 6
membrane   p. 4
metabolism   p. 4
molecule   p. 4
multicellular   p. 8
mutation   p. 6

natural causality   p. 9
natural selection   p. 6
nucleus   p. 8
nutrient   p. 4
observation   p. 9
organ   p. 4
organelle   p. 4
organic molecule   p. 4
organism   p. 4
organ system   p. 4
photosynthesis   p. 4
population   p. 4
prokaryotic   p. 8
scientific method   p. 9

species   p. 4
spontaneous generation   p. 10
subatomic particle   p. 4

theory   p. 11
tissue   p. 4
unicellular   p. 8

variable   p. 9

❆ THINKING THROUGH THE CONCEPTS

Multiple Choice

1. A requirement that is NOT involved in critical thinking is
   a. an intolerance for ambiguity
   b. questioning the facts
   c. examining the whole experimental system
   d. defining terms
   e. discovering biases

2. A scientist examines a group of cells under the microscope. She notices the presence of nuclei in these cells. Chemical tests reveal that each cell is surrounded by a cell surface covering composed of cellulose. She concludes that the cells come from an organism that is a member of the Kingdom
   a. Monera
   b. Protista
   c. Plantae
   d. Fungi
   e. Animalia

3. Which statement is correct?
   a. Genes are proteins that produce DNA.
   b. Each organism has its own unique DNA code.
   c. The DNA of humans is more similar to the DNA of birds than it is to that of apes.
   d. Differences among organisms reflect different nucleotide sequences in their DNA.
   e. Each DNA molecule is a single strand of nucleotides.

4. Choose the answer that best describes the scientific method.
   a. observation, hypothesis, experiment, absolute proof
   b. guess, hypothesis, experiment, conclusion
   c. observation, hypothesis, experiment, conclusion
   d. hypothesis, experiment, observation, conclusion
   e. experiment, observation, hypothesis, conclusion

5. Which of the following are characteristics of living things?
   a. They reproduce.
   b. They respond to stimuli.
   c. They are complex and organized.
   d. They acquire energy.
   e. all of the above

6. The three natural processes that form the basis for evolution are
   a. adaptation, natural selection, and inheritance
   b. predation, genetic variation, and natural selection
   c. mutation, genetic variation, and adaptation
   d. fossils, natural selection, and adaptation
   e. genetic variation, inheritance, and natural selection

Review Questions

1. What are the differences between a salt crystal and a tree? Which is living? How do you know? How would you test your "knowledge"? What controls would you use?

2. What is the difference between a theory and a hypothesis? Explain how each is used by scientists.

3. Define and explain the terms: natural selection, evolution, mutation, selective breeding, creationism, and population.

4. Starting with the cell, list the hierarchy of organization of life, briefly explaining each level.

5. Define homeostasis. Why must organisms continually acquire energy and materials from the external environment to maintain homeostasis?

6. Describe the scientific method. In what ways do you use the scientific method in everyday life?

7. What is evolution? Briefly describe how evolution occurs.

❆ APPLYING THE CONCEPTS

1. Natural occurrences in the heavily populated state of California, including earthquakes, heavy rains, and grass and forest wildfires have reduced the quality of life for many Californians. Look at it another way for a moment. What biomes are found in this state? What impacts have humans had on these natural ecosystems? Are changes in these areas the reason for fires and floods and mudslides? What needs to be done to alter the balance of "humans and nature"?

2. Design an experiment to test the effects of a new dog food, "Super Dog," on the thickness and water-shedding properties of the coats of golden retrievers. Include all the parts of a scientific experiment. Design objective methods to assess coat thickness and water-shedding ability.

3. Science is based on principles, including uniformity in space and time, and common perception. Assume humans one day encounter intelligent beings from another galaxy who evolved under very different conditions. Discuss the two principles mentioned above and how they would affect (1) the nature of scientific observations on the different planets and (2) communications about these observations.

▓ F O R   M O R E   I N F O R M A T I O N

Attenborough, D. *Life on Earth*. Boston: Little, Brown, 1979. Gorgeously illustrated and beautifully written introduction to the diversity of life on Earth; the inspiration for our title.

Leopold, A. *A Sand County Almanac*. New York: Oxford University Press, 1949 (reprinted in 1989). A classic by a natural philosopher; provides an eloquent foundation for the conservation ethic.

Swain, R. B. *Earthly Pleasures*. New York: Charles Scribner's Sons, 1981. Insightful essays stress the interrelatedness and diversity of life.

Thomas, L. *The Medusa and the Snail*. New York: Bantam Books, 1980. The late Lewis Thomas, physician, researcher, and philosopher, shares his awe of the living world in a series of delightful essays.

N E T   W A T C H

On-line resources for this chapter are on the World Wide Web at:
http://www.prenhall.com/~audesirk (click on the <u>table of contents</u> link and then select Chapter 1).

Unit 1
The Life of a Cell

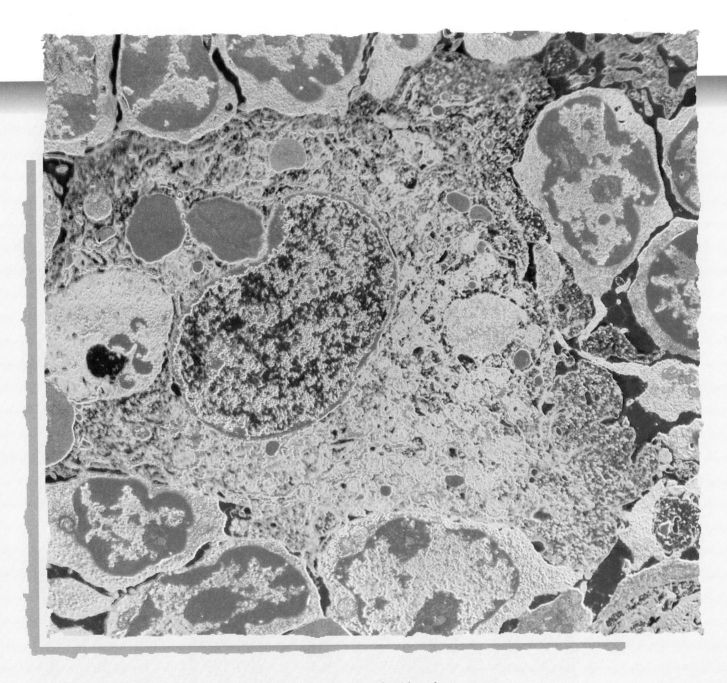

Two types of white blood cells are shown inside a human lymph node in this false-color transmission electron micrograph. A large macrophage, colored in green and yellow, is surrounded by many smaller lymphocytes.

2 Atoms, Molecules, and Life

"If there is magic on this planet, it is contained in water."

LOREN EISELEY in *The Immense Journey*

▦ AT A GLANCE

Matter and Energy
  Energy Exists in Several Forms That Can Be Interconverted

The Structure of Matter
  Atoms, the Smallest Units of Elements, Are Composed of
    Still Smaller Particles
  Atoms Will React with Other Atoms Only When There
    Are Vacancies in Their Outermost Electron Shells

Chemical Bonds: Joining Atoms to Make Molecules
  Atoms Can Become Stable by Gaining or Losing Electrons,
    Forming Charged Ions; the Attraction between
    Oppositely Charged Ions Produces Ionic Bonds
  Atoms Can Become Stable by Sharing Electrons with Other
    Atoms, Forming Covalent Bonds
  Hydrogen Bonds Are Weak Electrical Attractions between
    the Polar Parts of Molecules

Organic and Inorganic Molecules

Water and Life
  Water Interacts with Many Other Molecules
  The pH of a Solution Is a Measure of Its Concentration of
    Hydrogen Ions
  A Buffer Helps Maintain a Solution at a Relatively
    Constant pH
  Water Moderates the Effects of Temperature Changes
  Water Forms an Unusual Solid, Ice
  Water Molecules Tend to Stick Together

Michelangelo's David, Accademia, Florence, Italy

Why do biology texts begin with a description of chemistry? Aren't we trying to learn about the biological equivalent of Michelangelo's famous sculpture *David* rather than rocks? About the cathedral of Notre Dame rather than stone and mortar? Yes we are, but to understand those works of art and architecture we must first understand basic principles. Why did Michelangelo sculpt *David* from fine-grained marble, not from slate? Why are the arches of the cathedral roof shaped the way they are, and why are the walls studded with support structures called flying buttresses? A sculptor needs an intimate knowledge of his materials, or else his statue may crack apart midway through its creation. An architect must know a good deal about engineering, or else her roof may collapse and her walls may buckle.

Similarly, to understand the structure and function of living organisms, we need a basic understanding of the atoms and molecules of which organisms are composed and how interactions among those atoms and molecules produce growth, movement, and all the other things that organisms do. How does photosynthesis convert the energy of sunlight into the energy of sugar molecules? How do nerve cells in your brain communicate with one another? What causes cancer? What is acid rain, and why does it endanger the life of many freshwater lakes?

The evolution of life, in fact, is restricted to certain paths by the structure of matter and the laws of physics that govern interactions among pieces of matter. When you learn about the forces that allow water to move to the top of a tree, you also learn one reason why trees don't grow a thousand meters tall (the columns of water extending from root to leaf would break apart under their own weight). When you learn about energy and chemical reactions, you also learn why there can be huge herds of herbivores (plant-eating animals), such as bison or wildebeest, but not similar-sized herds of wolves or lions (most of the energy that herbivores get from eating plants is used up in their own life processes and is not available to the predators that eat them).

This chapter will introduce some of the fundamental properties of energy and matter, the structure of atoms, and how atoms interact with one another to form molecules. Later chapters in this section of the text will describe the molecules that form the bodies of all living organisms; the structure and function of cells, which are remarkably similar in everything from bacteria to elephants; and the chemical reactions that allow the acquisition and use of energy by individual cells and entire organisms.

Matter and Energy

Technically speaking, matter and energy are interchangeable, as expressed by Albert Einstein's famous equation $E = mc^2$: energy equals matter times the speed of light $(c)^2$. For the chemical reactions that occur within living organisms, however, we can treat matter and energy as quite distinct from one another: **Matter** is the physical material of the universe; **energy** is the capacity to do work, usually manifested by moving pieces of matter from place to place.

Energy Exists in Several Forms That Can Be Interconverted

Energy can exist in several forms and may be converted from one form to another. The two major categories of energy are kinetic energy and potential energy. **Kinetic energy** is energy of movement. This includes not only movement of large objects such as a football hurtling through the air but also other forms of movement such as electrical energy (movement of electrons) and heat (movement of atoms and molecules). **Potential energy** is "stored" energy that can be released as kinetic energy under the right conditions. A coconut high in a palm tree has potential energy owing to its location. Its potential energy is converted into kinetic energy when it falls. The food you eat has chemical potential energy, some of which is converted into kinetic energy of movement and heat when you run (Fig. 2-1).

The Structure of Matter

An **element**, such as carbon or oxygen, is a substance with specific properties; it can neither be broken down nor converted to different substances by ordinary chemical reactions. Elements are the building blocks of all living and nonliving matter. Although there are 92 naturally occurring elements in our universe, most are quite rare, and only a relative handful are essential to life on Earth. Table 2-1 lists the most common elements in the universe, Earth, and the human body. Atoms, described below, are the smallest unit of an element that retain the element's properties.

A **compound** is a substance composed of precise proportions of two or more elements, joined together in a specific geometric pattern. Water, for example, consists of one part oxygen and two parts hydrogen, in a precise arrangement (see Fig. 2-6).

A **mixture** is composed of two or more compounds or elements, in variable proportions. For example, a can of soda

Figure 2-1   The conversion of chemical potential energy into kinetic energy

When 20,000 people run the New York marathon each year, they do more than just make the bridges quake. They also collectively convert over 50,000,000 kilocalories of food into movement, heat, and sore muscles—that's equivalent to burning off over 6400 kilograms (14,000 pounds) of fat or eating almost 14,000 kilograms (about 30,000 pounds) of chocolate cake.

is a mixture that contains, among other things, water, sugar, and variable amounts of carbon dioxide (which makes it fizz), depending on how long ago you popped open the top.

Atoms, the Smallest Units of Elements, Are Composed of Still Smaller Particles

If you took a diamond (a form of carbon) and cut it into pieces, each piece would still be carbon. If you could continue to make finer and finer divisions, you would eventually produce a pile of carbon **atoms**, the smallest possible particles of the element carbon (each atom, of course, would be too small to see even in a powerful microscope). Atoms themselves, however, are composed of two parts: a central **nucleus** and one or more outer **electron shells** (Fig. 2-2). Within the nucleus and electron shells reside subatomic particles. The nucleus contains heavy, positively charged **protons** and equally heavy but uncharged **neutrons**. The electron shells contain light, negatively charged **electrons**. The positive charge of a proton is designated by a plus (+) sign, and the negative charge of an electron by a minus (−) sign. An atom by itself has an equal number of electrons and protons and is therefore electrically neutral.

Nuclei and electron shells play complementary roles in the structure and function of atoms. The nuclei provide stability, and the electron shells allow interactions with other atoms. Nuclei resist disturbance by outside forces. Ordinary sources of energy, such as heat, electricity, and light, hard-

TABLE 2-1 ❖ *Common Elements Important in Living Organisms*

Element	Symbol	Atomic Number[a]	Percent in Universe[b]	Percent in Earth[b]	Percent in Human Body[b]
Hydrogen	H	1	91	0.14	9.5
Helium	He	2	9	Trace	Trace
Carbon	C	6	0.02	0.03	18.5
Nitrogen	N	7	0.04	Trace	3.3
Oxygen	O	8	0.06	47	65
Sodium	Na	11	Trace	2.8	0.2
Magnesium	Mg	12	Trace	2.1	0.1
Phosphorus	P	15	Trace	0.07	1
Sulfur	S	16	Trace	0.03	0.3
Chlorine	Cl	17	Trace	0.01	0.2
Potassium	K	19	Trace	2.6	0.4
Calcium	Ca	20	Trace	3.6	1.5
Iron	Fe	26	Trace	5	Trace

[a]Atomic number = number of protons in the nucleus.

[b]Approximate percentage of atoms of this element, by weight, in the universe, in Earth's crust, and in the human body.

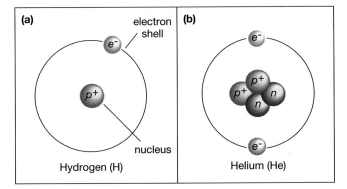

Figure 2-2   "Planetary" Atomic Models

Structural representations of the two smallest atoms, hydrogen **(a)** and helium **(b)**. In these "planetary" models, the electrons are imagined as miniature planets, circling in precisely defined orbits around a nucleus that contains protons and neutrons. Although not strictly accurate, planetary models are easy to visualize and understand.

ly affect them at all. The stability of its nucleus is the reason why a carbon atom remains carbon whether it is part of a diamond, carbon dioxide, or sugar. Electron shells, however, are dynamic; as you will soon see, atoms interact with one another by gaining, losing, or sharing electrons.

The Atomic Nucleus Contains a Fixed Number of Protons; This Number Is a Unique Characteristic of Each Element

The number of protons in the nucleus, called the **atomic number**, is characteristic of each element. For example, every hydrogen atom has one proton in its nucleus, every carbon atom has six, and every oxygen atom has eight (see Table 2-1). Different atoms of a given element may, however, have different numbers of neutrons. Atoms with the same num-

ber of protons (that is, atoms of the same element) but with different numbers of neutrons are called **isotopes**. All the isotopes of an element are virtually identical in their chemical reactions but may vary in their physical properties. The nuclei of some isotopes spontaneously disintegrate, releasing radioactivity as they do. Radioactive isotopes are extremely useful as "labels" in studying biological processes (see "Scientific Inquiry: The Use of Isotopes in Biology and Medicine").

Electrons Orbit the Nucleus at Fixed Distances, Forming Electron Shells That Correspond to Different Energy Levels

At one time, it was believed that electrons circle the nucleus in well-defined orbits, like tiny planets orbiting the sun. But electrons are much more unpredictable than planets. They zip about outside the nucleus in regions of space called **orbitals**. At different times a given electron may be found anywhere within its orbital. Although it is not strictly accurate, here we will use the "planetary" model of electrons orbiting at fixed distances from their central nucleus within electron shells because it helps us to visualize how they interact with electrons of other atoms to form chemical bonds (Fig. 2-2).

Electrons repel one another, owing to their negative electrical charge. As you may know from experimenting with a magnet, like charges repel and opposite charges attract. Electrons, with their negative charge, are drawn to the positively charged protons of the nucleus. Because of their mutual repulsion, however, only limited numbers of electrons can occupy the space closest to the nucleus. Large atoms accommodate many electrons by having several orbits at increasing distances from their nucleus. Because the electrons orbit through a three-dimensional space, their orbits are called electron shells (Fig. 2-3).

The electron shell closest to the nucleus is the smallest and can hold only two electrons. The second shell can hold up to eight electrons. The electrons in an atom normally occupy the shells closest to the nucleus. Thus a carbon atom, with six electrons, has two electrons in the first shell, closest to the nucleus, and four electrons in its second shell (see Fig. 2-3).

Electrons in shells closest to the nucleus have the least energy and those farthest away have the most. For this reason, electron shells are often called **energy levels**.

Why do we speak of energy *levels* and *shells*? Because electron shells are found only at specific distances from the nucleus—distances that correspond to discrete energy levels. Most people aren't used to thinking about energy as occurring in steps, but perhaps an analogy with gravity can help. Picture yourself climbing a flight of stairs. You must spend a certain amount of kinetic energy to climb each step, and no intermediate amount will suffice. In other words, you can't lift your foot half a step and stand between steps. The energy used in climbing each step is stored as the potential energy; the higher the position relative to the bottom of the stairs, the greater the potential energy. This potential energy can be transformed back into kinetic energy by jumping down the steps: the higher up you start, the harder you land.

Energy levels in atoms are analogous to the staircase (Fig. 2-4). It takes energy to force an electron out of a low-energy shell into a higher-energy shell. An electron cannot occupy an intermediate position between shells, so it takes a certain minimum amount of energy to boost an electron to the next shell. But an atom is unstable when electrons have been boosted from lower- to higher-energy shells. Eventually, the electron returns to the low-energy shell, releasing the stored energy as it does (as you do when you jump down stairs you've just climbed). When plants photosynthesize, the energy of sunlight kicks electrons in chlorophyll molecules out of their low-energy shells to higher-energy shells. As the electrons return to the low-energy shells, some of their "extra" energy is used to power the synthesis of sugar from carbon dioxide and water, as we will describe in Chapter 7.

Atoms Will React with Other Atoms Only When There Are Vacancies in Their Outermost Electron Shells

Just as an atom is the smallest possible particle of an element, so too a **molecule** is the smallest possible particle of a compound. A molecule consists of two or more atoms (of the same or different elements) held together by interactions among their outermost electron shells. The basic principles of atomic reactivity are these:

1. An atom is stable (that is, it will not react with other atoms) when its outermost electron shell is completely full or completely empty.
2. An atom is reactive when its outermost electron shell is only partially filled.

To illustrate these principles, consider the two smallest atoms, hydrogen and helium (see Fig. 2-2). Hydrogen has one proton in its nucleus, and one electron in its outermost electron shell, which can hold two. Helium has two protons in its nucleus and two electrons completely fill its outermost electron shell. Therefore, we predict that helium atoms, with a full shell, should be stable, and hydrogen atoms, with a half-empty shell, should be highly reactive. Our predictions are correct. Except for nuclear fusion reactions, helium is almost completely unreactive. Hydrogen, however, is highly flammable; that is, it reacts readily with oxygen in an explosive way. For example, the space shuttle and many other rockets use liquid hydrogen as fuel to power lift-off.

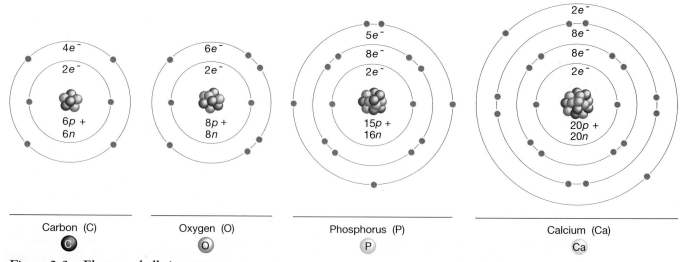

Figure 2-3 Electron shells in atoms

Most biologically important atoms have at least two shells of electrons. The first shell, closest to the nucleus, can hold two electrons; the next three shells usually hold a maximum of eight electrons each.

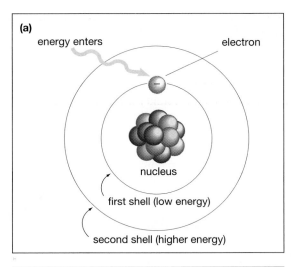

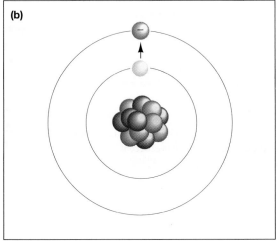

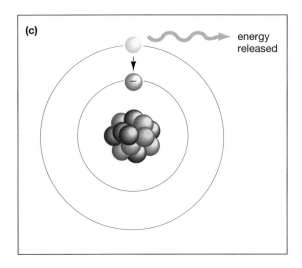

Figure 2-4   **Energy absorption and emission by an atom**

The electron shells of an atom are found at discrete energy levels; the lowest energy level is the one closest to the nucleus. **(a)** Energy (such as electrical energy or light) enters an atom and is absorbed by an electron. **(b)** Only the specific amounts of energy that boost the electron from a low-energy inner shell to a higher-energy outer shell can be absorbed. Intermediate amounts of energy are not absorbed. **(c)** The electron spontaneously returns to a low-energy shell. The "extra" energy is released. It may drive energy-requiring chemical reactions (for example, photosynthesis) or be emitted, often in the form of light.

Chemical Bonds: Joining Atoms to Make Molecules

An atom with an outermost electron shell that is partially full can gain stability by losing electrons (emptying the shell completely), gaining electrons (filling the shell), or sharing electrons with another atom (with both atoms behaving as though they had full shells). Losing, gaining, and sharing electrons result in forces called **chemical bonds** that hold atoms together in molecules. There are two major types of chemical bonds: ionic and covalent. Chemical bonds are not *things*, such as glue, but are *attractive forces* that hold atoms together. Each element has bonding properties that result from the configuration of electrons in its outer shell (see Table 2-2).

Atoms Can Become Stable by Gaining or Losing Electrons, Forming Charged Ions; the Attraction between Oppositely Charged Ions Produces Ionic Bonds

Some atoms have almost empty outermost electron shells, and some atoms have almost full outermost shells. Both types can become stable by losing electrons, completely emptying their outermost shells, or by gaining electrons, completely filling their outermost shells. The formation of table salt—sodium chloride—illustrates this principle. Sodium (Na) has an outermost electron shell containing only one electron, and chlorine (Cl) has an outer shell that lacks only one electron to be completely full (Fig. 2-5a). Sodium, therefore, can become stable by losing the electron from its outer shell, leaving that shell empty, and chlorine can fill its outer shell by gaining an electron. This is precisely what each of them does to form sodium chloride: Sodium loses an electron, becoming positively charged (Na$^+$), and chlorine picks up an electron, becoming a negatively charged chloride ion (Cl$^-$) (Fig. 2-5b). Atoms that have lost or gained electrons, altering the balance between protons and electrons, are charged. These charged atoms are called **ions**.

Opposite charges attract; therefore, sodium ions and chloride ions tend to stay near one another, forming crystals that contain repeating orderly arrangements of the two ions (Fig. 2-5c). The electrical attraction between oppositely charged ions that holds them together in crystals is called an **ionic bond**.

Atoms Can Become Stable by Sharing Electrons with Other Atoms, Forming Covalent Bonds

An atom with a partially full outermost electron shell can also become stable by sharing electrons with another atom in a **covalent bond**. To understand what this means, consider the hydrogen atom, which has one electron in a shell built for two. A hydrogen atom can become reasonably stable if it shares its single electron with another hydrogen atom, forming a molecule of hydrogen gas, H_2 (Fig. 2-6a). Because the two hydrogen atoms are identical, neither nucleus can exert more attraction and capture the other's electron. So both electrons spend part of the time in the shell of both hydrogen atoms, with the result that each hydrogen behaves almost as if it had two electrons in its shell.

The stability of a covalent bond depends on the atoms involved. Some covalent bonds, such as those in water (H_2O; Fig. 2-6c) and carbon dioxide (CO_2), are extremely stable—that is, it takes a lot of energy to break the bonds. Other bonds, such as those in oxygen gas (O_2; Fig. 2-6b) or gasoline, are less stable, coming apart more easily. When a chemical reaction occurs in which less stable bonds are broken and more stable bonds are formed (such as burning gasoline with oxygen to form carbon dioxide and water), energy is released (see Chapter 4).

Most Biological Molecules Utilize Covalent Bonding

The atoms in most biological molecules are joined by covalent bonds. Hydrogen, carbon, oxygen, nitrogen, phosphorus, and sulfur are the most common atoms found in biological molecules. Except for hydrogen, each of these atoms lacks two or more electrons to fill its outermost electron shell and can share electrons with two or more other atoms. Hydrogen can form a covalent bond with one other atom, oxygen and sulfur with two, nitrogen with three, and phosphorus and carbon with up to four other atoms. This diversity of bonding arrangements permits biological molecules to be constructed in almost infinite variety and complexity.

Sometimes, an atom will share two pairs of electrons with another atom, forming a **double covalent bond** (each atom contributes two electrons; see Fig. 2-6b). If atoms share three pairs of electrons, a **triple covalent bond** is formed. Double and triple bonds create further variety in the shapes and functions of biological molecules. Table 2-2 summarizes bonding patterns in biological molecules.

Polar Covalent Bonds Form When Atoms Share Electrons Unequally

Electron sharing in covalent bonds is not always equal. In hydrogen gas, the two nuclei are identical, and the shared electrons spend equal time near each nucleus. Therefore, not only is the molecule as a whole electrically neutral, but each end, or pole, of the molecule is also electrically neutral. Such an electrically symmetrical bond is called a **nonpolar covalent bond**. In many other molecules, one nucleus may attract the electrons more strongly than the other nucleus does because its nucleus has a larger positive charge. This situation produces a **polar covalent bond**. Although the molecule as a whole is electrically neutral, it has charged parts: The atom that attracts the electrons more strongly has a slightly negative charge (the negative pole of the molecule), and the other atom has a slightly positive charge (the positive pole). In water, for example, oxygen attracts electrons more strongly than hydrogen does, so the oxygen end of a water molecule is negative and each hydrogen is positive (Fig. 2-6c). Water with its charged ends can be called a polar molecule.

Hydrogen Bonds Are Weak Electrical Attractions between the Polar Parts of Molecules

Because of the polar nature of their covalent bonds, nearby water molecules attract one another, the partially negatively charged oxygens of some molecules attracting the

Atom	Capacity of Outer Electron Shell	Electrons in Outer Shell	Number of Covalent Bonds Usually Formed	Common Bonding Patterns
Hydrogen	2	1	1	—H
Carbon	8	4	4	—C— —C= —C≡ =C=
Nitrogen	8	5	3	—N— —N= N≡
Oxygen	8	6	2	—O— —O=
Phosphorus	8	5	5	—P=
Sulfur	8	6	2	—S—

TABLE 2-2 ⬡ *Bonding Patterns of Atoms Commonly Found in Biological Molecules*

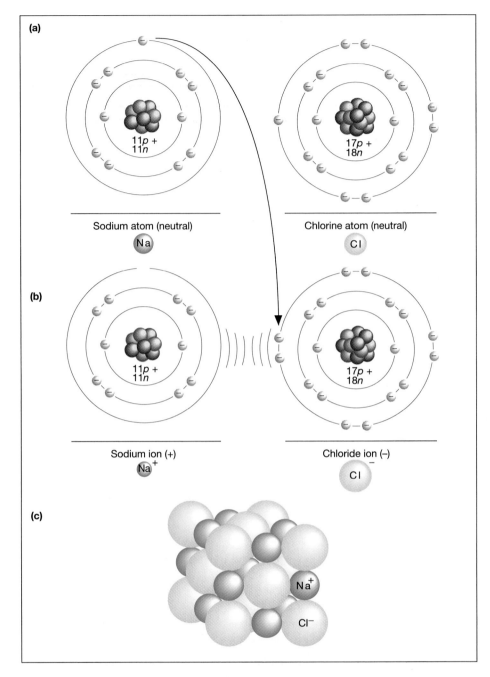

(a)

Sodium atom (neutral)
Na

Chlorine atom (neutral)
Cl

11p +
11n

17p +
18n

(b)

Sodium ion (+)
Na⁺

Chloride ion (−)
Cl⁻

11p +
11n

17p +
18n

(c)

Na⁺

Cl⁻

Figure 2-5 **The formation of ions and ionic bonds**

(a) Sodium has only one electron in its outer electron shell; chlorine has seven (lacking only one electron to fill the outer shell). **(b)** Sodium can become stable by losing an electron, and chlorine can become stable by gaining an electron. When this exchange happens, sodium has an empty and chlorine a full outer shell. Sodium becomes a positively charged ion, and chlorine a negatively charged ion. **(c)** Because oppositely charged particles attract one another, the resulting sodium and chloride ions nestle closely together in a crystal of salt, NaCl. Here we have used different-colored balls to represent entire ions: sodium (Na⁺) and chloride (Cl⁻). You will see balls representing atoms in several other figures as well.

partially positively charged hydrogens of other molecules. This electrical attraction is called a **hydrogen bond** (Fig. 2-7). As we will see shortly, hydrogen bonds between molecules give water several unusual properties that are essential to life on Earth.

Many biological molecules also form hydrogen bonds. Both nitrogen and oxygen atoms attract electrons more strongly than hydrogen atoms do. Therefore, the nitrogen or oxygen pole of a nitrogen-hydrogen or oxygen-hydrogen bond is slightly negative, and the hydrogen pole is slightly positive. The resulting polar parts of the molecules can form hydrogen bonds with water, with other biological molecules, or with distant, polar parts of the same molecule. While hydrogen bonds individually are quite weak, many of them working together are quite strong. As

we will see in Chapter 3, hydrogen bonds play crucial roles in shaping the three-dimensional structures of proteins and DNA.

Organic and Inorganic Molecules

Chemists classify molecules as organic or inorganic. The word *organic* originally signified that these molecules could be manufactured only within living organisms. Today, however, organic chemists can synthesize many of these molecules in the laboratory. Thus, the modern definition of an **organic molecule** is any molecule containing both carbon and hydrogen. Most organic molecules are large, with complex structures. **Inorganic molecules** include carbon dioxide and all molecules without carbon.

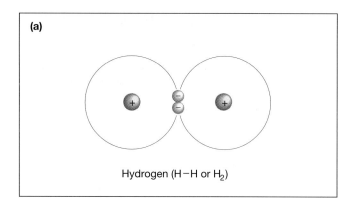

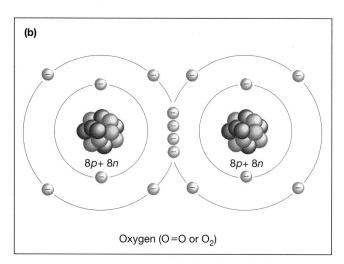

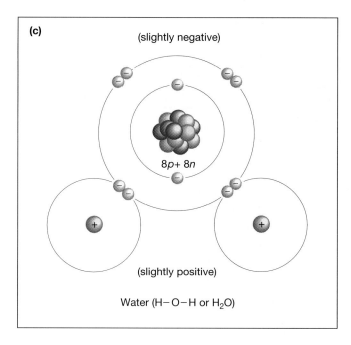

Figure 2-6   Covalent bonds

Electrons are shared between atoms to form covalent bonds. **(a)** In hydrogen gas, one electron from each hydrogen atom is shared, forming a single covalent bond. The resulting bond is usually represented by a single dash connecting the atomic symbols, and the molecule of hydrogen gas may be represented as H–H or H_2. **(b)** In oxygen gas, two oxygen atoms share four electrons, making a double bond, usually represented by a double dash or equal sign (O=O or O_2). **(c)** Oxygen lacks two electrons to fill its outer shell, so oxygen can make two single bonds, one with each of two hydrogen atoms to form water (H–O–H or H_2O). Oxygen exerts a greater pull on the electrons than does hydrogen, and so the oxygen end of the molecule has a slight negative charge, and the hydrogen end has a slight positive charge. This is an example of polar covalent bonding, and the water molecule with its slightly charged ends is called a polar molecule.

Many inorganic molecules are important to living organisms, including, for instance, minerals in the soil and carbon dioxide in the air, which are the raw materials that plants use to construct their bodies. One inorganic molecule, however, is extraordinarily abundant on Earth, has unusual properties, and is so essential to life that it merits special consideration. That molecule is water.

Water and Life

Life almost certainly arose in the waters of the primeval Earth. The first life form was probably a small sac enclosing water with an array of dissolved enzymes and simple genet-

ic material. Living organisms are still composed of about 60% to 90% water, and all life on Earth depends intimately on the properties of water. Why is water so crucial to life?

Water Interacts with Many Other Molecules

Water is an extremely good **solvent**—that is, it is capable of dissolving a wide range of substances, especially salts. Recall that a crystal of table salt is held together by the electrical attraction between positively charged sodium ions and negatively charged chloride ions (Fig. 2-5). Because water is a polar molecule, it has positive and negative ends. If a salt crystal is dropped into water, the positively charged hydrogen ends of water molecules will be attracted to and sur-

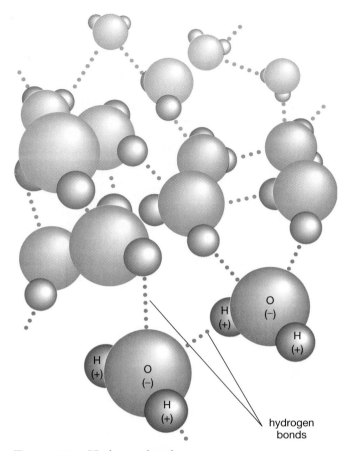

Figure 2-7   Hydrogen bonds

The partial charges on different parts of water molecules produce weak attractive forces called hydrogen bonds (dotted lines) between the hydrogens of one water molecule and the oxygens of other molecules.

round the negatively charged chloride ions, and the negatively charged oxygen ends of water molecules will surround the positively charged sodium ions. As water molecules enclose the sodium and chloride ions and shield them from interacting with each other, the ions separate from the crystal and drift away in the water—the salt dissolves (Fig. 2-8).

Water dissolves polar molecules in a similar fashion, as its positive and negative poles are attracted to oppositely charged regions of dissolving molecules. Charged and polar molecules are termed **hydrophilic** (Greek for "water-loving"), because of their electrical attraction for water molecules. Many biological molecules, such as sugars and amino acids, are hydrophilic and dissolve readily in water (Fig. 2-9). Water also dissolves gases such as oxygen and carbon dioxide.

Other liquids can dissolve some of these substances, but not all of them. Alcohol, for example, dissolves some sugars and proteins but not salts. By dissolving such a wide variety of molecules, the watery substance inside a cell provides a suitable environment for the countless chemical reactions essential to life on Earth.

Molecules that are uncharged and nonpolar, such as fats and oils, usually do not dissolve in water, and hence are

called **hydrophobic** ("water-fearing"). Nevertheless, water has an important effect on such molecules. Oils, for example, form globs when spilled into water. Why? In pure water, each water molecule can form hydrogen bonds with the maximum possible number of other water molecules. If two molecules of oil are added, at distant locations, each disrupts the hydrogen bonding of nearby water molecules. Any water molecule directly adjacent to the oil molecules cannot make hydrogen bonds in the direction of the oil.

Suppose now that the two oil molecules, randomly moving in the water, encounter one another. The instant they make contact, their nonpolar surfaces nestle closely together, surrounded by water molecules that form hydrogen bonds with one another but not with the oil. The two-molecule glob has a smaller surface area than the two separate molecules did, so there is less interference with hydrogen bonding among water molecules. To separate again, the oil molecules would have to shove apart the surrounding water molecules. Because it takes energy to break the hydrogen bonds interconnecting the water molecules, the water tends to prevent the oil molecules from moving apart. The tendency for hydrophobic molecules to clump together in watery solutions is often termed a **hydrophobic interaction**, although in fact it is the hydrogen bonding between water molecules that keeps the hydrophobic molecules together, rather than any special affinity among the hydrophobic molecules.

As we will see in Chapter 5, the membranes of living cells owe much of their structure to the organizing effect of water on hydrophobic molecules.

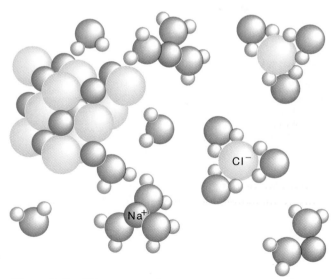

Figure 2-8   Water as a solvent

The polarity of water molecules allows water to dissolve polar and charged substances. When a salt crystal is dropped into water, the water molecules worm their way around the outsides of the sodium and chloride ions, surrounding them with oppositely charged ends of the water molecules. Thus insulated from the attractiveness of other molecules of salt, the ions float away, and the whole crystal gradually dissolves.

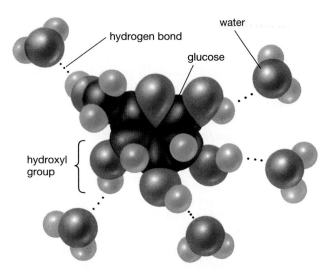

Figure 2-9 **Water dissolves many biological molecules**
Many biological molecules dissolve in water because they have polar parts, for example, OH⁻ (hydroxyl) groups, that can form hydrogen bonds with water molecules. This illustration depicts the hydrogen bonds that can form between the hydroxyl groups on a glucose molecule (a simple sugar) and surrounding water molecules.

Water Takes Part in Many Chemical Reactions

Water enters into many of the chemical reactions that occur in living cells. The oxygen that green plants release into the air is derived from water during photosynthesis. When your body manufactures a protein, fat, nucleic acid, or sugar, it produces water in the process, and, conversely, when you digest proteins, fats, and sugars in the foods you eat, water is used in the reactions.

The pH of a Solution Is a Measure of Its Concentration of Hydrogen Ions

Although water is generally regarded as a stable compound, individual water molecules constantly gain, lose, and swap hydrogen atoms. As a result, at any given time about two of every billion water molecules are ionized; that is, broken apart into hydrogen ions (H^+) and hydroxide ions (OH^-):

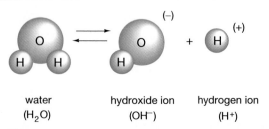

Pure water contains equal concentrations of hydrogen ions and hydroxide ions. In many solutions, though, the concentrations of H^+ and OH^- ions are not the same. If the concentration of H^+ ions exceeds the concentration of OH^- ions, the solution is **acidic**; if the concentration of OH^- ions is greater, the solution is **basic**. The degree of acidity is expressed on the **pH scale** (Fig. 2-10), in which neutrality (equal num-

bers of H^+ and OH^- ions) is assigned the number 7. Acids have a pH below 7; bases have a pH above 7. Pure water, with equal concentrations of H^+ and OH^-, has a pH of 7. Each unit on the pH scale represents a tenfold change in the concentration of H^+ ions. Thus, a cola drink (pH = 3) has 10,000 times the concentration of H^+ than does water (pH = 7).

An **acid** is a substance that gives off hydrogen ions. When hydrochloric acid (HCl), for example, is added to pure water, almost all of the HCl molecules separate into H^+ and Cl^- ions. Therefore, the concentration of H^+ ions greatly exceeds the concentration of OH^- ions, and the resulting solution is acidic. A **base** is a substance that combines with H^+ ions and thereby reduces their number. For instance, if sodium hydroxide (NaOH) is added to water, the NaOH molecules separate into Na^+ and OH^- ions. The OH^- ions combine with H^+ ions, reducing their number. The resulting solution is basic.

A Buffer Helps Maintain a Solution at a Relatively Constant pH

In most mammals, including humans, both the cell cytoplasm and the fluids that bathe the cells are nearly neutral (pH about 7.3 to 7.4). Small increases or decreases in pH may cause drastic changes in both structure and function, leading to the death of cells or entire organisms. Nevertheless, living cells seethe with chemical reactions that take up or give off H^+ ions. How, then, does the pH remain constant? The answer lies in the many buffers found in living organisms. A **buffer** is a compound that tends to maintain a solution at a constant pH by accepting or releasing H^+ ions in response to small changes in H^+ ion concentration. If the H^+ ion concentration rises, buffers combine with them; if the H^+ ion concentration falls, buffers release H^+. The result is that the concentration of H^+ ions is restored to its original level. Common buffers in living organisms include bicarbonate (HCO_3^-) and phosphate ($H_2PO_4^-$ and HPO_4^{2-}), both of which can accept or release H^+ ions, depending on the circumstances. If the blood becomes too acidic, for example, bicarbonate accepts H^+ ions to form carbonic acid:

$$HCO_3^- \quad + \quad H^+ \quad \rightarrow \quad H_2CO_3$$
(bicarbonate)   (hydrogen ion)   (carbonic acid)

If the blood becomes too basic, carbonic acid liberates H^+ ions that combine with the excess OH^- ions, forming water:

$$H_2CO_3 \quad + \quad OH^- \quad \rightarrow \quad HCO_3^- \quad + \quad H_2O$$
(carbonic acid) (hydroxide ion) (bicarbonate) (water)

In either case, the end result is that the blood pH remains near its normal value.

Water Moderates the Effects of Temperature Changes

Organisms can survive only within a limited temperature range: Both high and low temperatures can be fatal. High temperatures may damage the protein enzymes that guide the chemical reactions essential to life (see Chapter 4):

Most enzymes cease to function at temperatures well below boiling. Low temperatures are also dangerous, because enzyme action slows as the temperature drops. Subfreezing temperatures within the body are usually lethal, because cells are ruptured by spearlike ice crystals.

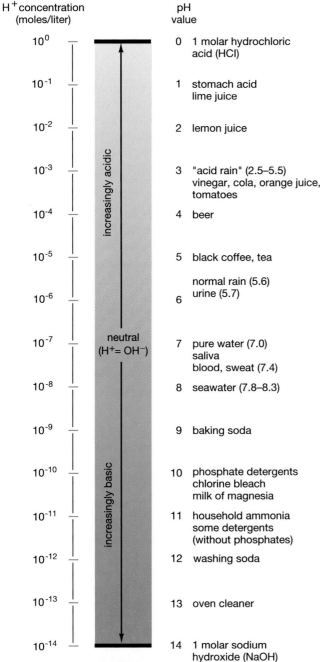

Figure 2-10   The pH scale

The pH scale expresses the concentration of hydrogen ions in a solution on a scale of 0 (very acidic) to 14 (very basic). Each unit of change in pH on the pH scale represents a tenfold change in the concentration of hydrogen ions. Lemon juice, for example, is about 10 times more acid than orange juice, and the most severe acid rains in the northeastern United States are almost 1000 times more acidic than normal rainfall. Except for the insides of your stomach, nearly all the fluids in your body are finely adjusted to a pH of 7.4.

Water has three properties that moderate the effects of temperature changes. First, water has a high **specific heat**, the amount of energy needed to raise the temperature of 1 gram of a substance by 1° C. Temperature reflects the speed of molecules: the higher the temperature, the greater the average speed. Generally speaking, if heat energy enters a system, the molecules of that system move more rapidly, and the temperature of the system rises. Individual water molecules, however, are weakly linked to one another by hydrogen bonds (see Fig. 2-7). When heat enters a watery system such as a lake or a living cell, much of the heat energy goes into breaking hydrogen bonds rather than speeding up individual molecules. Thus, 1 **calorie** of energy will heat 1 gram of water 1° C, whereas it takes only 0.6 calorie per gram to heat alcohol 1° C, 0.2 calorie for table salt, and 0.02 calorie for common rocks such as granite or marble. As a result, because the human body is mostly water, a sunbather can absorb a lot of heat energy without sending his body temperature soaring.

Second, water moderates the effects of high temperatures through its great **heat of vaporization**, the amount of heat required to convert liquid water to water vapor. Water has one of the highest heats of vaporization known, 539 calories per gram. Again, this is due to the hydrogen bonds interconnecting individual water molecules. For a water molecule to evaporate, it must move quickly enough to break all the hydrogen bonds holding it to the other water molecules in the solution. Only the fastest-moving water molecules, carrying the most energy, can break their hydrogen bonds and escape into the air as water vapor. The remaining liquid is cooler for the loss of these high-energy molecules. As a sunbather's body temperature begins to rise, she perspires, covering her body with a film of water. Heat energy is transferred from her skin to the water and from the water to the vapor as the water evaporates. Evaporating just 1 gram of water cools 539 grams of her body 1° C, and so a great loss of heat can occur without much loss of water.

Third, water moderates the effects of low temperatures because it has a high **heat of fusion**, the energy that must be removed from molecules of liquid water before they form the precise crystal arrangement of ice. Therefore, water, both in a living organism or in a lake, freezes more slowly than other liquids at a given temperature.

Water Forms an Unusual Solid, Ice

Water, of course, will become a solid after prolonged exposure to temperatures below its freezing point. But even solid water is unusual. Most liquids become denser when they solidify, and the solid sinks. Ice is less dense than liquid water, so when a pond or lake starts to freeze in winter, the ice stays on top, forming an insulating layer that delays the freezing of the rest of the water. If ice sank, ponds and lakes in much of North America would freeze solid during the winter, eliminating fish and making liquid drinking water far less available to animals.

S C I E N T I F I C   I N Q U I R Y

The Use of Isotopes in Biology and Medicine

As you read this text, you will encounter many statements that may cause you to wonder, "How do they know *that?*" How do biologists know that DNA is the genetic material of cells (Chapter 10)? How do paleontologists measure the ages of fossils (Chapter 16)? How do botanists know that sugars made in plant leaves during photosynthesis are transported to other parts of the plant in the sieve tubes of phloem (Chapter 26)? These discoveries, and many more, have been possible only through the use of isotopes.

Although all atoms of a particular element have the same number of protons, the number of neutrons may vary. Neutrons don't affect the chemical reactivity of an atom very much, but they do make their presence felt in other ways. First, neutrons add to the mass of an atom, which can be detected by sophisticated instruments, such as mass spectrometers. Second, nuclei with "too many" neutrons break apart spontaneously, often emitting **radioactive** particles in the process. Those particles can also be detected, for example with Geiger counters.

A particularly fascinating and medically important application of radioactive isotopes is positron emission tomography, more commonly known as PET scans (Fig. E2-1). In a common application of PET scans, a subject is given glucose that has been labeled with a harmless radioactive isotope of fluorine, ^{18}F. When the nucleus of ^{18}F

decays, it emits two bursts of energy that travel in opposite directions along the same line. Energy detectors arranged in a ring around the subject record the nearly simultaneous arrival of the two energy bursts. A powerful computer then calculates the location within the subject at which the decay must have occurred and generates a map of the frequency of ^{18}F decays. Because the ^{18}F is attached to glucose molecules, this map reflects the glucose concentrations within the subject. The brain uses prodigious amounts of this sugar for energy; the more active a brain cell is, the more glucose it uses. How can this information be used in biological research?

Let's suppose that a neuroscientist is trying to locate the areas of the brain that are involved in memory. She might give ^{18}F-labeled glucose to a few volunteer subjects and then ask them to memorize a list of words that she reads aloud. Because brain regions active during this process would need more energy and would take up more ^{18}F-glucose molecules than would inactive regions, they would have more ^{18}F decays. PET scans taken during the memorization would pinpoint brain regions active in storing memories of words.

Physicians also use PET scans in the diagnosis of brain disorders. For example, brain regions that are the origin of epileptic seizures generally have excessively high glucose utilization and show up in PET scans as "hot spots." Many brain

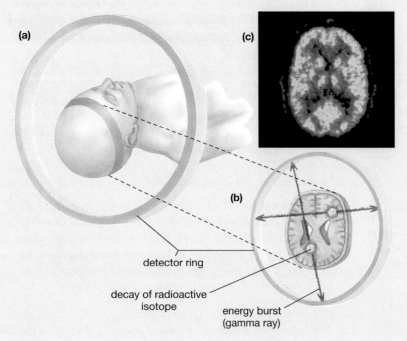

(a)

(c)

(b)

detector ring

decay of radioactive isotope

energy burst (gamma ray)

Figure E2-1   How positron emission tomography works

(a) A subject is given a substance containing a harmless radioactive isotope. The subject's head is inserted within a ring of detectors. Computerized analysis provides an image of the "slice" of the head that lies within the plane of the detector ring. **(b)** Inside the "head slice," radioactive decay releases energetic particles that activate the detectors. Computer analysis locates the site of the radioactive decay. **(c)** The number of decays at each location is color-coded and projected onto a television screen: Usually red and yellow mean many decays, green and blue mean few decays. The resulting image provides a visual representation of the activity and structure of the brain.

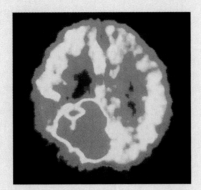

Figure E2-2 **A brain tumor**

A brain tumor consists of actively dividing, metabolically energetic cells. It usually has a rich blood supply and consumes both oxygen and glucose voraciously. Therefore, a brain tumor "lights up" in a PET scan. In this scan, a malignant brain tumor shows clearly as a red area.

tumors also light up in PET scans (Fig. E2-2). Abnormal metabolism of certain brain regions may also appear in patients with some mental disorders, such as schizophrenia.

Biology and medicine have profited immensely from close interactions with the other sciences, especially chemistry and physics. The development of PET scans required close cooperation with chemists (developing and synthesizing the radioactive probes), physicists (understanding positron-electron interactions and the geometry of the resulting energy emissions), and engineers (designing and building the electronic apparatus). Continued teamwork among scientists promises further advances in both fundamental understanding of biological processes and applications in medicine and agriculture.

Water Molecules Tend to Stick Together

Because of the hydrogen bonds that interconnect individual molecules, water has high **cohesion**—that is, water molecules tend to stick together. Cohesion among water molecules at the surface of a lake or pond produces **surface tension**, the tendency for the water surface to resist being broken. In general, objects that are denser than water sink. Because the water molecules at the surface of a pond stick to one another, however, the surface film acts almost as a solid, supporting relatively dense objects such as fallen leaves and water striders (Fig. 2-11a).

(a)

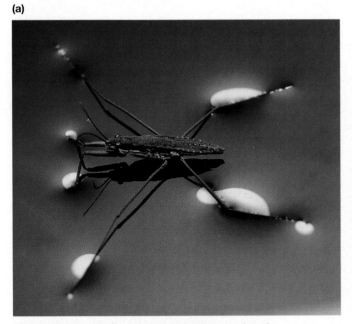

(b)

Figure 2-11 **Cohesion among water molecules**

(a) Cohesion among water molecules allows water striders to skate across the surface of still waters. **(b)** In giant redwoods, cohesion holds water molecules together in continuous strands from the roots to the topmost leaves even a hundred meters above the ground.

A more important role of cohesion occurs in the life of land plants (Fig. 2-11b). A plant absorbs water through its roots. How does the water reach the aboveground parts, especially if the plant is a hundred-meter-tall redwood? As we shall see in Chapter 26, the water is pulled up by the leaves. Water fills tiny tubes that connect the leaves, stem, and roots. Water molecules that evaporate from the leaves pull water up the tubes, much like a rope being pulled up from the top. The system works because the hydrogen bonds interconnecting water molecules are stronger than the weight of the water in the tubes, even a hundred meters' worth, and so the water "rope" doesn't break. Without the cohesion of water, there would be no land plants as we know them, and terrestrial life would undoubtedly have evolved quite differently.

✖ S U M M A R Y   O F   K E Y   C O N C E P T S

Matter and Energy
Matter is the physical material of the universe. Energy is the capacity to do work. Energy can exist in several forms that may be converted from one to another. The two major forms of energy are kinetic energy (the energy of motion) and potential energy (stored energy).

The Structure of Matter
An element is a substance that can neither be broken down nor converted to different substances by ordinary chemical means. The smallest possible particle of an element is the atom, which is itself composed of a central nucleus containing protons and neutrons and outer electron shells. All atoms of a given element have the same number of protons, which is different from the number found in the atoms of any other element.

Electrons are found in electron shells, at specific distances from the nucleus. Each shell can contain a fixed maximum number of electrons. The chemical reactivity of an atom depends on the number of electrons in its outermost electron shell; an atom is most stable, and therefore least reactive, when its outermost shell is either completely full or completely empty.

Chemical Bonds: Joining Atoms to Make Molecules
Atoms may combine to form molecules. The forces holding atoms together in molecules are called chemical bonds. There are two principal types of chemical bond, called ionic and covalent bonds. When one atom fills its outermost shell by acquiring electrons while another atom empties its shell by losing electrons, the results are negatively and positively charged particles called ions. Ionic bonds are electrical attractions between charged ions, holding them together in crystals.

Covalent bonds involve the sharing of electrons by two atoms, in which neither atom completely gains or loses an electron. In a nonpolar covalent bond, the two atoms share electrons equally. In a polar covalent bond, one atom may attract the electron more strongly than the other atom does; in this case, the strongly attracting atom bears a slightly negative charge, and the weakly attracting atom bears a slightly positive charge. Some polar covalent bonds give rise to hydrogen bonding, which is the attraction between charged regions of individual polar molecules or distant parts of a large polar molecule.

Organic and Inorganic Molecules
An organic molecule is one that contains both carbon and hydrogen; all other molecules are inorganic.

Water and Life
Properties of the water molecule that are important to the processes that occur within living organisms include its ability to dissolve many polar and charged substances, to force nonpolar substances to assume certain types of physical organization, to participate in chemical reactions, to maintain a fairly stable temperature in the face of wide temperature fluctuations in the environment, and to cohere to itself.

✖ K E Y   T E R M S

acid   p. 30
acidic   p. 30
atom   p. 22
atomic number   p. 23
base   p. 30
basic   p. 30
buffer   p. 30
calorie   p. 31
chemical bond   p. 25
cohesion   p. 33
compound   p. 22
covalent bond   p. 26
double covalent bond   p. 26

electron   p. 22
electron shell   p. 22
element   p. 22
energy   p. 22
energy level   p. 24
heat of fusion   p.31
heat of vaporization   p. 31
hydrogen bond   p. 27
hydrophilic   p. 29
hydrophobic   p. 29
hydrophobic interaction   p. 29
inorganic molecule   p. 27
ion   p. 25

ionic bond   p. 25
isotope   p. 23
kinetic energy   p. 22
matter   p. 22
mixture   p. 22
molecule   p. 24
neutron   p. 22
nonpolar covalent bond   p. 26
nucleus   p. 22
orbital   p. 23
organic molecule   p. 27
pH scale   p. 30
polar covalent bond   p. 26

potential energy p. 22
proton p. 22
radioactive p. 32
solvent p. 29
specific heat p. 31
surface tension p. 33
triple covalent bond p. 26

THINKING THROUGH THE CONCEPTS

Multiple Choice

1. What is the purest form of matter that cannot be separated into different substances by chemical means?
   a. compounds b. molecules
   c. hormones d. elements
   e. electrons
2. What phrase best describes molecular bonds?
   a. physical bridges b. energy links
   c. electronic glue d. electronic orbitals
   e. all of these phrases are equally descriptive
3. When an atom ionizes, what happens?
   a. It shares one or more electrons with another atom.
   b. It emits energy as it loses extra neutrons.
   c. It completely gives up or takes up one or more electrons.
   d. It shares a hydrogen atom with another atom.
   e. none of the above
4. If electrons in water molecules were equally attracted to hydrogen nuclei and oxygen nuclei, water molecules would be
   a. more polar b. less polar
   c. unchanged d. triple bonded
   e. unable to form
5. Where are hydrogen bonds important to life?
   a. between water molecules
   b. between hydrogen atoms
   c. in ionic substances
   d. all of the above
   e. none of the above
6. What is the defining characteristic of an acid?
   a. It donates hydrogen ions.
   b. It accepts hydrogen ions.
   c. It will donate or accept hydrogen ions depending on the pH.
   d. It has an excess of hydroxide ions.
   e. It has a pH greater than 7.

Review Questions

1. What are the six most abundant elements in living organisms?
2. Distinguish among atoms and molecules; elements, compounds, and mixtures; protons, neutrons, and electrons.
3. Compare and contrast covalent bonds and ionic bonds.
4. Why can water absorb a great amount of heat with little increase in its temperature?
5. Describe how water dissolves a salt. How does this compare with the effect of water on a hydrophobic substance such as corn oil?
6. Define acid, base, and buffer. How do buffers reduce changes in pH when hydrogen ions or hydroxide ions are added to a solution? Why is this important in living organisms?

APPLYING THE CONCEPTS

1. Many "over-the-counter" substances are sold to bring relief from "acid stomach" or "heartburn." What is the chemical basis for these compounds? Why do they work?
2. Fats and oils do not dissolve in water; polar and ionic molecules dissolve easily in water. Detergents and soaps help to clean dishes by dispersing fats and oils in water, so that they can be rinsed away. From your knowledge of the structure of water and the hydrophobic nature of fats, what general chemical structures (e.g., polar or nonpolar parts) must a soap or detergent have, and why?
3. What would the effects be for aquatic life if the density of ice were greater than that of liquid water? What would be the impact on all the rest of the living (terrestrial) organisms on Earth?
4. How does sweating help you to regulate your body temperature? Why do you feel hotter and more uncomfortable on a hot, humid day than on a hot, dry day?

FOR MORE INFORMATION

Atkins, P. W. *Molecules.* New York: Scientific American Library, 1987. A layperson's introduction to atoms and molecules, with superb illustrations.

Morrison, P., and Morrison, P. *Powers of Ten.* New York: W. H. Freeman and Co., 1982. A fascinating journey from the universe to the nucleus of an atom.

Storey, K. B., and Storey, J. M. "Frozen and Alive." *Scientific American*, December 1990. By triggering ice formation here, suppressing it there, and loading up their cells with antifreeze molecules, some animals, including certain lizards and frogs, can survive with 60% of their body water frozen solid.

NET WATCH

On-line resources for this chapter are on the World Wide Web at:
http://www.prenhall.com/~audesirk (click on the table of contents link and then select Chapter 2).

3 Biological Molecules

▓ AT A GLANCE

Synthesizing Organic Molecules: A Modular Approach
  Biological Molecules Are Joined Together or Broken Apart by Adding or Removing Water

The Principal Types of Biological Molecules
  Carbohydrates Are Sugars and Starches
  Lipids Are Diverse Molecules That Are Insoluble in Water
  Proteins Play Many Roles in Living Organisms
  Nucleic Acids Form DNA and RNA, the Molecules of Heredity

A computer-generated image of the shape of trypsin, a protein that serves as a digestive enzyme for protein

Living organisms are amazingly diverse: Bacteria, mushrooms, redwood trees, sea urchins, and people, at first glance, seem to have very little in common. However, this diversity is a veneer that covers an underlying unity. We have all evolved from common ancestors, a fact that can most readily be seen in the organic molecules of which our bodies are made. The organic molecules—the carbohydrates, lipids, proteins, and nucleic acids—are essentially the same in every organism. Over a couple of billion years of evolution, many differences have evolved between specific biological molecules in bacteria and bulldogs, but the similarities are striking. Why can people eat other organisms as diverse as chickens, corn, shrimp, and mushrooms? Because the building blocks that make up the biological molecules in these organisms are the same ones that we need to make our own molecules and because our bodies are able to break down many of their molecules to derive energy as well.

While the common structure and function of the types of organic molecules among organisms provides unity, the tremendous range of organic molecules accounts for the diversity of living organisms and the diversity of structures within single organisms and even within individual cells. This vast array of organic molecules, in turn, is possible because of the versatility of the carbon atom. A carbon atom has four electrons in its outermost shell, with room for eight. Therefore, carbon atoms are able to form many bonds. They become stable by sharing four electrons with other atoms, forming up to four single covalent bonds or smaller numbers of double or triple bonds. Molecules with many carbon atoms can assume complex shapes, including chains, branches, and rings.

Organic molecules are much more than just complicated skeletons of carbon atoms, however. Attached to the carbon backbone are groups of atoms, called **functional groups**, that determine the characteristics and chemical reactivity of the molecules. These functional groups are far less stable than the carbon backbone and are more likely to participate in chemical reactions. The common functional groups found in organic molecules are shown in Table 3-1.

The similarity among organic molecules found in all life on Earth is a consequence of two main features: the use

TABLE 3-1 ▪▪ *Important Functional Groups in Biological Molecules*

Group	Structure	Properties	Types of Molecules
Hydrogen (–H)		Polar or nonpolar, depending on what atom hydrogen is bonded to; involved in condensation and hydrolysis	Almost all organic molecules
Hydroxyl (–OH)		Polar; involved in condensation and hydrolysis	Carbohydrates, nucleic acids, alcohols, some acids, and steroids
Carboxyl (–COOH)		Acid; negatively charged when H^+ dissociates; involved in peptide bonds	Amino acids, fatty acids
Amino (–NH$_2$)		Basic; may bond an additional H^+, becoming positively charged; involved in peptide bonds	Amino acids, nucleic acids
Phosphate (–H$_2$PO$_4$)		Acid; up to two negative charges when H^+ dissociates; links nucleotides in nucleic acids; energy-carrier group in ATP	Nucleic acids, phospholipids
Methyl (–CH$_3$)		Nonpolar; tends to make molecules hydrophobic	Many organic molecules; especially common in lipids

of the same set of functional groups in virtually all organic molecules in all organisms and the use of the "modular approach" to synthesizing large organic molecules.

Synthesizing Organic Molecules: A Modular Approach

In principle, there are two ways to manufacture a large, complex molecule: One could synthesize the molecule atom by atom following an extremely detailed blueprint, or one could take preassembled smaller molecules and hook them together. Just as trains are made by coupling engines, boxcars, coal cars, and cabooses, so too life on Earth takes the modular approach. Small molecules (for example, sugars) are used as **subunits** with which to synthesize longer molecules (for example, starches), like cars in a train (Fig. 3-1a). The individual subunits are often called **monomers** (from Greek words meaning "one part"); long chains of monomers are called **polymers** ("many parts"). Most organic molecules in our bodies are in the form of either monomers or polymers.

Biological Molecules Are Joined Together or Broken Apart by Adding or Removing Water

Biological molecules almost always use the same type of chemical reaction, called a **dehydration synthesis** (literally "to form by removing water") to join subunits to one another. In a dehydration synthesis, a hydrogen (–H), removed from one subunit, and a hydroxyl (–OH), removed from a second subunit, combine to form a molecule of water (H_2O), as the remaining subunits are joined by a covalent bond (Fig. 3-1b). The reverse reaction, called **hydrolysis** (literally, "to break apart with water"), can split the molecule into individual subunits again (Fig. 3-1c). During hydrolysis, water is added back, a hydrogen to one subunit and a hydroxyl to the other.

The Principal Types of Biological Molecules

Although the bodies of living things often include thousands of different organic molecules, nearly all fall into one of four categories: carbohydrates, lipids, proteins, or nucleic acids (see Table 3-2).

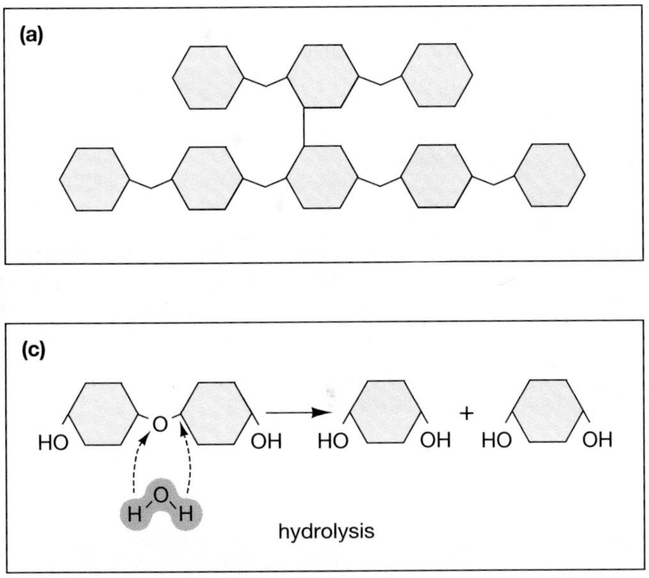

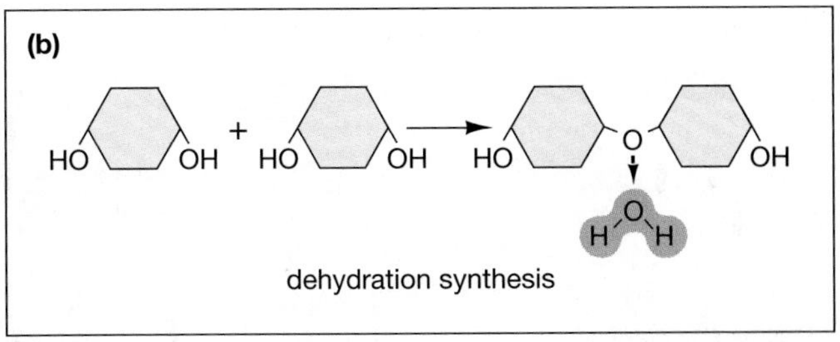

(a)

(b)

dehydration synthesis

(c)

hydrolysis

Figure 3-1   Synthesis and breakdown of organic molecules

(a) A typical organic molecule is composed of similar or identical subunits linked together by covalent bonds. Here we see sugar subunits joined to form a starch. **(b)** In a dehydration synthesis, two subunits are joined by a covalent bond. Simultaneously, a hydroxyl group is removed from one subunit and combines with a hydrogen removed from a second subunit to form water. **(c)** Hydrolysis is the reverse of dehydration synthesis. Hydrogen and hydroxyl from water are added to the subunits as the large organic molecule is broken apart into its subunits.

TABLE 3-2 ❖ The Principal Biological Molecules

Class of Molecule	Principal Subtypes	Example	Function
Carbohydrate: Usually contains carbon, oxygen, and hydrogen, in the approximate formula $(CH_2O)_n$	Monosaccharide: Simple sugar	Glucose	Important energy source for cells; subunit of which most polysaccharides are made
	Disaccharide: Two monosaccharides bonded together	Sucrose	Principle sugar transported throughout bodies of land plants
	Polysaccharide: Many monosaccharides (usually glucose) bonded together	Starch	Energy storage in plants
		Glycogen	Energy storage in animals
		Cellulose	Structural material in plants
Lipid: Contains high proportion of carbon and hydrogen; usually nonpolar and insoluble in water	Triglyceride: Three fatty acids bonded to glycerol	Oil, fat	Energy storage in animals, some plants
	Wax: Variable numbers of fatty acids bonded to long-chain alcohol	Waxes in plant cuticle	Waterproof covering on leaves and stems of land plants
	Phospholipid: Polar phosphate group and two fatty acids bonded to glycerol	Phosphatidylcholine	Common component of membranes in cells
	Steroid: Four fused rings of carbon atoms with functional groups attached	Cholesterol	Common component of membranes of eukaryotic cells; precursor for other steroids such as testosterone, bile salts
Protein: Chains of amino acids; contain carbon, hydrogen, oxygen, nitrogen, and sulfur		Keratin	Helical protein, principal component of hair
		Silk	Pleated sheet protein produced by silk moths and spiders
		Hemoglobin	Globular protein composed of four subunit peptides; transport of oxygen in vertebrate blood
Nucleic acid: Made of nucleotide subunits; may consist of a single nucleotide or long chain of nucleotides	Long-chain nucleic acids	Deoxyribonucleic acid (DNA)	Genetic material of all living cells
		Ribonucleic acid (RNA)	Genetic material of some viruses; in living cells, essential in transfer of genetic information from DNA to protein
	Single nucleotides	Adenosine triphosphate (ATP)	Principal short-term energy carrier molecule in cells
		Cyclic adenosine monophosphate (cyclic AMP)	Intracellular messenger

Carbohydrates Are Sugars and Starches

All **carbohydrates** are either small, water-soluble **sugars** (glucose or fructose, for example) or chains, such as starch or cellulose, that are made by stringing sugar subunits together. If a carbohydrate consists of just one sugar molecule, it is called a **monosaccharide** (Greek for "single sugar"). When two or more monosaccharides are linked together, they form a **disaccharide** ("two sugars") or a **polysaccharide** ("many sugars").

Carbohydrates such as sugars and starches are important energy sources for most organisms. Other carbohydrates, such as cellulose and similar molecules, provide structural support for individual cells or even for the entire bodies of organisms as diverse as plants, fungi, bacteria, and insects.

Single sugars (monosaccharides) usually have a backbone of three to seven carbon atoms (Fig. 3-2a, left). Most of the carbon atoms have both a hydrogen (–H) and a hydroxyl group (–OH) attached to them. Therefore, the general formula for a monosaccharide is $(CH_2O)_n$, where n is the number of carbons in the backbone. This formula explains the origin of the name carbohydrate, which literally means "carbon plus water." When dissolved in water, such as in the cytoplasm of a cell, the carbon backbone of a sugar usually "circles up" into a ring (Fig. 3-2a, right). It is in this ring form that sugars link together to make disaccharides and polysaccharides (see Fig. 3-3).

Most small carbohydrates are soluble in water. As in water molecules, the O–H bond in a hydroxyl group is polar, because oxygen attracts electrons more strongly than hydrogen does. Hydrogen bonds between water molecules and the polar hydroxyl groups of the carbohydrate keep the carbohydrate in solution (see Fig. 2-9).

There Are a Variety of Monosaccharides with Slightly Different Structures

Glucose (Fig. 3-2a) is the most common monosaccharide in living organisms and is the subunit of which most polysaccharides are made. Glucose has six carbons, and hence

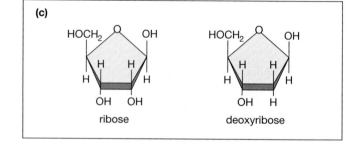

Figure 3-2   Monosaccharide structure

(a) The most common monosaccharides have "backbones" of either five or six carbon atoms (chain diagram of glucose, left). When dissolved in water, however, the chain bends upon itself to form a ring (center). The ring forms are usually drawn on paper as if you were looking at the edge of the ring (right). The thick edge projects out of the paper toward you, the thin edge recedes behind the paper, and the –OH, and –CH₂OH groups are perpendicular to the ring, in the plane of the paper. For convenience, carbon atoms at the corners of the polygons are omitted, but other atoms (such as oxygen) are usually shown. **(b)** Fructose and galactose have the same atomic composition as glucose but a different structure. **(c)** Ribose and deoxyribose are five-carbon monosaccharides that form parts of nucleic acids.

has the chemical formula $(CH_2O)_6$, or $C_6H_{12}O_6$. Many organisms also synthesize other monosaccharides that have the same chemical formula as glucose, but have slightly different structures (Fig. 3-2b). These include fructose (the "corn sugar" found in corn syrup) and galactose (part of lactose, or "milk sugar"). Some other common monosaccharides, such as ribose and deoxyribose, have five carbons (Fig. 3-2c). Ribose and deoxyribose are parts of the genetic molecules ribonucleic acid (RNA) and deoxyribonucleic acid (DNA), respectively.

Disaccharides Consist of Two Single Sugars Linked Together by Dehydration Synthesis

Monosaccharides, especially glucose and its relatives, have a short life span in a cell. Most are either broken down to free their chemical energy for use in driving needed cellular reactions or are linked together by dehydration synthesis to form disaccharides or polysaccharides (Fig. 3-3). Disaccharides are often used for short-term energy storage, especially in plants. Common disaccharides include **sucrose** (table sugar: glucose plus fructose), **lactose** (milk sugar: glucose plus galactose), and **maltose** (glucose plus glucose, formed during the digestion of starch). When energy is required, the disaccharides are broken apart again into their monosaccharide subunits by hydrolysis.

Polysaccharides Are Chains of Single Sugars

For long-term energy storage, monosaccharides, usually glucose, are joined together into polysaccharides, forming **starch** (in plants; Fig. 3-4) or **glycogen** (in animals). Starch, commonly found in roots and seeds, may occur as coiled, unbranched chains of up to 1000 glucose subunits, or, more frequently, as huge branched chains of up to half a million glucose monomers. Glycogen, stored in the liver and muscles of animals like ourselves, is usually much smaller than starch and has branches every 10 to 12 glucose subunits. Having many small branches probably makes it easier to split off the glucose subunits for quick energy release.

Many organisms use polysaccharides as structural materials. One of the most important structural polysaccharides is **cellulose**, which makes up a large part of the cell walls of plants and about half the bulk of a tree trunk (Fig. 3-5). When you picture the vast fields and forests blanketing much of our planet, you will not be surprised to learn that there is probably more cellulose on Earth than all other organic molecules put together. Ecologists estimate that about a *trillion tons of cellulose* is synthesized each year!

Like starch, cellulose consists of glucose subunits bonded together; however, whereas most animals can easily digest starch, only a few microbes, such as those in the digestive tracts of cows and termites, can digest cellulose. Why is this the case, given that starch and cellulose both consist of glucose? In cellulose the orientation of the bonds between subunits is different, so that every other glucose is "upside down" (compare Fig. 3-5 with Fig. 3-4). This difference in orientation prevents the digestive enzymes of animals from attacking the bonds between glucose subunits. Wholly different enzymes synthesized by certain microbes, though, can break these bonds. As a result, cellulose is food for these microbes, but for most animals cellulose is *roughage* or *fiber*, material that passes undigested through the digestive tract.

Polysaccharides are also the starting point for the synthesis of many other important molecules. The hard outer coverings (exoskeletons) of insects, crabs, and spiders are

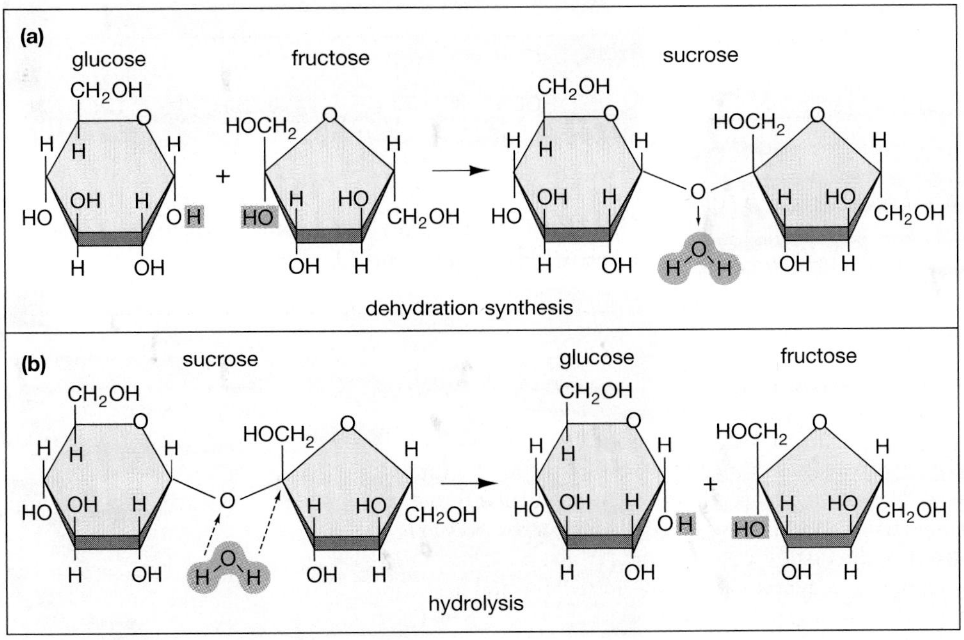

(a)

glucose   fructose   sucrose

dehydration synthesis

(b)

sucrose   glucose   fructose

hydrolysis

Figure 3-3 **Synthesis and breakdown of a disaccharide**

(a) The disaccharide sucrose is synthesized by a dehydration synthesis reaction in which a hydrogen (–H) is removed from glucose and a hydroxyl group (–OH) is removed from fructose, forming a water molecule and leaving the two monosaccharide rings joined by single bonds to the remaining oxygen atom. **(b)** Hydrolysis of sucrose is just the reverse of its synthesis, as water is split and added back to the monosaccharides.

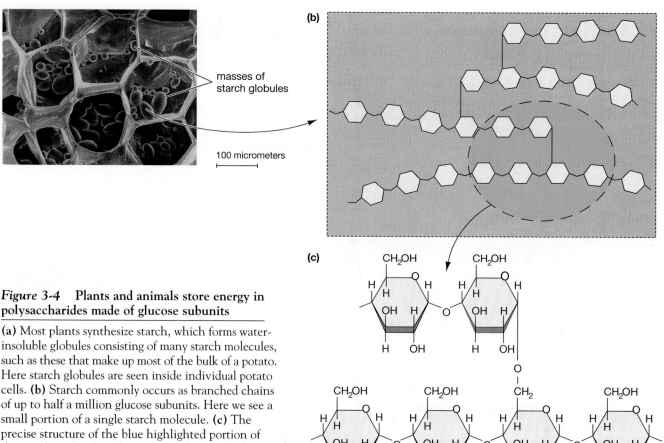

masses of
starch globules

100 micrometers

Figure 3-4   **Plants and animals store energy in polysaccharides made of glucose subunits**

(a) Most plants synthesize starch, which forms water-insoluble globules consisting of many starch molecules, such as these that make up most of the bulk of a potato. Here starch globules are seen inside individual potato cells. **(b)** Starch commonly occurs as branched chains of up to half a million glucose subunits. Here we see a small portion of a single starch molecule. **(c)** The precise structure of the blue highlighted portion of the starch molecule is shown here. Note the linkage between the individual glucose subunits for comparison with cellulose, shown in Figure 3-5.

made of **chitin**, a polysaccharide in which the glucose subunits have been chemically modified by the addition of a nitrogen-containing group (Fig. 3-6). Interestingly, chitin also stiffens the cell walls of many fungi. Bacterial cell walls contain still other types of modified polysaccharides, as do the lubricating fluids in our joints and the transparent corneas of our eyes.

Many other molecules are partly carbohydrate. Perhaps the most important of these molecules are the nucleic acids (discussed later), the carriers of hereditary information in all organisms. Other examples include mucus, some hormones, and many molecules in the cell membrane, including "identification molecules," such as those on red blood cells that determine blood type.

Lipids Are Diverse Molecules That Are Insoluble in Water

Lipids are a diverse assortment of molecules, all of which share two important features. First, lipids contain large regions composed almost entirely of hydrogen and carbon, with nonpolar carbon-carbon or carbon-hydrogen bonds. Second, these nonpolar regions make lipids hydrophobic and insoluble in water. The various types of lipids serve a

wide variety of functions. Some lipids are energy-storage molecules; some form waterproof coverings on both plant and animal bodies; some make up the bulk of all of the membranes of a cell; and still others are hormones.

Lipids are classified into three major groups: (1) oils, fats, and waxes, which are similar in structure and contain only carbon, hydrogen, and oxygen; (2) phospholipids, structurally similar to oils but also containing phosphorus and nitrogen; and (3) the fused-ring family of steroids.

Oils, Fats, and Waxes Are Lipids Containing Only Carbon, Hydrogen, and Oxygen

Oils, fats, and waxes are related in three ways: (1) They contain only carbon, hydrogen, and oxygen; (2) they contain one or more **fatty acid** subunits (long chains of carbon and hydrogen with a carboxyl group [–COOH] at one end; Fig. 3-7); and (3) they usually do not have ring structures. **Fats** and **oils** are formed by dehydration synthesis from one molecule of **glycerol** (a short, three-carbon molecule with one hydroxyl group [–OH] per carbon) and three fatty acid subunits. This structure of three fatty acids joined to one glycerol molecule gives fats and oils their chemical name, **triglyceride**. Fats and oils have a high

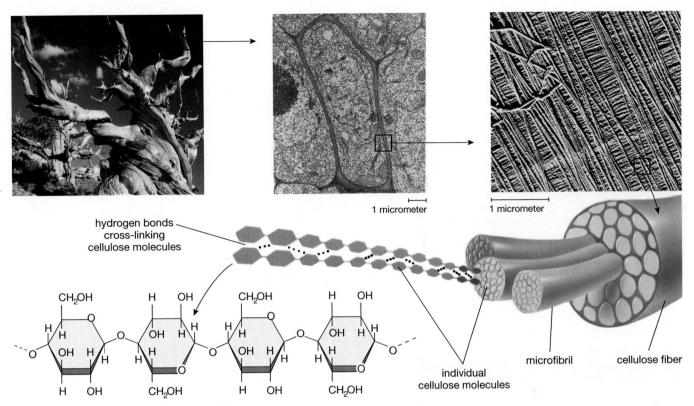

Figure 3-5   Cellulose structure and function

Cellulose, like starch, is composed of glucose subunits, but the orientation of the bond be-tween subunits in cellulose is different (compare with Fig. 3-4c) so that every other glucose molecule is "upside down." Unlike starch, cellulose has great structural strength, due partly to the difference in bonding and partly to the arrangement of parallel molecules of cellulose into long, cross-linked fibers. Plant cells often lay down cellulose fibers in layers that run at angles to each other, resulting in resistance to tearing in both directions. The final product can be incredibly tough, as this 3000-year-old bristlecone pine in California's White Moun-tains testifies.

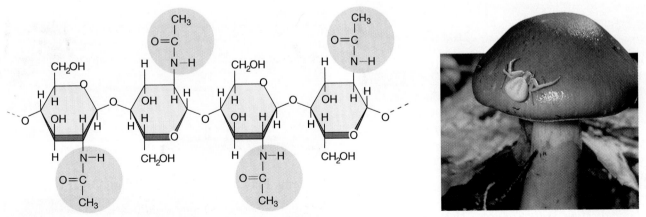

Figure 3-6   Chitin: a unique polysaccharide

Chitin has the same "alternating upside down" bonding of glucose molecules as cellulose; the difference is that the glucose subunits are modified by replacement of one of the hydroxyl groups with a nitrogen-containing functional group (yellow). Tough, slightly flexible chitin supports the otherwise soft bodies of arthropods (insects, spiders, and their relatives) and fungi.

Figure 3-7 **Triglycerides**

Triglycerides (fats and oils) are synthesized by a dehydration reaction linking three fatty acids to a single glycerol molecule. A double bond in the topmost fatty acid creates a kink in the chain.

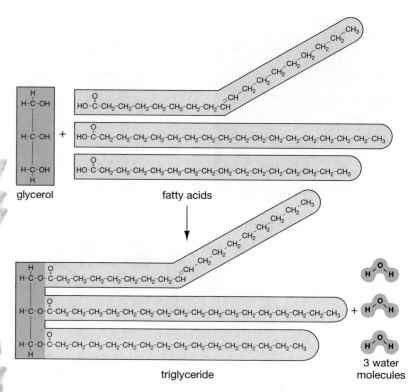

concentration of chemical energy, about 9.3 Calories per gram, compared with 4.1 for sugars and proteins (a Calorie with a capital C equals 1000 calories; the Calorie is used in measuring the energy content of foods). Fats and oils are used for semipermanent energy storage—for example, bears that feast during summer and fall put on fat to tide them over during their winter hibernation.

The difference between a fat (which is a solid at room temperature) and an oil (liquid at room temperature) lies in their fatty acids. Fats have fatty acids with all single bonds in their carbon chains. Hydrogens occupy all the other bond positions on the carbons. The resulting fatty acid (for example, stearic acid; Fig. 3-8a) is called **saturated** because it is "saturated" with hydrogens—that is, it has as many hydrogens as possible. If there are double bonds between some of the carbons, and consequently fewer hydrogens, the fatty acid is called **unsaturated** (oleic acid; Fig. 3-8b). Oils have mostly unsaturated fatty acids. The saturated fatty acids of fats can nestle closely together, forming solid lumps at room temperature (beef fat; Fig. 3-8a). The double bonds in the unsaturated fatty acids of oils, on the other hand, form kinks in the fatty acid chains (linseed oil; Fig. 3-8b). The kinks keep oil molecules apart, with the result that an oil is liquid at room temperature. An oil can be converted to a fat by breaking the double bonds between carbons, replacing them with single bonds, and adding hydrogens to the remaining bond positions. This is the "hydrogenated oil" listed in the ingredients on a box of margarine.

Waxes are similar to fats and oils except that the fatty acids are linked to large, long-chain alcohols rather than to glycerol. Waxes form a waterproof coating over the leaves and stems of land plants. Animals also synthesize waxes, as waterproofing for mammalian fur and insect exoskeletons, and, in a few cases, to build elaborate structures such as beehives.

Phospholipids Have Water-Soluble "Heads" and Water-Insoluble "Tails"

The cell membrane that separates the inside of a cell from the outside world contains several types of **phospholipids**. A phospholipid is similar to an oil, except that one of the fatty acids is replaced by a phosphate group with a short, polar, often

nitrogen-containing group attached to the end (Fig. 3-9). Unlike the fatty acid "tails," which are insoluble in water, the phosphate-nitrogen "head" is polar, or charged, and is water soluble. Thus a phospholipid has two dissimilar ends: a hydrophilic head attached to hydrophobic tails. As you will see in Chapter 6, this dual nature of phospholipids is crucial to the structure and function of the cell membrane.

Steroids Consist of Four Carbon Rings Fused Together

Steroids are structurally different from all the other lipids. All steroids are composed of four rings of carbon fused together, with various functional groups protruding from them (Fig. 3-10). Steroids include cholesterol, which is a vital component of the membranes of most eukaryotic cells and which is used by cells to synthesize other steroids. These other steroids include male and female sex hormones, hormones that regulate salt levels, hormones that assist in fat digestion, and insect hormones that stimulate the shedding of exoskeletons.

Proteins Play Many Roles in Living Organisms

Proteins are molecules composed of one or more chains of amino acids (see below). Proteins perform many functions. Protein **enzymes** guide almost all the chemical reactions that occur inside cells (see Chapter 4). Because each enzyme assists only one or a few specific reactions, cells usually contain hundreds of different enzymes. Other proteins are used for structural purposes, such as elastin, which gives skin its elasticity;

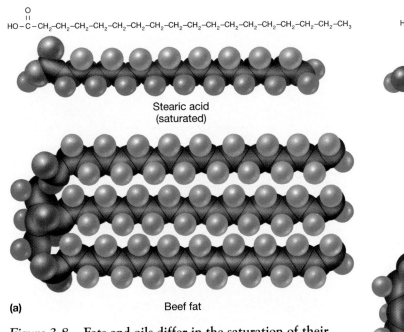

$$HO-C-CH_2-CH_2-CH_2-CH_2-CH_2-CH_2-CH_2-CH_2-CH_2-CH_2-CH_2-CH_2-CH_2-CH_2-CH_2-CH_2-CH_3$$

Stearic acid
(saturated)

(a)   Beef fat

Figure 3-8   **Fats and oils differ in the saturation of their fatty acids**

(a) A saturated fatty acid, such as stearic acid, has all single bonds between carbons. In beef fat, which is composed of triglycerides with a high proportion of stearic acids, the fatty acid chains pack closely together, forming a solid at room temperature. **(b)** An unsaturated fatty acid, such as oleic acid, has one or more double bonds between carbons, which cause kinks in the chain. These kinks in the fatty acids that make up linseed oil prevent close packing; as a result, the oil is liquid at room temperature.

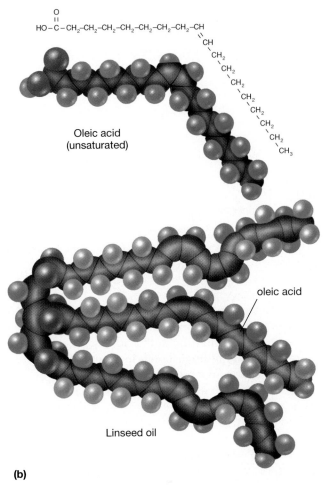

Oleic acid
(unsaturated)

oleic acid

Linseed oil

(b)

Figure 3-9   **Phospholipids**

Phospholipids are similar to fats or oils, except that only two fatty acid tails are attached to the glycerol backbone. The third position on the glycerol is occupied by a polar head composed of a phosphate group ($-PO_4^-$) to which is attached a second, often nitrogen-containing group. The phosphate group is negatively charged, and the nitrogen-containing group is positively charged.

polar head   glycerol backbone   fatty acid tails

Figure 3-10   Steroids

Steroids are synthesized from cholesterol, and all have almost the same molecular structure (colored rings; note that the carbon atoms at the corners of the rings and the hydrogen atoms attached to the rings have been omitted from these drawings). Great differences in steroid function are a result of the different functional groups attached to the rings. The male sex hormone testosterone, the female sex hormone estradiol (a type of estrogen), and the insect hormone ecdysone, which causes exoskeleton shedding, share a common cholesterol base but differ in functional group attachments.

keratin, the principal protein of hair, horns, and claws; and the silk of spider webs and silk moth cocoons (Fig. 3-11). Still other types of proteins are used for energy and material storage (albumin in eggs, casein in milk), transport (hemoglobin to carry oxygen in the blood), and cell movement (contractile proteins in muscle). Some hormones (insulin, growth hormone), antibodies that help to fight disease and infection, and many poisons (rattlesnake venom) are also proteins.

Proteins Are Formed from Chains of Amino Acids

Proteins are polymers of **amino acids** (Fig. 3-12). All amino acids have the same fundamental structure, consisting of a central carbon bonded to four different functional groups: a nitrogen-containing amino group (–NH₂); a carboxyl

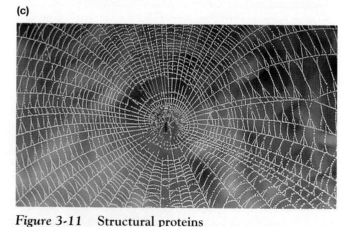

Figure 3-11   Structural proteins

Common structural proteins include those of **(a)** hair, **(b)** horn, and **(c)** spider web silk.

group (–COOH); a hydrogen; and a variable group (represented by the letter R):

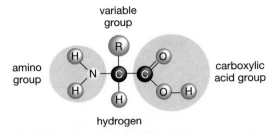

The R group differs among amino acids and gives each its distinctive properties (see Fig. 3-12).

The 20 amino acids commonly found in the proteins of living organisms have different properties based on the nature of the R group. Some amino acids are hydrophilic; their R groups are polar and soluble in water. Others are hydrophobic, with nonpolar R groups that are insoluble in water. Another type of amino acid, cysteine, has sulfur in its R group and can form bonds with other cysteines that can link protein chains together. These bonds between the R groups of cysteine are called **disulfide bridges** (see "A Closer Look: Protein Structure—A Hairy Subject").

Twenty amino acids may seem like a small number of subunits from which to construct thousands of different proteins, but perhaps an analogy with language will help you understand protein structure and function more easily. The English language uses 26 letters to construct thousands of words, each with a distinct meaning based on the exact sequence of letters. But only a small fraction of all possible letter combinations are used as words; for example, the combination "protein" has meaning, but "nteio-pr" does not. Similarly, living organisms construct thousands of different proteins—about 50,000 in your body—from an "alphabet" of 20 amino acids.

Amino acids differ in their chemical and physical properties—size, water solubility, electrical charge—because of their different R groups. Therefore, the exact sequence of amino acids dictates the function of each protein: whether it is water soluble or not, whether it is an enzyme or a hormone or a structural protein. Scrambled sequences of amino acids are useless. In some cases, just one wrong amino acid may cause a protein to function incorrectly.

Amino Acids Are Joined to Form Chains by Dehydration Synthesis

Like lipids and polysaccharides, proteins are formed by dehydration synthesis. The nitrogen of the amino group (–NH$_2$) of one amino acid is joined to the carbon of the carboxyl group (–COOH) of another amino acid by a single covalent bond (Fig. 3-13). This bond is called a **peptide bond**, and the resulting chain of two amino acids is called a **peptide**. More amino acids are added, one by one, until the protein is complete. Protein chains found in living cells vary in length from three to thousands of amino

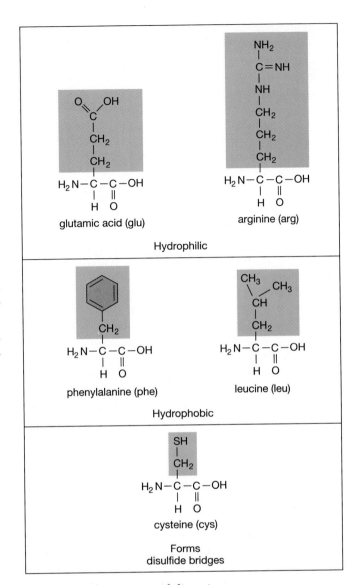

Figure 3-12   Amino acid diversity

The diversity of amino acid structures is a consequence of differences in the variable R group (colored blue). Some amino acids are classified according to the variable R group as hydrophilic, others as hydrophobic. Cysteine stands in a class by itself. Two cysteines in distant parts of a protein molecule can form a covalent bond between their sulfur atoms, making a disulfide bridge that brings the cysteines very close together and bends the protein chain (see Fig. 3-15).

acids. Often, the word *protein* is reserved for long peptides, say 50 or more amino acids in length. For simplicity, we will use both peptides and proteins to refer to amino acid chains.

Protein May Have Up to Four Levels of Structure

The phrase "amino acid chains" may evoke images of proteins as floppy, featureless structures, but this is not correct. Proteins are, instead, highly organized molecules that come in a variety of shapes. Biologists recognize four

A   C L O S E R   L O O K

Protein Structure—A Hairy Subject

A single strand of human hair, thin and not even alive, is nonetheless a highly organized, complex structure. Hair is composed mostly of a single, helical protein called keratin. If we look closely at the structure of hair, we can learn a great deal about biological molecules, chemical bonds, and why human hair behaves as it does.

In the molecular structure of hair, evolution anticipated the technique of sailors who made strong ropes out of weak individual fibers of hemp: A series of fibers are twisted about one another, with bundles of fibers making up the final product. A single hair consists of a hierarchy of structures (Fig. E3-1). The outermost layer is a set of overlapping shingle-like scales that protect the hair and keep it from drying out. Inside the hair lie closely packed, dead cells, each filled with long strands running lengthwise. These strands in turn are composed of thinner strands called microfibrils, embedded in a protein matrix. Each microfibril is a bundle of protofibrils, and each protofibril is composed of three or more helical keratin molecules twisted together. As a hair grows, living cells in the hair follicle embedded in the skin whip out new keratin at the rate of 10 turns of the protein helix every second.

Pull the ends of a hair and you will notice that it is rather strong. Hair gets its strength from three types of chemical bonds (see Fig. E3-1). First, the individual molecules of keratin are held in their helical shape by many hydrogen bonds. Before a hair will break, all the hydrogen bonds of all the keratin molecules in one cross-sectional plane of the strand must break to allow the helix to be stretched to its maximal extent. Second, each molecule is cross-linked to neighboring keratins by disulfide bridges between cysteines. Some of these bridges must break as the hair stretches. Finally, at least one peptide bond in each keratin molecule must break before the strand as a whole breaks.

Hair is also fairly stiff. The stiffness arises from hydrogen bonds within the individual helices of keratin and from disulfide bridges holding neighboring keratin molecules together. When hair gets wet, however, the hydrogen bonds between turns of the helices are replaced by hydrogen bonds between the amino acids and the water molecules surrounding them, and so the helices collapse. Wet hair is therefore very limp. If wet hair is rolled onto curlers and allowed to dry, the hydrogen bonds reform in slightly different places, holding the hair in a curve. The slightest moisture, even humid air, allows these hydrogen bonds to rearrange into their natural configuration, and the hair straightens out again.

Pull gently and you will discover still another property of hair. It stretches and then springs back into shape when you release the tension. When hair stretches, many of the hydrogen bonds within each keratin helix are broken, allowing the helix to be extended. Most of the disulfide bonds between different levels of the helices, in contrast, are distorted by stretching but do not break. When tension is released, these disulfide bridges contract, returning the hair to its normal length.

Finally, each hair has a characteristic shape: It may be straight, wavy, or curly. The curliness of hair is genetically specified and is determined biochemically by the arrangement of disulfide bridges (Fig. E3-2). Curly hair

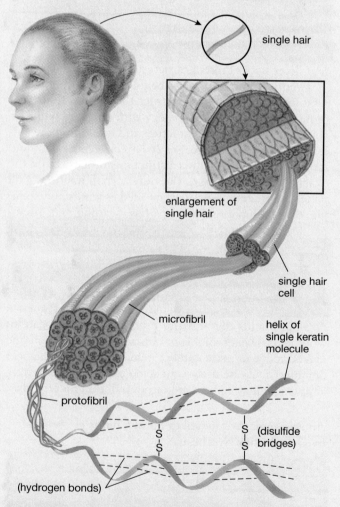

single hair

enlargement of single hair

single hair cell

microfibril

helix of single keratin molecule

protofibril

S—S
S—S   (disulfide bridges)

(hydrogen bonds)

Figure E3-1   The organization of hair

The microscopic organization of a single hair is one of bundles of fibers within further bundles of fibers. Hydrogen bonds and disulfide bridges between cysteines impart strength and elasticity to individual hairs.

Figure E3-2 **Straight or curly?**

If disulfide bridges join individual keratin molecules in a hair at the same level, the hair will be fairly straight. If disulfide bridges connect different levels within or between molecules, the hair will have a natural curl. The same effect can be obtained with a permanent wave that breaks and reforms the disulfide bridges of naturally straight hair. The new hair growing out, however, will be straight.

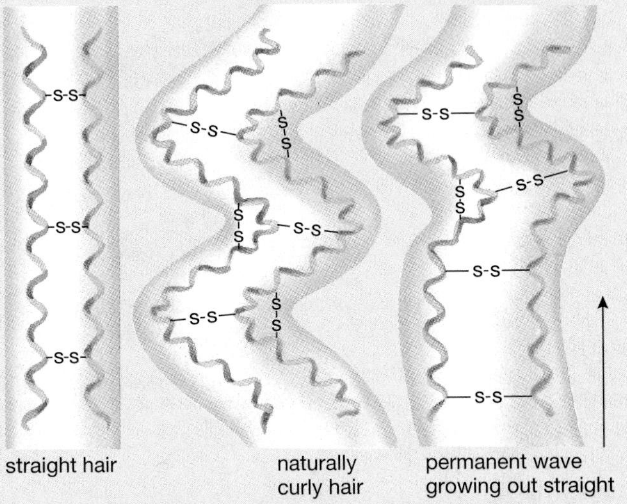

has disulfide bridges cross-linking the various keratin molecules at *different levels*, whereas straight hair has bridges mostly at the *same level*. When straight hair is given a "permanent wave," two lotions are applied. The first lotion breaks disulfide bonds between neighboring helices. The hair is then rolled tightly onto curlers, and a second solution, which reforms the bridges, is applied. The new disulfide bridges connect helices at different levels, holding the strands of hair in a curl. These new bridges are more or less permanent, and genetically straight hair can be transformed into biochemically curly hair. The new hair that grows in, of course, will have the genetically determined arrangement of bridges and will not be curly.

straight hair   naturally   permanent wave
                       curly hair   growing out straight

levels of organization in protein structure (Fig. 3-14). The **primary structure** is the sequence of amino acids that make up the protein (Fig. 3-15). This sequence is coded by the genes. Different types of proteins (for example, insulin versus growth hormone) have different sequences of amino acids. However, all molecules of the same type of protein, such as human insulin, have the same primary structure.

Hydrogen bonds cause many protein chains to form one of two simple, repeating **secondary structures**. Looking back at Figure 3-13, notice that every amino acid subunit in a peptide retains a –C=O from its carboxyl group and an –N–H from its amino group. Because oxygen attracts electrons more strongly than does carbon, the oxygen is relatively negative. Similarly, nitrogen attracts electrons more strongly than does hydrogen, leaving the hydrogen relatively positive. Therefore, hydrogen bonds

can form between the oxygen of the –C=O and the hydrogen of the –N–H groups of a protein chain. Many proteins, such as the hair protein keratin, have a coiled, springlike shape called a **helix** in which hydrogen bonds hold the turns of the coils together (Fig. 3-16a). Some proteins, such as silk, consist of many protein chains lying side by side, with hydrogen bonds holding adjacent chains in a **pleated sheet** arrangement (Fig. 3-16b).

It is rare for an entire protein to be a simple helix or pleated sheet. Most proteins assume complex, three-dimensional **tertiary structures** (see Fig. 3-14). Disulfide bridges formed between cysteine amino acids may bring otherwise distant parts of a protein close together. Internal stresses, owing to the size and properties of the functional groups on the particular amino acids present, also may contort a protein. For example, amino acids with very large R groups (such as phenylalanine; see Fig. 3-12) are

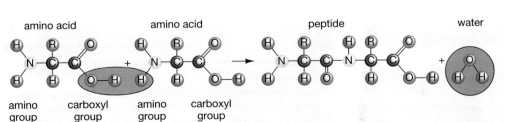

amino acid   amino acid   peptide   water

amino   carboxyl   amino   carboxyl
group   group   group   group

Figure 3-13 **Protein synthesis**

In protein synthesis, a dehydration synthesis joins the carbon of the carboxyl acid group of one amino acid to the nitrogen of the amino group of a second amino acid. The resulting covalent bond is called a peptide bond.

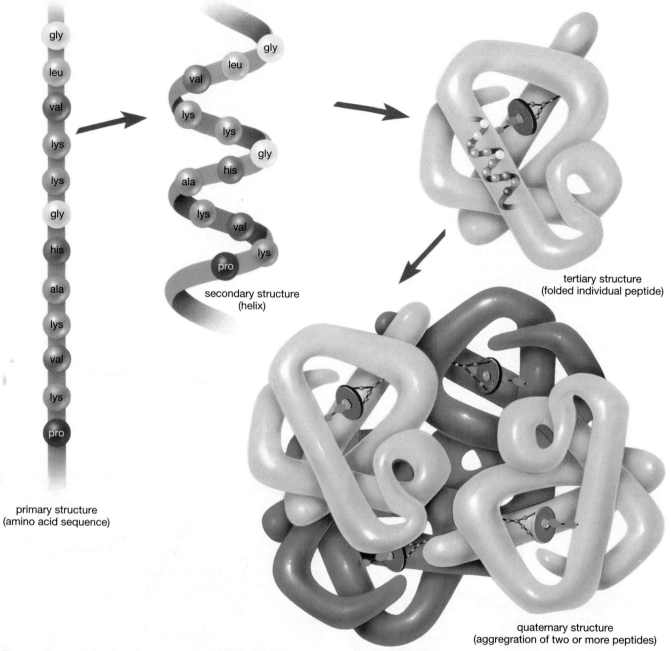

primary structure
(amino acid sequence)

secondary structure
(helix)

tertiary structure
(folded individual peptide)

quaternary structure
(aggregration of two or more peptides)

Figure 3-14   **The four levels of protein structure**

Levels of protein structure are represented here by hemoglobin, the oxygen-carrying protein in red blood cells. All levels of protein structure are determined by the amino acid sequence of the protein, interactions among the R groups of the amino acids (primarily hydrogen bonds and disulfide bridges between cysteines), and interactions between the R groups and their surroundings (usually water or lipids).

too bulky to fit side by side in a simple helix. As a result, the helix bends.

Probably the most important influence on the tertiary structure of a protein is its cellular environment, specifically whether the protein is dissolved in the water of the cytoplasm, in the lipids of the membranes, or half in one and half in the other. Hydrophilic amino acids can form hydrogen bonds with nearby water molecules, whereas hydrophobic amino acids cannot. Therefore, a protein dissolved in water folds into an irregular glob, with its hydrophilic amino acids facing the outside watery environment and its hydrophobic ones clustered in the center of the molecule.

Peptides may sometimes join into aggregations of two or more peptides to form the fourth level of protein organization, called **quaternary structure**. This fourth-level organization is maintained by the same types of

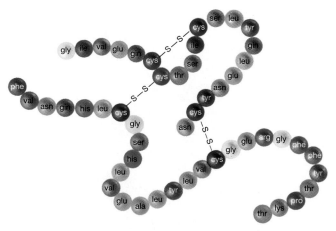

Figure 3-15   The primary structure of insulin

The insulin molecule actually consists of two small amino acid chains, whose sequences represent insulin's primary structure. Cysteines form disulfide bridges within and between the chains.

noncovalent interactions, such as hydrogen bonds, that produce secondary and tertiary structures. A single molecule of hemoglobin, the oxygen-carrying pigment in red blood cells, exhibits all four structural levels. He-

moglobin consists of two pairs of very similar peptides, held together by hydrogen bonds (see Fig. 3-14). Each peptide holds an iron-containing organic molecule called a heme that can bind one molecule of oxygen. Although no one knows precisely why hemoglobin contains four peptides, it is known that a single peptide binds oxygen more tightly than the four-peptide molecule does. This characteristic would be great for picking up oxygen in the lungs, but, because the single peptides would not release the oxygen again under normal conditions, the rest of the body's cells would die from lack of oxygen. Interactions among the four peptides allow hemoglobin to bind oxygen tightly enough to acquire oxygen in the lungs but loosely enough to give it up again to the body tissues.

The Functions of Proteins Are Intimately Linked to Their Three-Dimensional Structure

The exact type, position, and number of different R groups of the amino acids of a protein determine both the structure of the protein and its biological function. In any given protein, however, some R groups are more important than others. In hemoglobin, for example, certain amino acids

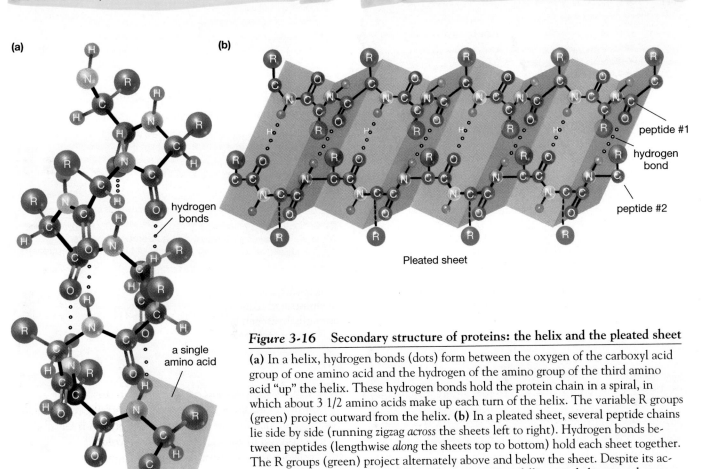

Pleated sheet

Figure 3-16   Secondary structure of proteins: the helix and the pleated sheet

(a) In a helix, hydrogen bonds (dots) form between the oxygen of the carboxyl acid group of one amino acid and the hydrogen of the amino group of the third amino acid "up" the helix. These hydrogen bonds hold the protein chain in a spiral, in which about 3 1/2 amino acids make up each turn of the helix. The variable R groups (green) project outward from the helix. **(b)** In a pleated sheet, several peptide chains lie side by side (running zigzag *across* the sheets left to right). Hydrogen bonds between peptides (lengthwise *along* the sheets top to bottom) hold each sheet together. The R groups (green) project alternately above and below the sheet. Despite its accordion-like appearance, each peptide chain is in its fully extended state and cannot easily be stretched further. For this reason, pleated sheet proteins such as silk are strong but not elastic.

must be present in precisely the right places to hold the iron-containing heme group that binds oxygen. Many of the other amino acids are interchangeable to some extent, if they are functionally equivalent. For example, the amino acids on the outside of a hemoglobin molecule mostly serve to keep it dissolved in the cytoplasm of a red blood cell. Therefore, as long as they are hydrophilic, it doesn't matter too much exactly which amino acids are where. As we will see in Chapter 15, however, replacing a hydrophilic amino acid with a hydrophobic amino acid can have catastrophic effects on the solubility of the hemoglobin molecule; in fact such a substitution is the molecular cause of a painful and sometimes life-threatening disorder called sickle-cell anemia.

Nucleic Acids Form DNA and RNA, the Molecules of Heredity

As you will learn in later chapters, the amino acid sequence of every protein in your body is specified by the genetic instructions residing in the nuclei of your cells. In fact, that's precisely what most genes are: a set of instructions spelling out the amino acid sequences of your body's proteins. Genes are composed of the fourth major category of biological molecule, **nucleic acids**. Nucleic acids are long chains of similar but not identical subunits called **nucleotides**. All nucleotides have a three-part structure: a five-carbon sugar (ribose or deoxyribose), a phosphate group, and a nitrogen-containing base that differs among nucleotides:

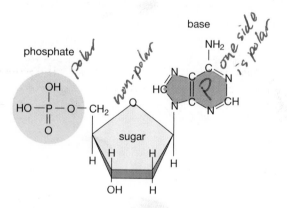

Deoxyribose nucleotide

There are two types of nucleotides, the ribose nucleotides (containing ribose sugar) and the deoxyribose nucleotides (containing deoxyribose sugar). Four different nitrogen-containing bases—adenine, guanine, cytosine, and thymine—are found bonded to deoxyribose nucleotides. Similarly, there are four types of ribose nucleotides, with adenine, guanine, cytosine, or uracil as the base (see Chapters 10 and 11).

Nucleotides may be strung together in long chains, with the phosphate of one nucleotide covalently bonded to the sugar of another:

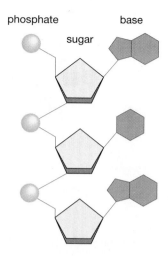

Nucleotide chain

Deoxyribose nucleotides form chains millions of units long called **deoxyribonucleic acid**, or **DNA**. DNA is found in the chromosomes of all living things, and its sequence of nucleotides, like the dots and dashes of a biological Morse code, spells out the genetic information needed to construct the proteins of each organism (see Chapter 10). Chains of ribose nucleotides, called **ribonucleic acid,** or **RNA**, are copied from the central repository of DNA in the nucleus of each cell. RNA carries DNA's genetic code into the cytoplasm and directs the synthesis of proteins (see Chapter 11).

Other Nucleotides Act as Intracellular Messengers, Energy Carriers, and Coenzymes

Not all nucleotides are part of DNA or RNA molecules. Some exist singly in the cell or occur as parts of other molecules. **Cyclic nucleotides**, such as cyclic adenosine monophosphate (cyclic AMP; Fig. 3-17a) are intracellular messengers that carry information from the cell membrane to other molecules in the cell. Cyclic AMP is synthesized when certain hormones come into contact with the cell membrane. Cyclic AMP then stimulates essential reactions in the cytoplasm or nucleus (see Chapter 35).

Some nucleotides have extra phosphate groups. These diphosphate and triphosphate nucleotides, such as **adenosine triphosphate** (ATP; Fig. 3-17b), are unstable molecules that carry energy from place to place within a cell. They capture energy where it is produced (during photosynthesis, for example) and give it up to drive energy-demanding reactions elsewhere (say, to synthesize a protein). Finally, certain nucleotides assist enzymes in their role of promoting and guiding chemical reactions. These nucleotides are called **coenzymes** and usually consist of a nucleotide combined with a vitamin (Fig. 3-17c). You will learn more about energy-carrier nucleotides, enzymes, and coenzymes in Chapter 4, where we discuss energy production and use in the cell.

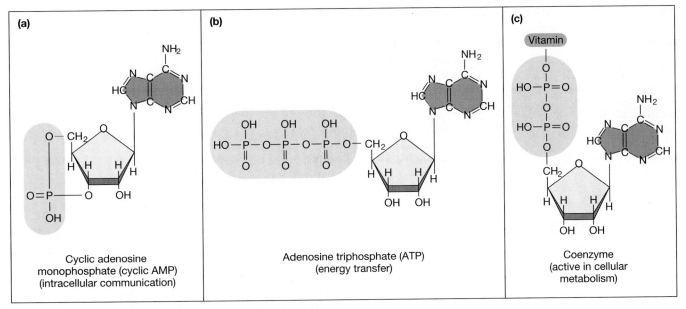

Figure 3-17   A sampling of the diversity of nucleotides

Individual nucleotides, constructed of a sugar, a phosphate group, and a nitrogen-containing base, are often modified by the addition of different functional groups and serve a variety of cellular functions.

✖ S U M M A R Y   O F   K E Y   C O N C E P T S

Synthesizing Organic Molecules: A Modular Approach

Most large biological molecules are polymers synthesized by linking together many smaller subunit monomers. Chains of subunits are connected by covalent bonds through dehydration synthesis; the chains may be broken apart again by hydrolysis reactions.

The Principal Types of Biological Molecules

The most important organic molecules fall into one of four classes: carbohydrates, lipids, proteins, and nucleic acids. Their major characteristics are summarized in Table 3-2.

Carbohydrates include sugars, starches, chitin, and cellulose. Sugars (monosaccharides and disaccharides) are used for temporary storage of energy and for the construction of other molecules. Starches and glycogen are polysaccharides that serve for longer-term energy storage in plants and animals, respectively. Cellulose and related polysaccharides form cell walls of bacteria, fungi, plants, and some microorganisms.

Lipids are nonpolar, water-insoluble molecules of diverse chemical structure that include oils, fats, waxes, phospholipids, and steroids. Lipids are used for energy storage (fats and oils), as waterproofing for the outside of many plants and animals (waxes), as the principal component of cellular membranes (phospholipids), and as hormones (steroids).

Proteins are chains of amino acids. The structure and function of a protein are determined by the sequence of amino acids in the chain. Proteins may be enzymes (which guide chemical reactions), structural molecules (hair, horn), hormones (insulin), or transport molecules (hemoglobin).

Nucleic acid molecules are chains of nucleotides. Each nucleotide is composed of a phosphate group, a sugar group, and a nitrogen-containing base. The two types of nucleic acids are deoxyribonucleic acid (DNA) and ribonucleic acid (RNA). Other nucleotides include intracellular messengers (cyclic AMP), energy-carrier molecules (ATP), and coenzymes.

✖ K E Y   T E R M S

adenosine triphosphate   p. 52
amino acid   p. 46
carbohydrate   p. 40
cellulose   p. 41
chitin   p. 42
coenzyme   p. 52

cyclic nucleotide   p. 52
dehydration synthesis   p. 38
deoxyribonucleic acid (DNA)   p. 52
disaccharide   p. 40
disulfide bridge   p. 47
enzyme   p. 44

fat   p. 42
fatty acid   p. 42
functional group   p. 37
glucose   p. 40
glycerol   p. 42
glycogen   p. 41

helix p. 49
hydrolysis p. 38
lactose p. 41
lipid p. 42
maltose p. 41
monomer p. 38
monosaccharide p. 40
nucleic acid p. 52
nucleotide p. 52
oil p. 42
peptide p. 47
peptide bond p. 47
phospholipid p. 44
pleated sheet p. 49
polymer p. 38
polysaccharide p. 40
primary structure p. 49
protein p. 44
quaternary structure p. 50
ribonucleic acid (RNA) p. 52
saturated p. 44
secondary structure p. 49
starch p. 41
steroid p. 44
subunit p. 38
sucrose p. 41
sugar p. 40
tertiary structure p. 49
triglyceride p. 42
unsaturated p. 44
wax p. 44

THINKING THROUGH THE CONCEPTS

Multiple Choice

1. What is the main function of polysaccharides in living cells?
   a. energy storage
   b. storage of hereditary information
   c. catalytic activity
   d. formation of membranes
   e. only b and c
2. Characteristics of carbon that contribute to its ability to form an immense diversity of organic molecules include its
   a. tendency to form covalent bonds
   b. ability to bond with up to four other atoms
   c. capacity to form single and double bonds
   d. ability to bond together to form extensive, branched or unbranched "carbon skeletons"
   e. all of the above
3. Foods that are high in fiber are most likely to be derived from
   a. plants
   b. dairy products
   c. meat
   d. fish
   e. all of the above
4. Proteins differ from one another because
   a. the peptide bonds linking amino acids differ from protein to protein
   b. the sequence of amino acids in the polypeptide chain differs from protein to protein
   c. each protein molecule contains its own unique sequence of sugar molecules
   d. the number of nucleotides found in each protein varies from molecule to molecule
   e. the number of nitrogen atoms in each amino acid differs from the number in all others
5. Which, if any, of the following choices does not properly pair an organic compound with one of its building blocks (subunits)?
   a. polysaccharide–monosaccharide
   b. fat–fatty acid
   c. nucleic acid–glycerol
   d. protein–amino acid
   e. all are paired correctly
6. Which one of the following statements is NOT true with regard to glycogen, starch, and cellulose? All are
   a. complex carbohydrates
   b. composed solely of glucose monomers
   c. easily digested by animals
   d. polysaccharides
   e. found in plants, except glycogen, which is found in animals

Review Questions

1. Which elements are commonly found in biological molecules?
2. List the four principal types of biological molecules and give an example of each.
3. What roles do nucleotides play in living organisms?
4. One way to convert corn oil to margarine (solid at room temperature) is to add hydrogen atoms, decreasing the number of double bonds in the molecules of oil. What is this process called? Why does it work?
5. Describe and compare dehydration synthesis and hydrolysis. Give an example of a substance formed using each chemical reaction, and describe the specific reaction in each instance.
6. Distinguish among the following: monosaccharide, disaccharide, and polysaccharide. Give two examples of each, and their functions.
7. Describe the synthesis of a protein from amino acids, then describe primary, secondary, tertiary, and quaternary structures of a protein.
8. Most structurally supportive materials in plants and animals are polymers of special sorts. Where would one find cellulose? Chitin? In what way(s) are these two similar? Different?
9. What kinds of bonds or bridges between keratin molcules are altered when hair is (a) wet and allowed to dry on curlers; (b) given a permanent wave?

❖ APPLYING THE CONCEPTS

1. A preview question for Chapter 6: In Chapter 2 you learned that hydrophobic molecules tend to cluster when immersed in water. In this chapter, you discovered that a phospholipid has a hydrophilic head and hydrophobic tails. What do you think would be the configuration of phospholipids that are immersed in water?
2. Many birds must store large amounts of energy to power flight during migration. Which type of organic molecule would be the most advantageous for energy storage? Why?
3. Remember the nuclear incident at Chernobyl? A scientist suspects that the food in a nearby ecosystem may have been contaminated with radioactive nitrogen over a period of months. Which substances in plants and animals could be examined for radioactivity to test his hypothesis?

❖ FOR MORE INFORMATION

Atkins, P. W. *Molecules*. New York: W. H. Freeman and Co., 1987. A fascinating, nontechnical introduction to the beauties of chemistry.

Doolittle, R. F. "Proteins." *Scientific American*, October 1985. Proteins are the molecular tools that, directly or indirectly, carry out almost all the functions of a cell.

Dushesne, L. C., and Larson, D. W. "Cellulose and the Evolution of Plant Life." *BioScience*, April 1989. Chemical and natural history aspects of the most plentiful organic molecule on Earth.

Goodsell, D. S. *The Machinery of Life*. New York: Springer, 1993. Goodsell depicts the molecules of a cell in all of their three-dimensional, interactive glory. A great way to gain a feel for the beauty and intricacy of the organic molecules of life.

Hill, J. W. *Chemistry for Changing Times*. 7th ed. Englewood Cliffs, NJ: Prentice Hall, 1995. A chemistry textbook that is both clearly readable and thoroughly enjoyable, for non-science majors.

Sharon, N., and Lis, H. "Carbohydrates in Cell Recognition." *Scientific American*, January 1993. Sugar-protein complexes on the surfaces of cells regulate cell identification and interaction between cells. See also Sharon's earlier article, "Carbohydrates," *Scientific American*, November 1980.

NET WATCH

On-line resources for this chapter are on the World Wide Web at:
http://www.prenhall.com/~audesirk (click on the table of contents link and then select Chapter 3).

4 Energy Flow in the Life of a Cell

> *"Disorder spreads through the universe, and life alone battles against it."*
> G. EVELYN HUTCHINSON

⊞ AT A GLANCE

Energy and the Ability to Do Work
The Laws of Thermodynamics Describe the Basic Properties of Energy
Living Things Use the Energy of Sunlight to Create the Low-Entropy Conditions Characteristic of Life

Energy Flow in Chemical Reactions
Exergonic Reactions Release Energy
Endergonic Reactions Require an Input of Energy
Coupled Reactions Link Exergonic with Endergonic Reactions
Chemical Reactions Are Reversible

Controlling the Metabolism of Living Cells
At Body Temperatures, High Activation Energy Makes Spontaneous Reactions Proceed Too Slowly to Sustain Life
Catalysts Reduce Activation Energy
Enzymes Are Biological Catalysts

Coupled Reactions and Energy-Carrier Molecules
ATP Is the Principal Energy Carrier in a Cell
Electron Carriers Also Transport Energy within Cells

A chameleon consumes a grasshopper in Kenya. Both predator and prey convert the chemical energy of their food molecules into the kinetic energy of movement.

Among the fundamental characteristics of all living organisms is the ability to guide chemical reactions within their bodies along certain pathways. The chemical reactions serve many functions, depending on the nature of the organism: to synthesize the molecules that make up the organism's body, to reproduce, to move, even to think. Chemical reactions either require or release **energy**, which can be defined simply as *the capacity to do work*, including synthesizing molecules, moving things around, and generating heat and light. In this chapter we discuss the physical laws that govern energy flow in the universe, how energy flow in turn governs chemical reactions, and how the chemical reactions within living cells are controlled by the molecules of the cell itself. Chapters 7 and 8 focus on photosynthesis, the chief "port of entry" for energy into the biosphere, and glycolysis and cellular respiration, the most important sequences of chemical reactions that release energy.

Energy and the Ability to Do Work

As you learned in Chapter 2, there are two types of energy: **kinetic energy** and **potential energy**. Both types of energy may exist in many different forms. Kinetic energy, or *energy of movement*, includes light (movement of photons), heat (movement of molecules), electricity (movement of electrically charged particles), and movement of large objects. Potential energy, or *stored energy*, includes chemical energy stored in the bonds that hold atoms together in molecules, electrical energy stored in a battery, and positional energy stored in a diver poised to spring (Fig. 4-1). Under the right conditions, kinetic energy can be transformed into potential energy, and vice versa. For example, the diver converted kinetic energy of movement into potential energy of position when she climbed the ladder up to the platform; when she jumps off, the potential energy will be converted back into kinetic energy.

To understand how energy flow governs interactions among pieces of matter, we need to know two things: (1) the quantity of available energy and (2) the usefulness of

Figure 4-1   Potential energy

The body of a diver perched atop the platform has potential energy, because of the different heights of the platform and the pool. The smacking sound and splashing water are manifestations of the conversion of this potential energy into kinetic energy when she dives into the water.

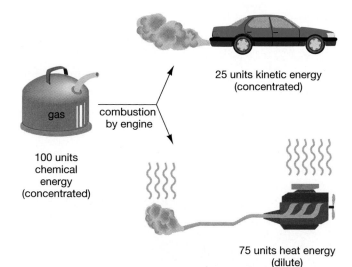

Figure 4-2   Thermodynamics of automobiles

According to the first law of thermodynamics, the quantity of energy originally present in the gasoline equals the energy in the resulting products: the molecules in the exhaust, the moving car, and heat. The form of the energy changes from chemical energy into heat or energy of movement. According to the second law, some of the energy in the products is in a less concentrated, less usable form than the chemical energy of the gasoline. In cars, about 75% of the chemical energy of the gas is given off as waste heat to the environment, contributing to random movement of molecules in the air.

the energy. These are the subjects of the laws of thermodynamics, which we will now examine.

The Laws of Thermodynamics Describe the Basic Properties of Energy

All interactions among pieces of matter are governed by the two **laws of thermodynamics**, physical principles that define the basic properties and behavior of energy. The laws of thermodynamics deal with "isolated systems," which are any parts of the universe that cannot exchange either matter or energy with any other parts. Probably no part of the universe is completely isolated from all possible exchange with every other part, but the concept of an isolated system is useful in thinking about energy flow.

The First Law of Thermodynamics States That Energy Can Neither Be Created nor Destroyed

The **first law of thermodynamics** states that within any isolated system, energy can neither be created nor destroyed, although it can be changed in form (for example, from chemical energy to heat energy). In other words, within an isolated system *the total quantity of energy remains constant.* The first law is therefore often called the law of conservation of energy. To use a famil-

iar example, let's see how the first law applies to driving your car (Fig. 4-2). We can consider that your car (with a full tank of gas), the road, and the surrounding air roughly constitute an isolated system. When you drive your car, you convert the potential chemical energy of gasoline into kinetic energy of movement and heat energy. The total amount of energy that was in the gasoline before it was burned is the same as the total amount of this kinetic energy and heat.

An important rule of energy conversions is this: Energy always flows "downhill," from places with a high concentration of energy to places with a low concentration of energy. This is the principle behind engines. As we described in Chapter 2, temperature is a measure of how fast molecules move. The burning gasoline in your car's engine consists of molecules moving at extremely high speeds: a high concentration of energy. The cooler air outside the engine consists of molecules moving at much lower speeds: a low concentration of energy. The molecules in the engine hit the piston harder than the air molecules outside the engine do, and so the piston moves upward, driving the gears that move the car. Work is done. When the engine is turned off, it cools down as heat is transferred from the warm engine to its cooler surroundings. The molecules on both sides of the piston move at the same speed, so the piston stays still. No work is done.

The Second Law of Thermodynamics States That the Amount of Useful Energy Always Decreases When Energy Is Converted from One Form to Another

The second law is related to the usefulness of energy. The **second law of thermodynamics** states that any change in an isolated system causes the quantity of concentrated, useful energy to decrease. Put another way, the second law states that all spontaneous changes result in a more uniform distribution of energy, reducing the energy differences that are essential for doing work; in other words, energy is always converted from more useful into less useful forms.

Let us continue our example of a car engine. The heat of the burning gasoline moves the piston, which moves the car. But the kinetic energy of the moving car is much less than the chemical energy that was originally contained in the gasoline (see Fig. 4-2). Where is the "missing" energy? Feel the engine and the exhaust: they are both hot. The burning gas not only moved the piston, it also heated up the engine, the exhaust system, and the air around the car. The friction of tires on pavement heated the road up a little, too. So, as the first law dictates, there really isn't any missing energy: the original quantity of energy in the gasoline still exists, only in different forms. However, some of the energy (the heat given off) merely increased the random movement of molecules in the engine block, the road, and the air. Such random movements aren't useful: You can't very well gather up the energy found in the warmed-up road and use it to drive your car a few more blocks.

The second law also tells us something about the organization of matter. Regions of concentrated energy are usually also regions of great orderliness. The eight carbon atoms in a single molecule of gasoline have a much more orderly arrangement than do the carbon atoms of the eight separate, randomly moving molecules of carbon dioxide that are formed when the gasoline burns.

Therefore, we can also phrase the second law in terms of the organization of matter: All processes in an isolated system result in an increase in randomness and disorder. This tendency toward loss of orderliness and high-level energy and an increase in randomness, disorder, and low-level energy is called **entropy**.

Eventually, several billion years from now, all the energy in the universe will be evenly dispersed as heat, and all the matter will be randomly distributed in small molecules. Without differences in energy, no further work will be possible. Life, therefore, will cease. (This rather gloomy prospect is explored in a marvelous science fiction short story by Isaac Asimov entitled "The Last Question": definitely recommended reading.)

Living Things Use the Energy of Sunlight to Create the Low-Entropy Conditions Characteristic of Life

If you stop to think about the second law of thermodynamics, you may wonder how life can exist at all. If chemical reactions, including those inside living cells, cause the amount of low-level, unusable energy to increase, and if matter tends toward increasing randomness and disorder, how can organisms accumulate the concentrated energy and precisely ordered molecules that characterize living things? The answer is this: The second law applies only to isolated systems. Living organisms, however, are not isolated systems. Nuclear reactions in the sun produce concentrated energy (sunlight) along with vast increases in entropy. Living things on Earth use that concentrated energy to synthesize complex molecules and maintain orderly structures. The small packets of low entropy that we call life do not violate the second law—they are achieved at the expense of an enormous loss of useful energy from sunlight. The entropy of the solar system as a whole constantly increases.

Energy Flow in Chemical Reactions

A chemical reaction begins with one set of substances, called the **reactants**, and converts them into another set, the **products**. All chemical reactions fall into one of two categories: exergonic and endergonic. A reaction is **exergonic** (Greek for "energy out"), if the reactants have more energy than the products. Consequently, the reaction releases energy:

Exergonic reaction

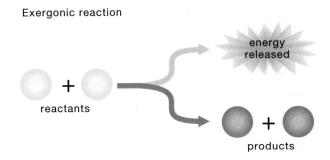

Conversely, a reaction is **endergonic** (Greek for "energy in"), if the products have more energy than the reactants. Because, according to the first law of thermodynamics, the extra energy of the products must come from somewhere, energy input from outside the system is required to drive the reaction:

Endergonic reaction

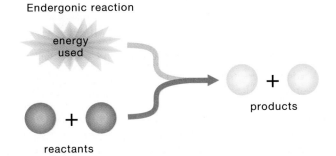

Let's look at two processes that illustrate these types of reactions: burning coal and photosynthesis.

Exergonic Reactions Release Energy

In an exergonic reaction, the reactants contain more energy than the products. For example, when coal (C) is burned it reacts with oxygen (O_2) to produce carbon dioxide (CO_2) and energy as described by this equation:

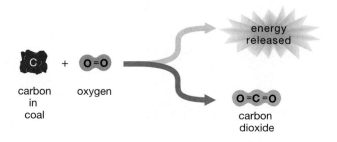

(To simplify things, we have assumed that coal is pure carbon.) This reaction illustrates two important concepts, diagrammed in Figure 4-3a. First, molecules of carbon and oxygen contain much more energy than molecules of carbon dioxide do, so the reaction releases energy. Energy release allows exergonic reactions to occur without a net input of energy. In our example, coal fires, once started, burn by themselves until the coal is consumed: You don't have to keep the fire going by adding outside energy from a flamethrower. (It may be helpful if you think of exergonic reactions as being "downhill," both in the graphical sense, as shown in Figure 4-3a, and in the everyday sense that cars and bicycles can coast downhill without engines or legs propelling them.)

Although burning coal releases energy, a lump of coal doesn't burst into flames by itself. This observation leads to the second point: All chemical reactions require an initial input of energy, called the **activation energy**, to get started. Chemical reactions require activation energy because atoms and molecules are surrounded by a cloud of negatively charged electrons. For two molecules to react with each other, their electron clouds must be forced together, despite their mutual electrical repulsion. Forcing them together requires energy. The usual source of activation energy is kinetic energy of movement. Only molecules moving with sufficient speed collide hard enough to force their electron clouds to mingle and react. Because molecules move faster at higher temperatures, most reactions occur more readily at high temperatures. The heat provided by setting coal on fire causes the carbon atoms to move faster; some atoms move so fast that they vaporize (become gaseous) and are no longer part of the solid lump of coal. These energetic carbon atoms collide with oxygen molecules and react to form carbon dioxide (CO_2), releasing heat. The heat causes more carbon atoms to vaporize and collide with oxygen molecules, producing more CO_2 and releasing more heat, which sustains the reaction until the coal is completely consumed.

Endergonic Reactions Require an Input of Energy

In contrast to what happens when coal is burned, many reactions in living systems result in products that contain more energy than the reactants. These reactions therefore require an input of energy into the low-energy reactants and a net input of energy into the reaction (Fig. 4-3b). As we shall see in Chapter 7, photosynthesis in green plants

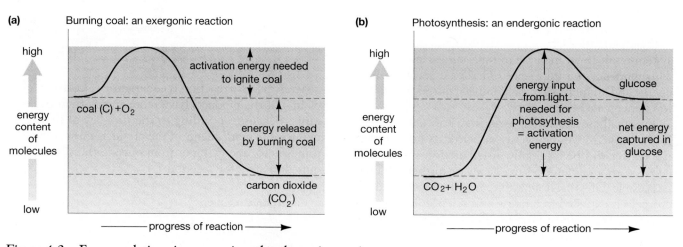

Figure 4-3   Energy relations in exergonic and endergonic reactions

(a) An exergonic ("downhill") reaction proceeds from high-energy reactants, such as coal, to low-energy products, such as CO_2. The energy difference between the chemical bonds of the reactants and products is released as heat. To start the reaction, however, an initial input of energy—the activation energy—is required. (b) An endergonic ("uphill") reaction, such as photosynthesis, proceeds from low-energy reactants such as CO_2 and H_2O to high-energy products, such as glucose, and therefore requires a net input of energy.

takes low-energy water and carbon dioxide and produces high-energy sugar and oxygen:

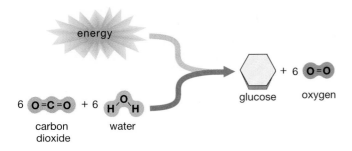

Photosynthesis requires energy, which plants obtain from sunlight. We might call endergonic reactions "uphill," in the same sense that cars and bicycles need inputs of energy to go uphill.

Coupled Reactions Link Exergonic with Endergonic Reactions

Because endergonic reactions require energy from other sources, they do not violate the rule that energy always flows downhill, from high concentrations of energy to lower concentrations of energy. Photosynthesis is an example of a **coupled reaction**. In a coupled reaction, an exergonic reaction provides the energy needed to drive an endergonic reaction.

In photosynthesis, the exergonic reaction occurs in the sun, and the endergonic reaction occurs in the plant:

Sun: Nuclear fusion of hydrogen → helium + large amount of solar energy released

Plants:

$$6\ CO_2 + 6\ H_2O + \underset{\substack{\text{solar} \\ \text{energy} \\ \text{input}}}{\text{small}} \rightarrow C_6H_{12}O_6\ \text{(glucose)} + 6\ O_2$$

The reactions occurring in the sun liberate energy as light. The sunlight captured by a plant possesses much more energy than is needed to drive photosynthesis. Therefore, the overall process, if we include the sun, is exergonic.

The fact that sunlight contains more energy than is needed to drive photosynthesis is an example of a general rule of coupled reactions. According to the second law of thermodynamics, the amount of useful energy decreases during every chemical reaction. This means that not all the energy released by an exergonic reaction can be used to drive an endergonic reaction; some energy is lost to the environment as heat and random movement of molecules. Therefore, in coupled reactions, the exergonic reaction must always release more energy than is required to drive the endergonic reaction (Fig. 4-4a).

Living organisms are the ultimate chemists, constantly using the energy given off by exergonic reactions (such as the chemical breakdown of food) to drive essential en-

dergonic reactions (such as brain activity, movement, or the synthesis of complex molecules; Fig. 4-4b). The exergonic and endergonic halves of coupled reactions often occur in different places, so there must be some way to transfer the energy from the exergonic reaction to the endergonic reaction. In photosynthesis, sunlight carries energy from exergonic reactions in the sun to the endergonic reactions in plants. In coupled reactions occurring within cells, energy is usually transferred from place to place by *energy-carrier* molecules, of which the most common is *adenosine triphosphate*, or *ATP*. We will examine the synthesis and use of ATP later in this chapter.

Chemical Reactions Are Reversible

So far, we have treated chemical reactions as if they always proceed in one direction, "forward" from reactants to products, and then are finished. This is usually not the case. Most chemical reactions are reversible: They may proceed spontaneously in either direction under the appropriate conditions of reactants, products, and energy. Furthermore, all reactions, if left to themselves, eventually reach a steady state called a **chemical equilibrium**, in which the reaction proceeds at an equal rate in both directions. At equilibrium, "forward" and "backward" reactions occur at exactly equal rates, with the result that, although *individual molecules* are constantly undergoing chemical reactions, the *net concentrations of molecules* do not change.

To understand the reversibility of chemical reactions and the nature of chemical equilibria, let's look at the reversible combination of oxygen with hemoglobin in the blood (see Chapter 3 for more about hemoglobin). As blood circulates through your lungs, hemoglobin picks up oxygen and becomes oxyhemoglobin. When the blood passes through the other organs of the body, the oxygen is released. Why?

Blood that enters the lungs has been depleted of oxygen during its journey through the body, so few of its hemoglobin molecules have oxygen bound to them. Therefore, in the high-oxygen environment of the lungs, oxygen binds to hemoglobin, forming oxyhemoglobin:

As more and more hemoglobin molecules become oxygenated, a few give up their oxygens. Finally, an equilibrium point is established: Just as many oxygen molecules

(a)

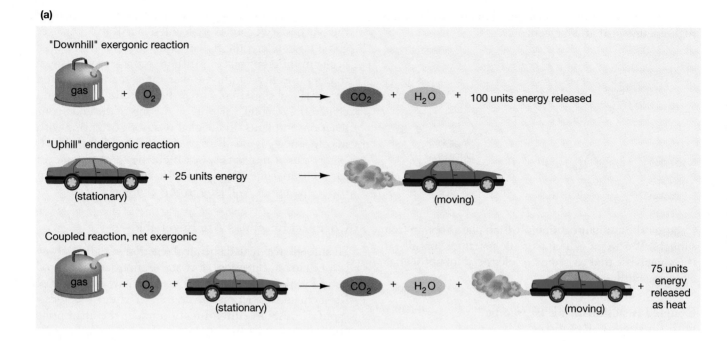

"Downhill" exergonic reaction

gas + O_2 ⟶ CO_2 + H_2O + 100 units energy released

"Uphill" endergonic reaction

(stationary) + 25 units energy ⟶ (moving)

Coupled reaction, net exergonic

gas + O_2 + (stationary) ⟶ CO_2 + H_2O + (moving) + 75 units energy released as heat

(b)

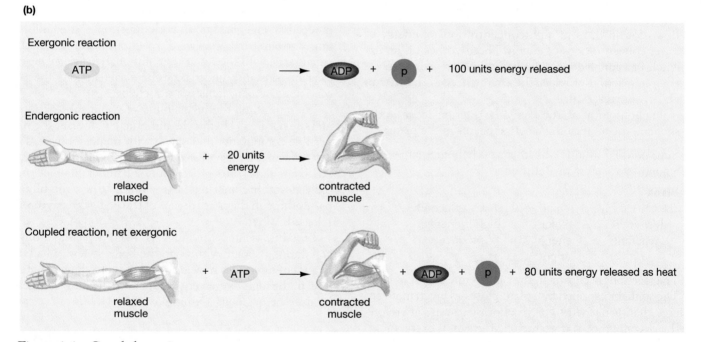

Exergonic reaction

ATP ⟶ ADP + p + 100 units energy released

Endergonic reaction

relaxed muscle + 20 units energy ⟶ contracted muscle

Coupled reaction, net exergonic

relaxed muscle + ATP ⟶ contracted muscle + ADP + p + 80 units energy released as heat

Figure 4-4   **Coupled reactions**

(a) Burning gasoline is an exergonic reaction that can be coupled to the endergonic reaction of moving a car. According to the second law of thermodynamics, the quantity of concentrated, useful energy decreases in the process. Therefore, the exergonic reaction must actually release more energy than the endergonic reaction requires. The excess energy is given off to the environment as heat. **(b)** Muscle movement is an endergonic reaction coupled to the exergonic reaction of ATP breakdown. The energy released by ATP as it is broken down into ADP plus phosphate (P) exceeds the energy put into muscle contraction, and the overall reaction is exergonic.

leave the air and bind to hemoglobin as leave the hemoglobin and reenter the air:

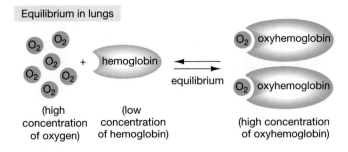

Equilibrium in lungs

At equilibrium, the reaction proceeds in both directions at the same rate.

Chemical equilibria are affected by the concentrations of the molecules involved: The high concentration of oxygen in the lungs produces a high concentration of oxyhemoglobin. In contrast, body tissues such as exercising muscles use up oxygen, so the tissue concentration of oxygen is very low. When oxygenated blood enters a muscle, the overall reaction reverses direction, with oxyhemoglobin in the blood losing oxygen to the muscle cells:

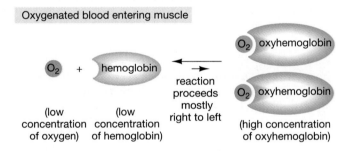

Oxygenated blood entering muscle

Because the muscle uses up oxygen about as fast as the hemoglobin releases it, almost all the oxygen leaves the he-

moglobin and enters the muscle cells. A new equilibrium is established, this time with most of the hemoglobin in the deoxygenated state:

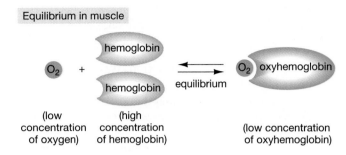

Equilibrium in muscle

Eventually the hemoglobin returns to the lungs, and the cycle repeats.

The hemoglobin-oxygen reaction illustrates an important principle of reactions occurring in living organisms. To remain alive, cells are constantly synthesizing some molecules and destroying others. By adding and removing the reactants and products, cells can control and automatically adjust the rates and directions of chemical reactions to meet their needs.

Controlling the Metabolism of Living Cells

Cells are miniature, incredibly complex chemical factories. The **metabolism** of a cell is the sum of all its chemical reactions. Many of these reactions are linked in sequences called **metabolic pathways** (Fig. 4-5). Photosynthesis is such a pathway. Glycolysis, the series of reactions that begin the digestion of glucose (see Chapter 8), is another. Different metabolic pathways may exchange molecules, with the re-

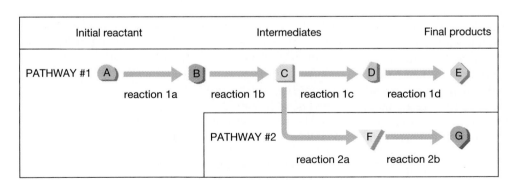

Figure 4-5   A simplified view of metabolic pathways

The original reactant molecule undergoes a series of reactions, each catalyzed by a specific enzyme. The product of each reaction serves as the reactant for the next reaction in the pathway. Metabolic pathways are often interconnected so that the product of a step in one pathway may serve as a reactant either for the next reaction in the same pathway or for a reaction in another pathway.

sult that all the reactions and all the molecules of a cell are actually interconnected in a single, enormously complicated metabolic pathway.

The chemical reactions within a cell are governed by the same laws of thermodynamics that control any other reactions. How, then, do orderly metabolic pathways arise? The biochemistry of cells is finely tuned in three ways:

1. Cells regulate chemical reactions through the use of protein catalysts called enzymes.
2. Cells couple reactions together, driving energy-requiring endergonic reactions with the energy released by exergonic reactions.
3. Cells synthesize energy-carrier molecules that capture energy from exergonic reactions and transport it to endergonic reactions.

At Body Temperatures, High Activation Energy Makes Spontaneous Reactions Proceed Too Slowly to Sustain Life

The laws of thermodynamics tell us that energy-releasing reactions can occur spontaneously, but they don't tell us how fast they will occur. In general, the rate of a reaction is determined by its activation energy, which is a measure of the initial amount of energy required to speed up reactant molecules enough to force their electron clouds together (Fig. 4-6). Reactions with low activation energies, such as the binding of oxygen to hemoglobin, can proceed swiftly at fairly low temperatures. Reactions with high activation energies, such as the reaction of coal with oxygen, where lots of energy is required to force the clouds together, will be very slow at low temperatures. Therefore, the rate of most reactions is limited by the speed of reactant molecules—the faster they move, the faster the reaction goes. Most reactions can be accelerated by raising the temperature, thereby increasing the speed of molecules.

Without high temperatures, many "spontaneous," exergonic reactions proceed extremely slowly. The reaction of sugar with oxygen to yield carbon dioxide and water is exergonic, but it has an enormous activation energy. Even quite high temperatures do not provide enough energy to overcome the activation energy barrier. As any cook knows, you can boil a sugar solution for hours and hardly any of the sugar breaks down. At the temperatures found in living organisms, sugar and many other energetic molecules would almost never break down spontaneously and give up their energy. Because of the catalytic enzymes produced by living cells, however, sugar is an important energy source for life on Earth. Let's see how enzymes and other catalysts influence chemical reactions.

Catalysts Reduce Activation Energy

Catalysts are molecules that speed up the rate of a reaction without themselves being used up or permanently altered. Catalysts speed up reactions by reducing the acti-

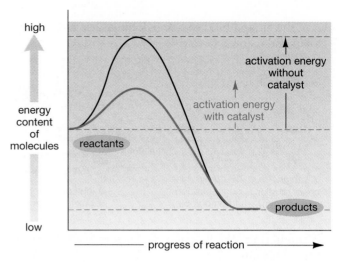

Figure 4-6   Activation energy controls the rate of chemical reactions

High activation energy (black line) means that reactant molecules must collide very forcefully in order to react. Only very fast-moving molecules will collide hard enough to react, so reactions with high activation energies proceed slowly at low temperatures, where most molecules move relatively slowly. Catalysts lower the activation energy of a reaction (red line), so a much higher proportion of molecules move fast enough to react when they collide. Therefore, the reaction proceeds much more rapidly.

vation energy (Fig. 4-6). As an example of catalytic action, let's consider the catalytic converters in car engines. When gasoline is burned completely, the final products are carbon dioxide and water:

$$2\,C_8H_{18}\text{ (octane)} + 25\,O_2 \rightarrow 16\,CO_2 + 18\,H_2O + \text{energy}$$

However, flaws in the combustion process generate other substances, including poisonous carbon monoxide (CO). Carbon monoxide reacts spontaneously but slowly with oxygen in the air to form carbon dioxide:

$$2\,CO + O_2 \rightarrow 2\,CO_2 + \text{energy}$$

In large cities, so many cars emit so much CO that the spontaneous reaction of CO with O_2 can't keep up, and unhealthy levels of carbon monoxide accumulate. Enter the catalytic converter. The platinum catalysts in the converter hasten the conversion of CO to CO_2, thereby reducing air pollution.

There are four important principles to note about all catalysts:

1. Catalysts speed up reactions.
2. Catalysts cannot cause energetically unfavorable reactions to occur; that is, they can speed up only those reactions that would occur spontaneously anyway, although at a much slower rate.
3. Catalysts do not change the equilibrium point of a reaction. Catalytic converters, for example, cannot reduce the concentration of CO below the naturally oc-

curring minimum level. If fresh mountain air is put through a converter, the concentration of CO in the air coming out is the same as the concentration going in.

4. Catalysts are not consumed in the reactions they promote. No matter how many reactions they accelerate, the catalysts themselves are not permanently changed. Converters in cars seldom have to be replaced unless they become contaminated by other materials.

Enzymes Are Biological Catalysts

Enzymes are catalysts synthesized by living organisms. Almost all enzymes are proteins. Just a few years ago, biology texts stated unequivocally that *all enzymes are proteins*. During the 1980s, however, Thomas Cech and Sidney Altman discovered that certain molecules of ribonucleic acid also function as enzymes. These molecules, dubbed **ribozymes**, catalyze reactions involved in processing genetic information for use by a cell and may have been crucial during the early evolution of life (see Chapter 19). Ribozymes notwithstanding, in the rest of this chapter we will discuss only the actions of protein enzymes.

Enzymes possess the four characteristics of catalysts just described. But enzymes have two additional attributes that set them apart from inorganic catalysts:

1. An enzyme is usually very specific, promoting only a few types of chemical reactions. An inorganic catalyst, in contrast, can usually speed up many different reactions. In most cases, an enzyme catalyzes a single reaction involving one or two specific molecules, while leaving even quite similar molecules untouched. You may recall from Chapter 3, for example, that animals have enzymes that break apart starch molecules but leave cellulose intact, despite the fact that starch and cellulose are both composed of glucose subunits.

2. Enzyme activity is regulated—that is, enhanced or suppressed—often by the very molecules whose reactions they promote.

The Structure of Enzymes Allows Them to Bind Specific Molecules and Catalyze Specific Reactions

Why are enzymes specific, and how are they regulated? Enzyme function is intimately related to enzyme structure. Enzymes are proteins with complex three-dimensional shapes (Fig. 4-7). Each enzyme has a dimple or groove, called the **active site**, into which reactant molecules, called **substrates**, can enter.

The active site of each enzyme has a distinctive shape and distribution of electrical charge that is complementary to its substrate, analogous to the fit between a lock and

(a)

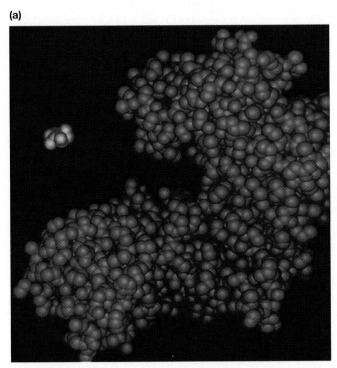

(b)

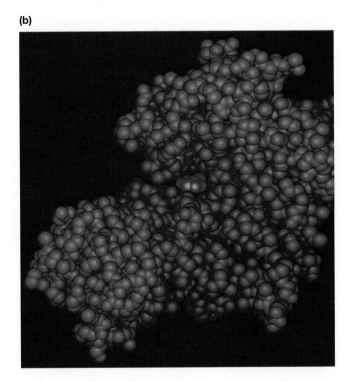

Figure 4-7 **How enzymes interact with substrates**

Computer-generated models of the enzyme hexokinase and its substrate glucose (hexokinase is the enzyme that catalyzes the first step in glucose metabolism). **(a)** The active site of hexokinase is a groove or dimple into which a glucose molecule may enter. **(b)** Entry of glucose into the active site causes both the glucose and hexokinase molecules to change shape (changes in the glucose molecule cannot be seen because it is too small and is buried) and is essential for the enzyme action to take place.

its key. The enzyme allows only a few extremely similar molecules to enter, much as the size and shape of a lock allow only one or a few keys to fit in. However, enzymes and substrates are actually much more flexible than locks and keys. For at least some enzymes, changes in the shape of both the enzyme and substrate are essential for enzyme action to occur (see Fig. 4-8).

The size, shape, and charge of the active site confer great specificity on an enzyme, allowing only certain molecules to enter and react, while rejecting even fairly similar molecules. For example, there are several protein-digesting enzymes in your intestines that break the peptide bonds that join amino acids. But each enzyme is quite specific. No single enzyme can digest every type of protein, because only a protein with the right amino acid sequence can fit into the active site of any individual enzyme. Other proteins, with amino acids that are too large or too small or have the wrong charge, cannot get in and consequently cannot be digested. Complete digestion of all the proteins in the human diet, therefore, requires several enzymes.

Conversely, some molecules may be able to enter the active site of an enzyme but do not have chemical bonds upon which the enzyme can act, so no reaction occurs. Many poisons, including some insecticides, enter the active site of enzymes essential to brain function and never leave again. The enzymes remain plugged up, and the reactions they normally promote cannot occur. Parts of the brain turn off, or become wildly hyperactive, sometimes with fatal consequences.

How does an enzyme promote a reaction? First, the shape and charge of the active site force substrates to enter the enzyme in specific orientations (Fig. 4-8, step 1). Second, when substrates enter the active site, both substrate and active site change shape (step 2). Certain amino acids that form the active site may make temporary chemical bonds to atoms of the substrates, or electrical interactions between active site and substrates may distort the chemical bonds within the substrates. The combination of substrate selectivity, substrate orientation, temporary chemical bonds, and bond strain promotes the specific chemical reaction catalyzed by a particular enzyme. When the final reaction between the substrates is finished, the product(s) no longer fit properly into the active site and are expelled (step 3). The temporary changes in shape, charge, and bonding patterns within the enzyme revert to their original configuration, and the enzyme is ready to accept another set of substrates (back to step 1).

Why does this sequence of events speed up the rate of chemical reactions? The interactions between substrate and enzyme, which are like minireactions with very low activation energies, allow the overall reaction to bypass its otherwise high activation energy barrier.

The Activity of Enzymes Is Influenced by Their Environment

Protein enzymes have very complex, three-dimensional structures that are sensitive to environmental conditions. Each enzyme has evolved to function optimally at a particular pH, temperature, and salt concentration. Some also require the presence of other molecules called coenzymes, often derived from water-soluble vitamins, in order to function.

Most enzymes function optimally at a pH between 6 and 8, the level found in most body fluids and maintained within cells. An exception is the protein-digesting enzyme pepsin. Pepsin is converted from an inactive to an active form by the highly acid (pH 2) conditions within the stomach. At this pH, the excess of hydrogen ions causes hydrogen to attach to certain locations on the protein, altering its configuration and exposing the active site. In proteins that function best at neutral pH (7), such interactions would distort the structure and destroy normal function.

Before refrigeration, foods such as meat were often preserved using concentrated salt solutions, which kill most bacteria, partly by interfering with enzyme function. Salts dissociate into ions, which form bonds with enzymes that interfere with the enzymes' normal three-dimensional structure. This process destroys enzyme activity, which requires a precise protein configuration. Dill pickles are very well-preserved in a vinegar-salt solution, which combines both salty and acidic conditions. Organisms that live in highly salty environments, as you might predict, have enzymes whose configuration depends on the presence of salt ions.

Temperature also affects the rate of enzyme-catalyzed reactions. Because molecules move more rapidly at higher temperatures, their random movements are more likely to bring them into contact with the active site of an appropriate enzyme. Thus these reactions are accelerated by higher temperatures and slowed by lower temperatures.

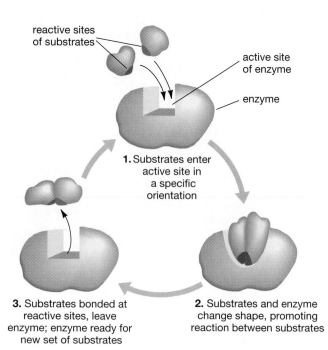

reactive sites of substrates

active site of enzyme

enzyme

1. Substrates enter active site in a specific orientation

3. Substrates bonded at reactive sites, leave enzyme; enzyme ready for new set of substrates

2. Substrates and enzyme change shape, promoting reaction between substrates

Figure 4-8   **The cycle of enzyme-substrate interactions**

A child who fell through the ice was rescued and survived unharmed after 20 minutes under water; the lowering of his body temperature slowed all his metabolic reactions and drastically reduced his need for oxygen. When temperatures rise too high, the bonds that regulate enzyme shape may be broken apart by the excessive molecular motion.

Some enzymes require helper molecules called **coenzymes** to function. These organic molecules bind to the enzyme and interact with the substrate molecule. Coenzymes help weaken the bonds of the substrate, allowing it to react with another molecule. Many water-soluble vitamins (such as the B vitamins) are essential because they are used by the body to synthesize coenzymes.

Cells Regulate the Amount and Activity of Their Enzymes, Thus Precisely Regulating Their Metabolic Reactions

Speeding up reactions is not always desirable. For example, you would not want your body to break down every glucose molecule you eat immediately. For one thing, you might starve to death overnight if you couldn't store any energy between supper and breakfast. What's more, glucose is an important ingredient in essential body chemicals, so some must be retained for synthesizing molecules such as hormones and components of cell membranes.

Cells have evolved the following three ways of regulating enzyme activity:

1. A cell may regulate the amounts of enzymes that it contains. Obviously, reactions will occur only if the necessary enzymes are available. As we will see in Chapter 11, cells often regulate the synthesis of enzymes to meet their changing needs.
2. A cell may synthesize an enzyme in an inactive form and activate it only when needed. Certain cells in your digestive system, for example, produce enzymes that digest food molecules such as proteins and lipids.

These enzymes could just as easily digest the proteins and lipids of the cells themselves. They don't because the enzymes are synthesized in an inactive form, with the active site blocked off. Within your stomach and small intestine, the interfering parts are cut off, thus activating the enzymes (see Chapter 32).

3. A cell can temporarily activate or inhibit enzymes, depending on the conditions at any given time. For example, the enzyme threonine deaminase begins the metabolic pathway that converts the amino acid threonine to the amino acid isoleucine. A cell needs suitable concentrations of both amino acids to manufacture proteins. The concentrations are regulated automatically by **feedback inhibition**, in which the activity of an enzyme is inhibited by its product or a later product of the metabolic pathway (Fig. 4-9). If enough isoleucine is present, it inhibits the activity of threonine deaminase, preventing further conversion of threonine.

On the molecular level, there are a variety of mechanisms of enzyme regulation. Here we will discuss two: allosteric regulation and competitive inhibition (Fig. 4-10). In **allosteric regulation**, enzyme action is inhibited or enhanced when molecules bind to a binding site on the enzyme that is distinct from the active site. This separate binding region is called the allosteric regulatory site (Fig. 4-10a). When the allosteric regulatory site is occupied, the enzyme changes shape and its activity may be enhanced or inhibited as a result (*allosteric* literally means "other shape"). Feedback inhibition frequently works through allosteric inhibition (Fig. 4-10c). The reaction product binds to the allosteric regulatory site, blocking enzyme activity and inhibiting further production of the product.

The second inhibitory mechanism is called **competitive inhibition**, a situation in which two or more molecules compete for entry into the enzyme's active site (Fig. 4-10b). Obviously, if one molecule occupies the active site, another cannot. Some poisons exert their effect be-

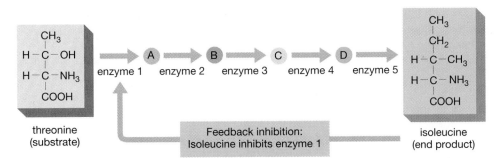

Figure 4-9 **Enzyme regulation by feedback inhibition**

In this example, the first enzyme in the metabolic pathway that converts threonine (substrate) to isoleucine (end product) is inhibited by high concentrations of isoleucine. If a cell lacks isoleucine, the first enzyme is not inhibited, and the pathway proceeds rapidly. As isoleucine concentrations build up, the isoleucine binds to the first enzyme to gradually shut down the pathway. When concentrations of isoleucine drop again and fewer molecules are around to inhibit the enzyme, the pathway resumes production.

(a) Enzyme structure

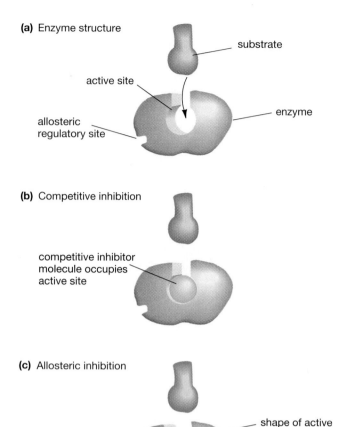

(b) Competitive inhibition

(c) Allosteric inhibition

Figure 4-10   **Enzyme regulation by competitive and allosteric inhibition**

Although they operate by very different mechanisms, both forms of inhibition prevent the substrate from entering the active site of the enzyme. **(a)** Many enzymes have an active site into which the substrate enters and an allosteric regulatory site on another part of the enzyme molecule. The enzyme is functional when the substrate can enter the active site. **(b)** In competitive inhibition, an inhibitor molecule competes with the substrate for entry into the active site: Whenever the inhibitor occupies the active site, the substrate cannot get in. **(c)** In allosteric inhibition, an inhibitor molecule binds to the allosteric regulatory site. This causes the shape or charge of the active site to change, thereby preventing the substrate from entering.

cause they resemble the normal substrate for an enzyme and bind to the active site, preventing normal metabolism. Two types of alcohol, methanol and ethanol, compete for the active site on the enzyme alcohol dehydrogenase. Methanol breakdown by this enzyme produces formaldehyde, which can cause blindness. Doctors administer ethanol intravenously to patients who have ingested methanol, because the ethanol competes with the methanol for the enzyme active site and blocks the production of formaldehyde.

Allosteric regulation and competitive inhibition can regulate concentrations of molecules in a cell because reactions are reversible. This includes the binding of inhibitor molecules (either allosteric or competitive) to enzymes. If inhibitors are scarce in the cell, the enzyme may be active most of the time. If the concentration of inhibitor molecules is high, then as soon as one leaves the enzyme, another is likely to take its place, and the enzyme will be inactive most of the time. In feedback inhibition, this means that when there is very little product in the cell there is very little enzyme inhibition, so the enzymes rapidly synthesize the product. As the product accumulates, it gradually shuts down the enzymes.

The ability of an enzyme to catalyze reactions is controlled by many factors, including the concentration of active enzyme, the concentration of inhibitor molecules, and the concentration of substrates. The interactions among these molecules usually maintain suitable concentrations of both substrates and products within a cell.

Coupled Reactions and Energy-Carrier Molecules

As we pointed out earlier, cells control energy flow by coupling reactions so that the energy released by exergonic reactions is used to drive endergonic reactions. Energy transfer is the role of **energy-carrier molecules**. Energy carriers work something like rechargeable batteries, picking up an energy charge at an exergonic reaction, moving to another location in the cell, and releasing the energy to drive an endergonic reaction.

ATP Is the Principal Energy Carrier in a Cell

The most common energy carrier molecule in living cells is **adenosine triphosphate**, or **ATP**. As you learned in Chapter 3, ATP is a nucleotide composed of a nitrogen-containing base, adenine; a sugar, ribose; and three phosphate groups (Fig. 4-11). Energy released in the cell through glucose metabolism is used to drive the synthesis of ATP from **adenosine diphosphate (ADP)** and inorganic phosphate:

ATP synthesis: Energy is stored in ATP

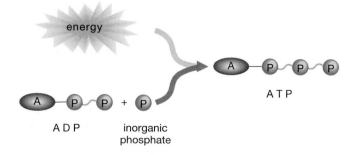

ATP: Principal energy carrier of the cell

	Adenosine diphosphate (ADP)	Adenosine triphosphate (ATP)
Shorthand representations	(A)—(P)—(P) or (ADP)	(A)—(P)—(P)—(P) or (ATP)
Energy content	low	high

Figure 4-11 **ADP and ATP structure**

A phosphate group is added to ADP (adenosine diphosphate) to make ATP (adenosine triphosphate). In most cases, only the last phosphate group and its high-energy bond are used to carry energy and transfer it to endergonic reactions within a cell. In the remainder of the text, we will usually employ the shorthand representations of ATP and ADP.

Most of the chemical bonds in ADP and ATP are "ordinary" covalent bonds, but the bonds joining the last two phosphate groups of ATP to the rest of the molecule are usually called high-energy bonds. Under most circumstances, only the last high-energy bond of ATP (the one that joins phosphate to ADP to make ATP) carries energy from exergonic to endergonic reactions. ATP is an unstable molecule, and its phosphate bond is easily broken, releasing energy that can be used to drive endergonic reactions.

The term *high-energy bond* is somewhat misleading, because, in fact, the bonds do not require an enormous amount of energy to form, nor do they release an enormous amount of energy when they are broken. Rather, a high-energy bond's relationship to a chemical reaction is a bit like Goldilocks's opinion of baby bear's porridge: It's just right. When your cells digest glucose, many of the reactions release just enough energy to add a phosphate to ADP to form ATP. Similarly, when ATP is broken down to ADP and inorganic phosphate, it releases an amount of energy that is just right to drive many essential cellular reactions. Nevertheless, the phrase "high-energy bond" is so common that we will continue to use it here.

ATP carries this energy to various sites in the cell that perform energy-requiring reactions, such as muscle contraction (see Fig. 4-4b). The ATP is then broken down to form ADP and inorganic phosphate:

ATP hydrolysis: Energy of ATP is released

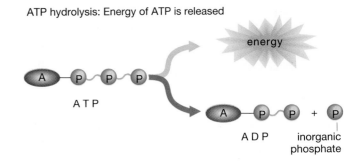

The energy released by ATP hydrolysis may drive the energy-requiring reaction. The ADP and inorganic phosphate are usually recycled back to energy-generating reactions that resynthesize ATP.

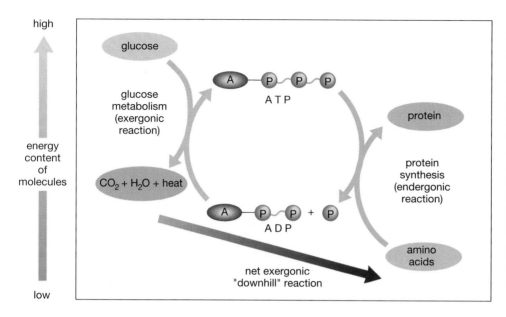

energy
content
of
molecules

low

glucose

glucose
metabolism
(exergonic
reaction)

$CO_2 + H_2O$ + heat

A — P — P — P
A T P

A — P — P + P
A D P

net exergonic
"downhill" reaction

protein

protein
synthesis
(endergonic
reaction)

amino
acids

Figure 4-12   Coupled reactions within living cells

Exergonic reactions (for example, glucose metabolism) drive the endergonic reaction of ATP synthesis from ADP. The ATP molecule moves to another part of the cell, where ATP hydrolysis liberates some of this energy to drive an essential endergonic reaction (such as protein synthesis). The ADP and inorganic phosphate are recycled back to the exergonic reactions, where they will be converted to ATP once again. The overall reaction is "downhill": More energy is produced by the exergonic reaction than is needed to drive the endergonic reaction. The extra energy is released as heat.

Most uses of energy in living cells involve pairs of coupled reactions linked by ATP (Fig. 4-12). In the first coupled reaction, energy released by an exergonic reaction drives ATP synthesis; in the second, ATP hydrolysis drives an energy-requiring endergonic reaction. Each coupled reaction is guided by a specific enzyme that positions the molecules and channels the ATP energy properly. Coupled reactions catalyzed by enzymes provide the energy and the specificity necessary to construct the different types of molecules needed by the cell.

So much ATP is used by living organisms that the life span of any given ATP molecule is very short. For example, a human runner can use up as much as half a kilogram (a pound) of ATP each minute (obviously, the ADP

thus produced must be quickly converted back to ATP, or it will be a very brief run). As a result, ATP is *not* a long-term energy-storage molecule. More stable molecules, such as sucrose, glycogen, starch, or fat, are used to store energy for hours, days, or months.

Electron Carriers Also Transport Energy within Cells

Energy may also be transported around a cell by other carrier molecules. In some exergonic reactions, including both glucose metabolism and the light-capture stage of photosynthesis, some energy is transferred to electrons. These energetic electrons (sometimes along with hydrogen atoms)

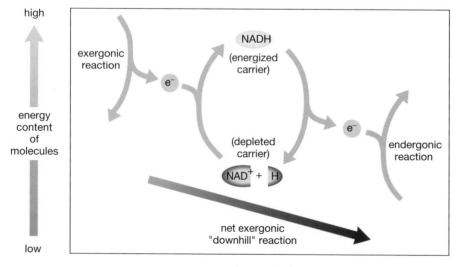

high

energy
content
of
molecules

low

exergonic
reaction

e^-

NADH
(energized
carrier)

(depleted
carrier)

NAD^+ + H

e^-

endergonic
reaction

net exergonic
"downhill" reaction

Figure 4-13   Electron carriers

Electron-carrier molecules such as nicotinamide adenine dinucleotide (NAD^+) pick up electrons generated by exergonic reactions and hold them in high-energy outer electron shells. Hydrogen atoms are often picked up simultaneously. The electron is then deposited, energy and all, with another molecule to drive an endergonic reaction, often the synthesis of ATP.

may be captured by **electron carriers** (Fig. 4-13), such as nicotinamide adenine dinucleotide (NAD⁺) and its relative flavin adenine dinucleotide (FAD). Loaded electron carriers then donate the electrons, along with their energy, to other molecules. You will see more about electron carriers and their role in cellular metabolism in Chapters 7 and 8.

✖ S U M M A R Y   O F   K E Y   C O N C E P T S

Energy and the Ability to Do Work

Kinetic energy is the energy of movement (light, heat, electricity, movement of large particles). Potential energy is stored energy (chemical energy, positional energy). The flow of energy among atoms and molecules obeys the laws of thermodynamics. The first law of thermodynamics states that within an isolated system the total amount of energy remains constant, although it may change in form. The second law of thermodynamics states that any change within an isolated system causes a decrease in the quantity of concentrated, useful energy, and an increase in the randomness and disorder of matter. Entropy is a measure of disorder within a system.

Energy Flow in Chemical Reactions

Chemical reactions fall into two categories. In exergonic reactions, the product molecules have less energy than the reactant molecules, so the reaction releases energy. In endergonic reactions, the products have more energy than the reactants, so the reaction requires an input of energy. Exergonic reactions can occur spontaneously, but all reactions, including exergonic ones, require an initial input of energy, called the activation energy, to overcome electrical repulsions between reactant molecules. Exergonic and endergonic reactions may be coupled together so that the energy liberated by an exergonic reaction drives the endergonic reaction. Living organisms couple exergonic reactions, such as light-energy capture or sugar metabolism, with endergonic reactions, such as the synthesis of organic molecules.

Both exergonic and endergonic reactions are reversible and can proceed in either direction, given suitable inputs of products, reactants, and energy. All reactions, if left to themselves, eventually reach a steady state called a chemical equilibrium. Living organisms provide the necessary inputs in controlled ways that drive reversible reactions in the directions necessary to maintain life.

Controlling the Metabolism of Living Cells

High activation energies slow many reactions, even exergonic ones, to an imperceptible rate under normal environmental conditions. Catalysts lower the activation energy and thereby speed up chemical reactions. Catalysts are not permanently altered during the reaction. Living organisms synthesize protein catalysts called enzymes that promote one or a few specific reactions. The reactants temporarily bind to the active site of the enzyme; the binding strains their original chemical bonds and makes it easier to form the new chemical bonds of the products. Enzyme action is regulated on a cellular level in three ways: (1) by altering the rate of enzyme synthesis, (2) by activating previously inactive enzymes, and (3) by inhibiting enzyme activity through allosteric and feedback inhibition.

Coupled Reactions and Energy-Carrier Molecules

Energy released by chemical reactions within a cell is captured and transported about the cell by energy-carrier molecules such as ATP and electron-carrier molecules. These molecules are the major means whereby cells couple exergonic and endergonic reactions occurring at different places within the cell.

✖ K E Y   T E R M S

activation energy   p. 60
active site   p. 65
adenosine diphosphate (ADP)   p. 68
adenosine triphosphate (ATP)   p. 68
allosteric regulation   p. 67
catalyst   p. 64
chemical equilibrium   p. 61
coenzyme   p. 67
competitive inhibition   p. 67
coupled reaction   p. 61

electron carrier   p. 71
endergonic   p. 59
energy   p. 57
energy-carrier molecules   p. 69
entropy   p. 59
enzyme   p. 65
exergonic   p. 59
feedback inhibition   p. 67
first law of thermodynamics   p. 58
kinetic energy   p. 57

laws of thermodynamics   p. 58
metabolic pathway   p. 63
metabolism   p. 63
potential energy   p. 57
product   p. 59
reactant   p. 59
ribozyme   p. 65
second law of thermodynamics   p. 59
substrate   p. 65

✖ T H I N K I N G   T H R O U G H   T H E   C O N C E P T S

Multiple Choice

1. According to the first law of thermodynamics, the total amount of energy in the universe
   a. is always increasing
   b. is always decreasing
   c. varies up and down
   d. is constant
   e. cannot be determined

2. What is predicted by the second law of thermodynamics?
   a. Energy is always decreasing.
   b. Disorder cannot be created or destroyed.
   c. Systems always tend toward greater states of disorder.
   d. All potential energy exists as chemical energy.
   e. all of the above

3. Which statement about exergonic reactions is true?
   a. The products have more energy than the reactants.
   b. The reactants have more energy than the products.
   c. They will not proceed spontaneously.
   d. Energy input reverses entropy.
   e. none of the above

4. ATP is important in cells because
   a. it transfers energy from exergonic reactions to endergonic reactions
   b. it is assembled into long chains which make up cell membranes
   c. it acts as an enzyme
   d. it accelerates diffusion
   e. all of the above

5. How does an enzyme increase the speed of a reaction?
   a. by changing an endergonic to an exergonic reaction
   b. by adding activation energy
   c. by lowering activation energy requirements
   d. by decreasing the concentration of reactants
   e. by increasing the concentration of products

6. Which of the following statements about enzymes is (are) true?
   a. They interact with specific reactants (substrates).
   b. Their three-dimensional shapes are closely related to their activities.
   c. They change the shape of the reactants.
   d. They have active sites.
   e. all of the above

Review Questions

1. Explain why living organisms do not violate the second law of thermodynamics; what is the ultimate energy source for most forms of life on Earth?

2. Define metabolism and explain how reactions can be coupled to one another.

3. Are chemical equilibria ever reached in living systems? Why or why not?

4. What is activation energy? How do catalysts affect activation energy? How does this change the rate of reactions?

5. Describe some exergonic and endergonic reactions that occur in plants and animals very regularly.

6. Describe the structure and function of enzymes, and compare them to inorganic catalysts. How is enzyme activity regulated?

✖ A P P L Y I N G   T H E   C O N C E P T S

1. A preview question for ecology: When a brown bear eats a salmon, does it acquire all the energy contained in the body of the fish? Why or why not? What implications do you think this would have for the relative abundance (by weight) of predators and their prey? Does the second law of thermodynamics help to explain the title of the book *Why Big Fierce Animals Are Rare?*

2. Many people in Sub-Saharan Africa have experienced the effects of malnutrition and starvation, but the very young are most severely affected. Some suffer permanent disability even if food is provided. How could a lack of food intake interfere with functions of individual cells and tissues? Which tissues are likely to suffer the most irreversible damage?

3. As you learned in Chapter 3, the subunits of virtually all organic molecules are joined by condensation reactions and can be broken apart by hydrolysis reactions. Why, then, does your digestive tract have separate enzymes to digest proteins, fats, and carbohydrates, and in fact several of each?

4. Suppose someone tried to disprove evolution with the following argument: "According to evolutionary theory, organisms have increased in complexity through time. However, such evolution of increased biological complexity contradicts the second law of thermodynamics. Therefore, evolution is impossible." If you were to support the theory of evolution, what would be your response to this argument?

✖ F O R   M O R E   I N F O R M A T I O N

Baker, J. J. W., and Allen, G. E. *Matter, Energy, and Life*. 4th ed. Reading, MA: Addison-Wesley, 1981. An excellent introduction to chemical-energy principles for those interested in biology.

Dickerson, R. E. "Cytochrome c and the Evolution of Energy Metabolism." *Scientific American*, March 1980. Cytochrome c is an electron carrier in mitochondria, and one of the best-studied proteins. Dickerson points out the evolutionary heritage of this molecule in diverse organisms and also discusses its role in energy acquisition by cells. See also Dickerson's earlier article on cytochrome c, "The Structure and History of an Ancient Protein." *Scientific American,* April 1972.

Fenn, J. *Engines, Energy, and Entropy.* New York: W. H. Freeman and Co., 1982. Elegantly simple introduction to the laws of thermodynamics and their relationship to everyday life.

Koshland, D. E., Jr. "Protein Shape and Biological Control." *Scientific American,* October 1973. Enzyme function is intimately related to structure. Koshland discusses enzyme specificity and regulation in terms of its protein structure.

Sackheim, G. *Introduction to Chemistry for Biology Students.* Redwood City, CA: Benjamin/Cummings Publishing Co., 1991. This book is designed specifically for biology students who don't have much chemistry background.

NET WATCH

On-line resources for this chapter are on the World Wide Web at:
http://www.prenhall.com/~audesirk (click on the <u>table of contents</u> link and then select Chapter 4).

5 Cell Structure and Function

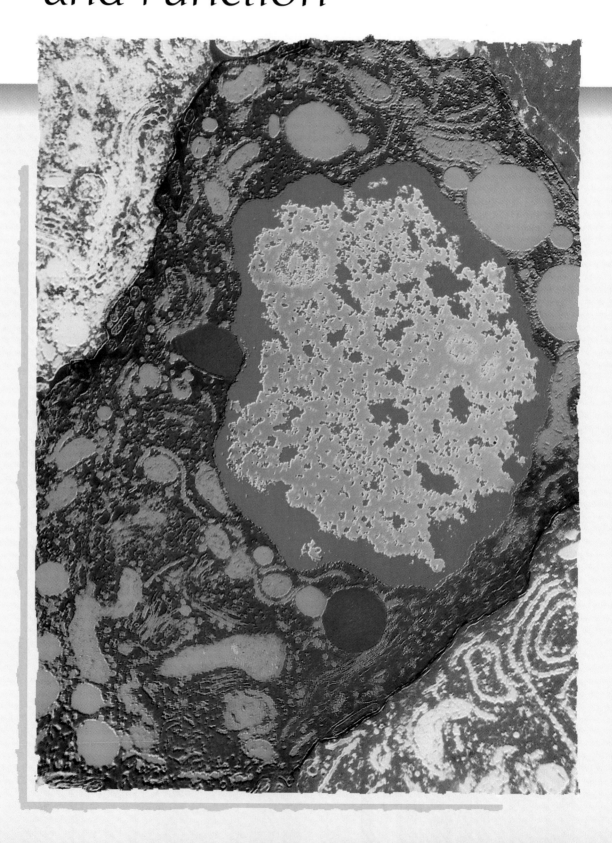

"It is all too easy now to underestimate cells. We have known about them for such large fractions of our lives that, for the most part, we cease being aware of how remarkable they really are."

BRUCE ALBERTS, DENNIS BRAY, JULIAN LEWIS, MARTIN RAFF, KEITH ROBERTS, AND JAMES WATSON from the Prologue to *Molecular Biology of the Cell*

❊ AT A GLANCE

The Development of Cell Theory

An Overview of Cell Structure and Function
The Plasma Membrane Isolates the Cell but Allows Interactions with the Environment
The Genetic Material Provides a Cellular "Blueprint"
Cytoplasm Fills Each Cell and Surrounds the Nucleus

Cell Function Limits Cell Size

Types of Cells: Prokaryotic and Eukaryotic
Prokaryotic Cells Are Relatively Simple
Eukaryotic Cells Are More Complex

The Nucleus: Control Center of the Cell
The Nuclear Envelope Allows Selective Exchange of Materials
Chromatin Consists of DNA and Its Associated Proteins
The Nucleolus Is the Site of Ribosome Assembly

The Membrane System of the Cell
Membranes Consist of a Double Layer of Phospholipids in Which Proteins Are Embedded
The Cell's Membrane System Includes the Plasma Membrane, Endoplasmic Reticulum, the Golgi Complex, and Lysosomes

Chloroplasts and Mitochondria: Energy Capture and Extraction
Chloroplasts Are the Site of Photosynthesis
Mitochondria Produce ATP Using Energy Stored in Food Molecules

Plastids and Vacuoles: Storage and Elimination
Many Types of Plastids Store Food in Plants
Vacuoles Serve Many Functions, Including Support and Storage and Elimination of Food and Wastes

The Cytoskeleton: Shape, Support, and Movement
Microfilaments Allow Cells to Change Shape and Guide the Movement of Organelles
Intermediate Filaments Provide a Supportive Framework
Microtubules Help Position, Anchor, and Move Organelles and Alter the Shapes of Cells
Cilia and Flagella Move the Cell or Move Fluid Past the Cell

This electron micrograph reveals the complexity of a single cell. Within the blue cytoplasm is a red and green nucleus; orange structures include round lipid granules, elongated mitochondria, and threadlike cut membranes of the Golgi complex and endoplasmic reticulum.

A vast period of evolutionary time separated the appearance of the first biological molecules from the first structures that could truly be considered alive. Evolution remained severely constrained until complex aggregates of these molecules became separated from their environment. Proteinlike molecules sometimes spontaneously form hollow balls, called microspheres, in water. Perhaps, several billion years ago, some of these microspheres trapped other biological molecules inside, making a sort of pre-cell. The first collections of molecules that could be considered alive surely possessed a thin covering that separated life from nonlife, a simple cell membrane.

The Development of Cell Theory

Human understanding of the cellular nature of life came slowly. In 1665, English scientist and inventor Robert Hooke reported observations with a primitive microscope. He aimed his instrument at an "exceeding thin...piece of Cork" and saw "a great many little Boxes" (Fig. 5-1). Hooke called the boxes "cells," because he thought they resembled the tiny rooms, or cells, occupied by monks. Cork comes from the dry outer bark of the Mediterranean oak. Hooke wrote that in the living oak and other plants, "These cells [are] fill'd with juices." What these cells were, and how they were related to the lives of entire plants, Hooke did not say.

In 1673, the Dutch inventor Anton van Leeuwenhoek reported to Britain's Royal Society on his observations of red blood cells, sperm, and myriad microscopic "animalcules" in pond water. But more than a century passed before biologists began to understand the role of cells in life on Earth. Microscopists first noted that many plants consist entirely of cells. The thick wall surrounding all plant cells made their observations easier. Animal cells, however, escaped notice until the 1830s, when German zoologist Theodor Schwann saw that cartilage contains cells that "exactly resemble—[the cells of] plants." In 1839, after studying cells for years, Schwann was confident enough to publish his theory, calling cells the elementary

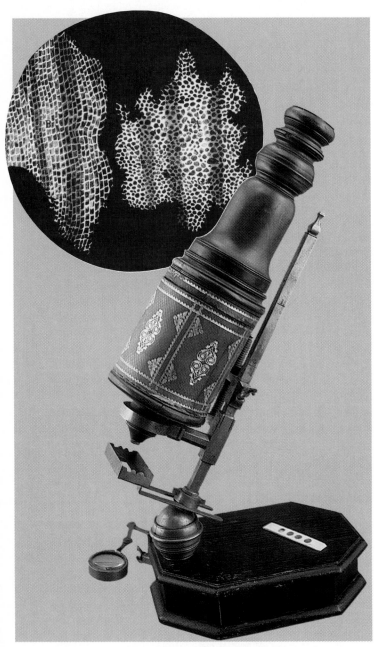

Figure 5-1   Early observations of cells

Robert Hooke's drawings of the cells of cork, as he viewed them with an early light microscope similar to the one shown here. The cells of cork are not living, and only the cell walls remain to outline the cell.

particles of both plants and animals. By the mid-1800s, German botanist Mattias Schleiden had further refined science's view of cells when he wrote: "It is...easy to perceive that the vital process of the individual cells must form the first, absolutely indispensable fundamental basis" of life.

Within a few years, several microscopists had noted that living cells could grow larger and divide into smaller but still living cells. In the late 1850s, the Austrian pathologist Rudolf Virchow wrote, "Every animal appears as a sum of vital units, each of which bears in itself the complete char-

acteristics of life." Furthermore, Virchow predicted, "All cells come from cells." The three principles of modern cell theory evolved directly from Virchow's statements.

1. Every living organism is made up of one or more cells.
2. The smallest living organisms are single cells, and cells are the functional units of multicellular organisms.
3. All cells arise from preexisting cells.

An Overview of Cell Structure and Function

In Chapter 1, we defined cells as the smallest units of life. Each cell contains genes that control the synthesis of molecules within the cell and allow it to reproduce and organelles that act as microscopic chemical factories that guide the chemical reactions within the cell, all enclosed within a thin plasma membrane that separates the cell from its environment. Most cells are small, ranging from about 1 to 100 micrometers (millionths of a meter) in diameter (Fig. 5-2). Living organisms may consist of just one cell (bacteria, protists) or aggregates of interacting, cooperating cells (fungi, plants, and animals). Because of the small size of nearly all cells, biologists use several types of microscopes to assist their studies of cell structure and function. Some of the diversity of images produced by modern microscopes is shown in "Scientific Inquiry: Viewing the Cell—A Gallery of Microscopic Images."

All cells have at least three components: a plasma membrane, genetic material, and cytoplasm (Fig. 5-3, p. 80).

The Plasma Membrane Isolates the Cell but Allows Interactions with the Environment

The **plasma membrane** (also called the *cell membrane*) is a double layer of phospholipids (see Chapter 3) in which are embedded a wide variety of proteins. The plasma membrane performs functions: (1) It isolates the cytoplasm from the external environment; (2) it regulates the flow of materials between the cytoplasm and its environment, for example, acquiring nutrients and expelling wastes; and (3) it allows interaction with other cells.

The Genetic Material Provides a Cellular "Blueprint"

Each cell has its own hereditary blueprint, in which is stored the instructions for making all the other parts of the cell and for producing new daughter cells. In all living cells, the genetic material is **deoxyribonucleic acid (DNA)**. In eukaryotic cells (plants, animals, fungi, and protists), the DNA is contained within a separate, membrane-bound structure called the **nucleus**. In prokaryotic

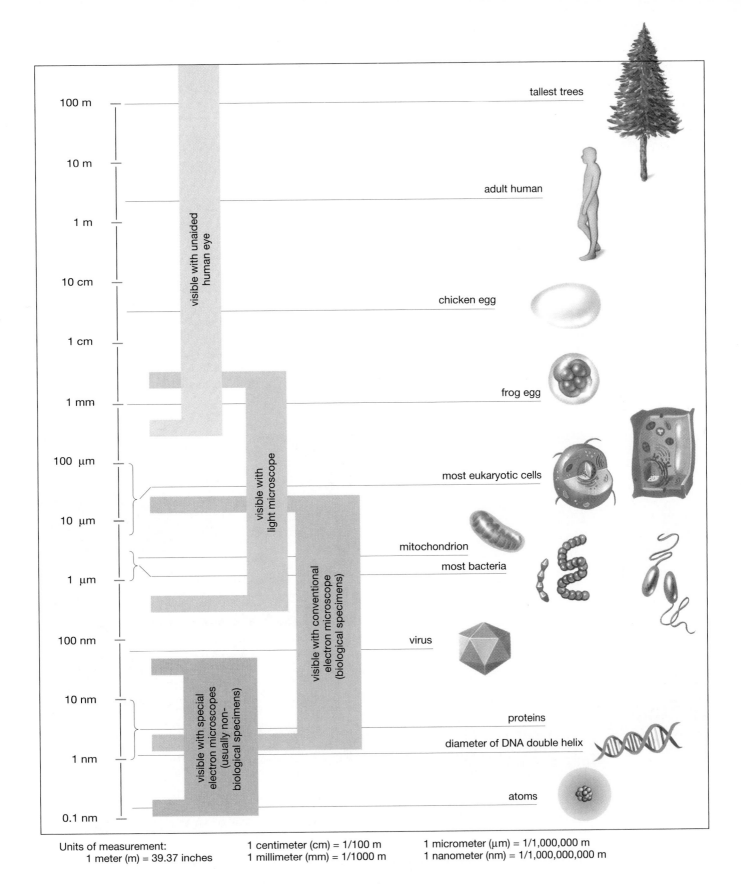

Figure 5-2 **Relative sizes**

Dimensions commonly encountered in biology range from about 100 meters (the height of the tallest redwoods) through a few micrometers (the diameter of most cells) to a few nanometers (the diameter of many large molecules). Note that, in the metric system (used almost exclusively in science), separate names are given to dimensions that differ by factors of 10, 100, and 1000.

S C I E N T I F I C   I N Q U I R Y

Viewing the Cell—A Gallery of Microscopic Images

Biologists, physicists, and engineers have collaborated in the development of a variety of microscopes to examine structures too small to see with the naked eye. In some cases, biologists use microscopes to look at living cells (Fig. E5-1a, b). In most instances, however, the specimen must be carefully prepared, usually sliced thin, and stained (Figs. E5-1c, d, E5-2, and E5-3).

Light microscopes use lenses, usually made of glass, to focus and magnify light rays that either pass through or bounce off a specimen. Light microscopes provide a wide range of images, depending on how the specimen is illuminated (for example, from the top [dark field] or the bottom [bright field]) and whether it has been stained (Fig. E5-1a–c). The resolving power of light microscopes—that

is, the smallest structure that can be seen—is about 1 micrometer (a millionth of a meter).

Electron microscopes use beams of electrons instead of light. The electrons are focused by magnetic fields rather than by lenses. Some types of electron microscopes can resolve structures as small as a few nanometers (billionths of a meter). *Transmission electron microscopes* (TEMs) pass electrons through a thin specimen and can reveal minute subcellular structures (Fig. E5-1d), including organelles and plasma membranes (Fig. E5-2). *Scanning electron microscopes* (SEMs) bounce electrons off specimens that have been coated with metals and provide three-dimensional images. SEMs can be used to view structures ranging in size from entire insects down to cells and even organelles (Fig. E5-3).

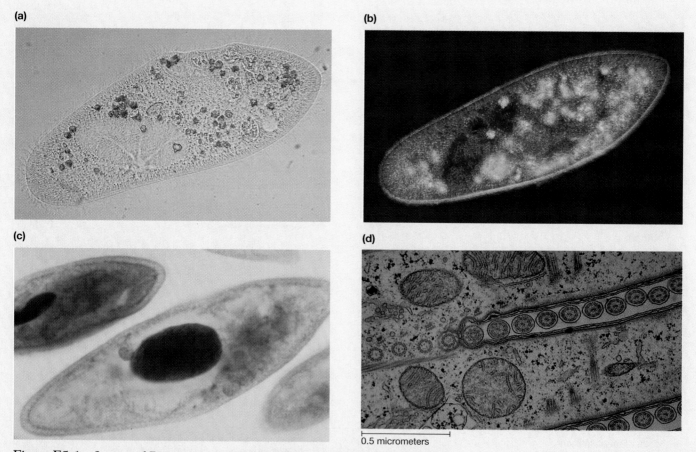

(a) **(b)** **(c)** **(d)**

0.5 micrometers

Figure E5-1   **Images of *Paramecium* seen through different types of microscopes**

Living *Paramecia* seen with **(a)** bright field and **(b)** dark field illumination. **(c)** This *Paramecium* has been stained to reveal some subcellular structures and photographed with a bright field microscope. **(d)** A transmission electron micrograph of the contractile vacuole of the *Paramecium*, the star-shaped organelle visible in (a) and (c). The roles of contractile vacuoles are discussed in Chapter 6. Note the increased detail compared with the specimens photographed through a light microscope. Because the magnification is greater, the entire *Paramecium* is not shown.

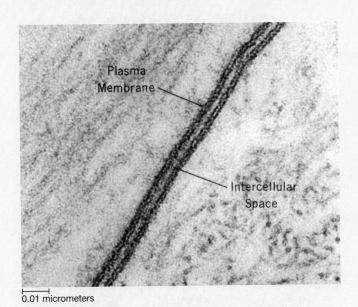

0.01 micrometers

Figure E5-2   **The transmission electron microscope**

Transmission electron microscopes can magnify much more than light microscopes. This photo shows a pair of plasma membranes that have been cut perpendicularly (like cutting a sheet of paper with scissors and looking at the paper edge on). In micrographs such as this one, membranes appear as a pair of dark lines (the phospholipid heads) separated by a thin layer of paler material (the phospholipid tails; see Fig. 5-11, p. 88). A plasma membrane is only about 7 billionths of a meter in thickness.

(b)

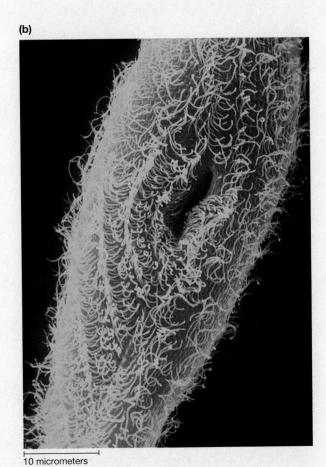

10 micrometers

(a)

(c)

5 micrometers

Figure E5-3 **Scanning electron microscopes provide images that look three-dimensional**

(a) A high-tech ant carries off a computer chip. **(b)** An SEM photograph of *Paramecium* shows the pattern of hairlike cilia that cover the surface of the cell. **(c)** This SEM photograph shows part of the inside of a cell. The spherical structures, many partially sliced open, are mitochondria (see Fig. 5-17, p. 95).

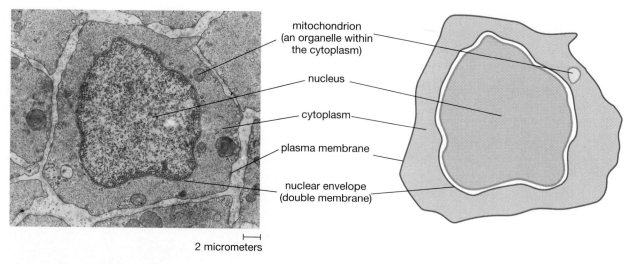

2 micrometers

Figure 5-3   An animal cell

An animal cell as seen with a transmission electron microscope. Note the plasma membrane (thin line surrounding cell), cytoplasm (material enclosed by the plasma membrane), and the nucleus, containing the genetic material, surrounded by the nuclear envelope.

cells (bacteria), the DNA, although localized to a particular place within the cytoplasm called the **nucleoid**, is not separated by membranes from the rest of the cytoplasm.

Cytoplasm Fills Each Cell and Surrounds the Nucleus

The **cytoplasm** consists of all the material inside the plasma membrane and, in eukaryotic cells, outside the nucleus, which is itself contained by a membrane. The cytoplasm includes water, salts, an assortment of organic molecules including many enzymes that catalyze reactions and other proteins that form an intracellular support system. The cytoplasm also contains a variety of discrete structures, called **organelles**, each of which performs a distinct cellular function.

Cell Function Limits Cell Size

Why are all living things composed of cells? Looking over the essential components of cells described in the previous section should enable you to make a pretty good guess at why microscopic organisms are cellular: The plasma membrane separates the complex life processes in the cytoplasm from the chaos of the outside world. The genetic material is the databank that directs those life processes.

What may be less obvious is why large organisms consist of many cells rather than one large cell. To answer that question, we must consider the physical constraints that limit cell size. All living organisms exchange nutrients and wastes with their external environment through the plasma membrane. These activities would cause two major problems if a large organism were composed simply of a

single cell (Fig. 5-4): Exchange would be limited both by the distance from the center of the cell to its surface and by the surface area of the cell.

First, as a cell becomes larger, its innermost regions become farther away from the membrane. Most nutrients and wastes move into, through, and out of cells by **diffusion**, the net transport of molecules from places of high concentration to places of low concentration (we will discuss diffusion more thoroughly in Chapter 6). Diffusion works well for moving molecules over very short distances (a few micrometers), but it is exceedingly slow for long-distance transport. Relying on diffusion alone, oxygen molecules would take more than *200 days* to reach the center of a cell 20 centimeters in diameter (8.5 inches; about the thickness of your chest). Clearly not much life could go on inside a cell that large!

The second difficulty arises from geometry: As a cell enlarges, its volume increases more rapidly than its surface area does. A cell that doubles its radius becomes eight times greater in volume but only four times greater in surface area. Thus, compared with their volume, large cells have a smaller surface area than small cells do. In general, as a cell increases in volume, more chemical reactions occur. Thus more nutrients and oxygen are needed, and more waste products must be eliminated—all through the plasma membrane. In a very large cell the surface area of the membrane would be too small to keep up with the cell's metabolic needs. The ratio of the cell surface to the cell volume thus sets the upper limit on individual cell size. Large organisms, therefore, consist of billions of coordinated cells, with structural and functional arrangements (such as respiratory and circulatory systems) that allow internal cells to exchange materials with the external environment.

distance to center (r)	1.0	2.0	3.0	1.0
surface area $(4\pi r^2)$	12.6	50.3	113.1	339.4
volume $(4/3\pi r^3)$	4.2	33.5	113.1	113.1
area/volume	3.0	1.5	1.0	3.0

Figure 5-4   Geometric considerations limit the size of cells

As a cell enlarges, the distance from the center of the cell to the outside world increases. Further, the volume increases much more rapidly than the surface area. As these spherical cells illustrate, doubling the radius halves the ratio of surface area to volume. Thus each unit volume of cytoplasm within the cell has only half the membrane area available to exchange nutrients and wastes with the external environment. However, if the volume of the largest sphere is divided among cells the size of the smallest sphere, the overall ratio of surface area to volume of the resulting "multicellular organism" remains large.

Types of Cells: Prokaryotic and Eukaryotic

As we briefly discussed in Chapter 1, there are two fundamentally different types of cells. The first type, represented by the bacteria, is called **prokaryotic**, Greek for "before the nucleus." The second type of cell, which evolved from the prokaryotic cell and is today found in protists, plants, fungi, and animals, is called **eukaryotic** ("true nucleus"). As their names imply, perhaps the most striking difference between prokaryotic cells and eukaryotic cells is that eukaryotic cells have their genetic material contained within a membrane-limited structure, the nucleus, whereas the genetic material of prokaryotic cells is not enclosed within a membrane. A variety of other features also differ between the two types of cells, making a closer look worthwhile.

Prokaryotic Cells Are Relatively Simple

Prokaryotic cells are usually very small (less than 5 micrometers long), with a relatively simple internal structure (Fig. 5-5; compare with the eukaryotic cell shown in Fig. 5-3). Most prokaryotic cells are surrounded by a relatively stiff cell wall. The substances that make up the cell wall are secreted by the cell itself. Many bacteria live in fluid environments, from ponds to the human bloodstream, from which water tends to enter the bacterial cell (see Chapter 6). Without a cell wall, and without specialized energy-intensive mechanisms to excrete water, water entering from the environment would cause the cell to swell and explode; in fact, penicillin and some other antibiotics fight bacterial infections by inhibiting cell wall synthesis, causing the bacteria to rupture.

In some bacteria, such as the one shown in Figure 5-5, the cell wall has a polysaccharide coating that prevents white blood cells (part of the human body's immune system) from engulfing the bacterium. Although it plays an important supportive role, the cell wall and its polysaccharide coat are quite porous. Movement of materials into and out of a prokaryotic cell is regulated by the plasma membrane that lies just beneath the cell wall.

The cytoplasm of most prokaryotic cells is relatively homogeneous in appearance (although some photosynthetic bacteria have elaborate internal membranes). The DNA is usually coiled, attached to the plasma membrane, and concentrated in a region of the cell called the nucleoid. It is not, however, physically separated from the rest of the cytoplasm by a membrane.

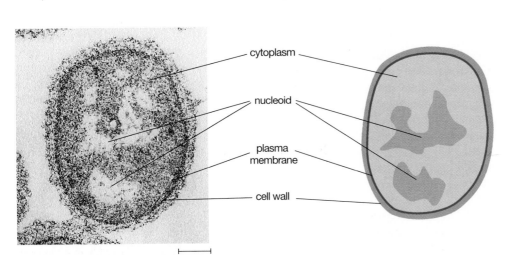

cytoplasm

nucleoid

plasma membrane

cell wall

0.2 micrometer

Figure 5-5   A prokaryotic cell

A transmission electron micrograph of a bacterium, a prokaryotic cell. The pale area occupying much of the cell, called the nucleoid, is the location of the cell's DNA; it is not separated from the rest of the cytoplasm by a membrane. In this bacterium, a cell wall surrounds the plasma membrane. This prokaryotic cell is far simpler in structure than the eukaryotic cell in Figure 5-3.

Chapter 21 provides a more detailed look at the diversity of prokaryotic cells.

Eukaryotic Cells Are More Complex

Eukaryotic cells differ from prokaryotic cells in many ways. In addition to being larger than prokaryotic cells (often more than 10 micrometers in diameter), eukaryotic cells contain a variety of membranous organelles that lend structural and functional organization to the cell. Technically, the material within the plasma membrane of a eukaryotic cell is divided into the nucleus, an organelle consisting of a double layer of membrane that encloses the genetic material, and the cytoplasm, which includes everything else. The cytoplasm in turn is composed of several types of organelles, occupying as much as half the volume of the cell, and a fluid matrix, the **cytosol** (literally, "cell solution"), in which the organelles reside. The cytosol is a watery solution of salts, sugars, amino acids, proteins, fatty acids, nucleotides, and other materials. Giving shape and organization to the cytoplasm is a network of protein fibers, the **cytoskeleton**. Many of the organelles and even individual molecules of the cytoplasm are thought to be attached to the cytoskeleton.

With a few rare exceptions, all eukaryotic cells contain the following organelles, each with its own structural and functional specialization:

1. **Nucleus.** Contains the cell's genetic information
2. **Endoplasmic reticulum (ER).** A network of membranous channels, often studded with small protein/ribonucleic acid (RNA) particles called **ribosomes**; ribosomes synthesize many different types of proteins, whereas protein enzymes in the membrane of the endoplasmic reticulum synthesize lipids
3. **Golgi complex.** Stacks of membranous sacs that modify proteins and lipids, synthesize carbohydrates, and package molecules for transport
4. **Vesicles.** Membranous sacs that store and transport molecules around the cell
5. **Mitochondrion.** Captures some of the chemical energy of food molecules as high-energy bonds of adenosine triphosphate (ATP)

Eukaryotic cells, however, are not all alike. Figures 5-6 and 5-7 illustrate the structures that are found in animal and plant cells, although few individual cells contain all the features shown in the drawings. Each type of cell possesses a few unique organelles not found in the other. You may want to refer to these illustrations as we describe the structures of the cell in more detail. As a reference and study guide, Table 5-1 lists the major characteristics of prokaryotic and eukaryotic cells, including a brief description of the functions of the organelles.

In the following pages, we will introduce you to the major components of eukaryotic cells. We will emphasize the union of *structure* and *function* that is crucial to understanding biology. In studying the anatomy of the cell, remember that each structure originally evolved, and per-

sists today, because it performs a specific function that is essential to the survival and reproduction of the cell.

The Nucleus: Control Center of the Cell

Deoxyribonucleic acid (DNA) is the genetic material of all living cells. A cell's DNA is the repository of information needed to construct the cell and direct the countless chemical reactions necessary for life and reproduction. Just as a blueprint is used selectively, one part at a time, to build a house, so too the hereditary information in DNA is used selectively by the cell, depending on its stage of development and its environmental conditions. In eukaryotic cells, the DNA is housed within the nucleus.

The nucleus consists of three readily distinguishable components (Fig. 5-8, p. 86). The **nuclear envelope** separates the nuclear material from the cytoplasm. Inside the nuclear envelope, the nucleus contains a granular-looking material called **chromatin** and a darker region called the **nucleolus**.

The Nuclear Envelope Allows Selective Exchange of Materials

The nucleus is isolated from the rest of the cell by a nuclear envelope that consists of two membranes riddled with pores (Fig. 5-8). Nuclear pores are complex structures consisting of at least eight protein subunits with a small channel through the middle. Water, ions, and small molecules such as ATP can pass freely through the central channel, but the passage of large molecules, particularly proteins and RNA, is regulated, presumably as the protein subunits move to enlarge and restrict the channel. Consequently, the pores help control the flow of information to and from the DNA.

Chromatin Consists of DNA and Its Associated Proteins

The bulk of the nucleus appears granular in an electron micrograph (see Figs. 5-3 and 5-8), with darker and lighter regions but no obvious structure. Since the nucleus is highly colored by several common stains used in light microscopy, early microscopists named the nuclear material chromatin, meaning simply "a colored substance." Biologists have since learned that chromatin consists of DNA associated with proteins. Although you can't see them in ordinary electron micrographs, eukaryotic DNA and its associated proteins form long strands called **chromosomes** ("colored bodies"). When cells divide, each chromosome coils upon itself, becoming thicker and shorter. The re-

Figure 5-6   A "typical" animal cell

In fact, the entire range of organelles depicted here seldom occurs in a single animal cell.

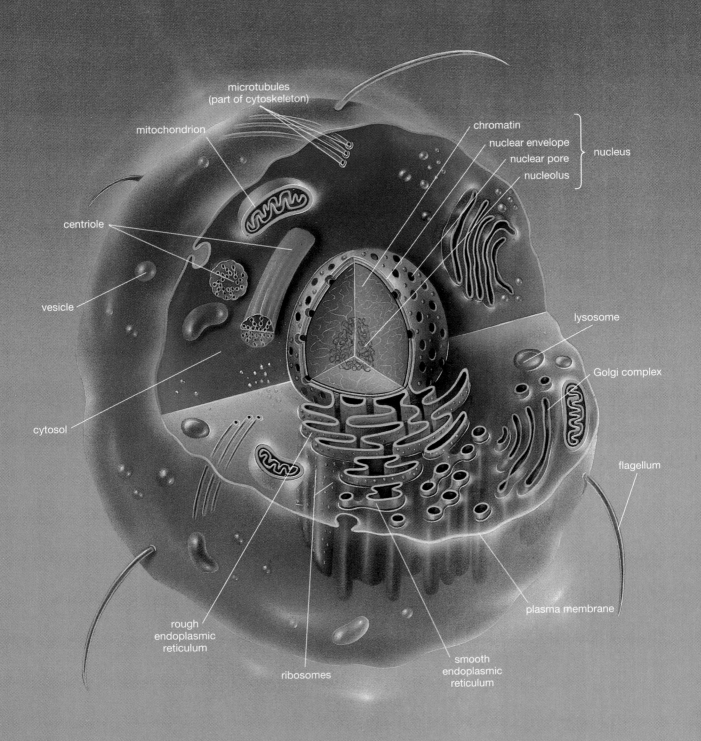

microtubules
(part of cytoskeleton)

mitochondrion

chromatin

nuclear envelope

nuclear pore

nucleolus

nucleus

centriole

vesicle

lysosome

Golgi complex

cytosol

flagellum

plasma membrane

rough
endoplasmic
reticulum

ribosomes

smooth
endoplasmic
reticulum

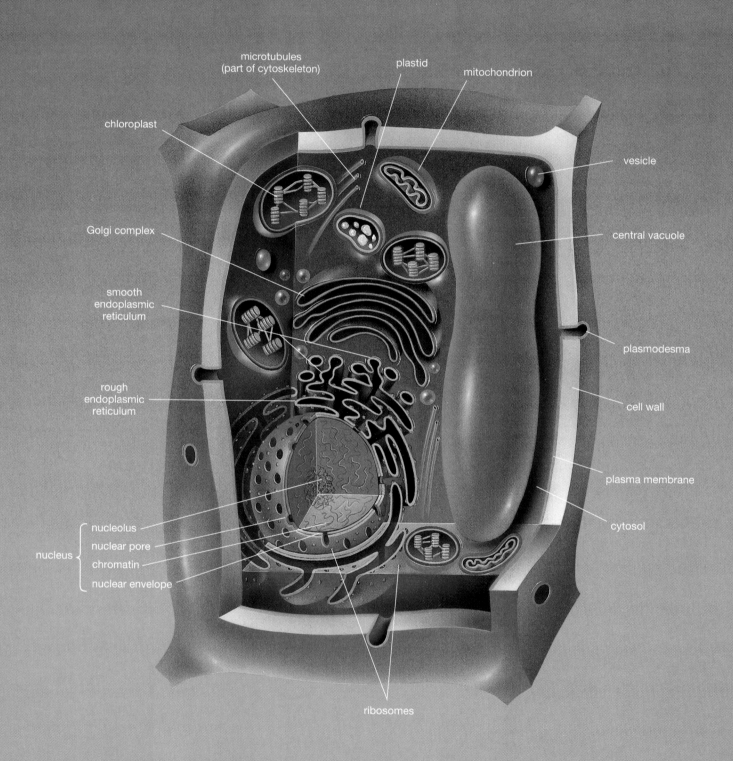

microtubules
(part of cytoskeleton)

plastid

mitochondrion

chloroplast

vesicle

Golgi complex

central vacuole

smooth
endoplasmic
reticulum

plasmodesma

rough
endoplasmic
reticulum

cell wall

plasma membrane

cytosol

nucleolus

nuclear pore

nucleus

chromatin

nuclear envelope

ribosomes

TABLE 5-1 ⊞ *Cell Structures, Their Functions, and Their Distribution in Living Cells*

Structure	Function	Prokaryotes[a]	Plants	Animals
Cell surface				
Cell wall	Protects, supports cell	Present	Present	Absent
Plasma membrane	Isolates cell contents from environment; regulates movement of materials into and out of cell; communicates with other cells	Present	Present	Present
Organization of genetic material				
Genetic material	Encodes information needed to construct cell and control cellular activity	DNA	DNA	DNA
Chromosomes	Contain and control use of DNA	Single, circular, no proteins	Many, linear, with proteins	Many, linear, with proteins
Nucleus	Membrane-bound container for chromosomes	Absent	Present	Present
Nuclear envelope	Encloses nucleus; regulates movement of materials into and out of nucleus	Absent	Present	Present
Nucleolus	Synthesizes ribosomes	Absent	Present	Present
Cytoplasmic structures				
Mitochondria	Produce energy by aerobic metabolism	Absent	Present	Present
Chloroplasts	Perform photosynthesis	Absent	Present	Absent
Ribosomes	Provide site of protein synthesis	Present	Present	Present
Endoplasmic reticulum	Synthesizes membrane components and lipids	Absent	Present	Present
Golgi complex	Modifies and packages proteins and lipids; synthesizes carbohydrates	Absent	Present	Present
Lysosomes	Contain intracellular digestive enzymes	Absent	Present	Present
Plastids	Store food, pigments	Absent	Present	Absent
Central vacuole	Contains water and wastes; provides turgor pressure to support cell	Absent	Present	Absent
Other vesicles and vacuoles	Contain food obtained through phagocytosis; contain secretory products	Absent	Present (some)	Present
Cytoskeleton	Gives shape and support to cell; positions and moves cell parts	Absent	Present	Present
Centrioles	Synthesize microtubules of cilia and flagella; may produce spindle in animal cells	Absent	Absent (in most)	Present
Cilia and flagella	Move cell through fluid or move fluid past cell surface	Absent[b]	Absent (in most)	Present

Handwritten annotation next to Prokaryotes column: "Bacteria blue/green algae"

[a]Many structures are listed as "absent" in prokaryotes; however, their functions are often essential to the life of any cell. In prokaryotes, these functions still occur, but not in discrete structures. For example, prokaryotes synthesize digestive enzymes on ribosomes attached to the cell membrane and immediately secrete the enzymes, where they digest food outside the cell. Therefore, although endoplasmic reticulum, secretory vesicles, and lysosomes are not found in prokaryotes, the functions of these organelles are still carried out.

[b]Many prokaryotes have structures called flagella, but these are not made of microtubules and move in a fundamentally different way than eukaryotic cilia or flagella do.

sulting "condensed" chromosomes are easily visible even with light microscopes (Fig. 5-9).

Cellular processes (the chemical reactions responsible for growth and repair, nutrient and energy acquisition and use, and reproduction) are governed by the information encoded in DNA. Because the DNA stays in the nucleus, while most of the chemical reactions that it controls occur in the cytoplasm, information molecules must be exchanged between the nucleus and the cytoplasm. Genetic information is copied from DNA into molecules of RNA, which move through the pores of the nuclear envelope into the cytoplasm. This information is then used to direct the synthesis of proteins (see below). These proteins include enzymes that regulate chemical reactions, membrane proteins that govern interactions between the cell and its environment, and a variety of structural proteins. Some of these proteins pass from the cytoplasm into the nucleus and regulate the transfer of information from DNA to RNA, depending on what is happening in the cytoplasm and the extracellular environment. We take a closer look at these processes in Chapter 11.

Figure 5-7 A "typical" plant cell

Not all of these structures occur in every plant cell. Compare this drawing with that of an animal cell in Figure 5-6.

The Nucleolus Is the Site of Ribosome Assembly

Most eukaryotic nuclei have one or more darkly staining regions called nucleoli ("little nuclei"; one nucleolus is shown

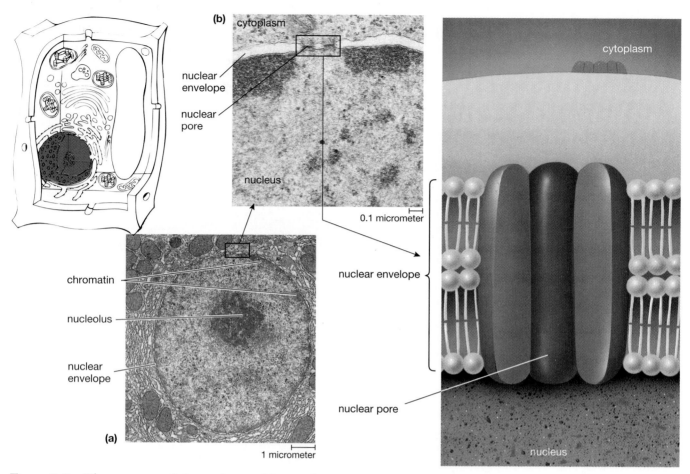

Figure 5-8   The structure of the nucleus and its envelope

(a) The eukaryotic nucleus consists of an outer double membrane, the nuclear envelope, enclosing the genetic material, DNA. The DNA is associated with proteins to form chromatin. One region of the chromatin, the nucleolus, stains much more darkly than the rest of the nucleus. **(b)** A transmission electron micrograph of the nuclear envelope clearly shows that it is composed of a double membrane that is fused around the lips of the nuclear pores. Micrographs such as this one can only hint at the structure of the pore, which is shown more clearly in the accompanying drawing.

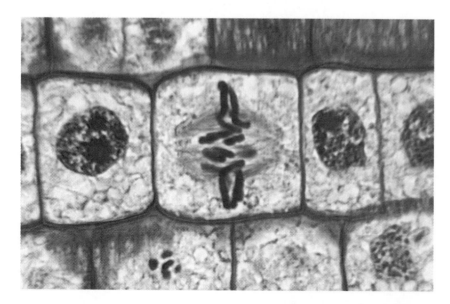

Figure 5-9   Chromosomes

Chromosomes, seen here in a light micrograph of a dividing cell in an onion root tip, are the same material as the chromatin seen in nondividing cells (DNA and proteins) but in a more compact state.

in Fig. 5-8a), which are the sites of ribosome synthesis. A ribosome is a small particle composed of RNA and protein that serves as a "workbench" for the synthesis of proteins (Chapter 13). Like many workbenches, ribosomes are non-specific—the same set of tools can be used to construct many different objects. Any ribosome can be used to synthesize any of the hundreds of proteins made by a cell. In electron micrographs, ribosomes appear as dark granules, either distributed in the cytosol (Fig. 5-10) or clustered along the membranes of the nuclear envelope and the endoplasmic reticulum (see Fig. 5-12).

The nucleolus consists of ribosomal RNA, proteins, ribosomes in various stages of synthesis, and DNA (bearing genes that specify the blueprint for ribosomal RNA). It stains darkly because of this accumulation of material. During cell division, ribosome synthesis slows. The chromosomes bearing the ribosomal RNA genes move away from one another. Consequently, the nucleolus disappears until after cell division is complete and the daughter cells begin synthesizing more ribosomes themselves.

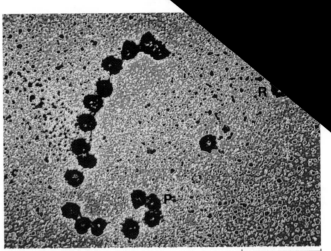

0.05 micrometers

Figure 5-10   **Ribosomes**

Ribosomes may be found free in the cytoplasm either singly (R) or in groups called polyribosomes (P). Free ribosomes synthesize proteins that will be used within the cell. Other ribosomes are attached to the endoplasmic reticulum (see Fig. 5-12).

The Membrane System of the Cell

All eukaryotic cells have an elaborate system of **membranes**. This membrane system is composed of the plasma membrane and several organelles within the cytoplasm, including the endoplasmic reticulum, nuclear envelope, Golgi complex, and a variety of membranous sacs such as lysosomes, which exchange membrane material with one another (see Figs. 5-6 and 5-7). All cellular membranes are composed of lipids and proteins that are synthesized in the endoplasmic reticulum. The membrane components are then sorted and modified in the Golgi complex and sent off to their appropriate destinations within the cell as small membranous sacs called vesicles. We will study this membrane system in three steps. First, we will provide a capsule summary of the structure of membranes, a topic covered in detail in Chapter 6. Second, we will examine the structure and function of individual components of the membrane system. Finally, we will trace the flow of membrane throughout the system.

Membranes Consist of a Double Layer of Phospholipids in Which Proteins Are Embedded

All the membranes of eukaryotic cells are composed of a double layer (bilayer) of phospholipids in which are embedded cholesterol and several kinds of proteins (Fig. 5-11). The exact types and amounts of phospholipids and proteins vary among membranes that enclose different types of organelles, according to the function of the organelle. In general, phospholipids and cholesterol form the structural framework of membranes, while proteins carry out specialized functions, such as transporting molecules across the membrane, catalyzing chemical reactions, or attaching one membrane to another.

The Cell's Membrane System Includes the Plasma Membrane, Endoplasmic Reticulum, the Golgi Complex, and Lysosomes

The Plasma Membrane Both Isolates the Cell and Allows Selective Interactions between the Cell and Its Environment

The plasma membrane forms the outer boundary of the living part of a cell. It is a marvelously complex structure that must perform the seemingly contradictory functions of separating the cytoplasm of the cell from the outside environment while simultaneously providing for transport of selected substances into or out of the cell. We will discuss plasma membrane structure and function in Chapter 6.

The Endoplasmic Reticulum Forms Membrane-Enclosed Channels within the Cytoplasm

The endoplasmic reticulum (ER) is a series of interconnected membranous tubes and channels in the cytoplasm (Fig. 5-12). Most eukaryotic cells have two forms of ER—rough and smooth. Numerous ribosomes stud the outside of **rough endoplasmic reticulum; smooth endoplasmic reticulum** lacks ribosomes.

The different structures of smooth and rough ER reflect different functions. Enzymes embedded in the membranes of the smooth ER are the major site of lipid synthesis, including the phospholipids of the ER and other membranes. In some cells the smooth ER synthesizes other types of lipids as well, such as the steroid hormones testosterone and estrogen produced in the reproductive organs

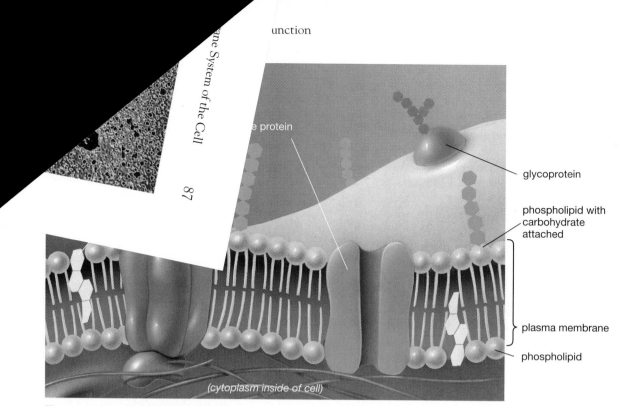

glycoprotein

phospholipid with carbohydrate attached

plasma membrane

phospholipid

...e protein

(cytoplasm inside of cell)

Figure 5-11 **A biological membrane**

The general structure of all biological membranes is similar: a phospholipid bilayer in which are embedded cholesterol and a variety of membrane proteins. Carbohydrates are sometimes attached to certain proteins to form glycoproteins and to phospholipids to form glycolipids.

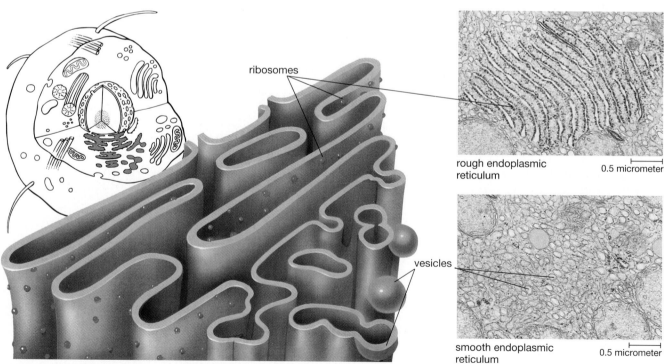

ribosomes

vesicles

rough endoplasmic reticulum   0.5 micrometer

smooth endoplasmic reticulum   0.5 micrometer

Figure 5-12 **Endoplasmic reticulum**

There are two types of endoplasmic reticulum: rough ER, coated with ribosomes, and smooth ER without ribosomes. Although the ER looks like a series of tubes and sacs in electron micrographs, it is actually a maze of folded sheets and interlocking channels. In many cells the rough and smooth ER are thought to be continuous, as depicted in the drawing. Ribosomes (red) stud the cytoplasmic face (purple) of the rough ER membrane.

of mammals. The ribosomes on the outside of rough ER synthesize proteins, including membrane proteins. Therefore, the ER can synthesize itself, both lipid and protein components. Although most of the membrane synthesized in the ER forms new or replacement ER membrane, some of it moves inward to replace nuclear membrane or outward to form the Golgi complex, lysosomes, and the plasma membrane (see Fig. 5-15).

Ribosomes on rough ER also manufacture the proteins that some secretory cells export into their surroundings, including digestive enzymes and protein hormones (for example, insulin secreted by cells of the pancreas). As these proteins are synthesized by the ribosomes on the outside of the ER, they are simultaneously transported into the channels inside. The proteins then move through the ER and accumulate in pockets at the ends, especially the ends near Golgi complexes. These pockets then "bud off," forming membrane-bound vesicles that migrate to the Golgi complex.

The Golgi Complex Sorts, Chemically Alters, and Packages Important Molecules in Membranous Sacs

The Golgi complex is a specialized set of membranous sacs derived from the endoplasmic reticulum. In fact, the Golgi looks very much like a stack of smooth ER that has been stepped on, flattening the middle and making the ends bulge out (Fig. 5-13). Vesicles that bud off from the smooth ER fuse with the sacs on one side of the Golgi stack, adding their membrane to the Golgi and emptying their contents into the Golgi sacs. Other vesicles bud off the Golgi on the opposite side of the stack, carrying away proteins, lipids, and other complex molecules. The Golgi complex performs the following three major functions:

1. It separates proteins and lipids received from the ER according to their destinations; for example, the Golgi separates digestive enzymes that are bound for lysosomes from hormones that will be secreted from the cell.
2. It modifies some molecules—for instance, adding sugars to proteins to make glycoproteins.
3. It packages these materials into vesicles that are then transported to other parts of the cell or to the plasma membrane for export.

Lysosomes Serve As the Cell's Digestive System

Some of the proteins manufactured in the ER and sent to the Golgi are intracellular digestive enzymes that can break down proteins, fats, and carbohydrates into their component subunits. In the Golgi, these enzymes are packaged in membranous vesicles called **lysosomes** (Fig. 5-14). The major function of lysosomes is to digest food particles, which range from individual proteins to complete microorganisms.

As we will see in Chapter 6, many cells "eat" by phagocytosis; that is, by engulfing extracellular particles with extensions of the plasma membrane. The food particles are then moved into the cytoplasm, enclosed within mem-

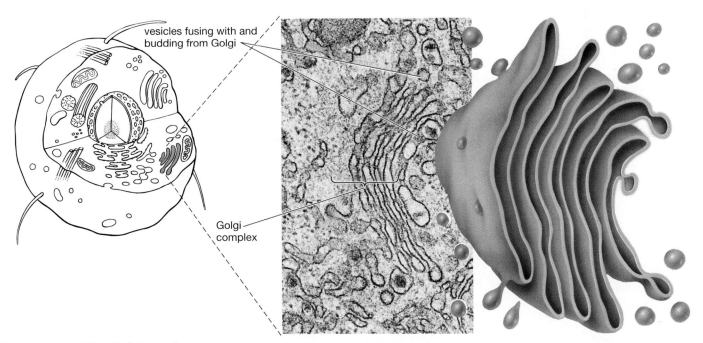

vesicles fusing with and budding from Golgi

Golgi complex

Figure 5-13   **The Golgi complex**
The Golgi complex is a stack of flat, membranous sacs derived from the endoplasmic reticulum. Vesicles constantly bud off from and fuse with the Golgi and ER, transporting material in several directions: from the ER to the Golgi and back again, and from the Golgi to plasma membrane, lysosomes, and vesicles.

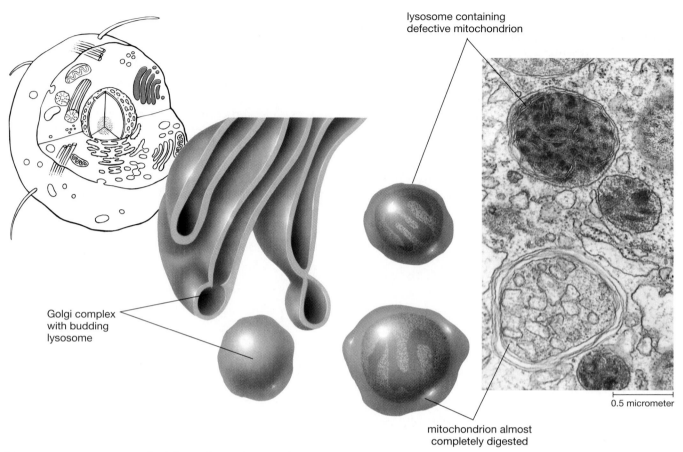

Figure 5-14   **Lysosomes bud from the Golgi complex**

Lysosomes are enzyme-filled vesicles that bud off from the Golgi complex. These enzymes digest food and worn-out organelles, such as mitochondria.

branous sacs, now called **food vacuoles**. Lysosomes recognize these food vacuoles and fuse with them. The contents of the two vesicles mix, and the lysosomal enzymes digest the food into amino acids, monosaccharides, fatty acids, and other small molecules. These simple molecules then diffuse out of the lysosome into the cytosol to nourish the cell. Cell biologists continue to search for the key to how lysosomes recognize these food vacuoles.

Lysosomes also digest defective or malfunctioning organelles, such as mitochondria or chloroplasts. After identifying these organelles, the cell encloses them in vesicles made of membrane from the ER. These vesicles fuse with lysosomes, and digestive enzymes within the lysosome enable the cell to recycle valuable materials from the defunct organelles. How the cell identifies organelles that have outlived their usefulness is a topic of continuing research.

Membrane Synthesized in the Endoplasmic Reticulum Flows through the Membrane System of the Cell in an Orderly Way

The nuclear envelope, rough and smooth ER, Golgi complex, lysosomes, food vacuoles, and plasma membrane all form an integrated membrane system. Membrane is synthesized in the ER and flows back and forth among these structures in an orderly way. As an example, let's look at the movement of materials destined for inclusion in the plasma membrane (Fig. 5-15). The ER synthesizes the phospholipids and proteins that make up the plasma membrane and simultaneously synthesizes proteins and lipids that belong in the ER itself.

How are these sorted out? The ER buds off a vesicle whose membrane is a mixture of ER and plasma membrane components. The vesicle fuses with the Golgi complex, adding its "mixed membrane" to the Golgi membrane. In the Golgi, ER and plasma membrane components are separated. ER material is removed to serve as "recycle" vesicles that pinch off and return to the ER. Plasma membrane material continues on through the Golgi, where it may be modified, for example by adding sugars to make glycoproteins or glycolipids (lipids to which a sugar is attached). Eventually the plasma membrane material becomes an "outward-bound" vesicle that buds off the far side of the Golgi and moves to the cell surface. The vesicle fuses with the plasma membrane, replenishing and enlarging the membrane.

The Golgi complex processes and packages all membrane-enclosed materials produced by the cell. Many of the vesicles pinched off from the Golgi contain secreto-

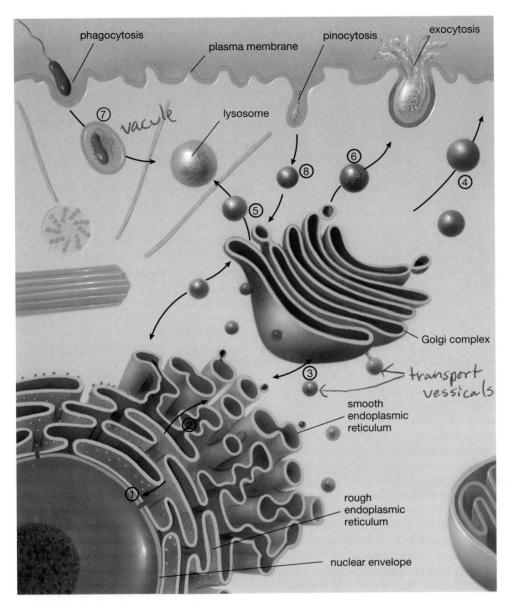

phagocytosis
plasma membrane
pinocytosis
exocytosis
⑦
vacule
lysosome
⑧
⑥
④
⑤
Golgi complex
③
transport vessicals
smooth endoplasmic reticulum
②
①
rough endoplasmic reticulum
nuclear envelope

Figure 5-15 **The flow of membrane and protein within a cell**

Membrane, regardless of its destination, is synthesized by the endoplasmic reticulum. ① Some of this membrane moves inward to form new nuclear envelope, while most of it moves outward to form ② smooth ER and ③ Golgi membrane. From the Golgi, membrane moves to form ④ new plasma membrane and ⑤ the membranes that surround other cell organelles such as lysosomes. Many proteins are synthesized in the rough ER and move through the smooth ER to the Golgi. Here the proteins are sorted according to function; some remain as parts of the Golgi membrane, some are returned to the ER, some are packaged in vesicles bound for the plasma membrane where they will be ⑥ secreted from the cell, and some are packaged in lysosomes. ⑦ Lysosomes may fuse with food vacuoles for intracellular digestion of food particles. ⑧ Plasma membrane brought into the cell is recycled through the Golgi complex. The processes by which plasma membrane moves from the surface to the interior of the cell (called phagocytosis and pinocytosis) and by which the membrane of vesicles fuses with the plasma membrane (called exocytosis) are described in Chapter 6.

ry products (for example, hormones) that are released outside the cell. Lysosomes contain digestive enzymes and often fuse with food vacuoles to carry out intracellular digestion. How these diverse materials are recognized, purified, modified properly, separated out, and individually packaged remains a challenge to cell biologists.

Chloroplasts and Mitochondria: Energy Capture and Extraction

Every cell has prodigious energy needs: to manufacture materials, to pick things up from the environment and throw other things out, to move, and to reproduce. As Lewis Thomas put it, "There are structures squirming inside each of our cells that provide all the energy for living." These structures are the chloroplasts and mito-

chondria, "essentially foreign creatures" thought to have evolved from bacteria that took up residence long ago within a fortunate eukaryotic cell (see "A Closer Look: The 'Essentially Foreign Creatures' within Us").

Chloroplasts and mitochondria are similar to each other in many ways (see Figs. 5-16 and 5-17, pp. 94 and 95). Both are usually oblong, about 1 to 5 micrometers long, and are surrounded by a double membrane. Both have enzyme assemblies that synthesize ATP, although the systems are used in a very different manner (see Chapters 7 and 8). Finally, both have many characteristics, including their own DNA, that seem to be remnants of their probable evolution from free-living organisms. However, they also have many differences, corresponding to their vastly different roles in cells: Chloroplasts capture the energy of sunlight during photosynthesis and store it in sugar, whereas mitochondria convert the energy of sugar into ATP for use by the cell.

The "Essentially Foreign Creatures" within Us—
The Evolution of Chloroplasts and Mitochondria

Many years ago botanists observed that the chloroplasts of some red algae bear a striking resemblance to certain photosynthetic prokaryotic bacteria. Both have separate, ribbonlike thylakoids, not at all like the stacked, disclike thylakoids found in most plants. Both also have somewhat unusual pigment granules attached to the outside of each thylakoid. Finally, most chloroplasts, even those in land plants, are about the size of an entire photosynthetic bacterium. Could these similarities be coincidence, or are they evidence of an evolutionary relationship?

Further study revealed that chloroplasts and mitochondria are very unusual organelles, in many respects resembling prokaryotic cells more than they resemble the other organelles of eukaryotic cells. First, chloroplasts and mitochondria contain a single circular chromosome composed of DNA without any proteins. Prokaryotic cells also have circular chromosomes of DNA without proteins. Second, chloroplasts and mitochondria contain their own ribosomes and synthesize proteins. The ribosomes are more similar to prokaryotic ribosomes than to the ribosomes found in the rest of a eukaryotic cell. Third, chloroplasts and mitochondria can grow, duplicate their DNA, and reproduce (Fig. E5-4). Fourth, they apparently cannot be manufactured by the cell. During cell division, occasionally one of the daughter cells receives all the chloroplasts or mitochondria, and the other receives none. When this happens, the cell that lacks these organelles never gets them back. Finally, mitochondria and chloroplasts are surrounded by two separate membranes, whereas most other organelles are bounded by one. The inner membrane of a mitochondrion, the thylakoid membranes of a chloroplast (derived from the inner membrane), and the plasma membrane of a prokaryotic cell are all about three-fourths protein and can carry out ATP synthesis. Eukaryotic membranes are usually only about half protein and play no role in ATP synthesis.

Why should chloroplasts and mitochondria resemble prokaryotic cells so strongly? Many years ago, a few biologists suggested that chloroplasts and mitochondria resemble prokaryotes because they might be the descendants of prokaryotic cells that took up residence within the cytoplasm of pre-eukaryotic cells. This **endosymbiotic hypothesis** has been persuasively argued by Lynn Margulis, and today most biologists agree that mitochondria and chloroplasts are, in Lewis Thomas's words, "essentially foreign creatures" within eukaryotic cells.

How could free-living bacteria become organelles in eukaryotic cells? Let's look first at a hypothetical scenario for the origin of mitochondria. Many single-celled

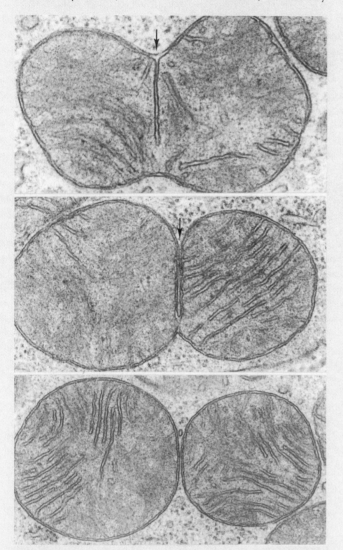

Figure E5-4   A mitochondrion divides

This series of micrographs illustrates the division of mitochondria. A membranous partition grows inward from one side of the mitochondrion, eventually reaching the other side and producing two daughter mitochondria from the parent. Not visible in these micrographs is the mitochondrial DNA, which duplicates before division, with one copy being incorporated into each daughter organelle.

organisms feed by phagocytosis (Fig. E5-5; see Chapter 6 for the details of phagocytosis). They engulf entire prey organisms in extensions of the plasma membrane, "swallowing" the prey in a vesicle of membrane. Usually the prey is doomed, as the vesicle fuses with enzyme-containing lysosomes and the prey inside is digested. According to the endosymbiotic theory, over a billion years ago the ancestors of eukaryotic cells had evolved, and some of them fed by phagocytosis. However, these cells lacked the enzymes needed to use oxygen in the metabolism of food molecules. As you shall see in Chapter 8, such a cell can extract only a small amount of usable energy as ATP from its food. Some bacteria, however, had evolved these enzymes, and these aerobic bacteria could produce much more ATP from the same amount of food.

Bacteria were probably favorite prey of pre-eukaryotic cells, and perhaps on occasion some bacteria were captured as prey but not digested. The bacteria took up residence in the cytoplasm of their new host. In most cases their presence must have harmed the host, for host and bacterium would compete for the same nutrients. The waste products of the bacterium may also have been poisonous. The host cell would likely have died—done in, as it were, by an indigestible bit of food.

But an aerobic bacterium in this situation would be in prokaryotic paradise. The cytoplasm of an anaerobic host cell would be rich in half-digested food molecules. Awash in food, the bacterium would probably generate large amounts of surplus ATP, some of which might pass into the host cytoplasm and be put to use by the host. If the bacterium did *not* kill the host, the relationship would be beneficial for both. The host-bacterium combination would survive better than either cell could alone. The "mitochondrion-bacterium" multiplied inside its host, and when the host cell divided, each daughter cell contained a few of these newfound powerhouses. The aerobic, eukaryotic cell—ancestor to all protists, plants, animals, and fungi—had been born.

Later, one of these new mitochondrion-containing cells may have similarly captured a photosynthetic bacterium. It, too, found a congenial home in its host's cytoplasm. It photosynthesized, grew, and multiplied. Surplus sugars synthesized by the "prechloroplast" may have passed to the host, providing some of the energy needed for the host to live and grow. The mitochondria already present in the cell metabolized these sugars, providing plenty of energy.

The plantlike protists and all true plants would be the descendants of this cell.

If mitochondria and chloroplasts descended from bacteria, this would explain why they have bacteria-like DNA and ribosomes and can grow and multiply. The endosymbiotic hypothesis also neatly explains the origin of the two membranes. Look again at Figure E5-5. When feeding by phagocytosis, a cell surrounds its prey with a vesicle made from its plasma membrane (Fig. E5-5, left). A live bacterium has its own plasma membrane. Therefore, the newly engulfed premitochondrion or prechloroplast would have two membranes, an outer one from the eukaryotic predator and an inner one from the bacterium (Fig. E5-5, right).

Needless to say, if chloroplasts and mitochondria arose from such interactions of predator and parasite, it was well over a billion years ago, and evolution has not stood still since. Chloroplasts and mitochondria are now fully incorporated into the lives of eukaryotic cells and cannot survive by themselves. Indeed, many of the proteins found in chloroplasts and mitochondria are not encoded by the organelle DNA but are specified by the nuclear DNA of the cell. This should not surprise us, because for over a billion years, natural selection has operated not on the "host" alone, nor on the "parasite" alone, but on the functional whole of the entire eukaryotic cell.

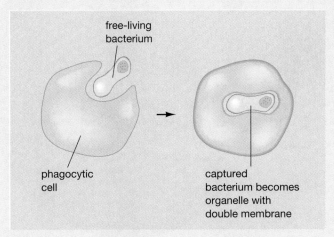

Figure E5-5   **The endosymbiotic hypothesis**

A model for the incorporation of bacteria into the cytoplasm of host cells, and the possible origin of the double membrane surrounding mitochondria and chloroplasts.

Chloroplasts Are the Site of Photosynthesis

Chloroplasts (Fig. 5-16) are found only in plants and certain protists, notably the unicellular algae. Chloroplasts are surrounded by two membranes, although there is very little space between them. The inner membrane encloses a semifluid material called the **stroma**. Embedded within the stroma are interconnected stacks of hollow membranous sacs. The individual sacs are called **thylakoids**, and a stack of sacs is a **granum** (plural, **grana**). During chloroplast development, thylakoids probably bud off from the inner membrane into the stroma; in mature chloroplasts, thylakoids are not connected to the inner membrane.

The thylakoid membranes contain the green pigment **chlorophyll** (which gives plants their green color) as well as other pigment molecules. During photosynthesis, chlorophyll captures the energy of sunlight and transfers it to other molecules in the thylakoid membranes. These molecules in turn transfer the energy to ATP and other energy-carrier molecules. The energy carriers diffuse into the stroma, where their energy is used to drive the synthesis of sugar from carbon dioxide and water. Photosynthesis is described in more detail in Chapter 7.

Mitochondria Produce ATP Using Energy Stored in Food Molecules

Whereas chloroplasts convert solar energy into chemical energy (some of which is stored in food molecules such as sugar and starch), mitochondria extract energy from food molecules and store it in the high-energy bonds of ATP. Almost all eukaryotic cells have mitochondria, which are often called the "powerhouses of the cell." As you will see in Chapter 8, different amounts of energy can be released from a food molecule, depending on how it is metabolized. The breakdown of food molecules begins in the cytosol, but the cytosol lacks the enzymes needed to use oxygen to break down food. This **anaerobic** (without oxygen) metabolism does not convert very much food energy into ATP energy. Mitochondria are the only places in a cell where oxygen can be used in food breakdown. **Aerobic** metabolism reactions are much more effective in generating energy than are the anaerobic reactions; 18 or 19 times more ATP is generated by aerobic metabolism in the mitochondria than by anaerobic metabolism in the cytosol. Not surprisingly, mitochondria are found in large numbers in metabolically active cells, such as muscle, and are less abundant in more inert cells, such as those of bone and cartilage.

Mitochondria are round, oval, or tubular sacs made of a pair of membranes (Fig. 5-17). Although the outer mitochondrial membrane is smooth, the inner membrane loops back and forth to form deep folds called **cristae** (singular, **crista**, meaning "crest"). As a result, the mitochondrial membranes enclose two fluid-filled spaces, the **intermembrane compartment** between the inner and outer membranes and the **matrix**, or inner compartment, within the inner membrane. Some of the reactions of food metabolism occur in the fluid matrix contained within the inner mem-

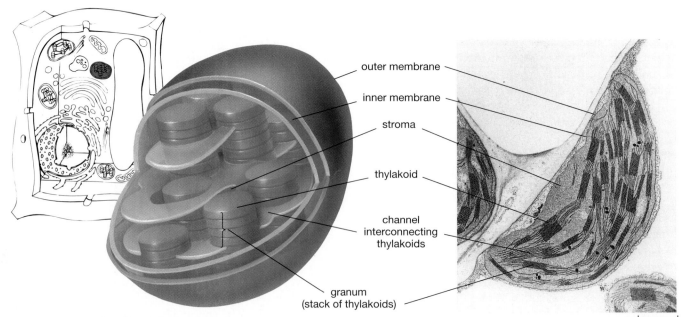

outer membrane
inner membrane
stroma
thylakoid
channel interconnecting thylakoids
granum (stack of thylakoids)

1 micrometer

Figure 5-16   **The chloroplast**
Chloroplasts are surrounded by a double membrane, although the inner membrane is not usually visible in electron micrographs. Enclosed by the inner membrane is the semifluid stroma, in which are embedded stacks of sacs collectively referred to as grana. The individual sacs of the grana are called thylakoids. Chlorophyll is embedded in the membranes of the thylakoids.

brane, while the rest are conducted by a series of enzymes attached to the membranes of the cristae.

Plastids and Vacuoles: Storage and Elimination

If a cell finds itself in a favorable environment, some of the molecules that it synthesizes and some of the food that it acquires and digests will be more than it needs at the moment. Cells have evolved organelles in which to store such valuable, if temporarily surplus, molecules. Other materials, such as the waste products of digestion, must be eliminated, and still other organelles serve that function.

Many Types of Plastids Store Food in Plants

Besides mitochondria and nuclei, plant cells have other organelles with double outer membranes. These are the **plastids**; we have already met the most important plastid, the chloroplast. Other types of plastids are used as storage containers for various types of molecules, including pigments that give ripe fruits their yellow, orange, or red colors. Especially important, particularly for perennial plants (which continue growing year after year), are plastids that store photosynthetic products from the summer for use during the following winter and spring. Plants usually convert the sugars made during photosynthesis into starch and store it in plastids (Fig. 5-18). Potatoes, for example, are masses of cells stuffed with starch-filled plastids.

Vacuoles Serve Many Functions, Including Support and Storage and Elimination of Food and Wastes

Most cells contain sacs, called **vacuoles**, that are bound by a single membrane. Eukaryotic cells may contain several types of vacuoles, both temporary and permanent, that serve a variety of functions. The central vacuole of many plant cells, for example, provides support for the cell and also serves as a storage site for metabolic wastes. Other vacuoles store food or assist the cell in eliminating excess water that diffuses inside. Because many of the functions of vacuoles are intimately connected with the movement of water across membranes, we will describe vacuoles in Chapter 6, which discusses membrane structure and function in more detail.

The Cytoskeleton: Shape, Support, and Movement

Organelles do not drift about the cytoplasm haphazardly. Rather, most organelles are attached to a network of protein fibers called the cytoskeleton (Fig. 5-19). Even indi-

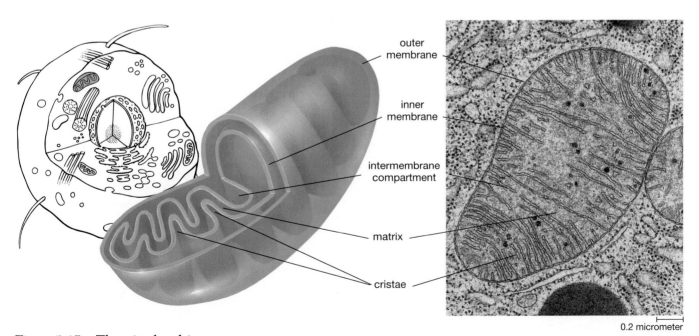

Figure 5-17   **The mitochondrion**

Mitochondria consist of a pair of membranes enclosing two fluid compartments, the intermembrane compartment between the outer and inner membranes and the matrix within the inner membrane. The outer membrane is smooth, but the inner membrane loops back and forth to form deep folds called cristae. Mitochondria are the site of aerobic metabolism.

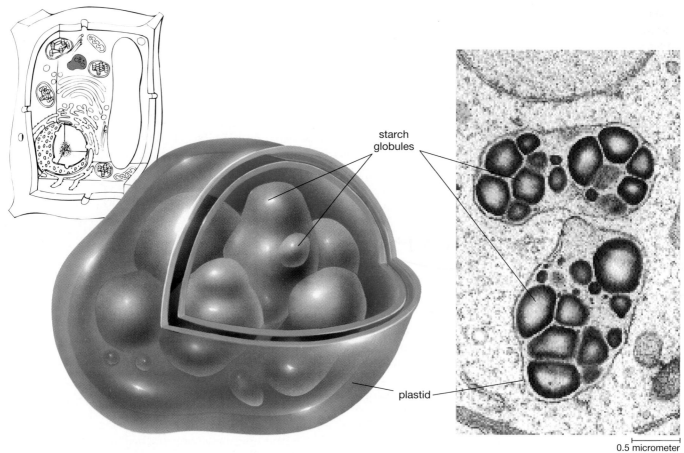

starch
globules

plastid

0.5 micrometer

Figure 5-18   **The plastid**

Plastids are organelles, found in the cells of plants and plantlike protists, that are surrounded
by a double outer membrane. Chloroplasts are the most familiar plastids; other types store
various materials, such as the starch filling these plastids in potato cells.

vidual enzymes, which are often parts of complex metabolic pathways, may be fastened in sequence to the cytoskeleton, so that molecules can be passed from one enzyme to the next. Several types of protein fibers, including
thin **microfilaments**, medium-sized **intermediate filaments**, and thick **microtubules** make up the cytoskeleton
(Table 5-2).

The cytoskeleton performs the following important
functions:

1. *Cell shape.* In cells without cell walls, the cytoskeleton, especially networks of intermediate filaments, determines the shape of the cell.
2. *Cell movement.* The assembly, disassembly, and sliding
   of microfilaments and microtubules causes cell movement. Cell movement includes both the familiar
   "crawling" of amoebae and white blood cells and the
   migration and shape changes that occur during the
   development of multicellular organisms.
3. *Organelle movement.* Microtubules and microfilaments move organelles from place to place within a
   cell. For example, microfilaments attach to vesicles

formed during endocytosis (see Chapter 6), when
large particles are engulfed by the plasma membrane
and pull the vesicles into the cell. Vesicles budded
off the ER and Golgi complex are probably guided by
the cytoskeleton as well.
4. *Cell division.* Microtubules and microfilaments are
   essential to cell division in eukaryotic cells (see
   Chapter 9). First, when eukaryotic nuclei divide, microtubules move the chromosomes into the daughter nuclei. Second, in animal cells, division of the
   cytoplasm of a single parent cell into two new daughter cells results from the contraction of a ring of microfilaments that pinch the "waist" of the parent cell
   around the middle.

Microfilaments Allow Cells to Change Shape and Guide the Movement of Organelles

Microfilaments are strands that consist mostly of the protein actin, sometimes in association with a second protein,
myosin. Myosin has small extensions that can grab onto

TABLE 5-2 ❖ *Components of the Cytoskeleton*

	Microfilaments	*Intermediate Filaments*	*Microtubules*
Structure	Solid strands about 7 nm in diameter; may be several cm long (muscle cells)	Solid strands 8 to 10 nm in diameter, 10 to 100 μm in length	Hollow tubes about 25 nm in diameter; may be more than 50 μm in length
Protein	Actin (most) and/or myosin	At least five different proteins	Tubulin
Function	Muscle contraction; changes in cell shape, including division of cytoplasm in dividing animal cells; cytoplasmic streaming; movement of pseudopodia	Maintenance of cell shape; attachments of microfilaments in muscle cells; support of nerve cell processes (axons)	Movement of chromosomes during cell division; movement of organelles within cytoplasm; movement of cilia and flagella

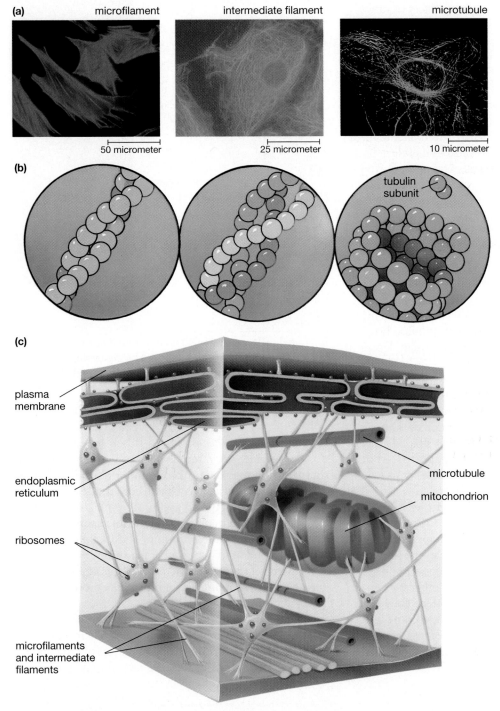

(a) microfilament — intermediate filament — microtubule

50 micrometer   25 micrometer   10 micrometer

(b) tubulin subunit

(c)

plasma membrane

endoplasmic reticulum

ribosomes

microfilaments and intermediate filaments

microtubule

mitochondrion

Figure 5-19 **The cytoskeleton**

Cells are given shape and organization by the cytoskeleton, which consists of three types of proteins: microfilaments, intermediate filaments, and microtubules. **(a)** Fluorescent substances that bind specifically to each of the cytoskeletal proteins are used to show the abundance, organization, and orientation of these proteins within living cells. **(b)** The structures of the cytoskeletal proteins. **(c)** An artist's interpretation of the cytoskeleton. Many organelles are probably attached to the cytoskeleton, which also reinforces the plasma membrane. Chemical reactions, movement of cell parts, and acquisition and release of materials are all coordinated by the cytoskeleton.

actin and flex, sliding the actin filament past the myosin, much like a sailor pulling up an anchor line hand over hand. The most familiar function of actin and myosin is muscle contraction (see Chapter 38), but actin and myosin in microfilaments also contribute to shape changes and organelle movement in most, if not all, eukaryotic cells.

Intermediate Filaments Provide a Supportive Framework

There are at least five types of intermediate filaments, each composed of a different protein. Each variety of intermediate filament is normally found in only one or a few cell types. Intermediate filaments serve a variety of functions, with one common feature: They are usually permanent frameworks that provide shape to cells and anchor various cell parts together. They do not assemble and disassemble. For example, the long axon of a nerve cell is supported by an internal skeleton of intermediate filaments in association with microtubules. In many cells, at least some of the proteins of the plasma membrane are attached to intermediate filaments. Finally, intermediate filaments anchor the microfilaments of actin in muscle cells, ensuring that the cells don't tear themselves apart during strong contractions.

Microtubules Help Position, Anchor, and Move Organelles and Alter the Shapes of Cells

Microtubules are hollow cylinders made from many dumbbell-shaped subunits of the protein tubulin (see Fig. 5-19). Some microtubules are relatively permanent features, such as those that make up the cilia and flagella by which many cells move (discussed shortly). Many microtubules, however, are transitory, being assembled and disassembled as required by the cell.

Both temporary and permanent microtubules are generated by **microtubule organizing centers** in various locations within the cell. All eukaryotic cells have a microtubule organizing center near the nucleus. During cell division, this center produces a football-shaped array of microtubules, called the **spindle apparatus**, that moves the chromosomes into the two daughter cells (see Chapter 9).

In animal cells, a prominent pair of **centrioles** is found at the microtubule organizing center near the nucleus (Fig. 5-20). Centrioles are short, barrel-shaped rings of microtubules. When a cell divides, the centrioles are duplicated, and one pair moves into each daughter cell. In ciliated cells, the centrioles multiply, and "offspring centrioles" move to the surface of the cell. Here they become anchored just beneath the plasma membrane and give rise to the microtubules of the cilia. Because of their location at the base of the cilia, these centrioles are often called **basal bodies**.

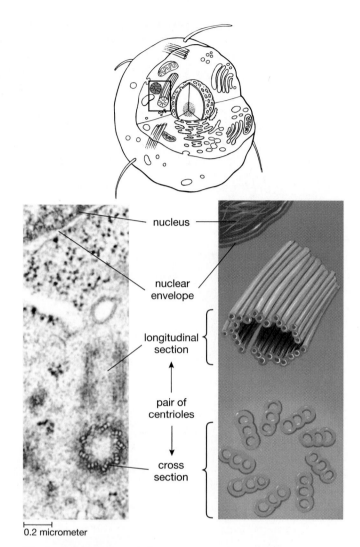

0.2 micrometer

nucleus

nuclear envelope

longitudinal section

pair of centrioles

cross section

Figure 5-20   Centrioles

In animal cells, a pair of centrioles is found at the microtubule organizing center near the nucleus. The centrioles, always lying at right angles to one another, are composed of nine triplets of short microtubules. A centriole is identical to the basal body of a cilium or flagellum (see Fig. 5-21).

Microtubules are often associated with small proteins that travel along them, like engines on a railroad track. These proteins attach to various organelles, drag them along the microtubule to appropriate places in the cell, then anchor them in place using the microtubule as a scaffold. The endoplasmic reticulum and Golgi complex are positioned and held in place by microtubules. Short-term movements of organelles and movements of chromosomes during cell division are also produced by proteins moving along microtubules.

Microtubules can also slide past one another, like escalators moving in opposite directions. Microtubule sliding is responsible for the whiplike action of cilia and flagella (see below). Microtubules can change position in

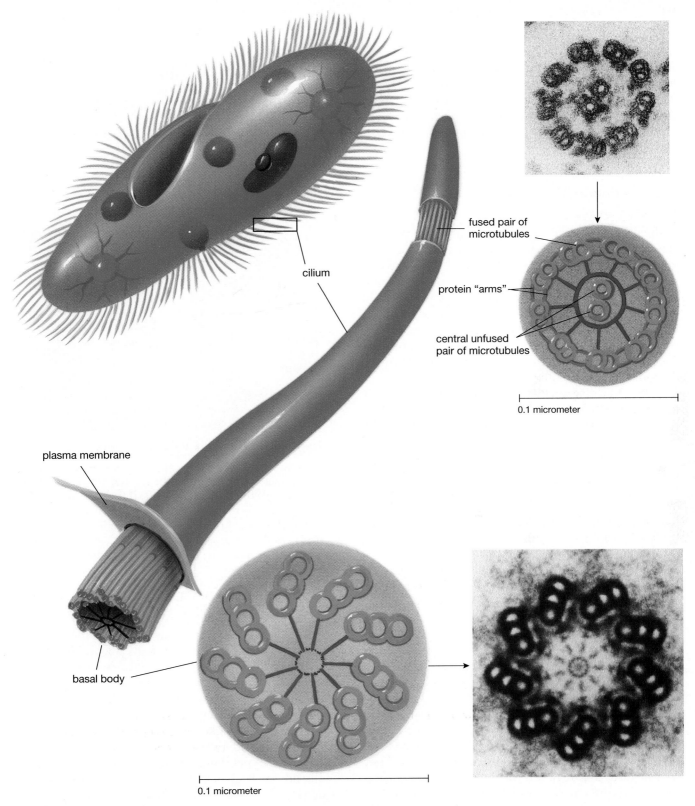

fused pair of
microtubules

protein "arms"

central unfused
pair of microtubules

0.1 micrometer

cilium

plasma membrane

basal body

0.1 micrometer

Figure 5-21 **Cilia and flagella**

Both cilia and flagella contain microtubules arranged in an outer ring of nine fused pairs of microtubules surrounding a central unfused pair (the 9 + 2 arrangement). The nine outer pairs have "arms" made of protein that interact with adjacent pairs to provide the force for bending (see Fig. 5-22). Cilia and flagella arise from basal bodies (centrioles) located just beneath the plasma membrane. Basal bodies have nine fused triplets, from which the fused pairs arise, and an indistinct central hub. It is not known how the basal body organizes the somewhat different structure of the cilium or flagellum.

still a third way: by adding tubulin subunits at one end and removing them from the other. This activity effectively moves the microtubule along in one direction. Assembly and disassembly of microtubules contribute to changes in the overall shapes of cells, for example when single nerve cells grow from your spinal cord out to the muscles of your legs.

Cilia and Flagella Move the Cell or Move Fluid Past the Cell

Both cilia and flagella are slender extensions of the plasma membrane. Each cilium and flagellum contains a ring of nine fused pairs of microtubules, with an unfused pair of microtubules in the center of the ring (forming what is called a "9 + 2" arrangement; Fig. 5-21). This pattern of microtubules is produced by a basal body (centriole) located just beneath the plasma membrane. A basal body consists of a ring of nine triplets of short microtubules. Two of the members of each triplet give rise to the pairs of microtubules in the cilium or flagellum.

Figure 5-22 diagrams how cilia and flagella bend. Tiny protein "arms" project out from each pair of microtubules in the outer ring. These arms attach to the neighboring pair of microtubules and flex, thereby moving the first pair along relative to the second. However, the basal body firmly anchors the "bottom" of all the microtubules in the entire cilium or flagellum. Therefore, adjacent microtubules can slide past one another only if the whole cilium or flagellum bends. (Think of a paint brush: When the bristles are straight, their tips are lined up evenly. When you flex the brush, the bristles slide a bit relative to one another, so that the tips of the bristles on the inside of the bend extend past the tips of the bristles on the outside of the bend.)

ATP energy powers the movement of the protein arms during microtubule sliding. Cilia and flagella often move almost continuously and consequently require enormous supplies of ATP, which are generated by mitochondria that are usually found in abundance near the basal bodies.

The main differences between cilia and flagella lie in their number, length, and the direction of the force they generate. In general, **cilia** (Latin for "eyelash") are short (about 10 to 25 micrometers long) and numerous. They provide force in a direction parallel to the plasma membrane, like the oars in a canoe. This is accomplished through a fairly stiff "rowing" motion during the power stroke, with most of the bending occurring at the base of the cilium, and a flexible return stroke (Fig. 5-23a). **Flagella** (Latin for "whip") are long (50 to 75 micrometers), usually few in number, and provide force perpendicular to the plasma membrane, like the motor on a motorboat. Flagella undulate with a continuous bending wavelike motion, without distinct power and return strokes (Fig. 5-23b).

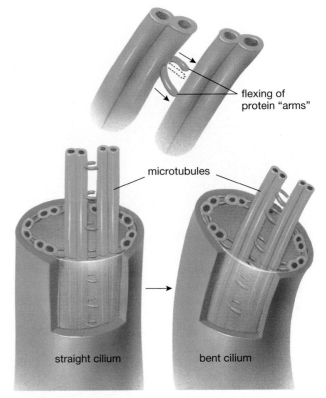

Figure 5-22   **How cilia and flagella work**

Cilia and flagella bend when the arms of one pair of microtubules temporarily attach to the shaft of an adjacent pair and flex, "walking" along the shaft. If the microtubules were free, rather than anchored, this sequence of bending and attaching would result in one pair sliding past the other pair. But because the microtubules are anchored in place by the basal body, the action causes the cilium or flagellum to bend.

Some unicellular organisms, such as *Paramecium* and *Euglena*, use cilia or flagella to move about. Most animal sperm also rely on flagella for movement. In multicellular animals, cilia are occasionally used for locomotion, moving the animal through a fluid. Many small aquatic invertebrates, for example, swim by the coordinated beating of rows of cilia, like the oars on a Roman galley. More often, however, the organism stays put, and its cilia move fluids and suspended particles past a surface. Ciliated cells line such diverse structures as the gills of oysters (moving food- and oxygen-rich water), the oviducts of female mammals (moving the eggs along from the ovary to the uterus), and the respiratory tracts of most land vertebrates (clearing debris and microorganisms from the windpipe and lungs).

Prokaryotic cells may also bear slender protrusions that undulate or spin, thereby enabling the cell to move about. However, the "flagella" of prokaryotic cells do not contain microtubules, and have no evolutionary relationship to eukaryotic flagella or cilia.

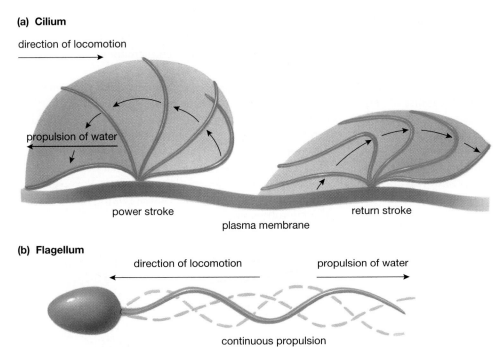

(a) Cilium

direction of locomotion

propulsion of water

power stroke

return stroke

plasma membrane

(b) Flagellum

direction of locomotion

propulsion of water

continuous propulsion

Figure 5-23 **Characteristic movement patterns of cilia and flagella**

(a) Cilia usually "row" along, providing a force of movement parallel to the plasma membrane, just as oars provide movement parallel to the sides of a rowboat. **(b)** Flagella often move in a wavelike motion, with a continuous bending that starts at the base and moves up to the tip. This motion provides a force of movement perpendicular to the plasma membrane. In this way a flagellum attached to a sperm can move the sperm straight ahead.

✖ S U M M A R Y   O F   K E Y   C O N C E P T S

The Development of Cell Theory
The principles of cell theory are:

1. Every living organism is made up of one or more cells.
2. The smallest living organisms are single cells, and cells are the functional units of multicellular organisms.
3. All cells arise from preexisting cells.

Cell Function Limits Cell Size
Cells are limited in size by two constraints. First, if a cell were too large, the rate of diffusion of essential materials from the outer surface of the cell to the center of the cell would be too slow to sustain life. Second, as a cell enlarges its volume increases more rapidly than its surface area. Therefore, the surface area of a very large cell would be too small to service the metabolic needs of the cell cytoplasm.

Types of Cells: Prokaryotic and Eukaryotic
All cells are either prokaryotic or eukaryotic. Prokaryotic cells lack membrane-bound organelles; specifically, the DNA of prokaryotic cells is not enclosed within a membrane-limited nucleus. Eukaryotic cells have several types of membrane-bound organelles, including a nucleus containing DNA. The cytoplasm of eukaryotic cells consists of a fluid cytosol and organelles located outside the nucleus. With rare exceptions, all eukaryotic cells contain a nucleus, endoplasmic reticulum, Golgi complex, various types of vesicles, and mitochondria (see Table 5-1, p. 85).

The Nucleus: Control Center of the Cell
Genetic material (DNA) is contained within the nucleus, which is bounded by the double membrane of the nuclear envelope. Pores in the nuclear envelope regulate the movement of molecules between nucleus and cytoplasm. The genetic material of eukaryotic cells is organized into linear strands called chromosomes, which consist of DNA and proteins. The nucleolus consists of the genes that code for ribosome synthesis, together with ribosomal RNA and ribosomal proteins. Ribosomes are particles of ribosomal RNA and protein that are the sites of protein synthesis.

The Membrane System of the Cell
The membrane system of a cell consists of the plasma membrane, endoplasmic reticulum (ER), Golgi complex, and vesicles derived from these membranes. Endoplasmic reticulum with ribosomes, called rough ER, manufactures many cellular proteins. Endoplasmic reticulum without ribosomes, called smooth ER, manufactures lipids. The ER is the site of all membrane synthesis within the cell. The Golgi complex is a series of membranous sacs derived from the ER. The Golgi processes and modifies materials synthesized in the rough and smooth ER. Some substances in the Golgi are packaged into vesicles for transport elsewhere in the cell. Lysosomes are vesicles budded off the Golgi that contain digestive enzymes, which digest food particles and defective organelles.

Chloroplasts and Mitochondria: Energy Capture and Extraction

All eukaryotic cells contain mitochondria, organelles that use oxygen to complete the metabolism of food molecules, capturing much of their energy as ATP. Cells of plants and some protists contain chloroplasts, which capture the energy of sunlight during photosynthesis, enabling the cells to manufacture organic molecules, particularly sugars, from simple inorganic molecules. Both mitochondria and chloroplasts probably originated from bacteria that were captured by pre-eukaryotic cells over a billion years ago, and have become incorporated into the normal functioning of the eukaryotic cell.

Plastids and Vacuoles: Storage and Elimination

Plastids are organelles with double membranes, including both chloroplasts and storage plastids that contain pigments or starch, that are found in plant cells. Many cells contain sacs bounded by a single membrane, called vacuoles, that may store food or wastes, excrete water, or support the cell. The size and function of many vacuoles change as a result of water movement across their membranes.

The Cytoskeleton: Shape, Support, and Movement

The cytoskeleton organizes and gives shape to eukaryotic cells and moves and anchors organelles. The cytoskeleton is composed of microfilaments, intermediate filaments, and microtubules (see Table 5-2, p. 97). Cilia and flagella are whiplike extensions of the plasma membrane that contain microtubules in a characteristic 9 + 2 pattern. Cilia and flagella bend as a result of flexing movements of the microtubules.

Study Note

Figures 5-6 and 5-7 illustrate the overall structure of animal and plant cells, respectively. Table 5-1 lists the principal organelles, their functions, and their occurrence in prokaryotic cells, animal cells, and plant cells.

✖ K E Y   T E R M S

aerobic p. 94	endosymbiotic hypothesis p. 92	nucleoid p. 80
anaerobic p. 94	eukaryotic p. 81	nucleolus p. 82
basal body p. 98	flagellum p. 100	nucleus p. 82
centriole p. 98	food vacuole p. 88	organelle p. 80
chlorophyll p. 94	Golgi complex p. 82	plasma membrane p. 76
chloroplast p. 94	granum p. 94	plastid p. 95
chromatin p. 82	intermediate filament p. 96	prokaryotic p. 81
chromosome p. 82	intermembrane compartment p. 94	ribosome p. 82
cilium p. 100	lysosome p. 89	rough endoplasmic reticulum p. 87
crista p. 94	matrix p. 94	smooth endoplasmic reticulum p. 87
cytoplasm p. 80	membrane p. 87	spindle apparatus p. 98
cytoskeleton p. 82	microfilament p. 96	stroma p. 94
cytosol p. 82	microtubule p. 96	thylakoid p. 94
deoxyribonucleic acid (DNA) p. 76	microtubule organizing center p. 98	vacuole p. 95
diffusion p. 80	mitochondrion p. 82	vesicle p. 82
endoplasmic reticulum (ER) p. 82	nuclear envelope p. 82	

✖ T H I N K I N G   T H R O U G H   T H E   C O N C E P T S

Multiple Choice

1. The outermost boundary of an animal cell is the
   a. plasma membrane   b. nucleus
   c. cytoplasm   d. cytoskeleton
   e. cell wall

2. Which organelle contains the cell's genetic material?
   a. Golgi complex   b. ribosomes
   c. nucleus   d. mitochondria
   e. chloroplast

3. Most of the cell's ATP is synthesized in the
   a. Golgi complex   b. ribosomes
   c. nucleus   d. mitochondria
   e. chloroplast

4. Which organelle sorts, chemically modifies, and packages newly synthesized proteins?
   a. Golgi complex   b. ribosomes
   c. nucleus   d. mitochondria
   e. chloroplast

5. Membrane-enclosed digestive organelles that contain enzymes are called
   a. lysosomes
   b. smooth endoplasmic reticulum
   c. cilia
   d. Golgi complex
   e. mitochondria

6. A series of membrane-enclosed channels studded with ribosomes are called
   a. lysosomes
   b. Golgi complex
   c. rough endoplasmic reticulum
   d. mitochondria
   e. smooth endoplasmic reticulum
7. Which of the following is NOT a true statement about the cytoskeleton?
   a. It helps support cells.
   b. Once formed, the cytoskeleton is permanent and unchanging.
   c. It is composed of microfilaments, microtubules, and intermediate filaments.
   d. It plays an important role in cell movement.
   e. It is composed of proteins.

Review Questions

1. Diagram "typical" prokaryotic and eukaryotic cells, and describe their important similarities and differences.
2. Which organelles are common to both plant and animal cells, and which are unique to each?
3. Define cytosol, stroma, and matrix.
4. Describe the plasma membrane and explain which structures give the membrane its unique properties.
5. Describe the nucleus, including the nuclear envelope, chromatin, chromosomes, DNA, and the nucleolus.
6. What are the functions of mitochondria and chloroplasts? In each case, explain how the internal structure of these organelles supports its function.
7. What is the function of ribosomes? Where in the cell are they typically found?
8. Describe the structure and function of the endoplasmic reticulum and Golgi apparatus.
9. How are lysosomes formed? What is their function?
10. Diagram the structure of cilia and flagella.

APPLYING THE CONCEPTS

1. If muscle biopsies (samples of tissue) were taken from the legs of a world-class marathon runner and a typical couch potato, which would you expect to have a higher density of mitochondria? Why? What about a muscle biopsy from the biceps of a weight lifter?
2. One of the functions of the cytoskeleton in animal cells is to give shape to the cell. Plant cells have a fairly rigid cell wall surrounding the plasma membrane. Does this mean that, for a plant cell, having a cytoskeleton is superfluous? Defend your answer in terms of other functions of the cytoskeleton.
3. Most cells are very small. What physical and metabolic constraints limit cell size? What problems would an enormous cell encounter? What adaptations might help a very large cell survive?
4. Lynn Margulis years ago proposed the "endosymbiotic hypothesis." Explain and critically evaluate this hypothesis.

FOR MORE INFORMATION

Glover, D. M., Gonzalez, C., and Raff, J. W. "The Centrosome." *Scientific American*, June 1993. The centrosome is the origin of much of a cell's cytoskeleton, and therefore regulates cell shape, movement, and cell division.

Goodsell, D. S. *The Machinery of Life*. New York: Springer, 1993. Wonderful, drawn-to-scale images of the organelles and molecules of the cell.

Kiester, E., Jr. "A Bug in the System." *Discover*, February 1991. Mitochondria are the descendants of bacteria that live within our cells. Although mitochondria are essential for human life, defects in mitochondrial genes can cause some devastating diseases.

Murray, M. "Life on the Move." *Discover*, March 1991. Many cells, including *Amoeba* and white blood cells, crawl about using tiny protein "motors" to manipulate their cytoskeleton.

Rothman, J. "The Compartmental Organization of the Golgi Apparatus." *Scientific American*, September 1985. Although they look homogeneous in electron micrographs, the sacs of the Golgi actually form three biochemically distinct processing units.

Symmons, M., Prescott, A., and Warn, R. "The Shifting Scaffolds of the Cell." *New Scientist*, February 18, 1989. Good overview showing the dynamic nature of the cytoskeleton.

Weber, K., and Osborn, M. "The Molecules of the Cell Matrix." *Scientific American*, October 1985. Biochemistry and microscopy are used to probe the structure and function of the cytoskeleton.

NET WATCH

On-line resources for this chapter are on the World Wide Web at:
http://www.prenhall.com/~audesirk (click on the table of contents link and then select Chapter 5).

6 Cell Membrane Structure and Function

"To stay alive, you have to be able to hold out against equilibrium, maintain imbalance, bank against entropy, and you can only transact this business with membranes in our kind of world."

LEWIS THOMAS in *The Lives of a Cell* (1974)

AT A GLANCE

Cell Walls

The Plasma Membrane
The Fluid Mosaic Model Describes the Structure of the Plasma Membrane
The Fluid Portion of Membranes Is Produced by a Double Layer of Phospholipids: The Phospholipid Bilayer
A Variety of Proteins in the Cell Membrane Form a Protein Mosaic

Transport across Membranes
Molecules in Fluids Move in Response to Gradients
Movement across Membranes Occurs by Both Passive and Energy-Requiring Transport

Passive Transport: Movement down Concentration Gradients
Molecules Diffuse from Areas of High Concentration to Areas of Low Concentration
Molecules Diffuse across Membranes down Their Concentration Gradients
Osmosis Is the Diffusion of Water across Membranes
Osmosis across the Cell Membrane Plays an Important Role in the Lives of Cells

Energy-Requiring Transport across Membranes
Active Transport Uses Energy to Move Molecules across Membranes
Cells Engulf Particles or Fluids by Endocytosis

Transport across Intracellular Membranes
Vacuoles Help Regulate Cell Volume

Cell Connections and Communication
Desmosomes Attach Cells Together
Tight Junctions Leakproof the Cell
Gap Junctions and Plasmodesmata Allow Communication between Cells

Diatoms, seen here under the light microscope. Different species of diatoms are found in both fresh and salt water. A glassy outer shell and a selective cell membrane help cells maintain relatively constant internal conditions.

Consider the challenge presented to a unicellular organism drifting about in a tide pool, stranded by the receding tide (Fig. 6-1). It inhabits an environment whose conditions change quickly and often. As the tide recedes, the sun warms the pool. As the day wears on, some of the water evaporates, boosting the concentration of dissolved salts. A sudden rain or the returning tide dilutes the water again. When night falls, the water temperature falls sharply. Just surviving requires a protective barrier between the inside and outside of the cell.

Merely withstanding environmental changes is not enough, however. The cell must absorb sunlight for photosynthesis or engulf other organisms to obtain the nutrients it needs to make new cellular components. To main-

Figure 6-1   A challenging environment
A tidepool contains countless life forms, including the photosynthetic single-celled diatom in the opening photograph. All cells face formidable challenges in coping with their external environment. The plasma membrane, which both isolates the cell from its environment and regulates interactions with that environment, is the focus of this chapter.

tain a consistent internal environment in which its components function best, the cell must shuttle salt, water, and other molecules in and out. It must excrete waste products. It must detect chemical clues to the location of food or to healthy and unhealthy environmental conditions. Finally, if it reproduces sexually, it must communicate with at least one other cell of its own species sometime during its life.

A lipid membrane only two molecules thick stands between the delicate inner workings of the cell and its hostile environment. The membrane also selectively permits the absorption and release of chemicals, sometimes using cellular energy. It allows chemical reactions to occur between one molecule outside the cell and another inside the cell. And it may deform to engulf food or enable the cell to move. All these processes rely on a variety of proteins embedded in the lipid membrane or attached to its surfaces.

Some single-celled organisms, such as the amoeba, survive with no barrier between life and the environment but a cell membrane. Cells of plants, bacteria, and many protists, such as diatoms, have walls outside the cell membrane that lend mechanical support. Cells of multicellular animals live surrounded by other cells and bathed in a bloodlike fluid. But all these cells possess very similar cell membranes and carry on many of the same processes across that membrane.

In this chapter, we explore the cell membrane structures that make defense, exchange of materials with the environment, and communication between cells possible.

Cell Walls

The outer surfaces of the cells of bacteria, plants, fungi, and some protists are covered with stiff, nonliving coatings called **cell walls**. Plant cell walls are composed of cellulose and other polysaccharides (see Fig. 3-5), while fungal cell walls are made of the modified polysaccharide **chitin** (see Fig. 3-6). Bacterial cell walls have a chitinlike framework to which short chains of amino acids and other molecules are attached.

Cell walls are produced by the cells they surround. In plants, membranous vesicles filled with sticky polysaccharides such as pectin (the ingredient that congeals grape juice into jelly) line up across the middle of a dividing cell. The vesicles fuse together, forming the new plasma membranes that separate the two daughter cells (see Chapter 9). The pectins formerly contained in the vesicles glue the daughter cells together, forming the **middle lamella** (Fig. 6-2). The two cells then secrete cellulose through their plasma membranes, underneath the middle lamella, forming the **primary cell wall**. Many plant cells later secrete more cellulose and other polysaccharides beneath the primary wall to form a

thick **secondary cell wall**. In some plant cells, the secondary wall may become thicker than the entire rest of the cell.

Cell walls support and protect otherwise fragile cells. For example, cell walls allow plants and mushrooms to resist the forces of gravity and wind and to stand erect on land. Tree trunks are the ultimate proof of the strength of cell walls, being composed almost entirely of cellulose and other materials laid down over the years and capable of supporting impressive loads (see Fig. 3-5 for more on the structure and properties of cellulose).

Although strong, cell walls are usually porous, permitting easy passage of small molecules such as minerals, water, oxygen, carbon dioxide, amino acids, and sugars (otherwise, the cell within would soon die). However, the structure that really governs the interactions between a cell and its external environment is the plasma membrane.

The Plasma Membrane

The **plasma membrane** of a cell can be thought of as a gatekeeper, allowing only specific substances in or out and passing messages from the external environment to the cell's interior. As gatekeeper, the plasma membrane must perform several specific functions:

1. Isolate the cell cytoplasm from the external environment.
2. Regulate the exchange of essential substances between the cytoplasm and the external environment.
3. Communicate with other cells.
4. Identify the cell as belonging to a particular species and a particular individual member of that species. In multicellular organisms, specific cell types often also bear unique molecular labels on their cell surfaces.

These are formidable tasks for a structure so thin that 10,000 stacked atop one another would scarcely equal the thickness of this page. The key to membrane function lies in membrane structure. Membranes are not simply homogeneous sheets: They are complex, heterogeneous structures, with different parts, indeed even individual molecules, performing very distinct functions.

As we pointed out in Chapter 5, all the membranes of a cell have a similar structure. Therefore, although this chapter focuses on the plasma membrane, the information presented here applies to other cellular membranes as well.

The Fluid Mosaic Model Describes the Structure of the Plasma Membrane

According to the **fluid mosaic model** of cellular membranes, developed by cell biologists S. J. Singer and

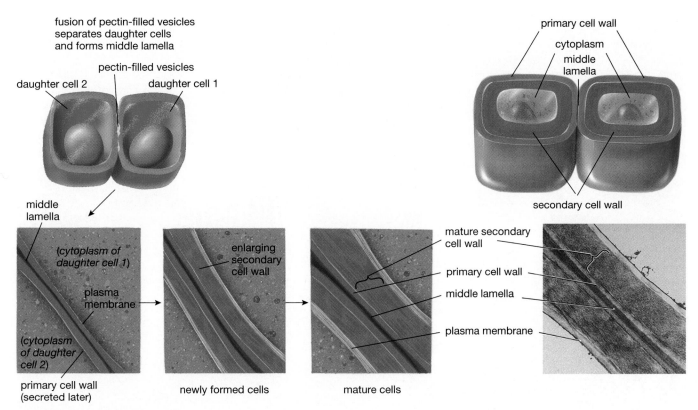

Figure 6-2   Plant cell walls

During cell division, the fusion of pectin-filled vesicles forms the middle lamella, which glues together the plasma membranes of the two daughter cells. Each cell then secretes cellulose and other carbohydrates to form its primary cell wall, between the middle lamella and the plasma membrane. Later, many cells may secrete additional cellulose and other carbohydrates beneath the primary wall, forming a secondary cell wall, pushing the primary cell wall and middle lamella farther away from the plasma membrane.

G. L. Nicolson in 1972, a membrane, when viewed from above, looks something like a lumpy, constantly shifting mosaic of tiles (Fig. 6-3). A bilayer of phospholipids forms a viscous, fluid "grout" for the mosaic, while an assortment of proteins are the "tiles," often sliding sluggishly about within the phospholipid bilayer. Thus, the overall image slowly changes over time, though the components remain relatively constant. As strange as this model may seem, it captures something of the dynamic quality of real membranes. With this model in mind, let's look more closely at the structure of membranes.

The Fluid Portion of Membranes Is Produced by a Double Layer of Phospholipids: The Phospholipid Bilayer

As you learned in Chapter 3, a phospholipid consists of two very different parts, a polar hydrophilic head and a pair of nonpolar hydrophobic tails:

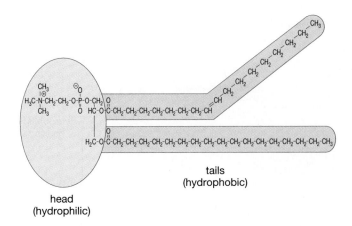

All living cells are surrounded by water, whether it is the pond in which an *Amoeba* spends its life or the extracellular fluid that bathes the cells of animals. The cell cytoplasm is also mostly water. Plasma membranes therefore separate a watery cytoplasm from a watery external envi-

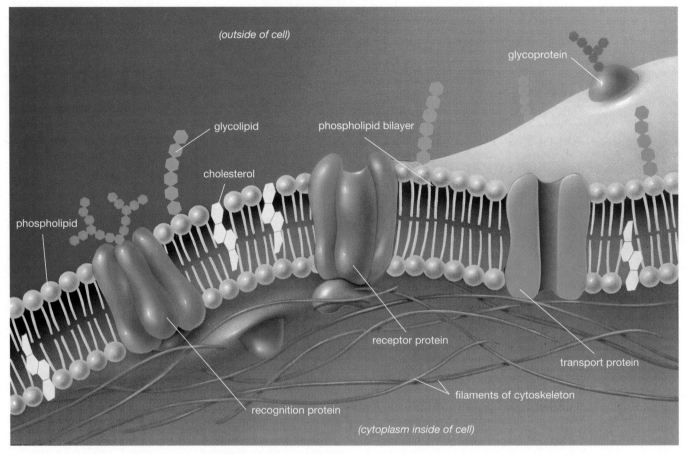

Figure 6-3   **The fluid mosaic model of the plasma membrane**

According to this model, the plasma membrane is a bilayer of phospholipids in which are embedded various proteins. Many proteins and lipids have carbohydrates attached to them, forming glycoproteins and glycolipids, respectively. The wide variety of membrane proteins fall mostly into three categories: transport proteins, receptor proteins, and recognition proteins.

ronment. Under these conditions, phospholipids spontaneously arrange themselves in a double layer called a **phospholipid bilayer:**

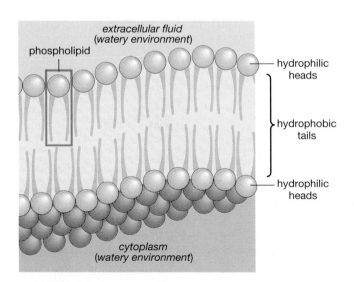

Hydrogen bonds can form between water and the phospholipid heads, so the heads face the cytoplasm or the extracellular fluid. Hydrophobic interactions (see Chapter 2) cause the phospholipid tails to hide inside the bilayer. Because individual phospholipid molecules are not bonded to one another, this double layer is quite fluid, and individual molecules move about easily.

Most biological molecules, such as salts, amino acids, and sugars, are polar and water soluble. They therefore cannot pass easily through the nonpolar, hydrophobic fatty acid tails of the phospholipid bilayer. Because most substances that contact a cell are water soluble, the phospholipid bilayer acts as a barrier to the entrance of these molecules. It is largely responsible for the first of the four membrane functions that we listed earlier—isolating the cell cytoplasm from the external environment.

The phospholipid bilayer of membranes also contains cholesterol. The membranes of mitochondria have just a few cholesterol molecules here and there, but some plasma membranes have as many cholesterols as they do phospholipids. Cholesterol affects membrane structure and function in several ways. It makes the bilayer stronger, more flexible but less fluid, and less permeable to water-soluble substances such as ions or monosaccharides.

The flexible, somewhat fluid nature of the bilayer is very important for membrane function. As you breathe, move your eyes, and turn the pages of this book, cells in your body change shape. If their plasma membranes were stiff instead of flexible, cells would break open and die. Further, as we saw in the previous chapter, vesicles of membrane constantly bud off from the endoplasmic reticulum, merge with the Golgi complex, bud off again, and fuse with the plasma membrane. Membranes can bud off vesicles and seal up again, or fuse with vesicles and smooth out again, because of the fluid nature of the bilayer.

A Variety of Proteins in the Cell Membrane Form a Protein Mosaic

A variety of proteins are embedded within or attached to the surface of a membrane's phospholipid bilayer. You will recall from Chapter 2 that some of the amino acids that make up proteins are hydrophilic whereas others are hydrophobic. Many membrane proteins are held in the lipid bilayer by the interactions between their hydrophobic amino acids and the hydrophobic tails of the phospholipids. Hydrophilic parts of the proteins may protrude either into the cytoplasm or into the extracellular fluid. Many of the proteins in plasma membranes also have carbohydrate groups attached to them, especially to the parts that stick outside the cell. These proteins are called **glycoproteins**.

Many proteins can move about within the relatively fluid phospholipid bilayer. Other membrane proteins, however, are anchored in place by a network of protein filaments connected to the cytoskeleton. In animal cells, which lack cell walls, the attachments between plasma membrane proteins and the underlying cytoskeleton produce the characteristic shapes of the cells, from the dimpled discs of red blood cells to the elaborate branching nerve cells.

Membrane Proteins Serve a Variety of Functions That Allow the Cell to Interact with Its Environment

There are three major categories of membrane proteins, each of which serves a different function: transport proteins, receptor proteins, and recognition proteins (see Fig. 6-3).

Transport proteins regulate the movement of water-soluble molecules through the plasma membrane. Some, called **channel proteins**, form pores or channels that allow small water-soluble molecules to penetrate through the membrane (see Fig. 6-3). These proteins form a structure something like a sleeve with a lining: Hydrophobic amino acids (the outer material of the sleeve) anchor the protein in the phospholipid bilayer and hydrophilic amino acids form the inside of the channel (the lining of the sleeve). Every plasma membrane bears a large assortment of protein channels, each lined with specific amino acids that allow certain molecules, usually ions such as potassium (K^+), sodium (Na^+), and calcium (Ca^{2+}), to pass through. Other transport proteins, called **carrier proteins**, have binding sites, much like the active sites of enzymes, that can grab onto specific molecules on one side of the membrane. The transport protein then changes shape, in some cases through the use of cellular energy, and moves the molecule across the membrane. We will discuss the mechanisms whereby transport proteins move molecules across the membrane in the next section.

Receptor proteins are molecular triggers that set off cellular responses when specific molecules in the extracellular fluid, such as hormones or nutrients, bind to them. Most cells bear dozens of different types of receptors on their plasma membranes. When activated by the appropriate molecule, some receptors set off elaborate sequences of cellular changes, such as increased metabolic rate, cell division, movement toward a source of nutrients, or secretion of hormones. Other receptors act like gates on channel proteins; activating the receptor opens the gates, allowing ions to flow through the channels. For example, these receptor-activated channels allow nerve cells in your brain to communicate with one another (see Chapter 36).

Recognition proteins and glycoproteins serve as identification tags and cell-surface attachment sites. The cells of your immune system, for example, recognize a *Salmonella* bacterium as a foreign invader and target it for destruction. These same immune cells ignore the trillions of cells of your own body because of different identification glycoproteins on their surfaces. During development, the growth of nerve fibers from your spinal cord down to the muscles in your feet is guided by attachments between recognition proteins on the nerve cell and the other cells it traverses on its way to the muscle.

As you can see from these brief descriptions, membrane proteins are largely responsible for the second, third, and fourth membrane functions in our list: moving substances across the membrane, communicating with other cells, and identifying the cell.

Transport across Membranes

Nearly all living cells are bathed in liquid, which may be the extracellular fluid of a human body, the pond water in which the single-celled amoeba crawls, or the water-saturated cell walls of a young plant. The plasma membrane separates the fluid cytoplasm of the cell from its fluid environment. Let's begin our study of membrane transport, then, with a brief look at the characteristics of fluids, starting with a few definitions.

1. A **fluid** is a liquid or a gas; that is, any substance that can move or change shape in response to external forces without breaking apart.
2. The **concentration** of molecules in a fluid is the number of molecules in a given unit of volume.
3. A **gradient** is a physical difference between two regions of space, so that molecules tend to move from one region to the other. Cells frequently encounter gradients of concentration, pressure, and electrical charge.

Molecules in Fluids Move in Response to Gradients

The individual molecules in a fluid move continually, bouncing off one another in random directions. Superimposed atop these random collisions, however, may be a very nonrandom *net movement* of particular types of molecules in the fluid, or even of the entire fluid, in response to gradients. Molecules move from regions of high concentration to low concentration (for example, sugar dissolving in coffee) or from high pressure to low pressure (such as air flowing out of a bicycle pump when you push down the handle). Charged molecules (ions) move in response to electrical gradients, toward unlike charges or away from like charges. The gradient of concentration, pressure, or electrical charge provides the potential energy that can drive the net movement of the molecules. By analogy with gravity, we will refer to such movements as going "down" the gradient.

Movement across Membranes Occurs by Both Passive and Energy-Requiring Transport

Because the cytoplasm of a cell is very different from the extracellular fluid, gradients of concentration, electrical charge, and occasionally pressure span the plasma membrane. In its role as gatekeeper of the cell, the plasma membrane provides for two types of movement (Table 6-1).

During **passive transport**, substances move down gradients of concentration, electrical charge, or pressure. This movement by itself requires no expenditure of energy. The gradients provide the potential energy that drives the movement and controls the direction of movement, into or out of the cell. The plasma membrane acts as a filter; the lipids and protein pores regulate which molecules can cross and may influence the rate and timing of movement, but they do not influence the direction of movement.

In energy-requiring transport, substances move *against* gradients of concentration, electrical charge, or pressure. In this case, transport proteins in the plasma membrane do control the direction of movement, using chemical energy from cellular metabolism to drive movement against the gradients.

A helpful analogy for understanding the difference between these types of transport is to consider what happens when you ride a bike. If you don't pedal, you can go only downhill, as in passive transport. However, if you put enough energy into pedaling, you can go uphill as well.

Passive Transport: Movement down Concentration Gradients

Although gradients of pressure (see Chapter 26) and electrical potential (see Chapter 36) are important in regulating the movement of certain substances in living organisms, most materials that move passively across plasma membranes do so in response to **concentration gradients**.

Molecules Diffuse from Areas of High Concentration to Areas of Low Concentration

Diffusion is the net movement of molecules in a fluid from regions of high concentration to regions of low concentration, driven by the concentration gradient. Diffusion can occur from one part of a fluid to another or across a membrane separating two fluid compartments. We will first examine the simplest case, the diffusion of molecules within a fluid.

To see how concentration gradients cause molecules to diffuse from one place to another, let's consider what happens when a drop of dye is placed in a glass of water (Fig. 6-4). With time, the drop will seem to become larger and paler, until eventually, even without stirring, the whole glass of water will be uniformly faintly colored. Why?

TABLE 6-1 ▓ *Transport across Membranes*	
Passive transport	Movement of substances across a membrane, going down a gradient of concentration, pressure, or electrical charge. Does not require the expenditure of energy by the cell.
Simple diffusion	Diffusion of water, dissolved gases, or lipid-soluble molecules through the phospholipid bilayer of a membrane.
Facilitated diffusion	Diffusion of (usually water-soluble) molecules through a membrane, assisted by membrane proteins.
Osmosis	Diffusion of water across a differentially permeable membrane—that is, a membrane that is more permeable to water than to dissolved solutes.
Energy-requiring transport	Movement of substances across a membrane, usually against a concentration gradient, using cellular energy.
Active transport	Movement of individual small molecules or ions through membrane-spanning proteins, using cellular energy, usually ATP.
Endocytosis	Movement of large particles, including large molecules or entire microorganisms, into a cell by a process in which the plasma membrane engulfs extracellular material, forming membrane-bound sacs that enter the cytoplasm.
Exocytosis	Movement of materials out of a cell by enclosing the material in a membranous sac that moves to the cell surface, fuses with the membrane, and opens to the outside, allowing its contents to diffuse away.

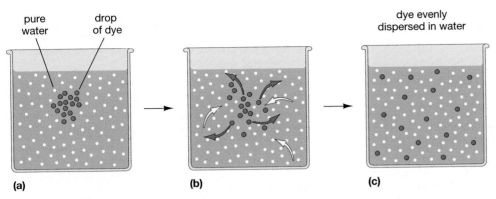

Figure 6-4   Diffusion of a dye in water

(a) A drop of pure dye (red dots) is placed in a glass of pure water (white dots in blue background). (b) Although individual molecules move at random, concentration gradients cause diffusion of dye molecules into the water and water molecules into the drop of dye. (c) Eventually, dye and water are both evenly dispersed. Individual molecules still move, but there is no longer any concentration gradient and therefore no further diffusion.

A drop of dye consists of many individual dye molecules moving in all directions. Those molecules that move but stay within the drop don't change its composition, which was all dye to begin with. But some molecules along the border of the drop, simply due to random motion, go out into the water. Thus, the net movement of dye molecules is from the region of high dye concentration (the drop) to the region of low dye concentration (the water). The same thing happens with water molecules. Random motion causes some to intersperse with the dye molecules, and the net movement of water is from the high water concentration outside the drop into the low water concentration inside the drop.

To the observer, dye molecules that move beyond the original border of the drop make the drop appear larger; water molecules that invade the drop dilute the dye, making the drop paler. At first, when the drop is pure dye and the water is pure water, there is a very steep concentration gradient, and the dye diffuses rapidly. As the concentration differences lessen, the dye diffuses more and more slowly. However, as long as the concentration of dye within the expanding drop is greater than the concentration of dye in the rest of the glass, the net movement of dye will be from drop to water, until the dye becomes uniformly dispersed in the water. Then, with no concentration gradient of either dye or water, diffusion stops. Individual molecules still move about, but there are no changes in concentration of either water or dye.

As you can appreciate from this imaginary experiment, diffusion cannot move molecules rapidly over long distances. Although the drop of dye immediately begins to diffuse into the water, uniform dispersion may take many minutes. As described in the previous chapter, the slow rate of diffusion over long distances is one of the reasons cells are small.

SUMMARY

THE PRINCIPLES OF DIFFUSION

1. Diffusion is the net movement of molecules down a gradient from high to low concentration.
2. The greater the concentration gradient, the faster the rate of diffusion.
3. If no other processes intervene, diffusion will continue until the concentration gradient is eliminated.
4. Diffusion cannot move molecules rapidly over long distances.

Molecules Diffuse across Membranes down Their Concentration Gradients

Many molecules cross plasma membranes by diffusion, driven by differences between their concentration in the cytoplasm and in the external environment. Molecules cross the plasma membrane at different locations and at different rates, depending on the properties of the molecule in question. Therefore, plasma membranes are said to be **differentially permeable**: They allow some molecules to pass through, or **permeate**, more rapidly than others.

Some Molecules Move across Membranes by Simple Diffusion

Water, dissolved gases such as oxygen and carbon dioxide, and lipid-soluble molecules such as ethyl alcohol and vitamin A easily diffuse across the phospholipid bilayer. This process is called **simple diffusion** (Fig. 6-5a). Generally, the rate of simple diffusion is a function of the concentration gradient across the membrane, the size of the molecule, and how easily it dissolves in lipids (its lipid solubility): Large concentration gradients, small molecule size, and high lipid solubility all increase the rate of simple diffusion.

Other Molecules Cross the Membrane by Facilitated Diffusion, with the Help of Membrane Proteins

Most water-soluble molecules, such as ions (K^+, Na^+, Ca^{2+}), amino acids, and monosaccharides, cannot move through the phospholipid bilayer. These molecules can diffuse across only with the aid of two types of transport proteins, channel proteins and carrier proteins, which we discussed earlier in this chapter. This process is called **facilitated diffusion**.

Channel proteins form permanent pores, or channels, in the lipid bilayer through which certain ions can cross the membrane (Fig. 6-5b). Usually, each channel protein has a specific interior diameter and distribution of electrical charges that allow only particular ions to pass through. Nerve cells, for example, have separate channels for sodium ions, potassium ions, and calcium ions. As we will see in Chapter 36, many of these channels have gates on one end, which open and close in response to electrical or chemical signals.

Carrier proteins bear groups of amino acids that bind specific molecules from the cytoplasm or extracellular fluid (Fig. 6-5c). Binding triggers a change in the shape of the carrier that allows the molecule to pass through the protein and thus across the plasma membrane. Carrier proteins that do not use cellular energy move molecules only down their concentration gradients (see the section on active transport, below).

Plasma membranes contain relatively few channels or carriers through which any given molecule can pass. The rate of facilitated diffusion is therefore a function of both the concentration gradient and the amount of transport

(a)

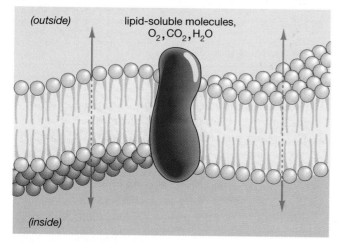

(b)

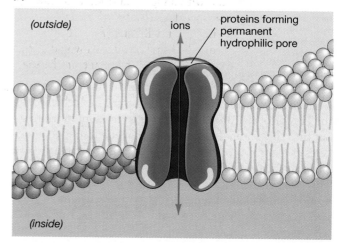

(c)

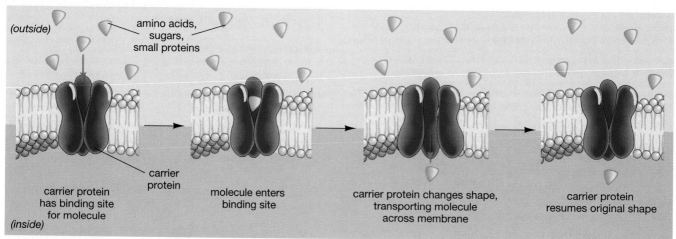

Figure 6-5   **Diffusion through the plasma membrane**

(a) Simple diffusion through the phospholipid bilayer: Water, gases such as oxygen and carbon dioxide, and lipid-soluble molecules can diffuse directly through the phospholipids. **(b)** Facilitated diffusion through a channel: Water-soluble molecules cannot pass through the phospholipid bilayer. Protein channels (pores) allow some water-soluble molecules, principally ions, to penetrate the cell. **(c)** Carrier-mediated facilitated diffusion: Carrier proteins may bind specific molecules, and, as a result, the carrier proteins change their shape, passing the molecule through the middle of the protein to the other side of the membrane.

proteins. Molecules that cross the membrane by facilitated diffusion usually do so more slowly than those that cross by simple diffusion through the lipid bilayer.

Osmosis Is the Diffusion of Water across Membranes

Water, like any other molecule, moves by diffusion from regions of high water concentration to regions of low water concentration. However, the diffusion of water across differentially permeable membranes has such dramatic consequences that it has been given a special name: **osmosis**. Let's investigate osmosis in another thought experiment.

A very simple kind of differentially permeable membrane consists of an impervious sheet perforated with tiny pores. The pores allow water molecules to pass through, but not larger molecules such as sugar. Suppose we make a bag out of such a membrane, fill it with a sugar solution, tie off the top, and place the bag in a glass of pure water. The bag will swell up, and, if it is weak enough, it will burst (Fig. 6-6). Why?

If you could see individual molecules, you would notice that there are two categories of water molecules in the sugar solution inside the bag (Fig. 6-6): "free" water molecules, well separated from the sugars; and "bound" water molecules, held to the sugars by hydrogen bonds (this is why sugars dissolve in water; see Fig. 2-9). In the pure water outside the bag, of course, there are only free water molecules. Now, free water molecules can diffuse through the pores in the membrane,

but the bound water molecules cannot, because they are attached, at least temporarily, to the bulky sugars.

Therefore, the concentration of free water molecules is lower inside the bag than in the pure water outside. This water concentration gradient favors the movement of free water molecules from the pure water outside the bag to the sugar solution inside the bag. The bag swells up as more water molecules enter the bag than leave it. The sugar can't escape at all, so the free water concentration inside the bag is always lower than in the pure water outside. Water continues to enter the bag until it bursts.

Now let's redo our experiment but with one change. Suppose we tie a piece of the same differentially permeable membrane across the end of a glass tube, suspend the tube in pure water, and put the sugar solution in the tube (Fig. 6-7). Water will diffuse across the membrane from the pure water into the solution in the tube. Because the upper end of the tube is open, as the water enters it will raise the level of the solution in the tube. Eventually, the solution in the tube will stop rising. Why?

As the solution in the tube rises above the level of water in the glass, its weight applies a pressure to the membrane across the bottom of the tube. Therefore, two opposing gradients are set up: a concentration gradient moving water *into* the tube and a pressure gradient pushing water *out* of the tube. When the two gradients are equal, no further net movement of water will occur across the membrane. The physical pressure that exactly balances the osmosis of water

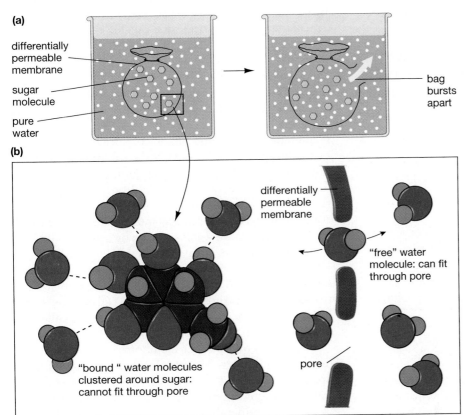

(a)

differentially permeable membrane

sugar molecule

pure water

bag bursts apart

(b)

differentially permeable membrane

"free" water molecule: can fit through pore

"bound" water molecules clustered around sugar: cannot fit through pore

pore

Figure 6-6   **Osmosis**

The membrane is differentially permeable to free water molecules (white dots) but not to larger molecules such as sugar (yellow hexagons) or water molecules held to the sugars by hydrogen bonds. **(a)** If a bag made of such a membrane is filled with a sugar solution and suspended in pure water, free water molecules will diffuse down their concentration gradient from the high concentration in the pure water outside the bag to the lower concentration in the sugar solution inside the bag. The bag swells up as water enters. If the bag is weak enough, the increasing water pressure will cause it to burst. **(b)** Membrane pores allow "free" water molecules to pass through; sugar molecules are too large. "Bound" water molecules, attracted to the sugars by hydrogen bonds, are thus also kept from passing through the pore.

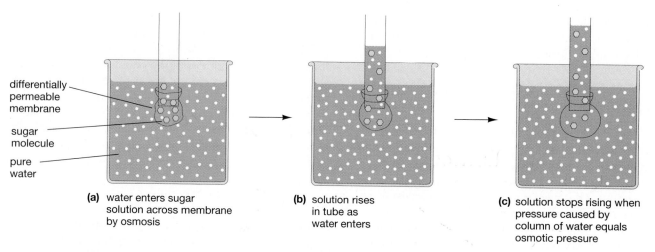

differentially
permeable
membrane

sugar
molecule

pure
water

(a) water enters sugar
solution across membrane
by osmosis

(b) solution rises
in tube as
water enters

(c) solution stops rising when
pressure caused by
column of water equals
osmotic pressure

Figure 6-7   **Osmotic pressure**

(a) A membrane permeable to free water molecules (white dots) but not to sugar (yellow hexagons) is fastened across the bottom of a glass tube. A sugar solution is placed in the tube and lowered into a glass of pure water until the level of the sugar solution is even with the surface of the water. **(b)** Water moves by osmosis across the membrane into the tube. As a result, the solution rises within the tube. **(c)** When the pressure of the column of solution in the tube forces water out through the membrane as fast as osmosis causes water to enter, the column stops rising. The resulting hydrostatic pressure equals the osmotic pressure of the solution.

due to the concentration difference between a solution and pure water is defined as the **osmotic pressure** of the solution. A solution with a high osmotic pressure has a low concentration of free water; a solution with a low osmotic pressure has a high concentration of free water.

As we have emphasized, osmosis is driven by differences in *water concentration* between solutions. Therefore, osmosis does not depend on the *type of dissolved molecules* in a solution, but only on their *concentration*. We can produce the same osmotic pressure by adding sucrose, glucose, amino acids, sodium ions, or a mixture of all of these to a solution, as long as the concentration of all types of particles that cannot permeate a membrane stays the same.

⠭ S U M M A R Y

THE PRINCIPLES OF OSMOSIS

1. Osmosis is the diffusion of water across a differentially permeable membrane.
2. Water moves across a membrane from a high concentration of free water molecules to a low concentration of free water molecules, or from high pressure to low pressure.
3. Dissolved substances, regardless of what they are, lower the concentration of free water molecules in a solution.

Osmosis across the Cell Membrane Plays an Important Role in the Lives of Cells

Most plasma membranes are highly permeable to water. Because all cells contain dissolved salts, proteins, sugars, and so on, the flow of water across the plasma mem-

brane depends on the concentration of water in the liquid that bathes the cells. The extracellular fluids of animals are usually **isotonic** ("having the same strength") to the insides of the body cells; that is, the concentration of water inside is the same as that outside the cells so there is no net tendency for water to either enter or leave the cells. Note that the *types of dissolved particles* are seldom the same inside and outside the cells, but the *total concentration of all dissolved particles* is equal, with the result that the water concentration inside is equal to that outside the cells.

If red blood cells, for example, are taken out of the body and immersed in salt solutions of varying concentrations, the effects of the differential permeability of the plasma membrane to water and dissolved particles become dramatically apparent (Fig. 6-8). If the solution has a higher salt concentration than the cytoplasm (that is, if the solution has a lower water concentration), water will leave the cells by osmosis. The cells will shrivel up until the concentrations of water inside and outside become equal. Solutions that cause water to leave by osmosis are called **hypertonic** ("having greater strength").

In the opposite situation, if the solution has little or no salt, water will enter the cells, causing them to swell. If the solution has little enough salt, the cells will burst. Solutions that cause water to enter by osmosis are called **hypotonic** ("having lesser strength").

Osmosis across plasma membranes is crucial to the functioning of many biological systems, including water uptake by plant roots (Chapter 26), absorption of dietary water from your intestines (Chapter 32), and reabsorption of water and minerals in your kidneys (Chapter 33).

Osmosis will occur across any membrane, including the membranes surrounding organelles within a cell, if the water concentration differs on the two sides of the membrane. Osmosis across internal membranes, such as those that surround vacuoles, can also be important in the lives of certain cells, as described later in this chapter.

Energy-Requiring Transport across Membranes

All cells need to move some materials "uphill" across their plasma membranes, against diffusion gradients. For example, every cell requires some nutrients that are less concentrated in the environment than in the cell cytoplasm. Therefore, diffusion would cause the cell to lose, not gain, these nutrients. Other substances, such as sodium and calcium ions in your brain cells, must be maintained at much lower concentrations inside the cells than in the extracellular fluid. When these ions diffuse into the cells, they must be pumped out again against their concentration gradients. Finally, many cells acquire or expel some substances, such as whole bacteria or large proteins, that are too large to diffuse across a membrane regardless of concentration gradients. Cells have evolved several processes that use cellular energy to move materials into or out of the cell.

Active Transport Uses Energy to Move Molecules across Membranes

In **active transport**, membrane proteins use cellular energy to move individual molecules across the plasma membrane, usually against their concentration gradient (Fig. 6-9). Active transport proteins span the membrane and have two active sites. One active site (which may be either on the face of the plasma membrane in contact with the cytoplasm or on the face in contact with the extracellular fluid, depending on the active transport protein) recognizes a particular molecule, say a calcium ion, and binds it. The second site (always on the inside of the membrane) binds an energy-carrier molecule, usually adenosine triphosphate (ATP). The ATP donates energy to the protein, causing it to change shape and move the calcium ion across the membrane. Active transport proteins are often called *pumps*, in analogy to water pumps, because they use energy to move molecules "uphill" against a concentration gradient. We will see that plasma membrane pumps are vital in mineral uptake by plants (Chapter 26), mineral absorption in your intestines (Chapter 32), and maintaining concentration gradients essential to nerve cell functioning (Chapter 36).

Cells Engulf Particles or Fluids by Endocytosis

Cells can acquire fluids or particles, especially large proteins or entire microorganisms such as bacteria, by a process called **endocytosis** (Greek for "into the cell"). Dur-

(a) Isotonic solution

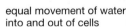
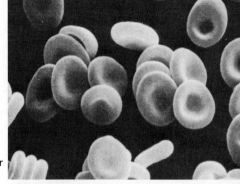

equal movement of water into and out of cells

(b) Hypertonic solution

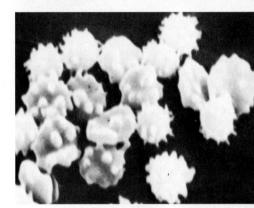

net water movement out of cells

(c) Hypotonic solution

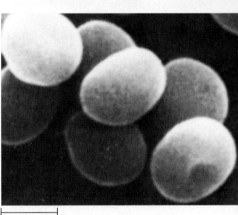

net water movement into cells

|— 10 micrometers —|

Figure 6-8   **The effects of osmosis**

Red blood cells are normally suspended in the fluid environment of the blood and lack the ability to regulate water flow across their plasma membranes. **(a)** If red blood cells are immersed in an isotonic salt solution, which has the same concentration of dissolved substances as the blood cells, there is no net movement of water across the cell membrane, and the red blood cells keep their characteristic dimpled disc shape. **(b)** A hypertonic solution, with too much salt, causes water to leave the cells, shriveling them up. **(c)** A hypotonic solution, with less salt than the cells, causes water to enter, and the cells swell.

ing endocytosis, the plasma membrane engulfs the particle or fluid droplet and pinches off a membranous sac called a **vesicle**, with the particle or fluid inside, into the

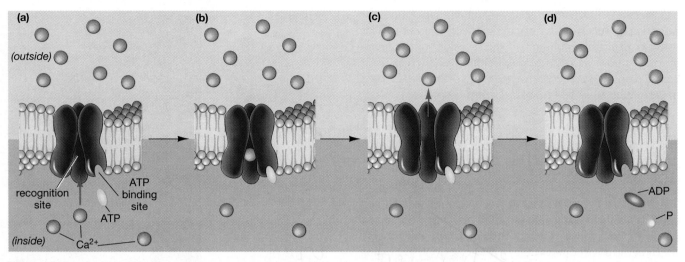

Figure 6-9   Active transport

Active transport uses cellular energy to move molecules across the plasma membrane, often against a concentration gradient. **(a)** A transport protein (blue) has a recognition site for the molecule to be transported, calcium ions (Ca^{2+}) in this case, and an ATP binding site. **(b)** The transport protein binds ATP and Ca^{2+}. **(c)** Energy from ATP changes the shape of the transport protein and moves the Ca^{2+} ion across the membrane. **(d)** The carrier releases the Ca^{2+} ion and the remnants of the ATP and resumes its original configuration.

cytoplasm (Fig. 6-10). Three types of endocytosis can be distinguished on the basis of the size of the particle acquired and the method of acquisition.

Pinocytosis Moves Liquids into the Cell

In **pinocytosis** ("cell drinking"), or **fluid-phase endocytosis**, a very small patch of plasma membrane dimples inward as it surrounds extracellular fluid and buds off into the cytoplasm as a tiny vesicle (Fig. 6-10a). Pinocytosis moves a droplet of extracellular fluid, contained within the dimpling patch of membrane, into the cell. Therefore, the cell acquires materials in proportion to their concentration in the extracellular fluid.

Receptor-Mediated Endocytosis Moves Specific Molecules into the Cell

Cells can take up certain molecules (cholesterol, for example) more efficiently by a process known as **receptor-mediated endocytosis** (Fig. 6-10b). Most plasma membranes bear many receptor proteins on their outside surfaces, each bearing a binding site for a particular nutrient molecule. In most cases, these receptors move through the phospholipid bilayer and accumulate in depressions of the plasma membrane called coated pits (Fig. 6-11). If the right molecule contacts a receptor protein in one of these coated pits, it attaches to the binding site. The coated pit deepens into a U-shaped pocket that eventually pinches off into the cytoplasm as a coated vesicle. Both the receptor-nutrient complex and a bit of extracellular fluid move into the cell in the coated vesicle.

Phagocytosis Moves Large Particles into the Cell

Phagocytosis ("cell eating") is used to pick up large particles, including whole microorganisms (see Fig. 6-10c).

When an *Amoeba*, for example, senses a tasty *Paramecium*, it produces extensions of its surface membrane, called **pseudopodia** (Latin for "false foot"; singular, **pseudopod**). The pseudopodia surround the luckless *Paramecium*, their ends fuse, and the prey is carried into the interior of the *Amoeba* for digestion. The resulting vesicle, called a food vacuole, fuses with lysosomes whose enzymes digest the prey (see Chapter 5). White blood cells also use phagocytosis and intracellular digestion to engulf and destroy bacteria that have invaded your body (Chapter 34).

Exocytosis Moves Material Out of the Cell

The reverse of endocytosis, called **exocytosis** (Greek for "out of the cell"), is often used by cells to dispose of unwanted materials, such as the waste products of digestion, or to secrete materials, such as hormones, into the extracellular fluid (Fig. 6-12). During exocytosis, a vesicle created by the Golgi complex moves to the cell surface, where the membrane of the vesicle fuses with the plasma membrane. The vesicle opens to the extracellular fluid and its contents diffuse out.

Transport across Intracellular Membranes

As we mentioned earlier, all the membranes of a cell, including those that surround such diverse organelles as chloroplasts, mitochondria, the nucleus, and vacuoles, are similar in structure and function to the plasma membrane. The transport processes that we have described in this chapter, from simple diffusion to active transport, also

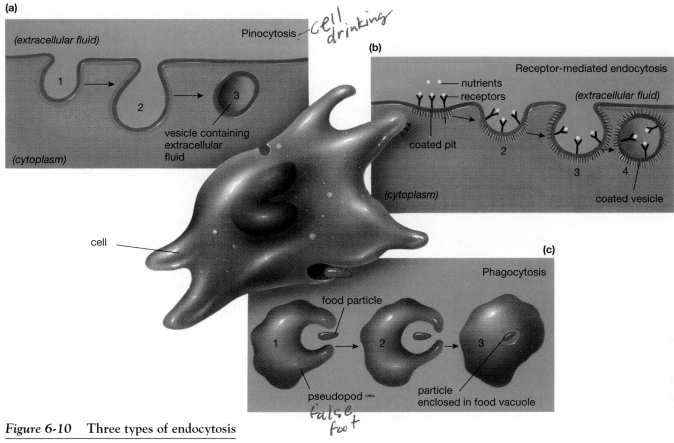

(a) *(extracellular fluid)* Pinocytosis — cell drinking

1 2 3 vesicle containing extracellular fluid

(cytoplasm)

(b) Receptor-mediated endocytosis

nutrients
receptors
(extracellular fluid)
coated pit
1 2 3 4
(cytoplasm)
coated vesicle

cell

(c) Phagocytosis

food particle
1 2 3
pseudopod — false foot
particle enclosed in food vacuole

Figure 6-10   **Three types of endocytosis**

(a) Pinocytosis: A dimple in the plasma membrane deepens and eventually pinches off as a fluid-filled vesicle. The vesicle contains a random sampling of the extracellular fluid. **(b)** Receptor-mediated endocytosis: Receptor proteins selectively bind molecules (for example, nutrients) in the extracellular fluid. The receptors migrate along the fluid lipid bilayer of the plasma membrane to dimpling sites (coated pit). The membrane dimples inward, carrying the receptor–captured molecule complexes with it. The end of the coated pit buds off a coated vesicle into the cytoplasm of the cell. The vesicle contains both extracellular fluid and a high concentration of the molecules that bind to the receptors. **(c)** Phagocytosis: Extensions of the plasma membrane, called pseudopodia, encircle an extracellular particle (for example, food). The ends of the pseudopodia fuse, forming a large vesicle (a food vacuole) containing the engulfed particle.

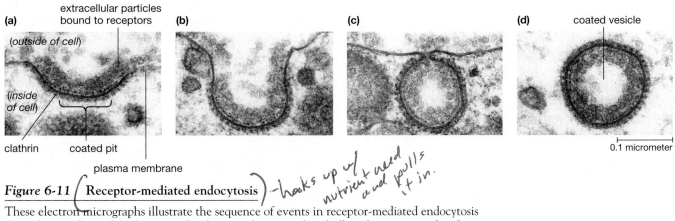

(a) extracellular particles bound to receptors
(outside of cell)
(inside of cell)
clathrin   coated pit
plasma membrane

(b) **(c)** **(d)** coated vesicle

0.1 micrometer

Figure 6-11 (**Receptor-mediated endocytosis**) — hooks up w/ nutrient need and pulls it in.

These electron micrographs illustrate the sequence of events in receptor-mediated endocytosis through coated pits. **(a)** This type of endocytosis begins with a shallow depression in the plasma membrane, coated on the inside with a protein called clathrin (dark fuzzy substance in the micrographs) and bearing receptor proteins on the outside (not visible in the micrographs). The pit deepens **(b)**, probably pulled in by elements of the cytoskeleton, and eventually pinches off as a coated vesicle **(c, d)**. The clathrin protein is eventually recycled back to the plasma membrane, restarting the cycle.

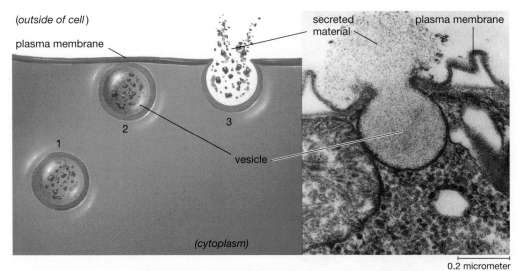

(outside of cell)

plasma membrane

secreted material

plasma membrane

1   2   3

vesicle

(cytoplasm)

0.2 micrometer

Figure 6-12   **Exocytosis** — the way unwanted stuff is removed
Exocytosis is functionally the reverse of endocytosis. The material to be ejected from the cell
is encapsulated into a membrane-bound vesicle that moves to the cell membrane and fuses
with it. As the vesicle opens to the outside, the material within leaves by diffusion.

occur across the internal membranes of a cell. For example, we will see in Chapters 7 and 8 that chloroplasts and mitochondria have specific protein channels that are permeable to hydrogen ions and that are crucial to ATP production in those organelles. Calcium ions are actively transported across certain internal membranes in muscle cells (Chapter 38). And if you look back at the section on membrane flow within cells in Chapter 5, you will see vesicles budding from and fusing with the membranes of the endoplasmic reticulum and Golgi complex, processes virtually identical to endocytosis and exocytosis. Transport across internal membranes is just as vital to the life of a cell as is transport across the plasma membrane, and similar mechanisms are used, as illustrated by transport in and out of cell vacuoles.

Vacuoles Help Regulate Cell Volume

Most cells contain one or more **vacuoles**, which are fluid-filled sacs surrounded by a single membrane (see Chapter 5). Some, such as the food vacuoles that form during phagocytosis, are temporary features of cells. Many cells, however, contain permanent vacuoles that have important roles in maintaining the integrity of the cell, especially by regulating the cell's water content.

Contractile Vacuoles Are Found in Freshwater Microorganisms

Freshwater protists such as *Paramecium* often contain complex **contractile vacuoles** composed of collecting ducts, a central reservoir, and a tube leading to a pore in the plasma membrane (Fig. 6-13). Because fresh water is hypotonic to the cytosol of these organisms, water constantly enters the

cell by osmosis. The increasing volume of incoming water might soon burst the fragile creature if it did not have a mechanism to excrete the water. Water is pumped into the collecting ducts and drains into the central reservoir (a process that requires cellular energy). When the reservoir is full, it contracts, squirting the water up the exit tube and out through the pore in the plasma membrane.

Central Vacuoles Are Found in Plant Cells

Three-quarters or more of the volume of many plant cells is occupied by a large **central vacuole** (see Fig. 5-7). The central vacuole has several functions. It provides a dump site for hazardous wastes, which plant cells often cannot excrete. Some plant cells store extremely poisonous substances, such as sulfuric acid, in their vacuoles; these substances deter animals from munching on the otherwise tasty leaves. Vacuoles may also store sugars and amino acids not immediately needed by the cell. Blue or purple pigments stored in central vacuoles are responsible for the colors of many flowers.

These dissolved substances make the vacuole contents hypertonic to the cell cytoplasm, which in turn is usually hypertonic to the extracellular fluid bathing the cells. Water therefore enters the vacuole by osmosis, which tends to make it swell. The pressure of the water within the vacuole, called **turgor pressure**, pushes the cytosol up against the cell wall with considerable force (Fig. 6-14). Primary cell walls are usually somewhat flexible, so the overall shape and rigidity of the cell depend on turgor pressure within the cell. Turgor pressure thus provides support for the nonwoody parts of plants. If you forget to water your houseplants, the central vacuoles and cytosol lose water and the cell shrinks away from its cell walls. Just as

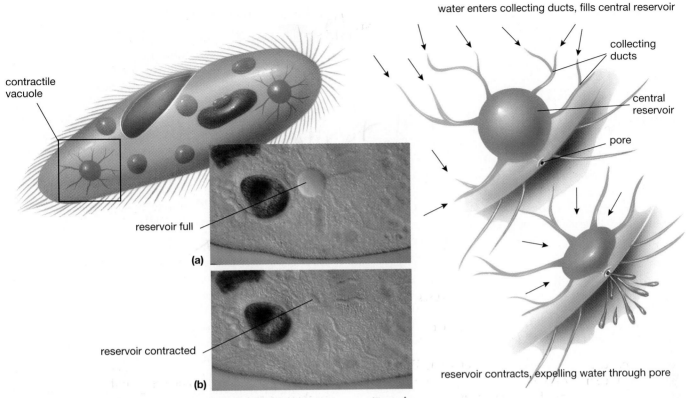

water enters collecting ducts, fills central reservoir

collecting ducts

central reservoir

pore

contractile vacuole

reservoir full

(a)

reservoir contracted

(b)

reservoir contracts, expelling water through pore

Figure 6-13 **Contractile vacuoles** *— get rid of exses water*

Many freshwater protists contain contractile vacuoles. **(a)** Osmosis causes water to constantly enter the cell, where it is taken up by collecting ducts and drains into the central reservoir of the vacuole. **(b)** When full, the reservoir contracts, expelling the water up the exit tube and outside the cell through a pore in the plasma membrane.

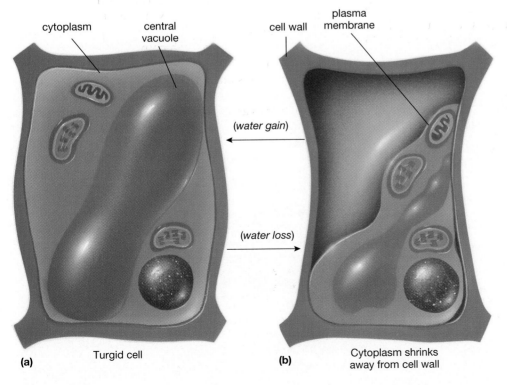

cytoplasm

central vacuole

cell wall

plasma membrane

(water gain)

(water loss)

(a) Turgid cell

(b) Cytoplasm shrinks away from cell wall

Figure 6-14 **The role of turgor pressure in plant cells**

(a) Water enters the central vacuole by osmosis, so pressure builds up within the vacuole. This pressure pushes the cytoplasm up against the cell wall, helping the cell to maintain its shape.
(b) When the central vacuole loses water and shrinks, a plant wilts. The cytoplasm pulls away from the cell wall, and the whole cell may become soft and shrunken.

a balloon goes limp when its air leaks out, so too the plant droops as its cells lose turgor pressure.

Cell Connections and Communication

In multicellular organisms, plasma membranes also function in holding together clusters of cells and in providing avenues through which cells can communicate with their neighbors. Depending on the organism and the cell type, one of four types of connection may occur between cells: desmosomes, tight junctions, gap junctions, and plasmodesmata.

Desmosomes Attach Cells Together

Animals, as you know, tend to be flexible, mobile organisms. Many of an animal's tissues are stretched, compressed, and bent as the animal moves about. If the skin, intestines, stomach, urinary bladder, and other organs are not to tear apart under the stresses of movement, their cells must adhere firmly to one another. Such animal tissues have junctions called **desmosomes** that hold adjacent cells together (Fig. 6-15). In a desmosome, the membranes of adjacent cells are glued together by proteins and carbohydrates. Intermediate filaments attached to the insides of the desmosomes extend into the interior of each cell, further strengthening the attachment.

Tight Junctions Leakproof the Cell

The animal body contains many tubes or sacs that must hold their contents without leaking: A leaky urinary bladder would spell disaster for the rest of the body. The spaces between the cells lining such sacs are sealed with strands of protein to form **tight junctions** (Fig. 6-16). The membranes of adjacent cells nearly fuse along a series of ridges, effectively forming waterproof gaskets between cells. Continuous tight junctions sealing each cell to its neighbors prevent molecules from escaping between cells.

Gap Junctions and Plasmodesmata Allow Communication between Cells

Multicellular organisms must coordinate the actions of their component cells. In animals, many cells, including heart muscle cells, most gland cells, some brain cells, and every cell of very young embryos, communicate through protein channels directly connecting the insides of adjacent cells (Fig. 6-17). These cell-to-cell channels are clustered in specialized regions called **gap junctions**. Hormones, nutrients, ions, and even electrical signals can pass through the channels at gap junctions.

Virtually all of the living cells of plants are connected to one another by much larger channels called **plasmodesmata** (Fig. 6-18). Each plasmodesma is a tube, lined with plasma membrane, that penetrates through the cell wall from the cytoplasm of one cell to the cytoplasm of its neighbor. Many plant cells have thousands of plasmodesmata. As a result, water, nutrients, and hormones pass quite freely from one cell to another.

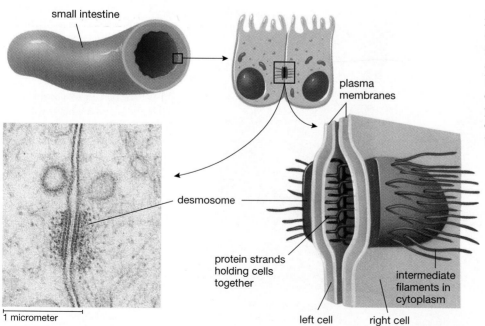

small intestine

plasma membranes

desmosome

protein strands holding cells together

intermediate filaments in cytoplasm

left cell

right cell

1 micrometer

Figure 6-15   **Desmosomes**

Cells lining the small intestine are firmly attached to one another by desmosomes. Intermediate filaments bound to the inside surface of each desmosome extend into the cytoplasm and attach to other elements of the cytoskeleton, strengthening the connection between cells.

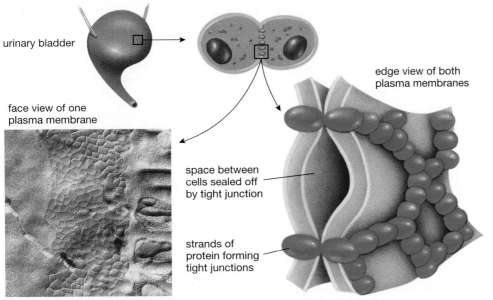

urinary bladder

face view of one
plasma membrane

edge view of both
plasma membranes

space between
cells sealed off
by tight junction

strands of
protein forming
tight junctions

Figure 6-16 Tight junctions
Leakage between cells of the urinary bladder is prevented by close-fitting tight junctions.

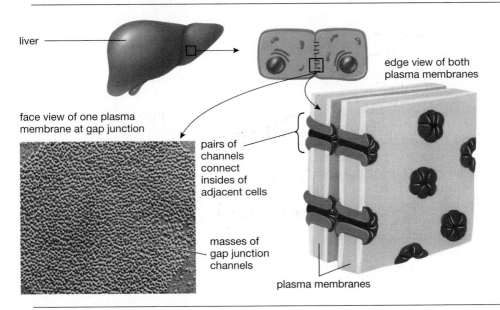

liver

face view of one plasma
membrane at gap junction

edge view of both
plasma membranes

pairs of
channels
connect
insides of
adjacent cells

masses of
gap junction
channels

plasma membranes

Figure 6-17 Gap junctions
Gap junctions contain cell-to-cell channels that interconnect the cytoplasm of adjacent cells.

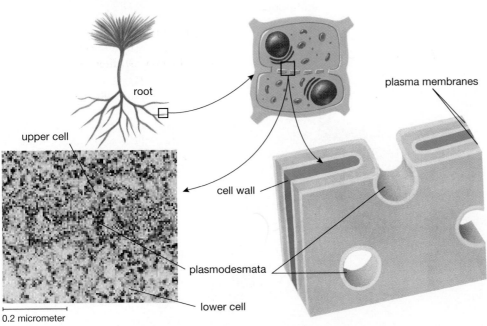

root

upper cell

plasma membranes

cell wall

plasmodesmata

lower cell

0.2 micrometer

Figure 6-18 Plasmodesmata
Plant cells are widely interconnected by rather large cell-to-cell pores called plasmodesmata.

E V O L U T I O N A R Y
C O N N E C T I O N S

Natural Selection Has Produced Cell Membranes with Diverse Compositions

Although the membranes of all cells have a similar structure, membrane function varies tremendously from organism to organism and from cell to cell within a single organism. This diversity arises largely from the different proteins and phospholipids in the membrane, which have evolved under differing selection pressures.

Our discussion of membranes emphasized the unique functions of the membrane proteins. Consequently, you may think that the phospholipids are just a waterproof place for the proteins to reside. This isn't quite true, as we can see by examining the plasma membrane phospholipids in the legs of caribou, animals that live in very cold regions of North America (Fig. 6-19). During the long arctic winters of these regions, temperatures plummet far below freezing, and for caribou to keep their legs and feet really warm would waste precious energy. These conditions have favored the evolution of specialized arrangements of arteries and veins in caribou legs that allow the temperature of their lower legs to drop almost to freezing (0° C). The upper legs and main trunk of the body, in contrast, remain at about 37° C. Further, the phospholipids in the membranes of cells in the upper legs of caribou are very different from those near the hooves.

Remember, the membrane of a cell needs to be somewhat fluid to allow the proteins to move to sites where they are needed. The fluidity of a membrane is a function of the fatty acid tails of its phospholipids: Unsaturated fatty acids remain more fluid at lower temperatures than saturated fatty acids do (see Chapter 3). Caribou have a range of phospholipids in the plasma membranes of the cells in their legs. The membranes of cells near the chilly hoof have lots of unsaturated fatty acids, whereas the membranes of cells near the warmer trunk have more saturated ones. This arrangement gives the plasma membranes throughout the leg the proper fluidity despite great differences in temperature.

As important as the phospholipids are, the membrane proteins probably play the major roles in determining cell function and in governing the interactions between a cell and its neighbors. The extreme complexity of the protein molecule makes it more susceptible than other types of molecules to mutations that alter its amino acid composition, its shape, and hence its function. Over billions of years, an incredible diversity of proteins has evolved. Every nerve cell in your body, for instance, has membrane proteins essential for producing electrical signals and conducting them along the nerves to various parts of the body. Other membrane proteins receive chemical messages from neighboring nerve cells or from hormones and other chemicals in the blood. Each cell in the brain has a specific set of membrane proteins, allowing it to respond to some stimuli while ignoring others. In fact, your ability to read this page depends on the proteins residing in the membranes of your brain cells.

As we progress through this book, we shall return many times to the concepts of membrane structure presented in this chapter. Understanding the diversity of membrane lipids and proteins is the key to understanding not just the isolated cell, but entire organs, which function as they do largely because of the properties of the membranes of their component cells.

Figure 6-19   **Caribou browse on the Alaskan tundra**

The lipid composition of the membranes in the cells of a caribou's legs varies with distance from the trunk. Unsaturated phospholipids predominate in the lower leg, while more saturated phospholipids are found in the upper leg.

✖ SUMMARY OF KEY CONCEPTS

Cell Walls
The outer surface of some protist cells and of each bacterial, plant, and fungal cell is surrounded by a rigid cell wall outside the plasma membrane. The cell wall is produced by the cell that it surrounds. It protects and supports the cell.

The Plasma Membrane
The plasma membrane has four major functions: (1) isolate the cytoplasm from the external environment; (2) regulate the flow of materials into and out of the cell; (3) communicate with other cells; and (4) identify the cell. The membrane, according to the fluid mosaic model, consists of a bilayer of phospholipids in which are embedded a variety of proteins. There are three major categories of membrane proteins: (1) transport proteins, which regulate the movement of most water-soluble substances through the membrane; (2) receptor proteins, which bind molecules in the external environment, triggering changes in the metabolism of the cell; and (3) recognition proteins, which identify cells as to species and cell type.

Transport across Membranes
Particles in a fluid tend to move in response to gradients of concentration, pressure, or electrical charge. In cells, passive transport is the movement of substances across the plasma membrane down concentration gradients. Cellular energy is not used. Energy-requiring transport processes move substances across the membrane, usually against concentration gradients, through the use of cellular energy.

Passive Transport: Movement down Concentration Gradients
Diffusion is the movement of particles from regions of higher concentration to regions of lower concentration. In simple diffusion, water, dissolved gases, and lipid-soluble molecules diffuse through the phospholipid bilayer. In facilitated diffusion, water-soluble molecules cross the membrane through protein channels or with the assistance of protein carriers. In both cases, molecules move down their concentration gradients and cellular energy is not required.

Osmosis is the diffusion of water across a differentially permeable membrane, down its concentration gradient. Dissolved solutes decrease the concentration of free water molecules. Osmosis does not require cellular energy.

Energy-Requiring Transport across Membranes
In active transport, protein carriers in the membrane use cellular energy (ATP) to drive the movement of molecules across the plasma membrane, usually against concentration gradients. Large molecules (for example, proteins), particles of food, microorganisms, and extracellular fluid may be acquired by endocytosis, either by pinocytosis, receptor-mediated endocytosis, or phagocytosis. The secretion of substances such as hormones and the excretion of wastes from a cell are accomplished by exocytosis.

Transport across Intracellular Membranes
The intracellular membranes that surround organelles such as the endoplasmic reticulum, Golgi complex, mitochondria, chloroplasts, and vacuoles are fundamentally similar in structure and function to the plasma membrane. Substances are transported across these membranes by most of the same mechanisms that are involved in transport across the plasma membrane.

Cell Connections and Communication
Cells may be connected to one another by a variety of junctions. Desmosomes attach cells firmly to one another, preventing tearing of a tissue during movement or stress. Tight junctions seal off the spaces between adjacent cells, leakproofing organs such as the urinary bladder. Gap junctions in animals and plasmodesmata in plants are the locations at which the cytoplasm of two adjacent cells are interconnected by pores through adjoining plasma membranes.

✖ KEY TERMS

active transport   p. 115
carrier protein   p. 109
cell wall   p. 106
central vacuole   p. 118
channel protein   p. 109
chitin   p. 106
concentration   p. 109
concentration gradient   p. 110
contractile vacuole   p. 118

desmosome   p. 120
differential permeability   p. 111
diffusion   p. 110
endocytosis   p. 115
exocytosis   p. 116
facilitated diffusion   p. 112
fluid   p. 109
fluid mosaic model   p. 106
fluid-phase endocytosis   p. 116

gap junction   p. 120
glycoprotein   p. 109
gradient   p. 109
hypertonic   p. 114
hypotonic   p. 114
isotonic   p. 114
middle lamella   p. 106
osmosis   p. 113
osmotic pressure   p. 114

passive transport   p. 110
permeate   p. 111
phagocytosis   p. 116
phospholipid bilayer   p. 108
pinocytosis   p. 116
plasma membrane   p. 106
plasmodesma   p. 120

primary cell wall   p. 106
pseudopod   p. 116
receptor-mediated endocytosis   p. 116
receptor protein   p. 109
recognition protein   p. 109
secondary cell wall   p. 106
simple diffusion   p. 111

tight junction   p. 120
transport protein   p. 109
turgor pressure   p. 118
vacuole   p. 118
vesicle   p. 115

✖ T H I N K I N G   T H R O U G H   T H E   C O N C E P T S

Multiple Choice

1. Active transport through the plasma membrane occurs through the action of
   a. diffusion
   b. membrane proteins
   c. DNA
   d. water
   e. osmosis

2. The following is a characteristic of a cell membrane.
   a. It separates the cell contents from its environment.
   b. It is permeable to certain substances.
   c. It is a lipid bilayer with embedded proteins.
   d. It contains sodium-potassium pumps.
   e. all of the above

3. If an animal cell is placed into a solution whose concentration of solutes is higher than that inside the cell
   a. the cell will swell
   b. the cell will shrivel
   c. the cell will remain the same size
   d. the solution is described as hypertonic
   e. both b) and d) are correct

4. Small, nonpolar hydrophobic molecules such as fatty acids
   a. pass readily through a membrane's lipid bilayer
   b. diffuse very slowly through the lipid bilayer
   c. require special channels to enter a cell
   d. are actively transported across cell membranes
   e. must enter the cell via endocytosis

5. Which of the following would be least likely to diffuse through a lipid bilayer?
   a. water
   b. oxygen
   c. carbon dioxide
   d. sodium ions
   e. the small, nonpolar molecule butane

6. Which of the following pieces of evidence would show that a substance enters a cell by active transport rather than diffusion?
   a. A carrier protein moves the substance across the membrane.
   b. The substance enters the cell when its concentration is higher outside than inside.
   c. The breakdown of ATP is required for the substance to move into the cell.
   d. The substance moves through a channel protein.
   e. all of the above

Review Questions

1. Describe the process of cell wall formation in plant cells.

2. Describe and diagram the structure of a plasma membrane. What are the two principal types of molecules in plasma membranes? What are the four principal functions of plasma membranes?

3. What are the three types of proteins commonly found in plasma membranes, and what is the function of each?

4. Define diffusion and compare it to osmosis. How do these two forces help plant leaves to remain firm? What is turgor pressure?

5. Define hypotonic, hypertonic, and isotonic. What would be the fate of an animal cell immersed in each of the three types of solution?

6. Describe the following types of transport processes: simple diffusion, facilitated diffusion, active transport, pinocytosis, receptor-mediated endocytosis, phagocytosis, and exocytosis.

7. Name four types of cell-to-cell junctions and the function of each. Which allow communication between the interiors of adjacent cells?

✖ A P P L Y I N G   T H E   C O N C E P T S

1. Different cells have somewhat different plasma membranes. The plasma membrane of a *Paramecium*, for example, is only about 1% as water permeable as the plasma membrane of a human red blood cell. Referring back to our discussion of the effects of osmosis on red blood cells and the role of contractile vacuoles in *Paramecium*, what do you think is the function of the low water permeability of *Paramecium*? What molecular differences do you think might account for this low water permeability?

2. A preview question for Chapter 34: The integrity of the plasma membrane is essential for cellular survival. Could this fact be utilized by the immune system to destroy foreign cells that have invaded the body? How might cells of

the immune system disrupt membranes of foreign cells (Two hints: Virtually all cells can secrete proteins, and some proteins form pores in membranes.)

3. A preview question for Chapter 27: Plant roots take up minerals (inorganic ions such as potassium) that are dissolved in the water of the soil. The concentration of such ions is usually much lower in the soil water than in the cytoplasm of root cells. Design the plasma membrane of a hypothetical mineral-absorbing cell, with special reference to mineral-permeable channel proteins and mineral-transporting active transport proteins. Justify your choice of channels and active transport proteins.

4. Red blood cells will swell up and burst when placed in a hypotonic solution like pure water. Why don't we swell up and burst when we go for a swim in water that is hypotonic to our cells and body fluids?

✕ F O R   M O R E   I N F O R M A T I O N

Bretscher, M. S. "The Molecules of the Cell Membrane." *Scientific American*, October 1985. Beautifully illustrated, this article explores the structure and function of cell membranes, with special attention to cell junctions.

Dautry-Varsat, A., and Lodish, H. "How Receptors Bring Proteins and Particles into Cells." *Scientific American*, May 1984. Receptor-mediated endocytosis is an important pathway both for cell nutrition and for cell-to-cell signaling.

McNeil, P. L. "Cell Wounding and Healing." *American Scientist*, May-June 1991. The fluidity of plasma membrane phospholipids makes cells able to withstand minor damage.

Sharon, N., and Lis, H. "Carbohydrates in Cell Recognition." *Scientific American*, January 1993. Carbohydrates, usually attached to proteins as part of glycoproteins, identify cells, serve as parts of receptors in hormone binding, and regulate the attachment and movement of cells.

N E T   W A T C H

On-line resources for this chapter are on the World Wide Web at: http://www.prenhall.com/~audesirk (click on the table of contents link and then select Chapter 6).

7 Capturing Solar Energy: Photosynthesis

> "... the cells of life
> Bound themselves into clans, a multitude of cells
> To make one being—as the molecules before
> Had made of many one cell. Meanwhile they had invented
> Chlorophyll and ate sunlight, cradled in peace
> On the warm waves."
>
> ROBINSON JEFFERS, *The Beginning and the End*

✸ AT A GLANCE

Photosynthesis: An Overview
Leaves and Chloroplasts Are Adaptations for Photosynthesis

The Light-Dependent Reactions: Converting Light to Chemical Energy
During Photosynthesis, Light Is First Captured by Pigments in Chloroplasts
The Light-Dependent Reactions Occur in Clusters of Molecules Called Photosystems

The Light-Independent Reactions: Securing Chemical Energy in Glucose Molecules
The C_3 Cycle Captures Carbon Dioxide
Carbon Fixed during the C_3 Cycle Is Used to Synthesize Glucose

The Relationship between the Light-Dependent and the Light-Independent Reactions

Water, CO_2, and the C_4 Pathway
When Stomata Are Closed to Conserve Water, Wasteful Photorespiration Occurs
C_4 Plants Reduce Photorespiration Using a Two-Stage Carbon Fixation Process
C_3 and C_4 Plants Are Adapted to Specific Environmental Conditions

A sunburst through the leaves of a fern depicts the harvesting of light energy by photosynthesis, the ultimate source of energy for nearly all life on Earth.

The flow of energy through life on Earth begins with the sun, but this was not always so. About 4.5 billion years ago, the Earth formed as chunks of matter collided and fused, transforming their energy of movement into heat. Storms and volcanic eruptions released still more energy on the newly formed planet, but no organisms existed to harness the enormous fluctuations in energy, nor is it likely that any could have withstood the violence of that time. Energy-rich organic molecules were formed, however, their synthesis driven by heat and lightning (see Chapter 19). As Earth cooled and calmed, living cells arose, and for the first billion years or so, they fed on the soup of organic molecules provided by the earlier chemical cauldron. But the cells gradually consumed the organic molecules, and the soup thinned. Sources of organic energy became scarce.

All the while, another source of energy bathed the planet: light from the sun. Through chance mutations in their molecules, some cells acquired the ability to harness the energy in sunlight. They combined simple inorganic molecules—carbon dioxide and water—into more complex organic molecules such as glucose. In a process called photosynthesis, these cells captured a small fraction of the sunlight's energy, storing it as chemical energy in these complex organic molecules (Fig. 7-1, left). Exploiting this new source of energy without competition, photosynthetic cells filled the seas with their progeny. As evolution continued, several types of photosynthesis arose. The most common released oxygen as a by-product. This new element in the atmosphere was harmful to many organisms because oxygen combines readily with many molecules and disrupts stable chemical arrangements. But the endless variation produced by random mutation eventually included some cells that could survive in oxygen and, later, cells that could use oxygen to break down glucose in a new, more efficient process.

Before oxygen was pumped into the atmosphere by photosynthetic organisms, cells (including photosynthetic cells) broke down glucose in a process, called glycolysis, that does not utilize oxygen (Fig. 7-1, upper right). But glycolysis is inefficient; it reclaims only a small fraction of the chemical energy of glucose to drive essential cellular reactions. When cells evolved the enzymes to utilize oxy-

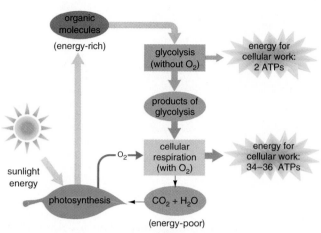

Figure 7-1   The relationship among photosynthesis, glycolysis, and cellular respiration

Chloroplasts in green plants use the energy of sunlight to synthesize high-energy carbon compounds such as glucose from low-energy molecules of water and carbon dioxide. Plants themselves, and other organisms that eat plants or one another, extract energy from these organic molecules by glycolysis and cellular respiration, yielding water and carbon dioxide once again. This energy in turn drives all the reactions of life. The orange arrows trace the flow of energy through a photosynthetic organism. Note that cellular respiration takes the products of glycolysis and extracts considerably more energy from them by combining them with oxygen.

gen to further break down the products of glycolysis in a process called cellular respiration, they were able to extract about 18 times more energy from each glucose molecule (Fig. 7-1, lower right). Cells that carried on cellular respiration grew faster and reproduced more rapidly than cells relying on glycolysis alone.

The complementary reactions of photosynthesis and cellular respiration are perhaps 2 billion years old; together they drive the flow of energy through individual organisms and ecosystems. This chapter investigates photosynthesis, the process by which almost all useful energy enters the biosphere. Chapter 8 examines glycolysis and cellular respiration, by which organisms extract the energy captured by photosynthesis, conveying it to adenosine triphosphate (ATP), the principal energy-carrier molecule in living cells.

Photosynthesis: An Overview

Photosynthesis uses the energy of sunlight to synthesize energy-rich products—glucose and oxygen—from energy-poor reactants—carbon dioxide and water. Photosynthesis thus converts the electromagnetic energy of sunlight into chemical energy stored in the bonds of glu-

cose and oxygen. The overall chemical reaction for photosynthesis is:

$$6\,CO_2 + 6\,H_2O + \text{light energy} \rightarrow C_6H_{12}O_6 + 6\,O_2$$

Photosynthesis occurs in plants, algae, and certain types of bacteria. In this chapter, we will restrict our discussion of photosynthesis to plants, with an emphasis on land plants. Photosynthesis in plants takes place within the chloroplasts, most of which are contained in leaf cells. Let's begin, then, with a brief look at the structures of leaves and chloroplasts. Chapter 27 will examine leaf structure and function more thoroughly.

Leaves and Chloroplasts Are Adaptations for Photosynthesis

The leaves of most land plants are only a few cells thick (Fig. 7-2a). The upper and lower surfaces of a leaf each consist of a layer of transparent cells, the epidermis. The outer surface of both epidermal layers is covered by a waxy, waterproof covering, the cuticle, that reduces the evaporation of water from the leaf. A leaf obtains CO_2 for photosynthesis from the air; adjustable pores in the epidermis, called **stomata**, open and close at appropriate times to admit CO_2. Inside the leaf are a few layers of cells collectively called **mesophyll** (which means simply "middle of the leaf"). The mesophyll cells contain the vast majority of a leaf's chloroplasts (Fig. 7-2b), and consequently photosynthesis occurs principally in these cells. Vascular bundles, or veins, supply water and minerals to the mesophyll cells and carry the sugars they produce to other parts of the plant.

As we discussed in Chapter 5, chloroplasts are organelles that consist of a double outer membrane enclosing a semifluid medium, the **stroma** (Fig. 7-2c, d). Embedded in the stroma are disc-shaped, interconnected membranous sacs called **thylakoids**. In most chloroplasts, the thylakoids are piled atop one another in stacks called **grana** (singular, **granum**).

The seemingly simple chemical reaction of photosynthesis actually involves dozens of enzymes catalyzing dozens of individual reactions. Conceptually, however, photosynthesis can be thought of as a pair of reactions coupled together by energy-carrier molecules (Fig. 7-2e). Each reaction occurs in a different site in the chloroplast.

1. In the **light-dependent reactions**, chlorophyll and other molecules in the membranes of the thylakoids capture sunlight energy and convert some of it into the chemical energy of energy-carrier molecules (ATP and NADPH).
2. In the **light-independent reactions**, enzymes in the stroma use the chemical energy of the carrier molecules to drive the synthesis of glucose or other organic molecules.

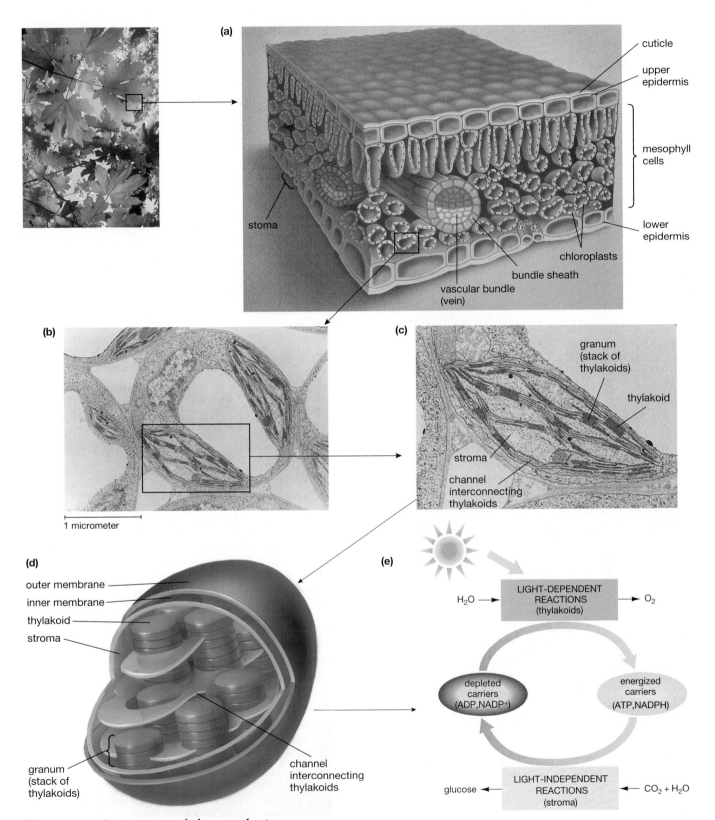

Figure 7-2 **An overview of photosynthesis**

(a) In land plants, photosynthesis usually occurs in the leaves. (b) Cells in the middle of the leaf, called mesophyll cells, contain chloroplasts, the organelles of photosynthesis (c, d). (e) Within the chloroplasts, sunlight energy is absorbed by pigments in the membranes of the thylakoids, and this energy is used to drive the synthesis of the energy carrier ATP and electron carrier NADPH (yellow). These are the light-dependent reactions. In the light-independent reactions, enzymes in the stroma of the chloroplast then use ATP and NADPH to synthesize carbohydrates from CO_2 and H_2O. In the process, ADP and $NADP^+$ (gray) are regenerated and pass back to the light-dependent reactions to be recharged.

The Light-Dependent Reactions: Converting Light to Chemical Energy

The light-dependent reactions convert the energy of sunlight into the chemical energy of two different carrier molecules: the familiar energy carrier ATP and the electron carrier nicotinamide adenine dinucleotide phosphate (NADPH).

During Photosynthesis, Light Is First Captured by Pigments in Chloroplasts

The sun emits energy in a broad spectrum of electromagnetic radiation, from short-wavelength gamma rays, through ultraviolet, visible, and infrared light, to very long-wavelength radio waves (Fig. 7-3a). As you may know, light and the other types of radiation are composed of individual packets of energy called **photons**. The energy of a photon corresponds to its wavelength: Short-wavelength photons are very energetic, whereas longer-wavelength photons have lower energies. Visible light consists of wavelengths with energies that are strong enough to alter the shape of certain pigment molecules but weak enough not to damage crucial molecules such as DNA.

Three processes may occur when light strikes an object such as a leaf: The light may be absorbed, reflected (bounced back again), or transmitted (passed through). Light that is absorbed can heat up the object or drive biological processes, such as photosynthesis. Light that is reflected or transmitted gives an object its color.

Chloroplasts contain several types of molecules that absorb different wavelengths of light. **Chlorophyll**, the key light-capturing molecule in thylakoid membranes, strongly absorbs violet, blue, and red light but reflects green, and therefore appears green (Fig. 7-3b).

Thylakoids also contain other molecules, called **accessory pigments**, that capture light energy and transfer it to chlorophyll. **Carotenoids** absorb blue and green light and appear yellow, orange, or red; **phycocyanins** absorb green and yellow and appear blue or purple. Because all wavelengths of light are absorbed to some degree by either chlorophyll, carotenoids, or phycocyanins, all wavelengths can drive photosynthesis to some extent (Fig. 7-3c).

The Light-Dependent Reactions Occur in Clusters of Molecules Called Photosystems

In the thylakoid membranes, chlorophyll, accessory pigment molecules, and electron-carrier molecules form highly organized assemblies called **photosystems**. Each thylakoid contains thousands of copies of two different kinds of photosystems, named photosystem I and photosystem II (Fig. 7-4a). Each consists of two major parts, a light-harvesting complex and an electron transport system.

The **light-harvesting complex** is composed of about 300 chlorophyll and accessory pigment molecules. These

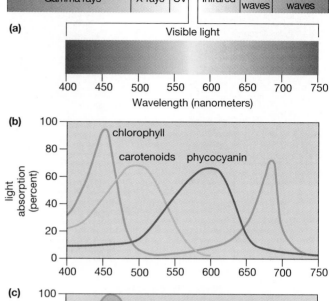

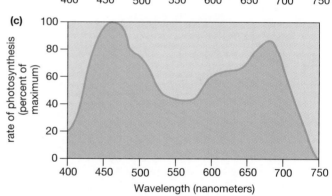

Figure 7-3   **Light, chloroplast pigments, and photosynthesis**

(a) Visible light, a small part of the electromagnetic spectrum, consists of wavelengths that correspond to the colors of the rainbow. **(b)** The first step in photosynthesis is the absorption of light by pigment molecules in the thylakoid membranes of chloroplasts. Different types of pigments selectively absorb certain colors; the height of each line represents the ability of each pigment to absorb light of each color. Chlorophyll (green line) strongly absorbs violet, blue, and red light and therefore looks green. The other pigments in chloroplasts absorb other colors of light. **(c)** Photosynthesis is driven to some extent by all colors of light, owing to the absorption of light by several thylakoid pigments.

molecules absorb light and pass the energy to a specific chlorophyll molecule called the **reaction center**. By analogy with television reception, the light-absorbing pigments are called antenna molecules, because they gather energy and transfer it to the energy-processing reaction center. The reaction center chlorophyll is located next to the **electron transport system**, which is a series of electron carriers embedded in the thylakoid membrane.

When the reaction center chlorophyll receives energy from the antenna molecules, one of its electrons absorbs the energy, leaves the chlorophyll, and jumps over to the electron transport system. This energetic electron moves

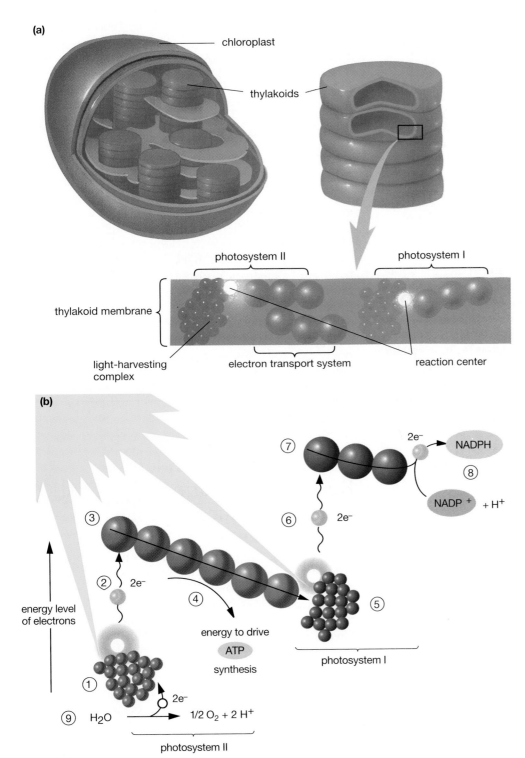

(a)

chloroplast

thylakoids

photosystem II

photosystem I

thylakoid membrane

light-harvesting complex

electron transport system

reaction center

(b)

energy level of electrons

② 2e⁻

③

④

energy to drive

ATP synthesis

⑤

photosystem I

⑦ 2e⁻ → NADPH

⑧

NADP⁺ + H⁺

⑥ 2e⁻

①

⑨ H₂O

2e⁻

1/2 O₂ + 2 H⁺

photosystem II

Figure 7-4 **Thylakoid structure and the light-dependent reactions of photosynthesis**

(a) The thylakoid membranes contain many copies of photosystems I and II. Each photosystem consists of a light-harvesting complex of pigment molecules and an adjacent electron transport system.
(b) A summary of the light-dependent reactions. ① Light is absorbed by the light-harvesting complex of photosystem II (light green), and the energy is passed to the reaction center chlorophyll molecule. ② This energy ejects electrons out of the reaction center. ③ The electrons pass to the adjacent electron transport system. ④ The transport system passes the energetic electrons along, and some of their energy is used to pump hydrogen ions into the thylakoid interior. The hydrogen ion gradient thus generated can drive ATP synthesis (see "A Closer Look : Chemiosmosis" for details). ⑤ Light strikes photosystem I (dark green), ⑥ causing it to emit electrons. ⑦ The electrons are captured by the photosystem I electron transport system. The electrons lost from the reaction center of photosystem I are replaced by those coming from the transport system of photosystem II. ⑧ The energetic electrons from photosystem I are captured in molecules of NADPH. ⑨ The electrons lost from the reaction center of photosystem II are replaced by electrons obtained from splitting water. This reaction also releases oxygen.

from one carrier to the next. At some of the transfers, the electron releases energy that drives reactions resulting in the synthesis of ATP from ADP or NADPH from NADP⁺ (NADP is the electron carrier NAD, described in Chapter 4, plus a phosphate group). This sequence of events is called **photophosphorylation**, because light energy (the "photo" in the name) is used to phosphorylate ("add a phosphate group to") ADP, thus forming ATP. With this overall scheme in mind, let's look a little more closely at the actual sequence of events in the light-dependent reactions (Fig. 7-4b).

Photosystem II Generates ATP

For historical reasons, the photosystems are numbered "backward," and the usual process of light-energy capture is most easily understood by starting with photosystem II. The light-dependent reactions begin when a photon of light is absorbed by an antenna molecule in photosystem II (step 1 in Fig. 7-4b). The photon's energy passes from molecule to molecule until it reaches the reaction center, where it boosts an electron completely out of the chlorophyll molecule (step 2, which shows two electrons being boosted by two photons of light, for reasons explained below). The first electron carrier of the adjacent electron transport system instantly accepts these energized electrons (step 3). The electrons move from carrier to carrier, releasing energy as they go. Some of the energy is used to produce a hydrogen ion gradient within the thylakoid. This gradient drives the synthesis of ATP by a process known as chemiosmosis (step 4). "A Closer Look: Chemiosmosis—The Mechanism of ATP Synthesis in Chloroplasts" describes chemiosmosis in more detail.

Photosystem I Generates NADPH

Meanwhile, light rays have also been striking the light-harvesting complex of photosystem I (step 5), each photon ejecting an electron from its reaction center chlorophyll (step 6, which shows two electrons being boosted by two photons of light). These electrons jump to photosystem I's electron transport system (step 7). Photosystem I's reaction center chlorophyll immediately obtains replacements for its lost electrons from the last electron carrier in photosystem II's electron transport system. Photosystem I's high-energy electrons move through its electron transport system to the electron carrier $NADP^+$. Each $NADP^+$ molecule picks up two energetic electrons and one hydrogen ion, forming NADPH (step 8). $NADP^+$ and NADPH are both water-soluble molecules dissolved in the chloroplast stroma.

Splitting Water Maintains the Flow of Electrons through the Photosystems

Overall, electrons flow from the reaction center of photosystem II, through the photosystem II electron transport system, to the reaction center of photosystem I, through the photosystem I electron transport system, and on to NADPH. To sustain this one-way flow of electrons, photosystem II's reaction center must continuously be supplied with new electrons to replace the ones it gives up. These replacement electrons come from water (step 9). In a series of reactions that scientists are still working to uncover, photosystem II's reaction center chlorophyll attracts electrons from water molecules within the thylakoid compartment, causing the water molecules to split apart:

$$H_2O \rightarrow 1/2\ O_2 + 2\ H^+ + 2\ e^-$$

For every two photons captured by photosystem II, two electrons are boosted out of the reaction center chlorophyll and are replaced by the two electrons obtained by splitting one water molecule. As water molecules are split, their oxygen atoms combine to form molecules of oxygen gas, O_2. The oxygen may be used directly by the plant in its own respiration or it may be given off to the atmosphere.

▓ S U M M A R Y

LIGHT-DEPENDENT REACTIONS

(1) The light-dependent reactions begin with the absorption of light by the light-harvesting complex of photosystem II. (2) The light energy energizes electrons from the reaction center of the complex, causing them to be ejected. (3) The electrons are transferred to photosystem II's electron transport system. As the electrons pass through the transport system, they release energy. (4) Some of the energy is used to create a hydrogen ion gradient that drives ATP synthesis. (5) Meanwhile, light is absorbed by the light-harvesting complex of photosystem I. (6) The light energy ejects electrons from the reaction center that (7) are picked up by photosystem I's electron transport system. Electrons lost from the reaction center are replaced by those coming from the transport system of photosystem II. (8) Some of this energy is captured as NADPH. (9) Finally, the "electron-deprived" chlorophyll of photosystem II attracts electrons from water molecules. A water molecule splits, donating electrons to the photosystem II chlorophyll and generating oxygen as a by-product.

The Light-Independent Reactions: Securing Chemical Energy in Glucose Molecules

The ATP and NADPH synthesized during the light-dependent reactions are dissolved in the stroma. Here, they provide the energy to power the synthesis of glucose from carbon dioxide and water. The reactions that eventually produce glucose are called the light-independent reactions, because they can occur independently of light as long as ATP and NADPH are available.

The C_3 Cycle Captures Carbon Dioxide

Carbon dioxide capture occurs in a set of reactions known as the **Calvin-Benson cycle** (after its discoverers), or the **C_3 (three-carbon) cycle**, because some of the important molecules in the cycle have three carbon atoms in them (Fig. 7-5). The C_3 cycle requires (1) CO_2 (normally from the air); (2) a CO_2-capturing sugar, ribulose bisphosphate (RuBP); (3) enzymes to catalyze all the reactions; and (4) energy in the form of ATP and NADPH (usually from the light-dependent reactions).

The C_3 cycle is most easily understood if we mentally divide it into three parts: carbon fixation, synthesis of phosphoglyceraldehyde (PGAL), and regeneration of ribulose bisphosphate. Keep track of the carbons as you follow along in Figure 7-5.

A CLOSER LOOK

Chemiosmosis—The Mechanism of ATP Synthesis in Chloroplasts

In the electron transport system of photosystem II, energetic electrons move from carrier to carrier. Until about 1960, it was thought that ATP synthesis was directly coupled to these electron transfers: Where the exergonic steps (reactions in which there is a net release of energy) were large enough, the energy given up by the electrons drove ATP synthesis. However, it turns out that ATP synthesis in chloroplasts is *not* a simple coupled reaction. The electron transfers do not directly drive ATP synthesis; rather, the energy released during the transfers is used to generate a concentration gradient of hydrogen ions across the thylakoid membrane. In a completely separate reaction, the energy stored in this gradient then powers ATP synthesis. Let's follow these reactions in some detail.

In the first step of the light-dependent reactions of photosynthesis, a photon strikes the light-harvesting complex of photosystem II, and the energy is absorbed by an antenna pigment molecule. The energy hops around from molecule to molecule within the complex until it reaches the reaction center chlorophyll. Here, an electron absorbs the energy and is ejected completely out of the chlorophyll molecule. Within a billionth of a second, the electron is captured by the first electron carrier of the adjacent electron transport system.

The electron moves from carrier to carrier, losing energy as it goes. The energy released from the exergonic reaction of electron movement is used for the active transport of hydrogen ions across the thylakoid membrane from the stroma into the thylakoid compartment:

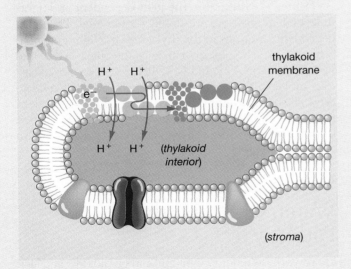

This transport raises the concentration of hydrogen ions (and therefore also the positive charge) inside the thy-

lakoid. The thylakoid membrane does not allow hydrogen ions to leak out, except at specific protein channels that are coupled to ATP-synthesizing enzymes. When hydrogen ions flow through these channels, down their gradients of charge and concentration, the energy released drives the synthesis of ATP:

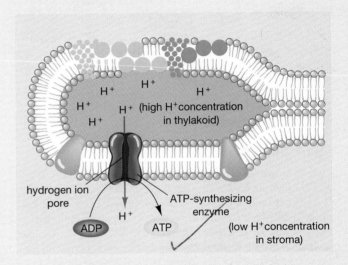

How does a gradient of hydrogen ions generate energy to drive ATP synthesis? Perhaps you can get a feeling for the process if we compare the hydrogen ion gradient across the thylakoid membrane to a dry-cell battery (the kind used in flashlights). In a battery, the negative pole (the flat end of the battery) tends to release electrons, and the positive pole (knobbed end) tends to absorb electrons and transfer them to chemical reactions going on inside the battery. The positive and negative poles, however, are insulated from each other. Electrons can flow from negative to positive poles only if wires connect the two. The flow of electrons can do work, such as lighting a bulb or running a motor. The chloroplast thylakoid operates in a similar way. Hydrogen ions in the thylakoid interior can move down their concentration and charge gradients out into the stroma only through the ATP-synthesizing enzyme channels. This flow of hydrogen ions can do work, namely drive ATP synthesis.

This mechanism of ATP synthesis was first proposed in 1961 by the British biochemist Peter Mitchell, who called it **chemiosmosis**. Chemiosmosis has been shown to be the mechanism of ATP generation in chloroplasts, mitochondria (see Chapter 8), and bacteria. For his brilliant hypothesis, Mitchell was awarded the Nobel Prize in 1978.

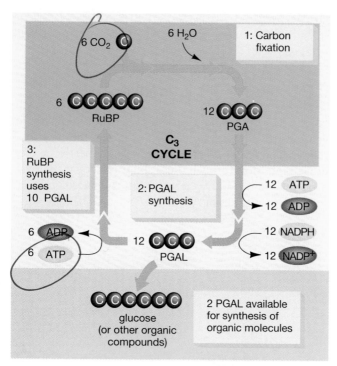

Figure 7-5   **The C₃ cycle of carbon fixation**
Carbon atoms are gray balls; depleted carrier molecules ADP and NADP⁺ are gray ovals; and energized carriers ATP and NADPH are yellow ovals. (1) Six molecules of ribulose bisphosphate (RuBP) react with 6 molecules of CO_2 and 6 molecules of H_2O to form 12 molecules of phosphoglyceric acid (PGA). This reaction is carbon fixation, the capture of carbon from CO_2 into organic molecules. (2) The energy of 12 ATPs and the electrons and hydrogens of 12 NADPHs are used to convert the 12 PGA molecules to 12 phosphoglyceraldehydes (PGALs). (3) Energy from 6 ATPs is used to rearrange 10 PGALs into 6 RuBPs, completing one turn of the C₃ cycle. The remaining 2 PGAL molecules are further processed into glucose or other organic molecules such as glycerol, fatty acids, or the carbon skeleton of amino acids, depending on the needs of the plant.

1. **Carbon fixation.** The C₃ cycle begins (and ends) with a five-carbon sugar, ribulose bisphosphate (RuBP). RuBP combines with CO_2 from the atmosphere to form an extremely unstable six-carbon compound. This compound spontaneously reacts with water to form two three-carbon molecules of phosphoglyceric acid (PGA), which gives the C₃ cycle its name. Capturing CO_2 is called carbon fixation because it "fixes" gaseous CO_2 into a relatively stable organic molecule (step 1 in Fig. 7-5).

2. **Synthesis of phosphoglyceraldehyde.** In a series of enzyme-catalyzed reactions, energy donated by ATP and NADPH (generated during the light-dependent reactions) is used to convert PGA to phosphoglyceraldehyde (PGAL; step 2).

3. **Regeneration of ribulose bisphosphate.** Through a complex series of reactions requiring ATP energy, 10 molecules of PGAL (10×3 carbons) can regenerate the 6 molecules of RuBP (6×5 carbons; step 3) used at the start of carbon fixation.

Carbon Fixed during the C₃ Cycle Is Used to Synthesize Glucose

Obviously, if the C₃ cycle starts with RuBP, adds carbon from CO_2, and ends one "turn" with RuBP again, there is carbon left over from the captured CO_2. Using the simplest "carbon accounting" numbers shown in Figure 7-5, if you start and end one turn of the cycle with six molecules of RuBP, there are two molecules of phosphoglyceraldehyde left over. These two PGAL molecules (three carbons each) are combined to form one molecule of glucose (six carbons). Later, glucose may be broken down during cellular respiration or linked together in chains to form starch (a storage molecule) or cellulose (a major component of cell walls), or modified further into amino acids, lipids, and other cellular constituents.

▦   S U M M A R Y

LIGHT-INDEPENDENT REACTIONS

For the synthesis of one molecule of glucose through one turn of the C₃ cycle: Six molecules of CO_2 are captured by six molecules of RuBP. A series of reactions driven by energy from ATP and NADPH produces 12 molecules of PGAL. Two PGAL molecules join to become one molecule of glucose. Further ATP energy is used to regenerate the 6 RuBP molecules from 10 PGAL molecules.

The Relationship between the Light-Dependent and the Light-Independent Reactions

As Figure 7-2e illustrates, the light-dependent and light-independent reactions are closely coordinated. The light-dependent reactions in the thylakoids use light energy to "charge up" the energy-carrier molecules ADP and NADP⁺ to form ATP and NADPH. These energized carriers move to the stroma, where their energy is used to drive glucose synthesis by means of the light-independent reactions. The depleted carriers, NADP⁺ and ADP, then return to the light-dependent reactions for recharging to ATP and NADPH.

Water, CO_2, and the C₄ Pathway

Photosynthesis requires light and carbon dioxide, and lack of either one reduces the rate of photosynthesis. The light-dependent reactions of photosynthesis cannot occur in the dark, regardless of CO_2 levels; the light-independent reactions of carbon fixation cannot occur without a supply of

CO$_2$, regardless of the light intensity. Therefore, you might think that an ideal leaf should have a large surface area to intercept lots of sunlight and be very porous to allow lots of CO$_2$ to enter the leaf from the air. For land plants, however, being porous to CO$_2$ also allows water vapor to be evaporated out of the leaf. Loss of water from the leaves is a prime cause of stress, and may be fatal, to land plants.

Many plants have evolved leaves that are a compromise between obtaining adequate CO$_2$ supplies and reducing water loss: large, waterproof leaves, with adjustable pores, the stomata, that admit carbon dioxide (see Fig. 7-2; see Chapter 26 for more details of leaf structure). When water supplies are adequate, the stomata open, letting in CO$_2$. If the plant is in danger of drying out, the stomata close. Closing the stomata reduces evaporation but at the cost of reducing CO$_2$ intake as well. Further, it also restricts the release of O$_2$, which is produced during photosynthesis.

When Stomata Are Closed to Conserve Water, Wasteful Photorespiration Occurs

What happens to carbon fixation when the stomata close? As it happens, the enzyme that catalyzes the reaction of RuBP with CO$_2$ is not very selective: It can cause *either* CO$_2$ *or* O$_2$ to combine with RuBP (Fig. 7-6). The reaction of O$_2$ with RuBP is the first step in a series of reactions, collectively called **photorespiration**, that generate CO$_2$, using carbon from RuBP. In photorespiration, oxygen instead of carbon dioxide becomes attached to RuBP. Thus photorespiration prevents carbon fixation by combining O$_2$ instead of CO$_2$ with RuBP. (Photorespiration, by the way, received its name because it uses up O$_2$ and produces CO$_2$. Unlike cellular respiration, it does not produce any useful cellular energy, such as ATP.)

Whether this matters very much depends on the relative amounts of carbon fixation versus photorespiration. The atmosphere contains much more O$_2$ (21%) than CO$_2$ (0.03%). In addition, the light reactions of photosynthesis release O$_2$ within the leaf. Therefore, some photorespiration occurs all the time, even under the best of conditions. During hot or dry weather, moreover, the stomata seldom open. The CO$_2$ that was initially present within the leaf is rapidly depleted by photosynthesis, new CO$_2$ from the air can't get in, and the O$_2$ generated by photosynthesis can't get out. With CO$_2$ levels low and O$_2$ levels very high, photorespiration dominates (Fig. 7-7a). Plants, especially seedlings, may die during hot, dry weather because respiration (both cellular respiration and photorespiration) exceeds photosynthesis, and the plants consume all their carbohydrate supplies by respiration.

C$_4$ Plants Reduce Photorespiration Using a Two-Stage Carbon Fixation Process

Some plants, described as C$_4$ plants, have evolved a way to reduce photorespiration and boost photosynthesis during dry weather. In the leaves of "regular" (C$_3$) plants, almost all the chloroplasts reside in the mesophyll cells, whereas the **bundle-sheath cells**, which surround the veins, lack chloroplasts. In plants adapted to hot, dry conditions, such as corn and crabgrass, both the mesophyll cells and the bundle-sheath cells contain chloroplasts (Fig. 7-7b). These plants use a two-stage carbon fixation pathway, the **C$_4$ pathway**.

The First Stage Forms a Four-Carbon Molecule That Gives C$_4$ Plants Their Name

In C$_4$ plants, the mesophyll cells contain a three-carbon molecule called phosphoenolpyruvate (PEP), instead of RuBP. CO$_2$ reacts with PEP to form a four-carbon molecule of oxaloacetic acid (hence the name, C$_4$ plants). This reaction is highly specific for CO$_2$ and is not hindered by high O$_2$ concentrations. This initial capture of carbon can continue for quite a while even when the stomata are closed and CO$_2$ concentrations within the leaf become very low. Oxaloacetic acid is used as a shuttle to transport carbon from the mesophyll to the bundle-sheath cells.

The Second Stage Shuttles Carbon into the C$_3$ Cycle

Oxaloacetic acid is transported into the bundle-sheath cells, where it breaks down, releasing CO$_2$ again. The high CO$_2$ concentration created in the bundle-sheath cells allows the RuBP + CO$_2$ reaction to fix carbon in the regular C$_3$ cycle. The remnant of the shuttle molecule returns to the mesophyll cells, where ATP energy is used to regenerate the PEP capture molecule.

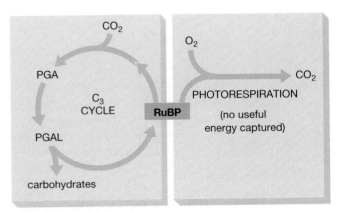

Figure 7-6 **Photorespiration**

Carbon fixation through the C$_3$ cycle and photorespiration are alternative pathways for RuBP. Some photorespiration always occurs. When the stomata of a leaf close during hot, dry weather, CO$_2$ levels drop while O$_2$ levels remain high. Under these conditions, photorespiration predominates and the plant loses carbon.

C_3 and C_4 Plants Are Adapted to Specific Environmental Conditions

The C_4 pathway is advantageous when lots of light and little water are available. The C_4 pathway is not *always* better, though, because it uses more energy as it regenerates PEP. In some cases, the straightforward C_3 carbon fixation pathway is more efficient. For example, if water is plentiful the stomata of C_3 plants can stay open and let in lots of CO_2. Or, if light levels are low, photosynthesis proceeds slowly and C_4 plants may not be able to generate the extra energy they require.

A plant with the structures and enzymes of the C_4 pathway always uses the C_4 method of carbon fixation, regardless of environmental conditions. C_4 plants thrive, therefore, in deserts and in midsummer in temperate climates, when light energy is plentiful but water is scarce. C_3 plants have the advantage, however, in cool, wet, cloudy climates, where their exclusive use of the C_3 pathway is more energy efficient. This is why your lawn of lush Kentucky bluegrass (a C_3 plant) may be taken over by spiky crabgrass (a C_4 plant) during a long, hot summer.

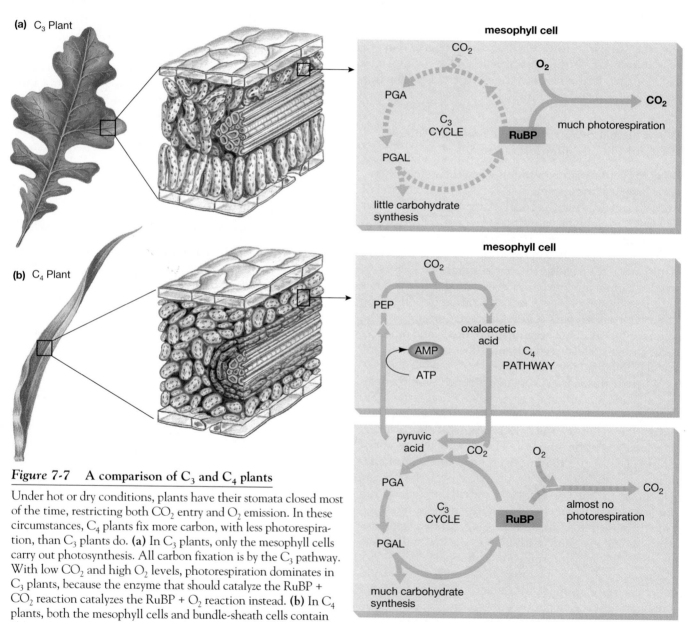

(a) C_3 Plant

(b) C_4 Plant

Figure 7-7   **A comparison of C_3 and C_4 plants**

Under hot or dry conditions, plants have their stomata closed most of the time, restricting both CO_2 entry and O_2 emission. In these circumstances, C_4 plants fix more carbon, with less photorespiration, than C_3 plants do. **(a)** In C_3 plants, only the mesophyll cells carry out photosynthesis. All carbon fixation is by the C_3 pathway. With low CO_2 and high O_2 levels, photorespiration dominates in C_3 plants, because the enzyme that should catalyze the RuBP + CO_2 reaction catalyzes the RuBP + O_2 reaction instead. **(b)** In C_4 plants, both the mesophyll cells and bundle-sheath cells contain chloroplasts and participate in photosynthesis. The initial carbon fixation step in the mesophyll cells is a reaction between phosphoenolpyruvic acid (PEP) and CO_2, with which O_2 does not compete. A four-carbon molecule of oxaloacetic acid is produced, giving the C_4 pathway its name. The oxaloacetic acid then releases CO_2 in the bundle-sheath cells, thus maintaining a high CO_2 concentration in their chloroplasts. Higher CO_2 levels allow efficient carbon fixation in the C_3 pathway of the bundle-sheath cells with little photorespiration. Notice that the regeneration of PEP requires energy: Two phosphates are removed from ATP to produce AMP (adenosine monophosphate).

❖ SUMMARY OF KEY CONCEPTS

Photosynthesis: An Overview

Photosynthesis uses the energy of sunlight to convert low-energy inorganic molecules of carbon dioxide and water into high-energy organic molecules such as glucose. In plants, photosynthesis takes place in the chloroplasts, in two major steps: the light-dependent and the light-independent reactions. Photosynthesis is summarized in Figure 7-8.

The Light-Dependent Reactions: Converting Light to Chemical Energy

In the light-dependent reactions, light excites electrons in chlorophyll molecules and transfers the energetic electrons to electron transport systems. The energy of these electrons drives three processes:

1. Some of the energy from the electrons is used to pump hydrogen ions into the thylakoid sacs. The hydrogen ion concentration is therefore higher inside the thylakoids than in the stroma outside. Hydrogen ions diffuse down this concentration gradient through ATP-synthesizing enzymes in the thylakoid membranes, providing the energy to drive ATP synthesis.
2. Some of the energy, in the form of energetic electrons, is added to electron-carrier molecules of $NADP^+$ to make the highly energetic carrier NADPH.
3. Some of the energy is used to split water, generating electrons, hydrogen ions, and oxygen.

The Light-Independent Reactions: Securing Chemical Energy in Glucose Molecules

In the stroma of the chloroplasts, ATP and NADPH provide the energy that drives the synthesis of glucose from CO_2 and H_2O. The light-independent reactions occur in a cycle of chemical reactions called the Calvin-Benson, or C_3, cycle. The C_3 cycle has three major parts:

1. In the carbon fixation step, CO_2 and water combine with ribulose bisphosphate (RuBP) to form phosphoglyceric acid (PGA).
2. PGA is converted to phosphoglyceraldehyde (PGAL), using energy from ATP and NADPH. PGAL may be used to synthesize organic molecules such as glucose.
3. 10 molecules of PGAL are used to regenerate 6 molecules of RuBP, again using ATP energy.

The Relationship between the Light-Dependent and the Light-Independent Reactions

The light-dependent reactions produce the energy carrier ATP and the electron carrier NADPH. The energy from these carriers is used in the synthesis of organic molecules during the light-independent reactions. The depleted carriers, ADP and $NADP^+$, return to the light-dependent reactions for recharging.

Water, CO_2, and the C_4 Pathway

The enzyme that catalyzes the reaction between RuBP and CO_2 is not very selective. It may also catalyze a reaction, called photorespiration, between RuBP and O_2. Photorespiration prevents carbon fixation and does not generate ATP. If CO_2 concentrations drop too low, or O_2 concentrations rise too high, photorespiration may exceed carbon fixation. Some plants have evolved an additional step for carbon fixation that minimizes photorespiration. In the mesophyll cells of these C_4 plants, CO_2 combines with phosphoenolpyruvic acid (PEP) to form the four-carbon molecule oxaloacetic acid. Oxaloacetic acid is transported into adjacent bundle-sheath cells, where it releases CO_2, thereby maintaining a high CO_2 concentration in the bundle-sheath cells. This CO_2 is then fixed in the C_3 cycle.

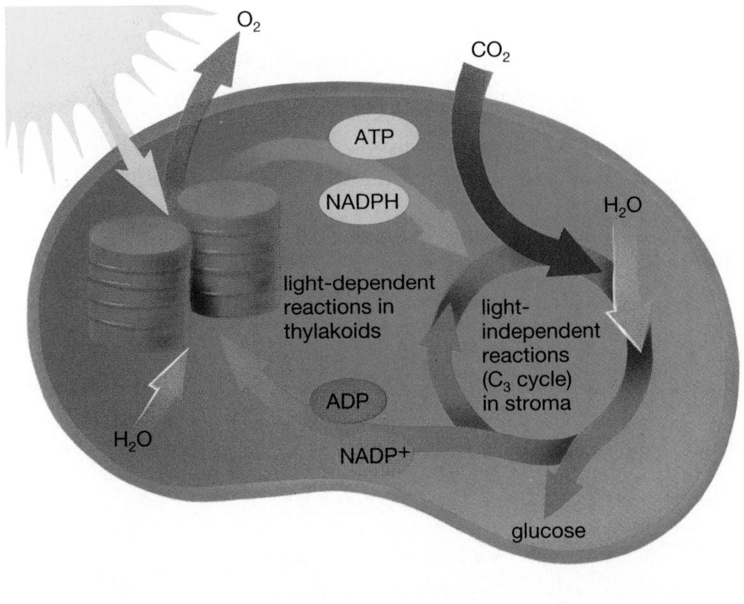

Figure 7-8   **A summary diagram for photosynthesis**

The light-dependent reactions in the thylakoids convert the energy of sunlight into the chemical energy of ATP and NADPH. Part of the sunlight energy is also used to split H_2O, forming O_2. In the stroma, the light-independent reactions (C_3 cycle) use the energy of ATP and NADPH to convert CO_2 and H_2O to glucose. The depleted carriers, ADP and $NADP^+$, return to the thylakoids to be recharged by the light-dependent reactions.

✖ K E Y   T E R M S

accessory pigments   p. 130
bundle-sheath cell   p. 135
Calvin-Benson cycle   p. 132
carbon fixation   p. 134
carotenoid   p. 130
C_3 cycle   p. 132
chemiosmosis   p. 133
chlorophyll   p. 130
C_4 pathway   p. 135

electron transport system   p. 130
granum   p. 129
light-dependent reactions   p. 129
light-harvesting complex   p. 130
light-independent reactions   p. 129
mesophyll   p. 129
photon   p. 130
photophosphorylation   p. 131
photorespiration   p. 135

photosynthesis   p. 129
photosystem   p. 130
phycocyanin   p. 130
reaction center   p. 130
stoma   p. 129
stroma   p. 129
thylakoid   p. 129

✖ T H I N K I N G   T H R O U G H   T H E   C O N C E P T S

Multiple Choice

1. Photosynthesis is measured in the leaf of a green plant exposed to different wavelengths of light. Photosynthesis is
   a. highest in green light   b. highest in red light
   c. highest in blue light   d. highest in red and blue light
   e. the same at all wavelengths

2. Where do the light-dependent reactions of photosynthesis occur?
   a. in the guard cells of the stomata
   b. in the chloroplast stroma
   c. within the thylakoid membranes of the chloroplast
   d. in the leaf cell cytoplasm
   e. in leaf cell mitochondria

3. The oxygen produced during photosynthesis comes from
   a. the breakdown of CO_2
   b. the breakdown of H_2O
   c. the breakdown of both CO_2 and H_2O
   d. the breakdown of oxaloacetic acid
   e. photorespiration

4. The role of accessory pigments is to
   a. provide an additional photosystem to generate more ATP
   b. allow photosynthesis to occur in the dark
   c. prevent photophosphorylation
   d. donate electrons to chlorophyll reaction centers
   e. capture additional light energy and transfer it to the chlorophyll reaction centers

5. The generation of ATP by electron transport in photosynthesis and cellular respiration depends upon
   a. a proton gradient across a membrane
   b. proton pumps driven by electron transport chains
   c. an ATP-synthesizing enzyme complex
   d. a, b, and c are all required for ATP generation
   e. none of the above are required for ATP generation

6. Where do the light-independent, carbon-fixing reactions occur?
   a. in the guard cell cytoplasm
   b. in the chloroplast stroma
   c. within the thylakoid membranes
   d. at night in the thylakoids
   e. in mitochondria

Review Questions

1. Write the overall equation for photosynthesis. Does it differ between C_3 and C_4 plants?
2. Draw a labeled diagram of a chloroplast. Explain specifically how chloroplast structure is related to its function.
3. Briefly describe the light-dependent and light-independent reactions. In what part of the chloroplast does each occur?
4. What is the difference between carbon fixation in C_3 and C_4 plants? Under what conditions does each mechanism of carbon fixation work most effectively?
5. Describe the process of chemiosmosis in chloroplasts, tracing the flow of energy from sunlight to ATP.

✖ A P P L Y I N G   T H E   C O N C E P T S

1. Many lawns and golf courses are planted with bluegrass, a C_3 plant. In the spring, the bluegrass grows luxuriously. In the summer, a weed called crabgrass, a C_4 plant, often appears and spreads rapidly. Explain this sequence of events, given the normal weather conditions of spring and summer and the characteristics of C_3 versus C_4 plants.

2. Suppose an experiment is performed in which Plant I is supplied with normal carbon dioxide but with water containing radioactive oxygen atoms. Plant II is supplied with normal water but with carbon dioxide containing ra-

dioactive oxygen atoms. Each plant is allowed to perform photosynthesis, and the oxygen gas and sugars produced are tested for radioactivity. Which plant would you expect to produce radioactive sugars, and which plant would you expect to produce radioactive oxygen gas? Why?

3. You continuously monitor the photosynthetic oxygen production from the leaf of a higher plant illuminated by white light. Explain what will happen (and why) if you place (a) red, (b) blue, and (c) green filters between the light source and the leaf.

4. A plant is placed in a CO_2-free atmosphere in bright light. Will the light-dependent reactions continue to generate ATP and NADPH indefinitely? Explain how you reached your conclusion.
5. You are called before the House Ways and Means Committee to explain why the U.S. Department of Agriculture should continue to fund photosynthesis research. How would you justify an expensive project to genetically engineer the enzyme that catalyzes the reaction of RuBP with CO_2 to prevent it reacting with oxygen as well as CO_2. What are the potential applied benefits of this research?

❋ F O R   M O R E   I N F O R M A T I O N

Bazzazz, F. A., and Fajer, E. D. "Plant Life in a CO_2-Rich World." *Scientific American*, January 1992. Burning fossil fuels is increasing CO_2 levels in the atmosphere (see Chapter 45). This increase could tip the balance between C_3 and C_4 plants.

Govindjee, and Coleman, W. J. "How Plants Make Oxygen." *Scientific American*, February 1990. The generation of oxygen during photosynthesis is just beginning to be understood.

Grodzinski, B. "Plant Nutrition and Growth Regulation by CO_2 Enrichment." *BioScience*, 1992. How higher CO_2 levels influence plant metabolism.

Hall D. O., and Rao, K. K. *Photosynthesis*. 5th ed. New Series in Biology. New York: Cambridge University Press, 1994. An excellent short book recommended to any student interested in finding out more about photosynthesis.

Hinkle, P. C., and McCarthy, R. E. "How Cells Make ATP." *Scientific American*, March 1978. A good explanation of chemiosmosis, which is a difficult concept for many students.

Mooney, H. A., Drake, B. G., Luxmoore, R. J., Oechel, W. C., and Pitelka, L. F. "Predicting Ecosystems' Responses to Elevated CO_2 Concentrations." *BioScience*, 1994. What effects will CO_2 enrichment of the atmosphere due to human activities have on ecosystems?

Youvan, D. C., and Marrs, B. L. "Molecular Mechanisms of Photosynthesis." *Scientific American*, June 1987. How photosynthesis works at the molecular level.

N E T   W A T C H

On-line resources for this chapter are on the World Wide Web at:
http://www.prenhall.com/~audesirk (click on the table of contents link and then select Chapter 7).

8 Harvesting Energy from Food: Glycolysis and Cellular Respiration

■■ A T A G L A N C E

Glucose Metabolism: An Overview

Glycolysis
  Glucose Must Be Activated by ATP before Its Energy Can
    Be Harvested

Fermentation
  Some Cells Ferment Pyruvic Acid to Lactic Acid
  Other Cells Ferment Pyruvic Acid to Alcohol

Cellular Respiration
  Pyruvic Acid Is Transported into the Mitochondrial Matrix
  Pyruvic Acid Is Broken Down by Reactions in the
    Mitochondrial Matrix
  Energetic Electrons Are Carried to Electron Transport
    Systems in the Inner Mitochondrial Membrane
  Chemiosmosis Captures Energy Stored in a Hydrogen Ion
    Gradient and Produces ATP
  ATP Is Transported from the Mitochondrial Matrix and
    Diffuses into the Cell Cytosol
  Cellular Metabolism Influences the Way Entire Organisms
    Function

An Anna's hummingbird feeds at a flower in California.

When a hummingbird sips nectar from a flower, it eats the products of photosynthesis. Photosynthesis in the plant's leaves had converted the energy of sunlight into the chemical energy of organic molecules such as the sugar in nectar. To supply its enormous energy needs—the highest of any animal—the hummingbird must eat its weight in nectar daily. Then its cells must efficiently extract energy from the glucose in the nectar. When the hummingbird's cells break down the high-energy glucose molecules in the presence of oxygen, they produce the low-energy molecules the plant started with—carbon dioxide and water—making energy available for muscles and growth (Fig. 7-1, lower right).

However, the hummingbird's cells cannot directly use the chemical energy derived from this process. Its wing muscles require energy stored in the molecule adenosine triphosphate (ATP). Its brain uses ATP in conducting nerve signals, and its ovaries use ATP in making eggs. The plant, for its part, uses ATP to make the petals, pigments, and fragrance that attract the bird, and the leaves and chlorophyll molecules that trap the sun's energy. In this chapter, we explore the metabolic processes by which organisms convert the energy of organic molecules into the usable energy of ATP, processes called glycolysis and respiration.

Most cells can metabolize a variety of organic molecules to produce ATP. We focus on the metabolism of the single molecule glucose for three reasons. First, virtually all single-celled organisms, and every cell of multicellular organisms, metabolize glucose for energy at least part of the time. Some, such as nerve cells in your brain, rely predominantly on glucose as a source of energy. Second, glucose metabolism is less complex than the metabolism of most other organic molecules. Finally, when cells use other organic molecules as energy sources, they usually convert the molecules to glucose or other compounds that enter the pathways of glucose metabolism (see "Health Watch: Metabolic Transformations—Why You Can Get Fat by Eating Sugar").

Glucose Metabolism: An Overview

During photosynthesis, photosynthetic organisms harvest and store the energy of sunlight in glucose. During glucose breakdown, that energy is released and converted to

ATP. The chemical equations for glucose formation by photosynthesis and the complete metabolism of glucose back to CO_2 and H_2O (the original reactants in photosynthesis) are almost perfectly symmetrical:

Photosynthesis

$$6 \; CO_2 + 6 \; H_2O + \text{sunlight energy} \rightarrow C_6H_{12}O_6 + 6 \; O_2$$

Complete Glucose Metabolism

$$C_6H_{12}O_6 + 6 \; O_2 \rightarrow 6 \; CO_2 + 6 \; H_2O + \text{energy}$$

This symmetry might lead you to think that a cell can convert all of the chemical energy contained in a glucose molecule to high-energy bonds of ATP. Unfortunately, according to the second law of thermodynamics, "you can't break even"; in other words, the conversion of energy into different forms always results in the decrease of the amount of concentrated, useful energy. In

fact, most of the "energy" listed on the right side of the second equation is heat, not ATP. Nevertheless, a cell can extract a great deal of ATP energy from glucose if the glucose molecule is completely broken down to CO_2 and H_2O.

Figure 8-1 summarizes the major steps of glucose metabolism in eukaryotic cells. The first stage, glycolysis, does not require oxygen and proceeds in exactly the same way under both aerobic (with oxygen) and anaerobic (without oxygen) conditions. Glycolysis splits apart a single glucose molecule (a six-carbon sugar) into two three-carbon molecules of pyruvic acid. This splitting releases a small fraction of the chemical energy stored in the glucose, some of which is used to generate two ATP molecules. The presence of oxygen becomes an issue only in the processes that follow glycolysis. Under anaerobic conditions, the pyruvic acid is usually converted by fermentation into lactic acid or ethanol. Fermentation does not produce any more ATP energy. Both glycolysis and fermentation occur in the cell cytosol.

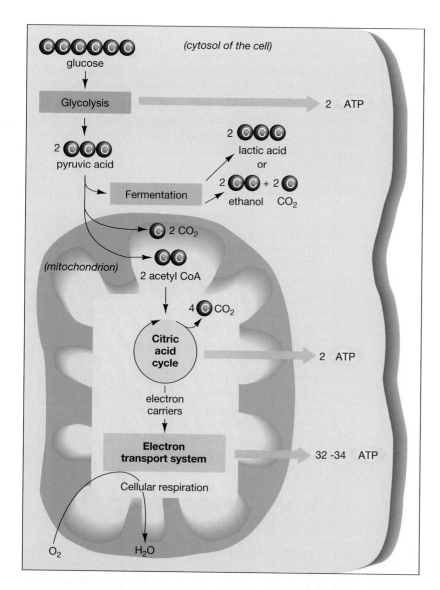

Figure 8-1   A summary of glucose metabolism

Refer to this diagram as we progress through the reactions of glycolysis (in the cell cytosol) and cellular respiration (in the mitochondria). Note that the breakdown of glucose occurs in stages, with various amounts of energy harvested as ATP along the way. The vast majority of the ATP is produced in the mitochondria, justifying their nickname, "powerhouse of the cell."

The pyruvic acid produced by glycolysis may also enter the mitochondria. Here, if oxygen is available, cellular respiration uses oxygen to break pyruvic acid down completely to carbon dioxide and water, generating an additional 34 or 36 ATP molecules (the amount differs among cells). The extra ATP produced by cellular respiration is so important to most organisms that anything that interferes with its production, such as lack of oxygen or the poison cyanide, quickly results in death.

Glycolysis

Glycolysis (in Greek, "to break apart a sweet") is a complex sequence of reactions in the cytosol of a cell in which a molecule of glucose is broken down into two molecules of pyruvic acid. This breakdown results in a small energy gain of two molecules of ATP and two molecules of the electron carrier NADH. Reduced to its essentials, glycolysis consists of two major steps: glucose activation and energy harvest (Fig. 8-2).

Glucose Must Be Activated by ATP before Its Energy Can Be Harvested

In glucose activation, a molecule of glucose undergoes two enzyme-catalyzed reactions, each of which uses ATP energy. These reactions convert a relatively stable glucose molecule into a highly reactive molecule of fructose diphosphate. Fructose is a sugar molecule similar (but not identical) to glucose; diphosphate refers to the two phosphate groups acquired from the ATP molecules. Forming fructose diphosphate costs the cell two ATP molecules. But this initial consumption of energy is necessary to ultimately produce much greater energy returns.

In the energy harvest steps, fructose diphosphate splits apart into two three-carbon molecules of phosphoglyceraldehyde (PGAL; in Chapter 7 we encountered PGAL in the C_3 cycle of photosynthesis). The two PGAL molecules then go through a series of reactions that culminate in the production of two molecules of pyruvic acid, one from each PGAL. Two of these reactions are coupled to ATP synthesis, generating two ATP molecules per PGAL, for a total of four ATPs. Because two ATPs were used to activate the glucose molecule in the first place, *there is a net gain of only two ATPs per glucose molecule.* At another step along the way from PGAL to pyruvic acid, two high-energy electrons and a hydrogen ion are added to the "empty" electron carrier NAD^+ to make the "energized" carrier NADH (note that these molecules are slightly different from $NADP^+$ and NADPH encountered in the light-dependent reactions of photosynthesis). Two PGAL molecules are produced per glucose molecule, so two NADH carrier molecules are formed when PGAL is converted to pyruvic acid. For a complete description of glycolysis, see "A Closer Look: Glycolysis."

▓ S U M M A R Y

GLYCOLYSIS

Each molecule of glucose is broken down to two molecules of pyruvic acid. During these reactions, two ATP molecules and two NADH electron carriers are formed.

Carrier molecules such as NAD^+ capture energy by accepting high-energy electrons, and they can carry these electrons to sites where their energy is used to form ATP.

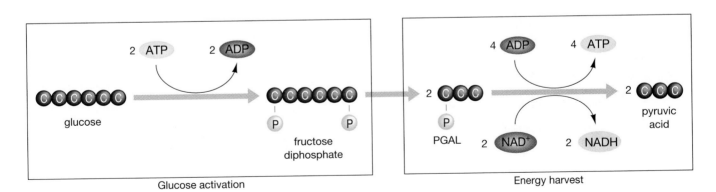

Figure 8-2   **The essentials of glycolysis**

(1) Glucose activation: The energy of two ATP molecules is used to convert glucose to the highly reactive fructose diphosphate. Fructose diphosphate splits into two smaller, but still reactive, molecules of phosphoglyceraldehyde (PGAL). (2) Energy harvest: Both PGAL molecules go through a series of reactions that generate four ATP and two NADH molecules. Therefore, glycolysis results in a net harvest of two ATP and two NADH molecules per glucose molecule.

A   C L O S E R   L O O K

Glycolysis

Glycolysis is a series of enzyme-catalyzed reactions that break down a single molecule of glucose into two molecules of pyruvic acid. To help you follow the reactions, we show only the "carbon skeletons" of glucose and the molecules produced during glycolysis as gray balls. Depleted carrier molecules of ADP and NAD⁺ are gray ovals; energized carriers ATP and NADH are yellow ovals; and phosphate groups attached to the carbons are yellow circles. Each colored arrow represents a reaction catalyzed by at least one enzyme. (A molecular model of hexokinase, the first enzyme in glycolysis, is shown in Fig. 4-7.)

1. A glucose molecule is energized by the addition of a high-energy phosphate from ATP.

2. The molecule is slightly rearranged,

3. then a second phosphate is added from another ATP.

4. The resulting molecule of fructose-1,6-diphosphate is split into two three-carbon molecules, one DHAP (dihydroxacetone phosphate) and one PGAL. Each has one phosphate attached.

5. DHAP rearranges into PGAL, so from now on there are two molecules of PGAL going through the identical reactions.

6. Each PGAL undergoes two almost simultaneous reactions. Two electrons and a hydrogen ion are donated to NAD⁺ to make the energized carrier NADH, and an inorganic phosphate (P) is attached to the carbon skeleton with a high-energy bond. The resulting molecules of 1,3-diphosphoglyceric acid have two high-energy phosphates.

7. One phosphate from each diphosphoglyceric acid is transferred to ADP to form ATP, for a net of two ATPs. This transfer compensates for the initial two ATPs used in glucose activation.

8. After another rearrangement, the second phosphate from each phosphoenolpyruvic acid is transferred to ADP to form ATP, leaving pyruvic acid as the final product of glycolysis. There is a net profit of two ATPs from each glucose molecule.

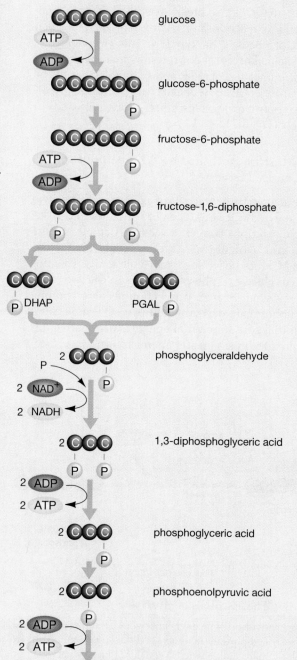

One major difference between anaerobic and aerobic glucose breakdown is the use that is made of these high-energy electrons. In the absence of oxygen, pyruvic acid acts as the electron acceptor during fermentation, producing molecules (ethanol or lactic acid) that the cell cannot use. During cellular respiration, oxygen becomes the electron acceptor, allowing the pyruvic acid to be fully broken down and its energy harvested as ATP.

Fermentation

The electrons carried in NADH are highly energetic, but their energy can be used to synthesize ATP only when oxygen is available (see "Cellular Respiration," below). Many organisms (particularly microorganisms) thrive in the guts of animals, deep in soil, in sediments underlying lakes and oceans, or in bogs and marshes where oxygen is nearly or entirely absent. As described below, even some of our own body cells must cope without oxygen for brief periods. In anaerobic conditions (under which life, and probably glycolysis, evolved), NADH production is *not* a method of energy capture; it is actually a way of getting rid of hydrogen ions and electrons produced during the metabolism of glucose to pyruvic acid. But this disposal method poses a problem for the cell because NAD⁺ is used up as it accepts electrons and hydrogen ions in becoming NADH. Without a way to regenerate NAD⁺ and to dispose of the electrons and hydrogen ions, glycolysis would have to stop once the supply of NAD⁺ was exhausted.

Fermentation solves this problem by enabling pyruvic acid to act as the final acceptor of electrons and hydrogen ions from NADH. NAD⁺ is thereby regenerated for use in further glycolysis. There are two main types of fermentation; one converts pyruvic acid to lactic acid, and the other converts it to carbon dioxide and ethanol.

Some Cells Ferment Pyruvic Acid to Lactic Acid

When you exercise vigorously, your muscles need lots of ATP as an energy source (Fig. 8-3a). Even though cellular respiration generates much more ATP than glycolysis alone, *cellular respiration is limited by organismal respiration*, that is, by the ability of the organism to provide oxygen (by breathing, for example). While you exercise vigorously, such as by sprinting to class after oversleeping, you may not be able to get enough air into your lungs and enough oxygen out of this air, into your blood, and delivered to your muscles to keep cellular respiration going fast enough. When deprived of oxygen, your muscles do not immediately stop working. Instead, glycolysis continues for a while, providing its meager two ATP molecules per glucose and generating both pyruvic acid and NADH.

To regenerate NAD⁺, muscle cells convert pyruvic acid molecules to lactic acid, using electrons and hydrogen ions from NADH:

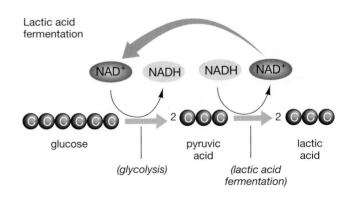

Figure 8-3   Fermentation

(a) A sprinter at the end of a race. The runner's respiratory and circulatory systems cannot supply oxygen to her leg muscles fast enough to keep up with the demand for energy, so glycolysis and lactic acid fermentation must provide the energy. Panting after the race brings in the oxygen needed to remove the lactic acid through cellular respiration. (b) Champagne spurts out of a newly opened bottle, driven out by CO₂ formed by fermenting yeast and trapped in the bottle by the cork.

Lactic acid
fermentation

NAD⁺   NADH   NADH   NAD⁺

glucose   →   2   pyruvic   →   2   lactic
                       acid   acid

(glycolysis)   (lactic acid
                    fermentation)

The regenerated NAD⁺ molecules produced by this conversion again are available to accept electrons during glycolysis, and energy production can continue. Lactic acid, however, is toxic in high concentrations. Soon, it causes intense discomfort and fatigue, causing you to stop or at least slow down. As you rest in the classroom, breathing rapidly after your sprint, oxygen once more becomes available and the lactic acid is reconverted to pyruvic acid. Interestingly enough, the conversion from lactic acid to pyruvic acid occurs not in the muscle cells, which lack the necessary enzymes, but in the liver. This pyruvic acid is then broken down through cellular respiration to carbon dioxide and water.

Various microorganisms also use lactic acid fermentation (some ferment even when oxygen is present, others are poisoned by oxygen), including the bacteria that produce yogurt, sour cream, and cheese. As you may know, acids taste sour, and lactic acid contributes to the distinctive tastes of these foods.

Other Cells Ferment Pyruvic Acid to Alcohol

Many microorganisms use another process to regenerate NAD^+ under anaerobic conditions: alcoholic fermentation. This series of reactions produces ethanol and CO_2 (rather than lactic acid) from pyruvic acid, using hydrogen ions and electrons from NADH:

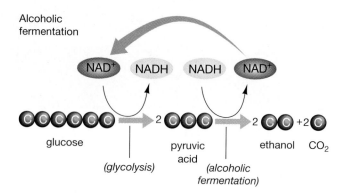

Alcoholic fermentation is somewhat risky for a microbe because the alcohol it generates is poisonous. If the alcohol concentration in its environment becomes too great, the microbe will die. Most yeasts (single-celled fungi), for example, die if the alcohol concentration exceeds 12%, although certain strains developed for wine production do not succumb until the alcohol concentration reaches 18%. (Under natural conditions, of course, yeasts are unlikely to find themselves confined, as they are in a wine bottle, in a small volume of water, with lots of sugar for energy production and no oxygen—selective pressure for alcohol tolerance was probably very slight before humans developed a taste for wine!)

Sparkling wines, such as champagne, are bottled while the yeasts are still alive and fermenting, trapping both the alcohol and the CO_2. When the cork is removed, the pressurized CO_2 is released, sometimes rather explosively (Fig. 8-3b). We add baker's yeast to bread dough so that the CO_2 it produces will make the bread rise; the alcohol the yeast generates evaporates while the bread is baking.

Cellular Respiration

Cellular respiration is a series of reactions in which the pyruvic acid produced by glycolysis is broken down to carbon dioxide and water and large amounts of ATP are produced. The final reactions of cellular respiration require oxygen because oxygen acts as the final acceptor of electrons.

In eukaryotic cells, cellular respiration occurs in the mitochondria. You may recall from Chapter 5 that a mitochondrion has two membranes that produce two compartments: an inner compartment enclosed by the inner membrane and containing the fluid **matrix**, and an **intermembrane compartment** between the two membranes (Fig. 8-4). Reactions involving enzymes in the matrix, electron transport proteins in the inner membrane, and the movement of hydrogen ions through ATP-synthesizing proteins in the inner membrane yield most of the ATP produced during cellular respiration.

Figure 8-4 summarizes the main events of cellular respiration.

Step 1: Glycolysis in the cell cytosol produces two molecules of pyruvic acid per molecule of glucose.

Step 2: The pyruvic acids are transported across both mitochondrial membranes into the matrix.

Step 3: Each pyruvic acid is split into CO_2 and a two-carbon acetyl group, which enters the citric acid cycle (also called the Krebs cycle). The citric acid cycle releases the remaining carbons as CO_2, produces one ATP, and donates energetic electrons to several electron-carrier molecules.

Step 4: The electron carriers donate their energetic electrons to the electron transport system of the inner membrane. The energy of the electrons transports H^+ from the matrix to the intermembrane compartment. At the end of the system, the electrons combine with O_2 and H^+ to form H_2O.

Step 5: In chemiosmosis, the hydrogen ion gradient created by the electron transport system discharges through ATP-synthesizing enzymes in the inner membrane, and the energy is used to produce large amounts of ATP.

Step 6: The ATP is transported back out of the mitochondrion into the cytosol, available to drive needed reactions in the rest of the cell.

We have already discussed glycolysis; now let's look a little more closely at the processes of cellular respiration in the mitochondria.

Pyruvic Acid Is Transported into the Mitochondrial Matrix

The outer membrane of mitochondria contains many large pores that are very permeable to pyruvic acid. Facilitated diffusion through proteins in the inner membrane allows pyruvic acid to pass down its concentration gradient from

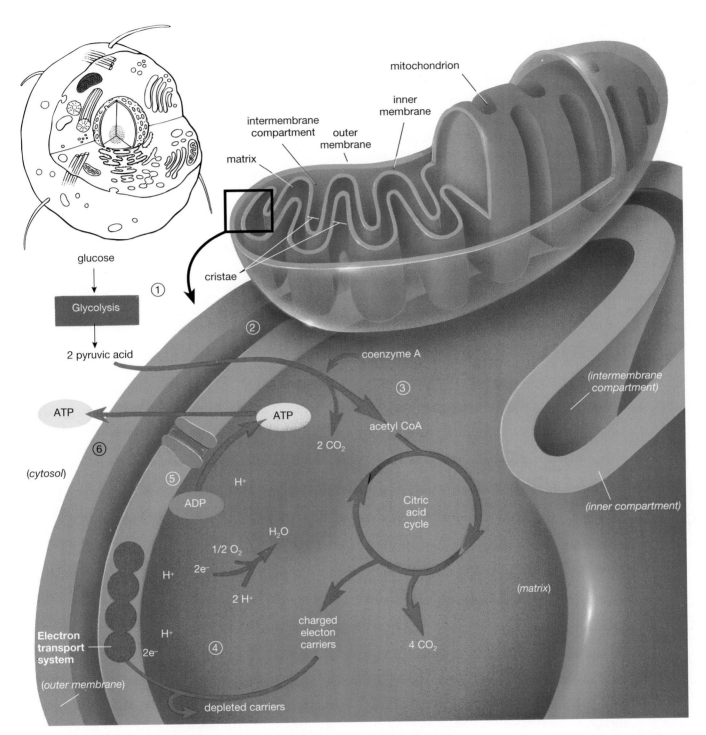

Figure 8-4   **Cellular respiration**

Cellular respiration takes place in the mitochondria, whose structure, like that of a chloroplast, reflects the compartmentalized reactions that occur there. The inner membrane separates the inner compartment, containing the soluble enzymes of the matrix, from the intermembrane compartment (between the inner and outer membranes). The lower part of the diagram summarizes the six essential steps in glucose metabolism, from the initial glycolysis in the cytosol to the final transport of ATP out of the mitochondrion and into the cytosol.

the cytosol, where it is synthesized as the end product of glycolysis, to the mitochondrial matrix, where it is used in cellular respiration.

Pyruvic Acid Is Broken Down by Reactions in the Mitochondrial Matrix

In the matrix, pyruvic acid reacts with a molecule called coenzyme A (Fig. 8-5, left). Each pyruvic acid is split into CO_2 and a two-carbon molecule called an acetyl group, which immediately attaches to coenzyme A to form the acetyl–coenzyme A complex (acetyl CoA for short). During this reaction, two energetic electrons and a hydrogen ion are transferred to NAD^+, forming NADH.

The two acetyl CoA molecules then enter a cyclic pathway known as either the **Krebs cycle**, named after its discoverer Hans Krebs, or the **citric acid cycle**, named after the first product in the reaction sequence, citric acid (Fig. 8-5, right). Each acetyl CoA briefly combines with a molecule of oxaloacetic acid. The two-carbon acetyl group is donated to the four-carbon oxaloacetic acid (which, as you may remember, also functions in the carbon fixation stage of the C_4 pathway) to form the six-carbon citric acid. Coenzyme A is released once again (coenzyme A, like an enzyme, is not permanently altered during these reactions, and is reused many times). Mitochondrial enzymes then lead each citric acid through a number of rearrangements that regenerate the oxaloacetic acid, give off two CO_2 molecules, and capture most of the energy of the acetyl group as one ATP and four electron carriers, one FADH$_2$ (flavin adenine dinucleotide) and three NADH.

The box "A Closer Look: The Mitochondrial Matrix Reactions" shows the complete set of reactions in the mitochondrial matrix, from acetyl CoA formation through the citric acid cycle.

▓ SUMMARY

THE MITOCHONDRIAL MATRIX REACTIONS

Synthesis of acetyl CoA produces one CO_2 and one NADH per pyruvic acid. The citric acid cycle produces two CO_2, one ATP, three NADH, and one FADH$_2$ per acetyl CoA. Therefore, at the conclusion of the matrix reactions, the two pyruvic acids produced from a single glucose molecule have been completely broken down by adding oxygen to form six CO_2 molecules. In the process, two ATPs, eight NADH, and two FADH$_2$ electron carriers have been produced.

Energetic Electrons Are Carried to Electron Transport Systems in the Inner Mitochondrial Membrane

At this point the cell has gained only four ATP molecules from the original glucose molecule: two during glycolysis and two during the citric acid cycle. The cell has, however, captured many energetic electrons in carrier molecules: 2 NADH during glycolysis and 8 more NADH and 2 FADH$_2$ from the matrix reactions, for a total of 10 NADH and 2 FADH$_2$. The carriers deposit their electrons in **electron transport systems** located in the inner mitochondrial

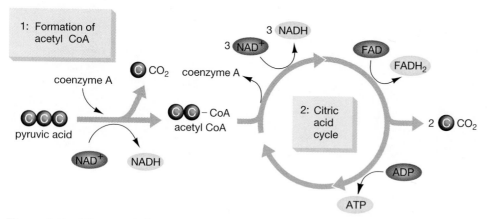

Figure 8-5   **The metabolic reactions occurring in the mitochondrial matrix**

The essentials of the metabolic reactions occurring in the mitochondrial matrix. Pyruvic acid reacts with coenzyme A to form CO_2 and acetyl CoA. During this reaction, an energetic electron is added to NAD^+ to form NADH. When acetyl CoA enters the citric acid cycle, the acetyl group combines with oxaloacetic acid to form citric acid, and coenzyme A is released. One turn through the reactions of the cycle produces three molecules of NADH, one of FADH$_2$, two of CO_2, and one of ATP for each acetyl CoA. Because each glucose molecule yields two pyruvic acids, the total energy harvest per glucose molecule in the matrix is two ATP, eight NADH, and two FADH$_2$.

A CLOSER LOOK

The Mitochondrial Matrix Reactions

Mitochondrial matrix reactions occur in two stages: the formation of acetyl coenzyme A and the citric acid cycle. Remember that glycolysis produces two pyruvic acids from each glucose molecule, so each set of matrix reactions occurs twice during the metabolism of a single glucose molecule.

First Stage: Formation of Acetyl Coenzyme A

Pyruvic acid is split to CO_2 and an acetyl group. The acetyl group attaches to coenzyme A to form acetyl CoA. Simultaneously, NAD^+ receives two electrons and a hydrogen ion to make NADH. The acetyl CoA enters the second stage of the matrix reactions.

Second Stage: The Citric Acid Cycle

1. Acetyl CoA donates its acetyl group to oxaloacetic acid to make citric acid.
2. Citric acid is rearranged to form isocitric acid.
3. Isocitric acid loses a carbon to CO_2, forming isoketoglutaric acid; NADH is formed from NAD^+.
4. Isoketoglutaric acid loses a carbon to CO_2, forming succinic acid; NADH is formed from NAD^+ and additional energy is stored in ATP. By this point, two molecules of CO_2 have been given off. (These two CO_2 molecules, along with the one that was released during the formation of acetyl CoA, account for the three carbons of the original pyruvic acid.)
5. Succinic acid is converted to fumaric acid, and the electron carrier FAD is charged up to $FADH_2$.
6. Fumaric acid is converted to malic acid.
7. Malic acid is converted to oxaloacetic acid and NADH is formed from NAD^+.

The citric acid cycle produces three molecules of CO_2 and NADH, one $FADH_2$, and one ATP per acetyl CoA. NADH and $FADH_2$ will donate their electrons to the electron transport system of the inner membrane, where the energy of the electrons will be used to synthesize ATP by chemiosmosis.

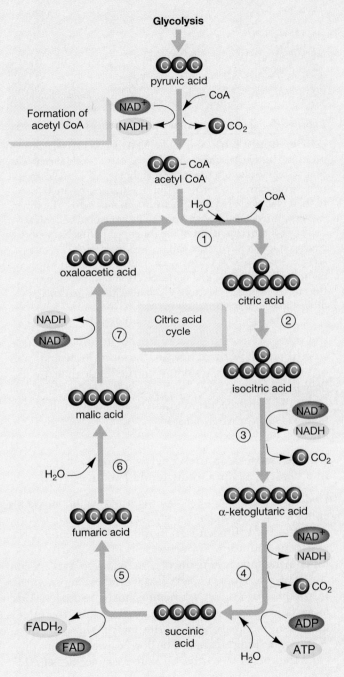

membrane (Fig. 8-6). These electron transport systems are similar in function to those embedded in the thylakoid membrane of chloroplasts. The energetic electrons move from molecule to molecule along the transport systems. Energy released by the electrons during these transfers is used to pump hydrogen ions from the matrix, across the

inner membrane, into the intermembrane compartment (see the section on chemiosmosis, below).

Finally, at the end of the electron transport system, oxygen and hydrogen ions accept the energetically depleted electrons: Two electrons, one oxygen atom, and two hydrogen ions combine to form water. This step clears out

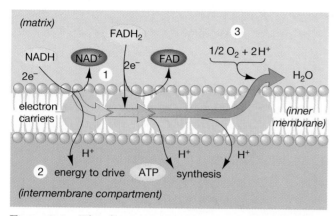

Figure 8-6   **The electron transport system of mitochondria**

① The electron carrier molecules NADH and FADH₂ deposit their energetic electrons with the carriers of the transport system located in the inner membrane. ② The electrons move from carrier to carrier within the transport system. Some of their energy is used to pump hydrogen ions across the inner membrane from the matrix into the intermembrane compartment. This movement out of the inner compartment creates a hydrogen ion gradient that can be used to drive ATP synthesis (for details, see "A Closer Look: Chemiosmosis in Mitochondria"). ③ At the end of the electron transport system, the energy-depleted electrons combine with oxygen and hydrogen ions in the matrix to form water.

the transport system, leaving it ready to run more electrons through. Without oxygen, the electrons would "pile up" in the transport system, with the result that the hydrogen ions would not be pumped across the inner membrane. The hydrogen ion gradient would soon dissipate and ATP synthesis would stop.

Chemiosmosis Captures Energy Stored in a Hydrogen Ion Gradient and Produces ATP

Hydrogen ion pumping across the inner membrane generates a large concentration gradient; that is, a high concentration of hydrogen ions in the intermembrane compartment and a low concentration in the matrix. The inner membrane is impermeable to hydrogen ions except at protein pores that are part of ATP-synthesizing enzymes. In the process known as **chemiosmosis**, hydrogen ions move down their concentration gradient from the intermembrane compartment to the matrix through these ATP-synthesizing enzymes. The flow of hydrogen ions provides the energy to synthesize 32 to 34 molecules of ATP from ADP and inorganic phosphate in the matrix. The box "A Closer Look: Chemiosmosis in Mitochondria" examines chemiosmosis in more detail.

ATP Is Transported from the Mitochondrial Matrix and Diffuses into the Cell Cytosol

The ATP that was synthesized from ADP and inorganic phosphate in the matrix during chemiosmosis is trans-

ported across the inner membrane from the matrix to the intermembrane compartment. It then diffuses out of the mitochondrion to the cytosol through the outer membrane, which is very permeable to ATP. These ATP molecules provide most of the energy needed by the cell. ADP simultaneously diffuses from the cytosol across the outer membrane and is transported across the inner membrane to the matrix, replenishing the supply of ADP.

▦ S U M M A R Y

ELECTRON TRANSPORT AND CHEMIOSMOSIS

Electrons from the electron carriers NADH and FADH₂ enter the electron transport system of the inner mitochondrial membrane. Here their energy is used to generate a hydrogen ion gradient. Movement of hydrogen ions down their gradient through the pores of ATP-synthesizing enzymes drives the synthesis of 32 to 34 molecules of ATP. At the end of the electron transport system, two electrons combine with one oxygen atom and two hydrogen ions to form water.

Cellular Metabolism Influences the Way Entire Organisms Function

Many students feel that the details of cellular metabolism are hard to learn and don't really help them to understand the living world around them. Have you ever read a murder mystery and wondered how cyanide could kill a person almost instantly? Cyanide reacts with one of the proteins in the electron transport system, immediately blocking the movement of electrons through the system and bringing cellular respiration to a screeching halt. Under more normal conditions, metabolic processes within individual cells have enormous impacts on the functioning of entire organisms. To take just two familiar examples, let's consider the migration of ruby-throated hummingbirds and track events in the Olympics.

Hummingbird Migration Requires Efficient Energy Storage and Use

As the days shorten in August, many birds in North America prepare to migrate to Central and South America, thereby escaping the cold weather and food shortages of the northern winter. For some, migration is a very dangerous undertaking. Ruby-throated hummingbirds, for example, fly across the Gulf of Mexico from the southeastern United States to Mexico and Central America. There is nowhere to stop and no food on a journey of over 1000 kilometers (621 miles) of open sea.

This trek presents real difficulties for a hummingbird. With its short wings, a hummer couldn't fly at all if it weighed too much. Yet flying 1000 kilometers re-

A CLOSER LOOK

Chemiosmosis in Mitochondria

ATP synthesis in mitochondria is similar to the process of chemiosmosis that we described in chloroplasts in Chapter 7. The inner membrane of a mitochondrion has an electron transport system that functions similarly to the one in the thylakoids. Further, the intermembrane compartment between the outer and inner membranes of a mitochondrion is analogous to the interior of a thylakoid.

Anatomically, the arrangement in mitochondria looks like this:

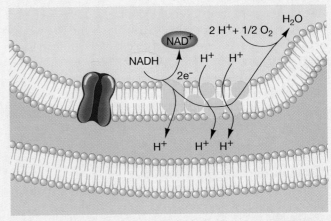

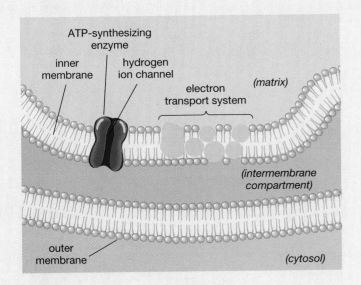

The electron carriers formed during glycolysis and the citric acid cycle—NADH and FADH$_2$—deposit their electrons with the electron transport system of the inner membrane (for clarity, FADH$_2$ is not shown in the illustration). As they pass through the electron transport system, the electrons provide the energy to pump hydrogen ions (H$^+$) across the inner membrane, from the matrix to the intermembrane compartment:

This pumping process increases the H$^+$ concentration in the intermembrane compartment and decreases the H$^+$ concentration in the matrix; therefore, a H$^+$ gradient is produced across the inner membrane. Like the thylakoid membrane of a chloroplast, the inner membrane of a mitochondrion is permeable to H$^+$ only at pores that are coupled with ATP-synthesizing enzymes. The movement of hydrogen ions down their concentration gradient through these pores drives ATP synthesis.

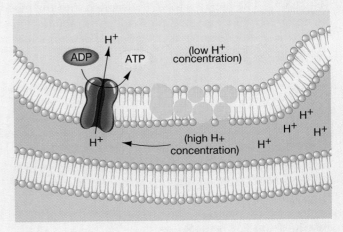

quires a lot of energy that must be stored in the bird's body. Hummers solve this dilemma with a twofold strategy of storing the highest-energy molecules possible—fat—and extracting the maximum usable energy during flight.

A ruby-throated hummingbird weighs 2 to 3 grams (0.11 to 0.16 ounces) before putting on weight for migration. It adds as much as 2 grams of fat in late summer, nearly doubling its weight. As you recall from Chapter 3, fats contain over twice as much energy per unit weight as do proteins or carbohydrates. If a hummer had to store 4 or 5 grams (0.21 to 0.26 ounces) of glycogen or protein, it would be too heavy to lift off.

Even so, the hummer must still squeeze every ATP molecule possible out of each fat molecule. The hummer that just makes it to Guatemala on 2 grams of fat using cellular respiration would collapse almost within sight of the Gulf Coast if it used lactic acid fermentation instead. The cells of a hum-

Metabolic Transformations—Why You Can Get Fat by Eating Sugar

As you know, humans do not live by glucose alone. Nor does the typical diet contain exactly the required amounts of each nutrient. Accordingly, the cells of the human body seethe with biochemical reactions, synthesizing one amino acid from another, making fats from carbohydrates, and channeling surplus organic molecules of all types into energy storage or release. Let's look at two examples of these metabolic transformations: producing ATP from fats and proteins and synthesizing fats from sugars.

Metabolizing Fats and Proteins

Even the leanest people have some fat in their bodies. During fasting or starvation, the body mobilizes these fat reserves for ATP synthesis, because even the bare maintenance of life requires a continuous supply of ATP, and seeking out new food sources demands even more energy. Fat metabolism flows directly into the pathways of glucose metabolism.

Chapter 3 illustrated the structure of a fat: three fatty acids connected to a glycerol backbone. In fat metabolism, the bonds between the fatty acids and glycerol are hydrolyzed (broken into subunits by the addition of water). The glycerol part of a fat, after activation by ATP, feeds directly into the middle of the glycolysis pathway (Fig. E8-1). The fatty acids are transported into the mitochondria, where enzymes in the inner membrane and matrix chop them up into acetyl groups. These attach to coenzyme A to form acetyl CoA, which enters the citric acid cycle.

In cases of severe starvation or of feeding almost exclusively on protein, amino acids too can be used to produce energy. First, they are converted to pyruvic acid, acetyl CoA, or the compounds of the citric acid cycle. These molecules then proceed through the remaining stages of cellular respiration, yielding various amounts of ATP depending on their point of entry into the pathway.

Synthesizing Fat from Sugar

The body has not only developed ways of coping with fasting or starvation; it has also evolved strategies for coping with situations in which food intake exceeds current energy needs. Just as fats can be funneled into the glucose metabolic pathway for energy production, the sugars and starches in cornflakes or candy bars can be converted into fats for energy storage. Complex sugars, such as starches and sucrose, are first hydrolyzed into their monosaccharide subunits (see Chapter 3). The

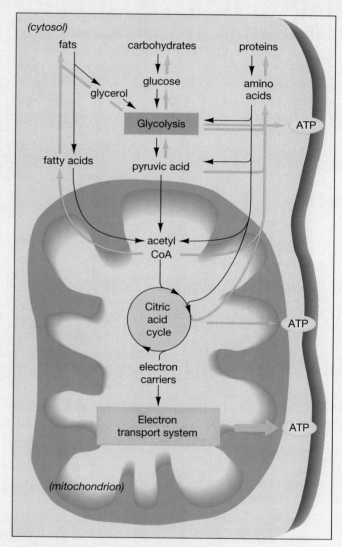

Figure E8-1   How other nutrients yield energy

Fats, carbohydrates such as starches, and proteins can all be broken down metabolically into molecules that enter glycolysis or the citric acid cycle, where they are used to generate ATP.

monosaccharides are broken down to pyruvic acid and converted to acetyl CoA. If the cell needs ATP, the acetyl CoA will enter the citric acid cycle. If the cell has plenty of ATP, acetyl CoA will be used to make fatty acids by a series of reactions that are essentially the reverse of fatty acid breakdown. In humans, the liver synthesizes fatty acids, but fat storage is relegated to fat cells, with their all too familiar distribution in the body, particularly around the waist and hips.

Energy use, fat storage, and eating are usually precisely balanced. Where the balance point lies, however, varies from person to person. Some people seem to eat nearly continuously without ever storing much fat, probably because their metabolic rates are high or adjust upward to compensate for excess food intake. Other people crave high-calorie foods even when they have a lot of fat already stored. As every chronic dieter knows, many people seem to have bodies that are "biologically set" to have large fat deposits. Constant restraint can

slim most people down, but as soon as the discipline is relaxed, the fat once again accumulates. Before the fashionably thin become too smug about their svelte condition, they should remember that overeating during times of easy food availability is highly adaptive behavior from an evolutionary point of view. If hard times come, the plump may easily (if hungrily) survive while the trim succumb to starvation. Only recently have many societies enjoyed regular agricultural surpluses, in which the drive to eat can lead to obesity.

mingbird's flight muscles are packed with mitochondria, so each cell is capable of producing large quantities of ATP.

Furthermore, the hummingbird's respiratory system is exquisitely designed to extract oxygen out of the air (Chapter 31) even while exhaling, enabling the lungs to provide a constant supply of oxygen to the cells. Therefore, even during strenuous flight, cellular respiration never falters for lack of oxygen.

Human Runners Face Similar Challenges

Humans, like hummingbirds, must regulate energy reserves and energy use. Why is the average speed of the 5000-meter run in the Olympics slower than that of the 100-meter dash? It's not because the distance runners could not, at any given time, run faster; the finishing kick at the end of the race testifies to that. The reason is that at top speed, their leg muscles consume ATP faster than their lungs can extract oxygen from the air to keep cellular respiration going. Glycolysis and lactic acid fermentation can keep the muscles functional for a short time, but soon the toxic effects of lactic acid buildup cause fatigue and cramps. Although runners can do a 100-meter

dash anaerobically, doing a 5000-meter run anaerobically is out of the question. Distance runners must therefore pace themselves, using cellular respiration to power their muscles for most of the race and saving the anaerobic sprint for the finish.

Marathon runners face somewhat the same dilemma that hummingbirds do. Nobody ever saw a fat marathoner: It's too much work to lug a lot of weight around for 26 miles. Nevertheless, a marathon may require 3000 kilocalories of stored energy, even with the most energy-efficient process, cellular respiration, supplying nearly all the ATP. Marathoners train by running 50 or 100 miles a week, not so much to build up their leg muscles (which are usually pretty stringy) as to build up the capacity of their respiratory and circulatory systems to deliver enough oxygen to their muscles. An efficient transport of oxygen to the cells is necessary to sustain the cellular respiration that such vigorous exercise demands.

As you can see, life on Earth depends on efficiently obtaining, storing, and using energy. Your understanding of the principles of cellular metabolism should enable you to fully appreciate the energy-related adaptations of living organisms.

❈ SUMMARY OF KEY CONCEPTS

Glucose Metabolism: An Overview
Cells produce usable energy by breaking down glucose to lower-energy compounds and capturing some of the released energy as ATP. In glycolysis, glucose is metabolized in the cytosol to two molecules of pyruvic acid, generating two ATP molecules. In the absence of oxygen, pyruvic acid is converted by fermentation to lactic acid or

ethanol and CO_2. If oxygen is available, the pyruvic acids are metabolized to CO_2 and H_2O through cellular respiration in the mitochondria, generating much more ATP.

Figure 8-7 and Table 8-1 summarize the locations, major mechanisms, and overall energy harvest for the complete metabolism of glucose from glycolysis through cellular respiration.

Glycolysis

During glycolysis, a molecule of glucose is activated by adding phosphates from two ATP molecules, to form fructose diphosphate. The fructose diphosphate then undergoes a series of reactions that break it down into two molecules of pyruvic acid. These reactions produce four ATP molecules and two NADH electron carriers. Because two ATPs were used in the activation steps, the net yield from glycolysis is two ATPs and two NADHs. Glycolysis, in addition to providing a small yield of ATP, uses up NAD^+ to produce NADH. Once the cell's supply of NAD^+ is consumed, glycolysis must stop. NADH may be regenerated by fermentation, with no additional ATP gain, or by cellular respiration, which also produces additional ATP.

Fermentation

If oxygen is absent, the pyruvic acids undergo either lactic acid fermentation, producing lactic acid, or alcoholic fermentation, producing ethanol and CO_2. In both cases, no new ATP is formed. However, both types of fermentation regenerate NAD^+ from NADH, thus replenishing the cell's supply of NAD^+.

Cellular Respiration

If oxygen is available, cellular respiration can occur. The pyruvic acids are transported into the matrix of the mitochondria. In the matrix, each pyruvic acid reacts with coenzyme A to form acetyl CoA plus CO_2. One NADH is also formed at this step. The two-carbon acetyl group of acetyl CoA enters the citric acid cycle, which releases the remaining two carbons as CO_2. One ATP, three NADHs, and one $FADH_2$ are also formed for each acetyl group that goes through the cycle. At this point, each glucose molecule has produced 4 ATPs (two from glycolysis and one from each acetyl CoA through the citric acid cycle), 10 NADHs (two from glycolysis, one from each pyruvic acid during the formation of acetyl CoA, and three from each acetyl CoA during the citric acid cycle), and 2 $FADH_2$s (one from each acetyl CoA during the citric acid cycle).

The NADHs and $FADH_2$s deposit their energetic electrons in the electron transport system embedded in the inner mitochondrial membrane. The energy of the electrons is used to pump hydrogen ions across the inner membrane from the matrix to the intermembrane compartment. At the end of the electron transport system, the depleted electrons combine with hydrogen ions and oxygen to form water. During chemiosmosis, the hydrogen ion gradient created by the electron transport system is used to produce ATP, as the hydrogen ions diffuse back across the inner membrane through pores in ATP-synthesizing enzymes. Electron transport and chemiosmosis yield 32 to 34 additional ATPs, for a net yield of 36 to 38 ATPs per glucose molecule.

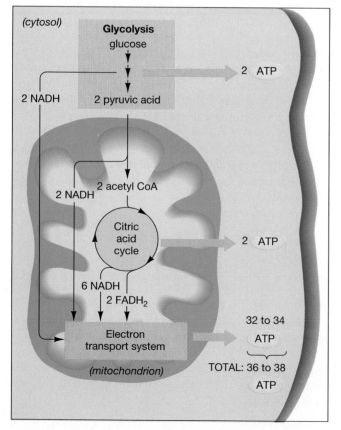

Figure 8-7   A summary of the energy harvest from the complete metabolism of one glucose molecule

Glycolysis and the citric acid cycle each produce two ATP molecules. The reactions within the mitochondrial matrix produce eight NADH molecules and two $FADH_2$ molecules. By donating its electrons to the electron transport system, each NADH molecule yields three ATP molecules, for a total of 24 ATPs. Each $FADH_2$ molecule yields two ATP molecules, for a total of 4 ATPs. The electrons from the two NADH molecules produced in the cytoplasm during glycolysis must be transported into the mitochondrion to reach the electron transport system. In heart and liver cells, this transport is "free"; in most cells, transport costs one ATP per NADH. The two "glycolytic NADH" molecules therefore yield either 4 or 6 ATP molecules, depending on the cell. Therefore, the energy harvest from electron transport is 32 to 34 ATPs. Including 2 ATPs from glycolysis and 2 ATPs from the citric acid cycle, the total energy yield from glucose metabolism is 36 to 38 ATPs.

TABLE 8-1 ■ *Summary of Glycolysis and Cellular Respiration*

Process	Location	Reactions	Electron Carriers Formed	ATP Yield
Glycolysis	Cytosol	Glucose broken down to two pyruvic acids	2 NADH	2 ATP
Acetyl CoA formation	Matrix of mitochondrion	Each pyruvic acid combined with coenzyme A to form acetyl CoA and CO_2	2 NADH	
Citric acid cycle	Matrix of mitochondrion	Acetyl group of acetyl CoA metabolized to two CO_2	6 NADH, 2 $FADH_2$	2 ATP
Electron transport	Inner membrane, intermembrane compartment	Energy of electrons from NADH, $FADH_2$ used to pump H^+ into intermembrane compartment, H^+ gradient used to synthesize ATP: three ATP per NADH, two ATP per $FADH_2$		32–34 ATP*

*Glycolysis produces two NADH molecules in the cytosol of the cell. Unlike the NADH and $FADH_2$ molecules generated in the matrix of the mitochondrion, the electrons from these two NADH molecules must be transported into the matrix before they can enter the electron transport system of the inner membrane. In most eukaryotic cells, the energy of one ATP molecule is used to transport the electrons from one NADH molecule into the matrix. Thus, the two "glycolytic NADH" molecules net only two ATPs, not the usual three, during electron transport. The heart and liver cells of mammals, however, use a different transport system, one that does not consume ATP. In these cells, then, the "glycolytic NADH" molecules net three ATPs apiece, just as the "mitochondrial NADH" molecules do.

■ KEY TERMS

cellular respiration p. 146
chemiosmosis p. 150
citric acid cycle p. 148

electron transport system p. 148
fermentation p. 145
glycolysis p. 143

intermembrane compartment p. 146
Krebs cycle p. 148
matrix p. 146

■ THINKING THROUGH THE CONCEPTS

Multiple Choice

1. Where does glycolysis occur?
   a. cytoplasm
   b. matrix of mitochondria
   c. inner membrane of mitochondria
   d. outer membrane of mitochondria
   e. chloroplast stroma
2. Where does respiratory electron transport occur?
   a. cytoplasm
   b. matrix of mitochondria
   c. inner membrane of mitochondria
   d. outer membrane of mitochondria
   e. chloroplast stroma
3. What are the economically important products of the fermentation of grape juice by yeast that are essential for the manufacture of champagne?
   a. lactic acid and NAD^+ b. ATP and CO_2
   c. ATP and ethanol   d. CO_2 and ethanol
   e. only the French know
4. The majority of ATP produced in aerobic respiration comes from
   a. glycolysis   b. the citric acid cycle
   c. the electron transport system   d. fermentation
   e. reverse osmosis
5. The process that converts glucose (6 C) into two molecules of pyruvate (3 C) is
   a. glycolysis   b. fermentation
   c. the citric acid cycle   d. respiratory electron transport
   e. the Calvin-Benson cycle

6. The process that converts pyruvate (3 C) into three molecules of CO_2 is
   a. glycolysis
   b. fermentation
   c. the citric acid cycle
   d. respiratory electron transport
   e. the Calvin-Benson cycle

Review Questions

1. Starting with glucose ($C_6H_{12}O_6$) write the overall reactions for (a) aerobic respiration and (b) fermentation in yeast.
2. Draw a labeled diagram of a mitochondrion and explain how its structure is related to its function.
3. What role do the following play in respiratory metabolism: (a) glycolysis, (b) mitochondrial matrix, (c) inner membrane of mitochondria, (d) fermentation, and (e) NAD^+?
4. Outline the major steps in (a) aerobic and (b) anaerobic respiration, indicating the sites of ATP production. What is the overall energy harvest (in terms of ATP molecules generated per glucose molecule) for each?
5. Describe the citric acid cycle. In what form is most of the energy harvested?
6. Describe the mitochondrial electron transport system and the process of chemiosmosis.
7. Why is oxygen necessary for cellular respiration to occur?

❈ A P P L Y I N G   T H E   C O N C E P T S

1. Some years ago a freight train overturned, spilling a load of grain. Because the grain was spoiled it was buried in the embankment. Although there is no shortage of other food locally, the local bear population has created a nuisance by continually uncovering the grain. What do you think has happened to the grain to make them do this, and how is it related to human cultural evolution?

2. In detective novels, "the odor of bitter almonds" is always the telltale clue to murder by cyanide poisoning. Cyanide works by attacking the enzyme that transfers electrons from respiratory electron transport to O_2. Why is it not possible for the victim to survive using anaerobic respiration? Why is cyanide poisoning almost immediately fatal?

3. Over a century ago, the French biochemist Louis Pasteur described a phenomenon, in the wine-making process using yeast, that we now call "the Pasteur effect." He observed that in a sealed container of grape juice containing yeast, the yeast will consume the sugar very slowly as long as oxygen remains in the container. As soon as the oxygen is gone, however, the rate of sugar consumption by the yeast increases greatly and the alcohol content in the container rises. Discuss the Pasteur effect based on what you know about aerobic and anaerobic cellular respiration.

4. Some species of bacteria that live at the surface of sediment on the bottom of lakes are facultative anaerobes that are capable of either aerobic or anaerobic respiration. How will their metabolism change during the summer when the deep water becomes anoxic (deoxygenated)? If the bacteria continue to grow at the same rate will glycolysis increase, decrease, or remain the same after the lake becomes anoxic? Explain why.

5. Dumping large amounts of raw sewage into rivers or lakes often leads to massive fish kills, although sewage itself is not toxic to fish. Similar fish kills also occur in shallow lakes that become covered in ice during the winter. What kills the fish? How might you reduce fish mortality after the accidental release of raw sewage into a small pond containing large bass?

6. Although respiration occurs in all living cells, different cells respire at different rates. Explain why. How could you predict the relative respiratory rates of different tissues in a fish by microscopic examination of cells?

❈ F O R   M O R E   I N F O R M A T I O N

Hinkle, P. C., and McCarty, R. E. "How Cells Make ATP." *Scientific American*, March 1978. A summary of cellular respiration, with an emphasis on electron transport in mitochondria.

McCarty, R. E. "H⁺-ATPases in Oxidative and Photosynthetic Phosphorylation." *BioScience*, January 1985. A description of the structure and function of the ATP-synthesizing enzymes in mitochondria and chloroplasts.

Nelson, M., Burgess, T. L., Alling, A., Alverez-Romo, N., Dempster, W. F., Walford, R. L., and Allen, J. P. "Using a Closed Ecological System to Study Earth's Biosphere." *BioScience*, 1993. An artificial ecosystem allows scientists to learn more about how natural ecosystems function.

N E T   W A T C H

On-line resources for this chapter are on the World Wide Web at:
http://www.prenhall.com/~audesirk (click on the <u>table of contents</u> link and then select Chapter 8).

Unit 11
Inheritance

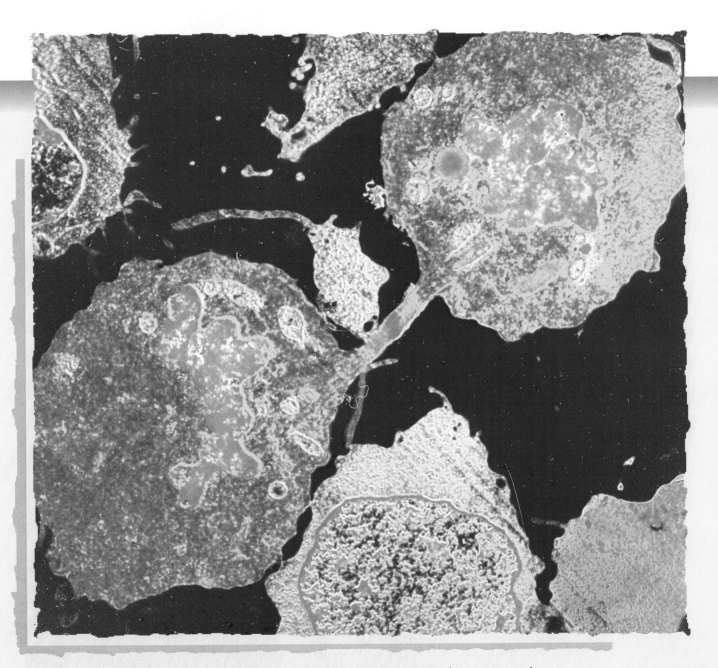

A human white blood cell (lymphocyte) undergoes mitosis in this false-color transmission electron micrograph. The two daughter cells, each of which contains a full set of genetic material (pink), are almost separated, but are still connected by a strand of cytoplasm. This is late telophase; the nuclear envelopes are about to re-form.

9 Cellular Reproduction

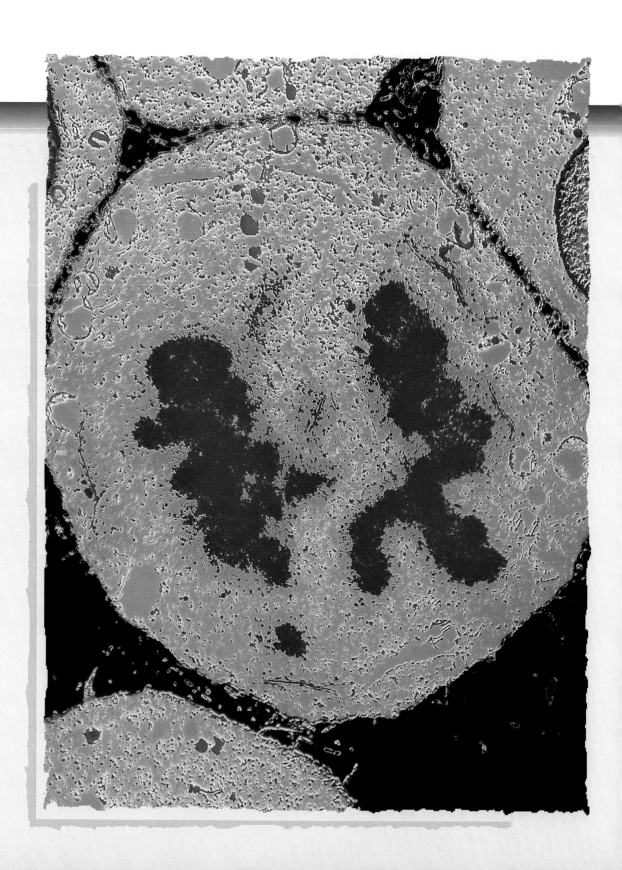

> *"All cells come from cells."*
>
> RUDOLF VIRCHOW
> German physician and early proponent of the cell theory (1858)

▓ A T   A   G L A N C E

The Essentials of Cellular Reproduction
Cell Division Transmits a Complete Set of Hereditary
    Information to Each Daughter Cell
Cell Division Transmits Essential Cytoplasmic Materials to
    Each Daughter Cell
The Activities of a Cell from One Cell Division to the Next
    Constitute the Cell Cycle

Prokaryotic and Eukaryotic Chromosomes Compared

The Prokaryotic Cell Cycle

The Eukaryotic Chromosome
The Eukaryotic Chromosome Consists of a Single Strand of
    DNA Complexed with Proteins
Eukaryotic Chromosomes Often Occur in Pairs Containing
    Similar Genetic Information

The Eukaryotic Cell Cycle
Growth, Replication of Chromosomes, and Most Cell
    Functions Occur during Interphase
Cells Vary in the Duration of Interphase
Cell Division Consists of Nuclear Division and Cytoplasmic
    Division

Mitosis
During Prophase, the Chromosomes Condense and the
    Spindle Apparatus Forms and Attaches to the
    Chromosomes
During Metaphase, the Chromosomes Are Aligned along
    the Equator of the Cell
During Anaphase, Sister Chromatids Separate and Are
    Pulled to Opposite Poles of the Cell
During Telophase, Nuclear Envelopes Re-form around Each
    Group of Chromosomes

Cytokinesis

Cell Division and Asexual Reproduction

A human cancer cell in anaphase of cell division. Two sets of chromosomes are seen in blue in this false-color transmission electron micrograph.

"All cells come from cells." With these words, Rudolf Virchow captured the crucial importance of cellular reproduction for both unicellular and multicellular organisms. Because all living organisms consist of one or more cells, and because all cells are descended from preexisting cells, cellular reproduction is absolutely essential for the continued existence of life on Earth.

How do cells reproduce? Although there are many variations, most cells reproduce according to the simple repeating phrases: enlarge, divide in two, enlarge, divide in two, and so on. This process, appropriately called **cell division,** produces two daughter cells that, if all goes well, have exactly the same genetic information as the parent cell, carried in chromosomes that are exact copies of the chromosomes of the parent cell. Each daughter cell also inherits about half of the parent cell's cytoplasm, including a full complement of organelles. Each round of growth and cell division is called a **cell cycle.**

In unicellular organisms such as bacteria or *Amoeba,* the cell cycle is synonymous with the **life cycle,** which consists of the events in the life of an organism from one generation to the next. In multicellular organisms, the cell cycle is usually only part of the life cycle. Multicellular organisms begin life as a fertilized egg. Repeated cell divisions produce the hundreds to trillions of cells that make up the adult organism. To complete the life cycle, most multicellular organisms produce a new fertilized egg, through the process of **sexual reproduction**. Cells in the reproductive organs of the adult organism undergo a specialized kind of cell division, called **meiosis,** which we will describe in Chapter 12. In animals, the daughter cells of meiosis are sex cells—sperm or eggs. When a sperm fertilizes an egg, the life cycle starts over again.

In this unit, we will study the mechanisms of inheritance: how a cell transmits copies of its chromosomes to its daughter cells; the structure and function of the molecules of DNA contained in those chromosomes that carry the genetic information; how multicellular parents transmit chromosomes to their offspring during meiosis; and how the function of DNA and the distributions of chromosomes during meiosis produce the observable patterns of inheritance that we see in life on Earth. We will begin, in this chapter, with cell cycles and cell division. We will see that cells have evolved elaborate mechanisms that ensure that each daughter cell inherits all the materials it needs to continue the flow of life.

The Essentials of Cellular Reproduction

When a cell divides, it must transmit to its offspring cells two essential requirements for life (Fig. 9-1): hereditary information to direct life processes, and materials in the cytoplasm that the offspring need to survive and to utilize their hereditary information.

Cell Division Transmits a Complete Set of Hereditary Information to Each Daughter Cell

The hereditary information of all living cells is **deoxyribonucleic acid (DNA)**. Like many large biological molecules, a molecule of DNA consists of a long chain of smaller subunits (see Chapter 3). DNA has four different types of subunits, called nucleotides. Segments of DNA a few hundred to many thousand nucleotides long are the units of inheritance—the genes—that carry the genetic information to produce specific traits. The sequence of nucleotides in a gene encodes information for the synthesis of the RNA and protein molecules that are needed to build a cell and carry out its metabolic activities. We will see in Chapters 10 through 13 how DNA encodes genetic instructions and how a cell regulates which genes it uses at any given time.

For any cell to survive, it must have a *complete set of genetic instructions*. Therefore, when a cell divides, it cannot simply split its set of genes in half and give each daughter cell half a set. Rather, the cell must first *duplicate its*

DNA, much like making a photocopy of an instruction manual. Each daughter cell then receives a complete "DNA manual" containing all the genes.

Cell Division Transmits Essential Cytoplasmic Materials to Each Daughter Cell

Like the blueprints for a house, the instructions encoded by DNA are useless without materials to work with. Each newly formed cell must receive the molecules it needs to read its genetic instructions and to keep it alive long enough to acquire new materials from the environment and to process them into new cellular components. Furthermore, as we described in Chapter 5, a cell cannot synthesize either mitochondria or chloroplasts from scratch: These organelles arise only by the division of previously existing mitochondria and chloroplasts. Usually, when a cell divides, its cytoplasm is divided about equally between the two daughter cells. This simple mechanism normally provides both daughter cells with all the organelles, nutrients, enzymes, and other molecules they need.

The Activities of a Cell from One Cell Division to the Next Constitute the Cell Cycle

Newly formed cells usually acquire nutrients from their environment, synthesize more of their own materials, and grow larger. After a variable amount of time, depending on the organism, the type of cell, and the nutrients available, the cell divides. This general description applies to both eukaryotic and prokaryotic cells. Prokaryotic cells are structurally and functionally very different from eukaryotic cells, however, and accordingly their cell cycle differs in many respects. We will therefore discuss prokaryotic and eukaryotic cycles separately.

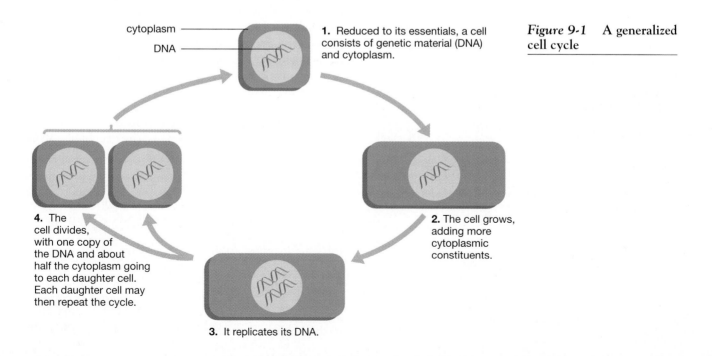

cytoplasm

DNA

1. Reduced to its essentials, a cell consists of genetic material (DNA) and cytoplasm.

2. The cell grows, adding more cytoplasmic constituents.

3. It replicates its DNA.

4. The cell divides, with one copy of the DNA and about half the cytoplasm going to each daughter cell. Each daughter cell may then repeat the cycle.

Figure 9-1   **A generalized cell cycle**

Prokaryotic and Eukaryotic Chromosomes Compared

The DNA of all living cells is packaged into one or more structures called **chromosomes** (see Chapter 5). A single, circular chromosome about a millimeter or two in circumference contains all the DNA of a prokaryotic cell (Fig. 9-2). One point on the chromosome is attached to the plasma membrane of the cell. A prokaryotic chromosome usually has relatively little protein associated with the DNA.

The chromosomes of eukaryotic cells are quite different. First, eukaryotic cells contain much more DNA than prokaryotic cells do. Second, in contrast to the single chromosome of a prokaryotic cell, a eukaryotic cell packages its enormous amounts of DNA, along with roughly equal amounts of protein, into many separate chromosomes. Even dividing up the DNA this way, eukaryotic chromosomes are often much longer than prokaryotic chromosomes. The chromosome of a bacterium is a millimeter or two in circumference, whereas the average human chromosome is over 40 millimeters long. (The total length of all the 46 chromosomes in a human cell is about *2000* millimeters, longer than the height of the average human.) Finally, the prokaryotic chromosome is not separated from the cytoplasm of the cell, whereas eukaryotic chromosomes are contained within a membrane-bound nucleus.

Differences in number, size, and location of chromosomes dictate considerable differences in the mechanisms of prokaryotic and eukaryotic cell division. Therefore, we will describe the prokaryotic cell cycle first. Then we will examine the eukaryotic chromosome in more detail, concentrating on those aspects of chromosome structure that

ruptured bacterium   bacterial chromosome

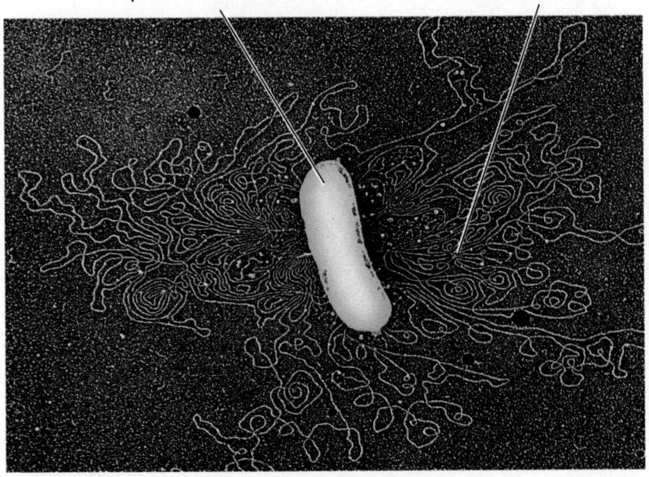

Figure 9-2   The bacterial chromosome

This bacterial cell (yellow central oblong) is about 1 × 3 micrometers in size. The cell was ruptured to free its chromosome. If you look carefully, you will see that there are no free ends to the chromosome: It is a single thin strand. If stretched out, the chromosome would be a circular molecule of DNA 1 or 2 millimeters in circumference.

are important in cell division. Finally, we will explore the eukaryotic cell cycle.

The Prokaryotic Cell Cycle

Under favorable conditions, the cell cycle of many prokaryotes proceeds rapidly. The common intestinal bacterium *Escherichia coli*, for example, can complete its cell cycle in 30 minutes or less. During most of this time, the cell absorbs nutrients from the environment, grows, and replicates its DNA. As soon as the DNA is replicated, the bacterium divides; in fact, in some bacteria the cell begins dividing even before DNA replication is quite finished.

With only a single chromosome and no separate nucleus, bacterial cells divide by **binary fission** (Fig. 9-3):

1. Before DNA replication, one point on the chromosome is attached to the plasma membrane.
2. The chromosome replicates (by a process that we will describe in Chapter 10). The resulting pair of identical chromosomes attach to the plasma membrane at nearby, but distinct, points.
3. The cell elongates, and new plasma membrane is added between the two attachment points, pushing them apart.
4. As the two chromosomes move toward opposite ends of the cell, the plasma membrane around the middle of the cell grows inward.
5. Finally, two new daughter cells are formed.

Each daughter cell thus receives one of the replicated chromosomes and about half the cytoplasm. With a complete set of genes and enough materials to work with, the two daughter cells begin to grow, starting the cycle over again.

The Eukaryotic Chromosome

The complicated events of eukaryotic cell division are largely an evolutionary solution for sorting out a large number of long chromosomes. To help you understand the mechanisms of eukaryotic cell division, we will begin by taking a closer look at the structure of the eukaryotic chromosome.

The Eukaryotic Chromosome Consists of a Single Strand of DNA Complexed with Proteins

Under the light microscope, interphase nuclei appear as large, darkly staining bodies without much internal structure. Early microscopists named the material in the nucleus **chromatin,** from the Greek word for "color," because chromatin is stained by some commonly used dyes. During cell division, dark threadlike structures form where

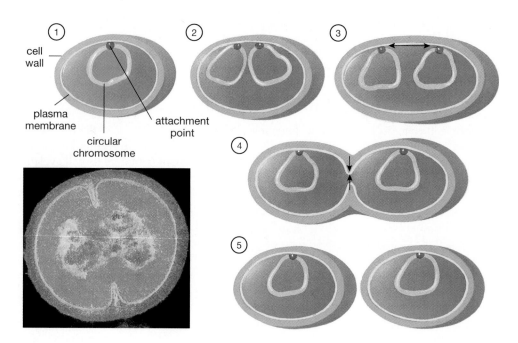

Figure 9-3   **Prokaryotic cells divide by binary fission**

① The circular chromosome is attached at one point to the cell membrane. ② The chromosome is replicated. The two copies are attached to the membrane at nearby points. ③ The cell elongates; new plasma membrane is added between the attachment points. ④ The plasma membrane grows inward at the middle of the cell. ⑤ The parent cell has divided into two daughter cells. (Inset) Cross section of a prokaryotic cell undergoing binary fission at a stage similar to ④.

the nucleus used to be. The microscopists called these threads chromosomes, meaning "colored bodies." With the advent of the electron microscope and advances in molecular biology, we now know that chromatin and chromosomes are actually the same thing (DNA and its associated proteins), but in different stages of condensation.

Each eukaryotic chromosome is a single long molecule of DNA, continuous from end to end, complexed with special proteins called histones. During most of its life (that is, interphase), a cell must be able to read the information contained in the chromosomes; it can do so only when the chromosomes are extended (see Chapter 11). In this extended state, individual chromosomes are too thin to be visible in light microscopes, or even in ordinary electron microscopes. During cell division, however, the chromosomes must be sorted out and moved into the daughter nuclei. Just as sewing thread is easier to organize when it is wound onto spools, sorting and transporting are easier if the chromosomes are condensed and shortened (Fig. 9-4).

The terms used by biologists to describe chromosomes during cell division often trouble beginning students. But you will find it difficult to understand cell division without understanding the terminology, so follow carefully. For simplicity, the drawings below illustrate chromosomes only in the condensed state.

Most chromosomes consist of two arms that extend out from a specialized region called the **centromere** (meaning "middle body"):

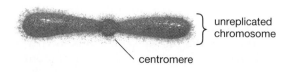

Before a cell divides, it duplicates its chromosomes. The two copies remain attached at their centromeres. As long as they are attached to one another, the copies are called sister **chromatids**. Each chromatid is a single DNA molecule identical to the DNA of the original chromosome before replication. The whole structure (two sister chromatids attached at their centromeres) is still considered to be a single chromosome:

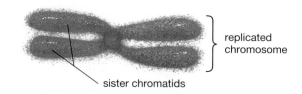

Figure 9-4 therefore shows several single chromosomes, each consisting of two sister chromatids.

During cell division, the two sister chromatids separate, with each chromatid becoming an independent daughter chromosome:

Eukaryotic Chromosomes Often Occur in Pairs Containing Similar Genetic Information

The chromosomes of each species have characteristic shapes, sizes, and staining patterns (Fig. 9-5). The nonreproductive cells of many organisms contain *pairs of chro-*

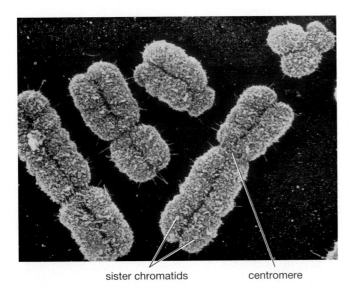

sister chromatids   centromere

Figure 9-4   **Human chromosomes during cell division**

In this scanning electron micrograph, the DNA and associated protein of the chromosomes, which are thin and extended during interphase, have coiled up into thick, short structures. The chromosomes were replicated before condensation; the two strands, called chromatids, are attached at the centromere. To get a sense of the how condensed they really are, look at the fuzzy texture of the chromosomes: Each strand of "texture" is a loop of chromosome, about as thin as the whole chromosome would look during interphase. The condensed chromosomes are about 5 to 20 micrometers long; the same chromosomes during interphase would relax into thin strands about 10,000 to 40,000 micrometers long.

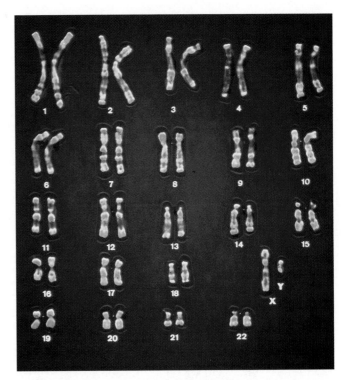

Figure 9-5   **A human male karyotype**

A karyotype is produced by staining and photographing the chromosomes of a dividing cell. Pictures of the individual chromosomes are cut out and arranged in descending order of size. Note that the chromosomes occur in pairs (homologues), which are similar in both size and staining patterns and have similar genetic material. If these chromosomes were from a female, there would be two X chromosomes. In humans, the Y chromosome is much smaller than the X chromosome.

mosomes that are the same length and that stain in the same pattern. Breeding and biochemical experiments show that the chromosomes of each pair also have similar, although usually not identical, genetic content. Therefore, the members of a pair are called **homologues,** meaning "to say the same thing" in Greek. Cells with pairs of homologous chromosomes are called **diploid.** The 46 chromosomes of the nonreproductive cells of human beings occur as 23 pairs of homologues.

Not all cells have pairs of homologous chromosomes. At some point in the life cycle of all sexually reproducing organisms, a special type of cell division called meiosis produces cells that have only one copy of each type of chromosome (see Chapter 12): These cells are called **haploid.** In animals, these haploid cells are the sex cells, or **gametes** (sperm and egg). Gametes cannot live independently; during sexual reproduction, male and female gametes fuse to re-form a diploid cell, the fertilized egg, or **zygote.** In plants, the haploid cells produced by meiosis are not gametes. They undergo ordinary cell division to produce multicellular, haploid organisms. Certain cells of these haploid organisms later become gametes and fuse to form a zygote. In both plants and animals one of each pair of chromosomes in

the diploid zygote is inherited from each parent. We will see how this happens in Chapter 12.

In biological shorthand, the number of different types of chromosomes is termed the haploid number and is designated n. For humans, $n = 23$; for other organisms, n ranges from 1 to several hundred. Diploid cells, with two homologues of each type, contain $2n$ chromosomes. Some organisms have more than two homologues of each type of chromosome in their cells and are called **polyploid.** Polyploid cells may be $3n$ (triploid), $4n$ (tetraploid), and so on.

The Eukaryotic Cell Cycle

The eukaryotic cell cycle consists of two major phases (Fig. 9-6). The first phase, **interphase,** is the period between cell divisions, during which the cell acquires nutrients from its environment, grows, and replicates its chromosomes. During the second phase, cell division (also called the M phase), one copy of each chromosome and approximately half the cytoplasm are parceled out into each of the two daughter cells.

Growth, Replication of Chromosomes, and Most Cell Functions Occur during Interphase

Most of the life of a cell (usually 90% or more) is spent in interphase. Many years ago, biologists studying the cell cycle noted that DNA synthesis, which occurs as a part of the replication of the chromosomes, is not continuous throughout interphase. Rather, interphase begins with a period in which DNA is not synthesized, then DNA synthesis occurs, and then a second period follows in which DNA is not synthesized. Cell biologists therefore divided interphase into three subphases, called G_1 (the "first gap" in DNA synthesis), S (when DNA synthesis occurs), and G_2 (the "second gap" in DNA synthesis; see Fig. 9-6).

The period that occurs after the most recent cell division and before duplication of the chromosomes is the G_1 phase. Although the designation as "first gap" in DNA synthesis implies that the cell is merely resting before getting on with the important business of DNA synthesis, this is not the case. In fact, most of the growth and activity of a cell occur during the G_1 phase. The cell acquires nutrients from its environment, carries out its specialized functions (for example, hormone synthesis and secretion), and grows.

Late in the G_1 phase, at a time called the restriction point, the cell undergoes a kind of "internal evaluation" of its ability to complete the cell cycle and produce two viable daughter cells. If the evaluation is negative, the cell does not go on to divide. If the evaluation is positive, the cell becomes committed to DNA replication and cell division. The commitment to cell division is much like launching a rocket: Once the "go-ahead" signal has been given, the cell cannot turn back and return to the G_1 phase.

Chromosome replication defines the **S, or synthesis, phase**, because this is the only time DNA synthesis normally occurs. Each chromosome is duplicated—just once. During the S phase, animal cells also duplicate their centrioles.

The period after DNA synthesis, but before the next cell division, is the **G_2 phase**. The cell is already committed to cell division before entering the G_2 phase. Most of G_2 seems to be spent in synthesizing molecules other than DNA that are required for cell division.

Cells Vary in the Duration of Interphase

Many types of mammalian cells progress steadily through interphase, spending about 5 hours in the G_1 stage, 7 hours replicating their DNA during the S phase, and 3 hours in G_2 preparing for division. Although cell division itself usually takes about an hour, certain cell types have very short cell cycles, whereas others may go for weeks or even an entire lifetime without dividing. These differences in cell cycle length usually arise from differences in the duration of G_1. The early cell divisions of an animal embryo, for instance, occur in rapid succession with virtually no G_1 phase at all and

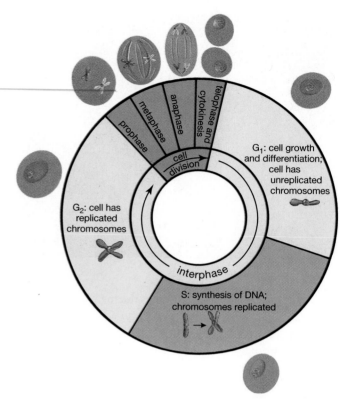

Figure 9-6 **The eukaryotic cell cycle**
The eukaryotic cell cycle consists of two major phases, interphase and cell division. Each is divided into several subphases, as described in the text.

therefore almost no growth between divisions (see Chapter 40). In contrast, nerve cells in the adult mammalian brain no longer divide but remain in the G_1 phase for life.

Cell Division Consists of Nuclear Division and Cytoplasmic Division

During nuclear division, identical, complete copies of all the chromosomes are packaged into two new nuclei. Nuclear division is called **mitosis**, from the Greek word for "thread," because the chromosomes appear as threadlike structures at this time (see Fig. 9-7). During cytoplasmic division, **cytokinesis**, the cytoplasm is split into two daughter cells, with each cell receiving one of the newly formed nuclei and (usually) roughly equal amounts of cytoplasm. Therefore, the two daughter cells are essentially identical both genetically and cytoplasmically. They are also genetically identical to the parent cell.

Although most cells adhere to this scheme, nuclear division and cytoplasmic division are potentially independent events. Some cells, including some fungi and the skeletal muscle cells of vertebrates, undergo mitosis without cytokinesis. This process produces single cells with many nuclei.

As we mentioned earlier, a multicellular organism grows from a single fertilized egg by repeated cell divisions. Therefore, with a few exceptions, such as sex cells and the

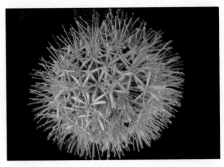

(a) The flower of the African blood lily.

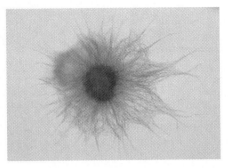

(b) Interphase in a cell of the endosperm (a food-storage organ in the seed) before mitosis begins. The chromosomes are in the thin, extended state and appear as an amorphous mass of stained material in light micrographs such as this one. The microtubules during interphase form a vaguely star-shaped array leading from the nucleus to all parts of the cell.

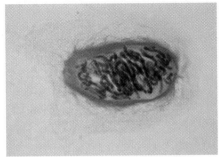

(c) Early prophase: The chromosomes have condensed into thick, separate strands.

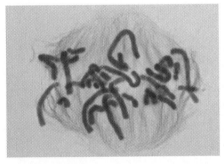

(d) Late prophase: The chromosomes are attached to microtubules of the almost fully developed spindle.

(e) Metaphase: The spindle has moved the chromosomes to the equator of the cell.

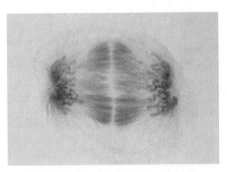

(f) Anaphase: The spindle is moving one set of chromosomes to each pole of the cell.

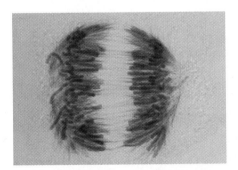

(g) Telophase: The chromosomes have been gathered into two clusters, one at the site of each future nucleus.

(h) Resumption of interphase: The chromosomes are relaxing again into their extended state. The spindle is disappearing and the microtubules are rearranging into the interphase array.

Figure 9-7   **Mitosis in a plant cell**

Mitosis as seen in the developing seed of the African blood lily, *Haemanthus katherinae*. In all of the micrographs, the chromosomes are stained bluish purple, and the microtubules of the spindle are stained pink to red. Compare these micrographs with the drawings of mitosis in an animal cell shown in Figure 9-8. (Each micrograph here is of a different single cell that has been fixed and stained at a particular stage of mitosis. The cells therefore have different sizes and shapes.)

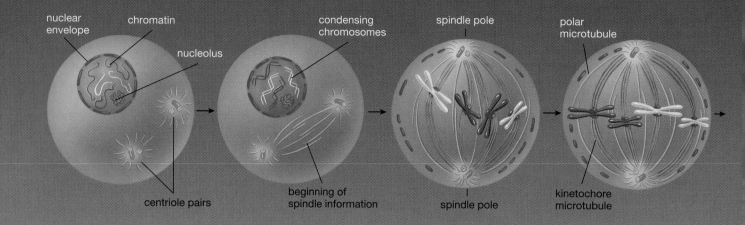

nuclear envelope chromatin nucleolus condensing chromosomes spindle pole polar microtubule

centriole pairs beginning of spindle information spindle pole kinetochore microtubule

(a) Late interphase: The chromosomes have been replicated but remain elongated and relaxed within the nucleus. The centrioles, within the microtubule organizing center, have also been replicated.

(b) Early prophase: The chromosomes condense and shorten, becoming visible with a light microscope. The centrioles begin to move apart, and the spindle apparatus begins to form.

(c) Late prophase: The nuclear envelope breaks down and the spindle microtubules invade the nuclear region. Some spindle microtubules (the kinetochore microtubules) attach to each chromosome at its kinetochore. Other microtubules (the polar microtubules) extend from each pole to the cell's equator, where their ends overlap.

(d) Metaphase: The spindle is fully formed. The kinetochore microtubules move the chromosomes to the cell's

Figure 9-8   Mitosis and cytokinesis in an animal cell

cells of the immune system (see Chapter 34), all cells of a multicellular organism are genetically identical. As incredible as it seems, a brain cell and a liver cell have the same genes. Their differences in structure and function occur because of differences in which genes are used during the life of the cell.

Much of biology, reduced to its essentials, is the study of cellular activities during interphase, which result in such diverse phenomena as photosynthesis, muscle movement, and thought. Most of the rest of this text will be devoted to those topics. Here, though, we focus on the events of cell division.

Mitosis

As we outlined earlier, eukaryotic cell division consists of mitosis (nuclear division) and cytokinesis (cytoplasmic division). These two processes usually occur together, but, in some cells, mitosis may occur without cytokinesis. Before we begin our discussion of mitosis, remember that the chromosomes are duplicated during the S phase of interphase. Therefore, when mitosis begins, each chromosome

already consists of two sister chromatids attached to one another at the centromere.

For convenience, mitosis is divided into four phases: prophase, metaphase, anaphase, and telophase (Figs. 9-7 and 9-8). As with most biological processes, however, these phases are not really discrete events. Rather, they form a continuum, each phase merging into the next.

During Prophase, the Chromosomes Condense and the Spindle Apparatus Forms and Attaches to the Chromosomes

The first phase of mitosis is called **prophase** (meaning "the stage before" in Greek). Prophase develops from the G_2 phase of interphase, as three major events occur: chromosome condensation, assembly of the spindle apparatus, and capture of the chromosomes by the spindle.

During Early Prophase, the Chromosomes Condense and the Nucleolus Disappears

The chromosomes have already been duplicated during interphase, and sister chromatids remain attached to each

| (e) ANAPHASE | (f) TELOPHASE | (g) CYTOKINESIS | (h) INTERPHASE OF DAUGHTER CELLS |

(e) Anaphase: Chromatids separate at the centromere, becoming independent chromosomes. The kinetochore microtubules shorten, pulling one former chromatid from each chromosome toward each pole of the cell. At about the same time, the polar microtubules slide past one another, moving poles farther apart.
(f) Telophase: One complete set of chromosomes has reached each pole of the cell. The chromosomes relax into their extended state, the spindle disappears, and the nuclear envelopes begin to re-form.

(g) Cytokinesis: Usually simultaneously with the end of telophase, the cytoplasm is divided along the equator of the parent cell, with each daughter cell receiving one nucleus and about half the original cytoplasm.
(h) Interphase of daughter cells: The daughter cells return to interphase. The spindle disappears, the nuclear envelope completely re-forms, the chromosomes finish decondensing, and the nucleolus reappears.

other at their centromeres. During interphase, the chromosomes are extended, allowing the information in their DNA to be used by the cell (Fig. 9-8a). During prophase, the chromosomes coil and condense, becoming visible in a light microscope (Fig. 9-8b). As the chromosomes condense, the nucleolus disappears. You may recall from Chapter 5 that the nucleolus contains an aggregation of the parts of several chromosomes that are involved in ribosome synthesis. As the chromosomes condense and separate from one another, this aggregation breaks up.

The Spindle Apparatus Consists of an Array of Microtubules That Will Separate the Sister Chromatids during Anaphase

Near the end of prophase, after the chromosomes are condensed, the **spindle apparatus** begins to form. The spindle apparatus is a football-shaped arrangement of microtubules that arises in the following way.

During interphase, a cell contains an array of microtubules radiating outward from the **microtubule organizing center** near the nucleus (see Chapter 5). During prophase, the interphase microtubules disintegrate into

individual subunits of the protein tubulin. The microtubule organizing center splits in two, and each daughter center begins to form a new array of microtubules from the tubulin subunits. Gradually the two microtubule organizing centers move to opposite sides of the nucleus (the poles of the cell). The microtubules extending from each pole almost surround the nucleus. This new microtubule array is called the spindle apparatus because of its resemblance to the thread wound around the spindle of a spinning wheel—fat in the middle, and tapering more or less to a point on either end.

Although the operation of the spindle during mitosis is similar in all eukaryotic cells, the spindles of animal and plant cells differ in two details. First, in animal cells, each microtubule organizing center contains a pair of **centrioles** (Fig. 9-9). During interphase, the two centrioles separate and a new centriole develops near the base of each "parent" centriole, resulting in a microtubule organizing center containing two pairs of centrioles. When the microtubule organizing center splits, each daughter center contains one pair of centrioles. Second, in animal cells, spindle microtubules extend not only across the nucleus toward the other pole, but also outward in all directions from the microtubule organizing center toward the plasma membrane, forming a star-shaped **aster.** Plant cells, in contrast, do not have centrioles and consequently do not form asters.

The Spindle Microtubules Attach to the Chromosomes at Their Centromeres

When the spindle is fully formed, the nuclear envelope abruptly disintegrates. The double membrane of the envelope breaks up into vesicles that resemble pieces of endoplasmic reticulum. The disruption of the nuclear envelope allows the microtubules of the spindle to invade the nuclear region (see Fig. 9-8c). Some of these microtubules will attach to the chromosomes and some will interact with one another.

Each sister chromatid has a structure called a **kinetochore** located at the centromere (Fig. 9-10). Therefore, each replicated chromosome has two kinetochores. Each kinetochore grabs some of the passing microtubules (now called **kinetochore microtubules**) as they invade the nuclear region. Although the exact mechanism is still a matter of investigation, the kinetochore of one sister chromatid of each chromosome captures microtubules leading to one pole, while the kinetochore of the other sister chromatid captures microtubules leading to the opposite pole. The sister chromatids of each chromosome therefore face opposite poles of the spindle (see Fig. 9-10). When the sister chromatids separate later in mitosis, the kinetochore microtubules will pull the newly independent chromosomes to opposite poles.

Other microtubules do not attach to chromosomes but retain free ends that overlap along the cell's equator. As we will see, these **polar microtubules** will push the two spindle poles apart.

During Metaphase, the Chromosomes Are Aligned along the Equator of the Cell

During **metaphase** (the "middle stage"), the kinetochore microtubules running to opposite spindle poles engage in a tug-of-war, each pulling toward its own pole. Apparently, long kinetochore microtubules pull harder than short ones do. Therefore, if a chromosome is farther from one pole than from the other pole, it is dragged toward the more distant pole. It is also possible that, at this stage, chromosomes are pushed away from the spindle poles; the closer a chromosome is to a spindle pole, the harder it is pushed away. By being pulled toward the more distant pole or being pushed away from the closer pole, or both, each chromosome ends up aligned along the cell's equator, midway between the spindle poles. Metaphase ends when all the chromosomes have lined up along the equator (see Fig. 9-8d.).

During Anaphase, Sister Chromatids Separate and Are Pulled to Opposite Poles of the Cell

At the beginning of **anaphase** (see Fig. 9-8e), the centromere of each chromosome divides, and the sister chromatids separate into two independent daughter chromosomes. The

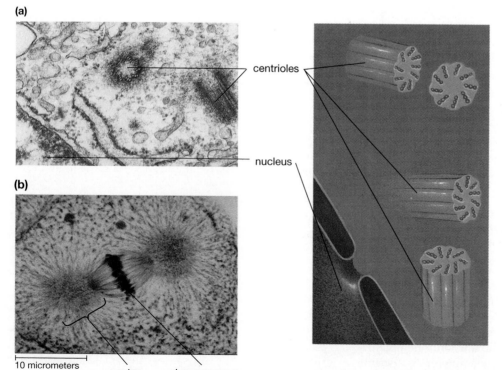

(a)

centrioles

nucleus

(b)

10 micrometers

aster   chromosomes

Figure 9-9   Details of the spindle apparatus in animal cells

(a) An electron micrograph of the microtubule organizing center of an animal cell. The centrioles were replicated just before cell division. The centrioles of each pair lie at right angles to each other. During mitosis, the two pairs migrate to opposite sides of the nuclear region (see Fig. 9-8b).
(b) Asters in embryonic whitefish cells. The asters consist of microtubules that extend both in toward the equator of the cell and out toward the plasma membrane. The centrioles, which are not visible at this magnification, lie in the center of the asters.

kinetochores now perform some fancy molecular moves; protein "motors" in the kinetochores, similar to those that cause bending of cilia and flagella, pull the kinetochores (and their attached chromosomes) poleward along the kinetochore microtubules (Fig. 9-11a). Simultaneously, the microtubules disassemble inside the kinetochores so that the microtubules shorten at about the same rate as the kinetochores move toward the poles.

An analogy might help you to understand these processes: Think of the kinetochores as a pair of sailors, pulling a small boat in which they are standing (the chromosome) to shore (the spindle pole) by pulling on a rope (the kinetochore microtubules) attached to the shore. One sailor (the motor) pulls the rope in, hand over hand, moving the boat to the shore. To keep from filling up the boat with rope, the other sailor cuts off the end of the rope as fast as the "motor sailor" pulls it in and throws the pieces overboard (into the cytoplasm).

At the same time as the kinetochores are towing their chromosomes toward the poles, the polar microtubules grow longer and push the spindle poles apart (Fig. 9-11b).

Note that one sister chromatid of each chromosome moves toward each pole of the cell. Because the sister chromatids are identical copies of the original chromosomes, each of the two clusters of chromosomes that form contains one copy of every chromosome.

The usefulness and accuracy of the mechanisms for sorting and separating chromosomes do not depend on the numbers of chromosomes in a cell. Mitosis works equally well in cells with only a handful of chromosomes, such as those of fruit flies (eight), and in those containing hundreds of chromosomes, such as the cells of horsetails and some ferns. Similarly, the same mechanisms are employed in haploid, diploid, and polyploid cells.

During Telophase, Nuclear Envelopes Re-form around Each Group of Chromosomes

When the chromosomes reach the poles of the spindle, **telophase** (the "end stage") has begun (see Fig. 9-8f). The spindle disintegrates. Vesicles that formed when the old nuclear envelope broke up during late prophase coalesce around each group of chromosomes, forming two new nuclear envelopes. The chromosomes relax into their extended state once again, and the nucleoli reappear. In most cells, cytokinesis occurs during telophase, enclosing each daughter nucleus into a separate cell.

Cytokinesis

In most cells, the division of the cytoplasm into nearly equal halves begins during telophase. In animal cells, microfilaments composed of the proteins actin and myosin form rings around the equator of the cell, surrounding the remnants of the spindle (Fig. 9-12). The microfilaments are attached to the plasma membrane. During cytokinesis, the rings contract and pull in the equator of the cell, much like pulling the drawstring around the waist of a pair of sweatpants. Eventually the "waist" contracts down to nothing, dividing the cytoplasm into two new daughter cells.

Cytokinesis in plant cells is quite different, perhaps because the stiff cell wall makes it impossible to divide one cell

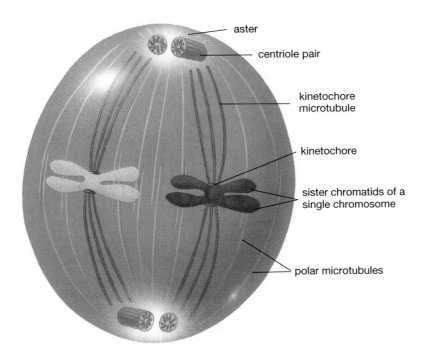

aster

centriole pair

kinetochore microtubule

kinetochore

sister chromatids of a single chromosome

polar microtubules

Figure 9-10 **A diagram of the spindle apparatus in an animal cell**

Note that there are three sets of spindle microtubules. First, the aster consists of microtubules radiating out from the pair of centrioles at the pole toward the plasma membrane (see Fig. 9-9b). Second, kinetochore microtubules run from one pole to one kinetochore of each chromosome. The two kinetochores of a single chromosome attach to kinetochore microtubules that lead to opposite poles of the cell. Finally, polar microtubules extend from each pole to, and slightly beyond, the equator of the cell. Polar microtubules overlap at the equator.

(a)

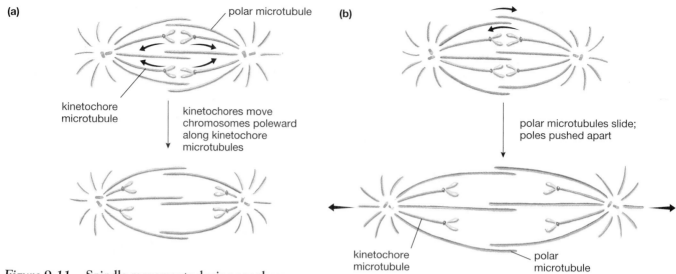

(b)

polar microtubule

kinetochore
microtubule

kinetochores move
chromosomes poleward
along kinetochore
microtubules

polar microtubules slide;
poles pushed apart

kinetochore
microtubule

polar
microtubule

Figure 9-11   **Spindle movements during anaphase**

These movements occur just after the separation of chromatids to form individual daughter chromosomes. **(a)** Movement of chromosomes toward the poles: Kinetochores of the newly formed daughter chromosomes move poleward along the kinetochore microtubules, while the microtubules simultaneously shorten. The kinetochores remain attached to the microtubules even as the microtubules shorten. **(b)** Separation of the poles: At their sites of overlap along the equator of the cell, polar microtubules slide past one another, probably by a mechanism similar to the bending of cilia or flagella (see Fig. 5-22). This mechanism pushes the poles of the spindle farther apart.

into two by pinching at the waist. Instead, the Golgi complex buds off carbohydrate-filled vesicles that line up along the equator of the cell between the two nuclei (Fig. 9-13). The vesicles fuse together, producing a structure called the **cell plate,** shaped much like a hot-water bottle filled with sticky carbohydrates. When enough vesicles have fused, the edges of the cell plate merge with the original plasma membrane around the circumference of the cell. Seen as if you were facing the dividing cell in Figure 9-13, the left membrane of the cell plate becomes part of the plasma membrane of the left daughter cell, while the right membrane of the cell plate becomes part of the plasma membrane of the right

daughter cell. The carbohydrate formerly contained in the vesicles remains between the plasma membranes as the middle lamella of the cell wall (see Chapter 6).

Cell Division and Asexual Reproduction

Through cell division and subsequent diversification of cells, a single cell, usually a fertilized egg, generates the multicellular bodies of plants and animals. Cell division is

(a)

(b)

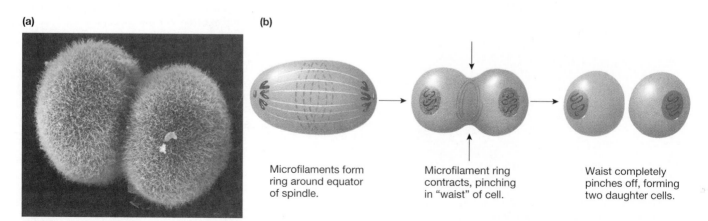

Microfilaments form
ring around equator
of spindle.

Microfilament ring
contracts, pinching
in "waist" of cell.

Waist completely
pinches off, forming
two daughter cells.

Figure 9-12   **Cytokinesis in animal cells**

(a) A ring of microfilaments just beneath the plasma membrane contracts around the equator of the cell, pinching it in two. **(b)** The mechanism of cytokinesis in animal cells.

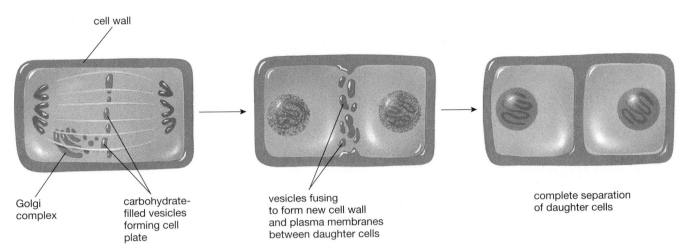

cell wall

Golgi complex

carbohydrate-filled vesicles forming cell plate

vesicles fusing to form new cell wall and plasma membranes between daughter cells

complete separation of daughter cells

Figure 9-13   **Cytokinesis in plant cells**

The mechanism of cell plate formation and cytokinesis in plant cells. Carbohydrate-filled vesicles produced by the Golgi complex congregate at the equator of the cell, forming the cell plate. The vesicles will fuse to form the two plasma membranes separating the daughter cells, while their carbohydrate contents form the middle lamella.

also the basis of **asexual reproduction,** in which offspring are formed from a single parent without the necessity of uniting male and female gametes. This is the usual mode of reproduction for many unicellular organisms, such as *Paramecium, Euglena,* and yeasts (Fig. 9-14a).

Many multicellular organisms can also reproduce asexually. Small replicas of the parent grow through cell division. A *Hydra,* for example, can reproduce by growing a miniature new *Hydra* as a bud (Fig. 9-14b). Eventual-

ly the bud separates from its parent, going off to live independently. Because mitosis produces genetically identical cells, these offspring are genetically identical to their parents.

Many plants reproduce both asexually and sexually. The beautiful aspen groves of Colorado, Utah, and New Mexico (Fig. 9-15) develop asexually from shoots growing up from the root system of a single parent tree. The entire grove, although seeming to be a population of separate

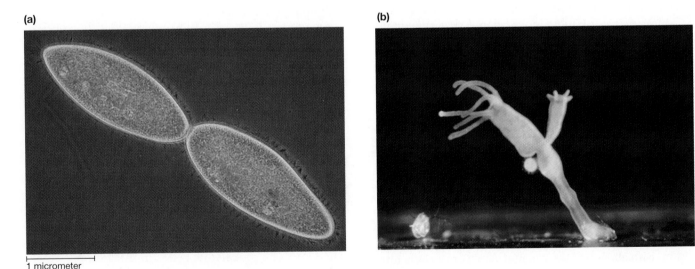

(a)

(b)

1 micrometer

Figure 9-14   **Modes of asexual reproduction**

(a) In unicellular microorganisms, such as the protist *Tetrahymena* shown here, cell division produces two new, independent organisms. **(b)** *Hydra,* a freshwater relative of jellyfish, grows a miniature replica of itself (a bud) protruding from its side. When fully developed, the bud breaks off and assumes independent life.

Figure 9-15   **The trees in aspen groves are genetically identical**

Each tree grows as a shoot from the roots of a single ancestral tree. This photo shows three separate groves near Aspen, Colorado. The genetic identity within a grove and genetic difference between groves are shown by their leaves in fall. One entire grove is still green, the second has turned bright gold, and the third has already lost its leaves.

trees to the admiring visitor, may actually be considered to be a single individual, with its multiple trunks interconnected by a common root system. Recently, biologists at the University of Colorado calculated that the single largest organism yet discovered on Earth is probably a huge aspen grove in Utah, covering 106 acres and including about 47,000 trunks with a total mass of perhaps almost 6 million kilograms (13 million pounds). Finally, although individual trunks may age and die, the grove lives on: Some groves are thought to be hundreds of thousands of years old. If you plant an aspen tree in your backyard, who knows what you may be starting!

❊ S U M M A R Y   O F   K E Y   C O N C E P T S

The Essentials of Cellular Reproduction
Every cell must begin life with (1) a complete set of genetic information (DNA) and (2) the materials necessary to survive and use the genetic information. Cellular reproduction involves replication of the DNA, with one copy packaged into each daughter cell, and cytoplasmic division, with about half the parental cytoplasm passing into each daughter cell.

Prokaryotic and Eukaryotic Chromosomes Compared
The DNA of a prokaryotic cell is contained in a single, circular chromosome, not separated from the cytoplasm of the cell. Eukaryotic cells contain much more DNA than prokaryotic cells do, packaged in many separate chromosomes that are contained within a membrane-bound nucleus.

The Prokaryotic Cell Cycle
The DNA of a prokaryotic cell is in the form of a single circular chromosome, attached at one point to the cell membrane. When a prokaryotic cell reproduces, it duplicates the chromosome, moves the two resulting copies to opposite ends of the cell, and divides the cell in two.

The Eukaryotic Chromosome

The chromosomes of eukaryotic cells consist of both DNA and protein. During interphase, the chromosomes are in an extended form, accessible for use in reading their genetic instructions. During cell division, the chromosomes condense to form short, thick structures. At some point in all eukaryotic life cycles, cells contain pairs of chromosomes called homologues. Homologues have virtually identical appearance and similar genetic information. Cells with pairs of homologous chromosomes are diploid. Those with only a single member of each pair are haploid, and those with more than two copies of each homologous chromosome are polyploid.

The Eukaryotic Cell Cycle

The eukaryotic cell cycle consists of two major phases, interphase and cell division. During interphase, the cell takes in nutrients, grows, and duplicates its chromosomes. Cell division consists of two processes, mitosis (nuclear division) and cytokinesis (cytoplasmic division). Mitosis parcels out one copy of each chromosome into two separate nuclei, and cytokinesis subsequently encloses each nucleus in a separate cell.

Mitosis

The chromosomes are duplicated during interphase, prior to mitosis. The two identical copies, called chromatids, are attached to one another at the centromere. During mitosis, the chromatids separate, one going to each daughter nucleus. The two daughter nuclei thus receive identical genetic information. Mitosis consists of four phases:

Prophase: The chromosomes condense and the spindle microtubules form. The nuclear membrane breaks apart into vesicles. The spindle microtubules invade the nuclear region. Some microtubules, now called kinetochore microtubules, attach to the chromosomes at kinetochores located at the centromeres. Other microtubules, called polar microtubules, extend from the spindle poles to the equator and overlap at the equator with polar microtubules extending from the opposite pole.

Metaphase: The chromosomes move to the equator of the cell.

Anaphase: The two chromatids of each chromosome separate; each former chromatid is now considered to be an individual chromosome. The spindle microtubules move the newly independent chromosomes to opposite poles of the cell.

Telophase: The chromosomes relax into their extended state, and nuclear envelopes re-form around each new daughter nucleus.

Cytokinesis

Mitosis may occur without cytokinesis, for example in mammalian muscle cells and many fungi. In most cells, however, cytokinesis roughly coincides with the end of telophase. Cytokinesis usually divides the cytoplasm into approximately equal halves. In animal cells, the plasma membrane is pinched in along the equator by a ring of microfilaments, separating the cytoplasm into two daughter cells, each containing one of the daughter nuclei. In plant cells, new plasma membrane forms along the equator by fusion of vesicles produced by the Golgi complex.

Cell Division and Asexual Reproduction

Some organisms reproduce asexually—that is, without the fusion of gametes. Asexual reproduction usually occurs by the growth of a new organism from the body of a single parent organism, accomplished by repeated cell divisions and differentiation of daughter cells. Therefore, the offspring are genetically identical to the parent.

❈ KEY TERMS

anaphase   p. 168
asexual reproduction   p. 171
aster   p. 168
binary fission   p. 161
cell cycle   p. 159
cell division   p. 159
cell plate   p. 170
centriole   p.167
centromere   p. 162
chromatid   p. 162
chromatin   p. 161
chromosome   p. 161

cytokinesis   p. 173
deoxyribonucleic acid (DNA)   p. 160
diploid   p. 163
G_1 phase   p. 164
G_2 phase   p. 164
gamete   p. 163
haploid   p. 163
homologue   p. 163
interphase   p. 163
kinetochore   p. 168
kinetochore microtubule   p. 168
life cycle   p. 159

meiosis   p. 159
metaphase   p. 168
microtubule organizing center   p. 167
mitosis   p. 173
polar microtubule   p. 168
polyploid   p. 163
prophase   p. 166
S phase   p. 164
sexual reproduction   p. 159
spindle apparatus   p. 167
telophase   p. 169
zygote   p. 163

✖ THINKING THROUGH THE CONCEPTS

Multiple Choice

1. Since most somatic (body) cells of a fruit fly have 8 chromosomes each, the zygote would have
   a. 2 chromosomes
   b. 4 chromosomes
   c. 8 chromosomes
   d. 16 chromosomes
   e. 32 chromosomes

2. At which stage of mitosis are chromosomes arranged along a plane at the midline of the cell?
   a. anaphase   b. telophase
   c. metaphase   d. G_2
   e. interphase

3. A diploid cell contains in its nucleus
   a. an even number of chromosomes
   b. an odd number of chromosomes
   c. one copy of each homologue
   d. either an even or an odd number of chromosomes
   e. two sister chromatids of each chromosome during G_1

4. How many chromosomes are in a human sperm cell?
   a. 8   b. 16
   c. 23   d. 46
   e. 92

5. Synthesis of new DNA occurs during
   a. prophase
   b. interphase
   c. mitosis.
   d. cytokinesis
   e. formation of the cell plate

6. What is the attachment site on the chromosome for pulling chromosomes apart during mitosis?
   a. the cell plate
   b. the aster
   c. the centriole
   d. the kinetochore
   e. none of the above

Review Questions

1. Define a cell cycle and a life cycle. What is the role of the cell cycle in a bacterium? In a human?

2. Diagram and describe the prokaryotic cell cycle.

3. Diagram and describe the eukaryotic cell cycle. Name the various phases and briefly describe the events that occur during each.

4. Define mitosis and cytokinesis. Do these two processes always occur together?

5. Diagram the stages of mitosis. How does mitosis ensure that each daughter nucleus receives a complete set of chromosomes?

6. Define the following terms: homologous chromosome; centromere; kinetochore; chromatid; diploid; haploid; polyploid.

7. Diagram the spindle apparatus. How do the two types of spindle microtubules operate during anaphase?

8. Describe and compare the process of cytokinesis in animal and plant cells.

9. Compare the roles of polar and kinetochore microtubules during mitotic cell division.

✖ APPLYING THE CONCEPTS

1. Neurons in the adult human central nervous system and heart muscle cells are postmitotic. In other words, they remain in interphase. By contrast, cells lining the lumen of the small intestine divide frequently. Discuss this in terms of why damage to the former (brain in stroke and heart in heart attack) is so dangerous. What do you think might happen to tissues like the intestinal lining if some disorder or drug blocked mitoses in all cells of the body?

2. Do you think that each type of cell in your body steadily progresses through one cell division after another, or do you think that the rate of division is likely to vary from time to time? Explain and justify your answer in terms of some common events in human life: growing up as a child; healing a wound in your skin; menstruation; replacing blood donated to the Red Cross; or the unfortunate case of cancer.

3. A drug called colchicine inhibits the polymerization of microtubules. What specific effects do you think colchicine would have on cell division? What effects would it have on the daughter cell(s)?

4. Cancer cells divide out of control. Chemotherapy and radiation therapy may interfere with cell division in other cells, with consequences such as loss of hair, and loss of the gastrointestinal lining, causing severe nausea. Are such therapies justified, in your opinion? If a doctor recommended such treatment to you, how might your answer differ depending on your age, or on the severity of the cancer?

5. Suppose you were a researcher using tissue from a deceased person who had donated his body to science, and you discovered a chromosomal abnormality that might affect the health of surviving offspring. Should you be allowed (or even required) to communicate this finding to relatives of the deceased?

✖ FOR MORE INFORMATION

Glover, D. M., Gonzalez, C., and Raff, J. W. "The Centrosome." *Scientific American*, June 1993. The structure and function of the centrioles that are essential to the motility of chromosomes during anaphase are presented in this up-to-date paper.

Grant, M. C. "The Trembling Giant." *Discover*, October 1993. Aspen groves are really single individuals; huge, slowly spreading from the roots of the original parent tree, and potentially almost immortal.

Mazia, D. "The Cell Cycle." *Scientific American*, January 1974. A clearly illustrated description of the essentials of cell division.

McIntosh, J. R., and McDonald, K. L. "The Mitotic Spindle." *Scientific American*, October 1989. Beautiful micrographs and lucid drawings describe the structure and function of the spindle apparatus.

Murray, A. W., and Kirschner, M. W. "What Controls the Cell Cycle." *Scientific American*, March 1991. The synthesis, processing, and degradation of a group of proteins seems to control the progression of a cell through the stages of the cell cycle.

NET WATCH

On-line resources for this chapter are on the World Wide Web at:
http://www.prenhall.com/~audesirk (click on the <u>table of contents</u> link and then select Chapter 9).

10

DNA:
The Molecule of Heredity

"... a structure this pretty just had to exist."
JAMES WATSON in *The Double Helix*

▓ A T A G L A N C E

The Composition of Chromosomes
DNA Is Composed of Four Types of Nucleotides
Is the Genetic Material DNA or Protein?

DNA: The Hereditary Molecule
Bacterial Transformation Indicates That the Hereditary
Molecule of Bacteria Is DNA
The Hereditary Molecule of Bacteriophages Is Also DNA
The Sequence of Bases in DNA Can Encode Vast Amounts
of Information

The Structure of DNA
The DNA of Chromosomes Is a Double Helix

DNA Replication: The Key to Constancy
The Steps of DNA Replication Are Catalyzed by the
Coordinated Action of Several Enzymes
Base Pairing and Proofreading Produce Almost Error-Free
Replication of DNA

A computer-generated model of the structure of DNA. In the words of James Watson, "... a structure this pretty just had to exist."

As you know, all living organisms on Earth are composed of cells. Each cell in existence today is the offspring of a previously existing cell, produced when the parent cell divided into two daughter cells. Furthermore, each daughter cell is, with extremely rare exceptions, genetically identical to its parent cell. The genetic information that guides the life of each cell is contained in its chromosomes. As we saw in Chapter 9, during mitosis, an exquisite, highly organized dance of the chromosomes is executed, which ensures that each daughter cell receives a full set of chromosomes. This information still leaves important questions unanswered: How is genetic information contained in the chromosomes? How are the chromosomes duplicated so precisely that each daughter cell receives a full set of chromosomes that are exact copies of the chromosomes of the parent cell?

Just 50 years ago, no one knew that **deoxyribonucleic acid,** or **DNA** (see the chapter opening art), is the molecule of heredity in all forms of life on Earth. Even after DNA was discovered and found to be part of chromosomes, it was not immediately recognized as the molecule that carries genetic information. Finally, when DNA was accepted as the molecule of heredity, questions still remained about how it encodes genetic information and how that information is faithfully copied when chromosomes are duplicated. These issues occupied many of the best minds of the mid-twentieth century. In this chapter, we will follow those minds along the paths of discovery that led to our present understanding of the structure and function of DNA. Along the way, we hope you will gain insight not only into the molecular nature of DNA but also into how scientists work and how they unraveled part of the complex fabric of inheritance. In Chapter 11, we will investigate the mechanisms whereby the genetic information carried by DNA is used by a cell to produce the structures and functions necessary for continued life.

The Composition of Chromosomes

In the early 1870s, the biochemist Friedrich Miescher extracted a previously unknown chemical substance from nuclei, an acidic material with an unusually high phosphorus content. Because of its location in the nucleus and its acidic properties, this material came to be known as nucleic acid. In the early 1900s, the discovery that chromosomes are the carriers of genetic information focused on their chemical composition. Biochemists found that a eukaryotic chromosome is composed only of protein and a specific kind of nucleic acid, deoxyribonucleic acid (DNA). One of these substances, therefore, must carry the cell's hereditary blueprint, but which one?

DNA Is Composed of Four Types of Nucleotides

DNA seemed to be a particularly simple molecule, composed of just four kinds of subunits, called **nucleotides,** strung together in a long chain. As we described in Chapter 3, each nucleotide consists of three parts: a phosphate group; deoxyribose, a five-carbon sugar; and a nitrogen-containing **base:**

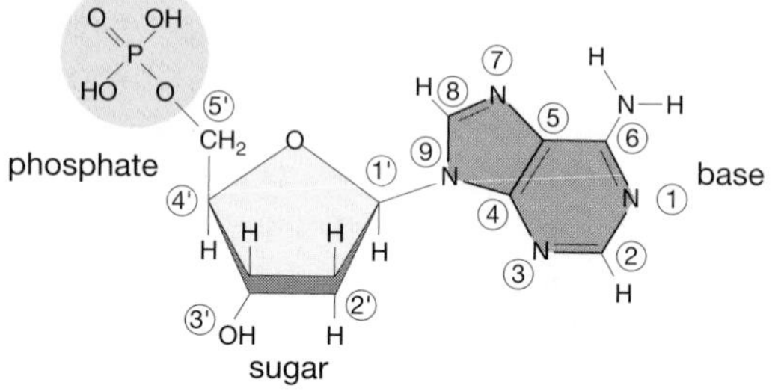

Biochemists have assigned numbers to the carbon and nitrogen atoms that make up the skeletons of the sugar and the base (the numbers for the sugar are given as 1' ["one prime"], 2', etc., to distinguish them from the numbers for the base).

The four different DNA nucleotides have the same phosphate and sugar but different bases. The bases come in two types: **pyrimidines,** thymine (abbreviated T) and cytosine (C), which consist of a single ring; and the two-ringed **purines,** adenine (A) and guanine (G):

Nucleotides with pyrimidine bases:

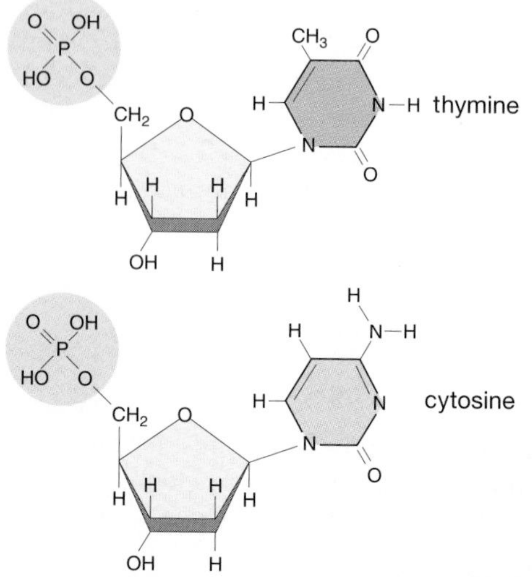

Nucleotides with purine bases:

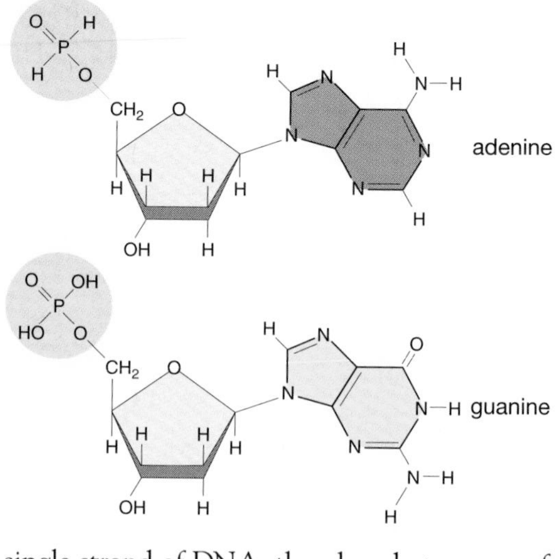

In a single strand of DNA, the phosphate group of one nucleotide bonds to the sugar of another. This arrangement forms a long strand consisting of a "backbone" of sugars and phosphates, with the bases protruding from the backbone:

Single strand of DNA:

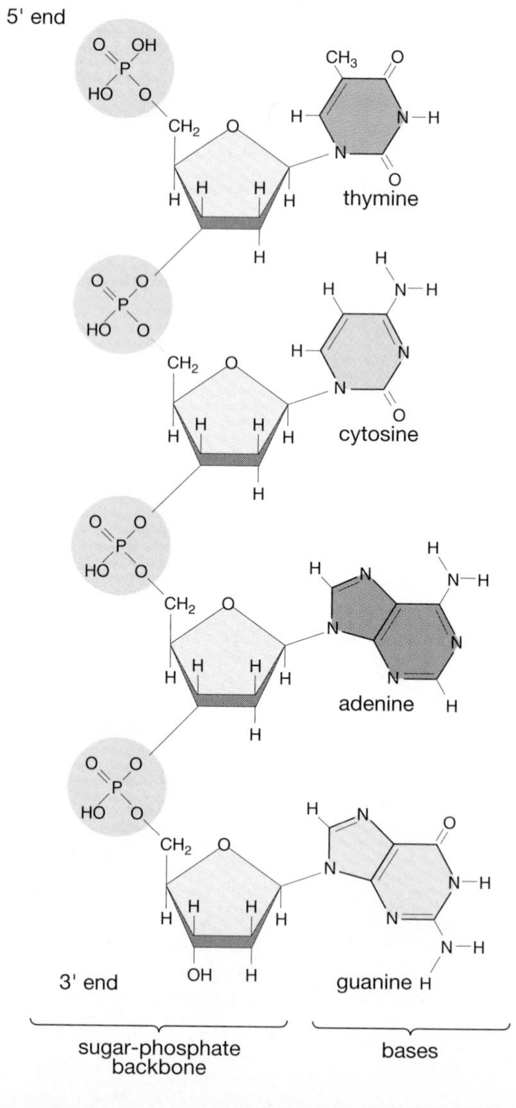

The phosphate is attached to the 5' carbon (the numbering system for the carbon atoms of the sugars comes in handy here). To form a single strand of DNA, the phosphate of one nucleotide undergoes a condensation reaction with the hydroxyl group attached to the 3' carbon of the next nucleotide, thus bonding the phosphate and sugar. This bonding arrangement orients all nucleotides in the strand in the same direction. Notice also that the DNA strand has a "free" sugar on one end (the 3' end of the strand) and a "free" phosphate on the opposite end (the 5' end of the strand).

Is the Genetic Material DNA or Protein?

Because DNA has only four subunits, each quite similar to the others, many biologists were skeptical that DNA could contain the information needed to construct all the molecules of complex organisms. Proteins, in contrast, are constructed of about 20 different amino acids. Intuitively, it would seem that having a greater number of building blocks offers much more opportunity for carrying the enormous amount of genetic information that each chromosome obviously contains. Besides that, most of the substance of living cells, aside from water, is protein or is synthesized through the action of enzymes, which are also proteins. An attractive hypothesis, therefore, was that the proteins of chromosomes are templates for the proteins of the rest of the cell, much like molds from which copies could be made.

Gradually, biologists accumulated information that indicated that proteins might not be the genetic material. For example, chromosomes in the sperm of some fish contain only DNA and one protein, called protamine. Like other animals, a fish receives half its genes from its male parent, through the male's sperm. Now, protamine is a very simple protein, mostly composed of a single amino acid. If protamine *were* the genetic material in fish sperm, how could it possibly serve as a template for all the other, more complex, proteins of the whole fish? Despite contradictory evidence such as this, many scientists continued to support the protein-as-gene hypothesis. Finally, two experiments with microorganisms showed that DNA must be the hereditary material.

DNA: The Hereditary Molecule

Throughout the history of biology, major advances have been made possible by using the "right" organism for a particular experiment. Often, a certain organism is especially suitable for a certain experiment because it is simple in structure or reproduces very rapidly. Such simple, rapidly reproducing organisms were the keys to proving that DNA, not protein, is the chemical of heredity.

Bacterial Transformation Indicates That the Hereditary Molecule of Bacteria Is DNA

In the mid-1920s, the bacteriologist Frederick Griffith tried to develop a vaccine to protect people against a bacterium, *Streptococcus pneumoniae*, that causes pneumonia.

He never produced a successful vaccine, so in one sense his experiments were failures. As it turned out, however, he discovered an intriguing phenomenon, which he called **transformation,** that eventually played a major role in the rise of molecular genetics.

Griffith found two varieties of *Streptococcus*. In one variety, each bacterium has a polysaccharide capsule; in the other form, the bacteria are naked. Griffith hypothesized that the capsule might affect the ability of the two varieties to cause disease. He ran four experiments with the two varieties of *Streptococcus* (Fig. 10-1).

1. **Injection with live encapsulated bacteria.** When he injected mice with encapsulated bacteria, the mice contracted pneumonia and died (Fig. 10-1a). Autopsy showed that the blood of infected mice contained hordes of encapsulated bacteria. Later it was found that when *Streptococcus* bacteria invade a mammalian host—for example, a mouse—the host's white blood cells try to ingest the invading microbes. Encapsulated bacteria seem to be "slippery," preventing the white blood cells from getting a grip on the bacterial surface. Consequently, white blood cells cannot ingest encapsulated bacteria very well. Encapsulated bacteria therefore multiply in the host and cause disease.
2. **Injection with live naked bacteria.** If mice were injected with the naked variety of *Streptococcus*, they remained healthy (Fig. 10-1b). No *Streptococcus* could be found in the mice's blood. Later discoveries showed that naked bacteria are easily destroyed by white blood cells, so naked bacteria do not cause disease.
3. **Injection with heat-killed encapsulated bacteria.** Griffith found that heating encapsulated *Streptococcus* to a high temperature killed the bacteria. Not surprisingly, dead bacteria did not cause pneumonia, and the mice remained free of living bacteria (Fig. 10-1c).
4. **Injection with a mixture of dead encapsulated bacteria and live naked bacteria.** Griffith injected mice with a mixture of heat-killed encapsulated bacteria and live naked bacteria (Fig. 10-1d). Although neither of these caused pneumonia when injected alone, the mixture did cause pneumonia. What's more, the infected mice teemed with live encapsulated bacteria, and these encapsulated bacteria bred true; that is, all of their offspring also were encapsulated.

A few years later, in 1933, J. L. Alloway obtained the same results as Griffith's fourth experiment without any mice at all. He killed encapsulated bacteria, filtered out the larger bacterial parts, and mixed the resulting solution with living naked bacteria. The mixture produced live, true-breeding, encapsulated bacteria.

What did Alloway's experiments mean? A likely hypothesis was that the living bacteria had acquired molecules of genetic information from the dead bacteria. The molecules encoded the instructions for making capsules, thereby *transforming* the formerly naked bacteria into the encapsu-

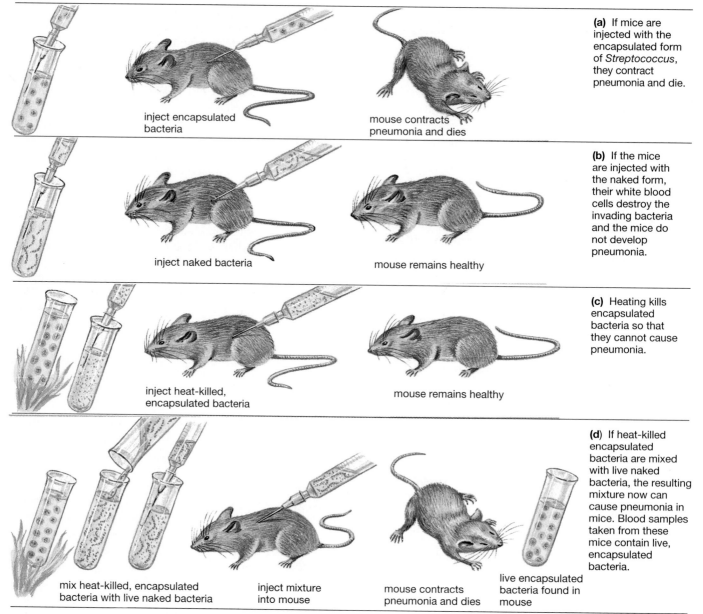

(a) If mice are injected with the encapsulated form of *Streptococcus*, they contract pneumonia and die.

inject encapsulated bacteria

mouse contracts pneumonia and dies

(b) If the mice are injected with the naked form, their white blood cells destroy the invading bacteria and the mice do not develop pneumonia.

inject naked bacteria

mouse remains healthy

(c) Heating kills encapsulated bacteria so that they cannot cause pneumonia.

inject heat-killed, encapsulated bacteria

mouse remains healthy

(d) If heat-killed encapsulated bacteria are mixed with live naked bacteria, the resulting mixture now can cause pneumonia in mice. Blood samples taken from these mice contain live, encapsulated bacteria.

mix heat-killed, encapsulated bacteria with live naked bacteria

inject mixture into mouse

mouse contracts pneumonia and dies

live encapsulated bacteria found in mouse

Figure 10-1   **The discovery of transformation**

Griffith's discovery of transformation in pneumonia bacteria.

lated form. As it turns out, transformation is an effective evolutionary strategy because it allows bacteria to acquire new genes, which might help them to colonize new hosts.

The Transforming Molecule Is DNA

In 1944, three researchers at Rockefeller University—Oswald Avery, Colin MacLeod, and Maclyn McCarty—discovered that the transforming molecule is DNA. They isolated DNA from encapsulated bacteria, mixed it with live naked bacteria, and produced live encapsulated bacteria. To show that transformation was caused by DNA, and not by small quantities of protein contaminating the DNA, they treated different samples with protein-destroying enzymes. These enzymes did not affect the transforming ability of

the DNA samples; however, treating samples with DNA-destroying enzymes prevented transformation.

Avery, MacLeod, and McCarty deduced that DNA is the molecule responsible for transformation and that a live bacterium can take up DNA from its environment and incorporate this DNA into its own chromosome (Fig. 10-2). When DNA carrying the genes for capsules becomes part of the chromosome of another bacterium, the recipient becomes capable of synthesizing capsules. On the basis of this evidence, they concluded that DNA is the genetic material of bacteria.

Despite this evidence, many biologists remained unconvinced that DNA was the universal hereditary molecule. Perhaps, some thought, DNA induces a mutation in naked bacteria, changing a (protein) gene that does not di-

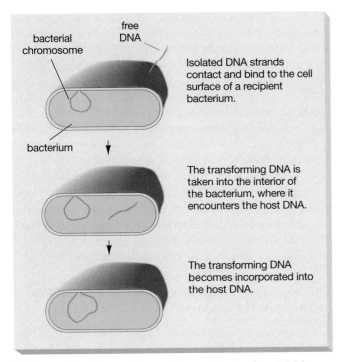

Figure 10-2   Bacterial transformation by free DNA

Transformation is not the "invasion" of a bacterium by free DNA. Rather, the recipient actively acquires the DNA and incorporates it into its chromosome. Transformation allows bacteria to acquire new genes and confers a selective advantage on the recipient bacteria. For example, transformation aids in the spread of antibiotic resistance from one type of bacterium to another.

rect capsule synthesis into one that does. Or perhaps DNA is the hereditary molecule of bacteria, but not of other organisms. A second experiment convinced most of the skeptics. This experiment used an even simpler system, a virus that infects bacteria.

The Hereditary Molecule of Bacteriophages Is Also DNA

Certain viruses infect only bacteria and are called **bacteriophages,** meaning "bacteria eaters" in Greek (Fig. 10-3). A bacteriophage (phage for short) depends on its host bacterium for every aspect of its life cycle (Fig. 10-4). When a phage encounters a bacterium, it attaches to the bacterial cell wall and injects its genetic material into the bacterium. The rest of the phage remains outside the bacterium. The phage genes subvert the bacterial metabolism into producing more phages. Finally, the genetic material of the phage directs the synthesis of an enzyme that ruptures the bacterium, liberating the newly manufactured phages.

Even though many bacteriophages have elaborate structures (see Fig. 10-3), they are chemically very simple, being composed only of DNA and protein. The question was, Which molecule is the phage genetic material?

In a stunning series of experiments published in 1952, Alfred Hershey and Martha Chase used the chemical simplicity of bacteriophages to deduce that the genetic material of phages is DNA (Fig. 10-5). Chemically, DNA and protein both contain carbon, oxygen, hydrogen, and

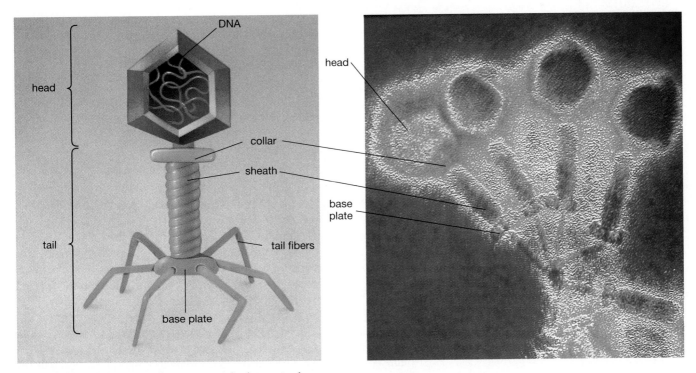

Figure 10-3   The general structure of a bacteriophage

The phage consists of a head region composed of a protein coat, containing a single DNA molecule, attached at the collar to a protein tail region (the sheath, base plate, and tail fibers) responsible for attachment to its host bacterium and injection of the genetic material.

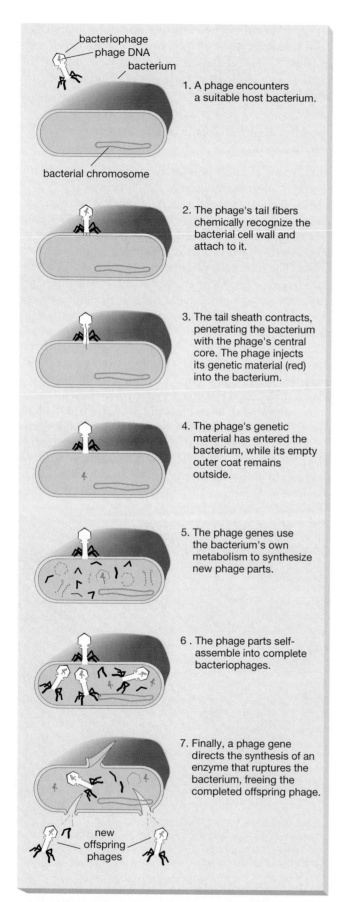

1. A phage encounters a suitable host bacterium.

2. The phage's tail fibers chemically recognize the bacterial cell wall and attach to it.

3. The tail sheath contracts, penetrating the bacterium with the phage's central core. The phage injects its genetic material (red) into the bacterium.

4. The phage's genetic material has entered the bacterium, while its empty outer coat remains outside.

5. The phage genes use the bacterium's own metabolism to synthesize new phage parts.

6. The phage parts self-assemble into complete bacteriophages.

7. Finally, a phage gene directs the synthesis of an enzyme that ruptures the bacterium, freeing the completed offspring phage.

Figure 10-4   **The life cycle of bacteriophages**

nitrogen. However, DNA also contains phosphorus but not sulfur, whereas protein contains sulfur (in the amino acids methionine and cysteine) but not phosphorus. Hershey and Chase forced one population of phages to synthesize DNA using radioactive phosphorus, thereby "labeling" their DNA. Another population was forced to synthesize protein using radioactive sulfur. When bacteria were infected by phages containing radioactively labeled protein, radioactivity did not appear inside the bacteria. Offspring phages were not radioactive either (Fig. 10-5a). When bacteria were infected by phages containing radioactive DNA, the radioactivity was subsequently found inside the bacteria. When the bacteria burst, some of the offspring phages also had radioactive DNA (Fig. 10-5b). These experiments showed that the genetic material that phages inject into their hosts is DNA, not protein.

The Sequence of Bases in DNA Can Encode Vast Amounts of Information

Although the experiments of Avery, MacLeod, and McCarty with bacterial transformation and those of Hershey and Chase with bacteriophage replication seemed to demonstrate that DNA must be the genetic material, you may still be wondering, "Can just four different types of bases in DNA encode all of the information needed to produce thousands of proteins, each with various combinations of 20 amino acids?" The answer is yes. The four different types of bases can be arranged in any linear order along a strand of DNA. Each sequence of bases represents a unique set of genetic instructions, like a biological Morse code (which, after all, can produce all of the English language with only two symbols). A stretch of DNA just 10 nucleotides long can exist in over a million different possible sequences of the four bases. Because an "average" chromosome has millions (in bacteria) to billions (plants and animals) of nucleotides, DNA molecules can encode a staggering amount of information.

The Structure of DNA

Even before the publication of Hershey and Chase's experiments in 1952, the demonstration that the bacterial transforming factor was DNA by Avery and his colleagues had stimulated a burst of research on the chemical nature of DNA. Alfred Mirsky studied the amount of DNA in the cells of various tissues of several organisms. He found that, although the quantity of DNA varies among species, it is constant in each nucleus of a given species no matter what tissue the nucleus comes from. Gametes are a notable exception in that they have half as much DNA as the other cells of the body. When we discuss meiosis and gamete formation in animals in Chapter 12, you will see why this is expected if DNA is the hereditary material.

What seemed equally significant, if only someone could figure out what it meant, were the results of Erwin Char-

(a)

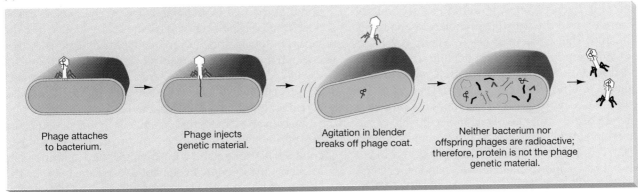

(b)

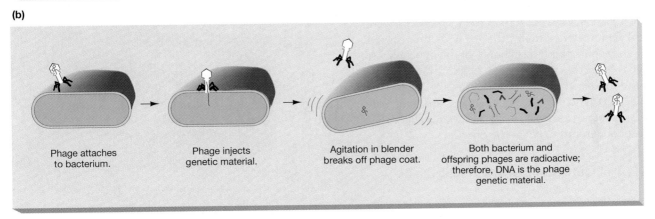

Figure 10-5   **The Hershey-Chase experiment**

To determine whether DNA or protein is the genetic material of bacteriophages, Hershey and Chase grew one population of phages in a medium containing radioactive sulfur, which radioactively labels the phage proteins. Another population of phages was grown in radioactive phosphorus, which labels the phage DNA. **(a)** The fate of protein labeled with radioactive sulfur (red). Phages attach to the cell walls of unlabeled bacteria and inject their genetic material. Whirling the mixture in a blender breaks off the phage heads and sheaths left on the outside of the bacteria. The radioactivity is found outside the bacteria, in the phage coats, but the bacteria, containing the phage genetic material, are not radioactive. New phages resulting from the infection are also not radioactive. The phage genetic material, therefore, is not protein. **(b)** The fate of DNA labeled with radioactive phosphorus (blue). When the phage-bacteria combinations are broken apart in a blender, the radioactivity is found inside the bacteria, while the phage pieces are unlabeled. Some of the phages synthesized inside the infected bacteria are labeled. The genetic material must be DNA, carrying the radioactivity inside the bacteria and to offspring phages.

gaff's studies. Chargaff analyzed the amounts of the four nucleotides of DNA in several species and found a curious consistency. Although the amounts of each of the four bases vary considerably from species to species, *the DNA of any given species contains equal amounts of adenine and thymine and equal amounts of cytosine and guanine.*

Let us summarize the data compiled up to 1952:

1. DNA is the molecule of heredity.
2. All cells in the body, gametes excepted, contain the same amount of DNA.

3. The amount of cytosine is the same as the amount of guanine; the amount of thymine is the same as the amount of adenine.

Several major questions still remained unanswered: How does DNA encode genetic information? How is it duplicated before mitosis, so that each daughter cell receives exactly the same genetic information? Biologists in the early 1950s agreed that the secrets of DNA function, and therefore of heredity itself, could be found only by understanding the structure of the molecule.

Finding out the structure of any biological molecule is no simple task. Even the most powerful electron microscopes cannot reveal the structure of molecules in atomic detail. To study the structure of DNA, Maurice Wilkins and Rosalind Franklin turned to X-ray diffraction. They bombarded crystals of purified DNA with X-rays and photographed the resulting diffraction patterns (Fig. 10-6). As you can see, the X-ray pattern of DNA does not provide a direct picture of the structure of the molecule. However, the diffraction pattern suggested to Wilkins and Franklin that a DNA molecule is a **helix** (that is, twisted like a corkscrew), has a uniform diameter of 2 nanometers, and consists of subunits, each of which is separated from neighboring subunits by 0.34 nanometer (a nanometer is a billionth of a meter). One full turn of the DNA helix occurs every 3.4 nanometers. Finally, the X-ray picture suggests that the sugar-phosphate "backbone" of the molecule (see Fig. 10-6a) is on the outside of the helix, while the bases are on the inside. This last point is also supported by the fact that DNA is acidic; phosphate groups are acidic, and the bases, of course, are basic, so logically one would expect the phosphate groups to be on the outside.

The DNA of Chromosomes Is a Double Helix

The chemical and X-ray diffraction data were not nearly enough information with which to work out the structure of DNA. Some good guesses were also needed (see the box "Scientific Inquiry: The Discovery of the Double Helix"). Combining a knowledge of how complex organic molecules bond with one another with an intuition that "important biological objects come in pairs," James Watson and Francis Crick proposed that the DNA molecule consists of two strands, twisted about each other into a **double helix,** much like a ladder twisted lengthwise into a corkscrew shape (Fig. 10-7). The sugar-phosphate backbones of the two DNA strands are on the outside of the double helix, like the uprights of the ladder. Notice that the sugar-phosphate uprights run in opposite directions, so that one upright has its free sugar "foot" (3' end) on one end of the DNA molecule and the other has its free sugar "foot" on the opposite end. The bases are packed into the middle, paired up to form the rungs of the ladder.

Watson and Crick proposed that each rung is composed of a purine and a pyrimidine, held together by hydrogen bonds. If adenine pairs with thymine, and guanine pairs with cytosine, then hydrogen bonds can form between the bases, holding the two halves of the rungs together:

Adenine and thymine are held together by two hydrogen bonds; cytosine and guanine are held together by three.

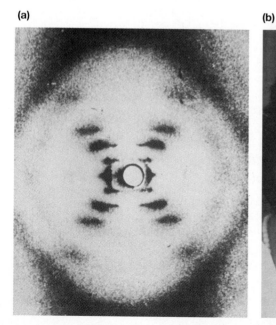

(a)

(b)

Figure 10-6   **The X-ray diffraction pattern of DNA, taken by Rosalind Franklin**

(a) The "cross" formed of dark spots is characteristic of helical molecules such as DNA. Measurements of various aspects of the pattern indicate the dimensions of the DNA helix; for example, the distance between spots in the cross corresponds to the distance between turns of the helix. **(b)** Rosalind Franklin, before her untimely death in 1958 at age 37, published about 40 scientific papers.

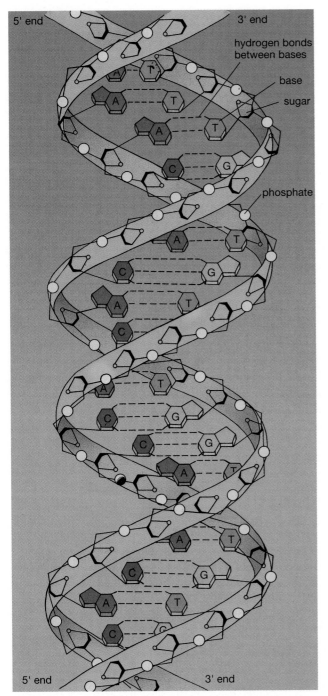

5' end   3' end

hydrogen bonds
between bases

base

sugar

phosphate

5' end   3' end

Figure 10-7   **The Watson-Crick model of DNA**

In this model, two strands of DNA are wound about each other in a double helix, like a ladder twisted about its lengthwise axis. Complementary bases pair together in the center of the helix. Three hydrogen bonds hold guanine to cytosine; two hold adenine to thymine. In this and the following figure, the "arrowheads" (with the oxygen at the point of the arrow) represent the sugars, and the yellow circles the phosphate groups. The two DNA strands run in opposite directions: Note that the free sugar (the 3' end of the sugar-phosphate backbone) is at the bottom of one strand, while the phosphate (the 5' end of the backbone) is at the bottom of the other strand. (The arrowheads point in the 3' to 5' direction for each strand.) As Figures 10-8 and 10-9 show, the directionality of the strands is important during DNA replication.

Adenine-thymine and guanine-cytosine pairs make sense of Chargaff's data, that the amount of adenine in DNA equals the amount of thymine and that the amount of guanine equals the amount of cytosine. In nucleic acids, bases that pair together by hydrogen bonds are called **complementary base pairs**. In DNA, adenine is complementary to thymine, and guanine is complementary to cytosine. This is called the base-pairing rule.

The structure of DNA was solved. Although further data would be needed to confirm its details, "a structure this pretty just had to exist," as Watson later put it. Watson and Crick published their double helix model for DNA in 1953, a model that changed all of genetics and practically all of biology in just a few years. In 1962, Watson, Crick, and Wilkins shared the Nobel Prize in recognition of their revolutionary work in deciphering the structure of DNA. (Rosalind Franklin, whose work was equally deserving of the Prize, died in 1958. Nobel Prizes are not awarded after someone's death, so her contributions sometimes do not receive the recognition they deserve.)

What remains for us to find out is how the structure *works*: How are identical copies formed during DNA replication, and how does the sequence of nucleotides in DNA direct synthesis of cellular components? In the remainder of this chapter and in the following chapter, we will explore how the Watson-Crick model of DNA helps provide answers to those questions.

DNA Replication: The Key to Constancy

In classically understated scientific style, Watson and Crick included the following comment in their paper describing the double helix: "It has not escaped our notice that the specific [base] pairing we have postulated immediately suggests a possible copying mechanism for the genetic material." In other words, the base-pairing rule offers a simple hypothesis for the replication of the DNA of chromosomes before mitosis and meiosis: The sequence of bases in one strand of the double helix accurately predicts the sequence of bases in the other, complementary strand. A chromosome could be replicated by separating the two DNA strands and synthesizing new strands using nucleotides with bases complementary to the parental strands. Each new chromosome would then consist of one parental strand and one newly synthesized, complementary daughter strand.

▦ S U M M A R Y

DNA REPLICATION

There are three fundamental steps to DNA replication in all living cells (Fig. 10-8).

1. The two DNA strands of the double helix of a parental chromosome unwind and separate.

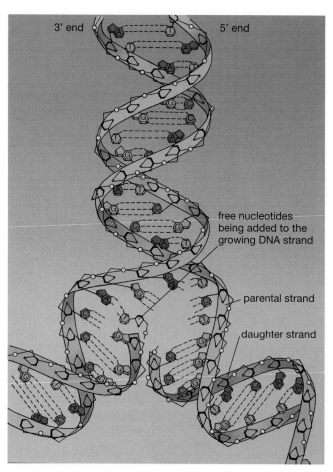

Figure 10-8   **Replication of DNA**
The sugar-phosphate backbones of the parental DNA strands are colored light tan and the backbones of the daughter DNA strands are colored pink. The parental DNA strands unwind, and the hydrogen bonds between complementary bases are broken. DNA polymerase enzymes move along the two DNA strands in opposite directions, traveling in the 3' to 5' direction (the direction that the "oxygen arrowheads" are pointing). The polymerase enzymes recognize the nucleotides in the parental DNA strands, pair each one with a complementary free nucleotide, and connect the new nucleotides together into daughter strands. This semiconservative replication process produces two new double helices of DNA, each composed of one parental strand and one daughter strand that is an exact copy of the other parental strand. Therefore, each new double helix is an exact copy of the parental DNA.

2. Each parental strand is used as a template for the formation of a complementary daughter strand of DNA. The daughter strand is formed by connecting nucleotides together in an order that is dictated by the nucleotide sequence of the parental strand: The base of each nucleotide added to the daughter strand is complementary to the base of the nucleotide in the corresponding location of the parental DNA.
3. Finally, one parental DNA strand and its newly synthesized, complementary daughter strand wind together into

one double helix, while the other parental strand and its complementary daughter strand wind together into a second double helix. Because each new double helix consists of one intact parental DNA strand and one newly synthesized complementary strand, the process of DNA replication is called **semiconservative replication** (one half of the parental DNA is "conserved," or inherited, by each daughter cell).

We will now examine these steps in more detail.

The Steps of DNA Replication Are Catalyzed by the Coordinated Action of Several Enzymes

In eukaryotic cells, the overall sequence of events of DNA replication is the following (Fig. 10-9):

DNA Replication Begins with the Unwinding of Small Sections of the Parental Double Helix

The first step in DNA synthesis is to separate the two DNA strands in one or more places, thereby unwinding the double helix (Fig. 10-9a). An enzyme appropriately named DNA helicase ("an enzyme that breaks apart the helix") wiggles its way between the two strands. According to one model, the helicase enzyme then "walks" along

Figure 10-9   **Molecular mechanisms of DNA replication**

(a) The enzyme DNA helicase separates the two DNA strands, unwinding a small portion of the double helix. **(b)** Two DNA polymerase molecules attach to the separated parental strands. DNA polymerase can synthesize new DNA strands only by proceeding in the 3' to 5' direction on the sugar-phosphate backbones. Therefore, the "branch point" polymerase attaches near the helicase and proceeds away from the helicase, while the "end" polymerase attaches at the other end of the unwound DNA region and proceeds toward the helicase (which is also the direction of future unwinding). **(c)** The two DNA polymerase molecules match up free nucleotides with the parental DNA strands, using the base-pairing rules, and synthesize daughter DNA strands (see Fig. 10-8). **(d)** The helicase advances, unwinding more parental DNA. The end DNA polymerase continues to synthesize additional daughter DNA, adding new nucleotides to the end of the strand already synthesized in part (c). The original branch point DNA polymerase, however, has run into the end of the unwound DNA and leaves the DNA, having synthesized only a short daughter DNA segment. Meanwhile, a new DNA polymerase molecule attaches at the new branch point and proceeds away from the helicase toward the first daughter DNA segment. **(e)** When the new branch point polymerase has synthesized daughter DNA that reaches to the end of the first daughter DNA segment, it leaves the DNA. Another enzyme, DNA ligase, attaches the second daughter DNA segment to the first daughter DNA segment.

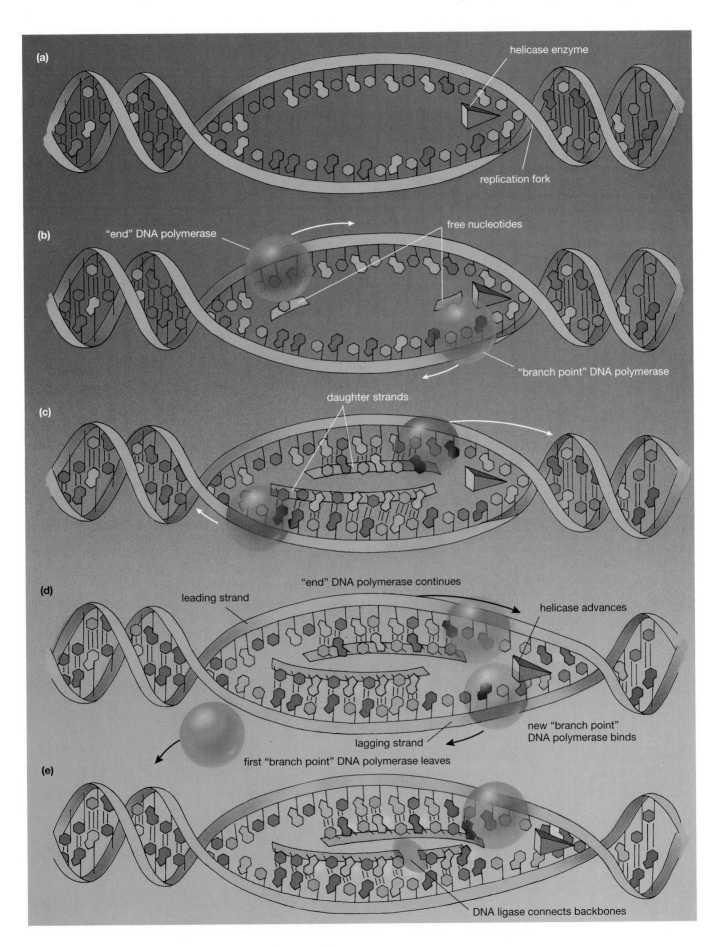

one strand, nudging the other strand out of the way as it goes. The result is that the two DNA strands separate, which exposes their bases. The junction between the separated strands and the remaining double helix is called the **replication fork.**

Daughter DNA Strands Are Synthesized That Are Complementary to Each Parental DNA Strand

DNA polymerase molecules then bind to the unwound strands, one to each strand. DNA polymerase performs a dual function. First, DNA polymerase recognizes bases exposed in a parental strand and matches them up with free nucleotides that have complementary bases. Individual nucleotides of all four types are synthesized by the cell before replication and are available within the nucleus. Second, DNA polymerase bonds together the sugars and phosphates of the free complementary nucleotides to form the daughter strand (Fig. 10-9b). In other words, when the DNA polymerase is in contact with an adenine in the parental strand, it adds thymine to the growing daughter DNA strand; when it encounters cytosine in the parental strand, it adds guanine to the daughter strand, and so on.

As far as DNA polymerase molecules are concerned, the parental DNA strands are one-way streets: DNA polymerase can travel in only one direction on a DNA strand, in the 3' to 5' direction (follow the "oxygen arrowheads" that point toward the 5' end of a DNA strand in Fig. 10-8). This means that the two DNA polymerase molecules, one on each strand, move in opposite directions. Their movement in opposite directions dictates the third step in DNA synthesis.

The Daughter DNA Pieces Are Joined into Continuous Strands

Let's look a little more closely at the process so far. Suppose that DNA helicase has wiggled its way into the helix and separated about 100 bases, forming a replication fork. Two DNA polymerase molecules then enter the fork. Now, the DNA polymerases will travel in opposite directions on the two strands of DNA, each moving toward the 5' end of its own strand. Let's assume that the two DNA polymerase molecules settle down on opposite ends of the separated DNA, one at the "end" of the fork, one at the "branch point." The end DNA polymerase builds a new daughter strand by moving along its parental strand, now called the leading strand, toward the branch point, and thus *toward the helicase.* Therefore, as the helicase continues to separate the parental DNA strands, the end polymerase simply follows along on the leading strand, synthesizing one long, continuous complementary daughter strand as it goes (Fig. 10-9c, d).

However, the polymerase that settled on the other strand, called the lagging strand, can't follow the heli-

case. Because it also moves in the 5' direction, it moves *away from the helicase* as it travels and soon reaches a "dead end" where the DNA is still intact (Fig. 10-9c). Therefore, as the helicase continues to separate the parental strands, more lagging strand immediately behind the enzyme is exposed, which the original polymerase can't reach (Fig. 10-9d). Soon, however, a new DNA polymerase molecule attaches to the lagging strand close behind the helicase, becoming a new branch point polymerase. Like the first branch point polymerase, it moves away from the helicase. What happens when this second branch point polymerase reaches the place where the first polymerase started? The two short DNA strands synthesized by the two DNA polymerases must be joined up to make a continuous DNA strand. An enzyme called DNA ligase bonds the two strands together (Fig. 10-9e). This process is repeated many times (perhaps 10 million or so for a human chromosome), until the entire lagging strand has been replicated. Because the original replication fork is seldom exactly at the end of the parental double helix, even the leading strand must be synthesized in several pieces and must also be stitched together by DNA ligase.

Base Pairing and Proofreading Produce Almost Error-Free Replication of DNA

A high degree of accuracy of DNA replication is provided by hydrogen bonding between correctly matched base pairs. Nevertheless, DNA replication is not perfect. Partly because of the speed of replication (50 to 500 nucleotides per second) and partly because of spontaneous chemical flip-flops in the bases, DNA polymerase occasionally incorporates incorrectly matched bases, perhaps one mistake for every 10,000 base pairs. In mammalian cells, however, the completed DNA strands contain only about one mistake for every billion base pairs. This phenomenal accuracy is ensured by several enzymes, including DNA polymerase itself, that "proofread" each daughter strand as it is synthesized. The one-way directionality of DNA structure allows the proofreading enzymes to recognize the parental strand, running in one direction, as the "right stuff," and to correct any mismatches by changing the daughter strand, which runs in the other direction.

During prophase of mitosis, each chromatid consists of a double helix of DNA, composed of one of the original strands of the parent DNA plus one new strand that is an exact copy of the other original strand. When sister chromatids separate at anaphase and are parceled out to the daughter cells, each cell receives an exact copy of each of the parental chromosomes. Thus, if there are no mistakes in the whole process, the integrity of the genetic information will be maintained from cell division to cell division and from parent to offspring.

SCIENTIFIC INQUIRY

The Discovery of the Double Helix

In the early 1950s, many biologists realized that the key to understanding inheritance lay in the structure of DNA. They also knew that whoever deduced the correct structure of DNA would receive recognition from fellow biologists, fame in the popular press, and very possibly the Nobel Prize. Less obvious were the best methods to use and who would be the person to do it.

The betting favorite in the race to discover the structure of DNA had to be Linus Pauling of Caltech. Pauling probably knew more about the chemistry of large organic molecules than any person alive, and he had hit upon the idea that accurate models could aid in deducing molecular structure. Like Rosalind Franklin and Maurice Wilkins, he was an expert in X-ray diffraction techniques. Finally, he was almost frighteningly brilliant. In 1950, he demonstrated these traits by showing that many proteins were coiled into single-stranded helices (see Chapter 3). Pauling, however, had two main handicaps. First, for years he had concentrated on protein research, and therefore he had little data about DNA. Second, he was active in the peace movement. During the early 1950s, some government officials, including Senator Joseph McCarthy, considered such activity to be potentially subversive and possibly dangerous to national security. This latter handicap may have proved decisive, as we shall see.

The second most likely competitors were Wilkins and Franklin, the English scientists who had set out to determine the structure of DNA by the most direct procedure, the careful study of the X-ray diffraction patterns of DNA. They were the only scientists who had really good data about the general shape of the DNA molecule. Unfortunately for Wilkins and Franklin, their methodical approach was also slow.

This left the door open for the eventual discoverers of the double helix, James Watson and Francis Crick, two young scientists with neither Pauling's tremendous understanding of chemical bonds nor Franklin and Wilkins's expertise in X-ray analysis. They did have three crucial advantages: the knowledge that models could be enormously helpful in studying molecular structure, a lesson learned from Pauling's work on proteins; access to the X-ray data; and a driving ambition to be first.

Watson and Crick did no experiments in the ordinary sense of the word; rather, they spent their time thinking about DNA, trying to construct a molecular model that made sense and fit the data. Because they were in England and because Wilkins was very open about his and Franklin's data, Watson and Crick were familiar with all the X-ray information relating to DNA. This information was just what Pauling lacked. Because of his presumed subversive tendencies, the U.S. State Department refused to issue Pauling a passport to leave the country, and consequently he could not attend meetings at which Wilkins presented the X-ray data, nor could he visit England to talk with Franklin and Wilkins directly.

Watson and Crick knew that Pauling was working on DNA structure and were terrified that he would beat them to it. In *The Double Helix*, Watson recounts his belief that, if Pauling could have seen the X-ray pictures, "in a week at most, Linus would have the structure."

About this time, you might be thinking, "But wait just a minute! That's not fair. If the goal of science is to advance knowledge, then everybody should have access to all the data. If Pauling was the best, he should have discovered the double helix first." Perhaps so. But science is an activity of scientists, who, after all, are people too. Virtually all scientists want to see the advancement and benefit of humanity, but each individual also wants to be the one responsible for the advancement and to receive the credit and the glory. We should not overlook the fact that the ambition to be first helps to inspire the intense concentration, the sleepless nights, and the long days in the laboratory that ultimately produce results.

At any rate, Pauling remained in the dark about the correct X-ray pictures of DNA and was beaten to the correct structure (Fig. E10-1). When Watson and Crick discovered the base-pairing rules that were the key to DNA structure, Watson sent off a letter about it to Max Delbruck, a friend and adviser at Caltech. He asked Delbruck not to reveal the contents of the letter to Pauling until their structure was formally published. Delbruck, perhaps more of a model scientist, firmly believed that scientific discoveries belong in the public domain and promptly told Pauling all about it. With the class of a great scientist and a great person, Pauling graciously congratulated Watson and Crick on their dazzling solution to the structure. The race was over.

Figure E10-1 DNA

James Watson and Francis Crick with a model of the structure of DNA.

❧ SUMMARY OF KEY CONCEPTS

The Composition of Chromosomes

Eukaryotic chromosomes are composed of DNA and protein. DNA is the molecule that carries genetic information. It is composed of subunits called nucleotides, linked together into long strands. Each nucleotide consists of a phosphate group, the five-carbon sugar deoxyribose, and a nitrogen-containing base. Four different bases occur in DNA: adenine, guanine, thymine, and cytosine.

DNA: The Hereditary Molecule

Transformation in bacteria and infection of bacteria by bacteriophages provided the first evidence that DNA is the hereditary molecule. Griffith discovered that living bacteria can acquire genetic material from dead bacteria, transforming the live bacteria. Avery and his colleagues showed that the genes taken up by living bacteria during transformation are composed of DNA.

A bacteriophage consists solely of DNA and protein. When it infects a bacterium, a phage injects its genetic material into the bacterium, where it directs the synthesis of more phages. Hershey and Chase demonstrated that phages inject DNA, but not protein, into bacteria. Therefore, DNA must be the genetic material of phages.

The Structure of DNA

The DNA of chromosomes is composed of two strands wound about one another in a double helix. The sugars and phosphates that link one nucleotide to the next form the backbone on each side of the double helix, while the bases from each strand pair up in the middle of the helix. Only specific pairs of bases, called complementary base pairs, can link together in the helix, held by hydrogen bonds: adenine with thymine and guanine with cytosine.

DNA Replication: The Key to Constancy

When a chromosome is replicated before mitosis or meiosis, the two DNA strands of each double helix unwind. DNA polymerase enzymes move along each strand, linking up free nucleotides into new DNA strands. The sequence of nucleotides in each newly formed strand is complementary to the sequence on a parental strand. As a result, two double helices are synthesized, each consisting of one parental DNA strand plus one newly synthesized, complementary strand that is an exact copy of the other parental strand. The two daughter DNA molecules are therefore duplicates of the parental DNA molecule.

❧ KEY TERMS

bacteriophage   p. 181
base   p. 178
complementary base pair   p. 185
deoxyribonucleic acid (DNA)
    p. 177

DNA polymerase   p. 188
double helix   p. 184
helix   p. 184
nucleotide   p. 178
purine   p. 178

pyrimidine   p. 178
replication fork   p. 188
semiconservative replication
    p. 186
transformation   p. 179

❧ THINKING THROUGH THE CONCEPTS

Multiple Choice

1. If you wanted to label amino acids but not DNA, which of the following radioactive isotopes would you use?
   a. ^{18}F b. ^{35}S c. ^{14}C
   d. ^{3}H e. ^{32}P

2. Whose pioneering work on the gene used the bacteriophage to show that DNA was the part of this virus which went into the bacterial cell for replication?
   a. Watson and Crick   b. Avery c. Hershey and Chase
   d. Griffith   e. Beadle and Tatum

3. Knowing what you now know about RNA, DNA, and DNA replication from this chapter and mitosis from the last chapter, sister chromatids
   a. each have one single strand of the two complementary DNA strands
   b. each have a fully formed double helix of DNA
   c. are the same as the homologous chromosomes of a diploid cell
   d. are present in bacteriophages, bacteria, and eukaryotes
   e. none of the above

4. How many different possible base sequences are there in a nucleotide chain 3 bases long?
   a. 1   b. 3
   c. 9   d. 64
   e. more than 64

5. Chargaff, in analyzing DNA, found that the amount of
   a. cytosine equalled that of guanine
   b. cytosine equalled that of thymine
   c. cytosine equalled that of adenine
   d. each nucleotide was unrelated
   e. each nucleotide was equal to each other

6. Semiconservative replication refers to the fact that
   a. genetic information is contained in the transforming molecule
   b. only the DNA of the bacteriophage enters the bacterial cell
   c. certain bases pair with specific bases
   d. one DNA strand remains while a new one is made
   e. point mutations sometimes interfere with accurate replication

Review Questions

1. Draw the structure of a nucleotide. Which parts are identical in all nucleotides, and which can vary?
2. Name the four types of nitrogen-containing bases found in DNA. Which bases are complementary to one another? How are they held together in the double helix of DNA?
3. What is transformation? In Griffith's experiments, why were the transformed bacteria able to synthesize capsules?
4. Diagram the life cycle of a bacteriophage.
5. Describe the structure of DNA in a chromosome. Where are the bases, sugars, and phosphates in the structure?
6. Describe the process of DNA replication.

✖ A P P L Y I N G   T H E   C O N C E P T S

1. Diagram and describe the Hershey-Chase experiment. Suppose they had found that *both radioactively labeled DNA and radioactively labeled protein* were injected into the bacteria. What would they have been able to conclude from this? (Note that if the bacteriophage chromosome was like the eukaryotic chromosome, that is, if it contained both DNA and protein, this is exactly what would have happened.)

2. A preview question for Chapter 11: Genetic information is encoded in the sequence of nucleotides in DNA. Let's suppose that the nucleotide sequence on one strand of a double helix encodes the information needed to synthesize a hemoglobin molecule. Do you think that the sequence of nucleotides on the other strand of the double helix would also encode useful information? Why or why not? An analogy to think about: Suppose that English were a "complementary language," with letters at opposite ends of the alphabet complementary to one another (for example, A is complementary to Z, B to Y, C to X, etc.). Would the complementary sentence to "All the world's a stage" make sense?

3. Suppose mammalian cells are grown and divide many times in a culture medium containing thymine made radioactive with a form of hydrogen (^3H). The cells are then removed from the radioactive medium and allowed to replicate several times in normal culture medium. Daughter chromosomes are tested each generation to determine whether they contain radioactive thymine. While in the radioactive medium, every strand of DNA of each chromosome has radioactive thymine. Assuming that each chromosome contains a single long double-stranded molecule of DNA, predict the radioactive status of daughter chromosomes after one, two, and three rounds of cell division in normal medium. Explain how your predictions are consistent with the Watson-Crick explanation of semiconservative DNA replication.

4. During the Inquisition, the church forced Galileo to disavow some of his scientific statements. Today, scientific advances are being made at an astounding rate, and nowhere is this more evident than in our understanding of the biology of heredity. Using DNA as a starting point, do you think there are limits to the knowledge people should acquire?

5. Historians debate the degree to which key people actually changed the course of history. From this chapter's historical perspective, discuss whether the pace or actual progress of knowledge is influenced by the individual scientists involved.

✖ F O R   M O R E   I N F O R M A T I O N

Crick, F. H. C. "The Structure of the Hereditary Material." *Scientific American*, October 1954. It is entertaining to consider a paper on one of science's most fundamental results while the finding was still new and before the author won the Nobel Prize for Physiology and Medicine.

Holliday, R. "A Different Kind of Inheritance." *Scientific American*, June 1989. Methyl groups attached to DNA may alter its use in a cell. What's more, patterns of DNA methylation may be passed down from parent to offspring.

Judson, H. F. *The Eighth Day of Creation: Makers of the Revolution in Biology*. New York: Simon and Schuster, 1979. A remarkably readable history of the beginnings of molecular biology.

Kimura, M. "The Neutral Theory of Molecular Evolution." *Scientific American*, November 1979. Kimura persuasively argues that some evolutionary changes can occur without selection.

Mirsky, A. E. "The Discovery of DNA." *Scientific American*, June 1968. The early history of DNA research.

Radman, M., and Wagner, R. "The High Fidelity of DNA Duplication." *Scientific American*, August 1988. Faithful duplication of chromosomes requires both reasonably accurate initial replication of DNA sequences and final proofreading.

Rennie, J. "DNA's New Twists." *Scientific American*, March 1993. Relevant to this and several subsequent chapters, this simple review covers some of the interesting modern contradictions which go beyond the standard textbook dogma.

Watson, J. D. *The Double Helix*. New York: Atheneum Publishers, 1968. If you still believe the Hollywood images, that scientists are either maniacs or cold-blooded, logical machines, be sure to read this book. Although scarcely models for the behavior of future scientists, Watson and Crick are certainly human enough!

N E T   W A T C H

On-line resources for this chapter are on the World Wide Web at:
http://www.prenhall.com/~audesirk (click on the <u>table of contents</u> link and then select Chapter 10).

11

Gene Expression and Regulation

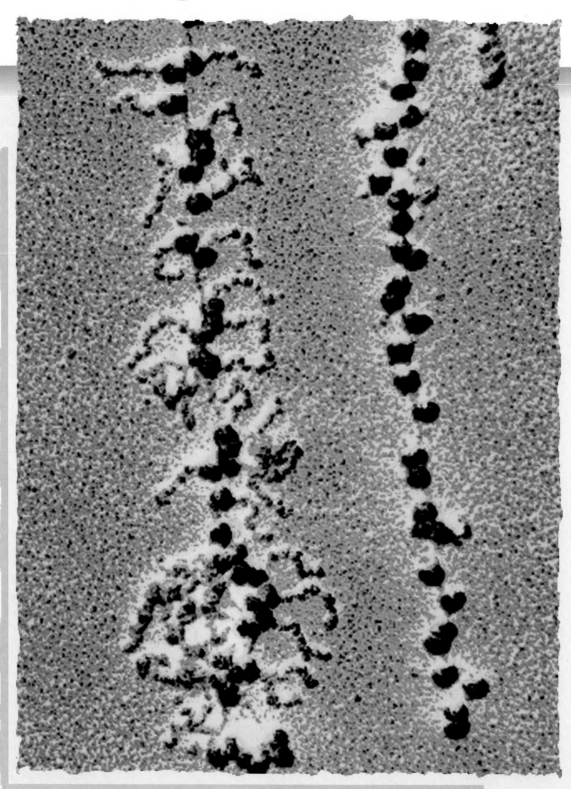

⊞ AT A GLANCE

The Relationship between Genes and Proteins
Most Genes Encode the Information for the Synthesis of a Protein
The Sequence of Bases in DNA Codes for the Sequence of Amino Acids in Proteins

Synthesizing Proteins from the Instructions in DNA
Genetic Information Flows in a Cell from DNA to RNA to Protein

The Genetic Code
The Genetic Code Uses Three Bases to Specify Each Amino Acid

RNA: Intermediary in Protein Synthesis
Transcription Produces RNA Molecules That Are Complementary Copies of One Strand of DNA
Three Types of RNA Cooperate in Protein Synthesis

Protein Synthesis

Mutations in DNA and Their Effects
Point Mutations Are Changes in a Single Base
Insertion and Deletion Mutations Result from Addition or Removal of Nucleotides
Mutations Differ in Their Effects on Protein Structure and Function
Mutations Provide the Raw Material for Evolution

Gene Regulation

Gene Regulation in Prokaryotes

Gene Regulation in Eukaryotes
Eukaryotic Genes Consist of DNA Segments That Code for the Amino Acid Sequence of Proteins Interrupted by Noncoding DNA Segments
Eukaryotic Cells May Regulate the Transcription of Individual Genes, Large Parts of Chromosomes, or Entire Chromosomes

Protein synthesis in action: The large beads are ribosomes, traveling along a central strand of messenger RNA. The smaller beaded strands emerging from the larger beads are the beginnings of proteins. The molecules involved in protein synthesis are described in this chapter.

Solving the structure of the double helix of DNA and discovering how DNA is replicated during cell division is only the first step in understanding the molecular basis of inheritance. In this chapter, we will take the next two steps.

First, how does a cell use its DNA to create the cell itself? You may recall from previous chapters that most of the organic molecules of a cell are proteins or are synthesized through the actions of protein enzymes. Therefore, to go from DNA to cell structure and function means synthesizing the appropriate proteins. How does DNA direct protein synthesis and function in a cell? We will see that the sequence of bases in DNA is a code that is translated into the sequence of amino acids in proteins. Genes are, in fact, a library containing the information needed to construct the proteins of a cell.

Second, how does an organism regulate the use of its DNA in the appropriate manner during its development and subsequent interactions with its environment? For example, if you begin life as a fertilized egg, and if all the cells of your body are derived from that egg through mitosis, then it follows that all the cells of your body contain the same genes. How, then, can different cells of the body have different structures and functions? Why do hair follicle cells synthesize hair proteins and not hormones? Further, does the environment influence how the genes are used? If so, how? The answer is that, within certain limits, cells can regulate which genes are used and which are not, depending on the function of the cell and the environment in which it lives.

The Relationship between Genes and Proteins

Long before anything was known about the molecular structure of genes, biologists tried to find out how genes work by studying how the functioning of an organism was affected by changes in its genes. In the early 1900s, the English physician Archibald Garrod studied the inheritance of human metabolic disorders. Garrod knew that the human body is a chemical caldron, churning with reactions that convert molecules of food into other molecules the body needs. These biochemical conversions proceed in steps, with each step catalyzed by a specific enzyme. The

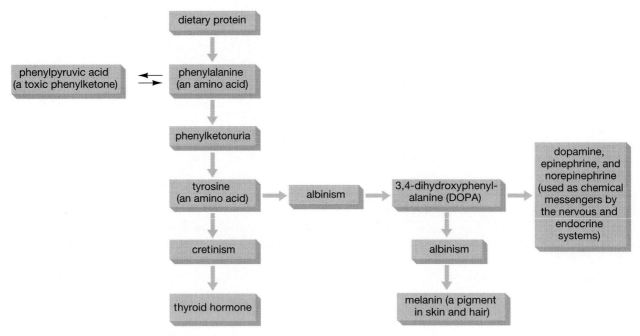

Figure 11-1   Enzymes catalyze biochemical conversions
Some of the pathways of phenylalanine and tyrosine metabolism in humans. The arrows represent enzyme-catalyzed chemical reaction(s) that produce a product (for example, tyrosine) from a substrate (for example, phenylalanine). If the enzyme that converts phenylalanine to tyrosine is defective, phenylalanine and phenylpyruvic acid build up, damaging developing brain cells. Tyrosine is a precursor for several other essential compounds, including thyroid hormone, the nervous system chemicals epinephrine and norepinephrine, and the pigment melanin. Hereditary disorders caused by single-gene mutations that lead to defective enzymes are shown.

metabolism of the amino acids phenylalanine and tyrosine is an instructive example (Fig. 11-1).

Dietary proteins usually contain both amino acids. Once in the body, the amino acids are used as building blocks for bodily proteins and for other functions, each catalyzed by a specific set of enzymes. If a single enzyme is defective, it may cause one or both of two types of effects. First, the product of the reaction catalyzed by that enzyme will not be synthesized. For example, one of the pathways that tyrosine may take in the human body is to be converted into the pigment melanin, which colors skin, hair, and the iris of the eye. Defects in the enzymes that produce melanin from tyrosine will result in a lack of melanin, and the affected person will be an albino, with extremely pale skin, white hair, and usually pink eyes.

Second, if a substance is not converted to a given product, either the substance itself or products of alternative pathways the substance can take are likely to accumulate in the body. In phenylketonuria (PKU), the enzyme that converts phenylalanine to tyrosine is defective. Therefore, phenylalanine and related substances called phenylketones accumulate. (*Phenylketonuria* means that **phenylk**etones are found in the **u**rine: hence, PKU.) In infants, high phenylalanine levels injure developing brain cells, causing severe mental retardation.

Garrod catalogued several of these "inborn errors of metabolism" and deduced, from their patterns of inheritance, that each defective enzyme was probably caused by a rare, defective version of a single gene. Given the technology of the time, this hypothesis was impossible to test in humans.

Most Genes Encode the Information for the Synthesis of a Protein

The common red bread mold, *Neurospora crassa*, proved to be an ideal organism for studying the relationship between genes and enzymes. Although we commonly see it on stale bread, *Neurospora* is an extremely independent organism that can synthesize almost all the organic compounds it needs. It can grow on a minimal medium containing an energy source, such as sucrose, a few minerals to supply essential elements, such as nitrogen and phosphorus, and a single vitamin, biotin.

Besides being easy to raise, *Neurospora* is genetically ideal as well. For most of its life cycle, *Neurospora* is haploid, with just one copy of each chromosome and, therefore, with just one copy of each gene. Most plants and animals, in contrast, are diploid, with two copies of each chromosome and thus two copies of each gene. Consequently, it may be difficult to tell if a plant or animal carries a defective copy of a gene, because there may be a normal copy that masks the effects of the defective copy. (An analogy might help: Let's imagine two workers who are both supposed to be making widgets, but only one actually works, while the other goofs off. If the productive worker makes enough widgets, the boss may never know that anything is wrong. The same thing can happen with genes; under some circumstances, the "boss" [your body] may never suffer from a defective copy of a gene. We will see a lot more about interactions between the two copies of each gene in the next chapter.) Because *Neurospora* is haploid, the effects of a defective gene cannot be masked, there being no other copy of that gene.

Geneticists George Beadle and Edward Tatum bombarded *Neurospora* with X-rays, which cause mutations in DNA. They produced mutant molds that lost the ability to synthesize some substances (Fig. 11-2). One particularly interesting mutant couldn't grow on minimal medium unless the amino acid arginine was added. Beadle and Tatum therefore concluded that the mutant must have lost the ability to synthesize arginine. By starting with minimal medium and adding one precursor molecule in the pathway of arginine synthesis at a time, Beadle and Tatum found that the mutant lacked only the enzyme that catalyzed one specific step in arginine synthesis. Genetic analysis showed that the mutant differed from normal molds in a single gene. From this and many other experiments, they concluded that a gene encodes the information needed for the synthesis of a specific enzyme. This conclusion became known as the "one-gene, one-enzyme" hypothesis. Garrod had reached the same conclusion on the basis of his studies of human metabolic disorders, but he couldn't prove it.

Geneticists have since learned that not all genes encode the information needed by a cell to produce enzymes. Some genes carry information for the synthesis of structural proteins, such as collagen in skin and keratin in hair, or hormones, such as insulin. For a few genes, the final product isn't protein at all, but ribonucleic acid (RNA), such as the RNA of ribosomes. Nevertheless, the ultimate cellular products encoded by most genes are usually proteins or parts of proteins. In general, then, each gene encodes the information for a single protein; this generalization might be called the **one-gene, one-protein hypothesis**.

The Sequence of Bases in DNA Codes for the Sequence of Amino Acids in Proteins

Knowing the relationship between genes and DNA, we can reinterpret the conclusions of Beadle and Tatum's experiments: A gene is a segment of DNA that contains the information needed to synthesize a protein. Different genes have different base sequences, and different proteins have different amino acid sequences. Therefore, the sequence of bases in DNA must encode the sequence of amino acids in a protein. How does a cell convert the message of its DNA into proteins?

Synthesizing Proteins from the Instructions in DNA

As you know, the DNA of eukaryotic cells is located in the nucleus. Protein synthesis, on the other hand, occurs on ribosomes in the cytoplasm. Therefore, DNA cannot directly guide protein synthesis. There must be an intermediate molecule that carries the information from DNA

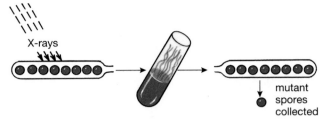

(a) Mold spores are bombarded with X-rays to induce mutations. They are germinated and grown on an enriched medium. When these molds reproduce, their spores are collected.

(b) Spores are germinated individually in enriched medium containing all the amino acids. A piece of each resulting mold is placed on minimal medium lacking amino acids. If the mold still grows on minimal medium (left), then no mutation has occurred and this mold is discarded. If the mold cannot grow on minimal medium (right), a mutation has occurred and the experiments are continued.

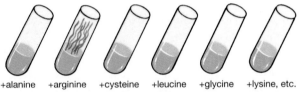

+alanine +arginine +cysteine +leucine +glycine +lysine, etc.

(c) Pieces of mutated mold are placed in tubes containing minimal medium plus one amino acid. The mutant can grow only if arginine is added. Therefore, the mutant cannot synthesize arginine.

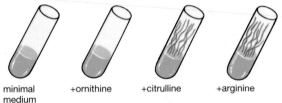

minimal medium   +ornithine   +citrulline   +arginine

(d) Mutants are tested in media to which precursors in the synthesis pathway of arginine have been added. Because the mold can grow if supplied with citrulline or arginine, but not ornithine, the mutation must have ruined a single enzyme, the one that normally catalyzes the conversion of ornithine to citrulline.

Figure 11-2   **Beadle and Tatum's experiments with the mold *Neurospora***

These experiments showed that a mutation of a single gene may produce a single defective enzyme.

in the nucleus to the ribosomes in the cytoplasm. This molecule is **ribonucleic acid**, or **RNA**.

RNA is similar to DNA but differs in three respects: (1) RNA is usually single stranded; (2) RNA has a different type of sugar in its backbone—ribose instead of deoxyribose; and (3) the base thymine in DNA is replaced by uracil in RNA:

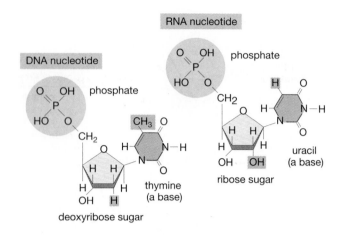

There are three functional types of RNA in a cell; all are involved in translating genetic information in DNA to the amino acid sequence of proteins. These types are messenger RNA, transfer RNA, and ribosomal RNA. We will examine their functions in more detail shortly.

Genetic Information Flows in a Cell from DNA to RNA to Protein

Information flows from DNA to proteins in a two-step process (Fig. 11-3).

1. In **transcription**, the information contained in the DNA of a specific gene is copied into RNA. All three types of RNA are transcribed from DNA. The sequence of bases in one particular type of RNA, appropriately called **messenger RNA (mRNA),** carries information from the nucleus to the ribosomes about the sequence of amino acids in the protein to be manufactured.
2. In **translation**, the two other types of RNA, **transfer RNA (tRNA)** and **ribosomal RNA (rRNA)**, convert the information of messenger RNA into the correct amino acids and help to synthesize the protein.

To understand the molecular mechanisms of this information flow from DNA to protein, geneticists first had to break the language barrier: How does the language of base sequences in DNA and messenger RNA translate into the language of amino acid sequences in proteins? This translation relies on a "dictionary" called the genetic code.

The Genetic Code

We have used the word *code* several times to refer to the information stored in DNA and ultimately translated into the amino acid sequence of proteins. This **genetic code** is conceptually similar to Morse code: One set of symbols (bases in nucleic acids, dots and dashes in Morse code) can be translated into another set of symbols (amino acids in proteins, letters of the alphabet). The question is, What combinations of bases stand for which amino acids?

The Genetic Code Uses Three Bases to Specify Each Amino Acid

There are four different bases in DNA: adenine (A), cytosine (C), guanine (G), and thymine (T). There are also four bases in RNA: adenine, cytosine, guanine, and uracil (U). But there are 20 different amino acids in proteins, so the bases cannot serve as a one-to-one code for amino acids. There are simply not enough of them. Perhaps, though, the genetic code might use a short sequence of bases to encode each amino acid, just as Morse code uses a short sequence of dots and dashes to encode the letters of the alphabet. If a sequence of two bases codes for an amino acid, then there will be 16 (4 × 4) possible combinations of bases. This isn't enough either. Three bases per amino acid, however, gives 64 (4 × 4 × 4) possible com-

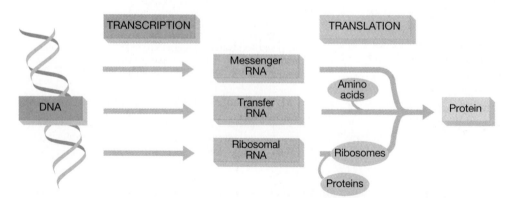

Figure 11-3   **Information flow from DNA to RNA to protein**
During transcription, a gene in DNA provides the information for the synthesis of RNA. During translation, the information in messenger RNA is used to specify the sequence of amino acids in a protein. Transfer RNA and ribosomal RNA are essential for decoding the information in messenger RNA and forming chemical bonds between amino acids to make the protein.

binations, which is more than enough. Under the assumption that nature operates as economically as possible, biologists hypothesized that the genetic code must be triplet: Three bases specify one amino acid. In 1961, Francis Crick and three co-workers demonstrated that this hypothesis is correct (see "Scientific Inquiry: Cracking the Genetic Code.")

A few requirements must be met for any language to be understood. The users must know what the words mean, where words start and stop, and where sentences start and stop. The Crick experiments demonstrated that the "words" of the genetic code are all three bases long and that a set of three bases "means" one amino acid. Shortly after this discovery, researchers began to decipher the code. They ground up bacteria and isolated the components needed to synthesize proteins. To this mixture, they added artificial messenger RNA, which allowed them to control what "words" were to be transcribed. Researchers could then see which amino acids were incorporated into the resulting proteins. For example, an RNA composed entirely of uracil (UUUUUU…) directed the mixture to synthesize a protein composed solely of phenylalanine. Therefore, the triplet specifying phenylalanine must be UUU. Because the genetic code was deciphered using these artificial RNAs, the code is usually written as the base triplets in messenger RNA that code for each amino acid (Table 11-1). These messenger RNA triplets are called **codons**.

What about punctuation? How does the cell recognize where codons start and stop and where codes for entire proteins, which are composed of many amino acids, start and stop? Research showed that the codon AUG signals "start"—that is, the beginning of a protein (**start codon**). Three codons, UAG, UAA, and UGA, signal "stop"—the end of a protein (**stop codon**). Now, if all the codons have three bases, and the beginning and end of a protein are specified, then punctuation between codons is unnecessary. To see why this is so, consider what would happen if English used only three-letter words: A sentence such as THEMANSAWTHECAT would be perfectly understandable, even without spaces between the words, as long as the reader knew where the sentence started and stopped. The genetic code doesn't need, and doesn't have, punctuation between codons.

Though there are only 20 amino acids for which to code, there are 60 codons, in addition to the start and stop codons. All 60 codons are used in the genetic code. The genetic code is thus highly redundant, or **degenerate**. In other words, a single amino acid may be specified by several codons. For example, six different codons all code for arginine (see Table 11-1). Even though the code is redundant, however, it is not ambiguous: Each codon specifies one, and only one, amino acid.

RNA: Intermediary in Protein Synthesis

Protein synthesis requires all three types of RNA: messenger RNA, transfer RNA, and ribosomal RNA. In the following sections, we will take a closer look at how these RNA molecules are synthesized, how they differ, and how they function in protein synthesis.

TABLE 11-1 ❖ The Genetic Code (Codons of mRNA)

First Base		Second Base			
	U	C	A	G	Third Base
U	UUU Phenylalanine UUC Phenylalanine UUA Leucine UUG Leucine	UCU Serine UCC Serine UCA Serine UCG Serine	UAU Tyrosine UAC Tyrosine UAA Stop UAG Stop	UGU Cysteine UGC Cysteine UGA Stop UGG Tryptophan	U C A G
C	CUU Leucine CUC Leucine CUA Leucine CUG Leucine	CCU Proline CCC Proline CCA Proline CCG Proline	CAU Histidine CAC Histidine CAA Glutamine CAG Glutamine	CGU Arginine CGC Arginine CGA Arginine CGG Arginine	U C A G
A	AUU Isoleucine AUC Isoleucine AUA Isoleucine AUG Start (Methionine)	ACU Threonine ACC Threonine ACA Threonine ACG Threonine	AAU Asparagine AAC Asparagine AAA Lysine AAG Lysine	AGU Serine AGC Serine AGA Arginine AGG Arginine	U C A G
G	GUU Valine GUC Valine GUA Valine GUG Valine	GCU Alanine GCC Alanine GCA Alanine GCG Alanine	GAU Aspartic Acid GAC Aspartic Acid GAA Glutamic Acid GAG Glutamic Acid	GGU Glycine GGC Glycine GGA Glycine GGG Glycine	U C A G

SCIENTIFIC INQUIRY

Cracking the Genetic Code

The hypothesis that three bases in DNA code for one amino acid in protein is an attractive and logical possibility. Fewer than three bases can't unambiguously code for all 20 amino acids; more than three are superfluous. But how could you prove that nature really uses a triplet code?

As it did in the Hershey-Chase experiments (see Chapter 10), the simplicity of the bacteriophage proved invaluable in deciphering a fundamental principle of genetics. Francis Crick and his co-workers exposed bacteriophages to a dye called acridine, which causes insertion mutations: One, two, or three nucleotides were inserted into the DNA molecule at random places near the beginning of a particular gene. Crick and his colleagues found that inserting one or two nucleotides into the DNA causes the synthesis of defective enzymes that prevent the phages from reproducing in their host bacteria. Inserting three nucleotides, however, sometimes produces phages that synthesize normal or nearly normal enzymes.

Crick concluded that during RNA synthesis the DNA of a gene is "read" in a linear order, starting at the beginning. Each set of three bases makes up a "word"; that is, three bases in DNA encode a single amino acid. The code must specify the beginning and end of a gene, but within a gene there are no spaces or punctuation between words. How did he arrive at these conclusions?

To understand Crick's reasoning, let's suppose that we have a gene with this repetitive sequence of bases:

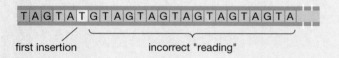

If DNA always has three-letter words, then this base sequence spells out the English word *tag* over and over again. You don't need spaces to separate the words if they all have three and only three letters.

Suppose that the acridine dye inserts another thymine somewhere near the beginning of the gene. The gene now reads:

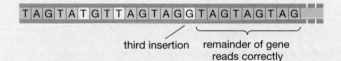

first insertion   incorrect "reading"

Protein synthesis is like a computer: garbage in, garbage out. From the point of the insertion to the end of the gene, the gene now reads GTA, GTA, GTA,..., which is nonsense.

Inserting another T nearby still results in nonsense, as most of the gene now reads AGT, AGT, AGT:

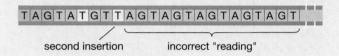

second insertion   incorrect "reading"

A third insertion, however, results in most of the gene reading TAG, TAG, TAG again:

third insertion   remainder of gene
reads correctly

How does this scenario explain Crick's results? One insertion near the beginning of the bacteriophage gene causes the rest of the triplets to encode the wrong amino acids, so the enzyme specified by that gene completely malfunctions. The second insertion still leaves most of the triplets calling for the wrong amino acids. If there are three insertions near the start of the gene, however, then the first few amino acids would be wrong, but all the triplets beyond the third insertion would be correct once again. Most of the enzyme would therefore be synthesized correctly. If the incorrect parts of the enzyme were relatively unimportant to its overall function, it may work well enough to allow the phages to reproduce.

Only a triplet code can account for the fact that three insertions, and not one or two, restore near-normal enzyme function. Further, only a code without punctuation or spaces between words will be confused by any of the insertions. Finally, a code without punctuation between words can work only if the start and stop points of protein synthesis are clearly marked. All three conclusions have been found to be correct, as we describe in the text.

Transcription Produces RNA Molecules That Are Complementary Copies of One Strand of DNA

All RNA molecules are synthesized using molecules of DNA as a template. As we saw earlier in this chapter, RNA nucleotides are chemically very similar to DNA nucleotides. Because the two "languages" are so much alike, RNA synthesis has been named transcription, meaning "the process of copying over."

Transcription of DNA into RNA is restricted in two major ways. First, in any cell, transcription normally copies only the DNA of selected genes into RNA. For example, hair follicle cells transcribe the DNA encoding keratin proteins, insulin-secreting cells of the pancreas transcribe the genes encoding insulin, and so on. Some genes, such as the genes for ribosomal RNA, are transcribed hundreds or thousands of times during interphase of most cells.

Second, when transcription of these selected genes is needed, transcription normally copies only one strand of DNA into RNA. This happens because the useful information of any given gene normally resides on only one strand of the DNA double helix. Why? Remember, the two strands of DNA are *complementary*, not *identical*. If the sequence of bases on the strand on which the gene lies codes for a sequence of amino acids that forms a functional protein, the complementary strand will have a different sequence of bases, which would probably not code for a useful protein. The DNA strand that actually contains the gene, and is transcribed into RNA, is called the **template strand** of the DNA, because it is the template from which the complementary RNA strand is made. A chromosome, which is one long DNA molecule, contains many genes. One strand may be the template strand of some genes, while the other strand may be the template strand for other genes.

With these constraints in mind, we can view transcription as a three-step process: initiation, elongation of the RNA molecule, and termination (Fig. 11-4). These three steps correspond to the three major parts of most genes, in both eukaryotes and prokaryotes: a promoter at the beginning of the gene; the "body" of the gene, which for most genes consists of the DNA bases that actually code for amino acids in the protein to be synthesized; and a termination signal at the end of the gene.

RNA Synthesis Begins at the Promoter of a Gene

RNA synthesis is carried out by an enzyme called **RNA polymerase.** The first step in transcription is for RNA polymerase to locate the beginning of the gene, causing initiation of transcription. The **promoter** region of a gene is a short sequence of DNA bases located just "upstream" (in the 3′ direction) of the body of the gene. RNA polymerase recognizes the promoter base sequence as marking the beginning of a gene and binds to the DNA at that site (Fig. 11-4a). Although RNA polymerase is shown occupying only a few bases in Figure 11-4, it is a fairly large protein that occupies at least 50 bases, including both the promoter and the first dozen or so bases in the body of the gene.

RNA Synthesis Proceeds from the Promoter to the End of the Gene

Once the RNA polymerase has bound to the promoter site, it changes shape, forcing the DNA double helix to open up at the beginning of the body of the gene. RNA polymerase then moves along the template strand of the DNA. RNA polymerase travels in the 3′ to 5′ direction along the template strand, just as DNA polymerase does (refer to Figs. 10-8 and 10-9). During elongation, using free RNA nucleotides present in the nucleus, RNA polymerase synthesizes a single strand of RNA that is complementary to the DNA of the template strand (Fig. 11-4b). The same base-pairing rules are used for RNA synthesis as for DNA replication, except that adenine in DNA is paired with uracil in RNA. The DNA-to-RNA base-pairing rules thus become:

Base in DNA	Complementary Base in RNA
adenine	uracil
cytosine	guanine
guanine	cytosine
thymine	adenine

Although RNA polymerase adds RNA nucleotides to the growing RNA strand according to base pairing with the DNA nucleotides of the gene, this base pairing does not persist. After about 10 nucleotides have been added to the growing RNA chain, the beginning of the RNA molecule separates from the DNA (Fig. 11-4c). As the RNA continues to elongate, it forms a long "tail" drifting away from the DNA, as can be seen in electron micrographs (Fig. 11-5).

At the End of RNA Synthesis, the RNA Molecule Separates from the DNA of the Gene

RNA polymerase continues along the template strand until it reaches the termination signal, a sequence of DNA bases that trigger two events (see Fig. 11-4d). First, the RNA molecule completely separates from both the DNA and the RNA polymerase. Second, the RNA polymerase leaves the template strand of the DNA. These events terminate transcription.

Three Types of RNA Cooperate in Protein Synthesis

Messenger RNA (mRNA) carries the code for the amino acid sequences of proteins from the genes in DNA to the ribosomes, the actual sites of protein synthesis. In contrast, ribosomal RNA (rRNA) and transfer RNA (tRNA) do not carry information to be translated into protein. Instead, these RNA molecules are the final products of certain genes and thus are an exception to the generalization that genes code for proteins.

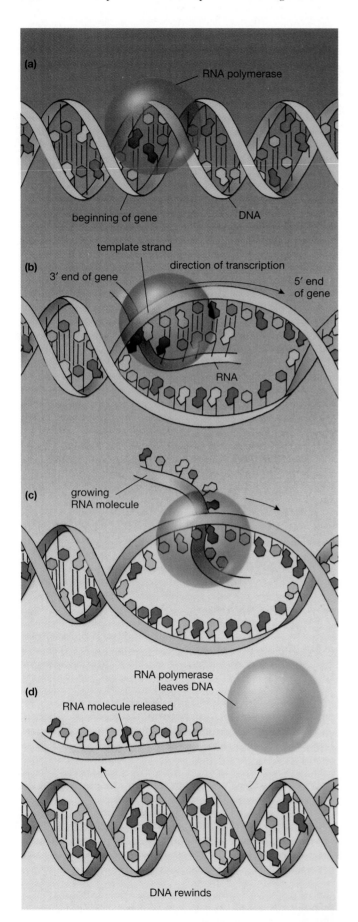

(a) RNA polymerase

beginning of gene   DNA

(b) template strand

direction of transcription

3' end of gene

5' end of gene

RNA

(c) growing RNA molecule

RNA polymerase leaves DNA

(d) RNA molecule released

DNA rewinds

Figure 11-4   A diagram of RNA transcription

(a) The enzyme RNA polymerase binds to the promoter region of DNA near the beginning of a gene (initiation). **(b)** The DNA double helix unwinds. The RNA polymerase travels along one of the DNA strands (the template strand), catalyzing the formation of a continuous strand of RNA from free RNA nucleotides (elongation). The nucleotides incorporated into the growing RNA strand are complementary to the nucleotides in the template strand of DNA. **(c)** The RNA polymerase continues to the end of the gene. **(d)** At the end of the gene, the RNA polymerase leaves the DNA. The DNA rewinds, and the RNA molecule is released (termination).

The Sequence of Bases in Messenger RNA Encodes the Sequence of Amino Acids in a Protein

Messenger RNA is a long, single-stranded molecule that includes the codons that will be translated into the amino acid sequence of a protein. In prokaryotes, mRNA is directly transcribed from the DNA of a gene, and translation into proteins often begins even before transcription is complete. In eukaryotes, things are a bit more complicated, because the RNA transcribed from DNA contains more nucleotides than will ultimately be translated into protein. We will examine the formation of eukaryotic mRNA in more detail a bit later in this chapter.

The bottom line in eukaryotic cells, however, is that mRNA molecules are synthesized in the nucleus and enter the cytoplasm through the pores in the nuclear envelope. In the cytoplasm, mRNA binds to ribosomes, where the codons of mRNA are translated into the language of amino acids in proteins. (You might think of mRNA as a "molecular photocopy" of the DNA of the gene. The gene itself remains safely stored in the nucleus, like a valuable document in a library, while copies are sent to the cytoplasm to be used in protein synthesis.)

Ribosomal RNA Forms an Important Part of the Protein-Synthesizing Machinery of a Ribosome

Ribosomes are composites of ribosomal RNA and a variety of proteins. Each ribosome is composed of two subunits (Fig. 11-6). In eukaryotic cells, the small subunit consists of one molecule of rRNA and about 30 proteins. It recognizes and binds mRNA and part of tRNA. The large ribosomal subunit consists of three rRNA molecules and 45 to 50 proteins. It contains an enzymatic region that catalyzes the addition of amino acids to the growing protein chain and two sites (designated P and A) that bind to tRNA. It is thought that rRNA plays an important part, perhaps the major part, both in recognizing mRNA and in catalyzing the formation of peptide bonds between amino acids of the protein.

Transfer RNA Molecules Decode the Sequence of Bases of mRNA into the Amino Acid Sequence of a Protein

Transfer RNA molecules bind amino acids and deliver them to the ribosome, where they are incorporated into protein chains. There are many different types of tRNAs, at least one type for each amino acid. Transfer RNAs are like "code books," the only molecules in the cell that can decipher the codons of mRNA and translate them into the amino acids of proteins. Transfer RNAs are complex molecules, twisted about into a shape something like a three-leaf clover with a stem (Fig. 11-7a). For our purposes, the stem and the central leaf are the important parts. Enzymes in the cytoplasm recognize each specific tRNA molecule and attach the correct amino acid to the stem (Fig. 11-7b). The energy of ATP is used to form the tRNA–amino acid bond. Some of the ATP energy is stored in the tRNA–amino acid bond, and this energy will be used to forge the peptide bond when the amino acid is added to a growing protein molecule.

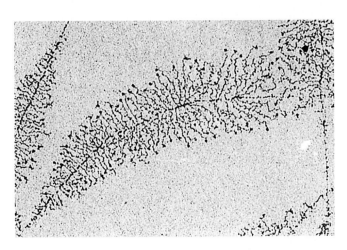

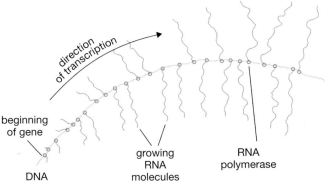

Figure 11-5   **A photograph of RNA transcription**

This electron micrograph shows the progress of RNA transcription in the egg of the African clawed toad (*Xenopus*). As shown in the diagram, in each "Christmas tree" structure, the central "trunk" is DNA and the "branches" are RNA molecules. A series of RNA polymerase molecules are traveling down the DNA, synthesizing RNA as they go. The beginning of the gene is on the left. Therefore, the short RNA molecules on the left have just begun to be synthesized, while the long RNA molecules on the right are almost finished.

The outside bend of the central tRNA leaf bears three exposed bases, called the **anticodon,** that actually decipher the mRNA code: The anticodon of each tRNA is complementary to the codon of mRNA that specifies the amino acid to which that tRNA is attached. For example, the mRNA codon GUA is complementary to the anticodon CAU of a tRNA bearing the amino acid valine (Fig. 11-7).

Protein Synthesis

Now that we have introduced all the actors involved in protein synthesis, let's look at the actual events. Once the amino acids have linked up with their appropriate transfer RNAs (Fig. 11-7b), protein synthesis can be considered to occur in two stages. First, during transcription, messenger RNA is transcribed from the DNA template of the genes in the nucleus (see Fig. 11-4). The mRNA travels to a ribosome in the cytoplasm. Second, during translation, the ribosome binds mRNA and the appropriate tRNAs. On the ribosome, the codons of mRNA are translated into the amino acid sequence of a protein.

We described transcription earlier in this chapter; here we will examine the second stage, translation (Fig. 11-8). Like transcription, translation has three steps: initiation of protein synthesis, elongation of the protein chain, and termination.

Protein Synthesis Begins When the Initiator tRNA and the Start Codon of mRNA Bind to a Ribosome

As you'll remember, the genetic code signals where protein synthesis is to begin with the start codon AUG on mRNA (see Table 11-1). In eukaryotic cells, the first step in translation is the binding of several protein "initiation factors" and a tRNA bearing the complementary "start anticodon" UAC (sometimes called the "initiator" tRNA) to the small subunit of a ribosome (Fig. 11-8a). (Unless they are actively synthesizing proteins, the two subunits of a ribosome remain apart.) The small subunit then binds to an mRNA molecule and moves along it until the start codon is encountered. At

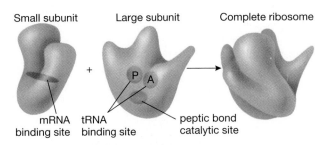

Figure 11-6   **Ribosome structure**

A ribosome has two subunits, each composed of protein and rRNA. The small subunit binds messenger RNA. The large subunit has three functional sites. Two, called the P and A sites, bind tRNA, while the third catalyzes the formation of the peptide bond between amino acids of the growing protein.

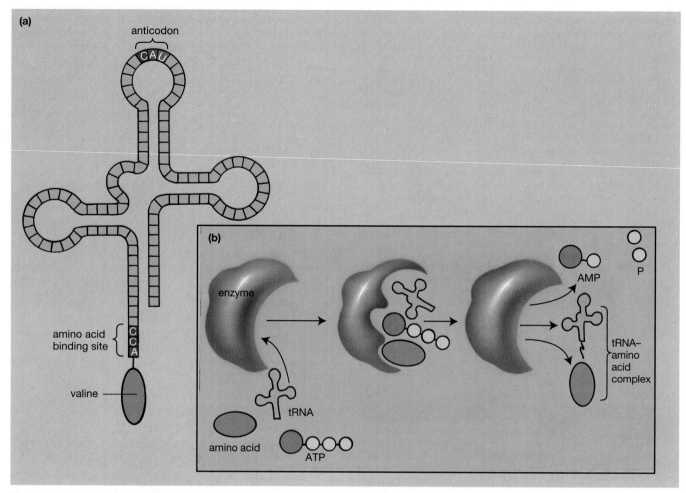

Figure 11-7   Transfer RNA

(a) Transfer RNA is a single RNA strand, folded back upon itself into loops like a three-leaf clover growing from a single stem. The loops are held together by hydrogen bonds between complementary bases. The central "leaf" of the tRNA clover bears three bases called the anticodon, which base pair with complementary bases in mRNA during protein synthesis (see Fig. 11-8). The anticodons differ among different tRNA molecules. The "stem" binds the amino acid encoded by the anticodon bases of the leaf. All tRNA stems end with the bases CCA. (b) Each type of tRNA binds to a specific amino acid. Enzymes in the cytoplasm, one for each amino acid, catalyze the bond between each tRNA and its corresponding amino acid, using the energy of ATP. Some of the ATP energy is stored in the tRNA–amino acid bond and will be used to drive the formation of the peptide bond when the amino acid is used in protein synthesis.

this point, the UAC anticodon on the initiator tRNA base pairs with the AUG of the start codon (Fig. 11-8b). The large ribosomal subunit then attaches to the small subunit. As it does so, the initiator tRNA simultaneously binds to the P site on the large subunit (Fig. 11-8c). The ribosome is now fully assembled and ready to begin translation.

Protein Synthesis Proceeds One Amino Acid at a Time

The completed ribosome is large enough to encompass two codons' worth of mRNA: Initially, this includes the start codon plus the codon coding for the next amino acid in the protein to be synthesized (GUU in our example). The an-

ticodon of a tRNA–amino acid complex (valine in our example) recognizes the second mRNA codon and moves into the A site on the large subunit (Fig. 11-8d). The two amino acids carried by the two tRNAs now lie next to one another. The catalytic site on the large subunit breaks the bond holding the "start" amino acid (methionine) to its tRNA and uses the released energy to form a peptide bond between the methionine and the valine carried by the second tRNA. At the end of this step, the initiator tRNA is now "empty," while the second tRNA bears a short, two-amino acid, protein chain (Fig. 11-8e). (Incidentally, because the start codon specifies the amino acid methionine, all protein chains are synthesized with methionine at the beginning. In most proteins, however, the methionine is soon removed.)

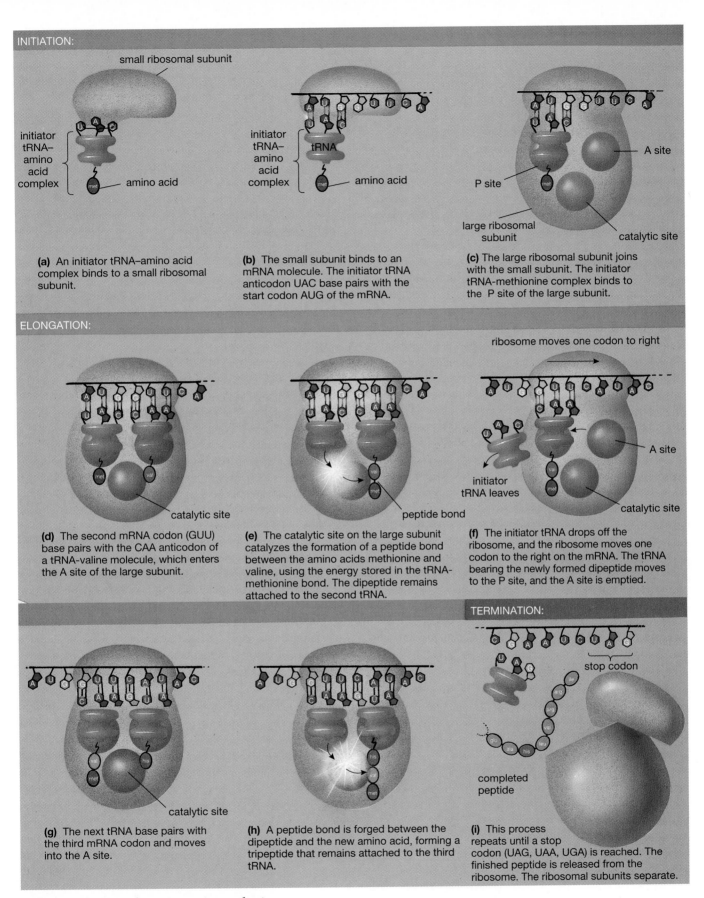

INITIATION:

small ribosomal subunit

initiator tRNA–amino acid complex

amino acid

(a) An initiator tRNA–amino acid complex binds to a small ribosomal subunit.

initiator tRNA–amino acid complex

tRNA

amino acid

(b) The small subunit binds to an mRNA molecule. The initiator tRNA anticodon UAC base pairs with the start codon AUG of the mRNA.

A site

P site

large ribosomal subunit

catalytic site

(c) The large ribosomal subunit joins with the small subunit. The initiator tRNA-methionine complex binds to the P site of the large subunit.

ELONGATION:

catalytic site

(d) The second mRNA codon (GUU) base pairs with the CAA anticodon of a tRNA-valine molecule, which enters the A site of the large subunit.

peptide bond

(e) The catalytic site on the large subunit catalyzes the formation of a peptide bond between the amino acids methionine and valine, using the energy stored in the tRNA-methionine bond. The dipeptide remains attached to the second tRNA.

ribosome moves one codon to right

A site

initiator tRNA leaves

catalytic site

(f) The initiator tRNA drops off the ribosome, and the ribosome moves one codon to the right on the mRNA. The tRNA bearing the newly formed dipeptide moves to the P site, and the A site is emptied.

catalytic site

(g) The next tRNA base pairs with the third mRNA codon and moves into the A site.

(h) A peptide bond is forged between the dipeptide and the new amino acid, forming a tripeptide that remains attached to the third tRNA.

TERMINATION:

stop codon

completed peptide

(i) This process repeats until a stop codon (UAG, UAA, UGA) is reached. The finished peptide is released from the ribosome. The ribosomal subunits separate.

Figure 11-8 **Translation: protein synthesis**

Protein synthesis is the translation of the sequence of bases in mRNA to the sequence of amino acids in the encoded protein.

At this point, the empty initiator tRNA drops off the ribosome, and the ribosome shifts to the next codon on the mRNA molecule (Fig. 11-8f). The tRNA holding the growing protein chain shifts too, from the A site to the P site on the ribosome. A new tRNA–amino acid complex binds to the emptied A site (Fig. 11-8g). The catalytic site on the large subunit breaks the bond between the dipeptide and its tRNA and links the dipeptide with the amino acid (histidine) in the A site (Fig. 11-8h). The "empty" tRNA in the P site leaves the ribosome, the ribosome shifts over another codon, and the process repeats.

Protein Synthesis Terminates When a Stop Codon Is Reached on mRNA

Near the end of the mRNA, a stop codon is reached. No tRNA recognizes a stop codon. Instead, "termination factors" cut the finished protein chain off the last tRNA, releasing it from the ribosome (Fig. 11-8i).

Messenger and Transfer RNAs Form a Decoding Chain from the Sequence of Bases in DNA to the Sequence of Amino Acids in Protein

We can now understand how a cell decodes the genetic information stored in its DNA to synthesize a protein (Fig. 11-9).

1. The DNA is organized into genes that are dozens to thousands of bases long. You might think of a gene as a sentence within the genetic information manual of the cell.
2. The "words" that make up the "gene sentences" are groups of three bases.
3. A codon of mRNA consists of three bases that are complementary to the three bases of a DNA "word."
4. An anticodon of tRNA, in turn, is complementary to a specific codon of mRNA.

5. The tRNA bears a specific amino acid, which is attached to the tRNA by enzymes that can "read" the anticodon.

This decoding chain, from bases in DNA to codon of mRNA to anticodon of tRNA to amino acid, results in the incorporation of the correct amino acid in the growing protein.

Mutations in DNA and Their Effects

So far, both in this chapter and in Chapter 10, we have emphasized the precision and fidelity of genes and their functions: DNA molecules are replicated exactly before the cell divides; mRNA molecules are exact complementary copies of the DNA of a gene; and the information in mRNA is used to produce a protein with exactly the correct amino acid sequence. Of course, no living thing is perfect. Mistakes can be made in any of these processes. A single faulty copy of mRNA or a single defective protein molecule doesn't usually affect a cell very much, because there are many correct molecules in the cell at the same time, carrying out the proper cellular functions. A single faulty copy of a gene, however, is extremely serious, because a cell may have only one (in haploid cells) or two (in diploid cells) copies of the gene and may synthesize many, or only, defective proteins as a result.

Changes in DNA are called **mutations.** Because the information in a gene is encoded in the specific sequence of bases, a mutation is a change in the sequence of bases. How can the base sequence change? One way for a mutation to occur is through a mistake in base pairing during replication (see Chapter 10). A few base-pairing mistakes occur spontaneously, in the best of circumstances: Even with proofreading, replication of several billion bases re-

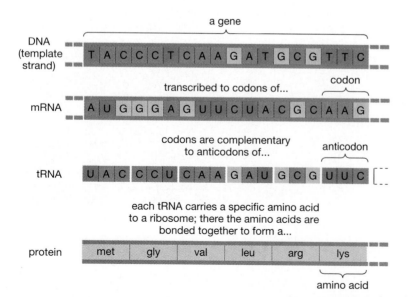

Figure 11-9   From DNA to protein

A summary of the chain of events by which a DNA sequence is decoded to produce a protein.

sults in some mistakes—not many, but a few. Certain chemicals (such as the aflatoxins synthesized by some molds that live on grain and peanuts) and some types of radiation (such as X-rays) increase errors of base pairing during replication or even induce changes in DNA composition between replications.

Random changes in DNA composition are unlikely to code for improvements in the functioning of the gene products, much as typing random words in the midst of a script of *Hamlet* will be unlikely to improve on Shakespeare's work. Therefore, cells have evolved mechanisms to monitor DNA and repair damaged regions (Fig. 11-10a). Nevertheless, mistakes do occur.

Point Mutations Are Changes in a Single Base

In a **point mutation,** a pair of bases becomes incorrectly matched (Fig. 11-10b). Repair enzymes recognize the mismatch, cut out the incorrect nucleotide, and replace it with a nucleotide bearing a complementary base. Occasionally, however, the enzymes replace the correct nucleotide instead of the incorrect one. This mistake produces a complementary pair of bases at that location once again but not the same pair that the chromosome originally had.

Insertion and Deletion Mutations Result from Addition or Removal of Nucleotides

As their names suggest, an **insertion** occurs when new nucleotide pairs are inserted in the midst of a gene (Fig. 11-11a), while a **deletion** occurs when nucleotide pairs are removed from a gene (Fig. 11-11b).

Mutations Differ in Their Effects on Protein Structure and Function

If a mutation occurs in cells whose offspring become gametes (sperm or eggs), it may be passed on to future generations. But how does a change in base sequence affect the organism that inherits the mutated DNA? As the box "Scientific Inquiry: Cracking the Genetic Code" points out, deletions and insertions can have catastrophic effects on a gene, because all the codons following the deletion or insertion will be misread. The protein synthesized from such misread directions is almost certain to be nonfunctional.

Four categories of effects may result from point mutations (Table 11-2). As a concrete example, let's consider possible mutations of the DNA sequence CTC, which codes for glutamic acid.

1. A mutation might not change the amino acid sequence of the encoded protein. Remember that the genetic code is degenerate, so one amino acid may be encoded by several different codons. If a mutation changes CTC to CTT, the new triplet still codes for glutamic acid. Therefore, the protein synthesized from the mutated gene remains the same.

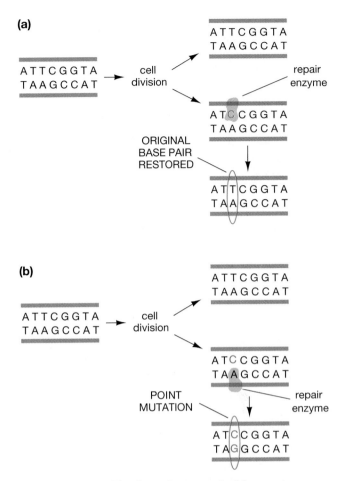

Figure 11-10 **The fate of mismatched base pairs**
Repair of damaged or mismatched regions of DNA sometimes results in restoration of the original base sequence and sometimes results in mutations. In this diagram and in Figure 11-11, the sugar-phosphate backbone of DNA is represented by solid lines, and the bases are represented by the letters A, C, G, and T. **(a)** Correction of mismatched bases. During DNA replication, a mistake in base pairing occurs, resulting in an incorrect A-C pair instead of the proper A-T pair. Repair enzymes in the daughter cell recognize the mismatch, cut out the incorrect base in the daughter DNA, and substitute the correct complementary base (T). No mutation has occurred. **(b)** Point mutation. With the same replication error as in part (a), if the repair enzymes cut out the mismatched base on the parental strand instead of on the daughter strand, then the new base pair differs from the original. A mutation has occurred.

2. A mutation might code for an amino acid that is functionally equivalent to the original amino acid. Many proteins have large "background" regions whose exact amino acid sequence is relatively unimportant. For example, in hemoglobin, the amino acids on the outside of the protein must be hydrophilic to keep the protein dissolved in the cytoplasm of red blood cells. Exactly *which* hydrophilic amino acids are on the outside doesn't seem to mat-

(a)

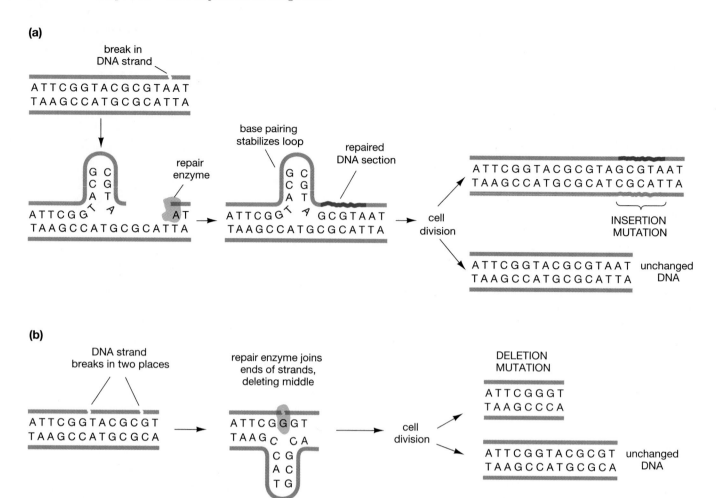

(b)

Figure 11-11   Insertion and deletion

(a) Insertion. Radiation or chemical agents cause a break in one of the DNA strands, and the freed DNA end loops out. If the free loop has certain sequences of bases, base pairing may stabilize the loop. Repair enzymes fill in the resulting gap with new nucleotides, making one DNA strand (with the loop) longer than the other (intact) strand. When DNA replication occurs, each strand is used as a template for synthesis of a new complementary strand. The long strand thus carries an insertion mutation to one of the daughter cells. **(b)** Deletion. If a DNA strand breaks in two places, a repair enzyme may stitch together the remaining pieces without filling in the missing nucleotides. When the DNA is replicated, one daughter cell receives the deletion mutation.

TABLE 11-2 ▪ *Examples of Functional Outcomes of Single Substitutions in the Glutamic Acid Codon of DNA*

	DNA	*mRNA*	*Amino Acid*	*Properties*	*Functional Effect*
Original sequence	CTC	GAG	Glutamic acid	Hydrophilic, acidic	—
Mutation 1	CTT	GAA	Glutamic acid	Hydrophilic, acidic	Neutral
Mutation 2	CTA	GAU	Aspartic acid	Hydrophilic, acidic	Neutral
Mutation 3	CAC	GUG	Valine	Hydrophobic, neutral	Lose water solubility; possibly catastrophic
Mutation 4	ATC	UAG	Stop codon	Ends translation	Synthesize only part of protein; catastrophic

ter too much. A mutation from CTC to CTA, replacing glutamic acid (hydrophilic) with aspartic acid (also hydrophilic), probably wouldn't affect the solubility of hemoglobin. Mutations that do not detectably change the function of the encoded protein are called **neutral mutations**.

3. A mutation might encode for a functionally different amino acid. A mutation from CTC to CAC replaces glutamic acid (hydrophilic) with valine (hydrophobic). This substitution, which is the genetic defect in sickle-cell anemia (see Chapter 15), causes hemoglobin molecules to stick together, clumping up and distorting the shape of the red blood cells. This is a potentially fatal mutation.

4. A mutation might produce a stop codon. An inappropriate stop codon will cut short the translation of mRNA before the protein is finished. This defect is almost certainly catastrophic to protein function; if the protein is essential to life, as hemoglobin is, the mutation will be lethal to the organism.

Mutations Provide the Raw Material for Evolution

Although most mutations are harmful or neutral, mutations are essential for evolution, because ultimately all genetic variation originates as these random changes in DNA sequence. Natural selection tests new sequences in the crucible of competition for survival and reproduction. Occasionally, a mutation proves beneficial in the organism's interactions with its environment. The mutant sequence may spread throughout the population and become common, as its possessors outcompete rivals bearing the original, unmutated DNA sequence.

Gene Regulation

Knowing how proteins are synthesized and the likely effects of gene mutations on protein structure and function does not provide a complete understanding of how an organism's genes are used to produce its structures and behaviors. For example, most of the cells of your body have the same DNA but don't use all the DNA all the time. Individual cells express (produce proteins encoded by) only a small fraction of their genes, those that are appropriate to the function of that particular cell type. As we mentioned earlier in the chapter, this constraint restricts the transcription of much of a cell's DNA. Muscle cells, for example, synthesize the contractile proteins actin and myosin but not insulin or hair proteins. Gene expression also changes over time, depending on the needs of the body from moment to moment. Both an organism's developmental stage and its environment can affect which genes are transcribed. Therefore, under-

standing gene function requires an understanding of how genes are regulated.

The use of genetic information by a cell is a multistep process, beginning with the transcription of DNA and often ending with an enzyme's catalyzing a needed reaction (Fig. 11-12). Regulation can occur at any of these steps.

1. The Rate of Transcription of Individual Genes Can Be Regulated

The rate of transcription of specific genes depends on the type of cell, its stage in the cell cycle, and the metabolic activity of both the cell and the whole organism. Some genes are never transcribed in certain types of cells; for example, the gene for insulin is not transcribed in mus-

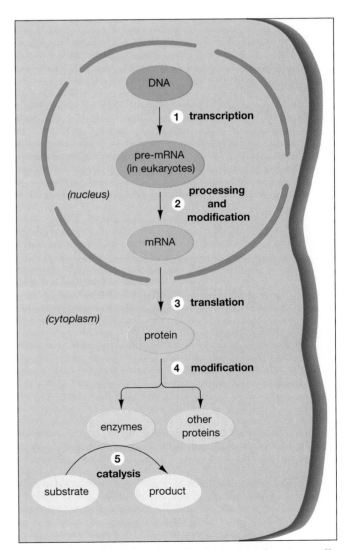

Figure 11-12 **An overview of information flow in a cell**
This simplified diagram shows the series of steps from DNA to protein to chemical reactions catalyzed by enzymes. Information flow can be regulated at any step.

cle cells. Transcription of other genes is turned on or off on demand.

2. RNA Molecules Transcribed from DNA Can Be Processed into Different Final mRNAs

As we will discuss shortly, the genes of eukaryotic cells are much larger than the final mRNA used to synthesize a protein. The genes are first transcribed into very long "pre-messenger RNA" molecules. Differential processing of these pre-mRNA molecules can produce different types of final "true mRNA," which are therefore translated into different proteins. Messenger RNA molecules must also be modified, for example by the addition or removal of nucleotides at the ends, before they can be used by the cell.

3. Messenger RNAs Can Be Translated at Different Rates

Messenger RNAs vary in their stability and the rate at which they are translated into protein. Some mRNAs are extremely stable and therefore offer the possibility of repeated translation; others are rapidly degraded. Further, a cell may block translation of certain mRNAs, depending on its metabolic requirements.

4. Proteins Might Require Modification before They Can Carry Out Their Functions in a Cell

Many proteins must be modified before they become active. For instance, the protein-digesting enzymes produced by cells of your stomach wall and pancreas are initially synthesized in an inactive form (this form keeps the enzymes from digesting the very cells that produce them). After these inactive forms are secreted into the digestive tract, portions of the enzymes are then snipped out to unveil the active site.

5. The Rate of Activity of Enzymes Can Be Regulated

Enzyme activity is often controlled by competitive or allosteric inhibition, as discussed in Chapter 4.

These methods of regulating gene activity are all important and are probably used to some extent by virtually all eukaryotic cells. We will restrict our discussion, however, to the first two steps: the transcription of DNA to RNA in prokaryotic and eukaryotic cells and the processing of pre-mRNA to true mRNA in eukaryotic cells.

Gene Regulation in Prokaryotes

Prokaryotic DNA is often organized in coherent packages called **operons,** in which the genes for related functions lie close to one another (Fig. 11-13). An operon consists of four parts: (1) a **regulatory gene,** which controls the timing or rate of transcription of other genes; (2) a **promoter,** which RNA polymerase recognizes as the place to start transcribing; (3) an **operator,** which governs access of RNA polymerase to the promoter; and (4) the **structural genes,** which actually encode for the related enzymes or structural proteins (Fig. 11-13a). Whole operons are regulated as units, so that functionally related enzymes are synthesized simultaneously when the need arises. Prokaryotic operons are regulated in different ways, depending on the functions they control. Some operons synthesize enzymes that are needed by the cell just about all the time, such as the enzymes that synthesize amino acids. These operons are usually transcribed continuously, except under unusual circumstances when the bacterium encounters a vast surplus of a particular amino acid. Other operons synthesize enzymes that are needed only occasionally, for instance to digest a relatively rare food substance. They are transcribed only when the bacterium encounters the rare food.

As an example of the latter type of operon, consider the common intestinal bacteria *Escherichia coli*. These bacteria must live on whatever types of nutrients their host eats, and they can synthesize a variety of enzymes to metabolize a potentially wide variety of foods. The genes that code for most of these enzymes are transcribed only when the enzymes are needed. The enzymes that metabolize lactose, the principal sugar in milk, are a case in point. The **lactose operon** contains three structural genes, each coding for an enzyme that aids in lactose metabolism (Fig. 11-13a).

The lactose operon is shut off, or **repressed,** unless specifically activated by the presence of lactose. The regulatory gene of the lactose operon directs synthesis of a protein, called a **repressor protein,** that binds to the operator site. RNA polymerase, although still able to bind to the promotor, cannot get past the repressor protein and transcribe the structural genes. Consequently, the lactose-metabolizing enzymes are not synthesized (Fig. 11-13b).

When *E. coli* colonize the intestines of a newborn mammal, however, they find themselves bathed in a sea of lactose whenever the host nurses from its mother. Lactose molecules enter the bacteria and bind to the repressor proteins, changing their shape (Fig. 11-13c). The lactose-repressor combination cannot attach to the operator site. Therefore, when RNA polymerase binds to the promoter of the lactose operon, it can transcribe the structural genes. Lactose-metabolizing enzymes are synthesized, allowing the bacterium to use lactose as an energy source.

After the young mammal is weaned, it usually never consumes milk again. The intestinal bacteria no longer encounter lactose, the repressor proteins are free to bind to the operator, and the genes for lactose metabolism are shut down.

Structure of the lactose operon

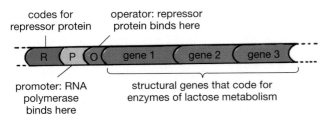

(a) The lactose operon consists of a regulatory gene, a promoter, an operator, and three structural genes that code for enzymes involved in lactose metabolism. The regulatory gene codes for a protein, called a repressor, which can bind to the operator site under certain circumstances.

Lactose absent

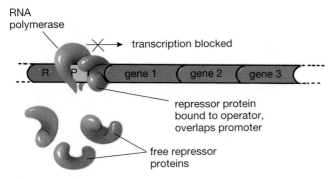

(b) When lactose is not present, repressor proteins bind to the operator of the lactose operon. When RNA polymerase binds to the promotor, the repressor protein blocks access to the structural genes, which therefore cannot be transcribed.

Lactose present

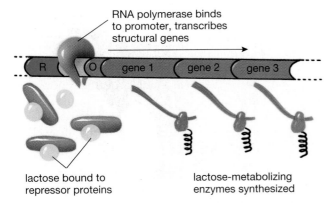

(c) When lactose is present, it binds to the repressor protein. The lactose-repressor complex cannot bind to the operator, so RNA polymerase has free access to the promoter. The RNA polymerase transcribes the three structural genes coding for the lactose-metabolizing enzymes.

Figure 11-13 **Structure and regulation of the lactose operon of *Escherichia coli***

Gene Regulation in Eukaryotes

Gene regulation is quite different in eukaryotes. Not only are genes for related functions sometimes found on entirely different chromosomes, but even many individual genes are split up on the chromosome. As a result, transcription and its regulation are more complex.

Eukaryotic Genes Consist of DNA Segments That Code for the Amino Acid Sequence of Proteins Interrupted by Noncoding DNA Segments

In the 1970s, molecular geneticists discovered that most eukaryotic structural genes have much more DNA than is needed to encode the amino acids of proteins. Each gene consists of two or more base sequences that encode the protein, interrupted by other base sequences that are not translated into protein. The coding segments are called **exons** because they are **ex**pressed in protein. The noncoding segments are called **introns** because they **int**ervene between the exons (Fig. 11-14a).

Each eukaryotic gene has its own promoter. A nearby region of the chromosome, called the **enhancer,** regulates binding of RNA polymerase to the promoter. When specific regulatory proteins bind to the enhancer, they facilitate the binding of RNA polymerase to the promoter, thus enhancing transcription.

When a eukaryotic gene is transcribed, a very long molecule of RNA is synthesized, starting before the first exon and ending after the last exon (Fig. 11-14b). The resulting RNA contains many more nucleotides than the true codons for the amino acids of the encoded protein. Two major steps convert this RNA molecule into true mRNA. First, RNA nucleotides are added at the beginning (the cap) and the end (the tail) of the molecule. Second, enzymes in the nucleus precisely cut the molecule apart, splice together the sections that code for the protein, and discard the rest.

Why are eukaryotic genes split up like this? There appear to be at least two functions served by gene fragmentation. The first function is to produce multiple proteins from a single gene. In rats, there is a gene that is transcribed in both the thyroid and the brain. In the thyroid, one splicing arrangement results in the synthesis of a hormone called calcitonin, which helps to regulate calcium concentrations in the blood. In the brain, another splicing arrangement results in the synthesis of a peptide that probably is used as a messenger molecule for communication among brain cells.

The second function is more speculative but has some good experimental evidence in its support: Fragmented genes may provide a quick and efficient way for eukaryotes to evolve new proteins with new functions. This possibility is explored in "A Closer Look: Rube Goldberg Genetics—Making New Proteins from Old Parts."

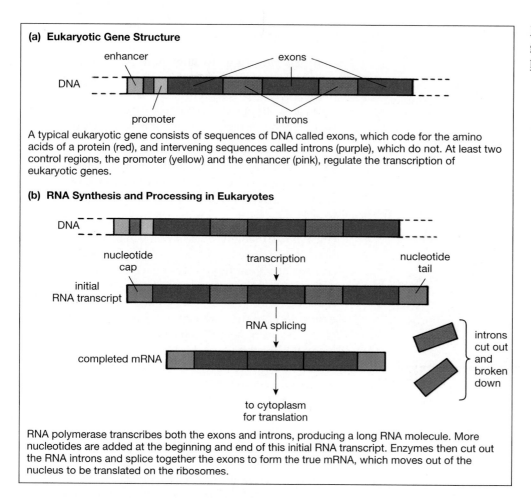

(a) Eukaryotic Gene Structure

A typical eukaryotic gene consists of sequences of DNA called exons, which code for the amino acids of a protein (red), and intervening sequences called introns (purple), which do not. At least two control regions, the promoter (yellow) and the enhancer (pink), regulate the transcription of eukaryotic genes.

(b) RNA Synthesis and Processing in Eukaryotes

RNA polymerase transcribes both the exons and introns, producing a long RNA molecule. More nucleotides are added at the beginning and end of this initial RNA transcript. Enzymes then cut out the RNA introns and splice together the exons to form the true mRNA, which moves out of the nucleus to be translated on the ribosomes.

Figure 11-14   Eukaryotic gene structure and the processing of RNA

Eukaryotic Cells May Regulate the Transcription of Individual Genes, Large Parts of Chromosomes, or Entire Chromosomes

As do prokaryotes, eukaryotes regulate the rate of transcription of genes. Transcriptional regulation can operate on three levels: the individual gene, large parts of chromosomes, or entire chromosomes.

Regulatory Proteins Alter the Rate of Transcription of Individual Genes

Some of the best-known examples of transcriptional regulation at the level of the individual gene are the cellular actions of steroid hormones (Chapter 3). The stimulation by estrogen of albumin (egg white) synthesis in female birds (Fig. 11-15) is one example of this type of regulation at work. Being lipid soluble, steroid hormones readily penetrate cell membranes and enter the interiors of cells. During the breeding season, estrogen is secreted into the bloodstream by the bird's ovaries and enters the cells of the oviduct. The estrogen binds to receptor proteins in the cytoplasm. The estrogen-protein complex enters the nucleus, where it binds to DNA, probably near the enhancer for the albumin gene. Proximity to the enhancer makes it easier for RNA polymerase to contact the promoter of the albumin gene. Rapid transcription occurs and albumin is syn-

thesized. Similar activation of genes by steroid hormones occurs in other animals, including humans.

Condensed Regions of Chromosomes Normally Are Not Transcribed

Certain parts of chromosomes are in a highly condensed, compact state, in which the DNA seems to be inaccessible to RNA polymerase. Some of these regions are structural parts of chromosomes that don't contain genes. For example, condensed DNA is usually found at the centromeres that hold sister chromatids together during cell division. In other cases, DNA can change from the condensed state to a looser configuration that allows genes to be transcribed, depending on the stage in the life cycle of the animal or the type of cell (Fig. 11-16).

Entire Chromosomes May Be Inactivated, Thereby Preventing Transcription

In rare cases, an entire chromosome may be condensed and inaccessible to RNA polymerase. This occurs with one of the X chromosomes of female mammals. As you probably know, female mammals have two homologous X chromosomes. But only one X chromosome is available for transcription in any given cell. The other entire X chromosome is condensed into a tight mass. In the light

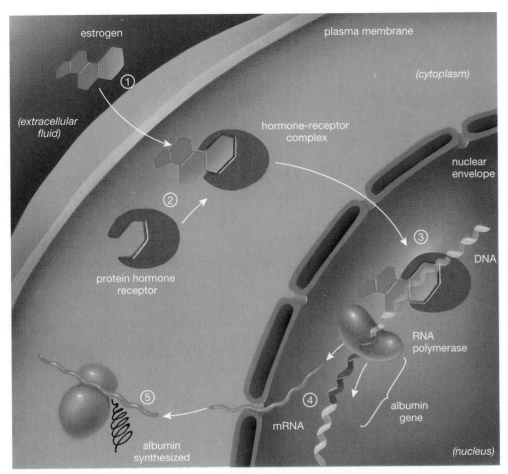

Figure 11-15   **Stimulation of transcription by estrogen, a steroid hormone**

① Estrogen diffuses through the plasma membrane into the cytoplasm. ② Estrogen combines with a receptor protein. ③ The hormone-receptor complex enters the nucleus and binds to the enhancer of the albumin gene. ④ RNA polymerase transcribes the albumin gene. ⑤ The mRNA leaves the nucleus and is translated into albumin.

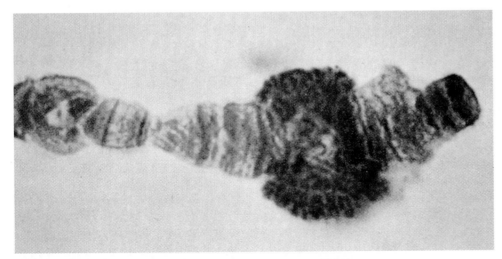

Figure 11-16   **A giant chromosome of a midge (a small fly)**

Most of the chromosome is in a tight, condensed state, but a few regions have puffed out in a much looser configuration. The purple stain shows the RNA, which is transcribed mostly from the loose, puffed regions of DNA. Changing patterns of puffing and condensation of the chromosomes reflect different genes turning on and off during development.

A CLOSER LOOK

Rube Goldberg Genetics—Making New Proteins from Old Parts

The cartoonist Rube Goldberg created marvelous fictional contraptions to perform simple functions, such as sharpening a pencil (Fig. E11-1). The beauty of a Goldberg "invention" was that you could have really built it from ordinary household items and scrap lumber. In many respects, evolution works a lot like Rube Goldberg, modifying ordinary, preexisting structures to perform new functions. Consider, if you will, an elephant's ears: huge flaps of skin laced with blood vessels. The selective advantage of "ordinary" external ears, like those on a wolf or a deer, is that they funnel sound into the ear canal, helping animals locate the source of sounds. From these humble beginnings, the enormous ear flaps on an elephant have been adapted for quite a different role: dissipating heat in the African savanna. They also have the added benefit of looking most impressive when an elephant threatens a rival.

In evolution, it is difficult to devise anything from scratch. Inventing a new protein, say of 200 amino acids, means putting together a string of 600 nucleotides in DNA in just the right sequence. Mutations cannot assemble a new, useful string of 600 nucleotides in one fell swoop. Instead, a very different, quick, and quite effective strategy has evolved: making new proteins from old parts.

Many proteins consist of several subunits, each with a completely different function. A protein that actively transports potassium across the cell membrane, for example, might have three subunits: one to anchor the protein in the membrane, one to bind potassium ions, and one to bind ATP to power the transport (Fig. E11-2a). The membrane anchor and ATP-binding subunits are modules that might be useful for other transport proteins as well. Exchange the potassium-binding subunit for a calcium-binding subunit, and presto! The cell has a calcium-transport protein (Fig. E11-2b).

How can an organism exchange subunits among various proteins? It turns out that chromosomes are not quite the stable, nearly perfectly replicating structures that geneticists pictured not very long ago. Chromosomes occasionally break in two, and one end may become attached to a completely different chromosome. Sometimes segments pop out of one chromosome and insert themselves in another. These DNA rearrangements provide a mechanism for rearranging protein subunits as well.

Because genes are segments of DNA on chromosomes, scrambling parts of a chromosome will sometimes scramble parts of genes. You might think that doing so

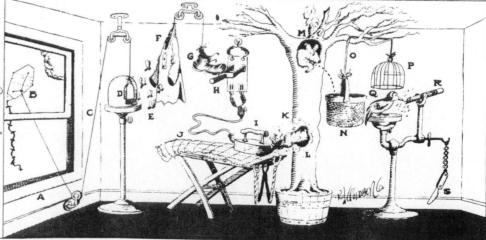

Figure E11-1   A Rube Goldberg contraption

Rube Goldberg™ property of and © Rube Goldberg Inc. Made available by United Media.

would completely ruin the genetic instructions in the rearranged genes. Remember, however, that a eukaryotic gene includes both expressed coding regions (exons) and intervening noncoding regions (introns). Exons often code for individual protein subunits (Fig. E11-2c). What if chromosomes break within introns, so that intact, functional exons can be moved from one chromosome to another (Fig. E11-2d)? By interchanging exons among genes, a eukaryotic organism can create new genes and thereby adapt more quickly to changing environmental conditions. Molecular biologists have recently found evidence that this is probably just what happened in the case of ATP-powered transport molecules and enzymes, several of which have essentially the same ATP-binding subunit.

Rube Goldberg would have appreciated that.

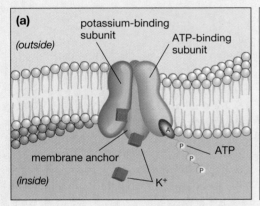

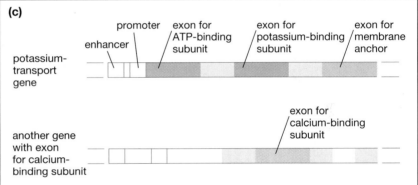

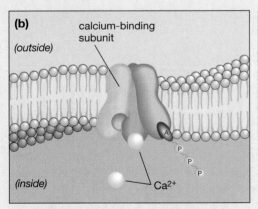

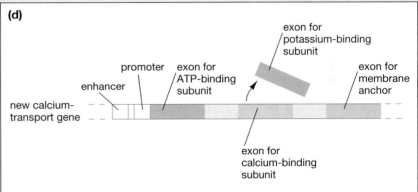

Figure E11-2   Shuffling exons to make new functional proteins

(a) Hypothetical protein subunits that might make up an active transport protein for potassium ions. (b) Substituting a calcium-binding subunit for the potassium-binding subunit creates a calcium-transporting protein. (c) The potassium-transport gene might have three exons, one coding for each subunit of the transport protein. Another gene, even on another chromosome, has an exon coding for a calcium-binding protein. (d) If the potassium-binding exon is cut out and replaced with the calcium-binding exon, a new gene is formed, now coding for a calcium-transporting protein.

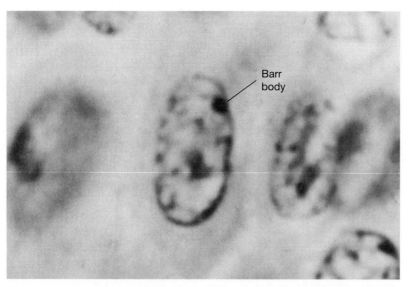

Figure 11-17   Nuclei of human cells, stained to show Barr bodies

In general, human nuclei have one fewer Barr body than the number of X chromosomes in the nucleus. Each nucleus from a normal XX woman, has one Barr body. Nuclei from men (XY) lack Barr bodies.

microscope, the **inactivated** X chromosome shows up as a dark spot in the nucleus, called a **Barr body** (Fig. 11-17), named after its discoverer, Murray Barr. Apparently, both X chromosomes are in the "loose" state in fertilized eggs. After a few cell divisions, one or the other condenses and forms a Barr body. (The cells in the ovaries that will give rise to eggs are an exception; here, both X chromosomes are in the loose, uncondensed state.) Which X chromosome is inactivated in any given cell is random, but all its daughter cells will then have the same condensed chromosome. As a result, female mammals (including human females) are mosaics: patches of cells with one X chromosome active are interspersed with patches in which the other X chromosome is active.

☒ SUMMARY OF KEY CONCEPTS

The Relationship between Genes and Proteins
Genes are segments of DNA on chromosomes. The ultimate cellular product encoded by a gene is usually a protein. Therefore, with a few exceptions (principally transfer RNA and ribosomal RNA), the specific base sequence of a gene encodes the amino acid sequence of a protein or a part of a protein.

Synthesizing Proteins from the Instructions in DNA
Ribonucleic acid (RNA) is a single-stranded nucleic acid involved in transcribing and translating genetic information in DNA to amino acid sequences. Information flows from DNA to proteins in a two-step process. (1) *Transcription*: The information contained in the DNA of a gene is copied into messenger RNA. (2) *Translation*: The sequence of bases in mRNA provides the information needed by transfer RNA and ribosomal RNA to synthesize a protein with the amino acid sequence specified by the base sequence of the DNA of the gene.

The Genetic Code
The sequence of bases in mRNA carries the genetic code for the amino acid sequence in a protein. Sequences of three bases in mRNA, called codons, specify the amino acids of the protein. There are also start and stop codons that signal the beginning and end of protein synthesis.

RNA: Intermediary in Protein Synthesis
Synthesizing proteins from the information in DNA requires RNA molecules as intermediates. RNA is transcribed from one DNA strand by the enzyme RNA polymerase. RNA polymerase recognizes the promoter region of DNA

at the beginning of a gene. Starting there, RNA polymerase uses free ribose nucleotides to synthesize an RNA strand that is complementary to the DNA of the gene.

There are three types of RNA. The sequence of bases in messenger RNA (mRNA) carries the information needed to determine the amino acid sequence of a protein. Ribosomal RNA (rRNA) and proteins form ribosomes. Ribosomes consist of large and small subunits. The small subunit recognizes mRNA. The large subunit bears two binding sites for transfer tRNA (tRNA) as well as a catalytic site that forges the peptide bond between amino acids as a protein is synthesized. Each tRNA binds a specific amino acid and transports it to a ribosome. A set of three bases in tRNA, called the anticodon, is complementary to the codon in mRNA that specifies the amino acid to which that tRNA is bound.

Protein Synthesis
Protein synthesis occurs in the following sequence:

1. Messenger RNA is transcribed from a gene. The mRNA leaves the nucleus and travels to a ribosome.
2. Two codons of mRNA bind to the small subunit of the ribosome. The first codon is the start codon, which signals where protein synthesis is to begin.
3. Transfer RNAs, carrying their amino acids, move to the mRNA. The anticodons of two tRNA molecules base pair with the two codons of mRNA, and the tRNAs bind to the large ribosomal subunit.
4. The large subunit of the ribosome catalyzes the formation of a peptide bond between the amino acids carried by the two tRNA molecules. The "first" amino acid detaches from its tRNA. The chain of two amino acids remains attached to the "second" tRNA.
5. The first tRNA leaves the ribosome. The ribosome moves one codon over on the mRNA. A third tRNA, with its attached amino acid, base pairs with the third codon on mRNA. A new peptide bond is formed between the amino acid of the third tRNA and the dipeptide still attached to the second tRNA.
6. This process continues until a stop codon is reached, whereupon the mRNA and the newly formed protein leave the ribosome.

Mutations in DNA and Their Effects
A mutation is a change in the sequence of bases in a gene. Mutations are caused by mistakes in base pairing during replication, by chemical agents, and by certain kinds of radiation. Common gene mutations include point mutations, insertions, and deletions. Although mutations are usually neutral or harmful, occasionally a mutation will promote better adaptation to the environment and thus will be favored by natural selection.

Gene Regulation
Which genes are transcribed in a cell at any given time is regulated by the function of the cell, the developmental stage of the organism, and the environment. In prokaryotic cells, DNA is often organized into operons. Whole operons are regulated as units. Access of RNA polymerase to the promoter of a gene may be either prevented or enhanced by other molecules in the cell, including nutrients and hormones.

Eukaryotic genes consist of exons (sequences of bases that encode the information for the amino acid sequence of a protein) and introns (bases that separate exons and do not themselves encode amino acid sequences). Therefore RNA transcribed from eukaryotic genes must be processed to form a true mRNA, by cutting out the introns and splicing together the exons. Regulatory proteins may alter the transcription rate of individual genes. Large parts of chromosomes may also be rendered inaccessible to RNA polymerase by changes in DNA structure.

✖ KEY TERMS

anticodon   p. 201
Barr body   p. 214
codon   p. 197
degeneracy   p. 197
deletion   p. 205
enhancer   p. 209
exon   p. 209
genetic code   p. 196
inactivation   p. 214
insertion   p. 205
intron   p. 209
lactose operon   p. 208

messenger RNA (mRNA)   p. 196
mutation   p. 204
neutral mutation   p. 207
one-gene, one-protein hypothesis   p. 195
operator   p. 208
operon   p. 208
point mutation   p. 205
promoter   p. 199
regulatory gene   p. 208
repression   p. 208
repressor protein   p. 208

ribonucleic acid (RNA)   p. 195
ribosomal RNA (rRNA)   p. 196
ribosome   p. 200
RNA polymerase   p. 199
start codon   p. 197
stop codon   p. 197
structural gene   p. 208
template strand   p. 199
transcription   p. 195
transfer RNA (tRNA)   p. 196
translation   p. 196

✖ T H I N K I N G   T H R O U G H   T H E   C O N C E P T S

Multiple Choice

1. "Anticodon" is the term applied to
   a. the list of amino acids corresponding to the genetic code
   b. the concept that multiple codons sometimes code for a single amino acid
   c. the part of the tRNA interacting with the codon
   d. the several three-nucleotide stretches which code for "stop"
   e. the control mechanism of prokaryotic genes

2. It is possible to take the cytoplasmic mRNA and reverse transcribe it into a complementary DNA (called cDNA). In contrast with genomic DNA (in the cell's nucleus) from a eukaryotic cell, the cDNA derived from the same gene
   a. would likely be shorter   b. would contain introns
   c. would have histones bound to it
   d. would have condensed chromatin
   e. would be identical

3. If ^3H (tritiated) thymidine were taken up by cells, what would that probably indicate?
   a. cells are dividing
   b. a high level of transcription
   c. a high level of translation
   d. the cell is in G_1 and is actively synthesizing mRNA
   e. introns are being excised in the processing of RNA

4. Eukaryotic DNA
   a. takes part directly in protein synthesis by leaving the nucleus and being translated on the ribosome
   b. takes part indirectly in protein synthesis; the DNA itself stays in the nucleus
   c. has nothing to do with protein synthesis; it is only involved in cell division
   d. is involved in protein synthesis which takes place in the nucleus
   e. codes for mRNA but not tRNA or rRNA

5. A single-stranded molecule that is assembled using DNA as a template and contains the information for assembly of a specific amino acid sequence in the cytoplasm is
   a. messenger RNA   b. complementary DNA
   c. transfer RNA   d. intron DNA
   e. ribosomal RNA

6. Making a protein from a template of messenger RNA
   a. is catalyzed by DNA polymerase
   b. is catalyzed by RNA polymerase
   c. is called translation
   d. is called transcription
   e. utilizes the RNA exactly as it is copied from the eukaryotic DNA template

Review Questions

1. What are the three types of RNA? What are their functions? Where and how are they synthesized?

2. Compare the chromosomal DNA from a specific eukaryotic gene and the mRNA derived from it with respect to the major differences.

3. Define the following terms: genetic code; codon; anticodon. What is the relationship between the bases in DNA, the codons of mRNA, and the anticodons of tRNA? What does it mean to say that the genetic code is degenerate?

4. Diagram and describe protein synthesis.

5. In general, all cells in your body have the same genes. How is it that different cells in your body differ? What is the name of the mechanism by which different cell-specific forms of one protein can be coded for by one "gene?" What starting molecule is converted to what final molecule? What components are pieced together to make the final molecule? What components are chopped out?

6. How would a radioactively labeled precursor of uracil (^3H-uridine) be used to localize sites of active gene transcription? Specifically, what molecule would take up the ^3H-uridine?

7. What is an operon? Are operons found in prokaryotes, eukaryotes, or both?

8. Describe the process of gene regulation in prokaryotes, using the lactose operon as an example.

9. Diagram the structure of a eukaryotic gene, including both the internal structure of the gene and the nearby control regions of the chromosome.

10. How is mRNA formed from a eukaryotic gene?

11. How do steroid hormones regulate eukaryotic genes?

12. Define mutation, and give one example of how a mutation might occur. Describe four functional consequences of point mutations. Would you expect most mutations to be beneficial or harmful? Explain your answer.

13. What amino acid has the same coding sequence as the start codon?

14. How would the genetic material on the two complementary DNA strands differ in terms of coding for proteins?

✖ A P P L Y I N G   T H E   C O N C E P T S

1. The same genetic code is used in most organisms, including bacteria and people. Suppose that the gene that encodes a human protein, insulin for example, were inserted into a bacterium. If the bacterium transcribes this gene into messenger RNA, would you expect that the amino acid sequence

   of the protein would be the same as the protein synthesized in a human being?

2. If you look at the genetic code (Table 11-1), you will notice that most of the codons that specify the same amino acid have the same first two bases, and differ only in the

third base. Now, most mutations that cause a change in a protein are harmful. If you sequenced the DNA for the hemoglobin gene from 100 people, in which codon position do you think you would find the most differences? Why?

3. If a single base pair deletion occurred toward the beginning of the coding sequence of a gene, why might transcription of the whole gene be interrupted by a stop codon?

4. As you have learned in this chapter, many factors influence gene expression. Use of hormones such as anabolic steroids or growth hormones among athletes has created quite a controversy recently. Hormones certainly affect gene expression, but, in the broadest sense, so do vitamins and food. What do you think are appropriate boundary lines for the use of hormones? Should athletes take growth hormone? Should a child be given growth hormone if he is unlikely to achieve normal height without it?

5. "Basic" science research refers to the quest for knowledge without any immediate practical application, sometimes using simple organisms such as harmless bacteria. In contrast, "applied" scientific research often has a specific goal, such as a cure for a disease. Given limitations in funding, it is often suggested that applied science be given priority. Thinking over the information in the last few chapters, argue for or against equal levels of funding for basic and applied scientific research.

✤ FOR MORE INFORMATION

Beadle, G. W. "The Genes of Men and Molds." *Scientific American,* September 1948. A classic paper by the winner of the 1958 Nobel Prize on the classic experiments on *Neurospora* referenced in this chapter.

Beardsley, T. "Smart Genes." *Scientific American,* August 1991. How genes are regulated is just as important as which genes are on the chromosomes. New findings are beginning to outline how gene expression is regulated during development.

Chambron, P. "Split Genes." *Scientific American,* May 1981. The segmented nature of eukaryotic genes is described.

Cohen, J. S., and Hogan, M. E. "The New Genetic Medicines." *Scientific American,* December 1994. Can pieces of DNA be administered which tie up the expression of targeted genes for the control of disease?

Crick, F. H. C. "The Genetic Code." *Scientific American,* October 1962. The determination of the triplet nature of the genetic code.

Crick, F. H. C. "The Genetic Code: III." *Scientific American,* October 1966. The genetic code is completely solved.

Grunstein, M. "Histones as Regulators of Genes." *Scientific American,* October 1992. Histones are proteins found associated with DNA in eukaryotic chromosomes. Once thought to be merely a scaffold for DNA, they are actually important in gene regulation.

Nirenberg, M. W. "The Genetic Code: II." *Scientific American,* March 1963. Nirenberg deciphered much of the genetic code. Here he describes some of those experiments.

Steitz, J. A. "Snurps." *Scientific American,* June 1988. "Snurps" are small nuclear ribonucleoproteins, which snip introns out of eukaryotic pre–messenger RNA. Snurps are probably examples in which the RNA part of the molecule is the real catalyst.

Tjian, R. "Molecular Machines That Control Genes." *Scientific American,* February 1995. Complexes of proteins regulate which genes are transcribed in a cell, and therefore help to determine the cell's structure and function.

NET WATCH

On-line resources for this chapter are on the World Wide Web at:

http://www.prenhall.com/~audesirk (click on the table of contents link and then select Chapter 11).

12 Gene Exchange, Meiosis, and Eukaryotic Life Cycles

AT A GLANCE

Genetic Variability, Genetic Exchange, and the Evolution of Sexual Reproduction

Mutations in DNA Are the Ultimate Source of Genetic Variability

The Effects of a Mutation Depend on the Nature of the Mutation, the Organism in Which It Occurs, and the Environment in Which the Organism Lives

Genetic Exchange May Combine Useful Mutations

Eukaryotic Organisms Use a Specialized Cell Division Process Called Meiosis to Combine Genetic Material from Two Separate Parents in a Single Offspring

"Permanent" Diploidy Protects against Some of the Harmful Effects of Mutations

Meiosis and Sexual Reproduction

Meiosis Separates Homologous Chromosomes in a Diploid Cell to Produce Haploid Daughter Cells Containing One Copy of Each Type of Chromosome

The Mechanisms of Meiosis

Meiosis I

Meiosis II

Mitosis, Meiosis, and Eukaryotic Life Cycles

In Haploid Life Cycles, the Majority of the Cycle Consists of Haploid Cells

In Diploid Life Cycles, the Majority of the Cycle Consists of Diploid Cells

In Alternation-of-Generations Life Cycles, There Are Both Diploid and Haploid Multicellular Stages

The Roles of Meiosis and Sexual Reproduction in Producing Genetic Variability

Male elephant seals fight for control of a harem, a group of females with which one of them will breed.

In Chapters 9 through 11, we discussed several important aspects of inheritance. First, DNA molecules carry the hereditary information of all living organisms; the sequences of bases in specific stretches of DNA, the genes, encode the information needed to specify the amino acid sequences of proteins. Second, chromosomes are long strands of DNA, which, in eukaryotic organisms, are found in the nucleus of each cell. Third, the cells of most plants and animals are diploid; that is, they have pairs of chromosomes that have similar, though usually not identical, DNA sequences. Finally, when cells divide, each daughter cell receives a full set of chromosomes containing one copy of all the chromosomes of the parent cell.

In multicellular organisms, cell division is the basis of asexual reproduction (see Chapter 9): groves of aspens that are clones of an ancestral parent or a new *Hydra* budding off from its parent. Asexual reproduction through mitosis seems to be so simple and efficient. In contrast, **sexual reproduction**, in which DNA from two parents is combined in a single offspring, is such a nuisance. Consider male elephant seals, fighting to obtain a harem (see the opening photograph). These immense beasts develop long, rubbery snouts as displays to show off in front of rival males. They bash each other with their massive necks and bite each other until their blood stains the beach. A defeated male may never reproduce at all, but victory is not pure bliss, either. The victor often spends so much energy defending his harem that, by the end of the breeding season, he is drained of all reserves, a mockery of his former magnificence. Many do not survive the winter.

Why not just reproduce asexually, as aspens and *Hydra* can? Or perhaps we should word the question the other way around—why, if asexual reproduction is so effective, do aspens nevertheless flower and set seed, reproducing sexually as well as asexually? Offspring produced by asexual reproduction are usually genetically identical to their parents. This duplication may be an advantage if the parent organism is well adapted to its environment, if that environment never changes, and if the organism and its offspring never move to a new location. But suppose the environment changes—the climate grows warmer or new predators appear. Neither the parent nor its asexually reproduced offspring may be well adapted any longer, and neither may survive. On the other hand, suppose that the offspring vary somewhat, both from the parent and from one another, because the genetic material from two unique individuals is combined. Perhaps some of them would be better adapted to the changing environment.

In this chapter, we will consider some of the advantages and disadvantages of sexual reproduction from evolutionary and genetic perspectives. We will see that a combination of overall constancy of genetic information together with limited genetic variability may have been the selective pressure behind the evolution of sexual reproduction, a specialized type of cell division called meiosis, and diploid cells with paired chromosomes. We will then examine the mechanisms of meiosis, whereby diploid cells with paired chromosomes give rise to haploid cells with unpaired chromosomes. Finally, we will see how meiosis, sexual reproduction, and mitosis interact in the life cycles of eukaryotic organisms.

Genetic Variability, Genetic Exchange, and the Evolution of Sexual Reproduction

Mutations in DNA Are the Ultimate Source of Genetic Variability

Although we have emphasized the fidelity of DNA replication and proofreading, we have also noted that changes in base sequence *do* occur. These changes in DNA are **mutations**. Often we think of mutations in a negative light, focusing on genetic disorders caused by defective genes, such as Huntington's disease or muscular dystrophy, or on science fiction fantasies, such as gigantic mutant ants destroying cities in old movies. Mutations, however, are the foundation of evolution. Without mutations in DNA, the only life on Earth would probably be extremely primitive single cells floating around in the primordial soup. Bacteria are different from bison, and you are different from your relatives, because of differences in DNA sequence that originally arose as mutations.

The Effects of a Mutation Depend on the Nature of the Mutation, the Organism in Which It Occurs, and the Environment in Which the Organism Lives

What are the effects of a mutation on a cell or an entire organism? As we described in Chapter 11, depending on the exact change in base sequence, the proteins synthesized from the directions encoded in mutated DNA may be completely unchanged, subtly altered (for better or for worse), or even rendered completely nonfunctional. Whether these effects are helpful or harmful to the organism, or matter at all, depends on many factors. One of the clearest examples of the ambiguous effects of mutations and the interaction between mutations and the environment is the development of antibiotic resistance in microbes. Many bacteria are resistant to some common antibiotics (see Chapter 21). In many cases, the resistant bacteria synthesize enzymes that break down the antibiotic; in other cases, they have altered cellular constituents, such as ribosome structure. In general, synthesizing antibiotic-metabolizing enzymes is a drain on a bacterium's energy and nutrient reserves; altered cellular components usually do not function quite as well as the original "wild-type" components do. Mutations causing altered cellular components or synthesis of antibiotic-metabolizing enzymes can therefore be harmful or beneficial, *depending on the presence or absence of the antibiotic in the environment*. Without antibiotics around, the mutations are harmful to the bacteria; when people use antibiotics, the mutations become beneficial.

Genetic Exchange May Combine Useful Mutations

Let's suppose that two different useful mutations occur in two separate organisms of the same species. It would clearly be useful for the two mutations to be combined in a single organism, which would then be genetically superior in that environment. How can this be accomplished? Asexual reproduction, of course, cannot combine the mutations. Every offspring of the organism having one of the mutations would have that same mutation, and every offspring of the organism having the second mutation would have just that mutation. It is possible that the second mutation might occur later in an organism, or its asexually reproduced offspring, that already had one of the useful mutations. But most mutations are rare, so the odds against the organism's having both are enormous. (For example, if the likelihood of each useful mutation occurring separately is one in a million, then the likelihood of both occurring in a single organism is one in a trillion.)

Enter genetic exchange. If organisms with the individual mutations could exchange some DNA, they might produce an organism with both mutations. Probably all living organisms have some mechanism for genetic exchange between individuals (Fig. 12-1). Genetic exchange is the principle underlying sexual reproduction. The potential advantages of combining useful mutations provide the evolutionary rationale for the development of the complex processes that create genetic exchange.

Eukaryotic Organisms Use a Specialized Cell Division Process Called Meiosis to Combine Genetic Material from Two Separate Parents in a Single Offspring

Most modern eukaryotic cells are either **haploid**, containing a single copy of each type of chromosome, or **diploid**, containing pairs of chromosomes. The two members of each pair of a single type of chromosome are

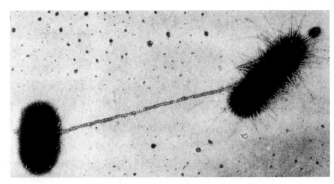

Figure 12-1 **Bacterial conjugation**

Bacteria exchange genetic material by a process called conjugation. Two bacterial cells become temporarily connected by a hollow tube, the pilus, through which DNA passes from one cell to the other.

called **homologues**, from a Greek word meaning "to say the same thing." Actually, homologous chromosomes have similar, but not identical, genetic material. As we will see later in this chapter and in Chapter 13, the fact that homologues are *not* exactly the same has extremely important consequences for inheritance. Haploid cells include sperm and eggs, of course, but also the "body cells" of organisms such as the bread mold *Neurospora*, the unicellular alga *Chlamydomonas*, and part of the plant life cycle called the gametophyte stage (discussed later in this chapter). Diploid cells include the body cells of most animals and the part of the plant life cycle called the sporophyte stage.

The first eukaryotic cells to evolve, about 1 to 1.5 billion years ago, were probably haploid. Relatively early on, single-celled eukaryotic organisms evolved two events that allowed them to shuffle and recombine genetic information. First, two haploid (parental) cells would fuse, resulting in a temporary, large, diploid cell. In this diploid cell, there would be two copies of each chromosome, one copy contributed by each of the parental cells. Second, this diploid cell would then soon undergo a type of cell division, called meiosis, that yields haploid daughter cells, each containing one copy of each type of chromosome—one of each of the pairs of homologues.

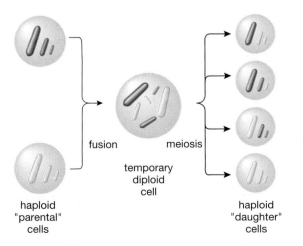

haploid "parental" cells — fusion — temporary diploid cell — meiosis — haploid "daughter" cells

We'll get to the actual mechanisms of meiosis shortly; for now just note that a single daughter cell formed from meiosis might contain some chromosomes from each parental cell, and perhaps combine two or more useful mutations that would allow it to survive and reproduce more successfully in its environment. This form of sexual reproduction is still used today by *Neurospora* and *Chlamydomonas* (see Fig. 12-8).

"Permanent" Diploidy Protects against Some of the Harmful Effects of Mutations

Going from haploid cells to a temporary diploid cell and immediately back to haploid cells may allow a single organism to combine two or more beneficial mutations. But, as you know, few mutations are beneficial; in fact, most are neutral or harmful. Let us use the analogy from Chapter 11: Randomly changing the amino acid sequence of a protein will usually make the protein function less effectively, just as randomly inserting words into Shakespeare's *Hamlet* will probably make the play worse. If a haploid cell had just one chromosome containing one gene encoding a vital protein, what would happen if that gene mutated and the resulting protein became nonfunctional? That's right: The cell would die or at least limp along much less successfully. But what would happen if that mutation occurred in a diploid cell with two homologous chromosomes, each bearing a copy of the vital gene? Now the same mutation, but in just one of the homologous chromosomes, would be much less harmful. There would still be some normal protein synthesized from the directions in the unmutated gene copy. Under the right conditions, for example if the protein is an enzyme not needed in huge quantities, a cell with one normal copy may function just as well as a cell with two normal copies (see Chapter 13).

A great advantage of diploidy is that it minimizes the potential negative effects of mutations. A diploid cell can have quite a few harmful mutations without suffering serious effects, whereas the same number of harmful mutations would almost certainly kill a haploid cell. On the other hand, a diploid cell might not benefit quite as much from a useful mutation, because it would produce both the original protein and the new, improved version, whereas a haploid cell would produce only the improved protein. Given the high frequency of harmful mutations compared with beneficial mutations, however, diploidy is probably advantageous.

Although we will never know for sure, diploidy probably evolved when a "temporary" diploid cell that formed from the fusion of two haploid cells delayed meiosis for a while, remaining diploid for a significant portion of its life cycle. Eventually, some of these diploid cells underwent some mitotic divisions as well, producing diploid daughter cells. From there, it is a relatively short evolutionary leap to multicellular, diploid, sexually reproducing organisms in which meiosis serves to produce haploid **gametes** (that is, sperm or eggs) that cannot live independently. Sperm fuses with egg to yield a diploid cell, the fertilized egg, or **zygote** (see Fig. 12-9), which, through repeated mitotic cell divisions, gives rise to a new, multicellular organism.

Meiosis and Sexual Reproduction

Although the details vary widely from organism to organism, there are three features typical of sexual reproduction in almost all multicellular eukaryotes.

1. Organisms engaging in sexual reproduction have diploid cells with pairs of homologous chromosomes at some stage in their life cycle.
2. At some point, the homologues are separated from one another through meiosis, which produces haploid cells. In animals, these haploid cells are gametes. In plants and many fungi, the haploid cells are spores that undergo mitosis to produce a multicellular, haploid body. Certain cells of this haploid form then differentiate into gametes at a later time (this life cycle is discussed later in this chapter).
3. The haploid gametes fuse to form a diploid cell once again (the zygote), with one copy of each homologous chromosome donated by each parent. This fusion reestablishes pairs of homologous chromosomes.

Meiosis Separates Homologous Chromosomes in a Diploid Cell to Produce Haploid Daughter Cells Containing One Copy of Each Type of Chromosome

The key to sexual reproduction in eukaryotic cells is **meiosis**, the production of haploid cells with unpaired chromosomes. The word *meiosis* comes from a Greek word meaning "to diminish," and that is just what meiosis does: It reduces the number of chromosomes by half. Meiosis is not, however, a random division. In meiosis, each daughter cell receives one member from each pair of homologous chromosomes.

Because meiosis evolved from mitosis, many of the structures and events in meiosis are similar or identical to those of mitosis. Therefore, you may want to review the essentials of mitosis in Chapter 9 before proceeding.

Let's begin with an overview of the primary events of meiosis (Fig. 12-2). Meiosis involves two nuclear divisions, termed **meiosis I** and **meiosis II**. The overall process of meiosis proceeds as follows:

1. All the chromosomes are replicated during interphase before meiosis I, with sister chromatids attached at the centromere of each chromosome.

2. During meiosis I, homologous chromosomes are separated; each daughter nucleus I receives one homologue of each pair of chromosomes. The sister chromatids do *not* separate during meiosis I, so at the completion of meiosis I each chromosome still consists of two sister chromatids. Remember that sister chromatids joined at a single centromere are *not* independent chromosomes (see Chapter 9). Therefore, because each daughter nucleus contains one homologue of each pair of chromosomes, the daughter nuclei are considered to be haploid. The chromosomes *do not replicate again* between meiosis I and meiosis II.
3. In meiosis II, the sister chromatids split into two independent daughter chromosomes, with one going into each daughter nucleus II.

Therefore, the reduction in the number of chromosomes from diploid to haploid occurs during meiosis I. Meiosis II is essentially the same as mitosis in a haploid cell, with the number of chromosomes remaining the same. Each nuclear division is usually accompanied by cytokinesis (division of the cytoplasm). Therefore, a cell that undergoes both meiosis I and meiosis II produces a total of four haploid cells.

Further, in most cases homologous chromosomes in the parental diploid cell have similar, but not identical, base sequences in their DNA. Therefore, because the haploid daughter cells produced by meiosis receive one homologue of each pair of chromosomes, the daughter cells are genetically similar, but not identical, to one another.

The Mechanisms of Meiosis

The phases of meiosis are given the same names as the roughly equivalent phases in mitosis, although, as we shall see, there are important differences, in meiosis I (Fig. 12-3).

Meiosis I

In multicellular organisms, meiosis occurs only in a relatively few, specialized cells in the reproductive organs. Meiosis begins in these cells with the replication of the chromosomes. (Because one function of meiosis is to *reduce* the number of chromosomes from diploid to haploid, you may think that to begin by duplicating the chromosomes is an unnecessary step. However, the cellular mechanisms of meiosis evolved from those of mitosis, and the replication of chromosomes is probably leftover evolutionary baggage. As we will see many times throughout this text, evolution is not an engineer; it's more like a tinker.) As in mitosis, the sister chromatids of each chromosome are attached to one another at the centromere. From this point on, meiosis I differs tremendously from mitosis.

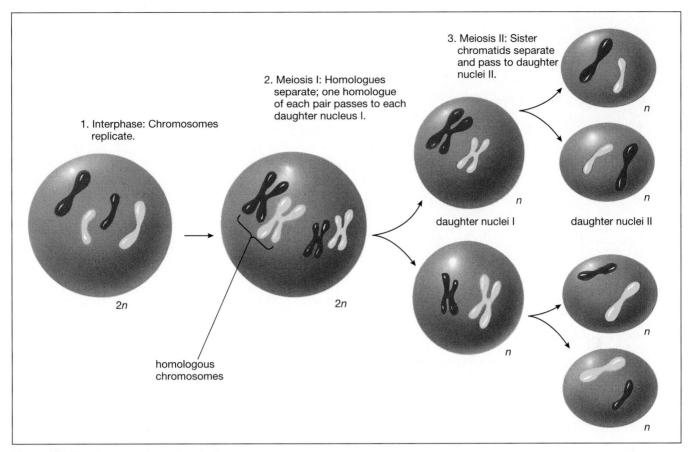

1. Interphase: Chromosomes replicate.

2. Meiosis I: Homologues separate; one homologue of each pair passes to each daughter nucleus I.

3. Meiosis II: Sister chromatids separate and pass to daughter nuclei II.

2n

2n

homologous chromosomes

daughter nuclei I

daughter nuclei II

n

n

n

n

n

n

Figure 12-2   **An overview of meiosis**

Meiosis consists of two nuclear divisions: meiosis I and meiosis II. Together, the two divisions separate homologous chromosomes and produce haploid nuclei that each contain one member of each pair of homologues. The number of different types of chromosomes in a cell is designated by the letter n. Haploid cells, with only one copy of each type of chromosome, are indicated by n ($1 \times n$). Diploid cells, with two copies of each type, are indicated by $2n$.

During Prophase I, Homologous Chromosomes Pair Up and Exchange DNA

During **prophase I**, homologous chromosomes line up side by side, exchange segments of DNA, and condense (Fig. 12-4). In some way that is not yet understood, homologous chromosomes find one another early in prophase I of meiosis (step 1 in Fig. 12-4). (To make it easier to keep track of the homologues, we'll call one homologue the "maternal chromosome" and the other the "paternal chromosome," because, as we will see, one was originally inherited from the organism's mother and the other was inherited from the organism's father.) The ends of the maternal and paternal chromosomes attach to the nuclear envelope right next to each other, and the two chromosomes are brought together, side by side, much like zipping the zipper on a jacket (step 2). Protein strands join the homologous chromosomes so that they match up exactly along their entire length.

Then large enzyme complexes assemble at random places along the paired chromosomes (step 3). The enzymes snip apart the chromosomes and graft the broken ends together again. However, *the enzymes almost always graft a maternal "stump" onto a paternal "stem," and vice versa* (step 4). Thus, the maternal and paternal chromosomes now intertwine, forming crosses, or **chiasmata** (singular, **chiasma**). Eventually the enzyme complexes leave the chromosomes, and the protein zippers that held homologues together disappear. The chromosomes coil and condense, becoming visible with a light microscope, with chiasmata connecting homologues (step 5 in Fig. 12-4, and Fig. 12-5, p. 226).

Chiasmata serve two functions. First, the maternal and paternal chromosomes have swapped some segments of their DNA, an event called **crossing over**. If, as is likely, they had slightly different DNA, then neither chromosome is quite the same as it was before the chiasmata formed. We will discuss the genetic consequences of cross-

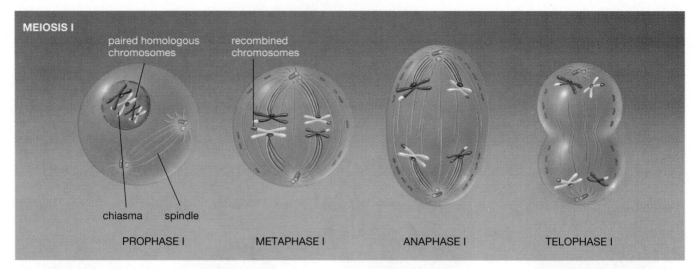

MEIOSIS I

paired homologous chromosomes

recombined chromosomes

chiasma   spindle

PROPHASE I   METAPHASE I   ANAPHASE I   TELOPHASE I

Prophase I. Chromosomes thicken and condense. Homologous chromosomes come together in pairs, and chiasmata occur as chromatids of homologues exchange parts. The nuclear envelope disintegrates and the spindle forms.

Metaphase I. Paired homologous chromosomes line up along the equator of the cell. One homologue of each pair faces each pole of the cell, with both chromatids of a given homologue becoming attached to microtubules leading to the same pole.

Anaphase I. Homologues separate, one member of each pair going to each pole of the cell. Sister chromatids do not separate.

Telophase I. Two clusters of chromosomes have formed, each containing one member of each pair of homologues. The daughter nuclei are therefore haploid. Cytokinesis often occurs during telophase I. There is usually little or no interphase between meiosis I and meiosis II.

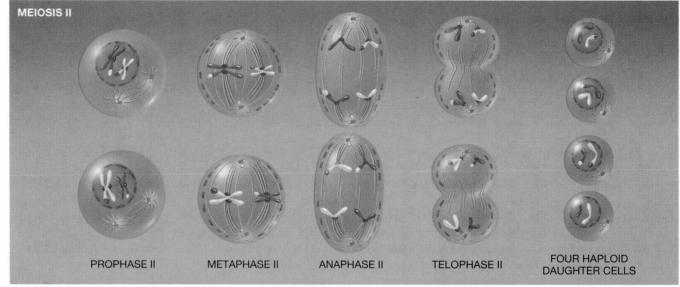

MEIOSIS II

PROPHASE II   METAPHASE II   ANAPHASE II   TELOPHASE II   FOUR HAPLOID DAUGHTER CELLS

Prophase II. If chromosomes have relaxed after telophase I, they recondense. Spindles re-form, and the spindle microtubules attach to the sister chromatids.

Metaphase II. Chromosomes line up along the equator, with sister chromatids of each chromosome attached to microtubules leading to opposite poles of the cell.

Anaphase II. Chromatids separate into independent daughter chromosomes, one former chromatid moving toward each pole.

Telophase II. Chromosomes finish moving to opposite poles.

Four haploid cells. Cytokinesis results in four haploid cells, each containing one member of each pair of homologous chromosomes.

Figure 12-3   **The details of meiosis**

Meiosis is the process by which the homologous chromosomes of a diploid cell are separated. Each of the four resulting haploid daughter cells contains one member of each pair of homologous chromosomes. In these diagrams, two pairs of homologous chromosomes are shown, large and small. The yellow chromosomes are from one parent (for example, father), and the violet chromosomes are from the other parent.

ing over in the next chapter. The second function of chiasmata is to hold the homologous chromosomes together during their attachment to the spindle microtubules. As we will see, this continued pairing is essential.

As in mitosis, the spindle begins to assemble outside the nucleus during prophase I. Near the end of prophase I, the nuclear envelope breaks up and the spindle microtubules capture the chromosomes by attaching to their kinetochores.

During Metaphase I, Paired Homologous Chromosomes Move to the Equator of the Cell

During **metaphase I**, the spindle microtubules move the chromosomes to the equator of the cell (see Fig. 12-3). Unlike in mitosis, where *individual chromosomes* line up along the equator, during metaphase I of meiosis *homologous pairs of chromosomes* line up along the equator. The

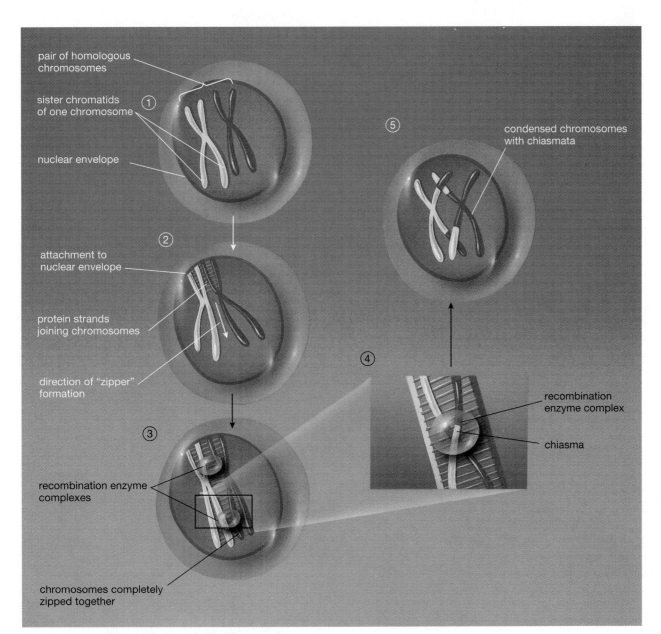

Figure 12-4   **The mechanism of crossing over**

See text for details. ① Homologous chromosomes pair up side by side. ② One end of each chromosome binds to the nuclear envelope. Protein strands "zip" homologous chromosomes together. ③ Homologous chromosomes are fully joined by protein strands. Recombination enzymes bind to the chromosomes. ④ Recombination enzymes snip chromatids apart and reattach the chromatids. Chiasmata are formed when one end of a chromatid of a paternal chromosome (yellow) is attached to the other end of a chromatid of a maternal chromosome (violet). ⑤ The protein strands and recombination enzymes leave as the chromosomes condense. The chiasmata remain as locations where homologous chromosomes are twisted around each other, helping to hold homologues together.

chiasmata   chiasmata

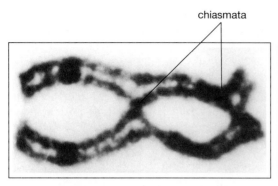
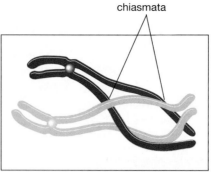

Figure 12-5 **Chiasmata**

This figure shows chiasmata in homologous chromosomes during prophase I in a salamander testis. Chiasmata are sites of DNA exchange between homologues.

two homologues of each pair are attached to kinetochore microtubules leading to opposite poles of the spindle.

Because the attachment of the chromosomes to the spindle is really the key to understanding the movement of chromosomes during meiosis I, let's compare the spindle attachments of mitosis with those of meiosis I (Fig. 12-6). In mitosis, the two sister chromatids of each chromosome attach to kinetochore microtubules leading toward opposite poles of the cell (Fig. 12-6a). Therefore, when the centromeres split during anaphase, the former sister chromatids are pulled to opposite poles. In meiosis I, both chromatids of one homologue attach to kinetochore microtubules extending toward one pole of the cell, while both chromatids of its paired homologue attach to kinetochore microtubules leading to the opposite pole (Fig. 12-6b). The kinetochore microtubules will thus separate the paired homologous chromosomes, but not pull apart the sister chromatids of any single chromosome.

Which member of a pair of chromosomes faces which pole of the cell is random. For some pairs, the maternal chromosome may face "north," while for other pairs, the maternal chromosome may face "south." We will look at the genetic consequences of this randomness, also called *independent assortment*, later in this chapter.

During Anaphase I, Homologous Chromosomes Separate

In **anaphase I**, the paired homologous chromosomes separate from one another and move to opposite poles of the cell (see Fig. 12-3). Anaphase I begins when the chiasmata holding homologous chromosomes together suddenly loosen up. The kinetochore microtubules tug the homologues apart, with the chiasmata apparently sliding along to the ends of the chromosomes, eventually letting the two homologues separate. The centromeres holding sister chromatids together, however, do *not* split, so the sister chromatids remain attached to one another. Therefore, one chromosome of each homologous pair (still consisting of two sister chromatids) moves to one pole of the dividing cell while its homologue moves to the other pole. This step is referred to as the "segregation" of the homologues.

At the end of anaphase I, there are two clusters of chromosomes. Each cluster contains one member of each pair of homologous chromosomes and is therefore haploid.

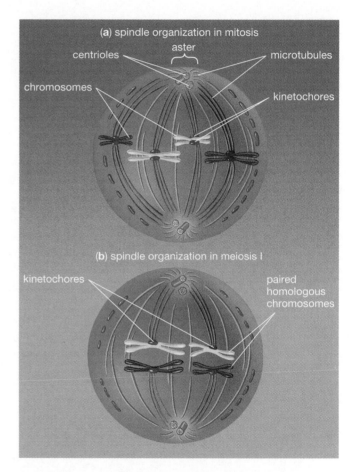

Figure 12-6 **A comparison of the spindles formed during mitosis and meiosis I**

(a) Homologous chromosomes are not paired. The kinetochores of sister chromatids are attached to kinetochore microtubules that lead to opposite poles. Therefore, when the sister chromatids separate during anaphase, the newly independent daughter chromosomes move to opposite poles of the cell.
(b) Homologous chromosomes are paired. The two kinetochores of the sister chromatids of a single chromosome are attached to kinetochore microtubules that lead to the same pole. Both kinetochores of the sister chromatids of the paired homologous chromosomes are attached to kinetochore microtubules that lead to the opposite pole. Therefore, during anaphase I, sister chromatids of each chromosome remain together, moving to the same pole, but homologous chromosomes separate and move to opposite poles.

*During Telophase I, Two Haploid Clusters
of Chromosomes Are Formed*

In **telophase I**, the spindle disappears. In many cases, nuclear envelopes may not re-form; particularly in plants, cytokinesis may not occur either. Meiosis I is usually followed immediately by meiosis II, with little or no intervening interphase. It is important to remember that *the chromosomes do not replicate between meiosis I and meiosis II*. Frequently, the chromosomes remain in their condensed, shortened state, so prophase II can be distinguished from telophase I only by the reappearance of the spindle.

Meiosis II

In meiosis II, the sister chromatids of each chromosome separate. Meiosis II is easy to follow, because it is virtually identical to mitosis in a haploid cell (see Fig. 12-3). During **prophase II** the spindle re-forms. The chromosomes attach to microtubules as they did in mitosis: *The two sister chromatids of each chromosome attach to microtubules that extend to opposite poles of the cell.* **Metaphase II** is marked by completion of the spindle apparatus and migration of the chromosomes to the equator of the cell. During **anaphase II**, the centromeres holding sister chromatids together split, and the chromatids separate from one another. The kinetochore microtubules pull each former chromatid, now an independent chromosome, to opposite poles of the cell. **Telophase II** and cytokinesis conclude meiosis II, as nuclear envelopes re-form, the chromosomes relax into their extended state, and the cytoplasm is divided. Because both daughter cells produced in meiosis I undergo meiosis II, a total of four haploid cells are formed by the end of meiosis.

Mitosis, Meiosis, and Eukaryotic Life Cycles

The **life cycles** of almost all eukaryotic organisms have a common overall pattern (Fig. 12-7). First, two haploid cells fuse during the process of **fertilization**, bringing together genes from different parent organisms and endowing the resulting diploid cell with new gene combinations. Second, at some point in the life cycle, meiosis occurs, recreating haploid cells. Third, at some time in the life cycle, mitosis

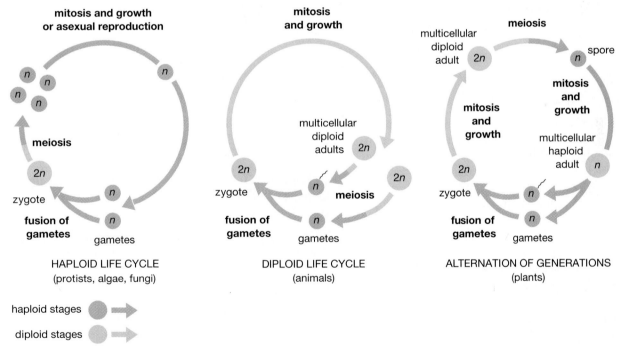

Figure 12-7   **The three major types of eukaryotic life cycles**

The lengths of the arrows correspond roughly to the proportion of the life cycle spent in each stage. **(a)** Haploid life cycle (many protists, algae, and fungi): Most of the life cycle is spent with haploid cells. Mitosis of haploid cells results in either asexual reproduction of unicellular organisms or growth of multicellular organisms. At some point, specialized reproductive haploid cells fuse, and the resulting diploid cell almost immediately undergoes meiosis. **(b)** Diploid life cycle (animals): Most of the life cycle is spent with diploid cells. When haploid cells are produced, they fuse to form a diploid zygote. Mitosis of the zygote produces multicellular bodies composed of diploid cells. **(c)** Alternation of generations (plants): Diploid and haploid stages are of roughly equal duration. Cells of the diploid stage undergo meiosis to form spores. The spores then undergo mitosis to produce a multicellular haploid body. Later, specialization of haploid cells produces gametes, which fuse to form a zygote. Through mitosis, the zygote develops into a multicellular diploid form.

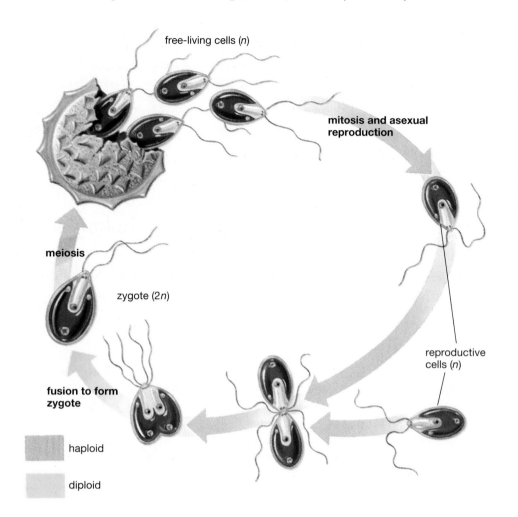

free-living cells (*n*)

mitosis and asexual reproduction

meiosis

zygote (2*n*)

reproductive cells (*n*)

fusion to form zygote

haploid

diploid

Figure 12-8   **The life cycle of the unicellular alga** *Chlamydomonas*

Chlamydomonas multiplies asexually by mitosis of haploid cells. When nutrients are scarce, specialized haploid cells (usually of different strains) fuse to form a diploid cell. Meiosis then immediately produces four haploid cells, usually with different genetic composition than either of the original parental strains.

of either haploid or diploid cells, or both, results in the growth of multicellular bodies and/or asexual reproduction.

The seemingly vast differences between the life cycles of, say, ferns and humans are due to variations in three aspects: (1) the interval between meiosis and the fusion of haploid cells, (2) at what points in the life cycle mitosis and meiosis occur, and (3) the relative proportions of the life cycle spent in the diploid and haploid states. These aspects of life cycles are interrelated, and we can conveniently label life cycles according to the relative dominance of diploid and haploid stages.

In Haploid Life Cycles, the Majority of the Cycle Consists of Haploid Cells

Some eukaryotes, such as many fungi and the unicellular alga *Chlamydomonas*, spend most of their life cycles in the haploid state, with single copies of each type of chromosome (Fig. 12-8). Asexual reproduction by mitosis produces a population of identical, haploid cells. Under certain environmental conditions, specialized "sexual" haploid cells are produced. Two of these sexual haploid cells fuse, forming a diploid cell. This cell immediately undergoes meiosis, producing haploid cells again. In organisms with a haploid life cycle, mitosis never occurs in diploid cells. As we mentioned at the beginning of this chapter, this life cycle is probably

fairly similar to the life cycle of eukaryotic organisms in which meiosis and sexual reproduction first evolved.

In Diploid Life Cycles, the Majority of the Cycle Consists of Diploid Cells

Most animals have life cycles that are just the reverse of the *Chlamydomonas* cycle. Virtually the entire animal life cycle is spent in the diploid state (Fig. 12-9). Haploid gametes (sperm in males and eggs in females) are formed through meiosis. These fuse to form a diploid fertilized egg, the zygote. Growth and development of the zygote into the adult organism is a result of mitosis in diploid cells.

In Alternation-of-Generations Life Cycles, There Are Both Diploid and Haploid Multicellular Stages

The life cycles of most plants include both multicellular diploid and multicellular haploid body forms. In the typical pattern (Fig. 12-10), a multicellular diploid body gives rise to haploid cells, called **spores**, through meiosis. The spores then undergo mitosis to create a multicellular haploid stage (the "haploid generation"). At some point, certain haploid cells differentiate into haploid gametes. Two gametes then may fuse to form a diploid zygote. The zygote grows by mitosis into a diploid multicellular body (the

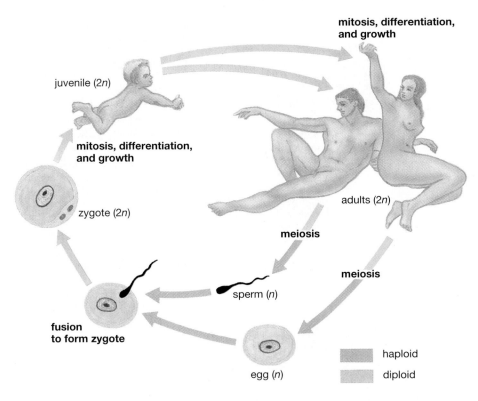

mitosis, differentiation, and growth

juvenile (2*n*)

mitosis, differentiation, and growth

zygote (2*n*)

fusion to form zygote

sperm (*n*)

meiosis

meiosis

adults (2*n*)

egg (*n*)

haploid

diploid

Figure 12-9   The human life cycle

Through meiosis, the two sexes produce different types of gametes—sperm in males and eggs in females—that fuse to form a diploid zygote. Mitosis and specialization of cells produce an embryo, child, and ultimately a sexually mature adult. The haploid state lasts only a few hours to a few days; the diploid state may survive for a century.

"diploid generation"). This life cycle is called **alternation of generations** because a multicellular diploid generation alternates with a multicellular haploid generation.

In "primitive" plants, such as ferns, both the haploid and diploid stages are free-living, independent plants. The flowering plants, however, have reduced the haploid stage to a minimum, represented only by the pollen grain and a small cluster of cells in the ovary of the flower.

The Roles of Meiosis and Sexual Reproduction in Producing Genetic Variability

As we outlined in the introduction to this chapter, genetic variability among organisms is essential for survival and reproduction in a changing environment, and therefore for evolution. In the long run of millions of years, mutation of DNA is the ultimate basis for genetic variability; on shorter time scales, down to just a single generation, shuffling of homologous chromosomes, crossing over, and fusion of gametes from separate parents during sexual reproduction provide most of the genetic variability among members of a species.

Shuffling Homologues Creates Novel Combinations of Chromosomes

Maternal and paternal homologues are randomly shuffled out into the daughter cells of meiosis I. Remember, at metaphase I the paired homologues line up at the cell's equator. For each pair of homologues, the maternal chro-

mosome faces one pole and the paternal chromosome faces the opposite pole, but which homologue faces which pole is random. Let's consider an organism with three pairs of homologous chromosomes, large, medium, and small. Further, let's color-code the maternal homologue of each pair yellow, the paternal homologue violet. The alignment of chromosomes at metaphase I can assume four configurations:

Anaphase I can therefore produce eight possible sets of chromosomes ($2^3 = 8$):

Each of these chromosome clusters will then undergo meiosis II to produce two gametes. Therefore, our hypothetical parent organism could produce gametes with eight different chromosome sets.

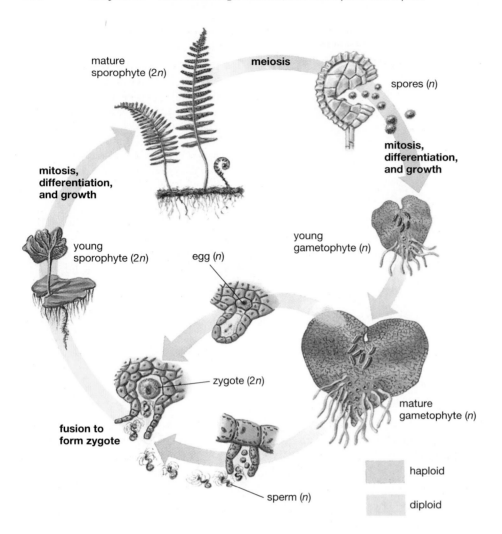

mature
sporophyte (2n)

meiosis

spores (n)

**mitosis,
differentiation,
and growth**

**mitosis,
differentiation,
and growth**

young
sporophyte (2n)

egg (n)

young
gametophyte (n)

zygote (2n)

mature
gametophyte (n)

**fusion to
form zygote**

sperm (n)

haploid

diploid

Figure 12-10   **Alternation of
generations**

The life cycle of a fern is
representative of alternation of
generations. The diploid stage is
called the *sporophyte*; this is the
fern usually seen in the woods.
Specialized cells in the sporophyte
undergo meiosis, producing
haploid spores. Spores differ from
gametes in that spores undergo
mitosis and grow into a
multicellular haploid body (the
gametophyte). In ferns, the haploid
stage is inconspicuous, a small
plant growing in moist places on
the forest floor. Sperm and eggs are
produced during the haploid stage,
and they fuse to make a diploid
zygote. Through mitosis, the
zygote develops into a new
multicellular diploid body, the
sporophyte.

Crossing Over Creates Chromosomes with Novel Combinations of Genes

Crossing over produces chromosomes that differ from
those of either parent. Consequently, offspring may receive new combinations of genes that may never before have occurred on single chromosomes. Each time
our hypothetical parent produces gametes, its homologous chromosomes cross over in new and different
places. Therefore, it can produce many more than eight
different types of gametes. *Because of crossing over, the
parent probably never produces any two gametes that are
completely identical!*

Fusion of Gametes Adds Further Genetic Variability to the Offspring

Finally, two gametes, possibly genetically unique in all the
world, fuse to form a diploid offspring. As an example of the
resulting variability, let's look at the potential for genetic
variability in human beings. Humans have 23 pairs of homologous chromosomes. A single person, therefore, can theoretically produce 2^{23} = about 8 million different gametes,
just on the basis of random separation of the homologues. Fusion of gametes from just two different people could produce
8 million × 8 million = 64 trillion genetically different children! When you add in the variability produced by crossing
over, is it any wonder that, with the exception of identical
twins, there is truly no one just like you?

❈ SUMMARY OF KEY CONCEPTS

Genetic Variability, Genetic Exchange, and the Evolution of Sexual Reproduction

For evolution to occur, there must be genetic differences
among organisms. These genetic differences originate as

mutations. Multiple mutations that arise in separate members of a species may be combined in single individuals
through genetic exchange, most often through sexual reproduction.

Meiosis and Sexual Reproduction

Sexual reproduction usually combines genetic contributions from two parental organisms. At some point in the life cycle of all sexually reproducing eukaryotic species, diploid cells undergo meiosis, which separates homologous chromosomes and produces haploid cells with only one homologue from each pair. These haploid cells or their descendants fuse to form diploid cells that receive one homologue of each pair from each parent, reestablishing pairs of homologous chromosomes.

The Mechanisms of Meiosis

During meiosis, a diploid cell undergoes two specialized cell divisions, meiosis I and meiosis II, to produce four haploid daughter cells. During interphase before meiosis, chromosomes are replicated. The cell then proceeds into meiosis I.

Meiosis I: During prophase I, homologous chromosomes, each consisting of two chromatids, pair up and exchange parts by crossing over. During metaphase I, paired homologues move together to the equator of the cell, with the two members of each pair facing opposite ends of the cell. Homologous chromosomes separate during anaphase I, and two nuclei form during telophase I. Each daughter nucleus receives only one member of each pair of homologues and is therefore haploid. However, sister chromatids do *not* separate during meiosis I, so each chromosome is still duplicated, consisting of two sister chromatids attached at the centromere. Chromosomes do *not* duplicate again between meiosis I and meiosis II.

Meiosis II: Meiosis II occurs in both of the daughter nuclei and resembles mitosis in a haploid cell. The chromosomes move to the equator of the cell during metaphase II. The two chromatids of each chromosome separate and are moved to opposite ends of the cell during anaphase II. This second division produces four haploid nuclei. Cytokinesis usually occurs during or shortly after telophase II, producing four haploid cells.

Meiosis separates pairs of homologous chromosomes in diploid cells to form unpaired chromosomes in haploid cells. Homologous chromosomes contain similar, but usually slightly different, genetic information. The separation of homologues during meiosis I is random: Either member of each pair of homologues ("maternal" or "paternal") may end up in either daughter nucleus. Therefore, the haploid cells produced by meiosis usually differ in their genetic composition.

Figure 12-11 compares mitosis with meiosis. The illustrations are arranged so that comparable stages are aligned. Note that there are no stages in mitosis that are comparable to meiosis I. Meiosis II is essentially identical to mitosis in a haploid cell.

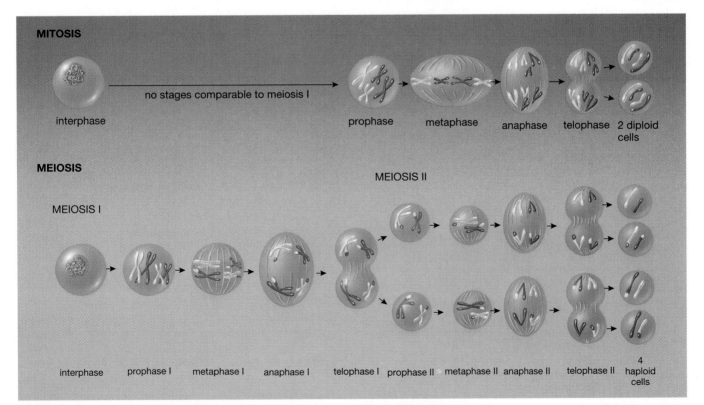

Figure 12-11   **A comparison of mitosis and meiosis**

In these diagrams, comparable phases of mitosis and meiosis are aligned. In both, chromosomes are replicated during interphase. Meiosis I, with pairing of homologous chromosomes, formation of chiasmata and exchange of chromosome parts, and separation of homologues to form haploid daughter nuclei, has no counterpart in mitosis. Meiosis II, however, is similar to mitosis occurring in a haploid cell.

Mitosis, Meiosis, and Eukaryotic Life Cycles

Most eukaryotic life cycles are similar in three respects: (1) Sexual reproduction combines haploid gametes to form a diploid cell; (2) at some point in the life cycle, diploid cells undergo meiosis to produce haploid cells once again; and (3) at some time in the life cycle, mitosis of either the haploid or diploid cells, or both, results in the growth of multicellular bodies and/or sexual reproduction. Three general types of life cycle can be distinguished on the basis of the proportion of the life cycle spent in diploid or haploid stages: (1) haploid cycles (mostly microbes and fungi), in which most of the life cycle is spent in the haploid state; (2) diploid cycles (animals), in which most of the life cycle is spent in the diploid state, with meiosis immediately preceding gamete formation; and (3) alternation of generations (plants), in which there are both haploid and diploid multicellular stages.

The Roles of Meiosis and Sexual Reproduction in Producing Genetic Variability

From one generation to the next, shuffling of homologous chromosomes, crossing over, and fusion of gametes from separate parents during sexual reproduction provides most of the genetic variability among members of a species. The random shuffling of homologous maternal and paternal chromosomes creates completely new chromosome combinations. Crossing over creates chromosomes with novel combinations of genes that may never before have occurred on single chromosomes. Because of crossing over, the parent probably never produces any two gametes that are completely identical. Finally fusion of two genetically unique gametes adds further genetic variability to the offspring. Fusion of gametes from just two different people could produce 64 trillion genetically different children!

✖ KEY TERMS

alternation of generations p. 229	**haploid** p. 220	**mutation** p. 220
anaphase p. 226	**homologue** p. 221	**prophase** p. 223
chiasma p. 223	**life cycle** p. 227	**sexual reproduction** p. 219
crossing over p. 223	**meiosis** p. 222	**spore** p. 228
diploid p. 220	**meiosis I** p. 222	**telophase** p. 227
fertilization p. 227	**meiosis II** p. 222	**zygote** p. 222
gamete p. 222	**metaphase** p. 225	

✖ THINKING THROUGH THE CONCEPTS

Multiple Choice

1. Which statement is most correct?
   a. All mutations are harmful.
   b. Mitosis and meiosis both add to genetic diversity.
   c. Crossing over helps each gamete get a different set of genes in meiosis.
   d. Mitosis always makes diploid daughter cells while meiosis always produces gametes.
   e. The only haploid cells are gametes.
2. When does crossing over occur?
   a. during translation   b. during fertilization
   c. during mitosis   d. during mutation
   e. when homologues have aligned and replicated
3. When do homologous chromosomes pair up?
   a. only in mitosis   b. only in meiosis I
   c. only in meiosis II   d. in both mitosis and meiosis
   e. in neither mitosis nor meiosis
4. Two homologues
   a. would be represented by maternal vs. paternal copies of the same chromosome
   b. would represent the two single strands of DNA which wrap into the double helix
   c. are still joined at the centromere between meiosis I and meiosis II
   d. are both present in haploid gametes
   e. none of the above

5. Curiously, there is no crossing over of any chromosome in the male fruit fly *Drosophila*, which has 4 pairs of chromosomes. How many different combinations of maternal vs. paternal chromosomes are possible in a male fruit fly's sperm?
   a. 2   b. 4
   c. 8   d. 16
   e. many more than the above
6. During alternation of generations
   a. the zygote produces a gametophyte
   b. the sporophyte is haploid
   c. spores give rise to gametophytes
   d. sporophytes produce sperm and eggs
   e. the gametophyte is diploid

Review Questions

1. Diagram the events of meiosis. At what stage do haploid nuclei first appear?
2. Describe the process of homologue pairing and crossing over. At what stage of meiosis does this occur? Name two functions of chiasmata.
3. In what ways are mitosis and meiosis similar? In what ways are they different?
4. Draw a generalized diagram of each of the three types of life cycles of eukaryotic organisms, and give an example of an organism using each type. Label the haploid and diploid stages, and indicate at which point in the cycle meiosis occurs.

5. What are probable advantages and disadvantages of sexual reproduction? Of diploidy?

6. Describe how meiosis provides for genetic variability. If an animal had a haploid number of 2 (no sex chromosomes), how many genetically different types of gametes could it produce (assume no crossing over)? If it had a haploid number of 5?

7. If there were no crossing over in human meiosis, how many different combinations of chromosomes (i.e., paternal vs. maternal centromeres, given this assumption) would be possible in the gametes (sperm and ova)? Express your answer as a number and an exponent. What does crossing over do to the number of possible different combinations of genes?

✖ APPLYING THE CONCEPTS

1. Crossing over apparently occurs at random locations along a pair of homologous chromosomes. Would you expect crossing over to recombine two genes more often if they are close to one another on the chromosome or if they are on opposite ends of the chromosome? Why?

2. In humans, meiosis is the first step in the formation of sperm and eggs. Normally, meiosis separates homologous chromosomes during meiosis I and separates sister chromatids during meiosis II. Occasionally, however, something goes awry. Suppose that during meiosis I, *both members of one pair of homologous chromosomes* moved to the same pole of the cell. If meiosis II occurred normally, what would be the chromosome complement of the resulting gametes? Suppose now that meiosis I occurred normally, but both sister chromatids of one chromosome migrated to the same pole during meiosis II. What would be the chromosome complement of resulting gametes? One relatively common genetic defect is Down syndrome, otherwise known as trisomy 21 (three copies of chromosome 21). Can abnormalities of meiosis explain trisomy 21? How?

3. In the bodies of humans and most other animal species, both mitosis and meiosis occur. Discuss the similar and different functions of mitosis and meiosis in our bodies and why it is important that both processes occur.

4. Some animal species can reproduce either asexually or sexually, depending on the state of the environment. Asexual reproduction tends to occur in stable, favorable environments, whereas sexual reproduction increases in unstable and/or unfavorable circumstances. Discuss the possible advantages or disadvantages this behavior might have on survival of the species in an evolutionary sense, or on survival of individuals in the population.

5. Some species, such as the cheetah, have had their populations reduced to such a small size that their genetic diversity has been reduced. Why might a lack of genetic diversity threaten a species? What can people do to help maintain genetic diversity, especially in light of captive breeding programs for endangered species in zoos? Could a lack of genetic diversity threaten a small, isolated, ethnically distinct human population?

6. Farm animals with desirable traits are sometimes produced by removing eggs from the female and fertilizing them in a dish using sperm from a male with desirable traits. After the zygote has undergone several divisions, the individual cells can be separated, returned to the same or different females, and they will develop into genetically identical animals. Human sperm banks already carry sperm from men with particular traits that a woman may choose. Human *in vitro* fertilization (see the Scientific Inquiry box in Chapter 39) could theoretically be used to create identical twins, triplets, etc. Argue for or against such human "cloning."

✖ FOR MORE INFORMATION

Amabile-Cuevas, C. F., and Chicurel, M. E. "Horizontal Gene Transfer." *American Scientist*, 81: 332–341, 1993. "Vertical" transfer of genetic material traverses generations as in sexual reproduction. More examples are being uncovered of transfer of genetic material horizontally, from one organism to another.

Kinoshita, J. "Swap Meet." *Discover*, April 1991. A discussion of hypotheses about the evolution of sex.

Lyttle, T. W. "Cheaters Sometimes Prosper: Distortion of Mendelian Segregation by Meiotic Drive." *Trends in Genetics* 9, June 1993. A gene usually perpetuates itself by survival and reproduction of the organism which expresses its coded protein. In a strikingly more direct survival of the fittest, some genes kill the gametes which do not possess the genes.

Stahl, F. "Genetic Recombination." *Scientific American*, February 1987. A look at the mechanisms of crossing over.

NET WATCH

On-line resources for this chapter are on the World Wide Web at:
http://www.prenhall.com/~audesirk (click on the <u>table of contents</u> link and then select Chapter 12).

13 Patterns of Inheritance

"In the mingling of the seed, sometimes the woman, with sudden force, overpowers the man, and then the children, born of maternal seed, will resemble more the mother; but if from paternal seed, the father. The children you see resembling both their parents, having the features of both, have been created from father's body and mother's blood, when the seeds course through the bodies excited by Venus, in harmony of mutual passion, breathing as one, with neither conquering and neither being conquered."

LUCRETIUS in *On the Nature of Things* (96–55 B.C.)

⚹ AT A GLANCE

Gregor Mendel and the Origin of Genetics
   Doing It Right: The Secrets of Mendel's Success

Inheritance of Single Traits: The Law of Segregation
   The Inheritance of Dominant and Recessive Alleles on
      Homologous Chromosomes Can Explain the Results of
      Mendel's Monohybrid Crosses
   Mendel's Hypothesis Can Be Used to Predict the Outcome
      of New Types of Single-Trait Crosses

Inheritance of Multiple Traits: Independent Assortment

Inheritance Patterns of Genes Located on the Same Chromosome
   Genes on the Same Chromosome Tend to Be Inherited
      Together
   Crossing Over May Separate Linked Genes
   The Frequency of Crossing Over between Linked Genes Can
      Be Used to Map the Location of Genes on a Chromosome

The Inheritance of Sex and Sex-Linked Genes
   Sex Linkage Is a Special Case of Linkage between Genes on
      the Same Chromosome

Variations on the Mendelian Theme
   In Incomplete Dominance, the Phenotype of Heterozygotes
      Is Intermediate between the Phenotypes of
      the Homozygotes
   There May Be Multiple Alleles of a Single Gene
   Many Traits Are Influenced by Several Genes
   A Single Gene May Have Multiple Effects on Phenotype
   The Environment Influences the Expression of All Genes

The Molecular Basis of Mendelian Genetics

Reflections on Genetic Diversity and Human Welfare

Inheritance provides for both similarity and difference. These animals are both tigers because of the vast majority of identical genes that they inherited from common ancestors; they differ in color because of differences in a single gene.

How can two brown-eyed parents produce a blue-eyed child? If you breed two black Labrador retrievers, why may the offspring be yellow, brown, or black (Fig. 13-1)? Although you may not realize it yet, you have already come a long way toward understanding the answers to questions such as these. First, you have learned that DNA molecules carry the hereditary information of all living organisms; the sequences of nucleotides in specific stretches of DNA, the **genes**, encode the information needed to specify the amino acid sequences of proteins. Second, DNA is physically part of the chromosomes, which, in eukaryotic organisms, are found in the nucleus of each cell. Third, in the cells of a diploid organism, chromosomes occur in pairs, called homologous chromosomes, that are similar in appearance and have the same genes. But the nucleotide sequences of the genes may differ on homologous chromosomes because of mutations in DNA that may have arisen in that organism or its ancestors. Fourth, during meiosis, homologous chromosomes separate, one member of each pair entering each haploid daughter cell. Therefore, when gametes are formed, either directly after meiosis or later during the organism's life cycle, each gamete contains one member of each pair of chromosomes. During sexual reproduction, each offspring receives one member of each pair of homologous chromosomes from each of its two parents.

What does this process have to do with blue-eyed children and brown dogs? Quite a bit. If two homologous chromosomes have different nucleotide sequences in some of their genes, then each homologue will direct the synthesis of slightly different proteins. Those protein differences can have profound, visible effects in an organism. For example, as we will see in Chapter 15, different eye colors in people are caused by differences in the amount and distribution of pigment molecules that are synthesized by protein enzymes. Small genetic differences produce enzyme differences, which in turn alter the amount and distribution of pigments in the iris, yielding blue, green, brown, or black eyes. Therefore, slightly different nucleotide sequences of genes on homologous chromosomes, separation of homologues during meiosis, and re-pairing of homologous chromosomes from two parents can produce offspring with genetic differences, enzyme differences, and consequently differences in appearance or physiology.

Figure 13-1   **Inheritance of color**

Labrador retrievers come in yellow (left), brown (middle), or black (right). Color is determined by interactions between two sets of genes.

In this chapter, we will examine the patterns of inheritance commonly observed in diploid organisms, starting with the work of Gregor Mendel.

Gregor Mendel and the Origin of Genetics

Gregor Mendel (Fig. 13-2) was a monk in the monastery of St. Thomas in Brünn (now Brno, in the Czech Republic). Before settling down in the monastery, Mendel tried his hand at several pursuits, including health care and teaching. In an effort to earn his teaching certificate, Mendel attended the University of Vienna for two years, where he studied botany and mathematics, among other subjects. This training proved crucial to his later experiments, which were the foundation for the modern science of genetics. At St. Thomas, in the mid-1800s, Mendel carried out both his monastic duties and a groundbreaking series of experiments on inheritance in the common edible pea. Although Mendel worked without knowledge of genes or chromosomes, it will be easier to follow his experiments after a brief look at the genetic differences that may occur on homologous chromosomes.

The specific physical location of a given gene on a chromosome is called its **locus** (Fig. 13-3). Homologous chromosomes carry the same genes, located at the same loci. Although the nucleotide sequence at a given gene locus is always *similar* on homologous chromosomes, it may not be *identical*. Different nucleotide sequences at the same gene locus on two homologous chromosomes produce alternative forms of the gene, called **alleles**. The human ABO blood types, for example, are produced by different alleles of the same gene (see Fig. 13-16).

If two homologous chromosomes in an organism have the same allele at a given gene locus, the organism is said to be **homozygous** for that gene locus. (*Homozygous* comes

Figure 13-2   **Gregor Mendel**

A portrait of Gregor Mendel, painted about the time of his pioneering genetics experiments. Mendel was born in Austria in 1822 and died in 1884.

from Greek words meaning "same pair.") If two homologous chromosomes have different alleles at a given gene locus, the organism is **heterozygous** ("different pair") at that

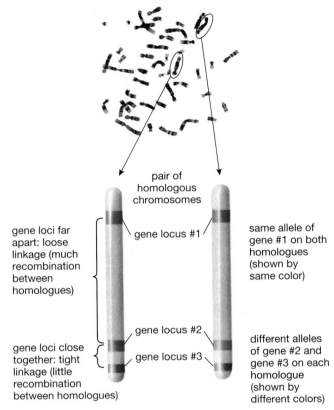

Figure 13-3 **The relationships among genes, alleles, and chromosomes**

The drawing illustrates a pair of homologous chromosomes. A given gene is a segment of DNA at the same location (locus) on both homologous chromosomes. A chromosome carries many genes; each gene occupies a different locus on the chromosome. Differences in nucleotide sequences at the same gene locus (depicted by different colors) produce different alleles of the gene.

locus. You will recall that, during meiosis, homologous chromosomes are separated. When gametes are formed, each gamete receives one member of each pair of homologous chromosomes. Therefore, an organism homozygous at a particular gene locus produces gametes that all contain the same allele. An organism heterozygous at the same gene locus produces gametes of two kinds; about half of the gametes contain one allele and half contain the other allele.

These facts, combined with some careful breeding experiments and a little "genetic bookkeeping," can be used to explain, and indeed to predict, the patterns of inheritance of many traits in organisms as diverse as pea plants and human beings. Interestingly, both the patterns of inheritance and many essential facts about genes, alleles, and the distribution of alleles in gametes and zygotes during sexual reproduction were deduced by Gregor Mendel before DNA, chromosomes, or meiosis had been discovered. Because you know much more about cellular and molecular genetics than Mendel did, you should be able to anticipate Mendel's results even before we present them in the text. However, because Mendel's experi-

ments are a succinct, elegant example of science in action, let's follow Mendel along his paths of discovery.

Mendel was not the first person to study inheritance. However, as Mendel himself wrote:

> Whoever surveys the work in this field will come to the conviction that among the numerous experiments, not one has been carried out to an extent or in a manner that would make it possible to determine the number of different forms in which hybrid progeny appear, permit classification of these forms in each generation with certainty, and ascertain their numerical interrelationships. . . . [H]owever, this seems to be the one correct way of finally reaching the solution to a question whose significance for the evolutionary history of organic forms must not be underestimated."

Doing It Right: The Secrets of Mendel's Success

Why did Mendel succeed where others before him had failed? There are three key steps to any successful experiment in biology: (1) choosing the right organism to work with, (2) designing and performing the experiment correctly, and (3) analyzing the data properly. Mendel was the first geneticist to complete all three steps.

Mendel's choice of the edible pea as an experimental subject was excellent. The structure of the pea flower normally prevents pollen from entering from the outside—that is, from another flower (Fig. 13-4). Instead, pea flowers usually self-pollinate. As a result, the egg cells in each flower are fertilized by sperm from the pollen of the same flower (this is called **self-fertilization**). Even in Mendel's time commercial seed dealers sold many different types of peas that were **true-breeding**. In true-breeding plants, all the offspring produced through self-fertilization are essentially identical to the parent plant.

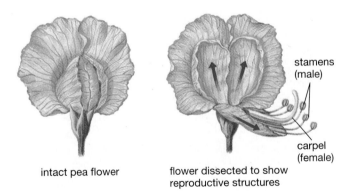

Figure 13-4 **The flower of the edible pea**

In the intact flower (left), the lower petals form a container enclosing the reproductive structures, the stamens (male) and carpel (female). Pollen normally cannot enter the flower from outside, and consequently peas usually self-fertilize. Plant breeders, however, can pull apart the petals (right) and remove the stamens so that self-fertilization cannot occur. Dusting the carpels with pollen of their choice results in controlled cross-fertilization.

Although peas normally self-fertilize, plant breeders can also mate two different plants by hand. By carefully picking the flowers apart, plant breeders can collect pollen from one plant and transfer it to a flower on another plant. The sperm from the "foreign" pollen will then fertilize the egg cells of the recipient flower (**cross-fertilization**). In this way, one can mate two different true-breeding plants and see what types of offspring they produce.

Mendel began by working with only one trait at a time, and he chose traits that had unmistakably different forms, such as white versus purple flowers (Fig. 13-5). Earlier scientists had often crossed plants that differed in several traits or that had different forms of a trait that might be confused with one another. Not surprisingly, the results were difficult to interpret. Studying single traits, with clearly different alternative forms, allowed Mendel to see through to the underlying principles of inheritance.

Perhaps most important, Mendel counted the numbers of offspring bearing the various traits and critically analyzed the numbers. The use of statistics as a tool for understanding underlying principles has since become an extremely important practice in biology.

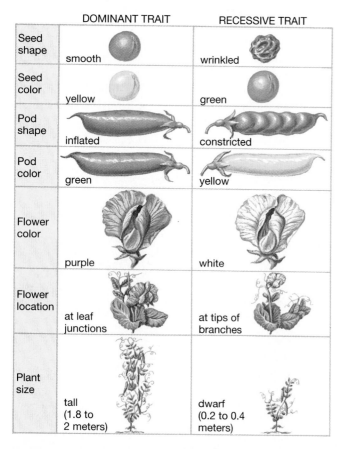

	DOMINANT TRAIT	RECESSIVE TRAIT
Seed shape	smooth	wrinkled
Seed color	yellow	green
Pod shape	inflated	constricted
Pod color	green	yellow
Flower color	purple	white
Flower location	at leaf junctions	at tips of branches
Plant size	tall (1.8 to 2 meters)	dwarf (0.2 to 0.4 meters)

Figure 13-5   **Traits of Mendel's peas**

Traits of pea plants used by Mendel in his studies of plant inheritance.

Inheritance of Single Traits: The Law of Segregation

Mendel started as simply as possible. He raised varieties of pea plants that were true-breeding for different forms of a single trait and cross-fertilized them. Such an experiment, involving organisms that differ in only one trait, is called a **monohybrid cross**. (In this usage, **hybrid** refers not to the offspring of parents of two different species, such as a mule, but to offspring of parents of the same species that differ in one or more inherited traits, such as hybrid corn.) Mendel saved the resulting hybrid seeds and grew them the next year.

In one of these experiments, Mendel cross-fertilized a white-flowered pea with a purple-flowered one. (In modern terminology, this was the **parental generation**, denoted by the letter **P.**) When he grew the resulting seeds, he found that all the first-generation offspring (the "first filial," or F_1 **generation**) produced purple flowers:

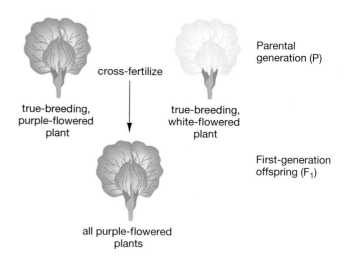

true-breeding, purple-flowered plant

cross-fertilize

true-breeding, white-flowered plant

Parental generation (P)

all purple-flowered plants

First-generation offspring (F_1)

What happened to the white color? The flowers of the hybrids were every bit as purple as the flowers of their purple parent. White seemed to have disappeared completely in the F_1 offspring.

Mendel allowed the F_1 flowers to self-fertilize, collected the seeds, and planted them the next spring. The second generation (F_2) had some plants with flowers of each color:

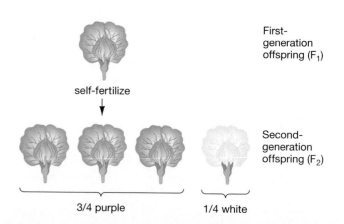

First-generation offspring (F_1)

self-fertilize

Second-generation offspring (F_2)

3/4 purple   1/4 white

Overall, about three-fourths of the plants had purple flowers and one-fourth had white flowers. The exact numbers were 705 purple and 224 white, or a ratio of 3.15 purple to 1 white.

Mendel allowed the F_2 plants to self-fertilize and produce yet a third (F_3) generation. He found that all the white-flowered F_2 plants produced white-flowered offspring; that is, they bred true. This result was true for as many generations as he had time and patience to raise: White-flowered parents always gave rise to white-flowered offspring. About one-third of the purple-flowered F_2 plants were also true-breeding. The remaining two-thirds were hybrids and produced both purple- and white-flowered offspring, again in the ratio of 3 to 1. Therefore, the F_2 generation included ¼ true-breeding purple plants, ½ hybrid purple, and ¼ true-breeding white.

The Inheritance of Dominant and Recessive Alleles on Homologous Chromosomes Can Explain the Results of Mendel's Monohybrid Crosses

Mendel formed a five-part hypothesis to explain these results.

1. Each trait is determined by pairs of discrete physical units (what modern geneticists call genes). Each individual organism has two genes that together control the expression of a given trait (for example, two genes for flower color). As you will no doubt have guessed, the two genes are located on homologous chromosomes. There may be two or more alternative forms (alleles) of a gene. White-flowered peas, for example, have a different allele of the gene controlling flower color than true-breeding purple-flowered peas do. Different alleles of the same gene are located at the same locus on a chromosome but have different nucleotide sequences.

2. Pairs of genes separate from each other during gamete formation. This part is Mendel's **law of segregation**: Each gamete receives only one of an organism's pair of genes. When a sperm fertilizes an egg, the resulting offspring receives one gene from the father and one from the mother. Now apply your knowledge of meiosis; homologous chromosomes separate during anaphase I of meiosis. When gametes form, each receives one member of each pair of homologous chromosomes, and therefore one member of each pair of genes.

3. Which member of a pair of genes becomes included in a gamete is determined by chance. If we assign letters to each member of a pair of genes, say G1 and G2, then each gamete is just as likely to receive a G1 as a G2. The gamete will not receive both, and there will not be a preference for one member of the pair. This randomness occurs because separation of pairs of homologous chromosomes during meiosis is random.

4. When two alternative forms of a gene are present, one (the **dominant** allele) may mask the expression of the other (the **recessive** allele). However, although the dominant allele of a gene masks the *expression* of the recessive allele, it doesn't alter the *physical nature* of the recessive allele, which can be passed unchanged into the individ-

ual's gametes. In Mendel's experiments with flower color, the allele for purple flowers is dominant to the allele for white flowers. At the end of this chapter we will discuss the molecular basis of dominance.

5. True-breeding organisms have two of the same alleles for the trait under study (that is, they are homozygous); hybrids have two different alleles (that is, they are heterozygous). A homozygous organism, with only one type of allele, can produce only one type of gamete:

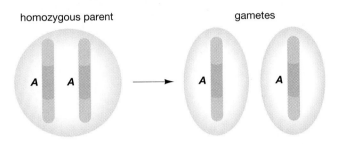

homozygous parent   gametes

A heterozygous organism, with two different alleles, produces equal numbers of gametes with each of the two alleles:

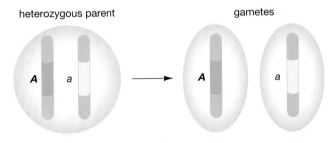

heterozygous parent   gametes

Let's see how Mendel's hypothesis explains the results of his experiments with flower color. Using letters to represent the different alleles, we will assign the uppercase letter P to the allele for purple (the dominant allele is usually represented by a capital letter) and the lowercase letter p to the allele for white (recessive). A homozygous purple-flowered plant has two alleles for purple (PP); a white-flowered plant has two alleles for white (pp). A PP plant produces sperm and eggs all of which carry the P allele; a pp plant produces sperm and eggs all of which carry the p allele:

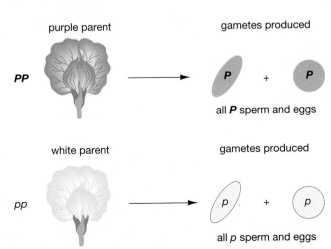

purple parent   gametes produced

PP   all **P** sperm and eggs

white parent   gametes produced

pp   all p sperm and eggs

The F_1 hybrid offspring are produced when P sperm fertilize p eggs or when p sperm fertilize P eggs. In either case, the F_1 offspring are Pp. Because P is dominant to p, all the offspring are purple:

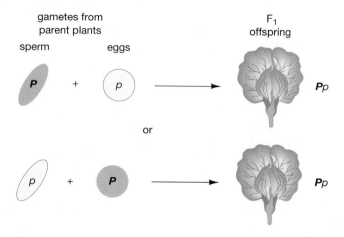

gametes from
parent plants

sperm   eggs F_1
offspring

P   +   p   →   Pp

or

p   +   **P**   →   Pp

All gametes produced by a heterozygous Pp plant have an equal chance of receiving either the P allele or the p allele (that is, the plant produces equal numbers of P and p sperm and equal numbers of P and p eggs). When a Pp plant self-fertilizes, the two types of sperm have an equal chance of fertilizing each type of egg:

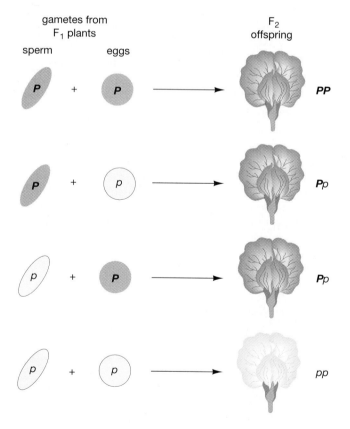

gametes from
F_1 plants

sperm   eggs F_2
offspring

P   +   **P**   →   **PP**

P   +   p   →   **Pp**

p   +   **P**   →   **Pp**

p   +   p   →   pp

Therefore, three types of offspring can be produced: PP, Pp, and pp. These occur in the approximate proportions of ¼ PP, ½ Pp, and ¼ pp.

The actual combination of alleles carried by an organism (for example, PP or Pp) is its **genotype**. An organism's traits, including its outward appearance, behavior, digestive enzymes, blood type, or any other observable feature, determine its **phenotype**. As we have seen, plants with genotypes PP and Pp both bear purple flowers. Thus, even though they have different genotypes, they have the same phenotype. Therefore, the F_2 generation consists of three genotypes (¼ PP, ½ Pp, and ¼ pp) but only two phenotypes (¾ purple and ¼ white).

Figures 13-6 and 13-7 present two methods of "genetic bookkeeping" for determining the expected proportions of offspring in a monohybrid cross. The first method uses a diagram called a Punnett square (Fig. 13-6), named after a famous geneticist of the early 1900s. The second method relies on probability theory (Fig. 13-7). The two methods yield the same results, so you can use whichever seems easier for you. Whichever method you choose, there is one crucial point to keep in mind: *These calculations give only the most probable proportions of offspring of different genotypes and phenotypes.* In a real experiment, one would expect only that the offspring will occur in *approximately* the predicted proportions. In the F_2 generation of Mendel's monohybrid cross, note that he didn't obtain *exactly* ¾ purple-flowered and ¼ white-flowered plants.

Mendel's Hypothesis Can Be Used to Predict the Outcome of New Types of Single-Trait Crosses

You probably have recognized that Mendel used the scientific method discussed in Chapter 1. He observed the results of his experiments and then formed a hypothesis. As you know, the scientific method has a third step: to use the hypothesis to predict the results of other experiments and to see if those experiments support or refute the hypothesis. Mendel did just that. If the hybrid F_1 flowers have one allele for purple and one for white (Pp), then we should be able to predict the outcome of cross-fertilizing these Pp plants with homozygous recessive (pp) white plants. (Can you?) As Figure 13-8a shows, Mendel's hypothesis predicts that there will be equal numbers of Pp (purple) and pp (white) offspring. This is precisely what happened.

This type of experiment has practical uses, too. By just looking at an organism that has the dominant phenotype, one usually cannot tell whether it is homozygous or heterozygous. Cross-fertilization of a phenotypically dominant individual with a homozygous recessive individual is called a **test cross**, because it can be used to test whether the dominant parent is homozygous or heterozygous. When crossed with a homozygous recessive, a homozygous dominant produces all phenotypically dominant offspring (Fig. 13-8b), whereas a heterozygous dominant yields offspring with both dominant and recessive phenotypes, in a 1:1 (one-to-one) ratio (Fig. 13-8a).

Inheritance of Multiple Traits: Independent Assortment

Having determined the mode of inheritance of single traits, Mendel then turned to the more complex question of multiple traits. He began with a **dihybrid cross**; that is, he cross-bred plants that differed in two traits, for example seed color (yellow or green) and seed shape (smooth or wrinkled). If he crossed a plant that was homozygous for smooth yellow seeds with one that was homozygous for wrinkled green seeds, the F_1 offspring all bore smooth yellow seeds. This result was no great surprise, because from separate monohybrid crosses of each of these traits he al-

ready knew that smooth (S) is dominant to wrinkled (s), and that yellow (Y) is dominant to green (y) (see Fig. 13-5). The F_1 offspring, therefore, are all genotypically $SsYy$. Allowing these F_1 plants to self-fertilize, Mendel found that the F_2 generation consisted of 315 smooth yellow seeds, 101 wrinkled yellow seeds, 108 smooth green seeds, and 32 wrinkled green seeds. These numbers work out to roughly $9/16$ smooth yellow seeds, $3/16$ wrinkled yellow seeds, $3/16$ smooth green seeds, and $1/16$ wrinkled green seeds, or a ratio of 9:3:3:1. The F_2 generations produced from other dihybrid crosses had similar phenotypic ratios.

Mendel realized that these ratios could be explained if he assumed that the genes for seed color and seed shape are inherited independently of each other and do not influ-

Figure 13-6   The Punnett square

The Punnett square method of determining likely genotypes and phenotypes of offspring. The Punnett square is intuitive and relatively nonmathematical, and many students prefer it to the probability theory. The example shown here analyzes the inheritance of flower color in peas, based on a cross of F_1 hybrids. Using the Punnett square requires the following six steps.

(1) Assign letters to represent the different alleles; dominant alleles are usually upper case (P), recessive alleles lower case (p). **Be sure that you can distinguish upper and lower case letters**: for example, W and w or S and s may be difficult to tell apart, especially in the heat of a test.

(2) Assuming that the genotypes of the parents are known, determine the types of equally likely, genetically different gametes that can be produced by the male and female parents. In the case of the gene pair here, $1/2$ of the gametes should receive the P allele, and $1/2$ should receive the p allele. (Of course, if a parent is homozygous, then all of its gametes will receive the same allele.)

(3) Draw the Punnett square. The number of rows in the square is the same as the number of different types of sperm, and the number of columns in the square is the same as the number of different types of eggs. Label the rows with the types of sperm, the columns with the types of eggs.

(4) The expected offspring are the "cells" of the square. Each offspring results from the fertilization of the egg in its column by the sperm in its row. Fill out the offspring genotypes systematically, for example, first using row 1 sperm to fertilize all types of eggs, then row 2 sperm to fertilize all types of eggs, and so on.

(5) Examine the offspring cells and add up the number of offspring with the same genotypes.

(6) **Remember that the numbers generated by a Punnett square do not mean that every mating will yield exactly four offspring.** Convert the numbers of offspring genotypes to expected fractions of offspring genotypes by dividing the number of each offspring genotype by the total number of offspring cells. If desired, add together those genotypes that produce the same phenotypes (PP and Pp in this example), to yield the expected fractions of offspring with each phenotype. Remember also that the fractions of different offspring generated by the square are only *expected* fractions. Real matings would be expected only to yield *approximately* these fractions of different offspring.

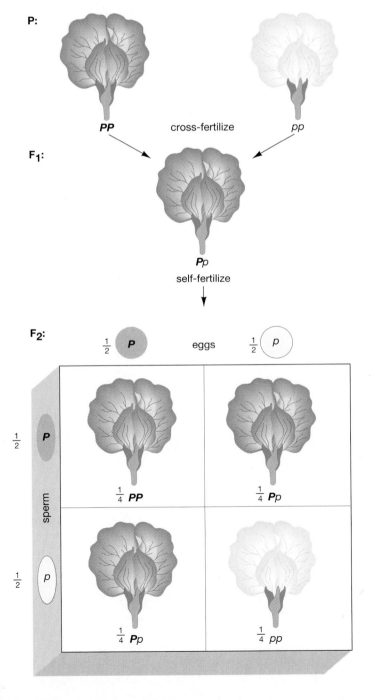

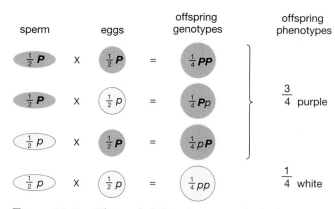

sperm		eggs		offspring genotypes	offspring phenotypes
$\frac{1}{2}$ **P**	X	$\frac{1}{2}$ **P**	=	$\frac{1}{4}$ **PP**	
$\frac{1}{2}$ **P**	X	$\frac{1}{2}$ **p**	=	$\frac{1}{4}$ **Pp**	$\frac{3}{4}$ purple
$\frac{1}{2}$ **p**	X	$\frac{1}{2}$ **P**	=	$\frac{1}{4}$ **pP**	
$\frac{1}{2}$ **p**	X	$\frac{1}{2}$ **p**	=	$\frac{1}{4}$ **pp**	$\frac{1}{4}$ white

Figure 13-7   **The probability theory method of determining likely genotypes and phenotypes of offspring**

The example is the F₁ cross illustrated in the Punnett square method in Figure 13-6. In probability theory, the probability that two independent events will occur together is the product (multiplication) of their independent probabilities. In this cross, each type of gamete (P or p) is equally likely to occur; that is, each has a probability of ½. Further, each type of sperm is equally likely to fertilize each type of egg. In terms of probability theory, the first event (which allele is in a particular sperm) does not depend on the second event (which allele is in the egg fertilized by that sperm). We can therefore obtain the probable ratio of offspring by multiplying the probability of obtaining each type of sperm by the probability of obtaining each type of egg.

To use probability theory to analyze genetic crosses, perform the following four steps:

(1) Assign letters to the alleles (the same as step 1 of the Punnett square method).

(2) Determine the types of equally likely, genetically different gametes that can be produced by the parents (the same as step 2 of the Punnett square method).

(3) Set up a multiplication table as shown below: ½ of the sperm are P, and ½ of the eggs are P; multiply these together to obtain ¼ PP offspring. Continue filling in the table; be sure that all types of sperm have the "opportunity" to fertilize all types of eggs. For a "genotype" table such as this one, the number of rows in the table (4) should equal the number of different types of sperm (2) multiplied by the number of different types of eggs (2).

(4) If desired, add together the genotypes that produce the same phenotypes (the same as step 6 of the Punnett square method).

ence each other during gamete formation (Fig. 13-9). (Imagine flipping two coins, a dime and a nickel. Whether the dime comes up heads or tails does not affect which side of the nickel comes up.) Thus the outcome for each trait could be regarded as a simple monohybrid cross, in which a 3:1 ratio of offspring would be expected. The laws of probability state that the independent combination of two 3:1 ratios yields a 9:3:3:1 ratio (Fig. 13-10b), and we can see from Mendel's results that this is just what happened. There were 423 smooth seeds (of either color) to 133 wrinkled ones (3.18:1) and 416 yellow seeds (of either shape) to 140 green ones (2.97:1). The Punnett square shows how two 3:1 ratios combine to form an overall 9:3:3:1 ratio (Fig. 13-10a).

The independence in the inheritance of two or more distinct traits is called the **law of independent assortment**: The distribution of alleles for one trait into the gametes does not affect the distribution of alleles for other traits. Think back to the events of meiosis in Chapter 12: How do you think that independent assortment of alleles for two different genes might occur? That's right: Independent assortment will occur if the gene loci are on different chromosomes. When paired homologous chromosomes line up at the equator of the cell during metaphase I, the orientation of the members of each pair is random, and the orientation of one pair of homologues does not influence the orientation of other pairs. In other words, if the paternal homologue of chromosome 1 faces the "north pole" of the cell, this orientation does not influence whether the paternal homologue of chromosome 2 faces north or south. Therefore, when the homologues separate during anaphase I, whether a given daughter cell receives the paternal homologue of chromosome 1 has no effect on the likelihood that it will receive the paternal or maternal homologue of chromosome 2.

In an Unprepared World, Genius May Go Unrecognized

In 1865, Gregor Mendel presented his theories of inheritance to the Brünn Society for the Study of Natural Science, and they were published the next year. It did *not* mark the beginning of genetics. In fact, it didn't make any impression at all on the biology of his time. Mendel's experiments, which eventually spawned one of the most important theories in all of biology, simply vanished from the scientific scene. Apparently, very few biologists read his paper, and those who did probably couldn't understand it.

It was not until 1900 that three biologists—Carl Correns, Hugo de Vries, and Erich Tschermak—working independently and knowing nothing of Mendel's work, rediscovered the principles of inheritance. No doubt to their intense disappointment, when they searched the scientific literature before publishing their results, they found that Mendel had scooped them over 30 years before.

Inheritance Patterns of Genes Located on the Same Chromosome

Gregor Mendel didn't know anything about the physical nature of genes or that genes are parts of chromosomes, because chromosomes weren't even discovered until long after his experiments were concluded. Once chromosomes were seen in microscopes, and their distribution during mitosis and meiosis was deduced, it became obvious that they were the vehicles of inheritance. It also became obvious that there are many more traits, and therefore many more genes, than there are chromosomes. The facts that genes are parts of chromosomes and that each chromosome bears many genes have important implications for inheritance.

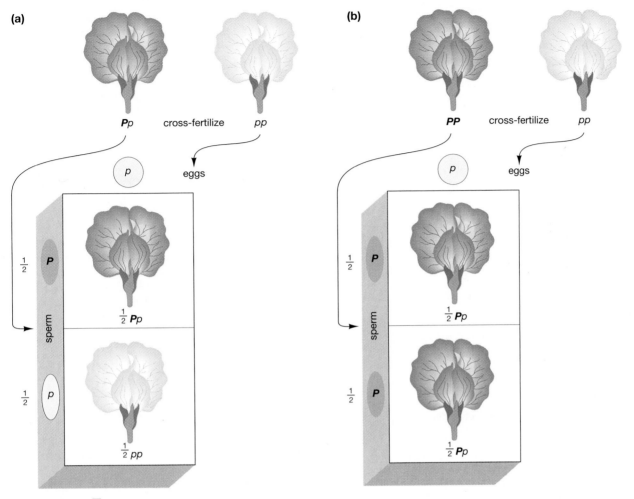

Figure 13-8*   Test crosses

Test crosses involve breeding a homozygous recessive organism (in these examples, *pp* plants with white flowers) with a phenotypically dominant organism (here, plants with purple flowers). If the phenotypically dominant parent is heterozygous *Pp* **(a)**, then half the offspring will be heterozygous purple and half white; if the parent is homozygous *PP* **(b)**, then all the offspring will be heterozygous purple.

Genes on the Same Chromosome Tend to Be Inherited Together

If *chromosomes* assort independently during meiosis I, then only genes located on *different chromosomes* will assort independently. Genes on the *same chromosome* are said to be **linked**, because they do not assort independently but tend to be inherited together. One of the first pairs of linked genes to be discovered were those for flower color and pollen grain shape in the sweet pea (a different type than the edible pea). Purple flower color (*P*) is dominant over red (*p*), and long pollen shape (*L*) is dominant over round (*l*). When a homozygous purple-flowered, long-pollen pea is crossed with a homozygous red-flowered, round-pollen pea, all the F₁ offspring have purple flowers and long pollen. However, the genes for flower color and for pollen shape are carried on the same chromosome, so the two genes tend to assort together during meiosis and to be inherited together. Thus, the phenotypes of the F₂ generation do not occur in a 9:3:3:1 ratio, as Mendel's law of independent assortment would predict. Instead, the F₂ generation has about ¾ purple-flowered, long-pollen plants and ¼ red-flowered, round-pollen plants (Fig. 13-11).

Crossing Over May Separate Linked Genes

There is one thing wrong with this tidy scheme: Genes on the same chromosome do not *always* stay together. In the sweet pea cross just described, for example, often the F₂ generation will include a few purple-flowered, round-pollen plants and red-flowered, long-pollen plants. How can this be?

As you know, during prophase I of meiosis, the nonsister chromatids of homologous chromosomes intertwine, producing X-shaped figures called **chiasmata** (see Fig. 12-5). At chiasmata, parts of nonsister chromatids may exchange with each other, a process called **crossing over**. In fact, there is almost always at least one exchange between homologues during any meiosis. Exchange of corresponding segments of DNA during crossing over forms new gene combinations on both homologous chromosomes. Therefore, when homologous chromosomes separate at anaphase I, the chro-

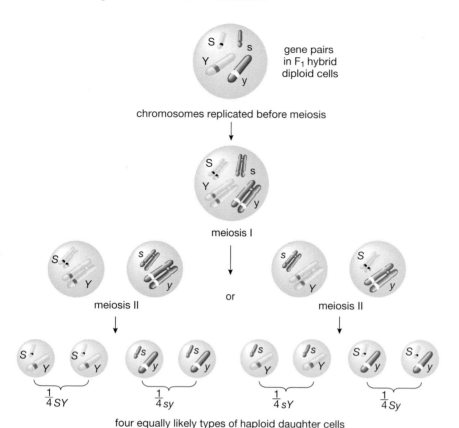

gene pairs in F₁ hybrid diploid cells

chromosomes replicated before meiosis

meiosis I

meiosis II   or   meiosis II

$\frac{1}{4}$ SY $\frac{1}{4}$ sy $\frac{1}{4}$ sY $\frac{1}{4}$ Sy

four equally likely types of haploid daughter cells produced by independent assortment during meiosis

Figure 13-9 **Independent assortment of chromosomes**

Independent assortment of alleles for two different genes arises from the mechanisms of chromosome movements during meiosis (see Chapter 12). At the end of meiosis I, two equally likely chromosome groupings are possible: SY and sy, or Sy and sY. At the end of meiosis II, the SY/sy grouping yields haploid cells, half of which are SY and half of which are sy. The Sy/sY grouping yields half Sy and half sY daughter cells. Because meiosis occurs in many reproductive cells in a plant, the two combinations are equally likely to occur, on the average. Therefore, an F₁ plant would produce gametes in the predicted proportions $\frac{1}{4}$ SY, $\frac{1}{4}$ sy, $\frac{1}{4}$ Sy, and $\frac{1}{4}$ sY.

(a)

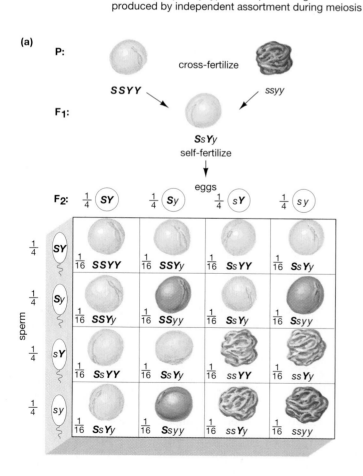

P: cross-fertilize

SSYY   *ssyy*

F₁:

SsYy
self-fertilize

eggs

F₂: $\frac{1}{4}$ (SY) $\frac{1}{4}$ (Sy) $\frac{1}{4}$ (sY) $\frac{1}{4}$ (sy)

sperm

$\frac{1}{4}$ (SY)
| $\frac{1}{16}$ **SSYY** | $\frac{1}{16}$ **SSYy** | $\frac{1}{16}$ **SsYY** | $\frac{1}{16}$ **SsYy** |

$\frac{1}{4}$ (Sy)
| $\frac{1}{16}$ **SSYy** | $\frac{1}{16}$ **SSyy** | $\frac{1}{16}$ **SsYy** | $\frac{1}{16}$ **Ssyy** |

$\frac{1}{4}$ (sY)
| $\frac{1}{16}$ **SsYY** | $\frac{1}{16}$ **SsYy** | $\frac{1}{16}$ ss**YY** | $\frac{1}{16}$ ss**Yy** |

$\frac{1}{4}$ (sy)
| $\frac{1}{16}$ **SsYy** | $\frac{1}{16}$ **Ssyy** | $\frac{1}{16}$ ss**Yy** | $\frac{1}{16}$ ssyy |

(b)

seed shape		seed color		F₂ offspring
$\frac{3}{4}$ smooth	x	$\frac{3}{4}$ yellow	=	$\frac{9}{16}$ smooth yellow
$\frac{3}{4}$ smooth	x	$\frac{1}{4}$ green	=	$\frac{3}{16}$ smooth green
$\frac{1}{4}$ wrinkled	x	$\frac{3}{4}$ yellow	=	$\frac{3}{16}$ wrinkled yellow
$\frac{1}{4}$ wrinkled	x	$\frac{1}{4}$ green	=	$\frac{1}{16}$ wrinkled green

Figure 13-10 **Calculating the expected offspring in a dihybrid cross**

(a) Punnett square analysis. One of the original parents is homozygous for smooth (SS) yellow (YY) seeds, and the other original parent is homozygous for wrinkled (ss) green (yy) seeds. The F₁ offspring therefore all have smooth yellow seeds (SsYy). If we allow the F₁ to self-fertilize, we obtain the offspring shown within the square. There are now 16 squares in the analysis, but the method is the same as in Figure 13-6. The expected proportions of offspring are 9/16 smooth yellow, 3/16 smooth green, 3/16 wrinkled yellow, and 1/16 wrinkled green, or 9:3:3:1. Note also that the Punnett square predicts 12 smooth to 4 wrinkled seeds and 12 yellow to 4 green seeds (both 3:1 ratios), just as would be expected from monohybrid crosses if we ignore the second trait. **(b)** Analysis by probability theory. *Remember that independent assortment means that each trait assorts independently of the others.* We know from monohybrid crosses that three-quarters of the F₂ offspring will be smooth and one-quarter will be wrinkled, and that three-quarters will be yellow and one-quarter will be green. Multiplying these independent probabilities produces the expected F₂ offspring. These are identical to the offspring ratios generated by the Punnett square.

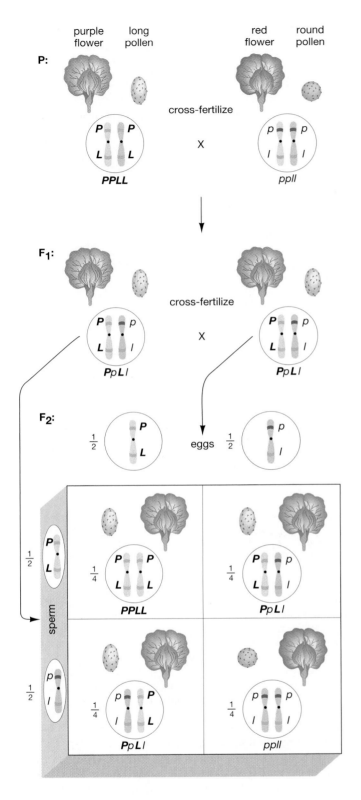

Figure 13-11   Linkage

Inheritance of linked genes for pollen grain shape and flower color in sweet peas. Because these genes are found on the same chromosome, the configurations *PL* and *pl* in the parental generation tend to be preserved in all subsequent generations. The F₂ offspring show a phenotypic ratio of 3 purple, long to 1 red, round rather than the 9:3:3:1 ratio one would expect from independently assorted genes.

mosomes that each haploid daughter cell receives are different from those of the parent cell.

Crossing over during meiosis explains why new combinations of traits occurred in the sweet pea cross. At the beginning of prophase I, all the F₁ peas had this pair of homologous chromosomes:

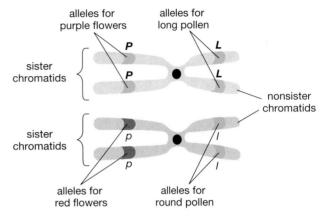

In a few reproductive cells, crossing over occurred between the locations of the genes for flower color and pollen shape, so that two nonsister chromatids exchanged alleles for flower color:

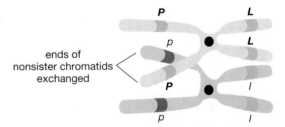

At anaphase I the separated homologous chromosomes had this gene composition:

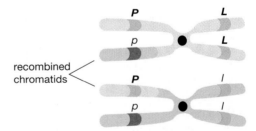

Four different types of chromosomes were distributed to the haploid daughter cells during meiosis II:

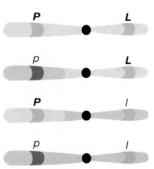

Therefore, some gametes were produced with each of the four chromosome configurations: *PL* and *pl* (the original parental types), and *Pl* and *pL* (recombined chromosomes).

In most of the reproductive cells, crossing over between these specific genes did not occur. Therefore, most of the F₂ offspring received chromosomes with the *PL* and *pl* allele combinations. However, fusion of sperm and eggs that formed from cells in which crossing over *did* occur gave rise to a few plants with purple flowers and round pollen and a few plants with red flowers and long pollen.

This situation is **genetic recombination**: generating new combinations of alleles by the exchange of DNA between homologous chromosomes. In most organisms, **sexual recombination** also occurs, in which offspring receive one homologous chromosome from each of two, genetically different parents. Together, genetic and sexual recombination provide great genetic variability among organisms.

The Frequency of Crossing Over between Linked Genes Can Be Used to Map the Location of Genes on a Chromosome

Crossing over occurs more often between genes that are far apart on a chromosome than between genes that are close together. In fact, if two genes are close enough together on a chromosome, crossing over almost never occurs between them; in contrast, if two genes are at opposite ends of a long chromosome, crossing over almost always occurs between them. Intermediate distances between genes result in intermediate amounts of crossing over. By swapping alleles between homologous chromosomes, crossing over produces offspring with recombined traits. Therefore, the proportion of offspring with recombined traits is a reflection of the spacing of genes along a chromosome.

This idea has been used to map chromosomes. The distance between two genes on a chromosome is defined as the percentage of recombination observed during a cross between one organism homozygous recessive for both genes and another organism heterozygous for both genes. (If no crossing over occurred, what would be the expected ratio of genotypes and phenotypes? Hint: This is a "two-gene" test cross.) If two genes recombine 10% of the time, they are placed 10 "map units" apart on the chromosome. In some organisms, such as the fruit fly *Drosophila*, it has been possible to correlate this type of recombination map with the physical appearance of certain chromosomes and locate the actual sites of genes (Fig. 13-12).

The Inheritance of Sex and Sex-Linked Genes

To say that the two members of a pair of homologous chromosomes carry the same genes is a bit of an oversimplification. In mammals and many insects, males have the same number of chromosomes as females do, but one "pair," the

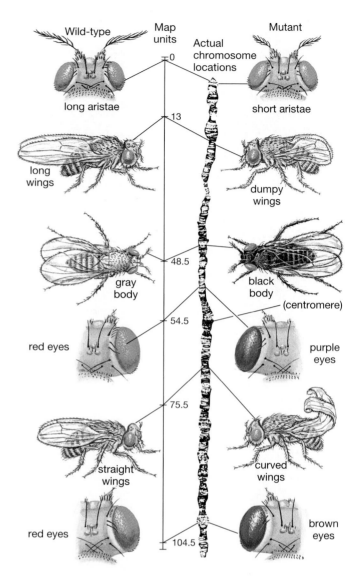

Figure 13-12   A chromosome map in the fruit fly
A map of chromosome 2 in the fruit fly, comparing the distances between genes obtained by recombination analysis with the physical locations of genes on the chromosomes. The order of the genes is the same in both maps, but the distances between genes are not. Crossing over apparently is hindered in some regions of the chromosome, especially around the centromere, distorting the recombination map.

sex chromosomes, are very different in appearance and genetic composition. Females have two identical sex chromosomes, called **X chromosomes**, and males have one X chromosome and one **Y chromosome**. Although the Y chromosome usually carries far fewer genes than the X chromosome does, a small part of both sex chromosomes is homologous, so X and Y chromosomes pair up during prophase of meiosis I and separate during anaphase I. The other chromosomes, which occur in pairs of identical appearance in both males and females, are called **autosomes**. Numbers of chromosomes vary tremendously, but there is always only one pair of sex chromosomes: For example, the

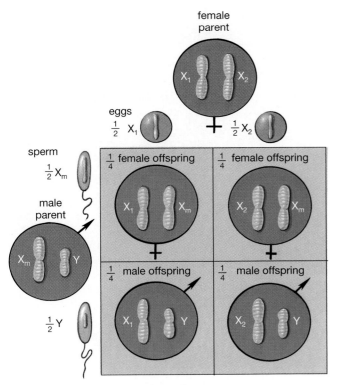

Figure 13-13   Sex determination

Sex determination in species such as humans and fruit flies, in which males carry two dissimilar sex chromosomes (XY) and females carry two similar sex chromosomes (XX). Only the distribution of sex chromosomes is illustrated. Male offspring receive the Y chromosome from the father; female offspring receive the father's X chromosome (labeled X_m in the drawing). The mother passes one of her X chromosomes (X_1 or X_2 in the drawing) to both male and female offspring.

fruit fly *Drosophila* has four pairs of chromosomes (three pairs of autosomes), humans have 23 pairs (22 pairs of autosomes), and dogs have 39 pairs (38 pairs of autosomes).

For organisms in which males are XY and females are XX, the sex of an offspring is determined by which sex chromosome is in the sperm that fertilizes the egg (Fig. 13-13). During sperm formation, the sex chromosomes segregate, and each sperm receives either the X or the Y chromosome (plus one member of each pair of autosomes). The sex chromosomes also segregate during egg formation, but because females have two X chromosomes, every egg receives one X chromosome. Therefore, an offspring is male if an egg is fertilized by a Y-bearing sperm, or female if an egg is fertilized by an X-bearing sperm.

Sex Linkage Is a Special Case of Linkage between Genes on the Same Chromosome

Genes that are found on one sex chromosome but not on the other are called **sex-linked**. In principle, the X chromosome might carry genes that are not found on the Y chromosome, and vice versa. However, in many animals, the Y chromosome carries relatively few genes other than

those determining maleness, whereas the X chromosome bears many genes that have nothing to do with specifically female traits. The human X chromosome, for example, contains genes for color vision, blood clotting, and certain structural proteins in muscles; those genes have no counterpart on the Y chromosome. Therefore, because they have two X chromosomes, females can be either homozygous or heterozygous for genes on the X chromosome. Normal dominant versus recessive relationships among alleles will be expressed. Males, in contrast, must fully express whatever alleles they have on their single X chromosome, whether those alleles are otherwise dominant or recessive.

How does sex linkage affect inheritance? Let's look at the first example of sex linkage to be discovered, the inheritance of eye color in the fruit fly *Drosophila*. Because of their small size, rapid reproductive rate, ease of rearing, and small number of chromosomes, *Drosophila* have been favored subjects for genetics studies for more than a century. Over the years, many mutations have been discovered in laboratory-bred *Drosophila* that produce significant phenotypic differences compared with wild-type *Drosophila*. Wild-type *Drosophila* have bright red eyes (Fig. 13-14). In the early 1900s, researchers in the laboratory of Thomas Hunt Morgan at Columbia University discovered a male fly with white eyes. This white-eyed male was mated to a virgin red-eyed female. The resulting offspring were all red-eyed flies (Fig. 13-14a), indicating that white eye color (*w*) is probably recessive to red (*W*). The F_2 generation, however, was a surprise. As expected, the total ratio of red- to white-eyed flies was about 3:1; however, *there were nearly equal numbers of red-eyed males and white-eyed males, but no females with white eyes at all* (Fig. 13-14b)! A test cross of the F_1 red-eyed females and the original white-eyed male yielded roughly equal numbers of red-eyed and white-eyed males and females.

From these data, could you figure out the mode of inheritance of eye color? Morgan made the brilliant hypothesis that *the gene for eye color must be located on the X chromosome, while the Y chromosome has no corresponding gene:*

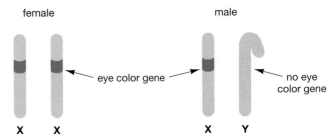

In the F_1 generation, both male and female offspring received an X chromosome, with its *W* allele for red eyes, from their mother. The F_1 males received a Y chromosome from their father, with no allele for eye color at all, so the males had a *WY* genotype (the use of *Y* here indicates that the Y chromosome does not contain the gene for eye color). The F_1 females received the father's X chromosome with its *w* allele, so the females had a *Ww* genotype. Thus, all male and female F_1 offspring had red eyes.

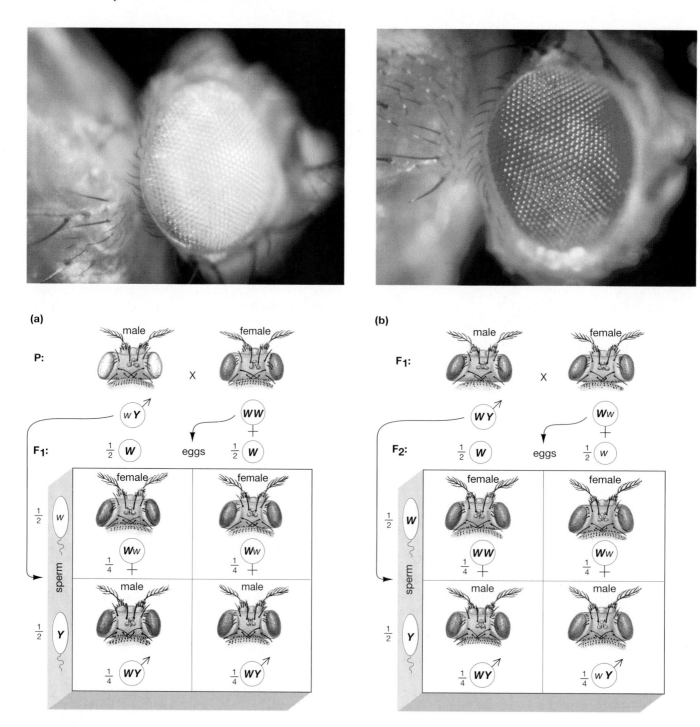

Figure 13-14 **Sex-linked inheritance**

Morgan's interpretation of the results of sex-linked inheritance of white eye color in fruit flies. The gene for eye color is carried on the X chromosome; there is no corresponding gene on the Y chromosome. The normal red eyes allele (W) is dominant to the mutant allele for white eyes (w). **(a)** The F_1 generation. Female offspring receive the w allele on their father's X chromosome, but phenotypically this is masked by the dominant W allele on their mother's X chromosome. Male offspring receive the W allele from their mother, and the Y chromosome with no eye color gene from their father. Therefore all the females are Ww and all the males are WY. Both males and females have red eyes. **(b)** The F_2 generation. All F_1 males carry the W allele, so all their F_2 female offspring receive the W allele as well. Therefore, the F_2 females are all red eyed. The F_1 females are all heterozygous Ww. Consequently, half their F_2 sons receive the W allele and half receive the w allele. The F_2 males also receive a Y chromosome from their fathers. Therefore half the F_2 males are WY (red eyes) and half are wY (white eyes).

Crossing two F₁ flies resulted in an F₂ generation with the chromosome distribution shown in Figure 13-14b. All the F₂ females received one X chromosome from their F₁ male parent with its *W* (red) allele. They therefore had red eyes. All the F₂ males, on the other hand, inherited their single X chromosome from their F₁ female parent. Because the F₁ females were heterozygous for eye color (*Ww*), the F₂ males had a 50-50 chance of receiving either an X chromosome with the *W* allele or one with the *w* allele. *With no corresponding gene on the Y chromosome, the F₂ males displayed the phenotype determined by the allele on the X chromosome.* Therefore, half the F₂ males had red eyes and half had white eyes.

Before you read on, try to answer this question: How much crossing over would you expect between X and Y chromosomes? If you concluded "very little" or "none," you're right. Most of the X and Y chromosomes are not homologous, so the normal mechanisms of crossing over do not apply.

Variations on the Mendelian Theme

In our discussion of patterns of inheritance thus far, we have made some major simplifying assumptions: that each trait is completely controlled by a single gene, that there are only two possible alleles of each gene, and that one allele is completely dominant to the other, recessive, allele. Most traits, however, are influenced in more varied and subtle ways.

In Incomplete Dominance, the Phenotype of Heterozygotes Is Intermediate between the Phenotypes of the Homozygotes

In his pea experiments, Mendel encountered a particularly simple situation: Heterozygotes and homozygous dominants had the same phenotype. This is often not the case. In snapdragons, for example, crossing homozygous red-flowered plants (*RR*) with homozygous white-flowered ones (*R´R´*) does not produce red-flowered F₁ hybrids. Instead, the F₁ flowers are pink. When the heterozygous phenotype is intermediate between the two homozygous phenotypes, the pattern of inheritance is called **incomplete dominance**. This apparent blending of the phenotype is not, however, the result of any alteration of the underlying alleles. The F₂ generation shows that the alleles for flower color have not changed—in homozygous flowers of the F₂ generation, the red and white colors are as strong as ever (Fig. 13-15). The F₂ offspring include about ¼ red, ½ pink, and ¼ white flowers. This distribution corresponds to the genotypic ratio of ¼ *RR*:½ *RR´*:¼ *R´R´*. *We will discuss the molecular basis for incomplete dominance later in this chapter.*

There May Be Multiple Alleles of a Single Gene

A single diploid organism can have at most only two different alleles for a given gene. Alleles, however, arise

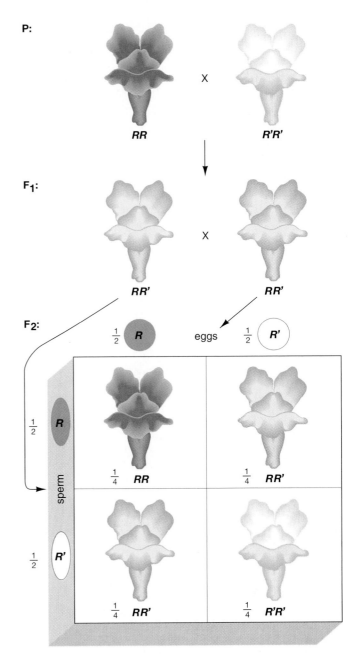

Figure 13-15 Incomplete dominance

Incomplete dominance in the inheritance of flower color in snapdragons. In cases of incomplete dominance, we will use capital letters for both alleles (here *R* and *R´*, "R prime") rather than upper- and lowercase letters. Hybrids (*RR´*) have pink flowers, and the homozygotes are red (*RR*) or white (*R´R´*). Because heterozygotes can be distinguished from homozygous dominants, the distribution of phenotypes in the F₂ generation (¼ red to ½ pink to ¼ white) is the same as the distribution of genotypes (¼ *RR* to ½ *RR´* to ¼ *R´R´*).

through mutation, and the genes of different organisms may have different mutations, each producing a new allele. If we could sample all the individuals of a species, we would often find several, sometimes even dozens, of alleles for every gene. The gene for eye color in fruit flies, for example, has many alleles, each recessive to normal red eyes and producing various shades of white, yellow, or pink when homozygous.

The ABO blood types in humans constitute a familiar system of such **multiple alleles**. The blood types A, B, AB, or O arise as a result of three different alleles of a single gene (usually designated I^A, I^B, and i). This gene directs the synthesis of glycoprotein "identification markers" that protrude from the surfaces of red blood cells. Alleles I^A and I^B direct the synthesis of glycoproteins A and B, respectively, whereas allele i produces no glycoproteins at all (Fig. 13-16). Individual humans may have one of six genotypes: $I^A I^A$, $I^B I^B$, $I^A I^B$, $I^A i$, $I^B i$, or ii. Alleles I^A and I^B are dominant to i. Therefore, individuals with genotypes $I^A I^A$ or $I^A i$ have type A glycoproteins on their red blood cells and have type A blood. Those with genotypes $I^B I^B$ or $I^B i$ synthesize type B glycoproteins and have type B blood. Homozygous recessive ii individuals lack these glycoproteins and have type O blood. However, alleles I^A and I^B are **codominant** to one another—that is, *both are phenotypically detectable in heterozygotes*. $I^A I^B$ individuals have red blood cells with both A and B glycoproteins and have type AB blood.

These red blood cell glycoproteins may react with antibodies in the blood plasma. If a patient with type B blood receives a transfusion of type A blood, the anti-A antibodies in the patient's serum cause the type A blood cells to clump (Fig. 13-16b).

(a)

Genotype	Blood Type	Red Blood Cells	Plasma Antibodies
$I^A I^A$ $I^A i$	A	A glycoprotein	anti-**B**
$I^B I^B$ $I^B i$	B	B glycoprotein	anti-**A**
$I^A I^B$	AB	A glycoprotein and B glycoprotein	none
$i\ i$	O	neither glycoprotein	anti-**A** anti-**B**

(b)

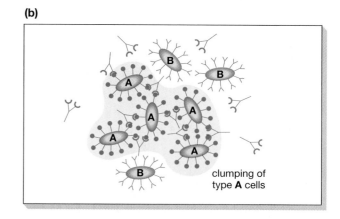

clumping of type **A** cells

Medical effects of blood transfusions			
Donor Type	Recipient Type	Effect on Recipient	Permissible Blood Donation?
A	A	–	Yes
	B	Clumping	No
	AB	–	Yes
	O	Clumping	No
B	A	Clumping	No
	B	–	Yes
	AB	–	Yes
	O	Clumping	No
AB (universal recipient)	A	Clumping	No
	B	Clumping	No
	AB	–	Yes
	O	Clumping	No
O (universal donor)	A	–	Yes
	B	–	Yes
	AB	–	Yes
	O	–	Yes

(a) Red blood cell glycoproteins and plasma antibodies found in individuals with each blood type. Type A individuals have the A glycoprotein on their red blood cell surfaces. Their plasma contains antibodies only against B glycoproteins (anti-B antibodies). Therefore, their plasma antibodies do not bind to their own red blood cell glycoproteins, and their own cells do not clump. Type B individuals have B glycoproteins and anti-A antibodies. Type AB blood contains red blood cells with both the A and B glycoproteins, but neither antibody. Type O individuals have both anti-A and anti-B antibodies but no reactive glycoproteins on their red blood cells. **(b)** The reaction between type A blood cells transfused into type B blood. Type B blood contains many anti-A antibodies. Because each antibody has two sites that can bind the A glycoprotein, the A cells become clumped, held together by the anti-A antibodies. These clumps can become large clots, with serious medical consequences. The permissible donors for each blood type are shown in the table.

Figure 13-16 **Human ABO blood group reactions**

Glycoproteins on the surfaces of the red blood cells determine blood type. If antibodies in the blood plasma bind to red blood cell glycoproteins, the red blood cells clump together. Each antibody can bind to only one specific glycoprotein; for example, anti-A antibody binds only to glycoprotein A and not to glycoprotein B.

Many Traits Are Influenced by Several Genes

If you look around your class, you are likely to see people of varied heights, skin colors, and body build, to take just a few obvious traits. Traits such as these are not governed by single genes but are influenced by interactions among two or more genes (plus, of course, interactions with the environment).

Continuous Variation in a Trait May Be Caused by Contributions from Several Genes

For many traits, such as skin color, there are more than just two or three phenotypes. There may be several phenotypes, or even seemingly continuous variation that cannot be split up into convenient, easily defined categories. In many of these cases, two or more genes may produce functionally equivalent contributions to the trait (e.g., roughly equal amounts of skin pigment), a situation called **polygenic inheritance**.

The simplest polygenic inheritance occurs when two gene pairs code for the same trait. In wheat, for example, there are two gene pairs for kernel color, which we might designate 1 and 2. Each gene has two alleles, R and R'. The R allele directs the synthesis of one "unit" of red pigment in the kernel, while the R' allele causes no pigment synthesis at all. If only gene pair 1 were present, then the inheritance of kernel color would follow simple incomplete dominance: R_1R_1 = red; $R_1R'_1$ = pink; $R'_1R'_1$ = white. But because kernel color is controlled by gene pairs 1 and 2, the color becomes more finely graded in intensity, depending on the number of R alleles of both genes. Kernels with the genotype $R_1R_1R_2R_2$ synthesize four units of pigment and therefore are dark red. Kernels with the genotype $R'_1R'_1R'_2R'_2$ produce no pigment and are white. Intermediate numbers of R alleles yield intermediate intensities of red.

As you can well imagine, the more genes that contribute to a single trait, the greater the number of categories of the trait, with increasingly fine gradation between categories. Continuing our example of color intensity, if three pairs of genes are involved, we would have seven phenotypic categories; with four pairs of genes, nine categories, and so on. With more than three pairs of genes, differences between phenotypes are small, and it is extremely difficult to classify the phenotypes reliably.

Polygenic inheritance leads to small, "quantitative" differences in traits (such as shades of color in wheat kernels) rather than the large "qualitative" differences that are caused by single-gene inheritance (such as the all-or-none, purple-or-white color of pea flowers). Therefore, polygenic inheritance is often referred to as **quantitative inheritance**.

The Action of One Gene May Modify the Action of Another Gene

In polygenic inheritance, several genes have functionally similar actions in controlling a trait. In **epistasis**, the expression of one gene depends on, or is modified by, the expression of another gene. One of the best understood cases of epistasis occurs in the inheritance of hair color in virtually all mammals, including, for example, the mouse (Fig. 13-17a). One gene controls the *synthesis* of melanin, the pigment in hair. The dominant allele of this gene (M) allows the synthesis of melanin, but the recessive allele (*m*) does not. A second gene controls the *distribution* of pigment in the hair. Agouti, as the normal fur of the wild house mouse is called, has individual hairs that are black with a yellow tip, giving the mouse an overall brownish gray appearance. Agouti (A) is dominant to plain black fur (*a*).

These two genes control hair color in the following way. If a mouse has the MM or Mm genotype, melanin will be produced. The fur will then be agouti if the mouse is AA or Aa for the melanin-distribution gene; the fur will be black if the mouse is aa. If a mouse is *mm*, then it cannot produce melanin. All mice with the mm genotype therefore have white hair and are albino. In an albino, the melanin-distribution gene cannot affect the phenotype, because there isn't any melanin to distribute. Therefore, the *melanin-synthesis gene* controls the expression of the *melanin-distribution gene*. A cross of two agouti mice can result in a variety of offspring, as the genes for melanin production and distribution assort independently (Fig. 13-17b).

The expression of almost all genes is influenced to some extent by other genes. To take an obvious, but usually overlooked, example, no genes can be expressed in an adult organism unless the genes that direct development operate properly. An organism is a cohesive, coordinated whole, and all its genes influence its ultimate anatomy, physiology, and behavior.

A Single Gene May Have Multiple Effects on Phenotype

We have just seen that a single phenotype may require the interaction of several genes. The reverse is also true: Single genes may have multiple phenotypic effects, a phenomenon called **pleiotropy**. Take the case of albino mice, for example. Albino mice lack pigment not only in their fur but also in their eyes. (Their eyes are bright pink because the blood circulating through capillaries inside their eyes shows through the transparent iris.) Without any eye pigment, they are extraordinarily sensitive to light. Albino mice can't see well in even fairly dim light, and normal daylight rapidly destroys the receptors in their eyes, causing blindness. Consequently, the single gene for pigment production actually can have multiple phenotypic effects: white fur, pink eyes, and blindness.

The Environment Influences the Expression of All Genes

An organism is not just the sum of its genes. Both the genotype and the environment in which an organism lives profoundly affect its phenotype, even for simple physical traits.

(a)

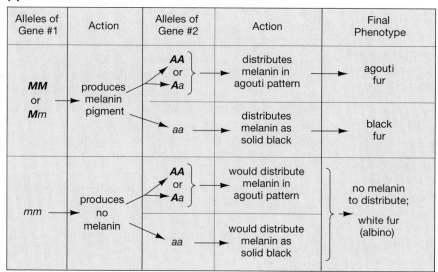

Alleles of Gene #1	Action	Alleles of Gene #2	Action	Final Phenotype
MM or **Mm**	produces melanin pigment	**AA** or **Aa**	distributes melanin in agouti pattern	agouti fur
		aa	distributes melanin as solid black	black fur
mm	produces no melanin	**AA** or **Aa**	would distribute melanin in agouti pattern	no melanin to distribute; white fur (albino)
		aa	would distribute melanin as solid black	

(b)

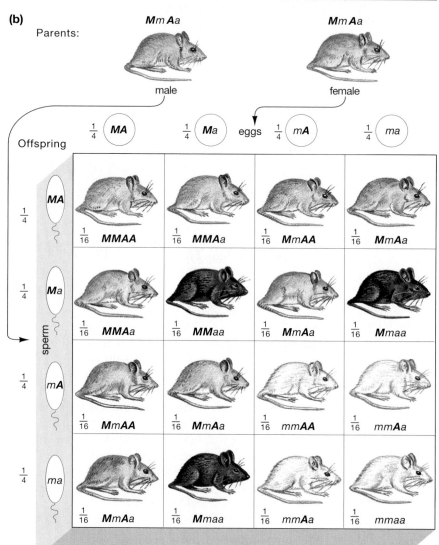

Figure 13-17   Gene interactions

Two genes interact in the inheritance of fur color in mice.
(a) To produce any melanin pigment at all, a mouse must have at least one dominant allele M for melanin production (MM or Mm genotypes). If, the melanin-distribution gene is present, the mouse's fur will be agouti (*AA* or *Aa* genotypes) or black (*aa* genotype). If a mouse is homozygous recessive *mm*, then it produces no melanin and will have white fur regardless of its genotype for the melanin-distribution gene.
(b) If two mice that are heterozygous for both genes are mated, their offspring will not have a 9:3:3:1 ratio of phenotypes because of the interaction between the genes. In this cross, the phenotypic ratio of offspring is 9 agouti to 3 black to 4 albino. Other types of crosses (for example, a heterozygote mated with a homozygous recessive) yield other modifications of the 9:3:3:1 ratio.

Figure 13-18   Environmental influence on genotype

A simple case of interaction between genotype and environment is the expression of the gene for black fur in the Himalayan rabbit. Cool areas (nose, ears, and feet) allow expression of the gene for black fur.

A striking example of environmental effects on gene action occurs in the Himalayan rabbit, which, like the Siamese cat, has pale body fur but black ears, nose, tail, and feet (Fig. 13-18). The Himalayan rabbit actually has the genotype for black fur all over its body. The enzyme that produces the black pigment, however, is temperature sensitive; above about 34° C (93° F), the enzyme is inactive. At the temperatures typical of rabbit hutches, extremities such as the ears and feet are cooler than the rest of the body, and black pigment can be produced there. The main body surface is warmer than 34° C, so this fur is pale.

Most environmental influences are more complicated and subtle. The interactions between complex genetic systems and varied environmental conditions can create a continuum of phenotypes that defies analysis into genetic and environmental components. This is particularly true of human characteristics. Genetic analysis of all but the simplest human traits is notoriously difficult: The human generation time is long, the number of offspring per couple is small, and in any case one can't very well kidnap a few thousand people and keep them for genetics experiments. Add to these factors the countless subtle ways in which people respond to their environments, and you can see that a precise determination of the genetic bases of complex traits such as intelligence or musical ability is, at best, extraordinarily difficult.

The Molecular Basis of Mendelian Genetics

You have now learned about the molecular basis of inheritance, the flow of information from DNA to protein, and the patterns of inheritance that can be observed in the traits of whole organisms. What is the connection among these facets of genetics? Although the precise molecular processes underlying most traits are still unknown, it has become clear that the various patterns of inheritance are outgrowths of the transcription and translation into protein of information stored in DNA. Here, we will focus on a few simple cases in which the molecular basis of relationships among alleles is fairly clear.

A Dominant Allele Usually Codes for a Functional Enzyme; a Recessive Allele Usually Codes for a Nonfunctional Enzyme

Consider the situation in which a single gene controls the expression of a single trait. There are two different alleles of the gene, each producing a different form of the trait, and one allele is completely dominant to the other allele. To see how such a system might work on the molecular level, let's look at the inheritance of body fat color in rabbits.

In domestic rabbits, body fat is usually white, but rabbits that are homozygous for a recessive allele have yellow fat. Why? As you know, rabbits eat plants. Most plants contain a yellow pigment called xanthophyll, which is fat soluble. If not broken down, xanthophyll dissolves in the rabbit's fat, coloring it yellow. The xanthophyll-metabolizing allele directs the synthesis of an enzyme that breaks down xanthophyll to a colorless compound. Rabbits that are homozygous for the xanthophyll-metabolizing allele synthesize more than enough functional enzyme to break down all of the xanthophyll in their diet, so these rabbits have white fat. The yellow-fat allele is a mutation that renders the xanthophyll-metabolizing enzyme nonfunctional. In rabbits with one xanthophyll-metabolizing allele and one mutant allele, the xanthophyll-metabolizing allele directs the synthesis of enough functional enzyme to break down dietary xanthophyll completely. Therefore, the body fat of heterozygotes will still be white. However, in rabbits homozygous for the mutant allele, there are no functional xanthophyll-metabolizing enzymes. The xanthophyll dissolves in the fat, coloring it yellow.

In general, dominant alleles direct the synthesis of functional enzymes and recessive alleles direct the synthesis of nonfunctional enzymes. An organism with one dominant allele synthesizes enough enzyme to produce a phenotype that is indistinguishable from the phenotype of organisms with two dominant alleles. This is an example of the protective effect of diploidy mentioned in Chapter 12: A functional allele on one homologous chromosome may completely compensate for a defective allele on the other homologue.

What about incomplete dominance? The simplest form of incomplete dominance is very similar to complete dominance: One allele directs the synthesis of a functional enzyme, and the other allele directs the synthesis of a nonfunctional enzyme. The difference between complete and incomplete dominance results from the difference in effects of one versus two copies of the functional allele. In incomplete dominance, the amount of enzyme synthesized under the direction of a single functional allele is not enough to catalyze all the reactions that would be needed to produce a dominant phenotype. For example, the red allele in snapdragon flowers (see Fig. 13-15) directs the synthesis of an enzyme that produces a red pigment. If a plant has two red alleles, its flowers produce a lot of red pigment and are colored red. If a plant has only one red allele, its flowers produce less red pigment and are colored pink.

Codominance Is Caused by Two or More Alleles, Each of Which Codes for a Different, Functional Enzyme

Mutations may occur anywhere in a gene. Therefore, different organisms may have quite different mutations in the same gene. These different mutations may all produce slightly different, though still functional, proteins, giving rise to multiple alleles of the same gene. In some circumstances, each of these multiple alleles may be detectable phenotypically, as is the case with the human ABO blood groups (see Fig. 13-16). The I^A and I^B alleles both code for slightly different, functional enzymes. The I^B enzyme attaches the monosaccharide galactose to a protein on red blood cell membranes, whereas the I^A enzyme attaches a slightly different compound, galactosamine. The i allele is a mutation producing a nonfunctional enzyme that can't attach anything to the red blood cell proteins. People with $I^B I^B$ or $I^B i$ genotypes have red blood cell proteins with galactose attached and have type B blood. Those with $I^A I^A$ or $I^A i$ genotypes have red blood cell proteins with galactosamine attached and have type A blood. The $I^A I^B$ genotype results in red blood cells with some galactose-bearing proteins and some galactosamine-bearing proteins (type AB blood). Finally, the red blood cells of ii individuals bear neither compound (type O blood). The I^A and I^B alleles are considered codominant because the phenotypes produced by both the I^A and I^B alleles can be detected in people with type AB blood.

Reflections on Genetic Diversity and Human Welfare

The science of genetics, which began in a monastery over a century ago, is critically important to humankind today. Genetics plays a role in many human activities, from possible new sources of energy to inherited human diseases, from paternity lawsuits to crop improvement. Research into crop and livestock improvement represents some of the greatest successes, as well as challenges, for geneticists. Virtually all the corn grown in the United States today, for example, is hybrid corn, the result of crossing parent strains that are themselves usually not very good as a crop. Faced with high prices, rising populations, and increasing demand for food, agricultural geneticists are constantly try-

ing to develop strains of plants or animals that produce more food per unit input of energy, labor, and money.

Scientists take three principal approaches to improving crops and livestock. First, breeders can search for individual plants or animals with the desired characteristics and use only those organisms as breeding stock. The individuals selected for breeding could arise from mutations or from chance recombinations of preexisting alleles. This approach relies on "good" alleles appearing spontaneously and on our ability to recognize them. Second, molecular geneticists can try to create superior genes in the laboratory or transplant them from one species to another. Gene transplantation is rapidly approaching commercial use, but creating superior genes is not yet feasible. The third approach is to look for desirable traits in wild populations of the same or closely related species. Breeders might then cross-breed or otherwise incorporate the desired genetic information into livestock or crop plants. This last approach shows great promise and, in fact, is how many crops and livestock breeds were developed in the first place, thousands of years ago. Wheat, for example, was a cross between varieties of wild grass, aided by irregularities in meiosis that resulted in a polyploid plant producing large edible kernels.

Today, our reserves of wild genes are diminishing at a frightening rate, largely because of the pressures of human population and development. An estimated 8 to 10 million species of organisms exist on Earth, and only about a million and a half of these have even been identified. What's worse, many, perhaps most, of the unidentified species may become extinct before science ever has a chance to study them. Large tracts of wilderness, both in the United States and in other countries, are diminishing rapidly. When these ecosystems go, countless species and varieties go with them.

Rare species and local varieties of otherwise widespread species have been important in crop development in the past and promise to be just as important in the future. In the 1960s, for example, plant geneticists cross-bred commercial strawberries with a wild variety growing in Cottonwood Canyon, Utah. The result: strawberries that produce fruit year-round. More recently, Florida breeders have developed a heat-tolerant blueberry bush by cross-breeding domestic blueberries with the rabbit-eye blueberry of the South. Geneticists at Oregon State University are developing a wildflower called meadowfoam into a source of high-quality commercial oil for the pharmaceutical and electronics industries. One of the most useful meadowfoam species is found in a 6-square-mile area near Medford, Oregon, and nowhere else in the world.

This is an important, but little recognized, reason to preserve wilderness: to preserve the genes of the plants and animals that live in them. This is also an important goal of efforts to save endangered species, because once a species becomes extinct, its genes are lost forever. Genes, with all their various alleles, have evolved over hundreds of millions of years and represent one of our most valuable and irreplaceable natural resources.

SUMMARY OF KEY CONCEPTS

Gregor Mendel and the Origin of Genetics

Homologous chromosomes carry the same genes located at the same loci, but the genes at a particular locus can exist in alternate forms called alleles. An organism whose homologous chromosomes carry the same allele at a given locus is homozygous for that particular gene. If the alleles differ, the organism is heterozygous. Gregor Mendel deduced many principles of inheritance in the mid-1800s, before the discovery of DNA, genes, chromosomes, or meiosis. He did this by choosing an appropriate experimental subject, designing his experiments carefully, and analyzing his data accurately.

Inheritance of Single Traits: The Law of Segregation

Breeding experiments that deal with the inheritance of a single trait at a time are called monohybrid crosses. The inheritance of each individual trait is determined by discrete physical units called genes. Each organism possesses a pair of similar, but not necessarily identical, genes that influence each trait. However, only one of each pair of genes is included in any individual gamete (the law of segregation). An offspring formed by the fusion of two gametes therefore receives pairs of genes, one of each pair inherited from each parent. Each gene may exist in alternative forms called alleles. Different alleles cause different forms of the trait (for example, purple or white flowers). If an individual possesses two different alleles of the same gene, one allele, called dominant, may completely mask the expression of the other, recessive allele. (For example, a pea with alleles both for purple and white flower color will have purple flowers.) In individuals with two different alleles of the same gene, which allele is included in any given gamete is determined by chance. Therefore, we can predict the relative proportions of offspring through the laws of probability.

The physical appearance of an organism (its phenotype) may not always be a reliable indicator of its alleles (the genotype) because of the masking of recessive alleles by dominant alleles. Organisms with two dominant alleles (homozygous dominant) have the same phenotype as organisms with one dominant and one recessive allele (heterozygous).

Inheritance of Multiple Traits: Independent Assortment

Genes are parts of chromosomes. The F_2 generation from a dihybrid cross (parental organisms differing in two traits) may have two fundamentally different outcomes:

If the genes for the two traits are on different chromosomes, then the F_2 offspring will appear in four different phenotypes, resulting from the independent assortment of the chromosomes (and hence the alleles) during meiosis. This is Mendel's law of independent assortment.

Inheritance Patterns of Genes Located on the Same Chromosome

If the genes are found on the same chromosome, they are said to be linked. In the absence of crossing over, the F_2 offspring will express only the two parental phenotypes. If crossing over does occur, then some of the F_2 offspring will express recombined phenotypes. The frequency of crossing over depends on the distance between the genes on the chromosome: the farther apart the genes, the higher the frequency of crossing over.

The Inheritance of Sex and Sex-Linked Genes

Sex linkage is a special and easily observed case of linkage of traits on the same chromosome. In many animals, females have two X chromosomes, whereas males have one X and one Y chromosome, with many fewer genes. Consequently, males have only one copy of most of the X chromosome genes, and recessive traits are more likely to be phenotypically expressed in males.

Variations on the Mendelian Theme

Not all inheritance follows the simple dominant-recessive pattern.

1. In incomplete dominance, heterozygotes have a phenotype intermediate between the two homozygous phenotypes.
2. Codominant alleles, such as those determining blood type, are both phenotypically detectable in heterozygotes. The heterozygotes are not intermediate in phenotype between the two parents, but have separately distinguishable features of both parental types.
3. Many traits are determined by several genes with similar actions, a phenomenon called polygenic inheritance. Traits that appear to exist in a continuum of finely graded forms are often determined polygenically.
4. The expression of some genes may depend on, or be modified by, the expression of other genes (epistasis).
5. Many genes have multiple phenotypic effects (pleiotropy).
6. The environment plays at least some role in the phenotypic expression of all traits.

The Molecular Basis of Mendelian Genetics

The observed patterns of inheritance in whole organisms arise from and can be explained on the molecular level. Different alleles of a given gene are slightly different nucleotide sequences of DNA that produce differences in proteins. A dominant allele usually codes for a functional enzyme, whereas a recessive allele usually codes for a nonfunctional enzyme. An organism with a single, completely dominant allele synthesizes enough enzyme to produce a phenotype indistinguishable from the phenotype of organisms with two dominant alleles. Codominant alleles both code for functional, but slightly different, enzymes that produce distinguishably different phenotypes.

✖ KEY TERMS

allele   p. 236
autosome   p. 246
chiasma   p. 243
codominance   p. 250
cross-fertilization   p. 238
crossing over   p. 243
dihybrid cross   p. 241
dominant   p. 239
epistasis   p. 251
F₁ generation   p. 238
F₂ generation   p. 238
F₃ generation   p. 239
gene   p. 235

genetic recombination   p. 246
genotype   p. 240
heterozygous   p. 236
homozygous   p. 236
hybrid   p. 238
incomplete dominance   p. 249
law of independent assortment   p. 243
law of segregation   p. 239
linkage   p. 243
locus   p. 236
monohybrid cross   p. 238
multiple alleles   p. 250
parental generation   p. 238

phenotype   p. 240
pleiotropy   p. 251
polygenic inheritance   p. 251
quantitative inheritance   p. 251
recessive   p. 239
self-fertilization   p. 237
sex chromosome   p. 246
sex linkage   p. 247
sexual recombination   p. 246
test cross   p. 240
true-breeding   p. 237
X chromosome   p. 246
Y chromosome   p. 246

✖ THINKING THROUGH THE CONCEPTS

Multiple Choice

1. An organism is described as *Rr* : red. The *Rr* is the organism's [A]; red is the organism's [B]; and the organism is [C].
   a. [A] phenotype; [B] genotype; [C] degenerate
   b. [A] karyotype; [B] hybrid; [C] recessive
   c. [A] genotype; [B] phenotype; [C] heterozygous
   d. [A] gamete; [B] linkage; [C] pleiotropic
   e. [A] zygote; [B] epistasis; [C] homozygous

2. The 9:3:3:1 ratio is a ratio of
   a. phenotypes in a test cross
   b. phenotypes in a monohybrid cross
   c. phenotypes in a dihybrid cross
   d. genotypes in a monohybrid cross
   e. genotypes in a dihybrid cross

3. A test cross is used to
   a. demonstrate whether two genes are linked
   b. test whether an organism expressing the dominant phenotype is homozygous or heterozygous
   c. show whether inheritance is polygenic or Mendelian
   d. determine whether a trait is sex-linked or autosomal
   e. find out whether the trait is characterized by incomplete dominance

4. Which of the following statements rewords Mendel's first law in terms of what you know about genetics and meiosis?
   a. Genes are arranged on sex chromosomes or autosomes.
   b. If two genes are on two different chromosomes, then how alleles of one gene distribute into gametes does not influence how alleles of the other gene distribute.
   c. If two genes are on the same chromosome, they are linked.
   d. Two alleles of a gene do not mix with each other while they are in a heterozygous organism; they can appear unaltered in subsequent generations.
   e. There would be no crossing over between X and Y chromosomes.

5. A man's lawyer tells him that he cannot use blood type in a paternity suit against him because the child could, in fact, be his according to blood type. Which of the following is the only possible combination supporting this hypothetical circumstance? (Answers in the order: mother:father:child)
   a. A:B:O   b. A:O:B

   c. AB:A:O   d. AB:O:AB
   e. B:O:A

6. A heterozygous red-eyed female *Drosophila* mated with a white-eyed male would produce
   a. red-eyed females and white-eyed males in the F₁
   b. white-eyed females and red-eyed males in the F₁
   c. half red- and half white-eyed females and all white-eyed males in the F₁
   d. all white-eyed females and half red- and half white-eyed males in the F₁
   e. half red- and half white-eyed females as well as males in the F₁

Review Questions

1. Define the following terms: gene, allele, dominant, recessive, monohybrid, dihybrid, true-breeding, homozygous, heterozygous, cross-fertilization, self-fertilization.

2. Explain the meaning of Mendel's law of segregation and law of independent assortment. Under what circumstances does the law of independent assortment apply? When is it violated?

3. Explain why genes located on one chromosome are linked during inheritance. Why do linked genes sometimes separate during meiosis?

4. Explain why human skin color does not occur in just two forms, e.g., black and white.

5. What is sex linkage? In mammals, which sex would be most likely to show recessive sex-linked traits?

6. What is the difference between a phenotype and a genotype? Does knowledge of an organism's phenotype always allow you to determine the genotype? What type of experiment would you perform to determine the genotype of a phenotypically dominant individual?

7. Define polygenic inheritance, epistasis, and pleiotropy. Describe an example of each.

8. If smooth versus wrinkled and yellow versus green pea seeds had happened to be close to each other on the same chromosome, what would Mendel have found?

9. In the white body with black extremities of the Himalayan rabbit, where does the enzyme work? Where doesn't it work? Can you think of an explanation for this?

✖ APPLYING THE CONCEPTS

1. There are some groups of insects in which females are XX and males are XO (the Y chromosome has not evolved in these species). In some other groups of animals, including moths, butterflies, and birds, females are ZW and males are ZZ for their sex chromosomes. Regarding chromosomal sex determination, discuss several basic ways in which these groups differ from mammals, in which females are XX and males are XY.

2. There are many hypotheses, but as yet no definitive answers, to the question, "Why does the Y chromosome carry so few genes?" Discuss the disadvantages of a Y chromosome lacking most of the genes carried on the X chromosome in the light of your knowledge of dominance and recessiveness, and the protective effect of diploidy.

3. Sometimes the term "gene" is used rather casually. Compare and contrast use of the terms "allele" and "locus" as alternatives to "gene."

4. As you read in this chapter, a person with type B blood has antibodies which would cause blood cells of a type A transfusion to clump. Expanding on this information, type A people have antibodies which attack type B glycoproteins, type O people have antibodies which attack both glycoproteins, and type AB people do not have antibodies which attack the blood group glycoproteins. Using this information along with the information in your chapter on which glycoproteins are in which blood group, explain why AB individuals are referred to as "universal recipients" in terms of blood transfusions and why people with type O blood are "universal donors."

5. Mendel's numbers seemed almost too perfect to be real; some believe he may have cheated a bit on his data. Perhaps he continued to collect it until the numbers matched his predicted ratios, then stopped. Recently, there has been much publicity over violations of scientific ethics including researchers plagiarizing other's work, using other scientists' methods to develop lucrative patents, or just plain fabricating data. How important an issue is this for society? What are the boundary lines of ethical scientific behavior? How should the scientific community or society "police" scientists? What punishments would be appropriate for violators of scientific ethics?

6. Although American society has often been described as a "melting pot," people often engage in "assortative mating," in which they marry others of similar height, socioeconomic status, race, and IQ. Discuss the consequences to society of assortative mating among humans. Would society be better off if people mated more randomly? Discuss why or why not.

✖ GENETICS PROBLEMS

(*Note: An extensive group of genetics problems, with answers, can be found in the Study Guide.*)

1. In certain cattle, hair color can be red (homozygous RR), white (homozygous $R'R'$), or roan (a mixture of red and white hairs; heterozygous RR').
   a. When a red bull is mated to a white cow, what genotypes and phenotypes of offspring could be obtained?
   b. If one of these offspring were mated to a white cow, what genotypes and phenotypes of offspring could be produced? In what proportion?

2. The palomino horse is golden in color. Unfortunately for horse fanciers, palominos do not breed true. In a series of matings between palominos, the following offspring were obtained:

   65 palominos, 32 cream-colored,
   34 chestnut (reddish brown)

   What is the probable mode of inheritance of palomino coloration?

3. In the edible pea, tall (T) is dominant to short (t), and green pods (G) are dominant to yellow pods (g). List the types of gametes and offspring that would be produced in the following crosses:
   a. $TtGg \times TtGg$
   b. $TtGg \times TTGG$
   c. $TtGg \times Ttgg$

4. In tomatoes, round fruit (R) is dominant to long fruit (r), and smooth skin (S) is dominant to fuzzy skin (s). A true-breeding round, smooth tomato ($RRSS$) was cross-bred with a true-breeding long, fuzzy tomato ($rrss$). All the F_1 offspring were round and smooth ($RrSs$). When these F_1 plants were bred, the following F_2 generation was obtained:

   Round, smooth: 43
   Long, fuzzy: 13

   Are the genes for skin texture and fruit shape likely to be on the same or on different chromosomes? Explain your answer.

5. In the tomatoes of question 4, an F_1 offspring ($RrSs$) was mated with a homozygous recessive ($rrss$). The following offspring were obtained:

   Round, smooth: 583
   Long, fuzzy: 602
   Round, fuzzy: 21
   Long, smooth: 16

   What is the most likely explanation for this distribution of phenotypes?

6. In humans, hair color is controlled by two interacting genes. The same pigment, melanin, is present in both brown-haired and blond-haired people, but brown hair has much more of it. Brown hair (B) is dominant to blond (b). Whether any melanin can be synthesized at all depends on another gene. The dominant form (M) allows melanin synthesis; the recessive form (m) prevents melanin synthesis. Homozygous recessives mm are albino. What will be the expected proportions of phenotypes in the children of the following parents?
   a. $BBMM \times BbMm$
   b. $BbMm \times BbMm$
   c. $BbMm \times bbmm$

7. In humans, one of the genes determining color vision is located on the X chromosome. The dominant form (C) pro-

duces normal color vision; red-green color blindness (*c*) is recessive. If a man with normal color vision marries a colorblind woman, what is the probability of their having a colorblind son? A colorblind daughter?

8. In the couple described in question 7, the woman gives birth to a colorblind daughter. The husband sues for a divorce, on the grounds of adultery. Will his case stand up in court? Explain your answer.

ANSWERS TO GENETICS PROBLEMS

1. a. A red bull (*RR*) is mated to a white cow (*R′R′*). The bull will produce all *R* sperm, while the cow will produce all *R′* eggs. All the offspring will be *RR′*, and will have roan hair (codominance).

   b. A roan bull (*RR′*) is mated to a white cow (*R′R′*). The bull produces half *R* and half *R′* sperm, while the cow produces *R′* eggs. Using the Punnett square method:

sperm \ eggs	*R′*
R	*RR′*
R′	*R′R′*

Using probabilities:

sperm	egg	offspring
1/2 *R*	*R′*	1/2 *RR′*
1/2 *R′*	*R′*	1/2 *R′R′*

The predicted offspring will be 1/2 *RR′* (roan) and 1/2 *R′R′* (white).

2. The offspring occur in three types, classifiable as dark (chestnut), light (cream), and intermediate (palomino). This distribution suggests incomplete dominance, with the alleles for chestnut (C) combining with the allele for cream (C′) to produce palomino heterozygotes (CC′). We can test this hypothesis by examining the offspring numbers. There are approximately 1/4 chestnut (CC), 1/2 palomino (CC′), and 1/4 cream (C′C′). If palominos are heterozygotes, we would expect the cross CC′ × CC′ to yield 1/4 CC, 1/2 CC′, and 1/4 C′C′. Our hypothesis is supported.

3. a. *TtGg* × *TtGg*. This is a "standard" dihybrid cross. Both parents produce *TG*, *Tg*, *tG*, and *tg* gametes. The expected proportions of offspring are 9/16 tall green, 3/16 tall yellow, 3/16 short green, 1/16 short yellow.

   b. *TtGg* × *TTGG*. In this cross, the heterozygous parent produces *TG*, *Tg*, *tG*, and *tg* gametes. However, the homozygous dominant parent can only produce *TG* gametes. Therefore, all offspring will receive at least one *T* allele for tallness and one *G* allele for green pods, and thus all the offspring will be tall with green pods.

   c. *TtGg* × *Ttgg*. The second parent will produce two types of gametes, *Tg* and *tg*. Using a Punnett square:

sperm \ eggs	*Tg*	*tg*
TG	*TTGg*	*TtGg*
Tg	*TTgg*	*Ttgg*
tG	*TtGg*	*ttGg*
tg	*Ttgg*	*ttgg*

The expected proportions of offspring are 3/8 tall green, 3/8 tall yellow, 1/8 short green, 1/8 short yellow.

4. If the genes are on separate chromosomes—that is, assort independently—then this would be a typical dihybrid cross with expected offspring of all four types (approximately 9/16 round smooth, 3/16 round fuzzy, 3/16 long smooth, and 1/16 long fuzzy). However, only the parental combinations show up in the F_2 offspring, indicating that the genes are on the same chromosome.

5. The genes are on the same chromosome and are quite close together. On rare occasions, crossing over occurs between the two genes, producing recombination of the alleles.

6. a. *BBMM* (brown) × *BbMm* (brown). The first parent can produce only *BM* gametes, so all offspring will receive at least one dominant allele for each gene. Therefore, all offspring will have brown hair.

   b. *BbMm* (brown) × *BbMm* (brown). Both parents can produce four types of gametes: *BM*, *Bm*, *bM*, and *bm*. Filling in the Punnett square:

sperm \ eggs	*BM*	*Bm*	*bM*	*bm*
BM	*BBMM*	*BBMm*	*BbMM*	*BbMm*
Bm	*BBMm*	*BBmm*	*BbMm*	*Bbmm*
bM	*BbMM*	*BbMm*	*bbMM*	*bbMm*
bm	*BbMm*	*Bbmm*	*bbMm*	*bbmm*

Remembering that all *mm* offspring are albino, the expected proportions are 9/16 brown-haired, 3/16 blond-haired, 4/16 albino.

   c. *BbMm* (brown) × *bbmm* (albino):

sperm \ eggs	*bm*
BM	*BbMm*
Bm	*Bbmm*
bM	*bbMm*
bm	*bbmm*

The expected proportions of offspring are 1/4 brown-haired, 1/4 blond-haired, 1/2 albino.

7. A man with normal color vision is CY (remember, the Y chromosome does not have the gene for color vision). His colorblind wife is *cc*. Their expected offspring will be:

sperm \ eggs	*c*
C	C*c*
Y	*c*Y

We therefore expect that all the daughters will have normal color vision, while all the sons will be colorblind.

8. The husband should win his case. All his daughters must receive one X chromosome, with the C allele, from him, and therefore should have normal color vision. If his wife gives birth to a colorblind daughter, her husband cannot be the father (unless there was a new mutation for color blindness in his sperm line, which is very unlikely).

✖ FOR MORE INFORMATION

Benzer, S. "Genetic Dissection of Behavior." *Scientific American*, December 1973. Not only physical traits, but behaviors too are under the influence of genes. The fruit fly once again proves a useful model system in which to study inheritance.

Crow, J. F. "Genes That Violate Mendel's Rules." *Scientific American*, February 1979. A close look at genetic recombination and its role in evolution.

Fisher, R. A. "Has Mendel's Work Been Rediscovered?" *Annals of Science*, 1:115–137, 1936. Reprinted on pages 139–172 of Stern and Sherwood's *The Origin of Genetics*, a famous geneticist and statistician argues that Mendel's ratios are too good to be true. Was Mendel's approach to science high on insight and low on honesty in data collection?

Hoyt, E. "Wild Relatives." *Living Wilderness*, Summer 1990. Undisturbed ecosystems are a valuable gene bank with great potential for improving crop plants.

Sapienza, C. "Parental Imprinting of Genes." *Scientific American*, October 1990. It is not quite true that all genes are equal, regardless of whether they have been inherited from mother or father. In some cases, which parent a gene comes from greatly alters its expression in the offspring.

Stern, C., and Sherwood, E. R. *The Origin of Genetics: A Mendel Source Book*. San Francisco: W. H. Freeman, 1966. There is no substitute for the real thing, in this case a translation of Mendel's original paper to the Brünn Society.

Wright, S. "Mendel's Ratios." Also in Stern and Sherwood's *The Origin of Genetics: A Mendel Source Book*, pages 173–175, another famous geneticist takes issue with Fisher's accusation against Mendel.

NET WATCH

On-line resources for this chapter are on the World Wide Web at:
http://www.prenhall.com/~audesirk (click on the table of contents link and then select Chapter 13).

14 Molecular Genetics and Biotechnology

"In the year 6565, ain't gonna need no husband, won't need no wife.
You'll pick your son, pick your daughter too
From the bottom of a long glass tube. Woh-woh"

From the song *"In the Year 2525"* by RICK EVANS

⊞ A T   A   G L A N C E

DNA Recombination in Nature
DNA Recombination Occurs Naturally through Processes
Such as Sexual Reproduction, Bacterial Transformation,
and Viral Infection
Recombination Provides Raw Material for Evolution
There Are Both Similarities and Differences between
Recombination in Nature and in the Laboratory

Recombinant DNA Technology
A DNA Library Consists of DNA from a Particular
Organism Inserted into Bacterial Plasmids
The Genes of Interest Must Be Identified in the DNA
Library
Selected DNA Sequences in the Library Can Be Amplified
Amplified Genes Can Be Used for Research, Industrial,
Medical, or Agricultural Purposes

Locating Genes
Restriction Enzymes Can Be Used to Provide Markers on a
Chromosome
Restriction Fragment Length Polymorphisms Can Be Used
to Locate a Gene

Sequencing Genes

Reflections on the Ethics of Biotechnology
Some Uses of Biotechnology Have Social Impacts
Bioengineered Organisms May Be Released into the
Environment
Biotechnology May Eventually Allow Alterations in Human
Genomes

*A plant that glows in the dark! Molecular genetic techniques
have been used to insert a gene from a firefly coding for the
enzyme luciferase into this tobacco plant. This enzyme breaks
down the chemical luciferin, releasing light in the process. This
tobacco plant has been provided with water containing luciferin.
The glow provides evidence that the plant's cell are expressing
the firefly gene.*

Biotechnology—virtually every week the newspaper features articles about the potential health benefits of some new medical advance in the works, based on biotechnology. Almost every state and large city in the United States tries frantically to lure "biotech" companies that promise high-paying, low-pollution jobs. Biotechnology often seems to be portrayed as both the medical and the economic miracle of our times. Yet furious debate rages about whether milk from cows treated with a recombinant growth hormone should be sold in supermarkets. What exactly *is* biotechnology? Is it boon or bane or both?

Biotechnology is the manipulation of the molecular basis of inheritance by methods collectively called **recombinant DNA technology**. Biotechnology is usually practiced to achieve one or more of three goals:

1. To understand more about the processes of inheritance and gene expression
2. To provide better understanding and treatment of various diseases, particularly genetic disorders
3. To generate economic benefits, including improved plants and animals for agriculture and efficient production of valuable biological molecules

The features of biotechnology that provide both vast promise and potential threat are the specificity with which biotechnology can, or will soon be able to, direct genetic changes; the speed with which genetic changes can be made; and the ability to transfer genetic material between species.

This chapter has three major themes. First, we look at a few important recombinant DNA technologies, briefly examining some of the methods that molecular biologists use to manipulate genes. Second, we discuss a few of the applications, both real and potential, of biotechnology. Third, we explore some of the practical and ethical issues that the use of biotechnology may raise.

Before we begin our discussion of human-directed biotechnology, however, it will be useful to provide a bit of background on what we might call "natural biotechnology." Let us start, then, with a survey of some naturally occurring methods of DNA recombination.

DNA Recombination in Nature

DNA recombination, whether by natural events or human intervention, involves two distinct processes: (1) changing the nucleotide composition of the DNA of a single cell, a few cells, or an entire organism; (2) selecting valuable DNA combinations.

DNA Recombination Occurs Naturally through Processes Such as Sexual Reproduction, Bacterial Transformation, and Viral Infection

You are already familiar with two major methods of "recombinant DNA," although we called them "genetic recombination" and "sexual recombination." Genetic recombination occurs through crossing over during meiosis I, when genes from a maternal chromosome and a paternal chromosome produce a chromosome with a new combination of alleles that may never have occurred before. In sexual recombination, chromosomes from two separate organisms combine to produce offspring that often are genetically unique. These natural recombinations, of course, usually occur within a single species. Many people have a tendency to consider single-species recombination as "natural" and therefore good, whereas they see between-species recombinations performed in the lab as "unnatural" and possibly intrinsically bad. However, between-species recombinations occur in nature, too.

Transformation May Combine DNA from Different Bacterial Species

Bacteria employ several methods of recombination that allow gene transfer between unrelated species. As you know, transformation allows bacteria to pick up free DNA from the environment (see Chapter 10). Sometimes, the free DNA acquired during transformation is part of the chromosome of another bacterium, including DNA from a bacterium of another species.

Transformation may also occur when bacteria pick up tiny circles of DNA called **plasmids** (Fig. 14-1). Plasmids, which range in size from about 1000 to 100,000 nucleotides, are self-replicating "parasites" normally found in the cytoplasm of many bacteria. A single bacterium may contain dozens or even hundreds of copies of a plasmid. Although the bacterium's "own" chromosome contains all the genes that it normally needs for survival, the genes carried by plasmids may be useful, too. For example, some plasmids contain genes that code for enzymes that digest certain antibiotics, such as penicillin. In environments where exposure to antibiotics is high, such as hospitals, these plasmids obviously are extremely valuable to their bacterial hosts (see Chapter 21). When a plasmid-containing bacterium dies, its plasmids may be liberated into the environment and transform other living bacteria. Plasmids may also be exchanged between living bacteria. The acquisition or exchange of plasmids is probably the most common form of bacterial transformation.

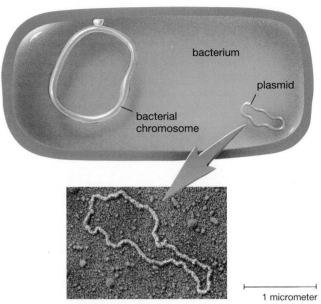

Figure 14-1   A bacterial plasmid

These tiny rings of DNA, only a few micrometers in diameter, have become important tools in recombinant DNA research.

Viruses May Transfer DNA among Eukaryotic Species

Recent evidence suggests that viruses sometimes transfer genes among eukaryotic organisms. The DNA of certain viruses can insert itself into a chromosome of its eukaryotic host and exist there quietly for days, months, or even years. Then, perhaps in response to a stimulus that is stressful to the host, the viral DNA leaves the chromosome, occasionally taking a bit of the eukaryotic DNA along with it. The viral DNA then takes over the host cell metabolism, replicates itself, and directs the synthesis of new viruses. The offspring viruses thus may include some host DNA. If one of these offspring infects a new host of a different species, it may insert itself, along with the piece of DNA from the former host, into a chromosome of its new host. In this way, the new host may acquire some genes that originally belonged to an unrelated species.

Recombination Provides Raw Material for Evolution

Recombination, by whatever mechanism, changes the genetic makeup of organisms. As is the case with mutations, natural recombination is random and undirected—a bacterium, for example, does not deliberately take up a plasmid that contains an antibiotic-resistance gene so that it can flourish in hospitalized patients. Naturally occurring DNA recombinations are tested, just as mutations are, by natural selection. Probably most recombinations prove to be harmful or neutral. A few, however, prove helpful in a particular environment. The organisms with the new DNA combinations thrive and pass on the new combinations to their offspring.

There Are Both Similarities and Differences between Recombination in Nature and in the Laboratory

Before we begin our discussion of recombinant DNA technologies, let's compare natural and laboratory DNA recombination.

1. Both natural and laboratory DNA recombination involve exchange of DNA between organisms, including between species.
2. Naturally occurring DNA recombinations are relatively random and undirected; in general, specific genes are not preferentially moved, and there is no "goal" that drives DNA movements. In laboratory DNA recombinations, specific pieces of DNA are moved between deliberately chosen organisms to achieve a specific goal.
3. The "usefulness" of naturally occurring DNA recombinations is determined by natural selection. Human interests determine the usefulness of laboratory DNA recombinations.

Recombinant DNA Technology

Within the past few years, the technologies and applications of recombinant DNA have mushroomed. Rather than attempting to summarize all the technologies, we will follow the sequence of procedures that might be used to solve a particular problem or produce a specific product. Although many other techniques are also commonly used, one logical sequence of activities would be the following: (1) produce a "DNA library" of an organism; (2) identify individual genes of interest; (3) produce a copy or (preferably) many copies of the gene; and (4) insert the gene into the desired organism and regulate the expression of the gene in a useful way.

A DNA Library Consists of DNA from a Particular Organism Inserted into Bacterial Plasmids

The first task in recombinant DNA technology is to produce a **DNA library**—a readily accessible, easily duplicable assemblage of all the DNA of a particular organism. Why build a DNA library? Think about your college library. It wouldn't be very useful if all the books were scattered about in random piles on the floor. Only consistent filing and a user-friendly cataloguing system allow you to find the books you need. To a geneticist, the chromosomes in a living cell are much like books in a pile: All the genes of the organism are there, but they are incredibly difficult to find and study. (The cell, of course, has its own "filing system," but geneticists are only beginning to decipher it.) A DNA library organizes the DNA in a way that researchers can use. As we will explain below, a DNA library also allows "molecular photocopying" of the DNA, so that researchers can easily obtain the thousands to billions of gene copies that they need for their experiments. Restriction enzymes, plasmids, and bacteria provide molecular geneticists with the filing systems, catalogues, and shelf space they need for their DNA libraries.

Restriction Enzymes Cut DNA at Specific Nucleotide Sequences

Many bacteria produce **restriction enzymes** that sever DNA at particular nucleotide sequences. In nature, restriction enzymes defend bacteria against bacteriophage invasion by cutting apart the phage DNA. The host bacteria protect their own DNA against being cut by the enzymes, probably by attaching methyl groups to some of the DNA bases. Restriction enzymes thus "restrict" phage infections to the types of phage whose DNA is not cut apart by the enzymes.

As shown in Figure 14-2, many restriction enzymes sever palindromic sections of DNA, which read the same in one direction on one strand as they read in the reverse direction on the other strand (a palindrome in English is a word that reads the same forward and backward, such as *madam*). Furthermore, the DNA is cut between the same two bases on the two strands, in this case between guanine and adenine. For the enzyme illustrated in Figure 14-2, this cut results in two pieces of DNA, one with a single-stranded end reading TTAA and one with a single-stranded end reading AATT. Complementary DNA regions like these, even from different organisms, can pair up, held together by hydrogen bonds between the bases. If the appropriate DNA repair enzymes are present, the two pieces can be rejoined.

Many restriction enzymes have been isolated from various species of bacteria. Each enzyme cuts DNA apart at a different

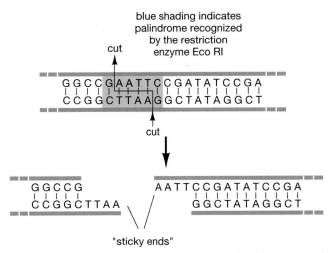

Figure 14-2   Restriction enzymes sever double-stranded DNA at specific nucleotide sequences

Many of the most useful restriction enzymes, such as the one called Eco RI illustrated here, sever DNA at palindromes, where the nucleotide sequence reads the same on one DNA strand (GAATTC) as it does in the opposite direction on the other strand. When a double-stranded DNA molecule is cut by Eco RI, two short, single-stranded "sticky ends" remain: AATT on one strand and TTAA on the other. Because these sticky ends are complementary, they can base pair, temporarily holding these strands, or any strands with the same sticky-end sequences, together again (see Fig. 14-3).

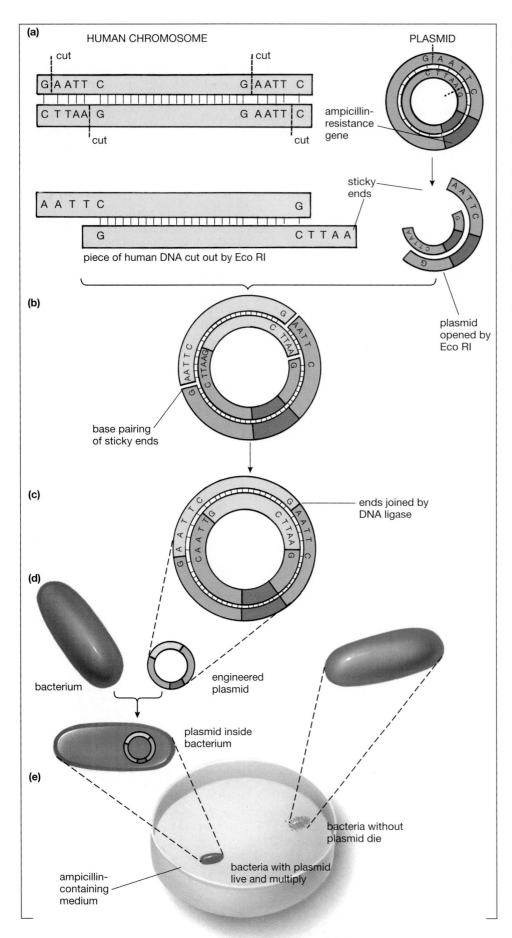

Figure 14-3 **The use of restriction enzymes in building a human DNA library**

(a) Human chromosomes and bacterial plasmids (with the ampicillin-resistance gene) are both cut with the Eco RI restriction enzyme. Both types of DNA therefore have AATT and TTAA sticky ends. **(b)** The severed human and plasmid DNA molecules are mixed. Plasmid-human combinations are temporarily held together by complementary base pairing of sticky ends. **(c)** DNA ligase bonds the backbones of the plasmid-human combinations together. **(d)** The new plasmid-human rings are mixed with bacteria that have been treated with calcium salts to make the bacteria permeable to DNA. Some bacteria take up a plasmid-human ring. **(e)** Bacteria are placed on culture dishes filled with medium containing the antibiotic ampicillin. Bacteria without plasmids succumb to the ampicillin; those with plasmids survive. Plasmids within the bacteria constitute a human DNA library.

nucleotide sequence. The specificity and variety of restriction enzymes have enabled molecular geneticists to identify and isolate segments of DNA from many organisms, including humans.

Restriction Enzymes Are Used to Insert DNA into Plasmids to Build a DNA Library

Suppose that DNA is isolated from a human source, say white blood cells, and is cut apart with the restriction enzyme Eco RI (Fig. 14-3a). Everywhere that the GAATTC-CTTAAG pairing occurs, the human DNA will be severed, leaving single-stranded AATT and TTAA "sticky ends" protruding. The next step is to isolate many copies of a bacterial plasmid that includes an easily identifiable "marker gene," such as a gene that confers resistance to an antibiotic called ampicillin (we'll see why this marker gene is useful in a minute). The plasmids are then exposed to the same restriction enzyme. The plasmids, too, will be cut open, with AATT and TTAA sticky ends protruding.

Now the human DNA fragments and opened plasmids are mixed together. The sticky ends will hydrogen-bond, forming human-human, plasmid-plasmid, and, most important, plasmid-human DNA combinations (Fig. 14-3b). DNA ligase enzymes are then added. DNA ligase bonds the sugar-phosphate backbones together, inserting human DNA into plasmids (Fig. 14-3c). Ideally, each plasmid receives only a relatively small piece of human DNA. Millions or billions of plasmids collectively would incorporate DNA from the entire human genome.

The new rings of plasmid-plus-human DNA are mixed with bacteria treated with calcium salts to make them permeable to DNA (Fig. 14-3d). The bacteria take up the plasmid-human DNA. Usually, 100 to 1000 times more bacteria than plasmids are used, so that no individual bacterium ends up with more than one plasmid-human DNA molecule. Of course, this procedure also ensures that most of the bacteria don't have any plasmid-human DNA combinations at all. Here's where the ampicillin-resistance marker gene becomes useful (Fig. 14-3e). The bacteria are placed on a culture dish containing medium with ampicillin. The antibiotic kills off all the bacteria that have not taken up a plasmid. The result: a population of bacteria, all with a plasmid, some with recombined plasmid-human DNA. This constitutes our human DNA library.

The Genes of Interest Must Be Identified in the DNA Library

The human **genome**, consisting of all the genes in an individual, contains about 6 billion nucleotides (3 billion pairs). The average protein has 300 to 400 amino acids, so at three nucleotides per amino acid, only about a thousand nucleotides encode the information for a typical protein. Identifying or finding a particular gene within the human genome, then, would appear to be a search for a very small needle within a very large haystack. Fortunately, there are several "tricks" that can be used to identify or locate genes.

It is fairly straightforward to find a gene if you first know the amino acid sequence of its encoded protein (Fig. 14-4). For example, the amino acid sequences of most of the major human hormones are known. From the amino acid sequence, one can then work backward through the genetic code to determine likely DNA sequences (remember, however, that a single amino acid may be encoded by several nucleotide sequences, which complicates things a little). Machines are now commercially available that can synthesize DNA chains from nucleotide sequences typed on a keyboard. The synthesizers cannot make really long chains, but they can accurately synthesize a chain a few dozen nucleotides in length. We'll see what to do with this artificial DNA shortly.

Another way of identifying a gene is to find a cell that you know synthesizes vast amounts of a particular protein. Immature red blood cells, for example, synthesize lots of hemoglobin. That means that they have lots of messenger RNA (mRNA) for hemoglobin. This mRNA is complementary to the DNA of the hemoglobin gene and can also serve to locate the gene.

Gene Probes Are Used to Search the DNA Library for Bacteria Containing Plasmids Bearing the Desired Gene

So far, our DNA library is pretty useless, because it lacks a card file or computer index that would tell us which bacterium contains the plasmid with the gene we want to study. This is where synthetic DNA or isolated mRNA, labeled with radioactive isotopes, comes in. Figure 14-5 explains how these "probes" are used to locate bacteria containing specific genes.

Selected DNA Sequences in the Library Can Be Amplified

Now that we have located the gene in the library, we can amplify it by making millions or billions of copies for further use, a process called **cloning**. One merely picks

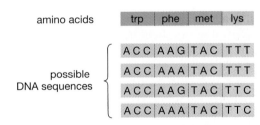

Figure 14-4 **Synthetic DNA can be used to locate a gene**

If the amino acid sequence of a protein is known, the likely nucleotide sequence of the DNA that encodes the protein can be deduced. The diagram shows DNA sequences that might code for a peptide with the amino acid sequence tryptophan-phenylalanine-methionine-lysine (trp-phe-met-lys). Note that, because phenylalanine and lysine may each be encoded by two DNA triplets, there are four possible DNA sequences for this single peptide.

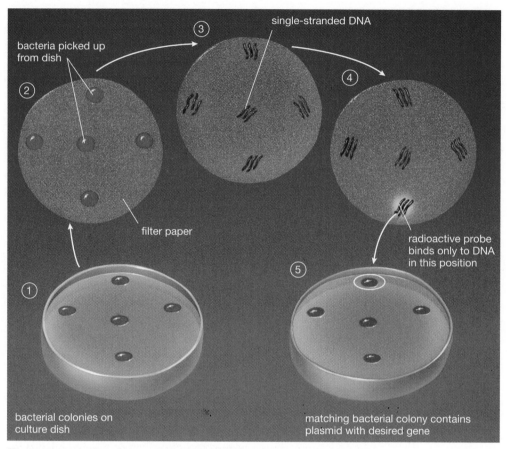

Figure 14-5   **Searching a human DNA library for a specific gene**
① Bacteria from the DNA library are sparsely spread out onto a culture dish. Each bacterium multiplies into a visible colony. If few enough bacteria were originally put in the dish, then each colony should consist of the descendants of a single bacterium, containing in turn a single type of plasmid-human combination. ② A sheet of special filter paper is pressed onto the culture dish. It picks up a few bacteria from each colony, faithfully preserving the colony positions. The original culture dish is saved. ③ The filter paper is placed in a basic solution. This solution breaks open the bacteria, freeing the plasmids, and separates the double-stranded DNA of the plasmids into single strands. ④ The paper is now bathed in solution of neutral pH containing a radioactive synthetic mRNA or DNA "probe" (yellow). The probe hydrogen-bonds only to plasmid DNA that is complementary to the nucleotide sequence of the probe; that is, to plasmids that contain the human gene complementary to the probe. ⑤ The locations of radioactivity on the paper are matched with bacterial colonies on the culture dish. Colonies in the same position consist of bacteria containing plasmids with the desired gene. Samples of these bacteria are cultured in their own dishes.

bacteria from the appropriate colonies located during the search and grows them in appropriate culture conditions. Biotechnology firms often use huge vats to produce many kilograms of bacteria, all containing a specific human gene (Fig. 14-6).

Another method for making copies of specific stretches of DNA, not only from DNA libraries, but even from single cells, is called the **polymerase chain reaction (PCR)**. In the polymerase chain reaction, billions of copies of selected genes can be made. The polymerase chain reaction has become so important in genetic research and biotechnology that *Science* magazine named the key enzyme involved in the procedure, a DNA polymerase from a specialized microbe, the "Molecule of the Year" in 1989 (see "A Closer Look: The Polymerase Chain Reaction—Hot Springs and Hot Science").

Amplified Genes Can Be Used for Research, Industrial, Medical, or Agricultural Purposes

At this point, procedures vary greatly, depending on what one wants to do with the gene. Further, the technologies tend to be quite complex and differ for each application. Therefore, we will discuss only a couple of possibilities, and generally we will describe only the applications, not the methodologies.

Amplified Genes Can Be Used to Produce Commercially Useful Amounts of Valuable Proteins

One prominent application of recombinant DNA is to produce large amounts of human proteins for medical uses. The genes for human and bovine (cow) growth hormone, insulin, blood clotting factors, and enzymes that dissolve blood clots have all been inserted into bacteria or eukaryotic cells for commercial production. The recombinant cells are grown in huge vats. The desired product is extracted from the culture medium, or the cells themselves, and purified.

Why is this process useful? Take the case of human growth hormone. Growth hormone, secreted by the pituitary gland, governs bodily growth, protein metabolism, and perhaps aging. A small percentage of the population does not produce enough growth hormone, remains very short, and ages prematurely. Formerly, the only source of human growth hormone was human pituitaries. Tens of thousands of pituitaries were removed from corpses and processed to purify the growth hormone, which would then be injected into growth hormone–deficient children. This approach, of course, was very expensive. To make matters worse, in 1985 a few recipients died from a rare infection, Creutzfeldt-Jakob disease, apparently caused by prions (see Chapter 21) that had contaminated the hormone preparation.

Meanwhile, a biotechnology company, appropriately called Genentech, inserted the gene for human growth hormone into bacteria and in late 1985 began to sell bacterially produced hormone. Because no human body parts are involved, no human diseases can be transmitted by the Genentech hormone. Similarly, blood clotting factors produced through recombinant DNA technology are now given to hemophiliacs rather than blood transfusions or blood clotting factors isolated from donated human blood. (Unfortunately, this development came too late to save hundreds of hemophiliacs from contracting AIDS from contaminated blood products; see Chapter 34.)

Viral Proteins May Be Synthesized for Use as Vaccines

Presently, vaccines are effective mostly against viral diseases, such as smallpox or measles. The vaccine itself usually consists of killed or weakened virus or of a strain of virus that is genetically incapable of causing disease. The killed, weakened, or genetically defective virus, although not disease causing, is recognized by the immune system. This recognition triggers the production of defenses that would protect against later encounters with disease-causing strains of the same virus (see Chapter 34).

A safer and potentially more effective vaccine might be synthesized using recombinant DNA techniques. The immune system usually recognizes not a whole virus, but one or more proteins on the viral surface. Recombinant DNA technologies might be used to find the genes for the appropriate viral proteins, insert them into bacteria, and produce huge quantities of the protein. A vaccine would then consist of pure protein and would be completely safe, because no living viruses would ever have been involved in its production. Recombinant protein vaccines would be especially attractive in the case of the HIV virus, which causes AIDS, because the virus is so lethal; in fact, one vaccine composed of weakened virus appears to have the potential to cause AIDS, at least in certain people. In late 1994, the World Health Organization decided to begin large-scale clinical trials of two AIDS vaccines based on fragments of one of the HIV surface proteins.

Genes May Be Inserted into or Deleted from Free-Living Organisms

Genetic engineers are inserting and deleting genes from bacteria that they intend to release into the outside world. For example, several laboratories are working to produce bacteria that selectively metabolize toxic substances, from crude oil to hazardous industrial wastes. In the 1980s, one of the first intentional releases of bioengineered bacteria—the so-called iceminus bacteria—raised a storm of controversy, even though a gene had been cut out of, rather than engineered into, the bacteria (see "Reflections on the Ethics of Biotechnology").

Genes can also be inserted into eukaryotic cells. Under the appropriate circumstances, entire multicellular or-

Figure 14-6   A gene-copying "factory"

A fermenting apparatus at a biotechnology company raises huge numbers of bacteria, each containing a plasmid that incorporates a valuable human gene.

A CLOSER LOOK

The Polymerase Chain Reaction—Hot Springs and Hot Science

One major problem with all the recombinant DNA technologies described in this chapter is that they require *lots* of DNA: millions or billions of copies. One DNA molecule, or even a few hundred, is usually not enough for even the most sensitive modern instruments to detect. In principle, lots of DNA can be produced by gene cloning in bacteria, but this takes time and money. For prenatal diagnosis of genetic defects, neither a lot of DNA nor a lot of time is often available. For decades, a dream of molecular geneticists has gone something like this: "If only someone could find a way to get me millions of copies of the gene I'm interested in, really quickly, and cheaply, too, of course." Well, Kary Mullis of the Cetus Corporation did just that: He invented a process, called the polymerase chain reaction, or PCR for short, that rapidly and selectively replicates specific parts of a DNA molecule. PCR is so elegant and so crucial to continued advances in molecular biology that it earned Mullis a share in the Nobel Prize in 1993.

The key to PCR lies in the nature of DNA polymerase. Remember that DNA polymerase works on single strands of DNA (other enzymes initially unwind the double helix and separate the two strands). If you add DNA polymerase and a batch of free nucleotides to a long single strand of DNA, the polymerase does not immediately begin synthesizing a new DNA strand. However, if you add a short piece of DNA that is complementary to part of the strand (this piece of DNA is called a primer), it will bind to its complementary region of the strand, and DNA polymerase will start adding new nucleotides at the end of the primer. Therefore, by adding a primer that is complementary to a specific place on the original DNA strand, you can effectively "tell" the DNA polymerase where to start copying. The polymerase chain reaction is so sensitive and so fast that it can actually produce billions of copies of a gene in a single afternoon, starting, if necessary, from a single molecule of DNA.

The polymerase chain reaction begins after two primers have been made in a DNA synthesizer: One primer is complementary to the template strand at the beginning of the gene (let's call this the "template primer"), and the second primer is complementary to the other strand at the end of the gene (the "nontemplate primer").

With a sample of DNA and these two primers in hand, PCR can begin (Fig. E14-1). In these diagrams, DNA is represented as long "railroad tracks" of sugar phosphates with notched "ties" of base pairs connecting the two rails. The primers and the gene that they will help to copy are colored blue.

The First Cycle
A DNA double helix is heated to just below boiling, about 98° C (208° F). The double helix separates into two single strands (steps 1 and 2 in Fig. E14-1). The separated strands are cooled and the primers, DNA polymerase, and free nucleotides are mixed in. The primers bind to their complementary regions of the DNA (step 3). DNA polymerase begins copying the DNA strand, beginning at a primer (the beginning of the gene) and continuing to the end of the strand. One DNA polymerase molecule binds at the "template primer" and synthesizes a new strand complementary to the DNA of the template strand (step 4, top). A second DNA polymerase molecule binds at the nontemplate primer and synthesizes a new strand complementary to the nontemplate strand (step 4, bottom). One complete copy of the gene, starting at the beginning of the gene, plus a lot of DNA that extends beyond the gene, has been made.

The Second Cycle
The DNA is heated up to 98° C again, separating the DNA into four single strands. The DNA is cooled down, and DNA polymerase, nucleotides, and primers are added (step 5). DNA synthesis proceeds as before, with both the original DNA strands and the first-cycle copies serving as templates. Note that the first-cycle copies begin at one end of a gene (let's call this end 1), but extend beyond the other end of the gene (end 2). When these first-cycle copies are used as templates, the primers bind to end 2 of the gene and DNA polymerase moves toward end 1. Therefore, DNA polymerase does not copy the "extra" nucleotides beyond end 2. When DNA polymerase reaches end 1, copies of the template and nontemplate strands have been synthesized, with no extraneous DNA (step 6).

The Third Cycle
The DNA is heated, cooled, and the ingredients added once again (step 7). Note that in this cycle, two of the eight DNA strands to be copied include *only* the gene. Therefore, this third cycle of PCR at last synthesizes two DNA double helices that are exact copies of just the gene, with no superfluous DNA at either end (step 8).

Further Cycles
As further cycles of PCR continue, the proportion of DNA copies that consist only of the gene increases rapidly. In the fourth cycle, 8 out of 16 DNA double helices include only the gene of interest; in the fifth cycle, 22 out of 32; and so on. Because each cycle takes only a few minutes to complete, within a few hours PCR produces many billions of copies of the gene, and most of these copies include just the gene, with no extra baggage.

Uses of PCR
PCR machines are now standard equipment in molecular biology labs. PCR has resulted in quicker and more accu-

rate prenatal diagnosis of genetic defects and certain diseases. It provides a more sensitive test for AIDS than the older antibody tests, by allowing a person's white blood cells to be examined for HIV virus genes. PCR is crucial in the Human Genome Project (see Chapter 15).

PCR promises to revolutionize other areas of biology and medicine as well. For example, a group of British and Russian sci-entists used PCR to identify the remains of the Russian royal family, killed by the Bolsheviks in 1917. Svante Paabo of the University of California at Berkeley used PCR to amplify bits of DNA remaining in Egyptian mummies. Even more remark-ably, in 1994 Scott Woodward of Brigham Young University claimed to have sequenced a bit of DNA from mitochondria in an 80-million-year-old bone, possibly from a dinosaur.

The First Cycle

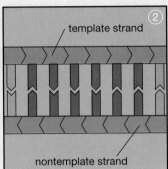

double-stranded DNA

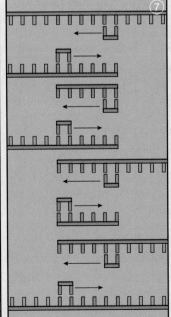

strands separated by heating

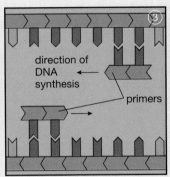

solution cooled; primers, nucleotides, and DNA polymerase added

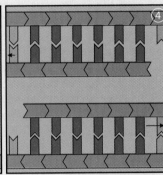

DNA copied from gene to end of strand

The Second Cycle

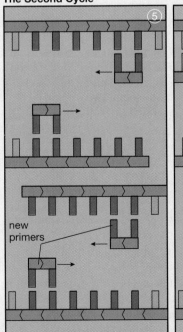

strands separated by heating; solution cooled, primers added

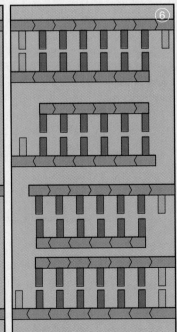

DNA copied

The Third Cycle

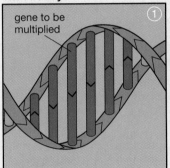

strands separated by heating; solution cooled, primers added

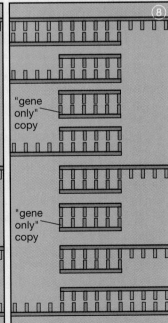

DNA copied; some copies consist only of the desired gene

Figure E14-1 **How the polymerase chain reaction copies genes**

The steps are described in the text.

Oh, yes—what about the hot springs mentioned in the title? Well, as you know, most proteins denature at high temperatures. If you heat DNA up to 98° C to separate the two strands, you irreversibly denature most DNA polymerase molecules as well. Therefore, early versions of PCR required the addition of new DNA polymerase in every cycle, after the DNA sample was cooled down. In 1987, someone at Cetus had a brainstorm: what about the bacteria that live in hot springs, such as those in Yellowstone National Park? Sure enough, the bacterium *Thermus aquaticus* (literally, "hot water") has evolved a DNA polymerase molecule that survives repeated heating and cooling cycles. Since then, even more stable DNA polymerase has been discovered in bacteria living in the deep ocean near volcanic vents in the ocean floor and is now commercially available. Because the free nucleotides and the short single-strand DNA primers are heat stable as well, a PCR machine is now a rather simple device that mostly just runs heating and cooling cycles over and over again. The investigator loads test tubes up with the ingredients, turns on the machine, and then goes away for a while.

ganisms can be genetically engineered to contain foreign genes. One drawback to this scenario is that many valuable traits may be polygenic—the desirable phenotype may be due to the interaction of several, perhaps many, genes, some of which may not yet be known.

Many genetically engineered plants have already been developed, for instance plants resistant to herbicides. Herbicides, more commonly called weed-killers at the local greenhouse, are used to kill undesirable plants that compete with flowers or crops for water, nutrients, and light. Unfortunately, most herbicides have the potential to harm desirable plants, too, so only limited amounts can be used. Biotechnology can potentially solve this dilemma. A gene that confers resistance to a common herbicide, glyphosate, has been identified and inserted into several types of crop plants. Once inserted, the glyphosate-resistance gene allows the crops to withstand heavier applications of herbicide. Higher concentrations of herbicide kill off more weeds that otherwise compete with the crops. Whether this application is beneficial or not depends on one's views about the safety of herbicides and how likely it is that the genes for herbicide resistance might be inadvertently transferred to other plants, including weeds (see the "Reflections" section).

In other examples, the FlavrSavr tomato, developed by the biotechnology company Calgene, has been engineered to resist rotting. It can therefore be allowed to ripen longer on the vine, perhaps at last producing a supermarket tomato that actually tastes like a tomato. French researchers at the Moet-Chandon champagne company have engineered grape vines to resist the grape fan-leaf virus. Genes for virus resistance have also been inserted into cantaloupe, cucumber, papaya, squash, potato, tomato, and tobacco.

One of the most valuable genetic modifications of plants would be the insertion of genes for nitrogen fixation into common crops such as corn or wheat. With a few exceptions, such as soybeans and alfalfa, our major crop plants cannot use atmospheric nitrogen as a nitrogen source. Consequently, they require nitrogen fertilizers, which are synthesized and spread using the energy of fossil fuels. Rain washes some of the fertilizer off into nearby streams and lakes, causing pollution. Further, the expense of fertilizers is burdensome to farmers, especially in the Third World. Several types of bacteria can extract nitrogen from the atmosphere and capture it in usable form, a process called nitrogen fixation (see Chapter 45). If the genes for nitrogen fixation could be inserted into wheat and corn, farmers could grow much more food at much lower cost. Especially in poor countries, an improvement in food supplies would have a tremendous impact on human health and well-being.

Recombinant DNA Techniques Can Be Used to Diagnose Genetic Disorders

A variety of human diseases are inherited, including sickle-cell anemia, Tay-Sachs disease, Huntington's disease, and muscular dystrophy. So far, there are no cures for these genetic disorders. Carriers (heterozygotes for recessive disorders who show no significant symptoms) or affected fetuses or both can sometimes be identified using recombinant DNA technology. Some of the techniques and the ethical concerns that they raise are discussed in Chapter 15.

Amplified Genes May Be Used to Treat Genetic Disorders

A few children are born each year with severe combined immune deficiency, in which the affected person has virtually no defense against disease (see Chapter 34). In about 25% of these children, the failure of the immune system can be traced to abnormal alleles of a single gene. In the late 1980s, R. Michael Blaese of the National Cancer Institute and W. French Anderson of the National Institutes of Health inserted normal alleles of this gene into white blood cells and obtained reasonably normal gene expression. At 12:52 P.M., September 14, 1990, the first clinical trial of human gene therapy began: A 4-year-old girl received a transfusion of her white blood cells that were genetically engineered to contain the normal allele. The treatment worked; today she has a functioning immune system and attends school. Unfortunately, so far only 2 out of 10 patients have responded to the treatment.

In 1994, experimental trials for cystic fibrosis, a usually fatal inherited lung disease, were begun with a "nasal spray" containing viruses engineered to infect lung cells and provide them with a crucial protein that helps prevent fluid buildup in the lungs.

As these examples illustrate, the most likely candidates for genetic engineering in humans are defects involving discrete structures, such as lungs or bone marrow, that are normally the only, or the most important, active sites of transcription of certain genes. These therapies would, of course, offer effective treatments only for the affected individual: The patient's reproductive cells would not be "fixed," so he or she could still pass the genetic defect on to future generations. It is likely to be many years before genetic engineers are in a position to modify the genetic composition of human sperm or eggs. Whether society will allow them to do so is also an open question, which we discuss in the "Reflections" section at the end of the chapter.

Locating Genes

The applications of biotechnology that we have discussed so far all rely on some prior knowledge: The desired genes are already known by either their protein products, their mRNA, or, in a few instances, at least part of their DNA sequences. Unfortunately, geneticists do not know much, if anything, about the molecular nature of most genes or their protein products. How can such information be obtained?

You already know something about methods for locating genes: for example, linkage analysis, which uses the frequency of crossing over to map chromosomes (see Chapter 13). To locate a gene by linkage analysis, one calculates the incidence of recombination between two traits during sexual reproduction. If recombination *never* occurs between the two traits, then either the genes controlling those traits lie very close together on a chromosome or, in fact, the traits are controlled by the same gene. If the location of the gene controlling one trait is already known, then it should be easy to find the other gene.

In principle, to locate an unidentified gene in a DNA library, you could make a probe for a known marker gene that linkage analysis tells you is close to the unidentified gene. Somewhere nearby should be the gene that you're interested in.

Using ordinary linkage analysis to find a gene in a DNA library, however, has two problems. First, geneticists usually know the location of relatively few genes on any given chromosome, and eukaryotic chromosomes are very large: The average human chromosome, for example, contains about 60 million nucleotides. Therefore, if linkage analysis reveals that an unknown gene is "close" to a known gene, say, only 10% of the chromosome's length away, that means that the few thousand nucleotides that make up the unknown gene might lie anywhere within about 6 million nucleotides. That informa-

tion isn't necessarily very useful. A second, related problem with linkage analysis is the relative scarcity of genes in a eukaryotic genome. Although there are tens of thousands of genes in most eukaryotic genomes, these constitute less than 5% of the genome: The rest is introns, regulatory sequences, or vast amounts of other, apparently nonfunctional, DNA. Therefore, classical linkage analysis usually cannot be expected to locate an unknown gene more closely than within a few hundred thousand nucleotides, at best. That still isn't very useful. What geneticists need are a lot more markers, a lot closer together.

Restriction Enzymes Can Be Used to Provide Markers on a Chromosome

You may recall that restriction enzymes cut DNA only at specific nucleotide sequences. Further, there are a host of known restriction enzymes, and each cuts DNA at a different sequence of nucleotides. Let's suppose that a pair of homologous chromosomes is exposed to a given restriction enzyme (Fig. 14-7). The DNA is cut up into pieces called **restriction fragments** (fragments of DNA produced by restriction enzymes). These fragments can be separated by size, using a method known as gel electrophoresis. Would the pattern of sizes be the same for both homologous chromosomes? Perhaps, or perhaps not, depending on the nucleotide sequences of the chromosomes and the recognition sequence of the enzyme. In the example diagrammed in Figure 14-7, the two homologous chromosomes have different nucleotide sequences near one end. The restriction enzyme Hind III can cut apart the upper chromosome in two places, but the lower chromosome in only one place (Fig. 14-7a). Gel electrophoresis therefore shows three pieces of DNA from the upper chromosome and two pieces of DNA from the lower chromosome (Fig. 14-7b). These differences in the size of the DNA pieces are called **restriction fragment length polymorphisms**, or **RFLPs** for short (*polymorphism* means "many forms" in Greek).

Restriction Fragment Length Polymorphisms Can Be Used to Locate a Gene

Gene location by RFLP analysis is conceptually similar to locating a gene by crossing over. In RFLP analysis of a human genetic disorder, researchers look for a restriction enzyme "cut site" that is always, or almost always, inherited along with the disorder. The cut site must therefore be within, or very close to, the gene causing the disorder. In the late 1980s, RFLP analysis was used to locate and sequence the gene for cystic fibrosis.

It is easy to understand why a unique restriction enzyme cut site might be located within a defective allele: If the altered nucleotide sequence that causes the genetic defect is also the one recognized and cut by the restriction enzyme, then the enzyme will cut the defective allele but not the normal allele. But why should a unique cut site be located *close to* a defective gene? To understand this, you must consider the inheritance of genetic defects and the nature of linkage. Defective alleles for inherited genetic disorders are usually quite rare. Therefore, all affected individuals may well be descended from a single, though distant, ancestor. Suppose that this person, strictly by chance, had a second altered nucleotide sequence on the same chromosome. How he or she got it, by inheritance or mutation, doesn't matter. If this second altered nucleotide sequence is close enough to the locus of the defective allele, then there will be little or no crossing over between the two sites.

This second nucleotide change can be used to find the locus of the defective allele if one more condition is met: The "normal" and "altered" nucleotide sequences must be affected differently by a restriction enzyme. Either the normal sequence must be cut by the enzyme while the altered sequence is not cut, or vice versa. Given the vast number of restriction enzymes, a "cut–no cut" difference can almost always be found relatively close to any given defective allele, although it may take years of work to find one close enough to be useful for DNA sequencing (described shortly).

Other Uses of RFLPs

Locating a gene is not the only application of RFLPs. For example, RFLP analysis can also be used for prenatal diagnosis of certain genetic defects, if the nucleotide sequence or at least the location of the gene is known. In Chapter 15, we describe how RFLPs are used to diagnose sickle-cell anemia. A variation on standard RFLP analysis, popularly known as DNA fingerprinting, is used increasingly in forensic medicine to help convict or clear a suspect; a class of RFLPs from blood or semen samples left at the scene of a crime is compared with RFLPs from the suspect's

tissues. DNA fingerprinting promises to play a crucial role in the O. J. Simpson murder trial and other well-publicized cases (see "Scientific Inquiry: DNA Fingerprinting in Forensics—The Case of the Telltale Palo Verde"). The methodology and some other applications of RFLP analysis are explained in a fascinating *Scientific American* article, "Chromosome Mapping with DNA Markers" (see "For More Information").

Sequencing Genes

Both basic research and medical/industrial applications of biotechnology are greatly enhanced if the sequence of nucleotides in a gene is known. During the 1980s, automated machines were developed that allow fairly large genes (up to a few tens of thousands of nucleotides long) to be sequenced (Fig. 14-8). The process isn't cheap, and even 50,000 nucleotides are less than a tenth of a percent of the 60 million nucleotides in an average human chromosome, which is why precise localization of genes is still very important.

Why is knowing the nucleotide sequence of a gene useful? For one thing, the nucleotide sequence determines the amino acid sequence of the encoded protein. In many instances, it is now easier to locate and sequence a gene than it is to isolate, purify, and sequence a protein. In recent years, biochemists have frequently "forecast" the amino acid sequence of a protein, its location within a cell (mem-

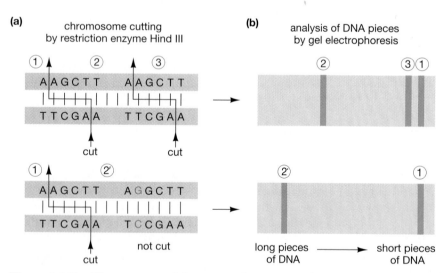

Figure 14-7   **The nature and detection of restriction fragment length polymorphisms**

(a) Two homologous chromosomes differ in a single base substitution. A restriction enzyme, Hind III, cuts DNA at AAGCTT/TTCGAA palindromes. The upper chromosome has two of these sequences and is cut into three pieces, two short (numbered 1 and 3 on the illustration) and one medium-long (number 2). The lower chromosome, because of the base substitution (a G-C pair substituted for an A-T pair), has only one sequence cut by Hind III. Therefore, the lower chromosome is cut into only two pieces, one short (same as number 1 on the upper chromosome) and one very long (numbered 2'). **(b)** The DNA pieces produced by Hind III are separated by the technique of gel electrophoresis. DNA migrates through the gel, pushed along by an electrical field. Small pieces (1 and 3) slip through the gel more easily than large pieces (2 and 2') can and migrate farther to the right. The size and number of DNA pieces differ for the two chromosomes.

brane-bound or dissolved in the cytoplasm), and even some of its probable functions, solely on the basis of the nucleotide sequence of a gene. The gene for cystic fibrosis was sequenced in late 1989. The gene sequence was the starting point for pinpointing the underlying genetic defect.

Further, if one knows the nucleotide sequence of a gene and the nucleotide sequence that is cut by a variety of restriction enzymes, one may be able to design a rapid, definitive test for genetic disorders—for example, the DNA of affected people will not be cut by a certain restriction enzyme but the DNA of everyone else will be cut. This type of test is now the basis for prenatal diagnosis of sickle-cell anemia.

In one of the most remarkable instances of cooperation and collaboration in modern biology, a huge group of scientists is undertaking the ultimate sequencing task—to sequence the entire human genome. The rationale, approach, and methodology of the Human Genome Project are explored in Chapter 15.

Reflections on the Ethics of Biotechnology

In many respects, biotechnology presents a microcosm of the benefits and dilemmas that science and technology have always offered to humankind. There are few fundamental advances in science that do not have both bright and dark sides—the first spear allowed hunters to nourish their families better, but it also provided a more effective means of killing their fellow humans. There are groups in society who would greatly restrict the uses of biotechnology and who might prefer that the techniques of recombinant DNA had never been developed. Others feel that the benefits greatly outweigh the liabilities and that any potential for harm can rather easily be overcome by fairly simple legal means. Many others take no stand on biotechnology in general but have strong opinions about specific applications. Here,

we will explore three aspects of biotechnology that together represent many of the ethical issues that surround this emerging field. In the next chapter, we will examine some ethical concerns in the application of biotechnology to medicine.

Some Uses of Biotechnology Have Social Impacts

Perhaps the easiest ethical issue surrounding biotechnology is the objection that some applications may be socially undesirable. For example, in the late 1980s, a biotech firm began to mass-produce bovine growth hormone by recombinant DNA techniques. Growth hormone, when injected into cows, enhances milk production. Some people are vigorously opposed to this practice, for three main reasons. First, will the recombinant hormone change the composition of the milk? Experts convened by the National Institutes of Health and the Food and Drug Administration have concluded that the changes are minor and pose no threat to human health. Second, increased milk production is likely to lead to increased incidence of udder infections in cows. This effect might not only be harmful to the cows but might also lead to increased use of antibiotics, which might in turn lead to increased antibiotic resistance among microbes (see Chapter 21). Third, there is already a milk glut in both America and Europe. Many dairy farmers oppose the use of recombinant growth hormone because they fear that greater milk production per cow will drive prices down even further and put some small farmers out of business.

These concerns are real. However, they do not have any direct bearing on biotechnology in and of itself. Millions of cattle and other mammals are slaughtered for meat every year; suppose that a cheap way was found to extract growth hormone from their pituitary glands. The same arguments would arise without biotechnology's being part of the issue at all. In instances such as this, society may wish to intervene in *a specific use of biotechnology* for reasons of public health, economic issues, or social justice. The outcome, not the methodology, is the heart of the matter.

Figure 14-8 **A commercially available DNA sequencing machine**

Machines such as this, only a dream a couple of decades ago, are routine equipment in many molecular biology labs.

SCIENTIFIC INQUIRY

DNA Fingerprinting in Forensics—The Case of the Telltale Palo Verde

The polymerase chain reaction (PCR) procedure has caused a revolution in many fields of biology, but nowhere has the impact of this technique been felt more than in forensics, the collection of information to be used as evidence in court proceedings. The technique that is proving most useful has been nicknamed "DNA fingerprinting," because it may generate patterns of DNA fragments unique to each individual. This procedure involves the amplification of a set of DNA fragments of unknown sequence, a set that is likely to differ from one individual to the next. When these fragments are amplified and separated from one another on a gel, the pattern of bands is referred to as a DNA fingerprint. In theory, it represents a characteristic "snapshot" of the unique genome of an individual, allowing it to be distinguished from that of every other individual in the population. The quantity of DNA needed to analyze using PCR is exceedingly small; this amount of DNA can easily be obtained from a single spot of dried blood, from the cells clinging to the base of a hair shaft, from skin fragments under a victim's fingernails, or other small tissue samples collected at a crime scene.

Here is an example of the forensic use of DNA fingerprints that did not involve the suspect's DNA, but rather that of plants at the crime scene. On the night of May 2, 1992, a Phoenix woman was strangled and her body was dumped near an abandoned factory in the desert. Police discovered a pager near the body, making its owner the prime suspect. When questioned, the man who owned the pager claimed that the woman must have stolen the pager from his pickup truck, since he had never been near the factory. A search of the pickup revealed two seed pods from a palo verde tree (Fig. E14-2), several of which grew near the factory. Could DNA fingerprinting demonstrate that the seed pods found in the suspect's truck had fallen from the tree near where the body was found? If it could, this evidence would place the suspect at the crime scene.

Homicide investigators turned to Dr. Timothy Helentjaris, then at the University of Arizona, in nearby Tucson. In his research on the evolution of maize and other important crop plants, he had been generating DNA fingerprints of parts of the maize genome. Helentjaris knew that any match between a seed pod and a tree would have significance only if he could first demonstrate that palo verdes show considerable genetic variation from one tree to the next. Otherwise, it would be impossible to unambiguously link a given pod to any particular tree. Fortunately, he found considerable differences in DNA between individual plants, allowing for discriminatory fingerprints.

Helentjaris then was given the two seed pods from the suspect's truck along with pods collected by investigators from 12 different palo verde trees near the factory. The investigators knew which of the 12 trees was near the body but did not tell Helentjaris. He extracted DNA from seeds taken from each pod and performed the PCR reaction, producing 10 to 15 distinct bands from each pod. The result of this experiment was unmistakable: The pattern from the pod found in the truck exactly matched the pattern from only one of the 12 trees, the one nearest the body. In an important additional test, Helentjaris found that this pattern was also different from that of 18 pods collected from trees at random sites around Phoenix. This result was admitted as evidence in the trial, the first time a DNA fingerprint of a plant had been used in a court case. This information was crucial in demolishing the suspect's alibi, and at the end of the 5-week trial the suspect was found guilty of first-degree murder.

Figure E14-2   The telltale palo verde
DNA fingerprints from the pods (inset) of the palo verde tree helped convict a murderer.

Bioengineered Organisms May Be Released into the Environment

Although bioengineered *products* are relatively simple to deal with, bioengineered *organisms* are more problematic. Two examples illustrate the potential difficulties.

Many crop plants are quite sensitive to frost: oranges and tomatoes, to name just two. Ice formation in the air, in water, or on plant surfaces is greatly facilitated by small particles of dust or other matter that act as "seeds" for ice-crystal formation. In the absence of "seed particles," ice crystals do not form even at temperatures somewhat below freezing. One of the primary sources of ice seeding on plants is a common bacterium, *Pseudomonas syringae*. A specific protein on the surface of the bacterium promotes ice formation. Therefore, an obvious use of genetic engineering would be to remove the gene encoding this protein from the bacterium and to try to replace the "wild-type" bacterium on crops with the new "ice-minus" bacterium. The genetic engineering part of this scenario was accomplished in the mid-1980s, but there were several years of controversy and litigation before field trials were permitted (Fig. 14-9).

What was the controversy about? What if the ice-minus strain, once released into the environment, drifted beyond the fields onto which it was sprayed? What if it became established in the wild and became widespread? Susceptibility to frost is a major determining factor in the distribution of plants across the planet—might ice-minus bacteria trigger massive changes in vegetation? Like the ice on leaves, raindrops also usually form around "seeds" of dust or debris in the atmosphere. What if wild-type bacteria contribute significantly to normal rain seeding—might droughts become frequent if ice-minus bacteria became the norm? Genetic engineers responded that the altered bacteria were actually less fit than the wild type and that a minuscule number of bacteria were to be released, compared with the vast hordes of wild-type bacteria in the environment. Therefore, they argued, the chances were infinitesimally small that the ice-minus bacteria could actually become established in the wild. At least so far, the genetic engineers appear to be right: A test field was sprayed in 1987, and no ecological effects seem to have occurred.

A similar controversy is brewing about herbicide-resistant plants. Some people are opposed to herbicide-resistant plants because they feel that extensive herbicide use is hazardous to human health and the environment, and herbicide-resistant crops would almost certainly lead to greatly increased herbicide use. Alternatively, what if herbicide-resistance genes were to be transferred from crops to weeds, for example by a plant virus? Such "lateral transfers" of plant genes may be possible. If such a transfer occurred, herbicides might ultimately become completely useless.

Ice-minus bacteria and herbicide-resistant plants are examples of larger issues surrounding releases of genetically engineered organisms into the environment: People are not omniscient; people make mistakes. Can we be sure that genetically released organisms, or the modified genes they carry, will not cause ecological harm? If we can determine only that the risk is "small," then how small is small enough?

Genetic engineers correctly assert that humans have practiced crude genetic engineering for millenia, breeding plants and animals with desired properties. Modern biotechnology is merely a faster, more precise version of standard agricultural practice. What's more, various forms of genetic recombination occur all the time in nature. Why should modern genetic engineering be singled out as a special threat?

Some ecologists reply that ecological disasters of varying magnitude do indeed happen: the invasion of pastureland in the western United States by cheatgrass and knapweed, the displacement of bluebirds by sparrows, the evolution of antibiotic-resistant bacteria, and the mutation of a monkey virus into the AIDS virus, to name a few modern examples. (On the other hand, bioengineering may offer the possibility of *correcting* one ecological disaster of major proportions, the destruction of American chestnut trees by chestnut blight; see "Earth Watch: Biotechnology Attacks the Chestnut Blight.") Genetically engineered organisms, such as the ice-minus bacteria, are likely to be genetically inferior to their wild-type counterparts, and therefore be outcompeted, but there may remain some risk. Although generally supportive of properly controlled releases of genetically engineered organisms, in 1989 a committee of prominent ecologists wrote the following warning in the journal *Ecology*: "The direct effects of self-replicating introduced organisms may not necessarily decrease with time or with distance from the point of introduction. The absence of an immediate negative effect does not ensure that no effect will ever occur."

Biotechnology May Eventually Allow Alterations in Human Genomes

The greatest concern of many observers is that genetic engineering offers the potential to change the human genome. As we pointed out, it is not yet possible to eliminate, say,

Figure 14-9   Field tests of "ice-minus" bacteria

Although the bacteria are completely harmless to humans, this worker spraying "ice-minus" bacteria onto a strawberry patch was required to wear protective clothing, because the bacteria were legally considered to be a pesticide that had not yet been safety-tested on people. The "frostbusters" logo, although just a publicity gimmick, was accurate: Treated plants showed about 80% less frost damage than nearby, untreated control plants.

Biotechnology Attacks the Chestnut Blight

Scarcely a century ago, the towering American chestnut tree made up a major portion of the eastern hardwood forest (Fig. E14-3). Today the splendor of the chestnuts is barely a memory; few notice the struggling sprouts emerging from the roots of the former giants. The American chestnut has been all but driven to extinction by the ravages of a fungus called the chestnut blight, inadvertently imported from Asia around 1900 on Asian chestnuts. However, there is hope that the American chestnut will reappear, thanks to chance and the research techniques of biotechnology.

In 1950, an Italian researcher noted that some European chestnuts, also attacked by chestnut blight, were recovering—had the fungus lost its virulence? In the 1960s, plant pathologists found that some strains of chestnut blight had been weakened, apparently by a new viral infection that left particles containing double-stranded RNA (dsRNA, which encodes genetic information in many fungal viruses) floating throughout the fungal cytoplasm. In an unknown way, the viral dsRNA decreases the virulence of the fungus, allowing the tree to fight it successfully (Fig. E14-3b). By spreading the viral dsRNA infection to virulent chestnut blight fungi, foresters are gradually controlling the blight in Europe. Unfortunately, the European blight's virus is not very useful against the strains of chestnut blight fungus found in the United States.

Although viral-infected strains of chestnut blight also occur in the United States, they do not successfully control the blight. Therefore, the tools of biotechnology are being applied toward weakening the parasitic fungus. In 1994, researchers at the Roche Institute of Molecular Biology in New Jersey found a way to introduce both viral dsRNA and a DNA "gene" encoding the dsRNA into chestnut blight fungi in such a way that the RNA or DNA or both are transferred from one fungus to another. Ultimately, they hope to produce a large supply of weakened fungus that can be introduced into the wild, where it would infect the deadly fungal variety with the viral nucleic acids, weakening the fungus so that it would no longer be able to overcome the defenses of the chestnut tree.

Biotechnology may eventually be used, not only to weaken the fungus, but to strengthen the chestnut. Asian chestnuts are not killed by the chestnut blight fungus, and there is hope that the gene that confers fungal resistance in the Asian chestnut may be identified and introduced into "germlings" of the American chestnut, producing a blight-resistant strain. Although you will never see the chestnut forests that delighted your great-great-grandparents, resistant chestnut trees infected by weakened fungi may yet shade the homes of your children.

(a)

(b)

Figure E14-3   Chestnut blight

(a) A healthy American chestnut, which escaped the blight because it had been transplanted far west of the normal range of the chestnut before the invasion of the blight fungus. **(b)** Chestnut blight fungus has produced a lesion on this chestnut tree. The lesioned area was treated with infected fungus, and the tree is walling off the infection with a growth of new tissue.

insulin-dependent diabetes by cutting the defective gene out of fertilized eggs and inserting a functional one. But suppose that it does become possible (few molecular geneticists would be willing to bet against it). Then what? Could humankind agree on what constitutes a "bad" gene? Even if we could, would we be right? Probably most people could agree on diabetes, cystic fibrosis, and muscular dystrophy. But what about more subtle changes? What about genes that might alter personality? Or length of life? Or predisposition to take risks? Are we wise enough to direct our own evolution—or, if we develop the capability, are we wise enough to refrain?

SUMMARY OF KEY CONCEPTS

DNA Recombination in Nature

There are many naturally occurring forms of DNA recombination, where the genetic makeup of an organism incorporates DNA from two or more other organisms. These include genetic recombination (crossing over), sexual recombination, transformation (in bacteria), and viral transfer of DNA between eukaryotic species. Bacterial transformation often involves exchange of circular pieces of DNA called plasmids, which are found in the cytoplasm of many bacteria. Bacteria may take up plasmids that have been released by the destruction of other bacteria, thereby acquiring plasmid DNA from different species.

Recombinant DNA Technology

Many common uses of recombinant DNA technology include the following steps: (1) production of a DNA library, (2) identification of the desired gene, (3) production of copies of the desired gene, and (4) insertion and regulation of the gene in specific organisms.

A DNA library is a complete set of all the DNA of a particular organism, usually cloned in plasmids. DNA libraries are produced through the use of restriction enzymes that cut DNA at specific nucleotide sequences, leaving single-stranded "sticky ends" at each side of the cut. A restriction enzyme is used to cut DNA of both source genome and plasmid. The two sets of DNA fragments are mixed together; the single-stranded sticky ends cause plasmid and source genome DNA to base pair. DNA repair enzymes are then used to bond the source genome DNA to the plasmid DNA. Bacteria then take up the plasmids and may be stored indefinitely as a DNA library.

Genes may be identified through a number of techniques. If the amino acid sequence of a protein is known, then artificial DNA may be synthesized that corresponds to the DNA sequence encoding all or part of the amino acid sequence. Alternatively, mRNA may be isolated from a source that is rich in the specific mRNA transcribed from the desired gene. In either case, radioactive mRNA or synthetic DNA can be used to highlight the desired DNA region from a DNA library by base pairing with the gene.

Copies of a specific gene may be produced in bacteria from the DNA library by culturing those bacteria that contain plasmids with the desired gene. Alternatively, the DNA containing the gene may be amplified using the polymerase chain reaction.

Uses of cloned genes include (1) producing large quantities of specific proteins; (2) synthesizing vaccines; (3) diagnosing genetic disorders; (4) modifying DNA in intact organisms, which may be bacteria, agricultural plants, or animals, and releasing the organisms into the environment for specific uses; and (5) modifying the DNA of cells of specific organs in human beings to correct genetic defects.

Locating Genes

Restriction fragment length polymorphisms (RFLPs) may be used to locate genes in chromosomes. Restriction enzymes cut DNA at specific nucleotide sequences. If the same chromosomes from two or more sources are cut with the same restriction enzymes, differences in nucleotide sequences will yield differences in restriction fragment lengths. Inheritance of phenotypic traits, such as genetic diseases, can be compared with the differences in RFLPs to locate genes quite precisely in chromosomes.

Sequencing Genes

DNA nucleotide sequences can be determined by automated processes in the laboratory. The nucleotide sequence of a gene predicts the amino acid sequence of the encoded protein. Knowledge of the amino acid sequence of the encoded protein can reveal the function and location of the protein and may assist in understanding genetic disorders and devising treatments for them.

Reflections on the Ethics of Biotechnology

Biotechnology has generated ethical controversies regarding its proper uses. These controversies mostly deal with the risks involved in the introduction of genetically engineered organisms into the environment and the potential for making significant changes in the human genome.

KEY TERMS

cloning p. 265
DNA library p. 263
genome p. 265
plasmid p. 262

polymerase chain reaction (PCR)
   p. 266
recombinant DNA technology p. 261
restriction enzyme p. 263

restriction fragment p. 271
restriction fragment length polymorphisms (RFLPs) p. 271

✖ T H I N K I N G   T H R O U G H   T H E   C O N C E P T S

Multiple Choice

1. During recombinant DNA techniques, how are the bacterial cells which take up the plasmids isolated from ones which do not?
   a. by screening for restriction fragment length polymorphisms
   b. by using antibiotic resistance plasmid genes and antibiotic-containing medium
   c. each of the plasmids must be sequenced
   d. by use of the polymerase chain reaction
   e. this is where you use mRNA or information on the protein sequence

2. How do you help ensure that each bacterium in a library has only one gene-containing plasmid in it?
   a. Genes with more than one plasmid do not survive the antibiotic.
   b. You need to screen using a radioactive RNA probe.
   c. It is a matter of numbers and probability—far more bacteria than plasmids are mixed together.
   d. Only a small amount of calcium salt is used to facilitate the incorporation of the plasmid into the bacterial cell.
   e. This is why it is so important to spread the bacteria sparsely onto the culture dish.

3. Knowing the DNA coding sequence of a gene would give you [A]; knowing the amino acid sequence of the coded protein would give you [B].
   a. [A] approximate information about the amino acid sequence; [B] approximate information about the DNA sequence
   b. [A] approximate information about the amino acid sequence; [B] precise information about the DNA sequence
   c. [A] precise information about the amino acid sequence; [B] approximate information about the DNA sequence
   d. [A] precise information about the amino acid sequence; [B] precise information about the DNA sequence
   e. none of the above

4. Why is *Thermus aquaticus* so useful?
   a. It is necessary for the mass production of bacteria containing a plasmid with an inserted gene.

   b. It is used in automated DNA sequencing.
   c. It is used in RFLP mapping.
   d. It facilitates the polymerase chain reaction.
   e. It is used to create recombinant plasmids.

5. How are RFLPs detected?
   a. by looking at the chromosome in the microscope
   b. by doing a standard Mendelian cross
   c. by observing DNA of different lengths on a gel
   d. by seeing with which plasmids they will combine
   e. by amplifying the DNA using PCR

6. How would you identify the bacterium in a library if you knew the sequence of the protein it coded for?
   a. You would put radioactive protein on the petri dish.
   b. You would put a synthetic radioactive nucleotide probe, designed from the protein sequence and the genetic code, onto the petri dish.
   c. Because of the antibiotic, only the one of interest would survive.
   d. You would check to see whether a protein of the desired sequence is being synthesized by each bacterium.
   e. You would use DNA polymerase.

Review Questions

1. Describe three natural forms of genetic recombination, and discuss the similarities and differences between recombinant DNA technology and these natural forms of recombination.
2. What is a plasmid? How are plasmids involved in transformation in bacteria?
3. What is a restriction enzyme? How can restriction enzymes be used to splice a piece of human DNA within a plasmid?
4. What is a DNA library? Briefly describe the steps involved in creating a DNA library of a mouse.
5. What is a restriction fragment length polymorphism? Describe the use of RFLPs in locating a gene on a human chromosome.
6. Describe the polymerase chain reaction. Why is it so useful in prenatal diagnosis of genetic defects?
7. In PCR, when primers at each end of the gene are used, what is so unique about the copy of the copy after the second round of DNA synthesis?

✖ A P P L Y I N G   T H E   C O N C E P T S

1. If you knew the entire sequence of a protein, what strategy would you use to design a short nucleotide sequence most likely to be complementary to the corresponding sequence in the actual gene?
2. In cloning, what two tricks (one at the level of plasmid insertion and one at the level of bacterial plating and growing) must be used to ensure that each bacterium (to grow into a colony) has one and only one inserted gene? Make the simplifying assumption that the restriction nuclease used chops into small enough fragments to have just one gene per fragment.

3. Traditionally, a "marker" in a genetic cross is a phenotype such as eye color. Now, it is possible to locate genes more directly by reference to an RFLP "marker" right in the DNA. Contrast the techniques and circumstances that would apply to the utilization of these different approaches.
4. The library described in this chapter, prepared from the nuclear DNA, is sometimes called a "genomic library." It is also possible to prepare a "cDNA library" made from cDNA, the complementary DNA copied from the mRNAs expressed in a particular tissue or structure (like *Drosophila*

head). How do you think such a cDNA library would differ from the genomic library?

5. Let's say that you are a citizen member of a National Institutes of Health advisory panel on the uses of recombinant DNA. Discuss the ethical issues that surround the release of bioengineered organisms into the environment. What levels of benefits and risks do you think would justify such a release?

6. Do you think that using recombinant DNA technologies to change the genetic composition of a human egg cell is ever justified? If so, what restrictions should be placed on such a use?

✖ F O R   M O R E   I N F O R M A T I O N

Barton, J. H. "Patenting Life." *Scientific American*, March 1991. Some of the legal and economic issues surrounding the new biotechnological advances are discussed.

Baskin, Y. "DNA Unlimited." *Discover*, July 1990. A clear, succinct description of the methods and applications of the polymerase chain reaction.

Capecchi, M. R. "Targeted Gene Replacement." *Scientific American*, March 1994. Changing the genetic makeup of mice provides a new way to understand the function of genes.

Elmer-Dewitt, P. "The Genetic Revolution." *Time*, January 17, 1994. A description of the science and ethics of biotechnology, written for a nonscience audience.

Gasser, C. S., and Fraley, R. T. "Transgenic Crops." *Scientific American*, June 1992. More and more commercially valuable plants are being engineered to resist pests while growing and spoilage after ripening.

Mestel, R. "It's a Tall, Tall World." *Discover*, January 1994. Should short, but otherwise healthy, children be given recombinant growth hormone, just to make them taller?

Mullis, K. B. "The Unusual Origin of the Polymerase Chain Reaction." *Scientific American*, April 1990. Mullis conceived the idea of PCR while on a drive through the mountains of California.

Paabo, S. "Ancient DNA." *Scientific American*, November 1993. The scientific concept underlying the science fiction in the film *Jurassic Park* is not so far-fetched, but, in actuality, it is applied to reconstructing evolutionary history rather than to recreating the dinosaur.

Roberts, L. "Ethical Questions Haunt New Genetic Technologies." *Science*, and "Ecologists Wary about Environmental Releases," *Science*, both March 1990. These two articles provide a brief primer on the debates concerning the social and ecological issues surrounding the uses of recombinant DNA technology.

Verma, I. M. "Gene Therapy." *Scientific American*, November 1990. A brief discussion of the promises and pitfalls of gene therapy.

White, R., and Lalovel, J.-M. "Chromosome Mapping with DNA Markers." *Scientific American*, February 1988. The authors describe the use of RFLPs to locate genes on human chromosomes.

N E T   W A T C H

On-line resources for this chapter are on the World Wide Web at:
http://www.prenhall.com/~audesirk (click on the <u>table of contents</u> link and then select Chapter 14).

15 Human Genetics

⠿ AT A GLANCE

Methods in Human Genetics

Single-Gene Inheritance
Most Human Genetic Disorders Are Determined
by Recessive Alleles
A Few Human Genetic Disorders Are Determined
by Dominant Alleles
Many Human Traits Are Either Sex-Linked
or Sex-Influenced

Complex Inheritance
Many Human Traits Are Influenced by Many Genes
The Environment Influences the Expression of Genes

Chromosomal Inheritance
Some Genetic Disorders Are Caused by Abnormal Numbers
of Sex Chromosomes
In Humans, the Y Chromosome Determines Maleness
Some Genetic Disorders Are Caused by Abnormal Numbers
of Autosomes

The Human Genome Project

**Medical Genetics Poses New Medical
and Ethical Dilemmas**
Phenylketonuria Illustrates a Biomedical Dilemma
Knowledge of Human Genetics Poses Ethical Dilemmas

*Nature and nurture: These twins are genetically identical, but
have also grown up in extremely similar environmental situations.
The interplay of both heredity and environment shapes the
characteristics of all organisms, but especially of people.*

Human genetics is similar to the genetics of peas or fruit flies, except that it is more interesting to most people, and vastly more complex. First, humans have long life spans and few children per couple. Second, people choose their own mates, so crosses that might be genetically informative may not occur very often. Finally, humans interact extensively with their environment, an environment that is incredibly diverse from culture to culture and even within a single culture. These interactions often obscure underlying genetic patterns. This is particularly true for personality traits, in which learning and experience play such a major role.

Nevertheless, human genetics is a flourishing field of study, and a great deal is known about the inheritance of human traits. From the viewpoint of genetics, two factors partially compensate for the difficulties humans present as experimental subjects: the enormous number of human beings on Earth and the extensive documentation of human families, ranging from names written down in family Bibles to the records of ancestry of royalty. Given enough time, ingenuity, and patience, geneticists often can find records of matings that help them determine how particular human traits are inherited, especially physical features and genetic diseases.

Methods in Human Genetics

Because experimental crosses are out of the question, human geneticists search medical, historical, and family records to find crosses that have already been made voluntarily. Records extending across several generations can be arranged in the form of family **pedigrees**, diagrams that show the genetic relationships among a set of individuals, such as those shown in Figure 15-1. Careful analysis of pedigrees shows that certain traits, such as an unattached earlobe, are inherited as simple dominants, whereas other traits, such as albinism, are inherited as recessives.

Since the mid-1960s, great strides have been made in understanding gene function on the molecular level. For instance, geneticists now know the molecular basis of several inherited diseases, such as sickle-cell anemia and cys-

(a) A pedigree for a dominant trait

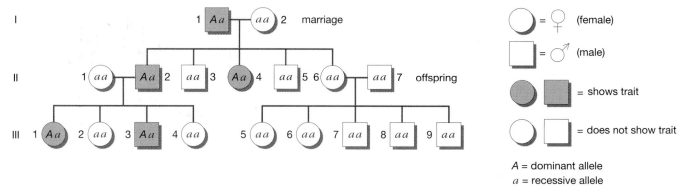

(b) A pedigree for a recessive trait

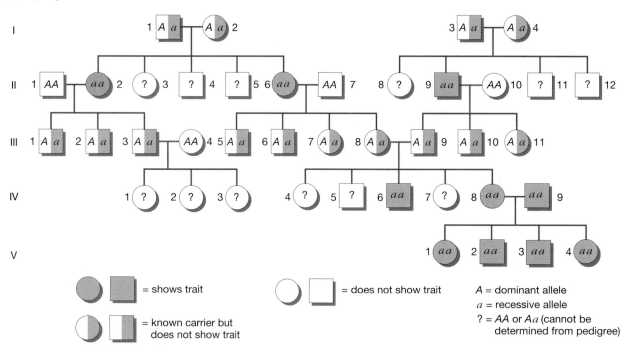

Figure 15-1   **Representative family pedigrees**

In pedigrees, squares usually represent males and circles females. Generations are indicated by the Roman numerals on the left side of the pedigree. **(a)** A pedigree for a dominant trait. Remember that all individuals bearing even one dominant allele will express the trait. Therefore, the trait will appear in every generation, in both males and females, and no child will possess the trait who does not have at least one parent with the trait. However, because heterozygotes show dominant traits, not all the children of phenotypically dominant parents will necessarily show the trait—that is, they may be homozygous recessive (for example, generation III, individuals 2 and 4). In this hypothetical pedigree, the genotypes of the individuals are given; in real pedigrees, genotypes must be deduced from inheritance patterns and cannot always be determined with certainty. **(b)** A pedigree for a recessive trait. Only homozygous recessives will show the trait. People with only one recessive allele will have the dominant phenotype. Parents with the dominant phenotype but who have a child who shows the recessive trait must themselves be heterozygotes. Therefore, the recessive phenotype may not appear in every generation (for example, generation III, families composed of individuals 5 through 8 and 9 through 11); the recessive allele may remain "hidden" in heterozygotes for several generations until two heterozygotes happen to marry (generation III, individuals 8 and 9) and have a homozygous child (generation IV, individuals 6 and 8). Two parents who both show the recessive phenotype (IV, 8 and 9) will have all homozygous recessive children (V, 1–4). In a pedigree for a rare trait, it is usually safe to assume that all individuals are homozygous normal (II, 1, 7, and 10) unless there is positive evidence of possession of the recessive allele (for example, an affected child).

tic fibrosis. Genetic engineering promises to increase our ability to predict genetic diseases and perhaps even to cure them (see Chapter 14 and "Scientific Inquiry: Collecting Fetal Samples and Analyzing Them Using Biotechnology" in this chapter).

Single-Gene Inheritance

Many "normal" human traits, such as freckles, long eyelashes, cleft chin, the ability to roll one's tongue (Fig. 15-2), are inherited in a simple Mendelian fashion; that is, they each appear to be controlled by a single gene with a dominant and recessive allele. In this chapter, rather than merely cataloguing inherited human traits, we will concentrate on a few examples of medically important genetic diseases and defects and explore their consequences for individuals and society.

Most Human Genetic Disorders Are Determined by Recessive Alleles

The human body depends on the integrated action of hundreds of enzymes and other proteins. A mutation in the gene coding for one of these enzymes almost always impairs or destroys enzyme function. As we described in the case of fat color in rabbits (Chapter 13), however, the presence of one normal allele often generates enough functional enzymes so that heterozygotes are phenotypically indistinguishable from homozygous normals. Therefore, normal alleles usually are dominant and mu-

tant alleles are recessive, although there are exceptions, such as polydactyly (more than five fingers on each hand), that are caused by dominant alleles. Most genetic diseases that are caused by lack of an essential enzyme are inherited as recessives.

If the metabolic pathway controlled by a particular gene is essential to survival, then individuals who are homozygous recessive will die. This outcome was especially certain before medical care became widely available in recent decades. In evolutionary terms, defective alleles are selected against. Consequently, even heterozygous **carriers** of serious genetic diseases (people who are phenotypically normal but can pass on their defective allele to their children) are usually rare. An unrelated man and woman will seldom both possess the same defective allele and produce a homozygous child. Related couples, however, especially first cousins or closer, have inherited some of their genes from recent common ancestors. Therefore, they are much more likely to carry the same allele and have an affected child (Fig. 15-3).

By their very nature, genetic diseases are an integral part of an affected person. They can neither be prevented with vaccines nor cured with antibiotics. Some, such as diabetes, can be treated, with varying degrees of success. As we pointed out in Chapter 14, genetic engineering offers the promise of a permanent genetic cure in certain cases. In instances in which the expression of the defective genes is localized to certain accessible organs, such as the liver or bone marrow, it may be possible to extract some of the patient's cells, insert the proper genes, and reimplant them in the patient. This procedure has been done for a few children suffering from severe combined

(a)

(b)

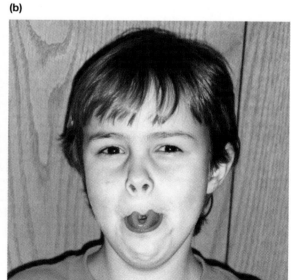

Figure 15-2   **Single-gene human traits**

These are controlled by dominant and recessive alleles of single genes. **(a)** Cleft chin seen in actor Kirk Douglas (left) and his son, Michael Douglas (right). **(b)** Tongue-rolling ability, demonstrated by Heather Audesirk (her father, Gerald Audesirk, also has this ability).

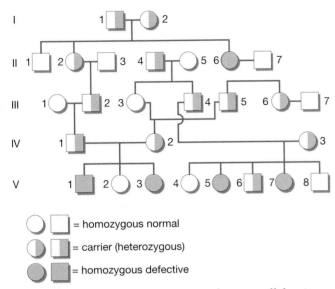

= homozygous normal

= carrier (heterozygous)

= homozygous defective

Figure 15-3   **Relatives may carry the same alleles**

Marriage between relatives increases the chance of genetic disorders. When related people marry, they often carry some of the same alleles, inherited from their common ancestors, and the likelihood that they will both carry the same defective recessive allele is much higher than normal. As a result, marriages between cousins or even closer relations are the cause of a disproportionate number of recessive diseases. In this family, marriages between cousins were common, including those between III, 3 and 5; III, 4 and IV, 3; and IV, 1 and 2.

immune deficiency (see Chapter 14). However, the genetic defect still remains in the reproductive cells that give rise to sperm or eggs, and the affected people can pass on the genes to their children.

For the present, then, the only way to prevent genetic diseases is to prevent the birth of affected babies. Reducing the incidence of recessive genetic diseases would require either identifying heterozygous carriers and preventing reproduction by couples who are both carriers or screening fetuses and aborting those that are homozygous recessive. Both courses of action are often both scientifically difficult and ethically questionable.

By definition, carriers of recessive genetic diseases cannot be identified by casual observation; if they can be identified at all, expensive medical procedures are usually required. With a few exceptions, it is impractical to screen the entire population of any country for a genetic disease. Screening tests can be useful, however, when the defective alleles are found almost exclusively within a readily identifiable group. Tay-Sachs disease and sickle-cell anemia are recessive genetic diseases, largely restricted to certain segments of the population, for which both carrier identification and prenatal diagnosis are possible. These two diseases illustrate the procedures used to diagnose recessive diseases and some possible courses of action. The practical and ethical dilemmas implicit in the diagnosis and treatment of genetic diseases are explored in the last section of the chapter.

Tay-Sachs Disease

Tay-Sachs disease occurs when nerve cells fail to synthesize an enzyme involved in the breakdown of lipids. Lipids accumulate in the brain of a child who is homozygous for the Tay-Sachs allele, causing progressive mental retardation, blindness, and failure of motor control (Fig. 15-4a). There is no cure, and death occurs in early childhood.

The Tay-Sachs allele is very rare in the United States population as a whole; only about one in 400 Americans is a carrier. However, about one in 30 American Jews is a carrier, so there is about one Tay-Sachs child in 3600 Jewish births. Heterozygotes for Tay-Sachs disease can easily be detected by a blood test (Fig. 15-4b), and many Jewish prospective parents take this test. If both husband and wife are identified as carriers, they have three choices: (1) forego reproduction entirely; (2) take a 25% chance of having an affected child who will face suffering and a certain early death; or (3) through **amniocentesis** or **chorionic villus sampling** determine whether each fetus is affected and abort homozygous recessives. The techniques of amniocentesis and chorionic villus sampling are described in "Scientific Inquiry: Collecting Fetal Samples and Analyzing Them Using Biotechnology." Genetic screening followed by abortion has reduced the incidence of Tay-Sachs disease by a factor of 20 since the 1970s.

Sickle-Cell Anemia

As with Tay-Sachs disease, almost all the carriers of **sickle-cell anemia** belong to one ethnic group, in this case Africans and African Americans. A single nucleotide substitution in DNA (adenine for thymine) causes the amino acid valine to be substituted for glutamic acid at one position on the outside of the hemoglobin molecule (see Chapter 11). Glutamic acid is highly charged and hydrophilic ("water loving"), whereas valine is neutral and hydrophobic ("water fearing"). Glutamic acid is crucially important in keeping hemoglobin molecules dissolved in the cytoplasm of the red blood cells. Substitution of valine at this position causes the hemoglobin molecules to clump together, forcing the red blood cells into the sickle-like shape that gives the disease its name (Fig. 15-5). During exercise or stress, the sickled cells break and clog the capillaries, cutting off circulation. In some instances, this lack of circulation may cause fatal strokes or heart attacks. Heterozygotes have about half normal and half abnormal hemoglobin, which you might expect would lead to partial sickle-cell symptoms. But heterozygotes usually have few sickled cells and show no symptoms whatsoever (in fact, many world-class black athletes are heterozygotes).

In the United States, about 8% of the African-American population is heterozygous; in parts of Africa, 15% to 20% of the population may be heterozygous. Given how severe the disease can be to homozygotes, why hasn't natural selection

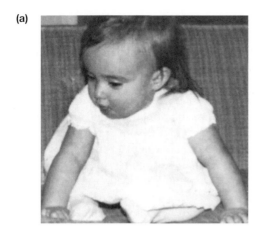

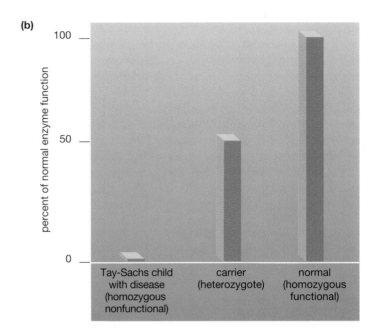

Figure 15-4   **Tay-Sachs disease**

(a) The progression of Tay-Sachs disease. Evelyn, beautiful and seemingly bright, happy, and normal for the first few months after birth, never managed simple tasks such as sitting upright (notice that at 9 months she is using her hands for balance). As the disease progressed, she became unable to respond to sounds or other stimuli and suffered from recurring infectious diseases. She died shortly before her sixth birthday. (b) The Tay-Sachs allele codes for a non-functional enzyme. Heterozygotes may appear phenotypically normal, but, because they have only one normal allele, they often have only about half as much of the normal enzyme as homozygous dominants do. In Tay-Sachs, this difference in the concentration of normal enzymes in the blood can be used to detect heterozygotes. This blood test is the basis for genetic counseling that can spare parents from having a child with what Evelyn's parents have called "this cruel . . . disease."

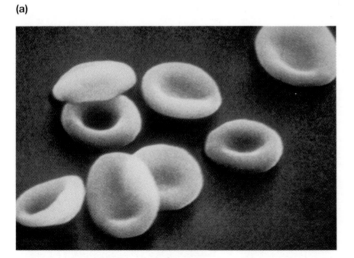

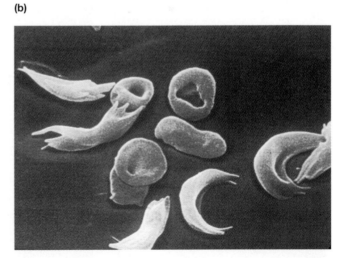

Figure 15-5   **Sickle-cell anemia**

Sickle-cell anemia is caused by abnormal hemoglobin molecules that clump together, especially in low oxygen concentrations such as occur in capillaries during exercise. The clumps force the red blood cell out of its normal disc shape (a) into a longer, sickle shape (b). The sickled cells are more fragile than normal red blood cells, making them likely to break or clog in capillaries. Tissues "downstream" from such a clogged capillary do not receive oxygen or have their wastes removed. This effect can cause pain, and, if it occurs in the brain or heart, serious injury.

eliminated the sickle-cell allele? The prevalence of the sickle-cell allele is probably an evolutionary compromise: People who are heterozygous for the sickle-cell allele enjoy some protection against malaria. In Africa, where malaria is still all too common, this protection may make the difference between life and death. In Chapter 17 we will explore how the dual selective effects of diminished hemoglobin function and protection against malaria may have favored the preservation of both the normal and sickle-cell alleles in African populations.

Heterozygous carriers of the sickle-cell allele can be detected by a blood test, but until fairly recently there was no way to find out if a fetus is homozygous or heterozygous. Recombinant DNA techniques have now been devised that can distinguish chromosomes with the normal hemoglobin allele from those with the sickle-cell allele (see "Scientific Inquiry: Collecting Fetal Samples and Analyzing Them Using Biotechnology"). Analyzing fetal cells collected by amniocentesis or chorionic villus sampling now allows medical geneticists to diagnose sickle-cell anemia in fetuses.

A Few Human Genetic Disorders Are Determined by Dominant Alleles

Many physical traits are inherited as dominants, including cleft chin and freckles. But, for two reasons, few people have serious genetic diseases caused by dominant alleles. First, as we have already explained, when a mutation strikes a normal allele, it usually produces a nonfunctional, recessive allele. Second, everyone bearing a dominant defective allele will develop the disease: There can be no phenotypically normal carriers. Before the advent of modern medicine, if the disease was serious, these people died without reproducing and did not pass on their defective allele to future generations.

An exception to this rule is Huntington's disease (Fig. 15-6). An incurable disease, Huntington's causes a slow, progressive deterioration of parts of the brain, resulting in loss of motor coordination, flailing movements, personality disturbances, and eventual death. Huntington's disease is particularly insidious because symptoms usually do not occur until 30 to 50 years of age. Therefore, a person usually has already had children before he or she ever suffers the first symptoms.

Because everyone with even one defective dominant allele shows the defective trait (if they live long enough), dominant diseases could be virtually eliminated in a single generation if all the affected people chose not to reproduce (except, of course, for new mutations, which are usually very rare). In the case of Huntington's, however, few people know for sure if they have the disease until after they have had children. Therefore, to eliminate Huntington's, all people with a parent suffering from the disease would themselves have to forego having children. Since half of these people would be homozygous normal and could not pass on the Huntington's gene, the decision not to reproduce would have to be based on statistics. For some, the choice is clear; for others, the dilemma is excruciating.

In 1983, painstaking pedigree analysis, restriction fragment length polymorphism (RFLP) analysis (see Chapter 14), and a lot of luck combined to localize the Huntington's gene to a relatively small part of chromosome 4 (see "Sci-

(a) (b)

 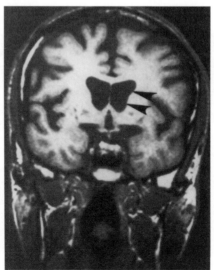 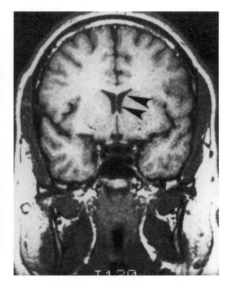

Figure 15-6   **Huntington's disease**

The public became aware of Huntington's disease when folk singer Woody Guthrie **(a)** developed symptoms in the 1950s. The reason for the behavioral and motor deficits that Huntington's patients develop is appallingly clear when we compare magnetic resonance images of the brains of a living patient with Huntington's disease **(b, left)** and a normal person of the same age **(b, right)**. Notice that the total area of brain tissue (light color) is much smaller and the darker fluid-filled spaces around it and inside it (see arrows) are much larger in the Huntington's victim's brain.

entific Inquiry: In Search of the Huntington's Gene"). No one knows what the protein encoded by the normal Huntington's gene is, or what it does in the brain, much less what's wrong with the protein encoded by the defective allele. Finding the gene and studying the proteins produced by the normal and defective alleles might enable researchers to devise therapies that slow or stop the progress of the disease. After 10 frustrating years of work, geneticists finally isolated the Huntington's gene in 1993. The next steps, finding out what the Huntington's gene does in healthy individuals and what goes wrong in afflicted people, are just beginning.

RFLP analysis has also succeeded in producing a test for Huntington's disease. Although the procedure is not infallible, geneticists can now sample a person's DNA and predict whether he or she has inherited the Huntington's allele *before any symptoms appear*. At first glance, this is a wonderful medical advance, which could result in the elimination of this devastating disease. For the individuals involved, however, the knowledge will be gained at enormous personal cost. What would *you* do if you were told that in 20 or 30 years you would develop Huntington's? Some people do not want to know.

Many Human Traits Are Either Sex-Linked or Sex-Influenced

As we described in Chapter 13, the X chromosome bears many genes that have no counterpart on the Y chromosome. With one X and one Y chromosome, males are effectively haploid for X-chromosome genes; that is, all alleles of X-chromosome genes are always expressed in men, a phenomenon called **sex-linked,** or **X-linked, inheritance**.

A son receives his X chromosome from his mother and his Y chromosome from his father. A man must therefore inherit X-chromosome genes from his mother, and he can pass them on only to his daughters. For rare, recessive alleles of X-chromosome genes, there is usually a striking pattern of inheritance (Fig. 15-7). Recessive traits appear most frequently in males and typically skip generations, with an affected male passing the trait on to a phenotypically normal, carrier daughter who in turn bears affected sons. The most familiar genetic defects due to recessive alleles of X-chromosome genes are red-green color blindness (Fig. 15-7), **hemophilia** (Fig. 15-8), and one form of muscular dystrophy, in which muscles of afflicted young males degenerate.

Each sex also has its own set of **sex-influenced traits** (in addition to the obvious ones of breast and genitalia development) that occur more commonly or more strongly in that sex but are not coded by genes on the sex chromosomes. The majority of sex-influenced traits affect males more often than females and seem to be enhanced by male sex hormones, such as testosterone. A familiar example is baldness, which appears as if it were dominant in men (heterozygotes become bald) but recessive in women (heterozygotes retain their hair). The hormonal

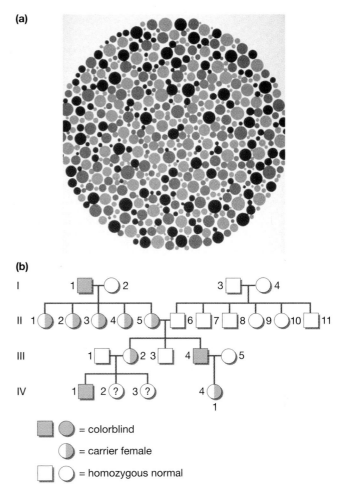

Figure 15-7   Color blindness

(a) This figure, called an Ishihara chart after its inventor, distinguishes between red-deficient and green-deficient vision in people with red-green color blindness (which is actually only color deficiency). People with normal color vision see the number 96, while those with red-deficient or green-deficient vision see either a 9 or a 6 depending on the type of deficiency. **(b)** Part of the family tree of Gerald Audesirk (III, 4), showing sex-linked inheritance of red-green color blindness. Audesirk and his maternal grandfather (I, 1) are both color deficient, but neither his mother and her sisters, nor any relative of his father, shows the defect. This pattern of skipping generations, a more common occurrence in males, and transmission from affected male to carrier female to affected male is typical of X-linked recessive traits.

connection in baldness is readily apparent in people with abnormal levels of sex hormones. Castrated men almost never become bald, but if castrated men with the allele for baldness are given high doses of testosterone, they lose their hair. Some women with tumors causing enlargement of the adrenal glands (sites of testosterone synthesis) produce large amounts of testosterone and may become bald. Another sex-influenced trait that is 20 times more common in men than in women is gout, which causes painful deposits of uric acid in joints.

SCIENTIFIC INQUIRY

In Search of the Huntington's Gene

As a young boy in 1860, George Huntington encountered an emaciated mother and daughter, both uncontrollably twitching, grimacing, twisting, and bowing: "I stared in wonderment, almost in fear," he later wrote. Huntington went on to study medicine at Columbia University and spent the rest of his life investigating this disease, now called Huntington's disease (HD) in his honor. Huntington showed that HD occurs in about half the children of an affected parent, evidence that it is inherited as a dominant allele of a gene.

Years later, in 1967, tragedy struck the family of Nancy Wexler, a young woman of 22, when her mother was diagnosed with HD. In response, Nancy's father, Milton Wexler, a clinical psychologist, founded the Hereditary Disease Foundation in 1968, hoping to find a cure or an effective treatment for the disease that eventually would kill his wife. Nancy Wexler was profoundly affected by her mother's suffering, and especially by her progressive loss of mental function. Trained as a clinical psychologist like her father, Nancy Wexler decided to devote her life to finding a cure for HD and now serves as president of the Hereditary Disease Foundation. She taught herself genetics by attending lectures, workshops, and seminars on the subject, by having many informal discussions with geneticists, and by reading extensively. During one of the seminars, Wexler learned that in several villages near Lake Maracaibo in Venezuela, an unusually high percentage of the villagers died of HD. Nancy visited these villages and found that about one in five persons there were stricken with the disease (Fig. E15-1).

By 1981, Wexler had begun assembling a large family tree showing the relationships of this huge, genetically interconnected Venezuelan population, where families with 13 or more children are common. Over 10,000 individuals have now been added to this extensive pedigree. Wexler also collected blood samples from these people, shipping them to the laboratory of Dr. James Guisella at Massachusetts General Hospital in Boston. In Guisella's lab, molecular genetic technology was used to identify a region of DNA near the tip of human chromosome 4 as the location of the gene for HD. Comparing the genetic in-

Figure E15-1   Tracking the gene for Huntington's disease

Nancy Wexler is shown here with one of the many inhabitants of the Lake Maracaibo region of Venezuela who suffer from or may carry the gene for Huntington's disease.

formation in the Venezuelan blood samples with the relationships depicted in the pedigree, Guisella's group devised tests that very accurately determine whether a person has inherited the gene for HD from an affected parent or is free of the "genetic time bomb," as Wexler describes the Huntington's gene.

Ironically, although Nancy Wexler was instrumental in the development of the screening test, she has decided not to find out whether she has inherited her mother's Huntington's gene. As she explains, each individual must make a deeply personal decision as to whether the pain of knowing the gene is present outweighs the joy of knowing the gene is absent, or whether living with the anxiety of not knowing is preferable. The availability of this test is of special importance to an individual who may have inherited the Huntington's gene and wants to have children. Genetic screening coupled with the voluntary decision not to risk passing the gene to the next generation could rapidly eliminate this tragic condition.

Is the Y chromosome totally devoid of genes other than those involved in determining maleness? No, but it turns out to be quite difficult to prove Y-linked inheritance. You might think that Y-linked traits should be easy to demonstrate, because *all* the males and *only* the males descended from a given affected male ancestor should show the trait. In practice, the situation is complicated by the influence of male sex hormones. Most traits that once were thought to be due to Y-chromosome genes are actually due to autosomal genes whose expression is strongly influenced by testosterone levels.

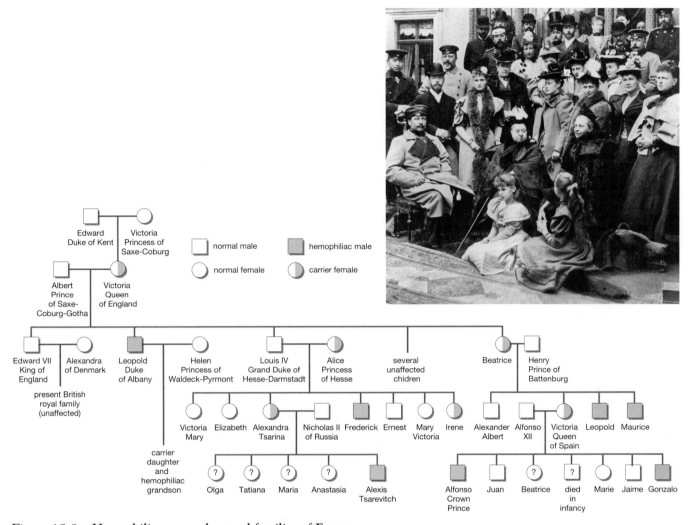

Figure 15-8 **Hemophilia among the royal families of Europe**

Hemophilia results from a deficiency of one of the substances causing blood clotting, a deficiency inherited as a recessive allele on the X chromosome. Affected males do not bleed to death from the first scratch or bruise; most hemophiliacs can clot off a minor wound. As a result, males often survive to pass on their allele to their daughters. Homozygous hemophiliac women are rare, but some are known. Surprisingly enough, they do not usually die from menstruation or even from bleeding following childbirth. In both of these instances, blood flow is stopped not by clotting, but by muscular contraction, shutting off circulation to the uterine wall.

The most famous genetic pedigree involves the transmission of sex-linked hemophilia from Queen Victoria of England (seated in right center, with crown) to her offspring, and eventually to virtually every royal house in Europe. Because all of Victoria's ancestors were free of hemophilia, the hemophilia allele must have arisen as a mutation either in Victoria herself when she was an embryo or in one of her parents.

Extensive intermarriage among royalty, who, after all, are not supposed to marry commoners, spread Victoria's hemophilia allele throughout Europe. Her most famous hemophiliac descendant was great-grandson Alexis, Tsarevitch (crown prince) of Russia. Bleeding episodes in the only son of the Tsar naturally distressed his parents greatly. The Tsarina Alexandra (Victoria's granddaughter) believed that the monk Rasputin, and no one else, could control Alexis's bleeding. Rasputin may actually have been able to do this through hypnosis, perhaps causing Alexis to cut off circulation to bleeding areas by muscular contraction. Although there were many causes underlying the Russian Revolution, which led to the rise of the Communist regime, the influence that Rasputin had over the imperial family may have contributed to the downfall of the Tsar. In any event, hemophilia was not to be the cause of Alexis's death; he was killed with the rest of his family by the Bolsheviks in 1918.

Complex Inheritance

Many human characteristics are not inherited in a simple either/or fashion. Even traits that are controlled by single genes, such as Huntington's disease, are often influenced by other genes and by environmental factors. For example, although everyone who has at least one allele for Huntington's contracts the disease, the age at which the first symptoms appear is extremely variable: Some people are affected in their teens; others do not develop symptoms until their fifties or even sixties.

Many Human Traits Are Influenced by Many Genes

Many, perhaps most, human traits are determined by the interaction of two or more genes. Examples of such polygenic inheritance in humans are the colors of eyes and skin.

The color of the iris in human eyes varies from very pale blue through green to almost black. Nevertheless, there are no blue, green, or black pigments in the human iris. Eye colors are actually caused by the distribution of a single yellowish brown pigment, melanin, the same pigment that colors skin and hair. The iris contains two layers of pigment, one at the back and one at the front. If there is little or no pigment at the front, the iris appears blue. The blue color arises from the scattering of light in the front layers, viewed against the dark background of melanin in the rear, just as the sky appears blue because of light scattered by the air, seen against the black background of space. In people with greater amounts of melanin in the front layers, the iris color may be green (blue plus yellowish brown), brown, or almost black. At least two, and probably more, genes direct the synthesis of melanin in the front of the iris, with each gene having two alleles showing incomplete dominance. The simplest scheme of two genes can create five shades of eye colors (Fig. 15-9).

Skin color is another case of different amounts of melanin. People don't really have white, yellow, red, or black skin: All are various shades of brown, with a pinkish tint from surface blood vessels showing through the paler tones. Skin color is inherited through the action of at least three pairs of genes, each with incompletely dominant pairs of alleles. As with eye color, polygenic inheritance explains both a more or less continuous gradation of skin colors and the occasional offspring whose skin color differs considerably from that of either parent.

The Environment Influences the Expression of Genes

Human physical traits are also affected by the environment. This fact is obvious to any Caucasian who spends much time in the summer sun: Skin color can be darkened by exposure to sunlight. At the same time, a person's genotype limits the range of environmental effects on phenotype. How fast we run, how high we jump, or how well we sing can all be greatly improved by training, but no amount of practice would ever enable most of us to soar above the rim like Michael Jordan.

Environmental factors influence the expression of every trait in all organisms, including behavioral traits. Humans, however, with our uniquely developed capacity for personality development, social interactions, and learning, are behaviorally much more flexible than any other animal. So great is the role of learning in human personality development that there are those who maintain that the human mind is almost infinitely changeable and that there are no genetic differences among people in personality or intelligence. However, increasing evidence indicates that at least some mental traits are influenced by both environmental factors and inherited tendencies. Some people are probably genetically predisposed to the development of certain mental diseases, such as schizophrenia, although environmental events may determine whether symptoms develop. Some of these disorders have been traced to imbalances in important chemicals in the brain. Affected people probably bear alleles coding for defective versions of the enzymes that regulate the synthesis, breakdown, and use of these chemicals.

Intelligence, too, has both genetic and environmental components. Dozens of studies have compared IQ levels in people of varying degrees of relatedness. Even when they have been separated at birth and raised in different environments, identical twins make similar scores on IQ tests, although not as similar as twins who have been raised together. Brothers and sisters who are not twins differ more than twins but are still fairly similar, while parents and children are less so. Thus, the more genetically related two people are, the more similar their IQ scores, indicating very strongly that intelligence, or at least whatever it is that IQ tests measure, is partially genetic. At the same time, unrelated people who have been raised together as children (for example, adoptees) show more similarity on IQ tests than unrelated people reared apart. Therefore, it is fair to say that *both heredity and environment play major roles in the development of intelligence*, and probably other personality traits as well.

The roles of environment and inheritance in human behavior are explored further in Chapter 42.

Chromosomal Inheritance

Like other diploid organisms, humans possess two copies of most genes, one on each of two homologous chromosomes. This rule is not true, of course, for sex-linked genes in men, whose Y chromosome lacks most of the genes borne on the X chromosome. It is also not true for people who inherit abnormal numbers of chromosomes. In Chapter 12 we examined the intricate mechanisms of meiosis,

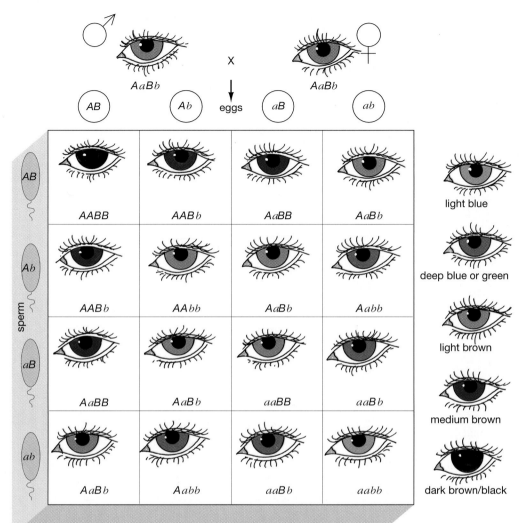

Figure 15-9   **Human eye color**

At least two separate genes, each with two incompletely dominant alleles, govern human eye color. A man and a woman, each heterozygous for both genes, could have children with five different eye colors, ranging from light blue (no dominant alleles) through light brown (two dominants) to almost black (all four alleles dominant).

which act to ensure that each sperm and each egg receive one homologous chromosome of each pair. Not surprisingly, this elaborate dance of the chromosomes occasionally misses a step, resulting in gametes that have too many or too few chromosomes. Such errors in meiosis, called **nondisjunction,** can affect the distribution of both sex chromosomes and autosomes. Most of the embryos that arise from the fusion of gametes with abnormal chromosome complements spontaneously abort, accounting for 20% to 50% of all miscarriages, but some survive.

Some Genetic Disorders Are Caused by Abnormal Numbers of Sex Chromosomes

Nondisjunction of the sex chromosomes occurs in both men and women. Nondisjunction in men produces sperm that are O (lacking any sex chromosome), XX, YY, or XY instead of the normal X or Y. In women, nondisjunction produces O or XX eggs instead of the normal X. When normal gametes fuse with these defective sperm or eggs, the zygotes have abnormal numbers of sex chromosomes (Table 15-1). The most common abnormalities are XO, XXX, XXY, and

XYY. (Genes on the X chromosome are absolutely essential to survival, and embryos with no X chromosome always spontaneously abort very early in development.)

Turner Syndrome (XO)

About one in every 5000 phenotypically female babies has only one X chromosome. At puberty, an XO female fails to menstruate or develop normal secondary sexual characteristics, such as breasts. These **Turner syndrome** women are sterile, usually short in stature, and often have folds of skin around their necks; under microscopic examination, their nuclei are shown to lack Barr bodies (see Fig. 11-17). Mentally, they are usually normal, except that they are frequently weak in mathematics and spatial perception. The differences between XO and XX women suggest that the theory of X chromosome inactivation in females (Chapter 11), which holds that in a normal female one X chromosome is condensed and "inactive," is oversimplified. Some genes on the inactivated X chromosome must be functional in XX females, preventing the Turner syndrome traits.

TABLE 15-1 ❖ Effects of Nondisjunction of the Sex Chromosomes during Meiosis

Nondisjunction in Father

Sex Chromosomes of Defective Sperm	Sex Chromosomes of Normal Egg	Sex Chromosomes of Offspring	Phenotype
O	X	XO	Female
XX	X	XXX	Female
YY	X	XYY	Male
XY	X	XXY	Male

Nondisjunction in Mother

Sex Chromosomes of Normal Sperm	Sex Chromosomes of Defective Egg	Sex Chromosomes of Offspring	Phenotype
X	O	XO	Female
Y	O	YO	Dies as embryo
X	XX	XXX	Female
Y	XX	XXY	Male

Trisomy X (XXX)

About 1 in 1000 women have three copies (trisomy) of X chromosomes. These women usually have no detectable defects at all, except for a relatively high incidence of below-normal intelligence. Unlike women with Turner syndrome, **trisomy X** women are fertile, and, interestingly enough, almost always bear normal XX and XY children. Some mechanism, presently unknown, must operate during meiosis to prevent the extra X chromosome from being included in the egg.

Klinefelter Syndrome (XXY)

About one male in 1000 is born with two X and one Y chromosomes. At puberty, these men show mixed secondary sexual characteristics, including partial breast development, broadening of the hips, and small testes. Men with **Klinefelter syndrome** are always sterile but usually not impotent. As is common in people with abnormal chromosome numbers, XXY males have an increased incidence of mental deficiency; in fact, about 1% of all people institutionalized for mental retardation are XXY males.

XYY Males

The last common type of sex chromosome abnormality is XYY, occurring in about one male in 1000. You might expect that having an extra Y chromosome, which presumably has few genes, would not make very much difference, and in most cases it doesn't seem to. However, XYY males may be affected in two ways: below-average intelligence and above-average height (about two-thirds of XYY males are over 6 feet tall, compared with the average male height of 5'9"). There is some debate about whether XYY males are geneti-

cally predisposed to violence. For instance, several studies have shown that a higher-than-expected percentage of men in prison are XYY. In several countries men accused of murder attempted to use their XYY constitution as a defense, like the insanity plea. They were not acquitted. The juries were probably right: Only a tiny percentage of XYY males ever commit any sort of crime, so an extra Y chromosome certainly doesn't force anyone into a life of violence.

In Humans, the Y Chromosome Determines Maleness

Studies of men and women with abnormal numbers of sex chromosomes lead to the inescapable conclusion that the Y chromosome determines maleness in humans. Simply having only one X chromosome, as both XY males and XO females do, does not automatically lead to maleness. In most respects, including external genitalia, XO individuals are clearly female. On the other hand, having a Y chromosome produces the male phenotype, no matter how many X chromosomes are present. Even the rare XXXY or XXXXY person is male. In computer terminology, femaleness is the "default" condition for human sex; explicit instructions encoded on the Y chromosome are required to produce a male. In fact, geneticists have found a specific gene on the Y chromosome that causes both the development of male sex organs and the regression of female sex organs during fetal development.

Some Genetic Disorders Are Caused by Abnormal Numbers of Autosomes

Nondisjunction of the autosomes may also occur, producing eggs or sperm with a missing autosome or two copies of an autosome. Fusion with a normal gamete (one

(a)

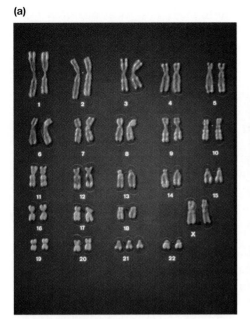

(b)

Figure 15-10   Trisomy 21, or Down syndrome

(a) Karyotype from a Down syndrome child reveals three copies of chromosome 21. **(b)** Children with abnormal numbers of chromosomes are almost always both physically and mentally defective, in ways that are characteristic of the particular chromosome abnormality. These girls have the typical relaxed mouth and distinctively shaped eyes usually seen in cases of Down syndrome.

copy of each autosome) leads to an embryo with either one or three copies of the affected autosome. With only one copy of any of the autosomes, the embryo aborts so early in development that the woman never knows she was pregnant at all. The presence of three copies of an autosome (trisomy) usually also causes a spontaneous abortion, but often later in pregnancy. Sometimes, however, trisomic babies are born, especially those with three copies of chromosomes 13, 18, or 21. Of these, trisomy 21 is the most common.

Trisomy 21, or Down Syndrome

In about one of every 900 births, the child has an extra copy of the twenty-first chromosome. Children with **trisomy 21**, also called **Down syndrome**, have several distinctive physical characteristics, including weak muscle tone, a small mouth held partially open because it cannot accommodate the tongue, and distinctively shaped eyelids (Fig. 15-10). Much more serious defects include low resistance to infectious diseases, heart malformations, and mental retardation so severe that only about 1 in 25 ever learns to read and only 1 in 50 learns to write.

The frequency of nondisjunction is influenced by the age of the parents, especially the mother (Fig. 15-11). As we will see in Chapter 39, meiosis begins in a woman's ovaries while she is still a fetus in her mother's womb, but it is suspended during prophase of meiosis I. When

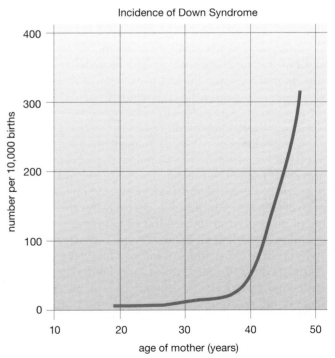

Figure 15-11   Down syndrome frequency increases with maternal age

This graph shows that nondisjunction of chromosome 21, producing children with Down syndrome, greatly increases after the mother reaches age 35. Geneticists have found that this relation is true even for women whose husbands are much younger.

a woman matures, meiosis resumes, a few cells at a time, during the monthly menstrual cycle. Therefore, a 40-year-old woman produces eggs that have been in meiosis for 40 years. Errors in chromosome distribution are more likely in old cells, perhaps just from aging itself, or perhaps from exposure to radiation, toxic chemicals, or viruses. In men, new sperm-producing cells are constantly being produced in the testes, and meiosis takes only a couple of weeks. Therefore, nondisjunction does not increase with age as rapidly in men as in women. Nevertheless, nondisjunction in sperm accounts for about 25% of the cases of Down syndrome, and there is a small age effect.

In the past two decades, it has become increasingly common for couples to delay having children, so that both husband and wife can establish careers. As a result, many women bear children when they are in their late 30s or early 40s, and in general they have husbands of similar age. Delaying pregnancy inevitably leads to more trisomic fetuses than if the same couples had reproduced earlier. Fetal trisomy can be diagnosed by examining the chromosomes of fetal cells collected through amniocentesis or chorionic villus sampling. Some couples may then choose a therapeutic abortion to prevent the birth of a trisomic child.

The Human Genome Project

Beginning with a few small meetings in 1985 and 1986, and now racing along with multimillion dollar budgets, the Human Genome Project is probably science's most ambitious response to the command of the ancient Oracle at Delphi: "Know thyself." The answers to a host of questions about humankind reside in the nucleotide sequences of our genes. Further, humanity is heir to about 4000 genetic disorders. If we knew the nucleotide sequence of the entire human genome, then, in theory at least, we could devise rapid screening and prenatal diagnosis of many genetic disorders, design better therapies for many diseases, and understand much more about the workings of the human brain.

Ever since DNA sequencing was first developed by Alan Maxam, Walter Gilbert, and Frederick Sanger, molecular geneticists have pondered the potential for sequencing the human genome, but the time, effort, and cost seemed daunting—$3 billion was the estimate in 1986. Recent developments in RFLP analysis, automated DNA sequencing, and polymerase chain reaction (PCR) technologies, however, have combined to make the project now seem quite reasonable. Although many meetings and much research went on before, the Human Genome Project officially got under way in 1988.

The project has many ultimate goals, but there are three of foremost importance:

1. Produce a genetic map. The locations of at least 3000 genes or other markers (such as restriction enzyme cut sites) will be sought. Remember, linkage maps locate genes by the relative frequency of crossing over (see Chapter 13). Linkage maps can be used in pedigree analysis to help locate other genes that are known only from phenotypic effects. Geneticists expect to complete this part of the project in 1995.

2. Generate a physical map. Each human chromosome will be cut apart with restriction enzymes. Unique DNA nucleotide sequences will be sought in each piece, with the unique sequences no farther apart than about 100,000 nucleotides. These unique nucleotide sequences can then be used as starting points for biochemically "walking" along the chromosomes, sequencing all the DNA or looking for previously unmapped genes.

3. Sequence the human genome. A complete set of haploid human chromosomes contains about 6 billion nucleotides (3 billion pairs on two complementary strands). This is an ambitious and expensive undertaking, but advocates point out that, if the entire genome sequence were known, and someone later pinpointed the gene locus of a genetic disorder, then one could obtain the gene sequence merely by looking it up in a computer. However, even with $100 million per year devoted to sequencing, it will probably be at least 10 more years before the entire genome is sequenced.

There is, of course, no one human genome—yours is different from everyone else's. By definition, the nucleotide sequence for a defective allele differs from that of the normal allele, and there is considerable variety among normal, functioning alleles, too. Therefore, at least for the DNA sequences that code for genes, it will be important to sequence the DNA from a fairly large sample of people.

The Human Genome Project has its detractors, mostly because of its cost. If the overall science budget remains constant, then the $50 million to $100 million slated to be spent each year on the Human Genome Project must mean decreased funding for other projects. Supporters, however, point to increased understanding of normal functioning, human evolution, and human genetic disorders that the project would produce. Certainly, the Human Genome Project has brought together many of the best molecular geneticists and is fostering cooperation in a field in which fierce competition is the norm.

Finally, the human genome is not the only one of interest. Plans are already being made to sequence the genomes of the bacterium *Escherichia coli*, the fruit fly *Drosophila melanogaster*, the roundworm *Caenorhabditis elegans*, the domestic dog, and at least one plant. These sequences would be used in fields as diverse as microbiology, developmental biology, veterinary medicine, and agriculture.

Medical Genetics Poses New Medical and Ethical Dilemmas

Genetics has opened up new vistas of understanding human nature. Analyses of family pedigrees, biochemical tests, amniocentesis, and recombinant DNA technologies have enabled geneticists to identify prospective parents who are carriers for some disorders, such as sickle-cell anemia and Tay-Sachs disease, and to detect fetuses who are homozygous for these and other disorders. These new powers demand a new set of decisions, both ethical and economic, by individuals, physicians, and society. Let us examine these new choices in one specific case, phenylketonuria.

Phenylketonuria Illustrates a Biomedical Dilemma

As you learned in Chapter 11, people with **phenylketonuria (PKU)** are homozygous for a recessive allele coding for a defective version of the enzyme that converts the amino acid phenylalanine to tyrosine (see Fig. 11-1). As is usual with recessive diseases, most homozygous children result from the mating of two heterozygous (and therefore phenotypically normal) parents. The level of phenylalanine in the developing fetus is regulated by its pregnant mother, so homozygous recessive infants are born with normal brain development. Once on their own, however, the infants' phenylalanine levels increase rapidly, damaging the brain. Fortunately, homozygous infants can be detected as early as 4 or 5 days after birth by a simple blood test that costs just a few dollars. Treatment consists of maintaining these children on a diet low in phenylalanine for at least the first 5 or 6 years of life, when most brain development occurs. Most naturally occurring proteins contain phenylalanine, so the low-phenylalanine diet is quite expensive and inconvenient.

This is not the end of the story, however. Women with PKU who were treated in infancy have normal mental development, but they are still deficient in phenylalanine metabolism and therefore have an elevated concentration of phenylalanine in their blood. Children of such women, fathered by homozygous normal men, will be heterozygous and should be phenotypically normal. The problem is that the phenylalanine levels of an otherwise normal fetus are raised by the high levels of its phenylketonuric mother. High phenylalanine levels during pregnancy, even just the first few months, cause mental retardation. The only protection for the fetus is if the mother adheres to the low-phenylalanine diet.

Thus, treating phenylketonuria in female infants brought with it new responsibilities. Many women now of childbearing age went off their low-phenylalanine diets years ago. Many have since forgotten that they have PKU or, not understanding genetics, think that they *used* to have PKU but have now outgrown it. Others planned to resume the low-phenylalanine diet when they wanted to become pregnant but became pregnant unintentionally. Therefore, the best medical advice for phenylketonuric women is to stay on the low-phenylalanine diet until menopause.

Knowledge of Human Genetics Poses Ethical Dilemmas

Insofar as we cannot cure most inherited diseases, or foresee a cure in the near future, what are the responsibilities and options open to individuals and societies? A hundred years ago, when no one understood the nature of inherited diseases, the choices were limited and the responsibilities few. A prospective parent could not know whether he or she carried an inherited defect, could not predict the likelihood of having an affected child, and could not tell whether a child was affected until the child was born or perhaps was even several years old.

The situation has changed dramatically. Today, many people *do* know that they carry a seriously defective gene, and genetic counselors can predict their chances of having an affected child. Should such an individual refrain from reproduction? In some cases, such as Tay-Sachs disease and the chromosomal abnormalities, prenatal diagnosis can detect the inherited defect in a fetus. Should the fetus be aborted? Should screening tests, at least for infants, be encouraged or required, as is now the case for PKU and, in some states, sickle-cell anemia? Would such a requirement constitute governmental interference in the rights of the individual? What about screening tests for prospective parents, at least in high-risk groups? Should these tests be required?

What, if any, are the rights and responsibilities of society in these decisions? In many Western countries, society pays most or all of the costs of medical care for affected children, often running to hundreds of thousands of dollars each. What, if any, are the obligations of society to children who might be born with genetic defects? Suppose a phenylketonuric woman goes off the low-phenylalanine diet against the advice of her physician and then bears a child. Would she be guilty of child abuse?

Some people argue that the human species is becoming genetically "loaded" with defective alleles: Whereas in earlier times people who had the alleles for diabetes, sickle-cell anemia, and many other genetically influenced diseases would have died, modern medicine now allows them to survive and reproduce. As a result, some believe that carriers of alleles for serious, incurable diseases should not reproduce, to eliminate not only homozygous recessives who are afflicted with the disease but also future carriers. Eventually, when we become knowledgeable enough, we may find that almost *all* of us carry recessive alleles for some inherited disease or other, yet few would argue that the majority of the population is morally obligated to remain childless. In a free society, probably the best solution is to give people the most accurate information possible about their genetic constitution and that of their future children. The choices, whatever their consequences for society as a whole, almost certainly will remain with the individual.

Collecting Fetal Samples and Analyzing Them Using Biotechnology

Prenatal diagnosis of a variety of genetic disorders, including sickle-cell anemia, Tay-Sachs disease, and Down syndrome, requires samples of fetal cells or chemicals produced by the fetus. Presently, there are two main techniques used to obtain these samples: amniocentesis and chorionic villus sampling. Once samples are collected, several tests, including some using recombinant DNA techniques, can be performed that allow prenatal diagnosis of many genetic disorders.

Amniocentesis

The human fetus, like all animal embryos, develops in a watery environment. A waterproof membrane called the amnion (see Chapter 40) surrounds the fetus and holds the amniotic fluid. As it develops, the fetus sheds some of its own cells into the amniotic fluid. When a fetus is 16 weeks or older, enough amniotic fluid has accumulated and there is enough space between the fetus and the amniotic membrane for the amniotic fluid to be collected by a procedure called amniocentesis. A physician determines the position of the fetus by ultrasound scanning (Fig. E15-2a) and inserts a sterilized needle through the abdominal wall of the pregnant woman and into the amniotic fluid (Fig. E15-2b). Ten to 20 milliliters of fluid is withdrawn. Biochemical analysis may be performed on the fluid immediately. However, there are very few cells in the fluid sample. For many analyses, such as karyotyping for Down syndrome, the cells must first be grown in culture. The cells multiply in culture, and after a week or two there are usually enough cells for karyotyping or other analyses.

Chorionic Villus Sampling

This is a newer procedure. The chorion is a membrane, produced by the fetus, that becomes part of the placenta (see Chapter 40). The chorion produces many small projections, called villi; the loss of a few villi seems to cause no harm. In chorionic villus sampling, a physician inserts a small tube into the uterus through the vagina and suctions off a few villi for analysis (see Fig. E15-2b).

The Two Techniques Compared

Chorionic villus sampling has two major advantages over amniocentesis. First, it can be done much earlier in pregnancy, perhaps as early as the eighth week. Early testing is especially helpful if the woman is contemplating a therapeutic abortion in case the fetus suffers from a major defect. Second, the sample contains a much higher concentration of fetal cells than amnio-

(a)

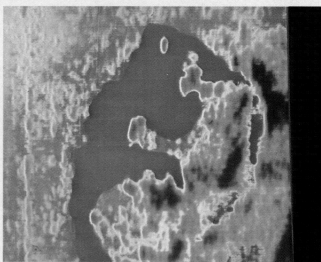

(b)

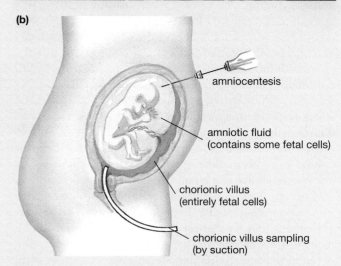

amniocentesis

amniotic fluid
(contains some fetal cells)

chorionic villus
(entirely fetal cells)

chorionic villus sampling
(by suction)

Figure E15-2   Prenatal tissue sampling techniques
(a) Ultrasound imaging reveals the position of the fetus. A needle can then be inserted into the amniotic sac to withdraw fluid without touching the fetus. **(b)** Two methods of obtaining fetal tissue samples.

centesis can obtain, so analyses can be performed immediately rather than waiting for a week or two. However, chorionic cells seem to be more likely than amniotic fluid to have abnormal numbers of chromosomes, even in a normal fetus, which complicates karyotyping. In most large metropolitan areas, both amniocentesis and chorionic villus sampling are available, and each has advantages and drawbacks.

Analyzing the Samples

Several analyses can be performed on the fetal cells or on the amniotic fluid. Biochemical analysis is used to determine the concentration of chemicals in the amniotic fluid. For example, Tay-Sachs disease and many other metabolic disorders can be detected by the low concentration of the enzymes that normally catalyze specific metabolic pathways, or by the accumulation of precursors or by-products. Analysis of the chromosomes of the fetal cells can show if all the chromosomes are present, if there are too many or too few of some, and if there are structural abnormalities of any of the chromosomes.

Analysis of the DNA of fetal cells using recombinant DNA techniques can detect some defective alleles, such as the Tay-Sachs and sickle-cell anemia alleles. Just a few years ago, prenatal diagnosis of sickle-cell anemia or other genetic disorders required thousands or millions of copies of the DNA. Therefore, fetal cells often had to be grown in culture for as long as 2 weeks before they had multiplied enough to be used. For sickle-cell anemia, the development of the polymerase chain reaction (PCR; see Chapter 14) has eliminated the wait.

As you may recall, PCR can produce virtually unlimited copies of a specific gene, starting with as little as one copy. The second step in prenatal diagnosis of sickle-cell anemia, therefore, is to extract the DNA from a few cells and to amplify the hemoglobin genes (both normal and sickle-cell versions) with PCR. This process takes only a few hours. The third step is to cut the PCR-amplified DNA with special restriction enzymes and sort the DNA according to size by gel electrophoresis (see Chapter 14). The hemoglobin genotype of the fetus can then easily be determined.

If the infant is homozygous for the sickle-cell allele, some therapeutic measures can be taken. In particular, regular doses of penicillin greatly reduce bacterial infections that otherwise kill about 15% of homozygous children. Further, knowing that a child has the disorder ensures correct diagnosis and rapid treatment, should "sickling crises" occur.

❖ S U M M A R Y   O F   K E Y   C O N C E P T S

Methods in Human Genetics

The genetics of humans is similar to the genetics of other animals, except that experimental crosses are not feasible. Analysis of family pedigrees and, more recently, molecular genetic techniques must be used to determine the mode of inheritance of human traits.

Single-Gene Inheritance

Many genetic disorders are inherited as recessives; therefore, only homozygous recessive persons show symptoms of the disease, whereas heterozygotes are phenotypically normal and usually cannot be detected by casual observation. Heterozygotes are called carriers, because the offspring of two carriers may be homozygous recessive and show the trait. Recessive genetic disorders include Tay-Sachs disease, sickle-cell anemia, and phenylketonuria.

Many normal traits, and a few diseases, are inherited as simple dominants. Both heterozygous and homozygous dominant individuals show the trait.

The sex chromosomes are alike in women (XX) but not alike in men (XY). The Y chromosome bears few genes other than those determining maleness. Therefore, women show normal dominant-recessive relationships among alleles of X chromosome genes, whereas men phenotypically display whatever allele they carry on their single X chromosome, a phenomenon called sex-linked, or X-linked, inheritance. Sex-linked conditions include red-green color blindness and hemophilia. Sex hormones help to determine the expression of sex-influenced traits. The majority of sex-influenced defects are more common in males than in females and include baldness and gout.

Complex Inheritance

Most human traits are polygenic—that is, they are influenced by the action of many genes. Examples include the color of skin and eyes and probably many personality traits. The environment affects the phenotypic expression of all genes in all organisms. This rule is particularly apparent in human beings, especially in intellectual and personality traits.

Chromosomal Inheritance

Gender in humans is determined by the presence of a Y chromosome. Individuals with one or more Y chromosomes are phenotypically male; individuals with no Y chromosome are phenotypically female. People with abnormal numbers of sex chromosomes often have mental and physical deficiencies. The most common defect is below-normal intelligence.

Abnormal numbers of autosomes usually lead to spontaneous abortion early in pregnancy. In rare instances, the fetus may survive to birth, but severe mental and physical deficiencies are always found. The likelihood of abnormal numbers of chromosomes increases with increasing age of the mother, and, to a lesser extent, the father.

The Human Genome Project
The largest single project in the history of biology, the Human Genome Project proposes to produce (1) a detailed linkage map; (2) a precise physical map; and (3) the complete nucleotide sequence of the "average" human genome.

✳ KEY TERMS

amniocentesis   p. 284
carrier   p. 283
chorionic villus sampling   p. 284
Down syndrome   p. 293
hemophilia   p. 287
Klinefelter syndrome   p. 292

nondisjunction   p. 291
pedigree   p. 281
phenylketonuria (PKU)   p. 295
sex-influenced trait   p. 287
sex-linked inheritance   p. 287
sickle-cell anemia   p. 284

Tay-Sachs disease   p. 284
trisomy 21   p. 293
trisomy X   p. 292
Turner syndrome   p. 287
X-linked inheritance   p. 287

✳ THINKING THROUGH THE CONCEPTS

Multiple Choice

1. Which one of the following is a lethal mutation that does not express itself until late in life in men and women?
   a. Down syndrome
   b. hemophilia
   c. Huntington's disease
   d. Tay Sachs disease
   e. sickle cell anemia

2. Which is NOT true of sickle cell anemia?
   a. It is most common in African Americans.
   b. It involves a one–amino acid change in hemoglobin.
   c. It involves red blood cells.
   d. It is lethal in heterozygotes since it is dominant.
   e. It confers resistance to malaria.

3. Which statement is NOT true?
   a. The frequency of some chromosome abnormalities goes up with maternal age.
   b. XYY individuals originate from two sperm cells fertilizing one ovum.
   c. Sperm are created throughout life.
   d. Meiosis to produce human eggs begins before birth.
   e. Chromosomal abnormalities are different from genetic abnormalities in that the former are usually not passed to the next generation.

4. Sex-linked disorders such as color blindness and hemophilia are
   a. on the X chromosome
   b. on the autosome
   c. on the Y chromosome
   d. the same as sex-influenced traits such as baldness; they are simply more likely to be expressed in men
   e. expressed only when two chromosomes are homozygous recessive

5. The improper disjoining (nondisjunction) of chromosome 21 in human beings leads to
   a. Tay-Sachs disease
   b. crossing over
   c. Klinefelter syndrome
   d. Turner syndrome
   e. Down syndrome

6. In human beings, red color blindness is modeled as a sex-linked recessive allele. A couple has a red-blind daughter. Which of the following statements is the only one which MUST be correct on the basis of this information?
   a. The father is red blind.
   b. The mother is red blind.
   c. Both father and mother are red blind.
   d. The daughter inherited all red-blind genes from the mother.
   e. Half of the mother's sons would be red blind.

Review Questions
(The answers to questions 1 and 2 are given at the end of the question list.)

1. If the frequency of heterozygous carriers for Tay-Sachs disease is 1 in 30 American Jews, why is the frequency of homozygous recessive babies of Jewish parents 1 in 3600?

2. If one parent of a couple has Huntington's disease (assume that this parent is heterozygous), calculate the fraction of their children that would be expected to develop the disease. What if both parents were heterozygous?

3. Why are most genetic diseases inherited as recessives rather than dominants?

4. How is sex determined in humans? What is the evidence for this?

5. Define polygenic inheritance. Why could polygenic inheritance allow parents to produce offspring that are notably different in eye or skin color than either parent?

6. What is sex-influenced inheritance? What is the evidence that hormonal levels control the expression of sex-influenced traits?

7. Define nondisjunction, and describe the common syndromes caused by nondisjunction of sex chromosomes and autosomes.

8. Describe amniocentesis and chorionic villus sampling, including the advantages and disadvantages of each. What are their medical uses? What technique can be used in conjunction with amniocentesis in order to expedite production of a sufficient sample of DNA material from a gene for analysis?

9. Referring to the material in Chapter 11, what type of mutation is sickle-cell anemia?

✖ ANSWERS TO QUESTIONS 1 AND 2

1. The probability of two or more independent events occurring simultaneously is the product of their individual probabilities. The probability of the husband being heterozygous is 1/30; the probability that the wife is heterozygous is also 1/30; the probability that two heterozygotes will produce a homozygous recessive child is 1/4 (see Chapter 13). Multiplying $1/30 \times 1/30 \times 1/4$ gives 1/3600.

2. Let H = the Huntington's allele (dominant) and h = the normal allele (recessive). The first set of parents would be Hh and hh. Half of their offspring would be expected to inherit Huntington's disease. If both parents are heterozygous, then the cross is $Hh \times Hh$. We would expect the offspring to be 1/4 HH, 1/2 Hh, and 1/4 hh; therefore, 3/4 would develop the disease.

✖ APPLYING THE CONCEPTS

1. Eugenics is the term applied to the notion that the human condition might be improved by improving the human genome. Do you think there are both good and bad sides to eugenics? What examples can you think of to back up your stand? What would an out-and-out eugenicist think of the medical advances which have ameliorated the problems of PKU?

2. Think about some of the personal, religious, and medical economic issues related to prenatal counseling and diagnosis. Would you avoid having children if you knew that both you and your spouse were heterozygous for a recessive disorder such as Tay-Sachs which is fatal at an early age and may involve considerable suffering? What would you do if you or your spouse were pregnant and learned that your offspring, if born, would be homozygous for such a disorder? Is the situation qualitatively different for Down syndrome? (Down syndrome children have a short life expectancy [20 to 30 years] with mental retardation but have a generally pleasant disposition.) If you were heterozygous for Huntington's disease, would you want to avail yourself

of the medical diagnostic tests being made available through the most recent technologies to find out? If the answer is "no" and you were a man, what would you think about your wife's potential decision to have amniocentesis or chorionic villus sampling?

3. Now that you know a bit more about human genetics and polygenic inheritance, rethink your answer to the following question from Chapter 12: Farm animals with desirable traits are sometimes produced by removing eggs from the female and fertilizing them in a dish using sperm from a male with desirable traits. After the zygote has undergone several divisions, the individual cells can be separated, returned to the same or different females, and they will develop into genetically identical animals. Human sperm banks already carry sperm from men with particular traits that a woman may choose. Human *in vitro* fertilization (see the Scientific Inquiry box in Chapter 39) could theoretically be used to create identical twins, triplets, etc. Argue for or against such human "cloning." Has your answer changed? Why or why not?

✖ FOR MORE INFORMATION

Collins, F., and Galas, D. "A New Five-Year Plan for the U.S. Human Genome Project." *Science*, 262:43–46, 1993.

Horgan, J. "Eugenics Revisited." *Scientific American*, June 1993. Determining patterns of inheritance, let alone finding suspect genes, for behavioral disorders is notoriously difficult. What's more, it isn't at all clear what one should do with such data, if they become available.

McGue, M. "Nature-Nurture and Intelligence." *Nature, 340:* 507–508, 1989. Nature works via nurture in producing the outward manifestations of intelligence.

Pines, M. "In the Shadow of Huntington's." *Science*, May 1984; Grady, D. "The Ticking of a Time Bomb in the Genes." *Discover,* June 1987; Roberts, L., "Huntington's Gene: So Near, Yet So Far." *Science,* 247:624–627, 1990; and Morell, V. "Huntington's Gene Finally Found." *Science,* 260:28-30, 1993. The story of the scientific detective work involved in the diagnosis of Huntington's disease using recombinant DNA techniques, the social dilemma it created, the frustrations of searching for the gene, and finally finding it. Particularly poignant, because one of the lead investigators may herself be a victim.

Roberts, L. "To Test or Not to Test." *Science, 247*:17-19, 1990. Some of the alleles causing cystic fibrosis cannot be detected —should the entire childbearing population be screened for carriers for this devastating disease?

Sutherland, G. R., and Richards, R. I. "Dynamic Mutations." *American Scientist*, March–April 1994. Several genetic disorders, including Huntington's disease, are characterized by large numbers of repeated nucleotide triplets.

Wivel, N. A., and Walters, L. "Germ-line Gene Modification and Disease Prevention: Some Medical and Ethical Perspectives." *Science, 262*: 533–538, 1993. If we can fix a defective allele in an egg or zygote, should we?

N E T   W A T C H

On-line resources for this chapter are on the World Wide Web at:

http://www.prenhall.com/~audesirk (click on the <u>table of contents</u> link and then select Chapter 15).

Unit III
Evolution

Beautifully preserved large and small fossil ammonites, which are prehistoric mollusks

16 Principles of Evolution

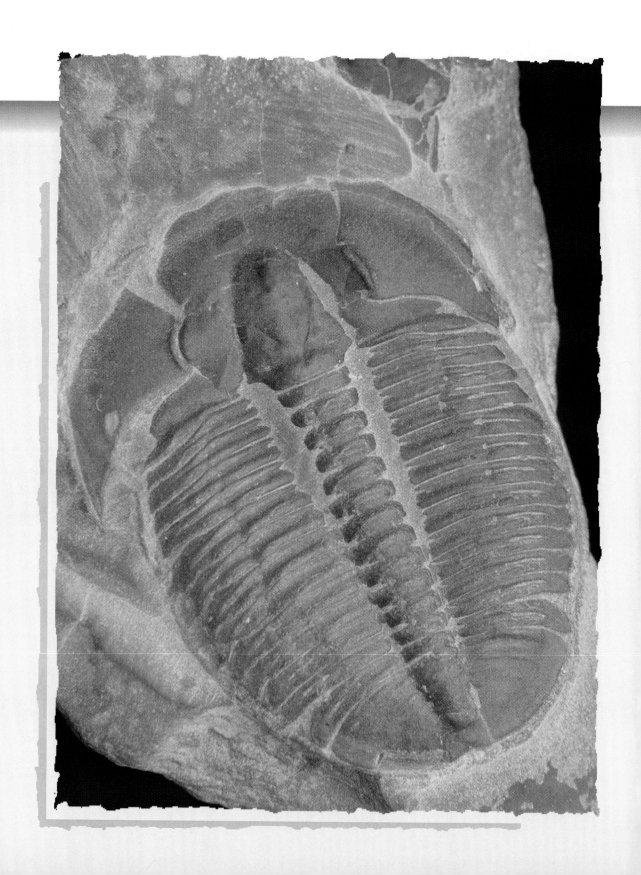

"When on board H.M.S. Beagle, as naturalist, I was much struck with…the distribution of the inhabitants of South America, and…the geological relations of the present to the past inhabitants of that continent. These facts seemed to me to throw some light on the origin of species—that mystery of mysteries, as it has been called by one of our greatest philosophers."

CHARLES DARWIN in *On the Origin of Species by Means of Natural Selection* (1859)

AT A GLANCE

The History of Evolutionary Thought
What Is a Species?
The Philosophies of Plato and Aristotle Influenced Early Christian Beliefs
Evidence Supporting Evolution Was Discovered Even before Darwin's Time
Charles Darwin's Voyage on the *Beagle* Sowed the Seeds for His Theory of Evolution
Darwin and Wallace Both Proposed That Evolution Occurs by Natural Selection
Evolutionary Theory Arises from Scientific Observations and Conclusions Based on Them

The Evidence for Evolution
The Fossil Record Provides Evidence of Evolutionary Change over Time
Comparative Anatomy Provides Structural Evidence for Evolution
Embryological Stages of Animals Can Provide Evidence for Common Ancestry
Modern Biochemical and Genetic Analyses Reveal Relatedness among Diverse Organisms
Artificial Selection Demonstrates That Organisms May Be Modified by Controlled Breeding
Evolution by Natural Selection Occurs Today

A Postscript by Charles Darwin

A fossil trilobite, an arthropod that was common 450 million years ago

The words above introduce what is perhaps the most important work in biology. In *On the Origin of Species*, Charles Darwin proposed that over thousands of millions of years, species arise from other, preexisting species through the process of "descent with modification," or **evolution**.

Before Darwin, how species originated remained the "mystery of mysteries" for a very simple reason. Over the time span of recorded human history, let alone the life of a single human being, no new species had been recognized (although undoubtedly many new species had appeared, especially plant species). It is quite difficult to determine how something happens if no one has ever observed it happening.

Nevertheless, one of the most striking features about our world is the remarkable variety of organisms inhabiting it. Why are there dozens of types of pine trees and scores of species of warblers? To explain the origin and diversity of life on Earth, nearly all peoples of the world historically turned to hypotheses of **creationism**. The most common of these hypotheses is that a supernatural being created each type of organism individually at the beginning of the world and that all modern organisms are essentially unchanged descendants of those ancestors.

Throughout history, however, scientists have sought natural, rather than supernatural, causes to explain natural events. As we pointed out in Chapter 1, one of the fundamental principles of science is that earthly phenomena are produced by natural, earthly causes that follow natural laws. Science cannot say whether or not divine power originally established those laws, but science firmly adheres to the principle that natural events have causes that arise from the operation of natural laws.

Therefore, throughout history, scientists have sought natural causes for the origin of species. But it was only in the nineteenth century that a truly coherent theory—evolution by descent with modification, driven by natural selection—was developed. This theory was published by two British naturalists, Charles Darwin and Alfred Russell Wallace, in 1858 and today still forms the foundation of our understanding of evolution. Darwin and Wallace did not work in a vacuum. Centuries of thought and observation preceded them and influenced their ideas. Before examining the principles of evolution, let us begin, then, with a brief survey of evolutionary thought.

The History of Evolutionary Thought

What Is a Species?

Before we can study the origin of species, we must first decide what a species is. Throughout most of human history, "species" was a poorly defined concept. When using the word *species*, most Europeans meant one of the originally created "kinds" referred to in the Bible. But because no one was present at the Creation to record the criteria of the Creator, one had to distinguish among species by visible differences in structure; in fact, the word *species* is Latin for "appearance." Clearly, warblers are different from eagles and ducks. But how do biologists distinguish among species of warblers? Today, biologists define a **species** as all the populations of organisms that are capable of interbreeding among themselves under natural conditions but that usually do not interbreed with members of other species. If interbreeding with another species does occur, the hybrid offspring are usually infertile or handicapped in some way (see Chapter 17). A **population** consists of all the individual members of a particular species within a given area.

Biologists have found that differences in appearance do not always mean that two populations belong to different species. For example, field guides published in the 1970s listed the Myrtle Warbler and Audubon's Warbler (Fig. 16-1) as distinct species; they differ in range and in the color of the throat feathers. More recently, scientists decided that they are, after all, merely local varieties of the same species. The reason: Where their ranges overlap, interbreeding between these warblers occurs, and the offspring are just as vigorous as the parents.

The Philosophies of Plato and Aristotle Influenced Early Christian Beliefs

Two principal lines of thought descended from the ancient Greeks to influence later Western ideas about the nature and origin of species. The philosophy of Plato (427–347 B.C.) rested on the foundation of the "ideal Form" that exists, perfect and forever unchanging, somewhere beyond Earth in the world of Forms. In this world view, each object on Earth, living or nonliving, rock, worm, or human, is a mere temporary reflection of its ideal Form. Plato's concept of unchanging Forms greatly influenced early Christian thought and came to be embodied in the idea that every species of living thing was created by God at the beginning of time. Although minor variations may occur among individual members of a given species, each species as a whole remains unchanged, very like a Platonic Form.

Plato's student Aristotle (384–322 B.C.), one of the first great naturalists, categorized all the living things that he encountered. Aristotle thought that all organisms fit into an orderly scheme, later called the *Scala Naturae*, or Ladder of Nature. The ladder stood, so to speak, upon nonliving matter; climbed rung by rung from fungi and mosses to higher

Figure 16-1 **Interbreeding blurs the distinction between species**

The Myrtle Warbler **(a)** and Audubon's Warbler **(b)** were formerly considered to be two separate species but are now considered to be merely local varieties of one widespread species.

plants, through primitive animals such as mollusks and insects; and finally culminated in human beings. Aristotle's ideas, even more than Plato's, were incorporated into Christian thought. The *Scala Naturae* was considered to be permanent and never changing: Each organism has its place on the ladder, ordained by God during creation.

Evidence Supporting Evolution Was Discovered Even before Darwin's Time

Creationism, the idea that each species was created individually by God and never changed thereafter, reigned unchallenged for nearly 2000 years. By the eighteenth century, however, new evidence arose that challenged this static view of creation.

Exploration of New Lands Revealed a Staggering Diversity of Life

As European naturalists explored the newly discovered lands of Africa, Asia, and America, they found that the diversity of living things was much greater than anyone had suspected. Some of these exotic species closely resembled one another yet also displayed variations in characteristics. This unpredicted expansion of information led some naturalists to consider that perhaps species could change after all and that some of the similar species might have developed from a common ancestor.

Fossils in Rocks Resembled Parts of Living Organims

At the same time, excavations for roads, mines, and canals revealed that rocks often occur in layers (Fig. 16-2). Sometimes, a few strangely shaped rocks were found embedded within one of these layers. These rocks, called **fossils** (from the Latin, meaning "dug up"), often resembled parts of living organisms. At first, fossils were thought to be ordinary rocks that wind, water, or people had worked into lifelike forms. As more and more fossils were discovered, however, it became obvious that they were in fact plants and animals that had died long ago and been changed into or preserved in rock (Fig. 16-3). After studying fossils carefully, William Smith (1769–1839) realized that certain fossils were always found in the same layers of rock. Further, the organization of fossils and rock layers was consistent: Fossil type A could always be found in a rock layer resting atop an older layer containing fossil B, which in turn rested atop a still older layer containing fossil C, and so on.

Fossil remains also showed a remarkable progression in form. Fossils found in the lowest (and therefore oldest) rock layers were usually very different from modern forms, with a gradual advancement to greater resemblance to

Figure 16-2   The Grand Canyon of the Colorado River
Layer upon layer of sedimentary rocks form the walls of the canyon, exposed in cliffs and mesas. The Grand Canyon strata cover more than a billion years of evolutionary history.

modern species in younger rocks, as if there were a *Scala Naturae* stretching back in time. Many of these fossils were the remains of plants and animals that no longer lived on Earth (Fig. 16-4). Putting these facts together, scientists came to the inescapable conclusion that different types of organisms had lived at various times in the past.

But what did this newfound richness of organisms, both living and extinct, mean? Was each organism produced by a separate act of creation? If so, why? And why bother to create so many, letting thousands become extinct? Georges Louis LeClerc, known by his title Comte de Buffon (1707–1788), suggested that perhaps the original creation provided a relatively small number of founding species and that some of the modern species had been "conceived by Nature and produced by Time"—that is, they had evolved through natural processes. Most people were not convinced. First, Buffon could not provide any mechanism whereby Nature could "conceive" new species. Second, no one thought that there was Time enough for their "production."

Geology Provided Evidence That Earth Is Exceedingly Old

In the early 1700s, few scientists suspected that Earth could be more than a few thousand years old. Counting generations in the Bible's Old Testament, for example, yields a maximum age of 4000 to 6000 years. By reading the descriptions of plants and animals from ancient writers such as Aristotle, it was clear that wolves, deer, lions, and other European organisms had not changed in over 2000 years. How, then, could whole new species arise, if Earth was created only a couple of thousand years before Aristotle?

To account for a multitude of species, both extinct and modern, while preserving creationism, the French paleontologist (a scientist who studies fossils) Georges Cuvier (1769–1832) proposed the theory of **catastrophism.** Cuvier hypothesized that a vast supply of species was created in the beginning. Successive catastrophes (such as the Flood described in the Bible) produced the layers of rock and destroyed many species, fossilizing some of their remains in the process. The reduced flora and fauna of the modern world, he theorized, are the species that survived the catastrophes. However, if modern *species* have survived from an original creation, then many *individuals* of those species should have died in the ancient catastrophes. Surely some of them would have been fossilized, and even the lowest and oldest rock layers should contain at least a few fossils of present-day species. Unfortunately for Cuvier's hypothesis, they do not. Louis Agassiz (1807–1873) proposed that there was a new creation after each catastrophe and that modern species result from the most recent creation. The fossil record forced Agassiz to postulate at least 50 separate catastrophes and creations!

Alternatively, perhaps Earth *is* old enough to allow for the production of new species. Geologists James Hutton (1726–1797) and Charles Lyell (1797–1875) contemplated the forces of wind, water, earthquakes, and volcanism.

(a) The bipedal carnivore *Allosaurus* confronts a *Stegosaurus*. A sudden volcanic eruption buries both under lava and ash.

(b) Minerals from the surrounding rocks seep into their skeletons.

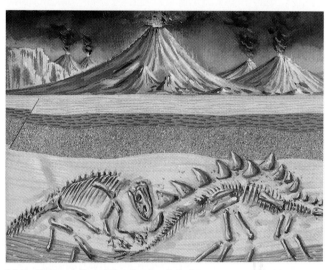

(c) Further eruptions bury the fossil skeletons more deeply, where they remain for tens of millions of years.

(d) Finally, erosion cuts through the overlying rock layers and exposes the fossils, where they may be found by paleontologists.

Figure 16-3   **Stages in fossilization**

Fossilization occurs when living organisms are buried beneath mud, silt, sand, or volcanic ash. This example might have occurred in Montana or Colorado.

They concluded that there was no need to invoke catastrophes to explain the findings of geology. Do not rivers in flood lay down layers of sediment? Do not lava flows produce layers of basalt? Why, then, should we assume that layers of rock are evidence of anything but ordinary natural processes, occurring repeatedly over long periods of time? This concept, called **uniformitarianism,** satisfies a scientific axiom often called Occam's Razor: The simplest explanation that fits the facts is probably correct. The implications of uniformitarianism were profound. If slow natural processes alone are enough to produce layers of rock thousands of feet thick, then Earth must be old indeed, many millions of years old. Hutton and Lyell, in fact, concluded that Earth was eternal: "No Vestige of a Beginning, no Prospect of an End," in Hutton's words. (Modern geologists estimate that Earth is about 4.5 billion years old; see Chapter 19, "Scientific Inquiry: Radiometric Dating Traces Fossils through Time.") Thus, Hutton and Lyell provided the time for evolution, but there was still no convincing mechanism.

Early Biologists Proposed Mechanisms for Evolution

One of the first to propose a mechanism for evolution was the French biologist Jean Baptiste Lamarck (1744–1829). Lamarck was impressed by the progression of forms in the fossil record. Older fossils tend to be simpler, while younger fossils are more complex and more like existing organisms. In 1801, Lamarck hypothesized that organisms evolved

(a)   **(b)**   **(c)**

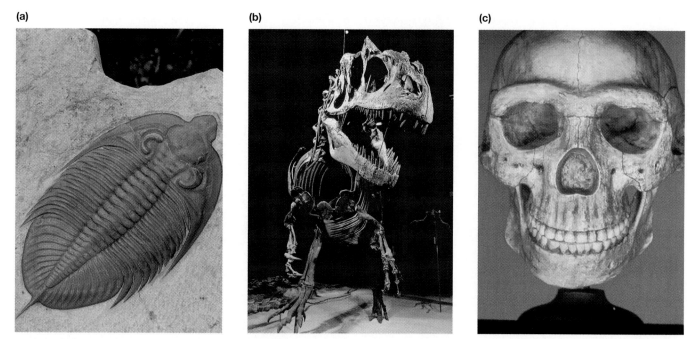

Figure 16-4   **Fossils of extinct animals**

(a) Trilobites were early arthropods (relatives of spiders and crabs) that flourished for scores of millions of years. Dozens of species of fossil trilobites are found in many parts of the world, but there have been no living trilobites for hundreds of millions of years. **(b)** *Allosaurus*, a predatory dinosaur similar to the more familiar *Tyrannosaurus rex*, preyed on large and formidable prey, including *Stegosaurus* (see Fig. 16-3). **(c)** The skull of *Australopithicus boisei*, an extinct relative of human beings. Clearly unlike either today's humans or apes, this species probably has no modern descendants.

through the **inheritance of acquired characteristics:** Living organisms can modify their bodies through use or disuse of parts, and these modifications can be inherited by their offspring. (As it turns out, the first part of his hypothesis is correct to some extent, but the second part is not.) Why would organisms modify their bodies? Lamarck proposed that all organisms possess an innate drive for perfection, an urge to climb the Ladder of Nature. In his best-known example, Lamarck hypothesized that ancestral giraffes stretched their necks to feed on leaves growing high up in trees, and as a result their necks became slightly longer. Their offspring inherited these longer necks and stretched even farther to reach still higher leaves. Eventually, this process might produce modern giraffes with very long necks indeed (Fig. 16-5).

Today, Lamarck's theory seems silly: The fact that a prospective father pumps iron doesn't mean that his children will look like Arnold Schwarzenegger. Remember, though, that in Lamarck's day no one had the foggiest idea how inheritance worked. Gregor Mendel wouldn't even be born for another 20 years, and the incorporation of his principles of inheritance into mainstream biology didn't happen until the early twentieth century.

Although Lamarck's theory fell by the wayside, by the mid-1800s some biologists were beginning to realize that the fossil record and the similarities between fossil forms and modern species could be best explained if present-day species had evolved from preexisting ones. The question

remained: *But how?* In 1858, Charles Darwin and Alfred Russell Wallace independently provided convincing evidence that the driving force behind evolutionary change was natural selection.

Charles Darwin's Voyage on the *Beagle* Sowed the Seeds for His Theory of Evolution

Like many modern students, Charles Darwin excelled only in subjects that intrigued him. Although his father was a physician, Darwin was uninterested in medicine and unable to stand the sight of surgery. He eventually obtained a degree in theology from Cambridge University, although theology too was of minor interest to him. What he really liked to do was to tramp over the hills, observing plants and animals, collecting new specimens, scrutinizing their structures, and categorizing them. As he himself later put it, "I was a born naturalist." Fortunately for Darwin (and for the development of biology), some of his professors at Cambridge, most notably the botanist John Henslow, had similar interests. So constant was their companionship in field studies that Darwin was sometimes called "the man who walks with Henslow."

In 1831, when Darwin was only 22 years old (Fig. 16-6), the British government sent Her Majesty's Ship *Beagle* on a 5-year surveying expedition that took them first along the coastline of South America and then around the world. As was common on such expeditions, the *Beagle* would carry along a naturalist to observe and collect geological

giraffe
ancestor

time

modern
giraffe

Figure 16-5   **Lamarck's hypothesis for the evolution of the giraffe**

Time flows from left to right. Short-necked ancestral giraffes stripped the leaves from the lower branches of trees and stretched and strained to reach leaves higher up, making their necks slightly longer in the process. Their offspring inherited these longer necks, did more stretching in their turn, and passed on still longer necks to their offspring. The modern giraffe on the right would be the outcome of this continual striving to feed on tree leaves high off the ground.

and biological specimens encountered along the route. Thanks to Henslow's recommendation to the captain, Robert FitzRoy, Darwin was offered the position of naturalist aboard the *Beagle*. (When they first met, FitzRoy almost rejected Darwin for the post because of the shape of Darwin's nose! Apparently FitzRoy felt that a person's personality could be predicted by the shape of the facial features, and Darwin's nose failed to measure up, as it were. Later, Darwin expressed the opinion that FitzRoy was "afterwards well satisfied that my nose had spoken falsely.")

The *Beagle* sailed to South America, making many stops along the coast. Here Darwin observed the plants and animals of the tropics and was stunned by the diversity of species compared with Europe. Although he boarded the *Beagle* convinced of the permanence of species, his experiences soon led him to doubt this. He discovered a snake with rudimentary hind limbs, calling it "the passage by which Nature joins the lizards to the snakes." Another snake vibrates its tail like a rattlesnake, but has no rattles and therefore makes no noise. Penguins use their wings to paddle through the water rather than fly through the air. If the Creator had individually created each animal in its present form, to suit its present environment, what could be the purpose behind these makeshift arrangements?

Perhaps the most significant stopover of the voyage was the month spent on the Galapagos Islands off the northwestern coast of South America. Here Darwin found huge tortoises (*galapagos* means "tortoise" in Spanish); different islands were home to distinctively different types of tortoises. On islands without tortoises, prickly pear cactus grew with their juicy (though spiny) pads and fruits spread out over the ground. On islands where tortoises lived, the prickly pears grew substantial trunks, bearing the fleshy pads and fruits high above the reach of the voracious and tough-mouthed

tortoises (Fig. 16-7). Darwin also found several varieties of mockingbirds and finches, and, as with the tortoises, different islands supported slightly different forms. Could the differences in these organisms have arisen after they became isolated from one another on separate islands? The diversity of tortoises and birds "haunted" him for years afterward.

In 1836, Darwin returned to England after 5 years on the *Beagle* and became established as one of the foremost naturalists of his time. But constantly gnawing on his mind was the problem of the origin of species. For over 20 years, Darwin pondered the problem, collecting evidence that might provide insight into its solution. When he finally published *On the Origin of Species* in 1859, his evidence had become truly overwhelming. Although its full implications would not be realized for decades, Darwin's theory of evolution by natural selection has become a unifying concept for virtual-

Figure 16-6   **A painting of Charles Darwin as a young man**

Figure 16-7 Tortoises act as agents of selection

Galapagos tortoises feed on prickly pear cactuses. On islands with tortoises, a young cactus quickly grows a tall trunk, which lifts the succulent pads beyond the reach of the tortoises.

ly all of biology. Events that change the world sometimes hinge on small details, even the shape of a nose!

Darwin and Wallace Both Proposed That Evolution Occurs by Natural Selection

Although their social and educational backgrounds were very different, Darwin and Wallace were quite similar in some respects. Both had traveled extensively in the tropics and had studied the staggering variety of plants and animals living there. Both found that some species differed only in a few fairly subtle, but ecologically important, features (Fig. 16-8). Darwin and Wallace were familiar with the fossil record, which showed a trend of increasing complexity through time. Finally, both were aware of the studies of Hutton and Lyell, who proposed that Earth is extremely ancient. These facts suggested to both men that species change over time; that is, they evolve. Both sought a mechanism that might direct change over many generations, causing new species to arise.

Part of the answer came to both men from an unlikely source: the writings of an English economist and clergyman, Thomas Malthus. In his *Essay on Population*, Malthus wrote, "It may safely be pronounced, therefore, that [human] population, when unchecked, goes on doubling itself every 25 years, or increases in a geometrical ratio." Darwin and Wallace realized that a similar principle holds true for plant and animal populations. In fact, most organisms can reproduce much more rapidly than humans (consider rabbits, dandelions, and houseflies) and consequently could produce overwhelming populations in short order. Nonetheless, the world is *not* chest-deep in rabbits, dandelions, or flies: Natural populations do *not* grow "unchecked" but tend to remain approximately constant in size. Clearly, vast numbers of individuals must die in each generation, and most must not reproduce. In fact, population growth is checked

by many environmental factors, including food supply, predators, diseases, and weather (see Chapter 43).

From their experience as naturalists, Darwin and Wallace realized that the individual members of a species often differ from one another in form and function. Further, *which individuals die in each generation is not arbitrary but depends to some extent on the structures and abilities of the organisms*. This observation was the source of the theory of evolution by natural selection. As Wallace put it, "those which, year by year, survived this terrible destruction must be, on the whole, those which have some little superiority enabling them to escape each special form of death to which the great majority succumbed." Here you see the origin of the expression "survival of the fittest." That "little superiority" that confers greater fitness might be better resistance to cold, more efficient digestion, or any of hundreds of other advantages, some very subtle. Everything now fell into place. Darwin wrote: "It at once struck me that under these circumstances favorable variations would tend to be preserved, and unfavorable ones to be destroyed." If the favorable variations were inheritable, then the entire species would eventually consist of individuals possessing the favorable trait. With the continual appearance of new variations (due, as we now know, to mutations), which in turn are subject to further selection, "the result . . . would be the formation of new species. Here, then, I had at last got a theory by which to work."

In 1858, remarkably similar papers from Darwin and Wallace were presented to the Linnaean Society in London. Like Gregor Mendel's manuscript on the principles of genetics, their papers made little impact. The secretary of the society, in fact, wrote in his annual report that nothing very interesting happened that year. Fortunately, the next year Darwin published his monumental *On the Origin of Species by Means of Natural Selection*, forcing everyone to take note of the new theory.

(a)

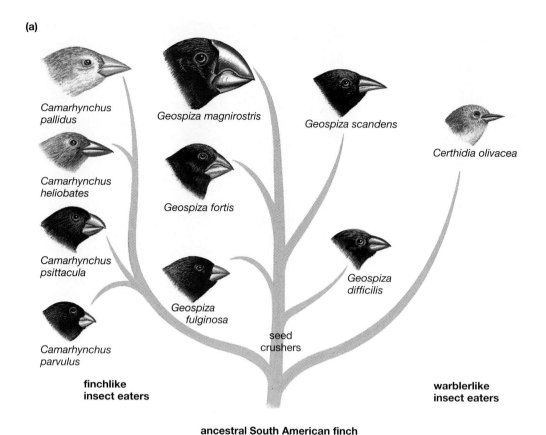

Camarhynchus pallidus

Geospiza magnirostris

Geospiza scandens

Certhidia olivacea

Camarhynchus heliobates

Geospiza fortis

Camarhynchus psittacula

Geospiza difficilis

Camarhynchus parvulus

Geospiza fulginosa

seed crushers

finchlike insect eaters

warblerlike insect eaters

ancestral South American finch

(b)

(c)

Figure 16-8   **Darwin's Finches, residents of the Galapagos Islands**

The Galapagos are a group of volcanic islands about 600 miles off the coast of Ecuador. Here Darwin found many species of finches, all similar to a species found on the South American mainland. Ordinarily, finches are seed-eating birds, with large bills adapted to crushing hard seeds. Apparently, thousands of years ago a finch or small flock of finches became lost during migration or were blown off course by a storm and arrived at the Galapagos. There they found few other birds. With few competitors, the finches found rich pickings of both seeds and insects. Over time, different groups of finches evolved adaptations to exploit different food sources. Modern biologists agree with Darwin that the variety of finch species on the Galapagos evolved from a common ancestor **(a)**. Particularly significant are the differences in beaks, which vary from a slender, warblerlike bill in the insect-eating *Certhidia olivacea* **(b)** to the massive seed-crushing bill of *Geospiza magnirostris* **(c)**. Beak differences are also important in species recognition. From behind, many Galapagos finches are very similar, and a male may approach a female of the wrong species. As soon as he catches sight of her beak, however, he immediately recognizes his mistake and looks elsewhere for a mate.

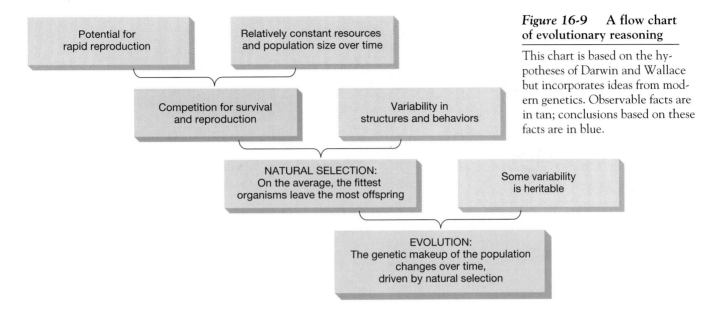

Figure 16-9 **A flow chart of evolutionary reasoning**

This chart is based on the hypotheses of Darwin and Wallace but incorporates ideas from modern genetics. Observable facts are in tan; conclusions based on these facts are in blue.

Evolutionary Theory Arises from Scientific Observations and Conclusions Based on Them

The essence of the Darwin-Wallace theory is very simple; it consists of three conclusions based on four observations. Below, we summarize their theory in modern terms (Fig. 16-9).

Observation 1: Natural populations of all organisms have the potential to increase rapidly, because organisms can produce far more offspring than are required merely to replace the parents.

Observation 2: Nevertheless, the sizes of most natural populations and the resources available to maintain them (such as food and appropriate habitat) remain relatively constant over time.

Conclusion 1: Therefore, there is competition for survival and reproduction. In each generation, many individuals must die young, fail to reproduce, produce few offspring, or produce less-fit offspring that fail to survive and reproduce in their turn.

Observation 3: Individual members of a population differ from one another in their ability to obtain resources, withstand environmental extremes, escape predators, and so on.

Conclusion 2: The most well-adapted (the "fittest") individuals in one generation will usually leave the most offspring. This is **natural selection**: the process by which the environment selects for those individuals whose traits best adapt them to that particular environment.

Observation 4: At least some of the variation in adaptive traits among individuals is due to genetic differences that may be passed on from parent to offspring.

Conclusion 3: Over many generations, differential, or unequal, reproduction among individuals with different genetic makeup changes the overall genetic composition of the population. This is evolution through natural selection.

As you know, the principles of genetics had not yet been discovered when Darwin wrote *On the Origin of Species*. Our observation 4, therefore, was an untested assumption for Darwin and a grave weakness in his theory. Although he could not explain how inheritance operated, Darwin's theory made an important prediction that we now know is correct. According to Darwin, the variations that appear in natural populations arise purely by chance. Unlike Lamarck, he proposed no internal drives for perfection or any other mechanisms that would ensure that variations would be favorable. Molecular genetics has shown that Darwin was correct: Variations arise because of chance mutations in DNA (see Chapters 10 and 11).

How might natural selection among chance variations change the makeup of a species? In *On the Origin of Species*, Darwin proposed the following example. "Let us take the case of a wolf, which preys on various animals, securing [them] by... fleetness.... The swiftest and slimmest wolves would have the best chance of surviving, and so be preserved or selected.... Now if any slight innate change of habit or structure benefited an individual wolf, it would have the best chance of surviving and of leaving offspring. Some of its young would probably inherit the same habits or structure, and by the repetition of this process, a new variety might be formed." The same argument would apply to the wolf's prey, in which the fastest or most alert would be the most likely to avoid predation and would pass on these traits to its offspring. Over generations, the population would change as a greater percentage of its individuals acquired adaptive traits. This interplay between predator and prey is an example of **coevolution,** an important source of **adaptations** which are traits that help organisms survive and reproduce in a particular environment: an environment that includes other organisms. Coevolution is discussed in more detail in Chapter 44.

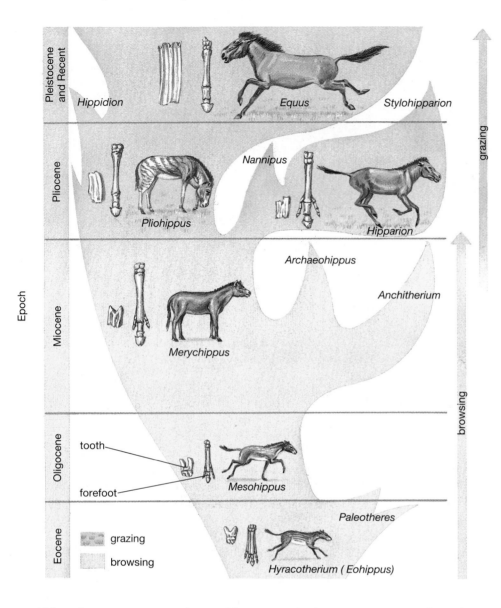

Epoch

Pleistocene and Recent

Hippidion   *Equus*   *Stylohipparion*

Pliocene

Nannipus

Pliohippus   *Hipparion*

Archaeohippus

Miocene   *Anchitherium*

Merychippus

Oligocene

tooth

forefoot

Mesohippus

Paleotheres

Eocene

grazing
browsing

Hyracotherium (Eohippus)

grazing

browsing

Figure 16-10   **The evolution of the horse**

Over the past 50 million years, horses evolved from small woodland browsers to large plains-dwelling grazers. Three major changes include size, leg anatomy, and tooth anatomy. No one is certain why horses became larger, but it may have been an anti-predator adaptation. In any case, a large body running over hard plains favored the evolution of large, hard hooves, attached by springlike, shock-absorbing joints to legs with stout bones. Finally, the teeth became larger, with more enamel. This change reflects a change in diet from the relatively soft leaves and buds of bushes to the abrasive blades of grasses, which contain silicon (glass). If a modern horse had teeth like *Hyracotherium*, its teeth would be ground away while it was still very young, leaving it to starve.

Although it is easiest to understand how natural selection would cause *changes within a species*, under the right circumstances, the same principles might produce *entirely new species*. In Chapter 18, we discuss the circumstances that give rise to new species. In the rest of this chapter, we briefly review some of the evidence for evolution.

The Evidence for Evolution

Virtually all biologists consider evolution to be a fact. Although debates still rage over the *mechanisms* of evolutionary change, exceedingly few biologists dispute that evolution occurs. Why? Because an overwhelming body of evidence permits no other conclusion.

The Fossil Record Provides Evidence of Evolutionary Change over Time

If fossils are the remains of members of species that are ancestral to modern species, then one would expect to find progressive series of fossils leading from an ancient, primitive organism, through several intermediate stages, and culminating in the modern form. (Just how fine the gradations between stages should be is currently a subject of debate among evolutionary biologists; see Chapter 18.) Probably the best-known of such series are the fossil horses (Fig. 16-10), but giraffes, elephants, and several mollusks all show a gradual evolution of body form over time, suggesting that species evolved from and replaced previous species. Certain sequences of fossil snails have such slight gradations in form between successive fossils that paleontologists cannot easily decide where one species leaves off and the next one begins.

Comparative Anatomy Provides Structural Evidence for Evolution

Through Convergent Evolution, Natural Selection Has Shaped Unrelated Organisms into Similar Forms in Similar Environments

Evolution by natural selection also predicts that, given similar environmental demands, unrelated organisms might independently evolve superficially similar struc-

(a)

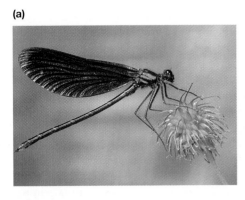

(b)

Figure 16-11   **Analogous structures**

Similar selective pressures acting on unrelated animals may result in the evolution of outwardly similar structures. The wings of insects **(a)** and birds **(b)** and the sleek, streamlined shapes of seals **(c)** and penguins **(d)** are examples of such analogous structures.

(c)

(d)

tures, a process called **convergent evolution.** Such outwardly similar body parts in unrelated organisms, termed **analogous structures,** are often completely different in internal anatomy, because the parts are not derived from common ancestral structures. The wings of flies and birds and the fat-insulated, streamlined shapes of seals (mammals) and penguins (birds), are two examples of analogous structures that have arisen through convergent evolution (Fig. 16-11).

Homologous and Vestigial Structures Provide Evidence for Relatedness of Organisms Adapted to Different Environments

Modern organisms are adapted for a wide variety of habitats and life-styles. The forelimbs of birds and mammals, for example, are variously used for flying, swimming, running over several types of terrain, and grasping objects such as branches and tools. Despite this enormous diversity of function, the internal anatomy of all bird and mammal forelimbs is remarkably similar (Fig. 16-12). It is inconceivable that nearly the same bone arrangements could be ideal for such different functions, as would be expected if each animal were created separately. Such similarity is exactly what we would expect, however, if bird and

mammal forelimbs evolved from a common ancestor. Through natural selection, each has been modified to perform a particular function. Such internally similar structures are called **homologous structures,** meaning that they have a similar evolutionary origin, despite possible differences in function. Studies of comparative anatomy have long been used to determine the relationships among organisms, on the grounds that the more similar the internal structures of two species, the more closely related the species must be; that is, the more recently they must have diverged from a common ancestor.

Evolution by natural selection also helps to explain the curious circumstance of **vestigial structures.** These are structures that serve no apparent purpose, including such things as molar teeth in vampire bats (which live on a diet of blood and therefore don't chew their food) and pelvic bones in whales and certain snakes (Fig. 16-13). Both of these vestigial structures are clearly homologous to important structures found in other vertebrates (animals with a backbone). Their continued existence in animals that have no use for them is best explained as a sort of "evolutionary baggage." For example, the ancestral mammals from which whales evolved had four legs and a well-developed set of pelvic bones. Whales do not have hind legs, yet they have small pelvic and leg bones embedded in their sides. During whale evolution, there was a selective advantage to the loss of the hind legs, the better to streamline the body for movement through water. Once mutation and selection reduced the pelvic bones so that they no longer interrupted the smooth line of the body, that selective pressure (selective forces exerted by the environment) diminished. The result is the modern whale with small, useless and unused pelvic bones.

Embryological Stages of Animals Can Provide Evidence for Common Ancestry

In the early 1800s, the German embryologist Karl von Baer noted that all vertebrate embryos look quite similar to one another early in development (Fig. 16-14). Fish, turtles, chickens, mice, and humans all develop tails and gill slits early in development. Only fish go on to develop gills, and only fish, turtles, and mice retain substantial tails. Why do such diverse vertebrates have similar developmental stages? The only plausible explanation is that ancestral vertebrates possessed genes that direct the development of gills and tails. All of their descendants still retain those genes. In fish, these genes are active throughout development, resulting in gill- and tail-bearing adults. In humans and chickens, these genes are active only during early developmental stages, and the structures are lost or inconspicuous in adults.

Modern Biochemical and Genetic Analyses Reveal Relatedness among Diverse Organisms

Biochemistry and molecular biology provide striking evidence for the evolutionary relatedness of all living organisms. At the most fundamental biochemical levels, all living cells are very similar. For example, all cells have DNA as the carrier of genetic information; all use RNA, ribosomes, and approximately the same genetic code to translate that genetic information into proteins; all use roughly the same set of 20 amino acids to build proteins; and all use ATP as an intracellular energy carrier.

Evolutionary relationships among species are reflected in similarities and differences in their proteins. The amino acid sequences of a few proteins, such as hemoglobin and cytochrome c (one of the electron carriers in the mitochondrion), have been determined for many different species. The sequences are remarkably similar across a huge spectrum of species. Further, an evolutionary tree comparing the degree of differences in amino acid sequence between species closely resembles the evolutionary trees that have been deduced from anatomical and embryological studies. Relatedness among organisms can also be evaluated by examining the morphology of their chromosomes. For example, the chromosomes of chimpanzees and humans are extremely similar, showing that these species are closely related.

Within the past decade, the techniques of molecular genetics have triggered a revolution in studies of evolutionary relationships. For the first time, DNA, rather than "secondary" features such as appearance, behavior, or even proteins, can be used to investigate relatedness among organisms. Some of these methods and findings are explored in the essay "Scientific Inquiry: Modern Molecular Genetic Techniques Shed Light on Evolutionary Relationships."

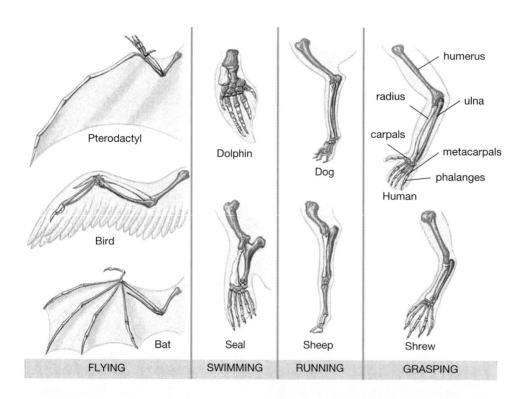

Figure 16-12   Homologous structures

The bones in the forelimbs of these animals are all similar, despite wide differences in function. The bones have been tinted different colors to highlight the similarities among the various species.

Pterodactyl

Dolphin

Dog

humerus

radius

ulna

carpals

metacarpals

phalanges

Human

Bird

Bat

Seal

Sheep

Shrew

FLYING   SWIMMING   RUNNING   GRASPING

salamander

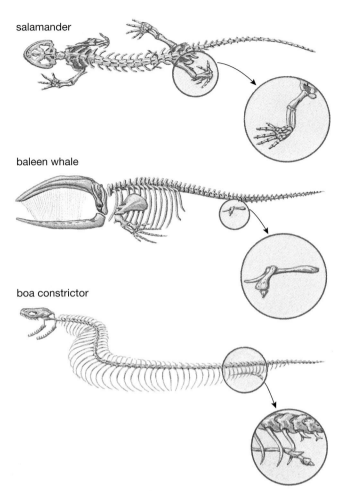

baleen whale

boa constrictor

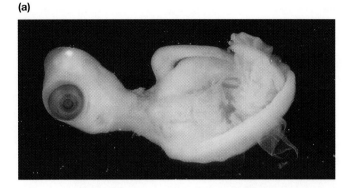

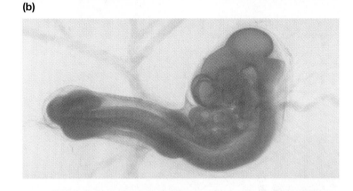

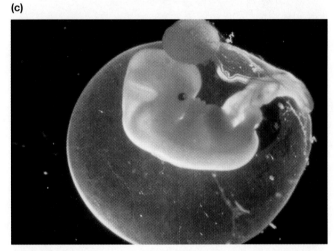

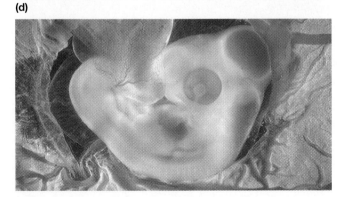

Figure 16-13   **Vestigial structures**

The functional hindlimb of a salamander (top) is probably similar to the limb of the common ancestor of all amphibians, reptiles, birds, and mammals. Baleen whales (center) and boa constrictors (bottom) have no functional legs but still develop vestigial pelvic bones and even miniature leg bones buried in their sleek sides.

Artificial Selection Demonstrates That Organisms May Be Modified by Controlled Breeding

One line of evidence supporting evolution that particularly impressed Charles Darwin was **artificial selection:** breeding domestic plants and animals to produce specific desirable features. The various breeds of dogs provide a striking example of artificial selection (Fig. 16-15). Dogs descended from wolves, and even today the two will readily cross-breed. With rare exceptions, however, few modern dogs resemble wolves. Some breeds, such as the Chihuahua and the Great Dane, are so different from one another that they would be considered separate species if they were found in the wild. Interbreeding would hardly be possible without a lot of human assistance. If humans could breed such radically different dogs in a few hundred to at most a few thousand years, Darwin reasoned, it seemed quite plausible that natural selection could produce the spectrum of living organisms in hundreds of millions of years.

Figure 16-14   **Embryological stages reveal evolutionary relationships**

Early embryonic stages of a turtle **(a)**, mouse **(b)**, human **(c)**, and chicken **(d)**, showing strikingly similar anatomical features.

(a)

(b)

Figure 16-15   **Dog diversity illustrates artificial selection**
A comparison of the ancestral dog (the gray wolf, *Canis lupus*) (a) and various breeds of dog (b). Artificial selection by humans has caused a great divergence in size and shape in only a few thousand years. Huge differences in size would make it extremely difficult for the large dogs in the rear of the photo to interbreed with the small terriers in the foreground.

Evolution by Natural Selection Occurs Today

One problem that many people have with the theory of evolution is that they think it all happened in the distant past. Darwin himself believed this; in *On the Origin of Species* he stated: "We see nothing of these slow changes in progress, until the hand of time has marked the lapse of ages." Modern evolutionary biologists, however, maintain that evolution is *not* merely a phenomenon of the past, but that it continues today. One example of present-day

evolution has been caused by the increasing industrial pollution in the nineteenth and twentieth centuries.

Tree trunks in British woodlands usually support a lush growth of mottled gray lichens (a symbiotic association of an alga and a fungus; see Chapter 22). Before the Industrial Revolution of the nineteenth century, most peppered moths, *Biston betularia*, were white with scattered specks of black pigment. This coloration matched the color and pattern of the lichens growing on the trees. Because the moths sat quietly on the lichens during the day, predatory birds could not easily see them (Fig. 16-16a). Occasionally, mutant black individuals appeared. These black moths, extremely conspicuous against the pale lichens, were easily spotted by birds and didn't live long.

During the Industrial Revolution, Britain's growing industries began to burn huge quantities of coal for fuel. With no pollution-control technology, soot from the smokestacks soon blanketed the countryside around the mills and factories (Fig. 16-17). The lichens on the trees became covered with pollutants and died off, leaving the trunks sooty black. The pale form of the peppered moth was no longer camouflaged while sitting on the blackened trunks (see Fig. 16-16b). Increasingly, pale moths fell prey to birds. Soot-covered trees, however, provided excellent camouflage for black moths. The black moths' color became, in Wallace's words, the "little superiority enabling them to escape" predation by birds. Black moths, once rare, survived and reproduced, passing on their genes for dark pigmentation to succeeding generations. As the years passed, ever-increasing numbers of black moths appeared, until by the end of the nineteenth century, about 98% of the moths around the industrial city of Manchester were black. Fortunately, pollution-control laws have now dramatically reduced emissions of soot and other pollutants, and lichens grow once again on the trees near British cities. As predicted by the principles of evolution, the pale moths are making a comeback in these areas, and the black form is becoming increasingly rare.

Evolution has also been documented more recently, and closer to home. In Florida, homeowners were dismayed to realize that roaches were ignoring a formerly effective poison bait called Combat. Researchers discovered that the bait had acted as an agent of natural selection—roaches that liked it were consistently killed. Those that survived had inherited a rare mutation that caused them to dislike glucose, a type of sugar found in the corn syrup used as bait in Combat. By the time researchers identified the problem in the early 1990s, the formerly rare mutation had become widespread in Florida's urban roach population.

In a meadow near Carson City, Nevada, biologists have closely monitored a rare subspecies of checkerspot butterfly. In 1983, about 80% of the eggs of the butterfly were laid on a native meadow plant *Collinsia*. But over the next decade, the meadow was invaded by a weed, *Plantago*. By 1993, over 70% of the checkerspots' eggs were being laid on *Plantago*. When eggs laid on *Plantago* were hatched in

(a)   **(b)**

Figure 16-16 **Protective colors differ in different environments**

The two color forms of the peppered moth, resting on a lichen-covered tree trunk **(a)** and a soot-blackened trunk **(b)**. The pale form is well camouflaged on the lichen but is conspicuous on the sooty trunk. The reverse is true of the black form.

the laboratory and the butterflies reared exclusively on *Collinsia*, the mature butterflies still sought out *Plantago* to lay their eggs, just as their mothers had. The preference for *Plantago* was clearly inherited and had rapidly increased within the population during a single decade.

Figure 16-17 **Pollution alters the environment**

The countryside around the industrial city of Manchester, England, is crisscrossed by black stone fences. If you break one of the stones, you can see that they are actually a bright golden sandstone. The black outer crust is the remnant of pollution from the Industrial Revolution of over a half-century ago.

It is important to remember that black coloration of British moths, distaste for glucose of Florida cockroaches, and preference for *Plantago* of Nevada butterflies were not *produced* by the pollution, poisoned corn syrup, or the invading weed. The mutations that produced these adaptations all arise spontaneously now and then. *The variations upon which natural selection works are produced by chance mutations.* A second point to remember is that selection does not necessarily produce well-adapted species. Some other species of moths in Britain lacked black mutants, and these became extinct in industrial areas. If the right raw material isn't there (black moth mutants in this case) then selection may drive a species to extinction. Finally, some people tend to regard evolution as a mechanism for producing the *Scala Naturae*, with ever-greater degrees of perfection appearing over time. It is not. *The processes of evolution select for organisms that are best adapted to a particular environment.* Under natural conditions, a distaste for glucose would put a cockroach at a disadvantage, causing it to avoid many rich food sources. In the presence of poisoned corn syrup, however, it becomes advantageous. Natural selection does not select for the "best" in any absolute sense, but only in the context of a particular environment, which varies from place to place, and which may change over time.

Other examples of modern-day evolution include antibiotic resistance in bacteria (see Chapter 21) and the appearance of the AIDS virus (HIV) (Chapter 34). Evolution is easiest to observe in rapidly reproducing, short-lived organisms such as insects or bacteria, but it occurs constantly in all the species on Earth.

S C I E N T I F I C   I N Q U I R Y

Modern Molecular Genetic Techniques Shed Light on Evolutionary Relationships

Evolution deals with inherited changes over time. Because DNA is the molecule of heredity, evolutionary changes must be reflected in changes in DNA. Chapter 17 will explore a few of the genetic mechanisms that contribute to evolution. In this essay, we will consider two methods of studying DNA and how those methods are applied to determining the evolutionary relatedness of organisms.

Cross-Species DNA Hybridization

The DNA in the chromosomes of all living organisms is double stranded, with the two strands held together by hydrogen bonds between complementary base pairs. The two strands of DNA can be separated by heating; when the temperature becomes high enough, the hydrogen bonds break and the two strands separate. The amount of separation can be measured by observing the absorption of ultraviolet (UV) light. Single DNA strands absorb UV light more effectively than double strands do, so the higher the proportion of single strands, the greater the UV absorption (Fig. E16-1).

Suppose now that you had two single strands of DNA in which only half of the bases were complementary. The complementary bases would still hold the two strands together, but with only half the normal number of hydrogen bonds. If you heated this "semidouble helix," you would find that, with fewer hydrogen bonds holding the strands together, the strands would separate at a lower temperature.

In the technique called DNA-DNA hybridization, single strands of DNA from two different species are allowed to form "hybrid" double helices (the procedures are fair-

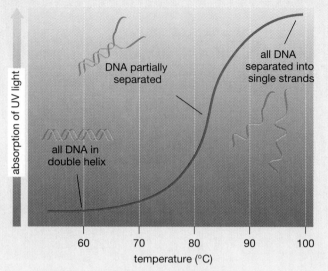

Figure E16-1   DNA-DNA hybridization

As DNA is heated, the two strands gradually separate. Separation is measured by UV light absorption; as more single strands are liberated, more UV light is absorbed. When DNA strands from two different species have been "hybridized," the lower their melting temperature, the poorer the match between them.

ly involved, but this is the final result). They are then heated up, and the temperature at which the two strands separate is measured. Separation temperatures for strands from different species will necessarily be less than those for the same species. The principle is this: The DNA of closely related species will have more-similar nucleotide sequences than does the DNA of distantly related species.

A Postscript by Charles Darwin

"It is interesting to contemplate an entangled bank, clothed with many plants of many kinds, with birds singing on the bushes, with various insects flitting about, and with worms crawling through the damp earth, and to reflect that these elaborately constructed forms . . . have all been produced by laws acting around us. These laws, taken in the highest sense, being Growth with Reproduction; Inheritance [and] Variability . . . ; a Ratio of Increase so high as to lead to a Struggle for Life, and as a con-

sequence to Natural Selection, entailing Divergence of Character and Extinction of less-improved forms . . . . There is grandeur in this view of life, with its several powers, having been originally breathed into a few forms or into one; and that, whilst this planet has gone cycling on according to the fixed law of gravity, from so simple a beginning endless forms most beautiful and most wonderful have been, and are being, evolved."

These are the concluding sentences of *On the Origin of Species* by Charles Darwin.

The more similar the nucleotide sequences, the greater the number of hydrogen bonds between them, and the higher the temperature required for the hybrid helices to separate. When the melting temperature for hybrid (two-species) DNA is very similar to that of the DNA for each single species, the two species are very similar genetically and are very closely related. Using this type of information, one can generate a tree of evolutionary relationships based on the separation temperature of the hybrid helices (Fig. E16-2).

Direct Comparison of Nucleotide Sequences

"Melting" DNA is, of course, a somewhat indirect way of measuring similarities in nucleotide sequences. With the advent of DNA sequencing machines, it is now possible to compare the nucleotide sequences of single genes from a variety of species. The number of nucleotide substitutions between genes encoding the same or similar proteins reflects the relatedness among species: the fewer the differences in nucleotide sequence, the more closely the organisms are related. One problem is that the rate of change in the nucleotide sequence may differ for different genes; for example, selection pressures may favor changes in certain proteins more than in others. Thus, comparisons of a *single* gene may not provide an accurate evolutionary tree. This is, in fact, an argument in favor of using DNA-DNA hybridization, because hybridization detects changes over millions of nucleotides, not just the few hundreds that make up any single gene. However, as gene sequencing becomes more automated and more economical, this technique will

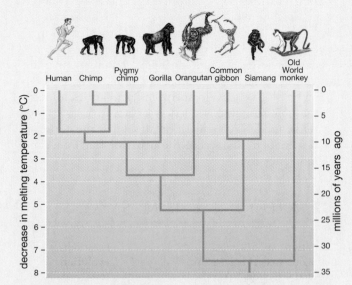

Figure E16-2   Relatedness can be determined by DNA melting temperatures

In this study by Charles Sibley and Jon Ahlquist, double helices of DNA were made by combining single strands from various pairs of species of primate. The melting temperatures were then determined and compared with temperatures of DNA from a single species (the difference is graphed here as the decrease in melting temperature). The greater the difference in melting temperature of hybrid versus "pure" DNA, the longer ago the two species diverged from a common ancestor.

provide increasing insight into the molecular changes underlying evolution.

❖ S U M M A R Y   O F   K E Y   C O N C E P T S

The History of Evolutionary Thought

A species consists of all the populations of organisms that can potentially interbreed and that do not usually interbreed with members of other species. Historically, the most common explanation for the origin of species has been creationism, that a divine being created each species in its present form, and that species have not significantly changed since the creation. Evidence provided by the diversity of living things, fossils, and geology challenged this view, though no convincing mechanism for the evolution of present-day species from previous ones was proposed. Since the middle of the nineteenth century, scientists have

concluded that species originate by the operation of natural laws, as a result of changes in the genetic makeup of the populations of organisms. This process is called evolution.

Charles Darwin and Alfred Russell Wallace independently proposed the theory of evolution by natural selection. Their theory can be concisely expressed as three conclusions based on four observations. These are summarized in modern biological terms in Figure 16-9.

The Evidence for Evolution

Many lines of evidence indicate that evolution has occurred and that natural selection is the chief mechanism driving changes in the characteristics of species over time. The evidence includes:

1. Fossils of ancient organisms are simpler in form than modern organisms. Sequences of fossils have been discovered that show a graded series of changes in form. Both of these facts would be expected if modern forms evolved from older forms.
2. Organisms thought to be related through evolution from a common ancestor show many similar anatomical structures. Examples include the limbs of amphibians, reptiles, birds, and mammals. Similarly, stages in embryological development and similarities in chromosome structure, sequences of amino acids in proteins, and DNA composition all support the notion of descent of related species through evolution from common ancestors.

3. Rapid, heritable changes have been produced in domestic animals and plants by selectively breeding organisms with desired features (artificial selection). The immense variations in species produced in a few thousand years of artificial selection by humans makes it seem likely that much larger changes could be wrought by hundreds of millions of years of natural selection.

4. Evolution can be observed today. Both natural and human activities drastically change the environment over short periods of time. Significant changes in the characteristics of species have been observed in response to these environmental changes. A well-studied example is the evolution of black coloration among moths in response to the darkening of their environment by industrial pollutants.

✖ KEY TERMS

adaptation   p. 311
analogous structures   p. 313
artificial selection   p. 315
catastrophism   p. 305
coevolution   p. 311
convergent evolution   p. 313

creationism   p. 303
evolution   p. 303
fossil   p. 305
homologous structures   p. 313
inheritance of acquired characteristics   p. 307

natural selection   p. 311
population   p. 304
species   p. 304
uniformitarianism   p. 306
vestigial structures   p. 313

✖ THINKING THROUGH THE CONCEPTS

Multiple Choice

1. Your arm is homologous with
   a. a seal flipper
   b. an octopus tentacle
   c. a bird wing
   d. a sea star arm
   e. both a and c
2. All organisms share the same genetic code. This commonality is evidence that
   a. evolution is occurring now
   b. convergent evolution has occurred
   c. evolution occurs gradually
   d. all organisms are descended from a common ancestor
   e. life began a long time ago
3. Which of the following are fossils?
   a. pollen grains buried in the bottom of a peat bog
   b. the petrified cast of a clam's burrow
   c. the impression a clam shell made in mud, preserved in mudstone
   d. an insect leg sealed in plant resin
   e. all of the above
4. In Africa, there is a species of bird called the Yellow-throated Longclaw. It looks almost exactly like the meadowlark found in North America, but they are not closely related. This is an example of
   a. uniformitarianism
   b. coevolution
   c. gradualism
   d. vestigial structures
   e. convergent evolution

5. Which of the following are examples of vestigial structures?
   a. your tailbone
   b. nipples on male mammals
   c. sixth fingers found in some humans
   d. your kneecap
   e. none of the above
6. Which of the following would *stop* evolution by natural selection from occurring?
   a. if humans became extinct because of a disease epidemic
   b. if a thermonuclear war killed most living organisms and changed the environment drastically
   c. if ozone depletion led to increased ultraviolet radiation, which caused many new mutations
   d. if genetic recombination, sexual reproduction, and mutation stopped, so that all offspring of all organisms were exact copies of their parents
   e. all of the above

Review Questions

1. Selection acts on individuals, but only populations evolve. Explain why this is true.
2. Describe what "survival of the fittest" really means by writing a clear explanation of evolutionary fitness.
3. Distinguish between catastrophism and uniformitarianism. How did these hypotheses contribute to the development of evolutionary theory?
4. Describe Lamarck's theory of inheritance of acquired characteristics. Why is it invalid?

5. What is natural selection? Describe how natural selection might have caused differential reproduction among the ancestors of a fast-swimming predatory fish, such as a barracuda.

6. Describe how evolution occurs through the interactions among the reproductive potential of a species, the normally constant size of natural populations, variation among individuals of a species, natural selection, and inheritance.

7. What is convergent evolution? Give an example.

8. How do biochemistry and molecular genetics contribute to the evidence that evolution occurred?

✖ A P P L Y I N G   T H E   C O N C E P T S

1. Does evolution through natural selection produce "better" organisms in an absolute sense? Are we climbing the *Scala Naturae*? Defend your answer.

2. Describe and compare the views of Plato and Darwin on the existence of variation in natural populations.

3. The idea of special creation and the study of fossils have each had an impact on evolutionary thought. Discuss why one is considered scientific endeavor and the other not scientific.

4. In evolutionary terms, "success" can be defined in many different ways. What are the most successful organisms you can think of in terms of (a) persistence over time, (b) sheer numbers of individuals alive now, (c) numbers of species (for a lineage), and (d) geographical range?

5. In what sense are humans currently acting as "agents of selection" on other species? Name some organisms that are *favored* by the environmental changes humans cause.

6. Darwin and Wallace's discovery of natural selection is one of the great revolutions in scientific thought. Some scientific revolutions spill over and affect the development of philosophy and religion. Is this true of evolution? Does (or should) Darwinism affect the way humans view their place in the world?

7. In your mind, what scientific question currently represents the "mystery of mysteries"?

✖ F O R   M O R E   I N F O R M A T I O N

Altman, S. A. "The Monkey and the Fig." *American Scientist*, May–June 1989. An amusing, informative discussion of many evolutionary themes, cast as a Socratic dialogue.

Bishop, J. A., and Cook, L. M. "Moths, Melanism and Clean Air." *Scientific American*, January 1975. This followup to H. B. D. Kettlewell's 1959 work shows that cleaning up the air reverses industrial melanism in moths—that is, becoming pigmented as a consequence of industrial activities—providing graphic evidence that evolution only adapts organisms to existing environments.

Darwin, C. *On the Origin of Species by Means of Natural Selection.* Garden City, N.Y.: Doubleday, 1960 (originally published in 1859). An impressive array of evidence amassed to convince a skeptical world.

Eiseley, L. C. "Charles Darwin." *Scientific American*, February 1956. An essay on the life of Darwin, by one of his foremost American biographers. Even if you need no introduction to Darwin, read this anyway, as an introduction to Eiseley, author of many marvelous essays.

Gould, S. J. *Ever Since Darwin*, 1977; *The Panda's Thumb*, 1980; and *The Flamingo's Smile*, 1985. New York: W. W. Norton. A series of witty, imaginative, and informative essays, mostly from *Natural History* magazine. Many deal with various aspects of evolution.

Grant, P. R. "Natural Selection and Darwin's Finches." *Scientific American*, October 1991. A drought in the Galapagos Islands provides dramatic evidence for natural selection as an agent of evolutionary change.

Kettlewell, H. B. D. "Darwin's Missing Evidence." *Scientific American*, March 1959. Industrial melanism as an example of modern-day evolution.

Weiner, J. "Evolution Made Visible." *Science*, January 6, 1995. A clear summary of modern evidence for evolution in action.

N E T   W A T C H

On-line resources for this chapter are on the World Wide Web at:
http://www.prenhall.com/~audesirk (click on the <u>table of contents</u> link and then select Chapter 16).

17 How Organisms Evolve

"What but the wolf's tooth whittled so fine
The fleet limbs of the antelope?
What but fear winged the birds, and hunger
Jewelled with such eyes the great goshawk's head?"

ROBINSON JEFFERS in *The Bloody Sire*

AT A GLANCE

Evolution and the Genetics of Populations
Genes, Influenced by the Environment, Determine the
    Traits of Each Individual
The Gene Pool Is the Sum of All the Genes Occurring in a
    Population
Evolution Is the Change of Gene Frequencies within a
    Population
Mutation and the Recombination of Alleles during Sexual
    Reproduction Provide Sources of Variability
The Equilibrium Population Is a Hypothetical Population in
    Which Evolution Does Not Occur

The Mechanisms of Evolution
Mutations Are the Ultimate Source of Genetic Variability
Migration Produces Gene Flow between Populations
Small Populations Are Subject to Random Changes in Allele
    Frequencies
Mating within a Population Is Almost Never Random
All Genotypes Are Not Equally Adaptive

Natural Selection
Natural Selection Acts on the Phenotype, Which Reflects
    the Underlying Genotype
Natural Selection Can Influence Populations in Three
    Major Ways
Natural Selection Takes Several Forms

Extinction
Localized Distribution and Extreme Specialization Make
    Species Vulnerable in Changing Environments
Interactions with Other Organisms May Drive Species to
    Extinction
Habitat Change and Destruction Are the Leading Causes of
    Extinction

Evolutionary Connections: Kin Selection and the Evolution of Altruism

A goshawk surveys its surroundings for prey. The exceptionally keen eyes of birds of prey evolved in response to selection pressures that favored those who could spot prey from far above. As Robinson Jeffers poetically suggests, those who could not see well went hungry and produced fewer offspring.

In Chapter 16 we discussed the history of the theory of evolution and presented some of the evidence that evolution actually happens. But what processes drive evolutionary change? Is natural selection the only cause of evolution? Does evolution always occur all the time in all populations of organisms? In this chapter, we examine evolutionary processes in more detail. As we do, you will see that *evolution is an inevitable consequence of the nature of living things*. It occurs as a direct result of the chemical structure of genes and the interactions between organisms and their environment.

Evolution and the Genetics of Populations

Individual organisms live, reproduce, and die. Individuals, however, do not evolve; populations do. Evolution is the change in gene frequency that occurs in a population over time. Inheritance, therefore, is the link between the lives of individual organisms and the evolution of **populations**, which are all the individuals of a species living in a given area. We will begin our discussion of the processes of evolution by reviewing the principles of genetics as they apply to individuals and then extend those principles to the genetics of populations. You may want to refer to Unit II to refresh your memory on specific points.

Genes, Influenced by the Environment, Determine the Traits of Each Individual

Each cell of every organism contains a repository of genetic information encoded in the DNA of its chromosomes. A gene is a segment of DNA located at a particular place on a chromosome. Its sequence of nucleotides encodes the sequence of amino acids of a protein, usually an enzyme that catalyzes one particular reaction in the cell. Slightly different sequences of nucleotides at a given gene's location, called alleles, generate different forms of the same enzyme. There are usually two or more alleles of a single gene. An individual having alleles of the same type is homozygous, and an individual having alleles of different types is heterozygous. The specific alleles borne on an organism's chromosomes (its genotype), interacting with the environment, determine its physical and behavioral traits (its phenotype).

Let's illustrate these principles with an example that should be familiar to you from Unit II. A pea flower is colored purple because a chemical reaction in its petals converts a colorless molecule to a purple pigment. When we say that a pea plant has the allele for purple flowers, we mean that a particular stretch of DNA on one of its chromosomes contains a sequence of nucleotides that codes for the enzyme catalyzing this reaction. A pea with the allele for white flowers has a different sequence of nucleotides at the corresponding place on one of its chromosomes. The resulting enzyme cannot produce purple pigment. If a pea is homozygous for the white allele, its flowers produce no pigment and are white.

As we will see later, natural selection operates on the phenotype and, in doing so, either favors or selects against the particular genotype that produced it. Thus natural selection alters the gene frequencies within a population.

The Gene Pool Is the Sum of All the Genes Occurring in a Population

A branch of genetics, called **population genetics**, deals with the frequency, distribution, and inheritance of alleles in populations. Because evolution is a change in the genetic makeup of populations over generations, you will need to learn the principles of population genetics to understand the mechanisms of evolution.

In population genetics, the **gene pool** is defined as all the genes that occur in a population. It is made up of all the alleles of all the genes found in all of the individuals. Each particular gene can also be considered to have a gene pool, which consists of all the alleles of that specific gene occurring in a population. For example, in a population of 100 pea plants, the gene pool for flower color would consist of 200 alleles (peas are diploid, so there are two color alleles per plant, multiplied by 100 plants). If we could analyze the genetic composition of every plant in the population, we might find that some have alleles for white flowers, some have alleles for purple flowers, and some have both alleles. If we added up the color alleles of all the plants in the population, we could determine the relative proportions of the different alleles, a number called the **allele frequency**. Let's say that the gene pool for flower color consisted of 140 alleles for purple and 60 alleles for white. The allele frequencies would then be purple, 0.7 (70%), and white, 0.3 (30%).

Evolution Is the Change of Gene Frequencies within a Population

What does all this have to do with evolution? Quite a bit. Suppose a flower-eating cow happens upon a field of purple flowers and, being enamored of purple flowers, eats all of them before they produce seeds. Because the allele for pur-

ple flowers (P) is dominant to the allele for white (p), all the purple alleles in the entire population are in the purple-flowered plants (PP or Pp). If none of these plants reproduce, while the white-flowered plants do reproduce, then the next generation will consist entirely of white-flowered peas (pp). The allele frequency for purple will drop to 0, while the allele frequency for white will rise to 1.0 (100%). As a result of the selective eating habits of the cow, *evolution will have occurred in that field*. The gene pool of the pea population will have changed, and **natural selection**, in the form of foraging by the cow, will have caused the change.

This simple example illustrates four important points about evolution.

1. *Natural selection does not cause genetic changes in individuals*. The alleles for purple or white flower color arose spontaneously, long before the cow ever found the pea field. The cow did not cause white alleles to appear. It merely favored the differential survival of white alleles compared with purple alleles.

2. *Natural selection befalls individuals, but evolution occurs in populations*. Individual pea plants either reproduced or did not, but it was the population as a whole that evolved as its gene frequencies changed.

3. *Evolution is a change in the allele frequencies of a population, owing to differential reproduction among organisms bearing different alleles*. In evolutionary terminology, the **fitness** of an organism is measured by its reproductive success: In our example, the white flowers had greater fitness than the purple flowers, because they produced more viable offspring.

4. *Evolutionary changes are not "good" or "progressive" in any absolute sense*. The white alleles were favored only because of the dietary preferences of this particular cow; in another environment, with other predators, the white allele may well be selected against.

Mutation and the Recombination of Alleles during Sexual Reproduction Provide Sources of Variability

Your observations of the life around you have undoubtedly made you aware that even within a species, most individuals are at least slightly different from one another. The number of different individuals that can be produced from the same set of thousands of genes, considering that most genes come in multiple alleles, is staggering. This variability is produced by genetic recombination and sexual reproduction, which recombine existing alleles. But where did the different genes and alleles come from in the first place? The ultimate source of new genes and new alleles is mutation. Together, these processes provide the raw material for evolution.

Mutations Are the Source of New Genes and Alleles

Cells have efficient mechanisms that protect the integrity of their genes. Enzymes constantly scan the DNA, repairing flaws caused by radiation, chemical damage, or mistakes in copying. Nevertheless, changes in nucleotide sequence can happen. These changes are mutations, and they vary tremendously in their impact. As we explained in Chapter 11, some changes in DNA have virtually no effect on the organism; many, perhaps most, are harmful; and a few may be beneficial or may aid the organism in coping with new or changed environments.

How significant is mutation in altering the gene pool of a population? Mutations are rare, occurring once in 10,000 to 1,000,000 genes per individual in each generation. Therefore, mutation is not a major force in evolution by itself by changing gene frequencies. However, *mutations are the source of new alleles*, new heritable variations upon which other evolutionary processes can work. As such, they are the foundation of evolutionary change. Without mutations there would be no diversity among life forms, and probably no life at all.

As we mentioned earlier, *mutations are not goal-directed*. A mutation does not arise as a result of, nor in anticipation of, environmental necessities (Fig. 17-1). A mutation simply happens and may in turn produce a change in the structure or function of the organism. Whether that

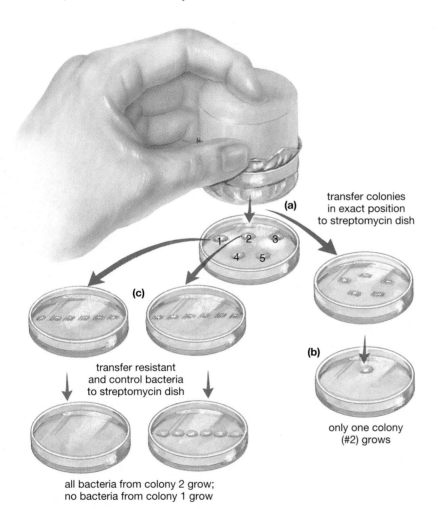

Figure 17-1   **Mutations occur spontaneously**

Experiment supporting the hypothesis that mutations occur spontaneously and not in response to specific selective pressures. **(a)** Several colonies of bacteria, each the offspring of a single individual and thus having the same genetic makeup, are grown on a solid nutrient medium in a dish. These bacteria have never been exposed to antibiotics. A piece of velvet the exact size of the dish is lightly pressed into the bacterial colonies and then touched to the surface of nutrient medium containing the antibiotic streptomycin in a second dish. Many bacteria from each original colony stick to the velvet and then come off the velvet into the second dish. Thus the exact positions of the "parent" colonies are duplicated in the second dish. **(b)** Only one daughter colony, in position #2, grows on the streptomycin-containing medium in the second dish. If mutations for streptomycin resistance were induced by the presence of streptomycin, then one would predict that any colony transplanted to the streptomycin-containing medium would develop these mutations. If, however, streptomycin resistance is an occasional spontaneous occurrence, then one would predict that the only colony grown on a normal medium that would survive is one that might already have been resistant. These alternatives are tested in **(c)**. Samples of the original colonies 1 and 2 are transferred to streptomycin-containing medium. All bacteria from colony 2 grow, but none from colony 1 grow, suggesting that the bacteria of colony 2 already possessed the mutation for streptomycin resistance prior to exposure and that the presence of streptomycin in the medium did not induce an adaptive mutation for streptomycin resistance.

change is helpful or harmful, now or in the future, depends on environmental conditions over which the organism has little or no control. The mutation provides *potential*; other forces, such as migration and especially natural selection, acting on that potential, may favor the spread of a mutation through the population or eliminate it.

Recombination during Sexual Reproduction Provides New Combinations of Existing Alleles

The production of new combinations of alleles occurs in three ways during sexual reproduction. During meiosis, homologous pairs of chromosomes are separated and parceled randomly into gametes. As a result of this segregation of homologues, each human (having 23 paired chromosomes) can produce 8 million different combinations of chromosomes in his or her gametes. Recombination of alleles on individual chromosomes also occurs regularly as a result of crossing over between pairs of homologous chromosomes, adding to the variability. Then, during fertilization, gametes join in random pairs. Together, these events guarantee that each fertilized egg bears a completely unique combination of alleles, or genotype. Many of these new allele combinations produce measurable differences in behavioral, anatomical, or physiological traits (new phenotypes) that in turn alter the fitness of the individual in a particular environment.

The Equilibrium Population Is a Hypothetical Population in Which Evolution Does Not Occur

It will be easier to understand the forces that cause populations to evolve if we first consider the characteristics of a population that would *not* evolve. In 1908, Godfrey H. Hardy and Wilhelm Weinberg defined an **equilibrium population** as one in which the allele frequencies and the distribution of genotypes remain constant with succeeding generations. In other words, the population remains in **genetic equilibrium** (see "A Closer Look: The Hardy-Weinberg Equilibrium Population"). If allele frequencies do not change, evolution does not occur. A population can remain in equilibrium only if several restrictive conditions are met:

1. *There must be no mutation.*
2. *There must be no **gene flow** between populations;* that is, there must be no net migration of alleles into the population (through immigration) or out of the population (through emigration).
3. *The population must be very large* (theoretically infinite).
4. *All mating must be random,* with no tendency for certain genotypes to mate with specific other genotypes.
5. *There must be no natural selection;* that is, all genotypes must be equally adaptive and reproduce equally well.

Under these conditions, allele frequencies within a population will remain the same indefinitely. If one or more of these conditions are violated, then allele frequencies will change: Evolution will occur.

As you might expect, few if any natural populations are truly in equilibrium. If so, then what is the importance of the Hardy-Weinberg principle? The Hardy-Weinberg conditions are useful starting points for studying the mechanisms of evolution. In the following sections, we will examine each condition, show why it is often violated by natural populations, and illustrate the consequences of its violation. In this way, you can better understand both the inevitability of evolution and the forces that drive evolutionary change.

The Mechanisms of Evolution

As the Hardy-Weinberg conditions predict, there are five major causes of evolutionary change within a population: mutation, migration, small populations, nonrandom mating, and natural selection.

Mutations Are the Ultimate Source of Genetic Variability

A population will remain in genetic equilibrium only if no mutations occur. Although mutations are not common, they are an inevitable result of the imperfections in the way DNA is copied as cells reproduce. When such a copying error occurs in a cell that produces gametes, the mutation may be passed to an offspring and enter the gene pool of a population. Over a sufficiently long time span, genetic change as a result of mutation is a certainty.

Migration Produces Gene Flow between Populations

In biology, the word *migration* has two distinct meanings. In the most familiar context, migration refers to the seasonal movement of many species between summer breeding grounds and winter refuges. In evolutionary biology, however, **migration** *is the flow of genes between populations.* Baboons, for example, live in social groupings called troops. Within each troop, all the females mate with a handful of dominant males. Juvenile males usually leave the troop. If they are lucky, they join and perhaps even become dominant in another troop. Thus the male offspring of one troop carry genes to the gene pools of other troops.

Migration has two significant effects.

1. *Gene flow spreads advantageous alleles throughout the species.* Suppose that a new allele arises in one population and that this new allele benefits the organisms that possess it. Migration can carry this new allele to other populations of the species.
2. *Gene flow helps to maintain all the organisms over a large area as one species.* If migrants constantly carry genes back and forth among populations, then the populations can never develop large differences in allele frequencies. Isolation of populations, with no gene flow

to or from other populations of the same species, is a key factor in the origin of new species, as will be discussed in Chapter 18.

Small Populations Are Subject to Random Changes in Allele Frequencies

To remain in genetic equilibrium, a population must be so large that chance events have no impact on its overall genetic makeup. Disaster may befall even the fittest organism. The maple seed that falls into a pond never sprouts; the deer and elk blasted away by the eruption of Mount St. Helens left no descendants. If a population is sufficiently large, chance events are unlikely to alter the overall gene frequency, since they would be expected to interfere equally with the reproduction of organisms of all genotypes. In a small population, however, certain alleles may be carried by only a few organisms. Chance events could reduce or even entirely eliminate such alleles from the population, altering its genetic makeup.

Genetic Drift Is An Example of Random Genetic Change in Small Populations

It is much more likely that chance events will change allele frequencies in a small population than in a large population, by a process called **genetic drift**. Consider, for ex-

ample, two hypothetical populations of haploid ladybugs in which the outer shell is either spotted or solid-colored, controlled by alternate alleles of a single gene. In each population, half the ladybugs are spotted and half are solid-colored (that is, the frequencies of both alleles are 0.5, or 50%), but one population has only four bugs and the other has 1000. Let us assume that each individual that survives to maturity produces two offspring identical to itself. For the population sizes to remain constant, exactly half the individuals must reproduce in each generation. Let us further assume that whether an individual ladybug reproduces is determined entirely by chance.

In the larger population, 500 bugs will be parents to the next generation. Though their survival is random, the odds against all 500 of the reproducing ladybugs being spotted are enormous. In fact, it would be extremely unlikely for even 300 parents to be spotted. In this large population, then, we would not expect a major change in allele frequencies to occur from generation to generation (Fig. 17-2). The effects of chance are minimized by the large population size. In the small population, on the other hand, only two individuals will reproduce. There is a 25% chance that both parents will be spotted (this is the same likelihood as flipping two coins and having both come up heads). If this happens, then the next generation will consist entirely of spotted ladybugs. Within a single generation, it is possible

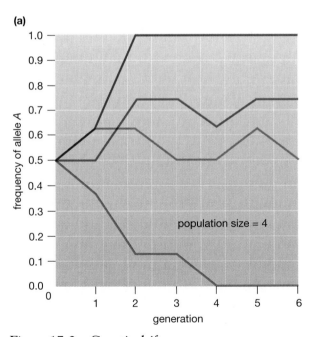

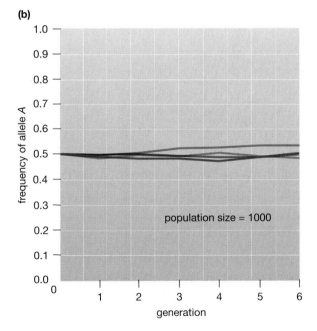

Figure 17-2 **Genetic drift**

Computer-generated graphs illustrating the effect of population size on genetic drift. In both graphs, the initial population was composed of half A and half *a* alleles, and six generations were simulated, with individuals chosen at random to contribute alleles to the next generation. Four simulations were run for each population size, producing the four lines on each graph. **(a)** With a population size of 4, one allele sometimes became "extinct" owing to chance. For example, in the top simulation run, the *a* allele became extinct by the second generation (therefore, the frequency of the A allele became 1.0). **(b)** With a population size of 1000, allele frequencies remained relatively constant.

A   C L O S E R   L O O K

The Hardy-Weinberg Equilibrium Population

The Hardy-Weinberg equilibrium model predicts that, if a large population undergoes no mutation, migration, or natural selection, and if all members of the population mate randomly, then the frequencies of alleles will not change from generation to generation. To see how this can be so, consider our familiar pea plants. Pea seeds can be round (R: dominant) or wrinkled (r: recessive). To determine the genotypes of the offspring of two individuals, for example two heterozygotes, we would draw a Punnett square:

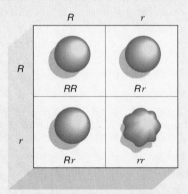

Each parent produces both R and r gametes. The expected offspring are ¼ RR, ½ Rr, ¼ rr. There are two ways to arrive at these frequencies. The first is to add up the offspring in all boxes of the square. Another way of doing it is by probabilities. Each gamete has an equal probability of containing either allele. Therefore we can assign probabilities to the gametes in the Punnett square: $R = 0.5$, $r = 0.5$. From the laws of probability, *the probability of two independent events occurring simultaneously is the product of their individual probabilities*. If you flip a coin, the probability of a head is ½. If you flip two coins simultaneously, the probability of two heads is ½ × ½ = ¼. Similarly, we can obtain the probability of obtaining each type of offspring by multiplying the relative proportions of each allele: We obtain 0.25 RR, 0.5 Rr, and 0.25 rr.

Let us suppose, now, that we have a population of 100 peas, that we collect sperm and egg cells from all of them, and we determine their genotypes. We may find, for example, that there are 60% R alleles and 40% r alleles in the gametes. The proportions of the two alleles R and r are identical to the probability that any given offspring will receive either R or r. We can thus draw a "population Punnett square":

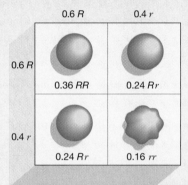

In the "population F$_1$ generation," we expect the following proportions of genotypes: 0.36 RR, 0.48 Rr, and 0.16 Rr. If the population remains the same size, 100 peas, then we will have 36 RR, 48 Rr, and 16 rr peas. What gametes would this F$_1$ generation in its turn produce? Under the Hardy-Weinberg conditions, all plants produce equal numbers of gametes, and by the principles of Mendelian genetics, all plants produce equal numbers of gametes with each of the two alleles for seed shape. To keep things simple, let's assume that each plant contributes two gametes, one with each of its two alleles. We therefore collect 72 R alleles from the homozygous dominants, 48 R alleles and 48 r alleles from the heterozygotes, and 32 r alleles from the homozygous recessives, for a total of 120 R and 80 r alleles. The allele frequencies of the gametes from the population F$_1$ generation, then, are 0.6 R and 0.4 r, just as we started out with. Therefore, the F$_2$ generation has the same distribution of genotypes as the F$_1$. If there are no disturbances, this process will go on indefinitely: The population remains in equilibrium.

Rather than going through Punnett squares, there is an easier way of calculating allele and genotype frequencies. The sum of all allele frequencies must equal 1. Let the frequency of the R allele be represented by p, and the frequency of the r allele by q. Then the sum of the frequencies $p + q = 1$. Just as we generated the genotype frequencies in the population Punnett square by multiplying allele frequencies, we can do the same with this equation:

$$(p + q) \times (p + q) = p^2 + pq + qp + q^2 = p^2 + 2\,pq + q^2 = 1$$

For our particular example, $p = 0.6$ and $q = 0.4$, so the genotypes of the population F_1 generation will be:

$$(0.6)^2 \, RR + 2 \, [(0.6) \times (0.4)] \, Rr + (0.4)^2 \, rr = 0.36 \, RR,$$
$$0.48 \, Rr, \text{ and } 0.16 \, rr$$

This is the same set of frequencies that we calculated with the "population Punnett square."

As these calculations show, in an equilibrium population, allele frequencies and the distribution of genotypes remain constant, generation after generation. In actual experiments, if measurements of allele frequencies in a population show significant changes over time, evolution is occurring in that population.

for the allele for solid-colored shell to completely disappear from the population.

Figure 17-2a illustrates two important points about genetic drift:

1. *Genetic drift tends to reduce genetic variability within a small population.* In extreme cases, all members of a population may become genetically identical (Fig. 17-2a, top line).
2. *Genetic drift tends to increase genetic variability between populations.* Purely as a result of chance, separate populations may evolve extremely different allele frequencies (Fig. 17-2a, top versus bottom lines).

Two special cases of genetic drift, called the population bottleneck and founder effect, further illustrate the enormous consequences that small population size may have on the allele frequencies of a species.

A Population Bottleneck Is an Example of Genetic Drift

In a **population bottleneck**, a species undergoes a drastic reduction in population size, so that only a few individuals contribute genes to the entire future population of the species. As our ladybug example showed, population bottlenecks may cause both *differences in allele frequencies* and *reductions in genetic variability* (Fig. 17-3a). Even if the population then rebounds and the species becomes common, these genetic effects of the bottleneck may remain for hundreds or thousands of generations.

Loss of genetic variability has been documented in the northern elephant seal and the cheetah (Fig. 17-3b). The elephant seal was hunted almost to extinction in the 1800s; by the 1890s only about 20 survived. Because elephant seals breed harem-style, with a single male mating with a stable group of females, one male may have fathered all the offspring at this extreme bottleneck point. The population today has expanded to about 30,000, but biochemical analysis shows that all northern elephant

seals are genetically almost identical. Other species of seals, whose populations have historically always remained large, are much more variable. The rescue of the northern elephant seal from extinction is rightly regarded as a triumph of conservation; however, with very little genetic variation, the elephant seal has much less potential to evolve in response to environmental changes. No matter how many elephant seals there are, the species must be considered to be threatened with extinction. Cheetahs are also genetically uniform, although the reason for the bottleneck is unknown. Consequently, cheetahs too could be gravely threatened by small changes in their environment.

A special case of a population bottleneck is the **founder effect**, which occurs when isolated colonies are founded by a small number of organisms. A flock of birds, for instance, may become lost during migration or may be blown off course by a storm (this is thought to have happened in the case of Darwin's Finches in the Galapagos Islands). Among humans, small groups may migrate for religious or political reasons (Fig. 17-4). Such a small group may have allele frequencies that are very different from the frequencies of the parent population because of chance inclusion of disproportionate numbers of certain alleles in the founders. If the isolation of the founders is maintained for a long period of time, a sizable new population may arise that differs greatly from the original population.

How much does genetic drift contribute to evolution? No one really knows. Only rarely are natural populations extremely small or completely cut off from gene flow from other populations. Populations occasionally do become very small, however, and it may be precisely these small populations that contribute most to major evolutionary changes. As we will see in the next chapter, biologists believe that new species often arise in small populations.

(a)

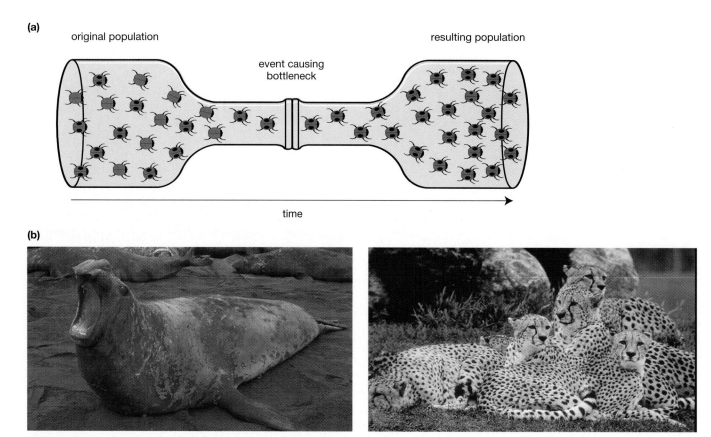

original population   resulting population

event causing
bottleneck

time

(b)

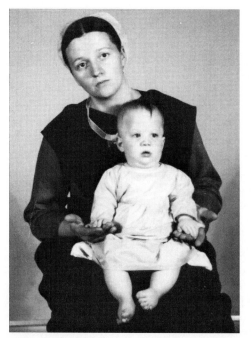

Figure 17-3   **Genetic bottlenecks reduce variability**

(a) If a population is reduced to a very small number of individuals, the gene pool is reduced and a population bottleneck occurs. The recovered population shows reduced genetic and phenotypic variability, because all are offspring of the few organisms that survived the bottleneck. **(b)** Both the northern elephant seal (left) and the cheetah (right) passed through a population bottleneck in the recent past, resulting in an almost total loss of genetic diversity. As a result, the ability of these populations to adapt to changing environments is very limited.

Figure 17-4   **A human example of the founder effect**

An Amish woman with her child, who suffers from a set of genetic defects known as the Ellis-van Creveld syndrome (short arms and legs, extra fingers, occasionally heart defects). Fleeing from religious persecution, about 200 members of the Amish religion migrated from Switzerland to Pennsylvania between 1720 and 1770. Since that time, virtually all the Pennsylvania Amish moved to Lancaster County and have remained reproductively isolated from non-Amish Americans. The population increased to about 8000 by 1964. In that year, geneticist Victor McKusick surveyed the Lancaster County Amish and discovered that they had an allele frequency for Ellis-van Creveld of about 0.07, compared with a frequency of less than 0.001 in the general population. Why? One couple who immigrated in 1744 carried the allele. Inbreeding among the Amish passed the allele along to their descendants: a clear example of a founder effect. In addition, by chance the Ellis-van Creveld carriers had more children than the Amish average, further increasing the allele frequency by genetic drift. The combination of an initially high frequency in the immigrants (1 or 2 out of 200) plus genetic drift has resulted in more cases of Ellis-van Creveld syndrome from Lancaster County than from the entire rest of the world.

Mating within a Population Is Almost Never Random

Organisms seldom mate strictly randomly. For example, most animals have limited mobility and are most likely to mate with nearby members of their species. Further, they may choose to mate with certain individuals of their species rather than with others. The White-crowned Sparrow is a case in point. Although all White-crowned Sparrows sing a fundamentally similar song, each local population has its own song dialect. A female usually chooses a mate that sings the same dialect that her father sang (Fig. 17-5). Among animals, there are three common forms of nonrandom mating: harem breeding, assortative mating, and sexual selection.

In some species, such as elephant seals, baboons, and bighorn sheep, only a few males fertilize all the females. Following some sort of contest, which may involve showing off with loud sounds or flashy colors, making threat-ening gestures, or actual combat, only certain males suc-ceed in gathering a harem and mating (Fig. 17-6).

Many animals mate assortatively—that is, they select mates that are similar to themselves. Humans, for exam-ple, tend to marry members of the opposite sex that are similar in height, race, intelligence, and social status.

Finally, in many mammal and bird species, mate selec-tion is primarily the prerogative of one sex, usually the fe-male. Males display their virtues, such as the bright plumage of a peacock (Fig. 17-7) or the rich territory of a songbird. A female evaluates the males and chooses her mate. This phenomenon, called **sexual selection**, is ex-plored in more detail later in this chapter.

All Genotypes Are Not Equally Adaptive

Genetic equilibrium requires that all genotypes must be equally adaptive—that is, none has any selective advan-tage over the others. It is probably true that some alleles

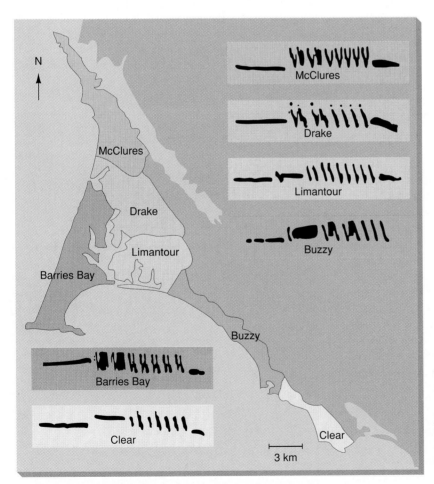

Figure 17-5   **Song-dialect preference illustrates nonrandom mating**

Song dialects among populations of White-crowned Sparrows at Point Reyes National Seashore north of San Francisco. As analysis of the sound patterns shows, the songs are fairly similar, but both birds and human listeners can recognize the different dialects. Male birds of each population learn their local dialect while in the nest and sing it when they mature. Fe-males preferentially mate with males that sing the dialect sung by the females' own fathers—that is, the females' own local dialect.

Figure 17-6   **Male competition promotes nonrandom mating**

Sparring contests between males result in extremely nonrandom mating among many animals, including deer, elk, seals, and many monkeys. Here, two male bighorn sheep square off against each other during the fall mating season. Although the horns are potentially deadly weapons, they are used in ritualized ways that minimize the danger of injury to either contestant.

are adaptively neutral, so organisms possessing any of several alleles will be equally likely to survive and reproduce. However, this is clearly not true of all alleles in all environments. Any time an allele confers, in Wallace's words, "some little superiority," natural selection will favor the enhanced reproduction of the individuals possessing it.

Natural selection is not the *only* evolutionary force. As we have seen, mutation provides initial variability in heritable traits. The chance effects of genetic drift may change allele frequencies, even spawning new species. Further, evolutionary biologists are just now beginning to appreciate the power of random catastrophe in shaping the history of life on Earth—mass extinctions that may exterminate flourishing and floundering species alike. Nevertheless, it is natural selection that shapes the evolution of adaptations that humankind has admired for millennia. For this reason, we will examine the mechanisms of natural selection in some detail.

Natural Selection

To most people, the words *natural selection* are synonymous with the phrase *survival of the fittest*. Natural selection evokes images of wolves chasing caribou, of lions snarling angrily in competition over a zebra carcass. Natural selection, however, is not really about *survival* but about *reproduction*. It is certainly true that an organism must survive at least for a while to live long enough to reproduce. In some cases, it may also be true that a longer-lived organism has more chances to reproduce. But no organism lives for-

ever, and the only way that its genes continue into the future is through successful reproduction. When an organism that fails to reproduce dies, its genes die with it. The organism reproduces, lives on, in a sense, through the genes that it has passed on to its offspring. Therefore, although evolutionary biologists often discuss survival, partly because survival is usually easier to measure than reproduction, natural selection is really an issue of **differential reproduction**: Individuals bearing certain alleles leave more offspring (who inherit those alleles) than other individuals with different alleles.

Natural Selection Acts on the Phenotype, Which Reflects the Underlying Genotype

The agents of natural selection cannot directly detect an organism's genotype. Rather, selection acts on phenotypes: the actual structures and behaviors that the organisms in a population display. Genotype and phenotype, however, are related in the following way. If you were to measure the phenotypes of a specific trait in all the individuals in a population, you would find a range of values (Fig. 17-8). This range of phenotypes would arise from differences both in the genotypes of the organisms and in the environments in

Figure 17-7   **The male peacock's showy tail has evolved through sexual selection**

Many male birds, including peacocks, attract mates by displaying their "wares." The features evolved for female attraction are often irrelevant, or even harmful, to the day-to-day survival of the males.

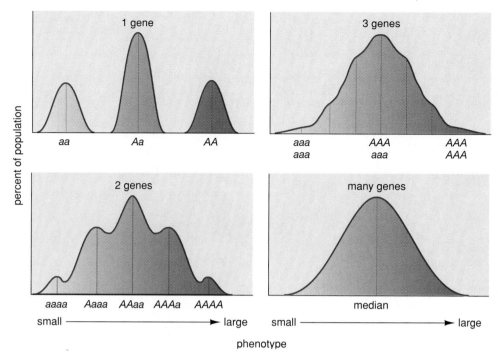

Figure 17-8   **Both genes and environment contribute to the phenotype**

This series of graphs illustrates the distribution of phenotypes that would be expected if one, two, three, or many genes, each with two incompletely dominant alleles (see Chapter 13), contributed to a particular body characteristic (for example, size). The vertical lines represent the precise size expected on the basis of genotype alone. In each case, environmental conditions (for example, amount of available food) create some variation in size, represented by the colored curves. As the number of genes contributing to the characteristic becomes large, the distribution of phenotypes approximates a smooth curve called a normal distribution (lower right-hand graph). The most common value for the phenotype is the middle value, also called the median.

which they live. However, environmental differences influencing phenotypes tend to average out in a large population. Thus, on the whole, genotype reflects phenotype: Most large plants will have genes promoting large size, whereas most small plants will have genes promoting small size. In our discussion of selection, therefore, we will ignore environmental causes of variability.

Natural Selection Can Influence Populations in Three Major Ways

Biologists recognize three major categories of natural selection based on its effect on the population over time (Fig. 17-9):

1. **Directional selection** favors individuals possessing values for a trait at one end of the distribution (representing a range of a particular trait) and selects against both average individuals and individuals at the opposite extreme of the distribution (for example, favors small size, selects against both average and large individuals).
2. **Stabilizing selection** favors individuals possessing an "average" value for a trait and selects against individuals with extreme values.

3. **Disruptive selection** favors individuals possessing relatively extreme values for a trait at the expense of individuals with average values. Disruptive selection favors organisms at both ends of the distribution of the trait (for example, favors both large and small body size).

Directional Selection Shifts Character Traits in a Specific Direction

If environmental conditions change in a consistent way—for example, if the climate becomes colder—then a species may evolve in a consistent direction in response, for example with thicker fur (Fig.17-9a). The evolution of long necks in giraffes was almost certainly due to directional selection: Ancestral giraffes with longer necks obtained more food and therefore reproduced more prolifically than their shorter-necked contemporaries did. Antibiotic resistance in bacteria is another example of directional selection (see Chapter 21).

How fast can directional selection change genotypes? That depends on both the genetic nature of the variability in the population and the strength of selection. The increased frequency of the black form of the peppered moth

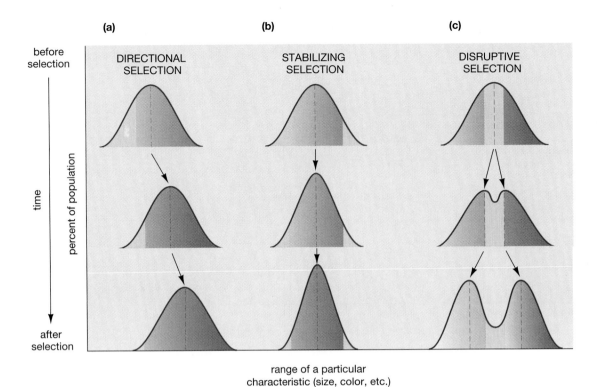

Figure 17-9   Three types of natural selection

A graphical illustration of three types of natural selection, acting on a normal distribution of phenotypes (in these examples, size). In all graphs, the pale areas represent individuals that are selected against—that is, do not reproduce as frequently. **(a)** In directional selection, phenotypes that are either larger or smaller than average (larger illustrated here) are favored. The average phenotype shifts position over the generations. **(b)** In stabilizing selection, the organisms most likely to reproduce are those with phenotypes close to the average for the population. The variability of phenotypes may decline, but the average value remains the same. **(c)** In disruptive selection, phenotypes that are both larger and smaller than average are favored. The population splits into two phenotypic groups.

in Britain during the Industrial Revolution was an extremely rapid case of directional selection (see Chapter 16). In that instance, the color did not vary in a finely graded manner but was either black or pale, controlled by two alleles of a single gene. Predation by birds was also a very strong selective force because pale moths were more visible on increasingly soot-covered tree trunks. Together, these two factors produced a dramatic change in the population in just a few years. If little variability exists in the population, or if the different alleles produce only slightly different phenotypes, then directional selection will drive much slower changes. In some instances, a population may not be able to respond fast enough to the selective forces and may become extinct.

Stabilizing Selection Acts against Individuals Who Deviate Too Far from the Average

Directional selection can't go on forever. Once a species is well adapted to a particular environment, and if the environment doesn't change, then most variations that appear through new mutations or recombination of old alleles will

be harmful. Therefore, the species will often undergo stabilizing selection, which favors the survival and reproduction of "average" individuals (see Fig. 17-9b). Stabilizing selection often occurs when a single trait is under opposing selective pressures from two different sources. Biologist M. K. Hecht, for example, studied lizards of the *Aristelliger* genus. He found that small lizards had a hard time defending territories, but large lizards were more likely to be preyed upon by owls. Therefore, *Aristelliger* lizards were under stabilizing selection favoring an "average" body size.

It is widely assumed, although difficult to prove, that many traits are under stabilizing selection. We have already mentioned several. Although the lengths of legs and necks of giraffes probably originated under directional selection for feeding on leaves high up in trees, they are almost certainly now under stabilizing selection, balancing the demands of eating and drinking. Similarly, female mate choice probably drove the evolution of elaborate sexual displays in many birds, but now increased vulnerability to predation may exert stabilizing selection: If a peacock's tail became so long that he couldn't fly, he would be unlikely to live long enough to woo a female.

Under certain circumstances, stabilizing selection may act not to eliminate variability, but to maintain it. Opposing selective pressures often give rise to **balanced polymorphism**, in which two or more alleles of a gene are maintained in a population because each is favored by a separate selective force. This seems to have occurred with the hemoglobin alleles in native Africans (see Chapter 14). The hemoglobin molecules of people who are homozygous for sickle-cell anemia (having two alleles for defective hemoglobin) clump up into long chains, distorting and weakening their red blood cells. This distortion causes severe anemia and potentially death. Before the advent of modern medicine, people homozygous for sickle-cell anemia were strongly selected against. Heterozygotes, who have one allele for defective hemoglobin and one allele for normal hemoglobin, suffer only mild anemia, though they may be adversely affected during strenuous exercise. Under these circumstances, you might wonder why natural selection has not eliminated the sickle-cell allele. Far from being eliminated, however, the sickle-cell allele is carried by nearly half the people in some areas of Africa. This distribution seems to result from the counterbalancing effects of anemia and malaria, a disease that formerly caused high death rates in equatorial Africa.

Malaria parasites multiply rapidly within the red blood cells of homozygous normal individuals. Before effective medical treatments were discovered, homozygous normals consequently often died of malaria. Heterozygotes, on the other hand, enjoy some protection against malaria. Malaria parasites inside a heterozygote's red blood cells use up oxygen, causing the sickle-cell hemoglobin to clump and the cells to become sickle shaped. Infected, sickled cells are destroyed by the spleen before the parasites can complete their development. Heterozygotes, therefore, have mild anemia but do not succumb to malaria. During the evolution of African populations, heterozygotes survived better than either type of homozygote and reproduced the most. As a result, both the normal hemoglobin allele and the sickle-cell allele have been preserved (Fig. 17-10).

Disruptive Selection Adapts Individuals within a Population to Different Habitats

Disruptive selection (see Fig. 17-9c) may occur when a population occupies an area that provides different types of resources that can be utilized by the species. In this situation, different characteristics best adapt individuals to use each type of resource. For example, an island, such as one of the Galapagos, may have several species of plants, some producing large, hard seeds and others small, soft seeds. Large seeds provide the most food per seed, but they can be cracked and eaten only by birds with large bills. Although large birds can easily eat small seeds, they would probably spend too much energy lugging their large bodies about looking for tiny seeds. If a single species of bird colonized such an island, what would happen? We would expect that larger-bodied, larger-beaked birds would spe-

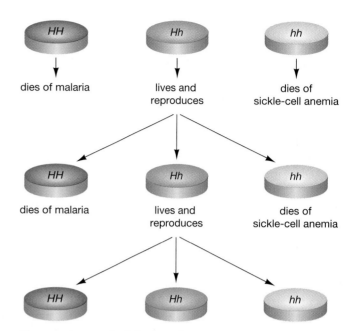

Figure 17-10   Stabilizing selection can produce balanced polymorphism

Sometimes two or more alleles, each producing a different phenotype, can be maintained in a population by opposing selection pressures. The alleles for normal (*H*) and sickle-cell (*h*) hemoglobin are maintained by selection against both homozygotes. Heterozygotes (*Hh*) reproduce the most, thereby keeping both alleles present in the population.

cialize on large seeds, while small-bodied, small-beaked birds would specialize on small seeds. Medium-sized birds might not be able to crack open the large seeds and might not get enough energy from small seeds, and so would be selected against. Disruptive selection would favor the survival and reproduction of both large and small, but not medium-sized, birds. Disruptive selection has not been extensively studied, although it has been documented or at least supported by studies of both butterflies and birds.

Natural Selection Takes Several Forms

Natural selection acts by eliminating individuals that do not have the characteristics needed for survival and reproduction in their environment. Characteristics that do help an individual to survive and reproduce in a particular environment are **adaptations**. The process of acquiring these characteristics is also called adaptation because *the end result of natural selection is adaptation to the environment.* An organism's environment can be divided into two components: the abiotic (nonliving) part and the biotic (living) part that consists of other organisms. Adaptations to both biotic and abiotic components occur through natural selection.

The abiotic environment includes physical factors such as climate, availability of water, and minerals in the soil. The abiotic environment provides the "bottom line" requirements that an organism must have to survive and reproduce. However, many of the adaptations that we see in modern organisms have arisen because of interactions with other organisms. As Darwin wrote, "the structure of every organic being is related . . . to that of all other organic beings, with which it comes into competition for food or residence, or from which it has to escape, or on which it preys." A simple example illustrates this concept.

A buffalo grass plant sprouts in a small patch of soil in the eastern Wyoming plains. Its roots must be able to take up enough water and minerals for growth and reproduction, and to that extent it must be adapted to its abiotic environment. Even in the dry prairies of Wyoming, this is a relatively trivial requirement *provided that the plant is alone and protected in its square meter of soil.* In reality, many plants, including other grasses, sagebrush bushes, and annual wildflowers also sprout in that same patch of soil. If our buffalo grass is to survive, it must compete for resources with the other plants. Its long, deep roots and efficient mineral uptake processes have evolved not so much because the plains are dry, but because it must share the dry prairies with other plants. Further, cattle (formerly bison) graze the prairies. Buffalo grass is extremely tough, with silica (glass) compounds reinforcing the blades, an adaptation that discourages grazing. Over millennia, tougher plants were harder to eat and so survived better and reproduced more—another adaptation to the biotic environment.

When two species or two populations of a single species interact extensively, each exerts strong selective pressures on the other. When one evolves a new feature or modifies an old one, the other often evolves new adaptations in response. As the Red Queen told Alice in *Through the Looking Glass,* "Here, you see, it takes all the running you can do to keep in the same place." This constant, mutual feedback between two species is called **coevolution**.

Competition for Scarce Resources Favors the Best-Adapted Individuals

One of the major selective forces in the biotic environment is **competition** with other members of the same species. As Darwin wrote in *On the Origin of Species:* "The struggle almost invariably will be most severe between the individuals of the same species, for they frequent the same districts, require the same food, and are exposed to the same dangers." In other words, no competing organism has such similar requirements for survival as another member of the same species. For example, both Lazuli Buntings and Western Bluebirds are brightly colored in blue, red, and white, and both nest and rear their young in the foothills of the Rocky Mountains in the summer. But they do not compete very much with each other, because they eat different foods: Bluebirds mostly catch insects, while buntings specialize in seeds. Each mosquito picked off by a bluebird makes little difference to a bunting, but makes it harder for other bluebirds to find enough to eat.

Different species may also compete for the same resources, although generally to a lesser extent. As we will

discuss more fully in Chapter 46, whether a particular plot of prairie is covered with grass, sagebrush, or trees is at least partly determined by competition among those plants for scarce soil moisture.

During Predation, Both Predator and Prey Act as Agents of Selection

Although we commonly think of predation as one animal preying upon another animal, predation actually includes any situation in which one organism eats another. In some instances, coevolution between predators and prey is a sort of "biological arms race," with each side evolving new adaptations in response to "escalations" by the other. Darwin used the example of wolves and deer: Wolf predation selects against slow or incautious deer, thus leaving faster, more alert deer to reproduce and continue the species. In their turn, alert, swift deer select against slow, clumsy wolves, because such predators cannot acquire enough food.

Symbiosis Produces Adaptations between Species That Live in Intimate Association with One Another

Symbiosis is any relationship in which individuals of different species closely interact for an extended time. Examples of symbiosis are parasitism, in which one species lives

Figure 17-11   An example of symbiosis

Several species of clownfish live in a symbiotic relationship with anemones, each species of fish favoring a particular species of anemone. The fish nestle within the stinging tentacles of the anemone, thus protected from being eaten by other fish. The clownfish evolved specialized skin secretions and behaviors, protecting it from being affected by the anemone. The fish may accidentally drop food onto the anemone once in a while, but the benefits are probably fairly one-sided.

and feeds on a larger species; commensalism, in which one species benefits and the other remains unharmed; and mutualism, in which both species benefit. The different types of symbiosis are described in Chapter 44. From an evolutionary perspective, symbiosis leads to the most intricate coevolutionary adaptations. Although a given predator usually preys on several species and may interact with a particular species only occasionally, partners in symbiosis often live together virtually their entire lives (Fig. 17-11). At least one of the partners, and usually both, must continually adjust to any evolutionary changes developed by the other.

Sexual Selection Sometimes Seems to Oppose Other Forms of Natural Selection

As mentioned earlier, in many species of birds and mammals and even some fish, one of the sexes, usually the female, selects the mate. Males compete for the attention of females through song, elaborate displays, the defense of large territories, or even by building elaborate structures such as that of the bowerbird (see Chapter 42). Choosing a male with a good territory is obviously advantageous, since good territories provide adequate food and shelter to raise young. However, females often also prefer elaborate "fashions" in their mates, such as bright colors and long feathers or fins that may make the male more vulnerable to predation. Why? A popular hypothesis is that structures and colors that do not serve any clear adaptive purpose actually provide the females with an outward sign of the males' fitness. Only vigorous, energetic males can survive when burdened with conspicuous coloration or large tails. Similarly, males sick or under parasitic attack

may be dull and frumpy compared with healthy males. Whatever the exact selective mechanisms, it is thought that many of the elaborate structures and behaviors found only in males have evolved through the selective pressure of female mate choice: Only the flashy males transmitted their genes to the next generation.

Darwin was so impressed with these structures that he coined the term *sexual selection* to designate the process of evolution through mate choice. Because conspicuous structures and bizarre behaviors make the males more vulnerable to predators, sexual selection often seems to work in opposition to other forms of natural selection. The trade-off between sexual selection and natural selection through predation has recently been demonstrated through observation and experiment in guppies (small freshwater fish). In streams where predation is a threat, male guppies are inconspicuous, blending with the sandy streambed. But in safer waters, male guppies show more conspicuous colors, apparently as a result of female preference for these markings. University of California biologist John Endler recently transplanted camouflage-colored male guppies from dangerous waters into safe waters. Within a year (about 20 guppy generations) their protected descendants had evolved conspicuous colors. However, nonsexual selective forces may also oppose one another; the height of a giraffe, for example, is a compromise between the advantage of reaching higher leaves for food and the disadvantage of vulnerability while drinking water (Fig. 17-12). In both sexual and nonsexual selection, then, some aspect of the environment (in sexual selection the "opinion" of the opposite sex, which is part of the social environment) influences reproductive success.

(a)

(b)

Figure 17-12 **A compromise between opposing selection pressures**

(a) The long neck and legs of a giraffe are a decided advantage in feeding on acacia leaves high up in trees. **(b)** But a giraffe has to get into an extremely awkward and vulnerable position to drink. Feeding and drinking thus place opposing selective pressures on the length of neck and legs.

Figure 17-13 **Altruism between mother and offspring**

A female killdeer lures a predator away from its nest by faking injury. The mother places herself in some small danger (she can always fly away if the predator comes too close) but saves her offspring from much greater danger.

Kin Selection Favors Altruistic Behaviors

Evolution is often portrayed in the popular press as being "red in tooth and claw." This image of bloody and vicious interaction, however, is not the complete picture. Although it is true that competitive and predatory interactions influence the evolution of most species, cooperation and even self-sacrifice can be important selective forces too. **Altruism** refers to any behavior that endangers an individual organism or reduces its reproductive success but benefits other members of its species. Altruistic behaviors are common in the animal kingdom. A mother killdeer flutters just out of reach of a predator, feigning an injured wing and luring the predator away from her nest (Fig. 17-13); female worker bees forego reproduction and devote their lives to raising the offspring of the hive queen (see Chapter 42); and young male baboons scout around the edges of the troop, even though doing so increases their danger from leopards.

You might think that altruism runs counter to natural selection: If altruism is encoded in an organism's genes, those genes are placed at risk every time the altruist performs one of its brave behaviors. But natural selection can indeed select for altruistic genes, if the altruistic individual helps relatives who possess the same alleles. This special case of natural selection is an example of **kin selection** and is explored in "Evolutionary Connections: Kin Selection and the Evolution of Altruism."

Extinction

Natural selection not only produces the fleet limbs of the antelope and the exquisite eyes of the goshawk. It may also lead to the death of all the members of a species, **extinction.** Trilobites, dinosaurs, saber-tooth cats—all are extinct, known

only from fossils. Paleontologists estimate that *at least* 99.9% of all the species that ever existed are now extinct. Why? The actual cause of extinction is probably always environmental change, either in the living or the nonliving parts of the environment. Two characteristics seem to predispose a species to extinction when the environment changes: localized distribution and overspecialization. Three major changes that drive species to extinction are competition among species, novel predators or parasites, and habitat destruction.

Localized Distribution and Extreme Specialization Make Species Vulnerable in Changing Environments

Species vary widely in their range, and hence in their susceptibility to extinction. Some species, such as herring gulls, white-tailed deer, and humans, inhabit entire continents or even the whole Earth, while others, such as the Devil's Hole pupfish (Fig. 17-14), have extremely limited ranges. Obviously, if a species occurs in only a very small area, any disturbance of that area could easily result in extinction. If Devil's Hole dries up from climatic change or well drilling nearby, its pupfish will immediately vanish. Wide-ranging species, on the other hand, usually do not succumb to local environmental catastrophes.

Another factor that may make a species vulnerable to extinction is extreme specialization. Each species evolves a set of genetic adaptations in response to pressures from its particular environment. Sometimes these adaptations limit the organism to a very specialized set of environmental conditions. The Everglades Kite, for example, feeds only on a certain freshwater snail (Fig. 17-15). As the swamps of the

Figure 17-14   **Very localized distribution can endanger a species**

The Devil's Hole pupfish is found in only one spring-fed water hole in the Nevada desert. During the last glacial period, the southwestern deserts received a great deal of rainfall, forming numerous lakes and rivers. As the rainfall decreased, pupfish populations were isolated in shrinking small springs and streams. Isolated small populations and differing environmental conditions caused the ancestral pupfish species to split up into several very restricted modern species, all of which swim on the brink of extinction.

Figure 17-15   **Extreme specialization places species at risk**

The Everglades Kite feeds exclusively on the apple snail, found in swamps of the southeastern United States. Such behavioral specialization renders the kite extremely vulnerable to any environmental change that may exterminate its single species of prey.

American Southeast are drained for farms and developments, the snail population shrinks. If the snail becomes extinct, the kite will surely go extinct along with it.

In the fossil record, such behavioral specialization is hard to recognize. Structural specializations, however, may be just as restrictive. A case in point is giantism. For poorly understood reasons, many animals evolved huge size, including certain amphibians, dinosaurs, and giant mammals, such as mammoths and ground sloths. To support their bulk, these animals must have consumed enormous amounts of food. If environmental conditions deteriorated, those giants may have been unable to find enough food and thus died out. Smaller animals that ate the same food but needed less of it survived.

Interactions with Other Organisms May Drive Species to Extinction

As described earlier, interactions such as competition, predation, and parasitism serve as forces of natural selection. In some cases, these same forces can lead to extinction, rather than adaptation.

Competition for limited resources occurs in all environments. If a species' competitors evolve superior adaptations, and it doesn't evolve fast enough to keep up, it may become extinct. A particularly striking example of extinction through competition occurred in South America 2 to 3 million years ago. For millions of years North and South America were isolated from one another, and each developed a distinctive array of animal life. When the Panamanian land bridge arose, connecting the two continents, massive migrations took place. In general, North American animals displaced their South American counterparts, and many South American species became extinct.

When formerly isolated populations encounter one another, it isn't only competitors that migrate between the

areas—predators and parasites do too. With the exception of humans, who have exterminated hundreds of species, predators probably cause few extinctions. Parasites, on the other hand, can be devastating. In North America, Dutch elm disease and chestnut blight are well-known instances of introduced parasites that almost completely destroyed widespread native species. We cannot tell much about prehistoric parasite invasions, but the extinction of South American animals, mentioned above, might have been at least partly due to diseases carried south by resistant North American migrants.

Habitat Change and Destruction Are the Leading Causes of Extinction

Habitat change, both contemporary and prehistoric is the single greatest cause of extinctions. Presently, habitat destruction due to human activities is proceeding at a frightening pace. Perhaps the most rapid extinction in the history of life will occur over the next 50 years, as tropical forests are cut for timber and to clear land for cattle and crops. As many as half the species presently on Earth may be lost because of tropical deforestation.

Prehistoric habitat alteration usually occurred over a longer time span but nevertheless had serious consequences. Climate changes, in particular, caused many extinctions. Several times, moist, warm climates gave way to drier, colder climates with more variable temperatures. Many plants and animals failed to adapt to the harsh new conditions and became extinct. One cause of climate change is continental drift (Fig. 17-16). As the continents flow about over the surface of Earth, they change latitudes. Much of North America was located around the equator many millions of years ago, an area characterized by consistently warm and wet tropical weather. But drift carried the continent up into temperate and arctic regions. As a result, the tropical weather was replaced by cooler temperatures, less rainfall, and seasonal changes.

An extreme, and very sudden, type of habitat destruction might be caused by catastrophic geological events, such as massive volcanic eruptions. Several prehistoric eruptions, which would make the Mount St. Helens explosion look like a firecracker by comparison, wiped out every living thing for dozens of miles around and probably caused global climatic changes as well.

The fossil record reveals episodes of extensive worldwide extinctions, especially among marine life (Fig. 17-17). Enormous meteorites, several kilometers in diameter, may have hit Earth at these times. If a huge meteorite struck land, it would kick up enormous amounts of dust. The dust might be thick enough, and spread widely enough, to block out most of the sun's rays. Fires started by the impact might be widespread, adding soot to the atmosphere. Many plants would die because they couldn't photosynthesize. Many animals, all of which ultimately depend on plants for food,

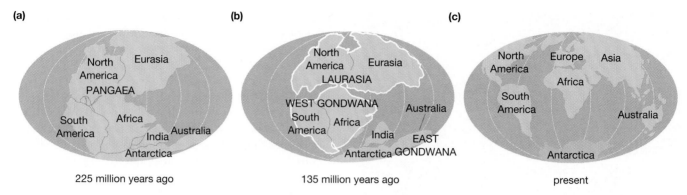

(a) 225 million years ago

(b) 135 million years ago

(c) present

Figure 17-16 **Continental drift has caused climate change**

Although slow, continental drift can cause tremendous environmental changes, as land masses are moved about on the surface of Earth. The solid surfaces of the continents slide about over a viscous, but fluid, lower layer. **(a)** About 225 million years ago, all the continents were fused together into one gigantic land mass, which geologists call Pangaea. **(b)** Gradually Pangaea broke up into Laurasia and West and East Gondwana. **(c)** Further drift eventually resulted in the modern positions of the continents. Continental drift continues today: The Atlantic Ocean, for example, widens by a few centimeters each year.

would also die. Smaller amounts of dust might still block out enough sunlight to cause global cooling, perhaps even triggering an ice age. Widespread extinctions would result.

Did such massive meteorite strikes really occur, and if so, would they cause extinctions? No one knows for sure, but considerable evidence points to meteorites as the causes of at least some major extinctions (Fig. 17-18). Recently, two groups of researchers have suggested that the Chicxulub crater near the Yucatan Peninsula of Mexico was the impact site of the meteorite that might have killed the dinosaurs.

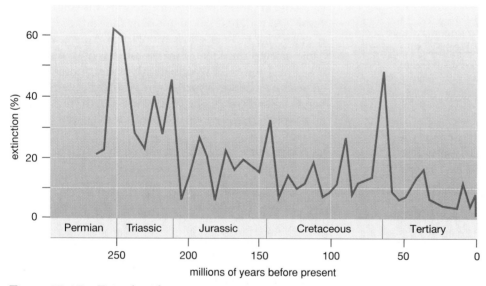

Figure 17-17 **Episodes of mass extinction**

This graph plots the percentage of genera of marine animals that have become extinct during geologic time. The higher the peak, the greater the extent of extinctions. Marine animals were chosen for this study because their fossils are abundant in sedimentary rocks and are easily dated. The large peak of extinctions near the boundary between the Cretaceous and Tertiary periods also approximately coincides with the end of the dinosaurs. Many paleontologists are convinced that the dinosaurs were going downhill for millions of years before this time, so there is hot debate over whether the Cretaceous-Tertiary extinction event (such as a meteorite impact) provided the final blow to the dinosaurs or whether the dinosaurs would have died out at that time anyway.

EARTH WATCH

Natural Selection, Genetic Diversity, and Endangered Species

Ever since the Endangered Species Act was passed in 1976, the United States has had an official policy of protecting rare species. In fact, the real goal of the act is not protection but recovery; as one U.S. Fish and Wildlife Service official put it, "the goal is to get species *off* the list." Wildlife biologists try to determine how large a population a species needs to have before it is no longer in danger of extinction from unpredictable events, such as a couple of years of drought or an epidemic of parasites. If a species reaches this critical population size, it is no longer legally "endangered" with extinction.

Does a "large enough" population (which usually is still very small by historical standards) really ensure a species' survival? From our discussion of genetic drift and population bottlenecks, you probably realize that the answer is no. If the population of a species has been reduced to the point where it is placed on the endangered species list, then it probably has lost much of its genetic diversity. As ecologist Thomas Foose aptly put it, loss of habitat and consequent reduction in population size mean that "gene pools are being converted into gene puddles." Even if the species recovers in numbers, its original gene pool has been lost. When the forces of natural selection change at some future time, the species may not have the necessary genetic variability to produce individuals adapted to the new environment, and it may become extinct.

What can be done? The best solution, of course, is to leave enough habitat of diverse types so that species never become endangered in the first place. The human population, however, has grown so large and appropriated so much of Earth's resources that this solution is not possible in many places (Fig. E17-1). For many species, the only so-lution is to preserve enough habitat so that the remaining population is large enough to retain all or most of the total genetic diversity of the species. If we are to be, as we imagine ourselves, the caretakers of the planet and not merely its ultimate consumers, then protection of other life forms and their genetic heritage will be a continuing responsibility as long as humankind exists.

Figure E17-1   Endangered by habitat destruction

This orangutan and its young, who live in the tropical rain forests of Borneo, are among the innumerable species whose continued survival is threatened by habitat destruction.

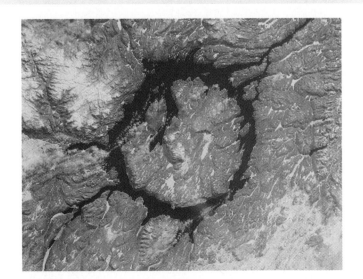

Figure 17-18   Meteorites may have triggered some mass extinctions

The Manicouagan crater in Quebec is about 45 miles in diameter. Giant impact craters such as this one often contain a central dome of rock that splashes up after the meteorite has buried itself in the ground, leaving a ring-shaped depression between the outer crater wall and the inner dome. In this satellite photo, water backed up behind a dam fills the crater ring. The Manicouagan meteorite struck a little over 200 million years ago; geologist Paul Olsen suggests that its impact triggered the mass extinction near the Triassic-Jurassic boundary (see Fig. 17-17).

E V O L U T I O N A R Y
C O N N E C T I O N S

Kin Selection and the Evolution of Altruism

Altruism is any behavior that is potentially harmful to the survival and future reproduction of an individual organism but enhances the reproductive potential of other organisms. Altruism includes worker bees' rearing the offspring of queen bees. Note that altruism does not imply conscious, voluntary decisions to engage in selfless behavior. Rather, most altruistic behaviors have a strong instinctive component; that is, many animals have altruism programmed in their genes.

From an evolutionary viewpoint, how can this be? Surely, if a mutation arose that caused altruistic behavior, and the bearers of that mutation lost their lives or failed to reproduce because of their self-sacrificing behaviors, their "altruistic alleles" would disappear from the population. Maybe, or maybe not. To understand the evolution of altruism, we will need to introduce a new concept: **inclusive fitness**. As formulated by W. D. Hamilton, the inclusive fitness of an allele is the fitness conferred on *all* organisms that have the allele. Therefore, *if an altruist benefits related members of its own species that bear the same altruistic allele, then the altruistic allele may be favored by natural selection.*

To see how altruism might increase the inclusive fitness of an allele, let's consider the Florida Scrub Jay. Year-old jays usually do not mate and reproduce. Instead, these yearlings remain at their parents' nest and help out with next year's brood. Let's assume, for simplicity, that this altruistic behavior is controlled by a single "altruistic" allele and that in the distant past helper jays had the altruistic allele, and nonhelpers had another, "selfish," allele.

At least for one year, altruistic yearlings do not reproduce; some probably die from predation or accidents and never reproduce at all. How, then, can this behavior be adaptive? It all has to do with relatedness and the probability of successful reproduction. First, an animal's offspring inherit 50% of its genes (the other 50% of the offspring's genes come from the other parent). On the average, an animal also shares 50% of its genes with its siblings. Therefore, *a scrub jay is just as related to its siblings as it would be to its own offspring.* The second factor influencing scrub jay reproduction is that ideal habitat for jays is limited. Inexperienced yearling jays would probably have a hard time acquiring a good nest site and would be hard put to feed their offspring. Their best "reproductive bet," then, is to put their energy into helping their parents. Selfish yearling jays that try to nest on their own will probably contribute fewer genes to the next generation than the altruistic yearlings do. This phenomenon, whereby the actions of an individual increase the survival or reproductive success of its relatives, is called kin selection.

As this example suggests, *kin selection can favor the evolution of altruism if the altruistic behavior benefits relatives that bear the same altruistic allele.* In most cases, an animal will not know if another carries the altruism allele, but the animal must at least be able to distinguish relatives from strangers: Relatives stand a good chance of possessing the altruism allele, but you never can tell with strangers. A yearling jay that helped out at the nest of unrelated adult jays would probably waste its time and effort.

Identification of relatives isn't too hard to imagine in the case of jays and their parents. Many biologists objected to other proposed instances of altruistic behaviors, however, arguing that animals cannot evaluate degrees of relatedness. Two findings seem to address this objection. First, many social groups, including wolf packs and baboon troops, are actually family groups. Therefore, an animal would not have to identify relatives in order for its altruistic behaviors to benefit them the most. Second, many animals, including birds, monkeys, tadpoles, bees, and even tunicate larvae, can indeed identify relatives (Fig. 17-19). Given the choice between relatives and strangers, these animals preferentially associate with their relatives, *even if they were separated at birth and have never seen those relatives before.* If animals selectively form related groups, then altruistic behaviors will most likely benefit relatives. Although it is not the only mechanism, kin selection has been a powerful selective force in the evolution of altruism in many species, probably including humans.

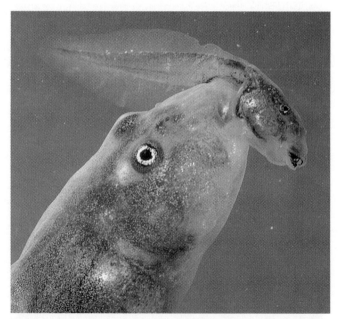

Figure 17-19 **Cannibalistic animals don't eat close relatives**

Spadefoot toad tadpoles, found in transient water holes of the Arizona desert, are cannibalistic. Many of their prey, however, are released unharmed after being tasted briefly. Researchers have discovered that the tadpoles can indeed distinguish, and spit out, their own brothers and sisters, preferring to eat unrelated members of their own species.

✖ S U M M A R Y   O F   K E Y   C O N C E P T S

Evolution and the Genetics of Populations

The gene pool of a population is the total of all the different alleles of all the genes carried by the members of a population. The sources of genetic variability within a population are mutation, which produces new genes and alleles, and recombination during sexual reproduction. In its broadest sense, evolution is a change in the frequencies of alleles in the gene pool of a population due to enhanced reproduction by individuals bearing certain alleles.

Allele frequencies in a population will remain constant over generations only if the following conditions are met: (1) There must be no mutation; (2) there must be no gene flow, no net migration of alleles into or out of the population; (3) the population must be very large; (4) all mating must be random; (5) all genotypes must reproduce equally well (that is, no natural selection). These conditions are rarely, if ever, met in nature. Understanding why they are not met leads to an understanding of the mechanisms of evolution.

The Mechanisms of Evolution

1. Mutations are random, undirected changes in DNA composition. Although most mutations are neutral or harmful to the organism, some prove advantageous in certain environments. Mutations are rare and do not change allele frequencies very much, but they provide the raw material for evolution.
2. Migration is the flow of genes between populations. If the alleles that migrants carry are different from those in the populations from which they come or to which they migrate, migration will cause changes in allele frequencies.
3. In any population, chance events kill or prevent reproduction by some of the individuals. If the population is small, chance events may eliminate a disproportionate number of individuals bearing a particular allele, thereby greatly changing the allele frequency in the population. This change is termed genetic drift.
4. Many organisms do not mate randomly. If only certain members of a population can mate, then the next generation of organisms in the population will all be offspring of this select group, whose allele frequencies may differ from those of the population as a whole. Population bottleneck and founder effect, two types of genetic drift, illustrate the consequences small populations have on frequency of alleles.
5. The survival and reproduction of organisms are influenced by their phenotype. Because phenotype depends at least partly on genotype, natural selection will tend to favor the reproduction of certain alleles at the expense of others.

Natural Selection

Natural selection is really an issue of differential, or unequal, reproduction. Three types of natural selection are:

1. *Directional selection*. Individuals with characteristics that are different from average in one direction (for example, smaller) are favored both over average individuals and over those that differ from average in the opposite direction.
2. *Stabilizing selection*. Individuals of the "average value" for a characteristic are favored over individuals of extreme values.
3. *Disruptive selection*. Individuals of extreme characteristics are favored over individuals with average values.

Natural selection occurs as a result of the interactions of organisms with both the biotic (living) and abiotic (nonliving) parts of their environments. Within a species, sexual selection and altruism are two types of natural selection. When two or more species interact extensively so as to exert mutual selective pressures on each other for long periods of time, they both evolve in response. Such coevolution can occur as a result of any type of relationship between organisms, including competition, predation, and symbiosis.

Extinction

Two factors that contribute to the likelihood of extinction of a species are localized distribution and overspecialization. Factors that actually cause extinctions include competition among species, novel predators or parasites, and habitat destruction.

✖ K E Y   T E R M S

adaptation   p. 335
allele frequency   p. 324
altruism   p. 338
balanced polymorphism   p. 335
coevolution   p. 336
competition   p. 336
differential reproduction   p. 332
directional selection   p. 333
disruptive selection   p. 333

equilibrium population   p. 326
extinction   p. 338
fitness   p. 324
founder effect   p. 329
gene flow   p. 326
gene pool   p. 324
genetic drift   p. 327
genetic equilibrium   p. 326
inclusive fitness   p. 342

kin selection   p. 338
migration   p. 326
natural selection   p. 324
population   p. 323
population bottleneck   p. 329
population genetics   p. 324
sexual selection   p. 331
stabilizing selection   p. 333
symbiosis   p. 336

✖ THINKING THROUGH THE CONCEPTS

Multiple Choice

1. Genetic drift is a _____ process.
   a. random
   b. directed
   c. selection-driven
   d. coevolutionary
   e. uniformitarian

2. Most of the 700 species of fruit flies found in the Hawaiian archipelago are each restricted to a single island. One hypothesis to explain this pattern is that each species diverged after a small number of flies had colonized a new island. This mechanism is called
   a. sexual selection
   b. genetic equilibrium
   c. disruptive selection
   d. the founder effect
   e. assortative mating

3. You are studying leaf size in a natural population of plants. The second season is particularly dry, and the following year you find that the average leaf size in the population is smaller than the year before. But the amount of overall variation is the same, and the population size hasn't changed. Also, you've done experiments that show that small leaves are better adapted to dry conditions. Which of the following has occurred?
   a. genetic drift
   b. directional selection
   c. stabilizing selection
   d. disruptive selection
   e. the founder effect

4. You have bacteria thriving in your gastrointestinal tract. This is an example of
   a. inclusive fitness
   b. balanced polymorphism
   c. symbiosis
   d. kin selection
   e. altruism

5. Lamarckian evolution could occur
   a. if each gene had only one allele
   b. if individuals had different phenotypes
   c. if the genotype was altered by the same environmental changes that altered the phenotype
   d. if the phenotype was altered by the environment
   e. under none of these conditions

6. Of the following possibilities, the best way to estimate the Darwinian fitness of an organism is to measure the
   a. size of its offspring
   b. number of eggs it produces
   c. number of eggs it produces over its lifetime
   d. number of offspring it produces over its lifetime
   e. number of offspring it produces over its lifetime that survive to breed

Review Questions

1. What is a gene pool? How would you determine the allele frequencies in a gene pool?

2. Define an equilibrium population, and outline the conditions that must be met for a population to remain in equilibrium.

3. How does population size affect the likelihood of changes in allele frequencies by chance alone? Can significant changes in allele frequencies (that is, evolution) occur because of genetic drift?

4. If you went out and measured the allele frequencies of a gene and found large differences from the proportions predicted by Hardy-Weinberg equilibrium, would that prove that natural selection is occurring in the population you are studying? Review the conditions that lead to Hardy-Weinberg equilibrium, and explain your answer.

5. People like to say that "you can't prove a negative." Study the experiment in Figure 17-1 again, and comment on what it demonstrates.

6. Describe the three types of natural selection. Which type(s) is (are) most likely to occur in stable environments and which type(s) in rapidly changing environments?

7. What is sexual selection? How is sexual selection similar to and different from other forms of natural selection?

8. Briefly describe competition, predation, symbiosis, and altruism, and give an example of each.

9. Define kin selection and inclusive fitness. Can these concepts help to explain the evolution of altruism?

✖ APPLYING THE CONCEPTS

1. In North America, the average height of human adults has been increasing steadily for decades. Is directional selection occurring? What data would you need to justify your answer?

2. Malaria is rare in North America. In populations of African Americans, what would you predict is happening to the frequency of the hemoglobin allele that leads to sickling in red blood cells? How would you go about determining if your prediction is true?

3. By the 1940s the Whooping Crane population had been reduced to under 50 individuals. Thanks to conservation measures, their numbers are now increasing. But what special evolutionary problems do Whooping Cranes have after passing through a population bottleneck?

4. In many countries, conservationists are trying to design national park systems so that "islands" of natural area (the big parks) are connected by thin "corridors" of undisturbed habitat. The idea is that this arrangement will allow animals and plants to migrate between refuges. Why would such migration be important?

5. Extinctions have occurred throughout the history of life on Earth. Why should we care if humans are causing a mass extinction event now?

6. A preview question for Chapter 18: A species is all the populations of organisms that potentially interbreed with one another but that are reproductively isolated from (cannot interbreed with) other populations. Using the five assumptions of the Hardy-Weinberg equilibrium population as a starting point, what factors do you think would be important in the splitting of a single ancestral species into two modern species?

✤ F O R   M O R E   I N F O R M A T I O N

Allison, A. C. "Sickle Cells and Evolution." *Scientific American*, August 1956. The story of the interaction between sickle-cell anemia and malaria in Africa.

Alvarez, W., and Asaro, F. "An Extraterrestrial Impact," and Courtillot, V. E. "A Volcanic Eruption." *Scientific American*, October 1990. Leading geologists debate the question, What caused the mass extinction of the dinosaurs at the end of the Cretaceous period?

Fellman, B. "To Eat or Not to Eat." *National Wildlife*, February–March 1995. How animals performing altruistic behaviors identify their relatives.

Gould, S. J. "The Evolution of Life on the Earth." *Scientific American*, October 1994. The importance of chance and catastrophe in shaping modern life.

May, R. M. "The Evolution of Ecological Systems." *Scientific American*, September 1979. Coevolution accounts for much of the structure of natural communities of plants and animals.

O'Brien, S. J., Wildt, D. E., and Bush, M. "The Cheetah in Peril." *Scientific American*, May 1986. According to molecular and immunological techniques, a population bottleneck has reduced the genetic variability of the world's cheetahs almost to zero.

Ryan, M. J. "Signals, Species, and Sexual Selection." *American Scientist*, January–February 1990. Ryan explores a variety of experiments on sexual selection, including the genetic basis of male characteristics and female choice.

Stebbins, G. L., and Ayala, F. "The Evolution of Darwinism." *Scientific American*, July 1985. A synthesis of molecular and classical evolutionary methodologies.

N E T   W A T C H

On-line resources for this chapter are on the World Wide Web at:

http://www.prenhall.com/~audesirk (click on the <u>table of contents</u> link and then select Chapter 17).

18 The Origin of Species

> *"… endless forms most beautiful and most wonderful have been, and are being, evolved."*
>
> CHARLES DARWIN in *On the Origin of Species by Means of Natural Selection* (1859)

⚏ AT A GLANCE

Speciation
Allopatric Speciation Occurs in Populations That Are
  Physically Separated
Sympatric Speciation Occurs in Populations That Live in
  the Same Area

Maintaining Reproductive Isolation between Species
Premating Isolating Mechanisms Prevent Mating between
  Species
Postmating Isolating Mechanisms Prevent Production of
  Vigorous, Fertile Offspring

Phyletic and Divergent Speciation

The Genetics of Speciation
One Model of Speciation Stresses Gradual Accumulation of
  Many Small Changes
A Second Model Stresses the Sudden Appearance of a Few
  Major Changes
Which Model Is Correct?

Rates of Speciation
During Adaptive Radiation One Species Gives Rise to Many
  under the Differing Selection Pressures of New Habitats

The Pattern of Evolution
Gradualism Explains Speciation as the Slow, Steady
  Accumulation of Small Changes over Time
Punctuated Equilibrium Explains Speciation as Occurring
  Relatively Rapidly amid Long Periods of Little Change
Gradualism and Punctuated Equilibrium Models Can Both
  Be Applied to the Evolution of the Horse
Many Evolutionary Biologists Accept a Synthesis of
  Gradualism and Punctuated Equilibrium

The great bird of paradise is one of almost 140 species of birds living on New Guinea, many of them unique to New Guinea and its neighboring islands. Why there are so many species, how they arose, and how they maintain reproductive isolation from other, often similar, species are major questions in evolutionary biology.

A tribe on New Guinea has 137 separate names for local birds, almost exactly the same as the number of species determined by Western ornithologists. The larvae of each of the 750 species of fig wasps on Earth can grow only in its own particular species of fig (Fig. 18-1). All three—New Guinea islanders, ornithologists, and fig wasps—perceive qualitative differences between members of different species; that is, differences in kind. Mere quantitative differences, for instance that this fig tree is a bit larger than that one, or this bird's feathers are a bit brighter blue than that one's, are usually ignored. To biologists, the major qualitative difference between species is the *ability to interbreed*. Thus, a species consists of all the populations of organisms that are potentially capable of interbreeding under natural conditions and that are reproductively isolated from (incapable of breeding with) members of other populations.

Our discussion of the mechanisms of evolution in the previous chapter dealt exclusively with quantitative changes within a population: the frequency of the sickle-cell anemia allele, the length of a giraffe's neck, or the size of a peacock's tail. At what point do these quantitative changes add up to a large enough total difference so that the population is a new species and cannot or will not interbreed with any other populations? Or is there a sudden jump, a quantum leap of speciation distinct from, and perhaps unrelated to, the accumulation of small, quantitative differences?

And what about the grand sweep of evolutionary history? Is all of evolution, from the first prokaryotic cell to the human brain, simply the accumulation of small changes over millions of years? Or are other processes at work?

These are some of the most contentious questions in biology. Virtually all biologists agree that evolution occurs and that the processes described in Chapter 17 produce quantitative changes in populations. But other questions remain. How do species arise? What is the relative importance of many small changes versus a few major changes? Is natural selection within populations a major force in generating the stunning diversity of living organisms? In this chapter, we will explore current hypotheses that try to explain the origin of species.

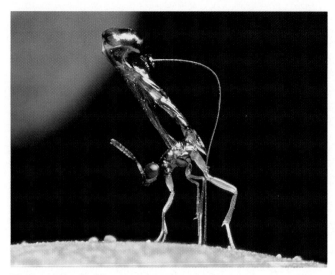

Figure 18-1 **Species specializations**
A female fig wasp uses a special organ at the tip of her abdomen, called an ovipositor, to inject her eggs into a fig. Each species of fig wasp is restricted to a single species of fig.

Speciation

Although we have described several evolutionary forces that lead to changes *within* species, we have not yet outlined a mechanism of **speciation**, whereby *new species* may be formed. To produce a new species, evolution must generate large enough genetic changes between populations so that either mating cannot occur or, if it does, the hybrid offspring are less fit. Speciation depends on the (1) isolation and (2) genetic divergence of two populations.

1. *Isolation of populations.* If two populations are to become sufficiently distinct, genetically, so that interbreeding is difficult or impossible, then there must be relatively little gene flow (migration) between them. If there is a great deal of gene flow, then genetic changes in one population will soon become widespread in the other as well.

2. *Genetic divergence.* It is not sufficient for two populations simply to be isolated—what if two formerly isolated populations are reunited? They will have become separate species only if, during the period of isolation, they had evolved sufficiently large genetic differences so that they cannot interbreed and produce vigorous, fertile offspring when they are reunited. If isolated populations are small, chance events may generate significant genetic differences by genetic drift (see Chapter 17). In both small and large populations, different selective pressures in separate environments may favor the evolution of large genetic differences.

Speciation has seldom been observed in the wild (with the exception of "instant speciation" in plants by poly-

ploidy, described below). However, based partly on theoretical considerations and partly on experiments and observations, evolutionary biologists have devised plausible hypotheses for the origin of new species in two different cases: **allopatric speciation** (two populations are geographically separated from one another) and **sympatric speciation** (two populations share the same geographical area) (Fig. 18-2).

At first glance, you might think that sympatric speciation violates our first principle of speciation, isolation of populations, because the speciating populations live in the same locale. However, it is *isolation from gene flow* that is crucial to speciation. Isolation from gene flow is simplest to envision if two populations are geographically separated by some kind of physical barrier, such as a river. Indeed, most cases of speciation, at least in animals, probably have occurred through such geographical isolation. As we will see below, however, even two populations living in the same area may have very little gene flow if they select different habitats within the area (for example, marshes as opposed to forests) or have differences in chromosome numbers so that they cannot form fertile hybrids. Therefore, the principle still holds: *Isolation from gene flow is the key to both allopatric and sympatric speciation.*

Allopatric Speciation Occurs in Populations That are Physically Separated

In Greek, the word *allopatric* means "having a different fatherland." Allopatric speciation occurs when two populations become **geographically isolated** from one another; that is, they are physically separated either by distance or by an impassible barrier (Fig. 18-3). As we described in Chapter 17, migration allows gene flow, which reduces genetic differences between populations and probably eliminates any possibility that two populations could ever become reproductively isolated from one another. Subsequently, if two populations become separated by a physical barrier, or even sheer distance, little or no gene flow can occur between them. If the pressures of natural selection differ in the two locations, or if the populations are small enough for genetic drift to occur, then the two populations may accumulate large genetic differences and become separate species. Founder events, in which a few members of a species become isolated from the main body of the species, may also be important in initiating genetic differences between populations. There is some debate as to whether genetic drift or natural selection normally plays a major role in allopatric speciation. Probably the majority opinion is that different selective pressures in the two geographical locales provide the major impetus to speciation. In either case, evolutionary biologists hypothesize that geographical isolation is involved in most cases of speciation, especially in animals.

(a) Allopatric speciation

(b) Sympatric speciation

time

geographical isolation

genetic divergence

reproductive isolation

ecological isolation

genetic divergence

reproductive isolation

Figure 18-2 **Models of allopatric and sympatric speciation**

(a) Allopatric speciation. (Top) A single species (white mice) occupies a relatively homogeneous habitat. (Middle) An impassable geographical barrier (here, a river changing course) splits the habitat into two parts, separating the species into two isolated populations. Genetic drift or different selective pressures cause the two populations to diverge genetically (tan vs. white mice). (Bottom) The barrier is removed (the river changes back to its original course), and the members of the two populations can share the same habitat. If the genetic differences between the two populations have become large enough so that interbreeding cannot occur (that is, they are reproductively isolated from one another), then the two populations constitute separate species (brown vs. white mice). **(b)** Sympatric speciation. (Top) A single species occupies a homogeneous habitat. (Middle) Climate change or other factors form two distinctly different habitats that are still physically part of the same general region; that is, there are no barriers to movement between habitats. Different selective pressures in the two habitats lead to genetic divergence of organisms living in each (tan vs. white mice). (Bottom) Sufficient genetic divergence causes reproductive isolation; former occupants of the two different habitats are now separate species (brown vs. white mice).

(a)   **(b)**

Figure 18-3   **Geographical isolation**

Geographical isolation that prevents gene flow is usually necessary for animal populations to split into two species. The tassel-eared squirrels that live on the rims of the Grand Canyon are a classic example. In the distant past, a single population of squirrels became separated as the canyon was carved into the plateau of northwestern Arizona. The two populations are split by the treeless desert of the canyon depths. Today, the Kaibab squirrel **(a)** is confined to the north rim, and the Abert squirrel **(b)** is found on the south rim and in forests throughout northern Arizona, Utah, New Mexico, and Colorado. It is not known whether the two are genetically distinct enough to constitute separate species.

Sympatric Speciation Occurs in Populations That Live in the Same Area

Sympatric means "having the same fatherland." As the name implies, sympatric speciation refers to speciation occurring within a single population and in a single geographical area. Sympatric speciation, like allopatric speciation, requires limited gene flow. There are two likely mechanisms whereby gene flow can be reduced between members of a single population: ecological isolation and chromosomal aberrations.

Ecological Isolation Restricts Different Populations to Different Habitats within the Same Area

If the same geographical area contains two distinct types of habitats (for example, food sources, nesting places, and so on), different members of a single species may begin to specialize in one habitat or the other. If conditions are right, natural selection for habitat specialization may cause the formerly single species to split into two species. Such a split seems to be occurring, right before biologists' eyes, so to speak, in the case of the fruit fly *Rhagoletis pomonella* (Fig. 18-4).

Rhagoletis is a parasite of the American hawthorn tree, laying its eggs in the hawthorn's fruit. When the maggots hatch, they eat the fruit. About 150 years ago, entomologists (scientists who study insects) noticed that *Rhagoletis* began to infest apple trees, which were introduced into North America from Europe. Now, it appears that *Rhagoletis* is splitting into two species, based on their preference

for apples or hawthorns. There are substantial genetic differences between apple-liking flies and hawthorn-liking flies. At least some of these genetic differences, such as the timing of emergence of the adult flies (described later in this chapter), are important for survival on a particular host plant. Since apples and hawthorns are often found quite close together, and flies, after all, can fly, why don't apple-flies and hawthorn-flies interbreed and cancel out any incipient genetic differences? There appear to be at least two reasons. First, female flies usually lay their eggs in the same type of fruit in which they themselves devel-

Figure 18-4   **Sympatric isolation**

The fruit fly *Rhagoletis pomonella* seems to be undergoing sympatric speciation through ecological isolation on two host trees: apples and hawthorns.

oped. Males also tend to rest on the same type of fruit in which they developed. Therefore, apple-liking males are likely to encounter and mate with apple-liking females. Second, apple fruits mature 2 or 3 weeks later than hawthorn fruits, and the two types of flies emerge with a timing appropriate for their chosen host fruits. Thus the two types of flies have very little chance of meeting. There is still some interbreeding between the two types of flies, but it looks as if they are well on their way to speciation. Will they make it? Entomologist Guy Bush suggests, "Check back with me in a few thousand years."

Changes in Chromosome Number Can Cause Immediate Reproductive Isolation of a Population

In some instances, new species can arise nearly instantaneously, either through changes in chromosome configuration or number or through irregularities during meiosis. A common speciation mechanism in plants is **polyploidy**, the acquisition of multiple copies of each chromosome. As you know, most plants and animals have paired chromosomes and are described as diploid. Occasionally, especially in plants, a fertilized egg duplicates its chromosomes but doesn't divide into two daughter cells. The resulting cell thus becomes tetraploid, with four copies of each chromosome. If all of the subsequent cell divisions are normal, this tetraploid zygote will develop into a plant that consists of tetraploid cells. Tetraploid plants are usually vigorous and healthy, and many can successfully complete meiosis to form viable gametes. The gametes, however, are diploid (meiosis normally produces haploid gametes from diploid cells). Therefore, if a sperm from a tetraploid plant fertilizes an egg cell of the diploid "parental" species, the resulting offspring will be triploid. The triploid offspring may develop normally, but when this triploid offspring begins meiosis, it will be unable to pair up all of its chromosomes, because there will be an odd number (three) of each. Meiosis fails, gametes are not formed, and so the triploid hybrids are sterile. The tetraploid plant is therefore reproductively isolated from its diploid parent.

Why is speciation by polyploidy common in plants but not in animals? Many plants can either self-fertilize or reproduce asexually, or both. If a tetraploid plant self-fertilizes, then its offspring will also be tetraploid. Asexual offspring, of course, are genetically identical to the parent, and are also tetraploid. In either case, the new tetraploid plant may perpetuate itself and form a new species. Animals, on the other hand, usually cannot self-fertilize or reproduce asexually. Therefore, if an animal produced a tetraploid offspring, the offspring would have to mate with a member of the diploid parental species and would produce all triploid offspring. As we pointed out, the triploid offspring would almost certainly be sterile. Speciation by polyploidy is extremely common in plants; in fact, nearly half of all species of flowering plants are polyploid, many of them tetraploid.

TABLE 18-1 ❖ Mechanisms of Reproductive Isolation

Premating Isolating Mechanisms: any structure, physiological function, or behavior that prevents organisms of two different populations from mating.

1. **Geographical isolation**: the separation of two populations by a physical barrier.
2. **Ecological isolation**: lack of mating between organisms belonging to different populations that occupy distinct habitats within the same general area.
3. **Temporal isolation**: the inability of organisms to mate if they have significantly different breeding seasons.
4. **Behavioral isolation**: lack of mating between species of animals that differ substantially in courtship and mating rituals.
5. **Mechanical incompatibility**: the inability of male and female organisms to exchange gametes, usually because of incompatibility of the reproductive structures.

Postmating Isolating Mechanisms: any structure, physiological function, or developmental abnormality that prevents organisms of two different populations, once mating has occurred, from producing vigorous, fertile offspring.

1. **Gametic incompatibility**: the inability of sperm from one species to fertilize eggs of another species.
2. **Hybrid inviability**: the failure of a hybrid offspring of two different species to survive to maturity.
3. **Hybrid infertility**: reduced fertility (often complete sterility) in hybrid offspring of two different species.

Maintaining Reproductive Isolation between Species

Once a species has formed, it can remain **reproductively isolated** from other species in two ways (Table 18-1). First, members of the species usually do not mate with members of other species. This strategy has a clear adaptive value. Separate species are usually genetically different in ways that adapt them to different environments. Any individual that mates with a member of another species will probably produce unfit or sterile offspring, thereby "wasting" its genes and contributing nothing to future generations. Thus there is strong selective pressure to avoid mating between different species. Incompatibilities between species that prevent mating are called **premating isolating mechanisms**.

Sometimes premating isolation fails or has not yet evolved, and members of different species do mate. Then the second isolating mechanism comes into play. If the resulting hybrid offspring die during development, then naturally the two species are still reproductively isolated from one another. In some cases, however, viable hybrid offspring are produced. Even so, if the hybrids are less fit or infertile, the two species may still remain separate, with little or no gene flow between them. Incompatibilities that prevent the formation of vigorous, fertile hybrids between species are called **postmating isolating mechanisms**.

Premating Isolating Mechanisms Prevent Mating between Species

Mechanisms that prevent mating between different species include geographical isolation, ecological isolation, temporal isolation, behavioral isolation, and mechanical incompatibility.

Geographical Isolation Prevents Members of Different Species from Meeting One Another

Members of different species obviously cannot mate if they never get near one another. As we have already seen, geographical isolation often provides the conditions for speciation in the first place. However, we cannot determine if geographically separated populations necessarily constitute distinct species. Should the barrier separating the two populations disappear (an intervening river changes course, for example), the reunited populations might interbreed freely and not be separate species at all. If they cannot interbreed, then other mechanisms, such as different courtship rituals, must have developed during their isolation. Geographical isolation, therefore, is usually considered to be a mechanism that *allows new species to form* rather than a mechanism that *maintains reproductive isolation between different species*.

Ecological Isolation Confines Members of Different Species to Different Habitats

If two populations have different resource requirements, they may use different local habitats within the same general area, and thus exhibit **ecological isolation**. White-crowned and White-throated Sparrows, for example, have extensively overlapping ranges. The White-throated Sparrow, however, frequents dense thickets, whereas the White-crowned Sparrow inhabits fields and meadows, seldom penetrating far into dense growth. The two species may coexist within a few hundred yards of one another and yet seldom meet during breeding season. The *Rhagoletis* fruit flies discussed earlier are at least partially isolated by habitat selection for apples versus hawthorns. Although ecological isolation may slow down interbreeding, it seems unlikely that it could prevent gene flow entirely. Other mechanisms usually also contribute to interspecific isolation.

Temporal Isolation Occurs between Species That Breed at Different Times

Even if two species occupy similar habitats, they cannot mate if they have different breeding seasons, a phenomenon called **temporal** (time-related) **isolation**. Bishop pines

Figure 18-5   Temporal isolation
Bishop pines, such as these, and Monterey pines coexist in nature. In the laboratory they produce fertile hybrids. In the wild, however, they do not interbreed because they release pollen at different times of the year.

and Monterey pines coexist near Monterey on the California coast (Fig. 18-5). Viable hybrids have been produced between these two species in the laboratory. In the wild, however, they release their pollen at different times: The Monterey pine releases pollen in early spring, the bishop pine in summer. Therefore, the two species never cross-breed under natural conditions. Hawthorn-liking and apple-liking *Rhagoletis* fruit flies are also partially isolated from one another because they emerge from their host fruits and breed at slightly different times of year.

Different Courtship Rituals Create Behavioral Isolation

Among animals, the elaborate courtship colors and behaviors that so enthrall human observers have evolved not only as recognition and evaluation signals between male and female; they may also aid in distinguishing among species. These different signals and behaviors create **behavioral isolation**. The striking colors and calls of male songbirds, for example, may attract females of their own species, but females of other species treat them with the utmost indifference. Among frogs, males are often impressively indiscriminate, jumping on every female in sight, regardless of the species, when the spirit moves them. Females, however, approach only male frogs croaking the "ribbet" appropriate to their species. If they do find themselves in an unwanted embrace, they utter the "release call," which causes the male to let go. As a result, few hybrids are produced.

Mechanical Incompatibility Occurs When Physical Barriers between Species Prevent Fertilization

In rare instances, ecological, temporal, and behavioral isolating mechanisms fail, and male and female of different species attempt to mate. Among different species of animals with internal fertilization (where the sperm is deposited inside the female's reproductive tract), in some cases the male and female sexual organs simply won't fit together. Among plants, differences in flower size or structure may prevent pollen transfer between species, if, for example, the species attract different pollinators (see Chapter 27 for a description of the interesting deceptions used by flowers to lure specific types of pollinators). Isolating mechanisms of this type are called **mechanical incompatibility**.

Postmating Isolating Mechanisms Prevent Production of Vigorous, Fertile Offspring

Sometimes premating isolation fails, and mating occurs between members of different species. However, if vigorous, fertile hybrids are not produced, there will still be little gene flow between the two species. Postmating isolating mechanisms include gametic incompatibility, hybrid inviability, and hybrid infertility.

Gametic Incompatibility Occurs When Sperm from One Species Are Unable to Fertilize Eggs of Another

Even though a male inseminates a female, his sperm may not fertilize her eggs, a situation called **gametic incompatibility**. For example, the fluids of the female reproductive tract may weaken or kill sperm of other species. Among plants, chemical incompatibility may prevent the germination of pollen from one species that lands on the stigma (pollen-catching structure) of the flower of another species.

Hybrid Inviability Occurs If Hybrid Offspring Survive Poorly

If fertilization does occur, the resulting hybrid may be weak or even unable to survive, a situation called **hybrid inviability**. The genetic programs directing development of the two species may be so different that hybrids abort early in development. Even if the hybrid survives, it may display behaviors that are mixtures of the two parental types. In attempting to do some things the way species A does them, and other things the way species B does, the hybrid may be hopelessly uncoordinated. Hybrids between certain species of lovebirds, for example, have great difficulty learning to carry nest materials during flight and probably could not reproduce in the wild (see Chapter 41).

Hybrid Infertility Occurs If Hybrid Offspring Are Unable to Produce Normal Sperm or Eggs

Animal hybrids, such as the mule (a cross between a horse and a donkey), are usually sterile. **Hybrid infertility** prevents hybrids from passing on their genetic material to offspring. A common reason is the failure of chromosomes to pair properly during meiosis, so that eggs and sperm never develop. Among plants that have speciated by polyploidy, any offspring produced by mating between the diploid "parent" species and the tetraploid "daughter" species will be triploid and sterile.

Phyletic and Divergent Speciation

The mechanisms of speciation and reproductive isolation that we have just described all apply only to "splitting," or **divergent speciation**. In divergent speciation, two populations of a single parent species become separate species as a result of isolation from gene flow followed by genetic divergence (Fig. 18-6a). Divergent speciation is often called "true" speciation, because it is possible (in principle at least) to determine if the two presumed species are in fact reproductively isolated.

In the fossil record, however, things are not so simple. Fossils, being dead and petrified, cannot breed, so it is difficult to determine whether they are reproductively isolated from other fossils. Therefore, paleontologists make distinctions among species purely on morphological (anatomical) grounds: If two fossils are structurally as different from one another as two modern species are, then the fossils are assigned to different species. Suppose that a whole species, under massive directional selection, changes over time so that the later specimens are very different from the earlier ones (Fig. 18-6b). If the differences are large enough, it may be assumed that the later organisms could not possibly interbreed with the earlier ones and so form a separate species. In other words, over time, a single species evolves into another species. In this situation, the two species never coexist and reproductive isolation between early and late forms cannot be directly tested. This process is called **phyletic speciation**.

Owing to its incompleteness, the fossil record does not reveal whether divergent or phyletic speciation dominates evolutionary history. If divergent speciation is most common, as many evolutionary biologists hypothesize, but few enough fossils are found, then most evolutionary lineages will appear to show phyletic speciation (Fig. 18-6c). Beginning with a modern species and working backward, one may be able to reconstruct a "phyletic evolutionary history" back to an ancient ancestor. This phyletic history may, however, simply be selective sampling of a much larger, divergent history in which most of the branches left no fossils.

The Genetics of Speciation

When plants speciate by polyploidy, it is clear that a sudden, major genetic change has taken place. But what about other speciation events, whether allopatric or sympatric?

There are two genetic models for speciation: the gradual accumulation of many small changes and the sudden appearance of a few major changes.

One Model of Speciation Stresses Gradual Accumulation of Many Small Changes

According to the first model, two populations, whether by genetic drift or through differential natural selection pressures, gradually accumulate many small genetic changes. Each mutation or different allele by itself has only a minor phenotypic effect. Over time, however, the accumulation of many small changes may result in reproductive isolation between the two populations, and they may become separate species.

A Second Model Stresses the Sudden Appearance of a Few Major Changes

The second model starts out with the idea that single regulatory genes, or small numbers of genes, control major developmental pathways, such as how originally similar cells become specialized to form different structures. In certain aspects of the development of animals,

this is known to be true. If so, then mutations in just a few regulatory genes might result in such significant changes in development that the mutants would immediately be reproductively isolated from their parental populations. A new species would therefore arise almost instantaneously.

Which Model Is Correct?

Because so few speciation events have been observed, and so little is known about the genetics of development, it is difficult to choose between these models. The extreme genetic similarity between certain species, for example chimps and humans, suggests to some observers that mutations in a few key regulatory genes may have been the cause of speciation. On the other hand, the "small change" model is supported by the finding that many phenotypic characteristics, including size, coloration, and various behavioral traits, are under the control of several to many genes (see Chapter 13 for a discussion of quantitative inheritance). Thus, one might expect that changes in many genes would be required to generate differences at the species level. Further, both theoretical modeling and common observation suggest that one-shot,

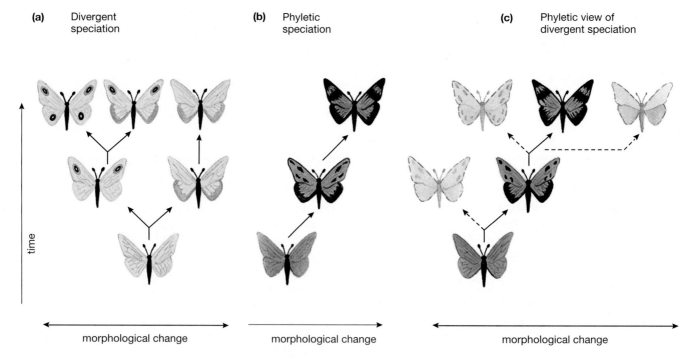

(a) Divergent speciation   **(b)** Phyletic speciation   **(c)** Phyletic view of divergent speciation

Figure 18-6   **Divergent speciation vs. phyletic speciation**

(a) In divergent speciation, a single species splits up into two or more daughter species that exist simultaneously and are reproductively isolated from one another. **(b)** In phyletic speciation, the whole population of a single species evolves over time. Sufficient genetic changes occur so that the later populations are considered to be a new species, different from earlier populations. **(c)** The fossil record is incomplete. Most species may become extinct without leaving either fossils or living descendants (dashed butterflies). Therefore, even if divergent speciation is most common, the fossil record may appear to show phyletic speciation (solid butterflies).

massive genetic changes are usually harmful. This debate is not likely to be settled until molecular developmental biology comes to the rescue with a better understanding of the genetic basis of development and of developmental differences between species.

Rates of Speciation

Over the course of evolutionary history, species continually form, exist for a time, and become extinct. Geographical isolation, genetic drift, mutations, the invasion of new habitats, and coevolution all ensure that speciation never stops. The rate of speciation varies considerably over evolutionary time, however, and bursts of speciation can be seen both in the fossil record and in the distribution of modern organisms.

During Adaptive Radiation One Species Gives Rise to Many under the Differing Selection Pressures of New Habitats

Sometimes a species gives rise to many new species in a relatively short time. This process, called **adaptive radiation**, occurs when populations of a single species invade different habitats and evolve in response to the differing selective pressures in those habitats. Adaptive radiation has occurred many times and in many groups of organisms. Adaptive radiation usually results from one of two causes. First, a species may encounter a wide variety of unoccupied habitats; for example, when the ancestors of Darwin's Finches colonized the Galapagos Islands, or when marsupial mammals first invaded Australia. With no competitors except other members of their own species, all the available habitats and food sources were rapidly utilized by new species that evolved from the original invaders. The ultimate in unoccupied habitats results from mass extinctions, such as those that might follow meteorite impacts (see Chapter 17). With most potential competitors exterminated by the effects of the impact, the survivors speciate prolifically, filling an array of empty habitats.

Adaptive radiation also occurs if a species develops a fundamentally new and superior **adaptation**, enabling it to displace less well adapted species from a variety of habitats. Such adaptations have apparently happened several times during evolutionary history, for example, when mammals diversified extensively at the expense of reptiles such as the dinosaurs.

The Pattern of Evolution

Although the processes that drive evolutionary change, such as genetic drift and natural selection, are well understood, there is considerable debate among evolutionary biologists concerning which processes actually play the major roles in shaping evolutionary history. Three subjects of particular interest are (1) the rate and timing of evolutionary change over time; (2) the role of speciation in producing morphological change; and (3) the relative importance of natural selection versus nonselective forces in determining structures and behaviors.

In both the popular press and scientific journals, these topics often are discussed within the framework of two opposing views of evolution, termed the gradualism and punctuated equilibrium models. These two models are actually at the opposite ends of a spectrum, as we will see. In all likelihood, the evolution of real organisms usually falls somewhere closer to the middle of the spectrum, sometimes evolving mostly according to "gradualistic" rules and sometimes according to "punctuated" rules. We will briefly outline the two extreme positions and then attempt to synthesize a coherent scheme.

Gradualism Explains Speciation as the Slow, Steady Accumulation of Small Changes over Time

The extreme **gradualism** model of speciation takes the following positions:

1. Speciation occurs from the accumulation of dozens or hundreds of small genetic differences between two populations. The two populations thus become separate species in a gradual fashion, as less and less interbreeding occurs and any hybrids that are formed become less and less fit. Speciation may take hundreds of thousands or even millions of years (Fig. 18-7a).
2. Morphological change within a lineage (e.g., horses; see Fig. 16-10) occurs continuously, although the rate may change somewhat from time to time.
3. Morphological change and speciation are not closely linked. Two populations of a species may undergo considerable morphological change without necessarily becoming reproductively isolated from one another.
4. Both continual morphological changes and speciation are driven largely by natural selection among individuals, with fitter organisms leaving more offspring than less fit organisms do.

Punctuated Equilibrium Explains Speciation as Occurring Relatively Rapidly amid Long Periods of Little Change

The extreme form of the **punctuated equilibrium** model of evolution takes contrasting stands on every issue:

1. Speciation is a relatively rapid event (in the context of geological time), driven by a relatively small number of genetic changes, such as mutations in a few developmentally important regulatory genes.

The rapidity of speciation is the "punctuated" part of the model (Fig. 18-7b).

2. Species remain morphologically the same for long periods of time; this is the "equilibrium" part of the model.

3. Morphological change and speciation are very tightly linked. Virtually all morphological change occurs during the brief period of speciation, and the newly formed species then remain essentially unchanged until the next speciation event.

4. Large-scale evolutionary changes are driven by selection among species, not among individuals. Further, "species selection" often may be driven not by natural selection but by random chance.

Gradualism and Punctuated Equilibrium Models Can Both Be Applied to the Evolution of the Horse

In the evolution of the horse, the fossil record shows a progression from *Hyracotherium*, the "dawn horse," through many intermediate steps to modern *Equus* (see Fig. 16-10). Extreme gradualism would propose that fossil horses are representative samples of changes that occurred gradually throughout the evolution of the horse. As environments changed from forest to open woodlands to prairies, natural selection favored larger, faster horses with strong, shock-absorbing legs, hard hooves, and large, grinding teeth. Divergent speciation occurred from time to time as genetic changes accumulated between, for example, woodland populations and prairie populations, but phyletic specia-tion also occurred, as whole populations changed into new species under selective pressures. During phyletic specia-tion, selection *between individuals within a population* drove the whole population to change, gradually becoming a new species.

One problem with this view, which provided the original stimulus for the development of the punctuated equilibrium model, is the nature of the fossil record. Every type of fossil horse is quite different from every other type; that is, the fossil record does not show a continuous accumulation of small changes, but rather a series of jumps from one fossil type to the next. Why haven't fossils been found that represent all the presumed intermediate stages? Gradualism replies that fossilization is a rare event; for every organism that becomes fossilized, and is found by modern humans, probably millions do not. Further, many thousands of years may pass during which suitable conditions for fossilization (for example, massive eruptions of volcanic ash, flash floods, buildup of sediment in shallow seas) do not occur. Therefore, many intermediate stages, perhaps even whole species, never appear in the fossil record.

The gaps in the fossil record are the starting point for the punctuated equilibrium model, championed by evolutionary biologists Stephen Jay Gould of Harvard and Niles Eldredge of the American Museum of Natural History and others. Punctuated equilibrium assumes that the gaps are not accidents: They are produced by such rapid evolutionary change that fossil remains of inter-

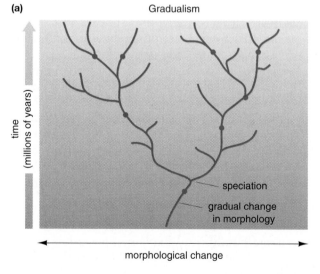

(a) Gradualism

time (millions of years) →

speciation

gradual change in morphology

morphological change

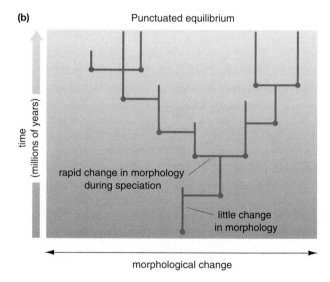

(b) Punctuated equilibrium

time (millions of years) →

rapid change in morphology during speciation

little change in morphology

morphological change

Figure 18-7   **Extreme views of gradualism vs. punctuated equilibrium**
Solid dots represent fossils considered to be separate species. Lines that do not reach the top of the graph represent extinctions. **(a)** Gradualism. Morphological change occurs gradually over time, although the rate of change may vary (more rapid during times of strong directional selection, much more slowly during times of stabilizing selection). Both phyletic and divergent speciation probably occur. **(b)** Punctuated equilibrium. Morphological change and speciation are simultaneous and very rapid, interrupted by long periods during which neither morphological change nor speciation occur. All speciation is divergent.

mediate forms would be extremely rare. Punctuated equilibrium hypothesizes that *Hyracotherium* underwent little change for millions of years (the "equilibrium"). Small fringe populations of *Hyracotherium* split off now and then. Some evolved very rapidly, perhaps mostly by genetic drift, to become entirely new species. At some point, perhaps because of a major change in the environment, one of these new species, *Mesohippus*, quickly replaced *Hyracotherium* as the dominant form of horse. The speciation of *Mesohippus* and its replacement of *Hyracotherium* may have taken place within a few thousand years—a mere instant, in the span of geological time (the "punctuation" between equilibria). Punctuated equilibrium assumes that evolution is largely driven by this frequent, rapid divergent speciation. Selection *between species* then results in the extinction of some species and the survival of others.

Many Evolutionary Biologists Accept a Synthesis of Gradualism and Punctuated Equilibrium

As we have already cautioned, the gradualism and punctuated equilibrium models just presented are extreme positions. Although there are certainly major disagreements among evolutionary biologists, something like a majority view could be summarized as follows.

Changes Too Subtle to be Observed between Generations Can Result in Rapid Evolution on a Geological Time Scale

The rate of evolution, as measured by morphological changes in fossils, varies greatly. In some organisms, there are indeed long periods of constancy—sharks, for example, have remained outwardly similar for scores of millions of years. Morphological constancy might arise from several causes. In a constant environment, stabilizing selection may limit the rate of evolutionary change. A rapidly fluctuating environment, with wide but frequent temperature changes, for instance, would promote equally rapid, fluctuating directional selection, with a net effect of preserving the status quo. Or even if the environment changes in a constant direction, species may simply "track" favorable habitats. During the ice ages, for example, the glaciers expanded from arctic regions, creating vast new areas of cold and severe conditions. Caribou and other cold-adapted species simply followed the vegetation that flourishes in this type of climate: south when the glaciers advanced, back north when the ice retreated. Great morphological changes were not required, because the organisms kept to the same habitat, even though its geographical location moved.

Rapid morphological changes, which may in some instances correspond with times of speciation, also occur. Genetic drift in small populations, directional selection among populations that do not track suitable habitats, coevolution owing to competition or cooperation with other

species, or sexual selection may all produce rapid evolutionary jumps.

Just how fast must evolution occur to be perceived as punctuated equilibrium? Returning to the fossil horses, let's suppose that a *Hyracotherium* 60 centimeters (24 inches) tall evolved into a *Mesohippus* 100 centimeters (40 inches) tall in 10,000 years and that the time between generations was 5 years. The height increase of 40 centimeters (16 inches) would have taken 2000 generations, or 0.2 millimeter per generation. This difference would be far too small to detect with the naked eye and is more than a hundred times less than the variability in size among modern horses, even of the same breed. To a geneticist, this rate of change doesn't seem very fast at all. In the fruit fly *Drosophila*, for example, mild selection pressure can cause evolutionary changes that are more than 100 times as fast as those required for *Hyracotherium* to evolve into *Mesohippus* in 10,000 years. In the span of geological time, however, 10,000 years is quite brief. No one would be surprised if intermediate fossils could not be found. Therefore, "punctuated" speciation would seem quite slow to a geneticist, but very fast to a geologist.

The Relationship between Speciation and Morphological Change Depends on Circumstances

Morphological change may occur whenever a small population is isolated from gene exchange with other populations and is subjected to new selective pressures. Considerable change may also occur by genetic drift or founder events or both, if the population is very small. Selection or genetic drift may also cause speciation, with the small, isolated population rapidly becoming a separate species from its larger parent population. Therefore, in these cases speciation may be linked with major morphological change.

Morphological change may occur in large populations subjected to widespread directional selection. However, if gene flow among populations is adequate, speciation is unlikely to occur. In this situation, morphological change will not be linked to speciation.

As we discussed previously, speciation could occur through the accumulation of many small genetic changes or through a few key mutations. Genetic data, although scanty, suggest that each mechanism may operate at different times in different populations. Which is more common cannot be decided with the evidence now available.

Natural Selection Occurs Both between Species and between Individuals within a Species

If two species are in direct competition for a particular resource in a particular habitat, one will probably move, evolve rapidly to use a different resource, or become extinct (see Chapters 17 and 44). In this case, species, not individuals, may be considered to be the unit of selection. However, the members of the species that "wins" in such a competition may well have attained their superior adaptations through natural selection acting on individual organisms. Further, the "winning" species may continue to evolve even after the extinction of the "losing" species, because of competition among individual members of the species. In these instances, the unit of selection is the individual. Most evolutionary biologists feel that selection occurs to some extent at both species and individual levels.

Both Natural Selection and Random Events Drive Evolution

Is natural selection the paramount force operating during evolution? Some evolutionary biologists emphasize the importance of nonselective, random events during evolution. They have two major objections to the concept that most features evolved as adaptations through natural selection. First, it is extremely difficult to prove that a particular feature, say the size of a bird's beak, was shaped by natural selection. If a biologist observes a bird with a large bill cracking a large, tough nut, it is tempting to assume that the large bill evolved because of the advantage it conferred in nutcracking. But maybe the bird acquired the large bill for other reasons, say for courtship or even through genetic drift, and eats nuts because it "accidentally" has a suitable bill. On one of the Galapagos Islands, however, zoologist Peter Grant of Princeton demonstrated quite conclusively that natural selection strongly influences beak size in a species of finch. When a prolonged wet spell altered the vegetation of the island, greatly increasing the availability of small seeds, the average bill size of breeding birds became longer and narrower over a period of just 4 years.

The second argument against the concept that natural selection is the primary force in evolution is that random forces do indeed operate in nature. In the previous chapter, we discussed the likelihood that gigantic meteorites strike the Earth from time to time, causing extinctions of many species. Species that were superbly adapted to warm, sunny climates might go extinct within a few years in the cold and gloom following the impact. The survival or extinction of species might have little to do with adaptations to their habitats, because the habitats would disappear. Smaller-scale catastrophes, such as floods, sea-level changes, or massive forest fires, could similarly cause the extinction of well-adapted but localized species.

The consensus is that both natural selection and random forces shape the evolutionary history of life on Earth. No one doubts that natural selection is the driving force behind the evolution of the magnificent structures and behaviors that we witness in the living world around us. And no one doubts that a 10-kilometer meteorite would drive many otherwise well-adapted species to extinction. Which processes dominate, over what time scales and over what fraction of species on Earth, is the real debate.

Concluding Thoughts on Evolution and Evolutionary Hypotheses

In the popular press, conflicts among evolutionary biologists are sometimes seen as conflicts about evolution itself. One occasionally reads statements to the effect that new theories are overthrowing Darwin and casting doubt on the reality of evolution. Nothing could be further from the truth. Proponents of punctuated equilibrium and gradualism, and everyone in between, unanimously agree that evolution occurred in the past and is still occurring today. The only argument is over the mechanisms of evolutionary change, their relative importance in the history of life on Earth, and which forces were most important in shaping the evolution of particular species. Meanwhile, wolves still tend to catch the slowest caribou, small populations still undergo genetic drift, habitats still change or disappear, and maybe somewhere in our galaxy another asteroid is swinging our way; evolution continues, still generating, in Darwin's words, "endless forms most beautiful."

✖ SUMMARY OF KEY CONCEPTS

Speciation

A species is defined as all the populations of organisms that are potentially capable of interbreeding under natural conditions and that are reproductively isolated from other populations. Speciation, the development of new species, requires that two populations be isolated from gene flow between them and develop significant genetic divergence. Allopatric speciation occurs by geographical isolation and subsequent divergence of the separated populations through genetic drift or natural selection. Sympatric speciation occurs by ecological isolation and subsequent divergence or by rapid chromosomal changes, such as polyploidy.

Maintaining Reproductive Isolation between Species

Reproductive isolation between species may be maintained by one or more of several mechanisms, collectively called premating isolating mechanisms and postmating isolating mechanisms. Premating isolating mechanisms include geographical isolation, ecological isolation, temporal isolation, behavioral isolation, and mechanical incompatibility. Postmating isolating mechanisms include gametic incompatibility, hybrid inviability, and hybrid infertility.

Phyletic and Divergent Speciation

In divergent speciation, a single species splits into two or more subpopulations that evolve into separate species. Daughter species therefore exist simultaneously. In phyletic speciation, the entire population of a species evolves in concert over time, so that the later population is recognized as being a new species. No one knows which form of speciation dominates evolutionary history, but it is likely that phyletic speciation is a selective sampling of divergent speciation as seen through a spotty fossil record.

The Genetics of Speciation

Separate species must have large enough genetic differences to prevent interbreeding. There are two models to explain the development of genetic differences between species: the gradual accumulation of many small genetic differences and the sudden appearance of a few major differences, probably in regulatory genes. It is likely that both mechanisms occur, and each may account for the genetic divergence of some species.

Rates of Speciation

Rates of speciation vary greatly over geological time. Invasion of new, unoccupied habitats or the development of fundamentally superior adaptations may trigger extremely rapid speciation, a process called adaptive radiation.

The Pattern of Evolution

There are two extreme models for the overall pattern of evolution. The gradualism model asserts that genetic and morphological changes accumulate slowly over time, with speciation being a relatively slow process not closely linked to observable morphological change. Natural selection among individuals is the prime driving force behind evolutionary change. The punctuated equilibrium model asserts that morphological change and speciation are very rapid, simultaneous events. Speciation events are separated by long periods during which species undergo little or no morphological change. Evolutionary change is driven by natural selection among simultaneously existing species, and also by random, chance events leading to speciation and extinction. Most evolutionary biologists agree that some aspects of both models probably occur during evolution and that evolutionary history may more closely resemble either model at different times or in the lineage of different organisms.

✖ KEY TERMS

adaptation   p. 355
adaptive radiation   p. 355
allopatric speciation   p. 345
behavioral isolation   p. 352
divergent speciation   p. 353
ecological isolation   p. 352
gametic incompatibility   p. 353

geographical isolation   p. 348
gradualism   p. 355
hybrid infertility   p. 353
hybrid inviability   p. 353
mechanical incompatibility   p. 353
phyletic speciation   p. 353
polyploidy   p. 351

postmating isolating mechanism   p. 352
premating isolating mechanism   p. 351
punctuated equilibrium   p. 355
reproductive isolation   p. 351
speciation   p. 348
sympatric speciation   p. 348
temporal isolation   p. 352

THINKING THROUGH THE CONCEPTS

Multiple Choice

1. Many closely related species of marine invertebrates exist on either side of the isthmus of Panama.They probably resulted from:
   a. premating isolation
   b. isolation by distance
   c. postmating isolation
   d. gametic incompatibility
   e. allopatric speciation
2. Many hybrids are sterile because their chromosomes don't synapse correctly at meiosis.Why aren't polyploid plants sterile?
   a. They backcross to the parental generation.
   b. Most are triploid.
   c. They cross-pollinate.
   d. They self-fertilize, using their diploid gametes.
   e. Their eggs develop directly, without fertilization.
3. In many species of fireflies, males flash to attract females. Each species has a different flashing pattern.This is probably an example of
   a. ecological isolation b. temporal isolation
   c. geographical isolation d. premating isolation
   e. postmating isolation
4. In phyletic speciation, fossil species are separated by
   a. time b. geography
   c. behavior d. hybrid inviability
   e. hybrid infertility
5. In terms of changes in gene frequencies, founder events result in
   a. gradual accumulation of many small changes
   b. large, rapid changes
   c. polyploidy
   d. hybridization
   e. mechanical incompatibility
6. After the demise of the dinosaurs, mammals evolved rapidly into many new forms because of
   a. the founder effect
   b. a genetic bottleneck
   c. adaptive radiation
   d. geological time
   e. genetic drift

Review Questions

1. Define the following terms: species, speciation, allopatric speciation, sympatric speciation. Explain how allopatric and sympatric speciation might work, and give a hypothetical example of each.
2. Many of the oak tree species found in central and eastern North America hybridize (interbreed). Are they "true species"?
3. Review the material on the possibility of sympatric speciation in *Rhagoletis* that breed on apples or hawthorns.What types of genetic, morphological, or behavioral data would really convince you that the two forms have actually become separate species?
4. A drug called colchicine affects the mitotic spindle and prevents cell division after the chromosomes have doubled at the start of meiosis. Describe how you would use colchicine to produce a new polyploid species of your favorite garden flower.
5. List and describe the different types of premating and postmating isolating mechanisms.
6. Define punctuated equilibrium and gradualism. Look up the evolutionary history of a lineage (e.g., dinosaurs, humans) and describe how each mechanism might account for the evolution of that lineage.
7. Describe phyletic and divergent speciation. Do you think you could distinguish between these two mechanisms in the fossil record?

APPLYING THE CONCEPTS

1. Do phyletic speciation and divergent speciation coincide with the gradualism and punctuated equilibrium models of evolution? Defend your answer.
2. Seedless varieties of fruits and vegetables have been created by breeders.They are triploid. Explain why they are seedless.
3. Why do you suppose there are so many endemic species on islands? (*Endemic* means "found nowhere else.") And why have the overwhelming majority of recent extinctions occurred on islands?
4. A contrarian biologist you've met says the fact that humans are pushing other species into small, isolated populations is actually good for biodiversity, because these are the conditions that theory says lead to new speciation events. Comment.
5. In southern Wisconsin there are several populations of gray squirrels (*Sciurus carolinensis*) with black fur. Design a study to determine if they are actually a separate species.
6. This chapter reviews a series of hypotheses that represent dichotomies: phyletic versus divergent speciation, punctuated equilibrium versus gradualism, species selection versus individual selection, and the importance of random events versus natural selection in the course of evolution.Think about the value of dichotomies in scientific (and political) debate.What are the advantages and disadvantages of taking extreme positions? Why do people do it?
7. It is difficult to gather data on speciation events in the past or to perform interesting experiments about the process of speciation. Does this difficulty make the study of speciation "unscientific"? Should we abandon the study of speciation?

✖ FOR MORE INFORMATION

Ayala, F. "The Mechanisms of Evolution." *Scientific American*, September 1979. Our increasing understanding of the molecular mechanisms of heredity and mutation reveals that organisms are extremely variable, providing tremendous amounts of material on which selection can operate.

Gould. S. J. "The Evolution of Life on the Earth." *Scientific American*, October 1994. The importance of chance and catastrophe in shaping modern organisms.

Gould, S. J., and Eldredge, N. "Punctuated Equilibrium Comes of Age." *Nature*, November 18, 1993. An overview of the history and evolution of the punctuated evolution hypothesis by two of its chief proponents.

Grant, P. R. "Natural Selection and Darwin's Finches." *Scientific American*, October 1991. A temporary climate change altering the food supply created selection pressures on finches that altered their average beak shape in several generations.

Smith, J. M. *Evolution Now: A Century after Darwin*. San Francisco: W. H. Freeman, 1982. The section titled "Evolution—Sudden or Gradual?" contains essays by Russell Lande and Smith himself on the "gradualist" side and by Stephen Jay Gould on the "punctuationist" side.

Stebbins, G. L., and Ayala, F. "The Evolution of Darwinism." *Scientific American*, July 1985. A synthesis of molecular and classical evolutionary methodologies.

NET WATCH

On-line resources for this chapter are on the World Wide Web at:
http://www.prenhall.com/~audesirk (click on the <u>table of contents</u> link and then select Chapter 18).

19 The History of Life on Earth

"[T]hence life was born,
Its nitrogen from ammonia, carbon from methane,
Water from the cloud and salts from the young seas . . . the cells of life
Bound themselves together into clans, a multitude of cells
To make one being—as the molecules before
Had made of many one cell.   ROBINSON JEFFERS in *The Beginning and the End*

⊞ AT A GLANCE

Origins
Prebiotic Evolution Was Controlled by the Early
   Atmosphere and Climate

The First Life
Eukaryotes Evolved Membrane-Enclosed Organelles and a
   Nucleus

Multicellularity

Multicellular Life in the Sea
Multicellular Plants Evolved Specialized Structures That
   Facilitated Their Invasion of Diverse Habitats
Multicellular Animals Evolved Specializations That
   Allowed Them to Capture Prey, Feed, and Escape More
   Efficiently

The Invasion of the Land
Some Plants Evolved Specialized Structures That Adapted
   Them to Life on Dry Land
Some Animals Evolved Specialized Structures That Adapted
   Them to Life on Dry Land

Human Evolution
Primate Evolution Has Been Linked to Grasping Hands,
   Binocular Vision, and a Large Brain
Hominids Evolved from Dryopithecine Primates
Australopithecines, the First True Hominids, Could Stand
   and Walk Upright
The Evolution of Human Behavior Is Highly Speculative

Evolutionary Connections: Why Do Humans Walk Upright?

The Great Nebula in Orion, a cloud of gases in which stars may be forming, as they did in the beginning of the Universe

Fifteen billion years ago, there was no universe as we know it: only a dense, minuscule mass in which energy and matter were melded into a single incomprehensibly ferocious state. Then the mass erupted. Particles of matter and antimatter formed, annihilating each other in great bursts of energy when they collided. So great was the energy released that astrophysicists can still detect its faint radiation, evidence to support this favored "Big-Bang" model of the formation of the universe.

According to this scenario, after the Big Bang, great clouds of matter formed. Gradually, gravity drew particle to particle, producing large aggregations. Many accumulations grew so large that their centers became dense and hot, triggering the thermonuclear reactions that fuel the light of the stars. Small, simple atoms of hydrogen fused to form helium, and these fused to give rise to still larger atoms. The nuclear forces of some stars grew so intense that they exploded, spewing their matter out into space.

Scientists theorize that our solar system originated about 5 billion years ago. At the far edge of an immense cluster of stars forming a spiral galaxy, a small cloud of matter began to condense, enriched with heavy elements provided by the self-destruction of ancient stars. The center of the cloud collapsed into the yellow dwarf star we call the sun. From the cloud of dust particles circling the newly formed sun, smaller aggregations formed the planets. Of the innermost four planets, Mercury and Venus, the two closest to the sun, became very hot, while the fourth, Mars, cooled into an eternal winter. The third settled into an orbit that would receive just the right intensity of sunlight to permit water to exist as a liquid. This became Earth, which the French oceanographer Jacques Cousteau has aptly described as the "Water Planet."

On this third planet, and so far as we know upon no other in our solar system, life evolved (Table 19-1). Although the mechanisms of evolution are known in considerable detail, they do not tell us how life first originated. Where did life come from? How did it develop into the myriad forms we see today? What is its future?

TABLE 19-1 ❊ *The History of Life on Earth*

Era	Period	Epoch	Years Ago* (millions)	Major Events
Precambrian			4600–3500	Origin of solar system and Earth.
			3500–590	Origin of first living cells; dominance of bacteria; origin of photosynthesis and evolution of oxygen atmosphere; origin of algae and soft-bodied marine invertebrates.
Paleozoic	Cambrian		590–505	Primitive marine algae flourish; origin of most marine invertebrate types.
	Ordovician		505–438	Invertebrates, especially arthropods and mollusks, dominant in sea; first fish, fungi.
	Silurian		438–408	Many fish, trilobites, mollusks in sea; first vascular plants; invasion of land by plants; invasion of land by arthropods.
	Devonian		408–360	Fishes and trilobites flourish in sea; origin of amphibians and insects.
	Carboniferous		360–286	Swamp forests of tree ferns and club mosses; dominance of amphibians; numerous insects; origin of reptiles.
	Permian		286–248	Origin of conifers; massive marine extinctions, including last of trilobites; flourishing of reptiles and decline of amphibians; continents aggregated into one land mass, Pangaea.
Mesozoic	Triassic		248–213	Origin of mammals and dinosaurs; forests of gymnosperms and tree ferns; breakup of Pangaea begins.
	Jurassic		213–144	Dominance of dinosaurs and conifers; origin of birds; continents partially separated.
	Cretaceous		144–65	Flowering plants appear and become dominant; mass extinctions of marine life and some terrestrial life, including last dinosaurs; modern continents well separated.
Cenozoic	Tertiary	Paleocene	65–54	Widespread flourishing of birds, mammals, insects, and flowering plants; drift brings continents into modern positions; mild climate at beginning of period, with extensive mountain building and cooling toward end.
		Eocene	54–37	
		Oligocene	37–24	
		Miocene	24–5	
		Pliocene	5–2	
	Quaternary	Pleistocene	2–0.01	Evolution of *Homo*; repeated glaciations in Northern Hemisphere; extinction of many giant mammals.
		Recent	0.01–present	

*From W. B. Harland, et al. *A Geologic Time Scale*. Cambridge, England: Cambridge University Press, 1982.

Origins

How and when did life first appear on Earth? Just a few centuries ago, this question would have been considered trivial. Although no one knew how life *first* arose, people thought that new living things appeared all the time, through **spontaneous generation** from both nonliving matter and other, unrelated forms of life. In 1609, a French botanist wrote, "There is a tree . . . frequently observed in Scotland. From this tree leaves are falling; upon one side they strike the water and slowly turn into fishes, upon the other they strike the land and turn into birds." Medieval writings abound with

similar observations and delightful recipes for creating life— even human beings. Microorganisms were thought to arise spontaneously from broth, maggots from meat, mice from mixtures of sweaty shirts and wheat.

In 1668 the Italian physician Francesco Redi disproved the maggots-from-meat hypothesis simply by keeping flies (whose eggs hatch into maggots) away from uncontaminated meat (see Chapter 1). Then in the mid-1800s Louis Pasteur in France and John Tyndall in England disproved the broth-to-microorganism idea as well (Fig. 19-1). Although their work effectively demolished the notion of spontaneous generation, it did

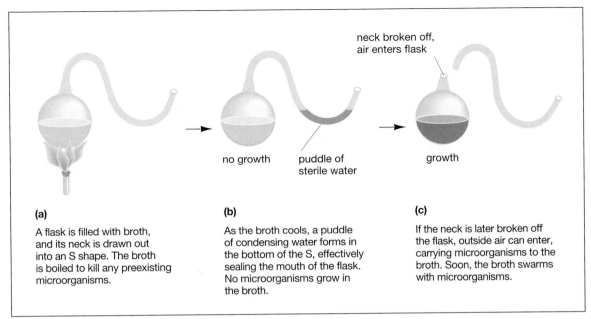

neck broken off,
air enters flask

no growth

puddle of
sterile water

growth

(a)

A flask is filled with broth,
and its neck is drawn out
into an S shape. The broth
is boiled to kill any preexisting
microorganisms.

(b)

As the broth cools, a puddle
of condensing water forms in
the bottom of the S, effectively
sealing the mouth of the flask.
No microorganisms grow in
the broth.

(c)

If the neck is later broken off
the flask, outside air can enter,
carrying microorganisms to the
broth. Soon, the broth swarms
with microorganisms.

Figure 19-1　**Spontaneous generation refuted**

Louis Pasteur's experiment disproving the spontaneous generation of microorganisms in broth.

not address the question of how life on Earth originated in the first place.

For almost half a century the subject lay dormant. Eventually, biologists returned to the question of the origin of life and began to seek answers. In the 1920s and 1930s, Alexander Oparin in Russia and John B. S. Haldane in England noted that the oxygen-rich atmosphere that we know would not have permitted spontaneous formation of the complex organic molecules necessary for life. Oxygen reacts readily with other molecules disrupting chemical bonds, thus tending to keep molecules simple. Oparin and Haldane speculated that the atmosphere of the young Earth was very low in oxygen and rich in hydrogen in the form of hydrogen gas (H_2), methane (CH_4), and ammonia (NH_3). Given these and other conditions, which we will discuss below, Oparin and Haldane proposed that life could have arisen from nonliving matter through ordinary chemical reactions. This process is called chemical evolution, or **prebiotic evolution**: that is, evolution before life existed.

Prebiotic Evolution Was Controlled by the Early Atmosphere and Climate

The primordial Earth differed greatly from the planet we now enjoy. As rock after rock smashed into the forming planet, their energies of motion were converted into heat. Radioactive atoms decayed, releasing still more heat. Soon the rock melted, and heavier elements such as iron and nickel sank to the center of the mass, where they remain molten still because of intense heat. Gradually Earth cooled, and elements combined to form compounds of

many sorts. Virtually all the oxygen combined with hydrogen to form water, carbon to form carbon dioxide, and heavier elements to form minerals. After millions of years, Earth cooled enough to allow water to exist as a liquid, and for thousands of years it must have rained, as water vapor condensed out of the cooling atmosphere. As the water struck the surface, it dissolved many minerals, forming a weakly salty ocean. Lightning from storms, heat from volcanoes, and intense ultraviolet light from the sun all poured energy into the young seas.

Judging from the chemical composition of the rocks formed at this time, geochemists have deduced that the primitive atmosphere probably contained substances such as carbon dioxide, methane, ammonia, hydrogen, nitrogen, hydrochloric acid, hydrogen sulfide, and water vapor. But because oxygen atoms were bound up in water, carbon dioxide, and minerals, there was virtually no free oxygen in the early atmosphere. The absence of free oxygen is an important factor in all hypotheses and experiments dealing with prebiotic evolution.

Organic Molecules Can Be Synthesized Spontaneously under Prebiotic Conditions

In 1953, inspired by the ideas of Oparin and Haldane, Stanley Miller, a graduate student, and his adviser Harold Urey of the University of Chicago set out to demonstrate prebiotic evolution in the laboratory. They mixed water, ammonia, hydrogen, and methane in a flask and provided energy with heat and electrical discharge (to simulate lightning). They found that simple organic molecules ap-

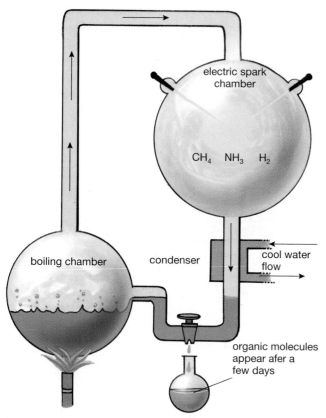

Figure 19-2 **The experimental apparatus of Stanley Miller**
Energy from heat and electrical discharge causes amino acids and other organic molecules to form from methane, ammonia, hydrogen, and water, all of which are thought to have been present in the atmosphere of the early Earth.

peared after just a few days (Fig. 19-2). In these and similar experiments, Miller and others have produced amino acids, short proteins, nucleotides, adenosine triphosphate (ATP), and other molecules characteristic of living things. Interestingly, the exact composition of the "atmosphere" used in these experiments is unimportant, provided that hydrogen, carbon, and nitrogen are available and that free oxygen is excluded. Similarly, a variety of energy sources, including ultraviolet light, electrical discharge, and heat, all seem about equally effective. Even though geochemists may never know exactly what the primordial atmosphere was like, it is certain that organic molecules were synthesized on the ancient Earth.

Prebiotic Conditions Would Allow Accumulation of Organic Molecules

Prebiotic synthesis would not have been very efficient or very fast; nonetheless, in a few hundred million years, large quantities of organic molecules could accumulate, especially since they didn't break down nearly as fast back then. On Earth today, most organic molecules have a short life; either they are digested by living organisms or they react

with atmospheric oxygen. Since the primeval Earth lacked both life and free oxygen, these sources of degradation were absent. However, the primordial atmosphere also lacked an ozone layer, which our present atmosphere has. The ozone layer is a region high in the atmosphere that is enriched with O_3 molecules that absorb some of the high-energy ultraviolet light of the sun before it reaches Earth. Consequently, the ultraviolet bombardment, which can break apart organic molecules, must have been fierce. Some places, however, such as those beneath rock ledges or at the bottoms of even fairly shallow seas, would have been protected from ultraviolet radiation. In these locations, organic molecules may have accumulated to relatively high concentrations.

These concentrations of organic molecules may have been crucial in the evolution of life. First, they may have provided the molecules that would form the first living organisms. Second, the chemical energy stored in these molecules would be food for the first cells. The rich accumulation of organic molecules in which scientists hypothesize that life first arose is often called the "primordial soup."

Was RNA the First Self-Reproducing Molecule?

In modern cells, proteins carry out most of the cellular functions, while DNA encodes the information the cell needs to synthesize those proteins. The old "chicken-and-egg" riddle applies here: Which came first—DNA or protein? If useful protein catalysts came first, how was the information needed to synthesize them passed from protocell to protocell? If nucleic acids came first, what would be the function of an information-storage molecule if there were no information to store?

In the 1980s, Thomas Cech of the University of Colorado and Sidney Altman of Yale offered an intriguing solution to this riddle. They found that some types of RNA molecules act as enzymes that, among other things, can cut apart RNA and synthesize more RNA molecules. During hundreds of millions of years of prebiotic chemical synthesis, RNA nucleotides may have occasionally bonded together to form short RNA chains. Let us suppose that, purely by chance, one of these RNA chains was a catalyst—dubbed a **ribozyme**—that could synthesize copies of itself from the free ribonucleotides in the surrounding waters. This first ribozyme probably wasn't very good at its job and made lots of mistakes. These mistakes, of course, were the first mutations. Like modern mutations, most undoubtedly ruined the catalytic abilities of the "daughter molecules," but a few may have been improvements. Molecular evolution could begin, as ribozymes with increased speed and accuracy of replication reproduced faster, making more and more copies of themselves.

Just as RNA-containing ribosomes are crucial to protein synthesis today, perhaps some early ribozymes began to bind amino acids and catalyze the synthesis of short proteins. Further ribozyme mutations might lead to the formation of the

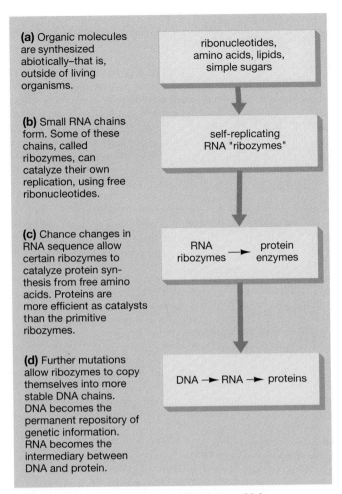

(a) Organic molecules are synthesized abiotically–that is, outside of living organisms.

ribonucleotides, amino acids, lipids, simple sugars

(b) Small RNA chains form. Some of these chains, called ribozymes, can catalyze their own replication, using free ribonucleotides.

self-replicating RNA "ribozymes"

(c) Chance changes in RNA sequence allow certain ribozymes to catalyze protein synthesis from free amino acids. Proteins are more efficient as catalysts than the primitive ribozymes.

RNA ribozymes → protein enzymes

(d) Further mutations allow ribozymes to copy themselves into more stable DNA chains. DNA becomes the permanent repository of genetic information. RNA becomes the intermediary between DNA and protein.

DNA → RNA → proteins

Figure 19-3   **A hypothesis for the origin of life**

In this scenario, RNA is the first "living molecule."

first protein enzymes. At the same time, these protein-synthesizing ribozymes were vulnerable to being cut up by other ribozymes, destroying the information so slowly accumulated over millions of years of random nucleotide substitutions. Further mutations might have allowed certain ribozymes to copy themselves over into DNA molecules that would be safe from destruction by their fellow ribozymes. In this hypothesis, then, RNA occupies center stage as the first living molecule, with both DNA and proteins evolving later (Fig. 19-3). RNA gradually receded into its present role as an intermediary between DNA and the protein enzymes that carry out most of the work of modern cells.

The First Living Cells May Have Consisted of Ribozymes within Microspheres

If water containing proteins and lipids is agitated to simulate waves beating against ancient shores, hollow structures called **microspheres** are formed (Fig. 19-4). These hollow balls resemble living cells in several respects. They have a well-defined outer boundary, separating internal contents from the external solution. If the composition of the microsphere is right, a "membrane" is formed that is remarkably similar in

appearance to a real cell membrane. Under certain conditions microspheres can absorb more material from the solution ("feed"), grow, and even divide (see Fig. 19-4).

If a microsphere happened to surround the right ribozymes, something very much like a living cell would have been formed. The ribozymes and their protein products would have been protected from free-roaming ribozymes in the primordial soup. Nucleotides and amino acids might have diffused across the membrane and have been used to synthesize new RNA and protein molecules. After sufficient growth, the microsphere may have divided, with a few copies of both ribozymes and proteins becoming incorporated into each daughter microsphere. If so, the first cells would have evolved.

But Did All This Happen?

The above scenario, although plausible and supported by many research findings, is by no means certain. One of the most striking aspects of origin-of-life research is the great diversity of assumptions, experiments, and contradictory hypotheses (see the suggested reading by Horgan, "In the Beginning . . ." for a taste of the controversies). Researchers disagree as to whether life arose in quiet pools, in the sea, in moist films on the surfaces of clay or iron pyrite (fool's gold), in furiously hot deep-sea vents. A few even argue that life may have arrived on Earth from space. Can any conclusions really be drawn from the research done so far? No one really knows the answer to that question, but we can offer a few observations.

First, the experiments of Miller and others show that amino acids, nucleotides, and other organic molecules would have been formed in abundance on the primordial Earth. Second, no one imagines that prebiotic evolution would have formed any molecules as large or sophisticat-

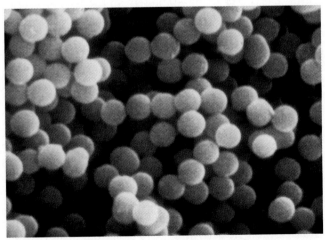

Figure 19-4   **Do microspheres resemble the earliest cells?**

Cell-like microspheres can be formed by agitating proteins and lipids in a liquid medium. Such microspheres can take in material from the surrounding solution, grow, and even "reproduce," as these are doing.

S C I E N T I F I C   I N Q U I R Y

Radiometric Dating Traces Fossils through Time

Early geologists could date rock layers and their accompanying fossils only in a *relative* way: Fossils found in deeper layers of rock were generally older than those found in shallower layers. With the discovery of radioactivity, it became possible to determine *absolute* dates, within certain limits of uncertainty. The nuclei of radioactive elements spontaneously break down, or decay, into other elements. For example, carbon-14 (usually written ^{14}C) emits an electron to become nitrogen-14 (^{14}N). Each radioactive element decays at a rate that is independent of temperature, pressure, or the chemical compound of which the element is a part. The time it takes for half the radioactive nuclei to decay at this characteristic rate is called the half-life. The half-life of ^{14}C, for example, is 5730 years.

How are radioactive elements used in determining the age of rocks? Knowing the rate of decay and measuring the proportion of decayed nuclei to undecayed nuclei, it is possible to estimate how much time has passed since these radioactive elements were trapped in rock; this process is called radiometric dating. A particularly straightforward dating technique uses the decay of potassium-40 (^{40}K) into argon-40 (^{40}Ar), which has a half-life of about 1.25 billion years. Potassium is a very reactive element and is a common constituent of volcanic rocks such as granite and basalt. Argon, on the other hand, is an unreactive gas. Let us suppose that a volcano, such as Hawaii's Kilauea volcano, erupts with a massive lava flow, covering the countryside. All the ^{40}Ar, being a gas, will bubble out of the molten lava, so when the lava solidifies into rock, it will start out with no ^{40}Ar. Potassium-40 present in the hardened lava will decay to ^{40}Ar, half the ^{40}K decaying every 1.25 billion years. The ^{40}Ar gas will be trapped in the rock. A geologist could take a sample of the rock and determine the pro-

portion of ^{40}K to ^{40}Ar (Fig. E19-1). If the analysis finds equal amounts of the two elements, the geologist will conclude that the lava hardened 1.25 billion years ago. With appropriate care, such age estimates are quite reliable. If a fossil is found beneath a lava flow dated at, say, 500 million years, then we know that the fossil is at least that old.

Some radioactive elements, as they decay, can even give an estimate of the age of the solar system. Analysis using uranium as it decays to lead has shown that the oldest Earth rocks, meteorites, and moon rocks collected by astronauts are all about 4.5 billion years old.

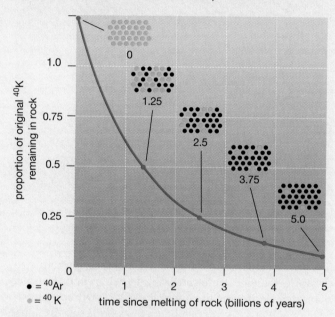

Figure E19-1   The relationship between time and the decay of radioactive ^{40}K to ^{40}Ar

ed as, say, hemoglobin. The first ribozymes, if they existed, may have been only a dozen nucleotides long. Third, ribozymes or other primitive catalysts need not have been very efficient to have accelerated the synthesis of important molecules. The stone tools of early humans weren't as handy as a Swiss army knife, either, but they were better than fingernails. Therefore, to a biologist, the origin of life means the origin of a few small molecules with terribly inefficient catalytic activities, perhaps surrounded by a film vaguely resembling a primitive membrane. Finally, several hundred million years probably elapsed between the appearance of organic molecules and the first primitive cell-like structures, and spontaneous synthesis must have proceeded over huge areas of the Earth. Given enough time

and space, even extremely rare events may in fact happen quite often. Most biologists accept that the origin of life is probably an inevitable consequence of the working of natural laws. We should emphasize, however, that *this proposition is not proved and never will be*. Biologists investigating the origin of life have neither millions of years nor trillions of liters of reaction solutions with which to work!

The First Life

The fossil record indicates that the earliest living cells arose about 3.5 billion years ago, based on radiometric dating techniques (see "Scientific Inquiry: Radiometric Dating Traces Fossils through Time"). What were these early cells like? The first cells were prokaryotic; that is, their genetic material was not sequestered from the rest of the cell within a membrane-limited nucleus. These cells probably obtained nutrients and energy by absorbing organic molecules from the primordial soup. There wasn't any free oxygen in the atmosphere, so the cells must have metabolized the organic molecules anaerobically. You will recall from Chapter 8 that anaerobic metabolism yields only small amounts of energy.

As you probably have already recognized, the earliest cells were primitive anaerobic bacteria. As these ancestral bacteria multiplied, however, they must have eventually used up the organic molecules produced by prebiotic synthesis. Simpler molecules, such as carbon dioxide and water, were still very abundant, as was energy, in the "dilute" form of sunlight. What was lacking, then, was not *materials* or *energy itself*, but *energetic molecules*: that is, molecules in which energy is stored in chemical bonds. Eventually some cells evolved the ability to use the energy of sunlight to drive the synthesis of their own complex, high-energy molecules from simpler molecules: Photosynthesis appeared.

Several kinds of photosynthetic bacteria evolved, but the ones that proved most important in the evolution of life were the cyanobacteria, sometimes called blue-green algae. Their photosynthetic reactions converted water and carbon dioxide to organic compounds, releasing oxygen as a by-product. (You will undoubtedly breathe in some oxygen molecules today that were excreted about 2 billion years ago by a cyanobacterium!) At first, the oxygen reacted with iron atoms in Earth's crust, or surface layer, forming huge deposits of iron oxide, also called rust. After all the accessible iron turned to rust, the concentration of free oxygen in the atmosphere increased. Chemical analysis of rocks suggests that significant amounts of free oxygen appeared in the atmosphere about 2 billion years ago.

Oxygen is potentially very dangerous to living things because it reacts with organic molecules, destroying them and releasing their stored energy. The accumulation of oxygen in the atmosphere provided the selective pressure for the next great advance in the Age of Microbes: the ability to use oxygen in metabolism, channeling its destructive power through aerobic respiration to generate useful energy for the cell. Because the amount of energy available to a cell is vastly increased when oxygen is used to metabolize food molecules, aerobic cells had a significant selective advantage.

Eukaryotes Evolved Membrane-Enclosed Organelles and a Nucleus

Hordes of bacteria would offer a rich food supply to any organism that could eat them. There are no fossil records of the first predatory cells, but paleobiologists speculate that once a suitable prey population (such as these bacteria) appeared, predation would have evolved quickly. These predators would have been specialized prokaryotic cells, lacking cell walls and consequently able to engulf whole bacteria as prey. According to the most widely accepted hypothesis, these predators were otherwise quite primitive, being capable of neither photosynthesis nor aerobic metabolism. Although they could capture large food particles, namely bacteria, they metabolized them inefficiently. About 1.4 billion years ago, however, one predator probably gave rise to the first eukaryotic cell.

Eukaryotic cells differ from prokaryotic cells in many ways, but perhaps most fundamental are the membrane-bound nucleus containing the genetic material and the inclusion of organelles for energy metabolism, mitochondria and (in plants) chloroplasts. How did these organelles evolve?

Mitochondria and Chloroplasts May Have Arisen from Engulfed Bacteria

The **endosymbiotic hypothesis**, championed most forcefully by Lynn Margulis of the University of Massachusetts, proposes that cells acquired the precursors of mitochondria and chloroplasts by engulfing certain types of bacteria. These cells and the bacteria trapped inside them (*endo* means "within") gradually entered into a symbiotic relationship, a close association between different organisms over an extended time. Let us suppose that an anaerobic predatory cell captured an aerobic bacterium for food, as it often did, but for some reason failed to digest this particular prey (Fig. 19-5a; see also "A Closer Look: The Evolution of Mitochondria and Chloroplasts" in Chapter 5). The aerobic bacterium remained alive and well. In fact, it was better off than ever, because the cytoplasm of its predator/host was chock-full of half-digested food molecules, the remnants of anaerobic metabolism. The aerobe absorbed these molecules and used oxygen to complete their metabolism, gaining enormous amounts of energy as it did so. So abundant were its food resources, and so bountiful its energy production, that the aerobe must have leaked energy, probably as ATP or similar molecules, back into its host's cytoplasm. The mitochondrion had been born. Interestingly, the amoeba *Pelomyxa palustris* may be a modern model for how this occurred. Unlike almost all other eukaryotic cells, it lacks mitochondria; however, a permanent population of aerobic bacteria carries out much the same role.

The predatory cell with its symbiotic bacteria could metabolize food aerobically, gaining a great selective advantage over its anaerobic compatriots. Soon its progeny filled the seas. One of these daughter cells managed a second feat: It captured a photosynthetic cyanobacterium and similarly failed to digest its prey (see Fig. 19-5a). The cyanobacterium flourished in its new host and gradually evolved into the first chloroplast. Some modern organisms resemble this hypothetical ancestral condition. A variety of corals, some clams, a few snails, and at least one species of *Paramecium*

(a) The probable origin of mitochondria and chloroplasts

An anaerobic, prokaryotic cell engulfs an aerobic bacterium. The bacterium is enclosed in a sac of the predatory cell's membrane. The resulting "proto-organelle" has a double membrane, one from the predator and one from the prey.

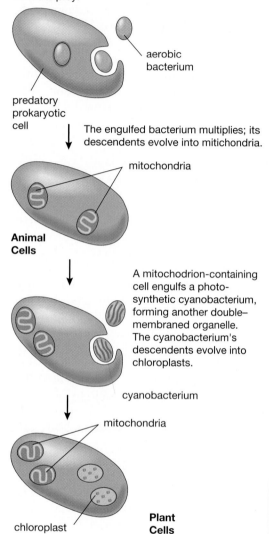

aerobic bacterium

predatory prokaryotic cell

The engulfed bacterium multiplies; its descendents evolve into mitichondria.

mitochondria

Animal Cells

A mitochodrion-containing cell engulfs a photo-synthetic cyanobacterium, forming another double–membraned organelle. The cyanobacterium's descendents evolve into chloroplasts.

cyanobacterium

mitochondria

chloroplast

Plant Cells

(b) Alternative possible origins of the membrane-enclosed nucleus

1. The nucleus may have originated as an infolding of the cell membrane that came to surround the DNA.

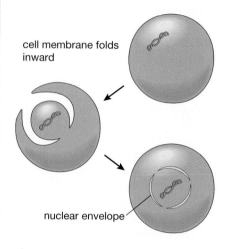

cell membrane folds inward

nuclear envelope

2. Or the nucleus may have originated when one bacterium engulfed another.

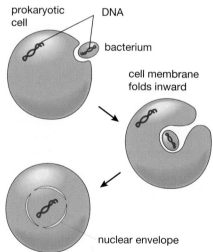

prokaryotic cell

DNA

bacterium

cell membrane folds inward

nuclear envelope

Figure 19-5 Possible mechanisms for the appearance of organelles in eukaryotic cells

---

harbor a permanent collection of algae in their cells (Fig. 19-6). These algae share some of their photosynthetically produced food molecules with the host cells.

Other eukaryotic organelles may have also originated through endosymbiosis. Many paleobiologists believe that cilia, flagella, centrioles, and microtubules may all have evolved from a symbiosis between a spirilla-like bacterium (a form of bacteria with an elongated corkscrew shape; see Chapter 21) and a primitive eukaryotic cell. Supporting this hypothesis was the discovery in 1990 that centrioles contain their own minute supply of DNA, which some interpret as a remnant of the DNA originally contained within the symbiotic bacterium.

The Origin of the Nucleus Is More Obscure

The evolution of the nucleus is more obscure. One possibility is that the plasma membrane folded inward, surrounding the DNA. This would create the nuclear membrane, and further infoldings could have produced the endoplasmic reticulum, which is continuous with the nuclear membrane (Fig. 19-5b, top). An alternative hypothesis is that, like many other eukaryotic organelles, the nucleus arose as a result of endosymbiosis. In this case, the engulfed bacterium took control of its host (Fig. 19-5b, bottom). However the nucleus originated, having the DNA sequestered within the nucleus seems to have con-

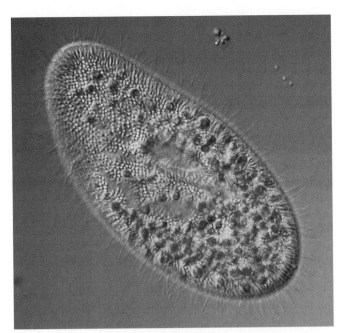

Figure 19-6   A modern intracellular symbiosis

Green unicellular algae of the genus *Chlorella* live within the cytoplasm of a *Paramecium*. A similar symbiotic relationship may have given rise to the ancestors of modern chloroplasts.

ferred great advantages in regulating the use of the genetic material. Today, organisms composed of eukaryotic cells are by far the most visible forms of life on Earth.

Multicellularity

Once predation had evolved, increased size became an advantage. A larger cell could more easily engulf a smaller cell, while in turn being more difficult for other predatory cells to ingest. Larger organisms can usually also move faster than small ones, making successful predation and escape more likely. But enormous single cells have problems. Oxygen and nutrients going in and waste products going out of the cell must diffuse through the plasma membrane. As we pointed out in Chapter 5, the larger a cell becomes, the less surface membrane is available per unit volume of cytoplasm. There are only two ways that an organism larger than a millimeter or so in diameter can survive. First, it can have a low metabolic rate so that it doesn't need much oxygen or produce much carbon dioxide. This seems to work for certain very large unicellular algae. Alternatively, an organism may be multicellular—that is, it may consist of many small cells packaged into a large unified body.

The fossil record reveals almost nothing about the origin of multicellularity, especially among animals. The first unicellular eukaryotic fossils occur in rocks about 1.4 billion years old, and the first signs of multicellular animals

occur as worm burrows and tracks in rocks 400 million years younger. The intervening animals almost certainly had no skeleton or other hard parts, and left few fossils. Consequently, we may never learn very much about them. Within another 500 million years, however, diverse types of multicellular animals appeared, leaving their fossilized remains in rocks in many parts of the world.

Multicellular Life in the Sea

The first multicellular organisms almost certainly evolved in the sea.

Multicellular Plants Evolved Specialized Structures That Facilitated Their Invasion of Diverse Habitats

During the Precambrian era, unicellular eukaryotic cells containing chloroplasts gave rise to the first multicellular plants. Multicellularity would have provided at least two advantages for plants. First, large, many-celled plants would have been difficult for unicellular predators to swallow. Second, specialization of cells would have conferred the potential for staying in one place in the brightly lit waters of the shoreline, as rootlike structures burrowed in sand or clutched onto rocks, while leaflike structures floated above in the sunlight. The profusion of green, brown, and red algae lining our shores today, some, such as the brown kelp, over 60 meters (200 feet) long, are the descendants of these early multicellular algae.

Multicellular Animals Evolved Specializations That Allowed Them to Capture Prey, Feed, and Escape More Efficiently

In a great burst of evolution, a wide variety of invertebrate animals (animals lacking backbones) appeared in the sea near the beginning of the Cambrian period, about 600 million years ago. For animals, one of the advantages of multicellularity is the potential for eating larger prey. The first animal capable of consuming large prey probably resembled a jellyfish. A single opening served both as a mouth, to take in food, and as an anus, to expel indigestible remains (Fig. 19-7a). A half-digested prey filling its digestive tract keeps such an animal from feeding again until it is finished with its first meal. Soon more efficient means of feeding evolved, employing a separate mouth and anus, found today in almost all animals (Fig. 19-7b). With this design, the animal can feed more or less continuously, as earthworms and sea cucumbers do today.

Coevolution of predator and prey led to increased sophistication in many kinds of animals; during the Silurian period (438 to 408 million years ago) mud-skimming, armored trilobites were preyed upon by ammonites and the chambered nautilus, which still survives almost unchanged in deep Pacific waters (Fig. 19-8). A major trend at this

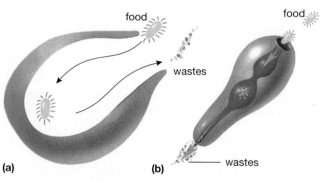

Figure 19-7   **Advances in digestive tracts**

(a) The earliest animals probably possessed a digestive system with only one opening. Feeding would have had to be periodic, because ingestion of food and expulsion of wastes would interfere with one another. **(b)** A major advance in feeding is to have separate openings for ingesting food and for ejecting wastes, allowing more or less continuous feeding.

time was toward greater mobility. Predators often need to travel over wide areas in search of suitable prey, while speedy escape is also an advantage for prey. Locomotion is usually accomplished by the contraction of muscles that move body parts through their attachments to some sort of skeleton. Most invertebrates at this time possessed either an internal hydrostatic skeleton, much like a water-filled tube (worms), or an external skeleton covering the body (arthropods such as trilobites). Greater sensory capabilities and more sophisticated nervous systems evolved along with the ability to move more efficiently. Senses for detecting touch, chemicals, and light became highly developed. The senses were usually concentrated in the head end of the animal, along with a nervous system capable of handling the sensory information and directing appropriate behaviors.

About 500 million years ago, one group of animals evolved a new form of support for the body: an internal skeleton. For a hundred million years these were inconspicuous members of the ocean community, but by 400 million years ago some had evolved into fishes. By and large, the fishes proved to be faster than the invertebrates, with more acute senses and larger brains. Eventually, they became the dominant predators of the open seas.

The Invasion of the Land

Between the period from 600 to 400 million years ago, both plants and animals evolved significantly, but they remained cushioned by the equanimity of the sea. Life in the ocean provides buoyant support against gravity and ready access to life-sustaining water. Reproduction is also simpler in the sea. At some point in their life cycles, both plants and animals produce sex cells that must fuse to form the beginnings of a new generation. Sea-dwelling organisms usually have mobile sperm or eggs, or both, that swim to each other through the water.

In contrast, on land an organism must bear its weight against the crushing force of gravity; it must find adequate water; and to reproduce, it must ensure that its gametes, particularly the sperm, are protected from drying out. Nevertheless, the land offered great potential, particularly for plants. Water strongly absorbs light, so even in the clearest water, photosynthesis is limited to the upper few hundred meters, and usually much less. Out of the water, the sun is dazzlingly bright, permitting rapid photosynthesis. Seawater also tends to be low in certain nutrients, particularly nitrogen and phosphorus. By comparison, terrestrial soils are rich storehouses of nutrients. Finally, by this time the sea swarmed with plant-eating animals. Because the land was devoid of animal life, plants that first colonized the land would have had no predators.

Some Plants Evolved Specialized Structures That Adapted Them to Life on Dry Land

In moist soils near the shore, a few small green algae began to grow, taking advantage of the sunlight and nutrients. They didn't have large bodies to support against the force of gravity, and living right in the film of water on the soil, they could easily obtain water. About 400 million years ago, some of these algae gave rise to the first multicellular land plants. Initially simple, low-growing forms, land plants rapidly evolved solutions to two of the main difficulties of plant life on land: obtaining and conserving water and staying upright despite gravity and winds. Root-like structures delved into the soil, mining water and minerals, while waterproof coatings on the aboveground parts reduced water loss by evaporation. Specialized cells formed tubes called vascular tissues to conduct water from roots to leaves. Extra-thick walls surrounding certain cells enabled stems to stand erect.

Primitive Land Plants Retained Swimming Sperm and Required Water to Reproduce

Reproduction out of water presented greater challenges. We will examine plant reproduction in more detail in Chapters 23 and 27. For now, the important point is that, like animals, plants produce sperm and eggs. Primitive marine plants have swimming sperm and sometimes swimming eggs as well. The first land plants retained swimming sperm. Consequently, they were restricted to swamps and marshes, where the sperm and eggs could be released into the water, or to areas with abundant rainfall, where the ground would occasionally be covered with water.

Conifers Encased Sperm in Pollen Grains, Allowing Them to Flourish in Dry Habitats

The reproductive strategy described above sufficed for millions of years. During the Carboniferous period, 360 to 286 million years ago, the climate was warm and moist,

(a)

(b)

(c)

(d)

Figure 19-8   **Diversity of ocean life during the Silurian period**

(a) Characteristic life of the oceans during the Silurian period, 438 to 408 million years ago. Among the most common fossils from that time are the trilobite **(b)** and its predators the ammonites **(c)** and the nautiloids. Although **(d)** illustrates a living *Nautilus*, the Silurian nautiloids were very similar in structure, showing that a successful body plan may exist virtually unchanged for hundreds of millions of years.

and great stretches of the land were covered with forests of giant tree ferns and club mosses that produced swimming sperm (see Fig. 19-9). The coal we mine today is derived from the fossilized remains of those forests. Meanwhile, some plants inhabiting drier regions had begun to evolve reproductive strategies that no longer depended on films of water. These early cone-bearing plants, called **conifers**, arose in the Permian period (286 to 246 million years ago). Their eggs were retained in the parent plant, while the sperm were encased in drought-resistant pollen grains that blew on the wind from plant to plant. Landing on a female cone near the egg, the pollen released sperm cells directly into living tissue, eliminating the need for a surface film of water. About 250 million years ago, mountains rose, swamps drained, and the moist climate dried up. Tree ferns and giant club mosses, with their swimming sperm, largely became extinct, while the conifers, which did not depend on water for reproduction, flourished and spread.

Flowering Plants Enticed Animals to Carry Pollen

About 130 million years ago, the flowering plants appeared, having evolved from a group of conifer-like plants. The initial advantage of the flowering plants seems to have been pollination by insects. The conifers are wind-pollinated, so they must produce an enormous amount of pollen because the vast majority of pollen grains fail to reach their target. Flower pollination by insects wastes far less pollen. Flowering plants also evolved other advantages, including more rapid reproduction and, in some cases, much more rapid growth. Today, flowering plants dominate the land, except in cold northern regions, where conifers still prevail.

Some Animals Evolved Specialized Structures That Adapted Them to Life on Dry Land

Soon after land plants evolved, providing potential food sources for animals, arthropods (probably early relatives of scorpions) emerged from the sea. Why arthropods? The

Figure 19-9   **The swamp forest of the Carboniferous period**

The treelike plants are tree ferns and giant club mosses, both now mostly extinct. Note the dragonfly at bottom center; some Carboniferous dragonflies had wingspans in excess of half a meter (about 1.5 feet)!

answer seems to be that they were **preadapted** for land life—that is, they already possessed structures, evolved under totally different selective pressures, that, purely by chance, were suited to life on land. Foremost among these was the external skeleton, or **exoskeleton**, a hard covering surrounding the body, such as the shell of a lobster or crab. Exoskeletons are both waterproof and strong enough to bear up a small animal under the stresses of gravity.

Land animals encounter another difficulty, also relatively easily solved by arthropods: breathing. To allow gas exchange, respiratory surfaces must be kept moist, and doing so is difficult in the dry air. Some arthropods, such as land crabs and spiders, evolved what amounts to an internal gill, kept moist within a waterproof sac (see Chapter 22). The insects developed tracheae, small branching tubes directly penetrating the body, with adjustable openings in the exoskeleton leading to the outside air.

For millions of years, arthropods had the land and its plants to themselves, and for tens of millions of years more, they were the dominant animals. Dragonflies with a wingspan of 70 centimeters (28 inches) flew among the Carboniferous tree ferns, while millipedes 2 meters (6.5 feet) long munched their way across the swampy forest floor. Eventually, however, the arthropods' splendid isolation came to an end.

Amphibians Evolved from Lobefin Fishes and Developed Lungs

About 400 million years ago, a group of fishes called the lobefins appeared, probably in fresh water. Lobefins had two important preadaptations to land-life: stout, fleshy fins with which they crawled about on the bottoms of shallow, quiet waters and an outpouching of the digestive tract that could be filled with air, like a primitive lung (Fig. 19-10). The coelacanth seen in Figure 24-33d is a lobefin that was believed to be long extinct before its discovery in 1939. In one group of lobefins, the lung evolved into a swim bladder, with which the fish could regulate their buoyancy and remain suspended in the water without active exertion. Many of these lobefins migrated back to the sea, where they evolved into the modern bony fishes. Another group of lobefins colonized very shallow ponds and streams, which shrank during droughts and whose water often became oxygen-poor. By taking air into their lungs, these lobefins could obtain oxygen anyway. Some of their descendants began to use their fins to crawl from pond to pond in search of prey or water, as some modern fish can do today (Fig. 19-11).

As the arthropods had discovered previously, the land is a rich source of food. Feeding on land and moving from pool to pool favored the evolution of fish that could stay out of water for longer periods and that could move about more effectively on land. With improvements in lungs and legs, the amphibians evolved from lobefins, first appearing in the fossil record about 350 million years ago. If an amphibian could have thought about such things, it would have thought that the Carboniferous swamp forests were heaven itself: no predators to speak of, abundant prey, and a warm, moist climate. As with the insects and millipedes, some amphibians evolved gigantic size, including salamanders over 3 meters (10 feet) long.

Despite their success, the early amphibians still lacked complete adaptation to life on land. Their lungs were

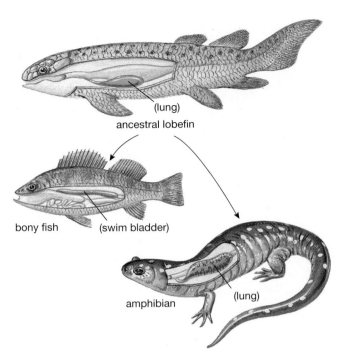

Figure 19-11   **A fish that walks on land**

Some modern fish resemble their lobefin ancestors in their ability to crawl about on land. The walking catfish was imported into Florida several years ago, escaped into local waters, and has spread across much of southern Florida by walking from pond to pond.

Figure 19-10   **From fish to amphibians**

A group of primitive fish, called the lobefins, gave rise both to the bony fish and the amphibians. Lobefins had a pair of lungs, which arose as outpocketings of the digestive tract. Lobefins probably used their lungs for breathing air when the pools in which they lived became stagnant, foul, and low in oxygen. In one group of lobefins, one lung evolved into a swim bladder, used by bony fishes to regulate buoyancy. The other lung disappeared. Another group of lobefins became further adapted for breathing air, evolving more complex, efficient lungs and sturdier fins for walking on land. These were the ancestors of the amphibians.

simple sacs without very much surface area, so they had to obtain some of their oxygen through their skins. Therefore, the skin had to be kept moist, restricting amphibians to swampy habitats where they wouldn't dry out. Further, amphibians deposit their sperm and eggs in water. As with the tree ferns and club mosses, when the Carboniferous climate turned dry at the beginning of the Permian period about 286 million years ago, amphibians were in trouble.

Reptiles, Which Arose from Amphibians, Evolved Several Adaptations to Dry Land

Meanwhile, just as the conifers had been evolving on the fringes of the swamp forests, so too a group of amphibians were evolving adaptations to drier conditions. These became the reptiles, which achieved four great advances over the amphibians. First, they evolved internal fertilization: The reptilian female provides the watery environment for the sperm inside her reproductive tract. Thus, sperm transfer could occur on land, without venturing back to the dangerous swamps full of fish and amphibian predators.

Second, they developed waterproof eggs enclosing their own supply of water for the developing embryo (see Chapter 24), again providing freedom from the swamps. Third, the ancestral reptiles developed scaly, waterproof skin. Finally, accompanying the evolution of waterproof skin came improved lungs that could provide the entire oxygen supply for an active animal. As the climate dried during the Permian period, reptiles became the dominant vertebrate land fauna, relegating the amphibians to swampy backwaters, where most remain today.

A few tens of millions of years later, the climate returned to more moist and equable conditions, providing for lush plant growth. Once again, gigantic size was selected for, as certain families of reptiles evolved into the dinosaurs (Fig. 19-12). These were among the most successful animals ever, if we consider persistence as a measure of success. They flourished for over a hundred million years, until about 65 million years ago when the last dinosaurs became extinct. No one is certain why they died, but a climate change, perhaps initiated by a gigantic meteorite impact, seems to have been the final blow (see Chapter 17). With less luxuriant plant growth, the herbivorous dinosaurs could not find enough food. Without large herbivores as prey, the carnivores were doomed too.

Even during the age of dinosaurs, many reptiles remained quite small. One major difficulty faced by small reptiles is keeping a high body temperature. Being active on land seems to require a rather warm body to maximize the efficiency of the nervous system and muscles. But a warm body loses heat to the environment unless the air is also warm. Small reptiles have a relatively large surface area through which heat is lost and a relatively small internal volume in which metabolic heat is generated. If the body is warmed metabolically when the air is cool, then an enormous amount of food must be consumed to provide sufficient energy. Apparently the food requirement is too high, for the naked-skinned small reptiles have a fairly

Figure 19-12   **A reconstruction of a Jurassic swamp**

In the foreground we see, standing in the swamp, the gigantic *Apatosaurus*, 20 meters (65 feet) long and weighing over 30 tons. On the right, a carnivorous *Allosaurus* uses its meter-long jaws to tear flesh from its prey. In the background grazes the 10-ton *Stegosaurus*, an astoundingly dim-witted herbivore with a brain the size of a walnut. *Stegosaurus* had a swelling of its spinal cord in the hip region that formed a second "brain" of sorts, 20 times larger than the one inside its skull. The plates on its back probably functioned in temperature regulation. Depending on air temperature, the animal's orientation to the sun, and the amount of blood flow across the surface of the plates, they could be used as radiators to cool the blood or as solar panels to pick up warmth.

low metabolism, not enough to keep their bodies warm in cool air, especially at night when there is no sun for radiant warmth. Two groups of small reptiles independently evolved insulation that minimizes heat loss: One group evolved feathers, while another group evolved hair.

Reptiles Gave Rise to Both Birds and Mammals

In the ancestral birds, insulating feathers retained body heat. Consequently, these animals could be active in cooler habitats and during the night, when their scaly relatives became sluggish. Later, some ancestral birds developed longer, stronger feathers on their forelimbs, perhaps allowing them to glide from trees or to assist in jumping after insect prey. From this point, the evolution of flight became possible.

The hair of ancestral mammals also provided insulation. Unlike the birds, which retained the reptilian habit of laying eggs, mammals evolved live birth and the ability to feed their young with secretions of the mammary (milk-producing) glands. Because these structures do not fossilize, we may never know when the uterus, mammary glands, and hair first appeared, or what their intermediate forms looked like.

The earliest mammals were small creatures, probably living in trees and being active mostly at night. When the dinosaurs became extinct, the mammals radiated out into the vast array of modern forms. Some stayed small and nocturnal, eating mostly seeds and insects; others evolved larger size and different habits, colonizing the habitats left empty by the extinction of the dinosaurs. One group remained in the trees, giving rise to the primates.

Human Evolution

Fossils of **primates** (lemurs, monkeys, apes, and humans) are relatively rare compared with those of many other animals. There are at least three reasons. First, most primates did not live in habitats that readily preserve fossils, such as swamps and shallow lagoons. Second, until recently primates were fairly small. Therefore predators and scavengers would be more likely to break up their bones into unrecognizable fragments than they would the bones of a bison or dinosaur. Third, ancestral primates may have had small populations, thus providing less material for fossilization in the first place.

Humans are intensely interested in their own origin and evolution, especially in trying to determine what conditions led to the evolution of the gigantic human brain. This obsession has led to sweeping speculations despite an often skimpy fossil record. Therefore, while the outline of human evolution that we will present is a synthesis of current thought on the subject, it is by no means as well understood as, say, the genetic code. Paleontologists disagree about the interpretation of the fossil evidence, and many ideas may have to be revised as new fossils are found.

Primate Evolution Has Been Linked to Grasping Hands, Binocular Vision, and a Large Brain

The first ancestral primates were the insect-eating tree shrews, whose fossils are found in rocks about 80 million years old. Nimble, probably nocturnal animals, tree shrews were smaller than all but the tiniest modern primates. Over the next 50 million years, the descendants of the tree shrews evolved forms similar to the modern tarsiers, lemurs, and monkeys (Fig. 19-13). These primates stayed in the trees. Some remained nocturnal, but many become active during the day. Primates, who fed on fruits and leaves, evolved several adaptations for life in the trees.

Grasping Hands in Early Primates Allowed Both Powerful and Precise Manipulations

The tree shrews already possessed handlike paws for holding on to small branches, and the primates further refined these appendages. Most primates are much larger than tree shrews, so only relatively large tree limbs could bear their weight. Long, grasping fingers that could wrap around and hold larger limbs made life in the trees a little safer. The primate line that led to humans evolved hands with the ability to perform both delicate maneuvers using a *precision grip* (manipulating small objects, writing, sewing) and powerful actions using a *power grip* (swinging a club, thrusting with a spear).

Binocular Vision Provides Accurate Depth Perception

One of the earliest primate adaptations seems to have been large, forward-facing eyes (see Fig. 19-13). Jumping from branch to branch is risky business unless an animal can accurately judge where the next branch is located. Accurate depth perception was made possible by binocular vision provided by forward-facing eyes with overlapping fields of view. Another adaptation was color vision. We cannot, of course, tell if a fossil animal had color vision, but modern primates have excellent color vision, and it seems reasonable to assume that earlier primates did too.

Many primates feed on fruit, and color vision helps in detecting ripe fruit among a welter of green leaves.

A Large Brain Facilitated Hand-Eye Coordination and Complex Social Interactions

Primates have brains that are larger, relative to their body size, than almost all other animals. No one really knows for certain what selective forces favored the evolution of large brains. It seems reasonable, however, that controlling and coordinating binocular, color vision, rapid locomotion through trees, and dexterous movements of the hands would be facilitated by increased brain power. Most primates also have fairly complex social systems, which probably would require relatively high intelligence. If sociality promoted increased survival and reproduction, then there would have been selective pressures for the evolution of larger brains.

Hominids Evolved from Dryopithecine Primates

Between 20 and 30 million years ago, in the moist tropical forests of Africa, a group of primates called the dryopithecines diverged from the monkey line. The dryopithecines appear to have been ancestral to the **hominids** (humans and their fossil relatives) and pongids (the great apes). Around 18 million years ago, global climatic cooling began to shrink the vast expanses of forest, splitting up the woodlands into isolated islands dotted upon a sea of grassland. Diversification of habitat and isolation of small populations led to the diversification of dryopithecines. One of these gave rise to the later hominids and pongids.

(a)

(b)

(c)

Figure 19-13   **Representative primates**

(a) Tarsier, (b) lemur, and (c) lion-tail macaque monkey. Note that all have relatively flat faces, with forward-looking eyes providing binocular vision. All also have color vision and grasping hands. These features served as preadaptations for tool and weapon use by early humans.

Figure 19-14   A possible evolutionary tree for humans

This hypothetical family tree shows skulls and facial reconstructions of representative specimens. Although many paleontologists consider this to be the most likely human family tree, there are several alternative interpretations of the known hominid fossils, each with its strong advocates. Fossils of the earliest australopithecine, *Australopithecus ramidus*, are too fragmentary to reconstruct a skull; the photo inset is of teeth, which reveal much about the specimen's relationship to other hominids.

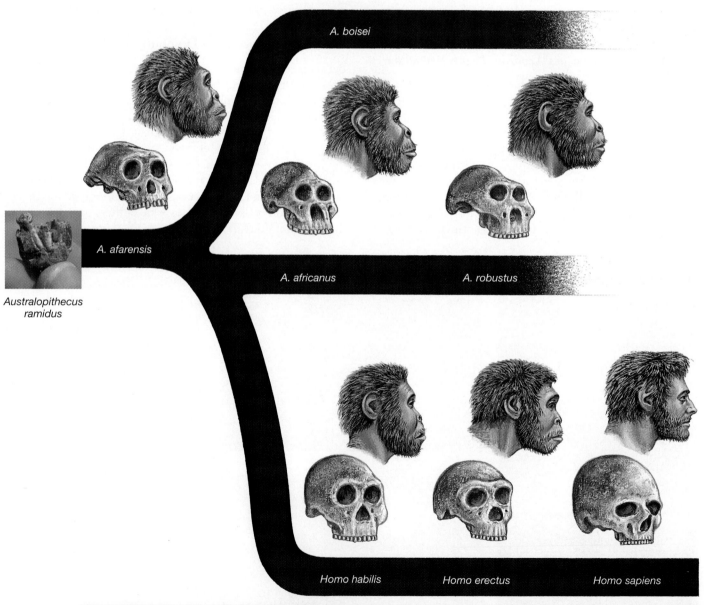

Australopithecus ramidus

A. afarensis

A. boisei

A. africanus

A. robustus

Homo habilis

Homo erectus

Homo sapiens

4   3   2   1   0

millions of years ago

Australopithecines, the First True Hominids, Could Stand and Walk Upright

A large gap ensues in the fossil record, until the first true hominids appear about 4.4 million years ago. These fossils are collectively called the australopithecines ("southern ape," after their original discovery site in southern Africa; Fig. 19-14). The brains of australopithecines were fairly large, although still much smaller than those of modern humans. Although the earliest australopithecines had legs that were shorter, for their height, than any modern humans, their knee joints allowed them to fully straighten their legs, permitting efficient bipedal (upright, two-legged) locomotion. Footprints almost 4 million years old, discovered in Tanzania by Mary Leakey, show that australopithecines could, and at least sometimes did, walk upright. Upright posture is an extremely important feature of hominids, because it freed their hands from use in walking. Later hominids thus were able to carry weapons, manipulate tools, and eventually achieve the cultural revolutions produced by modern *Homo sapiens*. What selective pressures led to the evolution of bipedalism? One interesting hypothesis is explored in "Evolutionary Connections: Why Do Humans Walk Upright?"

The oldest australopithecine species, represented by fossilized teeth, skull fragments, and arm bones, was unearthed in Ethiopia in 1993 from sediments that were dated, using radioactive isotopes, as 4.4 million years old (see "Scientific Inquiry: Radiometric Dating Traces Fossils through Time"). It was named *Australopithecus ramidus* by its discoverers (*ramid* means "root" in the local Ethiopian language). The teeth of these early fossils (although clearly hominid) have several features that are intermediate between ape and human. The second most ancient australopithecine, called *Australopithecus afarensis* was discovered in the Afar region of Ethiopia in 1974, a partial skeleton of a 3.5-foot tall female nicknamed Lucy. Fossil remains of this species 3.9 million years old have been unearthed. Later, the *A. afarensis* line apparently gave rise to at least two distinct forms: the small, omnivorous *A. africanus* (which in size and eating habits was similar to *A. afarenis*), and the large, herbivorous *A. robustus* and *A. boisei*. Most authorities agree that either *A. afarensis* or *A. africanus* was the ancestor of modern hominids, including humans, while the other forms became extinct, without leaving descendants, about a million years ago.

Homo habilis *Was the First Hominid to Use Tools*

By about 2 million years ago, certain African populations of *A. afarensis* or *A. africanus* seem to have given rise to a new form, *Homo habilis* (see Fig. 19-14). *Homo habilis* had a considerably larger body and brain than the australopithecines. *Homo habilis* was probably the first hominid that used stone and bone tools, mostly crude weapons of various sorts (Fig. 19-15). In a relatively short time, *H. habilis* spread throughout Europe, Asia, and Africa.

Homo erectus *Had a Larger Brain and Produced More Sophisticated Tools*

About 1.8 million years ago, *Homo erectus* appeared, probably a descendant of an African population of *H. habilis*. The brain of *H. erectus* was as large as the smallest modern adult human brains (around 1000 cubic centimeters). The face was notably different from that of modern humans (see Fig. 19-14), featuring large brow ridges (bony ridges above the eyes), a slightly protruding face, and no chin (the protruding tip of the lower jaw).

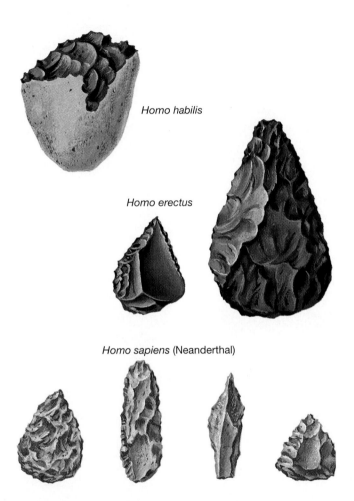

Homo habilis

Homo erectus

Homo sapiens (Neanderthal)

Figure 19-15   Representative hominid tools

Homo habilis produced only fairly crude chopping tools called hand axes, usually unchipped on one side to hold in the hand. *Homo erectus* manufactured much finer tools. The tools were often sharp all the way around the stone, so at least some of these blades were probably tied to spears rather than held in the hand. Neanderthal tools were works of art, with extremely sharp edges made by flaking off tiny bits of stone. In comparing these *H. habilis*, *H. erectus*, and *H. sapiens* weapons, note the progressive increase in the number of flakes taken off the blades and the corresponding decrease in flake size. Smaller, more numerous flakes produce a sharper blade and suggest more insight into tool making (perhaps passed down culturally from experienced tool makers to apprentices), more patience, finer control of hand movements, or perhaps all three.

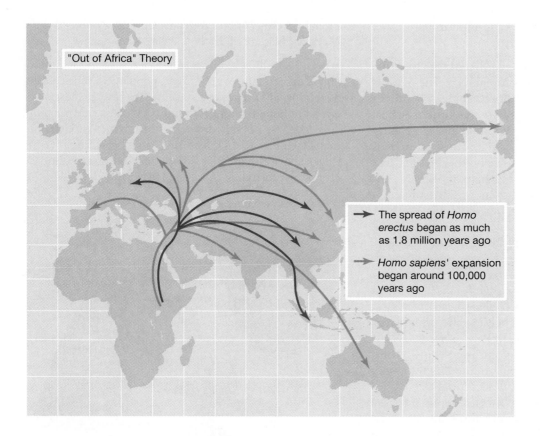

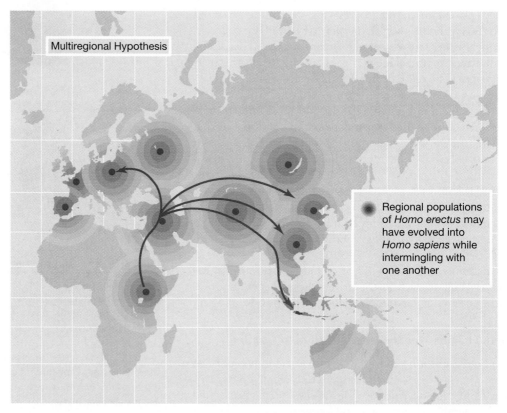

Figure 19-16   Competing hypotheses for the evolution of *Homo sapiens* from *Homo erectus*

The "out of Africa" advocates hypothesize that *H. sapiens* evolved in Africa, then migrated through-out the East, Europe, and Asia. The "multiregional" hypothesis states that populations of *H. sapiens* evolved in many regions simultaneously from the already widespread populations of *H. erectus*.

Behaviorally, *H. erectus* seems to have been more advanced than *H. habilis*. *Homo erectus* fashioned sophisticated stone tools, ranging from hand axes used for cutting and chopping to points probably used on spears (see Fig. 19-15). These weapons suggest that *H. erectus* ate animal food, probably both hunting on their own and scavenging for the remains of prey killed by lions and other predators. *Homo erectus*, like *H. habilis*, spread throughout most of Europe and Asia. In China, evidence from 500,000 years ago suggests that in addition to sophisticated stone tools, *H. erectus* used fire, although for what purpose (to cook food, keep warm, or ward off predators) no one knows.

Homo sapiens *Evolved from* Homo erectus

The first fossils of our own species, *Homo sapiens* (*sapiens* means "intelligent"), date from about 200,000 years ago. There are currently two hotly debated hypotheses as to the origin of *H. sapiens*. Some paleoanthropologists (scientists who study human origins) believe that *H. sapiens* evolved in Africa, then migrated into the Near East, Europe, and Asia, supplanting all other hominids, including *H. erectus*. The opposing hypothesis is that populations of *H. sapiens* evolved in many regions simultaneously from the already widespread populations of *H. erectus*. According to this hypothesis, continued migrations and interbreeding among *H. erectus* populations maintained them as a single species as they gradually evolved into *H. sapiens* (Fig. 19-16).

Although the earliest fossils are fragmentary, beginning about 100,000 years ago, Neanderthals, believed by many researchers to be a variety of *H. sapiens*, left many complete fossils in northern Africa, throughout Europe, and in parts of Asia. Contrary to the popular image of the hulking, stoop-shouldered "cave man," Neanderthals were quite similar to modern humans in many ways. Although more heavily muscled, Neanderthals walked fully erect, were dexterous enough to manufacture finely crafted stone tools, and had brains, on the average, slightly larger than those of modern humans. Although many of the European fossils show heavy brow ridges and a broad, flat skull, others, particularly from areas around the eastern shores of the Mediterranean Sea, were remarkably like ourselves. Neanderthal remains show evidence of modern behaviors, too, particularly ritualistic burial ceremonies (Fig. 19-17). Neanderthal skeletons have been discovered in clearly marked burial sites surrounded with stones, and often including offerings of flowers, bear skulls, and food. Altars found at other Neanderthal sites were probably used in "religious" rites associated with a bear cult.

Finally, about 90,000 years ago, fully modern humans appeared. Although the oldest specimens were discovered in Israel in 1987, these anatomically modern humans are called Cro-Magnon, after the district in France in which their remains were first discovered. Cro-Magnons had domed heads, smooth brows, and prominent chins. Their tools were precision instruments not very different from the stone tools used until recently in many parts of the world.

Where did the Cro-Magnons come from and where did the Neanderthals go? One group of paleoanthropologists believes that Cro-Magnon evolved from the modern-looking Neanderthals of the eastern Mediterranean region and spread into Europe from there. The contrary view is that Cro-Magnon evolved independently, from a separate population of *H. erectus*. Neanderthal and Cro-Magnon probably coexisted for a while, particularly in Europe. Some researchers believe that considerable interbreeding occurred between Neanderthals and Cro-Magnons and that their genes are a part of the European heritage. Others believe that the Neanderthals represent an evolutionary dead end. Whether the Neanderthals were eliminated by interbreeding, competition, or open warfare with Cro-Magnon may never be known.

Behaviorally, Cro-Magnon seems to have been similar to, if more sophisticated than, Neanderthal. Perhaps the most remarkable accomplishment of Cro-Magnon is the magnificent art left in caves such as Altamira in Spain and Lascaux in France (Fig. 19-18). Again, no one knows exactly why these paintings were made, but they attest to minds fully as human as our own.

The Evolution of Human Behavior Is Highly Speculative

Perhaps the most hotly debated subject in human evolution is the development of human behavior. Except in rare instances, such as the Neanderthal burials and the Cro-Magnon paintings, there is no direct evidence of the behavior of prehistoric hominids. A few hypotheses have been offered by biologists and anthropologists, on the basis of the fossil record and the behavior of modern humans and animals.

Figure 19-17   Neanderthal bear ceremony

Power and ferocity, and perhaps their ability to rise up on their hind legs like humans, made cave bears cult objects to Neanderthals. Some Neanderthals were buried with elaborate ceremonies, with bear skulls placed on the grave of the deceased.

***Figure 19-18*   The sophistication of Cro-Magnon people**

Cave paintings by Cro-Magnon people remarkably preserved by the relatively constant underground conditions of a cave in Lascaux, France.

The fossil record shows that the development of truly huge brains occurred within the last couple of million years. What were the selective advantages of large brains? This subject continues to stir controversy among paleoanthropologists. The enlarging brain may have been selected both because it provided for improved hand-eye coordination and because it facilitated complex social interactions.

As the early hominids such as *Australopithecus* descended from the trees into the savanna, they began to walk upright. Bipedal locomotion allowed them to carry things in their hands as they walked. Hominid fossils show shoulder joints capable of powerful throwing motions and an opposable thumb (a thumb that can be placed directly against each of the other fingers) that would enhance the ability to manipulate objects. These early hominids probably could see well and judge depth accurately with binocular vision. The brain expansion that occurred in the australopithecines may have been at least partly related to integration of visual input and control of hand and arm movements.

Homo erectus, and perhaps *H. habilis* before it, was social. So, of course, are many monkeys and apes. The later hominids, however, seem to have engaged in a new type of social activity: cooperative scavenging and hunting. *Homo erectus*, in particular, seems to have hunted extremely large game, which calls for cooperation in all phases of the hunt. Some paleoanthropologists believe that selection favored larger and more powerful brains as an adaptation for success in cooperative hunting. (Lest you become too impressed with such speculation, remember that lions and wolves also hunt cooperatively, yet are not noticeably more intelligent than their relatives, such as leopards and coyotes, who do not.)

The Cultural Evolution of Humans Now Far Outpaces Biological Evolution

In recent millennia, the biological evolution of humans has been far outpaced by our **cultural evolution**: learned behaviors passed down from previous generations. Careful estimates of the human population size since our species originated reveals three major surges of population growth. Each surge was initiated by a cultural-technological revolution: the development of new forms of technology that increased the resources available to sustain human life. The first revolution was the development of tools, beginning with early hominids and extending until 10,000 years ago, when the total human population numbered 5 million. Beginning then and extending over the next 8000 years, humans grew crops and domesticated animals, and the population gradually increased by 100-fold, to about 500 million. Following this agricultural revolution, we are now 300 years into the scientific-industrial revolution, and our population has reached 5.6 billion.

Human cultural evolution, and its accompanying increases in human population, have had profound effects on the continuing biological evolution of other life forms. Our agile hands and minds have transformed much of Earth's terrestrial habitats. People have become the single overwhelming agent of natural selection. In the words of evolutionary biologist Stephen Jay Gould of Harvard University: "We have become, by the power of a glorious evolutionary accident called intelligence, the stewards of life's continuity on Earth. We did not ask for this role, but we cannot abjure it. We may not be suited for it, but here we are."

EVOLUTIONARY CONNECTIONS

Why Do Humans Walk Upright?

Upright posture and bipedal locomotion, naked skin, and enlarged brains are hallmarks of humanity. Why did these features evolve in humans? The short answer is that no one knows, but British physiologist Peter Wheeler has an intriguing hypothesis: Our ancestors evolved the ability to walk upright because this posture exposes a minimum body surface to the blazing savanna sun. Wheeler constructed a small-scale model of "Lucy" (*Australopithecus afarensis*) and photographed it in bipedal and quadrupedal postures from a series of overhead locations that imitated the pathway of the sun. He found that a four-legged posture exposed the model to 60% more solar radiation than did the upright posture. Further, says Wheeler, standing upright exposes more of the body to cooler breezes a few feet above the ground. Together, these factors would increase the time the animal could spend foraging in the sun, allowing it to increase its access to food. Bipedalism may have also made possible our naked skin. Although hair is a shield against the sun in quadrupedal (four-legged) animals, in our bipedal ancestors this need was minimized, restricted to the shoulders and head. Naked skin then allowed our ancestors to dissipate even more heat. Wheeler carries the hypothesis a step further, to include the large human brain. The brain is one of the most metabolically active organs in the body and very easily damaged by overheating. Wheeler suggests that only after hominids had evolved upright posture and naked skin did they have enough heat-dissipating capacity to permit the evolution of their oversized brains.

SUMMARY OF KEY CONCEPTS

Origins
Before life arose, lightning, ultraviolet light, and heat formed organic molecules from water and the components of the primordial Earth's atmosphere. These molecules probably included nucleic acids, amino acids, short proteins, and lipids. By chance, some molecules of RNA may have had enzymatic properties, catalyzing the assembly of copies of themselves from nucleotides in Earth's waters. These may have been the forerunners of life. Protein-lipid microspheres enclosing these RNA molecules may have formed the first cell-like organisms.

The First Life
The first fossil cells are found in rocks about 3.5 billion years old. These cells were prokaryotes that fed by absorbing organic molecules that had been synthesized in the environment. Because there was no free oxygen in the atmosphere at this time, energy metabolism must have been anaerobic. As the cells multiplied, they depleted the organic molecules that had been formed by prebiotic synthesis. Some cells evolved the ability to synthesize their own food molecules using simple inorganic molecules and the energy of sunlight. These earliest photosynthetic cells were probably ancestors of today's cyanobacteria.

Photosynthesis releases oxygen as a by-product, and by about 2.2 billion years ago significant amounts of free oxygen had accumulated in the atmosphere. Aerobic metabolism, which generates more cellular energy than does anaerobic metabolism, probably arose about this time.

Eukaryotic cells evolved about 1.5 billion years ago. The first eukaryotic cells probably arose as symbiotic associations between predatory prokaryotic cells and bacteria. Mitochondria may have evolved from aerobic bacteria engulfed by predatory cells. Similarly, chloroplasts may have evolved from photosynthetic cyanobacteria.

Multicellularity
Multicellular organisms evolved from eukaryotic cells, first appearing about 1 billion years ago. Multicellularity offers several advantages, including increased speed of locomotion and greater size.

Multicellular Life in the Sea
The first multicellular organisms arose in the sea. In plants, increased size due to multicellularity offered some protection from predation. Specialization of cells allowed plants to anchor themselves in the nutrient-rich, well-lit waters of the shore. For animals, multicellularity allowed more efficient predation and more effective escape from predators. These in turn provided selection pressures for faster locomotion, improved senses, and greater intelligence.

The Invasion of the Land
The first land organisms were probably plants, appearing 400 to 600 million years ago. Although the land required special adaptations for support of the body, reproduction, and the acquisition, distribution, and retention of water, the land also offered abundant sunlight and protection from aquatic herbivores. Around 400 million years ago, arthropods invaded the land. Absence of predators and abundant land plants for food probably facilitated the invasion of the land by animals.

The earliest land vertebrates evolved from lobefin fishes, which had leglike fins and a primitive lung. A group of lobefins evolved into the amphibians about 350 million years ago. Reptiles evolved from amphibians, with several further adaptations for land-life: internal fertilization, waterproof eggs that could be laid on land, waterproof skin, and better lungs. Around 150 million years ago, birds and mammals evolved independently from separate groups of reptiles. Major advances included a high, constant body temperature and insulation over the body surface.

Human Evolution
One group of mammals evolved into the tree-dwelling primates. Primates show several preadaptations for human evolution: forward-facing eyes for binocular vision, color vision, grasping hands, and relatively large brains. Between 20 and 30 million years ago some primates descended from the trees; these were the ancestors of apes and humans. The australopithecines arose in Africa about 4 million years ago. These hominids walked erect, had larger brains than their forebears, and made primitive tools. One group of australopithecines evolved into true humans.

KEY TERMS

conifer p. 373
cultural evolution p. 382
endosymbiotic hypothesis p. 369
exoskeleton p. 374
hominid p. 377
microsphere p. 367
preadaptation p. 375
prebiotic evolution p. 365
primate p. 376
ribozyme p. 366
spontaneous generation p. 364

THINKING THROUGH THE CONCEPTS

Multiple Choice

1. Our best estimate for the age of Earth, 4.5 billion years, is supported by
   a. the assumption of uniformitarianism
   b. the Big Bang theory
   c. gradualism
   d. punctuated equilibrium
   e. radioactive dating of the oldest rocks found
2. There was no free oxygen in the early atmosphere because all of it was tied up in
   a. water
   b. ammonia
   c. methane
   d. rock
   e. radioactive isotopes
3. RNA became a candidate for the first information-carrying molecule when Tom Cech and Sidney Altman discovered that some RNAs can act as enzymes that
   a. degrade proteins
   b. turn light into chemical energy
   c. split water and release oxygen gas
   d. synthesize copies of themselves
   e. synthesize amino acids
4. Which three of the following observations support Lynn Margulis's endosymbiotic hypothesis for the origin of chloroplasts and mitochondria from ingested bacteria?
   a. Mitochondria are where aerobic respiration takes place.
   b. Mitochondria have their own DNA.
   c. Chloroplasts are where photosynthesis takes place.
   d. Chloroplasts have their own DNA.
   e. Bacterial plasma membranes are strikingly similar to the inner membrane of mitochondria.
5. The exoskeleton of early, marine-dwelling arthopods can be considered a preadaptation for life on land because it
   a. can support an animal's weight against the pull of gravity
   b. allows a wide diversity of body types
   c. resists drying
   d. absorbs light
   e. both a and c
6. The evolution of the shelled, amniotic (membrane-enclosed) egg was an important event in vertebrate evolution because it
   a. led to the Cambrian explosion
   b. was the first example of parents caring for their young
   c. allowed the colonization of freshwater environments
   d. freed organisms from having to lay their eggs in water
   e. allowed internal fertilization of eggs

Review Questions

1. What is the evidence that life might have originated from nonliving matter on the primordial Earth? What kind of evidence would you like to see before you would accept this hypothesis?
2. If they were so much more efficient at producing energy, why didn't the first cells with aerobic metabolism extinguish cells with only anaerobic metabolism?
3. Explain the endosymbiotic hypothesis for the origin of chloroplasts and mitochondria.
4. Name two advantages of multicellularity in plants and animals.
5. What advantages and disadvantages would terrestrial existence have had for the first plants to invade the land? For the first land animals?
6. Review the material on the evolution of flowering plants. How did flowering plants entice animals to carry pollen? (That is, what was in it for the animals?) Why do conifers "still prevail" in cold, northern regions?
7. Outline the general trends in the evolution of vertebrates, from fish to amphibians to reptiles to birds and mammals. Explain how these adaptations increased the fitness of the various groups for life on land.
8. Outline the evolution of humans from early primates. Include in your discussion such features as binocular vision, grasping hands, bipedal locomotion, social living, tool making, and brain expansion.

APPLYING THE CONCEPTS

1. What is cultural evolution? Is cultural evolution more or less rapid than biological evolution? Why?
2. Do you think that studying our ancestors can shed light on the behavior of modern humans? Why or why not?
3. A biologist would probably answer the age-old question "What is life?" by saying "the ability to self-replicate." Do you agree with this definition? If so, why? If not, how would you define life in biological terms?

4. Traditional definitions of humans have emphasized "the uniqueness of humans" because of language or tool use. But most animals can communicate with other individuals in sophisticated ways, and many vertebrates use tools to accomplish tasks. Pretend you are a biologist from Mars, and write a taxonomic description of the species *Homo sapiens*.

5. The "out of Africa" and "multiregional" hypotheses of *Homo sapiens* evolution make contrasting predictions about the extent and nature of genetic divergence among human races. One predicts that races are old and highly diverged genetically; the other predicts that races are young and little diverged genetically. What data would help you determine which hypothesis is closer to the truth?

6. In biological terms, what do you think was the most significant event in the history of life? Explain your answer.

❋ FOR MORE INFORMATION

Calvin, W. H. "The Emergence of Intelligence." *Scientific American*, October 1994. The evolution of several hallmarks of human intelligence may have been linked to the ability to coordinate rapid movements.

Diamond, J. "How to Speak Neanderthal." *Discover*, January 1990. Could the Neanderthals speak? How could we tell? Jared Diamond lucidly explains the arguments and methodology behind a scientific dispute.

Fenchel, T., and Finlay, B. J. "The Evolution of Life without Oxygen." *American Scientist*, January/February 1994. Clues to the origin of the first eukaryotic cells are provided by symbiotic relationships of organisms in oxygen-free environments.

Hay, R. L., and Leakey, M. D. "The Fossil Footprints of Laetoli." *Scientific American*, February 1982. The actual footprints of a hominid family were discovered by Hay and Leakey in volcanic ash 3.5 million years old.

Horgan, J. "In the Beginning . . . " *Scientific American*, February 1991. An exploration of the controversies surrounding research into the origin of life.

Morell, V. "Announcing the Birth of a Heresy." *Discover*, March 1987. Paleontologists Robert Bakker and Jack Horner speculate that dinosaurs were not the plodding beasts of monster flicks, but warm-blooded, advanced animals that may even have cared for their young.

Orgel, L. E. "The Origin of Life on Earth." *Scientific American*, October 1994. The evolution of self-reproducing molecules, of which RNA may have been the first, made life possible.

Waters, T. "Almost Human." *Discover*, May 1990. Artist John Gurche reconstructs fossil hominids by adding clay "muscle" and "skin" to replicas of hominid skulls.

NET WATCH

On-line resources for this chapter are on the World Wide Web at:
http://www.prenhall.com/~audesirk (click on the <u>table of contents</u> link and then select Chapter 19).

20 Taxonomy: Imposing Order on Diversity

> *"Systems of classification are not hat racks, objectively presented to us by nature. They are dynamic theories developed by us to express particular views about the history of organisms. Evolution has provided a set of unique species ordered by differing degrees of genealogical relationship. Taxonomy, the search for this natural order, is the fundamental science of history."*
>
> STEPHEN JAY GOULD, *Natural History* (1987)

❖ AT A GLANCE

Taxonomic Categories

The Origins of Taxonomy

Modern Criteria for Classification

The Five Kingdoms of Life

Taxonomy: An Inexact Science

Exploring Biodiversity: How Many Species Exist?

Evolutionary Connections: Classification Conundrums; or, Where Shall We Put the Algae?

Molecular biology provides new insights into the origin of species. Mitochondrial DNA sequencing has recently shown that the red wolf (top), currently listed as an endangered species, is almost certainly a hybrid between the gray wolf (bottom left) and the coyote (bottom right) and is probably not a separate species at all.

Taxonomic Categories

Taxonomy (from the Greek word *taxis*, meaning "arrangement") is the science by which organisms are classified and placed into categories based on their structural similarities and evolutionary relationships. Taxonomic categories form a hierarchy—that is, a series of levels each more inclusive than the last. There are seven major categories: **species, genus, family, order, class, division** or **phylum** (divisions are used for plants and plant-like microorganisms; phyla for animals and animal-like microorganisms), and **kingdom**. Each category from species to kingdom is increasingly more general and includes organisms whose common ancestor was increasingly remote in its evolutionary relationship. Some examples of classifications of specific organisms are given in Table 20-1.

The **scientific name** of an organism is actually formed from the two smallest of these categories: the genus and the species. For example, the scientific name for humans, *Homo sapiens*, places us in the genus *Homo* ("man") and the species *sapiens* ("thinking") within the genus *Homo*. The genus is a category that includes very closely related organisms that do not usually interbreed. The species is a category limited to organisms that will interbreed under natural conditions. Thus, the genus *Sialia* (bluebirds) includes the Eastern Bluebird (*Sialia sialis*), the Western Bluebird (*Sialia mexicana*), and the Mountain Bluebird (*Sialia currucoides*), very similar birds that do not normally interbreed (Fig. 20-1).

Scientific names are always underlined or italicized. The first letter of the genus name is always capitalized, and the first letter of the species name is always lowercased. These names are recognized by biologists worldwide, overcoming language barriers and allowing precise communication.

The Origins of Taxonomy

Aristotle (384–322 B.C.) was among the first to attempt to formulate a logical, standardized language for naming living things. Using characteristics such as structural complexity, behavior, and degree of development at birth, he

Figure 20-1   **Three species of bluebird**

The similarities among these species of bluebirds (genus *Sialia*) are obvious, but the species remain distinct because they do not interbreed.

Sialia sialis
(eastern bluebird)

Sialia mexicana
(western bluebird)

Sialia currucoides
(mountain bluebird)

classified about 500 organisms into 11 categories. Aristotle placed organisms into a hierarchy of categories each more inclusive than the one before it, a concept that is still used today.

Building on this foundation over 2000 years later, the Swedish naturalist Carolus Linnaeus (1707–1778) laid the groundwork for the modern classification system. He placed each organism into a series of hierarchically arranged categories based on its resemblance to other life forms, and he also introduced the scientific name based on genus and species. Nearly 100 years later, Charles Darwin (1809–1882) published *On the Origin of Species,* which added a new significance to these categories. Taxonomists then began to recognize that taxonomic categories reflect the evolution-

ary relatedness of organisms. The more categories two organisms share, the closer their evolutionary relationship.

Modern Criteria for Classification

As mentioned earlier, organisms are considered to belong to the same species if they will interbreed under natural conditions. However, this criterion can be applied only to organisms that reproduce sexually. Taxonomists must use other criteria to classify organisms that rarely engage in sexual reproduction (many bacteria and

	Human Being	Chimpanzee	Wolf	Fruit Fly	Sequoia Tree	Sunflower
TABLE 20-1 ❖ *Classification Reflects the Degree of Relatedness of Organisms*						
Kingdom	**Animalia**	**Animalia**	**Animalia**	**Animalia**	**Plantae**	**Plantae**
Phylum	**Chordata**	**Chordata**	**Chordata**	Arthropoda	Coniferophyta	Anthophyta
Class	**Mammalia**	**Mammalia**	**Mammalia**	Insecta	Coniferosida	Dicotyledoneae
Order	**Primates**	**Primates**	Carnivora	Diptera	Coniferales	Asterales
Family	Hominidae	Pongidae	Canidae	Drosophilidae	Taxodiaceae	Asteraceae
Genus	*Homo*	*Pan*	*Canis*	*Drosophila*	*Sequoiadendron*	*Helianthus*
Species	*sapiens*	*troglodytes*	*lupus*	*melanogaster*	*giganteum*	*annuus*

*Boldface categories are those that are shared by more than one of the organisms classified. Genus and species are always italicized or underlined.

(a)

(b)

Figure 20-2   **Convergent evolution molds similar body forms**

External similarities between organisms in completely different classes can be the result of convergent evolution (see Chapter 16) rather than evidence for a close evolutionary relationship. **(a)** Although dolphins (like humans) are mammals, the fish, such as this shark **(b)**, that share their habitat and many of their external features are in a different taxonomic class.

some fungi and plants) and to classify all types of organisms above the species level. Some of the most important criteria are anatomy, developmental stages, and biochemical similarities.

Historically, the most important and useful of these has been anatomy. In addition to obvious similarities in external body structure (see Fig. 20-1), taxonomists look carefully at details such as skeletons and tooth structure. For example, the presence of homologous structures, such as the finger bones of dolphins, bats, seals, and humans (Fig. 20-2; see also Fig. 16-12), provides evidence of an evolutionarily distant common ancestor. To distinguish between closely related species, taxonomists may use microscopes to discern finer details—the number and shape of the "teeth" on the tonguelike radula of a mollusk, the spines on a marine worm, or the fine structure of pollen grains of a flowering plant (Fig. 20-3).

Developmental stages also provide clues to common ancestry. For example, although the "sea squirt" spends its adult life permanently attached to rocks on the ocean floor, its free-swimming larva has a nerve cord, tail, and gills, placing it in the same phylum as the vertebrates: phylum Chordata (see Fig. 24-30).

Modern taxonomists are also aided by sophisticated biochemical techniques. One such technique is **electrophoresis.** During this procedure, molecules (such as proteins and nucleic acids) are separated according to their electrical charge and molecular weight. The presence and relative abundance of specific molecules in various species provides a way to assess their relatedness. Another technique is **DNA hybridization,** de-

scribed in detail in Chapter 16. When heated, the double-stranded DNA separates; as the DNA cools, it reforms. If DNA from two closely related species is combined and heated, hybrids of the two DNA types form upon cooling. Only complementary sequences can hybridize. Thus, the degree of hybridization is proportional to the degree of similarity between the DNA molecules from the two species. This similarity, in turn, reflects how closely the two species are related. Modern biochemical technology also allows scientists to determine the exact sequence of nucleotides in DNA and RNA and the amino acid sequences of proteins. The greater the similarity in the sequence of the subunits in genetic material or protein between two organisms, the closer their evolutionary relationship. The **polymerase chain reaction** (which can amplify minuscule quantities of DNA, allowing it to be analyzed; see Chapter 14) permits scientists to sequence DNA from the remains of organisms that are now extinct and compare it with that of their living relatives.

The Five Kingdoms of Life

Before 1970, taxonomists classified all forms of life into two kingdoms: Animalia and Plantae. Bacteria, fungi, and photosynthetic protists were considered plants, and the protozoa were classified as animals. In 1969, Robert H. Whittaker proposed the five-kingdom classification scheme that is widely used today, and which we follow in this text.

(a)   **(b)**   **(c)**

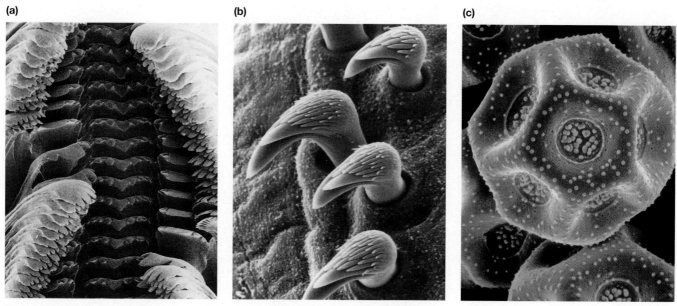

Figure 20-3   **Microscopic structures help taxonomists classify organisms**

(a) The "teeth" on the tonguelike radula of a snail or (b) the bristles on a marine worm are potential taxonomic criteria. (c) The shape and surface features of pollen grains help identify plants, particularly from fossil deposits and sediments.

Whittaker identified two kingdoms of primarily unicellular microorganisms based on whether they showed prokaryotic or eukaryotic cellular organization. The kingdom Monera consists of generally single prokaryotic cells, whereas the kingdom Protista consists of generally single eukaryotic cells. All of the organisms in the remaining three kingdoms (Plantae, Fungi, and Animalia) are eukaryotic, and most of them are multicellular. They may be classified further on the basis of their way of acquiring nutrients. Members of the kingdom Plantae photosynthesize, and members of the kingdom Fungi secrete enzymes outside their bodies and then absorb the externally digested nutrients. In contrast, members of the kingdom Animalia ingest their food and then digest it, either within an internal cavity or within individual cells. Comparisons among the five kingdoms are presented in Table 20-2.

Recently, a sixth kingdom has been proposed. Many microbiologists argue that the cell structure of the Archaebacteria (see Chapter 21) is so different from that of prokaryotic and eukaryotic cells that they should occupy their own kingdom. For the sake of simplicity, however, this text retains the five-kingdom classification system. The taxonomic categories described in the following chapters are listed in Table 20-3.

During evolution, the prokaryotic Monera gave rise to the eukaryotic Protista, which show all possible modes of nutrition. Different groups of protists then evolved into the multicellular plants, animals, and fungi, as illustrated in Figure 20-4.

TABLE 20-2 ⠿ *Some Characteristics of the Five Kingdoms*

Kingdom	Cell Type	Cell Number	Major Mode of Nutrition	Motility (Movement)	Cell Wall	Reproduction
Monera	Prokaryotic	Unicellular	Absorb or photosynthesize	Both motile and nonmotile	Present: peptidoglycan	Usually asexual, rarely sexual
Protista	Eukaryotic	Unicellular	Absorb, ingest or photosynthesize	Both motile and nonmotile	Present in algal forms: varies	Both sexual and asexual
Animalia	Eukaryotic	Multicellular	Ingest	All motile at some stage	Absent	Both sexual and asexual
Fungi	Eukaryotic	Most multicellular	Absorb	Generally nonmotile	Present: chitin	Both sexual and asexual
Plantae	Eukaryotic	Multicellular	Photosynthesize	Generally nonmotile	Present: cellulose	Both sexual and asexual

TABLE 20-3 ❊ *The Classification of the Major Groups of Organisms**

Kingdom	Division/Phylum	Common Name
Monera (unicellular, prokaryotic)	Division Eubacteria	"true" bacteria
	Division Archaebacteria	"ancient" bacteria
Protista (unicellular, eukaryotic)	Division Pyrrophyta	dinoflagellates
	Division Chrysophyta	diatoms
	Division Euglenophyta	euglenoids
	Division Myxomycota	plasmodial slime molds
	Division Acrasiomycota	cellular slime modes
	Phylum Sarcomastigophora	zooflagellates, amoebae
	Phylum Apicomplexa	sporozoans
	Phylum Ciliophora	ciliates
Animalia (multicellular, eukaryotic, heterotrophic, ingest nutrients)	Phylum Porifera	sponges
	Phylum Cnidaria	hydra, anemone, jellyfish, corals
	Phylum Platyhelminthes	flatworms
	Phylum Nematoda	roundworms
	Phylum Annelida	segmented worms
	Class Oligochaeta	earthworms
	Class Polychaeta	tube worms
	Class Hirudinea	leeches
	Phylum Arthropoda	"jointed legs"
	Class Insecta	insects
	Class Arachnida	spiders, ticks
	Class Crustacea	crabs, lobsters
	Phylum Mollusca	"soft-bodied"
	Class Gastropoda	snails
	Class Pelecypoda	mussels, clams
	Class Cephalopoda	squid, octopus
	Phylum Echinodermata	sea stars, sea urchins
	Phylum Chordata	chordates
	Subphylum Vertebrata	vertebrates
	Class Agnatha	lampreys, hagfish
	Class Chondrichthyes	sharks, rays
	Class Osteichthyes	bony fish
	Class Amphibia	frogs, salamanders
	Class Reptilia	turtles, snakes, lizards
	Class Aves	birds
	Class Mammalia	mammals
Fungi (multicellular, eukaryotic, heterotrophic, absorb nutrients)	Division Zygomycota	"zygote fungi"
	Division Ascomycota	"sac fungi"
	Division Deuteromycota	"imperfect fungi"
	Division Basidiomycota	"club fingi"
	Division Oomycota	"egg fungi"
Plantae (multicellular, eukaryotic, photosynthetic)	Division Rhodophyta	red algae
	Division Phaeophyta	brown algae
	Division Chlorophyta	green algae
	Division Bryophyta	mosses
	Division Pteridophyta	ferns
	Division Coniferophyta	evergreens
	Division Anthophyta	flowering plants

*There is no agreement on taxonomic classification systems. This table lists only those taxonomic categories described in this text.

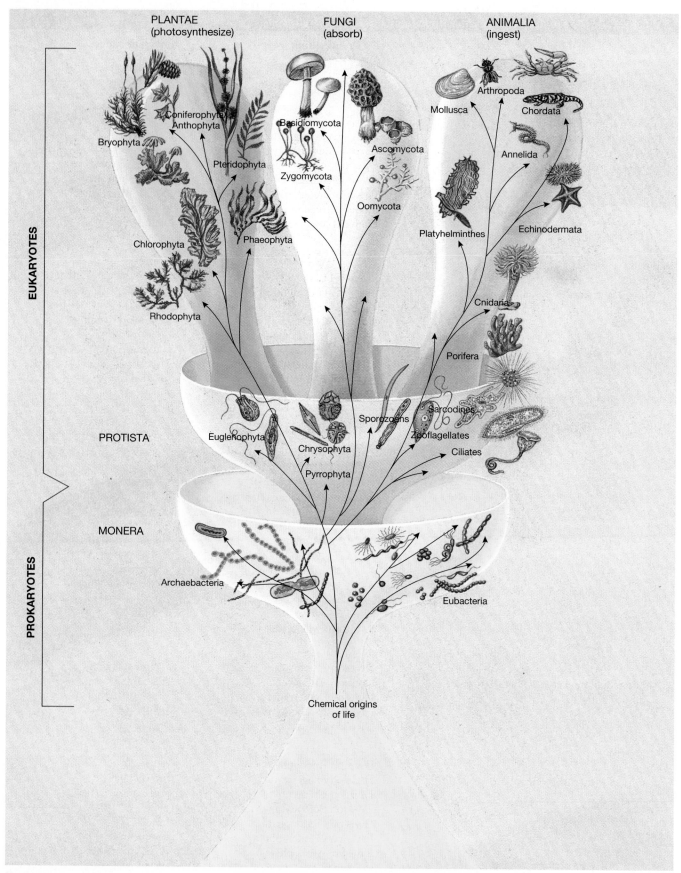

Figure 20-4   The tree of life

The five kingdoms and some of the major phyla or divisions within them are illustrated.

Taxonomy: An Inexact Science

As the Archaebacteria debate suggests, taxonomic categories are not the unchanging frameworks that we might wish them to be, even at the kingdom level (see "Evolutionary Connections: Classification Conundrums; or, Where Shall We Put the Algae?"). For example, in 1973 ornithologists officially declared the Baltimore Oriole and the Bullock's Oriole to be a single species, which they named the Northern Oriole. The reason: Where their ranges overlap, the two "species" of birds were interbreeding.

Another example is the red wolf from the southern United States, which is currently listed as an endangered species (see the photograph on the opening page of this chapter). Researchers recently analyzed the DNA from mitochondria taken from red wolves and found it to be identical to mitochondrial DNA from coyotes in some cases and to mitochondrial DNA from gray wolves in others. Mitochondrial DNA is inherited directly from the mother via the mitochondria present in the original egg cell. This mitochondrial DNA evidence, then, strongly suggests that red wolves are actually hybrids between gray wolves and coyotes and may not be a distinct species at all.

Asexually reproducing organisms pose a particular challenge to taxonomists, because the criterion of interbreeding cannot be used to distinguish species. For example, some taxonomists recognize 200 species of the parthenogenetic British blackberry (a plant that can produce seeds without fertilization); others recognize only 20 species. Most unicellular organisms reproduce asexually most of the time, making their classification difficult. For example, molecular biologists recently sequenced the DNA of two different strains of the bacterium *Legionella pneumophila,* which is responsible for Legionnaire's disease, and found only a 50% correspondence between the strains. This degree of genetic dissimilarity, which occurs within a single species of bacterium, is as great as the genetic difference between mammals and fishes! Because the taxonomy of bacteria is so controversial, we present only general descriptions of bacteria in Chapter 21.

Taxonomic categories are continuously debated and revised as taxonomists learn more about evolutionary relationships, particularly with the application of techniques derived from molecular biology. Although the precise evolutionary relationships of many organisms continue to elude us, taxonomy is enormously helpful in ordering our thoughts and investigations into the diversity of life on Earth.

Exploring Biodiversity: How Many Species Exist?

Scientists do not know even within an order of magnitude how many species share our world. Each year, between 7000 and 10,000 new species are named, most of them insects, many from the tropical rain forests. The total number of named species is currently around 1.4 million. However, many scientists believe that 7 to 10 million species may exist, and estimates range as high as 30 million. Of all the species that have been identified, about 5% are monerans and protists. An additional 22% are plants and fungi, and the rest are animals. This distribution has little to do with the actual abundance of these organisms and a lot to do with the size of the organisms, how easy they are to classify, how accessible they are, and the number of scientists studying them. An entire new "continent" of species lies largely unexplored on the deep-sea floor at depths of 1000 meters (3250 feet) and more. From the limited samples available, scientists estimate that hundreds of thousands of unknown species may reside there.

Although about 4000 species of bacteria have been named, bacterial diversity also remains largely unexplored. Consider a recent study by Norwegian scientists who used DNA hybridization to identify the different bacteria in a small sample of forest soil. To distinguish species, they arbitrarily defined bacterial DNA as coming from separate species if it differed by at least 30% from that of any other bacterial DNA. Using this criterion, they reported over 4000 types of bacteria in their soil sample and an equal number of new forms in a sample of shallow marine sediment!

Our ignorance of the full extent of life's diversity adds a new dimension to the tragedy of the destruction of the tropical rain forests, discussed in Chapter 46. Although these forests cover only about 6% of Earth's land area, they are believed to be home to two-thirds of the world's existing species, most of which have never been studied or named. Because the forests are being destroyed so rapidly, Earth is losing many species that people will never even know existed! For example, in 1990 a new species of primate, the black-faced lion tamarin, was discovered in a small patch of dense rain forest on an island just off the east coast of Brazil (Fig. 20-5). Only 12 individuals of this squirrel-sized monkey have ever been spotted, and captive breeding may be its only hope for survival. At current rates of destruction, most of the tropical rain forests, with their undescribed wealth of life, will be gone within the next century.

 E V O L U T I O N A R Y
C O N N E C T I O N S

Classification Conundrums; or, Where Shall We Put the Algae?

Classification schemes are used by biologists to place the huge number of organisms on Earth into natural groupings. Ideally these groupings are based on evolutionary relationships, which are inferred by taxonomists on the basis

Figure 20-5   **The black-faced lion tamarin**

This rare new species of monkey was discovered in 1990 in the vanishing Brazilian rain forest.

of the presence of shared distinguishing features. Early classification schemes placed all organisms into either the plant or the animal kingdom. Later, close examination of the unique structure of fungi and the diversity of single-celled organisms made it necessary to propose additional kingdoms that recognized the fundamental differences among plants, animals, fungi, and unicellular prokaryotes and eukaryotes. The current five-kingdom scheme was devised in response to this need.

The kingdom Protista, defined as comprising all single-celled eukaryotic organisms, is not a natural grouping, and scientists disagree about which organisms it should include. Plants, animals, and fungi all have close protistan relatives, and the separation of single-celled organisms from multicellular organisms is sometimes problematic. It is especially so for the algae, which have both single-celled and multicellular representatives within most smaller taxonomic groupings. Can closely related organisms be placed into separate kingdoms, Protista and Plantae, simply on the basis of multicellularity? If you look at different textbooks, you will see that the algae, photosynthetic organisms with simple reproduction, are some-

times placed entirely into Protista, and sometimes they are split between Protista and Plantae depending on whether they are single-celled or multicellular. Some taxonomists split the multicellular algae between two kingdoms, placing the multicellular brown and red algae with the protists and the multicellular green algae into the plant kingdom. These different attempts to classify closely related organisms are good examples of how difficult it is to develop standard criteria for grouping organisms, even at the kingdom level.

One approach to this problem, enthusiastically endorsed by Dr. Lynn Margulis, a biologist at the University of Massachusetts, is the creation of the kingdom Protoctista. This taxonomic category would include single-celled organisms and their close descendants (for example, the multicellular algae, but *not* the animals, fungi, and plants). Margulis describes the kingdom Protoctista as "the entire motley and unruly group of non-plant, non-animal, non-fungal organisms representative of lineages of the earliest descendants of the eukaryotes."

It is conceptually difficult to group one of the largest multicellular organisms in the world, the brown algae called giant kelps, with simple microscopic single-celled organisms. Kelps, some of which are up to 60 meters (almost 200 feet) long, possess a tissuelike level of organization that is relatively complex and can transport materials over long distances, as can the tissues of higher plants. The cells in kelps and some other algae are specialized and show division of labor. However, kelps reproduce like other algae and differently from plants. Dr. Thomas Cavalier-Smith of the University of British Columbia has proposed that brown algae merit their own kingdom (kingdom Chromista) based on ultra-

structural features and molecular comparisons of all algae. So, even among the algae, there are clear differences that some scientists believe are sufficient to justify the status of a separate kingdom.

As we learn more about the relationships between organisms and refine the criteria used to classify them, classification schemes will change. As the superficially simple question "In which kingdom should we place the algae?" illustrates, the taxonomic categories in textbooks are tentative and subject to revision as we continue to discover more about life on Earth.

✖ SUMMARY OF KEY CONCEPTS

Taxonomic Categories
Taxonomy is the science by which organisms are classified and placed into hierarchical categories that reflect their structural similarities and evolutionary relationships. The seven major categories, in order of increasing inclusiveness, are species, genus, family, order, class, division or phylum, and kingdom. The scientific name of an organism is composed of its genus name and species name.

The Origins of Taxonomy
A hierarchical concept was first used by Aristotle, but Linnaeus in the mid-1700s laid the foundation for modern taxonomy. In the 1860s, evolutionary theory proposed by Darwin provided an explanation for the observed similarities and differences between organisms, and modern taxonomists attempt to classify organisms according to their evolutionary relationships.

Modern Criteria for Classification
Today, taxonomists use features such as anatomy, developmental stages, and biochemical similarities to categorize or-

ganisms. Molecular biological techniques are used to determine the sequences of nucleotides in DNA and RNA and of amino acids in proteins. Biochemical similarities among organisms are a measure of evolutionary relatedness.

The Five Kingdoms of Life
A five-kingdom classification system consisting of the kingdoms Monera, Protista, Animalia, Fungi, and Plantae is widely used (see Table 20-2 for characteristics).

Taxonomy: An Inexact Science
Taxonomic categories are controversial and subject to revision, particularly in the case of asexually reproducing species. However, taxonomy is essential for precise communication and contributes to our understanding of the origins and diversity of species.

Exploring Biodiversity: How Many Species Exist?
Although only about 1.4 million species have been named, estimates of the total number of species range up to 30 million. New species are being identified at the rate of 7000 to 10,000 each year.

✖ KEY TERMS

class   p. 387
division   p. 387
DNA hybridization   p. 389
electrophoresis   p. 389
family   p. 387

genus   p. 387
kingdom   p. 387
order   p. 387
phylum   p. 387
polymerase chain reaction   p. 389

scientific name   p. 387
species   p. 387
taxonomy   p. 387

✖ T H I N K I N G   T H R O U G H   T H E   C O N C E P T S

Multiple Choice

1. It is possible to imagine the various levels of taxonomic classification as a kind of "family tree" for an organism. If the kingdom is analogous to the trunk of the tree, what taxonomic category would be analogous to the large limbs coming off that trunk?
   a. class
   b. family
   c. order
   d. phylum or division
   e. subfamily

2. The more taxonomic categories shared by two organisms, the more closely they are related in an evolutionary sense. Which biologist's work led to this insight?
   a. Aristotle
   b. Darwin
   c. Linnaeus
   d. Whittaker
   e. Pauling

3. Which of the following criteria could not be used to determine how closely related two different organisms are?
   a. similarities in the presence and relative abundance of specific molecules
   b. DNA hybridization
   c. the presence of homologous structures
   d. developmental stages
   e. occurrence of both organisms in the same habitat

4. Which of the following habitats appears to have the greatest number of species?
   a. the seafloor
   b. deserts
   c. tropical rain forests
   d. grasslands
   e. mountaintops

5. What organisms are considered plants by some taxonomists and protists by others?
   a. algae
   b. fungi
   c. mosses
   d. archaebacteria
   e. diatoms

6. An organism is described to you as having many nuclei-containing cells, each surrounded by a cell wall of chitin and absorbing its food. In what kingdom would you place it?
   a. Plantae
   b. Protista
   c. Animalia
   d. Fungi
   e. Monera

Review Questions

1. What contributions did Aristotle, Linnaeus, and Darwin each make to modern taxonomy?

2. What features would you study to determine whether a dolphin is more closely related to a fish or to a bear?

3. What techniques might you use to determine whether the extinct cave bear is more closely related to a grizzly bear or to a black bear?

4. Only a small fraction of the total number of species on Earth has been scientifically described. Why?

5. In England, "daddy long-legs" refers to a long-legged fly, but the same name refers to a spider-like animal in the United States. How do scientists attempt to avoid such confusion?

✖ A P P L Y I N G   T H E   C O N C E P T S

1. There are many areas of disagreement among taxonomists, for example, whether algae belong in the kingdom Protista or the kingdom Plantae, or whether archaebacteria should be in the kingdom Monera, or in an entirely new kingdom. What difference does it make whether biologists consider algae as plants or protists, or archaebacteria as monerans or something else? As Shakespeare put it, "What's in a name?"

2. The pressures created by human population growth and economic expansion place storehouses of biological diversity such as the tropics in peril. The seriousness of the situation is clear when we consider that probably only 1 out of every 20 tropical species is known to science. What sorts of arguments can you make for preserving biological diversity in poor and developing countries? Does such preservation require that these countries sacrifice economic development? Suggest some solutions to the conflict between the demand for resources and the importance of conserving diversity.

3. During major floods only the topmost branches of submerged trees may be visible above the water. If you were asked to sketch the branches below the surface of the water, based solely on the positions of the exposed tips, you would be attempting a reconstruction somewhat similar to the "family tree" by which taxonomists link various organisms according to their common ancestors (analogous to branching points). What sources of error do both exercises share? What advantages do modern taxonomists have?

4. The Florida panther, found only in the Florida Everglades, is currently classified as an endangered species, protecting it from human activities that could lead to its extinction. It has long been considered a subspecies of cougar (mountain lion), but recent mitochondrial DNA studies have shown that the Florida panther may actually be a hybrid between American and South American cougars. Should the Florida panther be protected by the Endangered Species Act?

✖ FOR MORE INFORMATION

Avise, J. C. "Nature's Family Archives." *Natural History*, March 1989. Shows how evolutionary relationships can be determined by analyzing differences in DNA contained in mitochondria.

Diamond, J. "The Search for Life on Earth." *Natural History*. A brief overview of the modern practice of taxonomy.

Gould, S. J. "What Is a Species?" *Discover*, December 1992. Discusses the difficulties of distinguishing separate species.

Lowenstein, J. M. "Molecular Approaches to the Identification of Species." *American Scientist*, 73: 541–547, 1985. A basic introduction to immunological methods of taxonomy.

Margulis, L., and Schwartz, K. *Five Kingdoms*. 2nd ed. New York: W. H. Freeman, 1988. An illustrated paperback guide to the diversity of life.

May, R. M. "How Many Species Inhabit the Earth?" *Scientific American*, October 1992. Although no one knows the answer to this question, an accurate count is crucial in an effort to manage the biological resources of this planet.

Moffett, M. W. *The High Frontier: Exploring the Tropical Rainforest Canopy*. Cambridge, MA: Harvard University Press, 1994. The tremendous diversity of life in the rainforest treetops is only now becoming known to us. This book documents the unexpected and spectacular diversity of animals in the upper reaches of this endangered habitat.

Wilson, A. C. "The Molecular Basis of Evolution." *Scientific American*, 1985. On using nucleotide and amino acid sequences to determine evolutionary relationships.

Wilson, E. O. *The Diversity of Life*. Cambridge, Mass. Harvard University Press, 1992. An outline of the processes that created the diversity of life; and a discussion of the threats to that diversity and the steps required to preserve it.

NET WATCH

On-line resources for this chapter are on the World Wide Web at:
http://www.prenhall.com/~audesirk (click on the <u>table of contents</u> link and then select Chapter 20).

21 The Hidden World of Microbes

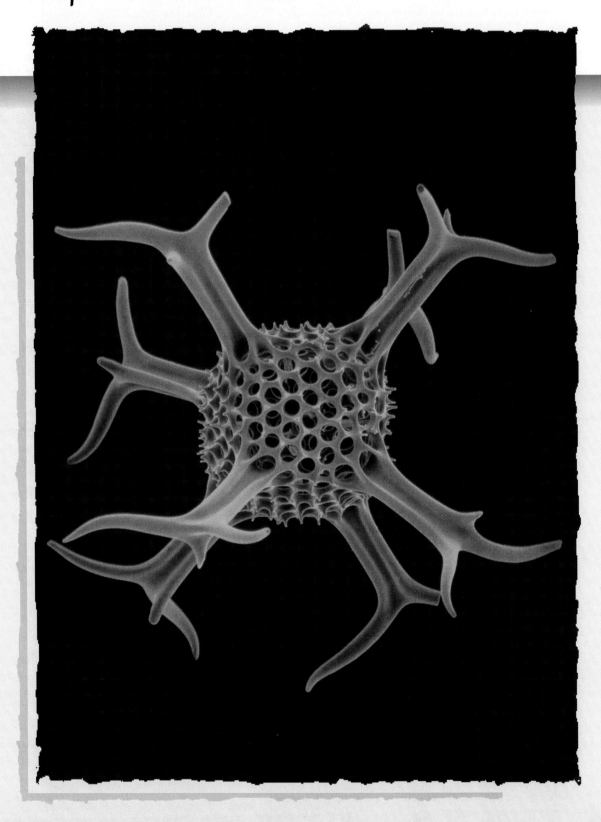

⊞ A T   A   G L A N C E

Viruses
  A Virus Consists of a Molecule of DNA or RNA Surrounded
    by a Protein Coat
  Viral Infections Cause Diseases That Are Difficult to Treat
  Some Infectious Agents Are Even Simpler than Viruses
  How Did These Infectious Particles Originate?

The Kingdom Monera
  Bacteria Are Difficult to Classify
  Bacteria Possess a Remarkable Variety of Shapes and
    Structures
  Bacterial Reproduction Is by Cell Division Called Binary
    Fission
  Bacteria Are Specialized for Specific Habitats
  Bacteria Perform Many Functions That Are Important to
    Other Forms of Life
  Some Bacteria Pose a Threat to Human Health
  Cyanobacteria Are Photosynthetic, and Many Can Also
    Capture Atmospheric Nitrogen
  Archaebacteria Are Sufficiently Different in Cellular
    Structure That Some Taxonomists Place Them in a
    Separate Kingdom

The Kingdom Protista
  Protists Are a Diverse Group Including Plantlike,
    Funguslike, and Animal-like Forms
  The Unicellular Algae, or Phytoplankton, Are Plantlike
    Protists
  The Slime Molds Are Funguslike Protists
  The Protozoa Are Animal-like Protists

**Evolutionary Connections: Our Unicellular
Ancestors**

*This scanning electron micrograph of a microscopic radiolarian
from the Indian Ocean reveals the intricate beauty of its glassy
shell.*

T hree and a half billion years ago, in warm, oxygen-free
primordial "soup," the first cells appeared, the forerunners
of all modern life forms. Their rodlike and filamentous
(threadlike) bodies, fossilized in rock dating back 3.5 billion years, appear little different from modern bacterial
rods and filaments. For the next 2 billion years, as prokaryotic cells spread and diversified, these early representatives of the kingdom Monera remained the only forms of
life on Earth. Finally, 1.4 billion years ago, a new type of
fossil appeared, single eukaryotic cells, the first protists.

This chapter introduces the hidden world of microbes,
beginning with viruses, puzzling parasitic particles that are
not alive and that defy classification. The kingdom Monera is composed of single prokaryotic cells, the simplest
forms of life. The chapter concludes with the kingdom
Protista: usually single, complex, eukaryotic cells. Figure
21-1 illustrates the relative sizes and complexity of these
three groups of microorganisms.

Viruses

Viruses possess no membranes of their own, no ribosomes
on which to make proteins, no cytoplasm, and no source
of energy. They cannot move or grow, and they can reproduce only inside a host cell. The simplicity of viruses
makes it impossible to call them cells, and indeed, seems
to place them outside the realm of living things.

A Virus Consists of a Molecule of DNA or RNA
Surrounded by a Protein Coat

A virus particle is so small (0.05–0.2 micrometer [μm: $\frac{1}{1000}$
of a millimeter] in diameter) that seeing it requires the
enormous magnification of an electron microscope. Viruses consist of two major parts: (1) a molecule of hereditary
material, either DNA or RNA, and (2) a coat of protein
surrounding the molecule. The protein coat may itself be
surrounded by an envelope formed from the membrane of
the host cell (Fig. 21-2a). Even if placed in a rich broth of
nutrients at optimal temperature, viruses remain inactive.
They are unable to grow or divide, because they lack the
complex cellular organization that these activities require.

The protein coat, however, is specialized to allow viruses to penetrate the cells of a specific host (the organism it infects). After a virus enters a host cell, its genetic material takes command. The hijacked host cell is forced to "read" the viral genes and to use the instructions encoded there to produce the components of new viruses. The pieces are rapidly assembled (Fig. 21-2b), and an army of new viruses bursts forth to invade and conquer neighboring cells (see "A Closer Look: Viral Replication").

Viral Infections Cause Diseases That Are Difficult to Treat

Each type of virus is specialized to attack a specific host cell (Fig. 21-3), and probably no organism is immune to all viruses. Even bacteria fall victim to viral invaders called **bacteriophages** (Fig. 21-4).

Within a particular organism, viruses specialize in attacking particular cell types. Those responsible for the common cold, for example, attack the membranes of the respiratory tract, those causing measles infect the skin, and the rabies virus attacks nerve cells. One type of herpes virus specializes in the mucous membranes of the mouth and lips, causing cold sores; a second type, transmitted through sexual contact, produces similar sores on or near the genitals. Herpes viruses take up permanent residence in the body, erupting periodically (often during times of stress) as infectious sores. The devastating disease AIDS (acquired immunodeficiency syndrome), which cripples the body's immune system, is caused by a virus that attacks a specific type of white blood cell that controls the body's immune response (see Chapter 34). Viruses have also been linked to specific types of cancer, such as T-cell leukemia, a cancer of the white blood cells. The recent identification of the papilloma virus, which causes genital warts, in 90% of cervical cancers sampled suggests that this virus may cause these cancers. Viruses are intracellular parasites that require the cellular machinery of their host, so the illnesses they cause are difficult to treat because antiviral agents may destroy host cells as well as virus. The antibiotics so effective against bacterial infections are useless against viruses, although some promising antiviral drugs are being developed (see Chapter 34).

Some Infectious Agents Are Even Simpler than Viruses

In the early 1970s, researchers discovered that some plant diseases are caused by particles only one-tenth the size of normal plant viruses. Called **viroids,** these particles are merely short strands of RNA, lacking even a protein coat. Like viruses, viroids apparently enter the nucleus of the infected cell, where they direct the synthesis of new viroids. About a dozen crop diseases have been attributed to viroids, including cucumber pale fruit disease, avocado sunblotch, and potato spindle tuber disease.

Prions are even more puzzling than viroids. Many years ago, doctors studying a primitive tribe in New Guinea were puzzled to observe frequent cases of a fatal degenerative disease of the nervous system, which the people called **kuru.**

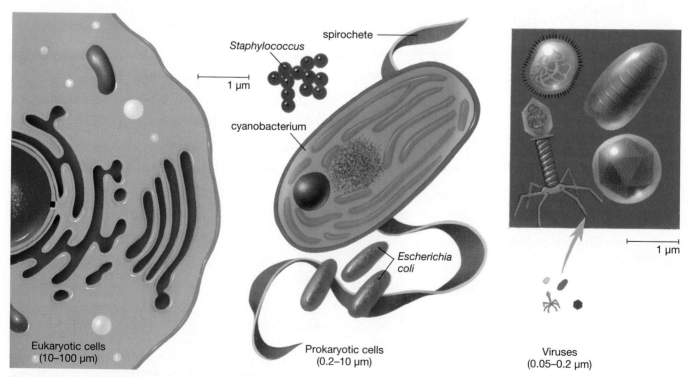

Staphylococcus
spirochete
1 μm
cyanobacterium
Escherichia coli

Eukaryotic cells
(10–100 μm)

Prokaryotic cells
(0.2–10 μm)

Viruses
(0.05–0.2 μm)

1 μm

Figure 21-1   **The sizes of microorganisms**
The relative sizes of eukaryotic cells, prokaryotic cells, and viruses (1 μm = 1/1000 millimeter).

(a)

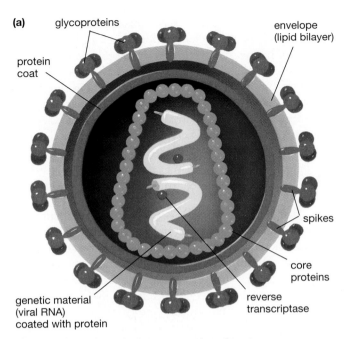

glycoproteins

protein coat

envelope (lipid bilayer)

spikes

core proteins

genetic material (viral RNA) coated with protein

reverse transcriptase

(b)

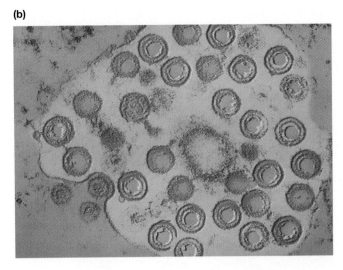

Figure 21-2   **Viral structure and replication**

(a) Cross section of the virus that causes AIDS. Inside is genetic material surrounded by a protein coat. Some viruses, including those causing herpes, rabies, and AIDS, have an outer envelope that may be formed from the membrane of the host cell. Spikes made of protein and carbohydrate may project from the envelope, and some viruses use these to attach to their host cell. **(b)** In this electron micrograph, herpes viruses are seen packed into an infected cell.

The victims had ritually consumed the brains of their tribespeople who had died from kuru. An infectious agent, transmitted in infected brain tissue, was clearly at work—but what was it? The symptoms of kuru showed a striking resemblance to those of scrapie, a disease of sheep that also causes degeneration of the brain. In 1982, Stanley Prusiner, a neurologist, isolated protein particles from hamsters infected with scrapie

and called them prions (Fig. 21-5). Prions are now known to be responsible for several neurodegenerative diseases, including kuru, Creutzfeldt-Jakob disease, scrapie, and mad-cow disease (a new form of scrapie that infects cattle).

How can a protein replicate itself, and be infectious? Recent research has shown that prions are a mutated form of a normal body protein, folded improperly. Researchers

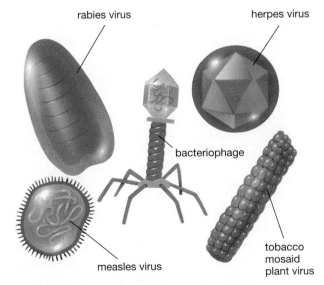

rabies virus

herpes virus

bacteriophage

measles virus

tobacco mosaic plant virus

Figure 21-3   **Viruses come in a variety of shapes**

Viral shapes are determined by their protein coats. The rabies and herpes viruses are surrounded by an extra envelope derived from membranes of the host cell.

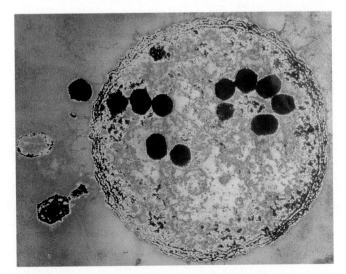

Figure 21-4   **Some viruses parasitize bacteria**

In this electron micrograph, bacteriophage viruses are seen attacking a bacterium. They have injected their genetic material inside, leaving their protein coats clinging to the bacterial cell wall. Black objects inside the bacterium are newly forming viruses.

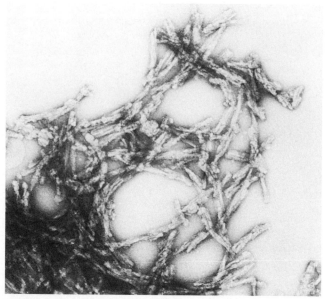

Figure 21-5   **Prions: puzzling proteins**

Prion proteins isolated from the brain of a hamster infected with scrapie, a degenerative disease of the nervous system.

believe that the mutant prion protein may interact with normal protein, forcing it to fold abnormally. The improperly folded protein thus increases in concentration over time. This hypothesis explains why diseases caused by prions may either be inherited (through inheritance of the mutated gene for prion protein) or be transmitted (by the abnormal protein that catalyzes its own production from normal protein). The discovery of prions has generated enormous interest among researchers, because prions are a totally new type of infectious particle—one that "reproduces" without DNA or RNA.

How Did These Infectious Particles Originate?

The origin of viruses, viroids, and prions is obscure. It is unlikely that these infectious particles are the forerunners of life, because they cannot reproduce by themselves. Some scientists believe that viruses and viroids originated from simple parasitic cells that evolved such complete dependence on their hosts that they lost the ability to perform the basic processes of life. Or perhaps viruses and viroids originated as loose fragments of genetic material that took up an independent existence. Whatever their origin, the success of these parasitic particles poses a continuing challenge to living things.

The Kingdom Monera

Members of the kingdom **Monera,** often called **bacteria** (singular, **bacterium**), are single, prokaryotic cells. They have changed very little in form from their fossil ancestors

of roughly 3.5 billion years ago. Prokaryotic cells lack organelles such as the nucleus, chloroplasts, and mitochondria (see Chapter 5 for a comparison of prokaryotic and eukaryotic cells). Bacterial cells are usually very small, ranging from about 0.2 to 10 micrometers in diameter, compared with eukaryotic cells, whose diameters range from about 10 to 100 micrometers. Around 250,000 average-sized bacteria could congregate on the period at the end of this sentence.

Bacteria Are Difficult to Classify

The classification of bacteria poses particular challenges to taxonomists. Because bacterial reproduction is usually asexual, bacterial species cannot be defined on the basis of their ability to interbreed. In addition, the fossil record of bacteria is quite sparse. Consequently, taxonomists classify bacteria using a variety of criteria: shape, means of locomotion, pigments, staining properties, nutrient requirements, and the appearance of bacterial colonies. As mentioned in the previous chapter, bacterial diversity is almost unexplored. Although about 4000 bacterial species have been described, there may be 1000 times that many yet undescribed.

Some microbiologists place bacteria in two major categories: a large division called the **eubacteria** (from the Greek words for "true bacteria") and a much smaller division, the **archaebacteria** (from the Greek words for "ancient bacteria"). These two types of bacteria have striking structural and biochemical differences. In this text, we organize our discussion of the kingdom Monera around the general features and diverse life-styles of bacteria, singling out two groups for special consideration: the cyanobacteria (members of the division Eubacteria) and the archaebacteria.

Bacteria Possess a Remarkable Variety of Shapes and Structures

A Cell Wall Gives Bacteria Their Characteristic Shapes

Nearly all bacteria are encased in a porous but rigid cell wall that protects them from osmotic rupture in watery environments and gives different types their characteristic shapes. The most common bacterial shapes are rodlike **bacilli,** spheres called **cocci,** and the corkscrew-shaped **spirilla** (Fig. 21-6). The cell wall contains a material called **peptidoglycan** that is unique to bacteria. Peptidoglycan is composed of chains of sugars cross-linked by peptides (short chains of amino acids). A staining technique, the **Gram stain,** distinguishes two types of cell wall construction in bacteria, classifying them as either gram positive or gram negative. The cell walls of gram negative bacteria include an additional outer membrane resembling a plasma membrane in structure. This outer membrane is sometimes toxic to mammals and is one mechanism by which some of these bacteria cause disease. The antibiotic penicillin works best on gram positive bacteria.

Outside the Cell Wall, Capsules, Slime Layers, and Pili Help Bacteria Survive in Specialized Environments

Surrounding the cell walls of some bacteria are sticky **capsules** or **slime layers,** composed of polysaccharide or protein. Capsules help certain disease-causing bacteria escape detection by their victim's immune system. Slime layers allow the bacteria that cause tooth decay to adhere in masses to the smooth surface of a tooth. This slime forms the basis of dental plaque (Fig. 21-7). Some bacteria cover themselves with a fuzz of hairlike projections called **pili** (singular, **pilus**). Pili are made of protein and generally serve to attach the bacterium to other cells. The pili of some infectious bacteria, such as those causing the venereal disease gonorrhea, attach to the cell membranes of their host, facilitating infection. Some bacteria produce special sex pili, described below (see Fig. 21-11).

Some Bacteria Can Move Using Simple Flagella That Are Attached with a Unique Wheel-like Structure

Some bacteria are equipped with **flagella** (singular, **flagellum**). These are simpler in structure than the flagella seen in some eukaryotic cells. Bacterial flagella, which may either cover the cell or form a tuft at one end (Fig. 21-8a, p. 406), can rotate rapidly, propelling the bacterium through its liquid environment. Recent research has revealed a unique

(a)

(b)

(c)

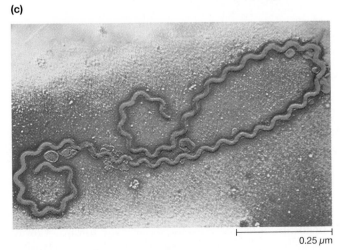

0.10 μm

4 μm

0.25 μm

Figure 21-6   **Three common bacterial forms**

(a) Spherical bacteria, also called cocci, of the genus *Micrococcus*; **(b)** rods, also called bacilli, shown here growing on the point of a pin; and **(c)** corkscrew-shaped spirilla; this species causes leptospirosis in humans.

Figure 21-7   **The cause of tooth decay**

Slime layers allow decay-causing bacteria to adhere in masses to the enamel of teeth, as shown in this scanning electron micrograph. The green bristles of a toothbrush can be seen sweeping the bacteria away.

A C L O S E R L O O K

Viral Replication

Viruses multiply, or "replicate," using their own genetic material, which consists of single- or double-stranded RNA or DNA, depending on the virus. This material serves as a template (a "blueprint") for the viral proteins and genetic material required to make new viruses. Sometimes, viral enzymes participate in replication as well, but the overall process depends on the biochemical machinery that the host cell uses to make its own proteins.

Viral replication follows a general sequence:

1. **Penetration.** Viruses may be engulfed by their host cell (endocytosis). Some viruses have surface proteins that bind to receptors on the host cell's membrane and stimulate endocytosis. Other viruses are coated with an envelope that can fuse with the host cell membrane. The viral genetic material is then released into the cytoplasm.

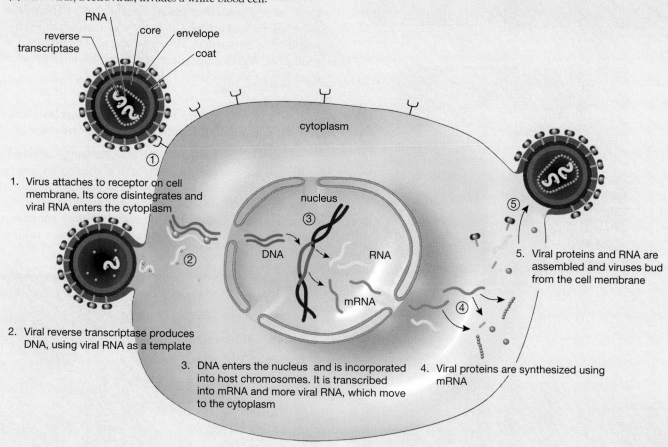

(a) HIV virus, a retrovirus, invades a white blood cell.

1. Virus attaches to receptor on cell membrane. Its core disintegrates and viral RNA enters the cytoplasm

2. Viral reverse transcriptase produces DNA, using viral RNA as a template

3. DNA enters the nucleus and is incorporated into host chromosomes. It is transcribed into mRNA and more viral RNA, which move to the cytoplasm

4. Viral proteins are synthesized using mRNA

5. Viral proteins and RNA are assembled and viruses bud from the cell membrane

Figure E21-1   How viruses replicate

wheel-like structure embedded in the bacterial membrane and cell wall that allows the flagellum to rotate (Fig. 21-8b). Flagella allow bacteria to disperse into new habitats, to migrate toward nutrients, and to leave unfavorable environments. Flagellated bacteria show orientation toward various stimuli, a behavior called a **taxis**. Some are **chemotactic**, moving toward chemicals given off by food, or away from toxic chemicals. Some are **phototactic**, moving toward or away from light, depending on the habitat they require. Other flagellated bacteria are **magnetotactic**. These detect Earth's magnetic field using magnets formed from iron crystals within their cytoplasm. They use their unique sensory system to direct their beating flagella to move them downward into aquatic sediments.

2. **Replication.** The viral genetic material is copied many times.

3. **Transcription.** Viral genetic material is used as a blueprint to make messenger RNA (mRNA).

4. **Protein synthesis.** In the host cytoplasm, viral mRNA is used to synthesize viral proteins.

5. **Viral assembly.** The viral genetic material and enzymes are surrounded by their protein coat.

6. **Release.** Viruses emerge from the cell by "budding" from the cell membrane or by bursting the cell.

Here we illustrate two types of viral life cycle. In Figure E21-1a, the *human immunodeficiency virus (HIV)*, which causes AIDS, is a retrovirus. Retroviruses use single-stranded RNA as a blueprint to make double-stranded DNA using a viral enzyme called reverse transcriptase. Many other retroviruses exist, and several cause cancers or tumors. In Figure E21-1b, the *herpes virus* contains double-stranded DNA that is transcribed into mRNA. The *influenza virus* (not illustrated) has a replication cycle somewhat like that of the herpes virus, but it uses single-stranded RNA as a template for mRNA and derives its envelope from the host cell membrane.

(b) Herpes virus, a double-stranded DNA virus, invades a skin cell.

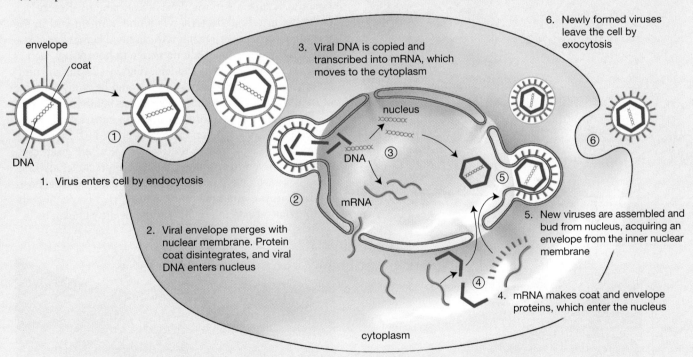

Protective Endospores Allow Some Bacteria to Withstand Adverse Conditions

When environmental conditions become inhospitable, many rod-shaped bacteria form protective resting structures called **endospores** (Fig. 21-9). The endospore (literally "inside spore") forms inside the bacterium. It contains genetic material and a few enzymes encased within a thick protective coat. Metabolic activity ceases. Endospores can survive extremely unfavorable conditions. Some can withstand boiling for an hour or more; others, still alive, have been found in the intestines of mummies 2000 years old. Endospores are important agents of bacterial dispersal because they can be carried for long distances in air or water and then produce new bacteria rapidly when they encounter favorable conditions.

Bacterial Reproduction Is by Cell Division Called Binary Fission

Most bacteria reproduce asexually by a simple form of cell division called **binary fission** (see Chapter 9), which produces genetically identical copies of the original cell (Fig. 21-10). Under ideal conditions, a bacterium may divide about once every 20 minutes, potentially giving rise to sextillions (1×10^{21}) of offspring in a single day! This rapid reproduction allows bacteria to exploit temporary habitats such as a mud puddle or warm pudding. Recall that many mutations, the source of genetic variability, occur as a result of mistakes in DNA replication during cell division (see Chapter 13). Thus, the rapid reproductive rate of bacteria provides ample opportunity for new forms to arise and also allows mutations that enhance survival to spread quickly (see "Health Watch: 'Unnatural Selection'—The Evolution of Drug-Resistant Pathogens").

Some bacteria transfer genetic material from a donor bacterium to a recipient during a process called **bacterial conjugation.** Some conjugating bacteria use specialized hollow sex pili to transfer genetic material (Fig. 21-11). The genetic material that is transferred during bacterial conjugation is located *outside* the single, circular bacterial chromosome, in a structure called a **plasmid**. A plasmid is a small circular DNA molecule that often carries genes for antibiotic resistance and sometimes carries alleles of genes also found on the main bacterial chromosome. Researchers in molecular genetics have made extensive use of bacterial plasmids, as described in Chapter 14. Conjugation produces new genetic combinations that may allow the resulting bacteria to survive under a greater variety of conditions.

Bacteria Are Specialized for Specific Habitats

Bacteria occupy a striking diversity of habitats. Recently, a colony of intestinal bacteria was found dormant within the intestinal contents of a mammoth that had lain in a peat bog for 11,000 years. Some bacteria thrive in hot springs such as those in Yellowstone National Park (Fig. 21-12). But the near-boiling temperatures of hot springs would be chilly for some of their relatives, found near deep ocean vents where superheated water is spewed through cracks in Earth's crust. These bacteria can grow at temperatures of up to 110°C (230°F). Bacteria are also found in the Dead Sea, where a salt concentration seven times that of the oceans precludes all other life. They are found in hot underground oil deposits and floating high in the atmosphere. Of course, rich bacterial communities are also found in and on the healthy human body.

(a)

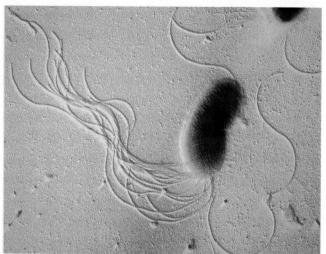

(b)

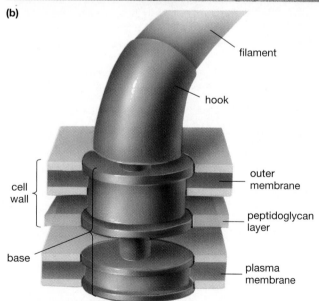

Figure 21-8   **The bacterial flagellum**

(a) A flagellated bacterium of the genus *Pseudomonas* uses its flagella to move toward favorable environments. **(b)** A unique "wheel-and-axle" arrangement anchors the bacterial flagellum within the cell wall and plasma membrane, allowing it to rotate rapidly.

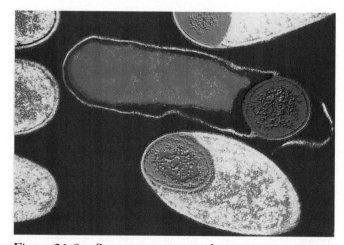

Figure 21-9   **Spores protect some bacteria**

Resistant spores, also called endospores, here colored red, have formed inside bacteria of the genus *Clostridium*, responsible for the potentially fatal food poisoning called botulism.

No single species of bacteria, however, is as versatile as these examples may suggest. In fact, bacteria are specialists; those found in hot springs, for example, could thrive nowhere else. Bacteria found on the human body are also specialized; different species colonize the skin, the mouth, and the large intestine.

Bacteria Perform Many Functions That Are Important to Other Forms of Life

Bacteria are able to colonize such diverse habitats partly because they are able to use a wide variety of nutrient sources. Blue-green bacteria engage in plantlike photosynthesis. Other bacteria are **chemosynthetic,** deriving energy through reactions that combine oxygen with inorganic molecules such as sulfur, ammonia, or nitrite. In the process, they release sulfates or nitrates, crucial plant nutrients, into the soil. Many bacteria, called **anaerobes,** do not depend on oxygen to extract energy. Some, such as the bacterium that causes tetanus, are poisoned by oxy-

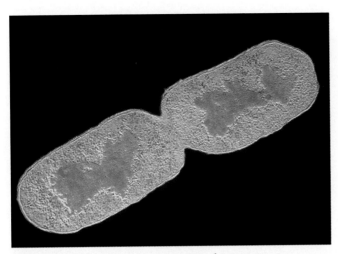

Figure 21-10   Reproduction in prokaryotes

Prokaryotic cells reproduce by a simple form of cell division called binary fission, as illustrated by this color-enhanced electron micrograph of a dividing *Escherichia coli,* found abundantly in the human intestine. Red areas are genetic material.

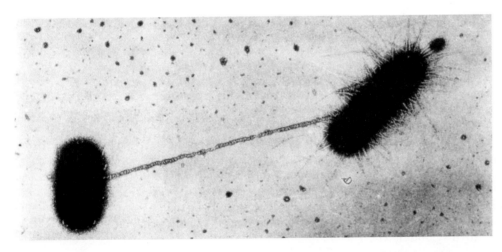

Figure 21-11   Conjugation: bacterial "mating"

Conjugation—the transfer of genetic material—occurs through a special large, hollow sex pilus, which is shown here connecting a pair of *Escherichia coli.* One bacterium (seen here on the top right) acts as a donor, transferring DNA to the recipient. In this photo, the donor bacterium is bristling with nonsex pili that probably help it attach to surfaces.

Figure 21-12   Bacteria may thrive in extreme conditions

Hot springs harbor heat- and mineral-tolerant bacteria. Some cyanobacteria can tolerate temperatures up to 85°C (186°F). Several species of cyanobacteria paint these hot springs in Yellowstone National Park with vivid colors, and each is confined to a specific area determined by temperature range. The bacterial pigments aid in photosynthesis. For scale, note the footpath in upper left.

(a)   **(b)**

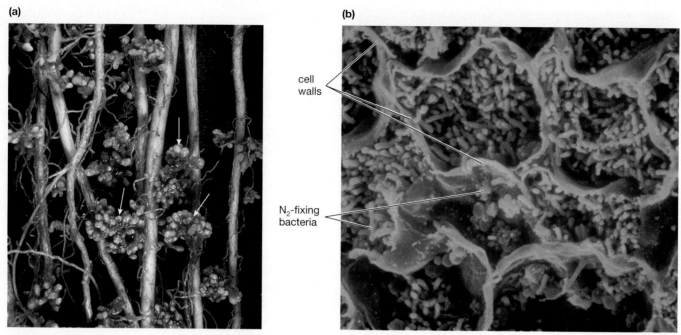

cell walls

N₂-fixing bacteria

Figure 21-13   **Nitrogen-fixing bacteria in root chambers**
(a) Special chambers called nodules on the roots of a legume (alfalfa) provide a protected and constant environment for nitrogen-fixing bacteria. **(b)** This scanning electron micrograph shows the nitrogen-fixing bacteria inside cells within the nodules.

gen. Others are opportunists, engaging in fermentation when oxygen is lacking and switching to cellular respiration (a more efficient process) when oxygen becomes available. Anaerobes such as the sulfur bacteria obtain energy from a unique type of bacterial photosynthesis. They use hydrogen sulfide (H_2S) instead of water (H_2O) in photosynthesis, releasing sulfur instead of oxygen.

Certain bacteria have the unusual ability to break down cellulose, the principal component of plant cell walls. Some of these have entered into a **symbiotic** (literally, "living together") relationship with a group of mammals called ruminants (including cows, sheep, and deer; see Chapter 32). These bacteria live in ruminant digestive tracts and help liberate nutrients from plant fodder that the animal is unable to break down itself. Symbiotic bacteria also inhabit your intestines. These feed on undigested food and synthesize nutrients such as vitamin K and vitamin B_{12}, which your body absorbs. Another form of bacterial symbiosis of enormous ecological and economic importance is the growth of **nitrogen-fixing** bacteria in specialized nodules on the roots of certain plants (**legumes,** which include alfalfa, soybeans, lupines, and clover; Fig. 21-13). These bacteria capture nitrogen gas (N_2, which the plant cannot use directly) from air trapped in the soil and combine it with hydrogen to produce ammonium (NH_4^+), an important nutrient for plants. Bacteria are also important in the production of human foods, including cheese, yogurt, and sauerkraut. The aging of meat tenderizes it through controlled bacterial digestion.

Most bacteria are heterotrophic, obtaining energy by breaking down complex organic (carbon-containing) molecules, and the range of compounds attacked by bacteria is staggering. Nearly anything that human beings can synthesize, including detergents and the poisonous solvent benzene, some bacteria can destroy. The term *biodegradable* (meaning "broken down by living things") refers largely to the work of bacteria. Even oil is biodegradable. Soon after an oil tanker dumped 11 million gallons of oil into Prince William Sound, Alaska, researchers from Exxon sprayed oil-soaked beaches with a fertilizer that encouraged the growth of natural populations of oil-eating bacteria. Within 15 days the oil deposits were noticeably reduced compared with unsprayed areas.

The appetite of some bacteria for nearly any organic compound is the key to their important role as decomposers in ecosystems. While feeding themselves, bacteria break down the waste products and dead bodies of plants and animals, freeing nutrients for reuse and allowing the recycling of nutrients that provides the basis for continued life on Earth (see Chapter 45).

Some Bacteria Pose a Threat to Human Health

The feeding habits of certain bacteria threaten our health and well-being, and bacterial infections are on the rise (see "Health Watch: 'Unnatural Selection'—The Evolution of Drug-Resistant Pathogens"). These **pathogenic** ("disease-producing") bacteria synthesize toxic substances that cause disease symptoms. Some bacteria, such as those

HEALTH WATCH

"Unnatural Selection"—The Evolution of Drug-Resistant Pathogens

In the early 1950s, several antibacterial drugs, including penicillin, streptomycin, and tetracycline, became available to treat bacterial infections. Penicillin was added to toothpaste, mouthwash, and chewing gum, and it was used to treat mild infections of all types. It soon became apparent that certain bacteria were becoming difficult to kill with these drugs—the pathogens had developed **antibiotic resistance**. By continuously exposing bacteria to antibiotics, humans have unwittingly introduced a strong agent of natural selection into the microbial world. Bacteria reproduce and mutate rapidly. Those bacteria whose mutations provide resistance to the effects of the drug survive, flourish, and soon dominate bacterial populations. When resistant bacteria cause disease, antibiotics are useless. In the late 1950s, researchers discovered that the resistant bacteria could transfer their genes for drug resistance directly to other bacteria, even to members of other bacterial species. This ability dramatically shortens the time necessary for drug resistance to spread within and between bacterial populations.

A penicillin-resistant strain of gonorrhea has developed as a result of the regular use of penicillin as a preventive measure by prostitutes in Southeast Asia. These resistant bacteria are now spreading through the United States. Both penicillin and tetracycline are added to animal feeds as growth promoters. Some farm animals harbor a strain of *Salmonella* that causes a virulent form of food poisoning. These bacteria have now developed antibiotic resistance, probably as a result of continued exposure. Tetracycline is losing its effectiveness against gonorrhea, meningitis, and urinary and respiratory tract infections. An ominous new development is the emergence of strains of tuberculosis bacterium resistant to most antibiotics. Once considered vanquished in developed countries, tuberculosis is now spreading among the homeless and AIDS patients; 28,000 cases were reported in the United States in 1992. Vigorous treatment of AIDS patients with antibiotics is believed to have hastened the emergence of these deadly new strains, which typically kill 50% of the people they infect.

Excessive use of antibiotics unnecessarily increases the exposure of bacteria to these powerful selective agents and encourages the spread of resistant strains. Antibiotics continue to be prescribed inappropriately, as in the treatment of viral infections such as colds. Researchers are working to develop new antibiotics, but drug development and testing take many years, while resistant bacterial strains spread rapidly. Unfortunately, most of the new drugs are more toxic to humans and far more expensive than those they replace. Clearly we must restrict use of antibiotics to situations in which they are urgently required, rather than relying on a steady flow of new ones. Awareness and restraint may yet enable our children to benefit from some of the same "miracle drugs" that protected our parents.

---

causing tetanus and botulism (a lethal form of food poisoning) produce toxins that attack the nervous system. These bacteria are anaerobes that survive as spores until introduced into a favorable environment. A deep puncture wound, through which tetanus bacteria enter the body, also protects the bacteria from contact with oxygen. As they multiply, the bacteria release their paralyzing poison into the bloodstream. A sealed container of canned food that has been improperly sterilized provides a haven for botulism bacteria. These produce a toxin so potent that a single gram could kill 15 million people.

The plague, or "Black Death," which killed 100 million people during the mid-fourteenth century, is caused by highly infectious bacteria, *Yersinia pestis*, spread by fleas carried by infected rats. In 1994, an outbreak of plague occurred in India for the first time in 30 years. Tuberculosis, a bacterial disease once almost vanquished from developed countries, is on the rise again in the United States and elsewhere. Two bacterial diseases, **gonorrhea** and **syphilis,** transmitted through direct sexual contact, have reached epidemic proportions in modern society (see Chapter 39: "Health Watch: Sexually Transmitted Diseases"). **Lyme disease,** named after the town of Old Lyme, Connecticut, where it was first described in 1975, is rapidly increasing in prevalence in the United States. This disease is caused by the spiral-shaped bacterium *Borrelia burgdorferi*. The bacterium is carried by the deer tick and transmitted to people who are bitten by the tick. At first, the symptoms resemble flu, with chills, fever, and body aches. If untreated, weeks or months later the victim may experience rashes, bouts of arthritis, and sometimes abnormalities of the heart and nervous system. Both physicians and lay people are becoming more familiar with the disease, so more victims are receiving treatment before serious symptoms develop.

The streptococcus bacterium comes in many forms that produce a variety of diseases. One type of streptococcus causes strep throat. Another, *Streptococcus pneumoniae,* causes pneumonia by stimulating an allergic reaction that clogs the lungs with fluid. Recently, another form seems to have developed unusual virulence. A small percentage of people who become infected with strep bacteria experience severe symptoms, described luridly in the headlines of a 1994 British tabloid newspaper which read: "Killer Bug Ate My Face." In Britain, 11 people died from these "flesh-eating" bacteria, and a few cases were also reported in the United States. When these streptococci enter broken skin, they spew out toxins that either destroy flesh directly or stimulate an overwhelming and misdirected attack by the immune system against the body's own cells. A limb can be destroyed in hours, and sometimes only amputation can halt the rapid tissue destruction. In other cases, these rare strep infections sweep through the body, causing death within a matter of days; Jim Hensen, creator of the muppets, died as a result of such an infection.

Although some bacteria assault the human body, most are harmless and many are beneficial. For example, the normal bacterial community in the female vagina creates an environment that is hostile to infections by parasites such as yeasts. Bacteria harmlessly inhabiting our intestines are an important source of vitamin K. As Lewis Thomas, the late physician, researcher, and author, so aptly put it, "Pathogenicity is, in a sense, a highly skilled trade, and only a tiny minority of all the numberless tons of microbes on the earth has ever been involved in it; most bacteria are busy with their own business, browsing and recycling the rest of life."

Cyanobacteria Are Photosynthetic, and Many Can Also Capture Atmospheric Nitrogen

Cyanobacteria, also known as blue-green bacteria (*cyan* is Greek for "dark blue"), are widespread and diverse. Depending on the amount and type of pigment they contain, they can also be purple, red, or yellow. They can be found on snowfields (where they dye the snow pink), in hot springs that may reach 85°C (186°F), in oceans, lakes, and in moist soil. Because most are aerobic and all are photosynthetic, they can exist only where light and oxygen are available. Like green plants, cyanobacteria possess chlorophyll and produce oxygen as a by-product of photosynthesis. Because bacteria lack organelles, including chloroplasts, chlorophyll is located on special membranes inside the cell (Fig. 21-14a). In addition to trapping solar energy, most cyanobacteria can acquire nitrogen from the atmosphere, making them extremely self-sufficient nutritionally. Some cyanobacteria form chains of cells that are unique among prokaryotes because they have a rudimentary division of labor. In these filamentous colonies (Fig. 21-14b), a few cells capture atmospheric nitrogen while the rest photosynthesize. When new islands are formed from volcanic eruptions, cyanobacteria are among the first colonizers of the bare rock; their activities prepare the way for more complex but less self-sufficient organisms.

Archaebacteria Are Sufficiently Different in Cellular Structure That Some Taxonomists Place Them in a Separate Kingdom

Recently, a unique group of bacteria—the archaebacteria—have come under close scientific scrutiny. Scientists have found such dramatic differences between the archaebacteria and other monerans that many believe the archaebacteria should be placed in their own kingdom. The lipids of archaebacterial cell membranes differ con-

(a)

(b)

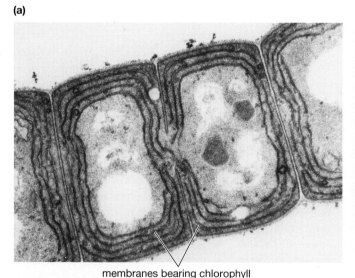

membranes bearing chlorophyll

N₂-fixing cells

Figure 21-14   **Cyanobacteria**

(a) Electron micrograph of a section through a cyanobacterial filament (genus *Oscillatoria*). Chlorophyll is located on the membranes visible within the cells. **(b)** Simple division of labor, rare among prokaryotic cells, is seen in this filamentous cyanobacterium (genus *Nostoc*). The larger cells are specialized for nitrogen fixation, while the rest of the cells photosynthesize.

siderably from those of both eukaryotic and other prokaryotic cells, as do the composition of their cell walls and the sequence of subunits in their ribosomal RNA.

Archaebacteria include **methanogens,** anaerobic bacteria that convert carbon dioxide to methane (sometimes called "swamp gas"). Methanogens are found in such diverse habitats as swamps, sewage-treatment plants, hot springs, deep-sea vent communities (see Chapter 46), and the stomachs of cows. Other archaebacteria include **halophiles,** bacteria that thrive in concentrated salt solutions such as the Dead Sea, and **thermoacidophiles,** which, as their name implies, thrive in hot, acidic environments such as hot sulfur springs. Archaebacteria were first discovered in these environments that are hostile to most other forms of life, and for many years researchers believed they were confined to such locations. Recently, however, microbioloogists have found thriving populations of archaebacteria in ocean water samples from all over the world. Why were they never detected before? Archaebacteria are almost impossible to grow in the laboratory, so when researchers attempted to identify marine bacteria from cultures grown from seawater samples, only eubacteria showed up. Only when researchers used molecular genetic techniques to look for archaebacterial RNA sequences in seawater did they find evidence of their presence and abundance.

The extreme environments in which many species of archaebacteria thrive, although rare now, were far more common when life first appeared on Earth. This observation has led some scientists to speculate that these organisms were among the first to evolve, hence the name *archaebacteria* (meaning "ancient bacteria"). But recent analysis of archaebacterial RNA nucleotide sequences has revealed that the archaebacteria are more similar to eukaryotic cells than are the eubacteria. Scientists now estimate that archaebacteria probably evolved about half a billion years after eubacteria.

The Kingdom Protista

The unicellular members of the kingdom **Protista** and all multicellular organisms are composed of eukaryotic cells that evolved roughly 2 billion years after the first prokaryotic cells. Eukaryotic cells contain many membrane-bound organelles that are absent in prokaryotic cells. Organelles such as mitochondria and chloroplasts are similar in size to typical bacteria and indeed probably evolved from bacteria (see Chapter 5). Within the kingdom Protista, we find some of the most complex eukaryotic cells in existence, with organelles taking on functions served by organs in multicellular organisms.

Since Anton van Leeuwenhoek first observed protists through his simple homemade microscope in 1674, at least 50,000 species have been described. Although some protists form colonies, most consist of a single eukaryotic cell. Most protists can reproduce asexually by mitotic cell division, but many are capable of a form of sexual reproduction, called conjugation as well (Fig. 21-15). All three major modes of nutrition are represented in this kingdom: The unicellular algae trap solar energy through photosynthesis; predatory protists ingest their food; and parasitic forms, some flagellates, and the versatile euglenoids can absorb nutrients from their surroundings.

(a)

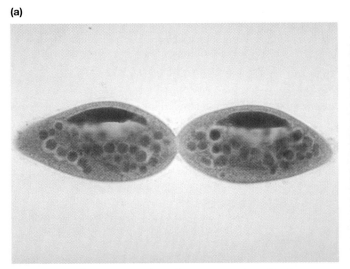

(b)

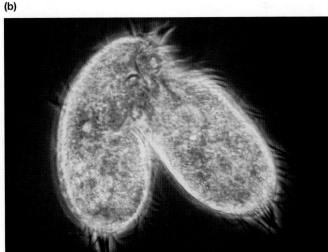

Figure 21-15   **Two modes of protistan reproduction**

(a) *Paramecium*, a ciliate, reproduces asexually by cell division that results in two daughters identical to the original parent. **(b)** Mating in a ciliate, *Euplotes*. Genetic material is exchanged across a cytoplasmic bridge. After the exchange occurs, new individuals formed by cell division will have gene combinations different from those of either parent cell.

Protists Are a Diverse Group Including Plantlike, Funguslike, and Animal-like Forms

The kingdom Protista is an extremely diverse group, including plantlike, funguslike, and animal-like forms, which have sometimes been classified as plants, fungi, or animals. The presence of both phyla (a taxonomic category of animals) and divisions (a taxonomic category of plants) within this kingdom reflects the difficulty of classifying protists. To the confusion of taxonomists, many protists (such as the euglenoids) fit equally well into animal-like or plantlike categories (see Fig. 21-19). Protistan taxonomy is still the subject of revision and controversy. Here we discuss members of the kingdom Protista in three categories: the plantlike unicellular algae, the funguslike slime molds, and the animal-like protozoa (Table 21-1).

The Unicellular Algae, or Phytoplankton, Are Plantlike Protists

Often called **phytoplankton** (literally, "floating plants"), these photosynthetic protists are widely distributed in oceans and lakes. Although they are microscopic, phyto-

plankton are immensely important. Marine phytoplankton account for nearly 70% of all the photosynthetic activity on Earth, absorbing carbon dioxide, recharging the atmosphere with oxygen, and supporting the complex web of aquatic life. There are three major divisions of plantlike protists: the division Pyrrophyta, also called dinoflagellates; the division Chrysophyta, the diatoms; and the division Euglenophyta, the euglenoids.

Dinoflagellates Swim Using Two Whiplike Flagella

Dinoflagellates are named for the motion created by their two whiplike flagella (*dino* is Greek for "whirlpool"). One flagellum encircles the cell, and the second projects behind it. Some dinoflagellates are covered only by a cell membrane; others have cellulose walls that resemble armor plates (Fig. 21-16). Although some live in fresh water, dinoflagellates are especially abundant in the ocean, where they are an important food source for larger organisms. Many dinoflagellates are bioluminescent, producing a brilliant blue-green light when disturbed. Specialized dinoflagellates, known as zooxanthellae, live

---

TABLE 21-1 ❖ *The Major Divisions of Protists*

General Category	Division/Phylum	Locomotion	Nutrition	Representative Features	Representative Genus
Plantlike protists: unicellular algae	Dinoflagellates (Division Pyrrophyta)	Swim with two flagella	Autotrophic; photosynthetic	Many bioluminescent; often have cellulose wall; most marine	*Gonyaulax* (causes red tide)
	Diatoms (Division Chrysophyta)	Glide along surfaces	Autotrophic; photosynthetic	Have silica shells; most marine	*Navicula* (glides toward light)
	Euglenoids (Division Euglenophyta)	Swim with one flagellum	Autotrophic; photosynthetic	Have an eyespot; all freshwater	*Euglena* (common pond-dweller)
Funguslike protists: slime molds	Acellular (plasmodial) slime molds (Division Myxomycota)	Sluglike mass oozes over surfaces	Heterotrophic	Form multinucleate plasmodium	*Physarum* (forms a large bright orange mass)
	Cellular slime molds (Division Acrasiomycota)	Amoeboid cells extend pseudopodia; sluglike mass crawls over surfaces	Heterotrophic	Form pseudoplasmodium with individual amoeboid cells	*Dictyostelium* (often used in laboratory studies)
Animal-like protists: protozoa	Zooflagellates	Swim with flagella	Heterotrophic	Inhabit soil or water or may be parasitic	*Trypanosoma* (causes African sleeping sickness)
	Sarcodines (Phylum Sarcomastigophora)	Extend pseudopodia	Heterotrophic	Both naked and shelled forms exist	*Amoeba* (common pond-dweller)
	Sporozoans (Phylum Apicomplexa)	Nonmotile	Heterotrophic; all parasitic	Form infectious spores	*Plasmodium* (causes malaria)
	Ciliates (Phylum Ciliophora)	Swim with cilia	Heterotrophic	Most complex single cells	*Paramecium* (fast-moving pond-dweller)

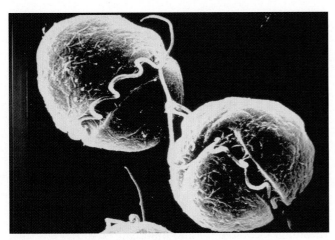

Figure 21-16   Dinoflagellates
Two dinoflagellates, covered with protective cellulose armor.
Two flagella lie within the grooves encircling the body.

Figure 21-17   A red tide
The explosive reproductive rate of certain dinoflagellates
under the right conditions can produce concentrations so
great that their microscopic bodies dye the sea red or brown,
as in this bay in Mexico.

within the tissues of corals and some clams, where they
provide photosynthetic nutrients and remove carbon
dioxide. Many coral are found only in shallow, well-lit
waters in which their zooxanthellae can survive.

The green chlorophyll in dinoflagellates is often masked
by red pigments that help trap light energy. Under certain
conditions, when the water is warm and rich in nutrients,
a population explosion occurs. Dinoflagellates can become
so numerous that the water is dyed red by the color of their
bodies, causing a "red tide" (Fig. 21-17). During red tides,
fish die by the thousands, suffocated by clogged gills or by
oxygen depletion resulting from the decay of the bodies of
billions of dinoflagellates. But oysters, mussels, and clams
have a feast, filtering millions from the water for food. In
the process, however, they concentrate a nerve poison produced by the dinoflagellates in their bodies. People who
eat these mollusks may be stricken with potentially lethal
paralytic shellfish poisoning.

Diatoms Encase Themselves within Glassy Walls

The photosynthetic **diatoms,** found in both fresh and salt
water, are so important to marine food webs that they
have been called the "pastures of the sea." They produce
glassy protective coverings, some of exceptional beauty
(Fig. 21-18). These consist of top and bottom halves that
fit together like a pillbox or petri dish. Accumulations of
the glassy walls of diatoms over thousands of years have
produced fossil deposits of "diatomaceous earth" that may
be hundreds of meters thick. This slightly abrasive substance is widely used in products such as toothpaste and
metal polish. Diatoms store reserve food as oil; their buoyancy in water helps their bodies float near the surface,
where light for photosynthesis is abundant. Prehistoric accumulations of diatoms and their stored oil may have contributed to today's petroleum reserves.

Figure 21-18   Some representative diatoms
This photomicrograph illustrates the intricate, microscopic
beauty and variety of the glassy walls of diatoms.

Euglenoids Lack a Rigid Covering and Swim Using Flagella

The **euglenoids** are named after the group's best-known representative, *Euglena* (Fig. 21-19), a complex single cell that moves about by whipping its flagellum through the water. Its simple light-sensing organelles consist of a photoreceptor, also called an eyespot, at the base of the flagellum and an adjacent patch of pigment. The pigment shades the photoreceptor only when light strikes from certain directions, enabling *Euglena* to determine the direction of the light source. Using information from the photoreceptor, the flagellum propels the protist toward light levels appropriate for photosynthesis. Most euglenoids live in fresh water. In contrast to diatoms or dinoflagellates, they lack a rigid outer covering, so some can move by wriggling as well as by whipping the flagellum. If *Euglena* is maintained in darkness, it loses its chloroplasts but can still absorb nutrients from its surroundings. In this state it closely resembles the animal-like zooflagellates described below.

The Slime Molds Are Funguslike Protists

The life cycle of the slime mold consists of two phases: a mobile feeding stage and a stationary reproductive stage called a **fruiting body** (see Fig. 21-20b). There are two major divisions of slime molds: the **acellular,** or **plasmodial, slime molds** of the division Myxomycota and the **cellular slime molds** of the division Acrasiomycota.

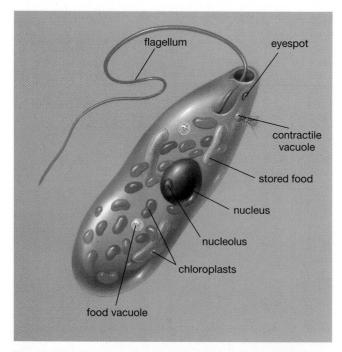

Figure 21-19   Euglena, a representative euglenoid
This diagram shows *Euglena*'s elaborate, single-celled structure. The cell is packed with green chloroplasts, which will disappear if the protist is kept in darkness.

The Acellular Slime Molds Form an Acellular Multinucleate Mass of Cytoplasm Called a Plasmodium

The acellular slime molds consist of a mass of cytoplasm that may spread thinly over an area of several square meters. Although the mass contains thousands of diploid nuclei, the nuclei are not confined in discrete cells surrounded by cell membranes, as in most multicellular organisms. This structure, called a **plasmodium,** explains why these protists are described as "acellular" (without cells). The plasmodium oozes through decaying leaves and rotting logs, engulfing food such as bacteria and particles of organic material. The mass may be bright yellow or orange—a large plasmodium can be rather startling (Fig. 21-20a). Dry conditions or starvation stimulate the plasmodium to form a fruiting body, on which haploid spores are produced (Fig. 21-20b). The spores are dispersed and germinate under favorable conditions, eventually giving rise to a new plasmodium.

(a)

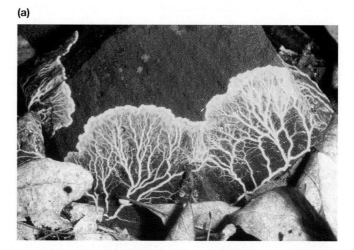

(b)

Figure 21-20   The acellular slime mold Physarum
(a) *Physarum* oozes over a stone on the damp forest floor.
(b) When food becomes scarce, the mass differentiates into black fruiting bodies in which spores are formed.

The Cellular Slime Molds Live as Independent Cells but Aggregate into a Pseudoplasmodium When Food Becomes Scarce

The cellular slime molds live in soil as independent haploid cells that move and feed by extending **pseudopods** (literally, "false feet"). They surround and engulf food such as bacteria. In the best-studied genus, *Dictyostelium*, individual cells release a chemical signal when food becomes scarce. This signal attracts nearby cells into a dense aggregation that forms a sluglike mass called a **pseudoplasmodium** ("false plasmodium") because it actually consists of individual cells (see Fig. 21-21). The pseudoplasmodium then behaves like a multicellular organism. After

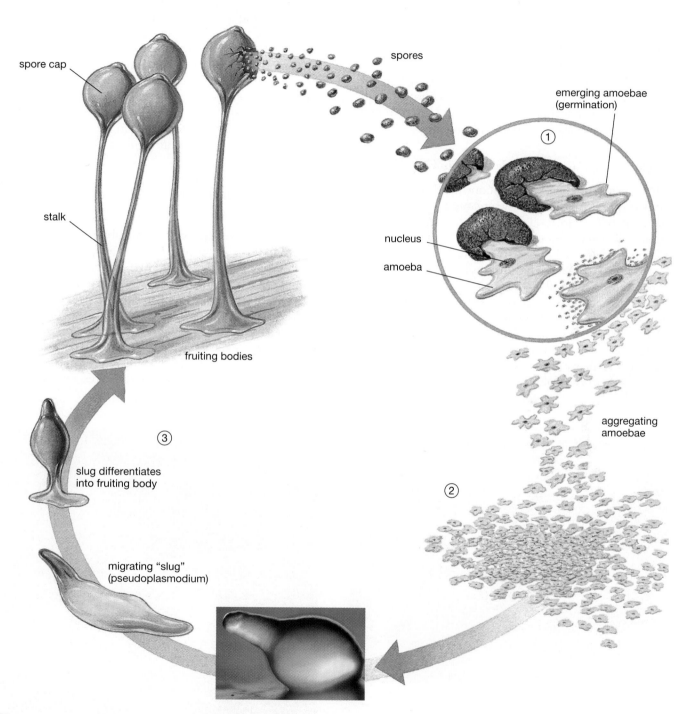

Figure 21-21   The life cycle of a cellular slime mold

① Single amoebalike cells emerge from spores, crawl, and feed. ② When food becomes scarce, they aggregate into a sluglike mass called a pseudoplasmodium (inset). ③ The pseudoplasmodium migrates toward the light and then forms a fruiting body, in which haploid spores are produced.

crawling toward a source of light, the cells in the aggregation take on specific roles, forming a fruiting body. Haploid spores formed within the fruiting body are dispersed by the wind and germinate directly into new amoeboid individuals.

The Protozoa Are Animal-like Protists

The **protozoa** (literally, "first animals") are described as animal-like because they can move and they obtain their food from other organisms. They are placed into three major phyla: phylum Sarcomastigophora, which includes the zooflagellates and sarcodines (amoebae); phylum Apicomplexa, the sporozoans; and phylum Ciliophora, the ciliates. All are unicellular, eukaryotic, and heterotrophic, but they differ in their methods of locomotion.

Zooflagellates Possess a Single Versatile Flagellum

All **zooflagellates** possess at least one flagellum. This versatile organelle may propel the organism, sense the environment, or ensnare food. The zooflagellates are a diverse group believed to be ancestral to the other protists. Many are free living, inhabiting soil and water; others are symbiotic, living inside other organisms in a relationship that may be either mutually beneficial or parasitic. One symbiotic form can digest cellulose and lives in the gut of termites, where it helps them extract energy from wood. A zooflagellate of the genus *Trypanosoma* is responsible for African sleeping sickness, a potentially fatal disease (Fig. 21-22). Like many parasites, this organism has a complex life cycle, part of which is spent in the tsetse fly, which transmits it to mammals while feeding on their blood. The parasite then develops in the host (which may be a person), entering the bloodstream. It may then be ingested by another tsetse fly that bites the host, thus beginning a new cycle of infection.

Another parasitic zooflagellate, *Giardia*, is an increasing problem in the United States, particularly to hikers who drink from apparently pure mountain streams. Cysts (tough, resting structures) of this flagellate are released in the feces of infected humans, dogs, or other animals (a single gram of feces may contain 300 million cysts) and enter freshwater streams and even community reservoirs. Cysts develop into the adult form (Fig. 21-23) in the small intestine of their mammalian host. In humans, infections may cause severe diarrhea, dehydration, nausea, vomiting, and cramps. Fortunately, these infections can be cured with drugs, and deaths from *Giardia* infections are uncommon.

Sarcodines, Including the Amoebae, Move Using Pseudopods

Sarcodines possess flexible cell membranes that they can extend in any direction to form pseudopodia, which are used for locomotion and for engulfing food (Fig. 21-24). One parasitic form causes amoebic dysentery, a particular problem in warm climates. Multiplying in the intestinal wall, this parasite causes severe diarrhea and may pierce the intestine, occasionally causing fatal infections.

Sarcodines called **amoebae** are common in freshwater lakes and ponds. Amoebae lack many of the specialized organelles found in flagellates and ciliates, but they have

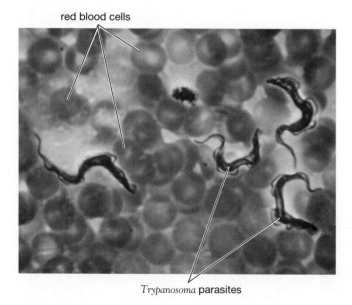

Figure 21-22   A disease-causing zooflagellate

This photomicrograph shows human blood that is heavily infested with the parasitic zooflagellate *Trypanosoma,* which causes African sleeping sickness.

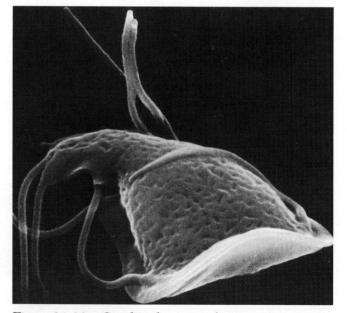

Figure 21-23   Giardia: the curse of campers

A zooflagellate (genus *Giardia*) that may infect drinking water, causing gastrointestinal disorders.

a complex internal structure and sophisticated ability to sense and capture prey. **Heliozoans** ("sun animals"), a striking form of freshwater sarcodine, may be found floating in ponds or attached by stalks to an underwater plant or rock (Fig. 21-25). They have stiff, needle-like pseudopodia, each of which is supported internally by a bundle of microtubules. Some heliozoans cover themselves with intricate and delicate shells of silica (glass). The **foraminiferans** and **radiolarians** are primarily marine sarcodines that also produce beautiful and elaborate shells. Shells of foraminiferans are constructed mostly of calcium carbonate (chalk); those of radiolarians are of silica (Fig. 21-26).

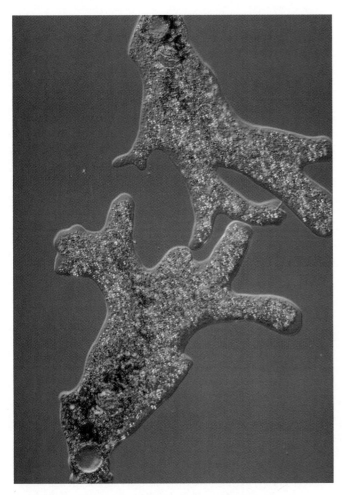

Figure 21-24   The amoeba

An amoeba uses cytoplasmic projections called pseudopodia to move about and to capture prey such as this *Paramecium*.

(a)

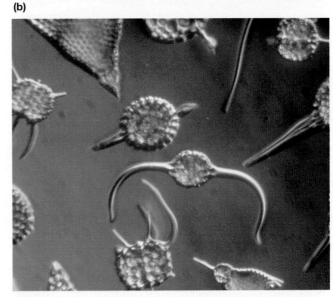

(b)

Figure 21-26   Foraminiferans and radiolarians

(a) The chalky shells of foraminiferans show numerous interior compartments. **(b)** The delicate, glassy shells of radiolarians. Pseudopodia, which sense the environment and capture food, extend out through the openings in the shells.

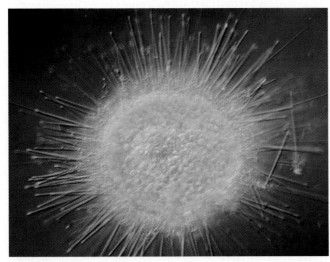

Figure 21-25   Heliozoans

Heliozoans are beautiful freshwater sarcodines; the needle-like pseudopodia are clearly visible in this specimen (genus *Acanthocystis*).

These elaborate shells are pierced by myriad openings through which pseudopods extend. The chalky shells of foraminiferans, accumulating over millions of years, have resulted in immense deposits of limestone such as form the famous white cliffs of Dover, England.

Sporozoans Are All Parasitic and Have No Means of Locomotion

All **sporozoans** are parasitic, living inside the bodies and sometimes inside the individual cells of their hosts. They are named after their ability to form infectious spores, resistant structures transmitted from one host to another through food, water, or the bite of an infected insect. As adults, sporozoans have no means of locomotion. Many have complex life cycles, a common feature of parasites. A well-known example is the malarial parasite *Plasmodium* (Fig. 21-27). Parts of its life cycle are spent in the stomach, and later the salivary glands, of the female *Anopheles* mosquito. When the mosquito bites a person, it passes the *Plasmodium* to the unfortunate victim. The sporozoan develops in the liver, then enters the blood, where it reproduces rapidly in red blood cells. The release of large quantities of spores through rupture of the blood cells causes the recurrent fever of malaria. Uninfected mosquitoes may acquire the parasite by feeding on the

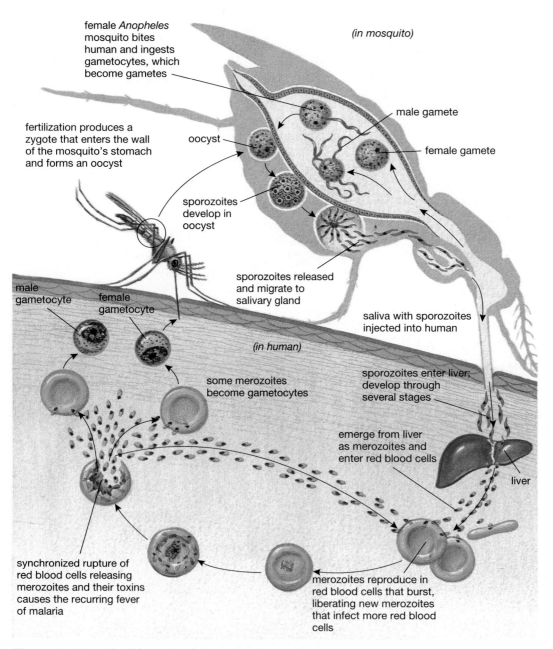

female *Anopheles* mosquito bites human and ingests gametocytes, which become gametes

(in mosquito)

fertilization produces a zygote that enters the wall of the mosquito's stomach and forms an oocyst

oocyst

male gamete

female gamete

sporozoites develop in oocyst

sporozoites released and migrate to salivary gland

male gametocyte

female gametocyte

(in human)

saliva with sporozoites injected into human

some merozoites become gametocytes

sporozoites enter liver; develop through several stages

emerge from liver as merozoites and enter red blood cells

liver

synchronized rupture of red blood cells releasing merozoites and their toxins causes the recurring fever of malaria

merozoites reproduce in red blood cells that burst, liberating new merozoites that infect more red blood cells

Figure 21-27   The life cycle of the malarial parasite

In the complex life cycle of the malarial parasite *Plasmodium*, humans and mosquitoes serve as alternate hosts.

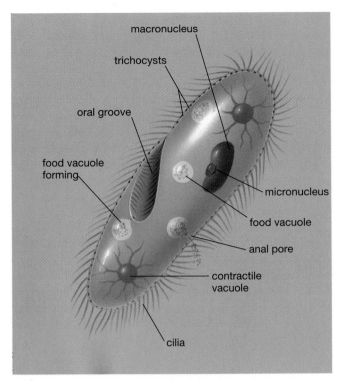

Figure 21-28   **The complexity of ciliates**

The ciliate *Paramecium* illustrates some important ciliate organelles. The oral groove acts as a mouth, and food vacuoles, miniature digestive systems, form at its apex; waste is expelled by exocytosis through an anal pore. The contractile vacuoles regulate water balance. Trichocysts help this predator stun its prey, while cilia propel it rapidly through the water.

blood of a malaria victim, spreading the parasite when it bites another person, thus continuing the infectious cycle.

Although the drug chloroquine kills the malarial parasite, unfortunately, drug-resistant populations of *Plasmodium* are rapidly spreading throughout Africa, where the disease is prevalent. Programs to eradicate mosquitoes have failed because the mosquitoes rapidly evolve resistance to pesticides. Researchers hope to use genetic engineering to breed strains of mosquito that kill, rather than transmit, the sporozoan parasite.

Ciliates, Named for Their Locomotory Cilia, Are the Most Complex of the Protozoa

Ciliates, which inhabit fresh or salt water, represent the peak of unicellular complexity. They possess many specialized organelles, including the **cilia**, the short hairlike outgrowths, after which they are named. Their cilia may cover the cell, or they may be localized, as in *Didinium* (see Fig. 21-29). In the well-known freshwater genus *Paramecium* (Fig. 21-28), rows of cilia cover the entire body surface. Their coordinated beating propels the cell through the water at a protistan speed record of a millimeter per second. Although only a single cell, *Paramecium* responds to its environment as if it had a well-developed nervous system. Confronted with some noxious chemical or with a physical barrier, the cell immediately backs up by reversing the beating of its cilia and then proceeds in a new direction.

Ciliates are accomplished predators (Fig. 21-29). Some, including *Paramecium* and *Didinium*, immobilize their prey

(a)   **(b)**   **(c)**

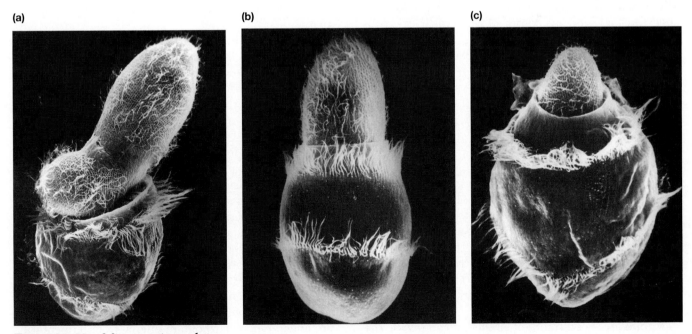

Figure 21-29   **Microscopic predator**

In this series of scanning electron micrographs, *Paramecium* is stung **(a)** and gradually engulfed **(b, c)** by another predatory ciliate, *Didinium*, whose cilia are confined to two bands encircling the egg-shaped body. This microscopic drama could occur on a pinpoint with room to spare.

with explosive darts called **trichocysts** embedded in the outer covering of the cell. Prey is escorted to a mouthlike opening, the oral groove. It is digested in a food vacuole, which forms a temporary "stomach," and is excreted by exocytosis. Excess water is accumulated in a contractile vacuole, which periodically contracts, emptying the fluid through a pore to the outside.

E V O L U T I O N A R Y C O N N E C T I O N S

Our Unicellular Ancestors

In the kingdoms Monera and Protista, we find persisting today the types of cellular organization that gave rise to the complex multicellular organisms with which we are most familiar. Modern monerans have changed little from their ancestors, whose fossilized remains date back 3.5 billion years. Life might still consist of prokaryotic single cells if the protists, with their radical eukaryotic design, had not appeared on the scene nearly 1.5 billion years ago. As you learned in Chapters 5 and 19, eukaryotic cells may have originated when one moneran, perhaps a bacterium capable of cellular respiration, took up residence inside a partner, forming the first "mitochondrion." A separate but equally crucial merger may have occurred when a photosynthetic bacterium (probably resembling a blue-green bacterium) took up residence within a nonphotosynthetic partner and became the first "chloroplast." The foundations of multicellularity were laid with the eukaryotic cell, whose intricacy allowed specialization of entire cells for specific functions within a multicellular aggregation. Thus primitive protists, some consuming their food in chunks, some photosynthesizing, and others absorbing nutrients from the environment, almost certainly followed divergent evolutionary paths that led to the three multicellular kingdoms predominating today: the fungi, plants, and animals, which are the subjects of the following three chapters.

✖ S U M M A R Y   O F   K E Y   C O N C E P T S

Microorganisms are classified in two kingdoms: Monera and Protista. Monerans consist of tiny prokaryotic cells lacking organelles such as nuclei, mitochondria, and chloroplasts. Protists are eukaryotic cells that possess the full range of organelles and resemble the cells of multicellular organisms. Without the photosynthetic, nitrogen-trapping, and decomposing abilities of the bacteria and the photosynthetic activities of protists, life as we know it would grind to a halt.

Viruses
Viruses are parasites consisting of a protein coat surrounding genetic material. They are noncellular and unable to move, grow, or reproduce outside a living cell. They invade cells of a specific host and use the host cell's energy, enzymes, and ribosomes to produce more virus particles, which are liberated when the cell ruptures. Many viruses are pathogenic to humans, including those causing colds and flu, herpes, AIDS, and certain forms of cancer.

Viroids are short strands of RNA that can invade the nucleus and direct the synthesis of new viroids. To date, viroids are only known to cause certain diseases of plants.

Prions have been implicated in neurodegenerative diseases such as kuru and scrapie. Prions are unique in that they lack genetic material. They are composed solely of mutated prion protein, which may act as an enzyme catalyzing the formation of more prions from normal prion protein.

The Kingdom Monera
Members of the kingdom Monera—the bacteria—are unicellular and prokaryotic. A cell wall of peptidoglycan determines their characteristic shape: coccus (round), bacillus (rodlike), or spiral. Many form spores that disperse widely and withstand inhospitable environmental conditions. Bacteria obtain energy in a variety of ways. Some, including the cyanobacteria, rely on photosynthesis. Others are chemosynthetic, breaking down inorganic molecules to obtain energy. Heterotrophic forms are capable of consuming a wide variety of organic compounds. Many are anaerobic, able to obtain energy from fermentation when oxygen is not available.

Some bacteria are pathogenic, causing disorders such as pneumonia, tetanus, botulism, and the venereal diseases gonorrhea and syphilis. Most, however, are harmless to humans and play important roles in natural ecosystems. Bacteria have colonized nearly every habitat on Earth. Some live in the digestive tracts of larger organisms such as cows and sheep, where they break down cellulose. Nitrogen-fixing bacteria enrich the soil and aid in plant growth, while many others live off the dead bodies and wastes of other organisms, liberating nutrients for reuse.

Cyanobacteria engage in plantlike photosynthesis, and many can also fix nitrogen, making them nutritionally very self-reliant.

The archaebacteria are a unique and diverse group that flourish under extreme conditions, including hot, acidic, very salty, and anaerobic environments. They differ in several ways from all other bacteria, including cell wall composition, ribosomal RNA sequence, and cell membrane lipid structure.

The Kingdom Protista

The kingdom Protista consists of organisms composed of single, highly complex eukaryotic cells. They are classified as plantlike, funguslike, and animal-like. The plantlike unicellular algae are important photosynthetic organisms in marine and freshwater ecosystems. They include dinoflagellates, diatoms, and the exclusively freshwater euglenoids.

The funguslike acellular (plasmodial) slime molds form a multinucleate plasmodium that crawls in amoeboid fashion, ingesting decaying organic matter. Drought or starvation stimulates the formation of a fruiting body on which spores are formed. Cellular slime molds exist as independent amoeboid cells. Under adverse conditions, they aggregate in response to a chemical signal and form a pseudoplasmodium that differentiates into a spore-forming fruiting body.

Protozoa are nonphotosynthetic protists that absorb or ingest their food. They are widely distributed in soil and water; some are parasitic. They include the zooflagellates, the amoeboid sarcodines, the parasitic sporozoans, and the predatory ciliates.

◪ KEY TERMS

acellular slime mold   p. 414
amoeba   p. 416
anaerobe   p. 407
antibiotic resistance   p. 409
archaebacteria   p. 402
bacillus   p. 402
bacterial conjugation   p. 406
bacteriophage   p. 400
bacterium   p. 402
binary fission   p. 406
capsule   p. 403
cellular slime mold   p. 414
chemosynthetic   p. 407
chemotactic   p. 404
ciliate   p. 419
cilium   p. 419
coccus   p. 402
cyanobacterium   p. 410
diatom   p. 413
dinoflagellate   p. 412
endospore   p. 405

eubacteria   p. 402
euglenoid   p. 414
flagellum   p. 403
foraminiferan   p. 417
fruiting body   p. 414
gonorrhea   p. 409
Gram stain   p. 402
halophile   p. 411
heliozoan   p. 417
kuru   p. 400
legume   p. 408
Lyme disease   p. 409
magnetotactic   p. 404
methanogen   p. 411
Monera   p. 402
nitrogen-fixing   p. 408
pathogen   p. 408
peptidoglycan   p. 402
phototactic   p. 404
phytoplankton   p. 412
pilus   p. 403

plasmid   p. 406
plasmodial slime mold   p. 414
plasmodium   p. 414
prion   p. 400
Protista   p. 411
protozoan   p. 416
pseudoplasmodium   p. 415
pseudopod   p. 415
radiolarian   p. 417
sarcodine   p. 416
slime layer   p. 403
spirillum   p. 402
sporozoan   p. 418
symbiotic   p. 408
syphilis   p. 409
taxis   p. 404
thermoacidophile   p. 411
trichocyst   p. 420
viroid   p. 400
virus   p. 399
zooflagellate   p. 416

✤ THINKING THROUGH THE CONCEPTS

Multiple Choice

1. Which of the following is true?
   a. Viruses cannot reproduce outside a host cell.
   b. Prions are infectious proteins.
   c. Viroids lack a protein coat.
   d. Some viruses cause cancer.
   e. all of the above
2. Which structure is used to transfer genetic material between bacteria?
   a. flagellum   b. pilus
   c. peptidoglycan   d. spore
   e. capsule
3. Most pathogenic bacteria cause disease by
   a. directly destroying individual cells of the host
   b. fixing nitrogen and depriving the host of this nutrient
   c. producing toxins that disrupt normal functions
   d. depleting the energy supply of the host
   e. depriving the host of oxygen
4. Cyanobacteria
   a. have chlorophyll   b. have chloroplasts
   c. are poisoned by oxygen d. are not widely distributed
   e. all of the above
5. Which organisms are sometimes called the "pastures of the sea?"
   a. dinoflagellates   b. foraminiferans
   c. diatoms   d. radiolarians
   e. amoebae
6. Which of the following pairs of organism and disease is incorrect?
   a. sporozoan: malaria   b. prion: kuru
   c. bacterium: syphilis   d. archaebacterium: gonorrhea
   e. virus: AIDS

Review Questions

1. Describe the structure of a typical virus. How do viruses replicate?
2. List the major differences between monerans and protists.
3. Describe some of the ways in which bacteria obtain energy and nutrients.
4. What are nitrogen-fixing bacteria and what role do they play in ecosystems?
5. What is an endospore? What is its function?
6. Why do bacteria readily become resistant to antibiotics, and what steps can be taken to prevent this situation?
7. Describe some examples of bacterial symbiosis.
8. What is the importance of dinoflagellates in marine ecosystems? What happens when they reproduce rapidly?
9. What is the major ecological role played by unicellular algae?
10. What protozoan group consists entirely of parasitic forms?
11. Describe the life cycle and mode of transmission of the malarial parasite.
12. What protozoan group's shells form the white limestone cliffs of Dover, England?
13. How do sarcodines feed? What disease is caused by a parasitic sarcodine?
14. In which protozoan group are the most complex single cells found? How do these cells feed?
15. Name three divisions of plantlike protists, three of animallike protists, and two of funguslike protists and give an example of each.

✤ APPLYING THE CONCEPTS

1. In some less-developed countries, antibiotics can be purchased without a prescription. Why do you think this is done? What biological consequences would you predict?
2. Before the discovery of prions, many (perhaps most) biologists would have agreed with the statement: "It is a *fact* that no infectious organism or particle can exist that lacks nucleic acid (such as DNA or RNA)." What lessons do prions have to teach us about nature, science, and scientific inquiry? You may wish to review Chapter 1 to help answer this question.
3. Recent research shows that ocean water off southern California has warmed by 2–3°F (1–1.5°C) over the past four decades, possibly due to the greenhouse effect. This has led indirectly to a depletion of nutrients in the water and thus a decline in photosynthetic protists such as diatoms. What effects is this warming likely to have for the life in the oceans? You may wish to review information on food chains in Chapter 45 to help answer this question.
4. Argue for and against the statement: "Viruses are alive."
5. The internal structure of many protists is much more complex than that of cells of multicellular organisms. Does this mean the protist is engaged in more complex activities than the multicellular organism? If not, why should the protistan cell be much more complicated?
6. Why would the lives of multicellular animals be impossible if moneran and protistan organisms did not exist?

✛ F O R   M O R E   I N F O R M A T I O N

Caldwell, M. "Mad Cows and Wild Proteins." *Discover*, April 1991. A discussion of the discovery and effects of prions.

Diamond, J. "The Return of Cholera." *Discover*, March 1992. The cholera bacterium is causing epidemics in some less-developed countries.

Dixon, B. "Overdosing on Wonder Drugs." *Science 86*, May 1986. The dangers of overuse of antibiotics are explored.

Greene, W. C. "AIDS and the Immune System." *Scientific American*, September 1993. Studies of the life cycle of the AIDS virus suggest possible avenues for treatment.

Knoll, A. "Life's Expanding Realm." *Natural History*, June 1994. The role of cyanobacteria in modifying Earth's surface and their impact on the evolution of life on Earth.

McEvedy, C. "The Bubonic Plague." *Scientific American*, February 1988. The story of the bacterium that killed nearly one third of the world's population in the fourteenth century.

Simons, K., Garoff, H., and Helenius, A. "How an Animal Virus Gets Into and Out of Its Host Cell." *Scientific American*, February 1982. (Offprint No. 1511).

Woese, C. "Archaebacteria." *Scientific American*, June 1981.

Zimmer, C. "Triumph of the Archaea." *Discover*, February 1995.

N E T   W A T C H

On-line resources for this chapter are on the World Wide Web at:
http://www.prenhall.com/~audesirk (click on the table of contents link and then select Chapter 21).

22 The Fungi

AT A GLANCE

Fungal Form and Function
Most Fungi Have Filamentous Bodies
Fungi Obtain Their Nutrients from Other Organisms
Most Fungi Can Reproduce Both Sexually and Asexually

Economic, Ecological, and Health Impacts of Fungi
Some Fungi Form Symbiotic Relationships with Certain
Algae and Plants

The Classification of Fungi
The Zygote Fungi (Division Zygomycota) Can Reproduce by
Forming Diploid Zygospores
The Sac Fungi (Division Ascomycota) Form Spores in a
Saclike Case Called an Ascus
The Club Fungi (Division Basidiomycota) Produce Club-
Shaped Reproductive Structures Called Basidia
The Imperfect Fungi (Division Deuteromycota) Seem to
Reproduce Entirely by Asexual Means
The Egg Fungi (Division Oomycota) Are Quite Different
from Other Fungi

Evolutionary Connections: Fungal Ingenuity

A ripe puffball touched by rain, a breeze, or a scurrying animal releases clouds of spores that are dispersed far and wide on air currents. Here, a rounded earthstar puffball emits spores when hit by a raindrop.

The soft bodies of fungi do not fossilize easily, and the evolutionary relationships of the fungi remain obscure. Fungal taxonomy is an inexact science and the subject of much debate. Like plants and many microorganisms, fungi have cell walls. Although once classified as plants, fungi are as different from plants as they are from animals. Fungi are heterotrophic, feasting on other organisms. Most have filamentous bodies that penetrate soil, animal wastes, or the dead or living bodies of plants and animals. Frequently, the only noticeable portion of a fungus is its reproductive structure (see the chapter-opening photograph). This chapter examines the "true fungi": those that have cell walls, feed by absorption, and usually have filamentous bodies. The cellular and acellular slime molds, sometimes classified as fungi, are covered in Chapter 21 as members of the kingdom Protista.

Fungal Form and Function

Most Fungi Have Filamentous Bodies

The fungal body, with rare exceptions, consists of microscopically thin, threadlike filaments called **hyphae** (singular, **hypha**) that grow in an interwoven mass called a **mycelium** (Fig. 22-1). The hyphae of some fungi consist of single elongated cells with numerous nuclei; the hyphae of others are subdivided by partitions called **septa** into many cells, each containing from one to many nuclei, depending on the species. Pores in the septa allow the cytoplasm to stream between cells, distributing nutrients.

In contrast to the nuclei found in animal cells, those of the fungal body are usually haploid (possessing only a single set of chromosomes). Like plant cells, fungal cells are surrounded by cell walls. Although some fungal cell walls contain cellulose, most are strengthened by chitin, the same substance found in the exoskeletons of arthropods (see Chapter 24).

Fungi are not motile, but their filaments can grow rapidly and in any direction within a suitable environment. The fungal mycelium insinuates itself into aging bread or cheese, beneath the bark of decaying logs, or into

(a)   **(b)**   **(c)**

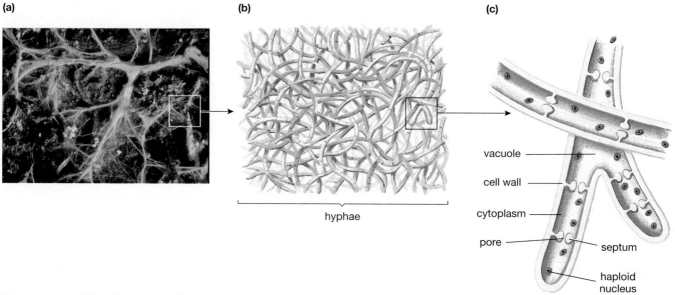

vacuole

cell wall

cytoplasm

pore

septum

haploid
nucleus

hyphae

Figure 22-1   **The filamentous body of a fungus**

(a) A fungal mycelium spreads over decaying vegation. The mycelium is composed of a tangle of microscopic hyphae, only one cell thick, drawn in **(b)** and enlarged in **(c)** to show their internal organization.

the soil. Periodically, the hyphae grow together and differentiate into reproductive structures that project above the surface beneath which the mycelium grows. These structures, including mushrooms, puffballs, or the powdery molds on food, which represent only a fraction of the complete fungal body, are often the only part of the fungus that is easily seen.

Fungi Obtain Their Nutrients from Other Organisms

Like animals, fungi are heterotrophic, surviving by breaking down nutrients stored in the bodies or wastes of other organisms. Some fungi are **saprobes;** these digest the bodies of dead organisms. Others are parasitic, feeding on living organisms and causing disease. Others, including lichens and mycorrhizae, live in mutually beneficial symbiotic relationships with other organisms. There are even a few predatory fungi, which attack tiny worms in the soil.

Fungal nutrient absorption resembles that of bacteria; both have cell walls that prevent them from ingesting food. Like bacteria, fungi must therefore secrete enzymes outside their bodies that digest complex molecules into smaller subunits that can then be absorbed. The fungal body, composed of long filaments only one cell thick, presents an enormous surface area through which to secrete enzymes and absorb nutrients.

Rapidly growing filaments penetrate deeply into a source of nutrients, such as a ripe apple, digesting the flesh and hastening its decomposition. Some of the resulting

small nutrient molecules diffuse through the fungal cell walls and into the cells, but some remain outside, providing nourishment for other organisms.

Most Fungi Can Reproduce Both Sexually and Asexually

Fungal reproduction is varied and often complex, involving both asexual and sexual processes, depending on the species of fungus and the local environment. During simple asexual reproduction, a mycelium simply breaks into pieces, each of which grows into a new individual. Many fungi reproduce both asexually and sexually through different types of spores, small resistant structures that can disperse and produce new fungi. These spores are formed on or within special structures that project above the mycelium. These special structures allow the spores to be dispersed by wind or water (see photograph on the opening page of this chapter). Haploid asexual spores are produced by mitotic divisions of haploid fungal cells. In favorable habitats, the asexual spores begin mitotic divisions that produce a new haploid mycelium that is a genetic replica of the parent cells.

Sexual spore formation begins when two haploid nuclei of compatible mating types fuse, resulting in a diploid zygote. This zygote then undergoes meiosis to form haploid sexual spores. These spores are dispersed, germinate, and divide mitotically to form a new haploid mycelium. The reproductive capacity of fungi is prodigious: a single giant puffball may contain 5 trillion sexual spores (see Fig. 22-9c).

Figure 22-2 **A helpful fungal parasite**

Although some parasitic fungi attack crops, others assist farmers by attacking insect crop pests. The body of this ant has been invaded by a parasitic fungus that is forming reproductive structures.

Economic, Ecological, and Health Impacts of Fungi

Although the sight of a brightly colored mushroom on the forest floor is a pleasant surprise, fungal bodies are far more extensive than their visible portions suggest. Microscopic fungal filaments can be found throughout moist, rich soils, penetrating decaying vegetation and the wastes and dead bodies of animals. As decomposers, fungi make an incalculable contribution to ecosystems. The extracellular digestive activities of many fungi liberate nutrients such as carbon, nitrogen, and phosphorus compounds and minerals that can be used by plants. Further, fungal bodies provide food for small insects and worms. If fungi and bacteria were suddenly to disappear, the consequences would be disastrous. Nutrients would remain locked in the bodies of dead plants and animals, the recycling of nutrients would grind to a halt, soil fertility would rapidly decline, and waste and organic debris would accumulate. In short, ecosystems would collapse.

Parasitic fungi, in contrast, cause disease. The majority of plant diseases are caused by fungi. American chestnut and elm tree populations have been drastically reduced by the fungi that cause chestnut blight and Dutch elm disease. Fungal parasites such as those that cause corn smut also result in billions of dollars in crop losses annually (see Fig. 22-9d). However, farmers have discovered that fungal parasites that attack insects and other arthropod pests (Fig.

22-2) can be an important ally in pest control. These "fungal pesticides" are being used in Florida to control citrus mites, and it is hoped they will make further contributions as farmers become more aware of the need to reduce their reliance on chemical pesticides. The human body can host several parasitic fungi. These cause a range of diseases such as ringworm and athlete's foot, which infect the skin; valley fever and histoplasmosis, which infect the lungs; and common vaginal yeast infections.

Several types of fungi are edible. Some, such as the truffle, are rare delicacies. Fungi are also important in the production of many staple foods and beverages such as bread, cheese, wine, and beer.

Some Fungi Form Symbiotic Relationships with Certain Algae and Plants

Lichens Are Formed by Fungi Living with Photosynthetic Algae or Bacteria

Lichens are a **symbiotic** (literally "living together") association between fungi, usually ascomycetes (described later in this chapter), and unicellular green algae or cyanobacteria (Fig. 22-3). Together, these organisms form a living unit so tough and undemanding of nutrients that it is among the first to colonize newly formed volcanic islands. The algal partner provides food for itself and the fungus

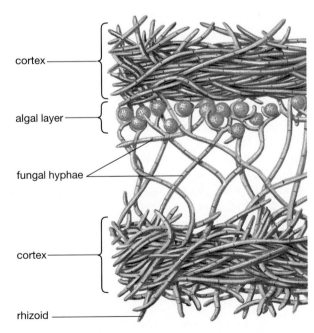

Figure 22-3 **The lichen: a symbiotic partnership**

Most lichens have a layered structure bounded on the top and bottom by an outer layer, or cortex, formed from fungal hyphae. Attachment structures called rhizoids, formed from fungal hyphae, emerge from the lower cortex and anchor the lichen to a surface, such as a rock or a tree. An algal layer in which the alga and fungus grow in close association is found beneath the upper cortex.

through photosynthesis; the fungus provides the algae with support and protection from dehydration. Brightly colored lichens can be found growing on bare rock and have invaded inhospitable habitats ranging from deserts to the Arctic. Their size and slow rate of growth suggest that some lichens in the Arctic are 4000 years old. About 20,000 species of lichen have been identified, each a combination of a different fungus and one of several types of green algae or blue-green bacteria, and with a unique color and growth pattern (Fig. 22-4).

Mycorrhizae Are Fungi Associated with the Roots of Many Plants

Mycorrhizae (singular, **mycorrhiza**) are symbiotic associations between fungi and plant roots. Over 5000 species of mycorrhizal fungi (mostly basidiomycetes or ascomycetes, described shortly) are found growing in intimate association with the roots of about 80% of all plants that have roots, including most trees. These associations benefit both the plant and its fungal partner. The hyphae of mycorrhizal fungi surround the plant root and frequently invade the root cells (Fig. 22-5). The fungus digests and absorbs organic nutrients from the soil, passing some of them directly into the root cells. The fungus also absorbs water and passes it to the plant, an advantage in dry, sandy soils. In return, sugar produced photosynthetically by the plant is passed from the root to the fungus. Plants that participate in this unique relationship, especially those in poor soils, tend to grow larger and more vigorously than those deprived of the fungus. Disturbing new data suggest that mycorrhizae and other types of fungi may be undergoing a dramatic decline that could threaten the health of forests and the communities of plants and animals that rely on them (see "Earth Watch: The Case of the Disappearing Mushrooms").

Some scientists believe that mycorrhizal associations may have been important in the invasion of land by plants over 400 million years ago. Such a relationship between an aquatic fungus such as a water mold (division Oomycota) and a green alga (ancestral to terrestrial plants) could have helped the alga acquire the water and mineral nutrients it needed to survive out of water.

The Classification of Fungi

Although nearly 100,000 species of modern fungi have been described, biologists have only begun to comprehend the diversity of these organisms—at least 1000 additional species are described each year. Like plants, fungi are grouped into divisions, which are comparable to animal phyla. The major divisions of fungi are the Zygomycota (zygote fungi), Oomycota (egg fungi), Ascomycota (sac fungi), Basidiomycota (club fungi), and Deuteromycota (imperfect fungi) (Table 22-1).

(a)

(b)

(c)

Figure 22-4   **Diverse lichens**

(a) Goat's beard, a lichen found hanging from tree limbs. (b) A colorful encrusting lichen, growing on dry rock, illustrates the tough independence of this symbiotic combination of fungus and alga. (c) A leafy lichen grows from a dead tree branch.

E A R T H   W A T C H

The Case of the Disappearing Mushrooms

Mycologists (scientists who study fungi) and gourmet cooks may seem to have little in common, but recently they have become united in a common concern: European mushrooms are rapidly declining in numbers, in average size, and in species diversity. Although the problem is most easily recognized in Europe, where people have been gathering wild mushrooms for centuries, American mycologists are also alarmed; the same decline may be occurring here as well. Why are mushrooms disappearing? Overhunting of edible mushrooms is not the culprit, because poisonous forms are equally affected. The loss is evident in all types of mature forests, which eliminates changing forest management practices as the cause. The most likely cause is air pollution, as the loss of mushrooms is greatest where the air contains the highest levels of ozone, sulfur, and nitrogen. Although mycologists have not yet determined exactly how air pollution harms mushrooms, the evidence is clear. In Holland, for example, over the past 20 years the average number of fungal species per thousand square meters has dropped from 37 down to 12. Twenty out of 60 fungal species surveyed in England are declining. Concern is intensified by the fact that the mushrooms most affected are those whose hyphae form mycorrhizal associations with tree roots. Trees with diminished mycorrhizae may have less resistance to periodic droughts or extreme cold spells. Because air pollution is also harming forests directly, the additional loss of mycorrhizae could be devastating.

The Zygote Fungi (Division Zygomycota) Can Reproduce by Forming Diploid Zygospores

The zygomycetes, also called the **zygote fungi**, include about 600 species. Familiar and annoying zygomycetes are those of the genus *Rhizopus,* which cause soft fruit rot and black bread mold. They are named for the ability of their haploid hyphae to "mate," fusing their nuclei to produce diploid **zygospores** (Fig. 22-6, inset). These resistant structures are dispersed through the air and can remain dormant until conditions are favorable for growth. Zygospores then undergo meiosis and germinate into structures bearing haploid spores. The spores then give rise to new hyphae. These hyphae may reproduce either asexually, by forming haploid spores in black spore cases called **sporangia** (Fig. 22-6, inset), or sexually, by fusing to produce more zygospores. The life cycle of the black bread mold, showing both sexual and asexual reproduction, is depicted in Figure 22-6.

The Sac Fungi (Division Ascomycota) Form Spores in a Saclike Case Called an Ascus

The 30,000 species of ascomycetes, also called **sac fungi,** are named after the saclike case, or **ascus,** in which spores form during sexual reproduction. Some ascomycetes live in decaying forest vegetation; these may form beautiful cup-shaped reproductive structures or corrugated, mushroomlike fruiting bodies called morels (Fig. 22-7).

This division also includes many of the colorful molds that attack stored food and destroy fruit and grain crops and other plants. Ascomycetes secrete enzymes that include both cellulase and protease, which in warm, humid climates cause significant damage to textiles such as cotton and wool. Ascomycetes cause both Dutch elm disease and chestnut blight, which have destroyed nearly all the American elm and American chestnut trees in the United States. Some ascomycetes are a boon to plants, forming mycorrhizal associations.

Claviceps purpurea, an ascomycete that attacks rye plants, produces several toxins, one of which is the active ingredient in LSD. If infected rye is made into flour and consumed, the toxins can produce convulsions, hallucinations, and death. This happened frequently in northern Europe in the Middle Ages, but modern agricultural techniques have essentially eliminated the disease. Another of its toxins, called ergot, is currently used in drugs that induce labor and control hemorrhaging after childbirth.

Figure 22-5   **Mycorrhizae enhance plant growth**
Here hyphae of mycorrhizae are seen entwining about the root of an aspen tree. Plants grow significantly better in a symbiotic association with these fungi, which help make nutrients and water available to the roots.

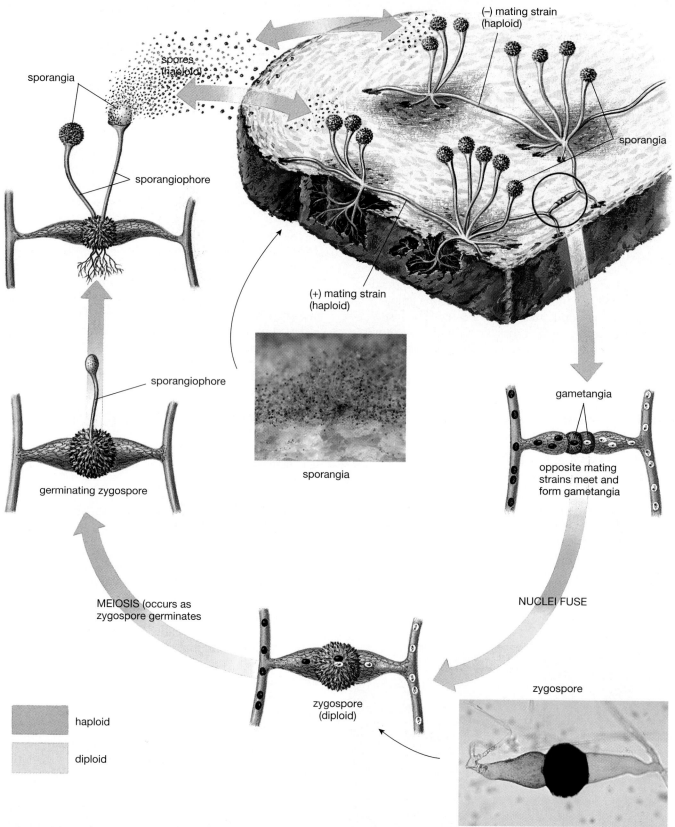

Figure 22-6   The life cycle of a zygomycete

Here we illustrate the black bread mold (genus *Rhizopus*). During asexual reproduction (shown on top), haploid spores are produced within sporangia. These disperse and germinate on food such as bread, producing haploid hyphae. These in turn may produce additional spores within sporangia by mitosis. During sexual reproduction hyphae that appear identical but are actually of different mating types (designated + and − on the slice of bread) contact one another and form specialized cells called gametangia at the point of contact. The gametangia (with haploid nuclei) then fuse, producing a diploid zygospore (bottom center and inset). The zygospore undergoes meiosis and germinates, producing a sporangiophore (center left), on which spore cases, or sporangia, are formed (upper left). These produce and liberate haploid spores that begin the asexual reproductive cycle described above.

TABLE 22-1 The Major Divisions of Fungi

Common Name (Division)	*Reproductive Structures*	*Cellular Characteristics*	*Economic and Health Impacts*	*Representative Genera*
Zygote fungi (Zygomycota)	Produce sexual diploid zygospores	Cell walls contain chitin; septa are absent	Cause soft fruit rot and black bread mold	*Rhizopus* (causes black bread mold); *Pilobolus* (dung fungus)
Sac fungi (Ascomycota)	Sexual spores formed in saclike ascus	Cell walls contain chitin; septa are present	Cause molds on fruit; can damage textiles; cause Dutch elm disease and chestnut blight; include yeasts and morels	*Saccharomyces* (yeast); *Ophiostoma* (causes Dutch elm disease)
Club fungi (Basidiomycota)	Sexual reproduction involves production of haploid basidiospores on club-shaped basidia	Cell walls contain chitin; septa are present	Cause smuts and rusts on crops; include some edible mushrooms	*Amanita* (poisonous mushroom); *Polyporus* (shelf fungus)
Imperfect fungi (Deuteromycota)	Not observed engaging in sexual reproduction	Cell walls contain chitin; septa are present	Cause athlete's foot, ringworm, histoplasmosis; source of penicillin	*Penicillium* (produces penicillin); *Arthrobotrys* (nematode predator)
Egg fungi (Oomycota)	Sexual reproduction involves fertilization of a large egg; flagellated asexual zoospores require water to swim	Cell walls contain cellulose; septa may be present or absent	Cause downy mildew of grapes and late blight of potatoes	*Achyla* (water mold); *Phytophthora* (late blight of potatoes)

In this group we also find the yeasts, one of the few unicellular fungi. Some yeasts form hyphae when nutrients are scarce; these can elongate and reach distant food sources. Although a parasitic yeast is a common cause of vaginal infections (Fig. 22-8), baker's and brewer's yeasts make possible the proverbial loaf of bread and jug of wine. Another gastronomic delicacy, the truffle, is also a member of this diverse division (see "Evolutionary Connections: Fungal Ingenuity;" Fig. 22-14).

The Club Fungi (Division Basidiomycota) Produce Club-Shaped Reproductive Structures Called Basidia

Basidiomycetes are called the **club fungi** because they produce club-shaped reproductive structures. The division Basidiomycota consists of about 25,000 species, including the familiar mushrooms, puffballs, and shelf fungi, sometimes called monkey-stools (Fig. 22-9a). Although several mushroom species are considered delicacies, mushrooms can be

(a)

(b)

Figure 22-7 **Diverse ascomycetes**

(a) The cup-shaped fruiting body of the scarlet cup fungus. **(b)** The morel, an edible delicacy. (But consult an expert before sampling any wild fungus—some are deadly!)

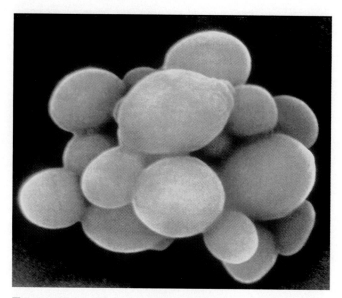

Figure 22-8   The unusual yeast

The yeast is an unusual, normally nonfilamentous ascomycete that reproduces most commonly by budding. The yeast shown here is *Candida*, a common cause of vaginal infections.

deadly. The good-tasting members of the genus *Amanita* (Fig. 22-9b) contain potent and deadly toxins. Among the basidiomycetes are some devastating plant pests descriptively called *rusts* and *smuts*, which cause billions of dollars worth of damage to grain crops each year (Fig. 22-9d). Some members of this group, however, enter into mutually beneficial mycorrhizal relationships with plants.

Basidiomycetes typically reproduce sexually; their life cycle is shown in Figure 22-10. Mushrooms and puffballs are actually reproductive structures: dense aggregations of hyphae that emerge under proper conditions from a massive underground mycelium. On the undersides of mushrooms are leaflike gills that produce specialized club-shaped diploid cells called **basidia.** Basidia form haploid reproductive **basidiospores** by meiosis. These are released by the billions from the gills of mushrooms or the inner surface of puffballs to be dispersed by wind and water.

Falling on fertile ground, a mushroom spore may germinate and produce an underground mycelium. These hyphae grow outward from the original spore in a roughly circular pattern as the older hyphae in the center die. The subterranean body periodically sends up numerous mush-

Figure 22-9   Diverse basidiomycetes

(a) Shelf fungi, the size of dessert plates, are conspicuous on trees. (b) A poisonous mushroom of the genus *Amanita*. (c) The giant puffball *Lycopedon giganteum* may produce up to 5 trillion spores. (d) Corn smut causes major losses of this crop each year.

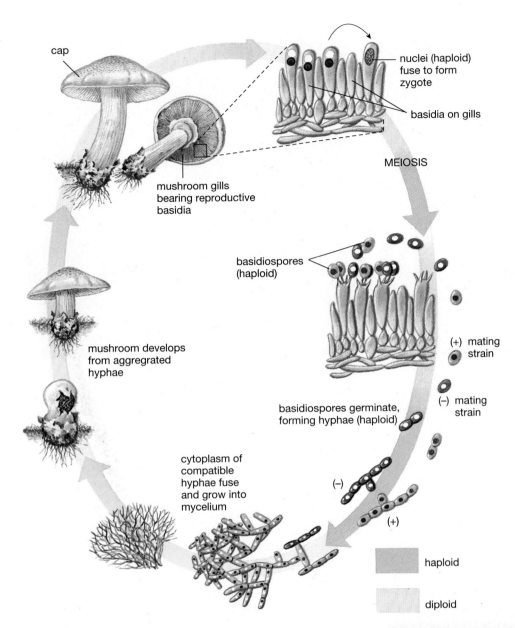

cap

nuclei (haploid) fuse to form zygote

basidia on gills

mushroom gills bearing reproductive basidia

MEIOSIS

basidiospores (haploid)

(+) mating strain

(−) mating strain

mushroom develops from aggregrated hyphae

basidiospores germinate, forming hyphae (haploid)

cytoplasm of compatible hyphae fuse and grow into mycelium

(−)

(+)

haploid

diploid

Figure 22-10   **The life cycle of a "typical" basidiomycete**

The mushroom (top left) is a reproductive structure, formed from aggregated hyphae made up of cells that each contain two haploid nuclei. Within the cap, leaflike gills bear numerous special cells called *basidia* (top right). Within each basidium, the two haploid nuclei fuse, producing a diploid zygote. The zygote then undergoes meiosis, forming haploid basidiospores that are released from the basidia (right). After being dispersed to new habitats by wind or water, the basidiospores germinate, forming haploid hyphae of two different mating types (+ and −). When hyphae of the different mating types meet, some of the cells fuse. These cells, each containing two haploid nuclei, produce an extensive underground mycelium (bottom). In the presence of adequate water and nutrients, portions of the mycelium aggregate, swell, and differentiate, poking up through the soil as mushrooms and completing the cycle.

rooms, which emerge in a ringlike pattern called a **fairy ring**. The diameter of the fairy ring can reveal the approximate age of the fungus—some are estimated to be 700 years old (Fig. 22-11). Recently, researchers have used genetic analysis to trace the extent of some basidiomycete mycelia. One, estimated to be at least 1500 years old, covers 38 acres in northern Michigan, and another, on the forested slopes of Mt. Adams in Washington, covers 1500 acres, although its age is only 400 to 1000 years.

The Imperfect Fungi (Division Deuteromycota) Seem to Reproduce Entirely by Asexual Means

Deuteromycetes are called the **imperfect fungi** because none have been observed forming sexual reproductive structures. In some species the sexual stage has been lost during evolution; in others it may exist but has not yet been observed. This large division includes about 25,000 described species

Figure 22-11   **A mushroom fairy ring**

Mushrooms emerge in a fairy ring from an underground fungal mycelium growing outward from a central point where a spore germinated, perhaps centuries ago.

(a)

(b)

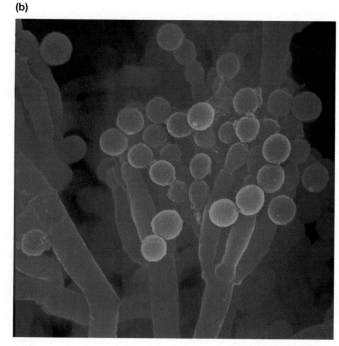

Figure 22-12   Penicillium, **an imperfect fungus**

(a) *Penicillium* growing on an orange. Reproductive structures, which coat the fruit's surface, are visible, while hyphae beneath draw nourishment from inside. **(b)** The spore-bearing structures of *Penicillium*, called conidiophores, are shown in this scanning electron micrograph. The spores, called conidia, are asexual.

of great diversity and considerable importance to humans. It was a member of this division that Alexander Fleming discovered contaminating and killing his bacterial cultures. His keen observations led to the isolation of penicillin, the first antibiotic, from the fungus *Penicillium* (Fig. 22-12). To these fungi we also owe the indescribable flavor and aroma of Roquefort and Camembert cheeses. Other imperfect fungi are human parasites, causing diseases such as ringworm and athlete's foot. Some are not content to live on dead organisms or even to parasitize live ones—they act as predators, laying deadly traps for unsuspecting roundworms (see "Evolutionary Connections: Fungal Ingenuity").

The Egg Fungi (Division Oomycota) Are Quite Different from Other Fungi

The oomycetes, also called the **egg fungi**, differ significantly from other true fungi. For example, oomycete cell walls often contain cellulose, the main component in plant cell walls, instead of (or in addition to) chitin. Sexual reproduction involves the fertilization of a large egg cell, hence the name egg fungi. Oomycetes reproduce asexually using actively swimming flagellated cells called **zoospores** that are not found in other fungal divisions. The swimming spores require very moist conditions. A final distinction is that oomycetes are diploid (instead of haploid) throughout most of their life cycle.

This rather small division of 475 species includes inoffensive water molds that live in water and damp soil, as well as some species with profound economic importance. For

example, an oomycete causes a disease called downy mildew of grapes (Fig. 22-13). Its inadvertent introduction into France from the United States in the late 1870s nearly destroyed the French wine industry. Another member of this division has destroyed millions of avocado trees in California, while still another is responsible for "late blight," a devastating disease of potatoes. When accidentally introduced into Ireland around 1845, this fungus destroyed nearly the entire potato crop, causing the devastating potato famine during which 1 million people in Ireland starved and many more emigrated to the United States.

Figure 22-13   **A parasitic oomycete**

An oomycete that causes downy mildew on grapes nearly destroyed the French wine industry during the 1870s.

E V O L U T I O N A R Y
C O N N E C T I O N S

Fungal Ingenuity

Natural selection, operating over thousands of years on the diverse forms of fungi, has produced some remarkable adaptations by which fungi disperse their spores and obtain nutrients. A few of these are highlighted here.

The Rare, Sexy Truffle

Although many fungi are prized as food, none are as avidly sought as the truffle. A single specimen resembling a small, shriveled, blackened apple (Fig. 22-14) may sell for $100. Truffles are the spore-containing structures of an ascomycete that forms a mycorrhizal association with the roots of oak trees. Although it develops underground, the truffle has evolved a fascinating mechanism to entice animals, especially wild pigs, to dig it up. The truffle releases a chemical that closely resembles the pig's sex attractant. As aroused pigs dig up and devour the truffle, millions of spores are scattered to the winds. Truffle collectors use muzzled pigs to hunt their quarry; a good truffle-pig can smell an underground truffle 50 meters (about 150 feet) away!

The Shotgun Approach to Spore Dispersal

The delicate structures shown in Figure 22-15 are actually fungal shotguns, the reproductive structures of the zygomycete *Pilobolus*. If you closely scrutinize piles of horse manure, you may be fortunate enough to observe this miniature beauty. Hyphae penetrating the dung send up clear bulbs capped with sticky black spore cases. As they mature, the sugar concen-

Figure 22-15   An explosive zygomycete

The delicate translucent reproductive structures of the zygomycete *Pilobolus* will literally blow their tops when ripe, dispersing the black caps with their payload of spores.

tration in the bulbs increases, drawing in water by osmosis. Meanwhile, the bulb begins to weaken just below its cap. Suddenly, like an overinflated balloon, the bulb bursts and blows its spore-carrying top up to a meter (3 feet) away. *Pilobolus* bends toward the light, thereby increasing the chances that its spores will land on open pasture when the bulb bursts. Here they adhere to grass until consumed by a grazing herbivore, perhaps a horse. Later (some distance away) the spores are deposited unharmed in a pile of their favorite food: manure. Growing hyphae penetrate this rich source of nutrients, sending up new projectiles to continue this ingenious cycle.

The Nematode Nemesis

Microscopic nematode roundworms abound in rich soil, and fungi have evolved several fascinating forms of nematode-nabbing hyphae that allow them to exploit this rich source of protein. Some produce sticky pods that adhere to passing nematodes, penetrate the nematode body with hyphae, and then begin digesting it from within. One species shoots a microscopic harpoonlike spore into passing nema-

Figure 22-14   The truffle

The truffle, a rare ascomycete (about the size of a small apple) is a gastronomic delicacy.

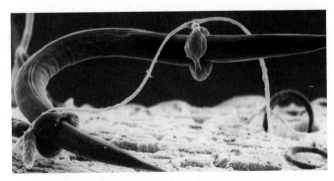

Figure 22-16   Nemesis of nematodes

Arthrobotrys, the nematode strangler, traps its nematode (roundworm) prey in a modified hypha that swells when the inside of the loop is contacted.

todes; the spore develops into a new fungus inside the worm. The fungal strangler *Arthrobotrys*, a member of the imperfect fungi, produces nooses formed from three hyphal cells. When a nematode blunders into the noose, its contact with the inner parts of the noose stimulates noose cells to swell with water (Fig. 22-16 on the previous page). In a fraction of a second the hole constricts, trapping the worm. Fungal hyphae then penetrate and feast on their prey.

❊ SUMMARY OF KEY CONCEPTS

Fungal Form and Function
Fungal bodies generally consist of filamentous hyphae, which are either multicellular or multinucleate and form large, intertwined networks called mycelia. Fungal nuclei are generally haploid. A cell wall of chitin (or, in a few cases, cellulose) surrounds fungal cells.

All fungi are heterotrophic, either parasitic or saprobic. They secrete digestive enzymes outside their bodies and absorb the liberated nutrients.

Fungal reproduction is varied and complex. Asexual reproduction can occur either through fragmentation of the mycelium or through asexual spore formation. Sexual spores form after fusion of compatible haploid nuclei to form a diploid zygote that undergoes meiosis to form haploid sexual spores. These produce haploid mycelia through mitosis.

Economic, Ecological, and Health Impacts of Fungi
Fungi are extremely important decomposers in ecosystems. Their filamentous bodies penetrate rich soil and decaying organic material, liberating nutrients through extracellular digestion.

The majority of plant diseases are caused by parasitic fungi. Some parasitic fungi can help control insect crop pests. Others can cause human diseases, including ringworm, athlete's foot, and common vaginal infections.

A lichen is a symbiotic association between a fungus and unicellular algae or cyanobacteria. This self-sufficient combination can colonize bare rock. Mycorrhizae are associations between fungi and the roots of most vascular plants. The fungus derives photosynthetic nutrients from the plant roots and in return carries water and nutrients into the root from the surrounding soil.

The Classification of Fungi
The major divisions of fungi and their characteristics are summarized in Table 22-1.

❊ KEY TERMS

ascus   p. 429
basidiospore   p. 432
basidium   p. 432
club fungi   p. 431
egg fungi   p. 434
fairy ring   p. 433
hypha   p. 425

imperfect fungi   p. 433
lichen   p. 427
mycelium   p. 425
mycorrhiza   p. 428
sac fungi   p. 429
saprobe   p. 426
septum   p. 425

sporangium   p. 429
symbiosis   p. 427
zoospore   p. 434
zygospore   p. 429
zygote fungi   p. 429

❊ THINKING THROUGH THE CONCEPTS

Multiple Choice
1. There are no fungi that are
   a. predators   b. photosynthetic
   c. decomposers   d. parasites
   e. symbiotic
2. With what plant organ do mycorrhizae interact?
   a. roots   b. leaves
   c. stems   d. fruits
   e. all parts of the plant

3. What term refers to the mass of threads that forms the body of most fungi?
   a. hyphae   b. mycelium
   c. sporangia   d. ascus
   e. basidium
4. Which of the following pairs is incorrect?
   a. downy mildew: egg fungi
   b. edible mushroom: club fungus
   c. black bread mold: sac fungus
   d. *Penicillium*: imperfect fungus
   e. yeast: sac fungus

5. Which of the following is true of both fungi and some types of bacteria?
   a. They both produce gametes.
   b. They both engulf microscopic animals.
   c. They both absorb materials across cell walls.
   d. They both fix nitrogen.
   e. They interact to form lichens.
6. Which of the following structures would you expect to find in the corn smut fungus?
   a. ascospores   b. basidiospores
   c. zoospores   d. zygospores
   e. fairy rings

Review Questions

1. Describe the structure of the fungal body. How do fungal cells differ from most plant and animal cells?
2. What portion of the fungal body is represented by mushrooms, puffballs, and similar structures? Why are these elevated above the ground?
3. What two plant diseases, caused by parasitic fungi, have had an enormous impact on U.S. forests? In what division are these found?
4. List some fungi that attack crops. In what division is each found?
5. Describe two different modes of asexual reproduction in fungi.
6. What is the major structural ingredient in the cell walls of fungi?
7. List the major divisions of fungi, describe the feature that gives each its name, and give one example of each.
8. Describe how a fairy ring of mushrooms is produced. Why is the diameter related to its age?
9. Describe two different symbiotic associations between plants and fungi. In each case, explain how each partner in these associations is affected.

APPLYING THE CONCEPTS

1. Dutch elm disease in the United States is caused by an exotic, an organism (in this case a fungus) introduced from another part of the world. What damage has this done? What other fungal pests fall into this category? Why are parasitic fungi particularly likely to be transported out of their natural habitat? What can governments do to limit this importation?
2. The discovery of penicillin revolutionized the treatment of bacterial diseases. However, penicillin is now rarely prescribed. Why is this? HINT: You might wish to refer back to Chapter 21.
3. The discovery of penicillin was the result of a chance observation by an observant microbiologist, Alexander Fleming. How would you search systematically for new antibiotics produced by fungi? Where would you go to look for these?
4. Fossil evidence suggests that mycorrhizal associations between fungi and plant roots existed in the late Paleozoic when the plant invasion of land began. This suggests an important link between mycorrhizae and the successful invasion of land by plants. Why might mycorrhizae have been important fungi in the colonization of terrestrial habitats by plants?
5. General biology texts in the 1960s included fungi in the plant kingdom. Why do biologists no longer consider fungi as legitimate members of the plant kingdom?
6. What ecological consequences would occur if all fungi on Earth were destroyed by humans using a new and deadly fungicide?

FOR MORE INFORMATION

Angier, N. "A Stupid Cell with All the Answers." *Discover*, November 1986. Fascinating description of the uses of yeasts in molecular biology, including beautiful illustrations.

Barron, G. "Jekyll-Hyde Mushrooms." *Natural History*, March 1992. Fungi include a variety of forms; some absorb decaying plant material, others prey on microscopic worms.

Kiester, E. "Prophets of Gloom." *Discover*, November 1991. Lichens can be used as indicators of air quality and provide evidence of deteriorating environmental conditions.

Radetsky, P. "The Yeast Within." *Discover*, March 1994. Describes the newly discovered ability of baker's yeast to form filaments, and the possible implications of this for the study of yeasts that cause vaginal infections.

Strobel, G. A., and Lanier, G. N. "Dutch Elm Disease." *Scientific American*, August 1981. A description of the life cycle of this parasitic fungus and of techniques that limit its spread.

Vogel, S. "Taming the Wild Morel." *Discover*, May 1988. Describes the research that has allowed these rare delicacies to be cultivated in the laboratory.

NET WATCH

On-line resources for this chapter are on the World Wide Web at:
http://www.prenhall.com/~audesirk (click on the <u>table of contents</u> link and then select Chapter 22).

23 The Plant Kingdom

◾ AT A GLANCE

Evolutionary Trends in Plants
   The Plant Body Increased in Complexity as Plants Made the
      Evolutionary Transition from Water to Dry Land
   The Invasion of Land Required Protection and a Means of
      Dispersal for Sex Cells and Developing Plants
   Plants Have Both a Sporophyte and a Gametophyte
      Generation

Watery Origins—The Algae
   The Red Algae (Division Rhodophyta) Are Found Primarily
      in Clear Tropical Oceans
   The Brown Algae (Division Phaeophyta) Dominate in Cool
      Coastal Waters
   The Green Algae (Division Chlorophyta), Found Mostly in
      Ponds and Lakes, Probably Gave Rise to the Land Plants

Land—The New Frontier
   The Liverworts and Mosses (Division Bryophyta) Are Only
      Partially Adapted to Dry Land
   The Vascular Plants, or Tracheophytes, Have Conducting
      Vessels That Also Provide Support
   The Seedless Vascular Plants Include the Club Mosses,
      Horsetails, and Ferns
   The Seed Plants Dominate the Land, Aided by Two
      Important Adaptations: Pollen and Seeds

A slight breeze releases a shower of pollen from the flowers, called catkins, of this oak.

Although there is no universally accepted classification system for plants, in this text we include in the kingdom Plantae organisms that share the following three characteristics: They are eukaryotic, photosynthetic, and (with few exceptions) multicellular. To date, nearly 300,000 species of multicellular plants have been identified.

Although eukaryotic cells resembling unicellular green algae have been found in fossil deposits dating back 1 billion years, the first clearly identifiable algae are found in deposits from the Cambrian period, 500 to 600 million years ago. Algae are relatively simple and structurally different from modern land plants. Nevertheless, several lines of evidence, discussed later in this chapter, point to green algae as the ancestors of all modern land plants. Terrestrial plants first appeared 430 to 500 million years ago. Two groups of plants arose from the ancestral green algae. One group, the liverworts and mosses (bryophytes), can be considered the amphibians of the plant kingdom, because they straddle the boundary between aquatic and terrestrial existence. Another group, the ferns, conifers, and flowering plants (vascular plants, or tracheophytes), evolved specialized vessels that transport water and nutrients and provide support for the plant body. These dominate Earth today. The general features of the major divisions of plants are summarized in Table 23-1.

Evolutionary Trends in Plants

The Plant Body Increased in Complexity as Plants Made the Evolutionary Transition from Water to Dry Land

Plant life arose in the sea, an environment that supports the plant body, provides a relatively constant temperature, and bathes the entire plant in nutrients. During their evolutionary history, plants underwent a series of changes under the selection pressures imposed by terrestrial environments, where plants are surrounded by dry, nonsupporting air. Adaptations to dry land include:

1. Roots or rootlike structures that anchor the plant and absorb water and nutrients from the soil

2. Conducting vessels that transport water and minerals upward from the roots and photosynthetic products from the leaves to the rest of the plant body

3. The stiffening substance **lignin**, which impregnates the water- and mineral-conducting vessels and supports the plant body, helping it to expose maximum surface area to the sunlight

4. A waxy **cuticle** covering the surfaces of leaves and stems that limits the evaporation of water

5. Pores, called **stomata**, in the leaves and stems that open to allow gas exchange but close when water is scarce, reducing the amount of water lost to evaporation

The Invasion of Land Required Protection and a Means of Dispersal for Sex Cells and Developing Plants

In addition to nourishing and supporting the plant, the surrounding water provides an ideal medium for reproduction. Gametes (sex cells) of some algae are carried passively by water currents; other algae commonly have flagellated, actively swimming gametes and spores, called **zoospores**. Not only do many algal gametes meet in the water, but **zygotes** (fertilized sex cells) and spores are dispersed by the water to suitable habitats.

To be fully successful on land, plants needed new methods of transporting sperm to eggs. The developing plant also required protection from drying and a means of dispersal independent of water. Gradually, through natural selection of random mutations, some plants evolved pollen, seeds, and later, the flower and the fruit. The dry, microscopic pollen grains produced by nonflowering seed plants allow wind, instead of water, to carry the male gametes. Later came the evolution of attractive flowers, which entice animal pollinators who deliver pollen more precisely than the wind. The seed provides waterproof protection and nourishment for the developing embryo, while some modifications of the seed coat enhance dispersal. Some fruits attract animal foragers, who consume them and incidentally disperse the indigestible seeds in their feces.

Plants Have Both a Sporophyte and a Gametophyte Generation

Plants generally have a "two-generation" life cycle. The bodies of animals are composed of diploid cells (containing two sets of chromosomes); their only haploid cells (containing one set of chromosomes) are sperm and eggs. In contrast, plants produce separate diploid and haploid generations that alternate with one another. The diploid generation is called the **sporophyte**, in which the plant body consists of diploid cells and produces haploid spores by meiosis. These haploid spores grow mitotically to produce the haploid generation, called the **gametophyte**. The gametophyte plant body is haploid and produces haploid gametes by mitosis. These gametes fuse to form a diploid zygote, which then develops into the diploid sporophyte,

continuing this complex life cycle, appropriately described as **alternation of generations** (see Chapter 27).

From the evolutionarily ancient algae to the more recent seed plants, we see a general trend toward increased prominence of the sporophyte generation and decreased size and duration of the gametophyte generation (see Table 23-1). Algal life cycles are quite diverse. However, in some types of green algae, whose ancestors may have given rise to terrestrial plants, the sporophyte generation is represented only by the diploid zygote. In nonvascular plants such as mosses, the sporophyte is present, but it is smaller than the gametophyte and remains attached to it (see Fig. 23-6). In the seedless vascular plants, such as ferns, the sporophyte is dominant, while the gametophyte is a much smaller, but still independent, plant. Finally, in seed plants the male and female gametophytes are microscopic and barely recognizable as an alternate generation (see Figs. 23-11 and 23-13). These small gametophytes, however, still produce the eggs and sperm that unite and form the zygote that develops into the diploid sporophyte.

Watery Origins—The Algae

The first plants—the **algae**—evolved in the sea, which provided many of their needs. Growing in this hospitable environment, modern algae remain simpler than terrestrial plants in several ways. Algae lack true roots, stems, and leaves, which have distinctive internal structures in vascular plants. Algae also lack complex reproductive structures such as flowers and cones. Algal gametes are usually shed directly into the water, where they unite and develop.

Algal life cycles are complex and vary considerably between, and even within, the algal divisions. In some genera, such as the green alga *Chara*, the haploid gametophyte is dominant. In *Fucus*, a brown alga, the diploid stage is dominant; the only haploid cells are gametes. In other genera, the sporophyte and gametophyte may appear nearly identical, as in the green alga *Ulva* (see Fig. 23-4). Alternatively, they may be very different in size and appearance, as in the giant kelp, whose sporophyte can be nearly 100 meters tall (328 feet; see Fig. 23-2b) but whose gametophyte is filamentous and microscopic.

Algae are colored by pigments that capture light energy for photosynthesis. These pigments are often red or brown, absorbing the green, violet, and blue light that most readily penetrates deep water. The combination of these pigments with green chlorophyll lends algae their distinctive colors and gives algal divisions their names: division Rhodophyta—the red algae; division Phaeophyta—the brown algae; and division Chlorophyta—the green algae.

The Red Algae (Division Rhodophyta) Are Found Primarily in Clear Tropical Oceans

The 4000 species of red algae, ranging from bright red to nearly black, derive their color from red pigments that mask

TABLE 23-1 ■ *Features of the Major Plant Groups*

Division	Relationship of Sporophyte and Gametophyte	Transfer of Reproductive Cells	Early Embryonic Development	Dispersal	Water and Nutrient Transport Structures	Typical Habitat
Red algae (Rhodophyta)	Highly variable	Gametes or asexual zoospores released into water	Occurs independently after zygote or zoospore settles to substrate	By water currents	Absent	Mostly marine, some freshwater
Brown algae (Phaeophyta)	Highly variable	Gametes or asexual zoospores released into water	Occurs independently after zygote or zoospore settles to substrate	By water currents	Absent (usually)	Almost entirely marine
Green algae (Chlorophyta)	Highly variable	Gametes or asexual zoospores released into water	Occurs independently after zygote or zoospore settles to substrate	By water currents	Absent	Mostly freshwater, some marine
Liverworts and mosses (Bryophyta)	Gametophyte dominant—sporophyte develops from zygote retained on gametophyte	Motile sperm swims to stationary egg retained on gametophyte	Occurs within archegonium of gametophyte	Haploid spores carried by wind	Absent	Moist terrestrial
Ferns (Pteridophyta)	Sporophyte dominant—develops from zygote retained on gametophyte	Motile sperm swims to stationary egg retained on gametophyte	Occurs within archegonium of gametophyte	Haploid spores carried by wind	Present	Moist terrestrial
Conifers (Coniferophyta)	Sporophyte dominant—microscopic gametophyte develops within sporophyte	Wind-dispersed pollen carries sperm to stationary egg in cone	Occurs within a protective seed containing a food supply	Seeds containing diploid sporophyte embryo dispersed by wind or animals	Present	Varied terrestrial habitats—dominate in dry, cold climates
Flowering plants (Anthophyta)	Sporophyte dominant—microscopic gametophyte develops within sporophyte	Pollen, dispersed by wind or animals, carries sperm to stationary egg within flower	Occurs within a protective seed containing a food supply; seed encased in fruit	Fruit, carrying seeds, dispersed by animals, wind, or water	Present	Varied terrestrial habitats—dominant terrestrial plant

Figure 23-1   Red algae

Red coralline algae from the Pacific Ocean off California provide an anchoring site for bright yellow hydroids (phylum Cnidaria). Coralline algae, which deposit calcium carbonate within their bodies, contribute to coral reefs in tropical waters.

their green chlorophyll (Fig. 23-1). Red algae are mostly marine and strictly multicellular. They dominate in deep, clear tropical waters, where their red pigments absorb the deeply penetrating blue-green light and transfer this light energy to chlorophyll, where it is used in photosynthesis.

Some red algae deposit calcium carbonate (limestone) in their tissues, and some of these contribute to the formation of reefs. Others are harvested as food in Asia, and from some the gelatinous substance carrageenan is extracted and used as a stabilizing agent in products such as paints, cosmetics, and ice cream. Agar, a material used to grow bacteria in laboratories, is also extracted from red algae. However, the major importance of these and other algae lies in their photosynthetic ability: The energy they capture and the food they synthesize help support the heterotrophic organisms in marine ecosystems.

Figure 23-2   Diverse red and brown algae

(a) The sporophyte generation of the giant kelp plant *Macrocystis* forms underwater forests off southern California. **(b)** The growing tip of the giant kelp. Gas-filled bladders are clearly visible at the base of leaflike blades. **(c)** *Fucus*, a genus found near shores, is shown here exposed at low tide. Notice the gas-filled bladders, which provide buoyancy in water.

The Brown Algae (Division Phaeophyta) Dominate in Cool Coastal Waters

The brown algae increase their light-gathering ability with brownish yellow pigments that (in combination with green chlorophyll) produce their brown to olive-green color. Like the red algae, the 1500 species of brown algae are almost entirely marine and strictly multicellular. This group includes the dominant "seaweed" found along rocky shores in the temperate (cooler) oceans of the world, including the east and west coasts of the United States. Brown algae are found in habitats ranging from near shore, where they cling to rocks exposed at low tide, to far offshore. Several types use gas-filled floats to support their bodies (Fig. 23-2a, c). The giant kelp plants found along the Pacific coast occasionally reach heights of 100 meters (328 feet) and may grow over 15 centimeters (6 inches) in a single day. With their dense growth and towering height, kelp plants form undersea forests that provide food, shelter, and breeding areas for a variety of marine animals (Fig. 23-2b).

The Green Algae (Division Chlorophyta), Found Mostly in Ponds and Lakes, Probably Gave Rise to the Land Plants

The green algae, a large and extremely diverse group of 7000 species, is of special interest because green algal ancestors are believed to have given rise to the terrestrial plants. Three lines of evidence support the hypothesis that land plants evolved from green algal ancestors:

1. Green algae use the same type of chlorophyll and **accessory pigments** in photosynthesis as do land plants.
2. Green algae store food as starch, as do land plants, and have cell walls made of cellulose, similar in composition to those of land plants.
3. Most green algae live in fresh water, where they were subjected to selection pressures that resulted in adaptations that helped them cope with the challenges of dry land, as described below.

In contrast to the nearly constant environmental conditions of the ocean, freshwater habitats (ponds, swamps, streams, and lakes) are highly variable. Dramatic fluctuations in temperature and rainfall over evolutionary time probably exerted intense selection pressure on ancient freshwater algae to withstand extremes of temperature and periods of dryness. The resulting adaptations served their descendants well as they invaded land.

Although most green algae are multicellular, there are a few unicellular forms. Some green algae form thin filaments and others form colonies that are intermediate between unicellular and multicellular forms. Colonial green algae form clusters of cells that are somewhat interdependent. These colonies range from a few cells to a few thousand cells, as in species of *Volvox* (Fig. 23-3a). Filamentous green algae, such as *Spirogyra*, form long chains of cells (Fig. 23-3b).

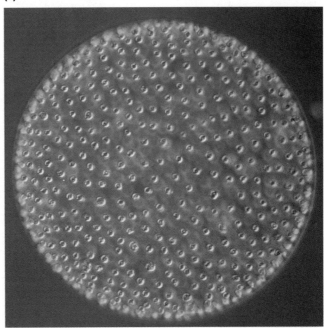

(a)

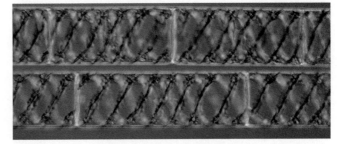

(b)

Figure 23-3 **Two forms of microscopic green algae**

(a) *Volvox* is a colonial green alga composed of cells embedded in a transparent, gelatinous green sphere. New daughter colonies can be seen developing inside the sphere. **(b)** *Spirogyra* is a filamentous green alga composed of strands only one cell thick. Inside these microscopic transparent cells, spiral chloroplasts are clearly visible.

Most green algae are quite small, but some large ones are found in the sea. The green alga *Ulva*, or sea lettuce, is similar in size to the leaves of its namesake (Fig. 23-4).

Land—The New Frontier

Moving onto land provided significant advantages for the descendants of the ancient green algae. The ponds and seas had grown crowded, teeming with hungry animals and with other plants competing for light and nutrients. The land, in contrast, offered abundant space and resources to the new colonists. For example, the raw materials for photosynthesis—carbon dioxide and sunlight—are present in far higher concentrations in the air than in water. But in other ways terrestrial environ-

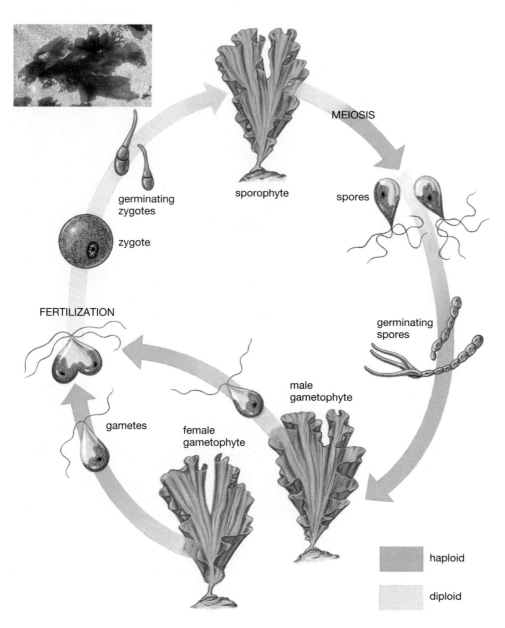

Figure 23-4   **Life cycle of *Ulva*, a large marine green alga**

In this case the sporophyte and gametophyte plants are equal in size and nearly indistinguishable. But the sporophyte is diploid and produces haploid spores through meiosis, whereas the male and female gametophytes are haploid and produce gametes by mitosis. The fusion of these gametes results in a diploid zygote that develops into the sporophyte plant. (Inset) A photograph of ulva. These could be either sporophytes or gametophytes.

MEIOSIS

germinating zygotes

sporophyte

spores

zygote

FERTILIZATION

germinating spores

gametes

female gametophyte

male gametophyte

haploid

diploid

ments are much less hospitable than aquatic environments. The supportive buoyancy of water is missing, and the air tends to dry things out. These conditions favored the evolution of structures that support the body and vessels that transport water and nutrients because the entire plant wasn't being bathed in them. The terrestrial plant body also needed waterproofing. Special reproductive adaptations, including pollen, seeds, and flowers, were favored by natural selection because land plants could no longer rely on water to carry their sex cells and disperse the fertilized eggs.

Two major groups of land plants arose from the ancient algal ancestors. One group, the **bryophytes**, straddles the boundary between aquatic and terrestrial life, much like the amphibians of the animal kingdom. The other group, the vascular plants, or **tracheophytes**, is completely adapted to life on land.

The Liverworts and Mosses (Division Bryophyta) Are Only Partially Adapted to Dry Land

The 16,000 species of bryophytes show some, but not all, of the adaptations necessary for a completely terrestrial existence. The bryophytes, like the algae, lack true roots, leaves, and stems. Rootlike anchoring structures called **rhizoids** bring water and nutrients into the plant body. Bryophytes lack well-developed structures for conducting water and nutrients. Because they must rely on slow diffusion or poorly developed conducting tissues to distribute water and other nutrients, their body size is limited; most are less than 2 centimeters (less than an inch) tall.

The major bryophyte representatives are the liverworts and mosses (Figs. 23-5 and 23-6). The liverworts and most mosses are confined to moist areas. Some mosses, however, possess a waterproof cuticle that retains moisture and stomata that can be closed, preventing water loss. Some mosses can survive in deserts, on bare rock, and in far northern and southern latitudes where humidity is low and liquid water is scarce for much of the year.

Bryophytes are adapted to terrestrial existence by their enclosed reproductive structures, which prevent the gametes from drying out. These are the **archegonia**, in which eggs develop, and the **antheridia**, where sperm are formed. They may be located on the same plant, or the entire plant may be either male or female, depending on the species. In all bryophytes the sperm must swim to the egg (which emits a chemical attractant) through a film of water. Bryophytes living in dry areas must time their reproduction to coincide with the infrequent rains.

The bryophyte life cycle is shown in Figure 23-6. As in many algae, the larger "leafy" plant body is the haploid gametophyte, which forms sperm and eggs by mitosis. Bryophytes retain the fertilized egg in the archegonium. Here the embryo grows and matures into a small diploid sporophyte that remains attached to the parent gametophyte plant. At maturity, the sporophyte produces haploid spores by meiosis within a capsule. When the capsule is opened, spores are released and dispersed by the wind. If a spore lands in a suitable environment, it may develop into another haploid gametophyte plant.

The Vascular Plants, or Tracheophytes, Have Conducting Vessels That Also Provide Support

Although the bryophytes were quite successful, they left vast areas of the land unoccupied. When the moister regions were covered with the short green fuzz of bryophytes, any plant that could stand taller would benefit by basking in sunlight while shading its short competitors. Natural selection thus favored adaptations that allowed some plants to become taller: support for the body and vessels that conduct water and nutrients absorbed by the roots into the upper portions of the plant. In the **vascular** (from the Latin word for "vessel-bearing") plants, specialized groups of conducting cells (which we will call **vessels**) impregnated with the stiffening substance lignin served both supportive and conducting functions (see Chapter 25).

The Seedless Vascular Plants Include the Club Mosses, Horsetails, and Ferns

The seedless vascular plants reached treelike proportions and dominated the landscape during the Carboniferous period (from 286 to 360 million years ago). Their bodies—transformed by heat, pressure, and time—are burned today as coal. Their modern representatives, the club mosses, horsetails, and ferns, have diminished in size and importance and have been largely replaced by the more versatile seed plants.

The club mosses (division Lycophyta) are now limited to representatives a few centimeters in height (Fig. 23-7a). Their leaves are small and scalelike, resembling the leaflike structures of mosses. Club mosses of the genus *Lycopodium*, commonly known as ground pine, form a beautiful ground cover in some temperate coniferous and deciduous forests.

Modern horsetails (division Sphenophyta) form a single genus, *Equisetum*, with only 15 species, most less than 1 meter tall (Fig. 23-7b). The bushy appearance of some species has given them the common name horsetails. They are also called scouring rushes, because they were used by early settlers to scour pots and floors. All species of *Equisetum* deposit large amounts of silica (glass) in their outer layer of cells, giving them an abrasive texture.

The ferns (division Pteridophyta), with 12,000 species, are far more successful (Fig. 23-7c). In the tropics, "tree ferns" still reach heights reminiscent of their ancestors in the Carboniferous period. Ferns are the only seedless vascular plants that have broad leaves. Broad leaves can capture more sunlight, and this advantage over the small-leaved club mosses and horsetails may account for the relative success of modern ferns.

The life cycle of a fern is depicted in Figure 23-8, p. 448. A major difference between vascular plants and bryophytes is that the diploid sporophyte is dominant in vascular plants. On special leaves of club mosses and ferns, and on conelike structures of horsetails, haploid spores are

Figure 23-5 **Liverworts**

Liverworts grow inconspicuously in moist, shaded areas. This is the female gametophyte plant, bearing umbrella-like archegonia, which hold the eggs. Sperm must swim up the stalks through a film of water to fertilize the eggs.

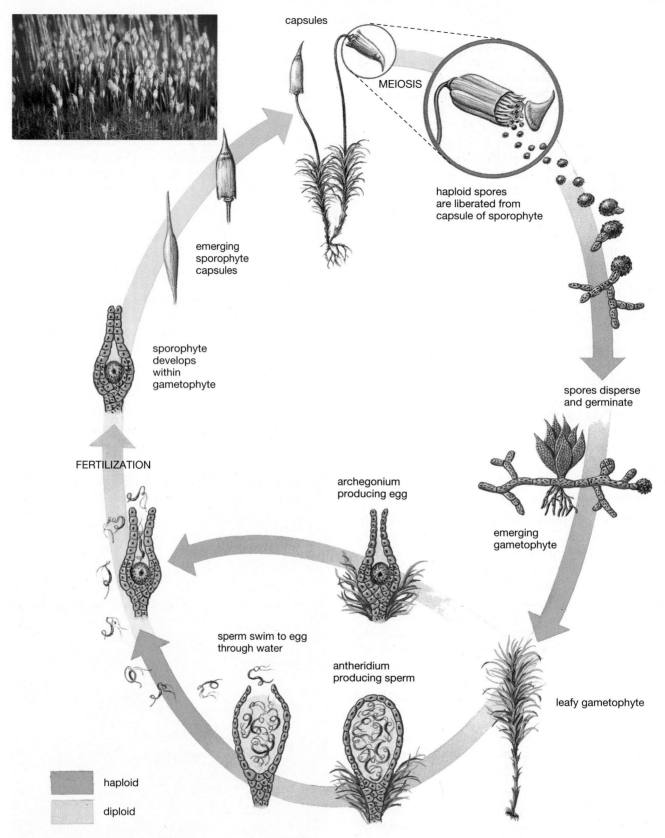

capsules

MEIOSIS

haploid spores
are liberated from
capsule of sporophyte

emerging
sporophyte
capsules

spores disperse
and germinate

sporophyte
develops
within
gametophyte

emerging
gametophyte

FERTILIZATION

archegonium
producing egg

sperm swim to egg
through water

antheridium
producing sperm

leafy gametophyte

haploid

diploid

Figure 23-6   Life cycle of a moss

The life cycle of a moss shows alternation of diploid and haploid generations. The leafy green gametophyte (lower right), which often grows in patches resembling a cushion, is actually the haploid gametophyte generation that produces sperm and eggs. The sperm develop in the antheridium and must swim through a film of water to the egg (which remains in the archegonium where it is formed). The zygote develops into a stalked, diploid sporophyte that emerges from the gametophyte plant. The sporophyte is topped by a brown capsule in which haploid spores are produced by meiosis. These are dispersed and germinate, producing another green gametophyte generation. (Inset) Moss plants showing both stages in the life cycle. The short, leafy green plants are the haploid gametophytes; the reddish brown stalks are the diploid sporophyte generation. The sporophytes are about 1 centimeter (less than half an inch) in height.

produced. These are dispersed by the wind and give rise to tiny, haploid gametophyte plants, which produce sperm and eggs. Two traits in the seedless vascular plants are reminiscent of the bryophytes. First, the small gametophytes lack conducting vessels. Second, as in bryophytes, the sperm must swim through water to reach the egg, so these plants still depend on the presence of water for sexual reproduction.

The Seed Plants Dominate the Land, Aided by Two Important Adaptations: Pollen and Seeds

Seed plants have dominated the land for the past 250 million years. Their success can be attributed to their reproductive versatility. Freed from the requirement of water for reproduction, seed plants have invaded nearly all terrestrial habitats, from swamps to deserts. Two major reproductive adaptations have given seed plants an edge over their seedless competitors: pollen and seeds.

Pollen is all that remains of the male gametophyte of seed plants. In seed plants, both male and female gametophytes (which produce the sex cells) are greatly reduced in size, while the sporophyte is large. The female gametophyte is a small group of haploid cells that produces the egg, and the male gametophyte is the pollen grain. Sperm-producing cells are carried within the pollen grains, which are dispersed by wind or by animal pollinators such as bees. Thus, seed plants are not limited in their distribution by the need for water in which sperm can swim to the egg; they are fully adapted to life on dry land.

(a)

(c)

(b)

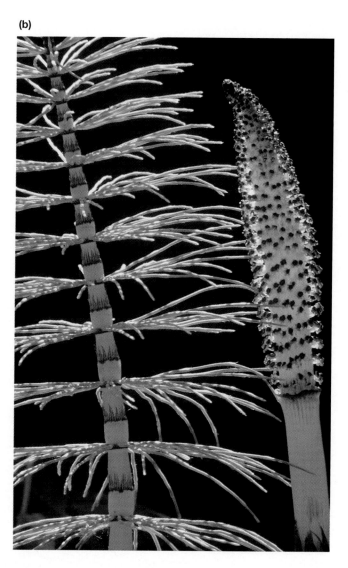

Figure 23-7   **Some seedless vascular plants**

All are found in moist woodland habitats. **(a)** The club mosses (sometimes called ground pines) grow in temperate forests. This specimen is liberating spores. **(b)** The giant horsetail (genus *Equisetum*) extends long, narrow branches in a series of rosettes. Its leaves are reduced to insignificant scales. On special stems, cone-shaped spore-forming structures called strobili are produced. **(c)** This deer fern, whose leaves emerge from coiled fiddleheads, was photographed in Redwood National Park, California.

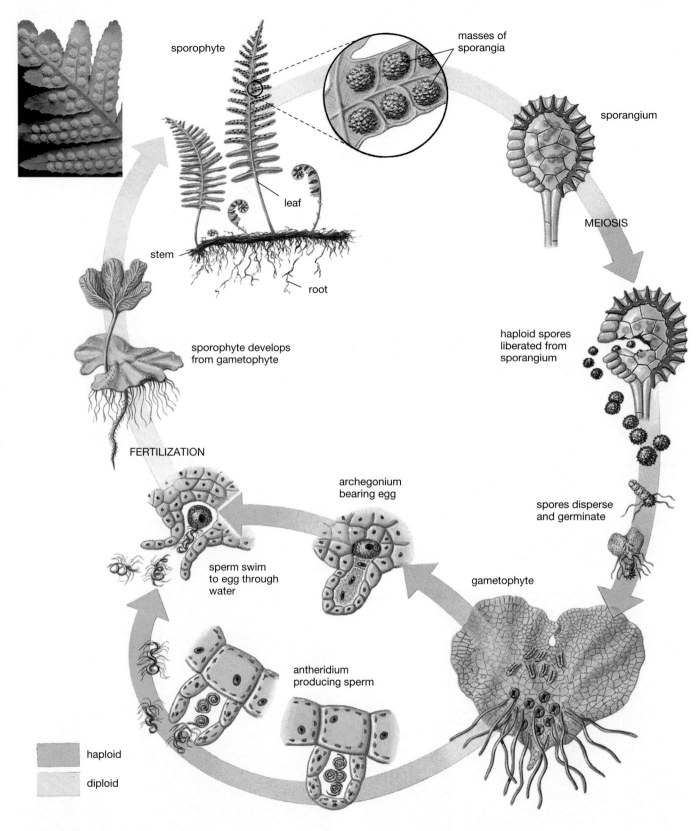

masses of sporangia

sporophyte

leaf

stem

root

sporangium

MEIOSIS

haploid spores liberated from sporangium

sporophyte develops from gametophyte

spores disperse and germinate

FERTILIZATION

archegonium bearing egg

gametophyte

sperm swim to egg through water

antheridium producing sperm

haploid

diploid

Figure 23-8 **Life cycle of a fern**

The fern's life cycle shows alternation of generations. The dominant plant body (upper left) is the diploid sporophyte. Haploid spores, formed in structures called sporangia located on the underside of certain leaves, are dispersed by the wind to germinate on the moist forest floor into inconspicuous haploid gametophyte plants. On the lower surface of these small, sheetlike gametophytes, male antheridia and female archegonia produce sperm and eggs. The sperm must swim to the egg, which remains in the archegonium. The zygote develops into the large sporophyte plant. (Inset) Underside of a fern leaf, showing clusters of sporangia.

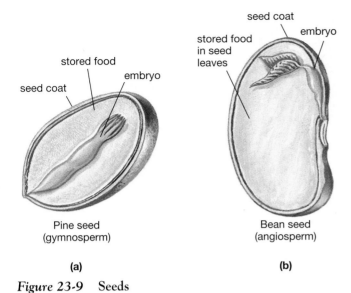

Figure 23-9   **Seeds**

Seeds from **(a)** a gymnosperm (pine) and **(b)** an angiosperm (bean). Both consist of an embryonic plant and stored food confined within a seed coat. In the angiosperm seed, food is stored within large seed leaves, or cotyledons, which take up most of the volume of the seed.

The second reproductive adaptation is the seed itself (seed structure is discussed in detail in Chapter 27). **Seeds** are somewhat similar to the eggs of birds and reptiles in that seeds consist of an embryonic plant, a supply of food for the embryo, and a protective outer coat (Fig. 23-9). The seed coat maintains the embryo in a state of suspended animation or dormancy until conditions are proper for growth. The stored food helps sustain the emerging plant until it develops roots and leaves and can make its own food by photosynthesis. Some seeds possess elaborate adaptations that make possible dispersal by wind, water, and animals. These adaptations have helped them invade nearly every habitat on Earth.

Seed plants may be considered in two general groups: gymnosperms, which lack flowers, and angiosperms, the flowering plants. Although these are not official taxonomic categories, they are useful in organizing our discussion of the seed plants.

Gymnosperms Are Nonflowering Seed Plants

Gymnosperms (whose name means "naked seed" in Greek), evolved earlier than the flowering plants. One group, the **conifers** (division Coniferophyta), with 500 species, still dominates large areas of the planet. Other gymnosperms, such as the cycads and ginkgos (divisions Cycadophyta and Ginkgophyta; Fig. 23-10) have declined to a small remnant of their former range and abundance.

The ginkgos were probably the first of the surviving seed plants to evolve, becoming widespread during the Jurassic period, which began 213 million years ago. Today they are represented by the single species *Ginkgo biloba*, the maidenhair tree. Ginkgo trees are either male or female; the female trees bear foul-smelling, fleshy seeds the size of cherries (see Fig. 23-10b). Ginkgos have been maintained by cultivation, particularly in Asia; if not for cultivation, they might be extinct today. Because they are more resistant to pollution than most other trees, ginkgos (usually male trees) have been extensively planted in American cities.

Cycads resemble large ferns, from which they probably evolved (see Fig. 23-10a). Today there are about 160 species, found mostly in tropical or subtropical climates. Most are

Figure 23-10   **Two uncommon gymnosperms**

(a) A cycad. Common in the age of dinosaurs, these are now limited to about 160 species living in warm, moist climates. **(b)** The ginkgo, or maidenhair tree, has been kept alive by cultivation in China and Japan. Relatively resistant to pollution, these trees have become popular in American cities. Both ginkgos and cycads have separate sexes. The ginkgo shown here is female and bears fleshy seeds the size of large cherries, which are noted for their foul smell when ripe.

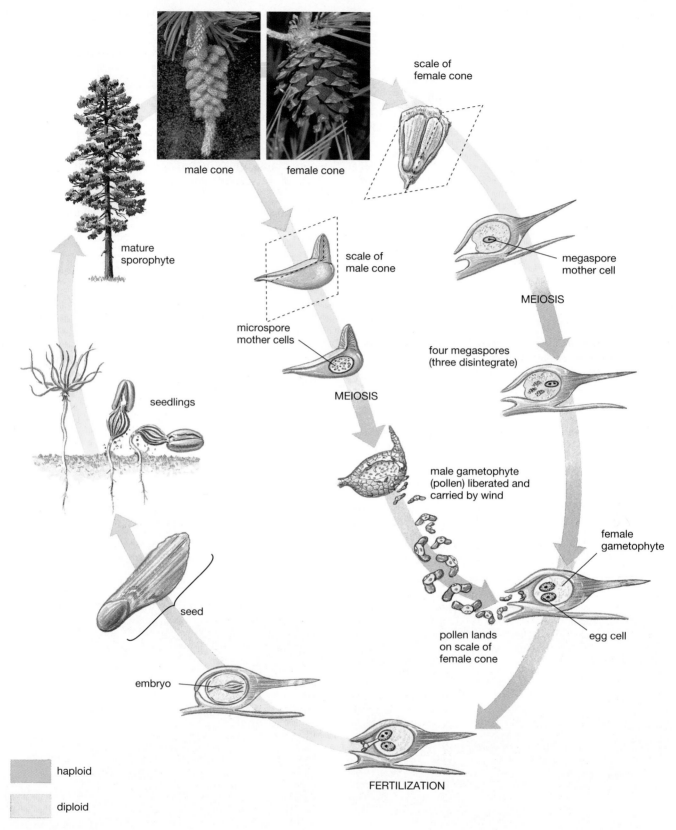

male cone

female cone

scale of
female cone

megaspore
mother cell

MEIOSIS

four megaspores
(three disintegrate)

mature
sporophyte

scale of
male cone

microspore
mother cells

MEIOSIS

male gametophyte
(pollen) liberated and
carried by wind

female
gametophyte

seedlings

pollen lands
on scale of
female cone

egg cell

seed

embryo

FERTILIZATION

haploid

diploid

Figure 23-11   **Life cycle of the pine**

The pine is a seed plant of the division Coniferophyta. The tree is the sporophyte generation (upper left), bearing both male and female cones. Within the ovules at the base of each scale of the female cone, special cells undergo meiosis to form megaspores. One megaspore then develops into the female gametophyte. The female gametophyte in turn produces egg cells. Meanwhile, in the male cones, special cells undergo meiosis to produce the male gametophytes: the pollen. Pollen grains, carrying sperm nuclei, are dispersed by the wind and land on the scales of the female cone. The pollen grows a pollen tube that penetrates the female gametophyte and conducts the sperm to the egg. The fertilized egg then develops into an embryonic plant enclosed in a seed formed from the ovule. The seed is eventually released from the cone, germinates, and grows into a sporophyte tree.

about 1 meter (3 feet) in height, although some species can reach 20 meters (65 feet). Cycads grow slowly and live for a long time; one Australian specimen is estimated to be 5000 years old. The fleshy seeds of cycads were once a staple food in Guam, but they contain a toxin that may cause a neurological disorder resembling Parkinson's disease.

Conifers spread widely as Earth became drier during the Permian period, which followed the Carboniferous period. Today they are most abundant in the cold latitudes of the far north and at high elevations where conditions are rather dry. Not only is rainfall limited in these areas, but soil water remains frozen and unavailable during the long winters. Conifers, including pines, firs, spruce, hemlocks, and cypresses, are adapted to withstand dry, cold conditions in several ways. First, conifers retain green leaves throughout the year, enabling them to continue photosynthesizing and growing slowly during times when most other plants become dormant. Conifers are often called **evergreens** for this reason. Second, conifer leaves are thin needles covered with a thick cuticle whose small waterproof surface minimizes evaporation. Finally, conifers produce a resinous "antifreeze" in their sap that enables them to continue transporting nutrients in subfreezing temperatures. This substance gives them their fragrant "piney" scent.

Reproduction is similar in all conifers, with pines serving as a good illustration (Fig. 23-11). The tree itself is the diploid sporophyte. It develops male and female cones. The male cones are relatively small (usually 2 centimeters, about ¾ of an inch, or less), delicate structures. They release clouds of pollen during the reproductive season and then disintegrate (Fig. 23-11, top). Each pollen grain is a male gametophyte, consisting of several specialized haploid cells, some of which form tiny winglike structures that allow the pollen to be carried by the wind for long distances. Immense clouds of pollen are released by the male cones, so inevitably some land by chance on the female cone. Each female cone consists of a series of woody scales arranged in a spiral around a central axis (Fig. 23-11, top). At the base of each scale are two ovules (immature seeds) within which haploid female gametophytes develop and produce egg cells. A pollen grain landing nearby sends out a pollen tube that slowly burrows into the female gametophyte. After nearly 14 months, the tube finally reaches the egg cell and releases sperm that fertilize it. The fertilized egg becomes enclosed in a seed as it develops into a tiny embryonic plant. The seed is liberated when the cone matures and its scales separate.

Angiosperms Are Flowering Seed Plants

Three major adaptations have contributed to the enormous success of **angiosperms**: flowers, fruits, and broad leaves. The oldest fossils of flowering plants (division Anthophyta) are estimated to be 127 million years old. Angiosperms are believed by some scientists to have evolved from gymnosperm ancestors that formed an association with animals (most likely insects), who carried their pollen from plant to plant. The insects benefited by eating some of the protein-rich pollen, while the plant no longer had to produce prodigious quantities of pollen and send it flying on the fickle winds to ensure fertilization. According to this hypothesis, this relationship between certain types of ancient gymnosperms and their animal pollinators was so beneficial that, through natural selection, plants evolved flowers that attract insects and other animals.

One hundred million years ago, flowering plants already dominated Earth, as they do today. Modern angiosperms are incredibly diverse, with over 230,000 species (Fig. 23-12). They range in size from the diminutive duckweed, a few millimeters in diameter, that floats on ponds to the mighty eucalyptus tree, over 100 meters (328 feet) tall. From desert cactus to tropical orchids to grasses to parasitic mistletoe, angiosperms dominate the plant kingdom. The **flower**, one of their most important adaptations, is discussed in detail in Chapter 27.

The angiosperm life cycle is shown in Figure 23-13. Like their gymnosperm ancestors, angiosperms have a dominant sporophyte plant that produces and nurtures tiny male and female gametophytes. These in turn produce the sex cells. In angiosperms, both male and female gametophytes are formed within the flower (see Chapter 27). Male gametophytes (pollen) are formed inside the anther. Within the ovary of the flower, an ovule produces the female gametophyte in which the egg is formed. Fertilization occurs within the ovule, and the zygote develops into an embryo enclosed in a seed formed from the ovule. The ovary surrounding the seed matures into a **fruit**. The word *angiosperm* (from the Greek for "enclosed seed") refers to the enclosure of the seed within a fruit.

The fruit is a second adaptation that has contributed to the success of angiosperms. Just as flowers encourage animals to transport pollen, so too, many fruits entice them to disperse seeds. These seeds pass through animal digestive tracts unharmed; examples can easily be observed in bird droppings. As dog owners are well aware, some fruits (called burs) disperse by clinging to animal fur. Others, such as the fruits of maples, form wings that carry the seed through the air. The variety of dispersal mechanisms made possible by the fruit has helped the angiosperms invade nearly all possible habitats.

A third feature that gives angiosperms an adaptive advantage in warmer, wetter climates is broad leaves that are shed under adverse conditions, such as periods of cold and drought. When water is plentiful, as it is during the long, warm growing season of temperate and tropical climates, broad leaves give trees an advantage by collecting more sunlight for photosynthesis. The extra energy gained during the spring and summer allows the trees to drop their leaves and enter a dormant period during the fall and winter of temperate climates (or during the dry season in certain tropical climates). In the north, the period of warmth and moisture is considerably shorter than in the tropics. Tropical angiosperm trees remain leafless and dormant for a much longer period than do northern trees, thereby avoiding water loss

Figure 23-12   **Diverse angiosperms**

Both grasses **(a)** and many trees, such as this birch **(b)**, in which flowers are shown as buds (green) and blossom (brown), have inconspicuous flowers and rely on wind for pollination. **(c)** Flowers, such as those shown on this hedgehog cactus and eucalyptus tree (inset in part e), entice insects and other animals that carry pollen between individual plants. **(d)** The smallest angiosperm is the duckweed, found floating on ponds. These specimens are about 3 millimeters (⅛ inch) in diameter. **(e)** The largest angiosperms are eucalyptus trees, which may reach 45 meters (150 feet) in height.

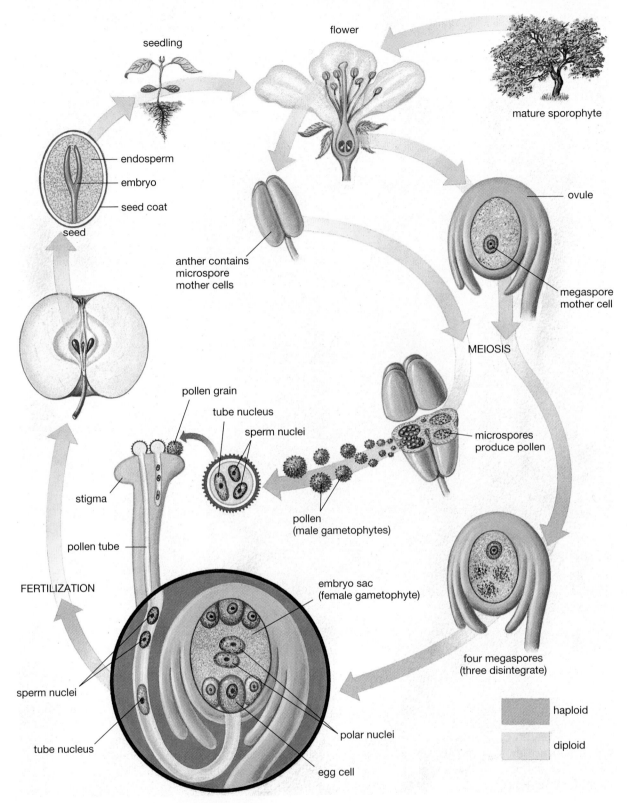

Figure 23-13 **Life cycle of a flowering plant**

The dominant plant (upper right) is the diploid sporophyte, whose flowers usually produce both male and female gametophytes. Male game-tophytes (pollen grains) are produced within the anthers, where special cells undergo meiosis to produce haploid microspores. These divide mitotically to produce pollen, in which a tube nucleus and two sperm nuclei are formed. The female gametophyte develops within the ovule of the ovary. Here special cells undergo meiosis to produce a haploid megaspore. The megaspore divides mitotically to produce the female ga-metophyte (also called the _embryo sac_), whose contents include one egg cell and two polar nuclei. After a pollen grain lands on the stigma of the carpel, it produces a pollen tube that burrows through the style to the ovary and into the female gametophyte. There it releases its two sperm nuclei. One sperm nucleus fuses with the egg to form a zygote. The second fuses with the polar nuclei and forms the endosperm, a source of food for the developing embryo, in a process called double fertilization. The ovule gives rise to the seed, which contains the en-dosperm and an embryo that develops from the zygote. The seed is dispersed, germinates, and develops into a mature sporophyte.

Figure 23-14 Two ways of coping with the dryness of winter

The evergreen (a conifer) retains its needles throughout the year. The small surface area and heavy cuticle of the needles slow loss of water through evaporation. In contrast, the aspen (an angiosperm) sheds its leaves each fall. The dying leaves turn brilliant shades of gold as pigments used to capture light energy for photosynthesis are exposed when the chlorophyll disintegrates.

through their broad leaves. The water-conserving conifers continue photosynthesizing and growing slowly during the winters, and for this reason they dominate in northern ecosystems (Fig. 23-14; see also Chapter 46).

The flowering plants may be grouped into two broad classes based on their internal and external structure, discussed in more detail in Chapter 25. The **monocots** (class Monocotyledoneae) are a group of about 65,000 species including grasses, corn and other grains, irises, lilies, and palms. The **dicots** (class Dicotyledoneae) are a considerably larger group with about 170,000 species, including most of the angiosperm trees, shrubs, and herbs.

❖ SUMMARY OF KEY CONCEPTS

The kingdom Plantae consists of eukaryotic, photosynthetic, usually multicellular organisms. The ability of plants and other photosynthetic organisms to capture the energy of sunlight in high-energy molecules provides nearly all other forms of life on Earth with a source of usable energy.

Evolutionary Trends in Plants

The first algae appeared between 500 and 600 million years ago. Green algal ancestors probably gave rise to two groups of terrestrial plants: bryophytes, restricted to moist environments and lacking many adaptations to land environments; and vascular plants with specialized vessels that also provide support, which allowed them to flourish on land.

In the course of evolution, plants became increasingly adapted to terrestrial existence. They developed (1) rootlike structures for anchorage and for absorption of water and nutrients; (2) conducting vessels to transport water and nutrients throughout the plant; (3) a stiffening substance, called lignin, to impregnate the vessels and support the plant body; (4) a waxy cuticle to slow loss of water through evaporation; (5) stomata that can open, allowing gas exchange, and can also close, preventing water loss.

As plants invaded the land, new reproductive adaptations evolved. Reduction of the male gametophyte to pollen allowed wind to replace water as a means of carrying sperm to eggs. Flowers attracted animals, who carry pollen more precisely and efficiently than wind, while fruit enticed animals to disperse seeds. The seed nourishes, protects, and helps disperse the developing embryo.

Plants exhibit alternation of generations in which a haploid gametophyte generation alternates with a diploid sporophyte generation. There has been a general evolutionary trend toward reduction of the haploid gametophyte, which is dominant in bryophytes but microscopic in seed plants.

Watery Origins—The Algae

The simplest plants are the aquatic algae. There are three major divisions of algae. The red algae (Rhodophyta) dominate in clear tropical waters, and the brown algae (Phaeophyta) populate temperate oceans. The green algae (Chlorophyta) are primarily small, freshwater forms believed to be ancestral to modern land plants. Algal divisions are named for their predominant colors, which result from a combination of green chlorophyll and light-trapping pigments. Algae lack true roots, stems, and leaves and rely on the water to carry sex cells.

Land—The New Frontier

Two major groups of plants arose from these ancient algal ancestors, the bryophytes and the tracheophytes. Bryophytes, which include the liverworts and mosses, are small, simple land plants that lack conducting vessels. Although some have adapted to dry areas, most are found in moist habitats. Reproduction in bryophytes requires water through which the sperm must swim to the egg.

In tracheophytes, or vascular plants, a system of vessels has evolved, stiffened by lignin, that conducts water and nutrients absorbed by the roots into the upper portions of the plant and supports the body as well. Owing to this support system seedless vascular plants, including the club mosses (division Lycophyta), horsetails (division Sphenophyta), and ferns (division Pterophyta), can grow larger than bryophytes. As in bryophytes, the sperm must swim to the egg for sexual reproduction to occur, and the gametophyte lacks conducting vessels.

Vascular plants with seeds have two major new adaptive features: pollen and seeds. They are often classified into two categories: gymnosperms and angiosperms. Gymnosperms include ginkgos, cycads, and the highly successful conifers. These plants were the first fully terrestrial plants to evolve. Their success on dry land is partially due to the evolution of the male gametophyte into the pollen grain. Pollen protects and transports the male gamete, eliminating the need for the sperm to swim to the egg. The seed, a protective resting structure containing an embryo and a supply of food, is a second important adaptation contributing to the success of seed plants.

Angiosperms, the flowering plants, dominate much of the land today. In addition to pollen and seeds, angiosperms also produce flowers and fruits. The flower allows angiosperms to utilize animals as pollinators. In contrast to wind, animals can carry pollen longer distances with greater accuracy and less waste. Fruits may attract animal consumers, who incidentally disperse the seeds in their feces.

✻ KEY TERMS

accessory pigment   p. 443
alga   p. 440
alternation of generations   p. 440
angiosperm   p. 451
antheridium   p. 445
archegonium   p. 445
bryophyte   p. 444
conifer   p. 449
cuticle   p. 440

dicot   p. 454
evergreen   p. 451
flower   p. 451
fruit   p. 451
gametophyte   p. 440
gymnosperm   p. 449
lignin   p. 440
monocot   p. 454
pollen   p. 447

rhizoid   p. 444
seed   p. 449
sporophyte   p. 440
stoma   p. 440
tracheophyte   p. 444
vascular   p. 445
vessel   p. 445
zoospore   p. 440
zygote   p. 440

✖ THINKING THROUGH THE CONCEPTS

Multiple Choice

1. Which of the following groups of organisms live primarily in freshwater habitats?
   a. red algae   b. brown algae
   c. green algae   d. liverworts
   e. bryophytes
2. In which of the following plants is the gametophyte the dominant generation?
   a. mosses   b. ferns
   c. pine trees   d. sunflowers
   e. The sporophyte is dominant in all of the above.
3. What is the function of a fruit?
   a. It attracts pollinators.
   b. It provides food for the developing embryo.
   c. It stores excess food produced by photosynthesis for later use.
   d. It helps ensure the dispersal of seeds from the parent plant.
   e. It evolved so that people would cultivate the plant, assuring its survival.
4. What is the function of lignin?
   a. to provide support for the plant
   b. to waterproof plant surfaces
   c. to store food
   d. to promote gas exchange in plant leaves
   e. to transport dissolved nutrients
5. Which of the following plants produce sperm that swim to the egg?
   a. sugar maple tree   b. Douglas fir tree, a conifer
   c. rattlesnake fern   d. common dandelion
   e. none of the above
6. Which of the following is the correct sequence during alternation of generations?
   a. sporophyte—diploid spores—gametophyte—haploid gametes
   b. sporophyte—haploid spores—gametophyte—haploid gametes
   c. sporophyte—haploid gametes—gametophyte—haploid spores
   d. sporophyte—haploid gametes—gametophyte—diploid spores
   e. sporophyte—diploid gametes—gametophyte—diploid spores

Review Questions

1. What is meant by "alternation of generations"? What two generations are involved? How does each reproduce?
2. Explain the evolutionary changes in plant reproduction that adapted plants to increasingly dry environments.
3. Describe evolutionary trends in the life cycles of plants, with emphasis on the relative sizes of the gametophyte and sporophyte.
4. Assuming that green algae have fewer accessory pigments than red or brown algae, would you expect to find them in shallow or deep water? Where would you find the red algae, and why?
5. From which algal division did green plants probably arise? Explain the evidence supporting this hypothesis.
6. List the various structural adaptations necessary for the invasion by plants of dry habitats on land. Which of these are possessed by bryophytes? By ferns? By gymnosperms and angiosperms?
7. The number of species of flowering plants is greater than the number of species in the rest of the plant kingdom. What feature(s) are responsible for the enormous success of angiosperms? Explain why.
8. List the adaptations of gymnosperms that have helped them become the dominant tree in dry, cold climates.
9. What is a pollen grain? What role has it played in helping plants colonize dry land?
10. The majority of all plants are seed plants. What is the advantage of a seed? How do plants that lack seeds meet the needs served by seeds?

✖ APPLYING THE CONCEPTS

1. If you were a geneticist working for a firm specializing in plant biotechnology, explain what *specific* parts (fruit, seeds, stems, roots, etc.) of the following plants you would try to alter by genetic engineering, what changes you would try to make, and why, on (a) corn, (b) tomatoes, (c) wheat, and (d) avocados.
2. Prior to the development of synthetic drugs, over 80% of all medicines were of plant origin. Even today, indigenous tribes in remote Amazonian rain forests can provide a plant product to treat virtually any ailment. Herbal medicine is also widely and successfully practiced in China. Most of these drugs are unknown to the Western world. Unfortunately, the forests from which much of this plant material is obtained are being converted to agriculture. We are in danger of losing many of these potential drugs before they can be discovered. What steps can you suggest to preserve these natural resources while also allowing nations to direct their own economic development?

3. Only a few hundred of the hundreds of thousands of species in the plant kingdom have been domesticated for human use. One example is the almond. The domestic almond is nutritious and harmless, but its wild precursor can cause cyanide poisoning. The oak makes potentially nutritious seeds (acorns) that contain very bitter-tasting tannins. If we could breed the tannin out of acorns, they might become a delicacy. Why do you suppose we have failed to domesticate oaks?

✖ FOR MORE INFORMATION

Cox, P. A., and Balick, M. J. "The Ethnobotanical Approach to Drug Discovery." *Scientific American*, June 1994. Biologists seek new pharmaceutical compounds by analyzing the plants used as drugs by indigenous cultures.

Diamond, J. "How to Tame a Wild Plant." *Discover*, September 1994. Cultivated plants have ecological and genetic properties that make them well suited for agriculture.

Doyle, J. "DNA, Phylogeny, and the Flowering of Plant Systematics." *BioScience*, June 1993. Chloroplast DNA is used in reconstructing evolutionary relationships, but there are difficulties with each of the methods in use.

Grant, M. C. "The Trembling Giant." *Discover* 1993. The largest living thing is a clone of 47,000 aspen trees covering 106 acres in the Wasatch Mountains of Utah.

Kaufman, P. B. *Plants—Their Biology and Importance*. New York: Harper and Row, 1989. Complete, readable coverage of all aspects of plant taxonomy, physiology, and evolution.

Milot, V. "Blueprint for Conserving Plant Diversity." *BioScience*, June 1989. Points out the importance of preserving genetic diversity in endangered plant species.

Nicholson, R. "Death and Taxus." *Natural History*, September 1992. The bark of the yew tree contains compounds that may help in the treatment of cancer.

NET WATCH

On-line resources for this chapter are on the World Wide Web at:

http://www.prenhall.com/~audesirk (click on the <u>table of contents</u> link and then select Chapter 23).

24 The Animal Kingdom

⚏ A T A G L A N C E

The Features of Animals

Evolutionary Trends in Animal Body Plans

Over Evolutionary Time, Animals Have Increased in
 Complexity

Animal Phyla Show Trends toward Increasing Cellular
 Organization

Animal Phyla Show General Trends in Body Symmetry

Cephalization Increased over Evolutionary Time

A Coelom Evolved in More Complex Animals

Segmentation Is First Seen in Annelid Worms

Digestive Systems Increased in Complexity

The Major Animal Phyla

The Sponges: Phylum Porifera

The Hydra, Anemones, and Jellyfish: Phylum Cnidaria

The Flatworms: Phylum Platyhelminthes

The Roundworms: Phylum Nematoda

The Segmented Worms: Phylum Annelida

**The Insects, Arachnids, and Crustaceans: Phylum
Arthropoda**

Insects Are the Most Diverse and Abundant Arthropods

Spiders, Scorpions, and Their Relatives Are Members of the
 Class Arachnida

Crabs, Shrimp, Crayfish, and Their Relatives Are Members
 of the Class Crustacea

The Snails, Clams, and Squid: Phylum Mollusca

Snails and Their Relatives Are Members of the Class
 Gastropoda

Scallops, Clams, Oysters, and Their Relatives Are Members
 of the Class Bivalvia

Octopuses, Squid, and Their Relatives Are Members of the
 Class Cephalopoda

**The Sea Stars, Sea Urchins, and Sea Cucumbers:
Phylum Echinodermata**

**The Tunicates, Lancelets, and Vertebrates: Phylum
Chordata**

The Invertebrate Chordates Include Lancelets and Tunicates

The Vertebrates Have a Backbone and Other Adaptations
 That Have Contributed to Their Success

Success on Dry Land Required Numerous Adaptations

**Evolutionary Connections: Are Humans a Biological
Success?**

*Resembling an undersea flower, a coral extends tentacles to
snare passing food. When disturbed, the coral can retract into a
limestone shell. These shells remain after the death of the coral,
forming the basis for coral reefs in tropical oceans.*

The earliest animal fossils are found in late Precambrian rock 630 million years old. Some resemble jellyfish, others, segmented worms, and still others resemble creatures from a science fiction writer's dreams. All clearly had been evolving as animals for a long time. The transition from protists to multicellular animals is still lost in time. The most widely accepted theory is that animals evolved from zooflagellate protists that formed simple colonies. The simplest animals, the sponges, resemble such colonies. Out of these loosely organized cells have evolved organisms of incredible variety and diversity, animals with a vast array of appearances, behaviors, and levels of complexity. In spite of these differences, all animals share common features, which we discuss below.

The Features of Animals

Animals have several features, which, taken together, distinguish them from all other kingdoms.

1. Animals are multicellular.
2. Animals are heterotrophic—that is, they obtain their energy by consuming the bodies of other organisms.
3. Animals can reproduce sexually. Although animal species exhibit a tremendous diversity of reproductive styles, all are capable of sexual reproduction.
4. Animal cells lack a cell wall.
5. Animals are motile during some stage of their life. Even the stationary sponges have a free-swimming larval stage.
6. Animals are usually able to make rapid responses to external stimuli as a result of the activity of nerve cells, muscle or contractile tissue, or both.

Evolutionary Trends in Animal Body Plans

Over Evolutionary Time, Animals Have Increased in Complexity

In our survey of the animal kingdom, we will present the major phyla in an order that approximates the sequence in which they evolved. As we progress from the

sponges through the cnidarians (jellyfish and their relatives) and through the three phyla of worms, you will see
a clear evolutionary trend toward increasing complexity.
This complexity is reflected in the following specific adaptations: increasing cellular organization and specialization;
bilateral symmetry (described below); **cephalization** (the
concentration of sensory organs and a brain in a defined
head region); the evolution of a body cavity; **segmentation**
(repeated, similar body parts); and the evolution of a digestive system in which material flows in a single direction.

Some of these features are used in the simple evolutionary
groupings of animals shown in Figure 24-1.

The trend toward greater complexity culminates in the
arthropods, mollusks, echinoderms, and chordates (see
Table 24-1). Although there are striking structural differences among these four phyla, the differences reflect adaptations to different environments and life-styles rather than
further increases in complexity. Within the chordates, however, another trend toward greater complexity is clear: an increase in the size and sophistication of the brain.

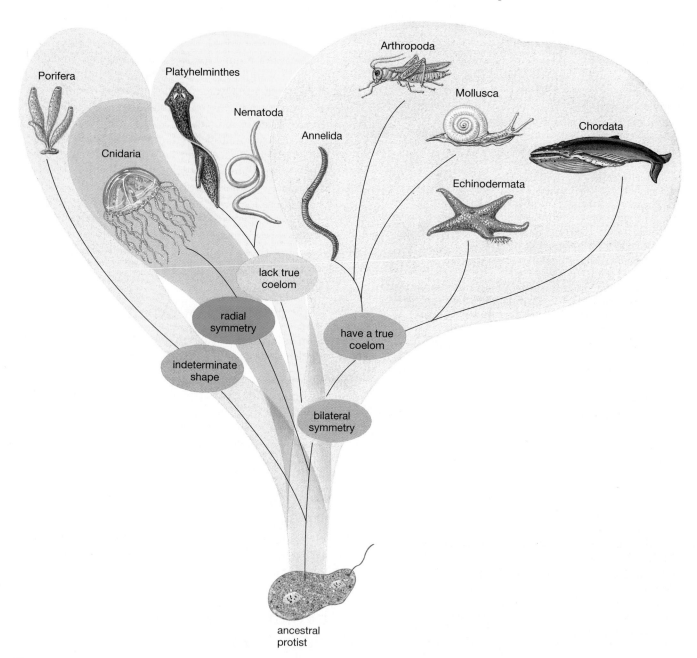

Figure 24-1   **A simple classification scheme for animals**

This classification scheme is based on anatomical features. Animals show evolutionary trends
from indeterminate shape to radial symmetry to bilateral symmetry. Bilaterally symmetrical
animals form two groups: The simplest lack a body cavity called a coelom, while more complex phyla have a coelom.

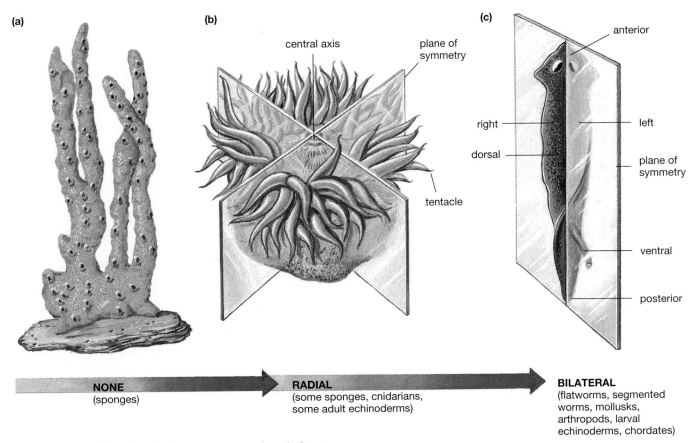

(a)

(b)

central axis

plane of symmetry

tentacle

(c)

anterior

right

left

dorsal

plane of symmetry

ventral

posterior

NONE
(sponges)

RADIAL
(some sponges, cnidarians, some adult echinoderms)

BILATERAL
(flatworms, segmented worms, mollusks, arthropods, larval echinoderms, chordates)

Figure 24-2   **Trends in body symmetry and cephalization**

(a) Sponges, the simplest animals, lack a head and most are asymmetrical. **(b)** Some sponges, cnidarians (anemones, hydra), and some adult echinoderms (sea urchins, sea stars) have bodies that are radially symmetrical. Any plane that passes through the central axis divides the body into mirror-image halves. Animals in these groups lack a well-defined head. **(c)** Nearly all the more complex animals, starting with flatworms (phylum Platyhelminthes) show bilateral symmetry. The body can be split into two mirror-image halves by a single plane running down the midline. Animals with bilateral symmetry also have an anterior head end, a posterior tail end, a dorsal upper surface, and a ventral underside.

Animal Phyla Show Trends toward Increasing Cellular Organization

In the course of animal evolution, specialized cells became organized into **tissues**—for example, groups of muscle cells. Various tissues combined to form **organs**: two or more tissues integrated to perform a specialized function, illustrated by the kidney and the eye. Organs, in turn, formed **organ systems**: two or more organs that work together to perform a specific function, as illustrated by the digestive or excretory system.

Although sponges have specialized types of cells, the individual cells act more or less independently and are not organized into true tissues or organs. Cnidarians show well-defined tissues, such as their nerve net, which coordinates movement and sensory information, but they lack organs. The flatworms possess not only organs, such as gonads and eyespots, but also an organ system, the reproductive system. Organ systems are found in all the more complex animals.

A second trend in cellular organization is in the number of tissue layers, called **germ layers**, that arise during embryonic development. Flatworms and all the more complex animals have three germ layers: an inner layer of **endoderm** (forming the lining of most hollow organs), a middle layer of **mesoderm** (forming muscle, and, when present, the circulatory and skeletal systems), and an outer layer of **ectoderm** (forming epithelial tissue that covers the body and lines its inner cavities and nervous tissue composed of nerve cells). The more primitive cnidarians, in contrast, have only two germ layers: endoderm and ectoderm.

Animal Phyla Show General Trends in Body Symmetry

The simplest animals, the sponges, have many representatives whose bodies are irregular and variable in shape, even within a single species. Other types of sponges show **radial symmetry**, in which any line through a central axis divides the animal into roughly equal halves (Fig. 24-2). This was the first type of symmetry to evolve in animals; it is also seen in cnidarians and in some adult echinoderms. **Bilateral symmetry** evolved later and is characteristic of all

TABLE 24-1 ▦ *Comparison of the Major Animal Phyla*

Common name (Phylum)		*Sponges* (*Porifera*)	*Hydra, Anemones, Jellyfish* (*Cnidaria*)	*Flatworms* (*Platyhelminthes*)	*Roundworms* (*Nematoda*)
Body Plan	**Level of organization**	Cellular—lack tissues and organs	Tissue—lack organs	Organ system	Organ system
	Germ layers	Absent	Two	Three	Three
	Symmetry	Absent	Radial	Bilateral	Bilateral
	Cephalization	Absent	Absent	Present	Present
	Body cavity	Absent	Absent	Absent	Pseudocoel
	Segmentation	Absent	Absent	Absent	Absent
Internal Systems	**Digestive system**	Intracellular	Gastrovascular cavity; some intracellular	Gastrovascular cavity	Separate mouth and anus
	Circulatory system	Absent	Absent	Absent	Absent
	Respiratory system	Absent	Absent	Absent	Absent
	Excretory system (fluid regulation)	Absent	Absent	Canals with flame cells	Excretory gland cells
	Nervous system	Absent	Nerve net	Head ganglia with longitudinal nerve cords	Head ganglia with dorsal and ventral nerve cords
	Reproduction	Sexual; asexual (budding)	Sexual; asexual (budding)	Sexual (some hermaphroditic); asexual (body splits)	Sexual (some hermaphroditic)
	Support	Endoskeleton of spicules	Hydrostatic skeleton	Absent	Hydrostatic skeleton

more complex animals, including larval echinoderms. A bilaterally symmetrical animal can be divided by a single line into roughly mirror-image right and left halves, with an upper, or **dorsal**, surface and a lower, or **ventral**, surface.

Cephalization Increased over Evolutionary Time

Bilateral symmetry is accompanied by cephalization, which produces an **anterior** (head) end and a **posterior** (tail) end (see Fig. 24-2). During the evolution of cephalization, sensory cells, sensory organs, clusters of nerve cells, and organs for ingesting food were increasingly con-

centrated at the anterior end of the animal. The flatworms are the simplest animals to show cephalization. Although they have a defined head with sensory organs, they still ingest food through a muscular tube located near the middle of their bodies. More complex animals have well-defined heads containing a brain and bearing important sensory structures.

A Coelom Evolved in More Complex Animals

As animals became more complex, they evolved a cavity between the gut, where food is digested and absorbed, and the body wall, called a **coelom** (Fig. 24-3). Simple animals

Segmented Worms (Annelida)	Insects, Arachnids, Crustaceans (Arthropoda)	Snails, Clams, Squid (Mollusca)	Sea Stars, Sea Urchins (Echinodermata)	Vertebrates (Chordata)
Organ system	Organ system	Organ system	Organ system	Organ system
Three	Three	Three	Three	Three
Bilateral	Bilateral	Bilateral	Bilateral larvae, radial adults	Bilateral
Present	Present	Present	Absent	Present
Coelom	Coelom	Coelom	Coelom	Coelom
Present	Present	Absent	Absent	Present (but reduced)
Separate mouth and anus	Separate mouth and anus	Separate mouth and anus	Separate mouth and anus (usually)	Separate mouth and anus
Closed	Open	Open	Absent	Closed
Absent	Tracheae, gills, or book lungs	Gills, lungs	Tube feet, skin gills, respiratory tree	Gills, lungs
Nephridia	Excretory glands resembling nephridia	Nephridia	Absent	Kidneys
Head ganglia with paired ventral cords; ganglia in each segment	Head ganglia with paired ventral nerve cords; ganglia in segments, some fused	Well-developed brain in some cephalopods; several paired ganglia, most in the head; nerve network in body wall	Head ganglia absent; nerve ring and radial nerves; nerve network in skin	Well-developed brain; dorsal nerve cord
Sexual (some hermaphroditic)	Usually sexual	Sexual (some hermaphroditic)	Sexual (some hermaphroditic); asexual by regeneration (rare)	Sexual
Hydrostatic skeleton	Exoskeleton	Hydrostatic skeleton	Endoskeleton of plates beneath outer skin	Endoskeleton of cartilage or bone

such as cnidarians and flatworms lack any internal space between their body wall and their gut. Roundworms have a space, called a **pseudocoel**, between the body wall and gut. The prefix *pseudo* (meaning "false") refers to the fact that this opening has a different embryological origin than a true coelom. In earthworms and nearly all of the more complex animals, a true coelom, which can serve a variety of functions, is present. In the earthworm, this fluid-filled cavity acts as a kind of skeleton, providing support for the body and a framework against which muscles can act. Internal organs bulge into the coelom, which serves as a protective buffer between them and the outside world. Its presence has allowed the internal organs such as the

heart and digestive tract to move independently of the body wall. Thanks to your coelom, you may remain externally inactive after a meal even though your digestive tract is churning energetically.

Segmentation Is First Seen in Annelid Worms

Segmentation, or the presence in the body of similar repeated units, is first seen in the annelids (earthworms and their relatives; see Fig. 24-13). Segmentation appears to be an evolutionary device for increasing body size using a minimum of new genetic information, since the genetic "blueprint" for each segment is similar. In more complex animals,

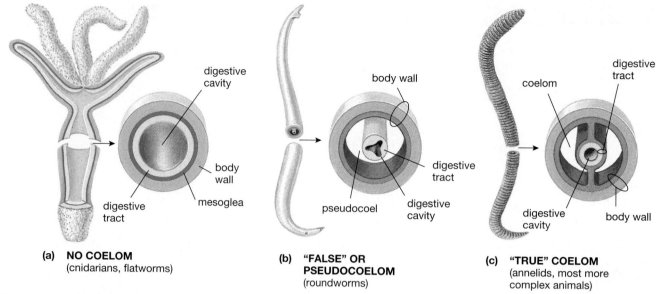

| (a) **NO COELOM**
(cnidarians, flatworms) | (b) **"FALSE" OR
PSEUDOCOELOM**
(roundworms) | (c) **"TRUE" COELOM**
(annelids, most more
complex animals) |

Figure 24-3   **Trends in body cavities**

(a) Cnidarians and flatworms lack the space between the body wall and digestive tract called a coelom. **(b)** Roundworms have a space, called a pseudocoel, between the body wall and digestive tract. The pseudocoel resembles a coelom, but has a different embryological origin. **(c)** Annelids and other complex animals have their digestive tracts and other internal organs suspended within a coelom.

the repeated segments have become specialized for specific functions or are visible only in certain body parts, such as the series of similar vertebrae in the vertebrate backbone.

Digestive Systems Increased in Complexity

The simplest form of digestion is found in the sponges, which lack any specialized digestive tract. In sponges, digestion is entirely intracellular; individual cells trap, ingest, and digest smaller unicellular organisms. The cnidarians and flatworms have a digestive system consisting of a sac with a single opening (see Figs. 24-7 and 24-9). This opening, politely called a mouth, also serves as an anus through which undigested material is expelled.

In the roundworms and all of the more complex animals, an efficient, tubular, one-way digestive system has evolved. A mouth, located at the anterior end near the sensory structures, allows food to be ingested as it is detected. The food is then processed in stages as it passes through a series of specialized regions. First it is physically broken down, then enzymatically digested, then absorbed. Wastes are voided through a separate anus, usually located near the posterior end of the animal.

The Major Animal Phyla

For convenience, biologists often place animals in one of two major categories: **vertebrates**, those with a backbone, or vertebral column, and **invertebrates**, those

lacking a backbone. You are probably most familiar with the vertebrates: fish, amphibians, reptiles, birds, and mammals. The invertebrates include everything else, from sponges to worms to snails to insects. Our human bias is clearly reflected in these categories. Over 97% of all the animal species on Earth are invertebrates, including 27 different phyla, while the vertebrates constitute only part of a single phylum, the phylum Chordata.

The first animals to evolve were invertebrates. They probably originated from colonies of protozoa whose members had become specialized to perform distinct roles within the colonial body. In our survey of the kingdom Animalia, we will begin with the sponges, whose body plan most closely resembles the probable ancestral protozoan colonies. Our discussion will roughly follow the order in which the various phyla of animals appeared on Earth, ending with the vertebrates in the phylum Chordata. The features of the major animal phyla are compared in Table 24-1.

The Sponges: Phylum Porifera

Sponges are the simplest multicellular animals, lacking true tissues and organs. They resemble colonies in which single-celled organisms live together for mutual benefit. The simple cellular organization of sponges was discovered in an experiment by the embryologist H. V. Wilson in 1907. He mashed a sponge through a piece of silk, microorganisms are filtered out and taken into individual cells where they are digested, and wastes are released. Sponges have three major cell types (see Fig. 24-5), each with a specialized role. Flattened **epithelial cells** cover their outer body surfaces. Some epithelial cells are modified into pore cells that surround pores, controlling their size and regulating the flow of water. The pores are closed when harmful substances are present. **Collar cells** maintain a flow of water through the sponge by beating a flagellum that extends into the inner canal. The collar that surrounds the flagellum acts as a fine sieve, filtering out microorganisms that are then ingested by the cell. Some of the food is passed to the **amoeboid cells.** These cells roam freely between the epithelial and collar cells, digesting and distributing nutrients, producing reproductive cells, and secreting small skeletal elements called **spicules.**

Sponges may grow to over a meter in height, and an internal skeleton, which is composed of spicules, provides support for the body (see Fig. 24-5). The spicules may be formed from calcium carbonate (chalk), silica (glass), or protein. The natural bath sponge, now rarely used, is a proteinaceous sponge skeleton.

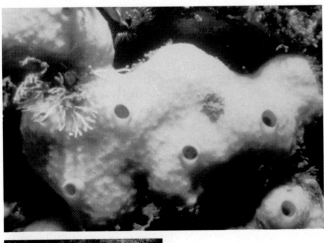

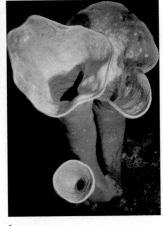

Figure 24-4   The diversity of sponges

Sponges come in a wide variety of sizes, shapes, and colors. Some are over a meter tall; others grow in free-form pattern over undersea rocks.

thereby dissociating it into single cells and cell clusters. He then placed them in seawater for 3 weeks, after which time the cells had reaggregated into a functional sponge. This experiment indicated that individual sponge cells were able to survive and function independently.

Although some sponges have a definite size and shape, others grow in free-form shape over rocks in their aquatic habitats (Fig. 24-4). All sponges have a similar general body plan (Fig. 24-5). The body is perforated by numerous tiny pores (called ostia), through which water enters, and by fewer, large openings (called oscula), through which it is expelled. Within the sponge, water travels through canals. During its passage, oxygen is extracted,

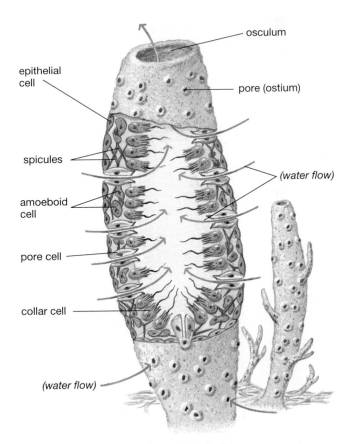

Figure 24-5   The body plan of sponges

Sponges all have a similar body plan. Currents created by collar cells draw water in through numerous tiny pores (ostia). Microscopic food particles are filtered out by collar cells and shared among the various cell types. Water exits through larger pores, the oscula. Spicules form a supportive internal skeleton.

Over 5000 species of sponges have been identified; all are aquatic and most are marine. All adult sponges are **sessile** (fixed to one spot), attaching themselves permanently to rocks or other underwater surfaces. Sponges may reproduce asexually by **budding**, in which the adult produces miniature versions of itself that drop off and assume an independent existence, or sexually through the fusion of sperm and eggs. Fertilized eggs develop inside the adult into active larvae that escape through the oscula. Water currents disperse the larvae to new areas, where they settle and develop into adult sponges.

The Hydra, Anemones, and Jellyfish: Phylum Cnidaria

The cnidarians are clearly a step above the sponges in complexity. Their cells are organized into distinct tissues, including contractile tissue that acts like muscle. The nerve cells are organized into tissue called a **nerve net** that branches through the body and controls the contractile tissue to bring about movement and feeding behavior. Cnidarians generally lack true organs, however, and have no brain.

Cnidarians come in a bewildering and beautiful variety of forms (Fig. 24-6), all of which are actually variations on two basic body plans: the **polyp** and the **medusa**, illustrated in Figure 24-7. The generally tubular polyp, with its foot attached and its **tentacles** reaching upward, is adapted to a life spent quietly attached to rocks, where it awaits prey like a predatory flower. Although the medusa ("jellyfish") swims weakly by contracting its bell-shaped body, it is primarily carried by ocean currents, trailing its tentacles like multiple fishing lines. Both polyp and medusa are radially symmetrical, with body parts arranged in a circle around an axis drawn through the mouth and digestive cavity (see Fig. 24-2). This arrangement of parts is well suited to these animals that are either sessile or carried randomly by water currents, because they are prepared to capture prey or defend themselves from any direction.

(a)

(b)

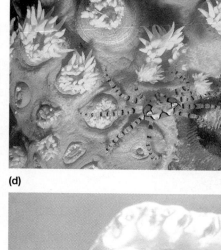

(c)

(d)

Figure 24-6   Cnidarian diversity

(a) A red-spotted anemone spreads its tentacles to capture prey. These large anemones are found in the ocean off Washington State. **(b)** A close-up of coral reveals bright yellow polyps in various stages of tentacle extension. At the lower right, areas where the coral has died expose the calcium carbonate skeleton that supports the polyps and forms the reef. A strikingly patterned crab (phylum Arthropoda: class Crustacea) sits atop the coral, holding tiny white anemones in its claws. Their stinging tentacles help protect the crab. **(c)** A small medusa from the ocean off southern California. **(d)** The Portuguese Man-of-War, a cnidarian whose stings are dangerous to human beings. A stunned fish is trapped in its tentacles.

Although all cnidarians are predatory, none hunt actively. Instead, they rely on their victims' blundering by chance into the grasp of their enveloping tentacles. Cnidarian tentacles are armed with **cnidocytes**, cells containing poisonous or sticky darts that are injected explosively into prey upon contact (Fig. 24-8). Stung and firmly grasped, the prey is forced through an expansible mouth into a digestive sac, the **gastrovascular cavity**. Digestive enzymes secreted into this cavity break down some of the food, and further digestion occurs within the cells lining the cavity. Because the gastrovascular cavity has only a single opening, when digestion is completed, undigested material is expelled through the mouth. Although this two-way traffic prevents continuous feeding, it is adequate to support the low energy demands of these animals.

Cnidarians can reproduce both asexually and sexually. Some medusae and some polyps, such as hydra and sea anemones, bud off miniature replicas of themselves (see Fig. 39-1). Sexual reproduction involves the fusion of sperm and eggs released into the water or retained within the parent. The fertilized egg often develops into a free-swimming ciliated larval stage that settles and becomes a tiny polyp.

Of the 9000 or more species of the phylum Cnidaria, all are aquatic and most are marine. One group, the corals, are of particular ecological importance (see Fig. 24-6b). These polyps secrete a hard protective "house" of limestone, a rock consisting mainly of calcium carbonate, that persists long after their death, serving as a base to which others

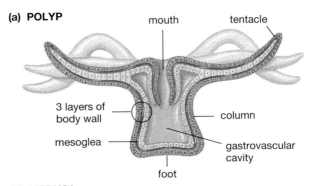

(a) POLYP

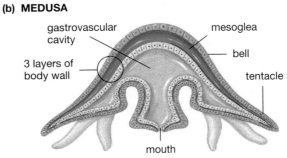

(b) MEDUSA

Figure 24-7   Polyp and medusa

The two basic body forms of cnidarians are actually variations on a single, simple theme. **(a)** The polyp form is seen in hydra (see Fig. 24-8), sea anemones (Fig. 24-6a), and the individual polyps within a coral (Fig. 24-6b). **(b)** The medusa form, seen in the jellyfish (Fig. 24-6c), resembles an inverted polyp. Both forms exhibit radial symmetry, with body parts arranged in a circle around a central axis.

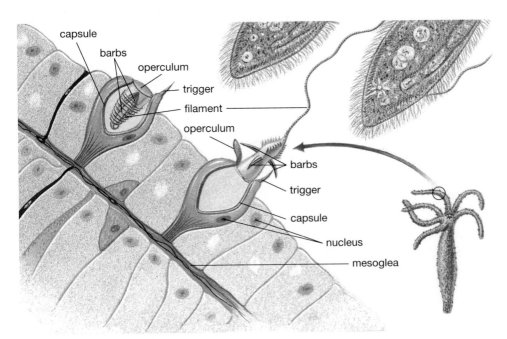

Figure 24-8   Cnidarian weaponry: the cnidocyte

At the slightest touch to the trigger, the cnidocyte of cnidarians, such as this hydra, violently expels the poisoned or sticky dart that lies coiled and inverted inside. During the process, the barbs and hollow filament actually turn inside out, impaling the prey and injecting a paralyzing venom. Cnidocytes are microscopic, and only a few species inject venom in sufficient quantity to harm a person.

may attach themselves. The cycle continues until, after thousands of years, massive coral reefs are formed. Corals are restricted to the warm, clear waters of the tropics, where their reefs form undersea habitats that are the basis of an ecosystem of stunning diversity and unparalleled beauty (see Chapter 46).

The Flatworms: Phylum Platyhelminthes

Although flatworms do not look anything like cnidarians, the two share some features that have led biologists to speculate that they evolved from a common ancestor. Both have a gastrovascular cavity with a single opening. Certain flatworms also show striking similarities to the larval stage of cnidarians.

Flatworms, however, are clearly more complex than cnidarians. First, they are bilaterally symmetrical rather than radially symmetrical (see Fig. 24-2). This body plan, found in all the more complex animals, is an adaptation to active movement. The anterior end, where the sense organs are concentrated, first encounters the environment ahead. On the basis of what is encountered, these sense organs inform the organism to feed, forge onward, or retreat. In **free-living** (nonparasitic) flatworms such as the freshwater planarians (Fig. 24-9), the sense organs consist of eyespots for detecting light and dark and cells responsive to chemical and tactile stimuli. To process information, flatworms have clusters of nerve cells called

ganglia in the head that form a simple brain. A pair of nerve cords conducts nervous signals to and from the head ganglia.

Flatworms are the simplest organisms with well-developed organs, in which tissues are grouped into functional units. When a free-living flatworm encounters food, usually smaller animals, it sucks up its prey using a muscular tube called a **pharynx** located in the middle of the ventral side of its body. The food is digested in an intricately branched gastrovascular cavity that distributes nutrients to all parts of the body (see Fig. 24-9a). Free-living flatworms have a simple system for excreting and regulating body fluids. It consists of a network of canals ending in bulbs containing beating cilia. The flickering motion of the cilia has led to the descriptive name: **flame cells** (see Fig. 24-9b). The beating cilia drive liquids through the system, emptying excess fluids to the outside through numerous tiny pores.

Flatworms lack both respiratory and circulatory systems. Nutrients are distributed by the branched digestive tract, from which they readily diffuse into nearby cells. Gas exchange by diffusion between the cells and the environment is aided by the flattened body, which ensures that all the cells are relatively close to the outside.

Flatworms can reproduce both sexually and asexually. Free-living forms may reproduce by cinching themselves around the middle until they separate into two halves, each of which regenerates its missing parts. All forms can reproduce sexually, and most are **hermaphroditic** (possessing both male and female sexual organs; Figs. 24-9c and 24-10). This feature is a great advantage to parasitic

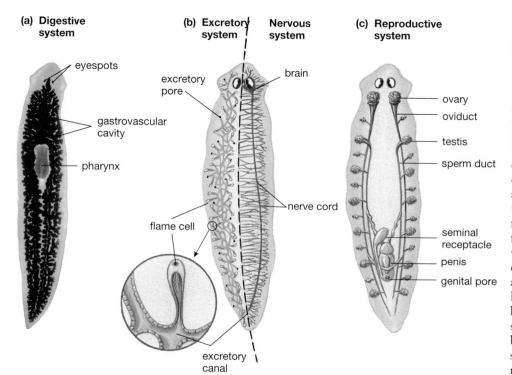

(a) Digestive system

- eyespots
- gastrovascular cavity
- pharynx

(b) Excretory system   Nervous system

- excretory pore
- flame cell
- excretory canal
- brain
- nerve cord

(c) Reproductive system

- ovary
- oviduct
- testis
- sperm duct
- seminal receptacle
- penis
- genital pore

Figure 24-9   **Flatworm organ systems**

Flatworms such as planarians have well-developed organ systems. **(a)** The elaborately branched digestive system, the centrally located ventral pharynx, and eyespots in the head are clearly visible. **(b)** (Left) The excretory system consists of branching tubes that conduct excess fluid to the outside through numerous pores. At intervals, flame cells, with their flickering tufts of cilia, keep the fluid moving. (Right) The nervous system of flatworms shows clear cephalization, with eyes and a brain composed of ganglia cells located within a well-defined head. Ladderlike nerve cords carry signals through the rest of the body. **(c)** Flatworm reproductive systems include both male and female reproductive organs, such as an ovary and testis.

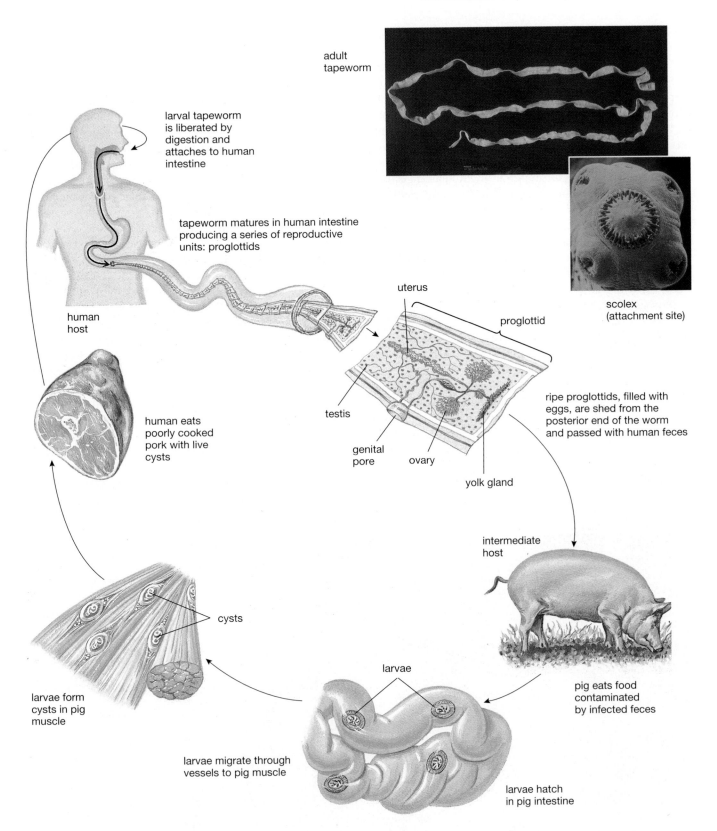

adult tapeworm

larval tapeworm is liberated by digestion and attaches to human intestine

tapeworm matures in human intestine producing a series of reproductive units: proglottids

scolex (attachment site)

uterus

proglottid

human host

human eats poorly cooked pork with live cysts

testis

genital pore

ovary

yolk gland

ripe proglottids, filled with eggs, are shed from the posterior end of the worm and passed with human feces

cysts

intermediate host

pig eats food contaminated by infected feces

larvae

larvae form cysts in pig muscle

larvae migrate through vessels to pig muscle

larvae hatch in pig intestine

Figure 24-10   **The life cycle of the human pork tapeworm**

Each reproductive unit, or proglottid, is a self-contained reproductive factory including both male and female sex organs.

forms because each worm can reproduce through self-fertilization, even if it is the only one present in its host.

Although many flatworms, such as the planarians, are free-living, those of major importance to humans are **parasites**. These include the tapeworms, several of which can infect humans. In most cases, people become infected by eating improperly cooked beef, pork, or fish that has been infected by the worms, whose larvae form encapsulated resting structures, called **cysts**, in the muscles of these animals. The cysts hatch in the human digestive tract, where they attach to the intestine and mature. There they may grow to a length of over 7 meters (20 feet), absorbing digested nutrients directly through their outer surface and releasing packets of eggs that are shed in the host's feces. If pigs eat grass contaminated with infected human feces, the eggs hatch in the pig's digestive tract, releasing larvae that burrow into its muscles and form cysts, thereby continuing the infective cycle (Fig. 24-10). Another group of parasitic flatworms are the flukes. Of these, the most devastating are liver flukes (common in Asia) and blood flukes, such as those of the genus *Schistosoma*, which cause schistosomiasis. Prevalent in Africa and parts of South America, this disease affects an estimated 200 million people; its symptoms include diarrhea, anemia, and possible brain damage. Like most parasites, flukes have a complex life cycle that includes an intermediate host, in this case a snail. The irrigation ditches filled by the Aswan Dam in Egypt have contributed to the spread of schistosomiasis by creating an extensive new habitat for the snail host.

The Roundworms: Phylum Nematoda

Nematodes, also called roundworms, have been enormously successful in colonizing nearly every habitat on Earth. Although only about 10,000 species of nematodes have been named, there may be as many as 500,000. Most are microscopic, such as that shown in Figure 24-11, but some parasitic forms reach a meter in length. Nematodes have a rather simple body plan consisting of a tubular gut running from mouth to anus. This one-way digestive tract shows greater complexity and a major evolutionary advance over flatworms. A fluid-filled pseudocoelom surrounds the organs and forms a **hydrostatic skeleton**, a framework against which muscles can act. A tough, flexible, nonliving cuticle encloses and protects the thin, elongated body (Fig. 24-11). Sensory organs in the head transmit information to a simple "brain," composed of a ring of ganglia. Nematodes lack both circulatory and respiratory systems. Because most are extremely thin and all have low energy requirements, diffusion suffices for gas exchange and the distribution of nutrients. Nematodes reproduce sexually, and the sexes are separate, with the male (who is usually smaller) fertilizing the female by placing sperm inside her body.

Although you may be blissfully unaware of their presence, roundworms are nearly everywhere outdoors, where they play an important role in breaking down organic matter. A single rotting apple may contain 90,000 worms. Billions thrive in each acre of topsoil. Nearly all plants and an-

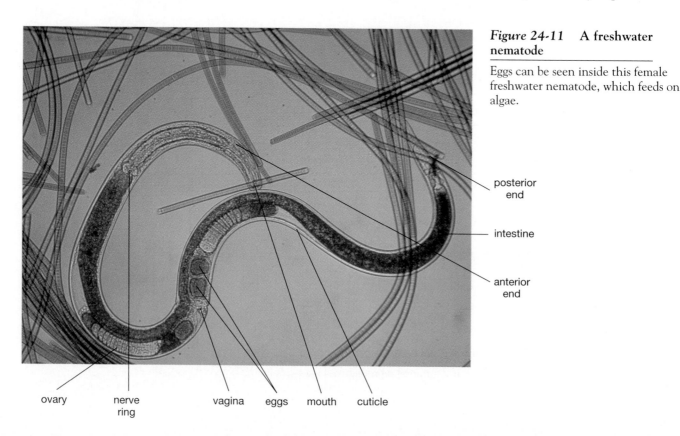

Figure 24-11   A freshwater nematode

Eggs can be seen inside this female freshwater nematode, which feeds on algae.

posterior end

intestine

anterior end

ovary   nerve ring   vagina   eggs   mouth   cuticle

(a)

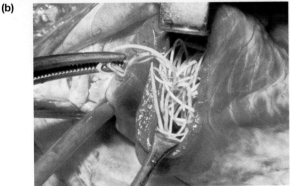

(b)

Figure 24-12   **Some parasitic nematodes**

(a) Encysted larva of the *Trichinella* worm in muscle tissue of a pig, where it may live for up to 20 years. **(b)** Adult heartworms in the heart of a dog. The juveniles are released into the bloodstream, where they may be ingested by mosquitoes and passed to another dog by the bite of an infected mosquito.

imals are host to several parasitic species. During your life you may be parasitized by one of the 50 species that specialize on humans, most of which are relatively harmless.

Although most roundworms are harmless, there are important exceptions. For example, hookworm larvae in soil may bore into human feet, enter the bloodstream, and travel to the intestine, where they cause continuous bleeding. The *Trichinella* worm, which causes trichinosis, may be ingested by eating improperly cooked pork. Infected pork may contain up to 15,000 larval cysts per gram (Fig. 24-12a). The cysts hatch in the human digestive tract and invade blood vessels and muscles, causing bleeding and muscle damage. Another dangerous nematode parasite, the heartworm of dogs, is transmitted by mosquitoes (Fig. 24-12b). In the southern United States, and increasingly in other parts of the country, it poses a severe threat to the health of unprotected pets.

The Segmented Worms: Phylum Annelida

As the name suggests, a prominent feature of the annelids (from the Latin word for "little ring") is the division of the body into a series of repeating segments. Externally, these appear as ringlike depressions on the surface. Internally, many of the segments contain identical copies of nerve ganglia, excretory structures, and muscles (Fig. 24-13). Segmentation is advantageous for locomotion because the body compartments, each of

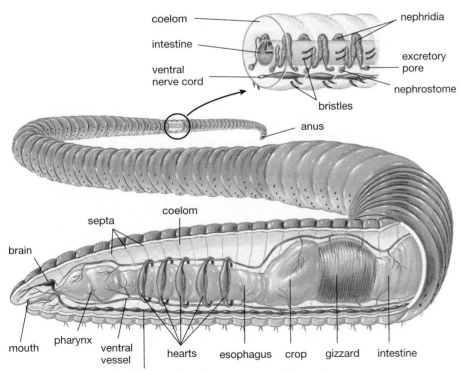

Figure 24-13   **An annelid, the earthworm**

This diagram shows an enlargement of segments, many of which are repeating similar units separated by partitions called septa. The digestive system, which has both a mouth and an anus, is divided into a series of compartments specialized to process food in an orderly sequence.

coelom

intestine

ventral nerve cord

nephridia

excretory pore

nephrostome

bristles

anus

septa

coelom

brain

mouth   pharynx   ventral vessel   hearts   esophagus   crop   gizzard   intestine

ventral nerve cord

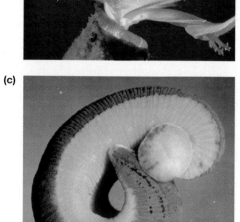

Figure 24-14   **Diverse annelids**

(a) A polychaete annelid projects brightly spiraling gills from a tube attached to rock. When the gills retract, the tube is covered by the trap door visible on the lower right. (b) The "fireworm" polychaete swims using paddles on each segment. The bristles on each paddle can deliver a fiery sting. (c) This leech, a freshwater annelid (class Hirudinea) found in a Georgia pond, shows numerous segments. The sucker encircles its mouth, allowing it to attach to its prey. Medicinal leeches were used by doctors up until the 1800s to suck the "tainted" blood from patients suffering from a variety of disorders.

which is controlled by separate muscles, collectively are capable of far greater complexity of movement than is seen in the nonsegmented worms. Another evolutionary advance that first appears in annelids is a fluid-filled true coelom between the body wall and the digestive tract (see Fig. 24-3c). The incompressible fluid in the coelom in many annelids is confined by the partitions separating the segments and serves as a hydrostatic skeleton, making possible such feats as burrowing through soil.

In contrast to nematodes, annelids have a well-developed **closed circulatory system** that distributes gases and nutrients throughout the body (see Chapter 30). In closed circulatory systems (yours is one also) blood remains confined to the heart and blood vessels. In the earthworm, for example, blood with oxygen-carrying hemoglobin is pumped through well-developed vessels by five pairs of "hearts." These hearts are actually short, expanded segments of specialized blood vessels that contract rhythmically. The blood is filtered and wastes are removed by excretory organs called **nephridia** that are found in many of the segments. Nephridia resemble the individual tubules of the vertebrate kidney (see Chapter 33). The annelid nervous system consists of a simple ganglionic brain in the head and a series of repeating paired segmental ganglia joined by a pair of ventral nerve cords traveling the length of the body.

Digestion in annelids occurs in a series of compartments, each specialized for a different phase of food processing (see Fig. 24-13). For example, in the earthworm, a muscular pharynx draws in food, consisting of bits of decaying plant and animal debris in soil. The food is conducted through the esophagus to a storage chamber, the crop, and then released slowly into the muscular gizzard, where it is ground into tiny particles by muscular contractions of the gizzard and the sharp-edged sand grains it contains. The food then passes into the intestine, where it is digested and nutrients are absorbed. Undigested food and soil exit through the anus.

The phylum Annelida includes about 9000 species, including the familiar earthworm and its relatives, the oligochaetes (meaning "few bristles"). In general, these exchange gas by diffusion through moist skin. The largest group of annelids, the polychaetes (meaning "many bristles") is found primarily in the ocean. Some have numerous bristles and paired fleshy paddles on most of their segments, used in locomotion. Others live in tubes from which they project feathery gills that both exchange gases and sift the water for microscopic food (Fig. 24-14a, b). A third group of annelids (class Hirudinea) consists of the leeches (Fig. 24-14c). These worms, found in freshwater or moist terrestrial habitats, are either parasitic or **carnivorous**, some preying on smaller invertebrates, others sucking the blood of larger animals.

The Insects, Arachnids, and Crustaceans: Phylum Arthropoda

Spread your picnic tablecloth in the shade of an oak beside a stream-fed pond and prepare to discover the diversity of arthropods. As you shoo flies from the potato salad, a yellowjacket may industriously attack your hamburger, flying off with a small piece of meat in its grasp. While a woolly caterpillar undulating up the tree trunk distracts you, ants will be discovering the cookie crumbs, and a spider, suspended by a gossamer thread, may lower itself onto your head. As you move a large sheltering stone in the stream, watch carefully for the sudden backward flipping of a crayfish. Dragonflies hover near the water's edge, their wings iridescent in the sunlight. Later, as dusk falls, you'll be glad you brought the mosquito repellent!

In numbers, both of individuals and of species, arthropods are the dominant animals on Earth. About 1 million species have been discovered, and scientists estimate that up to 9 million remain undescribed. The phylum Arthropoda includes many classes, three of which are particularly large and important: the insects (class Insecta), spiders and their relatives (class Arachnida), and crabs, shrimp, and their relatives (class Crustacea). In general, arthropod sexes are separate, and fertilization is internal. The success of this group can be attributed to several important adaptations that have allowed them to exploit nearly every habitat on Earth. These adaptations include an exoskeleton, segmentation, well-developed sensory and nervous systems, efficient gas-exchange mechanisms, and well-developed circulatory systems.

Segmentation in arthropods is evidence that they shared a common ancestor with annelids. Arthropod segments, however, tend to be fewer, less distinct from one another, and specialized for distinct functions such as sensing the environment, feeding, and movement (Fig. 24-15).

The **exoskeleton** (*exo* meaning "outside" in Greek) is an external skeleton that encloses the arthropod body like a suit of armor. In places it is thin and flexible, allowing movement of the paired, jointed appendages from which the phylum Arthropoda (Greek, "jointed foot") derives its name. The exoskeleton is secreted by the epidermis (the outer layer of skin) and composed chiefly of protein and a polysaccharide called **chitin**. It provides an important defense against small predators and is responsible for the greatly increased agility of arthropods over their annelid wormlike ancestors. By providing rigid attachment sites for muscles together with stiff but flexible appendages, the exoskeleton makes possible the flight of the bumblebee and the intricate, delicate manipulations of the spider as it weaves its web (Fig. 24-16). The exoskeleton also contributed enormously to the arthropod invasion of dry terrestrial habitats by providing a water-

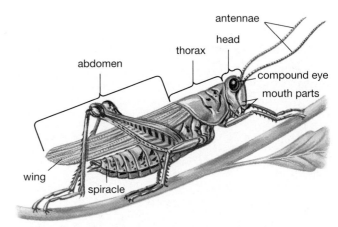

Figure 24-15   Segments are fused and specialized in insects

Insects such as this grasshopper show fusion and specialization of body segments into a distinct head, thorax, and abdomen. Segments are visible beneath the wings on the abdomen.

tight covering for delicate, moist tissues such as those used for gas exchange.

Like a suit of armor, the exoskeleton has some unique problems in addition to its advantages. First, because it cannot expand as the animal grows, the exoskeleton must be shed, or **molted**, periodically and replaced with one of a larger size (Fig. 24-17). Molting uses energy and leaves the animal temporarily vulnerable before the new skeleton hardens ("soft-shelled" crabs are eaten during this delicate period). The exoskeleton is also heavy; its weight

Figure 24-16   The exoskeleton allows precision movements

A garden spider, having immobilized its prey with a paralyzing venom, rapidly encases it in web. Such dexterous manipulations are made possible by the exoskeleton and jointed appendages characteristic of arthropods.

Figure 24-17 **The exoskeleton must be molted periodically**

A newly emerged praying mantis (a predatory insect) hangs beside its outgrown exoskeleton (left).

increases exponentially as the animal grows. It is no coincidence that the largest arthropods are crustaceans (crabs and lobsters), whose watery habitat supports much of their weight.

Most arthropods possess a well-developed sensory system, including **compound eyes** (Fig. 24-18) and acute chemical and tactile senses. The arthropod nervous system is similar in plan to that of annelids, but more complex. It consists of a brain composed of fused ganglia in the head, connected to a series of ganglia running down the length of the body and linked by a ventral nerve cord. The capacity for finely coordinated movement combined with sophisticated sensory abilities and a well-developed nervous system has allowed the evolution of complex behavior. In fact, the interactions among certain social insects, such as the honeybee, are more intricate than those of most vertebrate societies. Among social insects, communication and genetically programmed learning play important roles (see Chapter 42).

Efficient gas exchange is required to supply adequate oxygen to muscles that allow the rapid flight, swimming, or running displayed by many arthropods. Gas exchange is accomplished by **gills** in aquatic forms such as the crustaceans, and by either **tracheae** or **book lungs** in terrestrial forms (Fig. 24-19). Arthropods also have well-developed circulatory systems with a feature not seen in annelids: the **hemocoel**, or blood cavity. Blood not only travels through vessels but also empties into the hemocoel, where it bathes the internal organs directly. This arrangement, known as an **open circulatory system**, is also present in most mollusks (see Chapter 30).

Insects Are the Most Diverse and Abundant Arthropods

The number of insect species (members of the class Insecta; Fig. 24-20) is estimated at 800,000, roughly the same as the total number of species in all other classes of animals combined. Insects have three pairs of legs, usually supplemented by two pairs of wings. The capacity for flight distinguishes them from all other invertebrates and has contributed to their enormous success (Fig. 24-20c). As anyone who has unsuccessfully pursued a fly can testify, flight helps in escaping from predators. It also allows the insect to find widely dispersed food. Swarms of locusts (see Fig. 24-20d) have been traced all the way from Saskatchewan, Canada, to Texas on the trail of food. Flight requires rapid and efficient gas exchange. Insects use the network of narrow branching tubes called tracheae, which conduct air to all parts of the body (see Fig. 24-19c).

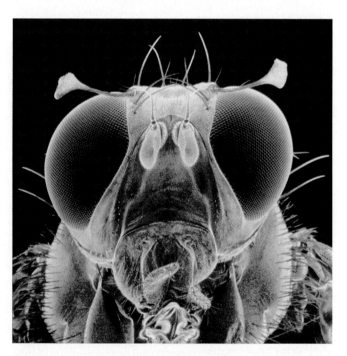

Figure 24-18 **Arthropods possess compound eyes**

This scanning electron micrograph shows the compound eye of a fruit fly. Compound eyes consist of an array of similar light-gathering and sensing elements whose orientation gives the arthropod a panoramic view of the world. Insects have reasonable image-forming ability and good color discrimination.

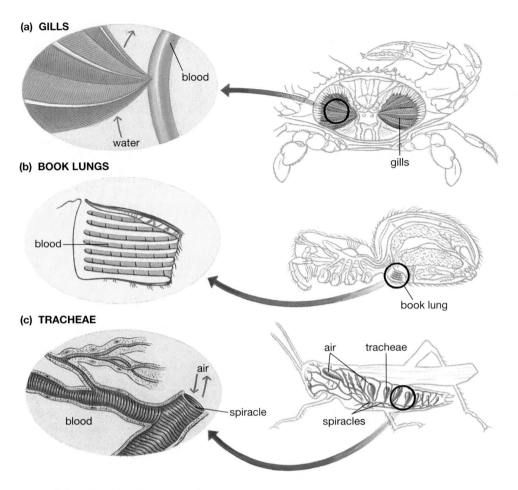

Figure 24-19   Arthropod respiratory structures

(a) Gills, adapted for life in water, expose a large surface area of tissue rich in blood vessels to the water for gas exchange. The drying effects of life on land demand protection of delicate, moist respiratory surfaces. These surfaces are, therefore, placed inside the body, with air entering through a small opening to minimize evaporation. **(b)** The book lungs of spiders resemble internal gills. **(c)** The internal tracheae of insects, elaborately branching networks of hollow air passages, carry air close to each cell.

Figure 24-20   The diversity of insects

(a) The rose aphid sucks sugar-rich juice from plants. **(b)** A mating pair of Hercules beetles. The large "horns" are found only on the male. **(c)** A June beetle displays its two pairs of wings as it comes in for a landing. The outer wings protect the abdomen and the inner wings, which are relatively thin and fragile. **(d)** Insects such as this locust can cause devastation of both crops and natural vegetation. **(e)** Caterpillars are larval forms of moths or butterflies. This caterpillar larva of the Australian fruit-sucking moth displays large eye-spot patterns that may frighten potential predators, who mistake them for eyes of a large animal.

During their development, insects undergo **metamorphosis**, which frequently involves a radical change in body form from juvenile to adult. In insects with complete metamorphosis, the immature form, called a **larva**, is wormlike in shape (for example, the maggot of a housefly or the caterpillar of a moth or butterfly; see Fig. 24-20e). The larva hatches from an egg, grows by eating voraciously and shedding its exoskeleton several times, then forms a nonfeeding **pupa**. Encased in an outer covering, the pupa undergoes a radical change in body form, emerging in its adult winged form. The adults mate and lay eggs, continuing the cycle. Metamorphosis may include a change in diet as well as in shape, thereby eliminating competition for food between adults and juveniles and in some cases allowing the insect to exploit different foods when they are most available. For instance, a caterpillar that feeds on new green shoots in springtime metamorphoses into a butterfly that drinks nectar from the summer's blooming flowers. Some insects undergo a more gradual metamorphosis (called incomplete metamorphosis), hatching as young that bear some resemblance to the adult, then gradually acquiring more adult features as they grow and molt.

Spiders, Scorpions, and Their Relatives Are Members of the Class Arachnida

The arachnids comprise about 50,000 species of terrestrial arthropods, including spiders, mites, ticks, and scorpions (Fig. 24-21). All have eight walking legs, and most are carnivorous, many subsisting on a liquid diet consisting of blood or predigested prey. For example, spiders, the most numerous arachnids, first immobilize their prey with a paralyzing venom. They then inject digestive enzymes into the helpless victim (often an insect) and suck in the resulting soup. Arachnids breathe using either tracheae or a specialized arachnid respiratory structure, book lungs (see Fig. 24-19b), or both. Arachnids have simple eyes, each with a single lens, in contrast to the compound eyes of insects and crustaceans. The eyes are particularly sensitive to movement, but in some species they probably can form images. Most spiders have eight eyes placed in such a way as to give them a panoramic view of predators and prey.

Crabs, Shrimp, Crayfish, and Their Relatives Are Members of the Class Crustacea

The roughly 30,000 species of crustaceans, including crabs, crayfish, lobster, shrimp, and barnacles, compose the only class of arthropods that is primarily aquatic (Fig. 24-22). Crustaceans range in size from the microscopic waterflea *Daphnia*, found in ponds, to the largest of all arthropods, the Japanese crab, with legs spanning nearly 4 meters (12 feet). Crustaceans have two pairs of sensory antennae, but the rest of their appendages are highly variable in form and number, depending on the habitat and life-style of the species. Most crustaceans have compound eyes similar to those of insects, and nearly all respire using gills (see Fig. 24-19a).

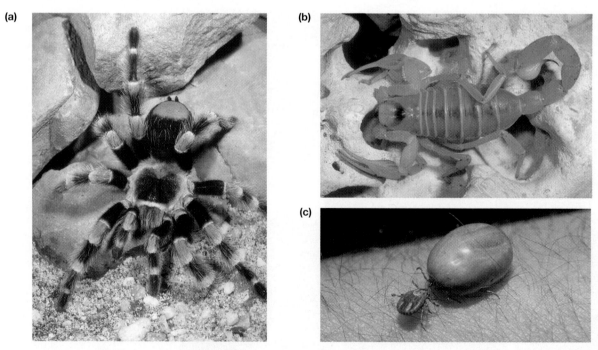

Figure 24-21 **The diversity of arachnids**

(a) The tarantula is among the largest spiders but is relatively harmless. (b) Scorpions, found in warm climates including deserts of the American Southwest, paralyze their prey with venom from a stinger at the tip of the abdomen. A few species can harm human beings. (c) Ticks before and after feeding on blood. The uninflated exoskeleton is flexible and folded, allowing the animal to become grotesquely bloated while feeding.

(a)

(b)

(c)

(d)

Figure 24-22 **The diversity of crustaceans**

(a) The microscopic waterflea, *Daphnia*, common in freshwater ponds. Notice the eggs developing within the body. **(b)** The sowbug, found in dark moist places such as under rocks, leaves, and decaying logs, is one of the few crustaceans to successfully invade the land. **(c)** The hermit crab protects its soft abdomen by inhabiting an abandoned snail shell. **(d)** The goose-neck barnacle anchors itself to rocks, boats, or even animals such as whales. These goose-neck barnacles attach using a tough, fleshy stalk. Other types of barnacles attach with shells resembling miniature volcanoes (see Fig. 24-25b). Early naturalists thought barnacles were mollusks until the jointed legs, seen extending into the water, were observed.

The Snails, Clams, and Squid: Phylum Mollusca

Mollusks, like arthropods, share a common ancestry with annelids. In number and variety (about 100,000 species have been described), they are second only to the arthropods. Mollusks (whose name comes from the Latin *mollis*, meaning "soft") have a moist, muscular body supported by a hydrostatic skeleton. Some protect their body with a shell of calcium carbonate; others escape predation by moving swiftly or, if caught, by tasting terrible. Mollusks have a tissue called the **mantle** that forms a chamber for the gills and secretes the shell in shelled species. Most mollusks have an open circulatory system, like that of arthropods, with blood directly bathing the organs in a hemocoel. The nervous system, like that of annelids and arthropods, consists of ganglia connected by nerves, but many more of the ganglia are concentrated in the brain. Reproduction is always sexual; some species have separate sexes and others are hermaphroditic. Although mollusks are enormously diverse, a simplified diagram of the body plan of a mollusk is shown in Figure 24-23.

Among the many classes of mollusks, three will be discussed in more detail: snails and their relatives (the class Gastropoda), clams and their relatives (the class Bivalvia), and octopuses and their relatives (the class Cephalopoda).

Snails and Their Relatives Are Members of the Class Gastropoda

Gastropods are by far the largest class of mollusks, with about 35,000 known species. They crawl on a muscular foot (*gastropoda* is from the Greek for "stomach foot"), and many have shells that vary widely in form and color. Some of the most beautiful gastropods, the sea slugs, lack shells, but their brilliant colors warn predators that they are poisonous or at least bad tasting (Fig. 24-24).

Gastropods feed with a **radula**, a flexible ribbon of tissue studded with spines that is used to scrape algae from rocks or grasp larger plants or prey (see Figs. 24-23 and 20-3a). Most gastropods use gills for respiration in addition to their moist skin, through which dissolved gases readily diffuse. The gills may be either enclosed in a cavity beneath the shell or exposed, as in the sea slugs. A few gastropods (including the destructive garden snails and slugs) live in moist terrestrial habitats. These terrestrial gastropods (and some freshwater forms that evolved from them) use a simple lung for breathing.

Scallops, Clams, Oysters, and Their Relatives Are Members of the Class Bivalvia

Included among the bivalves are scallops, oysters, mussels, and clams. Not only do bivalves lend exotic variety to the human diet, but they also are extremely important members of the marine community near shore, where they attach to rocks that are alternately covered and exposed by the tides (Fig. 24-25). Bivalves possess two shells connected by a flexible hinge

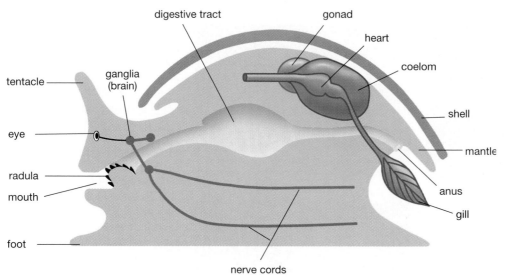

Figure 24-23   **A generalized mollusk**

The general body plan of a mollusk, showing the mantle, foot, gills, shell, and other features such as a radula that are seen in most (but not all) species of mollusk.

(a)

(b)

(a)

(b)

Figure 24-24   **The diversity of gastropods**

(a) A Florida tree snail displays a brightly striped shell, and eyes at the tip of stalks that are retracted instantly if touched. **(b)** Spanish shawl sea slugs prepare to mate. The brilliant colors of many sea slugs warn potential predators that they are distasteful.

Figure 24-25   **The diversity of bivalves**

(a) This swimming scallop from Vancouver parts its hinged shells. The upper shell is covered with an encrusting sponge. **(b)** Mussels attach to rocks in dense aggregations exposed at low tide. White barnacles are seen attached to the mussel shells and surrounding rock.

(hence their name, meaning "two shells"). A strong muscle clamps the shells closed in response to danger (this muscle is what you are served when you order scallops in a restaurant).

Clams use a muscular foot for burrowing in sand or mud. In mussels, which are sessile, the foot is smaller and is used to help secrete threads that anchor the animal to rocks. Scallops lack a foot and move by a sort of whimsical jet propulsion achieved by flapping their shells together. Bivalves are filter feeders, using their gills as both respiratory and feeding structures. Water is circulated over the gills, which are covered with a thin layer of mucus that traps microscopic food particles. Food is conveyed to the mouth by beating cilia on the gills. Probably because they filter-feed and do not move extensively, bivalves have "lost their heads" over the course of their evolution.

Octopuses, Squid, and Their Relatives Are Members of the Class Cephalopoda

The cephalapods (whose name means "head-foot"), a fascinating group that includes octopuses, squid, nautiluses, and cuttlefish (Fig. 24-26), are among the largest, swiftest, and smartest of all invertebrates. All cephalopods are predatory carnivores, and all are marine. In these mollusks, the foot has evolved into tentacles with well-developed chemosensory abilities and suction discs for detecting and grasping prey. Prey grasped by tentacles may be immobilized by a paralyzing venom in the saliva before being torn apart by beaklike jaws. The cephalopod eye resembles our own in complexity and exceeds it in efficiency of design (see Chapter 37).

Cephalopods move rapidly by jet propulsion, which is accomplished by the forceful expulsion of water from the mantle cavity. The octopus may also travel along the seafloor using its tentacles like multiple, undulating legs. Cephalopods are the only mollusk with a closed circulatory system. This is probably an adaptation to the very active life-style of the cephalopods, since it allows more efficient transport of oxygen and nutrients. The cephalopod brain, especially that of the octopus, is exceptionally large and complex for an invertebrate brain. It is enclosed in a skull-like case of cartilage and endows the octopus with highly developed capabilities to learn and remember. In the laboratory, octopuses can rapidly learn to associate certain symbols with food and to open a screw-cap jar to obtain food.

The Sea Stars, Sea Urchins, and Sea Cucumbers: Phylum Echinodermata

Although echinoderms have evolved a bewildering diversity of forms, they have never left their ancestral home on the ocean floor. Their descriptive common names re-

(a)

(b)

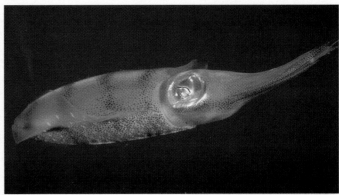

(c)

Figure 24-26   **The diversity of cephalopods**

(a) An octopus can crawl rapidly using its eight suckered tentacles. It can alter its color and skin texture to blend with its surroundings. In emergencies this mollusk can jet backward by vigorously contracting its mantle. Octopuses and squid can emit clouds of dark purple ink to confuse pursuing predators. **(b)** The squid moves entirely by jet propulsion by contracting its mantle, pushing the animal backward through the water. The giant squid is the largest invertebrate, reaching a length of 15 meters (almost 50 feet), including tentacles. **(c)** The chambered nautilus secretes a shell with internal, gas-filled chambers providing buoyancy in the water. Note the well developed eyes and the tentacles used to capture prey.

(a) **(b)** **(c)**

Figure 24-27 **The diversity of echinoderms**

(a) A sea cucumber off southern California feeds on debris in the sand. **(b)** The sea urchin's spines are actually projections of the internal skeleton. **(c)** The sea star has reduced spines and often has five arms.

flect this fact: sand dollar, sea urchin, sea star (or starfish), sea cucumber, and sea lily (Fig. 24-27).

Although their free-swimming embryos are bilaterally symmetrical, echinoderm adults have radial symmetry, an adaptation to a sluggish, or in some forms a sessile, existence. Most echinoderms lack a head and move very slowly in any direction as they feed on algae or small particles sifted from sand or water. The sea star, though, is a predator. It can slowly pursue prey (including bivalve mollusks) from any direction. Echinoderms move on numerous tiny **tube feet**, delicate cylindrical projections that extend from the lower surface of the body and terminate in a suction cup.

Tube feet are part of a unique echinoderm feature, the **water-vascular system**, which functions in locomotion, respiration, and food capture (Fig. 24-28). Seawater enters through an opening (the sieve plate) on the animal's upper surface and is conducted through a ring canal that encircles the esophagus, from which branch a number of radial canals. These conduct water to the tube feet, each of which is controlled by a muscular squeeze bulb (ampulla). Contraction of the bulb forces water into the tube foot, causing it to extend. The suction cup may be pressed against the seafloor or a food object, to which it adheres tightly until pressure is released.

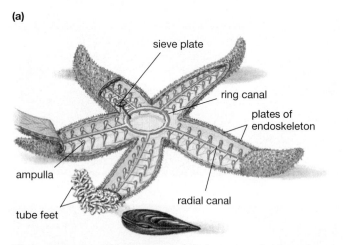

(a)

sieve plate

ring canal

plates of
endoskeleton

ampulla

radial canal

tube feet

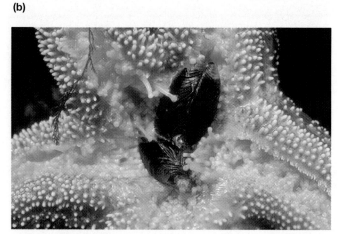

(b)

Figure 24-28 **The water-vascular system of echinoderms**

(a) Seawater enters through the sieve plate and is transported into the ring canal, from which it is distributed to each of the arms through radial canals. The water inflates squeeze bulb–like ampullae that expand and contract to extend or retract the tube feet. The plates of the endoskeleton can be seen embedded in the body wall. **(b)** The sea star often feeds on pelecypod mollusks such as this mussel. Numerous tube feet are attached to the shells, exerting a relentless pull. The sea star turns the delicate tissue of its stomach inside out, extending it through the centrally located ventral mouth. An opening in the bivalve shells of less than 1 millimeter is sufficient for the stomach tissue to insinuate between the shells, secreting digestive enzymes that weaken the mollusk, causing it to open further. Partially digested food is transported to the upper portion of the stomach, where digestion is completed.

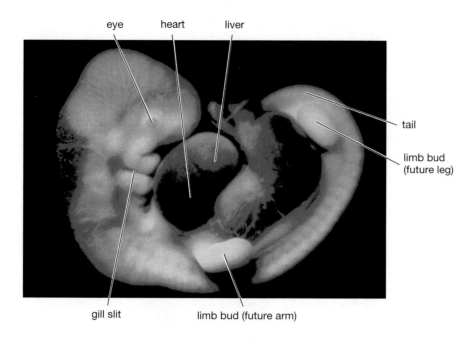

eye   heart   liver

tail

limb bud
(future leg)

gill slit   limb bud (future arm)

Figure 24-29   Vertebrate features in the human embryo

The 5-week-old human embryo is about 1 centimeter long and clearly shows external gill slits (more properly called grooves, since they do not penetrate the body wall) and a tail. Although the tail will disappear completely, the gill grooves contribute to the formation of the lower jaw.

Echinoderms have a relatively simple nervous system with no distinct brain. Movements are loosely coordinated by a system consisting of a nerve ring encircling the esophagus, radial nerves to the rest of the body, and a nerve network through the epidermis. In sea stars, simple receptors for light and chemicals are concentrated on the arm tips, and sensory cells are scattered over the skin. The echinoderms lack a circulatory system, although movement of the fluid in their well-developed coelom serves this function. Gas exchange occurs through the tube feet, and in some forms, numerous tiny "skin gills" project through the epidermis. Sea cucumbers possess an internal system of canals called a respiratory tree.

Most species reproduce by shedding sperm and eggs into the water, where fertilization occurs and a free-swimming larva develops. The sexes are usually separate. Sea stars have the ability to regenerate lost parts; new individuals may form from a single arm, provided that part of the central body is attached to it. When mussel fisherman tried to rid mussel beds of predatory sea stars by hacking them into pieces and throwing the pieces back, needless to say, the strategy backfired!

Echinoderms possess an **endoskeleton** (*endo* meaning "inside" in Greek) composed of plates of calcium carbonate beneath the outer skin (Fig. 24-28a). The name *echinoderm* (Greek, "hedgehog skin") comes from projections of the endoskeleton that extend as bumps or spines through the outer layer of skin. These spines are especially well developed in sea urchins and much reduced in sea stars and sea cucumbers.

The Tunicates, Lancelets, and Vertebrates: Phylum Chordata

The chordates are an extremely diverse group of animals united by four features that all possess at some stage of their lives:

1. A **notochord**. A stiff but flexible rod that extends the length of the body and provides an attachment site for muscles.
2. A **nerve cord**. Lying dorsal to the digestive tract, this hollow neural structure develops a thickening at its anterior end that becomes a brain.
3. **Gill slits**. Located in the pharynx (the cavity behind the mouth), these may form functional respiratory openings or may appear only as grooves during an early stage of development.
4. A **tail**. An extension of the body past the anus.

This list may seem puzzling because, although humans are chordates, at first glance we seem to lack every feature except the nerve cord. But evolutionary relationships are sometimes seen most clearly during early stages of development, and it is during our embryonic life that we develop, and lose, our notochord, our gill slits, and, finally, our tails (Fig. 24-29). Humans share these chordate features with two invertebrate chordate groups, the lancelets and the tunicates, and with all other vertebrates.

The Invertebrate Chordates Include Lancelets and Tunicates

The invertebrate chordates lack a head and, of course, the backbone that is the defining feature of vertebrates. The small (5 centimeters long) fishlike lancelet is an invertebrate chordate that spends most of its time half-buried in the sandy sea bottom, filtering tiny food particles from the water. As can be seen in Figure 24-30a, all the typical chordate features are present in the adult organism.

The tunicates form a larger group of marine invertebrate chordates. It is difficult to imagine a less likely relative of humans than this sessile, filter-feeding, vaselike animal (Fig 24-30b). Its ability to move is limited to a forceful contraction of the saclike body, which sends a

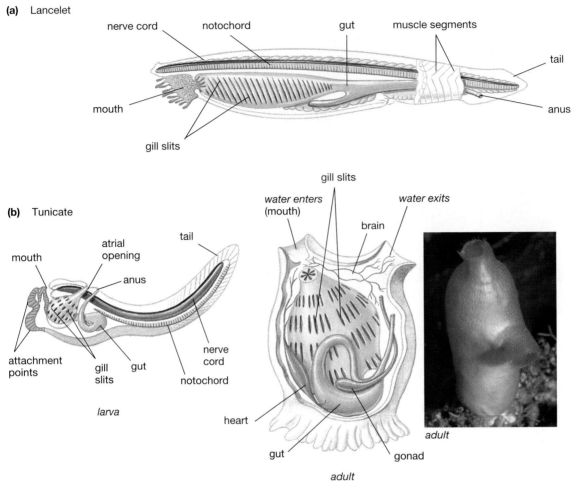

Figure 24-30   Invertebrate chordates

(a) A lancelet, a fishlike invertebrate chordate. The adult organism exhibits all the features of chordates. **(b)** The tunicate larva (left) has all the chordate features. The adult tunicate (middle) has lost the tail and notochord and has assumed a sedentary life as shown in the photo (right).

jet of seawater into the face of anyone who plucks it from its undersea home; hence the common name sea squirt. However, tunicates produce actively swimming tadpole-like larvae that possess all the proper chordate features (see Fig. 24-30b). Because fossils of intermediate forms have never been discovered, the exact sequence of evolutionary events that led from invertebrate chordates to the first, fishlike vertebrates remains shrouded in mystery.

The Vertebrates Have a Backbone and Other Adaptations That Have Contributed to Their Success

In the vertebrates (members of the subphylum Vertebrata), the embryonic notochord is usually replaced during development by a backbone, or **vertebral column**, composed of bone or cartilage, a tissue that resembles bone but is slightly softer and more flexible. This structure provides support for the body, an attachment site for muscles, and protection

for the delicate nerve cord and brain. The backbone is part of a living endoskeleton, capable of growth and self-repair. Because this internal skeleton provides support without the armorlike weight of the arthropod exoskeleton, it has allowed vertebrates to achieve great size and mobility and has contributed to their invasion of the land and the air.

Vertebrates show other adaptations that have contributed to their successful invasion of most habitats. One is the presence of paired appendages. These first appeared as fins in fish and served as stabilizers for swimming. Over millions of years, some fins were modified by natural selection into legs that allowed animals to crawl onto dry land, and later into wings that allowed some to take to the air. Another adaptation that has contributed to the success of vertebrates is an increase in size and complexity of their brains and sensory structures. These parallel adaptations allow vertebrates to perceive their environment in detail and to respond to it in a great variety of ways.

Today, vertebrates are represented by seven major classes: jawless fishes (class Agnatha), cartilaginous fishes (class Chondrichthyes), bony fishes (class Osteichthyes), am-

(a)

(b)

Figure 24-31 **Jawless fishes**

(a) The colorful but unattractive hagfishes live in communal burrows in the mud, feeding on polychaete worms. **(b)** Some lampreys are parasitic, attaching to fish (such as this carp) with suckerlike mouths lined with rasping teeth (inset).

phibians (class Amphibia), reptiles (class Reptilia), birds (class Aves), and mammals (class Mammalia).

Jawless Fishes Were the First Vertebrates to Evolve

Vertebrates arose in the sea; the earliest vertebrate fossils are those of strange jawless fishes protected by bony armor plates. Today, two groups of jawless fishes (members of the class Agnatha) survive: the hagfishes and the lampreys, which have many similar features but are not closely related. Both have skeletons of cartilage and are eel-like in shape. Both have unpaired fins located along the midline of the body. Both lack scales, and their smooth, slimy skin is perforated by circular gill openings.

Hagfishes are exclusively marine (Fig. 24-31a). Purple to pink, they live in communal burrows in the mud, feeding primarily on polychaete worms. They eagerly attack

dying fish, however, using pincerlike teeth that surround the tongue to burrow into the coelomic cavity and ingest the prey's soft internal organs. They are regarded with great disgust by fishermen because they produce large quantities of slime as a defense against predators.

Lampreys are found in both fresh and salt water, and even the marine forms return to fresh water to spawn. Lampreys include both parasitic and nonparasitic species. A parasitic lamprey has a suckerlike mouth lined with teeth, which it uses to attach itself to larger fish. With other rasping teeth on its tongue, the lamprey excavates a hole in the host's body wall through which it sucks blood and body fluids (Fig. 24-31b). Beginning in the 1920s, lampreys spread into the Great Lakes, where, lacking effective predators, they have multiplied prodigiously and greatly reduced commercial fish populations, including the lake trout. Vigorous measures to control the lamprey population have allowed some recovery of the other fish populations of the Great Lakes.

About 425 million years ago, in the mid-Silurian period, primitive jawless fishes that were ancestral to the lampreys and hagfishes gave rise to a group of fish that possessed an important new structure found in all the more advanced vertebrates: jaws. Jaws allowed fish to grasp and chew their food, permitting them to exploit a much wider range of food sources than could jawless fish. Although the first jawed fishes have been extinct for 230 million years, they are the ancestors of the two major classes of jawed fishes that survive today: the cartilaginous fishes (class Chondrichthyes) and the bony fishes (class Osteichthyes).

Cartilaginous Fishes Have an Endoskeleton Composed of Cartilage

The class Chondrichthyes, whose name is derived from Greek words meaning "cartilage fishes," includes 625 marine species, among them the sharks, skates, and rays (Fig. 24-32). These graceful predators lack any bone in their skeleton, which is formed entirely of **cartilage**, a tissue that resembles bone, but is less brittle and more flexible. The body is protected by a leathery skin roughened by tiny scales. Members of this group respire using gills. Although some must swim to circulate water through their gills, most can pump water across their gills. These and all other fish have a two-chambered heart (see Chapter 30).

Sharks may have several rows of razor-sharp teeth, the back rows moving forward as the front teeth are lost to age and use. Although a few species consider us potential prey, most sharks are shy of humans. Sharks include the largest fishes, such as the gentle whale shark, which can grow to over 15 meters (45 feet) in length. Skates and rays are also retiring creatures, although some can inflict dangerous wounds with a spine near their tail, and others produce a powerful electrical shock that can stun their prey.

Figure 24-32 **Cartilaginous fishes**

(a) The tropical blue-spotted sting ray swims by graceful un-dulations of lateral extensions of the body. **(b)** A sand tiger shark displaying several rows of teeth. As outer teeth are lost, they are replaced by new ones formed behind them. Both sharks and rays lack a swim bladder and tend to sink toward the bottom when they stop swimming.

Bony Fishes Have Invaded Nearly Every Aquatic Habitat

The name of the class Osteichthyes (derived from Greek words for "bone fish") refers to the skeleton of these animals, which is composed of bone rather than cartilage. From the snakelike moray eel to bizarre, luminescent deep-sea forms to the streamlined tuna, this enormously successful group has spread to nearly every possible watery habitat, both freshwater and marine (Fig. 24-33). Although about 17,000 species have been identified, scien-

tists predict that perhaps nearly twice this number may exist if the undescribed species from deep water and remote areas are considered. For example, a bony fish called a coelacanth, believed to have been extinct for 75 million years, was caught in deep water off the coast of South Africa in 1939 (Fig. 24-33d).

A feature found in early representatives of this group and retained in a few modern species is the presence of lungs that supplement the gills. Lungs are adaptations that allow life in fresh water, which can become foul and stagnant or dry up entirely. The swim bladder, a sort of internal balloon that allows most bony fishes to float effortlessly at any level, probably evolved from the lungs of freshwater ancestors. Some groups evolved another feature, modified fleshy fins that could be used (in an emergency) as legs, dragging the fish from a drying puddle to a deeper pool. From such ancestors arose a group that made the first tentative invasion of the land: the amphibians.

Success on Dry Land Required Numerous Adaptations

Land offered many advantages, including abundant food, shelter, and no predators, to those animals that first crawled from the water. But the price was high. Deprived of water's support, the body was heavy and clumsy to drag along on modified fins or weak, poorly adapted legs. Unsupported by water, gills collapse and become useless. The dry air and relentless sun suck vital water from unprotected skin and eggs, while temperature fluctuates dramatically compared with the relatively steady environmental conditions of the sea. The terrestrial habitat thus exerted strong selection pressure favoring the preservation of a number of mutations that happened to be adaptive on land. These included changes in skeletal structure that provided better support for the body, coverings for skin and eggs that reduced the evaporation of water, protection of the respiratory membranes, control of body temperature, and more efficient circulation.

Amphibian Means "Double Life"

The class Amphibia consists of about 2500 species, including frogs, toads, and salamanders, that straddle the boundary between aquatic and terrestrial existence (Fig. 24-34). The limbs of amphibians show varying degrees of adaptation to movement on land, from the belly-dragging crawl of salamanders to the efficient leap of frogs. Lungs replace gills in most adult forms, and a three-chambered heart (in contrast to the two-chambered heart of fishes) circulates blood more efficiently (see Chapter 30). However, the skin of frogs and salamanders must remain moist, since it serves as an additional respiratory organ that supplements poorly developed lungs. This constraint greatly restricts their habitats on land and explains why frogs are rarely found far from water.

(a)

(b)

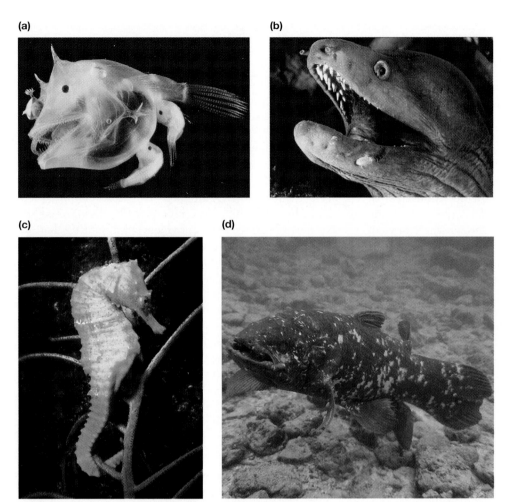

(c)

(d)

Figure 24-33 The diversity of the bony fishes

Bony fishes have colonized nearly every aquatic habitat. **(a)** This female deep-sea angler fish attracts prey with a living lure projecting just above her mouth. The fish is ghostly white because in the 2000-meter depth where anglers live, no light penetrates and thus colors are superfluous. The male deep-sea angler fish are extremely small and remain as permanent parasites attached to the female, always available to fertilize her eggs. Two parasitic males can be seen attached to this female. **(b)** This tropical green moray lives in rocky crevices. Look closely and you will see a small fish (a banded cleaner goby) on its lower jaw. This fish eats parasites that cling to the moray's skin. **(c)** The tropical seahorse may anchor itself with its prehensile tail, adapted for grasping, while feeding on small crustaceans. **(d)** A rare photo of a coelacanth, a fish once believed to be long extinct, in its natural habitat off the coast of South Africa.

(a)

(b)

(c)

Figure 24-34 Amphibian means "double life"

The double life of amphibians is illustrated by the transition from completely aquatic larval tadpole **(a)** to adult bullfrog leading a semiterrestrial life **(b)**. **(c)** The red salamander is restricted to moist habitats in the eastern United States. Salamanders hatch in a form that closely resembles the adult.

EARTH WATCH

Amphibians in Decline

The earliest amphibians appeared on Earth 350 million years ago, and frogs and toads have been around for nearly 150 million years. But recently herpetologists (biologists who study reptiles and amphibians) have been reporting an alarming trend: Worldwide, thousands of species of frogs, toads, and salamanders are experiencing a dramatic decline in numbers. Although the causes of the declines are diverse and poorly understood, they can be traced to a single source: human modification of the biosphere—the portion of Earth that sustains life.

Habitat destruction, particularly in the tropics, is one major cause of the decline. However, the unique biology of amphibians makes them vulnerable even where their habitat is not threatened. The double life of amphibians exposes them to toxic substances in a wide range of habitats, including water, air, and soil. Many amphibians' eggs develop in ponds and streams during the spring. Acid precipitation (see Chapter 45) has made spring a dangerous time for aquatic organisms. The melting of acid snow and ice causes a springtime "pulse" of intense acidity in freshwater ecosystems, just as many amphibian eggs are undergoing critical stages of development. Recent research by Andrew Blaustein, an ecologist at Oregon State University, demonstrated that amphibian eggs are also damaged by ultraviolet (UV) light. The intensity of UV light is increasing over parts of the globe because atmospheric pollutants have caused thinning of the protective ozone layer (see Chapter 45). Further, at all stages of life, amphibians are covered by a thin, permeable skin through which pollutants in air or water can penetrate. Some amphibians also feed on insects that have accumulated insecticides in their bodies.

Many scientists believe that the decline in the amphibian population signals an overall deterioration of Earth's ability to support life; and the decline is worldwide. Yosemite toads and yellow-legged frogs are disappearing from the mountains of California, while tiger salamanders have been nearly wiped out in the Colorado Rockies. Leopard frogs, eagerly chased by rural children throughout the United States, are suddenly becoming rare. While logging destroys the habitats of amphibians from the Pacific Northwest to the tropics (Fig. E24-1), even those in protected areas are dying. In the Monteverde Cloud Forest Preserve in Costa Rica, the golden toad (see Fig. 39-6) was common

in the early 1980s but has not been seen since 1989. The gastric brooding frog fascinated biologists by swallowing its eggs, brooding them in its stomach, and later regurgitating fully formed offspring. This species was abundant and seemed safe within a national park in Australia. Then suddenly, in 1980, the gastric brooding frog disappeared and has never been seen since. Evidence is mounting that setting aside small islands of nature in preserves amid surrounding environmental contamination and destruction is not enough to save species from extinction.

Amphibians are not just sensitive indicators of the health of the biosphere; they are a crucial component of many ecosystems. They may keep insect populations in check, while in turn serving as food for larger carnivores. Their decline will further disrupt the balance of these delicate communities. Margaret Stewart, an ecologist at the State University of New York, Albany, aptly summarized the problem: "There's a famous saying among ecologists and environmentalists: 'Everything is related to everything else.' . . . You can't wipe out one large component of the system and not see dramatic changes in other parts of the system."

Figure E24-1   Amphibians in danger

The corroboree toad, shown here with its eggs, is rapidly declining in its native Australia. Tadpoles can be seen developing within the eggs. The thin water- and gas-permeable skin of the adult, and the jellylike coating surrounding the eggs make them vulnerable to both air and water pollutants.

Fertilization of amphibian eggs is usually external and must therefore occur in water so that the sperm can swim to the eggs. The eggs are particularly vulnerable to loss of water by evaporation because they are surrounded by only a jellylike coating. The eggs must therefore be laid in water, where

they develop into aquatic larvae—the tadpoles of frogs and toads, for example. The dramatic transition from completely aquatic larva to semiterrestrial adult gives the class Amphibia its name, which means "double life." Their double life and the thin, permeable skin of amphibians have made

(a)

(b)

(c)

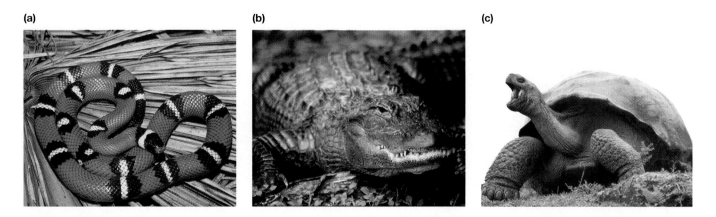

Figure 24-35   **The diversity of reptiles**

(a) The mountain king snake has evolved a color pattern very similar to that of the poisonous coral snake, which potential predators avoid. This mimicry helps the harmless king snake avoid predation. **(b)** The American alligator, found in swampy areas of the south, has survived for 150 million years with little change. **(c)** The tortoises of the Galapagos Islands, Ecuador, may live to be over 100 years old.

them particularly vulnerable to pollutants and ultraviolet light, as described in "Earth Watch: Amphibians in Decline."

Reptiles Have Adaptations That Suit Them to Life on Dry Land

The approximately 7000 species of reptiles (members of the class Reptilia) include the lizards and snakes (by far the most successful of the modern groups) and the turtles, alligators, and crocodiles, which have survived virtually unchanged from prehistoric times (Fig. 24-35). Reptiles evolved from an amphibian ancestor about 250 million years ago. Early reptiles—the dinosaurs—ruled the land for nearly 150 million years.

Some reptiles, particularly desert dwellers such as tortoises and lizards, have achieved complete independence from their aquatic origins. This independence was achieved through a series of adaptations, of which three are outstanding. First, reptiles evolved a tough, scaly skin that resists water loss and protects the body. Second, reptiles evolved internal fertilization, in which the male deposits sperm within the female's body. Third, reptiles evolved a shelled **amniotic egg** that can be buried in sand or dirt, far away from water with its hungry predators. The shell prevents the egg from drying, while an internal membrane, the **amnion**, encloses the embryo in the watery environment that all developing animals require (Fig. 24-36). In addition to these features, reptiles have

(a)

(b)

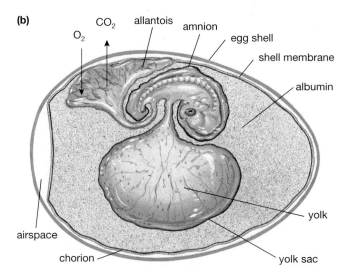

Figure 24-36   **The amniotic egg**

(a) An anole lizard from Carolina struggles free of its egg. **(b)** The amniotic egg of reptiles and birds is shown diagrammatically. In addition to the shell, which helps prevent dehydration, the egg contains several membranes. Enclosing the embryo in a watery "pond" is the amnion. The allantois stores the urinary wastes of the embryo and exchanges gases that can diffuse in and out through the shell. A yolk sac surrounds the fatty yolk, a high-energy food source. The albumin is a protein food source. The chorion encloses the albumin, the embryo and its associated membranes.

(a)   **(b)**   **(c)**

Figure 24-37   **The diversity of birds**

(a) The delicate hummingbird beats its wings about 60 times per second and weighs about 0.15 ounce (4 grams). **(b)** This young frigate bird, a fish-eater from the Galapagos Islands, has nearly outgrown its nest. **(c)** The ostrich is the largest of all birds, weighing over 136 kilograms (300 pounds) and producing eggs weighing over 1500 grams (3 pounds).

more efficient lungs, dispensing with the skin as a respiratory organ. The three-chambered heart became modified, allowing better separation of oxygenated and deoxygenated blood, and the limbs and skeleton evolved adaptations that provided better support and more efficient movement on land.

Birds Have Many Adaptations That Support the Demands of Flight

Having spread through the sea and over the land, vertebrates took to the air, a source of abundant insect food and a haven from predators. The 8600 species of birds

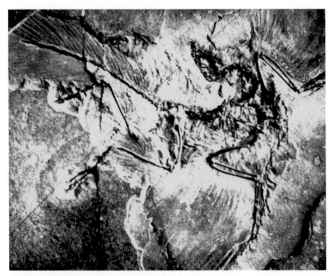

Figure 24-38   *Archaeopteryx,* **the "missing link" between reptiles and birds**

This 150-million-year-old fossil shows a remarkable animal possessing a beak with sharp teeth, a long jointed tail, clawed wings, and feathers.

(class Aves) attest to the advantages of flight (Fig. 24-37). The first birdlike creatures, reptiles modified for flight, appeared roughly 150 million years ago (Fig. 24-38). Body scales were dramatically modified to form feathers, while scales on the legs remained, testimony to the reptilian origin of these species.

Many aspects of bird anatomy and physiology help meet the rigorous demands of flight. For example, the body temperature of reptiles, fish, amphibians, and invertebrates varies with the temperature of their environment. Although these animals are often called cold blooded, they exert some control of their body temperature by behavioral means: basking in sun or seeking shade. Birds, in contrast, regulate their body temperature by physiological as well as behavioral mechanisms, maintaining a temperature that is usually higher than that of their surroundings. This elevated body temperature allows both muscles and metabolic processes to operate at peak efficiency, regardless of the outside temperature, supplying the power and the energy necessary to fly. Birds and mammals (which also have this capacity) are often described as warm blooded. These animals have a high metabolic rate that increases their demand for energy and requires efficient oxygenation of tissues. The heart of a bird, which has four chambers, helps meet this demand by completely separating oxygenated and deoxygenated blood. The respiratory system of birds is supplemented by air sacs that supply oxygenated air to the lungs even as the bird exhales.

Birds are exceptionally light for their size, another adaptation for flight. Feathers protect, streamline, and insulate the body; they also form lightweight extensions to the wings and the tail for the lift and control demanded by flight. Hollow bones reduce the weight of the skeleton to a fraction of that of other vertebrates. Reproductive organs are considerably reduced in size during nonbreeding periods, and female birds possess only a single ovary, further minimizing weight. The shelled egg that contributed to the reptiles' success on land frees the mother bird from carrying her developing offspring. Finally, the nervous system of birds accommodates the special demands of flight with extraordinary coordination and balance combined with acute eyesight.

Mammals Possess Hair and Produce Milk for Their Offspring

As one line of reptiles was developing feathers, a different group, the mammals (class Mammalia), was evolving hair. The mammals first appeared approximately 250 million years ago and came into prominence after the extinction of the dinosaurs roughly 70 million years ago. Today they are represented by some 4500 species. Like birds, mammals are "warm blooded" and have high metabolic rates. In most mammals, fur protects and insulates the warm body. Like birds, mammals have four-chambered hearts that increase the amount of oxygen delivered to the tissues. Legs designed for running rather than crawling make many mammals fast and agile. In contrast to birds, whose bodies are almost uniformly molded to the requirements of flight, mammals have evolved a remarkable diversity of form. The bat, mole, impala, whale, seal, monkey, and cheetah exemplify the radiation of mammals into nearly all habitats, with bodies finely adapted to their varied lifestyles (Fig. 24-39).

Mammals are named for the **mammary glands** used by all female members of this class to suckle their young (Fig. 24-39c). In addition to these unique milk-producing glands, the mammalian body is arrayed with sweat, scent, and sebaceous (oil-producing) glands, none of which are found in reptiles. With the exception of the egg-laying **monotremes**, such as the platypus and

(a)

(b)

Figure 24-39 **The diversity of mammals**

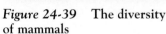

(c)

(d)

(a) A humpback whale gives its offspring a boost. (b) A bat, the only mammal capable of true flight, navigates at night using a kind of sonar. Large ears aid in detecting echoes as its high-pitched cries bounce off nearby objects. (c) Mammals are named after the mammary glands with which females nurse their young, as illustrated by this mother cheetah. (d) The male orangutan can reach 75 kilograms (165 pounds). These gentle, intelligent apes occupy swamp forests in limited areas of the tropics. They are endangered by hunting and habitat destruction.

(a)

(b)

Figure 24-40   **Nonplacental mammals**

(a) Monotremes, such as this platypus from Australia, lay leathery eggs resembling those of reptiles. The newly hatched young obtain milk from slitlike openings in the mother's abdomen. **(b)** Marsupials, such as the wallaby, give birth to extremely immature young who immediately grasp a nipple and develop within the mother's protective pouch (inset).

spiny anteater (Fig. 24-40a), mammalian embryos develop in the uterus. In one specialized group, the **marsupials** (including opossums, koalas, and kangaroos), the period of uterine development is short and the young are born at a very immature stage of development. Immediately after birth they crawl into a protective pouch (Fig. 24-40b), firmly grasp a nipple, and complete their development nourished by milk. Most mammals, called **placental** mammals, retain their young in the uterus for a much longer period. (*Placental* refers to the **placenta**, the uterine structure that functions in gas, nutrient, and waste exchange between circulatory systems of the mother and embryo.)

The mammalian nervous system has contributed significantly to the success of this group by making possible behavioral adaptation to changing and varied environments. The brain is more highly developed than in any other class, endowing mammals with unparalleled curiosity and learning ability. The highly developed brain allows mammals to alter their behavior on the basis of experience and helps them survive in a changing environment. Relatively long periods of parental care after birth allow some mammals to learn extensively under parental guidance; humans and other primates are good examples. In fact, the intellectual development of humans has led to their domination of the environment, as explored in the following section.

EVOLUTIONARY CONNECTIONS

Are Humans a Biological Success?

Physically, human beings are fairly unimpressive biological specimens. We are not very strong for such large animals, or very fast, and we lack the natural weapons of fang and claw. It is the human mind, with its tremendously developed cerebral cortex, that truly sets us apart from other animals. Our minds, in single bursts of brilliance and in the collective pursuit of common goals, have created wonders. No other animal could even appreciate the Parthenon, much less sculpt the graceful columns of this ancient Greek temple. We alone can eradicate smallpox and polio, domesticate other life forms, penetrate space with rockets, and fly to the stars in our imaginations.

And yet, are we, as it appears at first glance, the most successful of all living things? The duration of human existence is a mere instant in the 3.5-billion-year span of life on Earth. But during the last 300 years, the human population has increased from 0.5 billion to 5.5 billion and now grows by a million people every 4 days. Is this a measure of our success? As we have expanded our range over the globe, we have driven at least 300 major species to extinction. Within your lifetime, the rapid destruction of tropical rain forests and other diverse habitats may wipe out millions of species of plants, invertebrates, and vertebrates, most of which we will never know. Many of our activities have altered the environment in ways that are hostile to life, including our own. Acid from power plants and automobiles rains down on the land, threatening our forests and lakes—and eroding the marble of the Parthenon. Deserts spread as land is stripped by overgrazing and the demand for firewood.

Our aggressive tendencies, spurred by pressures of expanding wants and needs, their scope magnified by our technological prowess, have given us the capacity to destroy ourselves and most other life forms as well.

The human mind is the source of our most pressing problems—and our greatest hope for solving them. Will we now devote our mental powers to reducing our impact, controlling our numbers, and preserving the ecosystems that sustain us and other forms of life? Are we a phenomenal biological success—or a brilliant catastrophe? Perhaps the next few centuries will tell.

❖ SUMMARY OF KEY CONCEPTS

The Features of Animals
Animals are multicellular, sexually reproducing, heterotrophic organisms. Most can perceive and react rapidly to environmental stimuli and are motile at some stage in their lives. Their cells lack a cell wall.

Evolutionary Trends in Animal Body Plans
Several trends are apparent in the evolution of animals. There was a trend toward greater overall complexity, culminating with the arthropods, mollusks, echinoderms, and chordates. Cellular organization increased from specialized cells through tissues to organs to organ systems. Three separate germ layers arose. Body forms evolved from asymmetry through radial symmetry to bilateral symmetry. Sense organs and clusters of neurons became increasingly concentrated in the head, a process called cephalization. Body cavities were originally absent until animals with pseudocoels and later true coeloms evolved. Segmentation, first seen in annelids, is present to some degree in most of the more complex animals. In the simple sponges, digestion is intracellular; later, a saclike gastrovascular cavity evolved, and finally, a one-way digestive tract.

The Major Animal Phyla
Vertebrates, animals with backbones, constitute a single subphylum within the phylum Chordata. All other animals lack a backbone and are called invertebrates. Over 97% of the animal species on Earth are invertebrates.

The Sponges: Phylum Porifera
Sponge bodies are often free-form in shape and are always sessile. Sponges have relatively few types of cells, and although division of labor among the cell types is present, there is little coordination of activity. Sponges lack the muscles and nerves required for coordinated movement, and digestion occurs exclusively within the individual cells.

The Hydra, Anemones, and Jellyfish: Phylum Cnidaria
Far more coordination is evident among cells of cnidarians than occurs in sponges. A simple network of nerve cells directs the activity of contractile cells, allowing loosely coordinated movements. Digestion is extracellular, occurring in a central gastrovascular cavity with a single opening. Cnidarians exhibit radial symmetry, an adaptation to the free-floating life-style of the medusa or the sedentary existence of the polyp.

The Worms: Phylum Platyhelminthes, Phylum Nematoda, and Phylum Annelida
Flatworms (phylum Platyhelminthes) are the first to show a distinct head with sensory organs and a simple brain. A system of canals forming a network through the body aids in excretion. The roundworms (phylum Nematoda) are the first to possess a separate mouth and anus. The segmented worms (phylum Annelida) are the most complex, with a well-developed closed circulatory system and excretory organs resembling the basic unit of the vertebrate kidney. The segmented worms have a compartmentalized digestive system, like that of vertebrates, which processes food in a sequence. Annelids have a true coelom, a fluid-filled space between the body wall and the internal organs, found in most complex animals, including vertebrates.

The Insects, Arachnids, and Crustaceans: Phylum Arthropoda
Arthropods are the most diverse and abundant organisms on Earth. They have invaded nearly every available terrestrial and aquatic habitat. Jointed appendages and well-developed nervous systems make possible complex, finely coordinated behavior. The exoskeleton (which conserves water and provides support) and specialized respiratory structures (which remain moist and protected) enable the insects and arachnids to inhabit dry land. The diversifica-

tion of insects has been enhanced by their ability to fly. Crustaceans, which include the largest arthropods, are restricted to moist, usually aquatic habitats and respire using gills.

The Snails, Clams, and Squid: Phylum Mollusca

Mollusks lack a skeleton, sometimes protecting the soft, moist, muscular body with a single shell (many gastropods and a few cephalopods) or a pair of hinged shells (the bivalves). The lack of a waterproof external covering limits this phylum to aquatic and moist terrestrial habitats. Although the body plan of gastropods and bivalves limits the complexity of their behavior, the cephalopod's tentacles are capable of precisely controlled movements. The octopus has the most complex brain and the best-developed learning capacity of any invertebrate.

The Sea Stars, Sea Urchins, and Sea Cucumbers: Phylum Echinodermata

Echinoderms are an exclusively marine group. Like other complex invertebrates and chordates, echinoderm larvae are bilaterally symmetrical; however, the adults show radial symmetry. This, in addition to a primitive nervous system that lacks a definite brain, adapts them to a relatively sedentary existence. Echinoderm bodies are supported by a nonliving internal skeleton that sends projections through the skin. The water-vascular system, which functions in locomotion, feeding, and respiration, is a unique echinoderm feature.

The Tunicates, Lancelets, and Vertebrates: Phylum Chordata

The phylum Chordata includes two invertebrate groups, the lancelets and tunicates, as well as the familiar vertebrates. All members of this phylum possess a noto-chord, a dorsal nerve cord, gill grooves, and a tail at some stage in their development. Vertebrates are a subphylum of chordates that have a backbone, which is part of a living endoskeleton. Vertebrate evolution is believed to have proceeded from the fishes to amphibians to reptiles, which gave rise to both birds and mammals. The heart increases in complexity from the two chambers in fishes, to three in amphibians and most reptiles, to four in the warm-blooded birds and mammals. The evolution of efficient transport of oxygen supported active life-styles, high metabolic rates, and constant body temperatures.

During the progression from fishes to amphibians to reptiles, a series of adaptations evolved that helped vertebrates colonize dry land. Amphibians have legs and most have simple lungs for breathing in air rather than in water. Most are confined to relatively damp terrestrial habitats by their need to keep their skin moist, use of external fertilization, and requirement that their eggs and larvae develop in water. Reptiles, with well-developed lungs, dry skin covered with relatively waterproof scales, internal fertilization, and the amniotic egg with its own water supply, are well adapted to the driest terrestrial habitats. Birds and mammals are also fully terrestrial and have additional adaptations, such as an elevated body temperature, which allows the muscles to respond rapidly regardless of the temperature of the environment. The bird body is molded for flight, with feathers, hollow bones, efficient circulatory and respiratory systems, and well-developed eyes. Mammals have insulating hair and give birth to live young that are nourished with milk. The mammalian nervous system is the most complex in the animal kingdom, providing mammals with enhanced learning ability that helps them adapt to changing environments.

✖ KEY TERMS

amnion p. 487	coelom p. 462	gastrovascular cavity p. 467
amniotic egg p. 487	collar cell p. 465	germ layer p. 461
amoeboid cell p. 465	compound eye p. 474	gill p. 474
anterior p. 462	cyst p. 470	gill slit p. 481
bilateral symmetry p. 461	dorsal p. 462	hemocoel p. 474
book lung p. 474	ectoderm p. 461	hermaphroditic p. 468
budding p. 466	endoderm p. 461	hydrostatic skeleton p. 470
carnivorous p. 472	endoskeleton p. 481	invertebrate p. 464
cartilage p. 483	epithelial cell p. 465	larva p. 476
cephalization p. 460	exoskeleton p. 473	mammary gland p. 489
chitin p. 473	flame cell p. 468	mantle p. 477
closed circulatory system p. 472	free-living p. 468	marsupial p. 490
cnidocyte p. 467	ganglion p. 468	medusa p. 466

mesoderm   p. 461
metamorphosis   p. 476
molt   p. 473
monotreme   p. 489
nephridium   p. 472
nerve cord   p. 481
nerve net   p. 466
notochord   p. 481
open circulatory system   p. 474
organ   p. 461
organ system   p. 461

parasite   p. 470
pharynx   p. 468
placenta   p. 490
placental   p. 490
polyp   p. 466
posterior   p. 462
pseudocoel   p. 463
pupa   p. 476
radial symmetry   p. 461
radula   p. 477
segmentation   p. 460

sessile   p. 465
spicule   p. 465
tail   p. 481
tentacle   p. 466
tissue   p. 461
tracheae   p. 474
tube foot   p. 480
ventral   p. 462
vertebral column   p. 482
vertebrate   p. 464
water-vascular system   p. 480

✖ THINKING THROUGH THE CONCEPTS

Multiple Choice

1. Which of the following animals have radial symmetry?
   a. jellyfish   b. sea star
   c. sea anemone   d. sea urchin
   e. all of the above
2. Animals in which of the following phyla have collar cells?
   a. Porifera   b. Cnidaria
   c. Annelida   d. Gastropoda
   e. Chordata
3. What is a radula?
   a. a flexible supportive rod on the dorsal surface of chordates
   b. a stinging cell used by sea anemones to capture prey
   c. a gas-exchange structure in insects
   d. a spiny ribbon of tissue used for feeding in snails
   e. a locomotory structure of sea stars
4. Which of the following classes of animals includes the first vertebrates to appear on Earth?
   a. Agnatha, the jawless fishes
   b. Chondrichthyes, the sharks
   c. Osteichthyes, the bony fishes
   d. Tunicata, the sea squirts
   e. none of the above
5. Which of the following molts its exoskeleton, allowing it to grow larger?
   a. the blue crab
   b. the bat sea star
   c. the scallop
   d. the sea urchin
   e. the Venus clam
6. Which of the following has a gastrovascular cavity?
   a. sponges   b. sea stars
   c. nematodes   d. flatworms
   e. earthworms

Review Questions

1. List the distinguishing characteristics of each of the phyla discussed in this chapter and give an example of each.
2. Briefly describe each of the following adaptations and explain the adaptive significance of each: amniotic egg, bilateral symmetry, cephalization, closed circulatory system, coelom, placenta, radial symmetry, segmentation.
3. Describe and compare respiratory systems in the three major arthropod classes.
4. Describe the advantages and disadvantages of the arthropod exoskeleton.
5. State in which of the three major mollusk classes each of the following characteristics is found:
   a. two hinged shells
   b. a radula
   c. tentacles
   d. some sessile members
   e. the best-developed brains
   f. numerous eyes
6. Give three functions of the water-vascular system of echinoderms.
7. To what life-style is radial symmetry an adaptation? Bilateral symmetry?
8. List the vertebrate class (or classes) in which we find each of the following:
   a. a skeleton of cartilage
   b. a two-chambered heart
   c. the amniotic egg
   d. warm-bloodedness
   e. a four-chambered heart
   f. a placenta
   g. lungs supplemented by air sacs
9. Distinguish between vertebrates and invertebrates. List the major phyla found in each broad grouping.
10. List four distinguishing features of chordates.
11. Describe the ways in which amphibians are adapted to life on land, and in what ways they are still restricted to a watery or moist environment.
12. List the adaptations that distinguish reptiles from amphibians and help them adapt to life in dry terrestrial environments.
13. List the adaptations of birds that contribute to their ability to fly.
14. How do mammals differ from birds, and what adaptations do they share?
15. How has the mammalian nervous system contributed to the success of this group?

✖ APPLYING THE CONCEPTS

1. The class Insecta is the largest taxon of animals on Earth, and its greatest diversity is in the tropics, where habitat destruction and species extinction are occurring at an alarming rate. What biological, economic, and ethical arguments can you advance to persuade people and governments to preserve this biological diversity?

2. Animals, particularly laboratory rats, are used extensively in medical research. What advantages and disadvantages do lab rats have as research subjects in medicine? What are the advantages and disadvantages of using animals more closely related to humans?

3. Discuss at least three ways in which the ability to fly has contributed to the success and diversity of insects.

4. Discuss and defend what attributes you would use to define biological success among animals. Are humans a biological success by these standards? Why or why not?

✖ FOR MORE INFORMATION

Blaustein, A. "Amphibians in a Bad Light." *Natural History*, October 1994. Recent declines in amphibian population size and overall diversity are linked to possible harm from ultraviolet light penetrating a depleted ozone layer.

Diamond, J. "Stinking Birds and Burning Books." *Natural History*, October 1994. A description of a recently described species of bird (the pitohuis) and its peculiar chemical ecology.

Duellman, W. E. "Reproductive Strategies of Frogs." *Scientific American*, July 1992. Free-living tadpoles are only one way in which these amphibians progress from egg to adult.

Hamner, W. "A Killer Down Under." *National Geographic*, August 1994. Among the most poisonous animals in the world is the box jellyfish that lives off the coast of northern Australia.

Horridge, G. A. "The Compound Eye of Insects." *Scientific American*, July 1977. A close look at the multifaceted eyes of insects.

McMenamin, M. A. S. "The Emergence of Animals." *Scientific American*, May 1987.

Montgomery, S. "New Terror of the Deep." *International Wildlife*, July/August 1992. A description of the threat that overfishing by humans poses to shark populations.

Morell, V. "Life on a Grain of Sand." *Discover*, April 1995. The sand beneath shallow waters is home to an incredible range of microscopic creatures.

Phillips, K. "Frogs in Trouble." *International Wildlife*, November–December 1990. A documentation of the puzzling worldwide decline in amphibian populations.

Rahn, H., Ar, A., and Paganelli, C. V. "How Bird Eggs Breathe." *Scientific American*, February 1979. (Offprint No. 1420). A description of the remarkable adaptations of the amniotic egg for gas exchange.

Rennie, J. "Living Together." *Scientific American*, January 1992. Most parasites do not kill their hosts, and the interaction of parasite and host provides some fascinating insights into the evolution of life on Earth.

NET WATCH

On-line resources for this chapter are on the World Wide Web at:
http://www.prenhall.com/~audesirk (click on the <u>table of contents</u> link and then select Chapter 24).

Unit IV
Plant Anatomy and Physiology

A rosette of leaves on a storksbill plant found in California

25 The Structure of Land Plants

AT A GLANCE

An Overview of Plant Structure

Plant Development and Growth

Plant Tissues and Cell Types
The Dermal Tissue System Forms the Covering for the Plant
    Body
The Ground Tissue System Makes Up Most of the Young
    Plant Body
The Vascular Tissue System Consists of Xylem and Phloem

Roots: Anchorage, Absorption, and Storage
Primary Growth Causes Roots to Elongate
The Epidermis of the Root Is Very Permeable to Water
Cortex Makes Up Much of the Interior of a Young Root
The Vascular Cylinder Contains Conducting Tissues

Stems: Reaching for the Light
The Stem Has a Complex Organization Including Four
    Types of Tissue
Stem Branches Form from Lateral Buds Consisting of
    Meristem Cells
Secondary Growth Produces Thicker, Stronger Stems

Leaves: Nature's Solar Collectors
Leaves Have Two Major Parts: Blades and Petioles

Special Adaptations of Roots, Stems, and Leaves
Some Specialized Roots Store Food, Others Photosynthesize
Some Specialized Stems Produce New Plants, Store Water
    or Food, or Produce Thorns or Climbing Tendrils
Specialized Leaves May Conserve and Store Water, Store
    Food, or Even Capture Insects

Some of the diversity of plant forms is seen in these flowers and organ pipe cactus in Organ Pipe National Monument. The thick stems of cacti store water and carry out photosynthesis. Their leaves, reduced to spines, conserve water while providing protection.

Before the first animal crawled out of the sea onto the land, plants already stood there. When glaciers advanced and retreated across the land, plant communities sprung up in their wake, and again, animals followed the plants. The burst of human civilization that occurred about 10,000 years ago depended on the domestication of plants to provide a reliable food supply. As these examples suggest, terrestrial animals, including people, have always intimately depended on terrestrial plants.

Many people take plants for granted. We admire a field of wildflowers or a giant sequoia, but we seldom stop to think about the intricate structures and functions that allow even the most common plants to survive and prosper. Plants cannot move to escape enemies, to find food or water, to avoid the arrival of winter, or to locate a mate. Yet spruces perch atop the permafrost in Alaska, cypresses and mangroves stand immersed in swamps in Florida, and cacti bloom in the searing heat of Death Valley, California. Redwoods and eucalyptus trees tower a hundred meters above the forest floor; grasses survive grazing year after year; and bristlecone pines in the White Mountains of California may live to be 4000 years old. Clearly, plants have evolved many extremely successful adaptations that enable them to thrive in a wide variety of habitats.

By far the most abundant land plants are the tracheophytes, or vascular plants, a group that includes ferns, conifers, and flowering plants. Tracheophytes have specialized conducting cells, or vessels, that transport water and nutrients throughout their bodies and also provide support. Two groups within the tracheophytes are particularly widespread, the flowering plants (division Anthophyta, also called **angiosperms**) and the conifers (division Coniferophyta, part of a broader group, the **gymnosperms**). As discussed in Chapter 23, two major evolutionary advances distinguish these groups and have contributed to their success on land. One of these is pollen, a much reduced male gametophyte carried by animals, wind, or water. By eliminating the requirement that sperm swim to the egg through water, pollen has allowed conifers and flowering plants to invade much drier habitats. A second adaptation is the seed, which encases the embryonic plant and a supply of food within a protective outer coat, providing both protection and a means of dispersal. In this unit, we explore the adaptations that make these plants so successful on land, beginning with the tissues and structures that constitute the plant body. Because of their dominance in most terrestrial habitats and their importance to humans, we will focus on the flowering plants.

An Overview of Plant Structure

Flowering plants consist of two major regions, the root system and the shoot system (Fig. 25-1). The **root system** is usually below ground and serves five functions. Roots (1) anchor the plant in the ground; (2) absorb water and minerals; (3) store surplus sugars manufactured during photo-

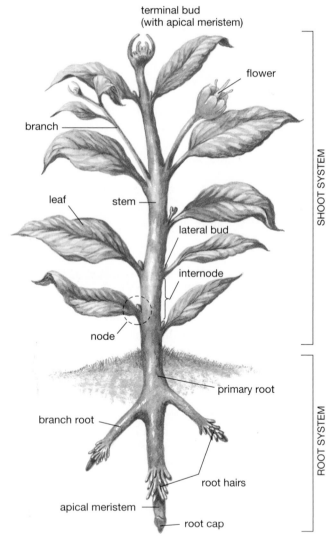

terminal bud
(with apical meristem)

flower

branch

leaf

stem

lateral bud

internode

node

primary root

branch root

root hairs

apical meristem

root cap

SHOOT SYSTEM

ROOT SYSTEM

Figure 25-1   **Flowering plant structure**

A flowering plant consists of the shoot and the root systems. The shoot system includes stems (often with branches), buds, and leaves. In the appropriate season, the shoot often bears flowers and fruit. The root system is usually highly branched, with many lateral roots growing from one or more main roots. Root hairs are drawn much larger than actual size.

synthesis; (4) transport water, minerals, sugars, and hormones to and from the shoot; and (5) produce some hormones.

The rest of the plant is the **shoot system,** usually found above ground. The shoot system consists of stems (with branches), leaves, buds, and (in season) flowers and fruits. The functions of shoots include: (1) photosynthesis, mainly in leaves and young green stems; (2) transport of materials between leaves, flowers, fruits, and roots; (3) reproduction; and (4) hormone synthesis.

Figure 25-2 illustrates the two broad groups of flowering plants: **monocots** (class Monocotyledoneae, which includes grasses, lilies, and orchids) and **dicots** (class Dicotyledoneae, which includes deciduous trees and bushes and many garden flowers). Monocots and dicots differ in the structure of their seeds, flowers, leaves, vascular tissue, and root pattern. Don't worry about terms that are not yet familiar to you; just look over the figure for now and refer to it later as we examine the parts of the flowering plants in more detail.

Plant Development and Growth

Animals and plants develop in dramatically different ways. One difference is very obvious: the timing and distribution of growth. As you grew from a baby to an adult, all parts of your body became larger. When you reached your adult height, you stopped growing (up, at least!). In contrast, flowering plants grow throughout their lives, never reaching a stable "adult" body form. What's more, most plants grow longer only at the tips of their branches and roots, and structures that developed earlier remain in exactly the same place; a swing tied to a tree branch does not move farther from the ground each year. Why do plants grow this way?

From the moment they sprout, plants are composites of two fundamentally different categories of cells: meristem cells and differentiated cells. Embryonic, undifferentiated **meristem cells** are capable of cell division. Mature, **differentiated cells** are specialized in structure and function, and usually do not divide. Normally, each time a meristem cell divides, one daughter cell develops specialized structures and becomes a differentiated cell, while the other daughter cell remains meristematic (Fig. 25-3). Continued divisions of meristem cells, then, keep a plant growing throughout its life, while their differentiated daughter cells form relatively permanent parts of the plant, such as mature leaves or the trunks of trees.

Plants grow through the division and differentiation of apical and lateral meristem cells. **Apical** ("tip") **meristems** are located at the ends of roots and shoots (including main stems and branches; see Figs. 25-10, 25-14, and 25-16). **Lateral** ("side") **meristems,** or **cambia** (singular, **cambium**), form cylinders that run parallel to the long axis of roots and stems (see Fig. 25-16).

Plant growth takes two forms: primary growth and secondary growth. **Primary growth** occurs through mitotic cell division of apical meristem cells followed by differen-

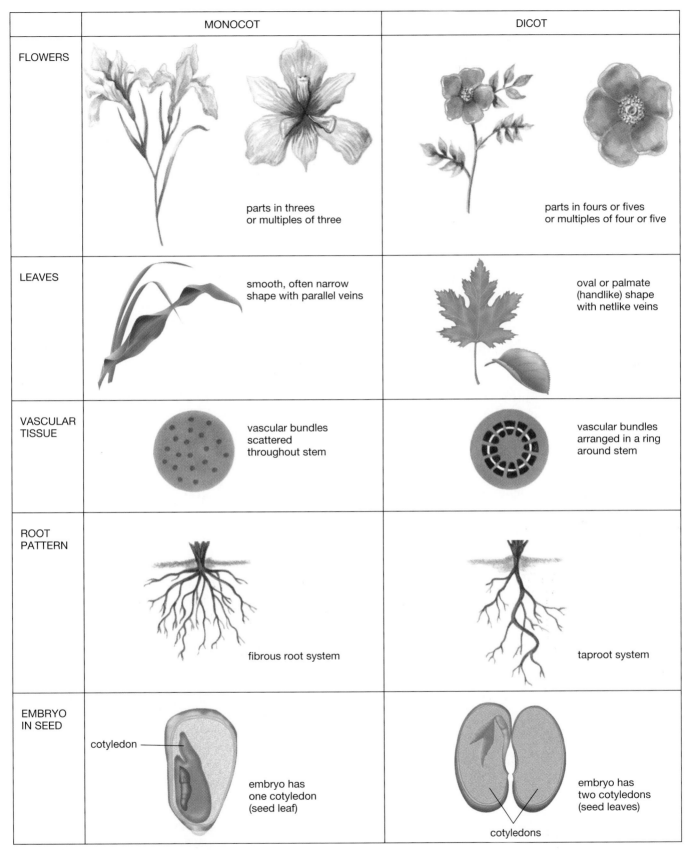

	MONOCOT	DICOT
FLOWERS	parts in threes or multiples of three	parts in fours or fives or multiples of four or five
LEAVES	smooth, often narrow shape with parallel veins	oval or palmate (handlike) shape with netlike veins
VASCULAR TISSUE	vascular bundles scattered throughout stem	vascular bundles arranged in a ring around stem
ROOT PATTERN	fibrous root system	taproot system
EMBRYO IN SEED	cotyledon — embryo has one cotyledon (seed leaf)	embryo has two cotyledons (seed leaves) cotyledons

Figure 25-2 **Monocots and dicots compared**

Distinguishing traits of the two major classes of flowering plants, the monocots and the dicots. The word *monocot* refers to the fact that the embryo in the seed has a single ("mono") cotyledon, or seed leaf; dicot means "two cotyledons."

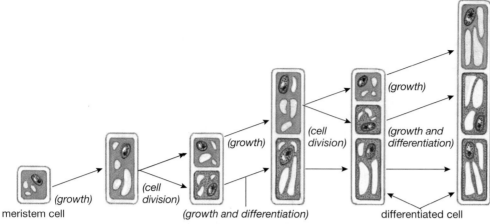

(growth)

meristem cell

(cell division)

(growth)

(growth and differentiation)

(cell division)

(growth)

(growth and differentiation)

differentiated cell

Figure 25-3   How meristem cells produce mature plant cells

Plant cell development occurs through divisions of meristem cells and differentiation of one of the daughter cells. A meristem cell (green) takes in nutrients, grows, and divides. One daughter cell remains meristematic, while the other differentiates into a particular specialized cell type (tan). You may have noticed that many plants (trees, for example) grow longer only at the tips of their branches. This diagram shows why: at each division, the lower daughter cell becomes differentiated while the upper cell remains meristematic. Therefore, the meristem cells are carried along at the tip of the shoot as the plant grows.

tiation of the resulting daughter cells. This type of growth occurs both in young plants and in the growing tips of roots and shoots of older plants. Primary growth is responsible both for increase in length and for the development of the specialized plant structures.

Secondary growth occurs through mitotic cell division of lateral meristem cells and differentiation of their daughter cells. Secondary growth is therefore responsible for growth in diameter. The stems and roots of most conifers and dicots become thicker and woodier as they age, for example, as a result of secondary growth. Although we later discuss secondary growth only in stems, keep in mind that secondary growth occurs in roots as well.

Plant Tissues and Cell Types

The major structures of land plants, including roots, stems, and leaves, consist of three tissue systems (Fig. 25-4). The **dermal tissue system** covers the outer surfaces of the plant body. The **ground tissue system,** which consists of all nondermal and nonvascular tissues, makes up most of the body of young plants. Its functions include photosynthesis, support, and storage. The **vascular tissue system** transports water, minerals, sugars, and plant hormones throughout the plant. Each tissue system arises from divisions of meristem cells and differentiation of the daughter cells, and each consists of one or (usually) several types of differentiated cells. Some flowering plants are called herbaceous. These are soft-bodied plants with flexible stems, such as lettuce, beans, and grasses, and usually live only one year. Other plants, such as trees

and bushes, are described as woody. They are usually perennial (living many years) and develop hard, thickened, woody stems as a result of secondary growth. As you will see, different types of tissue are present in herbaceous and woody plants.

The Dermal Tissue System Forms the Covering for the Plant Body

The dermal tissue system is the outer covering of the plant body. There are two types of dermal tissue: epidermal tissue and periderm (Fig. 25-5).

Epidermal tissue forms the **epidermis,** the outermost cell layer that covers the leaves, stems, and roots of all young plants (Fig. 25-5a). Epidermal tissue also covers flowers, seeds, and fruit. In herbaceous plants, the epidermis is retained as the outer covering of the entire plant body throughout its life. The epidermal tissue of the aboveground parts of a plant is generally composed of thin-walled cells packed tightly together and covered with a waterproof, waxy **cuticle,** secreted by the epidermal cells, that reduces evaporation of water from the plant. The epidermal cells of roots, in contrast, are not covered with cuticle; roots, of course, absorb water and minerals through the epidermis, and a waterproof cuticle would prevent absorption.

Some epidermal cells produce fine extensions, called hairs. Many root epidermal cells bear **root hairs,** which greatly increase the absorptive surface area of the root. Epidermal hairs on the stems and leaves of desert plants reduce evaporative water loss by reflecting sunlight and producing an unstirred layer of air near the plant's surface.

Periderm replaces epidermal tissue on the roots and stems of woody plants as they age. It is composed primar-

or secretion of hormones. Under the proper conditions, many parenchyma cells are capable of cell division.

Collenchyma tissue consists of elongated, polygonal (many-sided) cells with irregularly thickened cell walls (Fig. 25-6b). Collenchyma cells are alive at maturity but usually cannot divide. Although strong, the cell walls of collenchyma are still somewhat flexible. In herbaceous plants, and in the leaf stalks and young growing stems of all plants, collenchyma tissue is an important source of support. The strings in celery stalks, for example, are mostly collenchyma cells in association with vascular tissue.

Sclerenchyma tissue consists of cells with thick, hardened secondary cell walls (Fig. 25-6c), reinforced with a stiffening substance called lignin. Like collenchyma, sclerenchyma cells support and strengthen the plant body, but unlike collenchyma, they die after they differentiate. Their hardened cell

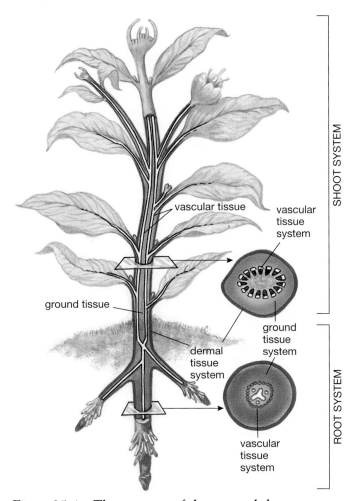

Figure 25-4 image labels: SHOOT SYSTEM, ROOT SYSTEM, vascular tissue, vascular tissue system, ground tissue, ground tissue system, dermal tissue system, vascular tissue system

Figure 25-4   **The structure of the root and shoot**

Both the root and shoot of a flowering plant consist of three tissue systems: dermal tissue system, ground tissue system, and vascular tissue system. (Notice that the structure of the leaves, vascular tissue system, and roots indicate that our sample plant is a dicot.)

ily of **cork cells**, which have thick, waterproofed walls and are dead at maturity (Fig. 25-5b). Cork cells form the protective outer layers of the bark of trees and woody shrubs and the woody covering of their roots.

The Ground Tissue System Makes Up Most of the Young Plant Body

The ground tissue system, which makes up the bulk of a young plant, consists of all nondermal and nonvascular tissues. There are three types of ground tissues: parenchyma, collenchyma, and sclerenchyma (Fig. 25-6).

Parenchyma tissue is the most abundant of the ground tissues. Parenchyma cells are thin-walled cells, alive at maturity, that typically carry out most of the metabolic activities of the plant (Fig. 25-6a). Depending on their location within the plant body, parenchyma cells have such diverse functions as photosynthesis, storage of sugars and starches,

(a)

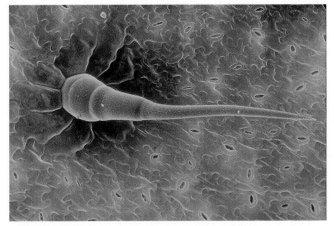

(b)

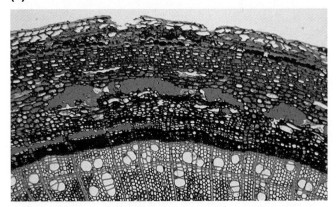

Figure 25-5   **Dermal tissues cover the surfaces of plants**

(a) The epidermis of a young root or shoot is a single layer of cells. In shoot epidermis, such as the epidermis of a zinnia leaf shown here, the outer surfaces of the cells are covered with a waxy, waterproof coating, the cuticle, that reduces the evaporation of water. The "leaf hair" protruding from the epidermis also reduces evaporation by slowing the movement of air across the surface of the leaf. **(b)** Woody stems and roots develop a thick, tough, waterproof periderm, which consists mostly of thick-walled cork cells.

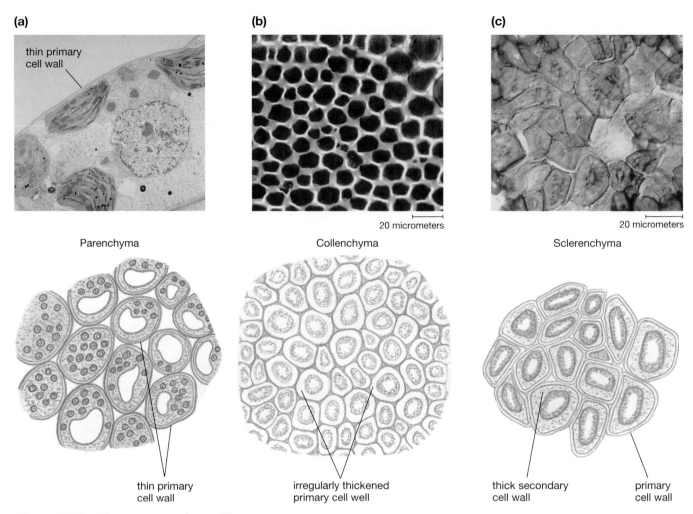

(a)

thin primary
cell wall

Parenchyma

thin primary
cell wall

(b)

20 micrometers

Collenchyma

irregularly thickened
primary cell well

(c)

20 micrometers

Sclerenchyma

thick secondary   primary
cell wall   cell wall

Figure 25-6   **The structure of ground tissue**

Ground tissue consists of three cell types: parenchyma, collenchyma, and sclerenchyma.
(a) Parenchyma cells are living cells with thin, flexible cell walls. These parenchyma cells have
abundant chloroplasts and carry out photosynthesis. Other parenchyma cells, particularly in
stems and roots, store starches (see the potato cells in Fig. 3-4a). **(b)** Collenchyma cells, such as
these from a celery stalk, are living cells with irregularly thickened, but still somewhat flexible,
walls. They help support the plant body. **(c)** Sclerenchyma cells, such as these "stone cells" that
give the meat of a pear its slightly gritty texture, have thick, rigid secondary cell walls.

walls then remain as a source of support. Sclerenchyma tissue
can be found in many parts of the plant body, including xylem
and phloem (described below). Sclerenchyma cells provide
the fibers of hemp and jute, which are used for making rope.
Other types of sclerenchyma cells form nut shells, the outer
covering of peach pits, and the gritty texture of pears.

The Vascular Tissue System Consists of Xylem and Phloem

The vascular tissue system consists of two complex conducting tissues: xylem and phloem. The major role of each
tissue is the transport of materials—water and minerals
up from the roots to the rest of the plant in xylem, and
water and sugars, amino acids, and hormones throughout
the plant body in phloem.

Xylem Conducts Water and Dissolved Minerals from the Roots to the Rest of the Plant Body

Xylem conducts water and minerals from roots to shoots in
tubes that are made from one of two types of cells: tracheids
and vessel elements (Fig. 25-7). Most conifers have only
tracheids; flowering plants usually have both tracheids and
vessel elements. As these cells differentiate, they develop
thick cell walls that help support the weight of the plant.
In some trees (pines, for example), the bulk of the tree
trunk consists of the thick cell walls of tracheids. The final
step in the differentiation of both tracheids and vessel elements is death: the cytoplasm and plasma membrane disintegrate, leaving behind a hollow tube of cell wall.

Tracheids are thin cells with slanted ends resembling
the tips of hypodermic needles. Tracheids are stacked

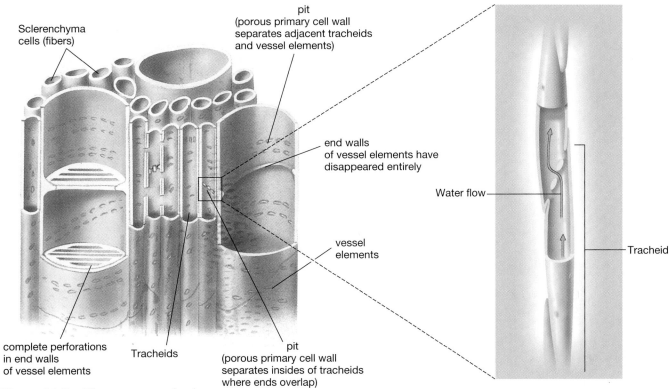

Sclerenchyma cells (fibers)

pit
(porous primary cell wall separates adjacent tracheids and vessel elements)

end walls
of vessel elements have disappeared entirely

vessel elements

Water flow

Tracheid

complete perforations in end walls of vessel elements

Tracheids

pit
(porous primary cell wall separates insides of tracheids where ends overlap)

Figure 25-7   **The structure of xylem**

Xylem is a mixture of cell types, including parenchyma and sclerenchyma fibers (ground tissue) and two types of conducting cells, tracheids and vessel elements. Tracheids are thin, with tapered, overlapping ends punctuated by pits. The pits are not holes; they have a water-permeable primary cell wall that separates the interiors of adjoining cells. Vessels consist of vessel elements stacked atop one another. The end walls of vessel elements are sometimes absent and sometimes have narrow openings that connect adjacent vessel elements. Both tracheids and vessel elements have pits in their side walls, allowing water and dissolved minerals to move sideways between adjacent conducting tubes.

atop one another with the slanted ends overlapping (see Fig. 25-7). The overlapping end walls contain **pits** where secondary cell walls are absent. These pits allow water and minerals to pass from one tracheid to the next by crossing only the thin, water-permeable primary cell wall.

Vessel elements also meet end to end, but they are larger in diameter than tracheids. Their ends may be either flat or overlapping and tapered. Open holes (pores) connect adjoining elements (cell walls remaining after the cells have died). In some cases, the ends of vessel elements completely disintegrate, leaving an open tube (see Fig. 25-7). Thus, vessel elements form wide-diameter, relatively unobstructed pipelines (called **vessels**) from root to leaf.

Phloem Conducts Water, Sugars, Amino Acids, and Hormones throughout the Plant Body

Phloem carries water with dissolved substances synthesized by the plant, including sugars, amino acids, and hormones through **sieve tubes** constructed of a single strand of cells

called **sieve-tube elements** (Fig. 25-8). As sieve-tube elements mature, most of their internal contents disintegrate, leaving behind only a thin layer of cytoplasm lining the plasma membrane. At the ends of sieve-tube elements, where adjacent cells meet, holes form in the primary cell walls, creating structures called **sieve plates.** The plasma membranes of the stacked sieve-tube elements are fused around the lips of the sieve plate pores, forming membrane-lined channels connecting the interiors of adjoining cells. A continuous conducting system is forged by many sieve-tube elements linking up end to end in this way.

A sieve-tube element has a plasma membrane, a few small mitochondria, and some endoplasmic reticulum, and is therefore considered to be alive, but it usually lacks ribosomes, Golgi complex, and a nucleus. How, then, can a sieve-tube element remain alive? Each sieve-tube element is nourished by a smaller, adjacent **companion cell.** Companion cells maintain the integrity of the sieve-tube elements by donating high-energy compounds and perhaps even by repairing the sieve-tube plasma membrane. As we will see in Chapter 26, companion cells also regulate the movements of sugars into and out of the sieve tubes.

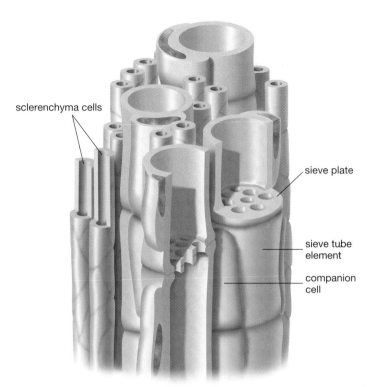

Figure 25-8   **The structure of phloem**

Phloem is a mixture of cell types, including sclerenchyma, sieve-tube elements, and companion cells. Sieve-tube elements have primary cell walls and have a thin layer of cytoplasm surrounding a fluid-filled core. Sieve-tube elements, stacked end to end, form the conducting system of phloem. Where they join, sieve-tube elements form sieve plates, where membrane-lined pores allow fluid to pass from cell to cell. Each sieve-tube element has a companion cell that nourishes it and regulates its function.

Roots: Anchorage, Absorption, and Storage

As a seed sprouts, the **primary root** grows down into the soil. Many dicots, such as carrots and dandelions, develop a taproot system. A **taproot system** consists of the primary root, which usually becomes longer and stouter with time, and many smaller roots that grow out from the sides of the primary taproot (Fig. 25-9a). In contrast, in monocots such as grasses, the primary root soon dies off, replaced by many new roots that emerge from the base of the stem. These secondary roots are nearly equal in size, forming a **fibrous root system** (Fig. 25-9b).

Primary Growth Causes Roots to Elongate

In young roots of both taproot and fibrous systems, divisions of the apical meristem give rise to four anatomically and functionally distinct regions (Fig. 25-10). At the very tip of the root, daughter cells produced on the "soil side" of the apical meristem differentiate into the **root cap**. The

root cap protects the apical meristem from being scraped off as the root pushes down between the rocky particles of the soil. Root cap cells have thick cell walls and secrete a slimy lubricant that helps to ease the way between soil particles. Nevertheless, root cap cells wear away and must be continuously replaced by new cells from the meristem.

Daughter cells produced on the "shoot side" of the apical meristem differentiate into three parts: an outer envelope of epidermis; a vascular cylinder at the core of the root; and, between the two, the cortex (Fig. 25-10).

The Epidermis of the Root Is Very Permeable to Water

The root's outermost covering of cells is the epidermis, which is in contact with the soil and any air or water trapped among the soil particles. The cell walls of the epidermal cells are highly water permeable. Therefore, water can penetrate into the interior of the root either by passing through the membranes of the epidermal cells or by passing between cells of the epidermis, through the porous cell walls. Many epidermal cells grow long projections, called root hairs, into the surrounding soil (Fig. 25-11). By increasing the surface area, root hairs increase the root's ability to absorb water and minerals. Root hairs may add dozens of square meters of surface area to the roots of even small plants.

Cortex Makes Up Much of the Interior of a Young Root

Cortex occupies most of the inside of a young root. The cortex consists of two very different regions: an outer mass of large, loosely packed parenchyma cells just beneath the

(a)   **(b)**

Figure 25-9   **Typical root systems in dicots and monocots**

(a) Dicots often have a taproot system, consisting of a long central root with many smaller, secondary roots branching from it. **(b)** Monocots usually have a fibrous system, with many roots of equal size.

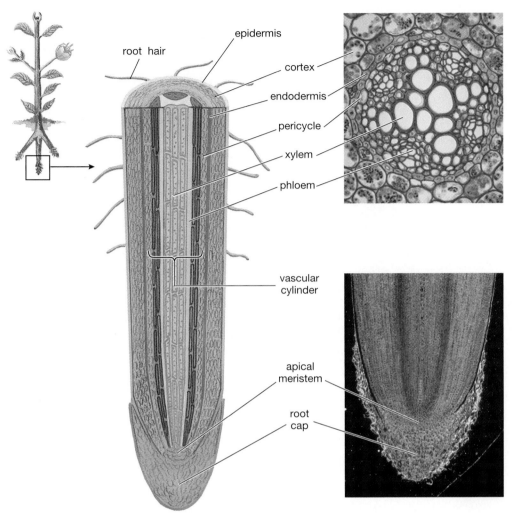

root hair

epidermis

cortex

endodermis

pericycle

xylem

phloem

vascular
cylinder

apical
meristem

root
cap

Figure 25-10   Primary growth in roots

Primary growth in roots results from cell divisions in the apical meristem, located near the tip of the root, and subsequent elongation of daughter cells as they differentiate into epidermis, cortex, and vascular cylinder. The root cap forms at the very tip of the root, also from divisions of cells in the apical meristem.

epidermis and an inner layer of smaller, close-fitting cells that form a ring around the vascular cylinder, called the **endodermis** (see Fig. 25-10). Sugars produced by photosynthesis in the shoot are transported down to the parenchyma cells of the cortex, where they are converted to starch and stored. These cells are particularly abundant in roots specialized for carbohydrate storage, such as the thick roots of carrots and dandelions.

The endodermis is a layer of cells with highly specialized cell walls (Fig. 25-12). Where endodermal cells contact each other, their cell walls are filled with a waxy material, forming the **Casparian strip.** The Casparian strip resembles the mortar in a brick wall: The waxy waterproofing covers the top, bottom, and sides of the endodermal cells. However, it does not cover the cell surfaces facing the rest of the cortex or those facing the vascular cylinder. Water and dissolved minerals can travel *around* both epidermal and cortex parenchyma cells by moving through their porous cell walls. But the waxy Casparian strip seals off the vascular cylinder cell walls, forcing water and minerals that enter the vascular cylinder to pass through the living membranes of the endodermal cells. These membranes regulate the types and

amounts of materials that the roots can absorb (see Fig. 25-12). We describe how this arrangement works in Chapter 26.

Figure 25-11   Root hairs

Root hairs, shown here in a sprouting radish, greatly increase a root's surface area for the absorption of water and minerals from the soil.

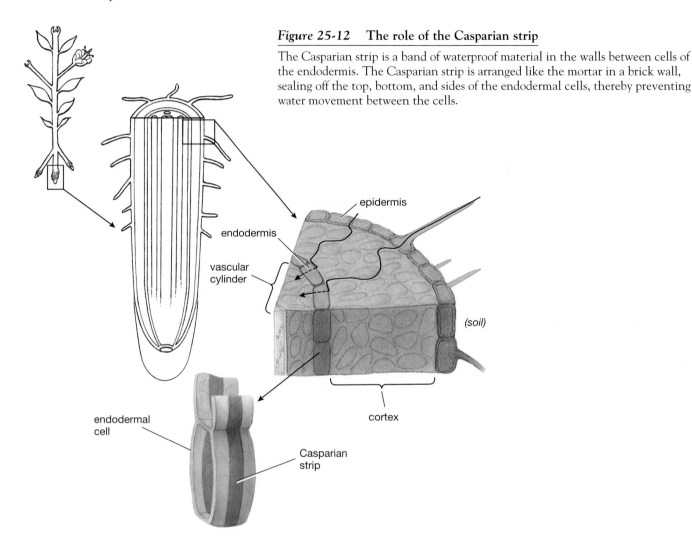

Figure 25-12   **The role of the Casparian strip**

The Casparian strip is a band of waterproof material in the walls between cells of the endodermis. The Casparian strip is arranged like the mortar in a brick wall, sealing off the top, bottom, and sides of the endodermal cells, thereby preventing water movement between the cells.

The Vascular Cylinder Contains Conducting Tissues

The **vascular cylinder** contains the conducting tissues of xylem and phloem that transport water and dissolved materials within the plant. The outermost layer of the vascular cylinder, called the **pericycle**, is a remnant of meristem that retains the ability to divide. Under the influence of plant hormones, pericycle cells divide and form the apical meristem of a **branch root** (Fig. 25-13). Branch root development is similar to primary root development except that the branch must first break out through the cortex and epidermis of the primary root. It does so partly by crushing the cells that lie in its path, and partly by secreting enzymes that digest them away. The vascular tissues of the branch root connect with the vascular tissues of the primary root.

Stems: Reaching for the Light

The Stem Has a Complex Organization Including Four Types of Tissue

Like roots, **stems** develop from a small group of actively dividing cells, the apical meristem, that lies at the tip of the

young shoot. The daughter cells of the apical meristem differentiate into the specialized cell types of stem, buds, leaves, and flowers.

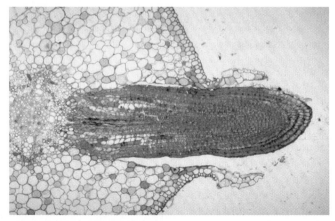

Figure 25-13   **Branch roots**

Branch roots emerge from the pericycle of a root. Notice that the central axis of the branch is already differentiating into vascular tissue.

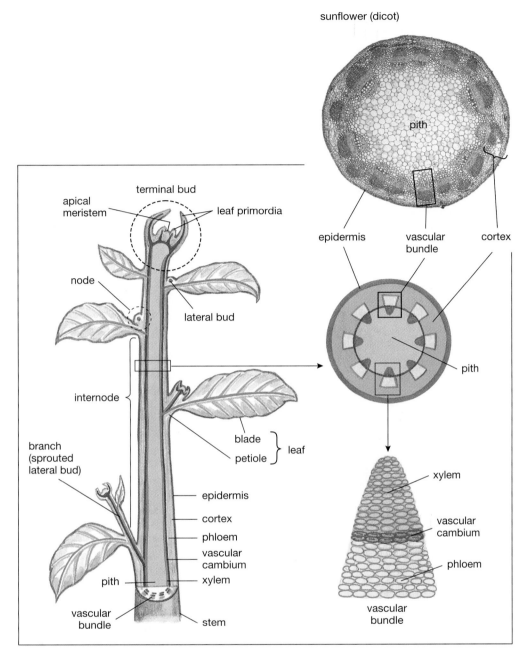

sunflower (dicot)

epidermis

vascular bundle

cortex

pith

terminal bud

apical meristem

leaf primordia

node

lateral bud

internode

branch (sprouted lateral bud)

blade

petiole } leaf

epidermis

cortex

phloem

vascular cambium

pith   xylem

vascular bundle

stem

xylem

vascular cambium

phloem

vascular bundle

Figure 25-14   The structure of a young dicot shoot

At the tip of the stem, the terminal bud includes the apical meristem and several leaf primordia, produced by the meristem. Other daughter cells of the apical meristem differentiate into epidermis, cortex, pith, and vascular tissues. As the young stem grows, the leaf primordia develop into mature leaves. Meanwhile, epidermis, cortex, pith, and vascular cells elongate between the points of attachment of leaf to stem, effectively pushing the leaves apart. A remnant of meristem tissue, called a lateral bud, remains in the crotch between each leaf and the stem. Lateral buds may sprout into branches (see Fig. 26-15). Points on the shoot where leaves and lateral buds are located are called nodes; the naked stem between nodes is an internode. In cross section, vascular tissue forms a ring of vascular bundles in dicots such as the sunflower shown in the photomicrograph inset. Vascular cambium produces secondary xylem toward the center of the stem and secondary phloem toward the outside of the stem.

As the shoot grows, small clusters of meristem cells are "left behind" at the surface of the stem. These meristem cells form the **leaf primordia** (singular, **primordium**) and **lateral buds** that appear at characteristic locations, called **nodes,** on the stem; regions of stem between these nodes are called **internodes** (Fig. 25-14). Leaf primordia develop into the mature leaves typical of the species of plant. Under appropriate conditions, lateral buds grow into branches. We will discuss the growth of branches shortly.

Most young stems are composed of four tissues: epidermis (dermal tissue), vascular tissues, cortex, and pith (the last two are both ground tissues). As Figure 25-2 illustrates, monocots and dicots differ somewhat in the arrangement of vascular tissues. We will discuss only dicot stems here.

The Epidermis of the Stem Is Specialized to Retard Water Loss While Allowing Entry of Carbon Dioxide

You will recall that the epidermis of a young root is the primary pathway for water absorption. In the stem (and leaves), the epidermis is exposed to dry air, making it therefore a potential pathway for water loss, not gain. Epidermal cells of the stem, unlike those of the root, thus secrete a waxy covering, the cuticle, that reduces evaporation of water. Unfortunately, the cuticle also reduces the diffusion of oxygen and carbon dioxide into and out of the plant. The epidermis, however, is often perforated with adjustable pores called **stomata** (singular, **stoma**) that regulate the diffusion of oxygen, carbon dioxide, and water vapor into and out of the stem. Stomata are discussed in more detail in the next chapter.

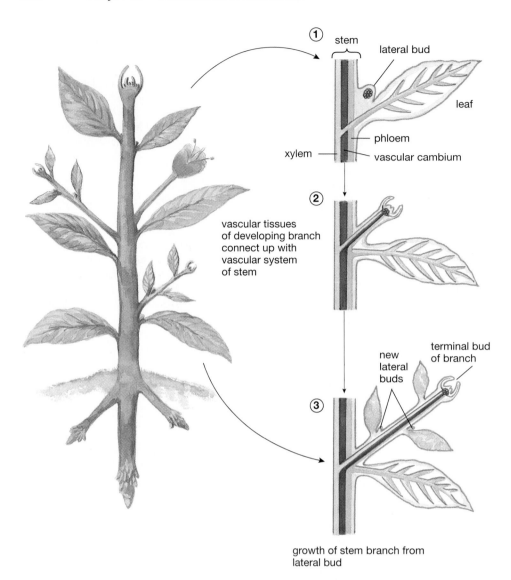

① stem
lateral bud
leaf
phloem
xylem
vascular cambium

②
vascular tissues
of developing branch
connect up with
vascular system
of stem

③
new
lateral
buds
terminal bud
of branch

growth of stem branch from
lateral bud

Figure 25-15   How branches form

Stem branches grow from lateral buds located at the outer surface of a stem. The bud apical meristem generates an outward-growing branch, replicating the pattern of nodes and internodes characteristic of the type of plant. Meanwhile, cortex cells beneath the sprouting bud differentiate into vascular tissues and connect up with the vascular system of the stem.

The Cortex and Pith Support the Stem, Store Food, and Sometimes Photosynthesize

Cortex (located between the epidermis and vascular tissues) and **pith** (inside the vascular tissues at the center of the stem) are similar in most respects; in fact, in some stems it is difficult to tell where cortex ends and pith begins. Cortex and pith perform three major functions: support, storage, and, in some cases, photosynthesis.

1. *Support.* In very young stems, water filling the central vacuoles of cortex and pith cells causes turgor pressure (see Chapter 6). Turgor pressure pushes the cytoplasm up against the cell wall, stiffening the cells much as air inflates a tire. Just as an underinflated tire goes flat, lack of water causes the cells to go limp; if you forget to water your houseplants, their drooping tips show the importance of turgor pressure in keeping young stems erect. Somewhat older rtems also have collenchyma or sclerenchyma cells with thickened cell walls. Because of their strong cell walls, these cells don't depend on turgor pressure for strength. That's why only the tips

of a wilted plant droop: Collenchyma and sclerenchyma cells support the older regions of the stem.

2. *Storage.* Parenchyma cells in both cortex and pith convert sugar into starch and store the starbh as a food reserve.

3. *Photosynthesis.* In many stems, the outer layers of cortex cells contain chloroplasts and carry out photosynthesis. In some desert plants, such as cacti, the leaves are reduced or absent, and the cortex of the stem is the only green photosynthetic part of the plant.

Vascular Tissues in Stems Transport Water, Dissolved Nutrients, and Hormones

As in roots, the vascular tissues of stems transport water, minerals, sugars, and hormones. Vascular tissues are continuous in root, stem, and leaf, interconnecting all the parts of the plant. The **primary xylem** and **primary phloem** found in young stems arise from the apical meristem. In young dicot stems, the primary xylem, **vascular cambium** (meristematic tissue that produces secondary xylem and phloem), and primary phloem either may form concentric cylinders or may appear as a ring of bundles running up the stem, with each bundle containing both phloem and xylem (see Fig. 25-14). Secondary growth in dicot stems, as we will discuss below, always results in concentric cylinders of xylem and phloem (see Fig. 25-16).

Stem Branches Form from Lateral Buds Consisting of Meristem Cells

Branches are formed from lateral buds. A lateral bud is a cluster of dormant meristem cells left behind by the apical meristem as the stem grows. Lateral buds are located on the surface of the stem, just above the attachment points of the leaves (see Fig. 25-14). When stimulated by the appropriate hormones (see Chapter 28), the meristem cells of a lateral bud break out of dormancy and the bud sprouts, growing into a branch (Fig. 25-15). As the meristem cells divide, they release hormones that change the developmental fate of the cells between the bud and the vascular tissues of the stem. Parenchyma cells of the cortex differentiate into xylem and phloem, ultimately connecting with the main vascular systems in the stem. As the branch grows, it duplicates the development of the stem: It has

an apical meristem at its tip and produces its own leaf primordia and lateral buds as it grows.

Secondary Growth Produces Thicker, Stronger Stems

In conifers and perennial dicots, stems last for several years, becoming thicker and stronger each year. This secondary growth in stem thickness results from cell division in lateral meristems known as the vascular cambium and cork cambium (Fig. 25-16).

Secondary Xylem and Phloem Are Produced by Vascular Cambium

The vascular cambium is a cylinder of meristem cells located between the primary xylem and primary phloem. Daughter cells of the vascular cambium produced toward the inside of the stem differentiate into **secondary xylem,** while those produced toward the outside of the stem differentiate into **secondary phloem** (see Fig. 25-16). Because the center of the stem is already filled with pith and primary xylem, newly formed secondary xylem pushes the vascular cambium and all

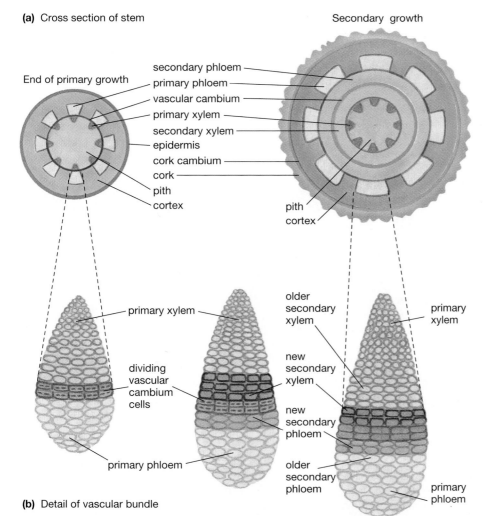

(a) Cross section of stem

End of primary growth

secondary phloem
primary phloem
vascular cambium
primary xylem
secondary xylem
epidermis
cork cambium
cork
pith
cortex

Secondary growth

pith
cortex

primary xylem

dividing vascular cambium cells

primary phloem

older secondary xylem

new secondary xylem

new secondary phloem

older secondary phloem

primary xylem

primary phloem

(b) Detail of vascular bundle

Figure 25-16 **Secondary growth in a dicot stem**

(a) Cross section of a dicot stem at the end of primary growth (left) and during early secondary growth (right). **(b)** Anatomical details of a vascular bundle during secondary growth. A vascular cambium forms between the primary xylem and primary phloem. When vascular cambium cells divide, daughter cells formed on the inside of the vascular cambium differentiate into secondary xylem, while cells formed on the outside of the cambium differentiate into secondary phloem. Note that, because xylem and pith already fill the inside of the stem, newly formed secondary xylem forces the cambium, phloem, and all outer tissues farther out, increasing the diameter of the stem. The cork cambium produces cork cells that cover the outside of the stem.

outer tissues farther out, increasing the diameter of the stem. This secondary xylem, with its thick cell walls, forms the wood that makes up most of the trunk of a tree. Young xylem (the sapwood just inside the vascular cambium) transports water and minerals; older xylem (the heartwood nearest the pith) only contributes to the strength of the trunk.

Phloem cells are much weaker than xylem. As they die with age, the sieve-tube elements and companion cells are crushed between the hard xylem on the inside of the trunk and the tough cork on the outside (see below). Only a thin strip of recently formed phloem remains alive and functioning.

In trees adapted to temperate latitudes, such as oaks and pines, cell division in the vascular cambium ceases during the cold of winter. In spring, the cambium cells divide to form new xylem and phloem. The young cells grow by absorbing water and swelling while the newly formed cell walls are still soft. As the cells mature, the cell walls thicken and harden, preventing further growth. Water ir readily available in spring; therefore, young xylem cells take up a lot of water, swell considerably, and thus are large when mature. As the summer progresses and water becomes scarcer, new xylem cells cannot absorb as much water and consequently are smaller when they mature. As a result, early (spring) wood is pale (because of a thin layer of dark cell wall surrounding a large pale interior) and late (summer) wood is dark (because of a thick cell wall and very little interior space; Fig. 25-17). This pattern of alternating light and dark xylem forms the familiar **annual rings** of growth in temperate trees.

Secondary Growth Causes the Epidermis to Be Replaced by Woody Cork

You will recall that epidermal cells are mature, differentiated cells that can no longer divide. Therefore, as new secondary xylem and phloem are added each year, enlarging the stem, the epidermis can't expand to keep up with the increasing circumference. The epidermis splits off and dies. Apparently prodded by hormones, some parenchyma cells in the cortex become rejuvenated and form a new lateral meristem, the **cork cambium** (see Fig. 25-16). These cells divide, forming daughter cells toward the outside of the stem. These daughter cells, called cork cells, or simply cork, develop tough, waterproof cell walls that protect the trunk from drying out and from abuse. Cork cells die as they mature and may form a protective layer half a meter thick on some tree species, such as the fire-resistant sequoia (Fig. 25-18). As the trunk expands from year to year, the outermost layers of cork split apart or peel off, accommodating the growth. (The cork in the neck of a wine bottle is part of the outermost layer of cork from a certain type of oak, carefully peeled off by harvesters.)

The common term **bark** includes all the tissues outside the vascular cambium: phloem, cork cambium, and cork cells. Complete removal of a strip of bark all the way around a tree, called girdling, is invariably fatal to a tree because it severs the phloem. With the phloem gone, sugars synthesized in the leaves cannot reach the roots. Hence, the roots die and no longer take up water and minerals, resulting in the death of the entire tree.

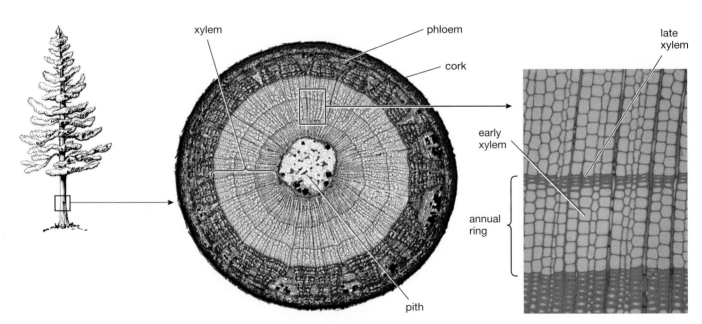

Figure 25-17   How tree rings are formed

Many trees, such as this pine, form annual rings of xylem. As this micrograph shows, secondary xylem cells formed during the wet spring are large, whereas secondary xylem cells formed during the hotter, drier summer are small. The ratio of cell wall to "hole" (the now-empty interior of the cell) determines the color of the wood: Early wood, formed during the spring, with lots of hole, is pale; late wood, formed during the summer, with lots of wall, is dark.

Figure 25-18 **Cork forms the outer layer of bark**

An ancient sequoia in the Sierra Nevada Mountains of California. The cork cambium of a sequoia produces new layers of cork each year, eventually producing a protective, fire-resistant outer covering half a meter or more thick. This massive cork layer contributes to a sequoia's great longevity; forest fires that kill lesser trees merely burn off a few inches of sequoia cork, leaving the living parts of the tree inside unharmed. The small blackened areas on the cork are from past fires.

Leaves: Nature's Solar Collectors

Leaves are the major photosynthetic structures of most plants. As you may remember from Chapter 7, photosynthesis uses the energy of sunlight to convert water and carbon dioxide to sugars, releasing oxygen as a by-product:

$$6\ H_2O + 6\ CO_2 + energy \rightarrow C_6H_{12}O_6\ (glucose) + 6\ O_2$$

Therefore, the cells of a leaf must be provided with water (H_2O) and carbon dioxide (CO_2). Water is obtained from the soil and transported to the leaf through the xylem, but CO_2 must diffuse into the leaf from the air. Thus you might think that an ideal leaf should have a large surface area for gathering light and should be porous to permit CO_2 to enter from the air for photosynthesis. However, a large, porous leaf would also lose large amounts of water through evaporation. Waterproofing the entire surface would reduce the diffusion of CO_2 into the leaf. The leaves of flowering plants represent an elegant compromise among these conflicting demands (Fig. 25-19).

Leaves Have Two Major Parts: Blades and Petioles

A typical angiosperm leaf consists of a broad, flat portion, the **blade**, connected to the stem by a stalk called the **petiole** (see Fig. 25-14). The petiole positions the blade in space, usually orienting the leaf for maximum exposure to the sun. Inside the petiole are vascular tissues of xylem and phloem that are continuous with those in the stem,

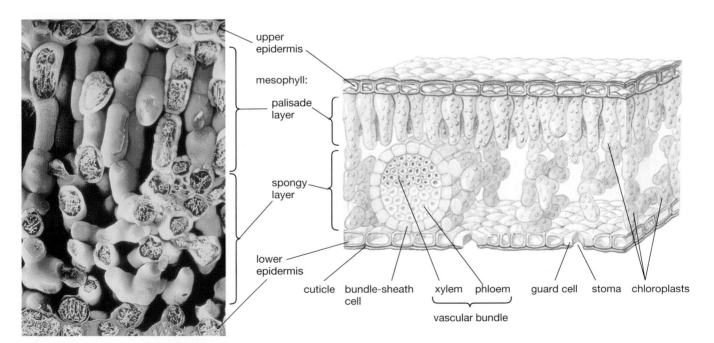

upper epidermis
mesophyll:
palisade layer
spongy layer
lower epidermis
cuticle
bundle-sheath cell
xylem
phloem
vascular bundle
guard cell
stoma
chloroplasts

Figure 25-19 **The structure of a typical dicot leaf**

The cells of the epidermis lack chloroplasts and are transparent, allowing sunlight to penetrate to the chloroplast-containing mesophyll cells beneath. The stomata that pierce the epidermis and the loose, open arrangement of the mesophyll cells ensure that CO_2 can diffuse into the leaf from the air and reach all the photosynthetic cells.

root, and blade. Within the blade, the vascular tissues branch into **vascular bundles,** or **veins.**

The leaf epidermis consists of a layer of nonphotosynthetic, transparent cells that secrete a waxy, waterproof cuticle on their outer surfaces. The epidermis and its cuticle are pierced by adjustable pores, the stomata, that regulate the diffusion of CO_2 and water into and out of the leaf. Each stoma is surrounded by two sausage-shaped **guard cells** that regulate the size of the opening into the interior of the leaf (see Fig. 25-19). Unlike the surrounding epidermal cells, guard cells contain chloroplasts and can carry out photosynthesis. As we will see in Chapter 26, photosynthesis in the guard cells contributes to their ability to adjust the size of the pore.

Beneath the epidermis lies the loosely packed parenchyma cells of the **mesophyll** ("middle of the leaf"). In many leaves, mesophyll cells are of two types: a layer of columnar **palisade cells** just beneath the upper epidermis and a layer of irregularly shaped **spongy cells** above the lower epidermis. Both palisade and spongy cells contain chloroplasts; these cells perform most of the photosynthesis of the leaf. The openness of the leaf interior (see Fig. 25-19) allows CO_2 to diffuse easily to all the mesophyll cells. Vascular bundles, each containing both xylem and phloem, are embedded within the mesophyll, with fine veins reaching very close to each photosynthetic cell. Thus, each mesophyll cell receives energy from sunlight transmitted through the clear epidermis; carbon dioxide from the air, diffusing through the stomata; and water from the xylem. The sugars it produces are carried away to the rest of the plant by the phloem.

Special Adaptations of Roots, Stems, and Leaves

Not all roots are sinuous fibers, not all stems are smooth and upright, and not all leaves are flat and fanlike. Just as evolution has changed the basic shape of the vertebrate forelimb to suit the demands of running, swimming, and flying, so too, plant parts have become modified in response to environmental demands. You may be surprised to learn that many familiar structures are derived from unlikely parts of a plant.

Although we will highlight unusual adaptations, don't forget that *all plants are adapted to their environments.* The "typical" leaf of an oak or maple is just as much a special adaptation as a cactus spine or daffodil bulb.

Some Specialized Roots Store Food, Others Photosynthesize

Roots have probably undergone fewer unusual modifications of their basic structure than either stems or leaves. Some roots have extreme specializations for storage, such as the familiar beet, carrot, or sweet potato. Some of the most bizarre root adaptations occur in certain orchids that grow perched on trees. A few of these aerial orchids have green, photosynthetic roots; in fact, for some orchids, the green roots are the only photosynthetic part of the plant.

Some Specialized Stems Produce New Plants, Store Water or Food, or Produce Thorns or Climbing Tendrils

Many plants have stems modified for functions very different from the original one of raising leaves up to the light. Strawberries, for example, grow horizontal **runners** that snake out over the soil, sprouting new strawberry plants where nodes touch the soil (Fig. 25-20). These new plants are connected with the "mother" plant, but once the plantlets form roots, they can live independently if the runner is severed.

Some plants, such as the saguaro cactus and the baobab tree (Fig. 25-21), store water in aboveground stems. Many other plants store carbohydrates in underground stems. The common white potato is actually a storage stem; each eye is a lateral bud, ready to send up a branch next year, using the energy stored as starch in the potato to power the growth of the branch. Irises have horizontal underground stems called **rhizomes** that store carbohydrates produced during the summer. Irises can be propagated by cutting up the rhizome; if it contains enough stored food, each piece with a node can generate a complete plant.

Many aboveground stems produce modified branches with special functions. One common branch adaptation is the **thorn,** usually growing from the normal branch location just above the site of attachment of a leaf (Fig. 25-22a). Hard, pointy thorns, of course, discourage animals from dining on the branches. Grapes and Boston ivy have some of their branches modified into grasping **tendrils** that coil around trees, trellises, or buildings, providing the otherwise prostrate plant better access to sunlight (Fig. 25-22b).

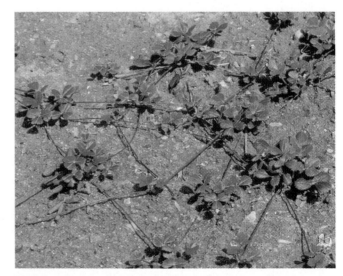

Figure 25-20 **Runner stems produce new plants**

The beach strawberry can reproduce with horizontal stems called runners that spread out over the surface of the sand. If a node of a runner touches the soil, it will sprout roots and develop into a complete plant.

Figure 25-21   **Some stems store water**

The baobab tree usually is not very tall but develops an enormously fat, water-storing trunk. The baobab grows in dry regions, and when it rains, it is to the tree's advantage to store all the water it can get. Some of these trees have trunks so large that people have hollowed small houses out in them, in one case even a jail cell!

Specialized Leaves May Conserve and Store Water, Store Food, or Even Capture Insects

The most important environmental factors that affect the growth of leaves are light, temperature, and water availability. For example, plants growing on the floor of a tropical rain forest have plenty of water year round but very little light owing to the deep shade cast by several layers of trees above them. Consequently, their leaves tend to be extremely large, an adaptation demanded by the low light level and permitted by the abundant water (Fig. 25-23a).

At the other extreme, deserts receive bright sunlight virtually every day of the year but have limited water and scorching temperatures. Desert plants have evolved two strikingly different adaptations to this situation. One group of plants, the succulents, have very thick leaves with large cells that store water from the infrequent rains against the inevitable long droughts (Fig. 25-23b). Succulent leaves are covered with a thick cuticle that greatly reduces the evaporation of

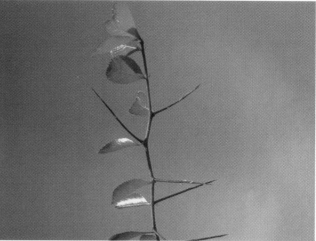

Figure 25-22   **Some branches stab, others grasp**

The branches growing from a single stem may differ in structure or function. **(a)** The hawthorn has both "regular" branches forming the crown of the tree and many smaller branches modified into sharp thorns. That these thorns are actually branches in disguise is clearly shown by their position in the normal branch location, the crotch between a leaf and its stem, and by the fact that the thorns themselves often branch. **(b)** Grape tendrils are long, soft, leafless branches. When a tendril contacts an object in the environment, it curls in the direction of the touching surface, wrapping itself firmly around the object.

water. The cacti display the opposite strategy, reducing the leaves into thin spines that protect the plant from herbivores and reduce water loss (Fig. 25-23c). Photosynthesis in cacti occurs in cortex cells of the green, water-storing stems.

Modified leaves in other plants function in ways having nothing to do with photosynthesis or water conservation. The common edible pea, for example, grasps fences, mailbox posts, or other plants with clinging tendrils (Fig. 25-23d). Unlike the tendrils of grapes, which are derived from branches, pea tendrils are slender, supple leaflets. Some plants, such as onions, daffodils, and tulips, use

(a)

(b)

(c)

(d)

(e)

Figure 25-23   **The many roles of leaves**

Differing environments have produced countless leaf types. **(a)** In tropical rain forests, light levels near the ground are very low. Consequently, plants closest to the ground often bear huge, dark green leaves for maximal light absorption. Many rainforest plants, such as this philodendron, have found a new habitat in modern houses, where they are grown because of their deep green foliage and tolerance of dim light. **(b)** Desert plants receive plenty of light but very little water. Succulents have evolved fleshy leaves that store water from the occasional rains, just as the baobab tree does in its trunk (Fig. 25-21). **(c)** The cactus takes the other extreme in desert adaptations: Reduce the leaves to vanishingly thin lines, which cannot store water but don't have much surface area for evaporation either. These spines also have very sharp points that discourage animals who might be tempted to munch on the tasty stem beneath. **(d)** Pea tendrils are modified leaflets that respond to contact with objects by coiling, much as stem tendrils do. **(e)** Daffodil bulbs are enormous buds, with short central stems surrounded by thick, water- and food-storing leaves. Like other monocots, these bulbs form roots as outgrowths of the base of the stem.

thick, fleshy leaves as storage organs (Fig. 25-23e). A daffodil bulb consists of a short stem bearing thick, overlapping leaves that store nutrients over the winter.

Finally, a few plants have turned the table on the animals and have become predators. For example, Venus flytraps (Fig. 25-24a) and sundews (Fig. 25-24b) both have leaves modified into snares for unwary insects. These plants live in nitrogen-poor swamps and derive most of their nitrogen supply from the bodies of their prey.

As varied and sometimes bizarre as these leaf specializations are, the most extreme and most important leaf modification is the flower. As we will see in the next chapter, these "reproductive leaves" enabled the flowering plants to become the dominant plants on land.

Figure 25-24   **Some predatory leaves**

Predatory leaves adorn the Venus flytrap and the sundew. **(a)** When an insect blunders into the fringed, hinged leaves of the flytrap, the leaves close up rapidly, trapping their victim. Digestive enzymes reduce the insect to nourishing molecules that are absorbed by the plant. **(b)** The sundew leaf bears glistening droplets that attract hungry insects. The droplets are incredibly sticky, and the unsuspecting insect becomes not the eater, but the eaten. As in the flytrap, enzymes secreted by the leaf digest the insect and the leaf absorbs the resulting nutrients.

☒ S U M M A R Y   O F   K E Y   C O N C E P T S

An Overview of Plant Structure

The body of a land plant consists of root and shoot. Roots are usually underground and have five functions. They (1) anchor the plant in the soil; (2) absorb water and minerals from the soil; (3) store surplus photosynthetic products; (4) transport water, minerals, photosynthetic products, and hormones; and (5) produce hormones. Shoots are usually aboveground and consist of stem, leaves, buds, and (in season) flowers and fruit. Shoot functions include: (1) photosynthesis, (2) transport of materials, (3) reproduction, and (4) hormone synthesis.

Plant Development and Growth

Plant bodies are composed of two main classes of cells. Meristem cells are undifferentiated cells that retain the capacity for cell division. Differentiated cells arise from divisions of meristem cells, become specialized for particular functions, and normally do not divide.

Most meristem cells are located in apical meristems at the tips of roots and shoots and in lateral meristems in the shafts of roots and shoots. Primary growth (growth in length and differentiation of parts) results from division and differentiation of cells from apical meristems; secondary growth (growth in diameter) results from division and differentiation of cells from lateral meristems.

Plant Tissues and Cell Types

Plant bodies consist of three tissue systems: the dermal, ground, and vascular systems. The dermal tissue system forms the outer covering of the plant body. The dermal tissue system of leaves and of primary roots and stems is usually a single cell layer of epidermis. Dermal tissue after

secondary growth is a multilayered covering of cork.

The ground tissue system consists of a variety of cell types, including parenchyma, collenchyma, and sclerenchyma. Most are involved in photosynthesis, support, or storage. Ground tissue makes up most of a young plant during primary growth. During secondary growth of stems and roots, ground tissue becomes an increasingly small part of the plant body.

The vascular tissue system consists of xylem, which transports water and minerals from the roots to the shoots, and phloem, which transports water, sugars, amino acids, and hormones throughout the plant body.

Roots: Anchorage, Absorption, and Storage

Primary growth in roots results in a structure consisting of an outer epidermis, an inner vascular cylinder of conducting tissues, and cortex between the two. The apical meristem near the tip of the root is protected by the root cap. Cells of the root epidermis absorb water and minerals from the soil. Root hairs are projections of epidermal cells that increase the surface area for absorption. Most cortex cells store surplus sugars produced through photosynthesis, usually as starch. The innermost layer of cortex cells is the endodermis, which controls the movement of water and minerals from the soil into the vascular cylinder. The vascular cylinder contains the conducting tissues—xylem and phloem.

Stems: Reaching for the Light

Primary growth in dicot stems results in a structure consisting of an outer, waterproof epidermis; supporting and photosynthetic cells of cortex beneath the epidermis; vascular tissues of xylem and phloem; and supporting and storage cells of pith at the center of the stem. Leaves and lateral buds are found at nodes along the surface of the stem. Under the proper hormonal conditions, lateral buds may sprout into a branch.

Secondary growth in stems results from cell divisions in the vascular cambium and cork cambium. Vascular cambium produces secondary xylem and secondary phloem, increasing the diameter of the stem. Cork cambium produces waterproof cork cells that cover the outside of the stem.

Leaves: Nature's Solar Collectors

Leaves are the main photosynthetic organs of plants. The blade of a leaf consists of a waterproof outer epidermis surrounding mesophyll cells that have chloroplasts and carry out photosynthesis, and vascular bundles of xylem and phloem that carry water, minerals, and photosynthetic products to and from the leaf. The epidermis is perforated by adjustable pores called stomata that regulate exchange of gases and water.

Special Adaptations of Roots, Stems, and Leaves

Through evolution, the roots, stems, and leaves of many plants have been modified into diverse structures such as spines, tendrils, thorns, and bulbs. These unusual structures are often involved in water or energy storage, support, or protection.

KEY TERMS

angiosperm   p. 497
annual ring   p. 510
apical meristem   p. 498
bark   p. 510
blade   p. 511
branch root   p. 506
cambium   p. 498
Casparian strip   p. 505
collenchyma   p. 501
companion cell   p. 503
cork cambium   p. 510
cork cell   p. 501
cortex   p. 508
cuticle   p. 500
dermal tissue system   p. 500
dicot   p. 498
differentiated cell   p. 498
endodermis   p. 505
epidermal tissue   p. 500
epidermis   p. 500
fibrous root system   p. 504
ground tissue system   p. 500
guard cell   p. 512
gymnosperm   p. 497
internode   p. 507

lateral bud   p. 507
lateral meristem   p. 498
leaf   p. 511
leaf primordium   p. 507
meristem cell   p. 498
mesophyll   p. 512
monocot   p. 498
node   p. 507
palisade cell   p. 512
parenchyma   p. 501
pericycle   p. 506
periderm   p. 500
petiole   p. 511
phloem   p. 503
pit   p. 503
pith   p. 508
primary growth   p. 498
primary phloem   p. 508
primary root   p. 504
primary xylem   p. 508
rhizome   p. 512
root cap   p. 504
root hair   p. 500
root system   p. 498
runner   p. 512

sclerenchyma   p. 501
secondary growth   p. 500
secondary phloem   p. 509
secondary xylem   p. 509
shoot system   p. 498
sieve plate   p. 503
sieve tube   p. 503
sieve-tube element   p. 503
spongy cell   p. 512
stem   p. 506
stoma   p. 507
taproot system   p. 504
tendril   p. 512
thorn   p. 512
tracheid   p. 502
vascular bundle   p. 512
vascular cambium   p. 508
vascular cylinder   p. 506
vascular tissue system   p. 500
vein   p. 512
vessel   p. 503
vessel element   p. 503
xylem   p. 502

THINKING THROUGH THE CONCEPTS

Multiple Choice

1. You and a friend carve your initials 5 feet above the ground on a tree on campus. The tree is now 40 feet tall. When you come back for your twenty-fifth reunion, the tree will be 100 feet tall. How high above the ground should you look for your initials?
   a. 5 feet   b. 40 feet
   c. 60 feet   d. 95 feet
   e. 100 feet
2. Most of a tree is made up of xylem cells, so it is safe to say that most of a tree is
   a. meristem   b. dead
   c. ground tissue   d. involved in carrying sugars
   e. dermal tissue

3. If you wanted to show a friend how much you had learned in biology class, which of the following would you NOT use to identify a plant as a dicot?
   a. leaves with veins arranged like a net
   b. hand-shaped leaves   c. six petals
   d. a taproot   e. a seed with two cotyledons
4. When water enters a plant root, it is forced to travel through the endodermal cells by the waxy
   a. cuticle   b. epidermis
   c. periderm   d. xylem
   e. Casparian strip
5. Which of the following is NOT a special adaptation of certain roots, stems, or leaves?
   a. roots–photosynthesis   b. stems–water storage

c. leaves–defense   d. roots–prey capture
e. stems–producing new plants

6. Which of the following is NOT true of the root cap?
   a. It is made of thick-walled cells.
   b. It is formed on the "soil side" of the apical meristem.
   c. It produces root hairs for absorption.
   d. It produces a slimy lubricant.
   e. It is constantly worn away.

Review Questions

1. Describe the locations and functions of the three tissue systems in land plants.
2. Distinguish between primary growth and secondary growth, and describe the cell types involved in each.
3. Compare primary and secondary growth in roots and stems. How do the tissue systems differ between a plant that has undergone only primary growth as compared with a plant that has undergone substantial secondary growth?
4. Distinguish between meristem cells and differentiated cells. What meristems cause primary growth? Which ones form secondary growth? Where is each type located?
5. Diagram the internal structure of a root after primary growth, labeling and describing the function of epidermis, cortex, endodermis, pericycle, xylem, and phloem. What tissues are located in the vascular cylinder?
6. How do xylem and phloem differ?
7. What are the main functions of roots, stems, and leaves?
8. What types of cells form root hairs? What is the function of root hairs?
9. Diagram the internal structure of leaves. What structures regulate water loss and CO_2 absorption by a leaf?

✳ APPLYING THE CONCEPTS

1. If you girdle a tree what will happen to it? Why does leaving a small portion of the bark uncut prevent this problem?
2. The tropical rain forest contains a large number of as-yet unidentified plants, many of which may have uses as medicines or food. If you were given the job of searching a particular portion of the rain forest for useful products, how would you use the information you gained from this chapter to help you narrow your search? What kinds of plant tissues or organs do you think would be most likely to contain such products?
3. In the desert there are typically two types of plants with respect to their root systems—small grasses or herbs, and shrubs or small trees. The grasses and herbs typically form fibrous root systems. The shrubs and trees form taproot systems. What advantages can you think of for each system? How does each type of root allow for survival in a desert environment?
4. Grasses (monocots) form their primary meristem near the ground surface rather than at the tips of branches the way dicots do. How does this feature allow you to grow a lawn and mow it every week in the summer? What would happen if you had a dicot lawn and tried to mow it?
5. Think about the meals that you have eaten in the past day. What plant foods did you eat? What plant parts (leaf, stem, root, etc.) did each of these come from? What parts of the world did these foods most likely come from? How does our varied diet affect politics and international affairs?
6. Discuss the structures and adaptations that might be found in the leaves of plants living in (a) dry, sunny habitats; (b) wet, sunny habitats; (c) dry, shady habitats; and (d) wet, shady habitats. Which of these habitats do you think would be most inhospitable (e.g., in which habitat would it be most difficult to design a functioning leaf)?

✳ FOR MORE INFORMATION

Baskin, Y. "Forests in the Gas." *Discover*, October 1994. As carbon dioxide increases in the atmosphere, plants will change in their relationships in the natural environment.

Frits, H. C. "Tree Rings and Climate." *Scientific American*, May 1972. The width of the annual rings of trees reflects the length of the growing season and the amount of rainfall and can be used to determine prehistoric climate.

Meyerowitz, E. M. "The Genetics of Flower Development." *Scientific American*, May 1994. An easy-to-read article about research in which flower parts are rearranged to determine how plant tissues develop.

Stuart, D. "Green Giants." *Discover*, April 1990. Sequoias, the most massive living organisms on the planet, are among the oldest as well. Stuart describes how they can grow so large and live so long.

Vogel, S. "When Leaves Save the Tree." *Natural History*, September 1993. The shape and arrangement of leaves on a tree allow it to withstand wind and storm.

Weiss, R. "When Plants Act like Animals." *National Wildlife*, December 1994–January 1995. A fun look at plant movements.

NET WATCH

On-line resources for this chapter are on the World Wide Web at:
http://www.prenhall.com/~audesirk (click on the <u>table of contents</u> link and then select Chapter 25).

26 How Land Plants Acquire and Transport Nutrients

▓ AT A GLANCE

A Comparison of Plant and Animal Nutrition

Plant Nutrition
  Roots Acquire Minerals Using a Four-Step Process
  Symbiotic Relationships Help Plants Acquire Nutrients

The Acquisition of Water

The Transport of Water and Minerals
  Water Movement in Xylem Is Explained by the Cohesion-
    Tension Theory
  Adjustable Stomata Control the Rate of Transpiration

The Transport of Sugars
  The Pressure Flow Theory Explains Sugar Movement in
    Phloem

These giant redwoods extract water and minerals out of the soil and transport them hundreds of feet to their topmost leaves.

In a very real sense, you are what you eat. Atoms from your food make up most of the molecules of your body. And energy derived from food powers your brain, your muscles, and all the cells of your body. Even the first cells must have obtained materials and energy from their environment to grow and carry on cellular processes. As organisms evolved, the ways they gained material and energy diversified dramatically.

In its very broadest sense, **nutrition** comprises all the activities by which organisms acquire materials, called **nutrients**, and energy from their environment and use them to carry on all the activities of life. Nutrition occurs in four steps (1) *acquisition* of nutrients and energy; (2) *digestion* (if required) of nutrients; (3) *distribution* of nutrients to all parts of the body; and (4) *synthesis* of the molecules that make up the organism's body, using the nutrients and energy from food.

A Comparison of Plant and Animal Nutrition

Plant and animal nutrition have both similarities and crucial differences (Table 26-1). Animals acquire most of their nutrients when they eat other organisms. They must then digest the complex prey molecules, freeing the simpler components for use in synthesizing their own molecules. Plants, in contrast, acquire only simple molecules dissolved in water and generally use these directly in synthesizing their own unique molecules. With rare exceptions such as the predatory plants described in Chapter 25, plants dispense with digestion.

Although both plants and animals use the high-energy molecule adenosine triphosphate (ATP) to carry energy, animal cells can synthesize ATP only by breaking down energy-rich organic molecules acquired from food. Plants, in contrast, harvest energy from sunlight. Photosynthesis produces ATP in the light-dependent reactions and produces carbohydrates in the light-independent reactions (see Chapter 7). Plants use some carbohydrate molecules to make other molecules of the plant body and metabolize others to produce ATP, just as animals do.

Both land plants and animals distribute nutrients to all their cells in fluid, but they move the liquid differently. In many animals, a muscular heart pumps nutrient-laden blood throughout the body. Plants, in contrast, utilize the combined forces of active transport, osmosis, diffusion, and cohesion between water molecules to move sugars produced in the photosynthetic portions of the plant to all other parts of the plant body. The same forces, combined with evaporation of water through leaves and stems, pull nutrients that are dissolved in water from the soil up into the plant body.

Plant Nutrition

Plants acquire nutrients as simple, inorganic compounds. Plants need a few elements in relatively large quantities; these elements are called **macronutrients**. Macronutrients such as potassium help maintain the ionic composition of the cytoplasm. Other macronutrients are combined into organic molecules that form the plant body and its food reserves. Most plants also need trace amounts of **micronutrients,** mostly elements that allow enzymes to function. Table 26-2 briefly describes the sources and functions of plant macro- and micronutrients.

Carbon dioxide and oxygen normally enter a plant by diffusion from the air into leaves, stem, and roots. Roots extract water and all other nutrients, collectively called **minerals**, from the soil.

Roots Acquire Minerals Using a Four-Step Process

Soil consists of bits of pulverized rock, air, water, and organic matter (Fig. 26-1). Although both the rock particles and the organic matter contain many essential nutrients, only minerals dissolved in the soil water are accessible to the roots. The concentration of minerals in the soil water is very low, usually much lower than the concentration within plant cells and fluids. For example, potassium (K^+) is the most abundant positive ion in the cytoplasm of most living cells. Its concentration within root cells is at least 10 times greater than in soil water. As a result, diffusion cannot move potas-

sium into the root. As a general principle, most minerals are moved into a root against their concentration gradients by active transport (movement of molecules from areas of low concentration to areas of high concentration requires energy; see Chapter 6). Because roots are below the ground, they cannot photosynthesize. Sugar synthesized in the leaves is transported in the phloem to the roots, where mitochondria in the root cells produce ATP by cellular respiration. Some of this ATP is used to drive the active transport of minerals.

Most mineral absorption by roots occurs in a four-step process, as diagrammed in Figure 26-2.

1. *Active transport into root hairs.* Root hairs projecting from the epidermal cells provide most of the surface area of the root and are in intimate contact with the soil water. The plasma membranes of the root hairs use the energy of ATP to transport minerals from the soil water, concentrating the minerals in the root hair cytoplasm (step 1 in Fig. 26-2).

2. *Diffusion through cytoplasm to pericycle cells.* You may remember from Chapter 6 that the cytoplasm of adjacent living plant cells is interconnected by pores called **plasmodesmata.** Therefore, minerals can diffuse through plasmodesmata from the epidermal cells into the cortex, endodermis, and pericycle cells (step 2).

3. *Active transport into the extracellular space of the vascular cylinder.* At the center of the vascular cylinder lies the xylem, into which the minerals must ultimately be transported. The tracheids and vessel elements of xylem are dead, without cytoplasm or plasma membrane—merely an outer skeleton of cell wall shot full of holes (see Fig. 25-7). Therefore, plasmodesmata do not connect pericycle cells with the insides of the xylem. However, any minerals that enter the extracellular space surrounding the xylem can easily diffuse into the xylem cells through the holes in their walls. Therefore, pericycle cells actively transport minerals out of their own cytoplasm into the extracellular space surrounding the xylem (step 3).

TABLE 26-1 *Plant and Animal Nutrition Compared*		
	Plant	*Animal*
Form in which nutrients are acquired	As individual inorganic molecules	Mostly in bulk as parts of prey
Digestion of nutrients	Usually none needed	Digest to component subunits and inorganic minerals
Transport of nutrients throughout body	Driven by evaporation and osmosis	Driven by hydrostatic pressure of pumping heart
Synthesis of molecules of own body	Synthesize all organic molecules from inorganic precursors	Synthesize some organic molecules from inorganic precursors; must obtain others in diet

TABLE 26-2 ■ *Essential Plant Nutrients*

Element	Molecular Form Absorbed	Source	Major Functions
Macronutrients (usually 0.1% or more of dry weight of plant)			
Carbon	CO_2	Air	Major element in organic molecules; photosynthesis
Oxygen	O_2	Air	Major element in organic molecules; cellular respiration
Hydrogen	H_2O	Soil	Major element in organic molecules
Nitrogen	NO_3^- or NH_4^+	Soil	Major element in proteins, nucleic acids, and chlorophyll
Potassium	K^+	Soil	Principal positive ion inside cells; control of stomatal opening and closing; enzyme activation
Phosphorus	$H_2PO_4^-$ or HPO_4^{2-}	Soil	Major element in nucleic acids, phospholipids, and electron carriers in chloroplasts and mitochondria
Calcium	Ca^{2+}	Soil	Component of adhesive compounds in cell walls; important in control of membrane permeability; enzyme activation
Magnesium	Mg^{2+}	Soil	Component of chlorophyll; enzyme activation; ribosome stability
Sulfur	SO_4^{2-}	Soil	Component of proteins and many coenzymes
Micronutrients (usually less than 0.01% of dry weight of plant)			
Iron	Fe^{2+} or Fe^{3+}	Soil	Needed for synthesis of chlorophyll; component of many electron carriers
Chlorine	Cl^-	Soil	Required for photosynthesis
Manganese	Mn^{2+}	Soil	Required for photosynthesis; enzyme activation
Molybdenum	MoO_4^{2-}	Soil	Required for nitrogen metabolism
Copper	Cu^{2+}	Soil	Enzyme activation; component of electron carriers in chloroplasts
Boron	BO_3^{3-} or $B_4O_7^{2-}$	Soil	Involved in sugar transport
Zinc	Zn^{2+}	Soil	Enzyme activation; protein synthesis; hormone synthesis

4. *Diffusion into the xylem.* The active transport of minerals into the extracellular space of the vascular cylinder increases the concentration of minerals in the extracellular space. This high concentration creates a gradient that promotes the diffusion of minerals from the extracellular space into the tracheids and vessel elements of the xylem (step 4).

You can now appreciate one of the functions of the waterproof Casparian strip that seals the spaces between the endodermal cells that surround the vascular cylinder (see Chapter 25). If water and dissolved minerals could flow through the extracellular space *between* endodermal cells, then minerals would leak back out of the extracellular

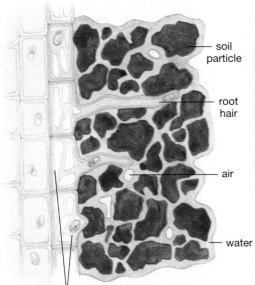

soil particle

root hair

air

water

cells of epidermis

Figure 26-1   Soil and roots

(Left) Soil has a complex structure, consisting mostly of rock particles, water, and air, along with some organic matter and living organisms. Roots can take up only those substances that are dissolved in the soil water. Root hairs penetrate between rock particles and increase the surface area of the root that is in contact with the soil water. (Right) Root hairs cover the root of a sprouting radish seedling.

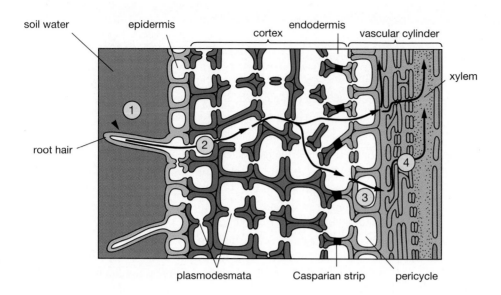

Figure 26-2   **Mineral uptake by roots**

The Casparian strip separates the extracellular space in the root into two compartments: an outer compartment (dark blue) that is continuous with the soil water and an inner compartment (light blue) that is continuous with the inside of the conducting cells of the xylem. Mineral uptake occurs in a four-step process. ① Active transport proteins in root hair membranes pump minerals into the root hair cytoplasm. ② Minerals diffuse inward from cell to cell through plasmodesmata that interconnect the cytoplasm of all living cells in the root. ③ The outermost living cells at the root interior, the pericycle cells, actively transport minerals out of their cytoplasm into the extracellular space surrounding the xylem. ④ Entry of the minerals raises the concentration of minerals in the extracellular space, so the minerals diffuse into the xylem cells through the pits in their walls.

space of the vascular cylinder as fast as they were pumped in. This leakage would both waste the energy that was used to actively transport the minerals into the root in the first place and reduce the concentration gradient that allows the minerals to diffuse into the conducting cells of the xylem. The Casparian strip, however, effectively leakproofs the vascular cylinder, retaining the concentrated mineral solution within the vascular cylinder.

Symbiotic Relationships Help Plants Acquire Nutrients

Many minerals are too scarce in soil water to support plant growth, although there may be plenty bound up in the surrounding rock particles. Further, one nutrient—nitrogen—is almost always in short supply both in rock particles and in soil water. Many plants have evolved beneficial relationships with other organisms that help them acquire these scarce nutrients. Examples include root-fungus relationships, called mycorrhizae, and root-bacteria relationships formed in nodules of legumes.

Fungal Mycorrhizae Help Plants Acquire Minerals

Under normal conditions, water-soluble minerals are released very slowly from rock particles. Furthermore, the

chemical forms of the minerals may not be suitable for uptake by the plasma membranes of plant root cells. Most land plants form symbiotic relationships with fungi to form root-fungus complexes called **mycorrhizae** (singular, **mycorrhiza**) that help the plant extract and absorb minerals (also described in Chapter 22). Fungal strands intertwine between the root cells and extend out into the soil (Fig. 26-3). In some way that is not yet understood, the fungus renders nutrients accessible for uptake by the roots, perhaps by converting rock-bound minerals into simple soluble compounds that root plasma membranes can transport. The fungus, in return, receives sugars and amino acids from the plant. In this way, both the fungus and the plant can grow in places where neither could survive alone, including deserts and high-altitude, rocky soils that are low in nutrients.

Nodules on the Roots of Legumes Help Plants Acquire Nitrogen

Amino acids, nucleic acids, and chlorophyll all contain nitrogen, so plants need large amounts of this element. Unfortunately, although nitrogen is abundant in the atmosphere, most of it is not readily available to plants. About 79% of the atmosphere is nitrogen gas, N_2, but plants can take up nitrogen only through their roots, in the form of ammonium ion (NH_4^+) or nitrate ion (NO_3^-).

(a)
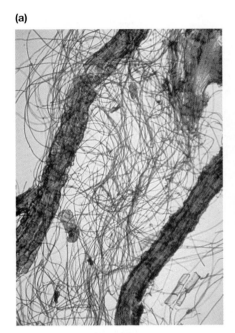

(b)

CK   GE

Figure 26-3 Mycorrhizae, a root-fungus symbiosis

(a) A tangled meshwork of fungal strands surrounds and penetrates into the root. (b) Seedlings growing under identical conditions with (on the right) and without (left) mycorrhizal fungi show the importance of mycorrhizae in plant nutrition. Plants that participate in this unique relationship tend to grow larger and more vigorously than those deprived of the fungus.

Although N_2 diffuses from the atmosphere into the air spaces in the soil, it cannot be used by plants because they don't have the enzymes needed to carry out **nitrogen fixation,** the conversion of N_2 into ammonium or nitrate. A variety of **nitrogen-fixing bacteria** do have these enzymes. Some of these bacteria are free-living in the soil. However, nitrogen fixation is very costly, energetically speaking, using at least 12 ATPs per ammonium ion synthesized. Consequently, bacteria don't routinely manufacture a lot of extra ammonium and liberate it into the soil.

Some plants, particularly the **legumes** (peas, clover, and soybeans), enter into a mutually beneficial relationship with specific species of nitrogen-fixing bacteria. By secreting chemicals into the soil, legumes attract nitrogen-fixing bacteria to their roots (Fig. 26-4a). Once there, the bacteria enter the root hairs. The bacteria then digest channels through the cytoplasm of the epidermal cells and into underlying cortex cells. As both bacteria and their host cortex cells multiply, a **nodule** housing the root-bacteria complexes is formed (Fig. 26-4b). A cooperative relationship ensues. The plant transports sugars from its leaves down to the cortex, just as it normally would for storage. The bacteria within the cortex cells take up the sugar and use its energy for all of their metabolic processes, including nitrogen fixation. The bacteria obtain so much energy that they produce more ammonium than they need. The surplus ammonium diffuses into the cytoplasm of their host cells, providing the plant with a steady supply of usable nitrogen.

The Acquisition of Water

Once a high concentration of minerals builds up in the vascular cylinder, water absorption becomes very straightforward (Fig. 26-5). Water moves by osmosis from regions of high water concentration to regions of low water concentration. Dissolved minerals bind water molecules and thus tie them up, lowering the concentration of free water molecules (see Chapter 6). Therefore, soil water (low in minerals) has a high free water concentration, while the vascular cylinder water (high in minerals) has a low free water concentration. This situation promotes the movement of water from the soil into the vascular cylinder.

On its way to the vascular cylinder, water first moves through the highly porous cell walls that occupy the extracellular space surrounding the epidermal and cortex cells. Thus, the soil water has an uninterrupted pathway through the extracellular space of the outer layers of the root (step 1 in Fig. 26-5). At the endodermis, the waterproof Casparian strip blocks further movement of water through the spaces between cells. The faces of the endodermal cells, however, are not waterproofed. Therefore, the plasma membranes of the endodermis form a pair of semipermeable membranes separating an outer solution of high free water concentration (in the cortex) from an inner solution of low free water concentration (within the vascular cylinder). Water moves across the membranes by osmosis from high to low concentration—that is, from the extracellular space outside the Casparian strip, through the cytoplasm of endodermal cells, and into the extracellular space of the vascular cylinder inside the Casparian strip (step 2). Water can then freely move from the extracellular space within the vascular cylinder into the tracheids and vessel elements of the xylem, through porous pits in their cell walls (step 3).

As you will see when we discuss transport in xylem, water is also pulled up the xylem, powered by the force of water evaporating from the leaves. This movement further lowers the water concentration within the vascular cylinder and promotes the entry of water across the endodermis.

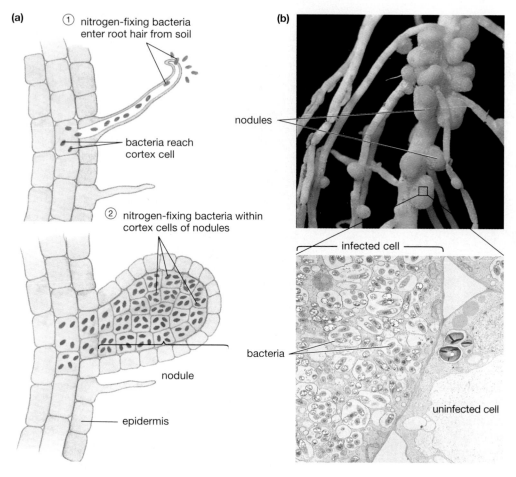

(a)
① nitrogen-fixing bacteria enter root hair from soil

bacteria reach cortex cell

② nitrogen-fixing bacteria within cortex cells of nodules

nodule

epidermis

(b)
nodules

infected cell

bacteria

uninfected cell

Figure 26-4   **Nitrogen fixation in legumes**

(a) Nitrogen-fixing bacteria penetrate a legume root through the root hairs. ① Bacteria in the soil are attracted to a root hair, and digest their way into the root hair, through the cytoplasm of epidermal cells, and into underlying cortex cells. ② The bacteria multiply and induce the cortex cells to multiply, forming a nodule. **(b)** Nodules in legume roots (top) consist of cortex cells filled with nitrogen-fixing bacteria, as shown in the photomicrograph (bottom). Cells containing nitrogen-fixing bacteria are described as "infected," even though the relationship is beneficial.

The Transport of Water and Minerals

Once water and minerals enter the root xylem, they still must be moved to the uppermost reaches of the plant (in redwood trees, this may be over 100 meters away!). The processes of active transport, diffusion, and osmosis, all of which suffice to move water and minerals from soil to root xylem, would be hopelessly slow for getting these substances to the top of a tree. Land plants accordingly move fluids up the xylem from root to stem and leaf by **bulk flow**. During bulk flow, molecules of fluids move together, rather than molecule by molecule, from areas of high pressure to areas of lower pressure. Bulk flow contrasts with diffusion and osmosis, in which molecules move individually from areas of high concentration to areas of lower concentration. Because minerals are dissolved in water, they are passively carried along as the water flows upward. So the question is: Why is water pressure sufficiently higher in the roots and lower in the leaves to overcome the force of gravity and force the water to flow upward? The cohesion-tension theory provides an explanation.

Water Movement in Xylem Is Explained by the Cohesion-Tension Theory

According to the **cohesion-tension theory,** water is pulled up the xylem, powered by the evaporation of water from

the leaves (Fig. 26-6). As its name suggests, this theory has two essential parts.

1. *Cohesion.* Attraction among water molecules holds water together, forming a solid ropelike column within the xylem tubes.
2. *Tension.* This "water rope" is pulled up the xylem, with evaporation providing the necessary energy.

Let's briefly examine both of these factors.

Hydrogen Bonds between Water Molecules Produce Cohesion

You may recall from Chapter 2 that water is a polar molecule, with the oxygen carrying a slight negative charge while the hydrogens carry a slight positive charge. As a result, nearby water molecules attract one another, forming weak hydrogen bonds. Just as individually weak cotton threads together make the strong fabric of your jeans, the network of individually weak hydrogen bonds within water collectively produces a very high cohesion, or tendency to resist being separated. Experiments have demonstrated that the column of water within the xylem is at least as strong—and as unbreakable—as a steel wire of the same diameter. This is the "cohesion" part of the theory: Hydrogen bonds among water molecules provide the cohesion that holds together a "rope" of water extending the entire height of the plant within the xylem. Supplement-

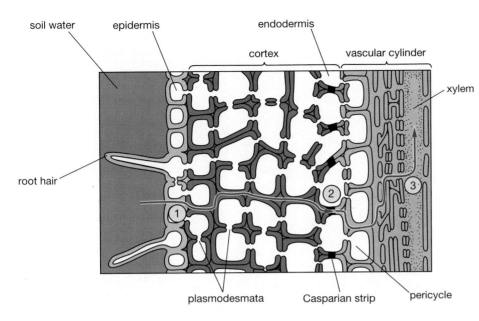

soil water   epidermis   endodermis   cortex   vascular cylinder   root hair   plasmodesmata   Casparian strip   pericycle   xylem

Figure 26-5   **Water uptake by roots**

① There is an uninterrupted pathway for water movement from the soil water, between cells of the epidermis and cortex, up to the waterproofing of the Casparian strip. Mineral uptake has created a higher concentration of minerals (and a lower concentration of water) in the extracellular space inside the Casparian strip (light blue) than in the extracellular space and soil water outside the Casparian strip (dark blue; see Fig. 26-2). ② Therefore water moves by osmosis across the endodermal cells into the extracellular space of the vascular cylinder. ③ Water moves into the tracheids and vessel elements of root xylem through porous pits in their cell walls.

ing the cohesion between water molecules is adhesion between water molecules and the walls of xylem. Attraction of water molecules to the cell walls of the thin xylem tubes help the water creep upward, just as water is pulled upward into a very narrow glass tube. This principle, called capillary action, helps water move upward within xylem.

Transpiration Produces the Tension That Pulls Water Upward

Evaporation of water through the stomata of a leaf, a process called **transpiration,** provides the force for water movement—the "tension" part of the theory. As a leaf transpires, the concentration of water in the mesophyll cells falls. This lower water concentration causes osmosis of water from the xylem in the nearby veins into the dehydrating mesophyll cells. Water molecules leaving the xylem are attached to other water molecules in the same xylem tube by hydrogen bonds. Therefore, when one water molecule leaves, it pulls adjacent water molecules up the xylem. As these water molecules move upward, other water molecules farther down move up to replace them. This process continues all the way to the roots, where water in the extracellular space around the xylem is pulled in through the holes in the walls of vessel elements and tracheids. This upward and inward movement of water finally causes soil water to move into the vascular cylinder by osmosis through the endodermal cells. The force generated by the evaporation of water from the leaves, transmitted down the xylem to the roots, is so strong that water can be absorbed from quite dry soils. Can this cohesion-tension theory explain the movement of water from soil to the topmost leaves of giant redwoods? Using a special apparatus, botanists have measured water tension in xylem and have found tensions strong enough to pull water up 200 meters (over 600 feet).

▓ SUMMARY

WATER TRANSPORT IN XYLEM

Transpiration from leaves removes water from the top of a xylem tube. This water is replaced by water from farther down the tube, so water moves by bulk flow up the xylem. This upward flow removes water from the root xylem and the extracellular space surrounding it, thereby promoting osmosis of water from the soil into the vascular cylinder of the root. *The flow of water is unidirectional, from root to shoot, because only the shoot can transpire.*

Adjustable Stomata Control the Rate of Transpiration

Transpiration has both positive and negative effects on a plant. On the plus side, transpiration provides the force that transports water and minerals to the leaves at the top of the plant. On the minus side, transpiration is by far the largest source of water loss—a loss that may threaten the very survival of the plant, especially in hot, dry weather.

Most water transpires through the stomata of leaves and stem, so you might think that a plant could prevent water loss by simply closing its stomata. Don't forget, however, that photosynthesis requires carbon dioxide from the air, which diffuses into the leaf mainly through open stomata. *Therefore, a plant, by opening and closing its stomata, must achieve a balance between carbon dioxide uptake and water loss.*

A Stoma Is an Opening Surrounded by Guard Cells

A **stoma** consists of a central opening surrounded by two sausage-shaped, photosynthetic **guard cells** that regulate the size of the opening (Fig. 26-7). With some exceptions, sto-

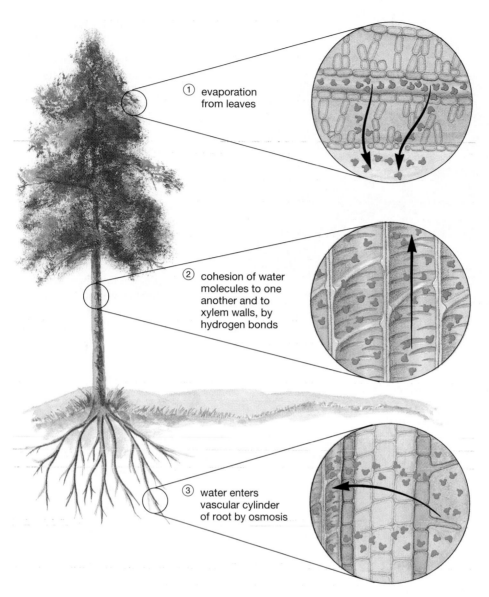

Figure 26-6   The cohesion-tension theory of water flow from root to leaf in xylem

① Water molecules evaporate out of the leaves, and other water molecules replace them from the xylem of the leaf veins. ② Within the xylem, hydrogen bonding holds nearby water molecules together so firmly that the column of water behaves very much like a rope. The top of the "water rope" is pulled up by evaporation, and the rest of the rope comes along as well, all the way down to the roots. ③ As the molecules of the water rope retreat up the xylem in the roots, the decreased water concentration within the root xylem and the surrounding extracellular space causes water to enter from the soil water by osmosis, thus steadily replenishing the bottom of the rope.

(labels in figure)
① evaporation from leaves
② cohesion of water molecules to one another and to xylem walls, by hydrogen bonds
③ water enters vascular cylinder of root by osmosis

mata open during the day and close at night, but they will also close if the leaf begins to dehydrate. The adaptive value of this arrangement is obvious: The stomata open only during the day, when sunlight allows photosynthesis. Even then, however, the stomata close if water loss becomes too great.

Guard cells change the size of the opening between them by changing their own shape. How do they do that? There are two levels to this question: First, how does changing the shape of the guard cells open and close the opening? Second, what physiological processes cause the change in shape?

Guard Cell Swelling Opens the Stomata

Stomata open when the guard cells take up water and swell, and they close when guard cells lose water and shrink. This might seem paradoxical, because it seems that swollen guard cells would block the central opening. The key lies in the construction of the cell wall of guard cells

(Fig. 26-8a). The tough cellulose fibers in the wall encircle the guard cells like a series of inelastic belts. Thus, when water enters the guard cells and their volume increases, they cannot become fatter but must become longer. Each pair of guard cells is attached at both ends, so the only way the cells can become longer is by bowing outward like a cooked sausage, opening a hole between them (Fig. 26-8b). Thickened guard cell walls on the sides that surround the central opening keep these sides from stretching and contribute to their arching away from one another as the guard cells elongate.

Guard Cells Change Shape by Regulating Their Potassium Content

Guard cells swell with water and open the stoma when they actively transport potassium inside. According to the principles of osmosis, water will enter a guard cell if its cytoplasm has a lower water concentration than the cyto-

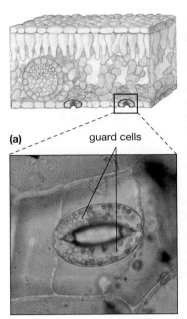

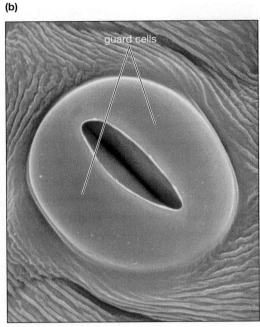

(b)

guard cells

(a)

guard cells

Figure 26-7   **Stomata**

Stomata seen through the light microscope **(a)** and scanning electron microscope **(b).** In the light micrograph, note that the guard cells contain chloroplasts (the green ovals within the cells), but the other epidermal cells do not.

plasm of surrounding cells and will leave the guard cell if it has a higher concentration of water. Large changes in potassium concentration within the guard cells cause correspondingly large changes in water concentration, driving the osmotic flows that open and close a stoma (Fig. 26-9). When potassium enters the guard cells, water follows by osmosis, opening the stoma; when potassium leaves the guard cells, water leaves again by osmosis, and the stoma closes.

Several factors regulate the potassium concentration inside guard cells. The three most important factors are

light, carbon dioxide (CO_2), and water levels within the leaf. These factors help the plant achieve a balance between the need to photosynthesize and the need to conserve water.

1. *Light reception.* Guard cells contain special pigments that absorb light. When light strikes these pigments, they trigger a series of reactions that cause potassium to be actively transported from the extracellular fluid into the guard cells. Water follows and the stomata open. At night, when light is not present to activate

(a)

Closed stoma

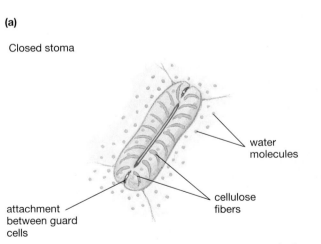

water molecules

cellulose fibers

attachment between guard cells

(b)

Open stoma

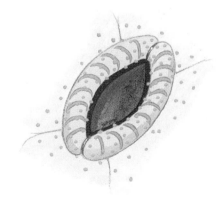

Figure 26-8   **Mechanics of stomatal opening and closing**

(a) Bands of cellulose fibers in guard cell walls enclose the cell like strong, inexpandable belts. When the guard cells are relatively dehydrated, they lie straight alongside one another, closing the central pore. **(b)** When the guard cells take up water, they swell. But the cellulose belts do not allow the cells to become fatter as they swell, so they must become longer. Because the cells are attached to each other at the ends, they can become longer only by bowing outward, thus opening up the central pore.

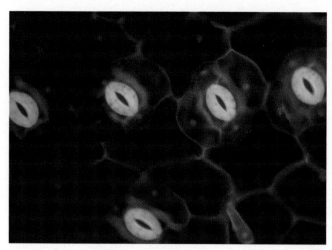

Figure 26-9   **The control of stomata**

The concentration of potassium ions drives osmosis of water into and out of guard cells. In this micrograph, the potassium concentration is indicated by the brightness of the yellow dye, clearly showing that the guard cells surrounding open stomata have a higher potassium concentration than the other epidermal cells.

the pigments, the potassium pumping stops, the "extra" potassium within the guard cells diffuses back out, and the stomata close, conserving water.

2. *Carbon dioxide concentration.* Guard cells contain specialized molecules responsive to CO_2 levels. Low CO_2 concentrations stimulate the active transport of potassium into the guard cells, causing stomata to open and allowing CO_2 to diffuse in. The concentration of CO_2 within the guard cells is regulated by the balance be-

tween photosynthesis, which uses up CO_2, and respiration, which generates CO_2. During the day, photosynthesis in guard cells uses up CO_2 faster than respiration produces it, so the CO_2 concentration drops, causing the stomata to open. At night, photosynthesis stops but respiration continues, raising the CO_2 level within the guard cells, halting transport of potassium inward, and allowing the guard cells to close.

3. *Water.* If a leaf loses water faster than it can be replaced from the xylem, it begins to wilt, and the mesophyll cells release a hormone called **abscisic acid.** The hormone strongly inhibits the active transport of potassium into the guard cells, overriding the stimulatory effects of light and low CO_2 levels, so potassium pumping stops. As potassium leaks out of the guard cells, water follows by osmosis, the guard cells shrink, and the stomata close.

The Transport of Sugars

Water and minerals transported into the leaves allow them to carry out photosynthesis, producing sugars from water and CO_2. These sugars must be moved to other parts of the plant, to nourish nonphotosynthetic structures such as roots or flowers, and to be stored in the cortex cells of the root and stem. Sugar transport is the function of phloem.

Botanists studying phloem contents employ a most unlikely lab assistant: the aphid. Aphids feed on the fluid contained in phloem sieve tubes. An aphid inserts a pointed, hollow tube, the stylet, through the epidermis and cor-

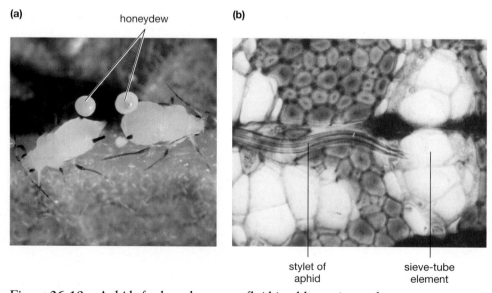

Figure 26-10   **Aphids feed on the sugary fluid in phloem sieve tubes**

(a) When an aphid pierces a sieve tube, pressure in the tube forces the fluid out of the plant and into the digestive tract of the aphid. Sometimes the pressure is so great that fluid is forced completely through the aphid and out its anus, as "honeydew." This fluid is collected by certain species of ants that act as "shepherds" to the aphids, defending them from predators in return for a diet of sweet honeydew. **(b)** The flexible stylet of an aphid, passing through many layers of cells to penetrate a sieve-tube element.

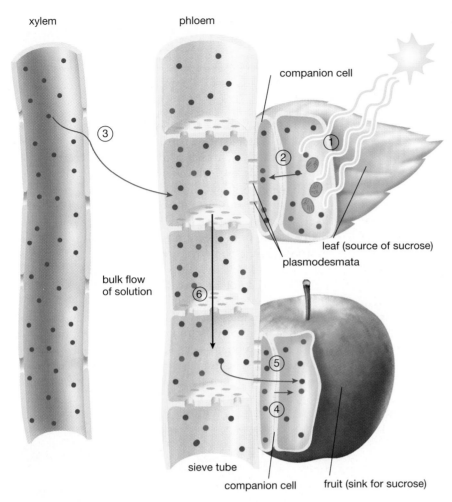

xylem

phloem

companion cell

③

bulk flow
of solution

②

①

⑥

leaf (source of sucrose)

plasmodesmata

⑤

④

sieve tube

companion cell

fruit (sink for sucrose)

● water ⟶ osmosis of water

● sucrose ⟶ active transport of sucrose

Figure 26-11   **The pressure flow theory**

This theory relies on differences in hydrostatic pressure to move fluid through phloem sieve tubes. ① A photosynthesizing leaf manufactures sucrose (red dots) that ② is actively transported (red arrow) into a nearby companion cell in phloem. The sucrose diffuses into the adjacent sieve-tube element through plasmodesmata, raising the concentration of sucrose in the sieve-tube element. ③ Water (blue dots) leaves nearby xylem and moves into the "leaf end" of the sieve tube by osmosis (blue arrow), raising the hydrostatic pressure as increasing numbers of water molecules enter the fixed volume of the tube. ④ The same sieve tube connects to a developing fruit. At the "fruit end" of the sieve tube, sucrose enters the companion cells by diffusion through plasmodesmata. It is then actively transported out of the companion cells and into the fruit cells. ⑤ Water moves out of the tube by osmosis, lowering the hydrostatic pressure within the tube. ⑥ High pressure in the leaf end of the phloem and low pressure in the fruit end cause water, together with any dissolved solutes, to flow in bulk from leaf to fruit (black arrow).

tex of a young stem into a sieve tube (Fig. 26-10). The aphid can then relax and let the plant do the work. The fluid in the sieve tubes is under pressure and flows through the stylet into the digestive tract of the aphid (sometimes with enough pressure to force its way out the other end!). By cutting off the aphid but leaving its stylet in place, botanists have collected sieve-tube fluid and found that it consists mostly of sucrose and water, as much as 25% sucrose by weight. How is this concentrated sugar solution moved about the plant?

The Pressure Flow Theory Explains Sugar Movement in Phloem

The movement of fluid in phloem is directed by sugar production and use. Any structure that actively synthesizes sugar will be a **source** away from which phloem fluid will be transported. Conversely, any structure that uses up sugar or converts sugar to starch will be a **sink** toward which phloem fluids will flow. A newly forming leaf will be a sink

as it develops, with phloem flow up into it from more mature leaves located farther down the plant. When the leaf matures, it will photosynthesize and produce sugar, becoming a source for phloem flow to other newly developing leaves above it, to flowers or fruits, or to the roots below it. Therefore, fluid in phloem can move either up or down the plant, depending on the metabolic demands of the various parts of the plant at any given time.

The most widely accepted mechanism for the transport of sugars in phloem is the **pressure flow theory** (Fig. 26-11), which relies on differences in hydrostatic pressure (water pressure) to move fluid through sieve tubes. Let's illustrate this theory by following sucrose movements from a mature leaf to a developing fruit.

1. *Sucrose source—photosynthesis.* When a leaf is photosynthesizing rapidly, it manufactures lots of glucose, much of which is converted to the larger molecule sucrose.

2. *Phloem sieve-tube loading.* Much of this sucrose is actively transported into companion cells of the phloem in the leaf veins. This movement raises the concen-

E A R T H   W A T C H

Plants Help Regulate the Distribution of Water

The land teems with a remarkable diversity of plant life. The distribution of plants over the planet is limited by environmental factors and the adaptations of the plants. Probably the most important environmental factor influencing plant distribution is water: Cacti and succulents inhabit deserts because they can withstand drought; orchids and mahogany trees need the frequent drenching rains of the rain forest. What people often overlook is the flip side of this plant-water relationship: Plants, through transpiration, help regulate the amount and distribution of rainfall, soil water, and even river flow.

Consider the Amazon rain forest (Fig. E26-1). An acre of soil supports hundreds of towering trees, each bearing millions of leaves. The surface area of the leaves dwarfs the surface area of the soil, so up to 75% of all the water evaporating from the acre of forest is from leaf transpiration. This transpiration raises the humidity of the air and causes rain to fall. In fact, about half of the water transpired from the leaves falls again as rain, with the overall result that about one-third of the total rainfall is water recycled by transpiration. Thus, in a very real sense, the high humidity and frequent showers that the rain forest needs to survive are partly *created by the forest itself!* If large areas of rain forest are cut down, less water evaporates in that area, so less rain falls, and new rainforest tree seedlings cannot grow. An entirely different plant community would probably become established on the disturbed land and might become a *permanent* new community.

Plant transpiration might even have a moderating influence on some aspects of the "greenhouse effect" brought about by increasing atmospheric CO_2 levels, mostly due to burning fossil fuels, such as coal, during industrial activity and to cutting down forests (see Chapter 45). According to predictions about the greenhouse effect, higher CO_2 levels will raise temperatures on Earth, and a warmer planet would increase water evaporation; increased evaporation in turn, should lead to drier soils and the expansion of deserts. As you just learned, however, stomata open partly in response to low CO_2 levels within the guard cells.

Elevated atmospheric CO_2 levels also raise CO_2 within the guard cells and cause partial stomatal closing. Further, a recent study found that plants grown in an atmosphere with high CO_2 levels have fewer stomata per unit area of leaf than plants grown at low, preindustrial CO_2 levels. If there are fewer stomata, and they are chronically partially closed, then one might expect less transpiration from plants living in a "greenhouse future" than from "preindustrial" plants. Some researchers suggest that soil moisture and river flow in southwestern states might actually *increase* in response to increased atmospheric CO_2, despite the greenhouse warming effect.

These examples show that plants wield an enormous influence on what we often consider to be nonliving aspects of the biosphere, such as humidity, rainfall, soil water, and river flow. The responses of plants to human activities, neither simple nor easily predictable, can have major impacts on ecosystems.

Figure E26-1   The Amazon rain forest
The rainforest community helps mold its own environment.

tration of sucrose within the companion cells so that sucrose then diffuses down its concentration gradient through plasmodesmata into an adjacent sieve-tube element. This diffusion in turn raises the sucrose concentration in the leaf sieve tube.

3. *Osmosis into the leaf sieve tube.* The high sucrose concentration in the leaf sieve tube lowers the water concentration in the sieve tube, causing water to enter the sieve tube by osmosis from nearby xylem and hy-

drostatic pressure to increase as more water molecules enter the tube.

4. *Sucrose sink—developing fruit.* Meanwhile, some distance away but connected by the same sieve tube, sucrose is actively transported out of the nearby sieve-tube elements and companion cells into the cells of a fruit. The concentration of sugar in the fruit is raised and the concentration of sugar in the sieve tubes leading to the fruit is lowered.

5. *Osmosis out of the fruit sieve tube.* Water leaves the sieve tube by osmosis and follows the sugar into the fruit. Hydrostatic pressure drops within the tube.

6. *Bulk flow, driven by a hydrostatic pressure gradient.* Water that follows the sucrose into the sieve tube near the leaf by osmosis causes hydrostatic pressure to build up in the leaf portion of the sieve tube. Meanwhile, water entering the fruit by osmosis causes reduced hydrostatic pressure in the sieve tube near the fruit. In response to this pressure gradient, water moves by bulk flow from the leaf portion of the phloem into the fruit portion of the phloem, carrying the dissolved sugar.

❋ SUMMARY OF KEY CONCEPTS

A Comparison of Plant and Animal Nutrition
An animal acquires both energy and materials from complex organic molecules in its diet, which must be digested. A plant absorbs inorganic molecules and uses the energy of sunlight captured during photosynthesis to synthesize its organic molecules from these inorganic starting materials. Plants acquire carbon dioxide and oxygen from the air and water, and all other essential nutrients from the soil.

Plant Nutrition
Most minerals are taken up from the soil water by active transport into the root hairs. These minerals diffuse into the root through plasmodesmata, to the pericycle just inside the vascular cylinder. Here they are actively transported into the extracellular space of the vascular cylinder. The minerals diffuse from the extracellular space into the tracheids and vessel elements of xylem.

Many plants have fungi called mycorrhizae associated with their roots that help them absorb soil nutrients. Nitrogen can be absorbed only as ammonium or nitrate, both of which are scarce in most soils. Legumes have evolved a cooperative relationship with nitrogen-fixing bacteria that invade legume roots. The plant provides the bacteria with sugars, and the bacteria use some of the energy to convert atmospheric nitrogen to ammonium that is then absorbed by the plant.

The Acquisition of Water
Because of the loose packing and porous walls of the cells of the root epidermis and cortex, water in the soil has a continuous, uninterrupted pathway through the outer layers of the root, up to the waterproof layer of the Casparian strip between endodermal cells. Both mineral uptake and the upward movement of water in xylem contribute to a concentration gradient across the endodermal cells, with a higher concentration of free water molecules in the extracellular space outside the endodermis than in the extracellular space inside the endodermis. Therefore, water moves by osmosis across the plasma membranes of the endodermal cells into the extracellular space of the vascular cylinder.

The Transport of Water and Minerals
The cohesion-tension theory explains xylem function: Cohesion of water molecules to one another by hydrogen bonds holds together the water within xylem tubes almost as if it were a solid rope. As water molecules evaporate from the leaves, the hydrogen bonds pull other water molecules up the xylem to replace them. This movement is transmitted down the xylem to the root, where water loss from the vascular cylinder promotes water movement across the endodermis from the soil water by osmosis.

The Transport of Sugars
The pressure flow theory explains sugar transport in phloem: Parts of the plant that synthesize sugar (for example, leaves) export sugar into the sieve tube. Increasing sugar concentrations attract water entry by osmosis, causing high hydrostatic pressure in that part of the phloem. Parts of the plant that consume sugar (for example, fruits) remove sugar from the sieve tube. Loss of sugar causes loss of water by osmosis, resulting in low hydrostatic pressure. Water and dissolved sugar move by bulk flow in the sieve tube from high to low pressure.

❋ KEY TERMS

abscisic acid   p. 528
bulk flow   p. 524
cohesion-tension theory   p. 524
guard cell   p. 525
legume   p. 523
macronutrient   p. 520
micronutrient   p. 520

mineral   p. 520
mycorrhiza   p. 522
nitrogen fixation   p. 523
nitrogen-fixing bacterium   p. 523
nodule   p. 523
nutrient   p. 519
nutrition   p. 519

plasmodesma   p. 520
pressure flow theory   p. 529
sink   p. 529
source   p. 529
stoma   p. 525
transpiration   p. 525

✖ T H I N K I N G   T H R O U G H   T H E   C O N C E P T S

Multiple Choice

1. Which of the following generally describes the relative amounts of mineral nutrients in the soil and in a plant growing in that soil?
   a. The plant has the same concentration of most minerals as does the soil.
   b. The plant has higher concentrations of most minerals than does the soil.
   c. The plant has lower concentrations of most minerals than does the soil.
   d. Plants do not take minerals out of the soil.
   e. none of the above

2. In a mycorrhizal relationship a plant root has a symbiotic relationship with
   a. a fungus that helps to obtain minerals from the soil
   b. a fungus that helps in the fixation of nitrogen from the air into a form usable by the plant
   c. bacteria that help to obtain minerals from the soil
   d. bacteria that help in the fixation of nitrogen from the air into a form usable by the plant
   e. an alga that helps the plant photosynthesize

3. Which of the following is NOT involved in the cohesion-tension theory of water movement in plants?
   a. the presence of hydrogen bonds holding water molecules together
   b. the attraction of water molecules to the walls of the xylem
   c. the diffusion of water from cells in the root to cells in the shoot
   d. the evaporation of water through the stomata
   e. the pulling of water molecules by other water molecules up through the plant

4. In the morning when the sun rises, plants open their stomata by
   a. pumping water out of the guard cells
   b. pumping water into the guard cells
   c. pumping potassium ions out of the guard cells
   d. pumping potassium ions into the guard cells
   e. all of the above may be involved at different times

5. Plants get all their nutrients from the soil, unlike animals, which get theirs primarily from eating and digesting other organisms.
   a. True
   b. False; plants get most of their food from the air as carbon dioxide.
   c. False; animals can manufacture many nutrients from inorganic sources.
   d. False; plants use the sun's energy as a source of nutrients.
   e. False; animals get their nutrients from cellular respiration.

6. The "greenhouse effect" is caused primarily by
   a. plant transpiration
   b. ozone depletion in the upper atmosphere
   c. acid rain
   d. carbon dioxide released during plant respiration
   e. increasing levels of carbon dioxide in the atmosphere due to human activities

Review Questions

1. Compare and contrast animal and plant nutrition, especially with respect to their sources of energy and materials.

2. What role does abscisic acid play in controlling the opening and closing of stomata? Describe the daily cycle of guard cells opening and closing. How are various environmental conditions involved in this process?

3. How are minerals and water taken up by roots? Diagram the structures involved, the pathways for water and minerals from soil water to xylem, and the transport processes at each step.

4. How does the pressure flow theory explain the movement of sugars through the plant?

5. Describe the cohesion-tension theory of water movement in xylem. What supplies the cohesion, and what is the source of the tension? How do these two interact to move water through a plant?

6. What occurs during nitrogen fixation? Describe the formation of a root nodule in a legume. How do both the bacteria and the legume benefit from this relationship?

✖ A P P L Y I N G   T H E   C O N C E P T S

1. A new mutant form of aphid, the klutzphid, has just appeared. Instead of feeding on phloem, it inserts its stylet into the vessel elements of xylem. What materials are found in the fluids of xylem? Would xylem fluid flow into the aphid? Justify your answer.

2. One of the foremost goals of plant molecular biologists is to insert the genes for nitrogen fixation, or the ability to enter into symbiotic relationships with nitrogen-fixing bacteria, into crop plants such as corn or wheat (see Chapter 14). Why would insertion of such genes be useful? What changes in farming practices would this technique allow?

3. We learned in Chapter 2 about the peculiar characteristics of water. Discuss several ways that the evolution of vascular plants has been greatly influenced by water's special characteristics.

4. When you buy cut flowers, the florist may advise you to do two things: Recut the stems while holding the stems under water, and add some table sugar to the water in which you place the arranged flowers so that they "stay fresh" longer. Discuss, in a technical manner, why both suggestions are good advice.

5. A major environmental problem is desertification, in which overgrazing by cattle or other animals results in too few plants in an area. Show how what you know about the

movement of water through plants enables you to understand this process, in which there is less water in the atmosphere, less rain, and thus dry, desertlike conditions.

6. Plants that grow in swamps and bogs have their roots in water that has a high concentration of dissolved materials. Such plants are often similar to plants that live in dry areas rather than wet ones. How could this be true? What effect does the dissolved material have on the movement of water into the plant?

✖ F O R   M O R E   I N F O R M A T I O N

Cochran, M. F. "Chestnuts: Back from the Brink." *National Geographic*, February 1990. Fungi girdle chestnuts, causing almost all trees in the United States to die. Recent discoveries are helping this tree come back.

Day, S. "A Shot in the Arm for Plants." *New Scientist*, January 9, 1993. Plants exposed to insects or disease may make chemicals that are transported throughout their body to protect them from future attack.

Mansfield, T. A., and Davies, W. J. "Mechanisms for Leaf Control of Gas Exchange." *BioScience*, March 1985. How stomata control gas exchange through the surface of a leaf.

Oaks, A. "A Re-evaluation of Nitrogen Assimilation in Plants." *BioScience*, February 1992. A somewhat technical comparison of nitrogen use comparing legumes with wheat. Legumes may change their release of nitrogen depending on the levels of nitrogen fertilizer in the soil.

Zimmerman, M. H. "How Sap Moves in Trees." *Scientific American*, March 1963. A delightful description of the use of aphids as research tools in botany.

N E T   W A T C H

On-line resources for this chapter are on the World Wide Web at:
http://www.prenhall.com/~audesirk (click on the table of contents link and then select Chapter 26).

27 Plant Reproduction and Development

> *"By the time I get to the wood, I am carrying all manner of seeds hooked in my coat or piercing my socks or sticking by ingenious devices to my shoestrings. I let them ride. After all, who am I to contend against such ingenuity? It is obvious that nature, or some part of it in the shape of these seeds, has intentions beyond this field, and has made plans to travel with me."*

LOREN EISELY, *The Immense Journey*

⊞ AT A GLANCE

Sexual versus Asexual Reproduction

Plant Life Cycles

The Evolution of Flowers
Complete Flowers Have Four Major Parts

Coevolution of Flowers and Pollinators
Some Flowers Provide Food for Pollinators
Sexy Deceptions Attract Pollinators
Some Plants Provide Nurseries for Pollinators

Gametophyte Development in Flowering Plants
Pollen Is the Male Gametophyte
The Embryo Sac Is the Female Gametophyte

Pollination and Fertilization

The Development of Seeds and Fruits
The Seed Develops from the Ovule and Embryo Sac
The Fruit Develops from the Ovary Wall
Seed Dormancy Helps Assure Germination at an
   Appropriate Time

The Germination and Growth of Seedlings
The Shoot Tip Is Protected by the Coleoptile
Cotyledons Nourish the Sprouting Seed
Controlling the Development of the Seedling

California poppies blanket the hillsides in California's Antelope Valley after early spring rains. Desert wildflowers are adapted to bloom and set seed rapidly, taking advantage of the brief spring moisture before summer brings drought and searing temperatures. The bright colors, shapes, and scents of flowers attract animal pollinators.

As you walk through a wildflower-strewn meadow, you may be tempted to think that the floral display was created just for your enjoyment. In fact, plants don't develop flowers for us, but for the birds and bees—and beetles, moths, and even bats. The flower is actually a sexual display that enhances the reproductive success of a plant. By enticing animals to transfer pollen from one plant to another, flowers enable stationary plants to "court" distant members of their own species. This critical selective advantage has allowed the flowering plants to become the dominant plants on land.

As you know, evolution commonly produces new structures by modifying old ones, and flowers are no exception. Flowers are not wholly new adaptations; their parts are actually highly modified leaves, shaped by mutation and natural selection into a form that enhances pollination. You will be able to understand the structure and function of flowers more easily if we begin, therefore, by briefly reviewing the essentials of plant reproduction and the plant life cycle.

Sexual versus Asexual Reproduction

Many plants can reproduce either sexually or asexually. Asexual reproduction in plants usually involves part of a single plant, say a stem, giving rise to a new plant. The cells that form an asexually produced offspring arise by mitosis from cells of the parent plant. Therefore, these offspring are genetically identical to the parent (see Chapter 9). In Chapter 25 we encountered several methods of asexual reproduction, including runners in strawberries and rhizome sprouting in irises. Asexual reproduction is often a highly effective reproductive strategy. For example, an offspring strawberry connected to its parent by a runner (see Fig. 25-20) draws nourishment from the parent until it grows large enough to fend for itself.

However, if an offspring is genetically identical to its parent, then the offspring is only as well adapted to the environment as its parent was. What if the environment changes? Sexually produced offspring usually combine genes from two different parents, and therefore they may be endowed with traits that differ from those of either parent. This new combination of traits may help the offspring cope with the environment better than either parent could. As a result, most organisms reproduce sexually, at least some of the time.

Plant Life Cycles

The sexual life cycle of plants is more complex than the familiar animal life cycle. In animals, cells in the reproductive organs (gonads) of a diploid adult undergo meiosis to produce haploid gametes, either sperm or eggs. Sperm and egg fuse to create a diploid fertilized egg, the **zygote.** Through repeated mitosis and differentiation of the daughter cells, the zygote develops into another diploid adult. Plants have two distinct, multicellular "adult" forms, one diploid and one haploid, that give rise to each other. For this reason, the plant life cycle is called **alternation of generations**, because diploid adults (called sporophytes) alternate with haploid adults (called gametophytes; see Figs. 27-1 and 27-2).

To illustrate the alternation of generations, let's examine the life cycle of a fern (Fig. 27-1), starting with the diploid adult form. This stage of the life cycle, the **sporophyte** ("spore plant" in Greek), bears reproductive cells that undergo meiosis to produce haploid cells that are **spores,** not gametes. The difference between spores and gametes is that spores do not fuse together to form a diploid cell. Instead, a fern spore is blown off the parent leaf by the wind and lands on the soil. There the spore germinates (begins to grow and develop), dividing repeatedly by mitosis to form a multicellular, haploid organism. This organism produces gametes and hence is called the **gametophyte** ("gamete plant" in Greek). Because its cells are haploid already, the "gonads" of the gametophyte can produce sperm and eggs without further meiosis. Usually a single gametophyte produces both sperm and eggs, but often at different times, thereby preventing self-fertilization. Sperm and egg fuse to form a zygote that develops into a new diploid sporophyte plant.

Alternation of generations occurs in all plants. In primitive land plants, including mosses and ferns, the gametophyte is an independent, although usually small, plant. It liberates mobile sperm cells that reach an egg by swimming through thin films of water covering adjacent gametophytes or by being splashed by raindrops from one plant to the next. Therefore, ferns and mosses can reproduce only in moist habitats.

But most terrestrial habitats are relatively dry. One way to reproduce in drier places is for the plant to surround its sperm in a watertight package that can be transported to

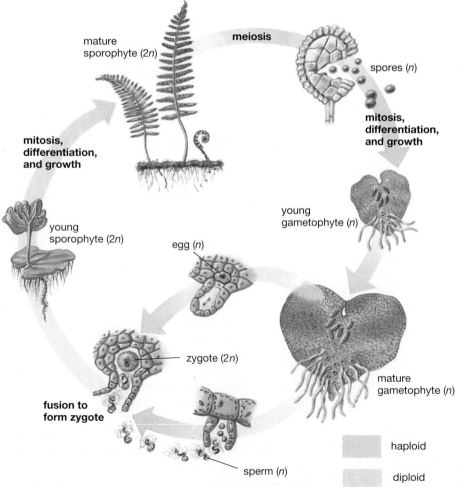

mature sporophyte (2n)

meiosis

spores (n)

mitosis, differentiation, and growth

mitosis, differentiation, and growth

young gametophyte (n)

young sporophyte (2n)

egg (n)

zygote (2n)

mature gametophyte (n)

fusion to form zygote

sperm (n)

haploid

diploid

Figure 27-1   The fern life cycle

Ferns typify the alternation-of-generations life cycle found in all plants, in which separate multicellular haploid and multicellular diploid "adult" organisms occur at different parts of the life cycle. The various stages of the life cycle are described in the text. The letter *n* refers to the haploid state, *2n* to the diploid state.

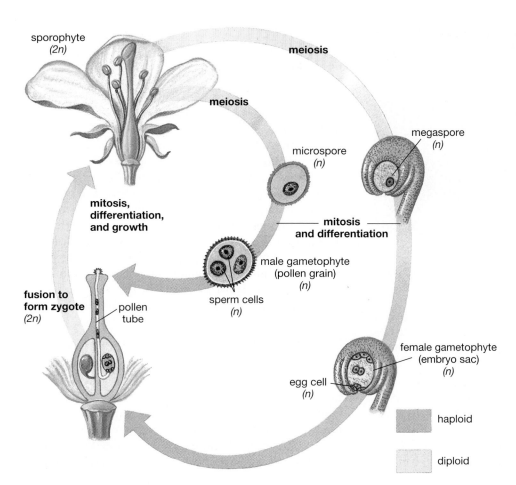

Figure 27-2   **The life cycle of a flowering plant**

Although this cycle shows the same stages as the life cycle of a fern (see Fig. 27-1), the haploid stages are much smaller and cannot live independently of the diploid plant.

another plant, where the sperm are liberated directly into the egg-bearing structures of the second plant. The seed plants (gymnosperms and flowering plants) do just that. In the flowering plants, two types of spores are formed by meiosis within the flowers borne by the sporophyte generation (Fig. 27-2). These haploid spores develop into gametophytes not in the soil, but *within the flower.* The gametophytes consequently are microscopic and do not live independently of the sporophyte. One type of spore, the megaspore, undergoes a few mitotic cell divisions and develops into the female gametophyte, a small cluster of cells permanently retained within the flower. The other type of spore, the microspore, develops into the male gametophyte: a tough, watertight **pollen grain** containing two sperm. The pollen grain drifts on the wind or is carried by an animal from one flower to another. On the recipient flower, the pollen grain elongates, burrowing through the flower tissues to the female gametophyte within. This miniature male gametophyte liberates its sperm inside the female gametophyte, where fertilization occurs. The zygote becomes enclosed in a drought-resistant **seed**. The seed, including an embryonic plant and a food reserve within a protective outer coating, may lie dormant for months or years, waiting for favorable conditions for growth.

In this chapter, we examine sexual reproduction in flowering plants, from the evolution of the flower through the formation of the seed and the development of the new seedling.

The Evolution of Flowers

The earliest seed plants were the gymnosperms, represented today mainly by conifers, a group that includes pines, firs, and spruces. As we described in Chapter 23, conifers bear male and female gametophytes on separate cones. During early spring, the small male cones release millions of pollen grains that float about on the breezes (Fig. 27-3). Most blow uselessly away, but so many grains are floating around that some enter the pollen chambers located on the scales of the female cones, where they are captured by sticky coatings of sugars and resins. The pollen grains germinate and tunnel to the female gametophytes at the base of each scale. Sperm are liberated and fertilize the eggs within the female gametophyte, and a new generation begins.

Clearly, wind pollination is an inefficient operation, since most of the pollen grains are lost. In a world of stationary plants and mobile animals, if a gymnosperm could entice an animal to carry its pollen from male to female cone, it would greatly enhance its reproductive rate and hence its evolutionary success. As it happens, gymnosperms and insects were poised to establish just such a relationship about 150 million years ago.

Insects, especially beetles, are among the most abundant animals on Earth. They exploit nearly every possible food resource on land, including the reproductive parts of gymnosperms. About 150 million years ago, some bee-

Figure 27-3   **Conifers are wind pollinated**

Even slight breezes blow thick clouds of pollen from ripe male cones.

tles fed on both the protein-rich pollen of male cones and the sugar-rich secretions of female cones. Beetles can make quite a mess when they feed, and pollen feeders often wind up with pollen dusted all over their bodies. If the same beetle were to visit one plant, eat pollen, and then wander over to another plant of the same species to dine on the sugary secretions of a female cone, some of the loose pollen would quite likely rub off on the female cone.

The stage was set for the evolution of flowering plants. Efficient pollination by insects requires that a given in-

sect visit several plants of the same species, pollinating them on the way. For the plants, two key adaptations were necessary. First, enough pollen or nectar (as the sugary secretions are called) must be produced within the reproductive structures so that insects will regularly visit them to feed. Second, the location and richness of these storehouses of pollen and nectar must be advertised to the insects, both to show them where to go and to entice them to specialize on that specific plant species. Any mutation that contributed to these adaptations would enhance the reproductive success of the plant carrying the mutation and would be favored by natural selection. By about 130 million years ago, flowers had evolved with exactly these adaptations. The advantages of flowers are so great that in today's temperate and tropical zones, flowering plants are overwhelmingly dominant, and an array of animals, including bees, moths, butterflies, hummingbirds, and even some mammals, feed almost exclusively at flowers.

Complete Flowers Have Four Major Parts

Flowers are the reproductive structures of flowering plants. Like leaves, they develop as outgrowths of stems and, in fact, the various flower parts have evolved from leaves. **Complete flowers,** such as those of crocuses, roses, and tomatoes, consist of a central axis upon which four successive sets of modified leaves are attached (Fig. 27-4). These modified leaves form the sepals, petals, stamens,

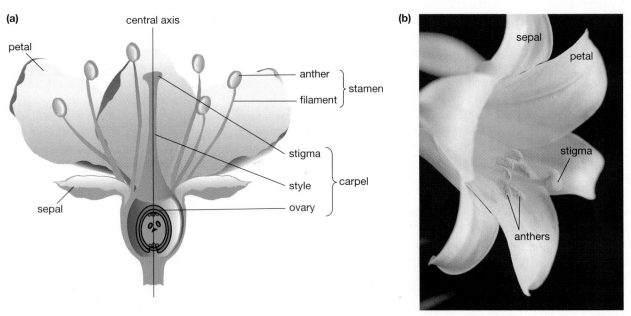

Figure 27-4   **A complete flower**

(a) A complete flower has four parts: sepals, petals, stamens (the male reproductive structures), and at least one carpel (the female reproductive structure). This drawing shows a complete dicot flower. **(b)** The lily is a complete monocot flower, with three sepals (virtually identical to the petals), three petals, six stamens, and three fused carpels. Each stamen consists of a filament bearing an anther at its tip. The carpels consist of an ovary hidden in the base of the flower, with a long style protruding out, ending in a sticky stigma. Note that the anthers are considerably below the stigma. This is probably an adaptation preventing self-pollination: Pollen cannot simply fall from the anther onto the stigma.

and carpels. The **sepals** are located at the base of the flower. In dicots, the sepals are often green and leaflike; in monocots they usually resemble the petals (see Fig. 27-4). In either case, sepals surround and protect the flower bud as the remaining three structures develop. Just above the sepals are the **petals,** which are usually brightly colored and fragrant, advertising the location of the flower.

The male reproductive structures, the **stamens,** are attached just above the petals. Stamens usually consist of a long slender **filament** bearing at its tip an **anther** that produces pollen. The female reproductive structures, the **carpels,** occupy the uppermost position in the flower. An idealized carpel is somewhat vase-shaped, with a sticky **stigma** for catching pollen mounted atop an elongated **style** connecting it with the bulbous **ovary.** Inside the ovary are one or more **ovules** in which the female gametophytes develop. When mature, the ovule will become the seed, and the ovary will develop into a protective, adhesive, or edible enclosure, the **fruit.**

As you may know from your own gardening experience, not all flowers are complete. **Incomplete flowers** lack one or more of the four floral parts. For example, many plants have separate male and female flowers, which may be borne on the same plant, as in cucumbers and squashes (Fig. 27-5). Alternatively, male and female flowers may be on different plants, as in the American holly: Male flowers lack carpels, and female flowers lack stamens. Incomplete flowers may also lack sepals or petals.

Coevolution of Flowers and Pollinators

Wind-pollinated flowers, such as those of grasses and oaks, are usually inconspicuous and unscented, often scarcely more than naked stamens that liberate pollen to the wind (Fig. 27-6). The beautiful flowers so much admired by humans, however, are pollinated by animals. The distinctive shapes, colors, and odors of animal-pollinated flowers as well as the sensory capabilities and life-styles of their pollinators are an example of **coevolution**—evolution in two species that interact extensively with one another, so that each acts as a major force of natural selection on the other. Animal-pollinated flowers must attract useful pollinators and frustrate undesirable visitors who might eat nectar or pollen without fertilizing the flower in return. The animals, in their turn, have been under selective pressures to locate flowers quickly, identify the ones that can provide them with adequate nutrition, and extract the nectar or pollen with a minimum expenditure of energy.

Animal-pollinated flowers can be loosely grouped into three categories, depending on the benefits (real or imagined) that they offer to potential pollinators: food, sex, or a nursery.

Figure 27-5   **Some plants have separate male and female flowers**

Plants of the squash family, such as these zucchinis, bear separate female (left) and male (right) flowers. Obviously, individual flowers cannot be self-pollinated, but the plant could still be self-pollinated if an insect carried pollen from a male flower to a female flower of the same plant. However, each plant initially produces only male flowers, so some cross-pollination between plants that flower at slightly different times is virtually assured.

Figure 27-6   **Pollination by wind**

The flowers of grasses and many deciduous trees are wind pollinated, with anthers (yellow structures hanging beneath flowers) exposed to the wind. Petals are usually reduced or absent.

Some Flowers Provide Food for Pollinators

Many flowers provide food for foraging animals such as bee-tles, bees, moths and butterflies, or hummingbirds. In return, the animals unwittingly distribute pollen from flower to flower.

(a)

human vision

violet   blue   green   yellow   orange   red   far-red

400   500   600   700
wavelength

(b)

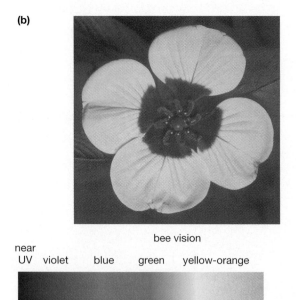

bee vision

near
UV   violet   blue   green   yellow-orange

400   500   600

Figure 27-7   **Ultraviolet patterns guide bees to nectar**

The spectra of color vision for humans **(a)** and bees **(b)** overlap considerably in the blue, green, and yellow ranges but differ on the edges. Humans are sensitive to orange and red, which bees do not perceive, whereas bees can see ultraviolet, which is in-visible to the human eye. Many flowers photographed under ordinary daylight (in part a) and under ultraviolet light (in part b) show striking differences in color patterns. Bees can see the ultraviolet patterns that presumably lead them to the nec-tar- and pollen-containing centers of the flowers.

Coevolution of Flowers and Insect Pollinators Has Produced Diverse Adaptations

Bees are culinary specialists, often feeding only on nectar and pollen. Bees use both scent and sight to locate and identify flowers from the air. We can thank the bees for most of the sweet-smelling flowers, because sweet "flowery" odors attract these pollinators. Bees also have good color vision, but they do not see exactly the same range of colors that humans do (Fig. 27-7). Although bees cannot distinguish red from gray or black, their color vision extends into the ultraviolet light range, which we cannot see. To attract a bee from afar, bee-pollinated flowers must look brightly colored *to a bee*. Typi-cally, these flowers are white, yellow, or blue and often have other markings, such as central spots or lines pointing toward the center, that reflect ultraviolet light (Fig. 27-7).

Bee-pollinated flowers have several structural adapta-tions that help to ensure pollen transfer. Many bee-polli-nated flowers, such as nasturtiums and foxgloves, produce nectar at the bottom of a tube (Fig. 27-8). Either pollen-laden stamens (usually in newly opened flowers) or the sticky stigma of the carpel (in older flowers) sticks out of the top of the tube. When a bee visits a young flower, she lands on the lip of the flower and thrusts her head into the tube to reach the nectar. Simultaneously, the stamens brush pollen onto her back. She may then visit an older

Figure 27-8   **Many bee-pollinated flowers are tubular**

In these foxgloves, nectar is produced at the base of the tube, and the anthers and stigmas extend into the tube. Often a lower "lip" on the tube serves as a landing platform for bees, because bees are not very good at hovering.

Figure 27-9 **The carrion flower is pollinated by flies**

Flies are attracted by its aroma of rotting meat, which gives this flower its name.

flower and repeat her foraging behavior. This time, she leaves pollen behind on the stigma.

Flowers adapted for moth and butterfly pollinators are often superficially similar to bee-pollinated flowers. However, their nectar tubes are usually deep and narrow, accommodating the long tongues of moths and butterflies.

Flowers pollinated by night-flying moths open only in the evening, are usually white, and give off strong, musky odors that help the moth locate the flower in the dark.

Many beetles prefer to feed on animal material, which has a high nutrient content. Therefore, beetle-pollinated flowers often smell like dung or rotting flesh, which attracts scavenging beetles. Flies have much the same taste in foods (or lack thereof!) as beetles, and most fly-pollinated flowers, such as the carrion flower (Fig. 27-9), also emit a powerful stench.

Hummingbird-Pollinated Flowers Have Deep, Nectar-Filled Tubes

Hummingbirds are one of the few vertebrates that are important pollinators, although several mammals also visit flowers (Fig. 27-10a, b). Birds have notoriously poor senses of smell, and hummingbird-pollinated flowers seldom synthesize fragrant chemicals. Instead, these flowers always produce much greater amounts of nectar than other flowers do because hummingbirds need lots of energy. If the flowers didn't produce lots of nectar, the hummers would go elsewhere and the flower would remain unpollinated. But a large supply of nectar would also attract insects. A bee, for example, might return again and again

(a)

(c)

(b)

Figure 27-10 **Bats and honey possums are unusual pollinators**

(a) A tropical bat feeds at a cluster of tubular flowers. Note the protruding stamens and stigma. As the bat hovers before the flower, the top of its head touches either the anthers or stigma or both, thus pollinating the flower. **(b)** As the honey possum stuffs its face into this flower, pollen adheres to its muzzle and whiskers. A visit to another flower may result in transfer of pollen. **(c)** A hummingbird hovers before a flower. Note that the flower has a deep, tubular shape and lacks the landing platform commonly found on bee-pollinated flowers.

to the same flower, never transferring pollen to another flower. Not surprisingly, hummingbird-pollinated flowers have evolved several adaptations that keep insects from drinking their nectar. These flowers are always tubular, matching the long bills and tongues of hummers (see Fig. 27-10c). The tube is much too deep for bees to reach the nectar at its base; the flower also lacks a lip upon which bees could rest while dining. In addition, most hummingbird-pollinated flowers are red or orange, colors that are brightly attractive to a bird but appear drab to a bee.

Sexy Deceptions Attract Pollinators

A few plants, most notably the orchids, take advantage of the mating drive and stereotyped behaviors of male wasps and flies to pollinate their flowers. Some orchid flowers mimic female wasps both in scent and shape (Fig. 27-11). The males land atop these "females" and attempt to copulate, but they get only a packet of pollen for their efforts. As they repeat their attempts on other orchids of the same species, the pollen packet is transferred.

Some Plants Provide Nurseries for Pollinators

Perhaps the most elaborate relationships between plants and pollinators occur in a few cases in which insects fertilize a flower and then lay their eggs in the flower's ovary. This arrangement is found between milkweeds and milkweed bugs, figs and certain wasps, and yuccas and yucca moths (Fig. 27-12). The yucca moth's remarkable behavior results in pollination of yuccas and a well-stocked pantry for its own offspring. A female moth visits a yucca flower, collects pollen, and rolls it into a compact ball. The moth flies off with the pollen ball to another yucca flower, drills a hole in the ovary wall, and lays its eggs inside the ovary. Then it takes its pollen ball and smears pollen all over the stigma of the flower, performing this

genetically programmed behavior flawlessly! By pollinating the yucca, the moth ensures that the plant will provide a supply of developing seeds for its offspring caterpillars. Since the caterpillars eat only a fraction of the seeds, the yucca also reproduces successfully. The mutual adaptation of yucca and yucca moth is so complete that neither can reproduce without the other.

(a)

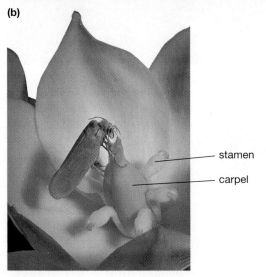

(b)

stamen

carpel

Figure 27-12   **A mutually beneficial relationship**

(a) Yuccas bloom on the dry plains of eastern Colorado in early summer.
(b) Within many of the yucca flowers, yucca moths carry out their part in one of nature's most unusual and most effective cooperative relationships between plant and animal.

Figure 27-11   **Sexual deception promotes pollination**

This male wasp is actually trying to copulate with an orchid flower. The result is successful reproduction, not for the wasp, but for the orchid.

Gametophyte Development in Flowering Plants

As Figure 27-2 illustrates, in the life cycle of flowering plants, the familiar plant of meadow, garden, and farm is the diploid sporophyte. The pollen grain (male) and the **embryo sac** (female) are the haploid gametophytes that develop within the flowers borne by the sporophytes. Both are much smaller than the gametophyte stages of ferns and mosses and cannot live independently of the sporophyte.

Pollen Is the Male Gametophyte

An anther consists of four chambers called pollen sacs (Fig. 27-13). Within each sac, hundreds to thousands of diploid **microspore mother cells** develop. Each microspore mother cell undergoes meiosis (see Chapter 12) to produce four haploid **microspores.** Each microspore divides once, by mitosis, to produce a male gametophyte, or pollen grain, consisting of only two cells: a large **tube cell** and a smaller **generative cell** that resides *within the cytoplasm of the tube cell* (Fig. 27-13). A tough surface coat develops

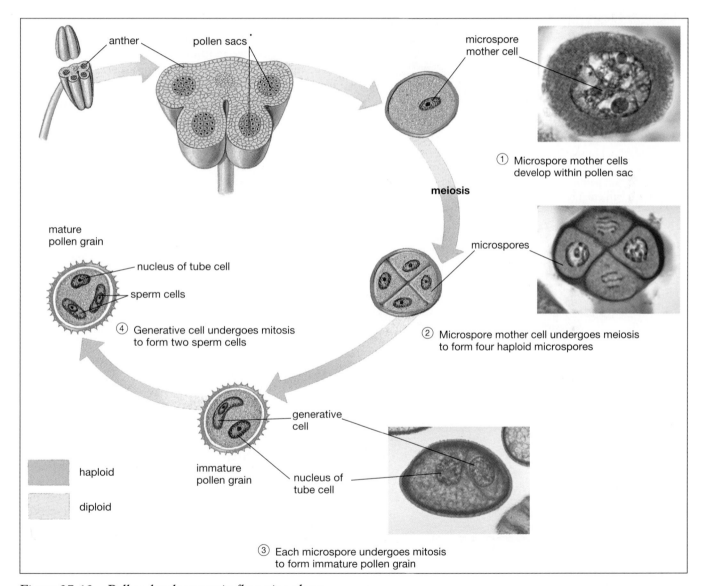

Figure 27-13   Pollen development in flowering plants

① Within the anthers, diploid microspore mother cells form. ② These cells undergo meiosis to produce four haploid microspores. ③ Through mitosis, each microspore divides into two (still haploid) cells, the tube cell and the generative cell. The entire generative cell resides within the cytoplasm of the tube cell. These two cells form the juvenile pollen grain (the male gametophyte stage of flowering plants). ④ Later (in some species, after pollination), the generative cell divides to form two sperm cells, which still remain within the tube cell cytoplasm. This three-celled "organism" is the mature male gametophyte.

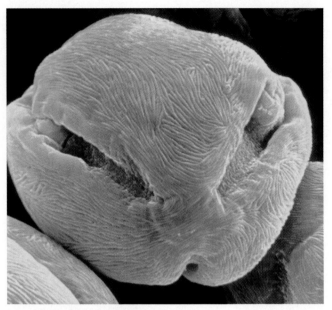

Figure 27-14 **Pollen grains**

The tough outer coverings of pollen grains are often elaborately sculptured in species-specific shapes and patterns. This pollen grain from a pear tree shows prominent furrows. When the pollen grain germinates on the stigma of a flower, the pollen tube will emerge through one of the furrows.

around the pollen grain, protecting the cells within during their journey to the carpel (Fig. 27-14).

When the pollen has matured, the pollen sacs of the anther split open. In wind-pollinated flowers such as those of grasses and oaks, the pollen grains spill out, a fortunate few to be carried by wind currents to other flowers of the same species. In animal-pollinated flowers, the pollen adheres weakly to the anther case until the pollinator comes along and brushes or picks it off.

The Embryo Sac Is the Female Gametophyte

Within an ovary, one or more dome-shaped masses of cells differentiate into ovules. Each ovule consists of outer layers of cells called **integuments** that surround a single, diploid **megaspore mother cell** (Fig. 27-15). The megaspore mother cell divides by meiosis to produce four large haploid **megaspores.** Three megaspores degenerate, and only one survives. This remaining megaspore undergoes an unusual set of mitotic divisions. Three nuclear divisions produce a total of eight haploid nuclei. Plasma membranes then divide up the cytoplasm into *seven*, not eight, cells: three small cells at each end, with one nucleus apiece, and one remaining large cell in the middle with

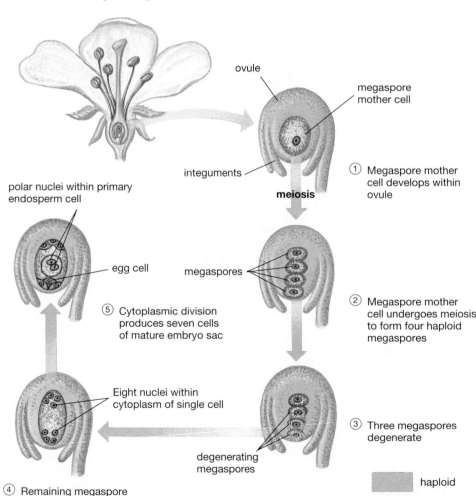

Figure 27-15 **Development of the female gametophyte**

① A single diploid megaspore mother cell matures within the integuments of an ovule.
② Through meiosis, it gives rise to four haploid megaspores.
③ Three of these degenerate.
④ The nucleus of the remaining megaspore then divides mitotically three times, producing eight haploid nuclei, four at each end of the as-yet undivided cell.
⑤ Cytoplasmic division then forms seven cells, six with one nucleus apiece, and one large primary endosperm cell with two nuclei, called polar nuclei. One of the small cells, near the opening in the integuments, is the egg cell. The other five cells degenerate soon after fertilization.

polar nuclei within primary endosperm cell

egg cell

⑤ Cytoplasmic division produces seven cells of mature embryo sac

Eight nuclei within cytoplasm of single cell

④ Remaining megaspore undergoes three mitotic divisions to form eight nuclei

degenerating megaspores

③ Three megaspores degenerate

② Megaspore mother cell undergoes meiosis to form four haploid megaspores

megaspores

① Megaspore mother cell develops within ovule

integuments

ovule

megaspore mother cell

meiosis

haploid

diploid

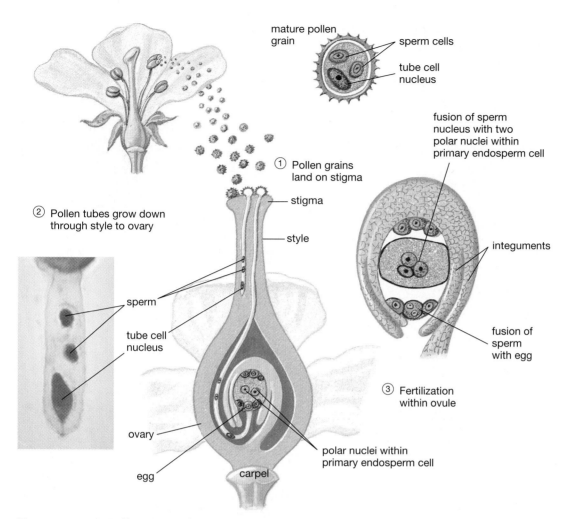

Figure 27-16 Pollination and fertilization of a flower

① The pollen grain lands on the sticky, often textured surface of the stigma and germinates. ② The tube cell elongates, burrowing down through the style. The sperm nuclei follow, inside the cytoplasm of the tube cell. ③ Upon reaching an ovule, the tube breaks through into the embryo sac, and the tube nucleus degenerates. One sperm enters the primary endosperm cell; its nucleus fuses with both endosperm nuclei to form a triploid cell. The other sperm fertilizes the egg cell to form the diploid zygote that will develop into the new embryonic plant. The other cells of the embryo sac degenerate.

two nuclei, called **polar nuclei**. This seven-celled organism, called the embryo sac (because that is where the embryo will develop), is the haploid female gametophyte. The large central cell containing two polar nuclei is the **primary endosperm cell.** The **egg** is the central small cell at the bottom of the embryo sac, located near an opening in the integuments of the ovule.

Pollination and Fertilization

When a pollen grain lands on the stigma of a flower of the same species of plant, a remarkable chain of events occurs (Fig. 27-16). The pollen grain absorbs water from the stigma. The tube cell elongates, growing down the style toward an ovule in the ovary. Meanwhile, the generative cell divides mitotically to form two sperm cells.

If all goes well, the pollen tube reaches the pore in the integument of an ovule and breaks into the embryo sac. Its tip ruptures, releasing the two sperm. One sperm fertilizes the egg cell to form the diploid zygote that will develop into a new sporophyte. The second sperm enters the primary endosperm cell. Its nucleus fuses with *both* polar nuclei, forming a triploid nucleus (having three sets of chromosomes). Through repeated mitotic divisions, the primary endosperm cell will develop into the triploid **endosperm,** a food-storage organ within the seed. The fusion of the egg with one sperm and the fusion of the polar nuclei of the primary endosperm cell with the second sperm is often called **double fertilization**, a process unique to flowering plants. The other five cells of the embryo sac degenerate soon after fertilization.

The distinction between pollination and fertilization is important. **Pollination** occurs when a pollen grain lands

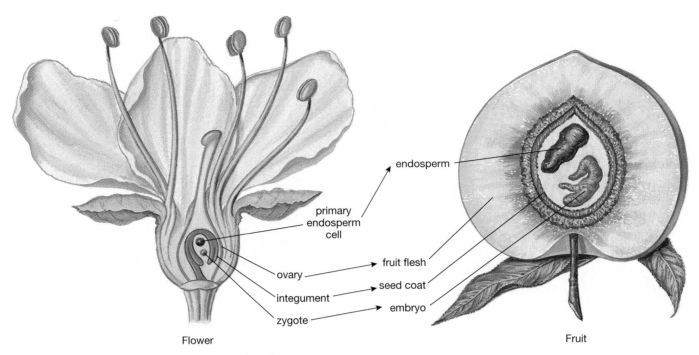

Flower   **Fruit**

Figure 27-17   **Development of fruit and seeds**

The fruit and the seed within it develop from various parts of the flower. Starting at the outside, the ovary wall ripens into the fruit flesh, which may be soft and tasty like a peach, or
variously hard, hooked, or tufted in other fruits. The integuments of the individual ovule,
which surround the embryo sac, harden and become waterproof, forming the seed coat. These
two fruit parts are derived from tissues of the parent sporophyte plant. Within the seed, two
structures develop from cells of the fertilized female gametophyte. The triploid endosperm
cell divides repeatedly, absorbs nutrients from the parent plant, and becomes the endosperm,
a food-storage structure within the seed. Finally, the zygote develops into the embryo.

on a stigma; **fertilization** is the fusion of sperm and egg. Although pollination is obviously a prerequisite for fertilization, these are two separate events. For example, pollination will not lead to fertilization if the tube cell fails to
grow properly, if the embryo sac is sterile, or if a sperm
from another pollen grain has already reached the egg.

The Development of Seeds and Fruits

Drawing upon the resources of the parent plant, the embryo sac and the surrounding integuments of the ovule
develop into a seed. The seed is surrounded by the accessory tissues of a fruit, which develops from the ovary
(Fig. 27-17).

The Seed Develops from the Ovule and Embryo Sac

The integuments of the ovule develop into the **seed coat,**
the thin, tough, and waterproof outer covering of the seed.
As we shall see, the characteristics of the seed coat play a
role in regulating when the seed will sprout.

Meanwhile, within the integuments, two distinct developmental processes occur (Fig. 27-18a). First, the

triploid endosperm cell divides rapidly. Its daughter cells
absorb nutrients from the parent plant, forming a large,
food-filled endosperm. Second, the zygote develops into
the embryo. Both dicot and monocot embryos consist of
three parts: the shoot, the root, and the **cotyledons,** or
seed leaves. The cotyledons absorb food molecules from
the endosperm and transfer them to other parts of the
embryo. In dicots ("two cotyledons"), the cotyledons usually absorb most of the endosperm during seed development, so that the mature seed is virtually filled with embryo (Fig. 27-18b). In monocots ("one cotyledon"), the
cotyledon absorbs some of the endosperm during seed development, but most of the endosperm remains in the
mature seed (Fig. 27-18c).

The future primary root develops at one end of the
main embryonic axis. The future shoot, at the other end,
is usually divided into two regions at the site of attachment of the cotyledons. Below the cotyledons, but above
the root, is the **hypocotyl** (*hypo* in Greek means "beneath"
or "lower"); above the cotyledons the shoot is called the
epicotyl (*epi* means "above"). At the tip of the epicotyl
lies the apical meristem of the shoot, the daughter cells
of which differentiate into the specialized cell types of
stem, leaves, and flowers. One or two developing leaves are
often already present.

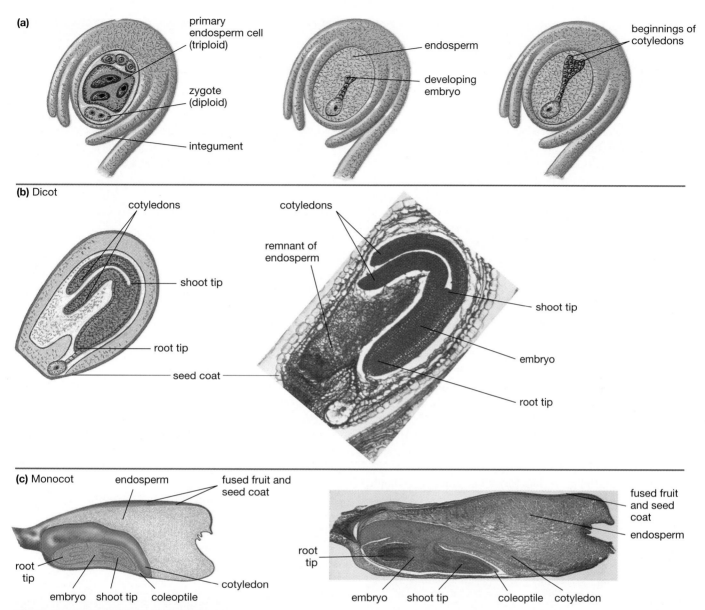

Figure 27-18   Seed development

(a) The endosperm develops first, absorbing nutrients from the parent plant. The embryo develops later, absorbing nutrients from the endosperm to fuel its growth. Monocot and dicot seeds differ in the number of cotyledons and the fate of the endosperm. (b) Dicot seeds, such as the shepherd's purse, have two cotyledons, and usually these absorb most of the endosperm as the seed develops, so that the mature seed is mostly cotyledon (the seed shown here is not quite mature and still has a remnant of endosperm left). (c) Monocot seeds, such as the corn kernel shown here, retain a large endosperm. (Cornmeal is the ground-up endosperm of corn seeds.) The embryo produces a single cotyledon. As the seed germinates, the cotyledon absorbs the food reserves of the endosperm and transfers them to the growing embryo.

The Fruit Develops from the Ovary Wall

The wall of the ovary develops into a fruit (see Fig. 27-17). There are a bewildering variety of fruits, with outer layers that are variously fleshy, hard, winged, or even spiked like a medieval mace. The selective forces favoring the evolution of all fruits, however, are similar: Fruits help to disperse the seeds to distant locations away from the parent plant (see "A Closer Look: Adaptations for Seed Dispersal").

Seed Dormancy Helps Assure Germination at an Appropriate Time

All seeds need warmth and moisture to germinate. But most newly matured seeds will not germinate immediately, even under ideal conditions. Instead, they enter a period of **dormancy** during which they will not sprout. Dormancy is usually marked by lowered metabolic activity and resistance to adverse environmental conditions. Seed dor-

A CLOSER LOOK

Adaptations for Seed Dispersal

Successful plant reproduction requires a suitable site for the seed to germinate and the young plant to grow. The growth of a seedling that sprouts right next to its parent will be inhibited by the larger plant's shade. If the offspring survives, it will compete with its parent for water and nutrients, to the detriment of both plants. Further, a plant species will become more widespread if its members send out some seeds to distant habitats. In flowering plants, seed dispersal is the function of fruits. A wide variety of fruits has evolved, each dispersing seeds in a different way.

Shotgun Dispersal

A few plants develop explosive fruits that eject their seed meters away from the parent plant. Mistletoes, common parasites of trees, produce fruits that shoot out sticky seeds. If one strikes a nearby tree, it sticks to the bark and germinates, sending rootlike fibers into the vascular tissues of its host, from which it draws its nourishment. Because the proper germination site for a mistletoe seed is not the ground, but a tree limb, it is clearly useful to shoot the seeds, not drop them.

Wind Dispersal

Dandelions and maples (Fig. E27-1) produce lightweight fruits with surfaces that catch the wind. (Yes, each indi-

vidual hairy tuft on a dandelion ball is a separate fruit!) Each fruit typically contains a single small seed; having only one seed reduces weight and lets the fruit remain aloft longer. These featherweight fruits aid the seed in traveling away from the parent plant, from a few meters for maples to kilometers for a milkweed or dandelion on a windy day.

Water Dispersal

Many fruits can float on water for a time and may be dispersed by streams and rivers. The coconut fruit, however, is the ultimate floater. Round, buoyant, and watertight, the coconut drops off its parent palm, rolls to the sea, and floats for weeks or months until it washes ashore on some distant isle (Fig. E27-2). There it germinates, perhaps establishing a new coconut colony on a formerly barren island.

Animal Dispersal

Perhaps the majority of fruits use animals as agents of seed dispersal. Two quite distinct strategies have evolved for dispersal by animals: Grab an animal as it passes by, or entice it to eat the fruit but not digest the seeds.

Anyone who takes a long-haired dog on a walk through an abandoned field knows about fruits that hitchhike on animal fur. Burdocks, burr clover, foxtails, and sticktights all develop fruits with prongs, hooks, spines, or adhesive hairs (Fig. E27-3). The parent plants hold these fruits very loosely, so that even slight contact with fur pulls the fruit free of

(a)   **(b)**

Figure E27-1   **Wind-dispersed fruits**

These fruits usually contain only one or two lightweight seeds. Some, such as dandelions **(a)**, have filamentous tufts that catch the breezes. Others, such as maple fruits **(b)**, are actually miniature glider-helicopters, silently whirling away from the tree as they fall. To see how the wings aid in seed dispersal, take two maple fruits and pluck the wing off one. Hold both fruits over your head and drop them. The wingless fruit will fall at your feet, while the winged one will glide some distance away.

Figure E27-2   **Dispersal by water**

After a long journey at sea, this coconut was washed high onto a beach by a storm. The large size and massive food reserves of coconuts are probably adaptations required for successful germination and seedling growth on barren, sandy beaches.

Figure E27-3   **The cocklebur seed uses hooked spines to hitch a ride on animal fur**

the plant and leaves it stuck on the animal. Some of these fruits don't hold on to the fur very tightly either and may fall off the next time the animal brushes against a tree or rock or come out when the animal grooms its fur.

Unlike hitchhiker fruits, edible fruits benefit both animal and plant. The plant stores sugars and tasty flavors in a fleshy fruit surrounding the seeds, enticing hungry animals to eat the fruit (Fig. E27-4). Some fruits, such as peaches and plums, contain large, hard seeds that animals usually do not eat. After eating the flesh of the fruit, the animals discard the seeds. Other fruits, including blackberries, raspberries, strawberries, and tomatoes, have small seeds that are swallowed along with the fruit flesh. The seeds then pass through the animal's digestive tract without harm. In some cases, passing through an animal's gut may even be essential to seed germination, by scraping or digesting away part of the

seed coat. Besides transport away from its parent, a seed that is swallowed and excreted benefits in another way: It ends up with its own supply of fertilizer!

Figure E27-4   **The color of ripe fruits attracts animals**
These bright red raspberry fruits have attracted a resplendent quetzal in Costa Rica. Only ripe fruits with mature seeds inside are sweet and brightly colored. Unripe fruits are usually unpalatable: green, hard, and bitter. This too is an evolutionary adaptation. The immature seeds within unripe fruit may not survive passage through an animal's gut.

mancy solves two problems, one intrinsic to the plant itself and one related to environmental factors. First, many seeds develop inside juicy fruits, such as apples, grapes, or oranges. If such a seed germinated while still enclosed in a fruit and hanging from the tree or vine, it might exhaust its food reserves before it ever touched the ground. When the fruit finally fell, the seedling would then lack the energy to burrow its roots into the soil and would die. Second, environmental conditions of temperature and precipitation suitable for seedling growth may not coincide with seed maturation. Seeds that mature in late summer in temperate climates, for example, face the harsh winter to come. Spending the winter as a dormant seed is clearly preferable to death by freezing as a tender young sprout.

Mechanisms for Maintaining and Breaking Dormancy Are Adaptations to Differing Environments

Plants have evolved many different mechanisms that produce seed dormancy, and the seeds of each species have their own set of requirements that must be met before germination can occur. Perhaps the three most common requirements are drying, exposure to cold, and disruption of the seed coat.

1. *Drying.* Many seeds must dry out before they are able to germinate. Drying prevents the seed from germinating while it is still within the fruit. Such seeds are often dispersed by animals that eat fruit but cannot digest the seeds, which are therefore excreted in their feces.

EARTH WATCH

Disrupting Coevolutionary Relationships Threatens Ecosystems

Flowering plants dominate terrestrial ecosystems largely because of the mutually beneficial relationships they forged with the animals that pollinate their flowers and disperse their seeds. Some plant and animal species, such as the yucca and yucca moths, have become totally dependent on one another. Absolute dependence is a risky proposition, however; if one partner becomes extinct, so might the other.

An example is the Calvaria tree and the dodo bird (Fig. E27-5a). Dodos were flightless, turkey-sized birds, found only on the island of Mauritius in the Indian Ocean. Dodos fed primarily on seeds and fruits, which they could gulp down whole, crunching them up in their powerful gizzards. Early sailors found the large, slow dodos to be easy prey, and by 1681 they had hunted the dodo to extinction.

Calvaria trees produce a large, edible fruit something like a peach, with a pulpy outside surrounding a stone-hard pit. Formerly one of the most common trees on Mauritius, only a few ancient specimens survive today. The ages of these patriarchs suggest that no Calvaria tree has successfully reproduced in about 300 years, even though the remaining trees still set normal fruit each year. Ecologist Stanley Temple has found out why.

Temple believes that the Calvaria tree is endangered because it depends exclusively on the dodo for seed dispersal and germination. Over thousands of years of coevolution with dodos, only Calvaria fruits with strong pits had any chance of surviving the trial-by-gizzard within a dodo and reproducing. Strong pit walls, however, also inhibit germination. In fact, the Calvaria pit came to *require* processing through the dodo before it could germinate: The pit walls must be partially ground away before the seed can sprout. By eliminating the dodo, humans inadvertently doomed the Calvaria forests as well. Now that we know the problem, a few Calvaria trees can be propagated in nurseries, by filing away the pit wall, or, as Temple discovered, by force-feeding Calvaria pits to domestic turkeys. But the forests of Calvaria trees that formerly clothed Mauritius, providing habitat for wildlife and lumber for people, are gone forever.

In many tropical forests, bats are the most important agents of seed dispersal (Fig. E27-5b). Bats may fly over 20 miles each night, consuming up to twice their weight in fruit and defecating the seeds in flight. Biologist Donald Thomas compared germination rates of seeds before and after passage through the bats' digestive tract. Nearly all the seeds germinated after passing through a bat, but seeds planted directly from fruit had only a 10% germination rate. Today in the tropical forests of southern Mexico, fruit-eating animals such as monkeys, deer, and tapir have been overhunted. Tropical fruits are rotting on the forest floor or sending up doomed sprouts under the shade of their parents; dispersal has stopped. As Alejandro Estrada of the University of Mexico put it, "The continued existence of tropical forests whose primates and . . . birds and bats have been shot is just as precarious as if their trees had been chain-sawed and bulldozed." The web of interdependent life forms linked by interactions forged over millennia of coevolution is fragile and easily disrupted. Only by understanding and preserving the complex and crucial interactions among plants and animals can we hope to conserve diverse, functioning ecosystems.

(a)   **(b)**

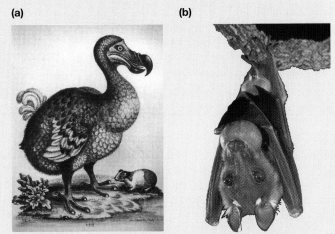

Figure E27-5   **Animal seed dispersers may be crucial to the survival of entire ecosystems**

(a) The extinction of the dodo in the late 1600s doomed the Calvaria forests of Mauritius. **(b)** A Wahlberg's epauleted bat in Kenya eats a ripe fig. Without bats and other seed-dispersing animals, tropical forests may not survive.

2. *Cold.* Seeds of many temperate and arctic plants will not germinate unless exposed to prolonged subfreezing temperatures, followed by warmth and moisture. This requirement ensures that they stay dormant during mild days in autumn and sprout only after winter yields to spring.

3. *Disruption of the seed coat.* The seed coat itself is often a barrier to seed germination. Many seed coats are imper-

meable to water and oxygen, or they bind the developing embryo so tightly that growth simply cannot occur; others contain chemicals that inhibit germination (see Chapter 28). In deserts, for example, rainfall is spotty and scarce. Years may go by without enough water for plants to germinate, grow, flower, and set more seed. Therefore, the seed must not sprout unless a rainfall is heavy enough to allow the plant to complete its life cycle, since anoth-

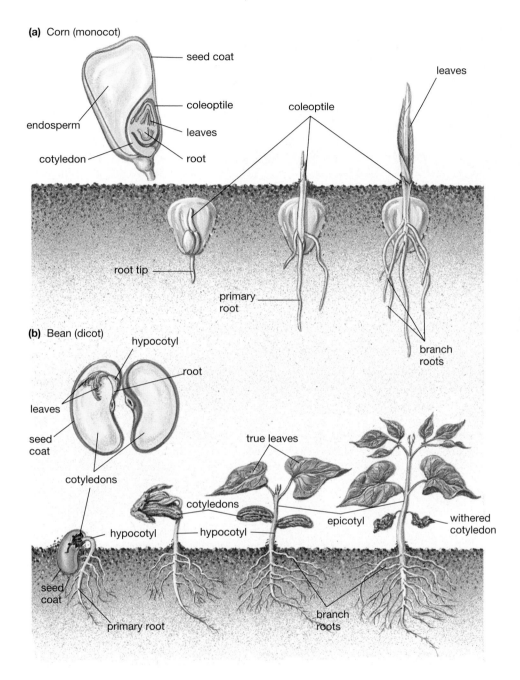

(a) Corn (monocot)

seed coat

coleoptile

endosperm

cotyledon

leaves

root

coleoptile

leaves

root tip

primary root

coleoptile

branch roots

(b) Bean (dicot)

hypocotyl

root

leaves

seed coat

cotyledons

hypocotyl

seed coat

primary root

true leaves

cotyledons

hypocotyl

epicotyl

withered cotyledon

branch roots

Figure 27-19   **Seed germination**

Seed germination in corn, a monocot **(a)**, and the common bean, a dicot **(b)**, differs in detail but involves the same three principles: (1) use of stored food to provide energy for seedling growth until photosynthesis can take over; (2) rapid growth and branching of the root to absorb water and nutrients; and (3) protection of the delicate shoot tip as it moves upward through the soil. In monocots (a), the delicate shoot tip is protected within a tough sheath, the coleoptile, which pushes up through the soil when the seed sprouts. In dicots (b), the hypocotyl (shown here) or the epicotyl forms a hook that emerges from the soil first, protecting the shoot tip.

er rain may not fall in time. The seeds of most desert plants have water-soluble chemicals in the seed coat that inhibit germination. Only a hard rainfall can wash away enough of the inhibitors to allow sprouting.

The Germination and Growth of Seedlings

During growth and development of a seed, **germination,** the formerly dormant embryo resumes growth and emerges from the seed. The embryo absorbs water, which makes it swell and burst open its seed coat. The root is usually the first structure to emerge from the seed coat, growing rapidly and absorbing water and minerals from the soil. Much

of the water is transported to cells in the shoot. As its cells elongate, the stem lengthens, pushing up through the soil.

The Shoot Tip Is Protected by the Coleoptile

The growing shoot faces a serious difficulty: It must push through the soil without scraping away the apical meristem and tender leaflets at its tip. A root, of course, must always contend with tip abrasion, and its apical meristem is protected by a root cap (see Fig. 25-10). Shoots spend most of their time in the air and do not develop permanent protective caps. Instead, germinating shoots have other mechanisms that cope with the abrasion of sprouting (Fig. 27-19). In monocots, a tough sheath, the **coleoptile,** encloses the shoot tip like a glove encloses a gardener's fin-

Figure 27-20   **Cotyledons nourish the developing plant**

In the squash family, a hypocotyl hook carries the cotyledons out of the soil. The cotyledons expand into photosynthetic leaves (the pair of smooth oval leaves). The first true leaf (crinkled single leaf) develops a little later. Eventually, the cotyledons shrivel up and die.

gers (Fig. 27-19a). The coleoptile "glove" pushes aside the soil particles as it grows. Once out in the air, the coleoptile tip degenerates, allowing the tender "finger" of shoot to emerge. Dicots do not have coleoptiles. Instead, the dicot shoot forms a hook in either the hypocotyl or epicotyl (Fig. 27-19b). The bend of the hook, encased in epidermal cells with tough cell walls, leads the way through the soil, clearing the way for the downward-pointing apical meristem with its delicate new leaves.

Cotyledons Nourish the Sprouting Seed

Food stored in the seed provides the energy for sprouting. You may recall that the cotyledons of dicots had already absorbed the endosperm while the seed was developing and are now swollen and full of food. In dicots with hypocotyl hooks, the elongating shoot carries the cotyledons out of the soil into the air. These aboveground cotyledons often become green and photosynthetic (Fig. 27-20) and transfer both previously stored food and newly synthesized sugars to the shoot. In dicots with epicotyl hooks, the cotyledons stay below the ground, shriveling up as the embryo absorbs their stored food. Monocots retain most of their food reserve in the endosperm until germination, when it is digested and absorbed by the cotyledon as the embryo grows. The cotyledon remains below the ground in the remnants of the seed.

Controlling the Development of the Seedling

Once out in the air, the shoot rapidly spreads its leaves to the sun. Simultaneously, the root system delves into the soil. The apical meristem cells of shoot and root divide, giving rise to the mature structures discussed in Chapter 25. Eventually this plant, too, will mature, flower, and set seed, renewing the cycle of life. How this cycle is regulated—why shoots grow upward while roots grow downward, and how plants produce flowers at the proper time of year—is the subject of the next chapter.

✷ S U M M A R Y   O F   K E Y   C O N C E P T S

Sexual versus Asexual Reproduction and Plant Life Cycles

The sexual life cycle of plants, called alternation of generations, includes both a multicellular diploid form (the sporophyte generation) and a multicellular haploid form (the gametophyte generation). In the seed plants, the gametophyte stage is greatly reduced. The male gametophyte is the pollen grain, a drought-resistant structure that can be carried from plant to plant by wind or animals. The female gametophyte is also reduced and is retained within the body of the sporophyte stage. In this way, seed plants can reproduce independently of liquid water.

The Evolution of Flowers

Flowering plants evolved from gymnosperms. In gymnosperms, pollen blows on the wind from male cones to female cones. Wind pollination is inefficient, though. In many habitats, flowering plants enjoy a selective advantage over gymnosperms because many types of flowers attract insects that carry pollen from plant to plant.

A complete flower consists of four parts: sepals, petals, stamens (male reproductive structures), and carpels (female reproductive structures). The sepals form the outer covering of the flower bud. The petals (and sometimes the sepals) are usually brightly colored and attract pollinators to the flower. The stamen consists of a filament that bears at its tip an anther in which pollen (the male gametophyte) develops. The carpel consists of the ovary in which one or more embryo sacs (the female gametophytes) develop, and a style that bears at its end a sticky stigma to which pollen adheres during pollination. Incomplete flowers lack one or more of the four floral parts.

Coevolution of Flowers and Pollinators

Most flowers are pollinated by the wind or by animals, usually insects or birds. Flowers show specific adaptations to pollination by particular pollinators, because of extensive coevolution between flowering plants and their pollinators.

Gametophyte Development in Flowering Plants

Pollen develops in the anthers. The diploid microspore mother cell undergoes meiosis to produce four haploid microspores. Each of these divides mitotically to form pollen grains. An immature pollen grain consists of two cells: the tube cell and the generative cell. The generative cell divides once to produce two sperm cells.

The embryo sac develops within the ovules of the ovary. A diploid megaspore mother cell undergoes meiosis to form four haploid megaspores. Three of these degenerate; the fourth undergoes three sets of mitotic divisions to produce the eight nuclei of the embryo sac. These eight nuclei come to reside in only seven cells. One of these cells, with a single nucleus, is the egg cell; another, with two nuclei, is the primary endosperm cell. These two cells are involved in seed formation; the rest of the cells degenerate.

Pollination and Fertilization

Pollination is the transfer of pollen from anther to stigma. When a pollen grain lands on a stigma, its tube cell grows through the style down to the embryo sac. The generative cell divides to form two sperm cells that travel down the style within the tube cell, eventually entering the embryo sac. One sperm fuses with the egg to form a diploid zygote, which will give rise to the embryo. The other sperm fuses with the binucleate primary endosperm cell to produce a triploid cell. This cell will give rise to the endosperm, a food-storage organ within the seed.

The Development of Seeds and Fruits

The embryo develops a root, shoot, and cotyledons. Cotyledons digest and absorb food from the endosperm and transfer it to the growing embryo. Monocot embryos have one cotyledon, and dicot embryos have two. The seed is enclosed within a fruit that develops from the ovary wall. The function of the fruit is to disperse the seeds away from the parent plant. Seeds often remain dormant for some time after fruit ripening. Environmental conditions involved in breaking dormancy may include an initial drying, exposure to cold, or disruption of the seed coat.

The Germination and Growth of Seedlings

Seed germination requires warmth and moisture. Energy for germination comes from food stored in the endosperm, transferred to the embryo by the cotyledons.

KEY TERMS

alternation of generations p. 536
anther p. 539
carpel p. 539
coevolution p. 539
coleoptile p. 551
complete flower p. 538
cotyledon p. 546
dormancy p. 547
double fertilization p. 545
egg p. 545
embryo sac p. 543
endosperm p. 545
epicotyl p. 546
fertilization p. 546
filament p. 539
flower p. 538
fruit p. 539
gametophyte p. 536
generative cell p. 543
germination p. 551
hypocotyl p. 546
incomplete flower p. 539
integument p. 544
megaspore p. 544
megaspore mother cell p. 544
microspore p. 543
microspore mother cell p. 543
ovary p. 539
ovule p. 539
petal p. 539
polar nucleus p. 545
pollen grain p. 537
pollination p. 545
primary endosperm cell p. 545
seed p. 537
seed coat p. 546
sepal p. 539
spore p. 536
sporophyte p. 536
stamen p. 539
stigma p. 539
style p. 539
tube cell p. 543
zygote p. 536

THINKING THROUGH THE CONCEPTS

Multiple Choice

1. When a bee gets nectar from a flower, it picks up pollen from the ___________ and carries it to another flower.
   a. stigma b. ovary
   c. sepals d. anthers
   e. filament
2. In plants, the gametophyte produces eggs and sperm by
   a. mitosis b. meiosis
   c. spores d. fertilization
   e. germination
3. If you found a dark, rotten-smelling flower near the ground, it would most likely be pollinated by
   a. bats b. bees
   c. butterflies d. moths
   e. beetles
4. Reproduction in flowering plants is known as double fertilization because
   a. one tube nucleus fuses with one egg and one sperm nucleus fuses with another egg
   b. two sperm nuclei fuse with one egg
   c. one sperm fuses with two eggs
   d. one sperm nucleus fuses with one egg and another sperm nucleus fuses with the two haploid nuclei of the primary endosperm cell
   e. two polar nuclei fuse with one egg nucleus
5. In a peach, the fruit is derived from the
   a. wall of the ovary b. endosperm
   c. megaspores d. sepals
   e. petals
6. The tassel on top of a corn plant is made up of male flowers that lack female structures. This flower is
   a. the sporophyte b. the gametophyte
   c. incomplete d. complete
   e. homologous

Review Questions

1. Diagram the plant life cycle, comparing ferns with flowering plants. Which stages are haploid and which are diploid? At which stage are gametes formed?
2. What are the advantages of the reduced gametophyte stages in flowering plants compared with the more substantial gametophytes of ferns?
3. Diagram a complete flower. Where are the male and female gametophytes formed? What are the male and female gametophytes called?
4. How does an egg develop within an embryo sac? How does this structure allow double fertilization to occur?
5. What does it mean when we say that pollen is the male gametophyte? How is pollen formed?
6. What are the parts of a seed, and how does each part help in the development of a seedling?
7. Describe the characteristics you would expect to find in flowers that are pollinated by the wind, beetles, bees, and hummingbirds, respectively, and explain why in each case.
8. What is the endosperm? From which cell of the embryo sac is it derived? Is endosperm more abundant in the mature seed of a dicot or a monocot?
9. Describe three mechanisms whereby seed dormancy is broken in different types of seeds. How are these mechanisms related to the normal environment of the plant?
10. How do monocot and dicot seedlings protect the delicate shoot tip during seed germination?
11. Describe three types of fruits and the mechanisms whereby these fruit structures help to disperse their seeds.

APPLYING THE CONCEPTS

1. A friend gives you some seeds that you want to grow in your yard. When you plant some, nothing happens. What things could you try to get the seed to germinate?
2. In areas where farms have been left uncultivated for several years it is often possible to see certain kinds of trees growing in straight lines, which mark old fences where birds sat and deposited seeds they had eaten. Why are such seeds more likely to germinate than those of the same species that have not passed through a bird's digestive tract? How might an anthropologist use such lines of trees to study past inhabitants of an area?
3. Charles Darwin once described a flower that produced nectar at the bottom of a tube 25 centimeters (10.5 inches) deep. He predicted that there must be a moth or other animal with a 25-centimeter-long tongue to match (he was right). Such specialization almost certainly means that this particular flower could be pollinated only by that specific moth. What are the advantages and disadvantages of such specialization?
4. Many plants that we call weeds are plants that were brought from another continent either accidentally or purposefully. In a new environment they have few competitors or animal predators, so they tend to grow in such large amounts that they come to be considered weeds. Think of as many ways as possible that humans become involved in plant dispersal. To what degree do you think humans have changed the distributions of plants? In what ways is this change helpful to humans? In what ways is it a disadvantage?
5. In the tropics there are a number of plant-animal coevolutionary relationships in which both are known to be completely dependent on the relationship. In light of the rapid rate of destruction of tropical ecosystems, how does this type of relationship leave both organisms particularly vulnerable to extinction? What political and economic problems may this rapid rate of extinction create?

FOR MORE INFORMATION

Fleming, T. H. "Cardon and the Night Visitors." *Natural History*, October 1994. A beautifully illustrated look at the way bats pollinate the world's largest cactus and may help to determine the numbers of each of the plant's three sexual types.

Handel, S. N., and Beattie, A. J. "Seed Dispersal by Ants." *Scientific American*, August 1990. Many plants depend on ants for seed dispersal, even producing fat deposits on the outside of the seed as a lure for the ants. Dispersal is not the only benefit to the seed: Being "planted" in an anthill seems to provide an ideal environment for germination and growth as well.

Heinrich, B. "Of Bedouins, Beetles, and Blooms." *Natural History*, May 1994. A colorful look at pollination strategies, especially at the large number of red flowers in Israel that use their color to attract beetles for sex.

Kearns, C., and Inouye, D. "Pistil-Packing Flies." *Natural History*, September 1993. A beautifully illustrated look at the wide variety of flowers that are pollinated by flies in alpine areas.

Murawski, D. A. "A Taste for Poison." *National Geographic*, December 1993. Certain passion vine butterflies form cyanide apparently from eating pollen from the passion vine. Other butterflies mimic the poisonous types.

Simons, P. "An Explosive Start for Plants." *New Scientist*, January 2, 1993. Seeds and spores spread and pollen is snapped onto bees by explosive movements in some plants.

Sunquist, F. "Blessed Are the Fruit Eaters." *International Wildlife*, May/June 1992. Especially in the tropics, fruit-eating mammals and birds are crucial to the dispersal and germination of seeds. Overhunting threatens not only the animals but the plants that depend on them.

NET WATCH

On-line resources for this chapter are on the World Wide Web at:
http://www.prenhall.com/~audesirk (click on the table of contents link and then select Chapter 27).

28 Plant Responses to the Environment

❖ AT A GLANCE

The Discovery of Plant Hormones
The Work of Several Researchers Revealed the Hormonal
Mechanisms of Phototropism

Plant Hormones and Their Actions

**The Plant Life Cycle: Reception, Response, and
Regulation**
Abscisic Acid Maintains Seed Dormancy; Gibberellin
Stimulates Germination
Auxin Controls the Orientation of the Sprouting Seedling
The Genetically Determined Shape of the Mature Plant Is
the Result of Interactions between Hormones
Hormones Control the Differentiation of Xylem and Phloem
Daylength Controls Flowering
Hormones Coordinate the Development of Seeds and Fruit
Senescence and Dormancy Prepare the Plant for Winter

*The brightly colored flesh of the saguaro cactus fruit attracts a
lesser long-nosed bat. Nearly all of the activities of plants, from
growth to fruit production and ripening, are under the control of
plant hormones.*

The time is late fall, the place a stretch of woods in Ohio.
Masses of acorns burden the oaks, and gray squirrels busi-
ly gather and bury the nuts, storing them against the coming winter. With so many acorns "squirreled away," some
are forgotten, others are not needed. They sprout next
spring as the sun warms the forest soil.

The squirrels, of course, haven't planted the acorns so
that new oaks will grow. The acorns are oriented randomly,
some with the future stem end down and root end up.
Buried beneath several inches of soil, how does the germinating seedling "know" which way is up? If it does sprout
and send its shoot up into the air, how does it develop into
a mature oak? How does it distinguish the seasons, flowering in spring and going dormant in fall?

Plants do not have specialized organs comparable to our
eyes or ears for perceiving the environment. Nevertheless,
plants *do* perceive many features of their world: the direction of gravity; the direction, intensity, and duration of
sunlight; the strength of the wind; and in some cases, even
the touch of a fly upon a leaf. Plants also respond to these
stimuli by regulating their growth and development in appropriate ways. Usually, plants control their bodily functions through the action of chemicals called hormones or
growth regulators. In this chapter we explore how plants
detect external stimuli and produce and distribute hormones in response. We will also look at how these hormones govern the development of the plant body.

The Discovery of Plant Hormones

Everyone who keeps houseplants on a windowsill knows
that the plants bend toward the window as they grow, in
response to the sunlight streaming in. Over a hundred
years ago Charles Darwin and his son Francis studied this
phenomenon of growth toward the light, or **phototropism.**

The Work of Several Researchers Revealed
the Hormonal Mechanisms of Phototropism

The Darwins illuminated grass coleoptiles from various
angles (a coleoptile is the protective sheath surrounding
a monocot seedling; see Fig. 27-19a). They noted that a re-

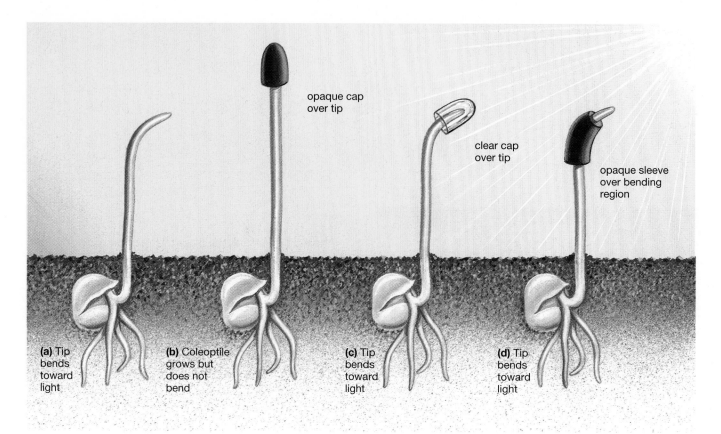

Figure 28-1   **The phototropism experiments of Charles and Francis Darwin**

(a) If light strikes a coleoptile from the side, a region a few millimeters down from the tip bends as it grows, until the tip points toward the light source. **(b)** If the tip is covered with an opaque cap, the coleoptile does not bend. **(c)** If a clear cap is placed over the tip, bending still occurs, proving that the mere presence of a cap does not prevent bending. **(d)** If a flexible opaque collar is placed around the bending region, the coleoptile still bends toward the light.

gion of the coleoptile a few millimeters below the tip bent toward the light until the tip faced directly into the light source (Fig. 28-1). If they covered the very tip of the coleoptile with an opaque (lightproof) cap, the coleoptile didn't bend. A clear cap placed over the coleoptile, however, did not prevent bending. Further, bending toward the light still occurred if they covered the bending region with an opaque sleeve. The Darwins concluded that (1) the tip of the coleoptile perceives the direction of light and (2) bending occurs farther down the coleoptile; therefore (3) the tip must transmit information about the light direction down to the bending region.

How does the coleoptile bend? Although the Darwins didn't know this, growth in the bending region is entirely due to elongation of preexisting cells. Therefore, the coleoptile must bend because of unequal elongation of cells (Fig. 28-2). The cells on the outside of the bend (away from the light) elongate more than the cells on the inside of the bend (toward the light). If one side of the coleoptile becomes longer than the opposite side, then the whole shaft must bend away from the longer side. On the basis of this information, we can reinterpret the Darwins' results and say that the information transmitted from the tip to the bending region causes

greater elongation of cells on the side of the coleoptile shaft away from the light.

About 30 years after the Darwins' experiments, Peter Boysen-Jensen showed that the information transferred down from the tip is chemical in nature (Fig. 28-3). He cut the tips off coleoptiles and found that the remaining stump neither elongated nor bent toward the light. If he replaced the tip and put the patched-together coleoptile in the dark, it elongated straight up. In the light, it showed normal phototropism. Placing a thin layer of porous gelatin between the severed tip and the stump still allowed elongation and bending, but an impervious barrier eliminated these responses. Boysen-Jensen concluded that a chemical is produced in the tip and moves down the shaft, causing cell elongation. In the dark, the chemical that causes the cells to elongate diffuses straight down from the tip and causes the coleoptile to elongate straight up. Presumably, light causes the chemical to become more concentrated on the "shady" far side of the shaft, so cells on the shady side elongate faster than cells on the "sunny" near side, causing the shaft to bend toward the light.

The next step was to isolate and identify the chemical. In the 1920s, Frits Went devised a way to collect the elongation-promoting chemical. He cut off the tips of oat

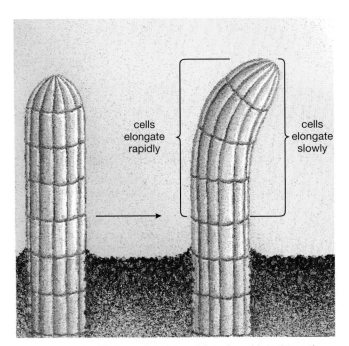

Figure 28-2   The mechanism of bending by coleoptiles
Coleoptiles bend by unequal elongation of cells on opposite sides of the coleoptile. If cells on one side elongate more rapidly than cells on the other side, the coleoptile bends away from the longer side.

coleoptiles and placed them on a block of agar (a porous, gelatinous material) for a few hours (Fig. 28-4a). Went hoped that the chemical would migrate out of the coleoptiles into the agar. He then cut up the agar, now presumably loaded with the chemical, and placed small pieces on the tops of coleoptile stumps. If he put a piece of agar squarely atop a stump, the stump elongated straight up (Fig. 28-4b). All the stump cells received equal amounts of the chemical and elongated at the same rate. If he placed a piece on one side of a cut stump, the stump would invariably bend away from the side with the agar (Fig. 28-4c). It was apparent that cells on the side under the agar received more of the chemical and were more stimulated to elongate. Went called the chemical **auxin,** from a Greek word meaning "to increase." Kenneth Thimann later purified auxin and determined its molecular structure.

Plant Hormones and Their Actions

Animal physiologists have long recognized that chemicals called **hormones** are produced by cells in one location and transported to other parts of the body, where they exert specific effects. By analogy, auxin and other plant-

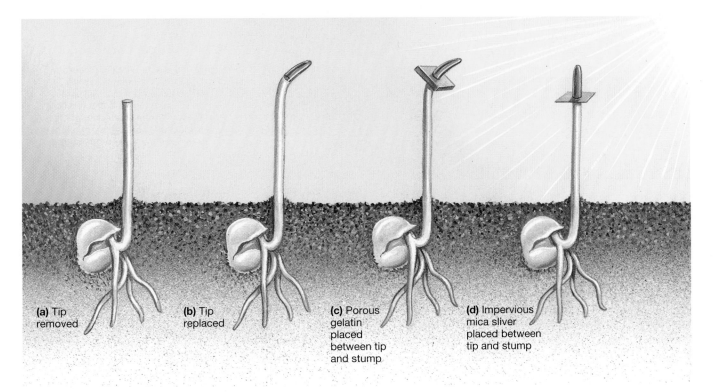

Figure 28-3   Boysen-Jensen's experiments
These experiments demonstrated the chemical nature of the signal passing from the tip to the bending region of the coleoptile during phototropism. **(a)** If the tip is cut off a coleoptile, the remaining stump neither elongates nor bends toward the light. **(b)** Replacing the tip restores both elongation and phototropism. **(c)** A permeable gelatin "filter" between the tip and stump does not prevent phototropism, but **(d)** an impenetrable slice of mica eliminates the response.

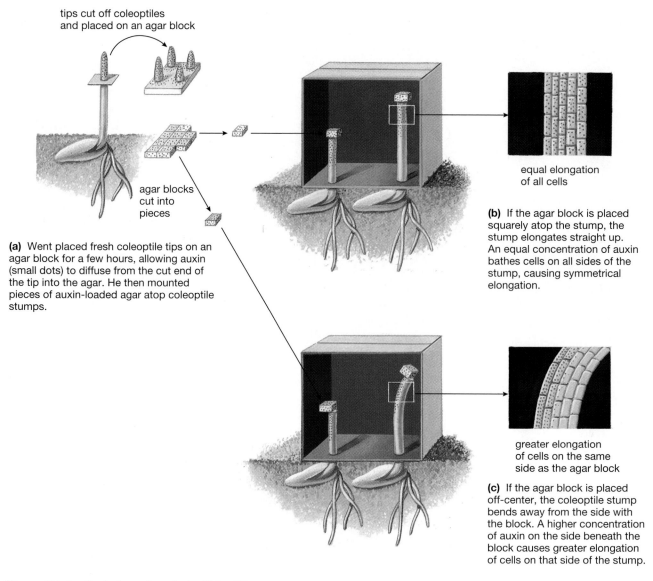

tips cut off coleoptiles
and placed on an agar block

agar blocks
cut into
pieces

(a) Went placed fresh coleoptile tips on an
agar block for a few hours, allowing auxin
(small dots) to diffuse from the cut end of
the tip into the agar. He then mounted
pieces of auxin-loaded agar atop coleoptile
stumps.

equal elongation
of all cells

(b) If the agar block is placed
squarely atop the stump, the
stump elongates straight up.
An equal concentration of auxin
bathes cells on all sides of the
stump, causing symmetrical
elongation.

greater elongation
of cells on the same
side as the agar block

(c) If the agar block is placed
off-center, the coleoptile stump
bends away from the side with
the block. A higher concentration
of auxin on the side beneath the
block causes greater elongation
of cells on that side of the stump.

Figure 28-4   **Isolation of auxin by Frits Went**

regulating chemicals are called **plant hormones.** So far,
plant physiologists have identified five major classes of
plant hormones: auxins, gibberellins, cytokinins, ethylene, and abscisic acid (Table 28-1). Several other types of
hormones are thought to exist but have not yet been isolated and identified. Each hormone can elicit a variety of
responses from plant cells, depending on the type of cell,
its physiological state, and the presence of other hormones.
Further, the roles of some plant hormones vary among
plant species. This should not be too surprising; after all,
in animals a single hormone can be involved in actions as
diverse as a salmon's transition from fresh to salt water, a
frog's metamorphosis from tadpole to adult, and a snake's
shedding its skin (see "Evolutionary Connections: The
Evolution of Hormones," Chapter 35).

Auxin, as we have seen, promotes elongation of cells in
coleoptiles and other parts of the shoot. In roots, low con-

centrations of auxin stimulate elongation; slightly higher
concentrations inhibit elongation. Both light and gravity
affect the distribution of auxin in roots and shoots, so
auxin plays a major role in both phototropism and **gravitropism** (directional growth with respect to gravity).
Auxin affects many other aspects of plant development,
too. It stimulates root branching, the differentiation of
conducting tissues (xylem and phloem), and the development of fruits. Auxin also prevents sprouting by lateral buds.

Gibberellins are a group of chemically similar molecules that, like auxin, promote elongation of cells in stems.
In some plants, gibberellins stimulate flowering, fruit development, seed germination, and bud sprouting.

Cytokinins promote cell division in many plant tissues;
consequently, they stimulate the sprouting of buds and the
development of fruit, endosperm, and the embryo. Cy-

TABLE 28-1 ⬥ *Hormone Actions in Plants*	
Hormone	*Functions*
Abscisic acid	Closing of stomata; seed dormancy; bud dormancy
Auxin	Elongation of cells in coleoptiles and shoots; phototropism; gravitropism in shoots and roots; root growth and branching; apical dominance; development of vascular tissue; fruit development; retarding senescence in leaves and fruit; ethylene production in fruit
Cytokinin	Promotion of sprouting of lateral buds; prevention of leaf senescence; promotion of cell division; stimulation of fruit, endosperm, and embryo development
Ethylene	Ripening of fruit; abscission of fruits, flowers, and leaves; inhibition of stem elongation; formation of hook in dicot seedlings
Gibberellin	Germination of seeds and sprouting of buds; elongation of stems; stimulation of flowering; development of fruit

tokinins also stimulate plant metabolism, preventing or at least delaying the aging of plant parts, especially leaves.

Ethylene is an unusual hormone in that it is a gas at normal environmental temperatures. Ethylene is best known, and most commercially valuable, for its ability to cause fruit to ripen. It also stimulates the separation of cell walls in abscission layers, allowing leaves, flowers, and fruit to drop off at the appropriate times.

Abscisic acid is a hormone that helps plants withstand unfavorable environmental conditions. As you learned in Chapter 26, it causes stomata to close when water availability is low. It inhibits the activity of gibberellin, thus helping to maintain dormancy in buds and seeds during times when germination would be dangerous.

In the rest of this chapter, we describe a year in the life of a plant, illustrating how hormones regulate its growth and development.

The Plant Life Cycle: Reception, Response, and Regulation

The life cycle of a plant results from a complex interplay between its genetic information and its environment. Hormones mediate many of the genetic determinants of growth and development, as well as nearly all responses to environmental factors. At each stage in its life cycle, a plant produces a distinctive set of hormones that interact with one another in directing the growth of the plant body.

Abscisic Acid Maintains Seed Dormancy; Gibberellin Stimulates Germination

As we pointed out in the previous chapter, a seed maturing within a juicy fruit on a warm autumn day has ideal conditions for germination, yet it remains dormant until the following spring. In many seeds, *abscisic acid enforces dormancy*. Abscisic acid slows down the metabolism of the embryo within the seed, preventing its growth. The seeds of some desert plants contain high concentrations of ab-

scisic acid; only a really hard rain can wash out the abscisic acid, freeing the embryo from its inhibitory effects and allowing the seed to germinate. Seeds of northern plants usually require a prolonged period of cold, such as occurs during winter, to break dormancy; in these seeds, chilling induces the destruction of abscisic acid.

Germination is stimulated by other hormones, especially gibberellin. The same environmental conditions that cause the breakdown of abscisic acid also promote the synthesis of gibberellin. Gibberellin induces the transcription of genes that code for the enzymes that digest the food reserves of the endosperm and cotyledons, making sugars, lipids, and amino acids available to the growing embryo.

Auxin Controls the Orientation of the Sprouting Seedling

When the growing embryo breaks out of the seed coat, it immediately faces a crucial problem: Which way is up? The roots must burrow downward, while the shoot must grow upward to emerge into the light. Auxin apparently controls the responses of both roots and shoots to light and gravity.

Auxin Stimulates Shoot Elongation Away from Gravity and toward the Light

Let's begin by looking at the growth of a shoot as it first emerges from the seed, buried underground. As we described earlier, auxin is synthesized in shoot tips, moves down the shaft of the stem, and stimulates cell elongation. If the stem is not exactly vertical, organelles in the cells of the stem detect the direction of gravity and somehow cause auxin to accumulate on the stem's lower side (Fig. 28-5a). Therefore the lower cells elongate rapidly, forcing the stem to bend upward. When the shoot tip is vertical, the auxin distribution becomes symmetrical. Now the stem grows straight up, emerging from the soil into the light.

Auxin also mediates phototropism. Ordinarily, the distribution of auxin caused by light is the same as the distribution caused by gravity, because the direction of brightest light (the sun) is roughly opposite that of gravity. For

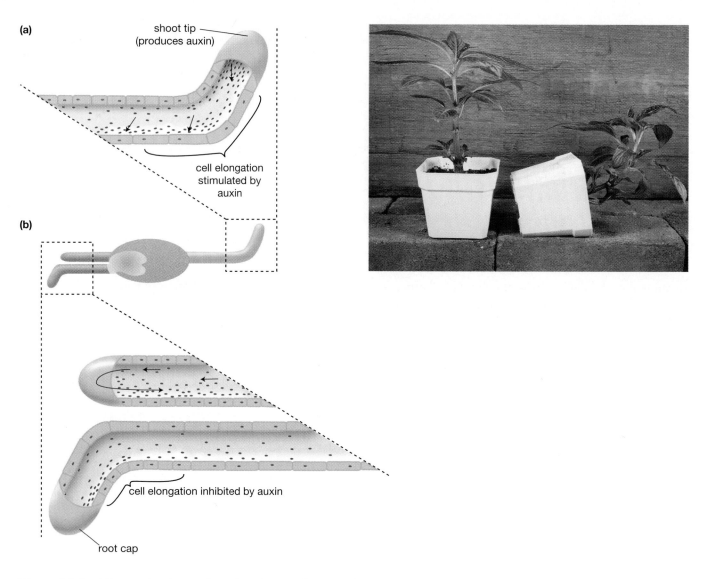

(a) shoot tip
(produces auxin)

cell elongation
stimulated by
auxin

(b)

cell elongation inhibited by auxin

root cap

Figure 28-5   **The mechanisms of gravitropism in shoots and roots**

(a) In shoots, auxin (blue) produced by the shoot tip accumulates on the lower side of the shoot. Auxin promotes rapid elongation of cells on the lower side, causing the shoot to bend upward. The photo shows two impatiens plants, one of which was placed on its side in the dark for 16 hours. In just this short time, faster elongation of cells on the lower side of the stem has brought the plant to a nearly vertical orientation. **(b)** In roots, auxin (blue) is transported to the root from the shoot. The root cap senses the direction of gravity and redirects the flow of auxin, so that auxin accumulates on the lower side of the root. Auxin inhibits cell elongation in roots. As cells on the top elongate faster than those on the bottom, the root bends downward.

example, if a young shoot still buried underground is close enough to the surface so that some light penetrates down to it, both light and gravity cause auxin to be transported to the lower side of the shoot and promote upward bending. Thus, under normal conditions gravitropism and phototropism add to each other's effects.

Auxin May Control the Direction of Root Growth

Gravitropism in roots is less well understood than it is in stems. According to one model, auxin controls the direction of root growth (Fig. 28-5b). Auxin is transported from the shoot down to the root. If the root is not vertical, the

root cap senses the direction of gravity and causes the auxin to accumulate on the lower side of the root. Unlike shoots, in which moderate concentrations of auxin *stimulate* cell elongation, in roots these same concentrations of auxin *inhibit* cell elongation. Therefore, cell elongation in the lower side of the root, where auxin accumulates, is inhibited, while cell elongation remains unaffected in the upper side of the root. The result is that the root bends downward. When the root tip points directly downward, the auxin distribution becomes equal on all sides, and the root continues to grow straight down. Note that *auxin slows down but does not eliminate root cell elongation:* A vertical root continues to grow.

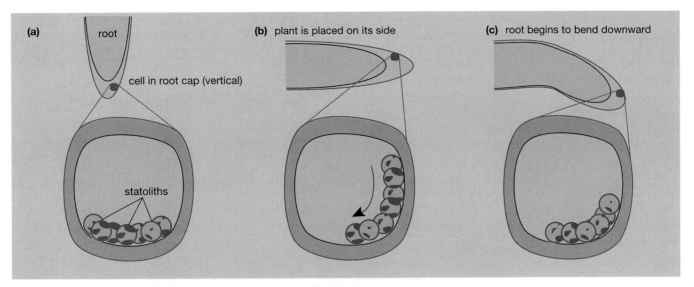

Figure 28-6   Gravity-sensing organelles in stems and root caps

These structures, called statoliths, fall to the downward side of the cell. **(a)** Normal orientation of statoliths in a root cap cell. **(b)** The plant is placed on its side (as in Fig. 28-5), and the statoliths begin tumbling downward, **(c)** coming to rest on the new lower surface. In some way, this change triggers the bending of the root downward.

Plants May Sense Gravity Using Organelles Called Statoliths

Although auxin plays a major role in the unequal growth rates of cells that cause bending toward gravity (in roots) and away from gravity (in shoots), auxin itself does not detect gravity. Instead, specialized cells in stems and in root caps contain starch-filled organelles called **statoliths**. By staining these statoliths and observing them under the microscope, plant physiologists have discovered that they settle to the downward side of the cell within minutes (Fig. 28-6). When a plant is laid on its side, causing a formerly vertical root to be horizontal, the time it takes the statoliths to settle is similar to the time it takes the root to begin its unequal cell elongation so that it is once more heading downward. Exactly how the falling statoliths initiate the response to gravity in stems and roots is still under investigation.

The Genetically Determined Shape of the Mature Plant Is the Result of Interactions between Hormones

As a plant grows, both its root and shoot develop branching patterns that are largely determined by its genetic heritage. For example, the stems of some plants, such as sunflowers, hardly branch at all; others, such as oaks and cottonwoods, branch profusely in seeming confusion; still others branch in a very regular pattern, producing the conical shapes of firs and spruces.

The amount of growth in shoot and root systems must also be kept in balance. The shoot must be large enough to supply the roots with sugars, while the roots must be large enough to provide the shoot with water and miner-als. Interactions between auxin and cytokinin regulate root and stem branching, thereby regulating the relative sizes of root and shoot systems.

Stem Branching Is Influenced by the Growing Tip of the Shoot

Gardeners know that pinching back the tip of a growing plant makes it become bushier. The botanical explanation for this practice is that the growing tip suppresses the sprouting of lateral buds to form branches, a phenomenon known as **apical dominance.** Although exactly how lateral bud sprouting is controlled is still a subject of research, there is some evidence that the proper levels of auxin and cytokinin must be present (Fig. 28-7). Auxin is produced by the shoot tip (its concentration is highest here) and transported down the stem, gradually decreasing in concentration. Cytokinin is produced by the roots and is transported up the stem. Therefore, the relative concentrations of these two hormones will vary along the length of the stem, so buds at different positions will experience differing hormonal influences.

Auxin by itself appears to inhibit the sprouting of lateral buds, whereas auxin and cytokinin together stimulate bud sprouting. The lateral buds closest to the shoot tip receive a great deal of auxin, probably enough to inhibit their growth, but receive very little cytokinin because they are so far from the roots. Therefore, they remain dormant. Lower buds receive less auxin while receiving much more cytokinin. They are stimulated by optimal concentrations of both hormones, so they sprout (Fig. 28-7). In many plants, this interaction between auxin and cytokinin produces an orderly progression of bud sprouting, from the bottom to the top of the shoot.

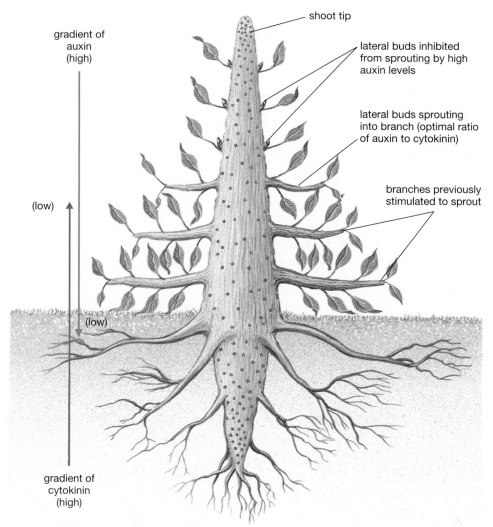

gradient of
auxin
(high)

shoot tip

lateral buds inhibited
from sprouting by high
auxin levels

lateral buds sprouting
into branch (optimal ratio
of auxin to cytokinin)

(low)

branches previously
stimulated to sprout

(low)

gradient of
cytokinin
(high)

Figure 28-7   The role of auxin and cytokinin in lateral bud sprouting

A simplified diagram of the interplay of auxin (blue dots) and cytokinin (red dots) in the control of sprouting of lateral buds. Auxin is produced by shoot tips and moves downward, while cytokinin is produced by root tips and moves upward. The auxin concentration therefore decreases with distance from the shoot tip, and the cytokinin concentration decreases with distance from the root tip. Sprouting of lateral buds depends on the ratio of cytokinin to auxin; the exact ratio needed for sprouting varies among species.

Auxin Stimulates Root Branching

Even in extremely low concentrations, auxin stimulates the branching of roots. As we described in Chapter 25, branch roots arise from the pericycle layer of the vascular cylinder. Auxin, transported down from the stem, stimulates pericycle cells to divide and form a branch root.

Hormone Gradients Create a Balance between the Root and Shoot Systems

Through the interaction of auxin and cytokinin, the root and shoot systems regulate each other's growth. This interaction is important because they supply complementary nutrients. An enlarging root system synthesizes large amounts of cytokinin, which stimulates lateral buds to break dormancy and sprout. If the root system isn't keeping up with the growing shoot system, less cytokinin is produced. The sprouting of lateral buds is delayed, slowing the growth of the shoot system. Simultaneously, as the

stem grows and branches, it produces lots of auxin, which stimulates root branching and growth. Thus, neither root nor shoot can outgrow the other, and the plant is adequately supplied with all its needs.

Hormones Control the Differentiation of Xylem and Phloem

As a plant grows, its parts must be interconnected by the vascular tissues of xylem and phloem. Differentiation of cells into vascular tissues appears to be yet another function of auxin and perhaps gibberellin as well. Apical meristems release auxin and gibberellin into cells that are maturing just behind them. High hormone levels stimulate these cells to differentiate into xylem and phloem. As leaves grow and lateral buds sprout, they too release auxin and gibberellin. These hormones cause cortex cells just beneath the leaves and buds to differentiate into strands of xylem and phloem that connect with the main vascular systems of the stem.

Daylength Controls Flowering

Ultimately, the plant matures enough to reproduce. The timing of flowering and seed production is finely tuned to the physiology of the plant and the rigors of its environment. In temperate climates, plants must flower early enough so that their seeds can mature before the killing frosts of autumn. Depending on how quickly the seed and fruit develop, flowering may occur in spring, as it does in oaks; in summer, as in lettuce; or even in autumn, as in asters.

What environmental cues do plants use to determine the season? Most cues, such as temperature or water availability, are quite variable: October can be warm, a late snow may fall in May, or the summer might be unusually cool and wet. *The only reliable cue is daylength:* Longer days always mean that spring and summer are coming; shorter days foretell the onset of autumn and winter.

With respect to flowering, plants are classified as day-neutral, long-day, or short-day plants (Fig. 28-8). A **day-neutral plant** is one that flowers as soon as it has grown and developed enough, regardless of the length of the day. Day-neutral plants include tomatoes, corn, and snapdragons. A **long-day plant** flowers when the daylength is *longer than some species-specific critical value*. A **short-day plant** flowers when the daylength is *shorter than some species-specific critical value*. Thus spinach is classified as a long-day plant because it flowers only if the day is *longer than* 13 hours, and cockleburs are short-day plants because they flower only if the day is *shorter than* 15.5 hours. Note that both will flower with 14 hours of light. Spinach, however, will also flower in much longer daylengths (for example, 16 hours of light), but cocklebur will not. Conversely, cocklebur will flower in much shorter daylengths (for example, 12 hours), but spinach will not.

Hormones Called Florigens Both Stimulate and Inhibit Flowering

Grafting a branch from a (short-day) cocklebur plant maintained on a short-day schedule onto a plant on a long-day schedule will induce flowering. In fact, plant physiologists have been able to induce flowering in the cocklebur by exposing a *single leaf* to short days in a special chamber,

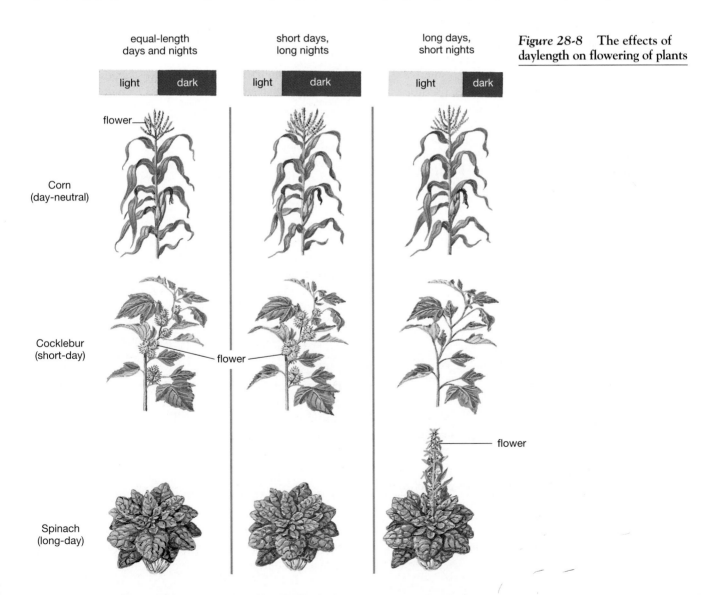

Figure 28-8   **The effects of daylength on flowering of plants**

while the rest of the plant was experiencing long days. Clearly a hormone must be traveling from the leaves to the bud. Studies on a variety of plants suggest that still unidentified hormones can both trigger and inhibit flowering. These substances, which may differ among different plant species, collectively are called **florigens** (literally "flower makers"). Daylength is a crucial stimulus for these hormones, but how do plants detect daylength?

Pigments Called Phytochromes Measure Daylength by Resetting the Biological Clock

To measure daylength, a plant needs some sort of metabolic *clock* to measure time (how long it has been light or dark) and a *light-detecting system* to set the clock. Virtually all organisms have an internal **biological clock** that in some way (usually poorly understood) measures time even without environmental cues. However, environmental cues, and particularly light, can reset the clock.

The light-detecting system of plants is a pigment in the leaves called **phytochrome** (meaning simply "plant color"). Phytochrome occurs in two interchangeable forms (Fig. 28-9). One form strongly absorbs red light and is called P_r; the other form absorbs far-red light (almost infrared) and is accordingly called P_{fr}. In most plants, P_{fr} is the active form of phytochrome; that is, a suitable concentration of P_{fr} stimulates or inhibits physiological processes, such as flowering or setting the biological clock. P_r has no effect on these same processes.

Phytochrome flips back and forth from one form to the other when it absorbs light of the appropriate color: When P_r absorbs red light, it is converted into P_{fr}, and when P_{fr} absorbs far-red light, it is transformed back into P_r. Daylight consists of all wavelengths of visible light, including both red and far-red. Therefore, during the day a leaf contains both forms of phytochrome. In the dark, P_{fr} rather rapidly breaks down or reverts to P_r.

Plants seem to use the phytochrome system and their internal biological clocks to detect daylength. Cockleburs, for example, flower under a lighting schedule of 8 hours of light and 16 hours of darkness. However, interrupting the middle of the dark period with just a minute or two of light prevents flowering. Thus, although cockleburs are usually classified as short-day plants, they might more accurately be called "long-night" plants, because what really matters is how long continuous darkness lasts. The color of the light used for the night flash is important. A midnight flash of red light inhibits flowering, but a far-red flash allows flowering. This observation, of course, implicates phytochrome in the control of flowering. Scientists are still researching, however, how the response of phytochrome to light determines whether or not a plant will flower. It seems likely that the biological clock measures the length of the night and that light reception by phytochrome tells the clock when sunrise and sunset have occurred, but this is not certain. How phytochromes influence the production of florigens is another area of active research.

Phytochromes Influence Other Responses of Plants to Their Environment

Phytochrome is involved in many plant responses. For example, P_{fr} inhibits elongation of seedlings, with profound and obviously adaptive results. Because P_{fr} breaks down or reverts to P_r in the dark, seedlings germinating in the darkness of the soil contain no P_{fr} and consequently elongate very rapidly, emerging from the soil. Seedlings growing beneath other plants will be exposed largely to far-red light, because the green chlorophyll of the leaves above them will absorb most of the red light but transmit the far-red. Far-red light converts P_{fr} to P_r, so shaded seedlings grow rapidly, which may bring them out of the shade. Once out in the sunlight, P_{fr} forms. P_{fr} slows down elongation, which prevents the seedlings from becoming too spindly.

Other plant responses that are stimulated by P_{fr} include leaf growth, chlorophyll synthesis, and straightening of the epicotyl or hypocotyl hook of dicot seedlings. As in the case of stem elongation, these responses are adaptations related to burial in the soil or shading by the leaves of other plants. For example, a newly germinating shoot needs to stay in its protective bend while still in the soil (that is, in the dark) and straighten out only in the open air, where sunlight converts P_r to P_{fr}.

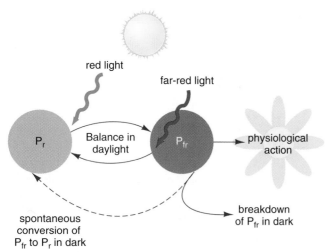

red light

far-red light

P_r Balance in daylight P_{fr} → physiological action

spontaneous conversion of P_{fr} to P_r in dark

breakdown of P_{fr} in dark

Figure 28-9   The light-sensitive pigment phytochrome

Phytochrome exists in two forms, inactive (P_r) and active (P_{fr}). P_r is converted to P_{fr} by red light. P_{fr} may then participate in physiological responses, be converted to P_r by far-red light, revert spontaneously to P_r, or break down to other, inactive compounds.

Hormones Coordinate the Development of Seeds and Fruit

When a flower is pollinated, auxin or gibberellin released by the pollen stimulates the ovary to begin developing into a fruit. If fertilization also occurs, the developing seeds release still more auxin or gibberellin or both into the surrounding ovary tissues. Cells of the ovary multiply and grow larger, often storing starches and other food materials, forming a mature fruit.

Seeds and fruits acquire nutrients for growth and development from their parent plant. If the seed is separated from the parent too soon, it may not complete its development. Not surprisingly, seed maturation and fruit ripening are closely coordinated. Unripe fruits are often inconspicuously colored (usually green like the rest of the plant), hard, bitter, and sometimes even poisonous. As a result, animals seldom eat unripe fruit. When the seeds mature, the fruit ripens: It becomes sweeter as starches are converted to sugar, softer, and more brightly colored, making it more noticeable and attractive to animals (Fig. 28-10). Interestingly, gibberellin sprayed on fruit such as grapefruit causes the peel to remain tough and green, although the inside continues to ripen. Citrus growers in Florida can now use this technique to discourage fruit flies, which are attracted to the yellow color and must penetrate the ripening peel to lay their eggs.

Ripening is stimulated by ethylene. Ethylene is synthesized by fruit cells in response to a surge of auxin that is released by the seeds. Because ethylene is a gas, a ripe fruit continually leaks ethylene into the air. In nature this probably doesn't make much difference. When people store fruit in closed containers, however, ethylene released from one fruit will hasten ripening in the rest, which is why "one rotten apple spoils the barrel." The discovery of the role of ethylene in ripening revolutionized modern fruit and vegetable marketing. The gibberellin-sprayed green (but ripe) grapefruit described above will turn its normal yellow when exposed to ethylene. Bananas, grown in Central America, can be picked green and tough and shipped to North American markets. By exposing them to ethylene at their destination, grocers can market perfectly ripe fruit. Unfortunately, not all fruits seem to ripen properly when separated from the plant. The pinkish tomato with all the flavor of wet cardboard that we buy in supermarkets is an example.

Senescence and Dormancy Prepare the Plant for Winter

The season is autumn. If animals haven't eaten the fruits yet, the time has come to let them drop to the ground. For perennial broadleaf plants, the leaves must be shed as well, because they will be a liability in winter, unable to photosynthesize but still allowing water to evaporate. Both leaves and fruits undergo a rapid aging called **senescence.** The culmination of senescence is the formation of the **abscission layer** at the base of the petiole, allowing the leaf or fruit to drop off.

Senescence and abscission are complex processes controlled by several different hormones. In most plants, healthy leaves and developing seeds produce auxin, which in turn helps to maintain the health of the leaf or fruit. Simultaneously, the roots synthesize cytokinin, which is transported up the stem and out the branches. Cytokinin also prevents senescence—a leaf plucked from a tree and floated in water in which cytokinin is dissolved stays green for weeks. As winter approaches, cytokinin production in the roots slows down, and fruits and leaves produce less auxin. Perhaps driven by these hormonal changes, much of the organic material in leaves is broken down to simple molecules that are transported to the roots for winter storage. Meanwhile, ethylene is released by both aging leaves and ripening fruit. Ethylene stimulates the production of enzymes that destroy the cell walls holding the abscission layers at the base of the petiole together. When the abscission layers weaken, leaves and fruits fall from the branches.

Other changes also occur that prepare the plant for winter. Buds, which developed into new leaves and branches during spring and summer, now become dormant, waiting out the winter tightly wrapped up. Dormancy in buds, as in seeds, is enforced by abscisic acid. Metabolism slows to a crawl, and the plant enters its long winter sleep, waiting for the signals of warmth and longer days in spring before awakening once again.

Figure 28-10   **Fruit ripening includes changes in color, texture, flavor, and sweetness**

The prickly pear cactus fruit is green, hard, and bitter before it ripens, so animals are discouraged from eating it. After the seeds mature, the fruit becomes soft, red, and tasty, attracting animals such as this desert tortoise. The mature seeds are not harmed by the animal's digestive tract and are dispersed in the animal's feces.

A CLOSER LOOK

Rapid-Fire Plant Responses

All plants are alive, but some are definitely more lively than others. Watch a fly brush against the sensory hairs in a Venus flytrap, and you will see a response that is almost animal-like in its purposefulness and speed of movement (Fig. E28-1). Why is it useful for a Venus flytrap to catch a fly, and how does it accomplish this task?

Many soils, and particularly those of bogs, are nitrogen-deficient. As an evolutionary response to chronic nitrogen shortages, several bog plants, including the Venus flytrap, pitcher plant, and sundew, have resorted to eating animals. By snaring an insect now and then, the plant obtains nitrogen from the chitin (in the exoskeleton), proteins, and nucleic acids of its prey. While the evolutionary advantage seems clear, the *mechanism* of movement is much less obvious. How does the plant perceive the touch of a fly, and how does it move its leaves rapidly enough to catch it?

In this chapter you have seen how hormones such as auxin can trigger expansion of cells, causing plants to move toward or away from light or gravity. But hormones don't carry signals fast enough to catch a fly. Venus flytraps have actually evolved a way to transmit signals that resembles the nerve impulse of animals and permits their animal-like predatory behavior.

Each of the fringed trapping leaves of a Venus flytrap bears three sensory "hairs" on its inside surface (Fig. E28-1a). Insects are attracted to nectar secreted by the leaves. In its foraging, if an insect touches one hair twice in rapid succession, or touches two different hairs, the hairs initiate an electrical potential change analogous to the action potential of animal nerve cells. The electrical potential sets off a rapid chain of events that causes the trap to close (Fig. E28-1b).

In a beautiful set of experiments, botanists Stephen Williams and Alan Bennett found that the flytrap leaf closes because of *irreversible, differential growth*. The flytrap leaves can be pictured most simply as two layers of cells, outer and inner (Fig. E28-2). The electrical potential triggered by hair movement stimulates cells of the outer layer to pump hydrogen ions (H^+) extremely rapidly into their cell walls. Enzymes in the cell walls are activated by acid conditions (created, as you'll remember from Chapter 2, by a high concentration of H^+) and loosen the cellulose fibers of the walls. As the walls weaken, the high osmotic pressure inside the cells causes them to absorb water from extracellular fluids, swiftly growing by about 25%. Because the outer layer expands while the inner layer does not, the leaf is pushed closed. Although reopening the trap takes several hours, the fundamental mechanism is similar: The cells on the inside of the leaf expand, pushing apart the lobes of the trap.

So much energy is used up by the hydrogen ion pumps that *closing the trap consumes nearly one-third of all the ATP within the entire leaf*. It is therefore very important that something digestible actually be in the leaf before it closes the trap!

Although much is known about the mechanisms producing movement in the Venus flytrap, mysteries still remain. How is touch transformed into an electrical stimulus by the sensory hairs? What is the nature of the electrical potential change? How does the electrical signal cause the cells to begin pumping hydrogen ions? As so often happens in biology, the answer to one question immediately poses several new, usually tougher, questions.

(a)

(b)

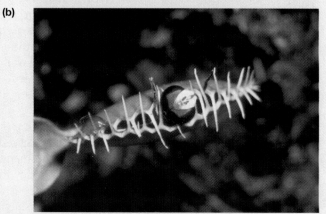

Figure E28-1   **A Venus flytrap captures its prey**

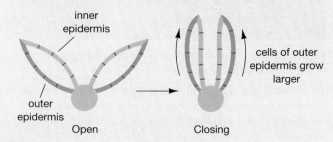

Figure E28-2   **How a Venus flytrap closes its leaf trap**

SUMMARY OF KEY CONCEPTS

The Discovery of Plant Hormones
Most responses of plants to their environment are produced by the actions of plant hormones. Early experiments on the mechanism of phototropism, the growth of plants toward the light, led to the first discovery of a plant hormone, auxin.

Plant Hormones and Their Actions
Plant hormones are chemicals that are produced by cells in one part of a plant body and transported to other parts of the plant, where they exert specific effects. The five major classes of plant hormones are: auxin, gibberellins, cytokinins, ethylene, and abscisic acid. The major functions of these hormones are summarized in Table 28-1.

The Plant Life Cycle: Reception, Response, and Regulation
Dormancy in seeds is enforced by abscisic acid. Falling levels of abscisic acid, and rising levels of gibberellin, trigger germination. As the seedling grows, it shows differential growth with respect to the direction of light (phototropism) and gravity (gravitropism). Auxin mediates phototropism and gravitropism in shoots, and gravitropism in roots. In shoots, auxin stimulates elongation of cells. In roots, similar concentrations of auxin inhibit elongation. Plants apparently detect gravity using organelles called statoliths.

Branching in stems results from the interplay of two hormones, auxin and cytokinin. High concentrations of auxin (produced in shoot tips and transported downward) inhibit the growth of lateral buds. An optimum concentration of both auxin and cytokinin (synthesized in roots and transported up the shoot) stimulates growth of lateral buds. Auxin also stimulates the growth of branch roots.

The timing of flowering is usually controlled by daylength. Flowering is both stimulated and inhibited by hormones called florigens. Plants appear to detect light and dark by changes in phytochrome, a pigment in the leaves. Plant processes influenced by phytochrome responses to light include flowering, straightening the epicotyl or hypocotyl hook, seedling elongation, leaf growth, and chlorophyll synthesis.

Developing seeds produce auxin, which diffuses into the surrounding ovary tissues and causes production of a fruit. A surge of auxin as the seed matures stimulates fruit cells to release another hormone, ethylene, which causes the fruit to ripen. Ripening includes the conversion of starches to sugars, softening of the fruit, development of bright colors, and often the formation of an abscission layer at the base of the petiole.

Several changes prepare perennial plants of temperate zones for winter. Leaves and fruits undergo a rapid aging process called senescence, including formation of an abscission layer. Senescence occurs because of a fall in levels of auxin and cytokinin and, perhaps, a rise in ethylene concentrations. Other parts of the plant, including buds, become dormant. Dormancy in buds is enforced by high concentrations of abscisic acid.

KEY TERMS

abscisic acid p. 561
abscission layer p. 567
apical dominance p. 563
auxin p. 559
biological clock p. 566
cytokinin p. 560
day-neutral plant p. 565

ethylene p. 561
florigen p. 566
gibberellin p. 560
gravitropism p. 560
hormone p. 559
long-day plant p. 565

phototropism p. 557
phytochrome p. 566
plant hormone p. 560
senescence p. 567
short-day plant p. 565
statolith p. 563

THINKING THROUGH THE CONCEPTS

Multiple Choice
1. If you put an underripe banana in a bag with an apple, it will quickly ripen because of the hormone _____ produced by the apple.
   a. auxin   b. cytokinin
   c. gibberellin   d. abscisic acid
   e. ethylene
2. Roots turn downward in a process known as
   a. reverse phototropism
   b. taxis
   c. apical dominance
   d. gravitropism
   e. mass flow
3. If you grow coleus plants, you will need to cut or pinch off the top bud frequently or the plant will become tall and spindly. Pinching off this bud will slow the production of _____ by the apical bud and allow the plant to become bushy.
   a. auxin   b. cytokinin
   c. gibberellin   d. abscisic acid
   e. ethylene

4. In your warm house, poinsettias act as short-day plants that turn red when the daylength is less than 12 hours. If you want to get a poinsettia ready for Christmas, which of the following would work best?
   a. Give it continuous light.
   b. Give it continuous dark.
   c. Keep it in the dark but shine a light on it for an hour every 13 hours.
   d. Keep it in the light except for turning off the light once a day for 1 hour.
   e. None of these would get it to turn red.

5. When a seed is first formed in the fall, it often will not germinate because _____ must first be washed from the seed by a hard rain or broken down by cycles of freezing and thawing.
   a. auxin   b. cytokinin
   c. gibberellin   d. abscisic acid
   e. ethylene

6. This same seed germinates as levels of _____ increase.
   a. auxin   b. cytokinin
   c. gibberellin   d. abscisic acid
   e. ethylene

Review Questions

1. What did the Darwins, Boysen-Jensen, and Went each contribute to our understanding of phototropism? Do their experiments truly prove that auxin is the hormone controlling phototropism? What other experiments would you like to see?
2. How do hormones interact to cause apical dominance? To control seed dormancy?
3. How can one hormone, an auxin, cause shoots to grow up and roots to grow down?
4. What is the phytochrome system? Why does this chemical exist in two forms? How do the two forms interact to help control the plant life cycle?
5. What hormone(s) causes fruit development? Where does this hormone come from? What hormone causes fruit ripening?
6. What is a biological clock?
7. Describe the role of phytochrome in stem elongation in seedlings growing in the shade of other plants. What is the likely adaptive significance of this response?
8. What is apical dominance? How do auxin and cytokinin interact in determining the growth of lateral buds?
9. What hormone(s) is involved in leaf and fruit drop? In bud dormancy?

✖ APPLYING THE CONCEPTS

1. Suppose you got a job in a greenhouse where the owner was trying to start the flowering of chrysanthemums (a short-day plant) for Mother's Day. You accidentally turned on the light in the middle of the night. Would you be likely to lose your job? Why or why not? What would happen if you turned on the lights in the day?
2. A student reporting on a project said that one of her seeds did not grow properly because it was planted upside down so that it got confused and tried to grow down. Do you think the teacher accepted this explanation? Why or why not? What plant hormone or hormones would be involved?
3. Agent Orange, a combination of two synthetic auxins, was used in Vietnam to defoliate the rain forest during the Vietnamese War. How can synthetic auxins be used to harm or kill plants when they are similar to natural growth hormones? What do you think would happen if natural auxins were used in excess quantities on plants?
4. Bean sprouts such as you might eat in a salad have to be grown in the dark in order to get them to form the long, yellowish stems that you see. We call such stems etiolated. If they are grown in the light, they will be short and green. Why do seedlings grow etiolated in the dark? Under what conditions does etiolation occur in nature? How do plant hormones enable these seeds to form this shape?
5. If you worked for a company that sold plant hormones, what uses would you suggest to your customers for each of the hormones we studied?
6. Suppose that on July 4, you discover that a short-day plant and a long-day plant, both growing in your garden, have bloomed. Discuss how it is possible for both to bloom.

✖ FOR MORE INFORMATION

Evans, M. L., Moore, R., and Hasenstein, K.-H. "How Roots Respond to Gravity." *Scientific American*, December 1986. Although botanists still dispute the mechanisms of root gravitropism, these authors convincingly argue that it is mediated by auxin.

Horton, T. "Longleaf Pine: A Southern Revival." *Audubon*, March–April 1995. The entire life cycle of this plant is centered around its adaptations to fire.

Marchand, P. J. "Waves in the Forest." *Natural History*, February 1995. Compares the life rhythms of forests in Japan and New England in which whole areas of the forest live and die in cycles.

Moffatt, A. S. "How Plants Cope with Stress." *Science*, November 1, 1994. A newly discovered hormone, "systemin," similar to animal hormones, enables plants to respond to stress.

Pennisi, E. "Plants Relay Signals Much As Animals Do." *Science News*, February 13, 1993. A short news article on the effects of ethylene that shows how plants respond to hormones.

NET WATCH

On-line resources for this chapter are on the World Wide Web at: http://www.prenhall.com/~audesirk (click on the <u>table of contents</u> link and then select Chapter 28).

Unit V
Animal Anatomy and Physiology

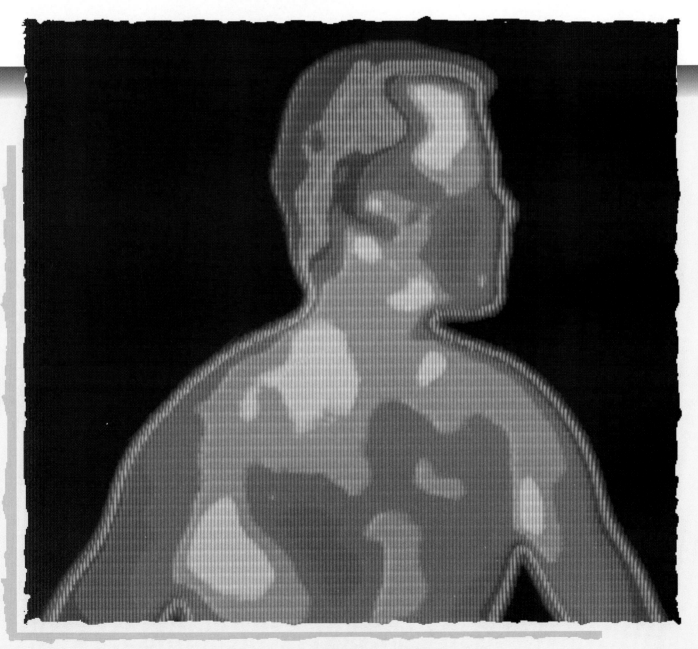

Human head and torso imaged using a thermogram, a device that maps heat distribution. Blue areas are cooler, yellow areas are warmest.

29 Homeostasis and the Organization of the Animal Body

"As a physiologist I am especially impressed by the common ignorance of bodily organs and their functions. It seems to me now, as it seemed to Robert Boyle nearly 300 years ago, that it is 'highly dishonorable for a Reasonable Soul to live in so Divinely built a Mansion as the Body [he or] she resides in, altogether unacquainted with the exquisite structure of it.'"

W. B. CANNON in *The Way of An Investigator* (1945)

☒ AT A GLANCE

Homeostasis
  Negative Feedback Tends to Maintain Constant Internal
    Conditions
  Positive Feedback Events Are Self-Limiting

The Organization of the Animal Body
  Animal Tissues Are Composed of Similar Cells That
    Perform a Specific Function
  Organs Include Two or More Interacting Tissue Types
  Organ Systems Consist of Two or More Interacting Organs

Through a variety of adaptations, animals are able to maintain relatively constant internal environments that differ substantially from their surroundings. This harp seal keeps warm on a Canadian ice floe with dense, waterproof fur and a thick layer of fat.

At the dawn of life, over 3.5 billion years ago, complex organic molecules arose within warm, primordial waters. But before assemblages of molecules could be considered alive, it was necessary for a boundary to form between the evolving cell and its environment. This enclosing membrane allowed the earliest life forms to maintain an internal environment distinct from their surroundings. As organisms evolved increasingly sophisticated mechanisms to regulate their internal environment, they were able to emerge from the nurturing waters. Today we find organisms thriving in extremely inhospitable regions of Earth because they are able to maintain internal environments that protect and promote the delicate processes of life. This ability is called homeostasis.

Homeostasis

In the mid-nineteenth century, the French physiologist Claude Bernard first described the "constancy of the interior milieu," the concept now known as homeostasis. The term **homeostasis** was coined in 1932 by Walter Cannon, who suggested that nearly all physiological processes operate in a manner that tends to maintain the internal constancy required for the maintenance of life. Although the term homeo*stasis* might imply a "static," or unchanging, environment, in fact, the "internal milieu" seethes with activity as the body continuously responds and adjusts to internal and external changes. The term *dynamic equilibrium* better describes this internal condition. Physical and chemical changes do occur, but only within the narrow range that cells require to function. The mechanisms by which organisms control the homeostatic conditions are called feedback systems. The continuous activation of negative feedback systems counteracts changes, while the less frequent use of positive feedback systems intensifies change for a specific purpose.

Negative Feedback Tends to Maintain Constant Internal Conditions

The most important principle governing homeostasis is **negative feedback,** in which a change in the internal environment triggers a response that tends to counteract the

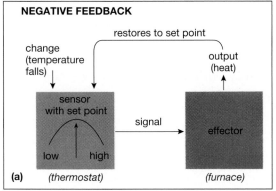

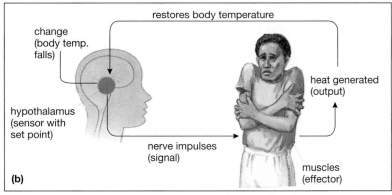

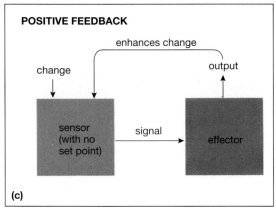

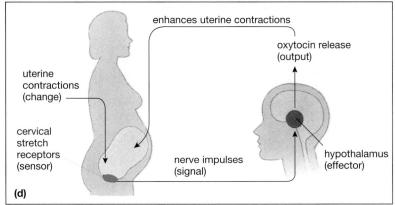

Figure 29-1   **Negative and positive feedback**

(a) Negative feedback controls the temperature in most homes. When the system changes (the house cools below a specific temperature), the change is sensed by a sensor with a specific set point (the thermostat, set to 68° F). The sensor sends a signal to an effector (the furnace) whose output (heat) returns the system to its set point. The signal is turned off, which turns off the effector. **(b)** Body temperature is both set by and sensed by neurons in the hypothalamus (the sensor with a set point). A drop in body temperature causes the hypothalamic neurons to send a signal to effectors such as muscles, which begin shivering. Shivering generates heat, which helps restore body temperature to its set point, shutting off the signal, and causing shivering to cease. **(c)** In positive feedback, a change in the system is sensed by a sensor with no set point, which then sends a signal to an effector. The output of the effector causes an even greater change in the system, rather than attempting to return it to previous conditions. **(d)** At the onset of labor, uterine contractions are sensed by receptor neurons in the cervix, which send a signal to the hypothalamus that results in secretion of a hormone (oxytocin) from the posterior pituitary gland. The hormone, in turn, stimulates more and stronger uterine contractions, ultimately leading to the expulsion of the baby and placenta from the uterus.

change and to restore and maintain the original conditions. This feedback is called "negative" because it opposes or resists the change.

A familiar example is your home thermostat (Fig. 29-1a). A drop in temperature from its setting is detected by a thermometer, which activates a switch that turns on a heating device. When the temperature is restored to the set point, the heater is turned off. Negative feedback requires a *set point* (the thermostat setting), a *sensor* (the thermometer), and an *effector* (the furnace), which accomplishes the change.

A biological example is the maintenance of your body at a temperature within about 1° of 98.6° F (37° C). The set point is established by neurons in the hypothalamus, a re-

gion of your brain that controls many homeostatic responses. Other neurons in the hypothalamus, abdomen, spinal cord, and large veins act as sensors for body temperature and transmit this information to the hypothalamus. When your body temperature drops, the hypothalamus activates various effector mechanisms including shivering (which generates heat through muscular activity), constriction of blood supply to the skin (which reduces heat loss), and elevation of metabolic rate (which generates heat). When normal body temperature is restored, the hypothalamus switches off these temperature control mechanisms (Fig. 29-1b).

As you read the following chapters, you will find many examples of homeostatic control using negative feedback. In Chapters 30, 31, and 35, you will encounter the mammalian hormones erythropoietin, ADH (antidiuretic hormone), and insulin, which regulate blood oxygen content, water balance, and blood sugar levels, respectively, all through negative feedback.

Positive Feedback Events Are Self-Limiting

A **positive feedback** system, in contrast to a negative feedback system, is one in which a change in the internal environment initiates a response that intensifies the original change (Fig. 29-1c). Instead of immediately returning conditions to a set point, this type of system encourages change under controlled circumstances to achieve a specific end. Positive feedback, as you can imagine, tends to create explosive events that must be carefully restricted. An example of positive feedback is nuclear fission, in which each of many particles that are split from one atom triggers the splitting of another atom, and so on. Controlled, the reaction supplies nuclear power. When deliberately set out of control, it produces the atomic bomb. Population growth is a familiar biological example of positive feedback, described in Chapter 43. The ecologist Paul Ehrlich coined the apt expression "population bomb" to describe unchecked population growth.

Consequently, positive feedback events in animal physiology are self-limiting and relatively infrequent. Positive feedback occurs, for example, during childbirth (see Fig. 29-1d). The early contractions of labor begin to force the baby's head against the cervix, located at the base of the uterus; this pressure causes the cervix to dilate (open). Stretch-receptive neurons in the cervix respond to this expansion by signaling the hypothalamus, which responds by triggering the release of a hormone (oxytocin) that stimulates more and stronger uterine contractions. Stronger contractions create further pressure on the cervix, which in turn prompts the release of more hormones. The feedback cycle is finally terminated by the expulsion of the baby and its placenta.

The Organization of the Animal Body

The animal body is made up of **organ systems** (for example, the digestive system and the excretory system), each made up of **organs** (the stomach and intestine of the digestive system; the kidney and urinary bladder of the excretory system). Each organ, in turn, consists of two or more of the four primary types of **tissue** (epithelial, connective, nerve, and muscle tissue). Each tissue type is composed of dozens to billions of structurally similar cells that act in concert to perform a similar function (Fig. 29-2).

Animal Tissues Are Composed of Similar Cells That Perform a Specific Function

A tissue is composed of cells that are similar in structure and designed to perform a specialized function. Tissue may also include extracellular components produced by these cells, as in the case of cartilage and bone. Here, we present a brief overview of the four major categories of animal tissue: epithelial tissue, connective tissue, muscle tissue, and nerve tissue.

Epithelial Tissue Forms Membranes That Cover the Body and Line Its Cavities

The cells of **epithelial tissues** (also called the epithelium) form continuous sheets called **membranes.** Epithelial membranes cover the body and line all the body cavities. This means that one surface of epithelial membranes is free; that is, it faces either the outside of the body or the inside of cavities within the body. Epithelial membranes create a barrier that either resists the movement of substances across it (such as the skin) or that allows movement of specific substances across it (such as the lining of the small intestine). Epithelial membranes can serve as effective barriers because their cells are packed closely together and connected to one another by tight junctions (see Chapter 6). No blood vessels penetrate epithelial tissue; it is nourished by diffusion from capillaries within the connective tissue that lies beneath it.

Another important property of epithelial tissues is that they are continuously lost and replaced by cell division. Consider the abuse suffered, for example, by the epithelium that lines your mouth. Scalded by coffee and scraped by corn chips, it would be destroyed within a few days if it were not continuously replacing itself. The stomach lining, abraded by food and attacked by acids and protein-digesting enzymes, is completely replaced every 2 to 3 days. Your skin's outer layer, the epidermis, is renewed about twice a month. Epithelial tissues are classified according to the shape of their cells and the number of cell layers present, as summarized in Table 29-1.

Some Epithelial Tissues Also Form Glands

During development, some epithelial tissues fold inward, and their cells change shape and function and form **glands,** clusters of cells that are specialized to secrete (release) substances. Glands are classified into two broad categories: exocrine glands and endocrine glands (Fig. 29-3, p. 578). **Exocrine glands** remain connected to the epithelium by a passageway or duct. Examples of exocrine glands are sweat and sebaceous (oil-secreting) glands, both found in the skin and derived from skin epithelium (see Fig. 29-7). Exocrine glands called salivary glands release saliva into the mouth, and still others line the stomach, where they secrete a protective layer of mucus. **Endocrine glands** become separated from the epithelium that produced them. The products of endocrine glands are mostly hormones, which are secreted into the extracellular fluid surrounding the glands and then diffuse into nearby capillaries. Endocrine glands and their hormones are covered in detail in Chapter 35.

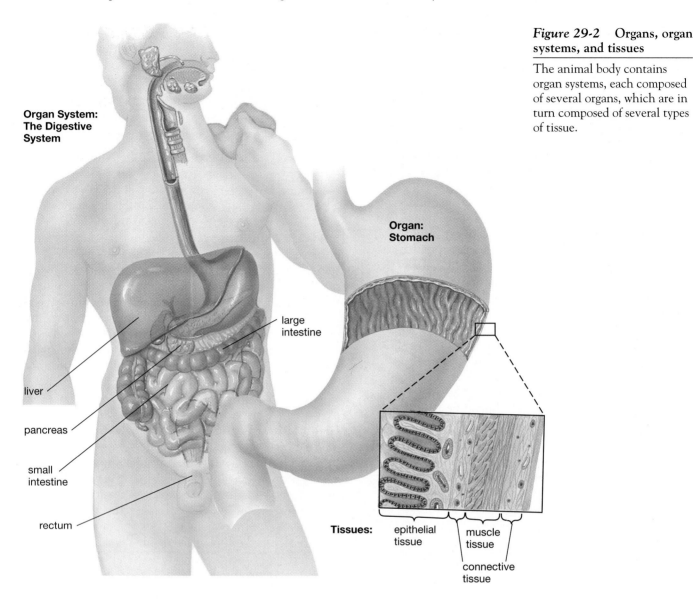

**Organ System:
The Digestive
System**

liver

pancreas

small
intestine

rectum

large
intestine

**Organ:
Stomach**

Tissues: epithelial
tissue

muscle
tissue

connective
tissue

*Figure 29-2   Organs, organ
systems, and tissues*

The animal body contains
organ systems, each composed
of several organs, which are in
turn composed of several types
of tissue.

Connective Tissues Have Diverse Structures and Functions

There are several diverse connective tissues, including the dermis of the skin, tendons, ligaments, cartilage, bone, fat, and blood. Most **connective tissues** share a common feature: Their cells are surrounded by large quantities of extracellular substances, often secreted by the connective tissue cells themselves. With the exception of blood, most connective tissue is interwoven with fibrous strands of an extracellular protein called **collagen** secreted by the cells. Connective tissue underlies all epithelial tissue, containing capillaries and fluid-filled spaces that nourish the epithelium. Underlying the epidermis of the skin, for example, is connective tissue, called the **dermis**, that is richly supplied with capillaries (see Fig. 29-7).

Tendons and **ligaments** are types of connective tissue that attach muscles to bones and bones to bones, respectively (see Fig. 38-10). They contain densely packed collagen fibers in an orderly parallel arrangement. **Cartilage** is a

flexible and resilient form of connective tissue. It consists of widely spaced cells surrounded by a thick, nonliving matrix of collagen, which they secrete (see Fig. 38-6). Cartilage covers the ends of bones at joints, provides the supporting framework for the respiratory passages, supports the ear and nose, and forms shock-absorbing pads between the vertebrae. **Bone** resembles cartilage that has been hardened by deposits of calcium phosphate. During bone formation, bone cells, or **osteocytes,** become embedded in concentric layers of hardened collagen matrix. These layers surround a central canal containing a blood capillary that nourishes the bone cells. These concentric circles of bone, each surrounding a canal, are called **Haversian systems** (Fig. 29-4). Both bone and cartilage are covered in detail in Chapter 38.

Blood and **lymph** are considered connective tissues because they are composed largely of extracellular fluid, the plasma and lymph, respectively. The cellular portion of blood consists of red and white blood cells and cell fragments called platelets. Blood, pumped by the heart and confined to vessels, carries dissoved oxygen and other nu-

TABLE 29-1 ❖ *Major Types of Epithelial Tissue*

Tissue Type	Diagram	Properties and Function	Location in Body
Simple squamous epithelium	Thin, flattened cells, one layer thick	Thin layer that allows movement of substances; found in areas where diffusion and filtration are occurring	Lines lung alveoli; forms capillary walls; lines blood vessels
Stratified squamous epithelium	Thin, flattened cells, many layers thick	Thick layer that is rapidly replaced by divisions of lower cells; has protective function; may secrete mucus	Upper layer of skin; lines mouth, anal canal, vagina
Simple cuboidal epithelium	Cube-shaped cells, one layer thick	Thin layer; functions in absorption and secretion	Lines kidney tubules; has secretory role in salivary glands, thyroid, pancreas, and liver
Simple columnar epithelium	Elongated cells, one layer thick; nuclei lined up	Forms a thick layer that functions in secretion and absorption; may secrete mucus	Lines the esophagus, stomach, intestines, and uterus
Pseudostratified columnar epithelium	Elongated cells, one layer thick; nuclei at different levels	Often possesses beating cilia; may secrete mucus; functions in trapping and transporting particles out of respiratory surfaces; moving sex cells	Lines the respiratory tract; lines the tubes of the reproductive system (oviducts, vas deferens)

Figure 29-3 **The development of exocrine and endocrine glands from epithelial tissue**

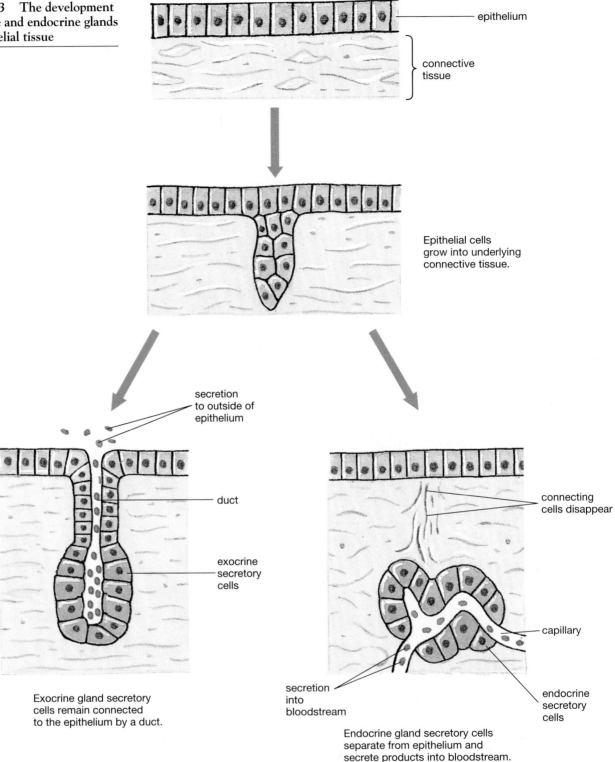

epithelium

connective tissue

Epithelial cells grow into underlying connective tissue.

secretion to outside of epithelium

duct

exocrine secretory cells

Exocrine gland secretory cells remain connected to the epithelium by a duct.

connecting cells disappear

capillary

secretion into bloodstream

endocrine secretory cells

Endocrine gland secretory cells separate from epithelium and secrete products into bloodstream.

trients to cells and carries away carbon dioxide and other cellular waste products. Blood distributes hormones within the body. White blood cells help destroy bacteria and other foreign invaders. Platelets aid in blood clotting. Lymph consists largely of fluid that has leaked out of blood capillaries and is carried back to the circulatory system within lymph vessels. Lymph contains some white blood cells and is responsible for transporting globules of fat from the small intestine to the blood. Blood and lymph are described in detail in Chapter 30.

Fat cells, collectively called **adipose tissue,** are specially modified to act as storage sacs for triglycerides, molecules used for long-term energy storage (see Chapter 3). Triglycerides may make up 90% of the volume of fat cells.

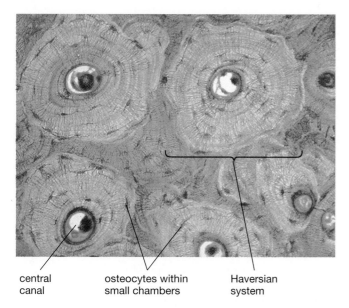

central
canal

osteocytes within
small chambers

Haversian
system

Figure 29-4   Bone is composed of Haversian systems

Haversian systems, concentric circles of bone deposited
around a central canal containing a blood vessel, are clearly
visible in this micrograph. Bone cells, or osteocytes, are
trapped within small chambers within the Haversian systems.

Muscle Tissue Has the Ability to Contract

Muscle tissue is a specialized tissue whose cells can contract (shorten) when stimulated, then relax passively. Muscle cells derive their elasticity from the arrangement of
actin (thin) and **myosin** (thick) protein filaments, which
can use energy to move relative to one another. There are
three types of muscle: skeletal (striated), cardiac, and
smooth. Muscle is discussed in more detail in Chapter 38.

Skeletal muscle is also called **striated muscle** because

its orderly arrangement of thin and thick protein filaments
give it a striped appearance (Fig. 29-5a). Skeletal muscle
is generally under voluntary, or conscious, control. As its
name implies, its main function is to move the skeleton,
as occurs when you walk or turn the pages of this text.
Cardiac muscle, found only in the heart, also has a striated appearance (Fig. 29-5b). Unlike skeletal muscle, it
is spontaneously active and not under conscious control.
Cardiac muscle cells are interconnected by gap junctions,
which allow electrical signals to spread rapidly through
the heart. **Smooth muscle** appears more uniform
("smooth") under the microscope (Fig. 29-5c) because it
lacks the orderly arrangement of thick and thin filaments
seen in cardiac and skeletal muscle. Smooth muscle is embedded in the walls of the digestive tract, the uterus, the
bladder, and the large blood vessels. It produces slow, sustained contractions that are usually involuntary.

Nerve Tissue Is Specialized to Transmit Electrical Signals

Nerve tissue makes up the brain, the spinal cord, and
the nerves that travel from them to all parts of the body.
Nerve tissue is composed of two types of cells: neurons
and glial cells. **Neurons** are specialized to generate electrical signals and to conduct these signals to other neurons, muscles, or glands. A neuron has four major parts,
each with a specialized function (Fig. 29-6). The **dendrites** receive signals from other neurons or from the external environment. The **cell body** directs maintenance
and repair of the cell. The **axon** conducts the electrical
signal to its target cell, and the **synaptic terminals** transmit the signal to the target cell at a specialized contact region called a **synapse. Glial cells** surround, support, and
protect neurons and regulate the composition of the ex-

(a)   **(b)**   **(c)**

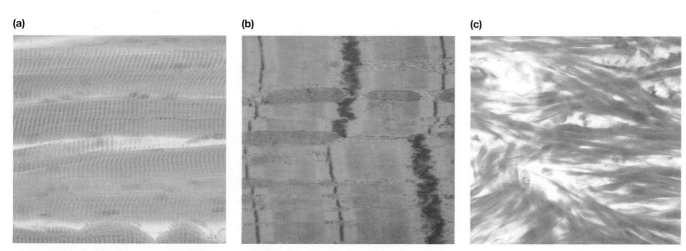

Figure 29-5   The three types of muscle tissue

(a) Skeletal muscle has a striped, or striated, appearance in microscopic cross section, due to
the regular alignment of thin and thick filaments containing actin and myosin. **(b)** Cardiac
muscle is also striated in appearance. Adjacent cardiac muscle cells are interconnected at
areas where their membranes are interlocked (dark, folded region in micrograph). Within
these regions, the adjacent membranes are packed with connecting gap junctions.
(c) Smooth muscle lacks the orderly internal structure of skeletal and cardiac muscle.

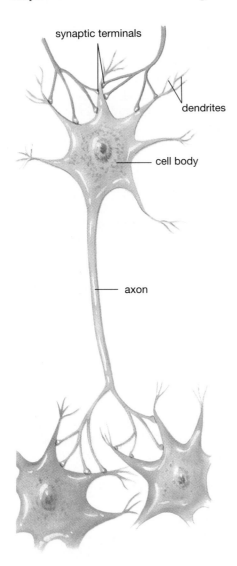

synaptic terminals

dendrites

cell body

axon

Figure 29-6   **A nerve cell**

A nerve cell, or neuron, has four major parts, each specialized for a specific function.

tracellular fluid, allowing neurons to function optimally. Neurons and the nervous system are covered in detail in Chapter 36.

Organs Include Two or More Interacting Tissue Types

Organs are formed from at least two different tissue types that function together; some organs, such as the skin, include all four of the tissue types described earlier. In this section, we examine the components and functions of a representative animal organ, the skin.

The Skin Illustrates the Properties of Organs

The **epidermis**, or outer layer of the skin, is a specialized epithelial tissue (Fig. 29-7). It is covered by a protective layer of dead cells produced by underlying living epidermal cells. These dead cells are packed with the protein **keratin**, which helps keep the skin both airtight and relatively waterproof.

Immediately beneath the epidermis lies a layer of connective tissue, the dermis. The loosely packed cells of the dermis are permeated by arterioles (small arteries), which feed blood pumped from the heart into a dense meshwork of capillaries that nourish both the dermal and epidermal tissue. Loss of heat through the skin is precisely regulated by neurons (nerve tissue) controlling the degree of dilation (expansion) of the arterioles.

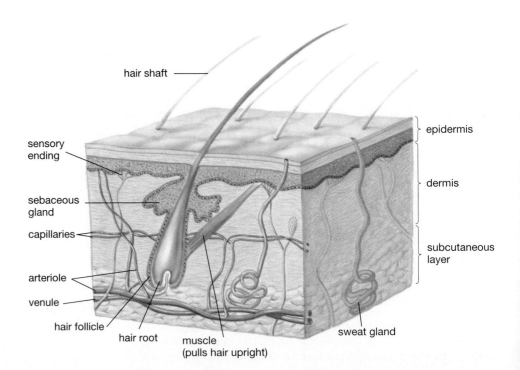

hair shaft

sensory ending

sebaceous gland

capillaries

arteriole

venule

hair follicle

hair root

muscle (pulls hair upright)

epidermis

dermis

subcutaneous layer

sweat gland

Figure 29-7   **Skin**

Mammalian skin, a representative organ, in cross section.

When cooling is required, the arterioles dilate and flood the capillary beds with blood, thus releasing excess heat; when heat conservation is required, the arterioles supplying the skin capillaries are constricted. Lymph vessels collect and carry off extracellular fluid within the dermis. A variety of sensory nerve endings responsive to temperature, touch, pressure, vibration, and pain are scattered throughout the dermis and epidermis and provide feedback to the nervous system.

The dermis is also packed with glands derived from epithelial tissue. Glands called **hair follicles** produce hair from protein-containing secretions. Sweat glands produce watery secretions that cool the skin and excrete substances, including salts and urea. **Sebaceous glands** secrete an oily substance (sebum) that lubricates the epithelium.

In addition to the epithelial, connective, and nerve tissues already mentioned, the skin also contains muscle tissue. Tiny muscles attached to the hair follicles can cause the hairs of the skin to "stand on end" in response to signals from motor neurons. Although this reaction is useless for heat retention in humans, most mammals are able to increase the thickness of their insulating fur in cold weather by erecting the individual hairs.

The structure of the skin is, in a general sense, representative of many organs. An outer epithelium is underlain by connective tissue containing a blood supply, a nerve supply, sometimes muscle, and glandular structures derived from the epithelium. If the organ is hollow, such as the bladder or blood vessels, its interior is also lined with epithelium underlain by connective tissue. Different organs have different types and proportions of glandular, muscular, and nervous tissue.

Organ Systems Consist of Two or More Interacting Organs

Organ systems consist of two or more individual organs (sometimes located in different regions of the body) that work together, performing a common function. An example is the digestive system, in which the mouth, esophagus, stomach, intestines, and other organs that supply digestive enzymes, such as the liver and pancreas, all function together to convert food into nutrient molecules (see Fig. 29-2). The major organ systems of the vertebrate body and their representative organs and functions are listed in Table 29-2. The structure and physiology of these organ systems are the subjects of the following chapters.

TABLE 29-2 ❖ *Major Vertebrate Organ Systems*

Organ System	Major Structures	Physiological Role
Circulatory system	Heart, vessels, blood	Transports nutrients, gases, hormones, metabolic wastes; also assists in temperature control
Lymphatic/immune system	Lymph, lymph nodes and vessels, white blood cells	Carries fat and excess fluids to blood; destroys invading microbes
Digestive system	Mouth, esophagus, stomach, small and large intestines, glands producing digestive secretions	Supplies the body with nutrients that provide energy and materials for growth and maintenance
Excretory system	Kidneys, ureters, bladder, urethra	Maintains homeostatic conditions within bloodstream; filters out cellular wastes, certain toxins, and excess water and nutrients
Respiratory system	Nose, trachea, lungs (mammals, birds, reptiles, amphibians), gills (fish and some amphibians)	Provides an area for gas exchange between the blood and the environment; allows oxygen acquisition and carbon dioxide elimination
Endocrine system	A variety of hormone-secreting glands including the hypothalamus, pituitary, thyroid, pancreas, and adrenals	Controls physiological processes, often in conjunction with the nervous system
Nervous system	Brain, spinal cord, peripheral nerves	Controls physiological processes in conjunction with the endocrine system; senses the environment, directs behavior
Muscular system	Skeletal muscle Smooth muscle Cardiac muscle	Moves the skeleton Controls movement of substances through hollow organs (digestive tract, large blood vessels) Initiates and implements heart contractions
Skeletal system	Bones, cartilage, tendons, ligaments	Provides support for the body, attachment sites for muscles, and protection for internal organs
Reproductive system	Male: testes, seminal vesicles, penis Female (mammal): ovaries, oviducts, uterus, vagina, mammary glands	Male: produces sperm, inseminates female Female (mammal): produces egg cells, nurtures developing offspring

❈ SUMMARY OF KEY CONCEPTS

Homeostasis

Homeostasis refers to the tendency of many physiological processes to maintain an organism's internal conditions within a narrow range that permits the continuation of life. These conditions are maintained through negative feedback, in which a change triggers a response that counteracts the change and restores conditions to a set level. Temperature control as well as many hormone systems use negative feedback to maintain homeostasis. Positive feedback, in which a change initiates events that intensify the change, occurs relatively rarely and is self-limiting. For example, the uterine contractions that lead to childbirth are driven by positive feedback.

The Organization of the Animal Body

The animal body is composed of organ systems consisting of two or more organs. Organs, in turn, are made up of tissues, which are composed of cells. A tissue is a group of cells and extracellular material that forms a structural and functional unit, specialized for a specific task. An-imal tissues include epithelial, connective, muscle, and nerve tissue.

Epithelial tissue forms membranous coverings over internal and external body surfaces and also gives rise to glands. Connective tissue usually contains considerable extracellular material and includes dermal tissue, bone, cartilage, tendons, ligaments, fat, and blood. Muscle tissue is specialized for movement, using sliding filaments of actin and myosin protein. There are three types of muscle tissue: skeletal, cardiac, and smooth. Nerve tissue is specialized for the generation and conduction of electrical signals.

Organs include at least two tissue types that function together. Mammalian skin is a representative organ. The epidermis, an epithelial tissue, covers and protects the dermis beneath it. The dermis contains blood and lymph vessels, a variety of glands, and tiny muscles that erect the hairs. Animal organ systems include the digestive, excretory, lymphatic/immune, respiratory, circulatory, nervous, muscular, skeletal, endocrine, and reproductive systems summarized in Table 29-2.

❈ KEY TERMS

actin p. 579	epithelial tissue p. 575	nerve tissue p. 579
adipose tissue p. 578	exocrine gland p. 575	neuron p. 579
axon p. 579	fat p. 578	organ p. 575
blood p. 576	gland p. 575	organ system p. 575
bone p. 576	glial cells p. 579	osteocyte p. 576
cardiac muscle p. 579	hair follicle p. 581	positive feedback p. 575
cartilage p. 576	Haversian system p. 576	sebaceous gland p. 581
cell body p. 579	homeostasis p. 573	skeletal muscle p. 579
collagen p. 576	keratin p. 580	smooth muscle p. 579
connective tissue p. 576	ligament p. 576	striated muscle p. 579
dendrite p. 579	lymph p. 576	synapse p. 579
dermis p. 576	membrane p. 575	synaptic terminal p. 579
endocrine gland p. 575	myosin p. 579	tendon p. 576
epidermis p. 580	negative feedback p. 573	tissue p. 575

❈ THINKING THROUGH THE CONCEPTS

Multiple Choice

1. The skin contains
   a. epithelial tissue   b. connective tissue
   c. nerve tissue   d. muscle tissue
   e. all of the above

2. Glands that become separated from the epithelium that produced them are called _____ glands.
   a. sebaceous   b. sweat
   c. exocrine   d. endocrine
   e. saliva

3. Epithelial membranes
   a. cover the body   b. line body cavities
   c. may create barriers that alter the movement of certain substances
   d. are continuously replaced by cell division
   e. all of the above

4. All of the following are examples of connective tissue EXCEPT
   a. tendons   b. ligaments
   c. blood   d. muscle
   e. adipose

5. Which of the following statements about muscle is true?
   a. Smooth muscle has a striated appearance.
   b. Skeletal muscle is not under conscious control.
   c. Cardiac muscle utilizes gap junctions.
   d. Smooth muscle is called voluntary muscle.
   e. Smooth muscle moves the skeleton.
6. All of the following are found in the dermis EXCEPT
   a. arteries
   b. sensory neuron endings
   c. hair follicles
   d. sebaceous glands
   e. cells packed with keratin

Review Questions

1. Define homeostasis and explain how negative feedback helps maintain it. Explain one example of homeostasis in the human body.
2. Explain positive feedback and provide one physiological example. Explain why this type of feedback is relatively rare in physiological processes.
3. Explain why body temperature cannot be maintained at *exactly* 37° C (98.6° F) at all times.
4. Describe the structure and functions of epithelial tissue.
5. What property distinguishes connective tissue from all other tissue types? List five types of connective tissue, and briefly describe the function of each type.
6. Describe the skin, a representative organ. Include the various tissues that compose it and the role of each tissue.

APPLYING THE CONCEPTS

1. Why does life on land present more difficulties in maintaining homeostasis than life in water? What made it evolutionarily advantageous for organisms to colonize dry land?
2. The majority of homeostatic regulatory mechanisms in animals are "autonomic," not requiring conscious control. Discuss several reasons why this is more advantageous to the animal than conscious regulation of homeostatic controls.
3. Third-degree burns are usually painless. Skin regenerates only from the edges of these wounds. Second-degree burns regenerate from cells located at the burn edges, in hair follicles and in sweat glands. First-degree burns are painful but heal rapidly from undamaged epidermal cells. From this information, draw the depth of first-, second-, and third-degree burns on Figure 29-7.
4. A coroner dictates the following description during an autopsy: "The tissue I am looking at forms part of the fetal skeleton. The matrix appears transparent. Fibers of collagen are present but are small and evenly dispersed in the matrix. Chondrocytes appear in tiny spaces, lacunae, within the matrix. Blood vessels have not yet penetrated the matrix." What tissue is the coroner describing?
5. The pancreas is both an exocrine and an endocrine gland. During embryonic development it appears as an outgrowth of the rudimentary digestive tube. Sketch the stages you might see as this compound organ develops.
6. Imagine you are a health-care professional teaching a prenatal class for fathers. Design a real-world analogy with sensors, electrical currents, motors, etc. to illustrate feedback relationships involved in the initiation of labor that could be understood by a layperson.

FOR MORE INFORMATION

Degabriele, R. "The Physiology of the Koala." *Scientific American*, June 1980 (Offprint No. 1476). Koalas eat poisonous eucalyptus leaves and almost never drink water. How can they survive?

Schmidt-Nielson, K. "The Physiology of the Camel." *Scientific American*, December 1959. Do camels really store water in their humps? How long can one go without drinking?

Schmidt-Nielson, K., and Schmidt-Nielson, B. "The Desert Rat." *Scientific American*, July 1953 (Offprint No. 1050). The ultimate desert dweller, the kangaroo rat never needs to drink water.

Storey, K. B., and Storey, J. M. "Frozen and Alive." *Scientific American*, December 1990. Some animals have special adaptations that allow them to withstand freezing

Wills, Christopher. "The Skin We're In." *Discover*, November 1994. A scientist looks at the controversial issue of melanism.

NET WATCH

On-line resources for this chapter are on the World Wide Web at:
http://www.prenhall.com/~audesirk (click on the table of contents link and then select Chapter 29).

30 Circulation

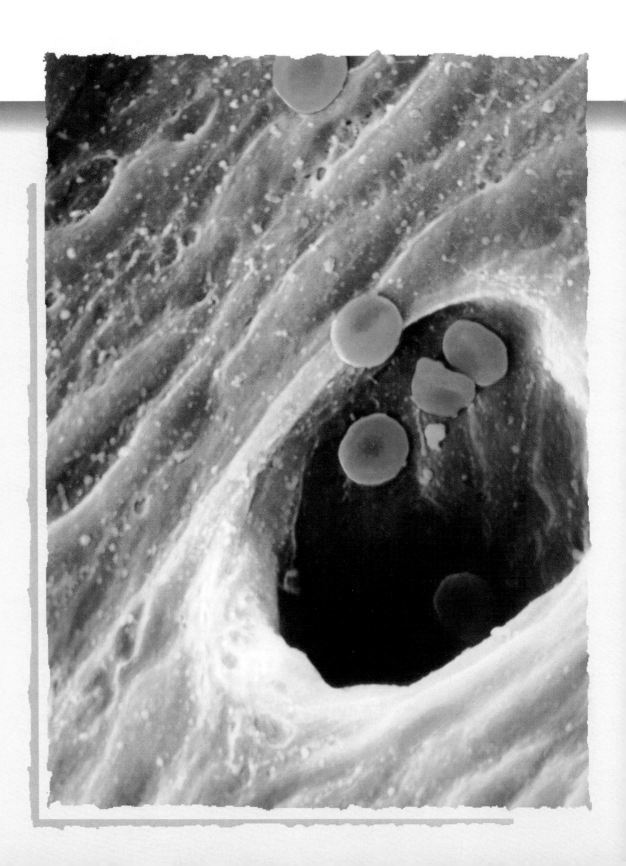

AT A GLANCE

Types of Circulatory Systems

The Vertebrate Circulatory System
  The Vertebrate Heart Consists of Muscular Chambers
    Whose Contraction Is Controlled by Electrical Impulses
  Blood Transports Dissolved Nutrients, Wastes, and Hormones
    throughout the Body
  The Blood Vessels Carry Blood to All Parts of the Body

The Lymphatic System
  Lymphatic Vessels Resemble the Veins and Capillaries of
    the Circulatory System
  The Lymphatic System Returns Fluids to the Blood
  The Lymphatic System Transports Fats from the Intestine to
    the Blood
  The Lymphatic System Helps Defend the Body against
    Disease

Red blood cells flow into a capillary from a larger vessel in this scanning electron micrograph. In its four-month life span, a red blood cell will travel over 900 miles through the circulatory system, ferrying oxygen to cells throughout the body.

Billions of years ago, the first living cells were nurtured by the sea in which they evolved. The sea brought them nutrients, which diffused into the cell, and washed away the wastes, which had diffused out. Today microorganisms and some simple multicellular animals still rely almost exclusively on diffusion for exchange of nutrients and wastes with the environment. Sponges, for example, circulate seawater through pores in their bodies, bringing the environment close to each cell (see Fig. 32-4a). As larger, more complex animals evolved, individual cells became increasingly distant from the outside world. The constant exchange demands of a living cell require, however, that diffusion distances be kept short. For a cell to avoid starving and stewing in its own wastes, a source of nutrients and a dumping place for wastes had to be brought within diffusing distance. With the evolution of the circulatory system, a sort of "internal sea" was created, serving the same purpose as the sea did for the first cells. This internal sea transports a fluid (blood) rich in food and oxygen close to each cell and carries away wastes produced by the cells. In this chapter we look at the different parts and functions of the vertebrate circulatory system, focusing on the human as a representative vertebrate.

Types of Circulatory Systems

All circulatory systems have three major parts:

1. A fluid, **blood**, that serves as a medium of transport
2. A system of channels, or **vessels**, that conduct the blood throughout the body
3. A pump, the **heart**, that keeps the blood circulating

Two major types of circulatory systems are found in animals: open and closed. **Open circulatory systems** include an open space within the body, the **hemocoel**, into which vessels empty and from which they pick up blood (Fig. 30-1a). Within this space, tissues are directly bathed in blood. Open circulatory systems are found in arthropods (insects, spiders, and crustaceans) and most mollusks (snails, clams).

In **closed circulatory systems**, blood is confined to the heart and a continuous series of vessels (Fig. 30-1b), allowing more rapid blood flow and more efficient transport than is possible in an open system. Closed systems are found in some invertebrates, such as earthworms and cephalopod mollusks (squids, octopuses), and in all vertebrates.

The Vertebrate Circulatory System

The circulatory system has many diverse roles and reaches its greatest development in the vertebrates. Among the most important functions of the vertebrate circulatory system are the following:

1. Transport of oxygen from the lungs to the tissues and of carbon dioxide from the tissues to the lungs
2. Distribution of nutrients from the digestive system to all body cells

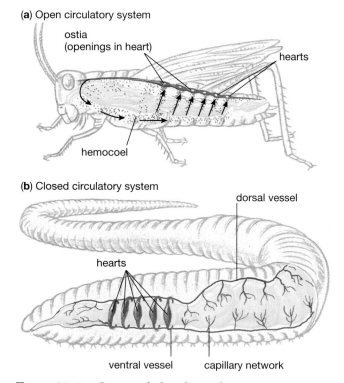

(a) Open circulatory system

(b) Closed circulatory system

Figure 30-1   **Open and closed circulatory systems**

(a) In the open circulatory system of insects and other arthropods, a series of hearts pumps blood through vessels into the hemocoel, where blood directly bathes the organs. When the hearts relax, blood is sucked back into the vessels through openings called ostia, guarded by one-way valves. When the hearts contract, the valves are pressed shut, forcing the blood to travel out through the vessels returning to the hemocoel. **(b)** In a closed circulatory system, blood remains confined to the heart and the blood vessels. In the earthworm, five contractile vessels serve as hearts and pump blood through major ventral and dorsal vessels from which smaller vessels branch off.

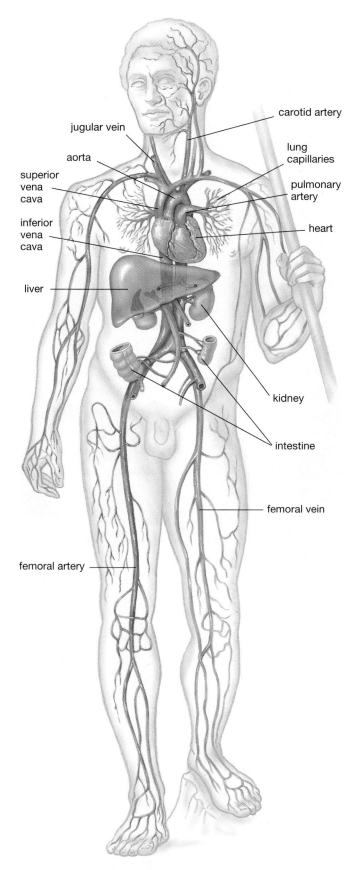

Figure 30-2   **The human circulatory system**

Oxygenated blood is shown in red, deoxygenated blood in blue.

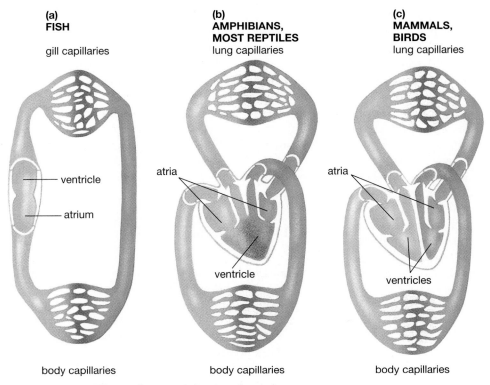

(a)
FISH

gill capillaries

ventricle

atrium

body capillaries

(b)
AMPHIBIANS,
MOST REPTILES
lung capillaries

atria

ventricle

body capillaries

(c)
MAMMALS,
BIRDS
lung capillaries

atria

ventricles

body capillaries

Figure 30-3   **The evolution of the vertebrate heart**

(a) The earliest vertebrate heart is represented by the two-chambered heart of fishes. Blood from the body tissues is collected in the atrium and transferred to the single ventricle. Contraction of the ventricle sends blood through the gill capillaries, where it picks up oxygen and gives off carbon dioxide, and then to the body capillaries, where it delivers oxygen to the tissues and picks up carbon dioxide. **(b)** In amphibians and most reptiles, the heart has two atria, the left receiving oxygenated blood from the lungs, the right receiving deoxygenated blood from body tissues. Although both empty into a single ventricle, the deoxygenated blood tends to remain on the right, where it is directed to the lungs, while most of the oxygenated blood stays on the left and is sent to the body tissues. In reptiles, there is often a partial wall down the middle of the ventricle, enhancing this separation. Hearts are drawn as if they were in a body facing you, so that right and left appear reversed. **(c)** The hearts of birds and mammals are actually two separate pumps, so any possible mixing of oxygenated and deoxygenated blood is prevented.

3. Transport of waste products and toxic substances to the liver, where many of them are detoxified, and to the kidney for excretion
4. Distribution of hormones from the organs that produce them to the tissues on which they act
5. Regulation of body temperature, which is achieved partly by adjustments in blood flow
6. Prevention of blood loss by means of the clotting mechanism
7. Protection of the body from bacteria and viruses by circulating antibodies and white blood cells

In the following sections we examine the three parts of the circulatory system: the heart, the blood, and the vessels, using the human as a representative vertebrate. The human circulatory system is shown in Figure 30-2. Finally, we describe the lymphatic system, which works closely with the circulatory system.

The Vertebrate Heart Consists of Muscular Chambers Whose Contraction Is Controlled by Electrical Impulses

Atria Collect Blood, and Ventricles Pump It throughout the Body

The vertebrate heart consists of muscular chambers capable of strong contractions that circulate blood through the body. During the course of vertebrate evolution, the heart has become increasingly complex, starting with the two-chambered heart of fish and culminating in the four-chambered hearts of birds and mammals (Fig. 30-3). The warm-blooded birds and mammals have high metabolic demands and require efficient delivery of oxygen to their tissues. The complete separation of oxygenated and deoxygenated blood, which is made possible by the four-chambered heart, assures that blood that reaches the tissues has the highest possible oxygen content.

Mammalian and bird hearts consist of two separate pumps, each with two chambers. In each pump, an **atrium** receives and briefly stores the blood, passing it to a **ventricle** that propels it through the body (Figs. 30-4 and 30-5). One pump is for **pulmonary circulation** and consists of the right atrium and ventricle. Oxygen-depleted blood from the body empties into the right atrium through a large vein, the superior vena cava. The right atrium contracts, transferring the blood to the right ventricle. Contraction of the right ventricle sends the oxygen-depleted blood to the lungs via pulmonary arteries. The other pump, consisting of the left atrium and ventricle, powers **systemic circulation**. Newly oxygenated blood from the lungs enters the left atrium through pulmonary veins and is passed to the left ventricle. Strong contractions of the left ventricle, the heart's most muscular chamber, send the oxygenated blood coursing out through the aorta to the rest of the body.

The Coordinated Contractions of Atria and Ventricles Produce the Cardiac Cycle

The alternating contraction and relaxation of the heart chambers is called the **cardiac cycle**. The two atria contract in synchrony, emptying their contents into the ventricles. A fraction of a second later, the two ventricles contract simultaneously, forcing blood into arteries leaving the heart. Both chambers then relax briefly before

the cycle is repeated. The period of ventricular contraction is called **systole**. The rest of the cycle, including relaxation of all the chambers followed by contraction of the atria, is called **diastole**. At normal resting heart rate, systole lasts about 0.3 second and diastole about 0.5 second. Systole and the last phase of diastole are illustrated in Figure 30-5. When blood pressure is measured (Fig. 30-6), the higher of the two readings is the systolic pressure, which is measured during ventricular contraction, and the lower reading is the diastolic pressure, measured between contractions.

Valves Maintain Directionality of Blood Flow, While Electrical Impulses Coordinate the Sequence of Contractions

Coordinating the activity of the four chambers to maintain proper blood flow through the heart and its vessels presents some challenges. First, when the ventricles contract, the blood must be directed out through the **arteries**, the large vessels that carry blood away from the heart, and not back up into the atria. Then, once blood has entered the arteries, it must be prevented from flowing back as the heart relaxes. These problems are solved by four simple one-way valves (see Fig. 30-4). Pressure in one direction opens them readily, while reverse pressure forces them tightly closed. **Atrioventricular valves** separate the atria from the ventricles; a **tricuspid** (from

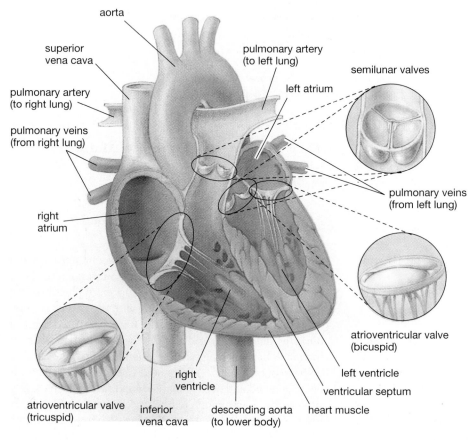

aorta
superior
vena cava
pulmonary artery
(to right lung)
pulmonary veins
(from right lung)
right
atrium
atrioventricular valve
(tricuspid)
inferior
vena cava
right
ventricle
descending aorta
(to lower body)
heart muscle
ventricular septum
left ventricle
atrioventricular valve
(bicuspid)
pulmonary veins
(from left lung)
semilunar valves
left atrium
pulmonary artery
(to left lung)

Figure 30-4   The human heart and its valves and vessels

The right atrium receives deoxygenated blood and passes it to the right ventricle, which pumps it to the lungs. Blood returning from the lungs enters the left atrium, which passes it to the left ventricle, which pumps oxygenated blood through the rest of the body. Note the thickened walls of the left ventricle, which must pump blood over a considerably longer distance. One-way valves are located between the aorta and the left ventricle, between the pulmonary artery and the right ventricle, and between the atria and ventricles.

(a) SYSTOLE

(b) DIASTOLE

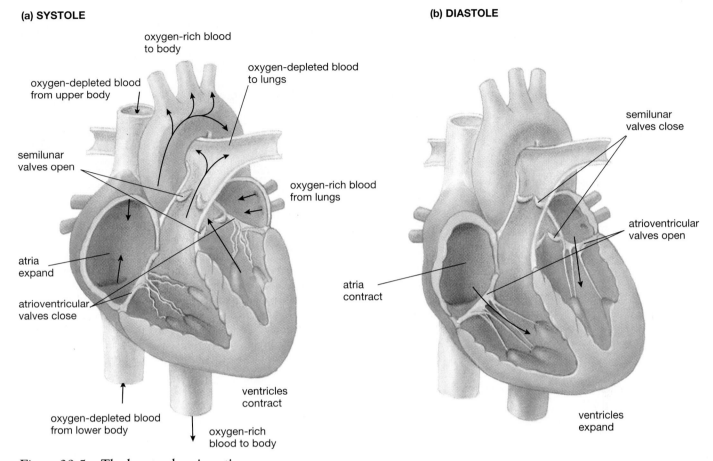

oxygen-rich blood
to body

oxygen-depleted blood
to lungs

oxygen-depleted blood
from upper body

semilunar
valves open

oxygen-rich blood
from lungs

atria
expand

atrioventricular
valves close

oxygen-depleted blood
from lower body

ventricles
contract

oxygen-rich
blood to body

semilunar
valves close

atrioventricular
valves open

atria
contract

ventricles
expand

Figure 30-5   **The heart valves in action**

(a) During ventricular contraction (systole), the pressure within the ventricles forces the semilunar valves open, allowing blood to flow into the aorta and pulmonary arteries. The atrioventricular valves are simultaneously pressed shut, preventing blood flow back into the atria. **(b)** As the ventricles reexpand (diastole), they would tend to draw blood back from the arteries, but this backward pressure forces the semilunar valves closed. Simultaneously, contraction of the atria forces open the atrioventricular valves, allowing blood to flow from the atria into the ventricles (arrows indicate the direction of blood flow).

Latin, meaning "three-pointed") **valve** separates the right ventricle and right atrium, and a **bicuspid** ("two-pointed") **valve** lies between the left atrium and the left ventricle. Two **semilunar** (Latin, "half-moon") **valves** allow blood to enter the pulmonary artery and the aorta when the ventricles contract but prevent it from returning as the ventricles relax.

A second challenge is to create smooth, coordinated contractions of the muscle cells that make up each chamber. When they are excited, muscle cells produce electrical signals that cause the muscle cells to contract. In the heart, the individual heart muscle cells communicate directly with one another through gap junctions in their adjacent membranes (Fig. 30-7). These connecting pores allow electrical signals that cause contraction to pass freely and rapidly between heart cells. The contraction of the heart is initiated and coordinated by a **pacemaker**, a cluster of specialized muscle cells that produce spontaneous electrical signals at

a regular rate. Although the nervous system can alter the rate of these signals, they are initiated by the pacemaker muscle cells themselves.   The heart's primary pacemaker is the **sinoatrial (SA) node**, located in the wall of the right atrium (Fig. 30-8). Signals from the SA node spread rapidly through both the right and left atria, causing the atria to contract in smooth synchrony.

A final challenge is to coordinate contractions of all four chambers. The atria must contract first and empty their contents into the ventricles in order to be able to refill while the ventricles contract. Thus, there must be a delay between the contractions of the atria and those of the ventricles. From the SA node, an electrical impulse creates a wave of contraction that sweeps through the muscles of the atria until it reaches a barrier of unexcitable tissue separating the atria from the ventricles. Here the excitation is channeled through a second small mass of specialized muscle cells, the **atrioventricular (AV) node**,

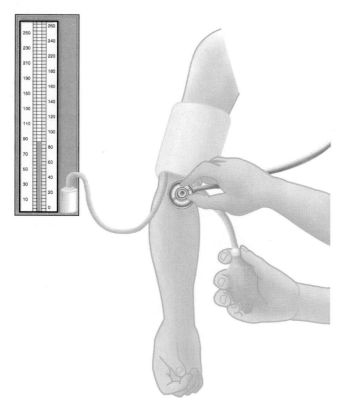

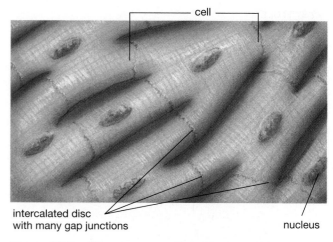

intercalated disc
with many gap junctions

nucleus

Figure 30-7   **The structure of cardiac muscle**

Cardiac muscle cells are branched. Adjacent cell membranes meet in folded areas called intercalated discs. These regions are densely packed with gap junctions (pores) that connect the interiors of adjacent cells. This arrangement allows direct transmission of electrical signals between the cells, coordinating their contractions.

Figure 30-6   **Measuring blood pressure**

Blood pressure is measured using an inflatable blood pressure cuff and a stethoscope. The cuff is placed around the upper arm and the stethoscope positioned over the artery just below the cuff. The cuff is inflated until its pressure closes off the main artery in the arm and blood ceases to flow. No pulse, therefore, can be detected below the cuff. Then the pressure is gradually reduced. When the pulse is first audible in the artery, this means that the pressure pulses created by the contracting left ventricle are just overcoming the pressure in the cuff and blood is flowing. This is the upper of the two readings: the systolic pressure. Cuff pressure is then further reduced until no pulse is audible. An inaudible pulse at this point indicates that the artery is no longer constricted and blood is flowing continuously through the artery; the pressure between ventricular contractions is just overcoming the cuff pressure. This is the lower of the two readings: the diastolic pressure. A blood pressure reading of 130 (systolic) over 80 (diastolic) is within the normal range. The numbers are in millimeters of mercury, a standard measure of pressure also used in barometers.

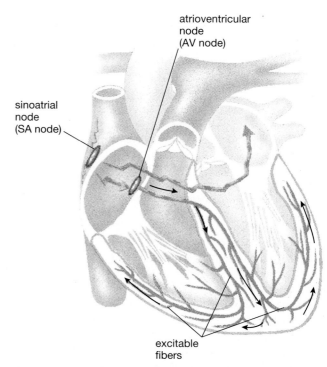

Figure 30-8   **The pacemaker of the heart**

A spontaneously active mass of modified muscle fibers in the right atrium called the sinoatrial (SA) node serves as the heart's pacemaker. The signal to contract spreads from the SA node through the muscle fibers of both atria, finally exciting the atrioventricular (AV) node in the lower right atrium. The AV node then transmits the signal to contract through bundles of excitable fibers that stimulate the ventricular muscle. The delay in transmission of the signals from the SA node through the AV node allows the atria to empty their contents completely into the ventricles before ventricular contraction begins.

located in the floor of the right atrium (see Fig. 30-8). The impulse is delayed at the AV node, postponing the ventricular contraction for about 0.1 second after contraction of the atria. This delay gives the atria time to complete the transfer of blood into the ventricles before ventricular contraction begins. From the AV node, the signal to contract spreads to the base of the two ventricles along tracts of excitable fibers. The impulse then travels rapidly from these fibers through the communicating muscle fibers, causing the ventricles to contract in unison. When the pacemaker fails to coordinate muscle contractions, uncoordinated, irregular contractions impair the heart's function, a condition known as **fibrillation**. Fibrillation of the ventricles is soon fatal because blood is not pumped out of the heart to the brain and other organs but is merely sloshed around.

The Nervous System and Hormones Influence Heart Rate

Left on its own, the SA node pacemaker would maintain a steady rhythm of about 100 beats per minute. However, the heart rate is significantly altered by the influence of nervous impulses and hormones. In the resting person, activity of the parasympathetic nervous system (which controls body functions during periods of rest; see Chapter 36) slows the heart rate to around 70 beats per minute. When exercise or stress creates a demand for greater blood flow to the muscles, the parasympathetic influence is reduced and the sympathetic nervous system (which prepares the body for emergency action) accelerates the heart rate. Likewise, the hormone epinephrine (also known as adrenaline), increases heart rate as it mobilizes the entire body for response to threatening or unfamiliar events. When astronauts were landing on the moon, their heart rates were over 170 beats per minute, even though they were sitting still!

Blood Transports Dissolved Nutrients, Wastes, and Hormones throughout the Body

Blood is the medium in which dissolved nutrients, gases, hormones, and wastes are transported through the body. It has two major components: (1) a fluid called **plasma** and (2) specialized cells (red blood cells, white blood cells, and platelets) that are suspended in the plasma (Table 30-1). On the average, the cellular components of blood account for 40% to 45% of its volume; the other 55% to 60% is plasma. The average person has 5 to 6 liters of blood, constituting about 8% of his or her total body weight.

Plasma Is Primarily Water in Which Proteins, Salts, Nutrients, and Wastes Are Dissolved

About 90% of the straw-colored plasma is water in which a number of substances are dissolved. Those substances include proteins, hormones, nutrients (glucose, vitamins, amino acids, lipids), gases (carbon dioxide, oxygen), salts

---

TABLE 30-1 ⠶ Blood Cells

Cell Type		Description	Average Number Present	Major Function
Red blood cell (erythrocyte)		Biconcave disc without nucleus, about one-third hemoglobin Approximately 8 μm[b] in diameter	5,000,000 per mm^{3a}	Transports oxygen and a small amount of carbon dioxide
White blood cells (leukocytes)			7500 per mm^3	
1. Neutrophil		About twice the size of red cells, nucleus with two to five lobes	62% of white cells	Destroys relatively small particles by phagocytosis
2. Eosinophil		About twice the size of red cells, nucleus with two lobes	2% of white cells	Inactivates inflammation-producing substances; attacks parasites
3. Basophil		About twice the size of red cells, nucleus with two lobes	Less than 1% of white cells	Releases anticoagulant to prevent blood clots and histamine, which causes inflammation
4. Monocyte		Two to three times larger than red cells, nuclear shape varies from round to lobed	3% of white cells	Gives rise to macrophage, which destroys relatively large particles by phagocytosis
5. Lymphocyte		Only slightly larger than red cell, nucleus nearly fills cell	32% of white cells	Functions in the immune response
Platelet		Cytoplasmic fragment of cells in bone marrow called megakaryocytes	250,000 per mm^3	Important in blood clotting

[a]mm^3 = cubic millimeter.
[b]μm = micrometer.

(sodium, calcium, potassium, magnesium), and wastes such as urea. Plasma proteins are the most abundant of the dissolved substances. The three major plasma proteins are (1) albumins, which help maintain the blood's osmotic pressure (which controls the flow of water across cell membranes); (2) globulins, which transport nutrients and also play a role in the immune system; and (3) fibrinogen, important in blood clotting, which is discussed later in this chapter.

Red Blood Cells Carry Oxygen from the Lungs to the Tissues

The oxygen-carrying red blood cells, also called **erythrocytes**, make up about 99% of the cells in the blood. They constitute about 40% of the total blood volume in females, and 45% in males. Each milliliter of blood contains about 5 billion erythrocytes. Proteins on the surface of red blood cells differ among individuals, thereby creating different blood types, as described in Chapter 13. The red blood cell resembles a ball of clay squeezed between thumb and forefinger (Fig. 30-9). Its biconcave (dented inward on both sides) shape provides a larger surface area than would a spherical cell of the same volume and increases the cell's ability to absorb and release oxygen through its membrane.

The red color of erythrocytes is caused by the pigment **hemoglobin** (Fig. 30-10). This large, iron-containing protein accounts for about one-third the weight of the blood cell. About 97% of the oxygen carried by the blood is bound to hemoglobin. The hemoglobin binds loosely to oxygen, picking up oxygen in the capillaries of the lungs, where the concentration is high, and releasing it where the concentration is low, in other tissues of the body. After releasing its oxygen, some of the hemoglobin picks up carbon dioxide from the tissues for transport back to the

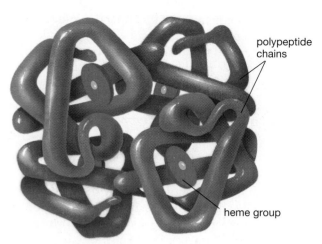

Figure 30-10   Hemoglobin

A molecule of hemoglobin is composed of four polypeptide chains (two pairs of similar chains), each surrounding a heme group. The heme group contains an iron atom and is the site of oxygen binding. When saturated, each hemoglobin molecule can carry four oxygen molecules (eight oxygen atoms).

lungs. The role of blood in gas exchange is discussed in more detail in Chapter 31.

Red Blood Cells Have a Relatively Short Life Span

Red blood cells are formed in the marrow, the soft interior portion of certain bones, including those of the chest, upper arms, upper legs, and hips. During their development, mammalian red blood cells lose their nuclei and their ability to divide. Without the ability to synthesize cellular materials, their lives are necessarily short; each cell lives about 120 days. Every second, over 2 million red blood cells die and are replaced by new ones from the bone marrow. Dead or damaged red blood cells are removed from circulation, primarily in the liver and spleen, and broken down to release their iron. The salvaged iron is carried in the blood to the bone marrow, where it is used to make more hemoglobin and packaged into new red blood cells. Although the recycling process is efficient, small amounts of iron are excreted daily and must be replenished by the diet. Bleeding from injury or menstruation also tends to deplete iron stores.

Negative Feedback Regulates Red Blood Cell Numbers

The number of red blood cells in the blood is maintained at an adequate level through a negative feedback system that involves a hormone called **erythropoietin**. Erythropoietin is produced by the kidneys in response to oxygen deficiency. This lack of oxygen may be caused by a loss of blood, insufficient production of hemoglobin, high altitude (where less oxygen is available), or lung disease that interferes with gas exchange in the lungs. The hormone stimulates rapid production of new red blood cells by the bone marrow. When adequate oxygen levels are restored,

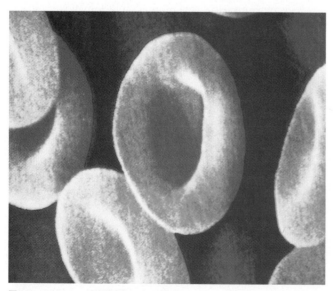

Figure 30-9   Red blood cells

Under the scanning electron microscope, the biconcave disc shape of red blood cells is clearly visible. These cells have been artificially colored.

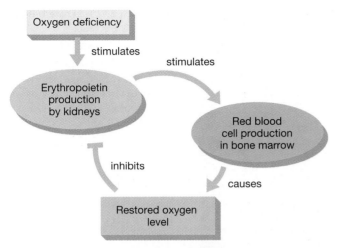

Figure 30-11   **Red blood cell regulation**

The production of red blood cells is regulated by a negative feedback system.

erythropoietin production is inhibited, and the rate of red blood cell production returns to normal. This negative feedback system is diagrammed in Figure 30-11.

Blood Type Is Determined by Specific Proteins on Red Blood Cell Membranes

Blood is classified as type A, B, AB, or O depending on the presence or absence of specific proteins (designated A and B) on the cell membranes of red blood cells. Type A blood has the A protein on its red blood cells, type B has the B protein, type AB has both, and type O has neither. Also, each blood type carries antibodies in the plasma to the proteins *not* present on its own red blood cells. Thus, a person with type A blood has antibodies to the B protein. If this person is transfused with type B blood, the antibodies to the B protein attack the transfused red blood cells, causing them to clump together and block small blood vessels, sometimes with fatal results. Blood type is inherited; the genetics and properties of these major blood types are discussed in detail in Chapter 13.

Another type of protein on red blood cells is the **Rh factor**. If it is present, blood is described as Rh-positive, and if the protein is absent, the blood is Rh-negative. Unlike A or B antibodies, antibodies to the Rh protein form only after massive exposure to the protein. The first exposure—for example, transfusion of Rh-positive blood into an Rh-negative individual—generally causes no ill effects but triggers the production of antibodies. Upon further exposure, the antibodies attack and destroy the Rh-positive red blood cells. If an Rh-negative woman marries an Rh-positive man, her children are likely to be Rh-positive, because Rh-positive blood is a dominant genetic trait. Her first Rh-positive child will trigger antibody production in her blood. Subsequent Rh-positive children will be in danger of being born with **erythroblastosis fetalis**, in which the mother's antibodies invade the fetus and attack

its red blood cells, causing the child to be born severely anemic. Fortunately this condition can now be easily prevented by injections of a substance that prevents formation of Rh antibodies by the pregnant woman.

White Blood Cells Help Defend the Body against Disease

There are five common types of white blood cells, or **leukocytes**, which together make up less than 1% of the total cellular component of the blood. These cells, described and illustrated in Table 30-1, are distinguished from one another by their staining characteristics, size, and the shape of their nuclei. All of these are derived from cells that originate in bone marrow. Most white blood cells function in some way to protect the body against foreign invaders and use the circulatory system to travel to the site of invasion. For example, **monocytes** and **neutrophils** travel through capillaries to wounds where bacteria have gained entry, then ooze out through narrow openings in the capillary walls. After leaving the capillaries, monocytes differentiate into **macrophages** ("big-eaters"), amoebalike cells that engulf foreign particles. Macrophages and neutrophils feed on bacterial invaders or other "foreign" cells including cancer cells (Fig. 30-12). They often die in the process, and their dead bodies accumulate and contribute to the white substance called pus, seen at infection sites.

Lymphocytes, described in detail in Chapter 34, are white blood cells responsible for the production of antibodies that help provide immunity against disease. Cells that give rise to lymphocytes migrate from bone marrow to tissues of the lymphatic system such as the thymus, spleen, and lymph nodes, described later in this chapter. Least abundant are the **eosinophils** and **basophils**. Eosinophil production is stimulated by parasitic infections. Eosinophils converge on the parasitic invaders, releasing substances that kill the parasite. Basophils release substances that inhibit blood clotting as well as chemicals, such as histamine, that participate in allergic reactions and responses to tissue damage and microbial invasion.

Platelets Are Cell Fragments That Aid in Blood Clotting

Platelets are not complete cells, but fragments of large cells called **megakaryocytes**. Megakaryocytes remain in the bone marrow, pinching off membrane-enclosed pieces of their cytoplasm, called platelets (Fig. 30-13). Platelets then enter the blood and play a central role in blood clotting. Like red blood cells, platelets lack a nucleus, and their life span is even shorter; about 10 to 12 days.

Clot formation is a complex process. It starts when platelets and other factors in the plasma contact an irregular surface, such as a damaged blood vessel. Platelets tend to stick to irregular surfaces, and they may build up and

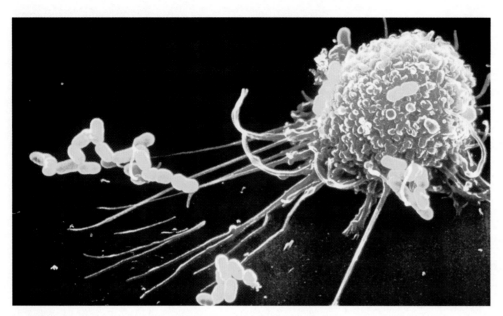

Figure 30-12   **A white blood cell attacks bacteria**

An amoebalike white blood cell is seen capturing bacteria. These bacteria are *Escherichia coli*, intestinal bacteria that can cause disease if they enter the bloodstream.

plug the damaged vessel if it is narrow enough. This mechanism is supplemented by blood coagulation, or clotting, which is the most important of the body's defenses against bleeding. The ruptured surface of an injured blood vessel not only causes platelets to adhere, but it also initiates a complex sequence of events among circulating plasma proteins. These events culminate in production of the enzyme **thrombin**.

Thrombin catalyzes the conversion of the plasma protein fibrinogen into stringlike molecules called **fibrin**. Fibrin molecules adhere to one another, end to end and side to side, forming a fibrous matrix. This protein web immobilizes the fluid portion of the blood, causing it to solidify in much the same way that gelatin does as it cools. The web traps red blood cells, further increasing the density of the clot (Fig. 30-14). Platelets then adhere to the fibrous mass and send out sticky projections that attach to one another. Within half an hour, the platelets contract, pulling the mesh tighter and forcing liquid out. This action creates a denser, stronger clot (on the skin it is called a scab) and also constricts the wound, pulling the damaged surfaces closer together in a way that promotes healing.

The Blood Vessels Carry Blood to All Parts of the Body

As blood leaves the heart, it travels from arteries to arterioles to capillaries to venules to veins, which return it finally to the heart. In this section we examine each type of blood vessel in more detail.

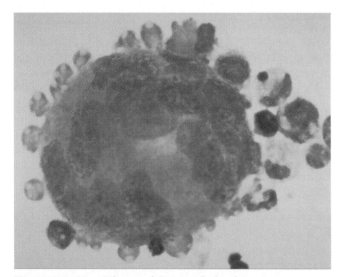

Figure 30-13   **The production of platelets**

Megakaryocytes are large cells in bone marrow that constantly manufacture proteins, enzymes, and cell membrane. But instead of growing larger, megakaryocytes bud off membrane-enclosed pieces of cytoplasm, called platelets, that enter the blood and help initiate and control blood clotting. In this photomicrograph, a single megakaryocyte is seen producing dozens of platelets.

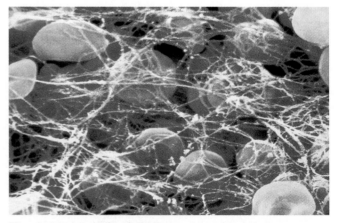

Figure 30-14   **Blood clotting**

In response to damage to a blood vessel, threadlike proteins called fibrin form in the blood. These produce a tangled sticky mass that traps red blood cells and eventually forms a clot.

Arteries and Arterioles Are Thick-Walled Vessels That Carry Blood Traveling Away from the Heart

After leaving the heart, blood first enters large vessels called arteries. These have thick walls containing smooth muscle and elastic tissue (Fig. 30-15). With each surge of blood from the ventricles, the arteries expand slightly, like thick-walled balloons. Between heartbeats they recoil, helping to pump the blood and maintain a steady flow through the smaller vessels. Arteries branch into vessels of even smaller diameter called **arterioles**, which play a major role in determining how blood is distributed within the body, as described later.

Capillaries Are Microscopic Vessels That Allow Exchange of Nutrients and Wastes between the Blood and Body Cells

The entire circulatory system is an elaborate device for providing each cell of a complex, multicellular organism with the type of exchange—diffusion—practiced by the simplest unicellular organisms. This exchange is accomplished at the level of the **capillaries**, the tiniest of all vessels. Here wastes, nutrients, gases, and hormones are exchanged between the blood and the body cells. Capillaries are finely adapted to their role of exchange. Their walls are only a single cell thick. Most nutrients, oxygen, and carbon dioxide diffuse readily through capillary cell membranes. Salts and small charged molecules (including some small proteins) move through fluid-filled spaces within the capillary cell membrane or between adjacent capillary cells (Fig. 30-16).

The pressure within capillaries causes a continuous leakage of fluid from the blood plasma into the spaces surrounding the capillaries and tissues. This fluid is known as the **interstitial fluid** and consists primarily of water in which are dissolved nutrients, hormones, gases, wastes, and small proteins from the blood. The large plasma proteins, red blood cells, and platelets are unable to leave the capillaries because of their size, although white blood cells can ooze through the openings between capillary cells. The exchange of materials between capillary blood and nearby cells occurs through this interstitial fluid, which bathes nearly all the cells of the body.

Capillaries are so narrow that red blood cells must pass through them in single file (Fig. 30-17). Consequently, all the blood is sure to pass very close to the capillary walls, where exchange occurs. In addition, capillaries are so nu-

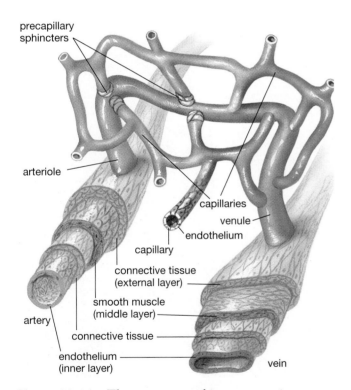

Figure 30-15   The structure and interconnections of the types of blood vessels

Arteries and arterioles are more muscular and maintain a greater muscular tension than veins and venules, whose walls are thinner and can stretch more. Capillaries have walls only a single cell thick, allowing movement of dissolved substances and white blood cells across the capillary wall. Oxygenated blood moves from arteries to arterioles to capillaries. Capillaries empty deoxygenated blood into venules, which empty into veins. The movement of blood from arterioles into capillaries is regulated by muscular rings called precapillary sphincters.

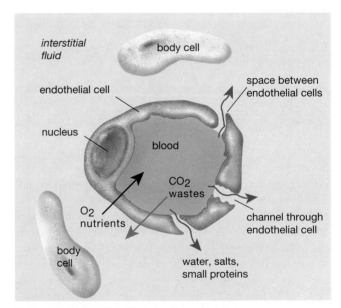

Figure 30-16   Capillary wall structure

The structure of capillary walls facilitates the exchange of dissolved gases, wastes, and nutrients. Capillary walls are only one endothelial cell thick. Narrow spaces separate adjacent endothelial cells, and channels penetrate through the cell membranes. Water containing salts and small proteins is forced out through the holes. Gases and most wastes and nutrients diffuse through the endothelial cell membranes along their concentration gradients.

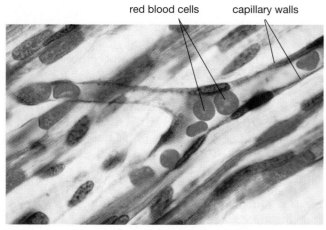

red blood cells   capillary walls

Figure 30-17   A capillary containing red blood cells

Capillaries are so narrow that red blood cells must pass through them single file, thereby facilitating exchange of gases by diffusion.

merous that no body cell is more than 100 micrometers (0.004 inch; about as thick as four pages of this book) from a capillary. These factors facilitate the exchange of materials by diffusion. It is estimated that the total length of capillaries in a human is over 80,600 kilometers (50,000 miles), enough to encircle the globe twice! The speed of blood flow drops very quickly as blood is forced through this narrow, almost endless network of capillaries, allowing more time for diffusion to occur.

Veins and Venules Carry Blood Traveling Back to the Heart

Blood from the capillaries, now carrying carbon dioxide and other cellular wastes, drains into larger vessels called **venules** that empty into still larger **veins** (see Fig. 30-15). Veins provide a low-resistance pathway by which blood can return to the heart. The walls of veins are much thinner and more expandable than those of arteries, although both contain a layer of smooth muscle. Because blood pressure in the veins is low, the contractions of skeletal muscle during exercise and breathing must assist in the return of blood to the heart. These muscular movements squeeze the veins, forcing blood through them. When veins are compressed, you might predict that blood would be forced away from the heart as well as toward it. To prevent this, veins are equipped with one-way valves that allow blood flow only toward the heart (Fig. 30-18). When you sit or stand for long periods, the lack of muscular activity allows blood to accumulate in the veins of the lower legs. This accumulation accounts for the swollen feet often experienced by airplane passengers. Long periods of inactivity can also contribute to varicose veins, in which the valves become stretched and weakened.

If blood pressure should fall, for instance after extensive bleeding, veins can help restore it. The sympathetic ner-

vous system (which prepares the body for emergency action) automatically stimulates contraction of the smooth muscles in the vein walls. This action decreases the internal volume of the veins and raises blood pressure, speeding up the return of blood to the heart.

Arterioles Control the Distribution of Blood Flow

The muscular walls of arterioles are under the influence of nerves, hormones, and chemicals produced by nearby tissues. They can therefore contract and relax in response to the changing needs of the tissues and organs they supply. For instance, as you read in your paperback thriller " . . . the blood drained from her face as she beheld the gruesome sight . . . ," keep in mind that the heroine is experiencing constriction of the arterioles that supply her skin with blood. In such threatening situations, the sympathetic nervous system is activated and stimulates the smooth muscle of the arterioles to contract. This contraction raises blood pressure overall, but selective constriction also redirects blood to the heart and muscles, where it may be needed for vigorous action, and away from the skin, where it is less essential.

On a hot summer day, however, you become flushed as the arterioles in your skin expand and bring more

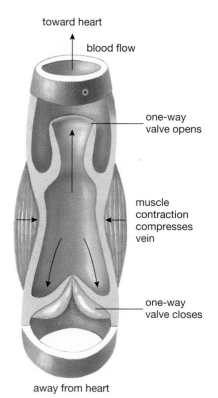

toward heart

blood flow

one-way valve opens

muscle contraction compresses vein

one-way valve closes

away from heart

Figure 30-18   Valves maintain directionality of blood flow in veins

Veins and venules have one-way valves that maintain flow in the proper direction. When the vein is compressed by nearby muscles, the valves allow blood to flow toward the heart but clamp shut to prevent backflow.

blood to the skin capillaries. Bringing this blood closer to the surface enables your body to dissipate excess heat to the outside and to maintain a proper internal temperature. In contrast, in extremely cold weather, fingers and toes can become frostbitten because the arterioles supplying the extremities constrict. The blood is shunted to vital organs such as the heart and brain, which cannot function properly if their temperature drops. By minimizing blood flow to the heat-radiating extremities, the body conserves heat.

The flow of blood in capillaries is regulated by tiny rings of smooth muscle called **precapillary sphincters** that surround the junctions between arterioles and capillaries (see Fig. 30-15). These sphincters open and close in response to local changes that signal the needs of nearby tissues. For example, accumulation of carbon dioxide, lactic acid, or other cellular wastes signals the need for increased blood flow to the tissues. These signals cause the precapillary sphincters as well as the muscles in nearby arterioles to relax, thus increasing blood flow through the capillaries.

The Lymphatic System

The **lymphatic system** consists of a network of lymphatic capillaries and larger vessels that empty into the circulatory system, numerous small lymph nodes, and two additional organs: the thymus and the spleen (Fig. 30-19). Although not strictly part of the circulatory system, the lymphatic system is closely associated with it. The lymphatic system has several important functions:

1. Removal of excess fluid and dissolved substances that leak from the capillaries
2. Transport of fats from the intestine to the bloodstream
3. Defense of the body by exposing bacteria and viruses to white blood cells

Lymphatic Vessels Resemble the Veins and Capillaries of the Circulatory System

Like blood capillaries, lymph capillaries form a complex network of microscopically narrow, thin-walled vessels into which substances can move readily. In contrast to those of blood capillaries, lymph capillary walls are composed of cells with openings between them that act as one-way valves. These openings allow relatively large particles, along with fluid, to be carried into the lymph capillary. Also unlike blood capillaries, which form a continuous connected network, lymph capillaries dead-end in the body's tissues (Fig. 30-20, p. 600). Materials collected by the lymph capillaries flow into larger lymph vessels, through which lymph is forced as blood is through veins. Large lymph vessels have somewhat muscular walls, but most of the impetus for lymph flow comes from the contraction of nearby muscles, such as those used in breath-

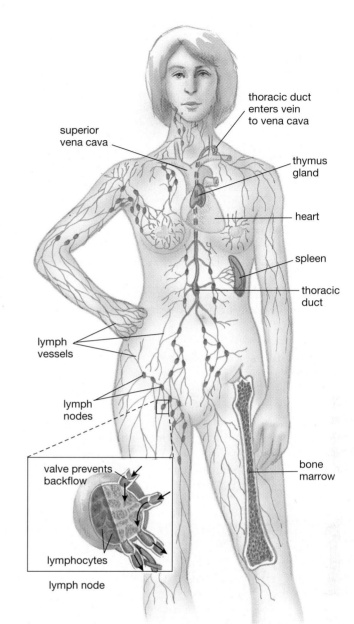

Figure 30-19   The human lymphatic system
This figure illustrates lymph vessels, lymph nodes, and two auxiliary lymph organs, the thymus and spleen. The lymph is returned to the circulatory system by way of the thoracic duct, which empties into the vena cava, a large vein. (Inset) A cross section of a lymph node. The node is filled with channels lined with white blood cells (lymphocytes) that attack foreign matter in the lymph.

ing and walking. As in blood veins, the direction of flow is regulated by one-way valves (Fig. 30-21, p. 600).

The Lymphatic System Returns Fluids to the Blood

As described earlier, dissolved substances are exchanged between the capillaries and body cells by means of interstitial fluid (derived from blood plasma), which bathes nearly all the cells of the body. In an average person, about 3 liters

H E A L T H   W A T C H

Cardiovascular Disorders

Cardiovascular disorders, including heart attacks, strokes, and congestive heart failure, are the leading cause of death in the United States, killing nearly 1 million Americans annually. Consider the stresses under which your circulatory system constantly operates. Your heart is expected to contract vigorously over 2.5 billion times during your lifetime without once stopping to rest. It is also expected to force blood through a series of vessels whose total length would encircle the globe twice. Add the possibility that the complex network of vessels may become constricted, weakened, or clogged for a number of reasons, and it's easy to see why the cardiovascular system is a prime candidate for malfunction.

Hidden Killers: High Blood Pressure and Atherosclerosis

High blood pressure, also called **hypertension**, is usually caused by constriction of the arterioles, which results in increased resistance to blood flow. In the majority of the 50 million Americans afflicted by this condition, the cause of this constriction is unknown. Heredity seems to play a role. For some individuals who are predisposed to hypertension, high salt intake in the diet, as well as obesity, may aggravate it. Although normal blood pressure tends to increase somewhat with age, an approximate borderline reading for high blood pressure is 140/90.

High blood pressure often gives few warning signals, but it undermines the cardiovascular system in several insidious ways. First, it causes strain on the heart by increasing resistance to blood flow. Although the heart may enlarge in response to this added demand, its own blood supply may not increase proportionately. The heart muscle is then inadequately supplied with blood, especially during exercise. Lack of sufficient oxygen to the heart can cause chest pain called **angina pectoris**. Second, high blood pressure contributes to "hardening of the arteries," or atherosclerosis, described below. Third, high blood pressure in conjunction with hardened arteries can lead to rupture of an artery and internal bleeding. Rupture of vessels supplying the brain causes **stroke**, in which brain function is lost in the area deprived of blood and the vital oxygen and nutrients it carries.

Hypertension can be treated in several ways. Mild hypertension may be alleviated by weight reduction, exercise, and sometimes by reduction of dietary salt. Stress-reduction therapies such as relaxation techniques, meditation,

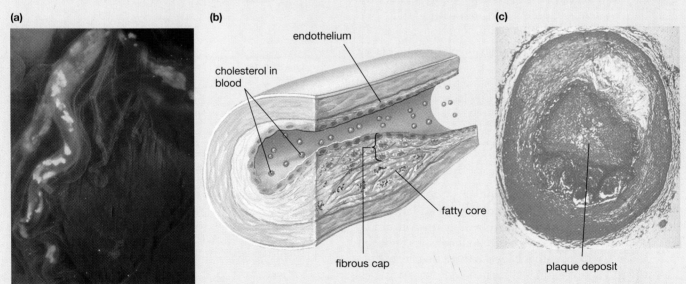

(a)   **(b)**   **(c)**

endothelium

cholesterol in blood

fatty core

fibrous cap

plaque deposit

Figure E30-1   **Plaque deposits clog arteries**

(a) In this remarkable photo of coronary arteries, plaque deposits are seen in glowing yellow. If they block a coronary artery, a heart attack will occur. **(b)** The anatomy of a plaque deposit is shown in this cross section of an artery. If the cap ruptures, a clot will form that may completely obstruct the artery, or the clot may break loose and clog a narrower artery "downstream." **(c)** This cross section of a coronary artery shows a large plaque deposit that has almost completely obstructed it. These deposits can become dislodged and be carried to a smaller artery that they may totally block, causing a heart attack or stroke.

and biofeedback may also be helpful. For more severe cases, drugs may be prescribed. These drugs include diuretics, which increase urination and reduce blood volume, drugs that reduce heart rate, and others that cause dilation (expansion) of the arteries and arterioles. Between 1970 and 1990, the number of deaths due to hypertension decreased by about half, as a result of increased public awareness, early detection, and successful treatment.

Atherosclerosis (derived from the Greek word *athero*, meaning "gruel" or "paste," and *scleros*, meaning "hard") causes a loss of elasticity in the large arteries and a thickening of the arterial walls. The thickening results from deposits called **plaques** composed of cholesterol and other fatty substances as well as calcium and fibrin. These plaques are deposited within the wall of the artery between the smooth muscle and the inner lining (Fig. E30-1). Occasionally the plaque ruptures through the lining into the interior of the vessel. This rupture stimulates the blood platelets to initiate blood clots. These clots further obstruct the artery and may completely block it. Arterial clots are responsible for the most serious consequences of atherosclerosis: heart attacks and strokes.

The American Heart Association estimates that about 1.5 million Americans suffered **heart attacks** in 1994, and about half a million people died from them. A heart attack occurs when one of the coronary arteries (arteries that supply the heart muscle itself; see Fig. E30-1a) is blocked. If a blood clot suddenly breaks loose, it may be carried to a narrower part of the artery, obstructing blood flow. Deprived of nutrients and oxygen, the heart muscle supplied by the blocked artery rapidly and painfully dies. If the damaged area is small, the victim may recover, but death of large areas of heart muscle is almost instantly fatal. Although heart attacks are the major cause of death from atherosclerosis, this disease causes plaques and clots to form in arteries throughout the body. If a clot or plaque deposit obstructs an artery supplying the brain, it can cause a stroke, with results similar to those caused by a ruptured artery.

As with hypertension, the exact cause of atherosclerosis is unclear, but several factors promote it. These factors include hypertension, cigarette smoking, genetic predisposition, obesity, diabetes, a sedentary life-style, and high blood levels of a type of cholesterol bound to a carrier molecule called low-density lipoprotein (LDL). If LDL-bound cholesterol levels are too high, cholesterol may be deposited in arterial walls. In contrast, cholesterol bound to HDL (high-density lipoprotein) is metabolized or excreted and hence is often called "good" cholesterol. By exercising regularly, controlling one's weight, not smoking, and reducing dietary cholesterol and saturated fat, one can greatly reduce the risk of developing atherosclerosis. Moderation in the consumption of animal and other saturated fats is also important because recent evidence links high dietary intake of these substances with high blood cholesterol levels.

Traditional treatment of atherosclerosis includes the use of drugs to lower blood pressure and blood cholesterol levels. In extreme cases, nitroglycerin is used to dilate blood vessels and ease the pain of angina caused by constriction of the coronary arteries. Coronary bypass surgery accounts for over 400,000 operations per year in the United States. This procedure consists of bypassing an obstructed coronary artery with a piece of vein, usually removed from the patient's leg.

Although prevention is by far the most cost-effective and successful strategy, a variety of other high-technology treatments are under development to fight atherosclerosis. Blood clots are often dissolved by injecting an enzyme, streptokinase, or another drug, TPA (tissue plasminogen activator), into the coronary artery. Both work by stimulating the production of an enzyme that breaks down fibrin, the protein that binds the clot together. When performed immediately after a heart attack, this treatment can significantly increase the victim's chances of survival. Another procedure involves squashing plaques flat against the artery walls by inserting a hollow tube with a tiny balloon attached to the end into the obstructed artery. The balloon is inflated, crushing the deposits and restoring blood flow. Alternatively, a tube bearing a tiny laser or drill is manipulated up the artery to the clot. The laser vaporizes the plaques; the drill shaves them away. It is hoped that these new technologies, in conjunction with changes in diet and life-style, will significantly reduce the rate of early deaths from atherosclerosis.

---

more fluid leaves the blood capillaries than is reabsorbed by them each day. One function of the lymphatic system is to return this excess fluid and its dissolved proteins and other substances to the blood. As interstitial fluid accumulates, its pressure forces the fluid through the openings in the lymph capillaries (see Fig. 30-20). The lymphatic system transports this fluid, now called **lymph**, back to the circulatory system.

The Lymphatic System Transports Fats from the Intestine to the Blood

As you will learn in Chapter 32, the small intestine is richly supplied with lymph capillaries. After absorbing digested fats, intestinal cells release fat globules into the interstitial fluid. These globules are too large to diffuse into blood

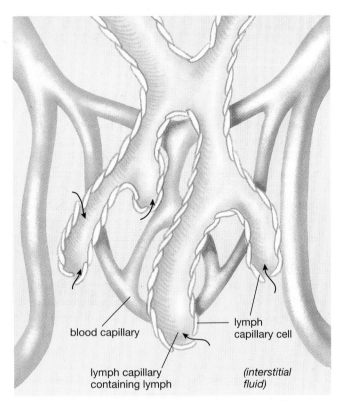

blood capillary

lymph
capillary cell

lymph capillary
containing lymph

*(interstitial
fluid)*

Figure 30-20   **Lymph capillaries**

Lymph capillaries end blindly in the body tissues, where pressure from the accumulation of interstitial fluid forces the fluid into the lymph capillaries through valvelike openings between lymph capillary cells.

capillaries but can move easily through the openings between lymph capillary cells. Once in the lymph, they are dumped into a large vein that enters the heart. After a fatty meal, these fat globules may make up 1% of the lymphatic fluid.

The Lymphatic System Helps Defend the Body against Disease

In addition to its other roles, the lymphatic system helps defend the body against foreign invaders such as bacteria and viruses. In the linings of the respiratory, digestive, and urinary tracts are patches of connective tissue containing large numbers of lymphocytes. The largest of these patch-

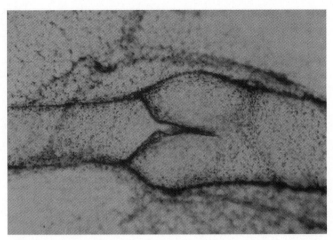

Figure 30-21   **A valve in a lymph vessel**

Like blood-carrying veins, lymph vessels have internal one-way valves that direct the flow of lymph toward the large veins into which they empty.

es are the **tonsils,** located in the cavity behind the mouth. The large lymph vessels are interrupted periodically by kidney bean–shaped structures about 2.5 centimeters (1 inch) long called **lymph nodes** (see Fig. 30-19). Lymph is forced through channels within the nodes that are lined with masses of macrophages. Lymphocytes are also produced in the nodes. Both of these white blood cells recognize and destroy foreign particles, such as bacteria and viruses, and are killed in the process. The painful swelling of lymph nodes that accompanies certain diseases, such as mumps, is largely a result of the accumulation of dead lymphocytes and macrophages and of the virus-infested cells they have engulfed.

The **thymus** and the **spleen** are often considered part of the lymphatic system (see Fig. 30-19). The thymus is located beneath the breastbone slightly above the heart and is responsible for production of lymphocytes. The thymus is particularly active in infants and young children but decreases in size and importance in early adulthood. The spleen is located in the left side of the abdominal cavity, between the stomach and diaphragm. Just as the lymph nodes filter lymph, the spleen filters blood, exposing it to macrophages and lymphocytes that destroy foreign particles and aged red blood cells.

✖ S U M M A R Y   O F   K E Y   C O N C E P T S

Types of Circulatory Systems

Circulatory systems transport blood rich in dissolved nutrients and oxygen close to each cell, where nutrients can be released and wastes absorbed by diffusion. All circulatory systems have three major parts: a fluid, blood; a system of channels, vessels, to conduct the blood; and a pump, a heart, to circulate the blood. Invertebrates have open or closed circulatory systems; all vertebrates have closed systems. In open systems, blood is pumped into a hemocoel, where it directly bathes internal organs. In closed systems, the blood is confined to the heart and blood vessels.

The Vertebrate Circulatory System

Vertebrate circulatory systems transport gases, hormones, and wastes; distribute nutrients; help regulate body temperature; and defend the body against disease.

The vertebrate heart evolved from two chambers in fishes, to three in amphibians and most reptiles, to four in birds and mammals. In the four-chambered heart, blood is pumped separately to the lungs and through the body, maintaining complete separation of oxygenated and deoxygenated blood. Deoxygenated blood is collected from the body in the right atrium and passed to the right ventricle, which pumps it to the lungs. Oxygenated blood from the lungs enters the left atrium, is passed to the left ventricles and pumped to the rest of the body.

The heart cycle consists of two stages: systole, during which the ventricles contract, and diastole, when the ventricles relax. The direction of blood flow is maintained by valves within the heart. The contractions of the heart are initiated and coordinated by the sinoatrial node, the heart's pacemaker. Heart rate can be modified by the nervous system and by hormones such as epinephrine.

Blood is composed of both fluid and cellular materials. The fluid plasma consists of water that contains proteins, hormones, nutrients, gases, and wastes. Red blood cells, or erythrocytes, are packed with a large iron-containing protein called hemoglobin, which carries oxygen. Their numbers are regulated by the hormone erythropoietin. Proteins on their cell membrane determine blood type. There are five types of white blood cells, or leukocytes, that fight infection (see Table 30-1). Platelets, which are fragments of megakaryocytes, are important for blood clotting.

Blood leaving the heart travels in sequence through arteries, arterioles, capillaries, venules, veins, and then back to the heart. Each vessel is specialized for its role. Elastic, muscular arteries help pump the blood along. The thin-walled capillaries are the sites of exchange of materials between the body cells and the blood. Veins provide a low-resistance path back to the heart, with one-way valves maintaining the direction of blood flow. The distribution of blood is regulated by the constriction and dilation of arterioles under the influence of the sympathetic nervous system and local factors such as the amount of carbon dioxide in the tissues. Local factors also regulate precapillary sphincters, which control blood flow in the capillaries.

The Lymphatic System

The human lymphatic system consists of lymphatic vessels resembling blood veins and capillaries, tonsils, lymph nodes, and the thymus and spleen. The lymphatic system removes excess interstitial fluid that leaks through blood capillary walls. It transports fats to the bloodstream from the intestine and fights infection by filtering the lymph through lymph nodes, where white blood cells ingest foreign invaders such as bacteria. The thymus, which is most active in young children, produces lymphocytes that function in immunity. The spleen filters blood past macrophages and lymphocytes, which remove bacteria and damaged blood cells.

✖ K E Y   T E R M S

angina pectoris p. 598
arteriole p. 595
artery p. 588
atherosclerosis p. 599
atrioventricular (AV) node p. 589
atrioventricular valve p. 588
atrium p. 588
basophil p. 593
bicuspid valve p. 589
blood p. 585
capillary p. 595

cardiac cycle p. 588
closed circulatory system p. 586
diastole p. 588
eosinophil p. 593
erythroblastosis fetalis p. 593
erythrocyte p. 592
erythropoietin p. 592
fibrillation p. 591
fibrin p. 594
heart p. 585
heart attack p. 599

hemocoel p. 585
hemoglobin p. 592
hypertension p. 598
interstitial fluid p. 595
leukocyte p. 593
lymph p. 599
lymphatic system p. 597
lymph nodes p. 600
lymphocyte p. 593
macrophage p. 593
megakaryocyte p. 593

monocyte   p. 593
neutrophil   p. 593
open circulatory system   p. 585
pacemaker   p. 589
plaque   p. 599
plasma   p. 591
platelet   p. 593
precapillary sphincter   p. 597

pulmonary circulation   p. 588
Rh factor   p. 593
semilunar valve   p. 589
sinoatrial (SA) node   p. 589
spleen   p. 600
stroke   p. 598
systemic circulation   p. 588
systole   p. 588

thrombin   p. 594
thymus   p. 600
tonsils   p. 600
tricuspid valve   p. 588
vein   p. 596
ventricle   p. 588
venule   p. 596
vessel   p. 585

✖ THINKING THROUGH THE CONCEPTS

Multiple Choice

1. Which of the following is NOT an important function of the vertebrate circulatory system?
   a. transport of nutrients and respiratory gases
   b. regulation of body temperature
   c. protection of the body by circulating antibodies
   d. removal of waste products for excretion from the body
   e. defense against blood loss, through clotting
2. What event initiates blood clotting?
   a. contact with an irregular surface by platelets and other factors in plasma
   b. production of the enzyme thrombin
   c. conversion of fibrinogen into fibrin
   d. conversion of fibrin into fibrinogen
   e. excess flow of blood through a capillary
3. The sites of waste, nutrient, gas, and hormone exchange between the blood and body cells are the
   a. arteries   b. arterioles
   c. capillaries   d. veins
   e. all blood vessels
4. What produces systolic blood pressure?
   a. contraction of the right atrium
   b. contraction of the right ventricle
   c. contraction of the left atrium
   d. contraction of the left ventricle
   e. the pause between heartbeats
5. Which of the following is NOT a component of plasma?
   a. water   b. globulins
   c. fibrinogen   d. albumins
   e. platelets

6. Lymph most closely resembles which of the following?
   a. blood   b. urine
   c. plasma   d. interstitial fluid
   e. water

Review Questions

1. Trace the flow of blood through the circulatory system, starting and ending with the right atrium.
2. List three types of blood cells and describe their principal functions.
3. What are five functions of the vertebrate circulatory system?
4. In what way do veins and lymph vessels resemble one another? Describe how fluid is transported in each of these vessels.
5. Describe three important functions of the lymphatic system.
6. Distinguish among plasma, interstitial fluid, and lymph.
7. Describe veins, capillaries, and arteries, noting their similarities and differences.
8. Trace the evolution of the vertebrate heart from two to four chambers.
9. Explain in detail what causes the vertebrate heart to beat.
10. Describe the cardiac cycle, and relate systole and diastole to the two readings taken during the measurement of blood pressure.
11. Describe how red blood cell number is regulated by a negative feedback system.
12. Describe the formation of an atherosclerotic plaque. What are the risks associated with atherosclerosis?

✖ APPLYING THE CONCEPTS

1. Discuss the steps you can take now and in the future to reduce your risks of developing heart disease.
2. Discuss why a four-chambered heart is much more efficient than a two-chambered heart in delivering oxygenated blood to the various body parts. What evolutionary changes in the life-styles of organisms selected for the evolution of the four-chambered heart?
3. Doctors have attempted to transplant baboon hearts into humans whose hearts were failing. Discuss the implications of this operation from as many angles as you can think of.

4. Considering the prevalence of cardiovascular disease and the high, increasing costs of treating it, certain treatments may not be available to all who might benefit. What factors would you take into account in rationing cardiovascular procedures, such as heart transplants?
5. Joe, a 45-year-old executive of a major corporation, has been diagnosed with mild hypertension. What treatments or lifestyle changes might Joe's doctor recommend? If Joe's hypertension becomes more severe, what treatments might Joe's doctor use? Should Joe be concerned about mild hypertension? Explain your answer.

⬔ F O R   M O R E   I N F O R M A T I O N

Katzir, A. "Optical Fibers in Medicine." *Scientific American*, May 1989. Describes the use of fiber optics in medicine, with emphasis on their role in cardiovascular surgery.

Lawn, R. M. "Lipoprotein(a) in Heart Disease" *Scientific American*, June 1992. Describes the interaction of this protein with blood fats and its role in heart disease.

Lillywhite, H. B. "Snakes, Blood Circulation, and Gravity." *Scientific American*, December 1988. Describes how the cardiovascular systems of snakes are adapted to their diverse life-styles.

Perutz, M. "Hemoglobin Structure and Respiratory Transport." *Scientific American*, December 1978. How hemoglobin plays the dual role of transporting oxygen to the tissues and carbon dioxide back to the lungs.

Radetsky, P. "The Mother of All Blood Cells." *Discover*, March 1995. Discusses the discovery by Irving Weissman of a "stem cell" that gives rise to all other types of blood cells.

Zimmer, C. "The Body Electric." *Discover*, February 1993. Describes computer imaging of the heart's electrical activity.

Zucker, M. "The Functioning of Blood Platelets." *Scientific American*, June 1980. Describes the complex role of platelets in blood clotting.

N E T   W A T C H

On-line resources for this chapter are on the World Wide Web at:
http://www.prenhall.com/~audesirk (click on the <u>table of contents</u> link and then select Chapter 30).

31 Respiration

❊ AT A GLANCE

The Evolution of Respiratory Systems
  Some Types of Animals Exchange Gases without Specialized Respiratory Structures
  A Variety of Respiratory Systems Have Evolved That Facilitate Gas Exchange by Diffusion

The Human Respiratory System
  The Conducting Portion of the Respiratory System Carries Air to the Lungs
  Gas Exchange Occurs in the Alveoli
  Oxygen and Carbon Dioxide Are Transported Using Different Mechanisms
  Air Is Inhaled Actively and Exhaled Passively
  Breathing Rate Is Controlled by the Respiratory Center of the Brain

Respiratory structures require large, moist surfaces for gas exchange. Nudibranch (literally "naked gill") mollusks, such as this rose nudibranch, extend numerous threadlike gills into the surrounding seawater. The flesh of nudibranches is distasteful, so these delicate respiratory structures are safe from nibbling predators.

Each cell in the animal body is a tiny factory that demands a continuous influx of energy to maintain itself. Most cells use a biochemical process called cellular respiration (see Chapter 8) to convert the energy in nutrients such as sugar into an energy form usable by the cell. The production of energy by cellular respiration requires a steady supply of oxygen and generates carbon dioxide as a waste product. Respiratory systems support cellular respiration by bringing a large, moist surface into intimate contact with both the external and the internal environments so that oxygen and carbon dioxide may be exchanged by diffusion between them. Diffusion of oxygen occurs rapidly enough to support metabolically active cells only over very short distances. Thus, as natural selection produced increasingly large, active animals, specialized respiratory systems became necessary to guarantee ample gas exchange throughout their bodies.

The Evolution of Respiratory Systems

Although animal respiratory systems are amazingly diverse, they all share two features that facilitate diffusion. First, the respiratory surface must remain moist, because gases must be dissolved in fluid when they enter or leave living cells. Second, the respiratory system must have a sufficiently large surface area in contact with the environment to allow adequate gas exchange by diffusion. As you may recall from Chapter 6, substances tend to diffuse from areas of high concentration to areas of low concentration: a situation described as a concentration gradient. Gas exchange in all organisms is ultimately a result of diffusion. In animals, carbon dioxide builds up and oxygen is depleted in cells as a result of their metabolic activities. This activity creates concentration gradients that favor diffusion of carbon dioxide out of cells and diffusion of oxygen into them.

In the following sections we examine some respiratory systems, each shaped by the environment in which it evolved. As we touch on the diverse types of respiratory systems (illustrated schematically in Fig. 31-1), notice how each meets the demand for a sufficiently large, moist surface area to allow gas exchange.

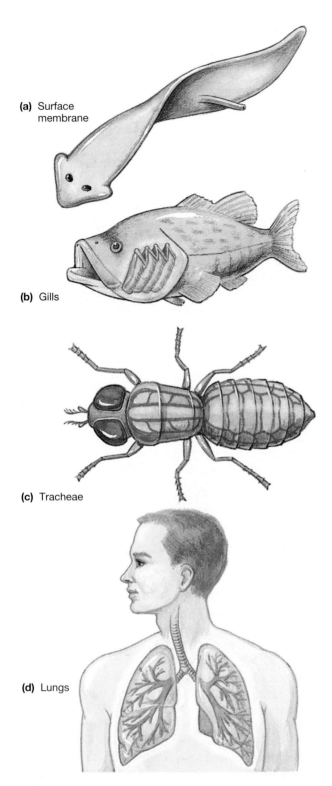

(a) Surface membrane

(b) Gills

(c) Tracheae

(d) Lungs

Figure 31-1   **Schematic illustration of respiratory systems**

(a) In the simplest case, gases are exchanged directly across the moist exterior of the body. **(b)** Gills evolved in larger animals that lived in water, bringing large surface areas richly supplied with blood capillaries into close contact with the surrounding water. As terrestrial animals evolved, their moist respiratory surfaces had to be protected within the body. **(c)** Insects evolved tracheae, and **(d)** vertebrates evolved lungs.

Some Types of Animals Exchange Gases without Specialized Respiratory Structures

Some animals that live in moist environments are able to exchange gases adequately without specialized respiratory systems. The size and shape of their bodies allow adequate diffusion for gas exchange. If the body is extremely small and elongated, as in microscopic nematode worms, for example, gases need to diffuse only a short distance to reach all parts of the body. Alternatively, the body may be thin and flattened, producing a large surface area for diffusion. In flatworms, most cells are close to the moist skin through which gases can diffuse.

If energy demands are sufficiently low, the relatively slow rate of gas exchange by diffusion may suffice even for a larger, thicker body. For example, the body of a jellyfish may be quite large, but the cells that are far from the surface are also relatively inert and require little oxygen.

Another adaptation for gas exchange is to bring the environment (usually water) close to all the body cells, making possible direct exchange of gases between body cells and water. Sponges, for example, circulate seawater through channels within their bodies, where it comes close to all their cells.

Alternatively, the body may combine low energy demands with a large skin surface through which diffusion occurs and a well-developed circulatory system. In the earthworm, for example, gases diffuse through the moist skin and are distributed throughout the body by an efficient circulatory system. Blood in skin capillaries rapidly carries off oxygen that has diffused through the skin, maintaining a concentration gradient that favors the diffusion of more oxygen inward. The worm's elongated shape assures a large skin surface relative to its internal volume, and the worm's sluggish metabolism demands relatively little oxygen. The skin must stay moist to remain effective as a gas-exchange organ; a dry earthworm will eventually suffocate.

A Variety of Respiratory Systems Have Evolved That Facilitate Gas Exchange by Diffusion

Most animals have evolved specialized respiratory systems that interface closely with their circulatory systems to exchange gases between the cells and the environment. The transfer of gases from the environment to the blood to the cells and back usually occurs in stages that alternate bulk flow with diffusion. During **bulk flow**, fluids or gases move in "bulk" through relatively large spaces, from areas of higher pressure to areas of lower pressure. Bulk flow contrasts with diffusion, in which molecules move individually from areas of high concentration to areas of lower concentration (see Chapter 6). In general, gas exchange in respiratory systems occurs in the following stages (see also Fig. 31-9):

1. Air or water containing oxygen is moved across a respiratory surface by bulk flow, often facilitated by muscular breathing movements.

2. Oxygen and carbon dioxide are exchanged through the respiratory surface by diffusion. Diffusion across the respiratory membranes carries oxygen into the capillaries of the circulatory system and removes carbon dioxide.

3. Gases are transported between the respiratory system and the tissues by bulk flow of blood as it is pumped throughout the body by the heart.

4. Gases are exchanged between the tissues and the circulatory system by diffusion. At the tissues, oxygen diffuses out of the capillaries and carbon dioxide diffuses into them along their concentration gradients.

Gills Facilitate Gas Exchange in Aquatic Environments

Gills are the respiratory structures of many aquatic animals. The simplest type of gill, found in certain mollusks and amphibians, consists of numerous projections of the body surface into the surrounding water (Fig. 31-2). In general, gills are numerous or elaborately branched, maximizing their surface area. In some animals, the size of the gill is determined by the availability of oxygen in the surrounding water. For example, salamanders living in stagnant water (which has little opportunity to mix with air) have larger gills than those that dwell in well-aerated water. Gills have a dense profusion of capillaries just beneath their delicate outer membrane. These bring blood close to the surface, where gas exchange occurs.

Some gills are not extensions of the skin but are protected inside the body. Crab gills are covered by a rigid exoskeleton (Fig. 31-3a). Fish gills are covered by a protective flap, the **operculum**, which keeps the delicate gill membranes from being nibbled off by predators and also streamlines the body, allowing the fish to swim faster. Fish create a continuous current over their gills by pumping water into their mouths and ejecting it through the opercular openings (see Fig. E31-1). They can increase the flow of water by swimming with their mouths open; some fast swimmers, such as the tuna and certain sharks, may rely exclusively on swimming to ventilate their gills.

The gills of fish contain an elaborate network of blood vessels whose organization enhances the exchange of gases across the gill membrane. This arrangement, called **countercurrent flow**, is described in "A Closer Look: Gills and Gas Exchange—Countercurrent Flow" and shown in Figure E31-1. Gills are useless out of water, because they collapse and dry out in the air. As terrestrial animals evolved, therefore, their respiratory organs required both support and protection from desiccation. Natural selection has produced three types of terrestrial respiratory structures: book lungs and tracheae in arthropods, and lungs in vertebrates.

Arthropods Respire Using Book Lungs or Tracheae

Although the external skeleton of terrestrial arthropods helps retain body water, it eliminates the skin as a respiratory surface. Spiders (and some other arachnids such as scorpions) enclose a series of moist, pagelike membranes within a chamber of the exoskeleton, forming a structure called a **book lung** (Fig. 31-3b). Insects use a system of elaborately branching tubes called **tracheae**, which convey air directly to the body cells. Tracheae subdivide into smaller channels that penetrate throughout the body tissues (Fig. 31-3c). Each body cell is close to a tracheal tube, minimizing diffusion distances. Tracheae usually communicate with the outside through openings called **spiracles** located along each side of the abdomen. In some large insects, muscular pumping movements of the abdomen enhance air movement through the tracheae.

Most Terrestrial Vertebrates Respire Using Lungs

Lungs are chambers containing moist, delicate respiratory surfaces that are protected within the body, where water loss is minimized and the body wall provides support. The first vertebrate lung probably appeared in a freshwater fish and consisted of an outpocketing of the digestive tract. Gas exchange in this simple lung helped the fish survive in stagnant water, in which oxygen is scarce. Amphibians, who straddle the boundary between aquatic and terrestri-

(a)   **(b)**

Figure 31-2 **Some aquatic animals respire using unprotected gills**

(a) This type of mollusk is called a nudibranch (literally "naked gill"). Its white-tipped gills extend gracefully into the surrounding water. **(b)** The aquatic larval stage of the tiger salamander uses gills whose frilly shape provides an enormous surface area for gas exchange.

A CLOSER LOOK

Gills and Gas Exchange—Countercurrent Flow

The gills of fish utilize countercurrent flow, a very effective mechanism for removing the maximum amount of oxygen from the water flowing over them (Fig. E31-1a). During countercurrent flow, two types of fluids (in this case blood and water) with different concentrations of one or more dissolved substances flow in opposite directions past one another. These fluids are separated by thin membranes. Countercurrent flow promotes diffusion of a substance (such as oxygen) down its concentration gradient from one fluid (water) to the other (blood).

Fish gills consist of a series of filaments supported by bony gill arches (Fig. E31-1b). Each filament is covered with thin folds of tissue called lamellae. Blood flows across each lamella within a dense network of capillaries. Within each lamel-

la, countercurrent flow enhances diffusion by maintaining a concentration gradient of oxygen between the water (which is relatively high in oxygen) and the blood (lower in oxygen). As shown in Figure E31-1c, water is deflected over the lamellae in a direction opposite the flow of blood in the capillaries. Thus, the most highly oxygenated blood is brought close to the water that is just entering the gills and that has an even higher oxygen content than the blood. As the water flows over the lamellae, gradually losing its oxygen to the blood, it encounters blood that is also increasingly low in oxygen. In this way, the gradient encouraging oxygen to move from the water into the blood is maintained across all the lamellae. Countercurrent flow is so effective that some fish extract 85% of the oxygen from the water that flows over their gills.

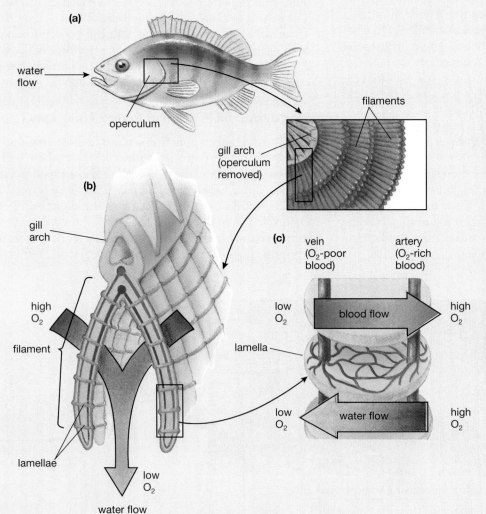

(a) water flow

operculum

filaments

gill arch (operculum removed)

(b)

gill arch

high O₂

filament

lamellae

low O₂

water flow

(c)

vein (O₂-poor blood)

artery (O₂-rich blood)

low O₂   blood flow   high O₂

lamella

low O₂   water flow   high O₂

Figure E31-1   The gill and countercurrent flow

(a) The gill reaches its greatest complexity in the fish, where it is protected under a bony flap, the operculum. A one-way flow of water is maintained over the gill by pumping water through the mouth and out the opercular opening. **(b)** Each filament is composed of a series of platelike lamellae densely supplied with capillaries. Water is deflected across the gill filaments so that it flows directly across each lamella, as shown in part (c). **(c)** Countercurrent flow occurs across each lamella. Water highest in dissolved oxygen encounters the most highly oxygenated blood. As the water loses its oxygen to the blood, it flows past blood that is increasingly low in oxygen, so the gradient encouraging oxygen to move into the blood is maintained.

(a) GILLS

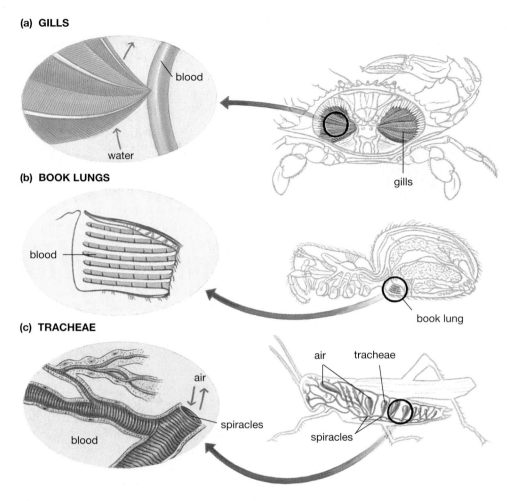

blood

water

(b) BOOK LUNGS

blood

gills

(c) TRACHEAE

book lung

air   tracheae

blood

spiracles

air

spiracles

Figure 31-3   **Arthropod respiratory structures**

(a) Crabs protect finely divided gills beneath a rigid exoskeleton. **(b)** The book lungs of spiders are sheetlike layers of thin tissue enclosed in a protective chamber. **(c)** The tracheae of insects open through spiracles in the abdominal wall and branch intricately throughout the body.

al life, may use gills in the larval stage and lungs in the more terrestrial adult form. For example, the purely aquatic tadpole exchanges its gills for lungs as it develops into a more terrestrial frog (Fig. 31-4). Frogs and salamanders use their moist skin as a supplemental respiratory surface.

The scales of reptiles (Fig. 31-5) reduce the loss of water through the skin and allow reptiles to survive in dry environments. But scales also reduce the diffusion of gases through the skin, so the lungs of reptiles are better developed than those of amphibians. Birds and mammals are

(a)

(b)

Figure 31-4   **From gills to lungs**

(a) The bullfrog, an amphibian, begins life as a fully aquatic tadpole with feathery external gills **(b)**. During metamorphosis into an air-breathing adult frog, the gills are lost and replaced by simple saclike lungs. In both tadpole and adult, gas exchange also occurs by diffusion through the skin, which must be kept moist to function as a respiratory surface.

***Figure 31-5*   Reptiles breathe with lungs**

The fully terrestrial reptile, illustrated by this mangrove snake, is covered with dry scales that restrict gas exchange through the skin. Reptilian lungs are more efficient than those of amphibians because reptiles cannot rely as much on their skin as an additional respiratory surface.

exclusively lung breathers. The bird lung has evolved special adaptations that allow extremely efficient gas exchange, which is necessary to support the enormous energy demands of flight (Fig. 31-6). Air sacs inflate during inhalation, storing fresh air. This air passes through the lung during exhalation, providing the bird with a continuous supply of fresh air even while exhaling.

The Human Respiratory System

The respiratory system in humans and other lung-breathing vertebrates can be divided into two parts: the **conducting portion** and the **gas-exchange portion**. The con-

ducting portion consists of a series of passageways that carry air into the gas-exchange portion, where gas is exchanged with the blood in tiny sacs in the lungs.

The Conducting Portion of the Respiratory System Carries Air to the Lungs

Air enters through the nose or the mouth, passes through a common chamber, the **pharynx**, and then travels through the **larynx** (Fig. 31-7). The opening to the larynx is guarded by the epiglottis, a flap of tissue supported by cartilage. During normal breathing, the epiglottis is tilted upward, allowing air to flow freely into the larynx (see Fig. 31-7). During swallowing, the epiglottis tilts downward and covers the larynx, directing substances into the esophagus. Within the larynx are the **vocal cords**, bands of elastic tissue controlled by muscles. Muscular contractions can cause the vocal cords to partially obstruct the opening within the larynx, so that exhaled air causes them to vibrate, giving rise to the tones of speech or song. The tones can be varied in pitch by stretching the cords and articulated into words by movements of the tongue and lips.

Inhaled air continues past the larynx into the **trachea**, a flexible tube whose walls are reinforced with semicircular bands of stiff cartilage. Within the chest, the trachea splits into two large branches called **bronchi** (singular, **bronchus**), one leading to each lung. Inside the lung, each bronchus branches repeatedly into ever-smaller tubes called **bronchioles**. These lead finally to the microscopic alveoli, tiny air sacs where gas exchange occurs (see Fig. 31-7).

During its passage through the conducting system, air is warmed and moistened. Much of the dust and bacteria it

(a)

(b)

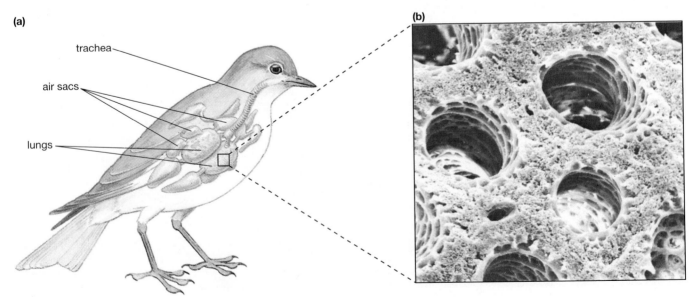

***Figure 31-6*   The respiratory system of the bird is extremely efficient**

(a) When air is inhaled, some fills the lungs while the rest travels past the lungs to fill air sacs. As air is exhaled, the fresh air that has been temporarily stored in the air sacs fills the lungs on its way out. **(b)** The bird uses tubular gas-exchange organs called parabronchi rather than saclike alveoli. The parabronchi allow the air to flow through the lung continuously.

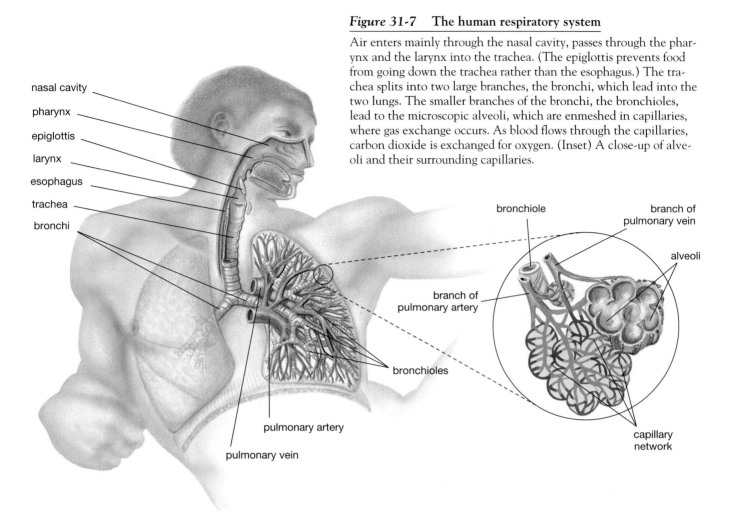

Figure 31-7   **The human respiratory system**

Air enters mainly through the nasal cavity, passes through the pharynx and the larynx into the trachea. (The epiglottis prevents food from going down the trachea rather than the esophagus.) The trachea splits into two large branches, the bronchi, which lead into the two lungs. The smaller branches of the bronchi, the bronchioles, lead to the microscopic alveoli, which are enmeshed in capillaries, where gas exchange occurs. As blood flows through the capillaries, carbon dioxide is exchanged for oxygen. (Inset) A close-up of alveoli and their surrounding capillaries.

carries is trapped in mucus secreted by cells lining the passages. The mucus, with its trapped debris, is continuously swept upward toward the pharynx by cilia lining the bronchioles, bronchi, and trachea. Upon reaching the pharynx, the mucus is coughed up or swallowed. Smoking interferes with this cleansing process by paralyzing the cilia (see "Health Watch: Smoking—A Life and Breath Decision").

Gas Exchange Occurs in the Alveoli

Each lung is packed with around 1.5 to 2.5 million **alveoli** (singular, **alveolus**). These tiny (0.2-millimeter diameter) chambers give magnified lung tissue the appearance of sponge cake (see Fig. E31-3). The thin-walled alveoli provide an enormous surface area for diffusion—about 75 square meters (800 square feet), similar in size to a tennis court and 80 times the total skin surface area of an adult human. The alveoli, which cluster about the end of each bronchiole like a bunch of grapes, are entirely enmeshed in capillaries (see Fig. 31-7). Because both the alveolar wall and the adjacent capillary walls are only one cell thick, the air is extremely close to the blood in the capillaries. The lung cells remain moist because they are coated by a thin layer of water lining each alveolus. Gases dissolve in this water and diffuse through the alveolar and capillary membranes (Fig. 31-8).

Blood is pumped to the lungs by the heart after circulating through the body tissues. The incoming blood surrounding the alveoli is low in oxygen (because the body cells have used it up) and high in carbon dioxide (which is released by the cells). In the alveoli, oxygen diffuses from the air, where its concentration is high, into the blood, where its concentration is low. Conversely, carbon dioxide diffuses out of the blood, where its concentration is high, into the air in the alveoli, where its concentration is lower (Fig. 31-9).

Blood from the lungs, oxygenated and purged of carbon dioxide, returns to the heart, which pumps it to the body tissues. In the tissues, the concentration of oxygen is lower than in the blood, and oxygen diffuses into the cells. Carbon dioxide, which has built up in the cells, diffuses into the blood.

Oxygen and Carbon Dioxide Are Transported Using Different Mechanisms

In the blood, oxygen binds loosely and reversibly with a large, iron-containing protein, **hemoglobin**, in the red blood cells, as noted in Chapter 30 (see Fig. 31-10). Each hemoglobin molecule can bind up to four oxygen molecules (eight oxygen atoms). Nearly all the oxygen carried by the blood is bound to hemoglobin. By removing oxy-

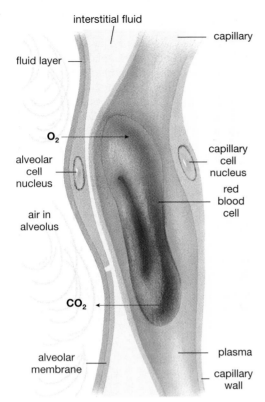

Figure 31-8   Gas exchange between alveoli and capillaries

The relationship between the alveolar membranes and the capillary walls facilitates the exchange of oxygen (O_2) and carbon dioxide (CO_2) by diffusion. Because the alveoli and capillaries are only one cell thick and because the cells are coated in a thin layer of water, gases dissolve and diffuse easily.

gen from solution in the plasma, hemoglobin maintains a high concentration gradient of oxygen from the air to the blood, enhancing diffusion into the blood. Thanks to hemoglobin, our blood can carry about 70 times as much oxygen as it could if the oxygen were simply dissolved in the plasma. As hemoglobin binds oxygen, the protein undergoes a slight change in shape, which alters its color. Deoxygenated blood is dark maroon-red and appears blue through the skin; oxygenated blood is a bright cherry-red. Carbon monoxide gas (CO) is highly toxic because it "fools" the hemoglobin, binding in place of oxygen and over 200 times as tenaciously. Hemoglobin containing CO is also bright red, but it is incapable of transporting oxygen. Most victims of asphyxiation have bluish lips and nail beds because their hemoglobin is deoxygenated, but the lips and nail beds of victims of carbon monoxide poisoning are brighter red than normal.

The transport of carbon dioxide (CO_2) is more complex. Most of the CO_2 carried in the blood (about 70%) reacts with water to form bicarbonate ion (HCO_3^-), which is carried in the blood plasma. About 20% of the CO_2 is returned to the lungs bound to hemoglobin that has released its O_2 to the tissues. The small amount remaining is carried dissolved in the plasma (Fig. 31-10). The pro-

duction of bicarbonate ion and the binding of CO_2 to hemoglobin both reduce the concentration of dissolved CO_2 in the blood and increase the gradient for CO_2 to flow from the body cells into the blood.

Air Is Inhaled Actively and Exhaled Passively

Outside the lungs, the chest cavity is airtight, bounded by neck muscles and connective tissue on top and the dome-shaped, muscular **diaphragm** on the bottom. Surrounding and protecting the lungs is the rib cage within the wall of the chest. Lining the chest and surrounding the lungs is a double layer of membranes called **pleural membranes**. These contribute to the airtight seal between the lungs and the chest wall.

Breathing occurs in two stages: **inhalation**, during which air is actively drawn into the lungs, and **exhalation**, during which it is passively expelled from the lungs. In-

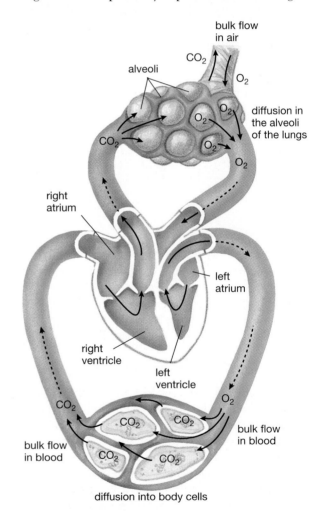

Figure 31-9   An overview of gas exchange

Oxygen is transported first by bulk flow as air is pumped in and out of the lungs. It moves by diffusion through the alveolar capillary walls and then is transported again by bulk flow in the bloodstream. Finally, diffusion carries oxygen into the cells of the body. Carbon dioxide follows the opposite path. Red blood is oxygenated, blue blood is deoxygenated.

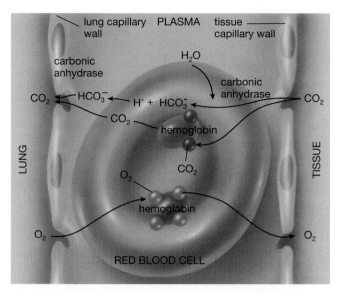

Figure 31-10   **The chemistry of gas exchange**

The chemical combination of water and carbon dioxide forming bicarbonate ion (HCO_3^-) occurs within red blood cells and is catalyzed by the enzyme carbonic anhydrase. The bicarbonate then diffuses back into the plasma. Carbonic anhydrase embedded in the capillary walls catalyzes the reverse reaction, allowing carbon dioxide (CO_2) to diffuse out of the capillaries and into the alveoli.

halation is accomplished by making the chest cavity larger. To do this, the diaphragm muscles contract, drawing the diaphragm downward, and the rib muscles also contract, lifting the ribs up and outward (Fig. 31-11). When the chest cavity is expanded, the lungs expand with it, because a vacuum holds them tightly against the inner wall of the chest. (If the chest is punctured and air leaks in between the pleural membranes and the lung, the lung will collapse.) As the lungs expand, their increased volume creates a partial vacuum that draws air into the lungs.

Exhalation occurs automatically when the muscles causing inhalation are relaxed. The relaxed diaphragm domes upward, and the ribs fall down and inward, decreasing the size of the chest cavity and forcing air out of the lungs. More air can be forced out by contracting the abdominal muscles. After exhalation, the lungs still contain air. This air prevents the thin alveoli from collapsing and fills the space within the conducting portion of the respiratory system. A normal breath moves only about 500 milliliters (1 pint) of air into the respiratory system. Of this, only about 350 milliliters reaches the alveoli for gas exchange. Deeper breathing during exercise causes several times this volume to be exchanged.

Breathing Rate Is Controlled by the Respiratory Center of the Brain

Breathing occurs rhythmically and automatically without conscious thought. But, unlike the heart, the muscles used in breathing are not self-activating; each contraction is stimulated by impulses from nerve cells. These impulses

originate in the **respiratory center**, which is located in the brainstem just above the spinal cord. Nerve cells in the respiratory center generate cyclic bursts of impulses that cause alternating contraction and relaxation of the respiratory muscles.

The respiratory center receives input from several sources and adjusts breathing rate and volume to meet the body's changing needs. The brainstem contains receptor neurons that monitor the concentration of carbon dioxide in the blood. The respiratory rate is regulated to maintain a constant level of carbon dioxide in the blood. An elevated level of carbon dioxide (caused, for example, by an increase in cellular activity) signals a need for more oxygen and causes the receptors to stimulate an increase in the rate and depth of breathing. These receptors are extremely sensitive; an increase in carbon dioxide of only 0.3% can double the breathing rate.

The respiratory rate is much less sensitive to changes in oxygen concentration, because even normal breathing supplies an overabundance of oxygen. But if blood oxygen levels fall drastically, receptors in the aorta and carotid arteries stimulate the respiratory center. Surprisingly, when a person begins strenuous activity, such as running, an increase in breathing rate actually *precedes* any changes in blood gas levels. Apparently, when higher brain centers activate muscles during heavy exercise, they simultaneously stimulate the respiratory center to increase breathing rate. Breathing activity is then "fine-tuned" by the receptors monitoring carbon dioxide concentrations.

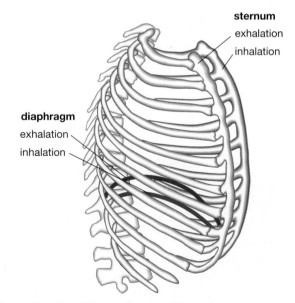

Figure 31-11   **The mechanics of breathing**

Rhythmic nerve impulses from the brain stimulate the diaphragm muscle to contract (pulling it downward) and the muscles surrounding the ribs to contract (moving them up and outward). The result is an increase in the size of the chest cavity, causing air to rush in. Relaxation of these muscles allows the diaphragm to dome upward, and the rib cage to collapse, forcing air out of the lungs.

H E A L T H   W A T C H

Smoking—A Life and Breath Decision

The American Lung Association reports that an estimated 419,000 Americans die of smoking-related diseases each year: 1 out of every 5 deaths in this country. Further, the cost of health care and loss of productivity due to smoking-related ailments costs the United States about $65 billion annually. The American Cancer Society estimates that 153,000 people died of lung cancer in 1994 and that 87% of those deaths were attributable to smoking. The rest of the deaths due to smoking result from a combination of smoking-induced emphysema, chronic bronchitis, heart disease, and other cancers including cancer of the mouth, larynx, esophagus, pancreas, bladder, and kidney.

Tobacco smoke has a dramatic impact on the human respiratory tract (Fig. E31-2). As smoke is inhaled through the nose, trachea, and bronchi, toxic substances such as nicotine and sulfur dioxide paralyze the cilia lining the respiratory tract; a single cigarette can inactivate them for a full hour. Because the function of these ciliary sweepers is to remove inhaled particles, smoking inhibits them just when they are most needed. The visible portion of cigarette smoke consists of billions of microscopic carbon particles. Adhering to these particles are toxic substances, of which a dozen or more are carcinogenic (cancer causing). With the cilia out of action, the particles stick to the walls of the respiratory tract or enter the lungs. Thus, smokers encounter a higher-than-normal risk of cancer in all areas of the respiratory tract touched by smoke.

Cigarette smoke also impairs the amoebalike white blood cells (macrophages) that defend the respiratory tract by engulfing foreign particles and bacteria. Consequently, still more bacteria, dust, and smoke particles enter the lungs. In response to the irritation of cigarette smoke, the respiratory tract increases production of mucus, a third method of trapping foreign particles. But without the cilia to sweep it along, the mucus builds up and can obstruct the airways; the familiar "smoker's cough" is an attempt to expel the mucus. Microscopic smoke particles find a secure lodging place in the tiny, moist alveoli deep within the lungs. There they accumulate over the years until the lungs of a heavy smoker are literally blackened (see Fig. E31-2). The longer the delicate tissues of the lungs are exposed to the carcinogens on the trapped particles, the greater the chance that cancer will develop.

Chronic bronchitis and emphysema together kill about 85,000 persons each year in the United States, and smoking is the leading contributing factor to each of these diseases, according to the American Lung Association. Chronic bronchitis is a persistent lung infection characterized by coughing, swelling of the lining of the respiratory tract, an increase in mucus production, and a decrease in the number and activity of cilia. The result: a decrease in air flow to the alveoli. **Emphysema** occurs when toxic substances in cigarette smoke, such as nitrogen oxides and sulfur dioxide, cause the body to produce substances that reduce the elasticity and increase the brittleness of lung tissue. As the brittle alveoli rupture, the lung gradually loses its normal sponge-cake appearance and more closely resembles blackened Swiss cheese (see Fig. E31-3). The loss of the alveoli, where gas exchange occurs, leads to oxygen deprivation of all body tissues. The emphysema victim's breathing becomes labored and grows increasingly worse until death.

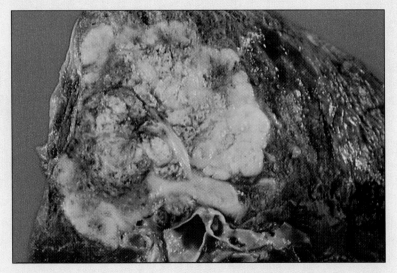

Figure E31-2   **Smoking causes lung cancer**

In this photo, lung cancer is visible as a large, pale mass; the lung tissue surrounding it is blackened by trapped smoke particles.

Carbon monoxide, present in high levels in cigarette smoke, binds tenaciously to red blood cells in place of oxygen. This reduces the blood's oxygen-carrying capacity and thereby increases the work the heart must do. Chronic bronchitis and emphysema compound this problem. As a result, smokers are 70% more likely than nonsmokers to die of heart disease. The carbon monoxide in cigarette smoke may also contribute to the reproductive problems experienced by pregnant women who smoke, including increased incidence of miscarriage, lower birth weight of their babies, and learning impairment of their young children.

There is convincing evidence that "passive smoking" (which occurs when nonsmokers are forced to breathe air polluted by cigarette smoke) poses real health hazards as well. Various studies have concluded that infants whose mothers smoke are more likely to suffer from bronchitis and pneumonia than are infants whose mothers don't smoke.

Decreased lung capacity has been reported in children whose mothers smoked, and abnormally thickened and stiffened heart walls have been reported in 11- and 12-year-old boys whose parents smoked. Angina sufferers are likely to experience chest pain after sitting in a smoke-filled room. Studies have concluded that nonsmoking spouses of smokers face a 30% higher risk of both heart attack and lung cancer than do spouses of nonsmokers. The Environmental Protection Agency estimates that passive smoking is responsible for 3000 deaths of nonsmokers each year, and it has classified environmental tobacco smoke as a known human carcinogen.

There is hope for people who stop smoking before irreversible lung damage has occurred. Healing begins immediately after smoking stops, and the chances of a heart attack or lung cancer gradually drop. After 10 to 15 smoke-free years, former smokers' mortality rates approach those of people who never smoked.

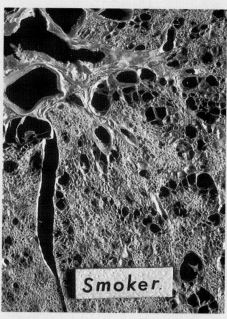

Figure E31-3   Smoking causes emphysema

A section through a normal lung appears almost opaque, but the lung of a smoker suffering from emphysema is full of large holes, each caused by the rupture of hundreds of alveoli as the lung tissue loses elasticity.

✖ S U M M A R Y   O F   K E Y   C O N C E P T S

The Evolution of Respiratory Systems

Respiration makes possible the exchange of oxygen and carbon dioxide between the body and the environment by diffusion of these gases across a moist surface. In moist environments animals whose bodies are very small or flattened may rely exclusively on diffusion through the body surface. Animals with low metabolic demands and/or well-developed circulatory systems may also lack specialized respiratory structures. Larger, more active animals have evolved specialized respiratory systems. Animals in aquatic environments have evolved gills, such as those of crustaceans, mollusks, fish, and many amphibians. On land, moist respiratory surfaces must be protected internally. This need has selected for the evolution of tracheae in insects, book lungs in arachnids, and lungs in terrestrial vertebrates.

The transfer of gases between respiratory systems and tissues occurs in a series of stages that alternate bulk flow with diffusion. Air or water moves by bulk flow past the respiratory surface, and gases in blood are carried by bulk flow. Gases move by diffusion across membranes between the respiratory system and the capillaries and between the capillaries and the tissues.

The Human Respiratory System

The human respiratory system consists of a conducting portion and a gas-exchange portion. Air passes first through the conducting portion, consisting of the nose and mouth, pharynx, larynx, trachea, bronchi, and bronchioles, and then into the gas-exchange portion, composed of microscopic sacs called alveoli. Blood within a dense capillary network surrounding the alveoli releases carbon dioxide and absorbs oxygen from the air.

Most of the oxygen in the blood is bound to hemoglobin within red blood cells. Hemoglobin binds oxygen at concentrations typical of those in alveolar capillaries and releases it at the lower oxygen concentrations found in the body tissues. Carbon dioxide diffuses into the blood from the tissues and is transported in three ways. Most is combined with water to form bicarbonate ion. Some is transported bound to hemoglobin, and a small amount is carried as dissolved carbon dioxide.

Breathing involves actively drawing air into the lungs by contracting the diaphragm and the rib muscles, which expand the chest cavity. Relaxing these muscles causes the chest cavity to collapse, expelling the air. Respiration is controlled by nerve impulses originating in the respiratory center of the brainstem, and its rate is modified by receptors, such as those in the brainstem that monitor carbon dioxide levels in the blood.

✖ K E Y   T E R M S

alveolus p. 611	emphysema p. 614	operculum p. 607
book lung p. 607	exhalation p. 612	pharynx p. 610
bronchiole p. 610	gas-exchange portion p. 610	pleural membrane p. 612
bronchus p. 610	gill p. 607	respiratory center p. 613
bulk flow p. 606	hemoglobin p. 611	spiracle p. 607
conducting portion p. 610	inhalation p. 612	trachea p. 610
countercurrent flow p. 607	larynx p. 610	tracheae p. 607
diaphragm p. 612	lung p. 607	vocal cord p. 610

✖ T H I N K I N G   T H R O U G H   T H E   C O N C E P T S

Multiple Choice

1. With which other system do specialized respiratory systems most closely interface in exchanging gases between the cells and the environment?
   a. the skin
   b. the excretory system
   c. the circulatory system
   d. the muscular system
   e. the nervous system
2. The gas-exchange portion of the human respiratory system is the
   a. larynx
   b. trachea
   c. bronchi
   d. pharynx
   e. alveoli
3. Which of the following animals use book lungs for respiration?
   a. spiders   b. snails
   c. insects   d. fish
   e. bookworms

4. How is most of the oxygen transported in the blood?
   a. dissolved in plasma
   b. bound to hemoglobin
   c. in the form of CO_2
   d. as bicarbonate
   e. dissolved in water
5. Which of the following statements regarding cigarette smoking is true?
   a. Cigarette smoke damages the alveoli.
   b. Cigarette smoke decreases the amount of oxygen in the blood.
   c. Cigarette smoke causes many types of cancer.
   d. Cigarette smoke can lead to heart damage.
   e. all of the above
6. Which of the following pairs of respiratory adaptations and animals are NOT correct?
   a. countercurrent flow: fish
   b. tracheae: birds
   c. gills: arthropods
   d. moist skin: snakes
   e. spiracles: insects

Review Questions

1. Describe three different arthropod respiratory systems and two different vertebrate respiratory systems.
2. Trace the route taken by air in the vertebrate respiratory system, listing the structures through which it flows and the point where gas exchange occurs.
3. Explain some characteristics of animals in moist environments that may supplement respiratory systems or make them unnecessary.
4. How are human respiratory movements initiated? How are they modified, and why are these controls adaptive?
5. What events occur during human inhalation? Exhalation? Which of these is always an active process?
6. Trace the pathway of an oxygen molecule in the human body starting with the nose and ending with a body cell.
7. Describe the effects of smoking on the human respiratory system.
8. Explain how bulk flow and diffusion interact to promote gas exchange between air and blood, and between blood and tissues.
9. Compare carbon dioxide and oxygen transport in the blood. Include in your answer the source and destination of each.
10. Explain how the structure and arrangement of alveoli make them well suited for their role in gas exchange.
11. Explain countercurrent flow and why it is so effective.

APPLYING THE CONCEPTS

1. Heart-lung transplants are sometimes performed, but donors are scarce. Based on your knowledge of the respiratory and circulatory systems and life-style factors that might damage them, what criteria would you use in selecting a recipient for such a transplant?
2. Nicotine is a drug in tobacco that is responsible for several of the effects that smokers crave. Discuss the advantages and disadvantages of low-nicotine cigarettes.
3. Discuss why a brief exposure to carbon monoxide is much more dangerous than a brief exposure to carbon dioxide.
4. Describe several adaptations that might evolve to help members of a species of mammal respire better if the population began living continuously for many generations at very high altitudes.
5. Mary, a strong-willed 3-year-old, threatens to hold her breath until she dies if she doesn't get her way. Can she carry out her threat? Explain.

FOR MORE INFORMATION

Adler, T. "Fish Dishes May Catch On among Smokers." *Science News*, July 30, 1994. Discusses chronic obstructive pulmonary disease.

Feder, M. E., and Burggren, W. W. "Skin Breathing in Vertebrates." *Scientific American*, May 1985. Describes how some vertebrates supplement the action of their lungs or gills to perform gas exchange better.

Harding, C. "Going to Extremes." *National Wildlife*, August/September 1993. Describes adaptations that allow diving animals to plunge to enormous depths without running out of oxygen.

Houston, C. "Mountain Sickness." *Scientific American*, October 1992. The mechanisms of potentially fatal altitude sickness are explained.

Schmidt-Nielsen, K. "How Birds Breathe." *Scientific American*, December 1971. Specializations of the bird respiratory system include additional air sacs and even hollow bones.

NET WATCH

On-line resources for this chapter are on the World Wide Web at:
http://www.prenhall.com/~audesirk (click on the table of contents link and then select Chapter 31).

32 Nutrition and Digestion

AT A GLANCE

Nutrition

The Primary Sources of Energy Are Carbohydrates and Fats
Lipids Include Fats, Phospholipids, and Cholesterol
Carbohydrates, Including Sugars and Starches, Are a Source of Quick Energy
Proteins, Composed of Amino Acids, Perform a Wide Range of Functions within the Body
Minerals Are Elements and Small Inorganic Molecules Required by the Body
Vitamins Are Required in Small Amounts and Play Many Roles in Metabolism
Nutritional Guidelines Help People Obtain a Balanced Diet

The Challenge of Digestion

Digestive Systems Are Adapted to the Life-Style of Each Animal

Human Digestion

The Mechanical and Chemical Breakdown of Food Begins in the Mouth
The Esophagus Conducts Food to the Stomach, Where Digestion Continues
Most Digestion Occurs in the Small Intestine
Most Absorption Occurs in the Small Intestine
Water Is Absorbed and Feces Are Formed in the Large Intestine
Digestion Is Controlled by the Nervous System and Hormones

A Brazilian toucan hungrily attacks an orange. Serrations on its long, colorful beak help it grip and cut the fruit.

In the primordial "soup" where life evolved, the earliest cells absorbed energy-rich nutrients directly through their cell membranes. But before simple life forms could invade the less hospitable environments that cover much of Earth, more complex ways of obtaining both energy and nutrients had to evolve. Over millions of years of natural selection, some became photosynthetic, harvesting the energy of sunlight and the abundant carbon dioxide of the early atmosphere to produce high-energy molecules for immediate use and storage. The dead or living bodies of photosynthetic organisms, with their rich supply of stored energy, offered an abundant food source for organisms that evolved ways to digest them.

The simplest form of digestion is utilized by organisms such as fungi and bacteria, which live in environments where potential food surrounds them. These organisms secrete digestive enzymes outside their bodies, where the enzymes break down complex molecules into simple subunits that can be absorbed. More complex cells such as those of ciliates or amoebae (see Chapter 21) take in particles of food (often smaller cells) and digest them within a vacuole into which enzymes are released. Such miniature intracellular digestive systems are also found in sponges, the simplest animals. More complex animals evolved digestive sacs, and still more complex animals have one-way digestive tubes running through the body. In this chapter we discuss the types of nutrients required by animals and the digestive systems through which they obtain those nutrients, with an emphasis on humans.

Nutrition

Animal nutrients fall into five major categories: lipids, carbohydrates, proteins, minerals, and vitamins. These substances provide the body with its basic needs: (1) energy to fuel cellular metabolism and activities; (2) the chemical building blocks, such as amino acids, to construct complex molecules unique to each animal; and (3) minerals and vitamins that participate in a variety of metabolic reactions.

The Primary Sources of Energy Are Carbohydrates and Fats

Each living cell in an animal's body relies on a continuous expenditure of energy to maintain its incredible complexity and perform its specific functions. Three nutrients provide energy for animals: fats, carbohydrates, and proteins. In a "typical" American diet, fats provide about 38%, carbohydrates about 46%, and protein about 16% of the energy. These molecules are broken down during cellular respiration, and the energy derived from them is used to produce adenosine triphosphate (ATP) (see Chapter 8).

The energy in nutrients is measured in **calories**. A calorie is the amount of energy required to raise the temperature of 1 gram of water by 1 degree Celsius. The calorie content of foods is measured in units of 1000 calories (kilocalories), also known as **Calories** (with a capital C). The human body at rest burns about 1550 Calories per day (somewhat more for males and less for females). Exercise significantly boosts caloric requirements: Well-trained athletes can temporarily raise their calorie consumption from a resting rate of about 1 Calorie per minute to nearly 20 Calories per minute during vigorous exercise (Table 32-1).

Lipids Include Fats, Phospholipids, and Cholesterol

Lipids are a diverse group of molecules that generally contain long chains of carbon atoms and are insoluble in water. The principal types of lipids are fats, or triglycerides, phospholipids, and cholesterol (see Chapter 3). Fats are used primarily as a source of energy. Phospholipids are important components of cell membranes and also provide the insulating covering of neurons (nerve cells). Cholesterol is used in the synthesis of cell membranes, bile (which aids in fat breakdown), and certain hormones. Some animals can synthesize all the specialized lipids they need. Others must acquire specific types of lipid building blocks, called **essential fatty acids**, from their food. For example, humans are unable to synthesize linoleic acid, which is required for the synthesis of certain phospholipids, so we need to obtain this essential fatty acid from our diet.

Humans and most other animals store energy primarily as fat. When an animal's diet provides more energy than is expended through its metabolic activities, most of the excess carbohydrate, fat, or protein is converted to fat for storage;

about 3600 Calories are stored in each pound of fat. Fats have two major advantages as energy-storage molecules. First, they are the most concentrated energy source, containing over twice the energy per unit weight of either carbohydrates or protein (about 9 Calories per gram for fats compared with about 4 per gram for proteins and carbohydrates). Second, lipids are hydrophobic, meaning they do not mix with water. Fat deposits, therefore, do not cause any extra accumulation of water in the body. For both these reasons, fats store more calories with less weight than do other molecules. Minimizing weight allows an animal to move faster (important for escaping predators and hunting prey) and to move using less energy (important when food supplies are limited).

Mammals, who maintain an elevated body temperature, often make their fat deposits do double duty by providing insulation as well as storing energy. Fat is typically stored in a layer beneath the skin. Here it insulates the body, because fat conducts heat at only one-third the rate of other body tissues. Mammals who live near the North or South Pole or in cold ocean waters are particularly dependent on this insulating layer, which reduces the amount of energy they must expend to keep warm (Fig. 32-1).

Carbohydrates, Including Sugars and Starches, Are a Source of Quick Energy

Carbohydrates consist of monosaccharide and disaccharide sugars as well as longer chains of sugars called polysaccharides (see Chapter 3). Polysaccharides include starches, the principal energy-storage material of plants; glycogen, a short-term energy-storage molecule in animals; and cellulose, the major structural component of plant cell walls. During digestion, carbohydrates are broken down into sugars and absorbed. For practical purposes, body cells obtain their energy from a single sugar: glucose. Glucose can be derived from fats, amino acids, and the carbohydrates consumed in the diet.

Animals, including humans, store a carbohydrate called **glycogen** (a large, highly branched chain of glucose mol-

Activity	Calories/ min	Time to "Work Off"		
		500 Cal Cheeseburger	300 Cal Ice Cream Cone	70 Cal Apple
Cross-country skiing (5 mph)	12	42 min	25 min	6 min
Rowing (machine, vigorous)	11	45 min	27 min	6 min
Jogging (5.5 mph)	11	45 min	27 min	6 min
Bicycling (10 mph)	7	1 hr 10 min	43 min	10 min
Dancing	7	1 hr 10 min	43 min	10 min
Gardening	6.5	1 hr 17 min	46 min	11 min
Swimming (crawl, 20 yd/min)	5	1 hr 40 min	1 hr	14 min
Walking (3 mph)	3.8	2 hr 12 min	1 hr 19 min	18 min
Sitting quietly	1.7	4 hr 54 min	2 hr 56 min	41 min

TABLE 32-1 ⊞ *Approximate Energy Consumed by Different Activities for a 150-Pound Person*

Figure 32-1   **Fat provides insulation**
This walrus can withstand the icy waters of the polar seas because it is insulated with a thick layer of fat beneath the skin.

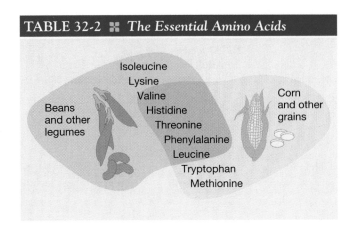

TABLE 32-2 ❖ *The Essential Amino Acids*

Beans and other legumes

Isoleucine
Lysine
Valine
Histidine
Threonine
Phenylalanine
Leucine
Tryptophan
Methionine

Corn and other grains

ecules) in the liver and muscles. Although humans can potentially store hundreds of pounds of fat, we store less than half a pound of glycogen. During exercise, such as running, the body draws on this store of glycogen as a source of quick energy. When the activity is prolonged, as in the case of a marathon runner, the stored glycogen can be totally depleted. The expression "hitting the wall" describes the extreme fatigue that distance runners may experience after exhausting their glycogen supply.

Proteins, Composed of Amino Acids, Perform a Wide Range of Functions within the Body

Each day, the body breaks down 20 to 30 grams of its own protein, which is metabolized for energy. This is replaced by protein taken in through the diet, and any excess amino acids are either broken down and used for energy or stored as fat. The breakdown of protein produces the waste product **urea**, which is filtered from the blood by the kidneys. Specialized diets in which protein is the major energy source place extra stress on the kidneys. The major role of dietary protein is as a source of amino acids to make new molecules (see also Chapter 3).

Amino acids are used to synthesize certain hormones, other amino acids, some neurotransmitters (chemicals used in communication between neurons), and new proteins. These proteins have diverse roles in the body, acting as enzymes, receptors on cell membranes, oxygen transport molecules (hemoglobin), structural components (hair and nails), hormones, antibodies, and muscle proteins.

In the digestive tract, dietary protein is broken down into its amino acid subunits. Then, in the body cells, the amino acids are linked in specific sequences to form new proteins. The human liver can synthesize (from other amino acids) 9 of the 20 different amino acids used in proteins. Those that cannot be synthesized, called **essential amino acids,** must be supplied by the diet in foods such as meat,

milk, eggs, corn, beans, and soybeans. Because many plant proteins are deficient in some of the essential amino acids, individuals on a vegetarian diet must include a variety of plants whose proteins together will provide all nine, or they risk protein deficiency. The nine essential amino acids and examples of how different plant proteins complement one another in providing them are illustrated in Table 32-2.

Minerals Are Elements and Small Inorganic Molecules Required by the Body

Animals require a wide variety of minerals, which are small inorganic molecules and elements (Table 32-3). Minerals must be obtained through the diet, either from food or dissolved in drinking water, because the body cannot manufacture the inorganic substances. Required minerals include calcium, magnesium, and phosphorus, which are major constituents of bones and teeth. Others, such as sodium and potassium, are essential for muscle contraction and the conduction of nerve impulses. Iron is used in the production of hemoglobin, and iodine is found in hormones produced by the thyroid gland. In addition, trace amounts of several other minerals, including zinc, copper, and selenium, are required, often as parts of enzymes.

Vitamins Are Required in Small Amounts and Play Many Roles in Metabolism

Vitamins are a diverse group of organic compounds that animals require in small amounts. In general, vitamins cannot be synthesized by the body (or cannot be synthesized in adequate amounts) and must be obtained from food. Although our skin can manufacture some vitamin D when it is exposed to sunlight, because we spend so much time indoors most of us do not synthesize enough and must supplement it through our diet. The vitamins considered essential in human nutrition are listed in Table 32-4.

Human vitamins are often grouped into two categories, water soluble and fat soluble. Water-soluble vitamins include vitamin C and the eleven different compounds that make up the B vitamin complex. These substances dissolve in the water of the blood plasma and are excreted by

TABLE 32-3 ☷ *Human Water and Mineral Requirements*[a]

Mineral	RDA[b] for Healthy Adult Male (milligrams)	Dietary Sources	Major Functions in Body	Deficiency Symptoms
Water	1.5 liters/day	Solid foods, liquids, drinking water	Transport of nutrients Temperature regulation Metabolic reactions	Thirst, dehydration
Calcium	800	Milk, cheese, green vegetables, legumes	Bone and tooth formation Blood clotting Nerve impulse transmission	Stunted growth Rickets, osteoporosis Convulsions
Phosphorus	800	Milk, cheese, meat, poultry, grains	Bone and tooth formation Acid-base balance	Weakness Demineralization of bone Loss of calcium
Potassium	2,500	Meats, milk, fruits	Acid-base balance Body water balance Nerve function	Muscular weakness Paralysis
Chlorine	2,000	Table salt	Formation of gastric juice Acid-base balance	Muscle cramps Apathy Reduced appetite
Sodium	2,500	Table salt	Acid-base balance Body water balance Nerve function	Muscle cramps Apathy Reduced appetite
Magnesium	350	Whole grains, green leafy vegetables	Activation of enzymes in protein synthesis	Growth failure Behavioral disturbances Weakness, spasms
Iron	10	Eggs, meats, legumes, whole grains, green vegetables	Constituent of hemoglobin and enzymes involved in energy metabolism	Iron-deficiency anemia (weakness, reduced resistance to infection)
Fluorine	2	Fluoridated water, tea, seafood	May be important in maintenance of bone structure	High frequency of tooth decay
Zinc	15	Widely distributed in foods	Constituent of enzymes involved in digestion	Growth failure Small sex glands
Iodine	0.14	Seafish and shellfish, dairy products, many vegetables, iodized salt	Constituent of thyroid hormones	Goiter
Copper Silicon Vanadium Tin Nickel Selenium Manganese	Not established (trace amounts)	Widely distributed in foods	Some unknown; some work in conjunction with enzymes	Unknown

[a]Modified from N. S. Scrimshaw and V. R. Vernon, "The Requirements of Human Nutrition," *Scientific American*, September 1976.
[b]Recommended Daily Allowance.

TABLE 32-4 :: Human Vitamin Requirements

Vitamin	RDA for Adult (milligrams)	Dietary Sources	Major Functions in Body	Deficiency Symptoms
Water soluble				
Vitamin B_1 (thiamin)	1.9	Milk, meat, bread	Coenzyme in metabolic reactions	Beriberi (muscle weakness, peripheral nerve changes, edema, heart failure)
Vitamin B_2 (riboflavin)	1.8	Widely distributed in foods	Constituent of coenzymes in energy metabolism	Reddened lips, cracks at corner of mouth, lesions of eye
Niacin	20	Liver, lean meats, grains, legumes	Constituent of two coenzymes in energy metabolism	Pellagra (skin and gastrointestinal lesions, nervous, mental disorders)
Vitamin B_6 (pyridoxine)	2	Meats, vegetables, whole-grain cereals	Coenzyme in amino acid metabolism	Irritability, convulsions, muscular twitching, dermatitis, kidney stones
Pantothenic acid	5–10	Milk, meat	Constituent of coenzyme A with a role in energy metabolism	Fatigue, sleep disturbances, impaired coordination
Folacin	0.4	Legumes, green vegetables, whole wheat	Coenzyme involved in nucleic and amino acid metabolism	Anemia, gastrointestinal disturbances, diarrhea, red tongue
Vitamin B_{12}	0.003	Meats, eggs, dairy products	Coenzyme in nucleic acid metabolism	Pernicious anemia, neurological disorders
Biotin	0.15–0.3	Legumes, vegetables, meats	Coenzymes required for fat synthesis, amino acid metabolism, and glycogen formation	Fatigue, depression, nausea, dermatitis, muscular pains
Choline	Not established Usual diet provides 500–900	Egg yolk, liver, grains, legumes	Constituent of phospholipids, precursor of the neurotransmitter acetylcholine	None reported in humans
Vitamin C (ascorbic acid)	45	Citrus fruits, tomatoes, green peppers	Maintenance of cartilage, bone, and dentine; collagen synthesis	Scurvy (degeneration of skin, teeth, blood vessels; epithelial hemorrhages)
Fat soluble				
Vitamin A (retinol)	1	Beta-carotene in green, yellow, red vegetables Retinol added to dairy products	Constituent of visual pigment Maintenance of epithelial tissues	Night blindness, permanent blindness
Vitamin D	0.01	Cod-liver oil, eggs, dairy products	Promotes bone growth and mineralization Increases calcium absorption	Rickets (bone deformities) in children; skeletal deterioration
Vitamin E (tocopherol)	15	Seeds, green leafy vegetables, margarines, shortenings	Antioxidant, prevents cellular damage	Possibly anemia
Vitamin K	0.03	Green leafy vegetables Intestinal bacteria produce	Important in blood clotting	Bleeding, internal hemorrhages

TABLE 32-5 ✖ *Dietary Changes Advocated by the U.S. Government*

Dietary Component	Percentage of Total Daily Energy Intake	
	Average American Diet	Dietary Goals
Carbohydrates	46	58
Lipids	38	30
Proteins	16	12

Summary of Recommendations:
1. Increase consumption of fruits, vegetables, and whole grains.
2. Decrease consumption of refined sugars.
3. Decrease consumption of fats, replace saturated with unsaturated fats.
4. Decrease consumption of animal fats by selecting lean meats, poultry, and fish.
5. Decrease consumption of high-cholesterol foods, such as butter and eggs.
6. Decrease consumption of salt and foods high in salt content.
7. Decrease caloric intake to maintain desirable weight.

the kidneys. They are not stored in the body in any appreciable amounts. Water-soluble vitamins generally work in the body in conjunction with enzymes to promote chemical reactions that supply energy or synthesize materials. Because each vitamin participates in several metabolic processes, a deficiency of a single vitamin can have wide-ranging effects (see Table 32-4).

The fat-soluble vitamins A, D, E, and K have even more varied roles. Vitamin K, for example, helps regulate blood clotting, and vitamin A is used to produce visual pigment (the light-capturing molecule in the retina of the eye). Fat-soluble vitamins can be stored in body fat and may accumulate in the body over time. For this reason, fat-soluble vitamins may be toxic if consumed in excessively high doses.

Nutritional Guidelines Help People Obtain a Balanced Diet

We are fortunate to live amid an abundance of food. However, the overwhelming diversity of food available in an American supermarket, and the easy availability of "fast food" can lead to poor nutritional choices. To help people make informed choices, the U.S. government has developed recommendations and goals for the "average American," which are summarized in Table 32-5. Further help is provided by the food pyramid designed by the U.S. Department of Agriculture that illustrates the relative abundances of different food groups in an optimal diet (Fig. 32-2). A third source of information is the nutritional labeling now required on commercially packaged foods. These provide complete information on calorie, fiber, fat, sugar, and vitamin content (Fig. 32-3). Some fast-food chains also make fliers available listing nutritional information about their products.

NUTRITION FACTS			
Serving Size			1 Cup (55g)
Servings Per Package			8
Amount Per Serving			
Calories 210		Calories from Fat 0	
			% Daily Value*
Total Fat 0g			0%
Saturated Fat 0g			0%
Cholesterol 0mg			0%
Sodium 20mg			1%
Total Carbohydrate 46g			15%
Dietary Fiber 6g			24%
Sugars 12g			
Protein 6g			
Vitamin A*	•	Vitamin C 2%	
Calcium 4%	•	Iron 18%	
Thiamin 38%	•		

*Percent Daily Values are based on a 2,000 calorie diet. Your Daily Values may be higher or lower depending on your calorie needs:

	Calories	2,000	2,500
Total Fat	Less than	65g	80g
Saturated Fat	Less than	20g	25g
Cholesterol	Less than	300mg	300mg
Sodium	Less than	2,400mg	2,400mg
Total Carbohydrate		300g	375g
Dietary Fiber		25g	30g

Calories per gram:
Fat 9 • Carbohydrate 4 • Protein 4

Figure 32-3   New food labels

The U.S. government recently required more complete nutritional labeling of foods, such as in this sample. The weight in grams of various nutrients (such as fat, cholesterol, and sodium) is converted into a percentage of the recommended daily allotment, assuming a 2000-Calorie diet. At the bottom of the label, the total recommended number of grams of these nutrients is listed for a 2000- and a 2500-Calorie diet.

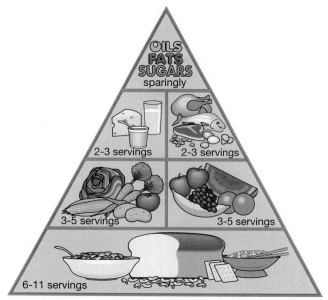

OILS FATS SUGARS
sparingly

2-3 servings   2-3 servings

3-5 servings   3-5 servings

6-11 servings

Figure 32-2   The dietary pyramid

Recommended by the U.S. Department of Agriculture, this chart shows suggested daily servings.

The Challenge of Digestion

The digestive systems of animals take in, and then **digest**, that is, physically grind up and chemically break down the complex molecules of their food into simpler molecules that can be absorbed. Material that cannot be used or broken down is then expelled from the body.

Animals eat the bodies of other organisms, bodies that may resist becoming food. The plant body, for example, armors each cell with a wall of indigestible cellulose. Animal bodies may be covered with equally indigestible fur, scales, or feathers. In addition, the complex lipids, proteins, and carbohydrates found in food are not in a form that can be used directly. They must be broken down before they can be absorbed and distributed to the cell of the animal that has consumed them, where they are recombined in unique ways. Different types of animals meet the challenge of acquiring nutrients with various types of digestive tracts, each finely tuned to a unique diet and life-style. Among this diversity, however, we find certain tasks that all digestive systems must accomplish:

1. *Ingestion.* The food must be brought into the digestive tract through an opening, usually called a **mouth**.
2. *Mechanical breakdown.* The food must be physically broken down into smaller pieces. This is accomplished by gizzards or teeth as well as by the churning action of the digestive cavity itself. The particles produced by mechanical breakdown provide a large surface area for attack by digestive enzymes.
3. *Chemical breakdown.* The particles of food must be exposed to digestive enzymes and other digestive fluids that cause large molecules to be broken down into smaller subunits.
4. *Absorption.* The small molecules must be transported out of the digestive cavity and into cells.
5. *Elimination.* Indigestible materials must be expelled from the body.

In the following section, we will explore briefly some of the diverse mechanisms by which animal digestive systems accomplish these functions.

Digestive Systems Are Adapted to the Life-Style of Each Animal

Digestion within Single Cells Occurs in the Sponges

Digestion *within* a single cell (called **intracellular digestion**) occurs after microscopic food particles have been engulfed by the cell. Once engulfed by a cell, the food is enclosed in a **food vacuole**, a space surrounded by membrane that serves as a temporary stomach. The vacuole fuses with small packets of digestive enzymes called **lysosomes**, and food is broken down within the vacuole into smaller molecules that can be absorbed into the cell cytoplasm. Undigested remnants remain in the vacuole, which eventually expels its contents outside the cell. Intracellular digestion is seen in the single-celled protists and in the simplest animals. Sponges, for example, rely entirely on intracellular digestion (Fig. 32-4). This limits their menu to microscopic food particles, such as protists, that they filter from the surrounding sea using the sieve-like "collars" of special collar cells.

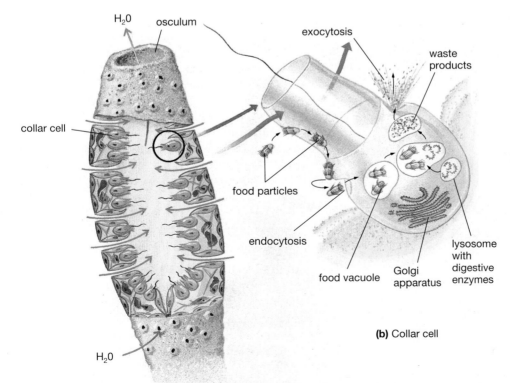

Figure 32-4   Intracellular digestion in a sponge

(a) Internal anatomy of a simple sponge showing the direction of water flow and the location of the collar cells. **(b)** Enlargement of a single collar cell. Water is filtered through the collar, and food particles (single-celled organisms) are trapped on the outside of the collar and conducted down to the cell membrane. Here the food is engulfed by endocytosis, then digested in lysosomes, and wastes are expelled.

H₂0   osculum   exocytosis

waste products

collar cell

food particles

endocytosis

food vacuole   Golgi apparatus   lysosome with digestive enzymes

(b) Collar cell

H₂0

(a) Sponge

A Sac with One Opening Forms the Simplest Digestive System

Larger, more complex organisms evolved a chamber within the body where chunks of food could be broken down by enzymes acting outside the cells. This process is called **extracellular digestion**. One of the simplest of these chambers is found in cnidarians, such as sea anemones, hydra, and jellyfish. These animals possess a digestive sac called a **gastrovascular cavity**, with a single opening through which food is ingested and wastes are ejected (Fig. 32-5). Although it is generally referred to as the mouth, this opening also serves as an anus. Food captured by stinging tentacles is escorted into the gastrovascular cavity, where enzymes break it down. Cells lining the cavity absorb the nutrients and engulf small food particles. Further digestion occurs using the intracellular digestive processes just described for the sponge. The undigested remains are eventually expelled through the same opening by which they entered.

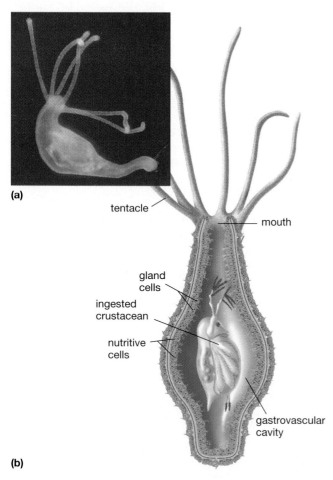

(a)

tentacle

mouth

gland cells

ingested crustacean

nutritive cells

gastrovascular cavity

(b)

Figure 32-5   Digestion in a sac

(a) A *Hydra* has just captured a waterflea (*Daphnia*, a microscopic crustacean) and ingested it. **(b)** Within the gastrovascular cavity, gland cells secrete enzymes that digest the prey into smaller particles and nutrients. Elongated cells lining the cavity ingest these particles (as described for the sponge; see Fig. 32-4), and digestion is completed intracellularly. Undigested waste is then expelled through the single opening.

While one meal is being digested, a second cannot be processed efficiently, because the same chamber is used. Thus, this type of digestive system is unsuited to active animals, who require frequent meals, or to animals whose food supplies so little nutrition that they must feed continuously. The needs of these animals are met by a digestive system consisting of a one-way tube with an opening at each end.

Digestion in a Tube Allows More Frequent Meals

Most animals, from nematode worms to earthworms, mollusks, arthropods, echinoderms, and vertebrates, have a digestive system that is basically a tube running through the body. A tubular digestive tract allows the animal to eat frequently. Further it consists of a series of specialized regions that process the food in an orderly sequence: first physically grinding it up, then enzymatically breaking it down, then absorbing the small nutrient molecules into body cells. Specialized tubular digestive tracts adapt different animals to eat a wide range of foods and to extract the maximum amount of nutrients from them.

In its simplest form, as seen in the threadlike nematode worm, the tube is relatively unspecialized along its length (Fig. 32-6a). In more complex organisms, such as the earthworm, the tube consists of a series of compartments, each with a specific role in the breakdown of food (Fig. 32-6b). The earthworm extracts nutrients from decaying organic material in soil. A tubular digestive system is essential to the earthworm, which continuously ingests soil as it burrows through the earth, passing it out at one end while taking it in at the other. The worm's muscular **pharynx** ingests soil and bits of vegetation, which are passed through the **esophagus** to the **crop**, a thin-walled storage organ. The crop collects the food and gradually passes it to the **gizzard**. Here, bits of sand and the contraction of muscles grind the food into smaller particles. The food then travels to the intestine, where enzymes break it down into simple molecules that can be absorbed by the cells lining the intestine.

Like the earthworm, humans and other vertebrates have tubular digestive tracts with several compartments in which food is first physically, then chemically, broken down before being absorbed by individual cells. Animals with tubular digestive systems thus use extracellular digestion to break down their food. Vertebrate digestive tracts, as illustrated by those of the bird, the cow (Fig. 32-7), and the human (Fig. 32-8) are specialized for the particular diet of the animal.

Ruminant Adaptations Help Them Digest Cellulose

Cellulose, like starch, consists of long chains of glucose molecules, but these molecules are linked together in a way that resists the attack of animal digestive enzymes (see Chapter 3). Because cellulose surrounds each plant cell, it is potentially one of the most abundant food energy sources on Earth. Ruminant animals—cows, sheep, goats, camels, and hippos, to name a few—have evolved elaborate digestive systems that can break down cellulose. Rumination, or cud-chewing, is the process of regurgitating food and rechew-

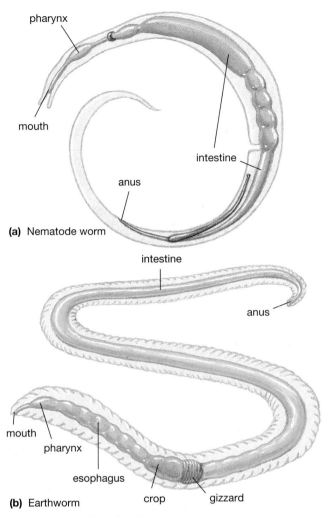

Figure 32-6   Tubular digestive tracts

(a) The nematode worm, an abundant, usually microscopic animal is among the simplest animals to have a one-way digestive system, but the digestive system is relatively unspecialized. Nematodes often live within their food source (several hundred thousand may be found in a rotting apple) and may eat almost continuously. **(b)** The earthworm has a one-way digestive system in which food is passed through a series of compartments. Each compartment is specialized to play a specific role in breaking down food and absorbing it.

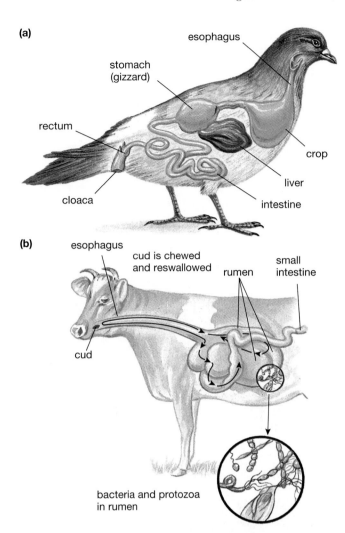

Figure 32-7   Adaptations of vertebrate digestive systems

(a) The digestive system of birds is adapted to the demands of flight. The expandable crop serves as a storage organ, allowing the bird to store food to meet the enormous caloric demands of flight. The gizzard replaces the teeth, using small stones that are stored in this organ and muscular action to break down the hard seeds and insect exoskeletons prevalent in the diet of many birds. Undigested food is expelled through the cloaca. **(b)** The stomach of the cow has four chambers. The largest is the rumen, which houses a flourishing population of microorganisms that digest the cellulose in the cow's vegetarian diet. Arrows trace the path of food through the digestive tract.

ing it, one of several adaptions that enable these animals to digest tough plant material. Ruminant stomachs consist of four chambers (Fig. 32-7b). The first chamber, the rumen, has evolved into a massive fermentation vat. Here, microorganisms—including many species of bacteria and ciliates—thrive in a mutually beneficial relationship with the ruminant. These microorganisms produce **cellulase,** the enzyme that breaks cellulose into its component sugars. After it is processed in the rumen, the plant material, now called *cud,* is regurgitated, chewed, and reswallowed to the rumen. The extra mechanical breakdown exposes more of the cellulose and cell contents to the enzymes of the microorganisms for further digestion. Gradually the cud is then released into the remaining three chambers for further breakdown before entering the intestine.

Human Digestion

The Mechanical and Chemical Breakdown of Food Begins in the Mouth

Both the mechanical and chemical breakdown of food begin in the mouth. In the adult human, 32 teeth of varying sizes and shapes cut and grind the food into small pieces. Human and other vertebrate teeth are specialized for the diet of the animal (Fig. 32-9). As the food is pulverized by the teeth, the first phase of chemical digestion occurs as three pairs of salivary glands (see Fig. 32-8) pour

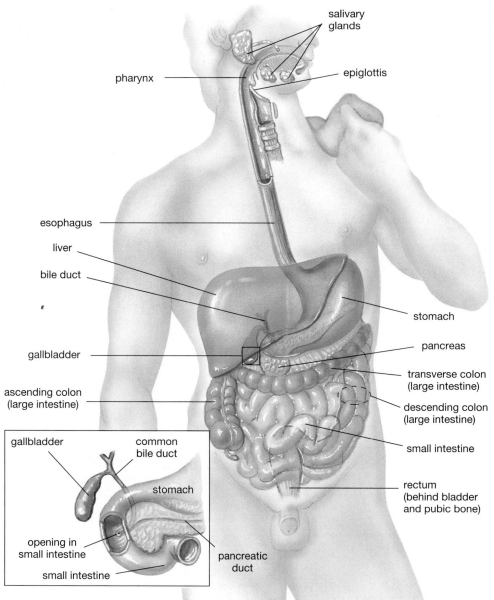

salivary
glands

epiglottis

pharynx

esophagus

liver

bile duct

stomach

gallbladder

pancreas

transverse colon
(large intestine)

ascending colon
(large intestine)

descending colon
(large intestine)

small intestine

rectum
(behind bladder
and pubic bone)

gallbladder

common
bile duct

stomach

opening in
small intestine

small intestine

pancreatic
duct

Figure 32-8   **The human
digestive tract**

Some organs that produce and
store digestive secretions, such as
the salivary glands, liver, gallbladder, and pancreas, are included.

out saliva in response to the smell, feel, taste, and (if you're
hungry) even the thought of food.

Saliva contains the digestive enzyme **amylase**, which
begins the breakdown of starches into sugar (Table 32-6).
Saliva has other functions as well. It contains a bacteria-killing enzyme and antibodies that help guard
against infection. It also lubricates the food to facilitate swallowing and dissolves some food molecules such
as acids and sugars, carrying them to taste buds on the
tongue. The taste buds help identify the type and quality of the food.

With the help of the muscular tongue, the food is manipulated into a mass and pressed backward into the **pharynx**, a cavity connecting the mouth with the esophagus
(Fig. 32-10a). The pharynx also connects, via the larynx,
the nose and mouth with the trachea, which conducts air
to the lungs. This arrangement occasionally causes problems, as anyone who has ever choked on a piece of food

can attest. Normally, however, the swallowing reflex ele-
vates the larynx so that it meets the **epiglottis**, a flap of tissue that blocks off the respiratory passages. Food is thus
directed into the esophagus (Fig. 32-10b).

The Esophagus Conducts Food to the Stomach, Where Digestion Continues

The esophagus is a muscular tube that propels food from the
mouth to the stomach. Circular muscles surrounding the
esophagus contract in sequence above the swallowed food
mass, squeezing it down toward the stomach (Fig. 32-11).
This muscular action, called **peristalsis**, also occurs in the
stomach and intestines, where it helps move food along
the digestive tract. Peristalsis is so effective that a person
can actually swallow when upside down. Mucus secreted by
cells lining the esophagus helps protect it from abrasion
and lubricates the food during its passage.

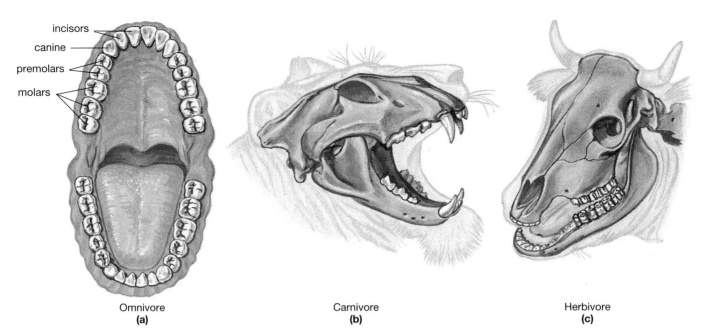

incisors
canine
premolars
molars

Omnivore
(a)

Carnivore
(b)

Herbivore
(c)

Figure 32-9   **Teeth are adapted to different diets**

(a) The upper and lower jaws of a human adult showing teeth specialized for cutting and grasping (incisors and canines) and grinding (molars and premolars). The shape of human teeth reflects our omnivorous diet, which includes both plant and animal material. **(b)** The teeth of a lion, a carnivore (meat eater), are specialized for grasping and tearing flesh and bone. **(c)** The teeth of the herbivorous (plant-eating) cow have large flat surfaces for grinding tough plant material.

The **stomach** is an expandable muscular sac capable of holding from 2 to 4 liters (as much as a gallon) of food and liquids. Food is retained in the stomach by a ring of circular muscle separating the lower portion of the stomach from the upper small intestine. This muscle, called the **pyloric sphincter**, regulates the passage of food into the small intestine, as described later. The stomach has three major functions. First, the stomach stores food and releases it gradually into the small intestine, at a rate suitable for proper digestion and absorption. Thus, the stomach enables us to eat large, infrequent meals. Carnivores carry this ability to an extreme. A lion, for instance, may consume about 18 kilograms (40 pounds) of meat at one meal, then spend the next few days quietly digesting it. A second function of the stomach is to assist in the mechanical breakdown of food. In addition to peristalsis, its

TABLE 32-6 ⬛ *Digestive Secretions*			
Site of Digestion	*Source of Secretion*	*Secretion*	*Role in Digestion*
Mouth	Salivary glands	Amylase	Breaks down starch into disaccharides
	Salivary glands	Mucus, water	Lubricates, dissolves food
Stomach	Cells lining stomach	Hydrochloric acid	Allows pepsin to work, kills bacteria, solubilizes minerals
	Cells lining stomach	Pepsin	Breaks down proteins into large peptides
	Cells lining stomach	Mucus	Protects stomach
Small intestine	Pancreas	Sodium bicarbonate	Neutralizes acidic chyme from stomach
	Pancreas	Amylase	Breaks down starch into disaccharides
	Pancreas	Peptidases	Split large peptides into small peptides
	Pancreas	Trypsin	Breaks down proteins into large peptides
	Pancreas	Chymotrypsin	Breaks down proteins into large peptides
	Pancreas	Lipase	Breaks down lipids into fatty acids and glycerol
	Liver	Bile	Emulsifies lipids
	Cells lining small intestine	Peptidases	Split small peptides into amino acids
	Cells lining small intestine	Disaccharidases	Split disaccharides into monosaccharides

(a)

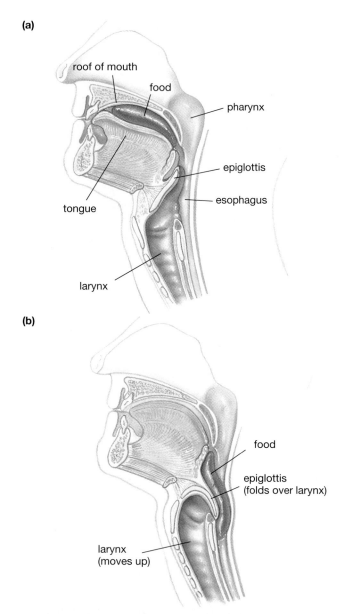

(b)

Figure 32-10   **The challenge of swallowing**

(a) Swallowing is complicated by the fact that both the esophagus (part of the digestive system) and the larynx (part of the respiratory system) open into the pharynx. **(b)** During swallowing, the larynx moves upward beneath a small flap of cartilage, the epiglottis. The epiglottis folds down over the larynx, sealing off the opening to the respiratory system and directing food down the esophagus.

muscular walls undergo a variety of contracting, churning movements that help break apart large pieces of food.

Finally, glands in the lining of the stomach secrete enzymes and other substances that facilitate the chemical breakdown of food. These include gastrin, hydrochloric acid (HCl), pepsinogen, and mucus. **Gastrin** (a hormone) stimulates secretion of hydrochloric acid by specialized stomach cells. Other cells release pepsinogen, an inactive form of the protein-digesting enzyme pepsin. Pepsin is a **protease,** an enzyme that helps break proteins into shorter chains of amino acids called peptides (see Table 32-6). Pepsin must

be secreted in an inactive form to prevent it from digesting the very cells that produce it. The highly acidic conditions in the stomach (pH 1 to 3), though, convert pepsinogen into pepsin, which functions best in an acidic environment.

As you may have noticed, the stomach produces all the ingredients necessary to digest itself, and indeed, this is what happens when a person develops ulcers (see "Health Watch: Ulcers—When the Digestive Tract Digests Itself"). However, cells lining the stomach normally produce a large quantity of thick mucus that coats the stomach lining and serves as a barrier to self-digestion. The protection is not perfect, however, and the cells lining the stomach are digested to some extent and must be replaced every few days.

Food in the stomach is gradually converted to a thick, acidic liquid called **chyme,** which consists of partially digested food and digestive secretions. Peristaltic waves then propel the chyme toward the small intestine. The sphincter at the base of the stomach allows only about a teaspoon of chyme to be expelled with each contraction, which occurs about every 20 seconds. It takes 2 to 6 hours, depending on the size of the meal, to empty the stomach completely. The continued churning movements of an empty stomach are felt as "hunger pangs."

Only a few substances, including water, some drugs, and alcohol, can enter the bloodstream through the stomach wall. Alcohol that is consumed when the stomach is empty is immediately absorbed into the bloodstream, with strong and rapid effects. Because food in the stomach slows alcohol absorption, the advice "never drink on an empty stomach" is based on sound physiological principles.

Most Digestion Occurs in the Small Intestine

The **small intestine** is a coiled, narrow tube about 3 meters (about 10 feet) long in a living person (lengths of over 20 feet are based on measurements made from cadavers in which the muscular intestinal walls have relaxed and stretched). Its diameter ranges from about 2 to 5 centimeters (1 to 2 inches) in an adult human. The small intestine has two major functions: to digest food into small molecules and to absorb these molecules, passing them to the bloodstream. The first role of the small intestine—digestion—is accomplished with the aid of digestive secretions from three sources: the liver, the pancreas, and the cells of the small intestine itself.

The Liver and Gallbladder Provide Bile, Important in Fat Breakdown

The **liver** is the largest and perhaps most versatile organ in the body. Its many functions include storage of fats and carbohydrates for energy, regulation of blood glucose levels, synthesis of blood proteins, storage of iron and certain vitamins, conversion of toxic ammonia (released by the breakdown of amino acids) into urea, and detoxification of other harmful substances such as nicotine and alcohol. The role of the liver in digestion is to produce bile, which is stored and concentrated in the **gallbladder** and released into the small intestine through the bile duct (see Fig. 32-8).

Bile is a complex mixture composed of **bile salts**, water, other salts, and cholesterol. Bile salts are synthesized in the liver from cholesterol and amino acids. Although they assist in the breakdown of lipids, bile salts are not enzymes. Rather, they act as detergents or emulsifying agents, dispersing globs of fat in the chyme into microscopic particles. These particles expose a large surface area for attack by **lipases**, lipid-digesting enzymes produced by the pancreas (see "A Closer Look: The Fate of Fats").

The Pancreas Supplies Several Digestive Secretions to the Small Intestine

The **pancreas** lies in the loop between the stomach and small intestine (see Fig. 32-8). It consists of two major types of cells. One type produces hormones involved in blood sugar regu-

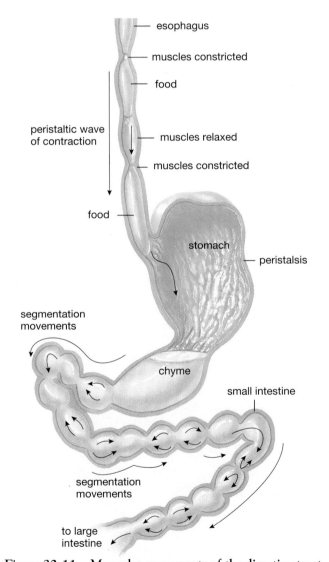

Figure 32-11   Muscular movements of the digestive tract
Food is propelled through the digestive system by the rhythmic peristaltic contractions of circular muscles. These contractions proceed downward through the digestive system, forcing the food along in front of them. In the stomach and small intestine, a variety of contractions, such as segmentation movement of the small intestine, help break down food particles and speed absorption.

lation (see Chapter 35), and the other produces a digestive secretion called **pancreatic juice**, which is released into the small intestine. The role of pancreatic juice is to neutralize the acidic chyme and to digest carbohydrates, lipids, and proteins. About 1 liter (1.06 quarts) of pancreatic juice is released into the small intestine each day. It contains water, sodium bicarbonate, and several digestive enzymes (see Table 32-6). Sodium bicarbonate (the active ingredient in baking soda) neutralizes the acidic chyme in the small intestine, producing a slightly basic pH. Pancreatic digestive enzymes require this basic pH for proper functioning, in contrast to the stomach's digestive enzymes, which require an acidic pH.

The pancreatic digestive enzymes break down three major types of food: Amylase breaks down carbohydrates, lipases digest lipids, and several proteases separate proteins and peptides. The pancreatic proteases include trypsin, chymotrypsin, and carboxypeptidase. Both trypsin and chymotrypsin break proteins and peptides into shorter peptide chains. Carboxypeptidase then completes protein digestion by separating the individual amino acids from the ends of the peptide chains. These proteases are secreted in an inactive form and become activated after they reach the small intestine.

The Intestinal Wall Completes the Digestive Process

The wall of the small intestine is studded with cells that are specialized to complete the digestive process and absorb the small molecules that result. These cells have various enzymes on their external membranes, which form the lining of the small intestine. These enzymes include proteases, which complete the breakdown of peptides into amino acids, and sucrase, lactase, and maltase, which break down disaccharides into monosaccharides (see Chapter 3). Small amounts of lipase digest lipids. Because these enzymes are actually embedded in the membranes of the cells lining the small intestine, this final phase of digestion occurs *as* the nutrient is being absorbed into the cell. Like the stomach, the small intestine is protected from digesting itself by large amounts of mucous secretions from specialized cells in its lining.

Most Absorption Occurs in the Small Intestine

The small intestine is not only the principal site of chemical digestion; it is also the major site of nutrient **absorption** into the blood. The small intestine has numerous foldings and projections that give it an internal surface area that is 600 times that of a smooth tube of the same length (Fig. 32-12). Covering the entire folded surface of the wall are minute, fingerlike projections called **villi** (literally, "shaggy hairs"). Villi, which range from 0.5 to 1.5 millimeters in length, give the intestinal lining a velvety appearance to the naked eye. They move gently back and forth in the chyme passing through the intestine. This movement increases their exposure to the molecules to be digested and absorbed. Further, each individual cell of the villi also bears a fringe of microscopic projections called **microvilli**. Taken together, these specializations of the

A CLOSER LOOK

The Fate of Fats

Because digestion occurs in a watery environment using water-soluble enzymes, the digestion of water-insoluble lipids poses a particular challenge: Fats in chyme tend to aggregate into globs that resist the attack of digestive enzymes. Bile salts, secreted by the liver and released by the gallbladder, provide the first level of attack. If you have ever poured detergent into a dishpan with oily water and watched the oil disperse, you've witnessed an effect very comparable to the action of bile salts on lipids. Like your dish detergent, bile salts are molecules with both hydrophilic (literally, "water-loving," or water-soluble) and hydrophobic ("water-fearing," or lipid-soluble) portions. The hydrophobic end dissolves in the lipid while the hydrophilic end dissolves in the surrounding watery fluid. As illustrated in Fig. E32-1, this structure enables bile salts to disperse the fat globs into microscopic particles ①. These particles expose a much greater surface area to the surrounding fluid, which is rich in pancreatic lipases (lipid-

digesting enzymes). As the lipases break triglycerides into fatty acids and monoglycerides, these subunits are absorbed into the lipid-soluble portion of tiny clusters of bile salts called **micelles** ②. The micelles carry the fatty acids and monoglycerides to cells lining the intestine. There, the fatty acids and monoglycerides diffuse through the cell membranes into the intestinal cells, leaving the bile salts behind ③.

Within the intestinal cells, the fat subunits are reassembled into triglycerides, mixed with cholesterol and phospholipids, and collected into droplets coated with protein ④. These droplets, called **chylomicrons**, are packaged into vesicles by the Golgi apparatus and expelled by exocytosis into the interstitial fluid ⑤. Too large to enter capillaries, chylomicrons diffuse into the lacteal, or lymph vessel, that projects into each intestinal villus. The fat is then carried in the lymph to large veins into which the lymphatic vessels empty their contents.

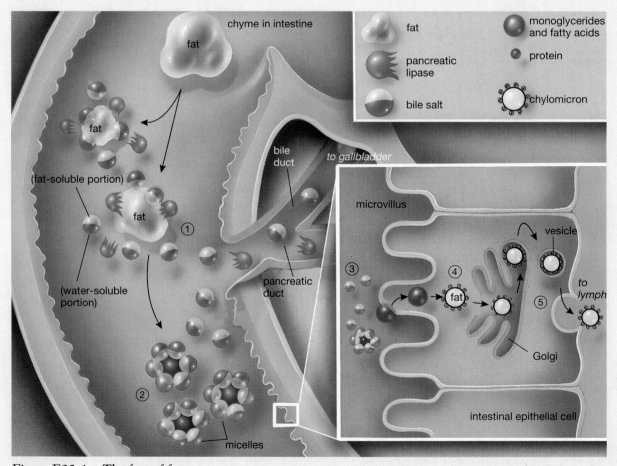

Figure E32-1   **The fate of fats**

small intestine wall give it a surface area of about 250 square meters, about the size of a tennis court.

Rhythmic, unsynchronized contractions of the circular muscles of the intestine, **segmentation movements**, slosh the chyme back and forth, bringing nutrients into contact with the absorptive surface of the small intestine (see Fig. 32-11). When absorption is complete, coordinated peristaltic waves conduct the leftovers into the large intestine.

Nutrients absorbed by the small intestine include water, monosaccharides, amino acids and short peptides, fatty acids produced by lipid digestion, vitamins, and minerals. The mechanisms by which this absorption occurs are varied and complex. In most cases, energy is expended to transport nutrients into the intestinal cells. (Water follows by osmosis as described in Chapter 6). The nutrients then diffuse out of the intestinal cells into the interstitial fluid, where they then enter the bloodstream.

Each villus of the small intestine is provided with a rich supply of blood capillaries and a single lymph capillary, called a **lacteal**, to carry off the absorbed nutrients and distribute them throughout the body (see Fig. 32-12). Most of the nutrients enter the bloodstream through the capil-laries, but fat subunits take a different route. After diffusing into the epithelial cells, they are resynthesized into fats, combined with other molecules, and then released as droplets into the interstitial fluid. By this means, they enter the lymph capillary and are eventually delivered to the bloodstream when the lymph vessels empty into the veins (see "A Closer Look: The Fate of Fats").

Water Is Absorbed and Feces Are Formed in the Large Intestine

The **large intestine** is about 1.5 meters (5 feet) long in the adult human and about 7.5 centimeters (3 inches) in diameter, which is considerably wider than the small intestine. The large intestine has two parts: For most of its length it is called the **colon**, but its final 15 centimeters (6 inches) is called the **rectum**. Into the large intestine flow the leftovers of digestion: a mixture of water, undigested fats and proteins, and indigestible fibers such as the cell walls of vegetables and fruits. The large intestine contains a flourishing population of bacteria that live on unabsorbed nutrients. These bacteria earn their keep by syn-

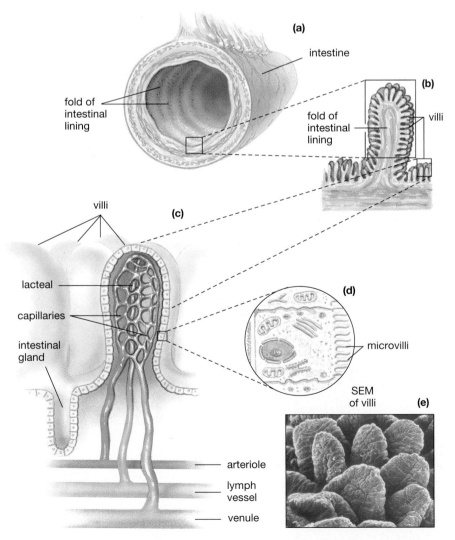

(a)

intestine

fold of intestinal lining

(b)

fold of intestinal lining

villi

villi

(c)

lacteal

capillaries

intestinal gland

(d)

microvilli

SEM of villi

(e)

arteriole

lymph vessel

venule

Figure 32-12   **The small intestine**
The surface area of the small intestine is greatly increased by **(a)** folds in the intestinal lining and **(b)** projections called villi (singular, villus) that extend from the folds, giving the lining a velvety appearance. The villi are enlarged in the micrograph **(e)** and the diagram **(c)**, showing the blood capillaries and lymph vessels that form a network within each villus where nutrients are absorbed. **(d)** Microvilli, which are microscopic projections of individual cell membranes, further increase the surface area for nutrient absorption.

thesizing vitamin B_{12}, thiamin, riboflavin, and, most important, vitamin K, which would otherwise be deficient in a normal diet. Cells lining the large intestine absorb these vitamins, as well as leftover water and salts.

After absorption is complete, the result is the semisolid feces consisting of indigestible wastes and the dead bodies of bacteria, which account for about one-third of the dry weight of feces. The feces are transported by peristaltic movements until they reach the rectum. Expansion of this chamber stimulates the desire to defecate. Although defecation is a reflex (as any new parents can attest), it is initiated voluntarily after around the age of two.

Digestion Is Controlled by the Nervous System and Hormones

Considerable coordination is required to break down a chef salad into amino acids and peptides, sugars, fatty acids, vitamins, minerals, and indigestible cellulose. As the mouth responds to the first bite, the stomach must be warned that food is on the way. In addition, the enzymes of the stomach and small intestine require different environments for proper functioning (highly acidic in the stomach, slightly basic in the small intestine), and secretions into various parts of the digestive tract must be coordinated with the arrival of food. Not surprisingly, the secretions and activity of the digestive tract are coordinated by both nerves and hormones (Table 32-7).

The initial phase of digestion is under the control of the nervous system and involves responses to signals originating in the head. These signals include the sight, smell, taste, and sometimes the thought of food, as well as the muscular activity of chewing. In response to these stimuli, saliva is secreted to the mouth, while nervous signals to the stomach walls initiate secretion of acid and the hormone gastrin, which stimulates further acid secretion. The concentration of acid is regulated by a negative feedback mechanism (see Chapter 29). When acid levels reach a certain point, they inhibit gastrin secretion, thereby inhibiting further acid production.

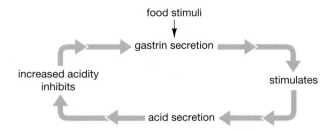

The arrival of food in the stomach triggers the second phase of digestion. Stimulation of the stomach wall causes production of a large quantity of mucus, which protects the stomach against self-digestion. The acidity of the stomach converts pepsinogen to its active form, pepsin, which begins protein digestion. However, the presence of protein in food tends to reduce the concentration of stomach acid. Thus, as protein is broken down, the acidity drops and the release of gastrin is no longer inhibited; gastrin is released again and stimulates further acid production. The cells secreting stomach acid are also activated by expansion of the stomach and by the presence of peptides produced by protein digestion.

As the liquid chyme is gradually released into the small intestine, its acidity stimulates the release of a second hormone, **secretin**, into the bloodstream by cells of the upper small intestine. Secretin causes the pancreas to pour sodium bicarbonate into the small intestine. Sodium bicarbonate neutralizes the acidity of the incoming chyme, creating an environment in which the pancreatic enzymes can function. A third hormone, **cholecystokinin**, is also produced by cells of the upper small intestine in response to the presence of chyme. This hormone stimulates the pancreas to release various digestive enzymes into the small intestine. It also stimulates the gallbladder to contract, squeezing bile through the bile duct to the small intestine. Bile assists in fat breakdown, as described earlier.

Gastric inhibitory peptide, a hormone secreted by the small intestine in response to the presence of fatty acids and sugars in chyme, inhibits acid production and peristalsis in the stomach. This inhibition slows the rate at which chyme is dumped into the small intestine, providing additional time for digestion and absorption to occur.

TABLE 32-7 ▪ *Some Important Digestive Hormones*			
Hormone	*Site of Production*	*Stimulus for Production*	*Effect*
Gastrin	Stomach	Food in mouth Distension of stomach Peptides in stomach	Stimulates acid secretion by cells in stomach
Secretin	Small intestine	Acid in small intestine	Stimulates bicarbonate production by pancreas and liver; increases bile output by liver
Cholecystokinin	Small intestine	Amino acids, fatty acids in small intestine	Stimulates secretion of pancreatic enzymes and release of bile by gallbladder
Gastric inhibitory peptide	Small intestine	Fatty acids and sugars in small intestine	Inhibits stomach movements and release of stomach acid

HEALTH WATCH

Ulcers—When the Digestive Tract Digests Itself

About 1 out of every 10 Americans eventually develops an ulcer. Ulcers occur when the mucus barriers of the stomach and upper small intestine break down, and the inner lining and deeper layers of the digestive tract are eroded by acid and the protein-digesting enzyme pepsin. The upper small intestine is the most common site for ulcers (Fig. E32-2), because it receives the highly acidic chyme but is less protected by mucus than the stomach is.

The causes of ulcers are complex and poorly understood. A tendency to develop ulcers can be inherited. Some, but not all, ulcer sufferers secrete excess stomach acid. In some cases, decreased secretion of sodium bicarbonate or mucus by cells lining the stomach and small intestine may contribute to the disease. There is a correlation between the development of ulcers and stress, which may stimulate acid and pepsin secretion. Smoking and chronic consumption of alcohol and aspirin all may reduce the resistance of the digestive tract lining to the effects of pepsin and acids and can aggravate ulcers. In 1983, B. J. Marshall, an Australian physician, identified a bacterium in stomach tissue from ulcer patients. When his claims that the bacterium caused the ulcer were met with skepticism, he swallowed a batch of them, and soon developed ulcer symptoms. The bacterium (*Helicobacter pylori*) is found in the digestive tracts of nearly all ulcer victims, and antibiotics that kill these bacteria have been very effective in reducing the recurrence of ulcers. Although an estimated half of the world's population harbors these bacteria, they cause ulcers in only 10% to 20% of those infected.

Ulcers can sometimes be relieved by stress-reduction programs, eliminating smoking and drinking, and dietary changes. Medications for ulcers include antibiotics, antacids, and drugs that decrease stomach acid production.

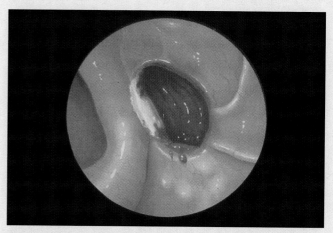

Figure E32-2 **A bleeding ulcer in the stomach**

An illustration of a bleeding ulcer in the stomach as it would be seen through a fiber-optic viewing device called an endoscope. The stomach lining has been digested away and blood seeps through the opening.

SUMMARY OF KEY CONCEPTS

Nutrition

Each type of animal has specific nutritional requirements. These requirements include: molecules that can be broken down to liberate energy, such as lipids, carbohydrates, and protein; chemical building blocks used to construct complex molecules, such as amino acids that can be linked together to form proteins; and minerals and vitamins that facilitate the diverse chemical reactions of metabolism.

The Challenge of Digestion

Digestive systems must accomplish five tasks: ingestion, mechanical followed by chemical breakdown of food, absorption, and elimination of wastes. Digestive systems convert the complex molecules of the bodies of other animals or plants that have been eaten into simpler molecules that can be utilized by the organism. Animal digestion at its simplest is intracellular, as occurs within the individual cells of a sponge. Extracellular digestion, utilized by all more complex animals, occurs in a body cavity. The simplest form is a saclike gastrovascular cavity in organisms such as flatworms and hydra. Still more complex animals utilize a tubular compartment with specialized chambers where food is processed in a well-defined sequence.

Human Digestion

In humans, digestion begins in the mouth, where food is physically broken down by chewing and chemical digestion is initiated by saliva. Food is then conducted to the stomach by peristaltic waves of the esophagus. In the acidic environment of the stomach, food is churned into smaller particles, and protein digestion begins. Gradually, the liquefied food, now called chyme, is released to the small intestine. Here it is neutralized by sodium bicarbonate from the pancreas. Secretions from the pancreas, liver, and the cells of the intestine complete the breakdown of proteins, fats, and carbohydrates. In the small intestine, the simple molecular products of digestion are absorbed into the bloodstream for distribution to the body cells. The large intestine absorbs the remaining water and converts indigestible material to feces.

Digestion is regulated by the nervous system and hormones. The smell and taste of food and the action of chewing causes the secretion of saliva and production of gastrin by the stomach. Gastrin stimulates stomach acid production. As chyme enters the small intestine, three additional hormones are produced by intestinal cells: secretin, which causes sodium bicarbonate production to neutralize the acid chyme; cholecystokinin, which stimulates bile release and causes the pancreas to secrete digestive enzymes into the small intestine; and gastric inhibitory peptide, which inhibits acid production and peristalsis by the stomach. This inhibition slows the movement of food into the intestine.

KEY TERMS

absorption p. 631
amylase p. 628
bile p. 631
bile salt p. 631
calorie p. 620
cellulase p. 627
cholecystokinin p. 634
chylomicron p. 632
chyme p. 630
colon p. 633
crop p. 626
digestion p. 625
epiglottis p. 628
esophagus p. 626
essential amino acid p. 621
essential fatty acid p. 620
extracellular digestion p. 626
food vacuole p. 625
gallbladder p. 630
gastric inhibitory peptide p. 634
gastrin p. 630
gastrovascular cavity p. 626
gizzard p. 626
glycogen p. 620
intracellular digestion p. 625
lacteal p. 633
large intestine p. 633
lipase p. 631
liver p. 630
lysosome p. 625
micelle p. 632
microvillus p. 631
mouth p. 625
pancreas p. 631
pancreatic juice p. 631
peristalsis p. 628
pharynx p. 628
protease p. 630
pyloric sphincter p. 629
rectum p. 633
secretin p. 634
segmentation movement p. 633
small intestine p. 630
stomach p. 629
urea p. 621
villus p. 631
vitamin p. 621

THINKING THROUGH THE CONCEPTS

Multiple Choice

1. An acidic mixture of partially digested food that moves from the stomach into the small intestine is called
   a. cholecystokinin b. bile
   c. lymph d. secretin
   e. chyme
2. The hormone responsible for stimulating secretion of hydrochloric acid by stomach cells is
   a. pepsin b. gastrin
   c. cholecystokinin d. insulin
   e. secretin
3. Chylomicrons are
   a. fatty particles that move from the small intestine into capillaries
   b. micelles produced by the action of bile
   c. droplets of protein coated with a layer of lipid
   d. molecules produced by the action of secretin on chyme
   e. structures made of triglycerides, cholesterol, and phospholipids
4. Humans lack digestive enzymes to attack chitin, a complex polysaccharide in the exoskeleton of lobster and crayfish. They also lack carbohydrases that degrade
   a. glycogen b. plant starch
   c. cellulose d. sucrose
   e. maltose
5. Which of the following is NOT characteristic of bile?
   a. It is produced in the gallbladder.
   b. It is a mixture of special salts, water, and cholesterol.
   c. It acts as a detergent or emulsifying agent.
   d. It helps expose a large surface area of lipid for attack by lipases.
   e. It works in the small intestine.
6. "Water-soluble compound that works primarily as enzyme helper" would be a good definition of
   a. vitamin C b. vitamin A
   c. vitamin B d. vitamin E
   e. vitamin D

Review Questions

1. List four general types of nutrients and describe the role of each in nutrition.
2. List and describe the function of the three principal secretions of the stomach.
3. List the substances secreted into the small intestine, and describe the origin and function of each.
4. Name and describe the muscular movements that usher food through the human digestive tract.
5. Vitamin C is a vitamin for humans but not for dogs. Certain amino acids are essential for humans but not for plants. Explain.
6. Name four structural or functional adaptations of the human small intestine that assure good digestion and absorption.
7. Describe protein digestion in the stomach and small intestine.

✖ A P P L Y I N G   T H E   C O N C E P T S

1. The food label on a soup can shows that the product contains 10 grams of protein, 4 grams of carbohydrate, and 3 grams of fat. How many Calories are found in this soup?

2. Stomach ulcers that resist antibiotic treatment are treated with several kinds of drugs. Anticholinergic drugs decrease nerve signals to the stomach walls that are produced by the sight, smell, and taste of food. Antacids neutralize stomach acid. Why would it be inadvisable to take anticholinergics and antacids together?

3. Small birds have high metabolic rates, efficient digestive tracts, and high-calorie diets. Some consume an amount of food equivalent to 30% of their body weight every day. They rarely eat leaves or grass but often eat small stones. The bird's duodenum has an attached pancreas and liver. Using this information, and Figure 32-7a, explain how a bird's digestive tract is adapted to its life-style (foods consumed, flight, habitat, etc.).

4. Control of the human digestive tract involves several feedback loops and messages that coordinate activity in one chamber with those taking place in subsequent chambers. List the coordinating events you discovered in this chapter, in order, beginning with tasting, chewing, and swallowing a piece of meat and ending with residue that enters the large intestine. What turns on and what shuts off each process?

5. Symbiotic protozoa in the digestive tracts of termites produce cellulase for them. In return termites provide protozoa with food and shelter. Imagine that the human species is gradually invaded, over many generations, by symbiotic protozoa capable of producing cellulase. What evolutionary adaptive changes in body structure and function might occur simultaneously?

6. Trace a ham and cheese with lettuce sandwich through the human digestive system, discussing what happens to each part of the sandwich as it passes through each region of the digestive tract.

7. One of the common remedies for constipation (difficulty eliminating feces) is a laxative solution containing magnesium salts. In the large intestine, magnesium salts are absorbed very slowly by the intestinal wall, remaining in the intestinal tract for long periods of time. This has an effect on water movement in the large intestine. Based on this information, explain the laxative action of magnesium salts.

✖ F O R   M O R E   I N F O R M A T I O N

Davenport, H. "Why the Stomach Does Not Digest Itself." *Scientific American*, January 1972. How the stomach protects itself from its own strongly acidic secretions, and what happens when the defenses fail.

Diamond, J. "Dining with the Snakes." *Discover*, April 1994. Follow food as it is consumed and digested by a snake.

Kretchmer, N. "Lactose and Lactase." *Scientific American*, April 1972. Discusses differences in tolerance to milk sugar among humans from different populations.

Livingston, E. H., and Guth, P. H. "Peptic Ulcer Disease." *American Scientist*, November–December 1992. Why and how ulcers develop.

Martin, R. J., White, B. D., and Hulsey, M. G. "The Regulation of Body Weight." *American Scientist*, November–December 1991. Describes some of the complex mechanisms that control eating, obesity, and the body's energy balance.

Mason, M. "Why Ulcers Run in Families." *Health*, September 1994. The story of the bacterium that causes ulcers.

Mayfield, E. "A Consumer's Guide to Fats." *FDA Consumer*, May 1994. Commonsense information on fat, cholesterol, diet, and health.

Moog, F. "The Lining of the Small Intestine." *Scientific American*, November 1981. A description of this intricate tissue, which is responsible for absorbing nutrients into the body.

Nuland, S. B. "The Beast in the Belly." *Discover*, February 1995. An interesting medical tale of bacterial disease, digestive enzymes, and food.

Willett, W. C. "Diet and Health: What Should We Eat?" *Science*, April 22, 1994. Summarizes studies that suggest that diet can play a major role in the prevention of disease.

N E T   W A T C H

On-line resources for this chapter are on the World Wide Web at:
http://www.prenhall.com/~audesirk (click on the <u>table of contents</u> link and then select Chapter 32).

33 The Urinary System and Homeostasis

⊞ AT A GLANCE

The Functions of the Urinary System

Some Simple Urinary Systems
Flame Cells Filter Fluids in Flatworms
Nephridia in Earthworms Resemble Parts of the
Vertebrate Kidney

The Human Urinary System
Urine Is Formed in the Kidneys
Blood Is Filtered by the Glomerulus
The Blood Filtrate Is Converted to Urine in the Nephron
The Loop of Henle Allows Urine to Become Concentrated
The Kidneys Are Important Organs of Homeostasis

A fountain in Belgium

Living cells have been compared to tiny factories, requiring a constant influx of raw materials and spewing out a constant stream of waste products, some of which are toxic at high levels. But unlike factories, living cells are delicate, requiring a precisely regulated environment. The extracellular fluid that bathes them must be close to neutral in pH; it must have precisely regulated concentrations of various salts; it must have appropriate levels of water and dissolved substances; and cellular waste products must not reach toxic levels.

The cells of the digestive system, through which most substances enter the body, are relatively unselective. Any molecule that can move into the body through the intestinal lining does so, including an excess of water, nutrients, salts, and minerals as well as nonnutritive substances such as drugs. However, even though we may load our digestive systems with pepperoni pizzas and hot fudge sundaes, our cells remain bathed in a precisely regulated solution of salts and nutrients. The burden of restoring and maintaining proper internal balance falls on the organs of homeostasis, particularly the kidneys of the urinary system.

The Functions of the Urinary System

All of the homeostatic functions of the urinary system are performed simultaneously as blood is filtered through the kidneys. Within the kidneys, the fluid portion of the blood is collected. From this fluid, water and important nutrients are reabsorbed into the blood, while toxic substances, cellular waste products, and excess vitamins, salts, hormones, and water are left behind to be excreted as urine. The rest of the urinary system channels and stores urine until it is released from the body, a process called **excretion**. The mammalian urinary system helps maintain homeostasis in several ways. These include:

1. Regulation of the blood levels of ions such as sodium, potassium, chloride, and calcium
2. Regulation of the water content of the blood
3. Maintenance of proper pH of the blood

4. Retention of important nutrients such as glucose and amino acids in the blood
5. Secretion of hormones, such as erythropoietin, which stimulates red blood cell production (see Chapter 30)
6. Elimination of cellular waste products such as urea

Urea is a product of amino acid metabolism. As you may recall, the digestive system breaks proteins into their amino acid building blocks, which are then absorbed. When amino acids are taken into cells, some are used directly to synthesize new proteins. Others have their amino ($-NH_2$) groups removed and are then used either as a source of energy or in the synthesis of new molecules. The amino groups are released as **ammonia** (NH_3), which is very toxic. In mammals, ammonia is carried in the blood to the liver, were it is converted to urea, a far less toxic substance (Fig. 33-1). Urea is filtered from the blood by the kidneys and excreted in **urine**, a fluid consisting of water and dissolved wastes and some excess nutrients.

In the following section, we see that even in relatively simple invertebrate animals, urinary systems perform functions similar to those of vertebrates: They collect and filter body fluids, retaining nutrients and releasing wastes.

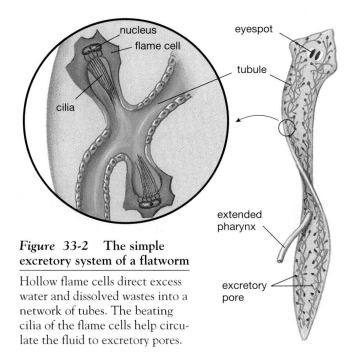

Figure 33-2   **The simple excretory system of a flatworm**

Hollow flame cells direct excess water and dissolved wastes into a network of tubes. The beating cilia of the flame cells help circulate the fluid to excretory pores.

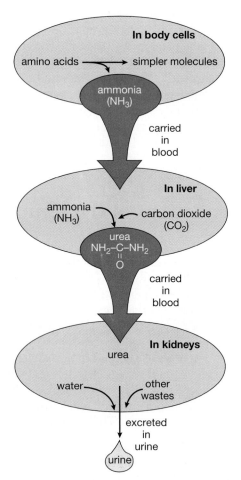

Figure 33-1   **A flow diagram showing the formation and excretion of urea**

Some Simple Urinary Systems

Flame Cells Filter Fluids in Flatworms

One of the simplest animals with an excretory system is the flatworm, which is found under rocks in streams. The flatworm's urinary system (often called an excretory system) consists of a network of tubes that branch throughout the body (Fig. 33-2). At intervals the tubes end blindly in single-celled bulbs called **flame cells**, which derive their name from the tuft of beating cilia extending into the hollow bulb. Under the microscope, the beating of the cilia resembles the flickering of a flame. Water and some dissolved wastes are filtered into the bulbs, where the beating cilia produce a current that conducts the fluid through the tubular network. Here, more waste products are added and nutrients withdrawn. Eventually, the waste liquid reaches one of numerous pores that release it to the outside. Flatworms also have a large skin surface through which wastes leave by diffusion.

Nephridia in Earthworms Resemble Parts of the Vertebrate Kidney

Earthworms, mollusks, and several other types of invertebrates have simple kidney-like structures called **nephridia** (singular, **nephridium**). In the earthworm, fluid fills the body cavity, or coelom, that surrounds the internal organs. This coelomic fluid collects both wastes and nutrients from the blood and tissues. The fluid is conducted into a funnel-shaped opening called the **nephrostome** and swept by cilia along a narrow, twisted tube (Fig. 33-3). Here, salts and other dissolved nutrients are absorbed back into the blood, leaving behind water and wastes. The resulting urine is stored in an enlarged bladderlike portion of the nephridium and is then excreted through an opening, the **excretory pore,** in the body

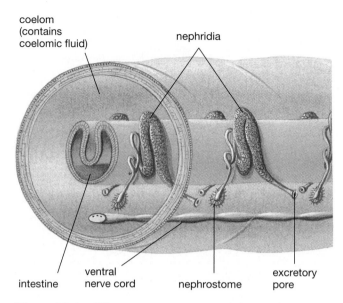

Figure 33-3   The excretory system of the earthworm

This system consists of structures called nephridia, one pair per segment. Coelomic fluid is drawn into the nephrostome, and urine is released through the excretory pore. Each nephridium resembles a vertebrate nephron.

wall. The earthworm body is composed of repeating segments, nearly every one of which contains its own pair of nephridia.

The Human Urinary System

Humans and other vertebrates filter their blood using **kidneys,** complex organs that in some ways resemble dense collections of nephridia. The kidneys are part of a larger group of structures collectively called the urinary system (Fig. 33-4). The kidneys actually produce the urine, and other portions of the system transport, store, and eliminate it. In the following sections we examine the major structures of the urinary system, tracing the pathway taken by waste products as they travel through the system. Then we explain kidney function in greater detail, and finally, we describe the role of the kidneys in homeostasis.

Human kidneys are paired, kidney bean-shaped organs located on either side of the spinal column and extending slightly above the waist. Each is approximately 13 centimeters (5 inches) long, 8 centimeters (3 inches) wide, and 2.5 centimeters (1 inch) thick. Blood carrying dissolved cellular wastes enters each kidney through a **renal artery.** After it has been filtered, the blood exits through the **renal vein** (Fig. 33-5). Urine leaves each kidney through a narrow, muscular tube called the **ureter.** Using peristaltic contractions, the ureters transport urine to the **bladder.** This hollow, muscular chamber collects and stores the urine.

The walls of the bladder, which contain smooth muscle, are capable of considerable expansion. Urine is retained in the bladder by two sphincter muscles located at its base, just above the junction with the urethra. When the bladder becomes distended, receptors in the walls sig-

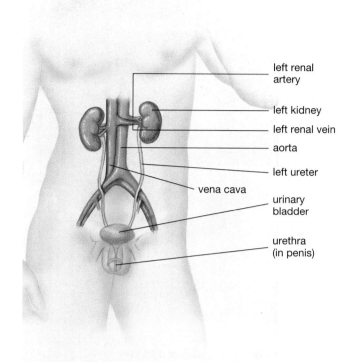

Figure 33-4   The human excretory system and its blood supply

nal its condition and trigger reflexive contractions. The sphincter nearest the bladder, the internal sphincter, is opened during this reflex. However, the lower, or external, sphincter, is under voluntary control, so the reflex can be suppressed by the brain unless bladder distension becomes acute. The average adult bladder will hold about 500 milliliters (approximately a pint) of urine, but the desire to urinate is triggered by considerably smaller accumulations. Urine completes its journey to the outside through the **urethra,** a single narrow tube about 3.8 centimeters (1.5 inches) long in the female and about 20 centimeters (8 inches) long in the male.

Urine Is Formed in the Kidneys

The kidney contains a solid outer layer where urine is formed and a hollow inner chamber called the **renal pelvis.** The renal pelvis is a branched collecting chamber that funnels urine into the ureter (see Fig. 33-5). The outer layer of the kidney is divided into a fan-shaped inner **renal medulla** and an overlying **renal cortex.** Microscopic examination of these structures reveals an array of tiny individual filters, or **nephrons.** Over 1 million nephrons are packed into the cortex of each kidney, with many extending into the renal medulla.

The nephron has three major parts: the **glomerulus,** a dense knot of capillaries from which fluid from the blood is filtered into a surrounding cuplike structure, called **Bowman's capsule,** and finally, a long, twisted **tubule** (from the

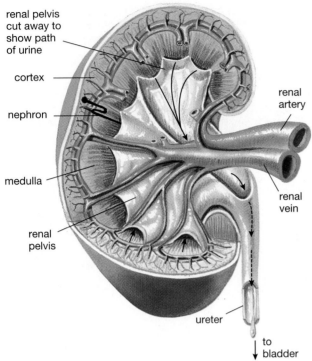

Figure 33-5   Cross section of a kidney
The cross section shows the blood supply and internal structure of a kidney. The renal artery, which brings blood to the kidney, and the renal vein, which carries the filtered blood away, branch extensively within the kidney. The two are joined by a highly permeable capillary network through which substances are exchanged between the blood and the nephrons. A nephron, considerably enlarged, is drawn to show its orientation in the kidney. The renal pelvis is the branched collecting chamber that funnels urine out of the kidney.

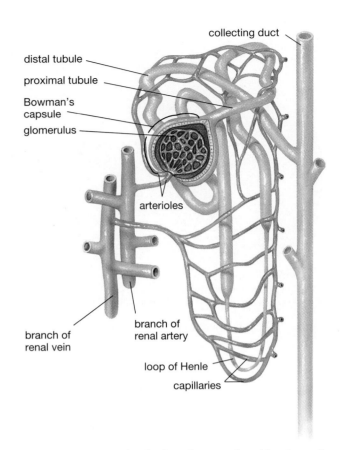

Figure 33-6   An individual nephron and its blood supply

Latin, "little tube"). The tubule is further subdivided into first the **proximal tubule,** then the **loop of Henle** (which extends into the renal medulla), and finally the **distal tubule,** which leads to the **collecting duct** (Fig. 33-6; see also Fig. E33-1). In the tubule, nutrients are selectively reabsorbed from the filtrate back into the blood, and wastes and some of the water are left behind to form urine. Additional wastes are also secreted into the tubule from the blood. Different portions of the tubule selectively modify the filtrate as it travels through (Fig. 33-7). These processes are examined in the following sections and covered in more detail in "A Closer Look: The Nephron and Urine Formation."

Blood Is Filtered by the Glomerulus

Blood is conducted to each nephron by an arteriole that branches from the renal artery. Within a cup-shaped portion of the nephron—Bowman's capsule—the arteriole branches into numerous microscopic capillaries that form an intertwined mass, the glomerulus (see Fig. 33-6). The walls of the glomerular capillaries are extremely permeable to water and small dissolved molecules, but they prevent the movement of most large proteins, such as the albumin found in blood. Beyond the glomerulus, the capillaries reunite to

form an arteriole whose diameter is smaller than that of the incoming arteriole. The differences in diameter between the incoming and outgoing arterioles create pressure within the glomerulus, driving water and many of the dissolved substances from the blood through the capillary walls. This process is called **filtration** (see Fig. 33-7), and the resulting fluid is called the **filtrate.** The watery filtrate, resembling blood plasma minus its plasma proteins, is collected in Bowman's capsule for transport through the nephron.

With the filtrate removed, the blood in the arteriole leaving the glomerulus is now very concentrated, having had much of its water removed but retaining substances too large to pass through the glomerular capillary walls, such as blood cells, large proteins, and fat droplets. Beyond the glomerulus, the arteriole branches into smaller, highly porous, capillaries. These capillaries surround the tubule, forming intimate contacts with it. At these points of contact, water and nutrients remaining in the filtrate after filtration are reabsorbed as the filtrate passes through the nephron and are returned to the blood; in addition, wastes remaining in the blood after filtration are passed into the filtrate for disposal.

The Blood Filtrate Is Converted to Urine in the Nephron

The blood filtrate collected in Bowman's capsule contains a mixture of both wastes and essential nutrients, including most of the blood's vital water. The nephron restores the nu-

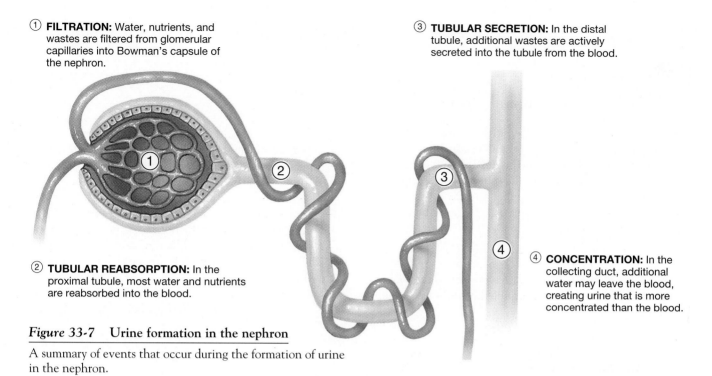

① **FILTRATION:** Water, nutrients, and wastes are filtered from glomerular capillaries into Bowman's capsule of the nephron.

② **TUBULAR REABSORPTION:** In the proximal tubule, most water and nutrients are reabsorbed into the blood.

③ **TUBULAR SECRETION:** In the distal tubule, additional wastes are actively secreted into the tubule from the blood.

④ **CONCENTRATION:** In the collecting duct, additional water may leave the blood, creating urine that is more concentrated than the blood.

Figure 33-7   Urine formation in the nephron

A summary of events that occur during the formation of urine in the nephron.

trients and most of the water to the blood, while retaining wastes for elimination. This is accomplished by two processes: tubular reabsorption and tubular secretion (see Fig. 33-7).

Tubular reabsorption is the process by which cells of the tubule remove water and nutrients from the filtrate within the tubule and pass them back into the blood. Reabsorption of salts and other nutrients, such as amino acids and glucose, generally occurs by active transport (that is, cells of the tubule expend energy to transport these substances out of the tubule). These nutrients then enter adjacent capillaries by diffusion. Water follows the nutrients out of the tubule by osmosis. Wastes such as urea remain in the tubule and become more concentrated as water leaves.

Tubular secretion is the process by which wastes and excess substances that were not initially filtered out into Bowman's capsule are removed from the blood for excretion. These wastes are actively secreted *into* the tubule by tubule cells. Secreted substances include hydrogen and potassium ions, ammonia, and many drugs.

The Loop of Henle Allows Urine to Become Concentrated

The kidneys of mammals and birds are able to produce urine that is more concentrated (has a higher osmotic pressure) than their blood. The ability to concentrate urine is a result of the structure of both the nephron and the collecting duct into which several nephrons empty. Urine can become concentrated because there is an osmotic concentration gradient of salts and urea in the interstitial fluid surrounding the loop of Henle. This gradient is produced by the loop of Henle within the renal medulla; the longer the loop, the greater the concentration gradient. The most concentrat-

ed fluid (with the greatest amount of dissolved substances and the least amount of water), which is far more concentrated than blood, surrounds the bottom of the loop. The collecting duct passes through this osmotic gradient. As the filtrate passes through the portion of the collecting duct surrounded by the osmotically concentrated fluid, additional water leaves the filtrate by osmosis, but wastes do not. Therefore, as it moves through the collecting duct, the filtrate, now called urine, can reach an osmotic equilibrium, becoming as concentrated as the surrounding fluid. Because the rest of the excretory system does not allow water to enter or urea to escape, the urine remains concentrated.

It is important to produce concentrated urine when water is scarce and to produce dilute watery urine when there is excess water in the blood. The degree of concentration of the urine is controlled by the amount of antidiuretic hormone, to be described shortly.

The Kidneys Are Important Organs of Homeostasis

Each drop of blood in your body passes through a kidney about 350 times a day; thus, the kidney is able to fine-tune the composition of the blood helping to maintain **homeostasis**. The importance of this task is illustrated by the fact that kidney failure results in death within a short time (see "Health Watch: When the Kidneys Collapse").

The Kidneys Regulate the Water Content of the Blood

One of the most important functions of the kidney is to regulate the water content of the blood (see Fig. 33-7). Human kidneys filter 125 milliliters (about half a cup) of

A C L O S E R L O O K

The Nephron and Urine Formation

The complex structure of the nephron is finely adapted to its function. In Figure E33-1, the nephron is presented in diagram form to illustrate the processing that occurs in each part. The numbers refer to the osmotic concentration of the filtrate (the higher the number, the greater the concentration of dissolved substances). Circled numbers refer to the following descriptions.

① *Filtration*. Water and dissolved substances are forced out of the glomerular capillaries into Bowman's capsule, from which they are funneled into the proximal tubule.

② *Tubular reabsorption*. In the proximal tubule, most of the important nutrients remaining in the filtrate are actively pumped out through the walls of the tubule and are reabsorbed into the blood. These include about 75% of the salts and water as well as amino acids, sugars, and vitamins. The proximal tubule is highly permeable to water, so water follows the nutrients, moving out of the tubule and back into the blood by osmosis.

③ The loop of Henle, which is unique to birds and mammals, is essential for urine concentration. It maintains a salt concentration gradient in the extracellular fluid surrounding it, with the highest concentration at the bot-

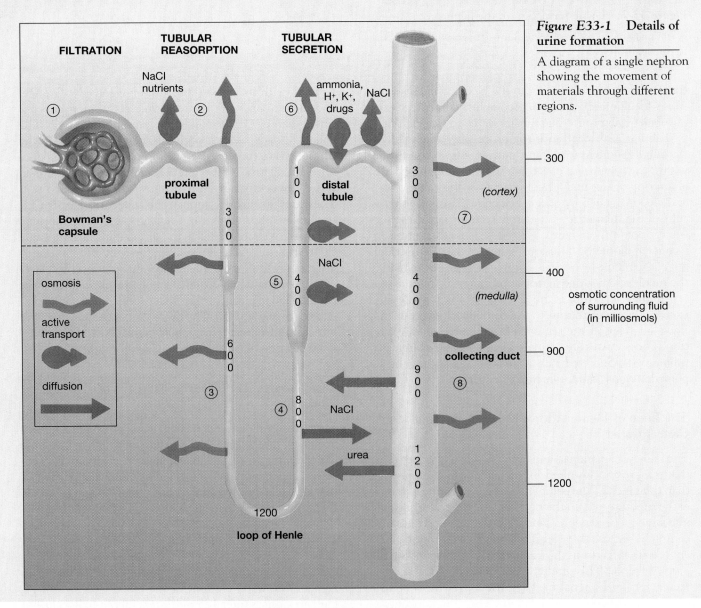

Figure E33-1   Details of urine formation

A diagram of a single nephron showing the movement of materials through different regions.

tom of the loop. The descending portion of the loop of Henle is very permeable to water but not to salt or other dissolved substances. As the filtrate passes through the descending portion, water leaves by osmosis as the concentration of the surrounding fluid increases.

④ The thin portion of the ascending loop of Henle is relatively impermeable to water and urea but is permeable to salt, which moves out of the filtrate by diffusion. Why? Although the osmotic concentrations inside and outside the tubule are about equal, at this stage the level of urea is higher outside and that of salt is higher inside. Thus, the diffusion gradient favors the movement of salt outward. Because water cannot follow it, the filtrate now becomes less concentrated than its surroundings.

⑤ The thick portion of the ascending loop of Henle is also impermeable to water and urea. Here, salt is actively pumped out of the filtrate, leaving water and wastes behind.

⑥ The watery filtrate, low in salt but retaining wastes such as urea, now arrives at the distal portion of the tubule, where more salt is pumped out. Because this portion is permeable to water, water follows by osmosis. Tubular secretion occurs throughout the tubule but is especially active in the distal portion. Here, substances such as K^+, H^+, NH_3, and some drugs and toxins are actively pumped into the tubule.

⑦ By the time the filtrate reaches the collecting duct, very little salt is left and about 99% of the water has been reabsorbed into the bloodstream. The collecting duct conducts the urine down through the increasing concentration gradient created by the loop of Henle. The collecting duct is very permeable to water when antidiuretic hormone (ADH) is present, so water moves out by osmosis as the concentration of the external fluid increases. If ADH is absent, the collecting duct remains impermeable to water, and the urine stays dilute and watery.

⑧ The lower portion of the collecting duct is also permeable to urea. Therefore, as the filtrate moves farther down the collecting duct, some urea diffuses out, contributing to the osmotic concentration of the surrounding fluid. As water (when ADH is present) and urea move out, the concentration of dissolved wastes in the urine in the collecting duct approaches equilibrium with the high osmotic concentration of the external fluid.

fluid from the blood each minute. This means that, without reabsorption of water, you would produce over 180 liters (45 gallons) of urine daily! Water reabsorption occurs passively by osmosis as the filtrate travels through the tubule and the collecting duct.

The amount of water reabsorbed into the blood is controlled by a negative feedback system (see Chapter 29) involving the amount of **antidiuretic hormone** (**ADH**; also called vasopressin) circulating in the blood. This hormone increases the permeability of the distal tubule and the collecting duct to water, thereby allowing more water to be reabsorbed from the urine. ADH is produced by cells in the hypothalamus and is released by the posterior pituitary gland (see Chapter 35).

ADH release is regulated by receptor cells in the hypothalamus that monitor the osmotic concentration of the blood and by receptors in the heart that monitor blood volume. For example, as the lost traveler staggers through the searing desert sun, dehydration occurs. With the loss of water, the osmotic concentration of his blood rises and his blood volume falls, triggering the release of more ADH (Fig. 33-8). Release of ADH increases water reabsorption and produces urine more concentrated than the blood. In contrast, a partygoer overindulging in beer will experience a decrease in blood concentration and an increase in blood volume, and her receptors will cause a decrease in ADH output. Reduced ADH concentration will make the distal tubule and collecting duct less permeable to water. When ADH is very low, little water is reabsorbed after the urine leaves the loop of Henle, and the urine produced will be more dilute than the blood. In extreme cases, urine flow may exceed 1 liter (about 1 quart) per hour. As the proper water level is restored, the increased osmotic concentration of the blood and decreased blood volume will stimulate increased ADH production, which will in turn stimulate increased water reabsorption.

Mammalian Kidneys Are Adapted to Diverse Environments

In addition to regulating water balance using hormones such as ADH, different mammals have kidneys whose structures are adapted to the availability of water in their natural habitats. Mammals that must conserve water do so by producing urine that is more concentrated (has a higher osmolarity) than their blood. The degree of concentration that can be achieved is determined by the length of the loop of Henle. The longer the loop, the higher the salt concentration produced in the fluid surrounding it. The higher the salt concentration, the more water moves out of the urine by osmosis as urine passes through the collecting duct, and the more concentrat-

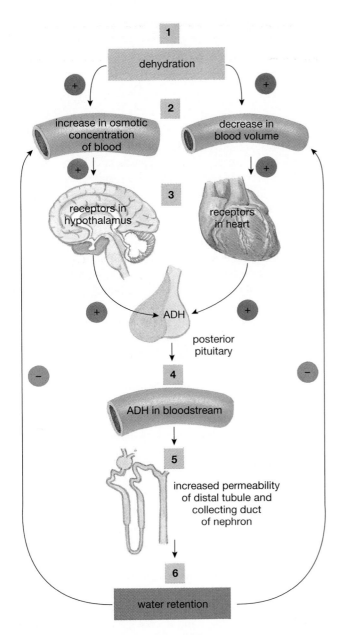

1

dehydration

2

increase in osmotic concentration of blood

decrease in blood volume

3

receptors in hypothalamus

receptors in heart

ADH

posterior pituitary

4

ADH in bloodstream

5

increased permeability of distal tubule and collecting duct of nephron

6

water retention

Figure 33-8   **Regulation of the water content of the blood**

Blood-water content is hormonally controlled by ADH through a negative feedback process. Stimulation is indicated by a plus sign (+), reduction by a minus sign (–). Dehydration, in addition to stimulating ADH release, triggers the sensation of thirst, leading to increased water intake. This response helps restore blood volume and diminish ADH secretion.

ed the urine becomes. As you might predict, animals living in very dry climates have the longest loops of Henle, while those living in watery environments have relatively short loops.

The beaver, for example, has only short-looped nephrons and is unable to concentrate its urine to more than twice its plasma concentration. Human kidneys have a mixture of long- and short-looped nephrons and can concentrate urine to about four times the plasma

concentration. The masters of urine concentration are desert rodents such as kangaroo rats, which can produce urine 14 times their plasma concentration (Fig. 33-9). Kangaroo rats (as you might predict) have only very long-looped nephrons. Because of their extraordinary ability to conserve water, they can completely dispense with drinking, relying entirely on water derived from their food.

Kidneys Help Regulate the Blood Pressure and Oxygen Content of the Blood

Two hormones produced by the kidneys are extremely important in regulating blood pressure and the blood's oxygen-carrying capacity. When blood pressure falls, the kidneys release **renin** into the bloodstream. Renin acts as an enzyme, catalyzing the formation of a second hormone, **angiotensin**, from a protein circulating in the blood. Angiotensin in turn causes arterioles to constrict, elevating blood pressure. Constriction of the arterioles carrying blood to the kidneys also reduces the rate of blood filtration, causing less water to be removed from the blood. Water retention causes an increase in blood volume and, consequently, an increase in blood pressure.

In response to low blood oxygen levels, the kidneys release a second hormone, **erythropoietin**, described in Chapter 30. Erythropoietin travels in the blood to the bone marrow, where it stimulates more rapid production of red blood cells, whose role is to transport oxygen.

Figure 33-9   **An adaptation to a dry environment**

The desert kangaroo rat of the southwestern United States can dispense with drinking partly because its long loops of Henle allow it to produce very concentrated urine.

H E A L T H   W A T C H

When the Kidneys Collapse

The kidney is a relatively rugged organ. Normally the need to urinate is our only evidence of its industrious activity. But if the kidneys fail, death occurs rapidly. The kidneys are vulnerable to attack from several sources. Some toxic substances can destroy kidney tubules. These include heavy metals (mercury, arsenic), organic compounds such as solvents (carbon tetrachloride), insecticides, and overdoses of some antibiotics and pain relievers. A second source of injury is through infection by bacteria, particularly intestinal bacteria that reach the kidney via the urethra. Still another source of injury is glomerulonephritis. In this disease, an abnormal immune response to an infection causes the antigen and antibody to fuse into insoluble particles, which are trapped in the glomeruli, resulting in inflammation and death of the tubules. Kidney damage is often treated using a remarkable machine: the artificial kidney (Fig. E33-2).

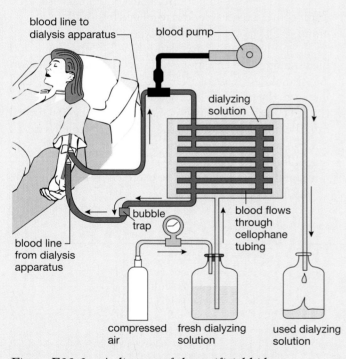

blood line to dialysis apparatus

blood pump

dialyzing solution

blood flows through cellophane tubing

bubble trap

blood line from dialysis apparatus

compressed air   **fresh dialyzing solution**   **used dialyzing solution**

Figure E33-2   **A diagram of the artificial kidney**

First used in 1945, the artificial kidney operates on the simple principle of diffusion of substances from areas of high concentration to areas of low concentration across an artificial permeable membrane. The passive diffusion of substances across an artificial semipermeable membrane is called **dialysis;** when blood is filtered using this principle, the process is called **hemodialysis.** During hemodialysis, the patient's blood is diverted from the body and pumped through narrow tubing made of a cellophane membrane suspended in dialyzing fluid. This membrane has pores too small to permit the passage of blood cells and large proteins, but large enough to pass small molecules such as water, sugar, salts, amino acids, and urea. The composition of the dialyzing fluid is adjusted to have normal blood levels of these substances so that important nutrients in the blood will be retained. As a result, only molecules whose concentration is in excess of normal will diffuse into the dialyzing fluid. Urea, for example, is present in relatively high concentration in the blood and absent in the dialyzing fluid, so it diffuses out of the blood and into the dialyzing fluid.

One challenge of dialysis is to eliminate excess water. Since the blood has a higher osmolarity than the dialyzing fluid, water tends to diffuse *into* the blood from the dialyzing fluid. This process is counteracted by applying pressure to the blood, which forces some of the small water molecules out of the blood against their concentration gradient.

Roughly 500 milliliters (1 pint) of blood are in the machine at any one time, flowing at a rate of several hundred milliliters per minute. Still, the patient must remain attached to the dialysis machine for 4 to 6 hours three times a week. Although patients requiring dialysis have remained alive for up to 20 years, their lives are far from normal. Their diet and fluid intake must be carefully regulated. Even so, their blood composition fluctuates and many more toxic substances accumulate than is normal. In spite of its drawbacks, dialysis keeps about 60,000 Americans alive. For those who are fortunate enough to find a compatible donor, kidney transplants are a far better solution; about 20,000 of these are performed yearly in the United States.

Kidneys Monitor and Regulate Dissolved Substances in the Blood

As the kidney filters the blood, it monitors and regulates blood composition in order to maintain a constant internal environment. Substances regulated by the kidney, in addition to water, include nutrients such as glucose, amino acids, vitamins, urea, and a variety of ions, including sodium, potassium, chloride, and sulfate. The kidney maintains a constant blood pH by regulating the amount of hydrogen and sodium bicarbonate ions. This remarkable organ also eliminates potentially harmful substances, including some drugs, food additives, pesticides, and toxic substances from cigarette smoke, such as nicotine.

✖ S U M M A R Y   O F   K E Y   C O N C E P T S

The Functions of the Urinary System

The urinary system plays a crucial role in homeostasis, the maintenance of a stable internal environment. The kidneys of mammalian urinary systems regulate water and ion content of the blood and blood pH. They help retain nutrients and eliminate cellular wastes and toxic substances, and they secrete the hormone erythropoietin.

Some Simple Urinary Systems

The simple excretory system of the flatworm consists of a network of tubules that branch through the body. Flame cells circulate body fluid through the tubules, where nutrients are reabsorbed. Wastes, including excess water, are excreted through numerous excretory pores. Many of the more complex invertebrates, including earthworms and mollusks, use nephridia. In the earthworm, these are paired structures resembling vertebrate nephrons that are found in most of the earthworm's segments. Coelomic fluid is drawn into a ciliated opening, the nephrostome, and nutrients and water are reabsorbed. Wastes and excess water are released through the excretory pore.

The Human Urinary System

The human urinary system consists of kidneys, ureters, bladder, and urethra. Kidneys produce urine, which is conducted by the ureters to a storage organ, the bladder. Distension of the muscular bladder wall triggers urination, during which urine passes out of the body through the urethra.

Each kidney consists of over a million individual nephrons in an outer renal cortex, with many extending into an inner renal medulla. Urine formed in the nephrons enters collecting ducts that empty into the renal pelvis, from which it is funneled into the ureter.

Each nephron is served by an arteriole that branches from the renal artery. The arteriole further branches into a mass of porous-walled capillaries called the glomerulus. Here water and dissolved substances are filtered from the blood by pressure. The filtrate is collected in the cup-shaped Bowman's capsule and conducted along the tubular portion of the nephron. During tubular reabsorption, nutrients are actively pumped out of the filtrate through the walls of the tubule. Nutrients then enter capillaries that surround the tubule, and water follows by osmosis. Some wastes remain in the filtrate, while others are actively pumped into the tubule by tubular secretion. The tubule forms the loop of Henle, which creates a salt concentration gradient surrounding it. After completing its passage through the tubule, the filtrate enters the collecting duct, which passes through the concentration gradient. Final passage of the filtrate through this gradient via the collecting duct allows concentration of the urine.

The kidneys are important organs of homeostasis. The water content of the blood is regulated by antidiuretic hormone (ADH), produced in the hypothalamus and released by the posterior pituitary gland. Low blood volume and high osmotic concentration of the blood signal dehydration and stimulate release of ADH into the bloodstream. ADH increases the permeability to water of the distal tubule and the collecting duct, allowing more water to be reabsorbed into the blood. In addition, the kidneys control blood pH, remove toxic substances, and regulate ions such as sodium, chloride, potassium, and sulfate. Excess glucose, vitamins, and amino acids are also excreted by the kidneys.

✖ K E Y   T E R M S

ammonia   p. 640
angiotensin   p. 646
antidiuretic hormone (ADH)   p. 645
bladder   p. 641
Bowman's capsule   p. 641
collecting duct   p. 642
dialysis   p. 647
distal tubule   p. 642
erythropoietin   p. 646
excretion   p. 639
excretory pore   p. 640
filtrate   p. 642

filtration   p. 642
flame cell   p. 640
glomerulus   p. 641
hemodialysis   p. 647
homeostasis   p. 643
kidney   p. 641
loop of Henle   p. 642
nephridium   p. 640
nephron   p. 641
nephrostome   p. 640
proximal tubule   p. 642
renal artery   p. 641

renal cortex   p. 641
renal medulla   p. 641
renal pelvis   p. 641
renal vein   p. 641
renin   p. 646
tubular reabsorption   p. 643
tubular secretion   p. 643
tubule   p. 641
urea   p. 640
ureter   p. 641
urethra   p. 641
urine   p. 640

✖ T H I N K I N G   T H R O U G H   T H E   C O N C E P T S

Multiple Choice

1. Which of the following is false?
   a. Urea is more toxic than ammonia.
   b. Ammonia is converted to urea in the liver.
   c. Ammonia is produced in body cells.
   d. The fluid collected in Bowman's capsule is called the filtrate.
   e. *Tubule* means "little tube."
2. The walls of the _____ are made more or less permeable to water, depending on the need to conserve water.
   a. ureter   b. urethra
   c. tubule   d. collecting duct
   e. glomerulus
3. Which of the following will cause a decrease in ADH production?
   a. dehydration
   b. drinking beer
   c. an increase in osmotic pressure of blood
   d. abnormally high blood sugar
   e. strenuous exercise
4. The function of the glomerulus and Bowman's capsule of the nephron is to
   a. reabsorb water into the blood
   b. eliminate ammonia from the body
   c. reabsorb salts and amino acids
   d. filter the blood and capture the filtrate
   e. concentrate the urine
5. What determines the ability of a mammal to concentrate its urine?
   a. the number of nephrons
   b. the length of the tubules
   c. the length of the collecting duct
   d. the size of the glomerulus
   e. the length of the loop of Henle
6. Which of the following processes does NOT occur in the nephron and collecting duct?
   a. filtration
   b. elimination of urea from the body
   c. reabsorption of nutrients
   d. tubular secretion
   e. concentration of urine

Review Questions

1. Explain the two major functions of excretory systems.
2. Trace a urea molecule from the bloodstream to the external environment.
3. What is the function of the loop of Henle? The collecting duct? Antidiuretic hormone?
4. Describe and compare the processes of filtration, tubular reabsorption, and tubular secretion.
5. Describe the role of the kidneys as organs of homeostasis.
6. Compare and contrast the excretory systems of humans, earthworms, and flatworms. In what general ways are they similar? Different?

✖ A P P L Y I N G   T H E   C O N C E P T S

1. Discuss the differences in function of the two major capillary beds in the kidneys: the glomerular capillaries and those surrounding the tubules.
2. Desert animals need to conserve water. These animals have larger kidneys than animals that live in moist environments and thus need not conserve water. The larger kidneys allow for a greater distance between the glomerulus and the bottom of the loop of Henle. Discuss why this anatomical difference assists water conservation in the desert animals.
3. Some "quick weight loss" diets require ingestion of much protein-rich food and elimination of carbohydrates. Two side effects of such diets are increased thirst and increased urination. Explain the connections between the diets and the side effects.
4. Your teenage son is dying from injuries sustained in a motorcycle accident. You have been asked to allow your son's kidneys to be donated to people currently relying on dialysis. Explain what benefits these kidneys would confer on a dialysis patient. How could two dialysis patients be helped by your son?
5. Some employers require their employees to submit to urine tests before they can be employed and at random intervals during their employment. Refusal to take the test or failure to "pass" the test could be grounds for termination. What is the purpose of the urine test? What employers might find such tests necessary? How do you feel about urine tests for obtaining or keeping a job? Explain your answers.

✖ F O R   M O R E   I N F O R M A T I O N

O'Brien, C. "Lucky Break for Kidney Disease Gene." *Science*, June 24, 1994. Discusses research on a dominant gene causing kidney disease.

N E T   W A T C H

On-line resources for this chapter are on the World Wide Web at:
http://www.prenhall.com/~audesirk (click on the <u>table of contents</u> link and then select Chapter 33).

34 Defenses Against Disease: The Immune Response

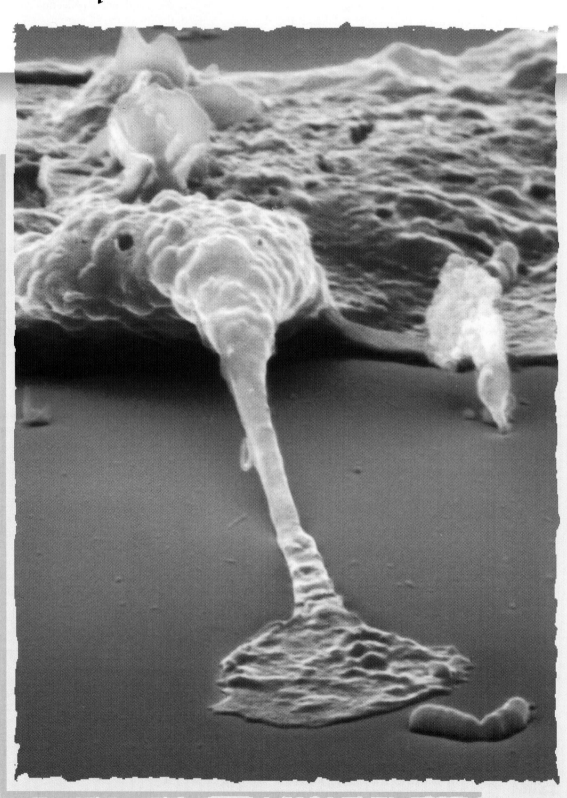

"If you wish to be well and keep well, take Braggs Vegetable Charcoal and Charcoal Biscuits. Absorbs all impurities in the stomach and bowels, effectively warding off cholera, smallpox, typhoid, and all malignant fevers. Eradicate worms in children. Sweeten the breath."

From an early 1900s newspaper ad

⠿ A T A G L A N C E

Defenses Against Microbial Invasion
The Skin and Mucous Membranes Form Barriers
Nonspecific Internal Defenses Combat Microbes

The Immune Response
A Successful Immune Response Recognizes, Overcomes, and Remembers

Recognition
Antibodies and T-Cell Receptors Recognize and Bind to Foreign Molecules, Triggering the Immune Response
An Antibody Contains Both Receptor and Effector Regions
T-Cell Receptors Bind Antigen and Trigger Responses
The Immune System Can Recognize Millions of Molecules
The Immune System Distinguishes "Self" from "Non-Self"

Attack
Humoral Immunity Is Produced by Antibodies in Blood
T Cells Produce Cell-Mediated Immunity

Memory

Medicine and the Immune Response
Antibiotics Slow Down Microbial Reproduction
Vaccinations Stimulate the Development of Memory Cells
Allergies Are Inappropriately Directed Immune Responses
An Autoimmune Disease Is an Immune Response against Some of the Body's Own Molecules
An Immune Deficiency Disease Results from the Inability to Mount an Effective Immune Response to Infection

AIDS
The Human Immunodeficiency Virus Is a Retrovirus That Infects and Destroys Helper T Cells
HIV Virus Is Transmitted by Exchange of Body Fluids
There Are Partially Effective Treatments, but No Cures, for AIDS
AIDS Is One of Many Widespread, Lethal Diseases

Cancer
Cancer Is Caused by Mutation, Activation, or Suppression of Genes That Control Cell Division

A macrophage, which is a type of white blood cell, sends out a pseudopod to capture a rod-shaped bacterium. The drama reproduced in this culture dish is played out many times each second in your body.

Every week, it seems, in newspapers, magazines, and on television, we encounter news of devastating, sometimes incurable, sometimes apparently new diseases: Lyme disease in the 1970s; AIDS in the 1980s; rodent-carried hantavirus in the Four Corners region of the American Southwest in 1993; Ebola virus outbreaks in Zaire in 1995; and new, more deadly forms of familiar old enemies, such as the bacterial infection that killed Muppets creator Jim Henson. It's enough to make you wonder how any of us survive more than a few years, at best. And the threat is real, too. Trillions of microbes lurk out there—in the air, in the water, in the foods we eat, and on the objects we touch—ready to invade our bodies. Nevertheless, most of us are healthy most of the time, not even aware of the constant combat between body and microbe. Even when an infection does occur, we almost always fend it off after a few days. How do we keep microbes out? If infected, how do we destroy the invaders?

Defenses Against Microbial Invasion

The human body has three lines of defense against microbial attack: (1) external barriers that keep microbes out of the body; (2) nonspecific internal defenses that combat all invading microbes; and (3) the immune system, that directs its assault against specific microbes (Fig. 34-1).

The Skin and Mucous Membranes Form Barriers

The first, and obviously best, defense is to keep microbes out in the first place. The human body has two surfaces exposed to the environment: the **skin** and the **mucous membranes** of the digestive and respiratory tracts. These surfaces are barriers to microbial invasion.

The Intact Skin Is Both a Barrier to Entry and an Inhospitable Environment for Microbial Growth

The outer surface of the skin consists of dry, dead cells filled with horny proteins similar to those in hair and nails. Consequently, most microbes that land on the skin cannot ob-

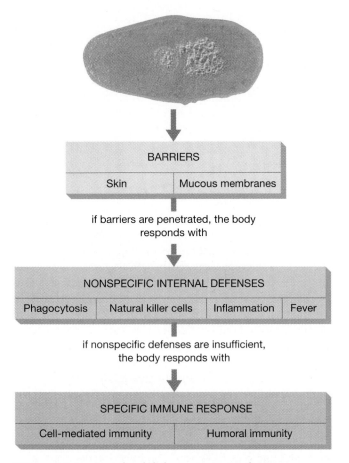

```
┌─────────────────────────────────────────┐
│              BARRIERS                    │
├──────────────────┬──────────────────────┤
│      Skin        │  Mucous membranes     │
└──────────────────┴──────────────────────┘
```

if barriers are penetrated, the body
responds with

```
┌──────────────────────────────────────────────────────────────┐
│              NONSPECIFIC INTERNAL DEFENSES                     │
├──────────────┬──────────────────┬──────────────┬─────────────┤
│ Phagocytosis │ Natural killer   │ Inflammation │   Fever     │
│              │     cells        │              │             │
└──────────────┴──────────────────┴──────────────┴─────────────┘
```

if nonspecific defenses are insufficient,
the body responds with

```
┌──────────────────────────────────────────────────┐
│             SPECIFIC IMMUNE RESPONSE               │
├────────────────────────┬───────────────────────────┤
│ Cell-mediated immunity │    Humoral immunity       │
└────────────────────────┴───────────────────────────┘
```

Figure 34-1   **Levels of defense against infection.**

these defenses, the warm, moist mucous membranes are much more vulnerable than the dry, oily skin, and many disease organisms enter the body through these membranes.

Nonspecific Internal Defenses Combat Microbes

Three nonspecific internal defenses are mustered against microbes that penetrate the skin or mucous membranes. These defenses are nonspecific because they attack a wide variety of microbes, rather than targeting specific invaders as the immune response does. First, the body has a standing army of **phagocytic cells** that destroy microbes and **natural killer cells** that destroy cells of the body that have been infected by viruses. The steady trickle of microbes that pass through the body's external barriers are mostly mopped up by these cells. Second, an injury, with its combination of tissue damage and relatively massive invasion of microbes, provokes an **inflammatory response.** The inflammatory response simultaneously recruits new members of the army of phagocytic cells and killer cells and walls off the injured area, isolating the infected tissue from the rest of the body. Third, if a population of microbes succeeds in establishing a major infection, the body often produces a **fever,** which both slows down microbial reproduction and enhances the body's own fighting abilities.

tain the water and nutrients they need. Secretions from sweat glands and sebaceous glands also cover the skin. These secretions contain acids and natural antibiotics, such as lactic acid, that inhibit the growth of bacteria and fungi. These multiple defenses make the unbroken skin an extremely effective barrier against microbial invasion.

Antimicrobial Secretions, Mucus, and Ciliary Action Defend the Mucous Membranes against Microbes

The membranes of the digestive and respiratory tracts are also well defended. First, they secrete mucus that contains antibacterial enzymes such as lysozyme, which destroys bacterial cell walls. Second, the mucus physically traps microbes that enter the body through the nose or mouth (Fig. 34-2). Cilia on the membranes sweep up the mucus, microbes and all, until it is either coughed or sneezed out of the body, or swallowed. If microbes are swallowed, they enter the stomach, where they encounter a combination of extreme acidity (pH 1 to 3) and protein-digesting enzymes, which can kill many types of microbes. Farther along in the digestive tract, the intestine is inhabited by bacteria that are harmless to the human body but secrete substances that destroy invading foreign bacteria or fungi. Despite

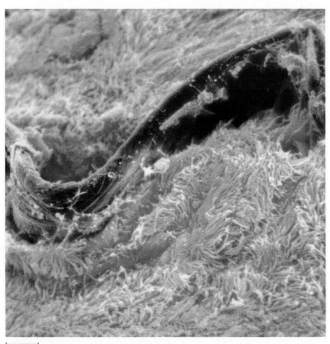

0.5 micrometer

Figure 34-2   **The protective function of mucus**
Mucus traps microbes and debris in the respiratory tract (a brown strand of dirt is shown caught in mucus atop the red cilia). The cilia lining the walls of the respiratory tract then sweep both mucus and foreign matter out of the body.

Phagocytic Cells and Natural Killer Cells Destroy Invading Microbes

The body contains several types of amoeboid cells that can engulf and digest microbes. The most important of these are the **macrophages** (literally, "big eaters"), white blood cells that crawl around in the extracellular fluid. Macrophages ingest microbes by phagocytosis. As we will see shortly, besides the immediate effect of destroying an individual microbe, macrophages also play a crucial role in the immune response by "presenting" parts of the microbe to other cells of the immune system (see "A Closer Look: Cellular Communication during the Immune Response").

Natural killer cells are another class of white blood cells. In general, natural killer cells do not directly attack invading microbes. Instead, natural killer cells strike at the body's own cells that have been invaded by viruses. Virus-infected cells usually bear some viral proteins on their surfaces. Natural killer cells recognize these proteins and kill the infected cell. In some way, natural killer cells also recognize and kill cancerous cells. Natural killer cells do not eat their victims; they strike from the outside. Their weapons are proteins that they secrete onto the plasma membrane of the infected or cancerous cell. Some of the proteins insert themselves into the target plasma membrane in a ring, much like making a barrel out of individual staves of wood, thereby opening up a large pore in the membrane (Fig. 34-3). Killer cells also secrete enzymes that break up some of the molecules of the target cell. Shot full of holes and chewed on by enzymes, the target cell soon dies.

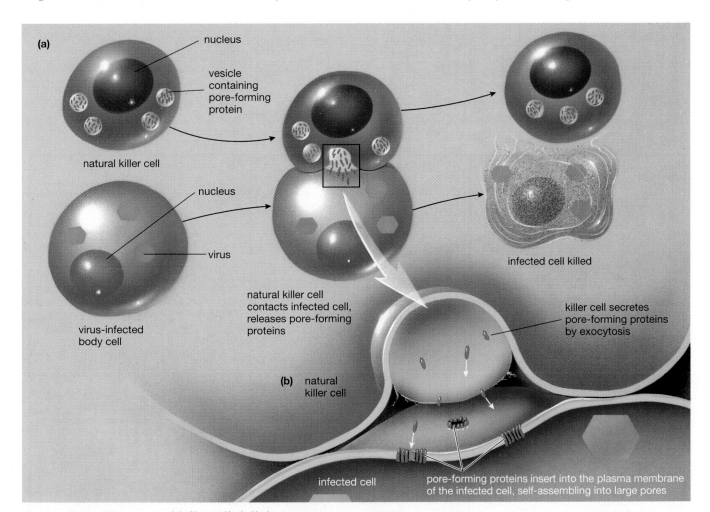

Figure 34-3   **How natural killer cells kill their targets**

(a) A natural killer cell contains vesicles full of pore-forming proteins. When it encounters a suitable target cell (here a virus-infected body cell), it briefly contacts the target and releases the pore-forming proteins by exocytosis. The pore-forming proteins self-assemble into large holes in the target cell's plasma membrane; the cell's contents leak out, and the cell dies. **(b)** A close-up view of the killing process. Pore-forming proteins leave the killer cell and insert themselves into the target cell's plasma membrane. The proteins line up next to one another like the staves in a barrel, forming large pores. The membrane of the killer cell is not affected; apparently it contains other proteins that bind to and inactivate any pore-forming proteins that try to insert themselves into the killer's membrane.

A CLOSER LOOK

Cellular Communication during the Immune Response

The immune system is a strange "system." Unlike the nervous system, for example, it is not composed of physically attached structures. Instead, as befits its mission of patrolling the entire body for microbial invaders, the immune system consists of an army of isolated cells. Nevertheless, the army is highly coordinated. Figure E34-1 presents an overview of just some of the complex communications among antigen, antibody, hormones, receptors, and cells that coordinate the immune response.

When a virus, for example, invades the body (step 1 in the figure), it sets off a cascade of events that can be loosely divided into three components.

Activation of Helper T Cells

One component of the immune response begins when macrophages ingest the virus (step 2 in Fig. E34-1) and digest it. Antigens "chewed off" the virus become attached to certain "self-identification" proteins of the macrophage's major

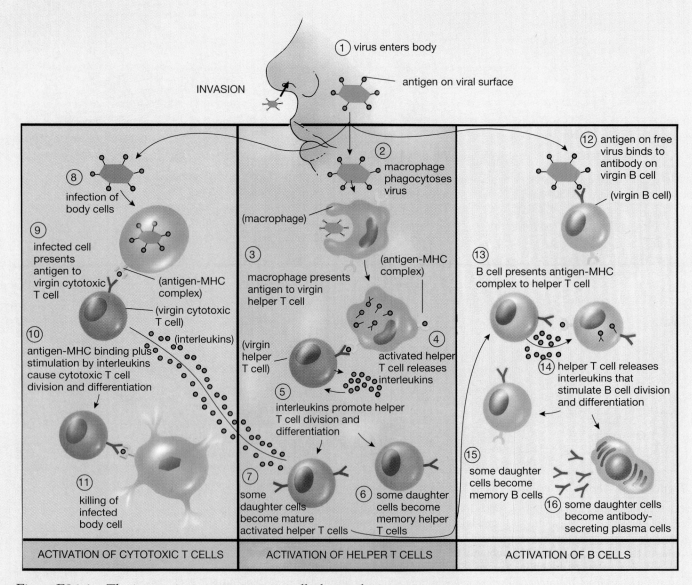

Figure E34-1   The interactions among immune cells during the immune response

The numbers are keyed to those used in the essay.

histocompatibility complex (MHC) and are displayed on the surface of the macrophage. These antigen-MHC complexes are recognized by virgin helper T cells (step 3); in immunology, *virgin* means that the T cell has never bound an antigen. When receptors on helper T cells bind to the antigen-MHC complex, the helper T cells release a hormone, called interleukin-2 (step 4; *interleukin* means "between white blood cells," which is an appropriate name for a hormone that allows communication among immune cells). This hormone stimulates cell division (step 5) in both the releasing cell and other T cells that have bound to antigen-MHC complexes (the antigens could be the same one, different antigens from the same microbe, or even antigens from an entirely different invading organism). Some daughter helper T cells become memory cells, providing future immunity (step 6); other daughter cells assist in stimulating the immune response of cytotoxic T cells and B cells (step 7).

Activation of Cytotoxic T Cells: Cell-Mediated Immunity

Meanwhile, the virus is infecting ordinary body cells, such as those lining the respiratory tract (step 8). Infected body cells display viral antigens on their surfaces, bound to another set of MHC molecules. Virgin cytotoxic T cells bind to the antigen-MHC complex (step 9) and are simultaneously stimulated by interleukin-2 released by the activated helper T cells. This dual binding-stimulation causes the cytotoxic T cells to multiply and become activated (step 10). When activated cytotoxic T cells then encounter infected cells displaying the antigen-MHC complex, they release toxic proteins that kill the infected cell (step 11).

Activation of B Cells: Humoral Immunity

Some B cells bear antibodies on their surfaces that bind antigens on the surface of free viruses that have not yet invaded a body cell (step 12). Antigen-antibody binding stimulates B cell division and maturation to some extent, but really turning B cells on requires a boost from helper T cells. B cells that have bound antigen ingest the antigen by receptor-mediated endocytosis (see Chapter 6), attach the antigen to MHC molecules, and display the antigen-MHC complex on their surfaces. The antigen-MHC complex is recognized by activated helper T cells (step 13), which then release several types of interleukin hormones that stimulate division and differentiation of antigen-binding B cells (step 14). Some of their progeny become memory cells (step 15); others become plasma cells that secrete antibodies into the bloodstream (step 16).

Summary of Immune Cell Communication

This interlocking communication network is quite complex. We can summarize its essentials, however, in five generalizations:

1. Macrophages bind to antigens, engulf them, and present them on their surfaces, along with "self-identification" MHC molecules, to helper T cells.
2. When helper T cells recognize the antigen-MHC complex, they multiply rapidly.
3. Meanwhile, cytotoxic T cells and B cells recognize the same antigen.
4. Hormones released by helper T cells stimulate cell division and maturation of only those cytotoxic T cells and B cells that have also been stimulated by antigen binding.
5. The stimulated cytotoxic T cells and B cells then provide cell-mediated and humoral immunity, respectively.

As you can see, helper T cells are essential in turning on both phases of the immune response. A loss of helper T cells, such as that caused by the AIDS virus, virtually eliminates the immune response to many diseases.

The Inflammatory Response Defends against Localized Injury

Large-scale breaches of the skin or mucous membranes, such as a cut, elicit an inflammatory response. Damaged cells release the chemical **histamine** into the wounded area. Histamine makes capillary walls leaky and relaxes the smooth muscle surrounding arterioles, leading to increased blood flow. With extra blood flowing through leaky capillaries, fluid seeps from the capillaries into the tissues around the wound. The wound becomes red, swollen, and warm (*inflammation* literally means "to set on fire"). Meanwhile, other chemicals released by injured cells initiate blood clotting (see Chapter 30), which both blocks damaged blood vessels, preventing microbes from escaping into the bloodstream, and seals off the wound from the outside world, limiting the entry of more microbes.

Still other chemicals, some released by wounded cells and others produced by the microbes themselves, attract macrophages and other phagocytic cells to the wound. The phagocytic cells squeeze out through the capillary walls, enter the wound, and engulf microbes, dirt, and tissue debris (Fig. 34-4). Unfortunately, each phagocyte can eat just so many microbes, and then it dies. The pus that collects around a wound consists largely of microbes, tissue debris, and living and dead white blood cells.

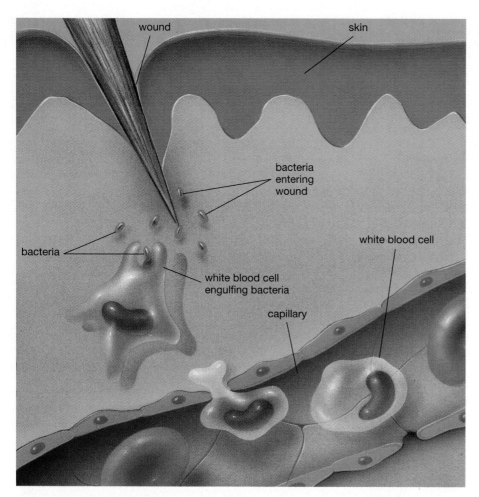

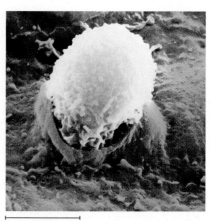

5 micrometers

Figure 34-4   **The inflammatory response**

During an inflammatory response triggered by a cut in the skin, the cells forming capillary walls separate slightly. Macrophages and other phagocytic cells squeeze through the gaps and enter the wounded tissue, where they engulf microbes, dirt, and debris from damaged cells. The micrograph shows a white blood cell leaving a capillary to join the fray against bacteria that have entered a cut.

Fever Is a Response by the Body That Combats Large-Scale Infections

If enough microbes make it past the inflammatory response, they may severely infect larger areas of the body, triggering a fever. Most Americans, having been trained by advertisements and TV dramas to regard fevers as dangerous and debilitating, believe that fevers are part of a microbe's assault on the body. The facts are just the opposite: A fever is part of the body's assault on the microbes. Severe fevers *are* dangerous, even fatal, but "average" fevers of 38° or 39°C (100° to 102°F) are beneficial.

The hypothalamus, a part of the brain, contains temperature-sensing nerve cells that are the body's thermostat. Normally, the thermostat is set at about 37°C (98.6°F). When disease organisms invade, however, the thermostat is turned up. Certain white blood cells, in responding to the infection, release hormones collective-ly called **endogenous pyrogens** ("self-produced fire-makers"). Pyrogens travel in the bloodstream to the hypothalamus and raise the thermostat's set point, triggering behaviors that increase body temperature: shivering, increased fat metabolism, or feeling cold so more clothing is put on. Pyrogens also cause other cells to reduce the concentration of iron in the blood.

Fever has both beneficial effects for the body's defenses and detrimental effects on the invading microbes. Many bacteria require more iron to reproduce at temperatures of 38° or 39°C than at 37°C, so fever and reduced iron in the blood combine to slow down their rate of reproduction. Simultaneously, fever increases the activity of phagocytic white blood cells that attack the bacteria, thereby producing a shorter and less serious infection.

Fever also helps fight viral infections. When certain cells of the body are invaded by viruses, they synthesize and release a protein called **interferon.** Interferon travels to other cells and increases their resistance to viral attack. Fevers increase the production of interferon. In one study, patients with colds were treated with aspirin (to reduce fever) or a placebo (an inactive substance that looks like the real drug, so the patients don't know whether they have been given the drug or not). The patients given aspirin had far more viruses in their noses and throats—and consequently sneezed and coughed out far more viruses—than the patients given the placebo did. This means that the immune systems of patients with lowered fevers weren't as effective at controlling their own infections and that these patients were much more infectious to other people.

As you can see, these defenses are all nonspecific; that is, their roles are to prevent or overcome any and all microbial invasions of the body. However, these nonspecific defenses are not impregnable. When they fail to do the job, the body then mounts a highly specific **immune response** directed against the particular organism that has successfully invaded the body.

The Immune Response

More than 2000 years ago, the Greek historian Thucydides recognized the essential features of the response of the immune system to infection. He observed that occasionally someone contracts a disease, recovers, and never catches that particular disease again—the person has become immune. With rare exceptions, however, immunity to one disease confers no protection against other diseases. Thus, the immune system attacks one type of microbe, overcomes

it, and provides future protection against that microbe but no others. This is why we refer to the immune response as a specific defense against invasion.

The immune system consists of about 2 trillion white blood cells, of a type called lymphocytes. Many are distributed throughout the body in the blood and lymph; others are clustered in specific organs, particularly the thymus, lymph nodes, and spleen. The immune response results from the interactions among several types of lymphocytes and the molecules that they produce. With such a large cast of characters, the theater of the immune response is difficult to follow without a program. Table 34-1 provides a brief summary of the major actors and their roles.

The immune response is extraordinarily complex in detail, but its essential features are conceptually simple. Therefore, we present the fundamentals of the immune response in the text; additional explanations outlining how the immune system performs some of its more complicated tasks are included in boxes entitled "A Closer Look . . ."

A Successful Immune Response Recognizes, Overcomes, and Remembers

The immune response is provided chiefly by two types of lymphocytes, called **B cells** and **T cells** (Fig. 34-5). As with all white blood cells, B and T lymphocytes arise from precursor cells in the bone marrow (see Chapter 30). Early in embryonic development, the newly forming T cells migrate to the thymus (hence the name T cell) and differentiate. B cells differentiate in the bone marrow itself (hence the name B cell). As we shall see shortly, B and T cells play quite different roles in the immune response. However, immune re-

---

TABLE 34-1 ▓ *The Major Molecules and Cells of the Immune Response*

Molecules

Antigens	Large organic molecules, usually proteins, polysaccharides, or glycoproteins, that can trigger an immune response; often found on the surface of cells.
Antibodies	Proteins produced by the cells of the immune system that bind to antigens and either neutralize the antigenic molecule itself or mark cells that bear the antigens for destruction.
Major histocompatibility complex (MHC)	A set of proteins found on the surface of cells that "label" the cell as belonging to a unique individual organism.
Effector molecules	A diverse group of molecules, including histamine and the cell-destroying proteins of killer cells and complement (soluble proteins found in blood).
Regulatory molecules	Hormonelike molecules produced by cells of the immune system that regulate the immune response.

Cells

Macrophages	Phagocytic white blood cells that both destroy invading microbes and help to alert other immune cells to the invasion.
B cells	Lymphocytes that produce antibodies; when stimulated, certain of their daughter cells (*plasma cells*) secrete large quantities of antibodies into the bloodstream.
T cells	A set of lymphocytes that regulate the immune response or kill certain types of cells. *Cytotoxic T cells* destroy specific targeted cells, usually either foreign eukaryotic cells, infected body cells, or cancerous body cells. *Helper T cells* stimulate immune responses by both B cells and killer T cells. *Suppressor T cells* inhibit immune responses by other lymphocytes.
Memory cells	A subset of the offspring of B and T cells that are long-lived and provide future immunity against a second invasion by the same antigen.

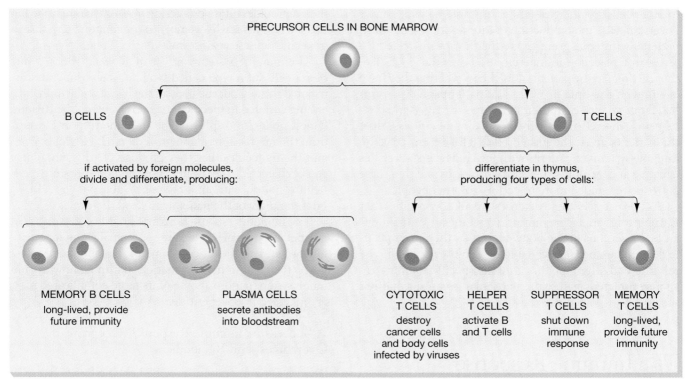

PRECURSOR CELLS IN BONE MARROW

B CELLS

if activated by foreign molecules,
divide and differentiate, producing:

T CELLS

differentiate in thymus,
producing four types of cells:

MEMORY B CELLS
long-lived, provide
future immunity

PLASMA CELLS
secrete antibodies
into bloodstream

CYTOTOXIC
T CELLS
destroy
cancer cells
and body cells
infected by viruses

HELPER
T CELLS
activate B
and T cells

SUPPRESSOR
T CELLS
shut down
immune
response

MEMORY
T CELLS
long-lived,
provide future
immunity

Figure 34-5   **The major cells of the immune system and their roles in the immune response**

sponses produced by both B and T cells consist of the same three fundamental steps: (1) recognizing the invader, (2) launching a successful attack, and (3) retaining the memory of the invader to ward off future infections.

Recognition

To understand how the immune system responds to invading microbes, we must answer three related questions: (1) How do the immune cells recognize foreign molecules? (2) How can they produce specific responses to so many different molecules? (3) How do they determine that a substance is "foreign" and not "self," thereby avoiding destruction of the tissues of their own body?

Antibodies and T-Cell Receptors Recognize and Bind to Foreign Molecules, Triggering the Immune Response

The key to the immune system's ability to attack invading microbes lies in the structure and function of large proteins called **antibodies** and **T-cell receptors**. Antibodies are found either attached to the surfaces of B cells or dissolved in the plasma of the blood, where they are called immunoglobulins (often abbreviated Ig). As their name implies, T-cell receptors are attached to the surfaces of T cells. They are never secreted from the cells into the bloodstream.

An Antibody Contains Both Receptor and Effector Regions

Antibodies are Y-shaped molecules composed of four peptide chains, one large (heavy) chain and one small (light) chain on each side of the Y (Fig. 34-6). Both heavy and light chains consist of a **constant region** that is similar in all antibodies of the same class (see Table 34-2) and a **variable region** that differs among antibodies. The combination of light and heavy chains results in an antibody with two functional parts: the "arms" and the "stem" of the Y.

The Arms of an Antibody Form Binding Sites That Recognize Specific Antigens

The variable regions that make up the arms form highly specific binding sites for large molecules. These binding sites are a lot like the active sites of enzymes (Chapter 4): Each binding site has a particular shape and electrical charge, so only certain molecules can fit in and bind. The binding sites at the tips of an antibody's arms are so specific that each antibody can bind just a few different types of molecules, perhaps only one.

In general, only large, complex molecules, such as proteins, polysaccharides, and glycoproteins, can bind to antibodies. These molecules are called **antigens** (short for "*anti*body response *gen*erating"). Antigens either may be attached to the surfaces of cells (for example, the body's own cells or invading microbes) or may be dissolved in the blood or extracellular fluid (such as snake venom or a chain constant regions that make up the stem of one type of antibody may function to attach the antibody to the plasma membrane of a cell; the constant regions of another type of antibody may bind to certain proteins in the blood, called **complement**, to promote destruction of microbes.

Antibodies Are Both Receptors for Foreign Molecules and Effectors That Help Destroy Invading Molecules and Microbes

Antibodies perform two very different functions in the immune response (Fig. 34-7). First, antibodies are receptors. In an antibody's role as a receptor, the stem attaches the antibody to the B cell's plasma membrane, while the two arms of the antibody protrude outward, sampling the blood and lymph for antigen. The binding of antigen to these "receptor antibodies" triggers responses in the lymphocytes bearing the antibodies.

Second, antibodies are effectors. As we will explain shortly, certain descendants of B cells (called plasma cells) secrete antibodies into the bloodstream. There the antibodies play several roles in neutralizing poisonous antigens or destroying microbes that bear antigens. Small differences in the stem—that is, the constant region—yield five classes of true antibodies (not including T-cell receptors; Table 34-2) that serve different roles in the defense of the body. The stems of some antibodies form a coating on microbes that targets the microbes for ingestion by macrophages (IgG); other stems promote secretion of the antibody into the respiratory and digestive tracts (IgA); still other stems bind to certain cells lining the respiratory and digestive tracts, which release histamine during allergic responses (IgE).

T-Cell Receptors Bind Antigen and Trigger Responses

T-cell receptors are somewhat different from antibodies in structure and function. First, they are found only on the surfaces of T cells, not in the bloodstream. Second, they consist of two peptide chains of about equal size. However, the ends of the two chains protrude out from the T cell, forming highly specific binding sites for antigen, just as the ends of the arms of an antibody molecule do. Third, T-cell receptors serve only a receptor function, recognizing antigen molecules and triggering a response in the T cell. They do not directly contribute to the destruction of invading microbes or toxic molecules.

The Immune System Can Recognize Millions of Molecules

A human being may encounter thousands of different types of invaders in a lifetime, from pollen grains and flu viruses to botulinus toxin (which causes botulism, a kind of food poisoning). The immune system recognizes and responds to this multiplicity of antigens because its cells produce millions of different antibodies and T-cell receptors, each capable of binding a different antigen.

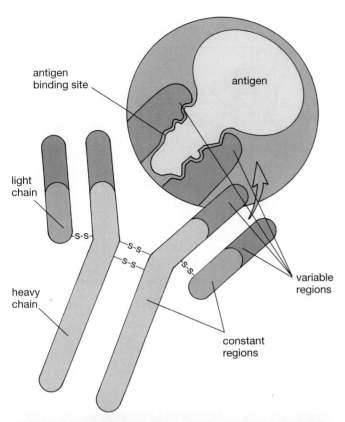

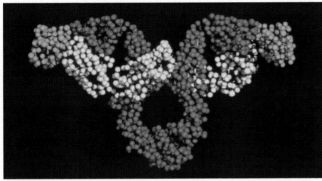

Figure 34-6   Antibody structure

Antibodies are proteins composed of two pairs of peptide chains, called light and heavy chains, arranged something like the letter Y. The variable regions on the two chains form a specific binding site at the end of each arm of the Y. Different antibodies have different variable regions, forming unique binding sites. The human body synthesizes millions of distinct antibodies, each binding a different antigen.

bacterial toxin). As we will discuss shortly, the binding of antigen to antibody triggers the immune response.

The Stem of an Antibody Determines Its Function

The stem of the Y-shaped antibody molecule consists of part of the constant regions of the two heavy chains. The stem determines the activity of an antibody. As we already mentioned, the constant regions are *similar* in different antibodies; they are not, however, the same. For example, the heavy

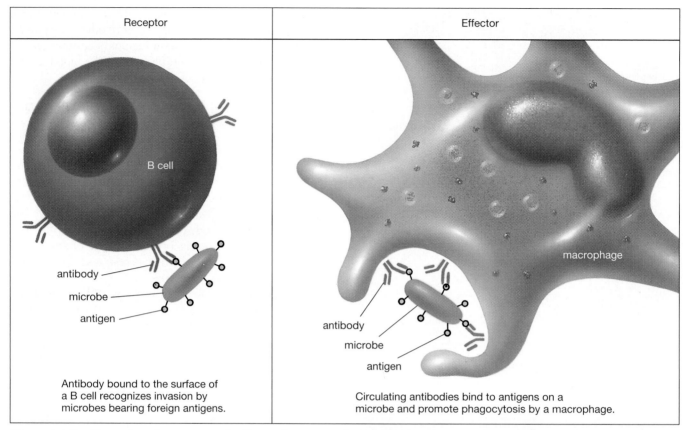

Receptor	Effector

antibody
microbe
antigen

Antibody bound to the surface of a B cell recognizes invasion by microbes bearing foreign antigens.

antibody
microbe
antigen

Circulating antibodies bind to antigens on a microbe and promote phagocytosis by a macrophage.

Figure 34-7   **Examples of receptor and effector functions of antibodies**
There are many other effector functions that are described in the text but not illustrated here.

This fact presented immunologists with a major problem: Because antibodies and T-cell receptors are proteins, and proteins are encoded by the genes, it would seem that a human being must have millions of antibody and T-cell receptor genes. But there are probably fewer than 100,000 genes in the entire human genome. How, then, can a fraction of these genes code for millions of antibodies and T-cell receptors? Two distinct but complementary mechanisms join forces to produce an enormous diversity of antibodies and T-cell receptors from a relative handful of genes. For simplicity, we will discuss only the generation of antibody diversity in B cells; similar phenomena produce a large diversity of T-cell receptors.

1. Genes encode parts of antibodies, not entire antibodies. These "antibody part" genes join together during immune cell development to form complete genes for the light and heavy chains of antibodies (Fig. 34-8). The progenitors of B cells contain a few genes for the constant regions of antibodies and several hundred genes for the variable regions (see Fig. 34-8a). During the development of each individual B cell, the chromosomes containing the "antibody part" genes are cut, sections are removed, and the remaining pieces are joined together, bringing one variable-region gene next to one constant-region gene (see Fig. 34-8b). Each cell comes

to possess one "constant + variable" light-chain gene, and one "constant + variable" heavy-chain gene. Which variable-region gene winds up alongside the constant-region gene is probably random. Only the constant and variable genes that are next to one another are transcribed and ultimately provide the genetic information for antibody molecules; the rest remain unused. Therefore, each B cell produces a single type of antibody, specified by the chance recombination of variable- and constant-region genes, that is different from the antibody produced by most or all other B cells, except of course its own daughter cells (see Fig. 34-8c).

It may help you to think of antibody gene formation in terms of card playing. Each B cell is dealt a "hand" of two variable-region genes, one for the light chain and one for the heavy chain, randomly chosen from two large "decks" of genes. With each deck containing hundreds of "cards" (genes), virtually every cell will synthesize its own unique antibody.

2. The genes for certain antibody parts are incredibly prone to mutate, thereby constantly generating new antibody genes. As B cells reproduce, some of their daughter cells accumulate mutations in their antibody genes. Thus, two sister cells may produce different antibodies.

TABLE 34-2 ⁞ *Antibodies and Their Roles*

IgM		IgM is composed of five Y-shaped antibody "monomers" held together at their stems. IgM is usually the first antibody secreted during an immune response. In the bloodstream it agglutinates antigens (because of the large number of antigen binding sites), activates complement proteins, and stimulates phagocytosis of bound microbes by macrophages.
IgG		IgG, the most common antibody in the blood, is composed of a single antibody "monomer." It activates both complement and macrophages. Special transport processes carry IgG across the placenta, where it protects the developing fetus against disease.
IgA		IgA is usually a dimer of two antibody monomers held together by their stems. An additional protein wound around the antibody stems helps IgA to be secreted from the bloodstream into saliva, tears, and mucus. Therefore, IgA is abundant on the surfaces of the respiratory and digestive tracts. It binds microbes on these surfaces, prevents them from entering the body, and allows them to be swept out of the body with mucus or other secretions. Thus, IgA provides a front line of defense for mucous membranes.
IgE		IgE, the "allergy antibody," is composed of a single antibody monomer. The IgE stem binds to mast cells in connective tissue and to certain white blood cells, including basophils and eosinophils. Its normal function appears to be protection against parasites, which are expelled by the sneezing and coughing induced by mast cell activation and weakened or killed by eosinophil activation. Allergic responses are produced when harmless substances bind to IgE. Allergy "shots" work by stimulating synthesis of IgG antibodies that bind the same antigen the IgE antibodies bind. If enough IgG antibodies are present, they bind up most of the allergy antigen and keep the allergy antigens from contacting IgE.
IgD		IgD is a single antibody monomer, usually found bound to the plasma membranes of B cells. Its function is unknown; it may serve mainly as an antigen receptor.

Kmart Versus Custom Tailoring: The Immune System Does Not Design Antibodies or T-Cell Receptors Expressly to Bind Invading Antigens

The end result of mutation and gene recombination is that each B cell has its own particular antibody genes, different from those of most other B cells. At any given time, the body may contain an army of perhaps 100 million different antibodies, so antigens almost always encounter antibodies that can bind them. A key fact about the antigen-antibody match, often misunderstood by students, is this: The immune system does not "design" antibodies to fit invading antigens. Instead, the immune system randomly synthesizes millions of different antibodies. The fact that virtually every possible invading antigen binds to at least a few antibodies is due purely to the immense numbers of different antibodies present in the body.

A more familiar example may help make this clear. When the Queen of England wants a dress, tailors go to Buckingham Palace, obtain the Queen's exact measurements, and custom-sew a dress to the Queen's specifications. Most of the rest of us go to a department store and look through the racks of ready-made clothes. If a store has a large enough selection, we will probably find something that fits reasonably well and is more or less the style we want.

The immune system is like a department store: The array of antibodies are simply there, waiting. They have *not* been designed to bind to any particular antigen. When an antigen enters the body, it may randomly bump into the variable regions of many antibodies. It cannot bind to most antibodies (in our clothes analogy, these antibodies don't "fit"), but, with millions of different antibodies around, the antigen inevitably encounters one that binds it pretty well. Antigen-antibody binding triggers changes in the B cells, usually leading to the destruction of microbes bearing that antigen. We will examine these changes in more detail in a moment.

The Immune System Distinguishes "Self" from "Non-Self"

The surfaces of the body's own cells bear large proteins and polysaccharides, just as microbes do. Some of these proteins, collectively called the **major histocompatibility complex (MHC** for short), are unique to each individual (except identical twins, which have the same genes and hence the same MHC proteins). Because your MHC proteins are different from those of everyone else, they act as antigens in other people's bodies (this is why transplants are rejected; the recipient's immune system recognizes MHC proteins on the donor's cells as foreign and destroys the transplant). Why doesn't your own immune system respond to these "self" antigens and destroy your own cells? The key seems to be the continuous presence of the body's antigens while the immune cells mature. As an embryo develops, some differentiating immune cells do indeed produce antibodies or T-cell receptors that can bind the body's own proteins and polysaccharides, not only the MHC proteins,

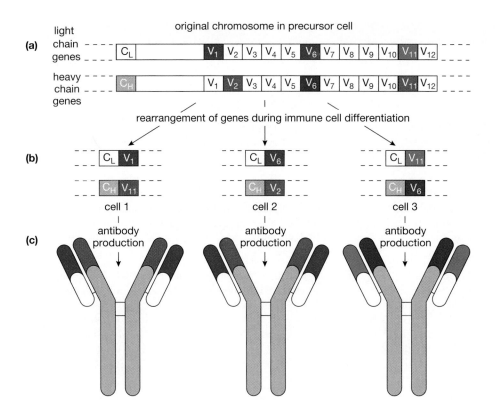

Figure 34-8 Recombination during the construction of antibody genes

(a) Each precursor cell of the immune system contains one or perhaps a few genes for the constant regions of the light and heavy chains of antibodies and many genes for the variable regions. (b) During the development of each B cell, these genes are rearranged, moving one of the variable-region genes next to a constant-region gene. Each cell thus generates a "recombined antibody gene" for each chain, which differs from the recombined antibody genes generated by other B cells. (c) Representations of the different antibodies synthesized by each B cell.

but lots of other molecules as well. However, if these *immature* immune cells contact molecules that bind to their antibodies or T-cell receptors, the immune cells are destroyed. In this way, potentially dangerous immune cells, which respond to the body's own cells as they would to foreign cells, are eliminated during immune system development. Thus the immune system distinguishes "self" from "non-self" by retaining only those immune cells that do not respond to the body's own molecules.

Attack

If the body is invaded by microbes, the immune system mounts two types of attack: B cells provide **humoral immunity,** which is provided by antibodies circulating in the bloodstream, while T cells produce **cell-mediated immunity,** as cytotoxic T cells destroy infected body cells (Fig. 34-9). The two types of responses are not as distinct as this description implies; however, the humoral and cellular responses are most easily understood when they are considered separately. The box "A Closer Look: Cellular Communication during the Immune Response" describes some of the interactions among B and T cells that regulate the immune response.

Humoral Immunity Is Produced by Antibodies in Blood

Each B cell bears a specific antibody on its surface (Fig. 34-10). When a microbial infection occurs, the antibodies borne by a few B cells will bind to antigens on the microbe (in reality, the process is considerably more complex than

this, but this is the end result). Antigen-antibody binding causes these B cells to divide rapidly (a type of T cell also helps stimulate cell division in B cells; see below and the box "A Closer Look: Cellular Communication during the Immune Response"). The resulting populations of cells are genetically identical to the parent B cells (a clone); they have been "selected" to multiply by the presence of particular invading antigens. The process is therefore called **clonal selection**. The daughter cells differentiate into two cell types: **plasma cells** and **memory cells.** Plasma cells become enlarged and packed with endoplasmic reticulum, churning out huge quantities of that cell type's specific antibody (Fig. 34-11). These antibodies are released into the bloodstream (hence the name "humoral" immunity; to the ancient Greeks, blood was one of the four body "humors"). Memory cells do not release antibodies. As we will see, memory cells play an important role in future immunity to this specific microbe.

Antibodies circulating in the bloodstream can affect only molecules or microbes that are exposed to them. Invaders that penetrate into the body's cells, as viruses do, are safe from antibody attack as long as they are within the cytoplasm of a body cell. Therefore, humoral immunity defends primarily against bacteria (most of which never enter the body's cells), bacterial toxins, and some fungi and protists. Viruses can be attacked during the times that they are outside of the body's cells, for example during the first stage of an infection or when they have finished replicating in one host cell, ruptured it, and have been released into the body fluids. As we will see shortly, however, cell-mediated immune responses are the primary defense against viral infections.

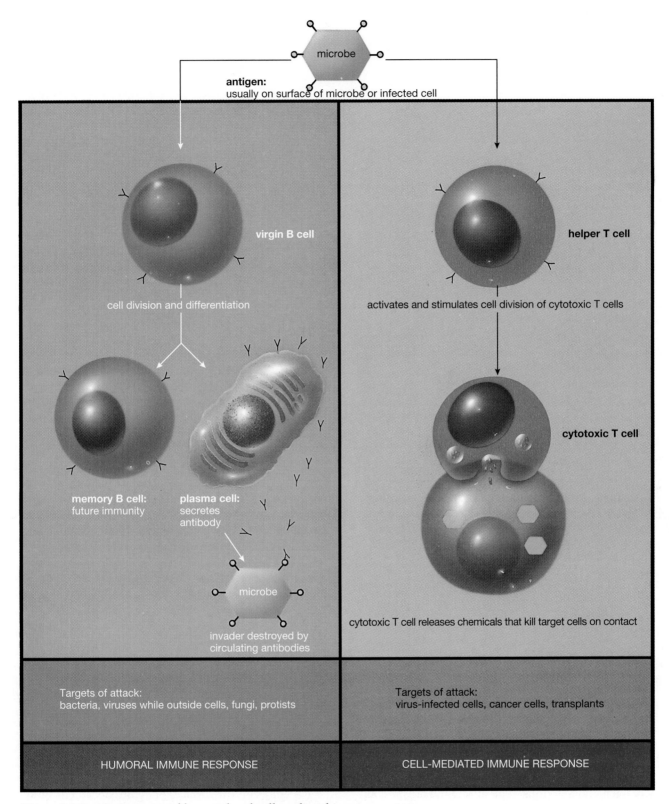

Figure 34-9   **A summary of humoral and cell-mediated immune responses**

See the text for details.

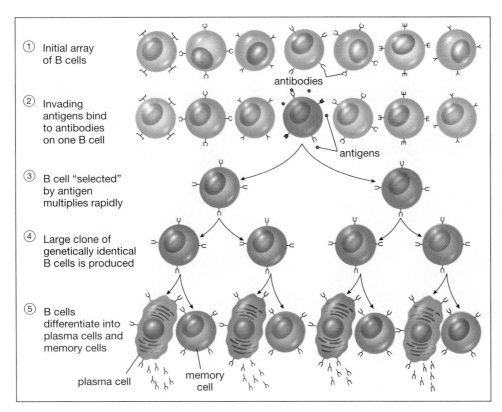

① Initial array of B cells

antibodies

② Invading antigens bind to antibodies on one B cell

antigens

③ B cell "selected" by antigen multiplies rapidly

④ Large clone of genetically identical B cells is produced

⑤ B cells differentiate into plasma cells and memory cells

plasma cell   memory cell

Figure 34-10 **Clonal selection among B cells by invading antigens**

① The body contains a large array of B cells, each of which synthesizes and bears on its surface a specific antibody, different from the antibodies borne by other B cells (see Fig. 34-8). ② When an antigen enters the bloodstream, it binds to the antibodies on one (or a few) B cells. ③ Antigen-antibody binding causes the selected B cell to multiply rapidly. ④ A clone of B cells, all synthesizing the same antibody, is produced. ⑤ The daughter B cells differentiate into plasma cells that secrete antibodies directed against the specific activating antigen or into long-lived memory cells.

Antibodies Destroy Extracellular Microbes and Molecules by Several Mechanisms

Antibodies in the blood may affect antigenic molecules, microbes bearing antigens, and infected body cells in four ways:

1. **Neutralization.** The antibody may combine with or cover up the active site of a toxic antigen such as a bacterial toxin, thereby preventing the toxin from harming the body.
2. **Promotion of phagocytosis.** The antibody may coat the surface of a microbe. The protruding antibody stems seem to identify the microbe as a target for phagocytic white blood cells to engulf.
3. **Agglutination.** Each antibody has two binding sites for antigen, one on each arm (the exception being IgM antibodies, which have 5 arms and thus 10 binding sites). These binding sites may attach to antigens on two different microbes, holding them together. As more and more antibodies link up with antigens on different microbes, the microbes clump together, or agglutinate. Agglutination seems to enhance phagocytosis.
4. **Complement reactions.** The antibody-antigen complex on the surface of an invading cell may trigger a series of reactions with blood proteins called the **complement system.** When these proteins bind to the antibody stems, they attract phagocytic white blood cells to the site, promote phagocytosis of the foreign cells, or in some instances directly destroy the invaders by creating holes in their plasma membranes, much as natural killer cells do (see Fig. 34-3).

T Cells Produce Cell-Mediated Immunity

Cell-mediated immunity is the primary defense against the body's own cells when they have become cancerous or have been infected by viruses. Cell-mediated immunity is also important in overcoming infection by fungi and protists. Several types of T cells contribute to cell-mediated immunity: helper cells, cytotoxic cells, and suppressor cells.

Three Types of T Cells Contribute to the Immune Response

Helper T cells, when their T-cell receptors bind an antigen, release chemicals that assist other immune cells in their defense of the body. These hormone-like chemicals stimulate cell division and differentiation in both B cells and cytotoxic T cells responding to the same microbial invasion. In fact, very little immune response, either cell-mediated or humoral, can occur without the boost provided by helper T cells (see "A Closer Look: Cellular Communication during the Immune Response"). This is why AIDS, which destroys helper T cells, is such a deadly disease. We will discuss AIDS more thoroughly shortly.

Receptors on the surfaces of **cytotoxic T cells** bind to antigens on the surface of an infected cell. The cytotoxic T cell then releases proteins that disrupt the infected cell's plasma membrane (Fig. 34-12). In at least some instances, the cytotoxic T cell's proteins are similar or identical to those of the natural killer cells that create giant holes in the target cell's membrane. After the infection has been conquered, **suppressor T cells** seem to shut off the immune response in both B and cytotoxic T cells.

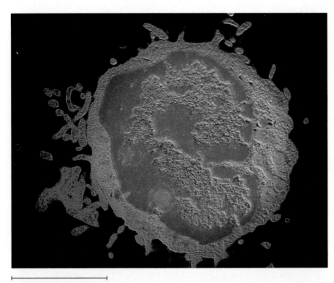

2 micrometers

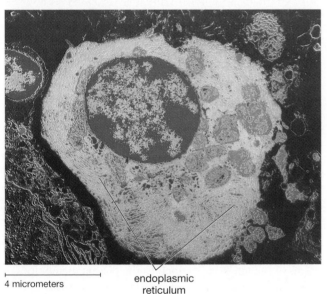

4 micrometers

endoplasmic
reticulum

Figure 34-11   A B cell becomes a plasma cell

False-color micrographs of B cells before (top) and after (bottom)
conversion to plasma cells. The plasma cell is much larger than
the B cell (note the difference in scale) and is virtually filled with
rough endoplasmic reticulum that synthesizes antibodies.

Some T cells formed during an immune response are
memory cells. Like memory B cells, these protect the body
against future infection.

Memory

As Thucydides observed 2000 years ago, a person who
overcomes a disease often remains immune to future encounters with that specific disease for many years. Retaining immunity is the function of memory cells. Plasma
cells and cytotoxic T cells do the immediate job of fighting disease organisms, but they usually live only a few days.
B and T memory cells, on the other hand, may survive for

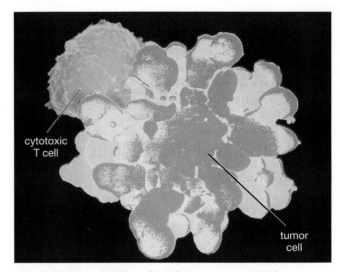

cytotoxic
T cell

tumor
cell

Figure 34-12   Cell-mediated immunity at work

A cytotoxic T cell (green) contacts a large cancer cell (pink
and yellow) in this false-color scanning electron micrograph.
Note the membrane bubbles forming on the tumor cell as it is
destroyed.

many years. If foreign cells bearing the same antigens reenter the body, they will be recognized by the appropriate
memory cells. These memory cells will multiply rapidly,
generate huge populations of plasma cells and cytotoxic T
cells, and produce a second immune response.

In the first encounter with a disease microbe, only a
few B and T cells respond. Each of these, however, leaves
behind hundreds or thousands of memory cells. Further,
memory cells respond to antigen much more rapidly than
their progenitor B and T cells could. Therefore, the second immune response is very rapid (Fig. 34-13). In most
instances, second or subsequent invasions by the same microbe are overcome so quickly that there are no noticeable

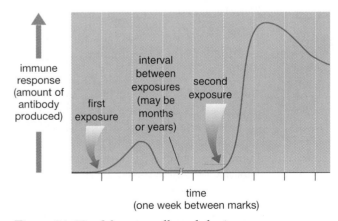

immune
response
(amount of
antibody
produced)

first
exposure

interval
between
exposures
(may be
months
or years)

second
exposure

time
(one week between marks)

Figure 34-13   Memory cells and the immune response

The immune response to the first exposure to a disease organism is fairly slow and not very large, as B and T cells are
selected and multiply. A second exposure activates memory
cells formed during the first response, and consequently the
second response is both faster and larger.

symptoms of infection at all. Although cold and flu viruses appear to be exceptions to this rule, the immune system does in fact provide lasting protection against these viruses. The problem, as we describe in "Health Watch: Flu—The Unbeatable Bug," is that one year's flu isn't caused by the same virus that you fought off the year before.

Medicine and the Immune Response

Antibiotics Slow Down Microbial Reproduction

Our description of the body's responses to infection may seem to suggest that nothing should harm us. As everyone knows, however, this is unfortunately not the case. If untreated, many diseases kill their victims, usually for a simple reason: The body provides ideal conditions for the growth and reproduction of disease microbes, which can multiply rapidly, sometimes dividing as fast as once an hour. The infection thus becomes a race between the invading microbes and the immune response. If the initial infection is massive, or if the microbes produce particularly toxic products, the full activation of the immune response may come too late.

Antibiotics help combat infection by slowing down the growth and multiplication of many microbes, including bacteria, fungi, and protists (but not viruses). Although antibiotics usually don't destroy every single microbe, they give the immune system enough time to finish the job. One problem with antibiotics, however, is that they are potent agents of natural selection. The occasional mutant microbe that is resistant to an antibiotic will, of course, pass on the gene(s) for resistance to its offspring. The result: Resistant mutants proliferate, while susceptible microbes die off. Eventually, many antibiotics become ineffective in treating diseases for which they formerly were the mainstay of medical care. This phenomena of antibiotic resistance is discussed more fully in Chapter 21.

Vaccinations Stimulate the Development of Memory Cells

As early as the year 1000, people in India, China, and Africa deliberately exposed themselves to mild cases of smallpox, to acquire immunity to the disease. In 1798, Edward Jenner discovered that being infected with cowpox conferred immunity to smallpox. This discovery initiated the modern practice of immunization. In the late 1800s, Louis Pasteur extended the use of immunization to several other diseases by injecting weakened or dead microbes into healthy people. The weakened microbes do not cause disease (or at least not a severe case) but bear antigens that elicit vigorous immune responses. As it would to a real infection, the immune system produces an army of memory cells, conferring immunity against subsequent exposure to the living, dangerous microbes. These injections of weak-

ened or killed microbes to confer immunity are called **vaccinations,** from the Latin word for "cow," in honor of Jenner's pioneering efforts with cowpox. Today, many diseases, including polio, diphtheria, typhoid fever, and measles, can be controlled through vaccination. Smallpox, one of the deadliest diseases of all, has been completely eradicated from Earth as a result of a vaccination program sponsored by the World Health Organization.

Through genetic engineering (see Chapter 14), we now enjoy the prospect of manufacturing tailor-made vaccines. One method is to synthesize the antigenic proteins from disease-causing microbes. These antigens can then be used as vaccines without having to raise, isolate, and weaken the disease microbes themselves or to inject people with microbes at all. A vaccine against anthrax, a severe disease of livestock, has been manufactured with this procedure. A second technique may be to insert the genes for the antigens of, for example, herpes into the genome of harmless microbes such as the cowpox virus. These "designer" microbes would produce herpes antigens without being able to cause the disease and could be used for vaccination.

Allergies Are Inappropriately Directed Immune Responses

More than 35 million Americans suffer from **allergies:** reactions to substances that are not harmful in themselves and to which many other people do not respond. Common allergies include those to pollen, dust, mold spores, and bee stings. Allergies are actually a form of immune response (Fig. 34-14). A foreign substance, such as a pollen grain, enters the bloodstream and is recognized as an antigen by a particular type of B cell. This B cell proliferates, producing plasma cells that pour out IgE antibodies against the pollen antigens. The stems of IgE antibodies attach to the plasma membranes of histamine-containing cells located in the respiratory and digestive tracts. When pollen grains encounter the attached IgE antibodies, they trigger the release of histamine, which causes increased mucus secretion, leaky capillaries, and other symptoms of inflammation. Because pollen grains most often enter the nose and throat, the major reactions occur in those locations, resulting in the runny nose, sneezing, and congestion typical of "hay fever." Antihistamine drugs block some of the effects of histamine, relieving the symptoms of allergies. Food allergies cause equivalent symptoms, including cramps and diarrhea, in the digestive tract.

IgE-Mediated Responses Probably Evolved as a Defense against Parasites

Why are some people allergic and not others? You can probably guess the answer. We are all exposed to the same antigens in pollens, molds, and foods. Therefore, people without allergies either must lack the genes for the allergy-causing antibodies or do not produce as much antibody as allergic individuals do. From an evolutionary perspective, a more interesting question is: Why are *any* people al-

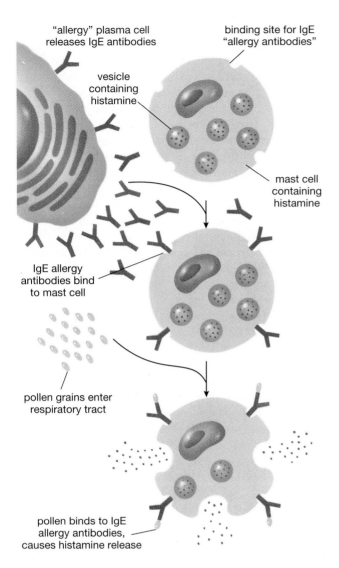

"allergy" plasma cell releases IgE antibodies

binding site for IgE "allergy antibodies"

vesicle containing histamine

mast cell containing histamine

IgE allergy antibodies bind to mast cell

pollen grains enter respiratory tract

pollen binds to IgE allergy antibodies, causes histamine release

Figure 34-14   **Allergic reactions**
Upon exposure to certain antigens, for instance pollen grains, some plasma cells synthesize antibodies that bind both to histamine-containing cells (middle) and to the antigen (bottom). When this dual binding occurs, the cells release histamine into their surroundings, causing local inflammation and the symptoms of allergy.

lergic? What useful function do IgE antibodies serve, so that the obvious disadvantages have not caused their elimination through natural selection? It turns out that IgE antibodies confer protection against parasites.

Many parasites invade the body through natural body openings—usually the mouth, nose, or anus. The initial stage of infection occurs when a parasite attaches to the lining of the nasal passages, throat, or intestine. It then usually burrows through into the body tissues (see Chapter 24). The typical symptoms of allergies—increased mucus secretions, sneezing, coughing, intestinal convulsions, and diarrhea—are apparently properly directed against parasites, not pollen, and help to dislodge and expel the parasites. The stems of IgE antibodies also attach to white blood cells

called eosinophils. When the variable-region arms of the attached IgE antibodies bind to a parasite, the eosinophil secretes toxic proteins that damage or kill the parasite.

An Autoimmune Disease Is an Immune Response against Some of the Body's Own Molecules

A person's immune system does not normally respond to the antigens borne on the body's own cells. Occasionally, however, something goes awry, and "anti-self" antibodies are produced. The result is an **autoimmune disease,** in which the immune system attacks some component of one's own body. Some types of anemia, for example, are caused by antibodies that destroy a person's red blood cells. Many cases of insulin-dependent (juvenile-onset) diabetes occur because the insulin-secreting cells of the pancreas are the victims of a misdirected immune response. Unfortunately, at present there is no way to cure autoimmune diseases. For some diseases, replacement therapy can alleviate the symptoms, for instance by administering insulin to diabetics or blood transfusions to anemics. Alternatively, the autoimmune response can be suppressed with drugs. Immune suppression, however, also reduces immune responses to the everyday assaults of disease microbes, so this therapy cannot be used except in the most life-threatening cases.

An Immune Deficiency Disease Results from the Inability to Mount an Effective Immune Response to Infection

Rarely, a child is born with a defect in which no immune cells, or very few, are formed. This condition is called **severe combined immune deficiency,** or **SCID.** Such a child may survive fetal life and even the first few months of postnatal life, protected by antibodies acquired from the mother during pregnancy or in her milk. Once these antibodies are lost, however, common bacterial infections may prove fatal. Some immune-deficient children have to live in a germproof "bubble," isolated from contact with every unsterilized object, including other people. One form of therapy is to transplant bone marrow (from which immune cells arise) from a normal donor into the child. In some children, marrow transplants have resulted in some antibody production, occasionally enough to confer normal immune responses. In 1990, researchers at the National Institutes of Health began clinical trials of injecting genetically engineered bone marrow cells into children with SCID (see Chapter 14). The therapy has been somewhat successful but is not yet in widespread use.

AIDS

Probably the most common, and most devastating, immune deficiency disease is **acquired immunodeficiency syndrome,** or **AIDS.** Two viruses, named human immunodeficiency viruses 1 and 2 (HIV-1 and HIV-2), cause AIDS by infecting and destroying helper T cells. If you refer to "A Closer

Look: Cellular Communication during the Immune Response," you will note that helper T cells potently stimulate both the cell-mediated and humoral immune responses.

AIDS does not directly kill its victims, but as the helper T cell population declines, the AIDS patient becomes increasingly susceptible to other diseases. In fact, it was the incidence of unusual diseases that led to the recognition of AIDS in the first place, in 1981. In that year, a man entered the UCLA Medical Center with a fungal infection in his throat. A few weeks later, he developed a rare form of pneumonia (*Pneumocystis carinii*), one almost never seen except in patients with cancer and people whose immune system is being suppressed to prevent rejection of organ transplants. After a series of infections, he soon died. At the same time, doctors across the country began encountering similar cases: Patients suffered debilitating effects from rare diseases or from common diseases that are not usually serious in normal adults. Although the particular diseases varied, all the patients had one feature in common: Their immune systems failed to ward off invading microbes because of a lack of helper T cells.

The Human Immunodeficiency Virus Is a Retrovirus That Infects and Destroys Helper T Cells

HIV-1 and HIV-2 are **retroviruses**—viruses that have RNA, not DNA, as their genetic material (see "A Closer Look: Viral Replication" in Chapter 21). As illustrated in Figure 34-15, HIV consists of an outer envelope, taken from an infected cell's plasma membrane as the virus leaves the cell, and two protein capsules, the innermost of which contains RNA and an enzyme called **reverse transcriptase.** The outer envelope binds to the plasma membrane of a helper T cell and allows the virus to invade the cell. Once inside, the reverse transcriptase copies the virus's RNA genome over into DNA; because the "normal" direction of transcription in living cells is DNA to RNA, the virus's direction, RNA to DNA, is dubbed "reverse" transcription. The infected cell's own metabolic machinery then uses the viral "DNA copy" to make more viruses and more reverse transcriptase. Multiplication of HIV eventually kills the infected helper T cell.

HIV seems to be transmitted only by direct exchange of body fluids, including blood and semen, and can't survive very long outside the body. However, once a person is infected, AIDS seems to be almost always fatal, and AIDS is now an extremely widespread disease (Fig. 34-16). By the time you read this, at least 17 million people, and probably many more, will have been infected with HIV. By the year 2000, an estimated 30 to 40 million people altogether will have been infected.

HIV Virus Is Transmitted by Exchange of Body Fluids

In the United States, AIDS patients are not an average cross section of society. Initially, almost all AIDS victims in the United States were homosexual men or intravenous drug users. Some victims were hemophiliacs. The disease spreads through the homosexual population by sexual encounters. Drug users spread AIDS by sharing unsterilized hypodermic needles. Many hemophiliacs acquired AIDS through contaminated blood transfusions (before it became standard practice to screen all donated blood for anti-HIV antibodies).

Initially, many people were somewhat complacent about AIDS. They reasoned that they were not personally at risk unless they were homosexual, intravenous drug users, or needed frequent blood transfusions. Although these are still the main avenues for infection in the United States, women and heterosexual men who are not drug users have also contracted AIDS. In fact, the number of AIDS cases among homosexuals has leveled off in the past few years, while the number of infected heterosexuals has continued to increase. It is now obvious that AIDS can be transmitted through heterosexual contact, although not as efficiently as through homosexual contact.

The pattern of AIDS infection is very different in Africa. The World Health Organization believes that heterosexual intercourse is probably the most common means of infection among Africans. Further, AIDS is frighteningly common in parts of Africa (see Fig. 34-16). For example, in Abidjan, the capital of Cote d'Ivoire (the Ivory Coast) in West Africa, AIDS-related infections are now the leading cause of death among young men and second only to complications of pregnancy and childbirth among young women.

There Are Partially Effective Treatments, but No Cures, for AIDS

For persons already infected with AIDS, there are two categories of therapy. Infections that are not actually caused by the HIV virus—such as Kaposi's sarcoma, a deadly form of cancer affecting the skin, or *Pneumocystis carinii* pneumonia—can be treated as they would in any patient. Within the past few years, more effective treatments for these diseases have improved the quality of life for AIDS patients and helped them to live longer. Second, the progress of AIDS can be slowed, but not stopped, by drugs such as ziduvidine (AZT) or dideoxyinosine (ddI). These drugs are "nucleotide mimics" that fool the viral reverse transcriptase. When incorporated into a growing DNA chain, they stop further DNA synthesis. Ideally, the viral RNA is thus never copied completely over into DNA, and therefore new virus cannot be synthesized. Unfortunately, these drugs are not completely successful in stopping reverse transcription; further, they also interfere to some extent with normal DNA replication, and in some patients they have very severe side effects.

What about normal immune responses to AIDS? As you probably know, AIDS patients produce anti-HIV antibodies (this was the original basis for most blood testing). For some reason, these antibodies do little to prevent the progress of the infection. One hypothesis is that HIV spends so much of its life cycle inside cells that it is seldom exposed to the antibodies.

Enormous resources are now being devoted to preventing AIDS infections. Two show some promise. First, HIV

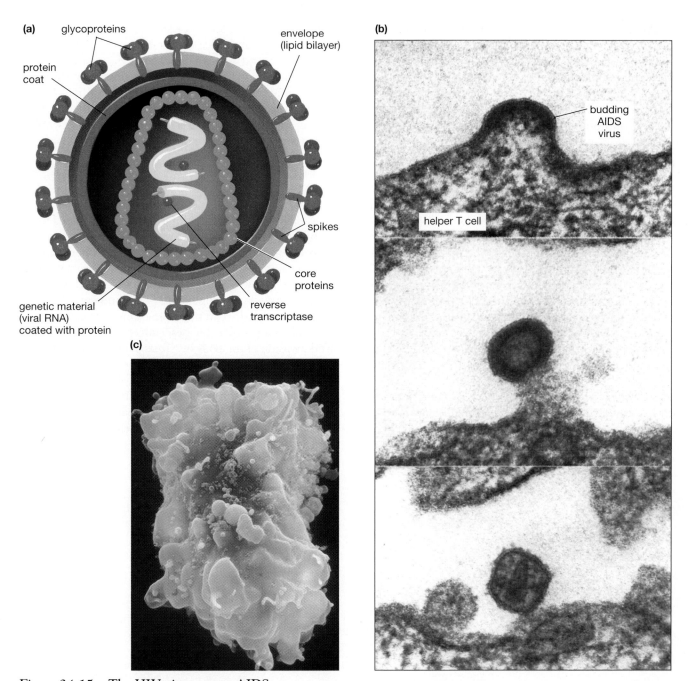

Figure 34-15   **The HIV virus causes AIDS**

(a) The structure of the AIDS virus. The virus consists of an outer envelope taken from the cells it infects and an inner protein capsule that contains RNA (the genetic material of HIV) and the enzyme reverse transcriptase (which copies the RNA over into DNA when the virus infects a cell). The proteins protruding through the envelope attach to the plasma membranes of helper T cells. These proteins are potential targets for AIDS vaccines. **(b)** An HIV virus emerges from an infected helper T cell, coating itself with a bit of the T cell's plasma membrane as it leaves. **(c)** The blue specks in the false-color scanning electron micrograph are HIV viruses that have just emerged from the large helper T cell.

gains entry into helper T cells only after binding to a specific receptor on the T cell's plasma membrane. Several researchers are exploring the possibility of flooding the body with other compounds that would bind to these receptors and prevent HIV from binding to them. Second, it may be possible to produce HIV vaccines. This is a tricky business.

As we just mentioned, the antibodies normally produced by the body do not prevent AIDS; therefore, vaccines would have to evoke a very different, and more effective, immune response than normal HIV infection does. Further, HIV has an incredible mutation rate, perhaps a thousand times faster than the flu virus. Nucleotide sequencing of viruses isolated from AIDS patients shows that different people have remarkably different strains of HIV. Even more surprising, HIV can be very different even when isolated from the same patient, but at different times. Nevertheless, a few parts of HIV appear to be encoded by genes that mutate less rapidly, and trials of AIDS vaccines are under way in several countries, but none appear particularly promising.

AIDS Is One of Many Widespread, Lethal Diseases

How should society respond to the AIDS epidemic? Perhaps of more immediate importance: How concerned should you be, personally? People in the high-risk categories should certainly take precautions against infection. Given the deadly nature of the disease, and the rising incidence of other sexually transmitted diseases, "safe-sex" practices are advisable for everyone. Health-care workers should exercise care when handling blood products and needles. Each year, thousands of health-care workers suffer "accidental sticks" from needles. These can usually be avoided. The odds that a patient will acquire AIDS from a physician or dentist are astronomically small, and no one should forego medical care for fear of AIDS.

From a global perspective, AIDS is a very widespread and serious disease, with more than 4 million full-blown AIDS cases worldwide over the past 15 years. On the other hand, many other diseases and disorders are much more common and cause many more deaths. For example, childbirth, malnourishment, and a host of infectious diseases cause far more deaths than AIDS in less-developed countries. In the United States, about one person in three will contract cancer, causing over 500,000 deaths each year. Thus, although AIDS garners a lot of publicity, it is only one of a host of public health threats that require our attention and resources.

Cancer

Cancer, along with *AIDS*, is one of the most dreaded words in the English language, and with good reason. As we already mentioned, nearly one out of three Americans will contract some form of cancer. For many, there will be no cure, only a slow wasting away to death. What *is* cancer? If we can prevent smallpox and polio, and cure dozens of other diseases, why can't we cure or prevent cancer?

Unlike most other diseases, cancer is not a straightforward invasion of the body by a foreign organism. Although some cancers may be triggered by viruses, in essence cancer is a malfunctioning of the growth controls of the body's own cells—a disease in which we destroy ourselves. Because it is a case of "self" fighting "self," most treatments designed to combat cancer also damage normal, healthy cells.

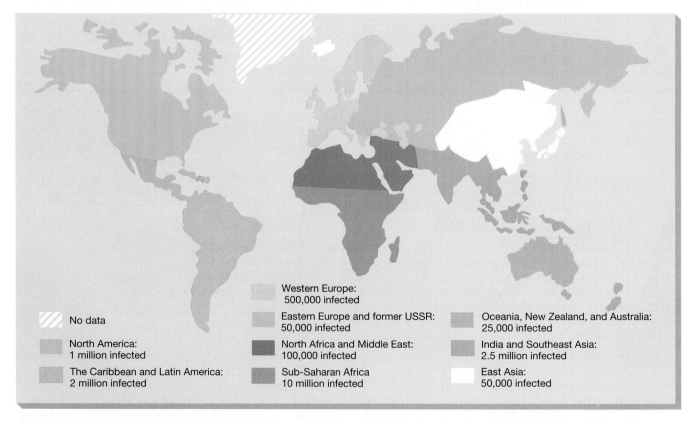

No data

North America:
1 million infected

The Caribbean and Latin America:
2 million infected

Western Europe:
500,000 infected

Eastern Europe and former USSR:
50,000 infected

North Africa and Middle East:
100,000 infected

Sub-Saharan Africa
10 million infected

Oceania, New Zealand, and Australia:
25,000 infected

India and Southeast Asia:
2.5 million infected

East Asia:
50,000 infected

Figure 34-16   **HIV infections worldwide**

The worldwide incidence of infection with HIV, estimated by the World Health Organization in 1994. The incidence figures are estimates for adults 15 to 49 years of age.

The usual development of any organ begins with rapid growth during embryonic life, slower growth as a juvenile, and finally maintenance of a constant size during adulthood. Individual cells may die and be replaced (as happens continually in the stomach lining), but most organs remain about the same size throughout adult life. A **cancer** is a population of cells that has escaped from normal regulatory processes and grows without control. As a cancer grows, it uses increasing amounts of the body's energy and nutrient supplies and literally squeezes out vital organs nearby.

Cancer Is Caused by Mutation, Activation, or Suppression of Genes That Control Cell Division

To learn the causes of cancer, we must answer two related but distinct questions: (1) What changes occur in a cancerous cell that allow it to escape normal growth controls? (2) What agents (genetic, viral, or environmental) initiate these cellular changes?

Cancer May Occur by Activation of Genes That Stimulate Cell Division or Inactivation of Genes That Suppress Cell Division

In the early 1980s, cancer researchers discovered **oncogenes:** genes that cause cancer. Potentially dangerous "proto-oncogenes" exist in all cells. These are genes that stimulate growth and cell division. These "growth" genes are active during embryonic development, but usually they are turned off, transcribed more slowly, or tightly controlled in mature organisms. However, a proto-oncogene may be activated or mutate into an oncogene. Consider a gene that normally directs the synthesis of a protein that promotes cell reproduction at "maintenance levels," such as those needed to replace cells lost through normal body wear and tear. A mutation in this proto-oncogene may change the protein so that it greatly accelerates the rate of cell division. This accelerated rate would probably create a cancer.

What causes activation of or mutations in proto-oncogenes? Some types of cancer are caused by viral infections

(Fig. 34-17). As in HIV, these viruses have genes composed of RNA, which the virus forces the cell to "reverse transcribe" into DNA. This new DNA is then inserted into chromosomes of the host cell, where it is usually transcribed continuously to synthesize new viral RNA. Nearby DNA on the host chromosome may also be incidentally transcribed. If this host DNA happens to include a previously silent proto-oncogene, the cell becomes cancerous.

Probably the most common causes of cancer are environmental insults, chiefly chemicals and radiation. We are besieged by cancer-causing chemicals (carcinogens), not only the eminently avoidable ones in cigarettes and various industrial processes but also those in the most innocent foods and even some synthesized in our own digestive tracts. Some chemicals and certain types of radiation induce point mutations in DNA. Others cause chromosomes to break in two and possibly rejoin in new and lethal combinations, by transferring oncogenes into actively transcribed regions of the chromosomes.

If these seemingly simple mechanisms cause cancer, why do cancers take so long to develop? No one knows the answer for certain, but most cancers seem to require two or more distinct steps (see "A Cancer Develops" below). Exposure to radiation early in life, perhaps, may mutate a proto-oncogene to a true oncogene. If this oncogene is located in a region of DNA that is not normally transcribed, no cancer will occur. Many years later, a viral infection or exposure to chemicals may move this oncogene to an active region of DNA, and cancer begins.

Cancer may also be caused by **tumor-suppressor genes**, which in many cells appear to regulate the rate of cell division. Perhaps the most well-studied tumor-suppressor gene, called *p53*, produces a protein that stimulates the production of a second protein that blocks cell division. In other cases, the protein product of a tumor-suppressor gene may block hormonal signals that would otherwise stimulate cell division. Many cancers, including retinoblastoma (a cancer on the retina of the eye) and cancers of the bladder, bone, brain, breast, cervix, lung, and ovaries, appear to arise when such tumor-suppressor genes are damaged

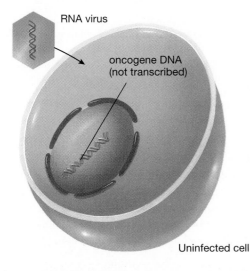

Uninfected cell

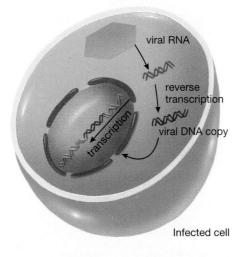

Infected cell

Figure 34-17 **Viral activation of an oncogene**

Certain RNA viruses invade animal cells and cause "reverse transcription" of their RNA genes into DNA. This DNA then inserts itself into the DNA of the host cell. The viral DNA copy includes nucleotides that promote transcription by host RNA polymerase. Nearby host genes, including oncogenes in some cells, will be transcribed along with the viral DNA copy, and the cell may become cancerous.

HEALTH WATCH

Flu—The Unbeatable Bug

Every winter, a wave of influenza sweeps across the world. Thousands of the elderly, the newborn, and those already suffering from illness succumb, while hundreds of millions more suffer the respiratory distress, fever, and muscle aches of milder cases. Occasionally, devastating flu varieties appear. In the great flu pandemic of 1918, the worldwide toll was 20 million dead in one winter. In 1968, the Hong Kong flu infected 50 million Americans, causing 70,000 deaths in 6 weeks. As recently as the winter of 1984–1985, the Centers for Disease Control estimated that about 57,000 Americans died from the flu.

The Flu Virus

Flu is caused by a virus that invades the cells of the respiratory tract, turning each one into a factory for manufacturing new viruses. The outer surface of the virus is studded with proteins, several of which serve as antigens recognized by the immune system. People survive the flu because their immune systems inactivate the viruses or kill off virus-infected body cells before the viruses finish reproducing. This is the same mechanism by which other viruses, such as mumps or measles, are conquered. So why don't people become immune to the flu, as they do to measles?

The answer lies in the flu virus's amazing ability to change. The genes of flu viruses are RNA, which lacks the proofreading mechanisms that reduce mutations in genes made of DNA (see Chapter 12). Therefore, flu RNA genes mutate rapidly: There are 10 mutations in every million newly synthesized viruses. A single mutation usually doesn't change the properties of the viral antigens very much. Four or five, however, may change them enough so that the immune system doesn't fully recognize it as the same old flu that was beaten off last year. Some of the memory cells don't recognize it at all, and the immune response produced by the rest doesn't work as well as it should. The virus, although slowed down somewhat, gets a foothold in the body and multiplies until a new set of immune cells recognizes the mutated antigens and starts up a new immune response. So you get the flu again this year.

Deadly New Strains

Far more serious are the dramatically new flu viruses that occasionally appear: the epidemic of 1918, the Asian flu of 1957, and the Hong Kong flu of 1968. In these, entirely new antigens seem to show up all at once. These are not just mutations of the old set, but novel antigens that the human immune system has never encountered before. Where do they come from? Believe it or not, they come from birds and pigs. Viruses strikingly similar to the human flu virus infect the intestinal tracts of birds, especially ducks, without causing any noticeable disease. The bird viruses don't infect peo-

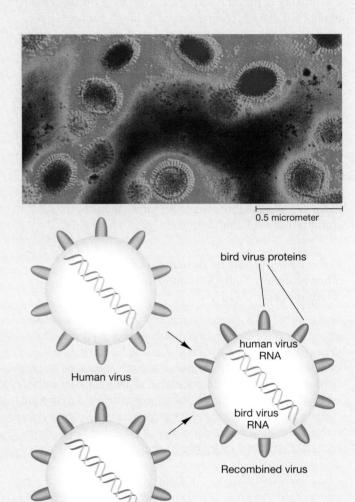

0.5 micrometer

Figure E34-2   **Formation of deadly new strains of influenza**

Rare recombination of genes from bird and human influenza viruses can result in deadly new strains of influenza. The photograph is a false-color electron micrograph of flu viruses. Note the protein "spikes" projecting from the virus coats. These attach to plasma membranes of cells in the human respiratory system, helping the virus gain entry into the cells.

ple, and human flu viruses don't infect birds. Both can infect pigs, however, with the result that both viruses may reproduce simultaneously in the same pig cell. Once in a great while (perhaps only three times this century), offspring viruses end up with a mixture of genes from human and bird viruses (Fig. E34-2). Some of these hybrid viruses combine the worst genes of both types (at least from the human perspective): from the human virus, the genes needed to subvert human cellular metabolism to produce new viruses; from the bird virus, genes for new surface antigens. Pigs live near people, and they cough when they have the flu just as people do, so the spread of virus from pig to person is virtually assured.

Have you ever wondered why flu strains are called "Asian" or "Hong Kong"? The reason is that Southeast Asia is usually the place where new strains crop up. Many farmers in Asia, especially in southern China, have "integrated" farms. Crops are grown to feed pigs and ducks, and the feces from the pigs and ducks are used to fertilize fish ponds. This is a very efficient farming practice, but, unfortunately, it also provides ideal mixing vessels for the flu virus (pigs) in close proximity to humans and ducks.

Effects of New Strains

If infected by a hybrid virus, the immune system must start from scratch, selecting out entirely new lines of B and T cells to attack the intruder. But the virus multiplies so rapidly in the meantime that many people die or become so weakened that they contract some other disease and die of that. The rest of the people recover, with immune systems now primed to resist any further assault from the new virus. Next year a few point mutations allow a slightly altered strain to infect millions of people, but with a partial immune response ready, few fatalities occur. Once again, for most of us, the flu becomes a routine annoyance.

Until next time, somewhere, the improbable happens again. Maybe this year.

---

or lost. Freed from the restraint of the protein product of the suppressor gene, the affected cell multiplies without control, starting a tumor.

A Cancer Develops: Oncogenes and Tumor-Suppressor Genes Interact in Colon Cancer

The events that produce most cancers are not well understood. In the late 1980s, however, the sequence of genetic accidents that cause colon cancer began to be determined. Apparently, no fewer than one oncogene and three tumor-suppressor genes must mutate, in a single cell or its progeny, to produce malignant colon cancer (Fig. 34-18). The final step, damage to the *p53* tumor-suppressor gene on chromosome 17, seems to be involved in dozens of different types of cancer.

The Immune System Defends against Cancerous Cells

Cancer cells form in our bodies every day, and not even the best of preventive measures can eliminate cancer completely. Gamma rays from the sun, radioactivity from the rocks beneath our feet, and naturally produced carcinogens in our food cannot be avoided. However, each of us can reduce his or her own chances of developing cancer. Some chemicals, including carotene and vitamins C and E, appear to offer protection against some forms of cancer. We can also avoid many well-known carcinogens. Cigarette smoking, for example, causes most of the lung cancers in the United States. Other chemicals, including those emitted from oil refineries and those used in certain industrial processes, can cause cancer in exposed workers. The ultraviolet rays from the sun that produce fashion-

able suntans are also a leading cause of skin cancer. Certain molds produce the most potent carcinogens known and can be reduced by storing food properly.

Fortunately, natural killer cells and cytotoxic T cells screen the body for cancer cells and destroy nearly all of them before they have a chance to proliferate and spread. If cancer cells are "self" cells, and the immune system does not respond to "self," how are cancer cells weeded out? Probably the very processes that cause cancer also cause new and slightly different proteins to appear on the surfaces of cancer cells. Cytotoxic T cells encounter these new proteins, recognize them as "non-self" antigens, and destroy the cancer cells. Without constant surveillance by the immune system, it is unlikely that any of us would survive more than a few years.

Medical Treatments for Cancer Depend on Distinguishing and Selectively Killing Cancerous Cells

Sometimes, however, the immune system does *not* recognize cancer cells as "non-self." Ignored by the immune system, the cancer grows and spreads. What can medical science do to cure cancer? The rate of cure is increasing, but it is still scarcely a third of all cancers (the suggested reading "A War Not Won" by T. Beardsley is instructive but depressing). The three main approaches taken are all quite crude: Burn the cancer out with radiation, cut it out with surgery, or poison it with drugs.

If a cancer is discovered when it is small enough, radiation or surgery may be able to eliminate it. Breast cancer, for example, can almost always be eliminated by surgery if detected early enough. However, while the body is rid of the cancer, surgery and radiation therapies may be traumatic, dangerous, and disfiguring.

In principle, chemotherapy might be able to destroy cancer cells without damaging normal cells, because cancer cells are, after all, different in many ways from normal cells. The most common chemotherapies involve drugs that are "nucleotide mimics" (different ones than are used to treat AIDS). These mimics are incorporated into DNA during chromosome replication. They then either prevent further replication or cannot be transcribed correctly. In either case, the cell dies or fails to reproduce. These drugs obviously have their main effect on dividing cells, and because cancer cells divide rapidly, the drugs kill cancer cells. Unfortunately, other cells of the body divide too, such as those in the hair follicles and intestinal lining. Chemotherapy drugs damage those cells, producing the well-known side effects of nausea, vomiting, and hair loss.

In the early 1990s, clinical trials of a very different therapy were begun. A special class of white blood cells, called tumor-infiltrating lymphocytes, seeks out and invades tumors. Through genetic engineering, genes for a protein called tumor necrosis factor, which destroys cancerous cells, are inserted into a patient's own tumor-infiltrating lymphocytes. These are then injected back into the patient, to seek and destroy the patient's cancer.

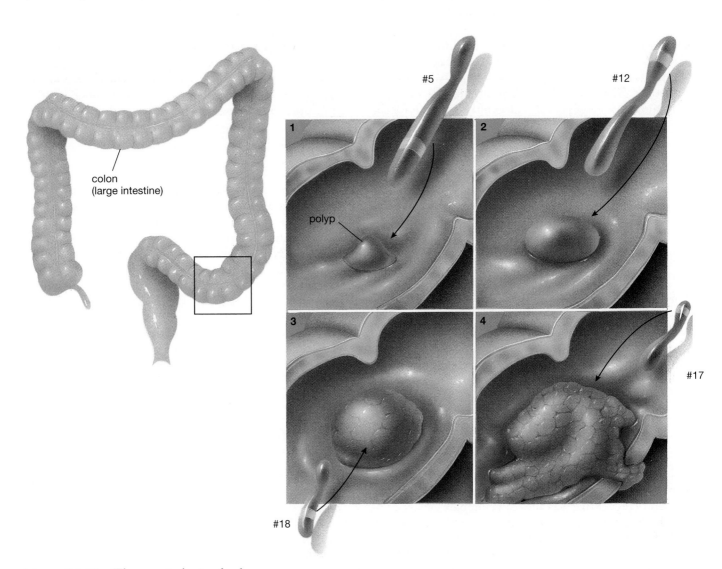

Figure 34-18   **The genetic basis of colon cancer**

The development of cancer of the colon involves four genetic changes. (1) A tumor-suppressor gene on chromosome 5 is mutated in one of the cells of the colon. The cell multiplies, forming a polyp, a small, projecting growth. (2) In one of the cells of the polyp, a proto-oncogene on chromosome 12 mutates into a true oncogene. This cell multiplies more rapidly, and the polyp becomes large. (3) A second tumor-suppressor gene, this time on chromosome 18, mutates in one of the cells that already has the first two mutations. This cell speeds up its multiplication rate, and the polyp becomes a tumor. (4) Finally, the *p53* tumor-suppressor gene on chromosome 17 mutates in a cell with all three previous mutations. Its daughter cells now multiply wildly, invading the wall of the colon and spreading cancer to other parts of the body.

✖ *S U M M A R Y   O F   K E Y   C O N C E P T S*

Defenses Against Microbial Invasion

The human body has three lines of defense against invasion by microbes: (1) the barriers of skin and mucous membranes; (2) nonspecific internal defenses, including phagocytosis, killing by natural killer cells, inflammation, and fever; and (3) the immune response.

The skin physically blocks the entry of microbes into the body. It is also covered with secretions from sweat and sebaceous glands that inhibit bacterial and fungal growth. The mucous membranes of the respiratory and digestive tracts secrete antibiotic substances, IgA antibodies, and mucus. Microbes are trapped in the mucus, swept up to the throat by cilia, and expelled or swallowed.

If microbes enter the body, white blood cells travel to the site of entry and engulf the invading cells. Natural killer cells secrete proteins that kill infected body cells or cancerous cells. Injuries stimulate the inflammatory response, in which chemicals are released that attract phagocytic white blood cells, increase blood flow, and make capillaries leaky. Later, blood clots wall off the injury site, preventing further invasion and spread of the microbes to other parts of the body. Fever is caused by chemicals called pyrogens released by white blood cells in response to infection. High temperatures inhibit bacterial growth and accelerate the immune response.

The Immune Response

The immune response involves two types of lymphocytes, B cells and T cells. Plasma cells, which are descendants of B cells, secrete antibodies into the bloodstream, causing humoral immunity. Cytotoxic T cells destroy some microbes, cancer cells, and virus-infected cells on contact, causing cell-mediated immunity. Helper T cells stimulate both the humoral and cell-mediated immune responses. Immune responses have three steps: recognition, attack, and memory.

Recognition

Antibodies (on B cells) and T-cell receptors (on T cells) recognize foreign molecules and trigger the immune response. Antibodies are Y-shaped proteins composed of a constant and a variable region. Antigens are molecules that generate an antibody response. Antibodies function both as receptors, detecting the presence of antigens, and as effectors, actively working to destroy antigens. Each B cell synthesizes only one type of antibody, unique to that particular cell and its progeny. The diversity of antibodies arises from gene shuffling and mutation of antibody genes during B-cell development. Each antibody has specific sites that bind one or a few types of antigen. Normally, only foreign antigens are recognized by the B cells.

Attack

Antigens bind to and activate only those B and T cells with the complementary antibodies or T-cell receptors. In humoral immunity, B cells with the proper antibodies, stimulated by the presence of particular antigens, divide rapidly to produce plasma cells that synthesize massive quantities of the antibody. The circulating antibodies destroy antigens and antigen-bearing microbes by four mechanisms: neutralization, promotion of phagocytosis by white blood cells, agglutination, and complement reactions. In cell-mediated immunity, T cells with the proper receptors bind antigen and divide rapidly. Cytotoxic T cells bind to antigens on microbes, infected cells, or cancer cells and kill the cells. Helper T cells stimulate, and suppressor T cells turn off, both the B-cell and cytotoxic-T-cell responses.

Memory

Some progeny cells of both B and T cells are long-lived memory cells. If the same antigen reappears in the bloodstream, these memory cells are immediately activated, divide rapidly, and cause an immune response that is much faster and more effective than the original response.

Medicine and the Immune Response

Antibiotics kill microbes or slow down their reproduction, thus allowing the immune system more time to respond and exterminate the invaders. Vaccinations are injections of antigens from disease organisms, often the weakened or dead microbes themselves. An immune response is evoked by the antigens, providing memory and a rapid response should a real infection occur. Allergies are immune responses to normally harmless foreign substances, such as pollen or dust. Certain cells respond to the presence of these substances by releasing histamine, which causes a local inflammatory response. Some diseases are caused by defective immune responses. Autoimmune diseases arise when the immune system destroys some of the body's own cells. Immune deficiency diseases occur when the immune system cannot respond strongly enough to ward off normally minor diseases.

AIDS

AIDS (acquired immunodeficiency syndrome) is caused by one of two viruses, called human immunodeficiency viruses 1 and 2. These viruses invade helper T cells and destroy them. Without helper T cells to stimulate the immune responses of B cells and cytotoxic T cells, the AIDS victim is extremely susceptible to a wide assortment of diseases. These infections eventually kill the patient.

Cancer

Cancer is a population of the body's cells that divides without control. Some cancers may be caused by activation of growth genes, called oncogenes, that cause cells to grow and multiply. These genes may be activated by viral infection or by mutations caused by chemicals or radiation. Other cancers may be caused by loss or inactivation of tumor-suppressor genes, which turn off or suppress cell division in normal cells.

✖ KEY TERMS

acquired immunodeficiency syndrome (AIDS)   p. 667
agglutination   p. 664
allergy   p. 666
antibody   p. 658
antigen   p. 658
autoimmune disease   p. 667
B cell   p. 657
cancer   p. 671
cell-mediated immunity   p. 662
clonal selection   p. 662
complement   p. 659
complement reaction   p. 664
complement system   p. 664
constant region   p. 658

cytotoxic T cell   p. 664
endogenous pyrogen   p. 656
fever   p. 652
helper T cell   p. 664
histamine   p. 655
humoral immunity   p. 662
immune response   p. 657
inflammatory response   p. 652
interferon   p. 656
macrophage   p. 653
major histocompatibility complex (MHC)   p. 661
memory cell   p. 662
mucous membrane   p. 651
natural killer cell   p. 652

neutralization   p. 664
oncogene   p. 671
phagocytic cell   p. 652
plasma cell   p. 662
retrovirus   p. 668
reverse transcriptase   p. 668
severe combined immune deficiency (SCID)   p. 667
skin   p. 651
suppressor T cell   p. 664
T cell   p. 657
T-cell receptor   p. 658
tumor-suppressor gene   p. 671
vaccination   p. 666
variable region   p. 658

✖ THINKING THROUGH THE CONCEPTS

Multiple Choice

1. In addition to the immune system, we are protected from disease by
   a. the skin
   b. mucous membranes
   c. natural secretions such as acids, protein-digesting enzymes, and antibiotics
   d. cilia
   e. all of the above

2. Body cells infected by viruses are destroyed mostly by
   a. IgA   b. phagocytes
   c. natural killer cells   d. histamines
   e. natural antibiotics

3. Fevers
   a. decrease interferon production
   b. decrease the concentration of iron in the blood
   c. decrease the activity of phagocytes
   d. increase the rate of reproduction of invading bacteria
   e. do all of the above

4. T and B cells are
   a. lymphocytes   b. macrophages
   c. natural killer cells d. red blood cells
   e. phagocytes

5. During allergic responses, IgE
   a. covers invaders and marks them for phagocytosis
   b. pokes holes in cell membranes of invading parasites
   c. binds to mast cells, causing them to release histamine
   d. triggers the production of antigens by lymphocytes
   e. promotes secretion of antibodies into the digestive and respiratory tracts

6. What shuts off the immune response in T and B cells after an infection has been conquered?
   a. IgA   b. histamine
   c. pyrogens   d. natural killer cells
   e. suppressor T cells

Review Questions

1. List the three lines of defense of the human body against invading microbes. Which are nonspecific (i.e., act against all types of invaders) and which are specific (i.e., act only against a particular type of invader)? Explain your answer.

2. How do natural killer cells and cytotoxic T cells destroy their targets?

3. Describe humoral immunity and cell-mediated immunity. Include in your answer the types of immune cells involved in each, the location of antibodies and receptors that bind foreign antigens, and the mechanisms by which invading cells are destroyed.

4. How does the immune system construct so many different antibodies?

5. How does the body distinguish "self" from "non-self"?

6. Diagram the structure of an antibody. What parts bind to antigens? Why does each antibody bind only to a specific antigen?

7. What are memory cells? How do they contribute to long-lasting immunity to specific diseases?

8. What is a vaccine? How does it confer immunity to a disease?

9. Compare and contrast an inflammatory response with an allergic reaction from the standpoint of cells involved, substances produced, and symptoms experienced.

10. Distinguish between autoimmune diseases and immune deficiency diseases, and give one example of each.

11. Describe the causes, progression, and eventual outcome of AIDS. How do AIDS treatments work?

12. What is cancer? How do oncogenes cause cancer? How do tumor-suppressor genes prevent cancer? How can environmental factors "turn on" oncogenes to cause cancer?

✣ APPLYING THE CONCEPTS

1. Some types of cancer are partly hereditary; that is, members of certain families are more likely than the general population to develop certain cancers, but not all family members develop the cancer. Given the multistep mechanism for colon cancer outlined in Figure 34-18, could you propose a genetic model that would account for predisposition to cancer while simultaneously not requiring that all people carrying, for example, a defective tumor-suppressor gene necessarily develop cancer?

2. Why is it essential that antibodies and T-cell receptors bind only relatively large molecules (like proteins) and not relatively small molecules (like amino acids)?

3. The text states that smallpox has been eradicated from the Earth. As of early 1995, this is not quite true: There are smallpox stocks in two laboratories, one in the United States, one in Russia. A debate is raging about whether these stocks should be destroyed (see *American Scientist*, November–December 1993, pp. 526–527, and *Science*, November 1993, pp. 1223–1226). In brief, one side argues that having smallpox around is too dangerous, particularly with terrorist groups proliferating. The other side argues that we may be able to learn things from smallpox, answers to questions we don't even know enough to ask yet, that may help us conquer future diseases. Do you think that the smallpox stocks should be destroyed? Why or why not?

4. The box "Health Watch: Flu—The Unbeatable Bug" states that the flu virus is different each year. If that is true, what good is it to get a "flu shot" each winter?

5. Pollen, house dust, or poison ivy provokes allergic reactions in some individuals. When these allergens cannot be avoided, a "hyposensitization" procedure is usually prescribed. Hyposensitization is most likely due to an opposite process—immunization. Explain how giving a patient gradually increasing doses of poison ivy extract over a period of time helps subdue allergic dermatitis.

6. Transplant patients often receive cyclosporine, a drug that inhibits the production of interleukin-2. How does cyclosporine prevent rejection of transplanted organs? Some patients who received successful transplants many years ago are now developing various kinds of cancers. Propose a hypothesis to explain this phenomenon.

7. Throughout pregnancy a mother may pass IgG antibodies across the placenta. After birth she may pass IgAs on through breast milk. The immunity provided to her offspring lasts only a few months. Explain.

✣ FOR MORE INFORMATION

Beardsley, T. "A War Not Won." *Scientific American*, January 1994. The war on cancer remains a standoff.

Caldwell, M. "The Long Shot." *Discover*, August 1993. To produce an effective vaccine against HIV turns out to be a lot more difficult than to produce one against smallpox or polio.

Goodenough, Ursula. "Deception by Pathogens." *American Scientist*, July–August 1991. Describes tactics disease organisms use to gain access to the body.

Greene, W. C. "AIDS and the Immune System." *Scientific American*, September 1993. We may know more about the AIDS virus than any other; can this knowledge produce effective treatment?

Janeway, C. A., Jr. "How the Immune System Recognizes Invaders." *Scientific American*, September 1993. All about antibodies, major histocompatibility complexes, antigen presentation and immune response.

Leder, P. "The Genetics of Antibody Diversity." *Scientific American*, May 1982. How only a few hundred genes can be used to make millions of antibodies.

Lemonick, Michael D. "The Killers All Around." *Time*, September 12, 1994. Are we losing the war against infectious diseases?

Levins, R., Awerbuch, T., Brinkmann, U., Eckardt, I., Epstein, P., Makhoul, N., de Possas, C. A., Puccia, C., Spielman, A., and Wilson, M. E. "The Emergence of New Diseases." *American Scientist*, January–February 1994. Diseases that appear to be new continue to challenge the human body and its physicians.

Lichtenstein, L. M. "Allergy and the Immune System." *Scientific American*, September 1993. Allergic responses, evolved as protection against parasites, turn against us when we respond violently to harmless pollen and foods.

Marrack, P., and Kappler, J. W. "How the Immune System Recognizes the Body." *Scientific American*, September 1993. How the immune system's potential weapons against the body are disarmed.

Nossal, G. J. V. "Life, Death, and the Immune System." *Scientific American*, September 1993. A nice introduction to the major players of the immune system.

Paul, W. E. "Infectious Diseases and the Immune System." *Scientific American*, September 1993. Microbes and the immune system engage in constant evolutionary warfare.

Radetsky, P. "The Roots of Cancer." *Discover*, May 1991. A lucid discussion of the genetics of cancer.

Steinman, L. "Autoimmune Disease." *Scientific American*, September 1993. An estimated 5 percent of Americans suffer from some type of autoimmune disorder.

Young, J. D–E., and Cohn, Z. A. "How Killer Cells Kill." *Scientific American*, January 1988. Cytotoxic T cells strike by secreting proteins that form large holes in the plasma membrane of their targets.

 ## NET WATCH

On-line resources for this chapter are on the World Wide Web at:
http://www.prenhall.com/~audesirk (click on the <u>table of contents</u> link and then select Chapter 34).

35 Chemical Control of the Animal Body: The Endocrine System

⌗ AT A GLANCE

Animal Hormone Structure and Function
Hormones Have a Variety of Chemical Structures
Hormones Function by Binding to Specific Receptors on
Target Cells
Hormones Are Regulated by Feedback Mechanisms

The Mammalian Endocrine System
Mammals Have Both Exocrine and Endocrine Glands
The Hypothalamus Controls the Secretions of the Pituitary
Gland
The Thyroid and Parathyroid Glands Are Located in the
Neck
The Pancreas Is Both an Exocrine and an Endocrine Gland
The Sex Organs Secrete Steroid Hormones
The Adrenal Glands Have Two Parts That Secrete Different
Hormones
Many Types of Cells Produce Prostaglandins
Other Endocrine Organs Include the Pineal Gland, Thymus,
Kidneys, Heart, and Digestive Tract

**Evolutionary Connections: The Evolution
of Hormones**

*The changes that occur throughout the life cycles of all animals
are under the control of hormones. One of the more dramatic
events during the life cycle of certain insects is metamorphosis,
illustrated here by an Indonesian birdwing butterfly emerging
from the pupal case it formed as a caterpillar.*

With the evolution of complex multicellular organisms came the need to coordinate the activities of cells in different parts of the body. Cell-to-cell communication is crucial to the control of movement, growth, reproduction, and the maintenance of homeostasis. Many different mechanisms have evolved by which cells within organisms communicate among themselves. One method is by direct contact. Molecules protruding from the surface membrane identify cells as belonging to an individual of a particular species, as parts of a unique individual organism, and as specific cell types, such as skin or liver. Surface contacts are important in the development of embryos, in which cells migrate around one another to arrive at their proper destination. Direct contact also plays a key role in defense against disease organisms, in which immune cells recognize invading foreign cells by their surface molecules (see Chapter 34).

Cells can also communicate with one another over distances. Cells of the nervous system convey electrical signals over long or short distances, then transmit these signals to other cells by releasing chemicals into the immediate vicinity of the nerve cell ending (Chapter 36). Cells of the **endocrine system**, in contrast, release chemicals into the bloodstream, where they are carried throughout the body and may have wide-ranging effects. These chemicals, or hormones, and the glands and organs that release them, are the subject of this chapter.

Animal Hormone Structure and Function

A **hormone** is a chemical secreted by cells in one part of the body that is transported in the bloodstream to other parts of the body, where it affects particular target cells. Hormones are released by the cells of major endocrine glands and endocrine organs located throughout the body (Fig. 35-1). There are four classes of chemicals used as hormones in the animal kingdom, illustrated in Table 35-1.

Hormones Have a Variety of Chemical Structures

Peptide Hormones Include Both Short and Longer Chains of Amino Acids

Most hormones are chains of amino acids ranging from a few to over a hundred amino acids in length. Technically, short amino acid chains are called peptides, whereas longer chains are called proteins. For convenience, however, all hormones composed of amino acid chains are described as **peptide hormones.** Insulin, antidiuretic hormone, all the hormones of the hypothalamus and the anterior and posterior pituitary, as well as many other hormones are peptide hormones.

Some Hormones Consist of Modified Single Amino Acids

A few hormones are modified amino acids. The amino acid tyrosine forms the basis for the hormones epinephrine (also called adrenaline), norepinephrine (also called noradrenaline), and the thyroid hormone thyroxine.

Steroid Hormones Resemble Cholesterol in Structure

Steroid hormones, also called **steroids**, all have a chemical structure resembling cholesterol, from which most of them are synthesized. Steroid hormones are secreted by the ovaries and placenta (estrogen and progesterone), the testes (testosterone), and the adrenal cortex (aldosterone).

Prostaglandins Are Modified Fatty Acids

Nearly every type of cell in the body has been found to produce **prostaglandins**. These substances consist of two fatty acid carbon chains attached to a five-carbon ring.

Hormones Function by Binding to Specific Receptors on Target Cells

Because nearly all cells have a blood supply, once hormones enter the bloodstream, they reach nearly every cell of the body. But in order to exert their precise control, hormones must act only on certain **target cells**. Hormone specificity is determined by receptors on target cells; if a cell lacks a specific receptor for a hormone, the hormone will not affect the cell. In addition, the same hormone may have several different effects depending on the nature of the target cell it contacts. Receptors for hormones are found in two general locations on target cells: on the cell membrane and inside the cell, usually within the nucleus (Fig. 35-2).

Some Hormones Bind to Surface Receptors

Most peptide hormones, as well as epinephrine and norepinephrine, are water soluble but not lipid soluble. Hence, these hormones cannot cross the phospholipid cell membranes. Instead they react with protein receptors protrud-

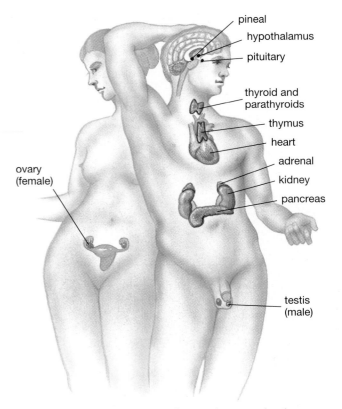

Figure 35-1   Major mammalian endocrine glands

The major mammalian endocrine glands discussed in the text are the hypothalamus-pituitary complex, the thyroid and parathyroid glands, the pancreas, the sex organs (ovaries in females, testes in males), and the adrenal glands. Other organs that secrete hormones include the pineal gland, thymus, kidneys, heart, and digestive tract.

ing from the outside surface of target cell membranes (see Chapter 6). In general, hormones that bind to surface receptors trigger rapid, short-term responses. Some receptors, such as those for norepinephrine, are directly linked to channels that are opened in response to the binding of the hormone. For example, epinephrine binding to receptors on heart muscle cells causes calcium channels to open, allowing more calcium to flow into the muscle cells, which in turn increases the strength of contraction of the heart.

More frequently, a **second messenger** system is used. In second messenger systems, when the hormone binds to the receptor, the shape of the receptor is altered, triggering a series of biochemical reactions that alter the activity of the cell (Fig. 35-2a). In many cases, the binding of the hormone to the receptor activates an enzyme. When activated, the enzyme catalyzes the conversion of ATP to **cyclic AMP** (see Chapter 3), a nucleotide that regulates many cellular activities. Cyclic AMP is often called a second messenger, because it transfers the signal from the first messenger, the hormone, to molecules within the cell. The formation of cyclic AMP initiates a series of reactions inside the cell. Each of these reactions involves an increasing number of molecules, amplifying the original signal. The end result varies with the

TABLE 35-1 ▪ *The Chemical Diversity of Vertebrate Hormones*

Chemical Type	Examples
Modified amino acids (synthesized from single amino acids)	Norepinephrine Thyroxine
Peptides and proteins (synthesized from multiple amino acids)	Oxytocin
Steroids (synthesized from cholesterol)	Testosterone Estradiol
Prostaglandins (synthesized from fatty acids)	Prostaglandin E₁

target cell. For example, channels may be opened in the cell membrane, or substances may be synthesized or secreted.

Another second messenger that may be activated by hormones is **calmodulin**. The binding of certain hormones to their receptors triggers the opening of calcium channels, allowing an influx of calcium into the cell. Calmodulin is a protein in the cytoplasm that binds calcium ions, changing shape as a result. The altered shape of the calcium-calmodulin complex allows it to activate enzymes, acting in a manner similar to cyclic AMP.

Other Hormones Bind to Intracellular Receptors

Steroid hormones and thyroid hormones are lipid soluble and are therefore able to diffuse into the cell membrane and bind to receptors inside the cell (Fig. 35-2b). Both steroid and thyroid hormones alter the activity of genes. It may take from minutes to days for these hormones to exert their full effects. Thyroid and most steroid hormones bind to protein receptors in the nucleus. The receptor-hormone complex binds to DNA and initiates the transcription of messenger RNA from specific genes. The messenger RNA then moves into the cytoplasm and directs the synthesis of new proteins, for example, enzymes involved in cell growth and metabolic activity.

Hormones Are Regulated by Feedback Mechanisms

Animals usually regulate the release of hormones through negative feedback (see Chapter 29). During negative feedback, the secretion of a hormone causes effects in target cells that inhibit further secretion of the hormone. Negative feedback is an important way of maintaining homeostasis—that is, keeping conditions within the body relatively constant over time.

Most hormones exert such powerful effects on the body that it would be harmful to have too much hormone working for too long. For example, suppose you have jogged several miles on a hot, sunny day and lost a pint of water through perspiration. In response to the loss of water from your bloodstream, your pituitary gland releases antidiuretic hormone (ADH), which causes your kidneys to reabsorb more water and produce a very concentrated urine (see Chapter 33). However, if you arrive home and drink a quart of Gatorade, you will more than replace the water you lost in sweat. Continued retention of this excess water could raise blood pressure and possibly damage your heart. Negative feedback ensures that when your blood water content returns to normal, ADH secretion is turned off and your kidneys begin eliminating the excess water (see Fig. 33-8). Negative feedback is also at work in the control of thyroxine secretion (see Fig. 35-9).

In a few cases, positive feedback controls hormone release. As mentioned in Chapter 29, contractions of the uterus early in childbirth cause the release of oxytocin by the posterior pituitary. The oxytocin stimulates stronger

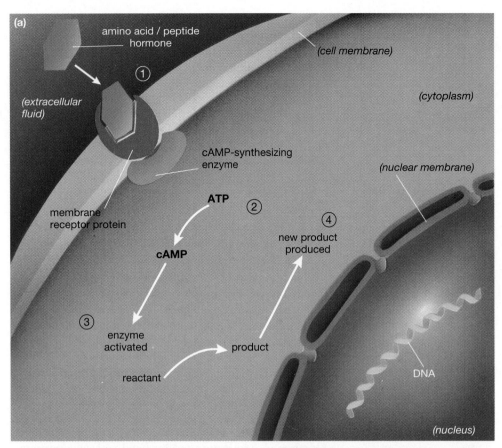

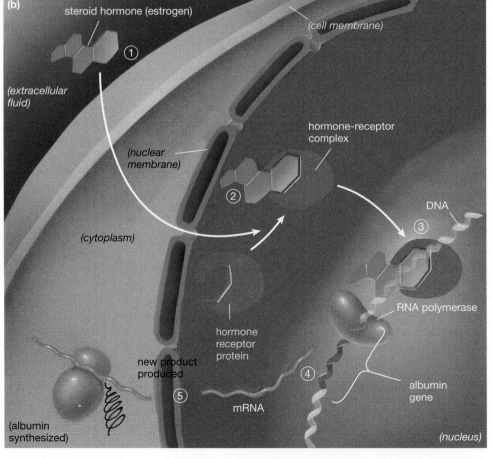

Figure 35-2   Modes of action of hormones

(a) Non-lipid-soluble peptide and amino acid hormones bind to a receptor on the outside of the target cell membrane ①. Hormone-receptor binding triggers synthesis of cyclic AMP (cAMP) ②. Cyclic AMP in turn activates specific enzymes ③ that promote specific cellular reactions that produce new products ④. This cyclic AMP "cascade" may generate a variety of responses. Examples include an increase in glucose synthesis induced by epinephrine and an increase in estrogen synthesis induced by luteinizing hormone.

(b) Lipid-soluble steroid hormones diffuse readily through the cell membrane into the target cell ① and then into the nucleus where they combine with a protein receptor molecule ②. The hormone-receptor complex binds to DNA and facilitates the binding of RNA polymerase to promoter sites on specific genes ③, accelerating transcription of DNA into messenger RNA (mRNA) ④. The mRNA then directs protein synthesis ⑤. In hens, for example, estrogen promotes transcription of the albumin gene, causing synthesis of albumin (egg white), which is packaged in the egg as a food supply for the developing chick.

contractions of the uterus, which cause more oxytocin release, creating a positive feedback cycle. Simultaneously, oxytocin causes uterine cells to release prostaglandins that further enhance uterine contractions, another example of positive feedback. Both positive and negative feedback are involved in the interactions between mother and child during breastfeeding (see Fig. 35-5).

The Mammalian Endocrine System

Endocrinologists (biologists who study the endocrine system) are still far from fully understanding hormonal control in mammals. New hormones, or new roles for previously known hormones, are discovered nearly every year. What we might call the major endocrine glands and endocrine organs, however, have been known for many years. These are the hypothalamus-pituitary complex, the thyroid and parathyroid glands, the pancreas, the sex organs, and the adrenal glands (Fig. 35-1). Table 35-2 lists these and other glands, their major hormones, and their principal functions.

Mammals Have Both Exocrine and Endocrine Glands

Mammals have two types of glands: exocrine glands and endocrine glands (Fig. 35-3). **Exocrine glands** produce secretions that are released outside the body (*exo* means "out of" in Greek) or into the digestive tract (which is actually a hollow tube continuous with the outside world). Exocrine gland secretions are released through tubes or openings called **ducts**. The exocrine glands include the sweat and sebaceous (oil-producing) glands of the skin, the lacrimal (tear-producing) glands of the eye, the mammary (milk-producing) glands, as well as glands producing digestive secretions.

Endocrine glands, sometimes called ductless glands, release their hormones within the body (*endo* means "inside of"). An endocrine gland generally consists of clusters of hormone-producing cells embedded within a network of capillaries. The cells secrete their hormones into the extracellular fluid surrounding the capillaries. The hormones then enter the capillaries by diffusion and are distributed throughout the body by the bloodstream.

The Hypothalamus Controls the Secretions of the Pituitary Gland

The **hypothalamus** is a part of the brain that contains clusters of specialized nerve cells, called **neurosecretory cells**. Neurosecretory cells synthesize peptide hormones, store them, and release them when stimulated. The **pituitary** is a pea-sized gland that dangles from the hypothalamus by a stalk (Figs. 35-1 and 35-4, p. 686). Anatomically, the pituitary consists of two distinct lobes, or parts: the **posterior pituitary** and the **anterior pituitary**. The hypothalamus con-

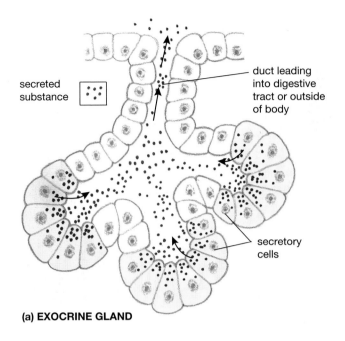

(a) EXOCRINE GLAND

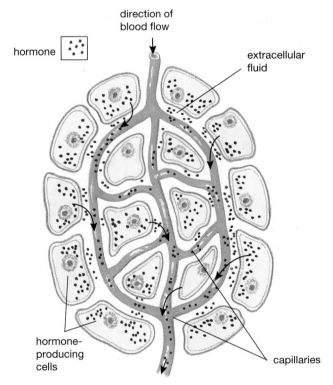

(b) ENDOCRINE GLAND

Figure 35-3   **The structure of exocrine and endocrine glands**

(a) The secretory cells of exocrine glands secrete substances into ducts that usually open outside the body (sweat glands, mammary glands) or into the digestive tract (pancreas, salivary glands). **(b)** Endocrine glands consist of hormone-producing cells embedded with a network of capillaries. The cells secrete hormones into the extracellular fluid from which they diffuse into the capillaries.

TABLE 35-2 ❖ *Mammalian Endocrine Glands and Hormones*

Endocrine Gland	Hormone	Type of Chemical	Principal Function
Hypothalamus (via posterior pituitary)	Antidiuretic hormone (ADH)	Peptide	Promotes reabsorption of water from kidneys; constricts arterioles
	Oxytocin	Peptide	In females, stimulates contraction of uterine muscles during childbirth, milk ejection, and maternal behaviors; in males, causes sperm ejection
Hypothalamus (to anterior pituitary)	Releasing and inhibiting hormones	Peptides	At least nine hormones; releasing hormones stimulate release of hormones from anterior pituitary; inhibiting hormones inhibit release of hormones from anterior pituitary
Anterior pituitary	Follicle-stimulating hormone (FSH)	Peptide	In females, stimulates growth of follicle, secretion of estrogen, and perhaps ovulation; in males, stimulates spermatogenesis
	Luteinizing hormone (LH)	Peptide	In females, stimulates ovulation, growth of corpus luteum, and secretion of estrogen and progesterone; in males, stimulates secretion of testosterone
	Thyroid-stimulating hormone (TSH)	Peptide	Stimulates thyroid to release thyroxine
	Growth hormone	Peptide	Stimulates growth, protein synthesis, and fat metabolism; inhibits sugar metabolism
	Adrenocorticotropic hormone (ACTH)	Peptide	Stimulates adrenal cortex to release hormones, especially glucocorticoids
	Prolactin	Peptide	Stimulates milk synthesis in and secretion from mammary glands
Thyroid	Thyroxine	Modified amino acid	Increases metabolic rate of most body cells; increases body temperature; regulates growth and development
	Calcitonin	Peptide	Inhibits release of calcium from bones
Parathyroid	Parathormone	Peptide	Stimulates release of calcium from bone; promotes absorption of calcium by intestines; promotes reabsorption of calcium by kidneys
Adrenal medulla	Epinephrine and norepinephrine (adrenaline and noradrenaline)	Modified amino acids	Increase levels of sugar and fatty acids in blood; increase metabolic rate; increase rate and force of contractions of the heart; constrict some blood vessels
Adrenal cortex	Glucocorticoids	Steroid	Increase blood sugar; regulate sugar, lipid, and fat metabolism; anti-inflammatory effects
	Aldosterone	Steroid	Increases reabsorption of salt in kidney
	Testosterone	Steroid	Causes masculinization of body features, growth

trols the release of hormones from both parts. The anterior pituitary is a true endocrine gland, composed of several types of hormone-secreting cells enmeshed in a network of capillaries. The posterior pituitary, on the other hand, is derived from an outgrowth of the hypothalamus.

The Posterior Pituitary Releases Hormones Produced by Cells in the Hypothalamus

The posterior pituitary contains the endings of two types of neurosecretory cells whose cell bodies are found in the hypothalamus. These neurosecretory cell endings are enmeshed in a capillary bed into which they release hormones to be carried into the bloodstream (Fig. 35-4). Two peptide hormones are synthesized in the hypothalamus and released from the posterior pituitary: **antidiuretic hormone (ADH)** and **oxytocin**.

Antidiuretic hormone, which literally means "hormone that prevents urination," helps prevent dehydration. As you learned in Chapter 33, ADH causes more water to be reabsorbed from the urine and retained in the body, by increasing the permeability to water of the collecting ducts of nephrons in the kidney. Interestingly, alcohol inhibits the release of ADH, greatly increasing urination, so a beer drinker may temporarily lose more fluid than he has taken in.

Other neurosecretory cells of the hypothalamus release oxytocin from their endings in the posterior pituitary. Oxytocin triggers the "milk ejection reflex" by causing contraction of muscle tissue within the breasts during lacta-

TABLE 35-2 ❖ *(Continued)*

Endocrine Gland	Hormone	Type of Chemical	Principal Function
Pancreas	Insulin	Peptide	Decreases blood glucose levels by increasing uptake of glucose into cells and converting glucose to glycogen, especially in liver; regulates fat metabolism
	Glucagon	Peptide	Converts glycogen to glucose, raising blood glucose levels
Ovaries[a]	Estrogen	Steroid	Causes development of female secondary sexual characteristics and maturation of eggs; promotes growth of uterine lining
	Progesterone	Steroid	Stimulates development of uterine lining and formation of placenta
Testes[a]	Testosterone	Steroid	Stimulates development of genitalia and male secondary sexual characteristics; stimulates spermatogenesis

Other Sources of Hormones

Digestive tract[b]	Secretin, gastrin, cholecystokinin, and others	Peptides	Control secretion of mucus, enzymes, and salts in digestive tract; regulate peristalsis
Thymus	Thymosin	Peptide	Stimulates maturation of cells of immune system
Pineal gland	Melatonin	Modified amino acid	Regulates seasonal reproductive cycles and sleep-wake cycles; may regulate onset of puberty
Kidney	Renin	Peptide	Acts on blood proteins to produce hormone (angiotensin) that regulates blood pressure
	Erythropoietin	Peptide	Stimulates red blood cell synthesis in bone marrow
Heart	Atrial natriuretic peptide (ANP)	Peptide	Increases salt and water excretion by kidneys; lowers blood pressure

[a]See Chapter 39, 40.
[b]See Chapter 32.

tion (breastfeeding). This reflex ejects milk from the saclike milk glands into the nipples (Fig. 35-5). Oxytocin also causes contractions of the muscles of the uterus during childbirth, helping to expel the infant from the womb.

Recent studies using laboratory animals indicate that oxytocin also has behavioral effects. In rats, for example, oxytocin injections cause virgin females to exhibit maternal behavior such as building a nest, licking pups, and retrieving pups that have strayed. Oxytocin may also have a role in male reproductive behavior. In several animals, oxytocin stimulates the contraction of muscles surrounding the tubes that conduct sperm from the testes to the penis, causing ejaculation.

The Anterior Pituitary Produces and Releases a Variety of Hormones

The anterior pituitary (see Fig. 35-4) produces six peptide hormones, four of which help regulate hormone production in other endocrine glands. Two of these, **follicle-stimulating hormone (FSH)** and **luteinizing hormone (LH)**, stimulate production of sperm and testosterone in males and eggs, estrogen, and progesterone in females. We will discuss the roles of FSH and LH in more detail in Chap-

ter 39. **Thyroid-stimulating hormone (TSH)** stimulates the thyroid gland to release its hormones, and **ACTH**, or **adrenocorticotropic hormone** ("hormone that stimulates the adrenal cortex"), causes the release of hormones from the adrenal cortex. We discuss the effects of thyroid and adrenal cortical hormones later in this chapter.

The remaining two hormones of the anterior pituitary, prolactin and growth hormone, do not act on other endocrine glands. **Prolactin**, in conjunction with other hormones, stimulates the development of the mammary glands (which are exocrine glands) during pregnancy. Suckling by the newborn infant then stimulates further release of prolactin, which in turn stimulates milk secretion. When the infant no longer suckles, prolactin secretion is turned off, and the ability to produce milk is lost within a few days.

Growth hormone regulates the growth of the body. Growth hormone acts on all the body's cells, increasing protein synthesis, fat utilization, and the storage of carbohydrates. During maturation, it has a stimulatory effect on bone growth, which influences the ultimate size of the adult organism. Much of the normal variation in human height is due to differences in secretion of growth hormone from the anterior pituitary. Too little growth hormone causes some cases of

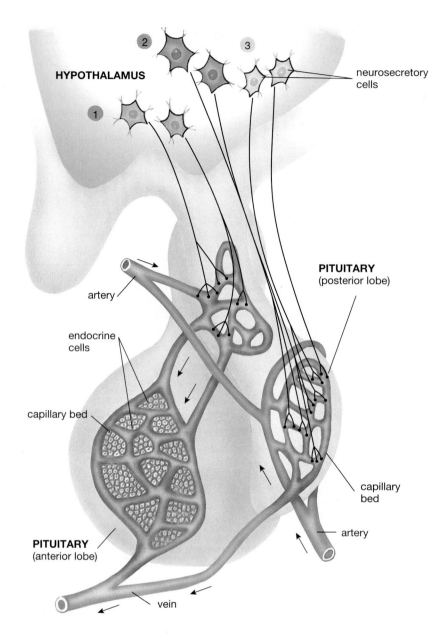

ANTERIOR LOBE	
Hormone	Target Organs
Follicle-stimulating hormone (FSH)	Gonads
Luteinizing hormone (LH)	
Thyroid-stimulating hormone (TSH)	Thyroid
Adrenocorticotropic hormone (ACTH)	Adrenal cortex
Prolactin	Mammary glands
Growth hormone	Most cells
POSTERIOR LOBE	
Hormone	Target Organs
Oxytocin	Uterus, mammary glands
Antidiuretic hormone (ADH)	Kidneys

Figure 35-4 **Anatomical relationships between the hypothalamus and pituitary**

The anterior lobe of the pituitary (left) consists of secretory cells enmeshed in a capillary bed. Release of hormones from these cells is controlled by releasing and inhibiting hormones produced by neurosecretory cells ① of the hypothalamus. These neurosecretory cells release their hormones into a capillary network directly "upstream" from the anterior pituitary (see Fig. 35-7). The posterior lobe of the pituitary (right) is an extension of the hypothalamus. Two types of neurosecretory cells, ② and ③, whose cell bodies are found in the hypothalamus, have cell endings on a capillary bed in the posterior lobe, where they release their hormones. The table shows the hormones of the anterior and posterior pituitary and their target organs.

dwarfism, whereas too much can cause gigantism (Fig. 35-6). Although in adulthood many bones lose their ability to grow, growth hormone continues to be secreted throughout life, helping to regulate protein, fat, and sugar metabolism.

A major advance in the treatment of dwarfism occurred when molecular biologists successfully inserted the gene for human growth hormone into bacteria, which churn out large quantities of the substance (see Chapter 14). Previously, tiny amounts were extracted from human cadavers at great cost. Now children with underactive pituitary glands who would previously have been dwarfs can achieve normal height.

Hypothalamic Hormones Exert Control over the Anterior Pituitary

Neurosecretory cells of the hypothalamus produce at least nine peptide hormones that regulate the release of hormones from the anterior pituitary. These peptides are called **releasing hormones** or **inhibiting hormones** depending on whether they stimulate or prevent the release of pituitary hormone (Fig. 35-7). Releasing and inhibiting hormones are synthesized in nerve cells in the hypothalamus, secreted into a capillary bed in the lower portion of the hypothalamus, and travel a short distance through blood vessels down the pituitary stalk to a second capillary bed surrounding the endocrine cells of the anterior pituitary (see Fig. 35-4). There, the releasers and inhibitors diffuse out of the capillaries and influence pituitary hormone secretion.

Because the releasing and inhibiting hormones are secreted very close to the anterior pituitary, they are produced only in minute amounts. Not surprisingly, they were extremely difficult to isolate and study. Andrew Schally and Roger Guillemin, American endocrinologists who shared the Nobel Prize in 1977 for characterizing several of these hormones, used the brains of millions of sheep

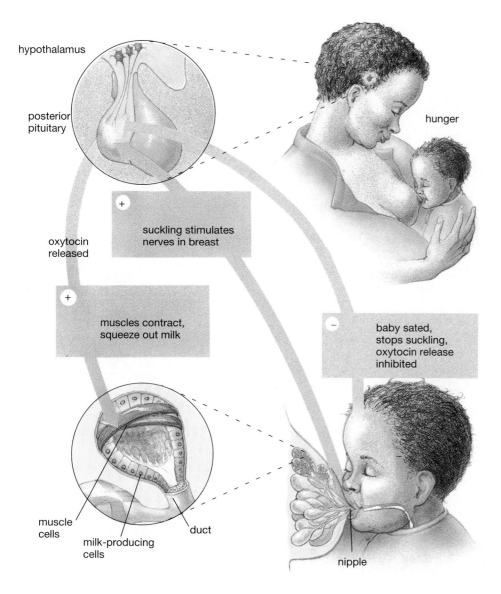

hypothalamus

posterior
pituitary

oxytocin
released

+ suckling stimulates
nerves in breast

+ muscles contract,
squeeze out milk

hunger

− baby sated,
stops suckling,
oxytocin release
inhibited

muscle
cells

milk-producing
cells

duct

nipple

Figure 35-5 **Hormones and breastfeeding**

The control of milk letdown by oxytocin during breastfeeding is regulated by positive (+) and negative (−) feedback between the baby and its mother. The breast, or mammary gland, is an exocrine gland. Here, clusters of milk-producing cells surround hollow bulbs, where milk collects in lactating women. The bulbs are surrounded by muscle that can expel the milk through the nipple. Milk is expelled when suckling by the baby stimulates nerve endings that send a signal to the mother's hypothalamus, causing secretion of oxytocin into the bloodstream by the posterior pituitary. When oxytocin reaches the muscles surrounding the milk ducts, it causes them to contract and expel milk through the nipple. This cycle continues until the infant is full and stops suckling. With the nipple no longer being stimulated, oxytocin release stops, the muscles relax, and milk flow ceases.

and pigs (obtained from slaughterhouses) to extract enough releasing hormone to analyze.

The Thyroid and Parathyroid Glands Are Located in the Neck

In the front of the neck, nestled around the larynx, lies the **thyroid gland** (Fig. 35-8a). The four small discs of the **parathyroid glands** are embedded in the back of the thyroid.

The thyroid produces two major hormones, **thyroxine** and **calcitonin**. Thyroxine is an iodine-containing modified amino acid that raises the metabolic rate of most body cells. Calcitonin is a peptide important in calcium metabolism.

In juvenile animals, including humans, thyroxine helps regulate growth, stimulating both metabolic rate and the development of the nervous system. Cretinism, caused by undersecretion of thyroid hormone from birth, results in mentally retarded dwarfs. Fortunately, early diagnosis and thyroxine supplementation can prevent this tragedy. Pre-

cocious development in vertebrate animals is triggered by oversecretion of thyroxine. In 1912, in one of the first demonstrations of hormone action, a physiologist discovered that thyroxine can induce early metamorphosis in tadpoles (see "Evolutionary Connections: The Evolution of Hormones").

Thyroxine influences most of the cells in the body, elevating their metabolic rate. Its effects include increasing oxygen consumption and heart rate and stimulating the synthesis of enzymes that break down glucose and provide energy. In adults, an elevated metabolic rate seems to be involved in regulating body temperature and stress reactions. Exposure to cold, for example, greatly increases thyroid hormone production.

Levels of thyroxine in the bloodstream are finely tuned by negative feedback loops. Thyroxine release is stimulated by thyroid-stimulating hormone (TSH) from the anterior pituitary, which in turn is stimulated by a releasing hormone from the hypothalamus. The amount of TSH released from

Figure 35-6   **When the anterior pituitary malfunctions**

An improperly functioning anterior pituitary can produce either too much or too little growth hormone. Too much can result in gigantism; too little causes dwarfism.

the pituitary is regulated by thyroxine levels in the blood (Fig. 35-9): High concentrations of thyroxine inhibit the secretion of both the releasing hormone and TSH, thus inhibiting further release of thyroxine from the thyroid.

Iodine deficiency causes a reduction in thyroxine that stimulates dramatic growth of the thyroid (increasing the number of thyroxine-producing cells), an example of a feedback mechanism acting to restore normal hormone levels. The enlarged gland bulges from the neck, producing a condition called **goiter** (Fig. 35-8b). Goiter was once common in some regions of the United States, but widespread use of iodized salt has now all but eliminated this condition in developed countries.

Calcitonin, along with **parathormone**, the hormone secreted by the parathyroids, controls the concentration of calcium in the blood and other body fluids. Calcium is essential for many processes, including nerve and muscle function, so the calcium concentration in body fluids must be kept within narrow limits. Calcitonin and parathormone regulate calcium absorption and release by the bones, which serve both as a skeleton and as a bank into which calcium can be deposited or withdrawn as necessary. In response to low blood calcium, the parathyroids release parathormone, which causes release of calcium from bones. The parathyroids increase in size in pregnant and lactating women, thereby enhancing parathormone output and allowing the mother's

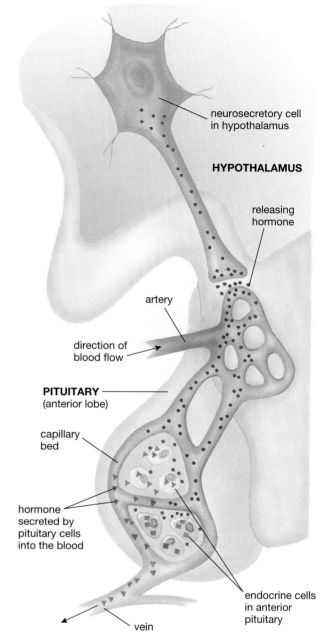

Figure 35-7   **The hypothalamus controls the anterior pituitary**

Hormone release from the anterior pituitary is under the control of releasing and inhibiting hormones from neurosecretory cells of the hypothalamus. Releasing hormones enter a capillary bed in the hypothalamus and travel downstream to capillaries in the anterior pituitary. There the releasing hormones contact the various endocrine cells of the pituitary. Only endocrine cells with matching cell membrane receptors respond to a given releasing hormone (see Fig. 35-2a), so each releasing hormone stimulates a particular type of endocrine cell to release its hormone while leaving other types unaffected.

body to meet the extra demands for calcium imposed by the developing fetus and, later, milk production. If blood calcium levels become too high, the thyroid releases calcitonin, which inhibits the release of calcium from bones.

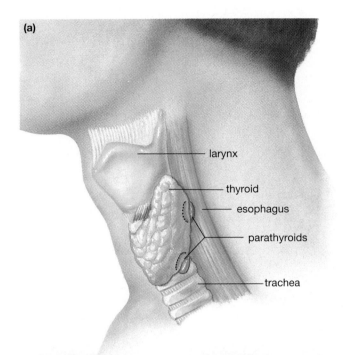

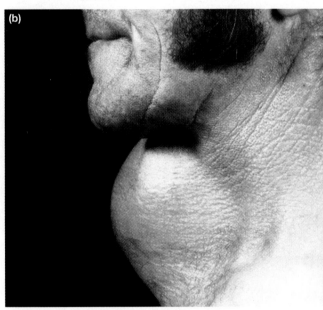

Figure 35-8   The thyroid and parathyroid glands

(a) The thyroid and parathyroid glands are located around the front of the larynx in the neck. (b) Individuals with iodine-deficient diets may suffer from goiter, a condition in which the thyroid becomes greatly enlarged.

The Pancreas Is Both an Exocrine and an Endocrine Gland

The **pancreas** is a double gland producing both exocrine and endocrine secretions (Fig. 35-10). The exocrine portion synthesizes digestive secretions that are released into the pancreatic duct and flow into the small intestine (see Chapter 32). The endocrine portion consists of clusters of cells, called **islet cells**, that produce peptide hormones.

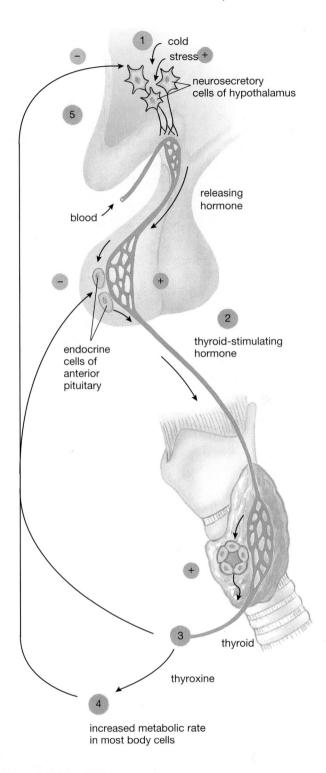

Figure 35-9   Negative feedback in thyroid function

Low body temperature or stress stimulates neurosecretory cells of the hypothalamus ①, whose releasing hormones trigger thyroid-stimulating hormone (TSH) release in the anterior pituitary ②. TSH then stimulates the thyroid to release thyroxine ③. Thyroxine causes increased metabolic activity in most cells of the body, generating ATP energy and heat ④. Both the raised body temperature and high thyroxine levels in the blood inhibit the releasing-hormone cells and the TSH-producing cells ⑤.

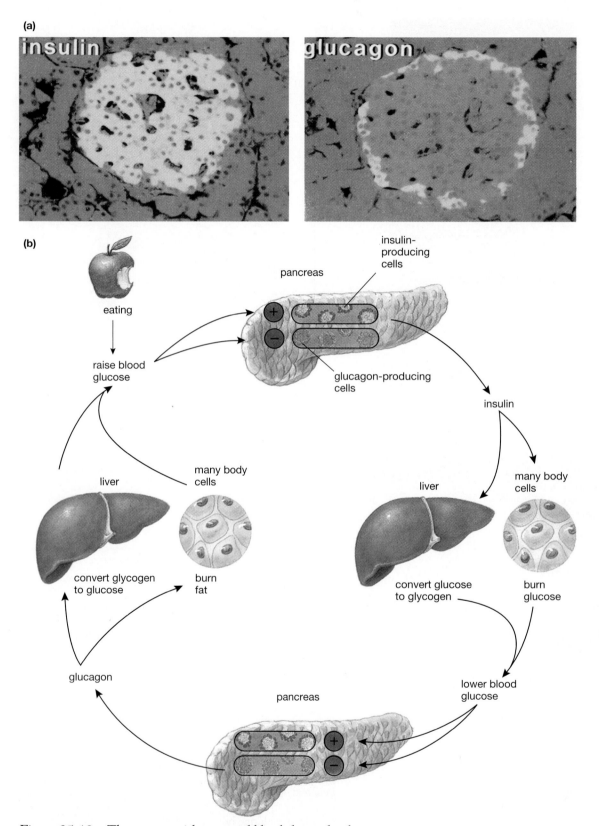

Figure 35-10   **The pancreas islets control blood glucose levels**

(a) The pancreatic islets contain two populations of hormone-producing cells, one producing insulin, the other glucagon. (b) These two hormones cooperate in a two-part negative feedback loop to control blood glucose concentrations. High blood glucose stimulates the insulin cells and inhibits the glucagon cells; low blood glucose stimulates the glucagon cells and inhibits the insulin cells. This dual control quickly corrects either high or low blood glucose levels.

One type of islet cell produces the hormone **insulin**; another type produces **glucagon**.

Insulin and glucagon work in opposite ways to regulate carbohydrate and fat metabolism. When blood glucose rises (for example, after a meal), insulin is released. Insulin causes most of the cells of the body to take up glucose and either metabolize it for energy or convert it to fat or glycogen (a starchlike molecule) for storage. By far the most important storage organ for glycogen is the liver. When blood glucose levels drop (for example, after a person skips breakfast or runs a 10-kilometer race), glucagon is released. Glucagon activates a liver enzyme that breaks down glycogen, releasing glucose into the blood. It also promotes lipid breakdown, releasing fatty acids that are metabolized for energy. Insulin, then, reduces blood glucose levels, whereas glucagon increases them; together they help keep blood glucose levels nearly constant.

Defects in insulin production, release, or reception by target cells result in **diabetes mellitus**, a condition in which blood glucose levels are high and fluctuate wildly with sugar intake. The lack of functional insulin in diabetics causes the body to rely much more heavily on fats as an energy source, leading to high circulating levels of lipids, including cholesterol. Severe diabetes causes fat deposits in the blood vessels, resulting in high blood pressure and heart disease; diabetes is an important cause of heart attacks in the United States. The fatty deposits in small vessels can also damage the retina of the eye, leading to blindness, and the kidneys, leading to kidney failure. Insulin supplements traditionally contained insulin extracted from the pancreases of cows and pigs, obtained from slaughterhouses. Recently, however, the gene for human insulin has been inserted into bacteria, allowing the production of large quantities of human insulin, which is now commercially available.

The Sex Organs Secrete Steroid Hormones

The male testes and female ovaries are important endocrine organs. The testes secrete several steroid hormones, collectively called **androgens**. The most important of these is **testosterone**. The ovary secretes two types of steroid hormones, **estrogen** and **progesterone**. The role of the sex hormones in development, the menstrual cycle, and pregnancy is discussed in Chapters 39 and 40.

The Adrenal Glands Have Two Parts That Secrete Different Hormones

Like the pituitary and pancreas, the adrenals (Latin for "on the kidney") are two glands in one: the adrenal medulla and the adrenal cortex (Fig. 35-11). The **adrenal medulla** is located in the center of each gland (*medulla* means "marrow" in Latin). It consists of secretory cells derived during development from nervous tissue, and its hormone secretion is controlled directly by the nervous

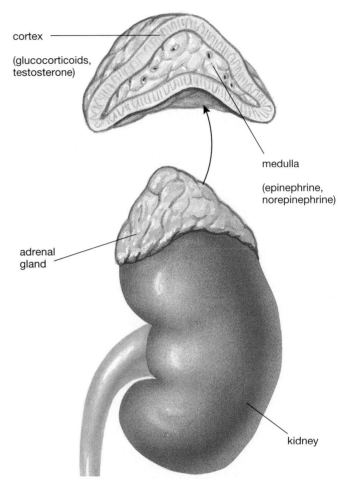

cortex
(glucocorticoids, testosterone)

medulla
(epinephrine, norepinephrine)

adrenal gland

kidney

Figure 35-11   **The adrenal glands**

Atop each kidney sits an adrenal gland, which is a two-part gland composed of very dissimilar cells. The outer cortex consists of ordinary endocrine cells that secrete steroid hormones. The inner medulla is derived from nervous tissue during development and secretes epinephrine and norepinephrine.

system. The adrenal medulla produces two hormones, **epinephrine** and **norepinephrine** (also called **adrenaline** and **noradrenaline**), in response to stress. These hormones, which are modified amino acids, prepare the body for emergency action. They increase the heart and respiratory rates, cause blood glucose levels to rise, and direct blood flow away from the digestive tract and toward the brain and muscles. The adrenal medulla is activated by the sympathetic nervous system, which prepares the body to respond to emergencies, as described in Chapter 36.

The outer layer of the adrenal gland forms the **adrenal cortex** (*cortex* is Latin for "bark"). The cortex secretes three types of steroid hormones synthesized from cholesterol, called **glucocorticoids**. These hormones help control glucose metabolism. Glucocorticoid release is stimulated by ACTH from the anterior pituitary; ACTH release in turn is stimulated by releasing hormones produced by the hypothalamus in response to stressful stimuli, including trauma, infection, or exposure to extremes of temperature. In some respects, the glucocorticoids act similarly to glucagon, raising blood glucose levels by stimulating glucose production and promoting the use of fats instead of glucose for energy production.

You may have noticed that many different hormones are involved in glucose metabolism: thyroxine, insulin, glucagon, epinephrine, and the glucocorticoids. Why? The reason can probably be traced to a metabolic requirement of the brain. Although most body cells can produce energy from fats and proteins as well as carbohydrates, brain cells can burn only glucose. Thus, blood glucose levels cannot be allowed to fall too low, or brain cells rapidly starve, leading to unconsciousness and death.

The adrenal cortex also secretes **aldosterone**, which regulates the sodium content of the blood. Sodium ions, derived from salt in the diet, are the major positive ions in blood and fluid surrounding cells. The sodium ion gradient across cell membranes (high outside, low inside) is used in many cellular events, including the production of electrical signals by nerve cells (see Chapter 36). If blood sodium falls, the adrenal cortex releases aldosterone, which causes the kidneys and sweat glands to retain sodium. Then salt and other sources of dietary sodium, combined with aldosterone-induced sodium conservation, raise blood sodium levels again, shutting off further aldosterone secretion. Too much salt increases the osmolarity of blood and extracellular fluid, drawing water out of cells by osmosis and increasing extracellular fluid and blood volume. The result is a decrease in aldosterone release that allows the kidneys to excrete more sodium.

Finally, the adrenal cortex produces the male sex hormone testosterone, although normally in much smaller amounts than the testes produce. Tumors of the adrenal medulla sometimes lead to excessive testosterone release, causing masculinization of women. Many of the "bearded ladies" who once appeared in circus sideshows probably suffered from this condition.

Many Types of Cells Produce Prostaglandins

Unlike most other hormones, which are synthesized by a limited number of cells, prostaglandins are produced by many, perhaps all, cells of the body. They are modified fatty acids, synthesized by the cell from membrane phospholipids. Several prostaglandins are known, and probably a great many more await discovery. One prostaglandin causes arteries to constrict and stops bleeding from the umbilical cords of newborn infants. Another prostaglandin works in conjunction with oxytocin during labor, stimulating uterine contractions. Prostaglandin-soaked vaginal suppositories are currently used to induce labor. Menstrual cramps are caused by the overproduction of uterine prostaglandins, stimulating uterine contractions.

Some prostaglandins cause inflammation (such as occurs in arthritic joints) and stimulate pain receptors. Drugs such as aspirin and ibuprofen, which inhibit prostaglandin synthesis, can provide relief. Some prostaglandins expand the air passages of the lungs, and one day asthma sufferers may benefit from research into this effect. Others stimulate production of the protective mucus that lines the stomach, a potential boon for ulcer patients. Research on this diverse and potent family of compounds is still in its infancy, but it promises many health benefits in the future.

Other Endocrine Organs Include the Pineal Gland, Thymus, Kidneys, Heart, and Digestive Tract

The **pineal gland** is located between the two hemispheres of the brain, just above and behind the brainstem (see Fig. 35-1). Named for its resemblance to a pine cone, the pineal is smaller than a pea. In 1646, the philosopher René Descartes described it as "the seat of the rational soul." Since then, scientists have learned more about this organ, although many of its functions are still poorly understood. The pineal produces the hormone **melatonin**, a modified amino acid. Melatonin is secreted in a daily rhythm, which in mammals is regulated by the eyes. In some vertebrates, such as the frog, the pineal itself contains photoreceptive cells, and the skull above it is thin, so the pineal can detect sunlight and thus daylength. By responding to daylengths characteristic of different seasons, the pineal appears to regulate the seasonal reproductive cycles of many mammals. Despite years of research, the function of the human pineal and of melatonin is still unclear. One hypothesis is that the pineal and melatonin secretion influence sleep-wake cycles. Improper pineal function may contribute to the depression that some people experience during the short days of winter.

The **thymus** is located in the chest cavity behind the breastbone, or sternum (see Fig. 35-1). In addition to producing white blood cells, the thymus produces the hormone **thymosin**, which stimulates the development of specialized white blood cells (T cells) that play an important role in the immune system (see Chapter 34). The thymus is extremely large in the infant, but begins decreasing in size after puberty.

The kidney, which plays a central role in maintaining body fluid homeostasis, has recently been recognized as an important endocrine organ as well. When the oxygen content of the blood drops, the kidney produces the hormone **erythropoietin**, which increases red blood cell production, as described in Chapter 30. The kidney also produces a second hormone, **renin**, in response to low blood pressure, such as may be caused by bleeding. Renin is an enzyme that catalyzes the production of another hormone, called **angiotensin**, from proteins in the blood. Angiotensin raises blood pressure by constricting arterioles. It also stimulates aldosterone release by the adrenal cortex, causing the kidneys to retain sodium, which in turn increases blood volume.

The heart seems an unlikely endocrine organ, but in 1981, a substance extracted from heart atrial tissue was found to cause an increase in the output of salt and water by the kidneys when injected into rats. Two years later, the active substance, **atrial natriuretic peptide (ANP)**, was described and its amino acid sequence determined. This peptide is released by cells in the atria of the heart when blood volume increases. Increased blood volume causes extra distension of the heart. Atrial natriuretic peptide causes reduction of blood volume by decreasing the release of both ADH and aldosterone. The pace of modern biomedical research is astonishing: Only 5 years after the factor was first detected in heart extract, clinical trials of synthetic ANP were started to test its value in treating high blood pressure and related problems.

The stomach and small intestine produce a variety of peptide hormones that help regulate digestion. These hormones include **gastrin**, **secretin**, and **cholecystokinin**, discussed in Chapter 32.

EVOLUTIONARY CONNECTIONS

The Evolution of Hormones

Not long ago, vertebrate endocrine systems were considered unique to our phylum, and the endocrine chemicals were thought to have evolved expressly for their role in vertebrate physiology. In recent years, however, physiologists have discovered that hormones are evolutionarily ancient. Insulin, for example, is found not only in vertebrates but also in protists, fungi, and bacteria, although research has not yet determined the function of insulin in most of those organisms. Protists also manufacture ACTH, although of course they have no adrenal glands to stimulate. Yeasts have receptors for estrogen but, of course, no ovaries.

Thyroid hormones have been found in certain invertebrates, such as worms, insects, and mollusks, as well as in vertebrates. Even among vertebrates, the effects of chemically identical hormones, secreted by the same glands, may vary dramatically from organism to organism. Let's look briefly at the diverse effects that the thyroid hormone thyroxine has on several different organisms.

Some fish undergo radical physiological changes during their lifetimes. A salmon, for example, begins life in fresh water, migrates to the ocean, and finally returns to fresh water to spawn. In the stream where it hatched, fresh water tends to enter the fish's tissues by osmosis; in salt water, the fish tends to lose water, becoming dehydrated. The fish's migrations, therefore, require complete revamping of salt and water control. In salmon, one of the functions of thyroxine is to produce the metabolic changes necessary to go from life in streams to life in the ocean.

In amphibians, thyroxine has the dramatic effect of triggering metamorphosis. In 1912, in one of the first demonstrations of the action of any hormone, tadpoles were fed minced horse thyroid. As a result, they metamorphosed prematurely into miniature adult frogs (Fig. 35-12). In high mountain lakes in Mexico, where the water is deficient in the iodine needed to synthesize thyroxine, natural selection has produced one species of salamander that has the ability to reproduce while still in its juvenile form.

Thyroxine regulates the seasonal molting of most vertebrates. From snakes to birds to the family dog, surges of thyroxine stimulate the shedding of skin, feathers, or hair. In people (who neither migrate, metamorphose, nor molt), thyroxine regulates growth and metabolism.

The use of chemicals to regulate cellular activity is extremely ancient. The diversity of life on Earth rests upon a conservative foundation: A relative handful of chemicals coordinate activities within single cells and among groups of cells. Life's diversity originated in part by changing the systems used to deliver the chemicals and by evolving new types of responses. Early in their evolution, animals developed a complement to hormonal communication that provides faster, more precise delivery of chemical messages: the nervous system. As we explain in the next chapter, the nervous system permits rapid responses to environmental stimuli, flexibility in response options, and ultimately consciousness itself.

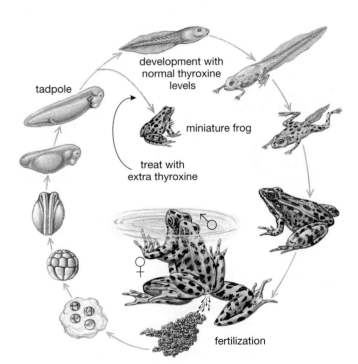

Figure 35-12 **Thyroxine controls metamorphosis in amphibians**

The life cycle of the frog begins with fertilization of the eggs (bottom); then development into an aquatic, fishlike tadpole; growth of the tadpole; and ultimately metamorphosis into an adult frog. Metamorphosis is triggered by a surge of thyroxine from the tadpole's thyroid gland. If a young tadpole is injected with extra thyroxine, it will metamorphose ahead of schedule into a miniature adult frog.

✖ SUMMARY OF KEY CONCEPTS

Animal Hormone Structure and Function
A hormone is a chemical secreted by cells in one part of the body that is transported in the bloodstream to other parts of the body, where it affects the activity of specific target cells.

Four types of molecules are known to act as hormones: peptides, modified amino acids, steroids, and prostaglandins.

Most hormones act on their target cells in one of two ways. Peptide hormones and modified amino acids bind to receptors on the surfaces of target cells and activate intracellular second messengers such as cyclic AMP and calmodulin. The second messengers then alter the metabolism of the cell. Steroid hormones diffuse through the cell membranes of the target cells and bind with receptor proteins in the cytoplasm. The hormone-receptor complex travels to the nucleus and promotes transcription of specific genes. Thyroid hormones also penetrate the cell membrane but diffuse into the nucleus, where they bind to receptors associated with the chromosomes and influence gene transcription.

Hormone action is often regulated through negative feedback, a process in which the hormone causes changes that inhibit further secretion of that hormone.

The Mammalian Endocrine System
Hormones are produced by endocrine glands, which are clusters of cells embedded within a network of capillaries. Hormones are secreted into the extracellular fluid and diffuse into the capillaries. The major endocrine glands of the human body are the hypothalamus-pituitary complex, the thyroid and parathyroid glands, the pancreas, the sex organs, and the adrenal glands. The hormones released by these glands and their actions are summarized in Table 35-2. Prostaglandins, unlike other hormones, are not secreted by discrete glands but are synthesized and released by many cells of the body. Other endocrine organs include the pineal gland, kidneys, heart, and the stomach and small intestine.

✖ KEY TERMS

adrenal cortex p. 691
adrenal medulla p. 691
androgen p. 691
anterior pituitary p. 683
calmodulin p. 681
cyclic AMP p. 680
diabetes mellitus p. 691
duct p. 683
endocrine gland p. 683
endocrine system p. 679
exocrine gland p. 683
goiter p. 688
hormone p. 679
hypothalamus p. 683
inhibiting hormone p. 686
islet cell p. 689
neurosecretory cell p. 683
pancreas p. 689
parathyroid gland p. 687
peptide hormone p. 680
pineal gland p. 692
pituitary p. 683
posterior pituitary p. 683
prostaglandin p. 680
releasing hormone p. 686
second messenger p. 680
steroid p. 680
target cell p. 680
thymus p. 692
thyroid gland p. 687

✖ THINKING THROUGH THE CONCEPTS

Multiple Choice

1. Steroid hormones
   a. alter the activity of genes
   b. trigger rapid, short-term responses in cells
   c. work via second messengers
   d. initiate open channels in cell membranes
   e. bind to cell surface receptors
2. Examples of posterior pituitary hormones are
   a. FSH and LH
   b. prolactin and parathormone
   c. secretin and cholecystokynin
   d. melatonin and prostaglandin
   e. ADH and oxytocin
3. Negative feedback to the hypothalamus controls the level of __________ in the blood.
   a. thyroxine b. estrogen
   c. glucocorticoids d. estrogen
   e. all of the above
4. The primary targets for FSH are cells in the
   a. hypothalamus b. ovary
   c. thyroid d. adrenal medulla
   e. pituitary
5. The kidney is a source of
   a. thyroxine and parathormone
   b. calcitonin and oxytocin
   c. renin and erythropoietin
   d. ANP and epinephrine
   e. glucagon and glucocorticoids

6. Hormones that are produced by many different body cells and cause a variety of localized effects are known as
   a. peptide hormones   b. parathormones
   c. releasing hormones   d. prostaglandins
   e. exocrine hormones

Review Questions

1. What are the four types of molecules used as hormones in vertebrates? Give an example of each.
2. What is the difference between an endocrine and an exocrine gland? Which type releases hormones?
3. When peptide hormones attach to receptors in target cells, what cellular events follow? How do steroid hormones behave by comparison?
4. Diagram the process of negative feedback, and give an example of negative feedback in the control of hormone action.

5. What are the major endocrine glands in the human body, and where are they located?
6. Describe the structure of the hypothalamus-pituitary complex. Which pituitary hormones are neurosecretory? What are their functions?
7. Describe how releasing hormones regulate the secretion of hormones by cells of the anterior pituitary. Name the hormones of the anterior pituitary and give one function of each.
8. Describe how the hormones of the pancreas act together to regulate the concentration of glucose in the blood.
9. Compare the adrenal cortex and medulla by answering the following questions. Where are they located within the adrenal gland? What are their embryological origins? What hormones do they produce? What organs do their hormones target? What homeostatic processes regulate blood levels of the respective hormones?

APPLYING THE CONCEPTS

1. An enterprising student decides to do a science project on the effect of the thyroid gland on frog metamorphosis. She sets up three aquaria with tadpoles. She adds thyroxine to the water of one, thiouracil to a second, and nothing to the third. Thiouracil reacts with thyroxine inside tadpoles to produce an ineffective compound. Assuming the student uses appropriate physiological concentrations, predict what will happen.
2. Diabetes mellitus is common diabetes. Two other forms, adrenal and pituitary diabetes, are also characterized by high levels of blood glucose. What specific parts of regulatory cycles are disrupted in these latter forms to cause hyperglycemia?
3. Suggest a hypothesis about the endocrine system to explain why many birds lay their eggs in the spring and why poultry farmers keep lights on at night in their egg laying operations.
4. Anabolic steroids, used by risk-taking athletes and bodybuilders, are chemically related to testosterone. They increase bone and muscle mass and do seem to improve athletic performance. But anabolic steroids can cause liver problems, heart attacks, strokes, testicular atrophy, and personality changes in males. Females on anabolic steroids also have liver and circulatory problems. In addition, their voices deepen, their bodies develop more hair, and their menstrual cycles are disturbed. Explain how the same compound can produce testicular atrophy in males but virilism in females.
5. Some parents, interested in college sports scholarships for their children, are asking physicians to prescribe growth hormone treatments. Farmers also have an economic incentive to treat cows with growth hormone, which can now be produced in large quantities with genetic-engineering techniques. What biological and ethical problems do you foresee for parents, children, physicians, coaches, college scholarship boards, food consumers, farmers, the U.S. Food and Drug Administration, and bio-tech companies?

FOR MORE INFORMATION

Atkinson, M., and MacLaren, N. "What Causes Diabetes?" *Scientific American*, July 1991.

Berridge, M. J. "The Molecular Basis of Communication within the Cell." *Scientific American*, October 1985. Both the "classical" cyclic AMP and more recently discovered second messengers convey information from cell surface receptors to DNA and cellular metabolism.

Guillemin, R., and Burgus, R. "The Hormones of the Hypothalamus." *Scientific American*, November 1972. The interaction between hypothalamus and pituitary is explored.

Sapolsky, R. "Stress in the Wild." *Scientific American*, January 1990. An interesting study of baboons showing the effects of hormones on stress responses.

Snyder, S. H. "The Molecular Basis of Communication between Cells." *Scientific American*, October 1985. Snyder describes the similarities and differences between neural and hormonal control systems in the body.

 # NET WATCH

On-line resources for this chapter are on the World Wide Web at:
http://www.prenhall.com/~audesirk (click on the <u>table of contents</u> link and then select Chapter 35).

36 Information Processing: The Nervous System

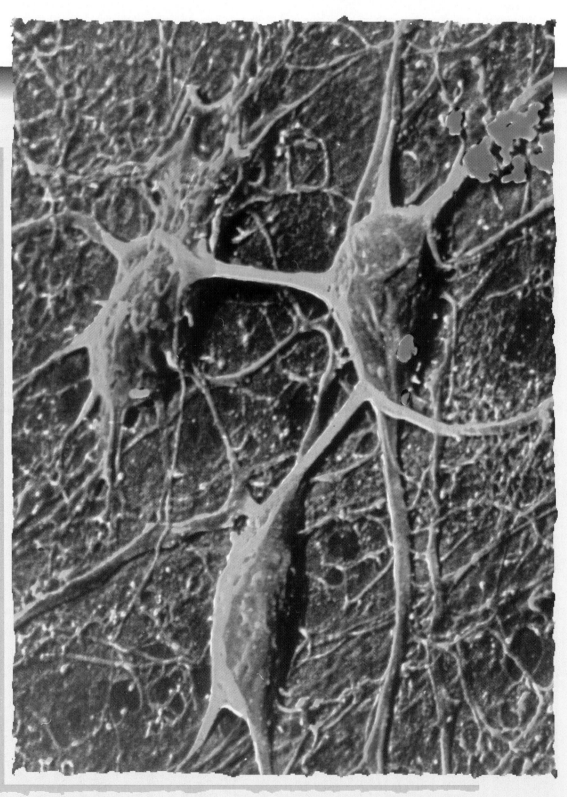

"Know thyself."

—*Inscribed above the entrance to the Temple of Apollo, home of the Oracle of Apollo at Delphi*

❖ AT A GLANCE

A Comparison of Nervous and Endocrine Communication

The Functions and Structure of Neurons
Dendrites Receive Signals from Other Neurons or the Environment
The Cell Body Maintains the Neuron and Integrates Electrical Signals from the Dendrites
The Axon Carries Electrical Signals from the Cell Body to Their Destination
Synaptic Terminals Communicate with Other Neurons, Muscles, or Glands

Mechanisms of Neural Activity
How Is the Resting Potential Generated?
Action Potentials Can Carry Messages Rapidly over Long Distances
Neurons Communicate at Synapses

Building and Operating a Nervous System
Information Processing Requires Four Basic Operations
Neural Pathways Direct Behavior
Increasingly Complex Nervous Systems Are Increasingly Centralized

The Human Nervous System
The Peripheral Nervous System Links the Central Nervous System to the Body
The Central Nervous System Consists of the Spinal Cord and Brain
The Spinal Cord Is a Cable of Axons Protected by the Backbone
The Brain Consists of Several Parts Specialized for Specific Functions
The Nervous System Uses Many Neurotransmitters and Neuromodulators

Brain and Mind
The "Left Brain" and "Right Brain" Are Specialized for Different Functions
The Mechanisms of Learning and Memory Are Poorly Understood
Insights on How the Brain Creates the Mind Come from Diverse Sources

Three neurons from the human cerebral cortex, the gray material that forms the outer layer of the brain, are seen in this scanning electron micrograph. The oval structures are cell bodies; the threadlike structures projecting out from them are dendrites and axons. Dendrites receive signals from other neurons and transmit them to the cell body. Axons carry electrical signals from the cell body to other neurons.

As you read this page, light reflected from it bombards your eyes, and sound waves from the radio assault your ears. Your senses of smell and taste are simultaneously stimulated as you drink coffee. Touch-sensitive receptors all over your body are activated as your clothes, the chair, and the pages of this book all push at your skin. You breathe, swallow, and shift positions in your chair. Untaxed by these marvelous but ordinary activities, your mind calls upon its reading skill, directs dexterous manipulations of a pen as you take notes, stores memories of what you have read, and still has the capacity to appreciate the music from the radio.

All this activity is controlled by your brain, an organ weighing about 3 pounds (about 1.5 kilograms) encased within your skull. Composed of billions of nerve cells, the brain is connected to the rest of the body by dozens of slender cables, or nerves. In this chapter, we first compare types of communication controlled by the endocrine and nervous systems. We explore how nerve cells create electrical signals, and how they use these signals to communicate with one another and to control the body. In Chapters 37 and 38, we examine the senses through which the brain experiences the outside world and the muscles that respond to its commands.

A Comparison of Nervous and Endocrine Communication

Although it is convenient to discuss hormonal control (Chapter 35) separately from nervous control, in some ways the two are quite similar. Both hormone-producing cells and nerve cells synthesize "messenger" chemicals that they release into extracellular spaces (Fig. 36-1). However, there are four main differences in how the nervous and endocrine systems use chemical messages. First, whereas hormone-producing and neurosecretory cells release hormones into the bloodstream, nerve cells usually release their chemical messengers (called neurotransmitters) very close to the cells they influence, often from less than a micrometer away. Second, whereas blood-borne hormones bathe millions of cells indiscriminately, a nerve cell re-

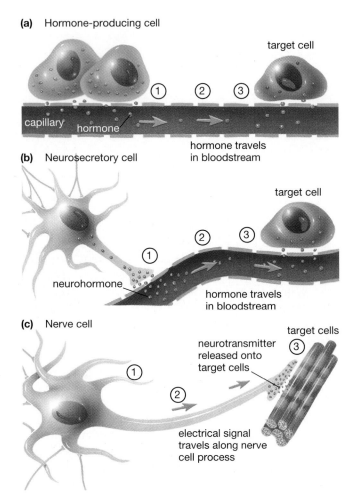

(a) Hormone-producing cell

target cell

capillary   hormone

hormone travels
in bloodstream

(b) Neurosecretory cell

target cell

neurohormone

hormone travels
in bloodstream

(c) Nerve cell

target cells

neurotransmitter
released onto
target cells

electrical signal
travels along nerve
cell process

Figure 36-1   **Three major types of "control cells"**

These all release chemicals that influence the activity of other cells of the body: **(a)** "classic" hormone-producing cells, **(b)** neurosecretory cells, and **(c)** ordinary nerve cells. Three common steps occur in each system: ① The chemical-releasing cell is stimulated. ② It sends a message to distant cells. ③ Selected target cells respond. Both classic hormone-producing cells and neurosecretory cells release their chemical messages into the bloodstream, which transports the chemicals to distant target cells. Regular nerve cells grow long cell extensions, called processes, to their target cells and release their chemical messages directly onto the target cell.

leases its neurotransmitter onto one or a small group of specific target cells. Third, a nerve cell speeds information from one part of the body to another by means of electrical signals traveling within the cell itself, and it releases its neurotransmitter only when it reaches its target. Hormones move much more slowly and are released at varying distances from the target cell. Finally, the effects of messages sent by neurons tend to be of much shorter duration than the effects of hormones.

Yet even these differences blur as we learn more about the hormonal and nervous systems of animals. Some hormones are in fact produced and released into the bloodstream by nerve cells, such as the neurosecretory cells of

the hypothalamus. Other chemicals that biologists formerly believed were strictly hormones, such as insulin, have recently been discovered in the brain, where they are synthesized and released by neurons and act more like neurotransmitters. As you read about the nervous system in this chapter, remember that no system in your body works alone. The hormonal and nervous systems are closely coordinated in their control of bodily functions.

The Functions and Structure of Neurons

Our study of the nervous system begins with the individual nerve cell, or **neuron.** As the fundamental unit of the nervous system, each neuron must perform five functions:

1. Receive information from the internal or external environment or from other neurons
2. Integrate the information it receives and produce an appropriate output signal
3. Conduct the signal to its terminal ending, which may be some distance away
4. Transmit the signal to other nerve cells, glands, or muscles
5. Coordinate the metabolic activities that maintain the integrity of the cell

Although neurons vary enormously in structure, a "typical" vertebrate neuron has four distinct structural regions that carry out the functions above. These regions are the dendrites, the cell body, the axon, and the synaptic terminals (Fig. 36-2).

Dendrites Receive Signals from Other Neurons or the Environment

Dendrites, branched tendrils that extend outward from the nerve cell body, are specialized to respond to signals from other neurons or from the external environment. Their branched form provides a large surface area to receive these signals. Dendrites of sensory neurons, described in the next chapter, have special membrane adaptations that allow them to respond to specific stimuli from the environment such as pressure, odorous molecules, light, or heat. In neurons of the brain and spinal cord, dendrites respond to the chemical neurotransmitters released by other neurons. These dendrites have protein receptors in their membranes that bind specific neurotransmitters and produce electrical signals as a result of this binding.

The Cell Body Maintains the Neuron and Integrates Electrical Signals from the Dendrites

Electrical signals travel down the dendrites and converge upon the **cell body** of the neuron, which serves as an integration center. In its integrating role, the cell body adds up the various signals from the dendrites and "decides"

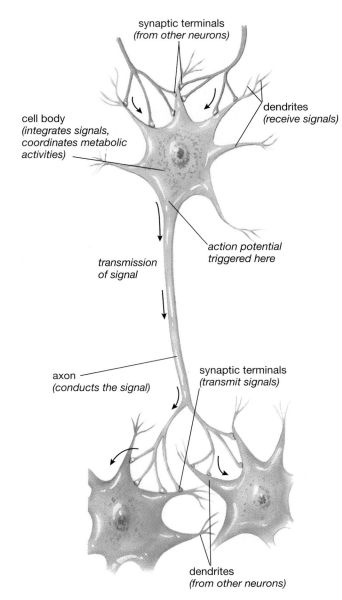

synaptic terminals
(from other neurons)

cell body
*(integrates signals,
coordinates metabolic
activities)*

dendrites
(receive signals)

*action potential
triggered here*

*transmission
of signal*

axon
(conducts the signal)

synaptic terminals
(transmit signals)

dendrites
(from other neurons)

***Figure 36-2*   A nerve cell showing its specialized parts
and their functions**

the cell body to the synaptic terminals, located at the far end of each axon. Axons are usually bundled together into **nerves,** much like the wires in an electrical cable. However, unlike electrical power distribution cables (in which energy is lost along the way from power station to customer), the cell membranes of axons are specialized to conduct action potentials undiminished from the cell body to their synaptic terminals. In vertebrates, nerves emerge from the brain and spinal cord and extend out to all regions of the body.

Synaptic Terminals Communicate with Other Neurons, Muscles, or Glands

Signals are transmitted to other cells at **synaptic terminals,** which appear as swellings at the branched endings of axons (see Fig. 36-2). Most synaptic terminals contain a specific type of chemical, a **neurotransmitter,** that they release in response to an action potential traveling down the axon. The synaptic terminals of one neuron may communicate with a gland, a muscle, or the dendrites or cell body of a second neuron, so that the output of the first cell becomes the input to the second.

Mechanisms of Neural Activity

About 40 years ago, using the giant axon of a mollusk—the squid—biologists developed ways to record electrical events inside individual neurons (Fig. 36-3). They found that unstimulated, inactive neurons maintain a constant electrical difference, or potential, across their cell membranes, similar to that found across the poles of a battery. As in a battery, the electrical potential across a neuron membrane stores energy. This potential, called the **resting potential,** is always negative inside the cell and ranges from –40 to –90 millivolts (thousandths of a volt). If the neuron is stimulated, either naturally or with an electrical current, the negative potential inside the neuron can be altered. Depending on the nature of the stimulus, the inside potential can be made either more or less negative. If the potential is made sufficiently less negative, it reaches a level called **threshold** (roughly 15 millivolts less negative than the resting potential), at which an action potential is triggered. The triggering point is near the junction of the cell body and the axon (see Fig. 36-2). During the action potential, the neuron suddenly becomes 20 to 50 millivolts positive inside. Action potentials last a few milliseconds (thousandths of a second) before the cell restores its negative resting potential. Let's look more closely at these electrical potentials, the language of the nervous system.

How Is the Resting Potential Generated?

How can a cell behave as a battery, storing energy in the form of electrical and chemical gradients across its cell membrane? To understand this phenomenon, recall two physical principles, *diffusion* and *electrical attraction,* and

whether to produce an **action potential,** the electrical output signal of the neuron. The cell body, containing the usual assortment of organelles, carries on activities common to most other body cells. These include the synthesis of complex molecules such as proteins, lipids, and carbohydrates and the coordination of the metabolic activities of the cell.

The Axon Carries Electrical Signals from the Cell Body to Their Destination

In a typical neuron, a long, thin fiber, called an **axon,** extends outward from the cell body, making neurons the longest cells in the body. Single axons, for example, stretch from your spinal cord to your toes, a distance of about a meter (3 feet). Axons are distribution lines, carrying action potentials from

(a)

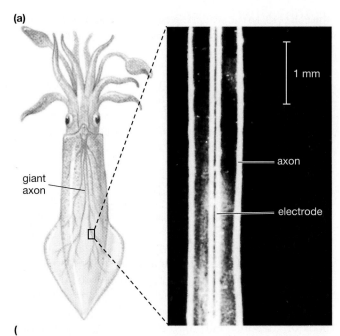

giant
axon

axon

electrode

1 mm

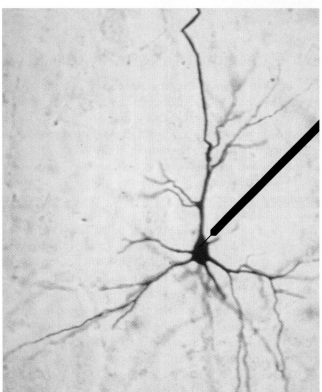

Figure 36-3   **Ways of studying neurons**

(a) Taking advantage of the giant axon of the squid, the British physiologists Bernard Katz, Alan Hodgkin, and Andrew Huxley pushed thin wires or narrow tubes down the inside of the axon. These electrodes were connected to voltmeters to record the electrical potential difference between the inside and outside of the axon, about −70 millivolts. **(b)** Modern neurobiologists use hollow glass electrodes drawn to a needlelike tip less than 1 micrometer in diameter and filled with a salt solution. The sharp tip penetrates the neuron without damaging it.

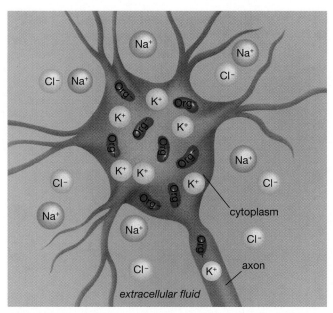

cytoplasm

axon

extracellular fluid

Figure 36-4   **The neuron maintains ionic gradients**

The ionic composition of the neuron cytoplasm is significantly different from that of the extracellular fluid. The neuron contains a high concentration of potassium ions (K$^+$) and large organic anions (Org$^-$), whereas the extracellular fluid is high in sodium chloride (Na$^+$ and Cl$^-$).

one property of cell membranes, *differential permeability* discussed in Chapter 6. These factors interact with concentration differences of ions inside and outside the cell to produce the resting potential, as described below.

The cell membrane of a neuron encloses cytoplasm with various ions dissolved in it. The neuron itself is immersed in a salt solution, the extracellular fluid (Fig. 36-4). The ions of the cytoplasm consist mainly of positively charged potassium ions (K$^+$) and large, negatively charged organic molecules, such as proteins. Outside the cell, the extracellular fluid contains more positively charged sodium ions (Na$^+$) and negatively charged chloride ions (Cl$^-$). These concentration differences are maintained by active transport using a specialized membrane protein called a **sodium-potassium pump**, described in more detail later.

As you have learned, a cell membrane resembles a lipid ocean in which protein icebergs are embedded. Because charged particles cannot pass through the lipids, they must travel through tunnel-shaped proteins, or channels, extending through the membrane. In an unstimulated neuron, shown below, only potassium ions can cross through the membrane. They travel through specific proteins called **potassium channels,** shown in yellow. Although **sodium channels** (shown in blue) are also present, in unstimulated neurons they remain closed. Only potassium ions can cross the membranes and potassium ions are in highest concentration inside the cell, so they tend to diffuse out of the cell. The large, negatively charged organic ions are left behind, as shown below.

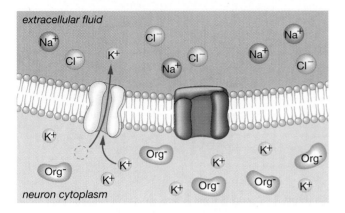

As more and more positively charged potassium ions leave, the inside of the cell becomes increasingly negative. But, because unlike charges attract one another, as potassium ions diffuse out, an electrical force develops that tends to pull them back inside. At some point, the diffusion of potassium ions out of the neuron due to concentration differences will be balanced by the electrical attraction tending to pull them back inside. At this point, there is no more net movement of potassium ions, and the cell reaches a stable resting potential—negative inside. This point is illustrated below.

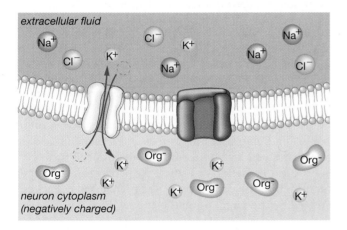

Notice that a resting potential is based on a balance between chemical and electrical gradients, maintained by active transport and a differentially permeable membrane. Both of these gradients represent a form of stored, or potential, energy that can be used to generate electrical signals as described below. Establishing a resting potential does not require significant changes in the potassium concentration inside and outside the cell. Only about 1/10,000 of the potassium ions initially inside our hypothetical cell must leave to set up a resting potential of –60 millivolts. It is interesting to note that, although neurons make special use of their electrical potentials, nearly all living cells maintain resting potentials.

Action Potentials Can Carry Messages Rapidly over Long Distances

An unchanging resting potential, like a single musical note, can't convey much information. Nervous information is therefore encoded in *transient changes in electrical potential* in nerve cells. An example is the action potential. Let's look at these signals in more detail.

If the potential inside the cell body of a neuron is made sufficiently less negative (such as by postsynaptic potentials, described later), the neuron membrane may reach threshold, triggering an action potential at the point where the axon leaves the cell body. An action potential is a sudden reversal of the electrical charge across the membrane, triggered by a temporary, localized increase in its permeability to sodium. This increased permeability allows a rapid influx of positively charged sodium ions. An action potential resembles a fast-moving wave of positive charge that travels, undiminished in size, along the axon to the synaptic terminal. Immediately after the action potential passes, the negative resting potential is restored within the axon. The events that occur during an action potential as recorded by an electrode inside the neuron are shown in Figure 36-5 and described below. A neuron at

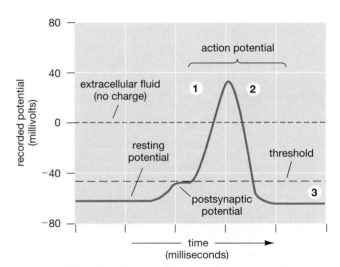

Figure 36-5 **The electrical events during an action potential**

These events are recorded inside a nerve cell. The resting potential is about 60 millivolts negative with respect to the outside. When the cell is stimulated to reach threshold by a postsynaptic potential, membrane channels permeable to sodium open up, and sodium enters the cell, powered both by diffusion and by electrical attraction; the inside of the cell becomes positively charged ①. Shortly thereafter, other membrane channels permeable to potassium open, and potassium leaves ②, driven by diffusion and electrical repulsion from the now-positive inside of the cell, until the resting potential is reestablished ③. Active transport molecules in the membrane, called the sodium-potassium pump, continuously pump sodium out and potassium in, maintaining the ionic gradient.

rest, diagrammed below, is something like a loaded musket, charged and ready to fire if the trigger is pulled. In a neuron, the "explosive charge" is the concentration gradient of sodium ions, which are highest in concentration outside the cell. The "trigger" is a set of membrane proteins, the sodium channels (shown in blue). These proteins are selectively permeable to sodium, are closed in a resting neuron, and are specialized to open suddenly when threshold is reached. The energy to pull the trigger is provided by postsynaptic potentials (described later) that bring the neuron to threshold.

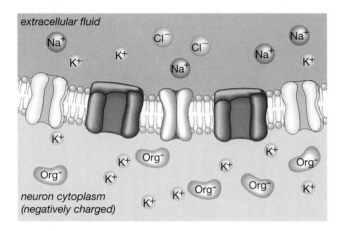

At threshold, sodium channels open, as shown below. Positively charged sodium ions flood into the cell, making the cell's interior momentarily positive.

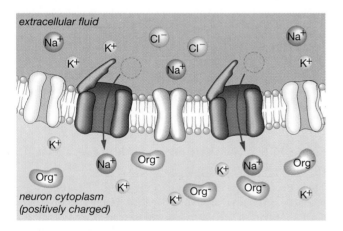

After a short time, the sodium channels spontaneously close, and a different set of potassium channels open, shown in orange below. Potassium ions now flow out of the cell through both types of potassium channels. The ions are driven out both by their diffusion gradient and by electrical repulsion from the positive sodium ions that recently entered.

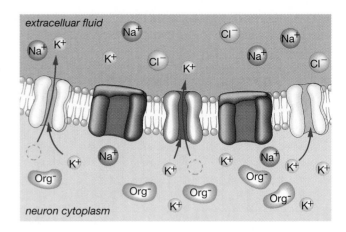

So many potassium ions leave that the inside once again becomes negative, reestablishing the resting potential. Thus, the action potential is a brief event; the neuron first becomes positive as sodium ions enter and then becomes negative again as potassium ions flow out.

Action potentials are "all-or-none"—that is, they do not vary in amplitude. If the neuron does not reach threshold, there will be no action potential at all, but if threshold is reached, then a full-sized action potential will occur and travel the entire length of the axon.

The Sodium-Potassium Pump Helps Maintain an Ionic Gradient across the Membrane

Only a tiny fraction of the total potassium and sodium in and around each neuron is exchanged during each action potential. But after a few thousand action potentials the sodium and potassium concentration gradients across the neuron membrane would be lost. This loss is prevented by a set of active transport molecules in the cell membrane called the sodium-potassium pump. The sodium-potassium pump uses energy from ATP to pump sodium out of the cell and potassium in, maintaining the concentration gradients of these ions across the cell membrane. Thus, the explosive charge we referred to earlier is created by the pump using the energy from ATP.

The Action Potential Is Conducted along the Axon without Diminishing in Size

The action potential is a signal. To be effective, it must be transmitted along the axon to cells specialized to receive the message, including other neurons, muscle, or glandular cells. If the neuron is to conduct an action potential along its axon to its synaptic terminal, the action potential must not diminish in magnitude and die out along the way. The cell maintains the magnitude of the action potential by renewing it at each successive point along the axon.

The action potential begins when threshold is reached, sodium channels open, and sodium ions enter the cell, making the inside of the cell positive at that point.

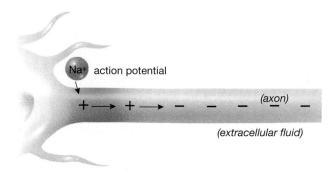

Although much of this positive charge leaks back out, some spreads passively and almost instantaneously along the inside of the axon, making the adjacent region less negative.

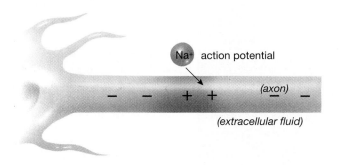

When the adjacent region of membrane reaches threshold, its sodium channels open, causing a further influx of sodium ions and an action potential in the adjacent membrane. This process continues along the entire length of the axon. Meanwhile, the sodium channels at the site of the original action potential close, and the resting potential is reestablished there. In this way, the action potential can travel undiminished down an axon several meters long (in giraffes and whales, for example).

Insulating Myelin Increases the Speed of Action Potential Conduction

Action potentials must travel rapidly. A giraffe couldn't run from a lion, for example, if it took 10 seconds for a signal to travel from brain to hoof. However, the opening and closing of ion channels during action potentials is relatively slow. Therefore, for a signal to travel as rapidly as possible, as few channels as possible should be opened and closed. In vertebrates, axons that need to conduct rapidly (such as those that carry signals to muscles used in pursuit or escape) are wrapped with insulating layers of membrane

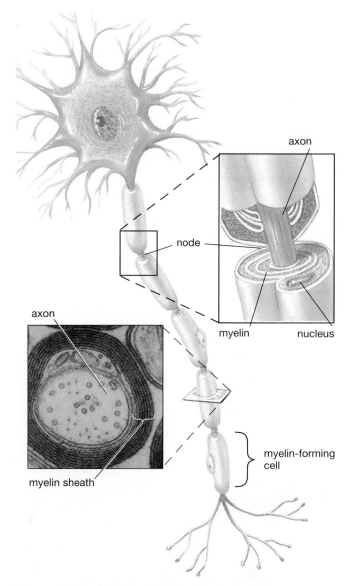

Figure 36-6   The myelin sheath

Some vertebrate axons are wrapped in a membrane "jellyroll" of myelin, formed from windings of membrane from specialized flattened cells. At intervals of around a millimeter, the wrapping is interrupted by bare places, called nodes, where action potentials occur.

called **myelin,** interrupted at intervals with naked areas of axon called **nodes** (Fig. 36-6). Myelin is formed from specialized nonneural cells that flatten and wrap themselves around the axon. These myelinated axons transmit signals much more rapidly than unmyelinated axons because ion channels are concentrated only at the nodes.

When an action potential occurs in a myelinated axon, the positive charge that enters the axon near the cell body can't leak back out through the myelin, but instead travels almost instantaneously to the next node. At the next node, channels open and a new action potential is initiated. Positive charges enter at the node and flow immediately to the next node, and so on. But the charge flowing beneath the myelin *does* diminish with distance, so myelinated axons maintain the signal by initiating a new action potential at each node. The transmission of an action potential along a myelinated axon is called **saltatory conduction** (literally, "jumping" conduction), because the action potential appears to jump from node to node.

Neurons Communicate at Synapses

Once an action potential has been conducted to the synaptic terminal of the neuron, the signal must be transmitted to another cell, usually another neuron. This transmission occurs at specialized regions called **synapses**, and the signals transmitted at synapses are called **postsynaptic potentials**.

The Synapse Includes Parts of Two Neurons with a Small Space Separating Them

When an action potential reaches the synaptic terminal of the axon, it encounters a region, called a synapse, where parts of two neurons are close together and are specialized to communicate with one another. A tiny gap, called a **synaptic cleft,** separates the synaptic terminal of the first neuron, the **presynaptic neuron,** from the second, or **postsynaptic neuron** (Fig. 36-7). Both the dendrites and the cell bodies of neurons are often covered with synapses.

When an action potential reaches a synaptic terminal, the inside of the synaptic terminal becomes positively charged. This charge triggers the synaptic terminal to release a chemical neurotransmitter into the synaptic cleft. The neurotransmitter molecules rapidly diffuse across the gap and bind to receptors in the membrane of the postsynaptic cell (see Fig. 36-7). The synapse includes

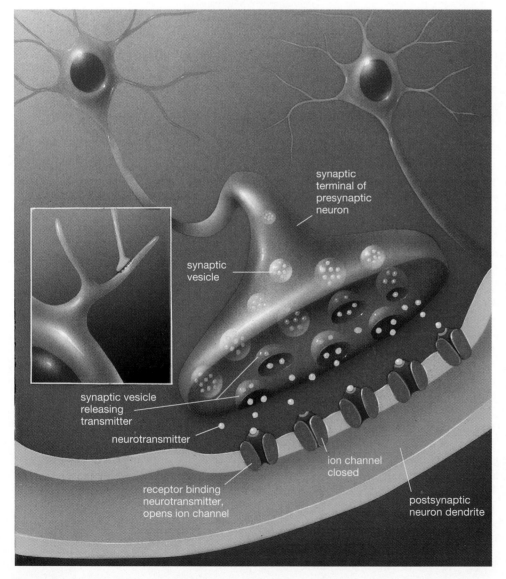

synaptic terminal of presynaptic neuron

synaptic vesicle

synaptic vesicle releasing transmitter

neurotransmitter

receptor binding neurotransmitter, opens ion channel

ion channel closed

postsynaptic neuron dendrite

Figure 36-7   The structure and function of the synapse

The ending of the presynaptic cell contains numerous membrane-bound spheres (called *synaptic vesicles*) that contain neurotransmitter. The postsynaptic cell has membrane receptors for the transmitter. When an action potential enters the synaptic terminal of the presynaptic cell, the vesicles dump their neurotransmitter into the space between the neurons. The neurotransmitter diffuses rapidly across the space, binds to the postsynaptic receptors, and causes ion channels to open. Ions flow through these open channels, causing a postsynaptic potential in the postsynaptic cell.

the synaptic terminal of the presynaptic cell, the synaptic cleft, and the specialized membrane of the postsynaptic cell just across the cleft that contains receptors for neurotransmitters.

Postsynaptic Potentials Are Produced by Neurotransmitter Binding to Receptors

The receptors, which are specialized proteins in the postsynaptic membrane, have two roles. First, they bind to a specific type of neurotransmitter. Second, after binding the neurotransmitter, the receptors cause specific types of ion channels in the membrane of the postsynaptic neuron to open. When ion channels are opened, ions flow across the cell membrane along their concentration gradients. The flow of ions in the postsynaptic neuron causes a small, brief change in electrical charge, called a postsynaptic potential, inside the postsynaptic neuron where the synapse occurs (Fig. 36-8). Depending on what type of channels are opened and what type of ions flow, postsyn-

aptic potentials can be either excitatory, making the neuron less negative inside and more likely to fire an action potential, or inhibitory, making it more negative and less likely to fire. An excitatory postsynaptic potential is called an EPSP; an inhibitory postsynaptic potential is called an IPSP. A synapse that produces EPSPs in the postsynaptic cell is called an **excitatory synapse,** and a synapse producing IPSPs is an **inhibitory synapse.** Postsynaptic potentials cannot travel far in a neuron; after a few millimeters, at most, the ions leak back across the membrane and the signal is lost. However, postsynaptic potentials travel far enough to reach the cell body, where they determine whether an action potential will be produced, as described below.

Postsynaptic Potentials Are Integrated in the Cell Body

The dendrites and cell body of a single neuron can receive EPSPs and IPSPs from the synaptic terminals of thousands of presynaptic neurons. The postsynaptic potentials produced by different presynaptic neurons are then "added up," or **integrated,** in the cell body of the postsynaptic neuron. The postsynaptic cell will produce an action potential only if the excitatory and inhibitory potentials, when added together, raise the electrical potential inside the neuron above threshold (see Fig. 36-8).

Neurotransmitters Are Rapidly Removed from the Synaptic Cleft

Neurotransmitters act only briefly on the postsynaptic cell. A few transmitters are destroyed by enzymes released into the synaptic cleft; others are removed from the synaptic cleft by active transport back into the presynaptic neuron, and still others simply diffuse away into the extracellular fluid. Some types of neurotransmitters and a few of their functions in the human nervous system are discussed later in this chapter.

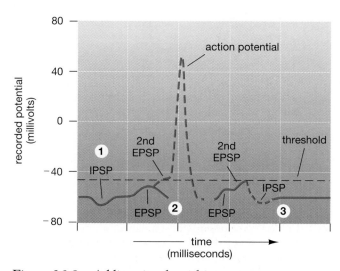

Figure 36-8   Adding signals within a neuron

This diagram shows electrical events recorded inside the cell body of a neuron that receives both excitatory and inhibitory postsynaptic potentials and integrates these signals. ① A presynaptic neuron making an inhibitory synapse on the recorded cell causes an inhibitory postsynaptic potential (IPSP). This brings the potential inside the cell body farther away from threshold. ② A different presynaptic neuron making an excitatory synapse on the recorded cell causes an excitatory postsynaptic potential (EPSP). This brings the potential closer to threshold. If a second EPSP from an excitatory synapse occurs at almost the same time, the two are added together and (in this case) bring the potential in the cell body above threshold, and an action potential is initiated in the axon. ③ If EPSPs from excitatory synapses and an IPSP from an inhibitory synapse occur at about the same time, they are all added together. The IPSP can prevent the EPSPs from bringing the potential inside the cell body to threshold, and so can prevent an action potential from occurring.

Building and Operating a Nervous System

The individual neuron uses a language of action potentials. Yet somehow this basic language allows even simple animals to perform a variety of complex behaviors. One key to the versatility of the nervous system is the presence of complex networks of neurons. These neural networks range from a few to billions of cells. As in computers, small, simple elements can perform amazing feats when connected properly.

Information Processing Requires Four Basic Operations

Before we delve into the basic anatomy of nervous systems, we should first examine the operating principles. At a minimum, a nervous system must be able to perform four operations:

1. Determine the type of stimulus
2. Signal the intensity of a stimulus
3. Integrate information from many sources
4. Initiate and direct the response

Let's examine each of these operations in more detail.

The Type of Stimulus Is Distinguished by Wiring Patterns in the Brain

The nervous system must be able to identify the type of stimulus (for example, light, touch, or sound). Action potentials are all pretty much alike; their properties tell us nothing about the kind of stimulus that elicited them. Instead, the nervous system monitors *which* neurons are firing action potentials. Thus, your brain interprets action potentials occurring in the axons of your optic nerves (originating in the eye and traveling to a specific area of the brain) as the sensation of light, action potentials in olfactory nerves (originating in receptors in the nose and traveling to a different region of the brain) as odors, and so on. This genetic wiring may occasionally yield false information. Being poked in the eye may cause action potentials in the optic nerve, because of slight trauma. Even though the stimulus is mechanical, your brain nevertheless interprets all action potentials occurring in the optic nerve as light, and you "see stars." For the same reason, a blow to the head can make your ears "ring."

The Intensity of a Stimulus Is Coded by the Frequency of Action Potentials

Because all action potentials are of roughly the same magnitude and duration, no information about the strength, or **intensity,** of a stimulus (for example, the loudness of a sound) can be encoded in a single action potential. Instead, intensity is coded in two other ways. First, intensity can be signaled by the frequency of action potentials in a single neuron. The more intense the stimulus, the faster the neuron fires (Fig. 36-9). Second, a nervous system usually has many neurons that can respond to the same input. Stronger stimuli tend to excite more of these neurons, whereas weaker stimuli excite fewer. Thus, intensity can also be signaled by the number of similar neurons firing at the same time (see Fig. 36-9).

The Nervous System Processes Information from Many Sources through Convergence

Your brain is continuously bombarded by sensory stimuli originating both inside and outside the body. The brain must filter all these inputs, determine which ones are important, and decide how to respond. Nervous systems integrate information much as do individual neurons, through **convergence.** In this process, many neurons fun-

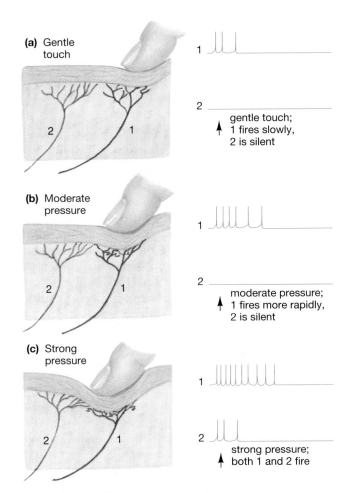

Figure 36-9   Signaling stimulus intensity
The intensity of a stimulus is signaled by the rate at which individual neurons produce action potentials and by the number of neurons firing. For example, two touch receptors may have endings in adjacent patches of skin. **(a)** A gentle touch elicits only a few action potentials and from only one of the sensory neurons. **(b)** Moderate pressure still stimulates only one receptor, but this receptor now fires faster, informing the brain that the touch is more intense than before. **(c)** Strong pressure activates both receptors, firing one very fast, and the other more slowly, thus signaling to the brain that the pressure is very intense.

nel their signals to fewer neurons. For example, many sensory neurons may converge onto a smaller number of brain cells (Fig. 36-10). The brain cells add up the postsynaptic potentials resulting from the synaptic activity of these sensory neurons, and, depending on their relative strengths (and other internal factors such as hormones or metabolic activity), they produce appropriate outputs.

Divergence of Signals Allows Complex Responses

The output of the integrating cells is responsible for initiating activity. The actions directed by the brain may involve many parts of the body and require **divergence,** the flow of electrical signals from a relatively small number of

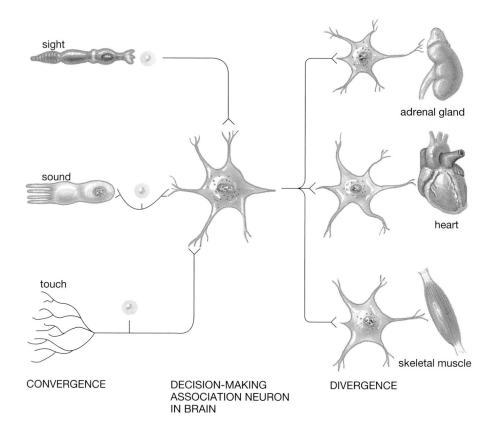

sight

sound

touch

adrenal gland

heart

skeletal muscle

CONVERGENCE

DECISION-MAKING
ASSOCIATION NEURON
IN BRAIN

DIVERGENCE

***Figure 36-10*   Convergence and divergence in the nervous system**

Integration of information and initiation of coordinated action involves convergence of inputs to and divergence of outputs from neurons. In this simplified example, inputs of sight, sound, and touch (say, being attacked by a swarm of bees) all converge on a "decision-making" neuron, which is strongly stimulated as a result. Its outputs then diverge to the adrenal glands ("pump out adrenaline"), the heart ("beat faster and stronger"), and skeletal muscles ("move, legs!").

decision-making cells onto many different neurons controlling muscle or glandular activity (see Fig. 36-10).

Neural Pathways Direct Behavior

Most behaviors are controlled by neuron-to-muscle pathways composed of four elements:

1. **Sensory neurons** that respond to a stimulus, either internal or external to the body
2. **Association neurons** that "decide" what to do, on the basis of input from many sensory neurons, stored memories, hormonal states, and other factors
3. **Motor neurons** that receive instructions from the association neurons and activate the muscles or glands
4. **Effectors**, usually muscles or glands, that perform the behavior or other response directed by the nervous system

The Simplest Behavior Is the Reflex

The simplest type of behavior is the **reflex,** a relatively involuntary movement of a body part in response to a stimulus. Examples of human reflexes include the familiar knee jerk and pain-withdrawal reflexes. The pain-withdrawal reflex (which moves a body part away from a painful stimulus) uses one neuron of each type (Fig. 36-11). Reflexes of this sort do not require the brain, although, as we know, other pathways inform the brain of pricked fingers and may in fact trigger other more complex behaviors (cursing, for example!).

Nearly all animals are capable of much more subtle and varied behavior than can be accounted for by simple reflexes. In principle, these more complex behaviors can be organized by *interconnected neural pathways* in which several types of sensory input (along with memories, hormones, and other factors) converge on a set of association neurons (see Fig. 36-10). By integrating the postsynaptic potentials from several sources, the association neurons can "decide" what to do and stimulate the motor neurons to direct appropriate activity in muscles and glands.

Increasingly Complex Nervous Systems Are Increasingly Centralized

In all the animal kingdom, there are really only two designs for nervous systems: diffuse nervous systems, found in the cnidarians (*Hydra,* jellyfish, and their relatives; Fig. 36-12), and more centralized nervous systems, found to varying degrees in more complex organisms. Not surprisingly, nervous system design is highly correlated with the life-style of the animal. In the radially symmetrical cnidarians, there is no "front end," and so there has been no evolutionary pressure to concentrate the senses in one place. A *Hydra* sits anchored to the bottom, and prey or danger are equally likely to come from any direction. Cnidarian nervous systems are composed of a network of neurons, often called a **nerve net,** woven through the tissues of the animal. Here and

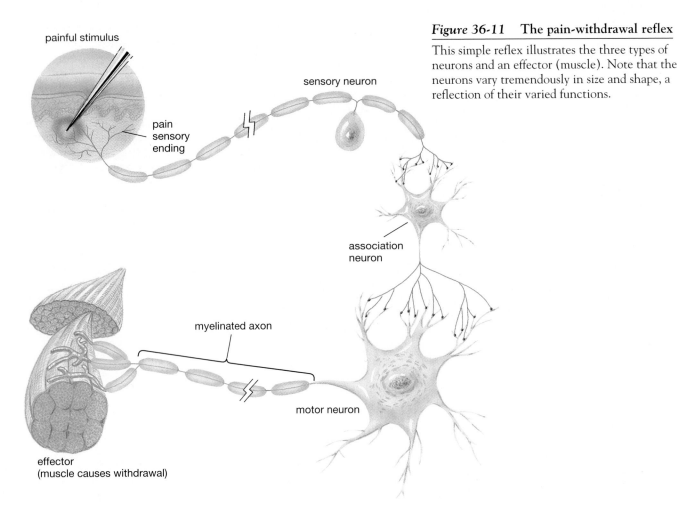

painful stimulus

pain
sensory
ending

sensory neuron

association
neuron

myelinated axon

motor neuron

effector
(muscle causes withdrawal)

Figure 36-11   The pain-withdrawal reflex

This simple reflex illustrates the three types of neurons and an effector (muscle). Note that the neurons vary tremendously in size and shape, a reflection of their varied functions.

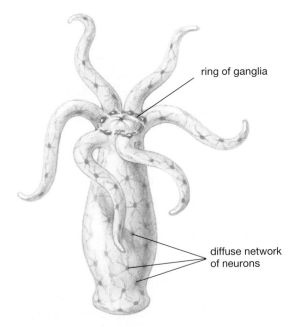

ring of ganglia

diffuse network
of neurons

Figure 36-12   A nerve net

The diffuse nervous system of *Hydra* contains a few concentrations of neurons, particularly at the bases of the tentacles, but no brain. Conduction of neural signals may occur in virtually any direction throughout the body.

there we can find a cluster of neurons, called a **ganglion** (plural, **ganglia**), but nothing resembling a real brain.

Almost all other animals are bilaterally symmetrical, with definite head and tail ends. Because the head is the first part of the body to encounter food, danger, and potential mates, it is advantageous to have sense organs concentrated there. Sizable ganglia evolved that integrate the information gathered by the senses and initiate appropriate action. Over evolutionary time, the sense organs gathered in the head and the ganglia became centralized into a brain. This trend is clearly seen in the mollusks (Fig. 36-13). Centralization reaches its peak in the vertebrates, where nearly all the cell bodies of the nervous system are localized in the brain and spinal cord. The organization of the vertebrate nervous system is shown in Figure 36-14.

The Human Nervous System

The human nervous system may be divided into two parts: central and peripheral. The **central nervous system** consists of a **brain** and a **spinal cord** that extends down the dorsal part of the torso. The **peripheral nervous system** consists of nerves connecting the central nervous system to the rest of the body (Fig. 36-14).

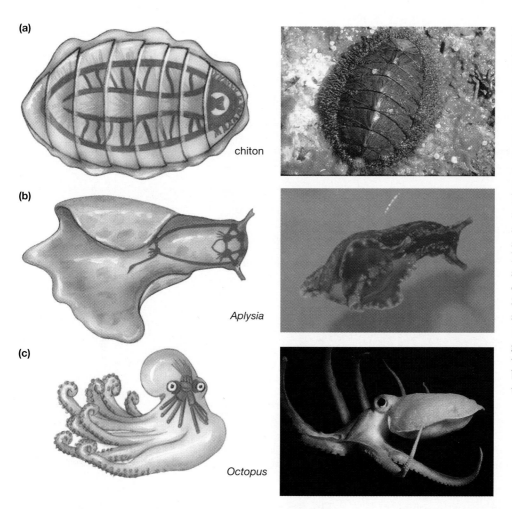

(a)

chiton

(b)

Aplysia

(c)

Octopus

Figure 36-13 **The brain has become localized to the head**

Bilaterally symmetrical animals usually have nervous systems concentrated in the head. The trend toward increasing concentration of the nervous system in the head is clearly illustrated by various mollusks (colored structures represent the nervous systems). **(a)** The chiton, although it does have a head end, seldom crawls in any direction and has had little selective pressure to concentrate sense organs and brain in the head. **(b)** Some marine snails, such as the shell-less *Aplysia*, can crawl quite rapidly or even swim. Still more of their neurons are aggregated into a brain. **(c)** Mollusk mobility and intelligence culminates in *Octopus*, with its large, complex brain and behavioral capabilities rivaling those of some mammals.

The Peripheral Nervous System Links the Central Nervous System to the Body

The peripheral nervous system consists of **peripheral nerves** that link the brain and spinal cord to the rest of the body, including the muscles, the sensory organs, and the organs of the digestive, respiratory, excretory, and circulatory systems. Within the peripheral nerves are axons of sensory neurons that bring sensory information *to* the central nervous system from all parts of the body. Peripheral nerves also contain the axons of motor neurons that carry signals *from* the central nervous system to the organs and muscles.

The motor portion of the peripheral nervous system can be subdivided into two parts: the **somatic nervous system** and the **autonomic nervous system.** Motor neurons of the somatic nervous system synapse on skeletal muscles and control voluntary movement. Their cell bodies are located in the gray matter of the spinal cord (see Fig. 36-16), and their axons go directly to the muscles they control. (Muscles and their control are discussed in Chapter 38.)

Motor neurons of the autonomic nervous system control involuntary responses. They synapse on the heart, smooth muscle, and glands. The autonomic nervous system is controlled both by the medulla and the hypothalamus of the brain, described later in the chapter. It con-

sists of two divisions, the **sympathetic nervous system** and the **parasympathetic nervous system** (Fig. 36-15). The two divisions of the autonomic nervous system generally make synaptic contacts with the same organs but usually produce opposite effects.

The sympathetic nervous system acts on organs in ways that prepare the body for stressful or highly energetic activity, such as fighting, escaping, or giving a speech. During such "fight-or-flight" activities, the sympathetic nervous system curtails activity of the digestive tract, redirecting some of its blood supply to be used by the muscles of arms and legs. Heart rate accelerates. The pupils of the eyes open wider, admitting more light, and the air passages in the lungs expand, accommodating more air. The parasympathetic nervous system, in contrast, dominates during maintenance activities that can be carried on at leisure, often called "rest and rumination." Under its control, the digestive tract becomes active, heart rate slows, and air passages in the lungs constrict.

Two differences in the organization of the sympathetic and parasympathetic nervous systems are evident in Figure 36-15. First, parasympathetic axons are found in nerves that originate from two separate locations, the brain (midbrain and medulla) and the base of the spinal cord. In contrast, sympathetic axons are found in nerves that origi-

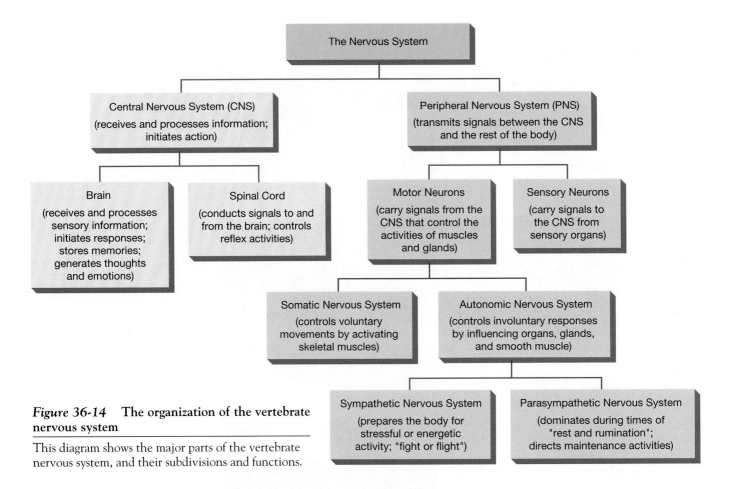

Figure 36-14   **The organization of the vertebrate nervous system**

This diagram shows the major parts of the vertebrate nervous system, and their subdivisions and functions.

nate from the middle and lower portions of the spinal cord. Second, in both the sympathetic and parasympathetic divisions, there are two neurons that carry messages in sequence from the central nervous system to each target organ, but they synapse at different locations. In the sympathetic nervous system, the synapse occurs in ganglia that are near the spinal cord. In the parasympathetic nervous system, the synapse occurs in smaller ganglia located at or very near each target organ.

The Central Nervous System Consists of the Spinal Cord and Brain

The central nervous system consists of the brain and spinal cord. It is the integrating portion of the nervous system, where sensory information is received and processed, thoughts are generated, and responses are directed. The central nervous system consists primarily of association neurons—somewhere between 10 and 100 billion of them!

The brain and spinal cord are protected in three ways. The first line of defense is a bony armor, consisting of the skull that surrounds the brain and the vertebral column that protects the spinal cord. Beneath the bones lies a triple layer of connective tissue called **meninges** (see Fig. 36-19). Between the layers of the meninges, a clear lymph-like liquid, the **cerebrospinal fluid,** cushions the brain and spinal cord.

The Spinal Cord Is a Cable of Axons Protected by the Backbone

The spinal cord is a neural cable about as thick as your little finger that extends from the base of the brain to the hips, protected by the bones of the vertebral column (Fig. 36-16). Between the vertebrae, nerves called dorsal roots, carrying axons of sensory neurons and ventral roots, carrying axons of motor neurons, arise from the dorsal and ventral portions of the spinal cord, respectively; these merge to form the peripheral nerves of the spinal cord, which are part of the peripheral nervous system. In the center of the spinal cord are neuron cell bodies, which form a butterfly-shaped area of **gray matter.** These are surrounded by bundles of axons called **white matter** owing to their white insulating myelin coating (see Fig. 36-16). The spinal cord relays signals between the brain and the rest of the body, and it contains the neural circuitry for certain behaviors, including reflexes.

To illustrate some of the functions of the parts of the spinal cord, let's examine a simple spinal reflex, the pain-withdrawal reflex, which involves neurons of both the central nervous system and the peripheral nervous system (Fig. 36-17; see also Fig. 36-11). The cell bodies of the sensory neurons from the skin (in this case signaling pain) are found just outside the spinal cord in a row of ganglia. Each of these **dorsal root ganglia** is located on a spinal

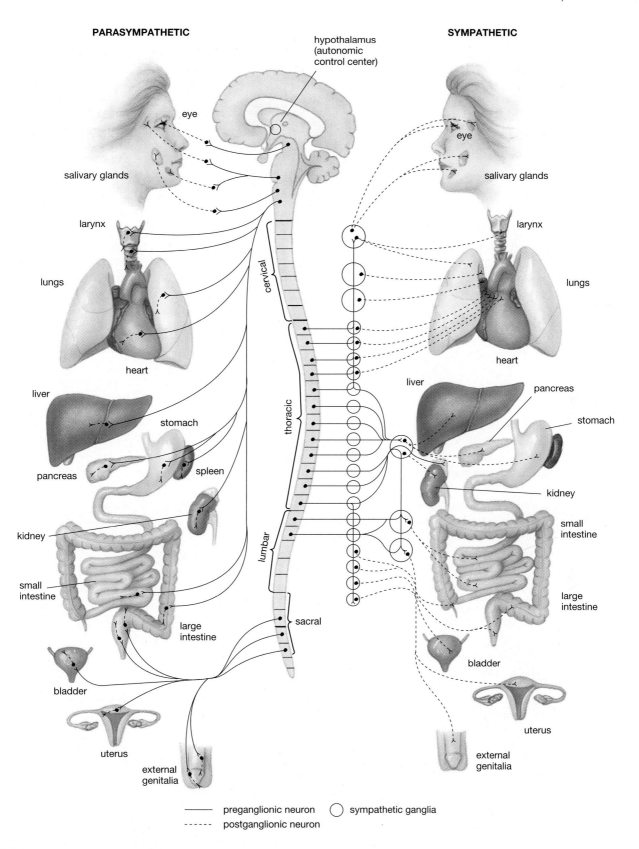

PARASYMPATHETIC

hypothalamus
(autonomic
control center)

SYMPATHETIC

eye

salivary glands

larynx

lungs

heart

liver

stomach

pancreas

spleen

kidney

small
intestine

large
intestine

bladder

uterus

external
genitalia

cervical

thoracic

lumbar

sacral

eye

salivary glands

larynx

lungs

heart

liver

pancreas

stomach

kidney

small
intestine

large
intestine

bladder

uterus

external
genitalia

——— preganglionic neuron   ◯ sympathetic ganglia

- - - - postganglionic neuron

Figure 36-15   **The autonomic nervous system**

The autonomic nervous system has two divisions, the sympathetic and parasympathetic. The two divisions supply nerves to many of the same organs but produce opposite effects. Activation of the autonomic nervous system is mostly involuntary, produced by nervous outputs from the hypothalamus.

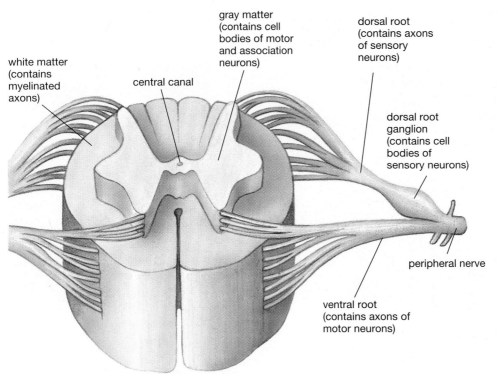

white matter (contains myelinated axons)

central canal

gray matter (contains cell bodies of motor and association neurons)

dorsal root (contains axons of sensory neurons)

dorsal root ganglion (contains cell bodies of sensory neurons)

peripheral nerve

ventral root (contains axons of motor neurons)

Figure 36-16   The spinal cord

The spinal cord runs from the base of the brain to the hips, protected by the vertebrae of the spine. Peripheral nerves emerge from between the vertebrae. A cross section of the spinal cord reveals an outer region of myelinated axons (white matter) traveling to and from the brain, surrounding an inner, butterfly-shaped region of dendrites and the cell bodies of association and motor neurons (gray matter). The cell bodies of the sensory neurons are located outside the cord in the dorsal root ganglion.

nerve and nestled close to the vertebral column. Both association and motor neuron cell bodies are found in the gray matter in the center of the spinal cord. The axons in the surrounding white matter communicate with the brain. Association neurons for the pain-withdrawal reflex, for example, not only synapse on motor neurons but also have axons extending up to the brain. Signals carried along these axons alert the brain to the painful event. The brain, in turn, sends impulses down axons in the white matter to cells in the gray matter. These signals can modify spinal reflexes. With sufficient motivation, you can suppress the pain-withdrawal reflex; to rescue a child from a burning building, for example, you could reach into the flames.

In addition to simple reflexes, the entire program for operating some fairly complex activities also resides within the spinal cord. All the neurons and interconnections needed to walk and run, for example, are found within the

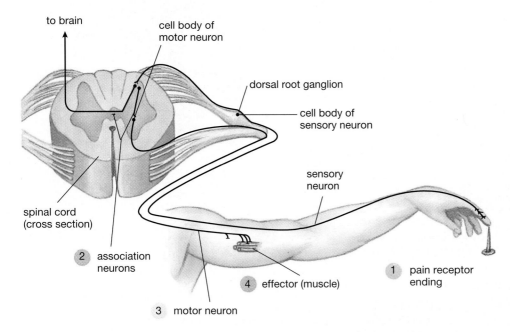

to brain

cell body of motor neuron

dorsal root ganglion

cell body of sensory neuron

sensory neuron

spinal cord (cross section)

2   association neurons

4   effector (muscle)

3   motor neuron

1   pain receptor ending

Figure 36-17   The vertebrate pain-withdrawal reflex

This simple reflex circuit includes one each of the four elements of a neural pathway. The sensory neuron has pain-sensitive endings in the skin ① and a long fiber leading to the spinal cord. The sensory neuron stimulates an association neuron in the spinal cord ②, which in turn stimulates a motor neuron, also in the cord. The axon of the motor neuron ③ carries action potentials to muscles ④, causing them to contract and withdraw the body part from the damaging stimulus. Note that the sensory neuron also makes a synapse on other association neurons not directly involved in the reflex, which carry signals to the brain, informing it of the danger below.

cord. In these cases, the role of the brain is to initiate and guide the activity of spinal neurons. The advantage of this semi-independent arrangement is probably an increase in speed and coordination, since messages do not have to travel all the way up the cord to the brain and back down again (in the case of walking) merely to swing forward one of your legs. The motor neurons of the spinal cord also control the muscles involved in conscious, voluntary activities such as eating, writing, or playing tennis. Axons of the brain cells directing these activities carry signals down the cord and stimulate the appropriate motor cells.

The Brain Consists of Several Parts Specialized for Specific Functions

All vertebrate brains have the same general structure, with major modifications corresponding to life-style and intelligence. Embryologically, the vertebrate brain begins as a simple tube, which soon develops into three parts: the hindbrain, midbrain, and forebrain (Fig. 36-18a). It is believed that in the earliest vertebrates, these three anatomical divisions were also functional divisions: The **hindbrain** governed automatic behaviors such as breathing and heart rate, the **midbrain** controlled vision, and the **forebrain** dealt largely with the sense of smell. In nonmammalian vertebrates, these three divisions remain prominent. However, in mammals, and particularly in humans, the brain regions are significantly modified. Some have been reduced in size, and others, especially the forebrain, greatly enlarged (Fig. 36-18b).

The Hindbrain Includes the Medulla, Pons, and Cerebellum

In humans, the hindbrain is represented by the medulla, the pons, and the cerebellum (Fig. 36-19). In both structure and function, the **medulla** is very much like an enlarged extension of the spinal cord. Like the spinal cord, the medulla has neuron cell bodies at its center, surrounded by a layer of myelin-covered axons. The medulla controls several automatic functions, such as breathing, heart rate, blood pressure, and swallowing. Certain neurons in the **pons,** located above the medulla, appear to influence transitions between sleep and wakefulness and between stages of sleep. Others influence the rate and pattern of breathing. The **cerebellum** is crucially important in coordinating movements of the body. It receives information from command centers in the higher, conscious areas of the brain that control movement and also from position sensors in muscles and joints. By comparing what the command centers ordered with information from the position sensors, the cerebellum guides smooth, accurate motions and body position. Not surprisingly, the cerebellum is largest in animals whose activities require fine coordination. It is best developed in birds (see Fig. 36-18b), who engage in the complex activity of flight.

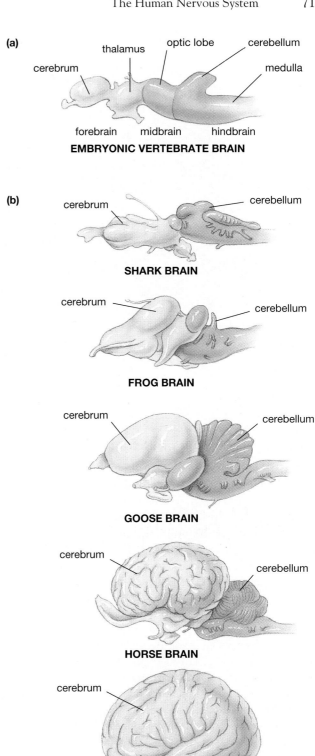

Figure 36-18   A comparison of vertebrate brains

(a) The embryonic vertebrate brain shows three distinct regions: the forebrain, midbrain, and hindbrain. (b) This basic structure persists in all adult brains, but the relative size and importance of the parts vary enormously.

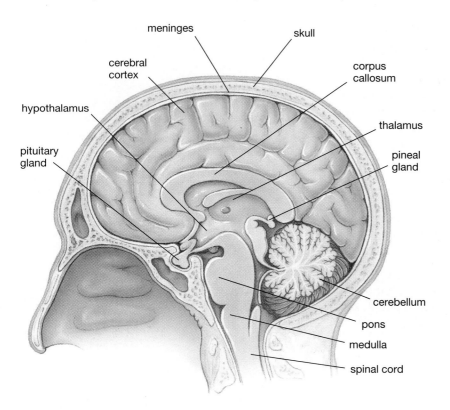

Figure 36-19   **The human brain**

A section taken through the midline of the human brain reveals some of its major structures.

The Midbrain Contains the Reticular Formation

The midbrain is extremely reduced in humans, but an important relay center, the **reticular formation,** passes through it (Fig. 36-20). The neurons of the reticular formation extend all the way from the central core of the medulla, through the pons, the midbrain, and on into lower regions of the forebrain. It receives input from virtually every sense and every part of the body and from many areas of the brain as well. The reticular formation plays a role in sleep and arousal, emotion, muscle tone, and certain movements and reflexes. It filters sensory inputs before they reach the conscious regions of the brain, although the selectivity of the filtering seems to be set by higher brain centers. Through a combination of genetically determined wiring and learning, the reticular formation "decides" which stimuli require attention. Important stimuli are forwarded to the conscious centers for processing, and unimportant stimuli are suppressed. The fact that a mother wakens upon hearing the faint cry of her infant but sleeps through loud traffic noise outside her window testifies to the effectiveness of the reticular formation in screening inputs to the brain and to the role of learning in determining the importance of sensory stimulation.

The Forebrain Includes the Thalamus, Limbic System, and Cerebral Cortex

The forebrain, also called the **cerebrum,** can be divided into three functional parts: the thalamus, the limbic system, and the cerebral cortex. In mammals, the cerebral cortex is much enlarged compared with that of fish, am-

phibians, and reptiles. This trend culminates in the human cerebral cortex, which is thrown into complex folds that increase its area (see Fig. 36-18).

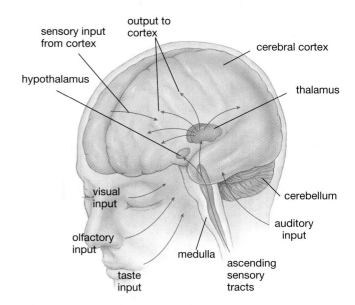

Figure 36-20   **The reticular formation**

The human reticular formation (shown here in blue) is a diffuse network of neurons running through the lower regions of the brain from the medulla in the hindbrain, through the midbrain, and up into the thalamus and hypothalamus of the forebrain. It receives input from most of the senses and sends outputs to many higher brain centers, filtering the sensory information that reaches the conscious brain.

The Thalamus   The **thalamus** (see Figs. 36-19 and 36-20) carries sensory information to the limbic system and cerebrum. This information includes sensory input from auditory and visual pathways, from the skin, and from within the body. Inputs from the cerebellum and limbic system are also channeled through this busy thoroughfare. Very little information processing goes on in the thalamus.

The Limbic System   Anatomically, the **limbic system** is a diverse group of structures located in an arc between the thalamus and the cerebrum (Fig. 36-21). These structures work together to produce our most basic and primitive emotions, drives, and behaviors, including fear, rage, tranquility, hunger, thirst, pleasure, and sexual responses. The limbic system includes the hypothalamus, portions of the thalamus, the amygdala, and the hippocampus.

The **hypothalamus** (literally "under the thalamus") contains many different clusters of neurons. Some of these are neurosecretory cells that release hormones (see Chapter 35). Through its hormone production and neural connections, the hypothalamus acts as a major coordinating center, controlling body temperature, hunger, the menstrual cycle, water balance, and the autonomic nervous system. In addition, stimulation of specific areas of the hypothalamus elicits emotions such as rage, fear, pleasure, and sexual arousal.

The **amygdala** is believed to be responsible for the production of appropriate behavioral responses to environmental stimuli. It receives input from many sources, including the auditory and visual areas of the cerebral cortex. Different clusters of neurons in the amygdala produce sensations of pleasure, punishment, or sexual arousal when stimulated. By stimulating different portions of the amygdala, researchers can either reduce or enhance aggressive behavior. Conscious humans whose amygdalas are electrically stimulated have reported feelings of rage or fear. Recent studies have revealed that damage to the human amygdala eliminates the person's ability both to feel fear and to recognize fearful facial expressions in other people.

The shape of the **hippocampus** as it curves around the thalamus inspired its name, which is derived from the Greek word meaning "sea horse." As in the amygdala and hypothalamus, behaviors that reflect a variety of emotions, including rage and sexual arousal, can be elicited by stimulating portions of the hippocampus. The hippocampus also plays an important role in the formation of long-term memory and is thus required for learning, discussed in more detail later in this chapter.

The Cerebral Cortex   In humans, by far the largest part of the brain is the **cerebral cortex**, the outer layer of the forebrain. The cerebral cortex and underlying parts of the forebrain are divided into two halves, called **cerebral hemispheres,** that communicate with each other by means of a large band of axons, the **corpus callosum.** The differences between the hemispheres is discussed a bit later in this chapter. The cerebral cortex is the most sophisticated in-

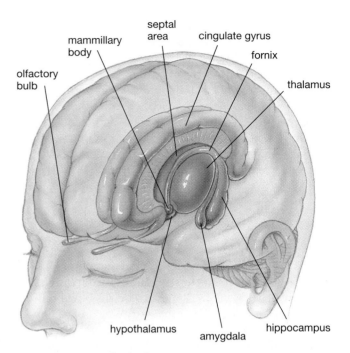

Figure 36-21   **The limbic system**

The limbic system extends through several brain regions. It seems to be the center of most unconscious, emotional behaviors, such as love, hate, hunger, sex, and fear.

formation processing center known, and it is also the area of the brain that scientists know the least about. Roughly *50 to 100 billion* neurons are packed into this thin surface layer. To accommodate this profusion of cells, the cortex is thrown into folds, called **convolutions,** that greatly increase its area. In the cortex, cell bodies of neurons predominate, giving this outer layer of the brain a gray appearance. These neurons receive sensory information, process it, store some in memory for future use, and direct voluntary movements.

The cerebral cortex is divided into four regions based on anatomical criteria: the frontal, parietal, occipital, and temporal lobes (Fig. 36-22). Functionally, the cortex contains primary sensory areas where signals originating in sensory organs such as the eyes and ears are received and converted into subjective impressions, for example, light and sound. Nearby association areas interpret the sounds, as speech, for example, and the visual stimuli as recognizable objects. Association areas also link the stimuli with previous memories stored in the cortex and generate commands to produce speech. Primary sensory areas in the parietal lobe interpret sensations of touch originating in all parts of the body, which is "mapped" in an orderly sequence. In an adjacent region of the frontal lobe, primary motor areas generate commands for movements in corresponding areas of the body (Fig. 36-22). The association area of the frontal lobe protected by bones of the forehead seems to be involved in complex reasoning such as decision making, predicting the consequences of actions, controlling aggression (see Fig. E36-3), and planning for the future, as discussed in the final section of the chapter. Damage to the cortex due to trauma, stroke, or a tumor results in specific deficits, such as problems with speech, difficulty reading, or the inability to sense or move specific parts of the body. Because brain cells cannot reproduce, once a brain region is destroyed it cannot be repaired or replaced, so these deficits are often permanent. Fortunately, however, in some cases diligent training can cause undamaged regions of the cortex to take control over and restore some of the lost functions.

The Nervous System Uses Many Neurotransmitters and Neuromodulators

Twenty-five years ago, neurobiologists thought that nervous systems operated with just a few excitatory and inhibitory neurotransmitters. Since that time, investigators have become increasingly aware that the brain is a teeming cauldron; its neurons synthesize and respond to a vast array of chemicals, including many of the hormones once thought unique to the endocrine system. For example, hormones that control digestive tract secretions are now known to be synthesized in the brain, where they influence appetite. At least 50 neurotransmitters and neuromodulators have been identified, and

more are added to the list each year. In the following sections, we first discuss a few "classic" neurotransmitters that have been recognized for many years and whose roles are partially understood. Then we describe a few recently recognized **neuromodulators,** substances released by neurons that alter the activity of groups of neurons over longer periods of time.

Acetylcholine

The neurotransmitter **acetylcholine** is found in many areas of the brain and is the only transmitter found at the synapses between motor neurons and skeletal muscles, where it is always excitatory. The drug curare, isolated from the South American poison arrow frog, blocks acetylcholine receptors on the postsynaptic membrane. This blockage prevents muscle contraction, causing paralysis and sometimes death. In contrast, many insecticides poison insects by inhibiting an enzyme (found in the synaptic cleft) that breaks down acetylcholine. This inhibition causes the muscles to contract uncontrollably, producing seizures and death. In the human central nervous system, breakdown of specific groups of acetylcholine-producing neurons is found in patients with Alzheimer's disease. In the peripheral nervous system, acetylcholine is the neu-

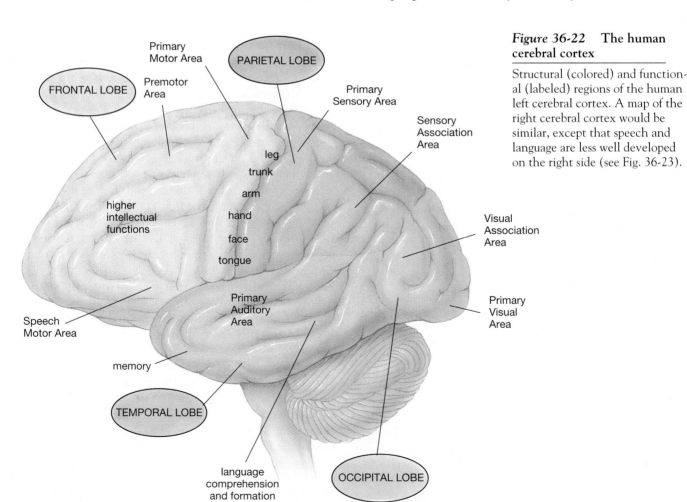

Figure 36-22   **The human cerebral cortex**

Structural (colored) and functional (labeled) regions of the human left cerebral cortex. A map of the right cerebral cortex would be similar, except that speech and language are less well developed on the right side (see Fig. 36-23).

rotransmitter released by neurons of the parasympathetic nervous system onto their target organs.

Dopamine

Dopamine is an important neurotransmitter in the brain, where its effects are largely inhibitory. The degeneration of dopamine-producing neurons causes Parkinson's disease, characterized by muscular rigidity (due to the continuous contraction of some muscles) and uncontrolled tremors. The drug levodopa (L-dopa), administered to Parkinson's patients, is used by unharmed neurons in the brain to synthesize dopamine, partially replacing that which has been lost. Victims of schizophrenia are often treated successfully with drugs that block dopamine receptors. Although the mechanisms of this disease remain obscure, one hypothesis is that schizophrenia is the result of excess dopamine receptors. Unfortunately, the drugs used to treat schizophrenia may induce symptoms of Parkinson's disease. Cocaine and amphetamines block the re-uptake of dopamine into the presynaptic neuron, causing it to remain in the synapse and prolonging its effects. Abuse of these drugs can sometimes produce symptoms resembling schizophrenia.

Serotonin

The neurotransmitter **serotonin** acts in the brain and spinal cord. It can inhibit pain sensory neurons in the spinal cord, and electrical devices that stimulate these neurons are sometimes implanted in patients suffering from chronic pain. Serotonin is also believed to affect sleep and mood. Animals in which serotonin production is blocked are unable to sleep normally. Too little serotonin (and norepinephrine, see below) may cause depression. Some drugs used to treat depression block either the enzymatic destruction of both serotonin and norepinephrine or their transport back into the presynaptic cell, thus prolonging the effects of these transmitters. The antidepressant Prozac, for example, selectively blocks re-uptake of serotonin into the presynaptic neuron, enhancing its effects. There is some evidence that the drug LSD, among many other effects, may block serotonin receptors.

Norepinephrine

Norepinephrine (also called noradrenaline) is chemically very similar to the hormone epinephrine (adrenaline) secreted by the adrenal glands (see Chapter 35). Noradrenaline is released by neurons of the sympathetic nervous system onto many organs (such as the heart, digestive system, and lungs; see Fig. 36-15). The effects of norepinephrine, which may be either inhibitory or excitatory, prepare the body to respond to stressful situations. Both amphetamines and cocaine block norepinephrine uptake into the presynaptic neuron, prolonging its effects. Some sudden deaths among cocaine users are caused by over-stimulation of the sympathetic nervous system by excess norepinephrine.

Neuromodulators

In addition to these and several other classic neurotransmitters, dozens of peptides are synthesized and released by neurons. Many may be released along with neurotransmitters, acting as neuromodulators. Neuromodulators, as their name suggests, modify the properties of synapses, making them more or less effective. They can also cause long-term changes in the excitability of neurons. Neuromodulators function over a longer time period than classic neurotransmitters and may influence many neurons at once.

Examples of peptide neuromodulators are the **opioids** (literally "opiate-like substances"), such as **endorphin.** For centuries the analgesic (pain-relieving) effects of opiates, such as morphine, opium, codeine, and heroin, have been recognized. In the early 1970s, neurobiologists discovered that these opiates bind to specific receptors on neurons of the central nervous system. Because opiates are plant products, it seemed unlikely that the human brain would have evolved receptors specifically for these chemicals. Perhaps, researchers reasoned, the opiates happen to resemble unknown substances produced by the brain that diminish pain perception. This hypothesis led to a search for such substances, which was rewarded in 1975 with the discovery of the opioids.

Some opioids act with serotonin in the spinal cord to block perception of pain. They are partially responsible for the suppression of pain in times of extreme stress, such as on a battlefield (or a football field). Opioids are released during strenuous exercise and may account for the well-known "runner's high." The analgesic effects of acupuncture are apparently caused by its ability to stimulate the release of opioids. Opioids and other peptide neuromodulators participate in the maintenance of body temperature and blood pressure. They have also been implicated in the regulation of behavioral states such as hunger, sexual excitement, anger, and depression, and they may even be involved in learning.

Brain and Mind

Historically, people have always had difficulty reconciling the physical presence of a few pounds of grayish material in the skull with the range of thoughts, emotions, and memories of the human mind. This "mind-brain problem" has occupied generations of philosophers and, more recently, neurobiologists. Beginning with observations of patients with head injuries and progressing to sophisticated surgical, physiological, and biochemical experiments, the outlines of how the brain creates the mind are beginning to emerge. Here, we will be able to touch upon only a few of the more fascinating features.

The "Left Brain" and "Right Brain" Are Specialized for Different Functions

The human brain appears bilaterally symmetrical, particularly the cerebrum, which consists of two extremely similar-looking hemispheres. However, it has been known since the early 1900s that this symmetry does not extend to brain function. Much of what is known of the differences in hemisphere function comes from two sources: studies of accident victims with localized damage to one hemisphere and studies of patients who have had the corpus callosum (which connects the two hemispheres) severed. This surgical procedure was performed in rare cases of uncontrollable epilepsy to prevent the spread of seizures through the brain.

Studies based on selective damage to the left cerebral hemisphere had led to the belief that the right hemisphere was relatively retarded, lacking the ability to speak, write, recognize words, or reason. For example, people suffering damage to localized areas of the left hemisphere, but not the right, often became unable to speak, read, or understand spoken language. In addition, the left hemisphere for most people is superior in mathematical ability and in logical problem-solving tasks (Fig. 36-23).

Roger Sperry, of the California Institute of Technology, worked with people whose hemispheres had been surgically separated by cutting the corpus callosum. In his studies, Sperry made use of the knowledge that axons within each optic nerve follow a pathway that causes the left half of each visual field to be projected on the right cerebral hemisphere, and vice versa (Fig. 36-23). Through an ingenious device that projected different images onto the left and right visual fields (thus sending different signals to each hemisphere), he and other investigators have gained more insight into the roles of the two hemispheres. If he projected an image of a nude figure onto the left visual field only, the patients would blush and smile but would claim to have seen nothing, because the image had

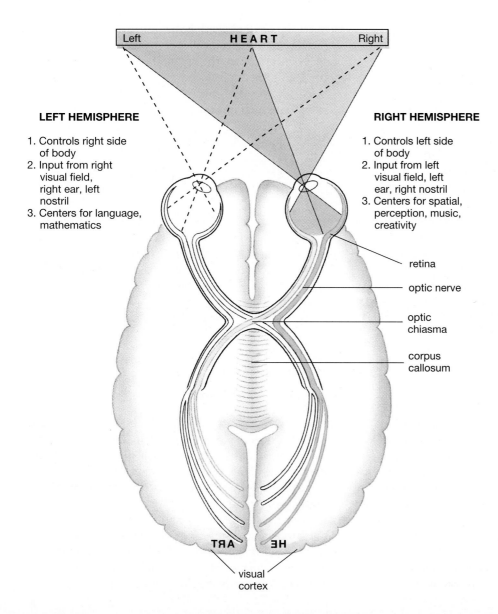

LEFT HEMISPHERE

1. Controls right side of body
2. Input from right visual field, right ear, left nostril
3. Centers for language, mathematics

RIGHT HEMISPHERE

1. Controls left side of body
2. Input from left visual field, left ear, right nostril
3. Centers for spatial, perception, music, creativity

retina

optic nerve

optic chiasma

corpus callosum

visual cortex

Figure 36-23 **Specialization of the two cerebral hemispheres**

In general, each hemisphere controls sensory and motor functions of the opposite side of the body. Further, the left side seems to predominate in rational and computational activities, whereas the right side governs creative and spatial abilities. Roger Sperry's split-brain experiments made use of the fact that images on the left half of the visual field are projected onto the right half of the retina and from there reach only the right visual cortex (orange), while the right half of the visual field is projected onto the left visual cortex (blue).

reached only the nonverbal right side of the brain! The same figure projected onto the right visual field was readily described verbally. These experiments, begun in the 1960s and refined since then, have revealed that the right side of the brain is actually superior to the left in several areas, including musical skills, artistic ability, recognition of faces, spatial visualization, and the ability to recognize and express emotions. For his pioneering work, Sperry was awarded the Nobel Prize in 1981.

Recent experiments indicate that the left-right dichotomy is not as rigid as was once believed. Patients who have suffered a stroke that disrupted blood supply to the left hemisphere typically show symptoms such as loss of speaking ability. Frequently, however, training can partially overcome these speech or reading deficits, even though the left hemisphere itself has not recovered. This fact suggests that the right hemisphere has some latent language capabilities. Interestingly, female stroke victims recover some lost abilities more often than males, and females also have a larger corpus callosum. These findings suggest a sex difference in the degree of specialization of the two hemispheres and the extent of their interconnections. Further evidence of this difference has recently been provided by sensitive techniques that allow imaging of neural activity in the brains of normal subjects performing various mental tasks. When subjects were asked to compare word lists for rhyming words, a specific region of the left cortex of male subjects became active, but in females, similar areas in *both* left and right hemispheres were activated (further brain-imaging studies are described in "Scientific Inquiry: Scanning the Brain").

The Mechanisms of Learning and Memory Are Poorly Understood

Although theories abound as to the cellular mechanisms of learning and memory, we are a long way from understanding these phenomena. In mammals, and particularly in humans, however, we do know a fair amount about two other aspects of learning and memory: the time course of learning and some of the brain sites involved in learning, memory storage, and recall.

Memory May Be Brief or Long Lasting

Experiments show that learning occurs in two phases: an initial **working memory** followed by **long-term memory.** For example, if you look up a number in the phone book, you will probably remember the number long enough to dial but forget it promptly thereafter. This is working memory. But if you call the number frequently, eventually you will remember the number more or less permanently. This is long-term memory.

Some working memory seems to be electrical in nature, involving the repeated activity of a particular neural circuit in the brain. As long as the circuit is active, the memory

stays. If the brain is distracted by other thoughts, or if electrical activity is interrupted, such as by electroconvulsive shock or by a concussion, the memory disappears and cannot be retrieved no matter how hard you try. In other cases, working memory involves temporary biochemical changes within neurons of a circuit, with the result that synaptic connections between them are strengthened.

Long-term memory, on the other hand, seems to be structural—the result, perhaps, of persistent changes in the expression of certain genes. It may require the formation of new, permanent synaptic connections between specific neurons or the permanent strengthening of existing but weak synaptic connections, for example, by increasing the area of synaptic contacts. These new or strengthened synapses last indefinitely, and the long-term memory persists unless certain brain structures are destroyed. Working memory can be converted into long-term memory, apparently by the hippocampus, which processes new memories then transfers them to the cerebral cortex for permanent storage.

Learning, Memory, and Retrieval May Be Controlled by Separate Regions of the Brain

Learning, memory, and the retrieval of memory seem to be separate phenomena, controlled by separate areas of the brain. Ample evidence shows that the hippocampus (part of the limbic system) is involved in learning. For example, intense electrical activity occurs in the hippocampus during learning. Even more striking are the results of hippocampal damage. A person whose hippocampus is destroyed retains most of his or her memories but is unable to learn anything that occurs after the loss. One victim was still unable to recall his address or find his way home after 6 years at the same residence. He could be entertained indefinitely by reading the same magazine over and over, and people whom he saw regularly required reintroduction at each encounter. People with extensive hippocampal damage can recall events momentarily, but the memory rapidly fades, as does the memory of a dream upon awakening. This phenomenon has led to the hypothesis that the hippocampus is responsible for transferring information from working into long-term memory.

Retrieval, or recall, of established long-term memories is localized in another area of the brain, the outer **temporal lobes** of the cerebral hemispheres. In a famous series of experiments in the 1940s, neurosurgeon Wilder Penfield electrically stimulated the temporal lobes of conscious patients undergoing brain surgery. The patients did not merely recall memories but felt that they were experiencing the past events right there in the operating room!

The site of storage of complex long-term memories is much less clear. The psychologist Karl Lashley spent many years training rats and subsequently damaging parts of their brains in an effort to locate the site of the memory, but failed. None of the injuries could erase a memory completely. In 1950, a frustrated Lashley wrote: "I sometimes

SCIENTIFIC INQUIRY

Scanning the Brain

For most of human history, the brain has been seen as a "black box" whose inputs and outputs were observable but whose internal workings were inherently unknowable. New techniques, however, are providing exciting insights into brain function. The most spectacular is the PET scan (positron emission tomography, described in Chapter 2), which allows a glimpse into the working

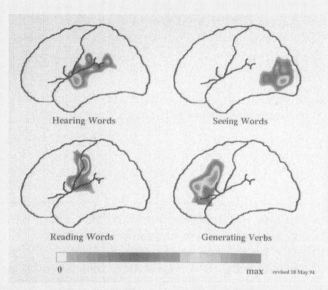

Figure E36-1   **Localization of language tasks**

Changes in glucose utilization reveal different cortex regions involved in different language-related tasks, based on research by Dr. Marcus Raichle of the Washington University School of Medicine in St. Louis. The scale ranges from white (lowest) through red (highest).

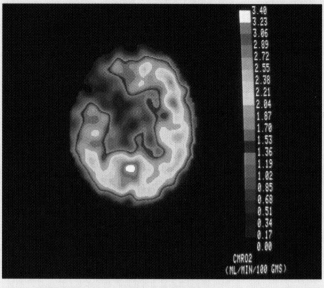

Figure E36-2   **Pinpointing a stroke**

As a result of stroke, part of the brain dies from lack of blood flow. The damaged region can be precisely localized by the lack of neural activity to the upper left region of this brain scan.

brain. Regions of the brain that are most active have higher energy demands and utilize more glucose than inactive areas. Scientists inject the subject with radioactive glucose then monitor levels of radioactivity (which reflect glucose consumption) throughout the brain. Different levels of glucose utilization are translated into colors on cross-sectional images of the brain. By monitoring radioactivity while a specific task is per-

feel, in reviewing the evidence on the localization of the memory trace, that the necessary conclusion is that learning just is not possible."

Some researchers suggest that each memory is stored in numerous distinct places in the brain. Or perhaps memories are stored like a hologram image, both everywhere and nowhere at the same time: The memory is more precise if the whole brain is intact, but each "bit" (probably several thousands of neurons) of cerebral hemisphere can store an essentially complete memory. Further research might provide definitive answers, but for now, the storage site of memories remains an unsolved mystery.

Insights on How the Brain Creates the Mind Come from Diverse Sources

Humans have always been intensely interested in the workings of their own minds. But until about 100 years ago, the mind was more appropriately a subject for philosophers than for scientists, because the tools to study the brain did not yet exist. Through the first half of the twentieth century, the mind was treated by psychologists as a "black box" whose internal workings could be deduced only through the investigation of how past and present experiences were interpreted and influenced behavior.

formed, scientists can identify parts of the brain involved in that task. Researchers can observe changes over a period of a few seconds as the brain responds to a specific odor or a visual or auditory stimulus or performs a type of thinking. Through PET scans, scientists have confirmed that different aspects of the processing of language occur in distinct areas of the cerebral cortex (Fig. E36-1). Pet scans can also precisely localize damaged portions of the brain, such as result from a stroke (Fig. E36-2). The contrast between normal and disturbed brain functioning can also be observed, for example, in some murderers and victims of Alzheimer's disease and schizophrenia (Fig. E36-3).

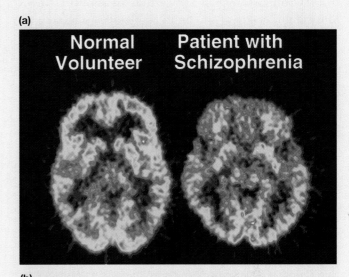

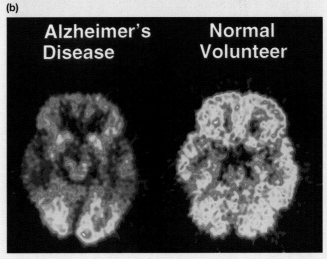

Figure E36-3   **Neural activity in normal and diseased brains**

PET scans can be used to compare neural activity in normal and diseased brains. These are images of sections of the brain taken so that the top of each image corresponds to the front of the brain, just behind the forehead, and the bottom to the back of the brain. Red areas are highest in metabolic activity, then yellow, then green. Blue areas show very low activity, while black areas exhibit little or no metabolic activity. **(a)** During a task requiring continuous attention, the frontal lobe (the region forming an arc across the top of the image) is strongly activated in a normal volunteer but much less active in a patient with schizophrenia. Schizophrenic individuals have difficulty focusing their attention on specific tasks. **(b)** The brain of a healthy elderly volunteer is compared to that of a victim of Alzheimer's disease, in which larger areas of blue and black indicate reduced neuronal activity in many brain regions.

New discoveries, however, are rapidly changing our views of the workings of the brain.

During recent decades, we have begun to understand the neural bases of at least some psychological phenomena. Many forms of mental illness, such as schizophrenia, manic depression, and autism, once thought to be due to childhood trauma or inept parenting, are now recognized as the result of biochemical imbalances in the brain. Studies are revealing a strong heritability factor (and hence, a biological basis) for traits that were once considered entirely learned, such as shyness and alcoholism.

A striking illustration of how the physical structure of the brain is related to personality was unwittingly provided by Phineas Gage in 1848. An explosion propelled a large metal rod through his skull, removing his left temporal lobe (Fig. 36-24). He miraculously survived, but his personality changed radically. Before the accident, Phineas was conscientious, industrious, and well liked. After his recovery, he became impetuous, profane, and incapable of planning or working toward a goal. Subsequent research has implicated the frontal lobe in emotional expression, control of aggression, and the ability to work for delayed rewards. Other sites of damage have revealed additional anatomical specializations. One patient with very localized damage to the left frontal lobe of the cerebral cortex was unable to name fruits and vegetables (although he could name everything else). Describing this patient, one science writer quipped: "Does the brain have a produce section?" Similarly, damage to certain areas of the cortex on the underside of the brain results in a selective inability to recognize faces.

In the past, much of our understanding of the human mind-brain connection came from the study of victims of brain damage such as that caused by a stroke, trauma, tumor, or surgical procedure. If the victim was cooperative, and if the case came to the attention of an interested researcher, tests were administered to define the change or loss of ability. Often, the exact extent of the damage remained unknown until revealed by autopsy.

Now new techniques, such as the PET and MRI scans, are permitting insight into the functioning of normal, as well as diseased, brains (see "Scientific Methods: Scanning the Brain"). These and increasingly sophisticated techniques of the future will create ever-larger windows into the "black box" that is the human brain. It is possible that the exact nature of consciousness will always remain obscure, but the next decades will see continued merging of the fields of psychology and neurophysiology, and a clearer understanding of how the human brain generates the human mind.

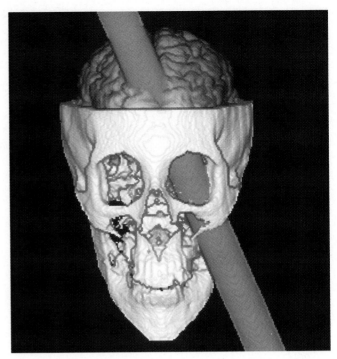

Figure 36-24   A famous accident

On the basis of studies of the skull of Phineas Gage, scientists have created this computer-generated reconstruction of the path taken by the steel rod that was blown completely through his head by an explosion, extensively damaging the frontal lobe of his cerebral cortex.

✖ S U M M A R Y   O F   K E Y   C O N C E P T S

A Comparison of Nervous and Endocrine Communication

There are four main differences between hormonal and neural control: (1) Hormones are released into the blood, but nerve cells release neurotransmitters directly onto their target cells; (2) a nerve cell releases neurotransmitter onto relatively few target cells; (3) neurons use electrical signals; and (4) neural signals last a much shorter time than most hormonal messages. These distinctions, however, are not absolute, and the hormonal and nervous systems are closely coordinated.

The Functions and Structure of Neurons

Nervous systems are composed of billions of individual cells called neurons. A neuron has five major functions, which are reflected in its structure: (1) Dendrites receive information from the environment or other neurons. (2) The cell body coordinates the cell's metabolic activities. (3) The cell body also adds together

electrical signals from the dendrites and from synapses on the cell body itself and "decides" whether to produce an action potential. (4) The axon conducts the signal to its output terminal: the synapse. (5) Synaptic terminals transmit the signal to other nerve cells, glands, or muscles.

Mechanisms of Neural Activity

The electrical activity in a neuron can be summarized as follows.

1. In an unstimulated neuron, the inside of the cell is negatively charged with respect to the extracellular fluid; this state is called the resting potential.
2. Signals received from other neurons are small, rapidly fading changes in potential called postsynaptic potentials. Inhibitory postsynaptic potentials (IPSPs) make the neuron more negative inside and less likely to produce an action potential. Excitatory postsynap-

tic potentials (EPSPs) make the neuron less negative inside and bring it closer to threshold.

3. If postsynaptic potentials, added together within the cell body, make the neuron substantially less negative, the neuron may reach threshold. Reaching threshold opens sodium channels and triggers an action potential. The action potential is a wave of positive charge that travels, undiminished in magnitude, to the synaptic terminals at the end of the axon. The speed at which the signal travels is fastest in neurons whose axons are insulated with myelin.

4. The resting potential is restored when the sodium channels close spontaneously and additional potassium channels open, allowing potassium to diffuse out and restore the negative resting potential.

A synapse, where two neurons communicate, consists of the synaptic terminal of the presynaptic neuron and a specialized region of the postsynaptic neuron, separated by the synaptic cleft. Neurotransmitter from the presynaptic neuron, released in response to an action potential, binds to receptors on the postsynaptic cell membrane. Ion channels then open, allowing ions to flow and producing either an EPSP or an IPSP, depending on the type of channels opened.

Building and Operating a Nervous System

Information processing in the nervous system requires four operations. The nervous system must: (1) signal the intensity of the stimulus, (2) determine the type of stimulus, (3) integrate information, and (4) initiate and direct the response. Nervous systems signal intensity by the frequency of action potentials in single neurons and by the number of similar neurons firing at the same time. The type of stimulus is determined by which neurons produce action potentials. The nervous system collects and processes sensory information from many sources. These sensory stimuli may come together (converge) on a few neurons whose activity in response to the inputs determines action. The "decision" to act may then be transmitted to many more neurons (divergence), which direct the activity.

Neural pathways normally have four elements: sensory neurons, association neurons, motor neurons, and effectors. Overall, nervous systems consist of numerous interconnected neural pathways, which may be either diffuse or centralized.

The Human Nervous System

The nervous system of humans and other vertebrates consists of the central nervous system (brain and spinal cord) and the peripheral nervous system (nerves leading from the central nervous system to the rest of the body).

The peripheral nervous system is further subdivided into sensory and motor portions. The motor portions consist of the somatic nervous system (which controls voluntary movement) and the autonomic nervous system (directing involuntary responses). The autonomic nervous system has both a sympathetic and parasympathetic component. The sympathetic nervous system prepares the body to "fight or take flight," whereas the parasympathetic nervous system promotes maintenance activities such as digestion.

Within the central nervous system, the spinal cord contains (1) neural pathways for reflexes and certain simple behaviors; (2) motor neurons controlling voluntary muscles; and (3) axons leading to and from the brain. The brain consists of three parts—the hindbrain, midbrain, and forebrain, each further subdivided into distinct structures.

The hindbrain in humans consists of the medulla and pons, which control involuntary functions (such as breathing), and the cerebellum, which coordinates muscular activities (such as walking). In humans, the small midbrain contains the reticular formation, a filter and relay for sensory stimuli. The forebrain includes the thalamus, a sensory relay station that shuttles information to and from higher, conscious centers in the forebrain. The diverse structures of the limbic system of the forebrain are involved in emotion, learning, and the control of instinctive behaviors such as sex, feeding, and aggression. Finally, the cerebral cortex of the forebrain is the center for information processing, memory, and initiation of voluntary actions. It includes primary sensory and motor areas and association areas that analyze sensory information and plan movements.

Neurotransmitters include acetylcholine, dopamine, norepinephrine, and serotonin. Neuromodulators modify synaptic transmission and may influence many neurons over a longer time span than classic transmitters. Examples are peptides called opioids that bind to the same receptors as do opiates.

Brain and Mind

The cerebral hemispheres are each specialized. In general, the left hemisphere is dominant in speech, reading, writing, language comprehension, mathematical ability, and logical problem solving. The right hemisphere specializes in recognizing faces and spatial relationships, artistic and musical abilities, and recognition and expression of emotions.

Memory takes two forms. Short-term memory is electrical or chemical. Long-term memory probably involves structural changes that increase the effectiveness of synapses. The hippocampus is an important site for learning and for the transfer of information into long-term memory. The temporal lobes are important for memory retrieval. The sites of memory storage are poorly understood.

✖ KEY TERMS

acetylcholine   p. 716
action potential   p. 699
amygdala   p. 715
association neuron   p. 707
autonomic nervous system   p. 709
axon   p. 699
brain   p. 708
cell body   p. 698
central nervous system   p. 708
cerebellum   p. 713
cerebral cortex   p. 715
cerebral hemisphere   p. 715
cerebrospinal fluid   p. 710
cerebrum   p. 714
convergence   p. 706
convolution   p. 715
corpus callosum   p. 715
dendrite   p. 698
divergence   p. 706
dopamine   p. 717
dorsal root ganglion   p. 710
effector   p. 707
endorphin   p. 717
excitatory synapse   p. 705
forebrain   p. 713

ganglion   p. 708
gray matter   p. 710
hindbrain   p. 713
hippocampus   p. 715
hypothalamus   p. 715
inhibitory synapse   p. 705
integration   p. 705
intensity   p. 706
limbic system   p. 715
long-term memory   p. 719
medulla   p. 713
meninges   p. 710
midbrain   p. 713
motor neuron   p. 707
myelin   p. 703
nerve   p. 699
nerve net   p. 707
neuromodulator   p. 716
neuron   p. 698
neurotransmitter   p. 699
node   p. 701
norepinephrine   p. 717
opioid   p. 717
parasympathetic nervous system   p. 709
peripheral nerve   p. 709

peripheral nervous system   p. 708
pons   p. 713
postsynaptic neuron   p. 704
postsynaptic potential   p. 704
potassium channel   p. 700
presynaptic neuron   p. 704
reflex   p. 707
resting potential   p. 699
reticular formation   p. 714
saltatory conduction   p. 704
sensory neuron   p. 707
serotonin   p. 717
sodium channel   p. 700
sodium-potassium pump   p. 700
somatic nervous system   p. 709
spinal cord   p. 708
sympathetic nervous system   p. 709
synapse   p. 704
synaptic cleft   p. 704
synaptic terminal   p. 699
temporal lobe   p. 719
thalamus   p. 715
threshold   p. 699
white matter   p. 710
working memory   p. 719

✖ THINKING THROUGH THE CONCEPTS

Multiple Choice

1. _____ are integration centers in neurons.
   a. dendrites   b. axons
   c. cell bodies   d. ion channels
   e. synapses

2. Which of the following helps establish the neuron resting potential?
   a. diffusion of ions and molecules
   b. electrical attraction of ions and molecules
   c. differential permeability of membranes
   d. concentration differences of ions and molecules
   e. all of the above

3. In a resting neuron, _____ channels remain open.
   a. all sodium   b. some potassium
   c. all sodium and chloride   d. some sodium
   e. all potassium

4. Which of the following statements about action potentials is NOT correct?
   a. Action potentials are "all-or-none" phenomena.
   b. During action potentials, the membrane potential becomes positive inside.
   c. Action potentials do not diminish in amplitude as they travel along an axon.
   d. An action potential is triggered when a neuron reaches threshold.
   e. An action potential begins when potassium ions enter a neuron.

5. Disturbances in body temperature, hunger perception, menstrual cycles, water balance, and sexual arousal are symptoms of defects in the
   a. cerebrum   b. thalamus
   c. medulla   d. hypothalamus
   e. cerebellum

6. Neurotransmitters can be eliminated from synapses by
   a. diffusing away
   b. enzymatic destruction
   c. active transport into a neuron
   d. all of the above
   e. none of the above

Review Questions

1. List four major parts of a neuron and explain what function each part is specialized for.

2. First describe in detail, then compare and contrast an action potential and a postsynaptic potential.

3. Diagram a synapse. How are signals transmitted from one neuron to another at a synapse?

4. How does the brain perceive the intensity of a stimulus? The type of stimulus?

5. What are the four elements of a simple nervous pathway? Describe how these elements function in the human pain reflex.

6. What is the difference between a diffuse and a concentrated nervous system? What types of animals possess each type?

7. Describe the autonomic nervous system. What are its two subdivisions, and what role does each play in daily life?

8. Draw a cross section of the spinal cord. What types of neurons are located in the spinal cord? Explain why severing the cord paralyzes the body below the level where it is severed.

9. Describe the functions of the following parts of the human brain: medulla, cerebellum, reticular formation, thalamus, limbic system, cerebrum.

10. What structure connects the two cerebral hemispheres? Describe the evidence that the two hemispheres are specialized for distinct intellectual functions.

11. Distinguish between long-term and working memory.

12. What is an opioid? What types of functions may opioids serve in the human brain?

✠ APPLYING THE CONCEPTS

1. Argue for or against the statement: "Consciousness by its nature is incomprehensible; the brain will never understand the mind."

2. In Parkinson's disease, which afflicts several million Americans, dopamine-producing cells degenerate in a small part of the brain important in the control of movement. Some physicians have reported improvement in patients after injecting cells taken from the same general brain region of an aborted fetus into appropriate parts of the Parkinson's victim's brain. Discuss this type of surgery from as many viewpoints as possible: ethical, financial, practical, etc. Based on your responses, is fetal transplant surgery the answer to Parkinson's disease?

3. A visually impaired student goes to his veterinarian to get flea powder for his seeing-eye dog. Unable to read the label precautions, he sits down near his dog and slowly works large quantities of the powder into the dog's coat. A few hours later the student ends up in the hospital emergency room with dizziness and breathing trouble. Hypothesizing that flea powder is the culprit, the physician draws blood for an enzyme analysis. What enzyme are they looking for? How might abnormally low levels of this enzyme produce the symptoms described?

4. Alpha-bungarotoxin, a snake toxin, binds to acetylcholine receptors in postsynaptic neuron membranes, preventing the attachment of acetylcholine. What symptoms would you expect to see in snake bite victims, and why?

5. If human spinal cord neuron axons were unmyelinated, would the spinal cord be larger or smaller? Would you move faster or slower? Explain.

✠ FOR MORE INFORMATION

Angier, N. "Storming the Wall." *Discover*, May 1990. Describes the blood-brain barrier and new methods of penetrating it with drugs.

Changeau, J-P. "Chemical Signaling in the Brain." *Scientific American*, November 1993. Acetylcholine receptors reveal information on communication between brain neurons.

Holloway, M. "Rx for Addiction." *Scientific American*, March 1991. By studying drug addiction, neurobiologists are learning more about the brain, and new ways to counteract addictions.

Montgomery, G. "The Mind in Motion." *Discover*, March 1989. Describes the use of the PET scan to view the workings of the living brain.

Radetsky, P. "The Brainiest Cells Alive." *Discover*, April 1991. What can you learn from culturing brain cells?

Scientific American special issue on Mind and Brain, September 1992. Articles describe the development of the brain, the biochemical basis of learning, major neurological disorders, and aging effects on the brain.

Souchurek, H. "Medicine's New Vision." *National Geographic*, January 1987. Striking photography and interesting case histories highlight this readable overview of noninvasive techniques for viewing inside the body.

NET WATCH

On-line resources for this chapter are on the World Wide Web at:
http://www.prenhall.com/~audesirk (click on the <u>table of contents</u> link and then select Chapter 36).

37 The Senses: Perception

> *"Nothing is understood by the intellect which is not first perceived by the senses."*
>
> ARISTOTLE (384–322 B.C.)

❖ AT A GLANCE

Receptor Mechanisms

Sensing Temperature: Thermoreception

Sensing Touch and Movement: Mechanoreception

Sound Perception: Hearing
  The Structure of the Ear Helps Capture, Transmit, and
    Transduce Sound
  Sound Transduction Is Aided by the Structure of the Cochlea

Gravity and Movement Perception: The Vestibular Apparatus

Sensing Light: Photoreception
  Animal Eyes Range from Simple to Complex
  The Mammalian Eye Collects, Focuses, and Transduces
    Light Waves
  Binocular Vision Allows Depth Perception

Smell and Taste: Chemoreception
  The Ability to Smell Arises from Olfactory Receptors
  Taste Receptors Are Found in Clusters on the Tongue

Sensing Pain

Evolutionary Connections: Natural Selection Shapes Perception

A star-nosed mole peeks up through a bed of moss, its 22-tentacled nose suggesting a grotesque, alien flower. The fleshy, writhing tentacles surrounding its nose are covered with nerve endings. Recent evidence suggests that they are actually electroreceptors, which the mole uses to detect the electrical fields produced by earthworms and other subterranean creatures that it hunts and eats.

Since birth, you have experienced your surroundings in ways specified by your nervous system and coded by your genes. The word *tree* may evoke a visual image of branching form, green and brown color, perhaps the sound of rustling leaves, the rough feel of bark, or the scent of pine needles. No doubt you feel secure in the belief that your senses provide accurate information about "the way things really are."

But think for a moment about the nature of the signals that the sense organs send to the brain. Sense organs don't transmit light, sound, or pressure. Instead, all sense organs transmit the same signal—action potentials—regardless of the stimulus. And these action potentials are virtually indistinguishable from one another. We must conclude, then, that the sensations we perceive as a result of these action potentials (color, light, sound, pain) are *purely a creation of the brain*.

The subjective sensations that the brain uses for light, sound, and so on, arise from different areas of the brain, which are stimulated by the action potentials carried by the different sensory nerves. For example, signals carried by the optic nerve eventually reach the visual cortex of the brain, which creates the sensations of light and form that we call vision. If action potentials from the optic nerve could be directed into the auditory cortex of the brain, we would presumably "hear" complex and different chords when we directed our eyes toward a tree or a sunset. Although our senses provide us with useful information about our environment, the *quality* of our perceptions arises strictly from the neural connections within the brain.

Because of the incredible complexity of the brain, scientists are a long way from understanding how these subjective perceptions are formed. If life has evolved independently on other planets, it is reasonable to suspect that those alien life forms have evolved entirely different ways of interpreting environmental stimuli. But what of life on Earth? Do all animals, or do all humans, for that matter, perceive the world in the same way? The basic anatomical similarity among human brains, and our ability to communicate easily with one another about our sensory perceptions, leads us to assume that human brains interpret the world in much the same way. Although scientists are less certain of what other vertebrates perceive, all vertebrates share a common ancestor. This evolutionary relationship has resulted in similarities in the structure of vertebrate sense organs and brains; those structural similarities suggest that the quality of all vertebrate perceptions may be somewhat similar. Keep in mind, however, that differing selective pressures on different animals have modified the sensitivity of their receptors and the range of stimuli to which they respond (see "Evolutionary Connections: Natural Selection Shapes Perception"). For example, dogs can hear extremely high-pitched sounds (which are not audible to the human ear), rhinos can produce and respond to sounds too low for us to hear, and bees and hawks can see light into the near ultraviolet range that is invisible to the human eye.

You needn't travel to another universe, however, to find modes of perception totally alien to your own. Place yourself in the sensory world of a bat, which emits high-pitched sounds that bounce back, enabling it to determine its surroundings as it hunts insects on a moonless night. As a bat, you "see" using sound waves! In fact, the auditory image may give rise to a sensation in the bat's brain that you cannot even imagine. Now imagine yourself as a migrating eel, orienting by means of Earth's magnetic field. How do you perceive that field? As a wavering note or a bluish tinge in the water? Again, more likely the sensation is impossible to imagine. If we could enter the brains of some of the creatures who share our world, we might feel as if we had been transported to an alien landscape!

Receptor Mechanisms

The word *receptor* is used in several contexts in biology. In the most general sense, a **receptor** is a structure that changes when it is acted upon by a stimulus from its surroundings, causing a signal to be produced. All receptors are **transducers**: structures that convert signals from one form to another. The receptor may be a membrane protein that changes configuration when it binds a specific hormone or neurotransmitter, as discussed in previous chapters. Alternatively, as described in this chapter, a **sensory receptor** may be an entire cell (often a neuron) that is specialized to produce an electrical response to particular stimuli in its environment—that is, it translates sensory stimuli into the language of the nervous system. All sensory receptors produce electrical signals, but each receptor type is specialized so that it produces its signal only in response to a particular type of environmental stimulus, as discussed below. Sensory receptors are often clustered into sensory organs, such as the eye, ear, or tongue. Their electrical activity gives rise to the subjective perceptions of light, sound, and taste that we describe as our "senses."

Stimulation of a sensory receptor causes a **receptor potential,** an electrical signal whose size is proportional to the strength of the stimulus. Receptor potentials may give rise to action potentials in the receptor neurons themselves. Alternatively, receptor potentials in small receptor cells cause release of neurotransmitter onto postsynaptic neurons, which in turn produce action potentials that travel to the central nervous system. In contrast to action potentials, receptor potentials vary according to the intensity of the stimuli—the stronger the stimulus, the larger the receptor potential. Intensity is then conveyed to the nervous system by the frequency, not the size, of action potentials (see Chapter 36).

Receptor cells are named after the stimulus to which they respond, as summarized in Table 37-1. **Thermoreceptors** ("heat" receptors) respond to fluctuations in temperature.

Mechanoreceptors respond to mechanical deformation of their cell membranes, which causes receptor potentials in one of two ways: by stretching the membrane of the receptor cell (as in receptors for touch or pressure) or by bending "hairs" that project from the receptor cell membrane. Receptor cells for sound, motion, and gravity bear hairlike structures and are called **hair cells.** Bending of these hairs produces a receptor potential.

Many other receptor cells produce receptor potentials in response to stimuli that alter specific receptor proteins embedded in their cell membranes. **Photoreceptors** (receptors for light), **chemoreceptors** (receptors for chemicals, which we perceive as tastes or odors), and pain receptors possess

TABLE 37-1 ❖ *Vertebrate Receptor Types*			
Type of Receptor	*Specific Sensory Cell Type*	*Stimulus*	*Location*
Thermoreceptor	Free nerve ending	Heat, cold	Skin
Mechanoreceptor	Hair cell	Vibration, motion, gravity	Inner ear
	Variety of specialized endings in skin (Pacinian corpuscle, Merkel's disc)	Vibration, pressure, touch	Skin
	Free nerve endings in muscles or joints (muscle spindle, Golgi tendon organ)	Stretch	Muscles, tendons
Photoreceptor	Rod, cone	Light	Retina of eye
Chemoreceptor	Olfactory receptor	Odor (airborne molecules)	Nasal cavity
	Taste receptor	Taste (waterborne molecules)	Tongue
Pain receptor	Free nerve ending	Chemicals released by tissue injury	Widespread in body

extensive areas of membrane studded with specialized receptor proteins. When light or a chemical contacts a receptor protein specialized to respond to that stimulus, the protein changes shape. This change alters the permeability of the cell membrane, generating a receptor potential.

Sensing Temperature: Thermoreception

Thermoreceptors sense temperature by responding to infrared radiation, or heat. Vertebrates have both "warm" receptors that respond to an increase in skin temperature and "cold" receptors that respond to a decrease in temperature. Scientists hypothesize that changes in the metabolic rate of the receptor—which is directly affected by temperature—cause the receptor potential. In human skin, warm and cold receptor cells end in many fine branches, called **free nerve endings,** located just beneath the epidermis (Fig. 37-1). Free nerve endings also respond to touch, pressure, and pain (described later in the chapter). Thermoreceptors are most responsive to temperature *changes*, which trigger a burst of action potentials that gradually decline in frequency if the temperature remains constant. This is why a swimming pool that feels uncomfortably cold or a hot tub that feels uncomfortably hot when you first step in often feels comfortable a few minutes later.

Thermoreceptors reach their highest degree of sensitivity in the **pit organs** of vipers (rattlesnakes, cottonmouths, and copperheads; Fig. 37-2). Within the pit organs are clus-

Figure 37-2   **Pit organs**

A pit viper, showing the right pit organ (located just below the eye), which is an exquisitely sensitive thermoreceptor. These are used to detect warm-blooded prey.

ters of free nerve endings so sensitive that they can respond to the body heat of a mouse up to 38 centimeters (16 inches) away. Because there is a pit organ on each side of the head, vipers can compare the intensity of signals from the two organs and determine the direction of their prey.

Sensing Touch and Movement: Mechanoreception

Mechanoreceptors are widespread within the body. From the variety of skin receptors, to the position sensors in joints, to the hair cells of the inner ear, mechanoreceptors show the greatest diversity of form and function.

The skin of humans and most other animals is exquisitely sensitive to touch. Embedded in and directly beneath human skin are several distinct types of mechanoreceptor neurons, each of which contains a sensory ending that produces a receptor potential when its membrane is deformed and stretched (Fig. 37-1). Some of these endings are enclosed in layers of connective tissue—for example, the **Pacinian corpuscle,** which responds to rapid changes in pressure such as may be produced by a poke or vibrations. Some touch receptor neurons end in free nerve endings, which can produce sensations of itching and tickling, as well as touch. The density of receptors in the skin varies tremendously over the surface of the body: Each square centimeter of fingertip has dozens of touch receptors, but on the back there may be fewer than one receptor per square centimeter.

Mechanoreceptors, such as those located in the walls of the stomach, rectum, and bladder, signal fullness by responding to stretch. Other mechanoreceptive endings in the joints and muscles sense the orientation and direction of movement of various body parts. These position sensors, collectively called **proprioceptors,** allow you to walk without

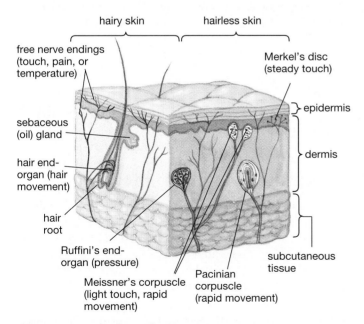

Figure 37-1   **Body surface receptors**

A sampling of receptors within the skin (consisting of the epidermis and dermis; see Chapter 29) and immediately beneath the skin (subcutaneous tissue) that respond to mechanical stimuli, temperature, and pain.

watching your feet or eat without watching the fork on its way to your mouth. Proprioceptors are concentrated in the ligaments and tissue surrounding the joints, in tendons, and in skeletal muscle. For example, Golgi tendon organs, found in tendons (which connect muscles to bone; see Chapter 38), and muscle spindles, found in skeletal muscle, respond to stretch (Fig. 37-3). They are activated by movement of the joints, which causes a sudden change in the degree of stretching of muscles and tendons, and also by the steady pull or tension that occurs in any position of the limb.

Specialized types of mechanoreceptors, called hair cells, detect sound and gravity in the mammalian inner ear (described below) and are also found in the **lateral line organs** of fish and the aquatic forms of amphibians. The lateral line organ, which detects water movement, consists of a series of clusters of hair cells located in pits or tubes that form a strip beginning in the head and extending along both sides of the body. As shown in Figure 37-4, the "hairs" of the hair cells are embedded in a gelatinous cap, the **cupula**, which is deflected by water currents, causing the hairs to bend. The receptor potentials caused by the bending produce action potentials that travel to the brain. Lateral line organs are used to detect water currents, water movements caused by prey or predators, and possibly low-frequency sounds.

In the following sections, we describe the inner ear of terrestrial vertebrates, using the human as an example. As you read about the mechanisms of sound, gravity, and movement detection, note the similarities to the lateral line organ. In each case, notice that the "hairs" of hair

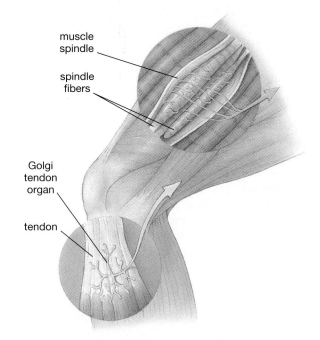

Figure 37-3   **Proprioceptors**
Proprioceptors are found in muscles, tendons, and tissues surrounding joints. They signal pressure and stretch and inform the brain of the position of the limbs.

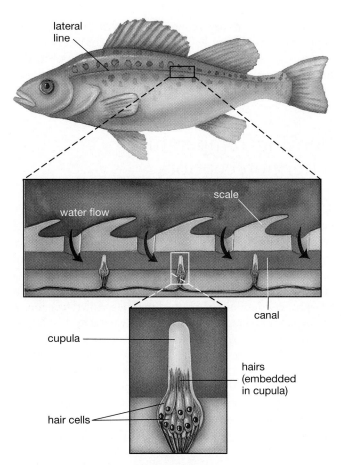

Figure 37-4   **The lateral line organ**
The lateral line organ of fish and aquatic amphibians consists of clusters of specialized hair cells whose hairs are embedded in a gelatinous structure, the cupula. These clusters are embedded at intervals within a canal along the length of each side of the body. Openings along the canal allow entry of water currents, which deflect the cupula. This deflection bends the hairs of the hair cells, causing a receptor potential.

cells are embedded in a gelatinous membrane that is moved by the stimulus and causes the hairs to bend.

Sound Perception: Hearing

Sound is produced by any vibrating object—a drum, a motor, or the vocal cords of a person or animal. These vibrations are transmitted through air in the form of sound waves and are intercepted by our ears. The ears of humans and other animals are elaborate and remarkable structures that detect the direction, intensity, and pitch of sound.

The Structure of the Ear Helps Capture, Transmit, and Transduce Sound

The ear of humans and most other vertebrates consists of three parts: the outer, middle, and inner ear (Fig. 37-5). The **outer ear** consists of the **external ear** and the **audi-**

tory canal. The external ear modifies sound waves with its fleshy folds. These modifications help the brain determine the location of the sound source. The air-filled auditory canal conducts the sound waves to the **middle ear**, consisting of the **tympanic membrane**, or eardrum; the **eustachian tube**; and three tiny bones called the hammer, anvil, and stirrup. The middle ear is surrounded by membranes and bone, but the eustachian tube connects the middle ear to the pharynx and equalizes the air pressure between the middle ear and the atmosphere.

Within the middle ear, sound first vibrates the tympanic membrane, which in turn vibrates the hammer, the anvil, and the stirrup. These bones transmit vibrations from the tympanic membrane to a much smaller membrane, the **oval window**. The oval window covers the opening of the **inner ear,** in which sound vibrations travel through fluid. The fluid-filled hollow bones of the inner ear form the spiral-shaped cochlea (*cochlea* means "snail" in Latin) and the vestibular apparatus (described later), where vibrations are translated into neural signals.

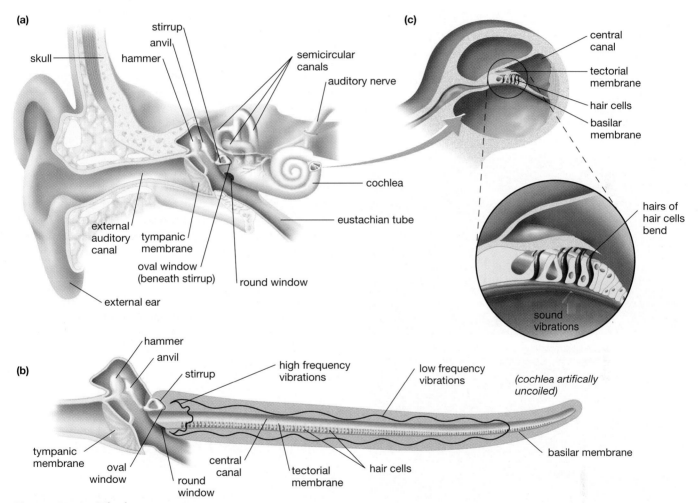

Figure 37-5   The human ear

The human ear contains structures that detect sound, gravity, and movement. **(a)** Sound waves enter the auditory canal and vibrate the tympanic membrane. These vibrations are transmitted through the bones of the middle ear to the membrane of the oval window connected to the fluid-filled cochlea, surrounded by thin bone. **(b)** Uncoiled, the cochlea consists of an outer tube surrounding the central canal. Vibrations of the fluid filling the cochlea cause the membranes of the central canal to vibrate, stimulating the hair cells, shown in part (c). The basilar membrane varies in width and tension along its length, causing lower-frequency vibrations (low-pitched sounds) to activate receptors toward the tip of the (uncoiled) cochlea and higher-frequency vibrations (high-pitched sounds) to activate receptors closer to the oval window. **(c)** The hairs of hair cells span the gap between the basilar and tectorial membranes in the central canal. Sound vibrations move the membranes relative to one another, bending the hairs and producing a receptor potential in the hair cells. The hair cells then cause action potentials in the auditory nerve.

Sound Transduction Is Aided by the Structure of the Cochlea

The receptor cells for hearing are located within a portion of the inner ear called the **cochlea** (see Fig. 37-5a). If we mentally unroll the cochlea, in lengthwise cross section it consists of two fluid-filled tubes, an outer U-shaped canal and a straight central canal (Fig. 37-5b). The central canal contains the **basilar membrane,** on top of which are located the receptors, or hair cells. Protruding into the central canal is another membrane, the **tectorial membrane,** a gelatinous structure in which the "hairs" of the hair cells are embedded.

How do these structures allow the perception of sound? When the bones of the middle ear vibrate the oval window, it passes these vibrations to the fluid in the cochlea, which in turn vibrates the basilar membrane, causing it to move relative to the tectorial membrane. This movement bends the hairs spanning the gap between the membranes (Fig. 37-5c), producing receptor potentials in the

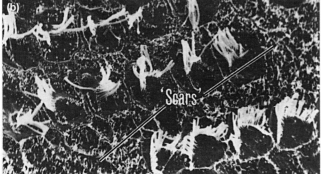

Figure 37-6   Loud sound damages hair cells

Scanning electron micrographs show the effect of intense sound on the hair cells of the inner ear. **(a)** Top view of the hairs of the hair cells in a normal guinea pig; hairs emerge from each receptor in a V-shaped pattern. **(b)** After 24-hour exposure to a sound level approached by loud rock music (2000 vibrations per second at 120 decibels), many of the hairs are damaged or missing entirely, leaving "scars." Because hair cells in humans do not regenerate, this type of hearing loss is permanent. (Scanning electron micrographs by Robert S. Preston, courtesy of Professor J. E. Hawkins, Kresge Hearing Research Institute, University of Michigan Medical School.)

hair cells. The hair cells then release transmitter onto neurons of the **auditory nerve.** Action potentials triggered in the auditory nerve axons travel to the brain.

The inner ear also allows us to perceive loudness (the magnitude of sound vibrations) and pitch (the frequency of sound vibrations). A weak sound causes small vibrations, which bend the hairs only slightly. This slight bending produces small receptor potentials in the hair cells and a low frequency of action potentials in the auditory nerve axons. A loud sound causes large vibrations, which cause greater bending of the hairs and a larger receptor potential. This leads to a high frequency of action potentials in the auditory nerve. Loud sounds sustained for a long time can actually damage the hairs (Fig. 37-6), resulting in hearing loss, a fate suffered by many prominent rock musicians and their fans.

The perception of pitch is a little more complex. Humans can detect vibration frequencies from about 30 vibrations per second (very low pitched) up to around 20,000 vibrations per second (very high pitched). The basilar membrane is stiff and narrow at the end near the oval window and more flexible and wider near the tip of the cochlea. This progressive change in structure causes different parts of the membrane to resonate best to particular frequencies of sound. High-frequency sound waves cause the greatest vibration toward the end of the basilar membrane near the oval window; increasingly lower frequencies produce vibration progressively farther toward the opposite end (see Fig. 37-5b). Thus, where the basilar membrane vibrates most, and consequently which receptors are stimulated most, varies with the frequency of sound. The brain interprets signals from receptors near the oval window as high-pitched sound, whereas signals from receptors farther along the uncoiled cochlea are interpreted as lower in pitch.

Gravity and Movement Perception: The Vestibular Apparatus

Besides hearing, the human inner ear is the site of two other modified mechanical senses, one for detecting motion and the other for detecting gravity. These are both found within the **vestibular apparatus** housed in a set of bony canals adjacent to the cochlea (Fig. 37-7).

Motion is detected by the **semicircular canals,** a set of three curving, fluid-filled, bone-covered tubes. In swellings at the base of each canal are clusters of hair cells similar to those of the inner ear, their hairs embedded in a gelatinous cupula very similar to that of the lateral line organ of fish (see Fig. 37-4). Sudden acceleration of the head causes the cupula and surrounding fluid to lag behind the movement of the canals, bending the hairs in one direction. Stopping the motion causes the fluid to slosh ahead,

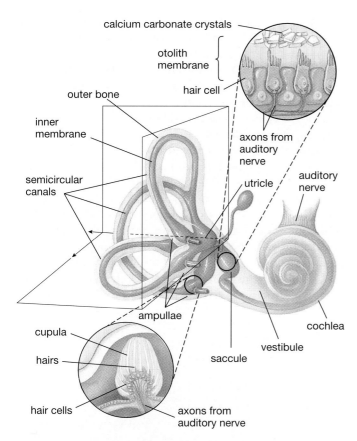

Figure 37-7   The vestibular apparatus

The vestibular apparatus, showing its relation to the cochlea. Both the semicircular canals and the utricle and saccule have clusters of hair cells whose hairs are embedded in gelatinous membranes. The cupula in each semicircular canal detects movement of the fluid, while the otolith membranes of the utricle and saccule (weighted with stones of calcium carbonate) exert pressure on the hairs in response to the pull of gravity.

bending the hairs in the opposite direction. Each of the canals is oriented at right angles to the others, so they allow us to sense movements in all directions (Fig. 37-7).

Gravity perception is the function of two chambers below the semicircular canals, the **utricle** and the **saccule,** each containing clusters of hair cells. The hairs of these receptors are embedded in a gelatinous mass, the **otolith membrane,** which contains crystals of calcium carbonate (*otolith* is from the Greek, meaning "ear stone"). The crystals make the membrane heavier than the surrounding fluid and thus more responsive to the downward pull of gravity, bending the hairs according to the orientation of the head (Fig. 37-7).

Sensing Light: Photoreception

Animal vision varies in its ability to provide a sharp, accurate representation of the real world, and several types of eyes have evolved independently. All forms of vision, how-

ever, use photoreceptors. These sensory cells contain receptor molecules called **photopigments** (because they are colored) that absorb light and chemically change in the process. This chemical change alters ion channels in the receptor cell membrane, producing a receptor potential.

Animal Eyes Range from Simple to Complex

The simplest animal light detector is the **eyespot,** found in flatworms (Fig. 37-8) and jellyfish. The eyespot has no lens and cannot focus light or form an image. Flatworms can distinguish light from dark and may also perceive the direction and intensity of the light. This information can be important to the animal: A passing shadow (which could be a predator) triggers withdrawal or escape responses.

The arthropods (insects, spiders, and crustaceans) evolved **compound eyes** that consist of a mosaic of many individual light-sensitive subunits called **ommatidia** (Fig. 37-9). Although much more complex than an eyespot, each ommatidium functions similarly as an on/off, bright/dim detector. Using a large number of individual units (up to 36,000 per eye in a dragonfly), most arthropods probably obtain a reasonably faithful, although grainy, image of the world. Compound eyes are excellent at detecting movements, an advantage in avoiding predators or in hunting. In addition, many arthropods, such as bees and butterflies, have good color perception.

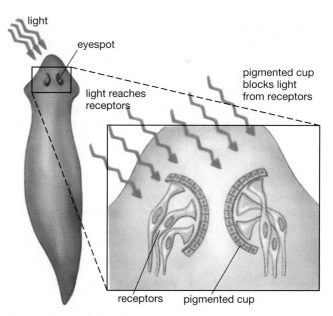

Figure 37-8   Eyespots

The most primitive eyes in the animal kingdom are eyespots, such as those found in flatworms. Eyespots consist of photoreceptors in a pigmented cup. Light entering the open end of the cup stimulates the photoreceptors, but the pigmented back side of the cup prevents light from other directions from reaching them. This arrangement allows detection of the direction of light, guiding simple behaviors such as finding a dark place to hide.

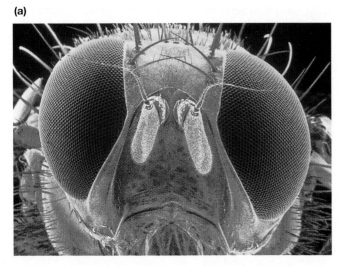

(a)

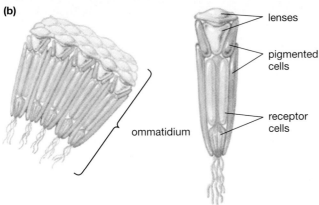

(b)

lenses

pigmented cells

receptor cells

ommatidium

Figure 37-9   **Compound eyes**

(a) Scanning electron micrograph of the head of a fruit fly showing a compound eye on each side of the head. **(b)** Each eye is made up of numerous individual light-receptive ommatidia. Within each ommatidium are several receptor cells, capped by a lens. Pigmented cells surrounding each ommatidium prevent light from passing through to adjacent receptors.

The cephalopod mollusks (see Chapter 24) and the vertebrates, including our ancestors, independently evolved a third type of eye, often called the **camera eye** (Fig. 37-10a). The camera eye consists of three basic parts: a light-sensitive layer (the retina), a lens for focusing light, and a set of muscles for adjusting focus by moving or changing the shape of the lens.

The Mammalian Eye Collects, Focuses, and Transduces Light Waves

Incoming light first encounters the **cornea,** a transparent covering over the front of the eyeball. Behind the cornea is a chamber filled with a watery fluid called **aqueous humor,** which provides nourishment for both the lens and cornea. The amount of light entering the eye is adjusted by the pigmented muscular tissue, the **iris,** whose circular

opening, the **pupil,** can be expanded or contracted. Light passing through the pupil encounters the **lens,** a structure resembling a flattened sphere and composed of transparent protein fibers. The lens is suspended behind the pupil by ligaments and muscles that regulate its shape. Behind the lens is another, much larger chamber filled with a clear jellylike substance, the **vitreous humor,** which helps maintain the shape of the eye.

After passing through the vitreous humor, light reaches the **retina,** a multilayered sheet of photoreceptors and neurons, where the light energy is converted into electrical nerve impulses that are transmitted to the brain (Fig. 37-10b). The retina is richly supplied with blood vessels and contains a layer of pigment that absorbs stray light rays that escape the photoreceptors. Behind the retina is a darkly pigmented tissue, the **choroid.** The choroid's rich blood supply helps nourish the cells of the retina. Its dark pigment also absorbs stray light whose reflection inside

(a)

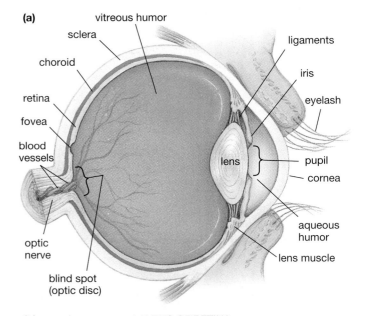

vitreous humor

sclera

choroid

retina

fovea

blood vessels

optic nerve

blind spot (optic disc)

ligaments

iris

eyelash

lens

pupil

cornea

aqueous humor

lens muscle

(b)   LAYERS OF RETINA

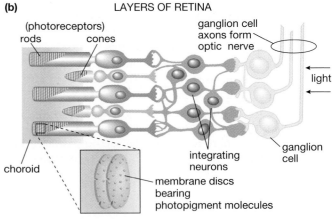

(photoreceptors)
rods   cones

choroid

ganglion cell axons form optic nerve

light

integrating neurons

ganglion cell

membrane discs bearing photopigment molecules

Figure 37-10   **The human eye**

(a) The anatomy of the human eye. **(b)** The human retina has rods and cones (photoreceptors), integrating cells, and ganglion cells. Each rod and cone bears a long extension packed with membranes in which the light-sensitive molecules are embedded.

the eyeball would interfere with clear vision. Surrounding the outer portion of the eyeball is a tough connective tissue layer, the **sclera**, visible as the white of the eye, which is continuous with the cornea.

In vertebrates (such as deer) that are most active at dusk, when little light is available, the choroid may be modified to reflect light, rather than absorb it. By reflecting light that escaped the photoreceptors during its initial passage, the choroid gives the receptors a second chance to detect it, maximizing the animal's ability to see in dim light. Reflective choroids give the eyes of these animals an eerie glow when bright light (such as from car headlights) is reflected back through the wide-open pupil.

The Adjustable Lens Allows Focusing of Both Distant and Nearby Objects

The visual image is focused most sharply on a small area of the retina called the **fovea**. Focusing is aided by the cornea, which contributes significantly to the bending of incoming light rays, producing an image of approximately the right size in the general vicinity of the retina. The shape of the cornea cannot be adjusted, however, and the lens is responsible for final, sharp focusing.

The shape of the lens can be adjusted so that it is either more rounded or more flattened when viewed from the side. This adjustment is accomplished by a circular muscle surrounding the lens. The size of the opening encircled by the muscle is largest when the muscle is relaxed. In its relaxed state, the muscle therefore applies maximum pull on the lens, which stretches and flattens it. In this configuration, the lens focuses distant objects. When focusing nearby, the muscle contracts, relaxing tension on the lens, and allowing the lens to resume a more rounded shape. The curved lens bends light rays more, allowing the eye to focus on closer objects (Fig. 37-11a).

Nearsighted people cannot focus on distant objects; farsighted people cannot focus on nearby objects. These conditions, which can be caused by abnormally long or short eyeballs, are corrected by external lenses of the appropriate shape (Fig. 37-11b, c). As humans age, the lens stiffens. By their mid-forties, most people have lost much of their ability to change the shape of the lens, and they become farsighted, requiring glasses for close work such as reading.

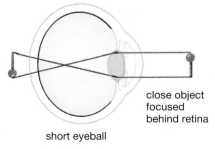

(a) NORMAL EYE

distant object, lens thin

close object, lens fat

retina

(b) NEARSIGHTED EYE

distant object focused in front of retina

long eyeball

concave lens diverges light rays, object focused on retina

(c) FARSIGHTED EYE

close object focused behind retina

short eyeball

convex lens converges light rays, object focused on retina

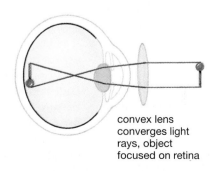

Figure 37-11   **Focusing in the human eye**

(a) (Left) To focus on a distant object, the lens is made thinner, causing relatively little bending of the light rays. (Right) To focus on a nearby object, the lens assumes a more nearly spherical shape, bending the light rays more sharply. **(b)** (Left) Nearsighted people usually have eyeballs that are too long. Light rays focus in front of the retina and are out of focus again by the time they strike the retina. (Right) Eyeglasses with a concave lens cause the light rays to diverge slightly so that the focal point falls on the retina. **(c)** (Left) Farsighted people have eyeballs that are too short. The lens cannot bend incoming light rays enough to focus on the retina, so the focal point falls behind the retina. (Right) Eyeglasses with a convex lens converge light rays, causing the focal point to fall on the retina.

Light Striking the Retina Is Captured by Photoreceptors and the Signal Is Processed by Layers of Overlying Neurons

The vertebrate eye provides the sharpest vision in the animal kingdom, even though the complex, multilayered retina is "built backward" from an engineering perspective (see Fig. 37-10b). There may be some unknown advantage to the structure, but more likely, it is a good example of how a complex structure that evolves as a result of the accumulation of chance mutations may not have the same features as a structure that had been designed "from scratch" for its purpose. The photoreceptors, called **rods** and **cones** after their shapes, have their light-gathering parts farthest away from the light, at the rear of the retina. Between the receptors and incoming light are several layers of neurons that process the signals from the photoreceptors. The outermost layer consists of **ganglion cells**, whose axons make up the **optic nerve.** The receptor potential from the photoreceptors is processed in complex ways by the retinal neurons. This processing results in an enhancement of our ability to detect edges, movement, dim light, and changes in light intensity.

The much modified signal from the photoreceptors is finally converted to action potentials carried by the ganglion cell axons in the optic nerve to the brain. Here, further processing ultimately results in the sensation of vision. Ganglion cell axons emerge from the outermost layer of the retina, then must pass back through the retina to reach the brain. The point where axons pass through the retina is called the **optic disc**, or **blind spot** (Fig. 37-12; see also Fig. 37-10a). This area lacks receptors, and objects focused here seem to disappear. You can locate your blind spot by closing your left eye and focusing on the star below with your right eye. Start with the book about a foot away and gradually move it closer. The spot will disappear when the image falls on the optic disc.

Interestingly, the camera eye of mollusks such as the octopus, which evolved independently from ours, is more efficient. Its photoreceptors are located in the outermost layer of the retina, so incoming light strikes the photoreceptors immediately. This design eliminates the blind spot and increases the availability of light to the receptors.

The Photoreceptor Cells Are Called Rods and Cones

Photoreception in both rods and cones begins with absorption of light by photopigment molecules embedded in the membranes of the photoreceptors. These membranes form flattened, hollow discs in rods and are deeply

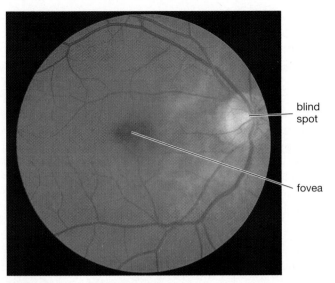

Figure 37-12   The human retina

A photograph of a portion of the human retina, taken through the cornea and lens of a living person. The blind spot (optic disc) and fovea are visible. Blood vessels supply oxygen and nutrients; notice that they are dense in the blind spot (where they won't interfere with vision) and there are fewer near the fovea.

folded in cones, giving them a large surface area on which to bear the photopigments (see Fig. 37-10b).

Light hitting the photopigment molecule causes the molecule to change shape. The altered shape of the molecule initiates a series of biochemical reactions inside the photoreceptor. These result in a change in the permeability of the receptor membrane to ions, producing a receptor potential in the photoreceptor cell. The receptor potential is then transmitted as described above.

Rods are far more numerous than cones (125 million rods versus 5 million cones) and dominate in the peripheral portions of the retina. Compared with cones, rods have much deeper stacks of pigment-bearing membrane (see Fig. 37-10b) and are about 100 times more sensitive to light. Studies have shown that a rod is capable of responding to a single photon, the smallest unit of light. Thus, rods are almost entirely responsible for our vision in dim light. Unlike cones, rods do not distinguish colors, so in moonlight, which is too dim to activate the cones, the world appears in shades of gray.

Although cones are found throughout the retina, they are concentrated in the fovea, where the lens focuses images most sharply (see Fig. 37-10a). The fovea, which consists entirely of densely packed cones, appears as a depression near the center of the retina because the layers of neurons normally covering the cones are pushed aside here while still retaining their synaptic connections. Because of this arrangement, light reaches the cones of the fovea with relatively little interference.

Unlike rods, cones respond to different wavelengths of light (colors). Human cones come in three varieties, each containing a slightly different photopigment. Each type of photopigment is most strongly stimulated by a particular wavelength of light, corresponding roughly to red, green, or blue. The brain distinguishes color according to the relative intensity of stimulation of different cones. For example, the sensation of yellow is caused by fairly equal stimulation of red and green cones. About 8% of all males lack normal color vision. Although described as "color- blind," they are actually only color deficient. The most common abnormality is red-green color deficiency due to a recessive allele on the X chromosome (see Chapter 15) that codes for a defective photopigment in the red cones. The altered red photopigment has about the same light-absorbing properties as the green photopigment, so the affected individual has trouble distinguishing red from green.

Not all animals have both rods and cones. Animals active almost entirely during the day (certain lizards, for example) may have all-cone retinas, whereas night-active animals (such as the ferret) or those dwelling in dimly lit habitats (such as deep-sea fishes) often have mostly rods.

Binocular Vision Allows Depth Perception

Most animals are bilaterally symmetrical and possess a pair of eyes. Having two eyes is useful in several ways—for example, two eyes are essential to allow a flatworm to determine the direction of a light source (see Fig. 37-8). Among vertebrates, we find two basically different eye placements: Herbivores usually have one eye on each side of the head; predators have both eyes facing forward (Fig. 37-13). The forward-facing eyes of predators and omnivores such as humans have slightly different but extensively overlapping visual fields. This **binocular vision** allows depth perception and accurate judgment of the size and distance of an object from the eyes. These abilities are important to a cat about to pounce on a mouse or to a monkey leaping from branch to branch.

In contrast, the widely spaced eyes of herbivores have little overlap in their visual fields; accurate depth perception is sacrificed in favor of a nearly 360 degree field of view. This allows these animals, who are frequently preyed upon, to spot a predator approaching from any direction.

Smell and Taste: Chemoreception

Through chemical senses, animals find food, avoid poisonous materials, and may locate homes or find mates (Fig. 37-14). Virtually all animals have chemoreceptors that sample the chemical composition of the environment. Ter-

(a)

(b)

Figure 37-13   **Eye position reflects life-style**

There are two usual placements for the eyes—one on each side of the head or both in front. The location of the eyes is related to life-style. **(a)** Herbivorous prey animals, such as rabbits, mice, horses, and deer, tend to have eyes placed at the sides, the better to scan all around for possible predators. **(b)** Predators, such as this owl, and primates tend to have eyes in front, and both can be brought to bear on a target. Each eye gets a slightly different view of the target, so that size and distance can be judged fairly accurately.

(a)

(b)

Figure 37-14   **Pheromones**

(a) Male moths find females not by sight, but by following air-borne scents (pheromones) released by the females. These odors are sensed by receptors on the male's huge antennae, whose enormous surface area maximizes the chances of detecting the female scent. **(b)** When dogs meet, they often sniff each other near the base of the tail. Scent glands there broadcast information about sex (both type and interest in) and status.

restrial vertebrates have two separate chemical senses: one for airborne molecules, called smell, or **olfaction**, and one for chemicals dissolved in water or saliva, the sense of **taste**.

The Ability to Smell Arises from Olfactory Receptors

In humans and most other vertebrates, receptors for smell, or olfaction, are nerve cells located in a patch of mucus-covered epithelial tissue in the upper portion of each nasal cavity (Fig. 37-15). The human olfactory epithelium is small compared with that of many other mammals (dogs are an obvious example) whose sense of smell is far more acute than ours. Olfactory neurons have hairlike dendrites that protrude into the cavity and lie embedded in a layer of mucus. Odorous molecules in the air diffuse into the mucus layer and bind with receptors on the dendrites.

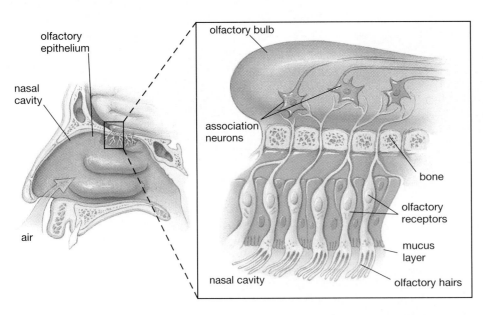

Figure 37-15   **Olfactory receptors**

The receptors for olfaction in humans are neurons that bear hairlike projections protruding into the nasal cavity. The projections are embedded in a mucus layer, in which odor molecules dissolve before contacting the receptors.

(a) The human tongue

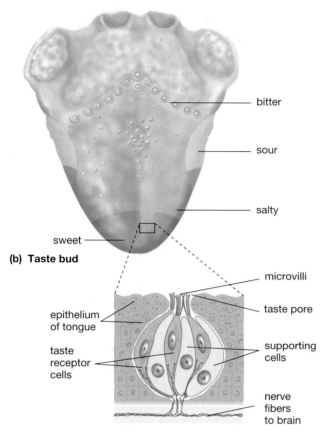

bitter

sour

salty

sweet

(b) Taste bud

microvilli

taste pore

epithelium
of tongue

taste
receptor
cells

supporting
cells

nerve
fibers
to brain

Figure 37-16   **Taste receptors**

(a) The human tongue bears numerous bumps in which are clustered masses of taste buds. **(b)** Each taste bud consists of supporting cells surrounding 60 to 80 taste receptor cells, whose microvilli bind the tasty molecules. Receptor cells synapse on nerve fibers that carry electrical signals to the brain.

Recent research suggests that there may be 1000 types of receptor proteins embedded in the olfactory dendrites. Each receptor protein is specialized to bind a particular type of molecule and stimulate the olfactory neuron to send a message to the brain. Researchers hypothesize that each neuron may bear only one type of receptor protein, and the brain may interpret the electrical signals coming from different neurons as distinct odors. Some chemicals may bind to more than one receptor type, and the brain may perceive *more* than 1000 different odors through its ability to interpret signals from particular *combinations* of olfactory neurons as distinct odors.

Taste Receptors Are Found in Clusters on the Tongue

The human tongue bears about 10,000 **taste buds**, embedded in small bumps covering the surface of the tongue.

Each taste bud consists of a cluster of about 60 to 80 taste receptors surrounded by supporting cells in a small pit that communicates with the mouth through a pore (Fig. 37-16). Microvilli (thin membrane projections) of taste receptor cells protrude through the pore. Dissolved chemicals enter the pore and bind to receptor molecules on the microvilli. There are only four types of taste receptors: sweet, sour, salty, and bitter. The great variety of apparent tastes we perceive is the result of two mechanisms. First, a particular substance may stimulate two or more receptor types to different degrees, making it taste "salty-sweet," for example. Second and more important, material being tasted usually also gives off molecules into the air inside the mouth. These odor molecules diffuse to the olfactory receptors, which contribute an odor component to the basic flavor (remember, the mouth and nasal passages are connected).

If you don't believe that what we call taste is really mostly smell, try eating different jelly beans while holding your nose (and closing your eyes). The tastes of grape, lime, and cherry jelly beans will be indistinguishable, like sweet sticky pastes. Likewise, when you have a bad cold, notice how normally tasty foods seem bland and lose much of their appeal.

Sensing Pain

Pain is an unusual sense in that it is somewhat nonspecific. Whether you burn, cut, or crush a fingertip, you will feel the same type of sensation: pain. This fact gives us a clue to the nature of pain perception: Most pain is produced by tissue damage, regardless of the cause. Over the past few years, researchers have found that pain perception is actually a special kind of chemical sense (Fig. 37-17).

When cells are broken open by a cut or a burn, for example, their contents flow into the extracellular fluid and blood. The cell contents include enzymes that convert certain blood proteins into a chemical called **bradykinin.** Pain receptors, which are dendrites of specific sensory neurons, have receptor molecules for bradykinin. Binding of bradykinin to these receptors results in action potentials that are interpreted as pain by the brain. Because each part of the body has a separate set of pain neurons that provide input to particular brain cells, the brain knows where the pain is occurring.

Drugs that provide pain relief, such as morphine or Demerol, block synapses in the pain pathways of the brain or spinal cord. In ways that we are just beginning to understand, the brain can modulate its perception of pain through its own narcotic-like **endorphins** (see Chapter 36). In critical situations, such as combat or during escape from a fire, endorphins may allow us to function by blocking our perception of pain until the emergency is over.

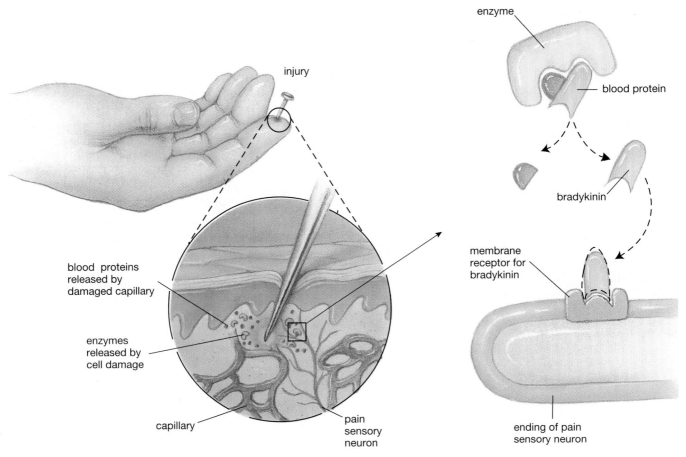

Figure 37-17   **Pain perception**

Pain perception is a specialized chemical sense. An injury, such as stabbing the finger with a needle, damages both cells and blood vessels. The cells release enzymes that convert certain blood proteins into bradykinin. Bradykinin stimulates pain-sensitive neurons by binding with membrane receptor molecules.

E V O L U T I O N A R Y
C O N N E C T I O N S

Natural Selection Shapes Perception

This chapter has reviewed the "common" senses of touch, sound, taste, smell, sight, and pain. But had this text been written by star-nosed moles, it undoubtedly would have included a large section on electroreception and nearly omitted the coverage of vision! Here we review a few of the "uncommon" senses that have evolved in response to different environments.

Echolocation

Some animals who hunt in darkness or murky water have evolved a type of sonar, similar to the navigational system used by ships. These animals transmit high-frequency sound waves (higher than can be detected by the human ear) then detect the vibrations reflected from objects to produce an "auditory image" of their nearby surroundings. This ability, called **echolocation,** is highly developed in bats and porpoises.

Bats can navigate and hunt insect prey in total darkness using echolocation. An echolocating bat emits pulses of noise at ultrasonic frequencies that bounce back from nearby objects. The intensity of this sound would make it unpleasantly loud if we could detect it. The patterns of returning sound convey accurate information about the size, shape, surface texture, and location of objects in the environment. Little brown bats can detect wires only 1 millimeter thick from a distance of 2 meters (over 6 feet). Several adaptations contribute to this remarkable sensitivity. The enormous, elaborately folded outer ears of the bat collect the returning echoes and help the bat locate their source (Fig. 37-18a). As the bat emits its cry, muscles attached to the bones of the middle ear contract briefly, reducing the bones' vibrations and preventing the bat from being deafened by its own calls. The tympanic membrane

(a)

(b)

Figure 37-18 **Echolocation**

Echolocation by bats and porpoises is based on very different anatomical structures. **(a)** A long-eared bat, showing the enormous size and elaborate folds of the external ears that help it localize returning echoes. **(b)** The bottlenose porpoise focuses ultrasonic clicks using the oil-filled bulge in the front of its head.

and bones of the middle ear are exceptionally light and easily vibrated by the faint returning echoes.

Porpoises produce ultrasonic clicks within their nasal passages and emit them through the front of their head (Fig. 37-18b). There, a large, flexible, oil-filled sac acts like an acoustic lens, directing the sound forward in a broad beam (for navigation) or a narrow beam (to locate prey). An echolocating porpoise can locate a pea-sized object on the floor of its tank and distinguish different species of fish. Porpoises may also use the narrowly focused beam to stun fish with a blast of sound, making them easier to capture.

Detecting Electrical Fields

Some fishes, called weak electric fishes, use electrical fields for electrolocation in much the same way that bats and porpoises use sound waves for echolocation. These fishes produce high-frequency electrical signals from an electric organ just in front of their tails, and they detect these signals using electroreceptor cells located inside the pores of their lateral lines (Fig. 37-19). Objects near the fish dis-

tort the electrical field surrounding the fish, and this distortion is detected by the electroreceptors, which send an altered pattern of action potentials to the brain. The fish is able to use this information to detect and localize nearby objects.

Recently, electroreceptors have been discovered in two mammals, the platypus and the star-nosed mole. Rather than sensing distortions in their own electrical fields, these mammals use their electroreceptors to detect electrical fields produced by other animals, such as potential prey. Both hunt under conditions where their eyes are nearly useless. The platypus, whose bill is covered with electroreceptors (as well as sensitive mechanoreceptors) hunts for crayfish and tadpoles at night in murky freshwater ponds and streams. The star-nosed mole (see the chapter-opening photograph) digs tunnels into streams, where it hunts worms in muddy water. The 22 writhing tentacles that surround the mole's nose had long puzzled zoologists. Researchers have now demonstrated that the moles can detect, and are attracted to, weak electrical fields such as those produced by earthworms, presumably by using their electroreceptor-studded noses.

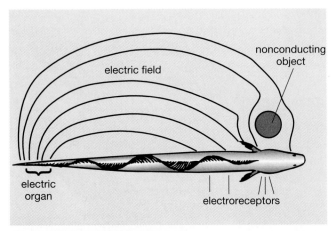

Figure 37-19   **Electroreception**

Weak electric fishes locate nearby objects by sensing distortions in their own electrical field that are produced by the objects. The fishes' fields are generated by electric organs near the tail and detected by electroreceptors in the pores of the lateral line system.

Detecting Magnetic Fields

Homing pigeons are famous for their ability to fly home after being released some distance away. They can accurately locate their home roost even under cloudy skies in terrain with few landmarks. A pigeon with a small magnet strapped to its back will still home successfully in sunny weather but will lose its way under overcast skies. Apparently, pigeons can navigate either by the sun or, if the sun is hidden, by magnetic fields. In cloudy weather, the magnets confuse the pigeon's magnetic compass. How do the pigeons detect magnetic fields? Pigeons have deposits of magnetite (a magnetic iron compound) located just beneath their skulls. These deposits may act as a built-in magnet that the pigeons use to tell direction.

Eels of eastern North America and western Europe swim out of streams and rivers into the Atlantic Ocean and migrate to the Sargasso Sea (near the West Indies) to spawn. Eels probably also use magnetic fields for navigation. You may recall from high school physics that moving an electrical conductor through a magnetic field induces an electrical current in the conductor (this is how we generate electricity commercially). Seawater, with its high salt concentration, is a fairly good conductor. The currents of the Gulf Stream, which flow from the Gulf of Mexico along the east coast of the United States, provide movement through Earth's magnetic field. The Gulf Stream generates an extremely weak electrical field, roughly equivalent to the field produced by a 1-volt battery with its poles more than 7 kilometers (12 miles) apart. At first, this field seemed far too weak to be detected. But researchers discovered that eels can detect electrical fields as weak as a 1-volt battery with poles *over 5000 kilometers* (3000 miles) apart. For an eel, finding the Gulf Stream must be a piece of cake!

✕ S U M M A R Y   O F   K E Y   C O N C E P T S

Receptor Mechanisms

Receptors are transducers, converting signals from one form to another. In response to external stimuli, sensory receptor cells produce receptor potentials that in turn produce action potentials, either in the sensory neuron itself or in a postsynaptic neuron. Receptor cells are named after the stimulus to which they respond. Thermoreceptors respond to heat. Mechanoreceptors respond to mechanical deformation of the cell membrane and produce sensations of touch, stretch, hearing, or gravity. Photoreceptors, chemoreceptors, and pain receptors respond to stimuli—light, taste and odor, and tissue damage, respectively—that influence receptor molecules embedded in the receptor cell membrane.

Sensing Temperature: Thermoreception

Thermoreceptors are free nerve endings located in the skin that respond to infrared radiation. Warm and cold receptors respond to increases and decreases in temperature, respectively. Pit vipers have thermoreceptors concentrated in pit organs near the eyes that allow detection of warm-blooded prey.

Sensing Touch and Movement: Mechanoreception

The skin contains dendrites of many different types of sensory neurons, which respond to mechanical stimuli such as pressure, touch, or vibration. Many internal organs, including the stomach, rectum, and bladder, have stretch receptors that signal fullness. Joints and muscles have similar receptors, collectively called proprioceptors, sensitive to stretch, that inform the brain about the position of the body. Hair cells in lateral line organs of fish, and in the mammalian inner ear are also mechanoreceptors.

Sound Perception: Hearing

Hearing is a modified mechanical sense specialized for reception of vibrations of the air or water. In the human ear, air vibrates the tympanic membrane, which transmits vibrations to the bones of the middle ear and then to the oval window of the fluid-filled cochlea. Within the cochlea, vibrations bend the hairs of hair cells, which are receptors located between the basilar and tectorial membranes. This bending produces receptor potentials in the hair cells that cause action potentials in the axons of the auditory nerve to the brain.

Gravity and Movement Perception: The Vestibular Apparatus

The inner ear also contains the vestibular apparatus, specialized to detect movement and gravity. Hair cells located in clusters at the base of each of three semicircular canals detect changes in movement in any direction. Below the semicircular canals are chambers, the utricle and saccule, containing hair cells that can detect the pull of gravity.

Sensing Light: Photoreception

All eyes have photoreceptor cells that contain a pigment that, upon absorption of light, causes a receptor potential. In the vertebrate eye, light enters the cornea and passes through the pupil to the lens, which focuses an image on the fovea of the retina. Two types of photoreceptor, rods and cones, are found deep in the retina. They produce receptor potentials in response to light. These signals are processed through several layers of neurons in the retina and finally are translated into action potentials in ganglion cells. Ganglion cell axons form the optic nerve, which carries action potentials to the brain. Rods are more abundant and more light-sensitive than cones and provide vision in dim light. Cones, which are concentrated in the fovea, provide color vision.

Smell and Taste: Chemoreception

Terrestrial vertebrates detect chemicals in the external environment either by olfaction (for airborne sources of chemicals) or by taste (chemicals dissolved in water or saliva). Each olfactory or taste receptor cell type responds to only one or a few specific types of molecules, allowing discrimination among tastes and odors. Olfactory neurons of vertebrates are found in a tissue lining the nasal cavity. Taste receptors are found in clusters called taste buds on the tongue.

Sensing Pain

Pain is a special type of chemical sense in which sensory neurons respond to chemicals released by damaged cells that produce the substance bradykinin. Binding of bradykinin to pain receptors results in action potentials that are interpreted as pain by the brain.

KEY TERMS

aqueous humor   p. 734
auditory canal   p. 730
auditory nerve   p. 732
basilar membrane   p. 732
binocular vision   p. 737
blind spot   p. 736
bradykinin   p. 739
camera eye   p. 734
chemoreceptor   p. 728
choroid   p. 734
cochlea   p. 732
compound eye   p. 733
cone   p. 736
cornea   p. 734
cupula   p. 730
echolocation   p. 740
endorphin   p. 739
eustachian tube   p. 731
external ear   p. 730
eyespot   p. 733

fovea   p. 735
free nerve ending   p. 729
ganglion cell   p. 736
hair cell   p. 728
inner ear   p. 731
iris   p. 734
lateral line organ   p. 730
lens   p. 734
mechanoreceptor   p. 728
middle ear   p. 731
olfaction   p. 738
ommatidium   p. 733
optic disc   p. 736
optic nerve   p. 736
otolith membrane   p. 733
outer ear   p. 730
oval window   p. 731
Pacinian corpuscle   p. 729
photopigment   p. 733
photoreceptor   p. 728

pit organ   p. 729
proprioceptor   p. 729
pupil   p. 734
receptor   p. 728
receptor potential   p. 728
retina   p. 734
rod   p. 736
saccule   p. 733
sclera   p. 735
semicircular canal   p. 732
sensory receptor   p. 728
taste   p. 738
taste bud   p. 739
tectorial membrane   p. 732
thermoreceptor   p. 728
transducer   p. 728
tympanic membrane   p. 731
utricle   p. 733
vestibular apparatus   p. 732
vitreous humor   p. 734

✖ T H I N K I N G   T H R O U G H   T H E   C O N C E P T S

Multiple Choice

1. Color vision is a function of _____ cells.
   a. cochlear   b. cupular
   c. corneal   d. choroid
   e. cone

2. *Basilar* and *tectorial* refer to types of
   a. membranes   b. eyes
   c. dendrites   d. proprioceptors
   e. canals

3. To go from the pharynx to the middle ear, a germ must travel through the
   a. oval window   b. vestibular apparatus
   c. eustachian tube   d. cochlea
   e. auditory canal

4. The fovea is
   a. the "blind spot"
   b. a clear area in front of the pupil and iris
   c. the area of the retina where an optic nerve exits the eye
   d. a point of attachment between the retina and choroid coat
   e. a central focal region of the vertebrate retina

5. Light entering the eye and striking the choroid would travel, in order, through the
   a. lens, vitreous humor, cornea, aqueous humor, and retina
   b. retina, aqueous humor, lens, vitreous humor, and cornea
   c. cornea, aqueous humor, retina, vitreous humor, and lens
   d. cornea, aqueous humor, lens, vitreous humor, and retina
   e. lens, aqueous humor, cornea, vitreous humor, and retina

6. Pain perception is carried out by a special type of
   a. mechanoreceptor   b. proprioceptor
   c. chemoreceptor   d. magnetoreceptor
   e. thermoreceptor

Review Questions

1. What are the names of the specific transducers used for taste, vision, hearing, smell, and touch?

2. Describe the anatomy of two types of receptor cells responsive to mechanical deformation; one found in the skin, the other in the inner ear.

3. Why are we apparently able to distinguish hundreds of different flavors, when we have only four types of taste receptors? How are we able to distinguish so many different odors?

4. What are proprioceptors? For what types of behavior are they important?

5. Describe the structure and function of the various parts of the human ear. Do this by tracing a sound wave from the air outside the ear to the cells causing action potentials in the auditory nerve.

6. How does the structure of the inner ear allow for the perception of pitch? Of sound intensity?

7. Diagram the overall structure of the human eye. Label the cornea, iris, lens, sclera, retina, and choroid. Describe the function of each structure.

8. How does the lens change shape to allow focusing of faraway objects? What defect makes focusing on faraway objects impossible, and what is this condition called? What type of lens can be used to correct it, and how does it do it?

9. List the similarities and differences between rods and cones.

10. Distinguish between taste and olfaction.

11. Describe how pain is signaled by tissue damage.

✖ A P P L Y I N G   T H E   C O N C E P T S

1. Describe an animal sense that humans lack altogether, and explain why it is needed by the animal but not by humans.

2. Explain the statement "Your sensory perceptions are purely a creation of your brain." Discuss the implications for communicating with other human beings, with other animals, and with intelligent life from another universe.

3. Bones in the middle ear help amplify sounds. Muscles attached to these bones help dampen sounds by a reflex arc through two different cranial nerves leading to and from the brain. Where do you think the receptor neurons are located to sense loud sounds that need to be dampened? Why is your hearing more likely to be damaged from a sudden loud sound than from a sound that gradually increases in volume?

4. Discuss the biological, ethical, and social implications of regulating noise as a form of pollution in the following places: rock concerts, aerobics classes, a neighborhood that was built near an existing airport, factories.

5. Why does a piece of hair in your mouth seem larger than it does when it is lying on your arm?

6. Corneal transplants can help restore vision and greatly improve the quality of life for the recipient. What are some properties of the cornea that make it an excellent candidate for transplantation? Suggest some ways in which society could improve the availability of corneal and other tissue for transplantation.

FOR MORE INFORMATION

Fenton, M. B., and Fullard, J. H. "Moth Hearing and the Feeding Strategies of Bats." *American Scientist,* May/June 1981. Explores evolutionary interactions between bats and moths.

Freedman, D. H. "In the Realm of the Chemical." *Discover,* June 1993.

Gutin, J. C. "Good Vibrations." *Discover,* June 1993.

Koretz, J. F., and Handelman, G. H. "How the Human Eye Focuses." *Scientific American,* July 1988. Describes focusing in the human eye with an emphasis on the loss of focusing ability that occurs with age.

McKean, K. "Pain." *Discover,* October 1986. Describes the discovery of bradykinin and its role in pain perception.

Montgomery, G. "Color Perception: Seeing with the Brain." *Discover,* December 1988. Describes, in layperson's terms, how the higher centers of the brain interpret signals from the receptors to arrive at a perception of color. Beautifully illustrated.

Nathans, J. "The Genes for Color Vision." *Scientific American,* February 1989. Good discussion of the physiology and genetics of color blindness.

Parker, D. E. "The Vestibular Apparatus." *Scientific American,* November 1980. Perception of gravity and motion requires not only the utricle, saccule, and semicircular canals but also input from other senses.

Zimmer, C. "The Electric Mole." *Discover,* August 1993. Describes the discovery of the electrolocation ability of the star-nosed mole.

NET WATCH

On-line resources for this chapter are on the World Wide Web at:
http://www.prenhall.com/~audesirk (click on the <u>table of contents</u> link and then select Chapter 37).

38 Action and Support: The Muscles and Skeleton

❖ AT A GLANCE

Muscle
The Intricate Structure of Skeletal Muscle Supports Its Active Function
Muscle Contraction Results from Thick and Thin Filaments Sliding Past One Another
Cardiac Muscle Powers the Heart
Smooth Muscle Produces Slow, Involuntary Contractions

The Skeleton
The Vertebrate Skeleton Serves Many Functions

Skeletal Tissues: Cartilage and Bone
Cartilage Provides Flexible Support and Connections
Bone Provides a Strong, Rigid Framework for the Body

Body Movement: Muscle-Skeleton Interactions
Muscles Move the Skeleton around Flexible Joints

The muscular and skeletal systems work in close harmony to provide the balance, strength, and flexibility shown by these dancers.

Muscles rippling, seemingly weightless bodies perform perfectly choreographed movements, a testimony to the functional design and precise interactions of the muscular and skeletal systems—and to years of hard work! This chapter begins by describing muscle tissue and the mechanisms underlying muscle contraction, then describes the tissues and structure of the skeletal system, ending with some of the muscle-skeleton interactions that give rise to movements, both ordinary and amazing.

Muscle

In single cells, movement is often the result of the sliding of two types of protein strands, **actin** and **myosin**, past one another (see Chapter 5). The activity of these strands, which make up microfilaments, contributes to shape changes and organelle movement in most eukaryotic cells. In animals, the evolution of muscles has been a remarkable elaboration of this preexisting system.

Any sort of rhythmic activity requires alternating muscle contractions and extensions, yet muscles are usually described only as contracting—for good reason. The only active phase of muscle movement is contraction. Muscles lengthen passively, by relaxing and allowing themselves to be stretched out. Muscles are extended by pushes and pulls from other sources; for example, contractions of opposing muscles, the weight of a limb, or pressure from blood or food that distends the muscular heart and stomach walls.

Animals have a variety of muscle types, each specialized to perform a particular function. Mammals have evolved three distinct types of muscle: skeletal, cardiac, and smooth. All work on the same basic principles but differ in function, appearance, and control (Table 38-1).

Skeletal muscle, so named because it is used to move the skeleton, is also called **striated muscle** because of its striped appearance under the microscope (*striated* means "striped"). Most skeletal muscle is under voluntary, or conscious, control. It can produce contractions ranging from quick twitches (as in blinking) to powerful, sustained tension (as in carrying an armload of textbooks). **Cardiac muscle** is so named because it is found only in the heart. It is spontaneously active, initiating its own contractions, but it is influenced by nerves and hormones. Like skeletal muscle, cardiac muscle has a striped appearance under the microscope. **Smooth muscle**, as its name suggests, lacks the orderly striped appearance of cardiac and skeletal muscle. Smooth muscle lines the walls of the digestive tract and large blood vessels and produces slow, sustained contractions. These contractions are primarily involuntary; that is, they are not under conscious control.

Our discussion of muscles begins with and emphasizes skeletal muscle, the most abundant form of muscle in the body, then provides an overview of cardiac and smooth muscle.

The Intricate Structure of Skeletal Muscle Supports Its Active Function

Individual muscle cells, called **muscle fibers,** are among the largest cells in the human body. Ranging from 10 to 100 micrometers in diameter (a bit smaller than the period at the end of this sentence), each muscle fiber runs the entire length of the muscle, which may be as much as 35 centimeters (about 14 inches) in a human thigh. Each muscle fiber, in turn, contains many individual contractile sub-

TABLE 38-1 ⬚ *Location, Characteristics, and Functions of the Three Muscle Types*

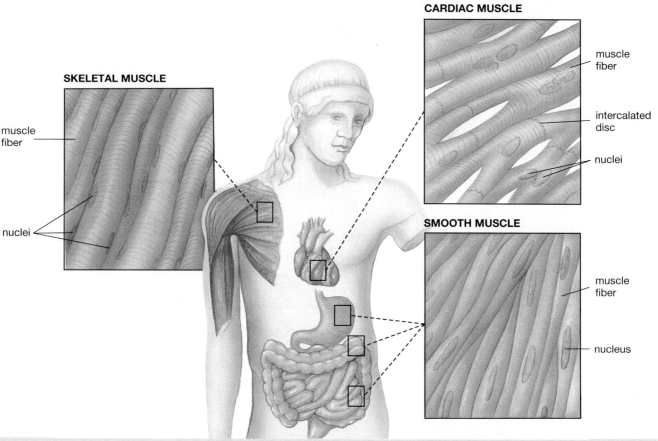

| | *Type of Muscle* | | |
Property	Smooth	Cardiac	Skeletal
Muscle appearance	Unstriped	Irregular stripes	Regular stripes
Cell shape	Spindle	Branched	Spindle or cylindrical
Number of nuclei	One per cell	Many per cell	Many per cell
Speed of contraction	Slow	Intermediate	Slow to rapid
Contraction caused by	Spontaneous, stretch, nervous system, hormones	Spontaneous	Nervous system
Function	Controls movement of substances through hollow organs	Pumps blood	Moves the skeleton
Voluntary control	Usually no[a]	Usually no[a]	Yes

[a]Smooth and cardiac muscle normally contract without conscious control. In some cases, however, their contractions may be initiated or modified voluntarily (for example, heart rate can be voluntarily slowed after biofeedback training, and bladder contractions are initiated consciously).

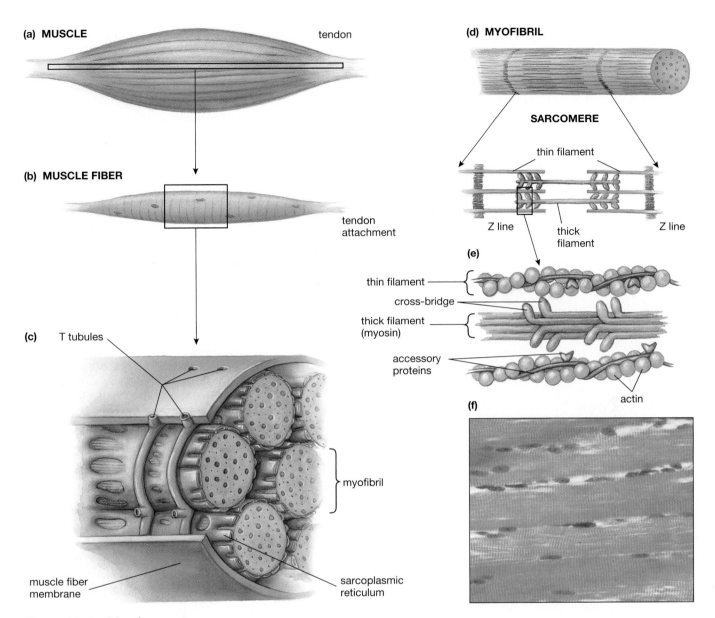

(a) MUSCLE

tendon

(b) MUSCLE FIBER

tendon attachment

(c)

T tubules

muscle fiber membrane

myofibril

sarcoplasmic reticulum

(d) MYOFIBRIL

SARCOMERE

thin filament

Z line

thick filament

Z line

(e)

thin filament

cross-bridge

thick filament (myosin)

accessory proteins

actin

(f)

Figure 38-1 **Muscle structure**

A muscle **(a)** is made up of individual muscle cells, called fibers **(b).** The sarcoplasmic reticulum subdivides each muscle fiber into smaller cylinders called myofibrils **(c).** The myofibril consists of a series of subunits called sarcomeres, attached end to end **(d).** Within each sarcomere are alternating thick and thin filaments **(e)** that can be connected by cross-bridges (projections of the myosin molecules that make up the thick filaments). In vertebrate skeletal muscles, the actin and myosin filaments are regularly arranged in the muscle fiber, giving it a striped appearance **(f).**

units, the **myofibrils,** extending from one end of the fiber to the other (Fig. 38-1a, b). Each cylindrical myofibril is surrounded by **sarcoplasmic reticulum.** Like the endoplasmic reticulum from which it is derived, the sarcoplasmic reticulum is a series of double sheets of membrane forming interconnected hollow tubes (Fig. 38-1c). The fluid within the sarcoplasmic reticulum stores high concentrations of calcium ions. Deep indentations of the muscle cell membrane, called **transverse tubules** or **T tubules,** extend down into the muscle fiber, passing very close to portions of the sarcoplasmic reticulum. This arrangement

of T tubules and sarcoplasmic reticulum is crucial to the control of muscle contraction, as described later.

Within each myofibril is a beautifully precise arrangement of filaments of actin and myosin, organized into subunits called **sarcomeres** (Fig. 38-1d). Sarcomeres are attached end to end throughout the length of the myofibril; their junction points are called **Z lines.** Attached to the Z lines are strands composed of actin plus two accessory proteins. These three proteins form the **thin filaments.** Suspended between the thin filaments are **thick filaments** composed of myosin protein. The thick and thin filaments are lined up in all the myofibrils, giving the cell its striped appearance (Fig. 38-1f). The strands of myosin extend small arms, called **cross-bridges,** contacting the thin filaments.

The complex structure of the thin filament is crucial to the regulation of muscle contraction. The actin protein is formed from a double chain of subunits resembling a twisted double strand of pearls. Each subunit has a binding site for a myosin cross-bridge. In a relaxed muscle, however, these sites are covered by thin strands formed by one type of accessory protein, held in place by a second, globular accessory protein (Fig. 38-1e). These accessory proteins prevent the myosin cross-bridges from attaching.

Muscle Contraction Results from Thick and Thin Filaments Sliding Past One Another

When a muscle contracts, the accessory proteins of the thin filament are moved aside, exposing the binding sites on the actin. As soon as the sites are exposed, myosin cross-bridges attach. Using energy from splitting adenosine triphosphate (ATP), the cross-bridges repeatedly bend, release, and reattach farther along, much like a sailor pulling in an anchor line hand over hand (Fig. 38-2a). The thin filaments are pulled past the thick filaments, shortening the sarcomere and contracting the muscle (Fig. 38-2b). Be-

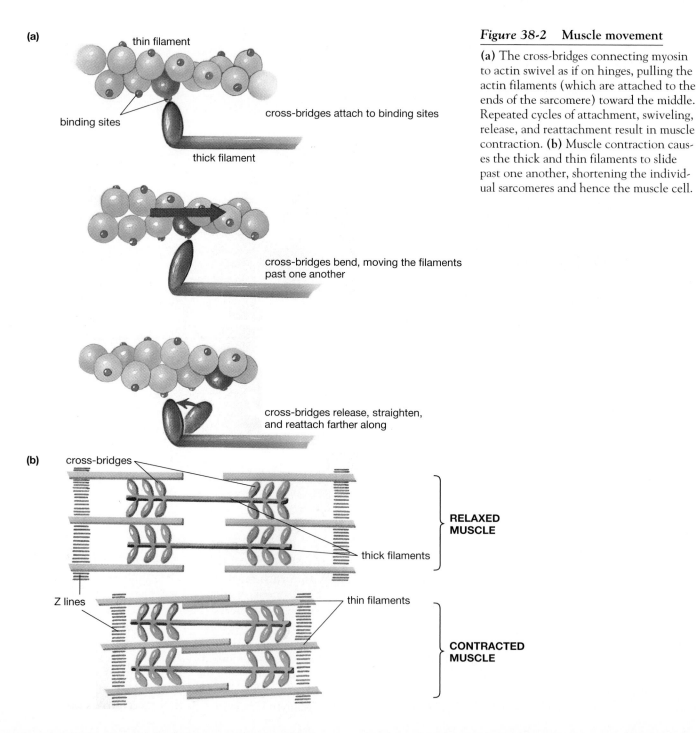

(a)
thin filament
binding sites
thick filament
cross-bridges attach to binding sites

cross-bridges bend, moving the filaments past one another

cross-bridges release, straighten, and reattach farther along

(b)
cross-bridges
thick filaments
Z lines
thin filaments
RELAXED MUSCLE
CONTRACTED MUSCLE

Figure 38-2   Muscle movement

(a) The cross-bridges connecting myosin to actin swivel as if on hinges, pulling the actin filaments (which are attached to the ends of the sarcomere) toward the middle. Repeated cycles of attachment, swiveling, release, and reattachment result in muscle contraction. **(b)** Muscle contraction causes the thick and thin filaments to slide past one another, shortening the individual sarcomeres and hence the muscle cell.

(a)

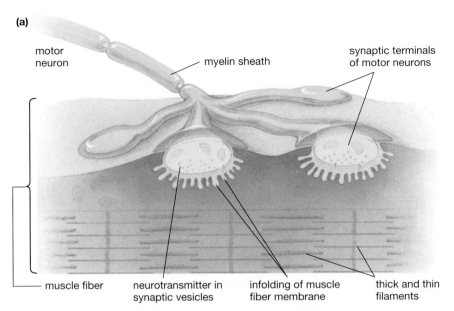

motor neuron

myelin sheath

synaptic terminals of motor neurons

muscle fiber

neurotransmitter in synaptic vesicles

infolding of muscle fiber membrane

thick and thin filaments

Figure 38-3   **The neuromuscular junction**

(a) Diagram of a neuromuscular junction in cross section. Action potentials in the motor neuron cause the synaptic terminals to release transmitter. Transmitter binds to receptors on the muscle cell membrane, which is folded beneath the terminal to allow more surface area for receptors. **(b)** A scanning electron micrograph shows motor neuron terminals synapsing on muscle fibers.

(b)

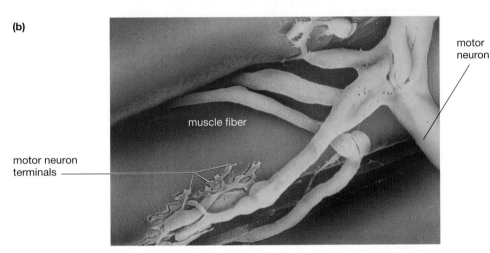

motor neuron

muscle fiber

motor neuron terminals

cause the thick and thin filaments slide past one another during contraction, muscle contraction is described as using a "sliding filament" mechanism.

How Is Muscle Contraction Controlled?

Muscle contracts when the binding sites on actin are exposed, and it relaxes when the binding sites are covered by thin filament accessory proteins. But what regulates the position of these accessory proteins? The answer is, the concentration of calcium ions around the filaments, which in turn is under the control of the nervous system.

In many respects, skeletal muscle cells are much like neurons. They have resting potentials, they are stimulated by synaptic contact with neurons, and they produce action potentials when stimulated above threshold. Skeletal muscles contract as a result of their own action potentials, which are caused by synaptic stimulation from motor neurons (neurons that control muscles). The cell bodies of the motor neurons (most of which reside in the spinal cord) send axons

out the spinal nerves to the muscles, where they form synapses on the muscle fibers (Fig. 38-3). These **neuromuscular junctions** work much like any other synapse: When an action potential reaches the synapse, the motor neuron releases a neurotransmitter (acetylcholine) that diffuses across the synaptic cleft to receptors on the muscle fiber membrane, producing an excitatory postsynaptic potential. Neuromuscular junctions differ from most other synapses in two important ways. First, they are always excitatory, never inhibitory. Second, every action potential in a motor neuron causes a large enough excitatory postsynaptic potential (EPSP) to evoke a muscle action potential; no "adding up" of EPSPs is required.

The muscle action potential travels into the interior of the muscle cell by passing down the T tubules (see Fig. 38-1c). The change in potential causes the sarcoplasmic reticulum to release calcium ions into the myofibrils. Calcium ions bind to the globular accessory proteins on the thin filament, altering their shape and causing them to pull the thin-stranded accessory proteins off the binding sites on the actin. This action allows the myosin cross-bridges to attach to the actin, initiating contraction. As soon as the action potential fades away, active transport proteins in the sarcoplasmic reticulum membrane pump the calcium back inside the reticulum. In the absence of calcium, the accessory proteins move back into place on the thin filaments, preventing further cross-bridge binding and ending the contraction.

If each action potential in a motor neuron elicits an action potential in a muscle fiber, causing all its sarcomeres to contract, how does the nervous system control the strength and degree of muscle contraction? The strength and degree of muscle contraction depend on the number of muscle fibers stimulated and on the frequency of action potentials in each fiber.

Most motor neurons innervate (make synapses onto) more than one muscle fiber. The group of fibers on which a single motor neuron synapses is called a **motor unit.** The number of muscle fibers in a motor unit varies from muscle to muscle. Muscles used for large-scale movement, such as those of the thigh or buttocks, may have hundreds of muscle fibers in each motor unit. In muscles used for fine control of small body parts, such as those of the lips, eyes, and tongue, only a few muscle cells may be stimulated by each motor neuron. Therefore, when a single motor neuron fires an action potential, it may cause contraction of a few muscle cells or of many, depending on the size of the motor unit.

The strength and extent of muscle contraction depends on the rate of action potentials fired in the motor neuron. A single action potential doesn't cause a muscle cell to contract fully, because calcium is pumped away too rapidly for many cross-bridge movement cycles to occur. Thus, many action potentials in rapid succession are required to fully contract a muscle. The strength of muscle contraction also depends on the frequency at which the motor neurons are firing: Higher-frequency firing produces stronger contractions, because the contractions produced by all the action potentials add to one another. If rapid firing is prolonged, the muscle produces a sustained maximal contraction called **tetany,** such as you experience when you carry an armful of books like this one.

Cardiac Muscle Powers the Heart

Cardiac muscle is found only in the heart. Cardiac muscle, like skeletal muscle, is striated owing to the regular arrangement of sarcomeres with their alternating thick and thin filaments. Action potentials spreading into the cell through the T tubules cause release of calcium from the sarcoplasmic reticulum, but, in contrast to skeletal muscle, calcium also enters from the extracellular fluid. Unlike skeletal muscle fibers, which contract in response to action potentials originating in motor neurons, cardiac muscle fibers can initiate their own contractions. This quality is particularly well developed in the specialized cardiac muscle fibers of the sinoatrial node, which serves as the heart's pacemaker (see Chapter 30). Action potentials originating in the pacemaker are spread rapidly throughout the heart by **intercalated discs.**

These discs are specialized areas in which membranes of adjacent muscle fibers are connected with numerous gap junctions. Intercalated discs allow electrical potentials to travel from one cell to the next, synchronizing their contractions (see Table 38-1). These connecting areas are absent in skeletal muscle, where each cell is individually stimulated by a branch of a motor neuron.

Smooth Muscle Produces Slow, Involuntary Contractions

Smooth muscle surrounds blood vessels and most hollow organs, including the uterus, bladder, and digestive tract. As its name suggests, smooth muscle lacks the regular arrangement of sarcomeres that characterizes skeletal and cardiac muscle (see Table 38-1). Smooth muscle generally produces either slow, sustained contractions (such as constriction of the arteries to elevate blood pressure during times of stress) or slow, wavelike contractions (such as the peristaltic waves that move food through the digestive tract). Like cardiac muscle cells, most smooth muscle cells are directly connected to one another by gap junctions, allowing synchronized contraction. Smooth muscle lacks sarcoplasmic reticulum; all the calcium needed for contraction flows in from the extracellular fluid during the action potential. Smooth muscle contraction may be initiated by stretch, by hormones, by nervous signals, or by some combination of these stimuli. Although contractions of the bladder can be initiated voluntarily, most smooth muscle contraction is under involuntary control.

The Skeleton

A **skeleton** can be broadly defined as a supporting framework. Within the animal kingdom, skeletons come in three radically different forms: hydrostatic skeletons (made of fluid), exoskeletons (on the outside of the animal), and endoskeletons (internal).

The **hydrostatic skeletons** of worms, mollusks, and cnidarians are the simplest, consisting of a fluid-filled sac (Fig. 38-4a). Fluid, which cannot be compressed, provides excellent support; but because it is formless, these animals rely on surrounding muscles in the body wall to determine their shape. The wavelike movements of a burrowing earthworm, alternately extending to stringlike thinness then fattening as it contracts, provide an excellent illustration of the flexibility of hydrostatic skeletons.

Exoskeletons (literally, "outside skeletons") encase the bodies of arthropods (such as spiders, crustaceans, and insects). Exoskeletons vary tremendously in thickness and rigidity, from the thin flexible covering of many insects and spiders to the armorlike covering of many crustaceans (Fig. 38-4b). All exoskeletons are thin and flexible at the joints, allowing complex and skillful movements such as those of a web-spinning spider. Exoskeletons and hydrostatic skeletons are discussed further in Chapter 24.

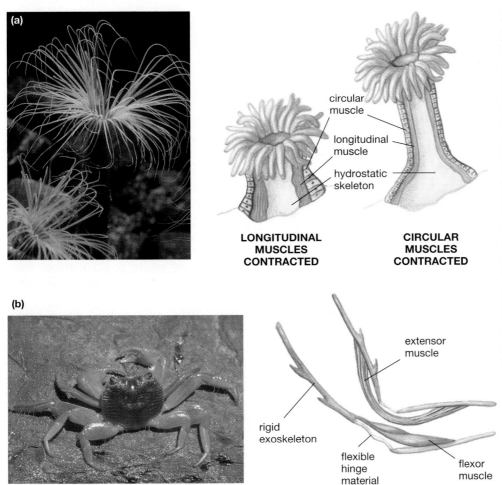

Figure 38-4 **Not all skeletons are made of bone**

(a) Hydrostatic skeletons. The skeleton of worms, many mollusks, and cnidarians such as this anemone is essentially a fluid-filled tube with soft walls. Two layers of muscles are arranged perpendicularly to one another: A layer of circular muscles forms a band around the circumference of the tube, and a layer of longitudinal muscles runs lengthwise. Because fluids are incompressible, if the circular muscles contract, the animal will become long and thin; if the longitudinal muscles contract, it will become short and fat. **(b)** Exoskeletons. Arthropods have armorlike skeletons on the outsides of their bodies. Joints allow movement, produced by pairs of antagonistic muscles spanning the joint.

circular muscle

longitudinal muscle

hydrostatic skeleton

LONGITUDINAL MUSCLES CONTRACTED

CIRCULAR MUSCLES CONTRACTED

extensor muscle

rigid exoskeleton

flexible hinge material

flexor muscle

Endoskeletons, the internal skeletons of humans and other vertebrates, are found only in echinoderms and chordates (see Chapter 24) and are actually the least common type of skeleton.

The Vertebrate Skeleton Serves Many Functions

The bony endoskeleton of humans and most vertebrates serves a wide variety of functions.

1. The skeleton provides a rigid framework that supports the body and protects the internal organs. The central nervous system, for example, is almost completely enclosed within the skull and vertebral column; the rib cage protects the lungs and the heart with its major blood vessels.
2. Bones produce red blood cells, white blood cells, and platelets (see Chapter 30). In adults, these cells of the circulatory system are produced by red bone marrow, found in porous areas of bone in the sternum (breastbone), ribs, upper arms and legs, and hips.
3. Bone serves as a storage site for calcium and phosphorus. Bone contains 99% of the calcium and 90% of the phosphorus in the human body. It absorbs and releases these minerals as needed, maintaining a constant concentration in the blood.

4. The skeleton even participates in sensory transduction. As you may recall from the previous chapter, three tiny bones of the middle ear (hammer, anvil, and stirrup) transmit the sound vibrations between the eardrum and the cochlea.

The 206 bones of the human skeleton can be placed in two categories: the axial skeleton and the appendicular skeleton. The **axial skeleton,** whose bones form the axis of the body, includes the bones of the head, vertebral column, and rib cage. The **appendicular skeleton,** whose bones form the appendages (extremities) and their attachments to the axial skeleton, includes the pectoral (shoulder) and pelvic (hip) girdles and the bones of the arms, legs, hands, and feet (Fig. 38-5).

Skeletal Tissues: Cartilage and Bone

The skeleton is composed primarily of two types of tissue: cartilage and bone. Both cartilage and bone are types of rigid connective tissue. Both consist of living cells embedded in a matrix of a protein called **collagen** (see Chapter 29).

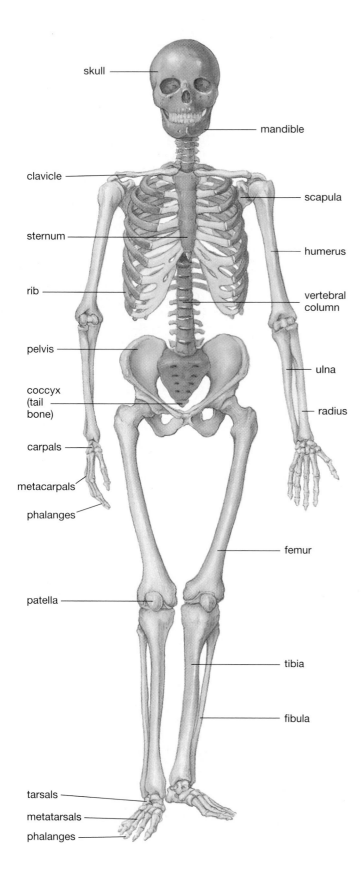

Figure 38-5   The human skeleton

The human skeleton, showing the axial skeleton (tinged in blue-gray) and the appendicular skeleton (bone color).

Cartilage Provides Flexible Support and Connections

Cartilage plays many roles in the human skeleton. It covers the ends of bones at joints, supports the flexible portion of the nose and external ears, connects the ribs to the sternum (breastbone), and provides the framework for the larynx, trachea, and bronchi of the respiratory system. It forms tough pads that act as shock absorbers. These pads are found in the knee joints and also form the **intervertebral discs** between the vertebrae of the backbone.

The living cells of cartilage are called **chondrocytes.** These cells secrete a flexible, elastic, nonliving matrix of collagen that surrounds the chondrocytes and forms the bulk of the cartilage (Fig. 38-6). No blood vessels penetrate into cartilage; chondrocytes must rely on gradual diffusion of materials through the collagen matrix to exchange wastes and nutrients. As you might predict, cartilage cells have a very slow metabolic rate, and damaged cartilage repairs itself very slowly, if at all.

During the development of the embryo, the skeleton is first formed from cartilage that is later replaced by bone (Fig. 38-7).

Bone Provides a Strong, Rigid Framework for the Body

Bone is the most rigid form of connective tissue. Although bone resembles cartilage, the collagen fibers of bone are hardened by deposits of calcium phosphate. Bones, such as those supporting your arms and legs, consist of an outer shell of **compact bone,** with **spongy bone** in the interior (Fig. 38-8). Compact bone is dense and strong and provides an attachment site for muscle. Spongy bone is lightweight, rich in blood vessels, and highly porous. Bone marrow, where blood cells are formed, is found in cavities of spongy bone. In contrast to cartilage, bone is well supplied with blood capillaries.

There are three types of cells associated with bone: **osteoblasts** (bone-forming cells), **osteocytes** (mature bone cells), and **osteoclasts** (bone-dissolving cells). Early in development, when bone is replacing cartilage,

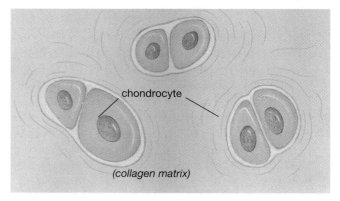

Figure 38-6   Cartilage

In cartilage, the cartilage cells, or chondrocytes, are embedded within an extracellular matrix of the protein collagen, which they secrete.

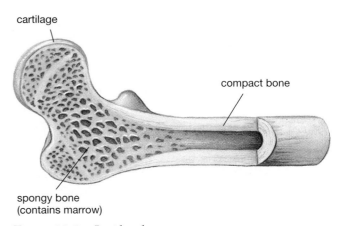

Figure 38-8   **Inside a bone**

A long bone such as is found in the arms and legs, showing the outer compact bone and inner spongy bone, where bone marrow is found.

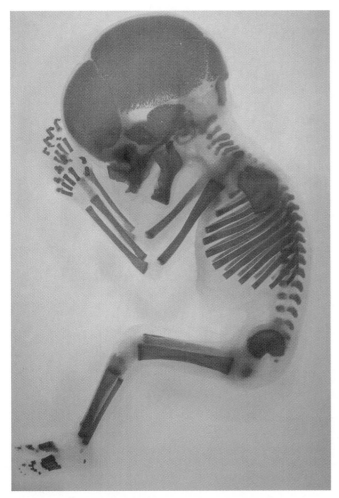

Figure 38-7   **Bone replaces cartilage during development**

Bone is stained magenta, and cartilage appears clear in this 16-week old human fetus. Clear areas of the wrists, knees, ankles, elbows, and breastbone show where cartilage will later be replaced by bone.

the osteoclasts invade and dissolve the cartilage; then osteoblasts replace it with bone.

As bones grow, osteoblasts form a thin layer covering the outside of the bone. The osteoblasts secrete a hardened matrix of bone and gradually become entrapped within it. They then stop secreting matrix and become osteocytes. Osteocytes are nourished by nearby capillaries and are connected to other osteocytes by thin extensions that these cells send out through narrow channels in the bone. Although unable to produce more bone, osteocytes may secrete substances that control the continuous remodeling of bone.

Bone Remodeling Allows Skeletal Repair and Adaptation to Stresses

Each year 5% to 10% of all the bone in your body is dissolved away and replaced, a process called bone remodeling. This allows your skeleton to subtly alter its shape in response to the demands placed upon it. For example,

bones that carry heavy loads or are subjected to extra stress become thicker to provide more strength and support. Archaeologists excavating skeletons from Pompeii were able to identify archers because the bones of their right and left arms differed considerably in thickness. Normal stresses are actually a major factor in maintaining bone strength. The bones of an arm or leg that are immobilized in a cast and protected from stresses rapidly lose significant amounts of calcium.

The process of bone remodeling replaces bone as it ages and becomes brittle. As the body gets older, however, the remodeling process slows, and bones tend to become more fragile as a result (see "Health Watch: Osteoporosis—The Hidden Crippler"). The continuous turnover of bone also allows the body to maintain constant levels of calcium in the blood; calcium from bones is retained in the blood if blood calcium drops, but it is returned to bone if blood calcium levels are adequate or high. This process is regulated by hormones; calcitonin and parathormone cause bones to absorb calcium from the blood and release calcium into the blood, respectively (see Chapter 35).

Bone remodeling is the result of the coordinated activity of two types of cells: osteoclasts that dissolve bone and osteoblasts that rebuild it. Osteoclasts cling to the bone surface, secreting acids and enzymes that dissolve the hard matrix. Working in small groups, osteoclasts tunnel into the bone, creating channels. These channels are invaded by capillaries and by osteoblasts. The osteoblasts fill the channel with concentric deposits of new bone matrix, leaving only a small opening for the capillary. As a result of this process, in cross section, hard bone is made up of tightly packed units, called **Haversian systems,** each consisting of concentric layers of bone with embedded osteocytes. The concentric deposits surround a central canal through which traverses a capillary (Fig. 38-9). Osteoclasts and osteocytes also play a crucial role in the repair of bone fractures, as described in "A Closer Look: Sticks and Stones—The Repair of Broken Bones."

A CLOSER LOOK

Sticks and Stones—The Repair of Broken Bones

With a thud accompanied by a sickening *crack*, a boy falls from high in a tree amid a crackling of small branches and a shower of leaves. The rapidly swelling, awkwardly bent forearm confirms his parents' fears: A bone is broken. Let's examine the sequence of events that occur during the next 6 weeks or so, as the bone heals.

First, blood from ruptured vessels forms a large clot surrounding the break. Phagocytic cells and osteoclasts in the blood ingest and dissolve the cellular debris and bone fragments (Fig. E38-1).

Second, bones are normally covered with a thin layer of connective tissue (the **periosteum**), rich in capillaries, osteoblasts, and osteoblast-forming cells. A fracture ruptures the periosteum and stimulates the production and release of numerous osteoblasts. These osteoblasts, in conjunction with cartilage-forming cells, secrete a porous mass of bone and cartilage called a **callus** surrounding the break. The callus replaces the original blood clot and holds the ends of the bones together while remodeling processes reform the original shape of the bone (Fig. E38-2).

Figure E38-1

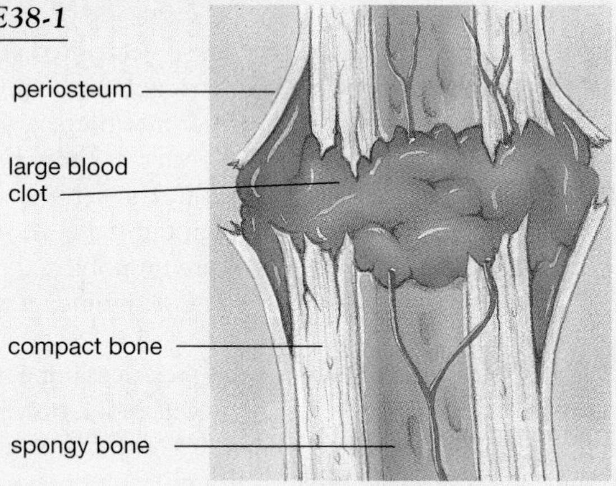

periosteum

large blood clot

compact bone

spongy bone

Figure E38-2

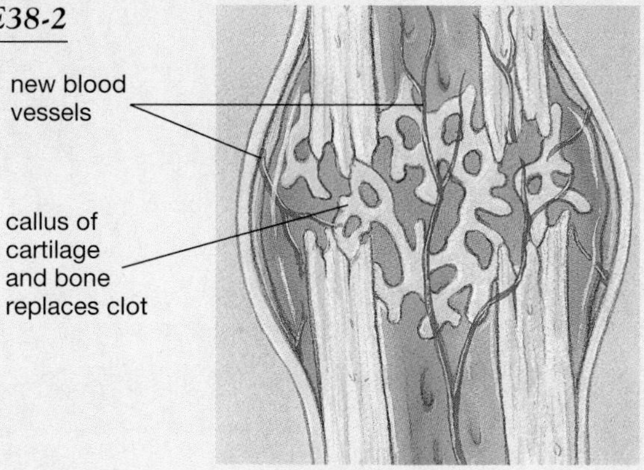

new blood vessels

callus of cartilage and bone replaces clot

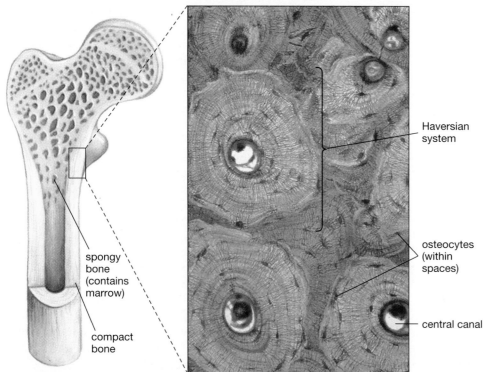

spongy bone (contains marrow)

compact bone

Haversian system

osteocytes (within spaces)

central canal

Figure 38-9   The fine structure of bone

Haversian systems are clearly visible in this micrograph. Each includes a central canal containing a capillary. The capillary nourishes the osteocytes, seen embedded in the concentric rings of bone material.

Third, osteoclasts, osteoblasts, and capillaries invade the callus. Nourished by the capillaries, osteoclasts break down the cartilage while osteoblasts replace it with bone (Fig. E38-3).

Finally, osteoclasts remove excess bone, restoring the original shape (Fig. E38-4).

Figure E38-3

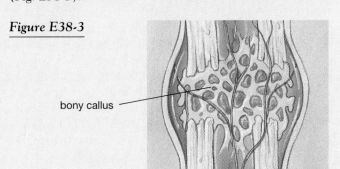

bony callus

Figure E38-4

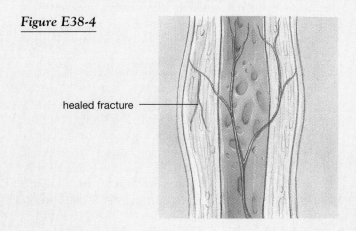

healed fracture

Body Movement: Muscle-Skeleton Interactions

In addition to providing support for the body, the skeleton facilitates movement by providing a framework that muscles can move. Movement of the skeleton is accomplished by the action of pairs of **antagonistic muscles**, one of which actively contracts, causing the other to be passively extended (Fig. 38-10; see also Fig. 38-4). Antagonistic muscles alter the configuration of the skeleton either by causing movement around joints (in exoskeletons and vertebrate endoskeletons), or by altering the shape of the internal fluid (in hydrostatic skeletons; see Fig. 38-4).

Muscles Move the Skeleton around Flexible Joints

The vertebrate skeleton allows movement by providing attachment points for skeletal muscles. Muscles move the skeleton around regions called **joints,** flexible attachment sites between adjacent bones. Skeletal muscles are attached to bones on either side of the joint by bands of tough, fibrous connective tissue called **tendons.** The portion of each bone that forms the joint is typically coated with a layer of cartilage, whose smooth, resilient surface allows the bone surfaces to slide past one another. Bones are joined to one another at joints by bands of fibrous connective tissue called **ligaments** (see Fig. 38-10).

Most skeletal muscles are arranged in antagonistic pairs, one called a **flexor** and the other an **extensor,** on opposite

sides of a joint (see Figs. 38-4b and 38-10). When one muscle contracts, it moves one bone with respect to the other and simultaneously stretches out the opposing mus-

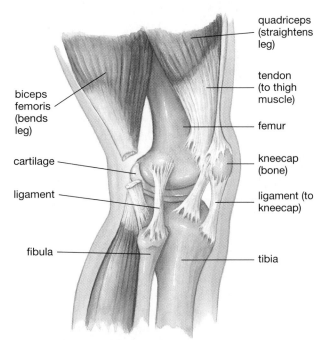

quadriceps (straightens leg)

tendon (to thigh muscle)

femur

kneecap (bone)

ligament (to kneecap)

tibia

biceps femoris (bends leg)

cartilage

ligament

fibula

Figure 38-10   A hinge joint

The human knee, a hinge joint, showing tendons, ligaments, and antagonistic muscles. The complexity of this joint, coupled with the extreme stresses placed on it during activities such as jumping, running, or skiing, make it very susceptible to injury.

HEALTH WATCH

Osteoporosis—The Hidden Crippler

As many as 20 million Americans, most of them women past menopause, are victims of a preventable but insidious form of bone loss called **osteoporosis** (literally, "porous bones"; Fig. E38-5a). Bone density increases steadily from birth through middle adulthood, reaching a peak at around age 35. From this point on, the activity of osteoclasts exceeds that of osteoblasts, and bone density begins a slow, natural decline.

Women are eight times as likely to suffer from osteoporosis as men. Why? One reason is that the bones of women are about 30% less massive than those of men to start with, so they can afford to lose less bone. Another factor is dietary calcium, which is lower in women's diets than in men's. Two-thirds of women between the ages of 18 and 30 get less than the RDA (Recommended Daily Allowance) of calcium, and women tend to consume even less calcium as they become older. As a result, when women reach the age when bone loss begins naturally, their bones may already be more fragile than they should be. Another factor unique to women is the role of the hormone estrogen. In women, estrogen stimulates osteoblasts and helps maintain bone density. After menopause, when estrogen production drops dramatically, women lose 3% to 5% of their bone mass each year for several years. Half of all women over age 65 are estimated to have some degree of osteoporosis.

Another important factor contributing to osteoporosis is lack of weight-bearing exercise such as walking, dancing, or running. Bones thrive on moderate stress, but as people age they tend to be less active. Being inactive or bedridden (or being weightless, as astronauts have discovered) results in rapid loss of bone calcium. In contrast, exercise, even in elderly people, can reverse bone loss and even increase bone mass. Alcoholism and smoking also contribute to bone loss and osteoporosis.

Although some bone loss is normal, in osteoporosis the loss is sufficient to weaken the bones, making them vulnerable to fractures and deformities. Frequently, the vertebrae of osteoporosis victims compress, causing a hunch-backed appearance (Fig. E38-5b). In extreme cases, simple activities such as lifting a shopping bag, opening a window, or sneezing can break a bone. Of women living to age 85, nearly one-third will fracture a hip weakened by osteoporosis. As a result of hip fractures, the elderly often become much less self-sufficient and active, and one quarter of them die within 6 months of complications such as pneumonia.

Supplemental hormones (estrogen or a combination of estrogen and progesterone) may slow the progress of osteoporosis in women past menopause. Sodium fluoride, in combination with supplements of calcium and vitamin D (which is crucial to the proper metabolism of calcium), is sometimes prescribed to increase bone deposition. Such treatments, however, may occasionally have undesirable side effects. Fortunately, much of the pain, incapacitation, and expense (estimated at $3.8 billion per year) caused by fractures due to osteoporosis can be prevented. A combination of regular exercise and adequate dietary or supplemental calcium and vitamin D will assure that bone mass is as high as possible before natural, age-related losses begin and will also minimize such losses in old age. In fact, easy measures taken to prevent osteoporosis can enhance the quality of life at any age.

(a)

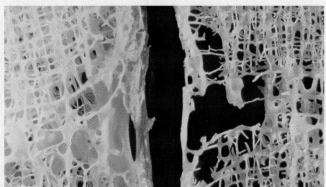

(b)

Figure E38-5   Osteoporosis

(a) Cross section of a normal bone (left) compared with a bone from a woman suffering from osteoporosis (right).
(b) The devastating effects of osteoporosis extend beyond the obvious deformities, such as a hunch-backed appearance. Its victims are also at high risk for bone fractures.

cle. In the most common types of joints, skeletal muscles span the joint; their contraction moves one bone while the other remains fixed. These joints, such as are found in the elbows, knees, and fingers, are called **hinge joints**. Like a hinged door, these joints are movable in only two dimensions. In hinge joints, pairs of muscles lie in roughly the same plane as the joint (see Fig. 38-10). One end of each muscle, called the **origin,** is fixed to a relatively immovable bone on one side of the joint; the other end, the **insertion,** is attached to a mobile bone on the far side of the joint. When the flexor muscle contracts, it bends the joint; when the extensor muscle contracts, it straightens the joint. Thus, alternate contractions of flexor and extensor muscles cause the movable bone to pivot back and forth at the joint.

Other joints, such as that of the hip or shoulder, are **ball-and-socket joints,** in which the rounded end of one bone fits into a hollow depression in another (Fig. 38-11). Ball-and-socket joints allow movement in several directions: Simply compare the wide-ranging swinging of your arm or upper leg with the limited bending of your knee. The range of motion in ball-and-socket joints is made possible by at least two pairs of muscles, oriented at right angles to each other, that provide flexibility of movement.

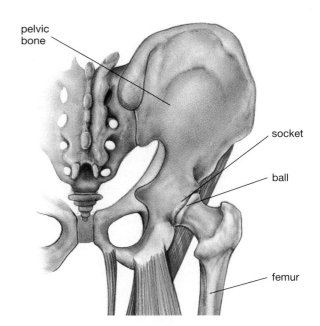

Figure 38-11   **A ball-and-socket joint**

The ball-and-socket joint, shown here in the human hip, consists of a rounded end (the ball; seen at the end of the femur) that fits into a cuplike depression (the socket; seen in the bones of the pelvis). This arrangement permits rotational movement.

SUMMARY OF KEY CONCEPTS

Muscle

Muscle can only actively contract. Skeletal muscle fibers consist of numerous subunits—myofibrils—surrounded by calcium-storing sarcoplasmic reticulum. Sarcomeres within each myofibril are formed by alternating thick filaments of myosin and thin filaments of actin and two accessory proteins.

During skeletal muscle contraction, the thick filaments form cross-bridges to exposed sites on the thin filaments. Using ATP, the bridges bend, release, and reattach in a way that slides the filaments past each other, shortening the fiber. Muscles produce action potentials when stimulated by motor neurons at neuromuscular junctions. Calcium, stored in the sarcoplasmic reticulum, is released when an action potential invades the muscle fiber. Calcium binds to one accessory protein, which pulls the other off the binding sites on the thin filaments, allowing contraction to occur. The strength and degree of muscle contraction are determined by the number of muscle fibers stimulated and the frequency of action potentials in each fiber. Rapid firing can result in maximum contraction, or tetany, if sustained.

Cardiac, or heart, muscle also consists of sarcomeres that contain alternating thick and thin filaments. Its cells tend to contract rhythmically and spontaneously, but these contractions are synchronized by electrical signals produced by specialized muscle fibers in the sinoatrial node. Cardiac muscle fibers are interconnected electrically by gap junc-

tions located in membrane regions called intercalated discs, allowing coordinated contraction.

Smooth muscle lacks organized sarcomeres, but like cardiac muscles, its cells are electrically coupled by gap junctions. Smooth muscle surrounds hollow organs (uterus, digestive tract, bladder) and blood vessels, producing slow sustained or rhythmic contractions, which are usually involuntary.

The Skeleton

Three types of skeletons are found in animals. Hydrostatic skeletons (found in cnidarians, mollusks, and worms) use fluid confined in a chamber and surrounded by muscle. Exoskeletons, found in arthropods, are outer coverings with flexible joints. Endoskeletons, including the bony vertebrate skeleton, are found in both echinoderms and chordates.

The vertebrate skeleton provides support for the body, attachment sites for muscles, and protection for internal organs. Red and white blood cells and platelets are formed in the marrow of bones. Bone acts as a storage site for calcium and phosphorus. The axial skeleton includes the skull, vertebral column, and rib cage. The appendicular skeleton consists of the pectoral and pelvic girdles and the bones of the arms, legs, hands, and feet.

Skeletal Tissues: Cartilage and Bone

Cartilage is found at the ends of bones and forms pads in the knee joints and the intervertebral discs. It also supports the nose, ears, and respiratory passages. During embryological development, cartilage is the precursor of bone. Cartilage is formed by chondrocytes, which surround themselves with a matrix of fibrous collagen.

Bone is formed by osteoblasts that secrete a collagen matrix that becomes hardened by calcium phosphate. A typical bone consists of an outer shell of compact, hard bone, to which muscles are attached, and inner spongy bone, which provides a space for bone marrow. Remodeling of bone occurs continuously. Osteoclasts tunnel through the bone using acids and enzymes. Nourishing capillaries invade the tunnels, and osteoblasts fill the space with concentric layers of new bone, leaving a small central space for the capillaries. This process produces Haversian systems. Osteoblasts trapped within the bone are called osteocytes.

Body Movement: Muscle-Skeleton Interactions

Skeletal muscles form antagonistic pairs that move the skeleton. In the vertebrate skeleton, movement occurs around joints, where bones are joined by ligaments. Muscles attach to bones on either side of the joint by tendons. Contraction of one muscle bends the joint and straightens its antagonistic muscle. At hinge joints, muscles are attached to the immovable bone at their origins. Their insertions attach to the mobile bone. Contraction of the flexor muscle bends the joint; contraction of its antagonistic extensor straightens it.

❒ K E Y   T E R M S

actin   p. 747
antagonistic muscles   p. 757
appendicular skeleton   p. 753
axial skeleton   p. 753
ball-and-socket joint   p. 759
callus   p. 756
cardiac muscle   p. 747
cartilage   p. 754
chondrocyte   p. 754
collagen   p. 753
compact bone   p. 754
cross-bridge   p. 750
endoskeleton   p. 753
exoskeleton   p. 752
extensor   p. 757
flexor   p. 757

Haversian system   p. 755
hinge joint   p. 759
hydrostatic skeleton   p. 752
insertion   p. 759
intercalated disc   p. 752
intervertebral disc   p. 754
joint   p. 757
ligament   p. 757
motor unit   p. 752
muscle fiber   p. 748
myofibril   p. 749
myosin   p. 747
neuromuscular junction   p. 751
origin   p. 759
osteoblast   p. 754
osteoclast   p. 754

osteocyte   p. 754
osteoporosis   p. 758
periosteum   p. 756
sarcomere   p. 749
sarcoplasmic reticulum   p. 749
skeletal muscle   p. 747
skeleton   p. 752
smooth muscle   p. 748
spongy bone   p. 754
striated muscle   p. 747
tendon   p. 757
tetany   p. 752
thick filament   p. 749
thin filament   p. 749
transverse (T) tubule   p. 749
Z line   p. 749

❒ T H I N K I N G   T H R O U G H   T H E   C O N C E P T S

Multiple Choice

1. Thin filaments in myofibrils consist of
   a. actin, tropomyosin, and troponin
   b. sarcomeres   c. cross-bridges
   d. Z lines   e. myosin
2. The deep infoldings of muscle fiber membranes that conduct action potentials are called
   a. sarcoplasmic reticula   b. Z lines
   c. myofilaments   d. T tubules
   e. sarcomeres
3. The force of muscle contraction depends on the
   a. number of muscle fibers stimulated
   b. number of motor units stimulated
   c. frequency of action potentials in each motor unit
   d. frequency of action potentials in each muscle fiber
   e. all of the above
4. Smooth muscle cells can be distinguished from striated ones because smooth fibers
   a. contract more rapidly
   b. lack regular arrangements of sarcomeres
   c. lack gap junctions

   d. contain only actin filaments
   e. contain sarcoplasmic reticulum
5. Bone-dissolving cells are called
   a. chondrocytes   b. osteoblasts
   c. osteoclasts   d. osteocytes
   e. erythroblasts
6. Haversian systems contain all of the following EXCEPT
   a. blood vessels   b. intervertebral discs
   c. osteocytes   d. calcium phosphate crystals
   e. concentric layers of bone

Review Questions

1. Sketch a relaxed myofiber containing a myofibril, sarcomeres, and thick and thin filaments. How would a contracted myofiber look by comparison?
2. Describe the process of skeletal muscle contraction, beginning with an action potential in a motor neuron and ending with the relaxation of the muscle. Your answer should include the following words: neuromuscular junction, T tubule, sarcoplasmic reticulum, calcium, thin filaments, binding sites, thick filaments, sarcomere, Z line, and active transport.

3. Explain the following two statements. Muscles can only actively contract. Muscle fibers lengthen passively.
4. What are the three types of skeletons found in animals? For one of these, describe how the muscles are arranged around the skeleton and how contractions of the muscles result in movement of the skeleton.
5. Compare the structure and function of the following: spongy and compact bone, smooth and striated muscle, and cartilage and bone.

6. Explain the functions of osteoblasts, osteoclasts, and osteocytes.
7. How is cartilage converted to bone during embryonic development? Where is cartilage found in the body, and what functions does it serve?
8. Describe a hinge joint and how it is moved by antagonistic muscles.

✖ APPLYING THE CONCEPTS

1. Discuss some of the problems that would result if the human heart were made of skeletal muscle instead of cardiac muscle.
2. Muscle fibers in Duchenne muscular dystrophy (DMD) patients lack a protein called dystrophin that helps control calcium release from sarcoplasmic reticulum. Lack of dystrophin leads to a constant leaking of calcium ions, which activates an enzyme that dissolves muscle fibers. DMD is inherited as a sex-linked recessive gene. Women with this gene have a 50% chance of passing the disease to their male children (and 50% chance of passing the gene to their female children). Afflicted children gradually become unable to walk and die of respiratory problems as young adults. Recently tests have been developed that allow a woman to determine (a) if she is a carrier, and (b) if her fetus has inherited the gene. What factor would make a woman a candidate for this test? If a woman discovers she carries this gene, what are her options with regard to having children? Discuss the ethical implications of these various options.
3. Myasthenia gravis is caused by abnormal production of antibodies that bind to acetylcholine receptors on muscle cells and eventually destroy the receptors. The disease causes muscles to become flaccid, weak, or paralyzed. Drugs, such as neostigmine, that inhibit the action of acetylcholinesterase, are used to treat myasthenia gravis. How does neostigmine restore muscle activity?
4. Some insects would have a tough time flying if one nerve impulse was required for each muscle contraction. Gnats, for example, may beat their wings 1000 times a second. At such high frequencies, contraction is "myogenic," originating from the stretching caused by the contraction of antagonistic muscles. Also insect flight muscle cells are filled with giant mitochondria. Suggest a mechanism to explain how myogenic contraction works inside cells. Explain the significance of giant mitochondria.
5. Human muscle cells contain a mixture of three types of muscle fibers: slow-twitch oxidative, fast-twitch oxidative, and fast-twitch glycolytic. Slow-twitch muscle cells break down ATP slowly. They contain many mitochondria and large amounts of myoglobin, a dark pigment that acts as a reservoir for oxygen. All fast-twitch muscle cells break down ATP rapidly. Fast-twitch oxidative fibers have moderate amounts of myoglobin and glycogen. Fast-twitch glycolytic fibers have little myoglobin and lots of glycogen. The relative numbers of these fibers in different muscles is under genetic control. Use this information to explain the location of dark and white meat in birds. Predict the relative numbers of slow- and fast-twitch fibers in muscle samples from the thighs of world-class marathon runners and world-class sprinters.
6. Some hours after death, skeletal muscle fibers run out of ATP and become rigid, although they are not contracted, a condition called rigor mortis. Based on your knowledge of the mechanisms of muscle contraction, including the role of calcium and ATP, explain what must be happening at the cellular level to cause this muscle stiffness.

✖ FOR MORE INFORMATION

Alexander, R. "Muscles Fit for the Job." *New Scientist*, April 15, 1989. Diverse muscles adapted for different functions.

Huyghe, P. "No Bone Unturned." *Discover*, December 1988. The story of a remarkable forensic anthropologist who uncovers the secrets hidden in unidentified skeletons.

Merton, P. A. "How We Control the Contraction of Our Muscles." *Scientific American*, May 1972. Useful muscular activity requires precise simultaneous adjustments of the contractions of many muscles.

Moore, K. "Little Is beyond His Scope." *Discover*, March 1985. A description of intricate arthroscopic surgery on joints.

Smith, K. K., and Kier, W. M. "Trunks, Tongues, and Tentacles: Moving with Skeletons of Muscle." *American Scientist*, January–February 1989. In certain organs of many animals, muscles provide support as well as movement.

Stossel, T. P. "The Machinery of Cell Crawling." *Scientific American*, September 1994. Cell movement relies on the orderly assembly and disassembly of scaffold proteins.

NET WATCH

On-line resources for this chapter are on the World Wide Web at:
http://www.prenhall.com/~audesirk (click on the <u>table of contents</u> link and then select Chapter 38).

39 Animal Reproduction

⊞ AT A GLANCE

Reproductive Strategies
Asexual Reproduction Does Not Involve the Fusion of
  Sperm and Egg
Sexual Reproduction Requires the Union of Sperm and Egg

Mammalian Reproduction
The Male Reproductive Tract Includes the Testes and
  Accessory Structures
The Female Reproductive Tract Includes Ovaries and
  Accessory Structures
The Menstrual Cycle Is Controlled by Complex Hormonal
  Interactions
Copulation Allows Internal Fertilization
During Pregnancy, the Developing Embryo Grows within
  the Uterus
Milk Secretion, or Lactation, Is Stimulated by Pregnancy
Reproduction Culminates in Labor and Delivery

On Limiting Fertility
Permanent Contraception Can Be Achieved through
  Sterilization
There Are Three Major Approaches to Temporary
  Contraception
Additional Contraceptive Methods Are under Development

A mother koala with her young reaches for eucalyptus leaves.
Most young mammals require considerable parental care,
feeding, and protection before they are able to face the
world alone.

The word *reproduction* may bring to mind images of romantic courtship and cute cuddly babies. From an evolutionary perspective, however, romance and the universal appeal of babies are frills that have evolved only because they further the real evolutionary goal: to pass on one's genes to another generation. From this viewpoint, an animal's life can be divided into three stages. First, it is born or hatched from an egg and grows to sexual maturity. Second, it gathers the resources needed to reproduce, which may include stores of food, impressive strength or weaponry, or a territory. Finally, it finds a mate (if necessary) and reproduces, which may include caring for its offspring until they can fend for themselves. The marvelous adaptations that we have discussed in the previous chapters, such as sophisticated sensory equipment or complex digestive systems, have evolved through millennia of mutation and natural selection because they have allowed survival and successful reproduction. Reproduction is the key to the continued existence of the species.

Reproductive Strategies

Animals reproduce either sexually or asexually. As you learned in Chapter 12, in **sexual reproduction** an animal produces haploid gametes through meiosis. Two gametes, usually from separate parents, fuse to form a diploid offspring. Because an offspring receives genes from two parents, it is genetically different from both of them. In most forms of **asexual reproduction,** on the other hand, a single animal produces offspring through repeated mitosis of cells in some part of its body. Therefore, the offspring are genetically identical to the parent.

We humans reproduce sexually, and we tend to regard sexual reproduction as the normal, best way to do it. From a biological standpoint, by bringing together genes from two different parental organisms, sexual reproduction allows for new gene combinations that may enhance the survival and reproduction of the offspring. Nevertheless, asexual reproduction is more efficient, because there is no need to find a mate, court, and fend off rivals, and there is no waste of sperm and eggs that never unite to form an offspring. Not surprisingly, most animals reproduce asexually, at least some of the time.

Figure 39-1   Budding

The offspring of some cnidarians, such as the anemone shown here, grow as buds that appear as miniature adults sprouting from the body of the parent. When sufficiently developed, the buds break off and assume independent existence.

Let's begin, then, with a brief survey of asexual reproduction among animals before moving on to sexual reproduction.

Asexual Reproduction Does Not Involve the Fusion of Sperm and Egg

Budding Produces a Miniature Version of the Adult

Many sponges and cnidarians, such as *Hydra* and some anemones, reproduce by **budding** (Fig. 39-1). A miniature version of the animal (a **bud**) grows directly on the body of the adult, drawing nourishment from its parent. When it has grown large enough, the bud breaks off and becomes independent.

Regeneration Produces a New Individual from Parts of Another

Regeneration from body fragments is a potential form of reproduction in some animals such as sea stars (Fig. 39-2a). If sea stars are cut up, fragments that contain part of the center part of the body can regrow the rest of the star. (Oyster "ranchers," attempting to rid their oyster beds of predatory sea stars, used to catch sea stars, hack them to pieces, and throw the parts back into the sea. Much to their dismay, this method merely resulted in more sea stars than ever, as the fragments regenerated entire animals.) A few brittle stars routinely reproduce in a similar fashion; they split apart and each half regenerates a complete animal. Despite these asexual capabilities, sea stars usually reproduce sexually, casting huge numbers of sperm and eggs into the sea.

During Fission, an Animal Divides, Producing Two New Individuals

Some animals reproduce by **fission.** A few corals can divide lengthwise to produce two smaller but complete individuals. Some flatworms and annelids divide across the

middle and regenerate the missing parts (Fig. 39-2b). Of course, the "tail half" of the animal must regenerate the head, including the brain!

During Parthenogenesis Eggs Develop without Fertilization

The females of some animal species can reproduce by a process known as **parthenogenesis,** in which haploid egg cells develop into adults without being fertilized. Parthenogenetically produced offspring of some species remain haploid. Male honey bees, for example, are haploid, developing from unfertilized eggs; their diploid sisters develop from fertilized eggs. On the other hand, some fish, amphibians, and reptiles regain the diploid number of chromosomes in parthenogenetically produced offspring by duplicating all the chromosomes either before or after meiosis. The resulting offspring are all females.

Some species of fish, including relatives of the mollies and platies found in tropical fish stores, and some lizards, such as the whiptail, have done away with males completely. Their populations consist entirely of parthenogenetically reproducing females. Still other animals, such as the aphid, can reproduce either sexually or parthenogenetically, depending on environmental factors such as the season of the year or the availability of food (Fig. 39-3).

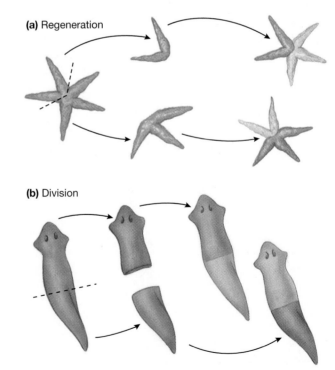

Figure 39-2   Other modes of asexual reproduction

(a) Many sea stars can regenerate new individuals from fragments if the fragment includes part of the center of the body. **(b)** Certain flatworms divide across the middle. At first, each offspring is missing half the adult body, but these are regrown from cells near the broken edge.

Figure 39-3   A female aphid gives live birth

In spring and early summer, when food is abundant, aphid females reproduce parthenogenetically. In fact, the development of the reproductive tract proceeds so rapidly that females are born pregnant! In fall, reproduction becomes sexual, as the females mate with males. Aphids have thus evolved the ability to exploit the advantages of asexual reproduction (rapid population growth during times of abundant food, no energy spent in seeking a mate, no wasted gametes) and sexual reproduction (genetic recombination).

Sexual Reproduction Requires the Union of Sperm and Egg

In animals, sexual reproduction occurs when a haploid sperm fertilizes a haploid egg, generating a diploid offspring. In most animal species, an individual is either male or female. These species are termed **dioecious** (Greek for "two houses"). The sexes are defined by the type of gamete that each produces. Females produce **eggs,** which are large, nonmotile cells containing substantial food reserves. Males produce small, motile **sperm** that have almost no cytoplasm and hence no food reserves.

In **monoecious** ("one house") species, such as earthworms and many snails, single individuals produce both sperm and eggs. Such individuals are commonly called **hermaphrodites** (after Hermaphroditos, a male Greek god whose body was merged with that of a female water nymph, producing a being half male and half female). Although most hermaphrodites exchange sperm with other individuals if they have the opportunity (retaining the advantages of genetic exchange), some hermaphrodites can fertilize their eggs with their own sperm if necessary. These animals, including tapeworms and many pond snails, are relatively immobile and may find themselves isolated from other members of their species. Obviously, the ability to fertilize oneself is advantageous under these circumstances.

For dioecious species and for hermaphrodites that cannot self-fertilize, successful reproduction requires that sperm and eggs from different animals be brought together for fertilization. The union of sperm and egg is accomplished in a variety of ways, depending on the mobility of the animals and on whether they breed in water or on land.

External Fertilization Occurs outside the Parent Bodies

In **external fertilization,** the union of the sperm and egg takes place outside the bodies of both parents. Usually, parents release sperm and eggs into water, through which the sperm swim to reach an egg. This procedure, called **spawning,** obviously is restricted to animals that breed in water. Because sperm and egg are relatively short lived, spawning animals must synchronize their reproductive behaviors, both *temporally* (male and female spawn at the same time) and *spatially* (male and female spawn in the same place). Animals employ a combination of environmental cues, chemical signals called pheromones, and behaviors to synchronize spawning.

Most spawning animals rely on environmental cues to some extent. Breeding usually occurs only during certain seasons of the year, but more precise synchrony is required to coordinate the actual release of sperm and egg. Grunion, fish that inhabit coastal waters off southern California, time their unusual reproductive rituals by the season, time of day, and phase of the moon (Fig. 39-4a). On fall nights during the highest tides (which occur during a full moon), grunion swim up onto sandy beaches. Writhing masses of males and females release their gametes into the wet sand and then swim back out to sea on the next wave. Many corals of Australia's Great Barrier Reef also synchronize spawning by the phase of the moon. On the fourth or fifth night after the full moons of November and December, all the corals of a particular species on an entire reef release a blizzard of sperm and eggs into the water (Fig. 39-4b).

Other animals communicate their sexual readiness to one another by releasing pheromones into the water. A **pheromone** is a chemical released from the body of one animal that affects the behavior of a second animal. Pheromones synchronize spawning of many immobile or sluggish invertebrates, such as mussels and sea stars. Usually, when a female is ready to spawn, she releases eggs and a pheromone into the water. Nearby males, detecting the mating pheromone, quickly release millions of sperm. The sperm themselves are lured by a chemical attractant released by the eggs in some, if not most, animals. Such "egg pheromones," which have been detected in animals as diverse as sea stars and humans, help ensure fertilization.

Synchronized timing alone does not guarantee efficient reproduction. Corals, sea stars, and mussels all waste enormous quantities of sperm and eggs because they are released too far apart. In species of mobile animals both temporal *and* spatial synchrony can be ensured by mating behaviors. Most fish, for example, have some sort of courtship ritual in which the male and female come very close together and release their gametes in the same place and at the same time (Fig. 39-5). Frogs carry this ritual one step further, by assuming a characteristic mating pose called **amplexus** (Fig. 39-6). At the edges of ponds and lakes, the male frog mounts the female and prods her in the side. This prod-

(a)

(b)

Figure 39-4   **Environmental cues may synchronize spawning**

(a) At the highest tides of fall, grunion swarm ashore on the few undeveloped beaches left in southern California. The fish burrow slightly into the sand and release sperm and eggs. The eggs hatch in the warm sand and develop over the following 2 weeks. When the next-highest tide comes, the young fish wash out of the sand back into the ocean. **(b)** Along the Great Barrier Reef of Australia, thousands of corals spawn simultaneously, creating this "blizzard" effect. The inset photo shows a package of sperm and eggs erupting from a spawning hermaphroditic coral. Spawning in these corals is linked to the phase of the moon.

ding stimulates her to release eggs, which he immediately fertilizes by releasing a cloud of sperm above them.

Internal Fertilization Occurs within the Female Body

In **internal fertilization,** sperm are taken into the body of the female, where fertilization occurs. This method has two advantages over external fertilization, especially in terrestrial environments. First, sperm require a direct fluid path to reach the eggs, which, on land, can be guaranteed only inside the body of the female. Second, even in aquatic environments, internal fertilization increases the like-

lihood that most eggs will be fertilized, because the sperm are confined in a small space with the eggs, rather than being left to thrash about in a large volume of water.

Internal fertilization usually occurs by **copulation,** in which the penis of the male is inserted into the body of the female, where it releases sperm (Fig. 39-7). In a variation of internal fertilization, males of some species package their sperm in a container called a **spermatophore** (Greek for "sperm carrier"). Males of some species of mites and scorpions simply drop the spermatophore on the ground. If a female finds the spermatophore, she fertilizes herself by inserting it into her reproductive cavity. The male squid picks up his spermatophore with a tentacle and inserts it into the female. In both cases, the sperm are then liberated inside the female's reproductive tract.

Simply depositing sperm into the body of the female does not guarantee fertilization. Fertilization can occur only if an egg is mature and released into the female reproductive tract during the limited time when sperm are present. Most mammals copulate only at certain seasons of the year or when the female signals readiness to mate. The season or signal often coincides with **ovulation,** or release of the egg cell from the ovary. Copulation itself triggers ovulation in a few animals, such as rabbits. An alternative strategy, employed by many female snails and

Figure 39-5   **Courtship rituals synchronize release of sperm and eggs**

Violent courtship rituals among Siamese fighting fish (*Betta splendens*) ensure fertilization of the female's eggs, as male and female curl about one another, releasing sperm and eggs together. The male retrieves the eggs as they fall, spits them into his bubble nest (seen here as bubbles floating on the surface above him), and cares for the offspring during their first few weeks of life.

Figure 39-6   Golden toads in amplexus

The smaller male rides atop the female and stimulates her to release eggs. Because the large eggs of frogs and toads are surrounded by a transparent coat of jelly, they are ideal subjects for observing and studying embryonic development.

insects, is to store sperm for days, weeks, or even months, thus assuring a supply of sperm whenever eggs are ready.

Mammalian Reproduction

Male and female sexes of mammals are separate. Mammals reproduce sexually, uniting sperm and eggs through internal fertilization. Many mammals reproduce only during certain seasons of the year and consequently produce sperm and eggs only at that time. Human reproduction is similar to that of other mammals, but it is not restricted by season. Men produce sperm more or less continuously, and women ovulate about once a month. Our discussion will focus on human reproduction.

The Male Reproductive Tract Includes the Testes and Accessory Structures

The male reproductive tract consists of the paired **gonads** (organs that produce sex cells), the **testes** (singular, **testis**) where sperm are produced, and accessory structures that store the sperm, produce secretions that activate and nourish them, and finally conduct them to the inside of the female reproductive tract (Fig. 39-8 and Table 39-1).

The Testes Are the Site of Sperm Production

The testes produce both sperm and male sex hormones. The testes are located in the **scrotum,** a pouch that hangs outside the main body cavity. This location keeps the testes about 4° C cooler than the core of the body and provides the optimal temperature for sperm development. (Tight jeans may look sexy, but they push the scrotum up against the

Figure 39-7   Internal fertilization is essential for reproduction on land

(a) Ladybugs mate on a dandelion flower. **(b)** South American tortoises must cope with confining shells. **(c)** King penguins mate comfortably in the snow.

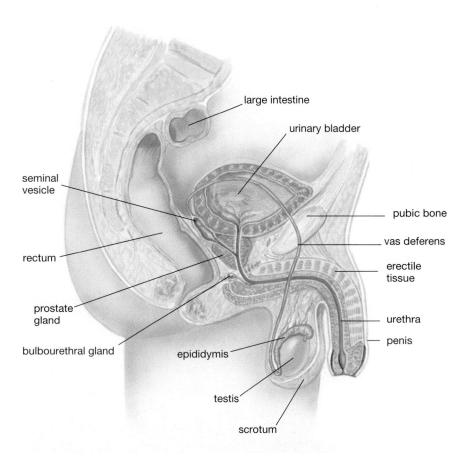

large intestine

urinary bladder

seminal
vesicle

rectum

prostate
gland

bulbourethral gland

pubic bone

vas deferens

erectile
tissue

urethra

penis

epididymis

testis

scrotum

Figure 39-8   **The human male
reproductive tract**

The male gonads, the testes, hang beneath
the abdominal cavity in the scrotum.
Sperm pass from the seminiferous tubules of
a testis to the epididymis, and from there
through the vas deferens and urethra to the
tip of the penis. Along the way, fluids are
added from three sets of glands: the seminal
vesicles, the bulbourethral glands, and the
prostate gland.

TABLE 39-1 ▪ *Structures and Functions of the Human Male Reproductive Tract*

Structure	Type of Organ	Function
Testis	Gonad	Produces sperm and testosterone
Epididymis and vas deferens	Ducts	Store sperm; conduct sperm from testes to penis
Urethra	Duct	Conducts semen from vas deferens and urine from urinary bladder to the tip of the penis
Penis	External "appendage"	Deposits sperm in female reproductive tract
Seminal vesicles	Glands	Secrete fluids that contain fructose (energy source) and prostaglandins (possibly cause "upward" contractions of vagina, uterus, and oviducts, assisting sperm transport to oviducts); fluids may wash sperm out of ducts of male reproductive tract into vagina
Prostate	Gland	Secretes fluids that are basic (neutralize acidity of vagina) and contain factors that enhance sperm motility
Bulbourethral glands	Glands	Secrete mucus (may lubricate penis in vagina)

body, raising the temperature of the testes. Some researchers think that wearing tight pants may reduce sperm counts and decrease fertility. This is not, however, a reliable means of birth control!) Coiled, hollow **seminiferous tubules,** in which sperm are produced, nearly fill each testis (Fig. 39-9a). In the spaces between the tubules are the **interstitial cells,** which synthesize the male hormone testosterone.

Just inside the wall of each seminiferous tubule lie the diploid cells, or **spermatogonia** (singular, **spermatogoni-**

um), from which all the sperm eventually will arise, and the much larger **Sertoli cells** (Fig. 39-9b, c). Each time a spermatogonium divides, it can take one of two developmental paths. First, it may undergo mitosis. Mitosis ensures that the male has a steady supply of new spermatogonia throughout his life. Second, it may undergo **spermatogenesis**—that is, the production of sperm by the process of meiosis followed by differentiation (Fig. 39-9d). Spermatogenesis begins with growth and differentiation

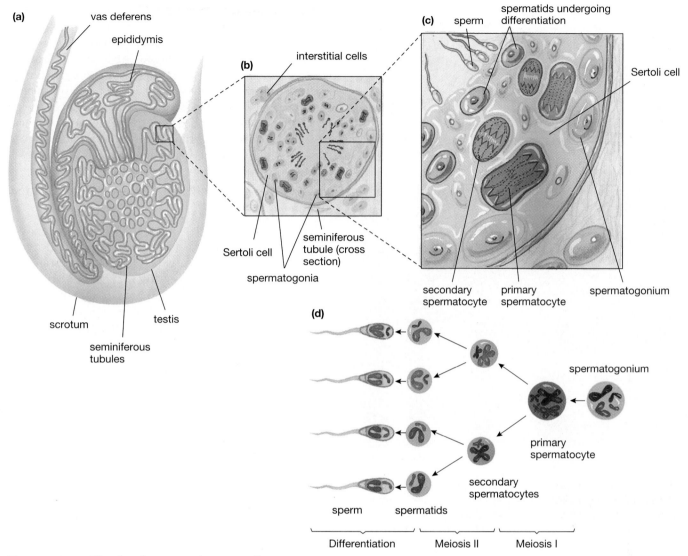

Figure 39-9   The development of sperm cells

(a) A lengthwise section of the testis, showing the location of the seminiferous tubules, epididymis, and vas deferens. (b) A cross section of a seminiferous tubule. The walls of the seminiferous tubules are lined with Sertoli cells and spermatogonia protruding into the central cavity (lumen) of the tubule. (c) As spermatogonia undergo meiosis, the daughter cells move inward, embedded in infoldings of the Sertoli cells. There they differentiate into sperm, drawing upon the Sertoli cells for nourishment. Mature sperm are finally freed into the lumen of the tubules for transport to the penis. Testosterone is produced by the interstitial cells found in the spaces between tubules. (d) Spermatogenesis is accomplished by meiotic divisions that produce haploid sperm (compare with the actual locations shown in part [c]). Although four chromosomes are shown for clarity, in humans the diploid number is 46 and the haploid number is 23.

of spermatogonia into **primary spermatocytes.** These are large diploid cells that will develop into sperm. The primary spermatocytes then undergo meiosis (see Chapter 12). At the end of meiosis I, each primary spermatocyte gives rise to two haploid **secondary spermatocytes.** Each secondary spermatocyte divides again during meiosis II to produce two **spermatids,** for a total of four spermatids per primary spermatocyte. Spermatids undergo radical rearrangements of their cellular components as they differentiate into sperm.

Sertoli cells regulate the process of spermatogenesis and nourish the developing sperm. The spermatogonia, spermatocytes, and spermatids are embedded in infoldings of the Sertoli cells. As spermatogenesis proceeds, they migrate up from the outermost edge of the seminiferous tubule to the central cavity of the tubule (Fig. 39-9c). The mature sperm, several hundred million a day, are finally liberated into the lumen.

A human sperm (Fig. 39-10) is unlike any other cell of the body. Most of the cytoplasm disappears, leaving a hap-

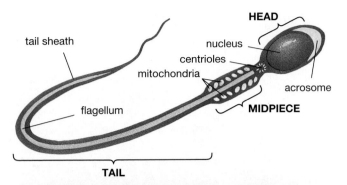

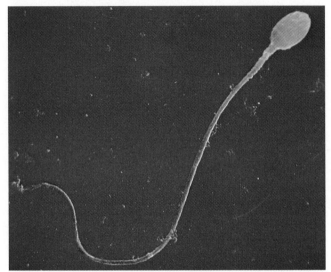

Figure 39-10   A human sperm cell

A mature sperm is a stripped-down cell equipped with only the essentials: a haploid nucleus containing the male genetic contribution to the future zygote, a lysosome (called the acrosome) containing enzymes that will digest the barriers surrounding the egg, mitochondria for energy production, and a tail (actually a long flagellum) for locomotion. The photo is a false-color electron micrograph of a human sperm.

loid nucleus nearly filling the head. Atop the nucleus lies a specialized lysosome, called the **acrosome.** The acrosome contains enzymes that will be needed to dissolve protective layers around the egg, enabling the sperm to enter and fertilize it. Behind the head is the midpiece, which is packed with mitochondria that provide the energy needed to move the tail that protrudes out the back. Whiplike movements of the tail, which is really a long flagellum, propel the sperm along inside the female reproductive tract.

At puberty, the time of rapid growth and sexual maturation, the hypothalamus releases **gonadotropin-releasing hormone (GnRH)**, which stimulates the anterior pituitary to produce **luteinizing hormone (LH)** and **follicle-stimulating hormone (FSH)**. Spermatogenesis begins as a result of the interplay of LH and FSH and **testosterone** secreted by the testes (Fig. 39-11). Luteinizing hormone stimulates the interstitial cells to produce testosterone. The combination of testosterone and FSH stimulates the Sertoli cells and spermatogonia, causing spermatogenesis.

Testosterone also stimulates the development of secondary sexual characteristics (such as the growth of facial hair in males and breast development in females), maintains sexual drive, and is required for successful intercourse (a term we will use for human copulation). Sperm, however, are not involved in these functions. Therefore, if one could suppress FSH release (blocking spermatogenesis) but not LH release (thereby allowing continued testosterone production), a man would be infertile but not impotent. Efforts are under way to develop a drug to do just that, as a form of male birth control.

Accessory Structures Produce Semen and Conduct the Sperm outside the Body

The seminiferous tubules merge to form a single convoluted tube, the **epididymis** (see Fig. 39-9). The epididymis becomes the **vas deferens,** which leaves the scrotum and enters the abdominal cavity. Most of the hundreds of millions of sperm produced each day are stored in the vas deferens and epididymis. The vas deferens joins the **urethra,** leading from the bladder to the tip of the penis. This final common path is shared, at different times, of course, by sperm (during ejaculation) and urine (during urination).

The fluid ejaculated from the penis, called **semen,** consists of sperm mixed with secretions from three glands that empty into the vas deferens or urethra: the **seminal vesicles,**

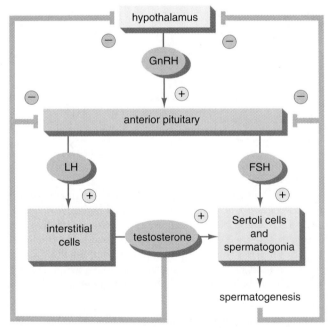

Figure 39-11   Hormonal control of spermatogenesis

Gonadotropin-releasing hormone (GnRH) from the hypothalamus stimulates the anterior pituitary to release LH and FSH. LH stimulates the interstitial cells to produce testosterone. Testosterone and FSH stimulate the Sertoli cells and the spermatogonia, causing spermatogenesis. Testosterone and chemicals produced during spermatogenesis inhibit further release of FSH and LH, forming a negative feedback loop that keeps the rate of spermatogenesis and the concentration of testosterone in the blood nearly constant. + stimulates; − inhibits.

the **prostate gland,** and the **bulbourethral glands.** The secretions activate swimming by the sperm, provide energy for swimming, and neutralize the acidic fluids of the vagina that normally inhibit bacterial growth (see Table 39-1).

The Female Reproductive Tract Includes Ovaries and Accessory Structures

The female reproductive tract is almost entirely contained within the abdominal cavity (Fig. 39-12 and Table 39-2). It consists of paired gonads, the **ovaries** (Fig. 39-13a), and accessory structures that accept sperm, conduct the sperm to the egg, and nourish the developing embryo.

The Ovaries Are the Site of Egg Production

The human female produces precursor egg cells, or **oogonia** (singular, **oogonium**), while still a fetus in her mother's womb, beginning the process of **oogenesis,** the formation of egg cells. The oogonia divide by mitosis and then grow into **primary oocytes.** No oogonia remain after the third month of fetal development, and no new ones form during the rest of her life. Still during the fetal stage, all the primary oocytes begin meiosis but then halt during prophase of meiosis I. At birth, the ovaries contain about 2 million primary oocytes; many die each day, until at puberty (usually 11 to 14 years of age) only about 400,000 remain. Because only a few oocytes resume meiosis during each month of a woman's reproductive span (from puberty to menopause at age 45 to 55), there is no shortage of oocytes.

Surrounding each oocyte is a layer of much smaller cells that both nourish the developing oocyte and secrete female sex hormones. Together, the oocyte and these accessory cells make up a **follicle** (Fig. 39-13b). Approximately once a month during a woman's reproductive years, she undergoes a menstrual cycle, which is described below. During the menstrual cycle, pituitary hormones stimulate

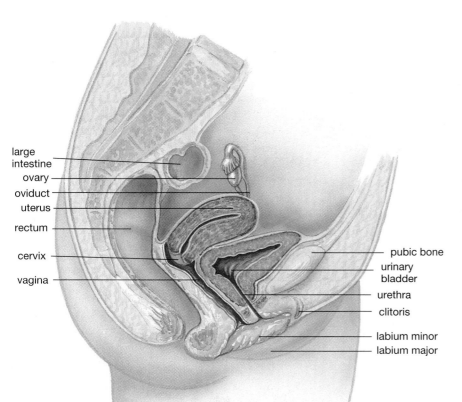

large
intestine
ovary
oviduct
uterus
rectum
cervix
vagina

pubic bone
urinary
bladder
urethra
clitoris
labium minor
labium major

Figure 39-12   **The human female reproductive tract**

Eggs are produced in the ovaries and swept by cilia into the oviduct. A male deposits sperm in the vagina, from which they move up through the cervix and uterus into the oviduct. Sperm and egg usually meet in the oviduct, where fertilization occurs. The fertilized egg attaches to the lining of the uterus, where the embryo develops.

Structure	Type of Organ	Function
Ovary	Gonad	Produces eggs, estrogen, and progesterone
Fimbria	Mouth of duct	Cilia sweep egg into oviduct
Oviduct	Duct	Conducts egg to uterus; site of fertilization
Uterus	Muscular chamber	Site of development of fetus
Cervix	Connective tissue ring	Closes off lower end of uterus, supports fetus, and prevents foreign material from entering uterus
Vagina	Large "duct"	Receptacle for semen; birth canal

TABLE 39-2 ⁘ *Structures and Functions of the Human Female Reproductive Tract*

(a)

fimbriae

oviduct

ovary

uterus

Figure 39-13   **Oogenesis in the human female**

(a) External view of the ovary and oviduct. **(b)** The development of follicles in an ovary, portrayed in a time sequence going clockwise from the lower right. ① A primary oocyte begins development within a follicle. ②, ③ The follicle grows, providing both hormones and nourishment for the enlarging oocyte. ④ At ovulation, the secondary oocyte, or egg, bursts through the ovary wall, surrounded by some follicle cells (now called the corona radiata). The remaining follicle cells develop into the corpus luteum, which secretes hormones. If fertilization does not occur, the corpus luteum breaks down after a few days. **(c)** The cellular stages of oogenesis. The oogonium enlarges to form the primary oocyte. At meiosis I, almost all the cytoplasm is included in one daughter cell, the secondary oocyte. The other daughter cell is a small polar body that contains chromosomes but little cytoplasm. At meiosis II, almost all the cytoplasm of the secondary oocyte is included in the egg, and a second small polar body discards the remaining "extra" chromosomes. The first polar body sometimes also undergoes the second meiotic division. In humans, meiosis II does not occur unless the egg is fertilized.

(b)

ruptured follicle

corpus luteum

degenerating corpus luteum

ovary

④ ovulated secondary oocyte (egg)

③ mature follicle with secondary oocyte

② developing follicles

① primary follicle containing primary oocyte

(c)

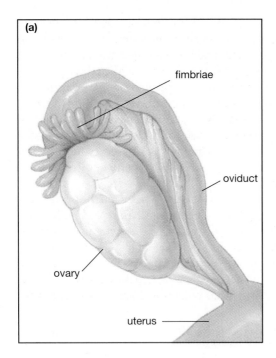

egg

polar body

polar body

polar body

polar body

secondary oocyte (egg)

primary oocyte

oogonium

Meiosis II (after fertilization)

Meiosis I

development of a dozen or more follicles, although usually only one completely matures. At this time, the primary oocyte completes the first meiotic division (which was halted during development) to become a single **secondary oocyte** and a **polar body,** which is little more than a discarded set of chromosomes (Fig. 39-13c). Meanwhile, the

small accessory cells of the follicle multiply and secrete **estrogen.** As it matures, the follicle grows, eventually erupting through the surface of the ovary and releasing the secondary oocyte (Fig. 39-14). The second meiotic division occurs, not in the ovary, but in the oviduct (the tube leading out of the ovary), and then only if the sec-

(a)

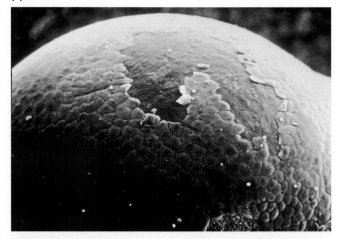

(b)

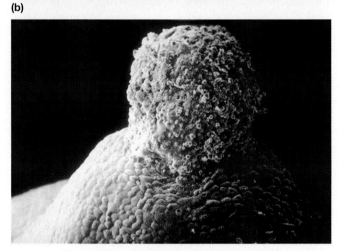

Figure 39-14   **A follicle erupts from the ovary**

The mature follicle grows so large, and is filled with so much fluid that it moves to the surface of the ovary **(a)** and literally bursts through the ovary wall like a miniature volcano **(b)**. It then releases the secondary oocyte into the oviduct.

ondary oocyte is fertilized. For convenience, we will refer to the ovulated secondary oocyte as the "egg."

Some of the follicle cells accompany the egg, but most remain behind in the ovary. These cells enlarge and become glandular, forming the **corpus luteum,** which secretes both estrogen and a second hormone, **progesterone.** If fertilization does not occur, the corpus luteum breaks down a few days later.

Accessory Structures Include the Oviducts, Uterus, and Vagina

Each ovary is adjacent to, but not continuous with, an **oviduct** (sometimes called the **fallopian tube** in humans; see Fig. 39-13a). The open end of the oviduct is fringed with ciliated "fingers" called **fimbriae** that nearly surround the ovary. The cilia create a current that sweeps the egg into the mouth of the oviduct. Fertilization usually occurs in the

oviduct, and the **zygote,** as the fertilized egg is now called, is swept down the oviduct by beating cilia and released into the pear-shaped **uterus,** or womb. Here it will develop for the next 9 months. The wall of the uterus has two layers that correspond to its dual functions of nourishment and childbirth. The inner lining, or **endometrium,** is richly supplied with blood vessels. (This lining will form the mother's contribution to the **placenta,** the structure that transfers oxygen, carbon dioxide, nutrients, and wastes between fetus and mother; see Chapter 40.) The outer muscular wall of the uterus, the **myometrium,** contracts strongly during delivery, expelling the infant out into the world.

Developing follicles secrete estrogen, which stimulates the uterine lining to grow an extensive network of blood vessels and nutrient-producing glands. After ovulation, estrogen and progesterone released by the corpus luteum promote continued growth of the endometrium. Thus, if an egg is fertilized, it encounters a rich environment for growth. If the egg is not fertilized, however, the corpus luteum disintegrates, estrogen and progesterone levels fall, and the overgrown endometrium disintegrates as well. The uterus contracts, squeezing out the excess endometrial tissue (and sometimes causing menstrual cramps in the process). The resulting flow of tissue and blood is called **menstruation** (from the Latin *mensis,* meaning "month").

The outer end of the uterus is nearly closed off by a ring of connective tissue, the **cervix.** The cervix holds the developing baby in the uterus, expanding only at the onset of labor to permit passage of the child. Beyond the cervix lies the **vagina,** which opens to the outside of the body. The vagina serves both as the receptacle for the penis during intercourse and as the birth canal.

The Menstrual Cycle Is Controlled by Complex Hormonal Interactions

The human male produces sperm continuously, thereby increasing the number of potential offspring that a man can father. In contrast, it is fruitless for a woman to ovulate unless her reproductive tract is properly prepared for pregnancy. The human female reproductive system goes through a complex **menstrual cycle,** in which hormonal interactions among the hypothalamus, pituitary gland, and ovary coordinate ovulation with the preparation of the uterus to receive and nourish the fertilized egg.

You may recall from Chapter 35 that hormone release by the anterior pituitary gland is controlled by neurosecretory cells in the hypothalamus. Some of these neurosecretory cells produce gonadotropin-releasing hormone (GnRH), which stimulates endocrine cells in the anterior pituitary to release FSH and LH. A key to understanding the menstrual cycle is that these neurosecretory cells spontaneously release GnRH all the time, unless actively prevented from doing so by other hormones, notably progesterone. We will begin our discussion of the menstrual cycle with the spontaneous release of GnRH.

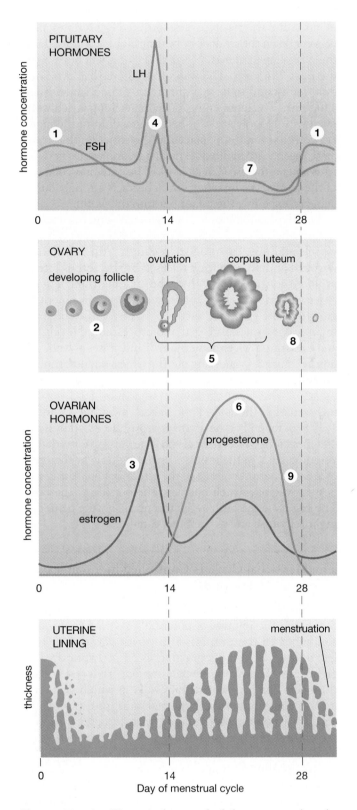

Figure 39-15   Hormonal control of the menstrual cycle
The menstrual cycle is generated by interactions among the hormones of the hypothalamus, anterior pituitary gland, and ovaries. The hormonal changes in turn drive cyclic changes in the uterine lining. The numbers on the graphs refer to the hormonal interactions discussed in the text. They zigzag back and forth among pituitary hormones, ovarian structures, and ovarian hormones, as each of these exerts effects on the others.

Gonadotropin-releasing hormone stimulates the anterior pituitary to release FSH and LH (step 1 in Fig. 39-15). FSH and LH circulate in the bloodstream and initiate the development of several follicles within the ovaries. The follicle cells surrounding the developing oocyte are stimulated by FSH and LH to secrete estrogen. Under the combined influences of FSH, LH, and estrogen, the follicles grow during the next two weeks (step 2). Simultaneously, the primary oocyte within each follicle enlarges, storing both food and regulatory substances (mostly proteins and messenger RNA) that will be needed by the fertilized egg during early development (see Chapter 40). For reasons that are not completely understood, only one, or rarely two, follicles complete development each month.

As the maturing follicle enlarges, it secretes ever greater amounts of estrogen (step 3). This estrogen has three effects. First, it promotes the continued development of the follicle itself and the primary oocyte contained within it. Second, it stimulates growth of the endometrium of the uterus. Third, high levels of estrogen stimulate both the hypothalamus and pituitary, resulting in a surge of LH and FSH at about the twelfth day of the cycle (step 4).

The function of the peak in FSH concentration is still a subject of investigation, but the surge of LH has three important consequences: (1) It triggers the resumption of meiosis I in the oocyte, resulting in the formation of the secondary oocyte and the first polar body; (2) it causes the final explosive growth of the follicle, culminating in ovulation (step 5); and (3) it transforms the remnants of the follicle that remain in the ovary into the corpus luteum.

The corpus luteum secretes both estrogen and progesterone (step 6). The combination of these hormones inhibits the hypothalamus and pituitary, shutting down the release of FSH and LH (step 7), thereby preventing the development of any more follicles. Simultaneously, estrogen and progesterone stimulate further growth of the endometrium, which eventually becomes about 5 millimeters thick. (Note that the effects of progesterone and estrogen depend on the target organ. Progesterone *stimulates* the endometrium but *inhibits* hormone release from the hypothalamus and pituitary.)

In menstrual cycles in which pregnancy does not occur, the corpus luteum disintegrates about 1 week after ovulation. The corpus luteum survives only while it is stimulated by LH (or by a similar hormone released by the developing embryo, described below). However, because progesterone secreted by the corpus luteum shuts off the LH secretion that sustained it, the corpus luteum essentially "self-destructs" around the twenty-first day of the cycle (step 8). With the corpus luteum gone, estrogen and progesterone levels plummet (step 9). Deprived of stimulation by estrogen and progesterone, the endometrium of the uterus also dies, and its blood and tissue are shed, forming the menstrual flow beginning about the twenty-seventh or twenty-eighth day of the cycle. Simultaneously, the reduced progesterone level no longer inhibits the hypothalamus and pituitary, and the spontaneous release of

GnRH from the hypothalamus resumes. Release of GnRH in turn stimulates release of FSH and LH (step 1), which initiates development of a new set of follicles, starting the cycle over again.

During pregnancy, the embryo itself prevents these changes from occurring. Shortly after the ball of cells formed by the dividing fertilized egg embed themselves in the endometrium, they start secreting an LH-like hormone called **chorionic gonadotropin (CG)**. This hormone travels in the bloodstream to the ovary, where it prevents breakdown of the corpus luteum. The corpus luteum continues to secrete estrogen and progesterone, and the uterine lining continues to grow, nourishing the embryo. So much CG is released by the embryo that it is excreted by the mother in her urine; in fact, most pregnancy tests use the presence of CG in a woman's urine to determine pregnancy.

Although negative feedback regulates the levels of most hormones, the hormones of the menstrual cycle are regulated by both positive and negative feedback. During the first half of the cycle, FSH and LH stimulate estrogen production by the follicles. High levels of estrogen then *stimulate* the midcycle surge of FSH and LH release (positive feedback). During the second half of the cycle, estrogen and progesterone together *inhibit* the release of FSH and LH (negative feedback). The early positive feedback causes hormone concentrations to reach high levels, and the later negative feedback shuts the system down again unless pregnancy intervenes.

Copulation Allows Internal Fertilization

As terrestrial mammals, humans use internal fertilization to deposit the sperm in the moist environment of the female reproductive tract. To do this, the penis is inserted into the vagina, where sperm are released during ejaculation. The sperm swim upward in the female reproductive tract, from the vagina through the opening of the cervix into the uterus, and on up into the oviducts. If the female has ovulated within the past day or so, the sperm will meet an egg in one of the oviducts. Only one sperm can succeed in fertilizing it, starting the development of a new human being.

During Copulation, Sperm Are Deposited inside the Female Vagina

The male role in copulation begins with erection of the penis. Before erection, the penis is relaxed (flaccid), because the arterioles supplying it are constricted, allowing little blood flow (Fig. 39-16a). Under the dual influences of psychological and physical stimulation, the arterioles dilate and blood flows into spaces in the tissue within the penis. As these tissues swell, they squeeze off the veins that drain the penis (Fig. 39-16b). Pressure builds up, causing an erection. After the penis is inserted into the vagina, movements further stimulate touch receptors on the penis, triggering ejaculation. Ejaculation occurs when muscles en-

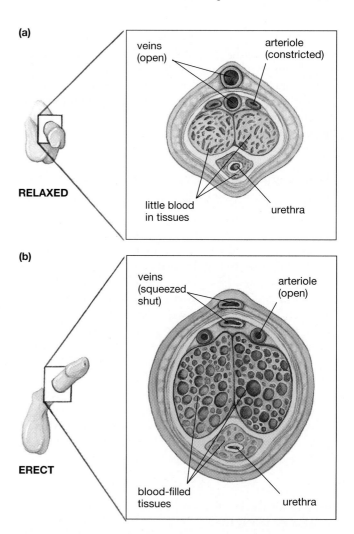

Figure 39-16   Changes in blood flow within the penis cause erection

(a) Normally, smooth muscles encircling the arterioles leading into the penis are contracted, limiting blood flow. (b) During sexual excitement, these muscles relax, and blood flows into spaces within the penis. The swelling penis squeezes off the veins leaving the penis, thereby increasing the pressure produced by fluids within the penis and causing it to become elongated and firm.

circling the epididymis, vas deferens, and urethra contract, forcing semen out of the penis and into the vagina. On average, 3 or 4 milliliters of semen, containing 300 to 400 million sperm, is ejaculated. Male orgasm causes both ejaculation and a feeling of intense pleasure and release.

Similar changes occur in the female. Sexual excitement causes increased blood flow to the vagina and external parts of the reproductive tract, including the labia and clitoris (see Fig. 39-12). The clitoris, which is derived from the same embryological tissue as the penis, becomes erect. Stimulation by the penis of the male often, but not always, results in female orgasm, a series of rhythmic contractions of the vagina and uterus accompanied by sensations of pleasure and release. Female orgasm is not necessary for fertilization.

The intimate contact involved in copulation creates a situation where disease organisms can readily be transmitted. Since the "sexual revolution," which began in the 1960s, many people have had multiple sexual partners, and the incidence of sexually transmitted diseases has greatly increased (see "Health Watch: Sexually Transmitted Diseases").

During Fertilization, the Sperm and Egg Nuclei Unite

Neither sperm nor egg lives very long. An egg may remain viable for a day, and sperm, under ideal conditions, may live for two. Therefore, fertilization can succeed only if copulation occurs within a couple of days before or after ovulation. You will recall that the egg leaves the ovary surrounded by follicle cells (Fig. 39-17a). These cells, now called the **corona radiata,** form a barrier between the sperm and the egg. A second barrier, the jellylike **zona pellucida** ("clear area"), lies between the corona radiata and the egg. Recent research suggests that the human egg releases a chemical attractant that lures the sperm toward it.

In the oviduct, hundreds of sperm reach the egg and encircle the corona radiata, each sperm releasing enzymes from its acrosome (Fig. 39-17b). These enzymes weaken both the corona radiata and the zona pellucida, allowing the sperm to wriggle through to the egg. If there aren't enough sperm, not enough enzymes are released, and none of the sperm will reach the egg. This may be the selective pressure for the ejaculation of so many sperm. Perhaps 1 in 100,000 reach the oviduct, and 1 in 20 of those find the egg, so only a few hundred of the 300 million sperm that were ejaculated join the attack on the barriers surrounding the egg.

When the first sperm finally contacts the surface of the egg, the cell membranes of egg and sperm fuse, and the sperm head is drawn into the cytoplasm of the egg. As the sperm enters, it triggers two vital changes in the egg. First, vesicles near the surface of the egg release chemicals into the zona pellucida, reinforcing it and preventing further sperm from entering the egg. Second, the egg undergoes its second meiotic division, producing a haploid gamete at last. Fertilization occurs as the haploid nuclei of sperm and egg fuse, forming a diploid nucleus that contains all the genes of a new human being.

Human males who have fewer than 20 million sperm per milliliter of semen (about one-fifth the normal amount) usually cannot fertilize a woman during intercourse because too few sperm reach the egg. If the sperm are otherwise normal, such men can father children by artificial insemination, in which a large quantity of their semen is injected directly into the oviduct. In other cases, a blocked oviduct may prevent sperm from reaching the egg. Today, some couples seek high-technology help in the form of *in vitro* fertilization (see "Scientific Inquiry: *In Vitro* Fertilization").

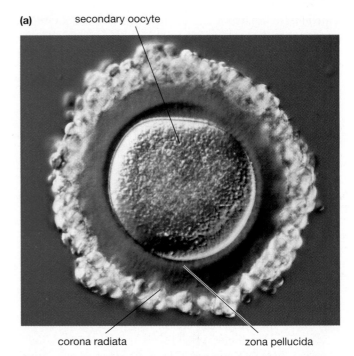

(a)   secondary oocyte

corona radiata   zona pellucida

(b)

Figure 39-17   The secondary oocyte and fertilization

(a) A human secondary oocyte shortly after ovulation. Sperm must digest their way through the small follicular cells of the corona radiata and the clear zona pellucida to reach the oocyte itself. **(b)** Sperm surround the oocyte, attacking its defensive barriers.

During Pregnancy, the Developing Embryo Grows within the Uterus

The zygote begins to divide while being carried by cilia down the oviduct to the uterus, a process that takes about 4 days (Fig. 39-18). By about 1 week after fertilization, the zygote has developed into a hollow ball of cells, the **blastocyst** (Fig. 39-18b). A thickened region of the blasto-

SCIENTIFIC INQUIRY

In Vitro Fertilization

Louise Brown, the first "test-tube baby," was born in England in 1978. Since that well-publicized event, more than 200 centers for *in vitro* fertilization (IVF) have been established in the United States; these have produced over 26,000 healthy babies. The demand for IVF is fueled by an epidemic of infertility. Twenty years ago one out of every six American couples was unable to conceive after trying for 1 year or more; that rate has now tripled. One reason for the increase in infertility is that modern couples often delay childbearing, and fertility declines with age. A second reason is a higher incidence of sexually transmitted diseases such as chlamydia and gonorrhea that can scar and block the oviducts or sperm ducts. Both blocked ducts and low sperm counts can be overcome with IVF, because oocytes are removed directly from the ovaries, and their meeting with sperm is guaranteed within the confines of a small glass dish (*in vitro* means "in glass" in Latin). The popularity of IVF, which costs around $6000 per attempt, is a testimony to the strong biological drive to have children. With an average success rate of 14%, a typical conception via IVF costs over $40,000.

Although the technique is simple in concept, the procedure is complex and delicate. First, the woman is given daily injections of drugs or hormones or both to stimulate multiple ovulation. Using blood tests and ultrasound imaging of the ovaries, doctors determine when the time is ripe for ovulation. Then the woman is injected with human chorionic gonadotropin, which begins the process of expelling the oocytes from the follicles. Just before the oocytes are ejected, surgeons insert a thin fiber-optic viewing device through a small abdominal incision to locate the mature follicles. Next, they insert a long, hollow needle into each ripe follicle and suck out the follicular cells and fluid, which are examined under a microscope. With luck, an oocyte is present (Fig. E39-1). Usually, at least four oocytes are harvested and incubated in a glass dish to which freshly collected sperm are then added. In 48 hours, about two-thirds of the oocytes will have been fertilized and have reached the eight-cell stage. A few of these early embryos (usually two to four) are sucked into a tube and expelled very gently into the uterus. Transplanting multiple embryos increases the success rate for implantation, but at the same time it increases the probability for multiple births, which carry considerably higher risks than single births.

Recently, IVF has become a weapon in the fight to save endangered species. The National Zoo in Washington, D.C., has been working since 1984 to adapt IVF technology to help endangered species reproduce. They have developed a mobile IVF laboratory that can travel to zoos where endangered species are housed. A tremendous advantage of IVF is that it will allow sperm from a male of an endangered species to be transported between continents, if necessary, to fertilize an appropriate female. This method eliminates the danger and trauma of transporting the animals themselves. IVF also overcomes the very real probability that, once together, the animals will refuse to mate. In April 1990, the first "test-tube tiger" was born (Fig. E39-2). Only 200 individuals of this rare Siberian subspecies remain in the wild. Use of IVF increases the chances that we may save this and other precious forms of life on Earth.

Figure E39-2   The world's first test-tube tiger
Born in 1990, this test-tube tiger is the first successful use of IVF in the fight to save endangered species.

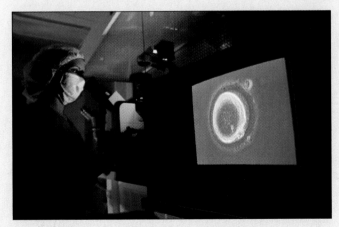

Figure E39-1   Examining the harvested egg
The microscopic egg cell is found in the follicular fluid, and its image is projected onto a screen for examination.

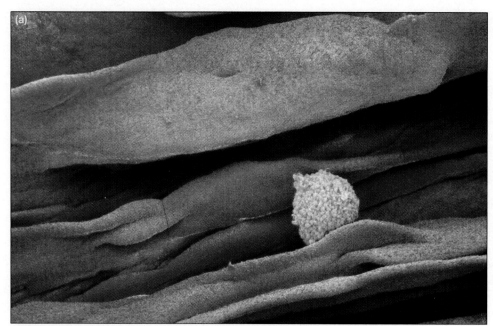

Figure 39-18   **The journey of the egg**

(a) The egg, surrounded by the corona radiata, travels down the oviduct toward the uterus. It emits chemicals that attract sperm, increasing its chances of being fertilized. **(b)** The egg is fertilized in the oviduct and slowly travels down to the uterus. Along the way, the zygote divides a few times, until a hollow blastocyst is formed. The inner cell mass will form the embryo, while the surrounding cells will adhere to the uterine endometrium, burrow in, and begin forming the placenta.

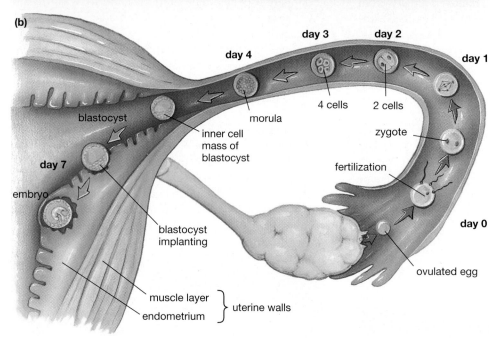

cyst, called the **inner cell mass,** will become the embryo itself, while the sticky outer ball will adhere to the uterus and burrow into the endometrium, a process called **implantation.** Blood from ruptured uterine vessels plus glycogen secreted by glands in the endometrium nourish the growing embryo.

Obtaining nutrients directly from the nearby cells of the endometrium suffices only for the first week or two of embryonic growth. During this time, the placenta, composed of interlocking tissues of the embryo and the endometrium, begins to form. Through the placenta, the embryo will receive nutrients and oxygen and dispose of wastes into the maternal circulation. The details of embryonic development are presented in Chapter 40.

Milk Secretion, or Lactation, Is Stimulated by Pregnancy

As the fetus grows, nourished by nutrients diffusing through the placenta, changes are occurring in the mother's breasts that prepare her to continue nourishing her child after it is born. The breast contains milk glands arranged in a circle around the nipple, each with a duct leading to the nipple. Although the breasts begin enlarging at puberty under the influence of estrogen, their glandular structure does not fully develop until pregnancy occurs (Fig. 39-19). Then, large quantities of estrogen and progesterone (acting together with several other hormones) stimulate the milk glands to grow, branch, and de-

During the first few days after birth, the milk glands secrete a thin, yellowish fluid called **colostrum.** Colostrum is high in protein and contains antibodies that are absorbed directly through the infant's intestine and help protect the newborn against some diseases. Colostrum is gradually replaced by mature milk, which is higher in fat and milk sugar (lactose) and lower in protein.

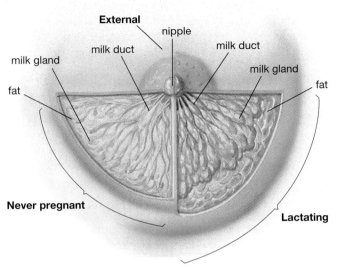

Figure 39-19   The structure of the mammary glands

During pregnancy, both fatty tissue and the milk-secreting glands and ducts increase in size.

velop the capacity to secrete milk. The actual secretion of milk is promoted by the pituitary hormone prolactin (see Chapter 35). The level of prolactin rises steadily from about the fifth week of pregnancy until birth.

Immediately after birth, estrogen and progesterone levels plummet, and prolactin, which stimulates milk secretion, takes over. Suckling stimulates nerve endings in the nipples. These signal the hypothalamus, which in turn triggers an extra surge of prolactin and oxytocin from the pituitary. Oxytocin causes muscles surrounding the milk glands to contract, ejecting the milk into the ducts leading to the nipples (see Chapter 35).

Reproduction Culminates in Labor and Delivery

Near the end of the ninth month, give or take a few weeks, the process of birth normally begins (Fig. 39-20). Birth is the result of a complex interplay between uterine stretching caused by the growing fetus and fetal and maternal hormones that finally trigger **labor** (contractions of the uterus that result in the birth).

Unlike skeletal muscles, uterine muscles can contract spontaneously, and these contractions are enhanced by stretching. As the baby grows, it stretches the uterine muscles, which contract occasionally weeks before delivery. Recent research suggests that the final trigger for labor is provided by the fetus. The near-term fetus produces steroid hormones that cause increased estrogen and prostaglandin production by the placenta and uterus. These hormones make the uterus even more likely to contract. When the combination of hormones and stretching activate the uterus beyond some critical point, strong contractions begin, signaling the onset of labor. As the contractions proceed, the baby's head pushes against the cervix, making it expand in diameter (dilate). Stretch receptors in the walls of the cervix send signals to the hypothalamus, triggering oxytocin release. Under the dual stimulation of prostaglandin and oxytocin, the uterus contracts even more strongly. This positive feedback cycle is finally halted when the baby emerges. After a brief rest, uterine contractions resume, causing the uterus to shrink remarkably. During these contractions, the placenta is

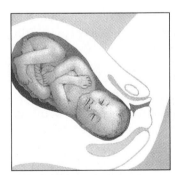 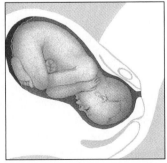 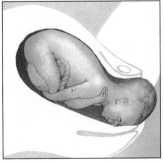 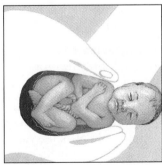

(a) The baby is oriented head downward, facing the mother's side. The cervix is beginning to thin (efface) and expand in diameter (dilate).

(b) The cervix is completely dilated to 10 centimeters (almost 4 inches) and the baby's head has entered the vagina, or birth canal. The baby has rotated to face the mother's back.

(c) The baby's head is emerging (crowning).

(d) The baby has rotated to the side once again as the shoulders emerge.

Figure 39-20   Delivery

sheared off from the uterus and is expelled through the vagina as the "afterbirth."

Further release of prostaglandins in the umbilical cord causes the muscles surrounding fetal blood vessels in the umbilical cord to contract, shutting off blood flow. (Tying off the cord is standard practice but is not usually necessary; if it were, other mammals would not survive birth!) Although he or she is still intimately dependent on the parents for survival, a new human being has been born.

On Limiting Fertility

Successful reproduction is essential if any species, including our own, is to endure. During most of human evolution, child mortality was high. Therefore, natural selection favored people who tended to produce many children. With a few tragic exceptions, people today enjoy low infant mortality and a life span triple that of ancient times. Although people today no longer need to have many children to ensure that a few survive to adulthood, we still have reproductive drives appropriate to prehistoric times. As a result, every 4 days a million new people are added to our increasingly overcrowded planet, and controlling human births has become a critical environmental issue. On the individual level, limiting reproduction allows people to plan their families and generally provide the best opportunities for themselves and their children.

Historically, limiting fertility has not been easy. In the past, some cultures have tried such inventive, if bizarre, techniques as swallowing froth from the mouth of a camel or placing crocodile dung in the vagina. Even 50 years ago there were no reliable methods of birth control. Since the 1970s, however, several effective techniques have been developed for **contraception,** or preventing pregnancy. Their mechanisms are described below, and their reliability and some possible side effects are summarized in Table 39-3. Of course, the choice of a contraceptive should be made only in consultation with a physician who can provide more complete information.

Permanent Contraception Can Be Achieved through Sterilization

The most effective, and, in the long run, most effortless method of contraception is **sterilization,** in which the pathways through which sperm or egg must travel are interrupted. In men, the vas deferens leading from each testis may be severed in an operation called a **vasectomy** (Fig. 39-21a). Sperm are still produced, but they cannot reach the penis during ejaculation. The surgery is performed under a local anesthetic, and vasectomy has no known physical side effects on health or sexual performance. The slightly more complex operation of **tubal ligation** renders a woman infertile by cutting her oviducts (Fig. 39-21b).

About 37% of women of childbearing age in the United States have chosen this form of birth control. Ovulation still occurs, but sperm cannot travel to the egg, nor can the egg reach the uterus. Sterilization is generally permanent. Sometimes, however, in a delicate and expensive operation, a surgeon can reconnect the vas deferens or oviducts.

There Are Three Major Approaches to Temporary Contraception

Temporary contraception techniques fall into three general categories: preventing ovulation, preventing sperm and egg from meeting when ovulation does occur, and preventing implantation of a fertilized egg in the uterus.

Synthetic Hormones Can Prevent Ovulation

As you learned earlier in this chapter, during a normal menstrual cycle, ovulation is triggered by a midcycle surge of LH. An obvious way to prevent ovulation is to suppress LH release by providing a continuing supply of estrogen and progesterone. Estrogen and progesterone (usually in synthetic form) are the components of **birth control pills.** "The Pill" is an extremely effective form of birth control, but it must be taken daily, usually for 21 days each menstrual period.

Two new long-term contraceptives, Norplant and Depo-Provera, have been approved for use in the United States since 1990. Both contain synthetic hormones resembling progesterone that prevent ovulation. Norplant consists of six slim, 1.3-inch long silicone rubber rods inserted under the skin of the upper arm. The rods provide gradual, steady diffusion of hormone into the bloodstream for 5 years. In extensive tests, Norplant has proved slightly more effective than the birth control pill. Women using Norplant usually become fertile within months after the capsule is removed. Depo-Provera is injected once every 3 months.

Abstinence and Barrier Methods Prevent Sperm from Reaching the Egg

Abstaining from sexual intercourse is the only completely effective means (other than sterilization) of preventing sperm and egg from meeting. Abstinence also has the advantage of complete safety from possible contraceptive side effects and from sexually transmitted diseases.

There are several effective barrier methods that prevent the encounter of sperm and egg. One is the **diaphragm,** a rubber cap that fits snugly over the cervix, preventing sperm from entering the uterus. In conjunction with a **spermicide,** diaphragms are very effective and have no serious known side effects. Alternatively, a **condom** may be worn over the penis, preventing sperm from being deposited in the vagina. A female condom is now available that completely lines the vagina. Diaphragms and condoms must be applied shortly before intercourse, often at a time when the participants would rather be

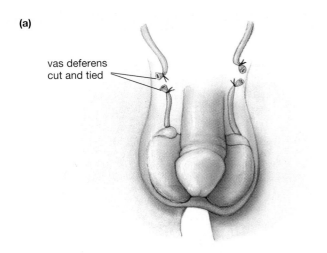

(a)

vas deferens
cut and tied

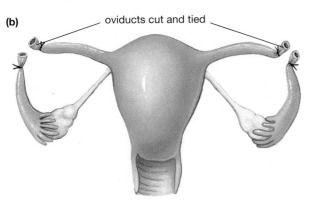

(b)

oviducts cut and tied

Figure 39-21   **Sterilization**

Sterilization is the most effective form of birth control, but it is not dependably reversible. **(a)** In the male, the vas deferens is reached through a small slit in the scrotum. It is cut and each end is tied off. **(b)** In the female, the oviduct is reached through an abdominal incision. It is cut, and each end is tied off.

thinking about something else. Furthermore, if the diaphragm or condom happens to have even a small hole, sperm may still enter the uterus and fertilize the egg. (It seems a cruel irony that a man ejaculating 50 million sperm will often be infertile, but if just a few drops of semen escape past a diaphragm or condom, pregnancy might result.)

Other, less effective, procedures include using spermicides alone, **withdrawal** (removal of the penis from the vagina just before ejaculation), and **douching** (washing sperm out of the vagina before, it is hoped, they have had a chance to enter the uterus). Spermicides have some contraceptive effect, but withdrawal and douching are essentially useless. A final method of preventing fertilization is the **rhythm method:** abstinence from intercourse during the ovulatory period of the menstrual cycle. In practice, rhythm usually has a high failure rate, because of lack of discipline on the part of the users or inaccuracies in determining the

menstrual cycle, which varies somewhat from month to month. Because a slight rise in body temperature and changes in the discharge of mucus from the cervix can be used to predict ovulation, the rhythm method can be made more effective if these indicators are monitored and sexual activity is regulated accordingly.

Intrauterine Devices Prevent Implantation

Even if an egg is fertilized, pregnancy will not occur unless the blastocyst implants in the uterus. The **intrauterine device (IUD)** is a small copper or plastic loop, squiggle, or shield that is inserted into the uterus and remains there until removed by a physician. Highly effective (if it stays in place), the IUD seems to work by irritating the uterine lining so that it cannot receive the embryo. A second method of preventing implantation is the "morning after" pill, which contains a massive dose of estrogen. For some women, preventing implantation has the major drawback that it is, in effect, an extremely early abortion.

Abortion Removes the Embryo from the Uterus

When contraception fails, pregnancy may be terminated by **abortion.** Abortion commonly involves dilating the cervix and removing the embryo and placenta by suction. The compound RU-486, taken as a pill, is now routinely used in France to terminate pregnancy up to about a month after conception and may be available in the United States by 1996. This substance binds to progesterone receptors and blocks the actions of progesterone, which are essential to the maintenance of pregnancy.

Because abortions are usually surgical procedures, they are potentially more dangerous to a woman's health than the contraceptive techniques described above. Although science can describe fetal development during pregnancy, it cannot provide judgments about when a fetus becomes a "person," or about the relative merits of fetal versus maternal rights. Therefore, abortion remains controversial.

Additional Contraceptive Methods Are under Development

Further advances in contraception are under development. For example, a vaginal ring that continuously releases synthetic progesterone has been tested in several countries but is not yet on the market. A pill that prevents ovulation by blocking receptors for gonadotropin-releasing hormone is in the testing stage. There are also once-a-month contraceptive pills under development, most of which prevent the uterus from becoming fully prepared for implantation. Another possibility is a contraceptive vaccine that would remain effective for one to several years. One such vaccine induces antibodies against human chorionic gonadotropin, which is essential for the implantation of the embryo in the uterus. Another in-

TABLE 39-3 ■ Birth Control Techniques[a]

	What Is It?	How Does It Work?
NO! (negative response)	Saying NO! is deciding not to be sexually active until you're ready to take on that heavy a commitment.	Say NO! as often as necessary to get your point across. Wear a button that says "NO!" Guaranteed stopper: "If you really loved me, you wouldn't ask."
Natural Family Planning	Combination of cervical mucus (Billings) and basal body temperature (BBT) methods to determine fertile period when you can get pregnant. Billings: based on recognizing changes in mucus discharge that occur just BEFORE ovulation; BBT: based on body temperature changes just AFTER ovulation.	You must learn from a professional nurse practitioner or doctor how to predict your fertile period, using changes in cervical mucus and body temperature.
Condom (for men) Condom (for women)	For men: Thin rubber or latex (latex is strongest) disposable sheath (usually lubricated with non-oxynol-9) worn over penis during sex. Never use Vaseline as a lubricant. ("Natural" condoms don't protect against STDs.) For women: Soft, loose-fitting prelubricated polyurethane pouch with two rings: one inserted deep in vagina, the other, when in place, remains just outside the vagina.	For both: Placed correctly, catches sperm so they can't enter vagina. Also shields you from exposure to AIDS or other STDs. Protects your partner too.
Foam (spermicides)	Sperm-killing foam inserted into vagina before having sex. Choose foam that contains non-oxynol-9 for best protection against pregnancy and STDs.	Inserted deep into vagina with plastic applicator, forms chemical barrier over uterine entrance. Sperm die when they hit foam.
The Sponge[c]	Soft disposable polyurethane sponge filled with one gram of sperm-killing non-oxynol-9 that is activated by moistening sponge with water. Can be put in vagina hours before having sex, is effective immediately, and offers contraceptive protection for up to 24 hours. Some clinics say it's number 1 choice for young lovers.	Sponge moistened with water is placed far back in vagina up to 18 hours before having sex. Sponge fits around cervix, keeps sperm from entering uterus. Traps sperm, killing them on contact.
Diaphragm/ Cervical Cap	The diaphragm and cervical cap are birth control devices that fit over the cervix and prevent pregnancy by preventing sperm from entering uterus. Both must be individually fitted by a medical professional. Both are made of rubber-like materials. Cap is half the size of diaphragm, and can be left in place twice as long.	Place one teaspoonful of spermicide containing non-oxynol-9 inside the dome and another around the edge of diaphragm or cap. Spermicide seals barrier and kills sperm.
The Pill	A pill made of a combination of synthetic hormones almost like those produced by the ovaries. Take at the same time every day whether or not you have sex. It's the day-by-day action of the pills that protects you from pregnancy.	Prevents ovary from releasing an egg. With no egg present for a sperm to fertilize, a woman cannot become pregnant. Caution: Use a back-up method plus the pill in first couple months of use.
IUD (intrauterine device)	Small plastic device treated with copper or hormones. Nylon thread attached for easy checking. Fitted by medical professional.	Prevents egg from being implanted in uterine wall. Copper IUD is replaced every 6 years; hormonal IUD, every year.
Norplant	New! Long-lasting! Doctor inserts 6 tiny flexible progestogen-filled capsules under skin of upper arm to provide protection up to 5 years. Fertility returns when Norplant is removed.	Progestogen is released in steady low doses, preventing pregnancy by blocking ovulations.
Depo-Provera (injectable)	An injectable form of contraception (given as a shot) that protects you from pregnancy for a full 3 months. Contains a chemical similar to the hormone (progesterone) that ovaries produce during the second half of the menstrual cycle.	Depo-Provera works by preventing egg cells from ripening. If an egg is not released from your ovaries during a menstrual cycle, it can't be fertilized by sperm, so you can't get pregnant. But you have to be very sure to get a shot every 3 months—no more, no less.

[a]Modified with permission from "No!" and Other Methods of Birth Control, Private Line, P.O. Box 31, Kenilworth, Illinois 60043.

[b]All of these birth control methods (except Norplant and to a lesser extent, Depo-Provera) take effort on your part. Nothing works if you leave it in the dresser drawer!

[c]The company manufacturing the sponge stopped distributing it in 1995 for business reasons. The sponge may or may not be available when you read this. Ask at your drugstore.

Does It Have Any Side Effects?	Are There Any Dangers?	How Reliable Is It?[b]
Only positive ones. If you're not loaded down with a relationship you can't handle, you'll have time to make a life for *yourself* and you won't have to worry about getting pregnant or getting a sexually transmitted disease (STD) such as chlamydia, genital warts, herpes, or AIDS.	Only that people who put pressure on you might make fun of you or drop you cold. But if they do, ask yourself what they really want from you—love and friendship, or sex?	100% effective.
No—but it doesn't offer any protection against sexually transmitted diseases (STDs).	You could become pregnant unless you avoid sex during your fertile period. Some women have trouble identifying mucus changes.	10 to 15 women out of 100 become pregnant in a given year using the Billings or BTT methods. The calendar method alone is not a reliable form of birth control.
For both: Next to not having sex, condoms or pouch used along with spermicidal foam containing non-oxynol-9 are the best available way to protect you against AIDS and other STDs—or pregnancy. (One or both of you could be allergic to spermicidal foam or lubricant.)	NO. AIDS is no joke. Condoms could save your life. Use one every time you have sex even if your partner is on the pill, has an IUD, is wearing a diaphragm—or has had a vasectomy. For women: A pouch, properly used, provides even better protection than a condom against STDs.	2 to 10 women out of 100 become pregnant in a given year when the man uses a condom correctly every time. The time period for clinical testing of the pouch has been too short for accurate statistics. The manufacturer reports that 2 to 3 women out of 100 become pregnant in a given year when the woman uses a condom every time as directed.
Occasionally the foam you're using may cause a mild vaginal irritation, cause bladder or yeast infection, or irritate your partner. Usually a change to a different brand solves the problem, but be sure the foam you use contains non-oxynol-9. If necessary, change to a birth control method that doesn't require a spermicide.	No, except your partner should use a condom when you use foam to give you the best protection, next to not having sex, against STDs and pregnancy. Just remember that protection is not 100% guaranteed.	4 to 29 women out of 100 become pregnant in a given year when using foam alone.
You don't have to worry about hormonal side effects. In clinical trials small number of women tested discontinued using because of itching, irritation, rash or allergic reactions.	Manufacturer's instructions stress use for ONLY 24-hour period. Should NOT be used during menstrual period. Slight danger of toxic shock syndrome (TSS). Sponge does not stop flow of vaginal secretions.	8 to 10 women out of 100 become pregnant in a given year when using the sponge as directed.
None unless the spermicide you're using causes bladder or yeast infection. If it does, change to another brand.	Slight danger of toxic shock syndrome (TSS). Should be used for birth control only, not to control vaginal secretions. Be sure you have inserted it correctly. Don't forget the spermicidal foam or jelly.	5 to 10 women out of 100 in a given year become pregnant when using a diaphragm or cap consistently and correctly. (If you gain or lose weight, your diaphragm may no longer fit correctly and should be checked by a medical professional.)
Positive: regular periods, less anemia, less cramping, less benign breast disease. May inhibit some forms of cancer. Negative: (normally disappear within 3 months) may include nausea, spotting, missed periods, headaches, mood changes, dark skin areas. Major but rare: blood clots, high blood pressure, gallbladder disease, heart attacks, liver tumors.	Some women should not take the pill because of other health problems. A health professional will choose the best pill for you. Read carefully all material in pill package. Follow doctor's advice. Smoking increases the chance of blood clots or stroke, even in young people on the pill.	1 woman out of 300 becomes pregnant in a given year when using the pill correctly. Always check with your doctor about possible interactions between the pill and any other medications you are taking.
Possible cramps and heavy menstrual flow caused by the body's effort to push out IUD. Expect heavy menstrual periods the first few months.	Risk of pelvic inflammatory disease (PID) or tubal pregnancy. Tell doctor if you have fever or stomach pain.	1 to 5 women out of 100 in a given year become pregnant while using an IUD.
Irregular menstrual cycle for first 6 months with longer periods, spotting between periods or skipped periods. Possible headaches, weight gain.	The method is too new for accurate statistics. Counseling to understand method is extremely important. If pregnancy does occur, have Norplant removed at once.	1 woman out of 300 becomes pregnant in a given year when using Norplant for the first 2 years. After 2 years, less protection for overweight women.
Irregular menstrual bleeding at first, then little or no menstruation, weight gain, but not much. Possibly headache, nervousness, abdominal pain, dizziness, fatigue. Check with doctor if you experience any unusual symptoms.	Important that injection be given during first 5 days of menstrual period to avoid possibility of receiving the shot during a pregnancy—which could result in ectopic pregnancy. If you should get pregnant, see your doctor immediately.	1 woman out of 300 in a given year becomes pregnant while receiving Depo-Provera injections as prescribed.

H E A L T H   W A T C H

Sexually Transmitted Diseases

Sexually transmitted diseases (STDs), caused by viruses, bacteria, protists, or arthropods that infect the sexual organs and reproductive tract, are a serious and growing health problem worldwide. The World Health Organization estimates that there are 250 million new cases of STD each year. As the name implies, these diseases are transmitted either exclusively or primarily through sexual contact. Here we discuss some of the more common of these diseases.

Bacterial Infections

Gonorrhea, an infection of the genital and urinary tract, is one of the most common of *all* infectious diseases in the United States, estimated to infect at least 2 million people each year. The causative bacterium, which cannot survive outside the body, is transmitted almost exclusively by intimate contact. It penetrates the membranes lining the urethra, anus, cervix, uterus, oviducts, and throat. In males, inflammation of the urethra results in a discharge of pus from the penis and painful urination. About 10% of infected males and 50% of infected females have symptoms that are mild or absent and do not seek treatment. They become carriers who can readily spread the disease. Gonorrhea can lead to infertility by blocking the oviducts with scar tissue. Treatment by penicillin was formerly highly successful, but penicillin-resistant strains now require use of other antibiotics. Infants born to infected mothers may acquire the bacterium during delivery. The bacterium attacks the eyes of newborns and was once a major cause of blindness. Today, most newborns are immediately given antibiotic eyedrops to kill the bacterium.

Syphilis is a far more dangerous, though less prevalent, disease than gonorrhea. It is caused by a spiral-shaped bacterium that enters the mucous membranes of the genitals, lips, anus, or breasts. Like gonorrhea, it is readily killed by exposure to air and is spread only by intimate contact. Syphilis begins with a sore at the site of infection. Syphilis may be cured with antibiotics, but if untreated, syphilis bacteria spread through the body, multiplying and damaging many organs, including the skin, kidneys, heart, and brain, sometimes with fatal results. About 4 out of every 1000 newborns in the United States have been infected with syphilis before birth. The skin, teeth, bones, liver, and central nervous system of such infants may be damaged.

Chlamydia causes inflammation of the urethra in males and the urethra and cervix in females, but in many cases there are no obvious symptoms, so the infection goes un-treated and is spread. Like the gonorrhea bacterium, *Chlamydia* can infect and sometimes block the oviducts, resulting in sterility. Chlamydial infection can cause eye inflammations in infants born to infected mothers.

Viral Infections

Acquired immunodeficiency syndrome, or **AIDS,** is caused by the HIV virus. Because the virus does not survive exposure to air, it is spread primarily by sexual activity and by contaminated blood and needles. The HIV virus attacks the immune system, leaving the victim vulnerable to a variety of infections, which almost invariably prove fatal. Children born to mothers with AIDS sometimes become infected before or during birth. There is no cure, although certain drugs, such as AZT, can prolong life. AIDS is discussed in detail in Chapter 34.

Genital herpes reached epidemic proportions during the 1970s and continues to spread to more than a million new victims yearly. It causes painful blisters on the genitals and surrounding skin and is transmitted primarily when blisters are present. The herpes virus never leaves the body but resides in certain nerve cells, emerging unpredictably, possibly in response to stress. The first outbreak is the most serious; subsequent outbreaks produce fewer blisters and may be quite infrequent. The drug acyclovir, which inhibits viral DNA replication, may reduce the severity of outbreaks. Pregnant women with an active case of genital herpes may transmit the virus to the developing child, causing severe mental or physical disability or stillbirth. Herpes may also be transmitted from mother to infant if the infant contacts blisters as it is born.

Protists and Arthropod Infections

Trichomoniasis is caused by *Trichomonas,* a flagellated protist that colonizes the mucous membranes lining the urinary tract and genitals of both males and females. The symptoms are a discharge caused by inflammation in response to the parasite. The protist is spread by intercourse but can also be acquired through contaminated clothing and toilet articles. Lengthy untreated infections can result in sterility.

Crab lice, also called pubic lice, are microscopic arachnids that live and lay their eggs in pubic hair. Their mouthparts are adapted for penetrating the skin and sucking blood and body fluids, a process that causes severe itching. "Crabs" are not only irritating; they may also spread infectious diseases. They can be controlled through careful hygiene and chemical treatments.

duces the formation of antibodies to a protein called SP-10 that is unique to sperm.

You may have noticed that most of the contraceptive techniques are directed at the woman, not the man. It is much easier to interfere with ovulation than with sperm formation. A woman ovulates only once a month, and ovulation itself does not influence a woman's sexual drives. In contrast, testosterone is essential both for sperm formation and sexual performance; early "male pills" caused not only infertility but also impotence. Nevertheless, there is a major research effort under way to develop male contraceptives equivalent to The Pill. A promising contraceptive is a daily dose of testosterone and a modified form of gonadotropin-releasing hormone, which together seem to block sperm production without affecting sexual performance. Clinical trials are now under way.

SUMMARY OF KEY CONCEPTS

Reproductive Strategies

Animals reproduce either sexually or asexually. Sexual reproduction involves the union of haploid gametes, usually from two separate parents, and produces an offspring that is genetically different from either parent. In asexual reproduction, offspring are usually genetically identical to the parent. Asexual reproduction may occur by budding, regeneration, fission, or parthenogenesis.

Among animals that engage in sexual reproduction, the female produces large, nonmotile eggs, and the male produces small, motile sperm. Animals may be either monoecious (a single animal produces both sperm and eggs) or dioecious (a single animal produces one type of gamete). The union of sperm and egg, called fertilization, may occur outside the bodies of the animals (external fertilization) or inside the body of the female (internal fertilization). External fertilization must occur in water so that the sperm can swim to meet the egg. Most internal fertilization is through copulation, in which the male deposits sperm directly into the female reproductive tract.

Mammalian Reproduction

The human male reproductive tract consists of paired testes that produce sperm and testosterone and accessory structures that conduct the sperm to the female's reproductive tract and secrete fluids that activate swimming by the sperm and provide energy. In human males, spermatogenesis and testosterone production are stimulated by FSH and LH, secreted by the anterior pituitary. Spermatogenesis and testosterone production are nearly continuous, beginning at puberty and lasting until death.

The human female reproductive tract consists of paired ovaries that produce eggs as well as the hormones estrogen and progesterone, and accessory structures that conduct sperm to the egg and receive and nourish the embryo during prenatal development. In human females, oogenesis, hormone production, and development of the lining of the uterus vary in a monthly menstrual cycle. The cycle is controlled by hormones from the hypothalamus (gonadotropin-releasing), anterior pituitary (FSH and LH), and ovaries (estrogen and progesterone).

During copulation, the male inserts his penis into the female's vagina and ejaculates semen. The sperm move through the vagina and uterus into the oviduct, where fertilization usually takes place. The unfertilized egg is surrounded by two barriers, the corona radiata and the zona pellucida. Enzymes released from the acrosomes at the tips of sperm digest these layers, permitting sperm to reach the egg. Only one sperm enters the egg and fertilizes it.

The fertilized egg undergoes a few cell divisions in the oviduct and then implants in the uterine lining. Implantation and subsequent release of chorionic gonadotropin by the embryo maintain the integrity of the corpus luteum and the endometrium during early pregnancy, preventing further menstrual cycles.

During pregnancy, milk glands in the mother's breasts enlarge under the influence of estrogen, progesterone, and other hormones. After birth, milk secretion is triggered by prolactin and oxytocin, whose release is triggered by suckling.

After about 9 months, uterine contractions are triggered by a complex interplay of uterine stretch and prostaglandin and oxytocin release. As a result, the uterus expels the baby and then the placenta.

On Limiting Fertility

Permanent contraception can be achieved by sterilization: severing the vas deferens in males (vasectomy) or the oviducts in females (tubal ligation). Temporary contraception techniques include those that prevent ovulation: birth control pills, Norplant, and Depo-Provera. Barrier methods prevent sperm and egg from meeting. These include the diaphragm, the contraceptive sponge, and the condom, accompanied by spermicide. Spermicide alone is less effective. Withdrawal and douching are poor techniques. The rhythm method involves sexual abstinence around the time of ovulation. Intrauterine devices prevent implantation of the blastocyst. Abortion causes the expulsion of the developing embryo.

✖ KEY TERMS

abortion   p. 781
acquired immunodeficiency syndrome
    (AIDS)   p. 784
acrosome   p. 770
amplexus   p. 765
asexual reproduction   p. 763
birth control pill   p. 780
blastocyst   p. 776
bud   p. 764
budding   p. 764
bulbourethral gland   p. 771
cervix   p. 773
chlamydia   p. 784
chorionic gonadotropin (CG)   p. 775
colostrum   p. 779
condom   p. 780
contraception   p. 780
copulation   p. 766
corona radiata   p. 776
corpus luteum   p. 773
crab lice   p. 784
diaphragm   p. 780
dioecious   p. 765
douching   p. 781
egg   p. 765
endometrium   p. 773
epididymis   p. 770
estrogen   p. 772
external fertilization   p. 765
fallopian tube   p. 773
fimbria   p. 773
fission   p. 764

follicle   p. 771
follicle-stimulating hormone (FSH)
    p. 770
genital herpes   p. 784
gonad   p. 767
gonadotropin-releasing hormone
    (GnRH)   p. 770
gonorrhea   p. 784
hermaphrodite   p. 765
implantation   p. 778
inner cell mass   p. 778
internal fertilization   p. 766
interstitial cell   p. 768
intrauterine device (IUD)   p. 781
labor   p. 779
luteinizing hormone (LH)   p. 770
menstrual cycle   p. 773
menstruation   p. 773
monoecious   p. 765
myometrium   p. 773
oogenesis   p. 771
oogonium   p. 771
ovary   p. 771
oviduct   p. 773
ovulation   p. 766
parthenogenesis   p. 764
pheromone   p. 765
placenta   p. 773
polar body   p. 772
primary oocyte   p. 771
primary spermatocyte   p. 769
progesterone   p. 773

prostate gland   p. 771
regeneration   p. 764
rhythm method   p. 781
scrotum   p. 767
secondary oocyte   p. 772
secondary spermatocyte   p. 769
semen   p. 770
seminal vesicle   p. 770
seminiferous tubule   p. 768
Sertoli cell   p. 768
sexual reproduction   p. 763
spawning   p. 765
sperm   p. 765
spermatid   p. 769
spermatogenesis   p. 768
spermatogonium   p. 768
spermatophore   p. 766
spermicide   p. 780
sterilization   p. 780
syphilis   p. 784
testis   p. 767
testosterone   p. 770
trichomoniasis   p. 784
tubal ligation   p. 780
urethra   p. 770
uterus   p. 773
vagina   p. 773
vas deferens   p. 770
vasectomy   p. 780
withdrawal   p. 781
zona pellucida   p. 776
zygote   p. 773

✖ THINKING THROUGH THE CONCEPTS

Multiple Choice

1. Budding, regeneration, and splitting apart are processes used by
    a. dioecious species
    b. hermaphroditic organisms
    c. organisms requiring new gene combinations for each generation
    d. sexually reproducing species
    e. asexually reproducing species

2. All of the following are barrier contraceptive devices EXCEPT
    a. IUD
    b. contraceptive sponge
    c. caps that fit over the cervix
    d. diaphragm
    e. male condom

3. In humans, spermatogenesis yields _____ sperm for each diploid germ cell, and oogenesis yields _____ secondary oocyte(s) for each germ cell.
    a. one, four   b. two, one
    c. one, two   d. four, one
    e. four, two

4. Which structure adds the *final* secretions to semen as it moves out of the human reproductive tract?
    a. epididymis
    b. bulbourethral gland
    c. seminal vesicle
    d. prostate gland
    e. interstitial cells

5. During fetal development, oogonia in human females halt meiosis at
    a. the secondary oocyte stage
    b. the polar body stage
    c. metaphase of meiosis II
    d. telophase of meiosis II
    e. prophase of meiosis I

6. The primary hormone that inhibits GnRH is
    a. FSH
    b. LH
    c. progesterone
    d. estrogen
    e. a hypothalamic releasing factor

Review Questions

1. List the advantages and disadvantages of asexual reproduction, sexual reproduction, external fertilization, and internal fertilization, including an example of an animal showing each type.
2. Compare the structures of the egg and sperm. What structural modifications are found in sperm to facilitate movement, energy use, and digestion?
3. What is the role of the corpus luteum in a menstrual cycle? In early pregnancy? What determines its survival after ovulation?
4. Construct a chart of common sexually transmitted diseases. List the disease's name, the cause (organism), symptoms, and treatment.
5. List the structures, in order, through which a sperm passes on its way from the seminiferous tubules of the testis to the oviduct of the female.
6. Name the three accessory glands of the male reproductive tract. What are the functions of the secretions they produce?
7. Diagram the menstrual cycle and describe the interactions among hormones produced by the pituitary gland and ovaries that produce the cycle.
8. Describe the changes in the breast that prepare a mother to nurse her child. How do hormones influence these changes and stimulate milk production?
9. Describe the events that lead to the expulsion of the baby and placenta from the uterus. Explain why this is an example of positive feedback.

APPLYING THE CONCEPTS

1. Identify and discuss some of the ethical issues involved in *in vitro* fertilization.
2. Discuss the most effective or appropriate method of birth control for each of the following couples: Couple A has intercourse three times a week but does not ever want to have children; Couple B has intercourse only once a month, and may want to have children someday; and couple C has intercourse three times a week and wants to have children someday.
3. Female condoms were recently introduced in the United States. What advantages and disadvantages can you think of for this form of birth control?
4. Pelvic endometriosis is a relatively common disease of women in which bits of the endometrial lining find their way onto abdominal organs and respond in typical ways to hormones during a menstrual month. When the uterine lining bleeds during menstruation, so do these implants. Common treatments are oral contraceptives, Danazol (a compound that inhibits gonadotropins), and synthetic GnRH analogues that are paradoxically powerful inhibitors of FSH and LH. How does each of these compounds provide relief?
5. Would contraceptive drugs that block cell receptors for FSH be useful in males and/or females? Explain. What side effects would such drugs have?
6. Think of all the choices a couple have to obtain a child, including: *in vitro* fertilization using the couple's eggs and sperm, *in vitro* fertilization using a donor's sperm or egg, and insemination of a surrogate mother with sperm from the couple's husband. Think of some more. Now describe what ethical issues these various options present?

FOR MORE INFORMATION

Crews, D. "Animal Sexuality." *Scientific American*, January 1994. Explores the wide range of mechanisms that control male and female sexual development among different types of animals.

Eberhard, W. G. "Runaway Sexual Selection." *Natural History*, December 1987. Interesting article on the evolution of male genitalia.

Riddle, J. M., and Estes, J. W. "Oral Contraceptives in Ancient and Medieval Times." *American Scientist*, May–June 1992. How did women control their fertility before modern medicine stepped in?

Ulmann, A., Teutsch, G., and Philibert, D. "RU-486." *Scientific American*, June 1990. Describes the abortion pill now widely used in France and other applications for this drug.

Wassarman, P. W. "Fertilization in Mammals." *Scientific American*, December 1988. Describes the events leading to the penetration of the egg by the sperm, with emphasis on the role and composition of the zona pellucida.

Wright, K. "The Sniff of Legend." *Discover*, April 1994. This article explores human pheromones, sex attractants, and a sixth sense organ in the nose.

NET WATCH

On-line resources for this chapter are on the World Wide Web at:
http://www.prenhall.com/~audesirk (click on the table of contents link and then select Chapter 39).

40 Animal Development

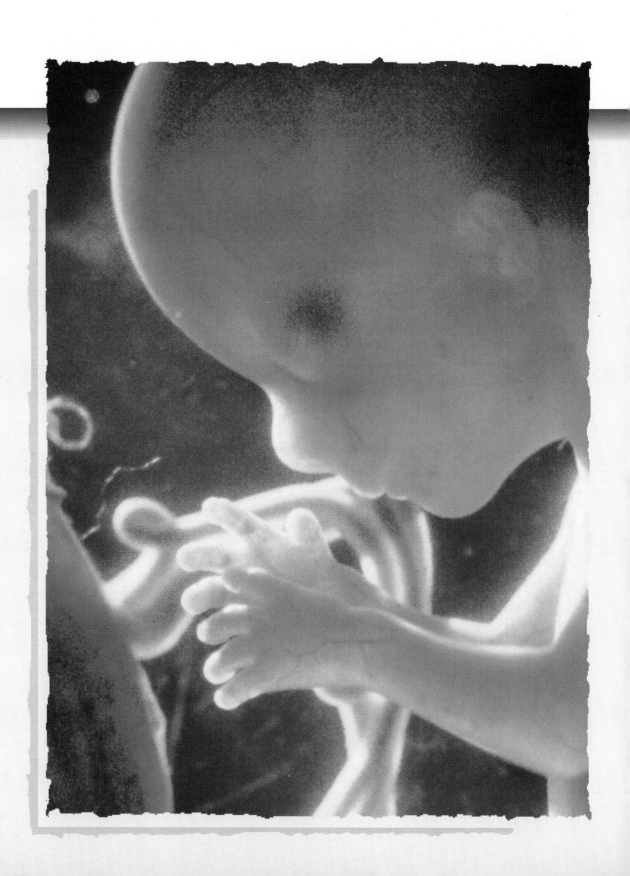

▓ AT A GLANCE

Differentiation
Gene Transcription Is Precisely Regulated during Development

Indirect and Direct Development
During Indirect Development, Animals Undergo a Radical Change in Body Form
Newborn Animals That Undergo Direct Development Resemble Miniature Adults

Stages of Animal Development
Cleavage Distributes Gene-Regulating Substances
Gastrulation Forms Three Tissue Layers
Adult Structures Develop during Organogenesis
Sexual Maturation Is Controlled by Genes and the Environment
Aging Seems to Be Genetically Programmed

Human Development
During the First Two Months, Rapid Differentiation and Growth Occur
The Placenta Secretes Hormones and Exchanges Materials between Mother and Embryo
Growth and Development Continue during the Last Seven Months
Development Culminates in Birth

A four-month-old human fetus showing the head, hands, and umbilical cord that connects the fetus to the placenta, where it obtains oxygen and nutrients from the mother's bloodstream

Out of millions of contestants, a single sperm fuses with the egg. The two haploid nuclei—one from the egg, contributed by the mother, and one from the sperm, contributed by the father—fuse to create a diploid cell, the beginning of a new generation. How does this single cell give rise to the trillions of cells of the adult body? It cannot be a simple matter of cell division after cell division, for that would merely give rise to a massive lump of identical cells. Rather, as the cells divide, they also must **differentiate**—that is, they must specialize to become particular cell types, such as liver, brain, or muscle.

How do the cells of the body become different from one another, when all are descended from the same fertilized egg? How do all the organs of the adult body assume their correct locations and connect with one another in the proper way? What governs the onset of puberty and reproductive behavior? Why do animals age? Is death inevitable? These questions inspire the study of **development,** the process by which an organism proceeds from fertilized egg through adulthood to eventual death. In this chapter, we examine animal development, especially the early embryonic stages, and discuss some of the mechanisms that control it. Finally, we describe the development of the human embryo from fertilization to birth.

Differentiation

How do cells become differentiated from one another during development? Because the characteristics of each cell are ultimately determined by its genes, one possibility might be that differentiation results from a progressive loss of genes. By this scenario, the zygote would contain all the genes needed to direct the construction of the whole organism. Each differentiated cell, though, would contain only those genes needed for its particular function in the body; unnecessary genes would be lost during the process of specialization. In a clever experiment, the British molecular biologist J. B. Gurdon showed that gene loss cannot be the mechanism of differentiation. Gurdon implanted nuclei from intestinal cells of tadpoles of the African clawed frog, *Xenopus*, into unfertilized eggs whose own nuclei had been destroyed (Fig. 40-1). Although the

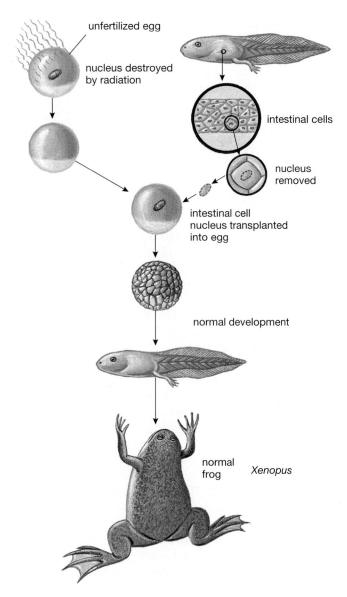

Figure 40-1 **Cells retain all of their genes during differentiation**

J. B. Gurdon's experiment proving that cells do not lose genes as they differentiate. Gurdon destroyed the nuclei of unfertilized frog eggs and then transplanted nuclei of intestinal cells from a tadpole into the egg. The resulting egg cells developed into normal tadpoles and eventually adult frogs, demonstrating that the intestinal cells retained all the genes necessary for normal development of an entire organism.

operation was very difficult and most of the "patients" died, some of the eggs with intestinal nuclei, when fertilized, developed into normal adult frogs.

Because the nucleus from an intestinal cell provided all the genetic information necessary to form a normal tadpole, these experiments led to the following conclusion: Each differentiated cell contains all the genetic information needed for the development of the entire organism. Cell types differ, then, not in which genes they

contain, but in which genes they make use of. In other words, cell types differ because different genes are transcribed to messenger RNA and translated into proteins.

Gene Transcription Is Precisely Regulated during Development

In Chapter 11, we discussed some of the mechanisms that control gene transcription, the production of messenger RNA using the gene as a blueprint. Although geneticists are just beginning to unravel the details of how transcription is controlled, the principles are straightforward. Cellular materials, often proteins or proteins combined with activating substances such as steroid hormones, travel to the nucleus and bind to the chromosomes. These proteins then block transcription of certain genes or promote the transcription of other genes (see "A Closer Look: Homeobox Genes and the Control of Body Form"). Which genes are transcribed largely determines the shape, structure, and activity of the cell.

There are two major sources of chemical substances that control gene transcription and therefore direct cell differentiation: (1) substances positioned in specific locations within the egg cytoplasm and (2) chemical messages received from other cells.

During the development of the egg (oogenesis), various gene-regulating substances become concentrated in different specific places in the egg cytoplasm (scientists are still studying how this is accomplished). The fertilized egg then divides in particular planes so that each of its daughter cells receives different gene-regulating substances (Fig. 40-2). Thus, the developmental fate of a daughter cell is determined by the part of the egg cytoplasm it receives and hence by the gene-regulating substances it inherits. Experiments that uncovered this principle are described later in the chapter.

During later embryonic development and continuing throughout adult life, cells constantly receive chemical messages, including nutrients, hormones, and neurotransmitters, from other cells of the body. These chemical messages can alter the transcription of genes and the ac-

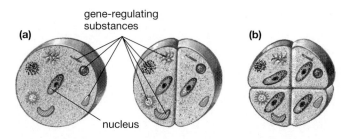

Figure 40-2 **Distribution of gene-regulating substances**

Gene-regulating substances are not distributed evenly during division of a fertilized egg. **(a)** Different gene-regulating substances (represented by the various symbols) are positioned in particular places in the egg cytoplasm during oogenesis. **(b)** As the egg divides, these materials remain in about the same positions, so the daughter cells inherit different substances.

tivity of enzymes within a cell, as described in Chapter 35. The effects of these two processes during development are described in more detail later in the chapter.

Indirect and Direct Development

All animal eggs contain food reserves of **yolk,** rich in lipids and protein. This food reserve is crucial to the **embryo,** as the organism is called in its early stages of development. The yolk provides nourishment for the early embryo until it develops into a form that can obtain food from outside sources. The amount of yolk in an egg corresponds closely with the way in which an animal develops. Animal development usually proceeds down either one of two paths: indirect development or direct development.

During Indirect Development, Animals Undergo a Radical Change in Body Form

In **indirect development,** the juvenile animal that hatches from the egg differs significantly from the adult, as a caterpillar differs from a butterfly. As it matures, the offspring undergoes radical changes in body form. Indirect development occurs in most of the invertebrates, including insects and echinoderms, and in a few vertebrates, notably the amphibians. Animals with indirect development typically produce huge numbers of eggs, and each egg has only a small amount of yolk. The yolk nourishes the developing embryo during a rapid transformation into a small,

sexually immature form called a **larva** (Fig. 40-3). Because the yolk is small and the time spent as an embryo is relatively short, indirect development does not place great demands on the mother, and many offspring can be produced.

During the larval feeding stage, some larval animals not only look very different from adult animals, but they also occupy entirely different habitats. In addition, most larvae feed on different organisms than they will as adults. For instance, the aquatic larva of the dragonfly feeds on aquatic organisms such as tadpoles, but the adult dragonfly, which is terrestrial, feeds on insects (Fig. 40-3b). Eventually the larvae undergo a revolution in body form, or **metamorphosis,** and become sexually mature adults.

Although people tend to regard the adult form as the "real animal" and larvae as "preparatory stages," most of the life span of some animals, especially insects, is spent as a larva. The adult may live for only a few days, reproducing frantically and in some cases not even eating. The mayfly, for example, metamorphoses from an aquatic larva that may have spent a year or more feeding and growing. Emerging in huge swarms from freshwater streams, ponds, and lakes, adult mayflies may live a few hours or, at most, a few days. The sole occupation of the adults is to mate and lay eggs; their fragile dead bodies then accumulate in piles to be swept away by the wind.

Newborn Animals That Undergo Direct Development Resemble Miniature Adults

Other animals, including such diverse groups as reptiles, birds, mammals, and land snails, show **direct development,** in which the newborn animal is a miniature, but sexually

Figure 40-3 **Indirect development**

Many animals, including many marine mollusks such as this common whelk **(a),** undergo indirect development. The larval stage is often very different from the adult in size, appearance, and life-style. The molluscan larva (upper left) is barely visible to the naked eye. **(b)** The larval dragonfly (left) is aquatic and feeds on small fish, whereas the adult form is terrestrial and eats other insects.

Figure 40-4 **Direct development**

The offspring of animals with direct development closely resemble their parents from the moment of birth, except of course in size. Lizards **(a)**, land snails **(b)**, and birds **(c)** hatch from large, yolk-filled eggs. Mammalian mothers **(d)** nourish their young within their bodies for weeks or months.

immature, version of the adult (Fig. 40-4). As it matures, it may grow much bigger, but it does not radically change its body form. These juveniles are typically much larger than larvae and consequently need much more nourishment before emerging into the world.

Two different strategies have evolved that meet the embryo's food requirement. Reptiles, birds, and land snails produce large eggs containing large amounts of

yolk. An ostrich egg, for example, weighs several pounds. Mammals, some snakes, and a few fish have relatively little yolk in their eggs but instead nourish the developing embryo within the body of the mother (Fig. 40-4d). Either way (in contrast to indirect development), providing food for directly developing embryos places great demands on the mother, and relatively few offspring are produced.

Reptiles, Birds, and Mammals Produce Similar Extraembryonic Membranes

Both birds and mammals evolved from reptiles. As an adaptation to direct development in a terrestrial environment, reptile and bird embryos produce four membranes, called **extraembryonic membranes.** The **chorion** lines the shell and exchanges oxygen and carbon dioxide through the shell. The **amnion** encloses the embryo in a watery environment, the **allantois** surrounds wastes, and the **yolk sac** contains the stored food. Although mammalian eggs contain almost no yolk, much of the reptilian genetic program for development still persists, including the four extraembryonic membranes. Table 40-1 compares the structures and functions of these extraembryonic membranes in reptiles and mammals.

Stages of Animal Development

Animal development (particularly in the vertebrates) may be loosely divided into several stages. The initial stages of cleavage, gastrulation, organogenesis, and growth occur during embryonic life (Fig. 40-5), during which nearly all the organs that will be present in the adult are formed. After birth, the animal typically undergoes further growth, achieves sexual maturity, reproduces, ages, and finally dies. Let's look at each of these events.

Cleavage Distributes Gene-Regulating Substances

Development begins with **cleavage,** which is the division of the fertilized egg without an increase in size. Cleavage reduces the cell size and distributes gene-regulating substances to the newly formed cells. As you know, an egg is a very large cell. Unlike most cell divisions that proceed through the cycle—divide, grow, duplicate genetic material, then divide again—embryonic cell divisions skip the growth phase during cleavage. Consequently, as cleavage progresses, the available cytoplasm is split up into ever-smaller cells whose sizes approach those of the adult organism. Finally, a solid ball of small cells, the **morula,** is formed. The morula is still about the same size as the zygote. Then a cavity opens within the morula, so that the cells become the outer covering of a hollow (often spherical) structure, the **blastula.**

To a considerable extent, the pattern of cleavage is controlled by the amount of yolk, because yolk hinders the division of the cytoplasm (cytokinesis). The almost yolkless eggs of sea urchins divide symmetrically, but eggs with extremely large yolks, such as a hen's egg, don't even divide all the way through. Nevertheless, a hollow blastula

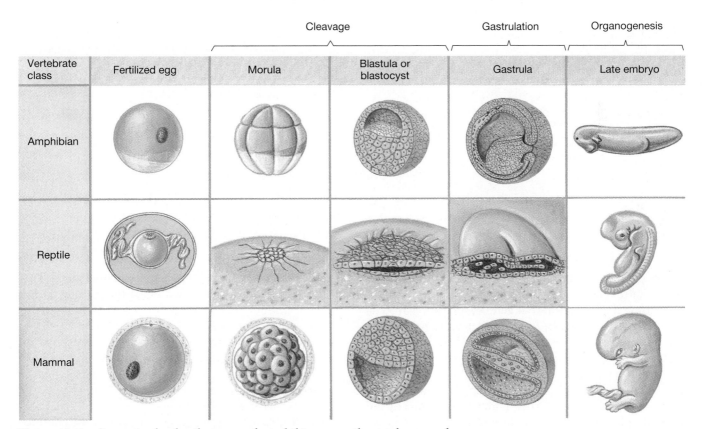

Figure 40-5 **Stages in the development of amphibians, reptiles, and mammals**

Use these illustrations as a guide to accompany the descriptions in the text.

TABLE 40-1 ❖ *Vertebrate Embryonic Membranes*

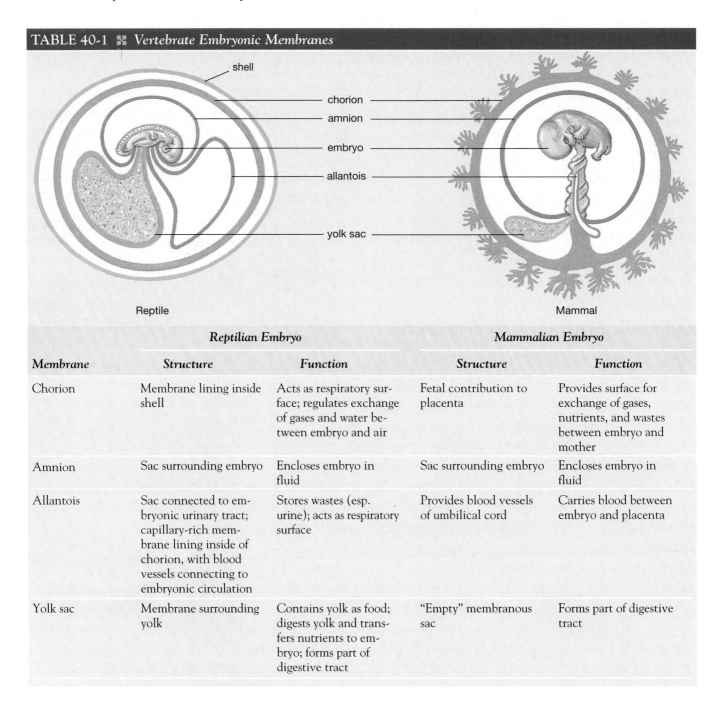

Reptile

Mammal

	Reptilian Embryo		**Mammalian Embryo**	
Membrane	**Structure**	**Function**	**Structure**	**Function**
Chorion	Membrane lining inside shell	Acts as respiratory surface; regulates exchange of gases and water between embryo and air	Fetal contribution to placenta	Provides surface for exchange of gases, nutrients, and wastes between embryo and mother
Amnion	Sac surrounding embryo	Encloses embryo in fluid	Sac surrounding embryo	Encloses embryo in fluid
Allantois	Sac connected to embryonic urinary tract; capillary-rich membrane lining inside of chorion, with blood vessels connecting to embryonic circulation	Stores wastes (esp. urine); acts as respiratory surface	Provides blood vessels of umbilical cord	Carries blood between embryo and placenta
Yolk sac	Membrane surrounding yolk	Contains yolk as food; digests yolk and transfers nutrients to embryo; forms part of digestive tract	"Empty" membranous sac	Forms part of digestive tract

is always produced, though in reptiles and birds the blastula is flattened rather than spherical.

During cleavage, different gene-regulating materials become incorporated into different daughter cells (see Fig. 40-2). This process was discovered decades ago in experiments on frog eggs (Fig. 40-6). The unfertilized frog egg has pale yolk on the "bottom," or vegetal, pole and pigmented cytoplasm on the "top," or animal, pole. At fertilization, some of the pigment shifts toward the animal pole, leaving behind a **gray crescent** of intermediate coloring. The first cleavage division of frog eggs normally passes through the center of the gray crescent, so that each daughter cell receives roughly half the crescent (Fig. 40-6a). If the two daughter cells are gently separated, each will develop into a normal tadpole. But if scientists experimentally force the first division to miss the gray crescent, one daughter cell receives the entire gray crescent, while the other receives none (Fig. 40-6b). If the cells are then separated, the cell with the gray crescent develops into a tadpole; the one without any crescent material merely forms a lump of cells that soon dies. Clearly, gene-regulating substances in the gray crescent region are required for the normal development of the tadpole.

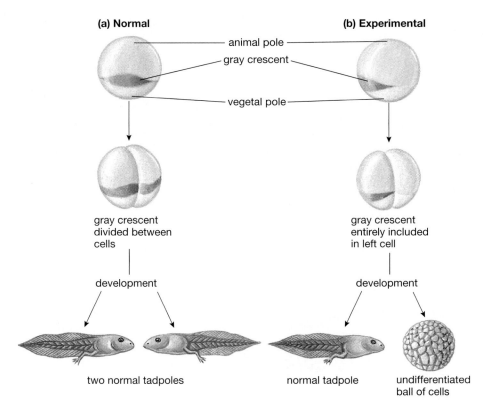

(a) Normal

animal pole
gray crescent
vegetal pole

gray crescent
divided between
cells

development

two normal tadpoles

(b) Experimental

gray crescent
entirely included
in left cell

development

normal tadpole   undifferentiated
ball of cells

Figure 40-6 **Distribution of gene-regulating substances during cleavage**

Different gene-regulating substances are distributed to different daughter cells during cleavage. In frog eggs, a pigmented region known as the gray crescent contains substances needed for normal embryonic growth and differentiation. **(a)** Normally, the first cleavage division cuts neatly through the center of the gray crescent. If the two daughter cells are separated in the lab, each cell contains gray crescent material, and each can subsequently develop into a normal tadpole. **(b)** Scientists can experimentally force the first cleavage division to miss the gray crescent, so that one daughter cell receives the entire gray crescent while the other receives none. The cell with the crescent material develops normally, but the cell lacking crescent substance cannot.

Gastrulation Forms Three Tissue Layers

In the next step of development, an indentation called the **blastopore** forms on one side of the blastula. Blastula cells migrate in a continuous sheet in through the blastopore, much as if you punched in an underinflated basketball (Fig. 40-7), to form three embryonic tissue layers. The enlarging dimple is destined to become the digestive tract; the cells that line its cavity are now called **endoderm** (Greek for "inner skin"). The cells remaining on the outside will form the epidermis of the skin and the nervous system and are called **ectoderm** ("outer skin"). Meanwhile, some cells migrate between the endoderm and ectoderm, forming a third and final layer, the **mesoderm** ("middle skin"). Mesoderm gives rise to muscles, skeleton, and the circulatory system (Table 40-2). This process of cell movement is called **gastrulation,** and the three-layered embryo that results is the **gastrula.**

The Cellular Environment Influences the Developmental Fate of Cells

During gastrulation, the developmental fate of most of the embryo's cells is determined by chemical messages received from other cells, a process called **induction.** In amphibian embryos, cells derived from the gray crescent region of the zygote form the site of dimpling as the blastula is transformed into the gastrula. This area, called the dorsal lip of the blastopore, controls the developmental fate of the cells around it, as Hans Spemann and Hilde Mangold demonstrated in the 1920s (Fig. 40-8a). They transplanted the

dorsal lip of the blastopore from one embryo to another. The transplanted dorsal lip then induced the cells of the host to form a second embryo, showing that dorsal lip tissue controls differentiation in the surrounding cells. One can also perform a control experiment, transplanting cells from regions of the gastrula other than the dorsal lip of the blastopore (Fig. 40-8b). These cells give rise to tissues appropriate to the region into which they were transplanted, rather than the region from which they were taken. These experiments clearly demonstrate that gene expression within differentiating cells (in this case nonblastopore cells) can be controlled by substances produced by other cells (in this case, cells of the dorsal lip of the blastopore).

TABLE 40-2 *Derivation of Adult Tissues from Embryonic Cell Layers*	
Embryonic Layer	**Adult Tissue**
Ectoderm	Epidermis of skin; lining of mouth and nose; hair; glands of skin (sweat, sebaceous, and mammary glands); nervous system; lens of eye; inner ear
Mesoderm	Dermis of skin; muscle, skeleton; circulatory system; gonads; kidneys; outer layers of digestive and respiratory tracts
Endoderm	Lining of digestive and respiratory tracts; liver; pancreas

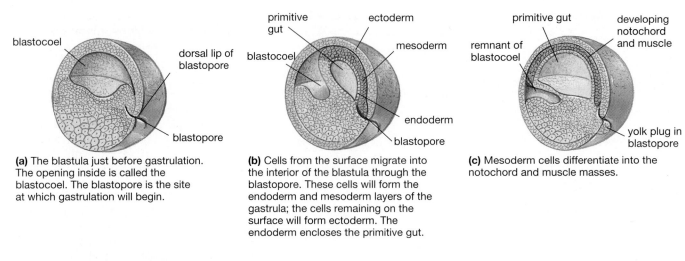

(a) The blastula just before gastrulation. The opening inside is called the blastocoel. The blastopore is the site at which gastrulation will begin.

(b) Cells from the surface migrate into the interior of the blastula through the blastopore. These cells will form the endoderm and mesoderm layers of the gastrula; the cells remaining on the surface will form ectoderm. The endoderm encloses the primitive gut.

(c) Mesoderm cells differentiate into the notochord and muscle masses.

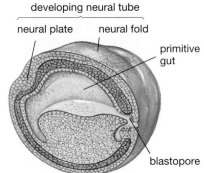

(d) The notochord induces ectoderm cells lying directly above it to form the neural tube.

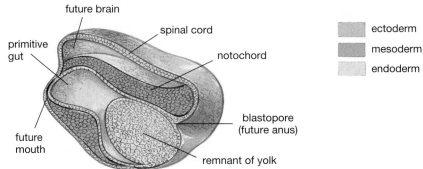

(e) The neural tube enlarges and differentiates into brain and spinal cord. A future mouth is produced when the opening formed by the primitive gut breaks through at the end of the embryo opposite the blastopore. The blastopore is the future anus.

Figure 40-7   **Gastrulation in the frog**

Adult Structures Develop during Organogenesis

Gradually, ectoderm, mesoderm, and endoderm rearrange themselves into the organs characteristic of the animal species (see Table 40-2). This process, called **organogenesis,** also usually occurs by induction.

In some cases, adult structures are, in effect, "sculpted" by the death of excess cells produced during embryonic development. Death of some cells is programmed to occur at a precise time during development. At least two mechanisms seem to be at work in different tissues. Some cells die during development unless they receive a "survival signal." Embryonic vertebrates, for example, have far more motor neurons in their spinal cords than adult animals do. Motor neurons are programmed to die unless they successfully synapse on a skeletal muscle, which releases a chemical that prevents the death of its own motor neuron.

For other cells, the situation is just the reverse: Some cells live unless they receive a "death signal" from other cells of the developing animal. Many embryonic structures disappear during development. All vertebrates pass through developmental stages with tails and webbed hands and feet. In humans, these stages can be seen clearly in the 5-week human embryo (see Fig. 40-13). Two weeks later, the webbing cells have died, revealing separate fingers, and the tail cells are dying, causing the tail to regress (Fig. 40-9). In frogs, the tail is lost during metamorphosis from its tadpole larva. Thyroid hormone, which triggers metamorphosis, stimulates cells in the tail to synthesize enzymes that digest the tail away. If the thyroid gland is surgically removed, the frog retains its tail.

Sexual Maturation Is Controlled by Genes and the Environment

Development does not stop at birth; animals continue to change throughout their lives. Animals become sexually mature at an age that is determined by both genes and environment. Many animals undergo months to years of growth and development before they can become sexually mature. This developmental process is largely regulated by their genes. Once the animal has

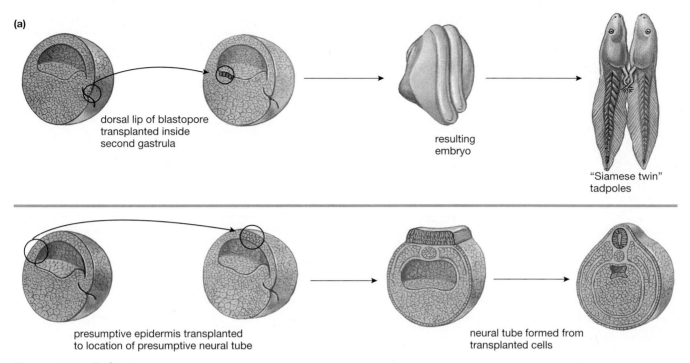

(a)

dorsal lip of blastopore
transplanted inside
second gastrula

resulting
embryo

"Siamese twin"
tadpoles

presumptive epidermis transplanted
to location of presumptive neural tube

neural tube formed from
transplanted cells

Figure 40-8   **Induction**

Experiments demonstrating the role of induction in the differentiation of frog embryos.
(a) The dorsal lip of the blastopore is removed from the gastrula of a heavily pigmented strain
of newt and transplanted into a different region of the gastrula from a lightly pigmented
strain of newt. The transplanted dorsal lip of the blastopore induced the adjacent regions of
the "host" gastrula to form a separate, nearly complete tadpole with the dark pigmentation of
the transplanted tissue. **(b)** As the first experiment suggests, many cells are "followers," guided
along certain developmental paths by their proximity to inducing tissues such as the dorsal lip
of the blastopore. Here, cells that would normally become skin from the gastrula of a darkly
pigmented strain of newt are transplanted to another site on a "host" gastrula of the lighter
strain. The host gastrula has induced cells that would normally have formed skin to produce a
neural tube. The developmental fate of follower cells is determined by the region into which
they are transplanted, and not by the region from which they were removed.

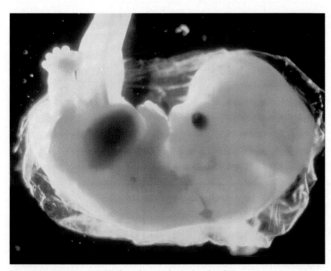

Figure 40-9   **Cell death sculpts body form**

By the seventh week, the human form has been more clearly
defined by selective death of cells that form the tail and connect
the fingers and toes. The tail has nearly disappeared, and fingers
and toes are separated from one another.

reached the appropriate age, the precise onset of sexual maturity is often triggered by environmental stimuli.
Songbirds, for example, almost always become sexually mature in the spring, stimulated by the increasingly
long days. Internal, and perhaps social, factors also influence maturation in many species. The age of puberty
among women, for instance, has dropped substantially
during the past few centuries. Puberty at a younger age
is due in part to improved nutrition, but social stimulation by early close contact between the sexes may also
be involved.

Aging Seems to Be Genetically Programmed

Most of your cells will function less efficiently, or divide more
slowly, as you age. Is death, for cells and for entire organisms, a programmed part of life? From an evolutionary standpoint, natural selection promotes only those mechanisms
that keep an organism alive and healthy during the time
that it is producing and nurturing its young. Repair mechanisms that extend longevity past this time are not favored

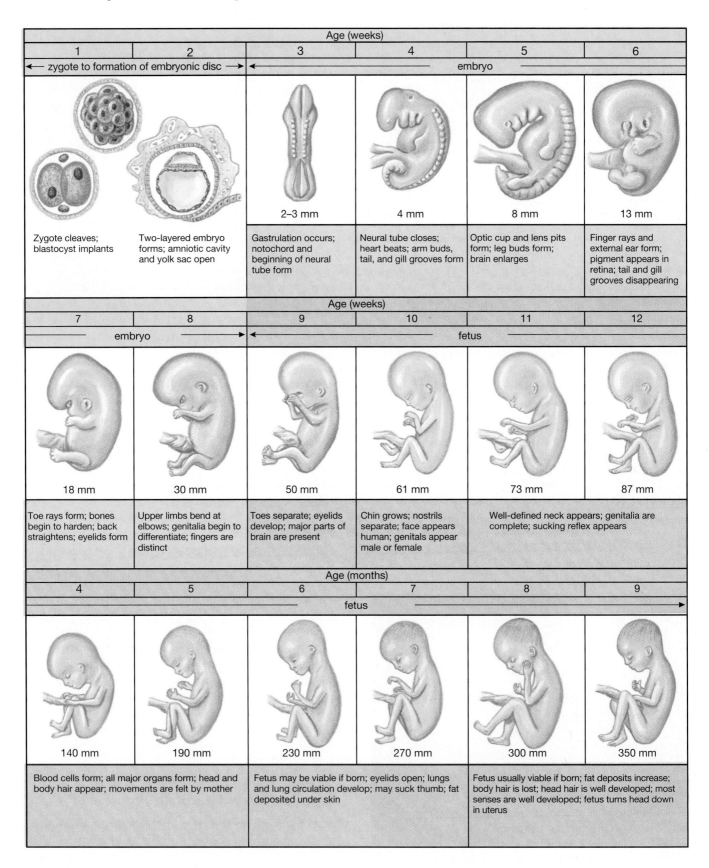

Figure 40-10 **Human embryonic development**

A calendar of human embryonic development, from blastula to birth.

and may even be harmful to the population as a whole. For example, older, nonreproductive individuals might compete with younger ones for limited resources, such as food.

Evidence for programmed cell death is seen in cells grown in dishes in the laboratory. These divide a relatively fixed number of times, then stop, and eventually die. Recent research has suggested that this maximum life span varies from species to species and that longevity depends on the ability of the cell to repair damage to its DNA. Longer-lived cells and longer-lived animals have more natural enzymes that protect against DNA damage and are better at repairing damaged DNA. Nevertheless, all normal cells seem to stop dividing and die eventually.

Not all cells, however, are "normal." Cancer cells survive and reproduce indefinitely in cell culture or in the body (which is what makes cancer so devastating). Some lines of cancer cells have been reproducing in culture for decades. Although scientists are still working to uncover the mechanism that regulates the life span of a cell, cancer cells provide evidence that this mechanism can be bypassed. However, cultured cancer cells frequently mutate and change their characteristics. Can we discover a method of defusing the self-destruct mechanism while retaining proper controls over the cells in our bodies? No one knows, but we can be certain it will not happen soon.

Human Development

Human development is controlled by the same mechanisms that control the development of other animals. In fact, our development strongly reflects our evolutionary heritage, a fact that we shall emphasize in the brief discussion to follow. Figure 40-10 summarizes the stages of human embryonic development. You may want to refer to this figure as we go along and, perhaps, later if you have a child of your own.

During the First Two Months, Rapid Differentiation and Growth Occur

A human egg is normally fertilized in a woman's oviduct and undergoes a few cleavage divisions on its way to the uterus. Just before being implanted in the uterine wall, the embryo, now called a **blastocyst** (the mammalian version of a blastula), consists of a thin-walled, hollow ball with a thicker **inner cell mass** on one side (Fig. 40-11). The thin outer wall becomes the chorion and will form the embryonic contribution to the placenta, while the inner cell mass develops into the embryo and the three other extraembryonic membranes.

After implantation, the inner cell mass grows and splits, forming two fluid-filled sacs that are separated by a double layer of cells called the **embryonic disc** (Fig. 40-12). One sac, bounded by the amnion, forms the amniotic cavity. The amnion eventually grows around the embryo, en-

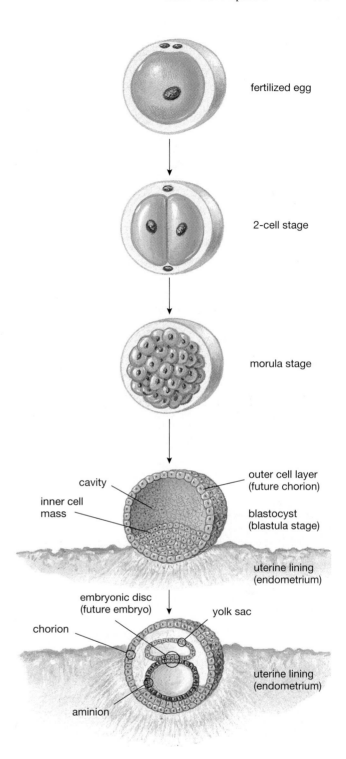

Figure 40-11 **Human development during the first and second week**

As it travels through the oviduct, the fertilized egg undergoes cleavage, forming a morula. In the uterus, the morula becomes a blastocyst (blastula) and implants in the uterine lining. At this stage it consists of an outer layer of cells surrounding an inner cell mass. As it burrows into the uterine lining, the outer cell layer forms the chorion, the embryonic contribution to the placenta. The inner cell mass forms the amnion, yolk sac, and the embryonic disc, which will become the embryo.

closing it in what one author has called its "private aquarium," providing the watery environment needed by all animal embryos. The yolk sac, corresponding to the yolk sac of reptiles and birds, forms the second cavity, although in humans it contains no yolk. At this stage, the embryonic disc consists of an upper layer of future ectoderm cells (on the side facing the amniotic cavity) and a lower layer of future endoderm cells (on the side facing the yolk sac).

Gastrulation begins about the fifteenth day after fertilization (Fig. 40-12a). The upper and lower layers split apart slightly, and a slit, the **primitive streak** (corresponding to the blastopore), appears in the center of the upper layer. Cells of the upper layer migrate through the primitive streak into the interior of the embryo, forming mesoderm. The upper layer is now called ectoderm and the lower layer is called endoderm. One of the earliest mesoderm structures to develop is the **notochord,** a supporting rod found at some stage in all chordates (Fig. 40-12b).

During the third week of development, the embryo and its amniotic sac begin to curl ventrally (toward the yolk sac; Fig. 40-12c). As the embryo grows, the endoderm pinches, forming a tube that will become the digestive tract (Fig. 40-12d). Simultaneously, the notochord induces formation of a groove in the overlying ectoderm, which folds inward and then closes over to become the **neural tube,** the forerunner of the brain and spinal cord. By the end of the fourth week, the amnion completely surrounds the embryo, punctured only by the umbilical cord connecting the embryo to the placenta.

By the end of the sixth week, the embryo clearly displays its chordate ancestry (see Chapter 24), having developed a notochord, primitive gill grooves, and a prominent tail (Fig. 40-13). These, of course, disappear as development continues. The embryo already has the rudimentary beginnings of the eyes, a beating heart, and separating fingers on its tiny hands. Especially notable at this

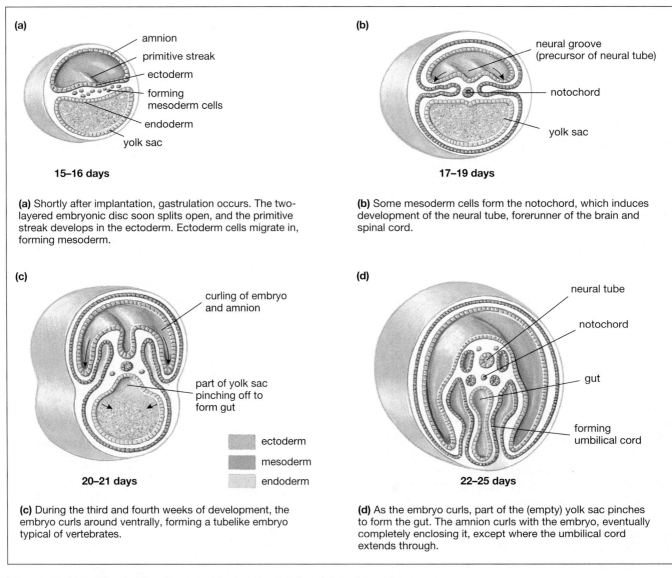

(a)
amnion
primitive streak
ectoderm
forming mesoderm cells
endoderm
yolk sac

15–16 days

(a) Shortly after implantation, gastrulation occurs. The two-layered embryonic disc soon splits open, and the primitive streak develops in the ectoderm. Ectoderm cells migrate in, forming mesoderm.

(b)
neural groove (precursor of neural tube)
notochord
yolk sac

17–19 days

(b) Some mesoderm cells form the notochord, which induces development of the neural tube, forerunner of the brain and spinal cord.

(c)
curling of embryo and amnion
part of yolk sac pinching off to form gut

ectoderm
mesoderm
endoderm

20–21 days

(c) During the third and fourth weeks of development, the embryo curls around ventrally, forming a tubelike embryo typical of vertebrates.

(d)
neural tube
notochord
gut
forming umbilical cord

22–25 days

(d) As the embryo curls, part of the (empty) yolk sac pinches to form the gut. The amnion curls with the embryo, eventually completely enclosing it, except where the umbilical cord extends through.

Figure 40-12   Human development during the third and fourth week

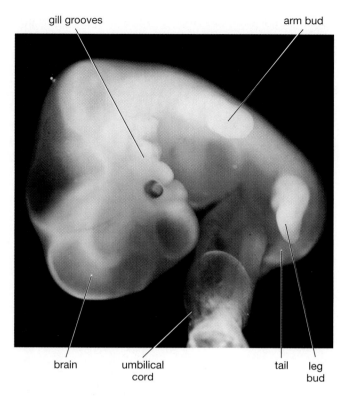

gill grooves   arm bud

brain   umbilical   tail   leg
         cord   bud

Figure 40-13   **The fifth week**

At the end of the fifth week, the human embryo is about half head. The feet and hands have begun to develop digits. A tail and gill grooves are clearly visible, evidence of our evolutionary relationship to other vertebrates.

stage is the rapid growth of the brain, which is nearly as large as the rest of the body. In fact, many of the structures of the adult brain are already recognizable.

As the second month draws to an end, nearly all the major organs have formed, and the embryo begins to look quite human (Fig. 40-14). The gonads appear and develop into testes or ovaries, depending on the presence or absence of the Y chromosome. Sex hormones are secreted, testosterone from the testes or estrogen from the ovaries, and these hormones affect the future development of the embryonic organs, including not only the reproductive organs but also certain regions of the brain. After the second month of development, the embryo is called a **fetus,** denoting that it has taken on a generally human appearance.

These first 2 months of pregnancy are times of extremely rapid differentiation and growth for the embryo and also times of considerable danger. Although the fetus is vulnerable throughout development, rapidly developing organs are the most sensitive to environmental insults, such as drugs or certain medications taken by the mother.

The Placenta Secretes Hormones and Exchanges Materials between Mother and Embryo

During the first few weeks of pregnancy, embryonic cells burrow into the thickened lining of the uterus (the endometrium). The outer cells of the embryo form the chorion, which penetrates the endometrium with fingerlike projections called **chorionic villi.** From this complex in-

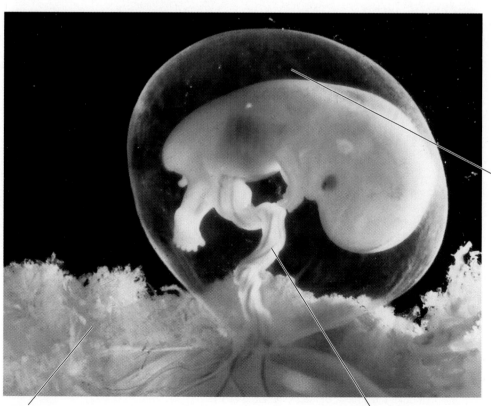

Figure 40-14   **The eighth week**

At the end of the eighth week, the embryo is clearly human in appearance and is now called a fetus. Most of the major organs of the adult body have begun to develop.

amniotic sac

placenta   umbilical cord

terweaving of tissues arises the **placenta,** a marvelously intricate organ. The placenta has two major functions: It secretes hormones, and it allows the selective exchange of materials between the mother and the fetus.

As the placenta develops during the first 2 months of pregnancy, it begins secreting estrogen and progesterone. Estrogen stimulates growth of the uterus and mammary glands, while progesterone also stimulates the mammary glands and inhibits premature contractions of the uterus.

The placenta also regulates the exchange of materials between the blood of the mother and the blood of the fetus without allowing the two to mix. The chorionic villi contain a dense network of fetal capillaries and are bathed in pools of maternal blood (Fig. 40-15). This arrangement permits diffusion of many small molecules between the fetal and maternal blood. Oxygen diffuses from maternal blood to fetal blood, and carbon dioxide from fetal blood to maternal blood. Nutrients travel from mother to fetus, some aided by active transport. Fetal urea diffuses into the mother's blood, to be filtered out by the mother's kidneys.

While allowing exchange by diffusion, the membranes of the capillaries and chorionic villi act as barriers to the passage of some large proteins and most cells. In spite of this barrier, some disease-causing organisms and many harmful chemicals can penetrate the placental barrier, as described in "Health Watch: The Placenta Provides Only Partial Protection."

Growth and Development Continue during the Last Seven Months

The fetus continues to grow and develop for another 7 months. Although the rest of the body is "catching up" with the head in size, the brain continues to develop rapidly, and the head remains disproportionately large. Nearly every nerve cell ever formed during the entire human life span develops during embryonic life. As the brain and spinal cord grow, they begin to generate noticeable behaviors. As early as the third month of pregnancy, the fetus begins to move about and respond to stimuli. Some instinctive behaviors appear, such as sucking,

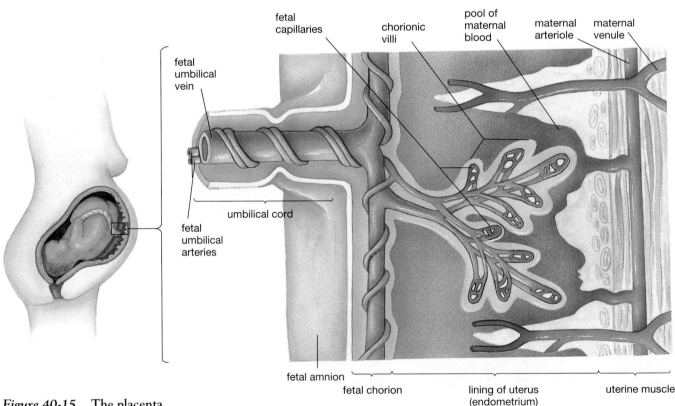

Figure 40-15   **The placenta**

The placenta is formed from both the chorion of the embryo and the endometrium of the mother. Capillaries of the endometrium break down, releasing blood to form pools within the placenta. Meanwhile, the chorion develops projections (the chorionic villi) extending into these pools of maternal blood. Blood vessels from the umbilical cord branch extensively within the villi. The resulting structure separates the maternal and fetal blood supplies, while generating a large surface area for diffusion of oxygen, carbon dioxide, nutrients, and wastes between the fetal capillaries and the maternal blood pools. Umbilical arteries carry deoxygenated blood from the fetus to the placenta, and umbilical veins carry oxygenated blood back to the fetus.

A CLOSER LOOK

Homeobox Genes and the Control of Body Form

In 1948, Edward Lewis of the California Institute of Technology began a systematic analysis of mutations in the fruit fly *Drosophila* that cause one body part to be replaced by another; for example, legs growing where antennae should be (Fig. E40-1). Lewis discovered that these organ substitutions could be caused by mutations in single genes, which he called "master genes," that control the activity of the many other genes necessary to produce the misplaced organ.

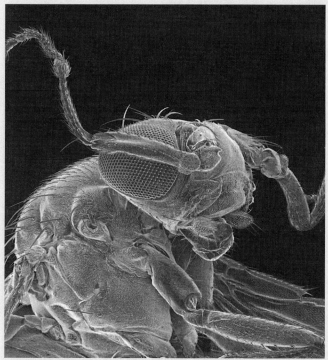

Figure E40-1

A mutation in a homeobox segment of a gene in this *Drosophila* has caused legs to grow where antennae should be.

In 1983, researchers found that the different master genes in *Drosophila* shared a nearly identical DNA sequence, now called a **homeobox.** Each homeobox segment of DNA codes for a similar 60-amino-acid protein. The proteins encoded by each homeobox have, however, small crucial differences that allow them to specify which body regions will develop into which organs. These proteins are found in the cell nucleus, where they bind to specific genes, turning them on or off. Homeoboxes, by coding for proteins that activate or inactivate other genes,

seem to directly specify the course of differentiation of embryonic cells. Homeoboxes determine the head-to-tail axis of the embryo and establish the appropriate sequence of structures along its length, thereby determining the overall shape of the body and the location and shape of its parts. Consequently, mutations in the homeobox can cause legs to replace antennae in the fruit fly.

Researchers studying diverse developmental questions, from the formation of color patterns on butterfly wings to the development of the forelimb in chickens and frogs, have found that chemical gradients often specify the fate of cells. In other words, during development, cells may take different forms in response to different concentrations of a regulatory substance. As a result, a gradient (such as occurs by diffusion from a concentrated source) of a single substance can produce an entire sequence of structural characteristics (such as the sequence of structures from the shoulder to the fingers of a forelimb). One source of such chemical gradients is the homeobox. For example, during the development of the clawed frog *Xenopus,* a forelimb bud forms from mesoderm that expresses an identified homeobox gene. The homeobox codes for a protein that forms a concentration gradient: high on the "thumb side" of the bud and low on the "pinkie side." As the limb lengthens, this protein forms a second gradient: high at the shoulder and lower toward the hand. Gradients of identical proteins have been found in the forelimbs of developing chick and mouse embryos as well.

Homeobox gene segments have a common evolutionary origin, and their discovery is merging the interests of evolutionary and developmental biologists. Researchers have discovered homeoboxes in humans and mice that are nearly identical to those of the fruit fly and that can be transplanted into the fruit fly and function normally. Because evolutionary lines leading to insects and mammals diverged 500 million years ago, the conservation of these short segments of DNA suggests that homeoboxes play a fundamental role in organizing the body plan of all multicellular animals. Recently, homeoboxes resembling those specifying the head end of the fruit fly have been discovered in the cnidarians (jellyfish), which are among the simplest of all multicellular animals.

The development of a single cell, the fertilized egg, into the fantastic complexity of a tadpole or a human infant is simultaneously one of life's greatest mysteries and one of biology's most exciting fields of research. The discovery of the homeobox has brought us one step closer to understanding this incredible journey.

H E A L T H   W A T C H

The Placenta Provides Only Partial Protection

When your grandparents were bearing children, doctors assumed that the placenta protected the developing fetus from most of the substances in maternal blood that could harm it. We know now that this is far from true. In fact, most medications and drugs and even some disease organisms readily penetrate the placental barrier and affect the fetus.

Infections May Cross the Placenta
The German measles virus can cross the placenta and attack the fetus, causing potentially severe retardation and other defects. As mentioned in Chapter 39, the virus causing genital herpes (during active outbreaks) and the bacterium causing syphilis can cause mental or physical defects in the developing fetus. The virus causing AIDS may also cross the placenta, so infants may be born with this incurable disease.

Drugs Readily Cross the Placenta
A tragic example of drugs crossing the placenta is the tranquilizer thalidomide, commonly prescribed in Europe in the early 1960s (Fig. E40-2). Thalidomide's devastating effects on embryos were discovered only when many babies were born with missing or extremely abnormal limbs. More recently, in the late 1980s, the anti-acne drug Accutane was found to cause gross deformities in babies born to women using it (Accutane contains retinoic acid, which has been shown to activate homeobox genes; see "A Closer Look: Homeobox Genes and the Control of Body Form"). Although these are extreme examples, *any drug, including aspirin, has the potential to harm the fetus, and any woman who thinks she may be pregnant should seek medical advice about any drugs she takes.* The use of so-called recreational drugs such as heroin and cocaine has devastating effects on the developing fetus. Babies of heroin addicts are often born addicted. In the United States, hundreds of thousands of infants have been born in recent years to users of "crack" cocaine. Research suggests that these children may be impaired, both behaviorally and emotionally.

The Effects of Smoking
Probably the most common toxic substances to which fetuses are exposed are those in cigarette smoke. Because many women who smoke are so addicted that they don't stop smoking during pregnancy, nearly a million human embryos are exposed to the poisons (including nicotine, carbon monoxide, and a host of carcinogens) each year in the United States alone. Women who smoke have a higher incidence of miscarriages than do nonsmokers and

Figure E40-2   Drugs interfere with development
Children born to mothers taking the tranquilizer thalidomide cope heroically with missing and deformed limbs. Drugs with no obvious ill effects on the mother may have devastating impacts on her developing fetus.

they give birth to smaller infants who are more likely to die shortly after birth than infants of nonsmokers. There is evidence that some children born to heavy smokers also suffer behavioral and intellectual impairment.

Fetal Alcohol Syndrome

The effects of alcohol on the developing fetus can be devastating. When a pregnant woman drinks, alcohol in the blood of her unborn child reaches levels as high as in her own blood. Children born to alcoholic mothers often exhibit **fetal alcohol syndrome (FAS).** Such children are retarded and may be hyperactive and irritable. FAS children have small heads and abnormally small, improperly developed brains, (Fig. E40-3), characteristic facial features, and a higher-than-normal incidence of defects of the heart and other organs. Their growth is also inhibited. FAS is the third most common cause of mental retardation in the United States, with one out of every 500 to 750 children being affected. It is likely that far higher numbers of children suffer from *fetal alcohol effect*, a milder form of FAS caused by more moderate alcohol consumption. A woman who takes one or two alcoholic drinks a day during the first 3 months of pregnancy also significantly increases her chances of having a miscarriage. *Researchers have not established any safe level of alcohol consumption during any phase of pregnancy. The U.S. Surgeon General advises pregnant women and those who are likely to become pregnant to avoid all alcohol consumption.*

In summary, a pregnant woman should assume that whatever chemicals she ingests will find their way into the bloodstream of her developing infant. Women who are likely to become pregnant need to consider that crucial stages of development occur before most women even realize that they are pregnant. The mother's choices during this critical 9-month period can strongly influence her child's future well-being.

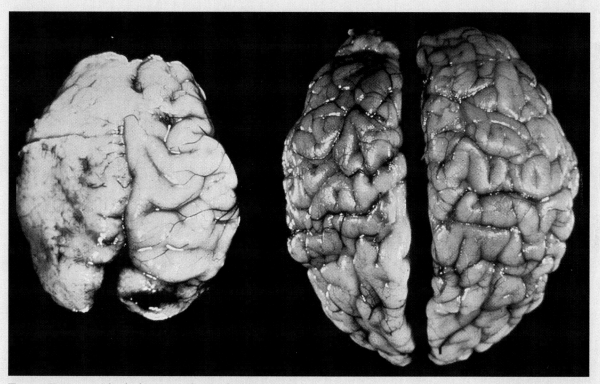

Figure E40-3   **Alcohol impairs brain development**

The devastating effects of alcohol on the developing brain are illustrated in these photos showing the brain of a child with fetal alcohol syndrome (left) and the brain of a normal child of the same age (right).

which will have obvious importance soon after birth. Structures that the fetus will need when it emerges from the womb, such as the lungs, stomach, intestine, and kidneys, enlarge and become functional, though they will not be used until after birth.

Fetuses 7 months or older can usually survive outside the womb, but larger and more mature fetuses have a much greater chance of survival. After about 9 months of development, the head is so large that it can barely fit through the mother's pelvis. The skull is compressed into a slightly conical shape as it passes through the birth canal.

Development Culminates in Birth

Normally, during the last months of pregnancy the fetus becomes positioned head downward in the uterus, with the crown of the skull resting against (and being held up by) the cervix (see Chapter 39). Usually delivered head first, the infant is in for a rude awakening. Formerly, all the world was soft, fluid-cushioned, and warm. Suddenly, the baby must obtain oxygen and eliminate carbon dioxide by breathing. It must regulate its own body temperature, and it must

Figure 40-16   On her own

suckle to obtain food. Considering that they have recently experienced what by any standard is a change for the worse, babies are often remarkably cheerful (Fig. 40-16)!

✖ S U M M A R Y   O F   K E Y   C O N C E P T S

Differentiation
All the cells of an animal body contain a full set of genetic information, yet each cell is specialized for a particular function. During development, cells differentiate by stimulating and repressing transcription of specific genes. Gene transcription is regulated in two ways: (1) The egg cytoplasm contains gene-regulating substances that are incorporated into specific daughter cells during the first few cleavage divisions. The developmental fate of each daughter cell is determined by which substances it receives. (2) Later in development, certain cells produce chemical messages that induce other cells to differentiate into particular cell types.

Indirect and Direct Development
Animals undergo either direct or indirect development. In indirect development, eggs (usually with relatively little yolk) hatch into larval feeding stages that later metamorphose into adults with body forms notably different from that of the larva. In direct development, the newborn animal is sexually immature but otherwise resembles a small adult. Animals with direct development usually either have large, yolk-filled eggs or nourish the developing embryo within the mother's body.

Stages of Animal Development
Animal development occurs in several stages. *Cleavage:* The zygote undergoes cell divisions with little intervening growth, so the egg cytoplasm is partitioned into small-

er cells. Cleavage divisions result in the formation of a solid ball of cells, the morula. A cavity then opens up within the morula, forming a hollow ball of cells, the blastula. *Gastrulation:* A dimple forms in the blastula, and cells migrate from the surface into the interior of the ball, eventually forming a three-layered gastrula. The three cell layers of ectoderm, mesoderm, and endoderm give rise to all the adult tissues (see Table 40-2). *Organogenesis:* The cell layers of the gastrula form organs characteristic of the animal species. *Sexual maturation:* The juvenile animal increases in size and achieves sexual maturity. *Aging:* Cells begin to function less efficiently, at least partly as a result of failure to repair their DNA, and eventually the animal dies.

Human Development
The human developmental program clearly reflects our reptilian ancestry. Reptilian embryos develop four extraembryonic membranes, the chorion, amnion, allantois, and yolk sac, that function in gas exchange, provision of the watery environment needed for development, waste storage, and yolk storage, respectively. These membranes are retained by the human embryo but are modified to suit development inside the mother's womb (see Table 40-1). The chorion invades the uterine lining (endometrium), and together these tissues form the placenta. Here the exchange of materials occurs by diffusion between maternal and fetal blood, while the blood supplies remain separate. By the end of the sec-

ond month, the major organs have formed, the embryo appears human and is called a fetus. In the next 7 months, until birth, the fetus continues to grow, and the lungs, stomach, intestine, and kidneys enlarge and be-come functional. Human embryonic development fol-lows the same principles as the development of other mammals. The stages of human development are sum-marized in Figure 40-10.

❋ KEY TERMS

allantois   p. 793
amnion   p. 793
blastocyst   p. 799
blastopore   p. 795
blastula   p. 793
chorion   p. 793
chorionic villus   p. 801
cleavage   p. 793
development   p. 789
differentiation   p. 789
direct development   p. 791
ectoderm   p. 795

embryo   p. 791
embryonic disc   p. 799
endoderm   p. 795
extraembryonic membrane   p. 793
fetal alcohol syndrome (FAS)   p. 805
fetus   p. 801
gastrula   p. 795
gastrulation   p. 795
gray crescent   p. 794
homeobox   p. 803
indirect development   p. 791
induction   p. 795

inner cell mass   p. 799
larva   p. 791
mesoderm   p. 795
metamorphosis   p. 791
morula   p. 793
neural tube   p. 800
notochord   p. 800
organogenesis   p. 796
placenta   p. 802
primitive streak   p. 800
yolk   p. 791
yolk sac   p. 793

❋ THINKING THROUGH THE CONCEPTS

Multiple Choice

1. Cells become differentiated through all of the following events EXCEPT
   a. changes in homeobox activity
   b. chemical messages received from other cells
   c. unequal distribution of gene-regulating substances dur-ing cleavage
   d. transcription of different genes
   e. progressive loss of genes as cells divide

2. Indirect development is characteristic of animals that produce
   a. few eggs
   b. eggs with large amounts of yolk
   c. young that are sexually immature versions of adults
   d. all of the above
   e. none of the above

3. In bird eggs, the allantois
   a. exchanges oxygen and carbon dioxide
   b. produces the shell
   c. stores wastes
   d. encloses the embryo in a watery environment
   e. contains stored food

4. The endoderm gives rise to the
   a. lining of the digestive tract
   b. epidermis of skin
   c. skeletal system
   d. muscles
   e. nervous system

5. In human development, ectoderm cells migrate through the primitive streak to form
   a. endoderm
   b. mesoderm
   c. the chorion
   d. the yolk sac
   e. the amnion

6. The process by which a tissue causes another tissue to dif-ferentiate is called
   a. gastrulation   b. metamorphosis
   c. cleavage   d. induction
   e. indirect development

Review Questions

1. Define differentiation. How do cells differentiate; that is, how is it that adult cells express some but not all the genes of the fertilized egg?

2. Describe the two processes of differentiation during devel-opment: the influence of the egg cytoplasm and induction by other cells.

3. Distinguish between indirect and direct development, and give examples of each.

4. What is yolk? How does it influence cleavage?

5. Name two structures derived from each of the three embry-onic germ layers—endoderm, ectoderm, and mesoderm.

6. What is gastrulation? Describe gastrulation in frogs and humans.

7. Describe the process of induction, and give two examples.

8. How does cell death contribute to development?

9. Describe the structure and function of four extraembryon-ic membranes found in reptiles and birds. Are these four present in placental mammals? Explain.

10. Explain how the structure of the placenta prevents mixing of fetal and maternal blood while allowing the exchange of substances between the mother and the fetus.

11. List and explain two very different functions of the placenta.

12. Is the placenta an effective barrier against substances that can harm the fetus? Describe two different types of harm-ful agents that can cross the placenta and their effects on the fetus.

13. Why are homeoboxes important in both developmental and evolutionary biology?

✖ APPLYING THE CONCEPTS

1. When fetal ectoderm tissue from the mouth of one embryo is transplanted into a region of mesoderm of an embryo of another species, the mesoderm induces the ectoderm to form mouth structures of the type of organism from which the ectoderm tissue was taken. When this type of experiment is done using chick ectoderm implanted into a mouse embryo, the chicken tissue produces teeth. Discuss what this experiment tells us about why the adage "scarce as hen's teeth" is true. Also, what does this experiment tell us about the evolutionary origin of chickens?

2. On the basis of your knowledge of genetics (Unit II) and evolution (Unit III) explain why the human embryo passes through a developmental stage in which it has gill grooves and a tail.

3. Embryologists have used embryo fusion to produce tetra-parental (four-parent) mice. They have also produced "geeps" using goat and sheep embryos. The resulting bodies are patchworks of cells from both animals. Why does fusion succeed with very early embryos (4- to 8-cell stages) and fail when much older embryos are used?

4. If the nuclei of frog intestinal cells can be transplanted into enucleated eggs (eggs from which the nucleus has been re-moved) to produce clones of the parent, is it theoretically possible to produce human clones? Would such clones yield offspring that are *exactly* identical to the parents who supplied the nuclei? Explain.

5. Estimates of the number of U.S. women who self-medicate with over-the-counter (OTC) drugs during the first trimester or throughout pregnancy range from 65% to 95%. Evidentally the public doesn't perceive nonprescription substances as drugs. Many subtle disorders, especially of the nervous system, will probably be connected to over-the-counter use in coming years. If you were engaged in pharmaceutical research, which OTC drugs would you focus on as possible culprits? What trimester of pregnancy would you study closely? Who would you chose as your study population?

6. At one stage in development, human embryos contain two ribbons of tissue capable of developing into breasts. They stretch from the armpits down to the lower abdomen. Sometimes children are born with supernumerary (extra) breasts located along these milk streaks. What normally happens to the process or timing of induction to limit the location of breasts? What must go wrong in supernumerary breast development? Why do dogs or cats have rows of mammary glands?

✖ FOR MORE INFORMATION

Beaconsfield, P., Birdwood, G., and Beaconsfield, R. "The Placenta." *Scientific American*, August 1980 (Offprint No. 1478). The placenta is one of the most remarkable of mammalian structures, allowing internal development of offspring within the mother.

Caldwell, M. "How Does a Single Cell Become a Whole Body?" *Discover*, November 1992. Explores the miracles of development in layperson's terms.

Cooke, J. "The Early Embryo and the Formation of Body Pattern." *American Scientist*, January–February 1988. Although the developmental patterns of worms, fruit flies, and humans are superficially very different, they are probably based on common mechanisms.

DeRobertis, E. M., Oliver, G., and Wright, C. V. E. "Homeobox Genes and the Vertebrate Body Plan." *Scientific American*, July 1990. Genes can be injected into amphibian eggs, allowing direct observation of the effects of specific gene products on development.

Dewitt, P. E. "Cloning: Where Do We Draw the Line?" *Time*, November 8, 1993. Does cloning of animals suggest that this technique will be applied to humans?

Rusting, R. "Why Do We Age?" *Scientific American*, December 1992. The underlying causes of human aging are being explored using experimental animals as diverse as fruit flies and roundworms.

NET WATCH

On-line resources for this chapter are on the World Wide Web at:
http://www.prenhall.com/~audesirk (click on the <u>table of contents</u> link and then select Chapter 40).

Unit VI
Animal Behavior

Young arctic foxes at play

41 The Foundations of Animal Behavior

"As I watched the geese, it appeared to me as little short of a miracle that a hard, matter-of-fact scientist should have been able to establish a real friendship with the wild, free-living animals, and the realization of this fact made me strangely happy. It made me feel as though man's expulsion from the Garden of Eden had thereby lost some of its bitterness."

KONRAD LORENZ, *King Solomon's Ring*

❖ AT A GLANCE

The Genetic Basis of Behavior

Innate Behavior
Kineses Are Changes in the Rate of Random Movements
Taxes Are Movements toward or away from a Stimulus
A Reflex Is an Instinctive Movement of Part of the Body
Fixed Action Patterns Are Complex Innate Behaviors
Innate Behaviors Are Usually Adaptive

Learning
Imprinting Is a Genetically Programmed Form of Learning
Habituation Is a Decline in Response to a Repeated
    Harmless Stimulus
Classical and Operant Conditioning Are More Complex
    Forms of Learning
Trial-and-Error Learning Occurs through Experience
Insight, or Reasoning, Is the Most Sophisticated Form of
    Learning

The Instinct to Learn and the Learning of Instincts

Evolutionary Connections: Why Do Animals Play?

This gentoo penguin feeds its chick in the Falkland Islands.

A mother bird alights on her nest, bug in beak, and drops it into the gaping mouth of one of her nestlings. Is her behavior a result of motherly love and concern, or is she responding reflexively to the stimuli of the open mouths? In this and the following chapter you will find evidence that animals' reactions to each other and to their surroundings are based on a foundation of simple or complex instinctive (genetically programmed) responses to stimuli. Depending on the complexity of the animal's brain, these responses may be modified to varying degrees by learning. In general, the simpler the nervous system of an organism, the greater the role of its genes in determining behavior. Those aspects of behavior that are under genetic control have evolved through natural selection as adaptive responses that enhance survival.

In this chapter, we introduce basic concepts of animal **behavior**, which can be defined as any observable response to external or internal stimuli. Our examples are drawn largely (but not exclusively) from the work of scientists in the field of **ethology**, the study of animal behavior as it occurs under natural or near-natural conditions. The work of early ethologists, including Konrad Lorenz, Niko Tinbergen, and Karl von Frisch provided insight into the fascinating diversity of animal behavior, its adaptiveness, and its causes. Modern ethologists continue to investigate the evolution of behavior as a response to specific selection pressures. In Chapter 42, we explore adaptive features of animal communication systems and their role in social behavior.

The Genetic Basis of Behavior

All behavior has some genetic basis, since the complexity of nervous system and physical structure of the body are genetically determined. One way that genes influence behavior can be seen in the radically differing body forms of different species. Imagine the streamlined body of a dolphin slicing through the water and the skillful web-making manipulations of a long-legged spider. Now try to imagine either of them performing the behavior of the other!

Much behavior consists of responses to environmental stimuli that are received and filtered by sense organs, then interpreted by the brain. The design of sensory organs, and the

(a)

(b)

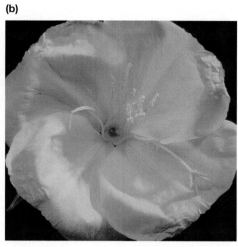

Figure 41-1   **Sensory abilities control perception**

Bees see ultraviolet wavelengths as color, but humans do not, so humans and bees perceive many flowers differently. **(a)** The ultraviolet-reflecting center of this flower (visible here under ultraviolet light using ultraviolet-sensitive film) guides bees toward the nectar and encourages them to pollinate the flower. **(b)** Humans are unable to see this pattern.

way the brain interprets their signals are genetically determined and differ among different species. Echolocating bats, for example, produce and respond to sounds far higher in pitch than the human ear can detect, and they use those sounds to produce some type of mental "image" of objects around them. A bee is guided toward nectar by patterns of reflected ultraviolet light on flowers that are invisible to the human eye (Fig. 41-1). The tick is exquisitely sensitive to the odor of butyric acid, produced by the skin of mammals. The scent will cause it to drop off its perch, usually landing on the passing animal. Clearly, behavior is influenced by perception.

Besides sending the impulses that directly control behavior, the nervous system, by its complexity and its specific neural connections, determines how much and what type of learning is possible. The relative contributions of heredity and learning vary among animal species and among behaviors within an individual. As you will see in the descriptions that follow, some behaviors are entirely genetically preprogrammed, or innate, whereas others can be modified by learning.

Innate Behavior

A fruit fly placed in a tunnel with light at one end and darkness at the other will fly toward the light; a hungry newborn human infant, touched on the side of her mouth, will turn her head and attempt to suckle. These are examples of **innate**, or **instinctive**, behaviors. Innate behavior is performed in reasonably complete form the first time an animal of the right age and motivational state encounters a particular stimulus. (The proper motivational state for feeding, for example, would be hungry and not frightened.) Innate behavior is programmed by the genes and passed from one generation to the next. Because innate behavior is performed without learning or prior experience, it tends to be highly stereotyped—that is, it is performed similarly each time. Innate behaviors fall into four categories: (1) kineses, (2) taxes, (3) reflexes, and (4) fixed action patterns.

Kineses Are Changes in the Rate of Random Movements

A **kinesis** is a behavior in which an organism changes the speed of its random movements in response to an environmental stimulus. It escapes hostile conditions by speeding up and stops moving when, by chance, it blunders into a favorable environment. For example, the pillbug (a land-dwelling crustacean) uses a kinesis to reach the moist areas it needs to survive. As the air surrounding it becomes drier, the pillbug moves faster (but in no particular direction) until it encounters a damper area, where it slows and eventually stops. This simple behavior results in congregations of pillbugs under damp leaves and rotting logs.

Taxes Are Movements toward or away from a Stimulus

In contrast to a kinesis, a **taxis** (plural, **taxes**) is a *directed* movement toward (a positive taxis) or away from (a negative taxis) a stimulus. A moth flying toward a light is one example (Fig. 41-2). Much of the behavior of very simple organisms, including single-celled protists, is taxes. For

Figure 41-2   **A taxis toward light**

Moths and other night-flying insects show a positive taxis toward light, resulting in congregations such as this.

example, the protist *Euglena* (see Chapter 21) is photosynthetic, but its chlorophyll can be damaged by intense light. As you might predict, *Euglena* shows a positive taxis toward dim light but a negative taxis toward intense light. Mosquito behavior also involves several taxes. Males orient toward the high-pitched whine of the female. Female mosquitoes (only females suck blood) show taxes to the warmth, humidity, and carbon dioxide exuded by their prey. Mosquito repellents probably act by blocking the receptors for these stimuli, rendering the insect incapable of sensing her victim. Taxes can be influenced by other stimuli. For example, the grayling butterfly shows a taxis toward bright light, but only when it is pursued. A grayling fleeing a predator flies directly toward the sun, temporarily blinding its pursuer. Taxes can be duplicated by relatively simple robots, suggesting correspondingly simple neural mechanisms, as explored in "Scientific Inquiry: Robot Cricket Finds Her Mate."

Vertebrates may also show taxes, although vertebrate taxes are more difficult to detect because they interact with other behaviors. For example, fish swim upright by orienting their back surface away from the force of gravity and toward light (normally the water surface). If the

gravity-detecting organ of a fish is removed, or the light enters the side of a fish tank, the animal becomes disoriented, as shown in Figure 41-3.

A Reflex Is an Instinctive Movement of Part of the Body

In contrast to taxes or kineses, which involve orientation of the entire body, a **reflex** is a movement of a body part, such as blinking an eye or withdrawing a hand from a hot stove. Because the conscious brain is not involved, reflexes are usually stereotyped and rapid. Take the knee jerk reflex, for example. Tapping the patellar (kneecap) ligament stretches the thigh muscle, to which the kneecap is attached, the same stretch as would occur if your knee suddenly buckled. A stretch receptor in the muscle carries an emergency message directly to a motor neuron in the spinal cord. This neuron commands the thigh muscle to contract, suddenly straightening the leg (Fig. 41-4). Had

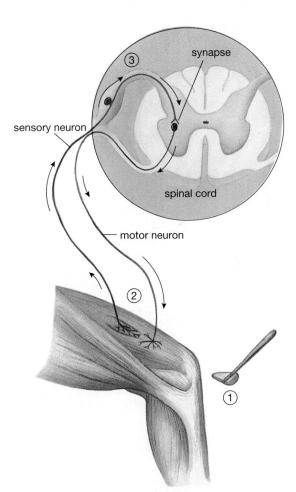

Figure 41-4   The knee jerk reflex

① Tapping the patellar ligament causes a sudden stretching of the thigh muscle. ② This stretching activates stretch receptors in the muscle that communicate to the spinal cord. ③ In the cord, the impulse is transmitted directly to a motor neuron, and the thigh muscle contracts, jerking the lower leg upward. The brain is not involved in this simple behavior.

NORMAL   INNER EAR REMOVED

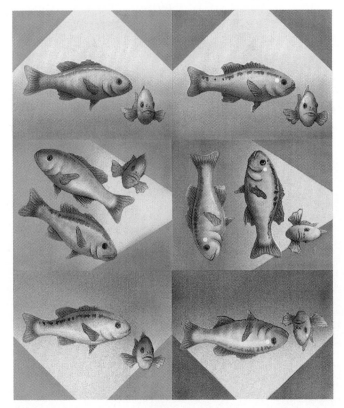

Figure 41-3   Taxes to light and gravity in the fish Crenilabrus

Fish on the left are normal; those on the right have had their inner ears removed, eliminating their response to gravity. These fish, unable to detect gravity, orient to the light alone.

S C I E N T I F I C   I N Q U I R Y

Robot Cricket Finds Her Mate

The cheerful chirping of a cricket is actually the "call song" of the male as he attempts to attract a female. The female follows the song unerringly, deftly detouring around obstacles and ignoring other sounds en route to her prospective mate. How intelligent is this apparently purposeful behavior? Barbara Webb, a psychologist at the University of Edinburgh, Scotland, attacked this problem in a novel way; she built a robot female cricket (Fig. E41-1). Her goal was to find out whether mate-finding behavior could be distilled down to relatively simple responses to stimuli, or taxes, such as could be wired into an electronic robot (and therefore wired into genetically predetermined neural connections). Although its tangle of wires appears bewildering, the circuitry of the robot is trivial when compared with the potential complexity of neural connections—even in a cricket brain.

On the laboratory bench, a loudspeaker "male cricket" broadcasts its species-specific call song: short, regularly repeating tones. As the robot rolls forward, microphonic ears conduct the song to electronic circuitry that filters it from other sounds and adds together the repeating syllables of the song that reach each ear. The summed sounds in the ear closest to the loudspeaker reach a critical threshold level first, activating a mechanism that turns the robot toward the sound. The turning halts when an equal intensity of sound hits both ears. Sensory "bumpers" help the robot detour around obstacles. The success of "robocricket" surpassed Barbara Webb's expectations; it not only found its "mate," it unexpectedly mimicked other cricket-searching behaviors. Placed between two loudspeakers broadcasting at equal volume, the robot, like a real cricket, arbitrarily chose one speaker. If the repeating syllables of the song were alternated between the two speakers, the robot (again like a real cricket) first positioned itself exactly between them, then made an arbitrary choice. The electronic circuitry provides insights into mechanisms that could be used by a simple nervous system to produce complex adaptive behavior.

Figure E41-1   **A model mimics cricket taxes**

The wiring in this model allows it to mimic several aspects of the mate-finding behavior of a real cricket (inset).

the knee actually buckled, this reflex could have prevented a dangerous fall. Your brain, which is informed of this activity by the eyes and spinal cord, does not participate in the reflex. Instead, it observes the reflex response to the doctor's hammer tap with detached interest and curiosity.

Fixed Action Patterns Are Complex Innate Behaviors

A **fixed action pattern** is a stereotyped and often complex series of movements. If the animal is at the right developmental stage and properly motivated, a fixed action pattern will be performed correctly the first time the appropriate stimulus, also called a **releaser,** is presented. Fixed action patterns may be so complex and so appropriate that they appear to be learned. How can we differentiate these complex instinctive acts from learned behaviors? Three approaches are suggested below.

Innate Behaviors Can Be Performed without Prior Experience

Scientists can demonstrate that a behavior is innate by depriving the animal of the opportunity to learn it. For example, red squirrels bury extra nuts in the fall for retrieval during the winter. Red squirrels can be raised from birth in a bare cage on a liquid diet, providing them with no experience with nuts, digging, or burying. Presented with nuts for the first time, such a squirrel (after eating several) will carry one to the corner of its cage, then make covering and patting motions with its forefeet. Nut bury-

ing is therefore an innate fixed action pattern, and the nut serves as a releaser.

Fixed action patterns can also be recognized by their occurrence immediately after birth, before there is any opportunity for learning. The cuckoo bird, for example, lays its egg in the nest of another bird species, to be raised by the unwitting adoptive parent. Immediately after hatching, the cuckoo chick shoves the nestowner's eggs (or baby birds) out of the nest, eliminating its competitors for food (Fig. 41-5).

Releasers Can Provide Evidence of a Fixed Action Pattern

Because a fixed action pattern is a programmed response to its releaser, scientists can elicit inappropriate behavior by providing the releaser in the wrong context. An example was inadvertently provided by bird banding. Parent birds clean their nests, throwing out light-colored objects such as broken egg shells or excrement. Biologists who had placed shiny metal identification bands on the legs of baby birds were dismayed to find the parents attempting to throw the band out of the nest, baby bird and all! The band was a releaser for the fixed action pattern of nest cleaning, regardless of the consequences.

Fixed action patterns are often released only by particular characteristics of the animal toward which they are directed. The female red-winged blackbird signals her readiness to mate by the angle of her tail, which serves as a releaser for mating by the male. He will even attempt to copulate with the stuffed tail of the female, provided the

(a)

(b)

Figure 41-5   **A fixed action pattern**

(a) The cuckoo chick, just hours after it hatches, and before its eyes have opened, evicts the eggs of its foster parents from the nest. **(b)** The parents feed the single, oversized cuckoo chick, unaware that it is not their own.

(a)

(b)

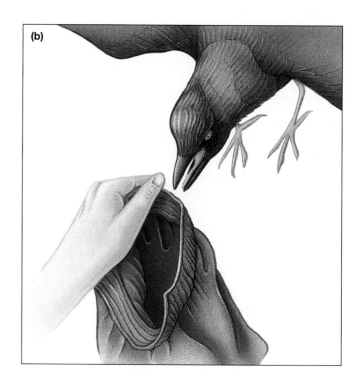

Figure 41-6   **Releasers**

(a) A male red-winged blackbird is stimulated to mate by the angle of the female's tail, whether she is attached to it or not. **(b)** Konrad Lorenz was attacked by his tame jackdaws as he carried his black bathing trunks. The trunks, resembling a dead jackdaw in the clutches of a predator, acted as a releaser for this defensive behavior.

feathers are angled appropriately (Fig. 41-6a). Konrad Lorenz described another interesting example. He was attacked by tame and normally friendly jackdaws (European relatives of crows) as he walked toward his swimming hole. The limp black bathing trunks he carried released the same defensive behavior as would a dead jackdaw in the clutches of a predator (Fig. 41-6b).

Scientists can use models to exaggerate releasers. Such exaggerated releasers, called **supernormal stimuli,** are far more effective than their natural counterparts. By selectively exaggerating specific features of a stimulus, ethologists can determine which features serve as releasers. For example, the herring gull chick, studied by Niko Tinbergen, instinctively pecks at a red spot on the parent gull's beak (Fig. 41-7). This fixed action pattern, which chicks perform immediately after hatching, causes the parent to regurgitate food that is then eaten by the chicks. Tinbergen found that the releasing features of the bill are its long, thin shape, the red color, and the presence of color contrasts. When he offered chicks a thin red rod with white stripes painted on it, they pecked at it more often than at a real beak (Fig. 41-8).

Adult herring gulls, which nest on the ground, show a fixed action pattern of egg retrieval released by the sight of an egg that has rolled out of the nest. Scientists placed various model eggs outside the nest and found that the most effective models are speckled, rounded, and very large. As shown in Figure 41-9, the gulls also preferred to

brood these monster egg models. This behavior is apparently common in ground-nesting birds. Tinbergen offered nesting geese a choice between their own egg and a volleyball, both placed just outside the nest. Invariably, the

Figure 41-7   **Supernormal stimuli are exaggerated releasers**

A herring gull chick pecks at the red spot on its mother's bill, causing her to regurgitate food for it. Niko Tinbergen and others have shown that the thin shape of the bill and the contrasting spot serve as releasers for pecking, which occurs immediately after hatching. The pecking then serves as a releaser for food regurgitation by the parent.

Model   Response
            (no. of pecks)

Figure 41-8   Experiments define releasers

Models are used to determine which features of the bill are releasers for pecking behavior. The length of the bar is proportional to the number of pecks delivered to the corresponding model. A brightly contrasting spot improves performance; the color red further increases pecking. But the red knitting needle with contrasting stripes is most effective; the long, thin shape, the red color, and the contrast are all exaggerated to produce a supernormal stimulus.

geese would attempt to retrieve the volleyball, whose size presented a truly supernormal stimulus.

Breeding Experiments Can Reveal Innate Behaviors

Sometimes scientists can produce hybrid species by mating closely related species that differ in a particular behavior. If the behavior is genetic, the hybrid offspring may try to perform a combination of the two behaviors. One excellent example was provided by behavioral biologist W. Dilger, who mated two different species of African lovebirds. One species carries nesting material to its nest site in its beak, one piece at a time. The second species tucks several pieces of nesting material into its tail feathers before flying to the nest (Fig. 41-10). Their hybrid offspring tried to place nesting material in their tail feathers but failed to release it from their beak, apparently trying to carry it in both their tails and their beaks.

Innate Behaviors Are Usually Adaptive

Innate behaviors are adaptive for a variety of reasons. First, survival may depend on the proper performance of behavior the first time the stimulus occurs. A good example is a camouflaged animal "freezing" at the sight of a predator. In such cases, even rapid learning is too slow; the animal may not have a second chance, so having this genetically programmed response is crucial to self-preservation. Second, animals with simple nervous systems may not have the neural capacity to learn behaviors important to survival. Thus, the simpler the organism, the more it relies on innate behavior. Third, social interactions important for survival may depend on the animal's rigidly performing specific roles. Complex insect societies such as those of termites and bees are possible only because insect behavior is almost entirely innate; any show of individuality could cause the society to collapse. Mating rituals provide another example. These innate rituals (described in Chapter

Figure 41-9   A supernormal egg

A herring gull will ignore her own eggs if offered a supernormal brooding stimulus.

Figure 41-10   A novel use for tail feathers

Lovebirds tearing off strips of nesting material and tucking them among their tail feathers for transport to the nest site.

42) allow animals of the same species to recognize and respond to one another automatically. The sexual rituals of different species are often incompatible, preventing the waste of genetic material and reproductive energy.

Learning

The eviction of banded baby birds by their nest-cleaning parents shows that instinctive behavior has its drawbacks, especially when the environment is not entirely predictable. Survival is enhanced if an animal can modify its behavior on the basis of experience. Thus, nervous systems have evolved with varying degrees of flexibility. In general, the more complex an animal's nervous system, the more its behavior can be modified by experience. This capacity to make *changes* in behavior on the basis of experience is called **learning.** Here we will discuss five categories of learning: imprinting, habituation, conditioning, trial-and-error learning, and insight.

Imprinting Is a Genetically Programmed Form of Learning

Learning always occurs within boundaries specified by the animal's genes. This fact is strikingly illustrated by the unique form of learning called **imprinting.** Imprinting refers to a strong association learned during a particular stage, called a **sensitive period**, in an animal's life. During this stage, the animal is primed to learn a specific type of information, which is then incorporated into an innate behavior. Imprinting is best known in birds such as geese, ducks, and chickens. These birds learn to follow the animal or object that they most frequently encounter during a sensitive period (for mallard ducks this period occurs about 13 to 16 hours after hatching). In nature, the mother is the object of imprinting. In the laboratory, however, these birds can be forced to imprint on a toy train or other moving object, although, if given a choice, they select a duck. Konrad Lorenz coined the word *imprinting* in the 1930s and demonstrated that the object of imprinting did not have to be appropriate (Fig. 41-11).

During imprinting, the animal is actually learning a releaser (the object followed) for a fixed action pattern (following) during a limited period in its development, the sensitive period. Parents may also imprint on their offspring. Herring gulls imprint on their young during the first 2 days after hatching. During this 2-day period, they will accept foster young placed in their nest. Thereafter, the parents will attack and kill baby chicks that are not their own.

Habituation Is a Decline in Response to a Repeated Harmless Stimulus

A more common form of simple learning is **habituation,** defined as a decline in response to a harmless, repeated stimulus. The ability to habituate prevents an animal from wasting its energy and attention on irrelevant stimuli. This

Figure 41-11   Konrad Lorenz

Konrad Lorenz, known as the "father of ethology," is followed by goslings that imprinted on him shortly after they hatched. They now treat him as they would their mother.

form of learning is shown by the simplest animals and has even been demonstrated in the single-celled protists. For example, the protist *Stentor* (Fig. 41-12a) contracts to touch but gradually stops retracting if touching is continued. The sea anemone, which also lacks a brain, shows a similar response to touch (Fig. 41-12b).

The ability to habituate is clearly adaptive. If an anemone contracted every time it was brushed by a strand of waving seaweed, it would waste a great deal of energy, and its retracted posture would prevent it from snaring food. Humans habituate to many stimuli: city dwellers to nighttime traffic sounds, and country dwellers to choruses of crickets and tree frogs. Each may initially find the other's habitat unbearably noisy at first but habituate after a time.

Classical and Operant Conditioning Are More Complex Forms of Learning

A more complex form of learning, most commonly seen in the laboratory, is called conditioning. During **classical conditioning** an animal learns to perform a response (normally caused by one stimulus) to a new stimulus. Ivan Pavlov, a Russian physiologist, is sometimes called the "father of conditioning." In the early 1900s, he performed his famous experiment on dogs. Pavlov first placed dried meat powder into a dog's mouth, causing reflex salivation. Then he rang a bell just before the meat powder was presented. After several repetitions, the dog would salivate to the bell alone. The

(a)

(b)

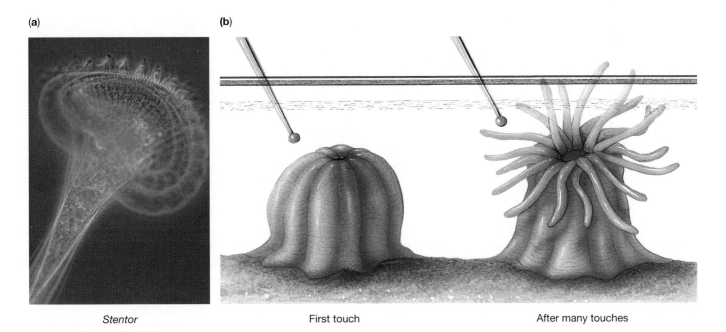

Stentor   First touch   After many touches

Figure 41-12   **Habituation**

(a) The protist *Stentor* is a complex single cell that can show habituation, a simple form of learning. **(b)** Habituation in the sea anemone. (Left) The animal, touched for the first time, withdraws. (Right) After many touches, the anemone habituates to this harmless stimulus. Learning occurs even though the anemone possesses only a simple network of neurons and lacks a brain.

dog who leaps up at the sight of her leash or drools on the floor at the sound of a can opener has been classically conditioned, however unintentionally. This form of learning can also be demonstrated in simple organisms. For example, some investigators have reported that flatworms can learn to associate a flash of light with an electrical shock, which causes them to contract. After several exposures to light paired with shock, they contract to the light alone.

During **operant conditioning** an animal learns to perform a behavior (such as pushing a lever or pecking a button) to receive a reward or to avoid punishment. This technique is most closely associated with the American comparative psychologist B. F. Skinner. Skinner designed the "Skinner box" in which an animal (often a white rat or pigeon) is isolated and allowed to train itself (Fig. 41-13). The box might contain a lever that ejects a food pellet when pressed. The animal in its explorations inevitably bumps the lever and is rewarded. Soon it will be pressing the lever repeatedly for food. Skinner boxes can be quite complex. A pigeon might learn that one lever will provide food only when a green light is shining, but when a red light is on, a specific spot on the wall must be pecked to avoid a shock.

Trial-and-Error Learning Occurs through Experience

In the natural environment, animals are faced with naturally occurring rewards and punishments and they learn by trial and error. (Operant conditioning is actually a special form

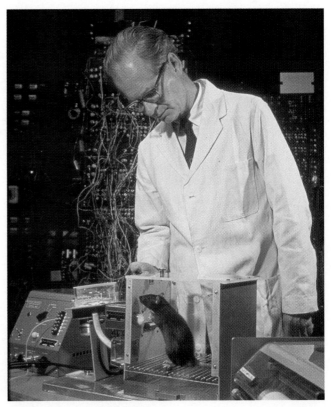

Figure 41-13   **B. F. Skinner**

B. F. Skinner, a pioneer in the study of operant conditioning, is shown here with a "Skinner box," in which a rat is trained to respond in certain ways by rewards or punishments.

of **trial-and-error learning** that occurs under artificial and carefully controlled laboratory conditions.) Through trial-and-error learning, animals acquire new and appropriate responses to stimuli through experience. Much learning of this type occurs during play and exploratory behavior in animals with complex nervous systems (see "Evolutionary Connections: Why Do Animals Play?"). Trial-and-error learning makes a major contribution to the behavior of young children, and adults as well. It is how a child learns which foods taste good or bad, that a stove may be hot, and not to pull the cat's tail.

In some instances, trial-and-error learning can modify the releaser for innate behavior, thus making it more adaptive. For example, to the hungry toad any flying insect is a releaser for feeding. Although feeding behavior usually results in a meal, occasionally the toad captures a mouthful of trouble: a bee. Having its tongue stung results in trial-and-error learning that takes only a single experience (Fig. 41-14). Learning modifies the releaser to exclude bees and even other insects that resemble them.

Insight, or Reasoning, Is the Most Sophisticated Form of Learning

The most complex form of learning is called **insight,** or reasoning. Insight involves manipulating concepts in the mind to arrive at adaptive behavior (Fig. 41-15). It can be considered a kind of mental trial-and-error learning. Insight arises from the ability to remember a variety of experiences and to apply these lessons in creative ways to new situations.

In 1917, animal behaviorist W. Kohler showed that a hungry chimpanzee, without any training, would stack boxes to reach a banana suspended from the ceiling (Fig. 41-15b). This type of mental problem solving was once believed to be limited to very intelligent animals such as primates. In 1984, however, R. Epstein and associates at Harvard observed pigeons under controlled conditions exhibiting similar insight: They moved a box beneath a suspended banana, then stood on the box to reach the banana. As investigators learn more about how to design appropriate experiments, we may find that insight is much more common than was originally suspected.

The Instinct to Learn and the Learning of Instincts

You have probably concluded by now that neither *innate* nor *learned* adequately describes the behavior of any given organism. Especially in an adult animal, every behavior is

(a) A naive toad is presented with a bee.

(b) While trying to eat the bee, the toad is stung painfully on the tongue.

(c) Presented with a harmless robber fly, which resembles a bee, the toad cringes.

(d) The toad is presented with a dragonfly.

(e) The toad immediately eats the dragonfly, demonstrating that the learned aversion is specific to bees and insects resembling them.

Figure 41-14   **Trial-and-error learning in the toad**

(a)

(b)

Figure 41-15   **Insight**
(a) A dog, lacking insight in this situation, fails a detour problem on the first try. It will eventually learn by trial and error.
(b) Unable to reach the bananas, the chimpanzee exhibits insight by stacking boxes beneath the bananas to extend its reach.

an intimate mixture of the two. In some cases, the nature and timing of learning are so rigidly programmed by the genes that we might use the term *innate learning*. An example of innate learning is imprinting: learning rigidly programmed to occur at a certain critical period of development. Adult bees also use a type of innate learning to allow them to locate their hive after a foraging trip. Ethologists discovered that bees memorize the location of their hive with respect to certain landmarks *only* on their first flight of the day. If the hive is moved even a short distance later in the day, the bees will have trouble finding it. A beekeeper who wishes to move the hive must do it during the night. The bees memorize the new location on their first flight and return to it later.

We have seen how genes influence learning, but learning may also modify innate behavior. Recall the example of the herring gull chicks' pecking their parent's beak for food. Immediately after hatching, the chicks prefer a striped knitting needle to the real beak. Within a few days, however, the chicks learn enough about the appearance of their parents that they begin pecking more frequently at models more closely resembling the parents. After 1 week, young gulls have learned enough about how their parents look to prefer models of their own species to models of a closely related species.

Habituation is another example of learning that can fine-tune an organism's innate responses to environmental stimuli. Early investigators were impressed to see young birds crouching when a hawk flew over but ignoring harmless birds such as geese. They hypothesized that crouching was released only by the very specific shape of predatory birds. Niko Tinbergen and Konrad Lorenz tested this hypothesis using an ingenious model. As shown in Figure

41-16, when moved in one direction the model resembled a goose, which the chicks ignored. When its movement was reversed, it resembled a hawk and elicited crouching. Further research revealed that naive chicks instinctively crouch when *any* object moves over their heads. Over time, their response habituates to things that soar by harm-

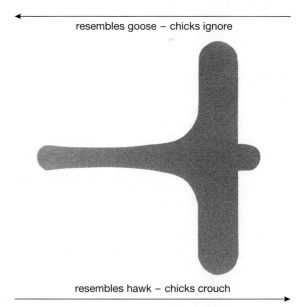

resembles goose – chicks ignore

resembles hawk – chicks crouch

Figure 41-16   **Habituation modifies innate responses**
The model used by Konrad Lorenz and his student Niko Tinbergen to investigate the response of chicks to the shape of objects flying overhead. The chicks' response depended on which direction the model moved. Moving toward the right, the model resembles a predatory hawk, but moving left it resembles a harmless goose.

lessly and frequently, such as leaves, songbirds, and geese. Predators are much less common, and the novel shape of a hawk still elicits instinctive crouching. Thus, learning modifies the innate response, making it more adaptive.

Trial-and-error learning can also result in more appropriate responses to releasers for fixed action patterns, as in the case of our bee-catching toad (see Fig. 41-14). The naive toad instinctively snaps at all flying insects of appropriate size, but from painful experience it learns to make certain exceptions!

E V O L U T I O N A R Y C O N N E C T I O N S

Why Do Animals Play?

Pigface, a giant 50-year-old African softshell turtle, spends hours each day batting a ball around his aquatic home in the National Zoo in Washington, D.C., to the delight of thousands of visitors and the puzzlement of behavioral biologists. Play has always been somewhat of a mystery. It has been observed in many birds and in most mammals, but, until zookeepers tossed Pigface a ball a few years ago, it had never been seen in animals as evolutionarily ancient as reptiles.

Animals at play are fascinating. Pygmy hippopotamuses push one another, shake and toss their heads, splash in the water, and pirouette on their hind legs. Otters delight in elaborate acrobatics. Bottlenose dolphins balance fish on their snouts, throw objects, and carry them in their mouths while swimming. Even baby vampire bats have been observed chasing, wrestling, and slapping each other with their wings. Solitary play often involves a single animal manipulating an object, like a cat with a ball of yarn, or the dolphin with its fish, or as seen on our cover, a macaque monkey making and playing with a snowball. Play may also be social. Often young of the same species play together, but parents may join them (Fig. 41-17a). Social play often includes chasing, fleeing, wrestling, kicking, and gentle biting (Fig. 41-17b, c).

What are the features of play? (1) Play seems to lack any clear goal. (2) Play is abandoned in favor of escaping from danger, feeding, and courtship. (3) Play seems to involve feelings of pleasure. (4) Young animals play more frequently than adults. (5) Play often involves movements borrowed from other behaviors (attacking, fleeing, stalking, etc.). (6) Play uses considerable energy. (7) Play is potentially dangerous. Young humans and other animals are frequently injured, and occasionally killed, during play. In addition, play may distract the animal from the presence of danger while making it conspicuous to predators. So why do animals play?

The only logical conclusion is that play must have survival value and that natural selection has favored those

Figure 41-17   Young animals at play

(a) A bonobo mother plays with her young. (b) Young polar bears wrestle on the ice of Hudson Bay, Canada. (c) Young warthogs in Kenya engage in a friendly shoving match.

individuals who enjoy playful activities. One of the best explanations for the survival value of play is the "practice theory," first proposed by K. Groos in 1898. He suggested that play allows young animals to gain experience in a variety of behaviors that they will use as adults. By performing these acts repeatedly in a lighthearted context, the animal gains skill that will later be important in hunting, fleeing, or social interactions.

More recent research supports and extends Groos's proposal. Play occurs most intensely early in life when the brain is developing and crucial neural connections are being formed. John Byers, a zoologist at the University of Idaho, has observed that animals with larger brains tend to be more playful than animals with smaller brains.

Because larger brains are generally linked to increased learning ability, this relationship supports the idea that adult skills are learned during juvenile play. Watch children roughhousing or playing tag and you will see how strength and coordination are fostered by play and how skills are developed that may have helped our hunting ancestors survive. Quiet play with other children, with dolls, blocks and other toys, helps children prepare to interact socially, nurture their own children, and deal with the physical world.

Although Shakespeare tells us "play needs no excuse," there is good evidence that the tendency to play has evolved as an adaptive behavior in animals capable of learning. Play is quite literally "serious fun"!

SUMMARY OF KEY CONCEPTS

The Genetic Basis of Behavior
All behavior has some genetic basis. Both the complexity of the nervous system that directs the behavior and the physical structure of the body that performs the behavior are genetically determined. The relative contributions of heredity and learning to behavior vary among animal species and among behaviors within an individual.

Innate Behavior
Innate behaviors can be placed into four categories: kineses, taxes, reflexes, and fixed action patterns. Animals showing kineses orient by varying the speed of essentially random movements, stopping when they encounter favorable conditions. In contrast, taxes are directed movements toward or away from specific stimuli. Reflexes are rapid movements of part of the body. Some, such as the knee jerk reflex, are directed by the spinal cord and do not involve higher brain centers.

A fixed action pattern is a complex innate behavior elicited by a specific stimulus called a releaser. Learning can sometimes modify the releasers for fixed action patterns. Fixed action patterns can be distinguished from learned behaviors (1) if they are performed without prior experience, (2) if manipulating the releaser can result in inappropriate behavior, and (3) if a genetic basis can be determined through breeding experiments.

Innate, or instinctive, behaviors allow the animal to respond properly the first time it encounters a stimulus. They allow animals with simple nervous systems to perform behaviors too complex for them to learn. Finally, social interactions important for survival, such as courtship and mating, may depend on animals' rigidly performing specific roles arising from innate behavior.

Learning
Learning is the ability to make changes in behavior as a result of experience. Learning is especially adaptive in changing and unpredictable environments and may modify innate behavior to make it more appropriate. Forms of learning include imprinting, habituation, conditioning, trial and error, and insight.

Imprinting occurs during a genetically programmed sensitive period. It often involves attachment between parent and offspring or learning the features of a future mate. Habituation is the decline in response to a harmless stimulus that is repeated frequently. It often modifies innate escape or defensive responses. During classical conditioning an animal learns to make a reflexive response, such as withdrawal or salivation, to a stimulus that did not originally elicit that response. During operant conditioning an animal learns to make a new response, such as pressing a button, to obtain a reward or to avoid punishment. Trial-and-error learning may modify innate behavior or may produce new behavior as a result of rewards and punishments provided by the environment. Insight, the most complex form of learning, can be considered a form of mental trial-and-error learning. An animal showing insight makes a new and adaptive response to an unfamiliar situation.

The Instinct to Learn and the Learning of Instincts
In animals with complex nervous systems, learning and instinct interact to produce adaptive behavior. Certain types of learning occur instinctively, during a rigidly defined time span. Imprinting and location of the hive by bee foragers are examples. Instinctive responses are often relatively unselective. Examples include baby gulls pecking at long thin objects, toads feeding on all flying insects, and chicks crouching whenever large birds fly overhead. Learning allows animals to modify these innate responses so that they occur only with appropriate stimuli.

✖ KEY TERMS

behavior   p. 811
classical conditioning   p. 818
ethology   p. 811
fixed action pattern   p. 815
habituation   p. 818
imprinting   p. 818

innate   p. 812
insight   p. 820
instinctive   p. 812
kinesis   p. 812
learning   p. 818
operant conditioning   p. 819

reflex   p. 813
releaser   p. 815
sensitive period   p. 818
supernormal stimulus   p. 816
taxis   p. 812
trial-and-error learning   p. 820

✖ THINKING THROUGH THE CONCEPTS

Multiple Choice

1. Which of the following is NOT an example of innate behavior?
   a. kineses   b. taxes
   c. reflexes   d. fixed action patterns
   e. habituation

2. What type of behavioral process is represented when an animal presses a lever to receive a food reward?
   a. operant conditioning   b. classical conditioning
   c. habituation   d. insight
   e. imprinting

3. The most complex form of learning is called
   a. habituation   b. trial and error
   c. insight or reasoning   d. fixed action patterning
   e. operant conditioning

4. Which of the following is NOT a feature of play?
   a. Young animals play more often than adults.
   b. Play often involves movements borrowed from other behaviors.
   c. Play uses considerable energy.
   d. Play always has a clear goal.
   e. Play is potentially dangerous.

5. Concerning the relationship between learning and innate behavior:
   a. the ability to learn does not have a genetic basis
   b. the limits of learning are defined by the animal's genetic makeup
   c. an innate behavior cannot be modified by learning
   d. innate behavior and learning are completely unrelated
   e. none of the above

6. A taxis is
   a. a learned behavior
   b. a change in the rate of movement in response to a stimulus
   c. a hired car
   d. an innate movement toward or away from a stimulus
   e. a result of operant conditioning

Review Questions

1. List four types of innate behavior and provide an example of each.
2. Distinguish between taxes and kineses.
3. What is the name for a stimulus that elicits a fixed action pattern? Give three examples of this type of stimulus.
4. Describe general types of experiments that might distinguish complex instinctive acts from learned behavior.
5. Compare classic conditioning, operant conditioning, and trial-and-error learning. Give an example of each.
6. Compare insight with trial-and-error learning.
7. Give examples in which habituation and trial-and-error learning make an innate response more appropriate to a particular situation.
8. Explain what imprinting is, and why it could be described as "genetically controlled learning."
9. Explain why neither "innate" nor "learned" adequately describes the behavior of any given organism.
10. Explain why animals play. Include the features of play in your answer.

✖ A P P L Y I N G   T H E   C O N C E P T S

1. You have moved to the country after growing up in the city. Although you are used to the sounds of traffic at night, the unfamiliar sounds of birds, crickets, and tree toads keep you awake at night. Should you move back to the city to get some sleep, or tough it out in the country? Explain your answer.

2. You are a police officer assigned to the K-9 squad. Your new partner is a German Shepherd dog, which you are assigned to train as a narcotics dog. You plan to use motivational techniques rather than negative reinforcement. What behavior principles (processes) would you use in your training? What are the advantages of positive over negative reinforcement?

3. Your 3-year-old child persistently approaches the hot stove even though you have told her not to touch it. What type of learning is she attempting to use? Is there any other learning process that might work that would not require the child to touch the hot stove?

4. Based on the information you learned about mosquito taxes, design a mosquito trap or killer that exploits their innate behaviors. Now design one for moths.

5. You have moved into a somewhat rough neighborhood, and your child wants to go out and play with the neighborhood children. You are concerned about what he might learn. Discuss the pros and cons of allowing him to play with these children.

✖ F O R   M O R E   I N F O R M A T I O N

Benzer, S. "Genetic Dissection of Behavior." *Scientific American,* December 1973. Mutant fruit flies are used to study the genetic bases of behavior.

Brown, S. L. "Animals at Play." *National Geographic,* December 1994. Clearly written description of why animals play accompanied by beautiful photographs that capture wild animals at play.

Gould, J. L., and Gould, C. G. "The Instinct to Learn." *Science 81,* May 1981. Birds, bees, and perhaps even humans are genetically programmed to learn specific things at particular stages in life.

Harbrecht, D. "Games Animals Play." *International Wildlife,* September/October 1993. Superb photographs and clear descriptions of the range and purpose of play.

Horgan, J. "Eugenics Revisited: Trends in Behavioral Genetics." *Scientific American,* June 1993. Explores the fascinating interface between genetics and behavior.

Lorenz, K. *King Solomon's Ring: New Light on Animal Ways.* New York: Thomas Y. Cromwell, 1952. Beautifully written and filled with interesting anecdotes; provides important insights into early ethology.

Phillips, K. "School Riddles." *International Wildlife,* March/April 1995. Explores research on fish schooling behavior and how it helps avoid predation.

N E T   W A T C H

On-line resources for this chapter are on the World Wide Web at:
http://www.prenhall.com/~audesirk (click on the <u>table of contents</u> link and then select Chapter 41).

42 The Social Behavior of Animals

> *"The elements of our own behavior are found in all organisms."*
>
> E. G. CONKLIN, 1944

▓ A T   A   G L A N C E

Communication
Visual Communication Includes Active and Passive Signals
Communication by Sound Has Many Advantages
Chemical Communication Uses Pheromones
Communication by Touch Helps Establish Social Bonds

Social Behavior
Competition for Resources Underlies Many Forms of
   Social Behavior
Sexual Reproduction Requires Social Interactions
   between Mates
Social Behavior within Animal Societies Requires
   Cooperative Interactions

Human Ethology
Newborn Infants Exhibit Some Innate Behaviors
Innate Tendencies Can Be Revealed by Exaggerating
   Human Releasers
Simple Behaviors Shared by Diverse Cultures May Be Innate
Do People Respond to Pheromones?
Comparisons of Identical and Fraternal Twins Reveal
   Genetic Components of Behavior

Evolutionary Connections: Altruism, Kin Selection, and the Selfish Gene

Elaborate instincts: The male satin bowerbird instinctively builds an intricate structure, or bower, that attracts females of the same species.

The Australian satin bowerbird male selects a prime location on the forest floor and picks it clean. Then, after covering the site with coarse grass and twigs, he constructs his elaborate bower, a prop for his courtship dance. Meticulously, he weaves two parallel walls of twigs a foot high and 4 inches thick. Then he chews a twig until its end bristles into a brush and uses it to paint the walls with a mixture of saliva and berry juice. Not satisfied, he decorates the sunny southern end of the avenue between the walls with blue objects: blue berries, blue feathers, even blue bits of glass or plastic. If his bower and treasure trove are sufficiently enticing, a female may come to watch him dance. At the sight of her, he leaps about, showing off each treasure. A successful dance ends with mating. The female then flies off to build a nest and tend her eggs alone.

The satin bowerbird's behavior seems incredibly complex and intelligent—until we look at other bowerbirds. Lauderbach's bowerbird collects only red and pale gray objects. The fawn-breasted bowerbird decorates his bower floor and walls with pale green berries. The great bowerbird favors white treasure piles with pale green borders. Each builds a particular shape of bower unique to its species, decorates it in species-specific colors, and performs a species-specific dance that clearly reveals the overwhelming role of heredity in this elaborate courtship ritual. The females, in turn, respond instinctively only to those bowers built by males of their species. Clearly, communicating his species is an important function of the male bowerbird's elaborate courtship behavior.

Social behavior is a necessity for nearly all animals, since at least minimal social interactions are required for sexual reproduction. Social interactions occur between parent and offspring in those species that nurture their young. In addition, most animals interact competitively in pursuit of resources such as food, living space, or mates. Some species, such as bees and ants, form complex societies based on instinctive social interactions. As we explore in Chapters 43 and 44, social interactions help regulate population size and density and govern some of the community interactions that provide the framework for ecosystem structure. Because many social behaviors are genetically coded, they provide raw material for natural selection, as we will examine more closely at the end of this chapter. All social behavior is based on the ability to communicate, so we start with an exploration of that phenomenon.

Communication

Social behavior is exhibited to some degree by all but the simplest organisms. The basis of all social behavior is communication, and the ultimate outcomes are survival and reproduction. In the context of animal behavior, **communication** is defined as the production of a signal by one organism that causes another to change its behavior in a way beneficial to one or both.

Although animals of different species may communicate (picture a cat, its tail erect and bushy, hissing at a strange dog), most communication occurs between members of the same species. Potential mates must communicate, as must parents and offspring. At the same time, members of the same species compete most directly with one another for food, space, and mates. Communication is often used to resolve such conflicts with minimal damage.

The mechanisms by which animals communicate are astonishingly diverse and use all the senses. In the following sections, we look at communication by visual displays, sound, chemicals, and touch.

Visual Communication Includes Active and Passive Signals

Animals with well-developed eyes, from insects to mammals, use vision to communicate. Visual signals may be **active,** in which a specific movement (such as baring fangs) or posture (lowering head, erecting fur) conveys a message (Fig. 42-1). Alternatively, visual signals may be **passive,** in which case the size, shape, or color of the animal conveys important information, often concerning its sex and reproductive state. For example, when female mandrills become sexually receptive, they develop a large, brightly colored swelling on their buttocks (Fig. 42-2). Active and passive signals may be combined, as illustrated by the lizard in Figure 42-3, the courtship display of the male peacock (see Fig. 42-14a), and the courtship behavior of the three-spined stickleback fish (see Fig. 42-15).

Like all forms of communication, visual signals have both advantages and disadvantages. On the plus side, they are instantaneous and can impart a great deal of information in a short time. The animal can convey the intensity of its response by varying its visual signals (Fig. 42-4). Visual communication is quiet and unlikely to alert distant predators, although the signaler does make itself conspicuous to those nearby. On the negative side, visual signals are usually ineffective in darkness or dense vegetation, though female fireflies signal potential mates using species-specific patterns of flashes. Finally, visual signals are limited to close-range communication.

Communication by Sound Has Many Advantages

The use of sound overcomes many of the shortcomings of visual displays. Like visual displays, sound communication is almost instantaneous. But unlike visual signals, sound can

(a)

(b)

Figure 42-1   Aggressive and submissive displays

These dogs were drawn by Charles Darwin, a perceptive student of animal behavior. **(a)** Upright stance, erect hair, ears, and tail, and a direct stare combine to make the aggressor appear formidable indeed. **(b)** Notice how those displays are reversed in the submissive pose.

Figure 42-2   A passive visual signal

The female mandrill's colorfully swollen buttocks serve as a passive visual signal that she is fertile and ready to mate.

Figure 42-3   An aggressive display

The South American *Anolis* lizard raises his head high in the air (an active visual signal), revealing a brilliantly colored throat pouch (a passive visual signal) warning others to keep their distance.

be transmitted through darkness, dense forests, and even water, as the intricate songs of humpback whales testify. If the animal is sufficiently energetic, its call may carry much farther than the eye can see. When the small kangaroo rat drums the Arizona desert floor with its hind feet, the sound can be heard more than 45 meters (150 feet) away. The howls of a wolf pack carry for miles on a still night, and the low, resonant song of the humpback whale can possibly be heard by other whales up to hundreds of miles away.

Auditory signals are similar to visual displays in that they can transmit rapidly changing information almost instantaneously (think of words and of the emotional nuances conveyed by the human voice during a conversation). Changes in motivation may be signaled by a change in the loudness or pitch of the sound. An individual may convey different messages by variations in the pattern, volume, and

pitch of the sound. Ethologist Thomas Struhsaker studied vervet monkeys in Kenya in the 1960s and found that they produced different calls in response to threats from each of their major predators: snakes, leopards, and eagles. In 1980, other researchers reported that the response of other vervet monkeys to each of these calls is appropriate to the particular predator. The "bark" that warns of a leopard or other four-legged carnivore causes monkeys on the ground to take to trees and those in trees to climb higher. The "rraup" call signaling an eagle or other hunting bird causes monkeys on the ground to look upward and take cover, while monkeys already in trees drop to the shelter of lower, denser branches. The "chutter" call indicates a snake and causes the monkeys to stand up and search the ground for this slower predator.

The use of sound is by no means limited to birds and mammals. Male crickets produce species-specific songs that attract female crickets of the same species (see "Scientific Inquiry: Robot Cricket Finds Her Mate" in Chapter 41). The annoying whine of the female mosquito as she prepares to bite alerts nearby males that she will soon have the blood meal necessary for laying eggs. Male water striders vibrate their legs, sending species-specific patterns of vibrations through the water that attract mates and repel other males (Fig. 42-5). From these rather simple signals to the virtuoso performance of human language, sound is one of the most important forms of communication.

Chemical Communication Uses Pheromones

Chemical substances produced by an individual that influence the behavior of others of its species are called **pheromones.** Chemicals may carry messages over long distances, and, unlike sound, take very little energy to produce. Pheromones may not even be detected by other species, including predators who might be attracted to visual or auditory displays. Like a signpost, a pheromone

(a)

(b)

Figure 42-4   Graded visual signals

(a) A relaxed, nonaggressive wolf. (b) The wolf signals aggression by lowering the head, ruffling the fur on its neck and along its back, facing the opponent with a direct stare, and exposing its fangs. These signals may vary in intensity, communicating different levels of aggression.

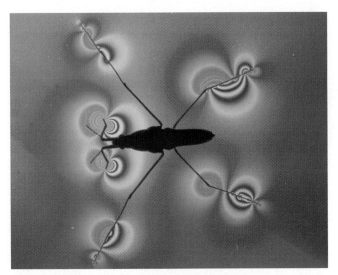

Figure 42-5   **Communication by vibration**
The light-footed water strider relies on the surface tension of water to support its weight. By vibrating its legs, the water strider sends signals radiating out over the surface of the water. These vibrations advertise the species and sex of the strider to others nearby.

persists over time and can convey a message after the animal has departed. Wolf packs, hunting over areas up to 1000 square kilometers (386 square miles), mark the boundaries of their travels with pheromones in urine that warn other packs of their presence. As anyone who has walked a dog can attest, the domesticated dog reveals its wolf ancestry by staking out its neighborhood with urine that carries a chemical message: "I live in this area."

This type of communication requires that an animal be able to synthesize as well as respond to a different chemical for each message. As a result, in general, fewer messages are communicated with chemicals than with sight or sound. In addition, pheromone signals lack the diversity and gradation of auditory or visual signals. Nonetheless, chemicals powerfully convey a few simple but critical messages.

Pheromones act in one of two ways. **Releaser pheromones** cause an immediate, observable behavior in the animal detecting them. They convey messages such as "this area is mine" or "I am ready to mate now." Foraging ants who discover food lay a trail of pheromones secreted by abdominal glands from the food back to the nest. Its message: "Follow to find food." **Primer pheromones**, in contrast, stimulate a physiological change in the animal detecting them, usually in its reproductive state. The queen honeybee produces a primer pheromone called **queen substance** that is eaten by her hivemates and prevents other females in the hive from becoming sexually mature. The urine of mature males of certain species of mice contains a primer pheromone that influences female reproductive hormones. This pheromone causes the newly mature female to become sexually receptive and fertile for the first time. It will also cause a female mouse newly pregnant by another male to abort her litter and become sexually receptive to the new male. There is indirect evidence

(discussed later in the chapter) that primer pheromones may even influence human reproductive cycles.

The sex attractant pheromones of some agricultural pests, such as the Japanese beetle and gypsy moth, have been synthesized. These synthetic pheromones can be used to confuse and disrupt mating or to lure these insects into traps. Pest control using pheromones has major environmental advantages. Pesticides kill beneficial as well as harmful insects and select for resistant insect strains. In contrast, pheromones are specific to one species, and insects resistant to the attraction of their own pheromones would not reproduce.

Communication by Touch Helps Establish Social Bonds

Physical contact between individuals is used in several ways, particularly to establish and maintain social bonds among group members. Primates, including humans, are "contact species" in which a variety of gestures including kissing, nuzzling, patting, petting, and grooming play an important social function (Fig. 42-6a). The greeting ceremony of wolves and dogs involves mutual licking, sniffing, and gentle nipping around the mouth (Fig. 42-6b). The bond between parent and offspring is often cemented by close physical contact, and sexual activity is frequently preceded by ritualized contact (Fig. 42-6c).

Touch can also influence human well-being. Recent research shows that when the limbs of premature human infants are stroked and moved for 45 minutes daily, the infants are more active, responsive, emotionally stable, and gain weight more rapidly than premature infants receiving the standard hospital treatment.

Social Behavior

Many social interactions use a combination of sight, sound, scent, and touch, as illustrated by the waggle dance of the honeybee (described later in this chapter). These complex behaviors play many different roles in the lives of animals. The following sections discuss the forms of communication animals use as they compete for limited resources, reproduce, and cooperate in complex societies.

Competition for Resources Underlies Many Forms of Social Behavior

Aggressive Displays Minimize Injury

One of the most obvious manifestations of competition for resources such as food, space, or mates is **aggression**, or antagonistic behavior, usually between members of the same species. Although the expression "survival of the fittest" evokes images of the strongest animal emerging triumphantly from among the dead bodies of its competitors, in reality most aggressive encounters between members of the same species are rather harmless. Natural selection has favored the evolution of symbolic displays or rituals for resolving conflicts. During fighting, even the victorious an-

Figure 42-6   **Communication by touch**

(a) An adult olive baboon grooms a juvenile. Grooming not only reinforces social relationships but also removes debris and parasites from the fur. (b) Wolves from the same pack engage in a greeting ceremony that involves nuzzling and licking. This behavior may indicate submission, as when a subordinate animal licks a dominant one. Wolf pups use muzzle licking to beg for food. (c) Touch is also important in sexual communication. These land snails *(Helix)* engage in courtship behavior that will culminate in mating.

imal may be injured, so serious fighters may not survive to pass on their genes. Aggressive *displays*, in contrast, allow the competitors to assess each other and acknowledge a winner on the basis of its size, strength, and motivation rather than on the wounds it can inflict.

During visual aggressive displays, animals exhibit weapons, such as claws and fangs (Fig. 42-7a), and often behave in ways that make them appear larger (Fig. 42-7b). Competitors may stand upright and erect their fur, feathers, ears, or fins (see Figs. 42-1a, 42-3, and 42-4). The displays may be accompanied by intimidating sounds (growls, croaks, roars, chirps) whose loudness can help decide the winner. Fighting is usually a last resort when displays fail to resolve the dispute.

In addition to visual and vocal displays of strength, many animal species engage in ritualized combat. Deadly weapons may clash harmlessly (Fig. 42-8) or may not be used at all. Frequently these encounters involve shoving rather than slashing. Again, the strength and motivation of the combatants are determined, and the loser slinks away in a submissive posture that minimizes the size of its body (see Fig. 42-1b).

Dominance Hierarchies Reduce Aggressive Interactions

Aggressive interactions use a great deal of energy, may cause injury, and can disrupt other important tasks such as finding food, watching for predators, courting a mate,

Figure 42-7   **Aggressive displays**

(a) Threat display of the male baboon. Despite the potentially lethal fangs so prominently displayed, aggressive encounters between baboons rarely cause injury. (b) The aggressive display of the male fighting fish includes elevating the fins and flaring the gill covers, thus making the body appear larger.

Figure 42-8   Displays of strength

Ritualized combat of male impalas, a type of African antelope. The deadly horns, which could stab a predator, clash harmlessly. Eventually one impala, sensing greater vigor in his opponent, will often retreat unharmed.

or raising young. So there are adaptive advantages to resolving conflicts with minimal aggression. In a **dominance hierarchy,** each animal establishes a rank that determines its access to resources. Domestic chickens, after a period of squabbling, sort themselves into a reasonably stable "pecking order." Thereafter, when competition for food occurs, all hens defer to the dominant bird, all but the dominant bird give way to the second, and so on. Conflict is minimized because each bird knows its place. Dominance among male bighorn sheep is reflected in the size of their horns (Fig. 42-9). In wolf packs, one member of each sex is the dominant, or "alpha," individual to whom all others are subordinate. Although aggressive encounters occur frequently while the dominance hierarchy is being established, after each animal learns its place, disputes are minimized. The dominant individuals obtain most access to the resources needed for reproduction, including food, space, and mates.

Figure 42-9   A dominance hierarchy

The dominance hierarchy of the male bighorn sheep is signaled by the size of the horns; these rams increase in status from right to left. These backward-curving horns, clearly not designed to inflict injury, are used in ritualized combat.

Territoriality Parcels Out Resources

Territoriality is the defense of an area where important resources are located. The defended resources may include places to mate, raise young, feed, or store food. Territorial animals generally restrict most or all of their activities to the defended area and advertise their presence there. Territories may be defended by males, females, a mated pair, or by entire social groups (as in the case of the defense of their nest by social insects). However, territorial behavior is most often seen in adult males, and territories are usually defended against members of the same species, those who compete most directly for the resources being protected. Territories are as diverse as the animals defending them. For example, they can be small depressions in the sand used as nesting sites by cichlid fish (Fig. 42-10), a hole in the sand used as a home by a crab, a tree where a woodpecker stores acorns (Fig. 42-11), or an area of forest providing food for a squirrel.

Acquiring and defending a territory require considerable time and energy, yet territoriality is seen in animals as diverse as worms, arthropods, fish, birds, and mammals. The fact that organisms as unrelated as worms and humans independently evolved similar behavior suggests that territoriality provides some important adaptive advantages. Although the benefits depend on the species and the type of territory it defends, some broad generalizations are possible. First (as with dominance hierarchies), once a territory

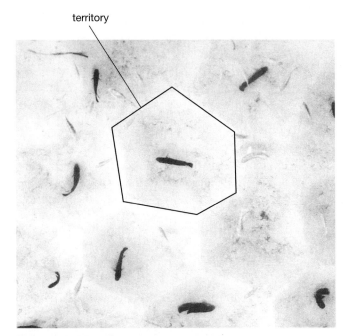

territory

Figure 42-10   Contiguous territories

Nesting males of the mouthbrooding cichlid fish (*Oreochromis mossambicus*) guard territories. When space is abundant, these territories are circular. When the population density rises, the territories approach hexagonal shapes because they are so closely packed. A female visits the territories and selects a male, who fertilizes the eggs she releases. She then gathers the eggs in her mouth (where they will develop) and departs.

is established through aggressive interactions, relative peace prevails as boundaries are recognized and respected. The saying "good fences make good neighbors" also applies to nonhuman territories. One reason for this respect is that an animal is highly motivated to defend its territory and will often defeat larger, stronger animals if they attempt to invade it. Conversely, an animal outside its territory is much less secure and more easily defeated. This principle was demonstrated by Niko Tinbergen's experiment using the stickleback fish, described in Figure 42-12.

Defending reproductive territories is advantageous for certain species. The reproductive success of these animals is enhanced by a high-quality breeding territory, which might have features such as large size, abundant food, and secure nesting areas. Males who successfully defend the most desirable territories have the greatest chance of mating and passing on their genes. For example, experiments have shown that male sticklebacks who defend large territories are more successful in attracting mates than are males who defend small territories. Females who select males with the best territories increase their reproductive success and pass their genetic traits (often including their mate-selection preferences) to their offspring.

Territoriality limits the population size of some species, helping to keep it within the limits set by the available resources. For example, the Scottish red grouse defends feeding and nesting territories. Scientists found that variations in territory size were correlated with the availability of food. In years when food was abundant, territories in the study area were smaller and the number of mating pairs was greater than in lean years.

Territories are advertised through sight, sound, and smell. If the territory is small enough, the owner's mere

Figure 42-11   A feeding territory

The acorn woodpecker excavates acorn-sized holes in dead trees, stuffing them with green acorns for dining during the lean winter months. He defends the trees vigorously against other acorn woodpeckers and against acorn-eating birds of other species, such as jays.

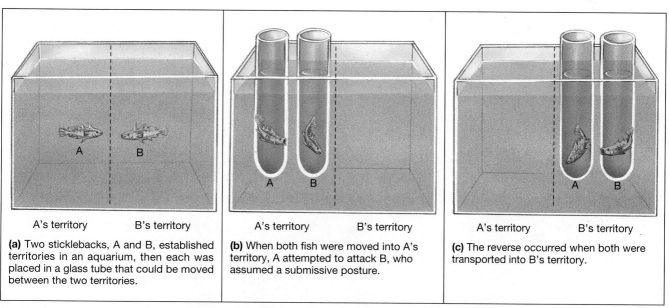

A's territory B's territory	A's territory B's territory	A's territory B's territory
(a) Two sticklebacks, A and B, established territories in an aquarium, then each was placed in a glass tube that could be moved between the two territories.	**(b)** When both fish were moved into A's territory, A attempted to attack B, who assumed a submissive posture.	**(c)** The reverse occurred when both were transported into B's territory.

Figure 42-12   Territory ownership and aggression

Niko Tinbergen's experiment demonstrating the effect of territory ownership on aggressive motivation.

Figure 42-13   Defense of a territory by song

A male meadowlark announces ownership of his territory to all listeners.

presence, reinforced by aggressive displays at intruders, may be a sufficient defense. A mammal that owns a territory but cannot always be present may scent-mark its terrestrial boundaries using pheromones. Male rabbits use pheromones secreted by chin and anal glands to mark their territories. Hamsters rub the areas around their dens with secretions from special glands in their flanks.

Vocal displays are a common form of territorial advertisement. Male sea lions defend a strip of beach by swimming up and down in front of it, calling continuously. Male crickets produce a specific pattern of chirps to warn other males away from their burrows. Birdsong is a striking example of territorial defense. The cheerful melody of the male meadowlark is part of an aggressive display, warning other males to steer clear of his territory (Fig. 42-13). The loudest songbird generally defends the largest territory and is most successful in attracting a mate and driving away intruders. In

ingenious experiments using the wood thrush, ethologist W. Dilger of Cornell University "invaded" the territories of male wood thrushes with a stuffed wood thrush accompanied by a loudspeaker through which he played recordings of the thrush's territorial song. When the volume was turned low, the resident thrushes attacked the stuffed bird, but they retreated rapidly when the volume was raised.

Sexual Reproduction Requires Social Interactions between Mates

Successful reproduction requires that several conditions be met. Animals must identify one another as members of the same species, as members of the opposite sex, and as being sexually receptive. Many animals resist the close approach of another individual; this resistance must be overcome before mating can occur. Some animals, such as frogs and many species of fish, must release eggs and sperm at precisely the same moment for fertilization to occur. The need to fulfill all these requirements has resulted in exceedingly complex, diverse, and fascinating courtship behavior.

Individuals who mate with members of other species, or members of the same sex, waste considerable energy and do not pass on their genes. Thus, animals have evolved elaborate ways to communicate their species and sex, often using vocalizations. The raucous, nighttime chirping that can keep campers awake is probably a chorus of male tree frogs, each singing a species-specific song. Male grasshoppers and crickets advertise species and sex by their calls, as does the female mosquito with her high-pitched whine. Male birds use song to attract a mate as well as defend a territory. For example, the male bellbird uses its deafening call to defend large territories and attract females from great distances. The females fly from one territory to another, alighting near the male in his tree. The male, beak gaping, leans directly over the flinching female and utters an earsplitting note. The female apparently endures this to compare volumes of the various males, choosing the loudest (who would also be the best defender of a territory) as a mate. Males and females of other bird species join

Figure 42-14   Sexual displays

(a) The extravagant tail of the male peacock is displayed during courtship. This oversized tail hampers flight and increases the male's vulnerability to predators. It probably evolved as females consistently selected the flashiest birds as mates, preferring this exaggerated releaser to a more practical tail. (b) The male frigate bird of the Galapagos Islands inflates a scarlet throat pouch to attract passing females.

in elaborate duets that help synchronize reproductive readiness and reinforce the bond between them.

Many species court using visual displays. The firefly, for example, flashes a message identifying its sex and species. Male fence lizards bob their heads in a species-specific rhythm, and females distinguish and prefer the rhythm of their own species. The tail of the male peacock and the scarlet throat of the male frigate bird serve as flashy advertisements of sex and species (Fig. 42-14). In contrast, the females are often quite drab. Because females are often in close association with their young, eye-catching (and predator-attracting) displays could be dangerous.

Species and sex recognition and the synchronization of reproductive behavior often require a complex series of signals, both active and passive, by both sexes. Such signals are beautifully illustrated by the complex underwater "ballet" executed by the male and female three-spined stickleback fish (Fig. 42-15).

(a) A male, inconspicuously colored, leaves the school of males and females to establish a breeding territory.

(b) As his belly takes on the red color of the breeding male, he displays aggressively at other red-bellied males, exposing his red underside.

(c) Having established a territory, the male begins nest construction by digging a shallow pit that he will fill with bits of algae cemented together by a sticky secretion from his kidneys.

(d) After he tunnels through the nest to make a hole, his back begins to take on the blue courting color that makes him attractive to females.

(e) An egg-carrying female displays her enlarged belly to him by assuming a head-up posture. Her swollen belly and his courting colors are passive visual displays.

(f) He leads her to the nest using a zigzag dance.

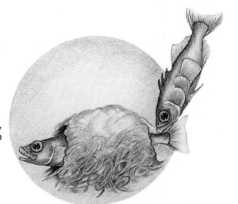

(g) After she enters, he stimulates her to release eggs by prodding at the base of her tail.

(h) He enters the nest as she leaves and deposits sperm that fertilize the eggs.

Figure 42-15   **Courtship of the three-spined stickleback**

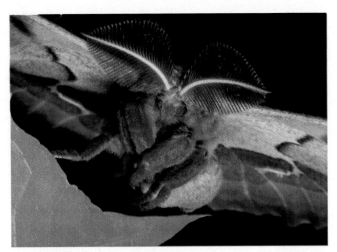

Figure 42-16   **Pheromone detectors**

The antennae of the male silk moth are plumelike structures specialized to detect the female's sex pheromone.

Pheromones can play an important role in reproductive behavior. The sexually receptive female silk moth, for example, sits quietly and releases a chemical message so powerful that it may be detected by males 4 to 5 kilometers (2.5 to 3 miles) away. The exquisitely sensitive and selective receptors on the antennae of the male respond to just a few molecules of the substance, allowing him to travel upwind along a concentration gradient to find the female (Fig. 42-16). Water is an excellent medium for dispersing chemical signals, and fish often use a combination of pheromones and elaborate courtship movements to ensure synchronous release of gametes. Mammals, with their highly developed sense of smell, often rely on pheromones released by the female during her fertile periods to attract males. The irresistible attraction of a female dog in heat to nearby males is one example. The primer pheromone in male mouse urine is another.

Most encounters between individuals of solitary species are competitive and aggressive. During the brief mating season, their reluctance to allow others to approach closely must be overcome for their mate but retained toward others. These conflicting needs introduce an element of tension into sexual encounters that may be overcome by submissive signals. The female Siamese fighting fish appeases the aggressive male with a submissive, head-down posture. Either the male or female of several species of birds defuses aggressive impulses by mimicking juvenile behavior such as begging. Courting male hamsters emit high-pitched cries like baby hamsters, eliciting a maternal response from the female.

Presenting "gifts" seems to inhibit aggression in some species. Finches present their mates with nesting material, terns give fish, and flightless cormorants offer one another gifts of seaweed (Fig. 42-17). The female of one species of fly sometimes devours the male when he makes sexual advances. The males of a closely related species avoid this fate by presenting the female with a gift: a dead insect. While she eats, the male mates and runs. The male of another species gains extra time to mate by "gift-wrapping" the insect in a silken web that she must remove before she can eat the gift.

Social Behavior within Animal Societies Requires Cooperative Interactions

Social groupings of animals are conspicuous but by no means universal. Group living has both advantages and disadvantages, and the relative weight of the positive and negative factors varies considerably among different animal species.

On the negative side, social animals may encounter:

1. Increased competition within the group for limited resources
2. Increased risk of infection from contagious diseases
3. Increased risk that offspring will be killed by other members of the group
4. Increased risk of being spotted by predators

For a species to have evolved social behavior, the benefits must have outweighed the costs. Benefits to social animals include:

1. Increased ability to detect, repel, and confuse predators
2. Increased hunting efficiency or increased ability to spot localized food resources
3. Advantages resulting from the potential for division of labor within the group
4. Conservation of energy
5. Increased likelihood of finding mates

The degree to which animals of the same species cooperate varies significantly from one species to the next.

Figure 42-17   **Courtship gifts defuse aggression**

A flightless cormorant from the Galapagos Islands returns to its nest bearing a "gift" of seaweed for its mate, who will aggressively snatch it away. A bird who fails to bring a gift will be driven away by its mate. The gift appears to allow the nesting bird to take out its aggressive impulses harmlessly.

Figure 42-18 **Cooperation in loosely organized social groups**

A herd of musk oxen functions as a unit when threatened by predators such as wolves. Males form a circle, horns pointed outward, around the females and young.

Some animals, such as the mountain lion, are basically solitary; interactions between adults consist of brief aggressive encounters and mating. Other animals cooperate on the basis of changing needs. For example, the coyote is solitary when food is abundant but hunts in packs in the winter when food becomes scarce. Loose social groupings, such as herds of musk oxen (Fig. 42-18), pods of dolphins, schools of fish, and flocks of birds provide a variety of benefits. For example, the characteristic spacing of fish in schools or the V pattern of geese in flight provides a hydro- or aerodynamic advantage for the individuals, reducing the energy required to swim or fly. Zoologists hypothesize that schooling fish confuse predators—their myriad flashing bodies make it difficult for the predator to focus on and pursue a single individual.

At the far end of the social spectrum are a few highly integrated cooperative societies, found primarily among the insects and mammals. As you read this section, you may notice that some cooperative societies are based on behavior that seems to sacrifice the individual for the good of the group. How could such behavior evolve? In "Evolutionary Connections: Altruism, Kin Selection, and the Selfish Gene," we explore the evolutionary basis for self-sacrificing behaviors that contribute to the success of cooperative societies. In the following section, we present examples of complex societies in insect, fish, and mammal species that vividly illustrate cooperative behavior.

Honeybees Form Complex Insect Societies

The most rigidly organized, most complex societies (humans excepted) are found among the social insects. In these communities, the individual is a mere cog in an intricate, smoothly running machine; it could not function by itself. Social insects are born into one of several castes within the society. These castes are groups of similar individuals genetically programmed to perform a specific function.

Honeybees emerge from their larval stage into one of three major preordained roles (Fig. 42-19). One role is that of queen. Only one queen is tolerated in a hive at

(a)

(b)

Figure 42-19   **Some stages in the life of a worker bee**

(a) Workers crowd around the queen (center), feeding her and licking the pheromone called queen substance from her body, which renders them sterile. **(b)** A forager collects pollen and nectar from a flower. Note the yellow pollen baskets on her legs.

any time. Her functions are to produce eggs (up to 1000 per day for a lifetime of 5 to 10 years) and to regulate the lives of the workers. Male bees, called drones, serve merely as mates for the queen. Lured by her sex pheromones, drones mate with the queen during her first week of life, perhaps as many as 15 times. This relatively brief "orgy" supplies her with sperm that will last a lifetime, enough to fertilize over 3 million eggs. Their sexual chore accomplished, the drones become superfluous and are eventually driven out of the hive or killed.

The hive is run by the third class of bees, sterile female workers. The tasks of the worker are determined by her age and conditions in the colony. The newly emerged worker starts as a waitress, carrying food such as honey and pollen to the queen, other workers, and developing larvae. As she matures, special glands begin wax production, and she becomes a builder, constructing perfectly hexagonal cells of wax where the queen will deposit her eggs and the larvae will develop. She will take a shift as maid, cleaning the hive and removing the dead, and as a guard, protecting the hive against intruders. Her final role in life is that of a forager gathering pollen and nectar, food

for the hive. She will spend nearly half of her 2-month life in this role. Acting as a forager scout, she will seek new and rich sources of nectar and, having found one, will return to the hive and communicate its location to other foragers using the **waggle dance,** an elegant form of symbolic communication (Fig. 42-20). Much of the meaning of the waggle dance was deciphered by Karl von Frisch during 35 years of research beginning in 1915.

Pheromones play a major role in regulating the lives of social insects. Honeybee drones are drawn irresistibly to the queen's sex pheromone (queen substance), which she releases during her mating flights. Back at the hive, she maintains her position as the only fertile female using the same substance (now acting as a primer pheromone). The queen substance is licked off her body and passed among all the workers, rendering them sterile. The queen's presence and health are signaled by her continuing production of queen substance; a decrease in production (which occurs normally in the spring) alters the behavior of the workers. Almost immediately they begin building extralarge "royal cells" and feeding the larvae that develop in them a special glandular secretion known as "royal jelly."

(a)

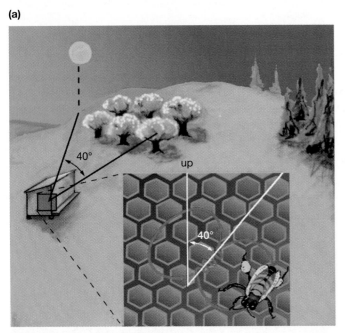

(b)

Figure 42-20   Bee language
The waggle dance. A forager, returning from a rich source of nectar, performs a waggle dance that communicates the distance and direction of the food source, as other foragers crowd around her, touching her with their antennae. The bee moves in a straight line while shaking her abdomen back and forth ("waggling") and buzzing her wings. The bee repeats this over and over in the same location, circling back in alternating directions. The richness of the food source is communicated by the vigor of the dance, and how long it continues. The rate at which she circles communicates the distance of the nectar source from the hive; the higher the rate, the closer the food source. The smell of the flowers on her body tells the other foragers what scent to search for. The direction of the food source is communicated by the direction of the waggling portion of the dance. (a) If the dance is performed on a wall inside the hive, the angle that the straight run deviates from the vertical represents the angle between the sun and the flowers.
(b) On a horizontal surface outside, the waggling run is aimed directly toward the flowers.

This unique food alters the development of the growing larvae so that, instead of a worker, a new queen emerges from the royal cell. The old queen will then leave the hive, taking a swarm of workers with her to establish residence elsewhere. If more than one new queen emerges, a battle to the death ensues, with the victorious queen taking over the hive.

Bullhead Catfish Illustrate a Simple Vertebrate Society

Vertebrates possess far more complex nervous systems than insects, and one might therefore expect vertebrate societies to be proportionately more complex. With the exception of human society, however, they are not. Perhaps because the vertebrate brain *is* more complex, vertebrate societies tend to be simpler than those of the social insects such as honeybees, army ants, and termites. Each individual is unique, and this uniqueness is enhanced because vertebrates exhibit more-flexible learned behavior. Although much social behavior has an innate component, vertebrates show a great deal more flexibility (and thus unpredictability) and less of the robotic precision that makes complex insect societies possible.

The social interactions of bullhead catfish, described by John Todd of the Woods Hole Oceanographic Institution in Massachusetts, provide a fascinating illustration of a relatively simple vertebrate in which complex social interactions are based almost entirely on pheromones. Todd observed these nocturnal fish in large aquariums in the laboratory. He discovered that when a group was housed together, territories were staked out, and a dominance hierarchy was established, with the dominant fish defending the largest and best-protected area of the tank. Contests between tankmates consisted of open-mouthed aggressive displays. Once a fish became dominant, its aggressive displays caused the subordinate fish to flee. Actual violence occurred only when a stranger was introduced into a tank with an established hierarchy. In this case, the established group exhibited cooperative behavior. The dominant fish allowed others to take refuge in his protected territory, then fought the intruder (Fig. 42-21). When the newcomer was defeated, the dominant fish chased the others back out of his territory.

Todd discovered that blinding the bullheads did not cause any appreciable change in their social interactions. When their sense of smell was temporarily blocked, however, the fish acted like permanent strangers, interacting aggressively for weeks until their sense of smell returned. Both the status of an individual and a change in status are communicated by scent. If a dominant fish was removed from his tank and later returned, usually both his territory and his status were remembered and respected by his tankmates. But if he was removed and subjected to defeat in the tank of a more aggressive fish, his pheromones were somehow altered. Upon return to his home tank, he was attacked by his former subordinates.

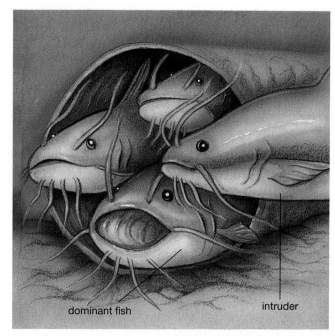

dominant fish   intruder

Figure 42-21   Cooperation among bullhead catfish

Three bullhead catfish occupy a section of pipe in a laboratory tank. The dominant fish is usually the exclusive occupant of the pipe, which is part of his territory. When an intruder is introduced into the tank, however, he allows two subordinate fish to seek refuge in the pipe and reacts aggressively to the intruder.

When many newly caught fish are placed in the same tank, bullheads may form a dense and peaceful community lacking territories or dominance hierarchies. Todd established such a community in one tank and placed a pair of aggressive rival fish in an adjacent tank. When water was pumped continuously from the "community tank" into the adjacent tank, the aggressive bullheads too became peaceful, resuming their fighting only when the flow was stopped. Under the crowded conditions of the community tank, an "anti-aggression pheromone" is apparently produced, minimizing conflict.

Naked Mole Rats Form a Complex Vertebrate Society

The most complex nonhuman society among mammals is formed by naked mole rats. These nearly blind, nearly hairless relatives of guinea pigs live in large underground colonies in southern Africa. Colonies ranging from 10 to 300 individuals inhabit a series of tunnels up to 2.5 kilometers (about 1.5 miles) long. The tunnels are interrupted by sleeping, nesting, and food-storage chambers and end in communal latrines. Mole rats eat fleshy roots they encounter while digging. By capturing colonies, permanently marking each individual, and maintaining the colonies in narrow glass-sided chambers (like ant farms), biologists have learned that mole rat societies have much in common with those of social insects such as the honeybee.

Figure 42-22   **The naked mole rat queen**

Encountering a lazy worker while patrolling the colony's underground tunnels, the queen threatens and shoves it, stimulating the worker to greater efforts.

The colony is dominated by a single reproducing female, the queen, to whom all other members are subordinate (Fig. 42-22). The queen is the largest individual in the colony and maintains her status by aggressive behavior, particularly shoving. The queen prods and shoves lazy workers, stimulating them to become more active. As in honeybee hives, there is a division of labor among the workers, in this case based on size. Small young rats clean the tunnels, gather food, and tunnel. Tunnelers line up head to tail and pass excavated dirt along the completed tunnel to an opening. Just below the opening, a larger mole rat flings the dirt into the air, adding it to a cone-shaped mound. Biologists observing this behavior from the surface dubbed it "volcanoing." In addition to volcanoing, large mole rats defend the colony against predators and members of other colonies.

If another female begins to become fertile, the queen apparently senses changes in the estrogen levels of the subordinate female's urine. The queen then selectively shoves the would-be breeder, causing stress that prevents her rival from ovulating. Large males are more likely to mate with the queen than are small ones, although all adult males are fertile. When the queen dies, a few of the females gain weight and begin shoving one another. Sometimes the aggression escalates and a rival is killed. Finally a single female becomes dominant; her body lengthens and she assumes the queenship and begins to breed. Litters averaging 14 pups are produced about four times a year. During the first month, the queen nurses her pups, and the workers feed the queen. Then the workers begin feeding the pups solid food.

Because the queen produces the pups who grow up to form the colony, all colony members are quite closely related. This genetic relatedness helps explain why behavior has evolved in which workers devote their lives to helping the queen reproduce rather than reproducing

themselves (see "Evolutionary Connections: Altruism, Kin Selection, and the Selfish Gene").

Human Ethology

Some scientists hypothesize that many human tendencies have a genetic basis, and they are attempting to study human ethology. However, human ethology is, and will remain, a less rigorous science than animal ethology. We cannot treat people as we do laboratory animals, nor can we control all the variables that influence our attitudes and actions. Nevertheless, scientists have attempted to isolate the genetic components of human behavior, and in the following sections we review some of these studies and their findings.

Newborn Infants Exhibit Some Innate Behaviors

Much of the behavior of very young infants is likely to have a large innate component, because there has been little time for learning to occur. The rhythmic movement of a baby's head in search of the mother's breast is a human fixed action pattern during the first days after birth. Suckling, which can be observed even in the human fetus, is also innate (Fig. 42-23). Other fixed action patterns seen in newborns or even premature infants include walking movements when the body is supported, and grasping with the hands and feet. Another example is smiling, which may occur soon after birth. Initially, smiling can be re-

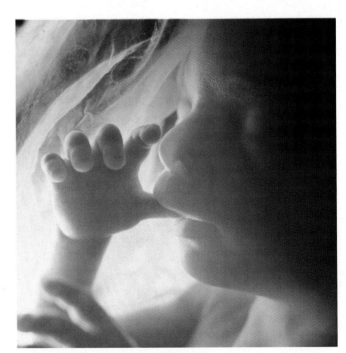

Figure 42-23   **A human instinct**

Thumb sucking is a difficult habit to discourage in young children, because sucking on appropriately sized objects is an instinctive, food-seeking behavior. This fetus sucks its thumb at about 4 1/2 months of development.

leased by almost any object looming over the newborn. Before an infant is 2 months old, supernormal stimulus (see Chapter 41) consisting of two dark, eye-sized spots on a light background will elicit smiling even more successfully than an accurate representation of a human face. As the child matures, learning and further development of the nervous system interact to limit the response to increasingly accurate representations of a face.

Another way to minimize the effects of learning is to observe children who are blind, deaf, or both, and have been unable to learn through sight and sound. Without ever having seen or heard them, these children produce normal smiles and laughter and expressions of frustration and anger.

As discussed in the previous chapter, some animals have a strong innate tendency to learn specific things during certain periods of development, a behavior called imprinting. The human fetus begins responding to sounds during the third trimester of pregnancy and, by 6 weeks after birth, is able to distinguish a variety of consonant sounds, strong evidence that the human brain is programmed to interpret language. Babies are notoriously difficult experimental subjects, and ingenious techniques have been devised to test them. One of the most successful uses the infant's ability to make sucking movements. The baby responds to the presentation of various consonant sounds by sucking on a pacifier containing a force transducer that records the sucking rate. After hearing one sound (such as "ba") repeatedly, the infant becomes bored with it and decreases her sucking rate. A new sound (such as "pa") causes an increase in sucking, revealing that the infant perceives these as different and is more excited by a new sound than by a familiar one.

The sucking-measurement technique has recently been used to demonstrate that newborns in their first 3 days of life can be conditioned to produce certain rhythms of sucking using their mother's voice as reinforcement. Infants clearly preferred their own mothers' voices over other female voices, as indicated by their sucking rhythm (Fig. 42-24). The ability of the infant to learn his or her mother's voice and respond positively to it within days of birth has strong parallels to imprinting and may help initiate bonding with the mother.

The ability of young children to acquire language rapidly and nearly effortlessly is almost certainly an example of developmentally programmed learning. Between ages one and eight, children typically acquire a vocabulary of 28,000 words whose meanings they recognize. If you are currently struggling with a foreign language class, you are probably painfully aware that your critical period for language acquisition is long past.

Innate Tendencies Can Be Revealed by Exaggerating Human Releasers

A behavior probably has an instinctive component if the stimulus that causes it (the releaser) can be exaggerated beyond the bounds of reality and elicit an even stronger re-

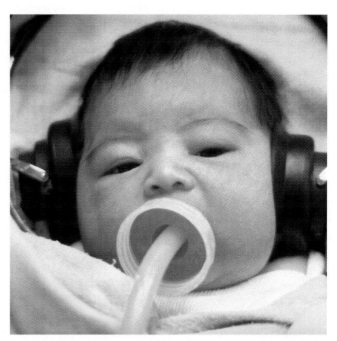

Figure 42-24   **Newborns prefer their mother's voices**
Using a nipple connected to a computer that plays audio tapes, researcher William Fifer of the New York State Psychiatric Institute demonstrated that newborns can be conditioned to suck at specific rates for the privilege of listening to their own mothers' voices through headphones. For example, if the infant sucks faster than normal, her mother's voice is played, if she sucks more slowly, another woman's voice is played. Researchers found that infants easily learned and were willing to work hard at this task just to listen to their own mothers' voices, presumably because they had become used to her voice in the womb. This response may be a human version of imprinting.

sponse. The eye-spots that cause young infants to smile are one example. In turn, the smile of an infant, along with certain characteristic baby features, may release protective feelings in adults. These features include a relatively large head with a domed forehead, chubby cheeks, small nose, short arms and legs, and a small, rounded body. Even 3-year-old children respond to these features with "mothering" behavior. The marketplace has exploited the releasing aspects of these features by exaggerating them in representations of baby animals and people and using their innate appeal to sell dolls, posters, calendars, and cards (Fig. 42-25).

One human signal with an innate, physiological basis is the involuntary enlargement of the pupil of the eye when viewing something pleasant, be it a loved one or a hot fudge sundae. We also react to this signal in others. To test this reaction, researchers showed male subjects identical photographs of smiling women that had been retouched to enlarge or contract the pupils. The subjects overwhelmingly preferred the women with large pupils, although none were consciously aware of the pupil size. We react positively to someone who gazes at us with dilated pupils (although recognition of this fea-

Figure 42-25   Human releasers
We instinctively respond to certain features associated with infants and very young children, such as big eyes, small noses, rounded faces, and large heads. These are sometimes exaggerated to produce supernormal stimuli, as in this example.

Figure 42-26   Common gestures
Gestures that have similar meanings in diverse and isolated cultures may be evidence of a common biological heritage. Here, motion pictures freeze the "eye flash" greeting (in which the eyes are widely opened and the eyebrows rapidly elevated) in a person from New Guinea (left) and Bali (right). Watch for this probably innate response in yourself when you encounter a friend. These photos were taken by I. Eibl-Eibesfeldt, who conducted this research.

ture is subconscious), because it implies interest and attraction. The positive response to enlarged pupils was recognized as long ago as the middle ages, when women sometimes artificially enlarged their pupils with eyedrops containing the drug belladonna (meaning "beautiful woman" in Italian).

Simple Behaviors Shared by Diverse Cultures May Be Innate

Another way to study the instinctive bases of adult human behavior is to compare simple acts performed by people from isolated and diverse cultures. This comparative approach, pioneered by the ethologist I. Eibl-Eibesfeldt, has revealed several gestures that seem to form a universal, and therefore probably innate, human language. Such gestures include a variety of facial expressions for pleasure, rage, and disdain and movements such as the "eye flash" and a hand upraised in greeting (Fig. 42-26).

Do People Respond to Pheromones?

Humans may have unconscious responses to pheromones. Our sense of smell is relatively poor, and the role of odor as a means of human communication is largely unknown. However, an interesting study by Martha McClintock of Harvard provided indirect evidence that primer pheromones may influence female reproductive physiolo-

gy. Using dorm-living college women as subjects, she found that the menstrual cycles of roommates and close friends became significantly more synchronous over a 6-month period. The cycles of women randomly chosen from the dormitory did not. Further, she found that college women who spent time with men frequently had significantly shorter menstrual cycles than those who didn't. Additional support comes from another study in which the timing of the menstrual cycle was altered by placing extract of male underarm sweat on the upper lip of women. Anatomist David Berliner believes he has isolated an odorless human pheromone from skin cells that produces a sense of well-being; he is currently marketing this discovery in the form of a perfume.

In most vertebrates, pheromones are detected by a small pitlike structure inside the nose called the vomeronasal organ (VNO), but the VNO was believed to be nonexistent in humans. In the mid-1980s, researchers finally took a closer look and were able to find a VNO, small but unmistakable, in nearly every human nose they examined. Preliminary studies suggest that certain odorless chemicals cause responses in the sensory cells lining the VNO. In other vertebrates, nerve fibers from the VNO travel to the hypothalamus and amygdala, brain structures important in producing unconscious and emotional responses. While far from conclusive, these findings are tantalizing and call for further investigation of the role of both primer and releaser pheromones in human behavior.

Comparisons of Identical and Fraternal Twins Reveal Genetic Components of Behavior

By studying identical and fraternal twins, investigators can come as close as possible to controlled breeding experiments in humans. Fraternal twins arise from two individual eggs and are no more similar genetically than other siblings. However, they are exactly the same age and share a very similar environment. In contrast, identical twins, arising from a single fertilized egg, have identical genes. The most fascinating twin findings are based on anecdotal observations of identical twins separated soon after birth, reared in different environments, and reunited for the first time as adults. They have been found to share nearly identical taste in jewelry, clothing, humor, food, and names for children and pets. Personal idiosyncrasies such as giggling, nail biting, drinking patterns, hypochondria, and mild phobias may be shared by these unacquainted twins.

More-rigorous studies are also supporting the heritability of many human behavioral traits. These studies have documented a significant genetic component for traits such as activity level, alcoholism, sociability, anxiety, intelligence, dominance, and even political attitudes. On the basis of tests designed to measure many aspects of personality, identical twins are about twice as similar in personality as fraternal twins. Further, identical twins reared apart were found to be as similar in personality as those reared together, indicating that the differences in their environments had little influence on their personality development.

The field of human behavioral genetics is controversial, because it challenges the long-held belief that environment is the most important determinant of human behavior. As discussed in Chapter 41, we now recognize that all behavior has some genetic basis and that complex behavior in nonhuman animals often combines elements of both learned and innate behavior. In the case of our own behavior, the debate over the relative importance of heredity and environment continues and is unlikely ever to be fully resolved. Human ethology is not yet recognized as a rigorous science, and it will always be hampered because we can neither view ourselves with detached objectivity nor treat each other as laboratory animals. In spite of these limitations, there is much to be learned about the interaction of learning and innate tendencies in people.

 E V O L U T I O N A R Y
C O N N E C T I O N S

Altruism, Kin Selection, and the Selfish Gene

Darwin's concept of the survival of individuals best adapted to their environment and able to leave the largest number of offspring remains the foundation of evolutionary theory. Social animals, however, often behave in ways that appear to endanger the individual, decreasing its chance to survive and reproduce. There are many examples. Worker honeybees and naked mole rats do not reproduce but care for the offspring of the queen. Worker ants die in defense of their nest. Young, mature Florida Scrub Jay males may remain at their parents' nest and help them raise subsequent broods instead of breeding themselves. Young adult jackals may also help their parents raise a new litter. In colonies of Belding ground squirrels, individuals may sacrifice their lives to warn the group of an approaching predator. These behaviors fulfill the biologist's definition of **altruism:** behavior that may decrease the reproductive success of one individual to the benefit of another.

How can such behavior be reconciled with the survival of the fittest? Why aren't the individuals that perform such self-sacrificing deeds rapidly eliminated from the population, taking the genes contributing to this behavior with them? In fact, the laws of natural selection operate in such a way that "selfish" behavior (that is, behavior that increases the chance of perpetuating itself) will always be most successful. So, in some way, altruism must be selfish, but how? A major insight into the selfishness of altruistic behavior is supplied by the theory that natural selection does not operate exclusively on the individual but operates at the level of the gene. From this theoretical viewpoint, individuals are short-lived carriers of genes. Genes, however, may persist for millions of years, passing from one generation to the next. One important way in which genes may be preserved is by individuals' performing innate behavior, even self-sacrificing behavior, that enhances the survival of others carrying the same genes. From this standpoint, genes promote the survival of copies of themselves in other individuals, a concept referred to as the **selfish gene.** This theory was proposed by the ethologist Richard Dawkins (a student of Niko Tinbergen) in his engaging book *The Selfish Gene.*

Statistically, closely related individuals are most likely to carry the same genes. Thus, an individual promotes survival of the types of genes it carries through inherited behaviors that maximize survival not only of that individual, but of its close relatives as well. This concept, called **kin selection,** is also discussed in Chapter 17. Kin selection helps to explain a variety of altruistic behaviors. Male honeybees, for example, are haploid, whereas females are diploid. Female workers share, on the average, 75% of their sisters' genes. If they were to mate and reproduce, they would share only 50% of the genes of their offspring. To maximize survival of their own genes, then, rather than reproducing, they are better off helping their mother (the queen) raise more of their sisters, who share a greater percentage of their genes than would their own offspring. In fact, this is what happens in the honeybee societies described in this chapter. The same is true for naked mole rats, in which colony members share about 80% of their genes.

Alarm calls of Belding ground squirrels (Fig. 42-27) may also be explained by kin selection. Paul Sherman, an

Figure 42-27   Altruistic behavior

A Belding ground squirrel sounds an alarm as danger approaches.

evolutionary ecologist at Cornell University, studied colonies of these rodents for several years, marking individuals so that they could be recognized. He observed that the squirrel calling an alarm often drew the attention of the predator and was sometimes killed, but it allowed other members of the colony to seek shelter. In ground squirrel colonies, males leave the burrow in which they were born and establish new burrows some distance away. Females, in contrast, stay close to home. So females in a given area are usually closely related, but males are not. Kin selection theory predicts that females should give more alarm calls than males, because their calls benefit relatives who share more of their genes, justifying the danger. It would also predict that females surrounded by close relatives would call more often than those without close relatives nearby. Sherman's data confirmed both of those predictions, supporting the theory of kin selection.

The concept of kin selection and the selfish gene provides important insights into how self-sacrificing, seemingly nonadaptive behaviors can evolve and persist within a species: Genetically programmed behaviors that sacrifice the individual will be favored by natural selection if they promote the survival of others carrying the same genes. The selfish gene concept helps explain the evolution of altruism.

✖ S U M M A R Y   O F   K E Y   C O N C E P T S

Communication

Communication, an action by one animal that alters the behavior of another, is the basis of all social behavior. It allows animals of the same species to interact effectively in their quest for mates, food, shelter, and other resources. Animals communicate through visual signals, sound, chemicals (pheromones), and touch.

Visual communication is quiet and can convey subtle, rapidly changing information. Visual signals may be active (body movements) or passive (body shape and color). Sound communication can also carry a wide range of rapidly changing information and is effective where vision is impossible. Although sound may attract predators, the animal may remain hidden while communicating.

Chemical signals take the form of releaser or primer pheromones. Releaser pheromones cause an immediate, observable behavior in the animal detecting them. Primer pheromones alter the physiological state of the recipient; releaser pheromones influence the recipient's behavior. Pheromones can be detected after the sender has departed, conveying a message over time. Physical contact reinforces social bonds and helps synchronize mating in a variety of animals, from mammals to snails.

Social Behavior

Although competitive interactions are often resolved through aggression, serious injuries are rare. Most aggressive encounters are settled using displays that communicate the motivation, size, and strength of the combatants.

Some species establish dominance hierarchies that minimize aggression and regulate access to resources. On the basis of initial aggressive encounters, each animal acquires a status in which it defers to more dominant individuals and dominates subordinates. When resources are limited, dominant animals obtain the largest share and are more likely to reproduce.

Territoriality, a behavior in which animals defend areas where important resources are located, also allocates resources and minimizes aggressive encounters. In general, territorial boundaries are respected, and the best-adapted individuals defend the richest territories and produce the most offspring.

Successful reproduction requires that animals recognize the species, sex, and sexual receptivity of potential mates. In some cases, they must also overcome a resistance to close approach by another individual. These requirements have resulted in the evolution of sexual displays that use all possible forms of communication.

Social living has both advantages and disadvantages, and species show a wide variation in the degree to which their members cooperate. Some form cooperative societies. The most rigid and highly organized are those of the social insects such as the honeybee, where the members follow rigidly defined roles throughout life. These roles are maintained through both genetic programming and the influence of primer pheromones. Nonhuman vertebrates also form complex, but usually less rigid, societies, such as are found among bullhead catfish. Naked mole rats exhibit the most complex and rigid vertebrate social interactions, resembling insect societies.

Human Ethology

The degree to which human behavior is genetically influenced is highly controversial. Because humans cannot be treated as laboratory animals, and because learning plays a major role in nearly all human behavior, investigators must rely on studies of newborn infants, observation of responses to exaggerated stimuli, comparative cultural studies, correlations between certain behaviors and physiology (which suggest a role for pheromones), and studies of identical and fraternal twins. Evidence is mounting that our genetic heritage plays a role in personality, intelligence, simple universal gestures, our responses to certain stimuli, and our tendency to learn specific things such as language at particular stages of development.

❈ KEY TERMS

active visual signal   p. 828
aggression   p. 830
altruism   p. 843
communication   p. 828
dominance hierarchy   p. 832

kin selection   p. 843
passive visual signal   p. 828
pheromone   p. 829
primer pheromone   p. 830
queen substance   p. 830

releaser pheromone   p. 830
selfish gene   p. 843
territoriality   p. 832
waggle dance   p. 838

❈ THINKING THROUGH THE CONCEPTS

Multiple Choice

1. Which of the following is a function of a territory?
   a. reduction of aggression   b. a site in which to nest
   c. a mating area   d. a food-storage area
   e. all of the above

2. Which of the following is an advantage of social groupings of animals?
   a. increased ability to detect, repel, or confuse predators
   b. increased hunting efficiency
   c. increased likelihood of finding mates
   d. energy conservation
   e. all of the above

3. Which of the following pairs of communication forms and advantages are NOT accurate?
   a. pheromones, long-lasting
   b. visual displays, instantaneous
   c. sound communication, effective at night
   d. pheromones, convey rapidly changing information
   e. touch, maintains social bonds

4. In an insect society, such as the honeybee society,
   a. the division of labor is based on biologically determined castes
   b. all adult members share labor equally
   c. all adult members have the opportunity to reproduce
   d. reproduction is altered seasonally among the adults
   e. the organization of the society is flexible and adaptable

5. A specific act of altruism will most likely occur between
   a. a male and a female
   b. a female and an unrelated neighbor
   c. a parent and its offspring
   d. two females
   e. two males

6. When a male rat defeats another male and takes over his harem, the new male often kills any current litters. What principle does this behavior represent?
   a. altruism   b. kin selection
   c. operant conditioning   d. sexual selection
   e. parental investment

Review Questions

1. List four senses through which animals communicate, and give one example of each form of communication. After each, present both advantages and disadvantages of that form of communication.

2. Distinguish between passive and active visual signals, providing an example of each. Which can convey the most rapidly changing information?

3. What are graded visual signals, and what is their purpose?

4. Define and distinguish between primer and releaser pheromones. Give an example of each.

5. A songbird will ignore a squirrel in its territory but act aggressively toward a member of its own species. Explain why.

6. Identify the criteria for successful reproduction. For each, provide an example of how a specific animal satisfies these criteria.

7. Why are most aggressive encounters among members of the same species relatively harmless?

8. Discuss advantages and disadvantages of group living.

9. What type of animal tends to form the most complex societies, and why?

10. Describe one of the experiments that reveal the importance of chemical communication in bullhead catfish. Suggest why visual communication is not highly developed in this species.

11. In what ways do naked mole rat societies resemble those of the honeybee?

✖ APPLYING THE CONCEPTS

1. You raise honeybees but are new at the job. Trying to increase honey production, you introduce several queens into the hive. What is the likely outcome? What different things could you do to increase production?

2. Describe and give an example of a dominance hierarchy. What role does it play in social behavior? Give a human parallel, and describe its role in human society. Are the two roles similar? Why or why not? Now repeat this exercise for territorial behavior, both in humans and in another species of animal.

3. Humans perform many altruistic behaviors, both for their relatives and for unrelated individuals. Discuss possible reasons why people might make sacrifices for unrelated individuals, and whether or not this behavior might be adaptive.

4. Remember "Cabbage Patch" dolls and their enormous popularity? Based on your readings in this chapter, suggest why they were so appealing.

5. You are manager of an airport, where planes are being endangered by large numbers of flying birds, which can be sucked into engines, disabling them. What might you do to discourage birds from nesting and flying near the airport and its planes, without harming the birds?

6. As parents, you decide you would like your child to speak a second language. When, ideally, would you start teaching the child this language, and why?

✖ FOR MORE INFORMATION

Dawkins, R. *The Selfish Gene*, 2nd ed. New York: Oxford University Press, 1989. Exceptionally clear explanation of the evolutionary basis of altruistic behavior, written for laypeople and scientists.

Eisner, T. and Wilson, E. O. *Animal Behavior*. New York: W. H. Freeman and Co., 1975. An outstanding collection of articles on animal behavior collected from *Scientific American*, written in a readable style by some of the field's best researchers.

Goldman, B. "The Essence of Attraction." *Health*, March/April 1994. Describes the discovery and marketing of a potential human pheromone.

Kirchner, W. H., and Towne, W. F. "The Sensory Basis of the Honeybee Dance Language." *Scientific American*, June 1994. Combines a historical look at the work of Karl von Frisch with a modern update on recent research that has almost fully elucidated the mechanisms of honeybee communication during the waggle dance.

Locke, J. L. "Phases in the Child's Development of Language." *American Scientist*, September/October 1994. Learning about speech begins before birth and continues to develop before any intelligible words are produced.

Macdonald, D., and Brown, R. "The Smell of Success." *New Scientist*, May 1985. Describes the amazing diversity of mammalian pheromones.

Radetsky, P. "Silence, Signs, and Wonder." *Discover*, August 1994. Discusses sign language as a means of communication and relates it to brain function.

Seeley, T. D. "The Honey Bee Colony as a Superorganism." *American Scientist*, November–December 1989. The honey bee society exemplifies the concept of kin selection in which the society, rather than the individual, serves as a "vehicle for the survival of genes."

Sherman, P. W., Jarvis, J. U. M., and Braude, S. H. "Naked Mole Rats." *Scientific American*, August 1992. Describes the newly investigated society of the naked mole rat, a vertebrate whose social behavior resembles that of some social insects.

Stevens, J. "The Biology of Violence." *BioScience*, May 1994. Discusses the genetic aspects of violence.

Wilson, E. O. "Empire of the Ants." *Discover*, March 1990. This entertaining article discusses the diversity of ant behavior that has contributed to their enormous success.

Wright, K. "The Sniff of Legend." *Discover*, April 1994. Discusses the discovery of human pheromones and a sixth sense organ that detects them.

NET WATCH

On-line resources for this chapter are on the World Wide Web at:
http://www.prenhall.com/~audesirk (click on the table of contents link and then select Chapter 42).

Unit VII
Ecology

The beauty and interdependence of Earth's biosphere is illustrated in Suzanne Duranceau's painting "Paradise."

43 Population Growth and Regulation

"There is no exception to the rule that every organic being naturally increases at so high a rate that, if not destroyed, the earth would soon be covered by the progeny of a single pair."

CHARLES DARWIN in *On the Origin of Species* (1859)

⊞ A T   A   G L A N C E

Introduction to Ecology

How Populations Grow
The Biotic Potential of a Population Produces Exponential Growth If Not Restrained
Exponential Growth Sometimes Leads to "Boom-and-Bust" Cycles
Environmental Resistance Limits Population Growth

Population Patterns in Space and Time
Populations Distribute Themselves in Different Ways
Populations Show Characteristic Patterns of Survivorship

The Human Population
The Human Population Is Growing Exponentially
The Age Structure of a Population Predicts Its Future Growth
The Population of the United States Is Growing Rapidly

A densely aggregated breeding population of king penguins on South Georgia Island in the Atlantic Ocean off Antarctica. Brown chicks are guarded and fed by their white parents.

Introduction to Ecology

In eastern Colorado lies a remnant of shortgrass prairie, dominated by buffalo grass and blue grama grass. In spring and summer it is ablaze with wildflowers: paintbrush, vetch, sunflower, and bladderpod (Fig. 43-1). Prairie dogs stand upright by their burrows, alerting the "town" with vigorous high-pitched cries as a single hawk soars lazily overhead. The prairie is an **ecosystem,** a complex, interrelated network of living organisms and their nonliving surroundings. An ecosystem can be as small as a puddle or as large as an ocean. Within our prairie ecosystem, wildflowers, prairie dogs, the grass on which they feed, the hawk that preys on them, and the myriad microscopic organisms that keep the soil fertile, constitute a **community** of interacting populations. A **population** consists of all the members of a particular species, be they hawks, grasshoppers, buffalo grass, or bacteria, living in a given area.

Just to the north, a farm was abandoned nearly 2 years ago. The community here is quite different, with Russian thistle, pigweed, amaranth, and cheatgrass invading the new habitat. A "For Sale" sign in the field foreshadows the high-density housing developments that will soon displace both farmland and prairie. The first human residents will be delighted to see a hawk, a rare and magnificent bird of prey, practically in their backyards. But as the prairie dog town is bulldozed, this large predator will also disappear.

How has the prairie (in contrast to the failed farm) sustained itself for centuries without artificial fertilizer or irrigation? Why are predators such as the hawk rare relative to their prey? What keeps prairie dogs from overpopulating their habitat and starving? What happens when two organisms compete for the same resources? Why is the abandoned farm community different from that of the untouched prairie? What will the abandoned farm look like in 40 years, if left to itself? Will it ever again resemble the prairie community? Why has the human population continued to increase, while other populations remain stable or decline in the face of human expansion? These are a few of the questions of **ecology,** the science that deals with the interrelationships among living things and their environment. The environment consists of an **abiotic** (nonliving) component, including soil, water, and weather, and a **biotic** (living) component, including all forms of life

Figure 43-1 **The prairie ecosystem**

This montage shows representatives of interacting populations, which together constitute the prairie community. Insets show **(a)** a fly on a lichen-covered rock, **(b)** an Indian Paintbrush wildflower, **(c)** a prairie dog, and **(d)** a hawk.

within the ecosystem. Ecology (a term derived from the Greek word *oikos*, meaning "a place to live") is a tremendously diverse, complex, and relatively young scientific discipline.

In preceding chapters we have studied the anatomy, physiology, and behavior of individual organisms. The science of ecology begins at the next level—the population, which we examine in this chapter. In the following three chapters, we will proceed from this starting point to levels of increasing complexity: first to communities and the interactions within them, then to the organization of ecosystems, and finally to an exploration of the diversity of ecosystems that make up the biosphere.

How Populations Grow

Studies of undisturbed ecosystems show that many populations tend to remain relatively stable over time. Yet we are vividly aware from the human example that populations can readily increase. Let's first examine how and why populations grow, then look at the factors that normally control this growth.

Three factors determine whether and how much the size of a population will change: births, deaths, and migration. Organisms join a population through birth or **immigration** (migration in) and leave it through death or **emigration** (migration out). A population remains stable if, on the average, as many individuals leave as join. Pop-

ulation growth occurs when the number of births plus immigrants exceeds the number of deaths plus emigrants.
Populations decline when the reverse occurs. A simple
equation for the change in population size is:

population change = (births – deaths) +
(immigrants – emigrants)

In many natural populations, organisms moving in and
out contribute relatively little to population change, leaving birth and death rates as the primary factors influencing population growth.

The ultimate size of any population (leaving out migration) is the result of a balance between two major opposing
factors. The first is **biotic potential,** or the maximum rate at
which the population could increase, assuming ideal conditions allowing a maximum birth rate and minimum death
rate. Opposing this potential for growth are limits set by
the living and non living environment. These limits include the availability of food and space, competition with
other organisms, and certain interactions among species,
such as predation and parasitism. Collectively, these limits
are called **environmental resistance.** Environmental resistance can both decrease the birth rate and increase the
death rate. The interaction between biotic potential and
environmental resistance usually results in a balance between population size and available resources. To understand how populations grow and how their size is regulated,
we must examine each of these forces in more detail.

The Biotic Potential of a Population Produces Exponential Growth If Not Restrained

Changes in population size (ignoring migration) are functions of the birth rate, the death rate, and the number of
individuals in the original population. Rates of change in
population size can be measured as the changes in each of
these variables for a given size of population during a given
unit of time. For example, birth and death rates may be expressed as the number of births or deaths per 1000 individuals per year.

The **growth rate** (r) of a population is determined by
subtracting the death rate (d) from the birth rate (b):

$$r \quad = \quad b \quad - \quad d$$
(growth rate)   (birth rate)   (death rate)

For example, the annual growth rate of a population of
10,000 in which 1500 births and 500 deaths occur each
year (the given unit of time) can be calculated as follows:

$$r = \text{birth rate} - \text{death rate}$$

$$r = 1500/10,000 - 500/10,000 = 0.10, \text{ or } 10\%$$

or

$$r = 0.15 - 0.05 = 0.10, \text{ or } 10\%$$

To determine the number of individuals added to a population in a given time period, multiply the growth rate

(r), which is a rate per individual, by the original population size (N):

$$\text{population growth } = rN$$

In this case, population growth (rN) equals 0.10 x 10,000
= 1000. If this growth rate is constant, then the following
year, r must be multiplied by an even larger population
size ($N + rN$ = 11,000). This means that in the following
year an additional 1100 individuals are added to the population size (N), which in turn increases the number of
individuals added in the third year, and so on.

This is **exponential growth.** During exponential growth,
the population grows (during a given time period) by a
fixed percentage of its size at the beginning of that time
period. Thus an increasing number of individuals is added
to the population during each succeeding time period, causing population size to grow at an ever-accelerating pace.
The curve formed by exponential population growth is
often called a J-shaped growth curve, or a **J-curve,** after its
shape, clearly illustrated in Figures 43-2 and 43-3.

Such growth occurs whenever births consistently exceed deaths, as they do if, on average, each individual produces more than one surviving offspring during its lifetime. Although the number of offspring produced by an
individual each year varies from millions for an oyster to
one or fewer for a human, each organism, whether working alone or as part of a sexually reproducing pair, has the
potential to replace itself many times over during its lifetime. This capacity, called biotic potential, has evolved
because natural selection has favored those individuals
that leave the most offspring. Several factors influence the
biotic potential of a particular species, including:

1. The age at which the average individual first reproduces
2. The average frequency with which reproduction occurs
3. The average number of offspring produced each time
4. The reproductive life span of the average organism
5. The average death rate under ideal conditions

We will use examples in which these factors differ to illustrate the concept of exponential growth (see Fig. 43-2).

The bacterium *Staphylococcus* is normally a harmless resident in and on the human body, where its population
growth is restricted by environmental resistance. But in an
ideal culture medium, such as warm custard, where it may
occasionally be introduced accidentally, each bacterial cell
can divide every 20 minutes, doubling the population three
times each hour (the by-products of bacterial metabolism
can result in food poisoning under these conditions). The
larger the population grows, the more cells there are to divide. The biotic potential of bacteria is so great that, were
nutrients unlimited, within 48 hours the offspring of a single bacterium would swamp Earth in a layer 7 feet deep!

In contrast, the bald eagle is a relatively long-lived,
rather slowly reproducing species. Let's assume that the
bald eagle can live 30 years, that it reaches sexual maturity at 4 years, and that each pair of eagles produces two off-

(a)

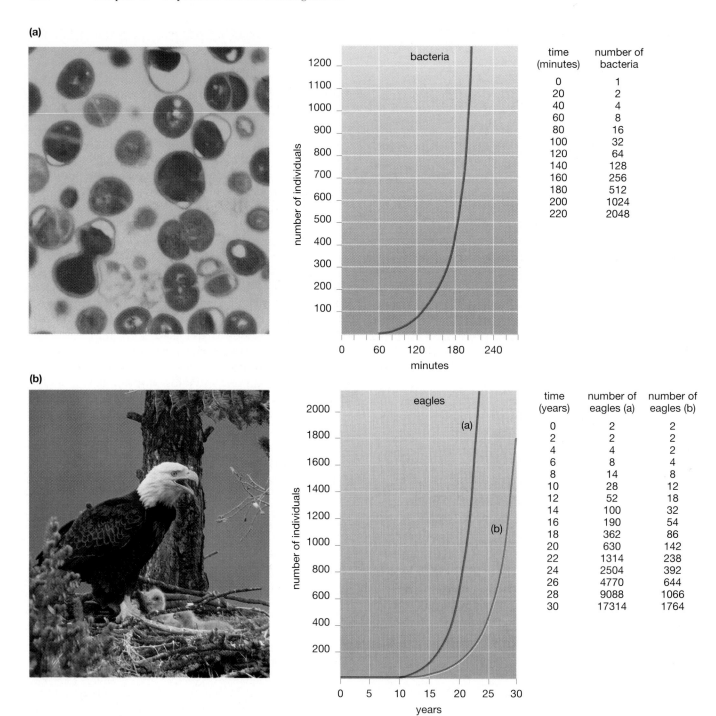

time (minutes)	number of bacteria
0	1
20	2
40	4
60	8
80	16
100	32
120	64
140	128
160	256
180	512
200	1024
220	2048

(b)

time (years)	number of eagles (a)	number of eagles (b)
0	2	2
2	2	2
4	4	2
6	8	4
8	14	8
10	28	12
12	52	18
14	100	32
16	190	54
18	362	86
20	630	142
22	1314	238
24	2504	392
26	4770	644
28	9088	1066
30	17314	1764

Figure 43-2   **Exponential growth curves**

All such curves share a similar J shape; the major difference is the time scale. **(a)** The growth of a population of bacteria, assuming one individual to start with and a doubling time of 20 minutes. **(b)** Growth of an eagle population, starting with a single pair of hatchlings, and assuming that the age at first reproduction is 4 years (red line) and 6 years (green line). Notice that after 26 years, the population of eagles that began reproducing at 4 years is seven times that of the eagles that began reproducing at 6 years.

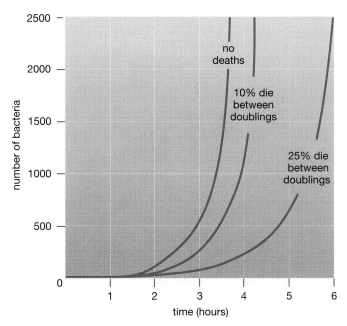

Figure 43-3 **The effect of death rates on bacterial population growth**

The graphs assume that the population doubles every 20 minutes. Notice that the population in which a quarter of the bacteria die every 20 minutes reaches 2500 only 2 hours and 20 minutes later than that in which no deaths occur.

spring per year for the remaining 26 years. Figure 43-2 compares the potential population growth of eagles with that of bacteria, assuming no deaths occur in either population during the time graphed. Notice that the shapes of the curves are virtually identical—both populations exhibit unrestricted exponential growth. Although the time scale differs, both population sizes eventually become astronomically large. Figure 43-2 also shows what happens if eagle reproduction begins at 6 years instead of 4. Under this condition, exponential growth still occurs, but the time required to reach a particular size increases considerably. This time factor has important implications for the human population: Delayed childbearing significantly slows population growth. If each woman bears three children in her early teens, the population will grow much faster than if each woman bears five children beginning at age 30.

So far we have looked only at birth rates. Even under ideal conditions, however, some mortality occurs. To illustrate the effect of differing death rates, we compare three bacterial populations in Figure 43-3. Again, the shapes of the curves are the same. In each case the population eventually approaches infinite size; but the time required to reach any given population size is increased by increasing mortality.

Exponential Growth Sometimes Leads to "Boom-and-Bust" Cycles

In nature, exponential growth occurs only under special circumstances and only for a limited time. For example, exponential growth is seen in populations that undergo

regular cycles, where rapid population growth is followed by a massive die-off. These **boom-and-bust cycles** occur in a variety of organisms for complex and varied reasons. Many short-lived, rapidly reproducing species, from algae to insects, have seasonal population cycles that are linked to predictable changes in rainfall, temperature, or nutrient availability (Fig. 43-4). In temperate climates, insect populations grow rapidly during the spring and summer, then crash with the killing hard frosts of winter. More-complex factors produce roughly 4-year cycles for small mammals, such as voles and lemmings, and much longer cycles in hare, muskrat, and grouse populations.

Lemming populations, for instance, may grow until they overgraze their fragile arctic tundra ecosystem. Lack of food, increasing populations of predators, and social stress caused by overpopulation may all contribute to a sudden high mortality. Many deaths occur as waves of lemmings emigrate from regions of high population density. During these dramatic mass movements, they are easy targets for predators, and many accidentally drown (some begin swimming when they encounter bodies of water, including the ocean). The reduced lemming population eventually contributes to a decline in predator numbers (see "A Closer Look: Cycles in Predator and Prey Populations") and recovery of the plant community. This recovery, in turn, sets the stage for another round of exponential growth in the lemming population (Fig. 43-5).

In populations that do not show boom-and-bust cycles, exponential growth may occur temporarily under special circumstances, for example, if population-controlling factors, such as predators or parasites, are eliminated or if the food supply is increased. Exponential growth can also occur when individuals invade a new habitat where conditions are favorable and competition or predation is scarce. Many such invasions have happened when people have introduced foreign, or **exotic,** species into ecosystems, often with tragic results (see "Earth Watch: Introduced Species Disrupt Community Interactions," Chapter

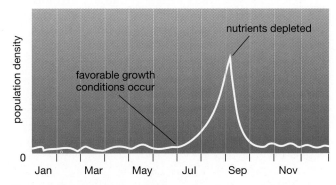

Figure 43-4 **A boom-and-bust population cycle**

Blue-green algae in a boom-and-bust cycle in Lake Erie. Algae survive at a low level through the fall, winter, and spring. Early in July, conditions become favorable for growth, and exponential growth occurs through August, after which nutrients become depleted and the population "goes bust."

A CLOSER LOOK

Cycles in Predator and Prey Populations

If we assume that prey, such as rabbits, are eaten exclusively by a particular predator, such as the lynx, it seems logical that both populations might show cyclic changes, with increases in the predator population lagging behind increases in prey population size. For example, a large rabbit population will provide abundant food for lynx and their offspring, which will survive in large numbers. The lynx population will in turn increase and thus so will predation on the rabbits. Increased predation will cause a decline in the rabbit population, which will allow fewer lynx to survive and reproduce, forcing a decline in the lynx population shortly thereafter.

Does this out-of-phase population cycle of predators and prey actually occur in nature? A classic example of such a cycle was observed using the ingenious method of counting all the pelts from northern Canada lynx and snowshoe hares purchased by the Hudson Bay Company between 1845 and 1935. The availability of pelts (which presumably reflects population size) showed dramatic, closely linked cycles between these predators and their prey (Fig. E43-1). Unfortunately, as with any field investigation, many variables could influence the relationship between hare and lynx. One problem is that hare populations have been shown to fluctuate even without lynx present, possibly owing to overshooting carrying capacity and reducing their food supply (in this case, the important predator-prey interaction is between the hares and the plants they prey upon). Further, lynx do not feed exclusively on hares but can eat a variety of small mammals. Environmental variables such as exceptionally severe winters could adversely affect both populations and produce similar cycles.

To test this hypothesis more scientifically, investigators turned to controlled laboratory studies on populations of small predators and their prey. The study illustrated here involved only braconid wasp predators and their bean weevil prey. As predicted, the two populations showed regular cycles, with the predator population rising and falling slightly later than the prey population (Fig. E43-2). The wasps lay their eggs on weevil larvae, which provide food for the newly hatched wasps. A large weevil population assures a high survival rate for wasp offspring, increasing the predator population. Then, under intense predation pressure, the weevil population plummets, reducing the population size of the next generation of wasps. Reduced predation then allows the weevil population to increase rapidly, and so on.

It is highly unlikely that such a straightforward example will ever be found in nature, but this type of predator-prey interaction clearly can contribute to the fluctuations observed in many natural populations.

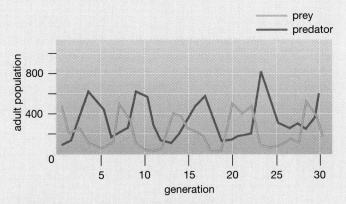

Figure E43-2   Experimental predator-prey cycles

Out-of-phase fluctuations in laboratory populations of the azuli bean beetle and its braconid wasp predator.

Figure E43-1 Population cycles

Snowshoe hares and their lynx predators are graphed on the basis of the number of pelts received by the Hudson Bay Company.

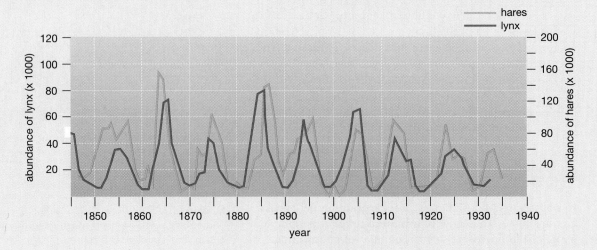

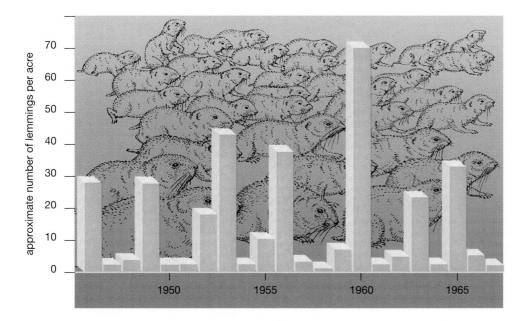

Figure 43-5 **Boom-and-bust lemming population cycles**

Lemming population density follows roughly a 4-year cycle, as illustrated by these data from Point Barrow, Alaska.

44). As you will learn in the next section, all populations showing exponential growth must eventually either stabilize or crash.

Environmental Resistance Limits Population Growth

Exponential growth carries with it the seeds of its own destruction. As individuals join the population, competition for resources intensifies. Predators may increase in number or they may make this newly abundant prey a larger part of their diet. Parasites and diseases spread more readily because of crowding and weakness resulting from lack of food or from stress caused by adverse social interactions. Consequently, after a period of exponential growth, populations tend to stabilize at or below the maximum size that the environment can sustain, called carrying capacity (see next section). The rate of growth gradually declines and reaches a long-term state of equilibrium, fluctuating around a growth rate of zero. In this equilibrium, the birth rate is balanced by the death rate and population size is stabilized. This type of population growth, which is typical of long-lived organisms colonizing a new area, is represented graphically by an S-shaped growth curve, or **S-curve** (Fig. 43-6).

Populations Usually Stabilize at or below Carrying Capacity

Populations may stabilize at a level called the **carrying capacity** of the ecosystem. The carrying capacity is the maximum size of a given population of organisms that a particular area can support indefinitely. It is determined primarily by the sustained availability of two types of resources: renewable resources, such as nutrients, water, and light, that are replenished by natural processes, and a nonrenewable resource, space, such as suitable nesting or denning areas.

Organisms will die if demands on renewable resources, such as food, water, and light (the energy source for plants), are too high. If space requirements are exceeded, animals may emigrate, often to less-suitable areas where their death rate will be higher. Reproduction will decline, because animals may not find adequate breeding sites or the seeds of plants may not reach a suitable place to germinate. Excess demands may damage ecosystems, reducing their carrying capacity. The result is either a population decline until the ecosystem recovers, or a permanently reduced population. For example, overgrazing by cattle on dry western grasslands has reduced the grass cover and allowed sagebrush (which cattle will not eat) to thrive. Once established, sagebrush replaces edible grasses and reduces the carrying capacity of the land for cattle. Other dramatic cases of overgrazing have occurred when herbivores, such as reindeer, have been introduced onto islands without large predators, as shown in Figure 43-7.

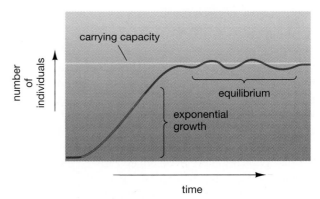

Figure 43-6 **The S-curve of population growth**

The population first grows exponentially, then fluctuates around carrying capacity.

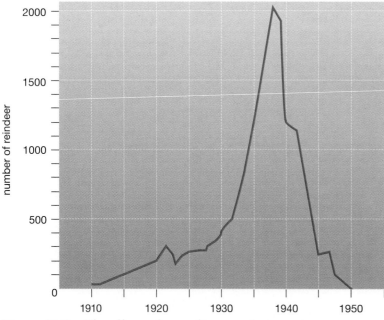

Figure 43-7   **The effects of exceeding carrying capacity**

Exceeding carrying capacity can damage an ecosystem, reducing its ability to support the population. In 1911, 25 reindeer were introduced onto one of the Pribilof Islands (St. Paul) in the Bering Sea off Alaska. Food was plentiful, and there were no predators of reindeer on the island. The herd grew exponentially (note the initial J shape) until it reached 2000 reindeer in 1938. At that point, the small island was seriously overgrazed, food was scarce, and the population declined dramatically. By 1950, only eight reindeer survived.

In nature, environmental resistance maintains populations at or *below* the carrying capacity of the environment. Factors of environmental resistance may be classified into two broad categories. **Density-independent** factors limit population size regardless of the density (number of individuals per given area) of the population. **Density-dependent** factors increase in effectiveness as the population density increases. In the following sections, we will look more closely at these factors and how they control population growth.

Density-Independent Factors Limit Populations Regardless of Their Density

Density-independent factors kill individuals or reduce reproductive rates regardless of population density. Perhaps the most important density-independent factor is weather. For example, many insects and annual plant populations are limited in size by the number of individuals that can be produced before the first hard freeze. Such populations typically do not reach carrying capacity, because density-independent factors intervene first. Weather is largely responsible for the boom-and-bust population cycles described earlier. Human activities can also limit the growth of natural populations in ways that are independent of population density. Pesticides and pollutants may cause drastic declines in natural populations, as does habitat destruction caused by construction of farms, roads, and housing developments.

Density-Dependent Factors Become More Effective As Population Density Increases

Organisms that live several years have evolved various mechanisms to compensate for seasonal weather changes, thus circumventing this form of density-independent population control. Many mammals, for example, develop thick coats and store fat for the winter; some also hibernate. Other animals, including many birds, migrate long distances to find food and a hospitable climate. Plants may survive the rigors of winter by entering a period of dormancy, dropping their leaves and drastically slowing their metabolic activities.

For these long-lived species, by far the most important elements of environmental resistance are density-dependent factors. Density-dependent factors are those that become increasingly effective as population density increases, thus exerting a negative feedback effect on population size. The larger the population grows, the more changes are triggered that counteract this growth. Density-dependent factors include community interactions, such as predation and parasitism, as well as competition within the species or with members of other species. These factors are discussed below and in Chapter 44.

Predators Help Control Their Prey Populations

Predation is the act of killing and eating another living organism. This broad definition includes the swift and violent capture of elk by wolves (Fig. 43-8) as well as the less

Figure 43-8   **Predators help control prey populations**

Gray wolves have brought down an elk, who may have been weakened by old age or parasites. Predators will often switch to the most abundant prey, limiting prey populations.

dramatic munching on a cactus by a moth (Fig. 43-9) or a bird eating a seed. Organisms that kill and eat other organisms are **predators**. Predation controls prey populations in a density-dependent manner. It can become an increasingly important factor in population control as prey populations increase, simply because the denser the prey population, the more predator-prey encounters there will be. Many predators will eat a variety of prey depending on what is most abundant and easiest to find. Coyotes may eat more mice when the mouse population is high but switch to eating more ground squirrels as the mouse population declines, incidentally allowing the mouse population to recover.

Predators may also exert density-dependent effects by increasing in number as their prey population grows. For instance, predators that feed heavily on lemmings, such as the arctic fox and snowy owl, regulate the number of offspring they produce according to the abundance of lemmings. The snowy owl may produce up to 13 chicks when lemmings are abundant but not reproduce at all in years when lemmings are scarce. In other cases, an increase in predators may cause a crash of the prey population, which is then followed by a decline in the predator population. This series of events can result in **out-of-phase population cycles** of both predators and prey (see "A Closer Look: Cycles in Predator and Prey Populations").

The influence of predators on prey populations varies considerably. Some predators feed primarily on prey made vulnerable because their populations have exceeded car-

(a)

(b)

Figure 43-9   **An introduced predator controls an introduced pest**

(a) A pasture in Queensland, Australia, is blanketed by the imported prickly pear cactus, whose spread was unchecked by predators. **(b)** The same site 3 years after the introduction of a predatory cactus moth appropriately named *Cactoblastis cactorum*. (Courtesy of the Department of Lands, Queensland, Australia.)

rying capacity. Such prey may be weakened by lack of food or may be exposed owing to lack of appropriate shelter. In such cases, predation may maintain prey populations near their sustainable density.

In other cases, predators may maintain the prey population at well below carrying capacity. A dramatic example is the prickly pear cactus, which was introduced into Australia. Lacking natural predators, it spread uncontrollably, destroying millions of acres of valuable pasture and range land. Finally, in the 1920s, a cactus moth (a predator of the prickly pear) was imported from Argentina and released to feed on the cacti. Within a few years, the cacti were virtually eliminated. Today the moth continues to maintain its prey at very low population densities (see Fig. 43-9).

Parasites Can Reduce Their Host Populations

In contrast to predation, **parasitism** is the process of feeding on another, usually larger, organism without killing it immediately or directly. **Parasites** often live on or in their prey, also called the **host**. Parasites include the familiar disease-causing microorganisms, mosquitoes, and intestinal worms, but also animals that graze on plants without killing them. Parasites may dramatically increase the death rate in a population in which individuals are already weakened by the effects of overcrowding. Further, infestation by parasites makes organisms more vulnerable to predators.

Parasitism is density dependent. Most parasites have limited motility and spread more readily between hosts at high population densities. For example, plant diseases spread readily through acres of densely planted crops, and childhood diseases spread rapidly through schools and day-care centers.

Both predators and parasites can have beneficial effects on the prey population as a whole. As you will learn in the next chapter, predators, parasites, and their prey co-evolve. Parasites and predators often keep host populations within carrying capacity. They also frequently destroy the least fit of the prey, leaving the better-adapted prey to reproduce. The result is usually a balance in which the prey population is regulated but not eliminated.

When a predator or parasite is introduced into an area where it did not evolve, however, the local prey species often are not adapted for it, and the consequences can be disastrous. The successful conquest of much of Earth by Europeans can be traced partly to the disease organisms they carried with them as they invaded continents whose populations had no previous exposure or resistance to these parasites. Smallpox, for example, imported by Europeans, ravaged the native population of Hawaii, the Amerindians of Argentina, and the aborigines of Australia. The population balance in ecosystems can be destroyed when species are introduced into new areas where they have no natural predators or parasites (see "Earth Watch: Introduced Species Disrupt Community Interactions," Chapter 44).

Competition for Resources Helps Control Populations

The resources that determine carrying capacity (space, nutrients, light) are limited relative to the demand for them. Further, use by one individual limits their availability to another. For this reason, **competition** limits population size in a density-dependent manner. There are two major forms of competition: **interspecific competition** (between members of different species; see Chapter 44) and **intraspecific competition** (between members of the same species). Because the needs of members of the same species for water and nutrients, shelter, breeding sites, light, and other resources are almost identical, intraspecific competition is more intense. Organisms have evolved several ways to deal with intraspecific competition. Some organisms, including most plants and many insects, engage in **scramble competition**. This is a kind of free-for-all with resources as the prize. For example, when a plant disperses its seeds in a small area, hundreds may germinate. As they grow, however, the larger ones begin to shade the smaller, those with the most extensive roots absorb most of the water, and the weaker individuals eventually wither and die.

Many animals (and even a few plants) have evolved **contest competition,** which helps regulate population size and reduce direct competition. Contest competition consists of social or chemical interactions used to limit access to important resources (also see Chapter 42). Territorial species—such as wolves, many fish, rabbits, and songbirds—defend an area containing important resources such as food or nesting sites. When the population begins to exceed the available resources, only the best-adapted individuals are able to defend adequate territories. Those without territories often do not reproduce and are also easy prey. The creosote bush, which secretes a chemical into the ground that prevents germination of seeds nearby, could be considered a territorial plant (see Fig. 43-11b).

Dominance hierarchies, or "pecking orders" (described in Chapter 42), are another form of contest competition observed in many species of social animals. High-ranking individuals have first access to food, breeding sites, and mates. When resources are limited, only dominant individuals obtain what they need to reproduce successfully. Thus, dominance hierarchies can limit population size.

As population densities increase and competition intensifies, some animals react by emigrating. Large numbers leave their homes to colonize new areas, and many, sometimes most, die in the quest. The massive movements of lemmings, which sometimes end in marches into the sea, are apparently in response to overcrowding. Migrating swarms of locusts plague the African continent, stripping all vegetation in their path (Fig. 43-10).

In laboratory studies, overcrowding of small mammals—such as mice, rats, and voles—causes **social stress.** Social stress results in a decrease in the size of reproduc-

Figure 43-10   **Emigration**

In response to overcrowding, emigration can reduce local populations. Locust swarms provide a dramatic example.

tive organs, reduced reproductive rate, slower growth, reduced resistance to disease, and cannibalism of young, all of which would reduce population size. In the wild, however, population densities rarely reach the levels achieved in the laboratory, so these findings may have limited applicability to natural populations.

The size of any population is a result of complex interactions between density-independent and density-dependent forms of environmental resistance. For example, a stand of pines weakened by drought (a density-independent factor) may more readily fall victim to the pine bark beetle (density dependent). Likewise, a caribou weakened by parasites (density dependent) is more likely to be killed by an exceptionally cold winter (density independent).

Population Patterns in Space and Time

Populations Distribute Themselves in Different Ways

Organisms may live in flocks, herds, pairs, or as solitary individuals, or they may cluster around resources such as water holes. The general pattern in which members of a population are dispersed within a given area is a population's distribution. Distribution may vary with time, changing with the breeding season, for example. Ecologists recognize three major types of spatial distribution: **clumped, uniform,** and **random** (Fig. 43-11).

Individuals in Many Populations Form Groups

There are many populations whose members live in groups and whose distribution can be described as clumped. Examples include family or social groupings, such as elephant

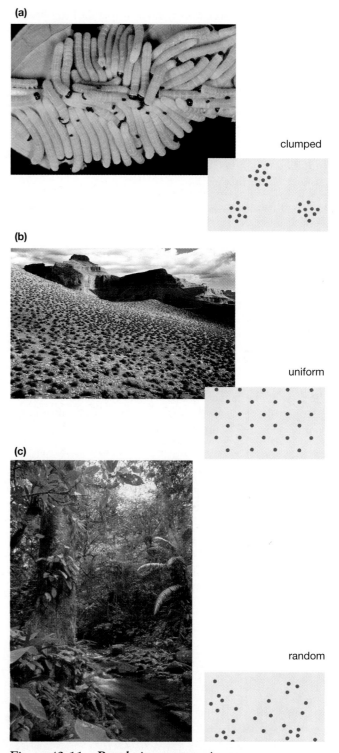

Figure 43-11   **Population patterns in space**

(a) Clumped, as illustrated by this gathering of caterpillars. **(b)** Uniform, seen in the spacing of these creosote bushes in the desert. **(c)** The random distribution of trees and plants in a rain forest.

herds, wolf packs, prides of lions, flocks of birds, or schools of fish (Fig. 43-11a). What are the advantages of clumping? Flocks provide many eyes that search for localized food, such as a tree full of fruit. Schooling fish avoid predation by confusing the predator with countless flashing bodies darting in all directions. Some species form temporary groups for mating and caring for their young. Other plant or animal populations cluster, not for social reasons, but because resources, such as nutrients, shelter, or water, are localized. Cottonwood trees, for example, aggregate along streams and rivers in grasslands. Animals also aggregate around water holes in the dry savanna of Africa.

Individuals of Some Populations Disperse Almost Uniformly

Uniformly distributed organisms maintain a relatively constant distance between individuals. This distribution occurs most frequently among animals that defend territories and exhibit territorial behaviors designed to protect scarce resources, such as breeding sites, nutrients, or water. Male Galapagos iguanas establish evenly spaced breeding territories. Shorebirds are also often found in evenly spaced nests, just out of reach of one another. Other territorial species, such as the Tawny Owl, mate for life and continuously occupy well-defined, relatively uniformly spaced territories (for breeding animals, the spacing refers to pairs, not individuals). Some plants growing in poor soil with limited water, such as the creosote bush (Fig. 43-11b), release chemicals that prevent germination of plants nearby. This mechanism helps assure adequate resources for each individual.

In a Few Populations, Organisms Are Randomly Distributed

In randomly distributed populations, which are relatively rare, individuals do not form social groups, the resources they need are more or less equally available throughout the area they inhabit, and resources are not scarce enough to require territorial spacing. Trees and other plants in rain forests come close to being randomly distributed (Fig. 43-11c).

There are probably no vertebrate species that maintain random distribution throughout the year because they must breed, a behavior that makes social interaction inevitable.

Populations Show Characteristic Patterns of Survivorship

Population patterns can be considered from the perspective of time as well as space. Over time, populations show characteristic patterns of deaths or (more optimistically) survivorship. These patterns, called **survivorship curves,** are revealed when the number of individuals of each age is graphed against time. Three different types of survivorship curve, which can be described as "early loss," "constant loss," and "late loss," based on the part of the life cycle during which most deaths occur, are shown in Figure 43-12.

Populations with convex survivorship curves (late loss) have relatively low infant death rates, and most individuals survive to old age. This curve is characteristic of humans and many other large animals, such as bighorn sheep. These species produce relatively few offspring, which are protected by the parents. Species with constant-loss survivorship curves have a fairly constant death rate; they have an equal chance of dying at any time during their life span. This phenomenon is seen in the American robin, the gull, and laboratory populations of organisms that reproduce asexually, such as *Hydra* and bacteria. The concave curve (early loss) is characteristic of organisms producing large numbers of offspring that have little parental care, being largely left to compete on their own. The death rate is very high among the offspring, but those that reach adulthood have a good chance of surviving to old age. Most invertebrates, most plants, and many fish exhibit concave survivorship curves. In addition, in some populations of black-tailed deer, 75% of the population dies within the first 10% of its life span, giving this mammalian population a concave curve as well.

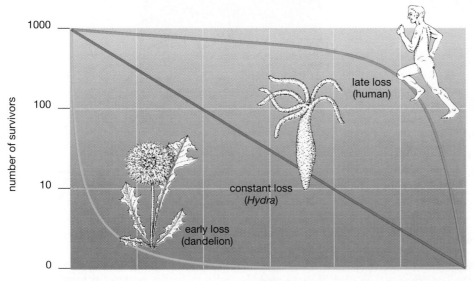

Figure 43-12 **Population patterns in time**

Three types of survivorship curve are shown. Because the life spans of these organisms differ so much, they can be compared on the same graph only if the percentages of survivors rather than the ages are used.

The Human Population

Your first child received free schooling and free medical care, and you were offered longer maternity leave and a larger pension if you signed a promise to have no more children. Pregnant with your second child, you face fines equivalent to all the benefits provided for the first child, up to a full year's pay. Each night, delegates come to your home to tell you of the harm you are doing to society. Your neighbors shun you, and the collective pressure finally drives you to the hospital and a state-sponsored abortion. A distant future scenario from a pessimistic science fiction novel? Hardly. Similar policies were instituted in China. With a population of over 1 billion and an increase of over 15 million people in 1983 alone, the government took drastic measures. Now, China's coercive policies have been eased somewhat, and birth rates are rising again. What is China's demographic future? What of the rest of the world? Let's examine the human population in light of what we know of exponential growth and carrying capacity.

The Human Population Is Growing Exponentially

Look at the graph of human population growth in Figure 43-13 and compare it with the exponential growth curves in Figure 43-2. The time span is different, but the growth curves are the same—exponential. It took over 1 million years for the human population to reach 1 billion, but the second billion was added in 100 years, the third in 30, the fourth in 15, and the fifth in 12 years. The population is projected to reach 6 billion in 1998. World population currently grows by about 88 million people yearly—a million people added every 4 days! Why hasn't environmental resistance put an end to our exponential growth? What is the carrying capacity of the world for humans?

Like all populations, humans have encountered environmental resistance, but unlike other populations, we have responded to environmental resistance by overcoming it. As a result, the human population has grown exponentially for an unprecedented time span; to accommodate our growing numbers, we have altered the face of the globe. Human population growth has been spurred by a series of "revolutions" that conquered various aspects of environmental resistance and increased Earth's carrying capacity for people.

Primitive people produced a **cultural revolution** when they discovered fire, invented tools and weapons, built shelters, and designed protective clothing. Weapons allowed more effective hunting and an increased food supply; clothing and shelter increased the habitable areas of the globe. Domesticated crops and animals supplanted hunting and gathering by about 8000 B.C. This **agricultural revolution** provided people with a larger, more dependable food supply and further increased Earth's carry-

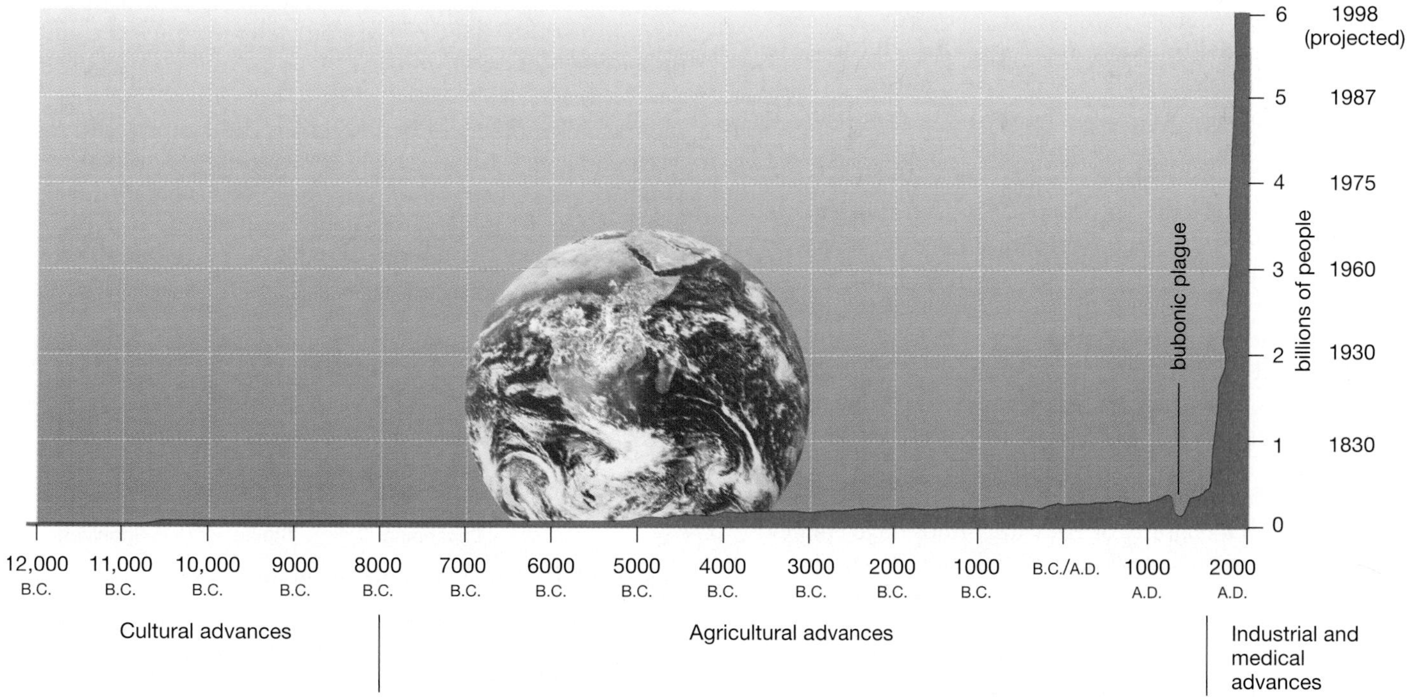

Figure 43-13   **Human population growth**

The human population from the Stone Age to the present has shown continued exponential growth. Note the dip in the fourteenth century caused by the bubonic plague. Note also the steadily decreasing intervals at which additional billions are added. (Inset) Earth is an island of life in a sea of emptiness; its space and resources are limited.

ing capacity for humans. Increased food resulted in longer lives and more childbearing years, but a high death rate from disease still restricted the population. Human population growth continued slowly for thousands of years until the **industrial-medical revolution** began in England in the mid-eighteenth century, spreading through Europe and North America in the nineteenth century. Medical advances dramatically decreased the death rate by reducing environmental resistance from disease. These advances included the discovery of bacteria and their role in infection, leading to control of bacterial disease through improved sanitation and the use of antibiotics, and the discovery of viruses, leading to the development of vaccines for diseases such as smallpox. The revolution continues today as research proceeds on vaccines against major killers, such as malaria and AIDS, and sophisticated medical advances, such as fertility drugs and *in vitro* fertilization, are developed.

In developed countries, such as those of western Europe, the industrial-medical revolution resulted in an initial rise in population due to decreased deaths, followed by a decline in birth rates. This decline can be attributed to many factors, including better education, increased availability of contraceptives, a shift to a primarily urban life-style, and more career options for women. In developed countries, such as Switzerland, Sweden, Austria, Germany, and England, populations have more or less stabilized. However, these countries are home to only about 4% of the world's population.

In less-developed countries, such as most of those in Central and South America, Africa, and Asia, medical advances have decreased death rates and increased the life span, but birth rates remain high. These countries have not experienced the increase in wealth that was partly responsible for the decline in birth rate of developed countries. Children serve as a form of social security in Third World nations because they may be the only support for parents in their old age. In agricultural societies, children are an important source of labor. Social traditions offer prestige to the man who fathers and the woman who bears many children. In Nigeria, the most crowded country in Africa, many men refuse to allow any form of birth control, and the average woman has more than six children. Nigeria is already suffering from loss of forests, the spread of deserts, soil erosion, and water pollution. With nearly half of its population of 93 million under the age of 15, continued growth is a certainty. Although a majority of African nations have concluded that their growth rates are too high and are working toward reducing them, lack of education and lack of access to contraceptives impede progress in curbing population growth. Of the projected 6.3 billion people on Earth in the year 2000, 5 billion will reside in less-developed countries. The prospects for population stabilization in the near future are nil, barring major catastrophes that might dramatically increase deaths. The reason can be seen clearly by looking at the age structures of the less-developed countries and comparing them with those of countries with stable populations, as described in the following section.

The Age Structure of a Population Predicts Its Future Growth

The **age structure** of a population is the distribution of males and females of various age groups, usually represented in a diagram. All age-structure diagrams come to a peak at the top, because relatively few people live into their nineties. The shape of the rest of the diagram, however, shows whether the population is expanding, stable, or shrinking (Fig. 43-14). If the number of children (age 0 to 14) exceeds the number of reproducing individuals (age 15 to 45), the population is expanding and the diagram resembles a pyramid. In the case of stable populations, the number of children is about equal to the number of reproducing adults, forming a straight-sided figure. In shrinking populations, there are fewer children than reproducing adults and the age-structure diagram is narrower at the base.

Figure 43-15 shows the average age structures of the populations of less-developed and developed countries. The outermost boundaries represent the projected population structure for the year 2025; the inner ones are for 1990. Each graph has been divided horizontally into three parts to show individuals who are prereproductive (0 to 14 years old), reproductive (15 to 45 years), and postreproductive (46 and older). In 1994, the less-developed countries (in Asia, Africa, India, and South and Central America) had an average annual growth rate of 1.9%, and the developed countries (in North America, Europe, Australia, Japan, and New Zealand) showed an average annual growth rate of 0.3%. In the developed countries, projections for the year 2025 are only slightly higher than present levels (see Fig. 43-15).

In contrast, each year in the less-developed countries, increasing numbers of people enter their reproductive years and give birth to an ever-increasing base of infants. Even if these countries were to reach **replacement-level fertility (RLF)** immediately (that is, if people of reproductive age had only sufficient children to replace them-

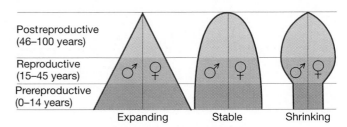

Figure 43-14   Age-structure diagrams

Here we show generalized figures for expanding, stable, and shrinking populations. The width of each figure is proportional to the number of individuals, with the male population on the left and the female population on the right. The vertical axis shows increasing age. The number of children (prereproductive, age 0 to 14 years) produced by the reproductive-age population determines whether the population is expanding (more children than reproductive-age adults), stable (number of children equal to number of reproductive-age adults), or shrinking (fewer children than reproductive-age adults).

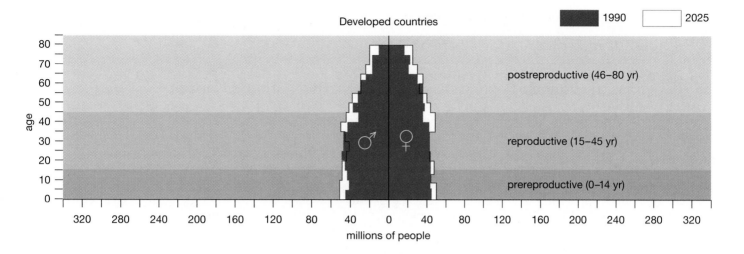

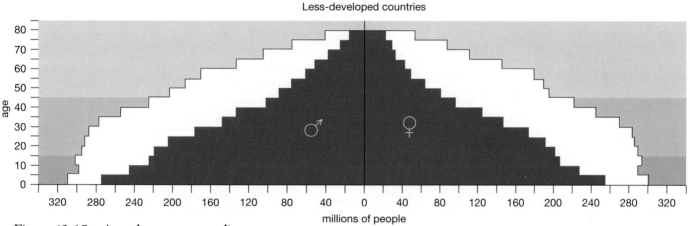

Figure 43-15   **Actual age-structure diagrams**

Here we show developed countries (top) and less-developed countries (bottom). Compare these with the idealized diagrams shown in Figure 43-14. The outer lines are projections to the year 2025. Notice how similar the current and projected age structures are. This similarity suggests that rapid population growth will continue in the less-developed countries for the foreseeable future.

selves), population growth would continue for decades because of built-in momentum produced by increasing numbers of people reaching reproductive age.

Notice, however, that predictions for these countries for the year 2025 show a structure very similar to the present one, the only difference being far more people in all age classes. Although several of these countries have recently initiated family-planning programs, parents in less-developed countries will probably continue to have larger than replacement-level families. The percent growth per year caused by natural increase (births – deaths) for various parts of the world is shown in Figure 43-16, p.866.

Ironically, population growth in these countries is helping to perpetuate the poverty and lack of education that in turn tend to sustain high birth rates. The relationship of income and education to birth rate has been documented in many countries, including the United States. Here, on the average, women who do not complete high school have twice as many children as those with more than 4 years of college.

The Population of the United States Is Growing Rapidly

As shown in Figure 43-17, the U.S. population is growing exponentially. In fact, at over 1% annually, our growth rate is over three times the average in the developed world. During the year ending in June 1995, for example, the U.S. population grew by over 2.5 million (nearly 300 people per hour), bringing the total over to 262 million. Recall the equation population change = (births - deaths) + (immigrants - emigrants). Let's examine each component of the equation to determine why the United States is growing so rapidly.

If each American woman had 2.1 children, we would have replacement-level fertility. (RLF is slightly higher than 2 because parents must replace both themselves and children who die before reaching maturity.) The U.S. birth rate in 1994 was exactly at replacement-level fertility, so why does our population continue to grow? Two factors

EARTH WATCH

Has the Human Population Exceeded Earth's Carrying Capacity?

A glance at the age structure of less-developed countries, where most of the world's population resides, shows a tremendous momentum for continued growth. World population in the year 2010 is predicted to be over 7 billion and growing. A modest United Nations (UN) projection is that the human population may stabilize in the year 2150 at 11.5 billion (Fig. E43-3). Can Earth support over twice its current population?

Earlier we defined carrying capacity as the maximum population that could be indefinitely sustained. Sustaining a population requires that the ecosystem not be damaged in ways that lower its ability to provide necessary resources. By this definition we have already exceeded Earth's carrying capacity for people. The upper limit of the planet's carrying capacity is determined by the ability of its plant life to harvest the energy from sunlight and produce high-energy molecules that organisms can use as food. Stanford University biologist Peter Vitousek estimates that human activities have already reduced the productivity of Earth's forests and grasslands by 12%. Each year, millions of acres of once-productive land are being turned into desert through overgrazing and deforestation, especially in less-developed countries. Earth's desert area is projected to increase by 20% by the year 2000 as a result of human activities. In a world where between 500 million and 1 billion people are chronically undernourished, 80% of the world's agricultural land is suffering moderate to severe erosion (Fig. E43-4).

In the years between 1990 and 2010, the world population is projected to increase by 33%. During this same period, the available fish, cropland, pastureland, and forests will all decline by 10% to 30% per person. The UN Food and Agricultural Organization reports that Africa's food production for each person has dropped by 20% since 1960; another 30% drop in the next 25 years is predicted. Each year the United States is losing 1 million acres of farmland to spreading urban development. The demand for wood in less-developed countries is far outstripping production, and large areas are being deforested each year. Deforestation in turn causes erosion of precious topsoil, runoff of much-needed fresh water, and the spread of deserts. The total world fish harvest peaked in 1989 and has been declining since, despite increased investments in fishing equipment and fish farming. Off the coast of New England, fish populations have plummeted since the early 1960s because of overfishing, and the fishing industry is collapsing. The destruction of tropical rain forests may exterminate 1 million species of animals and plants by the year 2000, as their habitats are destroyed. These are clear indications that our present population, at its present level of technology, is already "overgrazing" the world ecosystem and decreasing its ability to support all forms of life, including people.

Mass starvation, such as the famines in Somalia and Ethiopia, has occurred not because Earth lacks sufficient

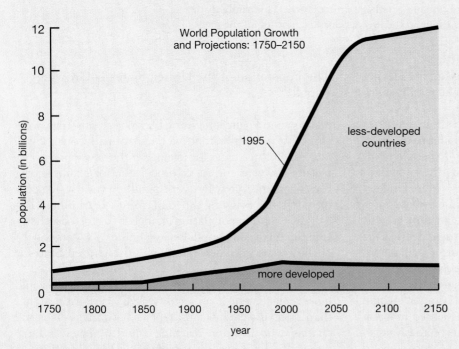

Figure E43-3   World population projected to the year 2150

The disparity in numbers between developed and less-developed countries is projected to increase even further than at present. Modified from Population Bulletin, Population Ref. Bureau, Vol. 50, March 1995.

(a)

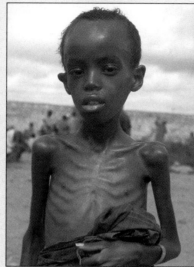

(b)

Figure E43-4
Desertification

(a) Human activities, including overgrazing livestock, deforestation, and poor agricultural practices, convert once-productive land into barren desert. (b) The loss of productive land, when combined with an expanding human population, can lead to tragedy.

food for its current population, but because resources are unequally distributed among, and even within, countries. Each natural population must live within the carrying capacity of the ecosystem it occupies. In contrast, human populations often expand well beyond the ability of their local ecosystem to support them, relying on wealth derived from other sources (such as minerals) to allow them to import the food they need. When human populations exceed the carrying capacity of the soil on which they live, and also lack the wealth to import and distribute adequate food, disaster strikes. An exploding population, coerced by poverty and ignorance into using destructive farming techniques that damage the land, is trapped in a downward economic spiral driven and perpetuated by the demands of population growth.

In estimating how many people Earth can support, we must keep in mind that humans desire more out of existence than a minimum caloric intake each day. It has been estimated that for everyone on Earth to live as Americans do, world population would have to be *reduced to 500 million, less than one-tenth the present population.* For all to achieve a standard of living similar to that in Europe, where the life-style is less extravagant, *the present population would have to be reduced by half.* Merely "redistributing the wealth" is obviously not an adequate answer.

There are some who predict that technological advances will continue to increase Earth's carrying capacity for humans indefinitely. In evaluating this forecast, keep in mind that technology takes time, wealth, and education

to develop and implement. Rapid population growth in less-developed countries increases poverty, overloads the educational system, and hampers technological development, while simultaneously making its need more urgent. Ironically, the countries that so desperately need these predicted technological advances are often unable even to take advantage of today's level of technology. The high-yield crops developed in the 1960s (the so-called Green Revolution) have helped considerably, but their impact has been lessened because they require expensive equipment, fertilizer, pesticides, and irrigation water to cultivate. Less-developed countries often cannot afford these inputs, and donated farm equipment has been left to rust in the fields for lack of fuel as well as the parts and know-how to maintain it.

Hope for the future lies in using the intelligence that has allowed us to overcome environmental resistance to see signs of "overgrazing" and act before we have irrevocably damaged our world ecosystem. Human population *will* stop its exponential growth. Either we will voluntarily reduce our birth rate or various forces of environmental resistance will increase our death rate; the choice is ours. Facing the problem of how to limit births is politically and emotionally difficult, but continued failure to do so will be disastrous. Our dignity, our intelligence, and our role as self-appointed stewards of life on Earth all demand that we make the decision to halt population growth ourselves before we have permanently reduced Earth's ability to support all life, including our own.

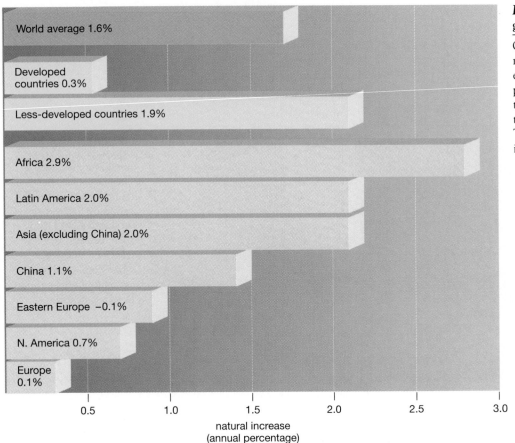

Figure 43-16 **Population growth by world regions**

Growth rates shown are due to natural increase (births - deaths) expressed as percent increase per year for various regions of the world. Europe now includes the former Soviet Union. These figures do not include immigration or emigration.

World average 1.6%

Developed countries 0.3%

Less-developed countries 1.9%

Africa 2.9%

Latin America 2.0%

Asia (excluding China) 2.0%

China 1.1%

Eastern Europe −0.1%

N. America 0.7%

Europe 0.1%

0.5   1.0   1.5   2.0   2.5   3.0

natural increase
(annual percentage)

are contributing to the rapid growth of the U.S. population: the "baby boom" and immigration.

Part of our current growth rate is a legacy of our recent past. Parents of the late 1940s through the 1960s had families that were larger than replacement-level, resulting in a "baby boom" and a momentum in population growth that has not yet subsided. For example, over a million more women entered their reproductive years in 1980 than did so in 1970. Even though these women are averaging only 2.1 children each, because there are more women having children, our population swells.

A second crucial component of the U.S. population equation is immigration, which contributes more to population growth here than in any other nation in the world. Legal immigration now averages 880,000 per year, causing about 30% of our total growth. Illegal immigration is estimated to add an additional 300,000 people each year.

According to Census Bureau projections, the U.S. population will not stabilize in the foreseeable future. In the past few years, there have been jumps in fertility as well as immigration that have caused dramatic increases in future

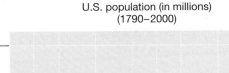

U.S. population (in millions)
(1790–2000)

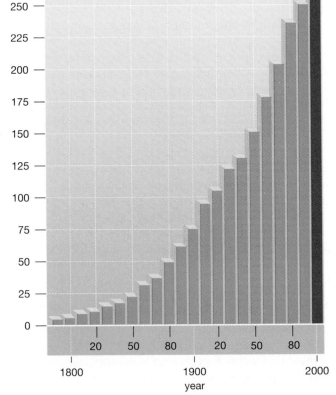

275
250
225
200
175
150
125
100
75
50
25
0

20   50   80   20   50   80

1800   1900   2000

year

Figure 43-17 **Population growth in the United States**

Since 1790, our growth has shown the J-shaped curve characteristic of an exponentially growing population. The figure for the year 2000 is a conservative projection based on a 1% annual growth rate.

population projections. The Census Bureau projects that by the year 2050 there will be 392 million Americans (about 130 million more than today), and the population will still be growing.

The rapid growth of the U.S. population has major environmental implications both for us and for the world as a whole. The average American uses six times as much energy as the world average. With 4.7% of the world's population, we are responsible for 14% of the total methane output and 22% of carbon dioxide and chlorofluorocarbon emissions. These gases are major contributors to the greenhouse effect, described in Chapter 45. Because people in less-developed countries use considerably less energy than the world average, the 2.6 million Americans added each year contribute more to climate change than the 34 million people added each year by China and India combined.

When and how will human numbers ever stabilize? How many people can Earth support? There are no certain answers to these questions, but in "Earth Watch: Has the Human Population Exceeded Earth's Carrying Capacity?" we explore them in more detail.

❊ S U M M A R Y   O F   K E Y   C O N C E P T S

Introduction to Ecology
Ecology is the study of the interrelationships between organisms and their environment. Ecology may be approached at the level of the population, the members of a particular species in a given area; the community with its complex interactions of populations; or an entire ecosystem, including communities and their abiotic environment.

How Populations Grow
Individuals join populations through births or immigration and leave through death or emigration. Thus

population change = (births – deaths) +
                    (immigrants – emigrants)

The ultimate size of a stable population is the result of interactions between biotic potential, the maximum possible growth rate, and environmental resistance, which limits population growth.

All organisms have the biotic potential to more than replace themselves over their lifetime, resulting in population growth. Populations tend to grow exponentially, with increasing numbers of individuals added during each successive time period. The maximum rate of growth of a given population is influenced by the age at first reproduction, frequency of reproduction, number of offspring produced each time, length of reproductive life span, and the death rate under ideal conditions. Populations cannot continue to grow exponentially indefinitely; they either stabilize or undergo periodic boom-and-bust cycles as a result of environmental resistance.

Environmental resistance restrains population growth by increasing the death rate or decreasing the birth rate. The maximum size at which a population may be sustained indefinitely by an ecosystem is called the carrying capacity, determined by limited resources such as space, nutrients, and light.

Populations are maintained at or below carrying capacity by density-independent forms of environmental resistance, such as weather, and density-dependent forms of resistance, which include predation, parasitism, and competition. Predators often eat the most abundant prey and may switch prey on the basis of the size of prey populations. When predators concentrate on a single prey, both predator and prey populations may cycle. Parasites spread most rapidly in dense populations and can increase death rates when animals are stressed because of overcrowding. Interspecific competition limits both the size and distribution of populations. Intraspecific competition may be resolved directly through scramble competition or indirectly through chemical interactions or social behaviors, collectively called contest competition. Emigration is another possible response to overcrowding.

Population Patterns in Space and Time
The distribution of populations can be classified as clumped or uniformly or randomly distributed. Clumping may occur for social reasons or around limited resources. Uniform distribution is usually the result of territorial spacing. Random distribution is rare, occurring only when individuals do not interact socially and when resources are abundant and evenly distributed.

Survivorship in populations shows specific survivorship curves that express the likelihood of survival at a given age. Convex (late-loss) curves are characteristic of long-lived species with few offspring and whose offspring must receive parental care. Species with constant curves have an equal chance of dying at any age. Concave (early-loss) curves are typical of organisms that produce numerous offspring, most of which die.

The Human Population
The human population has exhibited exponential growth for an unprecedented time by overcoming certain aspects of environmental resistance and increasing Earth's carrying capacity for people. These feats have been accomplished by the use of tools, agriculture, industry, and medical advances. Age-structure diagrams depict numbers of males and females in various age groups constituting a population. Expanding populations have pyramidal age structures, stable populations show rather straight-sided

figures, and decreasing populations are illustrated by figures that are constricted at the base.

Today, most of the world's people live in less-developed countries with rapidly expanding populations, where a variety of social and cultural conditions encourage large families. The United States is the fastest growing of the developed countries, because of high immigration rates and the "baby boom" of the 1940s through the 1960s. The carrying capacity of Earth for humans is unknown, but with a population of over 5.7 billion, resources are already too limited for all to be supported at a high living standard. A steady decline in productive land, wood, and fish harvests suggests that we are already damaging our world ecosystem and decreasing its ability to sustain us.

✖ K E Y   T E R M S

abiotic   p. 849
age structure   p. 862
agricultural revolution   p. 861
biotic   p. 849
biotic potential   p. 851
boom-and-bust cycle   p. 853
carrying capacity   p. 855
clumped distribution   p. 859
community   p. 849
competition   p. 858
contest competition   p. 858
cultural revolution   p. 861
density-dependent   p. 856
density-independent   p. 856

dominance hierarchy   p. 858
ecology   p. 849
ecosystem   p. 849
emigration   p. 850
environmental resistance   p. 851
exotic species   p. 853
exponential growth   p. 851
growth rate   p. 851
host   p. 858
immigration   p. 850
industrial-medical revolution   p. 862
interspecific competition   p. 858
intraspecific competition   p. 858
J-curve   p. 851

out-of-phase population cycle   p. 857
parasite   p. 858
parasitism   p. 858
population   p. 849
predation   p. 856
predator   p. 857
random distribution   p. 859
replacement-level fertility (RLF)   p. 862
scramble competition   p. 858
S-curve   p. 855
social stress   p. 858
survivorship curve   p. 860
uniform distribution   p. 859

✖ T H I N K I N G   T H R O U G H   T H E   C O N C E P T S

Multiple Choice

1. Why do many animals defend territories?
   a. to monopolize the resources within the territory
   b. to avoid predation
   c. to secure overwintering sites
   d. to advance within a dominance hierarchy
   e. to avoid each other

2. For exponential growth to occur, it is necessary that
   a. there is no mortality
   b. there are no density-independent limits
   c. the birth rate consistently exceed the death rate
   d. a species is very fast-reproducing
   e. the species is an exotic invader in an ecosystem

3. Which of the following currently contributes most to human population growth within the United States?
   a. the consequences of the "baby boom"
   b. immigration
   c. a birth rate above RLF
   d. average birth rate
   e. emigration

4. Which is the most common type of spatial distribution?
   a. logistic   b. uniform
   c. random   d. exponential
   e. clumped

5. Which continent has the highest rate of human population growth?
   a. North America   b. Africa
   c. Asia   d. Australia
   e. South America

6. If a population is above carrying capacity, what must happen?
   a. It must immediately crash.
   b. It can remain stable indefinitely.
   c. If the species is territorial, it can continue to increase.
   d. It must eventually decline.
   e. It must decline if it not territorial.

Review Questions

1. Define ecosystem, community, and population, making clear the differences between the terms.

2. Draw the growth curve of a population before it encounters significant environmental resistance. What is the name of this type of growth, and what is its distinguishing characteristic?

3. Define *biotic potential*. Explain natural selection in terms of biotic potential and environmental resistance.

4. What distinguishes density-dependent from density-independent limits to growth? Provide a specific example of each kind of limit. Also, explain why density-independent limits are less likely to affect long-lived species.

5. What are the two types of intraspecific competition? What are the behaviors that lead to each type? Which type is most common for plants?

6. Describe (or draw a graph illustrating) what is likely to happen to a population that far exceeds the carrying capacity of its ecosystem. Explain your answer.

7. Outline the technological and social "revolutions" that have successively removed limits to human population growth.

8. There are two aspects of the reproduction of individual females that determine how fast a population will increase. What are these two aspects of individual reproduction? Why does each affect population growth?

✳ APPLYING THE CONCEPTS

1. Red tree squirrels live in many of the same areas as do snowshoe hares. But squirrel populations hardly fluctuate from year to year, and they do not show cycles like those of the hare population at all. A crucial difference between the two species is that hares are not territorial, whereas squirrels fiercely defend territories throughout the year. Explain how such territoriality could help prevent a population from ever exceeding its environment's carrying capacity and how prevention of excess population could in turn prevent marked population fluctuations.

2. In conservation, the concern is often with how fast and what patterns of population decline may be occurring, since we wish to stop the decline of rare species. Saying that a population is declining is equivalent to saying that r is negative. What does this expression tell us about birth rate relative to death rate? Assume that $r = -0.20$. If we have an initial population size of 1000 animals, calculate the population sizes over the next 10 years, and plot the results against time (as in Figure 43-3). This curve shows exponential decline, an all too common pattern for endangered species.

3. The United States has a long history of accepting large numbers of immigrants. Discuss the implications of immigration for population stabilization.

4. What factors encourage rapid population growth in less-developed countries? What will it take to change this?

5. Contrast age structure in rapidly growing versus stable human populations. Why is there a momentum in population growth built into a rapidly growing population?

6. There is clear evidence not only that the world's human population is growing beyond sustainable limits, but also that some resources (e.g., topsoil and drinkable water) are actually decreasing because of overuse. Despite the evidence, some people still argue that it is possible to support many more people by producing foods differently, living with fewer luxuries, and so on. List the resources that are currently overused or appear to be close to overuse and then suggest ways to overcome those limitations. Do you believe that all of the limitations can be avoided by efficiency and technological innovation?

✳ FOR MORE INFORMATION

Bongaarts, J. "Can the Growing Human Population Feed Itself?" *Scientific American*, March 1994. A balanced account of the arguments between the optimists and pessimists concerning the future world food supply.

Brown, L. R., and others. *State of the World*. New York: W. W. Norton & Company, Inc., 1994. Annually updated collection of articles concerning global resources, pollution, and population.

Ehrlich, P. R., and Erlich, A. *Extinction*. New York: Random House, Inc. 1981. The causes and consequences of the disappearance of species.

Kunzig, R. "Twilight of the Cod." *Discover*, April 1995.

Ludwig, D., Hilborn, R., and Walters, C. "Uncertainty, Resource Exploitation, and Conservation: Lessons from History." *Science* 260:17–36, 1993. An influential and highly controversial article in which three ecologists argue that there is no evidence that humans can sustainably use natural resources.

Meadows, D. H., Meadows, D. L., and Randers, J. *Beyond the Limits: Confronting Global Collapse, Envisioning a Sustainable Future*. Post Mills, VT: Chelsea Green Publishing, 1992. An up-to-date effort to explain the causes of population growth as well as an attempt to project the future of population and world resources.

Myers, J. H., and Krebs, C. J. "Population Cycles in Rodents." *Scientific American*, June 1974. Natural populations of small rodents show periodic 3- to 4-year cycles in population size.

Sisk, T. D., Launerm, A. E., Switky, K. R., and Ehrlich, P. R. "Identifying Extinction Threats." *BioScience* 44:592–604, 1994. Using information on human population densities and growth to identify particularly threatened communities and ecosystems.

NET WATCH

On-line resources for this chapter are on the World Wide Web at:
http://www.prenhall.com/~audesirk (click on the table of contents link and then select Chapter 43).

44 Community Interactions

> "When we try to pick out anything by itself, we find it hitched to everything else in the universe."

JOHN MUIR

✖ AT A GLANCE

The Ecological Niche

Competition between Species
Interspecific Competition Helps Control Population Size

Predators and Their Prey
Predator-Prey Interactions Shape Evolutionary Adaptations
Keystone Predators Influence Community Structure

Symbiosis
Parasitism Harms, but Does Not Immediately Kill, the Host
Commensalism Benefits One Species without Affecting the Other
In Mutualistic Interactions, Both Species Benefit

Succession: Community Changes over Time
There Are Two Major Forms of Succession: Primary and Secondary
Succession Also Occurs in Ponds and Lakes
Succession Culminates in the Climax Community

This swallowtail butterfly larva shows an uncanny resemblance to a snake, discouraging would-be predators. The caterpillar's entire head is the "snake's nose," while the area behind the head is enlarged and bears markings that closely resemble eyes.

As winter wanes, spring rains create a small pond in the woods. Though it may last only a few months, the pond becomes a complete ecosystem. Its nonliving components define it physically: the depression in the forest floor holding the water, the fine grains of silt, dead leaves and twigs making up the complex bottom, and most of all the molecules of water. At first the ecosystem's living component may consist only of bacteria and fungi. But soon the bottom will teem with single-celled organisms; water-loving plants will sprout; and frogs, newts, and insects will lay eggs, populating the water with larvae. Snails will invade, and ducks may visit, leaving behind parasites that infect the snails. Soon dozens of species will inhabit the pond. The individuals of each species constitute a population. Although they each have their own way of life, the populations interact in three important ways. Populations may compete with other populations for limited resources. One population may prey upon another. And, in some cases, two species may live together in close association over an extended time. (This association may benefit one or both species and may work to the detriment of one or the other.) These three processes, competition, predation, and symbiosis, form the basis of a **community**.

Typically, populations within communities have evolved together, a process called **coevolution**. During coevolution, different species act as agents of natural selection on one another. Prey animals have evolved elaborate defenses that help them survive. Herbivores have digestive specializations that allow them to eat local plants. The plants in turn grow rapidly or defend themselves by chemical or physical means, keeping one step ahead of their predators. The whole community survives by a delicate balance between populations. This delicate balance may be toppled by the invasion of a single new species, as described in "Earth Watch: Introduced Species Disrupt Community Interactions."

The temporary pond quickly passes through a series of stages before reaching its mature late-summer form. The forest surrounding the pond, however, has remained stable for hundreds of years. Long-established communities are generally complex and self-sustaining and have interacting populations that remain relatively constant. Even these communities originate, however, over long periods of gradual change. In the case of the forest, this gradual change may have followed a fire, earthquake, or glacial scouring. During this development, one type of community gives way to another in a process called succession, which finally results in a stable, self-sustaining community.

In this chapter we focus on the complex web of community interactions that determine what species are found in stable ecosystems, and the community changes that occur during succession.

The Ecological Niche

Although the word *niche* may call to mind a small cubbyhole, in ecology it means much more. Each species occupies a unique **ecological niche** that encompasses all aspects of its way of life. The concept of the ecological niche includes the organism's physical home, or habitat. The habitat of a white-tailed deer, for example, is the eastern deciduous forest.

But the ecological niche goes far beyond habitat. It also specifies how the organism gets its supply of energy and materials—what might be called the organism's role or "occupation" within its ecosystem. An organism's predators, prey, and competitors as well as its behaviors and interactions are considered elements of its niche. In addition, the niche includes all the physical environmental factors necessary for survival, such as the range of temperatures under which the organism can survive, the amount of moisture it requires, the pH of the water or soil it may inhabit, the type of soil nutrients required, and the degree of shade it can tolerate. Although different types of organisms share many aspects of their niche with others, no two species ever occupy exactly the same ecological niche.

Competition between Species

Community interactions may be harmful or beneficial to the participating organisms (Table 44-1). During competition between different species, called **interspecific competition,** two or more species attempt to use the same limited resources, particularly food or space or both. In the case of interspecific competition, both species are harmed in the sense that access to resources is limited. The intensity of interspecific competition depends on how similar the requirements of the two species are. In other words, the degree of competition is proportional to the amount of niche overlap between the competing species.

Just as no two organisms can occupy exactly the same physical space at the same time, no two species can inhabit the same ecological niche. This important concept, often called the **competitive exclusion principle,** was formulated in 1934 by the Russian microbiologist G. F. Gause. If two species with the same niches are placed together and forced to compete for limited resources, inevitably one will outcompete the other, and the less well-adapted of the two will die out. Gause demonstrated this principle using two species of the protist *Paramecium: P. aurelia* and *P. caudatum.* In separate laboratory flasks, the two species thrived on bacteria and fed in the same parts of the flask. Both populations flourished. But when Gause placed the two species together, one always eliminated, or "competitively excluded," the other (Fig. 44-1). Gause then repeated the experiment, replacing *P. caudatum* with another species, *P. bursaria,* which tended to feed

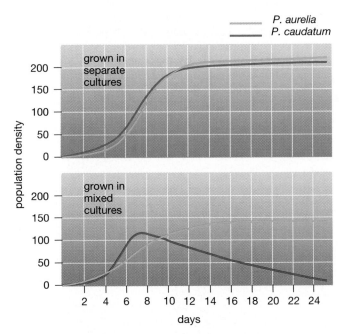

Figure 44-1   **Competitive exclusion**

(Top) Raised separately with a constant food supply, both *Paramecium aurelia* and *P. caudatum* show the S-curve typical of a population that initially grows rapidly, then stabilizes. (Bottom) Raised together and forced to occupy the same niche, *P. aurelia* consistently outcompetes *P. caudatum* and causes that population to die off. (Modified from G. F. Gause, *The Struggle for Existence.* Baltimore: Williams & Wilkins, 1934.)

TABLE 44-1	Interactions among Organisms	
Type of Interaction	*Effect on Organism A*	*Effect on Organism B*
Competition between A and B	Harms	Harms
Predation by A on B	Benefits	Harms
Symbiosis		
Parasitism by A on B	Benefits	Harms
Commensalism of A with B	Benefits	No Effect
Mutualism between A and B	Benefits	Benefits

at the bottom of the flask. In this case, the two species of *Paramecium* were able to coexist indefinitely, because they occupied slightly different niches.

The ecologist R. MacArthur tested Gause's laboratory findings under natural conditions by investigating five species of North American warbler. These birds all hunt for insects and nest in the same type of spruce tree. Although the niches of these birds appear to overlap considerably, MacArthur found that the five species concentrated their search in specific areas of the tree, employed different hunting tactics, and nested at slightly different times. By partitioning the resource, the warblers minimize the overlap of their niches and reduce competition among the different species (Fig. 44-2).

As MacArthur discovered, when species with similar requirements coexist, they often occupy a smaller niche than either would if it were by itself. This phenomenon, called **resource partitioning**, is an evolutionary adaptation that reduces the harmful effects of interspecific competition. Resource partitioning is the outcome of the coevolution of species with extensive (but not total) niche overlap. Coevolution (introduced in Chapter 16) occurs when two or more species act as agents of natural selection on one another. Because natural selection favors individuals with fewer competitors, over evolutionary time the competing species develop physical and behavioral adaptations that minimize their competitive interactions. A dramatic example of this type of adaptation was discovered by Darwin among finches of the Galapagos Islands. The finches had evolved different bill sizes and shapes and different feeding behaviors that reduced the competition among them (see Chapter 16).

Interspecific Competition Helps Control Population Size

Although coevolution tends to minimize niche overlap, closely related species still compete directly for limited resources. This competition may restrict the size and distribution of the competing populations. For example, barnacles of the genus *Chthamalus* share the rocky shores of Scotland with another genus of barnacle, *Balanus*, and their niches overlap considerably. The ecologist J. Connell found that *Chthamalus* dominates the drier upper shore and *Balanus* dominates the lower, which is more often submerged. When he scraped off *Balanus*, the *Chthamalus* population increased, spreading downward into the area its competitor had once inhabited. Because *Balanus* is intolerant of drying, removal of *Chthamalus* has little impact on the distribution of *Balanus*. Where the habitat is appropriate for both species, *Balanus* conquers, because it is larger and grows faster. But *Chthalamus* tolerates drier conditions, so on the upper shore, where only high tides submerge the barnacles, it has a competitive advantage. As this example illustrates, interspecific competition limits both the size and the distribution of the competing populations.

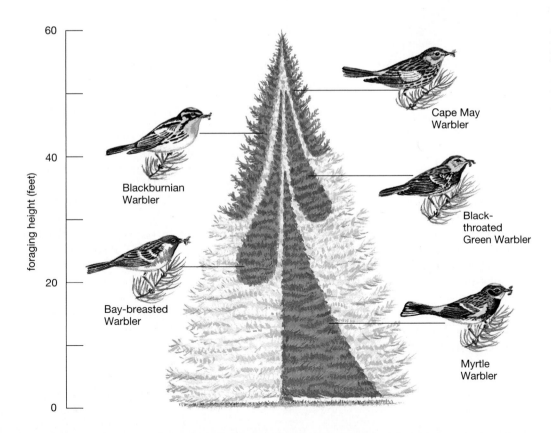

Figure 44-2　**Resource partitioning**

These five insect-eating species of North American warbler search for food in different regions of spruce trees.

Blackburnian Warbler

Bay-breasted Warbler

Cape May Warbler

Black-throated Green Warbler

Myrtle Warbler

foraging height (feet)

(a)

(b)

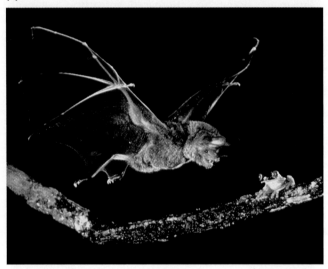

Figure 44-3   **Forms of predation**

(a) A predatory plant, the sundew, attracts and binds insects with its glistening sticky knobs. The insects provide supplemental nitrogen for the plant. **(b)** A few tropical bats feed on frogs or fish. Predatory bats home in on the mating calls of this tree frog.

Predators and Their Prey

Although predators are most commonly considered to be organisms that kill and eat other organisms, herbivorous animals are often included in this category even though they may kill only part of the plants they consume. For our purposes, then, predators include the antelope munching on sagebrush and the exotic sundew plant entangling and digesting its insect prey, as well as the bat swooping down on the frog and the bear catching a salmon (Fig. 44-3). Predators are usually either larger than their prey or hunt collectively, as wolves do when bringing down a moose. Predators are also less abundant than their prey, for reasons discussed in the next chapter.

Predator-Prey Interactions Shape Evolutionary Adaptations

To survive, predators must feed and prey must avoid becoming food. Therefore, these populations exert intense selective pressure on one another. As prey become more difficult to catch, predators must become more adept at hunting. Natural selection has endowed the cheetah with speed and camouflage spots and its zebra prey with speed and camouflage stripes. It has produced the keen eyesight of the hawk and the warning call of the Belding ground squirrel, the stealth of the jumping spider and the remarkable spider mimicry of the fly it stalks. In the following sections, we examine some of the results of predator-prey coevolution.

Bats and Moths Have Evolved Counteracting Strategies

Most bats are nighttime hunters that navigate and locate prey by echolocation. They emit extremely high-frequency and high-intensity pulses of sound and, by analyzing the returning echoes, create an "image" of their surroundings and nearby objects. In response to this unique prey-location system, certain moths (a favored prey of bats) have evolved simple ears that are particularly sensitive to the frequencies used by echolocating bats. When they hear a bat, these moths take evasive action, flying erratically or dropping to the ground. The bats may counter this defense by switching the frequency of their sound pulses away from the moth's sensitive range. Some moths have evolved a way to jam the bats' echolocation by producing their own high-frequency clicks. In response, when hunting a clicking moth, the bat may counter by turning off its own sound pulses temporarily and zeroing in on the moth by following the moth's clicks. These interactions dramatically illustrate the complexity of coevolutionary adaptations.

Camouflage Conceals Both Predators and Their Prey

An old saying of detective novels is that the best hiding place may be right out in plain sight! Both predators and prey have evolved colors and patterns that resemble their surroundings. Such coloration, called **camouflage**, renders animals inconspicuous even when they are in plain sight (Fig. 44-4). Some animals closely resemble specific (but uninteresting) objects such as leaves, twigs, or even bird droppings (Fig. 44-5). Camouflaged animals tend to remain motionless rather than fleeing their predators. (A fleeing bird dropping would be quite conspicuous!) While camouflaged animals often resemble plants, a few types of plants have evolved to resemble rocks, which are ignored by their herbivorous predators (Fig. 44-6).

Predators who ambush their prey rather than chasing it are also aided by camouflage. For example, the angler fish closely resembles the algae-covered rocks and algae on which it sits motionless, dangling a small lure from its upper lip (Fig. 44-7).

(a)

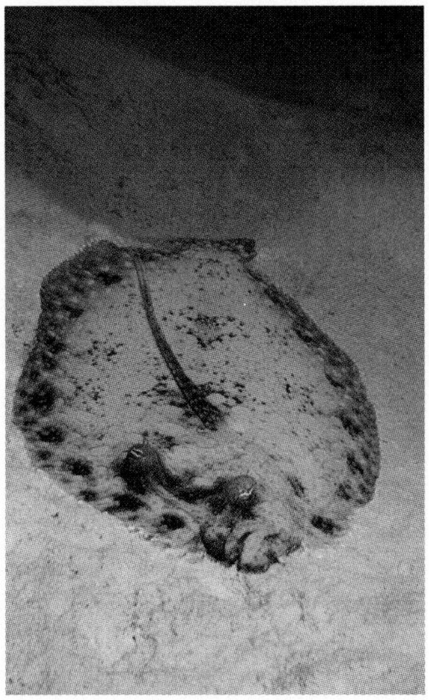

(b)

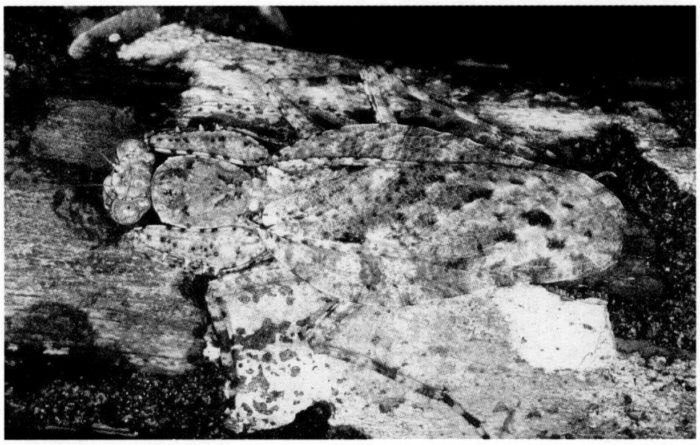

Figure 44-4 **Camouflage by blending in**

(a) The flounder's flat shape and its behavior (flipping sand over its back, then lying perfectly still) help the fish resemble a sandy bottom. The flounder can also alter its color and pattern somewhat to resemble its surroundings. **(b)** The tree bark mantis from Malaysia blends perfectly with the bark of this tree.

(a)

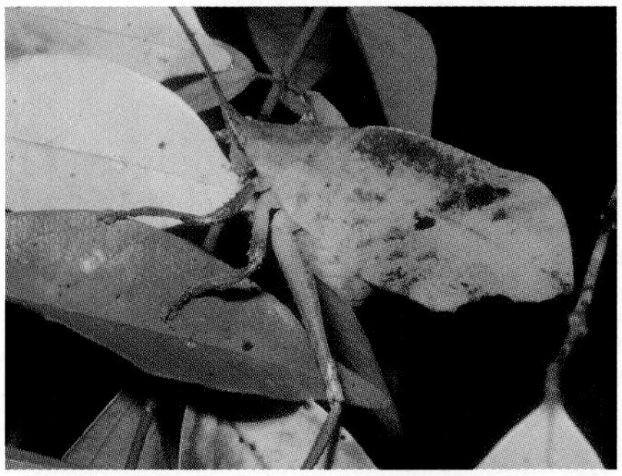

(b)

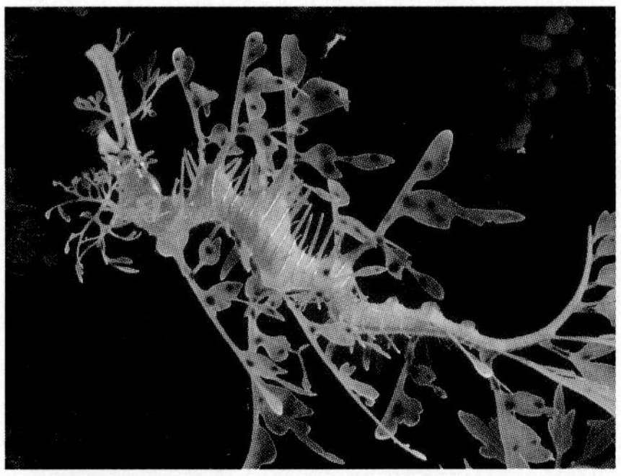

(c)

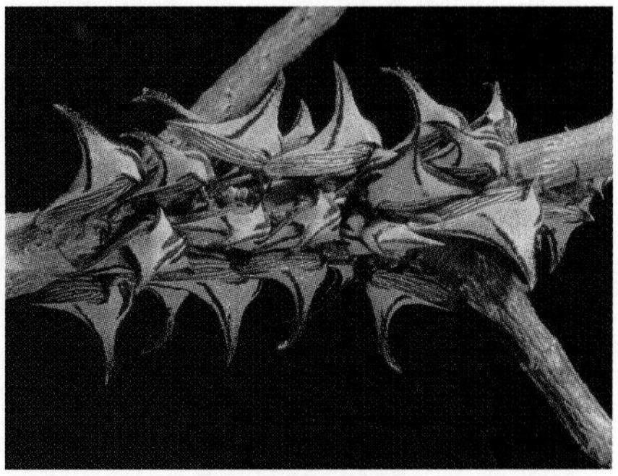

Figure 44-5 **Camouflage by resembling specific objects**

Resembling uninteresting parts of the environment allows some animals to avoid predation. **(a)** A leaf katydid from Costa Rica perfectly resembles a leaf, including an irregularity where the "leaf" has apparently been gnawed. **(b)** The leafy sea dragon (an Australian "seahorse" fish) has evolved extension of its body that precisely duplicates the algae in which it hides. **(c)** These Florida treehopper insects avoid detection by resembling thorns on a branch.

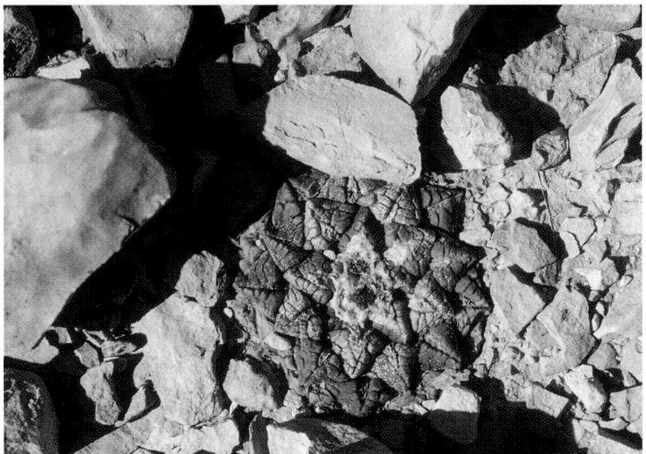

Figure 44-6 **A plant that resembles a rock**

This cactus of the American Southwest is appropriately called the living rock cactus.

Bright Colors Often Warn of Danger

Some animals have evolved very differently, exhibiting bright **warning coloration** (Fig. 44-8; see also Figs. 44-9, 44-13, and 44-14). These animals are usually distasteful and often extremely poisonous. Since poisoning your predator is small consolation if you have already been eaten, the bright colors declare: "Eat me at your own risk." A single unpleasant experience is enough to teach predators to avoid these conspicuous prey, as illustrated by the toad and bee in Figure 41-14.

Some Organisms Gain Protection through Mimicry

Mimicry refers to a situation in which a species evolves to resemble something else, often another type of organism. For example, once warning coloration evolved, there arose a selective advantage for tasty, harmless animals to resemble poisonous ones. The deadly coral snake has bril-

***Figure 44-7* Combining camouflage and aggressive mimicry**

The angler fish waits in ambush, its body camouflaged to match the algae-encrusted rock on which it normally rests. Above its mouth dangles a lure, closely resembling a small fish. This will attract other would-be predators, who will find themselves prey.

liant warning coloration, and the harmless mountain king snake avoids predation by resembling it (Fig. 44-9a, b). By resembling each other, two distasteful species may each benefit from predators' painful experience with the other. For example, predators rapidly learn to avoid the conspicuous stripes on bees, hornets, and yellowjackets. The monarch butterfly is poisonous and distasteful, and viceroy butterflies, which researchers have recently discovered are equally distasteful, have wing patterns strikingly similar to those of monarchs (Fig. 44-9c, d). A common color pattern results in faster learning by predators—and less predation on all similarly colored species.

Some predators have evolved **aggressive mimicry**, a "wolf-in-sheep's-clothing" approach to predation, in which they entice their prey to come close by resembling a harmless animal. For example, although the body of the angler fish is hidden by camouflage, it dangles a wriggling lure that resembles a small fish just above its mouth. A curious fish attracted to the lure is quickly swallowed (see Fig. 44-7). The saber-toothed blenny fish imitates a harmless fish, called a cleaner wrasse, that picks parasites from the skin of larger fish. Parasite-ridden fish welcome the attention of the cleaner wrasse and are fooled by the saber-toothed blenny, who bites them instead of their parasites (Fig. 44-10).

Certain prey species use still another form of mimicry called **startle coloration**. Several insects, including certain moths and caterpillars, have evolved patterns of color that closely resemble the eyes of a much larger, and possibly dangerous, animal (Fig. 44-11). If a predator gets too close, the prey suddenly flashes its eye-spots, startling the predator enough to make an escape.

A sophisticated variation on the theme of prey who mimic predators is seen in snowberry flies, who are hunted by territorial jumping spiders. When a fly spots an approaching spider, it spreads its wings, moving them back and forth in a jerky dance. Seeing this display, the spider is likely to flee from the harmless fly. Why? Researchers recently observed that the markings on the fly's wings closely resemble the legs of another jumping spider. The jerky movements of the fly mimic those made by a jumping spider when it drives another spider from its territory (Fig. 44-12). Natural selection has finely tuned both the behavior and the appearance of the fly to avoid predation by jumping spiders.

Some Predators and Prey Engage in Chemical Warfare

Both predators and prey have evolved a variety of toxic chemicals for attack and defense. Many plants produce defensive toxins. For example, lupines produce chemicals called alkaloids, which deter attack by the blue butterfly, whose larvae feed on the lupine's buds. In fact, different individuals of the same species of lupine produce different forms of this toxin, thus making the evolution of resistance to it more difficult. The venom of spiders and poisonous snakes such as the coral snake (see Fig. 44-9a) serves both to paralyze prey and to deter predators.

Certain mollusks, including squid, octopuses, and some sea slugs, emit clouds of ink when attacked. These colorful chemical "smoke screens" confuse their predators and mask their escape. A dramatic example of chemical defense is seen in the bombardier beetle. In response to the bite of an ant, the beetle releases secretions from special glands into an abdominal chamber. Here, enzymes catalyze an explosive chemical reaction that shoots a toxic, boiling-hot spray onto the attacking ant (Fig. 44-13).

Plants and Herbivores Have Many Coevolutionary Adaptations

Plants too have evolved chemical adaptations that deter predators. Many, such as the milkweed, synthesize toxic and distasteful chemicals. Animals rapidly learn not to eat foods

***Figure 44-8* Warning coloration**

The South American poison arrow frog, with its poisonous skin, advertises its unpleasant taste with bright and contrasting color patterns.

(a)

(b)

(c)

(d)

Figure 44-9   **Warning mimicry**

The warning coloration of the poisonous coral snake **(a)** is mimicked by the harmless king snake **(b)**. Mimicry of the warning coloration of the distasteful monarch butterfly **(c)** by the equally distasteful viceroy **(d)**. After attempting to eat either of these species, a bird will avoid both of them.

that make them sick, so milkweeds and other toxic plants suffer little nibbling. Consequently, such plants are often very abundant; any animal immune to the plant poisons enjoys a bountiful food supply. As plants evolved toxic chemicals for defense, certain insects evolved increasingly efficient ways to detoxify or even make use of the chemicals. The result is that nearly every toxic plant is eaten by at least one type of insect. For example, monarch butterfly caterpillars consume the toxic milkweed. The caterpillars not only tolerate the milkweed poison but also store it in their tissues as a defense against their own predators. The stored toxin is even retained in the metamorphosed monarch butterfly (see Fig. 44-9c). Although the desert camphor weed oozes a sticky, noxious resin that prevents most animals from eating it, the predatory assassin bug collects the substance and smears it on her abdomen (Fig. 44-14). The cam-

phor "glue" then coats her eggs as they emerge, making them distasteful to predators. Then, soon after hatching, young assassin bugs scrape the sticky substance from their discarded eggshells and transfer it to their forelegs, where it helps them capture their own insect prey. In this complex association, the assassin bug has evolved behaviors that allow it to utilize the defenses evolved by a plant against both its own predators and its prey.

Grasses have evolved tough silicon (glassy) substances in their blades, discouraging all predators except those with strong, grinding teeth and powerful jaws. Thus, grazing animals have come under selective pressure for longer, harder teeth. An example is the evolution of horses and the grass they eat. As grasses evolved tougher blades that reduce predation, horses evolved longer teeth with thicker enamel coatings that resist wear.

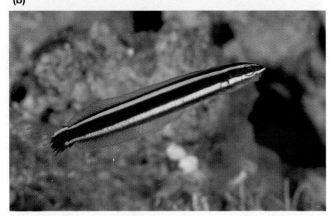

(a)

(b)

Figure 44-10   Aggressive mimicry

(a) Some fish, such as this marine wrasse, obtain food by eating parasites off the bodies of larger fish, a mutualistic relationship called a *cleaning symbiosis*. Fish welcome the approach of this predator of parasites. **(b)** The saber-toothed blenny has evolved to closely resemble the cleaner wrasse. Fish that allow the blenny to approach usually get a chunk bitten out of them—hence the term *aggressive mimicry*.

Keystone Predators Influence Community Structure

In some communities, a specific predator, described as a **keystone predator**, plays a major role in determining community structure. Removal of a keystone predator dramatically alters the community. For example, when the predatory sea star *Pisaster* was removed from sections of the rocky intertidal coast of the western United States, mussels (bivalve mollusks which are favored prey of *Pisaster*) became so abundant that they outcompeted algae and other invertebrates that normally coexist in intertidal communities. The African elephant is a keystone predator in the African savanna. By grazing on small trees and bushes, elephants prevent the encroachment of forests and help maintain the grassland community. It is difficult to identify keystone predators, since the predator must be selectively removed and the community studied for several years both before and after its removal. Therefore, ecologists have not yet determined whether communities dominated by keystone predators are rare or relatively common.

Symbiosis

Symbiosis, which literally means "living together," is defined as a close interaction between organisms of different species for an extended time. Considered in its broadest sense, symbiosis includes parasitism, commensalism, and mutualism. Although one species always benefits in symbiotic relationships, the second species may be harmed, not affected, or benefited (see Table 44-1).

Parasitism Harms, but Does Not Immediately Kill, the Host

Parasitism occurs when one species lives in or on its prey (also called host), often harming or weakening it but not

(a) 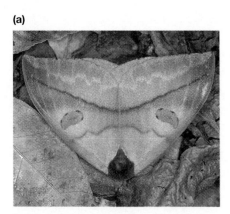   **(b)** 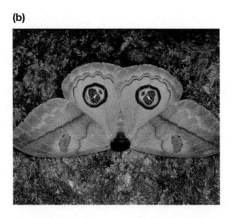   **(c)**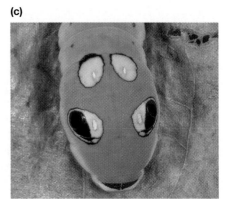

Figure 44-11   Startle coloration

(a) The peacock moth from Trinidad is well camouflaged, but should a predator approach too closely, it suddenly opens its wings to reveal spots resembling large eyes **(b).** The predator is startled, and the moth gets a chance to flee. **(c)** Would-be predators of this caterpillar larva of the swallowtail butterfly are deterred by its close resemblance to a snake. Note that the caterpillar's head is the snake's "nose."

immediately killing it. Although it is sometimes difficult to distinguish clearly between a predator and a parasite, parasites are generally much smaller and more numerous than their hosts. Familiar parasites include tapeworms, fleas, and numerous disease-causing protozoa, bacteria, and viruses. Many parasites, particularly worms and protozoa, have complex life cycles involving two or more hosts (see Chapter 24). There are few parasitic vertebrates; the lamprey eel, which attaches itself to a host fish and sucks its blood, is a rare example.

The variety of infectious bacteria and viruses and the precision of the immune system that counters their attacks are evidence of the powerful forces of coevolution between parasitic microorganisms and their hosts. Another example is the malaria parasite, which has provided strong selection pressure for a gene in humans that causes sickle-cell anemia. The parasites can't survive in the affected red blood cells. In certain areas of Africa where malaria is common, one-quarter of the human population carries the sickle-cell gene. *Trypanosoma*, a parasitic protozoan, causes both human sleeping sickness and a disease in cattle called nagana. African antelope, which coevolved with this parasite, are relatively unaffected by it. Cattle, a more recently introduced species, usually suffer but survive infection if they have been bred in an infested area for many generations. Newly imported cattle, however, generally die if not treated.

Commensalism Benefits One Species without Affecting the Other

Commensalism occurs when the relationship between two species benefits one without affecting the other. For example, as herds of bison graze, they disturb insects dwelling in the grass. Birds follow the bison, eating the insects. The

Figure 44-12   **Behavioral mimicry**

In response to the approach of a jumping spider (top), the snowberry fly spreads its wings, revealing a pattern resembling spider legs (bottom). The fly enhances the effect by performing a jerky, side-to-side dance that resembles the leg-waving display of another jumping spider defending its territory.

Figure 44-13   **Chemical warfare**

The bombardier beetle sprays a hot toxic brew in response to a leg pinch.

Figure 44-14   **A complex coevolutionary relationship**

A female assassin bug scrapes noxious resin from a camphor weed. Her warning coloration alerts potential predators of her painful sting.

(a)

(b)

Figure 44-15 **Commensal relationships**

(a) This tropical orchid clings to a rainforest tree. Its perch allows it access to sunlight without harming the tree. **(b)** Clusters of small barnacles hitchhike harmlessly on the hide of this manatee.

birds, which benefit, are considered commensal with the bison, who are unaffected. Birds and the trees in which they nest are involved in a commensal relationship. The birds obtain shelter and protection without affecting the trees. Many orchids attach to trees without harming them. The orchid obtains support and access to sunlight high above the dim tropical forest floor. Barnacles are harmless hitchhikers on marine animals, including whales and manatees, but benefit from a free ride through food-rich waters (Fig. 44-15).

In Mutualistic Interactions, Both Species Benefit

When two organisms interact in a way that benefits both, the relationship is called **mutualism**. The mutualistic interactions between flowering plants and their pollinators are discussed in Chapter 28. Mutualistic associations occur in the digestive tracts of cows and termites, where protists and bacteria find food and shelter while helping their hosts extract nutrients, and in our own intestines, where bacteria synthesize certain vitamins (see Chapter 32). The nitrogen-fixing bacteria inhabiting special chambers on the roots of legume plants are another important example (see Chapter 21). The bacteria obtain food and shelter from the plant and in return trap nitrogen in a form the plant can utilize. Some mutualistic partners have coevolved to the extent that neither can survive alone. An interesting example is the ant-acacia mutualism described in "Scientific Inquiry: Deciphering the Relationship between Ants and Acacias."

Mutualistic relationships involving vertebrates are rare and typically are less intimate and extended, as in the relationship of the cleaner wrasse (see Fig. 44-10) and the fish it cleans. Another is seen in the clown fish, which takes shelter among the venomous tentacles of the anemone. The fish derive shelter and protection and, at least occasionally, bring bits of food to their anemone host (Fig. 44-16).

Succession: Community Changes over Time

In a mature terrestrial ecosystem, the populations that make up the community interact with one another and with their nonliving environment in intricate ways. But this tangled web of life did not spring fully formed from bare rock or naked soil; rather, it emerged in stages over a long period, a process called succession. **Succession** is a change in a community and its nonliving environment over time. It is a kind of "community relay" in which assemblages of plants and animals replace one another in a sequence that is at least somewhat predictable.

Succession occurs under a variety of circumstances but is most easily observed in terrestrial and freshwater ecosys-

Figure 44-16 **Mutualism**

The clownfish snuggles unharmed among the stinging tentacles of the anemone. The fish obtains protection and sometimes brings food to the anemone in this mutualistic relationship.

SCIENTIFIC INQUIRY

Deciphering the Relationship between Ants and Acacias

Daniel Janzen, a doctoral student at the University of Pennsylvania, was walking down a road in Veracruz, Mexico, when he saw a flying beetle alight on a thorny tree, only to be driven off by an ant. Looking more closely, he saw that the tree, a bull's-horn acacia, was covered with ants. A large ant colony of the genus *Pseudomyrmex* made its home inside the enlarged thorns of the plant, whose soft pulpy interiors are easily excavated to provide shelter (Fig. E44-1).

To determine how important the ants are to the tree, Janzen began stripping the thorns by hand until he found and removed the thorn housing the ant queen, thus destroying the colony. He later turned to more efficient but dangerous methods, eliminating all the ants on a large stand of acacias with an insecticide. The acacias were unharmed by the poison, Janzen became ill from it, and the ants were all killed. Within a year of the spraying, Janzen found nearly all the acacia trees dead, consumed by insects and other herbivores and shaded out by competing plants. The ground surrounding the trees, which the ants normally kept trimmed, was overgrown. The trees were apparently dependent on their resident ants for survival.

Wondering if the ants could survive off the tree, Janzen painstakingly peeled the ant-inhabited thorns off 100 acacia trees, suffering multiple stings in the process. He housed each ant colony in a jar provided with local nonacacia vegetation and insects for food. The ant colonies all starved. Carefully examining the acacia, he found swollen structures filled with sweet syrup at the base of the leaves and protein-rich capsules on the leaf tips (Fig. E44-1, inset). Together, these provide a balanced diet for the ants.

Janzen's experiments strongly suggest that these species of ant and acacia have an obligatory mutualistic relationship: Neither can survive without the other. Of course, further observations were required to confirm this relationship. The fact that the ants starved in Janzen's jars did not rule out that they might survive successfully elsewhere, but, in fact, this species of ant is never found living independently of the acacia. Similarly, the bull's-horn acacia is never found without its resident ant colony. Thus, a chance observation followed by careful research led to the discovery of an important mutualistic association.

Figure E44-1   A mutualistic relationship

Yellow, protein-rich capsules are produced at the tips of certain acacia leaves. These provide food for the resident ants. (Inset) A hole in the enlarged thorn of the bull's-horn acacia provides shelter for members of the ant colony. The ant entering the thorn is carrying a food capsule produced by the acacia. As the ant colony grows, more thorns are invaded.

tems. Freshwater ponds and lakes tend to undergo a series of changes that transform them first into marshes and eventually to dry land. Shifting sand dunes are stabilized by creeping plants and may eventually support a forest. Volcanic eruptions may, as in the case of Mount St. Helens, wipe out previously existing ecosystems, or they may create new islands which are soon colonized. Forest fires

create a nutrient-rich environment that encourages rapid invasion of new life (Fig. 44-17).

The precise changes occurring during succession are as diverse as the environments in which succession occurs, but certain general stages can be recognized. In each case, succession is begun by a few hardy invaders called **pioneers** and ends with a diverse and relatively stable **climax com-**

(a)

(b)

(c)

Figure 44-17   **Succession in progress**

(a) (Left) On May 18, 1980, the explosion of Mount St. Helens in Washington devastated the pine forest ecosystem on its sides. (Right) Nine years later, pioneering evergreens were about 4 to 6 inches tall, and shrubs carpeted the nutrient-rich ash. **(b)** (Left) In the summer of 1988, extensive fires swept through the forests of Yellowstone National Park in Wyoming. (Right) Two years later, flowering plants had sprung up in the sunlight, and wildlife populations were rebounding. **(c)** (Left) The Hawaiian volcano Mount Kilauea has erupted repeatedly during the past 8 years, sending rivers of lava over the surrounding countryside. (Right) An early colonizer of a lava flow.

munity. As the community progresses from the pioneers to the climax community, the organisms gradually alter the nonliving environment. Ironically, these changes often favor competitors, which displace the existing populations. The climax community differs from earlier successional stages because it no longer alters the environment significantly. The climax community will persist unless external forces, such as human activities, natural events such as fires, hurricanes, or floods, or a gradual change in climate alter it.

During terrestrial succession certain general trends occur in ecosystem structure:

1. The soil increases in depth, and its content of organic material (derived from the bodies of organisms) also increases.

2. The overall productivity (the amount of organic material produced in a given area over a given time) increases.

3. The number of species increases, as does the number of interactions within the community.

4. Longer-lived species come to dominate the ecosystem, and the rate at which populations replace one another slows. The presence of longer-lived species combined with more complex community interactions frequently results in a community that is more stable and resistant to change.

5. In the climax community, the total mass of living organisms reaches a maximum, and the species present no longer alter the ecosystem in ways that encourage the growth of their competitors.

There Are Two Major Forms of Succession: Primary and Secondary

Succession on dry land takes two major forms: primary and secondary. During **primary succession,** an ecosystem is forged from bare rock, sand, or a clear glacial pool where there is no trace of a previous community. The for-

mation of an ecosystem "from scratch" is a process often requiring thousands or even tens of thousands of years. During **secondary succession,** a new ecosystem develops after an existing ecosystem is disturbed, as in the case of a forest fire or an abandoned farm field. It happens much more rapidly than primary succession because the previous community has left its mark in the form of soil and seeds. Succession in an abandoned farm field in the southeastern United States can reach its climax after two centuries. In the examples below, we examine these processes in more detail.

Primary Succession Can Begin on Bare Rock

Figure 44-18 illustrates primary succession as studied on Isle Royale, Michigan, an island in Lake Superior. Bare rock, such as that exposed by a retreating glacier, begins to liberate nutrients such as minerals by weathering. Cracks form as the rock alternately freezes and thaws, contracting and expanding. For lichens (symbiotic associations of fungi and algae; see Chapter 22), the weathered rock provides a place to attach where there are no competitors and plenty of sunlight. Lichens can photosynthesize, and they obtain minerals by dissolving some of the rock with an acid they secrete. As the pioneering lichens spread over the rock, drought-resistant, sun-loving mosses begin growing in the cracks. Fortified by nutrients liberated by the lichens, the moss forms a dense mat that traps dust, tiny rock particles, and bits of organic debris. The death of some of the moss adds to a growing nutrient base, while the moss mat itself acts as a sponge, trapping moisture. Within the moss, seeds of larger plants germinate. Eventually, their bodies contribute to a growing layer of soil. As woody shrubs such as blueberry and juniper take advantage of the newly formed soil, the moss and lichens may be shaded out and buried by decaying leaves and vegetation. Eventually,

Figure 44-18   Primary succession

Here, primary succession is shown occurring on bare rock in upper Michigan.

lichen on bare rock   bluebell, yarrow   blueberry, juniper   jack pine, black spruce, aspen   balsam fir, paper birch, white spruce, climax forest

time (years)

0 —————————————————————————— 1000

E A R T H   W A T C H

Introduced Species Disrupt Community Interactions

Exotic species, or species introduced into ecosystems where they did not evolve, sometimes find no predators or parasites in their new environment, and they find new prey that have little defense against them. The unchecked population growth of these invaders may seriously damage the ecosystem as they displace, outcompete, and prey on native species. Both starlings and English sparrows have spread dramatically since their deliberate introduction to the eastern United States in the 1890s. Their success threatens our native bluebird. Aggressive African bees, dubbed "killer bees," were introduced into Brazil in 1959 and began spreading northward. They have now made their way into Texas and California and are continuing to spread, oucompeting local bee species and threatening the U.S. beekeeping industry. A Japanese vine called kudzu was introduced to the South as a decorative plant in 1876 and then planted extensively in the 1940s to control erosion. Today kudzu is a major pest, overgrowing and killing trees and underbrush, and occasionally engulfing small houses while their owners are vacationing (Fig. E44-2b). The water hyacinth, introduced from South America as an ornamental plant, now clogs about 2 million acres of southern lakes and waterways, slowing boat traffic and displacing natural vegetation (Fig. E44-2c).

In 1986, a trading vessel bringing cargo from Europe up the St. Lawrence Seaway discharged fresh water containing millions of larvae of the zebra mussel into the St. Clair River near Detroit. By 1989, the invading mussels had spread to Lake Erie. Now they inhabit all the Great

Figure E44-2   **Exotic species**

(a) Workers blast jets of hot water at zebra mussels coating the interior of a Michigan water-treatment plant. **(b)** The Japanese vine kudzu will rapidly cover entire trees and houses. **(c)** The beauty that became a beast—water hyacinths, originally from South America, today clog waterways in our southern states.

Lakes and the Upper Mississippi, and they are spreading throughout eastern waterways. They cover piers, boats, machinery, and beaches (Fig. E44-2a). At water-treatment plants on the Great Lakes, zebra mussel populations reach 600,000 per square yard, clogging pipes and reducing water flow to a trickle (Fig. E44-2a). The mussels cover and suffocate other forms of shellfish, threatening many endan-

gered varieties with extinction. The U.S. Fish and Wildlife Service estimates that zebra mussel damage to fisheries, utilities, and industries dependent on the lakes they infest may cost $5 billion over the next 10 years.

By evading the checks and balances imposed by millennia of coevolution, exotic species are wreaking havoc on natural ecosystems throughout the world.

trees such as jack pine and blue spruce take root in the deeper crevices, and the sun-loving shrubs are shaded out. Within the depths of the forest, shade-tolerant seedlings of taller or faster-growing trees, such as balsam fir, paper birch, and white spruce, thrive. Eventually these overtower and replace the original trees, which are intolerant of shade. After a thousand years or more, a tall climax forest thrives on what was once bare rock.

An Abandoned Farm Field Will Undergo Secondary Succession

Figure 44-19 illustrates succession on an abandoned southeastern farm. The pioneers are fast-growing annual weeds such as crabgrass, ragweed, and sorrel, which root in the rich soil already present and thrive in direct sunlight. A few years later, perennial plants such as asters, goldenrod, broom sedge grass, and woody shrubs, such as blackberry, invade. These plants multiply rapidly and dominate for the next few decades. Eventually, they are replaced by

pines and fast-growing deciduous trees, such as tulip poplar and sweet gum, which sprout from windblown seeds. These trees become prominent after about 25 years, and a pine forest dominates the field for the rest of the first century. Meanwhile, shade-resistant, slow-growing hardwoods such as oak and hickory take root beneath the pines. After the first century, these begin to overtower and shade the pines, which eventually die from lack of sun. A relatively stable climax forest dominated by oak and hickory is present by the end of the second century.

Succession Also Occurs in Ponds and Lakes

In freshwater ponds or lakes, succession occurs both from changes within the pond or lake and as a result of an influx of nutrients from outside the ecosystem. Sediments and nutrients carried in by runoff from the surrounding land have a particularly large impact on small freshwater lakes, ponds, and bogs, which gradually undergo succession to dry land (Fig. 44-20).

Figure 44-19
Secondary succession

Here, secondary succession is illustrated on a plowed, abandoned southeastern farm field.

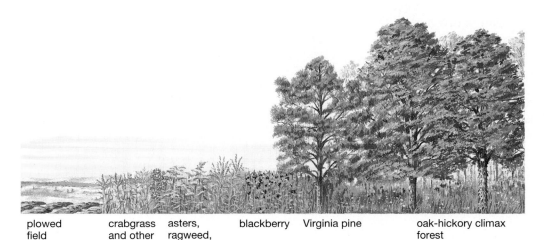

plowed field   crabgrass and other grasses   asters, ragweed, goldenrod   blackberry   Virginia pine   oak-hickory climax forest

time (years)

0 ——————————————————————————————— 200

(a)   (b)   (c)

Figure 44-20   **Succession in a freshwater pond**

In small ponds, succession is speeded by an influx of materials from the surroundings. **(a)** In this small pond, dissolved minerals carried by runoff from the surroundings support aquatic plants, whose seeds or spores were carried here by the winds or by birds and other animals. **(b)** Over time, the decaying bodies of aquatic plants build up soil that provides anchorage for more terrestrial plants. **(c)** Finally, the pond is entirely converted to dry land.

Succession Culminates in the Climax Community

Succession ends with a relatively stable climax community. In your travels, you have undoubtedly noticed that the type of climax community varies dramatically from one area to the next. For example, if you drive through Colorado, you will see a shortgrass prairie climax community on its eastern plains, spruce-pine forests in its mountains, tundra on their uppermost reaches, and sagebrush-dominated climax communities in its western valleys.

The exact nature of the climax community is determined by numerous geological and climatic variables, including temperature, rainfall, elevation, latitude, type of rock (which influences the type of nutrients available), exposure to sun and wind, and many more. Human activities may also dramatically alter the climax vegetation. Large stretches of grasslands in the West, for example, are now dominated by sagebrush because of overgrazing. The grass that normally outcompetes sagebrush is selectively eaten by cattle, allowing the sagebrush to prosper.

Some ecosystems are not allowed to reach the climax stage but are maintained in an earlier stage called a **subclimax**. The tallgrass prairie that once covered northern Missouri and Illinois is actually a subclimax of an ecosystem whose climax community is deciduous forest. The prairie was maintained by periodic fires, some set by lightning and others deliberately set by Native Americans to increase grazing land for buffalo. Forest now encroaches, and limited prairie preserves are maintained by carefully managed burning.

Agriculture also depends on the artificial maintenance of carefully selected subclimax communities. Grains are specialized grasses characteristic of the early stages of succession, and much energy goes into preventing competitors (weeds and shrubs) from taking over. The suburban lawn is also a painstakingly maintained subclimax ecosystem. Mowing destroys woody invaders, and selective herbicides kill pioneers such as crabgrass and dandelions.

To study succession is to study the variations in communities over *time*. The climax communities that form during succession are strongly influenced by climate and geology—the distribution of ecosystems in *space*. Deserts, grasslands, and deciduous forests are climax communities formed over broad geographical regions with similar environmental conditions; these regions of characteristic plant communities are called **biomes**. In Chapter 46 we explore some of the great biomes of the world. Although the communities that make up the various biomes differ radically in the types of populations they support, communities worldwide are structured according to general rules. These principles of ecosystem structure are described in Chapter 45.

⬥ S U M M A R Y   O F   K E Y   C O N C E P T S

The Ecological Niche

The ecological niche defines all aspects of a species' habitat and interactions with its living and nonliving environment. Each species occupies a unique ecological niche.

Competition between Species

Interspecific competition occurs when the niches of two populations within a community overlap and is proportional to the amount of niche overlap. When two species with the same niche are forced under laboratory conditions to occupy the same ecological niche, one species always outcompetes the other. Species within natural communities have evolved in ways that avoid excessive niche overlap, with behavioral and physical adaptations that allow resource partitioning. Interspecific competition limits both the size and the distribution of competing populations.

Predators and Their Prey

Predators eat other organisms and are usually both larger and less abundant than their prey. Predators and prey act as strong agents of selection on one another. Prey animals have evolved a variety of protective colorations that render them either inconspicuous (camouflage) or startling to their predators (startle coloration). Some prey have become poisonous and often exhibit warning coloration by which they are readily recognized and avoided. The situation in which one species has evolved to resemble another is called mimicry. Both predators and prey have evolved a variety of toxic chemicals for attack and defense. Plants that are preyed upon have evolved elaborate defenses ranging from poisons to thorns to overall toughness. These defenses, in turn, have selected for predators that can detoxify poisons, ignore thorns, and grind down tough tissues.

Symbiosis

Symbiotic relationships involve two different species that interact closely over an extended time. Symbiosis includes parasitism, in which the parasite feeds on a larger, less abundant host, usually harming it but not killing it immediately. In commensal symbiotic relationships, one species benefits, often by finding food more easily in the presence of another species, which is not affected by the association. Mutualism benefits both symbiotic species.

Succession: Community Changes over Time

A change in the types of populations inhabiting an ecosystem over time is called succession. Primary succession, which may take thousands of years, occurs where no remnant of a previous community exists. Primary succession could occur on rock scraped bare by a glacier or cooled from molten lava, a sand dune, or in a newly formed glacial lake. Secondary succession occurs much more rapidly, because it builds on the remains of a disrupted community, such as an abandoned field or the aftermath of a fire. During succession, pioneer organisms invade and alter the environment by their presence. Community changes occur as one group of organisms, particularly plants, alters the environment in ways that encourage growth of their competitors. These competitors replace the first colonizers and are subsequently replaced by others. Some ecosystems, including tallgrass prairie and farm fields, are maintained in relatively early stages of succession by periodic disruptions. Uninterrupted succession ends with a relatively stable group of organisms called the climax community.

⬥ K E Y   T E R M S

aggressive mimicry   p. 876
biome   p. 886
camouflage   p. 874
climax community   p. 881
coevolution   p. 871
commensalism   p. 879
community   p. 871
competitive exclusion principle   p. 872

ecological niche   p. 872
exotic species   p. 884
interspecific competition   p. 872
keystone predator   p. 878
mimicry   p. 875
mutualism   p. 880
parasitism   p. 878
pioneer   p. 881

primary succession   p. 883
resource partitioning   p. 873
secondary succession   p. 883
startle coloration   p. 876
subclimax   p. 886
succession   p. 880
symbiosis   p. 878
warning coloration   p. 875

✖ T H I N K I N G   T H R O U G H   T H E   C O N C E P T S

Multiple Choice

1. Which of the following is NOT usually true of climax communities? They have
   a. more species than early successional communities
   b. more organic matter in the soil
   c. more trophic levels
   d. more short-lived species
   e. more productivity

2. What were the differences in the niches of the *Paramecium* in Gause's second experiment that allowed them to coexist?
   a. food eaten
   b. body size
   c. feeding area
   d. preferred water temperature
   e. preferred pH

3. Which of the following is an example of a commensal relationship?
   a. orchids growing on tree limbs
   b. flowering plants and their pollinators
   c. lupines and blue butterflies
   d. bats and moths
   e. monarch and viceroy butterflies

4. What is the function of aggressive mimicry?
   a. to hide a prey from a predator
   b. to warn a predator that a prey is dangerous
   c. to warn a predator that a prey is distasteful
   d. to keep prey from recognizing a predator
   e. to startle a prey when it sees a predator

5. What is coevolution?
   a. two species selecting for different traits in each other
   b. individuals of two species living together
   c. the presence of two species in the same community
   d. two species evolving together through time

e. individuals of two species learning how to coexist with or to hunt with each other

6. The competitive exclusion principle implies that
   a. coexisting species can use the same resources
   b. coexisting species cannot eat exactly the same things
   c. coexisting species cannot have identical ecological interactions
   d. coexisting species cannot be exactly the same size
   e. coexisting species cannot be closely related to one another

Review Questions

1. Define an ecological community, and list three important types of community interactions.

2. An ecologist visiting an island finds two very closely related species of birds, one of which has a slightly larger bill than the other. Interpret this finding with respect to the competitive exclusion principle and the ecological niche, defining both terms.

3. List four general ways in which plants and animals protect themselves against being eaten, and provide a specific example of each.

4. List three different types of symbiosis; define and provide an example of each.

5. Introduced species are most often invasive when their new home lacks effective predators. What does this situation imply about the factors controlling populations of these species in their native ecosystems?

6. List two subclimax and two climax communities. How do they differ?

7. Describe succession, and explain why it occurs. Detail one of the successional sequences in this chapter.

8. What type of succession would occur on a clear cut in a national forest, and why?

✖ A P P L Y I N G   T H E   C O N C E P T S

1. Herbivorous animals that eat seeds are considered by some ecologists to be predators of plants, while herbivorous animals that eat leaves are considered to be parasites of plants. Discuss the validity of this classification scheme.

2. There are many more cases in which mutualisms seem to have evolved from parasite-host interactions than from predator-prey interactions. Ecologists think the reason is that it is often in the best interests of a parasite not to hurt its host. Try to think of why this would be true—when is what is good for a host good for its parasite? When not? Contrast this situation to predation.

3. Restoration ecologists try to rebuild more natural communities in areas damaged by humans, such as mine tailings, overgrazed pastures, and dredged wetlands. Essentially, these scientists try to help or direct natural succession to reclaim damaged lands. Think about a situation that restoration ecologists might face: restoring an abandoned rock quarry (which is almost pure gravel) to forest. What kinds of plant species

would you first seed into the site? What could you do to alter the physical environment to hasten the speed of succession?

4. The successional sequence of small ponds depends on the inflow and outflow of the pond. In particular, ponds with no outflowing stream quite rapidly undergo succession to a marsh and then to dry land. This sequence is much slower, and may not happen at all, in ponds with outflows. Explain why, using your understanding of the changes that occur during succession.

5. Would you expect in primary succession or secondary succession that late-successional species would most rely on the growth of the pioneer species to make the site suitable for them? Explain why, and describe how this explanation accounts for the differences in the speed of the two kinds of succession.

6. Many natural communities rarely or never reach a climax community because of natural disturbances. Indeed, human prevention of natural disturbances, especially for-

est fires, is often as destructive to communities as are more direct human impacts. Fires typically create a patchwork of areas in many different successional stages. Explain why this kind of patchwork can often support many more species than a uniform landscape of either climax community or early successional areas. Argue for or against programs that suppress fires—should Smokey the Bear retire? Why or why not?

7. Few regions of the United States are unaffected by introduced exotic species. Find out about an introduced species in your area (e.g., through news articles, by consulting an ecologist in your Biology Department, or calling a local Fish and Wildlife Service office). Learn where it came from, how it got here, and what impact it has had or might have on local native species. Could its introduction have been prevented? Suggest several ways to limit the introduction of exotic species.

✖ F O R   M O R E   I N F O R M A T I O N

Amos, W. H. "Hawaii's Volcanic Cradle of Life." *National Geographic*, July 1990. A naturalist explores succession on lava flows.

Daniels, P. "How Flowers Seduce the Bugs and the Bees." *International Wildlife*, November/December 1984. Beautifully illustrated description of plants and their mutualistic pollinators.

Horn, H. S. "Forest Succession." *Scientific American*, May 1975 (Offprint No. 1321). Describes succession in eastern deciduous forests.

Mills, E. L., Leach, J. H., Carlton, J. T., and Secor, C. L. "Exotic Species and the Integrity of the Great Lakes: Lessons from the Past." *BioScience* 44:666–676, 1994. The history of invasions by various species and their consequences.

Moore, J. "The Behavior of Parasitized Animals." *BioScience* 45:89–96, 1995. The ways in which parasites have evolved to manipulate the behavior of their hosts and hosts resist these effects.

Rinderer, T. E., Oldroyd, B. P., and Sheppard, W. S. "Africanized Bees in the U.S." *Scientific American*, 1993. What are the likely impacts of these aggressive invaders?

N E T   W A T C H

On-line resources for this chapter are on the World Wide Web at:
http://www.prenhall.com/~audesirk (click on the table of contents link and then select Chapter 44).

45 How Ecosystems Work

> "A thing is right, when it tends to preserve the integrity, stability, and beauty of the biotic community. It is wrong when it tends otherwise."
>
> ALDO LEOPOLD in *A Sand County Almanac* (1949)

✖ AT A GLANCE

The Flow of Energy
The Energy Captured by Photosynthetic Organisms Is Called Primary Productivity
Organisms Occupy Different Trophic Levels Based on How They Acquire Energy
Food Chains and Food Webs Describe the Feeding Relationships within Ecosystems
Energy Flows through Trophic Levels
Biological Magnification Occurs as Toxic Substances Are Passed through Trophic Levels

The Cycling of Nutrients
The Carbon Cycle Is an Atmospheric Cycle
The Nitrogen Cycle Is an Atmospheric Cycle
The Phosphorus Cycle Is a Sedimentary Cycle
Most Water Remains Chemically Unchanged during the Water Cycle

Human Intervention in Energy Flow and Nutrient Cycling
Acid Rain Is Caused by Overloading the Nitrogen and Sulfur Cycles
Interference with the Carbon Cycle Is Causing the Greenhouse Effect

Biodiversity and Ecosystem Stability

A microcosm is reflected in a dewdrop on the spore case of a moss. The sparkling energy of sunlight, captured by photosynthetic organisms, powers nearly all life on Earth.

Nearly all life on Earth is powered by the energy of sunlight, from the leaping of the jackrabbit to the active transport of molecules through a cell membrane. Each time this energy is used, some of it is lost as heat. But while solar energy continuously bombards Earth and is continuously lost as heat, nutrients remain. They may change in form and distribution, but nutrients do not leave the world ecosystem and are continuously recycled. Two basic laws thus underlie ecosystem function. First, energy moves *through* ecosystems in a continuous one-way flow. Energy needs constant replenishment from an outside source, the sun. Second, in contrast to energy, nutrients constantly cycle and recycle in a circular flow *within* ecosystems (Fig. 45-1). These laws shape the complex interactions within living communities.

In Chapters 7 and 8, you learned how energy is trapped by photosynthesis, released by cellular respiration, and used to construct the complex molecules of life. In Chapter 44, you learned some of the complex ways in which organisms interact within ecological communities. In this chapter, we relate some of these basic principles to the way ecosystems work.

The Flow of Energy

The Energy Captured by Photosynthetic Organisms Is Called Primary Productivity

Ninety-three million miles away, the sun fuses hydrogen into helium, releasing tremendous quantities of energy. A tiny fraction of this energy reaches Earth in the form of electromagnetic waves, including heat, light, and ultraviolet energy. Of the energy that reaches Earth, much is reflected by the atmosphere, clouds, and Earth's surface. Still more is absorbed as heat by Earth and its atmosphere, leaving only about 1% to power all life on Earth. Of this 1%, green plants capture 3% or less. The teeming life on Earth is thus supported by less than 0.03% of the energy reaching it from the sun. But how does this energy enter the biological community? The energy that powers ecosystems enters through the process of photosynthesis.

HEAT

producers

HEAT

NUTRIENTS

HEAT

primary consumers

detritus feeders
and decomposers

HEAT

higher-level
consumers

HEAT

nutrients

energy/heat

Figure 45-1   **Energy flow, nutrient cycling, and feeding relationships in ecosystems**

Note that nutrients (purple) neither enter nor leave the cycle, whereas energy (yellow) is lost at each level as heat and must be constantly replenished by sunlight.

In photosynthesis (performed by plants, plantlike protists, and cyanobacteria), pigments such as chlorophyll absorb certain wavelengths of sunlight. This solar energy is used to combine carbon dioxide and water into sugar, a compound that stores energy in chemical bonds. Respiration then uses the oxygen produced as a by-product of photosynthesis to break down glucose and capture the energy in its bonds in the form of ATP. Some of this energy

is used to power other chemical reactions, converting sugars into starches, cellulose, fats, vitamins, pigments, and proteins (Fig. 45-2). Photosynthetic organisms are called **autotrophs** (Greek, "self-feeders"), or **producers,** because they produce food for themselves. In addition, they directly or indirectly produce food for nearly all other life as well. Organisms that cannot photosynthesize cannot produce food for themselves; instead they must acquire ener-

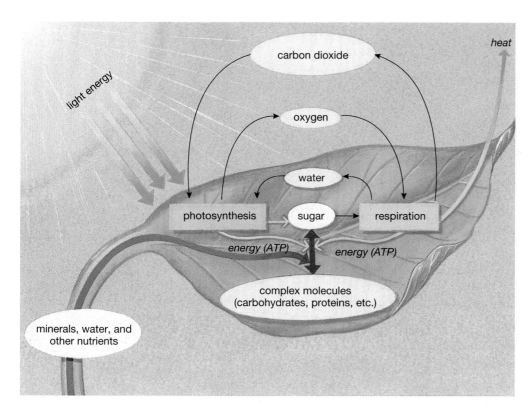

Figure 45-2 **Primary productivity: photosynthesis**

During photosynthesis, plants capture the energy of sunlight and store it in ATP, sugar, and other high-energy carbohydrates synthesized from carbon dioxide and water. Oxygen is released as a by-product. Some of the energy stored in sugar is liberated during respiration and used in the construction of more complex molecules. In the process, oxygen is used, water is released, and some energy is lost as heat.

gy prepackaged in the molecules of the bodies of other organisms. These organisms are called **heterotrophs** (Greek, "other-feeders"), or **consumers.**

The amount of life an ecosystem can support is determined by the energy captured by the producers. The energy that photosynthetic organisms store and make available to other members of the community over a given time period is called **net primary productivity.** Net primary productivity can be measured either in units of energy (calories) stored per unit area in a given time period or as the dry weight, also called **biomass,** of producers added to the ecosystem per unit area in a given time span. The productivity of an ecosystem is influenced by many environmental variables, including the amount of nutrients available to the producers, the amount of sunlight reaching them, the availability of water, and the temperature. In the desert, for example, lack of water limits productivity, while in the open ocean, light is limited in deep water and nutrients are limited in surface water. When resources are abundant, as in estuaries (regions where rivers meet the ocean) or in tropical rain forests, productivity is high. Some average productivities for a variety of ecosystems are shown in Figure 45-3.

Organisms Occupy Different Trophic Levels Based on How They Acquire Energy

Living things may be categorized according to their role in the flow of energy through communities. Energy flows through communities from photosynthetic producers through several levels of consumers. Each category of or-

ganism is called a **trophic** (literally "feeding") **level.** The producers, from redwood trees to cyanobacteria, form the first trophic level, obtaining their energy directly from sunlight (see Fig. 45-1). Consumers form several trophic levels, and certain consumers even switch trophic levels by eating organisms from different levels. Sparrows, for example, may eat either seeds or insects. Some consumers feed directly and exclusively on producers, the most abundant living energy source in the ecosystem. These **herbivores,** ranging from grasshoppers to giraffes, are called **primary consumers** and form the second trophic level. **Carnivores,** such as the spider, hawk, and wolf, are flesh-eaters, feeding primarily on herbivores; they are called **secondary consumers** and form the third trophic level. Some carnivores occasionally eat other carnivores, and when doing so they form the fourth trophic level: **tertiary consumers.** Depending on the nature of their prey, carnivores may also be **quaternary consumers** that occupy the fifth trophic level.

Food Chains and Food Webs Describe the Feeding Relationships within Ecosystems

To illustrate who feeds on whom in an ecosystem, ecologists commonly identify a representative of each trophic level that eats the representative of the level below it. This linear feeding relationship is called a **food chain.** As illustrated in Figure 45-4, different ecosystems have radically different food chains.

Natural communities rarely contain well-defined groups of primary, secondary, and tertiary consumers. A **food web** shows the many interconnecting food chains in an ecosystem, de-

Figure 45-3   **Ecosystem productivity compared**
Average net primary productivity in grams of organic material per square meter per year of some terrestrial and aquatic ecosystems.

scribing the actual feeding relationships within a given community much more accurately than does a food chain (Fig. 45-5). Some animals, such as raccoons, bears, rats, and humans, are omnivores (Latin, "eating all"), acting at different times as primary, secondary, and occasionally tertiary consumers. Many carnivores will eat either herbivores or other carnivores, thus acting either as secondary or tertiary consumers. An owl, for instance, is a secondary consumer when it eats a mouse, but a tertiary (third-level) consumer when it eats a shrew, which feeds on insects. If the shrew ate a carnivorous insect, it would be a tertiary consumer, and the owl that fed on it would then be a quaternary (fourth-level) consumer. When digesting a spider, a carnivorous plant, such as the sundew, can tangle the web hopelessly by serving simultaneously as a producer and a secondary consumer!

Detritus Feeders and Decomposers Release Nutrients for Reuse

Among the most important strands in the food web are the detritus feeders and decomposers. The **detritus feeders** are an army of small and often unnoticed animals and protists that live on the refuse of life: molted exoskeletons, fallen leaves, wastes, and dead bodies. The network of detritus feeders is extremely complex, including earthworms, mites, protists, centipedes, some insects and crustaceans, nematode worms, and even a few large vertebrates such as vultures. These organisms consume dead organic matter, extract some of the energy stored within it, and excrete it in a still further decomposed state. Their excretory products serve as food for other detritus feeders

and decomposers, until most of the stored energy has been utilized. The **decomposers** are primarily fungi and bacteria that digest food outside their bodies, absorb the nutrients they need, and release the remaining nutrients.

Through the activities of detritus feeders and decomposers, the bodies and wastes of living organisms are reduced to simple molecules, such as carbon dioxide, water, minerals, and organic molecules, that return to the atmosphere, soil, and water. By liberating nutrients for reuse, detritus feeders and decomposers form a vital link in the nutrient cycles of ecosystems. In some ecosystems, such as deciduous forests, more energy passes through the detritus feeders and decomposers than through the primary, secondary, and tertiary consumers. Though inconspicuous, this portion of the food web is absolutely essential to life on Earth. If the detritus feeders and decomposers were to disappear suddenly, communities would gradually be smothered by accumulated wastes and dead bodies. The nutrients stored in these bodies would be unavailable to enrich the soil. Without these nutrients, the quality of the soil would become poorer and poorer until plant life could no longer be sustained. With plants eliminated, energy would cease to enter the community, and the higher trophic levels would disappear as well.

Energy Flows through Trophic Levels

As discussed in Chapter 4, a basic law of thermodynamics is that energy use is never completely efficient. For example, as your car converts the energy stored in gasoline to the energy of movement, about 75% is immediately lost

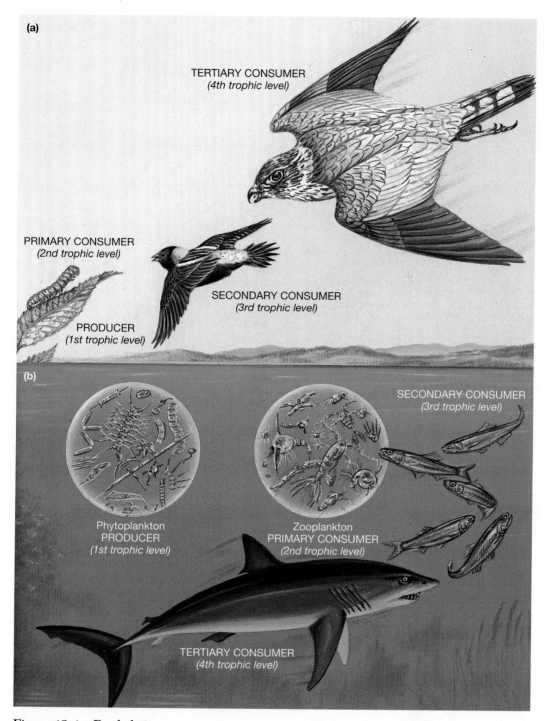

Figure 45-4   **Food chains**

(a) A simple terrestrial food chain. (b) A simple marine food chain.

as heat. So too in living systems; the splitting of the chemical bonds of ATP to power a muscular contraction also produces heat as a by-product, as anyone who has exercised vigorously is well aware. Small amounts of waste heat are also produced by the germination of a seed and the thrashing of the tail of a sperm.

The transfer of energy from one trophic level to the next is also quite inefficient. When a caterpillar (a primary consumer) eats the leaves of a tomato plant (a producer), only some of the solar energy originally trapped by the plant is available to the insect. Some was used by the plant to grow and maintain life. Some was converted into the chemical bonds of molecules such as cellulose, which the caterpillar cannot break down. Some energy remains in the plant. Therefore, only a fraction of the energy captured by the first trophic level is available to or-

Figure 45-5   A simple terrestrial food web

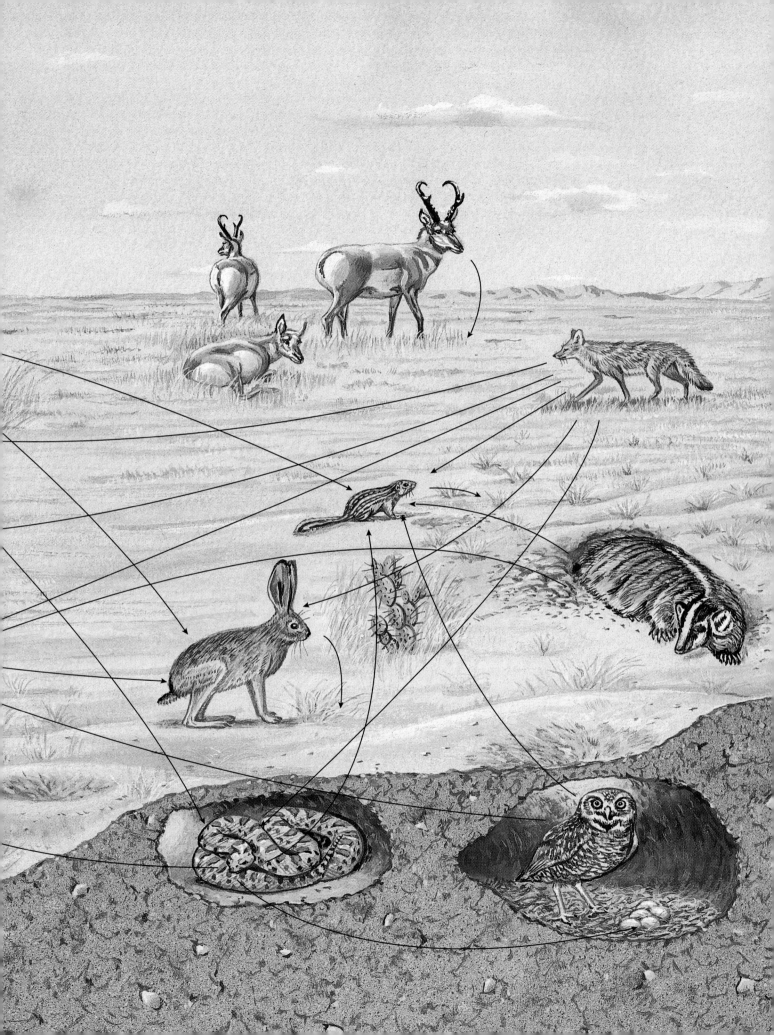

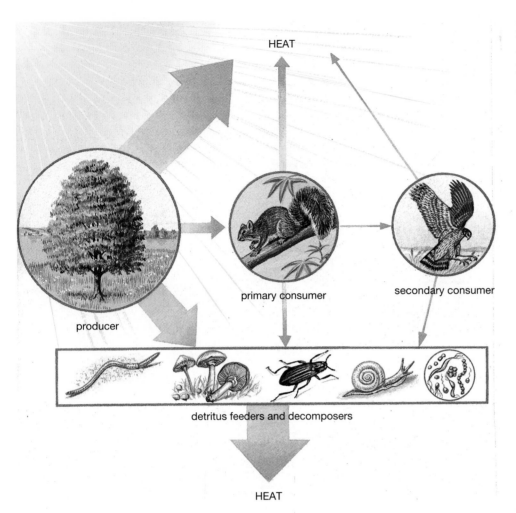

HEAT

producer

primary consumer

secondary consumer

detritus feeders and decomposers

HEAT

Figure 45-6   Energy loss

A diagram showing the loss of energy during its transfer between trophic levels in a forest community. The width of the arrows is roughly proportional to the quantity of energy transferred or lost.

ganisms in the second trophic level. The energy consumed by the caterpillar in turn is partially used to power crawling and the gnashing of mouthparts. Some is used to construct the indigestible exoskeleton, and much is given off as heat. All this energy is unavailable to the songbird in the third trophic level when it eats the caterpillar. The bird loses energy as body heat, uses more in flight, and converts a considerable amount into indigestible feathers, beak, and bone. All this energy will be unavailable to the hawk that catches it. A simplified model of energy flow through the trophic levels in a deciduous forest ecosystem is illustrated in Figure 45-6.

Energy Pyramids Illustrate Energy Transfer between Trophic Levels

Studies of a variety of ecosystems indicate that the net transfer of energy between trophic levels is roughly 10% efficient, although different ecosystems vary significantly. This means that, in general, the energy stored in primary consumers (herbivores) is only about 10% of the energy stored in the bodies of producers. The bodies of secondary consumers, in turn, possess roughly 10% of the energy stored in primary consumers. In other words, for every 100 calories of solar energy captured by grass, only about 10

calories are converted into herbivores, and only 1 calorie into carnivores. This inefficient energy transfer between trophic levels is called the "10% law." An **energy pyramid**, which shows maximum energy at the base and steadily diminishing amounts at higher levels, illustrates the energy relationships between trophic levels graphically (Fig. 45-7). Ecologists sometimes use biomass as a measure of the energy stored at each trophic level. Because the dry mass of the bodies of organisms at each trophic level is roughly proportional to the amount of energy stored in them, a **biomass pyramid** for a given ecosystem usually has the same general shape as its energy pyramid.

What does this mean for ecosystems? If you wander through an undisturbed ecosystem, you will notice that the predominant organisms are plants; these have the most energy available to them because they trap it directly from sunlight. The most abundant animals will be those feeding on plants, and carnivores will always be relatively rare. The inefficiency of energy transfer also has important implications for human food production. The lower the trophic level we utilize, the more food energy will be available to us; far more people can be fed on grain than on meat.

An unfortunate side effect of the inefficiency of energy transfer, coupled with human production and release of toxic chemicals, is the phenomenon of biological mag-

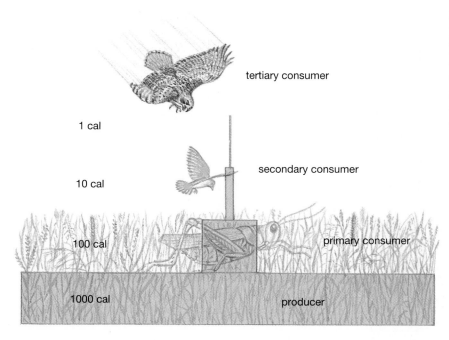

1 cal — tertiary consumer

10 cal — secondary consumer

100 cal — primary consumer

1000 cal — producer

Figure 45-7   An energy pyramid for a prairie ecosystem

Each trophic level from producer to tertiary consumer has less energy stored in it. The width of each rectangle is proportional to the energy stored at that trophic level. A pyramid of biomass for this ecosystem would look quite similar.

nification. This is the concentration of certain persistent toxic chemicals in the bodies of carnivores, including ourselves, as described below.

Biological Magnification Occurs as Toxic Substances Are Passed through Trophic Levels

In the 1940s, the properties of the new insecticide DDT seemed close to miraculous. In the tropics, DDT saved millions of lives by killing the mosquitoes that spread malaria. Increased crop yields resulting from DDT's destruction of insect pests saved millions more from starvation. DDT is long lasting, so a single application keeps on killing. The inventor was awarded the Nobel Prize, and people looked forward to a new age of freedom from insect pests. Little did they realize that the indiscriminate use of this pesticide was unraveling the complex web of life.

For example, in the mid-1950s, the World Health Organization sprayed DDT on the island of Borneo to control malaria. A caterpillar that fed on the thatched roofs of houses was relatively unaffected, while a predatory wasp that fed on it was destroyed. Thatched roofs collapsed, eaten by the uncontrolled caterpillars. Gecko lizards that ate the poisoned insects accumulated high concentrations of DDT in their bodies. Both they, and the village cats that ate the geckos, died of DDT poisoning. With the cats eliminated, the rat population exploded and the village was threatened with an outbreak of plague, carried by the uncontrolled rats. The outbreak was avoided by airlifting new cats to the villages.

Meanwhile, in the United States, wildlife biologists during the 1950s and 1960s witnessed an alarming decline in populations of several predatory birds, especially fish-eaters, such as Bald Eagles, cormorants, and Brown Pelicans. These top predators are never abundant, and the decline pushed some, such as the Brown Pelican, close to extinction. The aquatic ecosystems supporting these birds had been sprayed with relatively low amounts of DDT to control insects. Scientists were amazed to find concentrations of DDT in the bodies of predatory birds up to *one million* times the concentration present in the water. This finding led to the discovery of **biological magnification**, the process by which toxic substances accumulate in increasingly high concentrations in progressively higher trophic levels. In 1973, DDT was banned in the United States, although it is still used in some less-developed countries. In 1994, the Bald Eagle was removed from the endangered species list.

DDT and other substances that undergo biological magnification have two properties that make them dangerous: (1) They do not readily break down into harmless substances, and (2) they are fat soluble but not water soluble. Thus, they accumulate in the bodies of animals, particularly in the fat, rather than being broken down and excreted in the watery urine. Because the transfer of energy from lower to higher trophic levels is extremely inefficient, herbivores must eat large quantities of plant material (which may have been sprayed with DDT), carnivores must eat many herbivores, and so on. Because DDT is not excreted, the predator accumulates the substance from all its prey over many years. Thus, DDT reaches highest levels in top predators, such as predatory birds, where it interferes with the deposition of calcium into eggshells by mimicking and disrupting the functions of the natural sex hormone estrogen.

Today there is growing concern about the effects of a number of chlorinated compounds chemically related to DDT that also share its tendencies to persist, accumulate, and interfere with normal sex hormone functions. These chemi-

Figure 45-8   The price of pollution

Deformities such as the twisted beak of this double-crested cormorant from Lake Michigan have been linked to dioxin and related chemicals. Abnormalities of the reproductive and immune systems are also common in many different organisms exposed to these pollutants. Predatory animals are especially vulnerable because of bioaccumulation.

cals, which include dioxin and polychlorinated biphenyls (PCBs), are produced by a variety of manufacturing processes. Studies have focused on the Great Lakes, which until recently had relatively high levels of these chlorinated compounds. Here, populations of fish-eating river otters have declined sharply, and a variety of fish-eating birds, including the newly returned Bald Eagles, produce deformed offspring or eggs that never hatch (Fig. 45-8). Minute amounts of dioxin given to pregnant rats cause lowered sperm counts and genital deformities in their male offspring. In Florida's Lake Apopka, a spill of chlorinated chemicals including DDT in 1980 has recently been linked to a 90% decline in the birth rate of the lake's alligators and to abnormally small penis size in alligators reaching maturity in the lake. Some scientists suggest that human sperm counts have declined since the 1930s as a result of long-term exposure to pollution, although this effect is by no means proved.

A number of toxic substances are subject to biological magnification, including mercury and some radioactive compounds. In the 1960s, mercury from ocean pollution reached such high levels in swordfish (top predators) that the sale of swordfish meat was banned.

Understanding the properties of pollutants and the workings of food webs is crucial if we are to prevent both human health hazards and widespread loss of wildlife. Humans have considerable cause for concern, because we often feed high on the food chain. (When you eat a tuna sandwich, for example, you are a tertiary or even quaternary consumer.) In addition, our long life span provides more time for substances stored in the body to accumulate to toxic levels.

The Cycling of Nutrients

In contrast to the energy of sunlight, nutrients do not flow down onto Earth in a steady stream from above. For practical purposes, the same pool of nutrients has been supporting life for over 3 billion years. Nutrients are elements and small molecules that form all the chemical building blocks of life. Some, called **macronutrients,** are required by organisms in large quantities. These include water, carbon, hydrogen, oxygen, nitrogen, phosphorus, sulfur, and calcium. **Micronutrients,** including zinc, molybdenum, iron, selenium, and iodine, are required only in trace quantities. **Nutrient cycles,** also called **biogeochemical cycles,** describe the pathways these substances follow as they move from the living to the nonliving portions of ecosystems and back again to living tissues.

The major source, or **reservoir,** of important nutrients is generally in the nonliving environment. For example, the major reservoir of carbon and nitrogen is the atmosphere, so the cyclings of these elements are called **atmospheric cycles.** The reservoir of a **sedimentary cycle** is sediment—that is, soil or, as in the phosphorus cycle, rock. In the cycling of water, the major reservoir is the ocean. In the following section we briefly describe the cycles of carbon, nitrogen, phosphorus, and water.

The Carbon Cycle Is an Atmospheric Cycle

Chains of carbon atoms form the framework of all organic molecules. Carbon enters the living community through capture of carbon dioxide (CO_2) during photosynthesis. The major reservoir of this gaseous compound is the atmosphere, where it makes up 0.033% of the total gases.

Carbon enters food webs through producers, which trap CO_2 during photosynthesis. Some CO_2 is returned to the atmosphere through cellular respiration, and some that is incorporated into the plant body is later passed to herbivores. They, in turn, respire some of it and incorporate some into their tissues. All living things are eventually consumed by predators, detritus feeders, and decomposers, and ultimately most carbon is returned to the atmosphere as CO_2 (Fig. 45-9).

Some carbon is cycled much more slowly. For example, mollusks extract carbon dioxide dissolved in water and combine it with calcium to form calcium carbonate ($CaCO_3$), from which they construct their shells. Shells of dead mollusks collect in undersea deposits and may even-

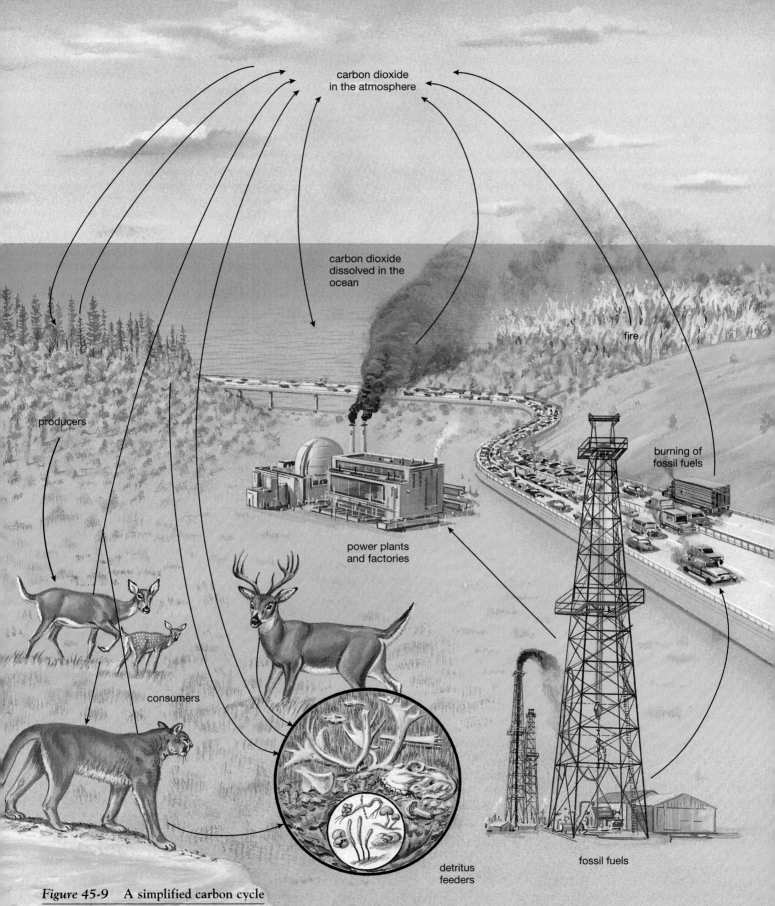

carbon dioxide
in the atmosphere

carbon dioxide
dissolved in the
ocean

fire

producers

burning of
fossil fuels

power plants
and factories

consumers

detritus
feeders

fossil fuels

Figure 45-9   A simplified carbon cycle

Carbon is captured from the atmosphere during photosynthesis and passed up through the
trophic levels. It is released during respiration from all trophic levels and by the burning of
forests and fossil fuels.

tually be converted to limestone. Limestone may dissolve gradually as it is exposed to water, making the carbon available to living organisms once more. Another long-term part of the carbon cycle is the production of fossil fuels. **Fossil fuels** are formed from the remains of ancient plants and animals. The carbon found in the organic molecules of these ancient organisms remains in these deposits, transformed by high temperature and pressure over millions of years into coal, oil, or natural gas. The energy of prehistoric sunlight is also trapped in fossil fuels. Burning fossil fuel releases its energy. Human activities, including the extensive use of fossil fuels and the cutting and burning of Earth's great forests (where much carbon is stored), are increasing the amount of carbon dioxide in the atmosphere, as described later in this chapter.

The Nitrogen Cycle Is an Atmospheric Cycle

The atmosphere is about 79% nitrogen gas (N_2), but neither plants nor animals can use this gas directly. Instead, plants must be supplied with nitrates (NO_3^-) or ammonia (NH_3). But how is atmospheric nitrogen converted to these molecules? Ammonia is synthesized by certain bacteria and cyanobacteria that engage in **nitrogen fixation**, a process that combines nitrogen with hydrogen. Some of these bacteria are found living in water and soil; others have entered a symbiotic association with plants called **legumes** (a group including soybeans, clover, and peas), where they live in special swellings on the roots (see Chapter 21). Decomposer bacteria can also produce ammonia from amino acids and urea found in dead bodies and wastes. Still other bacteria convert ammonia to nitrates. Nitrates are also produced by electrical storms and by other forms of combustion that cause nitrogen to react with atmospheric oxygen. In human-dominated ecosystems, such as farm fields, gardens, and suburban lawns, ammonia and nitrates are supplied by chemical fertilizers. These fertilizers are produced by using the energy in fossil fuels to artificially "fix" atmospheric nitrogen.

Plants incorporate the nitrogen from ammonia and nitrates into amino acids, proteins, nucleic acids, and vitamins. These nitrogen-containing molecules from the plant are eventually consumed, by primary consumers, detritus feeders, or decomposers. As it is passed through the food web, some of the nitrogen is liberated in wastes and dead bodies, which decomposer bacteria convert back to nitrates and ammonia. The cycle is balanced by a continuous return of nitrogen to the atmosphere by **denitrifying bacteria**. These residents of mud, bogs, and estuaries break down nitrates, releasing nitrogen gas back to the atmosphere (Fig. 45-10).

The Phosphorus Cycle Is a Sedimentary Cycle

Phosphorus is a crucial component of biological molecules, including the energy-transfer molecules ATP and NADP, nucleic acids, and the phospholipids of cell membranes. It is a major component of vertebrate teeth

and bones. The reservoir of phosphorus in ecosystems is rock, where it is found bound to oxygen in the form of phosphate. The phosphorus cycle is a sedimentary cycle, because unlike carbon and nitrogen, phosphorus does not enter the atmosphere. As phosphate-rich rocks are exposed and eroded, rainwater dissolves the phosphate. Dissolved phosphate is readily absorbed through the roots of plants and by other autotrophs such as photosynthetic protists and cyanobacteria, where it is incorporated into biological molecules such as ATP. From these producers, phosphorus is passed through food webs (Fig. 45-11). At each level, excess phosphate is excreted. Ultimately, decomposers return the phosphorus remaining in dead bodies back to the soil and water in the form of phosphate. Here it may be reabsorbed by autotrophs, or it may become bound to sediment and eventually reincorporated into rock. Some of the phosphate dissolved in fresh water is carried to the oceans. Although much of this phosphate ends up in marine sediments, some is absorbed by marine producers and eventually incorporated into the bodies of invertebrates and fish. Some of these, in turn, are consumed by seabirds, who excrete large quantities of phosphorus back onto the land. At one time, the guano (droppings) deposited by seabirds along the western coast of South America was mined and provided a major source of the world's phosphorus. Phosphate-rich rock is also mined, and the phosphate is incorporated into fertilizer. Soil that erodes from fertilized fields carries large quantities of phosphates into lakes, streams, and the ocean, where it stimulates the growth of producers.

Most Water Remains Chemically Unchanged during the Water Cycle

The water cycle, or **hydrologic cycle** (Fig. 45-12), differs from most other nutrient cycles because most water remains in the form of water throughout the cycle, and it is not used in the synthesis of new molecules. The major reservoir of water is the ocean, which covers about three-quarters of Earth's surface and contains over 97% of the available water. The hydrologic cycle is driven by solar energy, which evaporates water, and by gravity, which draws the water back to Earth in the form of precipitation (rain, snow, sleet, dew). Most evaporation occurs from the oceans, and much water returns directly to them by precipitation. Water falling on land takes more varied paths. Some is evaporated from the soil, lakes, and streams. A portion runs off the land back to the oceans, and a small amount enters underground reservoirs. Because the bodies of living things are roughly 70% water, some of the water in the hydrologic cycle enters the living communities of ecosystems. It is absorbed by the roots of plants, and much of this is evaporated back to the atmosphere from their leaves. A small amount is combined with carbon dioxide during photosynthesis to produce high-energy molecules. Eventually these are broken down

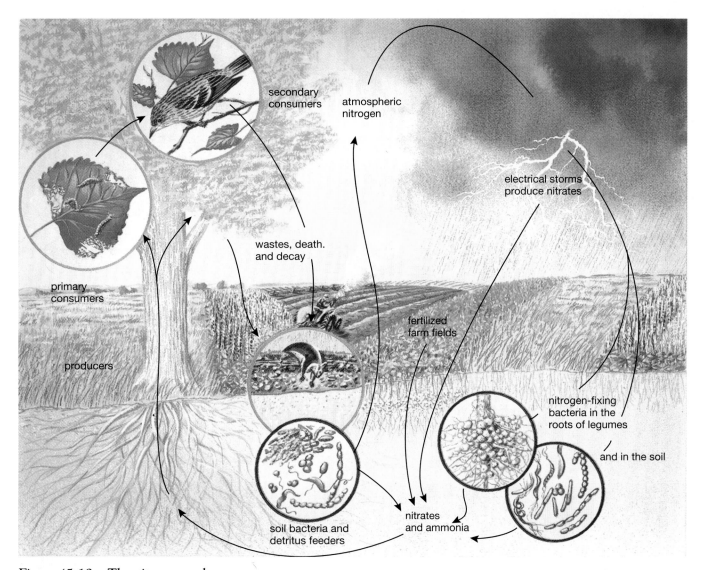

Figure 45-10 **The nitrogen cycle**

From its reservoir in the atmosphere, nitrogen gas is combined with oxygen to form nitrates by lightning or by the burning of forests or fossil fuels and carried to Earth dissolved in rain. Nitrogen-fixing bacteria produce ammonia. Nitrates and ammonia are also synthesized by humans for use in fertilizers. These are absorbed by plants and other producers and incorporated into biological molecules that are passed up through the trophic levels. Nitrates and ammonia are released by excretion or by decomposer bacteria. Other bacteria convert these molecules back to atmospheric nitrogen, completing the cycle.

during cellular respiration, releasing water back to the environment. Heterotrophs get water from their food or by drinking.

Although the hydrologic cycle would continue in the absence of living things, the distribution of life and the composition of biological communities depend on and, to a great extent are determined by, patterns of precipitation and evaporation. For example, the operation of the hydrologic cycle is considerably different over a desert and a tropical rain forest, and this difference is reflected in the composition of the communities in these ecosystems.

Human Intervention in Energy Flow and Nutrient Cycling

Many of the environmental problems that beset modern society are the result of human interference in the way ecosystems function. Primitive people were sustained solely by the energy flowing from the sun and produced wastes that were readily taken back into the nutrient cycles. But as the population grew and technology increased, humans began to act more and more independently of these natural processes. We have mined from Earth substances, such

Figure 45-11   **The phosphorus cycle**

Phosphate dissolves from phosphate-rich rocks or from fertilizers and enters plants and other producers, where it is incorporated into biological molecules. These molecules are passed through the trophic levels. Phosphate is excreted or returned to the soil and water by decomposer bacteria. It may then be reused by producers or eventually incorporated into rock.

Figure 45-12   **The hydrologic cycle**

The hydrologic cycle is the simplest of the nutrient cycles.

as lead, arsenic, cadmium, mercury, oil, and uranium, that are foreign to natural ecosystems and toxic to many of the organisms in them (Fig. 45-13). In our factories, we synthesize substances never before found on Earth: pesticides, solvents, and a wide array of other industrial chemicals harmful to many forms of life. The Industrial Revolution, which began in the mid-nineteenth century, resulted in a tremendous increase in our reliance on energy from fossil fuels (rather than sunlight) for heat, light, transportation, industry, and even agriculture.

In this section we describe two environmental problems of global proportions that are a direct result of human reliance on fossil fuels: acid rain and the greenhouse effect.

Acid Rain Is Caused by Overloading the Nitrogen and Sulfur Cycles

Each year, the United States alone discharges about 30 million tons of sulfur dioxide into the atmosphere, two-thirds of it from power plants burning coal or oil (Fig. 45-14). The rest is largely a by-product of industrial boilers, smelters, and refineries. Although volcanoes and hot springs also release sulfur dioxide, human industrial activities account for 90% of the sulfur dioxide in the atmosphere. Twenty-five million tons of nitrogen oxides are also released by

Figure 45-13   **A natural substance out of place**

An oil-soaked seabird is cleaned by rescuers after an oil spill off the coast of Alaska.

Figure 45-14   A source of acid rain

Power plants burning high-sulfur coal with inadequate emission controls are a major source of atmospheric sulfur dioxide, the prime contributor to acid rain.

Figure 45-15   Acid rain is corrosive

This marble statue shows the corrosive effects of acid rain.

the United States each year, from vehicles, power plants, and industry.

Excess production of these substances was identified in the late 1960s as the cause of a growing environmental threat: **acid rain**. Combined with water vapor in the atmosphere, nitrogen oxides are converted to nitric acid, and sulfur dioxide is converted to sulfuric acid. Days later, and often hundreds or thousands of miles from the source, the acids fall, either dissolved in rain or as microscopic dry particles. The acids are corrosive, eating away statues and buildings (Fig. 45-15), damaging trees and crops, and turning lakes lifeless. Although both sulfuric and nitric acids form solutions in water vapor, sulfuric acid may also form particles under dry conditions. These particles, which may visibly cloud the air, are deposited even in the absence of rain, snow, or fog. **Acid deposition** is the term used to describe both the wet and dry acid assaults on the environment.

Although sources of sulfur dioxide are widely distributed, in the United States they tend to be concentrated in the upper Ohio Valley, Indiana, and Illinois, where high-sulfur coal is burned in old power plants with few emission controls. As complex chemical reactions in the atmosphere

produce sulfuric and nitric acids, winds carry them toward New England, where lakes and forests are particularly vulnerable. Some areas of the country have rock rich in calcium carbonate, a natural buffer for acidity. In New England, however, hard rock, such as granite and basalt, predominates and cannot neutralize acid rain. Although the Northeast has been the most seriously damaged, it is not the only sensitive area; about 75% of the United States is at least moderately vulnerable to acid rain (Fig. 45-16).

Acid Rain Damages Life in Lakes, Farms, and Forests

In the Adirondack Mountains of New York, acid rain has made about 25% of all the lakes and ponds too acidic to support fish. But before the fish die, much of the food web that sustains them is destroyed. Clams, snails, crayfish, and insect larvae die first, then amphibians, and finally fish. The result is a crystal clear lake, beautiful—but dead. The impact is not limited to aquatic organisms. The loss of insect larvae and crustaceans has probably contributed to a dramatic decline in the population of black ducks, which feed on them.

Acid rain is not only a threat to lakes. Recent studies under laboratory conditions have shown that it also interferes with the growth and yield of many farm crops. Acid water draining through the soil washes out essential nutrients such as calcium and potassium. Acid rain may also kill decomposer microorganisms, thus preventing the return of nutrients to the soil. Plants, poisoned and deprived of nutrients, become weak and more vulnerable to infection and insect attack.

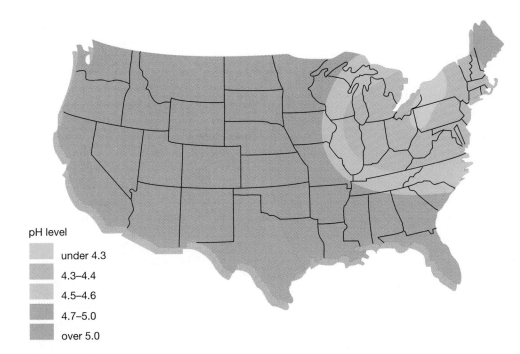

Figure 45-16   **Acid rain distribution**

The most acidic rain is concentrated in the Northeast, but thousands of lakes in the West are also vulnerable. (Source: National Acid Precipitation Assessment Program.)

pH level

- under 4.3
- 4.3–4.4
- 4.5–4.6
- 4.7–5.0
- over 5.0

High in the Green Mountains of Vermont, scientists have witnessed the death of about half of the red spruce and beech trees and one-third of the sugar maples since 1965. The snow, rain, and heavy fog that commonly cloak these eastern mountaintops is highly acidic. At a monitoring station atop Mount Mitchell in North Carolina, the pH of fog even on good days is 2.9, more acidic than vinegar. In this fragile and stressful environment, acid precipitation may well be tipping the balance against these forests (Fig. 45-17).

In the Black Forest of Germany, the devastation is even greater, with over half the trees affected. Many of the firs have turned yellow or brown, trunks have become gnarled, roots have shrunk, and plagues of insects and fungal diseases are attacking the weakened trees. German officials are convinced that acid rain from the burning of high-sulfur coal, in conjunction with other pollutants, is the cause. Vigorous pollution-control programs have begun, but many fear that the forests have been permanently damaged. In 1984, scientists confirmed that the growth rate of eastern U.S. forests had slowed dramatically over the past several decades. Germany's problems may foreshadow similar disasters for the forests of the United States.

Songbirds in the Netherlands are reproducing poorly, laying thin-shelled eggs in which develop embryos with bone deformities, evidence of calcium deficiency. Researchers have found that these birds get most of their calcium from eating snails, whose shells are a rich source of this nutrient. During the last 20 years, however, the snail populations in Dutch forests have nearly disappeared, a result, researchers believe, of acid rain. Songbird populations there may be the next to go.

Acid deposition increases the exposure of organisms to toxic metals, including aluminum, lead, mercury, and cadmium, which are extremely soluble in acidified water. Alu-

minum may inhibit plant growth. In acid lakes and streams, aluminum dissolved from the surrounding rock and soil causes mucus to accumulate in the gills of fish, suffocating them. Drinking water in some households has been found dangerously contaminated with lead dissolved by acid water from lead pipes. Boston's water supply must be chemically treated to counteract the acidity. Fish in

Figure 45-17   **Acid rain can destroy forests**

This forest atop Mount Mitchell in North Carolina was destroyed by acid deposition.

acidified water have been found to have dangerous levels of mercury in their bodies, which is subject to biological magnification as it is passed through trophic levels.

Interference with the Carbon Cycle Is Causing the Greenhouse Effect

Between 345 and 280 million years ago, under the unique conditions of the Carboniferous period, huge quantities of carbon were diverted from the carbon cycle when the bodies of plants and animals were buried in sediments, escaping decomposition. There, heat and pressure over time converted their bodies into fossil fuels such as coal, oil,

and natural gas. Without human intervention, this carbon would have remained essentially untouched for millions of years. Since the Industrial Revolution, however, modern cultures have increasingly relied on the energy stored in these fuels. As we burn them in our power plants, factories, and cars, we release CO_2 into the atmosphere. Since the start of the Industrial Revolution in 1850, the CO_2 content of the atmosphere has increased 25%. The burning of fossil fuels is the single greatest cause of this increase.

A second important source of additional CO_2 is global deforestation, the cutting of tens of millions of acres of forests each year. Deforestation is occurring principally in the tropics, where rain forests are rapidly being eliminat-

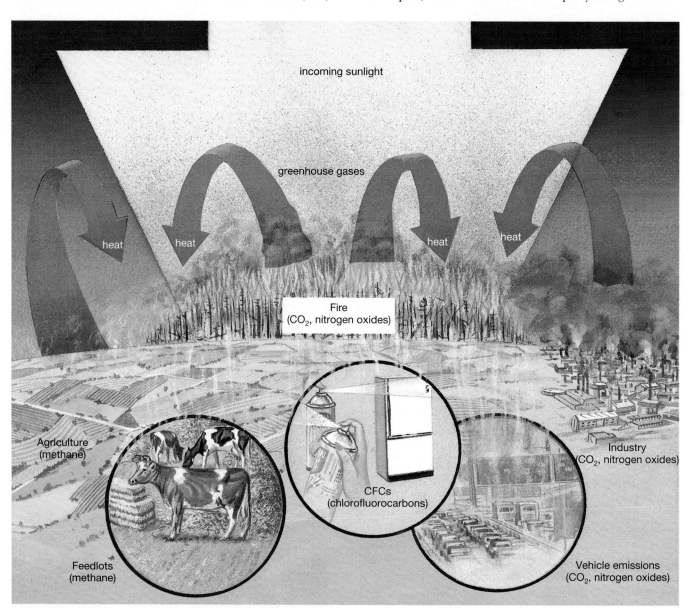

Figure 45-18   **The greenhouse effect**

Incoming sunlight warms the surface of Earth and is radiated back to the atmosphere. Greenhouse gases (such as CO_2 and nitrous oxides from combustion, chlorofluorocarbons from industrial activities, and methane from farms and feedlots) absorb some of this heat, trapping it in the atmosphere. The result is a gradual rise in average global temperatures.

ed to increase the availability of agricultural land. The carbon stored in the massive trees in these forests returns to the atmosphere after they are cut, either through burning or decomposition. During the past 200 years, human activities have released as much excess CO_2 as is currently stored in all life on Earth. Furthermore, the present CO_2 content of the atmosphere is projected to double within the next century—and possibly within the next 50 years.

Greenhouse Gases Trap Heat in the Atmosphere, with Unforeseen Consequences

Carbon dioxide still constitutes only a tiny fraction of Earth's atmosphere, about 0.035%. However, CO_2 has an important property that makes its buildup a cause for concern: It traps heat. Atmospheric CO_2 acts something like the glass in a greenhouse, allowing energy in the form of sunlight to enter but absorbing and holding that energy once it has been converted to heat (Fig. 45-18). Several other "greenhouse gases" share this property, including methane, chlorofluorocarbons, and nitrous oxide. Many interacting factors, such as changes in cloud cover caused by increased evaporation and possible increases in net primary productivity due to increased CO_2, make climate modeling difficult and uncertain. However, many scientists believe that the **greenhouse effect** is likely to cause a rise of 1.5° to 4.5° C in average global temperature by the end of the next century. Although this may not seem like much of an increase, average temperatures during the peak of the last ice age, which ended 10,000 years ago, were only about 5° C lower than at present. The temperature increase since the last ice age radically changed the species composition of forests throughout North America and caused ocean levels to rise so that the shore moved inland by about 160 kilometers (100 miles). Within the next century, Earth may be hotter than it has been at any time in the past million years, with the change occurring more rapidly than any change in Earth's history.

As the ice cap and glaciers melt in response to atmospheric warming, sea level will rise, threatening coastal cities and flooding coastal wetlands. These threatened ecosystems are the breeding grounds for many species of birds, fish, shrimp, and crabs, whose populations may be severely diminished.

Another serious consequence of the warming trend is a shift in the global distribution of temperature and rainfall. Even small temperature changes can dramatically alter the paths of major air and ocean currents, which would in turn change precipitation patterns in unpredictable ways. Some land that is currently cultivated only with the help of irrigation might become too dry for agriculture. Other areas might receive more rain. Agricultural disruption could be disastrous for human populations already on the brink of starvation.

The impact of warming on forests could be profound. The distribution of tree species is exquisitely sensitive to average annual temperature, and small changes could dra-

matically alter the extent and species composition of our forests. A recent study of eastern hemlock, yellow birch, beech, and sugar maple found that the temperature increases caused by a doubling of CO_2 concentration would result in the range of these species being shifted northward by hundreds of miles. Sugar maple could disappear from the United States, except for a remnant population in Maine. Beech, currently ranging from Georgia to southern Canada, might also be nearly eliminated from the United States by the warming climate. Although temperatures farther north would become hospitable for these trees, the ability of forests to move northward is limited by slow seed dispersal. In addition, the patterns of rainfall and the composition of the soil might prevent northern movement, causing some species to die out entirely. Meanwhile, the great forests of Mississippi and Georgia might be replaced by grassland.

Is the greenhouse effect merely a speculative future scenario? Recent NASA studies have revealed a global temperature increase of 0.5° to 0.7° C since 1860, paralleling the rise in atmospheric CO_2 and methane levels (Fig. 45-19). Much of this increase has occurred in the past decade; the 1980s included the six warmest years ever recorded, and 1990 set still another world record for high temperatures.

Although the scenario seems bleak, at the "Earth Summit" in Brazil in 1992, and again in Berlin in 1995, 155 countries pledged to curb greenhouse gas emissions, indicating global recognition and concern, the first step toward

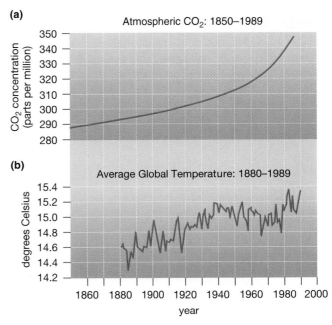

Figure 45-19   Atmospheric CO_2 and global temperatures

(a) The CO_2 concentration of the atmosphere has shown a steady increase since the Industrial Revolution. (b) Average global temperatures have also shown a gradual increase during the past century, paralleling the increasing atmospheric CO_2. (Data from NASA's Goddard Institute for Space Studies.)

change. As individuals, each of us can also make a difference. A car that gets 20 miles per gallon emits 1 pound of carbon dioxide per mile, so we can substantially reduce emissions by demanding and purchasing more fuel-efficient cars, and using car pools and public transportation. As electricity is generated in fossil fuel–fired power plants, so are tremendous quantities of carbon dioxide, sulfur dioxide, and nitrogen oxides. To conserve electricity, we can purchase more efficient appliances, replace incandescent with fluorescent light bulbs, and turn off unused lights. Insulating and weatherproofing our homes can significantly reduce fuel consumption, as will designing solar energy features into new homes. Recycling is also a tremendous energy saver—for example, 95% of the energy used to produce an aluminum can is conserved when the can is recycled.

Biodiversity and Ecosystem Stability

Biodiversity refers to the total number of different species within an ecosystem and the resulting complexity of interactions among them; in short, it defines the "richness" of an ecological community. As we discussed in Chapter 1, conservationists are increasingly concerned over the loss of biodiversity due to the expansion of the human population, and the resulting spread of agriculture and development and the extinction of species. An ecosystem with low biodiversity, such as a cornfield, supports relatively few species with few interactions among them. Adverse conditions such as drought or the invasion of a corn disease can devastate the entire ecosystem. In contrast, the diverse natural prairie that the cornfield replaced supported a variety of grasses, each tolerant of a slightly different range of conditions and resistant to different types of predators and parasites.

Although the concept that decreased diversity caused by human activities leads to a decrease in the productivity and stability of ecosystems seems logical, until recently it had never been tested in a rigorous manner. Now two different studies, one in the field and one in the laboratory, support the relationship. The field study was conducted for 11 years on over 200 small study plots within an oak-grassland ecosystem in Minnesota. Midway through the study, the area was hit with the worst drought in 50 years. Those plots with the highest biodiversity suffered the least loss of productivity and recovered faster than those with low diversity. In the controlled laboratory study, researchers in England created small artificial ecosystems with increasing numbers of organisms at four trophic levels. Again, those with the highest biodiversity had the highest productivity.

Some species have a much larger role than others in preserving the function of an ecosystem. What species are most crucial in each ecosystem? No one knows. Stanford ecologists Paul and Anne Ehrlich, in their book *Extinction*, compared the loss of biodiversity to the removal of rivets from the wing of an airplane. The rivet-removers continue to assume there are far more rivets than needed, until one day, upon takeoff, they are proved tragically wrong. As human activities drive species to extinction with little knowledge of the role each plays in the complex web of life, we run the risk of removing one rivet too many.

⊞ S U M M A R Y   O F   K E Y   C O N C E P T S

Ecosystems are sustained by a continuous flow of energy from sunlight and a constant recycling of nutrients.

The Flow of Energy

Energy enters the biotic portion of ecosystems when it is harnessed by autotrophs during photosynthesis. Net primary productivity is the amount of energy that autotrophs store in organic material in a given unit of area over a given period of time. It is extremely high in ecosystems such as estuaries and tropical rain forests, where light, nutrients, appropriate temperature, and adequate water are available. It is low in deserts and the open ocean, where the availability of some of those factors is restricted.

Trophic levels describe feeding relationships in ecosystems. Autotrophs are the producers, the lowest trophic level. Herbivores occupy the second level as primary consumers. Carnivores act as secondary consumers when they prey on herbivores or as tertiary or higher-level consumers when they eat other carnivores.

Feeding relationships in which each trophic level is represented by one organism are called food chains. In natural ecosystems, feeding relationships are far more complex and are described as food webs. Detritus feeders and decomposers, which digest dead bodies and wastes, use and release the energy stored in these substances and free nutrients for recycling.

The transfer of energy from one trophic level to the next is quite inefficient. In general, only about 10% of the energy captured by organisms at one trophic level is converted to the bodies of organisms in the next highest level. The higher the trophic level, the less energy is available to sustain it. As a result, plants are more abundant than herbivores, and herbivores are more common than carnivores. The storage of energy at each trophic level is illustrated graphically as an energy pyramid. The energy pyramid also explains biological magnification, the process by which toxic substances accumulate in increasingly high concentrations in progressively higher trophic levels.

The Cycling of Nutrients

A nutrient cycle depicts the movement of a particular nutrient from its reservoir (usually in the nonliving portion of the ecosystem) through the living portion of the ecosystem and back to its reservoir, where it is again available to the producers.

The reservoir of carbon is atmospheric carbon dioxide (CO_2). Carbon enters the producers through photosynthesis. From these autotrophs it is passed through the food web and released to the atmosphere as CO_2 during cellular respiration.

The reservoir of nitrogen is the atmosphere. Nitrogen gas is converted by bacteria and human industrial processes into ammonia and nitrates, which can be used by plants. Nitrogen passes from producers to consumers and is returned to the environment through excretion and the activities of detritus feeders and decomposers.

The reservoir of phosphorus is in rocks as phosphate, which dissolves in rainwater. Phosphate is absorbed by photosynthetic organisms, then passed through food webs. Some is excreted, and the rest is returned to the soil and water by decomposers. Some is carried to the ocean to be deposited in marine sediments. Humans mine phosphate-rich rock to produce fertilizer.

The major reservoir of water is the oceans. Water is evaporated by solar energy and returned to Earth as precipitation. The water flows into lakes, underground reservoirs, and in rivers to the ocean. Water is absorbed directly by plants and animals and is also passed through food webs. A relatively tiny amount is combined with CO_2 during photosynthesis to form sugars.

Human Intervention in Energy Flow and Nutrient Cycling

Environmental disruption occurs when people interfere with the natural functioning of ecosystems. Human industrial processes produce more nutrients than nutrient cycles can efficiently process and release toxic substances. Through massive consumption of fossil fuels, we have increased the flow of energy and disrupted the natural cycles of carbon, sulfur, and nitrogen, causing acid rain and greenhouse warming. Solutions to both acid rain and the greenhouse effect lie in reducing our use of fossil fuels and undertaking a major reforestation effort.

Biodiversity and Ecosystem Stability

Studies have shown that ecosystems with high biodiversity, or total number of different species, tend to be more productive, stable, and able to withstand disturbances than are ecosystems with lowered biodiversity due to human activities. Reduced biodiversity occurs with the spread of human agricultural-based ecosystems and with the extinction of species.

✖ K E Y   T E R M S

acid deposition   p. 906
acid rain   p. 906
atmospheric cycle   p. 900
autotroph   p. 892
biodiversity   p. 910
biogeochemical cycle   p. 900
biological magnification   p. 899
biomass   p. 893
biomass pyramid   p. 898
carnivore   p. 893
consumer   p. 893
decomposer   p. 894

denitrifying bacterium   p. 902
detritus feeder   p. 894
energy pyramid   p. 898
food chain   p. 893
food web   p. 893
fossil fuel   p. 902
greenhouse effect   p. 909
herbivore   p. 893
heterotroph   p. 893
hydrologic cycle   p. 902
legume   p. 902
macronutrient   p. 900

micronutrient   p. 900
net primary productivity   p. 893
nitrogen fixation   p. 902
nutrient cycle   p. 900
primary consumer   p. 893
producer   p. 892
quaternary consumer   p. 893
reservoir   p. 900
secondary consumer   p. 893
sedimentary cycle   p. 900
tertiary consumer   p. 893
trophic level   p. 893

✖ T H I N K I N G   T H R O U G H   T H E   C O N C E P T S

Multiple Choice

1. Why do scientists think that human-induced global warming will be more harmful to plants and animals than were past, natural climate fluctuations?
   a. because temperatures will change faster
   b. because the temperature changes will be larger
   c. because species now are less adaptable than species in the past
   d. because ecosystems are now more complicated than they used to be
   e. because the temperature changes will last longer

2. The reservoir for carbon is
   a. coal, oil, and natural gas
   b. plants
   c. CO_2 in the atmosphere
   d. methane (CH_4) in the atmosphere
   e. all of the above

3. Why would you expect there to be a smaller biomass of big predators (lions, leopards, hunting dogs, etc.) than of grazing mammals (gazelles, zebra, elephants, etc.) in the African savanna?
   a. too little cover for the predators to hide in
   b. the inefficiency of energy transfer between trophic levels
   c. like domestic cats, large predators are susceptible to hair balls
   d. many predators occur in social groups, thus limiting their numbers
   e. because grazers are better adapted to moving long distances, they can better follow the rains

4. What is the one group of organisms that is able to fix atmospheric nitrogen into forms usable by living organisms?
   a. plants   b. fungi
   c. insects   d. bacteria
   e. viruses

5. The biological process by which carbon is returned to its reservoir is
   a. photosynthesis   b. denitrification
   c. deamination   d. glycolysis
   e. cellular respiration

6. As a black widow spider consumes her mate, what is the *lowest* trophic level she could be occupying?
   a. the 3rd trophic level
   b. the 1st trophic level
   c. the 2nd trophic level
   d. the 4th trophic level
   e. the 5th trophic level

Review Questions

1. What makes the flow of energy through ecosystems fundamentally different from the flow of nutrients?

2. What is an autotroph? What trophic level does it occupy, and what is its importance in ecosystems?

3. Define primary productivity. Based on information presented here and in Chapter 44, would you predict higher productivity in a farm pond or an alpine lake? Defend your answer.

4. List the first three trophic levels. Among the consumers, which are most abundant? Why must there always be a greater biomass of plants than herbivores in any ecosystem? Relate your answer to the "10% law."

5. How do food chains and food webs differ? Which is the most accurate representation of actual feeding relationships in ecosystems?

6. Define detritus feeders and decomposers and explain their importance in ecosystems.

7. Trace the movement of carbon from its reservoir through the biotic community and back to the reservoir. How have human activities altered the carbon cycle, and what are the implications for future climate?

8. Explain how nitrogen gets from the air to a plant.

9. Trace a phosphorus molecule from a phosphate-rich rock into the DNA of a carnivore. What makes the phosphorus cycle fundamentally different from the carbon and nitrogen cycles?

10. Trace the movement of a water molecule from the moment it leaves the ocean until it eventually reaches a plant root, then a plant stoma, and then makes its way back to the ocean.

❖ APPLYING THE CONCEPTS

1. Introduced species often have their most devastating impacts on oceanic islands, such as the Hawaiian or Galapagos Islands. These islands frequently have very simple food webs, with few or no species in the highest trophic levels. Why would you expect species that naturally occur in simple food webs to be especially susceptible to introduced predators and herbivores?

2. What could your college or university do to reduce its contribution to acid rain and the greenhouse effect? Be specific, and if possible, offer practical alternatives to current practices.

3. Biological magnification has serious consequences not just for the species it directly hurts, such as predatory birds, but for other species in a community as well, because predators often control the populations of their prey (see Chapter 44). Consider the food web in Figure 45-5. If biological magnification of a pesticide killed all the predators in this community, what would the new food web look like? What would happen to competition between the remaining animal species? What effect might elimination of the predators have on the plant species?

4. Relate fossil fuel consumption to (a) the decline of forests in the Northeast and (b) the loss of aquatic life in lakes in the Northeast and Canada. Trace each step from the burning of gasoline in a car or power plant to the death of organisms.

5. Discuss the contribution of population growth and of consumption to (a) acid rain and (b) the greenhouse effect. How do population growth and consumption interact in developed and less-developed countries?

6. The disposal of household garbage is a major problem for American cities, with landfills overflowing and no place to start new ones. This problem of household waste could best be solved by decomposers and detritus feeders! But most of these species require oxygen to live, and deep landfills are usually anoxic (oxygen-free). Suggest a way of disposing of food, lawn, and paper wastes that would allow cities to take advantage of natural decomposition. What would be the best use of the decomposed remains of garbage?

7. Many wastes that humans produce cannot be processed by decomposers and detritus feeders. Give examples of some of these wastes and suggest alternatives to dumping them in landfills. Do any of your alternatives resemble (in a general way) processes that occur in natural ecosystems? Explain.

❖ FOR MORE INFORMATION

Bazzaz, F. A, and Fajer, E. D. "Plant Life in a CO$_2$-Rich World." *Scientific American*, January 1992. How increasing levels of CO$_2$ may alter the structure and function of ecosystems.

Cohn, J. P. "Restoring the Everglades." *BioScience* 44:579–583, 1994. Proposed changes in the ecosystem-wide water cycling of the Everglades could restore ecological health.

Graedel, T. E., and Crutzen, P. J. "The Changing Atmosphere." *Scientific American*, September 1989. Describes the harmful effects of human activities on the chemistry of the atmosphere.

Houghton, R. A., and Woodwell, G. M. "Global Climate Change." *Scientific American*, April 1989. An authoritative overview of the evidence that the greenhouse effect is altering the global climate.

Hutchinson, G. E. "The Biosphere." *Scientific American*, September 1970 (Offprint No. 11). Overview of Earth's thin film of life and how it is supported by the flow of energy and cycling of nutrients.

Matthews, S. W. "Under the Sun." *National Geographic*, October 1990. A comprehensive, readable exploration of the greenhouse effect, its causes, and its consequences.

Ojima, D. S., Galvin, K. A., and Turner, B. L., II. "The Global Impact of Land-Use Change." *BioScience* 44:300–304, 1994. An argument that to understand the effects of global change we must also consider social factors influencing human impacts. This article is part of a special feature of seven articles, all dealing with global climate change and its effects on ecosystems around the world.

Raloff, J. "EcoCancers: Do Environmental Factors Underlie a Breast Cancer Epidemic?" *Science News* 144:10–13, 1993. There is emerging evidence that some industrial pollutants are not only toxins but can also mimic vertebrate hormones, making their biological magnification especially worrisome.

Schneider, S. H. "The Changing Climate." *Scientific American*, September 1989. Discussion of how rising CO$_2$ levels threaten to dramatically change Earth's climate.

NET WATCH

On-line resources for this chapter are on the World Wide Web at:
http://www.prenhall.com/~audesirk (click on the <u>table of contents</u> link and then select Chapter 45).

46 Earth's Diverse Ecosystems

*"Living creatures press up against all barriers, they fill every possible niche all the world over. . . .
We see life persistent and intrusive—spreading everywhere, insinuating itself, adapting itself,
resisting everything, defying everything, surviving everything!"*

SIR JOHN ARTHUR THOMPSON, a Scottish biologist (1920)

⧉ A T   A   G L A N C E

Climate
    Both Climate and Weather Are Driven by the Sun
    Many Physical Factors Also Influence Climate

The Requirements of Life

Life on Land
    Terrestrial Biomes Support Characteristic Plant Communities
    Rainfall, Temperature, and Vegetation

Life in Water
    Freshwater Lakes Have Distinct Regions of Life and
       Temperature
    Freshwater Lakes Are Classified According to Their
       Nutrient Content
    Marine Ecosystems Cover Much of Earth

Rain forest surrounds the Segama River in Borneo.

In preceding chapters, we discussed the dynamics of population growth, community interactions, the flow of energy, and the cycling of nutrients, as well as the basic structure of ecosystems within which those processes occur. Ecosystems are extraordinarily diverse, yet clear patterns exist within this diversity. If you travel around the world, you may observe that, although the plant *species* may differ, very similar groups of plants are found wherever a particular climate exists. Deciduous forests, deserts, grasslands, tropical rain forests, coniferous forests, coral reefs, and estuaries all have distinguishing features that identify them no matter where they are found or what particular species inhabit them.

The reason we are able to place these communities into distinct categories is that each is dominated by organisms that are specifically adapted to the conditions of a particular environment. Variations in the availability of light, water, and in the range of temperatures shape the characteristic groupings of organisms that define the different types of ecosystems. The desert community, for example, is dominated by plants adapted to heat and drought. The cacti of the American Mojave Desert are strikingly similar to the euphorbia of South Africa, although these plants are in separate families and are only distantly genetically related. Their spinelike leaves and thick, green, water-storing stems are adaptations for water conservation (Fig. 46-1). Likewise, the plants of the arctic tundra and those of the alpine tundra of the Rocky Mountains show growth patterns clearly recognizable as adaptations to a cold, dry, windy climate. Thus, in regions of Earth where the environmental conditions are similar, we find similar types of organisms organized into similar types of communities.

Environments vary in the relative abundance of four basic resources that provide the requirements for life: nutrients, energy, water, and temperatures compatible with metabolic reactions. The availability of these basic resources shapes the different ecosystems that are the subject of this chapter.

Climate

Life on Earth, particularly life on land, is dramatically affected by both weather and climate. **Weather** refers to short-term fluctuations in temperature, humidity, cloud cover, wind, and precipitation over periods of hours or days.

Figure 46-1
Environmental demands mold physical characteristics
Evolution in response to similar environments has molded the bodies of **(a)** American cacti and **(b)** South African euphorbia into nearly identical shapes, although they are in completely different families. Both of these desert plants have thick, water-conserving stems, and leaves reduced to defensive spines that minimize evaporation.

Climate, in contrast, refers to overall patterns of weather that prevail from year to year and even century to century in a particular region. The amount of sunlight, water, and the range of temperatures determine the climate of a given region. Whereas weather affects individual organisms, climate influences and limits the overall distribution of entire species.

Both Climate and Weather Are Driven by the Sun

Both climate and weather are driven by a great thermonuclear engine: the sun. Solar energy reaches Earth in a range of wavelengths, from short, high-energy ultraviolet rays, through visible light, to the longer infrared wavelengths that produce heat. The solar energy that reaches Earth drives the wind, the ocean currents, and the global water cycle. Before it reaches Earth's surface, however, sunlight is modified by the atmosphere. A layer relatively rich in ozone (O_3) is located high in the upper atmosphere. This ozone layer absorbs much of the high-energy ultraviolet radiation from the sun that can damage biological molecules (see "Earth Watch: The Ozone Layer—We Have Punctured Our Protective Shield"). Dust, water vapor, and clouds scatter light, reflecting some of the energy back into space. Carbon dioxide, water vapor, ozone, methane, and other gases selectively absorb infrared wavelengths (heat), trapping warmth in the atmosphere. Human activities have intensified this natural greenhouse effect, as described in Chapter 45.

Only about half the solar energy reaching the atmosphere actually strikes Earth's surface. Of this, small fractions are immediately reflected directly back into space or used in photosynthesis, while the rest is absorbed as heat. Eventually, nearly all the incoming solar energy is sent back into space,

either as light or infrared radiation (heat). The solar energy temporarily absorbed and stored as heat by the atmosphere and Earth's surface maintains Earth's relative warmth.

Many Physical Factors Also Influence Climate

Many factors influence climate. Among the most important are latitude, air currents, ocean currents, and the presence of irregularly shaped continents.

Latitude Influences the Angle at Which Sunlight Strikes Earth

Latitude, measured in degrees, is the distance north or south from the equator, which is located at 0° latitude. The amount of sunlight that strikes a given area of Earth's surface has a major effect on average yearly temperatures. At the equator, sunlight hits Earth's surface nearly at a right angle, making the weather there consistently warm. Farther north or south, the sun's rays strike Earth's surface at a greater slant. This angle spreads the same amount of sunlight over a larger area, producing lower overall temperatures (Fig. 46-2).

Because Earth is tilted on its axis as it makes its yearly trip around the sun, the higher latitudes experience considerable variation in the directness of sunlight throughout the year, resulting in pronounced seasons. For example, in summer, the Northern Hemisphere is tilted toward the sun and thus receives sunlight more directly and experiences warmer weather; in our winter, the Southern Hemisphere is closer, as illustrated in Figure 46-2. The tilting hardly affects the angle of the equator to the sun's rays, so there is little seasonal variation at this latitude.

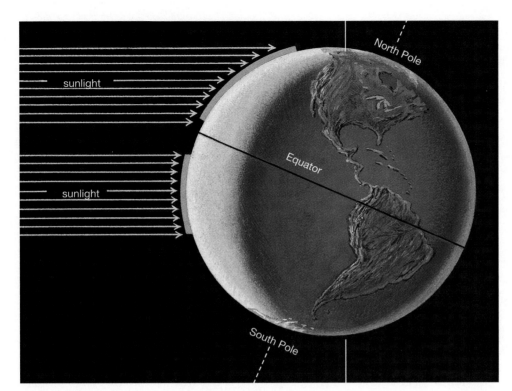

Figure 46-2   Earth's curvature and tilt produce seasons and climate

Seasons and climate are the result of an uneven distribution of solar energy. Near the equator, sunlight falls perpendicularly on Earth's surface, so its warmth is concentrated onto a relatively small area. Toward the poles, the same amount of sunlight falls over a much larger surface area. Temperatures are thus highest at the equator and lowest at the poles. The tilt of Earth on its axis causes seasonal variations in the directness of sunlight. In this figure, the Southern Hemisphere, which is tilted toward the sun, is receiving more direct sunlight than is the Northern Hemisphere, which is tilted away from the sun. Thus it is winter in the Northern Hemisphere and summer in the Southern Hemisphere.

Air Currents

Air currents are generated by Earth's rotation and differences in temperature between different masses of air. Because warm air is less dense than cold air, as the sun's direct rays fall on the equator, heated air rises. The warm air of the tropics is also laden with water evaporated by solar heat (Fig. 46-3a). As the water-saturated air rises, it cools somewhat. Because cool air cannot retain as much moisture as warm air, water condenses from the rising air and falls as rain. The direct rays of the sun and the rainfall produced when warm, moist air rises and is cooled create a band around the equator, called the tropics, which is both the warmest and the wettest region on Earth. The cooler dry air then flows north and south from the equator. At around 30° north and south latitudes, the cooled air begins to sink. As it sinks, the air is warmed by heat radiated from Earth. By the time it reaches the surface, it is both warm and very dry. Not surprisingly, the major deserts of the world are found at these latitudes (Fig. 46-3a, b). This air then flows back toward the equator. Farther north and south, this general circulation pattern is repeated, dropping moisture at around 60° north and south latitudes and creating extremely dry conditions at the North and South poles.

Ocean Currents Moderate Near-Shore Climates

Ocean currents are driven by Earth's rotation, the wind, and direct heating of the water by the sun. Continents interrupt the currents, breaking them into roughly circular patterns called **gyres,** which rotate clockwise in the Northern Hemisphere and counterclockwise in the Southern Hemisphere

(Fig. 46-4). Because water both heats and cools more slowly than land or air, ocean currents tend to moderate temperature extremes. Coastal areas, then, generally have less-variable climates than areas near the center of continents. For example, a gyre in the Atlantic Ocean brings warm water from equatorial regions north along the eastern coast of North America, creating a warmer, moister climate than is found farther inland. It then carries the still-warm water farther north and east, warming the western coast of Europe before returning south.

Continents and Mountains Complicate Weather and Climate

If Earth's surface were uniform, we would find climate zones occurring in bands corresponding to latitude. These zones would result from the interaction of temperature and rainfall, which are determined by the rise and fall of air masses (see Fig. 46-3a). The presence of irregularly shaped continents (which heat and cool relatively quickly) and oceans (which heat and cool more slowly) alters the flow of wind and water and contributes to the irregular distribution of ecosystems (see Fig. 46-7).

Variations in elevation within continents further complicate the picture. With increasing elevation, the atmosphere becomes thinner and retains less heat. The temperature drops approximately 2° C (3.5° F) for every 305 meters (1000 feet) elevation gain, so that snow-capped mountains may be found even in the tropics (Fig. 46-5).

Mountains also modify rainfall patterns. When water-laden air is forced to rise as it meets a mountain, it cools. Cooling reduces the air's ability to retain water, and the water

(a)

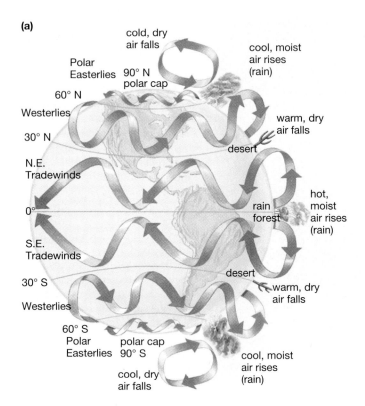

(b)

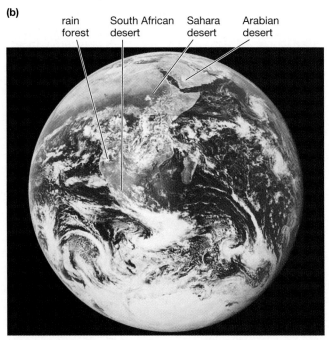

Figure 46-3   **The distribution of rainfall**

(a) Rainfall is determined primarily by the distribution of temperatures combined with Earth's rotation. The interaction of these two factors creates air currents that rise and fall predictably with latitude, producing broad general climatic regions. **(b)** Some of these regions are visible in this photograph of the African continent taken by *Apollo 11* astronauts. Along the equator, there are heavy clouds dropping moisture on the central African rain forests. Note the lack of clouds over the Sahara and Arabian deserts near 30° north latitude, and the South African desert near 30° south.

condenses as rain or snow on the windward (near) side of the mountain. The cool, dry air is warmed again as it travels down the far side of the mountain and absorbs water from the land, creating a local dry area called a **rain shadow.** For example, mountain ranges such as the Sierra Nevada of the western United States wring the moisture from the westerly winds coming off the Pacific Ocean, leaving deserts in the rain shadow on their eastern sides (Fig. 46-6).

The Requirements of Life

From the lichens on bare rock to the thermophilic (Greek for "heat-loving") algae in the hot springs of Yellowstone National Park to the bacteria thriving under the pressure-cooker conditions of a deep-sea vent, Earth teems with life. Underlying the diversity of habitats is the common ability to provide to varying degrees the four fundamental resources required for life on Earth. These resources are:

1. Nutrients from which to construct living tissue
2. Energy to power that construction
3. Liquid water to serve as a medium in which metabolic reactions occur
4. Appropriate temperatures in which to carry out these processes

As we shall see in the following sections, these resources are very unevenly distributed. Their availability limits the types of organisms that can exist within the various terrestrial and aquatic ecosystems on Earth.

Life on Land

Terrestrial organisms are restricted in their distribution largely by the availability of water and appropriate temperatures. Terrestrial ecosystems receive plenty of light, even on an overcast day, and the soil provides abundant nutrients. Water, however, is limited and very unevenly distributed, both in place and in time. Terrestrial organisms must be adapted to obtain water when it is available and conserve it when it is scarce.

Like water, favorable temperatures are very unevenly distributed in place and time. At the South Pole, even in summer, the average temperature is well below freezing, and (not surprisingly) life is scarce. Other places, such as central Alaska, have favorable temperatures only during certain seasons of the year, while the tropics have a uniformly warm, moist climate, and life abounds.

Terrestrial Biomes Support Characteristic Plant Communities

Terrestrial communities are dominated and defined by their plant life. Because plants can't escape from drought, sun, or winter weather, they tend to be precisely adapted

Figure 46-4 **Ocean circulation patterns are called gyres**

Gyres travel clockwise in the Northern Hemisphere and counterclockwise in the Southern Hemisphere. These currents tend to distribute warmth from the equator to northern and southern coastal areas.

to the climate of a particular region. Large land areas with similar environmental conditions and characteristic plant communities are called **biomes** (Fig. 46-7, p. 922). Biomes are usually named after the major type of vegetation found there. The predominant vegetation of each biome is determined by the complex interplay of rainfall and temperature (Fig. 46-8, p. 923). These factors determine the available soil moisture needed for plant growth as well as for compensation for water losses through evaporation. In addition to the total amount of rainfall and the overall yearly average temperature, the *variability* of rain and temperature over the year also determines which plants can grow in an area. Arctic tundra plants, for example, must be adapted to survive marshy conditions in the early summer but cold desertlike conditions for much of the rest of the year, when water is solidly frozen and unavailable. In the following sections we discuss the major biomes, beginning at the equator and working our way northward. We also discuss some of the effects human activities have on these biomes.

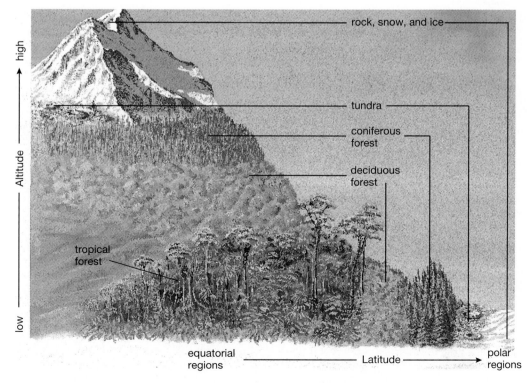

Figure 46-5 **Effects of elevation on temperature**

Climbing a mountain in some ways is like going north; in both cases, increasingly cool temperatures produce a similar series of biomes.

rock, snow, and ice

tundra

coniferous forest

deciduous forest

tropical forest

high

Altitude

low

equatorial regions ————— Latitude ————→ polar regions

EARTH WATCH

The Ozone Layer—We Have Punctured Our Protective Shield

A small fraction of the radiant energy produced by the sun, called ultraviolet, or UV, radiation, is so highly energetic that it can damage biological molecules. In small quantities, UV radiation helps human skin to produce vitamin D and causes tanning in fair-skinned people. But in larger doses, UV causes sunburn and premature aging of skin, skin cancer, and cataracts, a condition in which the lens of the eye becomes cloudy.

Fortunately, most of the UV radiation is filtered out by ozone in the stratosphere, a layer of atmosphere extending from 10 to 50 kilometers (6 to 30 miles) above Earth. In pure form, ozone (O_3) is a bluish, explosive, and highly poisonous gas. In the stratosphere, the normal concentration of ozone is about 0.1 part per million, compared with 0.02 part per million in the lower atmosphere. This ozone-enriched layer is called the **ozone layer.** Ultraviolet light striking ozone and oxygen causes reactions that break down as well as regenerate ozone. In the process, the UV radiation is converted to heat, and the overall level of ozone remains constant—normally.

In 1985, British atmospheric scientists published a startling discovery: The springtime levels of stratospheric ozone over Antarctica had declined by over 40% since 1977. A hole had been punctured in Earth's protective shield. In the localized "ozone hole" over Antarctica, ozone now dips to about one-third of its predepletion levels, a decline of 66% (Fig. E46-1). Although depletion of the ozone layer is severest over Antarctica, the ozone layer is somewhat reduced over most of the world, including nearly all of the continental United States. Parts of the Northern Hemisphere experience up to 6% reductions at some times of the year. Evidence of ozone thinning over the Arctic has led researchers to predict that we will soon have an Arctic, as well as an Antarctic, ozone hole.

On the basis of increased UV exposure due to ozone loss, the Environmental Protection Agency increased its estimates of skin cancer deaths over the next 50 years from 9300 up to 200,000. Studies using mice show that exposure to UV light in dosages similar to a day at the beach without sunscreen can depress their immune systems, making them more vulnerable to a variety of diseases. Human health effects are only one cause for concern. Photosynthesis by phytoplankton, the producers for marine ecosystems, is reduced under the ozone hole above Antarctica. Some types of trees and farm crops are also harmed by increased UV radiation.

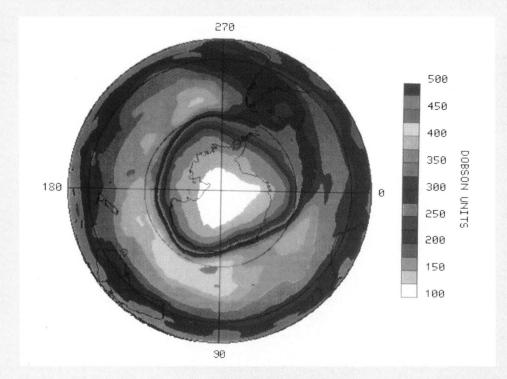

Figure E46-1 **Satellite image of the Antarctic ozone hole**

The ozone hole is seen in white and pink on this image recorded by a satellite in October 1993, when ozone levels dipped to less than one-third of normal. Dobson units measure the thickness of the ozone (in hundredths of millimeters) in a given layer of atmosphere if it were purified and held at standard temperature and pressure.

The decline in the thickness of the ozone layer is caused by increasing levels of chlorofluorocarbons (CFCs), which contain chlorine, fluorine, and carbon. Developed about 60 years ago, these gases were widely used as coolants in refrigerators and air conditioners, aerosol spray propellants, agents for producing polystyrene (Styrofoam), and cleansers for electronic parts. These chemicals are very stable and were considered safe. Their stability, however, proved to be a major problem, because they remain chemically unchanged as they gradually rise into the stratosphere. There, under intense bombardment by UV light, the CFCs break down, releasing chlorine atoms. Chlorine catalyzes the breakdown of ozone to oxygen gas (O_2) while remaining unchanged itself. Over the Arctic and Antarctic regions, the clouds are composed of ice particles that provide a surface on which the reaction can occur.

Although CFC production has not been halted, in a series of treaties beginning in 1987, industrialized nations throughout the world agreed to gradually phase out ozone-depleting chemicals, with the goal of eliminating their use entirely by 2030. Atmospheric CFCs are still increasing, but as a result of these treaties their *rate* of increase has dropped considerably. Because these highly stable compounds can persist 50 to 100 years and take decades to rise into the stratosphere, the millions of tons of CFCs already released will continue to erode our protective ozone shield, with possibly severe consequences.

Tropical Rain Forests

Large areas of South America and Africa lie along the equator. Here the temperature averages between 25° and 30° C (77° and 86° F) with little variation, and rainfall ranges from 250 to 400 centimeters (100 to 160 inches) each year. These evenly warm, evenly moist conditions combine to create the most diverse biome on Earth, the **tropical rain forest,** dominated by huge broadleaf evergreen trees (Fig. 46-9, p. 924).

Although rain forests cover only 6% of Earth's total land area, they are believed to be home to between 5 and 8 million species, representing half to two-thirds of the world's total. A recent survey of a 2.5-acre site in the upper Amazon basin revealed 283 species of trees, most of which were represented by a single individual.

Tropical rain forests typically have several layers of vegetation. The tallest trees reach 50 meters (150 feet) and tower above the rest of the forest. Below is a fairly continuous canopy of treetops at about 30 to 40 meters (90 to 120 feet). Below these is often another layer of shorter trees. Huge woody vines, often 100 meters (328 feet) or more in length, grow up the trees, reaching the sunlight far above. These layers of vegetation block out most of the sunlight. Plants that live in the dim green light that filters through to the forest floor often have enormous leaves to trap the little available energy.

Because edible plant material close to the ground is relatively scarce, much of the animal life in tropical rain forests is arboreal (living in the trees), including numerous birds, monkeys, and insects. Competition for the nutrients that do reach the ground is intense, among both

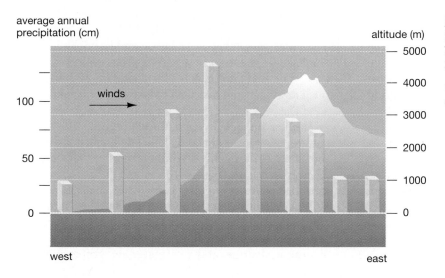

Figure 46-6   The rain shadow of the Sierra Nevada

Here, rainfall is graphed against altitude. Westerly (eastward-moving) winds deposit their moisture on the western slope, leaving a desert to the east.

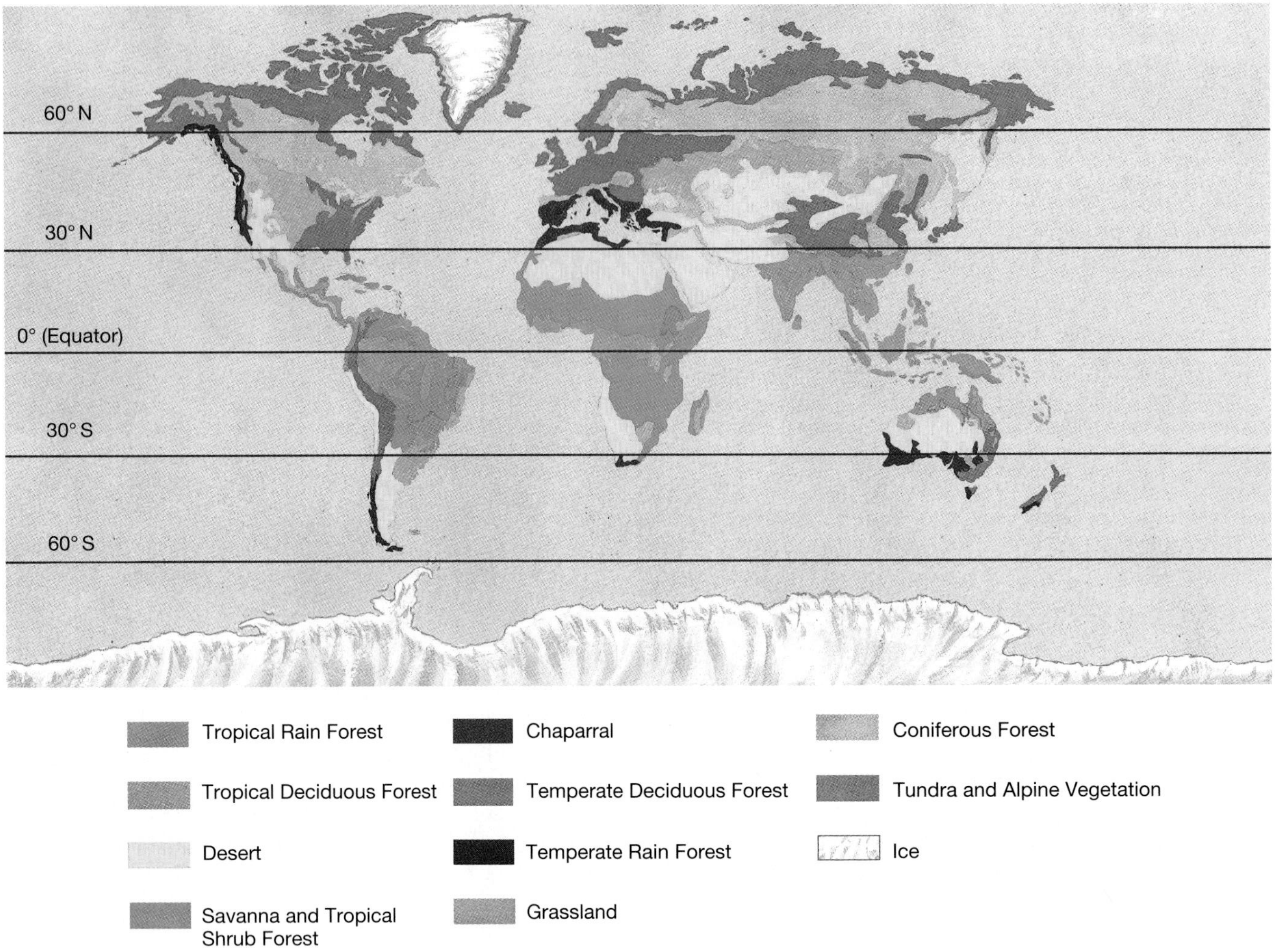
60° N
30° N
0° (Equator)
30° S
60° S
Tropical Rain Forest
Tropical Deciduous Forest
Desert
Savanna and Tropical Shrub Forest
Chaparral
Temperate Deciduous Forest
Temperate Rain Forest
Grassland
Coniferous Forest
Tundra and Alpine Vegetation
Ice

***Figure 46-7* The distribution of biomes**

Although mountain ranges and the sheer size of the continents complicate their pattern, note the overall consistencies. Tundra and coniferous forest always occur in the northernmost parts of the Northern Hemisphere, while the deserts of Mexico, the Sahara, Saudi Arabia, South Africa, and Australia are located around 20° to 30° north and south latitude.

animals and plants. Even such unlikely sources of food as the droppings of monkeys are in great demand. For example, dung beetles feed and lay their eggs on monkey droppings. When ecologists attempted to collect droppings of the South American howler monkeys to find out what the monkeys had been eating, they found themselves in a race with the beetles, hundreds of which would arrive within minutes after a dropping hit the ground!

Almost as soon as bacteria or fungi release any nutrients from dead plants or animals into the soil, rainforest trees and vines absorb the nutrients. This is one of the reasons why, despite the teeming vegetation, agriculture is very risky and usually destructive in rain forests. Virtually all the nutrients in a rain forest are tied up in the vegetation, so the soil is very infertile. If extensive land is cut down and the trees are carried away for lumber, few nutrients remain to support crops. Further, even if the nutrients are released by burning the vegetation, the heavy year-round rainfall quickly dissolves and erodes them away, leaving the soil infertile after a few seasons of cultivation. The exposed soil, which is rich in iron and aluminum, then takes on an impenetrable, bricklike quality as it bakes in the tropical sun. As a result, secondary succession on cleared rainforest land is slowed significantly. Even small forest cuttings take about 70 years to regenerate.

Human Impact In spite of their unsuitability for agriculture, rain forests are being felled for lumber and agriculture at a rate of roughly 38 million acres per year, or about *75 acres per minute* (Fig. 46-10). About 40% of the original rain forests are now gone. Harvard biologist Edward O. Wilson estimates that the destruction of tropical rain forests may be driving 27,000 species to extinction *each year*. With the loss of biodiversity, humanity loses access to a wealth of potential drugs and raw materials. For example, in

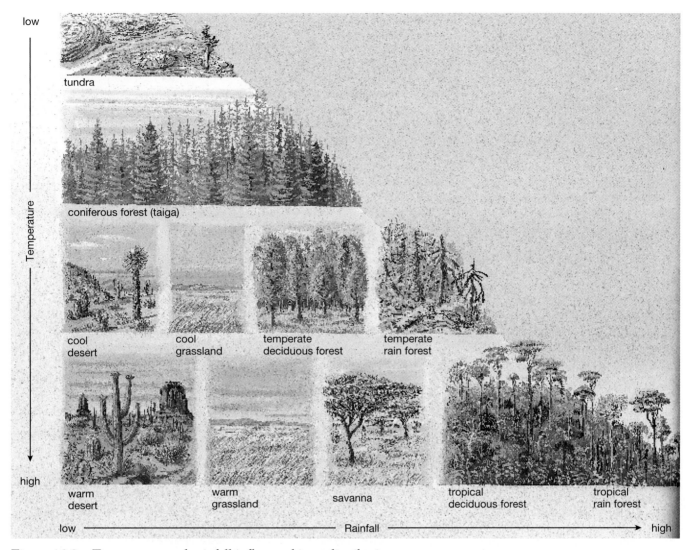

Figure 46-8   **Temperature and rainfall influence biome distribution**

Together, these factors determine the available soil moisture needed for plant growth.

Malaysia, the sap of a rare species of tree was discovered to yield a compound called calanolide A, which, in test tubes, completely blocks replication of the HIV virus. The removal of these forests is also contributing significantly to the buildup of carbon dioxide in the atmosphere, making the greenhouse effect more severe. It is estimated that one-quarter of the carbon released into the atmosphere during the past decade came from the loss of tropical rain forests. In West Africa, burning the forests is also causing acid rain, which is damaging the remaining trees.

Although disastrous losses are continuing, recognition of the problem is growing. Some areas have been set aside as protected preserves, and some reforestation efforts are under way. Local residents are becoming more involved in conservation efforts. In Brazil, a union of people who collect natural rubber from tree sap is fighting to preserve large tracts of land open only to rubber production and to the harvesting of fruits and nuts. This approach suggests the ultimate solution, which is tragically slow in coming—sustainable use,

the harvesting of products without permanently damaging the trees or the ecosystem they support.

Tropical Deciduous Forests

Slightly farther away from the equator, the rainfall is not nearly as constant, and there are pronounced wet and dry seasons. In these areas, which include India, much of Southeast Asia, and parts of South and Central America, **tropical deciduous forests** grow. During the dry season, the trees cannot get enough water from the soil to compensate for evaporation from their leaves. As a result, the plants have adapted to the dry season by shedding their leaves, thereby minimizing water loss. If the rains fail to return on schedule, the trees delay forming new leaves until the drought passes.

Savanna

Along the edges of the tropical deciduous forest, the trees gradually become more widely spaced, with grass-

(a)

(b)

(c)

(d)

Figure 46-9   **The tropical rain forest**

Towering trees draped with vines reach for the light in the dense tropical rain forest. Because food and light energy are found high in trees, amid their branches dwells the most diverse assortment of animals on Earth, including **(a)** a fruit-eating toucan, **(b)** a shimmering butterfly, **(c)** a tree-climbing orchid, and **(d)** a howler monkey.

es growing between them. Eventually, grasses become the dominant vegetation, with only scattered trees and thorny scrub forests here and there; this is the **savanna** (Fig. 46-11). Savanna grasslands typically have a rainy season during which virtually all of the year's precipitation falls—30 centimeters (12 inches) or less. When the dry season arrives, it comes with a vengeance. No rain may fall for months, and the soil becomes hard, dry, and dusty. Grasses are well adapted to this type of climate, growing very rapidly during the rainy season and dying back to drought-resistant roots during the dry times.

Only a few specialized trees, such as the thorny acacia or the water-storing baobab, can survive the devastating savanna dry seasons. In areas in which the dry season becomes even more pronounced, virtually no trees at all can grow, and the savanna imperceptibly grades into tropical grassland.

The African savanna has probably the most diverse and impressive array of large mammals on Earth, including numerous herbivores, such as antelope, wildebeest, buffalo, elephants, and giraffes, and carnivores, such as the lion, leopard, hyena, and wild dog.

Figure 46-10   **Fire engulfs a tropical rain forest in Brazil**

The cleared area will be converted to ranching or agriculture, doomed to failure because of the poor quality of the soil.

Figure 46-11   **The African savanna**

Elephants roam while a cheetah feasts on its prey **(a)**. Large herds of grazing animals, such as zebras **(c)** can still be seen on African preserves. The herds of herbivores provide food for the greatest assortment of large carnivores on the planet, including lions **(b)**, leopards, cheetahs, hyenas, and wild dogs.

Figure 46-12   Poaching threatens African wildlife
Rhinoceros horns, believed by some to have aphrodisiac properties, fetch staggering prices and encourage poaching. The black rhino is now nearly extinct.

Human Impact Africa's rapidly expanding human population threatens the wildlife of the savanna. Poaching has driven the black rhinoceros to the brink of extinction (Fig. 46-12) and also endangers the African elephant. The abundant grasses that make the savanna a suitable habitat for so much wildlife also make it suitable for grazing domestic cattle. As the human population of East Africa increases, so does the pressure of cattle grazing upon the savanna. Fences increasingly disrupt the migration of the great herds of herbivores in search of food and water. Ecologists have discovered that the native herbivores are much more efficient at converting grass into meat than are cattle. Perhaps the future African savanna may support herds of domesticated antelope and other large native grazers in place of cattle.

Deserts

Even drought-resistant grasses need at least 25 to 50 centimeters (10 to 20 inches) of rain a year, depending on its seasonal distribution and the average temperature. When less rain than this falls, many grasses fail and **desert** biomes occur. These biomes are found on every continent, often around 20° to 30° north and south latitude, and also in the rain shadows of major mountain ranges. As with all biomes, the designation *desert* includes a variety of environments. At one extreme are certain areas of the Sahara or Chile, where it virtually never rains and there is no vegetation at all (Fig. 46-13a). The more common deserts, however, are characterized by widely spaced vegetation and large areas of bare ground. The plants are often spaced evenly, as if planted by hand (Fig. 46-13b). Frequently, the perennial plants are bushes or cacti with large, shallow root systems. The shallow roots quickly soak up

the soil moisture after the infrequent desert storms, while the rest of the plant is typically covered with a waterproof, waxy coating to prevent evaporation of precious water. Water is stored in the thick stems of cacti and other succulents (Fig. 46-13c). The spines of cacti are leaves modified for protection and water conservation, presenting almost no surface area for evaporation. In many deserts, all the rain falls in just a few storms, and specialized annual wildflowers take advantage of the brief period of moisture to race through germination, growth, flowering, and seed production in a month or less (Fig. 46-14).

The animals of the deserts, like the plants, are specially adapted to survive on little water. Most deserts appear to be almost completely devoid of animal life during the day, because the animals seek relief from the sun and heat in cool underground burrows. After dark, when the deserts cool down considerably, horned lizards, snakes, and other reptiles emerge to feed, as do mammals, such as the kangaroo rat (see Fig. 46-13d), and birds, such as the burrowing owl. Most of the smaller animals survive without ever drinking at all, getting all the water they need from their food and that produced during cellular respiration in their tissues. Larger animals, such as desert bighorn sheep, are dependent on permanent water holes during the driest times of the year.

Human Impact While human activities are reducing the extent of many biomes, they are causing the spread of deserts, a process call **desertification.** A dramatic example is occurring in the Sahel, which borders the southern edge of the Sahara desert in Africa. Twenty-five years of below-average rainfall, coupled with rapid growth of the human population, have caused a steady southward spread of the desert. Western technology allowed the Sahelians to drill deep wells, abandon their nomadic way of life, and increase their livestock herds. The natural vegetation was destroyed as livestock competed for forage during the drought (Fig. 46-15). Native shrubs were uprooted by goats, preventing regeneration. The growing human population began cultivating increasingly fragile land without allowing time for the soil to recover between crops. As the desert spread southward, so did the people, with their livestock and crops, causing further deforestation, soil destruction, and desertification of once-productive land. The Sahel is an example of a human population's exceeding the carrying capacity of the land. The loss of productivity of the ecosystem is nearly irreversible, and massive famines, such as occurred in Ethiopia in the mid-1980s, are a tragic result.

Chaparral

In many coastal regions that border on deserts, such as southern California and much of the Mediterranean, a unique type of vegetation grows, called **chaparral.** The annual rainfall in these regions is similar to that of a desert, but the proximity of the sea provides a slightly longer rainy season in the winter and frequent fogs during the spring and fall, which re-

Figure 46-13   **Desert biomes**

(a) Under the most extreme conditions of heat and drought, deserts can be almost devoid of life, such as these sand dunes of the Sahara desert in Africa. **(b)** Throughout much of Utah and Nevada, the Great Basin Desert presents a monotonous landscape of widely spaced shrubs, such as sagebrush and greasewood. These shrubs often secrete a growth inhibitor from their roots, preventing germination of nearby plants and thus reducing competition for water. **(c)** The Sonoran Desert in southern Arizona and northern Mexico is typified by the saguaro cactus, prickly pear, and the spindly ocotillo, whose red flowers are pollinated by hummingbirds. The body of the saguaro is pleated and can expand after a rain to store water. **(d)** The kangaroo rat is an elusive inhabitant of the deserts of North America.

duce evaporation. Chaparral consists of small trees or large bushes with thick waxy or fuzzy evergreen leaves that conserve water. These shrubs are also able to withstand the frequent summer fires started by lightning (Fig. 46-16).

Human Impact   As is often the case with deserts, grasslands, and forests, the extent of chaparral vegetation is influenced by human activities. About 400 B.C., the Greek philosopher Plato wrote that "there are mountains in At-

tica [southern Greece] which can now keep nothing but bees, but which were clothed, not so very long ago, with timber suitable for roofing very large buildings." The original Greek forests were cut down not only for ceiling beams but also for the great Athenian naval fleets. Subsequently, heavy grazing by goats prevented regrowth of the forests. Only the chaparral plants, which could survive in the hot sun and were distasteful or poisonous to goats, regrew. The forests have never returned.

Figure 46-14   The Mojave Desert

In spring this southern California desert is carpeted with wildflowers. Through much of the year, annual wildflower seeds lie dormant waiting for the spring rains to fall.

Grasslands

In the temperate regions of North America, deserts occur in the rain shadows east of the mountain ranges, such as the Sierra Nevada and Rocky Mountains. Farther east, as the rainfall gradually increases, the land begins to support more and more grasses, giving rise to the prairies of the Midwest. These **grassland,** or prairie, biomes are usually concentrated in the centers of continents. You can see a large grassland in the center of the Eurasian continent in Figure 46-7. In general, they have a continuous cover of grass and virtually no trees at all except along the rivers. From the tallgrass prairies of Iowa and Illinois (Fig. 46-17)

to the shortgrass prairies of eastern Colorado, Wyoming, and Montana (Fig. 46-18), the North American grassland once stretched across almost half the continent.

Water and fire are the crucial factors in the competition between grasses and trees. The hot dry summers and frequent droughts of the shortgrass prairies can be tolerated by grass but are fatal to trees. Trees can grow in the more eastern prairies, but historically the trees were destroyed by frequent fires, often set by Native Americans to maintain grazing land for the bison. Although the tops of the grasses are destroyed by fire, their root systems usually survive; trees, on the other hand, are killed outright. The resulting grasslands of North America once supported huge

Figure 46-15   Overgrazing

In the Sahel region of Africa, overgrazing is degrading the already limited productivity of the land and contributing to the southward spread of the Sahara desert.

Figure 46-16   **The chaparral biome**

This biome is limited primarily to coastal mountains in dry regions, such as the San Gabriel Mountains in southern California. Chaparral is maintained by frequent fires set by summer lightning. Although the tops of the plants may be burned off, the roots send up new sprouts the next spring.

herds of bison; as many as 60 million in the early nineteenth century. Grasses growing and decomposing for thousands of years produced what may be the most fertile soil in the world.

Human Impact   With the elimination of the bison and the development of plows that could break the dense turf, the former prairie has become the "breadbasket" of North America, so named because enormous quantities of grain are cultivated in its fertile soil. The tallgrass prairie has been converted to agricultural land, except for tiny protected remnants.

On the western shortgrass prairie, cattle and sheep have replaced the bison and pronghorn antelope. As a result of their overgrazing the grasses, the boundary between the cool deserts and the grassland has often been altered in favor of desert plants. Much of the sagebrush desert of the American West is actually the result of overgrazing shortgrass prairie (Fig. 46-19). Cattle prefer grass to sagebrush, so heavy grazing destroys the grass. Consequently moisture that the grass would have absorbed is left in the soil, encouraging the growth of the woody sagebrush. Because the cattle will not eat sagebrush, it soon becomes the dominant vegetation; thus the prairie grasses are replaced by plants characteristic of the cool desert.

Figure 46-17   **Tallgrass prairie in Missouri**

In the central United States, moisture-bearing winds out of the Gulf of Mexico produce summer rains, allowing a lush growth of tall grasses and wildflowers such as these coneflowers.

Temperate Deciduous Forests

At their eastern edge the North American grasslands merge into the **temperate deciduous forest** biome (Fig. 46-20). Higher precipitation occurs there than in the grasslands (75 to 150 centimeters; 30 to 60 inches) and, in particular, more summer rainfall occurs. The soil therefore retains enough moisture for trees to grow, and the resulting forest shades out grasses. In contrast to the tropical forests, the temperate deciduous forest biome has cold winters, usually with at least several hard frosts and often long periods of below-freezing weather. Winter in this biome has an effect on the trees similar to that of the dry season in the tropical deciduous forests: During periods of subfreezing temperatures, liquid water is not available to the trees. To reduce evaporation when water is in short supply, the trees drop their leaves in the fall. They produce leaves again in the spring, when liquid water becomes available. During the brief time in spring when the ground has thawed but the trees have not yet blocked off all the sunlight, numerous wildflowers grace the forest floor.

Insects and other arthropods are numerous and conspicuous in deciduous forests. The decaying leaf litter on the forest floor also provides food and habitat for bacteria, earthworms, fungi, and small plants. Many arthropods feed on these or on each other. A variety of vertebrates, including mice, shrews, squirrels, raccoons, and many species of birds, dwell in the deciduous forests.

Human Impact   Large mammals such as black bear, deer, wolves, bobcats, and mountain lions were formerly abundant, but the predators have been largely wiped out

Figure 46-18 **Shortgrass prairie**

The lands east of the Rocky Mountains receive relatively little rainfall, and shortgrass prairie results, characterized by low-growing bunch grasses such as buffalo grass and grama grass, and a wealth of wildflowers such as this coneflower **(d)**. Pronghorn antelope **(a)**, prairie dogs **(b)**, and protected bison herds **(c)** occupy this biome.

Figure 46-19 **Sagebrush desert or shortgrass prairie?**

Biomes are influenced not only by temperature, rainfall, and soil but also by human activities. The shortgrass prairie field on the right has been overgrazed by cattle, causing the grasses to be replaced by sagebrush.

Figure 46-20   **Temperate deciduous forest**

In the eastern United States, the white-tailed deer **(a)** is the largest herbivore, and birds such as the Blue Jay **(c)** are abundant. In spring, a profusion of woodland wildflowers (such as these hepaticas **(b)** blooms briefly before the trees produce leaves.

by humans. Clearing for lumber and agriculture has dramatically reduced America's deciduous forests, and virgin forests are now almost nonexistent.

Temperate Rain Forests

On the Pacific coast, from the lowlands of the Olympic peninsula in Washington to southeast Alaska, is the **temperate rain forest** biome (Fig. 46-21). As in the tropical rain forest, there is no shortage of liquid water year-round. This abundance of water is due to two factors. First, there is a tremendous amount of rain. The Hoh River rain forest in Olympic National Park receives over 400 centimeters (160 inches) of rain each year, over 60 centimeters (24 inches) in the month of December alone. Second, the moderating influence of the Pacific Ocean prevents severe frost from occurring along the coast, so the ground seldom freezes.

The abundance of water means that the trees have no need to shed their leaves in the fall, and almost all the trees

are evergreens. In contrast to the broadleaf evergreen trees of the tropics, the temperate rain forests are dominated by conifers. The ground and often the trunks of the trees are covered with mosses and ferns. As in the tropical rain forests, so little light reaches the forest floor that tree seedlings usually cannot become established. Whenever one of the forest giants falls, however, it opens up a patch of light, and new seedlings quickly sprout, often right atop the fallen log. This event produces a "nurse log," as shown in Figure 46-21b.

Taiga

North of the grasslands and temperate forests, the **northern coniferous forest,** also called the **taiga** (Fig. 46-22), stretches horizontally across the entire continent of North America, mainly in southern Canada. Conditions in the taiga are harsher than those in the temperate deciduous forest. In the taiga, the winters are longer and colder, and the growing season is shorter. The few months of warm

Figure 46-21 **Temperate rain forest**

The Hoh River temperate rain forest in Olympic National Park. The coniferous trees do not block the light as effectively as do broadleaf trees, so ferns, mosses, and wildflowers grow in the pale green light of the forest floor. The dead feeds the living, as new trees grow from the decay of this fallen giant, called a "nurse log" **(b)**, and flowering foxglove **(a)** and fungi **(c)** find ideal conditions amid the moist, decaying vegetation.

weather are too short to allow trees the luxury of regrowing leaves in the spring. As a result, the taiga is populated almost entirely by evergreen coniferous trees with small, waxy needles. The waxy coating and small surface area of the needles reduce water loss by evaporation during the cold months, and the leaves remain on the trees year-round. Thus the tree is instantly ready to take advantage of good growing conditions when spring arrives, and can continue slow growth late into the fall.

Because of its harsh climate, the diversity of life in the taiga is much lower than in many other biomes. Vast stretches of central Alaska, for example, are covered by a somber forest composed almost exclusively of black spruce and an occasional birch. Large mammals, such as the wood bison, grizzly bear, moose, and wolf, which have mostly been wiped out in the southern regions of their original range, still roam the taiga, as do smaller animals, such as the wolverine, fox, snowshoe hare, and deer.

Human Impact Owing to the remoteness of the northernmost taiga and the severity of its climate, a greater percentage of the taiga remains in undisturbed condition than any other North American biome except the tundra. Nonetheless, the taiga is a major source of lumber for construction, and huge expanses of forest have been clearcut (Fig. 46-23).

Tundra

The last biome encountered before reaching the polar ice caps is the arctic **tundra**, a vast treeless region bordering the Arctic Ocean (Fig. 46-24). Conditions in the tundra are severe. Winter temperatures in the arctic tundra often reach –55° C (–40° F) or below, winds howl at 50 to 100 kilometers (30 to 60 miles) per hour, and precipitation averages 25 centimeters (10 inches) or less per year, making this a "freezing desert." Even during the summer, the temperatures can drop to freezing, and the growing season may last only a few weeks before a hard frost occurs. Somewhat less cold but similar conditions produce alpine tundra on mountaintops above the altitude where trees can grow.

The cold climate of the arctic tundra results in **permafrost,** a permanently frozen layer of soil often no more than half a meter (about 1.5 feet) below the surface. As a result, when summer thaws come, the water from melted snow and ice cannot soak into the ground and the tundra becomes a huge marsh. Trees cannot survive in the tundra; the permafrost limits root growth to the topmost meter or so of soil.

Nevertheless, the tundra supports a surprising abundance and variety of life. The ground is carpeted with small perennial flowers and dwarf willows no more than a few centimeters tall and often with a large lichen called reindeer moss, a favorite food of caribou. The standing pools provide superb mosquito habitat. The mosquitoes and other insects provide food for numerous birds, most of which migrate long distances to nest and raise their young during the brief summer feast. The tundra vegetation supports lemmings, which are eaten by wolves, snowy owls, arctic foxes, and even grizzly bears.

Human Impact The tundra is perhaps the most fragile of all the biomes because of its short growing season. A

willow 10 centimeters (4 inches) high may have a trunk 7 centimeters (3 inches) in diameter and be 50 years old. Human activities in the tundra leave scars that persist for centuries. Fortunately for the tundra inhabitants, the impact of civilization is localized around oil drilling sites, pipelines, mines, and military bases.

Rainfall, Temperature, and Vegetation

Terrestrial biomes are greatly influenced by both temperature and rainfall, whose effects interact. Temperature strongly influences the effectiveness of rainfall in providing soil moisture for plants and standing water for animals to drink. The hotter it is, the more rapidly water evaporates, both from the ground and from plants. As a result of this interaction of temperature with rainfall (and to a lesser extent, the distribution of rain throughout the year), areas that receive almost exactly the same rainfall can have startlingly different vegetation, all the way from desert to taiga. Take a trip with us from southern Arizona to northern Alaska as we visit ecosystems

Figure 46-22 **The taiga**
The small needles and pyramidal shape of conifers allows them to shed heavy snows. Winter is a challenge not only for the trees but also for animals such as this snowshoe hare and the bobcat that preys on it **(a)**. The hare is also prey for the great horned owl **(b)**. Taiga animals face diminished food supply but increased energy requirements during subfreezing weather.

Figure 46-23   Clearcutting

Clearcutting (the removal of all the trees in a given area) as seen in this Oregon forest, is relatively simple and cheap. But its environmental costs are high. Erosion will diminish the fertility of the soil, slowing new growth. Further, the dense stands of even-age trees that often regrow are more vulnerable to attack by parasites than a natural stand of trees of various ages would be.

that each receive around 28 centimeters (12 inches) of rain annually.

The Sonoran Desert near Tucson, Arizona (see Fig. 46-13c), has an average annual temperature of 20° C (68° F) and gets about 28 centimeters (a foot) of rain each

year. The landscape is dominated by giant saguaro cactus and low-growing, drought-resistant bushes. Fifteen hundred kilometers (931 miles) north, in eastern Montana, rainfall is about the same, but we have passed into the shortgrass prairie biome (see Fig. 46-18), largely because the average temperature is much lower, about 7° C (45° F). Central Alaska receives about the same annual rainfall (28 centimeters), yet is covered with taiga forest (see Fig. 46-22). As a result of the low average annual temperature (about –4° C, or 25° F), permafrost underlies much of the ground. During the summer thaw, the taiga earns its Russian name "swamp forest," although its rainfall is about the same as that of the Sonoran Desert.

Life in Water

Although this chapter has emphasized terrestrial biomes, the saltwater oceans and seas are the largest ecosystems on Earth, covering about 71% of its surface. Freshwater ecosystems, in contrast, cover less than 1%.

The unique properties of water lend some common features to aquatic ecosystems. First, because water is slower to heat and cool than air, temperatures in aquatic ecosystems are more moderate than in terrestrial ecosystems. Second, although water may appear quite transparent, it absorbs a considerable amount of the light

Figure 46-24   Tundra

Life on the tundra is adapted to cold. Plants such as dwarf willows and perennial wildflowers (such as this dwarf clover [c]) grow low to the ground, escaping the chilling tundra wind. Tundra animals, such as caribou (a) and arctic foxes (b), can regulate blood flow in their legs, keeping them just warm enough to prevent frostbite, while preserving precious body heat for the brain and vital organs.

energy that sustains life. Even in the clearest water, the intensity of light decreases rapidly with depth, so that at depths of 200 meters (600 feet) or more, little light is left to power photosynthesis. If the water is at all cloudy—for example, because of suspended sediment or microorganisms—the depth to which light can penetrate is greatly reduced. Third, nutrients in aquatic ecosystems tend to be concentrated near the bottom sediments where light levels often are too low to support photosynthesis. This separation of energy and nutrients limits aquatic life. Of the four requirements for life, aquatic ecosystems provide abundant water and appropriate temperatures. Thus, the major factors that determine the quantity and type of life in aquatic ecosystems are energy and nutrients.

Although they share some common features, aquatic ecosystems are extremely diverse. Freshwater ecosystems include rivers, streams, ponds, lakes, and marshes; marine (saltwater) ecosystems include estuaries, tide pools, the open ocean, and coral reefs. In the following sections, we look more closely at a few of these important aquatic ecosystems.

Freshwater Lakes Have Distinct Regions of Life and Temperature

Freshwater lakes vary tremendously in size, depth, and nutrient content. Although each lake is unique, moderate to large lakes in temperate climates share some common features, including distinct life zones and temperature stratification.

Life Zones Are Based on Access to Light and Nutrients

The distribution of life in lakes depends largely on access to light, to nutrients, and to a place for attachment (the bottom). The life zones of lakes, then, correspond to specific locations within the lake.

Near the shore is the **littoral zone.** In this zone, the water is shallow, and plants find abundant light, anchorage, and adequate nutrients from the bottom sediments. Not surprisingly, littoral-zone communities are the most diverse. Cattails and bulrushes abound nearest the shore, and water lilies and entirely submerged vascular plants and algae flourish at the deepest reaches of the littoral zone (Fig. 46-25). The plants of the littoral zone trap sediments carried in by streams and by runoff from the surrounding land, increasing the nutrient content in this region. Living among the anchored plants are microscopic organisms called **plankton.** There are two forms of plankton, **phytoplankton** (Greek, "drifting plants"), which includes photosynthetic protists, bacteria, and algae, and **zooplankton** (Greek, "drifting animals"), such as protozoa and tiny crustaceans. The greatest diversity of animals in the lake is also found in this zone. Littoral invertebrate animals include small crustaceans, insect larvae, snails, flatworms, and *Hydra*; vertebrates include frogs, minnows, and aquatic snakes and turtles.

As the water increases in depth farther from shore, plants are unable to anchor to the bottom and still collect enough light for photosynthesis. This open-water area is divided into two regions: the upper limnetic zone and the lower profundal zone (see Fig. 46-25).

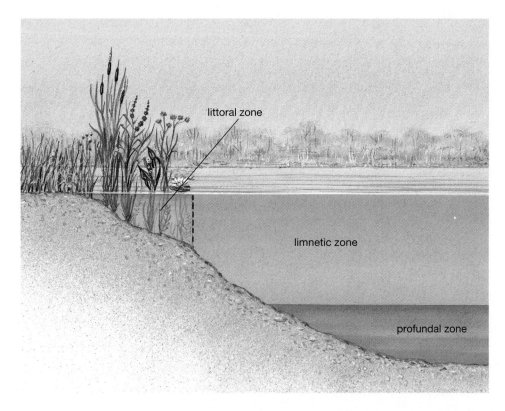

Figure 46-25   Lake life zones

There are three life zones in a "typical" lake: a near-shore littoral zone with rooted plants, an open-water limnetic zone, and a deep, dark profundal zone.

littoral zone

limnetic zone

profundal zone

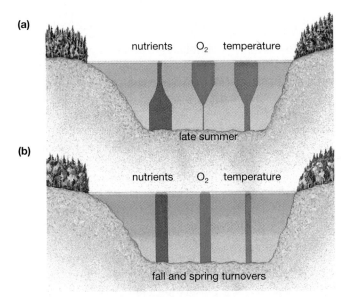

Figure 46-26   Temperature stratification in lakes

(a) Occurring during the summer, stratification results in high surface temperatures and oxygen levels, while surface nutrients become depleted because of a lack of mixing. (b) During the spring and fall turnovers, temperature stratification breaks down, and water from all levels mixes, resulting in uniform temperature, oxygen, and nutrient levels.

In the **limnetic zone,** enough light penetrates to support photosynthesis. Here phytoplankton including cyanobacteria (also called blue-green algae) serve as producers. These are eaten by protozoa and small crustaceans, which in turn are consumed by fish.

Below the limnetic zone lies the **profundal zone,** where light is insufficient to support photosynthesis. This area is nourished mainly by detritus that falls from the littoral and limnetic zones and by incoming sediment. It is inhabited primarily by decomposers and detritus feeders, such as snails and certain insect larvae and fish.

Temperature Stratification Occurs in Summer in Temperate Lakes

Lakes in regions that experience seasonal temperature changes become stratified during the summer, with a layer of relatively warm, less-dense water "floating" atop a layer of denser cold water (around 4° C, or 40° F; water reaches its highest density at 4° C; Fig. 46-26). The layers persist through the summer and restrict the mixing of surface and bottom waters. Because nutrients are concentrated in the bottom sediments, the upper region, where photosynthetic organisms flourish, becomes depleted of nutrients as the summer progresses. Simultaneously, the heterotrophic organisms near the bottom use up available oxygen, which is not replenished from the surface. So by late summer, the surface populations are limited by lack of nutrients, while those in the profundal zone are limited by lack of oxygen. (The boom-and-bust popula-

tion cycle of blue-green algae on the surface of Lake Erie, rapid growth followed by a die-off due to nutrient depletion, is shown in Figure 43-4.)

In the fall, surface temperatures drop and the stratification breaks down, allowing mixing of the nutrient-rich lower layer and the oxygen-rich upper layer. A similar situation occurs in the spring as the floating ice (at 0° C, or 32° F) melts, and surface waters warm to 4° C and mix with the deep water. Fall and spring turnovers maintain the lake's capacity to support its natural communities and cause spurts of growth among phytoplankton and zooplankton, particularly in the spring, when temperature and hours of sunlight are increasing.

Freshwater Lakes Are Classified According to Their Nutrient Content

Although each lake is unique, freshwater lakes may be classified on the basis of their nutrient content as either oligotrophic (Greek, "poorly fed") or eutrophic (Greek, "well fed").

Oligotrophic lakes are very low in nutrients. They often are formed by glaciers that scrape depressions in bare rock, and they are fed by mountain streams carrying little sediment. Because there is little sediment or microscopic life to cloud the water, oligotrophic lakes are clear, and light penetrates deeply. Therefore, photosynthesis (and by extension oxygenation, as oxygen is produced as a by-product of photosynthesis) in deeper water is possible, and the limnetic zone may extend all the way to the bottom. Oxygen-loving fish, such as trout, thrive in oligotrophic lakes.

Eutrophic lakes receive larger inputs of sediments, organic material, and inorganic nutrients (such as phosphorus) from their surroundings, allowing them to support dense communities. They are murkier, both from suspended sediment and dense phytoplankton populations, so the lighted limnetic zone is shallower. Dense "blooms" of algae occur seasonally in the limnetic zone. Their dead bodies fall into the profundal zone, where they are used as food by decomposer organisms. The metabolic activities of many of these decomposers use oxygen, depleting the oxygen content of the profundal zone.

Lakes are transient ecosystems, although very large lakes may persist for millions of years. Over time, they gradually fill with sediment, undergoing succession to dry land (see Chapter 44). As nutrient-rich sediment accumulates, oligotrophic lakes tend to become eutrophic, a process called eutrophication.

Human Impact   Human activities may greatly accelerate the process of eutrophication, because nutrients are carried into lakes from farms, feedlots, and sewage. Even if solid wastes are removed, water discharged from sewage-treatment plants is often rich in phosphates and nitrates dissolved from wastes and detergents. Excessively enriched water also washes off fertilized farm fields and feedlots, where the manure of thousands of cattle accumulates. These added nutrients support excessive growth of phytoplankton. These producers, particularly blue-green algae, form a scum on the lake surface, depriving the submerged plants of sunlight. The plant bodies are decomposed by bacteria, depleting the dissolved oxygen. Deprived of oxygen, fish, snails, and insect larvae die, and their bodies fuel more bacterial growth, further depleting oxygen. Even without oxygen, certain bacteria that produce foul-smelling gases thrive. Although it is full of life, the nutrient-polluted lake smells dead. Most of the trophic levels, including the fish, have been eliminated, and the community is dominated by bacteria and microscopic algae.

Acid rain, caused by the burning of fossil fuels, poses a very different threat to freshwater ecosystems. Few organisms can withstand the low pH of acidified lakes, which often appear clear and oligotrophic because of the absence of life. Acid rain is discussed in Chapter 45.

Marine Ecosystems Cover Much of Earth

In the oceans, the upper layer of water, where the light is strong enough to support photosynthesis, is called the **photic zone.** Below the photic zone lies the **benthic zone,** where the only energy comes from the excrement and bodies of organisms that sink or swim down.

As in lakes, most of the nutrients in the oceans are at or near the bottom where there is not enough light for photosynthesis. Nutrients dissolved in the water of the photic zone are constantly being incorporated into the bodies of living organisms. When these organisms die, some sink into the benthic zone, providing its organisms with nutrients. If no new nutrients entered the photic zone, life (in both the photic and the benthic zones) would eventually cease.

Fortunately, there are two sources of new nutrients: the land, from which rivers constantly remove nutrients and carry them to the oceans, and **upwelling,** which brings cold, nutrient-laden water from the ocean to the surface. Upwelling often occurs along coastlines, as in California, Peru, and West Africa, where prevailing winds displace surface water, causing it to be replaced by water from below. Upwelling also occurs around Antarctica. Not surprisingly, the major concentrations of life in the oceans are found where abundant light is combined with a source of nutrients, which happens most often in regions of upwelling and in shallow coastal waters, including coral reefs.

Coastal Waters Support the Most Abundant Marine Life

The most abundant life in the oceans is found in a narrow strip surrounding Earth's land masses, where the water is shallow and a steady flow of nutrients is washed off the land. Coastal waters include the **intertidal zone,** the area that is alternately covered and uncovered by water with the rising and falling of the tides, and the **near-shore zone,** relatively shallow but constantly submerged areas, in-

cluding bays and coastal wetlands such as salt marshes and estuaries (Fig. 46-27). The near-shore zone is the only part of the ocean where large plants can grow, anchored to the bottom. In addition, the abundance of nutrients and sunlight promotes the growth of a veritable soup of photosynthetic phytoplankton. Associated with these plants and protists are animals from nearly every phylum: annelid worms, sea anemones, jellyfish, sea urchins, sea stars, mussels, snails, fish, and sea otters, to name just a few. A large number and variety of organisms live permanently in coastal waters, but many that spend most of their lives in the open ocean come into the coastal waters to reproduce. Bays, salt marshes, and estuaries in particular are the breeding grounds for a wide variety of organisms, such as crabs, shrimp, and an array of fish, including most of our commercially important species.

Human Impact Coastal regions are of great importance not only to the organisms that live or breed there but also to humans who wish to use them for food sources, recreation, mineral and oil extraction, or living places. Like rain forests, wetlands of all types once covered 6% of Earth's surface, but nearly half of them have been dredged or filled in. In the United States, 100 million of our 215 million acres of wetlands have been lost. As populations increase in our coastal states and resources such as oil become increasingly scarce, the conflict between preservation of our coastal wetlands as wildlife and animal habitat and development of these areas for housing, marinas, and energy extraction will become increasingly intense. Because much of the life of the entire ocean is dependent upon the well-being of the coastal waters, it is essential that we protect these fragile, vital areas.

Coral Reefs

Coral reefs are actually created by animals and plants. In warm tropical waters, with just the right combination of bottom depth, wave action, and nutrients, specialized algae and corals (types of cnidarians) build reefs from their own calcium carbonate skeletons. Coral reefs are most abundant in tropical waters of the Pacific and Indian oceans, the Caribbean, and the Gulf of Mexico as far north as southern Florida, where the maximum water temperatures range between 22° C and 28° C (72° F and 82° F).

Reef-building corals are involved in a mutualistic relationship with unicellular algae called dinoflagellates, who live embedded in the coral tissue. These corals grow best at depths of less than 40 meters (130 feet), where light can penetrate and allow their algal partners to photosynthesize. The algae benefit from the high nitrogen, phosphorus, and carbon dioxide levels in the coral tissues. In return, algae provide food for the coral and help produce calcium carbonate, which forms the coral skeleton. Coral reefs provide an anchoring place for many other algae, a

Figure 46-27   **Near-shore ecosystems**

(a) A salt marsh in the eastern United States. Expanses of shallow water fringed by marsh grass (*Spartina*) provide excellent habitat and breeding grounds for many marine organisms and shorebirds. **(b)** Although the shifting sands present a challenge to life, grasses stabilize them, and animals such as this *Emerita* crab (inset) burrow in the sandy intertidal zone. **(c)** A rocky intertidal shore in Oregon, where animals and algae grip the rock against the pounding waves and resist drying during low tide. (Inset) Colorful sea stars cling to the rocks surrounded by a seaweed, fucus. **(d)** Towering kelp sway through the clear water off southern California, providing the basis for a diverse community of invertebrates, fishes, and an occasional sea otter (inset).

home for bottom-dwelling animals, and shelter and food for the most diverse collection of invertebrates and fish to be found in the oceans (Fig. 46-28). The Great Barrier Reef in Australia is home to more than 200 species of coral alone, and a single reef may harbor 3000 species of fish, invertebrates, and algae.

Human Impact Coral reefs are extremely sensitive to certain types of disturbance, especially silt caused by soil eroding from nearby land. As silt clouds the water, light is diminished and photosynthesis reduced, hampering growth of the corals. The reef may eventually become buried in mud, the corals smothered, and the entire marvelous community of diverse organisms destroyed. Ero-

sion from construction, roadways, and poor land management has produced enough silt to ruin several reefs near Honolulu, Hawaii. The reef inhabitants have been replaced by large numbers of sediment-feeding invertebrates such as sea cucumbers. In the Philippines, logging has dramatically increased erosion, so coral reefs (as well as rain forests) are being destroyed. Another hazard is sewage and runoff from agriculture, which fertilize near-shore ocean water, causing eutrophication and a dense growth of algae. This blocks sunlight from the coral's dinoflagellates, and decaying algal bodies deplete the water of oxygen, killing the coral.

Still another threat to the reef communities is over-fishing. In at least 80 countries, a variety of species, in-

Figure 46-28   **Coral reefs**

(a) Coral reefs, composed of the bodies of corals and algae, provide habitat for an extremely diverse community of extravagantly colored animals. Many fish, including this blue tang (b), feed on coral (note the bright yellow corals in background). A vast array of invertebrates such as this sponge (c) and blue-ringed octopus (d) live among the corals of Australia's Great Barrier Reef. This tiny (15 centimeters, or 6 inches, fully extended) octopus is one of the world's most venomous creatures, possessing enough venom to kill 10 adult humans.

cluding mollusks, turtles, fish, crustaceans, and even corals, are being harvested from reefs faster than they can replace themselves. Many of these are sold to shell collectors and aquarium owners in developed countries. In some tropical countries, dynamite is used to kill coral reef fish, destroying entire sections of the coral reef community in the process. Tropical fish collectors often use poison to stun the fish before collecting them, leaving most dead. Removal of predators from reefs may disrupt the ecological balance of the community, allowing an explosion in populations of coral-eating sea urchins, which are destroying reefs, for example, along the Kenyan coast in Africa.

As with rain forests, both protection and sustainable use are crucial to the survival of these fragile and diverse underwater ecosystems. Carefully regulated harvesting and tourism produce far more economic benefits than activities that destroy the reefs. Some countries have recognized this, and there are now many areas where reefs are protected.

The Open Ocean

Beyond the coastal regions lie vast areas of the ocean in which the bottom is too deep to allow plants to anchor and still receive enough light to grow. Most life in the open ocean is limited to the upper photic zone, in which

Figure 46-29   **The open ocean**

Porpoises skim the surface **(a)**, rare humpback whales leap clear of the water **(b)**, and fish such as this blue jack swim **(c)**. The photosynthetic phytoplankton **(d)** are the producers on which most other life ultimately depends. Phytoplankton are eaten by zooplankton **(e)**, represented by this microscopic crustacean, a copepod. The spiny projections on these planktonic creatures help keep them from sinking below the photic zone.

the life forms are **pelagic**—that is, free-swimming or floating—for their entire lives. The food web of the open ocean is dependent on phytoplankton consisting of microscopic photosynthetic protists, mainly diatoms and dinoflagellates (Fig. 46-29d). These organisms are consumed by zooplankton, such as tiny crustaceans that are relatives of crabs and lobsters (Fig. 46-29e). Zooplankton in turn serve as food for larger invertebrates, small fish, and even marine mammals such as the humpback whale (Fig. 46-29b).

One challenge faced by inhabitants of the open ocean is to remain afloat in the photic zone where sunlight and food are abundant. Many members of the planktonic community have elaborate flotation devices, such as oil droplets in their cells or long projections, to slow down their rate of sinking (see Fig. 46-29d,e). Most fish have swim bladders that can be filled with gas to regulate their buoyancy. Some animals, and even some of the phytoplankton, actively swim to stay in the photic zone. Many small crustaceans migrate to the surface at night to feed,

then sink into the dark depths during daylight, thus avoiding visual predators such as fish. The amount of pelagic life varies tremendously from place to place. The blue clarity of tropical waters is a result of a lack of nutrients, which limits the concentration of plankton in the water. Nutrient-rich waters that support a large plankton community are greenish and relatively murky.

Below the photic zone, the only available energy in most cases comes from the excrement and dead bodies drifting down from above. Nevertheless, a surprising quantity and variety of life exists in the benthic zone, including fishes of bizarre shapes, worms, sea cucumbers, sea stars, and mollusks.

Human Impact   Two major threats to the open ocean are pollution and overfishing. Open ocean pollution takes several forms. Ocean-going vessels dump millions of plastic containers overboard daily. Plastic six-pack holders, foam cups, and packing material washes and blows off the

Figure 46-30   Vent communities

Located in the ocean depths, vent communities include giant worms nearly 4 meters (12 feet) long. These worms are red with oxygen-trapping hemoglobin. They have no digestive tract but are host to sulfur bacteria, which provide the worms with energy by oxidizing hydrogen sulfide.

land, collecting on parts of the ocean surface. The plastic looks like food to unsuspecting sea turtles, gulls, porpoises, seals, and whales, who often die after trying to consume it. Until 1992, New York city barged its refuse and sewage sludge out to sea, creating a heavily contaminated area covering 40 square miles of open floor. The open ocean has served as a dumping ground for radioactive wastes. Oil contaminates the open ocean from many sources, including tanker spills, runoff from improper disposal on land, leakage from offshore oil wells, and natural seepage.

The increasing demand for fish to feed a growing human population, coupled with increasingly efficient fishing technologies, has resulted in depletion of many important fisheries. The northeastern U.S. cod fishery has collapsed due to overfishing. Populations of lobsters, salmon, haddock, king crab, and many other types of seafood have also declined dramatically because of overfishing.

Vent Communities

In 1977, a new and unusual source of nutrients, forming the basis of a spectacular undersea community, was discovered in the deep ocean. Geologists exploring the Galapagos Rift (an area of the Pacific floor where plates forming Earth's crust are separating) found vents spewing superheated water, black with sulfur and minerals. Surrounding these vents was a rich community of pink fish, blind white crabs, enormous mussels, white clams, sea anemones, and tube worms (Fig. 46-30). Twenty-two previously undescribed families and 284 new species of animals have been found in these novel communities.

In this unique ecosystem, sulfur bacteria serve as the primary producers, harvesting the energy stored in hydrogen sulfide spewed from beneath Earth's crust. This process, called chemosynthesis, replaces photosynthesis at these depths, over 2500 meters (7500 feet) below the surface. The bacteria proliferate in the warm water surrounding the vents, covering nearby rocks with thick, mat-like colonies. These colonies provide the food on which the animals of the vent community thrive. Many vent animals consume the bacteria directly. Others, such as the giant tube worm (which lacks a digestive tract), harbor the bacteria within cells in their bodies (see Fig. 46-30). In this mutualistic association, the bacteria provide high-energy carbon compounds, while the tube worm provides hydrogen sulfide, which it transports to the bacteria in its bloodstream using a unique form of hemoglobin.

The world still holds wonders and mysteries for those who seek them, and we have only begun to explore the versatility and diversity of life on Earth.

EARTH WATCH

Humans and Ecosystems

The expanding human population has left relatively few ecosystems undisturbed. Our impacts on natural ecosystems are so diverse and wide ranging that they far exceed the scope of this book. However, we can identify some general characteristics of ecosystems dominated by humans and contrast them to the characteristics of undisturbed ecosystems. Below are listed six characteristic differences between these two types of ecosystems and a few ideas for minimizing them.

First, ecosystems dominated by people tend to be simpler— that is, to have fewer species and fewer community interactions— than undisturbed ecosystems. Although a city street bustles with life and apparent complexity, count the number of species you encounter in an average block and compare it with the number you encounter while hiking in a wilderness for a similar distance. As humans enter an ecosystem, animals in the highest trophic levels are the first to go. Carnivores are always relatively rare, and their specialized needs are most easily disrupted. Many big carnivores, such as the wolf and mountain lion, require large undisturbed hunting territories. Human beings often selectively destroy large predators, believing them a threat to people or their livestock. As a result, even in relatively undisturbed areas they have often been eliminated. Grizzly bears and wolves no longer roam the Rocky Mountains. On the western short-grass prairie, prairie dog towns fall as human towns and ranches rise (Fig. E46-2). The black-footed ferret, a predator of prairie dogs, faces possible extinction. Farm fields have been deliberately simplified from the original prairie biome to eliminate competition and predation and allow the maximum productivity of a single crop species. Nowhere is the contrast between human and natural ecosystems greater than in the tropical rain forests, whose unparalleled diversity is replaced by failed attempts at farming.

There is probably no practical way to restore to human ecosystems the great diversity found in undisturbed areas, nor is doing so always desirable from a human standpoint. Agriculture, for example, demands a simplified ecosystem (although not all biomes are suited to it). Cities concentrate human activities and culture and may lessen the human impact on the surrounding countryside. However, as we recognize the benefits of artificially simplified ecosystems, we must be aware of the need to preserve intact as many natural communities as possible. The undisturbed forest traps and purifies water and can reduce air pollution. Swamps and estuaries, when not dredged or filled, contain a wealth of detritus feeders and decomposers that purify water. Coastal wetlands are breeding places for millions of birds and spawning sites for the majority of our commercially important fish and crustacean species. In

Figure E46-2   Habitat destruction
Loss of habitat due to human activities is a major threat to most of Earth's wildlife.

undisturbed, diverse ecosystems we find aesthetic pleasure, as well as a storehouse of species whose commercial or medicinal values are not yet recognized. For example, nearly half the medicines in use today were originally discovered in plants, and human beings have examined only a small fraction of existing plants for possible medical uses.

Second, whereas natural ecosystems run on sunlight, human ecosystems have become dependent on nonrenewable energy from fossil fuels. From the suburbanite pouring gas into a lawn mower to the farmer driving a tractor, managing a simplified ecosystem is an energy-intensive proposition. Energy must be expended to oppose the tendency of the natural system to restore complexity. Fertilizers also require large amounts of energy to produce, and farms must be heavily fertilized because nutrient cycles have been disrupted.

To counteract this trend, some farmers are returning to organic farming. Plant and animal waste and natural nutrient cycles are used to maintain soil fertility. Alternating legume crops, such as soybeans and alfalfa, with other crops helps maintain nitrogen in the soil. Mulching can help retain fertility, reduce water loss, and control weeds. The use of natural insect predators and limited pesticide spraying at critical times during a pest's life cycle can dramatically reduce the need for poisons. Organic farms can be as productive as more conventional farms while using 15% to 50% less energy from fossil fuels per quantity of food produced.

In our homes and commercial structures, better insulation and increased use of solar heat can result in dramatic energy savings. By increasing our use of renewable energy sources, such as wind and especially sunlight, we can conserve fossil fuels, dramatically reduce pollution, and move human ecosystems a step closer to those that occur naturally.

Third, natural ecosystems recycle nutrients, whereas human ecosystems tend to lose nutrients. Walk around a suburban neighborhood on trash-pickup day. Grass clippings and leaves are packed in plastic bags to be hauled away. To compensate for this loss of natural nutrients, suburbanites often heavily fertilize their lawns and gardens. A similar trend has occurred in modern farming. The exposed soil is eroded away by wind and water, removing crucial nutrients and requiring large inputs of fertilizer to replace them. Runoff from the field, carrying fertile topsoil and artificial fertilizers, may pollute nearby rivers, streams, and lakes. Pesticides may kill detritus feeders and decomposers, further disrupting natural nutrient cycles. Thus, whereas fertile soil accumulates in many natural ecosystems, it tends to be lost in those dominated by human beings. Some 3 billion tons of topsoil are eroded from farms each year in the United States. The Mississippi River alone carries off 40 tons each hour.

Again, organic farming can help reverse this trend, and we can apply the same principles in our own lawns and gardens. Farmers can counteract erosion by using contour planting, in which row crops are oriented so that they slow the flow of water instead of funneling it down a slope. Row crops, which tend to allow erosion, can be alternated in strips with dense, soil-catching crops such as wheat. Planting rows of trees as windbreaks helps prevent soil loss from blowing wind, and it creates a more diverse ecosystem and a nesting place for insect-eating birds. Farm fields need not be plowed in the fall and left unplanted to erode during the winter; in fact, an increasing number of farmers are planting crops in the stubble from the previous season, reducing erosion.

Fourth, natural ecosystems tend to store water and purify it through biological processes, whereas human ecosystems tend to pollute water and shed it rapidly. A thundershower strikes a forest, an adjacent city, and a farm. The rich soil of the forest sponges up the water, which gradually drains into the ground, filtered by the soil and purified by decomposers that break down organic contaminants. In the nearby city, water pours from sidewalks, rooftops, and streets, picking up soot, silt, oil, lead, and garbage. It races down gutters into storm sewers, and a weakly toxic soup gushes into the nearest stream or river. Farm runoff carries priceless topsoil, expensive fertilizer, and animal manure into rivers and lakes, where these resources become pollutants.

Preventing erosion will simultaneously conserve water and reduce water pollution from farm runoff. Manure from cattle feedlots, which is a significant source of both groundwater and surface-water pollution, could be placed on fields, where it would restore needed nutrients.

Although water will continue to run off our cities, the pollutants it carries can be reduced by minimizing our reliance on fossil fuels. We must eliminate leaded gasoline and tighten standards for emissions from diesel and gasoline engines and smokestacks. Efficient public transportation systems will reduce pollution and the massive frustration caused by congested traffic. Insulation will reduce power consumption in our homes and offices, as will increased reliance on solar heat.

Fifth, simple human ecosystems such as farms tend to be extremely unstable. Natural ecosystems have many species and tend to remain stable over time. Herbivorous insects are controlled by natural predators, including other insects, birds, and shrews. Insect populations are also limited because their preferred plants are scattered among many others rather than growing in a pure stand, as on a farm. On farms, both pest insects and their natural predators are exposed to pesticides. Unfortunately, the pests often develop resistance to the poison, while their predators are killed. In these simplified communities, unfavorable weather conditions or the introduction of an exotic species (see Chapter 44) can be disastrous.

Farmers can counteract this trend by planting smaller fields with a wider variety of crops. Alternating crops not only helps maintain soil fertility but also helps prevent the proliferation of disease and insect pests that are specialized for a particular crop. Populations of corn borers, for example, will die off during years when the field is planted in alfalfa. Use of biological controls such as natural insect predators and insect diseases can reduce reliance on pesticides.

Finally, human ecosystems are characterized by continuously growing populations, whereas nonhuman populations in natural ecosystems are relatively stable. As our population expands, the spread of human-dominated ecosystems presents a growing threat to the diversity of species and to the delicate balance that has evolved over the 3-billion-year history of life on Earth.

In summary, natural ecosystems tend to be complex, stable, and self-sustaining, powered by solar energy and nourished by recycled nutrients. They provide diverse habitats for wildlife, purify contaminants through the action of decomposers, and build up nutrient-rich soil. Modern human ecosystems are relatively simple and are sustained by large inputs of energy from fossil fuels. They tend to minimize wildlife habitat, contaminate soil and water, and lose nutrients and fertile soil. These problems are compounded by continued population growth, which is causing expansion of human-dominated ecosystems at the expense of undisturbed ones.

As we have pointed out, human ecosystems do not have to be as disruptive and alien to the operation of natural ecosystems as we have allowed them to become. Through understanding, education, and commitment, appropriate use of technology, and stabilization of our population, we can reverse many of these destructive trends.

❈ SUMMARY OF KEY CONCEPTS

Throughout the world, areas with similar climates have similar groups of organisms. Although species differ, adaptations to climate create recognizable distinguishing features, particularly among plants.

Climate

The availability of sunlight, water, and appropriate temperatures determines the climate of a given region. Sunlight that reaches Earth may be either reflected or absorbed by the atmosphere, land, and water, which radiates it back as heat, with a small amount captured by photosynthetic organisms. This heat maintains the temperature of Earth. Equal amounts of solar energy are spread over a smaller surface at the equator than farther north and south, making the equator relatively warm, while higher latitudes have lower overall temperatures. Earth's tilt on its axis causes dramatic seasonal variation at northern and southern latitudes.

Warm air currents at the equator and at 60° north and south latitudes rise and release moisture as they cool, increasing precipitation at these latitudes. Cooler, dry air sinks at around 30° north and south latitudes and at the poles, creating areas of very low moisture. These patterns are modified by the continents. Ocean gyres carry warm equatorial waters into northern and southern latitudes, moderating the climate of coastal areas.

Climate is modified by the rapid heating and cooling of continents as compared with the oceans, and by elevation. Air cools as it rises, so high elevations are cooler than low ones. Cool air holds less moisture, so rain falls on the windward side of mountain ranges, and rain shadows occur on the far sides. Deserts are often found in the rain shadows of mountain ranges. Because temperature decreases both with increasing latitude and increasing elevation, the vegetation changes observed going up a mountain are similar to those seen while traveling northward in the Northern Hemisphere.

The Requirements of Life

The requirements for life on Earth include nutrients, energy, liquid water, and a reasonable temperature. The differences in the form and abundance of living things in various locations on Earth are largely attributable to differences in the interplay of these four factors.

Life on Land

On land, the crucial limiting factors are temperature and liquid water. Large regions of the continents that have similar climates will have similar vegetation, determined by the interaction of temperature and rainfall (availability of water). These regions are called biomes.

Tropical forests are warm and wet, dominated by huge broadleaf evergreen trees. Most nutrients are tied up in vegetation, and most animal life is arboreal. Rain forests, home to at least 50% of all species, are rapidly being cut for agriculture, although the soil is extremely poor.

The African savanna is an extensive grassland with pronounced wet and dry seasons. It is home to the world's most diverse and extensive herds of large mammals.

Deserts, hot and dry, are found primarily between 20° and 30° of latitude and in the rain shadows of mountain ranges. Plants are widely spaced and have adaptations to conserve water. Animals tend to be small and nocturnal, also adapted to drought.

Chaparral exists in desertlike conditions, which are moderated by their proximity to a coastline, allowing small trees and bushes to thrive.

Grasslands, concentrated in the centers of continents, have a continuous grass cover and few trees. They produce the world's richest soils and have largely been converted to agriculture.

Temperate deciduous forests, whose broadleaf trees drop their leaves in winter to conserve moisture, dominate the eastern half of the United States. The wet temperate rain forests, dominated by evergreens, are found on the northern Pacific coast of the United States.

The taiga, or northern coniferous forest, covering much of the northern United States and southern Canada, is dominated by conifers whose small waxy needles are adapted for water conservation and year-round photosynthesis.

The tundra is a frozen desert where permafrost prevents the growth of trees and bushes remain stunted. Nonetheless, a diverse array of animal life and perennial plants flourishes in this fragile biome.

Life in Water

Energy and nutrients are the major limiting factors in the distribution and abundance of life in aquatic ecosystems. Nutrients are found in bottom sediments and are washed in from surrounding land, concentrating them near shore and in deep water.

Freshwater lakes have three life zones. The littoral zone, near shore, is rich in energy and nutrients and supports the most diverse community. The limnetic zone is the lighted region of open water where photosynthesis can occur. The profundal zone is the deep water, where light is inadequate for photosynthesis and the community is dominated by heterotrophic organisms.

Lakes in temperate climates become stratified during the summer and winter, with a layer of cold dense water at the bottom and less dense, warmer water (in summer) or colder water (in winter) in an upper layer. During the spring and fall turnovers when temperatures equalize, the layers mix, distributing oxygen and nutrients.

Oligotrophic lakes are clear, low in nutrients, and support sparse communities. Eutrophic lakes are rich in nutrients and support dense communities. During succession to dry land, lakes tend to go from an oligotrophic to a eutrophic condition.

Most life in the oceans is found in shallow water, where sunlight can penetrate, and is concentrated near the continents and in areas of upwelling, where nutrients are most plentiful.

Coastal waters, consisting of the intertidal zone and the near-shore zone, contain the most abundant life. Producers consist of aquatic plants anchored to the bottom

and photosynthetic protists called phytoplankton.

Coral reefs are confined to warm, shallow seas. The calcium carbonate reefs form a complex habitat supporting the most diverse undersea ecosystem.

In the open ocean, most life is found in the photic zone, where light supports phytoplankton. In the lower benthic zone, life is supported by nutrients that drift down from the photic zone. Specialized vent communities, supported by bacteria, thrive at great depths in the superheated waters where Earth's crustal plates are separating.

✖ KEY TERMS

benthic zone   p. 937
biome   p. 919
chaparral   p. 926
climate   p. 916
coral reef   p. 937
desert   p. 926
desertification   p. 926
eutrophic lake   p. 936
grassland   p. 928
gyre   p. 917
intertidal zone   p. 937
latitude   p. 916

limnetic zone   p. 936
littoral zone   p. 935
near-shore zone   p. 937
northern coniferous forest   p. 931
oligotrophic lake   p. 936
ozone layer   p. 920
pelagic   p. 940
permafrost   p. 933
photic zone   p. 937
phytoplankton   p. 935
plankton   p. 935
profundal zone   p. 936

rain shadow   p. 918
savanna   p. 924
taiga   p. 931
temperate deciduous forest   p. 929
temperate rain forest   p. 931
tropical deciduous forest   p. 923
tropical rain forest   p. 921
tundra   p. 932
upwelling   p. 937
weather   p. 915
zooplankton   p. 935

✖ THINKING THROUGH THE CONCEPTS

Multiple Choice

1. Most plant species fit into only a few morphological types. Which type seems most restricted by temperature and rainfall?
   a. trees
   b. shrubs
   c. grasses
   d. perennial herbs
   e. annual weeds
2. In which biome is the smallest fraction of carbon and nutrients present in the soil?
   a. tropical rain forest
   b. savanna
   c. tundra
   d. grassland
   e. coniferous forest
3. How do mountain ranges create deserts?
   a. by lifting land up into colder, drier air
   b. by completely blocking the flow of air into desert areas, thus preventing clouds from getting there
   c. by forcing air to first rise and then fall, thus causing rain on one side of the mountains and desert on the other
   d. by causing the world wind patterns that make certain latitudes very dry
   e. by causing very steep slopes that are subject to erosion
4. What is the primary reason that plants from distant, but climatically similar, places often look the same?
   a. common ancestry
   b. adaptation to the same physical conditions
   c. adaptation to similar herbivores
   d. continental drift
   e. effects of past climate change
5. Which of these biomes has been increased in area by human activities?
   a. savanna
   b. temperate rain forest
   c. grassland
   d. coniferous forest
   e. desert
6. What biome has the richest soil and has largely been converted to agriculture?
   a. tundra
   b. coniferous forest
   c. grassland
   d. tropical rain forest
   e. deciduous forest

Review Questions

1. Explain how air currents contribute to the formation of the tropics and the large deserts.
2. What are large, roughly circular ocean currents called? What effect do they have on climate, and where is that effect strongest?
3. What are the four major requirements for life? Which two are most often limiting in terrestrial ecosystems? In ocean ecosystems?

4. Explain why traveling up a mountain takes you through biomes similar to those you would encounter traveling north for a long distance.

5. Where are the nutrients of the tropical forest biome concentrated? Why is life in the tropical rain forest concentrated high above the ground?

6. Explain two different undesirable effects of agriculture in the tropical rain forest biome.

7. List some adaptations of (a) desert plants and (b) desert animals to heat and drought.

8. How and where does desertification occur?

9. How are trees of the taiga adapted to a lack of water and a short growing season?

10. How do deciduous and coniferous biomes differ?

11. What single environmental factor best explains why there is shortgrass prairie in Colorado, tallgrass prairie in Illinois, and deciduous forest in Ohio?

12. Where are the world's largest populations of large herbivores and carnivores found?

13. Where is life in the oceans most abundant, and why?

14. Why is the diversity of life so high in coral reefs? What human impacts threaten them?

15. Distinguish between the limnetic, littoral, and profundal zones of lakes in terms of their location and the communities they support.

16. Distinguish between oligotrophic and eutrophic lakes. Describe (a) a natural scenario and (b) a human-created scenario under which an oligotrophic lake might be converted to a eutrophic lake.

17. What is the reason and importance of the spring and fall overturn in temperate lakes?

18. Distinguish between the photic and benthic zones. How do organisms in the photic zone obtain nutrients? How are nutrients obtained in the benthic zone?

19. What unusual primary producer forms the basis for rift-vent communities?

20. On the basis of the location of the worst atmospheric ozone depletion, which biomes are likely to be most affected by in-

✖ APPLYING THE CONCEPTS

1. List at least six differences between human-dominated and undisturbed ecosystems, and discuss in some detail how these differences can be minimized.

2. In which terrestrial biome is your college or university located? Discuss similarities and differences between your location and the general description of that biome in the text. If you are living in a city, how has the urban environment modified your interaction with the biome?

3. During the 1960s and 1970s, many parts of the United States and Canada banned the use of detergents containing phosphates. Until that time, almost all laundry detergents and many soaps and shampoos had high concentrations of phosphates. What environmental concern do you think prompted these bans, and what ecosystem has benefited most from the bans?

4. Because ozone depletion is expected to get worse, not better, for decades to come, biologists have tried to assess which types of species will be most susceptible to increased UV penetration. Two of the groups that may be most vulnerable are quite different from each other: ocean plankton and long-lived birds and mammals. Try to think of what makes each of these groups so susceptible to increased UV radiation.

5. Although temperature and moisture are the primary determinants of terrestrial ecosystems, both natural and human-induced differences in soil types can strongly affect plant communities. Both naturally rocky soils (e.g., serpentine barrens and granitic outcrops) and the inorganic, gravelly soils produced by mining operations often support only grasses and shrubs, even in biomes that are largely dominated by trees. How could these different soils affect the ability of trees to grow? In particular, how could rocky soils affect the temperature or moisture experienced by plants growing on them, as opposed to the temperature and moisture experienced by plants growing nearby in soils with more organic matter?

6. Understanding the ways in which the four basic requirements for life determine where different biomes occur can help us predict the consequences of global warming. Global warming is expected to make most areas warmer, but it is also expected to change rainfall in ways that are hard to predict—some areas will get wetter and others drier. Our ignorance of how rainfall will change is not very important in understanding shifts in more northerly biomes, but it is very important for our understanding of changes in tropical areas. Look at Figure 46-8 and explain why this is true.

7. More northerly forests are far better able to regenerate after logging than are tropical rain forests. Try to explain why this would be true. HINT: The cold soils of northern climates greatly slow down decomposition rates.

✖ F O R   M O R E   I N F O R M A T I O N

Allan, J. D., and Flecker, A. S. "Biodiversity Conservation in Running Waters: Identifying the Major Factors That Threaten Destruction of Riverine Species and Ecosystems." *BioScience* 43:32–42, 1993. The major human impacts on river ecosystems and how to restore these communities.

Bell, R. H. V. "A Grazing Ecosystem in the Serengeti." *Scientific American,* July 1971. Describes the great migrations of the herbivores of the African savanna.

Dodd, J. L. "Desertification and Degradation in Sub-Saharan Africa: The Role of Livestock." *BioScience* 44:28–34, 1994. An argument that the process of desert formation is not as clearly linked to human-caused degradation as is usually portrayed.

Goreau, T. F., Goreau, N. I., and Goreau, T. J. "Corals and Coral Reefs." *Scientific American,* August 1979. Discusses the ecology of the great reefs built by tiny polyps harboring symbiotic protists.

Holloway, M. "Sustaining the Amazon." *Scientific American,* July 1993. Describes the threats to the Amazon rain forest and innovative ways to preserve tropical forests and their biodiversity.

Kusler, J. A., Mitsch, W. J., and Larson, J. S. "Wetlands." *Scientific American,* January 1994. Describes the characteristics and values of these threatened, but highly important ecosystems.

Mares, M. A. "Desert Rodents, Seed Consumption, and Convergence: Evolutionary Shuffling of Adaptations." *BioScience* 43:372–379, 1993. An example of how animal as well as plant species and communities living in similar climates can show convergence.

Milton, S. J., Dean, R. J., du Plessis, M. A., and Siegfried, W. R. "A Conceptual Model of Arid Rangeland Degradation." *BioScience* 44:70–76, 1994. Rangeland degradation proceeds in steps, each of them increasingly difficult and costly to reverse.

Toon, B., and Turco, R. P. "Polar Stratospheric Clouds and Ozone Depletion." *Scientific American,* June 1991. Describes the unique chemistry of the Antarctic clouds that causes ozone breakdown.

Tunnicliffe, V. "Hydrothermal-Vent Communities of the Deep Sea." *American Scientist,* July–August 1992. A comprehensive look at the structure and function of undersea vent communities, with excellent illustrations.

Wilson, E. O. "Threats to Biodiversity." *Scientific American,* September 1989. The elimination of habitats, particularly in the rain forest, is driving plants and animals to extinction in record numbers.

N E T   W A T C H

On-line resources for this chapter are on the World Wide Web at:
http://www.prenhall.com/~audesirk (click on the <u>table of contents</u> link and then select Chapter 46).

Answers to Multiple-Choice Questions

Chapter 1
1:a 2:c 3:d 4:c 5:e 6:d

Chapter 2
1:d 2:b 3:c 4:b 5:a 6:a

Chapter 3
1:a 2:e 3:a 4:b 5:c 6:c

Chapter 4:
1:d 2:c 3:b 4:a 5:c 6:e

Chapter 5
1:a 2:c 3:d 4:a 5:a 6:c 7:b

Chapter 6
1:b 2:e 3:e 4:a 5:d 6:c

Chapter 7
1:d 2:c 3:b 4:e 5:d 6:b

Chapter 8
1:a 2:c 3:d 4:c 5:a 6:c

Chapter 9
1:c 2:c 3:a 4:c 5:b 6:d

Chapter 10
1:b 2:c 3:b 4:d 5:a 6:d

Chapter 11
1:c 2:a 3:a 4:b 5:a 6:c

Chapter 12
1:c 2:e 3:b 4:a 5:d 6:c

Chapter 13
1:c 2:c 3:b 4:d 5:a 6:e

Chapter 14
1:b 2:c 3:c 4:d 5:c 6:b

Chapter 15
1:c 2:d 3:b 4:a 5:e 6:a

Chapter 16
1:e 2:d 3:e 4:e 5:a 6:d

Chapter 17
1:a 2:d 3:b 4:c 5:c 6:e

Chapter 18
1:e 2:d 3:d 4:a 5:b 6:c

Chapter 19
1:e 2:a 3:d 4:b,d,e 5:d 6:d

Chapter 20
1:d 2:b 3:e 4:c 5:a 6:d

Chapter 21
1:e 2:b 3:c 4:a 5:c 6:d

Chapter 22
1:b 2:a 3:b 4:c 5:c 6:b

Chapter 23
1:c 2:a 3:d 4:a 5:c 6:b

Chapter 24
1:e 2:a 3:d 4:a 5:a 6:d

Chapter 25
1:a 2:b 3:b 4:e 5:d 6:a

Chapter 26
1:b 2:a 3:c 4:d 5:a 6:d

Chapter 27
1:d 2:a 3:e 4:d 5:a 6:c

Chapter 28
1:e 2:d 3:a 4:c 5:d 6:c

Chapter 29
1:e 2:d 3:e 4:d 5:c 6:a

Chapter 30
1:d 2:a 3:c 4:d 5:e 6:d

Chapter 31
1:c 2:e 3:a 4:b 5:e 6:b

Chapter 32
1:e 2:b 3:e 4:c 5:a 6:c

Chapter 33
1:a 2:d 3:b 4:d 5:e 6:b

Chapter 34
1:e 2:c 3:b 4:a 5:c 6:e

Chapter 35
1:a 2:e 3e 4:b 5:c 6:d

Chapter 36
1:c 2e 3:b 4:e 5:d 6:d

Chapter 37
1:e 2:a 3:c 4:e 5:d 6:c

Chapter 38
1:a 2:d 3:e 4:b 5:c 6:b

Chapter 39
1:e 2:a 3:d 4:b 5:e 6:c

Chapter 40
1:e 2:e 3:c 4:a 5:b 6:d

Chapter 41
1:d 2:a 3:c 4:d 5:b 6:d

Chapter 42
1:e 2:e 3:d 4:a 5:c 6:b

Chapter 43
1:a 2:c 3:a 4:e 5:b 6:d

Chapter 44
1:d 2:c 3:a 4:e 5:a 6:c

Chapter 45
1:a 2:c 3:b 4:d 5:e 6:d

Chapter 46
1:a 2:a 3:c 4:b 5:e 6:c

Glossary

abiotic (ā-bī´-ah´-1tik): nonliving.

abortion: the procedure for terminating pregnancy in which the cervix is dilated and the embryo and placenta removed by suction.

abscisic acid (ab-sis´-ik): a plant hormone that generally inhibits the action of other hormones, enforcing dormancy in seeds and buds and causing closing of stomata.

abscission layer: a layer of thin-walled cells at the base of the petiole of a leaf that produces an enzyme that digests the cell wall holding leaf to stem, allowing the leaf to fall off.

absorption: the process by which nutrients are taken into cells.

accessory pigments: colored molecules other than chlorophyll a that absorb light energy and pass it to chlorophyll.

acetylcholine (ah-sēt´-il-kō´-lēn): a neurotransmitter found in the brain and in synapses of motor neurons onto skeletal muscles.

acid: a substance that releases hydrogen ions (H+) into solution; a solution with a pH of less than 7.

acid deposition: the deposition of nitric or sulfuric acid, either in dissolved in rain (acid rain) or in the form of dry particles, as a result of the production of nitrogen oxides or sulfur dioxide through burning, primarily of fossil fuels.

acrosome (ak´-rō-sōm): a vesicle located at the tip of an animal sperm, which contains enzymes needed to dissolve protective layers around the egg.

actin (ak'-tin): a major muscle protein, whose interactions with myosin produce contraction; found in the thin filaments of the muscle fiber.

action potential: a rapid change from a negative to a positive electrical potential in a nerve cell. This signal travels along an axon without change in size.

activation energy: in a chemical reaction, the energy needed to force the electron clouds of reactants together, prior to the formation of products.

active site: the region of an enzyme molecule that binds substrates and performs the catalytic function of the enzyme.

active transport: the movement of materials across a membrane through the use of cellular energy, usually against a concentration gradient.

active visual signal: a specific movement or posture that communicates information.

adaptation: a characteristic of an organism that helps it to survive and reproduce in a particular environment; also, the process of acquiring such characteristics.

adaptive radiation: The rise of many new species in a relatively short time as a result of a single species invading different habitats and evolving under different selective pressures in those habitats.

adenosine triphosphate (a-den´-ō-sen trīfos´-fāt; ATP): a molecule composed of ribose sugar, adenine, and three phosphate groups. The last two phosphate groups are attached by energy-carrier "high-energy bonds." ATP is the major energy carrier in cells.

adipose tissue (a´-di-pōse): tissue composed of fat cells.

adrenal cortex: the outer part of the adrenal gland, which secretes steroid hormones that regulate metabolism and salt balance.

adrenal medulla: the inner part of the adrenal gland, which secretes epinephrine (adrenalin) and norepinephrine (noradrenalin).

aerobic: using oxygen.

age structure: the distribution of males and females in a population according to age groups.

agglutination (a-glu-tin-ā´-shun): clumping of foreign substances or microbes, caused by binding with antibodies.

aggression: antagonistic behavior, usually between members of the same species, often resulting from competition for resources.

aggressive mimicry (mim´ik-ré): the evolution of a predatory organism to resemble a harmless one, thus gaining access to its prey.

algae (al´-gē; sing. alga): a general term for simple aquatic plants lacking vascular tissue.

allantois (al-an-tō´-is): one of the embryonic membranes of reptiles, birds, and mammals. In reptiles and birds, the allantois serves as a waste-storage organ. In mammals, the allantois forms most of the umbilical cord.

allele (al-ēl´): one of several alternative forms of a particular gene.

allele frequency: for any given gene, the relative proportion of each allele of that gene found in a population.

allergy: an inflammatory response produced by the body in response to invasion by foreign materials, such as pollen, which are themselves harmless.

allopatric speciation (al-ō-pat´-rik): speciation that occurs when two populations are separated by a physical barrier that prevents gene flow between them (geographical isolation).

allosteric regulation (al-ō-ster´-ik): enzyme regulation in which an inhibitor molecule binds to an enzyme at a site away from the active site, changing the shape or charge of the active site so that it can no longer bind substrate molecules.

alternation of generations: a life cycle typical of plants in which a diploid sporophyte (spore-producing) generation alternates with a haploid gametophyte (gamete-producing) generation.

altruism: a type of behavior that may decrease the reproductive success of the individual performing it but benefits that of other individuals.

alveolus (al-vē´-ō-lus; pl. alveoli): a tiny air sac within the lungs surrounded by capillaries where gas exchange with the blood occurs.

amino acid: the individual subunit of which proteins are made, composed of a central carbon atom to which is bonded an amino group ($-NH_2$), a carboxyl group ($-COOH$), a hydrogen atom, and a variable group of atoms denoted by the letter "R."

ammonia: NH_3; a highly toxic nitrogen-containing waste product of amino acid breakdown. In the mammalian liver, it is converted to urea.

amniocentesis (am-nē-ō-sen-tē´-sis): a procedure for sampling the amniotic fluid surrounding a fetus. A sterile needle is inserted through the abdominal wall, uterus, and amniotic sac of a preg-

nant woman, into the amniotic fluid. Ten to 20 milliliters of amniotic fluid is withdrawn. Various tests may be performed on the fluid and the fetal cells suspended in it to provide information on the developmental and genetic state of the fetus.

amnion (am´-nē-on): one of the embryonic membranes of reptiles, birds, and mammals, enclosing a fluid-filled cavity that envelops the embryo.

amniotic egg (am-nē-ot´-ik): the egg of reptiles and birds. It contains an amnion that encloses the embryo in a watery environment; this allows the egg to be laid on dry land.

amplexus (am-pleck´-sus): a form of external fertilization found in amphibians, in which the male holds the female during spawning and releases his sperm directly onto her eggs.

amygdala (am-ig´-da-la): part of the forebrain of vertebrates, involved in production of appropriate behavioral responses to environmental stimuli.

amylase (am´-ē-lās): an enzyme found in saliva and pancreatic secretions that catalyzes the breakdown of starch.

anaerobic: not using oxygen.

analogous structures: structures that have similar functions and superficially similar appearance but very different anatomy, such as the wings of insects and birds. The similarities are due to similar selective pressures rather than common ancestry.

anaphase (an´-a-fāz): the stage of mitosis in which the sister chromatids of each chromosome separate from one another and are moved to opposite poles of the cell. In meiosis I, the stage in which homologous chromosomes, consisting of two sister chromatids, are separated. In meiosis II, the stage in which the sister chromatids of each chromosome separate from one another and are moved to opposite poles of the cell.

androgens: the collective term for male sex hormones.

angina pectoris (an-jĭ´-na pek-tōr´-is): chest pain associated with reduced blood flow to the heart muscle caused by obstruction of coronary arteries.

angiosperm (an´-gē-ō-sperm): a flowering vascular plant.

annual ring: pattern of alternating light (early) and dark (late) xylem of woody stems and roots, formed because of unequal availability of water in different seasons of the year, usually spring and summer.

antagonistic muscles: a pair of muscles, one of which contracts and in so doing extends the other, an arrangement that makes possible movement of the skeleton at joints.

anterior (an-tēr´-ē-ur): the front, forward, or head end of an animal.

anterior pituitary: a lobe of the pituitary gland that produces prolactin and growth hormone as well as hormones that regulate hormone production in other glands.

anther (an´-ther): the uppermost part of the stamen in which pollen develops.

antheridium (an-ther-id´-ē-um): a structure in which male sex cells are produced, found in the bryophytes and certain seedless vascular plants.

antibiotic resistance: the ability of a mutated pathogen to resist the effects of an antibiotic that normally kills it.

antibody: a protein produced by cells of the immune system that combines with a specific antigen and usually facilitates its destruction.

anticodon: a sequence of three bases in transfer RNA that is complementary to the three bases of a codon of messenger RNA.

antidiuretic hormone (an-tē-dī-ūr-et´-ik): also called ADH; a hormone produced by the hypothalamus and released into the bloodstream by the posterior pituitary gland when blood volume is low. It increases the permeability of the distal tubule and the collecting duct to water, allowing more water to be reabsorbed into the bloodstream.

antigen: a complex molecule, usually protein or polysaccharide, that stimulates the production of a specific antibody.

apical dominance: the phenomenon whereby a growing shoot tip inhibits the sprouting of lateral buds.

apical meristem (āp´-i-kul mer´-i-stem): the cluster of meristematic cells found at the tip of a shoot or root (or one of their branches).

appendicular skeleton (ap-pen-dik´-ū-lur): that portion of the skeleton consisting of the bones of the extremities and their attachments to the axial skeleton; the pectoral and pelvic girdles, the arms, legs, hands, and feet.

aqueous humor (ā´-kwē-us): clear, watery fluid between the cornea and lens of the eye.

archegonium (ar-ke-gō´-nē-um): a structure in which female sex cells are produced, found in the bryophytes and certain seedless vascular plants.

arteriole (ar-tēr´-ē-ōl): a small artery that empties into capillaries. Contraction of the arteriole regulates blood flow to various parts of the body.

artery (ar´-tur-ē): a vessel with muscular, elastic walls that conducts blood away from the heart.

ascus (as´-kus): a saclike case in which sexual spores are formed by members of the fungal division Ascomycota.

asexual reproduction: reproduction that does not involve the fusion of haploid sex cells. The parent body may divide and new parts regenerate, or a new, smaller individual may be formed attached to the parent, to drop off when complete.

association neuron: In neural networks, a nerve cell that is postsynaptic to a sensory neuron and presynaptic to a motor neuron. In actual circuits, there maybe many association neurons between individual sensory and motor neurons.

aster: in cell division of animal cells, a star-shaped array of microtubules extending in all directions outward from the centriole.

atherosclerosis (ath´-er-ō-skler-ō´-sis): a disease characterized by obstruction of arteries by cholesterol deposits and thickening of the arterial walls.

atmospheric cycle: a nutrient cycle for a substance whose primary reservoir is the atmosphere.

atom: the smallest particle of an element that retains the properties of the element.

atomic number: the number of protons in the nuclei of all atoms of a particular element.

atrioventricular node (ā´-trē-ō-ven-trik´-ū-lar nōd): a specialized mass of muscle at the base of the right atrium through which the electrical activity initiated in the sinoatrial node is transmitted to the ventricles.

atrioventricular valve: heart valves between the atria and the ventricles that prevent backflow of blood into the atria during ventricular contraction.

atrium (ā´-trē-um): a chamber of the heart that receives venous blood and passes it to a ventricle.

auditory canal (aw´-dih-tōry): a canal within the outer ear that conducts sound from the external ear to the eardrum.

auditory nerve: the nerve leading from the mammalian cochlea to the brain, carrying information about sound.

autoimmune disease: a disorder in which the immune system produces antibodies against the body's own cells.

autonomic nervous system: part of the peripheral nervous system of vertebrates that synapses on glands, internal organs, and smooth muscle and produces largely involuntary responses.

autosome (aw´-tō-sōm): a chromosome found in homologous pairs in both males and females, and which does not bear the genes determining sex.

autotroph (aut´-ō-trof): a "self-feeder," usually meaning a photosynthetic organism.

auxin (awk´-sin): a plant hormone that influences many plant functions, including phototropism, apical dominance, and root branching. Auxin generally stimulates cell elongation and, in some cases, cell division and differentiation.

axial skeleton: the skeleton forming the body axis, including the skull, vertebral column, and rib cage.

axon: a long process of a nerve cell, extending from the cell body to synaptic endings on other nerve cells or on muscles.

B cell: a type of lymphocyte that participates in humoral immunity; gives rise to plasma cells that secrete antibodies into the circulatory system and to memory cells.

bacillus (buh-sil´-us; pl. bacilli): a rod-shaped bacterium.

bacterial conjugation: the exchange of genetic material between bacteria.

bacteriophage (bak-tir´-ē-ō-fāj): a virus specialized to attack bacteria.

bacterium (bak-tir´-ē-um; pl. bacteria): an organism consisting of a single prokaryotic cell surrounded by a complex polysaccharide coat.

balanced polymorphism: the prolonged maintenance of two or more alleles in a population, usually because each allele is favored by a separate selective force.

ball-and-socket joint: a joint in which the rounded end of one bone fits into a hollow depression in another, as in the hip; allows movement in several directions.

bark: the outer layer of a woody stem, consisting of phloem, cork cambium, and cork cells.

Barr body: an inactivated X chromosome found in cells of female mammals, which have two X chromosomes. The Barr body usually appears as a dark spot in the nucleus.

basal body: the organelle, structurally identical to a centriole, that gives rise to the microtubules of cilia and flagella.

base: a substance that is capable of combining with and neutralizing H+ ions. In molecular genetics, one of the nitrogen-containing, single- or double-ringed structures that distinguish one nucleotide from another. In DNA, the bases are adenine, guanine, cytosine, and thymine.

basidiospore (ba-sid´-ē-ō-spōr): a sexual spore formed by members of the fungal division Basidiomycota.

basidium (bas-id´-ē-um): a diploid cell, often club-shaped, formed on the basidiocarp of members of the fungal division Basidiomycota, which produces basidiospores by meiosis.

basilar membrane (bas-eh-lar): a membrane in the cochlea that bears hair cells that respond to the vibrations produced by sound.

behavior: as any observable response to external or internal stimuli.

behavioral isolation: lack of mating between species of animals that differ substantially in courtship and mating rituals.

bilateral symmetry: body plan in which only a single plane drawn through the central axis will divide the body into mirror-image halves.

bile (bīl): a liquid secretion of the liver stored in the gallbladder and released into the small intestine during digestion. Bile is a complex mixture of bile salts, water, other salts, and cholesterol.

bile salts: substances synthesized in the liver from cholesterol and amino acids, that assist in the breakdown of lipids by dispersing them into small particles on which enzymes may act.

binary fission: The process by which a single bacterium divides in half, producing two identical offspring.

binocular vision: The ability to see objects simultaneously through both eyes, providing greater depth perception and more accurate judgment of the size and distance of an object from the eyes.

biodiversity: all living things within a given geographical area and the interrelationships among them.

biological clock: a metabolic timekeeping mechanism found in most organisms, whereby the organism measures the approximate length of a day (24 hours) even without external environmental cues such as light and dark.

biological magnification: the increasing accumulation of a toxic substance in progressively higher trophic levels.

biome (bī´-ōm): a general type of ecosystem occupying extensive geographical areas, characterized by similar plant communities: for example, deserts.

biosphere (bī´-ō-sfēr): that part of the Earth inhabited by living organisms; includes both the living and nonliving components.

biotic (bī-ah´-tik): living.

biotic potential: maximum rate at which a population could increase, assuming a maximum birth rate and minimum death rate.

birth control pill: a temporary contraceptive method that prevents ovulation by providing a continuing supply of estrogen and progesterone, which in turn suppresses LH release; must be taken daily, usually for 21 days each menstrual cycle.

bladder: a hollow muscular storage organ for urine.

blade: the flat part of a leaf.

blastocyst (blas´-tō-sist): an early stage of human embryonic development, consisting of a hollow ball of cells, enclosing a mass of cells attached to its inner surface, which becomes the embryo.

blastopore: the site at which a blastula indents to form a gastrula.

blastula: in animals, the embryonic stage attained at the end of cleavage, in which the embryo usually consists of a hollow ball with a wall one or several cell layers thick.

blood: A fluid consisting of plasma in which blood cells are suspended, carried within the circulatory system.

Bohr effect (bor): the tendency of hemoglobin to release oxygen more readily when the blood is slightly more acidic than normal, as is caused by an increase in dissolved carbon dioxide.

book lungs: thin layers of tissue resembling pages in a book, enclosed in a chamber and used as a respiratory organ by certain types of arachnids.

boom-and-bust cycle: a population cycle characterized by rapid exponential growth followed by a sudden major decline in population size, seen in seasonal species and some populations of small rodents, such as lemmings.

Bowman's capsule: the cup-shaped portion of the nephron in which blood filtrate is collected from the glomerulus.

bradykinin (brā´-dē-kī´-nin): a chemical formed during tissue damage that binds to receptor molecules on pain nerve endings, giving rise to the sensation of pain.

brain: the part of the central nervous system of vertebrates enclosed within the skull.

branch root: a root that arises as a branch of a preexisting root, through divisions of pericycle cells and subsequent differentiation of the daughter cells.

bronchiole (bron´-kē-ōl): a narrow tube formed by repeated branching of the bronchi, which conducts air into the alveoli.

bronchus (bron´-kus): a tube that conducts air from the trachea to each lung.

bryophyte (brī´-ō-fīt): a division of simple nonvascular plants, including mosses and liverworts.

bud: in animals, a small copy of an adult that develops on the body of the parent. It eventually breaks off and becomes independent. In plants, an embryonic shoot, usually very short and consisting of an apical meristem with several leaf primordia.

budding: asexual reproduction by growth of a miniature copy, or bud, of the adult animal on the body of the parent. The bud breaks off to begin independent existence.

buffer: a compound that minimizes changes in pH by reversibly taking up or releasing H+ ions.

bulbourethral gland (bul-bō-ū-rē´-thrul): in male mammals, a gland that secretes a basic, mucus-containing fluid that forms part of the semen.

bulk flow: the movement of many molecules of a gas or fluid in unison from an area of higher pressure to an area of lower pressure.

bundle-sheath cells: cells that surround the veins of plants. In C_4 plants, but not C_3 plants, these cells contain chloroplasts.

C_3 cycle: the cyclic series of reactions whereby carbon dioxide is fixed into carbohydrates during the light-independent reactions of photosynthesis. Also called Calvin-Benson cycle.

C_4 pathway: the series of reactions in certain plants that fixes carbon dioxide into oxaloacetic acid, which is later broken down for use in the C_3 cycle of photosynthesis.

callus: a porous mass of bone and cartilage secreted by osteoblasts in response to a break in bone; it holds together the ends of the bones while repair takes place.

calmodulin (kal-mod´-ū-lin): a cytoplasmic protein that acts as a second messenger by binding calcium ions and changing configuration. The configuration change may activate enzymes, initiating a series of biochemical reactions.

calorie (kal´-ōr-ē): a measure of the energy derived from food. When capitalized (that is, Calorie), this unit is the amount of energy required to raise the temperature of 1 liter of water 1 degree Celsius. It represents 1000 calories (with a lowercase "c"). The energy content of foods is measured in Calories.

Calvin-Benson cycle: see C_3 cycle.

cambium (kam´-bē-um; pl. cambia): a lateral meristem that runs parallel to the long axis of roots and stems that causes secondary growth of woody plant stems and roots. See cork cambium; vascular cambium.

camera eye: the type of eye found in vertebrates and cephalopod mollusks, consisting of a light-sensitive retina, a lens that focuses images on the retina, and muscles that adjust the shape of the lens.

camouflage (cam´-a-flage): coloration and\or shape that renders an organism inconspicuous in its environment.

cancer: a disease in which some of the body's cells escape from normal regulatory processes and divide without control.

capillary: the smallest type of blood vessel, capillaries connect arterioles with venules. Capillary walls, through which exchange of nutrients and wastes occurs, are only one cell thick.

capsule: a polysaccharide or protein coating that surrounds the cell walls of some bacteria.

carbohydrate: a compound composed of carbon, hydrogen, and oxygen, with the approximate chemical formula (CH2O)n; includes sugars and starches.

carbon fixation: the initial steps in the C^3 cycle, in which carbon dioxide reacts with ribulose bisphosphate to form a stable organic molecule.

carbonic anhydrase (car-bon´-ik an-hī´-drās): an enzyme found in red blood cells that catalyzes the formation of bicarbonate ions from dissolved carbon dioxide and water.

cardiac cycle: the alternation of contraction (systole) and relaxation (diastole) of the heart chambers; systole and diastole.

cardiac muscle (kar´-dē-ak): specialized muscle of the heart, able to initiate its own contraction independent of the nervous system.

carnivore (kar´-neh-vōr): literally "meat eater," a predatory organism feeding on other heterotrophs; a secondary (or higher) consumer.

carnivorous (kar-niv´-e-rus): feeding on the bodies of other animals.

carotenoid (ka-rot´-en-oyd): red, orange, or yellow pigments found in chloroplasts that serve as accessory light-gathering molecules in thylakoid photosystems.

carpel (kar´pel): the female reproductive structure of a flower, composed of stigma, style, and ovary.

carrier: an individual who is heterozygous for a recessive condition. Carriers display the dominant phenotype but can pass on their recessive allele to their offspring.

carrier protein: a membrane protein that facilitates diffusion of specific substances across the membrane. The molecule to be transported binds to the outer surface of the carrier protein; the protein then changes shape, allowing the molecule to move across the membrane through the protein.

carrying capacity: the maximum population size that an ecosystem can maintain on a sustained basis. Determined primarily by the availability of space, nutrients, water, and light.

cartilage (kar´-teh-lij): a form of connective tissue forming portions of the skeleton, consisting of chondrocytes and their extracellular secretion of collagen. Cartilage resembles flexible bone.

Casparian strip (kas-par´-ē-an): a waxy, waterproof band in the cell walls between endodermal cells in a root, which prevents the movement of water and minerals in and out of the vascular cylinder through the extracellular space.

catalyst (cat´-a-list): a substance that speeds up a chemical reaction without itself being permanently changed in the process. Catalysts lower the activation energy of a reaction.

catastrophism: the hypothesis that Earth has experienced a series of geological catastrophes, probably imposed by a supernatural being, which accounts for the multitude of species, both extinct and modern, while preserving creationism.

cell: the smallest unit of life, consisting, at a minimum, of an outer membrane enclosing a watery medium containing organic molecules, including genetic material composed of DNA.

cell body: part of a nerve cell in which most of the common cellular organelles are located. Also often a site of integration of inputs to the nerve cell.

cell cycle: the sequence of events in the life of a cell, from one division to the next.

cell division: in eukaryotes, the process of reproduction of single cells, usually into two identical daughter cells, by mitosis accompanied by cytokinesis.

cell-mediated immunity: an immune response in which foreign cells or substances are destroyed by contact with T cells.

cell plate: in plant cell division, a series of vesicles that fuse to form the new plasma membranes and cell wall separating the daughter cells.

cell wall: a layer of material, usually made up of cellulose or celluloselike materials, found outside the plasma membranes of plants, fungi, bacteria, and some protists.

cellular respiration: the oxygen-requiring reactions occurring in mitochondria that break down the end products of glycolysis into carbon dioxide and water, while capturing large amounts of energy as ATP.

cellular slime mold: a funguslike protist consisting of individual amoeboid cells that can aggregate to form a sluglike mass, which in turn forms a fruiting body.

cellulase: an enzyme that catalyzes the breakdown of the carbohydrate cellulose into its component glucose molecules. Cellulase is almost entirely restricted to microorganisms.

cellulose: an insoluble carbohydrate composed of glucose subunits; forms the cell wall of plants.

central nervous system: in vertebrates, the brain and spinal cord.

central vacuole: a large, fluid-filled vacuole occupying most of the volume of many plant cells; performs several functions, including maintaining turgor pressure.

centriole (sen´-trē-ōl): in animal cells, a microtubule-containing structure found at the microtubule organizing center and the base of each cilium and flagellum. Gives rise to the microtubules of cilia and flagella and may be involved in spindle formation during cell division.

centromere (sen´-trō-mēr): the region of a replicated chromosome at which the sister chromatids are held together.

cephalization (sef-al-ī-zā´-shun): the increasing concentration over evolutionary time of sensory structures and nerve ganglia at the anterior end of animals.

cerebellum (ser-uh-bel´-um): part of the hindbrain of vertebrates, concerned with coordinating movements of the body.

cerebral cortex (ser-ē´-brel kōr´-tex): a thin layer of neurons on the surface of the vertebrate cerebrum, in which most neural processing and coordination of activity occurs.

cerebral hemisphere: one of two nearly symmetrical halves of the cerebrum, connected by a broad band of axons, the corpus callosum.

cerebrospinal fluid: a clear fluid produced within the ventricles of the brain that fills the ventricles and cushions the brain and spinal cord.

cerebrum (ser-ē´-brum): part of the forebrain of vertebrates concerned with sensory processing, direction of motor output, and coordination of most bodily activities. The cerebrum consists of two nearly symmetrical halves (the hemispheres) connected by a broad band of axons, the corpus callosum.

cervix (ser´-vicks): a ring of connective tissue at the outer end of the uterus, leading into the vagina.

channel protein: a membrane protein that forms a channel or pore completely through the membrane and that is usually permeable to one or a few water-soluble molecules, especially ions.

chemical bond: the force of attraction between neighboring atoms that holds them together in a molecule.

chemical equilibrium: the condition in which the "forward" reaction of reactants to products proceeds at the same rate as the "backward" reaction from products to reactants, so that no net change in chemical composition occurs.

chemiosmosis (kē-mē-os-mō´-sis): a process of ATP generation in chloroplasts and mitochondria. The movement of electrons down an electron transport system is used to pump hydrogen ions across a membrane, thereby building up a concentration gradient of hydrogen ions across the membrane. The hydrogen ions diffuse back across the membrane through the pores of ATP-synthesizing enzymes. The energy of their movement down their concentration gradient drives ATP synthesis.

chemoreceptor: a sensory receptor that responds to chemicals from the environment; used in the chemical senses of taste and smell.

chemosynthetic (kēm´-ō-sin-the-tic): capable of oxidizing inorganic molecules to obtain energy.

chiasma (kē-as´-ma; pl. chiasmata): during prophase I of meiosis, a point at which a chromatid of one chromosome crosses with a chromatid of the homologous chromosome. Exchange of chromosomal material between chromosomes takes place at a chiasma.

chitin (kī´-tin): a compound found in the cell walls of fungi and the exoskeletons of insects and some other arthropods, composed of chains of nitrogen-containing, modified glucose molecules.

chlorophyll (klōr´-ō-fil): a pigment found in chloroplasts that captures light energy during photosynthesis; absorbs violet, blue, and red light but reflects green.

chloroplast (klōr´-ō-plast): the organelle of plants and plantlike protists that is the site of photosynthesis; surrounded by a double membrane and containing an extensive internal membrane system bearing chlorophyll.

cholecystokinin (kō´-lē-sis-tō-ki´-nin): a digestive hormone produced by the small intestine that stimulates release of pancreatic enzymes.

chondrocyte (kon´-drō-sīt): the living cells of cartilage, which with its extracellular secretions of collagen, form cartilage.

chorion (kor´-ē-on): the outermost embryonic membrane in reptiles, birds, and mammals. In birds and reptiles, the chorion functions mostly in gas exchange. In mammals, the chorion forms most of the embryonic part of the placenta.

chorionic gonadotropin (CG): a hormone secreted by the chorion (one of the fetal membranes), which maintains the integrity of the corpus luteum during early pregnancy.

chorionic villi (kor-e-on-ik): in mammals, fingerlike projections of the chorion of the embryo that penetrate the uterine lining and form the embryonic portion of the placenta.

chorionic villus sampling: a procedure for sampling cells from the chorionic villi produced by a fetus. A tube is inserted into the uterus of a pregnant woman, and a small sample of villi are suctioned off for genetic and biochemical analysis.

choroid (kōr´-oyd): a darkly pigmented layer of tissue behind the retina that contains blood vessels and pigment that absorbs stray light.

chromatid (krō´-ma-tid): one of the two identical strands of DNA and protein forming a replicated chromosome. The two sister chromatids are joined at the centromere.

chromatin (krō´-ma-tin): the complex of DNA and proteins that makes up eukaryotic chromosomes.

chromosome (krō´-mō-sōme): in eukaryotes, a linear strand composed of DNA and protein, found in the nucleus of a cell, that contains the genes; in prokaryotes, a circular strand composed solely of DNA.

chylomicron (kī-lō-mī´-kron): a droplet consisting of triglycerides, cholesterol, and phospholipids, and coated with a thin layer of protein. Chylomicrons are formed in the cells lining the small intestine, enter the lymphatic system, and eventually reach the circulatory system.

chyme (kīme): an acidic, souplike mixture of partially digested food, water, and digestive secretions that is released from the stomach into the small intestine.

ciliate (sil´-ē-et): a category of protozoan characterized by cilia and a complex unicellular structure, including harpoonlike organelles called trichocysts. Members of the genus Paramecium are well-known ciliates.

cilium (sil´-ē-um; pl. cilia): a short, hairlike projection from the surface of certain eukaryotic cells, containing microtubules in a 9+2 arrangement. Movement of cilia may propel cells through a fluid medium or move fluids over a stationary surface layer of cells.

circadian rhythm (sir-kā´-dē-un): an event that recurs with a period of about 24 hours, even in the absence of environmental cues.

citric acid cycle: a cyclic series of reactions in which the acetyl groups from the pyruvic acids produced by glycolysis are broken down to CO2, accompanied by the formation of ATP and electron carriers. Occurs in the matrix of mitochondria.

class: the taxonomic category composed of related genera. Closely related classes form a division or phylum.

classical conditioning: a training procedure in which an animal learns to perform a response (such as salivation) to a new stimulus that did not elicit that response originally (such as a sound). This is accomplished by pairing a stimulus that elicits the response automatically (in this case, food) with the new stimulus.

cleavage: the early cell divisions of embryos, in which little or no growth occurs between divisions. Cleavage reduces the cell size and distributes gene-regulating substances to the newly formed cell.

climax community: a relatively stable community that forms the endpoint of succession.

cloning: the process of producing many identical copies of a gene; also used to denote producing many genetically identical copies of an organism.

closed circulatory system: the type of circulatory system found in certain worms and vertebrates in which the blood is always confined within the heart and vessels.

clumped distribution: characteristic distribution of populations in which individuals are clustered into groups. These may be social or based on the need for a localized resource.

coccus (ka´-kus; pl. cocci): a spherical bacterium.

cochlea (kōk´-lē-uh): a coiled, bony, fluid-filled tube found in the mammalian inner ear, which contains receptors (hair cells) that respond to the vibration of sound.

codominance: the relation between two alleles of a gene, such that both alleles are phenotypically expressed in heterozygous individuals.

codon: a sequence of three bases of messenger RNA that specifies a particular amino acid to be incorporated into a protein. Certain codons also signal the beginning and end of protein synthesis.

coelom (sē´-lōm): a space or cavity within the body separating the body wall from the inner organs.

coenzyme: an organic molecule, often derived from a water-soluble vitamin, that is bound to certain enzymes and is required for their proper functioning.

coevolution: the evolution of adaptations in two species due to their extensive interactions with one another, so that each species acts as a major force of natural selection upon the other.

cohesion: the tendency of the molecules of a substance to hold together.

cohesion-tension theory: a model for transport of water in xylem, which states that water is pulled up the xylem tubes, powered by the force of evaporation of water from the leaves (producing tension) and held together by hydrogen bonds between nearby water molecules (cohesion).

coleoptile (kō-lē-op´tīl): a protective sheath surrounding the shoot in monocot seeds, allowing the shoot to push aside soil particles as it grows.

collagen (kol´-uh-gen): a fibrous protein found in connective tissue such as bone and cartilage.

collar cells: specialized cells lining the inside channels of sponges. Flagella extend from a sievelike collar, creating a water current that draws microscopic organisms through the collar to be trapped.

collecting ducts: conducting tubes within the kidney that collect urine from many nephrons and conduct it through the medulla into the renal pelvis. Urine may become concentrated in the collecting ducts if ADH is present.

collenchyma (kōl-en´-ki-ma): an elongated, polygonal plant cell type with irregularly thickened primary cell walls that is alive at maturity and that provides support to the plant body.

colostrum (kō-los´-trum): a yellowish fluid high in protein and containing antibodies that is produced by the female breasts before milk secretion begins.

commensalism (kum-en´-sal-ism): a symbiotic relationship between two species in which one benefits while the other is neither harmed nor benefited.

communication: the act of producing a signal that causes another animal, usually of the same species, to change its behavior in a way beneficial to one or both participants.

community: all the interacting populations within an ecosystem.

compact bone: the hard and strong outer bone; composed of Haversian systems.

companion cell: a cell adjacent to a sieve-tube element in phloem, involved in control and nutrition of the sieve-tube element.

competition: interaction that occurs between individuals when both attempt to utilize a resource (for example, food or space) that is limited relative to the demand for it.

competitive exclusion principle: the concept that no two species can simultaneously and continuously occupy the same ecological niche.

competitive inhibition: in enzyme-catalyzed reactions, a condition in which two molecules (at least one a substrate for the enzyme) compete for entry into the active site of the enzyme, thus slowing the rate of reaction.

complement: a group of blood-borne proteins that participate in the destruction of foreign cells to which antibodies have bound.

complement reactions: interactions among foreign cells, antibodies, and complement proteins, resulting in the destruction of the foreign cells.

complementary base pair: in nucleic acids, bases that pair by hydrogen bonding; in DNA, adenine is complementary to thymine, and guanine is complementary to cytosine.

complete flower: a flower that has all four floral parts (sepals, petals, stamens, and carpels).

compound: a substance composed of two or more elements that can be broken into its constituent elements by chemical means.

compound eye: a type of eye found in arthropods, composed of numerous independent subunits, called ommatidia. Each ommatidium apparently contributes a single piece of a mosaic-like image perceived by the animal.

concentration: the number of particles of a dissolved substance in a given unit of volume.

concentration gradient: the difference in concentration of a substance between two parts of a fluid or across a barrier such as a membrane.

condom: a contraceptive sheath that may be worn over the penis during intercourse, preventing sperm from being deposited in the vagina.

cone: a cone-shaped photoreceptor cell in the vertebrate retina, not as sensitive to light as the rods. The three types of cones are most sensitive to different colors of light and provide color vision. See also rod.

conifer (kon´-eh-fer): a class of tracheophyte that reproduces using seeds formed inside cones and retains its leaves throughout the year.

connective tissue: a tissue type consisting of diverse tissues, including bone, fat, and blood, which generally contain large amounts of extracellular material.

constant region: part of an antibody molecule that is similar in all antibodies.

consumer: an organisms that eats other organisms; a heterotroph.

contest competition: a mechanism for resolving intraspecific competition using social or chemical interactions.

contractile vacuole: a fluid-filled vacuole found in certain protists that takes up water from the cell cytoplasm, contracts, and expels the water outside the cell through a pore in the plasma membrane.

control: that portion of an experiment in which all possible variables are held constant; in contrast to the "experimental" portion, in which a particular variable is altered.

convergence: a condition in which a large number of nerve cells provide input to a smaller number of cells.

convergent evolution: the independent evolution of similar structures among unrelated organisms, due to similar selective pressures. See analogous structures.

convolutions: foldings of the cerebral cortex of the vertebrate brain.

copulation: reproductive behavior in which the penis of the male is inserted into the body of the female, where it releases sperm.

cork cambium: a lateral meristem in woody roots and stems that gives rise to cork cells.

cork cell: a protective cell of the bark of woody stems and roots; at maturity, cork cells are dead, with thick, waterproofed cell walls.

cornea (kōr´-nē-uh): the clear outer covering of the eye in front of the pupil and iris.

corona radiata (ka-rō´-na rā-dē-a´-ta): the layer of cells surrounding an egg after ovulation.

corpus callosum (kōr´pus kal-ō´-sum): the band of axons that connect the two cerebral hemispheres of vertebrates.

corpus luteum (kor´-pus lū´-tē-um): in the mammalian ovary, a structure derived from the follicle after ovulation, which secretes the hormones estrogen and progesterone.

cortex: the part of a primary root or stem located between the epidermis and the vascular cylinder.

cotyledon (kot-ul-ē´don): also called a seed leaf; a leaflike structure within a seed that absorbs food molecules from the endosperm and transfers them to the growing embryo.

countercurrent flow: a structural arrangement that enhances diffusion between two fluids that differ in their concentration of dissolved substances by moving the fluids past one another in opposite directions, separated by semipermeable membranes.

coupled reaction: a pair of reactions, one exergonic and one endergonic, that are linked together so that the energy produced by the exergonic reaction provides the energy needed to drive the endergonic reaction.

covalent bond (ko-vā´-lent): a chemical bond between atoms in which electrons are shared.

creationism: the hypothesis that all species of organisms on Earth were created in essentially their present form by a supernatural being, and that significant modification of those species, specifically their transformation into new species, cannot occur through natural processes.

crista (kris´-ta; pl. cristae): a fold in the inner membrane of a mitochondrion.

crop: an organ found in both earthworms and birds in which ingested food is temporarily stored before being passed to the gizzard, where it is pulverized.

cross-bridge: in muscles, an extension of myosin that binds to and pulls upon actin to produce contraction of the muscles.

cross-fertilization: union of sperm and egg from two different individuals of the same species.

crossing over: the exchange of corresponding segments of the chromatids of two homologous chromosomes during meiosis.

cultural evolution: changes in the behavior of a population of animals, especially humans, by learning behaviors acquired by members of previous generations.

cupula (kūp´-ū-luh): a gelatinous structure found in the lateral line system of fish and the semicircular canals of the vestibular apparatus of other vertebrates. Hairs are embedded in the cupula, which is deflected by movement of the fluid filling the canals.

cuticle (kū´-ti-kul): a waxy or fatty coating on the exposed surfaces of epidermal cells of many land plants, which aids in the retention of water.

cyanobacteria: photosynthetic prokaryotic cells, utilizing chlorophyll and releasing oxygen as a photosynthetic by-product, sometimes called "blue-green algae."

cyclic AMP: a cyclic nucleotide formed within many target cells as a result of the reception of modified amino acid or peptide hormones, which causes metabolic changes in the cell; often called a second messenger.

cyclic nucleotide (sik´-lik nū´-klē-ō-tīd): a nucleotide in which the phosphate group is bonded to the sugar at two points, forming a ring. Cyclic nucleotides serve as intracellular messengers.

cyst (sist): an encapsulated resting stage in the life cycle of certain invertebrates, such as parasitic flatworms and roundworms.

cysteine amino acids: by attraction and repulsion among amino acid side groups, and by interaction between the cellular environment (water or lipids) and the amino acid side groups.

cytokinesis (sī-tō-ki-nē´-sis): division of the cytoplasm and organelles into two daughter cells during cell division. Usually cytokinesis occurs during telophase of mitosis.

cytokinin (sī-tō-kī´-nin): a plant hormone that promotes cell division, fruit growth, sprouting of lateral buds, and prevents aging of plant parts, especially leaves.

cytoplasm (sī´-tō-plazm): the material contained within the plasma membrane of a cell, exclusive of the nucleus.

cytoskeleton: a network of protein fibers in the cytoplasm that gives shape to a cell, holds and moves organelles, and is often involved in cell movement.

cytosol (sī´-tō-sol): the fluid part of the cytoplasm.

cytotoxic T cell: a type of T cell that directly destroys foreign cells upon contacting them.

day-neutral plant: a plant in which flowering occurs as soon as it has grown and developed, regardless of length of day.

decomposers: a group of organisms, mainly fungi and bacteria, that digest organic material by secreting digestive enzymes into the environment. In the process they liberate nutrients into the environment.

degeneracy: the property of the genetic code whereby several codons may specify the same amino acid.

dehydration synthesis: a chemical reaction in which two molecules are joined by a covalent bond with the simultaneous removal of a hydrogen from one molecule and a hydroxyl group from the other, forming water.

deletion: a mutation in which one or more pairs of nucleotides are removed from a gene.

dendrite (den´-drīt): branched tendrils that extend outward from the cell body, specialized to respond to signals from the external environment or from other neurons.

denitrifying bacteria (dē-nī´-treh-fī-ing): bacteria that break down nitrates, releasing nitrogen gas to the atmosphere.

density-dependent: description of any factor, such as predation, that limits population size more effectively as the population density increases.

density-independent: description of any factor, such as freezing weather, that limits a sensitive population's size and growth regardless of its density.

deoxyribonucleic acid (dē-ox-ē-rī-bō-nū- klā´-ik; DNA): a molecule composed of deoxyribose nucleotides; the genetic information of all living cells.

dermal tissue system: a plant tissue system that makes up the outer covering of the plant body.

dermis (dur´-mis): the layer of skin lying beneath the epidermis, composed of connective tissue and containing blood vessels, muscles, nerve endings, and glands.

desmosome (dez´-mō-sōm): a strong cell-to-cell junction that functions in attaching adjacent cells to one another.

detritus feeders (de-trī´-tus): a diverse group of organisms ranging from worms to vultures that live off the wastes and dead remains of other organisms.

development: the process by which an organism proceeds from fertilized egg through adulthood to eventual death.

diabetes mellitus (di-a-bē´-tes mel-ī´-tus): a disease characterized by defects in the production, release, or reception of insulin, characterized by high blood glucose levels that fluctuate with sugar intake.

diaphragm (dī´uh-fram): a dome-shaped muscle forming the floor of the chest cavity. Contraction of this muscle pulls it downward, enlarging the cavity and causing air to be drawn into the lungs. Also, a contraceptive rubber cap that fits snugly over the cervix, preventing the sperm from entering the uterus and thereby preventing pregnancy.

diastole (di-as´-tō-lē): the portion of the cardiac cycle during which the ventricles are relaxed; the lower of the two blood pressure readings.

diatom (dī´-e-tom): a category of protist that includes photosynthetic forms with two-part glassy outer coverings. Diatoms are important photosynthetic organisms in fresh and salt water.

dicot: short for dicotyledon; a type of flowering plant characterized by embryos with two cotyledons, or seed leaves, modified for food storage.

differential permeability: the property of a membrane by which some substances can permeate more readily than other substances.

differential reproduction: differences in reproductive output among individuals of a population, usually as a result of genetic differences.

differentiated cell: a mature cell specialized for a specific function; in plants, differentiated cells usually do not divide.

differentiation: the process whereby relatively unspecialized cells, especially of embryos, become specialized into particular tissue types.

diffusion: the net movement of particles from a region of high concentration of that particle to a region of low concentration, driven by the concentration gradient. May occur entirely within a fluid or across a barrier such as a membrane.

digestion: the process by which food is physically and chemically broken down into molecules that can be absorbed by cells.

dihybrid cross: a breeding experiment involving parents that differ in two distinct, genetically determined traits.

dinoflagellate (dī-nō-fla´-gel-et): a category of protist that includes photosynthetic forms in which two flagella project through armorlike plates. Abundant in oceans, these sometimes reproduce rapidly, causing "red tides."

dioecious (dī-ē´-shus): pertaining to organisms in which male and female gametes are produced by separate individuals.

diploid (dip´-loyd): referring to a cell with pairs of homologous chromosomes.

direct development: a developmental pathway in which the offspring is born as a miniature version of the adult and does not radically change its body form as it grows and matures.

directional selection: a type of natural selection in which one extreme phenotype is favored over all others.

disaccharide (dī-sak´-a-rīd): a carbohydrate formed by the covalent bonding of two monosaccharides.

disruptive selection: a type of natural selection in which both extreme phenotypes are favored over the average phenotype.

disulfide bridge: the covalent bond formed between the sulfur atoms of two cysteines in a protein; often causes the protein to fold, by bringing otherwise distant parts of the protein close together.

divergence: a condition in which a small number of nerve cells provide input to a larger number of cells.

divergent speciation: speciation in which a single parental species splits to form two or more new species that exist simultaneously. See phyletic speciation.

division: the taxonomic category contained within a kingdom and consisting of related classes of plants, fungi, bacteria, or plantlike protists.

DNA hybridization: a technique by which DNA from two species is separated into single strands and then allowed to reform. Hybrid double-stranded DNA from the two species can occur where the sequence of nucleotides is complementary. The greater the degree of hybridization, the closer the evolutionary relatedness of the two species.

DNA library: a readily accessible, easily duplicable complete set of all the DNA of a particular organism, usually cloned into bacterial plasmids.

DNA polymerase: an enzyme that bonds DNA nucleotides together into a continuous strand, using a preexisting DNA strand as a template.

dominance hierarchy: a social arrangement in which animals, usually through aggressive interactions, establish a rank for some or all of the members that determines access to resources.

dominant: an allele that can determine the phenotype of heterozygotes completely, so that they are indistinguishable from individuals homozygous for the allele. In the heterozygotes, the expression of the other (recessive) allele is completely masked.

dopamine (dōp´-uh-mēn): a transmitter in the brain whose actions are largely inhibitory. Loss of dopamine-containing neurons causes Parkinson's disease.

dormancy: a state in which an organism does not grow or develop; usually marked by lowered metabolic activity and resistance to adverse environmental conditions.

dorsal (dōr´-sul): the top, back, or uppermost surface of an animal oriented with its head forward.

dorsal root ganglion: a ganglion located on the dorsal (sensory) branch of each spinal nerve, containing the cell bodies of sensory neurons.

double covalent bond: a covalent bond that occurs when two atoms share two pairs of electrons.

double fertilization: in flowering plants, a phenomenon in which two sperm nuclei fuse with the nuclei of two cells of the female gametophyte. One sperm fuses with the egg to form the zygote, while the second sperm nucleus fuses with the two haploid nuclei of the primary endosperm cell to form a triploid endosperm cell.

douching: washing the vagina after intercourse in an attempt to remove sperm before they enter the uterus; an ineffective contraceptive method.

Down syndrome: a genetic disorder caused by the presence of three copies of chromosome 21. Common characteristics include mental retardation, distinctively shaped eyelids, a small mouth with protruding tongue, heart defects, and low resistance to infectious diseases.

duct: a tube or opening through which exocrine secretions are released.

echolocation: use of ultrasonic sounds, which bounce back from nearby objects, to produce an "auditory image" of nearby surroundings; used by bats and porpoises.

ecological isolation: lack of mating between organisms belonging to different populations that occupy distinct habitats within the same general area.

ecological niche (nitch): the role of a particular species within an ecosystem, including all aspects of its interaction with the living and nonliving environment.

ecology (ē-kol´-uh-gē): the study of the interrelationships of organisms with each other and with their nonliving environment.

eecosystem (ē′kō-sis-tem): all the organisms and their nonliving environment within a defined area.

ectoderm (ek′-tō-derm): the outermost embryonic tissue layer, which gives rise to structures such as hair, the epidermis of the skin, and the nervous system.

effector (ē-fek′-tōr): a part of the body (usually a muscle or gland) that carries out responses as directed by the nervous system.

egg: the haploid female gamete, usually large and nonmotile, containing food reserves for the developing embryo.

electron: a subatomic particle, found in the electron shells outside the nucleus of an atom, bearing a unit of negative charge and very little mass.

electron carrier: a molecule that can reversibly gain and lose electrons. Electron carriers generally accept high-energy electrons produced during an exergonic reaction and donate the electrons to acceptor molecules that use the energy to drive endergonic reactions.

electron shell: a region within which electrons orbit corresponding to a fixed energy level within which at a given distance from the nucleus of an atom.

electron transport system: a series of electron carrier molecules found in the thylakoid membranes of chloroplasts and the inner membrane of mitochondria, which extract energy from electrons and generate ATP or other energetic molecules.

electrophoresis: a biochemical technique that separates molecules according to their electrical charge and molecular weight.

element: a substance that cannot be broken down to a simpler substance by ordinary chemical means.

embryo: the stages of development of animals that begin with fertilization of the egg cell and end with hatching or birth. In mammals, embryo usually refers to the early stages in which the developing animal does not yet resemble the adult of the species.

embryo sac: the haploid female gametophyte of flowering plants.

embryonic disk: in human embryonic development, the flat, two-layered group of cells that separates the amniotic cavity from the yolk sac.

emergent property: an intangible attribute that arises as the result of complex ordered interactions among individual parts.

emigration (em-uh-grā′shun): movement of individuals out of an area.

emphysema (em-fuh-sē′-muh): a condition in which the alveoli become brittle and rupture, causing decreased area for gas exchange.

endergonic (en-der-gon′-ik): pertaining to a chemical reaction that requires an input of energy to proceed; an "uphill" reaction.

endocrine gland: a ductless, hormone-producing gland consisting of cells that release their secretions into the extracellular fluid from which the secretions diffuse into nearby capillaries.

endocytosis (en-dō-sī-tō′-sis): the movement of material into a cell by a process in which the plasma membrane engulfs extracellular material, forming membrane-bound sacs that enter the cytoplasm.

endoderm (en′-dō-derm): the innermost embryonic tissue layer, which gives rise to structures such as the lining of the digestive and respiratory tracts.

endodermis (en-dō-der′-mis): the innermost layer of small, close-fitting cells of the cortex of a root that form a ring around the vascular cylinder.

endogenous pyrogen: a chemical produced by the body that stimulates the production of a fever (elevated body temperature).

endometrium (en-dō-mē′-trē-um): the nutritive inner lining of the uterus.

endoplasmic reticulum (ER) (en-dō-plaz′-mik re-tik′-ū-lum): a system of membranous tubes and channels within eukaryotic cells; the site of most protein and lipid synthesis.

endorphin (en-dōr′-fin): one of a group of peptide neuromodulators in the vertebrate brain that, by reducing the sensation of pain, mimics some of the actions of opiates.

endoskeleton (en′-dō-skel′-uh-tun): a rigid internal skeleton with flexible joints to allow for movement.

endosperm: a triploid food storage tissue found in the seeds of flowering plants used to nourish the developing plant embryo.

endospore (spōr): a resistant or resting structure that disperses readily and withstands unfavorable environmental conditions.

endosymbiotic hypothesis: the hypothesis that certain organelles, especially chloroplasts and mitochondria, arose as mutually beneficial associations between the ancestors of eukaryotic cells and captured bacteria living within the cytoplasm of the pre-eukaryotic cell.

energy: the capacity to do work.

energy carrier: a molecule that stores energy in "high-energy" chemical bonds and releases the energy again to drive coupled endothermic reactions. ATP is the most common energy carrier in cells.

energy level: the specific amount of energy characteristic of a given electron shell in an atom.

energy pyramid: a graphical representation of the energy contained in succeeding trophic levels, with maximum energy at the base (primary producers) and steadily diminishing amounts at higher levels.

enhancer: a part of chromosome, usually located close to a gene, that binds proteins that alter the rate of gene transcription.

entropy (en′-trō-pē): a measure of the amount of randomness and disorder in a system.

environmental resistance: any factor that tends to counteract biotic potential, limiting population size.

enzyme (en′zīm): a protein catalyst that speeds up the rate of specific biological reactions.

epicotyl (ep′-ē-kot-ul): the part of the embryonic shoot located above the cotyledons but below the tip of the shoot.

epidermis (ep-uh-der′-mis): specialized epithelial tissue that forms the outer layer of skin. In plants, the outermost layer of cells of a leaf, young root, or young stem.

epididymis (e-pi-di′-dē-mus): tubes that connect with and receive sperm from the seminiferous tubules of the testis.

epiglottis (ep-eh-gla′-tis): a flap of cartilage in the lower pharynx that covers the opening to the larynx during swallowing. In doing so, the epiglottis directs the food down the esophagus.

epistasis (ep-i-stā′-sis): a pattern of inheritance in which the expression of one gene depends on, or is modified by, the expression of another gene.

epithelial tissue (eh-puh-thē′-lē-ul): a tissue type that forms membranes that cover the body surface and line body cavities, and that also gives rise to glands.

equilibrium population: a population in which allele frequencies and the distribution of genotypes do not change from generation to generation.

erythroblastosis fetalis (ē-rith′-rō-blas-tō′-sis fē-tal′-is): a condition in which the red blood cells of a newborn Rh-positive baby are attacked by antibodies produced by its Rh-negative mother, causing jaundice and anemia. Retardation and death are possible consequences if treatment is inadequate.

erythrocytes (ē-rith′-rō-sītes): red blood cells active in oxygen transport, which contain the red pigment hemoglobin.

erythropoietin (ē-rith′-rō-pō-ē′-tin): a hormone produced by the kidneys in response to oxygen deficiency that stimulates production of red blood cells by the bone marrow.

esophagus (eh-sof′-eh-gus): a muscular passageway connecting the pharynx to the next chamber of the digestive tract, the stomach in humans and other mammals.

essential amino acids: amino acids that are required nutrients. The body is unable to manufacture them, and so they must be supplied in the diet.

essential fatty acids: fatty acids that are required nutrients. The body is unable to manufacture them, and so they must be supplied in the diet.

estrogen: in vertebrates, a female sex hormone produced by follicle cells of the ovary, which stimulates follicle development, oogenesis, development of secondary sex characteristics, and growth of the uterine lining.

ethology (ē-thol′-ō-gē): the study of animal behavior under natural or near-natural conditions.

ethylene: a plant hormone that promotes ripening of fruits and leaf and fruit drop.

euglenoid (yū′-gle-nōyd): a category of protist characterized by one or more whiplike flagella used for locomotion and a photoreceptor for detecting light. Euglenoids are photosynthetic, but some are capable of heterotrophic nutrition if deprived of chlorophyll.

eukaryotic (ū-kar-ē-ot′-ik): referring to cells of organisms of the kingdoms Protista, Fungi, Plantae, and Animalia. Eukaryotic cells have their genetic material enclosed within a membrane-bound nucleus and contain other membrane-bound organelles.

eustachian tube (ū-stā′-shin): a tube connecting the middle ear with the pharynx that allow pressure between the middle ear and the atmosphere to equilibrate.

evolution: the descent of modern organisms from preexisting life forms; strictly speaking, any change in the proportions of different genotypes in a population from one generation to the next.

excitatory synapse: a synapse between two nerve cells in which the resting potential of the postsynaptic cell becomes less negative and more likely to fire an action potential due to the activity of the presynaptic cell.

excretion: Elimination of waste substances from the body. Excretion can occur from the digestive system, skin glands, urinary system, or lungs.

excretory pore: Opening in the body wall of certain invertebrates, such as the earthworm, through which urine is excreted.

exergonic (ex-er-gon´-ik): pertaining to a chemical reaction that liberates energy (either heat energy or in the form of increased entropy); a "downhill" reaction.

exhalation (ex-per-ā´-shun): the act of releasing air from the lungs, which results from relaxation of the respiratory muscles.

exocrine gland: a gland that releases its secretions into ducts that lead to the outside of the body or into the digestive tract.

exocytosis (ex-ō-sī-tō´-sis): the movement of material out of a cell by a process in which intracellular material is enclosed within a membrane-bound sac that moves to the plasma membrane and fuses with it, releasing the material outside the cell.

exon: a segment of DNA in a eukaryotic gene that codes for amino acids in a protein (see also intron).

exoskeleton (ex´-ō-skel´-uh-tun): a rigid external skeleton that supports the body, protects the internal organs and has flexible joints to allow for movement.

exotic species: A species introduced into an ecosystem where it did not evolve. Occasionally such species may flourish and outcompete native species.

exponential growth: a continuously accelerating increase in population size.

extensor muscle: a muscle that straightens a joint.

external ear: the fleshy portion of the ear that extends outside the skull.

external fertilization: union of sperm and egg outside the body of either parental organism.

extinction: the death of all members of a species.

extracellular digestion: the physical and chemical breakdown of food that occurs outside of a cell, usually in a digestive cavity.

extraembryonic membranes: in reptile, bird, and mammal embryonic development, the chorion, amnion, allantois, and yolk sac that function in gas exchange, provision of the watery environment needed for development, waste storage, and storage of the yolk, respectively.

eyespot: a simple, lensless eye found in various invertebrates, including flatworms and jellyfish. Eyespots can distinguish light from dark and sometimes the direction of light, but they cannot form an image.

facilitated diffusion: diffusion of molecules across a membrane, assisted by protein pores or carriers embedded in the membrane.

family: the taxonomic category contained within an order and consisting of related genera.

fat: a lipid composed of three saturated fatty acids covalently bonded to glycerol; fats are solid at room temperature. Also, fat-storing connective tissue whose cells are packed with triglycerides (fats); also called adipose tissue.

fatty acid: an organic molecule composed of a long chain of carbon atoms, with a carboxylic acid (—COOH) group at one end. Fatty acids may be saturated (all single bonds between the carbon atoms) or unsaturated (one or more double bonds between the carbon atoms).

feedback inhibition: in enzyme-mediated chemical reactions, the condition in which the product of a reaction inhibits one or more of the enzymes involved in synthesizing the product.

fermentation: anaerobic reactions that convert the pyruvic acid produced by glycolysis into lactic acid or alcohol and CO_2.

fertilization: the fusion of male and female haploid gametes to form a zygote.

fetal alcohol syndrome (FAS): a cluster of symptoms including retardation and physical abnormalities that occur in infants born to mothers who consumed large amounts of alcoholic beverages during pregnancy.

fetus: the later stages of mammalian embryonic development (after the second month for humans), when the developing animal has come to resemble the adult of the species.

fever: an elevation in body temperature caused by chemicals (pyrogens) released by white blood cells in response to infection.

fibrillation: rapid, uncoordinated, and ineffective contractions of heart muscle cells.

fibrin (fī´-brin): a clotting protein formed in the blood in response to a wound. Fibrin binds with other fibrin molecules and provides a matrix around which a blood clot forms.

fibrous root system: a root system, commonly found in monocots, characterized by many roots of approximately the same size arising from the base of the stem.

filament: in flowers, the stalk of a stamen, which bears an anther at its tip.

filtrate: Fluid produced by filtration; in the kidneys, fluid produced by filtration of the blood through the glomerular capillaries.

filtration: within Bowman's capsule in each nephron of a kidney, the process by which blood is pumped under pressure through permeable capillaries of the glomerulus, forcing out water, dissolved wastes, and nutrients.

fimbria (fim´-brē-a; pl. fimbriae): in female mammals, the ciliated, fingerlike projections of the oviduct that sweep the ovulated egg from the ovary into the oviduct.

first law of thermodynamics: a principle of physics that within any isolated system, energy can be neither created nor destroyed, but can be converted from one form to another.

fission: asexual reproduction by dividing the body into two smaller, complete organisms.

fitness: the reproductive success of an organism, usually expressed in relation to the average reproductive success of all individuals in the same population.

fixed action pattern: stereotyped, rather complex behavior that is genetically programmed (innate); often triggered by a stimulus called a releaser.

flagellum (fla-jel´-um; pl. flagella): a long, hairlike extension of the plasma membrane. In eukaryotic cells it contains microtubules arranged in a 9+2 pattern. Movement of flagella propel some cells through fluid media.

flame cell: in flatworms, a specialized cell containing beating cilia that conducts water and wastes through the branching tubes that serve as an excretory system.

flexor muscle: a muscle that flexes (decreases the angle of) a joint.

flower: the reproductive structure of an angiosperm plant.

fluid: a liquid or gas.

fluid mosaic model: a model of membrane structure. According to this model, membranes are composed of a double layer of phospholipids in which is embedded a variety of proteins. The phospholipid bilayer is a somewhat fluid matrix that allows movement of proteins within it.

fluid-phase endocytosis: nonselective movement of extracellular fluid into a cell, enclosed within a vesicle formed from the plasma membrane.

follicle: in the ovary of female mammals, the oocyte and its surrounding accessory cells.

follicle-stimulating hormone (FSH): a hormone produced by the anterior pituitary gland that stimulates spermatogenesis in males and development of the follicle in females.

food chain: an illustration of feeding relationships in an ecosystem using a single representative from each of the trophic levels.

food vacuole: a membrane-bound space within a single cell in which food is enclosed. Digestive enzymes are released into the vacuole, where intracellular digestion occurs.

food web: a relatively accurate representation of the complex feeding relationships within an ecosystem, including many organisms at various trophic levels, with many of the consumers occupying more than one level simultaneously.

forebrain: during development, the anterior portion of the brain. In mammals, the forebrain differentiates into the thalamus, the limbic system, and the cerebrum. In humans, the cerebrum contains about half of all the neurons in the brain.

fossil: the remains of an organism, usually preserved in rock. Fossils include petrified bones or wood; shells; impressions of body forms such as feathers, skin, or leaves; and markings made by organisms such as footprints.

fossil fuels: fuels such as coal, oil, and natural gas, derived from the bodies of prehistoric organisms.

founder effect: a type of genetic drift in which an isolated population founded by a small number of individuals may develop allele frequencies that are very different from those of the parent population, because of chance inclusion of disproportionate numbers of certain alleles in the founders.

fovea (fō´-vē-uh): the central region of the vertebrate retina, upon which images are focused. The fovea contains closely packed cones.

free-living: not parasitic.

free nerve endings: finely-branched endings of some receptor neurons, some of which respond to touch and pressure, others to heat and cold, and others to pain; also produce sensations of itching and tickling.

fruit: in flowering plants, the ripened ovary (plus, in some cases, other parts of the flower) that contains the seeds.

fruiting body: a spore-forming reproductive structure found in certain protists, bacteria, and fungi.

functional group: one of several groups of atoms commonly found in organic molecules, including hydrogen, hydroxyl, amino, carboxyl, and phosphate groups.

gallbladder: a small sac next to the liver in which the bile secreted by the liver is stored and concentrated. Bile is released from the gallbladder to the small intestine through the bile duct.

gamete (gam´-ēt): a haploid sex cell formed in sexually reproducing organisms.

gametic incompatibility: the inability of sperm from one species to fertilize eggs of another species.

gametophyte (ga-mēt´-ō-fīt): the multicellular haploid stage in the life cycle of plants.

ganglion (gang´-lē-un): a cluster of neurons.

ganglion cells: a type of cell comprising the innermost layer of the vertebrate retina whose axons form the optic nerve.

gap junction: a type of cell-to-cell junction in animals in which channels connect the cytoplasm of adjacent cells.

gastric inhibitory peptide: A hormone produced by the small intestine that inhibits the activity of the stomach.

gastrin: A hormone produced by the stomach that stimulates acid secretion in response to food.

gastrovascular cavity: a sac like chamber with digestive functions, found in simple invertebrates. A single opening serves as both mouth and anus, while the chamber provides direct access of nutrients to the cells.

gastrula (gas´-trū-la): in animal development, a three-layered embryo with ectoderm, mesoderm, and endoderm cell layers. The endoderm layer usually encloses the primitive gut.

gastrulation (gas-trū-lā´-shun): the process whereby a blastula develops into a gastrula, including the formation of endoderm, ectoderm, and mesoderm;

gene: a unit of heredity encoding the information needed to specify the amino acid sequence of proteins and hence particular traits. A gene is a segment of DNA located at a particular place on a chromosome.

gene flow: the movement of alleles from one population to another owing to migration of individual organisms.

gene pool: the total of all alleles of all genes in the population; for a single gene, the total of all the alleles of that gene that occur in a population.

generative cell: in flowering plants, one of the haploid cells of a pollen grain. The generative cell undergoes mitosis to form two sperm cells.

genetic code: the collection of codons of mRNA, each of which directs the incorporation of a particular amino acid into a protein during protein synthesis.

genetic drift: a change in the allele frequencies of a small population purely by chance.

genetic recombination: the generation of new combinations of alleles on homologous chromosomes, due to exchange of DNA during crossing over.

genotype (jēn´-ō-tīp): the genetic composition of an organism; the actual alleles of each gene carried by the organism.

genus (jē-nis): the taxonomic category consisting of very closely related species.

geographical isolation: the separation of two populations by a physical barrier.

germ layer: A tissue layer formed during early embryonic development.

germination: the growth and development of a seed, spore, or pollen grain.

gibberellin (jib-er-el´-in): a plant hormone that stimulates seed germination, fruit development, and cell division and elongation.

gills: in aquatic animals, a branched tissue richly supplied with capillaries around which water is circulated for gas exchange.

gizzard: a muscular organ found in earthworms and birds in which food is mechanically broken down prior to chemical digestion.

glomerulus (glō-mer´-ū-lus): a dense network of thin-walled capillaries located within the Bowman's capsule of each nephron. Here blood pressure forces water and dissolved nutrients through capillary walls for filtration by the nephron.

glucose: the most common monosaccharide, with the molecular formula $C6H12O6$. Most polysaccharides, including cellulose, starch, and glycogen, are made of glucose subunits covalently bonded together.

glycerol (glis´-er-ol): a three-carbon alcohol to which fatty acids are covalently bonded to make fats and oils.

glycogen (glī´-kō-jen): a long, branched polymer of glucose that is stored by animals in the muscles and liver and metabolized as a source of energy.

glycolysis (glī-kol´-i-sis): reactions carried out in the cytosol that break down glucose into two molecules of pyruvic acid, producing two ATP molecules. Glycolysis does not require oxygen but can proceed when oxygen is present.

glycoprotein: a protein to which a carbohydrate is attached.

goiter: a swelling of the neck caused by iodine deficiency, which affects the functioning of the thyroid gland and its hormones.

Golgi complex (gōl´-jē): a stack of membranous sacs found in most eukaryotic cells, which is the site of processing and separation of membrane components and secretory materials.

gonadotropin-releasing hormone (GnRH): a hormone produced by the neurosecretory cells of the hypothalamus, which stimulates cells in the anterior pituitary to release FSH and LH. GnRH is involved in the menstrual cycle and spermatogenesis.

gonads: organs where reproductive cells are formed; in males, the testes, and in females, the ovaries.

gonorrhea (gon-a-rē´-uh): a sexually transmitted bacterial infection of the reproductive organs. Untreated gonorrhea may result in sterility.

gradient: a difference in concentration, pressure, or electrical charge between two regions of space.

gradualism: a model of evolution, stating that morphological change and speciation are slow, gradual processes that are not necessarily simultaneous or linked.

Gram stain: A stain which is selectively taken up by the walls of ceratin types of bacteria (gram positive bacteria) and rejected by the cell walls of others called (gram negative bacteria). The stain is used to distinguish bacteria based on their cell wall construction.

granum (gra´-num; pl. grana): in chloroplasts, a stack of thylakoids.

gravitropism: growth with respect to the direction of gravity.

gray crescent: in frog embryonic development, an area of intermediate pigmentation in the fertilized egg containing gene-regulating substances required for the normal development of the tadpole.

gray matter: Outer portions of the brain and inner region of the spinal cord composed largely of neuron cell bodies which give these ares a gray color.

greenhouse effect: the ability of certain gases such as carbon dioxide and methane to trap sunlight energy in the atmosphere as heat. The glass in a greenhouse does the same, hence the name. This warming of the atmosphere is being enhanced by human production of these gases.

ground tissue system: a plant tissue system consisting of parenchyma, collenchyma, and sclerenchyma cells that makes up the bulk of a leaf or young stem, excluding vascular or dermal tissues. Most ground tissue cells function in photosynthesis, support, or carbohydrate storage.

guard cell: one of a pair of specialized epidermal cells surrounding the central opening of a stoma of a leaf, which regulates the size of the opening.

gymnosperms (jim´-nō-sperms): non-flowering seed plants such as conifers, cycads, and gingkos.

habituation (heh-bich-ū-ā´-shun): simple learning characterized by a decline in response to a harmless, repeated stimulus.

hair cell: the type of receptor cell found in the inner ear. Hair cells bear hairlike projections, the bending of which between two membranes causes the receptor potential.

hair follicle: a gland in the dermis of mammalian skin, formed from epithelial tissue, that produces a hair.

halophile (hā´-lō-fīl): leterally "salt-loving," a type of archaebacterium that thrives in concentrated salt solutions.

haploid (hap´-loyd): referring to a cell that has only one member of each pair of homologous chromosomes.

Haversian system (ha-ver´-sē-un): unit of hard bone consisting of concentric layers of bone matrix with embedded osteocytes surrounding a small central canal containing a capillary.

heart: A muscular organ within the circulatory system responsible for pumping blood throughout the body.

heart attack: a situation in which blood flow through a coronary artery is severaly reduced or blocked, depriving some of the heart muscle of its blood supply.

heat of fusion: the energy that must be removed from a compound to transform it from a liquid into a solid at its freezing temperature.

heat of vaporization: the energy that must be supplied to a compound to transform it from a liquid into a gas at its boiling temperature.

helix (hē´-licks): a spiral structure similar to a corkscrew or a spiral staircase; a type of secondary structure of a protein.

helper T cell: a type of T cell that aids other immune cells to recognize and act against antigens.

hemocoel (hē´-mō-sēl): a blood cavity within the bodies of certain invertebrates in which blood bathes tissues directly. A hemocoel is part of an open circulatory system.

hemoglobin (hē´mō-glō-bin): an iron-containing protein that gives red blood cells their color. Hemoglobin binds to oxygen in the lungs and releases it to the tissues.

hemophilia: a recessive, sex-linked disease in which the blood fails to clot normally.

herbivore (erb´-i-vōr): literally "plant-eater," an organism that eats feed directly and exclusively on producers; a primary consumer.

hermaphroditic (her-maf´-ruh-dit´-ik): possessing both male and female sexual organs. Some hermaphroditic animals can fertilize themselves; others must exchange sex cells with a mate.

heterotroph (het´-er-ō-trōf´): literally "other-feeder," meaning an organism that eats other organisms.

heterozygous (het-er-ō-zī´-gōt): carrying two different alleles of a given gene; sometimes called a hybrid.

hindbrain: the posterior portion of the brain, containing the medulla, pons, and cerebellum

hinge joint: a joint at which one bone is moved by muscle and the other bone remains fixed, such as in the knee, elbow, or fingers; allows movement in only two dimensions.

hippocampus (hip-ō-cam´-pus): part of the forebrain of vertebrates, important in emotion and especially learning.

histamine: a substance released by certain cells in response to tissue damage and invasion of the body by foreign substances. Histamine promotes dilation of arterioles, leakiness of capillaries and triggers some of the events of the inflammatory response.

homeobox (hō´-mē-ō-box): a sequence of DNA coding for special, 60 amino acid proteins. These proteins activate or inactivate genes that control development. Homeoboxes thus specify embryonic cell differentiation.

homeostasis (hōm-ē-ō-stā´sis): the maintenance of relatively constant environment required for optimal functioning of cells, maintained by the coordinated activity of numerous regulatory mechanisms, including the respiratory, endocrine, circulatory, and excretory systems. The precise regulation of the composition of fluid bathing the body cells is a homeostatic function performed largely by the urinary system.

hominid: A general term applied to all humans and their prehistoric relatives, beginning with the Australopithecines, whose fossils date back at least 4.4 million years.

homologous structures: structures that may differ in function but that have similar anatomy, presumably because of descent from common ancestors.

homologue (hō-´mō-log): a chromosome that is similar in appearance and genetic information to another chromosome with which it pairs during meiosis. Also called homologous chromosome.

homozygous (hō-mō-zī´-gōt): carrying two copies of the same allele of a given gene; also called true-breeding.

hormone: a chemical synthesized by one group of cells, secreted, and carried in the bloodstream to other cells, whose activity is influenced by reception of the hormone.

humoral immunity: an immune response in which foreign substances are inactivated or destroyed by antibodies circulating in the blood.

hybrid: an organism that is the offspring of parents differing in at least one genetically determined characteristic; also used to refer to the offspring of parents of different species.

hybrid infertility: reduced fertility (often complete sterility) in hybrid offspring of two different species.

hybrid inviability: the failure of a hybrid offspring of two different species to survive to maturity.

hydrogen bond: the weak attraction between a hydrogen atom bearing a partial positive charge (due to polar covalent bonding with another atom) and another atom, usually oxygen or nitrogen, bearing a partial negative charge. Hydrogen bonds may form between atoms of a single molecule or of different molecules.

hydrolysis (hī-drol´-i-sis): the chemical reaction that breaks a covalent bond through the addition of hydrogen to the atom forming one side of the original bond, and a hydroxyl group to the atom on the other side.

hydrophilic (hī-drō-fil´-ik): pertaining to a substance that dissolves readily in water, or to parts of a large molecule that form hydrogen bonds with water.

hydrophobic (hī-drō-fō´-bik): pertaining to a substance that does not dissolve in water.

hydrophobic interaction: the tendency for hydrophobic molecules to cluster together when immersed in water.

hydrostatic skeleton (hī-drō-stat´-ik): the use of fluid contained in body compartments to provide support for the body and mass against which muscles can contract.

hypertension: arterial blood pressure that is chronically elevated above the normal level.

hypertonic (hī-per-ton´-ik): referring to a solution that has a higher concentration of dissolved particles (and therefore a lower free water concentration) than the cytoplasm of a cell.

hypha (hī´-pha; pl. hyphae): a threadlike structure, many of which make up the fungal body, that consists of elongated cells, often with many haploid nuclei.

hypocotyl (hī´-pō-kot-ul): the part of the embryonic shoot located below the cotyledons but above the root.

hypothalamus (hī-pō-thal´-a-mus): a region of the brain that controls the secretory activity of the pituitary gland, and also synthesizes, stores, and releases certain peptide hormones, and directs autonomic nervous system responses.

hypothesis (hī-poth´-e-sis): in science, a supposition based on previous observations, which is offered as an explanation for an event, and used as the basis for further observations or experiments.

hypotonic (hī-pō-ton´-ik): referring to a solution that has a lower concentration of dissolved particles (and therefore a higher free water concentration) than the cytoplasm of a cell.

immigration (im-uh-grā´-shun): movement of individuals into an area.

immune response: a specific response by the immune system to invasion of the body by a particular foreign substance or microorganism, characterized by recognition of the foreign material by immune cells and its subsequent destruction by antibodies or cellular attack.

implantation: the process whereby the early embryo embeds itself within the lining of the uterus.

imprinting: the process by which an animal forms an association with another animal or object in the environment during a sensitive period.

inactivation: a process whereby certain chromosomes or parts of chromosomes are converted into a dense mass, preventing transcription.

inclusive fitness: the reproductive success of all organisms bearing a given allele, usually expressed in relation to the average reproductive success of all individuals in the same population. Compare with fitness.

incomplete dominance: a pattern of inheritance in which the heterozygous phenotype is intermediate between the two homozygous phenotypes.

incomplete flower: a flower that is missing one of the four floral parts (sepals, petals, stamens, or carpels).

independent assortment: a pattern of inheritance of multiple traits, in which the distribution of alleles for one trait into the gametes does not affect the distribution of alleles for other traits. Occurs with genes that are located on different chromosomes.

indirect development: a developmental pathway in which an offspring goes through radical changes in body form as it matures.

induction: the process by which a group of cells causes other cells to differentiate into a specific tissue type.

inflammatory response: a nonspecific, local response to injury to the body, characterized by phagocytosis of foreign substances and tissue debris by white blood cells, and "walling off" of the injury site by clotting of fluids escaping from nearby blood vessels.

inhalation: the act of drawing air into the lungs by enlarging the chest cavity.

inheritance of acquired characteristics: the hypothesis that organisms' bodies change during their lifetimes by use and disuse, and that these changes are inherited by their offspring.

inhibiting hormone: a hormone secreted by the neurosecretory cells of the hypothalamus that inhibits the release of specific hormones from the anterior pituitary gland.

inhibitory synapse: a synapse between two nerve cells in which the resting potential of the postsynaptic cell becomes more negative and less likely to fire an action potential due to the activity of the presynaptic cell.

innate (in-nāt´): inborn; instinctive; determined by the genetic makeup of the individual.

inner cell mass: in human embryonic development, the cluster of cells on one side of the blastocyst, which will develop into the embryo.

inner ear: the innermost part of the mammalian ear, composed of the bony, fluid-filled tubes of the cochlea and the vestibular apparatus.

inorganic molecule: any molecule that does not contain both carbon and hydrogen.

insertion: the site of attachment of a muscle to the relatively moveable bone on one side of a joint. In genetics, mutation in which one or more pairs of nucleotides are inserted within a gene.

insight: a complex form of learning that requires manipulation of mental concepts to arrive at adaptive behavior.

instinctive: innate; inborn; determined by the genetic makeup of the individual.

integration: in nerve cells, the process of adding up electrical signals from sensory inputs or other nerve cells, to determine the appropriate outputs.

integument (in-teg´-ū-ment): the outer layers of cells of the ovule surrounding the embryo sac; develops into the seed coat.

intensity: the strength of stimulation or response.

intercalated discs (in-tur´-cal-ā-ted): specialized interlocking membranes containing large numbers of gap junctions that connect adjacent cardiac muscle fibers.

interferon: a protein released by certain virus-infected cells that increases the resistance of other, uninfected, cells to viral attack.

intermediate filament: part of the cytoskeleton of eukaryotic cells, probably functioning mainly for support, composed of several types of proteins.

intermembrane compartment: the fluid-filled space between the inner and outer membranes of a mitochondrion.

internal fertilization: union of sperm and egg inside the body of the female.

internode: the part of a stem between two nodes.

interphase: the stage of the cell cycle between cell divisions. During interphase, chromosomes are replicated, and other cell functions occur, such as growth, movement, and acquisition of nutrients.

interspecific competition: competition between individuals of different species.

interstitial cells (in-ter-sti´-shul): in the vertebrate testis, testosterone-producing cells located between the seminiferous tubules.

interstitial fluid (in-tur-sti´-shul): fluid similar in composition to plasma (except lacking large proteins) which leaks from the capillaries and acts as a medium of exchange between the body cells and the capillaries.

intervertebral discs (in-tur-ver-tē´-brul): pads of cartilage found between the vertebrae that act as shock absorbers.

intracellular digestion: the chemical breakdown of food that occurs within single cells.

intraspecific competition: competition between individuals of the same species.

intrauterine device (IUD): a small copper or plastic loop, squiggle, or shield that is inserted in the uterus; a contraceptive method that works by irritating the uterine lining so that it cannot receive the embryo.

intron: a segment of DNA in a eukaryotic gene that does not code for amino acids in a protein.

invertebrate (in-vert´-uh-bret): a category of animals that never possess a vertebral column.

ion (ī´-on): an atom or molecule that has either an excess of electrons (and hence is negatively charged) or has lost electrons (and is positively charged).

ionic bond: a chemical bond formed by the electric attraction between positively and negatively charged ions.

iris: the pigmented muscular tissue of the vertebrate eye that surrounds and controls the size of a central opening, the pupil, through which light enters.

islet cells: clusters of cells in the endocrine portion of the pancreas that produce insulin and glucagon.

isotonic (ī-sō-ton´-ik): referring to a solution that has the same concentration of dissolved particles (and therefore the same free water concentration) as the cytoplasm of a cell.

isotope: one of several forms of a single element, the nuclei of which contain the same number of protons but different numbers of neutrons.

J-curve: the shape of the growth curve of an exponentially growing population, in which increasing numbers of individuals join the population with each succeeding time period.

joint: a flexible region between two rigid units of an exoskeleton or endoskeleton, to allow for movement between the units.

keratin (ker´-uh-tin): a fibrous protein found in hair, nails, and the epidermis of skin.

kidney: one of a pair of organs of the excretory system located on either side of the spinal column. Kidneys filter blood, removing wastes and regulating the composition and water content of the blood.

kin selection: selection favoring a certain allele because it increases the survival or reproductive success of relatives bearing the same allele.

kinesis (kin-nē´-sis): an innate behavior in which an organism changes the speed of its random movement in response to an environmental stimulus.

kinetic energy: the energy of movement; includes light, heat, mechanical movement, and electricity.

kinetochore (kī-nēt´-ō-kōr): a protein structure that forms at the centromere regions of chromosomes; attaches the chromosomes to the kinetochore microtubules of the spindle.

kinetochore microtubule: a microtubule of the spindle apparatus that connects the kinetochore of a chromosome with a spindle pole; involved in movement of the chromosome to the spindle pole during anaphase.

kingdom: the broadest taxonomic category, consisting of phyla or divisions. We recognize five kingdoms in this text: Monera, Protista, Fungi, Plantae, and Animalia.

Klinefelter syndrome: a set of characteristics typically found in individuals who have two X chromosomes and one Y chromosome. These individuals are phenotypically males, but sterile, and have several femalelike traits, including broad hips and partial breast development.

Krebs cycle: the citric acid cycle (in honor of Hans Krebs, who discovered many of its biochemical details).

lacteal (lak-t´ēl): a single lymph capillary that penetrates each villus of the small intestine.

lactose (lak´-tōs): a disaccharide composed of glucose and galactose; found in mammalian milk.

large intestine: The final section of the digestive tract consisting of the colon and the rectum, where feces are formed and stored.

larva (lar´-vuh): an immature form of an organism with indirect development prior to metamorphosis into its adult form. The caterpillars of moths and butterflies, and the maggots of flies, are larvae.

larynx (lār´-inx): that portion of the air passage between the pharynx and the trachea. The larynx contains the vocal cords.

lateral bud: a cluster of meristemic cells at the node of a stem; under appropriate conditions, it grows into a branch.

lateral line organ: a receptor organ in fish and aquatic amphibians that detects water movement. It utilizes clusters of hair cells whose hairs are bent by the deflection of a gelatinous cupula.

lateral meristem: also called cambium; a meristematic tissue that forms cylinders that run parallel to the long axis of roots and stems; usually found between the primary xylem and primary phloem (vascular cambium) and just outside the phloem (cork cambium).

leaf: an outgrowth of a stem, usually flattened and photosynthetic.

leaf primordium (pri-mōr´-dē-um; pl. primordia): a cluster of meristem cells located at the node of a stem that develops into a leaf.

learning: an adaptive change in behavior as a result of experience.

legume (leg´-yoom): a family of plants characterized by root swellings which nitrogen-fixing bacteria are housed. Includes soybeans. lupines, alfalfa, and clover.

lens: a clear object that bends light rays; in eyes, a flexible or movable structure used to focus light upon a layer of photoreceptor cells.

leukocyte (loo´-kō-sīt): any of the white blood cells circulating in the blood.

lichen (lī´-ken): a symbiotic association between an alga or cyanobacterium and a fungus, resulting in a composite organism.

life cycle: the events in the life of an organism from one generation to the next.

ligament: a tough connective tissue band connecting two bones.

light-dependent reactions: the first stage of photosynthesis, in which the energy of light is captured as ATP and NADPH; occurs in thylakoids of chloroplasts.

light-harvesting complex: in photosystems, the assembly of pigment molecules (chlorophyll and accessory pigments) that absorb light energy and transfer the energy to electrons.

light-independent reactions: the second stage of photosynthesis, in which the energy obtained by the light-dependent reactions is used to fix carbon dioxide into carbohydrates; occurs in the stroma of chloroplasts.

limbic system: a diverse group of brain structures, mostly in the lower forebrain, including the thalamus, hypothalamus, amygdala, hippocampus, and parts of the cerebrum, involved in basic emotion, drives, and behaviors, and in learning.

linkage: the inheritance of certain genes as a group because they are parts of the same chromosome. Linked genes do not show independent assortment.

lipase (lī´-pāse): an enzyme that catalyzes the breakdown of lipids such as fats.

lipid (li´-pid): one of a number of water-insoluble organic molecules, containing large regions composed solely of carbon and hydrogen. Lipids include oils, fats, waxes, phospholipids, and steroids.

liver: An organ with varied functions including bile production, glycogen storage, and detoxification of poisons.

locus: the physical location of a gene on a chromosome.

long-day plant: a plant that will flower only if the length of daylight is greater than some species-specific duration.

loop of Henle (hen´-lē): a specialized portion of the tubule of the nephron in birds and mammals that creates an osmotic concentration gradient in the fluid immediately surrounding it. This gradient in turn makes possible the production of urine more osmotically concentrated than blood plasma.

lungs: respiratory organs consisting of inflatable chambers within the chest cavity in which gas exchange occurs.

luteinizing hormone (LH): a hormone produced by the anterior pituitary gland that stimulates testosterone production in males and development of the follicle, ovulation, and production of the corpus luteum in females.

lymph (limf): pale fluid within the lymphatic system composed primarily of interstitial fluid and lymphocytes.

lymph nodes: small structures that act as filters for lymph. They contain lymphocytes and macrophages, which inactivate foreign particles such as bacteria.

lymphatic system: a system consisting of lymph vessels, lymph capillaries, lymph nodes, and the thymus and spleen. The system helps protect the body against infection, absorbs fats, and returns excess fluid and small proteins to the blood circulatory system.

lymphocyte (lim´-fō-sīt): a white blood cell type important in the immune response.

lysosome (lī´-sō-sōm): a membrane-bound organelle containing intracellular digestive enzymes.

macronutrient: a nutrient needed in relatively large quantities (often defined as composing more than 0.1% of an organism's body).

macrophage (mak´-rō-faj): a type of white blood cell that engulfs microbes. Macrophages destroy microbes by phagocytosis and also present microbial antigens to T cells, helping to stimulate the immune response.

major histocompatibility complex (MHC): proteins, usually located on the surfaces of body cells, that identify the cell as "self"; MHC proteins are also important in stimulating and regulating the immune response.

maltose (mal´-tōs): a disaccharide composed of two glucose molecules.

mammary glands (mam´-uh-rē): milk-producing organs used by female mammals to nourish their young.

mantle (man´-tul): an extension of the body wall in certain invertebrates, such as molluscs. It may secrete a shell, protect the gills, and, as in cephalopods, aid in locomotion.

marsupial (mar-sū´-pē-ul): a type of mammal whose young are born at an extremely immature stage and undergo further development in a pouch while they remain attached to a mammary gland. Includes kangaroos, opossums, and koalas.

matrix: the fluid contained within the inner membrane of a mitochondrion.

matter: the material of which the universe is made.

mechanical incompatibility: the inability of male and female organisms to exchange gametes, usually because of incompatibility of the reproductive structures.

mechanoreceptor: a receptor that responds to mechanical deformation, such as is caused by pressure, touch, or vibration.

medulla (med-oo´-la): part of the hindbrain of vertebrates that controls automatic activities such as breathing, swallowing, heart rate, and blood pressure.

medusa (meh-dū´-suh): a bell-shaped, often free-swimming stage in the life cycle of many cnidarians. Jellyfish are one example.

megakaryocyte (meg-a-kār´-ē-ō-sīt): a large cell type that remains in the bone marrow, pinching off pieces of itself. These fragments enter the circulation as platelets.

megaspore: a haploid cell formed by meiosis from a diploid megaspore mother cell. Through mitosis and differentiation, the megaspore develops into the female gametophyte.

megaspore mother cell: a diploid cell contained within the ovule of a flowering plant, which undergoes meiosis to produce four haploid megaspores.

meiosis (mī-ō´-sis): a type of cell division found in eukaryotic organisms, in which a diploid cell divides twice to produce four haploid cells.

membrane: continuous sheets of epithelial cells that cover the body and line body cavities. In cells, a thin sheet of lipids and proteins that surrounds the cell or its organelles, separating them from their surroundings.

memory cell: a long-lived descendant of a B or T cell that has been activated by contact with antigen. Memory cells are a reservoir of cells that rapidly respond to reexposure to the antigen.

meninges (men-in´-ges): three layers of connective tissue that surround the brain and spinal cord.

menstrual cycle: in human females, a complex 28-day cycle during which hormonal interactions among the hypothalamus, pituitary gland, and ovary coordinate ovulation and the preparation of the uterus to receive and nourish the fertilized egg. If pregnancy does not occur, the uterine lining is shed during menstruation.

menstruation: in human females, the monthly discharge of uterine tissue and blood from the uterus.

meristem cell (mer´-i-stem): an undifferentiated cell that remains capable of cell division throughout the life of a plant.

mesoderm (mes´-ō-derm): the middle embryonic tissue layer, lying between the endoderm and ectoderm, and usually the last to develop. Mesoderm gives rise to structures such as muscle and skeleton.

mesoglea (mez-ō-glē´-uh): a middle, jellylike layer within the body wall of cnidarians.

mesophyll (mez´-ō-fil): loosely packed parenchyma cells located beneath the epidermis of a leaf.

messenger RNA (mRNA): a strand of RNA, complementary to the DNA of a gene, that conveys the genetic information in DNA to the ribosomes to be used during protein synthesis. Sequences of three bases (codons) in mRNA specify particular amino acids to be incorporated into a protein.

metabolic pathway: a sequence of chemical reactions within a cell, in which the products of one reaction are the reactants for the next.

metabolism: the sum of all chemical reactions occurring within a single cell or within all the cells of a multicellular organism.

metamorphosis (met-a-mōr´-fō-sis): in animals with indirect development, a radical change in body form from larva to sexually mature adult, as seen in amphibians (tadpole to frog) and insects (caterpillar to butterfly).

metaphase (met´-a-fāz): the stage of mitosis in which the chromosomes, attached to kinetochore microtubules, are lined up along the equator of the cell.

methanogen (me-than´-ō-gen): a type of anaerobic archaebacterium capable of converting carbon dioxide to methane.

micelle (mī´-cēl): an aggregation of bile salts with its lipid-soluble ends facing inward. Fatty acids and monoglycerides dissolve in the hydrophobic portions and are carried to the cells of the lining of the small intestine.

microfilament: part of the cytoskeleton of eukaryotic cells, composed of the proteins actin and (sometimes) myosin; functions in the movement of cell organelles and in locomotion by extension of the plasma membrane.

micronutrient: a nutrient needed in relatively small quantities (often defined as composing less than 0.01% of an organism's body).

microsphere: a small, hollow sphere formed from proteins or proteins complexed with other compounds.

microspore: a haploid cell formed by meiosis from a microspore mother cell. Through mitosis and differentiation, the microspore develops into the male gametophyte.

microspore mother cell: a diploid cell contained within an anther of a flowering plant, which undergoes meiosis to produce four haploid microspores.

microtubule: a hollow, cylindrical strand found in eukaryotic cells, composed of the protein tubulin; part of the cytoskeleton used in movement of cell organelles, cell growth, and construction of cilia and flagella.

microtubule organizing center: the region(s) of a eukaryotic cell at which tubulin subunits are assembled into microtubules.

microvilli (mī-krō-vi´-lī): a fringe of microscopic projections of the cell membrane of each villus that serve to increase its surface area.

midbrain: during development, the central portion of the brain, which contains an important relay center, the reticular formation.

middle ear: part of the mammalian ear composed of the tympanic membrane, the eustachian tube, and three bones that transmit vibrations from the auditory canal to the oval window.

middle lamella: a thin layer of sticky polysaccharides, such as pectin, and other carbohydrates that separates and sticks together the primary cell walls of adjacent plant cells.

migration: in population genetics, the flow of genes between populations.

mimicry (mim´-ik-re): the situation in which a species has evolved to resemble another type of organism.

mineral: an inorganic substance, especially one found in rocks or soil.

mitochondrion (mī-tō-kon´-drē-un): an organelle, bounded by two membranes, that is the site of the reactions of aerobic metabolism.

mitosis (mī-tō´-sis): a type of nuclear division found in eukaryotic cells. Chromosomes are duplicated during interphase before mitosis. During mitosis, one copy of each chromosome moves into each of two daughter nuclei. The daughter nuclei are therefore genetically identical to each other.

mixture: a substance composed of two or more elements in variable proportions.

molecule (mol´-e-kūl): a particle composed of one or more atoms held together by chemical bonds. A molecule is the smallest particle of a compound that displays all the properties of that compound.

molt: to shed an external body covering, such as an exoskeleton, skin, feathers, or fur.

Monera (mō´-ne-ra): a taxonomic kingdom consisting of unicellular prokaryotic organisms, including bacteria, archaebacteria, and cyanobacteria.

monocot: short for monocotyledon; a type of flowering plant characterized by embryos with one seed leaf, or cotyledon.

monoecious (mon-ē´-shus): pertaining to organisms in which male and female gametes are produced in the same individual.

monohybrid cross: a breeding experiment in which the parents differ in only one genetically determined trait.

monomer (mo´-nō-mer): a small organic molecule, several of which may be bonded together to form a chain called a polymer.

monosaccharide (mo-nō-sak´-a-rīd): the basic molecular unit of all carbohydrates, usually composed of a chain of carbon atoms to which are bonded hydrogen and hydroxyl groups.

monotreme: a type of mammal that lays eggs; for example, the platypus.

morula (mor´-ū-la): in animals, an embryonic stage during cleavage, when the embryo consists of a solid ball of cells.

motor neuron: a neuron that receives instructions from the association neurons and activates effector organs such as muscles or glands.

motor unit: a single motor neuron and all the muscle fibers on which it synapses.

mouth: the opening of a tubular digestive system into which food is first introduced.

mucous membrane: the lining of the inside of the respiratory and digestive tracts.

muscle fiber: an individual muscle cell.

mutation: a change in the base sequence of DNA in a gene; usually refers to a genetic change that is significant enough to change the appearance or function of the organism.

mutualism (mū'-chū-al-ism): a symbiotic relationship in which both participating species benefit.

mycelium (mī-sēl´-ē-um): the body of a fungus, consisting of a mass of hyphae.

mycorrhiza (mi-ko-ri´za; pl. mycorrhizae): a symbiotic relationship between a fungus and the roots of a land plant that facilitates mineral extraction and absorption.

myelin (mī´-eh-lin): a wrapping of insulating membranes of specialized nonneural cells around the axon of a vertebrate nerve cell. Myelin increases the speed of conduction of action potentials.

myofibril (mī-ō-fī´-bril): a cylindrical subunit of each muscle cell, consisting of a series of sarcomeres. Myofibrils are surrounded by sarcoplasmic reticulum.

myometrium (mī-ō-mē´-trē-um): the muscular outer layer of the uterus.

myosin (mī´-ō-sin): one of the major proteins of muscle, whose interaction with actin produces contraction; found in the thick filaments of the muscle fiber. See also actin.

natural causality: the scientific principle that natural events occur as a result of preceding natural causes.

natural killer cell: a type of white blood cell that destroys some virus-infected cells and cancerous cells on contact. Part of the nonspecific internal defense against disease.

natural selection: the unequal survival and reproduction of organisms due to environmental forces, resulting in the preservation of favorable adaptations. Usually, natural selection refers specifically to differential survival and reproduction based on genetic differences among individuals.

negative feedback: a situation in which a change initiates a series of events that tend to counteract the change and restore the original state. Negative feedback in physiological systems maintains homeostasis.

nematocyst (nēm-āt´-ō-sist): a specialized cell found in cnidarians which, when disturbed, ejects a sticky or poisoned thread. Used by cnidarians to trap and sting prey.

nephridium (nef-rid´-ē-um): a type of excretory organ found in earthworms, mollusks, and certain other invertebrates. A nephridium somewhat resembles a single vertebrate nephron.

nephron (nef´-ron): the functional unit of the kidney, where blood is filtered and urine formed.

nephrostome: the funnel shaped opening of the nephridium of some invertebrates such as earthworms, into which coelomic fluid is drawn for filtration.

nerve: a bundle of axons of nerve cells, bound together in a sheath.

nerve cord: also called the spinal cord of vertebrates, a nervous structure lying along the dorsal side of the body of chordates.

nerve net: a simple form of nervous systems consisting of a network of neurons that extend throughout the tissues of an organism, such as a cnidarian.

net primary productivity: the energy stored in the primary producers of an ecosystem over a given time period.

neural tube: a structure derived from ectoderm during early embryonic development that later becomes the brain and spinal cord.

neuromodulator: a chemical, usually a peptide, released by neurons that alter the activity of groups of neurons over longer periods of time.

neuromuscular junction: the synapse formed between a motor neuron and muscle fiber.

neuron (nur´-on): a single nerve cell.

neurosecretory cell: a specialized nerve cell that synthesizes and releases hormones.

neurotransmitter: a chemical released by a nerve cell close to a second nerve cell, a muscle, or a gland cell, that influences the activity of the second cell.

neutral mutation: a mutation (change in DNA sequence) that has little or no effect on the function of the encoded protein.

neutralization: the process of covering up or inactivating a toxic substance with antibody.

neutron: a subatomic particle found in the nuclei of atoms, bearing no charge and having mass approximately equal to that of a proton.

nitrogen fixation: the process of combining atmospheric nitrogen with hydrogen to form ammonium (NH_4^+).

nitrogen-fixing: possessing the ability to remove nitrogen from the atmosphere and combine it with hydrogen to produce ammonium (NH_4^+).

nitrogen-fixing bacteria: bacteria that possess the ability to remove nitrogen (N_2) from the atmosphere and combine it with hydrogen to produce ammonium (NH_4^+).

node: a region of a stem at which leaves and lateral buds are located. Also, an interruption of the myelin on a vertebrate myelinated axon, exposing naked membrane at which action potentials are generated.

nodule: a swelling on the root of a legume or other plant that consists of cortex cells inhabited by nitrogen-fixing bacteria.

nondisjunction: an error in meiosis in which chromosomes fail to segregate properly into the daughter cells.

nonpolar covalent bond: a covalent bond with equal sharing of electrons.

norepinephrine(nor-epi-nef-rin´): also called noradrenaline, a neurotransmitter which is released by neurons of the parasympathetic nervous system and prepares the body to respond to stressful situations.

notochord (nōt´-ō-kōrd): a stiff but somewhat flexible, supportive rod found in all members of the phylum Chordata at some stage of development.

nuclear envelope: the double membrane system surrounding the nucleus of eukaryotic cells. The outer membrane is often continuous with the endoplasmic reticulum.

nucleic acid (nū-klā´-ik): an organic molecule composed of nucleotide subunits. The two common types of nucleic acids are ribonucleic acids (RNA) and deoxyribonucleic acids (DNA).

nucleoid (nū-klē-oyd): the location of the genetic material in prokaryotic cells; not membrane-enclosed.

nucleolus (nū-klē´-ō-lus): the region of the eukaryotic nucleus engaged in ribosome synthesis, consisting of the genes encoding ribosomal RNA, newly synthesized ribosomal RNA, and ribosomal proteins.

nucleotide: an individual subunit of which nucleic acids are composed. A single nucleotide consists of a phosphate group bonded to a sugar (deoxyribose in DNA), which is in turn bonded to a nitrogen-containing base (adenine, guanine, cytosine, or thymine). Nucleotides are linked together by bonds between the phosphate of one and the sugar of the next, to form a strand of nucleic acid.

nucleus: the membrane-bound organelle of eukaryotic cells that contains the cell's genetic material. Also, the central region of an atom, composed of protons and neutrons.

nutrient: a substance acquired from the environment and needed for survival, growth, and development of an organism.

nutrient cycle: a description of the movement of a specific nutrient (carbon, nitrogen, phosphorus, water, etc.) through the living and nonliving portions of an ecosystem.

nutrition: the process of acquiring nutrients from the environment and, if necessary, processing them into a form that can be used by the body.

oil: a lipid composed of three fatty acids, some of which are unsaturated, covalently bonded to a molecule of glycerol. Oils are liquid at room temperature.

olfaction (ōl-fak´-shun): a chemical sense, the sense of smell; in terrestrial vertebrates the result of detection of airborne molecules.

ommatidium (ōm-ma-tid´-ē-um): individual light-sensitive subunit of a compound eye. Each ommatidium consists of a lens and several receptor cells.

oncogene: a gene that, when transcribed, causes a cell to become cancerous.

one-gene, one-protein hypothesis: the proposition that each gene encodes the information for the synthesis of a single protein.

oogenesis: the process by which egg cells are formed.

oogonium (ō-ō-gō´-nē-um; plural, oogonia): a diploid cell in female animals that gives rise to a primary oocyte.

open circulatory system: a type of circulatory system found in some invertebrates, such as arthropods and molluscs, that includes an open space (the hemocoel) in which blood directly bathes body tissues.

operant conditioning: a laboratory training procedure in which an animal learns to perform a response (such as pressing a lever) through reward or punishment.

operator: in prokaryotes, a segment of DNA that governs access of RNA polymerase to the promoter and thus regulates transcription.

operculum: an external flap, supported by bone, that covers and protects the gills of most fish.

operon: a unit of organization of prokaryotic DNA, in which several genes that specify related functions (for example, enzymes in the same biosynthetic pathway) are grouped together on the chromosome, are transcribed at the same time, and are regulated together.

opioids (ōp´-ē-ōyd): a group of peptide neuromodulators found in the vertebrate brain that mimic some of the actions of opiates (such as opium) and also seem to influence many other processes, including emotion and appetite.

optic disk (op´-tik): the area of the retina at which the axons of the ganglion cell merge to form the optic nerve; the blind spot of the retina.

optic nerve (op´-tik): the nerve leading from the eye to the brain, carrying visual information.

orbital: the region of an atom, outside the nucleus, in which an electron is likely to be found.

order: the taxonomic category contained within a class and consisting of related families.

organ: a structure (such as the liver, kidney, or skin) composed of two or more distinct tissue types that function together.

organ system: two or more organs that work together to perform a specific function, i.e., the digestive system.

organelle (or-ga-nel´): a structure found in the cytoplasm of eukaryotic cells that performs a specific function; sometimes used to refer specifically to membrane-bound structures such as the nucleus or endoplasmic reticulum.

organic molecule: a molecule that contains both carbon and hydrogen.

organism (ōr´-ga-niz-em): an individual living thing.

organogenesis (or-gan-ō-jen´-i-sis): the process by which the layers of the gastrula (endoderm, ectoderm, mesoderm) rearrange themselves into organs.

origin: the site of attachment of a muscle to the relatively stationary bone on one side of a joint.

osculum (os´-kū-lum): a relatively large opening in the sponge body through which water is expelled.

osmosis (oz-mō´-sis): the diffusion of water across a differentially permeable membrane, usually down a concentration gradient of free water molecules. Water moves into the solution that has a lower free water concentra-tion from the solution with the higher free water concentration.

osmotic pressure: The pressure required to counterbalance the tendency of water to move from a solution with a higher concentra tion of free water molecules into one with a lower concentration of free water molecules.

osteoblast (os´-tē-ō-blast): a cell type that produces bone.

osteoclast: a cell type that dissolves bone.

osteocyte: a mature bone cell.

osteoporosis (os´-tē-ō-pōr-ō´-sis): a condition in which bones become porous, weak, and easily fractured; most common in elderly women.

otolith membrane (ō´-tō-lith): a gelatinous membrane in which are embedded the hairs of hair cells of the utricle and saccule of the inner ear. The otolith membrane is heavier than the surrounding fluid due to calcium carbonate crystals embedded in it and therefore responds to the pull of gravity.

outer ear: the outermost part of the mammalian ear, including the external ear and auditory canal leading to the tympanic membrane.

oval window: the membrane-covered entrance to the inner ear.

ovary: the gonad of female animals. Also, in flowering plants, a structure at the base of the carpel containing one or more ovules; develops into the fruit.

oviduct: in mammals, the tube leading from the ovary to the uterus.

ovulation: The release of a secondary oocyte, ready to be fertilized, from the ovary.

ovule: a structure within the ovary of a flower, inside which the female gametophyte develops. After fertilization, the ovule develops into the seed.

pacemaker: a cluster of specialized muscle cells in the upper right atrium of the heart that produce spontaneous electrical signals at a regular rate; the sino-atrial node.

Pacinian corpuscle (pas-in´-ē-an): a receptor of the skin and joints whose ending is wrapped in a layered capsule of connective tissue, and that responds to rapid movement.

palisade cells: columnar mesophyll cells containing chloroplasts just beneath the upper epidermis of a leaf.

pancreas (pan´-krē-as): a combined exocrine and endocrine gland located in the abdominal cavity next to the stomach. Its endocrine portion secretes the hormones insulin and glucagon, which regulate glucose concentrations in the blood. The exocrine portion secretes enzymes for fat, carbohydrate, and protein digestion into the small intestine and neutralizes the acidic chyme.

pancreatic juice: A mixtureof water, sodium bicarbonate, and enzymes released by the pancreas into the small intestine.

parasite (par´-uh-sīt): an organism that lives in or on the body of another organism, called a host, causing it harm as a result.

parasitism: a symbiotic relationship in which one organism (usually smaller and more numerous than its host) benefits by feeding on the other, which is usually harmed but not immediately killed.

parasympathetic nervous system: the division of the autonomic nervous system that produces largely involuntary responses related to maintenance of normal body functions, such as digestion.

parathyroids: a set of four small endocrine glands embedded in the surface of the thyroid gland that produce parathormone, which (with calcitonin from the thyroid) regulates calcium ion concentration in the blood.

parenchyma (par-en´-ki-ma): a plant cell type that is alive at maturity, usually with thin primary cell walls, that carries out most of the metabolism of a plant. Most dividing meristem cells in a plant are parenchyma.

parthenogenesis (par-the-nō-gen´-i-sis): a specialization of sexual reproduction, in which a haploid egg undergoes development without fertilization.

passive transport: movement of materials across a membrane down a gradient of concentration, pressure, or electrical charge without using cellular energy.

passive visual signal: a body shape, color, or size that communicates information, even in the absence of specific behaviors.

pathogen (path´-ō-gen): an organism capable of producing disease.

pedigree: a diagram showing genetic relationships among a set of individuals, usually with respect to a specific genetic trait.

peptide (pep´-tīd): a chain composed of two or more amino acids linked together by peptide bonds.

peptide bond: the covalent bond between the amino group nitrogen of one amino acid and the carboxyl group carbon of a second amino acid, which joins the two amino acids together in a peptide or protein.

peptide hormone: A hormone consisting of a chain of amino acids. Small proteins that function as hormones are included in this category.

peptidoglycan (pep-tid-ō-glī´-can): material found in prokaryotic cell walls consisting of chains of sugars cross-linked by short chains of amino acids called peptides.

pericycle (per´-i-sī-kul): the outermost layer of cells of the vascular cylinder of a root.

periderm: the outer cell layers of roots and a stem that have undergone secondary growth, consisting primarily of cork cambium and cork cells.

periosteum: a thin layer of connective tissue, rich in capillaries, osteoblasts, and osteoblast-forming cells that covers healthy bone. Rupture of the periosteum cause by a fracture stimulates production and release of osteoblasts.

peripheral nervous system: in vertebrates, that part of the nervous system connecting the central nervous system to the rest of the body.

peristalsis: rhythmic coordinated contractions of the smooth muscles of the digestive tract that move substances throughthe digestive tract.

permeate (per´-mē-āt): to pass through, as through pores in a membrane.

petal: part of a flower, often brightly colored and fragrant, serving to attract potential animal pollinators.

petiole (pet´-ē-ōl): the stalk that connects the blade of a leaf to the stem.

pH scale: a scale with values from 0 to 14, used for measuring the relative acidity of a solution. At pH 7 a solution is neutral, pH 0 to 7 is acidic, and pH 7 to 14 is basic. Each unit on the pH scale represents a tenfold change in the concentration of hydrogen ions.

phagocytosis (fa-gō-sī-tō´-sis): a type of endocytosis in which extensions of a plasma membrane engulf extracellular particles and transport them into the interior of the cell.

pharynx (far´-inx): in vertebrates, a chamber at the back of the mouth shared by the digestive and respiratory systems.

phenotype (fēn´-ō-tīp): the physical characteristics of an organism. Phenotype can be defined as outward appearance (such as flower color), as behavior, or in molecular terms (such as glycoproteins on red blood cells).

phenylketonuria (fen-ul-kē-tō-nū´-rē-a): a recessive disease in which the enzyme that catalyzes the conversion of the amino acid phenylalanine to tyrosine is defective.

pheromone (fār´-uh-mōn): a chemical produced by an organism that alters the behavior or physiological state of another of the same species.

phloem (flō´-um): a conducting tissue of vascular plants that transports a concentrated sugar solution up and down the plant.

phospholipid (fos-fō-li´-pid): a lipid consisting of glycerol to which two fatty acids and one phosphate group are bonded. The phosphate group bears another group of atoms, often containing nitrogen, and often bearing an electric charge. A double layer of phospholipids is found in all cell membranes.

phospholipid bilayer: a double layer of phospholipids that forms the basis of all cellular membranes; the phospholipid heads, which are hydrophilic, face the water of extracellular fluid or the cytoplasm, and the tails, which are hydrophobic, are buried in the middle of the bilayer.

photon (fō´-ton): The smallest unit of light energy.

photophosphorylation (fō-tō-fos-for-i-lā´-shun): ATP synthesis in the light-dependent reactions of photosynthesis; light energy ("photo") is used to add a phosphate group to ("phosphorylate") ADP to yield ATP.

photopigment (fō´-tō-pig-ment): a chemical substance in photoreceptor cells that changes molecular conformation when struck by light.

photoreceptors: receptor cells that respond to light; in vertebrates, rods and cones.

photorespiration: A series of reactions in plants in which O2 replaces CO2 during the C3 cycle, preventing carbon fixation. This wasteful process dominates when C3 plants are forced to close their stomata to prevent water loss.

photosynthesis: the complete series of chemical reactions in which the energy of light is used to synthesize high-energy organic molecules, usually carbohydrates, from low-energy inorganic molecules, usually carbon dioxide and water.

photosystem: in thylakoid membranes, a light-harvesting complex and its associated electron transport system.

phototropism: growth with respect to the direction of light.

phycocyanin (fī-kō-sī´-a-nin): a blue or purple pigment found in the membranes of chloroplasts and used as an accessory light-gathering molecule in thylakoid photosystems.

phyletic speciation: a form of evolution in which the gene pool of a species changes so greatly over time that a new species is formed. Old and new species do not exist simultaneously. See divergent speciation.

phylum (fī´-lum): the taxonomic category of animals and animallike protists contained within a kingdom and consisting of related classes.

phytochrome (fī´-tō-krōm): a light-sensitive plant pigment that mediates many plant responses to light, including flowering, stem elongation, and seed germination.

phytoplankton (fī´-tō-plank-ten): a general term describing photosynthetic protists that are abundant in marine and freshwater environments.

pili (pil´-ī; sing. pilus): hair-like projections made of protein and found on the surface of certain bacteria, often used to attach the bacterium to another cell.

pineal gland (pī-nē´-al): a small gland within the brain that secretes melatonin. The pineal controls the seasonal reproductive cycles of some mammals.

pinocytosis (pī-nō-sī-tō´-sis): nonselective movement of extracellular fluid into a cell, enclosed within a vesicle formed from the plasma membrane.

pioneer: an organism that is among the first to colonize an unoccupied habitat in the first stages of succession.

pit: an area in the cell walls between two plant cells in which secondary walls did not form, so that the two cells are separated only by a relatively thin and porous primary cell wall.

pit organs: thermoreceptive organs found in certain snakes (vipers) which consist of sensory cells protected within a pair of pits located between each eye and the nose; used to detect warm-blooded prey.

pith: cells forming the center of a root or stem.

pituitary: an endocrine gland located at the base of the brain that produces several hormones, many of which influence the activity of other glands.

placenta (pluh-sen´-ta): In mammals, a structure formed by a complex interweaving of the uterine lining and the embryonic membranes, especially the chorion; functions in gas, nutrient, and waste exchange between embryonic and maternal circulatory systems and secretes hormones.

plant hormone: the plant-regulating chemicals auxin, gibberellins, cytokinins, ethylene, and abscisic acid; plant hormones somewhat resemble animal hormones in that they are chemical produced by cells in one location that influence the growth or metabolic activity of other cells, often some distance away in the plant body.

plaque (plak): a deposit of cholesterol and other fatty substances within the wall of an artery.

plasma: the fluid, noncellular portion of the blood.

plasma cell: an antibody-secreting descendant of a B cell.

plasma membrane: the outer membrane of a cell, composed of a bilayer of phospholipids in which proteins are embedded.

plasmid (plaz´-mid): a small, circular piece of DNA found in the cytoplasm of many bacteria; usually does not carry genes required for the normal functioning of the bacterium but may carry genes that assist bacterial survival in certain environments, such as antibiotic resistance.

plasmodesma (plaz-mō-dez´-ma; pl. plasmodesmata): a cell-to-cell junction in plants that connects the cytoplasm of adjacent cells.

plasmodial slime mold: a funguslike protist consisting of a multinucleate mass or plasmodi-um that crawls and ingests decaying organic matter in amoeboid fashion and forms a fruiting body under adverse conditions. Also called acellular slime mold.

plasmodium (plas-mō´-dē-um): a sluglike mass of cytoplasm containing thousands of nuclei that are not confined within individual cells.

plastid (plas´-tid): in plant cells, an organelle bounded by two membranes that may be involved in photosynthesis (chloroplasts), pigment storage, or food storage.

platelets (plāt´-lets): cell fragments formed from megakaryocytes in bone marrow. Platelets, which lack nuclei, circulate in the blood and play a role in blood clotting.

pleated sheet: a type of secondary structure of a protein. In a pleated sheet, protein chains lie side-by-side, bonded to one another by hydrogen bonds.

pleiotropy (plē´-ō-trō-pē): a situation in which a single gene influences more than one phenotypic characteristic.

pleural membranes: Membranes that line the chest and surround the lungs.

point mutation: a mutation in which a single base pair in DNA has been changed.

polar body: in oogenesis, a small cell containing a nucleus but virtually no cytoplasm, produced by the first meiotic division of the primary oocyte.

polar covalent bond: a covalent bond with unequal sharing of electrons, so that one atom is relatively negative while the other is relatively positive.

polar microtubule: a microtubule of the spindle apparatus that extends from one spindle pole to the equator of the cell; involved in moving the spindle poles farther apart during anaphase.

pollen: the male gametophyte of a seed plant.

pollination: in flowering plants, when pollen lands on the stigma of a flower of the same species; in conifers, when pollen lands within the pollen chamber of a female cone of the same species.

polygenic inheritance: a pattern of inheritance in which the interactions of two or more functionally similar genes determine phenotype.

polymer (pa´-li-mer): a molecule composed of three or more (may be thousands) smaller subunits called monomers. The monomers that make up a single polymer may be identical (for example, the glucose monomers of starch) or different (for example, the amino acids of a protein).

polymerase chain reaction (PCR): a method of producing virtually unlimited numbers of copies of a specific piece of DNA, starting with as little as one copy of the desired DNA.

polyp (pol´-ip): the sedentary, vase-shaped stage in the life cycle of many cnidarians. Hydra and sea anemones are examples.

polyploid (pol-ē-ployd): having more than two homologous chromosomes of each type.

polysaccharide (pol-ē-sak´-a-rīd): a large carbohydrate molecule composed of branched or unbranched chains of repeating monosaccharide subunits, usually glucose or modified glucose molecules. Polysaccharides include starches, cellulose, and glycogen.

pons: a portion of the hindbrain just above the medulla containing neurons that influence sleep and the rate and pattern of breathing.

population: a group of individuals of the same species, found in the same time and place, and actually or potentially interbreeding.

population bottleneck: a form of genetic drift in which a population becomes extremely small, which may lead to differences in allele frequencies as compared with other populations of the species, and to a loss in genetic variability.

population genetics: the study of the frequency, distribution, and inheritance of alleles in a population of organisms.

positive feedback: a situation in which a change initiates events that tend to amplify the original change.

posterior (post-tēr´-ē-ur): the tail, hindmost, or rear end of an animal.

posterior pituitary: a lobe of the pituitary gland that is an outgrowth of the hypothalamus and releases antidiuretic hormone and oxytocin.

postmating isolating mechanism: any structure, physiological function, or developmental abnormality that prevents organisms of two different populations, once mating has occurred, from producing vigorous, fertile offspring.

postsynaptic neuron: the nerve cell at a synapse that changes its electrical potential in response to a chemical (the neurotransmitter) released by another (presynaptic) cell.

postsynaptic potential: an electrical signal produced in a postsynaptic cell by transmission across the synapse. It may be excitatory (EPSP), making the cell more likely to produce an action potential, or inhibitory (IPSP), tending to inhibit an action potential.

potassium channel: a protein that spans a nerve cell membrane, forming a pore, and which allows only potassium ions to flow through the pore.

potential energy: "stored" energy, usually chemical energy or energy of position within a gravitational field.

preadaptation: a feature evolved under one set of environmental conditions that, purely by chance, helps an organism adapt to new environmental conditions.

prebiotic evolution: evolution before life existed; especially abiotic synthesis of organic molecules.

precapillary sphincter (sfink´-ter): a ring of smooth muscle between an arteriole and a capillary that regulates the flow of blood into the capillary bed.

predation (pre-dā´-shun): the act of killing and eating another living organism.

predator: an organism that kills and eats other organisms.

premating isolating mechanism: any structure, physiological function, or behavior that prevents organisms of two different populations from exchanging gametes.

pressure flow theory: a model for transport of sugars in phloem, which states that movement of sugars into a phloem sieve tube causes water to enter the tube by osmosis, while movement of sugars out of another part of the same sieve tube causes water to leave by osmosis; the resulting pressure gradient causes bulk movement of water and dissolved sugars from the end of the tube into which sugar is transported toward the end of the tube from which sugar is removed.

presynaptic neuron: a nerve cell that releases a chemical (the neurotransmitter) at a synapse, which causes changes in the electrical activity of another (postsynaptic) cell.

primary cell wall: cellulose and other carbohydrates secreted by a young plant cell between the middle lamella and the plasma membrane.

primary consumer: an organism that feeds on producers; an herbivore.

primary endosperm cell: the central cell of the female gametophyte of a flowering plant, containing the polar nuclei (usually two). After fertilization, the primary endosperm cell undergoes repeated mitotic divisions to produce the endosperm of the seed.

primary growth: growth in length and development of initial structures of plant roots and shoots, due to cell division of apical meristems and differentiation of the daughter cells.

primary oocyte (ō´-ō-sīt): a diploid cell, derived from the oogonium by growth and differentiation, which undergoes meiosis to produce the egg.

primary phloem: phloem in young stems produced from an apical meristem.

primary root: the first root that develops from a seed.

primary spermatocyte (sper-ma´-tō-sīt): a diploid cell, derived from the spermatogonium by growth and differentiation, which undergoes meiosis to produce four sperm.

primary structure: the amino acid sequence of a protein.

primary succession: succession that occurs in an environment, such as bare rock, in which no trace of a previous community is present.

primary xylem: xylem in young stems produced from an apical meristem.

primer pheromone: a chemical produced by an organism that alters the physiological state of another of the same species.

primitive streak: in reptiles, birds,a nd mammals, the region of the ectoderm of the two-layered embryonic disk through which cells migrate to form mesoderm.

prion (prē´-on): a protein which, in mutated form acts as an infectious agent causing certain neurodegenerative diseases, including kuru and scrapie.

producer: a photosynthetic organism; an autotroph.

product: an atom or molecule resulting from a chemical reaction.

progesterone (prō-ge´-ster-ōn): a hormone produced by the corpus luteum that promotes development of the uterine lining.

prokaryotic (prō-kar-ē-ot´-ik): referring to cells of the kingdom Monera. Prokaryotic cells do not have their genetic material enclosed within a membrane-bound nucleus and also lack other membrane-bound organelles.

promoter: a specific sequence of DNA to which RNA polymerase binds, initiating gene transcription.

prophase (prō´-fāz): the first stage of mitosis, in which the chromosomes first become visible in the light microscope as thickened, condensed threads and the spindle begins to form. As the spindle is completed, the nuclear envelope breaks apart. The spindle invades the nuclear region and attaches to the kinetochores of the chromosomes. Also, the first stage of meiosis. In meiosis I, the homologous chromosomes pair up and exchange parts at chiasmata. In meiosis II, the spindle reforms and chromosomes attach to the microtubules.

proprioceptor (prō´-prē-ō-cep-tōr): a receptor that monitors the position of the parts of the body and their direction of movement by responding to stretch and pressure.

prostaglandin (pro-sta-glan´-din): a family of modified fatty acid hormones manufactured by many cells of the body.

prostate gland (prō´-stāt): a gland that produces part of the fluid component of semen. The prostate fluid is basic and contains a chemical that activates sperm movement.

protease (prō´-tē-ās): an enzyme that digests proteins.

protein: an organic molecule composed of one or more chains of amino acids.

protista (prō-tis´-tuh): a taxonomic kingdom including unicellular, eukaryotic organisms.

proton: a subatomic particle found in the nuclei of atoms, bearing a unit of positive charge and a relatively large mass roughly equal to the mass of the neutron.

protozoan (prō-te-zō´-an; pl. protozoa): a non-photosynthetic or animallike protist.

pseudocoel (sū-dō-sēl): "false coelom"; a body cavity with a different embryological origin than a coelom, but serving a similar function; found in roundworms.

pseudoplasmodium (sū-dō-plas-mō´-dē-um): an aggregation of individual amoeboid cells that form a sluglike mass.

pseudopod (sūd´-ō-pod): extension of the cell membrane by which certain cells, such as amoebae, locomote and engulf prey.

punctuated equilibrium: a model of evolution, stating that morphological change and speciation are rapid, simultaneous events (the "punctuation"), separated by long periods during which a species remains unchanged (the "equilibrium").

pupil: the adjustable opening in the center of the iris through which light enters the eye.

purine: a type of nitrogen-containing base found in nucleic acids that consists of two fused rings; includes adenine and guanine in both DNA and RNA.

pyloric sphincter (pī-lor´-ik sfink´-ter): a circular muscle at the base of the stomach that regulates the passage of chyme into the small intestine.

pyrimidine: a type of nitrogen-containing base found in nucleic acids that consists of a single ring; includes cytosine (both DNA and RNA), thymine (DNA only) and uracil (RNA only).

quaternary structure (kwat´-er-na-rē): the complex three- dimensional structure of a protein that is composed of more than one peptide chain.

queen substance: a chemical produced by a queen bee that can act as both a primer and a releaser pheromone.

radial symmetry: a body plan in which any plane drawn along a central axis will divide the body into approximately mirror-image halves. Cnidarians and many adult echinoderms show radial symmetry.

radioactive: pertaining to an atom with an unstable nucleus that spontaneously disintegrates with the emission of radiation.

radula (ra´-dū-luh): a ribbon of tissue in the mouth of gastropod mollusks that bears numerous teeth on its outer surface and is used to scrape and drag food into the mouth.

random distribution: spacing in which the probability of finding an individual is equal in all parts of an area.

reactant: an atom or molecule that is used up in a chemical reaction to form a product.

reaction center: in the light-harvesting complex of a photosystem, the chlorophyll molecule to which light energy is transferred by the antenna pigments. The captured energy ejects an electron from the reaction center chlorophyll, and the electron is transferred to the electron transport system.

receptor: (1) a cell that responds to an environmental stimulus (chemicals, sound, light, pH, etc.) by changing its electrical potential; (2) a protein molecule in a cell membrane that binds to another molecule (hormone, neurotransmitter) triggering metabolic or electrical changes in a cell.

receptor-mediated endocytosis: selective uptake of molecules from the extracellu-lar fluid, by binding to a receptor located at a coated pit on the plasma membrane, and pinching off the coated pit into a vesicle that moves into the cytosol.

receptor potential: an electrical potential change in a receptor cell produced in response to reception of an environmental stimulus (chemicals, sound, light, heat, etc.). The size of the receptor potential is proportional to the intensity of the stimulus.

receptor protein: a protein located on a membrane (or in the cytoplasm), that recognizes and binds to specific molecules. Binding by receptors often triggers a response by a cell, such as endocytosis, increased metabolic rate, or cell division.

recessive: an allele expressed only in homozygotes, and which is completely masked in heterozygotes.

recognition protein: a protein or glycoprotein protruding from the outside surface of a plasma membrane that identifies a cell as belonging to a particular species, to a specific individual of that species, and often to a specific organ within the individual.

reflex: a simple, stereotyped movement of part of the body that occurs automatically in response to a stimulus.

regeneration: (1) regrowth of a body part after loss or damage; (2) asexual reproduction by regrowth of an entire body from a fragment.

regulatory gene: a gene that controls the timing or rate of transcription of other genes.

releaser: a stimulus that triggers a fixed action pattern.

releaser pheromone: a chemical produced by one organism that alters the behavior of another of the same species.

releasing hormone: a hormone secreted by the hypothalamus that causes the release of specific hormones by the anterior pituitary gland.

renal artery: The artery carrying blood to each kidney.

renal cortex: the outer layer of the kidney where nephrons are located.

renal medulla: the layer of the kidney just inside the renal cortex where loops of Henle produce a highly concentrated interstitial fluid that is important for producing concentrated urine.

renal pelvis: The inner chamber of where urine from the collecting ducts accumulates before entering the ureters.

renal vein: The vein carrying cleansed blood away from each kidney

replacement-level fertility (RLF): the average birth rate at which a reproducing population exactly replaces itself during its lifetime.

replication fork: in DNA replication, a Y-shaped opening in the double helix of DNA, at which new DNA strands are synthesized.

repression: in the lactose operon, the condition that the genes of the operon are not transcribed unless specifically activated by the presence of lactose.

repressor protein: in the lactose operon, a protein that binds to the operator site, thereby preventing access of RNA polymerase to the promoter.

reproductive isolation: the failure of organisms of one population to breed successfully with members of another population; may be due to premating or postmating isolating mechanisms.

reservoir: the major source of any particular nutrient in an ecosystem, usually in the abiotic portion.

respiratory center: a cluster of neurons in the brainstem that sends rhythmic bursts of nerve impulses to the respiratory muscles, resulting in breathing.

resting potential: a negative electrical potential found in unstimulated nerve cells.

restriction enzyme: an enzyme, usually isolated from bacteria, that cuts double-stranded DNA at a specific nucleotide sequence; the cleavage sequence differs for different restriction enzymes.

restriction fragment: a piece of DNA that has been isolated by cleaving a larger piece of DNA with restriction enzymes.

restriction fragment length polymorphisms (RFLPs): differences in the lengths of restriction fragments, produced by cutting samples of DNA from different individuals of the same species with the same set of restriction enzymes; occurs because of differences in nucleotide sequences among different individuals of the same species.

reticular formation (reh-tik´-ū-lar): a diffuse network of neurons extending from the hindbrain, through the midbrain, and into the lower reaches of the forebrain, involved in filtering sensory input and regulating what information is relayed to higher brain centers for further attention.

retina (ret´-in-a): a multilayered sheet of nerve tissue at the rear of camera-type eyes, composed of photoreceptor cells plus associated nerve cells that refine the photoreceptor information and transmit it to the optic nerve.

retrovirus: a virus that uses RNA as its genetic material. When a retrovirus invades a eukaryotic cell, it "reverse transcribes" its RNA into DNA, which then directs the synthesis of more viruses, using the transcription and translation machinery of the cell.

reverse transcriptase: an enzyme found in retroviruses that catalyzes the synthesis of DNA from an RNA template.

Rh factor: a protein found on the red blood cells of some people (Rh-positive) but not others (Rh-negative). Exposure of Rh-negative individuals to Rh-positive blood triggers antibody production to Rh-positive blood cells.

rhizoid: (rī´-zoyd): a rootlike structure found in bryophytes that anchors the plant and absorbs water and nutrients from the soil.

rhizome (rī´-zōm): an undergound stem, usually horizontal, that stores food.

rhythm method: a contraceptive method involving abstinence from intercourse during ovulation.

ribonucleic acid (rī-bō-nū-klā´-ik; RNA): a molecule composed of ribose nucleotides, each of which consists of a phosphate group, the sugar ribose, and one of the bases adenine, cytosine, guanine, or uracil; transfers hereditary instructions from the nucleus to the cytoplasm; also the genetic material of some viruses.

ribosomal RNA (rRNA): a type of RNA that combines with proteins to form ribosomes.

ribosome: an organelle consisting of two subunits, each composed of ribosomal RNA and protein. Ribosomes are the site of protein synthesis, during which the sequence of bases of messenger RNA is translated into the sequence of amino acids in a protein.

ribozyme: an RNA molecule that can catalyze certain chemical reactions, especially those involved in synthesis and processing of RNA itself.

RNA polymerase: in RNA synthesis, an enzyme that catalyzes the bonding of free RNA nucleotides into a continuous strand, using RNA nucleotides that are complementary to those of a strand of DNA.

rod: a rod-shaped photoreceptor cell in the vertebrate retina, sensitive to dim light but not involved in color vision. See also cone.

root cap: a cluster of cells at the tip of a growing root, derived from the apical meristem. The root cap protects the growing tip from damage as it burrows through the soil.

root hair: a fine projection from an epidermal cell of a young root that increases the absorptive surface area of the root.

root system: the part of a plant, usually below ground, that anchors the plant in the soil, absorbs water and minerals, stores food, transports water, minerals, sugars, and hormones, and produces certain hormones.

rough endoplasmic reticulum: endoplasmic reticulum lined on the outside with ribosomes.

runner: a horizontally growing stem that may develop new plants at nodes that touch the soil.

S-curve: the growth curve that describes a population of long-lived organisms introduced into a new area. It consists of an initial period of exponential growth, followed by decreasing growth rate, and finally, relative stability around a growth rate of zero.

saccule (sak´-ūle): a portion of the vestibular apparatus responsible for detecting gravity.

saltatory conduction (sal´-ta-tōr-ē): literally "jumping" conduction; the transmission of an action potential along a myelinated axon, during which the action potential seems to jump from one node to the next.

saprobe (sap´-rōbe): an organism that derives its nutrients from the bodies of dead organisms.

sarcodine (sar-kō´-dīn): a category of nonphotosynthetic protist (protozoa) characterized by the ability to form pseudopodia. Some, such as amoebae, are naked, while others have elaborate shells.

sarcomere (sark´-ō-mēr): the unit of contraction of a muscle fiber; a subunit of the myofibril, consisting of actin and myosin filaments and bounded by Z-lines.

sarcoplasmic reticulum (sark´-ō-plas´-mik retik´-ū-lum): specialized endoplasmic reticulum found in muscle cells and forming interconnected hollow tubes. The sarcoplasmic reticulum stores calcium ions and releases them into the interior of the muscle cell to initiate contraction.

saturated: referring to a fatty acid with as many hydrogen atoms as possible bonded to the carbon backbone; a fatty acid with no double bonds in its carbon backbone.

sclera: a tough white connective tissue layer that covers the outside of the eyeball and forms the white of the eye.

sclerenchyma (skler-en´-ki-ma): a plant cell type with thick, hardened secondary cell walls, that usually dies as the last stage of differentiation and provides both support and protection for the plant body.

scramble competition: direct interactions between individuals attempting to acquire the same limited resources.

scrotum (skrō´-tum): the pouch of skin containing the testes of male mammals.

sebaceous gland (se-bā´-shus): a gland in the dermis of skin, formed from epithelial tissue, that produces an oily substance called sebum that lubricates the epidermis.

second law of thermodynamics: a principle of physics that any change in an isolated system causes the quantity of concentrated, useful energy to decrease, and the amount of randomness and disorder (entropy) to increase.

second messenger: a term applied to intracellular chemicals, such as cyclic AMP, that are synthesized or released within a cell in response to the binding of a hormone or neurotransmitter (the first messenger) to receptors on the cell surface. Second messengers bring about specific changes in the metabolism of the cell.

secondary cell wall: a thick layer of cellulose and other polysaccharides secreted by certain plant cells between the primary cell wall and the plasma membrane.

secondary consumer: an organism that feeds on primary consumers; a carnivore.

secondary growth: growth in diameter of a stem or root due to cell division in lateral meristems and differentiation of their daughter cells.

secondary oocyte: a large haploid cell derived from the first meiotic division of the diploid primary oocyte.

secondary phloem: phloem produced from the cells arising toward the outside of the vascular cambium.

secondary spermatocyte (ō´-ō-sīt): a large haploid cell derived by meiosis I from the diploid primary spermatocyte.

secondary structure: a repeated, regular structure assumed by protein chains, held together by hydrogen bonds; usually either a helix or a pleated sheet.

secondary succession: succession that occurs after an existing community is disturbed, in an environment modified by that community, for example, after a forest fire.

secondary xylem: xylem produced from cells arising at the inside of the vascular cambium.

secretin: hormone produced by the small intestine that stimulates production and release of digestive secretions by the pancreas and liver.

sedimentary cycle: a nutrient cycle for a substance whose primary reservoir is sediment, such as soil or rock.

seed: the reproductive structure of a seed plant. The seed is protected by a seed coat and contains an embryonic plant and a supply of food for it.

seed coat: the thin, tough, and waterproof outermost covering of a seed, formed from the integuments of the ovule.

segmentation (seg-men-tā´-shun): division of the body into repeated, often similar units.

segmentation movements: unsynchronized contractions of the small intestine that result in mixing of partially digested food and digestive enzymes. The movements also bring nutrients into contact with the absorptive intestinal wall.

segregation: a principle of inheritance, that the two alleles of a given gene separate from each other during gamete formation, with the result that each gamete receives one allele; occurs because of the separation of homologous chromosomes during meiosis.

self-fertilization: union of sperm and egg from the same individual.

selfish gene: the concept that genes promote the survival of themselves in individuals through innate self-sacrificing behavior that enhances the survival of others carrying the same genes; helps explain the evolution of altruism.

semen: the sperm-containing fluid produced by the male reproductive tract.

semicircular canals: three fluid-filled, semicircular tubes of the inner ear that function in the detection of movements of the head in all directions.

semiconservative replication: the process of replication of the DNA double helix; the two DNA strands separate, and each is used as a template for the synthesis of a complementary DNA strand. Each daughter double helix therefore consists of one parental strand and one new strand.

semilunar valve: the type of valve between the ventricles of the heart and the pulmonary artery and aorta. Semilunar valves prevent backflow of blood into the ventricles when they relax.

seminal vesicle: in male mammals, a gland that produces a basic, fructose-containing fluid that forms part of the semen.

seminiferous tubules (semi-i-ni´-fer-us): a series of tubes in the vertebrate testis in which sperm are produced.

senescence: in plants, a specific aging process, often including deterioration and dropping of leaves and flowers.

sensitive period: the particular stage in an animal's life during which it imprints.

sensory neuron: a nerve cell that responds to a stimulus, either from the internal or external environment.

sensory receptor: a cell specialized to respond to particular internal or external environmental stimuli by producing an electrical potential.

sepal (sē´-pul): set of modified leaves that surround and protect a flower bud, often opening into green, leaflike structures when the flower blooms.

septum (pl. septa): a partition that separates the fungal hypha into individual cells. Pores in the septa allow transfer of materials between cells.

serotonin (ser-uh-tō´-nin): a neurotransmitter in the central nervous system, which is involved in mood, sleep, and the inhibition of pain.

Sertoli cell: a large cell in the seminiferous tubule that regulates spermatogenesis and nourishes the developing sperm.

sessile (ses´-ul): not free to move about, usually permanently attached to a surface.

severe combined immune deficiency (SCID): a disorder in which no immune cells, or very few, are formed; the immune system is incapable of responding properly to invading disease organisms, and the victim is very vulnerable common infections.

sex chromosome: one of the pair of chromosomes that differ between the sexes and usually determine the sex of an individual; for example, human females have similar sex chromosomes (XX) while males have dissimilar ones (XY).

sex-influenced traits: a set of traits that occur more commonly or strongly in a single sex, often due to differing levels of sex hormones, but are not coded by genes on the sex chromosomes.

sex linkage: a pattern of inheritance characteristic of genes located on one type of sex chromosome (for example, X) and not found on the other type (for example, Y).

sex-linked inheritance: inheritance of traits controlled by genes carried on the X chromosome. Females show the dominant trait unless they are homozygous recessive, whereas males will express whatever allele is found on their single X chromosome.

sexual recombination: during sexual reproduction, the formation of new combinations of alleles in offspring, due to the inheritance of one homologous chromosome from each of two, genetically different parents.

sexual reproduction: a form of reproduction in which genetic material from two parental organisms is combined in the offspring. Usually, two haploid gametes fuse to form a diploid zygote.

sexual selection: a type of natural selection in which the choice of mates by one sex is the selective agent.

shoot system: all the parts of a vascular plant exclusive of the root. Usually aboveground, it consists of stem, leaves, buds, and (in season) flowers and fruits. Its functions include photosynthesis, transport of materials, reproduction, and hormone synthesis.

short-day plant: a plant that will flower only if the length of daylight is shorter than some species-specific duration.

sickle-cell anemia: a recessive disease caused by a single amino acid substitution in the hemoglobin molecule. Sickle-cell hemoglobin molecules tend to cluster together, distorting the red blood cell shape and causing them to break and clog the capillaries.

sieve plate: structues between two adjacent sieve-tube elements in phloem, where holes formed in the primary cell walls interconnect the cytoplasm of the elements.

sieve tube: in phloem, a single strand of sieve-tube elements, which transports sugar solutions.

sieve-tube element: one of the cells of a sieve tube, which form the phloem.

simple diffusion: diffusion of water, dissolved gases, or lipid- soluble molecules through the phospholipid bilayer of a membrane.

sinoatrial node (sī´-nō-āt´-rē-ul nōd): also called the SA node, a small mass of specialized muscle in the wall of the right atrium. It generates electrical signals rhythmically and spontaneously and serves as the heart's pacemaker.

skeletal (striated) muscle: the type of muscle that is attached to and moves the skeleton and is under the direct, usually voluntary, control of the nervous system.

skeleton: a supporting structure for the body, upon which muscles act to change the body configuration.

small intestine: portion of the digestive tract between the stomach and large intestine in which most digestion and absorption of nutrients occurs.

smooth endoplasmic reticulum: endoplasmic reticulum without ribosomes.

smooth muscle: type of muscle found around hollow organs, such as the digestive tract, bladder, and blood vessels, normally not under voluntary control.

sodium channel: a protein that spans a nerve cell membrane, forming a pore, and which allows only sodium ions to flow through the pore.

sodium-potassium pump: a set of active transport molecules in nerve cell membranes that use the energy of ATP to pump sodium ions out of the cell and potassium ions in, maintaining the concentration gradients of these ions across the cell membrane.

solvent: a liquid that is capable of dissolving (uniformly dispersing) other substances in itself.

somatic nervous system: that portion of the peripheral nervous system that controls voluntary movement by activating skeletal muscles.

spawning: a method of external fertilization in which male and female parents shed gametes into the water, and sperm must swim through the water to reach the eggs.

speciation: the process whereby two populations achieve reproductive isolation.

species (spē´-sēs): all of the organisms that are potentially capable of interbreeding under natural conditions or, if asexually reproducing, are more closely related to one another than to other organisms within the genus.

specific heat: the amount of energy required to raise the temperature of 1 gram of a substance 1° C.

sperm: the haploid male gamete, usually small, motile, and containing little cytoplasm.

spermatid: a haploid cell derived from the secondary spermatocyte by meiosis II. The mature sperm is derived from the spermatid by differentiation.

spermatogenesis: the process by which sperm cells are formed.

spermatogonium (pl. spermatogonia): a diploid cell lining the walls of the seminiferous tubules that gives rise to a primary spermatocyte.

spermatophore: in some animals, in a variation on internal fertilization, the males package their sperm in a container that can be inserted in the female reproductive tract.

spermicide: a sperm-killing chemical; used for contraceptive purposes.

spicule (spik´-ūle): subunits of the endoskeleton of sponges, made of protein, silica, or calcium carbonate.

spinal cord: part of the central nervous system of vertebrates, extending from the base of the brain to the hips, protected by the bones of the vertebral column; contains the cell bodies of motor neurons synapsing on skeletal muscles, the circuitry for some simple reflex behaviors, and axons communicating with the brain.

spindle apparatus: a football-shaped array of microtubules that moves the chromosomes to opposite poles of a cell during anaphase of meiosis and mitosis.

spiracles (spī´-re-kuls): openings in the abdominal segments of insects through which air enters the tracheae.

spirilla (spī´-rilla): a spiral shaped bacterium.

spleen: An organ of the lymphatic system in which lymphocytes are produced, and blood is filtered past lymphocytes and macrophages which remove foreign particles and aged red blood cells.

spongy cells: irregularly shaped mesophyll cells containing chloroplasts located just above the lower epidermis of a leaf.

spontaneous generation: the proposal that living organisms can arise from nonliving matter.

sporangium (spōr-an´-gē-um; pl. sporangia): a structure in which spores are produced.

spore: in the alternation of generation life cycle of plants, a haploid cell produced through meiosis, that then undergoes repeated mitotic divisions and differentiation of daughter cells to produce a multicellular, haploid organism called the gametophyte.

sporophyte (spōr´-ō-fīt): the diploid form of a plant that produces haploid, asexual spores through meiosis.

sporozoan (spōr-ō-zō´-en): a category of parasitic protist. Sporozoans have complex life cycles often involving more than one host, and are named for their ability to form infectious spores. A well-known member (genus Plasmodium) causes malaria.

stabilizing selection: a type of natural selection in which those organisms displaying extreme phenotypes are selected against.

stamen (stā´-men): the male reproductive structure of a flower, consisting of a filament and an anther in which pollen grains develop.

starch: a polysaccharide composed of branched or unbranched chains or glucose molecules, used by plants as a carbohydrate- storage molecule.

start codon: a codon in messenger RNA that signals the beginning of protein synthesis on a ribosome.

startle coloration: a form of mimicry in which a color pattern (often resembling large eyes) can be displayed suddenly by a prey organism when approached by a predator.

statoliths: In plants, starch-filled organelles in the stem and root cap that sink to the downward side of cells and act as a sensor for gravity.

stem: the normally vertical, aboveground part of a plant body that bears leaves.

sterilization: the method of contraception in which the pathways through which the sperm (vas deferens) or egg (oviducts) must travel are interrupted. Generally permanent, it is the the most form of contraception.

steroid: a class of hormone whose chemical structure resembles cholesterol that is secreted by the ovaries and placenta, the testes, and the adrenal cortex.

stigma (stig´-ma): the pollen-capturing tip of a carpel.

stoma (stō´-ma; pl. stomata): an adjustable opening in the epidermis of a leaf, surrounded by a pair of guard cells, that regulates the diffusion of carbon dioxide and water into and out of the leaf.

stomach: muscular sac between the esophagus and small intestine where food is stored and mechanically broken down, and in which protein digestion begins.

stop codon: a codon in messenger RNA that stops protein synthesis and causes the completed protein chain to be released from the ribosome.

stroke: an interruption of blood flow to part of the brain, caused by the rupture of an artery or the blocking of an artery by a blood clot. Loss of blood supply leads to rapid death of the area of the brain affected.

stroma (strō´-ma): the semi-fluid material inside chloroplasts in which the grana are embedded.

structural gene: a gene that codes for a protein used by the cell for purposes other than gene regulation. Structural genes code for enzymes or for proteins that are structural parts of a cell.

style: a stalk connecting the stigma of a carpel with the ovary at its base.

subatomic particle: the particles of which atoms are made: electrons, protons, and neutrons.

subclimax: a community where succession is stopped before the climax community is reached and is usually maintained by regular disturbance, for example, tallgrass prairie maintained by regular fires.

substrate: the atoms or molecules that are the reactants for an enzyme-catalyzed chemical reaction.

subunit: a small organic molecule, several of which may be bonded together to form a larger molecule. See also monomer.

succession (suk-se´-shun): the change in the community structure of an ecosystem over time. Community changes alter the ecosystem in ways that favor competitors, and species replace one another in a somewhat predictable manner until a stable, self-sustaining climax community is reached.

sucrose: a disaccharide composed of glucose and fructose.

sugar: a simple carbohydrate molecule, either a monosaccharide or a disaccharide.

supernormal stimulus: a stimulus that exaggerates crucial elements of the releaser, making it more effective than the normal releaser.

suppressor T cell: a type of T cell that depresses the response of other immune cells to foreign antigens.

surface tension: the property of a liquid to resist penetration by objects at its interface with the air, due to cohesion between molecules of the liquid.

survivorship curve: a curve resulting when the number of individuals of each age in a population is graphed against their age, usually expressed as a percentage of their maximum life span.

symbiosis (sim'-bī-ō'sis): a close association between different species over an extended period. Either or both may benefit from the association, or, in the case of parasitism, one of the participants is harmed. Symbiosis includes mutualism, commensalism, and parasitism.

sympathetic nervous system: the division of the autonomic nervous system that produces largely involuntary responses that prepare the body for stressful or highly energetic situations.

sympatric speciation (sim-pat´-rik): speciation of populations that are not physically divided; usually due to ecological isolation or chromosomal aberrations (such as polyploidy).

synapse (sin´-apz): the site of communication between nerve cells. One cell (presynaptic) usually releases a chemical (the neurotransmitter) that changes the electrical potential of the second (postsynaptic) cell.

synaptic cleft: a tiny space that separates the pre- and postsynaptic cells.

synaptic terminal: swellings at the branched endings of axons, where the axon forms a synapse.

syphilis (si´-ful-is): a sexually transmitted bacterial infection of the reproductive organs which, if untreated, can damage the nervous and circulatory systems.

systole (sis´-tō-lē): the portion of the heart cycle during which the ventricles contract; the higher of the two blood pressure readings.

T cell: a type of lymphocyte that recognizes and destroys specific foreign cells or substances, or that regulates other cells of the immune system.

T-cell receptor: a protein receptor located on the surface of a T cell, which binds a specific antigen and triggers the immune response of the T cell.

taproot system: a root system commonly found in dicots, consisting of a long, thick main root that develops from the primary root, and many smaller lateral roots that grow from the primary root.

target cell: a cell upon which a particular hormone exerts its effect.

taste: a chemical sense for substances dissolved in water or saliva; in mammals, perceptions of sweet, sour, bitter, or salt produced by stimulation of receptors on the tongue.

taste bud: a cluster of taste receptor cells and supporting cells located in a small pit beneath the surface of the tongue and communicating with the mouth through a small pore. The human tongue has about 10,000 taste buds.

taxis (tax´-is; pl. taxes): innate behavior that is a directed movement of an organism toward or away from a stimulus such as heat, light, or gravity.

taxonomy (tax-on-uh-mē): the science by which organisms are classified into hierarchically arranged categories that reflect their evolutionary relationships.

Tay-Sachs disease: a recessive disease caused by a deficiency in enzymes regulating lipid breakdown in the brain.

tectorial membrane (tek-tōr´-ē-ul): one of the membranes of the cochlea, in which the hairs of the hair cells are embedded. During sound reception, movement of the basilar membrane relative to the tectorial membrane bends the cilia.

telophase (tēl´-ō-fāz): the last stage of mitosis, in which a nuclear envelope reforms around each new daughter nucleus, the spindle disappears, and the chromosomes relax from their condensed form. In meiosis I, the stage during which the spindle disappears and the chromosomes usually relax from their condensed form. In meiosis II, the stage during which chromosomes relax into their extended state, nuclear envelopes reform, and cytokinesis occurs.

template strand: the strand of the DNA double helix from which RNA is transcribed.

temporal isolation: the inability of organisms to mate if they have significantly different breeding seasons.

temporal lobe: part of a cerebral hemisphere of the human brain, involved in recall of learned events.

tendon: a tough connective tissue band connecting a muscle to a bone.

tendril: a slender outgrowth of a stem that coils about external objects and supports the stem; usually a modified leaf or branch.

tentacle (ten´-te-kul): an elongate, extensible projection of the body of cnidarians and cephalopod mollusks that may be used for grasping, stinging, and immobilizing prey, and locomotion.

terminal bud: the bud at the extreme end of a stem or branch.

territoriality: the defense of an area in which important resources are located.

tertiary consumer (ter´-shē-ār-ē): a carnivore that feeds on other carnivores.

tertiary structure (ter´-shē-ār-ē): the complex three-dimensional structure of a single peptide chain. The tertiary structure is held in place by disulfide bonds between

test cross: a breeding experiment in which an individual showing the dominant phenotype is mated with an individual that is homozygous recessive for the same gene. The ratio of offspring with dominant versus recessive phenotypes can be used to determine the genotype of the phenotypically dominant individual.

testis (pl. testes): the gonad of male mammals.

testosterone: in vertebrates, a hormone produced by the interstitial cells of the testis; stimulates spermatogenesis and the development of male secondary sex characteristics.

tetany: smooth, sustained maximal contraction of a muscle in response to rapid firing by its motor neuron.

thalamus: part of the forebrain, the thalamus relays sensory information to many parts of the brain.

theory: in science, an explanation for natural events that is based on a large number of observations and is in accord with scientific principles, especially causality.

thermoacidophile (ther-mō-a-sid´-eh-fīl): a form of archaebacterium that thrives in hot, acidic environments.

thermoreceptor: a sensory receptor that responds to changes in temperature.

thick filaments: bundles of myosin protein within the sarcomere that interact with thin filaments to produce muscle contraction.

thin filaments: protein strands within the sarcomere that interact with thick filaments to produce muscle contraction. Composed primarily of actin, with accessory proteins.

thorn: a hard, pointed outgrowth of a stem; usually a modified branch.

threshold: the electrical potential (less negative than the resting potential) at which an action potential is triggered.

thrombin: an enzyme produced in the blood as a result of injury to a blood vessel. Thrombin catalyzes the production of fibrin, a protein that assists in blood clot formation.

thylakoid (thī´-la-koyd): a disk-shaped, membranous sac found in chloroplasts, the membranes of which contain the photosystems and ATP-synthesizing enzymes used in the light-dependent reactions of photosynthesis.

thymus (thī´-mus): a gland, considered part of the lymphatic system, located in the upper chest in front of the heart. The thymus secretes thymosin, which stimulates lymphocyte maturation. It begins to degenerate at puberty and has little function in the adult.

thyroid: an endocrine gland, located in front of the larynx in the neck, that secretes the hormones thyroxine (affecting metabolic rate) and calcitonin (regulating calcium ion concentration in the blood).

tight junction: a type of cell-to-cell junction in animals that prevents the movement of materials through the spaces between cells.

tissue: a group of (usually similar) cells that together carry out a specific function, i.e., muscle. The tissue may include extracellular material produced by its cells.

tonsils: patches of lymphatic tissue consisting of connective tissue containing many lymphocytes, located in the pharynx and throat.

trachea (trā´-kē-uh): a rigid but flexible tube supported by rings of cartilage, which conducts air between the larynx and the bronchi.

tracheae (trā´-kē): elaborately branching tubes that ramify through the bodies of insects and carry air close to each body cell. Air enters the tracheae through openings called spiracles.

tracheid (trā´-kē-id): an elongated xylem cell with tapering ends containing pits in the cell walls; forms tubes that transport water.

tracheophyte (trā´-kē-ō-fīt): a category of plant that has conducting vessels; a vascular plant.

transcription: the synthesis of an RNA molecule from a DNA template.

transducer: a device that converts signals from one form to another. Sensory receptors are transducers that convert environmental stimuli, such as heat, light, or vibration, into electrical signals (such as action potentials) recognized by the nervous system.

transfer RNA (tRNA): a type of RNA that (1) binds to a specific amino acid and (2) bears a set of three bases (the anticodon) complementary to the mRNA codon for that amino acid. Transfer RNA carries its amino acid to a ribosome during protein synthesis, recognizes a codon of mRNA, and positions its amino acid for incorporation into the growing protein chain.

transformation: a method of acquiring new genes, whereby DNA from one bacterium (usually released after the death of the bacterium) becomes incorporated into the DNA of another, living, bacterium.

translation: the process whereby the sequence of bases of messenger RNA is converted into the sequence of amino acids of a protein.

transpiration (trans´-per-ā-shun): evaporation of water through the stomata of a leaf.

transport protein: a protein that regulates the movement of water- soluble molecules through the plasma membrane.

transverse (T) tubules: deep infoldings of the muscle cell membrane that conduct the action potential inside the cell.

trial-and-error learning: a process by which adaptive responses are learned through rewards or punishments provided by the environment.

trichocyst (trik´-eh-sist): a stinging organelle of protists.

triglyceride (trī-gliss´-er-īd): a lipid composed of three fatty acid molecules bonded to a single glycerol molecule.

triple covalent bond: a covalent bond that occurs when two atoms share three pairs of electrons.

trisomy 21: see Down syndrome.

trisomy X: a condition of females who have three X chromosomes instead of the normal two. Most of these women are phenotypically normal and are fertile.

trophic level: literally "feeding level"; the position of an organism in a food chain, defined by the organisms's source of energy. trophic levels include producers, primary consumers, secondary consumers, and so on.

true-breeding: pertaining to an individual all of whose offspring produced through self-fertilization are identical to the parental type. True-breeding individuals are homozygous for a given trait.

tubal ligation: surgical procedure in which a woman's oviducts are cut so that the egg cannot reach the uterus, which makes her infertile.

tube cell: the outermost cell of a pollen grain; the tube cell digests a tube through the tissues of the carpel, ultimately penetrating into the female gametophyte.

tube feet: cylindrical extensions of the water-vascular system of echinoderms, used for locomotion, grasping food, and respiration.

tubular reabsorption: the process by which cells of the tubule of the nephron remove water and nutrients from the filtrate within the tubule and return them to the blood.

tubular secretion: the process by which cells of the tubule of the nephron remove additional wastes from the blood, actively secreting them into the tubule.

tubule (tūb´-ūle): the tubular portion of the nephron. It includes a proximal portion, the loop of Henle, and a distal portion. Urine is formed from the blood filtrate as it passes through the tubule.

tumor-suppressor gene: a gene that encodes information for a protein that inhibits cancer formation, probably by regulating cell division in some way.

turgor pressure: pressure developed within a cell (especially the central vacuole of plant cells) as a result of osmotic water entry.

Turner syndrome: a set of characteristics typical of a woman with only one X chromosome. These women are sterile, with a tendency to be very short and to lack normal female secondary sexual characteristics.

tympanic membrane (tim-pan´-ik): the eardrum; a membrane stretched across the opening of the ear, which transmits vibration of sound waves to bones of the middle ear.

uniform distribution: a relatively regular spacing of individuals within a population, often as a result of territorial behavior.

uniformitarianism: the hypothesis that Earth developed gradually through natural processes similar to those at work today occurring over long periods of time.

unsaturated: referring to a fatty acid with fewer than the maximum number of hydrogen atoms bonded to its carbon backbone; a fatty acid with one or more double bonds in its carbon backbone.

urea (ū-rē´-uh): a water-soluble, nitrogen-containing waste product of amino acid breakdown that is one of the principal components of mammalian urine.

ureter (ū´-re-tur): a tube that conducts urine from each kidney to the bladder.

urethra (ū-rē´-thra): the tube leading from the urinary bladder to the outside of the body; in males, the urethra also receives sperm from the vas deferens and conducts both sperm and urine (at different times) to the tip of the penis.

uric acid (ūr´-ik acid): a nitrogen-containing waste product of amino acid breakdown, which is a relatively insoluble white crystal. Uric acid is excreted by birds, reptiles, and insects.

urine: fluid produced and excreted by the urinary system containing water and dissolved wastes, such as urea.

uterus: in female mammals, the part of the reproductive tract that houses the embryo during pregnancy.

utricle (ū´-tri-cul): a portion of the vestibular apparatus responsible for detecting gravity.

vaccination: an injection into the body that contains antigens characteristic of a particular disease organism and that stimulates an immune response.

vacuole (vak´-ū-ōl): a vesicle, often large, consisting of a single membrane enclosing a fluid-filled space.

vagina: the passageway leading from the outside of the body to the cervix of the uterus.

variable: a condition, particularly in a scientific experiment, that is subject to change.

variable region: part of an antibody molecule that differs among antibodies; the ends of the variable regions of the light and heavy chains form the specific binding site for antigen.

vas deferens (vas de´-fer-ens): the tube connecting the epididymis of the testis with the urethra.

vascular bundle (vas´-kū-lar): a strand of xylem and phloem found in leaves and stems; in leaves, commonly called a vein.

vascular cambium: a lateral meristem located between the xylem and phloem of a woody root or stem and gives rise to secondary xylem and phloem

vascular cylinder: the centrally located conducting tissue of a young root, consisting of primary xylem and phloem.

vascular tissue system: a plant tissue system consisting of xylem (which transports water and minerals from root to shoot) and phloem (which transports water and sugars throughout the plant).

vasectomy: a surgical procedure in which a man's vas deferens are cut, preventing sperm from reaching the penis during ejaculation, which makes him infertile.

vein: a large-diameter, thin-walled vessel that carries blood from venules back to the heart.

ventral (ven´-trul): the lower, or underside of an animal whose head is oriented forward.

ventricle (ven´-tre-kul): the lower muscular chamber on each side of the heart, which pumps blood out through the arteries. The right ventricle sends blood to the lungs, and the left to the rest of the body.

venule (ven´-yul): a narrow vessel with thin walls that carries blood from capillaries to veins.

vertebrate: an animal that possesses a vertebral column.

vesicle (ves´-i-kul): a small membrane-bound sac within the cytoplasm.

vessel: a tube of xylem composed of vertically stacked vessel elements, with perforated or missing end walls, leaving a continuous, uninterrupted hollow cylinder.

vessel element: one of the cells of a xylem vessel; elongated, dead at maturity, with thick lignified lateral cell walls for support but with end walls either lacking entirely or heavily perforated.

vestibular apparatus (ves-tib´-ū-lar): the portion of the inner ear of mammals that responds to gravity and changes in the direction and speed of movement.

vestigial structures (ves-tij´-ē-ul): structures that serve no apparent purpose, but which are homologous to functional structures in related organisms and evidence of evolution.

villus (vi´-lus): fingerlike projections of the wall of the small intestine that increase its absorptive surface area.

viroid (vī´-rōyd): a particle of RNA capable of infecting a cell and directing the production of more viroids; responsible for certain plant diseases.

virus (vī´-rus): a noncellular parasitic particle consisting of a protein coat surrounding a strand of genetic material. Viruses multiply only within cells of living organisms.

vitamin: any one of a group of diverse chemicals that must be present in trace amounts in the diet to maintain health. Vitamins are used by the body in conjunction with enzymes in a variety of metabolic reactions.

vitreous humor (vit´-rē-us): a clear jellylike substance that fills the large chamber of the eye between the lens and retina.

vocal cords: folds of tissue that extend across the opening of the larynx and produce sound when air is forced between them. Muscles alter the tension on the vocal cords and control the size and shape of the opening, which in turn determines whether sound is produced and what its pitch will be.

waggle dance: symbolic communication used by honeybee foragers to communicate the location of a food source to their hivemates.

water-vascular system: a system in echinoderms consisting of a series of canals through which seawater is conducted and used to inflate tube feet for locomotion, grasping food, and respiration.

wax: a lipid composed of fatty acids covalently bonded to long- chain alcohols.

white matter: Portions of the brain and spinal cord consisting largely of myelin-covered axons that give these areas a white appearance

withdrawal: removal of the penis from the vagina just before ejaculation in an attempt to avoid pregnancy; an ineffective contraceptive method.

X-linked inheritance: see sex-linked inheritance.

xylem (zī´-lum): a conducting tissue of vascular plants that transports water and minerals from root to shoot.

yolk: protein- or lipid-rich substances contained in eggs as food for the developing embryo.

yolk sac: one of the embryonic membranes of reptile, bird, and mammalian embryos. In birds and reptiles, the yolk sac is a membrane surrounding the yolk in the egg. In mammals, the yolk sac is empty but forms part of the umbilical cord and the digestive tract.

Z lines: fibrous protein structures to which the thin filaments of skeletal muscle are attached, forming the boundaries of sarcomeres.

zona pellucida (pel-ū´-si-da): a clear, noncellular layer between the corona radiata and the egg.

zooflagellate (zō-ō-fla´-gel-et): a category of nonphotosynthetic protist that move using flagella.

zoospore (zō´-ō-spōr): a nonsexual reproductive cell that swims using flagella, such as is formed by members of the fungal division Oomycota.

zygospore (zī´-gō-spōr): produced by the division Zygomycota, a fungal spore surrounded by a thick, resistant wall, which forms from a diploid zygote.

zygote (zī´-gōt): in sexual reproduction, a diploid cell (the fertilized egg) formed by the fusion of two haploid gametes.

Photo Credits

Frontmatter
p. ii Dwight Kuhn Photography **p. v** Audesirk

Chapter 1
Opener NASA Headquarters **1-1a** Dr. Jeremy Burgess/Science Photo Library/Photo Researchers, Inc. **1-1b** Craig Tuttle/The Stock Market **1-1c** Kim Taylor/Bruce Coleman, Inc. **1-3** Biological Photo Service/Terraphotographics **1-4** Anua and Mahoj Shah/Animals Animals/Earth Science **1-5** Tom McHugh/Photo Researchers, Inc. **1-6** Will & Deni McIntyre/Photo Researchers, Inc. **1-7a** CNRI/Science Photo Library/Photo Researchers, Inc. **1-7b** Eric V. Grave/Photo Researchers, Inc. **1-7c** Richard L. Carlton/Photo Researchers, Inc. **1-7d** Animals Animals/Earth Scenes **1-7e** (p. xx) William Townsend/Photo Researchers, Inc. **1-8** John Durham/Science Photo Library/Photo Researchers, Inc. **1-9** Doug Perrine/DRK Photo **1-10** Tom Algire/Tom Stack & Associates **1-11** Chip Clark/Smithsonian Institution **E1-2** Luiz C. Marigo/Peter Arnold, Inc. **E1-2** (inset) Gunter Ziesler/Peter Arnold, Inc. **1-13** Audesirk

Unit 1
Opener Secchi, Iecaque, Roussel, Uclaf, Cnri/Photo Researchers, Inc.

Chapter 2
Opener Blaine Harrington III/The Stock Market **2-1** Jim Anderson/Woodfin Camp & Associates **E2-1c** Courtesy of Drs. M. J. Fulham & Giovanni Di Chiro, The Neuroimaging Section, NINDS, National Institutes of Health, Bethesda, MD. **E2-2** NIH/Science Source/Photo Researchers, Inc. **2-11a** David Dennis/Tom Stack & Associates **2-11b** Audesirk

Chapter 3
Opener Ken Edward/Biografix/SS/Photo Researchers, Inc. **3-4a** Dr. Jeremy Burgess/Science Photo Library/Photo Researchers, Inc. **3-5** (left) Larry Ulrich/DRK Photo **3-5** (middle) Jeremy Burgess, M.D./Photo Researchers, Inc. **3-5** (right) Biophoto Associates/Photo Researchers, Inc. **3-6** Richard Kolar/Animals Animals/Earth Science **3-11a** Robert Pearcy/Animals Animals/Earth Science **3-11b** Jeff Foott/DRK Photo **3-11c** Nuridsany et Perennou/Photo Researchers, Inc.

Chapter 4
Opener Bruce Davidson/Animals Animals/Earth Science **4-1** Focus on Sports **4-7a,b** T. A. Steitz/Yale University

Chapter 5
Opener Moredun Animal Health Ltd/Science Photo Library/Photo Researchers, Inc. **5-1** (upper) Biophoto Associates/Photo Researchers, Inc. **5-1** (lower) Cecil Fox/Science Source/Photo Researchers, Inc. **E5-1a,b,c** M. I. Walker/Photo Researchers, Inc. **E5-1d** David M. Phillips/Visuals Unlimited **E5-2** Dr. Don Fawcett/Photo Researchers, Inc. **E5-3a** Biophoto Associates/Photo Researchers, Inc. **E5-3b** Karl Aufderheide/Visuals Unlimited **E5-3c** Dr. K. Tanaka **5-3** Dr. Stanley L. Erlandsen/PH Archives **5-5** John Cardomona/Biological Photo Service/Terraphotographics **5-8a** Richard Rodewald/Biological Photo Service/Terraphotographics **5-8b** Don W. Fawcett/Photo Researchers, Inc. **5-9** Carolina Biological Supply Company **5-10** Omikron/Science Source/Photo Researchers, Inc. **5-12** Dr. Barry F. King/Biological Photo Service/Terraphotographics **5-13** Photo Researchers, Inc. **5-14** Zdenek Hruban **E5-4** Dr. Don Fawcett/T. Kanaseki/Photo Researchers, Inc. **5-16** W.P. Wergin/Biological Photo Service/Terraphotographics **5-17** Laboratory of Cell Biology University of Maryland **5-18** Biophoto Associates/Photo Researchers **5-19a** (left) Photo Researchers, Inc. **5-19a** (middle) Dr. Peter Dawson/Science Photo Library/Photo Researchers, Inc. **5-19a** (right) Francis LeRoy, Biocosmos/Science Photo Library/Photo Researchers, Inc. **5-20** Keith R. Porter/Laboratory of Cell Biology University of Maryland **5-21** (upper) William L. Dentler/Biological Photo Service/Terraphotographics **5-21** (lower) Dr. de Harven/Photo Researchers, Inc.

Chapter 6
Opener Jan Hinsch/Science Photo Library/Photo Researchers, Inc. **6-1** Farrell Grehan/Photo Researchers, Inc. **6-2** Biophoto Associates/Science Source/Photo Researchers, Inc. **6-8a,b,c** Dr. Joseph Kuranstin-Mills **6-11a–d** Perry, M. M. & Gilbert, A. B. *The Journal of Cell Science*, 39:257–272 (1979) **6-12** L. A. Hufnagel/PH Archives **6-13a,b** Thomas Eisner **6-15** Dr. Don Fawcett/G. Raviola; P. Elias; D. Friend/Photo Researchers, Inc. **6-16** L. Andrew Staehelin **6-17** Photo Researchers, Inc. **6-18** W.P. Wergin/Biological Photo Service/Terraphotographics **6-19** Steven J. Krasemann/DRK PHoto

Chapter 7
Opener Alfred Pasieka/Peter Arnold, Inc. **7-2a** Ken W. Davis/Tom Stack & Associates **7-2b,c** E. H. Newcome/Biological Photo Service/Terraphotographics

Chapter 8
Opener Francois Gohier/Photo Researchers, Inc. **8-3a** Focus on Sports **8-3b** Photo Researchers, Inc.

Unit II
Opener CNRI/Science Photo Library/Photo Researchers, Inc.

Chapter 9
Opener Dr. Gopal Murti/Science Photo Library/Photo Researchers, Inc. **9-2** Photo Researchers, Inc. **9-3** Photo Researchers, Inc. **9-4** Photo Researchers, Inc. **9-5** Photo Researchers, Inc. **9-7a–h** University of Oregon **9-9a** Biological Photo Service/Terraphotographics **9-9b** John D. Cunningham/Visuals Unlimited **9-12a** T. E. Schroeder/Biological Photo Service/Terraphotographics **9-14a** M. Abbey/Photo Researchers, Inc. **9-14b** (p.xxii) Carolina Biological Supply Co./Phototake NYC **9-15** Audesirk

Chapter 10
Opener Dr. Richard J. Feldman/Molecular Graphics/National Institutes of Health **10-3** Thomas Broker/CNRI/Phototake NYC **10-6a** Rosalind Franklin/Photo Researchers, Inc. **10-6b** CSHL Archives/Peter Arnold, Inc. **E10-1** Photo Researchers, Inc.

Chapter 11
Opener E. V. Kiseleva, 1989, *FEBS LETTS*, 25:251–253 **11-5** Dr. Oscar L. Miller, Jr./PH Archives **11-16** Science Source/Photo Researchers, Inc. **E11-1** Rube Goldberg/United Media **11-17** Dr. M. L. Barr

Chapter 12
Opener Minden Pictures **12-1** Brinton, C. C., Jr. & Carnahan, J. from *Principles of Population Genetics* 2/e, 19xx, edited by Daniel L. Hartl and Andrew C. C.ark, Sunderland, MA:Sinaver. **12-5** James Kezer

Chapter 13
Opener Ron Kimball Studios **13-1a,c** Animals Animals/Earth Scenes **13-1b** Tom Stack & Associates **13-2** Photo Researchers, Inc. **13-3** Audesirk **13-14** Carolina Biological Supply Co./Phototake NYC **13-18** (p. xxiii) Jane Burton/Bruce Coleman, Inc.

Chapter 14
Opener Visuals Unlimited **14-1** Professor Stanley N. Cohen/Photo Researchers, Inc. **14-6** Hank Morgan/Photo Researchers, Inc. **14-8** Applied Biosystems/Peter Arnold, Inc. **E14-2** Photo Researchers, Inc. **E14-2** (inset) Gilbert Grant/Photo Researchers, Inc. **E14-3a** Gary Braasch **E14-3b** Dr. Joseph Newhouse/PH Archives **14-9** Bettmann

Chapter 15
Opener David Stoecklein/The Stock Market **15-2a** Gamma-Liaison, Inc. **15-2b** Audesirk **15-4a** Marlene and Bob Lippman **15-5a,b** Photo Researchers, Inc. **15-6a** Bettmann **15-6b** From *Huntington's Disease: A Disorder of Families* by Susan E. Folstein. 1989. Fig. 5.6, p. 78. Johns Hopkins University Press. **15-7a** Hart-Davis/Science Photo Library/Photo Researchers, Inc. **E15-1** Peter Arnold, Inc. **15-8** Photo Researchers, Inc. **15-10a** CNRI/Science Photo Library/Photo Researchers, Inc. **15-10b** Lawrence Midgale **E15-2a** CNRI/Science Photo Library/Photo Researchers, Inc.

Unit III
Opener Martin Dohm/Photo Researchers, Inc.

Chapter 16
Opener Peter Arnold, Inc. **16-1a** Wayne Lankinen/Valan Photos **16-1b** Edgar T. Jones/Bruce Coleman, Inc. **16-2** Photo Researchers, Inc. **16-4a** DRK Photo **16-4b** Chip Clark/National Museum of Natural History, Smithsonian **16-4c** American Museum of Natural History **16-6** Bettmann **16-7** (left) Christian Grzimek/OKAPIA/Photo Researchers, Inc. **16-7** (right) Frans Lanting/Photo Researchers, Inc. **16-8b** Visuals Unlimited **16-8c** Bruce Coleman, Inc. **16-11a,b** Photo Researchers, Inc. **16-11c** Douglas T. Cheeseman, Jr./Peter Arnold, Inc. **16-11d** Minden Pictures **16-14a** Animals Animals/Earth Science **16-14b** Dan McCoy/Rainbow **16-14c** Photo Researchers, Inc. **16-14d** Carolina Biological Supply Company/Phototake NYC **16-15a** DRK Photo **16-15b** Timothy O'Keef/Tom Stack & Associates **16-16a,b** Bruce Coleman, Inc. **16-17** PH Archives

Chapter 17
Opener Phil Degginger/Bruce Coleman, Inc. **17-3b** (left) Peter Arnold, Inc. **17-3b** (right) Y. R. Tymstra/Valan Photos **17-4** Gregory Dimijian/Photo Researchers, Inc. **17-5** Visuals Unlimited **17-6** Photo Researchers, Inc. **17-7** DRK Photo **17-11** Robert & Linda Mitchell Photography **17-12a** Bruce Coleman, Inc. **17-12b** Photo Researchers, Inc. **17-13** Photo Researchers, Inc. **17-14** Tom McHugh/Steinhart Aquarium/Photo Researchers, Inc. **17-15** (p. xxiv) James A. Kern **E17-1** Minden Pictures **17-18** NASA Headquarters **17-19** Thomas A. Wiewandt, Ph.D.

Chapter 18
Opener Paul A. Johnsgard/Photo Researchers, Inc. **18-1** DRK Photo **18-3a,b** Photo Researchers, Inc. **18-4** Guy L. Bush **18-5** Bruce Coleman, Inc.

Chapter 19
Opener Royal Observatory, Edinburgh/aatb/Science Photo Library/Photo Researchers, Inc. **19-4** Visuals Unlimited **19-6** Visuals Unlimited **19-8a** The Field Museum **19-8b** James L. Amos/Photo Researchers, Inc. **19-8c** Carolina Biological Supply Co./Phototake NYC **19-8d** Photo Researchers, Inc. **19-9** Rudolph F. Zallinger/Peabody Museum of Natural History **19-11** Bruce Coleman, Inc. **19-12** Rudolph F. Zallinger/Peabody Museum of Natural History **19-13a** Tom McHugh/Photo Researchers, Inc. **19-13b** Minden Pictures, Inc. **19-13c** Tom Stack & Associates **19-14**

David L. Brill Photography **19-17** Chip Clark/National Museum of Natural History, Smithsonian **19-18** Gamma-Liaison, Inc.

Chapter 20

Opener (top) Tom Stack & Associates (left) Peter Arnold, Inc. (right) E.R. Digginger/Animals Animals/Earth Scenes **20-2a** Photo Researchers, Inc. **20-2b** Peter Arnold, Inc. **20-3a** Stephen Murphree/Dept. of Entomology, Auburn University/BPS **20-3b** Dr. Greg Rouse **20-3c** Dr. Jeremy Burgess/Science Photo Library/Photo Researchers, Inc. **20-5** Discovery Syndication

Chapter 21

Opener Peter Arnold, Inc. **21-2b** Photo Researchers, Inc. **21-4** Photo Researchers, Inc. **21-5** Dr. Stanley Pruisner **21-6a** Visuals Unlimited **21-6b,c** Photo Researchers, Inc. **21-7** Lennart Nilsson **21-8a** Dr. Tony Brain/Photo Researchers, Inc. **21-9** A. B. Dowsett/Photo Researchers, Inc. **21-10** Photo Researchers, Inc. **21-11** Brinton, C. C., Jr. & Carnahan, J. from *Principles of Population Genetics* 2/e, 19xx, edited by Daniel L. Hartl and Andrew C. C.ark, Sunderland, MA:Sinaver. **21-12** Yellowstone National Park **21-13a,b** Visuals Unlimited **21-14a** BioPhoto Associates/Photo Researchers, Inc. **21-14b** Photo Researchers, Inc. **21-15a** Carolina Biological Supply Co./Phototake NYC **21-15b** Photo Researchers, Inc. **21-16** Visuals Unlimited **21-17** Kevin Schafer/Peter Arnold, Inc. **21-18** Photo Researchers, Inc. **21-20a** Photo Researchers, Inc. **21-20b** Visuals Unlimited **21-21** Cabisco/Visuals Unlimited **21-22** Photo Researchers, Inc. **21-23** Drs. S. L. Erlandsen & D. E. Feely/PH Archives **21-24** Photo Researchers, Inc. **21-25** Photo Researchers, Inc. **21-26a,b** Photo Researchers, Inc. **21-29a,b,c** Photo Researchers, Inc.

Chapter 22

Opener Photo Researchers, Inc. **22-1a** Dwight Kuhn Photography **22-2** Dr. Charles Brewer-Carias **22-4a** Photo Researchers, Inc. **22-4b** Jeff Foott Productions **22-4c** Robert & Linda Mitchell Photography **22-5** Stanley L. Flegler/Visuals Unlimited **22-6** (middle) Breck P. Kent **22-6** (lower) Carolina Biological Supply/Phototake NYC **22-7a** Photo Researchers, Inc. **22-7** (p. xxv) D. Dvorak, Jr. **22-8** Photo Researchers, Inc. **22-9a,b,d** Tom Stack & Associates **22-9c** Photo Researchers, Inc. **22-11** Robert & Linda Mitchell Photography **22-12a** Audesirk **22-12b** Manfred Kage/Peter Arnold, Inc. **22-13** Dr. William Merrill **22-14** M. Viard/Jacana/Photo Researchers, Inc. **22-15** G. L. Baron/Biological Photo Service/Terraphotographics **22-16** N. Allin & G. L. Barron/Biological Photo Service/Terraphotographics

Chapter 23

Opener Dr. Jeremy Burgess/Photo Researchers, Inc. **23-1** Jeffrey L. Rotman **23-2a,c** Photo Researchers, Inc. **23-2b** Peter Arnold, Inc. **23-3a,b** Photo Researchers, Inc. **23-4** Photo Researchers, Inc. **23-5** Tom Stack & Associates **23-6** Carolina Biological Supply Co./Phototake NYC **23-7a** Dwight Kuhn Photography **23-7b** Tom Stack & Associates **23-7c** (p. xxvi) DRK Photo **23-8** Milton Rand/Tom Stack & Associates **23-10a** Audesirk **23-10b** A-Z Botanical Collection Ltd. **23-11** (left) Dr. William M. Harlow/Photo Researchers, Inc. **23-11** (right) Photo Researchers, Inc. **23-12a,d** Dwight Kuhn Photography **23-12b** Audesirk **23-12c** Photo Researchers, Inc. **23-12e** David Dare Parker/Auscape International **23-12e** (inset) Matt Jones/Auscape International **23-14** Tom Stack & Associates

Chapter 24

Opener Jeff Rotman/Peter Arnold, Inc. **24-4a** Audesirk **24-4b** Tom Stack & Associates **24-4c** Odyssey Productions **24-6a** Photo Researchers, Inc. **24-6b** Odyssey Productions **24-6c** Audesirk **24-6d** Visuals Unlimited **24-9a** Carolina Biological Supply Co./Phototake NYC **24-10** (upper) Martin Rotker/Phototake NYC **24-10** (lower) Stanley Flegler/Visuals Unlimited **24-11** Tom E. Adams/Peter Arnold, Inc. **24-12a** Carolina Biological Supply Co./Phototake NYC **24-12b** Howard Shiang, D.V.M. **24-14a** Kjell B. Sandved **24-14b** David Bull Photography **24-14c** Photo Researchers, Inc. **24-16** Audesirk **24-17** Dwight Kuhn Photography **24-18** David Scharf/Peter Arnold, Inc. **24-20a** Carolina Biological Supply Co./Phototake NYC **24-20b** Biological Photo Service/Terraphotographics **24-20c** Photo Researchers, Inc. **24-20d** Peter Arnold, Inc. **24-20e** DRK Photo **24-21a** Dwight Kuhn Photography **24-21b** Robert & Linda Mitchell Photography **24-21c** Audesirk **24-22a** Photo Researchers, Inc. **24-22b** Biological Photo Service/Terraphotographics **24-22c** Carolina Biological Supply Co./Phototake NYC **24-22d** Alex Kerstitch **24-24a** Photo Researchers, Inc. **24-24b** Alex Kerstitch **24-25a** Fred Bavendan/Peter Arnold, Inc. **24-25b** Ed Reschke/Peter Arnold, Inc. **24-26a** Peter Arnold, Inc. **24-26b** Photo Researchers, Inc. **24-26c** Alex Kerstitch **24-27a** Audesirk **24-27b** Jeff Foott Productions **24-27c** Bruce Coleman, Inc. **24-28b** Photo Researchers, Inc. **24-29** Photo Researchers, Inc. **24-30b** Tom McHugh/Photo Researchers, Inc. **24-31a** Photo Researchers, Inc. **24-31b** Tom Stack & Associates **24-31b** (inset) Breck P. Kent **24-32a** Photo Researchers, Inc. **24-32b** Jeffrey L. Rotman **24-33a** Peter David/Planet Earth Pictures **24-33b** Mike Neumann/Photo Researchers, Inc. **24-33c** DRK Photo **24-33d** Peter Scoones/Planet Earth Pictures **24-34a** Animals Animals/Earth Scenes **24-34b** Tom Stack & Associates **24-34c** Photo Researchers, Inc. **E24-1** DRK Photo **24-35a** D. B. Barker/Tom Stack & Associates **24-35b** Biological Photo Service/Terraphotographics **24-35c** Frans Lanting/Minden Pictures **24-36a** Carolina Biological Supply Co./Phototake NYC **24-37a,c** Photo Researchers, Inc. **24-37b** Carolina Biological Supply Co./Phototake NYC **24-38** Carolina Biologial Supply Co./Phototake NYC **24-39a** Flip Nicklin/Minden Pictures **24-39b,d** Photo Researchers, Inc. **24-39c** Peter Arnold, Inc. **24-40a** Photo Researchers, Inc. **24-40b** Mark Newman/Auscape International **24-40b** (inset) D. Parer and E. Parer-Cook/Auscape International

Unit IV

Opener Stephen Parker/Photo Researchers, Inc.

Chapter 25

Opener The Stock Market **25-5a** Photo Researchers, Inc. **25-5b** John D. Cunningham/Visuals Unlimited **25-6a** Biology Media/Photo Researchers, Inc. **25-6b,c** George Wilder/Visuals Unlimited **25-9a** Photo Researchers, Inc. **25-9b** Dwight Kuhn Photography **25-10** (upper) Ed Reschke/Peter Arnold, Inc. **25-10** (lower) E.R. Degginger/Animals Animals/Earth Scenes **25-11** Audesirk **25-13** Biological Photo Service/Terraphotographics **25-14** Ed Reschke/Peter Arnold, Inc. **25-17** (middle) Carolina Biological Supply Co. **25-17** (right) D. C. Cunningham/Visuals Unlimited **25-18** Visuals Unlimited **25-19** Dr. Jeremy Burgess/Photo Researchers, Inc. **25-20** Photo Researchers, Inc. **25-21** DRK Photo **25-22a** Audesirk **25-22b** Carolina Biological Supply Co./Phototake NYC **25-23a** Visuals Unlimited **25-23b,e** Audesirk **25-23c** C. Allan Morgan/Peter Arnold, Inc. **25-23d** Dwight Kuhn Photography **25-24a** Photo Researchers, Inc. **25-24b** Tom Stack & Associates

Chapter 26

Opener Jum Zipp/Photo Researchers, Inc. **26-1** Robert & Linda Mitchell Photography **26-3a** Photo Researchers, Inc. **26-3b** Visuals Unlimited **26-4bU** Photo Researchers, Inc. **26-4bW** E.H. Newcomb/S.R. Tandon/Biological Photo Service/Terraphotographics **26-7a,b** Photo Researchers, Inc. **26-9** P. Dayanandan/Photo Researchers, Inc. **26-10a** E. R. Degginger/Animals Animals/Earth Scenes **26-10b** Harvard University, The Biological Labs **E26-1** Randall Hyman

Chapter 27

Opener F. Stuart Westmorland/Photo Researchers, Inc. **27-3** Robert & Linda Mitchell Photography **27-4b** James L. Castner **27-5** PH Archives **27-6** PH Archives **27-7a,b** Thomas Eisner **27-8** John Gerlach/Tom Stack & Associates **27-9** James L. Castner **27-10a** Merlin D. Tuttle/Photo Researchers, Inc. **27-10b** Animals Animals/Earth Scenes **27-10c** Luiz C. Marigo/Peter Arnold, iNc. **27-11** Animals Animals/Earth Scenes **27-12a,b** DRK Photo **27-13** Carolina Biological Supply Co./Phototake NYC **27-14** Photo Researchers, Inc. **27-16** Carolina Biological Supply Co./Phototake NYC **27-18b,c** Carolina Biological Supply Co./Phototake NYC **E27-1a** PH Archives **E27-1b** Carolina Biological Supply Co./Phototake NYC **E27-2** PH Archives **E27-3** Scott Camazine/Photo Researchers, Inc. **E27-4** Michael & Patricia Fogden **E27-5a** George Edwards/Photo Researchers, Inc. **E27-5b** Merlin D. Tuttle/Photo Researchers, Inc. **27-20** PH Archives

Chapter 28

Opener Merlin D. Tuttle/Photo Researchers, Inc. **28-5a** Runk/Schoenberger/Grant Heilman Photography, Inc. **28-10** DRK Photo **E28-1a,b** Carolina Biological Supply Co./Phototake NYC

Unit V

Opener Alfred Pasieka/Photo Researchers, Inc.

Chapter 29

Opener Minden Pictures **29-4** Manfred Kage/Peter Arnold, Inc. **29-5a** John D. Cunningham/Visuals Unlimited **29-5b** Don Fawcett/Visuals Unlimited **29-5c** David M. Phillips/Visuals Unlimited

Chapter 30

Opener Professor P. Motta/Photo Researchers, Inc. **30-9** Photo Researchers, Inc. **30-12** (p. xxvii) Lennart Nilsson/Bolhrinoer Inoelheim International Gmbh **30-13** Discover Syndication **30-14** Photo Researchers, Inc. **30-17** Lennart Nilsson **E30-1a** Lou Lainey/Time Picture Syndicate **E30-1c** Biophoto Associates/Photo Researchers, Inc. **30-21** John D. Cuningham/Visuals Unlimited

Chapter 31

Opener Bob Evans/Western Marine Laboratory/Peter Arnold, Inc. **31-2a** Charles Seaborn/Odyssey Productions **31-2b** T. A. Wiewandt/DRK Photo **31-4a** Breck P. Kent **31-4b** E. R. Degginger/Bruce Coleman, Inc. **31-5** David M. Dennis/Tom Stack & Associates **31-6b** H. R. Duncker, Department of Anatomy, University of Giessen, Germany **E31-2** Bill Travis, M.D./National Cancer Institute **E31-3** O. Auerbach/Visuals Unlimited

Chapter 32

Opener Haroldo Palo Jr./Natural History Photographic Agency **32-1** Johnny Johnson/Tony Stone Images **32-3** Tobi Zausner **32-5a** E. R. Degginger/Photo Researchers, Inc. **32-12e** Lennart Nilsson **E32-2** St. Bartholomew's Hospital/Science Photo Library/Photo Researchers, Inc.

Chapter 33

Opener Douglas T. Mesney/The Stock Market **33-9** Tony Stone Images

Chapter 34

Opener Boehringer Ingelheim International GmbH/Lennart Nilsson **34-1** CNRI/Science Photo Library/Photo Researchers, Inc. **34-2** Boehringer Ingelheim International GmbH/Lennart Nilsson **34-4** ©NIBSC/SPL/Photo Researchers, Inc. **34-6** Leonard Lessin/Peter Arnold, Inc. **34-11** Photo Researchers, Inc. **34-12** Dr. Andrejs Liepins/Photo Researchers, Inc. **34-15b** Dr. Matthew A. Gonda, Ph.D./National Cancer

Institute 34-15c Lennart Nilsson E34-2 (p. xxviii)
CNRI/Science Photo Library/Photo Researchers, Inc.

Chapter 35

Opener E. R. Degginger/Photo Researchers, Inc. 35-6
Bettmann 35-8b Photo Researchers, Inc. 35-10a
From L. Orci, *Diabetes*, 31, 538–565 (1982).

Chapter 36

Opener Secci-Lecaque/Photo Researchers, Inc. 36-3a
Journal of Physiology 36-3b Carolina Biological Supply Co./Phototake NYC 36-6 C. Raines/Visuals Unlimited 36-13a Jeff Foott/Tom Stack & Associates 36-13b Dr. Eric Kandel/Peter Arnold, Inc. 36-13c
Tony Stone Images E36-1 M. Raichle/Washington University School of Medicine E36-2 Dr. William J. Powers E36-3a,b Monte S. Buchsbaum, M.D., Mount Sinai School of Medicine, New York, NY 35-24
Hanna Damasio, M.D./Science Magazines

Chapter 37

Opener Dwight Kuhn Photography 37-2 M. Fogden/Bruce Coleman, Inc. 37-6 Robeer S. Preston/Professor J.E. Hawkins 37-9a David Scharf/Peter Arnold, Inc. 37-12 Eichorst/PH Archives 37-13a L. West/Photo Researchers, Inc. 37-13b (p. xxix) Frans Lanting/Minden Pictures 37-14a Dwight Kuhn Photography 37-14b PH Archives 37-18a Charles E. Mohr/Photo Researchers, Inc. 37-18b Brian Parker/Tom Stack & Associates

Chapter 38

Opener Lois Greenfield 38-1f Dr. Brian Evden/Science Photo Library/Photo Researchers, Inc. 38-3b Photo Researchers, Inc. 38-4a Tom McHugh/Steinhart Aquarium/Photo Researchers, Inc. 38-4b Stanley Breeden/DRK Photo 38-7 Photo Researchers, Inc.
38-9 Manfred Kage/Peter Arnold, Inc. E38-5a
Michael Klein/Peter Arnold, Inc. E38-5b Larry Mulevhill/Science Source/Photo Researchers, Inc.

Chapter 39

Opener Hans & Judy Beste/Animals Animals/Earth Scenes 39-1 Fred Bavendam/Peter Arnold, Inc. 39-3
P. J. Bryant/University of California at Irvine/Biological Photo Service/Terraphotographics 39-4a Tom McHugh/Photo Researchers, Inc. 39-4b PH Archives 39-4b (inset) Peter Harrison/PH Archives 39-5 Anne et Jacques Six 39-6 Michael Fogden/Oxford Scientific Films/Animals Animals/Earth Scenes 39-7a Stephen J. Krasemann/DRK Photo 39-7b Tom McHugh/Photo Researchers, Inc. 39-7c Minden Pictures 39-10
Photo Researchers, Inc. 39-14a,b Gary Martin/Visuals Unlimited 39-17a Lennart Nilsson 39-17b Lennart Nilsson, *A Child is Born*, Dell Publishing Company. E39-1 Hank Morgan/Science Source/Photo Researchers, Inc. E39-2 AP/Wide World Photos 39-18a Lennart Nilsson

Chapter 40

Opener Petit Format/Nestle/Photo Researchers, Inc.
40-3a (left) Image Quest 3-D/Natural History Photographic Agency 40-3a (right) Jim Bain/Natural History Photographic Agency 40-3b (left) Stephen Dalton/Photo Researchers, Inc. 40-3b (right) L. West/Photo Researchers, Inc. 40-4a K. H. Switak/Photo Researchers, Inc. 40-4b J.A.L. Cooke/Animals Animals/Earth Science 40-4c Breck P. Kent 40-4d
Frans Lanting/Minden Pictures 40-9 Petit Format/Nestle/Science Source/Photo Researchers, Inc. 40-13
Lennart Nilsson 40-14 Lennart Nilsson E40-1
David Scharf/Peter Arnold, Inc. E40-2 Leonard McCombe/Life Magazine, Time Warner, Inc. E40-3 Sterling K. Clarren, M.D., Children's Hospital & Medical Center, University of Washington School of Medicine 40-16 Audesirk

Opener Minden Pictures, Inc.

Chapter 41

Opener Frans Lanting/Minden Pictures 41-1a,b
Terry Domico/Earth Images 41-2 G. I. Bernard/Oxford Scientific Films/Animals Animals/Earth Scenes
E41-1 Barbara Webb/Dept. of Artificial Intelligence, The University of Edinburgh E41-1 (inset) Gilbert Grant/Photo Researchers, Inc. 41-5a,b E. & D. Hosking/Frank Lane Picture Agency Ltd. 41-7 Frans Lanting/Minden Pictures 41-9 Thomas McAvoy/Life Magazine, Time Warner, Inc. 41-10 William Vandivert/James Carr 41-11 Thomas McAvoy/Life Magazine, Time Warner, Inc. 41-12a Manfred Kage/Peter Arnold, Inc.
41-13 Nina Leen/Life Magazine, Time Warner, Inc.
41-14a–e Lee Boltin Picture Library 41-17a Frans Lanting/Minden Pictures 41-17b (p. xxx) Daniel J. Cox/Gamma-Liaison, Inc. 41-17c Doug Cheeseman/Peter Arnold, Inc.

Chapter 42

Opener Norbert Wu/Peter Arnold, Inc. 42-1a,b New York Public Library Picture Collection 42-2 Ken Cole/Animals Animals/Earth Scenes 42-3 Richard K. LaVal/Animals Animals/Earth Scenes 42-4a Tom McHugh/Photo Researchers, Inc. 42-4b Jeff Lepore/Photo Researchers, Inc. 42-5 Nuridsany et Perennou/Photo Researchers, Inc. 42-6a Stephen J. Krasemann/Photo Researchers, Inc. 42-6b Jim Bandenburg/Minden Pictures, Inc. 42-6c Hans Pfletschinger/Peter Arnold, Inc. 42-7a M. P. Kahl/DRK Photo 42-7b
Anne et Jacques Six 42-8 Jeff Foott Productions 42-9
William Ervin/Natural Imagery 42-10 George W. Barlow 42-11 John D. Cunningham/Visuals Unlimited
42-13 Harold Hoffman/Photo Researchers, Inc. 42-14a
PH Archives 42-14b Frans Lanting/Minden Pictures, Inc.
42-16 J. L. Lepoore/Photo Researchers, Inc. 42-17
Bruce Coleman, Inc. 42-18 Fred Bruemmer/Peter Arnold, Inc. 42-19a Scott Camazine/Photo Researchers, Inc. 42-19b Hans Pfletschinger/Peter Arnold, Inc.
42-22 Raymond A. Mendez/Animals Animals/Earth Scenes 42-23 Lennart Nilsson, *A Child is Born*, Dell Publishing Company. 42-24 Dr. William Fifer 42-25
Todd Merritt Haiman/The Stock Market 42-26 Dr. I. Eibl-Eibesfeldt/Dr. Eibl-Eibesfeldt 42-27 Richard R. Hansen/Photo Researchers, Inc.

Opener Suzanne Duranceau/Illustrice, Inc.

Chapter 43

Opener Kin Heacox/Peter Arnold, Inc. 43-1 (background) W. Perry Conway 43-1a Jeff Foott/Tom Stack & Associates 43-1b Gary Milburn/Tom Stack & Associates 43-1c Rod Planck/Tom Stack & Associates 43-1d Gary R. Zahm/DRK Photo 43-2a G. Musil/Visuals Unlimited 43-2b Thomas Kitchin/Tom Stack & Associates 43-7 B & C Alexander 43-8
Tom McHugh/Photo Researchers, Inc. 43-9a,b Department of Lands 43-10 Gianni Tortoli/Photo Researchers, Inc. 43-11a Robert & Linda Mitchell Photography 43-11b Tom Bean/DRK Photo 43-11c
Gary Braasch 43-13 NASA Headquarters E43-4a
Minden Pictures E43-4b Stock Montage, Inc.

Chapter 44

Opener James L. Castner 44-3a Carolina Biological Supply Co./Phototake NYC 44-3b Merlin D. Tuttle/Bat Conservation International/Photo Researchers, Inc.
44-4a Marty Snyderman/Planet Earth Pictures 44-4b
Robert & Linda Mitchell Photography 44-5a Michael Fogden/DRK Photo 44-5b Paul A. Zahl/Photo Researchers, Inc. 44-5c Ray Coleman/Photo Researchers, Inc. 44-6 Robert & Linda Mitchell Photography 44-7
Andrew J. Martinez/Photo Researchers, Inc. 44-8

James L. Castner 44-9a Zig Leszczynski/Animals Animals/Earth Science 44-9b Breck P. Kent 44-9c,d
Dr. E. R. Degginger 44-10a,b Zig Leszczynski/Animals Animals/Earth Science 44-11a,b James L. Castner
44-11c Photo Researchers, Inc. 44-12 (upper) Robert P. Carr/Bruce Coleman, Inc. 44-12 (lower) Dr. Monica Mather/Simon Fraser University 44-13 Dan Aneshausley/Thomas Eisner 44-14 Thomas Eisner 44-15a Frans Lanting/Minden Pictures 44-15b Jeff Foott Productions 44-16 Charles Seaborn/Odyssey Productions 44-17a Gary Braasch 44-17b (left) Wendy Shattil/Bob Rozinski/Tom Stack & Associates 44-17b (right) Jeff Foott Productions 44-17c (left) John I. Kjargaard 44-17c (right) Krafft-Explorer/Photo Researchers, Inc. E44-1 Photo Researchers, Inc. E44-1 (inset) Carol Hughes/Bruce Coleman, Inc. E44-2a Ron Peplowski/The Detroit Edison Company E44-2b
Chuck Pratt/Bruce Coleman, Inc. E44-2c Robert & Linda Mitchell Photography 44-20a,b,c Robert & Linda Mitchell Photography

Chapter 45

Opener Dwight Kuhn Photography 45-8 Thomas A. Schneider 45-13 Bob Hallinen/Gamma-Liaison, Inc.
45-14 V. Leloup/Figaro/Gamma-Liaison, Inc. 45-15
Doug Plummer/Photo Researchers, Inc. 45-17 Will McIntyre/Photo Researchers, Inc.

Chapter 46

Opener Frans Lanting/Minden Pictures 46-1a PH Archives 46-1b Audesirk 46-3b NASA Headquarters E46-1 NASA/Science Photo Library/Photo Researchers, Inc. 46-9 (background) Loren McIntyre/Loren McIntyre 46-9a Peter Arnold, Inc. 46-9b
Bruce Lyon/Valan Photos 46-9c Michael Fogden/DRK Photo 46-9d Frans Lanting/Minden Pictures, Inc. 46-10 Peter May/Peter Arnold, Inc. 46-11 (background) W. Perry Conway 46-11a Aubry Lang/Valan Photos 46-11b Wendy Bass 46-11c Stephen J. Krasemann/DRK Photo 46-12 James Hancock/Photo Researchers, Inc. 46-13a Don & Pat Valantiss/DRK Photo 46-13b Brian Parker/Tom Stack & Associates
46-13c PH Archives 46-13d Tom McHugh/Photo Researchers, Inc. 46-14 Jeff Foott Productions 46-15
Gamma-Liaison, Inc. 46-16 Brian Parker/Tom Stack & Associates 46-17 Harvey Payne 46-18 (background) Tom Bean/DRK Photo 46-18a W. Perry Conway/Tom Stack & Associates 46-18b Nicholas Davore III/Photographers/Aspen 46-18c Bob Gurr/Valan Photos 46-18d Jim Brandenburg/Minden Pictures 46-19 PH Archives 46-20 (background) Gary Braasch 46-20a Thomas Kitchin/Tom Stack & Associates 46-20b Joe McDonald/Animals Animals/Earth Scenes 46-20c Stephen J. Krasemann/DRK Photo
46-21 (background) Robert A. Jureit/The Stock Market
46-21a Jeff Foott Productions 46-21b PH Archives
46-21c Richard Thom/Tom Stack & Associates 46-22 (background) PH Archives 46-22a Marty Stouffer Productions/Animals Animals/Earth Scenes 46-22b
Photo Researchers, Inc. 46-23 Gary Braasch 46-24 (background) S. J. Krasemann/Valan Photos 46-24a
Michael Giannechini/Photo Researchers, Inc. 46-24b
James Simon/Photo Researchers, Inc. 46-24c Jeff Foott/Tom Stack & Associates 46-27a T. P. Dickinson/Comstock 46-27b,c,d PH Archives 46-27b (inset)
Walter Dawn/National Audubon Society/Photo Researchers, Inc. 46-27c (inset) Thomas Kitchin/Tom Stack & Associates 46-27d (inset) Stan Wayman/Photo Researchers, Inc. 46-28a,c Charles Seaborn/Odyssey Productions 46-28b,d (b–p.xxxi) Alex Kerstitch 46-29 (background) Stephen J. Krasemann/DRK Photo 46-29a Jeffrey L. Rotman 46-29b
James D. Watt/Planet Earth Pictures 46-29c Fred McConnaughey/Photo Researchers, Inc. 46-29d Biophoto Associates/Science Source/Photo Researchers, Inc.
46-29e Robert Arnold/Planet Earth Pictures 46-30
Dr. Frederick Grassle/PH Archives E46-2 Stock Montage, Inc.

Index

Abiotic environment, 848
Abortion, 284, 781
  spontaneous, trisomy and, 293
Abscisic acid, 528, 560, 561, 567
Abscission layer, 567
Acacia, mutualism with ant, 881
Accessory pigments, photosynthesis and, 130, 443
Accutane, fetal effects of, 804
Acellular slime molds, 414
Acetylcholine, 716-717
Acetyl coenzyme A, 152
  in cellular respiration, 148
  synthesis of, 149
  in glycolysis, 155
Achyla, 431
Acid, 30
Acid deposition, 906, 907
Acid rain, 905-908, 923, 937
Acquired characteristics, inheritance of, 307
Acquired immunodeficiency syndrome (AIDS),
    401, 409, 667, 784
  human immunodeficiency virus and, 267, 317,
    404, 405, 669, 923
  treatments for, 668-670
  victims of, 668
Acrasiomycota, 391, 412, 414, 415
Acrosome, 770
ACTH. See Adrenocorticotropic hormone
Actin, 96-98, 579, 747
Action. See Movement; Muscles
Action potentials, 669, 701-704, 728, 752
  conduction of, 702-703
  muscle, 751-752
  saltatory conduction and, 704
  sodium-potassium pump and, 702
Activation energy, 60, 64
  catalysts and, 64-65
Active sites, active transport and, 65-66
Active transport, 110, 115
  plant nutrition and, minerals in, 520
Active visual signals, 828
Adaptation, 12, 311
  preadaptation and, 374
  speciation, 355
Adaptive radiation, speciation and, 355
Adenine
  in DNA, 52, 178, 184, 196, 199
  in RNA, 196, 199
Adenosine diphosphate (ADP), 68-70
Adenosine monophosphate (AMP), cyclic, 39, 52,
    53, 680-682
  as second messenger, 680-681
Adenosine triphosphate (ATP), 39, 52, 53, 61, 91-
    93, 100, 115, 129, 132, 133, 366, 519, 520,
    568, 620, 750
  active transport and, 115
  cilia and flagella and, 100
  as energy carrier molecule, 68-70
  hydrolysis of, 61, 6269
  synthesis of, 68-69
    cellular respiration and, 146-153
    from fats, 152
    glycolysis and, 141-143
    by photosynthesis, 129, 131, 132
  transport of, 150
ADH. See Antidiuretic hormone
Adipose tissue, 578
ADP. See Adenosine diphosphate
Adrenal glands, 691-692
  cortex of, 684, 691-692
  medulla of, 684, 691

Adrenalin, 591, 680, 684, 691
Adrenocorticotropic hormone (ACTH), 684, 685,
    686, 691, 693
Aerobic metabolism, 94, 95
African blood lily, mitosis in, 165
African clawed toad, differentiation in, 789, 790,
    803
African sleeping sickness, 416
Agar, 442
Agassiz, Louis, 305
Age-structure diagrams, human population growth
    and, 862-863
Agglutination, antibodies and, 664
Aggregated distribution of population, 859
Aggression, 828-832
  inhibiting with "gifts," 836
Aggressive mimicry, 876-878
Aging, 797, 799
Agnatha. See Fish, jawless
Agricultural revolution, human population growth
    and, 861
Agriculture
  climax community and, 883
Ahlquist, Jon, 319
AIDS. See Acquired immunodeficiency syndrome
    (AIDS)
Air currents, climate and, 917, 918
Alberts, Bruce, 75
Albumin, 46
  synthesis of, stimulation by estrogen, 210, 211, 682
Alcohol
  fetal effects of, 805
  ulcers and, 637
Alcoholic fermentation, 146
Aldosterone, 684, 692, 693
Alfalfa, 408
Algae, 440-444
  brown, 371, 391, 394-395, 440, 441, 442, 443
  coralline, 442
  green, 371, 428, 440, 441, 443, 444
  haploid life cycle in, 227, 228
  harbored by Paramecium, 369-370, 371
  life cycle of, 440
  red, 371, 391, 440-442
  taxonomic classification and, 394-395
Allantois, direct development of, 793, 794
Alleles, 239-240, 247, 323-324
  dominant, 239, 253
    genetic diseases and, 286
  frequency of, 324, 329
    in equilibrium population, 326, 328-329
    random changes and, 327, 329, 331-332
  genes and chromosomes and, 237
  inclusive fitness of, 342
  incomplete dominance and, 249, 253-254
  multiple, codominance and, 249-250, 254
  origin of, 249, 326
  recessive, 239, 253
    genetic diseases and, 283-284
Allergies, 666-667
Alligator, 487
Allopatric speciation, 348-349
Allosaurus, 306, 307, 376
Allosteric inhibition, 67-68
Alloway, A. L., 179
Altman, Sidney, 65, 366
Altruism
  natural selection and, 338, 342
  selfish gene concept and, 843, 844
Alveoli, 611, 612
Alzheimer's disease, 716, 721

Amanita, 431, 432
Amazon rainforest, 530
American chestnut tree blight, 275, 276, 339, 427,
    429
Amino acids, 44, 46-47, 366, 389
  as energy source, 152
  essential, 621
  gene identification and, 265
  modified, as hormones, 680-682
  in nutrition, 621
  specification of, 196-197
Amino group, 38, 47
Amish, founder effect among, 330
Ammonia, 365, 640, 903
Ammonites, 371, 373
Amniocentesis, 284, 286, 294, 296
Amnion
  direct development of, 793, 794
  human, 799, 800
  reptilian, 487
Amniotic egg, reptilian, 487
Amoeba, 107, 412, 416-418
  cell cycle of, 159
  pseudopodia of, 116
Amoeboid cells, of sponges, 465
AMP. See Adenosine monophosphate
Amphibians, 339, 391, 483, 485-487
  decline of, 486
  development of, 793
  origin of, 374-375
  respiration in, 609
  thyroxine in, 693
Amplexus, 765, 767
Amygdala, 715
Amylase, digestion and, 628, 629, 631
Anaerobic bacteria, 407-408
  primitive, 369, 370
Anaerobic metabolism, 94, 369
Analogous structures, evolution and, 313
Anaphase, 165, 167, 168-169
Anaphase I, 224, 226
Anaphase II, 224, 227
Anatomy
  comparative, evolution and, 312-315
  taxonomic classification and, 389
Anderson, W. French, 270
Androgens, 691
Anemones, 7, 336
Angina pectoris, 598, 599
Angiosperms. See Plants, flowering
Angiotensin, 646, 692
Angler fish, camouflage of, 485, 874, 876
Animal development, 788-807
  differentiation and, 789-791
  direct, 791-793
  gastrulation, 793, 795-796
  indirect, 791-793
  organogenesis, 793, 796
  sexual maturation, 796-797
  stages of, 793-799
    aging, 797, 799
    cleavage, 793-794
Animals, 6-8, 388-394
  behavior of. See Behavior; Individual behavior;
    Social behavior
  body organization of, 575-581
    organs and, 576, 580-581
    organ systems and, 576, 581
    tissues and, 575-580
  cells of, 83, 85
    see also Eukaryotic cells

Animals *continued*
  chemical communication in, 679
  coevolution of plants and, 550
  diploid life cycle in, 227, 228
  dispersal of seeds by, 548-550
  evolutionary trends in body plans of, 459-464
    body cavities and, 464
    cellular organization and, 461
    cephalization and, 460-462
    digestive tracts and, 464
    overall complexity and, 459-460
    segmentation and, 460, 463-464
    symmetry and, 461, 462
  excretion and, 645-646
  features of, 459
  nutrition of, plant nutrition compared with, 519-520
  origin of
    on land, 373-376
    in sea, 371-372, 373
  phyla of, 461-464
  see also specific animals
Annelida. *See* Segmented worms
Annual rings, tree growth and, 510
Anole lizard, 487
Anolis lizard, visual communication in, 828, 829
ANP. *See* Atrial natriuretic peptide
Ant, mutualism with acacia, 881
Antagonistic muscles, 757
Antenna molecules, 130
Anterior end, cephalization and, 462
Anther, 538, 539
Antheridia, 445
Anthophyta. *See* Plants, flowering
Anthrax, 666
Antibiotics
  bacteria resistant to, 66, 220, 409
  immune responses and, 666
Antibodies, 47, 657
  allergies and, 666-667
  blood types and, 250, 593
  sources and functions of, 658-663, 664
Antidiuretic hormone (ADH), 575, 681, 684, 686, 693
  urine concentration and, 643, 645
Antigens, 657, 658, 666
  recognizing "self" and "non-self" and, 661-662
  specific binding of, 659
Antiviral drugs, 400
Antocidon, 201
Anvil, 731
Apatosaurus, 376
Aphids
  phloem and, 528-529
  reproduction in, 764-765
Apical dominance, 563
Apical meristem, 498, 506
Apicomplexa, 391, 412, 416
Aplysia, nervous system of, 709
Appendicular skeleton, 753, 754
Apple maggot fly, ecological isolation of, 350-351
Aquatic ecosystems. *See* Ecosystems, freshwater lakes; Ecosystems, marine
Aqueous humor, 734
Arachnids, 391
  see also Scorpions; Spiders; Ticks
Archaebacteria, 390, 391, 393, 402, 410-411
Archaeopteryx, 488
Archegonia, 445, 446
Arginine, 47
Aristotle, 304, 387-388, 727
Arteries, 588, 589
Arthrobotrys, 391
Arthropods, 391
  compound eyes of, 733, 734

exoskeleton of, 752, 753
gas exchange in, 607, 609
infections caused by, sexually transmitted, 784
origin of, 373-374
see also Arachnids; Crustaceans; Insects
Artificial selection, evolution and, 315-316
Ascomycota, 391, 428, 431
Ascus, 429
Asexual reproduction. *See* Reproduction, asexual
Asimov, Isaac, 59
Aspen trees, 454
Assassin bug, coevolution with camphor weed, 877, 879
Atherosclerosis, 599
Atmosphere, prebiotic evolution and, 365-366
Atmospheric cycles, 900
Atomic number, 23
Atomic reactivity, 24
Atomic structure, 22-23
Atoms, 3, 4, 22-24
  bonding and, 25-27
ATP. *See* Adenosine triphosphate
Atrial natriuretic peptide (ANP), 685, 693
Atrioventricular node, 589-590
Atrioventricular valves, 588, 589
Atrium, of heart, 588, 589
Auditory canal, 730-731
Auditory communication, 829
Auditory nerve, 731-732
Auditory system, 730-732
  animal orientation and, 741
  human ear structure, 730-731
Audubon's warbler, 304
Australopithecines
  evolution of, 379
  evolution of human behavior and, 382
Australopithecus afarensis, 378, 379, 382
Australopithecus africanus, 378, 379, 382
Australopithecus boisei, 307, 378, 379
Australopithecus ramidus, 379
Australopithecus robustus, 378
Autoimmune diseases, 667
Autonomic nervous system, 709-711
Autosomes, 246, 247
  abnormal numbers of, 293-294
Autotrophs, 8, 892
  see also Plants, photosynthesis and
Auxin, 559-564, 563, 567
Avery, Oswald, 180, 182
Aves. *See* Birds
Axial skeleton, 753, 754
Axon, 579, 580, 699, 700
Azidothymidine (AZT; zidovudine), 668

Baboons, 342
  communication by touch in, 831
  migration of, 326
"Baby boom," U.S. population growth and, 866
Bacilli, 402, 403
Backbone. *See* Vertebral column; Vertebrates
Bacteria, 11, 402-411
  anaerobic, 407-408
    primitive, 369, 370
  bioengineered, 267, 274-275
  blue-green. *See* Cyanobacteria
  cell cycle of, 159
  cell walls of, 81, 106, 107, 402-403
  chemosynthetic, 407
  chemotaxic, 404
  classification of, 402
  community interactions of, 407-408
  conjugation of, 406, 407
  cyanobacteria. *See* Cyanobacteria
  denitrifying, 902
  DNA recombination between species of, 262

drug-resistant, 220, 409
habitats of, 406, 407
human health and, 408-410
"ice-minus," 267, 275
locomotion of, 403, 404
magnetotactic, 404
nitrogen-fixing, 408
  mutualism and, 880
nutrition of, 407-408
oil-eating, 408
photosynthetic, evolution of, 368
phototaxic, 404
polymerase chain reaction and, 270
reproduction of, 406, 407
restriction enzymes produced by, 263
sexually transmitted infections caused by, 784
structure of, 402-404
transformation, 179-181
see also Monerans; Prokaryotic cells
Bacteriophages, 400, 401
  DNA and, 181-183
  life of, 181, 182
Balanced polymorphism, 335
Balanus, interspecific competition of, 873
Baltimore oriole, taxonomic classification of, 393
Bananas, ripening of, 567
Baobab tree, water storage and, 512, 513, 514
Barnacles, 478
  commensalism with other marine animals, 880
  interspecific competition of, 873
Barr, Murray, 214
Barr body, 214
Base, 30
Basidia, 432
Basidiospores, 432
Basidomycota, 391, 428, 431-432
Basilar membrane, 731, 732
Basophils, 591, 593
Bats, 556
  coevolution with moths, 874
  dispersal of seeds and, 550
  orientation by sound, 740-741, 812
  as pollinators, 541
  vampire, play and, 822
B cells, 657, 658, 660-665
  cell-mediated immunity and, 655, 664-665
Beach strawberry, runners of, 512
Beadle, George, 195
Beagle, Darwin's voyage on, 307-308, 310
Bean weevils, wasps as predators of, 854
Beardsley, T., 673
Beaver, urine concentration in, 646
Bees. *See* Honeybees
Beetles, 475
Behavior
  genetic basis of, 811-812
  human
    environmental effects on, 290
    evolution of, 376-383
  see also Individual behavior; Social behavior
Behavioral isolation, speciation and, 352
Belding ground squirrel, 843-844
Bennett, Alan, 568
Benthic zone, 937
Berliner, David, 842
Bernard, Claude, 573
Betta splendens
  courting rituals and, 766
  spawning in, 831, 836
Bicuspid valve, 589
Big Bang theory, 363
Bighorn sheep, dominance hierarchies in, 832
Bilateral symmetry, 461-462
Bile, digestion and, 629, 630-631, 634
Bile salts, 631, 632

Binary fission, 161, 162
  prokaryotic cell cycle and, 161
  reproduction by, 764
    bacterial, 406, 407
Binocular vision, 737
  primate evolution and, 377
Biochemistry
  comparative, evolution and, 314, 318-319
  taxonomic classification and, 389
Biodiversity, 14, 910
Biogeochemical cycles, 900
Biological clock, daylength measurement and, 566
Biological magnification, 899-900
Biological molecules. See Organic molecules
Biology, 8-9, 11
  knowledge as illumination, 14-15
  scientific endeavor and, 11
  scientific method and, 9, 11
  scientific principles and, 8-9
  scientific theories and, 11
Biomass, 893
Biomass pyramid, 898
Biomes, 866, 919
  aquatic. See Ecosystems, freshwater lakes;
    Ecosystems, marine
  distribution of, 922, 923
  terrestrial. See Ecosystems, terrestrial biomes
Biosphere, 3, 4
Biotechnology, 261
  agricultural, 274-275
  bioengineered bacteria and, 267, 274-275
  chestnut blight and, 275, 276
  diagnosis of genetic disorders and, 270
  ethics of, 273-276
  human applications of, 266-267
  protein factories and, 266-267
  vaccines and, 267
  see also Molecular genetics; Recombinant DNA
    technology
Biotic environment, 848
Biotic potential, population growth and, 851
  boom and bust cycles of, 853, 855
Bipedalism, 379, 382-383
Birch trees, 452
Birds, 391
  albumin synthesis in, stimulation by estrogen,
    210, 211, 682
  analogous structures and, 314
  beak size in, 358
  behavioral isolation of, 352
  commensalism with bison, 879-880
  digestion in, 627
  direct development in, 791
  habituation in, 821
  homologous structures and, 314
  origin of, 376
  reproduction in, social behavior and, 834-835
  respiration in, 609, 610
  territoriality in, 834
  see also specific birds
Birth. See Childbirth
Birth control pills, 780, 782-783
  once-a-month, 781
Bison, 930
  commensalism with birds, 879-880
Biston betularia, evolution of, 316
"Black Death," 409
Black-faced lion tamarin, 393, 394
Bladder, 641
Blade, of leaf, 511
Blaese, R. Michael, 270
Blastocyst, 776, 778
  human, 799
Blastopore, 795
Blastula, 793, 795

Blaustein, Andrew, 486
Blenny fish, mimicry by, 878
Blind spot, 736
Blood, 576, 591-595
  clotting of, 592-594, 598-599
  gas transport in, 611-613
  osmosis and, 115
  pH of, 31
  plasma, 591-592
  platelets, 593-594
  red blood cells, 591-593
    life cycle of, 592
    regulation of number of, 592-593
  return of fluids to, 599-600
  white blood cells, 591, 593, 594
    tumor infiltrating, in cancer treatment, 674
  see also Hemoglobin; Sickle-cell anemia
Blood clots, treatment of, 599
Blood flow
  distribution of, 596-597
  see also Circulatory system
Blood flukes, 470
Blood pressure, 590
Blood types, ABO, 250, 593
Blood vessels, 594-597
Bluebirds, 387, 388
Body cavities, evolutionary trends in, 462-463
Bombardier beetle, toxic spray emitted by, 877,
  879
Bonds, 25-28
  covalent, 25, 26, 28
  high-energy, 69-70
  hydrogen, 25
  ionic, 25
  peptide, 47, 48
Bone, 576, 753-757
  broken, repair of, 756
  remodeling of, 755-757
  see also Skeleton
Bonobo, play and, 822
Bony fish, 375, 391, 482, 484, 485
Book lungs, in arthropods, 607, 609
Boom and bust cycles, exponential population
  growth and, 853, 855
Boron, in plant nutrition, 521
Borrelia burgdorferi, 409
Boston ivy, tendrils of, 512
Botulism, 409
Bowerbirds, 337, 826, 827
Bowman's capsule, 641, 642, 644
Boysen-Jensen, Peter, 588-589
Bradykinin, 739, 740
Brain, 713-716
  damage to, 718, 722
  evolution of human behavior and, 382
  "mind-brain problem" and, 717-722
    learning and memory and, 719-720
    left brain-right brain, 718-719
  PET scan of, 32-33, 720-721
  primate evolution and, 377
  see also Central nervous system; Nervous system
Branch root, 506
Bray, Dennis, 75
Bread mold, 194-195
Breathing, mechanics of, 612-613
British blackberry, species of, 393
Bronchi, 610, 611
Bronchioles, 610, 611
Bronchitis, smoking and, 614, 615
Brown, Louise, 777
Bryophyta. See Liverworts; Mosses
Budding, 764
  of sponges, 466
Buffalo grass, 336
Buffers, 30

Buffon, Georges-Louis, 305
Bulbourethral glands, 768, 771
Bulk flow, 606
Bullhead catfish, social behavior in, 839
Bullock's oriole, taxonomic classification of, 393
Bundle-sheath cells, 135
Burdocks, seed dispersal and, 548
Burning. See Combustion
Bush, Guy, 351
Butterflies, 7, 316, 317, 678, 813
  coevolution with milkweed, 877
  mimicry by Monarch, 877
  swallowtail, larva of, startle coloration in, 878
B vitamins, 621
Byers, John, 823

C_3 cycle, 132, 134-136
Cactus, 496, 915, 916
  adaptations of, 514
  camouflage of, 875
  moth as predator of, 857, 858
  photosynthesis in, 513
Calanolide A, 923
Calcitonin, 684, 687, 688, 755
Calcium, 23, 24
  in nutrition, human requirements for, 622
  osteoporosis and, 758
  in plant nutrition, 521
Callus, of bone, 756
Calmodulin, 681
Calories, 31, 620
Calvaria tree, coevolution with dodo bird, 550
Calvin-Benson cycle, 132, 134-136
Cambium, 498
Cambrian period, 364, 371
Camera eye, 734, 736
Camouflage, 874-876
Camphor weed, coevolution with assassin bug, 877,
  879
Cancer, 670-671, 673-674
  causes of, 671, 673
  destruction of ozone layer and, 920
  prevention of, 673
  treatment of, 670, 673-674
Cannon, Walter, 573
Capillaries, 595-596, 600
Capsules, bacterial, 403
Carbohydrates, 37-42, 619
  disaccharides, 39-41, 620
  as energy source, 620-621
  metabolism of, 152-153
  monosaccharides, 39-41, 620
  polysaccharides, 39-42, 620
Carbon, 23, 24
  bonding, 26
  in plant nutrition, 521
Carbon cycle, 900-902
  greenhouse effect, 908-909
Carbon-14 dating, 368
Carbon dioxide, 27, 365, 923
  plant transpiration in, 525
  potassium concentration in guard cells and, 526-
    528
  sources of, greenhouse effect and, 908-909
  transport in blood, 611-613
  see also Gas exchange
Carbon fixation, in photosynthesis, 134
  in C_4 plants, 135
Carbonic anhydrase, 613
Carboniferous period, 364, 372, 374, 375
Carboxylic acid group, 38, 47
Carboxypeptidase, digestion and, 631
Cardiac cycle, 588
Cardiac muscle, 579, 590, 747-748, 752
Cardiovascular disorders, 598-599

Caribou, 122, 357, 934
Carnivores, 629, 893
Carnivorous segmented worms, 472
Carotenoids, 130
Carpels, of complete flower, 538, 539
Carrier proteins, 109, 112
Carrion flower, pollination of, 541
Carrying capacity
    population growth and, 855-856
    world population and, 864-865
Cartilage, 753-755
Cartilaginous fish, 391, 482-484
Casein, 46
Casparian strip, 505, 506, 525
    plant nutrition and, 521, 522, 523
Catalysts, 64-65
    see also Enzymes
Catastrophism, 305, 306
Caterpillars. See Larvae
Cavalier-Smith, Thomas, 395
Cave paintings, 382
Cech, Thomas, 65, 366
Cell body, of neuron, 579, 580, 698, 699
Cell cycles, 159, 160
    eukaryotic, 163-166
        cell division and, 171-172
        interphase and, 163, 164
    prokaryotic, 161
Cell division, 159, 160
    asexual reproduction and, 170-172
Cell-mediated immunity, 655, 663-665
Cell membrane, 76, 85, 90-91, 104-123
    differentially permeable, 700
    endoplasmic reticulum of, 87-91
    fluid mosaic model of, 106-107
    Golgi complex of, 89-91, 97
    lysosomes of, 89-91, 97
    phospholipid bilayer of, 76, 87
    plasma membrane of, 87, 90, 91
    structure of, 87, 88
    transport across, 109-120
        energy-requiring, 15-116, 110
        intracellular membranes and, 116-120
        passive, 110-115
Cell plate, 170
Cells, 3, 4, 74-102
    connections and communication of, 120-121
    desmosomes and, 120
    gap junctions and plasmodesmata and, 120
    tight junctions and, 120
    cytoplasm of, 80, 81, 85, 87
    diploid. See Diploid cells
    first living, 368-370
    haploid. See Haploid
    metabolism of. See Metabolism
    plasma membrane of. See Cell membrane
    polyploid, 351
    sex. See Gametes
    size of, function limiting, 80, 81
    of sponges, 461
    theory of, development of, 75-76
    types of, 8
    see also Eukaryotic cells; Prokaryotic cells
Cellular reproduction, 158-173
    cell cycles and, 159
        eukaryotic, 163-166
        prokaryotic, 161
    cell division and asexual reproduction and, 171-172
    chromosomes and, eukaryotic, 161-163
    cytokinesis and, 169-170
    cytoplasmic materials and, 160
    hereditary information and, 160
    mitosis and, 166-169
Cellular respiration, 128, 129, 146-153

Cellular slime molds, 414, 415-416
Cellulase, 627
Cellulose, 39, 41-43
    digestion of, 626-627
Cell wall, 85
    bacterial, 81, 106, 107, 402-403
Cenozoic era, 364
Central nervous system, 708
    spinal cord, 710, 712-713
    See also Brain
Central vacuoles, 118
Centrioles, 85, 370
    during prophase, 167
Cephalization, evolutionary trends in, 460-462
Cephalopoda, 391, 479
Cerebellum, 713, 714
Cerebral cortex, 714-716
Cerebral hemisphere, 715, 718-719
Cerebrospinal fluid, 710
Cerebrum, 714
Cervix, 771, 773
CFCs. See Chlorofluorocarbons
CG. See Chorionic gonadotropin
Channel proteins, 109
Chaparral, 926-929
Chara, 440
Chargaff, Erwin, 182-183, 185
Chase, Martha, 181-183, 198
Cheetah, 489
    loss of genetic variability in, 329, 330
Chemical bonds. See Bonds
Chemical equilibrium, 61, 63
Chemical reactions, 59-63
    coupled, 6, 61
    endergonic, 59-62
    exergonic, 59-60, 62
    reversibility of, 61-63
    thermodynamics in, 58-59
Chemical warfare, between predators and prey, 876-877, 879
Chemiosmosis, 132, 133, 150-151
Chemoreceptors, 728, 737-739
Chemosynthetic bacteria, 407
Chemotaxic bacteria, 404
Chemotherapy, in cancer, 674
Chiasmata
    gene linkage and, 243
    during prophase I, 223, 225, 226
Chicxulub crater, 340
Childbirth, 779-780
Chitin, 42, 43
    in arthropod exoskeleton, 473
    in cell wall, 106
Chlamydia infection, 784
Chlamydomonas, haploid life cycle in, 221, 228
Chlorine, 23, 25
    in nutrition, human requirements for, 622
    in plant nutrition, 521
Chlorofluorocarbons (CFCs), destruction of ozone layer and, 921
Chlorophyll, photosynthesis and, 12, 94, 130
Chlorophyta. See Algae, green
Chloroplasts, 85, 91-94, 96
    origin of, 369-371
    photosynthesis and, 94, 128-130
Chloroquine, 419
Cholecystokinin, 685, 693
    digestion and, 634
Cholesterol, 39, 44, 46, 599
    in cell membranes, 108
Chondrichthyes. See Fish, cartilaginous
Chondrocytes, 754
Chordata
    invertebrate, 481-482

    see also Amphibians; Birds; Fish; Mammals; Reptiles; Tunicates; Vertebrates
Chorion, 296
    direct development of, 793, 794
Chorionic gonadotropin (CG), 775
Chorionic villous sampling, 284, 286, 294, 296
Choroid, 734-735
Chromatids, 162
Chromatin, 82, 161
Chromista, 395
Chromosomal inheritance
    abnormal numbers of autosomes and, 293-294
    abnormal numbers of sex chromosomes and, 291-292
Chromosomes, 82, 85, 86, 87
    capture during prophase, 168
    condensation during prophase, 167
    eukaryotic, 161-163
        daughter, 161
        during mitosis, 166-169
    genes, as parts of chromosomes, 242-246
    homologous, 235-237, 239
    locus of, 236-237
    prokaryotic cell cycle and, 161
    sex. See Sex chromosomes
    sex linkage and, 247-249
    see also Deoxyribonucleic acid; Genes; Genetics; Human genetics; Inheritance; Meiosis
Chrysophyta, 391, 412
Chthamalus, interspecific competition of, 873
Chylomicrons, digestion and, 632
Chyme, 630-634
Chymotrypsin, digestion and, 629, 631
Cilia, 85, 99-101, 370, 614
    protist, 419-420
Ciliates, 419-420
Ciliophora, 391, 412, 416
Circulatory system, 581, 584-601
    closed, of segmented worms, 472
    types of, 585-586
    vertebrate, 586-597
        blood and, 591-594
        blood vessels and, 594-597
        functions of, 586-587
        heart and, 587-591
    see also Blood; Cardiac muscle; Heart
Citric acid cycle, 146, 148
Clams, 477, 479
Class, 387
Classical conditioning, 818-819
Clathrin, 117
Claviceps purpurea, 429
Cleaner wrasse, mutualism and, 880
Cleaning symbiosis, 878
Clearcutting, in taiga, 934
Cleavage, 793-794
Climate, ecosystems and. See Ecosystems, climate and
Climax community, 881, 883, 886
Clitoris, 771, 775
Clonal selection, 662
Cloning, 265-266
Closed circulatory system, 472, 585, 586
Clown fish, 7, 336
    mutualism and, 880
Club mosses, 391, 445
Cnidarians, 391, 466-468
    cellular organization of, 461
    digestion in, 626
    digestive tract of, 464
    nervous system of, 707, 708
    symmetry of, 461
    see also Corals; Hydra; Sea anemones
Cnidocytes, 467
Coal, burning of, 60
Coastal waters, marine, 937

Cocci, 402, 403
Cochlea, 731, 732
Cockleburs
    control of flowering in, 566
    seed dispersal and, 549
Cockroaches, 316, 317
Coconuts, dispersal of, 548
Codominance, multiple alleles and, 249-250, 254
Codons, 197
Coelacanth, 374, 484, 485
Coelom, 462, 463, 464
Coenzymes, 52, 53, 66-67
Coevolution, 311, 336
    community interactions and, 871, 873
        predators and prey and, 874-878
    of flowers and pollinators, 538-543
        food and, 539-541
        nursery flowers and, 542
        sex and, 542
Cohesion-tension theory, 524-526
Cold
    seed dormancy and, 550
    see also Temperature
Coleoptiles, 551-552
    bending of, 557-559
Collagen, 753, 754
    in connective tissue, 576
Collar cells, of sponges, 465
Collecting duct, 642
Collenchyma tissue, of plants, 501, 502, 508
Colon, 633
Colon cancer, 673, 674
Color vision
    inheritance of colorblindness and, 287
    primate evolution and, 377
Colostrum, 779
Commensalism, 879-880
Common perception, principle of, 9
Communication. See Social behavior,
            communication
Community, 3, 4, 849, 871
Community interactions, 870-887
    commensalism and, 879-880
    competition and, 858, 872
        interspecific, 872-873
    ecological niche and, 872
    mutualism and, 880, 881
    parasitism and, 878-879
    predation and, 87-878
        coevolution of predators and prey and, 874-878
    succession and, 872, 880, 882-886
        climax community and, 881, 883, 886
        primary and secondary, 883, 885
    symbiosis and, 878-880
Compact bone, 754, 755
Companion cell, of phloem, 503
Comparative anatomy, evolution and, 312-315
Comparative biochemistry, evolution and, 314,
            318-319
Comparative cultural studies, 842
Comparative embryology, evolution and, 314, 315
Competition
    extinction and, 339
    natural selection and, 336
    population growth and, 858
    for resources
        community interactions and. See Community
            interactions, competition and
        social behavior and. See Social behavior,
            competition for resources
Competitive exclusion principle, 872
Competitive inhibition, 67-68
Complement, 659
Complement reactions, antibodies and, 664
Complement system, 664

Complete flowers, 538, 539
Compound, 22
Compound eyes
    of arthropods, 474, 733, 734
    of crustaceans, 476
Concentration, of molecules in fluid, 109
Concentration gradients, 110, 111
Conclusion, 9-11
Conditioning, classical and operant, 818-819
Condom, 780-783
Conducting portion, of human respiratory system,
            610-611
Cones, 734, 736-737
Coniferophyta, 449-451. see also Conifers
Conifers, 391, 441
    cones of, 537, 538
    origin of, 373
Conjugation, bacterial, 406, 407
Conklin, E. G., 827
Connective tissue, 576, 578
Connell, J., 873
Constant region, 658
Consumers, 893
Contest competition, population growth and, 858
Continents, climate and, 917, 918
Contraception
    permanent, 780
    temporary, 780-783, 785
Contraceptive sponge, 782-783
Contractile vacuoles, 118, 119
Convergence, 706, 707
Convolutions, cerebral, 715
Copper
    in human nutrition, requirements for, 621, 622
    in plant nutrition, 521
Copulation, 775-776
Coral reefs, 937-939
Corals
    reproduction in, 467-468, 764
    spawning in, 765, 766
Coral snake, warning coloration of, 877
Cork, of stems, 510, 511
Cork cambium, of stems, 510, 511
Cork cells, of plants, 501
Cornea, 734
Corn smut, 427, 432
Corona radiata, 776
Corpus callosum, 715, 718, 719
Corpus luteum, 773, 774
Correns, Carl, 242
Corroboree toad, 486
Cortex
    of roots, 504-505
    of stems, 508, 509
Cotyledons, 546, 547, 552
Countercurrent flow, 607, 608
Coupled reactions, 61, 62
    energy carrier molecules and, 68-70
Cousteau, Jacques, 363
Covalent bonds, 25, 26, 28
Crab lice, 784
Crabs, 476, 477
    molting by, 473
Creationism, 9, 303
Crenilabrus, taxes and, 813
Cretaceous period, 364
Cretinism, 687
Creutzfeld-Jakob disease, 401
Crick, Francis, 11, 184, 185, 189, 197, 198
Crickets, reproduction in, social behavior and, 834
Cristae, 94
Cro-Magnon Man, 381-382
Crop, of earthworm, 626
Cross-bridges, 750
Cross-fertilization, 237, 240, 243

Crossing over, 223, 230
    gene linkage and, 243, 245-246
Crustaceans, 391, 476, 477
    see also Crabs; Shrimp
Cuckoo, fixed action pattern in, 815
Cud, 627
Cultural evolution, 382
Cultural revolution, human population growth
            and, 861
Cupula, 730, 732, 733
Cuvier, Georges, 305
Cyanobacteria, 410
    chloroplast evolution from, 368
    photosynthetic, 368
Cycads, 49, 451
Cyclic adenosine monophosphate (cyclic AMP),
            39, 52, 53, 680-682
    as second messenger, 680-681
Cyclic nucleotides, 52, 53
Cystein, 47, 49
Cystic fibrosis, 271, 273
Cysts, of flatworms, 470
Cytokinesis, 166, 167, 169-170
Cytokinins, 560-561, 563, 564, 567
Cytoplasm, 4, 80, 81, 85, 87
Cytosine
    in DNA, 52, 178, 184, 196, 199
    in RNA, 196, 199
Cytoskeleton, 82, 85
    see also Eukaryotic cells, cytoskeleton of
Cytosol, 82
Cytoxic T cells, 655, 664, 665, 673

Daffodils, leaves and bulbs of, 513, 514
Dandelions, seed dispersal and, 548
Darwin, Charles, 11, 12, 303, 315, 316, 318, 336,
            337, 347, 358, 388, 557-558, 849, 873
    voyage on Beagle, 307-308, 310
Darwin, Francis, 557-558
Dawkins, Richard, 843
Daylength, control of flowering and, 565-566
Day-neutral plant, 565
ddI. See Dideoxyinosine
DDT, biological magnification and, 899-900
Decomposers, 894
Deer, 336
Defecation, 633
Defenses
    lymphatic, 600
    against microbial invasion, 651-657
    see also Immune response
Dehydration synthesis, 38, 39, 41
Delbruck, Max, 189
Deletions, 205, 206
Delivery. See Childbirth
Dendrites, 579, 580, 698, 699
Denitrifying bacteria, 902
Density-dependent limits to population growth, 856
Density-independent limits to population growth,
            856
Dental plaque, 403
Deoxyribonucleic acid (DNA), 5-6, 11, 12, 39, 41,
            52, 76, 80-82, 86, 87, 176-189
    bacterial transformation and, 179-180
    bacteriophage experiments and, 181-183
    cellular reproduction and, 160
    in centrioles, 370
    chemical nature of, 178-179
    differences from RNA, 195-196
    DNA hybridization and, 318-319
        taxonomic classification and, 389
    DNA sequencing and, 318-319
    fingerprinting and, 272, 274
    in interphase, 164
    mutations of. See Mutations

Deoxyribonucleic acid *continued*
  palindromic, severing, 263
  proteins and, 195-196
  protein synthesis and. *See* Transcription;
    Translation
  recombinant DNA technology and. *See*
    Biotechnology; Molecular genetics;
    Recombinant DNA technology
  replication of, 185-188
    faithfulness of inheritance and, 188
    molecular mechanisms of, 186, 187
    proofreading and, 188
    semiconservatism of, 186
  structure of, 183-185
    double helix and, 184-185, 189
  transcription and, 196, 199-201
  translation and, 196, 201-204
Deoxyribose, 40, 41
Depo-Provera, 780, 782-783
Dermis, 576, 580, 581
Desertification, 926
  world population and, 865
Deserts, 926-928
Desmosomes, 120
Detritus feeders, 894
Deuteromycota, 391, 431, 433-434
Developmental stages, taxonomic classification
    and, 389
Devil's Hole pupfish, localized distribution of, 338
Devonian period, 364
de Vries, Hugo, 242
Diabetus mellitus, 667, 691
Dialysis, 647
Diaphragm, contraceptive, 780-783
Diastole, 588, 589
Diatoms, 106, 412, 413
Dicots, 454, 498, 499
  growth of, 500
  stems of. *See* Plants, stems of
Dideoxyinosine (ddI), 668
Didinium, cilia of, 419
Differential permeability, 700
Differentiation
  in animal development, 789-791
  differentiated cells and, 498
Diffusion, 80, 699
  transport across cell membranes and, 110-114
Digestion, 619, 623-635
  of cellulose, 626-627
  extracellular
    in simple sac, 626
    in tube, 626-627
  human. *See* Human digestion
  intracellular, 625
  in segmented worms, 464
  ulcers, 635
Digestive tract, 581
  evolutionary trends in, 464
  hormones and, 634, 685
Dihybrid cross, 241
Dilger, W., 817, 834
Dinoflagellates, 412-413
Dinosaurs, 339
  evolution of, 375, 376
  extinction of, 340, 375
Dioecious species, reproduction in, 765
1,3-Diphosphoglyceric acid, in glycolysis, 144
Diploid cells, 163, 220-222
  life cycles and, 227-229
  see also Zygote
Direct development, 791-793
Directional selection, 333, 334
Disaccharides, 39-41, 620
Disruptive selection, 333, 334, 335
Distal tubule, 642

Disulfide bridges, 47, 48-49
Divergence, 706, 707
Divergent speciation, 353-354, 356
Division, 387
DNA. *See* Deoxyribonucleic acid
"DNA fingerprinting," 272, 274
DNA hybridization, 318-319
  taxonomic classification and, 389, 393
DNA library, 263
  building, 263, 265
  restriction enzymes and, 263-265
  searching, 265-266
DNA polymerase, 186, 188
  polymerase chain reaction and, 268-270
DNA sequencing, 318
Dobzhansky, Theodosius, 12
Dodo bird, coevolution with Calvaria tree, 550
Dogs, 738
  artificial selection and, 315-316
  classical conditioning of, 818
Dolphins, play and, 822
Dominance, incomplete, 249, 253-254
Dominance hierarchies, 831-832
  population growth and, 858
Dominant alleles, 253
Dopamine, 717
Dormancy, of seeds, 547, 549, 567
Dorsal root ganglia, 710, 712
Dorsal surface, 462
Double covalent bond, 27
Double fertilization, of flowering plants, 545
Double helix, of DNA, 184-185, 189
Douching, 781
Down syndrome, 293-294
Drosophila melanogaster
  evolution of, 357
  mutations in, 803
  sex chromosomes of, 247
  sex linkage in, 247-249
  taxonomy and, 388
Dryopithecines, 377
Ducts, 683
Dung fungus, 435
Dutch elm disease, 339, 427, 429
Dynamic equilibrium, 573

Eagles, 851-853
Ear. *See* Auditory system
Earthworms, 471, 472
  digestion in, 626, 627
  nephridia in, 640-641
Ebola virus, 651
Ecdysone, 46
Echinoderms, 391, 463, 479-481
  symmetry in, 462
  see also Sand dollars; Sea cucumbers; Sea stars;
    Sea urchins
Echolocation, 740-741
Ecological community, interactions in. *See*
    Community interactions
Ecological isolation, speciation and, 349, 350-351
  premating isolating mechanisms, 352
Ecology, 849-850
Ecosystems, 3, 4, 849, 914-945
  aquatic, acid rain and, 906
  carrying capacity of, population growth and,
    855-856
  climate and, 915-918
    factors influencing, 916-918
    solar radiation and, 916
  energy flow and
    energy pyramids and, 898-899
    food chains and food webs and, 893-895
    human intervention in, 903-910
    primary productivity and, 891-894

  trophic levels and, 893-895, 898
  freshwater lakes, 934, 935-937
    life zones on, 935-936
    succession in, 885-886
    temperature stratification and, 936
    types of, 936
  humans and, 942-943
  marine, 934, 937-941
    coastal waters, 937
    coral reefs, 937-939
    open ocean, 939-941
    vent communities, 941
  nutrient cycling and
    carbon cycle and, 900-902, 908-909
    human intervention in, 903-910
    nitrogen cycle and, 902
    phosphorus cycle and, 902, 904
    water cycle and, 902-903, 905
  requirements of life and, 918
  terrestrial biomes, 918-934
    chaparral, 926-929
    deserts, 926-928
    grasslands, 928-929, 930
    rainfall, temperature, and vegetation and,
      933-934
    savanna, 923-926
    taiga, 931-933
    temperate deciduous forests, 929, 931
    temperate rain forests, 931, 932
    tropical deciduous forests, 923
    tropical forests, 921-925
    tundra, 932-934
Ectoderm, 461
Eels
  moray, 485
  orientation by magnetic fields, 742
Ehrlich, Anne, 910
Ehrlich, Paul, 575, 910
Eibel-Eibesfeldt, I., 842
Einstein, Albert, 842
Eiseley, Loren, 21, 535
Ejaculation, 775
Elastin, 44
Eldredge, Niles, 356
Electrical attraction, 699
Electron carriers, 70-71
Electron microscopes, 78, 79
Electron orbitals, 23-24
Electrons, 22
Electron shells, 22-25
Electron transport systems, 130, 131, 148, 150, 151
Electrophoresis, taxonomic classification and, 389
Electroreception, 741-742
Elements, 2, 4, 22, 23
  radioactive dating and, 368
Elephants, 878
  ears of, 212
Elephant seal, 218, 219
  loss of genetic variability in, 329, 330
Elk, wolf as predator of, 856, 857
Ellis-van Creveld syndrome, 330
Elm tree, Dutch elm disease and, 339, 427, 429
Embryo, 791, 798-801
Embryology, comparative evolution and, 314, 315
Embryonic disk, human, 799, 800
Embryo sac, in flowering plants, development of,
    543-547
Emigration, population growth and, 850-851, 859
Emphysema, smoking and, 614, 615
Endangered Species Act, 341
Endergonic reactions, 59-62, 70
Endler, John, 337
Endocrine system, 578, 581, 678-694
  in animals, 679-683
    hormone actions on target cells and, 680

mammalian. *See* Mammals, endocrine system of
types of hormones and, 681
see also Hormones; *specific hormones*
Endocytosis, 110, 115-118
Endoderm, 461
Endodermis, of roots, 505
Endogenous pyrogens, 656
Endometrium, 763, 774
Endoplasmic reticulum, 82, 85, 87-91, 109
Endorphins, 717, 739
Endoskeleton, 753
of echinoderms, 481
Endosperm, of flowering plants, 545
Endospores, 405, 406
Endosymbiotic hypothesis, 92, 93
of origin of eukaryotic cells, 369-371
Energy
acquisition and use of, 4-5
activation, 64-65
kinetic, 22, 57, 59
potential, 22, 57, 58, 59
see also Cellular respiration; Glycolysis
Energy carrier molecules, 61, 68-70
Energy flow, 56-71
in chemical reactions, 59-63
coupled reactions and, 61, 62
endergonic reactions and, 59-62
exergonic reactions and, 59-60, 62
reversibility of reactions and, 61, 63
ecosystems and. *See* Ecosystems, energy flow and
energy carrier molecules and, 61, 68-70
entropy and, 59
laws of thermodynamics and, 58-59
metabolism and. *See* Metabolism
Energy levels, 24
Energy pyramids, 898-899
Energy-requiring transport, 110, 115-116
Enhancer region, 209, 210
Entropy, 59
Environmental resistance, population growth and.
See Population growth, limits to
Enzymes, 44, 65-68
regulation of, 67-68
structure and function of, 65-67
see also Coenzymes
Eocene epoch, 364
Eohippus, 13, 312
Eosinophils, 591, 593
Epicotyl, 546
Epidermal tissue, of plants, 500-501, 504
Epididymis, 769, 770
Epiglottis, 628, 630
Epilepsy, 718
Epinephrine, 591, 680, 684, 692
Epistasis, 251
Epithelial cells, 465
Epithelial tissue, 575-578
EPSP, 705
Epstein, R., 820
Equilibrium, chemical
Equilibrium population, 326, 328-329
Equisetum, 445
Equus, evolution of, 13, 356
Ergot, 429
Erythroblastosis fetalis, 593
Erythrocytes. *See* Blood, red blood cells;
Hemoglobin
Erythropoietin, 575, 592-593, 646, 685, 692
Escherichia coli
cell cycle of, 161
gene regulation in, 208-209
Esophagus
of earthworm, 626
human, digestion in, 628
Essay on Population (Malthus), 309

Essential amino acids, 621
Essential fatty acids, 620
Estrada, Alejandro, 550
Estradiol, 46, 681
Estrogen, 87, 685, 691, 758, 772
albumin synthesis and, 210, 211, 682
contraception and
evolution of, 693
menstrual cycle and, 774
Ethanol, 68
Ethics, of biotechnology, 273-276
genetic diseases and, 295-297
Ethology, 811
human, 840-845
comparative social studies and, 842
exaggerating human releasers and, 841-842
pheromones and, 842
studies of young children and, 840-841
twin studies and, 843
Ethylene, 561
fruit ripening, 567
Eubacteria, 391, 402
Euglena, 412, 414
asexual reproduction, 171
cilia and flagella and, 100
Euglenoids, 412, 414
Euglenophyta, 391, 412
Eukaryotic cells, 7, 8, 81-84
of animals, 83
chloroplasts, 85, 91-94, 96
chromatin of, 82
chromosomes of. *See* Chromosomes, eukaryotic
cytoskeleton of, 82, 95-100
cilia and flagella, 85, 99-101
intermediate filaments of, 96-98
microfilaments of, 96-98
microtubules of, 96-98, 100
differences from prokaryotic cells, 369
DNA recombination between species of, 262
gene regulation in, 209-211
gene structure and, 209, 210
transcription and, 210-211
glucose metabolism in. *See* Cellular respiration;
Glycolysis
meiosis and, 220-221, 222
mitochondria and, 79, 82, 86, 91-95
nuclear envelope of, 82, 86, 87
nucleolus of, 82, 85, 87
of plants, 84
plastids of, 85, 95, 96
rise of, 369-371
vacuoles of, 85, 95
see also Animals; Cell membrane; Chloroplasts;
Chromosomes; Fungi; Nucleus, of cell;
Plants; Protists
Euphorbs, 915, 916
Eustachian tube, 731
Eutrophic lakes, 936
Everglades kite, specialization of, 338, 339
Evergreens. *See* Conifers; Pine trees
Evolution, 11-14, 302-320
coevolution and. *See* Coevolution
convergent, 313
cultural, 382
evidence for, 312-317
artificial selection and, 315-316
comparative anatomy and, 312-315
comparative biochemistry and molecular
biology and, 314, 318-319
comparative embryology and, 314, 315
fossils and, 305, 306, 312
present-day evolution and, 316-317
of fungi, 435
history of thought on, 304-312
before Darwin, 304-307

Greek philosophers and, 304
species defined and, 304
of hormones, 693
of humans, 376-383
Australopithecines, 379
behavior and, 381-382
hominids and, 377-379
primates and, 376-377
mechanisms of, 323-342
equilibrium population and, 326, 328-329
equivalence of genotypes and, 331-332
migration and, 326-327
mutations and, 324-326
population genetics and, 323-326
random changes in allele frequencies and, 327
random mating and sexual slection and, 331
see also Natural selection
mutations and, 207
origin of life and. *See* Life on Earth
of plants, 439-440
prebiotic, 365-366
accumulation of organic molecules and, 367
first living cells and, 367
RNA and, 366-367
uncertainty regarding, 367-368
present-day, 316-317
speciation and, 355-358
gradualism and, 355-357
punctuated equilibrium and, 355-357
variation and inheritance and, 12
Excitatory synapse, 705
Excretion, 639
Excretory pore, 640
Excretory system. *See* Urinary system
Exergonic reactions, 59-60, 62, 70
Exhalation, 612-613
Exocrine glands, 575, 578, 683
Exocytosis, 91, 110, 116, 118
Exons, 209, 210, 213
Exoskeleton, 374, 752, 753
of arthropods, 473-474
Exotic species, population growth and, 853, 884-885
Experiments, 9, 10
Exponential population growth, 851-853, 855
boom and bust cycles and, 853, 855
Extensors, 757, 759
External ear, 730, 731
External fertilization, 765-766
Extinction, 14, 254, 491
competition and, 339
environment and, 338-340
habitat change and destruction and, 339-340
susceptibility to, 338
Extracellular digestion
in simple sac, 626
in tube, 626-627
Extraembryonic membranes, direct development
of, 793
Eye. *See* Vision
Eye color, inheritance of, 235, 290, 291
Eyespot, of flatworms, 733

Facilitated diffusion, 110, 112
FAD. *See* Flavin adenine dinuleotide
FADH$_2$. *See* Flavin adenine dinuleotide, reduced
form
Fairy ring, 433
Family, 387
FAS. *See* Fetal alcohol syndrome
Fat cells, 578
Fats, 42, 44, 620, 621
digestion of, 632, 633
metabolism of, 152-153
transport by lymphatic system, 600

Fatty acids, 42
  essential, 620
  saturated, 44, 45
  unsaturated, 44, 45
Feedback. See Negative feedback; Positive feedback
Female reproductive tract, human, 771-772
Fermentation, 142-143
  alcoholic, 146
  lactic acid, 145-146
Ferns, 391, 441
  alternation of generation life cycles in, 230
  life cycle of, 445, 448, 563
Fertility
  infertility and, 777
  limiting, 780-783
    permanently, 780
    temporarily, 780-783, 785
  replacement-level, human population growth
    and, 862, 863
Fertilization, 227
  external, 765-766
  in flowering plants, pollination versus, 545-546
  internal, 775
  self-fertilization, 237
  in vitro, 777
Fetal alcohol syndrome (FAS), 805
Fetus, 801
  development of, 802, 806
Fever, 652, 656-657
Fibrillation, 591
Fibrin, 594
Fibrous root system, 504
Filament, of complete flower, 538, 539
Filtrate, 642
Filtration, glomerular, 642, 643, 644
Fimbria, 773
Finches
  beak size in, 310
  Darwin's observations of, 310, 329, 873
Firefly, reproduction in, social behavior and, 835
Fish
  bony, 375, 391, 482, 484, 485
  cartilaginous, 391, 482-484
  gills, 606-609
  jawless, 391, 482, 483
  mutualism and, 880
  reproduction in, 764
  social behavior in, 839
  spawning in, 765, 766
  territoriality in, 833
  thyroxine in, 693
  see also specific fishes
FitzRoy, Robert, 308
Fixed action patterns, 815-817, 840
  breeding experiments and, 817
  performance without prior experience, 815
Flagella, 85, 99-101, 370, 412
  bacterial, 403-404, 406
Flame cells, in flatworms, 468, 640
Flatworms, 391, 462
  digestive tract of, 464
  eyespot of, 733
  flame cells in, 468, 640
  free-living, 468
  parasitic, 470
  reproduction in, 468, 470, 764
  symmetry in, 462, 468
Flavin adenine dinucleotide, reduced form
  (FADH$_2$), synthesis of, in cellular
  respiration, 148, 149
Flavin adenine dinucleotide (FAD), 71, 148
Flea, water, 2
Fleming, Alexander, 11, 434
Flexors, 757, 759
Florida scrub jays, altruism in, 342

Florigens, 566
Flounder, camouflage of, 875
Flowers. See Plant reproduction, flowers and
Flu, 672-673
Fluid mosaic model of membranes, 106-107
Fluid-phase endocytosis, 116, 117
Fluids
  cerebrospinal, 710
  interstitial, 595
  movement of molecules in, 109, 110
Flukes, 470
Fluorine, in nutrition, human requirements for,
  622
Fly
  apple maggot, ecological isolation of, 350-351
  fruit. See Drosophilia melanogaster; Rhagoletis
    pomonella
  as pollinator, 540, 541
  snowberry, coevolution with spiders, 876, 879
  Venus flytrap and, 568
Follicle, 771
Follicle-stimulating hormone (FSH), 684-686, 770,
  773-775
Food. See Nutrients; Nutrition; Plants, nutrition of
Food chains, 893, 895
Food labels, 625
Food vacuoles, 90, 116, 118, 625
Food web, 893-894, 896-897
Foraminiferans, 417
Forebrain, 713-716
Forensics, 272, 274
Forest fires, succession and, 881
Fossil fuels, 902, 905
Fossils, 305, 306
  evolution of horse and, 312, 356, 357
  radioactive dating of, 368
Founder effect, 329, 330
Fovea, 735, 736
Foxgloves, pollination of, 541
Franklin, Rosalind, 184, 185, 189
Free nerve endings, 729
Freshwater lakes. See Ecosystems, freshwater lakes
Frigate bird, 488
  reproduction in, social behavior and, 828, 834,
    835
Frogs
  brain of, 713
  differentiation in, 789, 790
  gastric breeding, 486
  leopard, 486
  respiration in, 609
  spawning in, 765-766, 767
  tree, reproduction in, social behavior and, 834
  warning coloration of, 876
  yellow-legged, 486
Frontal lobe, 716, 722
Fructose, 40, 41
Fructose diphosphate, in glycolysis, 143
Fructose-1,6-diphosphate, in glycolysis, 144
Fructose-6-phosphate, in glycolysis, 144
Fruit. See Plant reproduction, fruits and
Fruit fly. See Drosophilia melanogaster; Rhagoletis
    pomonella
Fruiting body, 414, 415
FSH. See Follicle-stimulating hormone
Fucus, 440, 442
Functional groups, 37-38
Fungi, 6-8, 388-392, 424-436
  body of, 425-426
  classification of, 428-429, 431-434
  club, 428, 431-433
  economic and ecological impacts of, 427-428
  egg, 428, 431, 434
  evolution of, 435
  imperfect, 431, 433-434

  nutrition of, 426
  parasitic, 427, 428
  reproduction of, 426, 430
  sac, 428, 431
  symbiotic relationships of, 427-429
  zygote, 428-431
Fusion, heat of, of water, 31

G$_1$ phase, 164
G$_2$ phase, 164
Gage, Phineas, 721-722
Galactose, 40, 41
Galapagos, 308-310
Gallbladder, in digestion, 630, 632, 634
Gametes, 163, 222
  incompatibility of, as postmating isolating
    mechanism, 351, 353
Gametophytes, 221, 230, 536, 537
  development in flowering plants, 543-545
Ganglion(a), 708
  of flatworms, 468
Ganglion cells, 736
Gap junctions, 120, 121
Garrod, Archibald, 194, 195
Gas exchange
  in arthropods, 474
  respiratory systems and, 605-610
Gas exchange portion, of human respiratory
    system, 611, 612
Gasoline, burning of, 62, 64
Gastric breeding frog, 486
Gastric inhibitory peptide, 634
Gastrin, 685, 693
  digestion and, 630, 634
Gastropoda, 391
Gastrovascular cavity
  of cnidarians, 467, 626
  of flatworms, 468
Gause, G. F., 872, 873
Geese, brain of, 713
Gene flow, 326
Genentech, 267
Gene pool, 324
Generative pool, 543, 545
Genes
  chromosomes and, 242-246
  copying, 265, 266, 268-270
  crossing over, 243, 245-246
  function in individual organisms, 323
  function in population, 324
  identifying, 265
  locating, 271-272
  Mendel's hypothesis and, 239-240
  mutations. See Mutations
  proteins and, 193-195
  regulation during development, 790-791
  regulation of, 207-211
    in eukaryotes, 209-211
    in prokaryotes, 208-209
  sequencing, 272-273
  structural, 209
  tumor-suppressor, 627
  see also Deoxyribonucleic acid; Ribonucleic acid;
    Chromosomes; Genetics; Human genetics
Genetic code, 196-198
  reading, 198
Genetic diseases
  diagnosis of, 270
    gene sequencing and, 272-273
  ethics and, 295-297
  see also Inheritance
Genetic divergence, speciation and, 348
Genetic drift, 327, 329, 357
Genetic engineering, 419, 666
Genetic recombination. See Crossing over

Genetics
  behavioral, 811-812
  molecular. *See* Biotechnology; Molecular
      genetics; Recombinant DNA technology
  of speciation, 353-355
  see also Human genetics; Population genetics
Genetic variation, 12
Genital herpes, 784
Genome, 265
  genetic engineering and, 275-276
  Human Genome Project and, 269, 273, 294
Genotypes, 240
  distribution of, in equilibrium population, 326
  equivalence of, 331-332
  phenotypes versus, natural selection and, 332-
      333
Genus, 387
Geographical isolation, speciation and, 348-351
  premating isolating mechanisms and, 352
Geology, 305-306
Germ layers, 461
Giardia, 416
Gibberellins, 560, 561, 567
Gigantism, 339
Gilbert, Walter, 294
Gills, 606-609
  in arthropods, 474, 475
Ginkgos, 449
Giraffe, height of, 337
Gizzard, of earthworm, 626
Glia, 579
Glomerulus, 641, 642
Glucagon, 685, 690, 691, 692
Glucocorticoids, 691, 692
Glucose, 39, 40-41, 63
  activation of, 143
  metabolism of, 140-155
  see also Glycolysis
Glucose-6-phosphate, in glycolysis, 144
Glutamic acid, 47, 284
Glycerol, 42
Glycogen, 39, 41, 620-621
Glycolysis, 63, 127, 128, 141-145, 153
Glycoproteins, 89
  in cell membranes, 109
GnRH. *See* Gonadotropin-releasing hormone
Goiter, 688
Goldberg, Rube, 212, 213
Golden toad, 4876
Golgi complex, 82, 85, 109
Gonadotropin-releasing hormone (GnRH), 770,
      773-775, 781
Gonads, 767, 768
Gonorrhea, 409, 784
Gonyaulax, 412
Gould, Stephen Jay, 356, 382
Gout, 288
Gradient, 109, 110
Gradualism, speciation and, 355-357
Gram stain, 402
Grant, Peter, 358
Granum(a), 94, 129
Grapes, tendrils of, 512, 513
Grasses, coevolution with horses, 877
Grasshoppers
  reproduction in, social behavior and, 834
  segmentation in, 473
Grasslands, 928-929, 930
Gravitropism, 560, 562
Gray matter, of spinal cord, 710, 712
Greenhouse effect, 530, 908-910, 923
  consequences of global warming and, 909
  need to address problem of, 910
  sources of additional carbon dioxide and, 908-909
Green Revolution, 865

Griffith, Frederick, 179, 180
Groos, K., 823
Growth hormone, 46, 273, 684, 685-686
  producing, 266-267
Grunion, spawning in, 765, 766
Guanine
  in DNA, 52, 178, 184, 196, 199
  in RNA, 196, 199
Guard cells
  of leaf, 512
  transpiration and, 525-527, 530
Guillermin, Roger, 686
Guisella, James, 288
Guppies, 337
Gurdon, J. B., 789, 790
Gymnosperms, 449
  see also Conifers
Gyres, 917, 919

Habitat destruction, 14
  amphibians and, 486
  extinction and, 339-341
Habituation, 818, 821
Hagfishes, 483
Hair
  color of, 253
  protein in, 47, 48-49
Hair cells, 728, 730
Hair follicles, 580, 581
Haldane, J. B. S., 365
Halophiles, 411
Hamilton, W. D., 342
Hammer, 731
Hantavirus, 651
Haploid cells, 163, 220-222
  life cycles and, 227-229
  see also Gametes; Meiosis
Hardy, G. H., 326, 328
Hare, snowshoe, lynx as predator of, 854
Haversian systems, 576, 579, 755, 756
Health. *See* Genetic diseases; Human health;
      specific disorders
Hearing. *See* Auditory system
Heart
  hormones and, 685, 692-693
  vertebrate, 587-591
    cardiac cycle and, 588
    coordination of activity of, 588-590
    outside influences on heart rate and, 591
    structure of, 587-588
  see also Cardiac muscle
Heart attacks, 599
Heartworm, 471
Heat
  of fusion, of water, 31
  specific, of water, 31
  of vaporization, of water, 31
  see also Temperature
Helentjaris, Timothy, 274
Helicobacter pylori, 635
Heliozoans, 416-417
Helium, 23
Helix, in proteins, 49-51
Helper T cells, 654-655, 658, 663, 664, 667-669
Hemispheres, 916, 917
Hemocoel, 474, 585
Hemodialysis, 647
Hemoglobin, 39, 46, 50, 592, 620
  oxygen binding with, 51, 61, 63, 611-612
  see also Sickle-cell anemia
Hemophilia, inheritance of, 287, 289
Henslow, John, 307, 308
Herbicides, resistance to, 270, 275
Herbivores, 629, 893
Hermaphrodites

flatworms, 468
  reproduction in, 765
Herpes virus, 400, 401, 405
  genital, 400
Herring gull, fixed action pattern in, 817, 821
Hershey, Alfred, 181-183, 198
Heterotrophs, 8, 893
Heterozygous organisms, 236
Hexokinase, 144
High-energy bonds, 69-70
Himalayan rabbit, hair color in, 253
Hindbrain, 713
Hinge joints, 757, 759
Hippocampus, 715
  learning and, 719
Hippopotamus, play and, 822
Hirudinea, 391
Histamine, 655
Histone, 162
HIV. *See* Acquired immunodeficiency syndrome
Hodgkin, Alan, 700
Homeobox, 803, 804
Homeostasis, 5, 639
  negative feedback and, 573-575
  positive feedback and, 574, 575
Homing pigeons, orientation by magnetic fields, 742
Hominids, evolution of, 377-379
Homo erectus
  evolution of, 378-381
  evolution of human behavior and, 382
Homo habilis
  evolution of, 378, 379, 381
  evolution of human behavior and, 382
Homologous structures
  evolution and, 313, 314
  taxonomic classification and, 389
Homologues, 163, 221
  during meiosis, 223-226
Homo sapiens, evolution of, 378-381
Homozygous organisms, 236
Honeybees
  dance of, 838
  innate learning and, 821
  as pollinators, 539-541
  queen substance and, 830, 837
  reproduction in, 764
  social behavior and, 837-839, 843
Honey possums, as pollinators, 541
Hooke, Robert, 75, 76
Hookworm, 471
Hormones
  actions on target cells, 680
    binding to intracellular receptors and, 681
    binding to surface receptors and, 680-681
  animal, types of, 681
  digestive, 634
  evolution of, 693
  menstrual cycle and, 773-775
  peptide, 680, 689
  of plants
    actions of, 559-561
    discovery of, 557-559
  steroid, 680, 681
  see also specific hormones
Horses
  brain of, 713
  coevolution with grasses, 877
  evolution of, 12, 13, 312, 356, 357
Human beings, taxonomy and, 388
Human development, 799-807
  birth and, 806
  see also Childbirth
  during first two months of pregnancy, 799-802
  during last seven months of pregnancy, 803
  placenta and, 801, 802, 804-805

Human digestion, 627-635
  control of, 634
  digestive secretions and, 629
  in esophagus and stomach, 628-630
  of fats, 632, 633
  in large intestine, 633-634
  mouth and, 627
  in small intestine, 629-631, 633
    absorption and, 631, 633
    intestinal wall and, 631
    liver and gallbladder and, 630-631
    pancreas and, 631
Human genetics, 280-298
  chromosomal inheritance and, 290-294
    abnormal numbers of autosomes and, 293-294
    abnormal numbers of sex hormones, 291-293
  complex inheritance and, 290
    environmental effects and, 290
    polygenic, 290
  dominant inheritance and, 286-287
  Human Genome Project and, 269, 273, 294
  molecular genetics and, 281
  pedigree analysis and, 281, 282
  sex-linked and sex-influenced inheritance and, 287-290
  single gene inheritance and, 283-290
    recessive, 283-286
    sickle-cell anemia and, 284-286
    Tay-Sachs disease and, 284, 285
  see also Genetic diseases
Human Genome Project, 269, 273, 294
Human health
  bacteria and, 408-410
  blood types and, 593
  broken bones and, repair of, 756
  cardiovascular disorders and, 598-599
  immune response and. See Immune response, medicine and
  osteoporosis and, 758
  sexually transmitted diseases and, 409, 776, 784
    see also Acquired immunodeficiency syndrome
  smoking and respiratory disease and, 614-615
  ulcers and, 635
  see also Genetic diseases; specific disorders
Human immunodeficiency virus. See Acquired immunodeficiency syndrome
Human population, growth of. See Population growth, human population and
Human reproduction, 767-776
  contraception and
    permanent, 780
    temporary, 780-783, 785
  copulation and, 775-776
  delivery and, 779-780
  female reproductive tract and, 771-773
  fertilization, 776
    in vitro, 777
  lactation and, 778-779
  male reproductive tract and, 767-771
  menstrual cycle and, 773-775
  pregnancy and, 776-778
Hummingbirds, 140, 141, 488
  migration of, 150-151, 153
  as pollinators, 14, 541
Humoral immunity, 662, 663
  antibody function and, 662
Huntington's disease, 270, 286-288
Hutchinson, G. Evelyn, 57
Hutton, James, 305-306, 309
Huxley, Andrew, 700
Hybrids
  infertility of, as postmating isolating mechanism, 351, 353

inviability of, as postmating isolating mechanism, 35, 351
Hydra, 466
  asexual reproduction in, 171
Hydrochloric acid, 365
  digestion and, 629
Hydrogen, 23, 365
  bonding and, 26-27
  in plant nutrition, 521
Hydrogen bonds, 26-29, 48, 50, 51
  in water, 524, 526
Hydrogen group, 38, 40, 47
Hydrogen sulfide, 365
Hydrologic cycle, 902-903, 905
Hydrolysis, 38, 39, 41
Hydrophilic amino acids, 50, 52
Hydrophilic head, of phospholipids, 107
Hydrophilic molecules, 29, 632
Hydrophobic amino acids, 50, 52
Hydrophobic interaction, 29
Hydrophobic molecules, 29, 632
Hydrophobic tail, of phospholipids, 107, 108
Hydrostatic skeletons, 752, 753
  of segmented worms, 470
Hydroxyl group, 38, 40, 43
Hypertension, 598-599
Hypertonic solutions, 114, 115
Hyphae
  dung fungus and, 435
  fungal, 425, 426, 431
  roundworms and, 435
Hypocotyl, 546
Hypothalamus, 714, 715, 774-775
  hormones of, 684
    control of pituitary by, 683-687
Hypothalamus-pituitary complex, 683-686, 688
Hypothesis, 9, 10, 240
Hypotonic solutions, 114, 115
Hyracotherium, 13, 312, 356, 357

Ice, 31
Ice-minus bacteria, 267, 275
IgA, 661
IgD, 661
IgE, 661, 667
  allergies and, 666-667
IgG, 661
IgM, 661, 664
Immigration, population growth and, 850-851
  in United States, 866
Immune response, 657-675
  attack and, 662-665
  cell-mediated immunity and, 655, 663-665
  humoral immunity and, 662-664
  medicine and, 666-667
    AIDS and, 667-670
    allergies and, 666
    antibiotics and, 666
    autoimmune diseases and, 667
    cancer and, 670-671, 673-674
    flu and, 672-673
    immune deficiency diseases and, 667
  memory and, 657, 665-666
  recognition and, 658-662
    antibodies and, 658-661
    distinguishing "self" from "non-self" and, 661-662
  specific, 658
Impalas, aggression in, 832
Imprinting, 818, 821
  human infants and, 841
Inclusive fitness, 342
Incomplete dominance, 249, 253-254
Incomplete flowers, 538, 539

Independent Assortment, Law of, 241-242, 244
Indirect development, 791-793
Individual behavior, 810-823
  genetic basis of, 811-812
  innate, 812-818
    fixed action patterns and, 815-817
    imprinting and, 818
    kineses and, 812
    learning and, 820-822
    reflexes and, 813, 814
    taxes and, 812-814
  learning, 719-720, 818-822
    classical and operant conditioning and, 818-819
    habituation and, 818, 821
    imprinting and, 818, 821
    insight and, 820, 821
    instinct and, 820-822
    memory and, 719-720
    trial-and-error, 819-820
  play, 822-823
Induction, gastrulation and, 795, 796
Industrial-medical revolution, human population growth and, 862
Infants
  communication by touch and, 830
  suckling in, 840, 841
Infections, viral. See Viruses, viral infections and
Infertility, 777
Inflammatory response, 652, 656
Influenza, 67-673
Influenza (type A) viruses, 405
Inhalation, 612-613
Inheritance, 12, 235-255
  of acquired characteristics, 307
  chromosomal, 290-294
    abnormal numbers of autosomes and, 293-294
    abnormal numbers of sex chromosomes and, 291-293
  codominance in, 249-250, 254
  complex, 290
    environmental effects on, 290
    polygenic, 290
  dominant, 286-287
  incomplete dominance in, 249, 253-254
  Mendelian principles of, 237-242
    multiple-trait experiments and, 241-242
    single-trait experiments and, 238-240
    variations in, 249-253
  polygenic, 251, 290
  sex-linked and sex-influenced, 246-249, 287-290
  single gene, 283-290
    recessive, 283-286
    sickle-cell anemia and, 284-286
    Tay-Sachs disease and, 284, 285
  see also Chromosomes; Deoxyribonucleic acid; Evolution; Genes; Genetics; Human genetics; Meiosis; Ribonucleic acid
Inhibiting hormones, hypothalamic, 684, 686
Inhibitory synapse, 705
Innate behavior. See Individual behavior, innate
Inner cell mass, 776, 799
Inorganic molecules, 27-28
Insects, 391, 463, 473-476
Insertion, of muscle, 759
Insertions (mutations), 205, 206
Insight, 820, 821
Instinctive behavior. See Individual behavior, innate
Insulin, 46, 575, 685, 690, 691, 692
  evolution of, 693
Integuments, of flowers, 544
Intelligence, genetics and environment and, 290
Intercalated discs, 752

Interconnected neural pathways, 707
Interferon, 656-657
Interleukin-2, 655
Intermediate compartment, of mitochondria, 94, 146, 147
Intermediate filaments, 96-98
Internal fertilization, 775
Internodes, of stems, 507
Interphase, 163-165
    of daughter cells, 167
    late, 166
Interspecific competition
    community interactions and, 872-873
    population growth and, 858
Interstitial fluid, 595
Intertidal zone, 937
Intervertebral discs, 754
Intestine, large
    cancer of, 673, 674
    digestion in, 633-634
Intestine, small.
        See Small intestine
Intracellular digestion, 625
Intracellular membranes, transport across, 116, 118-120
Intraspecific competition, population growth and, 858
Intrauterine device, 781-783
Introns, 209, 210
Invertebrates, 7, 464
In vitro fertilization, 777
Iodine, in nutrition, human requirements for, 621, 622
Ionic bonds, 25
IPSP, 705
Iris, 512, 734
Iron
    in human nutrition, requirements for, 622
    in plant nutrition, 521
Islet cells, 689-691
Isoleucine, 67
Isotonic fluids, 114, 115
Isotopes, 23
    radioactive, uses of, 32-33

Janzen, Daniel, 881
Jawless fishes, 483
J-curve, population growth and, 851-853, 866
Jeffers, Robinson, 127, 323, 363
Jenner, Edward, 666
Joints, movement around, 757, 759
Jurassic period, 341, 364, 376

Kangaroo rats, 829, 927
    urine concentration in, 646
Kaposi's sarcoma, 668
Katz, Bernard, 700
Kelps, 394-395
Keratin, 39, 46, 48, 49, 580
Keystone predator, 878
Kidneys, 639, 641-647
    filtration by glomerulus of, 642
    hormones and, 685
    as organs of homeostasis, 643-646
    structure of, 641-643
    urine formation in nephrons of, 641-645
    urine production and concentration in, 641-643
Killdeer, 338
Killer T cells, 654, 655, 673
Kineses, 812
Kinetic energy, 22
Kinetochore microtubules, 168
Kinetochores, 168, 169
Kingdoms, 6-8, 387

taxonomic classification and, 389-392
    see also Animals; Fungi; Monerans; Plants; Protists
King snake, mimicry by, 877
Kin selection
    altruism and, 342, 843-844
    natural selection and, 338
Klinefelter's syndrome, 292
Knee-jerk reflex, 813
Kohler, W., 820
Krebs, Hans, 148
Krebs cycle, 146, 148
Kudzu, population growth and, 884
Kuru, 400-401

Labia, 771, 775
Lactase, digestion and, 631
Lactation, 778-779
    oxytocin and, 684-685, 687
Lacteal, 633
Lactic acid fermentation, 145-146, 153
Lactose, 41
Lactose operon, 208-209
Ladder of Nature, 304, 317
Lakes, freshwater. See Ecosystems, freshwater lakes
Lamarck, Jean Baptiste, 306-308, 311
Lampreys, 483
Lancelet, 481, 482
Land. see also Ecosystems, terrestrial
    origin of life on, 373-376
Land crabs, evolution of
Large intestine
    cancer of, 673, 674
    digestion in, 633, 634
Larvae, 475, 476
    coevolution with milkweed, 877
    of swallowtail butterfly, startle coloration in, 878
Larynx, 610, 611
Lascaux caves, 382
Lashley, Karl, 719-720
Lateral buds, of stems, 507
Lateral line organs, 730
Lateral meristems, 498
Lateral nodes, of stems
Latitude, climate and, 916, 917
Law of Independent Assortment, 241-242, 244
Law of Segregation, 239-240
Leaf primordia, 507
Leakey, Mary, 379
Learning. See Individual behavior, learning
Leaves. See Plants, leaves of
Leeches, 472
Leeuwenhoek, Anton van, 75, 411
Legionella pneumophila, strains of, 393
Legumes, 523, 524, 902
Lemmings, population growth and, 853, 855
Lemurs, evolution of, 377
Lens, 734, 735
Leopard frogs, 486
Leopold, Aldo, 891
Leucine, 47
Leukocytes, 591, 593, 594
Lewis, Edward, 803
Lewis, Julian, 75
LH. See Luteinizing hormone
Lichens, symbiotic relationships of, 427-428
Life cycle, 159
Life on Earth, 362-383
    first life, 368-371
    human evolution and. See Evolution, of humans; Natural selection
    on land
        animals and, 373-376
        plants and, 372-373, 374

multicellularity and, 371
    origins of, 364-368
        prebiotic evolution and, 365-366
    in sea
        animals and, 371-372, 373
        plants and, 371
Ligaments, 576, 757
Light, photosynthesis and. See Plants, photosynthesis and
Light-dependent reactions, in photosynthesis, 129-132
    relation to light-independent reactions, 128, 134
Light-harvesting complex, 130
Light-independent reactions, in photosynthesis, 129, 132, 134
    relation to light-dependent reactions, 128, 134
Lignin, of vascular plants, 440
Limbic system, 714, 715
Linnaeus, Carolus, 388
Lipases, digestion and, 629, 631, 632
Lipids, 37-39, 42, 44, 619
    as energy source, 620
    oils, fats, and waxes, 42, 44
    phospholipids, 39, 44
    steroids, 39, 44, 46
    see also Fats
Littoral zone, 935
Liver, in digestion, 630, 632
Liver flukes, 470
Liverworts, 391, 439, 441, 444-445
Living organisms
    characteristics of, 1-6
        acquisition and use of materials and energy as, 4-5
        capacity to evolve as, 6
        complexity and organization as, 2, 4
        growth as, 5
        homeostasis as, 5
        reproduction as, 5-6
        responsiveness as, 5
    diversity of, 7-8
    see also Animals; Life on Earth; Multicellular organisms; Plants; Unicellular organisms; specific organisms
Lizards, 487
    reproduction in, social behavior and, 835
    visual communication in, 828, 829
Lobefins, evolution of, 374-375
Locusts, 474, 475
Long-day plant, 565
Long-term memory, 719
Loop of Henle, 642-645
Lorenz, Konrad, 811, 816, 818, 821
Lovebirds, behavioral genetics and, 817
Low-density lipoprotein (LDL), 599
Lucretius, 235
"Lucy" (Australopithecus afarensis), 379, 382
Lungs
    in bony fish
        book, in arthropods, 607, 609
    human, 610-615
Lupines, 14-15, 408
Luteinizing hormone (LH), 684-686, 770, 773-775
Lycopedon giganteum, 434
Lyell, Charles, 305-306, 309
Lyme disease, 409, 651
Lymph, 576, 599
Lymphatic system, 581, 597, 599-600
    body defense and, 600
    fat transport and, 600
    return of fluids to blood and, 599-600
    structure of vessels of, 597
Lymph nodes, 600

Lymphocytes, 591, 593, 657
    tumor infiltrating, in cancer treatment, 674
Lynx, as predator of snowshoe hare, 854
Lysosomes, 85, 87, 89-91, 625

MacArthur, R., 873
McCarty, Maclyn, 180, 182
McClintock, Martha, 842
McKusick, Victor, 330
MacLeod, Colin, 180, 182
Macronutrients, 900
    plant nutrition and, 520, 521
Macrophages, 653-657
Mad-cow disease, 401
Magnesium, 23
    in human nutrition, requirements for, 622
    in plant nutrition, 521
Magnetic fields, animal orientation and, 742
Magnetotactic bacteria, 404
Major histocompatibility complex (MHC), 654-
        655, 657, 661
    helper T cells and, 655
Malaria, 335, 418-419
    sickle-cell anemia and, 286
Male reproductive tract, human, 767-771
Maltase, digestion and, 631
Malthus, Thomas, 309
Maltose, 41
Mammals, 391, 489-490
    development of, 793
    endocrine system of, 683-693
        adrenal glands and, 691-692
        digestive tract and, 692
        gland structure and function and, 683
        heart and, 685, 692-693
        hypothalamus-pituitary complex and, 683-
            686, 688
        kidneys and, 692
        pineal gland and, 692
        prostaglandins and, 692
        sex organs and, 691
        thyroid and parathyroid glands and, 687-689
    origin of, 376
    see also specific mammals
Mammary glands, 489, 687
Manganese
    in nutrition, human requirements for, 622
    in plant nutrition, 521
Mantis
    praying, molting by, 474
    tree bark, camouflage of, 875
Mantle cavity, of cephalopods, 479
Maple trees, seed dispersal and, 548
Margulis, Lynn, 92, 369
Maricouagan crater, 341
Marine ecosystems. See Ecosystems, marine;
        Oceans
Marshall, B. J., 635
Marsupials, 490
Matrix, of mitochondria, 94, 146, 147, 149
Matrix reactions, in cellular respiration, 146, 148-
        150
Matter, structure of, 22-24
Maxam, Allam, 294
Mechanical incompatibility, speciation and, 351-
        353
Mechanoreceptors, 728, 729-730
Medical genetics. See Genetic diseases; Human
        genetics
Medulla, of hindbrain, 713, 714
Medusa, 466, 467
Megakaryocytes, 593, 594
Megaspore mother cell, 544
Megaspores, 544
Meiosis, 159

errors in, 291-294
    eukaryotic life cycles and, 227-229
    meiosis I, 222-227
    meiosis II, 222, 223, 224, 227
    principal events and outcomes of, 222
    sexual reproduction and, 222, 229-230
Melanin
    eye color and, 290
    hair color and, 253
    skin color and, 290
Melatonin, 685, 692
Membranes, 575
    see also Cell membrane; Plasma membrane
Memory, learning and, 719-720
Memory cells, 657, 658
    humoral immunity and, 662, 663, 665
Mendel, Gregor, 236, 307, 309
    genetic principles of, 237-242
Meninges, 710
Menstrual cycle, 771, 773-775, 842
Menstruation, 773
Mental illness, 721
Meristem, 498, 500, 509
Mesoderm, 461
Mesohippus, 357
Mesophyll, 129
    of leaf, 512
Mesozoic era, 364
Metabolic pathways, 63-64
Metabolism, 4
    aerobic, 94, 95
    anaerobic, 95, 369
    of carbohydrates, 152-153
    control of, 63-68
        activation energy and reaction rates and, 64-65
        catalysts and, 65-68
    of fats, 152-153
    of proteins, 152-153
Metamorphosis, 476, 791
Metaphase, 165, 166
Metaphase I, 224-226
Metaphase II, 224, 227
Meteorites, habitat destruction by, 339-341
Methane, 365
Methanogens, 411
Methanol, 68
Methyl group, 38
MHC. See Major histocompatibility complex
Micelles, digestion and, 632
Microbes. See Bacteria; Monera; Protists; Viruses
Microbial invasion, defenses against, 651-657
Microfilaments, 96-98
Micronutrients, 900
    plant nutrition and, 520, 521
Microphage, 650
Microscopes, 76, 78-79
Microspheres, 367
Microspore mother cells, 543
Microtubules, 96-98, 100, 370
Microvilli, of small intestine, 631, 633
Midbrain, 71, 714
Middle ear, 731
Middle lamella, 106, 107
Midges, 211
Miescher, Friedrich, 177
Migration, evolution and, 320-327
Milkweed
    coevolution with predators, 877
    seed dispersal and, 548
Miller, Stanley, 365-366, 367
Mimicry, 875-878
Minerals, in nutrition, 619, 621
    human requirements for, 622
    of plants, 520-528
Miocene epoch, 364

Mirsky, Alfred, 183
Mistletoes, seed dispersal and, 548
Mitchell, Peter, 79, 82, 85
Mites, 476
Mitochondria, 91-95
    chemiosmosis in, 150, 151
    origin of, 369-370
Mitosis, 165, 225, 226, 228
    anaphase, 165, 167, 168-169
    of apical meristem cells, 498, 506
    metaphase, 165, 166
    prophase, 165, 166
    telophase, 165, 167, 169
    see also Eukaryotic cells
Mixture, 22
Molds, slime
    acellular, 414
    cellular, 414, 415-416
    plasmodial, 414
Molecular biology, evolutionary, 314, 318-319
Molecular genetics, 260-277
    see also Biotechnology; Recombinant DNA
        technology
Molecules, 3, 4, 24
    bonding and, 25-27
    hydrophilic, 29
    hydrophobic, 29
    inorganic, 27-28
    nonpolar, 29
    organic. See Organic molecules
Mollusks, 463, 474, 477-479
    body plan of, 478
    camera eye of, 734, 736
    cephalopods, 479
    gas exchange in, 604, 607
    gastropods, 477
    nervous system of, 708, 709
    pelecypods, 391
    see also Clams; Octopuses; Scallops; Snails;
        Squid
Molting
    by arthropods, 473-474
    thyroxine and, 693
Molybdenum, in plant nutrition, 521
Monerans, 6-8, 390, 391
    see also Bacteria; Prokaryotic cells
Monkeys, evolution of, 377
Monocots, 455
Monocytes, 591, 593
Monoecious species, reproduction in, 765
Monohybrid cross, 238, 239
Monomers, 38
Monosaccharides, 39-41, 620
Monotremes, 490
Moray eel, 485
Morels, 431
Morgan, Thomas Hunt, 247, 248
Morphological change, speciation and, 355-357
Morula, 793
Mosquito, 418-419
    reproduction in, social behavior and, 829
    taxes and, 813
Mosses, 439, 441, 444-445
    club, 391, 445
    life cycle of, 446
Moths, 738
    coevolution with bats, 874
    peppered, evolution of, 316, 317, 333-334
    as predator of cactus, 857, 858
    reproduction in, social behavior and, 836
    taxis toward light, 752, 812
Motor unit, 752
Mountain king snake, 487
Mountains, climate and, 917-919
Mouth, digestion and, 625

MRI (magnetic resonance imaging), 722
Mucous membranes, as entry barrier, 651, 652
Muir, John, 871
Mullis, Kay, 268
Multicellular organisms, 3, 4, 7, 8
  origin of, 371
    on land, 373-376
    in sea, 371-372, 373
  see also specific organisms
Multiregional hypothesis, 380, 381
Muscle fibers, 748, 749
Muscles, 746-752
  anatomy of, 748-750
  contraction of
    control of, 751-752
    sliding filament model of, 750-751
  lactic acid fermentation and, 145-146, 152
  muscle-skeletal interactions and, 757, 759
Muscular dystrophy, 270
Muscular system, 581
Mushrooms, 429, 432-433
Musk oxen, cooperation in, 837
Mussels, 477, 478, 479
Mutations, 6, 204-207, 220-222
  effects of, 205, 207, 220
  evolution and, 207, 324-326
  insertions and deletions, 205
  neutral, 207
  point, 205
Mutualism, 880, 881
Mycelium, 425, 426
Mycorrhizae, 428, 429
  mineral accessibility and, 522, 523
  symbiotic relationships of, 428, 429
Myelin, 703-704
Myofibrils, 749, 751
Myometrium, 773
Myosin, 96-98, 579
Myrtle warbler, 304
Myxomycota, 391, 412, 414

NAD+. See Nicotinamide dinucleotide
NADH. See Nicotinamide adenine dinucleotide, reduced form
NADPH. See Nicotinamide adenine dinucleotide phosphate
Nagana, 879
Natural causality, principle of, 8-9
Natural killer cells, 652, 653
Natural selection, 6, 12, 122, 308-309, 312, 324, 333-338
  essentials of theory of, 311
  extinction and, 338-341
  forces of, 335-338
  genotype versus phenotype, 332-333
  population genetics and, 323-326
  speciation and, 358
  types of, 333-335
Nautiloids, 373
Nautilus, 371, 373
Navicula, 412
Neanderthal Man, 379, 381
Nearshore zone, 937.938
Negative feedback
  homeostasis and, 573-575
    hormonal maintenance of, 681
    red blood cell regulation and, 592-593
Nematoda. See Roundworms
Nephridia
  in earthworms, 640-641
  of segmented worms, 472
Nephrons, 641
  urine formation in, 641-645
Nephrostome, 640
Nerve cord, 481

Nerve net, cnidarian, 446, 707, 708
Nerves
  spinal, 710, 712
  see also Nervous system; Neurons
Nerve tissue, 579-580
Nervous system, 581, 696-723
  of arthropods, 474
  central, 708, 710, 712-713
    brain, 713-716
    spinal cord, 710, 712-713
    see also Brain
  centralized, 707-708
  design of, 707-708
  diffuse, 707-708
  functions of, 706
  human, 708-717
  information processing in, 705-707
  mechanisms of neural activity and, 699-705
    action potentials and, 701-704
    origin of resting potential and, 699-701
    synapses and postsynaptic potentials and, 704-705
  "mind-brain problem" and, 717-722
    learning and memory and, 719-720
    left brain-right brain and, 718-719
  of mollusks, 477
  neural pathways and, 707
  neurons and. See Neurons
  neurotransmitters and neuromodulators and, 716-717
  peripheral, 708-711
  vertebrate, 708, 710
Net primary productivity, 893, 894
Neural pathways, 707
Neural processing, in retina, 736
Neural tube, 800
Neuromodulators, 716, 717
Neuromuscular junctions, 751
Neurons, 579, 580
  functions and structure of, 698-699
  see also Nervous system
Neurosecretory cells, 683, 684, 686
Neurotransmitters, 621, 699, 705, 716-717, 751
Neutralization, antibodies and, 664
Neutral mutations, 207
Neutrons, 3, 27
Neutrophils, 591, 593
Nickel, in nutrition, human requirements for, 622
Nicolson, G. L., 107
Nicotinamide adenine dinucleotide, reduced form (NADH)
  in glycolysis, 143-145
  synthesis of, in cellular respiration, 148-151
Nicotinamide adenine dinucleotide phosphate (NADPH), 128-132, 134
Nicotinamide dinucleotide (NAD+), 70-71, 131
  in glycolysis, 143, 144
Nitrogen, 23, 365
  bonding and, 26
  in plant nutrition, 521
    acquisition of, 522-523
Nitrogen cycle, 902
Nitrogen fixation, 523, 524, 902
Nitrogen-fixing bacteria, 523, 524
Nodes, 703, 704
Nodules, of legumes, 523, 524
Nondisjunction, 291-294
Nonpolar covalent bonds, 26
Noradrenaline, 680, 684, 691, 717
Norepinephrine, 680, 681, 684, 691, 717
Norplant, 780, 782-783
North American warbler, competition for resources among, 873
Northern coniferous forest, 931-933
Notochord, of chordates, 481

Nuclear envelope, 82, 86, 87
Nucleic acids, 37-39, 42, 52
  nucleotides, 52
Nucleoid, 80, 81
Nucleolus, 82, 85, 87
Nucleotides, 52, 366, 368, 389
Nucleus of cell, 85-87, 92
  origin of, 370-371
  see also Cytokinesis; Mitosis
Nutrients, 4
  cycling of. See Ecosystems, nutrient cycling and
  plant nutrition and, 519
Nutrition, 618-624
  amino acids, 621
  carbohydrates, 619, 620-621
  fats in, 620, 621
  fungal, 426
  lipids in, 619, 620
  minerals in, 619, 621, 622
  of plants. See Plants, nutrition of
  protein in, 619, 620, 621
  vitamins in, 619, 621, 623, 624

Observation, 9, 10
Occam's razor, 306
Occipital lobe, 716, 722
Oceans
  currents in, climate and, 917, 919
  structure of, 2
  see also Ecosystems, marine
Octopuses, 477, 479, 811
  nervous system of, 709
Oils, 42, 44
  see also Lipids
O.J. Simpson trial, 272
Oleic acid, 45
Olfaction, 738-739
  see also Pheromones
Oligocene epoch, 364
Oligochaeta, 391, 472
Oligotrophic lakes, 936
Olsen, Paul, 341
Ommatidia, 733, 734
Oncogenes, 671, 673
One-gene, one-protein hypotheses, 194-195
Oogonia, 771, 772
Oomycota, 391, 428, 431, 434
Oparin, Alexander, 365
Open circulatory systems, 585, 586
  in arthropods, 474
Open ocean, 939-941
Operant conditioning, 819
Operator, 208
Operculum, 607
Operons, 208
  lactose, 208-209
Opioids, 717
Optic disk, 736
Optic nerve, 736
Orangutan, 341, 489
Orbitals, 23
Orchids
  commensalism with trees, 880
  pollination of, 542
Order, 387
Ordovician period, 364
Organ, 3, 4
Organelles, 4, 80, 82, 95
  origin of, 369, 370
  see also Chloroplasts; Endoplasmic reticulum; Golgi complex; Mitichondria; Nucleus; Ribosomes
Organic molecules, 4, 27, 36-53
  accumulation of, 366
  covalent bonding in, 25, 26, 28

Organic molecules *continued*
  functional groups of, 38
  prebiotic evolution of, 365-366
  principal types of, 39
  synthesizing, 38, 39
  see also Carbohydrates; Lipids; Nucleic acids;
      Proteins
Organism, 4
Organogenesis, 460, 461, 796
Organs, 580-581
  of flatworms, 468
Organ systems, 3, 4, 461, 462, 581
Origin, of muscle, 759
*Origin of Species by Means of Natural Selection, On
    the* (Darwin), 303, 308, 309, 311, 316, 318,
    336, 347, 388, 849
Oscula, of sponges, 465
Osmosis, 110
  guard cells and, 526-527
  transport across cell membranes and, 113-115
Osmotic pressure, 113-115
Osteichthyes. *See* Fish, bony
Osteoblasts, 754-757
Osteoclasts, 754-757
Osteocytes, 576, 579, 754, 755
Osteoporosis, 758
Ostrich, 488
Otolith membrane, 733
Otters, play and, 822
Outer ear, 730, 731
Out-of-Africa hypothesis, 380, 381
Oval window, 731
Ovaries
  of complete flower, 538, 539
  human, 771-774
  mammalian, 685
Overgrazing, 855, 856
Oviduct, 773
Ovulation, 766, 772
Ovules
  of complete flower, 538, 539
  of flowers, 544
Owls, 933
  vision and, 737
Oxaloacetic acid, in photosynthesis, 135, 136
Oxygen, 23, 24, 365
  anaerobic bacteria and, 369
  bonding and, 26, 28
  hemoglobin binding with, 51, 61, 63, 611-612
  in plant nutrition, 521
  see also Gas exchange
Oxytocin, 681, 683-687, 719
Ozone layer, 916, 920-921

Pacemaker, cardiac, 589, 590
Pacinian corpuscles, 729
Pain perception, 739-740
Pain receptors, 728
Pain-withdrawal reflex, 707, 708
Paleocene epoch, 364
Paleozoic era, 364
Palisade cells, of leaf, 512
Pancreas, 685, 689-691
  in digestion, 631, 634
Pancreatic juice, 631
Papilloma virus, 400
Paramecium, 7, 116, 412
  algae harbored by, 369-370, 371
  asexual reproduction in, 171
  cilia and flagella of, 100, 419
  competition for resources among, 872-873
  viewed through microscope, 78, 79
Parasitism, 470
  community interactions and, 878-879
  fungal parasites and, 427, 428

population growth and, 858
Parasympathetic nervous system, 591, 709, 711
Parathormone, 684, 688, 755
Parathyroid glands, 684, 687-689
Parenchyma tissue, of plants, 501, 502, 508
Parental generation, 238
Parietal lobe, 716, 722
Parthenogenesis, 764, 765
Passive smoking, 615
Passive transport, 110-115
Passive visual signals, 828
Pasteur, Louis, 11, 364, 365, 666
Pauling, Linus, 189
Pavlov, Ivan, 818
PCR. *See* Polymerase chain reaction
Pea
  flower color in, 324
  Mendel's experiments using, 237-242
  tendrils of, 513, 514
Peacock, courtship display and, 332, 828, 834, 835
Peacock moth, startle coloration in, 818
Pedigree analysis, 281, 282
Pelecypoda, 391
Pelomyxa palustris, 369
Penfield, Wilder, 719
Penguins, reproduction and, 767
Penicillin, 409
  discovery of, 11
Penicillium, 11, 431, 434
Penis, 768
  erection of, 775
PEP. *See* Phosphoenolpyruvate
Peppered moths, evolution of, 316, 317, 333-334
Pepsin, digestion and, 66, 629, 630
Pepsinogen, 630
Peptidases, digestion and, 629
Peptide bonds, 47, 48
Peptides, 47, 50, 684, 685
  digestion and, 630
  as hormones, 680, 689
Peptidoglycans, in cell walls, 402
Perception
  common, principle of, 9
  see also Sensory perception
Pericycle, 506
Periderm, of plants, 500-501
Periosteum, 756
Peripheral nervous system, 708-711
Peristalsis, 628, 629
Permafrost, 933, 934
Permian period, 364, 373, 375
PET. *See* Positron emission tomography scan
Petals, 538, 539
Petiole, of leaf, 511
Petrified forests, 12, 13
PGA. *See* Phosphoglyceric acid
PGAL. *See* Phosphoglyceraldehyde
Phaeophyta. *See* Algae, brown
Phagocytic cells, 652, 653
Phagocytosis, 91, 93, 116-118
  antibodies and, 664
Pharynx
  of earthworms, 626
  of flat worms, 468
  human, 610, 611, 628, 630
Phenotypes, 240
  genotypes versus, natural selection and, 332-333
Phenylalanine, 47, 49
Phenylketonuria (PKU), 194, 295
Pheromones, 738, 829-830
  honeybees and, 830, 838
  human, 842
  reproduction and, 830
  spawning and, 765
Phloem, 503-504

aphids and, 528-529
  primary, of stems, 508, 509
  secondary, of stems, 509, 510
  sugar transport in, 529-531
Phosphate group, 38
Phosphatidylcholine, 39
Phosphoenolpyruvate (PEP), in photosynthesis, 135
Phosphoenolpyruvic acid, in glycolysis, 144
Phosphoglyceraldehyde (PGAL)
  in glycolysis, 143, 144
  synthesis of, 132, 134
Phosphoglyceric acid (PGA)
  in glycolysis, 144
  in photosynthesis, 134
Phospholipid bilayer, 76, 87, 107-109, 112
Phospholipids, 39, 44, 45
Phosphorus, 23, 24
  bonding and, 26
  in nutrition, human requirements for, 622
  in plant nutrition, 521
Phosphorus cycle, 902, 904
Photic zone, 937
Photons, 130
Photophosphorylation, 131
Photopigments, 733
Photoreceptors, 728, 733-737
Photorespiration, 135
Photosynthesis, 4-5, 369
  chloroplasts and, 94, 128-130
  endergonic reaction of, 60-61
  energy flow and, 891-893
  exergonic reaction of, 61
  see also Plants, photosynthesis and; Protists
Photosystem I, 130-132
Photosystem II, 130-133
Phototaxic bacteria, 404
Phototropism, 557, 561
  hormonal mechanisms in, 558-559
pH scale, 30, 31, 66
Phycocyanins, 130
Phyletic speciation, 353-354, 356
Phylum(a), 387
Physarum, 412, 414
Phytochrome, 566
Phytophthora, 431
Phytoplankton, 412, 935
Pili, 403
Pilobolus, 435
Pineal gland, 685, 692
Pine trees, life cycle of, 450
Pinocytosis, 91, 116
Pioneers, 881, 883
Pith, of stems, 508, 509
Pit organs, 729
Pituitary gland, 774
  anterior, 683-686, 688, 691
  control by hypothalamic hormones, 683-687
  posterior, 683-685, 687
PKU. *See* Phenylketonuria
Placenta, 763, 779-780, 801-802
  protection provided by, 804
Placental mammals, 490
Plague, 409
Planarians, 468, 470
Plankton, 935
Plant hormones
  actions of, 559-561
  discovery of, 557-559
Plant life cycles, 561-568
  control of flowering and, daylength and, 656-666
  evolutionary trends in, 439-440
  ferns, 445, 448, 536
  mature plant development and, 563-564
    balance between root and shoot systems and,
      564

root branching and, 564
stem branching and, 563
plant hormones and
actions of, 559-561
phototropism and, 557-559
reproduction and, 535
seed and fruit development and, 556-557
seed dormancy and germination and, 561
seedling growth and
roots and, 562
shoot and, 561-562
senescence and dormancy and, 567
vascular tissue differentiation and, 564
Plant reproduction, 535-553
asexual, 535
flowers and, 537-546
coevolution with pollinators and, 538-543
control of flowering and, 565-566
evolution of, 537-538
gametophyte development and, 543-545
pollination and fertilization of, 545-546
structure of, 538
fruits and, 451
development of, 547, 566-567
dispersal of, 548-549
ripening of, 567
plant life cycle and, 536-537
seedlings, germination and growth of, 551-552
seeds and, 449, 497
development of, 546-547
dispersal mechanisms for, 548-550
dormancy of, 547, 549, 551
germination of, 551, 561
sexual, 535
Plants, 6-8, 388-394, 439-454
cells of, 84, 85
see also Eukaryotic cells
development and growth of, 498, 500
evolution of, 439-440
flowering, 7, 391, 451, 452, 454, 497
dicots, 498, 499
life cycle of, 450
monocots, 498, 499
origin of, 373
see also Plant reproduction, flowers and
fruits and. See Plant reproduction, fruits and
gymnosperms, 449, 497
see also Conifers
herbicide-resistant, 270, 275
leaves of, 511-512
adaptations of, 513-515
photosynthesis and, 129, 131, 132
structure of, 511-512
nutrition of
animal nutrition compared with, 519-520
minerals in, 520-522
sugar transport and, 529-531
transport of water and minerals and, 524-528
water in, 523
origin of
on land, 372-373, 374
in sea, 371
photosynthesis and, 126-137
chloroplasts and, 128-130
C_3 pathway and, 132, 134-136
C_4 pathway and, 134-136
leaves and, 129, 135, 512
light-dependent reactions in, 129-132
light-independent reactions in, 129, 132, 134
relations between light-dependent and light-independent reactions in, 128, 134
stems and, 508
predatory, 514, 515
relationship with water, 530
roots of, 498, 504-506

acquisition of minerals by, 520-522
acquisition of water by, 523
adaptations of, 512
balance between shoots and, 564
branching of, 564
growth of, 562
primary growth in, 504
seeds and. See Plant reproduction, seeds and
senescence and, 567
shoots of, 498
balance between roots and, 564
growth of, 561-562
stems of, 506-511
adaptations of, 512-513
branching of, 509, 563
primary, internal organization of, 508-509
primary growth of, 506-507
secondary growth in, 509-510
structure, overview of, 498, 499
tissues and cell types of, 500-504
dermal tissue system, 500-501
ground tissue system, 500-502
vascular tissue system, 500-503
vascular, 439, 444, 497
seedless, 445, 447
seed plants, 447, 449, 451
see also specific plants
Plaque
atherosclerotic, 598, 599
dental, 403
Plasma, 591-592
Plasma cells, humoral immunity and, 659, 662, 663, 665
Plasma membrane, 4, 76, 79, 80, 87, 90, 91, 106-109
Plasmids, 262, 265, 406
Plasmodesmata, 120, 121, 520
Plasmodium, 412, 414, 418-419
Plasmodial slime molds, 414
Plastids, 85, 95, 96
Platelets, 593-594
Plato, 304, 927
Platyhelminthes. See Flatworms
Platypus, 490, 741
Play, purpose of, 822-823
Pleated sheet, in proteins, 49, 51
Pleiotropy, 251
Pleistocene epoch, 364
Pleural menbranes, 612
Pliocene epoch, 364
Pneumonia, Pneumocystic carinii, 668
Poisons, 46
Polar body, 772
Polar covalent bonds, 26
Polar microtubules, 168, 169
Polar nuclei, 545
Pollen, 447, 497
development of, 543-545
Pollen grain, 537, 544, 545
Pollen sacs, 543
Pollination, 14-15
fertilization versus, 545-546
by insects, 538
by wind, 537, 538, 540
Pollution, 316, 317
Polychaeta, 391, 472
Polychlorinated biphenyls (PCBs), 900
Polygenic inheritance, 251, 290
Polymerase chain reaction (PCR), 266, 268-270, 389
prenatal diagnosis and, 297
Polymers, 38
Polymorphism, balanced, 335
Polyp, 466, 467
Polyploid cells, speciation and, 351

Polyribosomes, 87
Polysaccharides, 39-42, 620
Pongids, 377
Pons, 713, 714
Population, 304, 849
Population bottleneck, 329, 330
Population genetics, 323-326
equilibrium population and, 326, 328-329
evolution and, 323-326
gene function in individual organisms and, 323
genes in populations and, 324
Population growth, 309, 850-868
distribution of populations and, 859-860
aggregate, 859
random, 859, 860
uniform, 859, 860
exponential, 851-853, 855
boom and bust cycles and, 853, 855
human population and, 861-867
age structure and, 862-863
in United States, 863, 866-867
world population and carrying capacity and, 864-865
limits to, 855-859
carrying capacity as, 855-856
competition and population size and, 858
density-dependent, 856, 857
density-independent, 856
parasitism and host populations and, 858
predation and prey populations and, 856-858
rate of, 850-851
survivorship in populations and, 860
Populations, 3, 4
isolation of
behavioral, 351, 352
ecological, 349, 350-351
geographic, 348-351
temporal, 351, 352
Porifera. See Sponges
Porpoises, orientation by sound, 740-741
Positive feedback, homeostasis and, 574, 575
Positron emission tomography (PET) scan, of brain, 32-33, 720-721, 722
Posterior end, cephalization and, 462
Postmating isolating mechanisms, 351, 353
Postsynaptic neurons, 704, 705
Postsynaptic potentials, 702, 704, 705
Potassium
in human nutrition, requirements for, 621, 622
in plant nutrition, 521
concentration in guard cells and, 526-528
Potassium channels, 700-702
Potassium-40 dating, 368
Potato famine, 434
Potential energy, 22
Power grip, primate evolution and, 377
Prairie dogs, 849, 930
Praying mantis, 870, 875
molting by, 474
Prebiotic evolution. See Evolution, prebiotic
Precambrian era, 364, 371
Precapillary sphincters, 595, 597
Precision grip, primate evolution and, 377
Predation
community interactions and, 874-878
coevolution of predators and prey and, 874-878
natural selection and, 336
predator and prey populations and, 854, 856-858
Pregnancy, 776-778
Premating isolating mechanisms, 351-353
Pressure flow theory, of sugar movement in phloem, 529-531
Presynaptic neurons, 704, 705
Prickly pear cactuses, Darwin's observations of, 308-309

Primary cell wall, 106
Primary consumers, 893, 895
Primary endosperm cell, 545
Primary growth, of plants, 498, 500
Primary oocytes, 771
Primary phloem, of stems, 508, 509
Primary productivity, 891-894
Primary root, 504, 505
Primary spermatocytes, 769
Primary structure of proteins, 49, 50
Primary succession, on bare rock, 883, 885
Primary xylem, of stems, 508, 509
Primates
    evolution of, 376-377
        brain and, 377
        grasping hands and, 377
        vision of, 377
    see also specific primates
Primer pheromones, 830
Primitive streak, 800
"Primordial soup," 366, 619
Prions, 400-402
Proctoctista, 394
Producers, 892, 895
Products, of chemical reaction, 59
Progesterone, 685, 691, 758, 773
    menstrual cycle and, 774
Prokaryotic cells, 8, 81, 82
    cell cycle of, 161
    differences from eukaryotic cells, 369
    see also Bacteria; Monerans
Prolactin, 684-686, 779
Promoter, 208
Promoter region, 199
Pronghorn antelope, 930
Prophase, 165, 166
    early, 165, 166
    late, 165, 166
Prophase I, 223, 224, 225
Prophase II, 224, 227
Proprioceptors, 729, 730
Prostaglandins, 680, 681, 692, 779
Prostate gland, 768, 771
Proteases, digestion and, 630, 631
Proteins, 37-39, 44, 619, 620
    amino acids, 46-47
    in cell membranes, 109
        functions of, 109
    DNA and, 195-196
    as energy source, 621
    functions of, 51-52, 109
    genes and, 193-195
    metabolism of, 152-153
    repressor, 208
    structure of, 47-51
    synthesis of, 47, 49
        prebiotic evolution and, 367
        transcription and, 199-201
        translation and, 201-204
    transport, 109, 112, 115
Protists, 6-8, 390, 391, 394, 411-420
    animal-like, 416-420
    classification of, 412
    funguslike, 414-416
    habituation in, 818, 819
    infections caused by, sexually transmitted, 784
    plantlike, 412-414
    see also Amoeba; Diatoms; Euglena; Paramecium
Protons, 22, 23
Protozoa, 412, 416-420
Proximal tubule, 642
Prusiner, Stanley, 401
Pseudocoel, 463, 464
Pseudomanas syringae, 275
Pseudoplasmodium, of cellular slime molds, 415

Pseudopods, 116
    of cellular slime molds, 415
Pteridophyta. See Ferns
Pulmonary circulation, 588, 589
Pumps (transport proteins), 115
Punctuated equilibrium, speciation and, 355-357
Punnett square, 240, 241, 242
Pupa, 476
Pupil, 734
Purines, 178
Pygmy hippopotamus, play and, 822
Pyloric sphincter, 629
Pyrimidines, 178
Pyrrophyta, 391, 412
Pyruvic acid
    in cellular respiration, 146-149
    in fermentation, 145
    in glycolysis, 143, 144

Quaternary consumers, 893
Quaternary period, 364
Quaternary structure of proteins, 50
Queen substance, 830, 837

Rabbits
    body fat color in, 253
    Himalayan, hair color in, 253
    territoriality in
    vision and, 737
Rabies, 401
Radial symmetry
Radioactive dating, 368
Radioactive isotopes, uses of, 32-33
Radiolarians, 417
Radula, of gastropods, 477
Raff, Martin, 75
Rainfall, 917-918
    terrestrial biomes and, 933-934
Rain forests, temperate, 931, 932
Rain shadow, 918, 921
Random distribution of populations, 859, 860
Raven, Peter, 14
Reactants, 59-63
Reaction center, 130, 131
Reactions. See Chemical reactions
Receptor-mediated endocytosis, 116, 117
Receptor potentials, 728
Receptors, sensory, 728-729
Recessive alleles, 283-286
Recognition proteins, 109
Recombinant DNA technology, 261
    applications of. See Biotechnology
    copying genes and, 265, 266, 268-270
    DNA library and, 263
        building, 263, 265
        restriction enzymes and, 263-265
        searching, 265-266
    gene identification and, 265
    gene location and, 271-272
    gene sequencing and, 272
    goals of, 261
    naturally occuring DNA recombination, 262-263
        between bacterial species, 262
        between eukaryotic species, 262
        evolution and, 262
        laboratory recombination compared with, 263
Rectum, 633, 634, 771
Red blood cells, 584
Red bread mold, one-gene, one-protein
    hypotheses, 194-195
Redi, Francesco, 9, 10, 364
Red tide, 413
Redwinged blackbird, fixed action pattern in, 815-816
Reflexes, 707, 708, 813, 814

Regeneration, 764
Regulatory gene, 208
Releaser pheromones, 830
Releasers
    exaggerated, humans and, 841-842
    fixed action patterns and, 815-817
Releasing hormones, hypothalamic, 684, 685-687
Renal artery, 641, 642
Renal cortex, 641, 642
Renal medulla, 641, 642
Renal pelvis, 641, 642
Renal tubule, 641-642
Renal vein, 641, 642
Renin, 646, 685, 692
Replacement-level fertility, human population
    growth and, 862, 863
Repression, of lactose operon, 208, 209
Repressor protein, 208
Reproduction, 5-6, 762-807
    asexual, 219, 763-765
        cell division and, 170-172
        in cnidarians, 467
        of flatworms, 468, 470, 764
        fungal, 426, 430
    cellular. See Cellular reproduction
    differential, 332
    DNA and, 5-6
    human. See Human reproduction
    in plants. See Plant reproduction
    sexual, 159, 763
        external fertilization and, 765-766
        of flatworms, 468, 470
        fungal, 426
        internal fertilization and, 766
        mammalian. See Human reproduction
        meiosis and, 222, 229-230
        in plants, 535
    social behavior and, 834-836, 838
    taxonomic classification and, 388-389
Reproductive isolation, 351-355
Reproductive system
    animal, 581
    human. See Human reproduction
Reptiles, 391, 483, 487-488
    origin of, 375-376
    respiration in, 609, 610
    see also specific reptiles
Reservoir, of nutrients, 900
Resource partitioning, 873
Respiratory center, 613
Respiratory system, 581, 604-616
    evolution of, 605-610
    gas exchange and, 605-610
        book lungs and tracheae and, 607
        gills and, 606-609
        lungs and, 607
    human, 610-616
        blood transport of gases and, 611-612
        conducting portion of, 610-611
        control of respiration and, 613
        gas exchange portion of, 611, 612
        mechanics of breathing and, 612-613
        smoking and, 614-615
    respiration without specialized structures, 606
Resting potential, origin of, 699-701
Restriction enzymes, 263-265
Restriction fragment length polymorphisms
    (RLFPs), 271-272
Reticular formation, 714
Retina, 734, 735
    neural processing in, 736
Retroviruses, 404, 405, 668
Reverse transcriptase, 405, 668
Rhagoletis pomonella, ecological isolation of, 350-351, 352

Rh factor, 593
Rhinoceros, 926
Rhizomes, 512, 535
Rhizopus, 430, 431
Rhodophyta. *See* Algae, red
Rhythm method, 781, 782-783
Ribonucleic acid (RNA), 39, 41, 52, 82, 87
    differences from DNA, 194-195
    messenger, 196, 199, 200
    prebiotic evolution and, 366-367
    ribosomal, 196, 199, 200
    transcription and, 196, 199-201
    transfer, 196, 199, 201, 202
Ribose, 40, 41
Ribosomes, 82, 87, 200.85, 201
    in endoplasmic reticulum, 88
Ribozymes, 65
    prebiotic evolution and, 366-368
Ribulose biphosphate (RuBP), in photosynthesis,
    134-136
RLFPs. *See* Restriction fragment length
    polymorphisms
RNA. *See* Ribonucleic acid
RNA polymerase, 199-201
Roberts, Keith, 75
Rods, 634, 736-737
Root hairs, 500, 504, 505, 521
Roots. *See* Plants, roots of
Roundworms, 391, 462, 470-471
    digestion in, 464, 626, 627
    gas exchange in, 606
    hyphae and, 435
Ru-486, 781
RuBP. *See* Ribulose biphosphate
Ruby-throated hummingbirds. *See* Hummingbirds
Ruminants, digestion of cellulose by, 626-627
Runners, 512, 535

Saccharomyces, 431
Saguaro cactus, water storage and, 512, 927
Sahel, 926, 927
Salamanders, 374
    skin of, 484, 485
    tiger, 486
Salmonella, 109, 409
Salt, 23, 25
    bonds in, 26
    dissolving, 28-29
    structure of, 2
Sanger, Frederick, 294
Saprobes, 426
Sarcodines, 412, 416-418
Sarcomastigophora, 391, 412, 416
Sarcomeres, 749
Sarcoplasmic reticulum, 749, 752
Saturated fatty acids, 44, 45
Savanna, 923-926
Scala Naturae, 304, 317
Scallops, 477, 478, 479
Scanning electron microscopes, 78, 79
Scarlet cup fungus, 431
Schally, Andrew, 686
Schistosoma, 470
Schizophrenia, 721
Schleiden, Mattias, 76
Schwann, Theodor, 75
SCID. *See* Severe combined immunodeficiency
    disease
Science, as human endeavor, 11
Scientific method, 9, 11, 240
Scientific names, 387
    see also Taxonomy
Scientific principles
    of common perception, 9
    of natural causality, 8-9

of uniformity in space and time, 9
Sclera, 735
Sclerenchyma tissue, of plants, 501, 502, 508
Scorpions, 476
Scramble competition, population growth and, 858
Scrapie, 401
Scrotum, 767-768
"S-curve," population growth and, 855
Sea. *See* Ecosystem, marine; Oceans
Sea anemones, 467
    reproduction in, 467
Sea cucumbers, 480
    endoskeleton of, 481
Seahorse, 485
Sea slugs, 478
"Sea squirt," 389, 481-482
Sea stars, 463, 480, 878
    endoskeleton of, 481
    reproduction in, 764, 765
Sea urchins, 463, 480
    endoskeleton of, 481
Sebaceous glands, 580, 581
Secondary cell wall, 206
Secondary consumers, 893, 895
Secondary growth, of plants, 500
Secondary oocyte, 772, 776
Secondary phloem, of stems, 509, 510
Secondary spermatocytes, 769
Secondary structure of proteins, 49, 50, 51
Secondary succession, on abandoned field, 883,
    885
Secondary xylem, of stems, 509, 510
Second messenger system, 680-681
Secretin, 685, 693
    digestion in, 634
Sedimentary cycle, 900, 902
Seed coat, 546
    disruption of, seed dormancy and, 550-551
Seeds. *See* Plant reproduction, seeds and
Segmentation, evolutionary trends in, 460, 463-
    464
Segmentation movements, 633
Segmented worms, 391, 463, 471-472
    digestive tract and, 464
    reproduction in, 764
Segregation, Law of, 239-240
Selenium, in nutrition, human requirements for,
    621, 622
Self-fertilization, 237
Selfish gene concept, 843, 844
Semen, 770
Semicircular canals, 732, 733
Semiconservative replication model, DNA and,
    186
Semilunar valves, 589
Seminal vesicles, 768, 770
Seminiferous tubules, 768-769
Senescence, 567
Sensitive period, 818
Sensory perception, 726-743
    auditory system and, 730-732
    chemical senses and, 737-739
    mechanoreception and, 729-730
    natural selection shaping of, 740-742
    pain and, 739-740
    receptor mechanisms and, 728-729
    thermoreception and, 729
    vision and, 732-737
Sepals, 538, 539
Septa, fungal, 425
Sequoia tree, taxonomy and, 388
Severe combined immunodeficiency disease
    (SCID), 667
Sex, coevolution with flowers and pollinators and,
    538-543

Sex cells. *See* Gametes
Sex chromosomes, 246-249
    abnormal numbers of, 291-292
    sex linkage and, 247-249
Sex determination, chromosomal basis of, 246-249
Sex-influenced traits, 287
Sex linkage, 247-249, 287
Sexually transmitted diseases, 409, 776, 784
Sexual maturation, 796-797
Sexual reproduction, 219
    evolution of, 220-222
Sexual selection, 331, 332, 337
Sharks, 483, 484
    brain of, 713
    evolution of, 12, 13, 357
Sherman, Paul, 843-844
Short-day plant, 565
Shrimp, 476
Sialia, 387, 388
Sibley, Charles, 319
Sickle-cell anemia, 52, 270, 272, 284-286
    prenatal diagnosis of, 273, 295, 297-298
    stabilizing selection and, 335
Sieve plates, 503
Sieve-tube elements, 503
Silicon, in nutrition, human requirements for, 622
Simple diffusion, 110, 111-112
Simpson, O. J., 272
Singer, S. J., 106
Sinoatrial node, 589
Skeletal muscle, 579, 747
    anatomy of, 748-750
Skeletal system, 581
Skeleton
    appendicular, 753, 754
    axial, 753, 754
    endoskeleton, 753
    exoskeleton, 374, 752, 753
    hydrostatic, 752, 753
    muscle-skeletal interactions and, 757, 759
    osteoporosis and, 758
    vertebrate, 753
Skin, 580-581
    as entry barrier, 651, 652
Skin cancer, 920
Skin color, inheritance of, 290
Skinner, B. F., 819
"Skinner Box," 819
Sliding filament model of muscle contraction, 750-
    751
Slime layers, bacterial, 403
Slime molds, 412, 414-416
Small intestine, 629-631, 633
    absorption and, 631, 633
    intestinal wall and, 631
    liver and gallbladder and, 630-631
    pancreas and, 631
Smallpox, vaccination against, 267, 666
Smell, sense of. *See* Olfaction
Smith, William, 305
Smoking
    fetal effects of, 804
    passive, 615
    respiratory disease and, 614-615
    ulcers and, 637
Smooth muscle, 579, 748, 752
Snails, 338-339, 477, 478
Snakes, 877
    mountain king, 487
    pit organs of, 729
Snowberry fly, coevolution with spiders, 876, 879
Snowshoe hare, lynx as predator of, 854
Social behavior, 826-845
    communication, 828-830
        by chemicals, 829-830

Social behavior *continued*
    in honeybee, 837-839
        by sound, 828-829
        by touch, 830, 831
        visual, 828
    competition for resources, 830-834
        aggression and, 830-833
        dominance hierarchies and, 831-832
        territoriality and, 832-834
    cooperation and, 836-840
        in bullhead catfish, 839
        insect societies and, 837-839
        in naked mole rats, 839-840
    human ethology and. *See* Ethology, human
    reproduction and, 834-836
Social stress, overcrowding and, 858-859
Sodium, in nutrition, human requirements for, 621, 622
Sodium bicarbonate, digestion and, 629, 631, 634
Sodium channels, 700-703
Sodium chloride. *See* Salt
Sodium-potassium pump, 700, 702
Soil, 521
Solar radiation, climate and, 916
Solutions
    hypertonic, 114, 115
    hypotonic, 114, 115
    isotonic, 114, 115
Solvent, water as, 28-29
Somatic nervous system, 709
Somatostatin. *See* Growth hormone
Sound perception. *See* Auditory system
Spadefoot toads, 342
Sparrows, white-crowned, sexual selection among, 331
Spawning, 765, 766
Species, 3, 4, 387
    definition of, 304
    endangered
        see also Extinction
    number of, 393
    reproductive isolation of, 351-355
    speciation and, 348-358
        evolutionary progress and, 355-358
        genetics of, 353-355
        maintaining reproductive isolation between species and, 351-353
        morphological change and, 357
        phyletic versus divergent, 353-354, 356
        rates of, 355
        species selection versus individual selection and, 357-358
Specific heat, of water, 31
Sperm, 765-771, 775, 776
Spermatids, 769
Spermatogenesis, 768-770
Spermatogonia, 769, 770
Spermatophore, 766
Spermicide, 782, 783
Sperry, Roger, 718-719
Spicules, of sponges, 465
Spiders, 476
    coevolution with snowberry fly, 876, 879
    tarantula, 476
Spinal cord, 710, 712-713
Spindle apparatus, during prophase, 166-169
Spiracles, 607
Spirilla, 402, 403
Spirogyra, 443
Spleen, 600
Sponges, 391, 462, 464-466
    cellular organization of, 461
    digestion in, 625
    digestive tract of, 464
    gas exchange in, 606

symmetry in, 461
Spongy bone, 754, 755
Spongy cells, of leaf, 512
Spontaneous generation, 9, 10, 364, 365
Sporangia, 429, 431
Spores, 228, 405, 406
    fungal
        dispersal of, 435
        reproduction and, 426, 430
    of plants, 536, 537
Sporophyte, 230, 536, 537
Sporozoans, 412, 416
Squash, 539, 552
Squid, 479, 700
Squirrels, 350, 557, 843-844
Stabilizing selection, 333-335
Stamens, 538, 539, 540
Staphylococcus, population growth and, 851
Starch, 39, 41-43
Starfish. *See* Sea star
Star-nosed mole, 726, 741
Startle coloration, 876, 878, 879
Starvation, world population and, 864-865
Statoliths, 563
Stearic acid, 45
Stegosaurus, 306, 307, 376
Stems. *See* Plants, stems of
Stentor, habituation in, 818, 819
Sterilization, 780-781
Steroid hormones, 684-685
Steroids, 39, 44, 46
Stewart, Margaret, 486
Stickleback fish
    reproduction in, social behavior and, 835, 838
    territoriality in, 833
Stigma, of complete flower, 538, 539, 540
Stimulus, 5
Stimulus intensity, signaling, 706
Stirrup, 731
Stomach, human, digestion in, 628-630
Stomata, 129
    of leaf, 525-527, 530
    of stems, 507
Strawberries, runners of, 512
Streptococcus, 409-410
Streptococcus, transformation of, 179-180
Streptococcus pneumoniae, 409
Streptomycin, 409
Striated muscle, 579, 747
    anatomy of, 748-750
Stroke, 598
Stroma, 94, 129
Structural genes, 209
Struhsaker, Thomas, 829
Style, of complete flower, 538
Subatomic particles, 3, 4
Subclimax, 886
Substrates, 65-66
Subunits, 38
Succession
    primary, on bare rock, 883, 885
    secondary, on abandoned field, 883, 885
Succulent plants, leaves of, 513, 514
Sucking, in human infants, 840, 841
Sucrase, digestion and, 631
Sucrose, 39, 41
Sugars, 40, 41
    in plant nutrition, transport of, 529-531
Sulfur, 23, 47
    bonding and, 26
    in plant nutrition, 521
Sulfur cycle, acid rain and, 905-906
Sundews, 514, 515
Sunflower, taxonomy and, 388
Supernormal stimuli, 816

Suppressor T cells, 658, 664
Swallowing, 630
Swallowtail butterfly larva, startle coloration in, 878
Symbiosis
    bacterial, 408
    cleaning, 878
    community interactions and, 878-880
    fungal, 427-429
    natural selection and, 336-337
    in plant nutrition
        mineral accesibility and mycorrhizae and, 522, 523
        nitrogen acquisition and, 522-523
Symmetry, evolutionary trends in, 461, 462
Sympathetic nervous system, 709, 711
Sympatric speciation, 348-350
Synapse, 579, 704-705
Synaptic cleft, 704, 705
Synaptic potentials, 705
    integration of, 705
Synaptic terminals, 579, 580, 699
Synaptic vesicles, 704
Syphilis, 409, 784
Systemic circulation, 588
Systole, 588, 589

Taiga, 931-933
Tapeworms, 469, 470
Taproot system, 504
Tarantula, 476
Target cells, 680
Tarsiers, evolution of, 376, 377
Taste, 739
Tatum, Edward, 195
Taxes, 404, 812-814
Taxonomy, 386-395, 412
    inexact nature of, 393
    kingdoms and, 389-392
    modern criteria for classification and, 388-389
    number of species and, 393
    origins of, 387-388
    taxonomic categories and, 387
Tay-Sachs disease, 270, 284, 285, 295, 296-297
T cells, 657-662
    cytoxic, 655, 664, 665, 673
    helper, 654-655, 658, 663, 664, 667-669
    killer, 654, 655, 673
    suppressor, 658, 664
Tectorial membrane, 731, 732
Teeth
    dental plaque and, 403
    digestion and, 627, 629
Telophase, 165, 167, 169
Telophase I, 224, 227
Telophase II, 224, 227
Temperate deciduous forests, 929, 931
Temperate rain forests, 931, 932
Temperature
    terrestrial biomes and, 933-934
    water as moderator of, 30-31
Temperature stratification, in freshwater lakes, 936
Template strand, of DNA, 199
Temple, Stanley, 550
Temporal isolation, speciation and, 351, 352
Temporal lobe, 716, 719, 721, 722
Tendons, 576, 757
Tendrils, 512, 513
Tentacles, of polyp, 446, 447
Terrestrial biomes. *See* Ecosystems, terrestrial biomes
Territoriality, 832-834
Tertiary consumers, 893, 895
Tertiary period, 364
Tertiary structure of proteins, 49, 50

Test cross, 240, 243
Testes, 767-769
    mammalian, 685
Testosterone, 46, 87, 287, 681, 684, 685, 691, 697,
    769, 770, 785
Tetanus, 409
Tetany, 752
Tetracycline, 409
Tetrahymena, asexual reproduction in, 171
Thalamus, 714, 715
Thalidomide, 804
Theory, 11
Thermoacidophiles, 411
Thermodynamics, laws of, 58-59, 61, 62, 64, 94
Thermoreceptors, 728, 729
Thiamann, Kenneth, 559
Thick filaments, 749-751
Thin filaments, 749-751
Thomas, Donald, 550
Thomas, Lewis, 1, 15, 91, 92, 105, 410
Thorns, 512, 513
Threonine, 67
Threshold, 699
Thrombin, 594
Thucydides, 657, 665
Thylakoids, 92, 94, 129-131, 133
Thymine, in DNA, 52, 178, 184, 196, 199
Thymosin, 682, 685
Thymus gland, 600, 685, 692
Thyroid gland, 684, 687-689
Thyroid-stimulating hormone (TSH), 684-689
Thyroxine, 681, 684, 687-689, 692
    evolution of, 693
Ticks, 476
Tiger, test tube, 777
Tiger salamanders, 486
Tight junctions, 120, 121
Tin, in nutrition, human requirements for, 622
Tinbergen, Niko, 811, 816, 821, 833, 843
Tissue plasminogen activator (TPA), 599
Tissues, 3, 4, 575-580
    connective, 576, 578
    epithelial, 575-578
    nerve, 579-580
Toads
    African clawed, differentiation in, 789, 790, 803
    corroboree, 486
    golden, 486
    spadefoot, 342
    trial-and-error learning in, 820
    Yosemite, 486
Todd, John, 839
Tonsils, 600
Tools, 379, 381
Tortoises, 487
    Darwin's observations of, 308-309
    reproduction and, 767
Touch, communication by, 830, 831
TPA. *See* Tissue plasminogen activator
Tracheae
    in arthropods, 474, 607, 609
    human, 610, 611
Tracheids, 502-503
Tracheophytes. *See* Plants, vascular
Transcription, 199-201, 405
    elongation of RNA chain in, 199
    initiation of, 199
    regulation of, 207-208
        in eukaryotes, 209-211
    termination of, 199
Transformation, 179-181
Translation, 196, 201-204
Transmission electron microscopes, 78, 79
Transpiration, 525
    control of, 525

Transport proteins, 109, 112, 115
Transverse tubules, 749, 751
Tree bark mantis, camouflage of, 875
Tree frogs, reproduction in, social behavior and,
    834
Trees
    American chestnut, blight and, 275, 276, 427,
        429
    baobab, water storage and, 512, 513, 514
    Calvaria, coevolution with dodo bird, 550
    coniferous. *See* Conifers
    elm, Dutch elm disease and, 339, 427, 429
    growth of, annual rings and, 510
    maple, seed dispersal and, 548
    pine, life cycle of, 450
    sequoia, taxonomy and, 388
Tree shrews, evolution of, 376
Trial-and-error learning, 819-820
Triassic period, 341, 364
Trichinella, 471
Trichinosis, 471
Trichocysts, 420
Trichomonas infection, 784
Tricuspid valve, 588-589
Triglyceride, 39, 42, 44
Trilobites, 307, 371, 372, 373
Triple covalent bond, 26
Trisomy 21, 293-294
Trisomy X, 292
Trophic levels, 893
    energy flow through, 894-895, 898
Tropical forests, 14, 393, 921-925
Tropical rain forests, 514, 530
True-breeding organisms, 237, 239
Truffle, 427, 431, 435
Trypanosoma, 412, 416, 879
Trypsin, digestion and, 629, 631
Tschermak, Erich, 242
TSH. *See* Thyroid stimulating hormone
T tubules, 749, 751
Tubal ligation, 780
Tube cell, 543, 545
Tube feet, of echinoderms, 480
Tuberculosis, 409
Tubular resorption, 643-645
Tubular secretion, 643-645
Tumor infiltrating lymphocytes, in cancer
        treatment, 674
Tumor necrosis factor, 674
Tumor-suppressor genes, 672
Tundra, 932-933
Tunicates, 481-482
Turgor pressure, 118, 119
Turner's syndrome, 291-292
Twin studies, human ethology and, 843
Tympanic membrane, 731
Tyndall, John, 364
Tyrannosaurus rex, 307
Tyrosine, 680

Ulcers, 635
Ulva, 440, 443, 444
Unicellular organisms, 7, 8, 412
    see also Bacteria; Monera; Protists; *specific
        organisms*
Uniform distribution of populations, 859, 860
Uniformitarianism, 306
Uniformity in space and time, principle of, 9
United States, population growth in, 863, 866-867
Unsaturated fatty acids, 44, 45
Upwelling, 937
Uracil, 52
    in RNA, 196, 199
Urea, 621
    excretion of, 640

Ureter, 641, 642
Urethra, 641
    in male, 768, 770
Urey, Harold, 365
Urinary system, 581, 638-648
    functions of, 639
    human, 641-647
        kidney failure and, 647
        kidneys as organs of homeostasis and, 643-646
        kidney structure and function and, 641-643
        urine concentration and, 641-642
    simple, 640-641
Urine, 640
    concentration of, 641-643
    formation in nephron, 641-645
Uterus, 771, 773
Utricle, 733

Vaccines
    biotechnology and, 267, 660
    immune responses to, 666
Vacuoles, 85, 95
    central, 85
    food, 90, 116, 118, 625
Vagina, 771, 773, 775
Value systems, 9
Vampire bats, play and, 822
Vanadium, in nutrition, human requirements for,
    622
Vaporization, heat of, of water, 31
Variable region, 658
Variables, 9
Vascular bundles, of leaf, 512
Vascular cambium, of stem, 508, 509
Vascular cylinder, of roots, 506
Vascular tissues, of plants, 500-503
    differentiation of, 564
    stems, 508-509
Vas deferens, 768, 769, 770, 781
Vasectomy, 780, 781
Veins, 595, 596
    of leaf, 512
Vent communities, 941
Ventral roots, 712
Ventral surface, 462
Ventricles, of heart, 588, 589
Venules, 595, 596
Venus flytraps, 514, 515
    functioning, 568
Vertebral column, 482
    see also Vertebrates
Vertebrates, 391, 464, 482-484
    bony fishes, 375, 381, 482, 484, 485
    brain of, 713-716
    cartilage fishes, 391, 482-484
    circulatory system of. *See* Circulatory system,
        vertebrate
    jawless fishes, 483
    nervous system of, 708, 710
    skeleton of, 753
    vision in. *See* Vision
    see also Amphibians; Birds; Fish; Mammals;
        Reptiles; *specific vertebrates*
Vesicles, 82, 85, 89, 115, 116
Vessel elements, of xylem, 503
Vestibular apparatus, 732, 733
Vestigial structures, evolution and, 313, 315
Viceroy, mimicry by, 877
Villi, of small intestine, 631, 633
Vipers, pit organs of, 729
Virchow, Rudolf, 76, 159
Viroids, 400
Viruses, 399-402
    AIDS and. *See* Acquired immunodeficiency
        syndrome

Viruses *continued*
  flu, 672-673
  origins of, 402
  prions and, 400-402
  replication of, 404-405
  structure and reproduction of, 399-401
  viral infections and, 400-402
    fever and, 652, 656-657
    sexually transmitted, 784
    see also Acquired immunodeficiency syndrome
  viroids and, 400
Vision, 732-737
  binocular vision, 737
  camera eye, 734, 736
  compound eye of arthropods and, 474, 733, 734
  focusing of eye, 735
  neural processing in retina, 736
  photoreceptors, 736
  primate evolution and, 377
  structure of eye and, 734
  types of eyes and, 733-734
Visual communication, 828
Visual displays, reproduction and, 828, 835
Vitamins, in nutrition, 67, 619
  human requirements for, 621, 623, 624
Vitreous humor, 734
Vocal cords, 610
Vocalizations, reproduction and, 834
Volcanic eruptions
  habitat destruction and, 339
  succession and, 881
Volvox, 443
Vomeronasal organ (VNO), 842
von Baer, Karl, 314
von Frisch, Karl, 811

Waggle dance, 838
Walking catfish, 375
Wallaby, 490
Wallace, Alfred Russell, 11, 12, 303, 307, 309, 311, 316
Warblers
  North American, competition for resources among, 873
  species of, 304
Warning coloration, 875-877
Warthogs, play and, 822
Wasps
  as pollinators, 542

  as predators of bean weevils, 854
Water, 28-31, 33-34, 365
  acids, bases, and buffers and, 30
  bonding and, 26-27
  cohesion among molecules of, 33-34, 524
  dehydration synthesis, 38
  dispersal of seeds by, 548
  effects on other molecules, 28-29
  in human nutrition, requirements for, 622
  hydrolysis, 38
  in plant nutrition
    acquisition of, 523
    transport of, 524-528
    potassium concentration in guard cells and, 526-528
  relationship of plants with, 530
  as solid, 31
  as temperature moderator, 30-31
Water cycle, 902-903, 905
Water flea, 2
Water hyacinth, population growth and, 884
Water-vascular system, of echinoderms, 480
Watson, James, 11, 75, 177, 184, 185, 189
Waxes, 39, 42
Weather, 915-916
Webb, Barbara, 814
Weinberg, W., 326, 328
Went, Frits, 558-560
Wexler, Nancy, 288
Whales, 489
  evolution of, 313-314
Wheeler, Peter, 382-383
White blood cells, 591, 593, 594, 657
White-crowned sparrows, sexual selection among, 331
White matter, of spinal cord, 710, 712
Whittaker, Robert H., 389-390
Wilkins, Maurice, 184, 185, 189
Williams, Stephen, 568
Wilson, Edward O., 14, 922
Wilson, H. V., 464-465
Wind
  dispersal of seeds by, 548
  pollination by, 537, 538, 540
Withdrawal method of contraception, 781
Wolves, 336, 342
  communication by touch in, 831
  as predators of elk, 856, 857
  taxonomy and, 386, 388, 393

  visual communication in, 829
Working memory, 719
World population, carrying capacity and, 864-865
Worms
  heartworm, 471
  hookworm, 471
  tapeworms, 469, 470
  see also Earthworms; Flatworms; Roundworms; Segmented worms
Wrasse
  mimicry by, 878
  mutualism and, 880

X chromosomes, 214, 287
Xenopus, differentiation in, 789, 790
XO females, 291, 292
XXX females, 292
XXY males, 292
Xylem, 502, 503
  primary, of stems, 508, 509
  secondary, of stems, 509, 510
  water movement in, 524-526
XYY males, 292

Y chromosomes, 288, 292
Yeasts, 431, 432
  asexual reproduction in, 171
Yellow-legged frogs, 486
Yersinia pestis, 409
Yolk, 791
Yolk sac
  direct development of, 793, 794
  human, 799, 800
Yosemite toads, 486
Yuccas, pollination of, 542

Zebra mussel, 884-885
Zidovudine (Azidothymidine; AZT), 668
Zinc
  in human nutrition, requirements for, 621, 622
  in plant nutrition, 521
Z lines, 749
Zona pellucida, 776
Zooflagellates, 412, 416
Zooplankton, 935
Zoospores, 429, 434
Zygomycota, 391, 428-431
Zygote, 163, 773
  plant, 536